KEY OF ATOM COLORS USED IN MOLECULAR MODELS IN *HOLT*

Element	Color
Hydrogen, H	
Helium, He	
Carbon, C	
Nitrogen, N	
Oxygen, O	
Fluorine, F	
Neon, Ne	
Sodium, Na (similar color used for all Group 1 metals)	
Magnesium, Mg (similar color used for all Group 2 metals)	
Aluminum, Al	

Element	
Silicon, Si	
Phosphorus, P	
Sulfur, S	
Chlorine, Cl	
Argon, Ar	
Iron, Fe	
Copper, Cu	
Bromine, Br	
Silver, Ag	
Iodine, I	

COMMON IONS

Cation	Symbol	Cation	Symbol	Anion	Symbol	Anion	Symbol
Aluminum	Al^{3+}	Lead(II)	Pb^{2+}	Acetate	CH_3COO^-	Hydrogen sulfate	HSO_4^-
Ammonium	NH_4^+	Magnesium	Mg^{2+}	Bromide	Br^-	Hydroxide	OH^-
Arsenic(III)	As^{3+}	Mercury(I)	Hg_2^{2+}	Carbonate	CO_3^{2-}	Hypochlorite	ClO^-
Barium	Ba^{2+}	Mercury(II)	Hg^{2+}	Chlorate	ClO_3^-	Iodide	I^-
Calcium	Ca^{2+}	Nickel(II)	Ni^{2+}	Chloride	Cl^-	Nitrate	NO_3^-
Chromium(II)	Cr^{2+}	Potassium	K^+	Chlorite	ClO_2^-	Nitrite	NO_2^-
Chromium(III)	Cr^{3+}	Silver	Ag^+	Chromate	CrO_4^{2-}	Oxide	O_2^-
Cobalt(II)	Co^{2+}	Sodium	Na^+	Cyanide	CN^-	Perchlorate	ClO_4^-
Cobalt(III)	Co^{3+}	Strontium	Sr^{2+}	Dichromate	$Cr_2O_7^{2-}$	Permanganate	MnO_4^-
Copper(I)	Cu^+	Tin(II)	Sn^{2+}	Fluoride	F^-	Peroxide	O_2^{2-}
Copper(II)	Cu^{2+}	Tin(IV)	Sn^{4+}	Hexacyanoferrate(II)	$Fe(CN)_6^{4-}$	Phosphate	PO_4^{3-}
Hydronium	H_3O^+	Titanium(III)	Ti^{3+}	Hexacyanoferrate(III)	$Fe(CN)_6^{3-}$	Sulfate	SO_4^{2-}
Iron(II)	Fe^{2+}	Titanium(IV)	Ti^{4+}	Hydride	H^-	Sulfide	S^{2-}
Iron(III)	Fe^{3+}	Zinc	Zn^{2+}	Hydrogen carbonate	HCO_3^-	Sulfite	SO_3^{2-}

HOLT

Chemistry

TEACHER EDITION

R. Thomas Myers, Ph.D.
Kent, Ohio
Keith B. Oldham, D.Sc.
Peterborough, Ontario, Canada
Salvatore Tocci
East Hampton, New York

Teacher Edition WALK-THROUGH

Student Edition CONTENTS IN BRIEF

HOLT, RINEHART AND WINSTON
A Harcourt Education Company
Orlando • Austin • New York • San Diego • Toronto • London

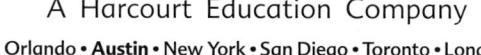

The content and skills your students need to succeed!

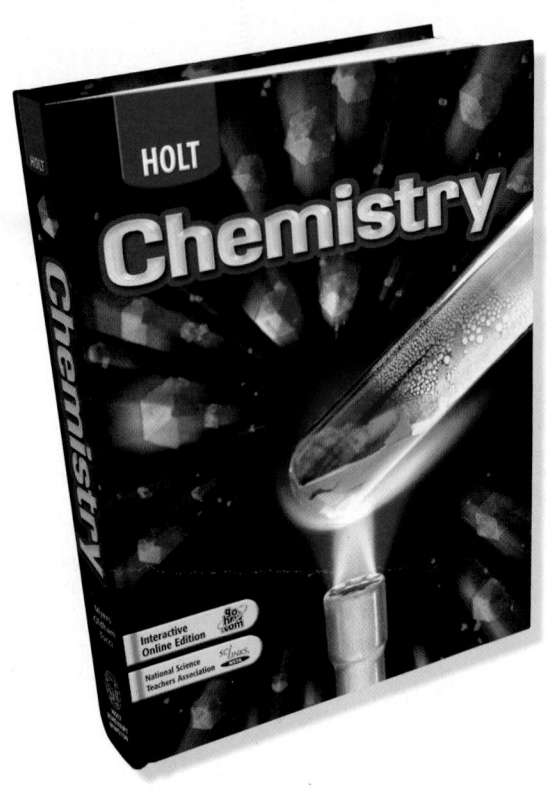

HOLT CHEMISTRY PRESENTS THE CONTENT YOU NEED

Twenty chapters allow you to finish the program in a standard one-year course, with critical content developed progressively throughout. The atom is presented in an "inside-out" approach—from atoms to the Periodic Table, to bonding, to molecules and reactions, and so on. The mole concept is introduced early on, expanded upon later in the text, and further developed throughout the rest of the program.

STUDENTS OF ALL ABILITY LEVELS DEVELOP SKILLS FOR SUCCESS

- **Pre-Reading** questions and objectives focus students' attention on upcoming topics.

- Outline-style head structure and non-intrusive graphics make the text material easy to navigate.

- Students can "see" chemistry through macro- to micro- visuals accompanied by the corresponding chemical equation.

- Activities, labs, and worksheets are graded by level of difficulty in the *Teacher Edition* to help you cater classroom instruction to the abilities of your students.

CONSISTENT PRACTICE DEVELOPS PROBLEM-SOLVING SKILLS

- **Sample Problems** and **Practice Hints** help students navigate through the problem-solving process.

- **Practice Problems** provide students with many opportunities to check their understanding with selected answers located in the **Appendix.**

- Unique **Skills Toolkits** present key processes in an easy-to-follow step-by-step manner.

A FLEXIBLE LABORATORY PROGRAM BUILDS INQUIRY AND CRITICAL-THINKING SKILLS

- The laboratory program includes in-text labs for each chapter. Additional labs are found in the *Chapter Resource Files.*

- A variety of labs—from **CBL™ Probeware Labs** to **Inquiry Labs**—help you meet curriculum and student needs.

- Additional **QuickLABs** and **Skill Practice Labs** help you select labs to fit your classroom schedule.

HOLT **Study Guide**

Chemistry

HOLT

Chemistry

Assessments

Includes
**Concept Review
Worksheets**

CHAPTERS
1 The Science of Chemistry
2 Matter and Energy
3 Atoms and Moles
4 The Periodic Table
5 Ions and Ionic Compounds
6 Covalent Bonds
7 The Mole and Chemical Compositions
8 Chemical Equations and Reactions

HOLT **Chemistry**
One-Stop Planner®
with Test Generator
CD-ROM for Macintosh® and Windows®

Printable
Teaching Resources

Customizable
Lesson Plans

Powerful
Test Generator

HOLT Chemistry

Chapter Resource File **1**

The Science of Chemistry

Skills Worksheets
Concept Reviews 1
Problem-Solving 7
Assessments
.... 17

Teacher Resources
Lesson Plans 73
Lab Notes and Answers 84
Answer Keys 124
Test Item Listing for
ExamView® Test Generator T1
S1

Teaching Transparencies

HOLT

Chemistry

includes
Bellringer Transparencies

Map Transparencies

ISBN 0-03-044441-0

Integrating
✓ Chemistry
✓ Physics
✓ Earth Science
✓ Space Science
✓ Mathematics

INTEGRATED TECHNOLOGY AND ONLINE RESOURCES EXPAND LEARNING BEYOND THE CLASSROOM

- Lighten the load with an interactive *Online Edition* or *CD-ROM Version* of the student text.

- **SciLinks,** a Web service developed and maintained by NSTA, contains current and prescreened links that engage students.

- *Holt ChemFile Interactive Tutor CD-ROM* gives students a fun way to explore chemistry concepts at their own pace.

- All the resources you need are on the *One-Stop Planner® CD-ROM with Test Generator,* with worksheets, customizable lesson plans, and a powerful test generator.

The Student Edition builds skills for success in science

Each section begins with a list of **Key Terms** and **Objectives** to enhance student comprehension of the material presented and improve reading skills.

Figures and tables in **Holt Chemistry** illustrate and clarify key concepts.

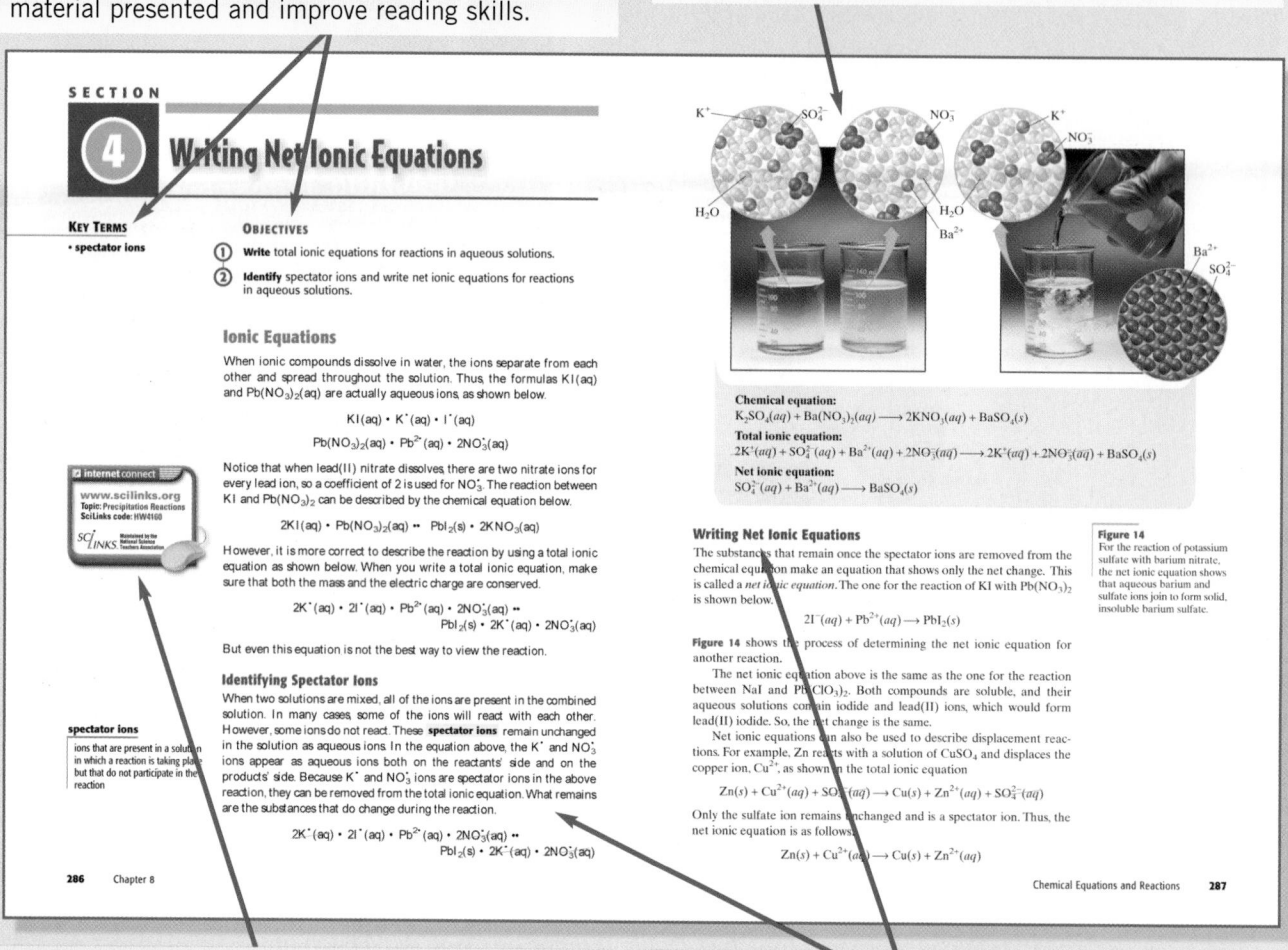

SciLinks, topical pre-screened links from NSTA, are called out whenever related material is available online.

Accessible navigation engages students with outline-style headings, content grouped into small chunks, and text that doesn't break between pages.

RELEVANT AND EXCITING FEATURES

Topic Link refers students to other parts of the textbook that contain material relating to the topic currently being studied.

Element Spotlight presents students with information about a particular element.

Consumer Focus uses real-world connections to make chemistry more relevant to students' lives.

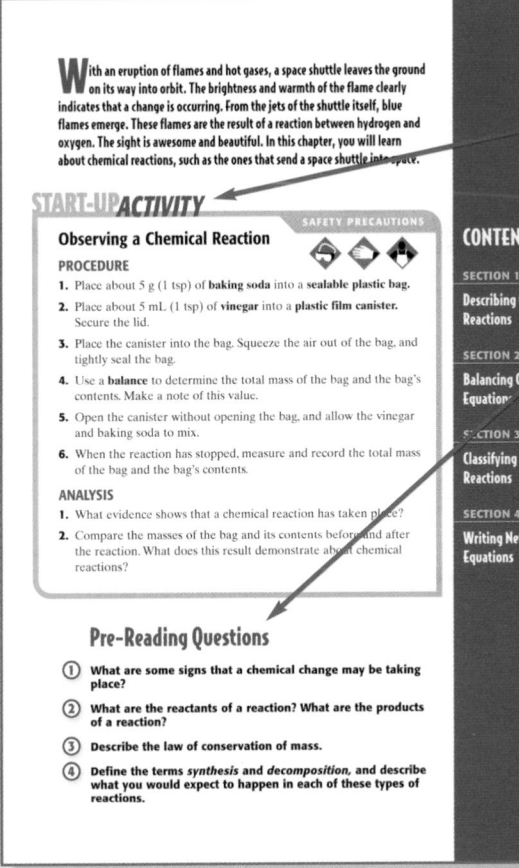

Start-up Activity piques student interest and draws them into the chapter content.

Pre-Reading Questions improve students' reading skills by helping them focus on upcoming text material.

LABS AND ACTIVITIES GRAB ATTENTION

QuickLABs give students a short, hands-on experiment requiring minimal materials.

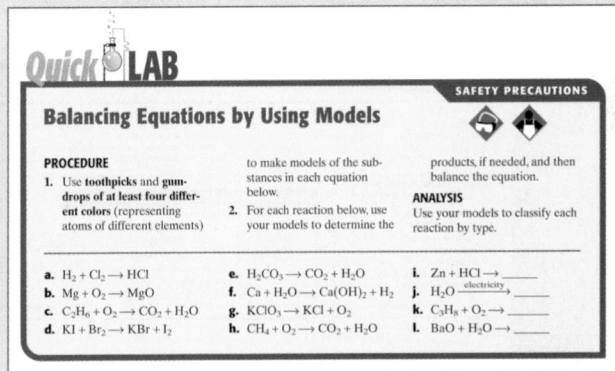

Chapter Labs are available in the back of the *Student Edition*:

- **Skill Builder Lab** provides the conceptual framework and the necessary laboratory techniques students need to practice their lab skills.

- **Inquiry Lab** gives students the opportunity to design their own experiments.

REVIEW FOR TEST-READINESS

Study Tip provides simple and practical ways for students to improve their study skills.

Chapter Highlights identifies key terms, concepts, and skills to help students study for the **Chapter Review**.

> **STUDY TIP**
> **WORKING PROBLEMS**
> If you have difficulty working practice problems, review the outline of procedures in **Skills Toolkit 2**. You may also refer back to the sample problems.

Each section and chapter ends with **Review** questions that enable students to check their understanding and apply problem-solving skills.

PROBLEM-SOLVING PRACTICE GETS STUDENTS READY FOR TESTING

Problem-Solving Skills are called out to give students the resources they need for additional help and practice.

Skills Toolkit provides students with practical problem-solving techniques in a step-by-step approach that students can apply to their chemistry studies.

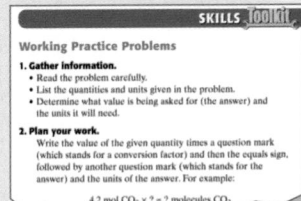

Sample Problem guides students through a step-by-step problem-solving process, with a **Practice Hint** to help students self-check their work.

Practice problems reinforce newly learned skills related to **Sample Problems**. Additional practice problems and answers for selected **Practice** problems are found in the **Appendix** of the student text, and detailed solutions are found in the *Chapter Resource File*.

A Teacher Edition
that makes planning easy

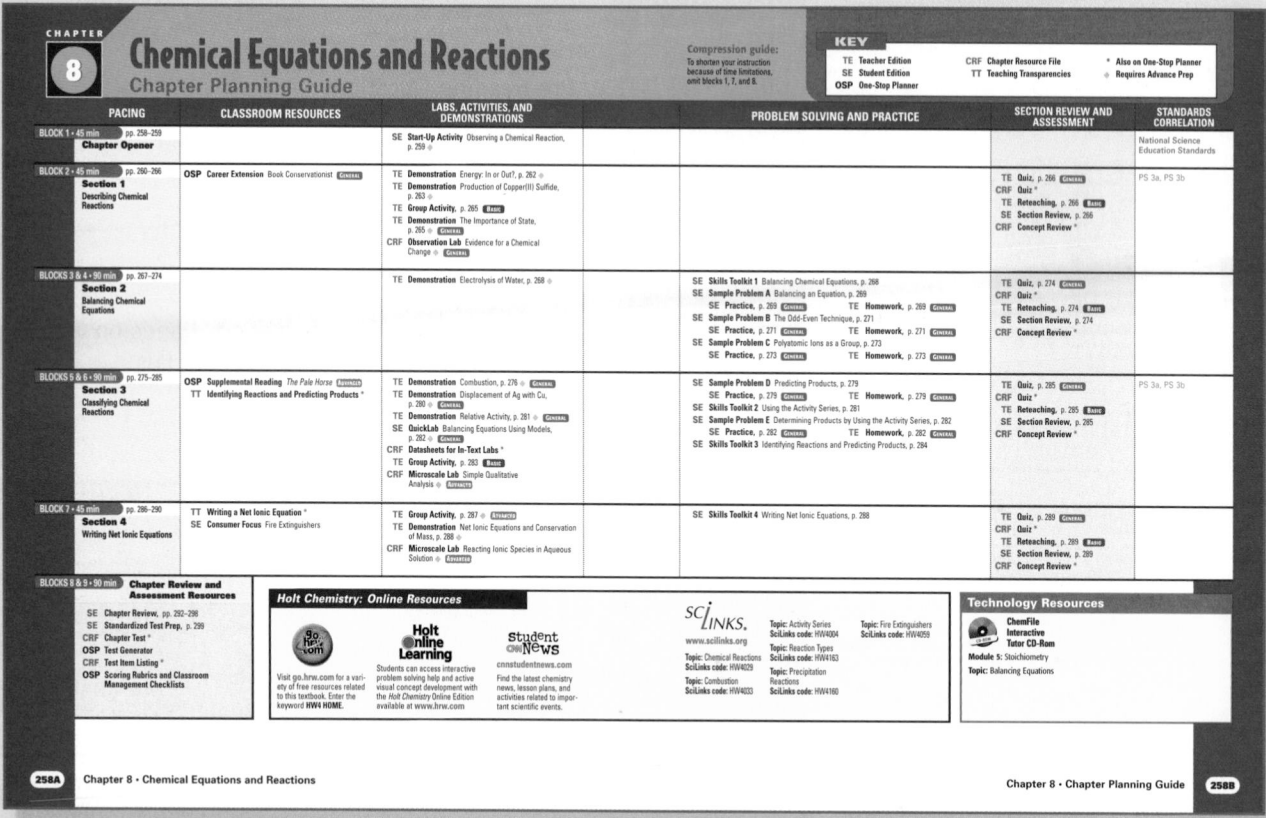

TEACHING RESOURCES
DESIGNED FOR CONVENIENCE

The **Chapter Planning Guide** breaks each chapter down into flexible 45-minute blocks and offers a full listing of activities and classroom resources available for that lesson. Look for guidance on:

- Pacing
- Labs and Demonstrations
- Classroom Resources
- Problem Solving and Practice
- Section Review and Assessment
- National Science Education Standards Correlations
- Online and Technology Resources

A complete **Lesson Cycle** builds structure around every lesson:

- **Focus** uses the objectives listed in the *Student Edition* to focus student attention on the upcoming content.

- **Motivate** uses demonstrations, discussions, and lively activities to get students excited about the material.

- **Teach** presents various teaching techniques including various typs of **Skill Builders, Using the Figure,** and more.

- Finally, **Close** with quiz questions to ensure students understand the information covered.

ACTIVITIES AND DEMONSTRATIONS FOR EVERY LEARNING LEVEL

Activities are leveled by ability level in the teacher's wrap—**Basic**, **General**, and **Advanced**—helping you choose appropriate activities for your students.

Learning styles are addressed throughout—**Interpersonal**, **Intrapersonal**, **Auditory**, **Kinesthetic**, **Logical**, **Visual**, and **Verbal**—so you can adapt material to different learning styles.

- **Bellringer** activities begin each section with an activity designed to get students thinking.

Bellringer

Have students list observations that they think indicate that a chemical reaction has taken place in the following situations: a cut apple turns brown, an egg changes when it cooks, a log burns, and a car rusts.

- **Demonstration** is a short activity that represents concepts in the text.

- **Activity** and **Group Activity** help motivate students to get involved in the material presented and share their ideas.

Group Activity —— BASIC

Provide students with a list of balanced equations that use the symbols from **Table 2**. Students should work in groups and read the equations in sentence form, expressing every item in the equation. For example, the equation $2KClO_3(s) \xrightarrow{heat} 2KCl(s) + 3O_2(g)$ could be read, "When solid potassium chlorate is heated, solid potassium chloride and oxygen gas form." **LS** Verbal

CREATING RELEVANCE AND UNDERSTANDING

On almost every page you will find exciting features to help ignite class discussion and keep students thinking.

- Connections
- Real-World Connections
- Did You Know?
- Teaching Tip
- Misconception Alert
- Skill Builder, for Reading, Graphing, Math, and Writing
- Homework

READING SKILL BUILDER —— BASIC

Paired Summarizing Have students read the rest of this section in pairs. Have one partner read each sentence and the other partner summarize the meaning of the sentence in his or her own words. Then have them change duties. If the partner who is to summarize

SKILL BUILDER —— ADVANCED

Graphing Skills Have students choose an element and list masses for various samples of the element. For example, they might choose nitrogen and list 7.00 g, 12.0 g, 16.0 g, and 20.0 g. Have them determine the number of moles of atoms for each mass. Ans. 0.500 mol, 0.857 mol, 1.14 mol,

REAL-WORLD CONNECTION

Barium Nitrate The compound used in **Sample Problem F**, barium nitrate, is extremely toxic, as are many barium compounds. This toxicity is more significant than it might be in other barium compounds because barium nitrate is much more soluble in water than other barium compounds. It can cause respiratory irritation, muscle spasms, and skin and eye irritation, as well as other symptoms. It is used in fireworks to produce a green flame.

MISCONCEPTION ALERT

Chemical Reaction or Not? Students might think that any change of a material involves a chemical reaction. Ask them to brainstorm a list of changes that a material can undergo. Then have them determine which of these changes do not involve permanent changes in the identity of the material. Such changes might include cutting paper or melting ice. Point out that these changes are physical changes and do not involve chemical reactions.

did you know? —— GENERAL

Pennies vary in composition and mass depending on when they were made. Pennies made between 1864 and 1962 are 95% copper and 5% tin and zinc. An exception are pennies made in 1943, which are made of steel coated with zinc. Pennies made between 1962 and 1982 are 95% copper and 5% zinc. Pennies made after 1982 are 97.5% zinc and 2.5% copper. Have interested students determine the masses and compositions of various coins and write a report about the history of U.S. coin composition. **LS** Verbal

Homework —— GENERAL

Additional Practice How many representative particles are present in each of the following?

1. 4.3 mol of tungsten Ans. 2.6×10^{24} atoms W
2. 2.45×10^{-6} mol of nickel(II) selenide Ans. 1.48×10^{18} formula units NiSe
3. 0.923 mol of selenium tetrabromide Ans. 5.56×10^{23} molecules SeBr$_4$

LS Logical

INCLUSION STRATEGIES MAKE MATERIAL ACCESSIBLE TO ALL

Written by professionals in the field of special needs education, **Inclusion Strategies** address many different learning exceptionalities in the classroom.

- Learning Disabled
- Developmentally Delayed
- Attention Deficit Disorder
- Behavior Control Issues
- Gifted and Talented
- English Language Learners

INCLUSION Strategies

- Learning Disabled
- English Language Learners
- Attention Deficit Disorder

Students can create a "Reaction Notebook" by labeling notebook pages with the different types of reactions discussed in the chapter. Students should write on each page specific examples of that reaction type, including the reactants and products of the reaction. Notes about patterns, trends, or similarities with other reactions can be written on each page. The notebook can be used as a study guide for individual study or discussion in small groups.

Assessment opportunities help you track students' progress

SECTION ASSESSMENT IN THE TEACHER EDITION

- **Homework** questions located throughout the *Teacher Edition* can be used for additional practice, or as chalkboard examples.

- **Quiz** provides additional questions to assess student progress.

- **Reteaching** activities help students understand section material by presenting a concept in a different way.

Homework ——— GENERAL

Additional Practice How many representative particles are present in each of the following?

1. 4.3 mol of tungsten Ans. 2.6×10^{24} atoms W

2. 2.45×10^{-6} mol of nickel(II) selenide Ans. 1.48×10^{18} formula units NiSe

3. 0.923 mol of selenium tetrabromide Ans. 5.56×10^{23} molecules $SeBr_4$

LS Logical

Close

Quiz ——— GENERAL

1. Balance the following equation. $NH_3 + O_2 \longrightarrow NO + H_2O$ **Ans.** $4NH_3 + 5O_2 \longrightarrow 4NO + 6H_2O$

2. Write a balanced equation for the reaction between chlorine gas and potassium bromide, forming potassium chloride and bromine. **Ans.** $Cl_2 + 2KBr \longrightarrow 2KCl + Br_2$

3. Why should you not balance an equation by changing subscripts? **Ans.** Changing subscripts

Reteaching ——— BASIC

Provide students with several simple unbalanced chemical equations and have them work in pairs to explain to each other step-by-step how to balance the equations. Then have the pairs compare their results to those of another group and discuss any differences.

LS Interpersonal

SECTION ASSESSMENT IN THE STUDENT EDITION

- **Pre-Reading Questions** assesses prior knowledge and serves as a reading warm-up.

- **Study Tip,** throughout the *Student Edition* and in the **Appendix,** offers advice students can use to create more effective study habits.

- **Practice** asks students to apply what they have learned in the **Sample Problems.**

- **Section Review** provides a thorough review and questions that include evaluate reading, writing, and critical-thinking skills.

1 Section Review

UNDERSTANDING KEY IDEAS

1. What is the definition of a mole?

2. How many particles are there in one mole?

3. Explain how Avogadro's number can give two conversion factors.

4. Which will have the greater number of ions, 1 mol of nickel(II) or 1 mol of copper(I)?

9. Find the mass in grams.
   a. 4.30×10^{16} atoms He, 4.00 g/mol
   b. 5.710×10^{23} molecules CH_4, 16.05 g/mol
   c. 3.012×10^{24} ions Ca^{2+}, 40.08 g/mol

10. Find the number of molecules or ions.
   a. 1.000 g I^-, 126.9 g/mol
   b. 3.5 g Cu^{2+}, 63.55 g/mol
   c. 4.22 g SO_2, 64.07 g/mol

11. What is the mass of 6.022×10^{23} molecules of ibuprofen (molar mass of 206.31 g/mol)

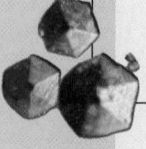

TEACHER EDITION CHAPTER ASSESSMENT

- **Assessing Prior Knowledge** provides a quick check of those topics students should be familiar with before they begin a new chapter.

- An **Alternative Assessments** guide in the *Teacher Edition* indexes options for activities, projects, and reports.

- The **Assignment Guide** in the *Teacher Edition* correlates all end-of-chapter test items to text sections for easy reference.

STUDENT EDITION CHAPTER ASSESSMENT

- **Chapter Highlights** is a handy guide to help students study key ideas, terms, and skills.

- The **Chapter Review** presents a variety of question types such as **Key Terms, Mixed Review, Technology and Learning,** and **Concept Mapping** to check students' understanding.

- **Test Prep** at the end of each chapter provides a selection of questions in standardized-test format to develop students' test-taking skills.

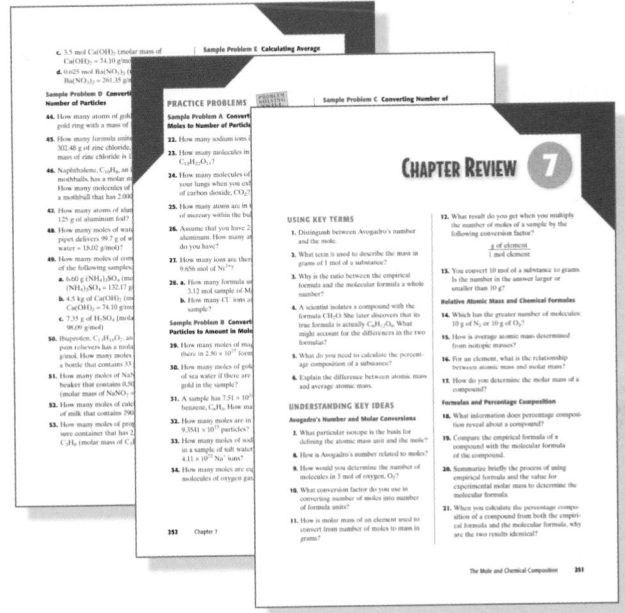

Assignment Guide

Section	Questions
1	6–13, 31–36, 50, 63, 73–76
2	2, 5, 14–18, 37–45, 51, 53, 56, 60, 77
3	1, 19–25, 46–49, 52, 54–55, 57, 61–62, 65–72
4	3, 4, 26–30, 58–59, 64

ADDITONAL RESOURCES IN THE CHAPTER RESOURCE FILES

- **Concept Review** contains review worksheets that reinforce the skills and concepts presented in the *Student Edition*.

- Blackline masters of **Chapter Tests** provide an objective-based assessment for the complete chapter. These tests include multiple choice, short answer, essay, and problem-solving questions.

- This resource also contains **Quizzes** and a complete test-item listing to accompany our **Test Generator.**

CUSTOM ASSESSMENT

One-Stop Planner CD-ROM with Test Generator
allows you to construct your own chapter tests from hundreds of items, correlated with text objectives. Questions can be edited and are coded by item type and level of comprehension. See page T12 for more information.

Resources that make teaching easier

CHAPTER RESOURCE FILES

A **Chapter Resource File** accompanies each chapter of **Holt Chemistry.** Everything you need to plan and manage your lessons in a convenient timesaving format is included in each chapter book. An introductory booklet that provides a guide to the resources in each chapter book is located in the **Chapter Resource File.** Each chapter book includes:

Skills Worksheets
• Problem Solving
• Concept Review

Labs and Activities
• Datasheets for In-Text Activities
• Quick Labs
• Skills Practice Labs
• Inquiry Labs
• CBL™ Probeware Labs
• Consumer Labs
• Design Your Own Labs
• Microscale Labs

Assessments
• Quizzes
• Chapter Test
• Test Item Listing
  (for ExamView® Test Generator)

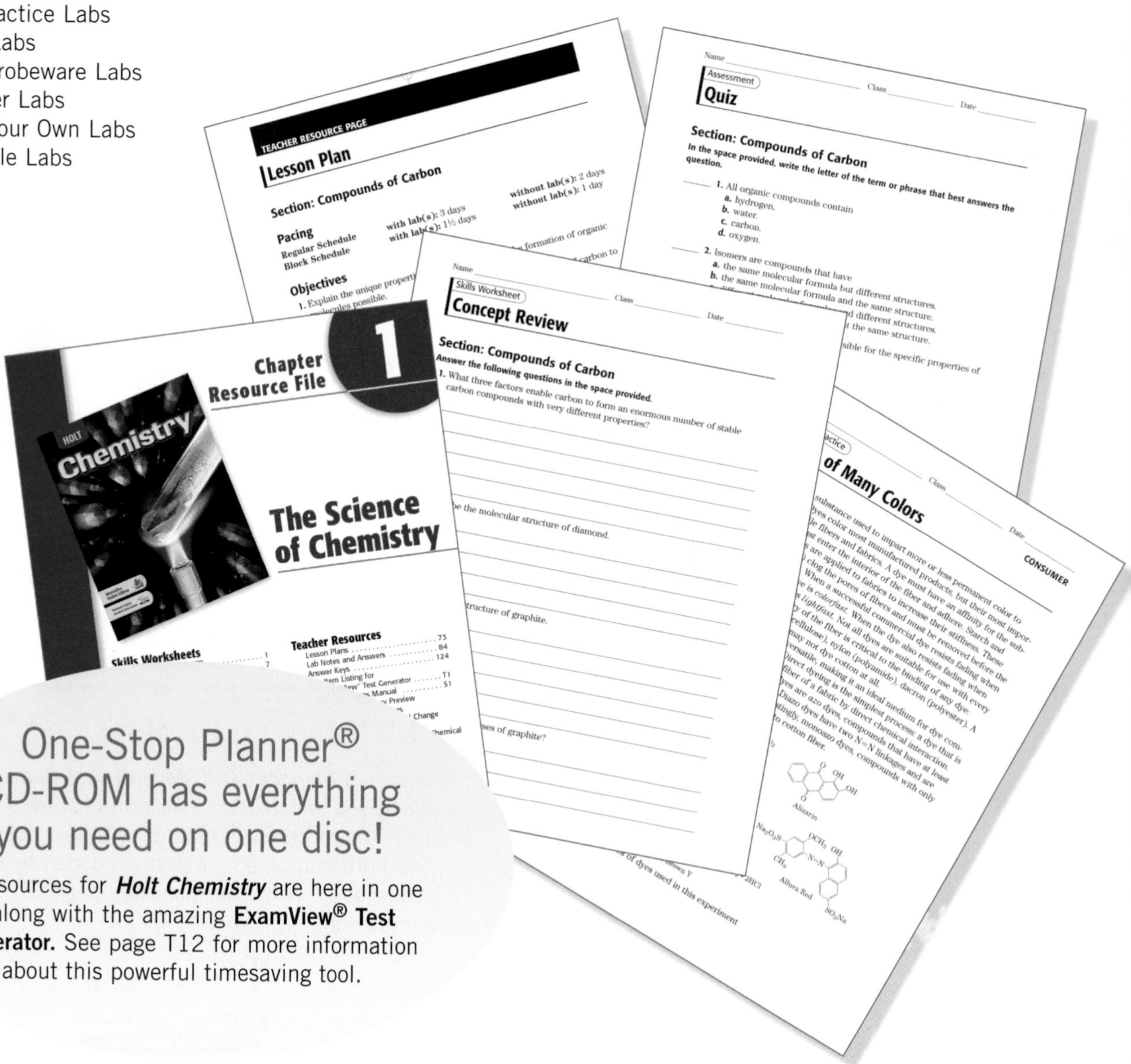

One-Stop Planner® CD-ROM has everything you need on one disc!

All the resources for **Holt Chemistry** are here in one place, along with the amazing **ExamView® Test Generator.** See page T12 for more information about this powerful timesaving tool.

ADDITIONAL RESOURCES REINFORCE AND EXTEND LESSONS

- *Holt Science Skills Workshop: Reading in the Content Area* contains exercises that target those reading skills that are specific to the comprehension of science texts.

- *Holt Science Laboratory Manager's Professional Reference* is a well-organized reference that was created to help teachers understand risk management and the hazards that can occur in the classroom.

- *Problem Solving Workbook* provides additional sample problems and practice exercises for study and review.

- *Study Guide* contains concept review worksheets that reinforce the skills and concepts presented in the *Student Edition*.

- Full-color **Teaching Transparencies** use graphics directly from the text to enhance classroom presentations.

SPANISH RESOURCES

- A **Spanish Glossary** is right at students' fingertips in the *Student Edition,* following the **Glossary.** It shows the English term, its Spanish equivalent, and a definition in Spanish.

- *Study Guide* in Spanish contains review worksheets that reinforce the skills and concepts presented in the *Student Edition* that have been translated into Spanish.

- *Assessments* in Spanish include: **Section Quizzes, Chapter Tests,** and **Concept Review.**

Technology that enhances teaching

One-Stop Planner CD-ROM®
with Test Generator

Planning and managing lessons has never been easier than with this convenient all-in-one CD-ROM with a variety of timesaving features, including:

Printable resources and worksheets

Resources available for *Holt Chemistry* are in one place, including science skills development, concept practice, math practice, vocabulary development, Spanish materials, and transparency masters.

Customizable lesson plans

Tailor your lessons to your classroom's specific needs. Includes block-scheduling lesson plans in several word-processing formats.

Powerful ExamView® Test Generator

Contains test items organized by chapter, plus hundreds of editable questions, so you can put together your own tests and quizzes.

HOLT **Chemistry**

One-Stop Planner®
with Test Generator
CD-ROM for Macintosh® and Windows®

Printable
Teaching Resources

Customizable
Lesson Plans

Powerful
Test Generator

STUDENT EDITION ON CD-ROM

Ideal for students who have limited access to the Internet, but who need to lighten the load of textbooks they carry home, the entire *Student Edition* is on one easy-to-navigate CD-ROM, page-for-page.

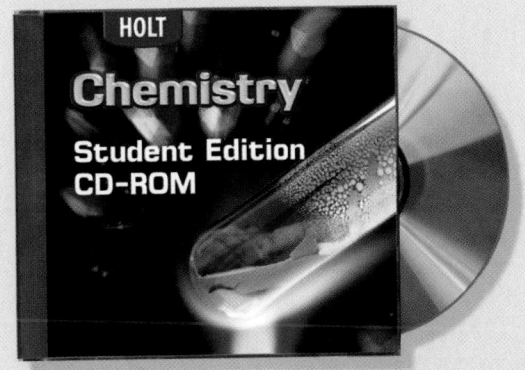

HOLT CHEMFILE INTERACTIVE TUTOR CD-ROM

This double CD-ROM program lets students explore chemistry at their own pace and then leads them step-by-step in visualizing and understanding ten key chemistry concepts.

With this individualized approach, students make macro-to-micro connections, practice math skills through animated solutions, and manipulate chemical systems to observe cause-and-effect relationships.

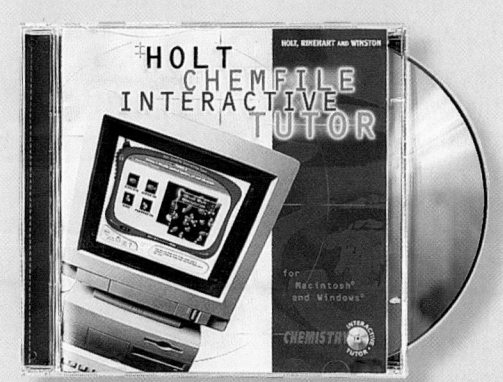

LESSON PRESENTATION CD-ROM

This CD-ROM is your guide to **Quick Concepts**— engaging, media-based presentations of core concepts. **Quick Concepts** are organized to match the content covered in each section of your textbook, and can be projected for large group instruction or viewed by individuals or groups of students with a computer. Because chapter content is distilled down to the core ideas, students know what is important to study.

CNN PRESENTS SCIENCE IN THE NEWS: CHEMISTRY CONNECTIONS

This CNN video collection includes broadcast news segments that bring the relevance of science directly into your classroom. Each news segment showcases useful applications of science concepts, often making cross-curricular connections to industry, careers, and a variety of other areas.

TECHNOLOGY RESOURCES

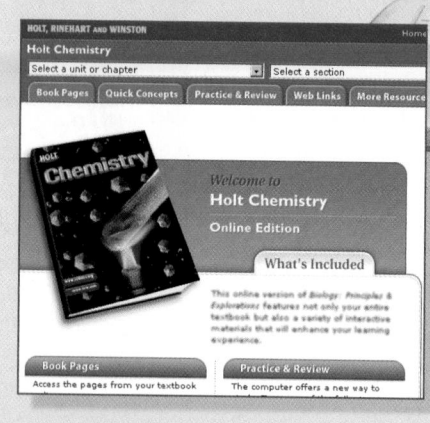

student CNN News ™

CNNStudentNews.com provides award-winning news and information for both teachers and students. You will find a wealth of helpful features, including:

- News as it happens
- Classroom Resources
- Student current-events activities
- Lesson plans
- Projects and activities

QUANTUM
INTELLIGENT
TUTORING
ENGINES

Six different artificial-intelligence-based online tutors help students learn key concepts in chemistry and physical science while working at their own pace. Imagine giving your students the help they need by offering them a tutor that coaches them through a problem and gives them the confidence to work through challenging subjects.

Try it out at www.hrw.com.

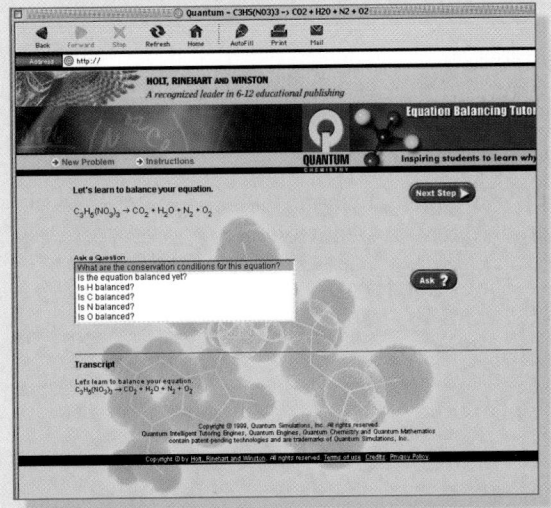

Holt's own award-winning Web site includes a variety of worksheets, activities, projects, research articles and ideas, interactive quizzes, review activities, and teacher resources.

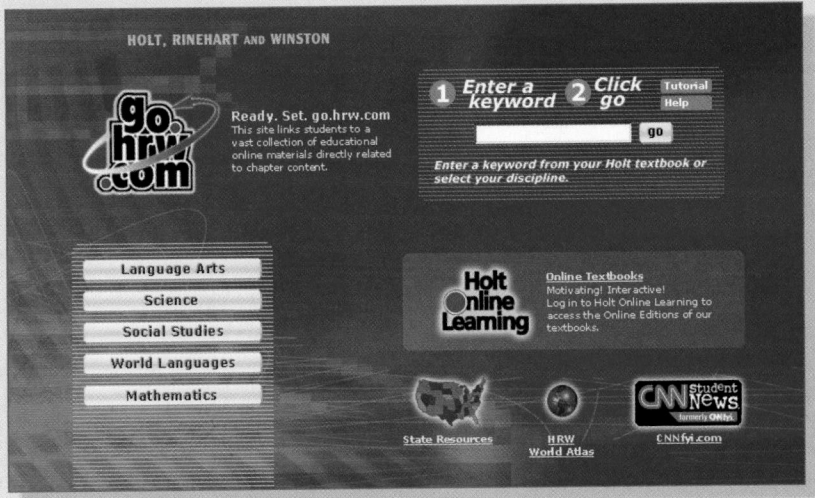

A variety of labs for every purpose

Holt Chemistry includes lab activities that meet the demands of your curriculum. This flexible laboratory program builds inquiry and critical-thinking skills.

All labs are **Bench-Tested,** and *Lab Ratings* help you select the labs that suit your students' abilities.

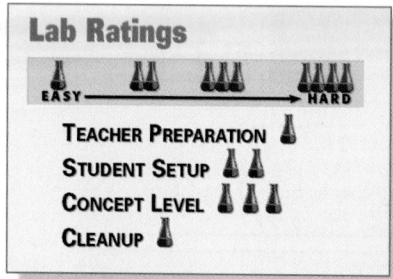

Lab Ratings

EASY ———————— HARD

TEACHER PREPARATION
STUDENT SETUP
CONCEPT LEVEL
CLEANUP

CHAPTER LABS

Start-up Activity is a simple activity at the beginning of each chapter. This activity helps introduce the content of the lesson.

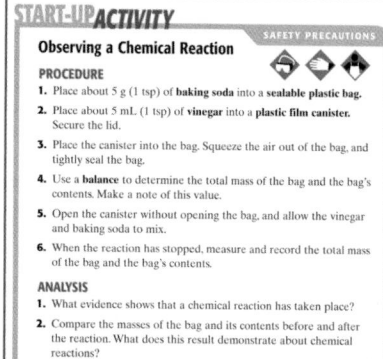

START-UP ACTIVITY

SAFETY PRECAUTIONS

Observing a Chemical Reaction

PROCEDURE

1. Place about 5 g (1 tsp) of **baking soda** into a **sealable plastic bag.**
2. Place about 5 mL (1 tsp) of **vinegar** into a **plastic film canister.** Secure the lid.
3. Place the canister into the bag. Squeeze the air out of the bag, and tightly seal the bag.
4. Use a **balance** to determine the total mass of the bag and the bag's contents. Make a note of this value.
5. Open the canister without opening the bag, and allow the vinegar and baking soda to mix.
6. When the reaction has stopped, measure and record the total mass of the bag and the bag's contents.

ANALYSIS

1. What evidence shows that a chemical reaction has taken place?
2. Compare the masses of the bag and its contents before and after the reaction. What does this result demonstrate about chemical reactions?

QuickLABs are easy activities that can be completed in less than one class period.

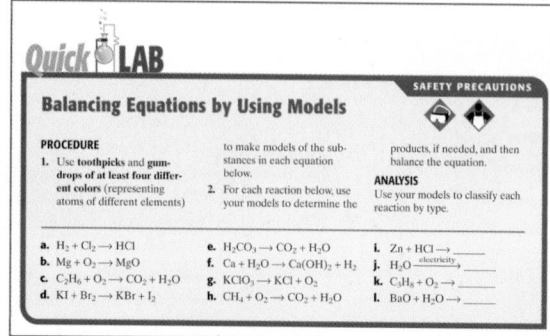

Quick LAB

SAFETY PRECAUTIONS

Balancing Equations by Using Models

PROCEDURE

1. Use **toothpicks** and **gumdrops of at least four different colors** (representing atoms of different elements) to make models of the substances in each equation below.
2. For each reaction below, use your models to determine the products, if needed, and then balance the equation.

ANALYSIS

Use your models to classify each reaction by type.

a. $H_2 + Cl_2 \rightarrow HCl$
b. $Mg + O_2 \rightarrow MgO$
c. $C_2H_6 + O_2 \rightarrow CO_2 + H_2O$
d. $KI + Br_2 \rightarrow KBr + I_2$
e. $H_2CO_3 \rightarrow CO_2 + H_2O$
f. $Ca + H_2O \rightarrow Ca(OH)_2 + H_2$
g. $KClO_3 \rightarrow KCl + O_2$
h. $CH_4 + O_2 \rightarrow CO_2 + H_2O$
i. $Zn + HCl \rightarrow \_\_\_\_$
j. $H_2O \xrightarrow{electricity} \_\_\_\_$
k. $C_3H_8 + O_2 \rightarrow \_\_\_\_$
l. $BaO + H_2O \rightarrow \_\_\_\_$

Skills Practice Lab

1 Laboratory Techniques

OBJECTIVES

- Demonstrate proficiency in using a Bunsen burner, a balance, and a graduated cylinder.
- Demonstrate proficiency in handling solid and liquid chemicals.
- Develop proper safety techniques for all lab work.
- Use neat and organized data-collecting techniques.
- Use graphing techniques to plot data.

MATERIALS

- balance
- beakers, 250 mL (2)
- Bunsen burner and related equipment
- copper wire
- crucible tongs
- evaporating dish
- graduated cylinder, 100 mL
- heat-resistant mat
- NaCl
- spatula
- test tube
- wax paper or weighing paper

Introduction

You have applied to work at a company that does research, development, and analysis work. Although the company does not require employees to have extensive chemical experience, all applicants are tested for their ability to follow directions, heed safety precautions, perform simple laboratory procedures, clearly and concisely communicate results, and make logical inferences.

The company will consider your performance on the test in deciding whether to hire you and determining what your initial salary will be. Pay close attention to the procedures and safety precautions because you will continue to use them throughout your work if you are hired by this company. In addition, you will need to pay attention to what is happening around you, make careful observations, and keep a clear and legible record of these observations in your lab notebook.

This laboratory orientation session will teach you some of the following techniques:

- how to use a Bunsen burner
- how to handle solids and liquids
- how to use a balance
- how to practice basic safety techniques in lab work

756   Skills Practice Lab 1

Chapter Labs are available in the back of the textbook:

- **Skill Practice Lab** provides the conceptual framework and the necessary laboratory techniques students need to practice their lab skills.

- **Inquiry Lab** gives students the opportunity to design their own experiments.

9 Stoichiometry and Gravimetric Analysis

Safety Procedures

- Wear safety goggles when working around chemicals, acids, bases, flames, or heating devices. Contents under pressure may become projectiles and cause serious injury.
- Never look directly at the sun through any optical device or use direct sunlight to illuminate a microscope.
- Avoid wearing contact lenses in the lab.
- If any substance gets in your eyes, notify your instructor immediately and flush your eyes with running water for at least 15 min.
- Secure loose clothing, and remove dangling jewelry. Don't wear open-toed shoes or sandals in the lab.
- Wear an apron or lab coat to protect your clothing when working with chemicals.
- If a spill gets on your clothing, rinse it off immediately with water for at least 5 min while notifying your instructor.
- Always use caution when working with chemicals.
- Never mix chemicals unless specifically directed to do so.
- Never taste, touch, or smell chemicals unless

specifically directed to do so.
- Add acid or base to water; never do the opposite.
- Never return unused chemicals to the original container.
- Never transfer substances by sucking on a pipette or straw; use a suction bulb.
- Follow instructions for proper disposal.
- Avoid wearing hair spray or hair gel on lab days.
- Whenever possible, use an electric hot plate as a heat source instead of an open flame.
- When heating materials in a test tube, always angle the test tube away from yourself and others.
- Glass containers used for heating should be made of heat-resistant glass.
- Know your school's fire-evacuation routes.
- Clean and decontaminate all work surfaces and personal protective equipment as directed by your instructor.
- Dispose of all sharps (broken glass and other contaminated sharp objects) and other contaminated materials (biological and chemical) in special containers as directed by your instructor.

Data Table 1

Material	Mass (g) step 11	Mass (g) step 12
empty beaker		
beaker and 50 mL of water		
50 mL of water		
beaker and 100 mL of water		
100 mL of water		
beaker and 150 mL of water		
150 mL of water		

Procedure

1. Copy Data Tables 1 and 2 in your lab notebook. Be sure that you have plenty of room for observations about each test.

Data Table 2

Material	Mass (g)
weighing paper	
weighing paper and NaCl	

2. Record in your lab notebook the location and use of the following emergency items: lab shower, eyewash station, and emergency telephone numbers.

3. Check to be certain that the gas valve at your lab station and at the neighboring lab stations are turned off. Notify your teacher immediately if a valve is on, because the fumes must be cleared before any work continues.

Laboratory Techniques **757**

- **CBL™ Probeware Lab** (Texas Instruments, Inc. Calculator Based Laboratory) is included for some of the chapters in the text.

- **Microscale Lab** is a cost-effective version of your favorite experiments that involves less time and employs lower-cost materials.

- **Skills Practice Lab** hones inquiry skills and develop scientific methods.

- **Design Your Own Lab** pushes students' problem-solving skills further as they create experiments in the lab.

TEACHER EDITION

Additional activities found in the *Teacher Edition* also help illustrate science concepts. Examples include **Demonstration, Activity, and Group Activity.**

Demonstration — BASIC
What's in a Mole? Measure 27 g (1 mol) of aluminum foil. You can fold the foil, if necessary, but do not crumple it. Tell students that there is 1 mol of aluminum present, and ask them how many particles are in the sheet. Ans. 6.022×10^{23} Then, crumple the aluminum foil and ask the students what has changed and what has not changed. Ans. The form has changed, but the number of particles is still the same. Tear the foil into four pieces and ask students what amount of aluminum is present in each piece. Ans. less than a mole
LS Visual

HOLT SCIENCE LABORATORY MANAGER'S PROFESSIONAL REFERENCE

This guide to risk management outlines the hazards that can occur in your classroom to help you learn how to prevent or eliminate them.

CHAPTER RESOURCE FILES LABS AND ACTIVITIES

- **Datasheets for In-text Activities** give students a template to guide them through text activities in a structured manner.

- **Inquiry Lab** is a problem-based experiment in which students design and carry out their own experimental procedures.

- **Consumer Lab** explores the chemical nature of common household products.

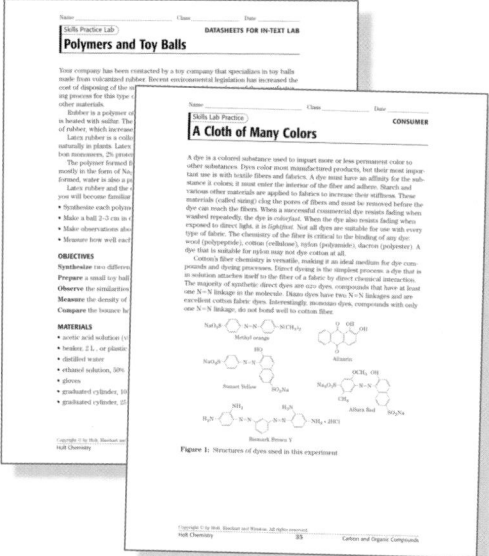

Meeting individual needs

Students have a wide range of abilities and learning exceptionalities. These pages show you how *Holt Chemistry* provides resources and strategies to help you tailor your instruction to engage every student in your classroom.

Learning exceptionality	Resources and strategies	
Learning Disabilities and Slow Learners Students who have dyslexia or dysgraphia, students reading below grade level, students having difficulty understanding abstract or complex concepts, and slow learners	• Inclusion Strategies labeled *Learning Disabled* • Activities labeled *Basic* • *Reteaching* activities	• Activities labeled *Visual* or *Kinesthetic* • Hands-on activities or projects • Oral presentations instead of written tests or assignments
Developmental Delays Students who are functioning far below grade level because of mental retardation, autism, or brain injury; goals are to learn or retain basic concepts	• Inclusion Strategies labeled *Developmentally Delayed* • Activities labeled *Basic*	• *Reteaching* activities • Project-based activities
Attention Deficit Disorders Students experiencing difficulty completing a task that has multiple steps, difficulty handling long assignments, or difficulty concentrating without sensory input from physical activity	• Inclusion Strategies labeled *Attention Deficit Disorder* • Activities labeled *Basic* • *Reteaching* activities • *Group Activities*	• Activities labeled *Visual* or *Kinesthetic* • Concepts broken into small chunks • Oral presentations instead of written tests or assignments
English as a Second Language Students learning English	• Activities labeled *English Language Learners* • Activities labeled *Basic*	• *Reteaching* activities • Activities labeled *Visual*
Gifted and Talented Students who are performing above grade level and demonstrate aptitude in crosscurricular assignments	• Inclusion Strategies labeled *Gifted and Talented* • Activities labeled *Advanced* • *Connection* activities	• Activities that involve multiple tasks, a strong degree of independence, and student initiative

General Strategies The following strategies can help you modify instruction to help students who struggle with common classroom difficulties.

A student experiencing difficulty with . . .	May benefit if you . . .	
Beginning assignments	• Assign work in small amounts • Have the student use cooperative or paired learning • Provide varied and interesting activities	• Allow choice in assignments or projects • Reinforce participation • Seat the student closer to you
Following directions	• Gain the student's attention before giving directions • Break up the task into small steps • Give written directions rather than oral directions • Use short, simple phrases • Stand near the student when you are giving directions	• Have the student repeat directions to you • Prepare the student for changes in activity • Give visual cues by posting general routines • Reinforce improvement in or approximation of following directions
Keeping track of assignments	• Have the student use folders for assignments • Have the student use assignment notebooks	• Have the student keep a checklist of assignments and highlight assignments when they are turned in
Reading the textbook	• Provide outlines of the textbook content • Reduce the length of required reading • Allow extra time for reading • Have the students read aloud in small groups	• Have the student use peer or mentor readers • Have the student use books on tape or CD • Discuss the content of the textbook in class after reading
Staying on task	• Reduce distracting elements in the classroom • Provide a task-completion checklist • Seat the student near you	• Provide alternative ways to complete assignments, such as oral projects taped with a buddy
Behavioral or social skills	• Model the appropriate behaviors • Establish class rules, and reiterate them often • Reinforce positive behavior • Assign a mentor as a positive role model to the student • Contract with the student for expected behaviors • Reinforce the desired behaviors or any steps toward improvement	• Separate the student from any peer who stimulates the inappropriate behavior • Provide a "cooling off" period before talking with the student • Address academic/instructional problems that may contribute to disruptive behaviors • Include parents in the problem-solving process through conferences, home visits, and frequent communication
Attendance	• Recognize and reinforce attendance by giving incentives or verbal praise • Emphasize the importance of attendance by letting the student know that he or she was missed when he or she was absent	• Encourage the student's desire to be in school by planning activities that are likely to be enjoyable, giving the student a preferred responsibility to be performed in class, and involving the student in extracurricular activities • Schedule a problem-solving meeting with parents, faculty, or both
Test-taking skills	• Prepare the student for testing by teaching ways to study in pairs, such as using flashcards, practice tests, and study guides, and by promoting adequate sleep, nourishment, and exercise • During testing, allow the student to respond orally on tape or to respond using a computer; to use	notes; to take breaks; to take the test in another location; to work without time constraints; or to take the test in several short sessions • Decrease visual distraction by improving the visual design of the test through use of larger type, spacing, consistent layout, and shorter sentences

Build critical reading skills

FEATURES HELP STUDENTS UNDERSTAND WHAT THEY READ

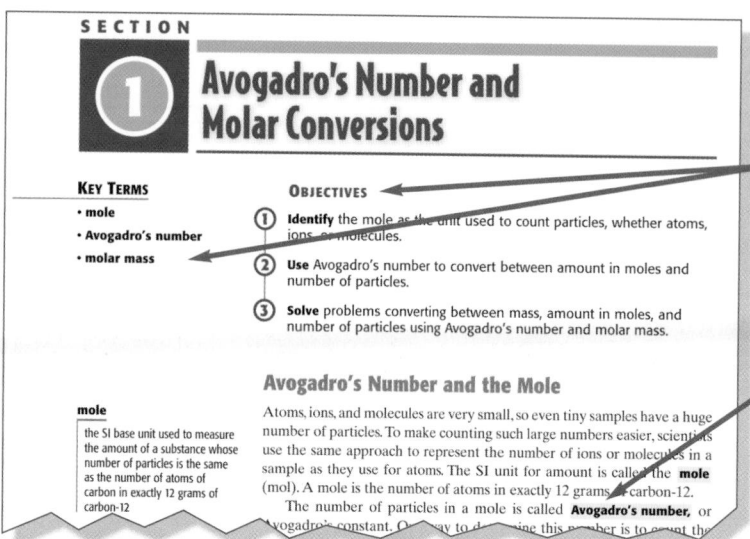

SECTION

1

Avogadro's Number and Molar Conversions

KEY TERMS
• mole
• Avogadro's number
• molar mass

OBJECTIVES

1. **Identify** the mole as the unit used to count particles, whether atoms, ions, or molecules.

2. **Use** Avogadro's number to convert between amount in moles and number of particles.

3. **Solve** problems converting between mass, amount in moles, and number of particles using Avogadro's number and molar mass.

Avogadro's Number and the Mole

mole
the SI base unit used to measure the amount of a substance whose number of particles is the same as the number of atoms of carbon in exactly 12 grams of carbon-12

Atoms, ions, and molecules are very small, so even tiny samples have a huge number of particles. To make counting such large numbers easier, scientists use the same approach to represent the number of ions or molecules in a sample as they use for atoms. The SI unit for amount is called the **mole** (mol). A mole is the number of atoms in exactly 12 grams of carbon-12.

The number of particles in a mole is called **Avogadro's number,** or Avogadro's constant. One way to determine this number is to count the

Pre-Reading Questions improve students' reading skills by helping them focus on upcoming text material.

Each section begins with **Objectives** and **Key Terms** that help focus student attention on the material students are about to read. Testing questions later in the section ensure understanding.

Vocabulary Words are highlighted in the text and defined in the margin when they are first used. A glossary at the back of the textbook includes all the **Key Terms.**

Reading Skill Builders throughout the teacher's wrap give tips and suggestions for improving key skills.

READING SKILL BUILDER — BASIC

Paired Summarizing Have students read the rest of this section in pairs. Have one partner read each sentence and the other partner summarize the meaning of the sentence in his or her own words. Then have them change duties. If the partner who is to summarize the sentence has trouble doing so, the student who read the sentence

Reading and Study Skills Appendix provides students with a variety of strategies to develop the skills they need to become better readers, including:

• Succeeding in Your Chemistry Class
• Making Concept Maps
• Making Power Notes
• Making Two-Column Notes
• Using the K/W/L Strategy
• Using Sequencing/Pattern Puzzles
• Other Reading Strategies
• Other Studying Strategies
• Cooperative Learning Techniques

Holt Science Skills Workshop: Reading in the Content Area targets reading skills that are specific to the comprehension of science. Students learn various skills that help them improve their reading.

Concept Review, in the *Chapter Resource Files,* includes straight recall and higher-order thinking questions to help reinforce what students have learned in a lesson.

Support for the development of critical math skills

MATHEMATICS IS INTEGRATED THROUGHOUT THE STUDENT EDITION

Skills Toolkit helps students decipher and understand important process skills, including balancing equations, setting up problems, and performing titrations.

Sample Problem provides a step-by-step approach to problem solving. Students can see the logical steps needed when approaching a chemistry problem. A **Practice Hint** helps guide students in the right direction.

Practice follows a **Sample Problem,** allowing students to immediately apply their new problem-solving skills.

Section Review provides additional problems to test student understanding of material presented in the lesson.

Chapter Review presents even more opportunities for students to practice their math skills with **Practice Problems** and **Mixed Review**.

Additional practice problems are found in the **Appendix.**

ADDITIONAL MATH PROBLEMS AND HINTS ARE INCLUDED IN THE TEACHER EDITION

Misconception Alert includes mathematical misconceptions. Student skills can be improved when this feature is used.

Skill Builder and **Teaching Tip** includes strategies to improve students' math skills.

Homework includes additional math problems that are grouped by difficulty level.

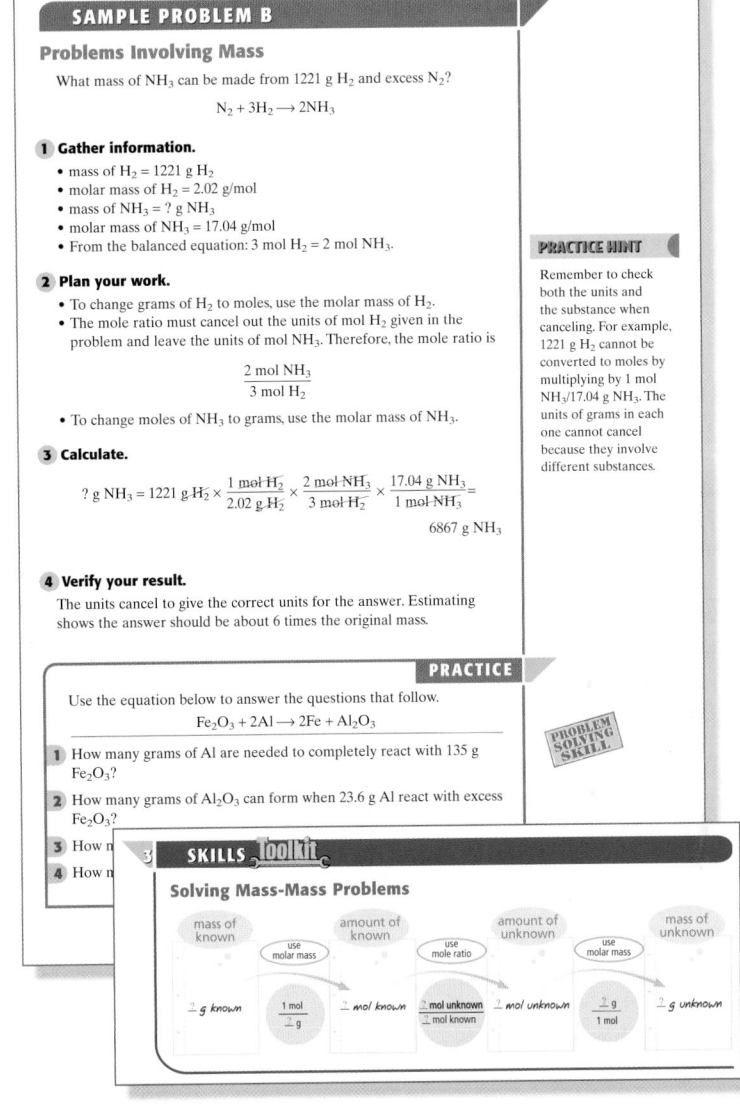

SAMPLE PROBLEM B

Problems Involving Mass

What mass of NH_3 can be made from 1221 g H_2 and excess N_2?

$$N_2 + 3H_2 \longrightarrow 2NH_3$$

1 Gather information.
- mass of H_2 = 1221 g H_2
- molar mass of H_2 = 2.02 g/mol
- mass of NH_3 = ? g NH_3
- molar mass of NH_3 = 17.04 g/mol
- From the balanced equation: 3 mol H_2 = 2 mol NH_3.

2 Plan your work.
- To change grams of H_2 to moles, use the molar mass of H_2.
- The mole ratio must cancel out the units of mol H_2 given in the problem and leave the units of mol NH_3. Therefore, the mole ratio is

$$\frac{2 \text{ mol } NH_3}{3 \text{ mol } H_2}$$

- To change moles of NH_3 to grams, use the molar mass of NH_3.

3 Calculate.

$$? \text{ g } NH_3 = 1221 \text{ g } H_2 \times \frac{1 \text{ mol } H_2}{2.02 \text{ g } H_2} \times \frac{2 \text{ mol } NH_3}{3 \text{ mol } H_2} \times \frac{17.04 \text{ g } NH_3}{1 \text{ mol } NH_3} =$$
$$6867 \text{ g } NH_3$$

4 Verify your result.
The units cancel to give the correct units for the answer. Estimating shows the answer should be about 6 times the original mass.

PRACTICE HINT

Remember to check both the units and the substance when canceling. For example, 1221 g H_2 cannot be converted to moles by multiplying by 1 mol NH_3/17.04 g NH_3. The units of grams in each one cannot cancel because they involve different substances.

PRACTICE

Use the equation below to answer the questions that follow.

$$Fe_2O_3 + 2Al \longrightarrow 2Fe + Al_2O_3$$

1 How many grams of Al are needed to completely react with 135 g Fe_2O_3?

2 How many grams of Al_2O_3 can form when 23.6 g Al react with excess Fe_2O_3?

3 How m

4 How m

SKILLS Toolkit

Solving Mass-Mass Problems

PROBLEM-SOLVING WORKBOOK

This workbook is full of detailed sample problems with extensive problem-solving strategies, as well as 630 additional practice problems.

This reference is a must for students who need extra help, reteaching, or more practice. Each portion of the workbook includes both individual sample problems with accompanying practice problems and sets of mixed review so students can practice and demonstrate their mastery of problem-solving skills.

Pacing guide

Today's chemistry classroom often requires a more flexible curriculum. *Holt Chemistry* can help you meet a variety of needs and challenges you and your students face in the classroom. The **Pacing Guide** below shows a number of ways to adapt the program to your teaching schedule.

This **Guide** can be further adapted, allowing you to mix and match or compress the material so you can spend more time on select topics, or to allow for special projects and activities.

- **General** provides the recommended course of study as indicated in the *Teacher Edition,* found in the individual chapter guides preceding each chapter.

- **Compressed** indicates how you can still cover the essentials, even in the face of drastic time constraints.

- **Basic** gives more time for the foundations of chemistry, especially mathematical problem-solving, with less emphasis on some advanced topics from later in the course.

- **Advanced** moves quickly through foundations of chemistry for students who may be comfortable with the basics, to provide additional time for advanced topics.

- **Heavy Lab/Activity** indicates ways to streamline "lecture" time to provide hands-on experience for more than a third of the blocks in the school year. (Note: even this approach does not cover all of the labs and activities that are available with *Holt Chemistry* and its *Chapter Resource Files.*)

Numbers indicate class periods recommended for the material within each chapter.	General	Compressed	Basic	Advanced	Heavy Lab/Activity
Chapter 1 *The Science of Chemistry*	10	6	11	7	10
Chapter Intro	1	–	1	–	1
1 What is Chemistry?	2	1	2	1	1
2 What is Matter?	2	2	3	2	2
3 How is Matter Classified?	2	1	2	1	1
Lab Experiment(s)	1	1	1	2	4
Chapter Review and Assessment	2	1	2	1	1
Chapter 2 *Matter and Energy*	9	5	10	6	9
Chapter Intro	1	–	1	–	1
1 Energy	2	1	2	1	1
2 Studying Matter and Energy	2	1	2	1	1
3 Measurements and Calculations in Chemistry	1	1	2	1	1
Lab Experiment(s)	1	1	1	2	4
Chapter Review and Assessment	2	1	2	1	1
Chapter 3 *Atoms and Moles*	9	6	12	9	12
Chapter Intro	1	–	1	–	1
1 Substances Are Made of Atoms	1	1	1	1	1
2 Structure of Atoms	1	1	2	1	1
3 Electron Configuration	2	2	3	2	2
4 Counting Atoms	1	–	2	1	1
Lab Experiment(s)	1	1	1	2	4
Chapter Review and Assessment	2	1	2	2	2
Chapter 4 *The Periodic Table*	10	6	10	9	11
Chapter Intro	1	–	1	–	1
1 How Are Elements Organized?	1	1	1	1	1
2 Tour of the Periodic Table	2	2	2	2	2
3 Trends of the Periodic Table	2	1	2	2	2
4 Where Did the Elements Come From?	1	–	1	1	1
Lab Experiment(s)	1	1	1	1	2
Chapter Review and Assessment	2	1	2	2	2

	General	Compressed	Basic	Advanced	Heavy Lab/Activity
Chapter 5 *Ions and Ionic Compounds*	8	4	9	6	9
Chapter Intro	1	–	1	–	1
1 Simple Ions	2	1	2	1	1
2 Ionic Bonding and Salts	2	1	2	2	2
3 Names and Formulas of Ionic Compounds	1	1	2	2	2
Lab Experiment(s)	–	–	–	–	2
Chapter Review and Assessment	2	1	2	1	1
Chapter 6 *Covalent Compounds*	8	4	10	5	8
Chapter Intro	1	–	1	–	1
1 Covalent Bonds	2	1	2	1	1
2 Drawing and Naming Molecules	2	1	4	2	3
3 Molecular Shapes	1	1	1	1	1
Lab Experiment(s)	–	–	–	–	1
Chapter Review and Assessment	2	1	2	1	1
Chapter 7 *The Mole and Chemical Composition*	9	5	12	7	7
Chapter Intro	1	–	1	–	1
1 Avogadro's Number and Molar Conversions	2	1	3	2	1
2 Relative Atomic Mass and Chemical Formulas	1	1	2	1	1
3 Formulas and Percentage Composition	2	1	3	2	1
Lab Experiment(s)	1	1	1	1	2
Chapter Review and Assessment	2	1	2	1	1
Chapter 8 *Chemical Equations and Reactions*	9	6	12	8	12
Chapter Intro	1	–	1	–	1
1 Describing Chemical Reactions	1	1	1	1	1
2 Balancing Chemical Equations	2	2	4	3	3
3 Classifying Chemical Reactions	2	2	3	2	2
4 Writing Net Ionic Equations	1	–	1	1	1
Lab Experiment(s)	–	–	–	–	3
Chapter Review and Assessment	2	1	2	1	1
Chapter 9 *Stoichiometry*	9	6	12	10	9
Chapter Intro	1	–	1	–	1
1 Calculating Quantities in Reactions	2	2	4	2	2
2 Limiting Reactants and Percentage Yield	2	1	3	2	2
3 Stoichiometry and Cars	1	1	1	2	1
Lab Experiment(s)	1	1	1	3	2
Chapter Review and Assessment	2	1	2	1	1
Chapter 10 *Causes of Change*	9	6	12	11	6
Chapter Intro	1	–	1	–	–
1 Energy Transfer	1	1	1	1	1
2 Using Enthalpy	1	1	2	2	1
3 Changes in Enthalpy During Chemical Reactions	2	2	3	3	2
4 Order and Spontaneity	1	–	2	3	–
Lab Experiment(s)	1	1	1	1	1
Chapter Review and Assessment	2	1	2	1	1
Chapter 11 *States of Matter and Intermolecular Forces*	9	4	11	10	7
Chapter Intro	1	–	1	–	–
1 States and State Changes	1	1	1	1	1
2 Intermolecular Forces	2	1	2	2	1
3 Energy of State Changes	2	1	3	3	1
4 Phase Equilibrium	1	–	2	3	–
Lab Experiment(s)	–	–	–	–	3
Chapter Review and Assessment	2	1	2	1	1

T23

	General	Compressed	Basic	Advanced	Heavy Lab/Activity
Chapter 12 *Gases*	8	4	12	8	9
Chapter Intro	1	–	1	–	1
1 Characteristics of Gases	1	1	2	1	1
2 The Gas Laws	2	1	4	3	2
3 Molecular Composition of Gases	2	1	3	3	2
Lab Experiment(s)	–	–	–	–	2
Chapter Review and Assessment	2	1	2	1	1
Chapter 13 *Solutions*	11	7	9	10	12
Chapter Intro	1	–	–	–	1
1 What Is a Solution?	1	1	1	1	1
2 Concentration and Molarity	2	2	4	2	2
3 Solubility and the Dissolving Process	2	1	1	2	2
4 Physical Properties of Solutions	2	1	1	3	2
Lab Experiment(s)	1	1	1	1	3
Chapter Review and Assessment	2	1	1	1	1
Chapter 14 *Chemical Equilibrium*	7	3	5	10	6
Chapter Intro	1	–	–	–	1
1 Reversible Reactions and Equilibrium	1	1	1	2	1
2 Systems at Equilibrium	2	–	2	4	–
3 Equilibrium Systems and Stress	1	1	1	3	1
Lab Experiment(s)	–	–	–	–	2
Chapter Review and Assessment	2	1	1	1	1
Chapter 15 *Acids and Bases*	11	7	12	13	10
Chapter Intro	1	–	1	–	1
1 What Are Acids and Bases?	2	2	2	2	2
2 Acidity, Basicity, and pH	2	2	3	3	2
3 Neutralization and Titrations	2	1	4	3	1
4 Equilibria of Weak Acids and Bases	1	–	–	3	–
Lab Experiment(s)	1	1	1	2	4
Chapter Review and Assessment	2	1	1	–	–
Chapters 16 *Reaction Rates*	8	5	5	8	7
Chapter Intro	1	–	–	–	1
1 What Affects the Rate of a Reaction?	2	1	1	3	1
2 How Can Reaction Rates be Explained?	2	2	2	3	2
Lab Experiment(s)	1	1	1	1	2
Chapter Review and Assessment	2	1	1	1	1
Chapter 17 *Oxidation, Reduction, and Electrochemistry*	10	6	7	13	10
Chapter Intro	1	–	–	–	1
1 Oxidation-Reduction Reactions	2	1	2	3	2
2 Introduction to Electrochemistry	1	1	1	2	1
3 Galvanic Cells	2	1	1	3	1
4 Electrolytic Cells	1	1	1	2	1
Lab Experiment(s)	1	1	1	2	3
Chapter Review and Assessment	2	1	1	1	1

	General	Compressed	Basic	Advanced	Heavy Lab/Activity
Chapter 18 *Nuclear Chemistry*	8	4	8	8	8
Chapter Intro	1	–	–	–	1
1 Atomic Nuclei and Nuclear Stability	1	1	1	1	1
2 Nuclear Change	2	1	3	3	2
3 Uses of Nuclear Chemistry	2	1	3	3	2
Lab Experiment(s)	–	–	–	–	1
Chapter Review and Assessment	2	1	1	1	1
Chapter 19 *Carbon and Organic Compounds*	9	0	0	10	9
Chapter Intro	1	–	–	–	1
1 Compounds of Carbon	1	–	–	2	1
2 Names and Structures of Organic Compounds	2	–	–	3	2
3 Organic Reactions	2	–	–	3	2
Lab Experiment(s)	1	–	–	1	2
Chapter Review and Assessment	2	–	–	1	1
Chapter 20 *Biological Chemistry*	8	0	0	11	8
Chapter Intro	1	–	–	1	1
1 Carbohydrates and Lipids	2	–	–	3	2
2 Proteins	1	–	–	2	1
3 Nucleic Acids	1	–	–	2	1
4 Energy in Living Systems	1	–	–	2	1
Lab Experiment(s)	–	–	–	–	1
Chapter Review and Assessment	2	–	–	1	1
TOTAL	179	94	179	179	179

Correlation to the National Science Education Standards

The following list shows the chapter correlation of **Holt Chemistry** with the National Science Education Standards (grades 9-12) for physical science content and Earth and space science content. For further detail, see the interleaf pages before each chapter.

UNIFYING CONCEPTS AND PROCESSES

Standard	Code
Systems, order, and organization	UCP 1
Evidence, models, and explanation	UCP 2
Change, consistency, and measurements	UCP 3
Evolution and equilibrium	UCP 4
Form and function	UCP 5

SCIENCE AS INQUIRY

Standard	Code
Abilities to do scientific inquiry	SAI 1
Understanding about scientific inquiry	SAI 2

SCIENCE AND TECHNOLOGY

Standard	Code
Abilities of technological design	ST 1
Understanding about science and technology	ST 2

HISTORY AND NATURE OF SCIENCE

Standard	Code
Science as a human endeavor	HNS 1
Nature of science	HNS 2
History of science	HNS 3

SCIENCE IN PERSONAL AND SOCIAL PERSEPECTIVES

Standard	Code
Personal health	SPSP 1
Populations, resources, and environments	SPSP 2
Natural hazards	SPSP 3
Risks and benefits	SPSP 4
Science and technology in society	SPSP 5

PHYSICAL SCIENCE CONTENT STANDARDS

Standard	Code	Chapter Correlation
Structure of Atoms		
Matter is made of minute particles called atoms, and atoms are composed of even smaller components. These components have measurable properties, such as mass and electrical charge. Each atom has a positively charged nucleus surrounded by negatively charged electrons. The electric force between the nucleus and electrons holds the atom together.	**PS 1a**	Chapter 1 Chapter 3
The atom's nucleus is composed of protons and neutrons, which are much more massive than electrons. When an element has atoms that differ in the number of neutrons, these atoms are called different isotopes of the element.	**PS 1b**	Chapter 3
The nuclear forces that hold the nucleus of an atom together, at nuclear distances, are usually stronger than the electric forces that would make it fly apart. Nuclear reactions convert a fraction of the mass of interacting particles into energy, and they can release much greater amounts of energy than atomic interactions. Fission is the splitting of a large nucleus into smaller pieces. Fusion is the joining of two nuclei at extremely high temperature and pressure, and is the process responsible for the energy of the sun and other stars.	**PS 1c**	Chapter 18
Radioactive isotopes are unstable and undergo spontaneous nuclear reactions, emitting particles and/or wavelike radiation. The decay of any one nucleus cannot be predicted, but a large group of identical nuclei decay at a predictable rate. This predictability can be used to estimate the age of materials that contain radioactive isotopes.	**PS 1d**	Chapter 18
Structure and Properties of Matter		
Atoms interact with one another by transferring or sharing electrons that are furthest from the nucleus. These outer electrons govern the chemical properties of the element.	**PS 2a**	Chapter 5 Chapter 6
An element is composed of a single type of atom. When elements are listed in order according to the number of protons (called the atomic number), repeating patterns of physical and chemical properties identify families of elements with similar properties. This "Periodic Table" is a consequence of the repeating pattern of outermost electrons and their permitted energies.	**PS 2b**	Chapter 1 Chapter 3 Chapter 4
Bonds between atoms are created when electrons are paired up by being transferred or shared. A substance composed of a single kind of atom is called an element. The atoms may be bonded together into molecules or crystalline solids. A compound is formed when two or more kinds of atoms bind together chemically.	**PS 2c**	Chapter 5 Chapter 7
The physical properties of compounds reflect the nature of the interactions among its molecules. These interactions are determined by the structure of the molecule, including the constituent atoms and the distances and angles between them.	**PS 2d**	Chapter 5 Chapter 6 Chapter 11 Chapter 13

PHYSICAL SCIENCE CONTENT STANDARDS, *continued*

Standard	Code	Chapter Correlation
Structure and Properties of Matter, *continued*		
Solids, liquids, and gases differ in the distances and angles between molecules or atoms and therefore the energy that binds them together. In solids the structure is nearly rigid; in liquids molecules or atoms move around each other but do not move apart; and in gases molecules or atoms move almost independently of each other and are mostly far apart.	**PS 2e**	Chapter 11 Chapter 12 Chapter 13
Carbon atoms can bond to one another in chains, rings, and branching networks to form a variety of structures, including synthetic polymers, oils, and the large molecules essential to life.	**PS 2f**	Chapter 19 Chapter 20
Chemical Reactions		
Chemical reactions occur all around us, for example in health care, cooking, cosmetics, and automobiles. Complex chemical reactions involving carbon-based molecules take place constantly in every cell in our bodies.	**PS 3a**	Chapter 1 Chapter 8 Chapter 9 Chapter 14 Chapter 17 Chapter 19 Chapter 20
Chemical reactions may release or consume energy. Some reactions such as the burning of fossil fuels release large amounts of energy by losing heat and by emitting light. Light can initiate many chemical reactions such as photosynthesis and the evolution of urban smog.	**PS 3b**	Chapter 1 Chapter 8 Chapter 10 Chapter 17
A large number of important reactions involve the transfer of either electrons (oxidation/reduction reactions) or hydrogen ions (acid/base reactions) between reacting ions, molecules, or atoms. In other reactions, chemical bonds are broken by heat or light to form very reactive radicals with electrons ready to form new bonds. Radical reactions control many processes such as the presence of ozone and greenhouse gases in the atmosphere, burning and processing of fossil fuels, the formation of polymers, and explosions.	**PS 3c**	Chapter 15 Chapter 17
Chemical reactions can take place in time periods ranging from the few femtoseconds (10^{-15} seconds) required for an atom to move a fraction of a chemical bond distance to geologic time scales of billions of years. Reaction rates depend on how often the reacting atoms and molecules encounter one another, on the temperature, and on the properties—including shape—of the reacting species.	**PS 3d**	Chapter 16
Catalysts, such as metal surfaces, accelerate chemical reactions. Chemical reactions in living systems are catalyzed by protein molecules called enzymes.	**PS 3e**	Chapter 16 Chapter 20
Motion and Forces	**PS 4**	
Conservation of Energy and the Increase in Disorder		
The total energy of the universe is constant. Energy can be transferred by collisions in chemical and nuclear reactions, by light waves and other radiations, and in many other ways. However, it can never be destroyed. As these transfers occur, the matter involved becomes steadily less ordered.	**PS 5a**	Chapter 2 Chapter 18

Standard	Code	Correlation
Conservation of Energy and the Increase in Disorder, *continued*		
All energy can be considered to be either kinetic energy, which is the energy of motion; potential energy, which depends on relative position; or energy contained by a field, such as electromagnetic waves.	**PS 5b**	Chapter 10 Chapter 12
Heat consists of random motion and the vibrations of atoms, molecules, and ions. The higher the temperature, the greater the atomic or molecular motion.	**PS 5c**	Chapter 2 Chapter 10 Chapter 12
Everything tends to become less organized and less orderly over time. Thus, in all energy transfers, the overall effect is that the energy is spread out uniformly. Examples are the transfer of energy from hotter to cooler objects by conduction, radiation, or convection and the warming of our surroundings when we burn fuels.	**PS 5d**	Chapter 10
Interactions of Energy and Matter		
Waves, including sound and seismic waves, waves on water, and light waves, have energy and can transfer energy when they interact with matter.	**PS 6a**	Chapter 3
Electromagnetic waves result when a charged object is accelerated or decelerated. Electromagnetic waves include radio waves (the longest wavelength), microwaves, infrared radiation (radiant heat), visible light, ultraviolet radiation, x-rays, and gamma rays. The energy of electromagnetic waves is carried in packets whose magnitude is inversely proportional to the wavelength.	**PS 6b**	Chapter 31
Each kind of atom or molecule can gain or lose energy only in particular discrete amounts and thus can absorb and emit light only at wavelengths corresponding to these amounts. These wavelengths can be used to identify the substance.	**PS 6c**	Chapter 3
In some materials, such as metals, electrons flow easily, whereas in insulating materials such as glass they can hardly flow at all. Semiconducting materials have intermediate behavior. At low temperatures some materials become superconductors and offer no resistance to the flow of electrons.	**PS 6d**	Chapter 5 Chapter 6

NATIONAL SCIENCE EDUCATION STANDARDS

Safety in Your Laboratory

Risk Assessment

MAKING YOUR LABORATORY A SAFE PLACE TO WORK AND LEARN

Concern for safety must begin before any activity in the classroom and before students enter the lab. A careful review of the facilities should be a basic part of preparation for each school term. You should investigate the physical environment, identify any safety risks, and inspect your work areas for compliance with safety regulations.

The review of the lab should be thorough, and all safety issues must be addressed immediately. Keep a file of your review, and add to the list each year. This will allow you to continue to raise the standard of safety in your lab and classroom.

Many classroom experiments, demonstrations, and other activities are classics that have been used for years. This familiarity may lead to a comfort that can obscure inherent safety concerns. Review all experiments, demonstrations, and activities for safety concerns before presenting them to the class. Identify and eliminate potential safety hazards.

1. **Identify the Risks**
   Before introducing any activity, demonstration, or experiment to the class, analyze it and consider what could possibly go wrong. Carefully review the list of materials to make sure they are safe. Inspect the equipment in your lab or classroom to make sure it is in good working order. Read the procedures to make sure they are safe. Record any hazards or concerns you identify.

2. **Evaluate the Risks**
   Minimize the risks you identified in the last step without sacrificing learning. Remember that no activity you perform in the lab or classroom is worth risking injury. Thus, extremely hazardous activities, or those that violate your school's policies, must be eliminated. For activities that present smaller risks, analyze each risk carefully to determine its likelihood. If the pedagogical value of the activity does not outweigh the risks, the activity must be eliminated.

3. **Select Controls to Address Risks**
   Even low-risk activities require controls to eliminate or minimize the risks. Make sure that in devising controls you do not substitute an equally or more hazardous alternative. Some control methods include the following:

   - Explicit verbal and written warnings may be added or posted.

   - Equipment may be rebuilt or relocated, have parts replaced, or be replaced entirely by safer alternatives.

   - Risky procedures may be eliminated.

   - Activities may be changed from student activities to teacher demonstrations.

4. **Implement and Review Selected Controls**
   Controls do not help if they are forgotten or not enforced. The implementation and review of controls should be as systematic and thorough as the initial analysis of safety concerns in the lab and laboratory activities.

SOME SAFETY RISKS AND PREVENTATIVE CONTROLS

The following list describes several possible safety hazards and controls that can be implemented to resolve them. This list is not complete, but it can be used as a starting point to identify hazards in your laboratory.

Identified risk	Preventative control
Facilities and Equipment	
Lab tables are in disrepair, room is poorly lighted and ventilated, faucets and electrical outlets do not work or are difficult to use because of their location.	Work surfaces should be level and stable. There should be adequate lighting and ventilation. Water supplies, drains, and electrical outlets should be in good working order. Any equipment in a dangerous location should not be used; it should be relocated or rendered inoperable.
Wiring, plumbing, and air circulation systems do not work or do not meet current specifications.	Specifications should be kept on file. Conduct a periodic review of all equipment, and document compliance. Damaged fixtures must be labeled as such and must be repaired as soon as possible.
Eyewash fountains and safety showers are present but no one knows anything about their specifications.	Ensure that eyewash fountains and safety showers meet the requirements of the ANSI standard (Z358.1).
Eyewash fountains are checked and cleaned once at the beginning of each school year. No records are kept of routine checks and maintenance on the safety showers and eyewash fountains.	Flush eyewash fountains for 5 min. every month to remove any bacteria or other organisms from pipes. Test safety showers (measure flow in gallons per min.) and eyewash fountains every 6 months and keep records of the test results.
Labs are conducted in multipurpose rooms, and equipment from other courses remains accessible.	Only the items necessary for a given activity should be available to students. All equipment should be locked away when not in use.
Students are permitted to enter or work in the lab without teacher supervision.	Lock all laboratory rooms whenever a teacher is not present. Supervising teachers must be trained in lab safety and emergency procedures.
Safety equipment and emergency procedures	
Fire and other emergency drills are infrequent, and no records or measurements are made of the results of the drills.	Always carry out critical reviews of fire or other emergency drills. Be sure that plans include alternate routes. Don't wait until an emergency to find the flaws in your plans.
Emergency evacuation plans do not include instructions for securing the lab in the event of an evacuation during a lab activity.	Plan actions in case of emergency: establish what devices should be turned off, which escape route to use, and where to meet outside the building.
Fire extinguishers are in out-of-the-way locations, not on the escape route.	Place fire extinguishers near escape routes so that they will be of use during an emergency.
Fire extinguishers are not maintained. Teachers are not trained to use them.	Document regular maintenance of fire extinguishers. Train supervisory personnel in the proper use of extinguishers. Instruct students not to use an extinguisher but to call for a teacher.

Identified risk	Preventative control
Safety equipment and emergency procedures, *continued*	
Teachers in labs and neighboring classrooms are not trained in CPR or first aid.	Teachers should receive training. The American Red Cross and other groups offer training. Certifications should be kept current with frequent refresher courses.
Teachers are not aware of their legal responsibilities in case of an injury or accident.	Review your faculty handbook for your responsibilities regarding safety in the classroom and laboratory. Contact the legal counsel for your school district to find out the extent of their support and any rules, regulations, or procedures you must follow.
Emergency procedures are not posted. Emergency numbers are kept only at the switchboard or main office. Instructions are given verbally only at the beginning of the year.	Emergency procedures should be posted at all exits and near all safety equipment. Emergency numbers should be posted at all phones, and a script should be provided for the caller to use. Emergency procedures must be reviewed periodically, and students should be reminded of them at the beginning of each activity.
Spills are handled on a case-by-case basis and are cleaned up with whatever materials happen to be on hand.	Have the appropriate equipment and materials available for cleaning up; replace them before expiration dates. Make sure students know to alert you to spilled chemicals, blood, and broken glass.
Work habits and environment	
Safety wear is only used for activities involving chemicals or hot plates.	Aprons and goggles should be worn in the lab at all times. Long hair, loose clothing, and loose jewelry should be secured.
There is no dress code established for the laboratory; students are allowed to wear sandals or open-toed shoes.	Open-toed shoes should never be worn in the laboratory. Do not allow any footwear in the lab that does not cover feet completely.
Students are required to wear safety gear but teachers and visitors are not.	Always wear safety gear in the lab. Keep extra equipment on hand for visitors.
Safety is emphasized at the beginning of the term but is not mentioned later in the year.	Safety must be the first priority in all lab work. Students should be warned of risks and instructed in emergency procedures for each activity.
There is no assessment of students' knowledge and attitudes regarding safety.	Conduct frequent safety quizzes. Only students with perfect scores should be allowed to work in the lab.
You work alone during your preparation period to organize the day's labs.	Never work alone in a science laboratory or a storage area.
Safety inspections are conducted irregularly and are not documented. Teachers and administrators are unaware of what documentation will be necessary in case of a lawsuit.	Safety reviews should be frequent and regular. All reviews should be documented, and improvements must be implemented immediately. Contact legal counsel for your district to make sure your procedures will protect you in case of a lawsuit.

Identified risk	Preventative control
Purchasing, storing, and using chemicals	
The storeroom is too crowded, so you decide to keep some equipment on the lab benches.	Do not store reagents or equipment on lab benches and keep shelves organized. Never place reactive chemicals (in bottles, beakers, flasks, wash bottles, etc.) near the edges of a lab bench.
You prepare solutions from concentrated stock to save money.	Reduce risks by ordering diluted instead of concentrated substances.
You purchase plenty of chemicals to be sure that you won't run out or to save money.	Purchase chemicals in class-size quantities. Do not purchase or have on hand more than one year's supply of each chemical.
You don't generally read labels on chemicals when preparing solutions for a lab, because you already know about a chemical.	Read each label to be sure it states the hazards and describes the precautions and first aid procedures (when appropriate) that apply to the contents in case someone else has to deal with that chemical in an emergency.
You never read the Material Safety Data Sheets (MSDSs) that come with your chemicals.	Always read the Material Safety Data Sheet (MSDS) for a chemical before using it and follow the precautions described. File and organize MSDSs for all chemicals where they can be found easily in case of an emergency.
The main stockroom contains chemicals that haven't been used for years.	Do not leave bottles of chemicals unused on the shelves of the lab for more than one week or unused in the main stockroom for more than one year. Dispose of or use up any leftover chemicals.
No extra precautions are taken when flammable liquids are dispensed from their containers.	When transferring flammable liquids from bulk containers, ground the container, and before transferring to a smaller metal container, ground both containers.
Students are told to put their broken glass and solid chemical wastes in the trash can.	Have separate containers for trash, for broken glass, and for different categories of hazardous chemical wastes.
You store chemicals alphabetically instead of by hazard class. Chemicals are stored without consideration of possible emergencies (fire, earthquake, flood, etc.), which could compound the hazard.	Use MSDSs to determine which chemicals are incompatible. Store chemicals by the hazard class indicated on the MSDS. Store chemicals that are incompatible with common fire-fighting media like water (such as alkali metals) or carbon dioxide (such as alkali and alkaline-earth metals) under conditions that eliminate the possibility of a reaction with water or carbon dioxide if it is necessary to fight a fire in the storage area.
Corrosives are kept above eye level, out of reach from anyone who is not authorized to be in the storeroom.	Always store corrosive chemicals on shelves below eye level. Remember, fumes from many corrosives can destroy metal cabinets and shelving.
Chemicals are kept on the stockroom floor on the days that they will be used so that they are easy to find.	Never store chemicals or other materials on floors or in the aisles of the laboratory or storeroom, even for a few minutes.

Safety Symbols and Safety Guidelines for Students

EYE PROTECTION

- Wear safety goggles, and know where the eyewash station is located and how to use it.
- Swinging objects can cause serious injury.
- Avoid directly looking at a light source, as this may cause permanent eye damage.

HAND SAFETY

- Wear latex or nitrile gloves to protect yourself from chemicals in the lab.
- Use a hot mitt to handle resistors, light sources, and other equipment that may be hot. Allow equipment to cool before handling it and storing it.

CLOTHING PROTECTION

- Wear a laboratory apron to protect your clothing.
- Tie back long hair, secure loose clothing, and remove loose jewelry to prevent their getting caught in moving parts or coming in contact with chemicals.

HEATING SAFETY

- When using a Bunsen burner or a hot plate, always wear safety goggles and a laboratory apron to protect your eyes and clothing. Tie back long hair, secure loose clothing, and remove loose jewelry.
- Never leave a hot plate unattended while it is turned on.
- If your clothing catches on fire, walk to the emergency lab shower, and use the shower to put out the fire.
- Wire coils may heat up rapidly during experiments. If heating occurs, open the switch immediately, and handle the equipment with a hot mitt.
- Allow all equipment to cool before storing it.

CHEMICAL SAFETY

- Do not eat or drink anything in the lab. Never taste chemicals.
- If a chemical gets on your skin or clothing or in your eyes, rinse it immediately with lukewarm water, and alert your teacher.
- If a chemical is spilled, tell your teacher, but do not clean it up yourself unless your teacher says it is OK to do so.

ELECTRICAL SAFETY

- Never close a circuit until it has been approved by your teacher. Never rewire or adjust any element of a closed circuit.
- Never work with electricity near water; be sure the floor and all work surfaces are dry.
- If the pointer of any kind of meter moves off the scale, open the circuit immediately by opening the switch.
- Light bulbs or wires that are conducting electricity can become very hot.
- Do not work with any batteries, electrical devices, or magnets other than those provided by your teacher.

GLASSWARE SAFETY

- If a thermometer breaks, notify your teacher immediately.
- Do not heat glassware that is broken, chipped, or cracked. Always use tongs or a hot mitt to handle heated glassware and other equipment because it does not always look hot when it is hot. Allow the equipment to cool before storing it.
- If a piece of glassware breaks, do not pick it up with your bare hands. Place broken glass in a specially designated disposal container.
- If a light bulb breaks, notify your teacher immediately. Do not remove broken bulbs from sockets.

WASTE DISPOSAL

- Use a dustpan, brush, and heavy gloves to carefully pick up broken glass, and dispose of it in a container specifically provided for this purpose.
- Dispose of any chemical waste only as instructed by your teacher.

HYGIENIC CARE

- Keep your hands away from your face and mouth.
- Always wash your hands thoroughly when you are done with an experiment.

Master Materials List

The following lists give the amounts of materials needed for each student group in a class to perform all the laboratory experiments in this book. **QuickLABs** are designated by the letter Q followed by the chapter in which they appear. **Start-Up Activities** are designated by the letter S followed by the chapter in which they appear. Experiments listed in the **_Holt Chemistry_ Laboratory**

Experiments section on pp. 746–827 are designated by the chapter to which they correspond. Visit **www.hrw.com/wards** for materials ordering information specific to _Holt Chemistry_ provided by WARD's and Sargent Welch (see catalog numbers listed below). Solution preparations can be found in the _Teacher Edition_ margin for that lab and on the **_One-Stop Planner CD-ROM_**.

Chemicals and Consumable Materials	Amount Needed per Lab Group	Chapter # for Lab	WARD's Catalog Number
2-L plastic soft-drink bottle	1	S12	Local supply
8-well microscale reaction strips	2	16	WLS70014-85A
Acetic acid solution, 5% (vinegar)	315 mL	S2, S8, S15, 19	WLC94811-06
Aluminum foil	1 sheet	1, Q1, Q7	Local supply
Antacid tablet	1	S15	Local supply
Apple	1 slice	S20	Local supply
Bag, plastic, sealable	1	S8	Local supply
Baking soda	5 g	S8	Local supply
Balloon	1	S12	Local supply
Blank seating chart	1	S4	Local supply
Blue litmus paper	1 strip	S15	WLS65265-A
$CaCl_2$	5 g	3	WLC94075-06
$CaCO_3$	approx. 10 g	9	WLC94195-06
Candy-coated chocolates	1 handful	Q13	Local supply
CH_3CH_2OH (denatured)	5 mL	Q20	WLC95063-06
Chromatography paper	1 sheet	13, Q13	WLS18864
Clay	5 g	S13	Local supply
Cooking oil	20 mL	S13	Local supply
Copper wire, 18 gauge	1 m	1, Q7	WLS85135-C
Cornstarch	1/2 tsp	20	Local supply
$Cu(NO_3)_2$	7 g	2	WLC98041-04
$CuSO_4 \cdot 5H_2O$	5 g	7	WLS3652-T
Detergent, dilute	100 mL	Q11	Local supply
Distilled water		throughout	Local supply
Effervescent tablet	2	Q16	Local supply
Eggshell	1	15	Local supply
Elemental samples of Ar, C, Cu, Sn, and Pb	20 g of each	4	See Catalog
Epsom salts	1 tsp	S5	Local supply
Ethanol solution, 50%	3 mL	19	Local supply
$FeSO_4 \cdot 7H_2O$	112 g/class	17	WLC94140-06
Filter paper	1	Q1, 2, 9	WLS33290-E
Flame test wire (nichrome)	2	3	WLS85125-C
Food coloring	2 mL	S13	Local supply
Graph paper	2 sheets	S10, S18	Local supply

Chemicals and Consumable Materials, _continued_	Amount Needed per Lab Group	Chapter # for Lab	WARD's Catalog Number
Graphite (carbon)	15 g	Q7	WL7073Y-25
Gumdrops, 4 different colors	10 of each color	Q8	Local supply
H_2O_2, 3%	5 mL	Q20	WLC94146-06
H_2SO_4 (conc.)	120 mL/class	15, 16, 17	WLC97092-06
Hard plastic	1 piece	19	Local supply
HCl (conc.)	300 mL/class	3, 10, 15, Q16	WLC97030-06
Ice	4 L	2, S11	Local supply
Iodine solution	5 drops	S20	Local supply
Iron filings	130 g	Q1, Q7	WLC94155-06
K_2SO_4	87 g/class	3	WLC94247-04
Kernels of popcorn of three brands	80	1	Local supply
KIO_3	1.07 g/class	16	WLC98082-02
$KMnO_4$	6.32 g/class	17	WLC98092-04
KNO_3	400 g/class	2	WLC98086-06
Li_2SO_4	64 g/class	3	AA36216-36
$Li_2SO_4 \cdot H_2O$	64 g/class	3	AA89817-14
Light stick	2	S16	Local supply
Liquid latex	10 mL	19	WLC94591-06
Liquid soap (detergent-free)	5 mL	S5	Local supply
Na_2CO_3	106 g/class	9	WLC94291-06
$Na_2S_2O_5$	0.05 g/class	16	WLC94322-06
Na_2SO_4	71 g/class	3	WLC94342-04
Na_4SiO_4	12 mL	19	WLC94341-06
NaCl	250 g	1, Q1, 2, 3, S3, S6, 13, Q13	WLC94298-06
NaOH pellets	112 g/class	10, 15	WLC97077-06
Note cards, 3 x 5	4	4	Local supply
Oil	5 mL	1	Local supply
Onion extract	5 mL	Q20	Local supply
Paper	1 piece	S3	Local supply
Paper bags	5	Q20	Local supply
Paper cups, 5 oz	2	19	Local supply
Paper plate	1	Q20	Local supply
Paper plate, small	3	20	Local supply
Paper towels	1 roll/class	9, 19	Local supply

Chemicals and Consumable Materials, *continued*	Amount Needed per Lab Group	Chapter # for Lab	WARD's Catalog Number
Paraffin wax	1/4 tsp	S6	WLC94639-06
Pepper	3 g	S3	Local supply
Periodic table, blank	1	4	Local supply
pH paper	1 strip	15	WLS65255-C
Phenolphthalein solution	12 g/class	15	WLC94654-02
Plastic bag, sealable	1	Q1	Local supply
Plastic bags	2	2	Local supply
Plastic pipets	15	7, 15, 16, Q20, S15	WLS69695
Plastic-foam cup	1	S6	Local supply
Plastic-foam cups (or calorimeter)	2	10	Local supply
Potato	4	Q17, S20, Q20	Local supply
Red cabbage	1	Q15	Local supply
Rock salt	40 g	2	Local supply
Rubber band	1	12	Local supply
Sand	1 cup	Q16	Local supply
Soft plastic	1 piece	S19	Local supply
Spoon, plastic	1	S3	Local supply
$SrCl_2 \cdot 6H_2O$	120 g	3, 9	WLC94365-06
Steel wool pad, one half	1	S2	Local supply
Sucrose	13 g	S13, S20	Local supply
Sulfur powder	80 g	Q1, Q7	WLC95145-06
Toothpicks	82	Q8, Q11	Local supply
Turkey	1 slice	S20	Local supply
Various household products	4	Q15	Local supply
Water	5 L	14	Local supply
Water bath, about 10°C	1	16	Local supply
Water bath, about 50°C	1	16	Local supply
Water-soluble starch	3.2 g	S13, 15	Local supply
Wooden splints (optional)	4	3	Local supply
Zinc strip, 1 cm x 5 cm	1	Q17	WLS8600-A

Equipment and Reusable Materials	Amount Needed per Lab Group	Chapter # for Lab	WARD's Catalog Number
Alligator clips	2	17	WLS31121-25B
Balance	1	1, Q1, 2, 7, 9, 10, 15, S7, S8, S9	WLS3448
Battery, 1.5 V	2	S17	Local supply
Beads, large	see instructions	S7	Local supply
Beads, small	200	S7	Local supply
Beaker tongs	1	9	WLS82105
Beaker, 100 mL	1	15	WLS4678-HH
Beaker, 150 mL	4	Q13, S15, S20	WLS4678-JJ

Equipment and Reusable Materials, *continued*	Amount Needed per Lab Group	Chapter # for Lab	WARD's Catalog Number
Beaker, 2 L (or plastic bucket)	1	19	WLS4684-K
Beaker, 250 mL	5	1, 2, 3, 9, 13, 15, Q15, Q16, S10, S11, S13	WLS4678-KK
Beaker, 400 mL	1	Q15, 17, Q20	WLS4678-LL
Beaker, 50 mL	3	Q1	WLS4678-GG
Beakers or other containers for holding beads	2	7	Local supply
Blender	1	Q15, 17, Q20	Local supply
Bolts	5	9	Local supply
Bottle, 50 mL (or small Erlenmeyer flask)	1	15	WLS8625-B
Bowl, large	1	Q16	Local supply
Büchner funnel	1	9	WLS35555-C
Bunsen burner and related equipment	1	1, Q1, 3, 7,	WLS11705, WLS13110-10, WLS1321-A
Buret	2	15, 17	WLS10627-C
Buret clamp	1	15, 17	WLS4727-D
CBL-2 pH probe (optional)	1	15	WLS13272-B
CBL-2 temperature probe (optional)	1	10	WLS13272-FF
CBL-2 unit (optional)	1	10, 15	WLS13272
Cobalt glass plate	1	3	WIS69945-B
Common household products	5	Q2	Local supply
Conductivity tester	1	S6	WLS29761-53A
Crucible and cover	2	7	WLS23687-E
Crucible tongs	1	1, 3, 7	WLS82115
Dessicator	1	7, 15 (optional)	WLS25137-A
Drying oven	1/class	9, 15	WLS64077-A
Earphone	1	Q17	Local supply
Erlenmeyer flask, 125 mL	4	17	WLS34107-DD
Erlenmeyer flask, 250 mL	1	15	WLS34107-FF
Evaporating dish	1	1	WLS25505-C
Eyedropper	1	S20	Local supply
Film canister, plastic with lid	1	S8, S10	Local supply
Filter flask	1	2	WLS34365-D
Flashlight	1	S13	Local supply
Forceps	1	Q11, 15	WLS35155-20
Funnel	1	2, 9	WLS35308-C
Funnels (small-, medium-, and large-bore)	3	Q16	WLS35308-A, WLS35308-C, WLS35308-E
Glass test plate or a microchemistry plate with wells, 7 cm x 15 cm	1	3	WLS69945-C
Graduated beakers	5	Q7	WLS4675-KK

T37

Equipment and Reusable Materials, *continued*	Amount Needed per Lab Group	Chapter # for Lab	WARD's Catalog Number
Graduated cylinder, 10 mL	1	13, 15, 19	WLS24667-BB
Graduated cylinder, 100 mL	1	1, 2, 9, 10, 17	WLS24667-EE
Graduated cylinder, 25 mL	1	19	WLS24667-CC
Gypsum sample	5 g	7	Local supply
Heat-resistant mat	1	1	Local supply
Hot plate	1	S11, 13, Q20	WLS41002
Hot plate or bunsen burner	1	2	WLS41002
Jar, large, with lid	1	2	Local supply
Light-emitting diode	1	17	Local supply
Magnet, pair	1	1	CP32950-10
Magnifying glass	1	19	Local supply
Markers, colored	1 set	13	Local supply
Meterstick	1	19	WLS44685
Mortar and pestle	1	7, 15, S15	WLS62250-B, WLS62251-B
Needle	1	Q11, 15	Local supply
Nuts	8	S9	Local supply
Paper clips	1	13	Local supply
Pencil	1	13	Local supply
Pennies	at least 20	S18	Local supply
Petri dish with lid	1	Q11	WLS26028-25
Pipe-stem triangle	1	7, 9	WLS82415-B
Pipet, micro-tip	7	15, Q20	WLS60684-33
Plaster of Paris sample	5 g	7	WLC95095-06
Plastic beaker, 250 mL	2	S14	WLS4677-03
Plastic cup	1	S18	Local supply
Plate, glass	1	Q11	Local supply
Plate, plastic	1	Q11	Local supply
Plate, steel	1	Q11	Local supply
Rectangular plastic bucket	2	S14	Local supply
Ring	2	1, 2, 7, 9, Q16	WLS73405-B
Ring stand	1	1, Q1, 2, 7, 9, 15, Q16, S11	WLS78306-A
Rubber band	1	S20	Local supply
Rubber policeman	1	2, 9	WLS73215-A
Ruler, metric	1	Q1, 13	WLS44625-10
Scissors	1	Q1, 13	WLS74367-C
Spatula	1	1, 2, 7, 9, 10, S6	WLS75289-A
Spectroscope	1	3	CP20105-00
Stirring rod, glass	1	Q1, 2, 7, 9, 10, Q20, S6, S11, S15	WLS40097-B

Equipment and Reusable Materials, *continued*	Amount Needed per Lab Group	Chapter # for Lab	WARD's Catalog Number
Stopwatch	1	Q16, 16, S7, S10, 15 (optional)	WLA5615
Teaspoon	1	S20	Local supply
Test tube	6	1, 15, Q20, S5, S6, S13	WLS79515-C
Test tube cork	1	S5	Local supply
Test tube rack	1	15, Q20, S5, S6, S13	WLS79065
Thermometer clamp	1	S11	WLS19425
Thermometer, nonmercury	1	2, 10, S2, S10	WLS880008-E
TI Graphing calculator with link cable	1	10, 15	WLS313283-P
Tongs	1	Q20	WLS82120
Towel	1	S3	Local supply
Tray, tub, or pneumatic trough	1	2	WLS82580
Tubing	1	2	WLS73575-F
Various objects	may vary	S1	Local supply
Wash bottle	1	9, 15, 17	WLS9486-07C
Watch glass	1	Q1, 10, S6	WLS83605-D
Weighing paper	1	1, 7, 15	WLS65223-10A
White paper or white background	1	15	Local supply
Wire gauze	1	1, 2	WLS85335-B
Wires, insulated copper with ends stripped	4	S17	Local supply
Wooden stick	1	19	WLS75943
Wool cloth	1	S3	Local supply

Safety Equipment	Amount Needed per Lab Group	Chapter # for Lab	WARD's Catalog Number
Safety gloves	1/person	throughout	WLS40273-H
Laboratory apron	1/person	throughout	WLS955-B
Safety goggles	1/person	throughout	WLS40380-03B
Gloves, nitrile	teacher use (solution preparation & cleanup)	throughout	WLS40305-05F

HOLT

Chemistry

AUTHORS

R. Thomas Myers, Ph.D.
Professor Emeritus of Chemistry
Kent State University
Kent, Ohio

Keith B. Oldham, D.Sc.
Professor Emeritus of Chemistry
Trent University,
Peterborough, Ontario, Canada

Salvatore Tocci
Science Writer
East Hampton, New York

Teacher Edition

HOLT, RINEHART AND WINSTON

A Harcourt Education Company

Orlando • **Austin** • New York • San Diego • Toronto • London

ABOUT THE AUTHORS

R. Thomas Myers, Ph.D.
Dr. Myers received his B.S. and Ph.D. in chemistry from West Virginia University in Morgantown, West Virginia. He was an assistant professor of chemistry and department head at Waynesburg College in Waynesburg, Pennsylvania, and an assistant professor at the Colorado School of Mines in Golden, Colorado. He then joined the chemistry faculty at Kent State University in Kent, Ohio, where he is currently a professor emeritus of chemistry.

Keith B. Oldham, D.Sc.
Dr. Oldham received his B.Sc. and Ph.D. in chemistry from the University of Manchester in Manchester, England and performed postdoctoral research at the Noyes Chemical Laboratory at the University of Illinois in Urbana, Illinois. He was awarded a D.Sc. from the University of Manchester for his novel research in the area of electrode processes. He was an assistant lecturer of chemistry at the Imperial College in London and a lecturer in chemistry at the University of Newcastle upon Tyne. Dr. Oldham worked as a scientist for the North American Rockwell Corporation where he performed research for NASA. After 24 years on the faculty, he is now a professor emeritus at Trent University in Peterborough, Canada.

Salvatore Tocci
Salvatore Tocci received his B.A from Cornell University in Ithaca, New York and a Master of Philosophy from the City University of New York in New York City. He was a science teacher and science department chairperson at East Hampton High School in East Hampton, New York, and an adjunct instructor at Syracuse University in Syracuse, New York. He was also an adjunct lecturer at the State University of New York at Stony Brook and a science teacher at Southold High School in Southold, New York. Mr. Tocci is currently a science writer and educational consultant.

CONTRIBUTING WRITERS

Inclusion Specialists

Joan A. Solorio
Special Education Director
Austin Independent School District
Austin, Texas

John A. Solorio
Multiple Technologies Lab Facilitator
Austin Independent School District
Austin, Texas

Lab Safety Consultant

Allen B. Cobb
Science Writer
La Grange, Texas

Lab Tester

Michelle Johnston
Trent University
Peterborough, Ontario, Canada

Teacher Edition Development

Ann Bekebrede
Science Writer
Sherborn, Massachusetts

Elizabeth M. Dabrowski
Science Department Chair
Magnificat High School
Cleveland, Ohio

Frances Jenkins
Science Writer
Sunburg, Ohio

Laura Prescott
Science Writer
Pearland, Texas

Matt Walker
Science Writer
Portland, Oregon

ACADEMIC REVIEWERS

Eric Anslyn, Ph.D.
Professor of Chemistry
Department of Chemistry and Biochemistry
The University of Texas
Austin, Texas

Paul Asimow, Ph.D.
Assistant Professor of Geology and Geochemistry
Division of Geological and Planetary Sciences
California Institute of Technology
Pasadena, California

Nigel Atkinson, Ph.D.
Associate Professor of Neurobiology
Institute for Cellular and Molecular Biology
The University of Texas
Austin, Texas

Scott W. Cowley, Ph.D.
Associate Professor
Department of Chemistry and Geochemistry
Colorado School of Mines
Golden, Colorado

Regina Frey, Ph.D.
Professor of Chemistry
Department of Chemistry
Washington University
St. Louis, Missouri

William B. Guggino, Ph.D.
Professor of Physiology
The Johns Hopkins University
Baltimore, Maryland

Joan Hudson, Ph.D.
Associate Professor of Botany
Sam Houston State University
Huntsville, Texas

Wendy L. Keeney-Kennicutt, Ph.D.
Associate Professor of Chemistry
Department of Chemistry
Texas A&M University
College Station, Texas

Samuel P. Kounaves
Associate Professor of Chemistry
Department of Chemistry
Tufts University
Medford, Massachusetts

Phillip LaRoe
Instructor
Department of Physics and Chemistry
Central Community College
Grande Island, Nebraska

Jeanne L. McHale, Ph.D.
Professor of Chemistry
College of Science
University of Idaho
Moscow, Idaho

Gary Mueller, Ph.D.
Associate Professor of Nuclear Engineering
Department of Engineering
University of Missouri
Rolla, Missouri

Brian Pagenkopf, Ph.D.
Professor of Chemistry
Department of Chemistry and Biochemistry
The University of Texas
Austin, Texas

Charles Scaife, Ph.D.
Chemistry Professor
Department of Chemistry
Union College
Schenectady, New York

Fred Seaman, Ph.D.
Research Scientist and Chemist
Department of Pharmacological Chemistry
The University of Texas
Austin, Texas

Peter Sheridan, Ph.D.
Associate Professor of Chemistry
Department of Chemistry
Colgate University
Hamilton, New York

Spencer Steinberg, Ph.D.
Associate Professor of Environmental Organic Chemistry
Department of Chemistry
University of Nevada
Las Vegas, Nevada

Continued on next page

Aaron Timperman, Ph.D.
Professor of Chemistry
Department of Chemistry
University of West Virginia
Morgantown, West Virginia

Richard S. Treptow, Ph.D.
Professor of Chemistry
Department of Chemistry
and Physics
Chicago State University
Chicago, Illinois

Martin VanDyke, Ph.D.
*Professor Emeritus of
Chemistry*
Front Range Community
College
Westminister, Colorado

Charles Wynn, Ph.D.
Chemistry Assistant Chair
Department of Physical
Sciences
Eastern Connecticut State
University
Willimantic, Connecticut

TEACHER REVIEWERS

David Blinn
Secondary Sciences Teacher
Wrenshall High School
Wrenshall, Minnesota

Robert Chandler
Science Teacher
Soddy-Daisy High School
Soddy-Daisy, Tennessee

Cindy Copolo, Ph.D.
Science Specialist
Summit Solutions
Bahama, North Carolina

Linda Culp
Science Teacher
Thorndale High School
Thorndale, Texas

Chris Diehl
Science Teacher
Belleville High School
Belleville, Michigan

Alonda Droege
Science Teacher
Seattle, Washington

Benjamen Ebersole
Science Teacher
Donnegal High School
Mount Joy, Pennsylvania

Jeffrey L. Engel
Science Teacher
Madison County High School
Athens, Georgia

Stacey Hagberg
Science Teacher
Donnegal High School
Mount Joy, Pennsylvania

Gail Hermann
Science Teacher
Quincy High School
Quincy, Illinois

Donald R. Kanner
*Physics and Chemistry
Instructor*
Lane Technical High School
Chicago, Illinois

Edward Keller
Science Teacher
Morgantown High School
Morgantown, West Virginia

Stewart Lipsky
Science Teacher
Seward Park High School
New York, New York

Mike Lubich
Science Teacher
Maple Town High School
Greensboro, Pennsylvania

Thomas Manerchia
*Environmental Science
Teacher, Retired*
Archmere Academy
Claymont, Delaware

Betsy McGrew
Science Teacher
Star Charter School
Austin, Texas

Jennifer Seelig-Fritz
Science Teacher
North Springs High School
Atlanta, Georgia

Dyanne Semerjibashian
Science Teacher
Star Charter School
Austin, Texas

Linnaea Smith
Science Teacher
Bastrop High School
Bastrop, Texas

Gabriela Waschesky, Ph.D.
*Science and Mathematics
Teacher*
Emery High School
Emeryville, California

(Credits and Acknowledgements continued on p. 908)

iv

Contents

In Brief

Contents

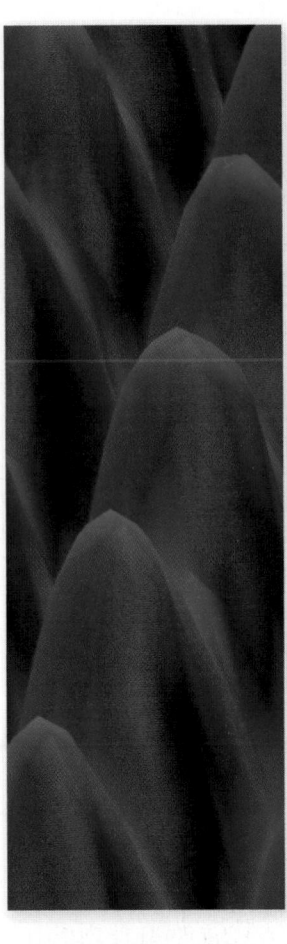

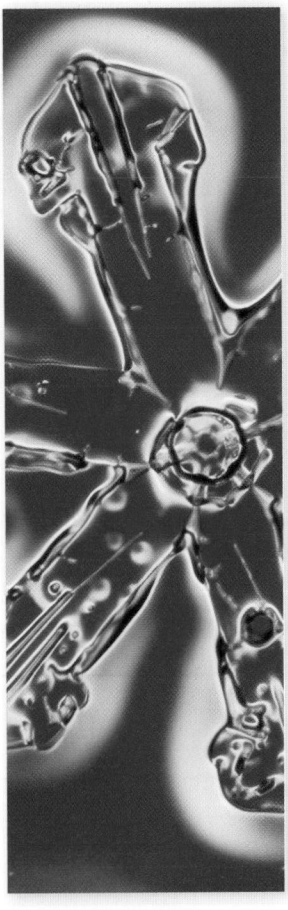

x

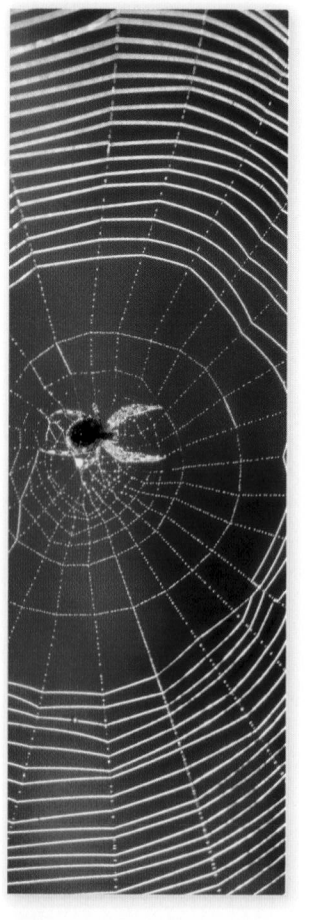

A P P E N D I C E S

LABORATORY EXPERIMENTS

 Safety in the Chemistry Laboratory **751**

SAMPLE PROBLEMS

SKILLS Toolkit

FEATURES

Science and Technology

Consumer Focus

Element Spotlight

HOW TO USE YOUR TEXTBOOK

Your Roadmap for Success with *Holt Chemistry*

Read the Objectives

Objectives tell you what you'll need to know.

STUDY TIP Reread the objectives when studying for a test to be sure you know the material.

Study the Key Terms

Key Terms are listed for each section. Learn the definitions of these terms because you will most likely be tested on them. Use the glossary to locate definitions quickly.

STUDY TIP If you don't understand a definition, reread the page where the term is introduced. The surrounding text should help make the definition easier to understand.

Take Notes and Get Organized

Keep a science notebook so that you are ready to take notes when your teacher reviews the material in class. Keep your assignments in this notebook so that you can review them when studying for the chapter test. Appendix B, located in the back of this book, describes a number of Study Skills that can help you succeed in chemistry, including several approaches to note taking.

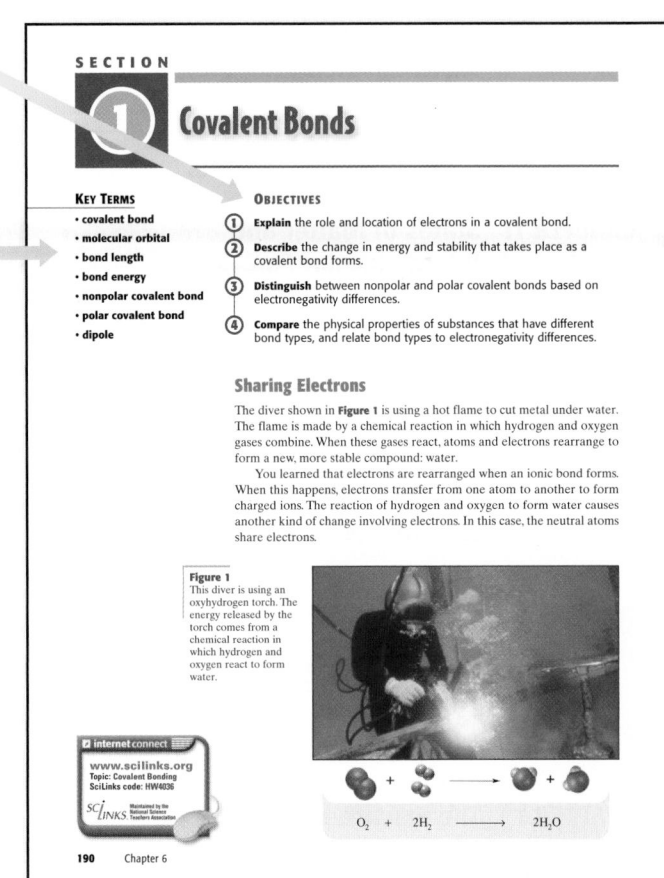

SECTION 1

Covalent Bonds

KEY TERMS
- covalent bond
- molecular orbital
- bond length
- bond energy
- nonpolar covalent bond
- polar covalent bond
- dipole

OBJECTIVES

1. **Explain** the role and location of electrons in a covalent bond.

2. **Describe** the change in energy and stability that takes place as a covalent bond forms.

3. **Distinguish** between nonpolar and polar covalent bonds based on electronegativity differences.

4. **Compare** the physical properties of substances that have different bond types, and relate bond types to electronegativity differences.

Sharing Electrons

The diver shown in **Figure 1** is using a hot flame to cut metal under water. The flame is made by a chemical reaction in which hydrogen and oxygen gases combine. When these gases react, atoms and electrons rearrange to form a new, more stable compound: water.

You learned that electrons are rearranged when an ionic bond forms. When this happens, electrons transfer from one atom to another to form charged ions. The reaction of hydrogen and oxygen to form water causes another kind of change involving electrons. In this case, the neutral atoms share electrons.

Figure 1
This diver is using an oxyhydrogen torch. The energy released by the torch comes from a chemical reaction in which hydrogen and oxygen react to form water.

internet connect
www.scilinks.org
Topic: Covalent Bonding
SciLinks code: HW4036

O_2 + $2H_2$ ⟶ $2H_2O$

190 Chapter 6

↗ Be Resourceful, Use the Web

Internet Connect boxes in your textbook take you to resources that you can use for science projects, reports, and research papers. Go to **scilinks.org,** and type in the SciLinks code to get information on a topic.

Visit go.hrw.com
Find worksheets and other materials that go with your textbook at **go.hrw.com.** Click on the textbook icon and the table of contents to see all of the resources for each chapter.

Next think about the bonds that form between carbon and chlorine and between aluminum and chlorine. The electronegativity difference between C and Cl is 0.6. These two elements form a polar covalent bond. The electronegativity difference between Al and Cl is 1.6. These two elements also form a polar covalent bond. However, the larger difference between Al and Cl means that the bond between these two elements is more polar, with greater partial charges, than the bond between C and Cl is.

Properties of Substances Depend on Bond Type
The type of bond that forms determines the physical and chemical properties of the substance. For example, metals, such as potassium, are very good electric conductors in the solid state. This property is the result of metallic bonding. Metallic bonds are the result of the attraction between the electrons in the outermost energy level of each metal atom and all of the other atoms in the solid metal. The metal atoms are held in the solid because all of the valence electrons are attracted to all of the atoms in the solid. These valence electrons can move easily from one atom to another. They are free to roam around in the solid and can conduct an electric current.

Table 3 Properties of Substances with Metallic, Ionic, and Covalent Bonds

Bond type	Metallic	Ionic	Covalent
Example substance	potassium	potassium chloride	chlorine
Melting point (°C)	63	770	–101
Boiling point (°C)	760	1500 (sublimes)	–34.6
Properties	• soft, silvery, solid • conductor as a solid	• crystalline, white solid • conductor when dissolved in water	• greenish yellow gas • not a good conductor

Covalent Compounds   197

Section Review

UNDERSTANDING KEY IDEAS

1. Describe the attractive forces and repulsive forces that exist between two atoms as the atoms move closer together.
2. Compare a bond between two atoms to a spring between two students.
3. In what two ways can two atoms share electrons when forming a covalent bond?
4. What happens in terms of energy and stability when a covalent bond forms?
5. How are the partial charges shown in a polar covalent molecule?
6. What information can be obtained by knowing the electronegativity differences between two elements?
7. Why do molecular compounds have low melting points and low boiling points relative to ionic substances?

CRITICAL THINKING

8. Why does the distance between two nuclei in a covalent bond vary?
9. How does a molecular orbital differ from an atomic orbital?
10. How does the strength of a covalent bond relate to bond length?
11. Compare the degree of polarity in HF, HCl, HBr, and HI.
12. Given that it has the highest electronegativity, can a fluorine atom ever form a nonpolar covalent bond? Explain your answer.
13. What does a small electronegativity difference reveal about the strength of a covalent bond?
14. Based on electronegativity values, which bond has the highest degree of ionic character: H—S, Si—Cl, or Cs—Br?

198   Chapter 6

Use the Illustrations and Photos

Art shows complex ideas and processes. Learn to analyze the art so that you better understand the material you read in the text.

Tables and graphs display important information in an organized way to help you see relationships.

A picture is worth a thousand words. Look at the photographs to see relevant examples of science concepts you are reading about.

Answer the Section Reviews

Section Reviews test your knowledge over the main points of the section. Critical Thinking items challenge you to think about the material in greater depth and to find connections that you infer from the text.

STUDY TIP When you can't answer a question, reread the section. The answer is usually there.

Do Your Homework

Your teacher may assign worksheets to help you understand and remember the material in the chapter.

STUDY TIP Don't try to answer the questions without reading the text and reviewing your class notes. A little preparation up front will make your homework assignments a lot easier. Answering the items in the Chapter Review will help prepare you for the chapter test.

Visit Holt Online Learning
If your teacher gives you a special password to log onto the **Holt Online Learning** site, you'll find your complete textbook on the Web. In addition, you'll find some great learning tools and practice quizzes. You'll be able to see how well you know the material from your textbook.

Visit CNN Student News
You'll find up-to-date events in science at **www.cnnstudentnews.com.**

The Science of Chemistry
Chapter Planning Guide

PACING	CLASSROOM RESOURCES	LABS, ACTIVITIES, AND DEMONSTRATIONS
BLOCK 1 · 45 min pp. 2–3 **Chapter Opener**		SE **Start-Up Activity** Classifying Matter, p. 3 ◆
BLOCKS 2 & 3 · 90 min pp. 4–9 **Section 1** What Is Chemistry?	TT **Water in Three States** * TT **Evidence of a Chemical Change** * OSP **Supplemental Reading** The Same and Not the Same "Chemistry and Industry" GENERAL OSP **Career Extension** Title GENERAL	TE **Demonstration** Changes in Matter, p. 4 ◆ TE **Demonstration** Physical Change Versus Chemical Change, p. 7 GENERAL TE **Demonstration** Evidence of Chemical Change I, p. 8 ◆ BASIC TE **Demonstration** Evidence of Chemical Change II, p. 8 ◆ GENERAL SE **Skills Practice Lab** Laboratory Techniques, p. 756 ◆ GENERAL CRF **Datasheets for In-Text Labs** * CRF **Observation Lab** Laboratory Procedures ◆ GENERAL
BLOCKS 4 & 5 · 90 min pp. 10–20 **Section 2** Describing Matter	TT **Using Conversion Factors** * TT **Comparison of Physical and Chemical Properties** * SE **Consumer Focus** Aspirin	TE **Demonstration** Measuring Volume, p. 11 GENERAL TE **Group Activity**, p. 12 BASIC TE **Group Activity** Gaining Perspective, p. 15 GENERAL TE **Demonstration** Mass, Volume, and Density, p. 17 ◆ SE **QuickLab** Thickness of Aluminum Foil, p. 18 ◆ GENERAL CRF **Datasheets for In-Text Labs** * SE **Inquiry Lab** Conservation of Mass—Percentage of Water in Popcorn, p. 760 ◆ GENERAL CRF **Datasheets for In-Text Labs** * CRF **CBL™ Probeware Lab** Analyzing Graphs and Establishing Relationships ◆ ADVANCED
BLOCKS 6 & 7 · 90 min pp. 21–29 **Section 3** How Is Matter Classified?	TT **Pure Substances** * TT **Models of a Molecule** * TT **Particle Models of Gold and Gold Alloy** * TT **Types of Mixtures** * TT **Classifying Matter** * OSP **Supplemental Reading** The Same and Not the Same "What Are You?" ADVANCED SE **Element Spotlight** Aluminum's Humble Beginnings	TE **Demonstration** Separating a Compound into Elements, p. 25 ◆ GENERAL TE **Demonstration** Separation by Distillation, p. 26 ◆ SE **QuickLab** Separating a Mixture, p. 27 ◆ GENERAL CRF **Datasheets for In-Text Labs** *

BLOCKS 8 & 9 · 90 min

Chapter Review and Assessment Resources

SE **Chapter Review,** pp. 31–34
SE **Standardized Test Prep,** p. 35
CRF **Chapter Test** *
OSP **Test Generator**
CRF **Test Item Listing** *
OSP **Scoring Rubrics and Classroom Management Checklists**

Holt Chemistry: Online Resources

Visit **go.hrw.com** for a variety of free resources related to this textbook. Enter the keyword **HW4 HOME**.

Students can access interactive problem solving help and active visual concept development with the *Holt Chemistry* Online Edition available at **www.hrw.com**.

student CNN News

cnnstudentnews.com

Find the latest chemistry news, lesson plans, and activities related to important scientific events.

PROBLEM SOLVING AND PRACTICE	SECTION REVIEW AND ASSESSMENT	STANDARDS CORRELATION
		National Science Education Standards
	TE Quiz, p. 9 GENERAL **CRF** Quiz * **TE** Reteaching, p. 9 BASIC **SE** Section Review, p. 9 **CRF** Concept Review *	PS 2e, PS 3a, PS 3b
SE Skills Toolkit 1 Using Conversion Factors, p. 13 **SE** Sample Problem A Converting Units, p. 14 **SE** Practice, p. 14 GENERAL **TE** Homework, p. 14 GENERAL **CRF** Problem Solving * ADVANCED **SE** Problem Bank, p. 858 GENERAL	**TE** Homework, p. 19 BASIC **TE** Quiz, p. 19 GENERAL **CRF** Quiz * **TE** Reteaching, p. 19 BASIC **SE** Section Review, p. 19 **CRF** Concept Review *	
	TE Homework, p. 28 BASIC **TE** Quiz, p. 28 GENERAL **CRF** Quiz * **TE** Reteaching, p. 28 BASIC **SE** Section Review, p. 28 **CRF** Concept Review *	PS 1a, PS 2b, PS 2c

www.scilinks.org

Topic: Chemicals
SciLinks code: HW4030

Topic: Chemical and Physical Changes
SciLinks code: HW4140

Topic: SI Units
SciLinks code: HW4114

Topic: Physical/Chemical Properties
SciLinks code: HW4097

Topic: Aspirin
SciLinks code: HW4012

Topic: Aluminum
SciLinks code: HW4136

Technology Resources

  Science in the NEWS

Each video segment is accompanied by a Critical Thinking Worksheet.

Segment 1
Chemical Industry Report

Segment 2
Chemicals Unknown and Untested

Segment 3
Chemicals in the Home

Segment 4
Student Superconductors

Segment 6
Cleaner Gas

 ChemFile Interactive Tutor CD-Rom

Module 1: States of Matter
Topic: Defining the States and Classes of Matter

Overview

This chapter introduces students to the study of chemistry. The students will learn about three states of matter. They will also learn about physical and chemical properties and changes in matter. The SI system of measurement is introduced along with conversion factors to convert one unit of measure to another. The concept of density is also presented. Finally, the classification of matter introduces the concepts of atoms, elements, molecules, compounds, and mixtures.

Assessing Prior Knowledge

Check for Content Knowledge

Students should be familiar with the following topics:

• solid, liquid, and gas

• evidence of changes in matter

• density, mass, and volume

Using the Figure

The art display in the photograph is a small portion of the work titled 100,000 POUNDS OF ICE AND NEON. It was constructed on the ice rink at the Tacoma Dome in the artist's hometown of Tacoma, Washington. Have students describe what they see in the image. Use their responses to introduce properties of matter. Point out that the solid water ice and the gaseous neon demonstrate two states of matter. As the ice melts, it turns into liquid water. Inform students that the properties and changes of matter are a large part of the material they will study throughout this course.

THE SCIENCE OF CHEMISTRY

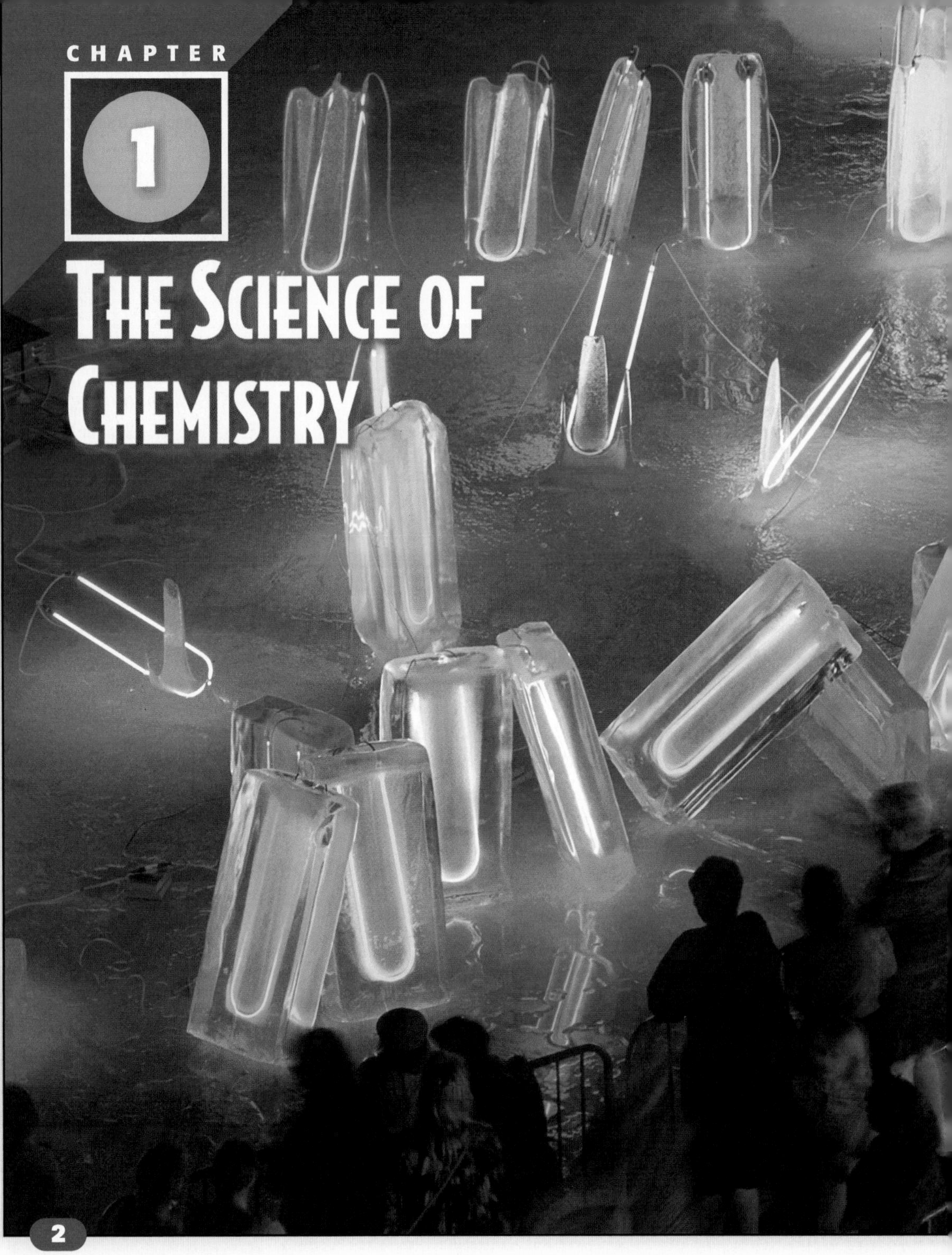

2

Standards Correlations

National Science Education Standards

PS 1a: Matter is made of minute particles called atoms. (Section 3)

PS 2b: An element is composed of a single type of atom. (Section 3)

PS 2c: A substance composed of a single kind of atom is called an element. The atoms may be bonded together into molecules or crystalline solids. A compound is formed when two or more kinds of atoms bind together chemically. (Section 3)

PS 2e: Solids, liquids, and gases differ in the distances and angles between molecules or atoms and therefore the energy that binds them together. In solids the structure is nearly rigid; in liquids molecules or atoms move around each other but do not move apart; and in gases molecules or atoms move almost independently of each other and are mostly far apart. (Section 1)

PS 3a: Chemical reactions occur all around us, for example in health care, cooking, cosmetics, and automobiles. Complex chemical reactions involving carbon-based molecules take place constantly in every cell in our bodies. (Section 1)

PS 3b: Chemical reactions may release or consume energy. Some reactions such as the burning of fossil fuels release large amounts of energy by losing heat and by emitting light. (Section 1)

For one weekend, an ice rink in Tacoma, Washington became a work of art. Thousands of people came to see the amazing collection of ice and lights on display. Huge blocks of ice, each having a mass of about 136 kg, were lit from the inside by lights. The glowing gas in each light made the solid ice shine with color. And as you can see, lights of many different colors were used in the display. In this chapter, you will learn about matter. You will learn about the properties used to describe matter. You will also learn about the changes matter can undergo. Finally, you will learn about classifying matter based on its properties.

START-UP ACTIVITY

SAFETY PRECAUTIONS

Classifying Matter

PROCEDURE

1. Examine the **objects** provided by your teacher.

2. Record in a table observations about each object's individual characteristics.

3. Divide the objects into at least three different categories based on your observations. Be sure that the objects in each category have something in common.

ANALYSIS

1. Describe the basis of your classification for each category you created.

2. Give an example that shows how using these categories makes describing the objects easier.

3. Describe a system of categories that could be used to classify matter. Explain the basis of your categories.

Pre-Reading Questions

① Do you think there are "good chemicals" and "bad chemicals"? If so, how do they differ?

② What are some of the classifications of matter?

③ What is the difference between a chemical change and a physical change?

CONTENTS 1

3

START-UP ACTIVITY

Skills Acquired:
- Classifying
- Collecting Data
- Identifying/Recognizing Patterns

Materials:

For each group of 2–3 students:
- objects, assorted (15)

Teacher's Notes:

- Present to the students a variety of objects that can be classified into at least 3 categories. Categories to consider when selecting objects include color, shape, use, material, and appearance.

- This activity will introduce students to an important aspect of chemistry: describing things and phenomena as belonging to categories.

Answers

1. Accept all reasonable answers.

2. Describing the assortment of objects would require describing every object individually. Classifying the objects into a few categories allows the objects to be described based on a few key characteristics.

3. Accept all reasonable responses. Answers may include properties such as the physical state, chemical makeup, toxicity, and electrical conductivity of the material.

Answers to Pre-Reading Questions

1. Sample answer: Chemicals can be helpful or harmful depending on their properties and on their use or misuse.

2. Matter is classified according to state, including solid, liquid, and gas. Matter is also classified according to its chemical makeup as an element, compound, or mixture.

3. A chemical change results in a change in the identity of matter. A physical change can result in a change in the appearance of the matter but not in its identity.

Overview

Before beginning this section, review with your students the Objectives listed in the Student Edition. This section introduces the study of chemistry and its presence in our everyday life. Students begin by learning about three states of matter—solid, liquid and gas. Students also learn about chemical and physical changes that matter undergoes.

🎵 Bellringer

Have the students make a list of any chemical that they can think of. Ask them which ones on that list they would like to live without and which ones they cannot live without.

Motivate

Identifying Preconceptions

Use the lists made in the Bellringer activity to identify if students think mainly of chemicals as being manufactured and hazardous. Be prepared to share examples of natural, helpful chemicals.

Demonstration

Changes in Matter

1. Put 8 drops of phenolphthalein solution in a 600 mL beaker before class. Add 200 mL of 0.25 M NaOH solution to the beaker containing the dye.

2. Now add 1 M HCl to the beaker, and stir until the solution turns clear again.

continued on next page

SECTION

What Is Chemistry?

KEY TERMS
- chemical
- chemical reaction
- states of matter
- reactant
- product

OBJECTIVES

① **Describe** ways in which chemistry is a part of your daily life.

② **Describe** the characteristics of three common states of matter.

③ **Describe** physical and chemical changes, and give examples of each.

④ **Identify** the reactants and products in a chemical reaction.

⑤ **List** four observations that suggest a chemical change has occurred.

Working with the Properties and Changes of Matter

Do you think of chemistry as just another subject to be studied in school? Or maybe you feel it is important only to people working in labs? The effects of chemistry reach far beyond schools and labs. It plays a vital role in your daily life and in the complex workings of your world.

Look at **Figure 1.** Everything you see, including the clothes the students are wearing and the food the students are eating, is made of chemicals. The students themselves are made of chemicals! Even things you cannot see, such as air, are made up of chemicals.

Chemistry is concerned with the properties of chemicals and with the changes chemicals can undergo. A **chemical** is any substance that has a definite composition—it's always made of the same stuff no matter where the chemical comes from. Some chemicals, such as water and carbon dioxide, exist naturally. Others, such as polyethylene, are manufactured. Still others, such as aluminum, are taken from natural materials.

chemical
any substance that has a defined composition

Figure 1
Chemicals make up everything you see every day.

4

📼 Videos

CNN. Presents Chemistry Connections

- **Segment 1** Chemical Industry Report
- **Segment 2** Chemicals Unknown and Untested

See the Science in the News video guide for more details.

Chapter Resource File

- **Lesson Plan**

You Depend on Chemicals Every Day

Many people think of chemicals in negative terms—as the cause of pollution, explosions, and cancer. Some even believe that chemicals and chemical additives should be banned. But just think what such a ban would mean—after all, everything around you is composed of chemicals. Imagine going to buy fruits and vegetables grown without the use of any chemicals at all. Because water is a chemical, the produce section would be completely empty! In fact, the entire supermarket would be empty because all foods are made of chemicals.

The next time you are getting ready for school, look at the list of ingredients in your shampoo or toothpaste. You'll see an impressive list of chemicals. Without chemicals, you would have nothing to wear. The fibers of your clothing are made of chemicals that are either natural, such as cotton or wool, or synthetic, such as polyester. The air you breathe, the food you eat, and the water you drink are made up of chemicals. The paper, inks, and glue used to make the book you are now reading are chemicals, too. You yourself are an incredibly complex mixture of chemicals.

Chemical Reactions Happen All Around You

You will learn in this course that changes in chemicals—or **chemical reactions**—are taking place around you and inside you. Chemical reactions are necessary for living things to grow and for dead things to decay. When you cook food, you are carrying out a chemical reaction. Taking a photograph, striking a match, switching on a flashlight, and starting a gasoline engine require chemical reactions.

Using reactions to manufacture chemicals is a big industry. **Table 1** lists the top eight chemicals made in the United States. Some of these chemicals may be familiar, and some you may have never heard of. By the end of this course, you will know a lot more about them. Chemicals produced on a small scale are important, too. Life-saving antibiotics, cancer-fighting drugs, and many other substances that affect the quality of your life are also products of the chemical industry.

chemical reaction

the process by which one or more substances change to produce one or more different substances

www.scilinks.org
Topic: Chemicals
SciLinks code: HW4030

SCiLINKS Maintained by the National Science Teachers Association

Rank	Name	Formula	Uses
1	sulfuric acid	H_2SO_4	production of fertilizer; metal processing; petroleum refining
2	ethene	C_2H_4	production of plastics; ripening of fruits
3	propylene	C_3H_6	production of plastics
4	ammonia	NH_3	production of fertilizer; refrigeration
5	chlorine	Cl_2	bleaching fabrics; purifying water; disinfectant
6	phosphoric acid (anhydrous)	P_2O_5	production of fertilizer; flavoring agent; rustproofing metals
7	sodium hydroxide	NaOH	petroleum refining; production of plastics
8	1,2-dichloroethene	$C_2H_2Cl_2$	solvent, particularly for rubber

Table 1   Top Eight Chemicals Made in the United States (by Weight)

Source: *Chemical and Engineering News.*

5

---

continued from previous page

3. Add more NaOH to the beaker until it turns pink again. Continue adding acid and base until the beaker is full.

4. Before class, gently mix 20 mL of dishwashing liquid with 100 mL of vinegar. Place 50 g of baking soda, $NaHCO_3$, into a 1000 mL graduated cylinder, and place the cylinder in a large container that will be able to catch the resulting bubbles. Add the vinegar-soap mixture, and watch the bubbles appear. Ask students what different properties they observed in each demonstration.

Safety Caution: Wear safety goggles and a lab apron. Students should be at least 3 m from the demonstration area.

Disposal: Neutralize the acid or base left over from each step, and pour it down the drain. Dissolve and flush any residue with plenty of water.

Teach

READING SKILL BUILDER — BASIC

Assimilating Knowledge Before students read this chapter, have them write a short list of all the things they know (or think they know) about the structure of matter and the changes in matter. Then have students list the things they might want to know about the nature of matter. For example, they might ask the following questions:

• What properties of liquid water change when it freezes? What properties do not change?

• How do you know that air is matter?

• How can you tell a block of aluminum from a block of steel if both are painted the same color?

• What differences account for the fact that iron rusts in air while gold does not?

Later, when students are reviewing for the chapter test, have them read over their list and correct or expand upon their answers. **LS Logical**

MISCONCEPTION ///ALERT\\\

Many students think that chemicals differ from substances and materials found in nature. Have students name their favorite foods and drinks. Point out that these foods are made up of carbohydrates, proteins, fats, water, and other components, all of which are chemicals.

One-Stop Planner CD-ROM

• **Supplemental Reading Projects**
Guided Reading Worksheet: The Same and Not the Same "Chemistry and Industry"
Assign this worksheet for cross-curricular connections to language arts.

Writing **Writing Skills** Have students research where the nearest chemical production facility is, what it produces, and why it is located where it is. Caution them not to include operations that are only storage or warehousing facilities. Have the students prepare a short essay on what they have learned. **LS Logical**

SKILL BUILDER — BASIC

Vocabulary Before introducing the states of matter, invite students to discuss their own ideas about the properties of solids, liquids, and gases. For example, ask them what they would say to a person who wanted to know the differences in the states of matter. This activity will help reveal students' misconceptions about matter. Students should then read the coverage in the text to see if their ideas match the descriptions found there. **LS Logical**

Using the Figure

Figure 2 Ask students how the models depicted in the circles help explain the different states of matter. If students express concern about the open appearance of the solid model, remind them that water is one of the few substances that expands when it freezes. In nearly any other substance, the particles would be closer together than they are in the liquid state.

Transparencies

TT Water in Three States

ChemFile **CHEMISTRY** *INTERACTIVE TUTOR*

- **Module 1: States of Matter**
  Topic: Defining the States and Classes of Matter

  This engaging tutorial reviews and reinforces the states and classes of matter through modeling and guided practice.

Physical States of Matter

states of matter

the physical forms of matter, which are solid, liquid, gas, and plasma

All matter is made of particles. The type and arrangement of the particles in a sample of matter determine the properties of the matter. Most of the matter you encounter is in one of three **states of matter:** solid, liquid, or gas. **Figure 2** illustrates water in each of these three states at the macroscopic and microscopic levels. *Macroscopic* refers to what you see with the unaided eye. In this text, *microscopic* refers to what you would see if you could see individual atoms.

The microscopic views in this book are models that are designed to show you the differences in the arrangement of particles in different states of matter. They also show you the differences in size, shape, and makeup of particles of chemicals. But don't take these models too literally. Think of them as cartoons. Atoms are not really different colors. And groups of connected atoms, or molecules, do not look lumpy. The microscopic views are also limited in that they often show only a single layer of particles whereas the particles are really arranged in three dimensions. Finally, the models cannot show you that particles are in constant motion.

Figure 2
a Below 0°C, water exists as ice. Particles in a solid are in a rigid structure and vibrate in place.

b Between 0°C and 100°C, water exists as a liquid. Particles in a liquid are close together and slide past one another.

c Above 100°C, water is a gas. Particles in a gas move randomly over large distances.

Water molecule, H_2O

Water molecule, H_2O

Water molecule, H_2O

6

Properties of the Physical States

Solids have fixed volume and shape that result from the way their particles are arranged. Particles that make up matter in the solid state are held tightly in a rigid structure. They vibrate only slightly.

Liquids have fixed volume but not a fixed shape. The particles in a liquid are not held together as strongly as those in a solid. Like grains of sand, the particles of a liquid slip past one another. Thus, a liquid can flow and take the shape of its container.

Gases have neither fixed volume nor fixed shape. Gas particles weakly attract one another and move independently at high speed. Gases will fill any container they occupy as their particles move apart.

There are other states that are beyond the scope of this book. For example, most visible matter in the universe is plasma—a gas whose particles have broken apart and are charged. Bose-Einstein condensates have been described at very low temperatures. A neutron star is also considered by some to be a state of matter.

Changes of Matter

Many changes of matter happen. An ice cube melts. Your bicycle's spokes rust. A red shirt fades. Water fogs a mirror. Milk sours. Scientists who study these and many other events classify them by two broad categories: *physical changes* and *chemical changes*.

Physical Changes

Physical changes are changes in which the identity of a substance doesn't change. However, the arrangement, location, and speed of the particles that make up the substance may change. Changes of state are physical changes. The models in **Figure 2** show that when water changes state, the arrangement of particles changes, but the particles stay water particles. As sugar dissolves in the tea in **Figure 3,** the sugar molecules mix with the tea, but they don't change what they are. The particles are still sugar. Crushing a rock is a physical change because particles separate but do not change identity.

internet connect

www.scilinks.org
Topic: Chemical and
        Physical Changes
SciLinks code: HW4140

SCiLINKS. Maintained by the National Science Teachers Association

Figure 3
Dissolving sugar in tea is a physical change.

7

Demonstration —— BASIC
Evidence of Chemical Change I

1. Place a few small marble chips in a Petri dish on an overhead projector.

2. Add enough 5% vinegar to cover the chips.

3. Have students note the evidence of chemical change. What chemical property of marble is demonstrated? **Ans.** Bubbles of gas are produced. The property demonstrated is reactivity with acid.

Safety Caution: Wear safety goggles and a lab apron. Students should stand back 3 m.

Disposal: Wearing safety goggles and a lab apron, decant the mixture, rinse the residue with tap water, decant the rinse water, and repeat the rinse. Pour the rinse water down the drain. Save the solid residue, wrap it in newspaper, and put it in the trash. **LS Visual**

Demonstration —— GENERAL
Evidence of Chemical Change II

1. Sand and/or polish a large sheet of copper foil, 30 cm × 30 cm.

2. Hold the foil at a 45° angle over a Bunsen burner flame.

3. Have students note evidence of chemical change. What is the dark stain? **Ans.** copper(II) oxide What are the reactants in this reaction? **Ans.** copper and oxygen

Safety Caution: Wear safety goggles and a lab apron. Tie back loose hair and clothing.

Disposal: Save the foil for reuse at a later time. Sweep up any flakes of copper oxide, and put them in the trash. **LS Logical**

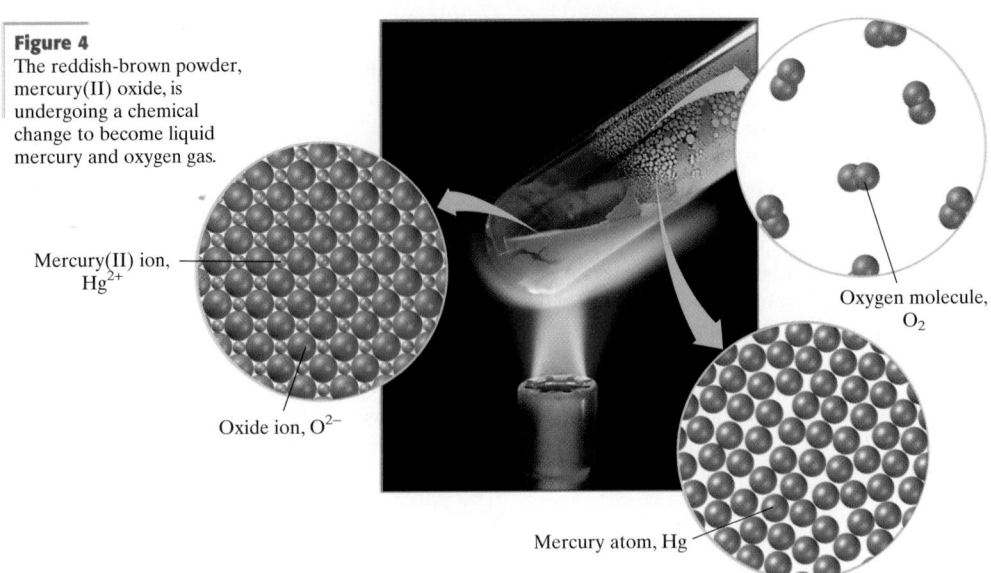

Figure 4
The reddish-brown powder, mercury(II) oxide, is undergoing a chemical change to become liquid mercury and oxygen gas.

Mercury(II) ion, Hg^{2+}

Oxide ion, O^{2-}

Oxygen molecule, O_2

Mercury atom, Hg

Chemical Changes

In a chemical change, the identities of substances change and new substances form. In **Figure 4,** mercury(II) oxide changes into mercury and oxygen as represented by the following word equation:

$$\text{mercury(II) oxide} \longrightarrow \text{mercury} + \text{oxygen}$$

In an equation, the substances on the left-hand side of the arrow are the **reactants.** They are used up in the reaction. Substances on the right-hand side of the arrow are the **products.** They are made by the reaction.

A chemical reaction is a rearrangement of the atoms that make up the reactant or reactants. After rearrangement, those same atoms are present in the product or products. Atoms are not destroyed or created, so mass does not change during a chemical reaction.

reactant

a substance or molecule that participates in a chemical reaction

product

a substance that forms in a chemical reaction

Evidence of Chemical Change

Evidence that a chemical change may be happening generally falls into one of the categories described below and shown in **Figure 5.** The more of these signs you observe, the more likely a chemical change is taking place. But be careful! Some physical changes also have one or more of these signs.

a. **The Evolution of a Gas** The production of a gas is often observed by bubbling, as shown in **Figure 5a,** or by a change in odor.

b. **The Formation of a Precipitate** When two clear solutions are mixed and become cloudy, a precipitate has formed, as shown in **Figure 5b.**

c. **The Release or Absorption of Energy** A change in temperature or the giving off of light energy, as shown in **Figure 5c,** are signs of an energy transfer.

d. **A Color Change in the Reaction System** Look for a different color when two chemicals react, as shown in **Figure 5d.**

Figure 5

a When acetic acid, in vinegar, and sodium hydrogen carbonate, or baking soda, are mixed, the solution bubbles as carbon dioxide forms.

b When solutions of sodium sulfide and cadmium nitrate are mixed, cadmium sulfide, a solid precipitate, forms.

c When aluminum reacts with iron(III) oxide in the clay pot, energy is released as heat and light.

d When phenolphthalein is added to ammonia dissolved in water, a color change from colorless to pink occurs.

① Section Review

UNDERSTANDING KEY IDEAS

1. Name three natural chemicals and three artificial chemicals that are part of your daily life.

2. Describe how chemistry is a part of your morning routine.

3. Classify the following materials as solid, liquid, or gas at room temperature: milk, helium, granite, oxygen, steel, and gasoline.

4. Describe the motions of particles in the three common states of matter.

5. How does a physical change differ from a chemical change?

6. Give three examples of physical changes.

7. Give three examples of chemical changes.

8. Identify each substance in the following word equation as a reactant or a product.

limestone $\xrightarrow{heat}$ lime + carbon dioxide

9. Sodium salicylate is made from carbon dioxide and sodium phenoxide. Identify each of these substances as a reactant or a product.

10. List four observations that suggest a chemical change is occurring.

CRITICAL THINKING

11. Explain why neither liquids nor gases have permanent shapes.

12. Steam is sometimes used to melt ice. Is this change physical or chemical?

13. Mass does not change during a chemical change. Is the same true for a physical change? Explain your answer, and give an example.

14. In beaker A, water is heated, bubbles of gas form throughout the water, and the water level in the beaker slowly decreases. In beaker B, electrical energy is added to water, bubbles of gas appear on the ends of the wires in the water, and the water level in the beaker slowly decreases.

    a. What signs of a change are visible in each situation?

    b. What type of change is happening in each beaker? Explain your answer.

Close

Quiz ──────────── GENERAL

1. What is a chemical? Ans. any substance with a definite composition

2. Which state of matter is characterized as having a definite shape and volume? Ans. solid

3. Which state of matter is characterized as having a definite volume but taking the shape of its container? Ans. liquid

4. What type of change does not change the identity of matter? Ans. physical

5. What term is used to identify the substances you begin with in a chemical reaction? Ans. reactants

6. List four changes that indicate a chemical change has probably occurred. Ans. gas evolution, precipitate formation, light emission, color change
LS Logical

Reteaching ────────── BASIC

Have the students create a graphic organizer or concept map that shows the differences between physical and chemical changes.
LS Visual

Chapter Resource File
• Concept Review
• Quiz

 Transparencies
TT Evidence of a Chemical Change

Answers to Section Review

1. Natural chemicals could include water, oxygen, carbon dioxide, sugar, or any other compound not manufactured in a laboratory. Artificial chemicals could include synthetic detergents, plastics, or any other chemically synthesized material.

2. Sample answer: Chemistry is involved in making the chemicals in soap, shampoo, hairspray, and toothpaste. Chemistry is important in the purification of the water used to shower. Cooking food for breakfast also involves chemistry.

3. Helium and oxygen are gases. Milk and gasoline are liquids. The others are solids.

4. Particles in solids vibrate. Particles in liquids remain in contact with each other but move past one another. Particles in gases move freely and independently of one another.

5. A physical change can change the form of a material but cannot change the identity of the substances involved. A chemical change always changes the identity of the substances involved.

6. Answers may vary, and may include boiling water, freezing water, dissolving sugar in water, chopping wood, crushing stone, and sharpening a pencil

Answers continued on p. 35A

Overview

Before beginning this section, review with your students the Objectives listed in the Student Edition. This section introduces the concepts of mass and volume. The SI System of measurement is introduced. Conversion factors are used to change from one unit of measure to another. Properties (both chemical and physical) can be used to characterize a substance.

Bellringer

Write the following words on the board:

peanut butter

water

fish

light

garbage

time

motion

the human brain

carbon dioxide

air

yourself

an idea

tree

energy

Ask students to sort the words into three categories: matter, not matter, or not sure.

Discuss responses. (Note that energy is not matter, although mass can be converted into energy.) Point out that all material things are matter. Finally, ask students why a clear definition of matter is important for studying chemistry.

Describing Matter

KEY TERMS

- matter
- volume
- mass
- weight
- quantity
- unit
- conversion factor
- physical property
- density
- chemical property

matter

anything that has mass and takes up space

volume

a measure of the size of a body or region in three-dimensional space

OBJECTIVES

① **Distinguish** between different characteristics of matter, including mass, volume, and weight.

② **Identify** and use SI units in measurements and calculations.

③ **Set up** conversion factors, and use them in calculations.

④ **Identify** and describe physical properties, including density.

⑤ **Identify** chemical properties.

Matter Has Mass and Volume

Matter, the stuff of which everything is made, exists in a dazzling variety of forms. However, matter has a fairly simple definition. **Matter** is anything that has mass and volume. Think about blowing up a balloon. The inflated balloon has more mass and more volume than before. The increase in mass and volume comes from the air that you blew into it. Both the balloon and air are examples of matter.

The Space an Object Occupies Is Its Volume

An object's **volume** is the space the object occupies. For example, this book has volume because it takes up space. Volume can be determined in several different ways. The method used to determine volume depends on the nature of the matter being examined. The book's volume can be found by multiplying the book's length, width, and height. Graduated cylinders are often used in laboratories to measure the volume of liquids, as shown in **Figure 6**. The volume of a gas is the same as that of the container it fills.

Figure 6
To read the liquid level in a graduated cylinder correctly, read the level at the bottom part of the *meniscus*, the curved upper surface of the liquid. The volume shown here is 73.0 mL.

Videos

CNN Presents Chemistry Connections

- **Segment 3** Chemicals in the Home

See the Science in the News video guide for more details.

Chapter Resource File

- **Lesson Plan**

Figure 7
A balance is an instrument that measures mass.

The Quantity of Matter Is the Mass

The **mass** of an object is the quantity of matter contained in that object. Even though a marble is smaller, it has more mass than a ping-pong ball does if the marble contains more matter.

Devices used for measuring mass in a laboratory are called *balances*. Balances can be electronic, as shown in **Figure 7,** or mechanical, such as a triple-beam balance.

Balances also differ based on the precision of the mass reading. The balance in **Figure 7** reports readings to the hundredth place. The balance often found in a school chemistry laboratory is the triple-beam balance. If the smallest scale on the triple-beam balance is marked off in 0.1 g increments, you can be certain of the reading to the tenths place, and you can estimate the reading to the hundredths place. The smaller the markings on the balance, the more decimal places you can have in your measurement.

Mass Is Not Weight

Mass is related to weight, but the two are not identical. Mass measures the quantity of matter in an object. As long as the object is not changed, it will have the same mass, no matter where it is in the universe. On the other hand, the weight of that object is affected by its location in the universe. The weight depends on gravity, while mass does not.

Weight is defined as the force produced by gravity acting on mass. Scientists express forces in *newtons,* but they express mass in *kilograms.* Because gravity can vary from one location to another, the weight of an object can vary. For example, an astronaut weighs about six times more on Earth than he weighs on the moon because the effect of gravity is less on the moon. The astronaut's mass, however, hasn't changed because he is still made up of the same amount of matter.

The force that gravity exerts on an object is proportional to the object's mass. If you keep the object in one place and double its mass, the weight of the object doubles, too. So, measuring weight can tell you about mass. In fact, when you read the word *weigh* in a laboratory procedure, you probably are determining the mass. Check with your teacher to be sure.

mass

a measure of the amount of matter in an object; a fundamental property of an object that is not affected by the forces that act on the object, such as the gravitational force

weight

a measure of the gravitational force exerted on an object; its value can change with the location of the object in the universe

11

did you know?

In practice, most mechanical balances depend on the presence of a gravitational field because the force exerted by an object is compared with the force exerted by standard masses placed on another pan or moved to positions on a beam.

Motivate

Identifying Preconceptions

Ask students if astronauts floating "weightless" in a space shuttle still have mass. Explain that mass is the amount of matter present and because matter cannot be changed just by changing location, the mass is unchanged.

Demonstration — GENERAL

Measuring Volume

1. Pour about 50 mL of colored water into a 100 mL beaker, and have a student measure the volume.

2. Pour the water into a 100 mL graduated cylinder, and have a student measure the volume.

3. Have students discuss why the values measured were different, and which instrument is the better one to use when measuring volume. **Ans.** The graduated cylinder is the better instrument for measuring volume. It has more subdivisions, so you can read the volume to more digits.
LS Visual

READING SKILL BUILDER — BASIC

Prediction Guide Write the following statements on the board before students read this section. Have students copy the statements and then decide if they "Strongly agree," "Agree somewhat," or "Strongly disagree," and give reasons. After students have read the section, they should determine if they have changed their minds and to give reasons why.

1. A sample of aluminum measuring 1 cm × 1 cm × 3 cm would have a different mass when measured at sea level than when measured on top of a mountain.

2. A professional basketball player may have a height greater than 2 meters.

3. Density is a property that can be used to identify a substance.
LS Verbal

Teaching Tip

SI Unit Review Students' familiarity with SI units will depend on what they were taught in previous courses. You may want to check their understanding of the metric system before proceeding. Having a firm idea of the size of a gram, a meter, and a liter is probably more important here than an ability to manipulate quantities mathematically.

Group Activity —— BASIC

Ask students to form groups. Assign different quantities to the different groups and ask what units would commonly be used to express the quantity they have been assigned. Unless students are accustomed to metric measurement, they are likely to list some of the following:

length or distance: inches, feet, yards, miles

weight: pounds, ounces, tons

volume: gallons, cups, teaspoons, tablespoons, fluid ounces

time: days, years, hours, minutes, seconds

temperature: degrees Fahrenheit

electric current: amperes

potential difference: volts

heat: British thermal units (BTU)
🄻🄢 Interpersonal

Figure 8
This graduated cylinder measures a *quantity*, the volume of a liquid, in a *unit*, the milliliter.

quantity

something that has magnitude, size, or amount

unit

a quantity adopted as a standard of measurement

📶 internet connect ≡
www.scilinks.org
Topic: SI Units
SciLinks code: HW4114
SCi*LINKS* Maintained by the National Science Teachers Association

Units of Measurement

Terms such as *heavy, light, rough,* and *smooth* describe matter qualitatively. Some properties of matter, such as color and texture, are usually described in this way. But whenever possible, scientists prefer to describe properties in quantitative terms, that is with numbers.

Mass and volume are properties that can be described in terms of numbers. But numbers alone are not enough because their meanings are unclear. For meaningful descriptions, units are needed with the numbers. For example, describing a quantity of sand as 15 kilograms rather than as 15 bucketfuls or just 15 gives clearer information.

When working with numbers, be careful to distinguish between a quantity and its unit. The graduated cylinder shown in **Figure 8,** for example, is used to measure the volume of a liquid in milliliters. *Volume* is the **quantity** being measured. *Milliliters* (abbreviated mL) is the **unit** in which the measured volume is reported.

Scientists Express Measurements in SI Units

Since 1960, scientists worldwide have used a set of units called the *Système Internationale d'Unités* or *SI*. The system is built on the seven base units listed in **Table 2.** The last two find little use in chemistry, but the first five provide the foundation of all chemical measurements.

Base units can be too large or too small for some measurements, so the base units may be modified by attaching prefixes, such as those in **Table 3.** For example, the base unit *meter* is suitable for expressing a person's height. The distance beween cities is more conveniently expressed in kilometers (km), with 1 km being 1000 m. The lengths of many insects are better expressed in millimeters (mm), or one-thousandth of a meter, because of the insects' small size. Additional prefixes can be found in **Appendix A.** Atomic sizes are so small that picometers (pm) are used. A picometer is 0.000 000 000 001 m. The advantage of using prefixes is the ability to use more manageable numbers. So instead of reporting the diameter of a hydrogen atom as 0.000 000 000 120 m, you can report it as 120 pm.

Table 2 SI Base Units

Quantity	Symbol	Unit	Abbreviation
Length	l	meter	m
Mass	m	kilogram	kg
Time	t	second	s
Thermodynamic temperature	T	kelvin	K
Amount of substance	n	mole	mol
Electric current	I	ampere	A
Luminous intensity	I_v	candela	cd

Table 3 SI Prefixes

Prefix	Abbreviation	Exponential multiplier	Meaning	Example using length
Kilo-	k	10^3	1000	1 kilometer (km) = 1000 m
Hecto-	h	10^2	100	1 hectometer (hm) = 100 m
Deka-	da	10^1	10	1 dekameter (dam) = 10 m
		10^0	1	1 meter (m)
Deci-	d	10^{-1}	1/10	1 decimeter (dm) = 0.1 m
Centi-	c	10^{-2}	1/100	1 centimeter (cm) = 0.01 m
Milli-	m	10^{-3}	1/1000	1 millimeter (mm) = 0.001 m

Refer to Appendix A for more SI prefixes.

Converting One Unit to Another

In chemistry, you often need to convert a measurement from one unit to another. One way of doing this is to use a **conversion factor.** A conversion factor is a simple ratio that relates two units that express a measurement of the same quantity. Conversion factors are formed by setting up a fraction that has equivalent amounts on top and bottom. For example, you can construct conversion factors between kilograms and grams as follows:

conversion factor

a ratio that is derived from the equality of two different units and that can be used to convert from one unit to the other

$$1 \text{ kg} = 1000 \text{ g} \text{ can be written as } \frac{1 \text{ kg}}{1000 \text{ g}} \text{ or } \frac{1000 \text{ g}}{1 \text{ kg}}$$

$$0.001 \text{ kg} = 1 \text{ g} \text{ can be written as } \frac{0.001 \text{ kg}}{1 \text{ g}} \text{ or } \frac{1 \text{ g}}{0.001 \text{ kg}}$$

SKILLS Toolkit 1

Using Conversion Factors

1. Identify the quantity and unit given and the unit that you want to convert to.

2. Using the equality that relates the two units, set up the conversion factor that cancels the given unit and leaves the unit that you want to convert to.

3. Multiply the given quantity by the conversion factor. Cancel units to verify that the units left are the ones you want for your answer.

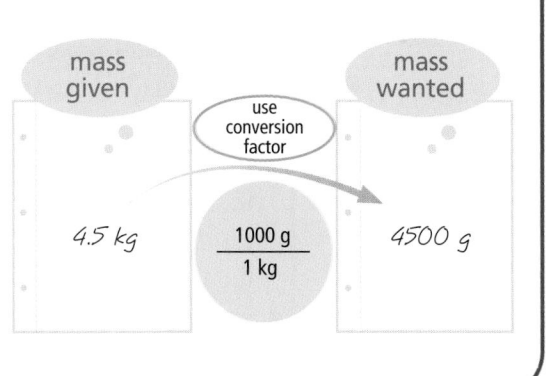

13

Transparencies

TT Using Conversion Factors

Answers to Practice Problems A

1. **a.** 0.000 765 kg
   b. 1340 mg
   c. 0.0342 g
   d. 23 745 000 000 mg
2. **a.** 1730 cm
   b. 0.002 56 km
   c. 56.7 m
   d. 5130 mm
3. shortest: 0.0128 km; longest: 17 931 mm

Homework ——— **GENERAL**

Additional Practice

1. Convert 253 mL to liters
   Ans. 0.253 L
2. Convert 1258 cm to meters
   Ans. 12.58 m
3. Convert 15 g to kilograms
   Ans. 0.015 kg
4. Convert 1254 kilocalories to calories Ans. 1 254 000 cal
5. How many seconds pass in 5.25 h? Ans. 18 900 s

LS Logical

SAMPLE PROBLEM A

Converting Units

Convert 0.851 L to milliliters.

1 Gather information.

- You are given 0.851 L, which you want to convert to milliliters. This problem can be expressed as this equation:

$$? \text{ mL} = 0.851 \text{ L}$$

- The equality that links the two units is 1000 mL = 1 L. (The prefix *milli-* represents 1/1000 of a base unit.)

2 Plan your work.

The conversion factor needed must cancel liters and leave milliliters. Thus, liters must be on the bottom of the fraction and milliliters must be on the top. The correct conversion factor to use is

$$\frac{1000 \text{ mL}}{1 \text{ L}}$$

3 Calculate.

$$? \text{ mL} = 0.851 \cancel{\text{ L}} \times \frac{1000 \text{ mL}}{1 \cancel{\text{ L}}} = 851 \text{ mL}$$

4 Verify your results.

The unit of liters cancels out. The answer has the unit of milliliters, which is the unit called for in the problem. Because a milliliter is smaller than a liter, the number of milliliters should be greater than the number of liters for the same volume of material. Thus, the answer makes sense because 851 is greater than 0.851.

> **PRACTICE HINT**
>
> Remember that you can cancel only those units that appear in both the top and the bottom of the fractions you multiply together. Be sure to set up your conversion factors so that the unit you want to cancel is in the correct place.

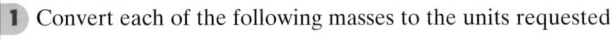
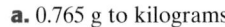
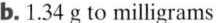

PRACTICE

PROBLEM SOLVING SKILL

1. Convert each of the following masses to the units requested.
   a. 0.765 g to kilograms
   b. 1.34 g to milligrams
   c. 34.2 mg to grams
   d. 23 745 kg to milligrams (Hint: Use two conversion factors.)

2. Convert each of the following lengths to the units requested.
   a. 17.3 m to centimeters
   b. 2.56 m to kilometers
   c. 567 dm to meters
   d. 5.13 m to millimeters

3. Which of the following lengths is the shortest, and which is the longest: 1583 cm, 0.0128 km, 17 931 mm, and 14 m?

14

Figure 9
The volume of water in the beaker is 1 L. The model shows the dimensions of a cube that is 10 cm on each side. Its volume is 1000 cm³, or 1 L.

Derived Units

Many quantities you can measure need units other than the seven basic SI units. These units are derived by multiplying or dividing the base units. For example, *speed* is distance divided by time. The derived unit of speed is meters per second (m/s). A rectangle's *area* is found by multiplying its length (in meters) by its width (also in meters), so its unit is square meters (m^2).

The volume of this book can be found by multiplying its length, width, and height. So the unit of volume is the cubic meter (m^3). But this unit is too large and inconvenient in most labs. Chemists usually use the liter (L), which is one-thousandth of a cubic meter. **Figure 9** shows one liter of liquid and also a cube of one liter volume. Each side of the cube has been divided to show that one liter is exactly 1000 cubic centimeters, which can be expressed in the following equality:

$$1 \text{ L} = 1000 \text{ mL} = 1000 \text{ cm}^3$$

Therefore, a volume of one milliliter (1 mL) is identical to one cubic centimeter (1 cm^3).

Properties of Matter

When examining a sample of matter, scientists describe its properties. In fact, when you describe an object, you are most likely describing it in terms of the properties of matter. Matter has many properties. The properties of a substance may be classified as physical or chemical.

Physical Properties

A **physical property** is a property that can be determined without changing the nature of the substance. Consider table sugar, or sucrose. You can see that it is a white solid at room temperature, so color and state are physical properties. It also has a gritty texture. Because changes of state are physical changes, melting point and boiling point are also physical properties. Even the lack of a physical property, such as air being colorless, can be used to describe a substance.

physical property

a characteristic of a substance that does not involve a chemical change, such as density, color, or hardness

Figure 10 Have students determine the slope of the line from two points that are not part of the data in the table. Have students share their results. Reinforce that the graph has a constant slope, so all values calculated should be close to 2.7 g/cm³.

density

the ratio of the mass of a substance to the volume of the substance; often expressed as grams per cubic centimeter for solids and liquids and as grams per liter for gases

Density Is the Ratio of Mass to Volume

The mass and volume of a sample are physical properties that can be determined without changing the substance. But each of these properties changes depending on how much of the substance you have. The **density** of an object is another physical property: the mass of that object divided by its volume. As a result, densities are expressed in derived units such as g/cm³ or g/mL. Density is calculated as follows:

$$density = \frac{mass}{volume} \quad \text{or} \quad D = \frac{m}{V}$$

The density of a substance is the same no matter what the size of the sample is. For example, the masses and volumes of a set of 10 different aluminum blocks are listed in the table in **Figure 10.** The density of Block 10 is as follows:

$$D = \frac{m}{V} = \frac{36.40 \text{ g}}{13.5 \text{ cm}^3} = 2.70 \text{ g/cm}^3$$

If you divide the mass of any block by the corresponding volume, you will always get an answer close to 2.70 g/cm³.

The density of aluminum can also be determined by graphing the data, as shown in **Figure 10.** The straight line rising from left to right indicates that mass increases at a constant rate as volume increases. As the volume of aluminum doubles, its mass doubles; as its volume triples, its mass triples, and so on. In other words, the mass of aluminum is directly proportional to its volume.

The slope of the line equals the ratio of mass (from the vertical y-axis) divided by volume (from the horizontal x-axis). You may remember this as "rise over run" from math class. The slope between the two points shown is as follows:

$$slope = \frac{rise}{run} = \frac{29.7 \text{ g} - 10.8 \text{ g}}{11 \text{ cm}^3 - 4 \text{ cm}^3} = \frac{18.9 \text{ g}}{7 \text{ cm}^3} = 2.70 \text{ g/cm}^3$$

As you can see, the value of the slope is the density of aluminum.

Figure 10
The graph of mass versus volume shows a relationship of direct proportionality. Notice that the line has been extended to the origin.

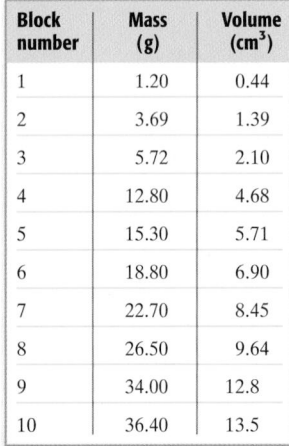

Block number	Mass (g)	Volume (cm³)
1	1.20	0.44
2	3.69	1.39
3	5.72	2.10
4	12.80	4.68
5	15.30	5.71
6	18.80	6.90
7	22.70	8.45
8	26.50	9.64
9	34.00	12.8
10	36.40	13.5

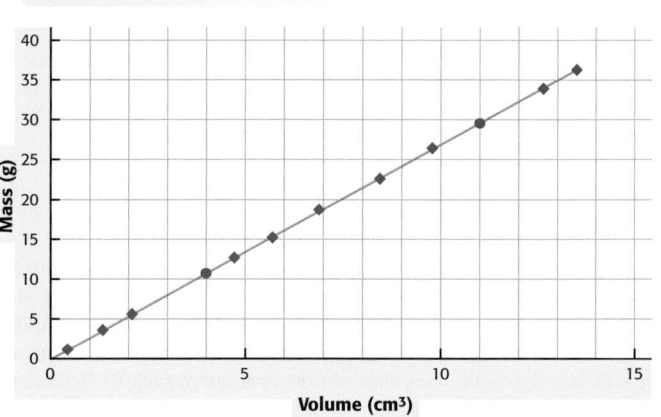

Mass Vs. Volume for Samples of Aluminum

16

did you know?

A football made of osmium would probably weigh over 200 lb. The density of osmium is 22.48 g/cm³. 1000 cm³ (1 L) of osmium would have a mass of 22.48 kg. The same quantity of water has a mass of 1 kg.

Table 4   Densities of Various Substances

Substance	Density (g/cm³) at 25°C
Hydrogen gas, H_2*	0.000 082 4
Carbon dioxide gas, CO_2*	0.001 80
Ethanol (ethyl alcohol), C_2H_5OH	0.789
Water, H_2O	0.997
Sucrose (table sugar), $C_{12}H_{22}O_{11}$	1.587
Sodium chloride, NaCl	2.164
Aluminum, Al	2.699
Iron, Fe	7.86
Copper, Cu	8.94
Silver, Ag	10.5
Gold, Au	19.3
Osmium, Os	22.6

*at 1 atm

Density Can Be Used to Identify Substances

Because the density of a substance is the same for all samples, you can use this property to help identify substances. For example, suppose you find a chain that appears to be silver on the ground. To find out if it is pure silver, you can take the chain into the lab and use a balance to measure its mass. One way to find the volume is to use the technique of water displacement. Partially fill a graduated cylinder with water, and note the volume. Place the chain in the water, and watch the water level rise. Note the new volume. The difference in water levels is the volume of the chain. If the mass is 199.0 g, and the volume is 20.5 cm³, you can calculate the chain's density as follows:

$$D = \frac{m}{V} = \frac{199.0 \text{ g}}{20.5 \text{ cm}^3} = 9.71 \text{ g/cm}^3$$

Comparing this density with the density of silver in **Table 4,** you can see that your find is not pure silver.

    Table 4 lists the densities of a variety of substances. Osmium, a bluish white metal, is the densest substance known. A piece of osmium the size of a football would be too heavy to lift. Whether a solid will float or sink in a liquid depends on the relative densities of the solid and the liquid. **Figure 11** shows several things arranged according to densities, with the most dense on the bottom.

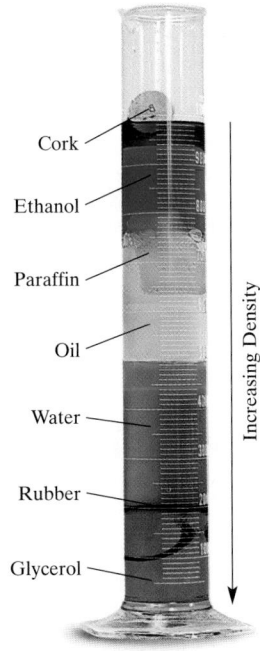

Cork

Ethanol

Paraffin

Oil

Water

Rubber

Glycerol

Increasing Density

Figure 11
Substances float in layers, and the order of the layers is determined by their densities. Dyes have been added to make the liquid layers more visible.

17

Teaching Tip

Discuss how the chemical properties of materials are indicated through safety precautions listed on product labels. Have some sample labels available including labels that mention flammability and drug interactions. Discuss how the effects of some drugs are heightened or lessened because of interaction with certain foods such as grapefruit.

Quick LAB

GENERAL

Skills Acquired:
- Collecting Data
- Measuring
- Organizing & Analyzing Data

Materials:
For each group of 2–3 students:
- aluminum foil, 2 brands, approx. 100 cm² of each
- balance
- metric ruler
- scissors

Teacher's Notes:
- Have students fold or wad the aluminum before placing it on the balance so that breezes do not affect the balance reading.
- Balances sensitive to 0.001 g will give good results. Less-sensitive balances will require larger pieces of aluminum foil.
- To calculate thickness, students will need to use the formula for density and the formula for the volume of a rectangular solid ($V = l \times w \times h$).

LS Visual

Answers

1. Volume will vary. Thickness will be near 0.0017 cm, 0.000 017 m, and 17 μm.
2. Answer will depend on brands used.

Chemical Properties

You cannot fully describe matter by physical properties alone. You must also describe what happens when matter has the chance to react with other kinds of matter, or the *chemical properties* of matter.

Whereas physical properties can be determined without changing the identity of the substance, **chemical properties** can only be identified by trying to cause a chemical change. Afterward, the substance may have been changed into a new substance.

For example, many substances share the chemical property of reactivity with oxygen. If you have seen a rusty nail or a rusty car, you have seen the result of iron's property of reactivity with oxygen. But gold has a very different chemical property. It does not react with oxygen. This property prevents gold from tarnishing and keeps gold jewelry shiny. If something doesn't react with oxygen, that lack of reaction is also a chemical property.

Not all chemical reactions result from contact between two or more substances. For example, many silver compounds are sensitive to light and undergo a chemical reaction when exposed to light. Photographers rely on silver compounds on film to create photographs. Some sunglasses have silver compounds in their lenses. As a result of this property of the silver compounds, the lenses darken in response to light. Another reaction that involves a single reactant is the reaction you saw earlier in this chapter. The formation of mercury and oxygen when mercury(II) oxide is heated, happens when a single reactant breaks down. Recall that the reaction in this case is described by the following equation:

$$\text{mercury(II) oxide} \longrightarrow \text{mercury} + \text{oxygen}$$

Despite similarities between the names of the products and the reactant, the two products have completely different properties from the starting material, as shown in **Figure 12.**

chemical property

a property of matter that describes a substance's ability to participate in chemical reactions

internet connect

www.scilinks.org
Topic: Physical/Chemical Properties
SciLinks code: HW4097

SCiLINKS Maintained by the National Science Teachers Association

Quick LAB

SAFETY PRECAUTIONS

Thickness of Aluminum Foil

PROCEDURE

1. Using **scissors** and a **metric ruler,** cut a rectangle of **aluminum foil.** Determine the area of the rectangle.
2. Use a **balance** to determine the mass of the foil.
3. Repeat steps **1** and **2** with each brand of aluminum foil available.

ANALYSIS

1. Use the density of aluminum (2.699 g/cm³) to calculate the volume and the thickness of each piece of foil. Report the thickness in centimeters (cm), meters (m), and micrometers (μm) for each brand of foil. (Hint: 1 μm = 10⁻⁶ m)
2. Which brand is the thickest?
3. Which unit is the most appropriate unit to use for expressing the thickness of the foil? Explain your reasoning.

18

Chapter Resource File
- Datasheets for In-Text Labs

Videos
CNN Presents Chemistry Connections
- **Segment 4** Student Superconductors
See the Science in the News video guide for more details.

Figure 12
The physical and chemical properties of the components of this reaction system are shown. Decomposition of mercury(II) oxide is a chemical change.

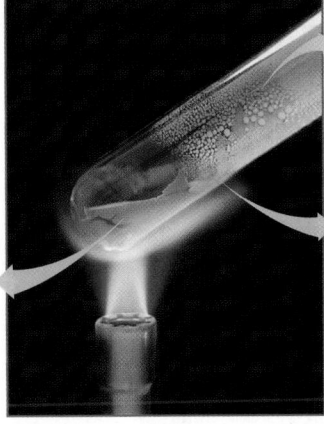

MERCURY(II) OXIDE

Physical properties: Bright red or orange-red, odorless crystalline solid

Chemical properties:
Decomposes when exposed to light or at 500°C to form mercury and oxygen gas; dissolves in dilute nitric acid or hydrochloric acid, but is almost insoluble in water

OXYGEN

Physical properties: Colorless, odorless gas

Chemical properties: Supports combustion; soluble in water

MERCURY

Physical properties: Silver-white, liquid metal; in the solid state, mercury is ductile and malleable and can be cut with a knife

Chemical properties: Combines readily with sulfur at normal temperatures; reacts with nitric acid and hot sulfuric acid; oxidizes to form mercury(II) oxide upon heating in air

 Section Review

UNDERSTANDING KEY IDEAS

1. Name two physical properties that characterize matter.
2. How does mass differ from weight?
3. What derived unit is usually used to express the density of liquids?
4. What SI unit would best be used to express the height of your classroom ceiling?
5. Distinguish between a physical property and a chemical property, and give an example of each.
6. Why is density considered a physical property rather than a chemical property of matter?
7. One inch equals 2.54 centimeters. What conversion factor is useful for converting from centimeters to inches?

PRACTICE PROBLEMS

8. What is the mass, in kilograms, of a 22 000 g bag of fertilizer?
9. Convert each of the following measurements to the units indicated. (Hint: Use two conversion factors if needed.)
   a. 17.3 s to milliseconds
   b. 2.56 mm to kilometers
   c. 567 cg to grams
   d. 5.13 m to kilometers
10. Convert 17.3 cm^3 to liters.
11. Five beans have a mass of 2.1 g. How many beans are in 0.454 kg of beans?

CRITICAL THINKING

12. A block of lead, with dimensions 2.0 dm × 8.0 cm × 35 mm, has a mass of 6.356 kg. Calculate the density of lead in g/cm^3.
13. Demonstrate that kg/L and g/cm^3 are equivalent units of density.
14. In the manufacture of steel, pure oxygen is blown through molten iron to remove some of the carbon impurity. If the combustion of carbon is efficient, carbon dioxide (density = 1.80 g/L) is produced. Incomplete combustion produces the poisonous gas carbon monoxide (density = 1.15 g/L) and should be avoided. If you measure a gas density of 1.77 g/L, what do you conclude?

19

Homework ——— **BASIC**

Graphic Organizer Have students create a graphic organizer or concept map that shows the relationship between matter, measurement, physical properties, and chemical properties. **LS** Visual

Close

Quiz ——— **GENERAL**

1. What two quantities characterize matter? **Ans.** mass and volume
2. What is the difference between mass and weight? **Ans.** Mass does not change with location in the universe. Weight changes depending on location in the universe.
3. What is density? **Ans.** the ratio of the mass to the volume of a substance
4. What are the basic SI units for mass? length? **Ans.** kilogram; meter
5. Name two chemical properties. **Ans.** Answers may vary. Sample answer: flammability and reactivity with water

LS Logical

Reteaching ——— **BASIC**

Have the students develop a graphic organizer or concept map that will help them understand and utilize conversion factors. **LS** Visual

Chapter Resource File
• Concept Review
• Quiz

Transparencies

TT Comparison of Physical and Chemical Properties

Answers to Section Review

1. mass and volume
2. Mass is a measure of the amount of matter in an object. Weight is a measure of the gravitational force exerted on an object. Mass does not depend on the object's location, but weight can vary with location in the universe.
3. grams per milliliter
4. meters
5. A physical property, such as mass, volume, color, or texture, is a property that can be determined without changing the nature of the substance. A chemical property, such as reactivity with oxygen, reactivity with acid, and flammability, is a property that can only be identified when the substance tries to undergo a chemical change.
6. Density can be measured without changing the substance.
7. $\dfrac{1 \text{ in.}}{2.54 \text{ cm}}$
8. 22 kg
9. a. 17 300 ms
   b. 0.000 002 56 km
   c. 5.67 g
   d. 0.005 13 km

Answers continued on p. 35A

Aspirin Although salicylates could be obtained from natural plant sources, such as from willow bark, the wintergreen plant, and the meadowsweet plant, they are expensive to isolate and produce from these sources. Fortunately, during the nineteenth century, German chemists had become expert in the development of synthetic forms of natural substances. German chemists were able to synthesize salicylic acid at one-tenth of the cost of producing the compound from natural sources. While natural products are held in high esteem, there are many instances where they are too costly for widespread consumption.

A more modern example of such a drug is taxol, which is used as an anti-cancer drug. It comes from a certain evergreen that is an endangered species. There was much controversy in the scientific community when this drug was discovered. The heart of the debate involved the problem of cutting down endangered trees to save cancer victims. Finally, the battle of words ended when a synthetic duplicate of the drug was developed.

CONSUMER FOCUS

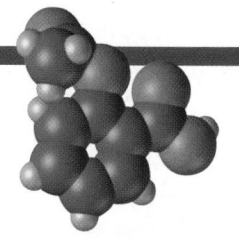

Aspirin

For centuries, plant extracts have been used for treating ailments. The bark of the willow tree was found to relieve pain and reduce fever. Writing in 1760, Edward Stone, an English naturalist and clergyman, reported excellent results when he used "twenty grains of powdered bark dissolved in water and administered every four hours" to treat people suffering from an acute, shiver-provoking illness.

The History of Aspirin

Following up on Stone's research, German chemists isolated a tiny amount of the active ingredient of the willow-bark extract, which they called *salicin*, from *Salix*, the botanical name for the willow genus. Researchers in France further purified salicin and converted it to salicylic acid, which proved to be a potent pain-reliever. This product was later marketed as the salt sodium salicylate. Though an effective painkiller, sodium salicylate has the unfortunate side effect of causing nausea and, sometimes, stomach ulcers.

Then back in Germany in the late 1800s, the father of Felix Hoffmann, a skillful organic chemist, developed painful arthritis. Putting aside his research on dyes, the younger Hoffmann looked for a way to prevent the nauseating effects of salicylic acid. He found that a similar compound, acetylsalicylic acid, was effective in treating pain and fever, while having fewer side effects. Under the name *aspirin*, it has been a mainstay in painkillers for over a century.

The FDA and Product Warning Labels

The Federal Drug Administration requires that all over-the-counter drugs carry a warning label. In fact, when you purchase any product, it is your responsibility as a consumer to check the warning label about the hazards of any chemical it may contain. The label on aspirin bottles warns against giving aspirin to children and teenagers who have chickenpox or severe flu. Some reports suggest that aspirin may play a part in Reye's syndrome, a condition in which the brain swells and the liver malfunctions.

Though side effects and allergic responses are rare, the label warns that aspirin may cause nausea and vomiting and should be avoided late in pregnancy. Because aspirin can interfere with blood clotting, it should not be used by hemophiliacs or following surgery of the mouth.

Questions

1. For an adult, the recommended dosage of 325 mg aspirin tablets is "one or two tablets every four hours, up to 12 tablets per day." In grams, what is the maximum dosage of aspirin an adult should take in one day? Why should you not take 12 tablets at once?

2. Research several over-the-counter painkillers, and write a report of your findings. For each product, compare the active ingredient and the price for a day's treatment.

3. Research Reye's syndrome, and write a report of your findings. Include the causes, symptoms, and risk factors.

internet connect

www.scilinks.org
Topic: Aspirin
SciLinks code: HW4012

SCI LINKS
Maintained by the National Science Teachers Association

20

Answers to Feature Questions

1. 3.90 g/day; Taking one large dose is likely to cause health problems due to an overdose of the drug.

2. Answers will vary.

3. Accept all reasonable answers. Reye's syndrome affects all body organs, especially the liver and the brain. Although the cause is unknown, symptoms often appear during recovery from a viral illness, such as influenza or chicken pox. Symptoms include persistent vomiting, drowsiness, personality changes, and irrational behavior. Research has shown an association between the development of Reye's syndrome and the use of medications containing salicylates during illnessess involving fever.

How Is Matter Classified?

KEY TERMS

- atom
- pure substance
- element
- molecule
- compound
- mixture
- homogeneous
- heterogeneous

OBJECTIVES

(1) **Distinguish** between elements and compounds.

(2) **Distinguish** between pure substances and mixtures.

(3) **Classify** mixtures as homogeneous or heterogeneous.

(4) **Explain** the difference between mixtures and compounds.

Classifying Matter

Everything around you—water, air, plants, and your friends—is made of matter. Despite the many examples of matter, all matter is composed of about 110 different kinds of **atoms.** Even the biggest atoms are so small that it would take more than 3 million of them side by side to span just one millimeter. These atoms can be physically mixed or chemically joined together to make up all kinds of matter.

atom

the smallest unit of an element that maintains the properties of that element

Benefits of Classification

Because matter exists in so many different forms, having a way to classify matter is important for studying it. In a store, such as the nursery in **Figure 13,** classification helps you to find what you want. In chemistry, it helps you to predict what characteristics a sample will have based on what you know about others like it.

Figure 13
Finding the plant you want without the classification scheme adopted by this nursery would be difficult.

21

Focus

Overview

Before beginning this section, review with your students the Objectives listed in the Student Edition. This section is devoted to classifying matter as elements, compounds, and mixtures. Students should be able to classify a sample of matter either when looking at the sample or when reading a description of the matter.

Bellringer

Have the students make a list of ways things are organized, and then have them write those methods on the board. Ask them to discuss which are the easiest to use with the greatest success.

Motivate

Identifying Preconceptions — GENERAL

To determine what students remember about different classifications of matter, ask students to describe the differences between a cup each of sugar water, muddy water, and pure water. Lead the discussion toward the particles that make up each one. **LS Verbal**

Chapter Resource File

- Lesson Plan

Discussion ——— GENERAL

Library Organization Have students consider the organization of materials in the school library in terms of the categories used, the physical layout of the materials, the method of searching for a particular item, etc. Have students discuss how the classification scheme is helpful and what difficulties would exist if no classification scheme were used. **LS** Interpersonal

Teach

READING SKILL BUILDER ——— GENERAL

K-W-L Ask the students to write down what they know about atoms, elements, compounds, and mixtures. They should make a second column telling what they want to know about these classes of matter. After reading the section they should write what they have learned about these things. **LS** Logical

Transparencies

TT Pure Substances

Figure 14
Copper, bromine, and dry ice are pure substances. Each is composed of only one type of particle.

Copper atom, Cu   Bromine molecule, Br$_2$ Carbon dioxide molecule, CO$_2$

pure substance

a sample of matter, either a single element or a single compound, that has definite chemical and physical properties

element

a substance that cannot be separated or broken down into simpler substances by chemical means; all atoms of an element have the same atomic number

Pure Substances

Each of the substances shown in **Figure 14** is a **pure substance.** Every pure substance has characteristic properties that can be used to identify it. Characteristic properties can be physical or chemical properties. For example, copper always melts at 1083°C, which is a physical property that is characteristic of copper. There are two types of pure substances: elements and compounds.

Elements Are Pure Substances

Elements are pure substances that contain only one kind of atom. Copper and bromine are elements. Each element has its own unique set of physical and chemical properties and is represented by a distinct chemical symbol. **Table 5** shows several elements and their symbols and gives examples of how an element got its symbol.

Table 5	Element Names, Symbols, and the Symbols' Origins	
Element name	**Chemical symbol**	**Origin of symbol**
Hydrogen	H	first letter of element name
Helium	He	first two letters of element name
Magnesium	Mg	first and third letters of element name
Tin	Sn	from *stannum*, the Latin word for "tin"
Gold	Au	from *aurum*, the Latin word meaning "gold"
Tungsten	W	from *Wolfram*, the German word for "tungsten"
Ununpentium	Uup	first letters of root words that describe the digits of the atomic number; used for elements that have not yet been synthesized or whose official names have not yet been chosen

Refer to Appendix A for an alphabetical listing of element names and symbols.

One-Stop Planner CD-ROM

• **Supplemental Reading Projects**
  Guided Reading Worksheet: The Same and Not the Same "What Are You?" *Assign this worksheet for cross-curricular connections to language arts.*

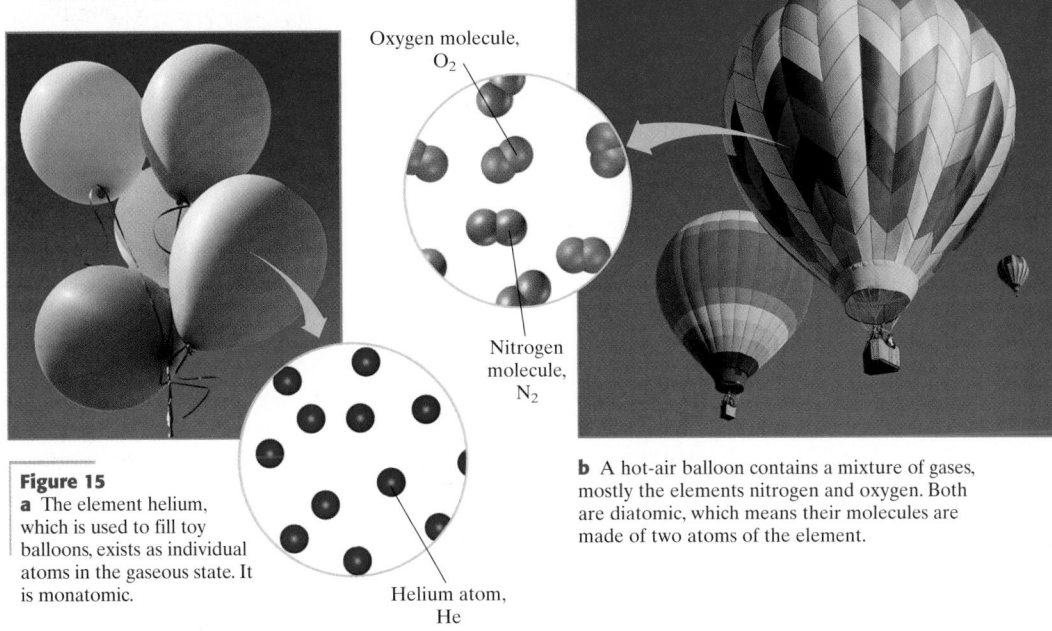

Oxygen molecule,
O_2

Nitrogen
molecule,
N_2

Figure 15
a The element helium, which is used to fill toy balloons, exists as individual atoms in the gaseous state. It is monatomic.

Helium atom,
He

b A hot-air balloon contains a mixture of gases, mostly the elements nitrogen and oxygen. Both are diatomic, which means their molecules are made of two atoms of the element.

Teaching Tip
Element Names Students who are interested in the origin of element names can find more information in reference sources such as the *CRC Handbook of Chemistry and Physics*.

Using the Figure
Figure 15 Tell students that knowing which elements are diatomic is important for writing correct formulas for these substances. Remind students that oxygen gas exists as diatomic molecules. Use the periodic table to show students the locations and identities of the diatomic elements. These are the halogens, hydrogen, oxygen, and nitrogen. Several mnemonic devices exist to help remember the diatomic elements. One such device is "HON Family," referring to H_2, O_2, N_2, and the fluorine family.

Teaching Tip
Molecular Elements Students should understand that a molecule may consist of two or more atoms of the same element or atoms of different elements. Besides diatomic elements, other molecular elements include sulfur, S_8, and phosphorus, P_4.

Elements as Single Atoms or as Molecules

Some elements exist as single atoms. For example, the helium gas in a balloon consists of individual atoms, as shown by the model in **Figure 15a.** Because it exists as individual atoms, helium gas is known as a *monatomic* gas.

Other elements exist as molecules consisting of as few as two or as many as millions of atoms. A **molecule** usually consists of two or more atoms combined in a definite ratio. If an element consists of molecules, those molecules contain just one type of atom. For example, the element nitrogen, found in air, is an example of a molecular element because it exists as two nitrogen atoms joined together, as shown by the model in **Figure 15b.** Oxygen, another gas found in the air, exists as two oxygen atoms joined together. Nitrogen and oxygen are *diatomic* elements. Other diatomic elements are H_2, F_2, Cl_2, Br_2, and I_2.

Some Elements Have More than One Form

Both oxygen gas and ozone gas are made up of oxygen atoms, and are forms of the element oxygen. However, the models in **Figure 16** show that a molecule of oxygen gas, O_2, is made up of two oxygen atoms, and a molecule of ozone, O_3, is made up of three oxygen atoms.

A few elements, including oxygen, phosphorus, sulfur, and carbon, are unusual because they exist as allotropes. An *allotrope* is one of a number of different molecular forms of an element. The properties of allotropes can vary widely. For example, ozone is a toxic, pale blue gas that has a sharp odor. You often smell ozone after a thunderstorm. But oxygen is a colorless, odorless gas essential to most forms of life.

molecule

the smallest unit of a substance that keeps all of the physical and chemical properties of that substance; it can consist of one atom or two or more atoms bonded together

Figure 16
Two forms of the element oxygen are oxygen gas and ozone gas.

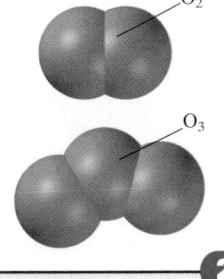

O_2

O_3

23

HISTORY
CONNECTION

In 1750, only 16 elements were known: antimony, arsenic, bismuth, carbon, cobalt, copper, gold, iron, lead, mercury, phosphorus, platinum, silver, sulfur, tin, and zinc. The next element to be identified was nickel, in 1751. Notice that none of the common gaseous elements, such as hydrogen, oxygen, and nitrogen, nor any of the halogens had been identified. As scientists learned more about the nature of gases and as laboratory techniques were developed and improved, additional elements were discovered at a much faster rate.

Teaching Tip

Definite Ratios If students do not understand what is meant by the phrase *combined in a definite ratio*, illustrate this idea with substances whose formulas are already familiar to students. All water exists in a ratio represented by the formula H_2O, two atoms of hydrogen to one atom of oxygen. It is a definite ratio because all water, regardless of its source, consists of these elements in the same ratio.

Using the Figure

Figure 18 Ask students to study and compare the models of aspirin. Point out that the shape of the space-filling model represents the best estimate of how the molecule would actually appear (but without the colors) if it could be seen. Stress that none of the models is exactly like the actual molecule. Instead, each model conveys information about the molecule in a slightly different way.

Transparencies

TT Models of a Molecule

---

compound

a substance made up of atoms of two or more different elements joined by chemical bonds

Compounds Are Pure Substances

Pure substances that are not elements are **compounds.** Compounds are composed of more than one kind of atom. For example, the compound carbon dioxide is composed of molecules that consist of one atom of carbon and two atoms of oxygen.

There may be easier ways of preparing them, but compounds can be made from their elements. On the other hand, compounds can be broken down into their elements, though often with great difficulty. The reaction of mercury(II) oxide described earlier in this chapter is an example of the breaking down of a compound into its elements.

Compounds Are Represented by Formulas

Because every molecule of a compound is made up of the same kinds of atoms arranged the same way, a compound has characteristic properties and composition. For example, every molecule of hydrogen peroxide contains two atoms each of hydrogen and oxygen. To emphasize this ratio, the compound can be represented by an abbreviation or *formula:* H_2O_2. *Subscripts* are placed to the lower right of the element's symbol to show the number of atoms of the element in a molecule. If there is just one atom, no subscript is used. For example, the formula for water is H_2O, not H_2O_1.

Molecular formulas give information only about what makes up a compound. The molecular formula for aspirin is $C_9H_8O_4$. Additional information can be shown by using different models, such as the ones for aspirin shown in **Figure 17.** A *structural formula* shows how the atoms are connected, but the two-dimensional model does not show the molecule's true shape. The distances between atoms and the angles between them are more realistic in a three-dimensional *ball-and-stick model*. However, a *space-filling model* attempts to represent the actual sizes of the atoms and not just their relative positions. A hand-held model can provide even more information than models shown on the flat surface of the page.

Figure 17
These models convey different information about acetylsalicylic acid (aspirin).

$C_9H_8O_4$

Molecular formula

Ball-and-stick model

Structural formula

Space-filling model

24

Compounds Are Further Classified

Such a wide variety of compounds exists that scientists classify the compounds to help make sense of them. In later chapters, you will learn that compounds can be classified by their properties, by the type of bond that holds them together, and by whether they are made of certain elements.

Mixtures

A sample of matter that contains two or more pure substances is a **mixture.** Most kinds of food are mixtures, sugar and salt being rare exceptions. Air is a mixture, mostly of nitrogen and oxygen. Water is *not* a mixture of hydrogen and oxygen for two reasons. First, the H and O atoms are chemically bonded together in H_2O molecules, not just physically mixed. Second, the ratio of hydrogen atoms to oxygen atoms is always exactly two to one. In a mixture, such as air, the proportions of the ingredients can vary.

mixture

a combination of two or more substances that are not chemically combined

Mixtures Can Vary in Composition and Properties

A glass of sweetened tea is a mixture. If you have ever had a glass of tea that was too sweet or not sweet enough, you have experienced two important characteristics of mixtures. A mixture does not always have the same balance of ingredients. The proportion of the materials in a mixture can change. Because of this, the properties of the mixture may vary.

For example, pure gold, shown in **Figure 18a,** is often mixed with other metals, usually silver, copper, or nickel, in various proportions to change its density, color, and strength. This solid mixture, or *alloy,* is stronger than pure gold. A lot of jewelry is 18-karat gold, meaning that it contains 18 grams of gold per 24 grams of alloy, or 75% gold by mass. A less expensive, and stronger, alloy is 14-karat gold, shown in **Figure 18b.**

Figure 18

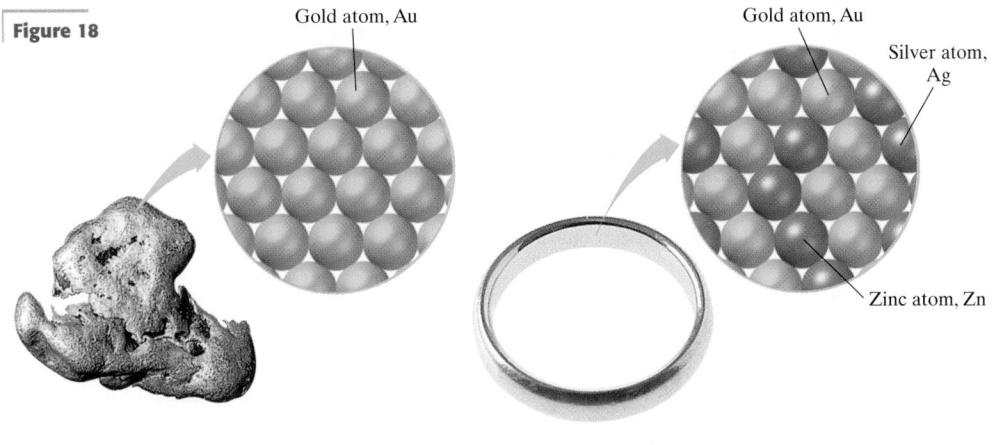

Gold atom, Au

Gold atom, Au

Silver atom, Ag

Zinc atom, Zn

a The gold nugget is a pure substance—gold. Pure gold, also called 24-karat gold, is usually considered too soft for jewelry.

b This ring is 14-karat gold, which is 14/24, or 58.3%, gold. This homogeneous mixture is stronger than pure gold and is often used for jewelry.

25

Demonstration

Separation by Distillation

1. Dissolve 25 g of $CuSO_4 \cdot 5H_2O$ in 100 mL of water, and show the blue solution to the class.

2. Pour the solution into a 250 mL Erlenmeyer flask that is clamped to a ring stand and resting on a hot plate. Add a boiling stone.

3. Insert into the mouth of the flask an appropriate one-hole rubber stopper fitted with a 10 cm piece of glass tubing.

4. Attach a 1 m piece of clear plastic or rubber tubing to the top of the glass tubing, and run the other end to another flask clamped so that it is submerged three-quarters of the way in a large beaker of cold water.

5. Gently heat the flask until boiling begins. Control the heat so that moderate boiling continues.

6. After a few minutes, remove the tube from the cold flask and then turn off the heat.

7. Show the cold flask to the class, and compare it with the hot solution. Ask students to explain the difference in color. Pour some condensate and solution into separate Petri dishes and compare the residues after evaporation.

Safety Caution: Wear safety goggles and a lab apron.

Disposal: Pour the colorless distillate down the drain. Let the blue solution evaporate, and save the crystals for reuse at a later time. Do not return them to the reagent bottle.

Water molecule, H_2O

Sugar molecule, $C_{12}H_{22}O_{11}$

Water molecule, H_2O

Silicon dioxide molecule, SiO_2

Figure 19
The mixture of sugar and water on the left is a homogeneous mixture, in which there is a uniform distribution of the two components. Sand and water, on the right, do not mix uniformly, so they form a heterogeneous mixture.

homogeneous

describes something that has a uniform structure or composition throughout

heterogeneous

composed of dissimilar components

Homogeneous Mixtures

Sweetened tea and 14-karat gold are examples of **homogeneous** mixtures. In a homogeneous mixture, the pure substances are distributed uniformly throughout the mixture. Gasoline, syrup, and air are homogeneous mixtures. Their different components cannot be seen—not even using a microscope.

Because of how evenly the ingredients are spread throughout a homogeneous mixture, any two samples taken from the mixture will have the same proportions of ingredients. As a result, the properties of a homogeneous mixture are the same throughout. Look at the homogeneous mixture in **Figure 19.** You cannot see the different materials that make up the mixture because the sugar is mixed evenly throughout the water.

Heterogeneous Mixtures

In **Figure 19** you can clearly see the water and the sand, so the mixture is not homogeneous. It is a **heterogenous** mixture because it contains substances that are not evenly mixed. Different regions of a heterogeneous mixture have different properties. Additional examples of the two types of mixtures are shown in **Table 6.**

Table 6	**Examples of Mixtures**
Homogeneous	Iced tea—uniform distribution of components; components cannot be filtered out and will not settle out upon standing
	Stainless steel—uniform distribution of components
	Maple syrup—uniform distribution of components; components cannot be filtered out and will not settle out upon standing
Heterogeneous	Orange juice or tomato juice—uneven distribution of components; settles out upon standing
	Chocolate chip pecan cookie—uneven distribution of components
	Granite—uneven distribution of components
	Salad—uneven distribution of components; can be easily separated by physical means

26

Videos

CNN. Presents Chemistry Connections

• **Segment 6** Cleaner Gas

See the Science in the News video guide for more details.

Transparencies

TT Types of Mixtures

Distinguishing Mixtures from Compounds

A compound is composed of two or more elements chemically joined together. A mixture is composed of two or more substances physically mixed together but not chemically joined. As a result, there are two major differences between mixtures and compounds.

First, the properties of a mixture reflect the properties of the substances it contains, but the properties of a compound often are very different from the properties of the elements that make it up. The oxygen gas that is a component of the mixture air can still support a candle flame. However, the properties of the compound water, including its physical state, do not reflect the properties of hydrogen and oxygen.

Second, a mixture's components can be present in varying proportions, but a compound has a definite composition in terms of the masses of its elements. The composition of milk, for example, will differ from one cow to the next and from day to day. However, the compound sucrose is always exactly 42.107% carbon, 6.478% hydrogen, and 51.415% oxygen no matter what its source is.

Separating Mixtures

One task a chemist often handles is the separation of the components of a mixture based on one or more physical properties. This task is similar to sorting recyclable materials. You can separate glass bottles based on their color and metal cans based on their attraction to a magnet. Techniques used by chemists include filtration, which relies on particle size, and distillation and evaporation, which rely on differences in boiling point.

Separating a Mixture

SAFETY PRECAUTIONS

PROCEDURE

1. Place the **mixture** of iron, sulfur, and salt on a **watchglass.** Remove the iron from the mixture with the aid of a **magnet.** Transfer the iron to a 50 mL **beaker.**

2. Transfer the sulfur-salt mixture that remains to a second 50 mL **beaker.** Add 25 mL of **water,** and stir with a **glass stirring rod** to dissolve the salt.

3. Place **filter paper** in a **funnel.** Place the end of the funnel into a third 50 mL **beaker.** Filter the mixture and collect the filtrate—the liquid that passes through the filter.

4. Wash the residue in the filter with 15 mL of water, and collect the rinse water with the filtrate.

5. Set up a **ring stand** and a **Bunsen burner.** Evaporate the water from the filtrate.

Stop heating just before the liquid completely disappears.

ANALYSIS

1. What properties did you observe in each of the components of the mixture?

2. How did these properties help you to separate the components of the mixture?

3. Did any of the components share similar properties?

27

REAL-WORLD CONNECTION — ADVANCED

Chromatography Have students research chromatographic techniques used to separate mixtures and the circumstances in which each type is best applied. Students could include paper chromatography, column chromatography, thin-layer chromatography, gas chromatography, gas-liquid chromatography, high-pressure liquid chromatography, ion-exchange chromatography, and others. Students can present their findings in a written report and a poster illustrating the technique. **LS Logical**

Skills Acquired:
• Experimenting

Materials:
• beaker, 50 mL (3)
• Bunsen burner
• filter paper
• funnel
• iron, 5 g
• magnet
• ring stand with ring clamp
• salt, 5 g
• stirring rod
• striker
• sulfur, 5 g
• watchglass
• water

Teacher's Notes: Place magnets in sealable plastic bags before use to allow the iron to be removed more easily.

Safety Caution: Safety goggles and a lab apron must be worn at all times.

Disposal: Dissolve the salt residue with water, and pour down the drain. The sulfur can be dried and remixed with the iron for reuse in this activity.

Answers

1. iron: dark gray solid, attracted to magnet; sulfur: yellow solid, not attracted to magnet, does not dissolve in water; salt: white crystalline solid, dissolves in water

2. Each component had a property that was different from the properties of the other components and allowed it to be separated from the mixture.

3. All components were solid, so they had the same physical state.

Chapter Resource File

• Datasheets for In-Text Labs

Close

Quiz ——————— GENERAL

1. What is a pure substance that is made of only one kind of atom?
Ans. an element

2. What classification of matter can be separated into two simpler parts using physical means?
Ans. a mixture

3. What type of mixture is a solution? Ans. homogeneous

4. What is a pure substance that is made up of more than one kind of element? Ans. compound

 Logical

Reteaching ——————— BASIC

Have the students create a flow-chart, concept map, or other graphic organizer that illustrates the connections between the terms *matter, element, compound, pure substance, mixture, homogeneous mixture,* and *heterogeneous mixture.* LS Visual

---

Chapter Resource File

• Concept Review

• Quiz

Transparencies

TT Classifying Matter

Matter

Pure substance
one kind of atom or molecule

H_2O
(water)

He
(helium)

Mixture
more than one kind
of atom or molecule

Element
a single kind of atom

Cl_2
(chlorine gas)

Compound
bonded atoms

CH_4
(methane)

Homogeneous mixture
uniform composition

Sugar   Water

Heterogeneous mixture
nonuniform composition

Water

Sand

③ Section Review

UNDERSTANDING KEY IDEAS

1. What are the two types of pure substances?

2. Define the term *compound.*

3. How does an element differ from a compound?

4. How are atoms and molecules related?

5. What is the smallest number of elements needed to make a compound?

6. What are two differences between compounds and mixtures?

7. Identify each of the following as an element, a compound, a homogeneous mixture, or a heterogeneous mixture.
   a. CH_4 d. salt water
   b. S_8 e. CH_2O
   c. distilled water   f. concrete

8. How is a homogeneous mixture different from a heterogeneous mixture?

CRITICAL THINKING

9. Why is a *monatomic compound* nonsense?

10. Compare the composition of sucrose purified from sugar cane with the composition of sucrose purified from sugar beets. Explain your answer.

11. After a mixture of iron and sulfur are heated and then cooled, a magnet no longer attracts the iron. How would you classify the resulting material? Explain your answer.

12. How could you decide whether a ring was 24-karat gold or 14-karat gold without damaging the ring?

13. To help a child take aspirin, a parent may crush the tablet and add it to applesauce. Is the aspirin-applesauce combination classified as a compound or a mixture? Explain your answer.

14. Four different containers are labeled C + O_2, CO, CO_2, and Co. Based on the labels, classify each as an element, a compound, a homogeneous mixture, or a heterogeneous mixture. Explain your reasoning.

28

Answers to Section Review

1. elements and compounds

2. a substance made up of atoms of two or more different elements joined by chemical bonds

3. An element is made up of one type of atom, but a compound must be made up of atoms of two or more different elements.

4. molecule consists of one or more atoms

5. two

6. Compounds have properties that differ from the properties of their components, but mixtures have properties that often reflect the properties of the components that make up

the mixture. A compound has a definite composition, so the components must be present in exact ratios. Mixtures can have varying compositions, so the proportions of the components can vary.

7. a. compound
   b. element
   c. compound
   d. homogeneous mixture
   e. compound
   f. heterogeneous mixture

Answers continued on p. 35A

Al
13
Aluminum
26.981 539
$[Ne]3s^23p^1$

Element Spotlight

Aluminum's Humble Beginnings

In 1881, Charles Martin Hall was a 22-year-old student at Oberlin College, in Ohio. One day, Hall's chemistry professor mentioned in a lecture that anyone who could discover an inexpensive method for making aluminum metal would become rich. Working in a wooden shed and using a cast-iron frying pan, a blacksmith's forge, and homemade batteries, Hall discovered a practical technique for producing aluminum. Hall's process is the basis for the industrial production of aluminum today.

Industrial Uses

- Aluminum is the most abundant metal in Earth's crust. However, it is found in nature only in compounds and never as the pure metal.
- The most important source of aluminum is the mineral bauxite. Bauxite consists mostly of hydrated aluminum oxide.
- Recycling aluminum by melting and reusing it is considerably cheaper than producing new aluminum.
- Aluminum is light, weather-resistant, and easily worked. These properties make aluminum ideal for use in aircraft, cars, cans, window frames, screens, gutters, wire, food packaging, hardware, and tools.

Real-World Connection Recycling just one aluminum can saves enough electricity to run a TV for about four hours.

Aluminum's resistance to corrosion makes it suitable for use outdoors in this statue.

A Brief History

1827: F. Wöhler describes some of the properties of aluminum.

1886: Charles Martin Hall, of the United States, and Paul-Louis Héroult, of France, independently discover the process for extracting aluminum from aluminum oxide.

1800	1900

1824: F. Wöhler, of Germany, isolates aluminum from aluminum chloride.

1854: Henri Saint-Claire Deville, of France, and R. Bunsen, of Germany, independently accomplish the electrolysis of aluminum from sodium aluminum chloride.

Questions

1. Research and identify at least five items that you encounter on a regular basis that are made with aluminum.
2. Research the changes that have occurred in the design and construction of aluminum soft-drink cans and the reasons for the changes. Record a list of items that help illustrate why aluminum is a good choice for this product.

internet connect

www.scilinks.org
Topic: Aluminum
SciLinks code: HW4136

SCILINKS. Maintained by the National Science Teachers Association

29

Element Spotlight
Aluminum's Humble Beginnings

Oberlin College has been declared a National Chemical Historic Landmark for the Hall Process. While Hall did his research in America, Paul-Louis Héroult simultaneously discovered the same process in France. The process they discovered is often called the Hall-Héroult process. The price of aluminum in 1852 was in excess of $540/lb. Just before Hall's discovery in 1886, the price was about $11/lb. Since Hall's discovery, the price of aluminum has been as low as $0.15/lb.

One of the most important compounds of aluminum is alumina, or aluminum oxide. Alumina occurs naturally as ruby and sapphire, and it is used extensively in glassmaking and related fields. Synthetic ruby and sapphire are used to produce laser light.

Aluminum is one of the most widely recycled metals. It is far less expensive to melt and reuse aluminum than to make cans from new aluminum. Aluminum is being used for some fuel cells that might be used to power cars in the future. It is also used in the manufacture of items including car engines, baseball bats, and aircraft parts.

did you know?

Outside the United States, the spelling *aluminium* is generally used. Before the development of the Hall-Héroult process, aluminum was so difficult to prepare from its ore that it was considered a precious metal. Aluminum was used in small quantities to decorate objects and to make novelty items, such as picture frames for the very wealthy.

Answers to Feature Questions

1. Answers could include cans, food packaging, tools, cars, window frames, and siding.
2. Accept all reasonable and factual answers.

1 CHAPTER HIGHLIGHTS

Alternative Assessments

Homework
• Pages 19, 28

SKILL BUILDER
• Page 6

Group Activity
• Page 14

CHAPTER REVIEW
• Item 43

Portfolio Assessments

SKILL BUILDER
• Page 14

CHAPTER REVIEW
• Item 44

Chapter Resource File

• **Observation Lab** Laboratory Procedures
• **CBL™ Probeware Lab** Analyzing Graphs and Establishing Relationships
• **Datasheets for In-Text Lab** Laboratory Techniques
• **Datasheets for In-Text Lab** Conservation of Mass—Percentage of Water in Popcorn

---

KEY IDEAS

SECTION ONE What Is Chemistry?
• Chemistry is the study of chemicals, their properties, and the reactions in which they are involved.
• Three of the states of matter are solid, liquid, and gas.
• Matter undergoes both physical changes and chemical changes. Evidence can help to identify the type of change.

SECTION TWO Describing Matter
• Matter has both mass and volume; matter thus has density, which is the ratio of mass to volume.
• Mass and weight are not the same thing. Mass is a measure of the amount of matter in an object. Weight is a measure of the gravitational force exerted on an object.
• SI units are used in science to express quantities. Derived units are combinations of the basic SI units.
• Conversion factors are used to change a given quantity from one unit to another unit.
• Properties of matter may be either physical or chemical.

SECTION THREE How Is Matter Classified?
• All matter is made from atoms.
• All atoms of an element are alike.
• Elements may exist as single atoms or as molecules.
• A molecule usually consists of two or more atoms combined in a definite ratio.
• Matter can be classified as a pure substance or a mixture.
• Elements and compounds are pure substances. Mixtures may be homogeneous or heterogeneous.

KEY SKILLS

Using Conversion Factors
Skills Toolkit 1 p. 13
Sample Problem A p. 14

KEY TERMS

chemical
chemical reaction
states of matter
reactant
product

matter
volume
mass
weight
quantity
unit
conversion factor
physical property
density
chemical property

atom
pure substance
element
molecule
compound
mixture
homogeneous
heterogeneous

USING KEY TERMS

1. What is chemistry?

2. What are the common physical states of matter, and how do they differ from one another?

3. Explain the difference between a physical change and a chemical change.

4. Identify the reactants and products in the following word equation:

iron oxide + aluminum ⟶
iron + aluminum oxide

5. During photosynthesis, light energy is captured by plants to make sugar from carbon dioxide and water. In the process, oxygen is also produced. Use this information to explain the terms *chemical reaction, reactant,* and *product.*

6. What units are used to express mass and weight?

7. How does a quantity differ from a unit? Give examples of each in your answer.

8. What is a conversion factor?

9. Explain what derived units are. Give an example of one.

10. Define *density,* and explain why it is considered a physical property rather than a chemical property of matter.

11. Write a brief paragraph that shows that you understand the following terms and the relationships between them: *atom, molecule, compound,* and *element.* **WRITING SKILLS**

12. What do the terms *homogeneous* and *heterogeneous* mean?

UNDERSTANDING KEY IDEAS

What Is Chemistry?

13. Your friend mentions that she eats only natural foods because she wants her food to be free of chemicals. What is wrong with this reasoning?

14. Determine whether each of the following substances would be a gas, a liquid, or a solid if found in your classroom.
a. neon
b. mercury
c. sodium bicarbonate (baking soda)
d. carbon dioxide
e. rubbing alcohol

15. Is toasting bread an example of a chemical change? Why or why not?

16. Classify each of the following as a physical change or a chemical change, and describe the evidence that suggests a change is taking place.
a. cracking an egg
b. using bleach to remove a stain from a shirt
c. burning a candle
d. melting butter in the sun

17. Over time, soap scum deposits build up around your bathtub. Unlike soap, the soap scum is a calcium-based compound that cannot be dissolved in water. What evidence do you have that indicates that the formation of soap scum is a chemical change rather than a physical change?

31

Assignment Guide

Section	Questions
1	1–5, 13–17, 33–35, 50
2	6–10, 18–25, 30–32, 36–38, 42–49
3	11, 12, 26–29, 39–41

REVIEW ANSWERS

Using Key Terms

1. Chemistry is a science that deals with matter and the changes it undergoes.

2. The common states of matter are solid, liquid, and gas. Solids have a fixed volume and shape. Liquids have a fixed volume but can flow and take the shape of their container. Gases have no fixed shape or volume and take the shape and volume of any container they occupy.

3. A physical change does not change the identity of a substance, while a chemical change causes a substance to change its identity as one or more new substances are made.

4. reactants: iron oxide and aluminum; products: iron and aluminum oxide

5. A chemical reaction is a process by which a substance or substances known as reactants are changed into one or more new substances known as products. During photosynthesis, a chemical reaction takes place that changes the reactants of carbon dioxide and water into two new products, sugar and oxygen.

6. Mass is expressed in kilograms. Weight is expressed in newtons.

7. A quantity is a measurable property, such as length, mass, time, or temperature. A unit is a standard used to measure a quantity, such as meters, kilograms, seconds, or kelvins.

8. A conversion factor is a ratio used to convert from one unit to another.

9. A derived unit is a combination of base units. Examples include square meter, cubic centimeter, and grams per cubic centimeter.

10. Density is the ratio of mass to volume. The density of a material may be observed and measured without changing the identity of the substance.

11. A molecule is the smallest unit of a substance that keeps all of the physical and chemical properties of that substance. An element is composed of molecules that can be single atoms or can be more than one of the same kind of atom. A compound is composed of two or more elements, so a molecule of a compound is made up of two or more different kinds of atoms.

12. Both consist of two or more different substances and can vary in composition. Homogeneous mixtures are uniform in composition because the molecules of the substances are mixed completely. The substances in heterogeneous mixtures are not evenly mixed.

Understanding Key Ideas

13. All matter, including natural foods, is made up of chemicals.

14. a. gas
    b. liquid
    c. solid
    d. gas
    e. liquid

15. Toasting bread is an example of a chemical change. The browning of the bread is a sign that a new substance is formed.

16. a. Cracking an egg is a physical change. The shell is broken into smaller pieces, but its chemical makeup has not changed.
    b. Using bleach is a chemical change. The color of the stain is removed.
    c. Burning a candle is a chemical change. As the candle burns, the wick becomes black, and gases are released along with energy in the form of light and heat.

Describing Matter

18. At the top of Mount Everest, the force of gravity is less than at sea level. Describe what would happen to your mass and to your weight if you traveled from sea level to the top of Mount Everest. Explain your answer.

19. Name the five most common SI base units used in chemistry. What quantity is each unit used to express?

20. What derived unit is appropriate for expressing each of the following?
    a. rate of water flow
    b. speed
    c. volume of a room

21. Pick an object you can see right now. List three physical properties of the object that you can observe. Can you also observe a chemical property of the object? Explain your answer.

22. Compare the physical and chemical properties of salt and sugar. What properties do they share? Which properties could you use to distinguish between salt and sugar?

23. What do you need to know to determine the density of a sample of matter?

24. For each pair below, indicate the substance that has the greater density. Explain your answer.
    a. penny and cork
    b. ice cube and liquid water

25. Substances A and B are colorless, odorless liquids that are nonconductors and flammable. The density of substance A is 0.97 g/mL; the density of substance B is 0.89 g/mL. Are A and B the same substance? Explain your answer.

How Is Matter Classified?

26. Is a compound a pure substance or a mixture? Explain your answer.

27. Iron (Fe) and sulfur (S_8) react to create iron(II) sulfide (FeS). Are iron and sulfur elements, compounds, or mixtures? Is iron(II) sulfide an element, a compound, or a mixture?

28. Which of the following items are homogeneous: a tree, a cheeseburger, flour, paint, and an ice cube?

29. Determine if each material represented below is an element, compound, or mixture, and whether the model illustrates a solid, liquid, or gas.

a. 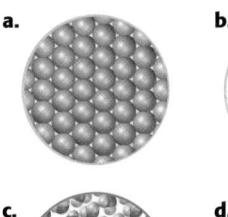 b.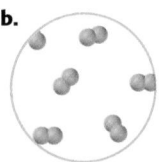

c. 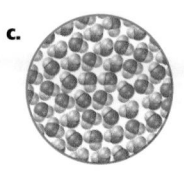 d.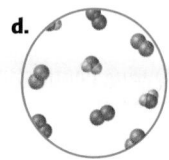

PRACTICE PROBLEMS

Sample Problem A Converting Units

30. Which quantity of each pair is larger?
    a. 2400 cm or 2 m
    b. 3 L or 3 mL
    c. 17 g or 1.7 kg

31. Using **Appendix A,** convert the following measurements to the units specified.
    a. 357 mL = ? L   e. 250 μg = ? g
    b. 25 kg = ? mg   f. 250 μg = ? kg
    c. 35 000 cm³ = ? L   g. 1.5 s = ? ms
    d. 2.46 L = ? cm³   h. 10.5 mol = ? mmol

32. What is the distance in meters between two points that are 150 km apart?

MIXED REVIEW

33. Use particle models to explain why liquids and gases take the shape of their containers.

34. When camping with your family, you boil a pot of water over a campfire. What physical and chemical changes take place during this process? What evidence of a chemical change do you observe?

35. You are given a sample of colorless liquid in a beaker. What type of information could you gather to determine if the liquid is water?

36. Calculate the density of a piece of metal if its mass is 201.0 g and its volume is 18.9 cm³.

37. The density of CCl_4 (carbon tetrachloride) is 1.58 g/mL. What is the mass of 95.7 mL of CCl_4?

38 What is the volume of 227 g of olive oil if its density is 0.92 g/mL?

CRITICAL THINKING

39. A white, crystalline material that looks like table salt releases gas when heated under certain conditions. There is no change in the appearance of the solid, but the reactivity of the material changes.
   a. Did a chemical or physical change occur? How do you know?
   b. Was the original material an element or a compound? Explain your answer.

40. A student leaves an uncapped watercolor marker on an open notebook. Later, the student discovers the leaking marker has produced a rainbow of colors on the top page.
   a. Is this an example of a physical change or a chemical change? Explain your answer.
   b. Should the ink be classified as an element, a compound, or a mixture? Explain your answer.

41. A student checks the volume, melting point, and shape of two unlabeled samples of matter and finds that the measurements are identical. From this he concludes that the samples have the same chemical composition. Is this a valid conclusion? What additional information might he collect to test his conclusion?

42. Equal amounts of three liquids—carbon tetrachloride ($D = 1.58$ g/cm³), mercury ($D = 13.546$ g/cm³), and water ($D = 1.00$ g/cm³)—are mixed together. After mixing, they separate to form three distinct layers. Draw and label a sketch showing the order of the layers in a test tube.

ALTERNATIVE ASSESSMENT

43. Your teacher will provide you with a sample of a metallic element. Determine its density. Check references that list the density of metals to identify the sample that you analyzed.

44. Make a poster showing the types of product warning labels that are found on products in your home.

CONCEPT MAPPING

45. Use the following terms to complete the concept map below: *volume, density, matter, physical property,* and *mass*.

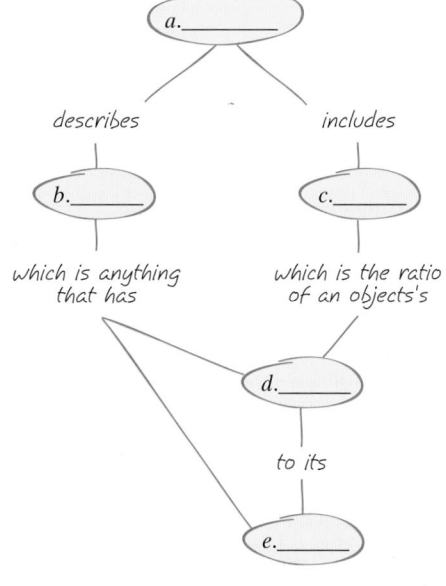

   d. Melting butter is a physical change. The state of the butter changes from solid to liquid, but the chemical makeup of the butter does not change.

17. Soap scum does not dissolve in water but soap does. The change in properties is a sign that a new substance has formed.

18. Mass stays the same because the amount of matter in your body has not changed. Weight would decrease because the force of gravity is less at the top of the mountain.

19. meter: length; kilogram: mass; second: time; kelvin: temperature; mole: amount of substance

20. a. Possible answers include L/min and L/sec.
   b. Possible answers include m/s, cm/s, km/h.
   c. m³

21. Answers will vary. Physical properties include shape, texture, state of matter, and taste. Chemical properties could not be observed unless a chemical change was attempted.

22. Both salt and sugar are white, crystalline solids at room temperature. Both dissolve in water. The melting point of salt is very high compared with that of sugar. Their densities are different. Either melting point or density could be used to distinguish between them.

23. You need to know the mass and volume of the sample.

24. a. The penny sinks in water, and the cork floats in water. The penny has the greater density.
   b. Ice floats in liquid water, so the liquid water has the greater density.

25. No, despite their similar properties, they must be different. The same substance would always have the same density under the same conditions.

26. A compound is a pure substance. It has a definite composition and a single set of properties that are different from the properties of the elements it is composed of.

27. Iron and sulfur are elements. Iron(II) sulfide is a compound.

28. The flour and the ice cube are homogeneous. All others are heterogeneous.

29. a. element, solid

   b. element, gas

   c. compound, liquid

   d. mixture, gas

Practice Problems

30. a. 2400 cm

   b. 3 L

   c. 1.7 kg

31. a. 0.357 L

   b. 2.5×10^7 mg

   c. 35 L

   d. 2460 cm^3

   e. 2.5×10^{-4} g

   f. 2.5×10^{-7} kg

   g. 1500 ms

   h. 10 500 mmol

32. 150 000 m

33. The particles of a liquid move past one another, and shift places as they are placed in a different container. The particles of a gas move independently of one another. When a gas is placed into a container, the particles eventually move to every part of the container, and the gas takes the shape of the container.

Mixed Review

34. Water boiling is a physical change. The fuel in the campfire burning is a chemical change. Evidence of this change include gas formed from the burning fuel, the color change of the wood, and the emission of energy as heat and light.

Answers continued on p. 35B

FOCUS ON GRAPHING

Study the graph below, and answer the questions that follow. For help in interpreting graphs, see Appendix B, "Study Skills for Chemistry."

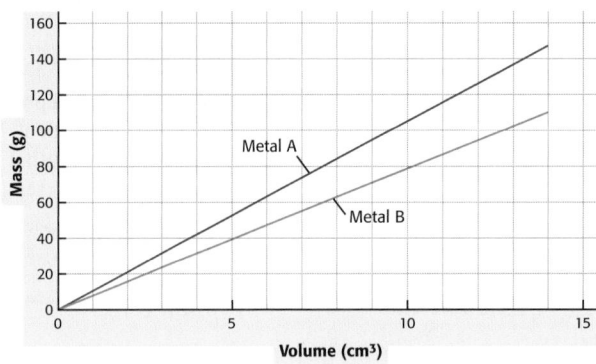

Mass Versus Volume for Two Metals

46. What does the straight line on the graph indicate about the relationship between volume and mass?

47. What does the slope of each line indicate?

48. What is the density of metal A? of metal B?

49. Based on the density values in **Table 4,** what do you think is the identity of metal A? of metal B? Explain your reasoning.

TECHNOLOGY AND LEARNING

50. Graphing Calculator

Graphing Tabular Data

The graphing calculator can run a program that graphs ordered pairs of data, such as temperature versus time. In this problem, you will answer questions based on a graph of temperature versus time that the calculator will create.

Go to Appendix C. If you are using a TI-83 Plus, you can download the program and data sets and run the application as directed. Press the **APPS** key on your calculator, and then choose the application **CHEMAPPS.** Press 1, then highlight **ALL** on the screen, press 1, then highlight **LOAD,** and press 2 to load the data into your calculator. Quit the application, and then run the program **GRAPH.** A set of data points representing degrees Celsius versus time in minutes will be graphed.

If you are using another calculator, your teacher will provide you with keystrokes and data sets to use.

   a. Approximately what would the temperature be at the 16-minute interval?

   b. Between which two intervals did the temperature increase the most: between 3 and 5 minutes, between 5 and 8 minutes, or between 8 and 10 minutes?

   c. If the graph extended to 20 minutes, what would you expect the temperature to be?

34

> **Chapter Resource File**
>
> • Chapter Test

1. Magnesium dissolves in hydrochloric acid to produce magnesium chloride and hydrogen gas. Which of the following represents the reactants in this reaction?
   a. magnesium and magnesium chloride
   b. hydrochloric acid and hydrogen gas
   c. magnesium and hydrochloric acid
   d. magnesium chloride and hydrogen gas

2. Matter that has a definite shape and a definite volume is
   a. an acid.
   b. a base.
   c. a solid.
   d. a gas.

3. Which of the following pairs of measurements are equal?
   a. 1.63 kg and 163 g
   b. 0.0704 m and 7.04 mm
   c. 0.015 mL and 1.5 L
   d. 325 mg and 0.325 g

4. A chemical property of a substance
   a. depends on how the substance reacts with other substances.
   b. can be described when the substance undergoes a physical change.
   c. cannot be quantitatively described.
   d. cannot be qualitatively described.

5. The SI base unit that is commonly used in chemistry to describe the amount of a substance is the
   a. degree Celsius.
   b. gram
   c. liter.
   d. mole.

6. Matter can be defined as anything that
   a. has weight.
   b. has mass and volume.
   c. can be described in SI units.
   d. exhibits both chemical and physical properties.

7. Which of the following is best classified as a homogeneous mixture?
   a. pizza
   b. blood
   c. hot tea
   d. copper wire

8. A compound differs from a mixture in that a compound
   a. contains only one element.
   b. varies in chemical composition depending on the sample size.
   c. has a definite composition by mass of the elements it contains.
   d. can be classified as either heterogeneous or homogeneous.

9. Which of the following is an element?
   a. He
   b. $BaCl_2$
   c. NaOH
   d. CO

10. Determine which of the following measurements is the largest.
    a. 0.200 L
    b. 0.020 kL
    a. 20.0 cL
    d. 2000 mL

35

Continuation of Answers

Answers to Section 1 Review, *continued*

7. Answers may vary, and may include burning a match, breaking down mercury(II) oxide, and photosynthesis

8. Limestone is the reactant. Lime and carbon dioxide are the products.

9. Carbon dioxide and sodium phenoxide are reactants. Sodium salicylate is the product.

10. evolution of a gas, formation of a precipitate, emission of light, change in color

11. The particles in liquids and gases are able to move past one another and are not held in place. Therefore, liquids and gases adopt the shape of the container in which they are placed.

12. physical

13. Yes; When ice melts, the same mass of water forms as the mass of ice that melted.

14. **a.** A gas is forming in each beaker.

    b. A physical change is occurring in beaker A. The bubbles forming throughout the liquid are the result of the water boiling as it is heated. A chemical change is occurring in beaker B. The gas is forming only at the wire, so it is probably not boiling. Instead, the electrical energy is breaking down the water.

Answers to Section 2 Review, *continued*

10. 0.0173 L

11. about 1081 beans

12. 11 g/cm^3

13. $\dfrac{1 \text{ kg}}{1 \text{ L}} \times \dfrac{1000 \text{ g}}{1 \text{ kg}} \times \dfrac{1 \text{ L}}{1000 \text{ mL}} \times \dfrac{1 \text{ mL}}{1 \text{ cm}^3} = \dfrac{1 \text{ g}}{1 \text{ cm}^3}$

14. Mostly complete combustion of carbon is happening. The density is close to the density of carbon dioxide, so mostly carbon dioxide is produced along with a small amount of carbon monoxide.

Answers to Section 3 Review, *continued*

8. In a homogeneous mixture, the components are distributed evenly throughout, so the properties of the mixture are the same throughout. In a heterogeneous mixture, the components are not evenly distributed, so there are regions of the mixture with different properties.

9. A compound must be composed of more than one type of atom, so it cannot be monatomic.

10. The compositions are identical. Sucrose is a compound, so it must have a definite composition no matter what its source is.

11. The material is a compound of the iron and sulfur. The material is not attracted to a magnet so it has different properties than the original materials had.

12. determine its density

13. a mixture; They are not chemically combined. Each material keeps its own properties and identity.

14. C + O_2 is a heterogeneous mixture because the materials are a solid and a gas. CO is a compound. CO_2 is a compound. Co is an element.

Answers to Chapter Review, *continued*

35. Answers should include seeing if the liquid has an odor, calculating its density, checking its boiling point and its freezing point.

36. 10.6 g/cm^3

37. 151 g

38. 2.5×10^2 mL

Critical Thinking

39. a. chemical; The physical appearance does not change, but the process of heating the material produces a gas and changes its reactivity.

  b. compound; The original material seems to break down to form a gas and a solid. A compound could be broken down by heating. An element cannot be broken down by heating.

40. a. This is a physical change because the dyes have only been separated, not changed in composition.

  b. The ink is a mixture because it was made up of different substances with different colors that were separated without changing their identities.

41. The conclusion is not valid as there was not enough information. Two different substances may have similar melting points. Volume is not a characteristic property as it changes with the amount of material. If mass is also measured, the density can be determined. Density is a characteristic property. The student should also compare the chemical properties of the sample to test his conclusion.

42. Sketches should show mercury on the bottom, carbon tetrachloride in the middle, and water on top.

Alternative Assessment

43. Students will have to devise a way of determining the volume of the object if it does not have a regular shape.

44. Grade posters based on the variety of products shown, organization, labeling, spelling, and overall artistic quality.

Concept Mapping

45. a. physical property

  b. matter

  c. density

  d. mass

  e. volume

Focus on Graphing

46. Mass increases as volume increases.

47. The slope of each line is the density of the metal.

48. Metal A: about 10.5 g/cm^3; Metal B: about 7.9 g/cm^3

49. Metal A: silver; Metal B: iron

Technology and Learning

50. a. about 48°C

  b. between 5 and 8 minutes

  c. about 53°C

Matter and Energy
Chapter Planning Guide

PACING	CLASSROOM RESOURCES	LABS, ACTIVITIES, AND DEMONSTRATIONS
BLOCK 1 · 45 min pp. 36–37 **Chapter Opener**		**SE Start-Up Activity** Chemical Changes and Energy, p. 37 ◆
BLOCKS 2 & 3 · 90 min pp. 38–45 **Section 1** Energy	**TT Heating Curve for Water** * **TT Conservation of Energy in a Chemical Reaction** *	**TE Group Activity** Units of Energy, p. 42 `GENERAL` **TE Group Activity**, p. 43 `GENERAL` **CRF Observation Lab** Specific Heat ◆ `GENERAL`
BLOCKS 4 & 5 · 90 min pp. 46–53 **Section 2** Studying Matter and Energy	**TT The Scientific Method** * **OSP Supplemental Reading** The Same and Not the Same "Whirligigs" `ADVANCED` **OSP Supplemental Reading** The Same and Not the Same "Thalidomide" `ADVANCED`	**SE QuickLab** Using the Scientific Method, p. 47 ◆ `GENERAL` **CRF Datasheets for In-Text Labs** * **TE Demonstration** Is Mass Conserved? p. 50 ◆ **TE Group Activity** The Black Box, p. 51 ◆ `BASIC` **SE Skills Practice Lab** Separation of Mixtures, p. 762 ◆ `GENERAL` **CRF Datasheets for In-Text Labs** * **SE Inquiry Lab** Separation of Mixtures—Mining Contract, p. 770 ◆ `GENERAL` **CRF Datasheets for In-Text Labs** *
BLOCK 6 · 45 min pp. 54–64 **Section 3** Measurements and Calculations in Chemistry	**TT Accuracy and Precision** * **SE Element Spotlight** Deep Diving with Helium	**TE Activity**, p. 54 ◆ `GENERAL` **TE Group Activity**, p. 57 `ADVANCED` **TE Group Activity**, p. 59 ◆ `GENERAL`

BLOCKS 7 & 8 · 90 min

Chapter Review and Assessment Resources

SE Chapter Review, pp. 66–70
SE Standardized Test Prep, p. 71
CRF Chapter Test *
OSP Test Generator
CRF Test Item Listing *
OSP Scoring Rubrics and Classroom Management Checklists

Holt Chemistry: Online Resources

Visit **go.hrw.com** for a variety of free resources related to this textbook. Enter the keyword **HW4 HOME**.

Students can access interactive problem solving help and active visual concept development with the *Holt Chemistry* Online Edition available at **www.hrw.com**.

student CNN **News**

cnnstudentnews.com

Find the latest chemistry news, lesson plans, and activities related to important scientific events.

KEY

TE Teacher Edition	**CRF** Chapter Resource File	***** Also on One-Stop Planner
SE Student Edition	**TT** Teaching Transparencies	**◆** Requires Advance Prep
OSP One-Stop Planner		

PROBLEM SOLVING AND PRACTICE	SECTION REVIEW AND ASSESSMENT	STANDARDS CORRELATION
		National Science Education Standards
	TE Homework, p. 44 `GENERAL` **TE** Quiz, p. 45 `GENERAL` **CRF** Quiz * **TE** Reteaching, p. 45 `BASIC` **SE** Section Review, p. 45 **CRF** Concept Review *	PS 5a, PS 5c
	TE Quiz, p. 53 `GENERAL` **CRF** Quiz * **TE** Reteaching, p. 53 `BASIC` **SE** Section Review, p. 53 **CRF** Concept Review *	
SE Skills Toolkit 1 Rules for Determining Significant Figures, p. 57 **SE Skills Toolkit 2** Rules for Using Significant Figures in Calculations, p. 58 **SE Sample Problem A** Determining the Number of Significant Figures, p. 59 　**SE Practice**, p. 59 `GENERAL`　　**TE Homework**, p. 59 `GENERAL` **SE Sample Problem B** Calculating Specific Heat, p. 61 　**SE Practice**, p. 61 `GENERAL`　　**TE Homework**, p. 61 `GENERAL` **SE Skills Toolkit 3** Scientific Notation in Calculations, p. 62 **SE Skills Toolkit 4** Scientific Notation with Significant Figures, p. 63 **CRF Problem Solving** * `ADVANCED` **SE Problem Bank**, p. 858 `GENERAL`	**TE** Quiz, p. 63 `GENERAL` **CRF** Quiz * **TE** Reteaching, p. 63 `BASIC` **SE** Section Review, p. 63 **CRF** Concept Review *	

SCLINKS®

www.scilinks.org

Topic: Matter and Energy
SciLinks code: HW4158

Topic: Conservation of Energy
SciLinks code: HW4035

Topic: Temperature Scales
SciLinks code: HW4124

Topic: Scientific Method
SciLinks code: HW4167

Topic: Chance Discoveries
SciLinks code: HW4139

Topic: Specific Heat
SciLinks code: HW4119

Topic: Helium
SciLinks code: HW4171

Technology Resources

  **Science in the NEWS**

Each video segment is accompanied by a Critical Thinking Worksheet.

Segment 7
Science Controversy:
Salt in the Diet

Overview

In Section 1 of this chapter, students learn that all changes in matter involve a change of energy but that the amount of energy in a system is conserved during any chemical or physical change. The difference between heat and temperature is also explained. The scientific method is described in Section 2 and the definitions for theory and scientific laws are given. Students also learn that experiments must be controlled in order to draw valid conclusions from the data produced. Finally, in Section 3, students are introduced to accuracy, precision, significant figures, and scientific notation.

Assessing Prior Knowledge
Check for Content Knowledge

Students should be familiar with the following topics:

• rounding

• using a scientific calculator

• solving problems with Algebra I skills

Using the Figure

Have students research how the chemical composition of a volcano's magma affects the nature of volcanic eruptions. For example, some types of magma lead to mild eruptions, forming lava that can flow great distances. Other types of magma are thicker and produce eruptions that alternate between lava flows and explosions.

Students should discover that magma that is rich in iron and magnesium tends to be very fluid. Magma that is rich in silicon, on the other hand, is thick enough to trap gases and erupt explosively.

CHAPTER

2

MATTER AND ENERGY

36

Standards Correlations

National Science Education Standards

PS 5a: The total energy of the universe is constant. Energy can be transferred by collisions in chemical and nuclear reactions, by light waves and other radiations, and in many other ways. However, it can never be destroyed. As these transfers occur, the matter involved becomes steadily less ordered. (Section 1)

PS 5c: Heat consists of random motion and the vibrations of atoms, molecules, and ions. The higher the temperature, the greater the atomic or molecular motion. (Section 1)

HNS 2b: Scientific explanations must meet certain criteria. First and foremost, they must be consistent with experimental and observational evidence about nature, and must make accurate predictions, when appropriate, about systems being studied. They should also be logical, respect the rules of evidence, be open to criticism, report methods and procedures, and make knowledge public. Explanations on how the natural world changes based on myths, personal beliefs, religious values, mystical inspiration, superstition, or authority may be personally useful and socially relevant, but they are not scientific. (Section 2)

HNS 2c: Because all scientific ideas depend on experimental and observational confirmation, all scientific knowledge is, in principle, subject to change as new evidence becomes available. The core ideas of science such as the conservation

The photo of the active volcano and the scientists who are investigating it is a dramatic display of matter and energy. Most people who view the photo would consider the volcano and the scientists to be completely different. The scientists seem to be unchanging, while the volcano is explosive and changing rapidly. However, the scientists and the volcano are similar in that they are made of matter and are affected by energy. This chapter will show you the relationship between matter and energy and some of the rules that govern them.

START-UP ACTIVITY

SAFETY PRECAUTIONS

Chemical Changes and Energy

PROCEDURE

1. Place a small **thermometer** completely inside a **jar**, and close the **lid**. Wait 5 min, and record the temperature.

2. While you are waiting to record the temperature, soak one-half of a **steel wool pad** in **vinegar** for 2 min.

3. Squeeze the excess vinegar from the steel wool. Remove the thermometer from the jar, and wrap the steel wool around the bulb of the thermometer. Secure the steel wool to the thermometer with a **rubber band.**

4. Place the thermometer and the steel wool inside the jar, and close the lid. Wait 5 min, and record the temperature.

ANALYSIS

1. How did the temperature change?

2. What do you think caused the temperature to change?

3. Do you think vinegar is a reactant or product? Why?

Pre-Reading Questions

1. **When ice melts, what happens to its chemical composition?**

2. **Name a source of energy for your body.**

3. **Name some temperature scales.**

4. **What is a chemical property? What is a physical property?**

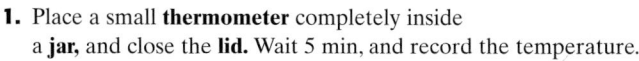

internet connect

www.scilinks.org
Topic: Matter and Energy
SciLinks code: HW4158

SCiLINKS Maintained by the National Science Teachers Association

START-UP ACTIVITY

Skills Acquired:
- Experimenting
- Measuring
- Collecting Data
- Interpreting
- Inferring

Materials:
For each group of 2–4 students:
- thermometer
- jar, large
- steel wool pad, one half
- vinegar, approx. 200 mL sample

Teacher's Notes:
- The thermometer must fit completely inside the jar so that the jar can be closed. Be sure to use a steel wool pad that does not contain soap.
- The vinegar removes any coating from the surface of the steel wool, exposing the iron in the steel. The iron comes in contact with oxygen and begins to react. The chemical reaction releases energy as heat and causes the temperature to rise.

Safety Caution: Remind students to use goggles, aprons, and gloves.

Answers
1. The temperature should have risen.
2. A chemical reaction released heat.
3. Vinegar is a reactant in this case, because it participated in a chemical reaction.

LS Visual

of energy or the laws of motion have been subjected to a wide variety of confirmations and are therefore unlikely to change in the areas in which they have been tested. In areas where data or understanding are incomplete, such as the details of human evolution or questions surrounding global warming, new data may well lead to changes in current ideas or resolve current conflicts. In situations where information is still fragmentary, it is normal for scientific ideas to be incomplete, but this is also where the opportunity for making advances may be greatest. (Section 2)

HNS 3d: The historical perspective of scientific explanations demonstrates how scientific knowledge changes by evolving over time, almost always building on earlier knowledge. (Section 2)

Answers to Pre-Reading Questions

1. It does not change.
2. The chemical energy stored in food.
3. Fahrenheit, Celsius, and Kelvin.
4. A chemical property describes the effects of chemical changes on matter. A physical property describes matter as it is and after it has undergone physical changes.

Overview

Before beginning this section, review with your students the Objectives listed in the Student Edition. In this section, students learn how energy is related to chemical and physical changes and learn energy is always conserved during such changes. The definition for heat is given and the difference between heat and temperature is explained. Finally, students are introduced to the Celsius and Kelvin temperature scales and practice converting between the two scales.

🔔 Bellringer

Have students work in small groups to brainstorm ideas relating to energy. They can list different types of energy, list why energy is important, and when energy is released or absorbed. After brainstorming for five minutes, ask the students to examine their lists and write their own definition for *energy*.

Motivate

Discussion

Sources of Energy Ask students where they would get the energy needed to perform the activities mentioned in the first paragraph. A common response is "from food." Then ask where the energy is stored in the food before it is released. Some students may already know that the energy involves chemical bonds. If so, ask how a chemical bond stores energy. Students may soon realize a need to learn more about the connection between energy and chemical substances.

KEY TERMS

- energy
- physical change
- chemical change
- evaporation
- endothermic
- exothermic
- law of conservation of energy
- heat
- kinetic energy
- temperature
- specific heat

OBJECTIVES

1. **Explain** that physical and chemical changes in matter involve transfers of energy.
2. **Apply** the law of conservation of energy to analyze changes in matter.
3. **Distinguish** between heat and temperature.
4. **Convert** between the Celsius and Kelvin temperature scales.

Energy and Change

If you ask 10 people what comes to mind when they hear the word *energy*, you will probably get 10 different responses. Some people think of energy in terms of exercising or playing sports. Others may picture energy in terms of a fuel or a certain food.

If you ask 10 scientists what comes to mind when they hear the word *energy*, you may also get 10 different responses. A geologist may think of energy in terms of a volcanic eruption. A biologist may visualize cells using oxygen and sugar in reactions to obtain the energy they need. A chemist may think of a reaction in a lab, such as the one shown in **Figure 1.**

The word *energy* represents a broad concept. One definition of **energy** is the capacity to do some kind of work, such as moving an object, forming a new compound, or generating light. No matter how energy is defined, it is always involved when there is a change in matter.

energy

the capacity to do work

Figure 1
Energy is released in the explosive reaction that occurs between hydrogen and oxygen to form water.

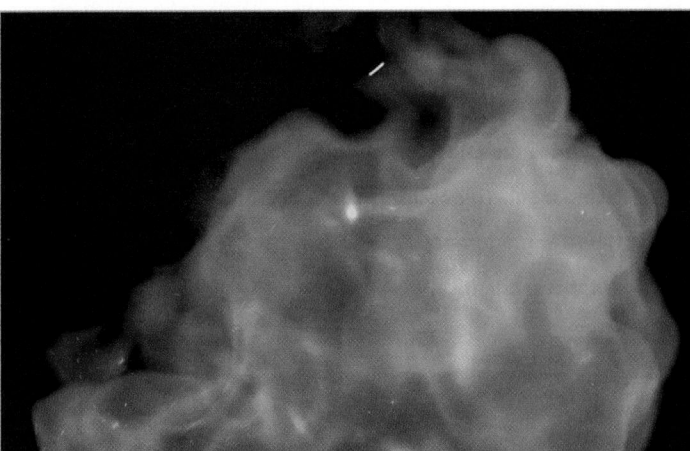

38

---

did you know?

Researchers at the Australian company Commonwealth Scientific and Industrial Research Organisation (CSIRO) are trying to develop artificial plants. They are attempting to use a mix of manufactured materials to mimic the role of chlorophyll in photosynthesis. CSIRO hopes their artificial plants will someday consume carbon dioxide from the atmosphere and produce fuels such as methane or food in the form of sugars.

Chapter Resource File

- Lesson Plan

Changes in Matter Can Be Physical or Chemical

Ice melting and water boiling are examples of **physical changes.** A physical change affects only the physical properties of matter. For example, when ice melts and turns into liquid water, you still have the same substance represented by the formula H_2O. When water boils and turns into a vapor, the vapor is still H_2O. Notice that in these examples the chemical nature of the substance does not change; only the physical state of the substance changes to a solid, liquid, or gas.

In contrast, the reaction of hydrogen and oxygen to produce water is an example of a **chemical change.** A chemical change occurs whenever a new substance is made. In other words, a chemical reaction has taken place. You know water is different from hydrogen and oxygen because water has different properties. For example, the boiling points of hydrogen and oxygen at atmospheric pressure are $-252.8°C$ and $-182.962°C$, respectively. The boiling point of water at atmospheric pressure is $100°C$. Hydrogen and oxygen are also much more reactive than water.

Every Change in Matter Involves a Change in Energy

All physical and chemical changes involve a change in energy. Sometimes energy must be supplied for the change in matter to occur. For example, consider a block of ice, such as the one shown in **Figure 2.** As long as the ice remains cold enough, the particles in the solid ice stay in place.

However, if the ice gets warm, the particles will begin to move and vibrate more and more. For the ice to melt, energy must be supplied so that the particles can move past one another. If more energy is supplied and the boiling point of water is reached, the particles of the liquid will leave the liquid's surface through **evaporation** and form a gas. These physical changes require an input of energy. Many chemical changes also require an input of energy.

Sometimes energy is released when a change in matter occurs. For example, energy is released when a vapor turns into a liquid or when a liquid turns into a solid. Some chemical changes also release energy. The explosion that occurs when hydrogen and oxygen react to form water is a release of energy.

physical change

a change of matter from one form to another without a change in chemical properties

chemical change

a change that occurs when one or more substances change into entirely new substances with different properties

evaporation

the change of a substance from a liquid to a gas

Figure 2
Energy is involved when a physical change, such as the melting of ice, happens.

Solid water, H_2O

Liquid water, H_2O

39

MISCONCEPTION ALERT

Some textbooks list *chemical energy* as a type of energy separate from kinetic energy and potential energy. However, *chemical energy* is an all inclusive term that generally refers to the total energy stored in matter as either kinetic energy or potential energy.

Teach

READING SKILL BUILDER — BASIC

Assimilating Knowledge Before students read this chapter, have them write a short, individual list of all the things they know about energy and energy transfer and about the process of scientific research, including the use of measurements in scientific calculations. Then have students list the things they might want to know about energy, scientific methods, and using measurements.

If students have trouble coming up with questions, here are some suggestions to get them on track.

• We eat food in order to have energy for life activities. So how is the energy stored in food?

• What are the major steps in scientific research?

• How do scientists record measurements in a way that reflects the precision of the measuring instrument?

LS Logical

Using the Figure

Figure 2 Exhibit a ball-and-stick model of a water molecule and manipulate it to show the ways the molecule as a whole can move and rotate. Then substitute springs for the O–H bonds and show the possible motions (vibrations) that can take place within the molecule.

Using the Figure

Figure 2 Explain to students that although water molecules in ice are held in a crystal lattice, they still vibrate in position, and therefore have kinetic energy. This is true of the molecules or atoms in any solid. Further explain that water molecules in the liquid phase are not held in a lattice and move around more than water molecules in ice and thus have more kinetic energy.

Teaching Tips

Identifying a System The concept of a system is better learned by example than by a verbal definition. To reinforce the idea, draw students' attention to (or have them identify) the system under consideration whenever they observe a demonstration or carry out an experiment.

Exothermic Reaction Instant heat packs usually employ one of two exothermic reactions. The first reaction is the crystallization of sodium acetate. Heat packs that contain sodium acetate are reusable because the crystals can be redissolved by placing the packs in boiling water. Other hot packs employ a chemical change in which fine iron filings are mixed with a solution containing a mild oxidizing agent. The iron is slowly oxidized, releasing energy as heat in the process.

|SKILL BUILDER — GENERAL

Vocabulary Tell students that the root words *therm* or *thermo* mean "heat" or "relating to heat." Ask them to brainstorm words that contain *therm* or *thermo* and to describe how those words relate to heat. **Ans.** Some possible words: thermometer, thermostat, Thermos®, thermocouple, isothermal, thermodynamics. Next, explain to students that although *therm* relates to heat, the words *exothermic* and *endothermic* refer to changes in total energy and that these changes can occur in other forms of energy. For example, combustion is an exothermic reaction because it releases energy in the form of light and in the form of heat. **LS** Verbal

endothermic

describes a process in which heat is absorbed from the environment

exothermic

describes a process in which a system releases heat into the environment

law of conservation of energy

the law that states that energy cannot be created or destroyed but can be changed from one form to another

Endothermic and Exothermic Processes

Any change in matter in which energy is absorbed is known as an **endothermic** process. The melting of ice and the boiling of water are two examples of physical changes that are endothermic processes.

Some chemical changes are also endothermic processes. **Figure 3** shows a chemical reaction that occurs when barium hydroxide and ammonium nitrate are mixed. Notice in **Figure 3** that these two solids form a liquid, slushlike product. Also, notice the ice crystals that form on the surface of the beaker. As barium hydroxide and ammonium nitrate react, energy is absorbed from the beaker's surroundings. As a result, the beaker feels colder because the reaction absorbs energy as heat from your hand. Water vapor in the air freezes on the surface of the beaker, providing evidence that the reaction is endothermic.

Any change in matter in which energy is released is an **exothermic** process. The freezing of water and the condensation of water vapor are two examples of physical changes that are exothermic processes.

Recall that when hydrogen and oxygen gases react to form water, an explosive reaction occurs. The vessel in which the reaction takes place becomes warmer after the reaction, giving evidence that energy has been released.

Endothermic processes, in which energy is absorbed, may make it seem as if energy is being destroyed. Similarly, exothermic processes, in which energy is released, may make it seem as if energy is being created. However, the **law of conservation of energy** states that during any physical or chemical change, the total quantity of energy remains constant. In other words, energy cannot be destroyed or created.

Accounting for all the different types of energy present before and after a physical or chemical change is a difficult process. But measurements of energy during both physical and chemical changes have shown that when energy seems to be destroyed or created, energy is actually being transferred. The difference between exothermic and endothermic processes is whether energy is absorbed or released by the substances involved.

Figure 3
The reaction between barium hydroxide and ammonium nitrate absorbs energy and causes ice crystals to form on the beaker.

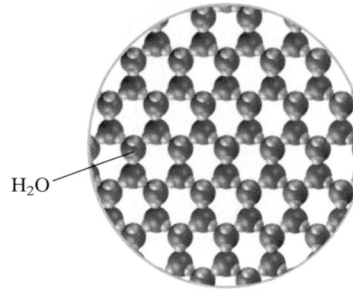

H_2O

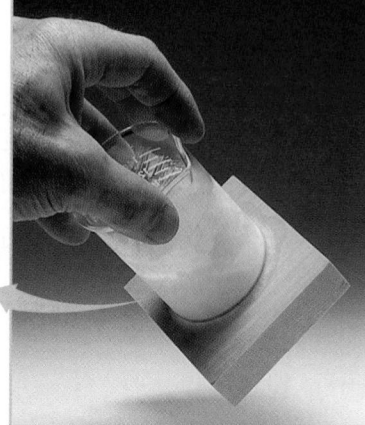

REAL-WORLD
CONNECTION — **GENERAL**

Glow-in-the-Dark Toys Glow-in-the-dark toys are a fun way to show an example of the conservation of energy. These toys contain phosphorescent materials. The phosphorescent materials absorb energy when exposed to light, and then release the energy at a later time. As a result the toys will glow for several minutes after the lights have been turned off. Encourage interested students to study phosphorescence to learn the theory behind it. Have them create a classroom activity that will help the other students learn how glow-in-the-dark toys work. **LS** Logical

Conservation of Energy in a Chemical Reaction

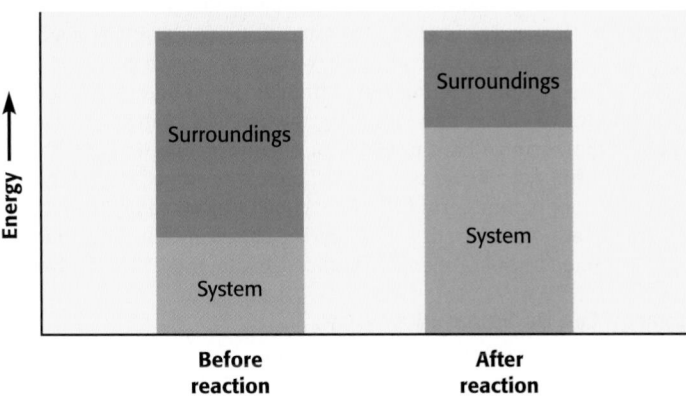

Before reaction

After reaction

Figure 4
Notice that the energy of the reactants and products increases, while the energy of the surroundings decreases. However, the total energy does not change.

Energy Is Often Transferred

Figure 4 shows the energy changes that take place when barium hydroxide and ammonium nitrate react. To keep track of energy changes, chemists use the terms *system* and *surroundings*. A system consists of all the components that are being studied at any given time. In **Figure 4,** the system consists of the mixture inside the beaker. The surroundings include everything outside the system. In **Figure 4,** the surroundings consist of everything else including the air both inside and outside the beaker and the beaker itself. Keep in mind that the air is made of various gases.

Energy is often transferred back and forth between a system and its surroundings. An exothermic process involves a transfer of energy from a system to its surroundings. An endothermic process involves a transfer of energy from the surroundings to the system. However, in every case, the total energy of the systems and their surroundings remains the same, as shown in **Figure 4.**

Energy Can Be Transferred in Different Forms

Energy exists in different forms, including chemical, mechanical, light, heat, electrical, and sound. The transfer of energy between a system and its surroundings can involve any one of these forms of energy. Consider the process of *photosynthesis*. Light energy is transferred from the sun to green plants. Chlorophyll inside the plant's cells (the system) absorbs energy—the light energy from the sun (the surroundings). This light energy is converted to chemical energy when the plant synthesizes chemical nutrients that serve as the basis for sustaining all life on Earth.

Next, consider what happens when you activate a light stick. Chemicals inside the stick react to release energy in the form of light. This light energy is transferred from the system inside the light stick to the surroundings, generating the light that you see. A variety of animals depend on chemical reactions that generate light, including fish, worms, and fireflies.

internet connect

www.scilinks.org
Topic: Conservation of Energy
SciLinks code: HW4035

sci LINKS. Maintained by the National Science Teachers Association

41

Teaching Tip
The Mass-Energy Equation
Toward the end of the 1800s, scientists believed that matter and energy were clearly distinct. Matter was perceived as something that consisted of particles. These particles had mass, and their exact location in space could be pinpointed. In contrast, energy was perceived as something very different from matter. However, by the beginning of the twentieth century, scientists had become to realize that there was no clear-cut distinction between matter and energy. In 1905, Albert Einstein proposed his special theory of relativity, a part of which implies that mass and energy are equivalent. From his theory, Einstein derived the equation: $E = mc^2$. This equation is called the mass-energy equation and shows the relationship between mass and energy.

Scientists now know that matter and energy can be transformed in to one another. In nuclear reactions such as fission and fusion, mass is transformed into energy. However, such transformations do not occur in normal chemical reactions such as the ones carried out in high school chemistry classes. This is why the statements of the laws of conservation of mass and conservation of energy are more correctly written as: "Mass (or energy) is conserved in normal chemical reactions."

Transparencies

TT Heating Curve for Water

NUTRITION

CONNECTION

Nutritionists use bomb calorimeters to measure the quantity of energy stored in different foods. A sample of food is placed inside the calorimeter and is ignited. The energy released by the combustion reaction is absorbed by a quantity of water that surrounds the combustion chamber. The amount of energy can be calculated by with the change in temperature of the water and the molar heat capacity of water. A calorimeter is an example of a system in which the total energy is conserved when the energy is transformed from one type (the potential energy stored in the food) to another (kinetic energy of the water molecules).

Units of Energy The SI unit for energy is the joule (J). However, students may encounter different units of energy in everyday life. Ask students to look up different energy units and research examples of when those units are used. For example, they might find that calories are used to measure the energy stored in foods and kilowatt-hours are used to measure the amount of electrical energy used by a household. Students should also find the conversion factors needed to convert these other units of energy into joules. Have students organize their findings in a poster-sized table to display in the classroom. Students may illustrate their poster if they wish to do so.
LS Interpersonal

Teaching Tip

Thermodynamics The second law of thermodynamics can be stated in a number of ways. One of the most popular ways is by using entropy. In this form, the second law is can be summarized as *the entropy of a system and its surroundings always increases.* However, the second law can also be stated in relationship to the transfer of energy as heat. This is often stated as, "Energy is transferred as heat spontaneously from an object at a high temperature to an object at a lower temperature, but will not be transferred spontaneously in the other direction."

heat

the energy transferred between objects that are at different temperatures; energy is always transferred from higher-temperature objects to lower-temperature objects until thermal equilibrium is reached

Figure 5
Billowing black smoke filled the sky over Texas City in the aftermath of the *Grandcamp* explosion, shown in this aerial photograph.

kinetic energy

the energy of an object that is due to the object's motion

Heat

Heat is the energy transferred between objects that are at different temperatures. This energy is always transferred from a warmer object to a cooler object. For example, consider what happens when ice cubes are placed in water. Energy is transferred from the liquid water to the solid ice. The transfer of energy as heat during this physical change will continue until all the ice cubes have melted. But on a warm day, we know that the ice cubes will not release energy that causes the water to boil, because energy cannot be transferred from the cooler objects to the warmer one. Energy is also transferred as heat during chemical changes. In fact, the most common transfers of energy in chemistry are those that involve heat.

Energy Can Be Released As Heat

The worst industrial disaster in U.S. history occurred in April 1947. A cargo ship named the *Grandcamp* had been loaded with fertilizer in Texas City, a Texas port city of 50 000 people. The fertilizer consisted of tons of a compound called *ammonium nitrate.* Soon after the last bags of fertilizer had been loaded, a small fire occurred, and smoke was noticed coming from the ship's cargo hold. About an hour later, the ship exploded.

The explosion was heard 240 km away. An anchor from the ship flew through the air and created a 3 m wide hole in the ground where it landed. Every building in the city was either destroyed or damaged. The catastrophe on the *Grandcamp* was caused by an exothermic chemical reaction that released a tremendous amount of energy as heat.

All of this energy that was released came from the energy that was stored within the ammonium nitrate. Energy can be stored within a chemical substance as chemical energy. When the ammonium nitrate ignited, an exothermic chemical reaction took place and released energy as heat. In addition, the ammonium nitrate explosion generated **kinetic energy,** as shown by the anchor that flew through the air.

Energy Can Be Absorbed As Heat

In an endothermic reaction, energy is absorbed by the chemicals that are reacting. If you have ever baked a cake or a loaf of bread, you have seen an example of such a reaction. Recipes for both products require either baking soda or baking powder. Both baking powder and baking soda contain a chemical that causes dough to rise when heated in an oven.

The chemical found in both baking powder and baking soda is sodium bicarbonate. Energy from the oven is absorbed by the sodium bicarbonate. The sodium bicarbonate breaks down into three different chemical substances, sodium carbonate, water vapor, and carbon dioxide gas, in the following endothermic reaction:

$$2NaHCO_3 \rightarrow Na_2CO_3 + H_2O + CO_2$$

The carbon dioxide gas causes the batter to rise while baking, as you can see in **Figure 6.**

42

MISCONCEPTION
**//// ALERT **

Transfer of Energy as Heat People know that they can add ice to water or another beverage to lower the temperature of the liquid. Be sure your students understand that the temperature of the liquid is lowered because energy was transferred in the form of heat from the liquid to the ice. In other words, the ice did not cool the liquid—the liquid warmed the ice!

Figure 6
Baking a cake or bread is an example of an endothermic reaction, in which energy is absorbed as heat.

Heat Is Different from Temperature

You have learned that energy can be transferred as heat because of a temperature difference. So, the transfer of energy as heat can be measured by calculating changes in temperature. Temperature indicates how hot or cold something is. **Temperature** is actually a measurement of the average kinetic energy of the random motion of particles in a substance.

For example, imagine that you are heating water on a stove to make tea. The water molecules have kinetic energy as they move freely in the liquid. Energy transferred as heat from the stove causes these water molecules to move faster. The more rapidly the water molecules move, the greater their average kinetic energy. As the average kinetic energy of the water molecules increases, the temperature of the water increases. Think of heat as the energy that is transferred from the stove to the water because of a difference in the temperatures of the stove and the water. The temperature change of the water is a measure of the energy transferred as heat.

Temperature Is Expressed Using Different Scales

Thermometers are usually marked with the Fahrenheit or Celsius temperature scales. However, the Fahrenheit scale is not used in chemistry. Recall that the SI unit for temperature is the *Kelvin*, K. The zero point on the Celsius scale is designated as the freezing point of water. The zero point on the Kelvin scale is designated as *absolute zero*, the temperature at which the minimum average kinetic energies of all particles occur.

In chemistry, you will have to use both the Celsius and Kelvin scales. At times, you will have to convert temperature values between these two scales. Conversion between these two scales simply requires an adjustment to account for their different zero points.

$$t(°C) = T(K) - 273.15\ K \qquad T(K) = t(°C) + 273.15°C$$

The symbols t and T represent temperatures in degrees Celsius and in kelvins, respectively. Also, notice that a temperature change is the same in kelvins and in Celsius degrees.

temperature
a measure of how hot (or cold) something is; specifically, a measure of the average kinetic energy of the particles in an object

internet connect
www.scilinks.org
Topic: Temperature Scales
SciLinks code: HW4124

SCi LINKS Maintained by the National Science Teachers Association

Discussion

Pose the following question to students: Suppose you are working in a laboratory that has some thermometers graduated in Kelvins and others graduated in degrees Celsius. You are assigned the task of trying to replicate the results of an experiment performed by another chemist in the lab. You have the chemist's laboratory notebook and are going over the details of the experiment. At one point you find the notation, "In the next 10 minutes, the temperature of the reaction rose 38." There is no unit indicated. Will you have to ask which type of thermometer the chemist was using? Explain your answer.

Group Activity ── GENERAL

Explain to students that particles (atoms and molecules) in matter are constantly in motion. This is true for solids as well as for gases and liquids. Further explain that the temperature of a substance is proportional to the kinetic energy of the particles in that substance. Challenge students to work in small groups to design models showing particles in the three common states of matter. Their models should show the differences in arrangement of particles and the differences in the motion of the particles for each state of matter. Their model of particles in a gas should also demonstrate the motion of the particles at two different temperatures. **LS** Interpersonal

43

PHYSICS
CONNECTION ── ADVANCED

The temperature of an ideal gas is related to the average translational kinetic energy of the particles by the equation:

$$\frac{1}{2}M\bar{v}^2 = \frac{3}{2}RT$$

Where M is the molar mass, $\bar{v}$ is the average velocity of the particles, R is the gas constant, and T is the temperature in kelvins. Encourage interested students to use a high school or college physics textbook to derive this equation. **LS** Logical

Homework ——— GENERAL

Have students work out the following temperature conversions:

1. 25°C to K Ans. 298 K

2. 300 K to °C Ans. 27°C

3. 37°C, normal body temperature, to K Ans. 310 K

4. 150 K to °C Ans. –123°C

LS Logical

Using the Figure

Figure 7 Emphasize that a kelvin and a degree Celsius have the same magnitude. In other words, they represent the same temperature interval. The only difference is the temperature at which the zero point occurs.

MISCONCEPTION //// ALERT \\\\

Some people think that the reason why the kelvin temperature scale is not widely used by nonscientists is because one would need a very long thermometer to measure temperature. Explain to students that a Celsius thermometer is also a kelvin thermometer—just with different numbers. Further explain that thermometers do not have to measure down to the zero point of a temperature scale. As an example, show a mercury or an alcohol thermometer that is used to measure body temperature. Those thermometers only measure a range from about 93°F to 105°F.

Transfer of Heat May Not Affect the Temperature

The transfer of energy as heat does not always result in a change of temperature. For example, consider what happens when energy is transferred to a solid such as ice. Imagine that you have a mixture of ice cubes and water in a sealed, insulated container. A thermometer is inserted into the container to measure temperature changes as energy is added to the ice-water mixture.

As energy is transferred as heat to the ice-water mixture, the ice cubes will start to melt. However, the temperature of the mixture remains at 0°C. Even though energy is continuously being transferred as heat, the temperature of the ice-water mixture does not increase.

Once all the ice has melted, the temperature of the water will start to increase. When the temperature reaches 100°C, the water will begin to boil. As the water turns into a gas, the temperature remains at 100°C, even though energy is still being transferred to the system as heat. Once all the water has vaporized, the temperature will again start to rise.

Notice that the temperature remains constant during the physical changes that occur as ice melts and water vaporizes. What happens to the energy being transferred as heat if the energy does not cause an increase in temperature? The energy that is transferred as heat is actually being used to move molecules past one another or away from one another. This energy causes the molecules in the solid ice to move more freely so that they form a liquid. This energy also causes the water molecules to move farther apart so that they form a gas.

Figure 7 shows the temperature changes that occur as energy is transferred as heat to change a solid into a liquid and then into a gas. Notice that the temperature increases only when the substance is in the solid, liquid, or gaseous states. The temperature does not increase when the solid is changing to a liquid or when the liquid is changing to a gas.

Figure 7
This graph illustrates how temperature is affected as energy is transferred to ice as heat. Notice that much more energy must be transferred as heat to vaporize water than to melt ice.

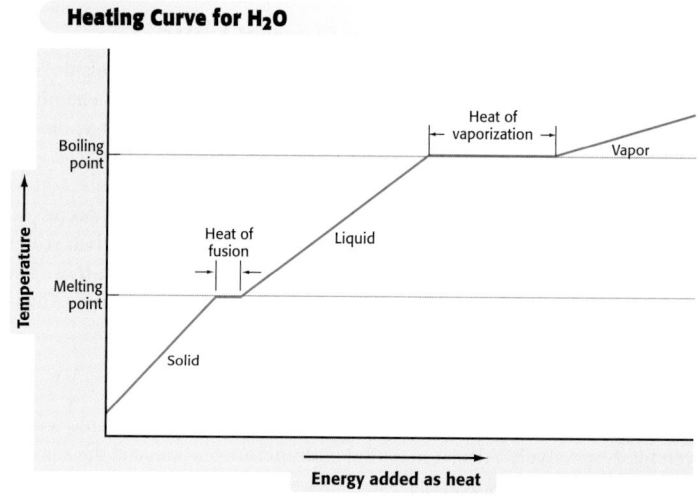

Heating Curve for H₂O

44

MATHEMATICS
CONNECTION — ADVANCED

Encourage students with strong math skills to derive the equations to convert Celsius temperatures to Fahrenheit temperatures and vice versa. Give them the following information: Water boils at 100°C and 212°F. Water freezes at 0°C and 32°F. Ans. °C = (5/9)*(°F–32) and °F = ((9/5)* °C)+32

LS Logical

Transparencies

TT Conservation of Energy in a Chemical Reaction

Transfer of Heat Affects Substances Differently

Have you ever wondered why a heavy iron pot gets hot fast but the water in the pot takes a long time to warm up? If you transfer the same quantity of heat to similar masses of different substances, they do not show the same increase in temperature. This relationship between energy transferred as heat to a substance and the substance's temperature change is called the **specific heat.**

The specific heat of a substance is the quantity of energy as heat that must be transferred to raise the temperature of 1 g of a substance 1 K. The SI unit for energy is the *joule* (J). Specific heat is expressed in joules per gram kelvin (J/g·K).

Metals tend to have low specific heats, which indicates that relatively little energy must be transferred as heat to raise their temperatures. In contrast, water has an extremely high specific heat. In fact, it is the highest of most common substances.

During a hot summer day, water can absorb a large quantity of energy from the hot air and the sun and can cool the air without a large increase in the water's temperature. During the night, the water continues to absorb energy from the air. This energy that is removed from the air causes the temperature of the air to drop quickly, while the water's temperature changes very little. This behavior is explained by the fact that air has a low specific heat and water has a high specific heat.

specific heat

the quantity of heat required to raise a unit mass of homogeneous material 1 K or 1°C in a specified way given constant pressure and volume

 Section Review

UNDERSTANDING KEY IDEAS

1. What is energy?
2. State the law of conservation of energy.
3. How does heat differ from temperature?
4. What is a system?
5. Explain how an endothermic process differs from an exothermic process.
6. What two temperature scales are used in chemistry?

PRACTICE PROBLEMS

7. Convert the following Celsius temperatures to Kelvin temperatures.
   - **a.** 100°C
   - **b.** 785°C
   - **c.** 0°C
   - **d.** −37°C

8. Convert the following Kelvin temperatures to Celsius temperatures.
   - **a.** 273 K
   - **b.** 1200 K
   - **c.** 0 K
   - **d.** 100 K

CRITICAL THINKING

9. Is breaking an egg an example of a physical or chemical change? Explain your answer.
10. Is cooking an egg an example of a physical or chemical change? Explain your answer.
11. What happens in terms of the transfer of energy as heat when you hold a snowball in your hands?
12. Why is it impossible to have a temperature value below 0 K?
13. If energy is transferred to a substance as heat, will the temperature of the substance always increase? Explain why or why not.

Chapter 2 • Matter and Energy **45**

Focus

Overview

Before beginning this section, review with your students the Objectives listed in the Student Edition. Students are introduced to the scientific method in this section. They learn that controlled experiments are an important part of scientific research, and that science sometimes progresses through unexpected discoveries. Finally, the definitions for scientific theories and scientific laws are given.

🔊 Bellringer

Have students write a brief paragraph describing their concept of the scientific method. After students finish, ask some of them to read their paragraphs to the class.

Motivate

Discussion

The Scientific Method Introduce students to the scientific method by asking them what part of a candle burns. **Ans.** the wax Ask them what observations they would make to show that it is not the wick that is burning. **Ans.** Observe the flame closely for a time. Ask students what state the wax is in when it burns. Elicit a statement (hypothesis) similar to, "The wax is a liquid when it burns." Ask students what observations (experiment) they would make to verify this statement. **Ans.** Close observation of the flame will reveal that the flame begins at a small distance from the wick, suggesting that the wax is gaseous when it burns. Have students form a statement (theory) that explains how a candle burns.

Studying Matter and Energy

KEY TERMS

- scientific method
- hypothesis
- theory
- law
- law of conservation of mass

scientific method

a series of steps followed to solve problems, including collecting data, formulating a hypothesis, testing the hypothesis, and stating conclusions

Figure 8
Each stage of the scientific method represents a number of different activities. Scientists choose the activities to use depending on the nature of their investigation.

OBJECTIVES

(1) **Describe** how chemists use the scientific method.

(2) **Explain** the purpose of controlling the conditions of an experiment.

(3) **Explain** the difference between a hypothesis, a theory, and a law.

The Scientific Method

Science is unlike other fields of study in that it includes specific procedures for conducting research. These procedures make up the **scientific method,** which is shown in **Figure 8.** The scientific method is not a series of exact steps, but rather a strategy for drawing sound conclusions.

A scientist chooses the procedures to use depending on the nature of the investigation. For example, a chemist who has an idea for developing a better method to recycle plastics may research scientific articles about plastics, collect information, propose a method to separate the materials, and then test the method. In contrast, another chemist investigating the pollution caused by a trash incinerator would select different procedures. These procedures might include collecting and analyzing samples, interviewing people, predicting the role the incinerator plays in producing the pollution, and conducting field studies to test that prediction.

No matter which approach they use, both chemists are employing the scientific method. Ultimately, the success of the scientific method depends on publishing the results so that others can repeat the procedures and verify the results.

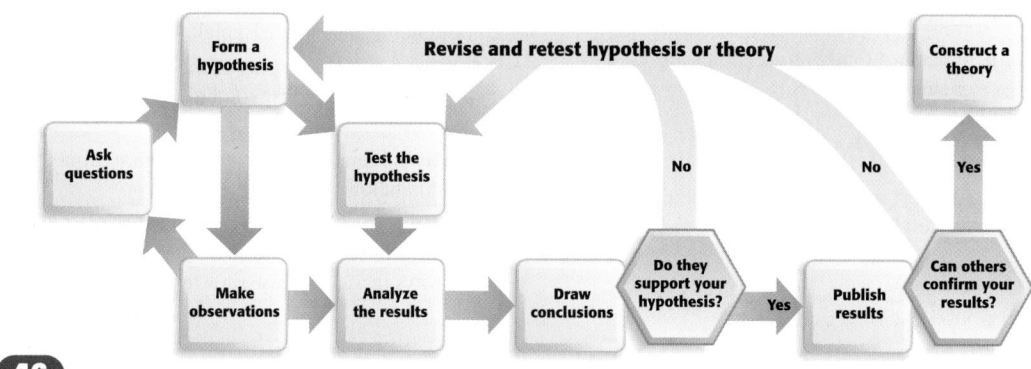

Videos

CNN. Presents Chemistry Connections

- **Segment 7** Science Controversy: Salt in the Diet ADVANCED

See the Science in the News video guide for more details.

Chapter Resource File

- **Datasheets for In-Text Labs**

Transparencies

TT The Scientific Method

Quick LAB

Using the Scientific Method

PROCEDURE

1. Have someone prepare **five sealed paper bags,** each containing **an item commonly found in a home.**

2. Without opening the bags, try to determine the identity of each item.

3. Test each of your conclusions whenever possible. For example, if you concluded that one of the items is a refrigerator magnet, test it to see if it attracts small metal objects, such as paper clips.

ANALYSIS

1. How many processes that are part of the scientific method shown in **Figure 8** did you use?

2. How many items did you correctly identify?

Experiments Are Part of the Scientific Method

The first scientists depended on rational thought and logic. They rarely felt it was necessary to test their ideas or conclusions, and they did not feel the need to experiment. Gradually, experiments became the crucial test for the acceptance of scientific knowledge. Today, experiments are an important part of the scientific method.

An experiment is the process by which scientific ideas are tested. For example, consider what happens when manganese dioxide is added to a solution of hydrogen peroxide. Tiny bubbles of gas soon rise to the surface of the solution, indicating that a chemical reaction has taken place. Now, consider what happens when a small piece of beef liver is added to a solution of hydrogen peroxide. Tiny gas bubbles are produced. So, you might conclude that the liver contains manganese dioxide. To support your conclusion, you would have to test for the presence of manganese dioxide in the piece of liver.

Experiments May Not Turn Out As Expected

Your tests would reveal that liver does not contain any manganese dioxide. In this case, the results of the experiment did not turn out as you might have expected. Scientists are often confronted by situations in which their results do not turn out as expected. Scientists do not view these results as a failure. Rather, they analyze these results and continue with the scientific method. Unexpected results often give scientists as much information as expected results do. So, unexpected results are as important as expected results.

In this case, the liver might contain a different chemical that acts like manganese dioxide when added to hydrogen peroxide. Additional experiments would reveal that the liver does in fact contain such a chemical. Experimental results can also lead to more experiments. Perhaps the chemical that acts like manganese dioxide can be found in other parts of the body.

internet connect

www.scilinks.org
Topic: Scientific Method
SciLinks code: HW4167

SC*LINKS.* Maintained by the National Science Teachers Association

47

Chapter Resource File

• Datasheets for In-Text Labs

Teach

Using the Scientific Method

Materials:

For each group of 2–4 students:
• items commonly found at home (5)
• paper bags (5)

Teacher's Notes: The items may include any common household object that will not break, leak, or otherwise cause a potential safety hazard. Avoid objects with sharp corners, as students will be handling the bags to feel the objects.

Answers

1. Answers will vary.

2. Answers will vary.

READING SKILL BUILDER ━ BASIC

Paired Summarizing Group students into pairs and have them read silently about the Scientific Method in this section. Then have one student summarize the ideas and the steps of the Scientific Method. The other student should listen to the retelling and point out any inaccuracies or ideas that were left out. Have students work with partners during this clarification process and refer to the text as needed. Next, ask students to repeat the process for the subsection titled Scientific Explanations. The students should switch summarizing and listening roles for the second half. **LS Interpersonal**

Research Skills The discovery of Teflon is an example of a serendipitous discovery. The invention of Silly Putty and the discovery that microwaves could be used to cook food are other serendipitous events. Have students research other serendipitous discoveries in the fields of science and technology. They should learn the history leading up to the researchers surprising observations, and how their discoveries have been put to use. Students may work in small groups if they wish. The students should summarize their findings in a poster to share with the class.
LS Logical

MISCONCEPTION ///ALERT\\\

The idea that a hypothesis is an "educated guess" is persistent in many science textbooks. However, philosophers of science believe that that is a very poor definition of *hypothesis*. In fact, some philosophers of science believe that the use of the word *hypothesis* is too loose and the definitions of it too varied that the word should be abandoned all together. In this book, the word *hypothesis* is better defined as a "prediction."

Scientific Discoveries Can Come from Unexpected Observations

Not all discoveries and findings are the results of a carefully worked-out plan based on the scientific method. In fact, some important discoveries and developments have been made simply by accident. An example in chemistry is the discovery of a compound commonly known as Teflon®. You are probably familiar with Teflon as the nonstick coating used on pots and pans, but it has many more applications.

Teflon is used as thermal insulation in clothing, as a component in wall coverings, and as a protective coating on metals, glass, and plastics. Teflon's properties of very low chemical reactivity and very low friction make it valuable in the construction of artificial joints for human limbs. As you can see in **Figure 9,** Teflon is also used as a roofing material.

Teflon was not discovered as a result of a planned series of experiments designed to produce this chemical compound. Rather, it was discovered when a scientist made a simple but puzzling observation.

Teflon Was Discovered by Chance

In 1938, Dr. Roy Plunkett, a chemist employed by DuPont, was trying to produce a new coolant gas to use as a refrigerant. He was hoping to develop a less expensive coolant than the one that was being widely used at that time. His plan was to allow a gas called *tetrafluoroethene* (TFE) to react with hydrochloric acid. To begin his experiment, Plunkett placed a cylinder of liquefied TFE on a balance to record its mass.

He then opened the cylinder to let the TFE gas flow into a container filled with hydrochloric acid. But no TFE came out of the cylinder. Because the cylinder had the same mass as it did when it was filled with TFE, Plunkett knew that none of the TFE had leaked out. He removed the valve and shook the cylinder upside down. Only a few white flakes fell out.

Curious about what had happened, Plunkett decided to analyze the white flakes. He discovered that he had accidentally created the proper conditions for TFE molecules to join together to form a long chain. These long-chained molecules were very slippery. After 10 years of additional research, large-scale manufacturing of these long-chained molecules, known as Teflon or polytetrafluoroethene (PTFE), became practical.

Figure 9
Teflon was used to make the roof of the Hubert H. Humphrey Metrodome in Minneapolis, Minnesota.

48

did you know?

Teflon Facts

• Teflon o-rings and tape are used to seal the protective case that holds the original Emancipation Proclamation.

• Teflon is listed as the world's most slippery substance by the *Guinness Book of World Records.*

• Teflon can be applied to fabrics and carpets to make them stain resistant.

Synthetic Dyes Were Also Discovered by Chance

If you have on an article of clothing that is colored, you are wearing something whose history can be traced to another unexpected chemistry discovery. This discovery was made in 1856 by an 18-year-old student named William Perkin, who was in his junior year at London's Royal College of Chemistry.

At that time, England was the world's leading producer of textiles, including those used for making clothing. The dyes used to color the textiles were natural products, extracted from both plants and animals. Only a few colors were available. In addition, the process to get dyes from raw materials was costly. As a result, only the wealthy could afford to wear brightly colored clothes for everyday use.

Mauve, a deep purple, was the color most people wanted for their clothing. In ancient times, only royalty could afford to own clothes dyed a mauve color. In Perkin's time, only the wealthy people could afford mauve.

Making an Unexpected Discovery

At first, Perkin had no interest in brightly colored clothes. Rather, his interest was in finding a way to make quinine, a drug used to treat malaria. At the time, quinine could only be made from the bark of a particular kind of tree. Great Britain needed huge quantities of the drug to treat its soldiers who got malaria in the tropical countries that were part of the British Empire. There was not enough of the drug to keep up with demand.

The only way to get enough quinine was to develop a synthetic version of the drug. During a vacation from college, Perkin was at home experimenting with ways of making synthetic quinine. One of his experiments resulted in a product that was a thick, sticky, black substance. He immediately realized that this attempt to synthesize quinine did not work. Curious about the substance, Perkin washed his reaction vessel with water. But the sticky product would not wash away. Perkin next decided to try cleaning the vessel with an alcohol. What he saw next was an unexpected discovery.

Analyzing an Unexpected Discovery

When Perkin poured an alcohol on the black product, it turned a mauve color. He found a way to extract the purple substance from the black product and determined that his newly discovered substance was perfect for dyeing clothes. He named his accidental discovery "aniline purple," but the fashionable people of Paris soon renamed it mauve.

Perkin became obsessed with his discovery. He left the Royal College of Chemistry and decided to open a factory that could make large amounts of the dye. Within two years, his factory had produced enough dye to ship to the largest maker of silk clothing in London. The color mauve quickly became the most popular color in the fashion industry throughout Europe. Perkin expanded his company and soon started producing other dyes, including magenta and a deep red. As a result of his unexpected discovery, Perkin became a very wealthy man and retired at the age of 36 to devote his time to chemical research. His unexpected discovery also marked the start of the synthetic dye industry.

STUDY TIP

LEARNING TERMINOLOGY

Important terms and their definitions are listed in the margins of this book. Knowing the definitions of these terms is crucial to understanding chemistry. Ask your teacher about any definition that does not make sense.

To determine your understanding of the terms in this chapter, explain their definitions to another classmate.

Figure 10
Through his accidental discovery of aniline purple, William Perkin found an inexpensive way to make the color mauve. His discovery brought on the beginning of the synthetic dye industry.

internet connect

www.scilinks.org
Topic: Chance Discoveries
SciLinks code: HW4139

SCiLINKS. Maintained by the National Science Teachers Association

49

MISCONCEPTION ALERT

A widely circulated myth states that the Teflon coating that often flakes off of damaged cookware is poisonous. This is not true. Teflon is chemically inert and nontoxic and any accidentally ingested pieces will not harm a person. In fact, the U.S. Food and Drug Administration has deemed Teflon coatings on cookware safe for kitchen use. Furthermore, extensive research has been conducted to demonstrate that Teflon coatings do not emit toxic fumes when heated for cooking.

Teaching Tip
Adhering Teflon to Metal
Students may wonder how "non-stick" Teflon stays on the metal surfaces of cookware. According to DuPont, Teflon finishes are applied in layers similar to the way paint is layered on a surface. The first layer applied is a "primer" layer that can adhere to roughened metal surfaces. Show students a pan with a Teflon coating. Point out that the inner surface of the pan is textured to help the Teflon stick to the metal. Also explain that no metal utensils should be used on Teflon cookware because they can scratch the Teflon finish. Once a Teflon finish is scratched, the Teflon has a tendency to flake or peel off and the pan will lose its nonstick properties. Consumers are advised to use plastic or nylon cooking utensils when using Teflon coated cookware. Furthermore, SilverStone, a cookware company owned by DuPont, is now marketing Teflon coated utensils.

MEDICINE
CONNECTION

A company called Comfort Socks manufactures socks made from Teflon fibers. The socks are intended to reduce friction and blisters on the feet of the person who wears them. Diabetics, pregnant women, and people with sensitive skin or circulatory problems use the socks.

One-Stop Planner CD-ROM

• **Supplemental Reading Projects**
Guided Reading Worksheet: The Same and Not the Same "Whirligigs"
Assign this worksheet for cross-curricular connections to language arts.

Demonstration

Is Mass Conserved?

(Approximate time: 10 minutes)

Present the following demonstration of an apparent contradiction to the law of conservation of mass.

1. Place a burner in the center of a metal tray or pan that is about 12 in. wide.

2. Take a small wad of medium steel wool, measure its mass on a balance, and record the mass on the board. A mechanical balance is best for the demonstration.

3. Ask students to think about things that burn and, have them formulate hypotheses on how the mass of the steel wool will change when it is burned. Some will think that it will decrease in mass or stay the same.

4. Pick up the steel wool with crucible tongs, and hold it in the burner flame. Ensure that very little of the material falls onto the tray. It is not important that all of the wool reacts.

5. After the steel wool has cooled sufficiently, place it back on the balance pan. Include material that fell on the tray. Measure and record the new mass. Students will notice that the mass has increased.

6. Ask the class to devise a hypothesis and an experiment that would determine whether or not the law of conservation of mass was violated and explain the increase.

Safety Caution: Wear safety goggles and a laboratory apron. Avoid long, loose sleeves, and tie back long hair. Keep students at least 3 m from the demonstration.

Disposal: Place cooled material in the trash.

Scientific Explanations

Questions that scientists seek to answer and problems that they hope to solve often come after they observe something. These observations can be made of the natural world or in a laboratory. A scientist must always make careful observations, not knowing if some totally unexpected result might lead to an interesting finding or important discovery. Consider what would have happened if Plunkett had ignored the white flakes or if Perkin had overlooked the mauve substance.

Once observations have been made, they must be analyzed. Scientists start by looking at all the relevant information or data they have gathered. They look for patterns that might suggest an explanation for the observations. This proposed explanation is called a **hypothesis.** A hypothesis is a reasonable and testable explanation for observations.

hypothesis

a theory or explanation that is based on observations and that can be tested

Chemists Use Experiments to Test a Hypothesis

Once a scientist has developed a hypothesis, the next step is to test the validity of the hypothesis. This testing is often done by carrying out experiments, as shown in **Figure 11.** Even though the results of their experiments were totally unexpected, Plunkett and Perkin developed hypotheses to account for their observations. Both scientists hypothesized that their accidental discoveries might have some practical application. Their next step was to design experiments to test their hypotheses.

To understand what is involved in designing an experiment, consider this example. Imagine that you have observed that your family car has recently been getting better mileage. Perhaps you suggest to your family that their decision to use a new brand of gasoline is the factor responsible for the improved mileage. In effect, you have proposed a hypothesis to explain an observation.

Figure 11
Students conduct experiments to test the validity of their hypotheses.

did you know?

Experiments as we understand them in chemistry are not always possible in some other branches of science, such as paleontology, historical geology, and archaeology. In these fields, experiments often take the form of numerous and systematic studies of existing fossils, formations, and chemical compositions of rocks. A hypothesis might state, "Because these layers of rock exist in this mountain, we will find the same layers in another mountain that is 50 miles away." The experiment then becomes the search for the expected layers in order to test the hypothesis.

One-Stop Planner CD-ROM

• **Supplemental Reading Projects**
Guided Reading Worksheet: The Same and Not the Same "Thalidomide"
Assign this worksheet for cross-curricular connections to language arts.

Figure 12
Any number of variables may be responsible for the improved mileage that a driver notices. A controlled experiment can identify the variable responsible.

Scientists Must Identify the Possible Variables

To test the validity of your hypothesis, your next step is to plan your experiments. You must begin by identifying as many factors as possible that could account for your observations. A factor that could affect the results of an experiment is called a *variable*. A scientist changes variables one at a time to see which variable affects the outcome of an experiment.

Several variables might account for the improved mileage you noticed with your family car. The use of a new brand of gasoline is one variable. Driving more on highways, making fewer short trips, having the car's engine serviced, and avoiding quick accelerations are other variables that might have resulted in the improved mileage. To know if your hypothesis is right, the experiment must be designed so that each variable is tested separately. Ideally, the experiments will eliminate all but one variable so that the exact cause of the observed results can be identified.

Each Variable Must Be Tested Individually

Scientists reduce the number of possible variables by keeping all the variables constant except one. When a variable is kept constant from one experiment to the next, the variable is called a *control* and the procedure is called a *controlled experiment*. Consider how a controlled experiment would be designed to identify the variable responsible for the improved mileage.

You would fill the car with the new brand of gasoline and keep an accurate record of how many miles you get per gallon. When the gas tank is almost empty, you would do the same after filling the car with the brand of gasoline your family had been using before. In both trials, you should drive the car under the same conditions. For example, the car should be driven the same number of miles on highways and local streets and at the same speeds in both trials. You then have designed the experiment so that only one variable—the brand of gasoline—is being tested.

51

Teaching Tip

Identifying Theories Ask students to name any theories they have already learned. Possibilities include the theory of plate tectonics, of evolution by punctuated equilibrium, and of evolution by natural selection. Ask them to explain how these theories satisfy the scientific definition of a theory.

MISCONCEPTION
///**ALERT**\\\

Many people believe in a hierarchy of hypotheses, theories, and laws. In this supposed hierarchy, hypotheses will eventually "grow up" to be theories when enough supporting evidence is collects, and theories will "grow up" to be laws when they have been "proven" to be true. This is untrue. A hypothesis may be a precursor to a theory or a law, or it may simply be a prediction that can never become something else. Theories and laws, on the other hand, cannot be transformed into one another because they serve two different functions in science. Laws are generalizations of observations and theories attempt to explain those generalizations.

Students should also note that a theory is not any less valid than a law. A theory can be generally accepted to be true (such as atomic theory) just as laws are. It is not acceptable to dismiss a theory (such as the theory of evolution) because it is "just a theory," because a theory cannot be anything else.

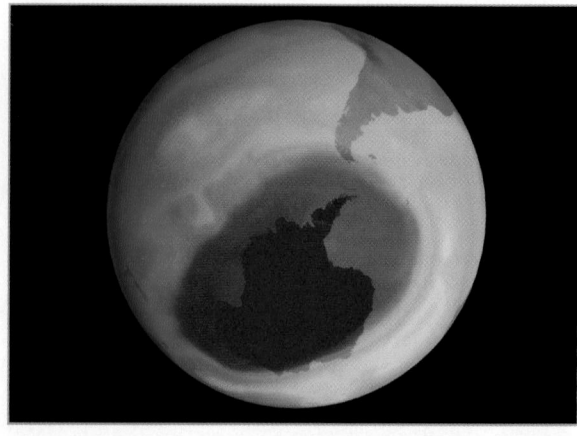

Figure 13
In 1974, scientists proposed a theory to explain the observation of a hole in the ozone layer over Antarctica, which is shown in purple. This hole is about the size of North America.

Data from Experiments Can Lead to a Theory

As early as 1969, scientists observed that the ozone layer was breaking down. Ozone, O_3, is a gas that forms a thin layer high above Earth's surface. This layer shields all living things from most of the sun's damaging ultraviolet light. In 1970, Paul Crutzen, working at the Max Planck Institute for Chemistry, showed the connection between nitrogen oxides and the reduction of ozone in air. In 1974, F. Sherwood Rowland and Mario Molina, two chemists working at the University of California, Irvine, proposed the hypothesis that the release of chlorofluorocarbons (CFCs) into the atmosphere harms the ozone layer. CFCs were being used in refrigerators, air conditioners, aerosol spray containers, and many other consumer products.

Repeated testing has supported the hypothesis proposed by Rowland and Molina. Any hypothesis that withstands repeated testing may become part of a **theory.** In science, a theory is a well-tested explanation of observations. (This is different from common use of the term, which means "a guess.") Because theories are explanations, not facts, they can be disproved but can never be completely proven. In 1995, Crutzen, Rowland, and Molina were awarded the Nobel Prize in chemistry in recognition of their theory of the formation and decomposition of the ozone layer.

theory

an explanation for some phenomenon that is based on observation, experimentation, and reasoning

law

a summary of many experimental results and observations; a law tells how things work

law of conservation of mass

the law that states that mass cannot be created or destroyed in ordinary chemical and physical changes

Theories and Laws Have Different Purposes

Some facts in science hold true consistently. Such facts are known as laws. A **law** is a statement or mathematical expression that reliably describes a behavior of the natural world. While a theory is an attempt to explain the cause of certain events in the natural world, a scientific law describes the events.

For example, the **law of conservation of mass** states that the products of a chemical reaction have the same mass as the reactants have. This law does not explain why matter in chemical reactions behaves this way; the law simply describes this behavior. In some cases, scientific laws may be reinterpreted as new information is obtained. Keep in mind that a hypothesis *predicts* an event, a theory *explains* it, and a law *describes* it.

52

did you know?

People have known for centuries that when a hot object and a cool object are placed near each other, the cool object becomes warmer and the hot object becomes cooler. This observation led people to believe that heat was an invisible fluid, called caloric, that flowed from a warm object to a cool object. Materials that gave off a lot of heat when burned were thought to contain more caloric than materials that gave off only a little heat. We still have the word *calorie* today to remind us of the caloric theory of heat.

In the late eighteenth century an American-British scientist named Benjamin Thompson (1753–1814), also known as Count Rumford, was in Germany overseeing the manufacture of cannons. He noticed that a tremendous amount of energy was generated as heat when the cannon barrels were bored. This observation led to research that replaced the caloric theory with the modern kinetic theory of heat.

 +  →

Hydrogen molecule   Oxygen atom   Water molecule

Figure 14
Models can be used to show what happens during a reaction between a hydrogen molecule and an oxygen atom.

Models Can Illustrate the Microscopic World of Chemistry

Models play a major role in science. A *model* represents an object, a system, a process, or an idea. A model is also simpler than the actual thing that is modeled. In chemistry, models can be most useful in understanding what is happening at the microscopic level. In this book, you will see numerous illustrations showing models of chemical substances. These models, such as the ones shown in **Figure 14**, are intended to help you understand what happens during physical and chemical changes.

Keep in mind that models are simplified representations. For example, the models of chemical substances that you will examine in this book include various shapes, sizes, and colors. The actual particles of these chemical substances do not have the shapes, sizes, or brilliant colors that are shown in these models. However, these models do show the geometric arrangement of the units, their relative sizes, and how they interact.

One tool that is extremely useful in the construction of models is the computer. Computer-generated models enable scientists to design chemical substances and explore how they interact in virtual reality. A chemical model that looks promising for some practical application, such as treating a disease, might be the basis for the synthesis of the actual chemical.

② Section Review

UNDERSTANDING KEY IDEAS

1. How does a hypothesis differ from a theory?
2. What is the scientific method?
3. Do experiments always turn out as expected? Why or why not?
4. What is a scientific law, and how does it differ from a theory?
5. Why does a scientist include a control in the design of an experiment?
6. Why is there no single set of steps in the scientific method?
7. Describe what is needed for a hypothesis to develop into a theory.

CRITICAL THINKING

8. Explain the statement "No theory is written in stone."
9. Can a hypothesis that has been rejected be of any value to scientists? Why or why not?
10. How does the phrase "cause and effect" relate to the formation of a good hypothesis?
11. How would a control group be set up to test the effectiveness of a new drug in treating a disease?
12. Suppose you had to test how well two types of soap work. Describe your experiment by using the terms *control* and *variable*.
13. Why is a model made to be simpler than the thing that it represents?

53

Answers to Section Review

1. A hypothesis is a testable explanation for events. A theory is also an explanation of observations, but it is the result of repeated testing and revision of hypotheses.

2. The scientific method is a series of steps followed to solve problems, including collecting data, formulating a hypothesis, testing a hypothesis, and stating conclusions.

3. No, because scientists do not account for all the variables during an experiment or scientists do not have enough information to formulate a hypothesis that predicts the outcome.

4. A scientific law reliably describes the observed behavior of the natural world but does not explain it. A theory explains the observed behavior.

5. A control is used to isolate the variable a scientist wants to study.

6. Scientists seldom follow a strict series of steps in their work. Rather they choose the processes they will use based on the nature of their investigation.

Answers continued on p. 71A

Overview

Before beginning this section, review with your students the Objectives listed in the Student Edition. In this section, students learn to distinguish between accuracy and precision in measurements. Students practice working with significant figures and scientific notation.

 **Bellringer**

Have students divide a sheet of paper into three columns. In the first column, they should list measuring devices and instruments. In the second column, they should list what the devices in the first column measure. Finally, in the last column, they should list the units in which the devices report their measurements.

Motivate

Group Activity —— GENERAL

Before class, prepare several measuring sticks (one for each group of three students) of various lengths. Cover the sticks so that only the markings at certain intervals are left uncovered, for example at 1 cm, 5 cm, 10 cm, 20 cm, 50 cm, and 1 m.

Divide the class into groups of three. Pick a common object with a height and width that can be easily measured. Ask each group to measure the object as accurately as possible with the measuring stick. Encourage each group to estimate the last digit of the measurement. Have each group record its measurement on the board. Compare the results. Who has the correct answer? Which group is entitled to use the greatest number of digits in its measurement? **LS Interpersonal**

SECTION

3 Measurements and Calculations in Chemistry

KEY TERMS

• accuracy
• precision
• significant figure

OBJECTIVES

(1) **Distinguish** between accuracy and precision in measurements.

(2) **Determine** the number of significant figures in a measurement, and apply rules for significant figures in calculations.

(3) **Calculate** changes in energy using the equation for specific heat, and round the results to the correct number of significant figures.

(4) **Write** very large and very small numbers in scientific notation.

Accuracy and Precision

When you determine some property of matter, such as density, you are making calculations that are often not the exact values. No value that is obtained from an experiment is exact because all measurements are subject to limits and errors. Human errors, method errors, and the limits of the instrument are a few examples. To reduce the impact of error on their work, scientists always repeat their measurements and calculations a number of times. If their results are not consistent, they will try to identify and eliminate the source of error. What scientists want in their results are *accuracy* and *precision*.

Measurements Must Involve the Right Equipment

Selecting the right piece of equipment to make your measurements is the first step to cutting down on errors in experimental results. For example, the beaker, the buret, and the graduated cylinder shown in **Figure 15** can be used to measure the volume of liquids. If an experimental procedure calls for measuring 8.6 mL of a liquid, which piece of glassware would you use? Obtaining a volume of liquid that is as close to 8.6 mL as possible is best done with the buret. In fact, the buret in **Figure 15** is calibrated to the nearest 0.1 mL.

Even though the buret can measure small intervals, it should not be used for all volume measurements. For example, an experimental procedure may call for using 98 mL of a liquid. In this case, a 100 mL graduated cylinder would be a better choice. An even larger graduated cylinder should be used if the procedure calls for 725 mL of a liquid.

The right equipment must also be selected when making measurements of other values. For example, if the experimental procedure calls for 0.5 g of a substance, using a balance that only measures to the nearest 1 g would introduce significant error.

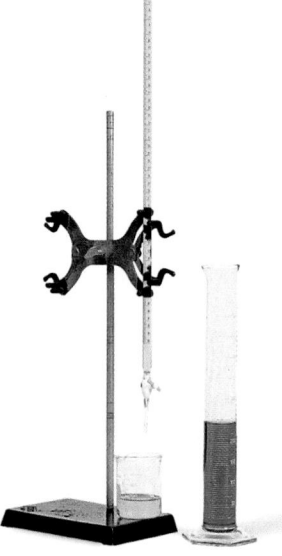

Figure 15
All these pieces of equipment measure volume of liquids, but each is calibrated for different capacities.

54

Chapter Resource File

• Lesson Plan

Figure 16
a Darts within the bull's-eye mean high accuracy and high precision.

b Darts clustered within a small area but far from the bull's-eye mean low accuracy and high precision.

c Darts scattered around the target and far from the bull's-eye mean low accuracy and low precision.

Accuracy Is How Close a Measurement Is to the True Value

When scientists make and report measurements, one factor they consider is accuracy. The **accuracy** of a measurement is how close the measurement is to the true or actual value. To understand what accuracy is, imagine that you throw four darts separately at a dartboard.

The bull's-eye of the dartboard represents the true value. The closer a dart comes to the bull's-eye, the more accurately it was thrown. **Figure 16a** shows one possible way the darts might land on the dartboard. Notice that all four darts have landed within the bull's-eye. This outcome represents high accuracy.

Accuracy should be considered whenever an experiment is done. Suppose the procedure for a chemical reaction calls for adding 36 mL of a solution. The experiment is done twice. The first time 35.8 mL is added, and the second time 37.2 mL is added. The first measurement was more accurate because 35.8 mL is closer to the true value of 36 mL.

accuracy

a description of how close a measurement is to the true value of the quantity measured

Precision Is How Closely Several Measurements Agree

Another factor that scientists consider when making measurements is precision. **Precision** is the exactness of a measurement. It refers to how closely several measurements of the same quantity made in the same way agree with one another. Again, to understand how precision differs from accuracy, consider how darts might land on a dartboard.

Figure 16b shows another way the four darts might land on the dartboard. Notice that all four darts have hit the target far from the bull's-eye. Because these darts are far from what is considered the true value, this outcome represents low accuracy. However, notice in **Figure 16b** that all four darts have landed very close to one another. The closer the darts land to one another, the more precisely they were thrown. Therefore, **Figure 16b** represents low accuracy but high precision. In **Figure 16c,** the four darts have landed far from the bull's-eye and each in a different spot. This outcome represents low accuracy and low precision.

precision

the exactness of a measurement

55

Discussion————— BASIC

Accuracy and Precision Discuss the following hypothetical lab situation with the class:

Suppose you give students a mixture of metals and ask them to determine the percentage of copper in the mixture. They experiment to find a way to separate copper from the other metals. Then they take three samples of the mixture and carry out the procedure on each. They obtain the following results:

Sample A: % Cu = 23.53

Sample B: % Cu = 23.47

Sample C: % Cu = 23.55

Students report these results to you. You tell them that they have done good lab work, but the mixture actually contained 23.94% copper.

Why did you compliment their lab work? **Ans.** Their results are consistent, indicating good lab technique. Discuss the outcome of the analysis in terms of accuracy and precision. **Ans.** The results had good precision but were not very accurate. What should students look for to account for their results? **Ans.** They should check their measuring equipment for accuracy. **LS Verbal**

Discussion————— GENERAL

Ask students to describe a scenario in which a measurement device is precise but not accurate. **Ans.** Sample answer: an analytical balance may be incorrectly zeroed and give consistent— i.e. precise—measurements that are not close to the true value—i.e. it's not accurate. **LS Verbal**

Prediction Guide

Before reading Section 3, have students decide whether the following statements are true or false:

- The terms *accuracy* and *precision* mean the same thing. Ans. false

- It is important to write out all the digits of a number when answering chemistry problems. Ans. false

- The number of digits written in an answer to a chemistry problem is important. Ans. true

- Scientific notation can be used for convenience. Ans. true

- Water heats up faster than copper. Ans. false

- Some values have an infinite number of significant figures. Ans. true

LS Verbal

Discussion ———— GENERAL

Suppose you have three balances. You can weigh an object to the nearest 0.01 g on the most sensitive of the three. The second balance weighs to the nearest 0.1 g, and the third can be read to the nearest 1 g. Give an example of the mass of an object known to weigh a little less than 100 g as read from each balance. Examples of readings could be 96.56 g, 96.6 g, and 97 g. Ask students which balance will measure the mass with the largest number of significant figures? Ans. The mass read from the balance that measures to the nearest 0.01 g has the most significant figures. Suggest a way to check the accuracy and another way to check the precision of the three balances. **LS** Logical

Significant Figures

When you make measurements or perform calculations, the way you report a value tells about how you got it. For example, if you report the mass of a sample as 10 g, the mass of the sample may be between 8 g and 12 g or may be between 9.999 g and 10.001 g. However, if you report the mass of a sample as 10.0 g, you are indicating that you used a measuring tool that is precise to the nearest 0.1 g. The mass of the sample can only be between 9.95 g and 10.05 g.

Scientists always report values using significant figures. The **significant figures** of a measurement or a calculation consist of all the digits known with certainty as well as one estimated, or uncertain, digit. Notice that the term *significant* does not mean "certain." The last digit or significant figure reported after a measurement is uncertain or estimated.

significant figure

a prescribed decimal place that determines the amount of rounding off to be done based on the precision of the measurement

Significant Figures Are Essential to Reporting Results

Reporting all measurements in an experiment to the correct number of significant figures is necessary to be sure the results are true. Consider an experiment involving the transfer of energy as heat. Imagine that you use a thermometer calibrated in one-degree increments. Suppose you report a temperature as 37.5°C. The three digits in your reported value are all significant figures. The first two are known with certainty, but the last digit is estimated. You know the temperature is between 37°C and 38°C, and you estimate the temperature to be 37.5°C.

Now assume that you use the thermometer calibrated in one-tenth degree increments. If you report a reading of 36.54°C, the four digits in your reported value are all significant figures. The first three digits are known with certainty, while the last digit is estimated. Using this thermometer, you know the temperature is certainly between 36.5°C and 36.6°C, and estimate it to be 36.54°C.

Figure 17 shows two different thermometers. Notice that the thermometer on the left is calibrated in one-tenth degree increments, while the one on the right is calibrated in one-degree increments.

Figure 17
If the thermometer on the left is used, a reported value contains three certain figures, whereas the thermometer on the right can measure temperature to two certain figures.

did you know?

Sir Edmund Hillary measured the height of Mount Everest to be exactly 29000 feet. However, simply writing this height as 29000 implied that he only made the measurement to the nearest 1000 feet (between 28500 and 29500 feet). Therefore, Hillary listed Everest's height at 29002 to emphasize the degree of precision in his measurements.

Rules for Determining Significant Figures

1. Nonzero digits are always significant.
- For example, 46.3 m has three significant figures.
- For example, 6.295 g has four significant figures.

2. Zeros between nonzero digits are significant.
- For example, 40.7 L has three significant figures.
- For example, 87 009 km has five significant figures.

3. Zeros in front of nonzero digits are not significant.
- For example, 0.0095 87 m has four significant figures.
- For example, 0.0009 kg has one significant figure.

4. Zeros both at the end of a number and to the right of a decimal point are significant.
- For example, 85.00 g has four significant figures.
- For example, 9.070 000 000 cm has 10 significant figures.

5. Zeros both at the end of a number but to the left of a decimal point may not be significant. If a zero has not been measured or estimated, it is not significant. A decimal point placed after zeros indicates that the zeros are significant.
- For example, 2000 m may contain from one to four significant figures, depending on how many zeros are placeholders. For values given in this book, assume that 2000 m has one significant figure.

Calculators Do Not Identify Significant Figures

When you use a calculator to find a result, you must pay special attention to significant figures to make sure that your result is meaningful. The calculator in **Figure 18** was used to determine the density of isopropyl alcohol, commonly known as rubbing alcohol. The mass of a sample that has a volume of 32.4 mL was measured to be 25.42 g. Remember that the mass and volume of a sample can be used to calulate its density, as shown below.

$$D = \frac{m}{V}$$

The student in **Figure 18** is using a calculator to determine the density of the alcohol by dividing the mass (25.42 g) by the volume (32.4 mL). Notice that the calculator displays the density of the isopropyl alcohol as 0.7845679012 g/mL; the calculator was programmed so that all numbers are significant.

However, the volume was measured to only three significant figures, while the mass was measured to four significant figures. Based on the rules for determining significant figures in calculations described in **Skills Toolkit 1,** the density of the alcohol should be rounded to 0.785 g/mL, or three significant figures.

Figure 18
A calculator does not round the result to the correct number of significant figures.

Teach, *continued*

Teaching Tip

Defining Standards This text uses the statistical method of rounding back quantities that end in 5. Values ending in 5 preceded by an even digit are not rounded up. For example, 4.065 kg is rounded to 4.06 kg. Values ending in 5 preceded by an odd digit are rounded up to the next (even) digit. For example, 0.0235 m is rounded to 0.024 m. Note that a calculated result of 18.8501 g, when rounded back to three significant figures, is 18.9 g. Because the number must end in a 5 with no following digits (other than zero), this rule is invoked infrequently when dealing with real measurements.

SKILL BUILDER — GENERAL

Math Skills Students should learn to recognize situations in which the rounding of numbers to a smaller number of significant figures is essential and may not follow the generally accepted rounding rules. To illustrate this, pose the following question: 34 students are traveling to a music competition in a neighboring town. The school owns several vans that can carry 14 students each. How many vans should the music teacher reserve? Ans. 3, answer to correct number of significant figures: 2.4, answer rounded to the nearest whole number following conventional rules: 2.
LS Logical

Rules for Using Significant Figures in Calculations

1. In multiplication and division problems, the answer cannot have more significant figures than there are in the measurement with the smallest number of significant figures. If a sequence of calculations is involved, do not round until the end.

$$12.257 \text{ m}$$
$$\times\ 1.162 \text{ m} \leftarrow \text{four significant figures}$$
$$14.2426234 \text{ m}^2 \xrightarrow{\text{round off}} 14.24 \text{ m}^2$$

$$8.472 \text{ mL}\overline{)3.05\text{g}} \quad 0.36000944 \text{ g/mL} \xrightarrow{\text{round off}} 0.360 \text{ g/mL}$$
$$\uparrow$$
$$\leftarrow \text{three significant figures}$$

2. In addition and subtraction of numbers, the result can be no more certain than the least certain number in the calculation. So, an answer cannot have more digits to the right of the decimal point than there are in the measurement with the smallest number of digits to the right of the decimal. When adding and subtracting you should not be concerned with the total number of significant figures in the values. You should be concerned only with the number of significant figures present to the right of the decimal point.

$$\begin{array}{r} 3.95 \text{ g} \\ 2.879 \text{ g} \\ +\ 213.6 \text{ g} \\ \hline 220.429 \text{ g} \xrightarrow{\text{round off}} 220.4 \text{ g} \end{array}$$

Notice that the answer 220.4 g has four significant figures, whereas one of the values, 3.95 g, has only three significant figures.

3. If a calculation has both addition (or subtraction) and multiplication (or division), round after each operation.

Exact Values Have Unlimited Significant Figures

Some values that you will use in your calculations have no uncertainty. In other words, these values have an unlimited number of significant figures. One example of an exact value is known as a *count value*. As its name implies, a count value is determined by counting, not by measuring. For example, a water molecule contains exactly two hydrogen atoms and exactly one oxygen atom. Therefore, two water molecules contain exactly four hydrogen atoms and two oxygen atoms. There is no uncertainty in these values.

Another value that can have an unlimited number of significant figures is a *conversion factor*. There is no uncertainty in the values that make up this conversion factor, such as 1 m = 1000 mm, because a millimeter is defined as exactly one-thousandth of a meter.

You should ignore both count values and conversion factors when determining the number of significant figures in your calculated results.

HISTORY CONNECTION

German mathematician Johann Karl Friedrich Gauss (1777–1855) studied the mathematics of error analysis. His work lead to development of the rules (listed in the **Skills Toolkit** above) for determining the number of significant figures to maintain after performing calculations.

Determining the Number of Significant Figures

A student heats 23.62 g of a solid and observes that its temperature increases from 21.6°C to 36.79°C. Calculate the temperature increase per gram of solid.

1 **Gather information.**

- The mass of the solid is 23.62 g.
- The initial temperature is 21.6°C.
- The final temperature is 36.79°C.

2 **Plan your work.**

- Calculate the increase in temperature by subtracting the initial temperature (21.6°C) from the final temperature (36.79°C).

$$\text{temperature increase} = \text{final temperature} - \text{initial temperature}$$

- Calculate the temperature increase per gram of solid by dividing the temperature increase by the mass of the solid (23.62 g).

$$\frac{\text{temperature increase}}{\text{gram}} = \frac{\text{temperature increase}}{\text{sample mass}}$$

3 **Calculate.**

$$36.79°C - 21.6°C = 15.19°C = 15.2°C$$

$$\frac{15.2°C}{23.62 \text{ g}} = 0.644\frac{°C}{g} \text{ rounded to three significant figures}$$

4 **Verify your results.**

- Multiplying the calculated answer by the total number of grams in the solid equals the calculated temperature increase.

$$0.644\frac{°C}{g} \times 23.62 \text{ g} = 15.2°C$$
rounded to three significant figures

PRACTICE HINT

Remember that the rules for determining the number of significant figures in multiplication and division problems are different from the rules for determining the number of significant figures in addition and subtraction problems.

PRACTICE

1 Perform the following calculations, and express the answers with the correct number of significant figures.

  a. 0.1273 mL − 0.000008 mL

  b. (12.4 cm × 7.943 cm) + 0.0064 cm²

  c. (246.83 g/26) − 1.349 g

2 A student measures the mass of a beaker filled with corn oil to be 215.6 g. The mass of the beaker is 110.4 g. Calculate the density of the corn oil if its volume is 114 cm³.

3 A chemical reaction produces 653 550 kJ of energy as heat in 142.3 min. Calculate the rate of energy transfer in kilojoules per minute.

PROBLEM SOLVING SKILL

59

did you know?

Some of the definitions for the SI base units are given with a large number of significant figures. For example, a meter is defined as the distance traveled by light in a vacuum in 1/299792458 of a second!

Chapter Resource File

• **Problem Solving**

---

1. a. .1273 mL

  b. 98.5 cm²

  c. 8.2 g

2. 0.923 g/cm³

3. 4593 kJ/min

Homework

Additional Practice

1. Determine the number of significant figures in each of the following quantities:

  a. 218 kPa Ans. 3

  b. 0.025 L Ans. 2

  c. 200. m² Ans. 3

  d. 1.05 g Ans. 3

2. Round the following quantities to the number of significant figures indicated in parentheses:

  a. 32.068 km (3) Ans. 32.1 km

  b. 155.8 g (3) Ans. 156 g

  c. 0.02274 cm (2) Ans. 0.023 cm

LS Logical

Group Activity —— BASIC

Give each group two sets of pennies (pre-1982 and post-1983), a 100 mL graduated cylinder, and a balance. Tell students that they must find the densities of each group of pennies to three significant figures. **Ans.** Students must measure the volume and mass of the pennies to three significant figures and then calculate the densities using the equation D = M/V. Students should find that the pre-1982 pennies have a density of approximately 8.96 g/cm³ and the post-1983 pennies have a density of approximately 7.13 g/cm³. Ask students to compare their density calculations with the known densities of copper and other metals. **Ans.** Students should find that the pre-1982 pennies have the same density as copper and the post-1983 pennies have a density close to the density of zinc. Explain to students that pennies made after 1983 have zinc cores covered with a thin layer of copper. **LS** Kinesthetic

---

Designing a Model Ask students to devise a way to demonstrate the differences in the specific heat capacity of several metals. If you wish, tell them that the rate of the temperature increase depends upon the heat capacity of the metal. Be sure students' plans address the question of which factors must be kept constant. A satisfactory method is to place thermometers in blocks of the various metals. The blocks must have equal masses. All of the blocks are then simultaneously lowered into hot water. The specific heat capacity of each sample should be inversely proportional to the rate of the temperature increase.
LS Logical

SKILL BUILDER — ADVANCED

Research Skills Encourage interested students to look up heat capacities of various materials in the *CRC Handbook of Chemistry and Physics.* Suggested materials to search for: air, water, steam, lithium, diamond, and copper(II) chloride. Students will find that they need to use several different tables of heat capacities and that some of the tables are dependent on pressure and temperature. They may also find tables that list only molar heat capacities rather than specific heat capacities. **LS** Intrapersonal

Specific Heat Depends on Various Factors

Recall that the specific heat is the quantity of energy that must be transferred as heat to raise the temperature of 1 g of a substance by 1 K. The quantity of energy transferred as heat during a temperature change depends on the nature of the material that is changing temperature, the mass of the material, and the size of the temperature change.

For example, consider how the nature of the material changing temperature affects the transfer of energy as heat. One gram of iron that is at 100.0°C is cooled to 50.0°C and transfers 22.5 J of energy to its surroundings. In contrast, 1 g of silver transfers only 11.8 J of energy as heat under the same conditions. Iron has a larger specific heat than silver. Therefore, more energy as heat can be transferred to the iron than to the silver.

Calculating the Specific Heat of a Substance

Specific heats can be used to compare how different materials absorb energy as heat under the same conditions. For example, the specific heat of iron, which is listed in **Table 1,** is 0.449 J/g·K, while that of silver is 0.235 J/g·K. This difference indicates that a sample of iron absorbs and releases twice as much energy as heat as a comparable mass of silver during the same temperature change does.

Specific heat is usually measured under constant pressure conditions, as indicated by the subscript p in the symbol for specific heat, c_p. The specific heat of a substance at a given pressure is calculated by the following formula:

$$c_p = \frac{q}{m \times \Delta T}$$

In the above equation, c_p is the specific heat at a given pressure, q is the energy transferred as heat, m is the mass of the substance, and ΔT represents the difference between the initial and final temperatures.

internet connect

www.scilinks.org
Topic: Specific Heat
SciLinks code: HW4119

SCiLINKS. Maintained by the National Science Teachers Association

Table 1	Some Specific Heats at Room Temperature		
Element	**Specific heat (J/g•K)**	**Element**	**Specific heat (J/g•K)**
Aluminum	0.897	Lead	0.129
Cadmium	0.232	Neon	1.030
Calcium	0.647	Nickel	0.444
Carbon (graphite)	0.709	Platinum	0.133
Chromium	0.449	Silicon	0.705
Copper	0.385	Silver	0.235
Gold	0.129	Water	4.18
Iron	0.449	Zinc	0.388

60

REAL-WORLD CONNECTION — BASIC

The thermal conductivity of a substance partly depends upon its specific heat capacity. If a substance has a low specific heat capacity, it is a better conductor of thermal energy than a substance with a higher specific heat capacity. For example, metals have low specific heat capacities and are therefore conduct heat very well. Because of this, metals are used to make cooking pots and pans. However, plastics generally have high specific heat capacities. Handles on pots are often made of plastic so that their temperature won't increase very much and the pots can be moved after they are heated. Challenge students to think of other instances in their every day lives where the specific heat capacities of substances are important. **LS** Interpersonal

Calculating Specific Heat

A 4.0 g sample of glass was heated from 274 K to 314 K and was found to absorb 32 J of energy as heat. Calculate the specific heat of this glass.

1 Gather information.
- sample mass (m) = 4.0 g
- initial temperature = 274 K
- final temperature = 314 K
- quantity of energy absorbed (q) = 32 J

2 Plan your work.
- Determine ΔT by calculating the difference between the initial and final temperatures.
- Insert the values into the equation for calculating specific heat.

$$c_p = \frac{32\ \text{J}}{4.0\ \text{g} \times (314\ \text{K} - 274\ \text{K})}$$

3 Calculate.

$$c_p = \frac{32\ \text{J}}{4.0\ \text{g} \times (40\ \text{K})} = 0.20\ \text{J/g}\cdot\text{K}$$

4 Verify your results.

The units combine correctly to give the specific heat in J/g·K. The answer is correctly given to two significant figures.

PRACTICE HINT

The equation for specific heat can be rearranged to solve for one of the quantities, if the others are known. For example, to calculate the quantity of energy absorbed or released, rearrange the equation to get $q = c_p \times m \times \Delta T$.

PRACTICE

1 Calculate the specific heat of a substance if a 35 g sample absorbs 48 J as the temperature is raised from 293 K to 313 K.

2 The temperature of a piece of copper with a mass of 95.4 g increases from 298.0 K to 321.1 K when the metal absorbs 849 J of energy as heat. What is the specific heat of copper?

3 If 980 kJ of energy as heat are transferred to 6.2 L of water at 291 K, what will the final temperature of the water be? The specific heat of water is 4.18 J/g·K. Assume that 1.0 mL of water equals 1.0 g of water.

4 How much energy as heat must be transferred to raise the temperature of a 55 g sample of aluminum from 22.4°C to 94.6°C? The specific heat of aluminum is 0.897 J/g·K. Note that a temperature change of 1°C is the same as a temperature change of 1 K because the sizes of the degree divisions on both scales are equal.

Answers to Practice Problems B
1. 0.069 J/g•K
2. 0.385 J/g•K
3. 329 K
4. 3.6 kJ

Homework

Additional Practice Have students solve the following problems:

1. A 5.00 g sample of a metal was heated from 25.0°C to 40.0°C and it absorbed 17.6 J of energy. What is its specific heat capacity? What was the identity of the metal? Ans. 0.235 J/g•K, silver

2. A 1.6 g sample of a metal was heated from 273 K to 300 K and it absorbed 5.57 J of energy. What is the metals specific heat capacity? Ans. 0.13 J/g•K

3. Air has a heat capacity of 1.007 J/g•K. The density of air is 1.161 g/L. How much energy is needed to heat 2.00 liters of air from 293 K to 298 K? Ans. 11.7 J

Chapter Resource File

- **Problem Solving**

REAL-WORLD CONNECTION

Thermos® Bottles Good insulators are materials that have very high heat capacities. Air and plastic make good insulators because of their high heat capacities. However, a vacuum is the best insulator because it contains no molecules and therefore cannot transfer energy as heat except radiatively. High quality Thermos bottles take advantage of a vacuum's insulating ability. The interior of a Thermos is a glass envelope that surrounds a vacuum. If a hot soup is placed in the Thermos, the vacuum prevents thermal energy from escaping to the surroundings. Conversely, if a cold drink is placed in the same Thermos, the vacuum prevents thermal energy from entering and increasing the temperature of the liquid.

Math Skills When students first encounter scientific notation, they may find it difficult to recognize the magnitude of the numbers. The following card game will help struggling students learn how to read numbers in scientific notation. Write numbers in scientific notation on note cards. Be sure to include negative numbers and numbers with negative exponents. Give each student 6–10 cards and ask them to arrange them in order from smallest number to largest without using their calculators. If students are having difficulty starting, tell them to group their cards into positive and negative numbers and then divide each group into numbers greater than 1 (or –1) and less than 1 (or –1).

After completing this exercise, you can quiz the students by writing numbers on the board such as Avogadro's number and Planck's constant and asking the students if they are very big numbers or very small numbers. **LS** Logical

Teaching Tip
Why Use Scientific Notation?
If students ask why scientific notation is important, you can list several examples of when it is very useful. For example, the population of the United States and the world are easier to write by scientific notation. Also, interplanetary and interstellar distances are normally expressed in scientific notation.

Scientific Notation

Chemists often make measurements and perform calculations using very large or very small numbers. Very large and very small numbers are often written in *scientific notation*. To write a number in scientific notation, first know that every number expressed in scientific notation has two parts. The first part is a number that is between 1 and 10 but that has any number of digits after the decimal point. The second part consists of a power of 10. To write the first part of the number, move the decimal to the right or the left so that only one nonzero digit is to the left of the decimal. Write the second part of the value as an exponent. This part is determined by counting the number of decimal places the decimal point is moved. If the decimal is moved to the right, the exponent is negative. If the decimal is moved to the left, the exponent is positive. For example, 299 800 000 m/s is expressed as 2.998×10^8 m/s in scientific notation. When writing very large and very small numbers in scientific notation, use the correct number of significant figures.

SKILLS Toolkit

Using Scientific Notation

1. **In scientific notation, exponents are count values.**

2. **In addition and subtraction problems, all values must have the same exponent before they can be added or subtracted.**

   • $6.2 \times 10^4 + 7.2 \times 10^3 = 62 \times 10^3 + 7.2 \times 10^3 = 69.2 \times 10^3 =$
   $$69 \times 10^3 = 6.9 \times 10^4$$

   • $4.5 \times 10^6 - 2.3 \times 10^5 = 45 \times 10^5 - 2.3 \times 10^5 = 42.7 \times 10^5 =$
   $$43 \times 10^5 = 4.3 \times 10^6$$

3. **In multiplication problems, the first factors of the numbers are multiplied and the exponents of 10 are added.**

   • $(3.1 \times 10^3)(5.01 \times 10^4) = (3.1 \times 5.01) \times 10^{4+3} =$
   $$16 \times 10^7 = 1.6 \times 10^8$$

4. **In division problems, the first factors of the numbers are divided and the exponent of 10 in the denominator is subtracted from the exponent of 10 in the numerator.**

   • $7.63 \times 10^3/8.6203 \times 10^4 = 7.63/8.6203 \times 10^{3-4} =$
   $$0.885 \times 10^{-1} = 8.85 \times 10^{-2}$$

62

did you know?

An entertaining and visual way of illustrating how scientific notion works is to view the nine-minute movie "The Powers of Ten." This movie was made in 1977 and zooms in and out—by powers of 10—from a park in California.

Scientific Notation with Significant Figures

1. Use scientific notation to eliminate all placeholding zeros.

- $2400 \longrightarrow 2.4 \times 10^3$ (both zeros are not significant)
- $750\,000. \longrightarrow 7.50000 \times 10^5$ (all zeros are significant)

2. Move the decimal in an answer so that only one digit is to the left, and change the exponent accordingly. The final value must contain the correct number of significant figures.

- $5.44 \times 10^7 / 8.1 \times 10^4 = 5.44/8.1 \times 10^{7-4} = 0.6716049383 \times 10^3 = 6.7 \times 10^2$ (adjusted to two significant figures)

Section Review

UNDERSTANDING KEY IDEAS

1. How does accuracy differ from precision?
2. Explain the advantage of using scientific notation.
3. When are zeros significant in a value?
4. Why are significant figures important when reporting measurements?
5. Explain how a series of measurements can be precise without being accurate.

PRACTICE PROBLEMS

6. Perform the following calculations, and express the answers using the correct number of significant figures.

   a. $0.8102 \text{ m} \times 3.44 \text{ m}$ **c.** $32.89 \text{ g} + 14.21 \text{ g}$

   b. $\dfrac{94.20 \text{ g}}{3.167\,22 \text{ mL}}$ **d.** $34.09 \text{ L} - 1.230 \text{ L}$

7. Calculate the specific heat of a substance when 63 J of energy are transferred as heat to an 8.0 g sample to raise its temperature from 314 K to 340 K.

8. Express the following calculations in the proper number of significant figures. Use scientific notation where appropriate.

   a. 129 g/29.2 mL

   b. (1.551 mm)(3.260 mm)(4.9001 mm)

   c. 35 000 kJ/0.250 s

9. A clock gains 0.020 s/min. How many seconds will the clock gain in exactly six months, assuming 30 days are in each month? Express your answer in scientific notation.

CRITICAL THINKING

10. There are 12 eggs in a carton. How many significant figures does the value 12 have in this case?

11. If you measure the mass of a liquid as 11.50 g and its volume as 9.03 mL, how many significant figures should its density value have? Explain the reason for your answer.

Answers to Section Review

1. Accuracy indicates how close a measurement is to the true value. Precision indicates how close repeated measurements are to each other.

2. Scientific notation makes it easier to write very large and very small values.

3. Zeros are significant when they are between nonzero digits, after a nonzero digit and before a decimal point, and following a nonzero digit after a decimal point.

4. The number of significant figures indicates the degree of uncertainty to which a measurement is known.

5. Measurements of the same thing can be very close to one another but still differ significantly from the true value.

6. **a.** 2.79 m^2
   b. 29.74 g/mL
   c. 47.10 g
   d. 32.86 L

7. 0.30 J/g•K

Answers continued on p. 71A

Element Spotlight

Deep-sea Diving with Helium

During the solar eclipse of 1868, Pierre Janssen (1824–1907) obtained the first evidence of the existence of helium when he detected a new line in the solar spectrum. Helium has since been detected spectroscopically in great abundance. It is the second most abundant element in the universe, and it is an important component in the processes by which stars get energy.

Although helium was probably present early in Earth's history, most of it drifted into space because of helium's low mass. All of the helium on Earth is the product of radioactive decay. It has collected in cavities, such as oil domes and gas domes, in Earth's crust. Commercial supplies of helium are extracted from natural gas, the bulk of which comes from Texas, Oklahoma, and Kansas.

did you know?

Helium has the lowest melting point of any element. It is the only liquid that cannot be solidified at atmospheric pressure by lowering the temperature. Near 0 K helium takes on a superfluid property, and it flows upward and covers all surfaces that it can reach. For this reason, helium has extensive applications in cryogenic research, and it is used as a coolant in superconductor research and in nuclear reactors.

When cooled to its superfluid state, helium can be used to measure the precise rotational speed of Earth and the exact movements of Earth's tectonic plates. Scientists hope that superfluid helium will

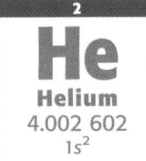

HELIUM

2
He
Helium
4.002 602
$1s^2$

Element Spotlight

Where Is He?
Universe:
about 23% by mass
Earth's crust:
0.000001% by mass
Air:
0.0005% by mass

Deep-sea Diving with Helium

Divers who breathe air while at great undersea depths run the risk of suffering from a condition known as nitrogen narcosis. Nitrogen narcosis can cause a diver to become disoriented and to exercise poor judgment, which leads to dangerous behavior. To avoid nitrogen narcosis, professional divers who work at depths of more than 60 m breathe heliox, a mixture of helium and oxygen, instead of air.

The greatest advantage of heliox is that it does not cause nitrogen narcosis. A disadvantage of heliox is that it removes body heat faster than air does. This effect makes a diver breathing heliox feel chilled sooner than a diver breathing air.

Breathing heliox also affects the voice. Helium is much less dense than nitrogen, so vocal cords vibrate faster in a heliox atmosphere. This raises the pitch of the diver's voice, and makes the diver's voice sound funny. Fortunately, this effect disappears when the diver surfaces and begins breathing air again.

Industrial Uses

- Helium is used as a lifting gas in balloons and dirigibles.
- Helium is used as an inert atmosphere for welding and for growing high-purity silicon crystals for semiconducting devices.
- Liquid helium is used as a coolant in superconductor research.

Real-World Connection Helium was discovered in the sun before it was found on Earth.

In Florida, divers on the Wakulla Springs project team breathed heliox at depths greater than 90 m.

internet connect
www.scilinks.org
Topic: Helium
SciLinks code: HW4171
SC*LINKS* Maintained by the National Science Teachers Association

A Brief History

1888: William Hillebrand discovers that an inert gas is produced when a uranium mineral is dissolved in sulfuric acid.

1908: Ernest Rutherford and Thomas Royds prove that alpha particles emitted during radioactive decay are helium nuclei.

| 1600 | 1700 | 1800 | 1900 |

1868: Pierre Janssen, studies the spectra of a solar eclipse and finds evidence of a new element. Edward Frankland, an English chemist, and Joseph Lockyer, an English astronomer, suggest the name helium.

1894: Sir William Ramsay and Lord Rayleigh discover argon. They suspect that the gas Hillebrand found in 1888 was argon. They repeat his experiment and find that the gas is helium.

Questions

1. Research the industrial, chemical, and commercial uses of helium.
2. Research properties of neon, argon, krypton, and xenon. How are these gases similar to helium? Are they used in a manner similar to helium?

some day help in the prediction of earthquakes. Superfluid helium may also be used in navigational gyroscopes on spacecrafts.

Answers to Feature Questions
1. Answers will vary.
2. Answers will vary.

CHAPTER HIGHLIGHTS

KEY TERMS

energy
physical change
chemical change
evaporation
endothermic
exothermic
law of conservation
   of energy
heat
kinetic energy
temperature
specific heat

scientific method
hypothesis
theory
law
law of conservation
   of mass

accuracy
precision
significant figure

KEY IDEAS

SECTION ONE Energy

- Energy is the capacity to do work.
- Changes in matter can be chemical or physical. However, only chemical changes produce new substances.
- Every change in matter involves a change in energy.
- Endothermic processes absorb energy. Exothermic processes release energy.
- Energy is always conserved.
- Heat is the energy transferred between objects that are at different temperatures. Temperature is a measure of the average random kinetic energy of the particles in an object.
- Specific heat is the relationship between energy transferred as heat to a substance and a substance's temperature change.

SECTION TWO Studying Matter and Energy

- The scientific method is a strategy for conducting research.
- A hypothesis is an explanation that is based on observations and that can be tested.
- A variable is a factor that can affect an experiment.
- A controlled experiment is an experiment in which variables are kept constant.
- A theory is a well-tested explanation of observations. A law is a statement or mathematical expression that describes the behavior of the world.

SECTION THREE Measurements and Calculations in Chemistry

- Accuracy is the extent to which a measurement approaches the true value of a quantity.
- Precision refers to how closely several measurements that are of the same quantity and that are made in the same way agree with one another.
- Significant figures are digits known with certainty as well as one estimated, or uncertain, digit.
- Numbers should be written in scientific notation.

KEY SKILLS

Rules for Determining Significant Figures
Skills Toolkit 1 p. 57

Rules for Using Significant Figures in Calculations
Skills Toolkit 2 p. 58
Sample Problem A p. 59

Calculating Specific Heat
Sample Problem B p. 61

Scientific Notation in Calculations
Skills Toolkit 3 p. 62

Scientific Notation with Significant Figures
Skills Toolkit 4 p. 63

65

Alternative Assessments

Group Activity
- Pages 42, 51, 59

Discussion
- Pages 43, 56

Portfolio Assessments

Homework
- Pages 44, 59, 61

Reteaching
- Pages 45, 63

- Page 47

SKILL BUILDER

- pages 48, 60, 62

Chapter Resource File

- **Datasheets for In-Text Lab**
  Separation of Mixtures
- **Datasheets for In-Text Lab**
  Separation of Mixtures–Mining Contract
- **Observation Lab** Specific Heat

REVIEW ANSWERS

1. Students' answers may vary. Sample answer is energy transferred as heat and light.

2. the law that states that energy cannot be created or destroyed but can be changed from one form to another

3. A chemical change is a change that occurs when one or more substances change into entirely new substances with different properties.

4. Temperature is a measure of the average kinetic energy of the particles of an object.

5. Heat is the energy transferred between objects that are at different temperatures.

6. A theory is an explanation based on observation, experimentation, and reasoning, and a law is a summary of many experimental results and observations.

7. Accuracy is a description of how close a measurement is to the true value of the quantity measured. Precision is a description of how close measurements are to each other.

8. Significant figures consist of all digits known with certainty as well as one estimated or uncertain digit.

9. On the hot sunny day, the water is warmer and the sun supplies a continuous source of light, some of which changes to heat when it strikes an object. As a result, the motion of the water molecules remains high and the water evaporates more rapidly than it would on a cold, overcast day.

10. Temperature is a measure of the average kinetic energy of the particles of an object. Therefore, the water molecules in the beaker at 37°C have higher average kinetic energy.

USING KEY TERMS

1. Name two types of energy.

2. State the law of conservation of energy.

3. Define chemical change.

4. What is temperature?

5. What is the difference between heat and temperature?

6. What is the difference between a theory and a law?

7. What is accuracy? What is precision?

8. What are significant figures?

UNDERSTANDING KEY IDEAS

Energy

9. Water evaporates from a puddle on a hot, sunny day faster than on a cold, cloudy day. Explain this phenomenon in terms of interactions between matter and energy.

10. Beaker A contains water at a temperature of 15°C. Beaker B contains water at a temperature of 37°C. Which beaker contains water molecules that have greater average kinetic energy? Explain your answer.

11. What is the difference between a physical change and a chemical change?

Studying Matter and Energy

12. What does a good hypothesis require?

13. Classify the following statements as observation, hypothesis, theory, or law:
    a. A system containing many particles will not go spontaneously from a disordered state to an ordered state.

b. The substance is silvery white, is fairly hard, and is a good conductor of electricity.
    c. Bases feel slippery in water.
    d. If I pay attention in class, I will succeed in this course.

14. What is a control? What is a variable?

15. Explain the relationship between models and theories.

16. Describe the scientific method.

17. Why is the conservation of energy considered a law, not a theory?

Measurements and Calculations in Chemistry

18. Why it is important to keep track of significant figures?

19. a. If you add several numbers, how many significant figures can the sum have?
    b. If you multiply several numbers, how many significant figures can the product have?

20. How many digits are not known with certainty in a number written with the proper number of significant figures?

21. Perform the following calculations, and express the answers with the correct number of significant figures.
    a. $2.145 + 0.002$
    b. $(9.8 \times 8.934) + 0.0048$
    c. $(172.56/43.8) - 1.825$

22. Which of the following statements contain exact numbers?
    a. There are 12 eggs in a dozen.
    b. Some Major League Baseball pitchers can throw a ball over 140 km/h.
    c. The accident injured 21 people.
    d. The circumference of the Earth at the equator is 40000 km.

Assignment Guide	
Section	Questions
1	1–5, 9–11, 65, 69–75
2	6, 12–17, 56, 58, 60–62, 67–68
3	7–8, 18–55, 57, 59, 63–64, 66

e. The tank was filled with 54 L of gas.

f. A nickel has a mass of 5 g.

23. Express 743 000 000 in scientific notation to the following number of significant figures:
a. one significant figure
b. two significant figures
c. four significant figures
d. seven significant figures

PRACTICE PROBLEMS

Sample Problem A Determining the Number of Significant Figures

24. How many significant figures are there in each of the following measurements?
a. 0.4004 mL   **c.** 1.000 30 km
b. 6000 g   **d.** 400 mm

25. Calculate the sum of 6.078 g and 0.3329 g.

26. Subtract 7.11 cm from 8.2 cm.

27. What is the product of 0.8102 m and 3.44 m?

28. Divide 94.20 g by 3.167 22 mL.

29. How many grams are in 882 µg?

30. The density of gold is 19.3 g/cm^3.
a. What is the volume, in cubic centimeters, of a sample of gold with mass 0.715 kg?
b. If this sample of gold is a cube, how long is each edge in centimeters?

31. a. Find the number of kilometers made up of 92.25 m.
b. Convert the answer in kilometers to centimeters.

Sample Problem B Calculating Specific Heat

32. Determine the specific heat of a material if a 35 g sample of the material absorbs 48 J as it is heated from 298 K to 313 K.

33. How much energy is needed to raise the temperature of a 75 g sample of aluminum from 22.4°C to 94.6°C? Refer to **Table 1.**

34. How much energy is needed to raise the temperature of 75 g of gold by 25°C? Refer to **Table 1.**

35. Energy in the amount of 420 J is added to a 35 g sample of water at a temperature of 10.0°C. What is the final temperature of the water? Refer to **Table 1.**

36. How much energy will be transferred as heat to a 15.3 g sample of cadmium when its temperature is raised from 322 K to 363 K?

Skills Toolkit 3 Scientific Notation in Calculations

37. Write the following numbers in scientific notation.
a. 0.000 673 0
b. 50 000.0
c. 0.000 003 010

38. The following numbers are written in scientific notation. Write them in ordinary notation.
a. 7.050×10^{-3} g
b. $4.000\ 05 \times 10^7$ mg
c. 2.3500×10^4 mL

39. Perform the following operation. Express the answer in scientific notation and with the correct number of significant figures.

$$\frac{(6.124\ 33 \times 10^6 \text{m}^3)}{(7.15 \times 10^{-3} \text{m})}$$

Skills Toolkit 4 Scientific Notation with Significant Figures

40. Use scientific notation to eliminate all placeholding zeros.
a. 7500
b. 92 002 000

41. How many significant figures does 7.324×10^{-3} have?

42. How many significant figures does the answer to $(1.36 \times 10^{-5}) \times (5.02 \times 10^{-2})$ have?

MIXED REVIEW

43. Why can a measured number never be exact?

44. Determine the specific heat of a material if a 78 g sample of the material absorbed 28 J as it was heated from 298 K to 345 K.

67

11. A substance that undergoes a physical change does not change into an entirely new substance. However, if that substance were to undergo a chemical change, it would change into a new substance or substances.

12. The hypothesis must explain an observation and must be testable.

13. a. law
b. observation
c. observation
d. hypothesis

14. A control is a variable that is kept constant. A variable is a factor that can affect the results of an experiment.

15. A theory can be interpreted in the form of a model, which can be a physical representation, a mathematical description, or a computer simulation.

16. The scientific method is a series of steps followed to solve problems, including collecting data, formulating a hypothesis, testing a hypothesis, and stating conclusions.

17. The law is an observation that is found to be universally true in ordinary chemical and physical changes.

18. The number of significant figures conveys the degree of certainty within a measurement.

19. a. The sum can have the number of significant figures in the value with the fewest decimal places.
b. The product can have the number of significant figures in the value with the least number of significant figures.

20. one

21. a. 2.147
b. 88
c. 2.12

22. a. exact
b. not exact
c. exact
d. not exact
e. not exact
f. not exact

23. a. 7×10^8

  b. 7.4×10^8

  c. 7.430×10^8

  d. $7.430\,000 \times 10^8$

24. a. four

  b. one

  c. six

  d. one

25. 6.411 g

26. 1.1 cm

27. 2.79 m^2

28. 29.74 g/ mL

29. 8.82×10^{-4} g

30. a. 37.0 cm^3

  b. 3.33 cm

31. a. 9.225×10^{-2} km

  b. 9.225×10^3 cm

32. 0.091 J/g•K

33. 4.9×10^3 J

34. 242 J

35. 13°C

36. 146 J

37. a. 6.730×10^{-4}

  b. $5.000\,00 \times 10^4$

  c. 3.010×10^{-6}

38. a. 0.007 050 g

  b. 40 000 500 mg

  c. 23 500. mL

39. 8.57×10^8 m^2

40. a. 7.5×10^3

  b. 9.2002×10^7

41. four

42. three

43. All measurements are subject to limits and errors, including human errors, errors in the method of measurement, and the limitations of the measuring instrument.

44. 0.0076 J/g•K

45. A piece of copper alloy with a mass of 85.0 g is heated from 30.0°C to 45.0°C. During this process, it absorbs 523 J of energy as heat.

  a. What is the specific heat of this copper alloy?

  b. How much energy will the same sample lose if it is cooled to 25°C?

46. A man finds that he has a mass of 100.6 kg. He goes on a diet, and several months later he finds that he has a mass of 96.4 kg. Express each number in scientific notation, and calculate the number of kilograms the man has lost by dieting.

47. A large office building is 1.07×10^2 m long, 31 m wide, and 4.25×10^2 m high. What is its volume?

48. An object has a mass of 57.6 g. Find the object's density, given that its volume is 40.25 cm^3.

49. A student measures the mass of some sucrose as 0.947 mg. Convert that quantity to grams and to kilograms.

50. Write the following measurements in scientific notation.

  a. 65 900 000 m

  b. 0.0057 km

  c. 22 000 mg

  d. 0.000 003 7 kg

51. Write the following measurements in long form.

  a. 4.5×10^3 g

  b. 6.05×10^{-3} m

  c. 3.115×10^6 km

  d. 1.99×10^{-8} cm

52. Write the following measurements in scientific notation.

  a. 800 000 000 m

  b. 0.000 95 m

  c. 60 200 L

  d. 0.0015 kg

53. Write the following measurements in long form.

  a. 9.8×10^4 mm

  b. 2.38×10^{-7} m

  c. 1.115×10^3 g

  d. 1.5×10^{-10} kg

54. Do the following calculations, and write the answers in scientific notation.

  a. 37 000 000 × 7 100 000

  b. 0.000 312/ 486

  c. 4.6×10^4 cm × 7.5×10^3 cm

  d. 8.3×10^6 kg/ 2.5×10^9 cm^3

55. Do the following calculations, and write the answers with the correct number of significant figures.

  a. 15.75 m × 8.45 m

  b. 5650 L/ 27 min

  c. 6271 m/ 59.7 s

  d. 0.0058 km × 0.228 km

56. Explain why the observation that the sun sets in the west could be called a *scientific law.*

57. Calculate the volume of a room with walls that are 3.125 m tall, 4.25 m wide, and 5.75 m long. Write the answer with the correct number of significant figures.

58. You have decided to test the effects of five garden fertilizers by applying some of each to five separate rows of radishes. What is the variable you are testing? What factors should you control? How will you measure the results?

CRITICAL THINKING

59. Suppose a graduated cylinder was not correctly calibrated. How would this affect the results of a measurement? How would it affect the results of a calculation using this measurement? Use the terms *accuracy* and *precision* in your answer.

60. Your friend says that things fall to Earth because of the law of gravitation. Tell what is wrong with your friend's statement, and explain your reasoning.

68

61. a. The table below contains data from an experiment in which an air sample is subjected to different pressures. Based on this set of observations, propose a hypothesis that could be tested.

b. What theories can be stated from the data in the table below?

c. Are the data sufficient for the establishment of a scientific law. Why or why not?

The Results of Compressing an Air Sample

Volume (cm³)	Pressure (kPa)	Volume × pressure (cm³ × kPa)
100.0	33.3	3330
50.0	66.7	3340
25.0	133.2	3330
12.5	266.4	3330

62. What components are necessary for an experiment to be valid?

63. Is it possible for a number to be too small or too large to be expressed adequately in the SI system? Explain your answer.

64. Around 1150, King David I of Scotland defined the inch as the width of a man's thumb at the base of the nail. Discuss the practical limitations of this early unit of measurement.

65. How does Einstein's equation $E = mc^2$ seem to contradict both the law of conservation of energy and the law of conservation of mass?

ALTERNATIVE ASSESSMENT

66. Design an experimental procedure for determining the specific heat of a metal.

67. For one week, practice your observation skills by listing chemistry-related events that happen around you. After your list is compiled, choose three events that are especially interesting or curious to you. Label three pocket portfolios, one for each event. As you read the chapters in this textbook, gather information that helps explain these events. Put pertinent notes, questions, figures, and charts in the folders. When you have enough information to explain each phenomenon, write a report and present it in class.

68. Make a poster of scientific laws that you have encountered in previous science courses. What facts were used to support each of these laws?

69. Energy can be transformed from one form to another. For example, light (solar) energy is transformed into chemical energy during photosynthesis. Prepare a list of several different forms of energy. Describe transformations of energy that you encounter on a daily basis. Try to include examples that involve more than one transformation, e.g., light → chemical → mechanical. Select one example, and demonstrate the actual transformation to the class.

CONCEPT MAPPING

70. Use the following terms to complete the concept map below: *energy, endothermic, physical change, law of conservation of energy,* and *exothermic.*

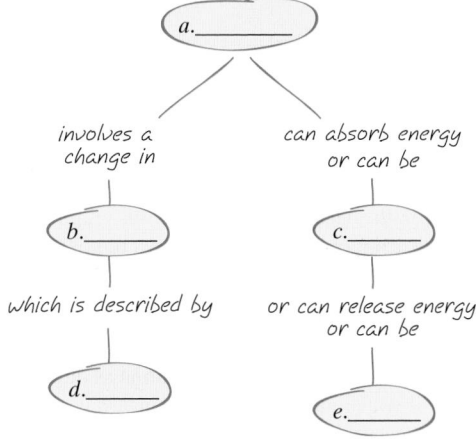

45. a. 0.41 J/g•K
b. 697 J
46. 1.006×10^2 kg, 9.64×10^1 kg; 4.2 kg
47. 1.4×10^6 m³
48. 1.43 g/cm³
49. 9.47×10^{-4} g; 9.47×10^{-7} kg
50. a. 6.59×10^7 m
b. 5.7×10^{-3} km
c. 2.2×10^4 mg
d. 3.7×10^{-6} kg
51. a. 4500 g
b. 0.006 05 m
c. 3 115 000 km
d. 0. 000 000 019 9 cm
52. a. 8×10^8 m
b. 9.5×10^{-4} m
c. 6.02×10^4 L
d. 1.5×10^{-3} kg
53. a. 98 000 mm
b. 0.000 000 238 m
c. 1115 g
d. 0.000 000 000 15 kg
54. a. 2.6×10^{14}
b. 6.42×10^{-7}
c. 3.4×10^8 cm²
d. 3.3×10^{-3} kg/cm³
55. a. 133 m²
b. 210 L/min
c. 105 m/s
d. 0.0013 km²
56. It has been observed repeatedly and it does not attempt to explain why the sun sets in the west.
57. $V = l \times w \times h = 5.75$ m $\times$ 4.25 m $\times$ 3.125 m = 76.4 m³
58. The type of fertilizer is the variable you are testing. Control factors are the types of radishes, the amount of water, the amount of sunshine, etc. There are at least four things that could be used to determine the results: size, quantity, appearance, and taste.
59. The results of the measurement would be less accurate, but they would not be less precise; The calibration error would introduce error into any calculations.

Chapter 2 • Matter and Energy **69**

60. The law of gravitation states that objects fall to Earth and how to calculate the force. It does not explain why.

61. a. Raising the pressure on a sample of air decreases its volume in a way that the product of the volume and the pressure is a constant value.

 b. None. The volume-pressure phenomenon is observed, but the available data provide no explanation of the observation.

 c. No. If this gas volume-pressure relationship is observed many times with differing gases and at varying pressures, then the description of this phenomenon is accepted into law.

62. One factor is allowed to vary while the others are kept constant. A control must be used for all variables under all reaction conditions.

63. No. As measuring technology advances and new magnitudes of measurement are possible, scientists add new prefixes to accommodate the new measurements.

64. not every thumb has the same size

65. The equation indicates that mass and energy are interconvertible. However, the law is not demonstrated in chemical and physical changes.

66. Designs may vary but should generally involve heating the object and allowing its heat to warm a known mass of water. Then they can use the specific heat of water to calculate the heat absorbed by the water.

Answers continued on p. 71A

FOCUS ON GRAPHING

Study the graph below, and answer the questions that follow.
For help in interpreting graphs, see Appendix B, "Study Skills for Chemistry."

71. What is the value for the slope of the curve during the period in which the temperature is equal to the melting point temperature?

72. Is there another period in the graph where the slope equals the value in question 71?

73. Draw the cooling curve for water. Label the axes and the graph.

74. Suppose water could exist in four states of matter at some pressure. Draw what the heating curve for water would look like. Label the axes and the graph.

Heating Curve for H$_2$O

Temperature →

Boiling point

Melting point

Heat of vaporization →

Vapor

Liquid

Heat of fusion

Solid

Energy added as heat →

TECHNOLOGY AND LEARNING

75. Graphing Calculator

Graphing Celsius and Fahrenheit Temperatures

The graphing calculator can run a program that makes a graph of a given Fahrenheit temperature (on the *x*-axis) and the corresponding Celsius temperature (on the *y*-axis). You can use the **TRACE** button on the calculator to explore this graph and learn more about how the two temperature scales are related.

Go to Appendix C. If you are using a TI-83 Plus, you can download the program **CELSIUS** and run the application as directed. If you are using another calculator, your teacher will provide you with keystrokes and data sets to use. After the graph is displayed, press **TRACE**. An X-shaped cursor on the graph line indicates a specific point. At the bottom of the screen the values are shown for that point. The one labeled X= is the Fahrenheit temperature and the one labeled Y= is the Celsius temperature. Use the right and left arrow keys to move the cursor along the graph line to find the answers to these questions.

a. What is the Fahrenheit temperature when the Celsius temperature is zero? (This is where the graph line crosses the horizontal *x*-axis.) What is the significance of this temperature?

b. Human internal body temperature averages 98.6°F. What is the corresponding value on the Celsius scale?

c. Determine the Fahrenheit temperature in your classroom or outside, as given in a weather report. What is the corresponding Celsius temperature?

d. At what temperature are the Celsius and Fahrenheit temperatures the same?

70

Chapter Resource File

• **Chapter Test**

STANDARDIZED TEST PREP 2

Answers

1. b
2. c
3. d
4. c
5. b
6. c
7. a
8. b
9. a
10. b
11. b

1. The equation $E = mc^2$ shows that
   a. chemical reactions are either exothermic or endothermic.
   b. mass and energy are related.
   c. a hypothesis may develop into a theory.
   d. the kinetic energy of an object relates to its motion.

2. Which of the following measurements contains three significant figures?
   a. 200 mL
   b. 0.02 mL
   c. 20.2 mL
   d. 200.0 mL

3. A control in an experiment
   a. is often not needed.
   b. means that the scientist has everything under control.
   c. is required only if the hypothesis leads to the development of a theory.
   d. allows the scientist to identify the cause of the results in an experiment.

4. A theory differs from a hypothesis in that the former
   a. cannot be disproved.
   b. always leads to the formation of a law.
   c. has been subjected to experimental testing.
   d. represents an educated guess.

5. All measurements in science
   a. must be expressed in scientific notation.
   b. have some degree of uncertainty.
   c. are both accurate and precise.
   d. must include only those digits that are known with certainty.

6. If the temperature outside is 26°C, the temperature would be _____ K.
   a. 26
   b. 273
   c. 299
   d. −247

7. When numbers are multiplied or divided, the answer can have no more
   a. significant figures than there are in the measurement with the smallest number of significant figures.
   b. significant figures than there are in the measurement with the largest number of significant figures.
   c. digits to the right of the decimal point than there are in the measurement with the smallest number of digits to the right of the decimal point.
   d. digits to the right of the decimal point than there are in the measurement with the largest number of digits to the right of the decimal point.

8. Which of the following is not part of the scientific method?
   a. making measurements
   b. introducing bias
   c. making an educated guess
   d. analyzing data

9. The accuracy of a measurement
   a. is how close it is to the true value.
   b. does not depend on the instrument being used to measure the object.
   c. indicates that the measurement is also precise.
   d. is something that scientists rarely achieve.

10. A measurement of 23 465 mg converted to grams equals
    a. 2.3465 g.
    b. 23.465 g.
    c. 234.65 g.
    d. 0.23465 g.

11. A metal sample has a mass of 45.65 g. The volume of the sample is 16.9 cm^3. The density of the sample is
    a. 2.7 g/cm^3.
    b. 2.70 g/cm^3.
    c. 0.370 g/cm^3.
    d. 0.37 g/cm^3.

71

Continuation of Answers

Answers to Section 1 Review, *continued*

8. **a.** 0°C

   b. 927°C

   c. −273°C

   d. −173°C

9. Breaking an egg is a physical change because the chemical nature of the egg has not been affected.

10. Cooking an egg is an example of a chemical change because the chemical properties of the egg are changed by the transfer of energy as heat.

11. Energy is transferred as heat from the hand, which is at a higher temperature, to the snowball, which is at a lower temperature.

12. Reducing the kinetic energy of particles to zero is impossible.

13. The temperature of the substance will not increase if it is undergoing a change of state.

Answers to Section 2 Review, *continued*

7. Many experiments that support the hypothesis, albeit with refinements and revisions, are usually required.

8. As new information is obtained, a theory may be revised or even discarded in favor of a new theory that provides a better explanation of the observed behavior.

9. Yes. Not only does it provide a record of what was tested, it may be the basis for the design of new experiments.

10. Many hypotheses predict a "cause and effect" relationship, most often by taking the form of an "if-then" statement.

11. The control group is unknowingly given a placebo instead of the drug.

12. Answers may vary. Each trial in the experiment should involve using the two types of soap under the same set of conditions.

13. A model is created to illustrate certain important aspects of the thing it represents. An attempt to create a model as complex as the thing it represents, just to demonstrate *some* characteristics of that thing, would generally require too much effort to be practical.

Answers to Section 3 Review, *continued*

8. **a.** 4.42 g/mL

   b. 24.78 mm^3

   c. 1.4×10^5 kJ/s

9. 5.2×10^3 s

10. Because 12 is a count value, there are an unlimited number of significant figures in this value.

11. The density value should have three significant figures because 9.03 mL has only three significant figures.

Answers to Chapter Review, *continued*

67. Events that students choose should generally involve clear-cut chemical or physical changes involving specific substances. Collect the folders, and use them as a basis for assigning reports throughout the year by selecting one topic from each student's folder.

68. Students should not overlook laws that they have learned in biology, physics, or Earth science.

69. Many energy chains that students will think of involve devices that are powered by fuel, battery, or current. Consider having students trace energy back to its ultimate origin, usually the sun.

70. **a.** physical change

   b. energy

   c. endothermic

   d. law of conservation of energy

   e. exothermic

71. 0

72. during the period in which the temperature is equal to the boiling point

73. The graph should have axes of "Temperature" and "Energy lost as heat." The curve should be the mirror image of the curve shown.

74. Students' answers may vary but should be similar to the graph shown.

75. **a.** 32°F; the freezing point of water

   b. 37.0°C

   c. Answers may vary.

   d. −40°

Atoms and Moles
Chapter Planning Guide

PACING	CLASSROOM RESOURCES	LABS, ACTIVITIES, AND DEMONSTRATIONS
BLOCK 1 · 45 min pp. 72–73 **Chapter Opener**		**SE** Start-Up Activity Forces of Attraction, p. 73 ◆
BLOCK 2 · 45 min pp. 74–78 **Section 1** Substances Are Made of Atoms	**TT** Law of Conservation of Mass * **TT** Law of Multiple Proportions *	**TE** Activity, p. 75 **BASIC**
BLOCK 3 · 45 min pp. 79–89 **Section 2** Structure of Atoms	**TT** Gold-Foil Experiment * **TT** Gold-Foil Experiment on the Atomic Level * **TT** Properties of Subatomic Particles * **OSP** Career Extension Recycling Engineer **GENERAL**	**TE** Demonstration Magnets and Cathode Rays, p. 80 ◆ **TE** Demonstration, p. 81 ◆ **ADVANCED** **TE** Group Activity, p. 87 **GENERAL**
BLOCKS 4 & 5 · 90 min pp. 90–99 **Section 3** Electron Configuration	**TT** Electromagnetic Spectrum * **TT** Wavelength and Frequency * **TT** Hydrogen's Line-Emission Spectrum * **TT** Shapes of *s, p,* and *d* Orbitals *	**TE** Group Activity, p. 87 ◆ **GENERAL** **TE** Demonstration Emission Spectra, p. 94 ◆ **SE** Skills Practice Lab Flame Tests, p. 772 ◆ **GENERAL** **CRF** Datasheets for In-Text Labs * **SE** Inquiry Lab Spectroscopy and Flame Tests— Identifying Materials, p. 776 ◆ **GENERAL** **CRF** Datasheets for In-Text Labs *
BLOCK 6 · 45 min pp. 100–105 **Section 4** Counting Atoms	**TT** Determining the Mass from the Amount in Moles * **OSP** Supplemental Reading Linus Pauling: In His Own Words, "What Is Chemistry" **ADVANCED** **SE** Element Spotlight Beryllium: An Uncommon Element	**TE** Group Activity, p. 101 ◆ **BASIC** **TE** Group Activity, p. 102 **GENERAL**

BLOCKS 7 & 8 · 90 min **Chapter Review and Assessment Resources**

SE Chapter Review, pp. 107–112
SE Standardized Test Prep, p. 113
CRF Chapter Test *
OSP Test Generator
CRF Test Item Listing *
OSP Scoring Rubrics and Classroom Management Checklists

Holt Chemistry: Online Resources

Visit **go.hrw.com** for a variety of free resources related to this textbook. Enter the keyword **HW4 HOME**.

Students can access interactive problem solving help and active visual concept development with the *Holt Chemistry* Online Edition available at **www.hrw.com**.

student CNN News

cnnstudentnews.com

Find the latest chemistry news, lesson plans, and activities related to important scientific events.

Compression guide:
To shorten your instruction because of time limitations, omit blocks 1, 6, and 7.

PROBLEM SOLVING AND PRACTICE	SECTION REVIEW AND ASSESSMENT	STANDARDS CORRELATION
		National Science Education Standards
	TE Homework, p. 77 BASIC **TE** Quiz, p. 78 GENERAL **CRF** Quiz * **TE** Reteaching, p. 78 BASIC **SE** Section Review, p. 78 **CRF** Concept Review *	PS 1a, PS 1b
SE Sample Problem A Determining the Number of Particles in an Atom, p. 86 **SE** Practice, p. 86 GENERAL **TE** Homework, p. 86 GENERAL **SE** Sample Problem B Determining the Number of Particles in Isotopes, p. 89 **SE** Practice, p. 89 GENERAL **TE** Homework, p. 88 GENERAL **SE** Problem Bank, p. 858 GENERAL	**TE** Homework, p. 82 BASIC **TE** Quiz, p. 89 GENERAL **CRF** Quiz * **TE** Reteaching, p. 89 BASIC **SE** Section Review, p. 89 **CRF** Concept Review *	PS 1a, PS 1b, PS 2b
SE Sample Problem C Writing Electron Configurations, p. 98 **SE** Practice, p. 98 GENERAL **TE** Homework, p. 98 GENERAL	**TE** Quiz, p. 99 GENERAL **CRF** Quiz * **TE** Reteaching, p. 99 BASIC **SE** Section Review, p. 99 **CRF** Concept Review *	PS 6c
SE Skills Toolkit 1 Determining the Mass from the Amount in Moles, p. 101 **SE** Sample Problem D Converting from Amount in Moles to Mass, p. 102 **SE** Practice, p. 102 GENERAL **TE** Homework, p. 102 GENERAL **SE** Skills Toolkit 2 Determining the Number of Atoms from the Amount in Moles, p. 103 **SE** Sample Problem E Converting from Amount in Moles to Number of Atoms, p. 103 **SE** Practice, p. 103 GENERAL **TE** Homework, p. 103 GENERAL **SE** Problem Bank, p. 858 GENERAL	**TE** Quiz, p. 104 GENERAL **CRF** Quiz * **TE** Reteaching, p. 104 BASIC **SE** Section Review, p. 104 **CRF** Concept Review *	

www.scilinks.org
Topic: Atoms and Elements
SciLinks code: HW4017

Topic: Development of Atomic Theory
SciLinks code: HW4148

Topic: Current Atomic Theory
SciLinks code: HW4038

Topic: Subatomic Particles
SciLinks code: HW4121

Topic: J. J. Thomson
SciLinks code: HW4156

Topic: Atomic Nucleus
SciLinks code: HW4014

Topic: Atomic Structures
SciLinks code: HW4015

Topic: Electromagnetic Spectrum
SciLinks code: HW4048

Topic: Light and Color
SciLinks code: HW4075

Topic: Producing Light
SciLinks code: HW4099

Topic: Energy Levels
SciLinks code: HW4051

Topic: Beryllium
SciLinks code: HW4021

Technology Resources

 **Science in the NEWS**

Each video segment is accompanied by a Critical Thinking Worksheet.

Segment 8
The Top Quark

Segment 9
Atom Lasers

 ChemFile Interactive Tutor CD-Rom

Module 2: Models of Atoms
Topic: Atomic Structure

Module 2: Models of Atoms
Topic: Electronic Structure

Overview

The chapter begins with historical evidence for the existence of the atom and Dalton's model for the structure of the atom. The chapter continues with descriptions of the experiments of Thomson, Rutherford, and Bohr. These scientists and other scientists revised and refined the model of the atom to the quantum model of today. The nuclear structure of the atom is also explained in the chapter and students learn to write electron configurations. Finally, the mole is introduced as a way of counting particles.

Assessing Prior Knowledge

Check for Content Knowledge

Students should be familiar with the following topics:

- theories
- models
- matter
- elements
- compounds
- physical and chemical properties

Using the Figure

Let students know that a scanning tunneling microscope traces the topography of a surface. The traces are used to make an image of the surface; therefore, the figure is not a visual image of actual atoms. Instead, it is a computer-generated map of a surface.

CHAPTER
3
ATOMS AND MOLES

72

Standards Correlations

National Science Education Standards

PS 1a: Matter is made of minute particles called atoms, and atoms are composed of even smaller components. These components have measurable properties, such as mass and electrical charge. Each atom has a positively charged nucleus surrounded by negatively charged electrons. The electric force between the nucleus and electrons holds the atom together. (Sections 1, 2)

PS 1b: The atom's nucleus is composed of protons and neutrons, which are much more massive than electrons. When an element has atoms that differ in the number of neutrons, these atoms are called different isotopes of the element. (Sections 1, 2)

PS 2b: An element is composed of a single type of atom. When elements are listed in order according to the number of protons (called the atomic number), repeating patterns of physical and chemical properties identify families of elements with similar properties. This "Periodic Table" is a consequence of the repeating pattern of outermost electrons and their permitted energies. (Section 2)

PS 6c: Each kind of atom or molecule can gain or lose energy only in particular discrete amounts and thus can absorb and emit light only at wavelengths corresponding to these amounts. These wavelengths can be used to identify the substance. (Section 3)

Until recently, if you wanted to see an image of atoms, the best you could hope to see was an artist's drawing of atoms. Now, with the help of powerful microscopes, scientists are able to obtain images of atoms. One such microscope is known as the scanning tunneling microscope, which took the image of the nickel atoms shown on the opposite page. As its name implies, this microscope scans a surface, and it can come as close as a billionth of a meter to a surface to get an image. The images that these microscopes provide help scientists understand atoms.

START-UP ACTIVITY

Forces of Attraction

SAFETY PRECAUTIONS

PROCEDURE

1. Spread some **salt** and **pepper** on a piece of **paper** that lies on a flat surface. Mix the salt and pepper but make sure that the salt and pepper are not clumped together.

2. Rub a **plastic spoon** with a **wool cloth.**

3. Hold the spoon just above the salt and pepper.

4. Clean off the spoon by using a **towel.** Rub the spoon with the wool cloth and bring the spoon slowly toward the salt and pepper from a distance.

ANALYSIS

1. What happened when you held your spoon right above the salt and pepper? What happened when you brought your spoon slowly toward the salt and pepper?

2. Why did the salt and pepper jump up to the spoon?

3. When the spoon is brought toward the paper from a distance, which is the first substance to jump to the spoon? Why?

Pre-Reading Questions

1. What is an atom?

2. What particles make up an atom?

3. Where are the particles that make up an atom located?

4. Name two types of electromagnetic radiation.

CONTENTS 3

internet connect
www.scilinks.org
Topic: Atoms and Elements
SciLinks code: HW4017
SCI LINKS. Maintained by the National Science Teachers Association

73

START-UP ACTIVITY

Skills Acquired:
• Collecting data
• Interpreting
• Identifying/Recognizing patterns

Materials:
For each group of 2–3 students:
• salt
• pepper
• paper
• plastic spoon
• wool cloth
• towel

Teacher's Notes:
• Make sure that students rub the spoon with the wool cloth in one direction only.
• If the salt or pepper is not attracted to the spoon, rub the spoon with the cloth again or spread the salt and pepper out so that they are not clumped together.
• If you cannot attract only pepper when the spoon is brought slowly toward the salt and the pepper, spread the salt and pepper out more so that they are not clumped together.

Safety Caution: Be sure that students do not eat the salt or pepper. Also make sure that students do not spill the salt or pepper on the floor which may cause someone to slip.

Answers
1. The salt and pepper flew up to the spoon. Only the pepper flew up to the spoon.
2. The spoon is (negatively) charged after being rubbed with the wool. The spoon's charge attracted the salt and pepper and made them stick to the spoon.
3. Pepper; Pepper particles have less mass than the salt particles do. The force caused by the charge is greater than the force of gravity on the pepper.

Answers to Pre-Reading Questions
1. the basic unit of matter
2. electrons, protons, and neutrons
3. Electrons are located outside of an atom's nucleus. Protons and neutrons make up the nucleus.
4. Students' answers may vary. Sample answer: visible and gamma radiation

Overview

Before beginning this section, review with your students the Objectives listed in the Student Edition. In this section, students will learn about the law of conservation of mass, the law of definite proportions, and the law of multiple proportions. Students will also learn how these laws led John Dalton to propose one of the first models for the atom.

 **Bellringer**

Have students pass around a sealed shoebox containing a rattling object. Ask them to write down their inferences about any properties of the object that they can detect. Ask your students to suggest other ways to learn about the object without opening the box. Relate the exercise to the way scientists began to learn about the atom.

Motivate

Discussion ——— GENERAL

Ask students for an alternative to the model of matter in which atoms are a building block. Students might respond that a piece of matter could be divided for an infinite number of times without finding a basic unit. Greek philosophers Democritus and Aristotle disagreed about which model was correct. Democritus believed in atoms. Aristotle disagreed with Democritus and because Aristotle was one of the eminent philosophers in history, his opinion prevailed for over 2000 years. **LS** Interpersonal

KEY TERMS

- law of definite proportions
- law of conservation of mass
- law of multiple proportions

OBJECTIVES

① **State** the three laws that support the existence of atoms.

② **List** the five principles of John Dalton's atomic theory.

Atomic Theory

As early as 400 BCE, a few people believed in an *atomic theory,* which states that atoms are the building blocks of all matter. Yet until recently, even scientists had never seen evidence of atoms. Experimental results supporting the existence of atoms did not appear until more than 2000 years after the first ideas about atoms emerged. The first of these experimental results indicated that all chemical compounds share certain characteristics.

What do you think an atom looks like? Many people think that an atom looks like the diagram in **Figure 1a.** However, after reading this chapter, you will find that the diagram in **Figure 1b** is a better model of an atom.

Recall that a compound is a pure substance composed of atoms of two or more elements that are chemically combined. These observations about compounds and the way that compounds react led to the development of the law of definite proportions, the law of conservation of mass, and the law of multiple proportions. Experimental observations show that these laws also support the current atomic theory.

☑ internet connect

www.scilinks.org
Topic: Development of Atomic Theory
SciLinks code: HW4148

SCILINKS. Maintained by the National Science Teachers Association

☑ internet connect

www.scilinks.org
Topic : Current Atomic Theory
SciLinks code: HW4038

SCILINKS. Maintained by the National Science Teachers Association

Figure 1
a Many people believe that an atom looks like this diagram.

b This diagram is a better model of the atom.

HISTORY CONNECTION —— ADVANCED

Chemical Forerunners Before there were chemists, there were alchemists who wanted to find methods to convert common metals to gold. The alchemists practiced, often in secret, throughout the world during the Middle Ages. Dalton's theory of the atom, particularly the atom's immutability, helped bring an end to the practice of alchemy, but much chemical information that alchemists recorded helped establish the new science of chemistry. Have interested students investigate the work of the alchemists and present their findings to the class. **LS** Verbal

Chapter Resource File

- Lesson Plan

The Law of Definite Proportions

The **law of definite proportions** states that two samples of a given compound are made of the same elements in exactly the same proportions by mass regardless of the sizes or sources of the samples. Notice the composition of ethylene glycol, as shown in **Figure 2**. Every sample of ethylene glycol is composed of three elements in the following proportions by mass:

51.56% oxygen, 38.70% carbon, and 9.74% hydrogen

The law of definite proportions also states that every molecule of ethylene glycol is made of the same number and types of atoms. A molecule of ethylene glycol has the formula $C_2H_6O_2$, so the law of definite proportions tells you that all other molecules of ethylene glycol have the same formula.

Table salt (sodium chloride) is another example that shows the law of definite proportions. Any sample of table salt consists of two elements in the following proportions by mass:

60.66% chlorine and 39.34% sodium

Every sample of table salt also has the same proportions of ions. As a result, every sample of table salt has the same formula, NaCl.

As chemists of the 18th century began to gather data during their studies of matter, they first began to recognize the law of definite proportions. Their conclusions led to changes in the atomic theory.

law of definite proportions

the law that states that a chemical compound always contains the same elements in exactly the same proportions by weight or mass

STUDY TIP

USING THE ILLUSTRATIONS

The illustrations in the text will help you make the connection between what you can see, such as a beaker of chemicals, and what you cannot see, such as the atoms that make up those chemicals. Notice that the model in **Figure 2** shows how the atoms of a molecule of ethylene glycol are arranged.

•To practice thinking at the particle level, draw pictures of water molecules and copper atoms.

Figure 2
a Ethylene glycol is the main component of automotive antifreeze.

b Ethylene glycol is composed of carbon, oxygen, and hydrogen.

c Ethylene glycol is made of exact proportions of these elements regardless of the size of the sample or its source.

Ethylene Glycol Composition by Mass

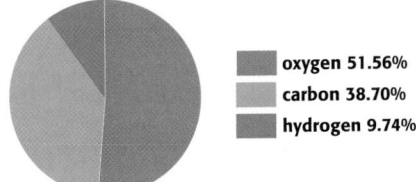

oxygen 51.56%
carbon 38.70%
hydrogen 9.74%

75

Teach

READING SKILL BUILDER — BASIC

Assimilating Knowledge Before reading this chapter, have students write a short list of all the things they already know or think they know about the following:

• the particle composition of matter

• the structure of an atom

• the major subatomic particles

• protons, neutrons, and electrons

• the characteristics, location, and arrangement of these particles

Then ask students to make a drawing of what they think an atom looks like. Also, have students discuss the things they might want to know about atoms and their structure. If students have trouble coming up with questions, the following is a suggestion to get them on track.

• Describe the structure of matter in common objects. If the objects could become very small, what would the inside of the matter look like?

• How would an iron nail and a piece of sulfur differ?
LS Interpersonal

Activity — BASIC

Have a beaker of water boiling in front of the class. Ask students to think about what is happening as water boils and if there is any evidence from which they might infer that matter is particulate in nature. Students should respond that the aggregate liquid water separates into particles of water as the water boils off into the air. Make sure students do not think that the particles of water break into smaller particles. **LS** Interpersonal

Discussion ——— GENERAL

Ask students how a brownie recipe is an analogy for the law of definite proportions. Students should see that a recipe calls for a certain measure of each of the ingredients. If the ingredients are mixed according to the recipe, the brownies should turn out the same every time they are made, no matter who makes them. The same theory applies to compounds. Each compound is made according to a "recipe" and contains the same proportions of its constituent elements no matter where the compound is found or who makes it. Make sure that students understand that this analogy is not perfect because nuts and chips can be added to brownies and the baked good is still considered brownies.
LS Interpersonal

Teaching Tip

Thinking Like Dalton The laws of definite proportions, conservation of mass, and multiple proportions re-create some of the reasoning that Dalton used in devising his atomic theory. After students have studied Dalton's theory, you may want to review how the theory explains these laws.

law of conservation of mass

the law that states that mass cannot be created or destroyed in ordinary chemical and physical changes

Figure 3
The total mass of a system remains the same whether atoms are combined, separated, or rearranged. Here, mass is expressed in kilograms (kg).

The Law of Conservation of Mass

As early chemists studied more chemical reactions, they noticed another pattern. Careful measurements showed that the mass of a reacting system does not change. The **law of conservation of mass** states that the mass of the reactants in a reaction equals the mass of the products. **Figure 3** shows several reactions that show the law of conservation of mass. For example, notice the combined mass of the sulfur atom and the oxygen molecule equals the mass of the sulfur dioxide molecule.

Also notice that **Figure 3** shows that the sum of the mass of the chlorine molecule and the mass of the phosphorus trichloride molecule is slightly smaller than the mass of the phosphorus pentachloride molecule. This difference is the result of rounding off and of correctly using significant figures.

Conservation of Mass

Hydrogen molecule
3.348×10^{-27} kg

Oxygen atom
2.657×10^{-26} kg

Water molecule
2.992×10^{-26} kg

$$H_2 \quad + \quad \tfrac{1}{2}O_2 \quad \longrightarrow \quad H_2O$$

Sulfur atom
5.325×10^{-26} kg

Oxygen molecule
5.314×10^{-26} kg

Sulfur dioxide molecule
1.064×10^{-25} kg

$$S \quad + \quad O_2 \quad \longrightarrow \quad SO_2$$

Phosphorus
pentachloride molecule
3.458×10^{-25} kg

Phosphorus
trichloride molecule
2.280×10^{-25} kg

Chlorine molecule
1.177×10^{-25} kg

$$PCl_5 \quad \longrightarrow \quad PCl_3 \quad + \quad Cl_2$$

76

HISTORY ——
CONNECTION

John Dalton based his atomic theory on the work of his contemporaries. Antoine-Laurent Lavoisier described the law of conservation of mass in 1782. In 1797, Joseph Proust proposed the law of definite proportions. A law of multiple proportions had been proposed by Claude-Louis Berthollet in the 1790s, but it was not accepted by many scientists. Dalton, however, saw that the law of multiple proportions supported his idea of atoms and restated it along with his atomic theory.

Transparencies

TT Law of Conservation of Mass

One-Stop Planner CD-ROM

• **Career Extension**
Real-World Connections Worksheet 5: Recycling Engineer
Assign this worksheet to emphasize relevant applications of text concepts.

Table 1 Compounds of Nitrogen and Oxygen and the Law of Multiple Proportions

Name of compound	Description	As shown in figures	Formula	Mass O (g)	Mass N (g)	Mass O(g) / Mass N(g)
Nitrogen monoxide	colorless gas that reacts readily with oxygen		NO	16.00	14.01	$\dfrac{16.00\,\text{g O}}{14.01\,\text{g N}} = \dfrac{1.14\,\text{g O}}{1\,\text{g N}}$
Nitrogen dioxide	poisonous brown gas in smog		NO_2	32.00	14.01	$\dfrac{32.00\,\text{g O}}{14.01\,\text{g N}} = \dfrac{2.28\,\text{g O}}{1\,\text{g N}}$

The Law of Multiple Proportions

Table 1 lists information about the compounds nitrogen monoxide and nitrogen dioxide. For each compound, the table also lists the ratio of the mass of oxygen to the mass of nitrogen. So, 1.14 g of oxygen combine with 1 g of nitrogen when nitrogen monoxide forms. In addition, 2.28 g of oxygen combine with 1 g of nitrogen when nitrogen dioxide forms. The ratio of the masses of oxygen in these two compounds is exactly 1.14 to 2.28 or 1 to 2. This example illustrates **the law of multiple proportions:** If two or more different compounds are composed of the same two elements, the ratio of the masses of the second element (which combines with a given mass of the first element) is always a ratio of small whole numbers.

The law of multiple proportions may seem like an obvious conclusion given the molecules' diagrams and formulas shown. But remember that the early chemists did not know the formulas for compounds. In fact, chemists still have not actually seen these molecules. Scientists think that molecules have these formulas because of these mass data.

law of multiple proportions

the law that states that when two elements combine to form two or more compounds, the mass of one element that combines with a given mass of the other is in the ratio of small whole numbers

Dalton's Atomic Theory

In 1808, John Dalton, an English school teacher, used the Greek concept of the atom and the law of definite proportions, the law of conservation of mass, and the law of multiple proportions to develop an atomic theory. Dalton believed that a few kinds of atoms made up all matter.

According to Dalton, elements are composed of only one kind of atom and compounds are made from two or more kinds of atoms. For example, the element copper consists of only one kind of atom, as shown in **Figure 4.** Notice that the compound iodine monochloride consists of two kinds of atoms joined together. Dalton also reasoned that only whole numbers of atoms could combine to form compounds, such as iodine monochloride. In this way, Dalton revised the early Greek idea of atoms into a scientific theory that could be tested by experiments.

iodine monochloride   **copper**

Figure 4
An element, such as copper, is made of only one kind of atom. In contrast, a compound, such as iodine monochloride, can be made of two or more kinds of atoms.

Homework ——— BASIC

The following statements match the five principles of Dalton's atomic theory. Scramble the order, and ask students to determine which of Dalton's principles explains each observation.

1. Matter can never really be thrown away. That is one reason that recycling is important.

2. There is no difference between copper found in an ancient Mayan necklace and copper wire freshly made from copper ore.

3. Zinc is a softer metal than iron, and it reacts more readily with acid than iron does.

4. The formula for ethanol is C_2H_6O, and the formula for acetic acid in vinegar is $C_2H_4O_2$.

5. When methane, CH_4, burns, it combines with oxygen, O_2, in the air to form molecules of water, H_2O, and carbon dioxide, CO_2.

LS Logical

INCLUSION Strategies

• *Learning Disabled*   • *English Language Learners*

Ask students to write the five principles of Dalton's Theory on five individual index cards. For each principle, have the students draw a model or describe an example that explains or defines each principle. The cards may be used for individual or small group study or displayed on a classroom bulletin board. Students can show their understanding of the concept by presenting their examples to the class or to a small group of other students.

Transparencies

TT Law of Multiple Proportions

ChemFile   CHEMISTRY INTERACTIVE TUTOR

• **Module 2: Models of the Atom Topic: Atomic Structure**

This engaging tutorial reviews the discoveries that form the basis of our current atomic model.

Close

Reteaching ———— BASIC

Divide students into groups of four and have them create a concept map or graphic organizer that includes all the principal ideas in this section. Reproduce the best map or organizer on the chalkboard and use it for further discussion. **LS Interpersonal**

Quiz ———————— GENERAL

1. Which statement in Dalton's atomic theory is supported by the law of conservation of mass? **Ans.** The statement that atoms cannot be created or destroyed

2. Do the statements of Dalton's atomic theory still apply today? Explain. **Ans.** Yes, except for statements 1 and 2. Recent technology allows atoms to be created and destroyed, atoms can be divided into smaller particles, and different isotopes of a given element have different atomic weights.

3. The following data is for two compounds (a and b) made up of lead (Pb) and oxygen (O). Do the data support the law of multiple proportions?

compound a: 0.773 g O for 10.0 g Pb

compound b: 1.546 g O for 10.0 g Pb

Ans. 1.546 g O/0.773 g O = 2.00; the ratio is 1:2, so the data support the law.

LS Intrapersonal

Dalton's Theory Contains Five Principles

Dalton's atomic theory can be summarized by the following statements:

1. All matter is composed of extremely small particles called *atoms*, which cannot be subdivided, created, or destroyed.

2. Atoms of a given element are identical in their physical and chemical properties.

3. Atoms of different elements differ in their physical and chemical properties.

4. Atoms of different elements combine in simple, whole-number ratios to form compounds.

5. In chemical reactions, atoms are combined, separated, or rearranged but never created, destroyed, or changed.

Dalton's theory explained most of the chemical data that existed during his time. As you will learn later in this chapter, data gathered since Dalton's time shows that the first two principles are not true in all cases.

Today, scientists can divide an atom into even smaller particles. Technology has also enabled scientists to destroy and create atoms. Another feature of atoms that Dalton could not detect is that many atoms will combine with like atoms. Oxygen, for example, is generally found as O_2, a molecule made of two oxygen atoms. Sulfur is found as S_8. Because some parts of Dalton's theory have been shown to be incorrect, his theory has been modified and expanded as scientists learn more about atoms.

 Section Review

UNDERSTANDING KEY IDEAS

1. What is the atomic theory?

2. What is a compound?

3. State the laws of definite proportions, conservation of mass, and multiple proportions.

4. According to Dalton, what is the difference between an element and a compound?

5. What are the five principles of Dalton's atomic theory?

6. Which of Dalton's five principles still apply to the structure of an atom?

CRITICAL THINKING

7. What law is described by the fact that carbon dioxide consists of 27.3% carbon and 72.7% oxygen by mass?

8. What law is described by the fact that the ratio of the mass of oxygen in carbon dioxide to the mass of oxygen in carbon monoxide is 2:1?

9. Three compounds contain the elements sulfur, S, and fluorine, F. How do the following data support the law of multiple proportions?

compound A: 1.188 g F for every 1.000 g S

compound B: 2.375 g F for every 1.000 g S

compound C: 3.563 g F for every 1.000 g S

78

Answers to Section Review

1. the theory that all matter is composed of atoms

2. A compound is a substance made from two or more kinds of atoms.

3. law of definite proportions: each sample of a given compound contains the same elements in the same proportions by mass regardless of the size or source of the sample; law of conservation of mass: mass is neither created nor destroyed during a chemical reaction; law of multiple proportions: when the same elements combine to form different compounds, they do so in mass ratios that can be expressed by small, whole numbers

Answers continued on p. 113A

Chapter Resource File

• Concept Review

• Quiz

Structure of Atoms

KEY TERMS

- electron
- nucleus
- proton
- neutron
- atomic number
- mass number
- isotope

OBJECTIVES

(1) **Describe** the evidence for the existence of electrons, protons, and neutrons, and describe the properties of these subatomic particles.

(2) **Discuss** atoms of different elements in terms of their numbers of electrons, protons, and neutrons, and define the terms *atomic number* and *mass number*.

(3) **Define** *isotope,* and determine the number of particles in the nucleus of an isotope.

Subatomic Particles

Experiments by several scientists in the mid-1800s led to the first change to Dalton's atomic theory. Scientists discovered that atoms can be broken into pieces after all. These smaller parts that make up atoms are called *subatomic particles.* Many types of subatomic particles have since been discovered. The three particles that are most important for chemistry are the electron, the proton, and the neutron.

internet connect

www.scilinks.org
Topic : Subatomic Particles
SciLinks code: HW4121

SC*LINKS* Maintained by the National Science Teachers Association

Electrons Were Discovered by Using Cathode Rays

The first evidence that atoms had smaller parts was found by researchers who were studying electricity, not atomic structure. One of these scientists was the English physicist J. J. Thomson. To study current, Thomson pumped most of the air out of a glass tube. He then applied a voltage to two metal plates, called *electrodes,* which were placed at either end of the tube. One electrode, called the *anode,* was attached to the positive terminal of the voltage source, so it had a positive charge. The other electrode, called a *cathode,* had a negative charge because it was attached to the negative terminal of the voltage source.

Thomson observed a glowing beam that came out of the cathode and struck the anode and the nearby glass walls of the tube. So, he called these rays *cathode rays.* The glass tube Thomson used is known as a *cathode-ray tube* (CRT). CRTs have become an important part of everyday life. They are used in television sets, computer monitors, and radar displays.

An Electron Has a Negative Charge

Thomson knew the rays must have come from the atoms of the cathode because most of the atoms in the air had been pumped out of the tube. Because the cathode ray came from the negatively charged cathode, Thomson reasoned that the ray was negatively charged.

Figure 5
The image on a television screen or a computer monitor is produced when cathode rays strike the special coating on the inside of the screen.

79

REAL-WORLD CONNECTION

Laws and Theories Dalton's atomic theory is a perfect example to use to clarify the concepts of scientific laws and theories. Many students and adults have the idea that all theories are unproven notions because of the way the word is used in everyday speech. Ask students for examples. A theory explains a law and is supported by scientific evidence. Laws describe the universe. They can be demonstrated to be true at any time or place. Dalton was confident in proposing his atomic theory because his theory accounted for the laws of conservation of mass, multiple proportions, and definite proportions.

Demonstration

Magnets and Cathode Rays

If you do not have a demonstration cathode-ray tube, you can perform this simple demonstration with a television and a bar magnet. Do not use a computer monitor for this demonstration.

1. Use a freeze-frame function on a laser disc player or a VCR to get a picture with contrasting color boundaries on the screen.

2. Bring a bar magnet toward the screen. The colors will shift as the electron beams are deflected toward phosphors of different colors. Move the magnet around to see how it affects other parts of the picture.

3. Reverse the magnet, and bring the other pole near the screen. The colors should shift in the opposite direction.

4. Find a picture with thin, straight, horizontal or vertical lines, and repeat steps 2 and 3.

Safety Caution: Do not hold the magnet near the screen for longer than is necessary to observe the effects because the metal grid just behind the screen can be magnetized and the colors will remain distorted. If this occurs, you can turn the set on and off several times to relieve the problem. Modern TVs have built-in demagnetizers.

Figure 6
A magnet near the cathode-ray tube causes the beam to be deflected. The deflection indicates that the particles in the beam have a negative charge.

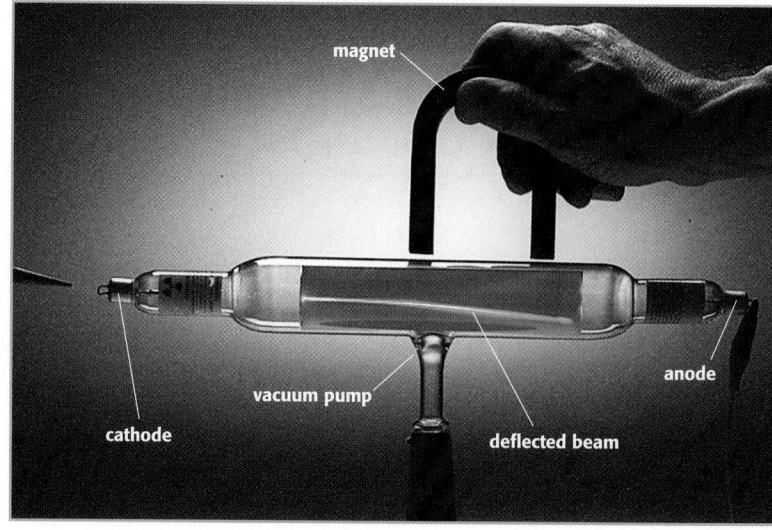

www.scilinks.org
Topic : J. J. Thomson
SciLinks code: HW4156

SCiLINKS Maintained by the National Science Teachers Association

electron

a subatomic particle that has a negative electric charge

He confirmed this prediction by seeing how electric and magnetic fields affected the cathode ray. **Figure 6** shows what Thomson saw when he placed a magnet near the tube. Notice that the beam is deflected by the magnet. Other researchers had shown that moving negative charges are deflected this way.

Thomson also observed that when a small paddle wheel was placed in the path of the rays, the wheel would turn. This observation suggested that the cathode rays consisted of tiny particles that were hitting the paddles of the wheel.

Thomson's experiments showed that a cathode ray consists of particles that have mass and a negative charge. These particles are called **electrons. Table 2** lists the properties of an electron. Later experiments, which used different metals for cathodes, confirmed that electrons are a part of atoms of all elements.

Electrons are negatively charged, but atoms have no charge. Therefore, atoms must contain some positive charges that balance the negative charges of the electrons. Scientists realized that positive charges must exist in atoms and began to look for more subatomic particles. Scientists also recognized that atoms must have other particles because an electron was found to have much less mass than an atom does.

Table 2	Properties of an Electron				
Name	**Symbol**	**As shown in figures**	**Charge**	**Common charge notation**	**Mass (kg)**
Electron	$e, e^-,$ or $_{-1}^{0}e$		-1.602×10^{-19} C	-1	9.109×10^{-31} kg

Rutherford Discovered the Nucleus

Thomson proposed that the electrons of an atom were embedded in a positively charged ball of matter. His picture of an atom, which is shown in **Figure 7,** was named the *plum-pudding model* because it resembled plum pudding, a dessert consisting of a ball of cake with pieces of fruit in it. Ernest Rutherford, one of Thomson's former students, performed experiments in 1909 that disproved the plum-pudding model of the atom.

Rutherford's team of researchers carried out the experiment shown in **Figure 8.** A beam of small, positively charged particles, called *alpha particles,* was directed at a thin gold foil. The team measured the angles at which the particles were deflected from their former straight-line paths as they came out of the foil.

Rutherford found that most of the alpha particles shot at the foil passed straight through the foil. But a very small number of particles were deflected, in some cases backward, as shown in **Figure 8.** This result greatly surprised the researchers—it was very different from what Thomson's model predicted. As Rutherford said, "It was almost as if you fired a 15-inch shell into a piece of tissue paper and it came back and hit you." After thinking about the startling result for two years, Rutherford finally came up with an explanation. He went on to reason that only a very concentrated positive charge in a tiny space within the gold atom could possibly repel the fast-moving, positively charged alpha particles enough to reverse the alpha particles' direction of travel.

Rutherford also hypothesized that the mass of this positive-charge containing region, called the *nucleus,* must be larger than the mass of the alpha particle. If not, the incoming particle would have knocked the positive charge out of the way. The reason that most of the alpha particles were undeflected, Rutherford argued, was that most parts of the atoms in the gold foil were empty space.

This part of the model of the atom is still considered true today. The **nucleus** is the dense, central portion of the atom. The nucleus has all of the positive charge, nearly all of the mass, but only a very small fraction of the volume of the atom.

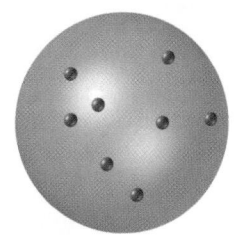

Figure 7
Thomson's model of an atom had negatively charged electrons embedded in a ball of positive charge.

nucleus

an atom's central region, which is made up of protons and neutrons

Figure 8

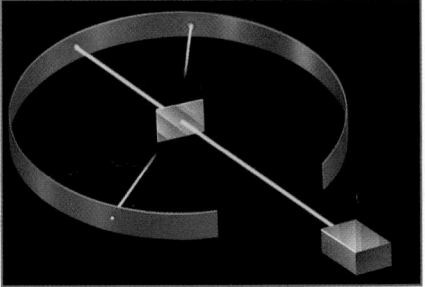

a In the gold foil experiment, small positively charged particles were directed at a thin foil of gold atoms.

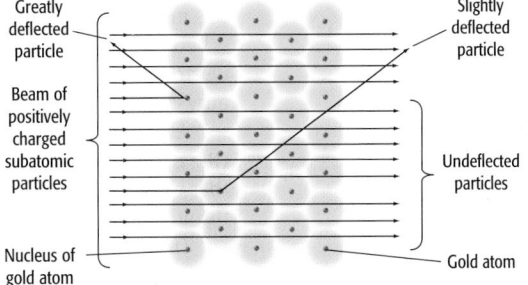

Greatly deflected particle

Beam of positively charged subatomic particles

Nucleus of gold atom

Slightly deflected particle

Undeflected particles

Gold atom

b The pattern of deflected alpha particles supported Rutherford's hypothesis that gold atoms were mostly empty space.

81

Transparencies

TT Gold-Foil Experiment

TT Gold-Foil Experiment on the Atomic Level

Demonstration

If you have a cathode ray tube like the one pictured in **Figure 6,** use it to demonstrate the rays and their deflection by a magnet. Follow the instructions that accompany the tube and the power supply. Keep students 3 m away from the apparatus, wear goggles, and have students wear goggles. Do not touch the terminals or the tube while the power supply is on. Keep magnets away from the ends of the tube. Demonstrate a Crooke's tube, which has a mica paddlewheel and a Maltese cross tube if they are available.

READING SKILL BUILDER — **BASIC**

Reading Organizer Students will become more familiar with the parts of the cathode-ray tube if they draw a labeled diagram showing its various parts and characteristics. Have them refer to **Figure 6** and the text to obtain the information they need. **LS Visual**

Discussion

Ask students to speculate on why Thomson inferred that electrons were embedded in a ball of positive charge. Get students to tell you that Thomson knew that the atom had no net charge, so he assumed that positive charge must be present to balance the negatively-charged electrons.

Homework ——— BASIC

Have students create a graphic organizer with three columns and three rows. The headings on the columns should be Question, Discovery, Inferences. The first column contains the following questions. Does the atom have a substructure? What is the arrangement of subatomic particles? How can all of the atom's mass be accounted for? Students should fill in the missing data in columns 2 and 3. **LS** Logical

Discussion ——— GENERAL

Present the following analogy to the gold-foil experiment. Have students imagine that 50 billiard balls are hung by strings at various heights throughout the front half of the classroom. A student is then blindfolded and given five tennis balls to throw toward the front of the room. Have students report what they think they would observe in this situation. Ask them what the billiard balls and the tennis balls symbolize. **Ans.** Some tennis balls would be deflected. The billiard balls represent gold nuclei, and the tennis balls represent the lighter alpha particles. **LS** Logical

Figure 9
If the nucleus of an atom were the size of a marble, then the whole atom would be about the size of a football stadium.

proton

a subatomic particle that has a positive charge and that is found in the nucleus of an atom; the number of protons of the nucleus is the atomic number, which determines the identity of an element

neutron

a subatomic particle that has no charge and that is found in the nucleus of an atom

82

Protons and Neutrons Compose the Nucleus

By measuring the numbers of alpha particles that were deflected and the angles of deflection, scientists calculated the radius of the nucleus to be less than $\frac{1}{10\,000}$ of the radius of the whole atom. **Figure 9** gives you a better idea of these sizes. Even though the radius of an entire atom is more than 10 000 times larger than the radius of its nucleus, an atom is still extremely small. The unit used to express atomic radius is the picometer (pm). One picometer equals 10^{-12} m.

The positively charged particles that repelled the alpha particles in the gold foil experiments and that compose the nucleus of an atom are called **protons.** The charge of a proton was calculated to be exactly equal in magnitude but opposite in sign to the charge of an electron. Later experiments showed that the proton's mass is almost 2000 times the mass of an electron.

Because protons and electrons have equal but opposite charges, a neutral atom must contain equal numbers of protons and electrons. But solving this mystery led to another: the mass of an atom (except hydrogen atoms) is known to be greater than the combined masses of the atom's protons and electrons. What could account for the rest of the mass? Hoping to find an answer, scientists began to search for a third subatomic particle.

About 30 years after the discovery of the electron, Irene Joliot-Curie (the daughter of the famous scientists Marie and Pierre Curie) discovered that when alpha particles hit a sample of beryllium, a beam that could go through almost anything was produced.

The British scientist James Chadwick found that this beam was not deflected by electric or magnetic fields. He concluded that the particles carried no electric charge. Further investigation showed that these neutral particles, which were named **neutrons,** are part of all atomic nuclei (except the nuclei of most hydrogen atoms).

HISTORY ——— CONNECTION

Progress in Steps The gold-foil experiment was undertaken by Rutherford and his graduate students with a routine "let's see what happens" attitude and with little expectation of exceptional results. However, the results caused great excitement among scientists. When Rutherford announced the results of the 1909 experiments in 1911, he postulated only the existence of an elementary, positively charged particle. He discovered the actual particle in 1918.

Table 3   Properties of a Proton and a Neutron

Name	Symbol	As shown in figures	Charge	Common charge notation	Mass (kg)
Proton	p, p^+, or $_{+1}^{1}p$	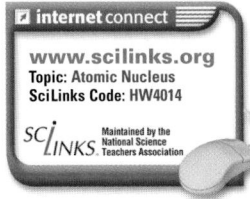	$+1.602 \times 10^{-19}$ C	$+1$	1.673×10^{-27} kg
Neutron	n or $_{0}^{1}n$		0 C	0	1.675×10^{-27} kg

Protons and Neutrons Can Form a Stable Nucleus

Table 3 lists the properties of a neutron and a proton. Notice that the charge of a neutron is commonly assigned the value 0 while that of a proton is +1. How do protons that are positively charged come together to form a nucleus? In fact, the formation of a nucleus with protons seems impossible if you just consider *Coulomb's law*. Coulomb's law states that the closer two charges are, the greater the force between them. In fact, the force increases by a factor of 4 as the distance is halved. In addition, the larger the two charges are the greater the force between them. If the charges are opposite, they attract one another. If both charges have the same sign, they repel one another.

If you keep Coulomb's law in mind, it is easy to understand why—with the exception of some hydrogen atoms—no atoms have nuclei that are composed of only protons. All protons have a +1 charge. So, the repulsive force between two protons is large when two protons are close together, such as within a nucleus.

Protons, however, do form stable nuclei despite the repulsive force between them. A strong attractive force between these protons overcomes the repulsive force at small distances. Because neutrons also add attractive forces without being subject to repulsive charge-based forces, some neutrons can help stabilize a nucleus. Thus, all atoms that have more than one proton also have neutrons.

internet connect

www.scilinks.org
Topic: Atomic Nucleus
SciLinks Code: HW4014

SciLINKS. Maintained by the National Science Teachers Association

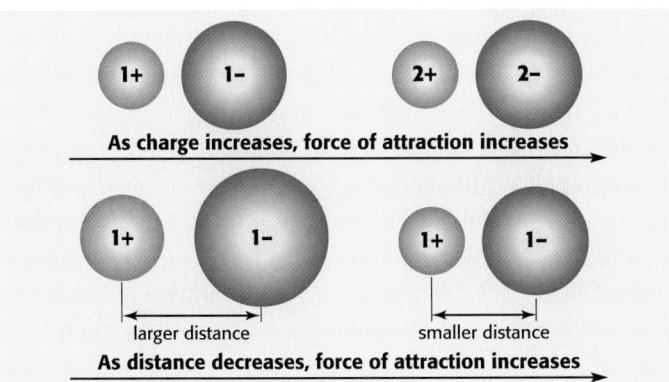

Figure 10
This figure shows that the larger two charges are, the greater the force between the charges. In addition, the figure shows the smaller the distance between two charges, the greater the force between the charges.

did you know?

The concept atomic number was proposed by Henry Moseley in 1914. At that time, there was still no explanation for the arrangement of the elements in the periodic table, and scientists continued to try to link periodicity to atomic mass. Moseley found that the progression of the elements in the table corresponded to an increase of one fundamental unit in the nucleus and that each element could be assigned an atomic number equal to the number of these units. In 1920, Rutherford announced the existence of Moseley's fundamental unit, the proton.

Transparencies

TT Properties of Subatomic Particles

Using the Table

Call students' attention to **Table 3.** Have them note that each of the three ways of designating the proton provides successively more information. Ask them to state in words the information in the symbol $_{+1}^{1}p$. Have them compare the masses of the proton and neutron. They should see that the masses are the same to three significant figures. Finally, explain Coulombs and reinforce that $+1 = +1.602 \times 10^{-19}$ Coulombs.

Teaching Tip

The Strong Force The strong force is one of the four fundamental forces of nature. The others are electromagnetic, weak, and gravitational. The strong force is 100 times stronger than the electrostatic force that repels the protons in the nucleus. However, the strength of the strong force dwindles with distance so that when the distance between two protons is equal to the combined diameters of a few particles, the strong force is much weaker than the electrostatic force of repulsion. As an analogy, have students think about the force between two magnets. When the magnets are close, the force of attraction or repulsion is great. As the magnets are separated, the force dwindles.

Teaching Tip

Atomic Numbers Point out atomic numbers on the periodic table so students realize that the numbers on the table are not just a sequencing of the elements, but the number of protons of an element's nucleus.

Videos

CNN Presents Science in the News

• **Segment 8** *The Top Quark*

See the Science in the News video guide for more details.

Atomic Number and Mass Number

All atoms consist of protons and electrons. Most atoms also have neutrons. Protons and neutrons make up the small, dense nuclei of atoms. The electrons occupy the space surrounding the nucleus. For example, an oxygen atom has protons and neutrons surrounded by electrons. But that description fits all other atoms, such as atoms of carbon, nitrogen, silver, and gold. How, then, do the atoms of one element differ from those of another element? Elements differ from each other in the number of protons their atoms contain.

Atomic Number Is the Number of Protons of the Nucleus

The number of protons that an atom has is known as the atom's **atomic number.** For example, the atomic number of hydrogen is 1 because the nucleus of each hydrogen atom has one proton. The atomic number of oxygen is 8 because all oxygen atoms have eight protons. Because each element has a unique number of protons in its atoms, no two elements have the same atomic number. So an atom whose atomic number is 8 must be an oxygen atom.

To date, scientists have identified 113 elements, whose atomic numbers range from 1 to 114. The element whose atomic number is 113 has yet to be discovered. Note that atomic numbers are always whole numbers. For example, an atom cannot have 2.5 protons.

The atomic number also reveals the number of electrons in an atom of an element. For atoms to be neutral, the number of negatively charged electrons must equal the number of positively charged protons. Therefore, if you know the atomic number of an atom, you immediately know the number of protons and the number of electrons found in that atom. **Figure 11** shows a model of an oxygen atom, whose atomic number is 8 and which has 8 electrons surrounding a nucleus that has 8 protons. The atomic number of gold is 79, so an atom of gold must have 79 electrons surrounding a nucleus of 79 protons. The next step in describing an atom's structure is to find out how many neutrons the atom has.

atomic number

the number of protons in the nucleus of an atom; the atomic number is the same for all atoms of an element

Figure 11
The atomic number for oxygen, as shown on the periodic table, tells you that the oxygen atom has 8 protons and 8 electrons.

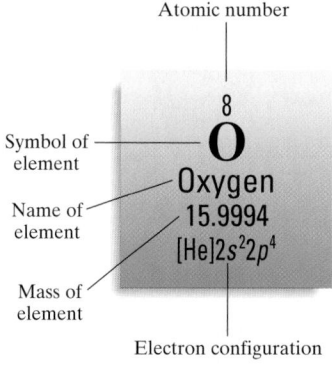

Atomic number

8

O

Oxygen

15.9994

$[He]2s^2 2p^4$

Symbol of element

Name of element

Mass of element

Electron configuration

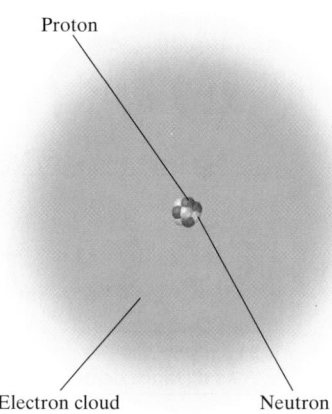

Proton

Electron cloud

Neutron

Mass Number Is the Number of Particles of the Nucleus

Every atomic nucleus can be described not only by its atomic number but also by its mass number. The **mass number** is equal to the total number of particles of the nucleus—that is, the total number of protons and neutrons. For example, a particular atom of neon has a mass number of 20, as shown in **Figure 12.** Therefore, the nucleus of this atom has a total of 20 protons and neutrons. Because the atomic number for an atom of neon is 10, neon has 10 protons. You can calculate the number of neutrons in a neon atom by subtracting neon's atomic number (the number of protons) from neon's mass number (the number of protons and neutrons).

mass number – atomic number = number of neutrons

In this example, the neon atom has 10 neutrons.

number of protons and neutrons (mass number) = 20
– number of protons (atomic number) = 10
──────────────────────────────
number of neutrons = 10

Unlike the atomic number, which is the same for all atoms of an element, mass number can vary among atoms of a single element. In other words, all atoms of an element have the same number of protons, but they can have different numbers of neutrons. The atomic number of every hydrogen atom is 1, but hydrogen atoms can have mass numbers of 1, 2, or 3. These atoms differ from one another in having 0, 1, and 2 neutrons, respectively. Another example is oxygen. The atomic number of every oxygen atom is 8, but oxygen atoms can have mass numbers of 16, 17, or 18. These atoms differ from one another in having 8, 9, and 10 neutrons, respectively.

mass number
the sum of the numbers of protons and neutrons of the nucleus of an atom

Figure 12
The neon atom has 10 protons, 10 neutrons, and 10 electrons. This atom's mass number is 20, or the sum of the numbers of protons and neutrons in the atom.

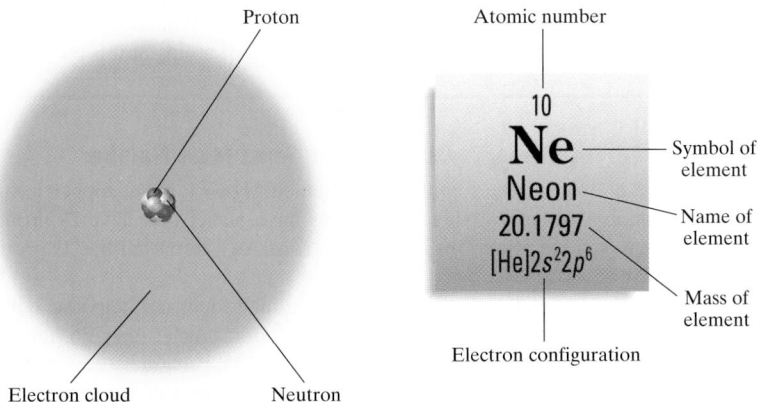

Proton

Atomic number

10

Ne — Symbol of element

Neon — Name of element

20.1797 — Mass of element

$[He]2s^2 2p^6$

Electron configuration

Electron cloud

Neutron

did you know?

Mass spectrometry is a technique for precisely determining the masses of charged particles. In a mass spectrometer, a charged particle is accelerated by an electric field into a magnetic field where the magnetic field deflects the particle along a curved path. The radius of the curve is determined by the particle's speed and its charge-to-mass ratio. In 1919, F. W. Aston used a mass spectrometer to demonstrate for the first time that isotopes existed. In a sample of air, he found a heavy form of neon with a mass of 22.

Additional Practice To master the concepts of atomic number and mass number, have students fill in a table for three atoms a, b, and c. Students should create a table with the headings Atomic symbol, Atomic number, Mass number, Number of protons, Number of neutrons, and Number of electrons. Students should enter the data given below and provide the remaining data.

Atom a: number of protons = 5,
         mass number = 11

Atom b: symbol = O,
         mass number = 16

Atom c: atomic number = 7,
         number of neutrons = 8

LS Intrapersonal

Answers to Practice Problems A

1. 11 protons and 11 electrons

2. 27

3. 80

4. The atomic number of this element is 54. The mass number of the atoms that have 77 neutrons is 131, while the mass number of the other atoms is 133.

SAMPLE PROBLEM A

Determining the Number of Particles in an Atom

How many protons, electrons, and neutrons are present in an atom of copper whose atomic number is 29 and whose mass number is 64?

1 Gather information.
- The atomic number of copper is 29.
- The mass number of copper is 64.

2 Plan your work.
- The atomic number indicates the number of protons in the nucleus of a copper atom.
- A copper atom must be electrically neutral, so the number of electrons equals the number of protons.
- The mass number indicates the total number of protons and neutrons in the nucleus of a copper atom.

> **PRACTICE HINT**
>
> Check that the atomic number and the number of protons are the same. Also check that adding the numbers of protons and neutrons equals the mass number.

3 Calculate.
- *atomic number* (29) = *number of protons* = 29
- *number of protons* = *number of electrons* = 29
- *mass number* (64) − *atomic number* (29) = *number of neutrons* = 35

4 Verify your results.
- *number of protons* (29) + *number of neutrons* (35) = *mass number* (64)

PRACTICE

1 How many protons and electrons are in an atom of sodium whose atomic number is 11?

2 An atom has 13 protons and 14 neutrons. What is its mass number?

3 Calculate the mass number for an atom that has 45 neutrons and 35 electrons.

4 An atom of an element has 54 protons. Some of the element's atoms have 77 neutrons, while other atoms have 79 neutrons. What are the atomic numbers and mass numbers of the two types of atoms of this element?

Different Elements Can Have the Same Mass Number

The atomic number identifies an element. For example, copper has the atomic number 29. All copper atoms have nuclei that have 29 protons. Each of these atoms also has 29 electrons. Any atom that has 29 protons must be a copper atom.

In contrast, knowing just the mass number does not help you identify the element. For example, some copper atom nuclei have 36 neutrons. These copper atoms have a mass number of 65. But zinc atoms that have 30 protons and 35 neutrons also have mass numbers of 65.

---

did you know? ——— ADVANCED

When J. J. Thomson was young, he was not interested in theoretical science. He wanted to become a railroad engineer. But at age 14, he was offered a scholarship to study chemistry at Owen's College in Manchester, England. Thus, Thomson began an illustrious career as chemist and physicist that continued into the twentieth century and embraced the new physics of the quantum era. Have interested students investigate the life and work of Thomson and present their findings to the class. **LS** Intrapersonal

Atomic Structures Can Be Represented by Symbols

Each element has a name, and the same name is given to all atoms of an element. For example, sulfur is composed of sulfur atoms. Recall that each element also has a symbol, and the same symbol is used to represent one of the element's atoms. Thus, S represents a single atom of sulfur, 2S represents two sulfur atoms, and 8S represents eight sulfur atoms. However, chemists write S_8 to indicate that the eight sulfur atoms are joined together and form a molecule of sulfur, as shown in the model in **Figure 13.**

Atomic number and mass number are sometimes written with an element's symbol. The atomic number always appears on the lower left side of the symbol. For example, the symbols for the first five elements are written with atomic numbers as follows:

$$_1H \quad _2He \quad _3Li \quad _4Be \quad _5B$$

Note that these subscript numbers give no new information. They simply indicate the atomic number of a particular element. On the other hand, mass numbers provide information that specifies particular atoms of an element. Mass numbers are written on the upper left side of the symbol. The following are the symbols of stable atoms of the first five elements written with mass numbers:

$$^1H \quad ^2H \quad ^3He \quad ^4He \quad ^6Li \quad ^7Li \quad ^9Be \quad ^{10}B \quad ^{11}B$$

Both numbers may be written with the symbol. For example, the most abundant kind of each of the first five elements can be represented by the following symbols:

$$_1^1H \quad _2^4He \quad _3^7Li \quad _4^9Be \quad _5^{11}B$$

An element may be represented by more than one notation. For example, the following notations represent the different atoms of hydrogen:

$$_1^1H \quad _1^2H \quad _1^3H$$

Sulfur, S_8

Hydrogen, H_2

Helium, He

■ internet connect ■

www.scilinks.org
Topic : Atomic Structures
SciLinks code: HW4015

SCI
LINKS. Maintained by the
National Science
Teachers Association

Figure 13
In nature, elemental sulfur exists as eight sulfur atoms joined in a ring, elemental hydrogen exists as a molecule of two hydrogen atoms, and elemental helium exists as single helium atoms.

87

Teaching Tip

Vocabulary The word *isotope*, coined around 1910, is derived from the Greek roots *iso*, which means "the same or equal," and *topos*, which means "place." Isotope was used to describe atoms of the same element (the same place on the periodic table) that had different masses.

Homework ———— **GENERAL**

Additional Practice

1. Calculate the number of protons, electrons, and neutrons in potassium-39 and potassium-41. The atomic number of potassium is 19. Ans. potassium-39: protons = 19, electrons = 19, neutrons = 20; potassium-41: protons = 19, electrons = 19, neutrons = 22

2. Lithium has two stable isotopes, lithium-6 and lithium-7. The atomic number of lithium is 3. Explain how the two are the same and how they are different. Ans. Both have three protons and three electrons; lithium-7 has four neutrons but lithium-6 has three neutrons.

LS Logical

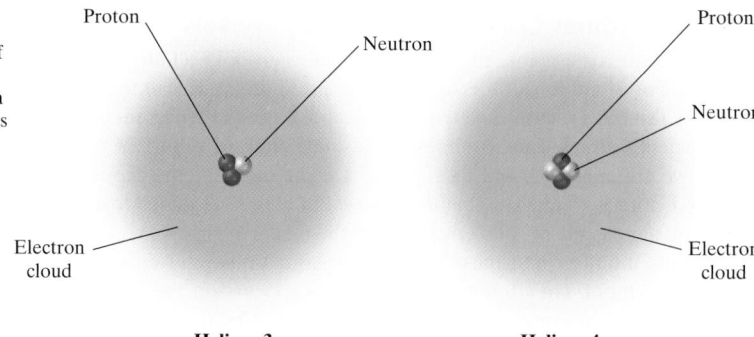

Figure 14
The two stable isotopes of helium are helium-3 and helium-4. The nucleus of a helium-4 atom is known as an *alpha particle*.

Helium-3   Helium-4

isotope

an atom that has the same number of protons (atomic number) as other atoms of the same element but has a different number of neutrons (atomic mass)

www.scilinks.org
Topic : Atoms and Elements
SciLinks code: HW4017

Isotopes of an Element Have the Same Atomic Number

All atoms of an element have the same atomic number and the same number of protons. However, atoms do not necessarily have the same number of neutrons. Atoms of the same element that have different numbers of neutrons are called **isotopes.** The two atoms modeled in **Figure 14** are stable isotopes of helium.

There are two standard methods of identifying isotopes. One method is to write the mass number with a hyphen after the name of an element. For example, the helium isotope shown on the left in **Figure 14** is written helium-3, while the isotope shown on the right is written as helium-4. The second method shows the composition of a nucleus as the isotope's nuclear symbol. Using this method, the notations for the two helium isotopes shown in **Figure 14** are written below.

$$^3_2\text{He} \quad ^4_2\text{He}$$

Notice that all isotopes of an element have the same atomic number. However, their atomic masses are not the same because the number of neutrons of the atomic nucleus of each isotope varies. In the case of helium, both isotopes have two protons in their nuclei. However, helium-3 has one neutron, while helium-4 has two neutrons.

Table 4 lists the four stable isotopes of lead. The least abundant of these isotopes is lead-204, while the most common is lead-208. Why do all lead atoms have 82 protons and 82 electrons?

Table 4   The Stable Isotopes of Lead

Name of atom	Symbol	Number of neutrons	Mass number	Mass (kg)	Abundance (%)
Lead-204	$^{204}_{82}\text{Pb}$	122	204	203.973	1.4
Lead-206	$^{206}_{82}\text{Pb}$	124	206	205.974	24.1
Lead-207	$^{207}_{82}\text{Pb}$	125	207	206.976	22.1
Lead-208	$^{208}_{82}\text{Pb}$	126	208	207.977	52.4

88

MISCONCEPTION ALERT

Students may think that not all atoms have isotopes and that those that do must exhibit different properties because they have different atomic masses. Some students may not appreciate that isotopes are forms of the same element. Stress that isotopes are common to all the elements, but that some isotopes do not occur naturally. Make it clear that only the number of protons determines the identity of an atom, and isotopes of an element have the same number of protons. The number of neutrons does not affect how an atom bonds with other atoms, so isotopes of the same element are indistinguishable in how they bond to other atoms.

Determining the Number of Particles in Isotopes

Calculate the numbers of protons, electrons, and neutrons in oxygen-17 and in oxygen-18.

1 Gather information.
- The mass numbers for the two isotopes are 17 and 18.

2 Plan your work.
- An oxygen atom must be electrically neutral.

3 Calculate.
- *atomic number = number of protons = number of electrons = 8*
- *mass number − atomic number = number of neutrons*
- For oxygen-17, 17 − 8 = 9 neutrons
- For oxygen-18, 18 − 8 = 10 neutrons

4 Verify your results.
- The two isotopes have the same numbers of protons and electrons and differ only in their numbers of neutrons.

PRACTICE HINT

The only difference between the isotopes of an element is the number of neutrons in the atoms of each isotope.

PRACTICE

1 Chlorine has two stable isotopes, chlorine-35 and chlorine-37. The atomic number of chlorine is 17. Calculate the numbers of protons, electrons, and neutrons each isotope has.

2 Calculate the numbers of protons, electrons, and neutrons for each of the following isotopes of calcium: $^{42}_{20}Ca$ and $^{44}_{20}Ca$.

PROBLEM SOLVING SKILL

Section Review

UNDERSTANDING KEY IDEAS

1. Describe the differences between electrons, protons, and neutrons.

2. How are isotopes of the same element alike?

3. What subatomic particle was discovered with the use of a cathode-ray tube?

PRACTICE PROBLEMS

4. Write the symbol for element X, which has 22 electrons and 22 neutrons.

5. Determine the numbers of electrons, protons, and neutrons for each of the following:
   a. $^{80}_{35}Br$ b. $^{106}_{46}Pd$ c. $^{133}_{55}Cs$

6. Calculate the atomic number and mass number of an isotope that has 56 electrons and 82 neutrons.

CRITICAL THINKING

7. Why must there be an attractive force to explain the existence of stable nuclei?

8. Are hydrogen-3 and helium-3 isotopes of the same element? Explain your answer.

89

Chapter Resource File
- Concept Review
- Quiz

Answers to Section Review

1. The mass of an electron is much smaller than the mass of a proton or the mass of a neutron. Protons have a +1 charge, electrons have a −1 charge, and neutrons have no charge. Protons and neutrons make up the atomic nucleus, while electrons surround the nucleus.

2. All isotopes of an element have the same number of protons and electrons.

3. electron

4. $^{44}_{22}X$

Answers continued on p. 113A

---

1. Both isotopes have 17 protons and 17 electrons. Chlorine-35 has 18 neutrons, and chlorine-37 has 20 neutrons.

2. Both have 20 protons and 20 electrons. Calcium-42 has 22 neutrons. Calcium-44 has 24 neutrons.

Close

Quiz ——— GENERAL

1. Explain how Rutherford's experiment resulted in a revision of Dalton's atomic model. **Ans.** Rutherford reasoned that because a few alpha particles were strongly repulsed, there must be a tiny highly-charged nucleus within the atom.

2. Compare the properties of electrons, protons, and neutrons. **Ans.** electrons and protons are charged, but neutrons are neutral. An electron charge is −1. A proton's charge is +1. Protons and neutrons are almost equal in mass. The mass of an electron is approximately 1/2000 of the mass of a proton.

3. Determine the number of protons, electrons, and neutrons for the following:
   a. $^{41}_{20}Ca$ **Ans.** protons, 20; electrons, 20; neutrons, 21
   b. $^{108}_{47}Ag$ **Ans.** protons, 47; electrons, 47; neutrons, 61

LS Logical

Reteaching ——— BASIC

Have students draw models of the nuclei of two isotopes carbon-12 and carbon-13. Have them create a way to distinguish the protons and neutrons. **LS** Visual

Focus

Overview

Before beginning this section, review with your students the Objectives listed in the Student Edition. Students will see how the study of the spectra of atoms led to the new quantum model in which four quantum numbers assign electrons to positions in the energy levels of the atom. Students will learn how to write electron configurations.

Bellringer

Ask students to draw a sketch of their concept of the atom. Their drawings, at this time, should show a tiny nucleus with protons and neutrons at the center of a large circular space. Use the sketches to correct any misconceptions.

Motivate

Identifying Preconceptions

Ask students if they think that their view of the atom as having protons and neutrons at the center of a sphere of almost empty space is correct. Tell them that the Rutherford model was an important step, but more research led to changes in the model.

Discussion

Rutherford envisioned the electrons outside the nucleus orbiting in the same way that the planets orbit the sun. The problem with this model is that the electrons and the nucleus are oppositely charged. The tug of the electrostatic force between electron and nucleus would inevitably pull the electrons into the nucleus.

SECTION 3
Electron Configuration

KEY TERMS

- orbital
- electromagnetic spectrum
- ground state
- excited state
- quantum number
- Pauli exclusion principle
- electron configuration
- aufbau principle
- Hund's rule

OBJECTIVES

(1) **Compare** the Rutherford, Bohr, and quantum models of an atom.

(2) **Explain** how the wavelengths of light emitted by an atom provide information about electron energy levels.

(3) **List** the four quantum numbers, and describe their significance.

(4) **Write** the electron configuration of an atom by using the Pauli exclusion principle and the aufbau principle.

Atomic Models

Soon after the atomic theory was widely accepted by scientists, they began constructing models of atoms. Scientists used the information that they had about atoms to build these models. They knew, for example, that an atom has a densely packed nucleus that is positively charged. This conclusion was the only way to explain the data from Rutherford's gold foil experiments.

Building a model helps scientists imagine what may be happening at the microscopic level. For this very same reason, the illustrations in this book provide pictures that are models of chemical compounds to help you understand the relationship between the macroscopic and microscopic worlds. Scientists knew that any model they make may have limitations. A model may even have to be modified or discarded as new information is found. This is exactly what happened to scientists' models of the atom.

Rutherford's Model Proposed Electron Orbits

The experiments of Rutherford's team led to the replacement of the plumpudding model of the atom with a nuclear model of the atom. Rutherford suggested that electrons, like planets orbiting the sun, revolve around the nucleus in circular or elliptical orbits. **Figure 15** shows Rutherford's model. Because opposite charges attract, the negatively charged electrons should be pulled into the positively charged nucleus. Because Rutherford's model could not explain why electrons did not crash into the nucleus, his model had to be modified.

The Rutherford model of the atom, in turn, was replaced only two years later by a model developed by Niels Bohr, a Danish physicist. The Bohr model, which is shown in **Figure 16,** describes electrons in terms of their energy levels.

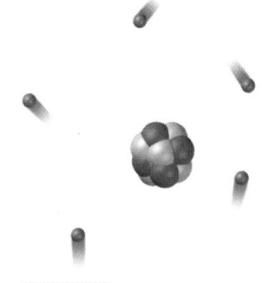

Figure 15
According to Rutherford's model of the atom, electrons orbit the nucleus just as planets orbit the sun.

90

Chapter Resource File

- Lesson Plan

Bohr's Model Confines Electrons to Energy Levels

According to Bohr's model, electrons can be only certain distances from the nucleus. Each distance corresponds to a certain quantity of energy that an electron can have. An electron that is as close to the nucleus as it can be is in its lowest energy level. The farther an electron is from the nucleus, the higher the energy level that the electron occupies. The difference in energy between two energy levels is known as a *quantum* of energy.

The energy levels in Bohr's model can be compared to the rungs of a ladder. A person can go up and down the ladder only by stepping on the rungs. When standing on the first rung, the person has the lowest potential energy. By climbing to the second rung, the person increases his or her potential energy by a fixed, definite quantity. Because the person cannot stand between the rungs on the ladder, the person's potential energy cannot have a continuous range of values. Instead, the values can be only certain, definite ones. In the same way, Bohr's model states that an electron can be in only one energy level or another, not between energy levels. Bohr also concluded that an electron did not give off energy while in a given energy level.

Electrons Act Like Both Particles and Waves

Thomson's experiments demonstrated that electrons act like particles that have mass. Although the mass of an electron is extremely small, electrons in a cathode ray still have enough mass to turn a paddle wheel.

In 1924, Louis de Broglie pointed out that the behavior of electrons according to Bohr's model was similar to the behavior of waves. For example, scientists knew that any wave confined in space can have only certain frequencies. The frequency of a wave is the number of waves that pass through a given point in one second. De Broglie suggested that electrons could be considered waves confined to the space around a nucleus. As waves, electrons could have only certain frequencies. These frequencies could correspond to the specific energy levels in which electrons are found.

Other experiments also supported the wave nature of electrons. Like light waves, electrons can change direction through diffraction. Diffraction refers to the bending of a wave as the wave passes by the edge of an object, such as a crystal. Experiments also showed that electron beams, like waves, can interfere with each other.

Figure 17 shows the present-day model of the atom, which takes into account both the particle and wave properties of electrons. According to this model, electrons are located in **orbitals,** regions around a nucleus that correspond to specific energy levels. Orbitals are regions where electrons are likely to be found. Orbitals are sometimes called *electron clouds* because they do not have sharp boundaries. When an orbital is drawn, it shows where electrons are most likely to be. Because electrons can be in other places, the orbital has a fuzzy boundary like a cloud.

As an analogy to an electron cloud, imagine the spinning blades of a fan. You know that each blade can be found within the spinning image that you see. However, you cannot tell exactly where any one blade is at a particular moment.

Figure 16
According to Bohr's model of the atom, electrons travel around the nucleus in specific energy levels.

orbital
a region in an atom where there is a high probability of finding electrons

Figure 17
According to the current model of the atom, electrons are found in orbitals.

91

PHYSICS CONNECTION ADVANCED

Niels Bohr (1885–1962), Nobel laureate, was a leader in the development of quantum physics for a half century. Educated in Copenhagen, Bohr subsequently went to England to work with J. J. Thomson and Ernest Rutherford. He eventually moved to the United States where he took part in the development of the atomic bomb. His concern about the destructive power of the bomb motivated his efforts to find a peaceful means for controlling it. He was awarded the first Atoms for Peace Award in 1957. Ask interested students to further research the work of Bohr. **LS** Intrapersonal

Teach, continued

Teaching Tip

Wave Motion Use a piece of rubber tubing or rope about 3 m long. Have a student hold one end while you shake the other end up and down at a steady rate. Speed up the motion to produce an increase in frequency. Have a second student use a meterstick to measure the wavelength as you double the frequency of the wave. Students should see that the wavelength is cut in half as the frequency doubles. Students can also see the tubing vibrate in two parts at one frequency, four parts at double the frequency, and eight parts at quadruple the frequency. Vary the energy of the vibration to demonstrate differences in amplitude. You and your students should wear goggles during the demonstration.

Using the Figure

Use **Figure 18** to reinforce the idea that the visible spectrum is only a small part of the electromagnetic spectrum. Point out that there is no qualitative difference between radio waves and visible light. The human eye contains structures that respond to electromagnetic waves of wavelengths 10^{-8} times the wavelengths of radio waves. Make sure that students appreciate that in the diagram, the units of wavelength change from pm to km across the spectrum and the unit of frequency changes from Hz to kHz.

internet connect

www.scilinks.org
Topic : Electromagnetic Spectrum
SciLinks code: HW4048

SC*LINKS* Maintained by the National Science Teachers Association

electromagnetic spectrum

all of the frequencies or wavelengths of electromagnetic radiation

Figure 18
The electromagnetic spectrum is composed of light that has a broad range of wavelengths. Our eyes can detect only the visible spectrum.

Electrons and Light

By 1900, scientists knew that light could be thought of as moving waves that have given frequencies, speeds, and wavelengths.

In empty space, light waves travel at 2.998×10^8 m/s. At this speed, light waves take only 500 s to travel the 150 million kilometers between the sun and Earth. The *wavelength* is the distance between two consecutive peaks or troughs of a wave. The distance of a wavelength is usually measured in meters. The wavelength of light can vary from 10^5 m to less than 10^{-13} m. This broad range of wavelengths makes up the **electromagnetic spectrum,** which is shown in **Figure 18.** Notice in **Figure 18** that our eyes are sensitive to only a small portion of the electromagnetic spectrum. This sensitivity ranges from 700 nm, which is about the value of wavelengths of red light, to 400 nm, which is about the value of wavelengths of violet light.

In 1905, Albert Einstein proposed that light also has some properties of particles. His theory would explain a phenomenon known as the *photoelectric effect.* This effect happens when light strikes a metal and electrons are released. What confused scientists was the observation that for a given metal, no electrons were emitted if the light's frequency was below a certain value, no matter how long the light was on. Yet if light were just a wave, then any frequency eventually should supply enough energy to remove an electron from the metal.

Einstein proposed that light has the properties of both waves and particles. According to Einstein, light can be described as a stream of particles, the energy of which is determined by the light's frequency. To remove an electron, a particle of light has to have at least a minimum energy and therefore a minimum frequency.

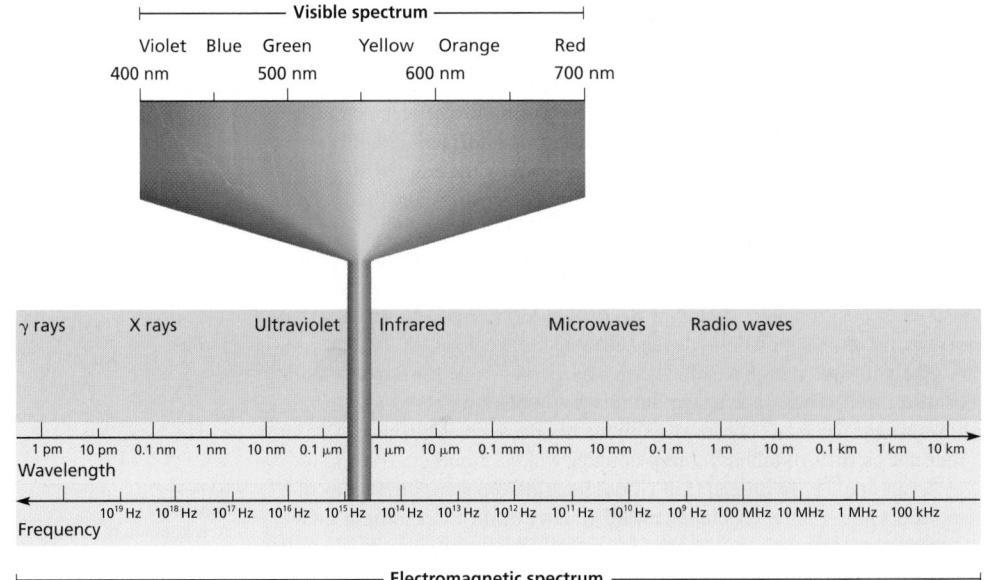

HISTORY CONNECTION

Electric current exists in the space between metal plates when light shines on one plate, but no current exists when there is no light. In 1899, Thomson found that the particles passing across the space were the same as the particles in cathode rays—electrons. Thus, light energy was "knocking" electrons from atoms of the metal plate. From these observations, it was clear to scientists that light could interact with electrons. In 1905, Einstein explained the photoelectric effect by means of his theory that light has both wave and particle characteristics.

Transparencies

TT Electromagnetic Spectrum

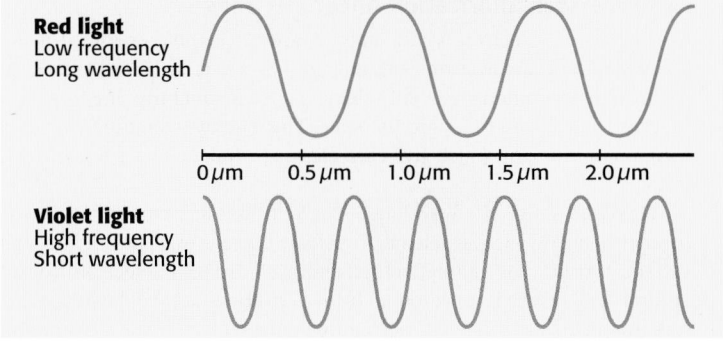

Red light
Low frequency
Long wavelength

0 μm   0.5 μm   1.0 μm   1.5 μm   2.0 μm

Violet light
High frequency
Short wavelength

Figure 19
The frequency and wavelength of a wave are inversely related. As frequency increases, wavelength decreases.

Light Is an Electromagnetic Wave

When passed through a glass prism, sunlight produces the visible spectrum—all of the colors of light that the human eye can see. You can see from **Figure 18** on the previous page that the visible spectrum is only a tiny portion of the electromagnetic spectrum. The electromagnetic spectrum also includes X rays, ultraviolet and infrared light, microwaves, and radio waves. Each of these electromagnetic waves is referred to as *light,* although we cannot see these wavelengths.

Figure 19 shows the frequency and wavelength of two regions of the spectrum that we see: red and violet lights. If you compare red and violet lights, you will notice that red light has a low frequency and a long wavelength. But violet light has a high frequency and a short wavelength. The frequency and wavelength of a wave are inversely related.

Light Emission

When a high-voltage current is passed through a tube of hydrogen gas at low pressure, lavender-colored light is seen. When this light passes through a prism, you can see that the light is made of only a few colors. This spectrum of a few colors is called a *line-emission spectrum.* Experiments with other gaseous elements show that each element has a line-emission spectrum that is made of a different pattern of colors.

In 1913, Bohr showed that hydrogen's line-emission spectrum could be explained by assuming that the hydrogen atom's electron can be in any one of a number of distinct energy levels. The electron can move from a low energy level to a high energy level by absorbing energy. Electrons at a higher energy level are unstable and can move to a lower energy level by releasing energy. This energy is released as light that has a specific wavelength. Each different move from a particular energy level to a lower energy level will release light of a different wavelength.

Bohr developed an equation to calculate all of the possible energies of the electron in a hydrogen atom. His values agreed with those calculated from the wavelengths observed in hydrogen's line-emission spectrum. In fact, his values matched with the experimental values so well that his atomic model that is described earlier was quickly accepted.

internet connect

www.scilinks.org
Topic: Light and Color
SciLinks code: HW4075

SciLINKS Maintained by the National Science Teachers Association

internet connect

www.scilinks.org
Topic : Producing Light
SciLinks code: HW4099

SciLINKS Maintained by the National Science Teachers Association

93

Using the Figure

Refer students to **Figure 19.** Ask them to think of the diagram as a snapshot that took a specific period of time. Ask them how they can tell that red light has a lower frequency than violet light. **Ans.** Fewer waves of red light were found during that time than waves of violet light. Ask students how they can tell that violet light has a shorter wavelength than red light. **Ans.** The crests are closer together. Finally, ask students to draw a conclusion about the relationship between frequency and wavelength. **Ans.** the longer the wavelength, the lower the frequency

INCLUSION Strategies

• *Learning Disabled*
• *Developmentally Delayed*
• *English Language Learners*

Have the students draw and color a rainbow on a poster board. Have the students label each color of the visible spectrum with its approximate frequency or range of frequencies. In addition, have the students give examples of living things that can detect parts of the electromagnetic spectrum that humans cannot detect.

did you know?

Ernest Rutherford's principal interest was radioactivity. In his gold foil experiments, Rutherford was studying the character and behavior of the alpha particle, one of the types of radiation produced by radioactive processes. Rutherford characterized and named alpha, beta, and gamma radiation. He published the first artificial disintegration of an element in 1919 when he engineered the collision of an alpha particle with a nitrogen atom. An oxygen atom and a hydrogen atom were the products. He also wrote several books on radioactivity.

Transparencies

TT Wavelength and Frequency

Demonstration
Emission Spectra

1. Obtain several gas spectrum tubes (Geissler tubes) (for example, H_2, He, O_2, Ne, Hg vapor, and CO_2) and a power supply made especially for the tubes.

2. Give students spectroscopes or diffraction gratings.

3. Have students observe the continuous spectrum of white light, such as from a halogen-bulb flashlight. Cover the flashlight lens with foil or black construction paper with a slit cut in it.

4. Have students observe the emission spectrum from the the center section of the hydrogen tube to demonstrate the spectrum shown in **Figure 20.** Darken the room as much as possible. The dark blue lines are faint and hard to see.

5. Have students view the remaining tubes in order of increasing complexity of the atom. Neon tubes show many lines and provide an impressive contrast to the hydrogen spectrum. Remind students that each line represents a possible electron transition from a higher energy state to a lower one.

Safety Caution: Wear safety goggles and a lab apron. Keep students at least 3 m away and have students wear goggles. Do not touch or allow anyone to touch any part of the apparatus while the power supply is on.

ground state

the lowest energy state of a quantized system

excited state

a state in which an atom has more energy than it does at its ground state

internet connect
www.scilinks.org
Topic : Energy Levels
SciLinks code: HW4051
SCLINKS. Maintained by the National Science Teachers Association

Figure 20
An electron in a hydrogen atom can move between only certain energy states, shown as $n = 1$ to $n = 7$. In dropping from a higher energy state to a lower energy state, an electron emits a characteristic wavelength of light.

Light Provides Information About Electrons

Normally, if an electron is in a state of lowest possible energy, it is in a **ground state.** If an electron gains energy, it moves to an **excited state.** An electron in an excited state will release a specific quantity of energy as it quickly "falls" back to its ground state. This energy is emitted as certain wavelengths of light, which give each element a unique line-emission spectrum.

Figure 20 shows the wavelengths of light in a line-emission spectrum for hydrogen, through which a high-voltage current was passed. The high-voltage current may supply enough energy to move an electron from its ground state, which is represented by $n = 1$ in **Figure 20,** to a higher excited state for an electron in a hydrogen atom, represented by $n > 1$. Eventually, the electron will lose energy and return to a lower energy level. For example, the electron may fall from the $n = 7$ energy level to the $n = 3$ energy level. Notice in **Figure 20** that when this drop happens, the electron emits a wavelength of infrared light. An electron in the $n = 6$ energy level can also fall to the $n = 2$ energy level. In this case, the electron emits a violet light, which has a shorter wavelength than infrared light does.

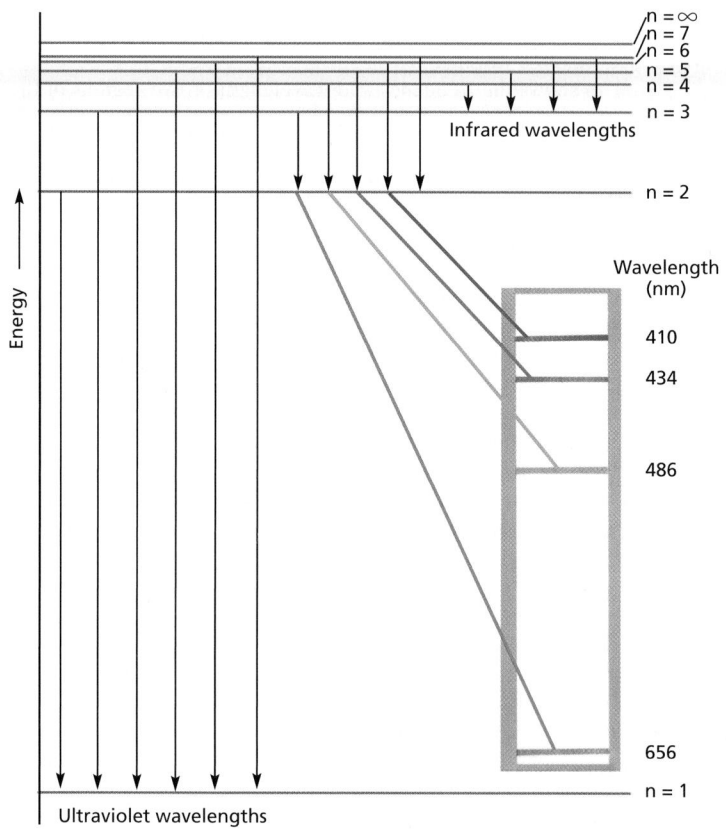

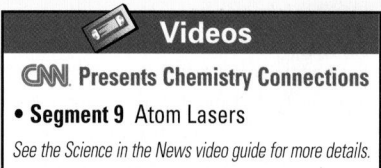

Transparencies

TT Hydrogen's Line-Emission Spectrum

Videos

CNN. **Presents Chemistry Connections**

• **Segment 9** Atom Lasers

See the Science in the News video guide for more details.

Table 5 Quantum Numbers of the First 30 Atomic Orbitals

n	*l*	*m*	Orbital name	Number of orbitals
1	0	0	1*s*	1
2	0	0	2*s*	1
2	1	−1, 0, 1	2*p*	3
3	0	0	3*s*	1
3	1	−1, 0, 1	3*p*	3
3	2	−2, −1, 0, 1, 2	3*d*	5
4	0	0	4*s*	1
4	1	−1, 0, 1	4*p*	3
4	2	−2, −1, 0, 1, 2	4*d*	5
4	3	−3, −2, −1, 0, 1, 2, 3	4*f*	7

Quantum Numbers

The present-day model of the atom, in which electrons are located in orbitals, is also known as the *quantum model.* According to this model, electrons within an energy level are located in orbitals, regions of high probability for finding a particular electron. However, the model does not explain how the electrons move about the nucleus to create these regions.

To define the region in which electrons can be found, scientists have assigned four **quantum numbers** to each electron. **Table 5** lists the quantum numbers for the first 30 atomic orbitals. The *principal quantum number,* symbolized by *n,* indicates the main energy level occupied by the electron. Values of *n* are positive integers, such as 1, 2, 3, and 4. As *n* increases, the electron's distance from the nucleus and the electron's energy increases.

quantum number
a number that specifies the properties of electrons

The main energy levels can be divided into sublevels. These sublevels are represented by the *angular momentum quantum number, l.* This quantum number indicates the shape or type of orbital that corresponds to a particular sublevel. Chemists use a letter code for this quantum number. A quantum number $l = 0$ corresponds to an *s* orbital, $l = 1$ to a *p* orbital, $l = 2$ to a *d* orbital, and $l = 3$ to an *f* orbital. For example, an orbital with $n = 3$ and $l = 1$ is called a 3*p* orbital, and an electron occupying that orbital is called a 3*p* electron.

The *magnetic quantum number,* symbolized by *m,* is a subset of the *l* quantum number. It also indicates the numbers and orientations of orbitals around the nucleus. The value of *m* takes whole-number values, depending on the value of *l.* The number of orbitals includes one *s* orbital, three *p* orbitals, five *d* orbitals, and seven *f* orbitals.

The *spin quantum number,* symbolized by $+\frac{1}{2}$ or $-\frac{1}{2}$ ($\uparrow$ or $\downarrow$), indicates the orientation of an electron's magnetic field relative to an outside magnetic field. A single orbital can hold a maximum of two electrons, which must have opposite spins.

95

Discussion

Ask students to think of examples of devices that emit visible light as a result of heating. Light bulbs and flames are obvious responses, as well as electric range burners and hot or molten metal. If students know that there is a relationship between visible light and electrons, ask them to speculate about why heated materials can emit light.

Using the Figure

Students may wonder why there are many more energy transitions shown in this diagram than the number of lines they saw in the visible hydrogen spectrum. The most energetic transitions to the first energy level ($n = 1$) produce light in the ultraviolet region of the spectrum. The very small drops in energy produce light in the infrared region. The only transitions that produce visible light are those to the second energy level ($n = 2$) from higher levels.

Teaching Tip

In 1885, Johann Jacob Balmer wrote an equation to calculate the wavelengths of the then-known spectral lines of hydrogen. The equation is $\frac{1}{\lambda} = R_H(\frac{1}{2^2} - \frac{1}{n^2})$. R_H is a constant known as the Rydberg constant. For the equation to correctly calculate the wavelengths, the values of *n* must be three or greater. Although Balmer did not know it at the time, the equation applied to those electronic transitions from energy levels 3, 4, 5, and so on to energy level 2, the spectral lines in the visible range. Upon studying the equation, other scientists surmised that additional lines existed and that their wavelengths could be calculated by changing $\frac{1}{2^2}$ to $\frac{1}{1^2}$, with the values of $n = 2$ or greater. This series, called the Lyman series, is found in the ultraviolet range. Similarly, $\frac{1}{2^2}$ could be changed to $\frac{1}{3^2}$ with the values of $n = 4$ or greater. This series is called the Paschen series and is found in the infrared range.

Math Skills Have interested students calculate the wavelength of each of the first three transitions in the Lyman series (ultraviolet) as described in the Teaching Tip on the previous page. The Rydberg constant is $1.097 \times 10^7 \text{m}^{-1}$. Ans. 1.215×10^{-7}m, 1.025×10^{-7}m, 0.972×10^{-7}m. **LS** Logical

Teaching Tip

Spectral Notation The orbital letters *s, p, d,* and *f* are derived from the words *sharp, principal, diffuse,* and *fundamental*. These words were first used by spectroscopists to describe the qualities of certain spectral lines. Later, it was discovered that these qualities were related to the sublevel structure of the electron arrangement, so their letters were retained.

MISCONCEPTION ///ALERT\\\

Because students first learn about the electron configuration of the hydrogen atom in which hydrogen's single electron occupies the first energy level, they sometimes come away with the mistaken idea that any atom in its ground state has all of its electrons in the first energy level.

Electron Configurations

Figure 21 shows the shapes and orientations of the *s, p,* and *d* orbitals. Each orbital that is shown can hold a maximum of two electrons. The discovery that two, but no more than two, electrons can occupy a single orbital was made in 1925 by the German chemist Wolfgang Pauli. This rule is known as the **Pauli exclusion principle.**

Pauli exclusion principle

the principle that states that two particles of a certain class cannot be in the exact same energy state

Another way of stating the Pauli exclusion principle is that no two electrons in the same atom can have the same four quantum numbers. The two electrons can have the same value of *n* by being in the same main energy level. These two electrons can also have the same value of *l* by being in orbitals that have the same shape. And, these two electrons may have the same value of *m* by being in the same orbital. But these two electrons cannot have the same spin quantum number. If one electron has the value of $+\frac{1}{2}$, then the other electron must have the value of $-\frac{1}{2}$.

electron configuration

the arrangement of electrons in an atom

The arrangement of electrons in an atom is usually shown by writing an **electron configuration.** Like all systems in nature, electrons in atoms tend to assume arrangements that have the lowest possible energies. An electron configuration of an atom shows the lowest-energy arrangement of the electrons for the element.

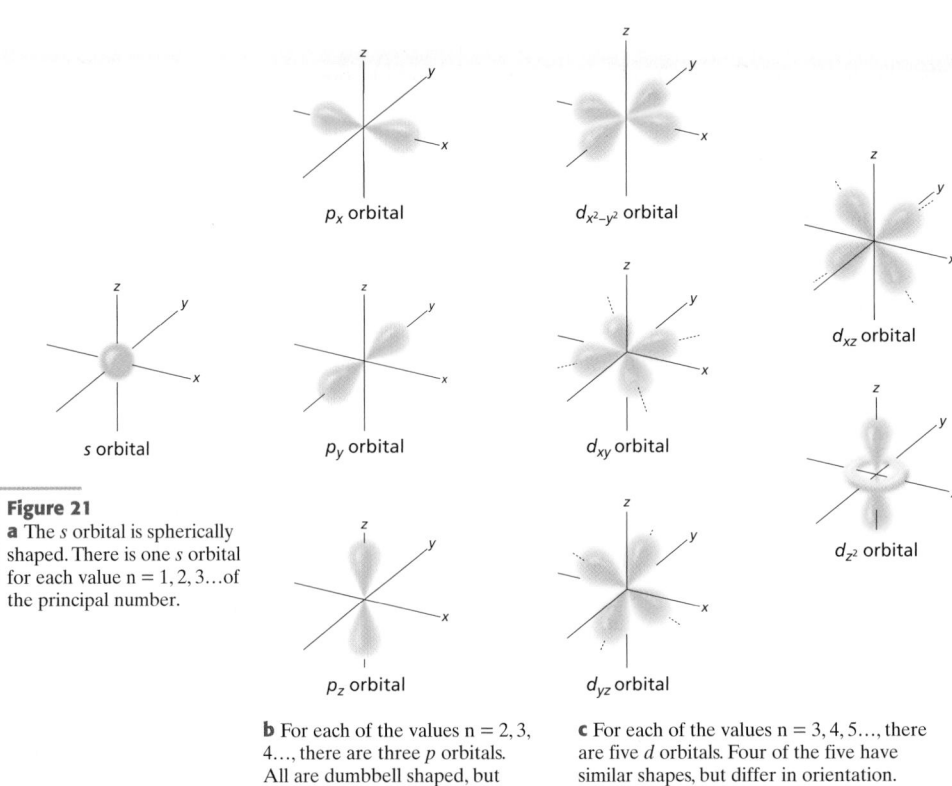

Figure 21
a The *s* orbital is spherically shaped. There is one *s* orbital for each value n = 1, 2, 3...of the principal number.

b For each of the values n = 2, 3, 4..., there are three *p* orbitals. All are dumbbell shaped, but they differ in orientation.

c For each of the values n = 3, 4, 5..., there are five *d* orbitals. Four of the five have similar shapes, but differ in orientation.

did you know?

Light Is Quantized Max Planck began to formulate the quantum theory in 1900. He theorized that energy is given off or absorbed in fundamental amounts called *quanta.* The energy of a quantum is given by the equation $E = h\nu$, where *h* is a constant called Planck's constant and ν is the frequency of light. Bohr based his atomic model on Planck's theory. This theory of quantization of energy paralleled Dalton's theory that matter is composed of discrete, fundamental units.

 Transparencies

TT Shapes of *s, p,* and *d* Orbitals

An Electron Occupies the Lowest Energy Level Available

The Pauli exclusion principle is one rule to help you write an electron configuration for an atom. Another rule is the aufbau principle. *Aufbau* is the German word for "building up." The **aufbau principle** states that electrons fill orbitals that have the lowest energy first.

Recall that the smaller the principal quantum number, the lower the energy. But within an energy level, the smaller the l quantum number, the lower the energy. Recall that chemists use letters to represent the l quantum number. So, the order in which the orbitals are filled matches the order of energies, which starts out as follows:

$$1s < 2s < 2p < 3s < 3p$$

After this point, the order is less obvious. **Figure 22** shows that the energy of the $3d$ orbitals is slightly higher than the energy of the $4s$ orbitals. As a result, the order in which the orbitals are filled is as follows:

$$1s < 2s < 2p < 3s < 3p < 4s < 3d$$

Additional irregularities occur at higher energy levels.

Can you determine which orbitals electrons of a carbon atom occupy? Two electrons occupy the $1s$ orbital, two electrons occupy the $2s$ orbital, and two electrons occupy the $2p$ orbitals. Now try the same exercise for titanium. Two electrons occupy the $1s$ orbital, two electrons occupy the $2s$ orbital, six electrons occupy the $2p$ orbitals, two electrons occupy the $3s$ orbital, six electrons occupy the $3p$ orbitals, two electrons occupy the $3d$ orbitals, and two electrons occupy the $4s$ orbital.

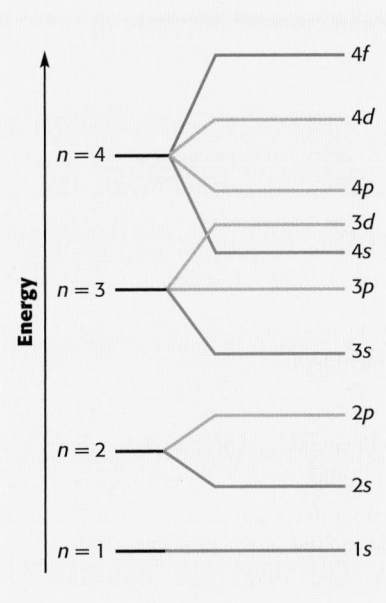

Figure 22
This diagram illustrates how the energy of orbitals can overlap such that $4s$ fills before $3d$.

aufbau principle

the principle that states that the structure of each successive element is obtained by adding one proton to the nucleus of the atom and one electron to the lowest-energy orbital that is available

Discussion

Figure 22 diagrams the normal order of orbital filling. The sequence is flawless for the elements in the first three periods. In the fourth period, there are two anomalies, chromium and copper. Even more anomalies occur among the remaining elements. By strictly following the orbital order, the following electron configuration for chromium is obtained.

Cr: $1s^2 2s^2 2p^6 3s^2 3p^6 4s^2 3d^4$

Experiments show, however, that the actual configuration is different. Anomalies occur because the $4s$ and $3d$ sublevels are close in energy. In this case, an electron is "borrowed" from the $4s$ orbital in order to achieve a set of five half-filled $3d$ orbitals. The true configuration is the following.

Cr: $1s^2 2s^2 2p^6 3s^2 3p^6 4s^1 3d^5$

An atom that has this configuration is less reactive than an atom that has the configuration that follows the rules. Similarly, the copper atom has a configuration of $1s^2 2s^2 2p^6 3s^2 3p^6 4s^1 3d^{10}$ rather than $1s^2 2s^2 2p^6 3s^2 3p^6 4s^2 3d^9$ because the atom is more stable with a complete set of five full $3d$ orbitals. In the fifth period, Mo, Ru, Rh, Pd, and Ag also have anomalous configurations.

Teaching Tip

The f orbitals are not shown nor are extensively discussed in this text because students often find these orbitals to be complicated. However, do introduce these orbitals to your advanced students. You can show these students pictures of the orbitals, help them create orbital diagrams such as the one in **Figure 22,** and ask them to write electron configurations that have f orbitals.

ChemFile **CHEMISTRY** TUTOR INTERACTIVE
• Module 2: Models of the Atom
  Topic: Electronic Structure

Paired Reading The material in this section will seem abstract to many students. Have pairs of students read the remainder of this section together. Remind them to study the diagrams carefully. Give them self-stick notes, and have them mark areas that are unclear. When students are finished, have two pairs discuss what they did not understand. Assign an element to each pair, and have them write its electron configuration.
LS Interpersonal

Teaching Tip

The Aufbau Principle According to the aufbau principle, electrons always fill a lower energy orbital before one of higher energy. Thus, when anomalies occur and an electron enters an orbital of supposedly higher energy than an unfilled orbital, it is probably because the energy of the orbital that the electron enters is actually lower in energy.

Hund's rule

the rule that states that for an atom in the ground state, the number of unpaired electrons is the maximum possible and these unpaired electrons have the same spin

An Electron Configuration Is a Shorthand Notation

Based on the quantum model of the atom, the arrangement of the electrons around the nucleus can be shown by the nucleus's electron configuration. For example, sulfur has sixteen electrons. Its electron configuration is written as $1s^2 2s^2 2p^6 3s^2 3p^4$. This line of symbols tells us about these sixteen electrons. Two electrons are in the $1s$ orbital, two electrons are in the $2s$ orbital, six electrons are in the $2p$ orbitals, two electrons are in the $3s$ orbital, and four electrons are in the $3p$ orbitals.

Each element's configuration builds on the previous elements' configurations. To save space, one can write this configuration by using a configuration of a noble gas. The noble gas electron configurations that are often used are the configurations of neon, argon, krypton, and xenon. The neon atom's configuration is $1s^2 2s^2 2p^6$, so the electron configuration of sulfur is written as shown below.

$$[Ne]3s^2 3p^4$$

Does an electron enter the first $3p$ orbital to pair with a single electron that is already there? Or does the electron fill another $3p$ orbital? According to Hund's rule, the second answer is correct. **Hund's rule** states that orbitals of the same n and l quantum numbers are each occupied by one electron before any pairing occurs. For example, sulfur's configuration is shown by the *orbital diagram* below. Electrons are represented by arrows. Note that an electron fills each $3p$ orbital before an electron enters an orbital that is already occupied.

$$\underset{1s}{\uparrow\downarrow} \quad \underset{2s}{\uparrow\downarrow} \quad \underset{2p}{\uparrow\downarrow\;\uparrow\downarrow\;\uparrow\downarrow} \quad \underset{3s}{\uparrow\downarrow} \quad \underset{3p}{\uparrow\downarrow\;\uparrow\;\uparrow}$$

SAMPLE PROBLEM C

Writing Electron Configurations

Write the electron configuration for an atom whose atomic number is 20.

PRACTICE HINT

Remember that an s orbital holds 2 electrons, three p orbitals hold 6 electrons, and five d orbitals hold 10 electrons.

1 Gather information.
• The atomic number of the element is 20.

2 Plan your work.
• The atomic number represents the number of protons in an atom.
• The number of protons must equal the number of electrons in an atom.
• Write the electron configuration for that number of electrons by following the Pauli exclusion principle and the aufbau principle.
• A noble gas configuration can be used to write this configuration.

98

MISCONCEPTION ALERT

Some students may think that orbitals are defined as part of an atom's structure. Be sure students understand that an orbital exists as a physical entity only when it is occupied by an electron. Students may also wonder about the shapes of the orbitals. Tell them that the shapes have much to do with the fact that electrons repel each other. Have them study the spatial arrangement of the orbitals around the x, y, and z axes in the p and d sublevels and note that the orbitals tend not to overlap each other.

3 **Calculate.**

- *atomic number = number of protons = number of electrons = 20*
- According to the aufbau principle, the order of orbital filling is 1*s*, 2*s*, 2*p*, 3*s*, 3*p*, 4*s*, 3*d*, 4*p*, and so on.
- The electron configuration for an atom of this element is written as follows:

$$1s^2 2s^2 2p^6 3s^2 3p^6 4s^2$$

- This electron configuration can be abbreviated as follows:

$$[Ar]4s^2$$

4 **Verify your results.**

- The sum of the superscripts is $(2 + 2 + 6 + 2 + 6 + 2) = 20$. Therefore, all 20 electrons are included in the electron configuration.

PRACTICE

1 Write the electron configuration for an atom of an element whose atomic number is 8.

2 Write the electron configuration for an atom that has 17 electrons.

PROBLEM SOLVING SKILL

Section Review

UNDERSTANDING KEY IDEAS

1. How does Bohr's model of the atom differ from Rutherford's?

2. What happens when an electron returns to its ground state from its excited state?

3. What does *n* represent in the quantum model of electrons in atoms?

PRACTICE PROBLEMS

4. What is the atomic number of an element whose atom has the following electron configuration: $1s^2 2s^2 2p^6 3s^2 3p^6 3d^2 4s^2$?

5. Write the electron configuration for an atom that has 13 electrons.

6. Write the electron configuration for an atom that has 33 electrons.

7. How many orbitals are completely filled in an atom whose electron configuration is $1s^2 2s^2 2p^6 3s^1$?

CRITICAL THINKING

8. Use the Pauli exclusion principle or the aufbau principle to explain why the following electron configurations are incorrect:

a. $1s^2 2s^3 2p^6 3s^1$

b. $1s^2 2s^2 2p^5 3s^1$

9. Why is a shorter wavelength of light emitted when an electron "falls" from $n = 4$ to $n = 1$ than when an electron "falls" from $n = 2$ to $n = 1$?

10. Calculate the maximum number of electrons that can occupy the third principal energy level.

11. Why do electrons fill the 4*s* orbital before they start to occupy the 3*d* orbital?

99

Homework ——— GENERAL

Additional Practice

1. Determine the electron configuration of boron.
Ans. $[He]2s^2 2p^1$

2. Determine the electron configuration of magnesium.
Ans. $[Ne]3s^2$

3. Determine the electron configuration of yttrium.
Ans. $[Kr]4d^1 5s^2$

LS Logical

Answers to Practice Problems C

1. $1s^2 2s^2 2p^4$ or $[He]2s^2 2p^4$

2. $1s^2 2s^2 2p^6 3s^2 3p^5$ or $[Ne]3s^2 3p^5$

Close

Reteaching ——— BASIC

Have students review the heads and subheads in this section and write down at least one question about an area that is unclear. Use the questions as the basis for a class discussion. **LS** Intrapersonal

Quiz ——— GENERAL

Determine the electron configurations for the following atoms: Sc, K, and P:

Sc: $1s^2 2s^2 2p^6 3s^2 3p^6 3d^1 4s^2$

K: $1s^2 2s^2 2p^6 3s^2 3p^6 4s^1$

P: $1s^2 2s^2 2p^6 3s^2 3p^3$

LS Logical

Chapter Resource File

- Concept Review
- Quiz

Answers to Section Review

1. Rutherford proposed that electrons were particles that travel in orbits and have continuous energy. Bohr's model describes electrons as particles but having quantized energy.

2. The electron emits energy in the form of electromagnetic radiation that has a particular wavelength, perhaps as visible light.

3. a principal energy level

4. 22

5. $1s^2 2s^2 2p^6 3s^2 3p^1$ or $[Ne]3s^2 3p^1$

6. $1s^2 2s^2 2p^6 3s^2 3p^6 3d^{10} 4s^2 4p^3$ or $[Ar]3d^{10} 4s^2 4p^3$

Answers continued on p. 113A

Focus

Overview

Before beginning this section, review with your students the Objectives listed in the Student Edition. This section deals with atomic mass units and grams. Students will learn to count atoms using the mole.

 **Bellringer**

Ask students to write out the number of copper atoms calculated on this page in regular notation with all the zeros. Use this exercise to emphasize the miniscule size of an atom.

Motivate

Discussion

An atom of copper has a mass of 1.0552×10^{-25} kg. Ask students if they think this is an inconvenient number to work with. Tell them that chemists like to make things easier for themselves, so they decided to invent a new unit for mass. The atomic mass unit, amu, is defined as one-twelfth of the atomic mass of the carbon-12 isotope. That mass is $1.6605402 \times 10^{-27}$ kg. Thus, 1 amu = $1.6605402 \times 10^{-27}$ kg. Show students on the chalkboard, that if you multiply the mass of a copper atom, 1.0552×10^{-25} kg, by 1 amu/$1.6605402 \times 10^{-27}$ kg, you will obtain 63.546 amu as the atomic mass of copper.

KEY TERMS

- atomic mass
- mole
- molar mass
- Avogadro's number

OBJECTIVES

1. **Compare** the quantities and units for atomic mass with those for molar mass.

2. **Define** *mole*, and explain why this unit is used to count atoms.

3. **Calculate** either mass with molar mass or number with Avogadro's number given an amount in moles.

Atomic Mass

You would not expect something as small as an atom to have much mass. For example, copper atoms have an average mass of only 1.0552×10^{-25} kg.

Each penny in **Figure 23** has an average mass of 3.13×10^{-3} kg and contains copper. How many copper atoms are there in one penny? Assuming that a penny is pure copper, you can find the number of copper atoms by dividing the mass of the penny by the average mass of a single copper atom or by using the following conversion factor:

$$1 \text{ atom Cu}/1.0552 \times 10^{-25} \text{ kg}$$

$$3.13 \times 10^{-3} \text{ kg} \times \frac{1 \text{ atom Cu}}{1.0552 \times 10^{-25} \text{ kg}} = 2.97 \times 10^{22} \text{ Cu atoms}$$

atomic mass

the mass of an atom expressed in atomic mass units

Figure 23
These pennies are made mostly of copper atoms. Each copper atom has an average mass of 1.0552×10^{-25} kg.

Masses of Atoms Are Expressed in Atomic Mass Units

Obviously, atoms are so small that the gram is not a very convenient unit for expressing their masses. Even the picogram (10^{-12} g) is not very useful. A special mass unit is used to express **atomic mass.** This unit has two names—the atomic mass unit (amu) and the Dalton (Da). In this book, *atomic mass unit* will be used.

But how can you tell what the atomic mass of a specific atom is? When the atomic mass unit was first set up, an atom's mass number was supposed to be the same as the atom's mass in atomic mass units. Mass number and atomic mass units would be the same because a proton and a neutron each have a mass of about 1.0 amu.

For example, a copper-63 atom has an atomic mass of 62.940. A copper-65 atom has an atomic mass of 64.928. (The slight differences from exact values will be discussed in later chapters.)

Another way to determine atomic mass is to check a periodic table, such as the one on the inside cover of this book. The mass shown is an average of the atomic masses of the naturally occurring isotopes. For this reason, copper is listed as 63.546 instead of 62.940 or 64.928.

100

HISTORY
CONNECTION

Establishing the Mole Amadeo Avogadro first proposed what has become known as Avogadro's hypothesis in 1811. It states that equal volumes of different gases at the same temperature and pressures contain the same number of molecules. Avogadro did not undertake measuring the number of particles in equal volumes of gases, but his hypothesis led eventually to the number 6.022×10^{23}, which bears his name. The volume of gas that contains Avogadro's number of particles is 22.41 L at STP.

Chapter Resource File

- Lesson Plan

Introduction to the Mole

Most samples of elements have great numbers of atoms. To make working with these numbers easier, chemists created a new unit called the *mole* (mol). A **mole** is defined as the number of atoms in exactly 12 grams of carbon-12. The mole is the SI unit for the amount of a substance.

Chemists use the mole as a counting unit, just as you use the dozen as a counting unit. Instead of asking for 12 eggs, you ask for 1 dozen eggs. Similarly, chemists refer to 1 mol of carbon or 2 mol of iron.

To convert between moles and grams, chemists use the molar mass of a substance. The **molar mass** of an element is the mass in grams of one mole of the element. Molar mass has the unit grams per mol (g/mol). The mass in grams of 1 mol of an element is numerically equal to the element's atomic mass from the periodic table in atomic mass units. For example, the atomic mass of copper to two decimal places is 63.55 amu. Therefore, the molar mass of copper is 63.55 g/mol. **Skills Toolkit 1** shows how to convert between moles and mass in grams using molar mass.

Scientists have also determined the number of particles present in 1 mol of a substance, called **Avogadro's number.** One mole of pure substance contains $6.022\ 1367 \times 10^{23}$ particles. To get some idea of how large Avogadro's number is, imagine that every living person on Earth (about 6 billion people) started counting the number of atoms of 1 mol C. If each person counted nonstop at a rate of one atom per second, it would take over 3 million years to count every atom.

Avogadro's number may be used to count any kind of particle, including atoms and molecules. For calculations in this book, Avogadro's number will be rounded to 6.022×10^{23} particles per mole. **Skills Toolkit 2** shows how to use Avogadro's number to convert between amount in moles and the number of particles.

mole

the SI base unit used to measure the amount of a substance whose number of particles is the same as the number of atoms in 12 g of carbon-12

molar mass

the mass in grams of 1 mol of a substance

Avogadro's number

6.022×10^{23}, the number of atoms or molecules in 1 mol

SKILLS Toolkit 1

Determining the Mass from the Amount in Moles

amount

$\underline{?}\ mol$

$\dfrac{\underline{?}\ g}{1\ mol}$

use molar mass

$\dfrac{1\ mol}{\underline{?}\ g}$

mass

$\underline{?}\ g$

101

---

did you know?

The Mole Is for the Tiny It makes no sense to talk of moles of macroscopic objects. Even a mole of periods, like the one at the end of this sentence, placed side by side would equal the radius of our galaxy. A mole of marbles stacked over the United States would cover the country to a depth of about 113 km (70 mi).

Transparencies

TT Determining the Mass by Using the Amount in Moles

Teach

Discussion

Ask students how they buy eggs or donuts. If they see that a dozen is a count of 12 objects, it will help them to understand the mole as a counting unit for 6.022×10^{23} particles. Ask about other units of counting. Paper is sold by the ream (500), pencils by the gross (144), boots and gloves by the pair. There are trios, quartets, quintets, and more. Ask students to think of other counting units. Introduce the mole. Have them visualize putting 6.022×10^{23} particles in a giant carton like an egg box but with 6.022×10^{23} depressions. Make sure they realize that any particle can be measured in moles—atoms, molecules, electrons, baseballs, and even people!

Teaching Tip

Appreciating 6.022×10^{23} Ask students to calculate how many dollars each person in the world would get if a mole of dollars were to be distributed evenly among them. Assume a world population of about 6 billion (6×10^9). Students will be surprised to find that the amount is incomprehensibly large (1×10^{14} dollars).

Group Activity —— BASIC

The mole will seem less abstract if students can see what a mole looks like. Seal a mole of any or all of the following pure substances in laboratory bottles or jars: H_2O, Na_2CO_3, $NaCl$, I_2, S, Cu, Zn, Fe, Mg, CH_3CH_2OH, $CuSO_4 \cdot 5H_2O$, $KMnO_4$, KI. Label the containers as One Mole with the name and formula of the substance. Allow students time to pick the containers up and feel the difference in masses and note the difference in volumes. Have them list each substance and classify it as either an element or a compound. Remember to tell students that each jar contains the same number of particles. **LS** Visual

---

Paired Reading Have pairs of students read Section 4 together and devise a flow chart for how they will solve two mole problems: calculating the mass of a substance from a given number of moles and calculating the number of atoms from a given number of moles.
LS Intrapersonal

Group Activity ——— GENERAL

Divide your class into groups of three or four and ask them to devise an experiment to determine the number of beans in a package without counting them. Their familiarity with the relationship between the mole and mass should suggest an experiment in which they would count a small number of beans and determine their mass. Then, by dividing this mass into the mass of the entire package, they can arrive at the total number of beans. Other grains such as rice or lentils could be used. **LS** Visual

Homework ——— GENERAL

Additional Practice

1. Calculate the mass in grams of 5.50 mol of sodium. Sodium has a molar mass of 22.99 g/mol.
Ans. 126 g

2. How many moles of helium are contained in 0.255 g of He?
Ans. 0.0638 mol

3. What is the mass in grams of 10.5 mol of nickel. Nickel has a molar mass of 58.69 g/mol.
Ans. 616 g

4. Calculate the number of moles of boron in 1.58×10^5 kg of B.
Ans. 1.46×10^7 mol
LS Logical

SAMPLE PROBLEM D

Converting from Amount in Moles to Mass

Determine the mass in grams of 3.50 mol of copper.

1 Gather information.
- amount of Cu = 3.50 mol
- mass of Cu = ? g Cu
- molar mass of Cu = 63.55 g

2 Plan your work.
- First, make a set-up that shows what is given and what is desired.

$$3.50 \text{ mol Cu} \times ? = ? \text{ g Cu}$$

- Use a conversion factor that has g Cu in the numerator and mol Cu in the denominator.

$$3.50 \text{ mol Cu} \times \frac{? \text{ g Cu}}{1 \text{ mol Cu}} = ? \text{ g Cu}$$

3 Calculate.
- The correct conversion factor is the molar mass of Cu, 63.55 g/mol. Place the molar mass in the equation, and calculate the answer. Use the periodic table in this book to find mass numbers of elements.

$$3.50 \text{ mol Cu} \times \frac{63.55 \text{ g Cu}}{1 \text{ mol Cu}} = 222 \text{ g Cu}$$

4 Verify your results.
- To verify that the answer of 222 g is correct, find the number of moles of 222 g of copper.

$$222 \text{ g of Cu} \times \frac{1 \text{ mol Cu}}{63.55 \text{ g Cu}} = 3.49 \text{ mol Cu}$$

The amount of 3.49 mol is close to the 3.50 mol, so the answer of 222 g is reasonable.

> **PRACTICE HINT**
>
> For elements and compounds, the mass will always be a number that is greater than the number of moles.

PRACTICE

1 What is the mass in grams of 1.00 mol of uranium?

2 What is the mass in grams of 0.0050 mol of uranium?

3 Calculate the number of moles of 0.850 g of hydrogen atoms. What is the mass in grams of 0.850 mol of hydrogen atoms?

4 Calculate the mass in grams of 2.3456 mol of lead. Calculate the number of moles of 2.3456 g of lead.

102

Answers to Practice Problems D
1. 238 g
2. 1.2 g
3. 0.84 mol; 0.86 g
4. 486.0 g; 0.01132 mol

Determining the Number of Atoms from the Amount in Moles

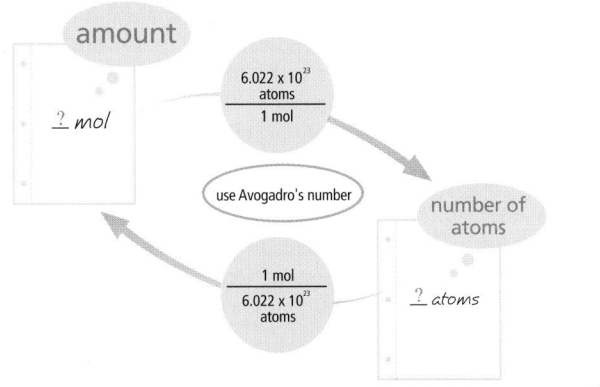

SAMPLE PROBLEM E

Converting from Amount in Moles to Number of Atoms

Determine the number of atoms in 0.30 mol of fluorine atoms.

1 Gather information.
- amount of F = 0.30 mol
- number of atoms of F = ?

2 Plan your work.
- To determine the number of atoms, select the conversion factor that will take you from the amount in moles to the number of atoms.

   amount (mol) × 6.022 × 10²³ atoms/mol = number of atoms

   amount (mol) $\times$ 6.022 $\times$ 10^{23} atoms/mol = number of atoms

3 Calculate.

$$0.30 \; \cancel{mol} \; F \times \frac{6.022 \times 10^{23} \; F \; atoms}{1 \; \cancel{mol} \; F} = 1.8 \times 10^{23} \; F \; atoms$$

4 Verify your results.
- The answer has units that are requested in the problem. The answer is also less than 6.022×10^{23} atoms, which makes sense because you started with less than 1 mol.

PRACTICE HINT

Make sure to select the correct conversion factor so that units cancel to give the unit required in the answer.

SKILL BUILDER — GENERAL

Math Skills Tell students that a book is 4.0 cm thick. If a mole of such books were stacked one on top of the other, ask students how high the stack would be. **Ans.** 2.4×10^{24} cm How many light years would the height of the stack correspond to if one light year = 9.4×10^{15} m? **Ans.** 2.5×10^6 light years Tell students this distance corresponds to the distance to the nearest galaxy. **LS** Logical

PRACTICE

1 How many atoms are in 0.70 mol of iron?

2 How many moles of silver are represented by 2.888×10^{23} atoms?

3 How many moles of osmium are represented by 3.5×10^{23} atoms?

103

Answers to Practice Problems E
1. 4.2×10^{23} atoms
2. 0.4796 mol
3. 0.58 mol

Homework — GENERAL

Additional Practice

1. How many moles are represented by 8.0×10^{13} atoms of Ca? **Ans.** 1.3×10^{-10} mol

2. How many atoms are present in 4.80 mol of iron? **Ans.** 2.9×10^{24} atoms

3. Determine the number of moles of potassium in 5.7×10^{25} atoms. **Ans.** 95 mol

4. Calculate the number of atoms in 7.50 mol of sulfur. **Ans.** 4.5×10^{24} atoms

LS Logical

Have small groups of students make up questions and problems like the sample problems in this section. Have them answer their own questions and solve their problems and then exchange their questions and problems with another group. **LS Interpersonal**

Quiz ──────── **GENERAL**

1. Why is the unit gram used in chemistry instead of amu? **Ans.** The amu is so small that ordinary-sized masses require cumbersome numbers.

2. What is the definition of 1 amu? **Ans.** 1/12 the atomic mass of the carbon-12 isotope

3. How is a dozen similar to a mole? **Ans.** Both are counting units for objects. There are 12 objects for a dozen and 6.022×10^{23} objects for a mole.

4. How many atoms are in 15.0 mol of zinc? **Ans.** 9.03×10^{24} atoms

5. Calculate the number of moles in 935 g of gold. **Ans.** 4.75 mol

6. How many moles of aluminum is 1.86×10^{25} atoms of Al? **Ans.** 30.9 mol

LS Logical

Figure 24
Carbon, which composes diamond, is the basis for the atomic mass scale that is used today.

Chemists and Physicists Agree on a Standard

The atomic mass unit has been defined in a number of different ways over the years. Originally, atomic masses expressed the ratio of the mass of an atom to the mass of a hydrogen atom. Using hydrogen as the standard turned out to be inconvenient because hydrogen does not react with many elements. Early chemists determined atomic masses by comparing how much of one element reacted with another element.

Because oxygen combines with almost all other elements, oxygen became the standard of comparison. The atomic mass of oxygen was defined as exactly 16, and the atomic masses of the other elements were based on this standard. But this choice also led to difficulties. Oxygen exists as three isotopes. Physicists based their atomic masses on the assignment of 16.0000 as the mass of the most common oxygen isotope. Chemists, on the other hand, decided that 16.0000 should be the average mass of all oxygen isotopes, weighted according to the abundance of each isotope. So, to a physicist, the atomic mass of fluorine was 19.0044, but to a chemist, it was 18.9991.

Finally, in 1960, a conference of chemists and physicists agreed on a scale based on an isotope of carbon. Carbon is shown in **Figure 24.** Used by all scientists today, this scale defines the atomic mass unit as exactly one-twelfth of the mass of one carbon-12 atom. As a result, one atomic mass unit is equal to $1.600\ 5402 \times 10^{-27}$ kg. The mass of an atom is indeed quite small.

④ Section Review

UNDERSTANDING KEY IDEAS

1. What is atomic mass?

2. What is the SI unit for the amount of a substance that contains as many particles as there are atoms in exactly 12 grams of carbon-12?

3. Which atom is used today as the standard for the atomic mass scale?

4. What unit is used for molar mass?

5. How many particles are present in 1 mol of a pure substance?

PRACTICE PROBLEMS

6. Convert 3.01×10^{23} atoms of silicon to moles of silicon.

7. How many atoms are present in 4.0 mol of sodium?

8. How many moles are represented by 118 g of cobalt? Cobalt has an atomic mass of 58.93 amu.

9. How many moles are represented by 250 g of platinum?

10. Convert 0.20 mol of boron into grams of boron. How many atoms are present?

CRITICAL THINKING

11. What is the molar mass of an element?

12. How is the mass in grams of an element converted to amount in moles?

13. How is the mass in grams of an element converted to number of atoms?

Answers to Section Review

1. the mass of an atom expressed in atomic mass units

2. mole

3. carbon-12

4. grams per mole

5. 6.022×10^{23} particles, or Avogadro's number

6. 0.500 mol Si

7. 2.4×10^{24} atoms

8. 2.00 mol

9. 1.3 mol

Answers continued on p. 113A

Chapter Resource File

• Concept Review
• Quiz

Be ⁴
Beryllium
9.012 182
[He]2s²

Element Spotlight

Where Is Be?
Earth's crust:
0.005% by mass

Beryllium: An Uncommon Element

Although it is an uncommon element, beryllium has a number of properties that make it very useful. Beryllium has a relatively high melting point (1278°C) and is an excellent conductor of energy as heat and electrical energy. Beryllium transmits X rays extremely well and is therefore used to make "windows" for X-ray devices. All compounds of beryllium are toxic to humans. People who experience prolonged exposure to beryllium dust may contract berylliosis, a disease that can lead to severe lung damage and even death.

Industrial Uses

- The addition of 2% beryllium to copper forms an alloy that is six times stronger than copper is. This alloy is used for nonsparking tools, critical moving parts in jet engines, and components in precision equipment.

- Beryllium is used in nuclear reactors as a neutron reflector and as an alloy with the fuel elements.

Real-World Connection Emerald and aquamarine are precious forms of the mineral beryl, $Be_3Al_2(SiO_3)_6$.

Crystals of pure beryllium look very different from the combined form of beryllium in an emerald.

A Brief History

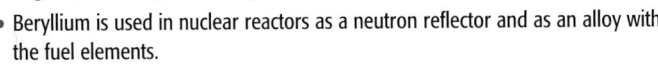

1800

1900

1828: F. Wöhler of Germany gives beryllium its name after he and W. Bussy of France simultaneously isolate the pure metal.

1926: M. G. Corson of the United States discovers that beryllium can be used to age-harden copper-nickel alloys.

1798: R. J. Haüy, a French mineralogist, observes that emeralds and beryl have the same optical properties and therefore the same chemical composition.

1898: P. Lebeau discovers a method of extracting high-purity beryllium by using an electrolytic process.

1942: A Ra-Be source provides the neutrons for Fermi's studies. These studies lead to the construction of a nuclear reactor.

Questions

1. Research how the beryllium and copper alloy is made and what types of equipment are made of this alloy.

2. Research how beryllium is used in nuclear reactors.

3. Research berylliosis and use the information to make a medical information brochure. Be sure to include symptoms, causes, and risk factors in your report.

☑ internet connect
www.scilinks.org
Topic : Beryllium
SciLinks code: HW4021
SCiLINKS Maintained by the National Science Teachers Association

105

Answers to Feature Questions

1. Students' answers may vary.

2. Students' answers may vary.

3. Students' answers may vary.

Alternative Assessments

HISTORY CONNECTION
• Page 74

• Pages 81, 98

did you know?
• Page 86

Group Activity
• Pages 87, 101

SKILL BUILDER
• Page 96

Portfolio Assessments

SKILL BUILDER
• Page 85

Chapter Resource File

• **Datasheets for In-Text Labs**
  Flame Tests

• **Datasheets for In-Text Labs**
  Spectroscopy and Flame
  Tests–Identifying Materials

KEY IDEAS

SECTION ONE   Substances Are Made of Atoms

• Three laws support the existence of atoms: the law of definite proportions, the law of conservation of mass, and the law of multiple proportions.

• Dalton's atomic theory contains five basic principles, some of which have been modified.

SECTION TWO   Structure of Atoms

• Protons, particles that have a positive charge, and neutrons, particles that have a neutral charge, make up the nuclei of most atoms.

• Electrons, particles that have a negative charge and very little mass, occupy the region around the nucleus.

• The atomic number of an atom is the number of protons the atom has. The mass number of an atom is the number of protons plus the number of neutrons.

• Isotopes are atoms that have the same number of protons but different numbers of neutrons.

SECTION THREE   Electron Configuration

• The quantum model describes the probability of locating an electron at any place.

• Each electron is assigned four quantum numbers that describe it. No two electrons of an atom can have the same four quantum numbers.

• The electron configuration of an atom reveals the number of electrons an atom has.

SECTION FOUR   Counting Atoms

• The masses of atoms are expressed in atomic mass units (amu). The mass of an atom of the carbon-12 isotope is defined as exactly 12 atomic mass units.

• The mole is the SI unit for the amount of a substance that contains as many particles as there are atoms in exactly 12 grams of carbon-12.

• Avogadro's number, 6.022×10^{23} particles per mole, is the number of particles in a mole.

KEY TERMS

law of definite
  proportions
law of conservation of
  mass
law of multiple
  proportions

electron
nucleus
proton
neutron
atomic number
mass number
isotope

orbital
electromagnetic
  spectrum
ground state
excited state
quantum number
Pauli exclusion principle
electron configuration
aufbau principle
Hund's rule

atomic mass
mole
molar mass
Avogadro's number

KEY SKILLS

Determining the Number of Particles in an Atom
Sample Problem A p. 86

Determining the Number of Particles in Isotopes
Sample Problem B p. 89

Writing Electron Configurations
Sample Problem C p. 98

Converting Amount in Moles to Mass
Skills Toolkit 1 p. 101
Sample Problem D p. 102

Converting Amount in Moles to Number of Atoms
Skills Toolkit 2 p. 103
Sample Problem E p. 103

USING KEY TERMS

1. Define *isotope*.

2. What are neutrons?

3. State the Pauli exclusion principle.

4. What is a cathode?

5. Define *mass number*.

6. What is a line-emission spectrum?

7. Define *ground state*.

8. Define *mole*.

9. State the law of definite proportions.

10. What is an orbital?

11. What is an electron configuration?

12. Which has less mass: an electron or a proton?

UNDERSTANDING KEY IDEAS

Substances Are Made of Atoms

13. What law is illustrated by the fact that ice, water, and steam consist of 88.8% oxygen and 11.2% hydrogen by mass?

14. What law is shown by the fact that the mass of carbon dioxide, which forms as a product of a reaction between oxygen and carbon, equals the combined masses of the carbon and oxygen that reacted?

15. Of the five parts of Dalton's atomic theory, which one(s) have been modified?

16. What is the atomic theory?

Structure of Atoms

17. Which of Dalton's principles was contradicted by the work of J. J. Thomson?

18. What information about an atom is provided by the atom's atomic number? the atom's mass number?

19. How were atomic models developed given that no one had seen an atom?

20. Why are atomic numbers always whole numbers?

21. If all protons have positive charges, how can any atomic nucleus be stable?

22. What observation did Thomson make to suggest that an electron has a negative electric charge?

Electron Configuration

23. Why does the Pauli exclusion principle include the word *exclusion*?

24. How do you use the aufbau principle when you create an electron configuration?

25. Explain what is required to move an electron from the ground state to an excited state.

26. Why can a *p* sublevel hold six electrons while the *s* sublevel can hold no more than two electrons?

27. What did Einstein's explanation about the photoelectric effect state about the nature of light?

28. What do electrons and light have in common?

29. Of a set of four quantum numbers, how many of these numbers can be shared by more than one electron of the same orbital?

107

Assignment Guide

Section	Questions
1	9, 13–16, 85–87
2	1, 2, 5, 12, 17–22, 39–46, 57, 62, 65–67, 73, 74, 81, 84, 88, 89, 93
3	3, 4, 6, 7, 10, 11, 23–32, 47–50, 60, 61, 79, 80, 82, 83, 90, 91, 95–98
4	8, 34–38, 51–56, 58, 59, 63, 64, 68–72, 75–78, 92, 94

REVIEW ANSWERS

1. atoms of the same element that have different masses

2. the particles that make up the nucleus and that have mass but no charge

3. No more than two electrons can occupy a single orbital, and the electrons must have opposite spins.

4. the electrode from which electrons leave

5. The mass number represents the total number of protons and neutrons that make up the nucleus of an atom.

6. distinct lines of colored light that are produced when the light emitted by excited atoms is passed through a prism

7. the lowest energy state of a quantized system

8. the SI base unit used to measure the amount of a substance that contains the same number of particles as the number of atoms of 12g of C-12

9. This law states that every sample of a compound has the same composition.

10. a region in an atom where there is a high probability of finding electrons

11. the arrangement of electrons in an atom

12. electron

13. law of definite proportions

14. law of conservation of mass

15. Atoms are not indivisible, and atoms of a given element are not identical.

16. The atomic theory states that all matter is composed of atoms.

17. the principle that atoms are indivisible

18. the number of protons, neutrons, and electrons that make up the atom

19. Models were developed by observing the behavior of matter at the macroscopic level and then explaining these observations. Such observations included the law of definite proportions, the law of conservation of mass, and the law of multiple proportions.

20. No atom can have only a part of a proton in its nucleus. Therefore, an atomic number must be a whole number that represents the number of protons in the nucleus.

21. A strong attractive force between the subatomic particles overcomes the repulsive force to keep the nucleus intact.

22. The cathode beam made up of electrons is deflected by a magnetic field.

23. Electrons are excluded from any orbital that already contains its maximum of two electrons.

24. The electrons must be placed in orbitals sequentially, starting with the lowest orbital that is available.

25. Energy must be supplied.

26. The p orbitals have three different orientations, each of which can hold two electrons for a maximum of six. In contrast, the s orbital has only one orientation and therefore can only hold 2 electrons.

27. Einstein proposed that light had properties of particles as well as properties of waves. Each particle of light has a certain quantity of energy.

28. Both have properties of waves and particles.

29. Of the four, three may be shared by more than one electron. The only quantum number that cannot be shared is the spin quantum number, $+\frac{1}{2}$ or $-\frac{1}{2}$.

30. When does an electron in an atom have the lowest energy?

31. How are the frequency and wavelength of light related?

32. Why does an electron occupy the $4s$ orbital before the $3d$ orbital?

33. The element sulfur has an electron configuration of $1s^2 2s^2 2p^6 3s^2 3p^4$.
 a. What does the superscript 6 refer to?
 b. What does the letter s refer to?
 c. What does the coefficient 3 refer to?

Counting Atoms

34. What is a mole? How is a mole related to Avogadro's number?

35. What significance does carbon-12 have in terms of atomic mass?

36. If the mass of a gold atom is 196.97 amu, what is the atom's molar mass?

37. How many water molecules are present in 1.00 mol of pure water?

38. What advantage is gained by using the mole as a unit when working with atoms?

PRACTICE PROBLEMS

Sample Problem A Determining the Number of Particles in an Atom

39. Calculate the number of neutrons of the atom whose atomic number is 42 and whose mass number is 96.

40. How many electrons are present in an atom of mercury whose atomic number is 80 and whose mass number is 201?

41. Calculate the number of protons of the atom whose mass number is 19 and whose number of neutrons is 10.

42. Calculate the number of electrons of the atom whose mass number is 75 and whose number of neutrons is 42.

Sample Problem B Determining the Number of Particles in Isotopes

43. Write nuclear symbols for isotopes of uranium that have the following numbers of neutrons. The atomic number of uranium is 92.
 a. 142 neutrons
 b. 143 neutrons
 c. 146 neutrons

44. Copy and complete the following table concerning the three isotopes of silicon, whose atomic number is 14.

Isotope	Number of protons	Number of electrons	Number of neutrons
Si-28			
Si-29			
Si-30			

45. Write the symbol for two isotopes of carbon. Both isotopes have six protons. One isotope has six neutrons, while the other has seven neutrons.

46. All barium atoms have 56 protons. One isotope of barium has 74 neutrons, and another isotope has 81 neutrons. Write the symbols for these two isotopes of barium.

Sample Problem C Writing Electron Configurations

47. Write the electron configuration for nickel, whose atomic number is 28. Remember that the $4s$ orbital has lower energy than the $3d$ orbital does and that the d sublevel can hold a maximum of 10 electrons.

48. Write the electron configuration of germanium whose atomic number is 32.

49. How many orbitals are completely filled in an atom that has 12 electrons? The electron configuration is $1s^2 2s^2 2p^6 3s^2$.

50. How many orbitals are completely filled in an atom of an element whose atomic number is 18?

108

Sample Problem D Converting Amount in Moles to Mass

51. How many moles are represented by each of the following.

  a. 11.5 g Na which has an atomic mass of 22.99 amu

  b. 150 g S which has an atomic mass of 32.07 amu

  c. 5.87 g Ni which has an atomic mass of 58.69 amu

52. Determine the mass in grams represented by 2.50 mol tellurium.

53. What is the mass in grams of 0.0050 mol of hydrogen atoms?

Sample Problem E Converting Amount in Moles to Number of Atoms

54. Calculate the number of atoms in 2.0 g of hydrogen atoms. The atomic mass of hydrogen is 1.01 amu.

55. Calculate the number of atoms present in each of the following:

  a. 2 mol Fe

  b. 40.1 g Ca, which has an atomic mass of 40.08 amu

  c. 4.5 mol of boron-11

56. How many mol of potassium are represented by 7.85×10^{23} potassium atoms?

MIXED REVIEW

57. In the diagram below, indicate which subatomic particles would be found in areas a and b.

58. What is the approximate atomic mass of an atom if the mass of the atom is

  a. 12 times that of carbon-12?

  b. one-fourth that of carbon-12?

59. What mass of silver, Ag, which has an atomic mass of 107.87 amu, contains the same number of atoms contained in 10.0 g of boron, B, which has an atomic mass of 10.81 amu?

60. Hydrogen's only electron occupies the 1s orbital but can be excited to a 4p orbital. List all of the orbitals that this electron can occupy as it "falls."

61. What is the electron configuration of zinc?

62. Identify the scientists who proposed each of the models illustrated below.

a. 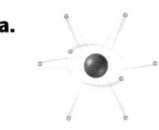 **c.**

b 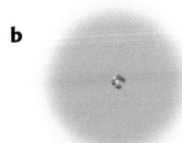 **d.**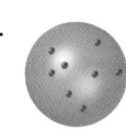

63. Determine the number of grams represented by each of the following:

  a. 3.5 mol of the element carbon, which has an atomic mass of 12.01 amu

  b. 5.0×10^9 atoms of neon, which has an atomic mass of 20.18 amu

  c. 2.25×10^{22} atoms of carbon, which has an atomic mass of 12.01 amu

64. How many atoms are in 0.75 moles of neptunium?

65. How did the results of the gold foil experiment lead Rutherford to recognize the existence of atomic nuclei?

66. Explain why atoms are neutral.

67. Explain Coulomb's law.

30. An electron will have the lowest energy when it is in the ground state.

31. There is an inverse relation between the two properties.

32. The 4s orbital has a lower energy than the 3d orbital does.

33. a. the number of electrons in the 2p orbital

  b. a type of orbital or sublevel

  c. the energy level

34. The mole is the SI unit for the amount of a substance that contains as many particles as there are atoms in exactly 12 grams of carbon-12. Avogadro's number is the number of particles in exactly one mole of a pure substance.

35. Carbon-12 is used as the standard for determining the atomic masses of all of the other elements.

36. 196.97 g/mol

37. 6.022×10^{23} molecules

38. Because the mole is a unit that represents a large number of particles, working with it is easy. For example, it is easier to work with 1 mol of oxygen than to use 6.022×10^{23} particles in calculations.

39. 54

40. 80

41. 9

42. 33

43. a. $^{234}_{92}\text{U}$

  b. $^{235}_{92}\text{U}$

  c. $^{238}_{92}\text{U}$

44.

Isotope	Number of protons	Number of electrons	Number of neutrons
Si-28	14	14	14
Si-29	14	14	15
Si-30	14	14	16

45. $^{12}_{6}\text{C}$ and $^{13}_{6}\text{C}$

46. $^{130}_{56}\text{Ba}$ and $^{137}_{56}\text{Ba}$

47. $1s^2 2s^2 2p^6 3s^2 3p^6 3d^8 4s^2$

48. $1s^2 2s^2 2p^6 3s^2 3p^6 3d^{10} 4s^2 4p^2$

49. 6

50. 9

REVIEW ANSWERS
continued

51. a. 0.500 mol
  b. 4.7 mol
  c. 0.100 mol
52. 319 g
53. 0.005 g
54. 1.2×10^{24} atoms
55. a. 1.2×10^{24} atoms
  b. 6.03×10^{23} atoms
  c. 2.7×10^{24} atoms
56. 1.30 mol
57. a. protons and neutrons
  b. electrons
58. a. 144 amu
  b. 3 amu
59. 99.8 g
60. $3d$, $4s$, $3p$, $3s$, $2p$, $2s$, and $1s$
61. $1s^2 2s^2 2p^6 3s^2 3p^6 3d^{10} 4s^2$
62. a. Bohr
  b. De Broglie
  c. Dalton
  d. Thomson
63. a. 42 g
  b. 1.7×10^{-13} g
  c. 0.449 g
64. 4.5×10^{23} atoms
65. Because some of the alpha particles bounced back, he concluded that an atom must have a very small but very dense nucleus.
66. Atoms have equal numbers of protons and electrons.
67. Coulomb's law states that the closer two charges are, the greater the electrostatic force between them.
68. a. 6.00 g
  b. 72.1 g
69. 0.307 kg
70. 6.022×10^{23} particles/mole
71. 0.39 mol; 2.3×10^{23} atoms
72. 89 mol; 5.4×10^{25} atoms
73. 5

68. What is the mass in grams of each of the following?
  a. 0.50 mol of carbon, C
  b. 1.80 mol of calcium

69. Determine the mass in kilograms of 5.50 mol of iron, Fe.

70. What is Avogadro's number?

71. How many moles are present in 11 g of silicon? how many atoms?

72. How many moles are present in 620 grams of lithium? how many atoms?

73. Suppose an atom has a mass of 11 amu and has five electrons. What is this atom's atomic number?

74. Explain why different atoms of the same element always have the same atomic number but can have different mass numbers.

75. What does an element's molar mass tell you about the element?

76. A pure gold bar is made of 19.55 mol of gold. What is the mass of the bar in grams?

77. How does halving the amount of a sample of an element affect the sample's mass?

78. James is holding a balloon that contains 0.54 g of helium gas. How many helium atoms is this?

79. Write the electron configuration of phosphorus.

80. Write the electron configuration of bromine.

81. What are the charges of an electron, a proton, and a neutron?

82. An advertising sign gives off red and green light.
  a. Which light has higher energy?
  b. One of the colors has a wavelength of 680 nm and the other has a wavelength of 500. Which color has which wavelength?

83. Can a stable atom have an orbital which has three electrons? Explain your answer.

CRITICAL THINKING

84. Predict what Rutherford might have observed if he had bombarded copper metal instead of gold metal with alpha particles. The atomic numbers of copper and gold are 29 and 79, respectively.

85. Identify the law that explains why a water molecule in a raindrop falling on Phoenix, Arizona, and a water molecule in the Nile River in Egypt are both made of two hydrogen atoms for every oxygen atom.

86. How would you rewrite Dalton's fourth principle to account for elements such as O_2, P_4, and S_8?

87. Which of Dalton's principles is contradicted by a doctor using radioisotopes to trace chemicals in the body?

88. What would happen to poisonous chlorine gas if the following alterations could be made?
  a. A proton is added to each atom.
  b. A neutron is added to each atom.

89. For hundreds of years, alchemists searched for ways to turn various metals into gold. How would the structure of an atom of $^{202}_{80}Hg$ (mercury) have to be changed for the atom to become an atom of $^{197}_{79}Au$ (gold)?

90. How are quantum numbers like an address? How are they different from an address?

91. The following electron configurations for electrons in the ground state are incorrect. Explain why each configuration is incorrect by using the Pauli exclusion principle or the aufbau principle, and then write the corrected electron configuration.
  a. $1s^2 2s^2 2p^5 3s^2 3p^3$
  b. $1s^2 2s^2 2p^8 3s^1$

92. Which has more atoms: 3.0 g of iron, Fe, or 2.0 grams of sulfur, S?

93. Predict which isotope of nitrogen is more commonly found, nitrogen-14 or nitrogen-15.

94. Suppose you have only 1.9 g of sulfur for an experiment and you must do three trials using 0.030 mol of S each time. Do you have enough sulfur?

95. How many orbitals in an atom can have the following designation?
a. $4p$
b. $7s$
c. $5d$

96. Explain why that if $n = 2$, l cannot be 2.

97. Write the electron configuration of rubidium.

98. Write the electron configuration of tin.

99. The magnetic properties of an element depend on the number of unpaired electrons it has. Explain why iron, Fe, is highly magnetic but neon, Ne, is not.

100. Many elements exist as polyatomic molecules. Use atomic masses to calculate the molecular masses of the following:
a. O_2
b. P_4
c. S_8

101. What do the electron configurations of neon, argon, krypton, xenon, and radon have in common?

102. Answer the following regarding electron configurations of atoms in the fourth period of the periodic table.
a. Which orbitals are filled by transition metals?
b. Which orbitals are filled by nonmetals?

ALTERNATIVE ASSESSMENT

103. So-called neon signs actually contain a variety of gases. Research the different substances used for these signs. Design your own sign on paper, and identify which gases you would use to achieve the desired color scheme.

104. Select one of the essential elements. Check your school library or the Internet for details about the role of each element in the human body and for any guidelines and recommendations about the element.

105. Research several elements whose symbols are inconsistent with their English names. Some examples include silver, Ag; gold, Au; and mercury, Hg. Compare the origin of these names with the origin of the symbols.

106. Research the development of the scanning tunneling microscope, which can be used to make images of atoms. Find out what information about the structure of atoms these microscopes have provided.

CONCEPT MAPPING

107. Use the following terms to complete the concept map below: *proton*, *atomic number*, *atomic theory*, *orbital*, and *electron*.

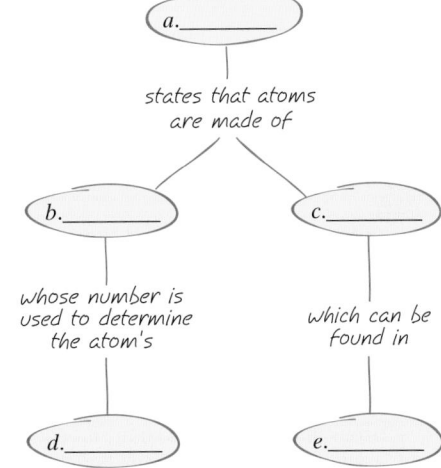

74. Atoms of the same element must have the same number of protons (same atomic number) but can have different numbers of neutrons. Because the mass number is the number of protons plus the number of neutrons, atoms of the same element can have different mass numbers.

75. the mass in grams of 1 mol of a substance

76. 3851 g

77. The mass is reduced by half because only half of the amount is present.

78. 8.1×10^{22} atoms

79. $1s^2 2s^2 2p^6 3s^2 3p^3$

80. $1s^2 2s^2 2p^6 3s^2 3p^6 3d^{10} 4s^2 4p^5$

81. −1, +1, 0

82. a. green
b. 680 nm, red; 500 nm green

83. No, an orbital can only have two electrons according to the Pauli exclusion principle.

84. He might have observed that the alpha particles were not deflected as much. The atomic nuclei of copper atoms have fewer protons; thus, the nuclei are smaller and have less total charge to deflect the alpha particle.

85. law of definite proportions

86. Atoms of the same element or of different elements combine in simple whole-number ratios to form compounds.

87. Radioisotopes show that atoms are divisible and that not all atoms of a given element are identical in their physical properties.

88. a. The atomic numbers would change, and a different element would form.
b. The atoms' mass numbers would change, and the atoms would become an isotope of chlorine.

89. The mercury atom would have to lose one proton (and one electron to remain neutral). It would also have to lose four neutrons from its nucleus.

111

90. Like an address that describes the location of a person's residence, the quantum numbers of an electron describe the location of an electron. However, unlike an address that gives the exact location of a person's home, the quantum numbers cannot pinpoint the exact location of the electron.

91. **a.** The aufbau principle states that the $2p$ orbital must be filled before electrons occupy any orbital that has a higher energy; $1s^2 2s^2 2p^6 3s^2 3p^2$.

    b. According to the Pauli exclusion principle, three p orbitals can hold a maximum of six electrons; $1s^2 2s^2 2p^6 3s^2 3p^1$.

92. 2 g S

93. nitrogen-14

94. 1.9 g S = 0.059 mol S, the experiment requires 0.09 mol, so there is not enough

95. **a.** 3

    b. 1

    c. 5

96. because $l = n - 1$

97. $1s^2 2s^2 2p^6 3s^2 3p^6 3d^{10} 4s^2 4p^6 5s^1$

98. $1s^2 2s^2 2p^6 3s^2 3p^6 4s^2 3d^{10} 4p^6 4d^{10} 5s^2 5p^2$

99. Iron atoms have unpaired electrons while neon atoms only have paired electrons.

100. **a.** 32.00 amu

     b. 123.9 amu

     c. 256.6 amu

101. These atoms have no unpaired electrons and have full s- and p-orbitals in the highest occupied energy levels.

102. **a.** $4s$, $3d$

     b. $4s$, $3d$, $4p$

Answers continued on p. 113A

FOCUS ON GRAPHING

Study the graph below, and answer the questions that follow.
For help in interpreting graphs, see Appendix B, "Study Skills for Chemistry."

108. What represents the ground state in this diagram?

109. Which energy-level changes can be detected by the unaided eye?

110. Does infrared light have more energy than ultraviolet light? Why or why not?

111. Which energy levels represent a hydrogen electron in an excited state?

112. What does the energy level labeled "$n = \infty$" represent?

113. If an electron is beyond the $n = \infty$ level, is the electron a part of the hydrogen atom?

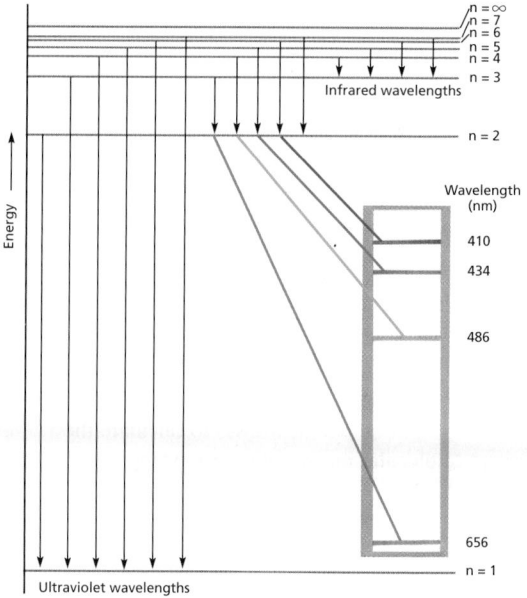

TECHNOLOGY AND LEARNING

114. **Graphing Calculator**

     Calculate Numbers of Protons, Electrons, and Neutrons.

     A graphing calculator can run a program that calculates the numbers of protons, electrons, and neutrons given the atomic mass and numbers for an atom. For example, given a calcium-40 atom, you will calculate the numbers of protons, electrons, and neutrons in the atom.

     Go to Appendix C. If you are using a TI-83 Plus, you can download the program

     NUMBER and data and can run the application as directed. If you are using another calculator, your teacher will provide you with keystrokes and data sets to use. After you have run the program, answer the questions below.

     a. Which element has the most protons?

     b. How many neutrons does mercury-201 have?

     c. Carbon-12 and carbon-14 have the same atomic number. Do they have the same number of neutrons? Why or why not?

112

Chapter Resource File

• Chapter Test

STANDARDIZED TEST PREP 3

Answers

1. b
2. d
3. c
4. a
5. b
6. a
7. c
8. c
9. d
10. b
11. d

1. Which of the following represents an electron configuration for a calcium atom, whose atomic number is 20?
   a. $1s^2 2s^2 2p^6 3s^1 3p^6 4s^2 3d^1$
   b. $1s^2 2s^2 2p^6 3s^2 3p^6 4s^2$
   c. $1s^2 2s^2 2p^6 3s^1 3p^6 4s^3$
   d. $1s^2 2s^2 2p^6 3s^2 3p^6 4s^2 3d^1$

2. Isotopes of an element differ in their
   a. atomic numbers.
   b. electron configurations.
   c. numbers of protons.
   d. numbers of neutrons.

3. The number of protons of an atom equals
   a. the number of neutrons.
   b. the mass number.
   c. the number of electrons.
   d. the number of isotopes.

4. The law of multiple proportions states that
   a. when two elements combine to form more than one compound, the masses of one element that combine with a given mass of the other element are in a ratio of small whole numbers.
   b. elements combine in the same mass ratio in a compound regardless of the quantity of the sample or the source.
   c. elements have different physical and chemical properties.
   d. atoms of elements cannot be created, destroyed, or subdivided when they participate in a chemical reaction.

5. How many neutrons are present in an atom of tin that has an atomic number of 50 and a mass number of 119?
   a. 50   c. 119
   b. 69   d. 169

6. As an electron in an excited state returns to its ground state,
   a. light energy is emitted.
   b. energy is absorbed by the atom.
   c. the atom is likely to undergo spontaneous decay.
   d. the electron configuration of the atom changes.

7. According to quantum theory, an electron
   a. remains in a fixed position.
   b. can replace a proton in the nucleus.
   c. occupies the space around the nucleus only in certain, well-defined orbitals.
   d. has neither mass nor charge.

8. The unit for molar mass is
   a. atoms per mole.
   b. grams per atoms.
   c. g/mol.
   d. mol/g.

9. Which of the following could represent a pair of isotopes?
   a. $^{20}_{8}X$ and $^{20}_{9}X$ c. $^{32}_{15}X$ and $^{34}_{16}X$
   b. $^{44}_{23}X$ and $^{44}_{23}X$ d. $^{40}_{19}X$ and $^{42}_{19}X$

10. Among the phenomena that enabled scientists to infer the existence of atoms was the
    a. Pauli exclusion principle.
    b. law of conservation of mass.
    c. relationship between mass and energy shown by the equation $E = mc^2$.
    d. observation that each element differed in the number of protons in one of its atoms.

11. An important result of Rutherford's work was to establish that
    a. atoms have mass.
    b. electrons have a negative charge.
    c. gold is an element.
    d. the atom is mostly empty space.

113

Continuation of Answers

Answers to Section 1 Review, *continued*

4. An element is composed of atoms, all of which have identical properties. A compound consists of two or more kinds of atoms, each of which has different properties, combined in simple, whole-number ratios.

5. 1. All matter is composed of extremely small particles called atoms, which cannot be subdivided, created, or destroyed.

   2. Atoms of a given element are identical in their physical and chemical properties.

   3. Atoms of different elements differ in their physical and chemical properties.

   4. Atoms of different elements combine in simple, whole-number ratios to form compounds.

   5. In chemical reactions, atoms are combined, separated, or rearranged but never created, destroyed, or changed.

6. Principles 3, 4, and 5 still apply.

7. the law of definite proportions

8. the law of multiple proportions

9. The three compounds have masses of fluorine combined with 1 g of sulfur in a ratio of small, whole numbers. The ratio is 1:2:3.

Answers to Section 2 Review, *continued*

5. **a.** 35 electrons, 35 protons, 45 neutrons

   b. 46 electrons, 46 protons, 60 neutrons

   c. 55 electrons, 55 protons, 78 neutrons

6. The atomic number of this element is 56, and the mass number is 138.

7. If the repulsive force between positively charged protons were the only force, a nucleus could not contain multiple protons.

8. No, they are isotopes of different elements, because they have different atomic numbers. All isotopes of an element have the same atomic number, not the same mass number.

Answers to Section 3 Review, *continued*

7. five

8. **a.** The Pauli exclusion principle states that a single orbital can hold a maximum of two electrons, so $2s^3$ is not correct.

   b. The aufbau principle states that electrons fill orbitals that have the lowest energy available, so $2p^5 3s^1$ is not correct.

9. The shorter wavelength represents the emission of more energy when the electron drops from $n = 4$ to $n = 1$.

10. 18

11. The $4s$ orbital is lower in energy than a $3d$ orbital.

Answers to Section 4 Review, *continued*

10. 2.2 g; 1.2×10^{23} atoms

11. the mass in grams of one mole of an element

12. by multiplying by the conversion factor, 1 mol/molar mass

13. First, you convert the mass to moles by multiplying by the conversion factor, 1 mole/molar mass; then you multiply the number of moles by Avogadro's number.

Answers to Chapter Review, *continued*

103. Students' answers may vary.

104. Students' answers may vary.

105. Students' answers may vary.

106. Students' answers may vary.

107. **a.** atomic theory

   b. proton

   c. electron

   d. atomic number

   e. orbital

108. $n = 1$

109. any of the falls of an electron to $n = 2$ level

110. No, because infrared energy has a longer wavelength.

111. $n > 1$

112. There are an infinite number of energy levels that an electron can theoretically occupy.

113. No, the electron is not considered a part of the atom.

114. **a.** fermium-257

   b. 121

   c. No. The atomic masses are different, and the number of neutrons can be found by subtracting the atomic number from the mass number. Carbon-12: 12 – 6 = 6 neutrons. Carbon-14: 14 – 6 = 8 neutrons.

The Periodic Table
Chapter Planning Guide

PACING	CLASSROOM RESOURCES	LABS, ACTIVITIES, AND DEMONSTRATIONS
BLOCK 1 • 45 min pp. 114–115 **Chapter Opener**		**SE** **Start-Up Activity** What Is a Periodic Table? p. 115
BLOCK 2 • 45 min pp. 116–123 **Section 1** How Are Elements Organized?	**TT** **Teaching Transparency** Blocks of the Periodic Table **OSP** **Supplemental Reading** The Periodic Kingdom `ADVANCED` **SE** **Consumer Focus** Good Health Is Elementary	**TE** **Demonstration** Families (Groups) of the Periodic Table, p. 119 **TE** **Group Activity,** p. 120 `GENERAL` **SE** **Skills Practice Lab** The Mendeleev Lab of 1869, p. 778 ◆ `GENERAL` **CRF** **Datasheets for In-Text Labs** *
BLOCKS 3 & 4 • 90 min pp. 124–131 **Section 2** Tour of the Periodic Table		**TE** **Demonstration** Orange-Yellow Light of Sodium, p. 125 ◆ **TE** **Demonstration** Reactivities of Mg and Ca, p. 126 ◆ **TE** **Group Activity,** p. 129 `GENERAL` **CRF** **Microscale Lab** Reactivity of Halide Ions ◆ `GENERAL`
BLOCKS 5 & 6 • 90 min pp. 132–141 **Section 3** Trends in the Periodic Table	**TT** **Teaching Transparency** Periodic Trends of Radii **TT** **Teaching Transparency** Additional Periodic Trends	**TE** **Demonstration** Size of an Oleic Acid Molecule, p. 135 ◆ `ADVANCED` **TE** **Group Activity,** p. 137 `GENERAL` **TE** **Demonstration** Ionic Size, p. 139 **CRF** **Observation Lab** Exploring the Periodic Table ◆ `GENERAL`
BLOCK 7 • 45 min pp. 142–148 **Section 4** Where Did the Elements Come From?	**TT** **Teaching Transparency** Nuclear Fusion **TT** **Teaching Transparency** Nuclear Fusion: Stellar Formation of Carbon-12 **SE** **Science and Technology** Superconductors	**TE** **Demonstration** Cloud Chamber, p. 144 ◆ `ADVANCED`

BLOCKS 8 & 9 • 90 min **Chapter Review and Assessment Resources**

- **SE** **Chapter Review,** pp. 150–154
- **SE** **Standardized Test Prep,** p. 155
- **CRF** **Chapter Test** *
- **OSP** **Test Generator**
- **CRF** **Test Item Listing** *
- **OSP** **Scoring Rubrics and Classroom Management Checklists**

Holt Chemistry: Online Resources

Visit **go.hrw.com** for a variety of free resources related to this textbook. Enter the keyword **HW4 HOME**.

Students can access interactive problem solving help and active visual concept development with the *Holt Chemistry* Online Edition available at **www.hrw.com**

student **CNN News**

cnnstudentnews.com

Find the latest chemistry news, lesson plans, and activities related to important scientific events.

KEY

TE	Teacher Edition	**CRF**	Chapter Resource File	*	Also on One-Stop Planner
SE	Student Edition	**TT**	Teaching Transparencies	◆	Requires Advance Prep
OSP	One-Stop Planner				

PROBLEM SOLVING AND PRACTICE	SECTION REVIEW AND ASSESSMENT	STANDARDS CORRELATION
		National Science Education Standards
	TE Quiz, p. 122 GENERAL **CRF** Quiz * **TE** Reteaching, p. 122 BASIC **SE** Section Review, p. 122 **CRF** Concept Review *	PS 2b
	TE Quiz, p. 131 GENERAL **CRF** Quiz * **TE** Reteaching, p. 131 BASIC **SE** Section Review, p. 131 **CRF** Concept Review *	PS 2b
	TE Quiz, p. 141 GENERAL **CRF** Quiz * **TE** Reteaching, p. 141 BASIC **SE** Section Review, p. 141 **CRF** Concept Review *	PS 2b
	TE Quiz, p. 147 GENERAL **CRF** Quiz * **TE** Reteaching, p. 147 BASIC **SE** Section Review, p. 147 **CRF** Concept Review *	

www.scilinks.org

Topic: The Periodic Table
SciLinks code: HW4094

Topic: Alkali Metals
SciLinks code: HW4007

Topic: Alkaline-Earth Metals
SciLinks code: HW4008

Topic: Halogens
SciLinks code: HW4065

Topic: Noble Gases
SciLinks code: HW4083

Topic: Metals
SciLinks code: HW4079

Topic: Transition Metals
SciLinks code: HW4168

Topic: Origin of Elements
SciLinks code: HW4093

Topic: Alchemy
SciLinks code: HW4006

Topic: Superconductors
SciLinks code: HW4170

Technology Resources

 Science in the NEWS

Each video segment is accompanied by a Critical Thinking Worksheet.

Segment 12
Alloy Technology

Segment 13
Atom Builders

Segment 11
Superconductivity

 ChemFile Interactive Tutor CD-Rom

Module 3: Periodic Properties

Overview

This chapter traces the historical development of the periodic table, including the modern periodic table. Students will learn how the periodic table orders elements by increasing atomic number and properties of elements. Certain groups of elements with distinctive characteristics will be examined. Students will learn how to use the periodic table to predict properties and trends of groups and periods. Students will also learn how one element can change to another element when nuclear changes occur.

Assessing Prior Knowledge

Check for Content Knowledge

Students should be familiar with the following topics:

- chemical and physical properties
- composition of matter
- nuclear structure
- atomic number
- electron configurations

Using the Figure

In the periodic table, the metals in Group 11 are known as the coinage metals. Throughout history, these metals—copper, silver, and gold—have been used separately and in combination with other metals for making coins. Because they are in the same group in the periodic table, they have similar chemical properties. For example, none of these metals readily reacts with most acids. Gold is the least reactive of the three metals; so unreactive that it is found as a pure element in nature. The other two metals are found as compounds in ores.

THE PERIODIC TABLE

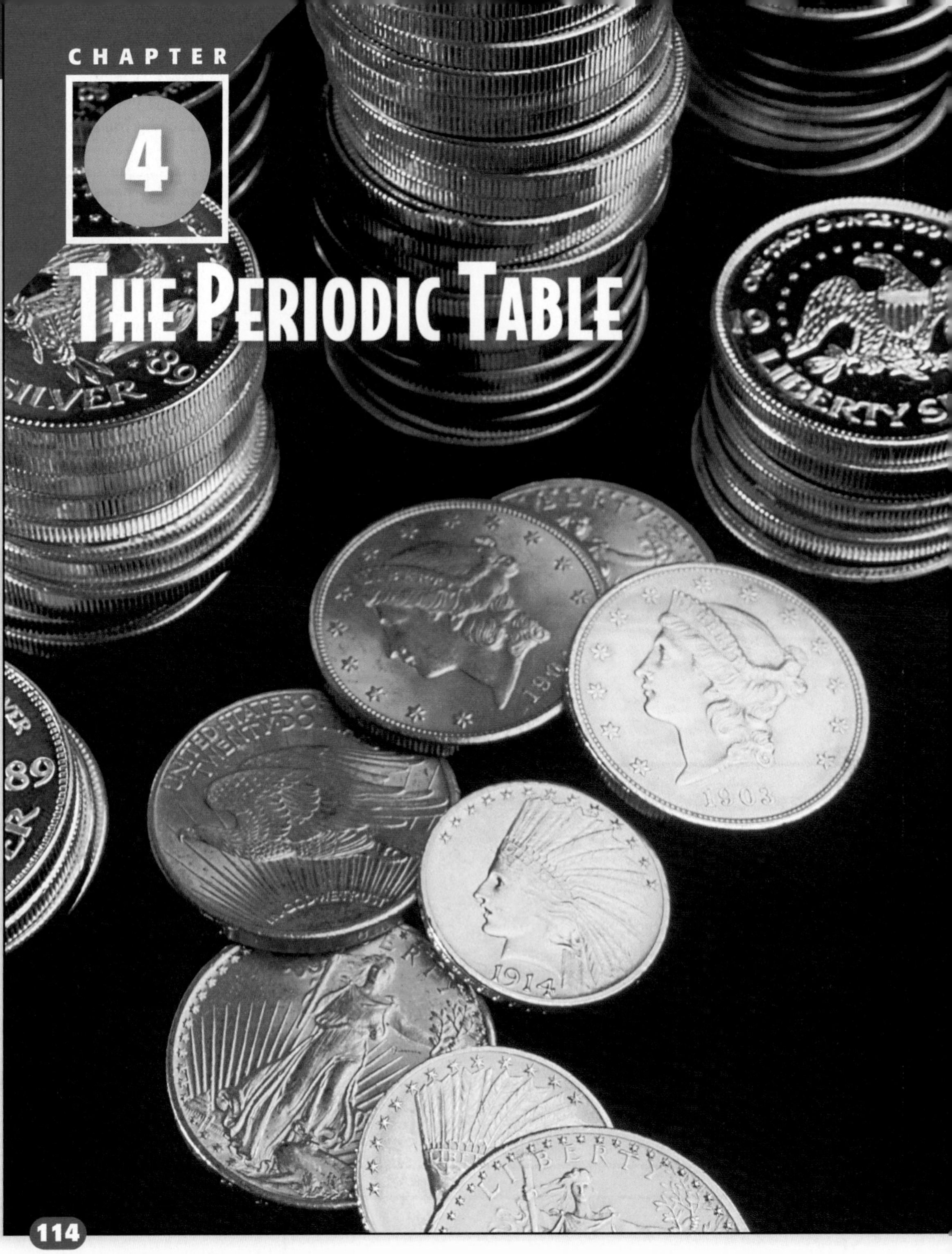

Standards Correlations

National Science Education Standards

PS 2b: An element is composed of a single type of atom. When elements are listed in order to the number of protons (called the atomic number), repeating patterns of physical and chemical properties identies of elements with similar properties. This "Periodic Table" is a consequence of the repeating pattern of outermost electrons and their permitted energies. (Sections 1, 2, 3)

The United States established its first mint to make silver and gold coins in Philadelphia in 1792. Some of these old gold and silver coins have become quite valuable as collector's items. An 1804 silver dollar recently sold for more than $4 million. A silver dollar is actually 90% silver and 10% copper. Because the pure elements gold and silver are too soft to be used alone in coins, other metals are mixed with them to add strength and durability. These metals include platinum, copper, zinc, and nickel. Metals make up the majority of the elements in the periodic table.

START-UP ACTIVITY

SAFETY PRECAUTIONS

What Is a Periodic Table?

PROCEDURE

1. Sit in your assigned desk according to the seating chart your teacher provides.

2. On the blank chart your teacher gives you, jot down information about yourself—such as name, date of birth, hair color, and height—in the space that represents where you are seated.

3. Find out the same information from as many people sitting around you as possible, and write that information in the corresponding spaces on the seating chart.

ANALYSIS

1. Looking at the information you gathered, try to identify patterns that could explain the order of people in the seating chart. If you cannot yet identify a pattern, collect more information and look again for a pattern.

2. Test your pattern by gathering information from a person you did not talk to before.

3. If the new information does not fit in with your pattern, reevaluate your data to come up with a new hypothesis that explains the patterns in the seating chart.

Pre-Reading Questions

① Define *element*.

② What is the relationship between the number of protons and the number of electrons in a neutral atom?

③ As electrons fill orbitals, what patterns do you notice?

CONTENTS [4]

internet connect

www.scilinks.org
Topic: The Periodic Table
SciLinks code: HW4094

SCILINKS. Maintained by the National Science Teachers Association

START-UP ACTIVITY

Skills Acquired:
• Collecting data
• Identifying and recognizing patterns
• Interpreting

Materials:
For each student:
• blank seating chart
• pencil or pen

Teacher Notes:
• Before class, make a seating chart sorting students by two characteristics, one in rows and the other in columns; for instance, alphabetical by last name in columns, and by date of birth in rows.

• You might want to list specific types of information students are to gather to avoid students asking one another uncomfortable questions.

• If individual students have difficulty coming up with a pattern, have pairs of students combine information and collectively come up with a pattern.

LS Interpersonal

Answers to Pre-Reading Questions

1. An element is any substance that cannot be separated by ordinary chemical means into simpler substances.

2. They are equal.

3. Answers should include the idea that electrons fill orbitals sequentially in repeating patterns by orbital and energy level. If students have any prior knowledge of the periodic table, they might answer that the elements in a group on the periodic table contain the same number of valence electrons.

Focus

Overview

Before beginning this section, review with your students the Objectives listed in the Student Edition. This section describes the historical development of the periodic table. It also explains how the modern periodic table is based on the periodic law, which states that properties of elements are a periodic function of their atomic numbers.

🔊 Bellringer

Have students list things in their classroom that they think are made from single elements. Encourage them to include things they can't see, such as the nitrogen and oxygen in the air. Other items might include objects made from metal, such as steel items (iron mixed with at least one other element) or students' coins.

Motivate

Identifying Preconceptions

Students might think that elements with similar chemical properties also have similar physical properties. As an example, have students investigate and compare the properties of chlorine and bromine. Although the chemical properties are quite similar, chlorine is a yellowish-green gas, and bromine is a reddish-brown liquid. They also differ in other physical properties, such as density, melting point, and boiling point.

SECTION

1

How Are Elements Organized?

KEY TERMS
- periodic law
- valence electron
- group
- period

OBJECTIVES

1. **Describe** the historical development of the periodic table.

2. **Describe** the organization of the modern periodic table according to the periodic law.

Patterns in Element Properties

Pure elements at room temperature and atmospheric pressure can be solids, liquids, or gases. Some elements are colorless. Others, like the ones shown in **Figure 1,** are colored. Despite the differences between elements, groups of elements share certain properties. For example, the elements lithium, sodium, potassium, rubidium, and cesium can combine with chlorine in a 1:1 ratio to form LiCl, NaCl, KCl, RbCl, and CsCl. All of these compounds are white solids that dissolve in water to form solutions that conduct electricity.

Similarly, the elements fluorine, chlorine, bromine, and iodine can combine with sodium in a 1:1 ratio to form NaF, NaCl, NaBr, and NaI. These compounds are also white solids that can dissolve in water to form solutions that conduct electricity. These examples show that even though each element is different, groups of them have much in common.

Topic Link

Refer to the chapter "The Science of Chemistry" for a definition and discussion of elements.

Figure 1
The elements chlorine, bromine, and iodine, pictured from left to right, look very different from each other. But each forms a similar-looking white solid when it reacts with sodium.

116

HISTORY
CONNECTION

Law of Triads In the early 1800s, the German chemist Johann Dobereiner attempted to classify several elements according to their properties. He noticed that certain groups of three elements had similar properties. He also noticed that the mass of the middle element was approximately the average of the other two masses. He called these groups of three elements triads. Examples of Dobereiner's triads are calcium, strontium, and barium; sulfur, selenium, and tellurium; and chlorine, bromine, and iodine.

Chapter Resource File

- Lesson Plan

John Newlands Noticed a Periodic Pattern

Elements vary widely in their properties, but in an orderly way. In 1865, the English chemist John Newlands arranged the known elements according to their properties and in order of increasing atomic mass. He placed the elements in a table.

As he studied his arrangement, Newlands noticed that all of the elements in a given row had similar chemical and physical properties. Because these properties seemed to repeat every eight elements, Newlands called this pattern the *law of octaves*.

This proposed law met with some skepticism when it was first presented, partly because chemists at the time did not know enough about atoms to be able to suggest a physical basis for any such law.

Dmitri Mendeleev Invented the First Periodic Table

In 1869, the Russian chemist Dmitri Mendeleev used Newlands's observation and other information to produce the first orderly arrangement, or periodic table, of all 63 elements known at the time. Mendeleev wrote the symbol for each element, along with the physical and chemical properties and the relative atomic mass of the element, on a card. Like Newlands, Mendeleev arranged the elements in order of increasing atomic mass. Mendeleev started a new row each time he noticed that the chemical properties of the elements repeated. He placed elements in the new row directly below elements of similar chemical properties in the preceding row. He arrived at the pattern shown in **Figure 2.**

Two interesting observations can be made about Mendeleev's table. First, Mendeleev's table contains gaps that elements with particular properties should fill. He predicted the properties of the missing elements.

ПЕРИОДИЧЕСКАЯ СИСТЕМА ЭЛЕМЕНТОВ

РЯДЫ	ГРУППЫ ЭЛЕМЕНТОВ									
	I	II	III	IV	V	VI	VII	VIII		
I	H 1									
II	Li 7	Be 9,4	B 11	C 12	N 14	O 16	F 19			
III	Na 23	Mg 24	Al 27,4	Si 28	P 31	S 32	Cl 35,5			
IV	K 39	Ca 40	? 45	Ti 50	V 51	Cr 52	Mn 55	Fe 56	Co 59	Ni 59
V	Cu 63,4	Zn 65,2	? 68	? 70	As 75	Se 79,4	Br 80			
VI	Rb 85,4	Sr 87,6	Yt ? 88	Zr 90	Nb 94	Mo 96	? 100	Ru 104,4	Rh 104,4	Pd 106,6
VII	Ag 108	Cd 112	In 113	Sn 118	Sb 122	Te 128?	J 127			
VIII	Cs 133	Ba 137	Di? 138	Ce? 140						
IX										
X			Er? 178	La? 180	Ta 182	W 186		Pt 197,4	Ir 198	Os 199
XI	Au 197?	Hg 200	Tl 204	Pb 207	Bi 210					
XII				Th 231		U 240				

Figure 2
Mendeleev's table grouped elements with similar properties into vertical columns. For example, he placed the elements highlighted in red in the table—fluorine, chlorine, bromine, and iodine—into the column that he labeled "VII."

HISTORY CONNECTION

Atomic Mass and Classification of Elements Remind students that neither Newlands nor Mendeleev had knowledge of atomic numbers or electron configuration. Because they knew that the elements had specific atomic masses, they logically based their tables on increasing atomic mass. The elements were arranged in families solely because of their observed properties, such as the ratios by which they combined with other elements, particularly oxygen. It is remarkable that an arrangement based primarily on chemical properties is now known also to represent an element's electron structure.

Teach

READING SKILL BUILDER — **BASIC**

Assimilating Knowledge Before reading this chapter, have each student write a short list of all the things they know about the arrangement of elements on the periodic table, and have them write what that arrangement reveals about atomic structure. When students have finished their list, ask them to contribute their entries to a group list on the board. Then have students list the things they want to know about the periodic table and its arrangement. If students have trouble coming up with questions, offer the following suggestions.

- What do the electron configurations of lithium and sodium have in common?
- What do the electron configurations of fluorine and chlorine have in common?
- What metallic elements can you name?
- Where are these elements located on the periodic table?

LS Verbal

Using the Figure

Refer students to **Figure 1.** Be sure students realize that the elements are in the top photo and the salts are arranged below them.

Using the Figure

Refer students to **Figure 2.** The illustration shows one representation of Mendeleev's original organization of the elements. When the predictions of the existence of yet undiscovered elements proved true, the attention of the scientific world was drawn to Mendeleev's table. Some names and symbols have changed over time. Also, Mendeleev used German symbols for elements, which resulted in J for iodine. Its name in German is Jod, pronounced "yote."

Mendeleev's Periodic Arrangement Ask students what they would tell Mendeleev about the tellurium/iodine discrepancy in atomic mass. Inform them that they should be able to deduce the reason from what they have already learned (differences in numbers of neutrons). In fact, iodine has only one stable isotope, I-127, whereas tellurium is a mixture of seven stable isotopes. Te-128 and Te-130 make up more then 65% of all atoms of tellurium, so the average atomic mass is higher than 127.

— BASIC

Reading Organizer If possible, supply students with a large blank copy of the periodic table with columns for each group of elements. Students can use it to make notes about the divisions of the table and the groups they will study in this chapter. Encourage students to write only major facts and concepts to avoid filling all of the space with notes. Caution students to save space for general information and not to copy the data from **Figure 4.** LS Logical

Homework — ADVANCED

Determining Triads To show how Mendeleev predicted the presence of elements unknown at the time, have students choose groups of three elements and average the atomic masses of the first and last elements. Have them find at least three sets of elements for which this average is close to the actual atomic mass of the middle element. LS Logical

Table 1   Predicted Versus Actual Properties for Three Elements

Properties	Ekaaluminum (gallium, discovered 1875)		Ekaboron (scandium, discovered 1877)		Ekasilicon (germanium, discovered 1886)	
	Predicted	Observed	Predicted	Observed	Predicted	Observed
Density	6.0 g/cm^3	5.96 g/cm^3	3.5 g/cm^3	3.5 g/cm^3	5.5 g/cm^3	5.47 g/cm^3
Melting point	low	30°C	*	*	high	900°C
Formula of oxide	Ea_2O_3	Ga_2O_3	Eb_2O_3	Sc_2O_3	EsO_2	GeO_2
Solubility of oxide	*	*	dissolves in acid	dissolves in acid	*	*
Density of oxide	*	*	*	*	4.7 g/cm^3	4.70 g/cm^3
Formula of chloride	*	*	*	*	$EsCl_4$	$GeCl_4$
Color of metal	*	*	*	*	dark gray	grayish white

He also gave these elements provisional names, such as "Ekaaluminum" (the prefix *eka-* means "one beyond") for the element that would come below aluminum. These elements were eventually discovered. As **Table 1** illustrates, their properties were close to Mendeleev's predictions. Although other chemists, such as Newlands, had created tables of the elements, Mendeleev was the first to use the table to predict the existence of undiscovered elements. Because Mendeleev's predictions proved true, most chemists accepted his periodic table of the elements.

Second, the elements do not always fit neatly in order of atomic mass. For example, Mendeleev had to switch the order of tellurium, Te, and iodine, I, to keep similar elements in the same column. At first, he thought that their atomic masses were wrong. However, careful research by others showed that they were correct. Mendeleev could not explain why his order was not always the same.

The Physical Basis of the Periodic Table

About 40 years after Mendeleev published his periodic table, an English chemist named Henry Moseley found a different physical basis for the arrangement of elements. When Moseley studied the lines in the X-ray spectra of 38 different elements, he found that the wavelengths of the lines in the spectra decreased in a regular manner as atomic mass increased. With further work, Moseley realized that the spectral lines correlated to atomic number, not to atomic mass.

When the elements were arranged by increasing atomic number, the discrepancies in Mendeleev's table disappeared. Moseley's work led to both the modern definition of atomic number, and showed that atomic number, not atomic mass, is the basis for the organization of the periodic table.

118

HISTORY CONNECTION

Atomic Number Henry Moseley proposed the concept of atomic number in 1914. At that time, there was no explanation for the arrangement of the elements on the periodic table, and scientists continued to try to link periodicity to atomic mass. Moseley found that the progression of the elements in the table corresponded to an increase of one fundamental unit in the nucleus and that each element could be assigned an atomic number equal to the number of these units. In 1920, Rutherford announced the existence of the proton, Moseley's fundamental unit.

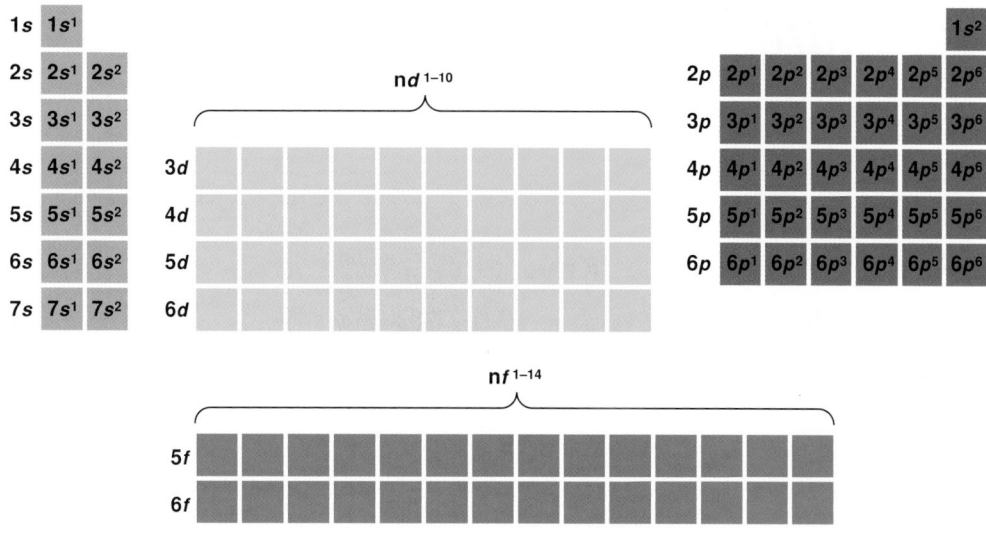

Figure 3
The shape of the periodic table is determined by how electrons fill orbitals. Only the s and p electrons are shown individually because unlike the d and f electrons, they fill orbitals sequentially.

The Periodic Law

According to Moseley, tellurium, whose atomic number is 52, belongs before iodine, whose atomic number is 53. Mendeleev had placed these elements in the same order based on their properties. Today, Mendeleev's principle of chemical periodicity is known as the **periodic law,** which states that when the elements are arranged according to their atomic numbers, elements with similar properties appear at regular intervals.

Organization of the Periodic Table

To understand why elements with similar properties appear at regular intervals in the periodic table, you need to examine the electron configurations of the elements. As shown in **Figure 3,** elements in each column of the table have the same number of electrons in their outer energy level. These electrons are called **valence electrons.** It is the valence electrons of an atom that participate in chemical reactions with other atoms, so elements with the same number of valence electrons tend to react in similar ways. Because s and p electrons fill sequentially, the number of valence electrons in s- and p-block elements are predictable. For example, atoms of elements in the column on the far left have one valence electron. Atoms of elements in the column on the far right have eight valence electrons (except for helium, which has two). A vertical column on the periodic table is called a **group.** A complete version of the modern periodic table is shown in **Figure 4** on the next two pages.

periodic law

the law that states that the repeating physical and chemical properties of elements change periodically with their atomic number

valence electron

an electron that is found in the outermost shell of an atom and that determines the atom's chemical properties

group

a vertical column of elements in the periodic table; elements in a group share chemical properties

Topic Link

Refer to the chapter "Atoms and Moles" for a discussion of electron configuration.

119

Teach, *continued*

Using the Figure

Figure 4 Inform students of the range of information shown in this table. Also point out that the lanthanide and actinide elements occur between Groups 3 and 4 in Periods 6 and 7 but are conventionally shown below the table so that the printed table is not too wide.

Group Activity —— GENERAL

Number pieces of paper 1 through 36, and place them in a can. Have each student draw a number. Each student should research and write a descriptive paragraph about the element whose atomic number he or she drew. The next day, collect the paragraphs. Divide the class into groups of three and distribute blank cards that have 25 boxes in a 5 × 5 in. square to each group. Have students fill the squares with any 25 of the 36 elements in any order. Read the student-made clues and allow each group of students to mark an X through each element they identify. The first group to correctly complete a vertical, horizontal, or diagonal row is the winner. **LS** Interpersonal

Teaching Tip

Electron configurations of transition metals Astute students may notice that many transition metals, such as chromium, niobium, and molybdenum, contain singly occupied *s* orbitals, and may wonder why. Explain that as *d* orbitals become populated, they can actually drop to a lower energy level than the *s* orbitals, so an electron which would have been in an *s* orbital drops to the *d* orbital instead.

Periodic Table of the Elements

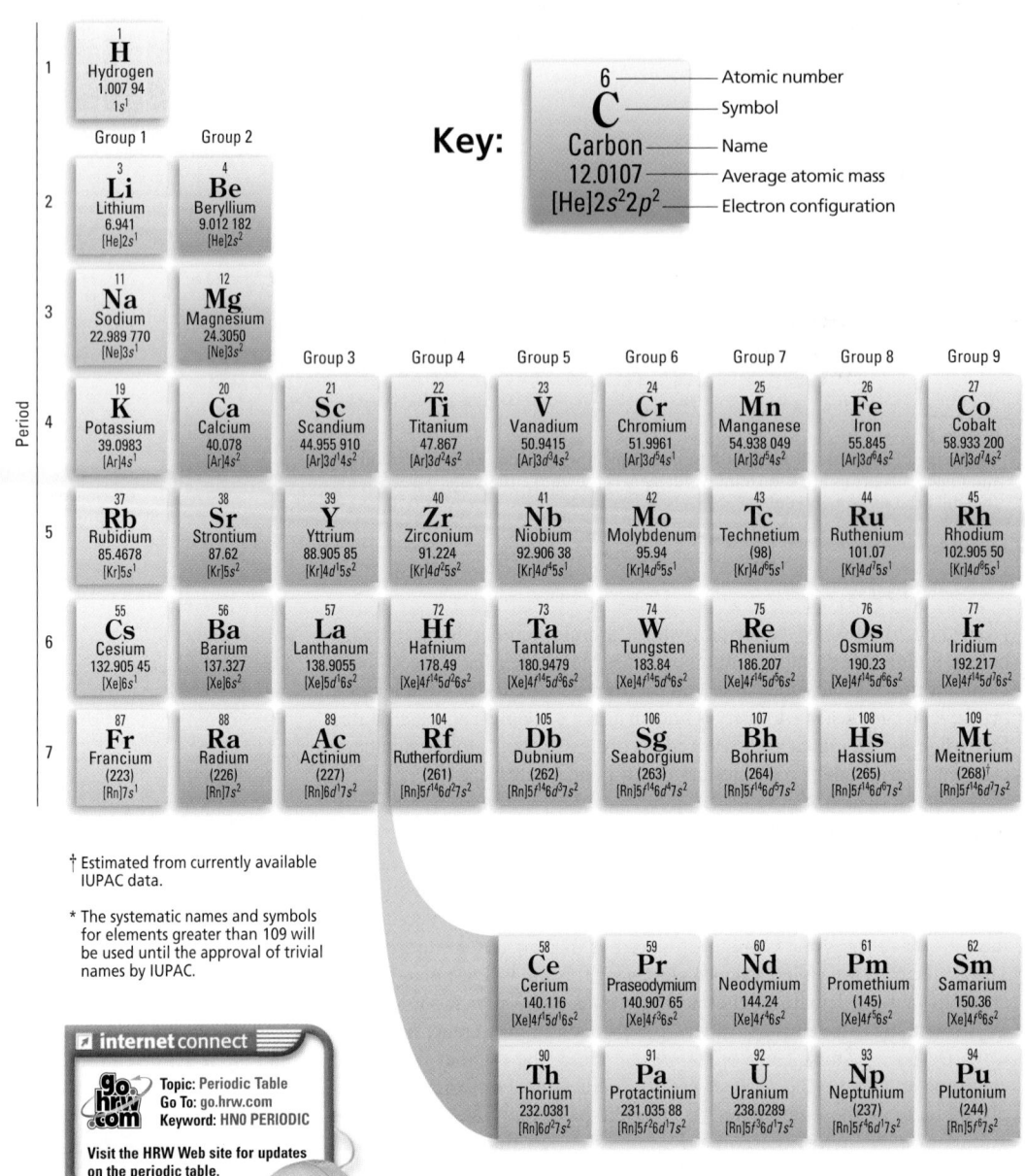

Key:

† Estimated from currently available IUPAC data.

* The systematic names and symbols for elements greater than 109 will be used until the approval of trivial names by IUPAC.

internet connect

Topic: Periodic Table
Go To: go.hrw.com
Keyword: HN0 PERIODIC

Visit the HRW Web site for updates on the periodic table.

120

MISCONCEPTION ALERT

Periodic Table Information Students sometimes think that they will have to memorize the information in the table. Point out that the table exists as a scientific tool and that its purpose is to give information. The student's job is to learn to read and interpret the information on the table, *not* to memorize it.

Metals

- Alkali metals
- Alkaline-earth metals
- Transition metals
- Other metals

Nonmetals

- Hydrogen
- Semiconductors (also known as *metalloids*)
- Other nonmetals
- Halogens
- Noble gases

Figure 4

Group 18

| 2 |
| He |
| Helium |
| 4.002 602 |
| $1s^2$ |

| Group 13 | Group 14 | Group 15 | Group 16 | Group 17 |

5	6	7	8	9	10
B	C	N	O	F	Ne
Boron	Carbon	Nitrogen	Oxygen	Fluorine	Neon
10.811	12.0107	14.006 74	15.9994	18.998 4032	20.1797
$[He]2s^22p^1$	$[He]2s^22p^2$	$[He]2s^22p^3$	$[He]2s^22p^4$	$[He]2s^22p^5$	$[He]2s^22p^6$

13	14	15	16	17	18
Al	Si	P	S	Cl	Ar
Aluminum	Silicon	Phosphorus	Sulfur	Chlorine	Argon
26.981 538	28.0855	30.973 761	32.066	35.4527	39.948
$[Ne]3s^23p^1$	$[Ne]3s^23p^2$	$[Ne]3s^23p^3$	$[Ne]3s^23p^4$	$[Ne]3s^23p^5$	$[Ne]3s^23p^6$

| Group 10 | Group 11 | Group 12 |

28	29	30	31	32	33	34	35	36
Ni	Cu	Zn	Ga	Ge	As	Se	Br	Kr
Nickel	Copper	Zinc	Gallium	Germanium	Arsenic	Selenium	Bromine	Krypton
58.6934	63.546	65.39	69.723	72.61	74.921 60	78.96	79.904	83.80
$[Ar]3d^84s^2$	$[Ar]3d^{10}4s^1$	$[Ar]3d^{10}4s^2$	$[Ar]3d^{10}4s^24p^1$	$[Ar]3d^{10}4s^24p^2$	$[Ar]3d^{10}4s^24p^3$	$[Ar]3d^{10}4s^24p^4$	$[Ar]3d^{10}4s^24p^5$	$[Ar]3d^{10}4s^24p^6$

46	47	48	49	50	51	52	53	54
Pd	Ag	Cd	In	Sn	Sb	Te	I	Xe
Palladium	Silver	Cadmium	Indium	Tin	Antimony	Tellurium	Iodine	Xenon
106.42	107.8682	112.411	114.818	118.710	121.760	127.60	126.904 47	131.29
$[Kr]4d^{10}5s^0$	$[Kr]4d^{10}5s^1$	$[Kr]4d^{10}5s^2$	$[Kr]4d^{10}5s^25p^1$	$[Kr]4d^{10}5s^25p^2$	$[Kr]4d^{10}5s^25p^3$	$[Kr]4d^{10}5s^25p^4$	$[Kr]4d^{10}5s^25p^5$	$[Kr]4d^{10}5s^25p^6$

78	79	80	81	82	83	84	85	86
Pt	Au	Hg	Tl	Pb	Bi	Po	At	Rn
Platinum	Gold	Mercury	Thallium	Lead	Bismuth	Polonium	Astatine	Radon
195.078	196.966 55	200.59	204.3833	207.2	208.980 38	(209)	(210)	(222)
$[Xe]4f^{14}5d^96s^1$	$[Xe]4f^{14}5d^{10}6s^1$	$[Xe]4f^{14}5d^{10}6s^2$	$[Xe]4f^{14}5d^{10}6s^26p^1$	$[Xe]4f^{14}5d^{10}6s^26p^2$	$[Xe]4f^{14}5d^{10}6s^26p^3$	$[Xe]4f^{14}5d^{10}6s^26p^4$	$[Xe]4f^{14}5d^{10}6s^26p^5$	$[Xe]4f^{14}5d^{10}6s^26p^6$

110*	111*	112*		114*
Uun	Uuu	Uub		Uuq
Unununium	Unununium	Ununbium		Ununquadium
(269)†	(272)†	(277)†		(285)†
$[Rn]5f^{14}6d^97s^1$	$[Rn]5f^{14}6d^{10}7s^1$	$[Rn]5f^{14}6d^{10}7s^2$		$[Rn]5f^{14}6d^{10}7s^27p^2$

A team at Lawrence Berkeley National Laboratories reported the discovery of elements 116 and 118 in June 1999. The same team retracted the discovery in July 2001. The discovery of element 114 has been reported but not confirmed.

63	64	65	66	67	68	69	70	71
Eu	Gd	Tb	Dy	Ho	Er	Tm	Yb	Lu
Europium	Gadolinium	Terbium	Dysprosium	Holmium	Erbium	Thulium	Ytterbium	Lutetium
151.964	157.25	158.925 34	162.50	164.930 32	167.26	168.934 21	173.04	174.967
$[Xe]4f^76s^2$	$[Xe]4f^75d^16s^2$	$[Xe]4f^96s^2$	$[Xe]4f^{10}6s^2$	$[Xe]4f^{11}6s^2$	$[Xe]4f^{12}6s^2$	$[Xe]4f^{13}6s^2$	$[Xe]4f^{14}6s^2$	$[Xe]4f^{14}5d^16s^2$

95	96	97	98	99	100	101	102	103
Am	Cm	Bk	Cf	Es	Fm	Md	No	Lr
Americium	Curium	Berkelium	Californium	Einsteinium	Fermium	Mendelevium	Nobelium	Lawrencium
(243)	(247)	(247)	(251)	(252)	(257)	(258)	(259)	(262)
$[Rn]5f^77s^2$	$[Rn]5f^76d^17s^2$	$[Rn]5f^97s^2$	$[Rn]5f^{10}7s^2$	$[Rn]5f^{11}7s^2$	$[Rn]5f^{12}7s^2$	$[Rn]5f^{13}7s^2$	$[Rn]5f^{14}7s^2$	$[Rn]5f^{14}6d^17s^2$

The atomic masses listed in this table reflect the precision of current measurements. (Values listed in parentheses are those of the element's most stable or most common isotope.) In calculations throughout the text, however, atomic masses have been rounded to two places to the right of the decimal.

121

Using the Figure

If you have a large periodic table on display in your classroom, you may want to have students compare it with the one on this page, especially if the information on the two tables differs in some way. Even if the tables differ in appearance, remind students that the arrangement of elements is the same in both tables. Keep in mind that the answers to problems in this book depend on the atomic masses from the table shown here.

SKILL BUILDER

Vocabulary Have students examine the key on the student page. Point out that words can indicate function. Chemistry texts generally apply one of three different words—*semimetal*, *metalloid*, or *semiconductor*—to the indicated elements along the stairstep line. All these terms describe the elements present in this area. We have chosen to use the word *semiconductor* to characterize these elements because it describes the property for which they are best known.

One-Stop Planner CD-ROM

- **Supplemental Reading Projects Guided Reading Worksheet:** *The Periodic Kingdom* by P.W. Atkins.

REAL-WORLD CONNECTION

Jewelry Make students aware that jewelry is made from alloys of gold, silver, or platinum because the pure metals are too soft to be durable. Sterling silver is an alloy containing 7.5% copper. Gold and platinum will not corrode in the air, but sterling silver will corrode slowly.

INCLUSION Strategies

- *Developmentally Delayed*   • *Learning Disabled*

Using blank note cards, create flash cards with the element name on the front and a key on the back including the atomic number, symbol, property, and a example of a common substance containing the element. Each member of a small group could be responsible for one or two of the groups or periods. Students may use the cards to study the elements or line them up on a desk to represent groups and periods of the periodic table.

Reteaching ━━━━━━ **BASIC**

Ask students to explain why the periodic table is such a valuable tool for scientists. Responses might include that the table organizes elements in such a way that scientists can predict properties of elements they might not be familiar with.
LS Logical

Quiz ━━━━━━━━━━ **GENERAL**

1. Mendeleev arranged the elements in his periodic table according to what property? Ans. atomic mass

2. Write the symbols and atomic numbers for silver, radon, and zinc. Ans. Ag, 47; Rn, 86; Zn, 30

3. Which of the elements from question 2 has the greatest atomic mass? Which has the lowest? Ans. radon, zinc

4. Give the group and period number for each of the elements in question 2. Ans. in order: Period 5, Group 11; Period 6, Group 18; Period 4, Group 12
LS Logical

Chapter Resource File

• Concept Review
• Quiz

---

period

a horizontal row of elements in the periodic table

A horizontal row on the periodic table is called a **period.** Elements in the same period have the same number of occupied energy levels. For example, all elements in Period 2 have atoms whose electrons occupy two principal energy levels, including the $2s$ and $2p$ orbitals. Elements in Period 5 have outer electrons that fill the $5s$, $5d$, and $5p$ orbitals.

This correlation between period number and the number of occupied energy levels holds for all seven periods. So a periodic table is not needed to tell to which period an element belongs. All you need to know is the element's electron configuration. For example, germanium has the electron configuration $[Ar]3d^{10}4s^24p^2$. The largest principal quantum number it has is 4, which means germanium has four occupied energy levels. This places it in Period 4.

The periodic table provides information about each element, as shown in the key for **Figure 4.** This periodic table lists the atomic number, symbol, name, average atomic mass, and electron configuration in shorthand form for each element.

In addition, some of the categories of elements are designated through a color code. You may notice that many of the color-coded categories shown in **Figure 4** are associated with a certain group or groups. This shows how categories of elements are grouped by common properties which result from their common number of valence electrons. The next section discusses the different kinds of elements on the periodic table and explains how their electron configurations give them their characteristic properties.

---

① Section Review

UNDERSTANDING KEY IDEAS

1. How can one show that elements that have different appearances have similar chemical properties?

2. Why was the pattern that Newlands developed called the *law of octaves*?

3. What led Mendeleev to predict that some elements had not yet been discovered?

4. What contribution did Moseley make to the development of the modern periodic table?

5. State the periodic law.

6. What do elements in the same period have in common?

7. What do elements in the same group have in common?

CRITICAL THINKING

8. Why can Period 1 contain a maximum of two elements?

9. In which period and group is the element whose electron configuration is $[Kr]5s^1$?

10. Write the outer electron configuration for the Group 2 element in Period 6.

11. What determines the number of elements found in each period in the periodic table?

12. Are elements with similar chemical properties more likely to be found in the same period or in the same group? Explain your answer.

13. How many valence electrons does phosphorus have?

14. What would you expect the electron configuration of element 113 to be?

Answers to Section Review

1. Reacting different elements with the same reactant and observing that they produce similar compounds shows that they have similar chemical properties.

2. The prefix *oct-* refers to "eight." The law of octaves deals with groups of eight elements.

3. In order to place elements in groups having similar properties, Mendeleev left gaps in his periodic table where no known element could be placed according to both atomic mass and chemical properties. He predicted that these gaps would be filled by elements yet to be discovered.

4. Moseley arranged elements by atomic number instead of atomic mass.

5. When elements are arranged according to their atomic numbers, elements with similar properties appear at regular intervals.

6. They have the same number of occupied energy levels.

7. They have the same number of valence electrons.

8. The first energy level of an atom can contain only two electrons.

9. Period 5, Group 1

Answers continued on p. 155A

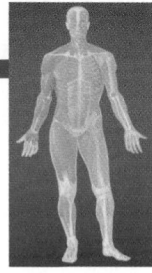

CONSUMER FOCUS

Essential Elements

Four elements—hydrogen, oxygen, carbon, and nitrogen—account for more than 99% of all atoms in the human body.

Good Health Is Elementary

Hydrogen, oxygen, carbon, and nitrogen are the major components of the many different molecules that our bodies need. Likewise, these elements are the major elements in the molecules of the food that we eat.

Another seven elements, listed in **Table 2,** are used by our bodies in substantial quantities, more than 0.1 g per day. These elements are known as *macronutrients* or, more commonly, as *minerals*.

Some elements, known as *trace elements* or *micronutrients*, are necessary for healthy human

life, but only in very small amounts. In many cases, humans need less than 15 nanograms, or 15×10^{-9} g, of a particular trace element per day to maintain good health. This means that you need less than 0.0004 g of such trace elements during your entire lifetime!

Table 2	Macronutrients	
Element	**Symbol**	**Role in human body chemistry**
Calcium	Ca	bones, teeth; essential for blood clotting and muscle contraction
Phosphorus	P	bones, teeth; component of nucleic acids, including DNA
Potassium	K	present as K⁺ in all body fluids; essential for nerve action
Sulfur	S	component of many proteins; essential for blood clotting
Chlorine	Cl	present as Cl⁻ in all body fluids; important to maintaining salt balance
Sodium	Na	present as Na⁺ in all body fluids; essential for nerve and muscle action
Magnesium	Mg	in bones and teeth; essential for muscle action

Questions

1. What do the two macronutrients involved in nerve action have in common?
2. You may recognize elements such as arsenic and lead as toxic. Explain how these elements can be nutrients even though they are toxic.

Periodic table with legend:
- Elements in organic matter
- Macronutrients
- Trace elements

123

CONSUMER FOCUS

Essential Elements

In addition to the elements listed in the table, other essential trace elements are iodine, cobalt, nickel, tin, silicon, fluorine, zinc, iron, copper, manganese, molybdenum, chromium, and selenium. The exact function of some of these elements is not known.

- An adequate supply of zinc is needed for proper growth and for wound healing. It is also needed in molecules that help your body eliminate carbon dioxide and digest protein. Women who are lactating require almost twice the normal amount of zinc. Sources of zinc include meat, yeast, sunflower seeds, and especially oysters.

- Copper is known to be needed by infants and is assumed to be necessary for adults. It is found in materials needed for cellular respiration.

- Chromium works with insulin to metabolize glucose and is thought to be beneficial to diabetics.

- Cobalt is part of vitamin B-12, cyanocobalamin, and is needed for the development of red blood cells.

- Iron is contained in hemoglobin molecules. Hemoglobin carries oxygen in the blood throughout the body. When the hemoglobin is low in the blood, a person might be given an iron supplement so that the body can produce more hemoglobin.

Answers to the Feature Questions

1. Sodium and potassium are both members of Group 1 of the periodic table.
2. Trace elements are considered nutrients in amounts that are too small for them to be toxic.

HEALTH

CONNECTION — **GENERAL**

Have students examine the label on a bottle of common multivitamins. Have them identify the elements listed, the amount present in the multivitamin, and what percent of the amount of the element recommended daily is present in a tablet. Ask students under what situations they might need to take a multivitamin instead of ingesting all of the needed element from their diet. Answers might include dietary restrictions on certain foods, allergies to certain foods, or that certain elements are not found in many different foods.

LS Intrapersonal

INCLUSION Strategies

• *Developmentally Delayed* • *Learning Disabled*

Have students research a variety of product labels to find nutritional information. List products in a chart indicating which elements are present in the "Nutritional Facts" portion of the label. Students may rank the products according to their nutritional daily value. Students may further plan meals that would meet one hundred percent of the daily requirements for some of the elements.

Overview

Before beginning this section, review with your students the Objectives listed in the Student Edition. This section examines the different families of main-group elements on the periodic table. Students will learn to relate the properties of these elements to their electron configurations. The characteristic properties and electron configurations of metals are also discussed.

 **Bellringer**

Provide students with a blank periodic table and ask them to label each group by the configuration of its valence electrons, assuming the configuration follows the pattern given by the aufbau principle, covered in the chapter "Atoms and Moles." Emphasize that this pattern applies to all the main-group elements, but there are many exceptions in the transition metals in the center of the table.

Motivate

Discussion

Point out that the main group elements contain some metals, some nonmetals, and some semiconductors. Have students examine the periodic table. Use the key to point out the metals, the nonmetals, and the semiconductors. Discuss why some elements along the stairstep line are semiconductor, and some are not. For example, aluminum is a good conductor of electric current and heat. Therefore, it is best classified as a metal even though it does exhibit some nonmetallic chemical properties.

KEY TERMS

- main-group element
- alkali metal
- alkaline-earth metal
- halogen
- noble gas
- transition metal
- lanthanide
- actinide
- alloy

main-group elements

an element in the *s*-block or *p*-block of the periodic table

OBJECTIVES

① **Locate** the different families of main-group elements on the periodic table, describe their characteristic properties, and relate their properties to their electron configurations.

② **Locate** metals on the periodic table, describe their characteristic properties, and relate their properties to their electron configurations.

The Main-Group Elements

Elements in groups 1, 2, and 13–18 are known as the **main-group elements.** As shown in **Figure 5,** main-group elements are in the *s*- and *p*-blocks of the periodic table. The electron configurations of the elements in each main group are regular and consistent: the elements in each group have the same number of valence electrons. For example, Group 2 elements have two valence electrons. The configuration of their valence electrons can be written as ns^2, where *n* is the period number. Group 16 elements have a total of six valence electrons in their outermost *s* and *p* orbitals. Their valence electron configuration can be written as ns^2np^4.

The main-group elements are sometimes called the *representative elements* because they have a wide range of properties. At room temperature and atmospheric pressure, many are solids, while others are liquids or gases. About half of the main-group elements are metals. Many are extremely reactive, while several are nonreactive. The main-group elements silicon and oxygen account for four of every five atoms found on or near Earth's surface.

Four groups within the main-group elements have special names. These groups are the *alkali metals* (Group 1), the *alkaline-earth metals* (Group 2), the *halogens* (Group 17), and the *noble gases* (Group 18).

Figure 5
Main-group elements have diverse properties and uses. They are highlighted in the groups on the left and right sides of the periodic table.

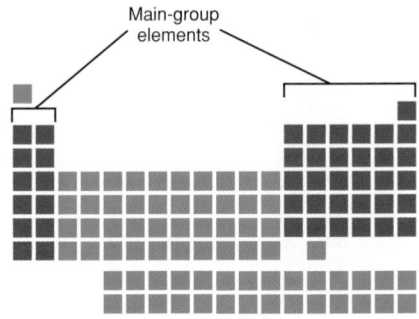

Main-group elements

124

did you know?

Although mercury is the only metal that is liquid at room temperature, three other metals—Rb, Cs, and Ga—have melting points not far above room temperature.

Chapter Resource File

- Lesson Plan

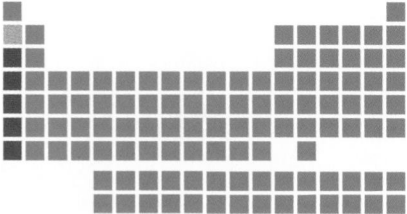

Figure 6
The alkali metals make up the first group of the periodic table. Lithium, pictured here, is an example of an alkali metal.

The Alkali Metals Make Up Group 1

Elements in Group 1, which is highlighted in **Figure 6,** are called **alkali metals.** Alkali metals are so named because they are metals that react with water to make alkaline solutions. For example, potassium reacts vigorously with cold water to form hydrogen gas and the compound potassium hydroxide, KOH. Because the alkali metals have a single valence electron, they are very reactive. In losing its one valence electron, potassium achieves a stable electron configuration.

Alkali metals are usually stored in oil to keep them from reacting with the oxygen and water in the air. Because of their high reactivity, alkali metals are never found in nature as pure elements but are found combined with other elements as compounds. For instance, the salt sodium chloride, NaCl, is abundant in sea water.

Some of the physical properties of the alkali metals are listed in **Table 3.** All these elements are so soft that they can be easily cut with a knife. The freshly cut surface of an alkali metal is shiny, but it dulls quickly as the metal reacts with oxygen and water in the air. Like other metals, the alkali metals are good conductors of electricity.

alkali metal

one of the elements of Group 1 of the periodic table (lithium, sodium, potassium, rubidium, cesium, and francium)

internet connect
www.scilinks.org
Topic: Alkali Metals
SciLinks code: HW4007

SciLINKS Maintained by the National Science Teachers Association

Table 3	Physical Properties of Alkali Metals					
Element	**Flame test**	**Hardness (Mohs' scale)**	**Melting Point (°C)**	**Boiling Point (°C)**	**Density (g/cm³)**	**Atomic radius (pm)**
Lithium	red	0.6	180.5	1342	0.53	134
Sodium	yellow	0.4	97.7	883	0.97	154
Potassium	violet	0.5	63.3	759	0.86	196
Rubidium	yellowish violet	0.3	39.3	688	1.53	(216)
Cesium	reddish violet	0.2	28.4	671	1.87	(233)

Refer to Appendix A for more information about the properties of elements, including alkali metals.

125

did you know?

Sodium as a Conductor Because sodium is a lightweight electrical conductor, it has been incorporated into high-voltage power lines by encasing it completely with aluminum or another metal. Sodium's low density reduces the weight of the cables without impairing their conductivity. The light weight reduces sag and permits longer cable spans.

Teach

Teaching Tips
Main-Group Elements Inform students that the main-group elements are elements whose outer-level electrons are in *s* orbitals or in *s* and *p* orbitals.

Alkali Metals If you have stock bottles containing Li, Na, and K, exhibit them to students and identify the oil that the metals are submerged in. The metals are too reactive to be exposed to air for any length of time. It is likely that the metal pieces have already become encrusted with oxidation products formed slowly over time.

Demonstration
Orange-Yellow Light of Sodium
Darken the room, and use a pair of crucible tongs to hold a piece of calcium sulfate in the burner flame. Calcium sulfate is gypsum, the mineral filler of drywall board. Note the low color production. Next, use the tongs to hold a piece of rock salt in the flame. Have students note the bright orange-yellow flame typical of sodium. As an option, place an opaque shield with a slit between the class and the burner, and pass out replica diffraction gratings to students. Hold the salt in the flame again while students view sodium's emission spectrum through the gratings. Students should note the pair of strong yellow lines typical of sodium.

Safety: Wear safety goggles and a lab apron. Students should be at least 3 m away from the demonstration because the rock salt could splatter.

Disposal: Put cooled specimens in the trash.

Teach, continued

Demonstration

Reactivities of Mg and Ca Place two Petri dishes half-full of room-temperature water on the overhead projector. Add two drops of phe-nolphthalein indicator to each dish. Explain to students that phenolph-thalein changes from colorless to pink in the presence of a base. Polish a 2-cm piece of magnesium ribbon with sandpaper, and imme-diately drop it into one of the dishes. Drop a pea-sized (or smaller) piece of calcium into the other dish. Be sure that both the Mg and Ca are submerged. Students should observe bubbles and a pink color begin to form around the calcium. Inform stu-dents that calcium reacts with water, but much more slowly than an alkali metal would. The prod-ucts are hydrogen gas and calcium hydroxide, a base. Students may or may not observe activity around the Mg sample. If nothing is observed, replace the room tem-perature water with boiling water. Add phenolphthalein. Magnesium should be about as reactive in hot water as calcium is in cold.

Safety: Wear safety goggles and a lab apron while handling chemicals.

Disposal: Set out a disposal con-tainer for any used indicator solu-tions that are left over at the end of the procedure. Unused indicators should be tightly covered and re-turned to the storage shelf. Save all metal strips for reuse next time, but if they are too corroded, put them in the trash. Put all other solids in the trash. Combine all liquids, and pour them down the drain.

Figure 7
The alkaline-earth metals make up the second group of the periodic table. Magnesium, pictured here, is an example of an alkaline-earth metal.

alkaline-earth metal

one of the elements of Group 2 of the periodic table (beryllium, magnesium, calcium, strontium, barium, and radium)

internet connect
www.scilinks.org
Topic: Alkaline-Earth Metals
SciLinks code: HW4008
SCiLINKS Maintained by the National Science Teachers Association

halogen

one of the elements of Group 17 of the periodic table (fluorine, chlorine, bromine, iodine, and astatine); halogens combine with most metals to form salts

internet connect
www.scilinks.org
Topic: Halogens
SciLinks code: HW4065
SCiLINKS Maintained by the National Science Teachers Association

126

The Alkaline-Earth Metals Make Up Group 2

Group 2 elements, which are highlighted in **Figure 7,** are called **alkaline-earth metals.** Like the alkali metals, the alkaline-earth metals are highly reactive, so they are usually found as compounds rather than as pure ele-ments. For example, if the surface of an object made from magnesium is exposed to the air, the magnesium will react with the oxygen in the air to form the compound magnesium oxide, MgO, which eventually coats the surface of the magnesium metal.

The alkaline-earth metals are slightly less reactive than the alkali metals. The alkaline-earth metals have two valence electrons and must lose both their valence electrons to get to a stable electron configuration. It takes more energy to lose two electrons than it takes to lose just the one electron that the alkali metals must give up to become stable. Although the alkaline-earth metals are not as reactive, they are harder and have higher melting points than the alkali metals.

Beryllium is found in emeralds, which are a variety of the mineral beryl. Perhaps the best-known alkaline-earth metal is calcium, an impor-tant mineral nutrient found in the human body. Calcium is essential for muscle contraction. Bones are made up of calcium phosphate. Calcium compounds, such as limestone and marble, are common in the Earth's crust. Marble is made almost entirely of pure calcium carbonate. Because marble is hard and durable, it is used in sculptures.

The Halogens, Group 17, Are Highly Reactive

Elements in Group 17 of the periodic table, which are highlighted in **Figure 8** on the next page, are called the **halogens.** The halogens are the most reactive group of nonmetal elements because of their electron con-figuration. Halogens have seven valence electrons—just one short of a stable configuration. When halogens react, they often gain the one elec-tron needed to have eight valence electrons, a filled outer energy level. Because the alkali metals have one valence electron, they are ideally suited to react with the halogens. For example, the alkali metal sodium easily loses its one valence electron to the halogen chlorine to form the compound sodium chloride, NaCl, which is table salt. The halogens react with most metals to produce salts. In fact, the word *halogen* comes from Greek and means "salt maker."

did you know?

Lightweight Alloys Magnesium and magne-sium alloys are vital to the aerospace industry because of their low densities. Machining mag-nesium parts can be hazardous because the magnesium can catch fire and burn with an intensely hot, blinding white light. Conse-quently, magnesium is often machined in an inert atmosphere.

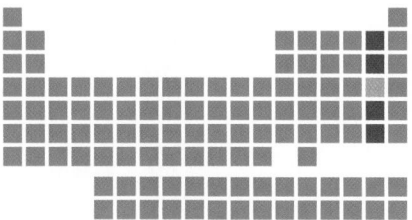

Figure 8
The halogens make up Group 17 of the periodic table. Bromine, one of only two elements that are liquids at room temperature, is an example of a halogen.

The halogens have a wide range of physical properties. Fluorine and chlorine are gases at room temperature, but bromine, depicted in **Figure 8,** is a liquid, and iodine and astatine are solids. The halogens are found in sea water and in compounds found in the rocks of Earth's crust. Astatine is one of the rarest of the naturally occurring elements.

The Noble Gases, Group 18, Are Unreactive

Group 18 elements, which are highlighted in **Figure 9,** are called the **noble gases.** The noble gas atoms have a full set of electrons in their outermost energy level. Except for helium ($1s^2$), noble gases have an outer-shell configuration of ns^2np^6. From the low chemical reactivity of these elements, chemists infer that this full shell of electrons makes these elements very stable. The low reactivity of noble gases leads to some special uses. Helium, a noble gas, is used to fill blimps because it has a low density and is not flammable.

The noble gases were once called *inert gases* because they were thought to be completely unreactive. But in 1962, chemists were able to get xenon to react, making the compound $XePtF_6$. In 1979, chemists were able to form the first xenon-carbon bonds.

internet connect
www.scilinks.org
Topic: Noble Gases
SciLinks code: HW4083

noble gas

an unreactive element of Group 18 of the periodic table (helium, neon, argon, krypton, xenon, or radon) that has eight electrons in its outer level (except for helium, which has two electrons)

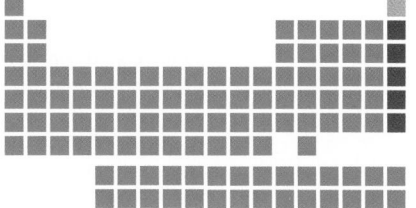

Figure 9
The noble gases make up Group 18 of the periodic table. Helium, whose low density makes it ideal for use in blimps, is an example of a noble gas.

127

did you know?

Noble Colors Excited neon atoms produce only an orange-red light. Other gases or mixtures of gases are used to produce other colors in lights for signs.

HISTORY
CONNECTION

Discovery of Noble Gases Argon was the first noble gas to be identified. Its existence was proposed in 1785 because chemists could not account for all of the major constituents of air. Argon makes up about 1% of air. It was not until 1894 that British chemist William Ramsay identified it. Because of its chemical inertness, it was given the name argon from the Greek *argos*, which means "lazy" or "inactive."

Teaching Tip

Hydrogen Hydrogen's atomic structure explains the difficulty in classifying it. Like the halogens, it can gain one electron to have the electronic configuration of a noble gas, helium. When it does so, it forms H⁻, called the hydride ion. Hydrogen is unlike the alkali metals in that it cannot form a true 1+ ion: H⁺, a bare proton, is not stable in the presence of other matter. Although we sometimes say that acids form H⁺ ions in water solution, what is really formed is a hydrated proton, usually written as H_3O^+, the hydronium ion.

Demonstration

Properties of Metals As an overview of the properties of metals, exhibit objects that incorporate metals. Ask students to contribute to the collection. Try to include samples representing as many metals as possible. Include electronic devices that have motors, magnets, or heating elements. Also include a mirror. Most modern mirrors are coated with aluminum on the back. Ask students to list the properties of metals that are exemplified by these objects.

SKILL BUILDER

Vocabulary The word *ductile* is derived from the Latin verb *ducere*, which means "to lead." Thus, a ductile metal is one that can be led, or pulled, to a smaller diameter. The word *malleable* derives from the Latin noun *malleus*, meaning "hammer." A malleable metal can be hammered until it is thinner. Most metal today is not hammered but rather is rolled and pressed.

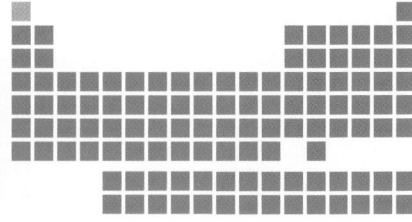

Figure 10
Hydrogen sits apart from all other elements in the periodic table. Hydrogen is extremely flammable and is used as fuel for space shuttle launches.

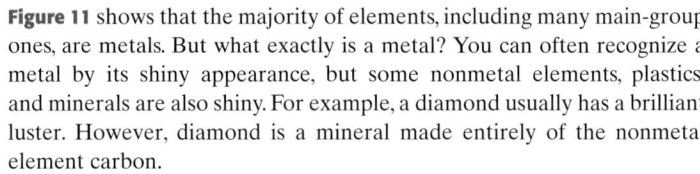

internet connect
www.scilinks.org
Topic: Metals
SciLinks code: HW4079

Hydrogen Is in a Class by Itself

Hydrogen is the most common element in the universe. It is estimated that about three out of every four atoms in the universe are hydrogen. Because it consists of just one proton and one electron, hydrogen behaves unlike any other element. As shown in **Figure 10,** hydrogen is in a class by itself in the periodic table.

With its one electron, hydrogen can react with many other elements, including oxygen. Hydrogen gas and oxygen gas react explosively to form water. Hydrogen is a component of the organic molecules found in all living things. The main industrial use of hydrogen is in the production of ammonia, NH_3. Large quantities of ammonia are used to make fertilizers.

Most Elements Are Metals

Figure 11 shows that the majority of elements, including many main-group ones, are metals. But what exactly is a metal? You can often recognize a metal by its shiny appearance, but some nonmetal elements, plastics, and minerals are also shiny. For example, a diamond usually has a brilliant luster. However, diamond is a mineral made entirely of the nonmetal element carbon.

Conversely, some metals appear black and dull. An example is iron, which is a very strong and durable metal. Iron is a member of Group 8 and is therefore not a main-group element. Iron belongs to a class of elements called *transition metals*. However, wherever metals are found on the periodic table, they tend to share certain properties.

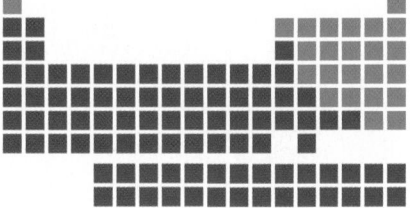

Figure 11
The regions highlighted in blue indicate the elements that are metals.

128

MISCONCEPTION ///ALERT

The Meaning of *Metal* Make sure students distinguish between the meaning of *metal* as applied to individual elements and as applied to metallic objects that we encounter every day, which are almost always alloys: homogeneous mixtures of metals with other metals and nonmetals.

Metals Share Many Properties

All metals are excellent conductors of electricity. Electrical conductivity is the one property that distinguishes metals from the nonmetal elements. Even the least conductive metal conducts electricity 100 000 times better than the best nonmetallic conductor does.

Metals also exhibit other properties, some of which can also be found in certain nonmetal elements. For example, metals are excellent conductors of heat. Some metals, such as manganese and bismuth, are very brittle. Other metals, such as gold and copper, are ductile and malleable. *Ductile* means that the metal can be squeezed out into a wire. *Malleable* means that the metal can be hammered or rolled into sheets. Gold, for example, can be hammered into very thin sheets, called "gold leaf," and applied to objects for decoration.

Transition Metals Occupy the Center of the Periodic Table

The **transition metals** constitute Groups 3 through 12 and are sometimes called the *d*-block elements because of their position in the periodic table, shown in **Figure 12.** Unlike the main-group elements, the transition metals in each group do not have identical outer electron configurations. For example, nickel, Ni, palladium, Pd, and platinum, Pt, are Group 10 metals. However, Ni has the electron configuration $[Ar]3d^8 4s^2$, Pd has the configuration $[Kr]4d^{10}$, and Pt has the configuration $[Xe]4f^{14}5d^9 6s^1$. Notice, however, that in each case the sum of the outer *d* and *s* electrons is equal to the group number, 10.

A transition metal may lose different numbers of valence electrons depending on the element with which it reacts. Generally, the transition metals are less reactive than the alkali metals and the alkaline-earth metals are. In fact, some transition metals are so unreactive that they seldom form compounds with other elements. Palladium, platinum, and gold are among the least reactive of all the elements other than the noble gases. These three transition metals can be found in nature as pure elements.

Transition metals, like other metals, are good conductors of heat and electricity. They are also ductile and malleable, as shown in **Figure 12.**

transition metal

one of the metals that can use the inner shell before using the outer shell to bond

www.scilinks.org
Topic: Transition Metals
SciLinks code: HW4168

SciLINKS Maintained by the National Science Teachers Association

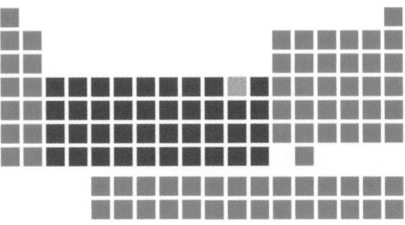

Figure 12
Copper, a transition metal, is used in wiring because it conducts electricity well. Because of its ductility and malleability, it can be formed into wires that bend easily.

129

TECHNOLOGY CONNECTION

Conduction Although silver and copper are the best conductors, gold is also a good conductor but has the advantage of not corroding or tarnishing under ordinary conditions. For this reason, gold is widely used on connectors in computers and other electronic devices.

Uses of Lanthanides The lanthanide elements are increasingly being used in cutting-edge technology. In addition to their use in TV phosphors, lanthanides are being used in microwave technology, new high-strength permanent magnets, and in improved types of lasers.

MISCONCEPTION ALERT

Different Periodic Tables

Students might think that all periodic tables are identical. They should not be concerned if they find a slightly different table in another book. The layout of the table printed in this book is the form most commonly used. For example, chemists have disagreed over the proper way to show the lanthanides and actinides. If the energy order of sublevels were followed strictly, the lanthanum configuration would end in $6s^2 4f^1$, and the actinium configuration would end in $7s^2 5f^1$. For these reasons, some tables show lanthanum and actinium with the other lanthanide and actinide elements. However, the next electron after $6s^2$ is in a $5d$ orbital, and the next two electrons after $7s^2$ are in $6d$ orbitals. Accordingly, cerium is the first lanthanide to have f electrons, and protactinium is the first actinide to have f electrons. The f orbitals become full in Yb and No, so the elements Lu and Lr are sometimes placed in the main body of the table.

---

lanthanide

a member of the rare-earth series of elements, whose atomic numbers range from 58 (cerium) to 71 (lutetium)

actinide

any of the elements of the actinide series, which have atomic numbers from 89 (actinium, Ac) through 103 (lawrencium, Lr)

alloy

a solid or liquid mixture of two or more metals

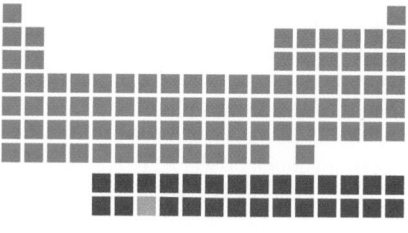

Figure 13
The lanthanides and actinides are placed at the bottom of the periodic table. Uranium, an actinide, is used in nuclear reactors. The collection of uranium-238 kernels is shown here.

Lanthanides and Actinides Fill *f*-orbitals

Part of the last two periods of transition metals are placed toward the bottom of the periodic table to keep the table conveniently narrow, as shown in **Figure 13**. The elements in the first of these rows are called the **lanthanides** because their atomic numbers follow the element lanthanum. Likewise, elements in the row below the lanthanides are called **actinides** because they follow actinium. As one moves left to right along these rows, electrons are added to the 4*f* orbitals in the lanthanides and to the 5*f* orbitals in the actinides. For this reason, the lanthanides and actinides are sometimes called the *f*-block of the periodic table.

The lanthanides are shiny metals similar in reactivity to the alkaline-earth metals. Some lanthanides have practical uses. Compounds of some lanthanide metals are used to produce color television screens.

The actinides are unique in that their nuclear structures are more important than their electron configurations. Because the nuclei of actinides are unstable and spontaneously break apart, all actinides are radioactive. The best-known actinide is uranium.

Other Properties of Metals

The melting points of metals vary widely. Tungsten has the highest melting point, 4322°C, of any element. In contrast, mercury melts at –39°C, so it is a liquid at room temperature. This low melting point, along with its high density, makes mercury useful for barometers.

Metals can be mixed with one or more other elements, usually other metals, to make an **alloy.** The mixture of elements in an alloy gives the alloy properties that are different from the properties of the individual elements. Often these properties eliminate some disadvantages of the pure metal. A common alloy is brass, a mixture of copper and zinc, which is harder than copper and more resistant to corrosion. Brass has a wide range of uses, from inexpensive jewelry to plumbing hardware. Another alloy made from copper is sterling silver. A small amount of copper is mixed with silver to produce sterling silver, which is used for both jewelry and flatware.

130

REAL-WORLD CONNECTION

Mercury poisoning can occur not only from contact with the metal itself but also by breathing mercury vapors or contact with mercury compounds. Methyl mercury compounds caused a large outbreak of poisonings in Japan in the 1950s among people who ate fish and shellfish containing high levels of mercury compounds. Absorbed mercury is concentrated in the kidneys and ultimately results in kidney dysfunction. Death occurs from the accumulation of toxic wastes in the blood.

Videos

 **Presents Physical Science**
- **Feature Story 12** Alloy Technology

See the Science in the News video guide for more details.

Figure 14
Steel is an alloy made of iron and carbon. When heated, steel can be worked into many useful shapes.

Many iron alloys, such as the steel shown in **Figure 14,** are harder, stronger, and more resistant to corrosion than pure iron. Steel contains between 0.2% and 1.5% carbon atoms and often has tiny amounts of other elements such as manganese and nickel. Stainless steel also incorporates chromium. Because of its hardness and resistance to corrosion, stainless steel is an ideal alloy for making knives and other tools.

 Section Review

UNDERSTANDING KEY IDEAS

1. Which group of elements is the most unreactive? Why?

2. Why do groups among the main-group elements display similar chemical behavior?

3. What properties do the halogens have in common?

4. Why is hydrogen set apart by itself?

5. How do the valence electron configurations of the alkali metals compare with each other?

6. Why are the alkaline-earth metals less reactive than the alkali metals?

7. In which groups of the periodic table do the transition metals belong?

8. Why are the nuclear structures of the actinides more important than the electron configurations of the actinides?

9. What is an alloy?

CRITICAL THINKING

10. Noble gases used to be called *inert gases.* What discovery changed that term, and why?

11. If you find an element in nature in its pure elemental state, what can you infer about the element's chemical reactivity?

12. Explain why the transition metals are sometimes referred to as the *d*-block elements.

13. Can an element that conducts heat, is malleable, and has a high melting point be classified as a metal? Explain your reasoning.

131

Focus

Overview

Before beginning this section, review with your students the Objectives listed in the Student Edition. This section examines trends in properties that can be predicted by the element's location on the periodic table, and thus its atomic structure. Such properties include ionization energy, atomic radius, electronegativity, ionic size, electron affinity, and melting and boiling points.

Bellringer

Have students draw atomic models of lithium, magnesium, and fluorine. From the models, have them predict whether the ions of these elements will be larger or smaller than the atoms. Be sure they can justify their predictions. Ans. Lithium and magnesium will lose all the electrons in their valence levels, so the ions will be smaller. Fluorine will gain an electron. Students should predict that the fluoride ion will be either the same size or larger.

LS Visual

Motivate

Discussion

Refer students to **Figure 15.** Inform students that rubidium and cesium react violently with water, exploding upon contact. Very little is known of francium, which is radioactive, but presumably its reaction with water would be even more violent. Ask students which alkali metal is the most active. Even if they cannot answer the question, it sets the stage for the section.

Trends in the Periodic Table

KEY TERMS

- **ionization energy**
- **electron shielding**
- **bond radius**
- **electronegativity**

OBJECTIVES

1 **Describe** periodic trends in ionization energy, and relate them to the atomic structures of the elements.

2 **Describe** periodic trends in atomic radius, and relate them to the atomic structures of the elements.

3 **Describe** periodic trends in electronegativity, and relate them to the atomic structures of the elements.

4 **Describe** periodic trends in ionic size, electron affinity, and melting and boiling points, and relate them to the atomic structures of the elements.

Periodic Trends

The arrangement of the periodic table reveals trends in the properties of the elements. A *trend* is a predictable change in a particular direction. For example, there is a trend in the reactivity of the alkali metals as you move down Group 1. As **Figure 15** illustrates, each of the alkali metals reacts with water. However, the reactivity of the alkali metals varies. At the top of Group 1, lithium is the least reactive, sodium is more reactive, and potassium is still more reactive. In other words, there is a trend toward greater reactivity as you move down the alkali metals in Group 1.

Understanding a trend among the elements enables you to make predictions about the chemical behavior of the elements. These trends in properties of the elements in a group or period can be explained in terms of electron configurations.

Figure 15
Chemical reactivity with water increases from top to bottom for Group 1 elements. Reactions of lithium, sodium, and potassium with water are shown.

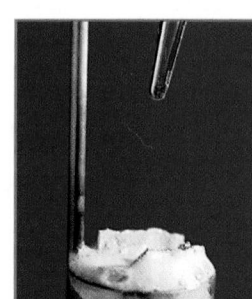
Lithium

Sodium

Potassium

132

REAL-LIFE ─
CONNECTION ── **BASIC**

Periodicity Have students name events that occur periodically, that is, in a regular, repeating pattern. Examples might include the days of the week, seasons, time of day, or a regular music lesson that occurs once a week. You could use a calendar to point out how each day of the week occurs at the same time in relation to the other days and how each month occurs at the same time during each year.

LS Logical

Chapter Resource File

• Lesson Plan

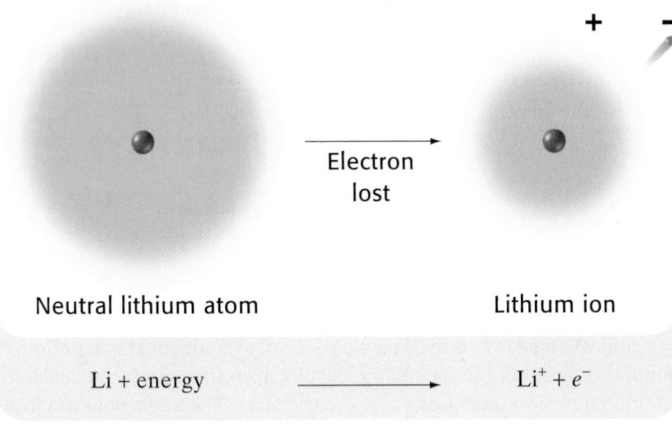

Electron
lost

Neutral lithium atom   Lithium ion

Li + energy $\longrightarrow$ Li$^+$ + e^-

Figure 16
When enough energy is
supplied, a lithium atom
loses an electron and
becomes a positive ion.
The ion is positive because
its number of protons now
exceeds its number of
electrons by one.

Ionization Energy

When atoms have equal numbers of protons and electrons, they are electrically neutral. But when enough energy is added, the attractive force between the protons and electrons can be overcome. When this happens, an electron is removed from an atom. The neutral atom then becomes a positively charged ion.

Figure 16 illustrates the removal of an electron from an atom. The energy that is supplied to remove an electron is the **ionization energy** of the atom. This process can be described as shown below.

A + ionization energy →   A$^+$ + e^-

neutral atom   ion   electron

ionization energy

the energy required to remove
an electron from an atom or ion

Ionization Energy Decreases as You Move Down a Group

Ionization energy tends to decrease down a group, as **Figure 17** on the next page shows. Each element has more occupied energy levels than the one above it has. Therefore, the outermost electrons are farthest from the nucleus in elements near the bottom of a group.

Similarly, as you move down a group, each successive element contains more electrons in the energy levels between the nucleus and the outermost electrons. These inner electrons shield the outermost electrons from the full attractive force of the nucleus. This **electron shielding** causes the outermost electrons to be held less tightly to the nucleus.

Notice in **Figure 18** on the next page that the ionization energy of potassium is less than that of lithium. The outermost electrons of a potassium atom are farther from its nucleus than the outermost electrons of a lithium atom are from their nucleus. So, the outermost electrons of a lithium atom are held more tightly to its nucleus. As a result, removing an electron from a potassium atom takes less energy than removing one from a lithium atom.

electron shielding

the reduction of the attractive
force between a positively
charged nucleus and its
outermost electrons due to
the cancellation of some of the
positive charge by the negative
charges of the inner electrons

(133)

Reading Organizer If students have been keeping notes on a blank periodic table, this would be a good time to conduct a short review of the relevant things that should be included. Many students may want to condense their notes on a fresh copy of the table or add the notes from this section on a new copy. **LS** Verbal

Using the Figure — GENERAL

Ask students whether they know of any trends in chemical or physical properties of the elements that are related to the arrangement of the periodic table. Students might discover that they already know about the progression from metals to semiconductors to nonmetals moving from left to right across the table. Have them look at Groups 13 through 16 to infer the trend in metallic properties from top to bottom in a group. (Metallic properties increase.) The combination of these two trends from right to left and top to bottom gives the dividing line between metals and nonmetals its stair-step shape.

Ask students to speculate about what might happen if the lithium atom in **Figure 16** were in the presence of a fluorine atom. **Ans.** Lithium would lose an electron, which would be added to the fluorine. **LS** Logical

Teaching Tip

Ionization Energy Before having the students read, ask them to consider how ionization energy might change from top to bottom in a group and from left to right across a period. Ask them to suggest reasons for their conclusions. Remind students of how atomic radii change in a group of elements. Note that all of the ionization energies given here are first ionization energies.

PHYSICS
CONNECTION — **BASIC**

Students are aware of how certain metals, particularly iron, are attracted to a magnet. Show the students a situation that is analogous to the shielding effect shown in atoms. Attach a magnet to a piece of iron. The magnet represents the nucleus and the iron represents electrons. Take another magnet and place several pieces of iron on top of it, on top of each other. Try this activity before class so that you can choose the number of pieces that will reduce the attraction of the outer piece of iron and the magnet, but the attraction is strong enough to be noticed. Then have a student attempt to pull the outer pieces of iron away from the magnets. Students should notice that the attraction is still present when the iron is shielded from the magnet, but it is less and it takes less energy to remove the iron. **LS** Kinesthetic

First Ionization Energy For multielectron atoms, more than one ionization energy exists. The ionization energy discussed in the text is the first ionization energy, which is the energy necessary to remove the first electron from a neutral atom. Once one electron is removed, the atom is then positively charged, and it is more difficult to remove a second electron. Thus, ionization energies become progressively greater as more than one electron is removed from an atom. For example, the first ionization energy of Be is 900 kJ/mol, but the second ionization energy is 1760 kJ/mol.

Homework ────── **ADVANCED**

Ionization Energies Provide students with the first five ionization energies for carbon. They are, in order, 1086 kJ/mol, 2353 kJ/mol, 4621 kJ/mol, 6223 kJ/mol, and 37 831 kJ/mol. Have each student write an explanation of why the fifth ionization energy of carbon is so much greater than the previous figures. **Ans.** After losing the first four electrons, carbon has a noble gas configuration, after which losing an extra electron takes extra energy. From their explanations, have students predict which ionization energy for calcium will show a large jump from the previous value. **Ans.** The third ionization energy will show a large jump because calcium has a noble gas configuration after losing two electrons. **LS Logical**

Figure 17
Ionization energy generally decreases down a group and increases across a period, as shown in this diagram. Darker shading indicates higher ionization energy.

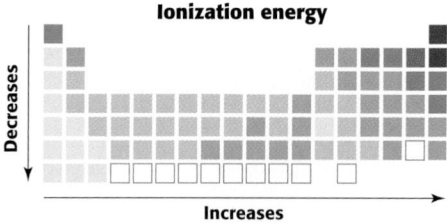

Ionization Energy Increases as You Move Across a Period

Ionization energy tends to increase as you move from left to right across a period, as **Figure 17** shows. From one element to the next in a period, the number of protons and the number of electrons increase by one each. The additional proton increases the nuclear charge. The additional electron is added to the same outer energy level in each of the elements in the period. A higher nuclear charge more strongly attracts the outer electrons in the same energy level, but the electron-shielding effect from inner-level electrons remains the same. Thus, more energy is required to remove an electron because the attractive force on them is higher.

Figure 18 shows that the ionization energy of neon is almost four times greater than that of lithium. A neon atom has 10 protons in its nucleus and 10 electrons filling two energy levels. In contrast, a lithium atom has 3 protons in its nucleus and 3 electrons distributed in the same two energy levels as those of neon. The attractive force between neon's 10 protons and 10 electrons is much greater than that between lithium's 3 protons and 3 electrons. As a result, the ionization energy of neon is much higher than that of lithium.

Figure 18
Ionization energies for hydrogen and for the main-group elements of the first four periods are plotted on this graph.

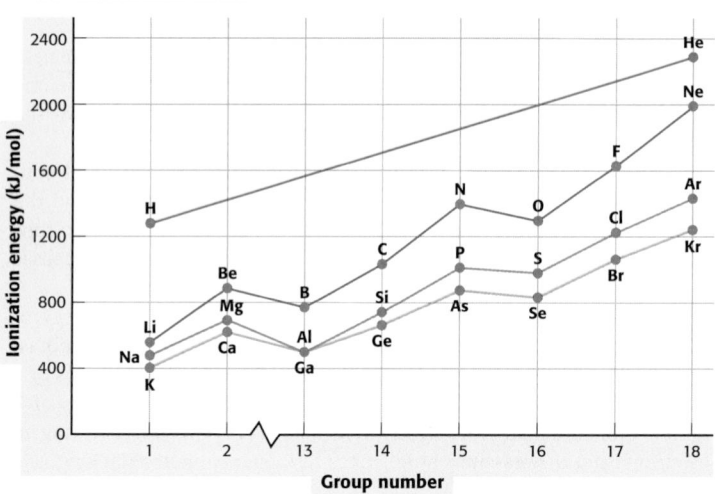

MISCONCEPTION ALERT

Losing Electrons and Charge Students are used to values decreasing when items are taken away from a whole quantity, or increasing when items are added. They might mistakenly think atoms should become more negative as they lose electrons and more positive as they gain them. *Remind students that charge convention considers an electron to have a* negative *charge.*

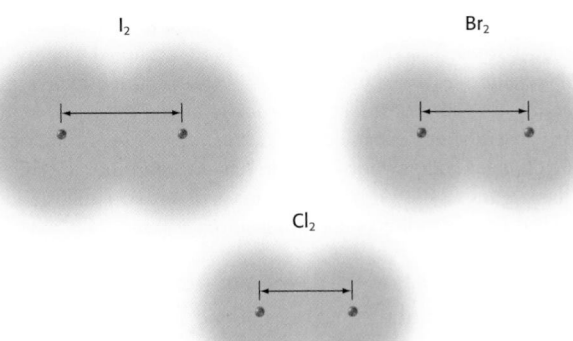

I₂ · Br₂ · Cl₂

Atomic Radius

The exact size of an atom is hard to determine. An atom's size depends on the volume occupied by the electrons around the nucleus, and the electrons do not move in well-defined paths. Rather, the volume the electrons occupy is thought of as an electron cloud, with no clear-cut edge. In addition, the physical and chemical state of an atom can change the size of an electron cloud.

Figure 19 shows one way to measure the size of an atom. This method involves calculating the **bond radius,** the length that is half the distance between the nuclei of two bonded atoms. The bond radius can change slightly depending on what atoms are involved.

bond radius

half the distance from center to center of two like atoms that are bonded together

Atomic Radius Increases as You Move Down a Group

Atomic radius increases as you move down a group, as **Figure 20** shows. As you proceed from one element down to the next in a group, another principal energy level is filled. The addition of another level of electrons increases the size, or atomic radius, of an atom.

Electron shielding also plays a role in determining atomic radius. Because of electron shielding, the effective nuclear charge acting on the outer electrons is almost constant as you move down a group, regardless of the energy level in which the outer electrons are located. As a result, the outermost electrons are not pulled closer to the nucleus. For example, the effective nuclear charge acting on the outermost electron in a cesium atom is about the same as it is in a sodium atom.

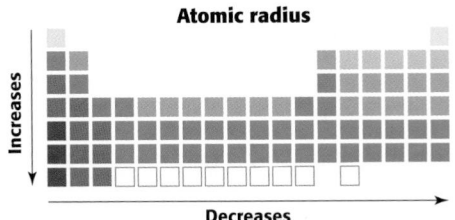

Atomic radius

Increases

Decreases

Figure 20
Atomic radius generally increases down a group and decreases across a period, as shown in this diagram. Darker shading indicates higher atomic radius.

HISTORY
CONNECTION

Size of an Atom The French physicist Jean Baptiste Perrin (1870–1942) published the first close estimate of the size of an atom in 1908. His calculations were based on an equation suggested by Albert Einstein in 1905.

Teaching Tip

The Structure of an Atom Ask students how atoms differ in structure from the top of a group toward the bottom. In addition to increased atomic mass and atomic number, the number of electrons also increases. Specifically, each atom has one more energy level of electrons than the atom above it. Ask students if that pattern matches observed trends in atomic radius. **Ans.** It does.

Trends in Atomic Radii Before discussing the reasons for the decrease in atomic radii from left to right through a period, project a copy of **Figure 21** from an overhead projector and ask students to study the trends illustrated. Ask them to describe the trend in radii from top to bottom in a group and to recall the reason for that trend. Next, point out the decrease in radii across a period, and ask students to use their knowledge of atomic structure and electron configurations to propose reasons for this counterintuitive trend.

Reading Organizer Students should update any notes they have been keeping on blank periodic tables by adding the trends discussed in this section as well as short notes about the reasons for those trends. They might want to color code each trend. **LS** Logical

As a member of Period 6, cesium has six occupied energy levels. As a member of Period 3, sodium has only three occupied energy levels. Although cesium has more protons and electrons, the effective nuclear charge acting on the outermost electrons is about the same as it is in sodium because of electron shielding. Because cesium has more occupied energy levels than sodium does, cesium has a larger atomic radius than sodium has. **Figure 21** shows that the atomic radius of cesium is about 230 pm, while the atomic radius of sodium is about 150 pm.

Atomic Radius Decreases as You Move Across a Period

As you move from left to right across a period, each atom has one more proton and one more electron than the atom before it has. All additional electrons go into the same principal energy level—no electrons are being added to the inner levels. As a result, electron shielding does not play a role as you move across a period. Therefore, as the nuclear charge increases across a period, the effective nuclear charge acting on the outer electrons also increases. This increasing nuclear charge pulls the outermost electrons closer and closer to the nucleus and thus reduces the size of the atom.

Figure 21 shows how atomic radii decrease as you move across a period. Notice that the decrease in size is significant as you proceed across groups going from Group 1 to Group 14. The decrease in size then tends to level off from Group 14 to Group 18. As the outermost electrons are pulled closer to the nucleus, they also get closer to one another.

Topic Link

Refer to Appendix A for a chart of relative atomic radii of the elements.

Figure 21
Atomic radii for hydrogen and the main-group elements in Periods 1 through 6 are plotted on this graph.

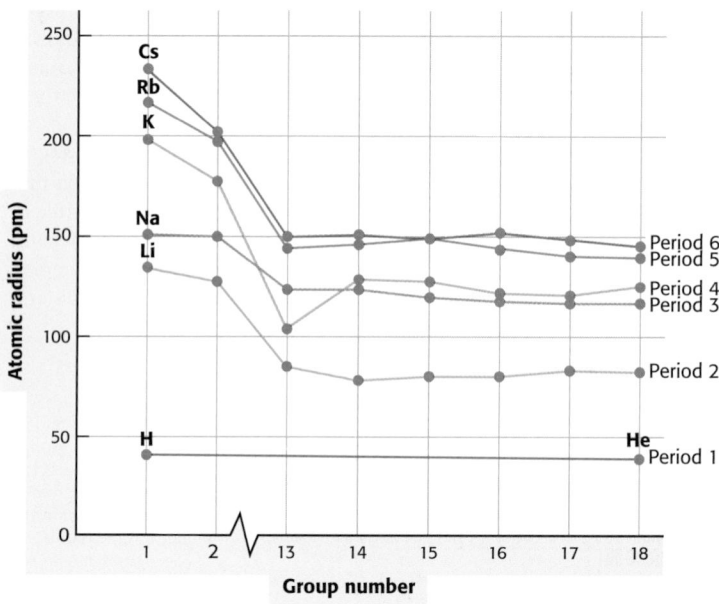

Atomic Radii of Main-Block Elements

MATH
CONNECTION

Many students perceive numbers abstractly and not as representations of measurements or counts of real objects. They can solve problems such as $6 - (-2) = 8$ but fail to understand why losing an electron makes an atom more positive and why gaining an electron makes an atom more negative. List the numbers of protons and electrons for several atoms and their ions. Show how the charges of the ions arise from the gain or loss of electrons.

ChemFile

• **Module 3: Periodic Properties**
This engaging tutorial reviews trends found in in atomic and ionic radii across a period and down a group.

Transparencies

TT Periodic Trends of Radii

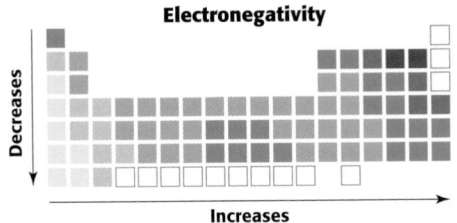

Electronegativity

Decreases ↓ / Increases →

Figure 22
Electronegativity tends to decrease down a group and increase across a period, as shown in this diagram. Darker shading indicates higher electronegativity.

Repulsions between these electrons get stronger. Finally, a point is reached where the electrons will not come closer to the nucleus because the electrons would have to be too close to each other. Therefore, the sizes of the atomic radii level off as you approach the end of each period.

Electronegativity

Atoms often bond to one another to form a compound. These bonds can involve the sharing of valence electrons. Not all atoms in a compound share electrons equally. Knowing how strongly each atom attracts bonding electrons can help explain the physical and chemical properties of the compound.

Linus Pauling, one of America's most famous chemists, made a scale of numerical values that reflect how much an atom in a molecule attracts electrons, called **electronegativity** values. Chemical bonding that comes from a sharing of electrons can be thought of as a tug of war. The atom with the higher electronegativity will pull on the electrons more strongly than the other atom will.

Fluorine is the element whose atoms most strongly attract shared electrons in a compound. Pauling arbitrarily gave fluorine an electronegativity value of 4.0. Values for the other elements were calculated in relation to this value.

electronegativity

a measure of the ability of an atom in a chemical compound to attract electrons

Electronegativity Decreases as You Move Down a Group

Electronegativity values generally decrease as you move down a group, as **Figure 22** shows. Recall that from one element to the next one in a group, the principal quantum number increases by one, so another principal energy level is occupied. The more protons an atom has, the more strongly it should attract an electron. Therefore, you might expect that electronegativity increases as you move down a group.

However, electron shielding plays a role again. Even though cesium has many more protons than lithium does, the effective nuclear charge acting on the outermost electron is almost the same in both atoms. But the distance between cesium's sixth principal energy level and its nucleus is greater than the distance between lithium's third principal energy level and its nucleus. This greater distance means that the nucleus of a cesium atom cannot attract a valence electron as easily as a lithium nucleus can. Because cesium does not attract an outer electron as strongly as lithium, it has a smaller electronegativity value.

137

Teaching Tip
Forming Compounds Advise students that they will better understand the role of electronegativity after learning about ionic and covalent bonding. An understanding of electronegativity is essential in knowing what type of bond can form between two atoms.

Group Activity ——— GENERAL
Divide students into groups of four. Have each group prepare 20 cards, each of which has a different symbol of an element. Ten of the cards should be metals, and the other half should be nonmetals. Do not include helium, neon, or argon; any element with atomic number greater than 104; or any of the lanthanides or actinides. Place the cards upside down, spread out on a desk. Have each student draw a card, and have the group use a periodic table to determine which element is most electronegative. The student with the most electronegative element earns a point. Replace the cards, mix them up, and repeat the process. Have a chart of electronegativities available for students to use when the trend is not decisive, such as if oxygen and chlorine were drawn.
LS Interpersonal

did you know?

Atoms with No Electronegativity Because electronegativity refers to the attraction of atoms for electrons in a compound, elements that do not form compounds are assigned no electronegativity values. Two types of elements fit this description. Of the noble gases, He, Ne, and Ar have no electronegativity values. Although noble gases are generally unreactive, heavier noble gases form compounds with active nonmetals. The other elements with no values assigned are the recently discovered elements that follow the actinide series. These elements generally are synthetic and have very short half-lives. They do not exist long enough to form compounds and thus, they have no electronegativity values.

✎ **Writing Skill** Have students write a paragraph that summarizes the trends in electronegativity shown in **Figure 23.** Be sure paragraphs note elements within the listed periods that are not included and explain why these elements are not included in the graph. **LS** Verbal

READING SKILL BUILDER — BASIC

Paired Reading Have pairs of students read the pages related to electronegativity. They should take turns reading and make a list of questions that arise about the reading. After completing reading the section on electronegativity, have the pairs of students discuss any concepts either of the students does not understand. **LS** Interpersonal

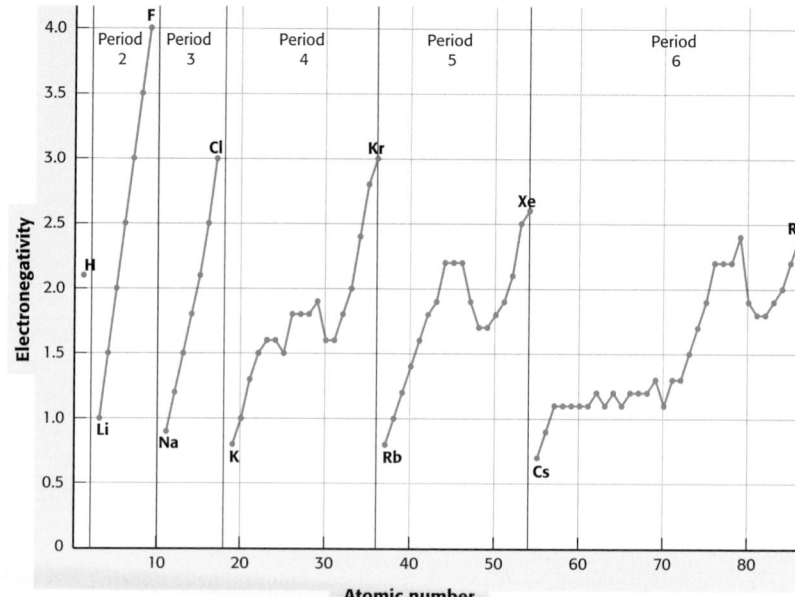

Electronegativity Versus Atomic Number

Figure 23
This graph shows electronegativity compared to atomic number for Periods 1 through 6. Electronegativity tends to increase across a period because the effective nuclear charge becomes greater as protons are added.

Electronegativity Increases as You Move Across a Period

As **Figure 23** shows, electronegativity usually increases as you move left to right across a period. As you proceed across a period, each atom has one more proton and one more electron—in the same principal energy level—than the atom before it has. Recall that electron shielding does not change as you move across a period because no electrons are being added to the inner levels. Therefore, the effective nuclear charge increases across a period. As this increases, electrons are attracted much more strongly, resulting in an increase in electronegativity.

Notice in **Figure 23** that the increase in electronegativity across a period is much more dramatic than the decrease in electronegativity down a group. For example, if you go across Period 3, the electronegativity more than triples, increasing from 0.9 for sodium, Na, to 3.2 for chlorine, Cl. In contrast, if you go down Group 1 the electronegativity decreases only slightly, dropping from 0.9 for sodium to 0.8 for cesium, Cs.

This difference can be explained if you look at the changes in atomic structure as you move across a period and down a group. Without the addition of any electrons to inner energy levels, elements from left to right in a period experience a significant increase in effective nuclear charge. As you move down a group, the addition of electrons to inner energy levels causes the effective nuclear charge to remain about the same. The electronegativity drops slightly because of the increasing distance between the nucleus and the outermost energy level.

MATH — CONNECTION — ADVANCED

Although the most common electronegativity values are determined by an equation developed by Linus Pauling, several other mathematicians proposed methods of determining these values. Pauling based his calculations on the bond energy of molecules that were not symmetrical. Robert Millikan developed an equation based on the average of the electron affinity and first ionization energy of an element. Allred and Rochow's equation is based on the charge of the nucleus and the electron and covalent radii. Have interested students investigate these equations in more detail. **LS** Logical

Other Periodic Trends

You may have noticed that effective nuclear charge and electron shielding are often used in explaining the reasons for periodic trends. Effective nuclear charge and electron shielding also account for two other periodic trends that are related to the ones already discussed: ionic size and electron affinity. Still other trends are seen by examining how melting point and boiling point change as you move across a period or down a group. The trends in melting and boiling points are determined by how electrons form pairs as *d* orbitals fill.

Periodic Trends in Ionic Size and Electron Affinity

Recall that atoms form ions by either losing or gaining electrons. Like atomic size, ionic size has periodic trends. As you proceed down a group, the outermost electrons in ions are in higher energy levels. Therefore, just as atomic radius increases as you move down a group, usually the ionic radius increases as well, as shown in **Figure 24a.** These trends hold for both positive and negative ions.

Metals tend to lose one or more electrons and form a positive ion. As you move across a period, the ionic radii of metal cations tend to decrease because of the increasing nuclear charge. As you come to the nonmetal elements in a period, their atoms tend to gain electrons and form negative ions. **Figure 24a** shows that as you proceed through the anions on the right of a period, ionic radii still tend to decrease because of the anions' increasing nuclear charge.

Neutral atoms can also gain electrons. The energy change that occurs when a neutral atom gains an electron is called the atom's *electron affinity.* This property of an atom is different from electronegativity, which is a measure of an atom's attraction for an electron when the atom is bonded to another atom. **Figure 24b** shows that electron affinity tends to decrease as you move down a group. This trend is due to the increasing effect of electron shielding. In contrast, electron affinity tends to increase as you move across a period because of the increasing nuclear charge.

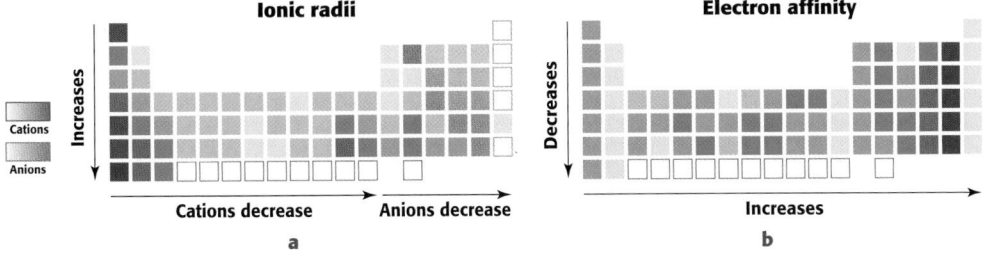

Figure 24
Ionic size tends to increase down groups and decrease across periods. Electron affinity generally decreases down groups and increases across periods.

139

Teaching Tip

Melting and Boiling Points Be sure students specifically understand what melting and boiling points are. Melting point is the temperature at which a solid becomes a liquid. Boiling point is the temperature at which the vapor pressure of the liquid equals atmospheric pressure. Liquids can evaporate below their boiling points.

Using the Figure — GENERAL

Have students look at a periodic table that indicates the state of the elements. From the states listed, have students predict which elements have the lowest melting points, moderate melting points, and the highest melting points. Have them check their predictions by using the trends of melting points and confirm them by using a table of melting points. Ans. Gases have low melting points. Liquids and low-melting solids, such as gallium, have moderate melting points. Transition elements have the highest melting points. **LS** Visual

Periodic Trends in Melting and Boiling Points

The melting and boiling points for the elements in Period 6 are shown in **Figure 25**. Notice that instead of a generally increasing or decreasing trend, melting and boiling points reach two different peaks as *d* and *p* orbitals fill.

Cesium, Cs, has low melting and boiling points because it has only one valence electron to use for bonding. From left to right across the period, the melting and boiling points at first increase. As the number of electrons in each element increases, stronger bonds between atoms can form. As a result, more energy is needed for melting and boiling to occur.

Near the middle of the *d*-block, the melting and boiling points reach a peak. This first peak corresponds to the elements whose *d* orbitals are almost half filled. The atoms of these elements can form the strongest bonds, so these elements have the highest melting and boiling points in this period. For Period 6, the elements with the highest melting and boiling points are tungsten, W, and rhenium, Re.

Figure 25
As you move across Period 6, the periodic trend for melting and boiling points goes through two cycles of first increasing, reaching a peak, and then decreasing.

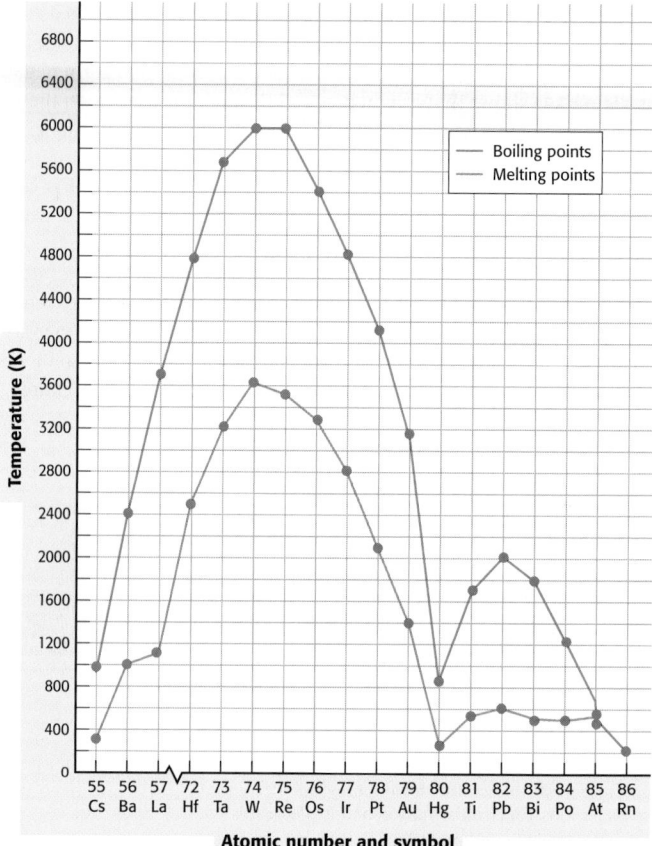

Melting Points and Boiling Points of Period 6 Elements

REAL-LIFE
CONNECTION

Melting Points Many practical uses of elements are based on their melting points. For example, metals that are used in engine parts that might reach high temperatures must not melt before these temperatures are reached. They also must be hard and durable. A transition metal, such as titanium or one of its alloys, might be used in an engine rather than a soft metal with a lower melting point. However, metals used in fuses and fire alarms must melt at a lower temperature.

As more electrons are added, they begin to form pairs within the *d* orbitals. Because of the decrease in unpaired electrons, the bonds that the atoms can form with each other become weaker. As a result, these elements have lower melting and boiling points. The lowest melting and boiling points are reached at mercury, whose *d* orbitals are completely filled. Mercury, Hg, has the second-lowest melting and boiling points in this period. The noble gas radon, Rn, is the only element in Period 6 with a lower boiling point than that of mercury.

As you proceed past mercury, the melting and boiling points again begin to rise as electrons are now added to the *p* orbital. The melting and boiling points continue to rise until they peak at the elements whose *p* orbitals are almost half filled. Another decrease is seen as electrons pair up to fill *p* orbitals. When the noble gas radon, Rn, is reached, the *p* orbitals are completely filled. The noble gases are monatomic and have no bonding forces between atoms. Therefore, their melting and boiling points are unusually low.

 Section Review

UNDERSTANDING KEY IDEAS

1. What is ionization energy?

2. Why is measuring the size of an atom difficult?

3. What can you tell about an atom that has high electronegativity?

4. How does electron shielding affect atomic size as you move down a group?

5. What periodic trends exist for ionization energy?

6. Describe one way in which *atomic radius* is defined.

7. Explain how the trends in melting and boiling points differ from the other periodic trends.

8. Why do both atomic size and ionic size increase as you move down a group?

9. How is electron affinity different from electronegativity?

10. What periodic trends exist for electronegativity?

11. Why is electron shielding not a factor when you examine a trend across a period?

CRITICAL THINKING

12. Explain why the noble gases have high ionization energies.

13. What do you think happens to the size of an atom when the atom loses an electron? Explain.

14. With the exception of the noble gases, why is an element with a high ionization energy likely to have high electron affinity?

15. Explain why atomic radius remains almost unchanged as you move through Period 2 from Group 14 to Group 18.

16. Helium and hydrogen have almost the same atomic size, yet the ionization energy of helium is almost twice that of hydrogen. Explain why hydrogen has a much higher ionization energy than any element in Group 1 does.

17. Why does mercury, Hg, have such a low melting point? How would you expect mercury's melting point to be different if the *d*-block contained more groups than it does?

18. What exceptions are there in the increase of ionization energies across a period?

141

Answers to Section Review

1. Ionization energy is the amount of energy needed to remove an electron from an atom.

2. In addition to being extremely small, an atom has an electron cloud that has no definite boundary.

3. It will strongly attract other electrons in a compound.

4. By keeping the effective nuclear charge constant, electron shielding contributes to the increase in atomic size as you move down a group.

5. Ionization energy decreases down a group and increases from left to right across a period.

6. half the distance between the nuclei of two bonded atoms

7. They are not as consistent. Across a period, they may increase, then decrease, then repeat this pattern.

8. In both cases, another principal energy level is added as you move from one element to the next, resulting in an increase in size.

9. Electron affinity refers to the attraction an atom has when it is not bonded. Electronegativity refers to the attraction an atom has when it is bonded to another atom.

Answers continued on p. 155A

Where Did the Elements Come From?

Overview

Before beginning this section, review with your students the Objectives listed in the Student Edition. This section describes how naturally occurring elements are formed and shows how natural and artificial transmutations change one element into another. It discusses how artificial transmutations take place in particle accelerators, and how synthetic elements can be produced.

Bellringer

Provide students with a list of elements that are synthetic (technetium, promethium, and any element with atomic number equal to or greater than 93). Have them note the location of each of these elements on a blank periodic table.

Motivate

Identifying Preconceptions

Remind students that certain words have different meanings according to how and in what context they are used. Have students list common meanings of the words *synthetic* and *synthesis*. The word *synthetic* as it is used in this section does not refer to *synthesis* by chemical means but rather by bombardment of nuclei with neutrons or other nuclei.

Discussion

Ask students to comment on the fact that all elements on Earth, including those that make up the molecules of the body, originated by fusion in stars.

KEY TERMS
• nuclear reaction
• superheavy element

OBJECTIVES

1 **Describe** how the naturally occurring elements form.

2 **Explain** how a transmutation changes one element into another.

3 **Describe** how particle accelerators are used to create synthetic elements.

Natural Elements

Of all the elements listed in the periodic table, 93 are found in nature. Three of these elements, technetium, Tc, promethium, Pm, and neptunium, Np, are not found on Earth but have been detected in the spectra of stars. The nebula shown in **Figure 26** is one of the regions in the galaxy where new stars are formed and where elements are made.

Most of the atoms in living things come from just six elements. These elements are carbon, hydrogen, oxygen, nitrogen, phosphorus, and sulfur. Scientists theorize that these elements, along with all 93 natural elements, were created in the centers of stars billions of years ago, shortly after the universe formed in a violent explosion.

Figure 26
Three natural elements—technetium, promethium, and neptunium—have been detected only in the spectra of stars.

internet connect

www.scilinks.org
Topic: Origin of Elements
SciLinks code: HW4093

*SCi*LINKS Maintained by the National Science Teachers Association

HISTORY
CONNECTION

The First Fusion Reaction The first fusion reaction caused by humans occurred in 1952 with the detonation of the first fusion bomb (hydrogen bomb). The bomb is an example of uncontrolled fusion. Scientists have still not found a reliable way to control and maintain fusion reactions because of the extremely high temperatures required, more than 20 million degrees Celsius.

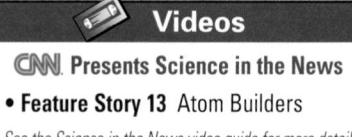
Videos

CNN Presents Science in the News
• **Feature Story 13** Atom Builders
See the Science in the News video guide for more details.

Chapter Resource File

• Lesson Plan

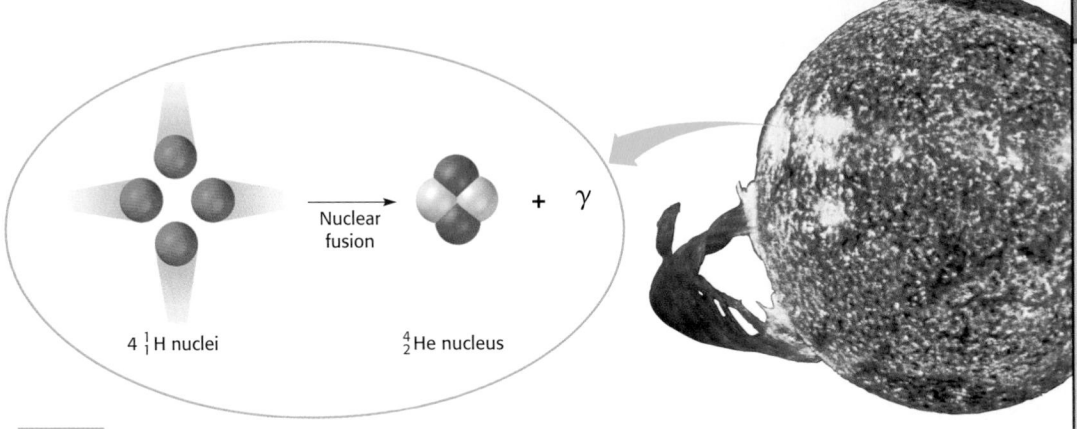

Nuclear fusion

$4\,{}^{1}_{1}\text{H}$ nuclei ${}^{4}_{2}\text{He}$ nucleus $+\ \gamma$

Figure 27
Nuclear reactions like those in the sun can fuse four hydrogen nuclei into one helium nucleus, releasing gamma radiation, γ.

Hydrogen and Helium Formed After the Big Bang

Much of the evidence about the universe's origin points toward a single event: an explosion of unbelievable violence, before which all matter in the universe could fit on a pinhead. This event is known as the *big bang*. Most scientists currently accept this model about the universe's beginnings. Right after the big bang, temperatures were so high that matter could not exist; only energy could. As the universe expanded, it cooled and some of the energy was converted into matter in the form of electrons, protons, and neutrons. As the universe continued to cool, these particles started to join and formed hydrogen and helium atoms.

Over time, huge clouds of hydrogen accumulated. Gravity pulled these clouds of hydrogen closer and closer. As the clouds grew more dense, pressures and temperatures at the centers of the hydrogen clouds increased, and stars were born. In the centers of stars, **nuclear reactions** took place. The simplest nuclear reaction, as shown in **Figure 27**, involves fusing hydrogen nuclei to form helium. Even now, these same nuclear reactions are the source of the energy that we see as the stars' light and feel as the sun's warmth.

nuclear reaction

a reaction that affects the nucleus of an atom

Other Elements Form by Nuclear Reactions in Stars

The mass of a helium nucleus is less than the total mass of the four hydrogen nuclei that fuse to form it. The mass is not really "lost" in this nuclear reaction. Rather, the missing mass is converted into energy. Einstein's equation $E = mc^2$ describes this mass-energy relationship quantitatively. The mass that is converted to energy is represented by m in this equation. The constant c is the speed of light. Einstein's equation shows that fusion reactions release very large amounts of energy. The energy released by a fusion reaction is so great it keeps the centers of the stars at very high temperatures.

143

did you know?

Age of a Star The sun is estimated to be 4.5–5.0 billion years old, with approximately 6 billion years left in its life span. The age of a star can be estimated from its color and the percentage of helium in its mass. Both can be determined by analyzing the light from the star with a spectrometer.

Rare Elements Francium and astatine occur in nature as products of nuclear decay in uranium ores. Scientists have estimated that at any given time, less than 25 g of each element exists in Earth's crust.

Teach

Using the Figure

Refer students to **Figure 27**. The reaction shown is simplified and is one of several possible fusion reactions believed to take place in the sun. Be sure that students note the composition of the alpha particle produced by fusion. It does not consist of the original four protons but instead has two protons and two neutrons. The neutrons have been formed from the protons by the emission of positrons.

Teaching Tip

Stable Nuclei Explain to students that iron and nickel nuclei are more stable than the nuclei of most other elements. For this reason, these elements tended to accumulate and are abundant in the universe today.

Heavy Elements Point out to students that all of the zinc, silver, gold, platinum, and mercury, as well as all of the bromine and iodine, came from explosions of red giant stars or from supernovae. Elements heavier than nickel also came from these sources.

──── BASIC

Reading Organizer Have students list each head in the section as either a main or secondary head in an outline. Have them leave room between each secondary head. As they read the section, have them summarize the main points under the appropriate head. **LS** Logical

Transparencies

TT Nuclear Fusion

Demonstration

Cloud Chamber Demonstrate the operation of a cloud chamber to students. Cloud chambers are available for purchase from most supply houses. However, to make a simplified Wilson cloud chamber, get a large glass jar with a lid, a piece of black velvet, a piece of blotter paper or felt, some alcohol, and a flat piece of dry ice. Cut the velvet to line the inside of the lid. Cut the blotter paper, and glue it to the bottom of the jar. Add rubbing alcohol to the blotter paper drop by drop until it is saturated but not dripping. Place the jar lid on the dry ice. Place a packaged nuclear source (obtained from a scientific supply house) in the lid, and screw on the jar. (The jar should be sitting upside down on the dry ice.) The jar should become saturated with alcohol vapor. Shine the beam of a small projector or a bright halogen flashlight through the chamber. The emitted nuclear particles should leave white streaks that are visible against the black velvet.

Safety: Wear safety goggles and a lab apron. Students must stand back at least 3 m. Use cloth or leather gloves when handling the dry ice. To break the dry ice, wrap it in a towel, place it on a firm surface, and strike it with a hammer. Ensure that there are no flames in the room. Have no more than 100 mL of alcohol in the room. Do not touch the radioactive source.

Disposal: Return leftover alcohol to the stock bottle. Allow the dry ice to sublime in a restricted location not accessible to students.

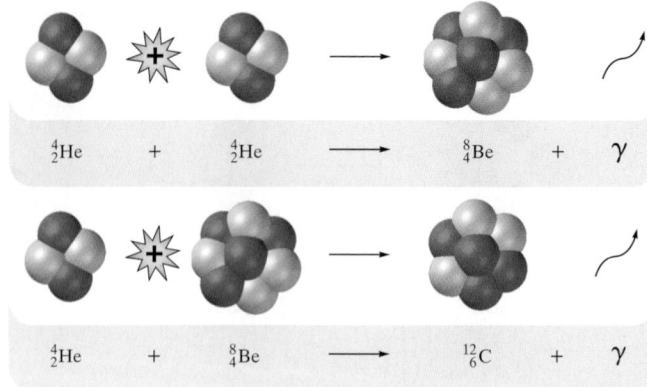

Figure 28
Nuclear reactions can form a beryllium nucleus by fusing helium nuclei. The beryllium nucleus can then fuse with another helium nucleus to form a carbon nucleus.

$$^{4}_{2}\text{He} + {}^{4}_{2}\text{He} \longrightarrow {}^{8}_{4}\text{Be} + \gamma$$

$$^{4}_{2}\text{He} + {}^{8}_{4}\text{Be} \longrightarrow {}^{12}_{6}\text{C} + \gamma$$

The temperatures in stars get high enough to fuse helium nuclei with one another. As helium nuclei fuse, elements of still higher atomic numbers form. **Figure 28** illustrates such a process: two helium nuclei fuse to form a beryllium nucleus, and gamma radiation is released. The beryllium nucleus can then fuse with another helium nucleus to form a carbon nucleus. Such repeated fusion reactions can form atoms as massive as iron.

Very massive stars (stars whose masses are more than 100 times the mass of our sun) are the source of heavier elements. When such a star has converted almost all of its core hydrogen and helium into the heavier elements up to iron, the star collapses and then blows apart in an explosion called a *supernova*. All of the elements heavier than iron on the periodic table are formed in this explosion. The star's contents shoot out into space, where they can become part of newly forming star systems.

Transmutations

internet connect
www.scilinks.org
Topic: Alchemy
SciLinks code: HW4006

SCI
LINKS. Maintained by the
National Science
Teachers Association

In the Middle Ages, many early chemists tried to change, or transmute, ordinary metals into gold. Although they made many discoveries that contributed to the development of modern chemistry, their attempts to transmute metals were doomed from the start. These early chemists did not realize that a *transmutation*, whereby one element changes into another, is a nuclear reaction. It changes the nucleus of an atom and therefore cannot be achieved by ordinary chemical means.

Transmutations Are a Type of Nuclear Reaction

Although nuclei do not change into different elements in ordinary chemical reactions, transmutations can happen. Early chemists such as John Dalton had insisted that atoms never change into other elements, so when scientists first encountered transmutations in the 1910s, their results were not always believed.

While studying the passage of high-speed alpha particles (helium nuclei) through water vapor in a cloud chamber, Ernest Rutherford observed some long, thin particle tracks. These tracks matched the ones caused by protons in experiments performed earlier by other scientists.

144

HISTORY
CONNECTION

Alchemy and Transmutation Remind students that one of the goals of the ancient alchemists was to turn lead and other "base" metals into gold. They attempted this feat by reacting different chemical substances with the metal. Today we realize that such a transmutation is not possible by chemical means. However, it is possible through nuclear reactions. Unfortunately, the cost of making gold in this way far exceeds the cost of obtaining it from ore. Although such reactions do not take place naturally on Earth, they occur constantly in stars.

Transparencies

TT Nuclear Fusion: Stellar Formation of Carbon-12

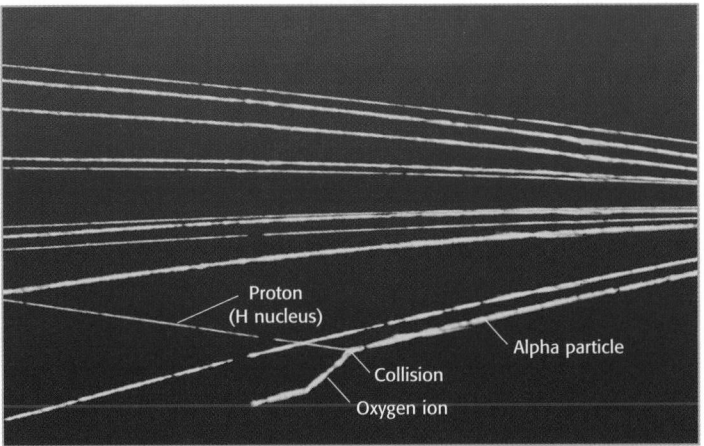

Proton
(H nucleus)

Alpha particle

Collision

Oxygen ion

Rutherford reasoned correctly that the atomic nuclei in air were disintegrating upon being struck by alpha particles. He believed that the nuclei in air had disintegrated into the nuclei of hydrogen (protons) plus the nuclei of some other atom.

Two chemists, an American named W. D. Harkins and an Englishman named P.M.S. Blackett, studied this strange phenomenon further. Blackett took photos of 400 000 alpha particle tracks that formed in cloud chambers. He found that 8 of these tracks forked to form a **Y**, as shown in **Figure 29.** Harkins and Blackett concluded that the **Y** formed when an alpha particle collided with a nitrogen atom in air to produce an oxygen atom and a proton, and that a transmutation had thereby occurred.

Synthetic Elements

The discovery that a transmutation had happened started a flood of research. Soon after Harkins and Blackett had observed a nitrogen atom forming oxygen, other transmutation reactions were discovered by bombarding various elements with alpha particles. As a result, chemists have synthesized, or created, more elements than the 93 that occur naturally. These are *synthetic elements.* All of the transuranium elements, or those with more than 92 protons in their nuclei, are synthetic elements. To make them, one must use special equipment, called *particle accelerators,* described below.

The Cyclotron Accelerates Charged Particles

Many of the first synthetic elements were made with the help of a cyclotron, a particle accelerator invented in 1930 by the American scientist E.O. Lawrence. In a cyclotron, charged particles are given one pulse of energy after another, speeding them to very high energies. The particles then collide and fuse with atomic nuclei to produce synthetic elements that have much higher atomic numbers than naturally occurring elements do. However, there is a limit to the energies that can be reached with a cyclotron and therefore a limit to the synthetic elements that it can make.

HISTORY

CONNECTION

Transuranium Elements All transuranium elements (elements with atomic number greater than 92) have been discovered since 1940. Many were discovered under the leadership of the American physicist Glenn Theodore Seaborg. Element 106 was named seaborgium in his honor. Much of Seaborg's research was spurred by the defense needs of the Allies in World War II. Seaborg discovered that the isotope U-235 could undergo fission, and his laboratory produced the Pu-239 used in the first atomic bomb.

145

Particle Accelerators Because even the largest particles accelerated are far too small to see, students might think of particle accelerators as being small also. Emphasize to them that the accelerators are quite large. The Tevatron, the synchrotron mentioned in the text, has a diameter of more than 0.8 km and a circumference of about 6.4 km. This facility has been in use since 1999.

Teaching Tip

Superheavy Elements Superheavy elements must not only be detected when they are created, but the results must also be duplicated by other scientists. For example, in 1999, scientists thought they had produced three atoms of element number 118 by fusing krypton-86 with lead-208. However, neither the scientists that originally made the claim nor other scientists have been able to reproduce it. As a result, the existence of this element and also that of element 116, which is produced by the decay of element 118, was never confirmed. In fact, the original claim was retracted (see notes on elements 116 and 118 in **Figure 4**).

The Synchrotron Is Used to Create Superheavy Elements

As a particle reaches a speed of about one-tenth the speed of light, it gains enough energy such that the relation between energy and mass becomes an obstacle to any further acceleration. According to the equation $E = mc^2$, the increase in the particle's energy also means an increase in its mass. This makes the particle accelerate more slowly so that it arrives too late for the next pulse of energy from the cyclotron, which is needed to make the particle go faster.

The solution was found with the synchrotron, a particle accelerator that times the pulses to match the acceleration of the particles. A synchrotron can accelerate only a few types of particles, but those particles it can accelerate reach enormous energies. Synchrotrons are now used in many areas of basic research, including explorations into the foundations of matter itself. The Fermi National Accelerator Laboratory in Batavia, IL has a circular accelerator which has a circumference of 4 mi! Subatomic particles are accelerated through this ring to 99.9999% of the speed of light.

Synthetic Element Trivia

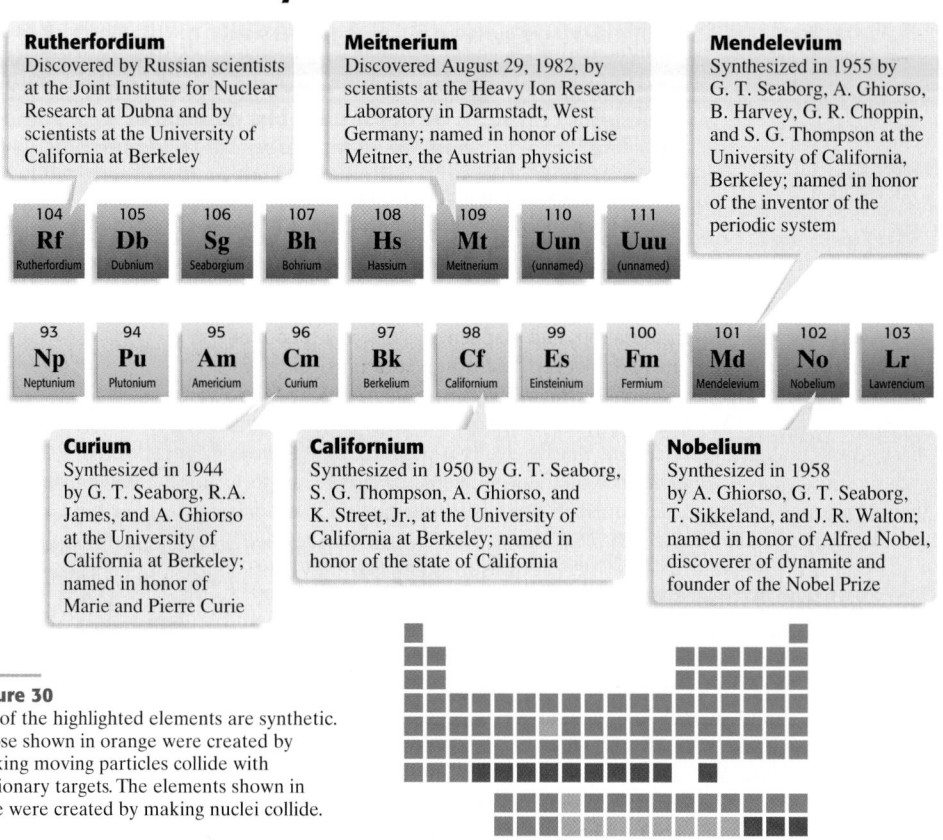

Rutherfordium
Discovered by Russian scientists at the Joint Institute for Nuclear Research at Dubna and by scientists at the University of California at Berkeley

Meitnerium
Discovered August 29, 1982, by scientists at the Heavy Ion Research Laboratory in Darmstadt, West Germany; named in honor of Lise Meitner, the Austrian physicist

Mendelevium
Synthesized in 1955 by G. T. Seaborg, A. Ghiorso, B. Harvey, G. R. Choppin, and S. G. Thompson at the University of California, Berkeley; named in honor of the inventor of the periodic system

Curium
Synthesized in 1944 by G. T. Seaborg, R.A. James, and A. Ghiorso at the University of California at Berkeley; named in honor of Marie and Pierre Curie

Californium
Synthesized in 1950 by G. T. Seaborg, S. G. Thompson, A. Ghiorso, and K. Street, Jr., at the University of California at Berkeley; named in honor of the state of California

Nobelium
Synthesized in 1958 by A. Ghiorso, G. T. Seaborg, T. Sikkeland, and J. R. Walton; named in honor of Alfred Nobel, discoverer of dynamite and founder of the Nobel Prize

Figure 30
All of the highlighted elements are synthetic. Those shown in orange were created by making moving particles collide with stationary targets. The elements shown in blue were created by making nuclei collide.

did you know?

Particle Accelerators Both cyclotrons and synchrotrons are types of particle accelerators. Both use magnetic and pulsed electric fields to accelerate charged particles. The fields are turned on and off so that the attraction of the charged particle for the field is always in the direction the particle is traveling, thereby increasing the acceleration of the particle.

Once the particles have been accelerated, they are made to collide with one another. **Figure 30** shows some of the **superheavy elements** created with such collisions. When a synchrotron is used to create an element, only a very small number of nuclei actually collide. As a result, only a few nuclei may be created in these collisions. For example, only three atoms of meitnerium were detected in the first attempt, and these atoms lasted for only 0.0034 s. Obviously, identifying elements that last for such a short time is a difficult task. Scientists in only a few nations have the resources to carry out such experiments. The United States, Germany, Russia, and Sweden are the locations of the largest such research teams.

One of the recent superheavy elements that scientists report is element 114. To create element 114, Russian scientists took plutonium-244, supplied by American scientists, and bombarded it with accelerated calcium-40 atoms for 40 days. In the end, only a single nucleus was detected. It lasted for 30 seconds before decaying into element 112.

Most superheavy elements exist for only a tiny fraction of a second. Thirty seconds is a very long life span for a superheavy element. This long life span of element 114 points to what scientists have long suspected: that an "island of stability" would be found beginning with element 114. Based on how long element 114 lasted, their predictions may have been correct. However, scientists still must try to confirm that element 114 was in fact created. The results of a single experiment are never considered valid unless the experiments are repeated and produce the same results.

superheavy element

an element whose atomic number is greater than 106

④ Section Review

UNDERSTANDING KEY IDEAS

1. How and where did the natural elements form?

2. What element is the building block for all other natural elements?

3. What is a synthetic element?

4. What is a transmutation?

5. Why is transmutation classified as a nuclear reaction?

6. How did Ernest Rutherford deduce that he had observed a transmutation in his cloud chamber?

7. How are cyclotrons used to create synthetic elements?

8. How are superheavy elements created?

CRITICAL THINKING

9. Why is the following statement *not* an example of a transmutation? Zinc reacts with copper sulfate to produce copper and zinc sulfate.

10. Elements whose atomic numbers are greater than 92 are sometimes referred to as the transuranium elements. Why?

11. Why must an extremely high energy level be reached before a fusion reaction can take place?

12. If the synchrotron had not been developed, how would the periodic table look?

13. What happens to the mass of a particle as the particle approaches the speed of light?

14. How many different kinds of nuclear reactions must protons go through to produce a carbon atom?

147

Answers to Section Review

1. They are produced by nuclear fusion reactions in stars.

2. hydrogen

3. A synthetic element is made by the bombardment of another element with neutrons, protons, or other atomic nuclei, in a particle accelerator.

4. The process by which one nucleus changes into another by spontaneous disintegration or bombardment with other particles.

5. Like any nuclear reaction, a transmutation involves changes in the nucleus of an atom.

6. Protons formed the tracks in the chamber. The only way free protons could have been present was if they were released from nuclei during a transmutation.

7. A cyclotron accelerates charged particles such as ions or electrons to very high energies. The particles then collide with target nuclei, changing them to other nuclei.

8. A synchrotron is used to accelerate particles to extremely high energies. These particles then collide with massive nuclei to create superheavy elements.

Answers continued on p. 155A

SCIENCE AND TECHNOLOGY

Superconductors

Superconductors are being used in many different fields. They are being used to construct devices that can detect and measure the very faint magnetic fields of the heart and brain. The results allow physicians to diagnose and study heart disease and neurological disorders such as epilepsy. Similar devices can be used to detect flaws in the structure of an aircraft and prevent aircraft failure due to structural problems.

SKILL BUILDER — ADVANCED

Writing Skills Hydrogen is one element that has mostly nonmetallic properties. However, scientists predict that it has superconductive properties under unusual conditions. Have students research hydrogen as a superconductor. Research should reveal that some scientists predict that at pressures approximately two million times atmospheric pressure, hydrogen loses much of its resistance and becomes a conductor. At twice that pressure, it becomes a superconductor because its electrons are forced out of their energy levels and become delocalized. **LS Logical**

Answers to the Feature Questions

1. As temperature increases, so does resistance. As resistance increases, conductivity decreases.
2. The mercury lost all resistance and became a superconductor.
3. Answers might include high-efficiency electric motors and generators and new modes of transportation.

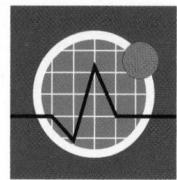

SCIENCE AND TECHNOLOGY

CAREER APPLICATION

Materials Scientist

A materials scientist is interested in discovering materials that can last through harsh conditions, have unusual properties, or perform unique functions. These materials might include the following: a lightweight plastic that conducts electricity; extremely light but strong materials to construct a space platform; a plastic that can replace iron and aluminum in building automobile engines; a new building material that expands and contracts very little, even in extreme temperatures; or a strong, flexible, but extremely tough material that can replace bone or connective tissue in surgery. Materials engineers develop such materials and discover ways to mold or shape these materials into usable forms. Many materials scientists work in the aerospace industry and develop new materials that can lower the mass of aircraft, rockets, and space vehicles.

internet connect

www.scilinks.org
Topic: Superconductors
SciLinks code: HW4170

SC*LINKS* Maintained by the
National Science
Teachers Association

148

Superconductors

Superconductivity Discovered

It has long been known that a metal becomes a better conductor as its temperature is lowered. In 1911, Heike Kamerlingh Onnes, a Dutch physicist, was studying this effect on mercury. When he used liquid helium to cool the metal to about −269°C, an unexpected thing happened—the mercury lost all resistance and became a superconductor. Scientists were excited about this new discovery, but the use of superconductors was severely limited by the huge expense of cooling them to near absolute zero. Scientists began research to find a material that would superconduct at temperatures above −196°C, the boiling point of cheap-to-produce liquid nitrogen.

The strong magnetic field produced by these superconducting electromagnets can suspend this 8 cm disk.

"High-Temperature" Superconductors

Finally, in 1987 scientists discovered materials that became superconductors when cooled to only −183°C. These "high-temperature" superconductors were not metals but ceramics; usually copper oxides combined with elements such as yttrium or barium.

High-temperature superconductors are used in building very powerful electromagnets that are not limited by resistance or heat. These magnets can be used to build powerful particle accelerators and high-efficiency electric motors and generators. Engineers are working to build a system that will use superconducting electromagnets to levitate a passenger train above its guide rail so that the train can move with little friction and thus save fuel.

Questions

1. How does temperature normally affect electrical conductivity in metals?
2. What happened unexpectedly when mercury was cooled to near absolute zero?
3. How might consumers benefit from the use of superconducting materials?

---

did you know?

Conductivity Clarify the relationship between heat and conductivity for students. When electric current passes along a wire, some of the energy of the flowing electrons is converted to heat as the electrons interact with the metal atoms in the wire. This effect, called *resistance,* is useful in a toaster, an electric heater, or a light bulb filament. In most cases, though, as in an electric motor, heat is an unwanted result of the resistance to flow of electrons and is wasted energy. Superconductivity is the result of minimal resistance.

Videos

CNN. Presents Science in the News
• **Feature Story 11** Superconductivity
See the Science in the News video guide for more details.

CHAPTER HIGHLIGHTS

KEY TERMS

periodic law
valence electron
group
period

main-group element
alkali metal
alkaline-earth metal
halogen
noble gas
transition metal
lanthanide
actinide
alloy

ionization energy
electron shielding
bond radius
electronegativity

nuclear reaction
superheavy element

KEY IDEAS

SECTION ONE   How Are Elements Organized?

• John Newlands, Dmitri Mendeleev, and Henry Moseley contributed to the development of the periodic table.

• The periodic law states that the properties of elements are periodic functions of the elements' atomic numbers.

• In the periodic table, elements are ordered by increasing atomic number. Rows are called *periods*. Columns are called *groups*.

• Elements in the same period have the same number of occupied energy levels. Elements in the same group have the same number of valence electrons.

SECTION TWO   Tour of the Periodic Table

• The main-group elements are Group 1 (alkali metals), Group 2 (alkaline-earth metals), Groups 13–16, Group 17 (halogens), and Group 18 (noble gases).

• Hydrogen is in a class by itself.

• Most elements are metals, which conduct electricity. Metals are also ductile and malleable.

• Transition metals, including the lanthanides and actinides, occupy the center of the periodic table.

SECTION THREE   Trends in the Periodic Table

• Periodic trends are related to the atomic structure of the elements.

• Ionization energy, electronegativity, and electron affinity generally increase as you move across a period and decrease as you move down a group.

• Atomic radius and ionic size generally decrease as you move across a period and increase as you move down a group.

• Melting points and boiling points pass through two cycles of increasing, peaking, and then decreasing as you move across a period.

SECTION FOUR   Where Did the Elements Come From?

• The 93 natural elements were formed in the interiors of stars. Synthetic elements (elements whose atomic numbers are greater than 93) are made using particle accelerators.

• A transmutation is a nuclear reaction in which one nucleus is changed into another nucleus.

149

Alternative Assessments

Homework

• Pages 118, 134, 146

HEALTH CONNECTION

• Page 123

SKILL BUILDER

• Pages 138, 148

Chapter Review

• Items 66–70

Portfolio Assessments

READING SKILL BUILDER

• Pages 118, 127, 133, 136, 143

SKILL BUILDER

• Page 145

Chapter Resource File

• **Microscale Lab** Reactivity of Halide Ions

• **Datasheets for In-Text Labs** The Mendeleev Lab of 1869

• **Observation Lab** Exploring the Periodic Table

4 CHAPTER REVIEW

REVIEW ANSWERS

1. alkaline-earth metals

2. halogens

3. metals

4. electron affinity

5. actinides

6. answers should involve the transmutation of one element to another by a change in the number of protons in its nucleus.

7. alkali metals

8. electron shielding

9. electronegativity and ionization energy

10. Answers should discuss particle accelerators and the colliding of particles at very high speeds to produce elements of higher atomic numbers.

11. noble gases

12. seven

13. Answers may include brass, steel, sterling silver, pig iron, etc.

14. Moseley arranged the elements according to their atomic numbers, while Mendeleev had arranged them based on their atomic mass.

15. He left gaps for the elements that he predicted would be discovered and which would have certain properties.

16. His success in predicting the properties of elements that had not yet been discovered gave him credibility.

17. The completion of an energy level determines the points at which the elements begin new periods.

18. It has two valence electrons and six occupied energy levels.

USING KEY TERMS

1. What group of elements do Ca, Be, and Mg belong to?

2. What group of elements easily gains one valence electron?

3. What category do most of the elements of the periodic table fall under?

4. What is the term for the energy released when an atom gains an electron?

5. What are elements 90–103 called?

6. Give an example of a nuclear reaction. Describe the process by which it takes place.

7. What are elements in the first group of the periodic table called?

8. What atomic property affects periodic trends down a group in the periodic table?

9. What two atomic properties have an increasing trend as you move across a period?

10. Write a paragraph describing in your own words how synthetic elements are created. Discuss what modification has to be made to the equipment in order to synthesize super-heavy elements. **WRITING SKILLS**

11. Which group of elements has very high ionization energies and very low electron affinities?

12. How many valence electrons does a fluorine atom have?

13. Give an example of an alloy.

UNDERSTANDING KEY IDEAS

How Are Elements Organized?

14. How was Moseley's arrangement of the elements in the periodic table different from Mendeleev's?

15. What did the gaps on Mendeleev's periodic table represent?

16. Why was Mendeleev's periodic table accepted by most chemists?

17. What determines the horizontal arrangement of the periodic table?

18. Why is barium, Ba, placed in Group 2 and in Period 6?

Tour of the Periodic Table

19. Why is hydrogen in a class by itself?

20. All halogens are highly reactive. What causes these elements to have similar chemical behavior?

21. What property do the noble gases share? How do the electron configurations of the noble gases give them this shared property?

22. How do the electron configurations of the transition metals differ from those of the metals in Groups 1 and 2?

23. Why is carbon, a nonmetal element, added to iron to make nails?

24. If an element breaks when it is struck with a hammer, could it be a metal? Explain.

25. Why are the lanthanides and actinides placed at the bottom of the periodic table?

26. Explain why the main-group elements are also known as representative elements.

150

Assignment Guide

Section	Questions
1	12, 14–18, 41, 43–45, 55, 62, 70
2	1–3, 5, 7, 13, 19–26, 42, 48, 50–52, 54, 57–60, 65, 67
3	4, 8, 9, 11, 27–33, 47, 53, 56, 61, 64, 71–77
4	6, 10, 34–40, 46, 49, 68, 69

Trends in the Periodic Table

27. What periodic trends exist for ionization energy? How does this trend relate to different energy levels?

28. Why don't chemists define atomic radius as the radius of the electron cloud that surrounds a nucleus?

29. How does the periodic trend of atomic radius relate to the addition of electrons?

30. What happens to electron affinity as you move across a period beginning with Group 1? Why do these values change as they do?

31. Identify which trend diagram below describes atomic radius.

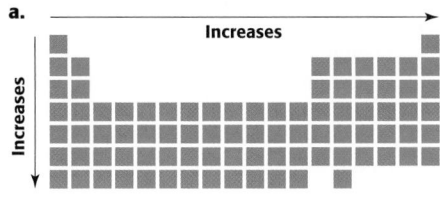

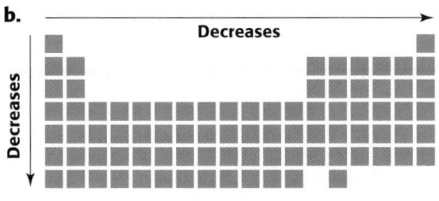

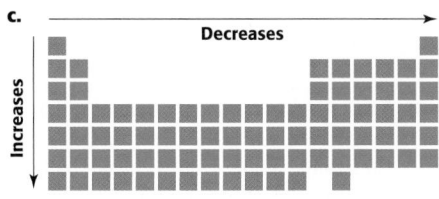

32. What periodic trends exist for electronegativity? Explain the factors involved.

33. Why are the melting and boiling points of mercury almost the lowest of the elements in its period?

Where Did the Elements Come From?

34. How does nuclear fusion generate energy?

35. What happens in the nucleus of an atom when a transmutation takes place?

36. Why are technetium, promethium, and neptunium considered natural elements even though they are not found on Earth?

37. Why must a synchrotron be used to create a superheavy element?

38. What role did supernovae play in creating the natural elements?

39. What is the heaviest element that can be made by normal processes inside a star?

40. What two elements make up most of the matter in a star?

MIXED REVIEW

41. Without looking at the periodic table, identify the period and group in which each of the following elements is located.
  a. $[Rn]7s^1$
  b. $[Ar]4s^2$
  c. $[Ne]3s^2 3p^6$

42. Which of the following ions has the electron configuration of a noble gas: Ca^+ or Cl^-? (Hint: Write the electron configuration for each ion.)

43. When 578 kJ/mol of energy is supplied, Al loses one valence electron. Write the electron configuration of the ion that forms.

44. The electron affinity value of oxygen is −146.1 kJ/mol. This value is the quantity of energy released when an oxygen atom gains an electron. Write the electron configuration of the ion that forms.

45. Without looking at the periodic table, write the complete electron configuration of the element in Group 2, Period 5. (Hint: The 5s orbital has a lower energy than the 4d orbital does.)

151

REVIEW ANSWERS
continued

19. With only one proton and one electron, hydrogen does not exhibit the same properties shared by the elements making up any of the groups in the periodic table.

20. All halogens have seven valence electrons and are therefore one electron short of having a full valence shell. As a result, they readily react to acquire this one electron.

21. Noble gases do not normally react with other elements because of the stability resulting from their electron configurations.

22. The outer electrons of Groups 1 and 2 are in s orbitals. The outer electrons of the transition metals are in d orbitals and s orbitals.

23. Pure iron is too soft to make nails. Adding carbon produces a harder alloy.

24. Some metals are brittle. Therefore, this element may be a metal if it is shown to be an excellent conductor of electricity.

25. This arrangement keeps the periodic table conveniently narrow.

26. They exhibit all the properties characteristic of the elements.

27. Ionization energy slightly decreases as you move down a group and increases significantly as you move across a period. Increased distance from the nucleus allows removal of an electron using less energy.

28. Because an electron cloud has no definite boundary, an exact radius is impossible to measure.

29. As electrons are added across a period, they are entering the same principal energy level. The increasing nuclear charge pulls these electrons closer, making the atoms progressively smaller across a period. As electrons are added down a group, they are entering another principal energy level, thus increasing the size of the atom.

REVIEW ANSWERS
continued

30. Electron affinity generally increases because the effective nuclear charge increases. Electron shielding does not play a role because electrons are not added to inner energy levels.

31. c

32. Electronegativity decreases as you move down a group because of the increasing electron shielding and increases as you move across a period because of the increasing effective nuclear charge.

33. Because the *d* orbitals are completely filled, mercury forms weaker bonds than most of the other elements in Period 6.

34. The mass of the fused nuclei is less than that of the mass of the nuclei that created it. This mass is converted to energy during fusion.

35. The number of protons either increases or decreases during a nuclear reaction, producing a different element.

36. These elements can be found elsewhere in the universe and are therefore natural.

37. Only a synchrotron can accelerate nuclei to energies high enough to result in fusion involving massive nuclei.

38. Elements heavier than iron were formed in supernovae.

39. iron

40. hydrogen and helium

41. a. Period 7, Group 1
b. Period 4, group 2
c. Period 3, group 18

42. Cl^-

43. $[Ne]3s^2$

44. $[He]2s^2 2p^5$

45. $1s^2 2s^2 2p^6 3s^2 3p^6 3d^{10} 4s^2 4p^6 5s^2$

46. In the middle ages, many alchemists tried to transform lead into gold. Could gold be made from lead by using the nuclear processes described in this chapter?

47. Name three periodic trends you encounter in your life.

48. Describe some differences between the *s*-block metals and the *d*-block metals.

49. Irène Joliot-Curie created the first artificial radioactive isotope, phosphorus-30, in 1934. She created phosphorus-30 by bombarding aluminum-27, a shiny metal with conductive properties, with helium nuclei. The resulting product was a nonmetal with completely different properties from aluminum. What caused the change in properties?

50. How do the electron configurations of the lanthanide and actinide elements differ from the electron configurations of the other transition metals?

51. Use the periodic table to describe the chemical properties of the following elements:
a. iodine, I
b. krypton, Kr
c. rubidium, Rb

52. The electron configuration of argon differs from those of chlorine and potassium by one electron each. Compare the reactivity of these three elements, and relate them to their electron configurations.

53. What trends were first used to classify the elements? What trends were discovered after the elements were classified in the periodic table?

54. Among the main-group elements, what is the relationship between group number and the number of valence electrons among group members?

55. Why is iodine placed just after tellurium on the periodic table even though its atomic mass is less than the atomic mass of tellurium?

152

CRITICAL THINKING

56. Consider two main-group elements, A and B. Element A has an ionization energy of 419 kJ/mol. Element B has an ionization energy of 1000 kJ/mol. Which element is more likely to form a cation?

57. Argon differs from both chlorine and potassium by one proton each. Compare the electron configurations of these three elements to explain the reactivity of these elements.

58. Why is it highly unlikely for calcium to form a Ca^+ cation and highly unlikely for sodium to form a Na^{2+} cation?

59. How would you prove that an alloy is a mixture and not a compound?

60. While at an amusement park, you inhale helium from a balloon to make your voice higher pitched. A friend says that helium reacts with and tightens the vocal cords to make your voice have the higher pitch. Could he be correct? Why or why not?

61. A scientist may measure the radius of a bonded atom five different times and may get five different results. The method used is correct, and the instrumentation is working correctly. How are different results possible?

62. In his periodic table, Mendeleev placed Be, Mg, Zn, and Cd in one group and Ca, Sr, Ba, and Pb in another group. Examine the electron configurations of these elements, and explain why Mendeleev grouped the elements this way.

63. The atomic number of yttrium, which follows strontium in the periodic table, exceeds the atomic number of strontium by one. Barium is 18 atomic numbers after strontium but it falls directly beneath strontium in the periodic table. Does strontium share more properties with yttrium or barium? Explain your answer.

64. Examine the following diagram.

Explain why the structure shown on the right was drawn to have a smaller radius than the structure on the left.

65. As of this writing, element 117 has yet to be created. Predict some of the properties that this element might have if it is synthesized.

ALTERNATIVE ASSESSMENT

66. Design an experiment to show that the electrical conductivity of a metal increases as the temperature decreases and decreases as the temperature increases. Test various metals to compare their conductivities at different temperatures. If your teacher approves your design, report your findings to the class after completing your experiments.

67. Select an alloy. You can choose one mentioned in this book or find another one by checking the library or the Internet. Obtain information on how the alloy is made. Obtain information on how the alloy is used for practical purposes.

68. The announcement of a new chemical element is sometimes claimed by two different groups of scientists. Research the history of one such element. What methods were used to create the element? How was the dispute settled so that only one group was recognized as the first to have synthesized the element?

69. Construct a model of a synchrotron. Check the library and Internet for information about synchrotrons. You may want to contact a synchrotron facility directly to find out what is currently being done in the field of synthetic elements.

70. In many labeled foods, the mineral content is stated in terms of the mass of the element, in a stated quantity of food. Examine the product labels of the foods you eat. Determine which elements are represented in your food and what function each element serves in the body. Make a poster of foods that are good sources of minerals that you need.

CONCEPT MAPPING

71. Use the following terms to complete the concept map below: *atomic number, atoms, electrons, periodic table,* and *protons.*

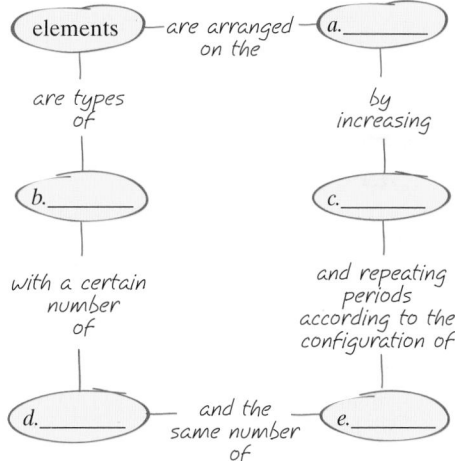

46. Although the conversion of lead into gold by nuclear means is possible, it is not possible by the processes discussed here, which involve increasing the atomic number of atoms. Gold has a smaller atomic number than lead has.

47. Answers will vary but could include, for example, the pattern of seasons, the repeating school year, the progression of weeks, or the pattern of daily meals.

48. The *d*-block metals are harder, denser, and less reactive, and, except for mercury, they have higher melting points than *s*-block metals.

49. Phosphorus-30 was produced by reactions that add two protons to the aluminum nucleus. The atom also acquires two more electrons to balance the protons. The atom has the electron configuration of phosphorus, and electron configurations determine chemical and many physical properties.

50. The lanthanide and actinide elements have partially filled *f*-electron shells.

51. a. A nonmetal in Group 17, it needs one electron to achieve a noble-gas configuration, which means it is likely to react with elements with which it can obtain one electron, such as an alkali metal. It has a high ionization energy, and thus the formation of positive ions is not likely.

b. A nonmetal of low reactivity in Group 18, it has a filled outer-energy level, so there is little tendency to lose, gain, or share electrons.

c. A reactive metal in Group 1, it has a low ionization energy because the loss of an electron to form a positive ion gives it a noble-gas configuration.

52. Argon has a complete energy level of electrons, and therefore is not reactive. Chlorine's electron configuration is one short of a full shell, so it has a tendency to react with elements from which it can gain one electron. Potassium has only a single valence electron, so it readily reacts to lose that electron.

53. Elements were first classified by reactivity and atomic mass. The modern periodic table has trends of atomic size, electron configuration, ionization energy, and electron affinity, among other factors not covered in this chapter.

54. Groups 1 and 2 have the same number of valence electrons as their group number. Groups 13–18 have n–10 valence electrons, n being the group number.

55. Iodine has an atomic *number* of 53, which is one more than tellurium's 52. This places it after tellurium.

56. Element A is more likely to lose an electron and become a cation because it has a lower ionization energy.

57. As a noble gas with a full valence shell, $3s^2 3p^6$, argon is unreactive. Chlorine has one less electron than argon, $3s^2 3p^5$, so it tends to react by gaining one electron to form an anion with a 1– charge. Potassium has one more electron than argon, $3s^2 3p^6 4s^1$, so it tends to react by losing one electron to form a cation with a 1+ charge.

58. With two valence electrons, calcium is unlikely to lose just a single electron because doing so does not achieve a stable electron configuration. Sodium achieves a full valence shell by losing one electron, so it is highly unlikely to lose two electrons.

Answers continued on p. 155A–B

FOCUS ON GRAPHING

Study the graph below, and answer the questions that follow.
For help in interpreting graphs, see Appendix B, "Study Skills for Chemistry."

72. What relationship is represented in the graph shown?

73. What do the numbers on the *y*-axis represent?

74. In every Period, which Group contains the element with the greatest atomic radius?

75. Why is the axis representing group number drawn the way it is in going from Group 2 to Group 13?

76. Which period shows the greatest change in atomic radius?

77. Notice that the points plotted for the elements in Periods 5 and 6 of Group 2 overlap. What does this overlap indicate?

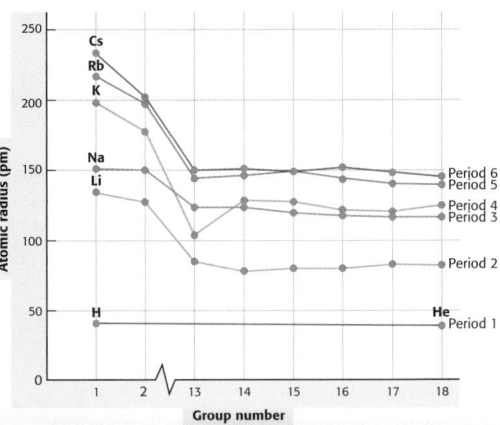

Atomic Radii of Main-Block Elements

TECHNOLOGY AND LEARNING

78. Graphing Calculator

Graphing Atomic Radius Vs. Atomic Number

The graphing calculator can run a program that graphs data such as atomic radius versus atomic number. Graphing the data within the different periods will allow you to discover trends.

Go to Appendix C. If you are using a TI-83 Plus, you can download the program and data sets and run the application as directed. Press the **APPS** key on your calculator, then choose the application **CHEMAPPS**. Press **8**, then highlight **ALL** on the screen, press **1**, then highlight **LOAD** and press **2** to load the data into your calculator. Quit the application, and then run the program **RADIUS**. For

L_1, press **2nd** and **LIST**, and choose **ATNUM**. For L_2, press **2nd** and **LIST** and choose **ATRAD**.

If you are using another calculator, your teacher will provide you with keystrokes and data sets to use.

a. Would you expect any atomic number to have an atomic radius of 20 pm? Explain.

b. A relationship is considered a function if it can pass a vertical line test. That is, if a vertical line can be drawn anywhere on the graph and only pass through one point, the relationship is a function. Does this set of data represent a function? Explain.

c. How would you describe the graphical relationship between the atomic numbers and atomic radii?

154

Chapter Resource File

• **Chapter Test**

1. In the modern periodic table, elements are arranged according to
   a. decreasing atomic mass.
   b. Mendeleev's original model.
   c. increasing atomic number.
   d. when they were discovered.

2. Which of the following elements is formed in stars?
   a. curium
   b. gold
   c. einsteinium
   d. americium

3. Group 17 elements, the halogens, are the most reactive of the nonmetal elements because they
   a. require only one electron to fill their outer energy level.
   b. have the highest ionization energies.
   c. have the largest atomic radii.
   d. are the farthest to the right in the periodic table.

4. The periodic law states that
   a. the chemical properties of elements can be grouped according to periodicity.
   b. the properties of the elements are functions of their atomic mass.
   c. all elements in the same group have the same number of valence electrons.
   d. all elements with the same number of occupied energy levels must be in the same group.

5. The energy it takes to remove an electron from an atom _____ as you move left to right across a period, from magnesium to chlorine.
   a. generally increases
   b. generally decreases
   c. does not change
   d. varies unpredictably

6. Which of the following is *not* a main-group element?
   a. neon
   b. bromine
   c. oxygen
   d. hydrogen

7. _____ is the element with the highest electronegativity.
   a. Oxygen
   b. Hydrogen
   c. Fluorine
   d. Carbon

8. To be classified as a metal, an element must
   a. be shiny.
   b. have a full outer energy level of electrons.
   c. conduct electricity.
   d. be malleable and ductile.

9. Because the noble gases have a stable electron configuration, they have
   a. high ionization energies.
   b. high electron affinities.
   c. large atomic radii.
   d. a tendency to form both cations and anions.

10. All elements in Group 2
    a. are also members of Period 2 in the periodic table.
    b. are known as the alkaline-earth metals.
    c. have the same atomic mass.
    d. have a stable electron configuration.

11. As electrons are added to the outer energy levels of atoms within a period, the atoms generally have
    a. increasing radii and increasing ionization energies.
    b. increasing radii and decreasing ionization energies.
    c. decreasing radii and decreasing ionization energies.
    d. decreasing radii and increasing ionization energies.

155

Continuation of Answers

Answers to Section 1 Review, *continued*

10. $6s^2$

11. The number of electrons that can be contained in that energy level, plus the number of *d*-electrons in the previous energy level, plus the number of *f*-electrons in the next previous energy level. Energy levels are defined by the principal quantum number, *n*.

12. They are more likely to be found in the same group because elements in the same group have the same number of valence electrons. This number determines the chemical properties of the elements.

13. 5

14. $[Rn]5f^{14}6d^{10}7s^27p^1$

Answers to Section 2 Review, *continued*

10. *Inert* indicates that the elements are totally unreactive. Some of the heavier Group 18 elements have been made to form compounds with active nonmetals, so they are not actually inert.

11. The element is relatively unreactive; if it were active, it would have reacted with other elements, such as oxygen in the air, to form compounds.

12. Going left to right across a period, transition elements add one electron in a *d* sublevel to the configuration of the previous metal.

13. It is possible that such an element is a metal. Knowing that it is a good conductor of electricity would confirm the designation.

Answers to Section 3 Review, *continued*

10. Electronegativity decreases down a group and increases from left to right across a period.

11. Each atom across a period has the same number of energy levels and therefore the same amount of shielding.

12. The electron configurations of the noble gases are stable. They have no tendency to lose electrons during a chemical reaction, so it takes a lot of energy to remove an electron from a noble gas.

13. The atom then has more protons than electrons, so the attraction between protons and electrons is more than the repulsion of the electrons and the atom becomes smaller.

14. A high ionization energy means that the atom has a strong attraction for its electrons, which is also shown by a high electron affinity.

15. The repulsion of electrons becomes equal to the attraction of the nucleus for the electrons.

16. The electron configuration of helium involves a completed valence level, which is stable and requires a high amount of energy to remove an electron. Compared to atoms in Group 1, hydrogen has a high ionization energy because its electron is not shielded from the nucleus by other electrons.

17. The fact that mercury has a filled *d*-shell gives it very weak inter-molecular forces, which contributes toward its being a liquid at room temperature. If the *d*-block were longer, it would have unpaired outer-shell electrons with which to bond, as with elements that come before it in its period, and would probably be a solid at room temperature.

18. Exceptions occur between groups 2 and 13 and between groups 15 and 16.

Answers to Section 4 Review, *continued*

9. No atom in the reaction changes to another kind of atom.

10. Element 92 is uranium. The prefix *trans-* indicates "beyond," so those beyond element 92 are transuranium elements.

11. Enough energy must be present to overcome the repulsion of the nuclei.

12. There would be no elements with atomic numbers above 106.

13. It increases.

14. Three.

Answers to Chapter Review, *continued*

59. You would find a way to separate it into its components by some physical process.

60. Helium is a noble gas and unreactive.

61. A bonded atom vibrates, constantly changing its radius. In addition, the atom being measured may be bonded to different atoms in each trial.

62. Mendeleev placed Be, Mg, Zn, and Cd in one group because he noticed that these elements had similar chemical reactivities. The same was true for Ca, Sr, Ba, and Pb.

63. Barium and strontium are in the same group, which means they have the same number of valence electrons. Therefore, they are likely to be chemically similar, while both will be chemically different from yttrium.

64. The atom on the right is shown losing an electron. The protons can pull the remaining electrons closer to the nucleus.

65. This element would fall in Group 17 and should have properties similar to those of the halogens.

66. Designs should include methods for varying and controlling the temperature of the conductor as well as measuring the current in amperes that is conducted at a constant voltage. An alternative method is to determine the change in voltage that will produce the same current in all trials.

67. There are a great number of alloys, especially of iron, aluminum, and magnesium. Some of the more interesting ones are used in aerospace applications. Even lithium is used in some lightweight, high-strength magnesium alloys. Students should discuss the applications for the alloys they choose.

68. Some of the superheavy elements fit these descriptions, particularly those from atomic numbers 104 to 109. Students should also take note of the controversies over naming these elements.

69. Models should be circular, and include representations of accelerator electromagnets and a particle beam.

70. Students' work should demonstrate a comprehensive listing of the minerals found in foods and their functions.

71. a. periodic table

   b. atoms

   c. atomic number

   d. protons

   e. electrons

72. The graph shows how atomic radius changes as you move across the elements in periods 2 through 6 and from hydrogen to helium.

73. Atomic radius, in units of picometers (pm).

74. Group 1.

75. The break indicates that the atomic radii of the transition elements in Groups 3 through 11 are not included in this graph.

76. Period 6.

77. This indicates that the atomic radii for these two elements are almost the same.

78. a. No. The graph seems to be approaching a limit of approximately 65–70.

   b. Yes. Each x-value (atomic number) on the graph corresponds to one and only one y-value (atomic radius), making it a function.

   c. As the atomic number increases, the atomic radius decreases.

Ions and Ionic Compounds
Chapter Planning Guide

PACING	CLASSROOM RESOURCES	LABS, ACTIVITIES, AND DEMONSTRATIONS
BLOCK 1 • 45 min pp. 156–157 **Chapter Opener**		**SE** **Start-Up Activity** Hard Water, p. 157 ◆
BLOCKS 2 & 3 • 90 min pp. 158–165 **Section 1** Simple Ions	**TT** Some Ions with Noble-Gas Configurations * **TT** Stable Ions Formed by the Transition Elements and Some Other Metals * **OSP** **Career Extension** Water Quality Manager GENERAL	**TE** **Demonstration** Achieving a Stable Octet, p. 159 ◆ **TE** **Group Activity,** p. 160 GENERAL
BLOCKS 4 & 5 • 90 min pp. 166–175 **Section 2** Ionic Bonding and Salts	**TT** Formation of Sodium Chloride * **TT** Sodium Chloride in Three Phases *	**TE** **Demonstration** Electrostatic Attraction, p. 167 ◆ BASIC **TE** **Group Activity** Interpreting Visuals, p. 168 GENERAL **TE** **Group Activity** Interpreting Visuals, p. 169 GENERAL **CRF** **Microscale Lab** Conductivity as an Indicator of Bond Type ◆ ADVANCED **TE** **Demonstration** Electrical Conduction in Ionic Compounds, p. 172 ◆ ADVANCED
BLOCK 6 • 45 min pp. 176–181 **Section 3** Names and Formulas of Ionic Compounds	**SE** **Element Spotlight** A Major Nutritional Mineral	**CRF** **Microscale Lab** Tests for Iron(II) and Iron(III) ◆ ADVANCED

BLOCKS 7 & 8 • 90 min **Chapter Review and Assessment Resources**

- **SE** **Chapter Review,** pp. 183–186
- **SE** **Standardized Test Prep,** p. 187
- **CRF** **Chapter Test** *
- **OSP** **Test Generator**
- **CRF** **Test Item Listing** *
- **OSP** **Scoring Rubrics and Classroom Management Checklists**

Holt Chemistry: Online Resources

Visit **go.hrw.com** for a variety of free resources related to this textbook. Enter the keyword **HW4 HOME**.

Holt Online Learning

Students can access interactive problem solving help and active visual concept development with the *Holt Chemistry* Online Edition available at **www.hrw.com**.

student CNN News

cnnstudentnews.com

Find the latest chemistry news, lesson plans, and activities related to important scientific events.

Compression guide:
To shorten your instruction because of time limitations, omit blocks 1, 3, 5, and 7.

KEY

TE Teacher Edition	**CRF** Chapter Resource File	***** Also on One-Stop Planner
SE Student Edition	**TT** Teaching Transparencies	**◆** Requires Advance Prep
OSP One-Stop Planner		

PROBLEM SOLVING AND PRACTICE	SECTION REVIEW AND ASSESSMENT	STANDARDS CORRELATION
		National Science Education Standards
	TE Quiz, p. 165 (GENERAL) **CRF** Quiz * **TE** Reteaching, p. 165 (BASIC) **SE** Section Review, p. 165 **CRF** Concept Review *	PS 2a
SE **Skills Toolkit 1** How to Identify a Compound as Ionic, p. 173 **TE** Homework, p. 173 (ADVANCED)	**TE** Homework, p. 170 (BASIC) **TE** Quiz, p. 175 (GENERAL) **CRF** Quiz * **TE** Reteaching, p. 175 (BASIC) **SE** Section Review, p. 175 **CRF** Concept Review *	PS 2a, 2c, 2d
SE **Skills Toolkit 2** Writing the Formula of an Ionic Compound, p. 177 **TE** Homework, p. 177 (BASIC) **SE** **Skills Toolkit 3** Naming Compounds with Polyatomic Ions, p. 179 **TE** Homework, p. 179 (BASIC) **SE** **Sample Problem A** Formula of a Compound with a Polyatomic Ion, p. 179 **SE** Practice, p. 180 (GENERAL)	**TE** Quiz, p. 180 (GENERAL) **CRF** Quiz * **TE** Reteaching, p. 180 (GENERAL) **SE** Section Review, p. 180 **CRF** Concept Review *	

SCILINKS

www.scilinks.org

Topic: Crystalline Solids
SciLinks code: HW4037

Topic: Inert Gases
SciLinks code: HW4070

Topic: Ionic Bonds
SciLinks code: HW4071

Topic: Salt Formations
SciLinks code: HW4112

Topic: Ionic Compounds
SciLinks code: HW4072

Topic: Salts
SciLinks code: HW4166

Topic: Salt Properties
SciLinks code: HW4113

Topic: Sodium
SciLinks code: HW4173

Technology Resources

  Science in the NEWS

Each video segment is accompanied by a Critical Thinking Worksheet.

Segment 14
Harvesting Salt

 ChemFile Interactive Tutor CD-Rom

Module 4: Chemical Bonding
Topic: Ionic Bonding

Overview

This chapter introduces students to ions and ionic compounds. Students will learn how ions are formed from their atoms, how ions bond to form ionic compounds, how to identify ionic compounds by their properties and how to name and write the formulas for ionic compounds.

Assessing Prior Knowledge

Check for Content Knowledge

Before beginning this chapter, be sure your students have had exposure to the following concepts:

1. atomic structure and electron configurations

2. the relationship between electron configurations and groups on the periodic table

Using the Figure

This picture shows many beautifully symmetrical crystals of sodium chloride. A crystal is a structure that is made up of repeating unit cells. When rock salt is spread on an icy road or table salt is sprinkled on food, regular cubes are observed. Most people do not know that these cubes may come from an enormous crystal that can extend for miles in an underground salt deposit. Microscopy demonstrates that a crystal's structure is uniform even on a very small scale.

Ionic crystals differ from covalently bonded crystals such as diamond because of the forces that hold the crystals together. The crystals are bonded because of the attraction between ions of opposite charge.

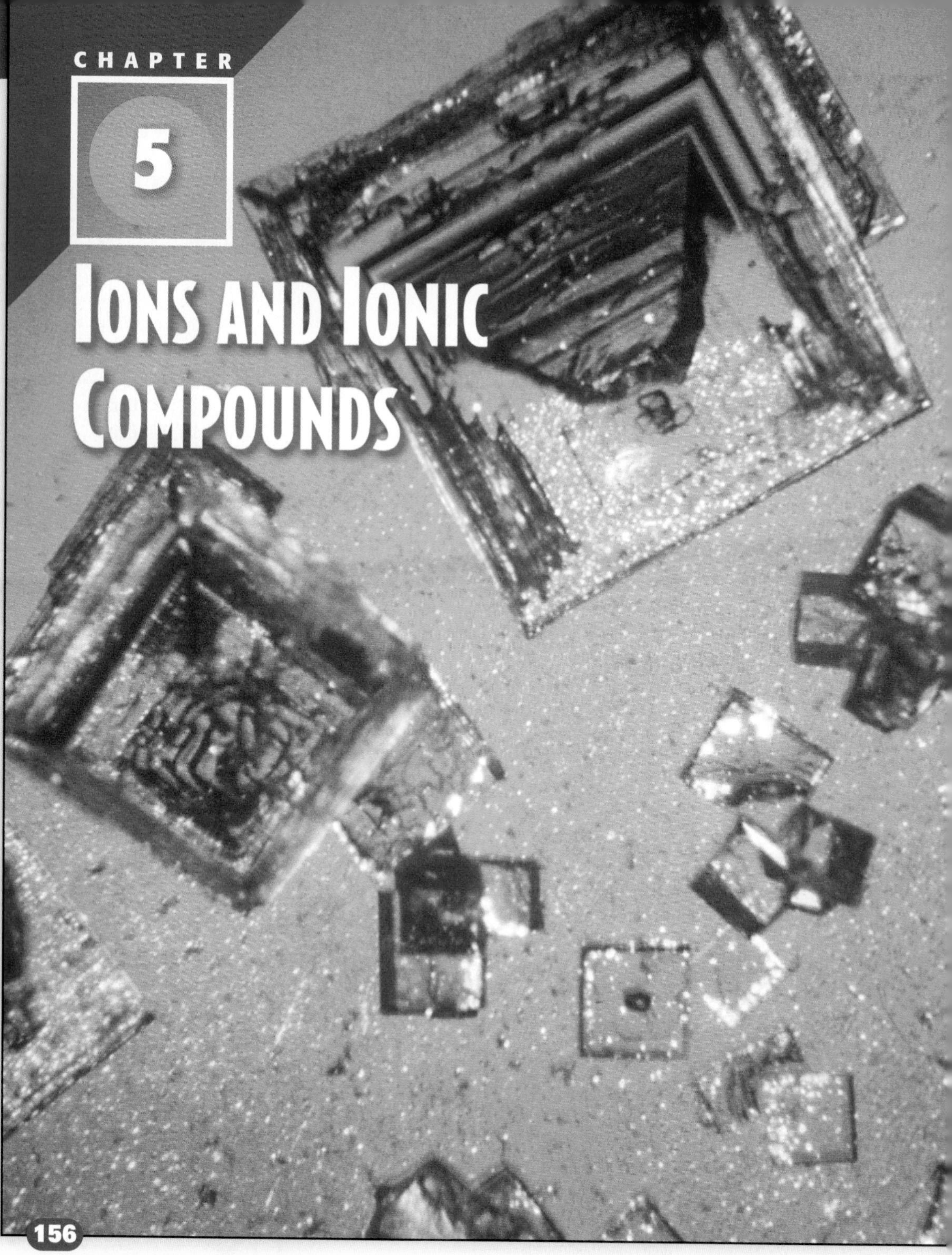

CHAPTER

5

IONS AND IONIC COMPOUNDS

156

The photograph provides a striking view of an ordinary substance—sodium chloride, more commonly known as table salt. Sodium chloride, like thousands of other compounds, is usually found in the form of crystals. These crystals are made of simple patterns of ions that are repeated over and over, and the result is often a beautifully symmetrical shape.

Ionic compounds share many interesting characteristics in addition to the tendency to form crystals. In this chapter you will learn about ions, the compounds they form, and the characteristics that these compounds share.

START-UP ACTIVITY

Hard Water

SAFETY PRECAUTIONS

PROCEDURE

1. Fill two **14 × 100 test tubes** halfway with **distilled water** and a third test tube with **tap water.**

2. Add about 1 tsp **Epsom salts** to one of the test tubes containing distilled water to make "hard water." Label the appropriate test tubes "Distilled water," "Tap water," and "Hard water."

3. Add a squirt of **liquid soap** to each test tube. Take one test tube, stopper it with a **cork,** and shake vigorously for 15 s. Repeat with the other two test tubes.

4. Observe the suds produced in each test tube.

ANALYSIS

1. Which water sample produces the most suds? Which produces the least suds?

2. What is meant by the term "hard water"? Is the water from your tap "hard water"?

Pre-Reading Questions

1. **What is the difference between an atom and an ion?**

2. **How can an atom become an ion?**

3. **Why do chemists call table salt sodium chloride?**

4. **Why do chemists write the formula for sodium chloride as NaCl?**

CONTENTS [5]

internet connect

www.scilinks.org
Topic: Crystalline Solids
SciLinks code: HW4037

SCI LINKS. Maintained by the National Science Teachers Association

157

START-UP ACTIVITY

Materials:

For each group:
- test tubes (3)
- test tube rack
- epsom salts
- liquid soap (one that does not contain detergents)

Teacher's Notes: Remind the students that they do not need to add very much Epsom salts to the distilled water. The same thing is true of the single squirt of liquid soap. Depending on how hard or soft the tap water is in your region, you may or may not get foaming in the test tube with the tap water. Because the soaps are designed to work in all kinds of water, students may get suds even with hard water, but the degree of foam formed will not be as great as in the distilled water test tube.

Safety Caution: Remind students to use goggles and aprons for this activity.

Answers

1. The distilled water produces the most suds. The Epsom salt/distilled water or the tap water will produce the least amount of suds.

2. "Hard water" is water that contains certain ions, specifically calcium and magnesium ions. Soap does not work as effectively in hard water, and scale can be deposited in pipes from hard water.

Answers to Pre-Reading Questions

1. An atom is neutral, and an ion is a charged particle.

2. An atom can become an ion by either losing or gaining electrons.

3. Sodium chloride is the systematic name for this compound, which is formed by the reaction of sodium with chlorine.

4. Ionic compounds are not composed of molecules, so the formula indicates the simplest whole number ratio of cations to anions. The compound sodium chloride has one Na^+ ion for each Cl^- ion, so the formula is written as NaCl.

Focus

Overview

Before beginning this section, review with your students the Objectives listed in the Student Edition. This section introduces students to ions. Students learn how valence shell electron configurations determine ion formation and reactivity. They will also learn how the properties of ions differ from those of their parent atoms.

Bellringer

Have students review the properties and atomic structures of the noble gas (Group 18) elements. Ask them what the electron configurations of these elements have in common.

Motivate

Identifying
Preconceptions — GENERAL

1. How many valence electrons do elements in Group 18 have? **Ans.** Helium has 2; all the rest have 8. Students may not notice that helium has a completed outer shell without having an octet. Noticing this will help the students to see why elements like lithium can exist as a 1+ ion and not have an octet.

2. How can you determine how many electrons are in the outermost energy level of an atom? **Ans.** By adding up the superscripts of the energy sublevels with the highest coefficient (n value). Students often see "p" sublevel electrons as belonging to a different energy level than "s" electrons, even if they have the same n value.

LS Verbal

Simple Ions

KEY TERMS
- octet rule
- ion
- cation
- anion

OBJECTIVES

1. **Relate** the electron configuration of an atom to its chemical reactivity.

2. **Determine** an atom's number of valence electrons, and use the octet rule to predict what stable ions the atom is likely to form.

3. **Explain** why the properties of ions differ from those of their parent atoms.

Chemical Reactivity

Some elements are highly reactive, while others are not. For example, **Figure 1** compares the difference in reactivity between oxygen and neon. Notice that oxygen reacts readily with magnesium, but neon does not. Why is oxygen so reactive while neon is not? How much an element reacts depends on the electron configuration of its atoms. Examine the electron configuration for oxygen.

$$[O] = 1s^2 2s^2 2p^4$$

Notice that the $2p$ orbitals, which can hold six electrons, have only four. The electron configuration of a neon atom is shown below.

$$[Ne] = 1s^2 2s^2 2p^6$$

Notice that the $2p$ orbitals in a neon atom are full with six electrons.

Figure 1
Because of its electron configuration, oxygen reacts readily with magnesium (**a**). In contrast, neon's electron configuration makes it unreactive (**b**).

a   **magnesium in oxygen**   b   **magnesium in neon**

158

HISTORY
CONNECTION

The independent existence of ions was proposed by the Swedish chemist Svante Arrhenius in 1884. In 1916, Walther Kossel proposed that atoms lose or gain electrons to achieve the configuration of a noble gas and that the compounds formed consist of ions.

Chapter Resource File
- **Lesson Plan**

Noble Gases Are the Least Reactive Elements

Neon is a member of the noble gases, which are found in Group 18 of the periodic table. The noble gases show almost no chemical reactivity. Because of this, noble gases have a number of uses. For example, helium is used to fill balloons that float in air, which range in size from party balloons to blimps. Like neon, helium will not react with the oxygen in the air. The electron configuration for helium is $1s^2$. The two electrons fill the first energy level, making helium stable.

The other noble gases also have filled outer energy levels. This electron configuration can be written as ns^2np^6 where n represents the outer energy level. Notice that this level has eight electrons. These eight electrons fill the s and p orbitals, making these noble gases stable. In most chemical reactions, atoms tend to match the s and p electron configurations of the noble gases. This tendency is called the **octet rule.**

Alkali Metals and Halogens Are the Most Reactive Elements

Based on the octet rule, an atom whose outer s and p orbitals do not match the electron configurations of a noble gas will react to lose or gain electrons so the outer orbitals will be full. This prediction holds true for the alkali metals, which are some of the most reactive elements. **Figure 2** shows what happens when potassium, an alkali metal, is dropped into water. An explosive reaction occurs immediately, releasing heat and light.

As members of Group 1, alkali metals have only one electron in their outer energy level. When added to water, a potassium atom gives up this electron in its outer energy level. Then, potassium will have the s and p configuration of a noble gas.

$$1s^2 2s^2 2p^6 3s^2 3p^6 4s^1 \longrightarrow 1s^2 2s^2 2p^6 3s^2 3p^6$$

The halogens are also very reactive. As members of Group 17, they have seven electrons in their outer energy level. By gaining just one electron, a halogen will have the s and p configuration of a noble gas. For example, by gaining one electron, chlorine's electron configuration becomes $1s^2 2s^2 2p^6 3s^2 3p^6$.

Figure 2
Alkali metals, such as potassium, react readily with a number of substances, including water.

Topic Link

Refer to the "Periodic Table" chapter for a discussion of the stability of the noble gases.

octet rule

a concept of chemical bonding theory that is based on the assumption that atoms tend to have either empty valence shells or full valence shells of eight electrons

159

MISCONCEPTION ALERT

Students sometimes make the mistake of thinking that all noble gases have "full" outer energy levels. Helium and neon have filled energy levels, but the remainder of the noble gases have unfilled d or f orbitals. For example, argon has a stable octet of $3s^2 3p^6$ in its outer level, and the third level still has five d orbitals available. These $3d$ orbitals fill with electrons in the transition elements in the fourth period.

Teach, continued

Teaching Tip

Ions may have the same electron configuration as noble gases, but ask students whether the ions have become noble gases. Also discuss the difference between sodium metal and sodium ions and their effect on the human body.

Teaching Tip

Students may need ways to remember which ions are cations and which are anions. Some different methods are: (1) negative and positive are in the same alphabetical order as anion and cation, (2) the "t" in cation is similar to the plus sign in a cation and the word anion can be thought of as an abbreviation for "*a negative ion.*"

Group Activity —— GENERAL

Write the electron configurations of the noble gases on the board as column headings. Divide the class into groups of three students each. Call out an element by name. Have the groups work out the electron configuration for the ion of that element and identify the noble gas with which the ion is isoelectronic. Pick a group, and have a student go to the board and write the symbol for the ion (including its charge) under the proper noble gas. Be sure to choose among only those elements listed in **Figure 5**. Other elements are unlikely to have ions isoelectronic with noble gases because of the presence of *d* or *f* orbitals. **LS** Logical

Valence Electrons

Topic Link

Refer to the "Periodic Table" chapter for more about valence electrons.

You may have noticed that the electron configuration of potassium after it loses one electron is the same as that of chlorine after it gains one. Also, both configurations are the same as that of the noble gas argon.

$$[Ar] = 1s^2 2s^2 2p^6 3s^2 3p^6$$

After reacting, both potassium and chlorine have become stable. The atoms of many elements become stable by achieving the electron configuration of a noble gas. These electrons in the outer energy level are known as valence electrons.

Periodic Table Reveals an Atom's Number of Valence Electrons

It is easy to find out how many valence electrons an atom has. All you have to do is check the periodic table.

For example, **Figure 3** highlights the element magnesium, Mg. The periodic table lists its electron configuration.

$$[Mg] = [Ne]3s^2$$

This configuration shows that a magnesium atom has two valence electrons in the 3*s* orbital.

Now check the electron configuration of phosphorus, which is also highlighted in **Figure 3.**

$$[P] = [Ne]3s^2 3p^3$$

This configuration shows that a phosphorus atom has five valence electrons. Two valence electrons are in the 3*s* orbital, and three others are in the 3*p* orbitals.

Figure 3
The periodic table shows the electron configuration of each element. The number of electrons in the outermost energy level is the number of valence electrons.

Group 1							Group 18
Hydrogen **H**							Helium **He**
	Group 2	Group 13	Group 14	Group 15	Group 16	Group 17	
Lithium **Li**	Beryllium **Be**	Boron **B**	Carbon **C**	Nitrogen **N**	Oxygen **O**	Fluorine **F**	Neon **Ne**
Sodium **Na**	Magnesium **Mg**	Aluminum **Al**	Silicon **Si**	Phosphorus **P**	Sulfur **S**	Chlorine **Cl**	Argon **Ar**

160

did you know?

Table salt as it is usually sold is not pure NaCl. It usually contains small amounts of finely divided insoluble substances such as silicates or complex salts of aluminum. Particles of these materials stick to the cubic NaCl crystals and keep them from clumping together in high humidity. In addition, iodized table salt contains small amounts of potassium iodide, which provides dietary iodine. For these reasons, NaCl solutions made with table salt are cloudy and generally are not suitable for use in the chemistry laboratory.

Atoms Gain Or Lose Electrons to Form Stable Ions

Recall that potassium loses its one valence electron so it will have the electron configuration of a noble gas. But why doesn't a potassium atom gain seven more electrons to become stable instead? The reason is the energy that is involved. Removing one electron requires far less energy than adding seven more.

When it gives up one electron to be more stable, a potassium atom also changes in another way. Recall that all atoms are uncharged because they have equal numbers of protons and electrons. For example, a potassium atom has 19 protons and 19 electrons. After giving up one electron, potassium still has 19 protons but only 18 electrons. Because the numbers are not the same, there is a net electrical charge. So the potassium atom becomes an **ion** with a 1+ charge, as shown in **Figure 4.** The following equation shows how a potassium atom forms an ion.

$$K \longrightarrow K^+ + e^-$$

An ion with a positive charge is called a **cation.** A potassium cation has an electron configuration just like the noble gas argon.

$$[K^+] = 1s^2 2s^2 2p^6 3s^2 3p^6 \qquad [Ar] = 1s^2 2s^2 2p^6 3s^2 3p^6$$

In the case of chlorine, far less energy is required for an atom to gain one electron rather than give up its seven valence electrons. By gaining an electron to be more stable, a chlorine atom becomes an ion with a 1− charge, as illustrated in **Figure 4.** The following equation shows the formation of a chlorine ion from a chlorine atom.

$$Cl + e^- \longrightarrow Cl^-$$

An ion with a negative charge is called an **anion.** A chlorine anion has an electron configuration just like the noble gas argon.

$$[Cl^-] = 1s^2 2s^2 2p^6 3s^2 3p^6 \qquad [Ar] = 1s^2 2s^2 2p^6 3s^2 3p^6$$

ion

an atom, radical, or molecule that has gained or lost one or more electrons and has a negative or positive charge

cation

an ion that has a positive charge

anion

an ion that has a negative charge

Figure 4
A potassium atom can lose an electron to become a potassium cation (**a**) with a 1+ charge. After gaining an electron, a chlorine atom becomes a chlorine anion (**b**) with a 1− charge.

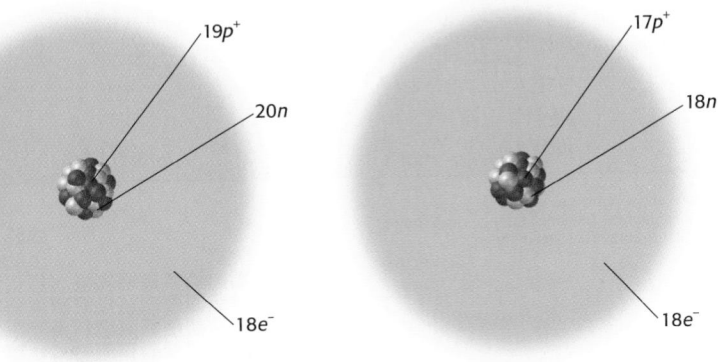

a potassium cation, K$^+$ b chloride anion, Cl$^-$

161

Teach, *continued*

SKILL BUILDER — BASIC

Interpreting Visuals Ask students to explain what all of the elements of the same color code in **Figure 5** have in common. Ans. They all form ions that are isoelectronic with the noble gas with the same color. Pick out random elements from the table, and ask students whether each forms an ion by gaining or losing electrons. Students should discover that elements preceding the noble gases (Groups 15–17) gain electrons to form anions. Those that follow the noble gases (Groups 1, 2, and 13) lose electrons to form cations.

LS Visual

Characteristics of Stable Ions

How does an atom compare to the ion that it forms after it loses or gains an electron? Use of the same name for the atom and the ion that it forms indicates that the nucleus is the same as it was before. Both the atom and the ion have the same number of protons and neutrons. When an atom becomes an ion, it only involves loss or gain of electrons.

Recall that the chemical properties of an atom depend on the number and configuration of its electrons. Therefore, an atom and its ion have different chemical properties. For example, a potassium cation has a different number of electrons from a neutral potassium atom, but the same number of electrons as an argon atom. A chlorine anion also has the same number of electrons as an argon atom. However, it is important to realize that an ion is still quite different from a noble gas. An ion has an electrical charge, so therefore it forms compounds, and also conducts electricity when dissolved in water. Noble gases are very unreactive and have none of these properties.

Figure 5
These are examples of some stable ions that have an electron configuration like that of a noble gas.

Some Ions with Noble-Gas Configurations

Each color denotes ions and a noble gas that have the same electron configurations. The small table at right shows the periodic table positions of the ions listed above.

Transparencies

TT Some Ions with Noble-Gas Configurations

Many Stable Ions Have Noble-Gas Configurations

Potassium and chlorine are not the only atoms that form stable ions with a complete octet of valence electrons. **Figure 5** lists examples of other atoms that form ions with a full octet. For example, examine how calcium, Ca, forms a stable ion. The electron configuration of a calcium atom is written as follows.

$$[Ca] = 1s^2 2s^2 2p^6 3s^2 3p^6 4s^2$$

By giving up its two valence electrons in the $4s$ orbital, calcium forms a stable cation with a 2+ charge that has an electron configuration like that of argon.

$$[Ca^{2+}] = 1s^2 2s^2 2p^6 3s^2 3p^6$$

Some Stable Ions Do Not Have Noble-Gas Configurations

Not all atoms form stable ions with an electron configuration like those of noble gases. As illustrated in **Figure 6**, transition metals often form ions without complete octets. Notice that, with the lone exception of rhenium, Re, these stable ions are all cations. Also notice in **Figure 6** that some elements, mostly transition metals, form stable ions with more than one charge. For example, copper, Cu, can give up one electron, forming a Cu^+ cation. It can also give two electrons, forming a Cu^{2+} cation. Both the Cu^+ and Cu^{2+} cations are stable even though they do not have noble-gas configurations.

Topic Link

Refer to the "Atoms" chapter for more about electron configuration.

Figure 6
Some stable ions do not have electron configurations like those of the noble gases.

Teaching Tip

Interpreting Visuals The octet rule is useful for predicting the ionic charges of only a few elements, mostly among the representative groups. Transition elements and a few others, as shown in **Figure 6,** typically can have several different ionic charges because electrons in d orbitals are very close in energy to the s and p valence electrons. As a result, these d electrons sometimes participate in ion formation.

Stable Ions Formed by the Transition Elements and Some Other Metals

Group 4	Group 5	Group 6	Group 7	Group 8	Group 9	Group 10	Group 11	Group 12	Group 13	Group 14
Ti^{2+} Ti^{3+}	V^{2+} V^{3+}	Cr^{2+} Cr^{3+}	Mn^{2+} Mn^{3+}	Fe^{2+} Fe^{3+}	Co^{2+} Co^{3+}	Ni^{2+}	Cu^+ Cu^{2+}	Zn^{2+}	Ga^{2+} Ga^{3+}	Ge^{2+}
		Mo^{3+}	Tc^{2+}			Pd^{2+}	Ag^+ Ag^{2+}	Cd^{2+}	In^+ In^{2+} In^{3+}	Sn^{2+}
Hf^{4+}			Re^{4+} Re^{5+}			Pt^{2+} Pt^{4+}	Au^+ Au^{3+}	Hg_2^{2+} Hg^{2+}	Tl^+ Tl^{3+}	Pb^{2+}

The small table at left shows the periodic table positions of the ions listed above.

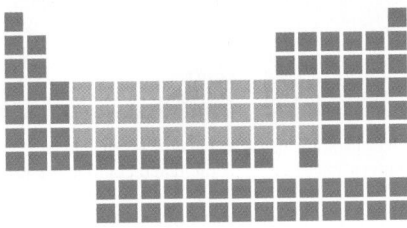

Transparencies

TT Stable Ions Formed by the Transition Elements and Some Other Metals

Interpreting Visuals Ask students which portion of **Figure 7** shows the particles in a higher state of energy. They should be able to see that the atomic form of sodium and chlorine are highly reactive because they are forming ions, while the ions formed are inert. **LS** Visual

Atoms and Ions

Many atoms form stable ions that have noble-gas configurations. It is important to remember that these elements do not actually *become* noble gases. Having identical electron configurations does not mean that a sodium cation is a neon atom. The sodium cation still has 11 protons and 12 neutrons, like a sodium atom that has not reacted to form an ion. But like a noble-gas atom, a sodium ion is very unlikely to gain or lose any more electrons.

Ions and Their Parent Atoms Have Different Properties

Like potassium and all other alkali metals of Group 1, sodium is extremely reactive. When it is placed in water, a violent reaction occurs, producing heat and light. Like all halogens of Group 17, chlorine is extremely reactive. In fact, atoms of chlorine react with each other to form molecules of chlorine, Cl_2, a poisonous, yellowish green gas. In nature, however, chlorine is almost always found in its anion form as part of ionic compounds.

Because both sodium and chlorine are very reactive, you might expect a violent reaction when these two are brought together. This is exactly what happens. If a small piece of sodium is lowered into a flask filled with chlorine gas, there is a violent reaction that releases both heat and light. After the reaction is complete, all that remains is a white solid. Even though it is formed from two dangerous elements, it is something you probably eat every day—table salt.

Chemists call this salt *sodium chloride*. Sodium chloride is made from sodium cations and chloride anions. As illustrated in **Figure 7,** these ions have very different properties than those of their parent atoms. That is why salt is not as dangerous to have around your house as the elements that make it up. It does not react with water like sodium metal does because salt contains stable sodium ions, not reactive sodium atoms.

Figure 7

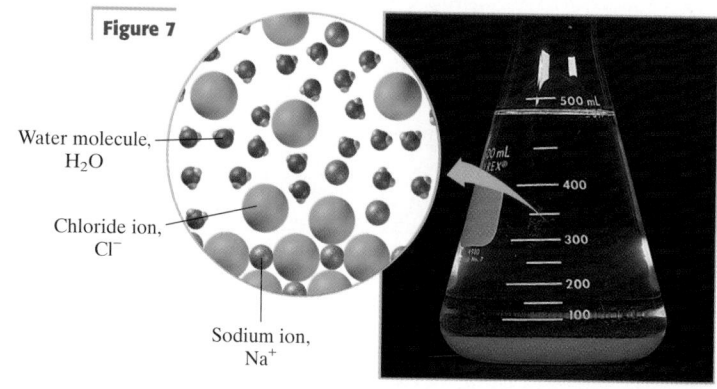

Water molecule, H_2O

Chloride ion, Cl^-

Sodium ion, Na^+

a Sodium chloride dissolves in water to produce unreactive sodium cations and chlorine anions.

b In contrast, the elements sodium and chlorine are very reactive when they are brought together.

Videos

CNN Presents Chemistry Connections
• **Segment 14** Harvesting Salt

See the Science in the News video guide for more details.

Atoms of Metals and Nonmetal Elements Form Ions Differently

Nearly all metals form cations, as can be seen by examining their electron configuration. For example, consider the configuration for the Group 2 metal magnesium, Mg.

$$[Mg] = 1s^2 2s^2 2p^6 3s^2$$

To have a noble-gas configuration, the atom must either gain six electrons or lose two. Losing two electrons requires less energy than gaining six. Similarly, for all metals, the energy required to remove electrons from atoms to form ions with a noble-gas configuration is always less than the energy required to add more electrons. As a result, the atoms of metals form cations.

In contrast, the atoms of nonmetal elements form anions. Consider the example of oxygen, whose electron configuration is written as follows.

$$[O] = 1s^2 2s^2 2p^4$$

To have a noble-gas configuration, an oxygen atom must either gain two electrons or lose six. Acquiring two electrons requires less energy than losing six. For other nonmetals, the energy required to add electrons to atoms of nonmetals so that their ions have a noble-gas configuration is always less than the energy required to remove enough electrons. As a result, the atoms of nonmetal elements form anions.

 # Section Review

UNDERSTANDING KEY IDEAS

1. Explain why the noble gases tend not to react.

2. Where are the valence electrons located in an atom?

3. How does a cation differ from an anion?

4. State the octet rule.

5. Why do the properties of an ion differ from those of its parent atom?

6. Explain why alkali metals are extremely reactive.

7. How can you determine the number of valence electrons an atom has?

8. Explain why almost all metals tend to form cations.

9. Explain why, as a pure element, oxygen is usually found in nature as O_2.

CRITICAL THINKING

10. How could each of the following atoms react to achieve a noble-gas configuration?

   a. iodine

   b. strontium

   c. nitrogen

   d. krypton

11. Write the electron configuration for each of the following ions.

   a. Al^{3+}

   b. Se^{2-}

   c. Sc^{3+}

   d. As^{3-}

12. In what way is an ion the same as its parent atom?

13. To achieve a noble-gas configuration, a phosphorus atom will form a P^{3-} anion rather than forming a P^{5+} cation. Why?

Answers to Section Review

1. The octet of electrons in their outer energy levels makes them stable.

2. in the outermost energy level

3. A cation is formed by an atom that loses one or more electrons. An anion is formed by an atom that gains one or more electrons.

4. Atoms lose or gain electrons to match the valence-electron configuration of a noble gas.

5. different electron configurations and charge

6. Their atoms can easily give up their one valence electron.

7. number of electrons in the outer energy level, listed in the periodic table

8. Metals have few enough valence electrons that the energy required to remove electrons from their atoms and form ions with a noble-gas configuration is less than the energy required to add more electrons.

9. Oxygen is extremely reactive, so its atoms react with each other to form molecules of O_2.

10. a. gain one electron

   b. lose two electrons

   c. gain three electrons

   d. no change

Answers continued on p. 187A

Overview

Before beginning this section, review with your students the Objectives listed in the Student Edition. This section introduces students to ionic compounds. Students learn about the steps atoms undergo in order to form an ionic compound, the characteristic properties of ionic compounds and the crystalline structure of these compounds.

 **Bellringer**

Have students review the concepts of ionization energy, electron affinity, anion formation, cation formation, and vaporization. Have them write down whether each of these processes involves the gain or loss of energy.

Motivate

Identifying Preconceptions — GENERAL

1. Name some substances that you think are ionic compounds. **Ans.** Answers will vary. Students should be led to name compounds containing the elements found in figures 5 and 6 in section 1.

2. Does the fact that the reaction of sodium with chlorine releases energy mean that no energy is needed for the formation of sodium chloride? **Ans.** While this topic is developed thoroughly in this section, remind students that although the overall reaction releases energy, individual steps in the process require energy.

LS Logical

Ionic Bonding and Salts

KEY TERMS
• salt
• lattice energy
• crystal lattice
• unit cell

OBJECTIVES

① **Describe** the process of forming an ionic bond.

② **Explain** how the properties of ionic compounds depend on the nature of ionic bonds.

③ **Describe** the structure of salt crystals.

Ionic Bonding

You may think that the material shown in **Figure 8** is very valuable. If you look closely, you will see what appear to be chunks of gold. The object shown in **Figure 8** is actually a mineral called *pyrite*, which does not contain any gold. However, the shiny yellow flakes make many people believe that they have discovered gold. All they have really discovered is a mineral that is made of iron cations and sulfur anions.

Because opposite charges attract, cations and anions should attract one another. This is exactly what happens when an ionic bond is formed. In the case of pyrite, the iron cations and sulfur anions attract one another to form an ionic compound.

Figure 8
The mineral pyrite is commonly called *fool's gold*. Unlike real gold, pyrite is actually quite common in Earth's crust.

HISTORY
CONNECTION

In 1912, the German physicist Max von Laue discovered that X rays passed through a crystal are scattered, or diffracted, in a pattern that depends on the arrangement of atoms, ions, or molecules in the crystal. This technique, called X-ray crystallography, was used in the early 1950s by British scientist Rosalind Franklin to confirm the double-helix structure of DNA.

Chapter Resource File
• Lesson Plan

Ionic Bonds Form Between Ions of Opposite Charge

To understand how an ionic bond forms, take another look at what happens when sodium and chlorine react to form sodium chloride. Recall that sodium gives up its only valence electron to form a stable Na^+ cation. Chlorine, with seven valence electrons, acquires that electron. As a result, a chlorine atom becomes a stable Cl^- anion.

The force of attraction between the 1+ charge on the sodium cation and the 1− charge on the chloride anion creates the ionic bond in sodium chloride. Recall that sodium chloride is the scientific name for table salt. Chemists call table salt by its scientific name because the word **salt** can actually be used to describe any one of thousands of different ionic compounds. Other salts that are commonly found in a laboratory include potassium chloride and calcium iodide.

All these salts are ionic compounds that are electrically neutral. They are made up of cations and anions that are held together by ionic bonds in a simple, whole-number ratio. For example, sodium chloride consists of sodium cations and chloride anions bonded in a 1:1 ratio. To show this 1:1 ratio, chemists write the formula for sodium chloride as NaCl.

However, the attractions between the ions in a salt do not stop with a single cation and a single anion. These forces are so far reaching that one cation attracts several different anions. At the same time, each anion attracts several different cations. In this way, many ions are pulled together into a tightly packed structure. The tight packing of the ions causes any salt, such as sodium chloride, to have a distinctive crystal structure. The smallest crystal of table salt that you could see would still have more than a billion billion sodium and chloride ions.

Transferring Electrons Involves Energy Changes

Recall that ionization energy is the energy that it takes to remove the outermost electron from an atom. In other words, moving a negatively charged electron away from an atom that will become a positively charged ion requires an input of energy before it will take place. In the case of sodium, this process can be written as follows.

$$Na + energy \longrightarrow Na^+ + e^-$$

Recall that electron affinity is the energy needed to add an electron onto a neutral atom. However, some elements, such as chlorine, easily accept extra electrons. For elements like this, energy is released when an electron is added. This process can be written as follows.

$$Cl + e^- \longrightarrow Cl^- + energy$$

But this energy released is less than the energy required to remove an electron from a sodium atom. Then why does an ionic bond form if these steps do not provide enough energy? Adding and removing electrons is only part of forming an ionic bond. The rest of the process of forming a salt supplies more than enough energy to make up the difference so that the overall process releases energy.

salt

an ionic compound that forms when a metal atom or a positive radical replaces the hydrogen of an acid

www.scilinks.org
Topic: Ionic Bonds
SciLinks code: HW4071

Maintained by the
National Science
Teachers Association

Demonstration
Electrostatic Attraction

Blow up a small rubber balloon, and tie a 40 cm string to it. Rub the balloon with some fur or wool to give it a negative charge. The fur or wool will now be positively charged. Allow the balloon to dangle near the positively charged fur or wool. A strong attraction between opposite charges will be evident.

Blow up a second balloon, and tie it to the other end of the string. Rub both balloons, and show that they repel each other strongly because they have like charges. Ask the students what this demonstration might have to do with ions in a crystal of an ionic compound.

Safety: None

Disposal: Put balloons in the trash.

Teach

READING SKILL BUILDER — BASIC

K-W-L Have the students draw three columns on their papers. Label column one K, column two W, and column three L. In the K column the students should write down what they already know about ionic compounds, ionic bonds, and crystals. In the W column they should write down what they want to learn about ionic compounds, bonds and crystals. They should fill in the L column with what they have learned after they read the section. **LS Verbal**

Teaching Tip

Salt Formation Point out that binary ionic compounds form easily when an atom of fairly low ionization potential meets an atom of high electron affinity. (Such atoms also have a large electronegativity difference.) Then one or more electrons are transferred and ions are formed.

Point out that when atoms become ions, their properties are dramatically altered. In forming the white ionic compound sodium chloride, NaCl, sodium changes from a silvery-gray soft solid to colorless ions, and chlorine changes from a yellowish-green gas to colorless ions.

Group Activity —— GENERAL

Interpreting Visuals Have the students separate into groups, and assign each group one of the steps in **Figure 9.** Have each group draw a graphic showing what has occurred in that step. For example, for the formation of ions, the students should use circles of the appropriate size to indicate the neutral atom versus the cation (smaller than the atom) or anion (larger than the atom) formed. For the bond breaking in chlorine gas, they should show the reactants as diatomic molecules and the products as separate atoms. For the formation of the sodium chloride crystal, they should indicate the crystal structure and the alternating positive and negative ions.
LS Visual

internet connect

www.scilinks.org
Topic: Salt Formations
SciLinks code: HW4112

SciLINKS Maintained by the National Science Teachers Association

lattice energy

the energy associated with constructing a crystal lattice relative to the energy of all constituent atoms separated by infinite distances

Salt Formation Involves Endothermic Steps

The process of forming the salt sodium chloride can be broken down into five steps as shown in **Figure 9** on the following page. Keep in mind that these steps do not really take place in this order. However, these steps, do model what must happen for an ionic bond to form between sodium cations and chloride anions.

The starting materials are sodium metal and chlorine gas. Energy must be supplied to make the solid sodium metal into a gas. This process takes energy and can be written as follows.

$$Na(solid) + energy \longrightarrow Na(gas)$$

Recall that energy is also required to remove an electron from a gaseous sodium atom.

$$Na(gas) + energy \longrightarrow Na^+(gas) + e^-$$

No energy is required to convert chlorine into the gaseous state because it is already a gas. However, chlorine gas consists of two chlorine atoms that are bonded to one another. Therefore, energy must be supplied to separate these chlorine atoms so that they can react with sodium. This third process can be written as follows.

$$Cl-Cl(gas) + energy \longrightarrow Cl(gas) + Cl(gas)$$

To this point, the first three steps have all been endothermic. These steps have produced sodium cations and chlorine atoms.

Salt Formation Also Involves Exothermic Steps

As **Figure 9** illustrates, the next step adds an electron to a chlorine atom to form an anion. This is the first step that releases energy. Recall that this step cannot supply enough energy to remove an electron from a sodium atom. Obviously, this step cannot produce nearly enough energy to drive the first three steps.

The chief driving force for the formation of the salt is the last step, in which the separated ions come together to form a crystal held together by ionic bonds. When a cation and anion form an ionic bond, it is an exothermic process. Energy is released.

$$Na^+(gas) + Cl^-(gas) \longrightarrow NaCl(solid) + energy$$

The energy released when ionic bonds are formed is called the **lattice energy.** This energy is released when the crystal structure of a salt is formed as the separated ions bond. In the case of sodium chloride, the lattice energy is greater than the energy needed for the first three steps. Without this energy, there would not be enough energy to make the overall process spontaneous. Lattice energy is the key to salt formation.

The value of the lattice energy is different if other cations and anions form the salt. For example, Na^+ ions can form salts with anions of any of the halogens. The lattice energy values for each of these salts are about the same. However, when magnesium cations, Mg^{2+}, form salts, these values are much higher than the values for salts of sodium. This large difference in lattice energy is due to the fact that ions with greater charge are more strongly attracted to the oppositely charged ions in the crystal. The lattice energy value for magnesium oxide is almost five times greater than that for sodium chloride.

If energy is released when ionic bonds are formed, then energy must be supplied to break these bonds and separate the ions. In the case of sodium chloride, the needed energy can come from water. As a result, a sample of sodium chloride dissolves when it is added to a glass of water. As the salt dissolves, the Na^+ and Cl^- ions separate as the ionic bonds between them are broken. Because of its much higher lattice energy, magnesium oxide does not dissolve well in water. In this case, the energy that is available in a glass of water is significantly less than the lattice energy of the magnesium oxide. There is not enough energy to separate the Mg^{2+} and O^{2-} ions from one another.

Figure 9
The reaction between $Na(s)$ and $Cl_2(g)$ to form sodium chloride can be broken down into steps. More energy is released overall than is absorbed.

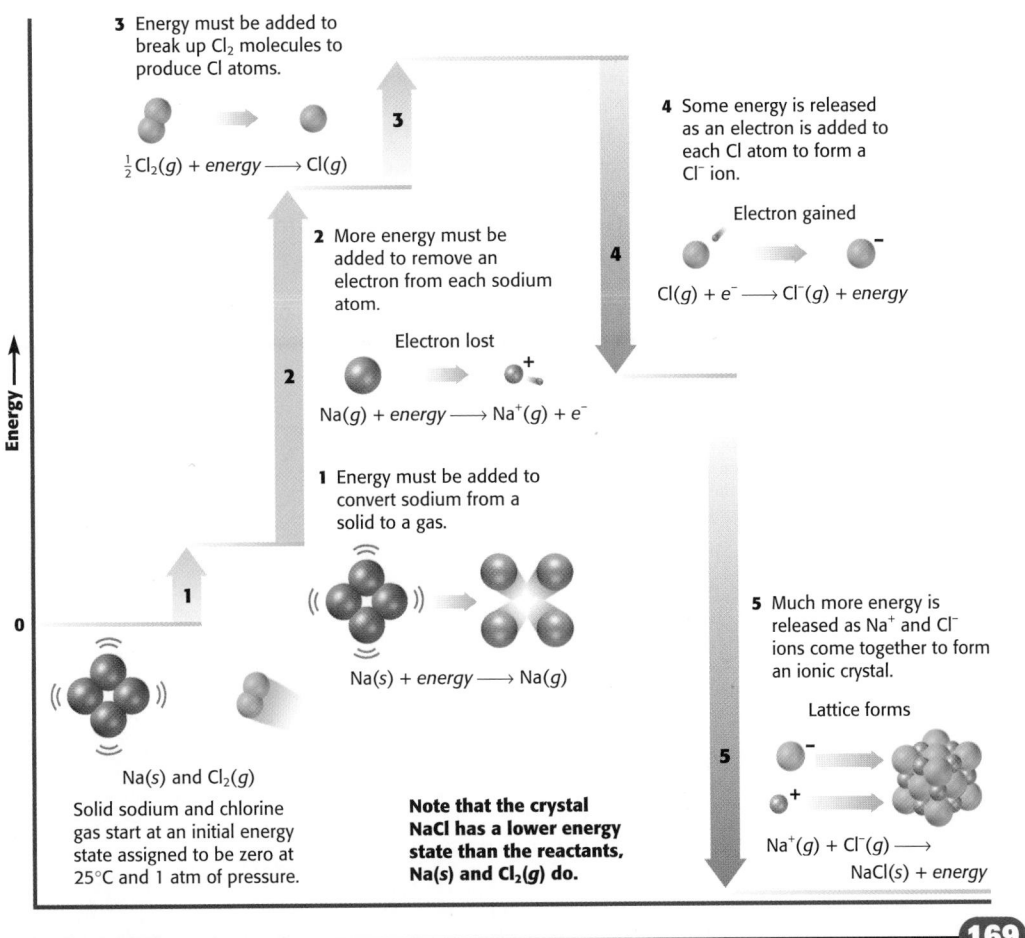

3 Energy must be added to break up Cl_2 molecules to produce Cl atoms.

$\frac{1}{2}Cl_2(g) + energy \longrightarrow Cl(g)$

4 Some energy is released as an electron is added to each Cl atom to form a Cl^- ion.

Electron gained

$Cl(g) + e^- \longrightarrow Cl^-(g) + energy$

2 More energy must be added to remove an electron from each sodium atom.

Electron lost

$Na(g) + energy \longrightarrow Na^+(g) + e^-$

1 Energy must be added to convert sodium from a solid to a gas.

$Na(s) + energy \longrightarrow Na(g)$

5 Much more energy is released as Na^+ and Cl^- ions come together to form an ionic crystal.

Lattice forms

$Na^+(g) + Cl^-(g) \longrightarrow NaCl(s) + energy$

Na(s) and $Cl_2(g)$

Solid sodium and chlorine gas start at an initial energy state assigned to be zero at 25°C and 1 atm of pressure.

Note that the crystal NaCl has a lower energy state than the reactants, Na(s) and $Cl_2(g)$ do.

169

Group Activity —— GENERAL
Interpreting Visuals Have the students determine the length or width or height of the classroom. Have the students separate into groups and calculate the length of an arrow showing the energy involved in their assigned step in **Figure 9**. The students should turn in a copy of their calculations showing use of factor labels and metric units. They should then cut pieces of paper to length and post them on a wall or floor. They should be careful to make sure that the arrows for endothermic reactions are pointing in the opposite direction of those for exothermic reactions. The students can then walk off the energy calculations.
LS Kinesthetic

MISCONCEPTION
/// **ALERT** \\\
Students should understand that even exothermic reactions require an initial input of energy to get them started. A pile of dry leaves or a gas burner may give off an abundant amount of heat once it is lit. Getting the fire started, though, requires an input of activation energy.

did you know?

The idea of breaking down the formation of crystals of compounds into a series of discrete changes was devised in order to calculate lattice energies, which cannot be measured directly. This series, diagrammed in **Figure 9,** is known as the Born-Haber cycle. It was devised by German physicist Max Born and German chemist Fritz Haber. This cycle is a specific application of Hess's law, which is discussed in the "Causes of Change" chapter.

Transparencies
TT Formation of Sodium Chloride

1. What are the properties that are characteristic of an ionic compound? **Ans.** They are solid at room temperature, are hard and brittle, have high melting and boiling points, are nonconductors as solids but good conductors in the liquid state and when dissolved in water.

2. You have been given a crystalline substance in an unmarked beaker in the chemistry laboratory and asked to determine if it is an ionic compound or not. You tap the crystals gently and they shatter but still retain their sharp edges. You heat the substance gently at first and then harder and after two to three minutes of heating it does not melt. It dissolves in water and the water solution conducts electricity. Is the substance an ionic compound? **Ans.** Yes.

LS Verbal

Ionic Compounds

Recall that salts are ionic compounds made of cations and anions. Many of the rocks and minerals in Earth's crust are made of cations and anions held together by ionic bonds. The ratio of cations to anions is always such that an ionic compound has no overall charge. For example, in sodium chloride, for every Na^+ cation, there is a Cl^- anion to balance the charge. In magnesium oxide, for every Mg^{2+} cation, there is an O^{2-} anion. Ionic compounds also share certain other chemical and physical properties.

Ionic Compounds Do Not Consist of Molecules

Figure 10 shows sodium chloride, an ionic compound, being added to water, a molecular compound. If you could look closely enough into the water, you would find individual water molecules, each made of two hydrogen atoms and one oxygen atom. The pot would be filled with many billions of these individual H_2O molecules.

Recall that the smallest crystal of table salt that you could see contains many billions of sodium and chloride ions all held together by ionic bonds. However, if you could look closely enough into the salt, all you would see are many Na^+ and Cl^- ions all bonded together to form a crystal. There are no NaCl molecules.

Elements in Groups 1 and 2 reacting with elements in Groups 16 and 17 will almost always form ionic compounds and not molecular compounds. Therefore, the formula CaO likely indicates an ionic compound because Ca is a Group 2 metal and O is a Group 16 nonmetal. In contrast, the formula ICl likely indicates a molecular compound because both I and Cl are members of Group 17.

However, you cannot be absolutely sure that something is made of ions or molecules just by looking at its formula. That determination must be made in the laboratory.

Figure 10
Salt is often added to water for flavor when pasta is being cooked.

Ionic Bonds Are Strong

Both repulsive and attractive forces exist within a salt crystal. The repulsive forces include those between like-charged ions. Within the crystal, each Na^+ ion repels the other Na^+ ions. The same is true for the Cl^- ions. Another repulsive force exists between the electrons of ions that are close together, even if the ions have opposite charges.

The attractive forces include those between the positively charged nuclei of one ion and the electrons of other nearby ions. In addition, attractive forces exist between oppositely charged ions. These forces involve more than a single Na^+ ion and a single Cl^- ion. Within the crystal, each sodium cation is surrounded by six chloride anions. At the same time, each chloride anion is surrounded by six sodium cations. As a result, the attractive force between oppositely charged ions is significantly greater in a crystal than it would be if the sodium cations and chloride anions existed only in pairs.

Overall, the attractive forces are significantly stronger than the repulsive forces, so ionic bonds are very strong.

Ionic Compounds Have Distinctive Properties

All ionic compounds share certain properties because of the strong attraction between their ions. Compare the boiling point of sodium chloride (1413°C) with that of water, a molecular compound (100°C). Similarly, most other ionic compounds have high melting and boiling points, as you can see in **Table 1.**

To melt, ions cannot be in fixed locations. Because each ion in these compounds forms strong bonds to neighboring ions, considerable energy is required to free them. Still more energy is needed to move ions out of the liquid state and cause boiling.

As a result of their high boiling points, ionic compounds are rarely gaseous at room temperature, while many molecular compounds are. Ice, for example, will eventually melt and then vaporize. In contrast, salt will remain a solid no matter how long it remains at room temperature.

internet connect
www.scilinks.org
Topic: Ionic Compounds
SciLinks code: HW4072

SC*LINKS* Maintained by the National Science Teachers Association

SKILL BUILDER — GENERAL

Interpreting Visuals Draw the students' attention to **Table 1.** Ask them to draw some conclusions about the general trends of the boiling points and melting points of ionic compounds. **LS** Logical

Table 1   Melting and Boiling Points of Compounds

Compound name	Formula	Type of compound	Melting point °C	Melting point K	Boiling point °C	Boiling point K
Magnesium fluoride	MgF_2	ionic	1261	1534	2239	2512
Sodium chloride	NaCl	ionic	801	1074	1413	1686
Calcium iodide	CaI_2	ionic	784	1057	1100	1373
Iodine monochloride	ICl	covalent	27	300	97	370
Carbon tetrachloride	CCl_4	covalent	−23	250	77	350
Hydrogen fluoride	HF	covalent	−83	190	20	293
Hydrogen sulfide	H_2S	covalent	−86	187	−61	212
Methane	CH_4	covalent	−182	91	−164	109

171

did you know?

Many ionic compounds decompose or sublime before they reach a boiling point or, in some cases, a melting point. Even so, these high transition temperatures demonstrate the strength of ionic bonds. $MgSO_4$ decomposes at 1124°C without melting; KCl melts at 772°C and sublimes at 1500°C; NH_4Cl sublimes at 350°C without melting; and $CaCO_3$ decomposes at 825°C.

Demonstration

Electrical Conduction in Ionic Compounds

(Approximate time: 15 minutes)

1. Construct the following assembly for use as a conductivity tester. Insert two straightened steel paper clips through a cork (not a rubber stopper) as shown. Spinning the paper clips will help drill them through the cork. Attach the battery, LED, resistor, and clips as shown below.

2. Half-fill a small uncracked porcelain crucible with crystalline lithium chloride, LiCl.

3. Insert the leads of the conductivity tester into the LiCl, and demonstrate that the LED does not light because the salt does not conduct electricity.

4. Remove the tester.

5. Support the crucible on a clay triangle resting on an iron ring on a ring stand. (See **Figure 11.**) Heat the crucible until the LiCl melts.

6. Insert the leads of the tester into the melted salt, and show that the LED lights because the melted salt now conducts. Remove heat, and let the crucible cool. Wash the leads of the tester with water to remove any LiCl.

7. Place 100 mL of distilled water into a clean 250 mL beaker. Place the clean leads into the water, and show that it does not conduct.

Continued on next page

Liquid and Dissolved Salts Conduct Electric Current

To conduct an electric current, a substance must satisfy two conditions. First, the substance must contain charged particles. Second, those particles must be free to move. Because ionic compounds are composed of charged particles, you might expect that they could be good conductors. While particles in a solid have some vibrational motion, they remain in fixed locations, as shown by the model in **Figure 11a.** Therefore, ionic solids, such as salts, generally are not conductors of electric current because the ions cannot move.

However, when the ions can move about, salts are excellent electrical conductors. This is possible when a salt melts or dissolves. When a salt melts, the ions that make up the crystal can freely move past each other, as **Figure 11b** illustrates. Molten salts are good conductors of electric current, although they do not conduct as well as metals. Similarly, if a salt dissolves in water, its ions are no longer held tightly in a crystal. Because the ions are free to move, as shown by the model in **Figure 11c,** the solution can conduct electric current.

As often happens in chemistry, there are exceptions to this rule. There is a small class of ionic compounds that can allow charges to move through their crystals. The lattices of these compounds have an unusually open structure, so certain ions can move past others, jumping from one site to another. One of these salts, zirconium oxide, is used in a device that controls emissions from the exhaust of automobiles.

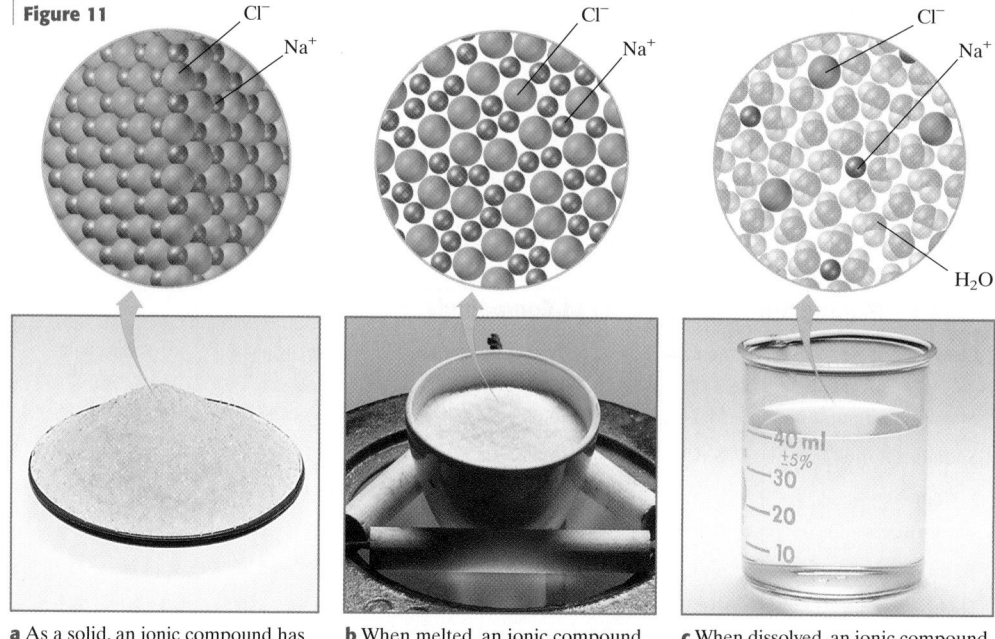

Figure 11

a As a solid, an ionic compound has charged particles that are held in fixed positions and cannot conduct electric current.

b When melted, an ionic compound conducts electric current because its charged particles move about more freely.

c When dissolved, an ionic compound conducts electric current because its charged particles move freely.

Red lead to battery cap — Long LED lead — LED

Black lead to battery cap

Short LED lead

9V

9V-transistor battery with battery cap

1 kV (¼ W) resistor with tape around it

Paper clips

Cork

Transparencies
TT Sodium Chloride in Three Phases

Salts Are Hard and Brittle

Like most other ionic compounds, table salt is fairly hard and brittle. *Hard* means that the crystal is able to resist a large force applied to it. *Brittle* means that when the applied force becomes too strong to resist, the crystal develops a widespread fracture rather than a small dent. Both of these properties can be attributed to the patterns in which the cations and anions are arranged in all salt crystals.

The ions in a crystal are arranged in a repeating pattern, forming layers. Each layer is positioned so that a cation is next to an anion in the next layer. As long as the layers stay in a fixed position relative to one another, the attractive forces between oppositely charged ions will resist motion. As a result, the ionic compound will be hard, and it will take a lot of energy to break all the bonds between layers of ions.

However, if a force causes one layer to move slightly, ions with the same charge will be positioned next to each other. The cations in one layer are now lined up with other cations in a nearby layer. In the same way, anions from one layer are lined up with other anions in a nearby layer. Because the anions are next to each other, the like charges will repel each other and the layers will split apart. This is why all salts shatter along a line extending through the crystal known as a cleavage plane.

internet connect
www.scilinks.org
Topic: Salt Properties
SciLinks code: HW4113

SCiLINKS. Maintained by the National Science Teachers Association

SKILLS Toolkit 1

How to Identify a Compound as Ionic

You can carry out the following procedures in a laboratory to determine if a substance is an ionic compound.

- Examine the substance. All ionic compounds are solid at room temperature. If the substance is a liquid or gas, then it is not an ionic compound. However, if it is a solid, then it *may* or *may not* be an ionic compound.

- Tap the substance gently. Ionic compounds are hard and brittle. If it is an ionic compound, then it should not break apart easily. If it does break apart, the substance should fracture into tinier crystals and not crumble into a powder.

- Heat a sample of the substance. Ionic compounds generally have high melting and boiling points.

- If the substance melts, use a conductivity apparatus to determine if the melted substance conducts electric current. Ionic compounds are good conductors of electric current in the liquid state.

- Dissolve a sample of the substance in water. Use a conductivity apparatus to see if it conducts electric current. Ionic compounds conduct electric current when dissolved in water.

Continued from previous page

8. While keeping the leads in the water, add about 5 g of LiCl and stir gently with a plastic or glass stirring rod. The LED should glow almost immediately and will reach maximum brightness in a few seconds.

Variations

You may substitute NaCl, NaBr, KCl, or KBr for the LiCl. Do not use nitrates.

Safety: Wear safety goggles and a laboratory apron. Keep students at least 10 ft away from the crucible once it is hot. Use only the battery-powered tester.

Disposal: Remove the battery from the tester, and put tape over the terminals. The tester circuit will be sturdier if all electrical connections are made using very small plastic wire nuts, available at any electronic supply store. Keep the crucible with its solidified LiCl to perform the same demonstration in the future.

Teaching Tip

Crystal Display Show large crystals of various shapes. Include NaCl and quartz. Attractive crystals can be purchased at science museums and from science supply companies. Consider borrowing some samples from mineral collectors or amateur geologists.

Homework ——— **ADVANCED**

Have students draw a flowchart based on the Skills Toolkit "How to Identify an Ionic Compound" to determine if an unknown compound that is found in the laboratory is ionic or not. **LS** Visual

173

Teaching Tip

Designing a Model Na^+ and Cl^- ions have a size ratio of about 1 to 1.8 (5:9) when they are in a crystal of NaCl. If you can obtain spheres of this approximate size ratio, you can create a space-filling model of the lattice. Use the model to demonstrate that each ion is surrounded by six ions of opposite charge. Plastic-foam spheres from a craft shop can be used. Constructing four of these models will allow the students to get a better idea of the extended three-dimensional nature of ionic crystals.

crystal lattice

the regular pattern in which a crystal is arranged

Figure 12
The crystal structure of sodium chloride (**a**) is not the same as that of calcium fluoride (**b**) because of the differences in the sizes of their ions and the cation-anion ratio making up each salt.

Salt Crystals

The ions in a salt crystal form repeating patterns, with each ion held in place because the attractive forces are stronger than the repulsive ones. The way the ions are arranged is the same in a number of different salts. Not all salts, however, have the same crystal structure as sodium chloride. Despite their differences, the crystals of all salts are made of simple repeating units. These repeating units are arranged in a salt to form a **crystal lattice.** These arrangements of repeating units within a salt are the reason for the crystal shape that can be seen in most salts.

Crystal Structure Depends on the Sizes and Ratios of Ions

As the formula for sodium chloride, NaCl, indicates, there is a 1:1 ratio of sodium cations and chlorine anions. Recall that the attractions in sodium chloride involve more than a single cation and a single anion. **Figure 12a** illustrates the crystal lattice structure of sodium chloride. Within the crystal, each Na^+ ion is surrounded by six Cl^- ions, and, in turn, each Cl^- ion is surrounded by six Na^+ ions. Because this arrangement does not hold for the edges of the crystal, the edges are locations of weak points.

The arrangement of cations and anions to form a crystal lattice depends on the size of the ions. Another factor that affects how the crystal forms is the ratio of cations to anions. Not all salts have a 1:1 ratio of cations to anions as found in sodium chloride. For example, the salt calcium fluoride has one Ca^{2+} ion for every two F^- ions. A Ca^{2+} ion is larger than an Na^+ ion, and an F^- ion is smaller than a Cl^- ion. Because of the size differences of its ions and their ratio in the salt, the crystal lattice structure of calcium fluoride is different from that of sodium chloride. As illustrated in **Figure 12b,** each calcium ion is surrounded by eight fluoride ions. At the same time, each fluoride ion is surrounded by four calcium ions. This is very different from the arrangement of six oppositely charged ions around any given positive or negative ion in a crystal of NaCl.

a   sodium chloride   **b**   calcium fluoride

Salts Have Ordered Packing Arrangements

Salts vary in the types of ions from which they are made. Salts also vary in the ratio of the ions that make up the crystal lattice. Despite these differences, all salts are made of simple repeating units. The smallest repeating unit in a crystal lattice is called a **unit cell.**

The ways in which a salt's unit cells are arranged are determined by a technique called X-ray diffraction crystallography. First, a salt is bombarded with X rays. Then, the X rays that strike ions in the salt are deflected, while X rays that do not strike ions pass straight through the crystal lattice without stopping. The X rays form a pattern on exposed film. By analyzing this pattern, scientists can calculate the positions that the ions in the salt must have in order to cause the X rays to make such a pattern. After this work, scientists can then make models to show how the ions are arranged in the unit cells of the salt.

Analysis of many different salts show that the salts all have ordered packing arrangements, such as those described earlier for NaCl and CaF_2. Another example is the salt cesium chloride, where the ratio of cations to anions is 1:1 just as it is in sodium chloride. However, the size of a cesium cation is larger than that of a sodium cation. As a result, the structure of the crystal lattice is different. In sodium chloride, a sodium cation is surrounded by six chloride anions. In cesium chloride, a cesium cation is surrounded by eight chloride anions. The bigger cation has more room around it, so more anions can cluster around it.

unit cell

the smallest portion of a crystal lattice that shows the three-dimensional pattern of the entire lattice

 ## Section Review

UNDERSTANDING KEY IDEAS

1. What force holds together the ions in a salt?
2. Describe how an ionic bond forms.
3. Why are ionic solids hard?
4. Why are ionic solids brittle?
5. Explain why lattice energy is the key to the formation of a salt.
6. Why do ionic crystals conduct electric current in the liquid phase or when dissolved in water but do not conduct electric current in the solid phase?

CRITICAL THINKING

7. Crystals of the ionic compound calcium fluoride have a different structure from that of the ionic compound calcium chloride. Suggest a reason for this difference.

8. Explain why each of the following pairs is not likely to form an ionic bond.
   a. chlorine and bromine
   b. potassium and helium
   c. sodium and lithium

9. The lattice energy for sodium iodide is 700 kJ/mol, while that for calcium sulfide is 2775 kJ/mol. Which of these salts do you predict has the higher melting point? Explain.

10. The electron affinity for chlorine has a negative value, indicating that the atom readily accepts another electron. Why does a chlorine atom readily accept another electron?

11. Use **Figure 9** on page 169 to describe how the formation of calcium chloride would differ from that of sodium chloride. (Hint: Compare the electron configurations of each atom.)

Answers to Section Review

1. forces of attraction between ions of opposite charges

2. The attraction between a cation and an anion forms an ionic bond.

3. Attractive forces result in strong bonds that make the solid hard.

4. Initially, cations are aligned with anions, so the attractive forces are very strong. When a force causes like-charged ions to align, the repulsive forces exceed the attractive forces, and the solid breaks apart.

5. Enough energy is released as lattice energy to overcome the energy-absorbing steps involved in the formation of a salt.

6. Ions must be able to move freely for a substance to conduct electric current.

7. The fluoride ion is much smaller than the chloride ion. More calcium cations can fit around the larger chloride ion.

8. a. Both are halogens that form anions.
   b. Helium is a noble gas and therefore unreactive.
   c. Both are metals that form cations.

Answers continued on p. 187A

Focus

Overview

Before beginning this section, review with your students the Objectives listed in the Student Edition. This section introduces students the naming of ionic compounds. Students will also learn how to write the formula of an ionic compound from its name. Students will learn these concepts for both simple binary compounds and those containing polyatomic ions.

 **Bellringer**

Write the following scrambled names on the board: Einstein Albert, Marie Madame Curie, and Carver Washington George. Have students put the names in the correct order. Discuss the importance of correct order and spelling to accurately denote the person. Extend the discussion to include the naming of ionic compounds. Cations and anions can be thought of as the first and last names, respectively, of ionic compounds.

Motivate

Identifying Preconceptions — GENERAL

At the beginning of section 2, you named some ionic compounds. Look at that list, and see if you notice any patterns in the names of the ionic compounds. **Ans.** Students may notice that the metallic ions are named like the elements and some may even notice the *-ide* ending on the binary compounds. **LS Logical**

Names and Formulas of Ionic Compounds

OBJECTIVES

1 **Name** cations, anions, and ionic compounds.

2 **Write** chemical formulas for ionic compounds such that an overall neutral charge is maintained.

3 **Explain** how polyatomic ions and their salts are named and how their formulas relate to their names.

Naming Ionic Compounds

You may recall that chemists call table salt *sodium chloride*. In fact, they have a name for every salt. With thousands of different salts, you might think that it would be hard to remember the names of all of them. But naming salts is very easy, especially for those that are made of a simple cation and a simple anion. These kinds of salts are known as binary ionic compounds. The adjective *binary* indicates that the compound is made up of just two elements.

Rules for Naming Simple Ions

Simple cations borrow their names from the names of the elements. For example, K^+ is known as the potassium ion, and Zn^{2+} is known as the zinc ion. When an element forms two or more ions, the ion names include roman numerals to indicate charge. In the case of copper, Cu, the names of the two ions are written as follows.

$$Cu^+ \text{ copper(I) ion} \qquad Cu^{2+} \text{ copper(II) ion}$$

When we read the names of these ions out loud, we say "copper one ion" or "copper two ion."

The name of a simple anion is also formed from the name of the element, but it ends in *-ide*. Thus, Cl^- is the chloride ion, O^{2-} is the oxide ion, and P^{3-} is the phosphide ion.

The Names of Ions Are Used to Name an Ionic Compound

Naming binary ionic compounds is simple. The name is made up of just two words: the name of the cation followed by the name of the anion.

NaCl sodium chloride	$CuCl_2$ copper(II) chloride
ZnS zinc sulfide	Mg_3N_2 magnesium nitride
K_2O potassium oxide	Al_2S_3 aluminum sulfide

Chapter Resource File

• Lesson Plan

Writing Ionic Formulas

Ionic compounds never have an excess of positive or negative charges. To maintain this balance the total positive and negative charges must be the same. Because both ions in sodium chloride carry a single charge, this compound is made up of equal numbers of the ions Na^+ and Cl^-. As you have seen, the formula for sodium chloride is written as NaCl to show this one-to-one ratio. The cation in zinc sulfide has a 2+ charge and the anion has a 2− charge. Again there is a one-to-one ratio in the salt. Zinc sulfide has the formula ZnS.

Compounds Must Have No Overall Charge

You must take care when writing the formula for an ionic compound where the charges of the cation and anion differ. Consider the example of magnesium nitride. The magnesium ion, Mg^{2+}, has two positive charges, and the nitride ion, N^{3-}, has three negative charges. The cations and anions must be combined in such a way that there are the same number of negative charges and positive charges. Three Mg^{2+} cations are needed for every two N^{3-} anions for electroneutrality. That way, there are six positive charges and six negative charges. Subscripts are used to denote the three magnesium ions and two nitride ions. Therefore, the formula for magnesium nitride is Mg_3N_2.

SKILLS Toolkit 2

Writing the Formula of an Ionic Compound

Follow the following steps when writing the formula of a binary ionic compound, such as iron(III) oxide.

- Write the symbol and charges for the cation and anion. Refer to **Figures 5** and **6** earlier in the chapter for the charges on the ions. The roman numeral indicates which cation iron forms.

  symbol for iron(III): Fe^{3+} symbol for oxide: O^{2-}

- Write the symbols for the ions side by side, beginning with the cation.

  $$Fe^{3+}O^{2-}$$

- To determine how to get a neutral compound, look for the lowest common multiple of the charges on the ions. The lowest common multiple of 3 and 2 is 6. Therefore, the formula should indicate six positive charges and six negative charges.

For six positive charges, you need two Fe^{3+} ions because $2 \times 3+ = 6+$.

For six negative charges, you need three O^{2-} ions because $3 \times 2- = 6-$.

Therefore the ratio of Fe^{3+} to O^{2-} is 2Fe:3O. The formula is written as follows.

$$Fe_2O_3$$

177

Discussion

Display samples of ionic compounds such as potassium chloride, potassium chlorate, sodium nitrate, sodium nitrite, sodium carbonate, sodium hydrogen carbonate, copper(I) chloride, and copper(II) chloride. Label each sample container with both the name and the formula of the compound. Ask the students to discuss any relationships they might see between the names and the formulas. Ask the students why they have to be precise in writing the formulas of these compounds.

Teach

Teaching Tip

Cation-anion Pairs Reinforce the idea of electroneutrality in the following drill. Make a vertical list of the symbols of six cations on the board, and make a parallel list of the symbols of six anions. The ions should have a variety of charges. Point to a cation-anion pair, and ask students to determine how many of each will produce electroneutrality. You may want to ask students to write or state the correct formula, but the main purpose of the exercise is to help students understand how to construct a neutral combination.

Homework ——— BASIC

1. Write the formulas for the following ionic compounds:
   - copper(II) oxide **Ans.** CuO
   - sodium fluoride **Ans.** NaF
   - zinc chloride **Ans.** $ZnCl_2$
   - aluminum sulfide **Ans.** Al_2S_3
   - potassium nitride **Ans.** K_3N

2. Write the names of the following compounds:
   - Ca_3N_2 **Ans.** calcium nitride
   - FeI_3 **Ans.** iron(III) iodide
   - Na_2O **Ans.** sodium oxide
   - $AlCl_3$ **Ans.** aluminum chloride
   - SrO **Ans.** strontium oxide

Vocabulary Have students read the topic "The Names of Polyatomic Ions Can Be Complicated" and make a chart indicating the difference between *-ate*, *-ite*, and *-ide* compounds. **LS** Visual

Teaching Tip

Students have learned that formulas for binary ionic compounds should represent the smallest possible ratio of ions. Some compounds of polyatomic ions may seem to violate this rule, especially compounds with mercury(I) ions, Hg_2^{2+}, or oxalate ions, $C_2O_4^{2-}$. Examples include Hg_2F_2 and $Ag_2C_2O_4$. Emphasize to students that these ions have particular structures that are reflected in their formulas and that their subscripts cannot be changed.

---

polyatomic ion

an ion made of two or more atoms

Table 2
Some Polyatomic Ions

Ion name	Formula
Acetate	CH_3COO^-
Ammonium	NH_4^+
Carbonate	CO_3^{2-}
Chromate	CrO_4^{2-}
Cyanide	CN^-
Dichromate	$Cr_2O_7^{2-}$
Hydroxide	OH^-
Nitrate	NO_3^-
Nitrite	NO_2^-
Permanganate	MnO_4^-
Peroxide	O_2^{2-}
Phosphate	PO_4^{3-}
Sulfate	SO_4^{2-}
Sulfite	SO_3^{2-}
Thiosulfate	$S_2O_3^{2-}$

Polyatomic Ions

Fertilizers have potassium, nitrogen, and phosphorus in a form that dissolves easily in water so that plants can absorb them. The potassium in fertilizer is in an ionic compound called *potassium carbonate*. Two ionic compounds in the fertilizer contain the nitrogen—ammonium nitrate and ammonium sulfate. The phosphorus supplied is in another ionic compound, calcium dihydrogen phosphate.

These compounds in fertilizer are made of cations and anions in a ratio so there is no overall charge, like all other ionic compounds. But instead of ions made of a single atom, these compounds contain groups of atoms that are ions.

Many Atoms Can Form One Ion

The adjective *simple* describes an ion formed from a single atom. A simple ion could also be called *monatomic*, which means "one-atom." Just as the prefix *mon-* means "one," the prefix *poly-* means "many." The term **polyatomic ion** means a charged group of two or more bonded atoms that can be considered a single ion. A polyatomic ion as a whole forms ionic bonds in the same way that simple ions do.

Unlike simple ions, most polyatomic ions are made of atoms of several elements. However, polyatomic ions are like simple ones in that their charge is either positive or negative. Consider the polyatomic ion ammonium, NH_4^+, found in many fertilizers. Ammonium is made of one nitrogen and four hydrogen atoms. These atoms have a combined total of 11 protons but only 10 electrons. So the ammonium ion has a 1+ charge overall. This charge is not found on any single atom. Instead, it is spread across this group of atoms, which are bonded together.

The Names of Polyatomic Ions Can Be Complicated

Naming polyatomic ions is not as easy as naming simple cations and anions. Even so, there are rules you can follow to help you remember how to name some of them.

Many polyatomic ions contain oxygen. The endings *-ite* and *-ate* indicate the presence of oxygen. Examples include sulfite, nitrate, and acetate. Often there are several polyatomic ions that differ only in the number of oxygen atoms present. For example, the formulas for two polyatomic ions made from sulfur and oxygen are SO_3^{2-} and SO_4^{2-}. In such cases, the one with less oxygen takes the *-ite* ending, so SO_3^{2-} is named *sulfite*. The ion with more oxygen takes the *-ate* ending, so SO_4^{2-} is named *sulfate*. For the same reason, NO_2^- is named *nitrite,* and NO_3^- is named *nitrate*.

The presence of hydrogen is often indicated by an ion's name starting with *hydrogen*. The prefixes *mono-* and *di-* are also used. Thus, HPO_4^{2-} and $H_2PO_4^-$ are monohydrogen phosphate and dihydrogen phosphate ions, respectively. The prefix *thio-* means "replace an oxygen with a sulfur" in the formula, as in potassium thiosulfate, $K_2S_2O_3$, compared with potassium sulfate, K_2SO_4. **Table 2** lists the names and formulas for some common polyatomic ions. Notice that some are made of more than one atom of the same element, such as peroxide, O_2^{2-}.

MISCONCEPTION ALERT

Students sometimes believe that the Roman numeral in the name of a compound represents the number of ions of that element present in the formula of that compound. Emphasize that Roman numerals represent the charge on the cation, whereas subscript Arabic numerals represent numbers of ions.

SKILLS Toolkit

Naming Compounds with Polyatomic Ions

Follow these steps when naming an ionic compound that contains one or more polyatomic ions, such as K_2CO_3.

- Name the cation. Recall that a cation is simply the name of the element. In this formula, K is potassium that forms a singly charged cation, K^+, of the same name.

- Name the anion. Recall that salts are electrically neutral. Because there are two K^+ cations present in this salt, these two positive charges must be balanced by two negative charges. Therefore, the polyatomic anion in this salt must be CO_3^{2-}. You may find it helpful to think of the formula as follows, although it is not written this way.

$$(K^+)_2(CO_3^{2-})$$

If you check **Table 2,** you will see that the CO_3^{2-} polyatomic ion is called carbonate.

- Name the salt. Recall that the name of a salt is just the names of the cation and anion. The salt K_2CO_3 is potassium carbonate.

SAMPLE PROBLEM A

Formula of a Compound with a Polyatomic Ion

What is the formula for iron(III) chromate?

1 **Gather information.**

- Use **Figure 6,** found earlier in this chapter, to determine the formula and charge for the iron(III) cation.

$$Fe^{3+}$$

- Use **Table 2,** found earlier in this chapter, to determine the formula and charge for the chromate polyatomic ion.

$$CrO_4^{2-}$$

2 **Plan your work.**

- Because all ionic compounds are electrically neutral, the total charges of the cations and anions must be equal. To balance the charges, find the least common multiple of the ions' charges. The least common multiple of 2 and 3 is 6. To get six positive charges, you need two Fe^{3+} ions.

$$2 \times 3 = 6+$$

To get six negative charges, you need three CrO_4^{2-} ions.

$$3 \times 2 = 6-$$

continued on next page

179

PRACTICE HINT

Sometimes parentheses must be placed around the polyatomic cation, as in the formula $(NH_4)_2CrO_4$.

MISCONCEPTION ALERT

Students may expect to find a single, systematic, numerical rule that allows them to write the formula of any oxyanion when given its name. The system of naming these ions dates back to the nineteenth century and is related to the names of acids containing these oxyanions. Ions ending in -ite are derived from acids ending in *-ous*. Ions ending in -ate are derived from oxygen-containing acids ending in *-ic*. Aside from the progression of *hypo . . . ite* to *ite* to *ate* to *per . . . ate*, students will have to memorize the number of oxygen atoms in the oxyanions of other nonmetal elements.

Close

Reteaching ———— BASIC

Have the students create a flow-chart, concept map or other graphic organizer that describes the steps in writing the formula of an ionic compound and one that describes the steps in naming an ionic compound.
LS Visual

Quiz ———————— GENERAL

1. What does a Roman numeral after the name of a cation indicate? **Ans.** The charge on the cation.

2. Why must the number of positive charges equal the number of negative charges in all ionic compounds? **Ans.** Because all compounds must be neutral in charge.

3. What is the name of $SrCrO_4$? **Ans.** Strontium chromate

4. How many oxygen atoms are there in $Mg(NO_3)_2$? **Ans.** six

5. What is the formula for rubidium phosphate? **Ans.** Rb_3PO_4
LS Logical

Answers to Practice Problems A

a. $Ca(CN)_2$
b. $Rb_2S_2O_3$
c. $Ca(CH_3COO)_2$
d. $(NH_4)_2SO_4$

Chapter Resource File

• Concept Review

• Quiz

---

3 Calculate.

• The formula must indicate that two Fe^{3+} ions and three CrO_4^{2-} ions are present. Parentheses are used whenever a polyatomic ion is present more than once. The formula for iron(III) chromate is written as follows.

$$Fe_2(CrO_4)_3$$

Notice that the parentheses show that everything inside the parentheses is tripled by the subscript *3* outside.

4 Verify your result.

• The formula includes the correct symbols for the cation and polyatomic anion.

• The formula reflects that the salt is electrically neutral.

PRACTICE

1 Write the formulas for the following ionic compounds.

   a. calcium cyanide   **c.** calcium acetate

   b. rubidium thiosulfate   **d.** ammonium sulfate

---

3 Section Review

UNDERSTANDING KEY IDEAS

1. In what ways are polyatomic ions like simple ions? In what ways are they different?

2. Why must roman numerals be used when naming certain ionic compounds?

3. What do the endings *-ite* and *-ate* indicate about a polyatomic ion?

4. Explain how calcium, Ca^{2+}, and phosphate, PO_4^{3-}, can make a compound with electroneutrality.

CRITICAL THINKING

5. Name the compounds represented by the following formulas.

   a. $Ca(NO_2)_2$ **c.** $(NH_4)_2Cr_2O_7$

   b. $Fe(OH)_3$ **d.** $CuCH_3COO$

6. Write the formulas for the following ionic compounds made of simple ions.

   a. sodium oxide

   b. magnesium phosphide

   c. silver(I) sulfide

   d. niobium(V) chloride

7. Name the following binary ionic compounds. If the metal forms more than one cation, be sure to denote the charge.

   a. Rb_2O **b.** FeF_2 **c.** K_3N

PRACTICE PROBLEMS

8. Write formulas for the following compounds.

   a. mercury(II) sulfate

   b. lithium thiosulfate

   c. ammonium phosphate

   d. potassium permanganate

Answers to Section Review

1. Both polyatomic and simple ions are electrically charged. Simple ions are monatomic whereas polyatomic ions are made of two or more atoms that function as a unit.

2. If an atom forms more than one cation, then a Roman numeral must be used to indicate which cation is present in the compound.

3. They indicate the presence of oxygen as part of the polyatomic ion.

4. Three Ca^{2+} ions and two PO_4^{3-} ions combine to form a compound that is electrically neutral.

5. **a.** calcium nitrite

   b. iron(III) hydroxide

   c. ammonium dichromate

   d. copper(I) acetate

6. **a.** Na_2O **c.** Ag_2S

   b. Mg_3P_2 **d.** $NbCl_5$

7. **a.** rubidium oxide

   b. iron(II) fluoride

   c. potassium nitride

8. **a.** $HgSO_4$ **c.** $(NH_4)_3PO_4$

   b. $Li_2S_2O_3$ **d.** $KMnO_4$

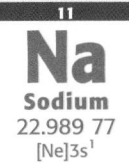

| Na |
| Sodium |
| 22.989 77 |
| [Ne]3s^1 |

11

Element Spotlight

Where Is Na?

Earth's Crust
2.36% by mass
Seventh most abundant element
Fifth most abundant metal

Sea Water
30.61% of all dissolved materials
1.03% by mass, taking the water
into account

A Major Nutritional Mineral

Sodium is important in the regulation of fluid balance within the body. Most sodium in the diet comes from the use of table salt, NaCl, to season and preserve foods. Sodium is also supplied by compounds such as sodium carbonate and sodium hydrogen carbonate in baked goods. Sodium benzoate is a preservative in carbonated beverages. Sodium citrate and sodium glutamate are used in packaged foods as flavor additives.

In ancient Rome, salt was so scarce and highly prized that it was used as a form of payment. Today, however, salt is plentiful in the diet. Many people must limit their intake of sodium as a precaution against high blood pressure, heart attacks, and strokes.

Industrial Uses

- Common table salt is the most important commercial sodium compound. It is used in ceramic glazes, metallurgy, soap manufacture, home water softeners, highway de-icing, herbicides, fire extinguishers, and resins.

- The United States produces about 42.1 million metric tons of sodium chloride per year.

- Elemental sodium is used in sodium vapor lamps for lighting highways, stadiums, and other buildings.

- Liquid sodium is used to cool liquid-metal fast-breeder nuclear reactors.

Food labels list the amount of sodium contained in each serving.

Real-World Connection For most people, the daily intake of sodium should not exceed 2400 mg.

A Brief History

1807: Sir Humphry Davy isolates sodium by the electrolysis of caustic soda (NaOH) and names the metal.

1990: The Nutrition Labeling and Education Act defines a Daily Reference Value for sodium to be listed in the Nutrition Facts portion of a food label.

| 1200 | 1800 | 1900 | 2000 |

1251: The Wieliczka Salt Mine, located in Krakow, Poland, is started. The mine is still in use today.

1930: Sodium vapor lamps are first used for street lighting.

1940: The Food and Nutrition Board of the National Research Council develops the first Recommended Dietary Allowances.

Questions

1. What is one possible consequence of too much sodium in the diet?

2. What is the chemical formula of the most important commercial sodium compound?

internet connect

www.scilinks.org
Topic: Sodium
SciLinks code: HW4173

SCiLINKS Maintained by the National Science Teachers Association

181

Element Spotlight

A Major Nutritional Mineral

In the United States, 32% of salt is mined as rock salt from large underground deposits that were left when ancient seas dried up and were later covered by geological layers. Another 55% comes from the evaporation of brine. The brine is usually obtained by pumping water through underground salt deposits. Mined rock salt contains impurities and is used for purposes such as melting ice. Table salt is produced by dissolving the mined salt, purifying the solution, and recrystallizing the salt in a way that produces small crystals.

Answers to Feature Questions
1. Answers may include high blood pressure (hypertension), heart attack, and stroke.
2. NaCl

Alternative Assessments

Group Activity
• Pages 5, 13, 14

SKILL BUILDER
• Pages 7, 16

Portfolio Assessments

Reteaching
• Pages 10, 20, 25

SKILL BUILDER
• Page 23

5 CHAPTER HIGHLIGHTS

KEY IDEAS

SECTION ONE   Simple Ions
• Atoms may gain or lose electrons to achieve an electron configuration identical to that of a noble gas.
• Alkali metals and halogens are very reactive when donating and accepting electrons from one another.
• Electrons in the outermost energy level are known as valence electrons.
• Ions are electrically charged particles that have different chemical properties than their parent atoms.

SECTION TWO   Ionic Bonding and Salts
• The opposite charges of cations and anions attract to form a tightly packed substance of bonded ions called a *crystal lattice*.
• Salts have high melting and boiling points and do not conduct electric current in the solid state, but they do conduct electric current when melted or when dissolved in water.
• Salts are made of unit cells that have an ordered packing arrangement.

SECTION THREE   Names and Formulas of Ionic Compounds
• Ionic compounds are named by joining the cation and anion names.
• Formulas for ionic compounds are written to show their balance of overall charge.
• A polyatomic ion is a group of two or more atoms bonded together that functions as a single unit.
• Parentheses are used to group polyatomic ions in a chemical formula with a subscript.

KEY TERMS

octet rule
ion
cation
anion

salt
lattice energy
crystal lattice
unit cell

polyatomic ion

KEY SKILLS

How to Identify a Compound as Ionic	Writing the Formula of an Ionic Compound	Naming Compounds with Polyatomic Ions	Formula of a Compound with a Polyatomic Ion
Skills Toolkit 1 p. 173	Skills Toolkit 2 p. 177	Skills Toolkit 3 p. 179	Sample Problem A p. 179

182

Chapter Resource File

• **Microscale Lab** Conductivity as an Indicator of Bond Type
• **Microscale Lab** Tests for Iron(II) and Iron(III)
• **Datasheets for In-Text Lab** Percent Composition of Hydrates
• **Datasheets for In-Text Lab** Hydrates—Gypsum and Plaster of Paris

USING KEY TERMS

1. How is an ion different from its parent atom?

2. What does a metal atom need to do in order to form a cation?

3. What does a nonmetal element need to do to form an anion?

4. Explain how the octet rule describes how atoms form stable ions.

5. Explain how you can tell from an element's number of valence electrons whether the element is more likely to form a cation or an anion.

6. Why is lattice energy the key to forming an ionic bond?

7. Explain why it is appropriate to group a polyatomic ion in parentheses in a chemical formula, if more than one of that ion is present in the formula.

UNDERSTANDING KEY IDEAS

Simple Ions

8. The electron configuration for arsenic, As, is $[Ar]3d^{10}4s^24p^3$. How many valence electrons does an As atom have? Write the symbol for the ion it forms to achieve a noble-gas configuration.

9. Explain why magnesium forms Mg^{2+} cations and not Mg^{6-} anions.

10. Explain why the properties of an ion differ from its parent atom.

11. How does the octet rule help predict the chemical reactivity of an element?

12. Why are the halogens so reactive?

13. If helium does not obey the octet rule, then why do its atoms not react?

14. How do the stable ions of the transition metals differ from those formed by the metals in Groups 1 and 2?

15. Explain why metals tend to form cations, while nonmetals tend to form anions.

16. Which of the following diagrams illustrates the electron diagram for a potassium ion found in the nerve cells of your body? (Hint: potassium's atomic number is 19.)

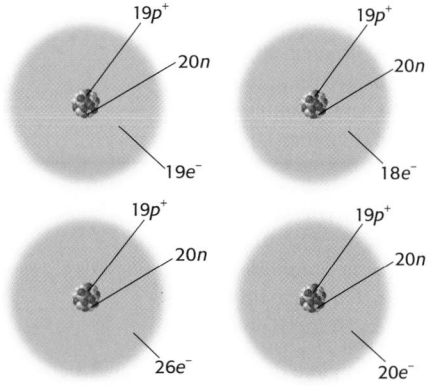

Ionic Bonding and Salts

17. Why do most ionic compounds have such high melting and boiling points?

18. Explain the importance of lattice energy in the formation of a salt.

19. Explain why a salt usually dissolves slowly.

20. What is the relationship between lattice energy and the strength of an ionic bond?

21. Why can't an ionic bond form between potassium and magnesium?

(183)

Assignment Guide

Section	Questions
1	1–5, 10–19, 38–41, 43, 51
2	6, 7, 20–28, 42, 44, 45, 48, 57–64
3	8, 9, 29–37, 46, 47, 49, 50, 52–56

REVIEW ANSWERS

Using Key Terms

1. It has a different number of electrons.

2. It needs to lose one or more electrons.

3. It needs to gain one or more electrons.

4. Atoms gain or lose electrons to fill their outer *s* and *p* orbitals with eight electrons.

5. An element with one, two, or three valence electrons is usually more likely to form a cation; an element with five, six, or seven valence electrons is usually more likely to form an anion.

6. Enough energy is released to make the overall process spontaneous.

7. The parentheses reflect the fact that the polyatomic ion functions as a single unit.

Understanding Key Ideas

8. five; As^{3-}

9. Less energy is required to remove magnesium's two valence electrons so that it forms a stable ion as opposed to adding six electrons to its outermost energy level.

10. Chemical properties depend on electron configuration. By either gaining or losing electrons, an atom changes its electron configuration and therefore its chemical properties also change.

11. The atoms of an element will react to achieve a noble-gas configuration. The atoms will either gain or lose electrons to achieve such a configuration.

12. With seven valence electrons, a halogen needs only one additional electron to form a stable ion.

13. Helium is inert because its one energy level, 1*s*, is filled with the maximum number of electrons.

14. The transition metals do not form stable ions that have noble gas configurations.

15. With one, two, or three valence electrons, metals generally lose electrons to achieve a noble gas configuration. More energy would be required for them to gain the necessary electrons. With five, six, or seven valence electrons, nonmetal elements tend to gain electrons.

16. b. 18 electrons

17. The electrical attraction between cations and anions in the crystal lattice is strong. Therefore, a high temperature is required to break down the lattice and change the solid crystal into a liquid. Ions still have strong attractions in the liquid state. Therefore, an even higher temperature is required to separate the ions into a gas.

18. Lattice energy provides enough energy to drive all the endothermic steps, such as the formation of cations, that are involved in the formation of a crystal lattice.

19. Only the ions at the surface of the crystal can go into solution, so ions are stripped away one layer at a time.

20. The greater the lattice energy, the stronger the ionic bond.

21. Both are metals and form cations. An ionic bond forms only between ions of opposite charges.

Names and Formulas of Ionic Compounds

22. What is the difference between the chlorite ion and the chlorate ion?

23. How are simple cations and anions named?

24. Why is peroxide, O_2^{2-}, considered a polyatomic ion and not a simple ion?

25. Identify and name the cations and anions that make up the following ionic compounds and indicate the charge on each ion.
   a. $NaNO_3$
   b. K_2SO_3
   c. $(NH_4)_2CrO_4$
   d. $Al_2(SO_4)_3$

26. Name the compounds represented by the following formulas.
   a. $Cu_3(PO_4)_2$
   b. $Fe(NO_3)_3$
   c. Cu_2O
   d. CuO

PRACTICE PROBLEMS

27. Write formulas for the following ionic compounds.
   a. lithium sulfate
   b. strontium nitrate
   c. ammonium acetate
   d. titanium(III) sulfate

28. Write formulas for the following ionic compounds.
   a. silver nitrate
   b. calcium thiosulfate
   c. barium hydroxide
   d. cobalt(III) phosphate

29. Complete the table below, and then use it to answer the questions that follow.

Element	Ion	Name of ion
Barium	Ba^{2+}	
Chlorine		chloride
Chromium	Cr^{3+}	
Fluorine	F^-	
Manganese		manganese(II)
Oxygen		oxide

Write the formula for the following substances:
   a. manganese chloride
   b. chromium(III) fluoride
   c. barium oxide

MIXED REVIEW

30. Write formulas for the following ionic compounds.
   a. barium nitrate
   b. calcium fluoride
   c. potassium iodide
   d. ammonium hydroxide

31. Name the following polyatomic ions.
   a. O_2^{2-}
   b. CrO_4^{2-}
   c. NH_4^+
   d. CO_3^{2-}

32. Complete the table below.

Atom	Ion	Noble-gas configuration of ion
S		
Be		
I		
Rb		
O		
Sr		
F		

33. Write formulas for the following polyatomic ions.
   a. cyanide
   b. sulfate
   c. nitrite
   d. permanganate

34. Name the following ionic compounds.
   a. $AlPO_4$
   b. $SrSO_4$
   c. KCN
   d. CuS

35. Determine the number of valence electrons in the following atoms.
   a. Al
   b. Rb
   c. Si
   d. F

CRITICAL THINKING

36. Why are most metals found in nature as ores and not as pure metals?

184

37. Why can't sodium gain a positive charge by acquiring a proton in its nucleus?

38. The electron configuration for a lithium atom is $1s^2 2s^1$. The configuration for an iodine atom is $1s^2 2s^2 2p^6 3s^2 3p^6 4s^2 3d^{10} 4p^6 5s^2 4d^{10} 5p^5$.

Write the electron configurations for the ions that form lithium iodide, a salt used in photography.

39. Why are there no rules for naming Group 18 ions?

40. Determine the ratios of cations to anions that are most likely in the formulas for ionic substances of the following elements:
a. an alkali metal and a halogen
b. an alkaline-earth metal and a halogen
c. an alkali metal and a member of Group 16
d. an alkaline-earth metal and a member of Group 16

41. Sodium chloride is prepared by reacting sodium with chlorine gas. Another way to prepare it is to react sodium hydroxide with hydrochloric acid and allow the water to evaporate. Will the method of preparation affect the type of unit cell and crystalline structure of sodium chloride? Explain.

42. Compound B has lower melting and boiling points than compound A does. At the same temperature, compound B vaporizes faster and to a greater extent than compound A. If only one of these compounds is ionic, which one would you expect it to be? Why?

ALTERNATIVE ASSESSMENT

43. Ions play an important physiological role in your body, such as nerve impulse conduction and muscle contraction. Select one such ion, and prepare a report detailing its function. Be sure to include recent medical information, including any diseases associated with a deficiency of the ion.

44. A number of homes have "hard water," which, as you learned in the Start-Up

Activity, does not produce as many soap suds as water that contains fewer ions. Such homes often have water conditioners that remove the ions from the water, making it "softer" and more likely to produce soapsuds. Research how such water softeners operate by checking the Internet or by contacting a company that sells such devices. Design an experiment to test the effectiveness of the softener in removing ions from water.

45. Devise a set of criteria that you can use to classify the following substances as ionic or nonionic compounds: $CaCO_3$, Cu, H_2O, NaBr, and C (graphite). Show your criteria to your teacher. If your plan is approved, carry out the analysis and report your results to the class.

CONCEPT MAPPING

46. Build a concept map about ionic bonding. Be sure to include the following terms in your concept map: *atoms, valence electrons, ions, cations, anions,* and *ionic compounds.*

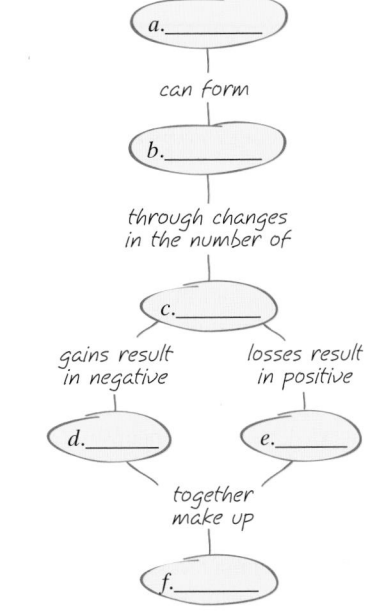

22. The chlorate ion, ClO_3^-, contains three O atoms while the chlorite ion, ClO_2^-, contains only two O atoms.

23. Cations are named directly after their parent atoms. The names of anions are also from the names of their parent atoms but end in *-ide*.

24. Peroxide is made of two O atoms as its formula indicates. Therefore, it is a polyatomic ion.

25. a. sodium (Na^+) and nitrate (NO_3^-)
b. potassium (K^+) and sulfite (SO_3^{2-})
c. ammonium (NH_4^+) and chromate (CrO_4^{2-})
d. aluminum (Al^{3+}) and sulfate (SO_4^{2-})

26. a. copper(II) phosphate
b. iron(III) nitrate
c. copper(I) oxide
d. copper(II) oxide

27. a. Li_2SO_4
b. $Sr(NO_3)_2$
c. NH_4CH_3COO
d. $Ti_2(SO_4)_3$

28. a. $AgNO_3$
b. CaS_2O_3
c. $Ba(OH)_2$
d. $CoPO_4$

29. barium; Cl^-; chromium(III); fluoride; Mn^{2+}; O^{2-}; **a.** $MnCl_2$; **b.** CrF_3; **c.** BaO

30. a. $Ba(NO_3)_2$
b. CaF_2
c. KI
d. NH_4OH

31. a. peroxide
b. chromate
c. ammonium
d. carbonate

32. S^{2-}; $[Ne]3s^2 3p^6$
Be^{2+}; $1s^2$
I^-; $[Kr]4d^{10}5s^2 5p^6$
Rb^+; $[Ar]3d^{10}4s^2 4p^6$
O^{2-}; $[He]2s^2 2p^6$
Sr^{2+}; $[Ar]3d^{10}4s^2 4p^6$
F^-; $[He]2s^2 2p^6$

33. a. CN^-

   b. SO_4^{2-}

   c. NO_2^-

   d. MnO_4^-

34. a. aluminum phosphate

   b. strontium sulfate

   c. potassium cyanide

   d. copper(II) sulfide

35. a. 3; **b.** 1; **c.** 4; **d.** 7

36. Most metals are active elements that lose electrons to form ionic compounds that are found in ores.

37. Nuclear processes do not occur in chemical reactions. Chemical reactions involve the rearrangement of electrons, not protons or neutrons.

38. $Li^+: 1s^2;$ $I^-:$ $1s^2 2s^2 2p^6 3s^2 3p^6 4s^2 3d^{10} 4p^6 5s^2 4d^{10} 5p^6$

39. Elements in this group do not normally form ions. They all have stable outer energy levels.

40. a. 1:1

   b. 1:2

   c. 2:1

   d. 1:1

41. No. NaCl has the same unit cell that makes up the crystal lattice no matter how the salt has been prepared.

42. Compound A is probably ionic because it has the higher melting and boiling points. In addition, compound A does not vaporize as readily. These properties are the result of strong ionic bonds.

Answers continued on p. 187A

FOCUS ON GRAPHING

Study the graph below, and answer the questions that follow. For help in interpreting graphs, see Appendix B, "Study Skills for Chemistry."

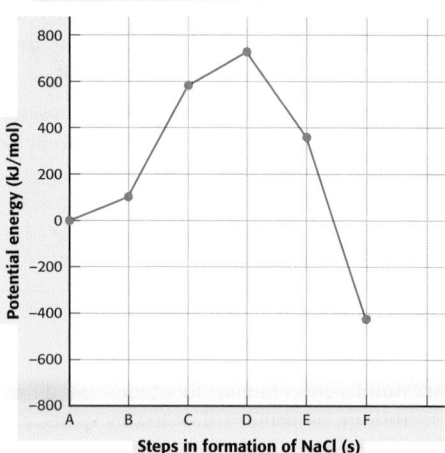

Potential Energy in the Formation of NaCl

The graph shows the changes in potential energy that occur when an ionic bond forms between $Na(s)$ and $Cl_2(g)$. The reactants, solid sodium and chlorine gas, start at an initial energy state that is assigned a value of zero at 25°C and 1 atm of pressure.

47. In terms of energy, what do the steps from point A to point D have in common?

48. What do the steps from point D to point F have in common?

49. What is occurring between points D and E?

50. Write the word equation to show what happens between points B and C when electrons are removed from 1 mol of sodium atoms.

51. Which portion of this graph represents the lattice energy involved in the formation of an ionic bond between sodium and chlorine?

52. Calculate the quantity of energy released when 2.5 mol of NaCl form.

TECHNOLOGY AND LEARNING

53. Graphing Calculator

Calculating the Number of Valence Electrons

The graphing calculator can run a program that can determine the number of valence electrons in an atom, given its atomic number.

Go to Appendix C. If you are using a TI-83 Plus, you can download the program **VALENCE** and run the application as directed. If you are using another calculator,

your teacher will provide you with keystrokes to use. After you have run the program, answer these questions.

How many valence electrons are there in the following atoms?

a. Rutherfordium, Rf, atomic number 104

b. Gold, Au, atomic number 79

c. Molybdenum, Mo, atomic number 42

d. Indium, In, atomic number 49

186

Chapter Resource File

• Chapter Test

STANDARDIZED TEST PREP

1. The correct formula for a copper atom that has lost two electrons is
   a. Co^+.
   b. Co^{2+}.
   c. Cu^{2-}.
   d. Cu^{2+}.

2. Which of the following atoms can achieve the same electron configuration as a noble gas when the atom forms an ion?
   a. argon
   b. potassium
   c. nickel
   d. iron

3. In forming NaCl, energy is required to
   a. change chlorine to a gas.
   b. add an electron to a chlorine atom.
   c. remove an electron from a sodium atom.
   d. bring together the sodium ions and chloride ions.

4. Which of the following is not a characteristic of a salt?
   a. hardness
   b. a high melting point
   c. an ability to conduct electric current in the molten state
   d. an ability to conduct electric current in the solid state

5. Which elements on the periodic table are most likely to form two or more ions with different charges?
   a. halogens
   b. alkali metals
   c. transition metals
   d. noble gases

6. Which of the following compounds does not contain a polyatomic ion?
   a. sodium carbonate
   b. sodium sulfate
   c. sodium sulfite
   d. sodium sulfide

7. The correct formula for ammonium phosphate is
   a. $(NH_4)_3PO_4$.
   b. $(NH_4)_2PO_4$.
   c. NH_4PO_4.
   d. $NH_4(PO_4)_2$.

8. When writing the formula for a compound that contains a polyatomic ion,
   a. write the anion's formula first.
   b. use superscripts to show the number of polyatomic ions present.
   c. use parentheses if the number of polyatomic ions is greater than 1.
   d. always place the polyatomic ion in parentheses.

9. Compounds that are electrically neutral must have equal numbers of
   a. anions and cations.
   b. positive and negative charges.
   c. molecules.
   d. ionic bonds.

10. The correct name for NH_4CH_3COO is
    a. ammonium carbonate.
    b. ammonium hydroxide.
    c. ammonium acetate.
    d. ammonium nitrate.

11. Which of the following is the correct formula for iron(III) sulfate?
    a. Fe_3SO_4
    b. $Fe_3(SO_4)_2$
    c. $Fe_2(SO_4)_3$
    d. $3FeSO_4$

12. How many valence electrons are indicated by the following electron configuration?
    $1s^2 2s^2 2p^6 3s^2 3p^5$
    a. two
    b. five
    c. three
    d. seven

13. Metals tend to lose one or more electrons to form
    a. cations.
    b. polyatomic ions
    c. anions.
    d. unit cells.

Answers

1. d.
2. b.
3. c.
4. d.
5. c.
6. d.
7. a.
8. c.
9. b.
10. c.
11. c.
12. d.
13. a.

187

Continuation of Answers

Answers to Section 1 Review, *continued*

11. a. $[He]2s^22p^6$;
    b. $[Ar]3d^{10}4s^24p^6$
    c. $[Ne]3s^23p^6$;
    d. $[Ar]3d^{10}4s^24p^6$

12. same number of protons and neutrons

13. Less energy is required to add three electrons than to remove five electrons.

Answers to Section 2 Review, *continued*

9. Calcium sulfide has the higher melting point because of its stronger ionic bond.

10. With its outer electrons in the $n = 3$ level, the chlorine atom is relatively small. The outer level is close enough to the nucleus that the nucleus has a strong attraction for an additional electron.

11. Calcium chloride requires two chlorine atoms for each calcium atom.

Answers to Chapter Review, *continued*
Alternative Assessment

43. Possible choices include Na^+, K^+, Ca^{2+}, and HCO_3^-.

44. Proposals may include using a water-testing kit to test the levels of certain ions both before and after treatment.

45. Testing criteria should include electrical conductivity (with a battery-powered tester), hardness, and crystalline structure. Care should be taken if the student plans on checking melting and boiling points.

Concept Mapping

46. a. atoms
    b. ions
    c. valence electrons
    d. anions
    e. cations
    f. ionic compounds

Focus on Graphing

47. These steps represent processes that absorb energy.

48. These steps represent processes that release energy.

49. An electron is being added to a chlorine atom, releasing energy in the process.

50. sodium atoms(g) + energy $\longrightarrow$ sodium ions(g) + electrons

51. the step between points E and F

52. 400 kJ/mol $\times$ 2.5 mol = 1000 kJ

Technology and Learning

53. a. 4; b. 2; c. 6; d. 3

Covalent Compounds
Chapter Planning Guide

PACING	CLASSROOM RESOURCES	LABS, ACTIVITIES, AND DEMONSTRATIONS
BLOCK 1 • 45 min pp. 188–189 **Chapter Opener**		**SE** **Start-Up Activity** Ionic Versus Covalent, p. 189 ◆
BLOCKS 2 & 3 • 90 min pp. 190–198 **Section 1** Covalent Bonds	**TT** **Formation of a Covalent Bond** * **OSP** **Supplemental Reading** Linus Pauling: In His Own Words "Mind and Molecules" (ADVANCED) **OSP** **Supplemental Reading** Consumer's Good Chemical Guide "Sweetness and Light" (ADVANCED) **TT** **Potential Energy Curve for H_2** * **TT** **Predicting Bond Character from Electronegativity Difference** * **TT** **Properties of Substances with Metallic, Ionic, and Covalent Bonds** *	**TE** **Activity** Modeling Covalent Bonds, p. 191 (GENERAL) **TE** **Activity**, p. 192 (BASIC) **TE** **Group Activity** Classifying Bonds, p. 196 (GENERAL) **TE** **Demonstration** Paper Chromatography, p. 197 ◆ (ADVANCED) **CRF** **Microscale Lab** Chemical Bonds ◆ (ADVANCED)
BLOCKS 4 & 5 • 90 min pp. 199–207 **Section 2** Drawing and Naming Molecules		
BLOCK 6 • 45 min pp. 208–214 **Section 3** Molecular Shapes	**OSP** **Career Extension** Occupational Applications Worksheet 9: Air-Quality Specialist (GENERAL) **TT** **Properties of Substances with Metallic, Ionic, and Covalent Bonds** * **TT** **Molecular Shapes with Three Electron Pairs Around Central Atom** * **TT** **Molecular Shape Affects Polarity** * **SE** **Element Spotlight** Silicon and Semiconductors	**TE** **Demonstration** Balloon Models, p. 209 ◆ **TE** **Group Activity** Modeling Molecular Shapes, p. 209 ◆ (BASIC) **TE** **Activity** Unshared Pairs and Shape, p. 210 (GENERAL) **TE** **Demonstration**, p. 212

BLOCKS 7 & 8 • 90 min **Chapter Review and Assessment Resources**

- **SE** Chapter Review, pp. 216–220
- **SE** Standardized Test Prep, p. 221
- **CRF** Chapter Test *
- **OSP** Test Generator
- **CRF** Test Item Listing *
- **OSP** Scoring Rubrics and Classroom Management Checklists

Holt Chemistry: Online Resources

go.hrw.com

Visit **go.hrw.com** for a variety of free resources related to this textbook. Enter the keyword **HW4 HOME**.

Holt Online Learning

Students can access interactive problem solving help and active visual concept development with the *Holt Chemistry* Online Edition available at **www.hrw.com**

student CNN News

cnnstudentnews.com

Find the latest chemistry news, lesson plans, and activities related to important scientific events.

Compression guide:
To shorten your instruction because of time limitations, omit blocks 1, 3, 5, and 7.

KEY

TE Teacher Edition	**CRF** Chapter Resource File
SE Student Edition	**TT** Teaching Transparencies
OSP One-Stop Planner	

* Also on One-Stop Planner
◆ Requires Advance Prep

PROBLEM SOLVING AND PRACTICE	SECTION REVIEW AND ASSESSMENT	STANDARDS CORRELATION
		National Science Education Standards
	TE **Homework**, p. 193 GENERAL TE **Quiz**, p. 198 GENERAL CRF **Quiz** * TE **Reteaching**, p. 198 BASIC SE **Section Review**, p. 198 CRF **Concept Review** *	PS 2a PS 2c
SE **Skills Toolkit 1** Drawing Lewis Structures with Many Atoms, p. 201 SE **Sample Problem A** Drawing Lewis Structures with Single Bonds, p. 202 　　SE **Practice**, p. 202 GENERAL　　TE **Homework**, p. 202 GENERAL SE **Sample Problem B** Drawing Lewis Structures for Polyatomic Ions, p. 203 　　SE **Practice**, p. 203 GENERAL　　TE **Homework**, p. 203 GENERAL SE **Sample Problem C** Drawing Lewis Structures with Multiple Bonds, p. 205 　　SE **Practice**, p. 205 GENERAL　　TE **Homework**, p. 205 GENERAL CRF **Problem Solving** * ADVANCED	TE **Quiz**, p. 207 GENERAL CRF **Quiz** * TE **Reteaching**, p. 207 BASIC SE **Section Review**, p. 207 CRF **Concept Review** *	
SE **Sample Problem D** Predicting Molecular Shapes, p. 211 　　SE **Practice**, p. 211 GENERAL　　TE **Homework**, p. 211 GENERAL CRF **Problem Solving** * ADVANCED	TE **Homework**, p. 211 BASIC TE **Homework**, p. 212 ADVANCED TE **Quiz**, p. 213 GENERAL CRF **Quiz** * TE **Reteaching**, p. 213 BASIC SE **Section Review**, p. 213 CRF **Concept Review** *	PS 2d

www.scilinks.org
Topic: Rubber
SciLinks code: HW4128

Topic: Covalent Bonding
SciLinks code: HW4036

Topic: Naming Compounds
SciLinks code: HW4081

Topic: VSEPR Theory
SciLinks code: HW4169

Topic: Molecular Shapes
SciLinks code: HW4080

Topic: Silicon
SciLinks code: HW4116

Technology Resources

ChemFile Interactive Tutor CD-Rom

Module 4: Chemical Bonding
Topic: Covalent Bonding

Module 4: Chemical Bonding
Topic: Molecular Geometry

Overview

This chapter introduces covalent compounds. Students will learn how covalent bonds form, how ionic and covalent substances differ, how to draw and name covalent compounds, and how the shape of a molecule affects its properties.

Assessing Prior Knowledge

Check for Content Knowledge

Students should be familiar with the following topics:

• electron configuration
• periodic trends
• ionic bonds
• polyatomic ions

Using the Figure —— BASIC

Have students brainstorm other items made of rubber, such as tires, rubber bands, and shoe soles. Discuss reasons why vulcanized rubber is a more useful manufacturing material than natural rubber. Reasons include its strength and flexibility. Tell students that it is covalent bonds that give vulcanized rubber the properties that make it a useful substance.

Have students research the life and career of Charles Goodyear (1800–1860). He invented what might be the most commonly known polymer, vulcanized rubber. He spent five years experimenting with ways to make a useful material out of natural rubber. After his discovery, many of his business dealings failed as he was forced to fight patent infringements. **LS** Visual

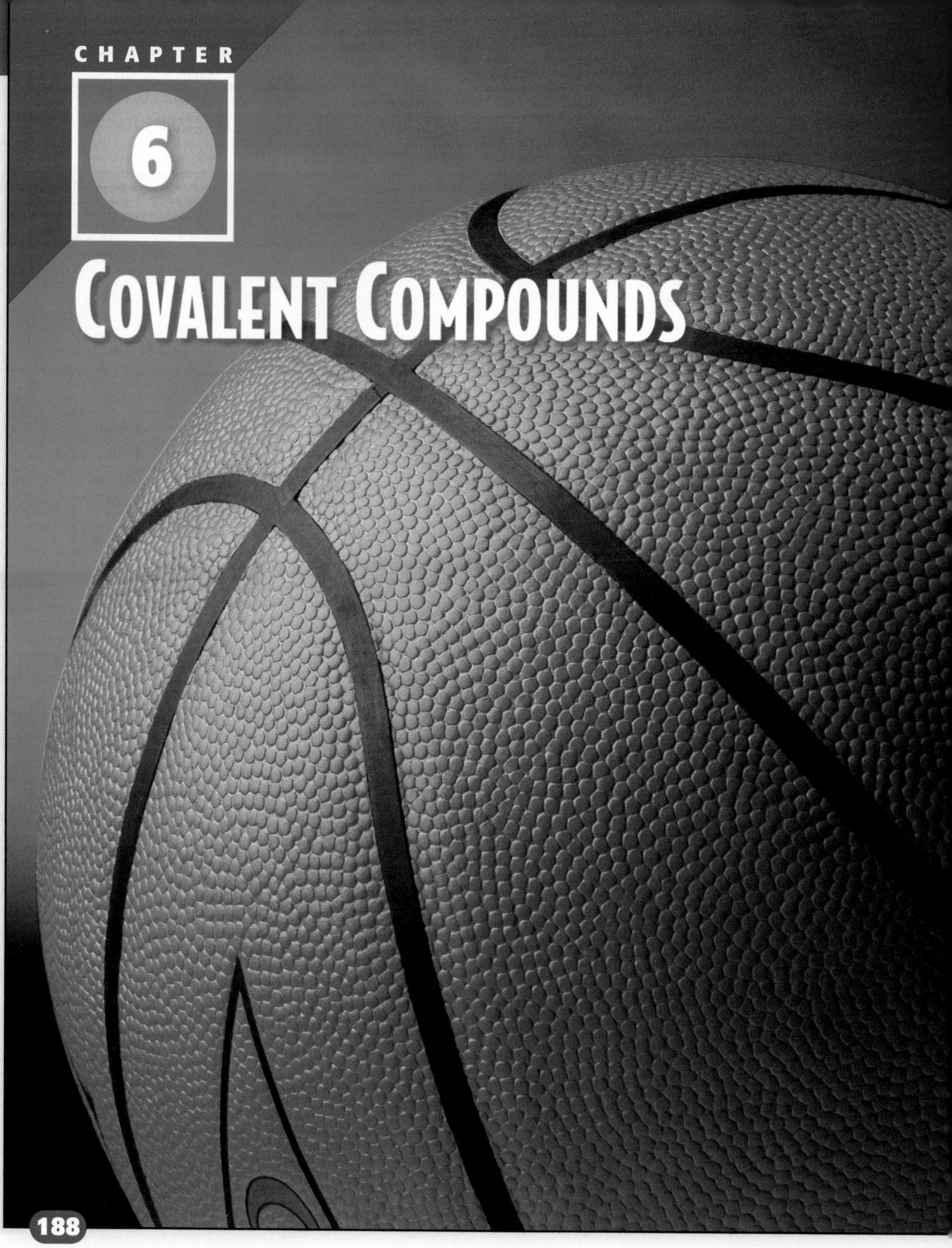

C H A P T E R

6

COVALENT COMPOUNDS

188

Standards Correlations

National Science Education Standards

PS 2a: Atoms interact with one another by transferring or sharing electrons that are furthest from the nucleus. These outer electrons govern the chemical properties of the element. (Section 1)

PS 2c: Bonds between atoms are created when electrons are paired up by being transferred or shared. A substance composed of a single kind of atom is called an element. The atoms may be bonded together into molecules or crystalline solids. A compound is formed when two or more kinds of atoms bind together chemically. (Section 1)

PS 2d: The physical properties of compounds reflect the nature of the interactions among its molecules. These interactions are determined by the structure of the molecule, including the constituent atoms and the distances and angles between them. (Section 3)

Natural rubber comes from tropical trees. It is soft and sticky, so it has little practical use. However, while experimenting with rubber in 1839, Charles Goodyear dropped a mixture of sulfur and natural rubber on a hot stove by mistake. The heated rubber became tough and elastic because of the formation of covalent bonds. The resulting compound was *vulcanized rubber*, which is strong enough to make up a basketball that can take a lot of hard bounces.

START-UP*ACTIVITY*

SAFETY PRECAUTIONS

Ionic Versus Covalent

PROCEDURE

1. Clean and dry **three test tubes.** Place a small amount of **paraffin wax** into the first test tube. Place an equal amount of **table salt** into the second test tube. Place an equal amount of **sugar** into the third test tube.

2. Fill a **plastic-foam cup** halfway with **hot water.** Place the test tubes into the water. After 3 min, remove the test tubes from the water. Observe the contents of the test tubes, and record your observations.

3. Place a small amount of each substance on a **watch glass.** Crush each substance with a **spatula.** Record your observations.

4. Add **10 mL deionized water** to each test tube. Use a **stirring rod** to stir each test tube. Using a **conductivity device** (watch your teacher perform the conductivity tests), record the conductivity of each mixture.

ANALYSIS

1. Summarize the properties you observed for each compound.

2. Ionic bonding is present in many compounds that are brittle, have a high melting point, and conduct electric current when dissolved in water. Covalent bonding is present in many compounds that are not brittle, have a low melting point, and do not conduct electric current when mixed with water. Identify the type of bonding present in paraffin wax, table salt, and sugar.

Pre-Reading Questions

① What determines whether two atoms will form a bond?

② How can a hydrogen atom, which has one valence electron, bond with a chlorine atom, which has seven valence electrons?

③ What happens in terms of energy after a hydrogen atom bonds with a chlorine atom?

CONTENTS 6

🔗 internet connect

www.scilinks.org
Topic: Rubber
SciLinks code: HW4128

SC*LINKS* Maintained by the National Science Teachers Association

189

Skills Acquired:
- Collecting Data
- Interpreting
- Inferring

Materials:

For each group of 2–3 students:
- conductivity device
- glass stirring rod
- paraffin wax, $\frac{1}{4}$ tsp.
- plastic-foam cup
- sodium chloride, NaCl, $\frac{1}{4}$ tsp.
- spatula
- sugar, $\frac{1}{4}$ tsp.
- test tubes (3)
- watch glass
- water, deionized, 10 mL
- water, hot, 1 cup

Teacher's Notes: The conductivity device is an open circuit containing a small battery connected with an insulated wire to a small light bulb or LED. A lead wire should come off of the battery, and a second lead wire should come off the light bulb. Some physics classroom kits contain clip-together electric circuit components, or you can solder together each component. To test for conductivity, place both lead-wire ends in the liquid. The ends should not touch. If the light bulb lights up, then the liquid conducts electric current. Rinse lead wires with deionized water between tests. Some devices cannot be used to test melted wax, as the wax may solidify around delicate components.

Safety Caution: Remind students to use goggles, aprons, and gloves. If students operate the conductivity devices, they should use caution when doing so in or near aqueous solutions.

Answers to Pre-Reading Questions

1. Two atoms will form a chemical bond if transferring or sharing electrons allows them to form a more stable arrangement of electrons having lower energy.

2. Hydrogen and chlorine can form a covalent bond by sharing electrons so that the hydrogen atom has two valence electrons and the chlorine atom has eight valence electrons. Sharing electrons in this way gives both atoms the electron configurations of noble gas atoms.

3. Energy is released when hydrogen and chlorine atoms form a bond.

Answers to Start-Up Activity

1. Paraffin wax melts in a hot water bath, is soft, and does not conduct electric current when mixed with water. Salt does not melt in a hot water bath, is brittle, and conducts electric current when mixed with water. Sugar does not melt in a hot water bath, is brittle, and does not conduct electric current when mixed with water.

2. Paraffin wax and sugar contain covalent bonds. Salt contains ionic bonds.

1 Covalent Bonds

Before beginning this section, review with your students the Objectives listed in the Student Edition. In this section, students will learn that some atoms form chemical bonds by sharing electrons. The relationship between bond energy, bond length, and bond stability is explored. Students will recognize that the polarity of a covalent bond depends upon the electronegativity difference between the bonding atoms and that the polarity of a molecule or compound affects its physical properties.

 **Bellringer**

Have students make a list of the elements that form ionic bonds. They may note that most ionic bonds contain a metal and a nonmetal.

Motivate

Identifying Preconceptions — GENERAL

Ask students what would be the formula for the carbon ion in CCl_4? A likely conclusion is C^{4+}. Make the point that under ordinary conditions the C^{4+} ion does not exist because too much energy is required to remove four electrons from the atom. Even though energy would be released as the hypothetical C^{4+} ion bonded ionically with four chloride ions, that amount of energy is less than the energy required to make the carbon ion. Thus, there must be another way in which atoms such as carbon can achieve the stability of a noble gas atom. **LS Logical**

KEY TERMS

- covalent bond
- molecular orbital
- bond length
- bond energy
- nonpolar covalent bond
- polar covalent bond
- dipole

OBJECTIVES

1. **Explain** the role and location of electrons in a covalent bond.
2. **Describe** the change in energy and stability that takes place as a covalent bond forms.
3. **Distinguish** between nonpolar and polar covalent bonds based on electronegativity differences.
4. **Compare** the physical properties of substances that have different bond types, and relate bond types to electronegativity differences.

Sharing Electrons

The diver shown in **Figure 1** is using a hot flame to cut metal under water. The flame is made by a chemical reaction in which hydrogen and oxygen gases combine. When these gases react, atoms and electrons rearrange to form a new, more stable compound: water.

You learned that electrons are rearranged when an ionic bond forms. When this happens, electrons transfer from one atom to another to form charged ions. The reaction of hydrogen and oxygen to form water causes another kind of change involving electrons. In this case, the neutral atoms share electrons.

Figure 1
This diver is using an oxyhydrogen torch. The energy released by the torch comes from a chemical reaction in which hydrogen and oxygen react to form water.

$$O_2 \ + \ 2H_2 \ \longrightarrow \ 2H_2O$$

internet connect
www.scilinks.org
Topic: Covalent Bonding
SciLinks code: HW4036
SCI LINKS. Maintained by the National Science Teachers Association

MISCONCEPTION /// ALERT ///

As students learn about the covalent bond, they may think that all compounds are held together by either ionic or covalent bonds. Point out that some compounds contain both kinds of bonds. For example, ionic compounds that contain polyatomic ions, such as Na_2SO_4 or K_3PO_4, are ionically bonded by the attraction between the Na^+ or K^+ ions and the PO_4^{3-} and SO_4^{2-} ions, but the bonds within the polyatomic ions are covalent.

Chapter Resource File

- **Lesson Plan**

One-Stop Planner CD-ROM

- **Supplemental Reading Projects**
Guided Reading Worksheets: Linus Pauling: In His Own Words "Mind and Molecules" and Consumer's Good Chemical Guide "Sweetness and Light" *Assign these worksheets for cross-curricular connections to language arts.*

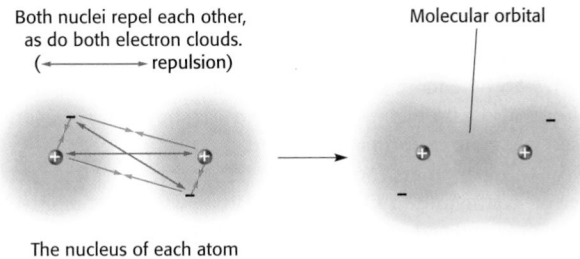

Both nuclei repel each other, as do both electron clouds.
(⟵⟶ repulsion)

Molecular orbital

Figure 2
The positive nucleus of one hydrogen atom attracts the electron of the other atom. At the same time, the two atoms' positive nuclei repel each other. The two electron clouds also repel each other.

The nucleus of each atom attracts both electron clouds.
(⟶⟵ attraction)

A covalent bond is formed.

Forming Molecular Orbitals

The simplest example of sharing electrons is found in diatomic molecules, such as hydrogen, H_2. **Figure 2** shows the attractive and repulsive forces that exist when two hydrogen atoms are near one another. When these forces are balanced, the two hydrogen atoms form a bond. Because both atoms are of the same element, the attractive force of each atom is the same. Thus, neither atom will remove the electron from the other atom. Instead of transferring electrons to each other, the two hydrogen atoms share the electrons.

The result is a H_2 molecule that is more stable than either hydrogen atom is by itself. The H_2 molecule is stable because each H atom has a shared pair of electrons. This shared pair gives both atoms a stability similar to that of a helium configuration. Helium is stable because its atoms have filled orbitals.

The sharing of a pair of electrons is the bond that holds the two hydrogen atoms together. When two atoms share electrons, they form a **covalent bond.** The shared electrons move in the space surrounding the nuclei of the two hydrogen atoms. The space that these shared electrons move within is called a **molecular orbital.** As shown in **Figure 2,** a molecular orbital is made when two atomic orbitals overlap. Sugar and water, shown in **Figure 3,** have molecules with covalent bonds.

covalent bond

a bond formed when atoms share one or more pairs of electrons

molecular orbital

the region of high probability that is occupied by an individual electron as it travels with a wavelike motion in the three-dimensional space around one of two or more associated nuclei

Figure 3
The sugar, $C_{12}H_{22}O_{11}$, and water, H_2O, in the tea are examples of covalent, or molecular, compounds.

191

Transparencies

TT Formation of a Covalent Bond

ChemFile • **CHEMISTRY** INTERACTIVE TUTOR

• **Module 4: Chemical Bonding Topic: Covalent Bonding**
This engaging tutorial reviews and reinforces the concept of covalent bonding through modeling and guided practice.

Chapter 6 • Covalent Compounds   191

Using the Figure

Point out in **Figure 4** that the *x*-axis tells how far two H nuclei are from each other. The *y*-axis represents the potential energy of the two H atoms. Be sure students understand that potential energy is energy of position. Starting from the right side, two H atoms meet as separated atoms. Each nucleus has an attraction for the electrons of the other. This attractive force draws the atoms closer together and their energy decreases. The nuclei are also repulsed because of their like charges, but a position is found at which the combined effect of attractive and repulsive forces reduces the energy to a minimum—a covalent bond has been formed. If the distance between the two nuclei decreases further, the potential energy rises steeply and the molecule becomes unstable.

Activity ———— BASIC

Using magnets, model the attractive forces between H atoms. The forces are similar in effect to the attractive force between opposite poles of magnets. Let students see that when magnets are close enough to each other, they clamp together in a position of lower energy. Work is required to separate magnets, so separated magnets have higher energy. Make the point that, in a similar way, energy is released when bonds are formed and absorbed when bonds are broken. **LS** Kinesthetic

---

Transparencies

TT Potential Energy Curve for H_2

Energy and Stability

Most individual atoms have relatively low stability. (Noble gases are the exception.) They become more stable when they are part of a compound. Unbonded atoms also have high potential energy, as shown by the energy that is released when atoms form a compound.

After two hydrogen atoms form a covalent bond, each of them can have an electron configuration like that of helium, which has relatively low potential energy and high stability. Thus, bonding causes a decrease in energy for the atoms. This energy is released to the atoms' surroundings.

Energy Is Released When Atoms Form a Covalent Bond

Figure 4 shows the potential energy changes that take place as two hydrogen atoms come near one another. In part (a) of the figure, the distance between the two atoms is large enough that there are no forces between their nuclei. At this distance, the potential energy of the atoms is arbitrarily set at zero.

In part (b) of the figure, the potential energy decreases as the attractive electric force pulls the two atoms closer together. As the potential energy goes down, the system gives off energy. In other words, energy is released as the attractive force pulls the atoms closer. Eventually, the atoms get close enough that the attractive forces between the electrons of one atom and the nucleus of the other atom are balanced by the repulsive force caused by the two positively charged nuclei as they are forced closer together. The two hydrogen atoms are now covalently bonded. In part (c) of the figure, the two atoms have bonded, and they are at their lowest potential energy. If they get any closer, repulsive forces will take over between the nuclei.

Potential Energy Determines Bond Length

Part (c) of **Figure 4** shows that when the two bonded hydrogen atoms are at their lowest potential energy, the distance between them is 75 pm. This distance is considered the length of the covalent bond between two hydrogen atoms. The distance between two bonded atoms at their minimum potential energy is known as the **bond length.**

bond length

the distance between two bonded atoms at their minimum potential energy; the average distance between the nuclei of two bonded atoms

Figure 4
Two atoms form a covalent bond at a distance where attractive and repulsive forces balance. At this point, the potential energy is at a minimum.

Potential Energy Curve for H_2

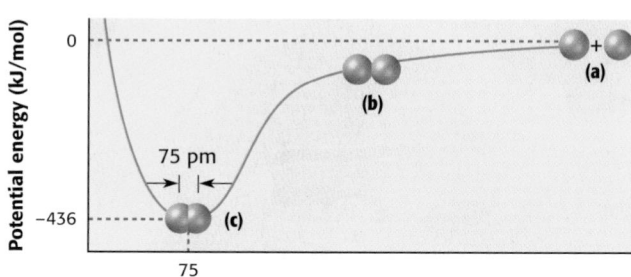

MISCONCEPTION ALERT

Students may not immediately recognize that bond energy—the amount of energy needed to break a bond—is the same as the potential energy released when the bond formed. Refer students to **Figure 4.** Point out that the curve, which extends from the stable H_2 molecule at the lowest part of the curve to the separated atoms at the top right, depicts not only the energy released as the separated atoms are brought together but also the energy required to separate the atoms after they have bonded. Point out to students that the difference between the potential energy of one mole of hydrogen molecules and two moles of hydrogen atoms is 436 kJ, the same energy listed as the bond energy of H_2 in **Table 1.**

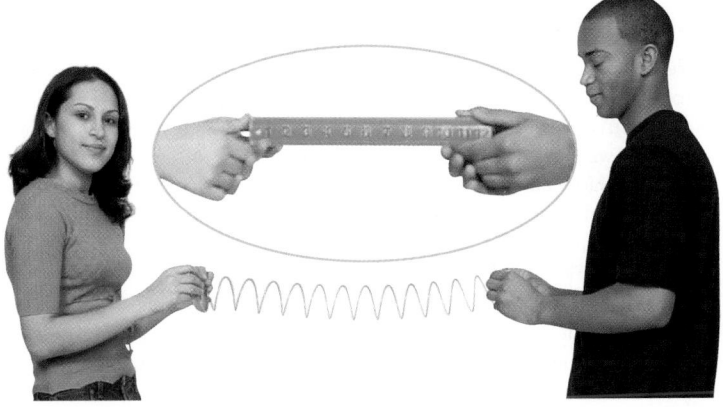

Figure 5
A covalent bond is more like a flexible spring than a rigid ruler, because the atoms can vibrate back and forth.

Bonded Atoms Vibrate, and Bonds Vary in Strength

Models often incorrectly show covalent bonds as rigid "sticks." If these bonds were in fact rigid, then the nuclei of the bonded atoms would be at a fixed distance from one another. Because the ruler held by the students in the top part of **Figure 5** is rigid, the students are at a fixed distance from one another.

However, a covalent bond is more flexible, like two students holding a spring. The two nuclei vibrate back and forth. As they do, the distance between them constantly changes. The bond length is in fact the average distance between the two nuclei.

At a bond length of 75 pm, the potential energy of H_2 is –436 kJ/mol. This means that 436 kJ of energy is released when 1 mol of bonds form. It also means that 436 kJ of energy must be supplied to break the bonds and separate the hydrogen atoms in 1 mol of H_2 molecules. The energy required to break a bond between two atoms is the **bond energy**. **Table 1** lists the energies and lengths of some common bonds in order of decreasing bond energy. Note that the bonds that have the highest bond energies (the "strongest" bonds) usually involve the elements H or F. Also note that stronger bonds generally have shorter bond lengths.

bond energy

the energy required to break the bonds in 1 mol of a chemical compound

Table 1 Bond Energies and Bond Lengths for Single Bonds

	Bond energy (kJ/mol)	Bond length (pm)		Bond energy (kJ/mol)	Bond length (pm)
H—F	570	92	H—I	299	161
C—F	552	138	C—Br	280	194
O—O	498	121	Cl—Cl	243	199
H—H	436	75	C—I	209	214
H—Cl	432	127	Br—Br	193	229
C—Cl	397	177	F—F	159	142
H—Br	366	141	I—I	151	266

193

did you know? ———— BASIC

Reactivity Series The bond energy of a diatomic compound provides a means of predicting the reactivity of the compound. If the bond energies of two similar compounds are compared, the compound with the lower bond energy is more likely to decompose or react with another compound. Have students examine the bond energies of the hydrogen halides (HF, HCl, HBr, and HI) shown in **Table 1** and predict the order of increasing reactivity. Remind students that because the energy of the two atoms separated by a large distance is defined as zero energy and because the energy of the atoms decreases as they are attracted together, bond energies stand for negative energies. Therefore, a large bond energy corresponds to a lower energy state. **LS** Logical

Teaching Tip
Units of Bond Length Many chemists still use the angstrom (Å) for describing bond length.

1 Å = 100 pm

Homework ———— GENERAL

Have students answer the following questions.

1. Why is the hydrogen molecule, H_2, more stable than the individual atoms that bond to form it? **Ans.** Free H atoms have no bonding energy, while a H_2 molecule has considerable bonding energy. Greater bonding energy means more stability. Also, each H atom now has a shared pair of electrons. A noble-gas configuration is more stable than a configuration with an unpaired electron.

2. What is created when two atomic orbitals overlap? **Ans.** a molecular orbital

3. What happens to the potential energy of two atoms as they approach each other to form a covalent bond? **Ans.** The potential energy decreases.

4. What name is given to the distance between two atoms in a covalent bond at which the potential energy is minimum? **Ans.** the bond length

5. List four examples of substances that have covalent bonds. **Ans.** Student answers will vary but can include H_2, CO, CO_2, O_2, H_2O, and sugar.
LS Logical

Using the Table ——— GENERAL

Have students study the data in **Table 1**. Ask them to make a generalization about how the magnitude of bond energy is related to bond length. Students should be able to determine that, in general, bond length is inversely related to bond strength. **LS** Verbal

Using the Figure

Figure 6 In general, metals are less electronegative than non-metals. Electronegativity can also be related to atomic size. For most representative families of the periodic table, the smaller the atom, the greater its electronegativity. Because its electrons are closer to the nucleus, a smaller atom has a stronger attraction for its own electrons—as well as for the electrons of other atoms—than a larger atom has. Point out to students that the noble gases are not shown in the figure because they do not generally form bonds.

Teaching Tip

Microwave Ovens When you heat food in a microwave oven, you are taking advantage of the fact that some molecules in the food are polar. Water molecules interact with the microwaves of a specific frequency that course through the oven. Microwaves are a type of electromagnetic radiation and differ from visible light and radio waves in the amount of energy they carry. In a microwave oven, polar molecules absorb the incident microwaves which cause the molecules to rapidly rotate, thus increasing their kinetic energy. It is good practice to allow food heated in a microwave to rest for a few moments before eating it. This allows time for the kinetic energy generated in the water molecules to be transferred to nonpolar molecules that did not absorb the microwaves.

Topic Link

Refer to the "Periodic Table" chapter for more about electronegativity.

nonpolar covalent bond

a covalent bond in which the bonding electrons are equally attracted to both bonded atoms

polar covalent bond

a covalent bond in which a shared pair of electrons is held more closely by one of the atoms

Figure 6
Fluorine has an electronegativity of 4.0, the highest value of any element.

Electronegativity and Covalent Bonding

The example in which two hydrogen atoms bond is simple because both atoms are the same. Also, each one has a single proton and a single electron, so the attractions are easy to identify. However, many covalent bonds form between two different atoms. These atoms often have different attractions for shared electrons. In such cases, electronegativity values are a useful tool to predict what kind of bond will form.

Atoms Share Electrons Equally or Unequally

Figure 6 lists the electronegativity values for several elements. In a molecule such as H_2, the values of the two atoms in the bond are equal. Because each one attracts the bonding electrons with the same force, they share the electrons equally. A **nonpolar covalent bond** is a covalent bond in which the bonding electrons in the molecular orbital are shared equally.

What happens when the electronegativity values are not the same? If the values differ significantly, the two atoms form a different type of covalent bond. Think about a carbon atom bonding with an oxygen atom. The O atom has a higher electronegativity and attracts the bonding electrons more than the C atom does. As a result, the two atoms share the bonding electrons, but unequally. This type of bond is a **polar covalent bond.** In a polar covalent bond, the shared electrons, which are in a molecular orbital, are more likely to be found nearer to the atom whose electronegativity is higher.

If the difference in electronegativity values of the two atoms is great enough, the atom with the higher value may remove an electron from the other atom. An ionic bond will form. For example, the electronegativity difference between magnesium and oxygen is great enough for an O atom to remove two electrons from a Mg atom. **Figure 7** shows a model of how to classify bonds based on electronegativity differences. Keep in mind that the boundaries between bond types are arbitrary. This model is just one way that you can classify bonds. You can also classify bonds by looking at the characteristics of the substance.

Electronegativities

H 2.2																	
Li 1.0	Be 1.6											B 2.0	C 2.6	N 3.0	O 3.4	F 4.0
Na 0.9	Mg 1.3											Al 1.6	Si 1.9	P 2.2	S 2.6	Cl 3.2
K 0.8	Ca 1.0	Sc 1.4	Ti 1.5	V 1.6	Cr 1.7	Mn 1.6	Fe 1.9	Co 1.9	Ni 1.9	Cu 2.0	Zn 1.7	Ga 1.8	Ge 2.0	As 2.2	Se 2.5	Br 3.0
Rb 0.8	Sr 1.0	Y 1.2	Zr 1.3	Nb 1.6	Mo 2.2	Tc 1.9	Ru 2.2	Rh 2.3	Pd 2.2	Ag 1.9	Cd 1.7	In 1.8	Sn 1.9	Sb 2.0	Te 2.1	I 2.7
Cs 0.8	Ba 0.9	La 1.1	Hf 1.3	Ta 1.5	W 2.4	Re 1.9	Os 2.2	Ir 2.2	Pt 2.3	Au 2.5	Hg 2.0	Tl 1.8	Pb 2.1	Bi 2.0	Po 2.0	At 2.2
Fr 0.7	Ra 0.9	Ac 1.1														

INCLUSION Strategies

- *Learning Disabled*
- *Developmentally Delayed*
- *English Language Learners*

Refer students to **Figure 6,** and ask them to make a list ranking the electronegativity of the elements from highest to lowest. Ask students to list the name of each element from the symbols given in the figure. By referring to the list, students will be able to explain which elements are most and least likely to form an ionic bond.

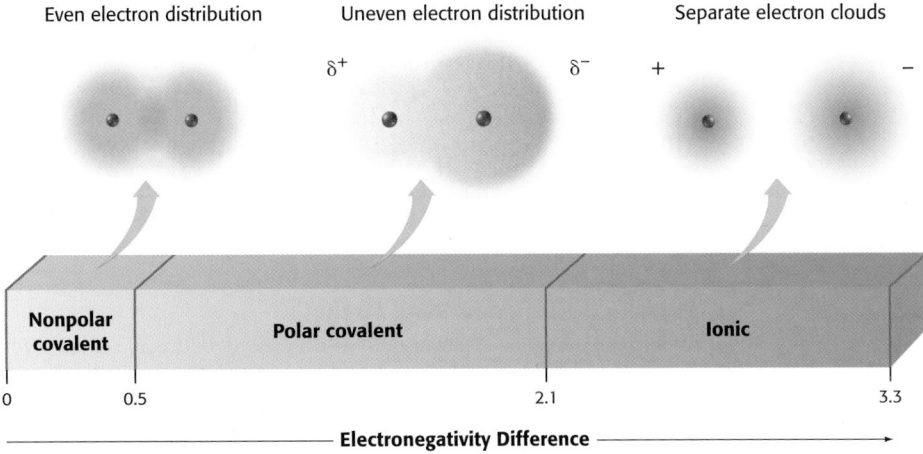

Even electron distribution   Uneven electron distribution   Separate electron clouds

δ⁺   δ⁻   +   −

| Nonpolar covalent | Polar covalent | Ionic |

0   0.5   2.1   3.3

◄──────── **Electronegativity Difference** ────────►

Polar Molecules Have Positive and Negative Ends

Hydrogen fluoride, HF, in solution is used to etch glass, such as the vase shown in **Figure 8**. The difference between the electronegativity values of hydrogen and fluorine shows that H and F atoms form a polar covalent bond. The word *polar* suggests that this bond has ends that are in some way opposite one another, like the two poles of a planet, a magnet, or a battery. In fact, the ends of the HF molecule have opposite partial charges.

The electronegativity of fluorine (4.0) is much higher than that of hydrogen (2.2). Therefore, the shared electrons are more likely to be found nearer to the fluorine atom. For this reason, the fluorine atom in the HF molecule has a partial negative charge. In contrast, the shared electrons are less likely to be found nearer to the hydrogen atom. As a result, the hydrogen atom in the HF molecule has a partial positive charge. A molecule in which one end has a partial positive charge and the other end has a partial negative charge is called a **dipole**. The HF molecule is a dipole.

To emphasize the dipole nature of the HF molecule, the formula can be written as $H^{\delta+}F^{\delta-}$. The symbol δ is a lowercase Greek *delta,* which is used in science and math to mean *partial.* With polar molecules, such as HF, the symbol δ^+ is used to show a partial positive charge on one end of the molecule. Likewise, the symbol δ^- is used to show a partial negative charge on the other end.

Although δ^+ means a positive charge, and δ^- means a negative charge, these symbols do not mean that the bond between hydrogen and fluorine is ionic. An electron is not transferred completely from hydrogen to fluorine, as in an ionic bond. Instead, the atoms share a pair of electrons, which makes the bond covalent. However, the shared pair of electrons is more likely to be found nearer to the fluorine atom. This unequal distribution of charge makes the bond polar covalent.

Figure 7
Electronegativity differences can be used to predict the properties of a bond. Note that there are no distinct boundaries between the bond types—the distinction is arbitrary.

dipole

a molecule or part of a molecule that contains both positively and negatively charged regions

Figure 8
Hydrogen fluoride, HF, is an acid that is used to etch beautiful patterns in glass.

195

PHYSICS
CONNECTION

Partial Charge Have students compare full positive and negative charges, as they occur in electrons and protons, with partial charges as they occur in polar molecules. Be certain that students recognize that a charge of +1 or −1 is a set amount of charge equal to $+1.6 \times 10^{-19}$ or -1.6×10^{-19} coulombs respectively.

A partial charge, designated δ+ or δ−, can be any fractional amount of charge between zero and $+1.6 \times 10^{-19}$ or zero and -1.6×10^{-19} coulombs respectively. The amount represented by partial charges δ+ or δ− depends on electronegativity differences and will vary, depending upon the elements involved.

Teaching Tip
Note that many other factors
besides polarity affect bond
strength. For example, atomic size
also affects bond strength.

SKILL BUILDER — GENERAL

Graphing Skills Have students
use the data in **Table 2** to graph
electronegativity difference versus
bond energy of the hydrogen
halides. Have them explain the
relationship between electronegativity difference and bond energy
as interpreted from the graph. **Ans.**
The graph shows that as electronegativity difference increases, bond
energy increases. **LS** Logical

Group Activity —— GENERAL

Classifying Bonds Divide your
class into teams of three or four
students. Assign a different element
from Group 13 to each of the teams.
Ask them to determine electronegativity differences between their
assigned element and all the elements in another group, for example, Groups 14, 15, 16, and 17.
Assign one of these groups to each
team. Students should arrange, in
order of increasing predicted polarity, the bonds that might form
between their Group 13 element
and the elements of their other
group. Have students classify each
bond as nonpolar, polar, or ionic.
Be sure students are aware that the
bonds they propose may not actually form. **LS** Interpersonal

Table 2	**Electronegativity Difference for Hydrogen Halides**	
Molecule	**Electronegativity difference**	**Bond energy**
H—F	1.8	570 kJ/mol
H—Cl	1.0	432 kJ/mol
H—Br	0.8	366 kJ/mol
H—I	0.5	298 kJ/mol

Polarity Is Related to Bond Strength

When examining the electronegativity differences between elements, you
may notice a connection between electronegativity difference, the polarity
of a bond, and the strength of that bond. The greater the difference
between the electronegativity values of two elements joined by a bond,
the greater the polarity of the bond. In addition, greater electronegativity
differences tend to be associated with stronger bonds. Of the compounds
listed in **Table 2,** H—F has the greatest electronegativity difference and
thus the greatest polarity. Notice that H—F also requires the largest input
of energy to break the bond and therefore has the strongest bond.

Electronegativity and Bond Types

You have learned that when sodium and chlorine react, an electron is
removed from Na and transferred to Cl to form Na^+ and Cl^- ions. These
ions form an ionic bond. However, when hydrogen and oxygen gas react,
their atoms form a polar covalent bond by sharing electrons. How do you
know which type of bond the atoms will form? Differences in electronegativity values provide one model that can tell you.

Bonds Can Be Classified by Bond Character

Figure 7 shows the relationship between electronegativity differences and
the type of bond that forms between two elements. Notice the general rule
that can be used to predict the type of bond that forms. If the difference in
electronegativity is between 0 and 0.5, the bond is probably nonpolar covalent. If the difference in electronegativity is between 0.5 and 2.1, the bond
is considered polar covalent. If the difference is larger than 2.1, then the
bond is usually ionic. Remember that this method of classifying bonds is
just one model. Another general rule states that covalent bonds tend to form
between nonmetals, while a nonmetal and a metal will form an ionic bond.

You can see how electronegativity differences provide information
about bond character. Think about the bonds that form between the
ions sodium and fluoride and between the ions calcium and oxide. The
electronegativity difference between Na and F is 3.1. Therefore, they
form an ionic bond. The electronegativity difference between Ca and O is
2.4. They also form an ionic bond. However, the larger electronegativity
difference between Na and F means that the bond between them has a
higher percentage of ionic character.

196

REAL-WORLD
CONNECTION

Washing Up If dirt and grime were polar, it
would be possible to wash ourselves and our
clothing using only pure water. But nonpolar
dirt, especially greasy dirt, does not mix well
with polar water and won't wash away with
just a rinse. Soap is also needed. A typical soap
is a long chain of carbon atoms bonded to
hydrogen atoms with a carboxylate group
($-COO^-$) at one end. The negative carboxylate
groups are attracted to polar water molecules,
and the long carbon chains form aggregates
that latch onto nonpolar molecules of the
substances in dirt, allowing the dirt to be
washed away.

Next think about the bonds that form between carbon and chlorine and between aluminum and chlorine. The electronegativity difference between C and Cl is 0.6. These two elements form a polar covalent bond. The electronegativity difference between Al and Cl is 1.6. These two elements also form a polar covalent bond. However, the larger difference between Al and Cl means that the bond between these two elements is more polar, with greater partial charges, than the bond between C and Cl is.

Properties of Substances Depend on Bond Type

The type of bond that forms determines the physical and chemical properties of the substance. For example, metals, such as potassium, are very good electric conductors in the solid state. This property is the result of metallic bonding. Metallic bonds are the result of the attraction between the electrons in the outermost energy level of each metal atom and all of the other atoms in the solid metal. The metal atoms are held in the solid because all of the valence electrons are attracted to all of the atoms in the solid. These valence electrons can move easily from one atom to another. They are free to roam around in the solid and can conduct an electric current.

Table 3   Properties of Substances with Metallic, Ionic, and Covalent Bonds

Bond type	Metallic	Ionic	Covalent
Example substance	potassium	potassium chloride	chlorine
Melting point (°C)	63	770	−101
Boiling point (°C)	760	1500 (sublimes)	−34.6
Properties	• soft, silvery, solid • conductor as a solid	• crystalline, white solid • conductor when dissolved in water	• greenish yellow gas • not a good conductor

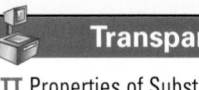

197

Transparencies

TT Properties of Substances with Metallic, Ionic, and Covalent Bonds

Using the Table

After students have studied **Table 3,** make a list of compounds similar to those in the table. Ask students to predict the properties of these substances.

Demonstration
Paper Chromatography

1. Cut a strip about 4 cm wide from a coffee filter (or paper towel).
2. Lay a pencil across a transparent plastic cup.
3. Tape the paper to the middle of the pencil so that the unattached end of the paper falls just above the bottom of the cup.
4. Using another pencil, draw a line across the paper 1 cm above the unattached end. This line will serve as a reference point. Use a water-soluble black marker to make a dot just above the middle of the line.
5. Add water to the cup to a depth of 1 cm.
6. Return the pencil and paper to the cup. The dot should be just above the water level.
7. Have students observe the result before the water reaches the top of the paper. Have them discuss the result in terms of polarity.

Teacher's Notes: Consider variations such as using food dyes or comparing different brands or colors of markers. Some will not produce separations. The separation of colors is based on the polarity of covalently bonded molecules. The process can be used to separate mixtures of covalent compounds based on their polarities. Water, or some other liquid solvent, carries the components of the mixture through a porous paper. The more polar the compound, the more the compound is attracted to the paper and the shorter the distance the component will rise up the paper.

Safety: Wear safety goggles and a lab apron.

Disposal: Throw the paper in the trash. Pour the water down the drain. Clean and save the pencil and cup.

Reteaching ——————— BASIC

Have students make a concept map with the title *Covalent Bonds.* Have them include all the key terms in this section. **LS** Visual

Quiz ——————————— GENERAL

1. Explain the difference between an ionic and a covalent bond. **Ans.** A covalent bond is formed by the sharing of one or more pairs of electrons. An ionic bond involves the attraction between oppositely charged ions, which are created by the transfer of one or more electrons.

2. What energy change occurs when two atoms form a covalent bond? **Ans.** Their potential energy decreases to a minimum.

3. In general, how is bond energy related to bond length? **Ans.** They are inversely related; the shorter the bond, the higher the energy.

4. How does electronegativity difference between two atoms affect the polarity of a covalent bond? **Ans.** The greater the electronegativity difference between two bonding atoms, the greater the polarity of the bond.

LS Verbal

Chapter Resource File

• Concept Review

• Quiz

In ionic substances, the overall attraction between all the cations and anions is very strong. Ionic compounds, such as potassium chloride, KCl, are made up of many K^+ and Cl^- ions. Each ion is held into place by many oppositely charged neighbors, so the forces—the ionic bonds—that hold them together are very strong and hard to break.

In molecular substances, such as Cl_2, the molecules are held together by sharing electrons. The shared electrons are attracted to the two bonding atoms, and they have little attraction for the atoms of other nearby molecules. Therefore, the attractive forces between separate Cl_2 molecules are very small compared to the attractive forces between the ions in KCl.

The difference in the strength of attraction between the basic units of ionic and molecular substances gives rise to different properties in the two types of substances. For example, the stronger the force between the ions or molecules of a substance in a liquid state, the more energy is required for the substance to change into a gas. **Table 3** shows that the strong forces in ionic substances, such as KCl, account for the high melting and boiling points they have compared to molecular substances, such as Cl_2. You will learn more about this relationship in a later chapter. The table also compares the conductivity of each substance.

① Section Review

UNDERSTANDING KEY IDEAS

1. Describe the attractive forces and repulsive forces that exist between two atoms as the atoms move closer together.

2. Compare a bond between two atoms to a spring between two students.

3. In what two ways can two atoms share electrons when forming a covalent bond?

4. What happens in terms of energy and stability when a covalent bond forms?

5. How can the partial charges be shown in a polar covalent molecule?

6. What information can be obtained by knowing the electronegativity differences between two elements?

7. Why do molecular compounds have low melting points and low boiling points relative to ionic substances?

CRITICAL THINKING

8. Why does the distance between two nuclei in a covalent bond vary?

9. How does a molecular orbital differ from an atomic orbital?

10. How does the strength of a covalent bond relate to bond length?

11. Compare the degree of polarity in HF, HCl, HBr, and HI.

12. Given that it has the highest electronegativity, can a fluorine atom ever form a nonpolar covalent bond? Explain your answer.

13. What does a small electronegativity difference reveal about the strength of a covalent bond?

14. Based on electronegativity values, which bond has the highest degree of ionic character: H—S, Si—Cl, or Cs—Br?

Answers to the Section Review

1. The positive nucleus of each atom attracts the electrons of the other atom. At the same time, the nuclei repel each other, as do the electron clouds.

2. The two nuclei vibrate back and forth, like students connected by a spring, coming closer together and then stretching farther apart.

3. The two atoms may share electrons equally, forming a nonpolar covalent bond, or unequally, forming a polar covalent bond.

4. Potential energy decreases, while stability increases.

5. The symbol δ+ is written as a superscript on the element with the partial positive charge. The symbol δ− is written as a superscript on the element with the partial negative charge.

6. Knowing the electronegativity difference suggests what type of bond will form between the atoms of the two elements.

7. Generally, the attractive forces between individual molecules are weak, accounting for the low melting point and boiling point of molecular compounds.

Answers continued on p. 221A

Drawing and Naming Molecules

KEY TERMS
- valence electron
- Lewis structure
- unshared pair
- single bond
- double bond
- triple bond
- resonance structure

OBJECTIVES

(1) **Draw** Lewis structures to show the arrangement of valence electrons among atoms in molecules and polyatomic ions.

(2) **Explain** the differences between single, double, and triple covalent bonds.

(3) **Draw** resonance structures for simple molecules and polyatomic ions, and recognize when they are required.

(4) **Name** binary inorganic covalent compounds by using prefixes, roots, and suffixes.

Lewis Electron-Dot Structures

Both ionic and covalent bonds involve **valence electrons,** the electrons in the outermost energy level of an atom. In 1920, G. N. Lewis, the American chemist shown in **Figure 9,** came up with a system to represent the valence electrons of an atom. This system—known as electron-dot diagrams or **Lewis structures** —uses dots to represent valence electrons. Lewis's system is a valuable model for covalent bonding. However, these diagrams do not show the actual locations of the valence electrons. They are models that help you to keep track of valence electrons.

Lewis Structures Model Covalently Bonded Molecules

A Lewis structure shows only the valence electrons in an atom or molecule. The nuclei and the electrons of the inner energy levels (if any) of an atom are represented by the symbol of the element. With only one valence electron, a hydrogen atom has the electron configuration $1s^1$. When drawing hydrogen's Lewis structure, you represent the nucleus by the element's symbol, H. The lone valence electron is represented by a dot.

H·

When two hydrogen atoms form a nonpolar covalent bond, they share two electrons. These two electrons are represented by a pair of dots between the symbols.

H:H

This Lewis structure represents a stable hydrogen molecule in which both atoms share the same pair of electrons.

valence electron

an electron that is found in the outermost shell of an atom and that determines the atom's chemical properties

Lewis structure

a structural formula in which electrons are represented by dots; dot pairs or dashes between two atomic symbols represent pairs in covalent bonds

Figure 9
G. N. Lewis (1875–1946) not only came up with important theories of bonding but also gave a new definition to acids and bases.

199

HISTORY
CONNECTION

Shared Discovery Gilbert Newton Lewis (1875–1946) and Irving Langmuir (1881–1957) independently developed the idea of two chlorine atoms sharing an electron to stabilize the orbits of each atom. Both men were versatile and productive scientists. Lewis also studied thermodynamics, developed an acid-base theory, and co-discovered deuterium oxide, D_2O, or heavy water. Langmuir's work in the field of surface chemistry earned him a Nobel Prize in 1932.

MISCONCEPTION
ALERT

Ask students how the electrons in Cl differ from the electrons in H. Students may think that because the two elements are different, the elements' electrons are different. Have them keep this in mind as they learn to model electrons as dots in Lewis structures. Some teachers keep track of electrons by using different symbols (dot, x, or square) for the valence electrons of different atoms. Be sure students know that all electrons are exactly the same and that using different symbols does not denote different kinds of electrons.

Focus

Overview

Before beginning this section, review with your students the Objectives listed in the Student Edition. Students will learn to draw Lewis structures to represent the single, double, and triple bonds in covalent molecules and ions. They will recognize when one Lewis structure is not enough to represent a structure and draw resonance structures. They will learn to name binary inorganic covalent compounds.

💿 Bellringer

List a number of diatomic compounds, such as NO, CO, HF, NaCl, HBr, and NaI, on the board. Have students classify the compounds according to the type of bonds they contain.

Motivate

Discussion

Comparing Models Write the electron configurations of a few elements on the board. Have the students decide which of the electrons shown are inner-shell electrons and which are valence electrons. Remind students that a shorthand way of writing electron configurations involves referencing the configuration of the noble gas that precedes the element in the periodic table. For example, the electron configuration for chlorine can be written either as $1s^2 2s^2 2p^6 3s^2 3p^5$ or as $[Ne]\ 3s^2 3p^5$. The second configuration represents the inner-shell electrons with the symbol [Ne] and lists the valence electrons. Contrast this type of shorthand representation with that of Lewis structures.

Chapter Resource File

- Lesson Plan

Table 4   Lewis Structures of the Second-Period Elements

Teach

Teaching Tip

The terms *Lewis structure, Lewis diagram, Lewis dot structure, electron dot structure,* and *electron dot diagram* are used to mean similar things in different texts. Some texts use different terms to distinguish between structures of atoms and structures of compounds.

SKILL BUILDER — BASIC

Vocabulary Refer students to the list of key terms, and ask students to find definitions for the terms. Ask them to use the definitions to write a brief paragraph about what the section is about. Have them compare their work with that of a partner. **LS** Verbal

Teaching Tip

There are many elements besides hydrogen, beryllium, and boron that do not always obey the octet rule. However, hydrogen and beryllium do bond to form a noble gas configuration, that of helium. Note that helium does not have an octet. Boron, aluminum, nitrogen, and phosphorus are among the elements that often do not bond to form a noble-gas configuration. Examples of compounds with atoms that do not form octets are BeS, BF_3, $SnCl_2$, AlN, NO_2, and PF_5.

Figure 10
An electron configuration shows all of the electrons of an atom, while the Lewis structure, above, shows only the valence electrons.

Element	Electron configuration	Number of valence electrons	Lewis structure (for bonding)
Li	$1s^2 2s^1$	1	Li·
Be	$1s^2 2s^2$	2	Be·
B	$1s^2 2s^2 2p^1$	3	·B·
C	$1s^2 2s^2 2p^2$	4	·C·
N	$1s^2 2s^2 2p^3$	5	:N·
O	$1s^2 2s^2 2p^4$	6	:O·
F	$1s^2 2s^2 2p^5$	7	:F·
Ne	$1s^2 2s^2 2p^6$	8	:Ne:

Lewis Structures Show Valence Electrons

The Lewis structure of a chlorine atom shows only the atom's seven valence electrons. Its Lewis structure is written with three pairs of electrons and one unpaired electron around the element's symbol, as shown below and in **Figure 10**.

$$:\ddot{C}l· $$

Table 4 shows the Lewis structures of the elements in the second period of the periodic table as they would appear in a bond. Notice that as you go from element to element across the period, you add a dot to each side of the element's symbol. You do not begin to pair dots until all four sides of the element's symbol have a dot.

An element with an octet of valence electrons, such as that found in the noble gas Ne, has a stable configuration. When two chlorine atoms form a covalent bond, each atom contributes one electron to a shared pair. With this shared pair, both atoms can have a stable octet. This tendency of bonded atoms to have octets of valence electrons is called the *octet rule.*

$$:\ddot{C}l:\ddot{C}l: $$

unshared pair

a nonbonding pair of electrons in the valence shell of an atom; also called *lone pair*

single bond

a covalent bond in which two atoms share one pair of electrons

Each chlorine atom in Cl_2 has three pairs of electrons that are not part of the bond. These pairs are called **unshared pairs** or *lone pairs.* The pair of dots that represents the shared pair of electrons can also be shown by a long dash. Both notations represent a **single bond.**

$$:\ddot{C}l:\ddot{C}l: \quad \text{or} \quad :\ddot{C}l—\ddot{C}l: $$

200

HISTORY

CONNECTION — ADVANCED

Defining Acids and Bases G. N. Lewis played a prominent role in defining the groups of compounds called acids and bases. His definition includes more compounds than other definitions that had come before. Ask students to use another text or reference book to find out what Lewis's definition is and how it differs from other ways of viewing acids and bases. **LS** Logical

For Lewis structures of bonded atoms, you may want to keep in mind that when the dots for the valence electrons are placed around the symbol, each side must contain an unpaired electron before any side can contain a pair of electrons. For example, see the Lewis structure for a carbon atom below.

$$\cdot \overset{\displaystyle\cdot}{C} \cdot$$

The electrons can pair in any order. However, any unpaired electrons are usually filled in to show how they will form a covalent bond. For example, think about the bonding between hydrogen and chlorine atoms.

$$H\cdot \ + \ \cdot \overset{\displaystyle\cdot\cdot}{\underset{\displaystyle\cdot\cdot}{Cl}}: \ \longrightarrow \ H—\overset{\displaystyle\cdot\cdot}{\underset{\displaystyle\cdot\cdot}{Cl}}:$$

SKILLS Toolkit 1

Drawing Lewis Structures with Many Atoms

1. Gather information.
- Draw a Lewis structure for each atom in the compound. When placing valence electrons around an atom, place one electron on each side before pairing any electrons.
- Determine the total number of valence electrons in the compound.

2. Arrange the atoms.
- Arrange the Lewis structure to show how the atoms bond in the molecule.
- Halogen and hydrogen atoms *often* bind to only one other atom and are *usually* at an end of the molecule.
- Carbon is often placed in the center of the molecule.
- You will find that, with the exception of carbon, the atom with the lowest electronegativity is often the central atom.

3. Distribute the dots.
- Distribute the electron dots so that each atom, except for hydrogen, beryllium, and boron, satisfies the octet rule.

4. Draw the bonds.
- Change each pair of dots that represents a shared pair of electrons to a long dash.

5. Verify the structure.
- Count the number of electrons surrounding each atom. Except for hydrogen, beryllium, and boron, all atoms must satisfy the octet rule. Check that the number of valence electrons is still the same number you determined in step 1.

201

HISTORY
CONNECTION — ADVANCED

 Linus Pauling In his career, Linus Pauling earned two Nobel Prizes, one for chemistry in 1954 and the other for peace in 1962. The peace prize was for his efforts to alert the world to the dangers of testing nuclear weapons in the atmosphere, helping to bring about a worldwide ban on aboveground testing. In chemistry, Pauling wrote about the nature of the chemical bond and the structures of molecules and crystals. Later in his career, Pauling applied his knowledge of molecular structure to proteins and other large biochemical molecules and investigated the mechanism of sickle cell anemia. Have students research the life of Linus Pauling and write a paper or prepare an oral presentation for the class. **LS Verbal**

Teaching Tip
Bonding Structures The Lewis structure for carbon shown in the text models the valence electrons of a carbon atom when it bonds with other atoms. The same is true for the Lewis structures in **Table 4.** Isolated atoms can have different structures than bonded atoms. As an isolated atom, the Lewis structure for carbon is $\overset{\displaystyle\cdot\cdot}{C}\cdot$.

Teaching Tip
Checking Octets Show students that they can check their Lewis structures by making sure that each atom is surrounded by four pairs of electrons (or one pair in the case of hydrogen). After completing a structure, draw a circle around each atom enclosing in the circle the eight electrons that complete the octet for that atom. The circles will overlap as the shared pair of electrons is enclosed in the circles surrounded each bonding atom.

MISCONCEPTION ALERT
As students begin to draw Lewis structures, they may think that all electrons are shared equally because the pair of dots representing two electrons are placed midway between the two symbols for the bonding atoms. Point out that a Lewis structure is only a way of accounting for the location of the bonds. Lewis structures do not describe the location of the electrons. A pair of electrons could be shared equally as in Cl_2 or it could be shared unequally as in HCl. Remind students that unequal sharing means that a bond is polar.

Teach, continued

Answers to Practice Problems A

1.

H
|
:S—H   :Cl—C—Cl:
·· |
··   H

H   H   H
| | |
H—N—H   H—C—C—H
·· | |
               H   H

2.

H
|
H—C—Ö—H
|
H

Homework —

Additional Practice Draw Lewis structures for each compound:

1. sulfur dichloride, SCl_2

Ans. :S̈—C̈l:
          |
       :Cl:
          ··

2. arsenic trifluoride, AsF_3

Ans. :F̈—Äs—F̈:
                 |
              :F:
              ··

3. silicon tetrahydride, SiH_4

Ans.   H
        H:Si:H
           H

4. trifluoromethane, CHF_3

Ans.   H
           |
    :F̈—C—F̈:
           |
        :F:
        ··

Drawing Lewis Structures with Single Bonds

Draw a Lewis structure for CH_3I.

1 Gather information.

Draw each atom's Lewis structure, and count the total number of valence electrons.

·Ċ· H· H· H· ·Ï: number of dots: 14

2 Arrange the atoms.

Arrange the Lewis structure so that carbon is the central atom.

H
H C I
H

PRACTICE HINT

You may have to try several Lewis structures until you get one in which all of the atoms, except hydrogen, beryllium, and boron, obey the octet rule.

3 Distribute the dots.

Distribute one bonding pair of electrons between each of the bonded atoms. Then, distribute the remaining electrons, in pairs, around the remaining atoms to form an octet for each atom.

H
··
H:C:Ï:
··
H

4 Draw the bonds.

Change each pair of dots that represents a shared pair of electrons to a long dash.

H
|
H—C—Ï:
| ··
H

5 Verify the structure.

Carbon and iodine have 8 electrons, and hydrogen has 2 electrons. The total number of valence electrons is still 14.

PRACTICE

1 Draw the Lewis structures for H_2S, CH_2Cl_2, NH_3, and C_2H_6.

2 Draw the Lewis structure for methanol, CH_3OH. First draw the CH_3 part, and then add O and H.

Lewis Structures for Polyatomic Ions

Lewis structures are also helpful in describing polyatomic ions, such as the ammonium ion, NH_4^+. An ammonium ion, shown in **Figure 11,** forms when ammonia, NH_3, is combined with a substance that easily gives up a hydrogen ion, H^+. To draw the Lewis structure of NH_4^+, first draw the structure of NH_3. With five valence electrons, a nitrogen atom can make a stable octet by forming three covalent bonds, one with each hydrogen atom. Then add H^+, which is simply the nucleus of a hydrogen atom, or a proton, and has no electrons to share. The H^+ can form a covalent bond with NH_3 by bonding with the unshared pair on the nitrogen atom.

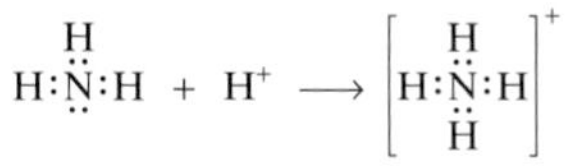

$$\overset{\text{H}}{\underset{..}{\text{H:N:H}}} + \text{H}^+ \longrightarrow \left[\overset{\text{H}}{\underset{\text{H}}{\text{H:N:H}}}\right]^+$$

The Lewis structure is enclosed in brackets to show that the positive charge is distributed over the entire ammonium ion.

$$\left[\overset{\text{H}}{\underset{\text{H}}{\text{H:N:H}}}\right]^+$$
Ammonium ion

Figure 11
Smelling salts often have an unstable ionic compound made of two polyatomic ions: ammonium and carbonate.

SAMPLE PROBLEM B

Drawing Lewis Structures for Polyatomic Ions

Draw a Lewis structure for the sulfate ion, SO_4^{2-}.

1 Gather information.

When counting the total number of valence electrons, add two additional electrons to account for the 2– charge on the ion.

$$:\overset{..}{\underset{..}{S}}:\quad :\overset{..}{\underset{..}{O}}:\quad :\overset{..}{\underset{..}{O}}:\quad :\overset{..}{\underset{..}{O}}:\quad :\overset{..}{\underset{..}{O}}:\qquad \text{number of dots: } 30 + 2 = 32$$

2 Arrange the atoms. Distribute the dots.

Sulfur has the lowest electronegativity, so it is the central atom. Distribute the 32 dots so that there are 8 dots around each atom.

$$\begin{array}{c}:\overset{..}{O}:\\ :\overset{..}{O}:\overset{..}{\underset{..}{S}}:\overset{..}{O}:\\ :\overset{..}{O}:\end{array}$$

3 Draw the bonds. Verify the structure.

- Change each bonding pair to a long dash. Place brackets around the ion and a 2– charge outside the bracket to show that the charge is spread out over the entire ion.
- There are 32 valence electrons, and each O and S has an octet.

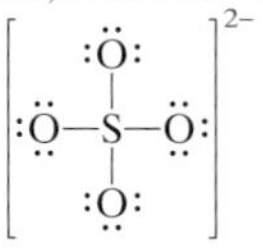

$$\left[\begin{array}{c}:\overset{..}{O}:\\ :\overset{..}{O}-\underset{\underset{:\overset{..}{O}:}{|}}{\overset{|}{S}}-\overset{..}{O}:\end{array}\right]^{2-}$$

PRACTICE

1 Draw the Lewis structure for ClO_3^-.

2 Draw the Lewis structure for the hydronium ion, H_3O^+.

PROBLEM SOLVING SKILL

203

SKILL BUILDER —ADVANCED

Writing Skills Have students research and write a report about the uses of and reactions involving ammonia and ammonium-containing salts. Student findings may include smelling salts, explosives, and fertilizers. **LS Verbal**

Answers to Practice Problems B

1.
$$\left[:\overset{..}{\underset{..}{O}}-\underset{\underset{:\overset{..}{O}:}{|}}{Cl}-\overset{..}{\underset{..}{O}}:\right]^-$$

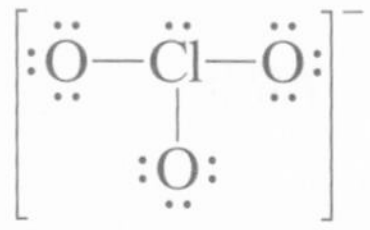

2.
$$\left[\begin{array}{c}\text{H}\\ |\\ \text{H}-\underset{..}{O}-\text{H}\end{array}\right]^+$$

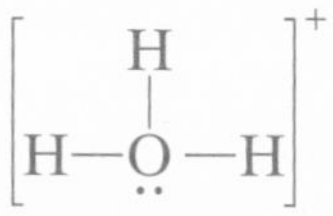

Homework ——GENERAL

Additional Practice Draw Lewis structures for each of the following polyatomic ions:

1. HS^-

Ans. $\left[\text{H}-\overset{..}{\underset{..}{S}}:\right]^-$

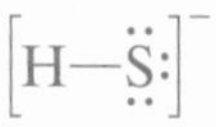

2. NF_4^+

Ans.
$$\left[\begin{array}{c}:\overset{..}{\underset{..}{F}}:\\ |\\ :\overset{..}{\underset{..}{F}}-\text{N}-\overset{..}{\underset{..}{F}}:\\ |\\ :\overset{..}{\underset{..}{F}}:\end{array}\right]^+$$

3. IO^-

Ans. $\left[:\overset{..}{\underset{..}{I}}-\overset{..}{\underset{..}{O}}:\right]^-$

4. PCl_4^+

Ans.
$$\left[\begin{array}{c}:\overset{..}{\underset{..}{Cl}}:\\ |\\ :\overset{..}{\underset{..}{Cl}}-\text{P}-\overset{..}{\underset{..}{Cl}}:\\ |\\ :\overset{..}{\underset{..}{Cl}}:\end{array}\right]^+$$

5. CH_3^+

Ans.
$$\left[\begin{array}{c}\text{H}\\ |\\ \text{H}-\text{C}-\text{H}\end{array}\right]^+$$

LS Visual

PRACTICE HINT

- If the polyatomic ion has a negative charge, add the appropriate number of valence electrons. (For example, the net charge of 2– on SO_4^{2-} means that there are two more electrons than in the neutral atoms.)
- If the polyatomic ion has a positive charge, subtract the appropriate number of valence electrons. (For example, the net charge of 1+ on H_3O^+ means that there is one fewer electron than in the neutral atoms.)

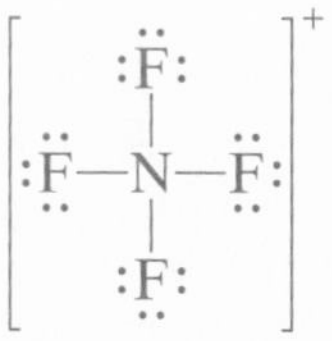

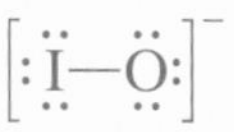

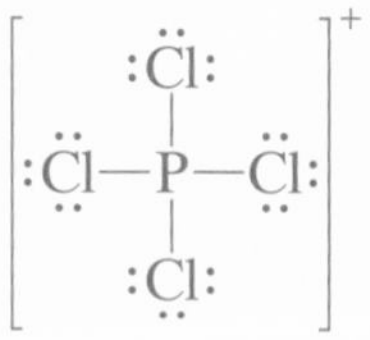

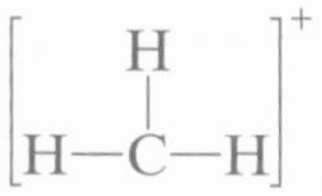

Teaching Tip

Remind students that like charges repel each other. In a multiple bond, two or three pairs of electrons must occupy the same small region near two nuclei. Electrons in such close proximity repel each other and cause stress in the bond. This stress is one reason that most multiple bonds are reactive.

Using the Figure

Ripening Fruit The role of ethene, C_2H_4, in ripening fruit is illustrated in **Figure 12.** Ethene is produced in many fruits in addition to tomatoes, such as apples, pears, and oranges. Such plants are called *climacteric plants*. Ethene is a plant hormone whose role is to trigger ripening, a process that involves the destruction of the green substance called chlorophyll. In unripe fruit, chlorophyll masks other colors which are always present in the skins of the fruit, for example, the red of apples and tomatoes, the orange of oranges, and the yellow of lemons. Ethene is used by merchants to rapidly ripen green bananas to prepare them for market. Tomatoes can be ripened more rapidly at home by placing them in a paper bag with apples, preferably bruised apples, which produce ethene faster than bruised apples do. The flowers of a broccoli plant will turn yellow more quickly if apples are stored along with it in the refrigerator.

double bond

a covalent bond in which two atoms share two pairs of electrons

Figure 12
Most plants have a hormone called *ethene*, C_2H_4. Tomatoes release ethene, also called *ethylene,* as they ripen.

Multiple Bonds

Atoms can share more than one pair of electrons in a covalent bond. Think about a nonpolar covalent bond formed between two oxygen atoms in an O_2 molecule. Each O has six valence electrons, as shown below.

$$:\ddot{O}\cdot \quad \cdot\ddot{O}:$$

If these oxygen atoms together shared only one pair of electrons, each atom would have only seven electrons. The octet rule would not be met.

Bonds with More than One Pair of Electrons

To make an octet, each oxygen atom needs two more electrons to be added to its original six. To add two electrons, each oxygen atom must share two electrons with the other atom so that the two atoms share four electrons. The covalent bond formed by the sharing of two pairs of electrons is a **double bond,** shown in the Lewis structures below.

$$:\ddot{O}::\ddot{O}: \quad \text{or} \quad :\ddot{O}=\ddot{O}:$$

Atoms will form a single or a multiple bond depending on what is needed to make an octet. While two O atoms form a double bond in O_2, an O atom forms a single bond with each of two H atoms in a water molecule.

$$\ddot{:O}:H \quad \text{or} \quad :\ddot{O}-H$$

Another example of a molecule that has a double bond is ethene, C_2H_4, shown in **Figure 12.** Each H atom forms a single bond with a C atom. Each C atom below has two electrons that are not yet part of a bond.

$$H:\ddot{C}\cdot \quad \cdot\ddot{C}:H$$

With only six electrons, each C atom needs two more electrons to have an octet. The only way to complete the octets is to form a **double bond.**

$$H:\overset{H}{\underset{}{\ddot{C}}}::\overset{H}{\underset{}{\ddot{C}}}:H \quad \text{or} \quad H-\overset{H}{\underset{}{C}}=\overset{H}{\underset{}{C}}-H$$

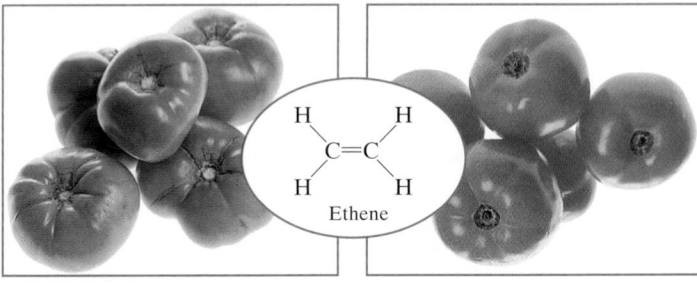

Ethene

204

Paramagnetism In order for two electrons to form a pair in an atomic or molecular orbital, they must have opposite spins. Atoms or molecules in which all electrons are paired are *diamagnetic*. That is, they are not affected much by an external magnetic field because the spins of the electrons nearly cancel each other. If an atom or a molecule contains one or more unpaired electrons, the unbalanced spin of the unpaired electron does react with a magnetic field. Such a substance is called *paramagnetic*. The O_2 molecule is known to be paramagnetic, which leads to the conclusion that a Lewis structure is not adequate for describing oxygen's structure. In a Lewis structure of O_2, all the electrons are represented as paired. Have students verify this by drawing the structure.

Carbon, oxygen, and nitrogen atoms often form double bonds by sharing two pairs of electrons. Carbon and nitrogen atoms may even share three pairs of electrons to form a **triple bond.** Think about the molecule N_2. With five valence electrons, each N atom needs three more electrons for a stable octet. Each N atom contributes three electrons to form three bonding pairs. The two N atoms form a triple bond by sharing these three pairs of electrons, or a total of six electrons. Because the two N atoms share the electrons equally, the triple bond is a nonpolar covalent bond.

triple bond

a covalent bond in which two atoms share three pairs of electrons

$$:N::N: \quad \text{or} \quad :N \equiv N:$$

SAMPLE PROBLEM C

Drawing Lewis Structures with Multiple Bonds

Draw a Lewis structure for formaldehyde, CH_2O.

1 Gather information.

Draw each atom's Lewis structure, and count the total dots.

$$\cdot \ddot{C} \cdot \quad H\cdot \quad H\cdot \quad \cdot \ddot{O} \cdot \qquad \text{total dots: 12}$$

2 Arrange the atoms. Distribute the dots.

- Arrange the atoms so that carbon is the central atom.
- Distribute one pair of dots between each of the atoms. Then, starting with the outside atoms, distribute the rest of the dots, in pairs, around the atoms. You will run out of electrons before all of the atoms have an octet (left structure). C does not have an octet, so there must be a multiple bond. To obtain an octet for C, move one of the unshared pairs from the O atom to between the O and the C (right structure).

incorrect:
$$\begin{array}{c} :\ddot{O}: \\ H:\ddot{C}:H \end{array}$$
correct:
$$\begin{array}{c} :O: \\ \overset{..}{H:C:H} \end{array}$$

3 Draw the bonds. Verify the structure.

- Change each pair of dots that represents a shared pair of electrons to a long dash. Two pairs of dots represent a double bond.
- C and O atoms both have eight electrons, and each H atom has two electrons. The total number of valence electrons is still 12.

$$\begin{array}{c} :O: \\ \| \\ H-C-H \end{array}$$

PRACTICE HINT

- Begin with a single pair of dots between each pair of bonded atoms. If no arrangement of single bonds provides a Lewis structure whose atoms satisfy the octet rule, the molecule might have multiple bonds.
- N and C can form single bonds or combinations of single and double or triple bonds.

PRACTICE

1. Draw the Lewis structures for carbon dioxide, CO_2, and carbon monoxide, CO.

2. Draw the Lewis structures for ethyne, C_2H_2, and hydrogen cyanide, HCN.

PROBLEM SOLVING SKILL

205

MISCONCEPTION ///ALERT\\\

Triple bonds are sometimes represented horizontally, $H:C:::C:H$. This representation may lead students to think that three electrons are involved in each of two bonds. Remember, Lewis dot structures are just models. They are not intended to show spatial orientation of electrons or electron clouds.

Teach, continued

Teaching Tip
Resonance and the Octet Rule

The molecule O_3 has a bond whose bond strength is between a single bond and a double bond. Molecules that have resonance structures are more stable than either of the Lewis structures indicates. For example, benzene, C_6H_6, has resonance structures and displays significant stability. The effect also explains how certain molecules with odd numbers of electrons do not follow the octet rule. One example is NO_2, which takes part in a series of reactions that lead to smog and can be drawn with the following resonance structures.

$$\ddot{O}=\ddot{N}-\ddot{O}: \longleftrightarrow :\ddot{O}-\ddot{N}=\ddot{O}$$

A molecule of NO_2 is a resonance hybrid of the two Lewis structures shown. Each oxygen atom in both Lewis structures follows the octet rule. However, the nitrogen atom in both structures does not because it has only seven valence electrons. This odd number of valence electrons surrounding the nitrogen atom gives nitrogen an unpaired electron and thus makes NO_2 a very reactive molecule.

Teaching Tip

It is common practice to omit the prefix *mono-* when naming the second element of a binary molecular compound. For example, nitrogen monoxide, NO, is frequently referred to as nitrogen oxide. One exception to this practice is carbon monoxide. In this case, the prefix is never omitted.

Figure 13
You can draw resonance structures for sulfur dioxide, SO_2, a chemical that can add to air pollution.

$$\ddot{O}=\ddot{S}-\ddot{O}: \longleftrightarrow :\ddot{O}-\ddot{S}=\ddot{O}$$

resonance structure

in chemistry, any one of two or more possible configurations of the same compound that have identical geometry but different arrangements of electrons

Topic Link

Refer to the "Ions and Ionic Compounds" chapter for more about naming ionic compounds.

Resonance Structures

Some molecules, such as ozone, O_3, cannot be represented by a single Lewis structure. Ozone has two Lewis structures, as shown below.

$$\ddot{O}=\ddot{O}-\ddot{O}: \longleftrightarrow :\ddot{O}-\ddot{O}=\ddot{O}$$

Each O atom follows the octet rule, but the two structures use different arrangements of the single and double bonds. So which structure is correct? Neither structure is correct by itself. When a molecule has two or more possible Lewis structures, the two structures are called **resonance structures.** You place a double-headed arrow between the structures to show that the actual molecule is an average of the two possible states.

Another molecule that has resonance structures is sulfur dioxide, SO_2, shown in **Figure 13.** Sulfur dioxide released into the atmosphere is partly responsible for acid precipitation. The actual structure of SO_2 is an average, or a *resonance hybrid,* of the two structures. Although you draw the structures as if the bonds change places again and again, the bonds do not in fact move back and forth. The actual bonding is a mixture of the two extremes represented by each of the Lewis structures.

Naming Covalent Compounds

Covalent compounds made of two elements are named by using a method similar to the one used to name ionic compounds. Think about how the covalent compound SO_2 is named. The first element named is usually the first one written in the formula, in this case *sulfur.* Sulfur is the less-electronegative element. The second element named has the ending *-ide,* in this case *oxide.*

However, unlike the names for ionic compounds, the names for covalent compounds must often distinguish between two different molecules made of the same elements. For example, SO_2 and SO_3 cannot both be called *sulfur oxide.* These two compounds are given different names based on the number of each type of atom in the compound.

Prefixes Indicate How Many Atoms Are in a Molecule

The system of prefixes shown in **Table 5** is used to show the number of atoms of each element in the molecule. SO_2 and SO_3 are distinguished from one another by the use of prefixes in their names. With only two oxygen atoms, SO_2 is named *sulfur dioxide.* With three oxygen atoms, SO_3 is named *sulfur trioxide.* The following example shows how to use the system of prefixes to name P_2S_5.

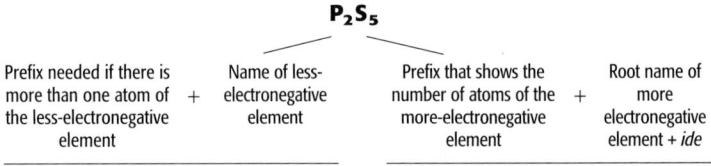

$$\mathbf{P_2S_5}$$

Prefix needed if there is more than one atom of the less-electronegative element	+	Name of less-electronegative element	Prefix that shows the number of atoms of the more-electronegative element	+	Root name of more electronegative element + *ide*
diphosphorus			**pentasulfide**		

MISCONCEPTION ALERT

Students often think that, like an ionic formula, a correctly written molecular formula should show the simplest ratio of atoms in the compound. However, a molecular formula represents the actual numbers of individual atoms in a molecule. Thus, the subscripts in the formula should not be reduced to the smallest possible ratio. For example, the molecular formulas for ethyne (acetylene) and

benzene are C_2H_2 and C_6H_6, respectively. Ethyne actually contains two carbon atoms and two hydrogen atoms. Similarly, benzene actually contains six of each of the atoms. Reducing the subscripts to the smallest possible ratios would produce CH, which is not the molecular formula for either of these molecules. Students will learn about empirical formulas, such as CH, in a later chapter.

206

Table 5 Prefixes for Naming Covalent Compounds

Prefix	Number of atoms	Example	Name
mono-	1	CO	carbon monoxide
di-	2	SiO_2	silicon dioxide
tri-	3	SO_3	sulfur trioxide
tetra-	4	SCl_4	sulfur tetrachloride
penta-	5	$SbCl_5$	antimony pentachloride

Refer to Appendix A for a more complete list of prefixes.

🌐 internet connect
www.scilinks.org
Topic: Naming Compounds
SciLinks code: HW4081

SCiLINKS. Maintained by the National Science Teachers Association

Prefixes are added to the first element in the name only if the molecule contains more than one atom of that element. So, N_2O is named *dinitrogen oxide*, S_2F_{10} is named *disulfur decafluoride*, and P_4O_6 is named *tetraphosphorus hexoxide*. If the molecule contains only one atom of the first element given in the formula, the prefix *mono-* is left off. Both SO_2 and SO_3 have only one S atom each. Therefore, the names of both start with the word *sulfur*. Note that the vowels *a* and *o* are dropped from a prefix that is added to a word begining with a vowel. For example, CO is carbon monoxide, not carbon monooxide. Similarly, N_2O_4 is named *dinitrogen tetroxide*, not *dinitrogen tetraoxide*.

 ② Section Review

UNDERSTANDING KEY IDEAS

1. Which electrons do a Lewis structure show?
2. In a polyatomic ion, where is the charge located?
3. How many electrons are shared by two atoms that form a triple bond?
4. What do resonance structures represent?
5. How do the names for SO_2 and SO_3 differ?

PRACTICE PROBLEMS

6. Draw a Lewis structure for an atom that has the electron configuration $1s^2 2s^2 2p^6 3s^2 3p^3$.
7. Draw Lewis structures for each compound:
   a. BrF   c. Cl_2O
   b. $N(CH_3)_3$ d. ClO_2^-

8. Draw three resonance structures for SO_3.
9. Name the following compounds.
   a. SnI_4 c. PCl_3
   b. N_2O_3 d. CSe_2
10. Write the formula for each compound:
    a. phosphorus pentabromide
    b. diphosphorus trioxide
    c. arsenic tribromide
    d. carbon tetrachloride

CRITICAL THINKING

11. Compare and contrast the Lewis structures for krypton and radon.
12. Do you always follow the octet rule when drawing a Lewis structure? Explain.
13. What is incorrect about the name *monosulfur dioxide* for the compound SO_3?

207

Chapter Resource File

• Concept Review
• Quiz

👆 **One-Stop Planner CD-ROM**

• **Career Extension**
  Occupational Applications Worksheet 9:
  Air-Quality Specialist
  Assign this worksheet to emphasize relevant applications of text concepts.

SKILL BUILDER — BASIC

Vocabulary Have students brainstorm examples of words containing prefixes that denote a number. Examples might include *bicycle, tricycle, octopus,* or *decade.* Other examples might include terms from mathematics, such as *triangle, pentagon, quadrilateral,* or *hexagon.* Review the meaning of each of the prefixes in the students' list of words. Tell them that such prefixes are also used to name molecular compounds. **LS** Verbal

Close

Reteaching — BASIC

Have students create a table like **Table 4** for the eight elements in the third row of the periodic table. Have them predict one compound that each atom would form and draw its Lewis structure. **LS** Logical

Quiz — GENERAL

1. Explain what is meant by the octet rule. **Ans.** Most atoms attain a greater degree of stability when they have eight electrons in their outer orbitals. Hydrogen is one exception.

2. Draw a Lewis structure for H_2S.
   Ans. H
           |
         :S—H

3. Explain why carbon can form only four single covalent bonds. **Ans.** Carbon has four valence electrons. It can pair with only four other atoms before its outer energy level is filled.

4. Draw resonance structures for ozone, O_3.
   Ans.
   $\ddot{O}=\ddot{O}-\ddot{O}: \longleftrightarrow :\ddot{O}-\ddot{O}=\ddot{O}$

5. Name the following compounds: SiF_3, PtF_6, P_2O_3, and V_4O_{10}
   Ans. silicon trifluoride, platinum hexafluoride, diphosphorus trioxide, tetravanadium decoxide
   LS Logical

SECTION

Focus

Overview

Before beginning this section, review with your students the Objectives listed in the Student Edition. Students will use Lewis structures and the valence shell electron pair repulsion theory (VSEPR) to predict the molecular geometry of molecules. After learning about molecular shape, students will be able to predict the polarity of a molecule and many of the compound's properties.

🔊 Bellringer

Without knowing about the theory in advance, have students write a short paragraph telling what they think the valence shell electron pair repulsion theory has to do with molecular shape.

Motivate

Identifying Preconceptions

Ask students what they think a molecule of BH_3 or CH_4 might look like. They may take Lewis structures too literally and form the impression that all molecules are flat, planar structures. Emphasize that although some atoms form molecules that lie in one plane, other atoms form molecules that are three-dimensional and that a Lewis structure does not reveal molecular geometry.

SECTION

3 Molecular Shapes

KEY TERM
• VSEPR theory

OBJECTIVES

① **Predict** the shape of a molecule using VSEPR theory.

② **Associate** the polarity of molecules with the shapes of molecules, and relate the polarity and shape of molecules to the properties of a substance.

Determining Molecular Shapes

Lewis structures are two-dimensional and do not show the three-dimensional shape of a molecule. However, the three-dimensional shape of a molecule is important in determining the molecule's physical and chemical properties. Sugar, or sucrose, is an example. Sucrose has a shape that fits certain nerve receptors on the tongue. Once stimulated, the nerves send signals to the brain, and the brain interprets these signals as sweetness. Inside body cells, sucrose is processed for energy.

People who want to avoid sucrose in their diet often use a sugar substitute, such as sucralose, shown in **Figure 14.** These substitutes have shapes similar to that of sucrose, so they can stimulate the nerve receptors in the same way that sucrose does. However, sucralose has a different chemical makeup than sucrose does and cannot be processed by the body.

Figure 14
Sucralose is chemically very similar to sucrose. Both have the same three-dimensional shape. However, three Cl atoms have been substituted in sucralose, so the body cannot process it.

Chapter Resource File

• Lesson Plan

Transparencies

TT Molecular Shapes with Three Electron Pairs Around Central Atom

---

CO is linear.

SO₂ is bent.

CO₂ is linear

a Molecules made up of only two atoms, such as CO, have a linear shape.

b Although SO₂ and CO₂ have the same numbers of atoms, they have different shapes because the numbers of electron groups surrounding the central atoms differ.

Figure 15
Molecules with three or fewer atoms have shapes that are in a flat plane.

A Lewis Structure Can Help Predict Molecular Shape

The shape of a molecule made of only two atoms, such as H_2 or CO, is easy to determine. As shown in **Figure 15,** only a linear shape is possible when there are two atoms. Determining the shapes of molecules made of more than two atoms is more complicated. Compare carbon dioxide, CO_2, and sulfur dioxide, SO_2. Both molecules are made of three atoms. Although the molecules have similar formulas, their shapes are different. Notice that CO_2 is linear, while SO_2 is bent.

Obviously, the formulas CO_2 and SO_2 do not provide any information about the shapes of these molecules. However, there is a model that can be used to predict the shape of a molecule. This model is based on the **valence shell electron pair repulsion (VSEPR) theory.** Using this model, you can predict the shape of a molecule by examining the Lewis structure of the molecule.

Electron Pairs Can Determine Molecular Shape

According to the VSEPR theory, the shape of a molecule is determined by the valence electrons surrounding the central atom. For example, examine the Lewis structure for CO_2.

$$\ddot{O}=C=\ddot{O}$$

Notice the two double bonds around the central carbon atom. Because of their negative charge, electrons repel each other. Therefore, the two shared pairs that form each double bond repel each other and remain as far apart as possible. These two sets of two shared pairs are farthest apart when they are on opposite sides of the carbon atom. Thus, the shape of a CO_2 molecule is linear. You'll read about SO_2's bent shape later.

Now think about what happens when the central atom is surrounded by three shared pairs. Look at the Lewis structure for BF_3, which has boron, an example of an atom that does not always obey the octet rule.

$$:\ddot{F}-B-\ddot{F}:$$
$$|$$
$$:\ddot{F}:$$

Notice the three single bonds around the central boron atom. Like three spokes of a wheel, these shared pairs of electrons extend from the central boron atom. The three F atoms, each of which has three unshared pairs, will repel each other and will be at a maximum distance apart. This molecular shape is known as *trigonal planar*, as shown in **Figure 16.**

internet connect
www.scilinks.org
Topic: VSEPR Theory
SciLinks code: HW4169

SCILINKS Maintained by the National Science Teachers Association

valence shell electron pair repulsion (VSEPR) theory

a theory that predicts some molecular shapes based on the idea that pairs of valence electrons surrounding an atom repel each other

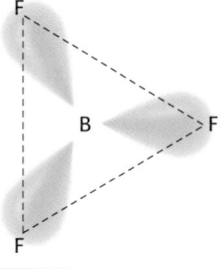

Figure 16
Trigonal planar molecules, such as BF_3, are flat structures in which three atoms are evenly spaced around the central atom.

209

did you know?

Hybridization The *s* orbital is spherical and thus has no specific orientation. The *p* orbitals are directed at 90° to one another. Bonds using *s* orbitals are different from those using *p* orbitals. The one *s* and three *p* bonds cannot be directed to the four corners of a tetrahedron. How is it that CH_4 indeed has a tetrahedral geometry?

One model explains that as bonding occurs, the *s* and three *p* orbitals mix together to form four hybrid orbitals called sp^3 orbitals. These four orbitals are exactly alike and are directed toward the corners of a tetrahedron.

Unshared Pairs and Shape
Students may wonder how it can be that BH_3 is a trigonal planar molecule, whereas NH_3 is trigonal pyramidal, and that H_2O is a bent molecule, whereas $BeCl_2$ is linear. Have students construct models for molecules such as BeF_2, CCl_4, PH_3, SO_2, and O_3. Also have them compare structures such as NO_2^- and NO_2. Give pairs of students a supply of toothpicks and gum-drops. Have them draw Lewis structures for the molecules. Then, students should arrange the pairs of electrons (toothpicks), both bonding and nonbonding, around the central atom (gumdrops) so that each pair has the maximum amount of space. Toothpicks representing bonding pairs should have a gum-drop at the end. Those representing nonbonding pairs should not. Extend the exercise by having students model PF_5 (trigonal bipyramidal) and SF_6 (octahedral). **LS** Visual

SKILL BUILDER — ADVANCED

Math Skills The angles of the geometric figures that model molecular shapes are the same or close to the bond angles in molecules. For example, BH_3 is a trigonal planar molecule. The angles between the B—H bonds are 120°. Have students draw the molecule and measure the angles. Refer students to an appropriate math textbook, and ask them to find the angles between bonds in tetrahedral CCl_4, trigonal bipyramidal PF_5, and octahedral SF_6. **LS** Visual

🔲 internet connect ≡≡≡
www.scilinks.org
Topic: Molecular Shapes
SciLinks code: HW4080

*sci*LINKS Maintained by the National Science Teachers Association

Next, think about what happens when the central atom is surrounded by four shared pairs of electrons. Examine the Lewis structure for methane, CH_4, shown below.

$$
\begin{array}{c}
\quad H \\
| \\
H - C - H \\
| \\
H
\end{array}
$$

Notice that four single bonds surround the central carbon atom. On a flat plane the bonds are not as far apart as they can be. Instead, the four shared pairs are farthest apart when each pair of electrons is positioned at the corners of a tetrahedron, as shown in **Figure 17.** Only the electron clouds around the central atom are shown.

In CO_2, BF_3, and CH_4, all of the valence electrons of the central atom form shared pairs. What happens to the shape of a molecule if the central atom has an unshared pair? Tin(II) chloride, $SnCl_2$, gives an example. Examine the Lewis structure for $SnCl_2$, shown below.

$$
\begin{array}{c}
:Sn - \overset{..}{\underset{..}{Cl}}: \\
| \\
:\overset{}{\underset{..}{Cl}}:
\end{array}
$$

Notice that the central tin atom has two shared pairs and one unshared pair of electrons. In VSEPR theory, unshared pairs occupy space around a central atom, just as shared pairs do. The two shared pairs and one unshared pair of the tin atom cause the shape of the $SnCl_2$ molecule to be bent, as shown in **Figure 17.**

The unshared pairs of electrons *influence* the shape of a molecule but are not visible in the space-filling model. For example, the shared and unshared pairs of electrons in $SnCl_2$ form a trigonal planar geometry, but the molecule has a bent shape. The bent shape of SO_2, shown in **Figure 15** on the previous page, is also due to unshared pairs. However, in the case of SO_2, there are two unshared pairs.

Figure 17
The electron clouds around the central atom help determine the shape of a molecule.

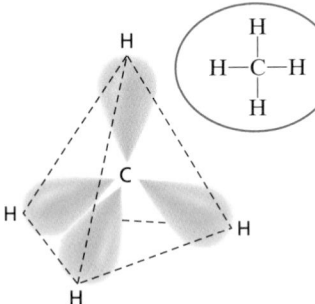

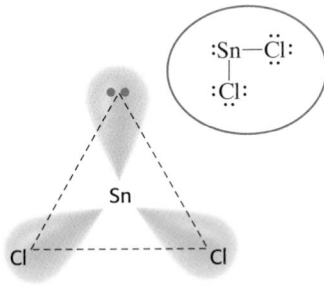

a A molecule whose central atom is surrounded by four shared pairs of electrons, such as CH_4, has a tetrahedral shape.

b A molecule whose central atom is surrounded by two shared pairs and one unshared pair, such as $SnCl_2$, has a bent shape.

210

INCLUSION Strategies

• *Tactile Learners*
Use molecular models to help tactile learners to recognize molecular shapes. Have students build the models, or build the models yourself and pass them around for students to handle.

Transparencies
TT Molecular Shapes with Four Electron Pairs Around Central Atom

ChemFile CHEMISTRY INTERACTIVE TUTOR

• **Module 4: Chemical Bonding**
  Topic: Molecular Geometry
  This engaging tutorial reviews and reinforces how to determine molecular geometry.

SAMPLE PROBLEM D

Predicting Molecular Shapes

Determine the shapes of NH_3 and H_2O.

1 Gather information.

Draw the Lewis structures for NH_3 and H_2O.

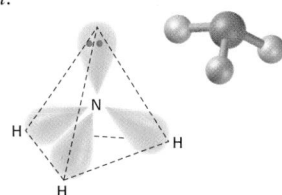

2 Count the shared and unshared pairs.

Count the number of shared and unshared pairs of electrons around each central atom.

NH_3 has three shared pairs and one unshared pair.

H_2O has two shared pairs and two unshared pairs.

3 Apply VSEPR theory.

- Use VSEPR theory to find the shape that allows the shared and unshared pairs of electrons to be spaced as far apart as possible.
- The ammonia molecule will have the shape of a pyramid. This geometry is called *trigonal pyramidal*.

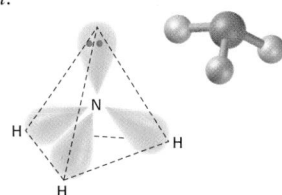

- The water molecule will have a bent shape.

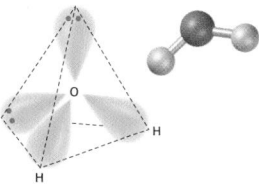

4 Verify the Structure.

For both molecules, be sure that all atoms, except hydrogen, obey the octet rule.

PRACTICE

Predict the shapes of the following molecules and polyatomic ions.

1. **a.** NH_2Cl **c.** NO_3^-
   b. $NOCl$ **d.** NH_4^+

PROBLEM SOLVING SKILL

211

- Keep in mind that the geometry is difficult to show on the printed page because the atoms are arranged in three dimensions.
- If the sum of the shared and unshared pairs of electrons in each molecule is four, the electron pairs have *tetrahedral* geometry. However, the shape of the molecule is based on the number of shared pairs of electrons present. That is, the shape is based only on the position of the atoms and not on the position of the unshared pairs of electrons.

did you know? ———— ADVANCED

Repulsion between pairs of electrons in a molecule keeps the pairs as far apart as possible. Just how far is measured by the bond angle. If the geometry of a molecule is tetrahedral and all the atoms bonded to the central atom are the same, then the angle between the bonds is the angle between the sides of a tetrahedron, 109.5°. Methane, CH_4, is an example. However, bond angles are less than 109.5° for a molecule that has four pairs of electrons but one is nonbonding, for example, NH_3. Because lone pairs are attracted by only one nucleus, their electron clouds tend to spread out further around the central nucleus and take up more space. This produces added repulsion on the other bonding pairs, causing them to squeeze together at a smaller angle. Ask students to research the bond angles of NH_3 and H_2O and report on how much they differ from the tetrahedral angle. **Ans.** Because lone pairs occupy more space than shared pairs, bond angles decrease as you go from tetrahedral to pyramidal to bent. **LS Visual**

Homework ———— BASIC

Have students make and fill out a graphic organizer entitled *VSEPR and Molecular Geometry*. The table should have five columns entitled *Molecular shape*, *Number of atoms*, *Number of lone pairs*, *Example*, and *Lewis structure*. There should be six rows. The first row, under the head *Molecular shape*, should include the following shapes with a sketch of each: linear, bent, trigonal planar, tetrahedral, and trigonal pyramidal.
LS Logical

Teaching Tip

It may help students to note that for both ammonia and water, the lone pairs and bonding pairs form a tetrahedron. However, lone pairs and not atoms occupy some corners of the tetrahedron, so the molecular shape is not tetrahedral. This text discusses molecular geometry and not electronic geometry.

Answers to Practice Problems D

1. trigonal pyramidal
2. bent
3. trigonal planar
4. tetrahedral

Homework ———— GENERAL

Additional Practice Predict the molecular geometry of the following molecules:

1. CCl_4 **Ans.** tetrahedral
2. HCN **Ans.** linear
3. $SiBr_4$ **Ans.** tetrahedral
4. PCl_3 **Ans.** trigonal pyramidal
5. H_2S **Ans.** bent **LS Visual**

Vocabulary Have students examine the terms *linear, bent, trigonal planar, tetrahedral,* and *trigonal pyramidal*. Using a dictionary, students should explain what the roots of each word mean. Then, have students describe each shape using what they have learned about the meanings of the terms.  Verbal

Homework —————— **ADVANCED**

Have students examine BF_3 and NH_3 and determine how shape affects the polarity of the two molecules. **Ans.** Because all three B–F bonds in BF_3 are on the same plane and radiate out at three equal angles, the dipoles cancel each other out. In NH_3, however, the three N–H bonds are not on the same plane, so there is a net dipole on the molecule. The N atom and lone pair have a partial negative charge, and the three H atoms collectively have a partial positive charge. **LS** Visual

Demonstration

To help students visualize the effects of bond angles and dipoles on molecular polarity, attach ropes to a small, wheeled cart. Have students pull on the ropes at various angles and in various combinations and observe which direction the cart moves. The net force on the cart models the direction of molecular polarity.

Transparencies

TT Molecular Shape Affects Polarity

Figure 18
Molecules of both water and carbon dioxide have polar bonds. The symbol ⟷ shows a dipole.

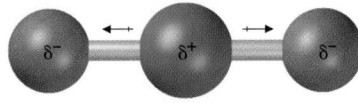

a Because CO_2 is linear, the molecule is nonpolar.

Carbon dioxide, CO_2
(no molecular dipole)

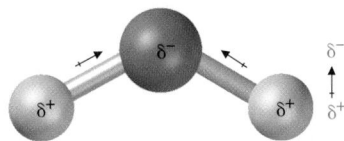

b Because H_2O has a bent shape, the molecule is polar.

Water, H_2O
(overall molecular dipole)

Molecular Shape Affects a Substance's Properties

A molecule's shape affects both the physical and chemical properties of the substance. Recall that both sucrose and sucralose have a shape that allows each molecule to fit into certain nerve endings on the tongue and stimulate a sweet taste. If bending sucrose or sucralose molecules into a different shape were possible, the substances might not taste sweet. Shape determines many other properties. One property that shape determines is the polarity of a molecule.

Shape Affects Polarity

The polarity of a molecule that has more than two atoms depends on the polarity of each bond and the way the bonds are arranged in space. For example, compare CO_2 and H_2O. Oxygen has a higher electronegativity than carbon does, so each oxygen atom in CO_2 attracts electrons more strongly. Therefore, the shared pairs of electrons are more likely to be found near each oxygen atom than near the carbon atom. Thus, the double bonds between carbon and oxygen are polar. As shown in **Figure 18**, each oxygen atom has a partial negative charge, while the carbon atom has a partial positive charge.

Notice also that CO_2 has a linear shape. This shape determines the overall polarity of the molecule. The polarities of the double bonds extend from the carbon atom in opposite directions. As a result, they cancel each other and CO_2 is nonpolar overall even though the individual covalent bonds are polar.

Now think about H_2O. Oxygen has a higher electronegativity than hydrogen does, so oxygen attracts the shared pairs more strongly than either hydrogen does. As a result, each covalent bond between hydrogen and oxygen is polar. The O atom has a partial negative charge, while each H atom has a partial positive charge. Notice also that H_2O has a bent shape. Because the bonds are at an angle to each other, their polarities do not cancel each other. As a result, H_2O is polar.

212

BIOLOGY ———————————————
CONNECTION

Lock-and-Key Model The shapes of complex biological molecules such as proteins is the subject of considerable study. Often these large molecules are folded into complex shapes and held together by both covalent and ionic bonds. The particular shape each molecule assumes affects its function. Some contain cavities or openings of a definite shape that fit just one particular smaller molecule. An action by the protein is initiated when that smaller molecule is attracted by the large molecule and drawn into the cavity. This is called the *lock-and-key model*. The lock is the large molecule with the shaped opening, which is called a *receptor*. The key is the small molecule that fits into the receptor site and initiates a reaction.

You can think of a molecule's overall polarity in the same way that you think about forces on a cart. If you and a friend pull on a wheeled cart in equal and opposite directions—you pull the cart westward and your friend pulls the cart eastward—the cart does not move. The pull forces cancel each other in the same way that the polarities on the CO_2 molecule cancel each other. What happens if the two of you pull with equal force but in nonopposite directions? If you pull the cart northward and your friend pulls it westward, the cart moves toward the northwest. Because the cart has a net force applied to it, it moves. The water molecule has a net partial positive charge on the H side and a net negative charge on the O side. As a result, the molecule has an overall charge and is therefore polar.

Polarity Affects Properties

Because CO_2 molecules are nonpolar, the attractive force between them is very weak. In contrast, the attractive force between polar H_2O molecules is much stronger. The H atoms (with partial positive charges) attract the O atoms (with partial negative charges) on other water molecules. The attractive force between polar water molecules contributes to the greater amount of energy required to separate these polar molecules. The polarity of water molecules also adds to their attraction to positively and negatively charged objects. Other properties realted to polarity and molecular shape will be discussed in a later chapter.

 # Section Review

UNDERSTANDING KEY IDEAS

1. In VSEPR theory, what information about a central atom do you need in order to predict the shape of a molecule?

2. What is the only shape that a molecule made up of two atoms can have?

3. Explain how Lewis structures help predict the shape of a molecule.

4. Explain how a molecule that has polar bonds can be nonpolar.

5. Give one reason why water molecules are attracted to each other.

PRACTICE PROBLEMS

6. Determine the shapes of Br_2 and HBr. Which molecule is more polar and why?

7. Use VSEPR theory to determine the shapes of each of the following.
   a. SCl_2
   b. PF_3
   c. NCl_3
   d. NH_4^+

8. Predict the shape of the CCl_4 molecule. Is the molecule polar or nonpolar? Explain your answer.

CRITICAL THINKING

9. Can a molecule made up of three atoms have a linear shape? Explain your answer.

10. Why is knowing something about the shape of a molecule important?

11. The electron pairs in a molecule of NH_3 form a tetrahedron. Why does the NH_3 molecule have a trigonal pyramidal shape rather than a tetrahedral shape?

213

Answers continued on p. 221A

Answers to Section Review

1. the number of shared and unshared pairs of electrons

2. linear

3. The Lewis structures show the number of bonded electron pairs and unshared electron pairs around the central atom.

4. The polarity of the bonds can be oriented so that they cancel each other, resulting in a nonpolar molecule as is the case with CO_2.

5. Water is a polar molecule. There is a strong attractive force between the O atom with its partial negative charge and the H atoms with their partial positive charges on another water molecule.

6. Both are linear. Br_2 is nonpolar, while HBr is a polar molecule because Br has a higher electronegativity than H. (The electronegativity difference for HBr is not zero.)

7. a. bent
   b. trigonal pyramidal
   c. trigonal pyramidal
   d. tetrahedral

8. The shape is tetrahedral. Because the polar bonds are oriented so that they cancel each other, the molecule is nonpolar.

Answers continued on p. 221A

Close

Reteaching ─────── BASIC

Divide your class into two groups and pair each student with a student in the other group. Give each group a different list of molecules having the geometries discussed in this section. Have each group draw the Lewis structures for the assigned molecules and exchange papers with the other group. Students will then sketch the shape of the molecules. Each pair of students should decide whether the shapes are correct and whether or not the molecules are polar. **LS** Visual

Quiz ─────── GENERAL

1. Why do electron pairs repel each other? **Ans.** They have like charges.

2. Explain how the VSEPR theory allows you to predict molecular structure. **Ans.** Nonbonding pairs of electrons or groups of electrons in bonds experience mutual repulsion and take positions as far from each other as possible

3. What are the molecular shapes of the following molecules.
   a. SiH_4 **Ans.** tetrahedral
   b. AsH_3 **Ans.** trigonal pyramidal
   c. CS_2 **Ans.** linear
   d. SCl_2 **Ans.** bent

4. Are the following compounds polar or nonpolar?
   a. CF_4 **Ans.** nonpolar
   b. NF_3 **Ans.** polar
   c. SF_2 **Ans.** polar
   d. $BeCl_2$ **Ans.** nonpolar

5. Explain why some molecules that contain more than one polar bond are not polar substances. **Ans.** The angles of the bonds cause the bond dipoles to cancel each other.

LS Logical

Chapter Resource File

- Concept Review
- Quiz

SILICON

14
Si
Silicon
28.0855
$[Ne]3s^2 3p^2$

Element Spotlight

Where Is Si?
Earth's crust
27.72% by mass

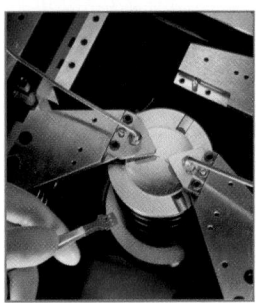

Many integrated circuit chips can be made on the same silicon wafer. The wafer will be cut up into individual chips.

📶 internet connect
www.scilinks.org
Topic: Silicon
SciLinks code: HW4116
SCi*LINKS* Maintained by the National Science Teachers Association

Silicon and Semiconductors

Silicon's most familiar use is in the production of microprocessor chips. Computer microprocessor chips are made from thin slices, or wafers, of a pure silicon crystal. The wafers are doped with elements such as boron, phosphorus, and arsenic to confer semiconducting properties on the silicon. A photographic process places patterns for several chips onto one wafer. Gaseous compounds of metals are allowed to diffuse into the open spots in the pattern, and then the pattern is removed. This process is repeated several times to build up complex microdevices on the surface of the wafer. When the wafer is finished and tested, it is cut into individual chips.

Industrial Uses

• Silicon and its compounds are used to add strength to alloys of aluminum, magnesium, copper, and other metals.

• When doped with elements of Group 13 or Group 15, silicon is a semiconductor. This property is important in the manufacture of computer chips and photovoltaic cells.

• Quartz (silicon dioxide) crystals are used for piezoelectric crystals for radio-frequency control oscillators and digital watches and clocks.

Real-World Connection Organic compounds containing silicon, carbon, chlorine, and hydrogen are used to make silicone polymers, which are used in water repellents, electrical insulation, hydraulic fluids, lubricants, and caulks.

A Brief History

1854: Henri S. C. Deville prepares crystalline silicon.

1943: Commercial production of silicone rubber, oils, and greases begins in the United States.

1800

1900

2000

1811: Joseph Louis Gay-Lussac and Louis Thenard prepare impure amorphous silicon from silicon tetrafluoride.

1824: Jöns Jacob Berzelius prepares pure amorphous silicon and is credited with the discovery of the element.

1904: F. S. Kipping produces the first silicone compound.

1958: Jack Kilby and Robert Noyce produce the first integrated circuit on a silicon chip.

Questions

1. Research and identify five items that you encounter on a regular basis and that are constructed by using silicon.

2. Research piezoelectric materials, and identify how piezoelectric materials that contain silicon are used in science and industry.

214

KEY TERMS

covalent bond
molecular orbital
bond length
bond energy
nonpolar covalent
  bond
polar covalent bond
dipole

valence electron
Lewis structure
unshared pair
single bond
double bond
triple bond
resonance structure

VSEPR theory

KEY IDEAS

SECTION ONE   Covalent Bonds

- Covalent bonds form when atoms share pairs of electrons.
- Atoms have less potential energy and more stability after they form a covalent bond.
- The greater the electronegativity difference, the greater the polarity of the bond.
- The physical and chemical properties of a compound are related to the compound's bond type.

SECTION TWO   Drawing and Naming Molecules

- In a Lewis structure, the element's symbol represents the atom's nucleus and inner-shell electrons, and dots represent the atom's valence electrons.
- Two atoms form single, double, and triple bonds depending on the number of electron pairs that the atoms share.
- Some molecules have more than one valid Lewis structure. These structures are called *resonance structures*.
- Molecular compounds are named using the elements' names, a system of prefixes, and *-ide* as the ending for the second element in the compound.

SECTION THREE   Molecular Shapes

- VSEPR theory states that electron pairs in the valence shell stay as far apart as possible.
- VSEPR theory can be used to predict the shape of a molecule.
- Molecular shapes predicted by VSEPR theory include linear, bent, trigonal planar, tetrahedral, and trigonal pyramidal.
- The shape of a molecule affects the molecule's physical and chemical properties.

KEY SKILLS

Drawing Lewis Structures with Single Bonds	**Drawing Lewis Structures for Polyatomic Ions**	**Drawing Lewis Structures with Multiple Bonds**	**Predicting Molecular Shapes**
Skills Toolkit 1 p. 201	Sample Problem B p. 203	Sample Problem C p. 205	Sample Problem D p. 211
Sample Problem A p. 202			

215

Alternative Assessments

SKILL BUILDER
- Pages 196, 203, 210

Reteaching
- Page 198

HISTORY CONNECTION
- Pages 200, 201

Group Activity
- Page 209

Homework
- Page 211

did you know?
- Page 211

CHAPTER REVIEW
- Items 55, 56, 57, 58, 59

Portfolio Assessments

SKILL BUILDER
- Page 195

Activity
- Page 210

Homework
- Page 212

Chapter Resource File
- **Microscale Lab** Chemical Bonds

Using Key Terms

1. The distance between two bonded atoms at their minimum potential energy is the bond length.

2. A shared pair is two electrons that are involved in forming a bond. An unshared pair is two electrons that are not involved in forming a bond.

3. In a Lewis structure, atomic symbols represent nuclei and inner-shell electrons.

4. Some molecules are represented by two or more Lewis structures that are called *resonance structures*. The molecule is a resonance hybrid of the two structures.

5. VSEPR theory helps predict molecular shapes.

6. A molecular dipole is a molecule with an uneven charge distribution, or a positive region and a negative region.

7. Electronegativity values are useful in determining the bond character.

8. If six electrons are shared between two atoms, they form a triple bond.

9. In a polar covalent bond two atoms share electrons unequally, while in a nonpolar covalent bond two atoms share electrons equally.

10. Bond energy is the energy required to separate two atoms that have formed a covalent bond.

---

6 CHAPTER REVIEW

USING KEY TERMS

1. How are bond length and potential energy related?

2. Describe the difference between a shared pair and an unshared pair of electrons.

3. How are the inner-shell electrons represented in a Lewis structure?

4. What term is used to describe the situation when two or more correct Lewis structures represent a molecule?

5. How is VSEPR theory useful?

6. Describe a molecular dipole.

7. Why is the electronegativity of an element important?

8. What type of bond results if two atoms share six electrons?

9. Contrast a polar covalent bond and a nonpolar covalent bond.

10. Describe how bond energy is related to the breaking of covalent bonds.

UNDERSTANDING KEY IDEAS

Covalent Bonds

11. How does a covalent bond differ from an ionic bond?

12. How are bond energy and bond strength related?

13. Why is a spring a better model than a stick for a covalent bond?

14. Describe the energy changes that take place when two atoms form a covalent bond.

15. Predict whether the bonds between the following pairs of elements are ionic, polar covalent, or nonpolar covalent.
    a. Na—F
    b. H—I
    c. N—O
    d. Al—O
    e. S—O
    f. H—H
    g. Cl—Br
    h. Sr—O

16. Where are the bonding electrons between two atoms?

17. How do the attractive and repulsive forces between two atoms compare when the atoms form a covalent bond?

18. Arrange the following diatomic molecules in order of increasing bond polarity.
    a. I—Cl
    b. H—F
    c. H—Br

19. What determines the electron distribution between two atoms in a bond?

20. Explain why the melting and boiling points of covalent compounds are usually lower than those of ionic compounds.

Drawing and Naming Molecules

21. Draw the Lewis structures for boron, nitrogen, and phosphorus.

22. Describe a weakness of using Lewis structures to model covalent compounds.

23. What do the dots in a Lewis structure represent?

Assignment Guide

Section	Questions
1	1–3, 7, 9–20, 42–44, 47–49, 55, 58–67
2	4, 8, 21–26, 32–34, 37, 39, 40, 50, 51, 53, 54
3	5, 6, 27–31, 35, 36, 38, 41, 45, 46, 52, 56, 57

24. How does a Lewis structure show a bond between two atoms that share four electrons?

25. Why are resonance structures used to model certain molecules?

26. Name the following covalent compounds.
   a. SF_4
   b. XeF_4
   c. PBr_5
   d. N_2O_5
   e. Si_3N_4
   f. PBr_3
   g. Np_3O_8

Molecular Shapes

27. Why do electron pairs around a central atom stay as far apart as possible?

28. Name the following molecular shapes.

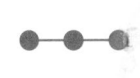

29. Two molecules have different shapes but the same composition. Can you conclude that they have the same physical and chemical properties? Explain you answer.

30. Why is VSEPR theory *not* needed to predict the shape of HCl?

31. a. What causes H_2O to have a bent shape rather than a linear shape?
   b. How does this bent shape relate to the polarity of the water molecule?

PRACTICE PROBLEMS

Sample Problem A Drawing Lewis Structures with Single Bonds

32. Draw Lewis structures for the following molecules. Remember that hydrogen can form only a single bond.
   a. NF_3 **d.** CCl_2F_2
   b. CH_3OH **e.** $HOCl$
   c. ClF

Sample Problem B Drawing Lewis Structures for Polyatomic Ions

33. Draw Lewis structures for the following polyatomic ions.
   a. OH^-
   b. O_2^{2-}
   c. NO^{2-}
   d. NO^{2+}
   e. AsO_4^{3-}

Sample Problem C Drawing Lewis Structures with Multiple Bonds

34. Draw Lewis structures for the following molecules.
   a. O_2
   b. CS_2
   c. N_2O

Sample Problem D Predicting Molecular Shapes

35. Determine the shapes of the following compounds.
   a. CF_4
   b. Cl_2O

36. Draw the shapes of the following polyatomic ions.
   a. NH_4^+
   b. OCl^-
   c. CO_3^{2-}

MIXED REVIEW

37. Draw Lewis structures for the following molecules.
   a. a refrigerant that has one C atom, one H atom, one Br atom, and two F atoms
   b. a natural-gas ingredient that has two C atoms and six H atoms

38. a. Determine the shapes of SCl_2, PF_3, and NCl_3.
   b. Which of these molecules has the greatest polarity?

39. Draw three resonance structures for NO_3^-.

217

Understanding Key Ideas

11. An ionic bond results from the transfer of electrons. A covalent bond results from the sharing of electrons.

12. As bond energy generally increases, bond strength generally increases.

13. Like a spring, a covalent bond can stretch and compress.

14. As two atoms come together to form a bond, their potential energy decreases until it is at a minimum. The distance between the atoms at the minimum potential energy is defined as the bond length.

15. a. ionic
   b. polar covalent
   c. nonpolar covalent
   d. polar covalent
   e. polar covalent
   f. nonpolar covalent
   g. nonpolar covalent
   h. ionic

16. The bonding electrons are found in a molecular orbital that is formed by the overlap of two atomic orbitals.

17. The attractive and repulsive forces balance.

18. from least to most polar: I—Cl, H—Br, H—F

19. The electronegativity difference between the two atoms determines the bond's electron distribution. The more electronegative atom holds electrons more closely than the less electronegative atom.

20. Attractive forces between molecules are generally much weaker than those between ions in an ionic solid. As a result, more energy must be supplied to separate the ions, giving ionic compounds generally higher melting and boiling points.

21. $\dot{B}\cdot$

   $:\overset{\cdot}{\underset{\cdot}{N}}\cdot$

   $:\overset{\cdot}{\underset{\cdot}{P}}\cdot$

22. Placing the dots to show valence electrons might give the impression that the exact location of an electron at any particular time can be determined. Students may also answer that for some molecules, a single Lewis structure cannot adequately explain the molecule's structure.

23. The dots represent valence electrons.

24. Two long dashes, representing a double bond, are used.

25. A single Lewis structure cannot account for how the electrons are arranged in the molecule.

26. **a.** sulfur tetrafluoride
   b. xenon tetrafluoride
   c. phosphorus pentabromide
   d. dinitrogen pentoxide
   e. trisilicon tetranitride
   f. phosphorus tribromide
   g. trineptunium octoxide

27. Electrons are negatively charged and therefore repel each other. Electron pairs are arranged as far apart as possible to minimize the repulsion between them.

28. tetrahedral
   linear
   trigonal planar

29. No, shape can determine both physical and chemical properties.

30. With only two atoms, HCl must have a linear shape.

31. **a.** The unshared pairs on the O atom are arranged so that they are as far apart as possible from the two shared pairs, giving the molecule a bent shape.
   b. The bent molecule has an overall polarity, because the dipoles of the two bonds are oriented at an angle.

40. Draw Lewis structures for the following polyatomic ions.
   a. CO_3^{2-}
   b. O_2^{2-}
   c. PO_4^{3-}

41. Name the following compounds, draw their Lewis structures, and determine their shapes.
   a. $SiCl_4$
   b. BCl_3
   c. NBr_3

42. **a.** How do ionic and covalent bonding differ?
   b. How does an ionic compound differ from a molecular compound?

43. How does the bond that forms between Ba and Br demonstrate that the boundaries between bond types as determined by electronegativity differences are not clear cut?

44. Explain why a halogen is unlikely to form a double bond with another element.

45. **a.** Draw the Lewis structure for $BeCl_2$.
   b. What is unusual about the Be atom in this compound?
   c. Label your Lewis structure to show the partial charge distribution.
   d. What type of bond do Be and Cl form?
   e. What is the shape of the $BeCl_2$ molecule based on VSEPR theory?
   f. Is $BeCl_2$ a polar or nonpolar molecule? Explain your choice.

46. According to VSEPR theory, what molecular shapes are associated with the following types of molecules?
   a. AB
   b. AB_2
   c. AB_3
   d. AB_4

47. What types of atoms tend to form the following types of bonding?
   a. ionic
   b. covalent
   c. metallic

218

CRITICAL THINKING

48. What is the difference between a dipole and electronegativity difference?

49. Why does F generally form covalent bonds with great polarity?

50. Explain what is wrong with the following Lewis structures, and then correct each one.

   a. H—H—S̈:

   b.
   ```
        :O:
        ‖
   H—C̈=Ö—H
   ```

   c.
   ```
       :Cl̈:
        |
        N
      :Cl   Cl̈:
   ```

51. Unlike other elements, noble gases are relatively inert. When noble gases do react, they do not follow the octet rule. Examine the following Lewis structure for the molecule XeO_2F_2.

   ```
          :F:
           |
    :Ö—Xe·
           |  ··
          :F:  Ö:
   ```

   a. Explain why the valence electrons of Xe do not follow the octet rule.
   b. How many unshared pairs of electrons are in this molecule?
   c. How many electrons make up all of the shared pairs in this molecule?

52. Ionic compounds tend to have higher boiling points than covalent substances do. Both ammonia, NH_3, and methane, CH_4, are covalent compunds, yet the boiling point of ammonia is 130°C higher than that of methane. What might account for this large difference?

53. Draw the Lewis structure of NH_4^+. Examine this structure to explain why this five-atom group exists only as a cation.

54. The length of a covalent bond varies depending on the type of bond formed. Triple bonds are generally shorter than double bonds, and double bonds are generally shorter than single bonds. Predict how the lengths of the C—C bond in the following molecules compare.

  a. C_2H_6
  b. C_2H_4
  c. C_2H_2

ALTERNATIVE ASSESSMENT

55. Natural rubber consists of long chains of carbon and hydrogen atoms covalently bonded together. When Goodyear accidentally dropped a mixture of sulfur and rubber on a hot stove, the energy joined these chains together to make vulcanized rubber. Vulcan was the Roman god of fire. The carbon-hydrogen chains in vulcanized rubber are held together by two sulfur atoms that form covalent bonds between the chains. These covalent bonds are commonly called *disulfide bridges.* Explore other molecules that have such disulfide bridges. Present your findings to the class.

56. Searching for the perfect artificial sweetener—great taste with no Calories—has been the focus of chemical research for some time. Molecules such as sucralose, aspartamine, and saccharin owe their sweetness to their size and shape. One theory holds that any sweetener must have three sites that fit into the proper taste buds on the tongue. This theory is appropriately known as the *triangle theory.* Research artificial sweeteners to develop a model to show how the triangle theory operates.

57. Computer Skill RasMol™ is a free software program for examining the shapes of molecules. Use a search engine on the Internet to locate the RasMol site. Download a copy of the program, which includes several models of large molecules. Present your findings to the class.

58. Devise a set of criteria that will allow you to classify the following substances as covalent, ionic, or metallic: $CaCO_3$, Cu, H_2O, NaBr, and C (graphite). Show your criteria to your teacher.

59. Identify 10 common substances in and around your home, and indicate whether you would expect these substances to contain covalent, ionic, or metallic bonds.

CONCEPT MAPPING

60. Use the following terms to complete the concept map below: *valence electrons, nonpolar, covalent compounds, polar, dipoles,* and *Lewis structures.*

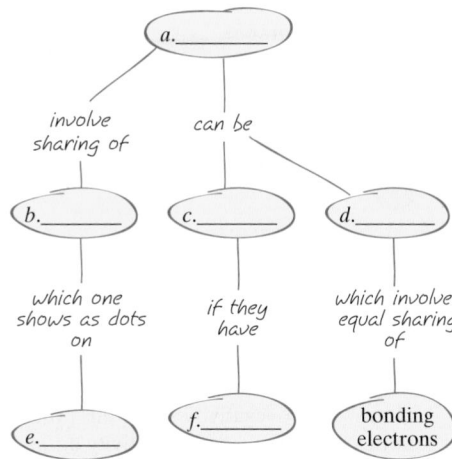

Practice Problems

32. a. :F̈—N̈—F̈:
            ‥
           :F̈:

  b.   H
              |
         H—C—Ö—H
              |
              H

  c. :C̈l—F̈:

  d.   :C̈l:
              |
        :C̈l—C—F̈:
              |
             :F̈:

  e. H—Ö—C̈l:

33. a. [H—Ö:]⁻

  b. [:Ö—Ö:]²⁻

  c. [·N̈—Ö:]²⁻

  d. [·N≡O:]²⁺

  e. [:Ö:]³⁻
              |
         :Ö—As—Ö:
              |
             :Ö:

34. a. :Ö=Ö:

  b. :S̈=C=S̈:

  c. :Ö=N=N̈:

35. a. tetrahedral

  b. bent

36. a.

  b. ●—●

  c.

Chapter 6 · Covalent Compounds   219

Mixed Review

37. a.

```
        H
        |
  :F—C—Br:
        |
       :F:
```

b.

```
    H   H
    |   |
  H—C—C—H
    |   |
    H   H
```

38. a. SCl_2 is bent. Both PF_3 and NCl_3 are trigonal pyramidal.

b. The bond angles of all three molecules are similar. However, the electronegativity difference is greatest between P and F, so PF_3 is the most polar.

39.

```
[ :O:  ]        [  O  ]        [ :O:  ]
[  N   ]  ←→    [  N  ]  ←→    [  N   ]
[:O. .O:]       [:O. .O:]      [.O. .O:]
```

40.

a.

```
[ :O: ]2-     [  O  ]2-     [ :O: ]2-
[  C  ]   ←→  [  C  ]   ←→  [  C  ]
[:O. .O:]     [:O. .O:]     [:O. O:]
```

b.

```
[ .. .. ]2-
[:O—O:]
[ .. .. ]
```

c.

```
[    :O:    ]3-
[    ||     ]
[:O—P—O:]
[   |   ]
[  :O:  ]
```

Answers continued on p. 221A

FOCUS ON GRAPHING

Study the graph below, and answer the questions that follow.
For help in interpreting graphs, see Appendix B, "Study Skills for Chemistry."

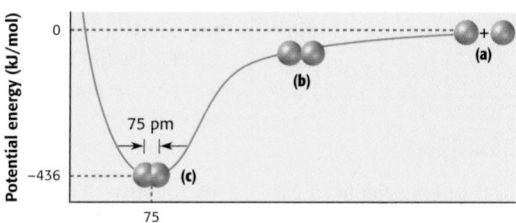

Potential Energy Curve for H₂

Potential energy (kJ/mol)

0

−436

75 pm

75

(a) (b) (c)

Distance between hydrogen nuclei (pm)

61. What do the blue spheres represent on this graph?

62. What are the coordinates of the minimum (the lowest point) of the graph?

63. What relationship does the graph describe?

64. What is significant about the distance between the hydrogen nuclei at the lowest point on the graph?

65. When the distance between the hydrogen nuclei is greater than 75 pm is the slope positive or is it negative?

66. Miles measures the energy required to hold two magnets apart at varying distances. He notices that it takes less and less energy to hold the magnets apart as the distance between them increases. Compare the results of Miles's experiment with the data given in the graph.

TECHNOLOGY AND LEARNING

67. Graphing Calculator

Classifing Bonding Type

The graphing calculator can run a program that classifies bonding between atoms according to the difference between the atoms' electronegativities. Use this program to determine the electronegativity difference between bonded atoms and to classify bonding type.

Go to Appendix C. If you are using a TI-83 Plus, you can download the program

BONDTYPE and data sets and run the application as directed. If you are using another calculator, your teacher will provide you with the keystrokes and data sets to use. After you have graphed the data sets, answer the questions below.

a. Which element pair or pairs have a pure covalent bond?

b. What type of bond does the pair H and O have?

c. What type of bond does the pair Ca and O have?

220

Chapter Resource File

• Chapter Test

STANDARDIZED TEST PREP 6

Answers

1. a
2. a
3. c
4. c
5. b
6. c
7. d
8. a
9. b
10. d
11. b
12. c

1. A chemical bond results from the mutual attraction of the nuclei for
   a. electrons.   **c.** protons.
   b. neutrons.   **d.** dipoles.

2. A polar covalent bond is likely to form between two atoms that
   a. differ in electronegativity.
   b. are of similar size.
   c. differ in state of matter.
   d. have the same number of electrons.

3. The Lewis structure of methane, CH_4, has
   a. two double bonds.
   b. one triple bond.
   c. four single bonds.
   d. two single bonds and two double bonds.

4. In a polar covalent bond, the electronegativity of the two atoms is
   a. exactly the same.
   b. so different that one atom takes an electron away from the other atom.
   c. different, but the electrons are still shared by both atoms.
   d. zero.

5. Because ammonia, NH_3, has an unshared pair of electrons, its molecular shape is
   a. tetrahedral.
   b. trigonal pyramidal.
   c. bent.
   d. linear.

6. The energy released in the formation of a covalent bond is the difference between the potential energy of the separated atoms (zero) and the
   a. maximum potential energy.
   b. kinetic energy of the atom.
   c. minimum potential energy.
   d. bond length expressed in picometers.

7. Multiple covalent bonds may be found in atoms that contain carbon, nitrogen, or
   a. helium.   **c.** hydrogen.
   b. argon.   **d.** oxygen.

8. VSEPR theory states that the repulsion between electron pairs surrounding an atom causes
   a. these pairs to be as far apart as possible.
   b. an electron sea to form.
   c. a bond to break.
   d. a covalent bond to form.

9. In writing a Lewis structure for a negative polyatomic ion, you must add one electron for each unit of
   a. positive charge.
   b. negative charge.
   c. mass.
   d. electronegativity.

10. To draw a Lewis structure, you do not need to know
    a. the number of valence electrons for each atom.
    b. the types of atoms in the molecule.
    c. the number of atoms in the molecule.
    d. bond energies.

11. The polarity of a bond is related to the
    a. Lewis structure of each atom.
    b. electronegativity difference between the two atoms.
    c. number of shared pairs that are present.
    d. number of unshared pairs that are present.

12. In Lewis structures, the atomic symbol represents
    a. valence electrons.
    b. atomic number.
    c. the nucleus and inner-shell electrons.
    d. a stable octet of electrons.

221

Continuation of Answers

Section 1 Review Answers, *continued*

8. As the atoms vibrate back and forth, the distance between them continually changes.

9. A molecular orbital is the region where bonding electrons between two atoms are likely to be found. An atomic orbital is where the electrons of an isolated atom are likely to be found.

10. Generally, the higher the bond energy, the shorter the bond.

11. HF is the most polar, followed by HCl, then HBr, and finally HI.

12. Yes, two fluorine atoms can equally share a pair of electrons to form a nonpolar covalent bond.

13. For single bonds, the smaller the electronegativity difference, the weaker the bond.

14. Cs—Br has the highest ionic character because the electronegativity difference is higher than that of H—S (nonpolar covalent bond) and Si—Cl (polar covalent bond).

Section 2 Review Answers, *continued*

7. a. $:\overset{..}{\underset{..}{Br}}—\overset{..}{\underset{..}{F}}:$

b.
```
      H       H
      |       |
  H—C—N—C—H
      |   |   |
      H   |   H
      H—C—H
          |
          H
```

c. $:\overset{..}{\underset{..}{O}}—\overset{..}{\underset{..}{Cl}}:$
   $:\overset{..}{\underset{..}{Cl}}:$

d. $\left[:\overset{..}{\underset{..}{Cl}}—\overset{..}{\underset{..}{O}}:\right]^-$
   $:\overset{..}{\underset{..}{O}}:$

8. $\overset{..}{\underset{..}{O}}{=}\overset{..}{S}{—}\overset{..}{\underset{..}{O}}: \longleftrightarrow :\overset{..}{\underset{..}{O}}{—}\overset{..}{S}{—}\overset{..}{\underset{..}{O}}: \longleftrightarrow :\overset{..}{\underset{..}{O}}{—}\overset{..}{S}{=}\overset{..}{\underset{..}{O}}$
   $:\overset{..}{\underset{..}{O}}:$ $:\overset{..}{\underset{..}{O}}:$ $:\overset{..}{\underset{..}{O}}:$

9. a. tin tetraiodide
   b. dinitrogen trioxide
   c. phosphorus trichloride
   d. carbon diselenide

10. a. PBr_5
    b. P_2O_3
    c. $AsBr_3$
    d. CCl_4

11. Both show a stable octet surrounding the symbol of each atom. They have different symbols representing the nucleus and inner-shell electrons.

12. No, an atom in a molecule may have an odd number of electrons and therefore may have an unpaired valence electron.

13. The correct name is sulfur (no prefix) trioxide (three oxygen atoms are present).

Section 3 Review Answers, *continued*

9. Yes, the atoms can be arranged in a linear shape, as is the case with BeF_2.

10. The molecule's shape can reveal information about the compound's physical and chemical properties.

11. The unshared pair of electrons repels other electrons just as would a shared pair of electrons. However, because there is no atom associated with the unshared pair, the fourth corner of the tetrahedron doesn't show up as part of the molecule.

Chapter Review Answers, *continued*

41. a. silicon tetrachloride, tetrahedral
    $:\overset{..}{\underset{..}{Cl}}:$
    $:\overset{..}{\underset{..}{Cl}}—Si—\overset{..}{\underset{..}{Cl}}:$
    $:\overset{..}{\underset{..}{Cl}}:$

    b. boron trichloride, trigonal planar
    $:\overset{..}{\underset{..}{Cl}}—B—\overset{..}{\underset{..}{Cl}}:$
    $:\overset{..}{\underset{..}{Cl}}:$

    c. nitrogen tribromide, trigonal pyramidal
    $:\overset{..}{\underset{..}{Br}}—\overset{..}{N}—\overset{..}{\underset{..}{Br}}:$
    $:\overset{..}{\underset{..}{Br}}:$

42. a. Ionic bonds involve a transfer of electrons, while covalent bonds involve sharing electrons.

    b. Ionic compounds are composed of positive and negative ions that are strongly attracted to one another, while covalent compounds are composed of separate molecules that are generally less strongly attracted to one another. Ionic substances tend to have higher melting and boiling points than covalent substances.

43. Ba and Br have an electronegativity difference of 2.1, which is on the border between the polar covalent and ionic ranges. Ba is a metal, and Br is a nonmetal, which can indicate that they form an ionic bond.

44. When a halogen forms a single bond, it achieves a full octet. If the halogen formed a double bond, it would no longer have an octet of valence electrons.

45. a. $:\overset{..}{\underset{..}{Cl}}—Be—\overset{..}{\underset{..}{Cl}}:$

    b. Be does not follow the octet rule.

    c. $:\overset{..}{\underset{..}{Cl}}—Be—\overset{..}{\underset{..}{Cl}}:$
       $\delta- \quad \delta+ \quad \delta-$

    d. polar covalent

    e. linear

    f. $BeCl_2$ is nonpolar because the linear orientation of the dipoles causes the opposite partial charges to cancel one another.

46. a. linear

    b. linear and bent

    c. trigonal planar and trigonal pyramidal

    d. tetrahedral

47. a. metals and nonmetals

    b. nonmetals

    c. metals

Critical Thinking

48. A dipole is a molecule with a partial positive charge at one end and a partial negative charge at the other end. An electronegativity difference represents the difference in two atoms' tendencies to attract bonding electrons when bonding with each other. Electronegativity difference allows you to determine whether a bond is a dipole.

49. F has the highest electronegativity and therefore has the greatest attraction for the electrons it shares in a covalent bond.

50. a. Hydrogen is never a central atom. The second hydrogen atom has two single bonds and can have only one single bond. The total number of valence electrons is wrong.

$$
\begin{array}{c}
\text{H} \\
| \\
\text{:}\overset{..}{\text{S}}\text{—H} \\
\end{array}
$$

    b. The octet rule is not followed for the carbon atom or for the central oxygen atom.

$$
\begin{array}{c}
\text{:O:} \\
\| \\
\text{H—C—}\overset{..}{\text{O}}\text{—H} \\
\end{array}
$$

    c. The top chlorine atoms contain too many electrons.

$$
\begin{array}{c}
\text{:}\overset{..}{\text{Cl}}\text{—}\overset{..}{\text{N}}\text{—}\overset{..}{\text{Cl}}\text{:} \\
| \\
\text{:}\overset{..}{\text{Cl}}\text{:} \\
\end{array}
$$

51. a. Xe is surrounded by 10 electrons—4 are bonding with O atoms, 2 are bonding with F atoms, and 2 are in a lone pair.

    b. 13

    c. In four single bonds, there are 8 electrons forming the shared pairs.

52. Because of its shape (tetrahedral), methane is a nonpolar molecule. As a result, CH_4 molecules are not strongly attracted to one another. In contrast, ammonia is a polar molecule. As a result, NH_3 molecules are attracted to one another. Therefore, it takes more energy to separate ammonia molecules and change ammonia to a gas than to separate methane molecules.

53. Eight valence electrons are needed for N to have an octet and for each H to have two valence electrons. The N atom has five electrons, and each H has one electron, for a total of nine electrons. Therefore, there must be one less electron (eight and not nine), resulting in the 1+ charge on the cation.

54. The triple bond in C_2H_2 is the strongest, making this bond the shortest of the three. The single bond in C_2H_6 is the weakest, making this bond the longest of the three. C_2H_4 forms a double bond.

Alternative Assessment

55. Answers will vary. Proteins are another type of substance that can have disulfide bridges.

56. Answers will vary.

57. Presentations will vary.

58. Answers will vary. Sample answer: $CaCO_3$ and NaBr have both a metal and nonmetals, so they are ionic substances. H_2O and C contain only nonmetals, so they are covalent substances. Cu has only metallic atoms, so it is metallic.

59. Answers will vary. Sample Answer: Aluminum foil (Al), copper wire (Cu), and gold jewelry (Ag) have metallic bonds. Table salt (NaCl) and baking soda ($NaHCO_3$) have ionic bonds. Water (H_2O), sugar ($C_{12}H_{22}O_{11}$), natural gas (CH_4), pencil lead (C), and hydrogen peroxide (H_2O_2) have covalent bonds.

Concept Mapping

60. a. covalent compounds

    b. valence electrons

    c. polar

    d. nonpolar

    e. Lewis structures

    f. dipoles

Focus on Graphing

61. hydrogen atoms

62. 75 pm, –436 kJ/mol

63. PE as a function of nuclear separation

64. It corresponds to the bond energy.

65. positive

66. Miles's experiment is similar to the PE curve to the right of the minimun. However, to the left, nuclei repel each other, whereas there is no repulsion between opposite poles of magnets.

The Mole and Chemical Composition
Chapter Planning Guide

PACING	CLASSROOM RESOURCES	LABS, ACTIVITIES, AND DEMONSTRATIONS
BLOCK 1 • 45 min pp. 222–223 **Chapter Opener**		SE **Start-Up Activity** Counting Large Numbers, p. 223
BLOCKS 2 & 3 • 90 min pp. 224–233 **Section 1** Avogadro's Number and Molar Conversions	TT **Converting Between Amount in Moles and Number of Particles** * TT **Converting Between Mass, Amount, and Number of Particles** *	SE **QuickLab** Exploring the Mole, p. 225 ◆ GENERAL CRF **Datasheets for In-Text Labs** * TE **Activity** Interpreting Flowcharts, p. 226 BASIC TE **Demonstration** What's in a Mole? p. 227 BASIC TE **Demonstration** Particle Mass Varies, p. 227 ◆ TE **Demonstration** Finding the Mass of One Item, p. 231
BLOCK 4 • 45 min pp. 234–240 **Section 2** Relative Atomic Mass and Chemical Formulas	TT **Understanding Formulas for Polyatomic Ionic Compounds** * TT **Calculating Molar Mass for Ionic Compounds** * OSP **Supplemental Reading** The Same and Not the Same "In Praise of Synthesis" ADVANCED	TE **Activity** Grade-Point Averages, p. 234 GENERAL TE **Group Activity** Concept Mapping, p. 237 GENERAL
BLOCKS 5 & 6 • 90 min pp. 241–249 **Section 3** Formulas and Percentage Composition	TT **Percentage Composition of Iron Oxides** * TT **Empirical and Actual Formulas** * TT **Comparing Empirical and Molecular Formulas** * SE **Element Spotlight** Get the Lead Out	TE **Demonstration** Modeling Composition Analysis, p. 241 TE **Demonstration** Determine an Empirical Formula, p. 242 ◆ TE **Group Activity,** p. 247 BASIC SE **Skills Practice Lab** Percent Composition of Hydrates, p. 780 ◆ GENERAL CRF **Datasheets for In-Text Labs** * SE **Inquiry Lab** Gypsum and Plaster of Paris, p. 784 ◆ GENERAL CRF **Datasheets for In-Text Labs** *

BLOCKS 7 & 8 • 90 min
Chapter Review and Assessment Resources

SE **Chapter Review,** pp. 251–256
SE **Standardized Test Prep,** p. 257
CRF **Chapter Test** *
OSP **Test Generator**
CRF **Test Item Listing** *
OSP **Scoring Rubrics and Classroom Management Checklists**

Holt Chemistry: Online Resources

Visit **go.hrw.com** for a variety of free resources related to this textbook. Enter the keyword **HW4 HOME**.

Students can access interactive problem solving help and active visual concept development with the *Holt Chemistry* Online Edition available at **www.hrw.com**.

student CNN News
cnnstudentnews.com

Find the latest chemistry news, lesson plans, and activities related to important scientific events.

KEY

TE	Teacher Edition	**CRF**	Chapter Resource File	*	Also on One-Stop Planner
SE	Student Edition	**TT**	Teaching Transparencies	◆	Requires Advance Prep
OSP	One-Stop Planner				

PROBLEM SOLVING AND PRACTICE	SECTION REVIEW AND ASSESSMENT	STANDARDS CORRELATION
		National Science Education Standards

SE Skills Toolkit 1 Converting Between Amount in Moles and Number of Particles, p. 226
SE Skills Toolkit 2 Working Practice Problems, p. 227
SE Sample Problem A Converting Amount in Moles to Number of Particles, p. 228
  SE Practice, p. 228 GENERAL   **TE Homework**, p. 228 GENERAL
SE Sample Problem B Converting Number of Particles to Amount in Moles, p. 229
  SE Practice, p. 229 GENERAL   **TE Homework**, p. 229 GENERAL
SE Skills Toolkit 3 Converting Between Mass, Amount, and Number of Particles, p. 230
SE Sample Problem C Converting Number of Particles to Mass, p. 231
  SE Practice, p. 231 GENERAL   **TE Homework**, p. 231 GENERAL
SE Sample Problem D Converting Mass to Number of Particles, p. 232
  SE Practice, p. 232 GENERAL   **TE Homework**, p. 232 GENERAL
CRF Problem Solving * ADVANCED
SE Problem Bank, p. 858 GENERAL
	TE Homework, p. 227 ADVANCED
	TE Homework, p. 231 BASIC
	TE Homework, p. 232 ADVANCED
	TE Quiz, p. 233 GENERAL
	CRF Quiz *
	TE Reteaching, p. 233 BASIC
	SE Section Review, p. 233
	CRF Concept Review *
		PS 2c

SE Sample Problem E Calculating Average Atomic Mass, p. 235
  SE Practice, p. 236 GENERAL   **TE Homework**, p. 235 GENERAL
SE Sample Problem F Calculating Molar Mass of Compounds, p. 239
  SE Practice, p. 239 GENERAL   **TE Homework**, p. 239 GENERAL
CRF Problem Solving * ADVANCED
SE Problem Bank, p. 858 GENERAL
	TE Homework, p. 236 ADVANCED
	TE Homework, p. 238 BASIC
	TE Quiz, p. 240 GENERAL
	CRF Quiz *
	TE Reteaching, p. 240 BASIC
	SE Section Review, p. 240
	CRF Concept Review *
		PS 2c

SE Sample Problem G Determining an Empirical Formula and Percentage Composition, p. 242
  SE Practice, p. 243 GENERAL   **TE Homework**, p. 242 GENERAL
SE Sample Problem H Determining a Molecular Formula from an Empirical Formula, p. 245
  SE Practice, p. 245 GENERAL   **TE Homework**, p. 245 GENERAL
SE Sample Problem I Using a Chemical Formula to Determine Percentage Composition, p. 247
  SE Practice, p. 248 GENERAL   **TE Homework**, p. 247 GENERAL
CRF Problem Solving * ADVANCED
SE Problem Bank, p. 858 GENERAL
	TE Quiz, p. 248 GENERAL
	CRF Quiz *
	TE Reteaching, p. 248 BASIC
	SE Section Review, p. 248
	CRF Concept Review *	

www.scilinks.org

Topic: Galaxies
SciLinks code: HW4062

Topic: Avogadro's Constant
SciLinks code: HW4019

Topic: Significant Figures
SciLinks code: HW4115

Topic: Isotopes
SciLinks code: HW4073

Topic: Chemical Formulas
SciLinks code: HW4028

Topics: Percentage Composition
SciLinks code: HW4131

Topic: Lead
SciLinks code: HW4074

Overview

This chapter introduces the mole and uses Avogadro's number to convert between the amount in moles and the number of particles. Students will solve problems using moles, particles, and molar mass. They will relate moles to chemical formulas and determining molar mass. Empirical and molecular formulas are also determined from percentage composition and formula mass, and percentage composition is derived from empirical and molecular formulas.

Assessing Prior Knowledge

Check for Content Knowledge
Students should be familiar with the following topics:

• conversion factors

• measurements

• significant figures

• atomic mass units

• naming compounds

Using the Figure

Astronomical Numbers To emphasize how large a galaxy is, explain that our galaxy, the Milky Way, has a diameter of about 1.30×10^5 light years, which is about 1.23×10^{18} km. Our sun has a diameter of only 1.39×10^6 km, which means that it would take 8.85×10^{11} suns to span our galaxy. Because one sun diameter equals 109 Earth diameters, it would take 9.64×10^{13} Earth diameters to equal the diameter of the Milky Way. This huge number refers only to the galaxy's diameter and does not take into account the enormity of its volume.

CHAPTER

7

THE MOLE AND CHEMICAL COMPOSITION

222

Standards Correlations

National Science Education Standards

PS 2c: Bonds between atoms are created when electrons are paired up by being transferred or shared. A substance composed of a single kind of atom is called an element. The atoms may be bonded together into molecules or crystalline solids. A compound is formed when two or more kinds of atoms bind together chemically. (Sections 1, 2)

Galaxies have hundreds of billions of stars. The universe may have as many as sextillion stars—that's 1000 000 000 000 000 000 000 (or 1×10^{21}) stars. Such a number is called *astronomical* because it is so large that it usually refers only to vast quantities such as those described in astronomy. Can such a large number describe quantities that are a little more down to Earth? It certainly can. In fact, it takes an even larger number to describe the number of water molecules in a glass of water! In this chapter, you will learn about the *mole*, a unit used in chemistry to make working with such large quantities a little easier.

START-UP ACTIVITY

Counting Large Numbers

PROCEDURE

1. Count out exactly **200 small beads.** Using a **stopwatch,** record the amount of time it takes you to count them.

2. Your teacher will tell you the approximate number of small beads in 1 g. Knowing that number, calculate the mass of 200 small beads. Record the mass that you have calculated.

3. Use a **balance** to determine the mass of the 200 small beads that you counted in step 1. Compare this mass with the mass you calculated in step 2.

4. Using the mass you calculated in step 2 and a balance, measure out **another 200 small beads.** Record the amount of time it takes you to count small beads when using this counting method.

5. Count the number of **large beads** in 1 g.

ANALYSIS

1. Which method of counting took the most time?

2. Which method of counting do you think is the most accurate?

3. In a given mass, how does the number of large beads compare with the number of small beads? Explain your results.

Pre-Reading Questions

① What are some things that are sold by weight instead of by number?

② Which would need a larger package, a kilogram of pencils or a kilogram of drinking straws?

③ If you counted one person per second, how many hours would it take to count the 6 billion people now in the world?

CONTENTS 7

🖥 **internet** connect

www.scilinks.org
Topic : Galaxies
SciLinks code: HW4062

SC*i*LINKS Maintained by the National Science Teachers Association

223

START-UP ACTIVITY

Skills Acquired:
- Collecting Data
- Interpreting
- Identifying/Recognizing Patterns

Materials:
For each group of 2–3 students:
- balance
- beads, small (200)
- beads, large (same total mass as 200 small beads)
- beakers or other containers for holding beads
- stopwatch

Teacher's Notes:
- Beads are available in bulk in craft stores. Be sure the beads are uniform in size and lightweight.
- For the purpose of significant figures, all numbers given can be treated as exact.

Safety Caution: Be sure that students immediately pick up any beads they drop or see on the floor. Students might slip if they step on a bead.

Answers

1. Answers may vary, but counting the beads in step 1 probably took the longest time.

2. Answers may vary, but counting the beads in step 4 is probably the most accurate.

3. There are fewer large beads in the same mass. A large bead has a larger mass than a small bead has, so fewer large beads make up a given mass.

Answers to Pre-Reading Questions

1. Answers will vary and may include: bulk items, delicatessen foods, produce, and landscaping material.

2. A kilogram of drinking straws would need a larger package. Each straw has less mass than a pencil, yet takes up about the same amount of space. It takes more straws to make up 1 kg.

3. $(6 \times 10^9$ people$) \times (1$ s/person$) \times (1$ h/3600 s$) = 2 \times 10^6$ h (about 200 years)

Avogadro's Number and Molar Conversions

Overview

Before beginning this section, review with your students the Objectives listed in the Student Edition. This section defines a mole and shows how it is a counting unit for large numbers of items. Avogadro's number, which is 6.022×10^{23}, is used to convert between amount in moles and number of particles. Molar mass is used to convert between amount in moles and mass.

 **Bellringer**

Have students list as many common counting units as they can. Lists might include pair, dozen, score, and gross. Write the following amounts on the board: 500 goldfish, 150 unicycles, and 50 jet planes. Ask students to determine how many groups of each unit are present in each amount.

Motivate

Discussion ── BASIC

Conversion Factors Ask students to brainstorm conversion factors, such as 1 deck/52 cards, 1 week/7 days, or 16 ounces/1 pound. Students may categorize the conversion factors or compile their lists for class reference. Ask students what all conversion factors have in common. **Ans.** All conversion factors are equal to one and will change one unit mentioned in the factor to the other unit. **LS Logical**

KEY TERMS
- mole
- Avogadro's number
- molar mass

OBJECTIVES

1. **Identify** the mole as the unit used to count particles, whether atoms, ions, or molecules.

2. **Use** Avogadro's number to convert between amount in moles and number of particles.

3. **Solve** problems converting between mass, amount in moles, and number of particles using Avogadro's number and molar mass.

Avogadro's Number and the Mole

mole

the SI base unit used to measure the amount of a substance whose number of particles is the same as the number of atoms of carbon in exactly 12 grams of carbon-12

Avogadro's number

6.022×10^{23}, the number of atoms or molecules in 1.000 mol

Atoms, ions, and molecules are very small, so even tiny samples have a huge number of particles. To make counting such large numbers easier, scientists use the same approach to represent the number of ions or molecules in a sample as they use for atoms. The SI unit for amount is called the **mole** (mol). A mole is the number of atoms in exactly 12 grams of carbon-12.

The number of particles in a mole is called **Avogadro's number,** or Avogadro's constant. One way to determine this number is to count the number of particles in a small sample and then use mass or particle size to find the amount in a larger sample. This method works only if all of the atoms in the sample are identical. Thus, scientists measure Avogadro's number using a sample that has atoms of only one isotope.

Figure 1
The particles in a mole can be atoms, molecules, or ions. Examples of a variety of molar quantities are given. Notice that the volume and mass of a molar quantity varies from substance to substance.

MOLAR QUANTITIES OF SOME SUBSTANCES

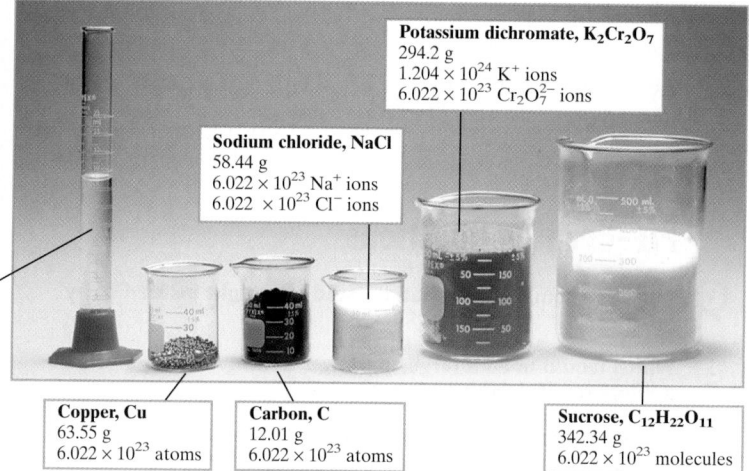

Potassium dichromate, $K_2Cr_2O_7$
294.2 g
1.204×10^{24} K^+ ions
6.022×10^{23} $Cr_2O_7^{2-}$ ions

Sodium chloride, NaCl
58.44 g
6.022×10^{23} Na^+ ions
6.022×10^{23} Cl^- ions

Water, H_2O
18.02 g
6.022×10^{23} molecules

Copper, Cu
63.55 g
6.022×10^{23} atoms

Carbon, C
12.01 g
6.022×10^{23} atoms

Sucrose, $C_{12}H_{22}O_{11}$
342.34 g
6.022×10^{23} molecules

HISTORY CONNECTION

Avogadro's Number Avogadro did not determine the number known as Avogadro's number. He worked with gases, attempting to prove that equal volumes of gases under the same conditions contain the same number of particles. His work laid the groundwork for further investigations by other scientists, such as Jean Baptiste Perrin (1870–1942). Perrin measured the displacement of colloidal particles that exhibit Brownian movement, and he used the results from these experiments to calculate the first value of Avogadro's number.

INCLUSION Strategies

• *Developmentally Delayed*

Using **Table 1,** ask students to give examples of objects that are counted by the units listed. Students can draw pictures of the items or find pictures in magazines. Ask the students to label each picture with its name, the unit used to count the objects, and the amount of individual parts equivalent to one unit.

Table 1	Counting Units
Unit	**Example**
1 dozen	12 objects
1 score	20 objects
1 roll	50 pennies
1 gross	144 objects
1 ream	500 sheets of paper
1 hour	3600 seconds
1 mole	6.022×10^{23} particles

Figure 2
You can use mass to count out a roll of new pennies; 50 pennies are in a roll. One roll has a mass of about 125 g.

The most recent measurement of Avogadro's number shows that it is $6.02214199 \times 10^{23}$ units/mole. In this book, the measurement is rounded to 6.022×10^{23} units/mol. Avogadro's number is used to count any kind of particle, as shown in **Figure 1.**

The Mole Is a Counting Unit

Keep in mind that the mole is used to count out a given number of particles, whether they are atoms, molecules, formula units, ions, or electrons. The mole is used in the same way that other, more familiar counting units, such as those in **Table 1,** are used. For example, there are 12 eggs in one dozen eggs. You might want to know how many eggs are in 15 dozen. You can calculate the number of eggs by using a conversion factor as follows.

$$15 \text{ dozen eggs} \times \frac{12 \text{ eggs}}{1 \text{ dozen eggs}} = 180 \text{ eggs}$$

Figure 2 shows another way that you can count objects: by using mass.

 **LAB**

Exploring the Mole

SAFETY PRECAUTIONS

PROCEDURE
1. Use a **periodic table** to find the atomic mass of the following substances: **graphite (carbon), iron filings, sulfur powder, aluminum foil,** and **copper wire.**
2. Use a **balance** to measure out 1 mol of each substance.

3. Use **graduated beakers** to find the approximate volume in 1 mol of each substance.

ANALYSIS
1. Which substance has the greatest atomic mass?
2. Which substance has the greatest mass in 1 mol?

3. Which substance has the greatest volume in 1 mol?
4. Does the mass of a mole of a substance relate to the substance's atomic mass?
5. Does the volume of a mole of a substance relate to the substance's atomic mass?

225

Teach, continued

Activity — BASIC

Interpreting Flowcharts Refer students to **Skills Toolkit 1.** Have students interpret the flow chart in their own words. Ask them to explain to a partner how to determine the number of particles when given the amount in moles, and vice versa. **LS Visual**

READING SKILL BUILDER — BASIC

Assimilating Knowledge Before students read this chapter, have them write a brief list of all the things they know about how particles are counted and how masses are measured. Then have students list what they want to find out. For example, they might ask the following questions:

- Why might one count the number of atoms in a sample?
- How do scientists find the number of particles in a sample?
- If you know the mass of 12 identical items, how can you find the mass of one item without measuring it directly?

Later, when students are reviewing for the chapter test, have them correct or expand upon their answers. **LS Verbal**

SKILL BUILDER — BASIC

Math Skills As students read through the chapter, have them make a list of conversion factors that can be used for converting among amount in moles, number of particles, and mass. Encourage them to keep the lists for when they study stoichiometry. Students may benefit from using their lists during a test. **LS Visual**

SKILLS Toolkit

Converting Between Amount in Moles and Number of Particles

1. Decide which quantity you are given: amount (in moles) or number of particles (in atoms, molecules, formula units, or ions).

2. If you are converting from amount to number of particles (going left to right), use the top conversion factor.

3. If you are converting from number of particles to amount (going right to left), use the bottom conversion factor.

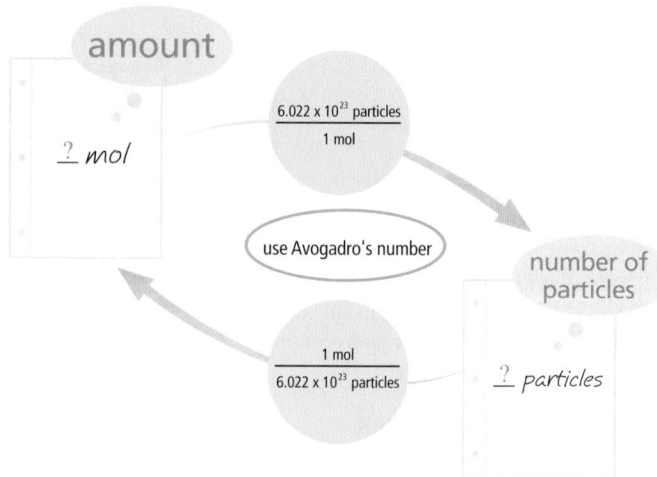

Amount in Moles Can Be Converted to Number of Particles

A conversion factor begins with a definition of a relationship. The definition of one mole is

$$6.022 \times 10^{23} \text{ particles} = 1 \text{ mol}$$

If two quantities are equal and you divide one by the other, the factor you get is equal to 1. The following equation shows how this relationship is true for the definition of the mole.

$$\frac{6.022 \times 10^{23} \text{ particles}}{1 \text{ mol}} = 1$$

The factor on the left side of the equation is a conversion factor. The reciprocal of a conversion factor is also a conversion factor and is also equal to one, so the following is true.

$$\frac{6.022 \times 10^{23} \text{ particles}}{1 \text{ mol}} = \frac{1 \text{ mol}}{6.022 \times 10^{23} \text{ particles}} = 1$$

Because a conversion factor is equal to 1, it can multiply any quantity without changing the quantity's value. Only the units are changed.

These conversion factors can be used to convert between a number of moles of substance and a corresponding number of molecules. For example, imagine that you want to convert 2.66 mol of a compound into the corresponding number of molecules. How do you know which conversion factor to use? **Skills Toolkit 1** can help.

internet connect

www.scilinks.org
Topic : Avogadro's Constant
SciLinks code: HW4019

SCLINKS Maintained by the National Science Teachers Association

226

Transparencies

TT Converting Between Amount in Moles and Number Particles

Choose the Conversion Factor That Cancels the Given Units

Take the amount (in moles) that you are given, shown in **Skills Toolkit 1** on the left, and multiply it by the conversion factor, shown in the top green circle, to get the number of particles, shown on the right. The calculation is as follows:

$$2.66 \ \cancel{mol} \times \frac{6.022 \times 10^{23} \ molecules}{1 \ \cancel{mol}} = 1.60 \times 10^{24} \ molecules$$

You can tell which of the two conversion factors to use, because the needed conversion factor should cancel the units of the given quantity to give you the units of the answer or the unknown quantity.

Working Practice Problems

1. Gather information.
- Read the problem carefully.
- List the quantities and units given in the problem.
- Determine what value is being asked for (the answer) and the units it will need.

2. Plan your work.
Write the value of the given quantity times a question mark (which stands for a conversion factor) and then the equals sign, followed by another question mark (which stands for the answer) and the units of the answer. For example:

$$4.2 \ mol \ CO_2 \times ? = ? \ molecules \ CO_2$$

3. Calculate.
- Determine the conversion factor(s) needed to change the units of the given quantity to the units of the answer. Write the conversion factor(s) in the order you need them to cancel units.
- Cancel units, and check that the units that remain are the same on both sides and are the units desired for the answer.
- Calculate and round off the answer to the correct number of significant figures.
- Report your answer with correct units.

4. Verify your result.
- Verify your answer by estimating. One way to do so is to round off the numbers in the setup and make a quick calculation.
- Make sure your answer is reasonable. For example, if the number of atoms is less than one, the answer cannot possibly be correct.

STUDY ▶ TIP

WORKING PROBLEMS
If you have difficulty working practice problems, review the outline of procedures in **Skills Toolkit 2**. You may also refer back to the sample problems.

227

Teach, *continued*

Answers to Practice Problems A

1. 1.13×10^{23} ions Na^+
2. 8.73×10^6 atoms As
3. 2.544×10^{24} molecules $C_2H_4O_2$
4. 3.6×10^{24} formula units NaOH

Homework ─── **GENERAL**

Additional Practice How many representative particles are present in each of the following?

1. 4.3 mol of tungsten Ans. 2.6×10^{24} atoms W

2. 2.45×10^{-6} mol of nickel(II) selenide Ans. 1.48×10^{18} formula units NiSe

3. 0.923 mol of selenium tetrabromide Ans. 5.56×10^{23} molecules $SeBr_4$

LS Logical

Teaching Tip ─── **BASIC**

Estimation Remind students how to estimate. Choose several simple mathematical problems and remind them to round off each number and use these rounded numbers to estimate the answer. For example, when multiplying 42×88, round the numbers to 40 and 90 and multiply these to obtain an estimate of 3600. This estimate is close to the actual answer of 3696. Emphasize the importance of using estimation to verify the reasonableness of an answer. **LS** Logical

---

Chapter Resource File

• **Problem Solving**

SAMPLE PROBLEM A

Converting Amount in Moles to Number of Particles

Find the number of molecules in 2.5 mol of sulfur dioxide.

1 Gather information.
• amount of SO_2 = 2.5 mol
• 1 mol of any substance = 6.022×10^{23} particles
• number of molecules of SO_2 = ? molecules

2 Plan your work.
The setup is: 2.5 mol SO_2 × ? = ? molecules SO_2

3 Calculate.
You are converting from the unit *mol* to the unit *molecules*. The conversion factor must have the units of *molecules/mol*. **Skills Toolkit 1** shows that this means you use 6.022×10^{23} molecules/1 mol.

$$2.5 \text{ mol } SO_2 \times \frac{6.022 \times 10^{23} \text{ molecules } SO_2}{1 \text{ mol } SO_2} = 1.5 \times 10^{24} \text{ molecules } SO_2$$

4 Verify your result.
The units cancel correctly. The answer is greater than Avogadro's number, as expected, and has two significant figures.

PRACTICE HINT

Take your time, and be systematic. Focus on units; if they are not correct, you must rethink your preliminary equation. In this way, you can prevent mistakes.

PRACTICE

1 How many ions are there in 0.187 mol of Na^+ ions?

2 How many atoms are there in 1.45×10^{-17} mol of arsenic?

3 How many molecules are there in 4.224 mol of acetic acid, $C_2H_4O_2$?

4 How many formula units are there in 5.9 mol of NaOH?

Number of Particles Can Be Converted to Amount in Moles

Notice in **Skills Toolkit 1** that the reverse calculation is similar but that the conversion factor is inverted to get the correct units in the answer. Look at the following problem. How many moles are 2.54×10^{22} iron(III) ions, Fe^{3+}?

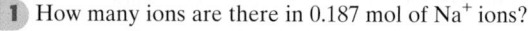

$$2.54 \times 10^{22} \text{ ions } Fe^{3+} \times ? = ? \text{ mol } Fe^{3+}$$

Multiply by the conversion factor that cancels the unit of *ions* and leaves the unit of *mol*. (That is, you use the conversion factor that has the units that you want to get on top and the units that you want to get rid of on the bottom.)

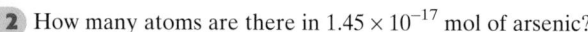
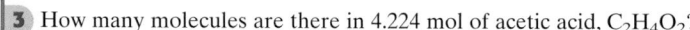

$$2.54 \times 10^{22} \text{ ions } Fe^{3+} \times \frac{1 \text{ mol } Fe^{3+}}{6.022 \times 10^{23} \text{ ions } Fe^{3+}} = 0.0422 \text{ mol } Fe^{3+}$$

This answer makes sense, because you started with fewer than Avogadro's number of ions, so you have less than one mole of ions.

SAMPLE PROBLEM B

Converting Number of Particles to Amount in Moles

A sample contains 3.01×10^{23} molecules of sulfur dioxide, SO_2.
Determine the amount in moles.

1 Gather information.
- number of molecules of $SO_2 = 3.01 \times 10^{23}$ molecules
- 1 mol of any substance $= 6.022 \times 10^{23}$ particles
- amount of $SO_2 = ?$ mol

2 Plan your work.
The setup is similar to the calculation in **Sample Problem A.**

$$3.01 \times 10^{23} \text{ molecules } SO_2 \times ? = ? \text{ mol } SO_2$$

3 Calculate.
The conversion factor is used to remove the unit of *molecules* and introduce the unit of *mol.*

$$3.01 \times 10^{23} \text{ molecules } SO_2 \times \frac{1 \text{ mol } SO_2}{6.022 \times 10^{23} \text{ molecules } SO_2} = 0.500 \text{ mol } SO_2$$

4 Verify your result.
There are fewer than 6.022×10^{23} (Avogadro's number) of SO_2 molecules, so it makes sense that the result is less than 1 mol. Three is the correct number of significant figures.

> **PRACTICE HINT**
>
> Always check your answer for the correct number of significant figures.

PRACTICE

1. How many moles of xenon do 5.66×10^{23} atoms equal?

2. How many moles of silver nitrate do 2.888×10^{15} formula units equal?

3. A biologist estimates that there are 2.7×10^{17} termites on Earth. How many moles of termites is this?

4. How many moles do 5.66×10^{25} lithium ions, Li^+, equal?

5. Determine the number of moles of each specified atom or ion in the given samples of the following compounds. (Hint: The formula tells you how many atoms or ions are in each molecule or formula unit.)

   a. O atoms in 3.161×10^{21} molecules of CO_2
   b. C atoms in 3.161×10^{21} molecules of CO_2
   c. O atoms in 2.222×10^{24} molecules of NO
   d. K^+ ions in 5.324×10^{16} formula units of KNO_2
   e. Cl^- ions in 1.000×10^{14} formula units of $MgCl_2$
   f. N atoms in 2.000×10^{14} formula units of $Ca(NO_3)_2$
   g. O atoms in 4.999×10^{25} formula units of $Mg_3(PO_4)_2$

229

HISTORY CONNECTION

Avogadro's Number Jean Baptiste Perrin made one famous experimental determination of Avogadro's number. Currently, several methods are used to determine this value. Measuring the decay rates and number of alpha particles from radioactive materials is one method. Another method combines measuring the charge on an electron and the electrolytic deposition of a certain amount of silver. Probably the most accurate determinations are made using density data combined with X-ray diffraction determination of atom spacing in crystals.

Answers to Practice Problems B

1. 0.940 mol Xe
2. 4.796×10^{-9} mol $AgNO_3$
3. 4.5×10^{-7} mol termites
4. 94.0 mol Li^+
5. a. 1.050×10^{-2} mol O
   b. 5.249×10^{-3} mol C
   c. 3.690 mol O
   d. 8.841×10^{-8} mol K^+
   e. 3.321×10^{-10} mol Cl^-
   f. 6.642×10^{-10} mol N
   g. 6.641×10^{2} mol O

Homework — GENERAL

Additional Practice How many moles are equivalent to each of the following?

1. 7.95×10^{24} copper(II) chloride formula units **Ans.** 13.2 mol $CuCl_2$
2. 6.93×10^{23} thallium atoms **Ans.** 1.15 mol Tl
3. 1.974×10^{14} sodium chloride formula units **Ans.** 3.278×10^{-10} mol NaCl

LS Logical

SKILL BUILDER — ADVANCED

Writing **Research Skills** Have students do research to find out any historical aspects of the mole. Have them investigate such items as who first named the quantity a *mole*, why this term was chosen, and whether there were intermediate steps in the development of the concept as we know it today. Have them organize their findings in a report. **LS** Verbal

> **Chapter Resource File**
> - Problem Solving

Teach, *continued*

Teaching Tip

Molar mass is the term used in this text to describe the mass of one mole. However, the terms *molecular weight, molecular mass, formula weight, formula mass, gram molecular weight,* or *gram molecular mass* might be used by some sources. These terms may have slightly different meanings yet still refer to similar values depending on the units used.

Teaching Tip ——— GENERAL

Remind students that the periodic table gives mass in *amu,* which is the mass per atom. Have students perform the conversion from *amu/atom* to *g/mol.* **LS Logical**

SKILL BUILDER — ADVANCED

Graphing Skills Have students choose an element and list masses for various samples of the element. For example, they might choose nitrogen and list 7.00 g, 12.0 g, 16.0 g, and 20.0 g. Have them determine the number of moles of atoms for each mass. **Ans.** 0.500 mol, 0.857 mol, 1.14 mol, and 1.43 mol Have them plot mass (*y*-axis) versus number of moles (*x*-axis) and determine the slope of the resulting line and compare this slope to the molar mass of the element. **Ans.** They are equal. **LS Visual**

Using the Figure

Emphasize that **Skills Toolkit 1** shows the same conversions as the left hand side of **Skills Toolkit 3.** Point out that *amount* is to the left of *number of particles* in **Skills Toolkit 1** and to the right in **Skills Toolkit 3.**

Molar Mass Relates Moles to Grams

In chemistry, you often need to know the mass of a given number of moles of a substance or the number of moles in a given mass. Fortunately, the mole is defined in a way that makes figuring out either of these easy.

Amount in Moles Can Be Converted to Mass

molar mass
the mass in grams of one mole of a substance

The mole is the SI unit for amount. The **molar mass,** or mass in grams of one mole of an element or compound, is numerically equal to the atomic mass of monatomic elements and the formula mass of compounds and diatomic elements. To find a monatomic element's molar mass, use the atomic mass, but instead of having units of *amu,* the molar mass will have units of *g/mol.* So, the molar mass of carbon is 12.01 g/mol, and the molar mass of iron is 55.85 g/mol. How to find the molar mass of compounds and diatomic elements is shown in the next section.

You use molar masses as conversion factors in the same way you use Avogadro's number. The right side of **Skills Toolkit 3** shows how the *amount* in moles relates to the *mass* in grams of a substance. Suppose you must find the mass of 3.50 mol of copper. You will use the molar mass of copper. By checking the periodic table, you find the atomic mass of copper, 63.546 amu, which you round to 63.55 amu. So, in calculations with copper, use 63.55 g/mol.

The Mole Plays a Central Part in Chemical Conversions

You know how to convert from number of particles to amount in moles and how to convert from amount in moles to mass. Now you can use the same methods one after another to convert from *number of particles* to *mass.* **Skills Toolkit 3** shows the two-part process for this conversion. One step common to many problems in chemistry is converting to amount in moles. **Sample Problem C** shows how to convert from number of particles to the mass of a substance by first converting to amount in moles.

3 SKILLS Toolkit

Converting Between Mass, Amount, and Number of Particles

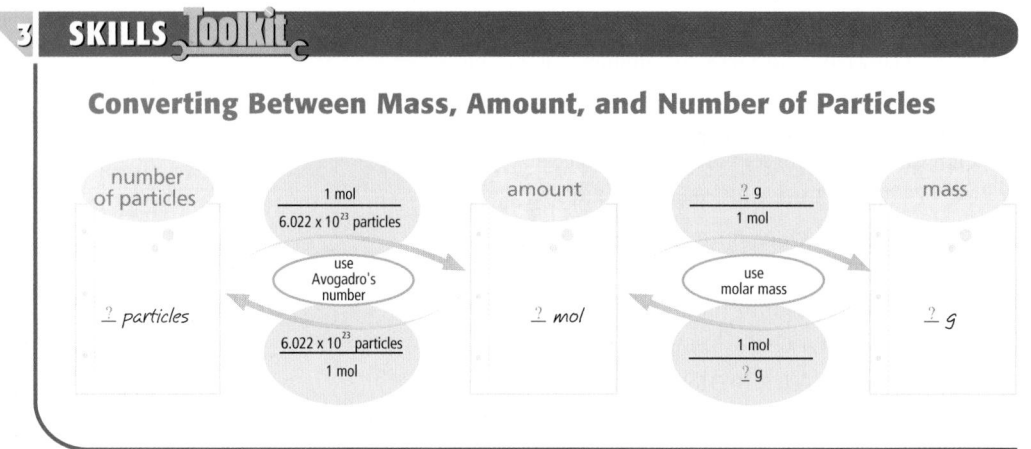

230

Transparencies

TT Converting Between Mass, Amount, and Number of Particles

SAMPLE PROBLEM C

Converting Number of Particles to Mass

Find the mass in grams of 2.44×10^{24} atoms of carbon, whose molar mass is 12.01 g/mol.

1 Gather information.
- number of atoms C = 2.44×10^{24} atoms
- molar mass of carbon = 12.01 g/mol
- amount of C = ? mol
- mass of the sample of carbon = ? g

2 Plan your work.
- **Skills Toolkit 3** shows that to convert from number of atoms to mass in grams, you must first convert to amount in moles.
- To find the amount in moles, select the conversion factor that will take you from number of atoms to amount in moles.

$$2.44 \times 10^{24} \text{ atoms} \times ? = ? \text{ mol}$$

- Multiply the number of atoms by the following conversion factor:

$$\frac{1 \text{ mol}}{6.022 \times 10^{23} \text{ atoms}}$$

- To find the mass in grams, select the conversion factor that will take you from amount in moles to mass in grams.

$$? \text{ mol} \times ? = ? \text{ g}$$

- Multiply the amount in moles by the following conversion factor:

$$\frac{12.01 \text{ g C}}{1 \text{ mol}}$$

3 Calculate.
Solve and cancel identical units in the numerator and denominator.

$$2.44 \times 10^{24} \text{ atoms} \times \frac{1 \text{ mol}}{6.022 \times 10^{23} \text{ atoms}} \times \frac{12.01 \text{ g C}}{1 \text{ mol}} = 48.7 \text{ g C}$$

4 Verify your result.
The answer has the units requested in the problem.

PRACTICE HINT

Make sure to select the correct conversion factors so that units cancel to get the unit required in the answer.

PRACTICE

Given molar mass, find the mass in grams of each of the following substances:

1. 2.11×10^{24} atoms of copper (molar mass of Cu = 63.55 g/mol)
2. 3.01×10^{23} formula units of NaCl (molar mass of NaCl = 58.44 g/mol)
3. 3.990×10^{25} molecules of CH_4 (molar mass of CH_4 = 16.05 g/mol)
4. 4.96 mol titanium (molar mass of Ti = 47.88 g/mol)

231

MISCONCEPTION ALERT

Moles and Mass Students will quickly associate moles with mass. Emphasize that moles are a measure of the number of particles, not the mass. For example, 16.00 g of oxygen is the mass of 1 mol of oxygen atoms. It is not 1 mol of oxygen. One mole of oxygen is 6.022×10^{23} atoms of oxygen.

Chapter Resource File
• Problem Solving

Homework — BASIC

Review For students who have difficulty with multistep problems, revisit simple mole-mass conversions, such as those in the "Atoms and Moles" chapter. For review, have students determine the mass of each of the following amounts:

1. 2.64 mol tellurium Ans. 337 g Te
2. 2.00×10^{-3} mol uranium Ans. 0.476 g U
3. 39 mol protons, 1.00742 g/mol Ans. 39 g protons

LS Logical

Answers to Practice Problems C
1. 223 g Cu
2. 29.2 g NaCl
3. 1063 g CH_4
4. 237 g Ti

Homework — GENERAL

Additional Practice Determine the mass in grams of each of the following quantities:

1. 6.12×10^{14} formula units of rhenium dioxide, 218.21 g/mol Ans. 2.22×10^{-7} g ReO_2
2. 5.3×10^{23} atoms of molybdenum Ans. 84 g Mo
3. 1.299×10^{26} ions of nitrite, 46.01 g/mol Ans. 9925 g NO_2^-

LS Logical

Demonstration
Finding the Mass of One Item
Obtain a handful of identical items, such as paper clips, buttons, or dimes. Count how many items you have. Find the mass of a paper bag. Place the items in the bag. Then find the mass of the bag and the items, and calculate the mass of the items by the difference between the mass of the bag and the mass of the bag and the items. From the mass and the number of items, calculate the mass of one item. Use a sensitive balance to confirm your answer. Compare this demonstration to determining the mass of a single atom, and assign the Homework on the next page.

Answers to Practice Problems D

1. 2.25×10^{24} atoms Cu
2. 3.00×10^{23} ions Ca^{2+}
3. 9.33×10^{25} atoms As

Homework ——— GENERAL

Additional Practice Determine the number of particles represented by each of the following masses:

1. atoms in 54.3 g cobalt **Ans.** 5.55×10^{23} atoms Co

2. molecules in 245 g diatomic oxygen gas, 32.00 g/mol **Ans.** 4.61×10^{24} molecules O_2

3. molecules in 0.0923 g hydrogen fluoride gas, 20.01 g/mol **Ans.** 2.78×10^{21} molecules HF

LS Logical

Teaching Tip

Remind students that it is not actually possible to isolate a sample of pure cations or pure anions. For example, you cannot isolate a 20 g sample of Ca^{2+} ions without also including a corresponding anion.

Homework ——— ADVANCED

Finding the Mass of One Atom
Have students determine what information they need to find the mass of a single atom. Have them describe the procedure they would use. **Ans.** Divide the molar mass by Avogadro's number. Have them find the mass of a single atom of each of the following elements:

1. helium **Ans.** 6.647×10^{-24} g
2. tungsten **Ans.** 3.053×10^{-22} g
3. potassium **Ans.** 6.493×10^{-23} g

LS Logical

Mass Can Be Converted to Amount in Moles

Converting from mass to number of particles is the reverse of the operation in the previous problem. This conversion is also shown in **Skills Toolkit 3**, but this time you are going from right to left and using the bottom conversion factors.

Sample Problem D shows how to convert the mass of a substance to amount (mol) and then convert amount to the number of particles. Notice that the problem is the reverse of **Sample Problem C**.

SAMPLE PROBLEM D

Converting Mass to Number of Particles

Find the number of molecules present in 47.5 g of glycerol, $C_3H_8O_3$. The molar mass of glycerol is 92.11 g/mol.

1 Gather information.
- mass of the sample of $C_3H_8O_3$ = 47.5 g
- molar mass of $C_3H_8O_3$ = 92.11 g/mol
- amount of $C_3H_8O_3$ = ? mol
- number of molecules $C_3H_8O_3$ = ? molecules

2 Plan your work.
- **Skills Toolkit 3** shows that you must first find the amount in moles.
- To determine the amount in moles, select the conversion factor that will take you from mass in grams to amount in moles.

$$47.5 \text{ g} \times ? = ? \text{ mol}$$

- Multiply mass by the conversion factor $\dfrac{1 \text{ mol}}{92.11 \text{ g } C_3H_8O_3}$

- To determine the number of particles, select the conversion factor that will take you from amount in moles to number of particles.

$$? \text{ mol} \times ? = ? \text{ molecules}$$

- Multiply amount by the conversion factor $\dfrac{6.022 \times 10^{23} \text{ molecules}}{1 \text{ mol}}$

3 Calculate.

$$47.5 \text{ g } C_3H_8O_3 \times \frac{1 \text{ mol}}{92.11 \text{ g } C_3H_8O_3} \times \frac{6.022 \times 10^{23} \text{ molecules}}{1 \text{ mol}} = 3.11 \times 10^{23} \text{ molecules}$$

4 Verify your result.
The answer has the units requested in the problem.

PRACTICE HINT

Because no elements have a molar mass less than one, the number of grams in a sample of a substance will always be larger than the number of moles of the substance. Thus, when you convert from grams to moles, you will get a smaller number. And the opposite is true for the reverse calculation.

PRACTICE

1. Find the number of atoms in 237 g Cu (molar mass of Cu = 63.55 g/mol).
2. Find the number of ions in 20.0 g Ca^{2+} (molar mass of Ca^{2+} = 40.08 g/mol).
3. Find the number of atoms in 155 mol of arsenic.

232

Chapter Resource File

- Problem Solving

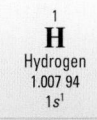 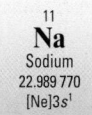 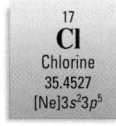

1	11	17
H	**Na**	**Cl**
Hydrogen	Sodium	Chlorine
1.007 94	22.989 770	35.4527
$1s^1$	$[Ne]3s^1$	$[Ne]3s^23p^5$
1.01 g/mol	22.99 g/mol	35.45 g/mol

Figure 3
Round molar masses from the periodic table to two significant figures to the right of the decimal point.

Remember to Round Consistently

Calculators may report many figures. However, an answer must never be given to more figures than is appropriate. If the given amount has only two significant figures, then you must round the calculated number off to two significant figures. Also, keep in mind that many numbers are exact. In the definition of the mole, the chosen amount is *exactly* 12 grams of the carbon-12 isotope. Such numbers are not considered when rounding. **Figure 3** shows how atomic masses are rounded in this text.

internet connect
www.scilinks.org
Topic : Significant Figures
SciLinks code: HW4115
SCi LINKS. Maintained by the National Science Teachers Association

 # Section Review

UNDERSTANDING KEY IDEAS

1. What is the definition of a mole?

2. How many particles are there in one mole?

3. Explain how Avogadro's number can give two conversion factors.

4. Which will have the greater number of ions, 1 mol of nickel(II) or 1 mol of copper(I)?

5. Without making a calculation, is 1.11 mol Pt more or less than 6.022×10^{23} atoms?

PRACTICE PROBLEMS

6. Find the number of molecules or ions.
  a. 2.00 mol Fe^{3+} **c.** 0.25 mol K^+
  b. 4.5 mol BCl_3 **d.** 6.022 mol O_2

7. Find the number of sodium ions, Na^+.
  a. 3.00 mol Na_2CO_3
  b. 3.00 mol $Na_4P_2O_7$
  c. 5.12 mol $NaNO_3$

8. Find the number of moles.
  a. 3.01×10^{23} molecules H_2O
  b. 1.000×10^{23} atoms C
  c. 5.610×10^{22} ions Na^+

9. Find the mass in grams.
  a. 4.30×10^{16} atoms He, 4.00 g/mol
  b. 5.710×10^{23} molecules CH_4, 16.05 g/mol
  c. 3.012×10^{24} ions Ca^{2+}, 40.08 g/mol

10. Find the number of molecules or ions.
  a. 1.000 g I^-, 126.9 g/mol
  b. 3.5 g Cu^{2+}, 63.55 g/mol
  c. 4.22 g SO_2, 64.07 g/mol

11. What is the mass of 6.022×10^{23} molecules of ibuprofen (molar mass of 206.31 g/mol)?

12. Find the mass in grams.
  a. 4.01×10^{23} atoms Ca, 40.08 g/mol
  b. 4.5 mol boron-11, 11.01 g/mol
  c. 1.842×10^{19} ions Na^+, 22.99 g/mol

13. Find the number of molecules.
  a. 2.000 mol H_2, 2.02 g/mol
  b. 4.01 g HF, 20.01 g/mol
  c. 4.5 mol $C_6H_{12}O_6$, 180.18 g/mol

CRITICAL THINKING

14. Why do we use carbon-12 rather than ordinary carbon as the basis for the mole?

15. Use **Skills Toolkit 1** to explain how a number of atoms is converted into amount in moles.

(233)

Answers to Section Review

1. the number of atoms in exactly 12 g of carbon-12

2. 6.022×10^{23} particles

3. There are Avogadro's number or 6.022×10^{23} particles in 1 mol. The relationship $6.022 \times 10^{23}/1$ mol can be used to convert amount in mole to number of particles. The inverse, 1 mol/6.022×10^{23} can be used to convert number of particles to amount in moles.

4. Both will have the same number, 6.022×10^{23}. (Students may wish to first compare more familiar quantities, such as a ton of feathers and a ton of bricks.)

5. It is more.

6. a. 1.20×20^{24} ions Fe^{3+}
  b. 2.7×10^{24} molecules BCl_3
  c. 1.5×10^{23} ions K^+
  d. 3.626×10^{24} molecules O_2

7. a. 3.61×10^{24} Na^+ ions
  b. 7.23×10^{24} Na^+ ions
  c. 3.08×10^{24} Na^+ ions

8. a. 0.500 mol H_2O
  b. 0.1661 mol C
  c. 0.09316 mol Na^+

Answers continued on p. 257A

Using the Figure

Refer students to **Figure 3**. After they are rounded to the hundredths place, molar masses of elements have three to five significant digits. Remind students that when they are adding numbers, decimals are aligned and the numbers are added. When determining significant figures, the number of places to the right of the decimal point in the sum should equal the least number of places to the right of the decimal point in the addends.

Close

Quiz ——— GENERAL

1. What is the conversion factor that can be used to change moles to number of particles?
Ans. 6.022×10^{23} particles/1 mol

2. How many atoms are present in 2.3 mol of titanium atoms?
Ans. 1.4×10^{24} atoms Ti

3. How many moles are present in 1.24×10^{22} ions Fe^{3+}?
Ans. 0.0206 mol Fe^{3+}

4. What is the mass in grams of 3.40 mol of thallium atoms?
Ans. 695 g Tl

5. How many moles of copper are present in 75.92 g of the element? Ans. 1.195 mol Cu

6. What is the mass in grams of 3.011×10^{23} atoms of indium?
Ans. 57.41 g In
LS Logical

Reteaching ——— BASIC

Ask students to explain in their own words how to convert from moles to number of particles and from number of particles to moles. Have them use examples in their explanations. **LS** Verbal

Chapter Resource File

• **Concept Review**
• **Quiz**

Overview

Before beginning this section, review with your students the Objectives listed in the Student Edition. This section investigates further molar masses, which are derived from chemical formulas and average atomic masses. Students will calculate average atomic mass and learn what information can be determined from a compound's chemical formula.

🔊 Bellringer

Have students compare the masses of a roll of pennies and a roll of dimes. Both contain 50 coins. Ask students why the masses of the rolls are different when both rolls contain the same number of coins. **Ans.** The masses of the coins are different. If given a roll of mixed coins, ask students what information they would need to determine the mass of the roll? **Ans.** the number of each type of coin in the roll and the mass of each coin

Motivate

Activity ——— GENERAL

Grade-Point Averages To give an example of a weighted average, have student's determine grade-point averages. For example, assume a student has 14 grades: 4 A's, 6 B's, 3 C's, and 1 D. An A is worth 4.0 points, a B is 3.0, a C is 2.0, and a D is 1.0.

The total number of points is:
(4 A's)(4.0 points/A) + (6 B's)(3.0 points/B) + (3 C's)(2.0 points/C) + (1 D)(1.0 point/D) = 41 points.

The GPA is:
41 points/14 grades = 2.9 points/grade. 🅛 Logical

Relative Atomic Mass and Chemical Formulas

KEY TERM

- average atomic mass

OBJECTIVES

① **Use** a periodic table or isotopic composition data to determine the average atomic masses of elements.

② **Infer** information about a compound from its chemical formula.

③ **Determine** the molar mass of a compound from its formula.

Average Atomic Mass and the Periodic Table

You have learned that you can use atomic masses on the periodic table to find the molar mass of elements. Many of these values on the periodic table are close to whole numbers. However, most atomic masses are written to at least three places past the decimal.

Why are the atomic masses of most elements on the periodic table not exact whole numbers? One reason is that the masses reported are *relative* atomic masses. To understand relative masses, think about the setup in **Figure 4.** Eight pennies have the same mass as five nickels do. Thus, you could say that a single penny has a relative mass of 0.625 "nickel masses." Just as you can find the mass of a penny compared with the mass of a nickel, scientists have determined the masses of the elements relative to each other. Remember that atomic mass is given in units of *amu*. This means that it reflects an atom's mass relative to the mass of a carbon-12 atom. So, now you may ask why carbon's atomic mass on the periodic table is not exactly 12.

Most Elements Are Mixtures of Isotopes

You remember that *isotopes* are atoms that have different numbers of neutrons than other atoms of the same element do. So, isotopes have different atomic masses. The periodic table reports **average atomic mass,** a weighted average of the atomic mass of an element's isotopes. A *weighted average* takes into account the relative importance of each number in the average. Thus, if there is more of one isotope in a typical sample, it affects the average atomic mass more than an isotope that is less abundant does.

For example, carbon has two stable isotopes found in nature, carbon-12 and carbon-13. The average atomic mass of carbon takes into account the masses of both isotopes and their relative abundance. So, while the atomic mass of a carbon-12 atom is exactly 12 amu, any carbon sample will include enough carbon-13 atoms that the average mass of a carbon atom is 12.0107 amu.

Topic Link

Refer to the "Atoms and Moles" chapter for a discussion of atomic mass and isotopes.

average atomic mass

the weighted average of the masses of all naturally occurring isotopes of an element

Figure 4
You can determine the mass of a penny relative to the mass of a nickel; eight pennies have the same mass as five nickels.

234

HISTORY ———
CONNECTION

Atomic Masses The first set of atomic masses was published as integers by John Dalton in the early 1800s. Hydrogen was given a mass of 1. Because the atomic masses of all elements were integers and hydrogen had a mass of 1, it was theorized by the English chemist Joseph Proust that all elements are actually made of hydrogen. This theory was accepted by many scientists until more sophisticated methods of measuring atomic mass were developed, and it was determined that some atomic masses were not whole numbers.

Like carbon, most elements are a mixture of isotopes. In most cases, the fraction of each isotope is the same no matter where the sample comes from. Most average atomic masses can be determined to several decimal places. However, some elements have different percentages of isotopes depending on the source of the sample. This is true of *native* lead, or lead that occurs naturally on Earth. The average atomic mass of lead is given to only one decimal place because its composition varies so much from one sample to another.

If you know the abundance of each isotope, you can calculate the average atomic mass of an element. For example, the average atomic mass of native copper is a weighted average of the atomic masses of two isotopes, shown in **Figure 5**. The following sample problem shows how this calculation is made from data for the abundance of each of native copper's isotopes.

Figure 5
Native copper is a mixture of two isotopes. Copper-63 contributes 69.17% of the atoms, and copper-65 the remaining 30.83%.

SAMPLE PROBLEM E

Calculating Average Atomic Mass

The mass of a Cu-63 atom is 62.94 amu, and that of a Cu-65 atom is 64.93 amu. Using the data in **Figure 5**, find the average atomic mass of Cu.

1 Gather information.
- atomic mass of a Cu-63 atom = 62.94 amu
- abundance of Cu-63 = 69.17%
- atomic mass of Cu-65 = 64.93 amu
- abundance of Cu-65 = 30.83%
- average atomic mass of Cu = ? g

2 Plan your work.
The average atomic mass of an element is the sum of the contributions of the masses of each isotope to the total mass. This type of average is called a *weighted average*. The contribution of each isotope is equal to its atomic mass multiplied by the fraction of that isotope. (To change a percentage into a fraction, divide it by 100.)

Isotope	Percentage	Decimal fraction	Contribution
Copper-63	69.17%	0.6917	62.94×0.6917
Copper-65	30.83%	0.3083	64.93×0.3083

3 Calculate.
Average atomic mass is the sum of the individual contributions:
$$(62.94 \text{ amu} \times 0.6917) + (64.93 \text{ amu} \times 0.3083) = 63.55 \text{ amu}$$

4 Verify your results.
- The answer lies between 63 and 65, and the result is closer to 63 than it is to 65. This is expected because the isotope 63 makes a larger contribution to the average.
- Compare your answer with the value in the periodic table.

Practice problems on next page

internet connect
www.scilinks.org
Topic: Isotopes
SciLinks code: HW4073

SCiLINKS Maintained by the National Science Teachers Association

PRACTICE HINT

In calculating average atomic masses, remember that the resulting value must be greater than the lightest isotope and less than the heaviest isotope.

235

Teach

READING SKILL BUILDER — BASIC

Reading Organizer Have students read about average atomic mass and go over **Sample Problem E.** Then, have them develop a graphic organizer that tells them how to find the average atomic mass of an element from the mass of each isotope and the relative abundance of that isotope. Graphic organizers might include units made up of the mass of an isotope times the percentage of the isotope, in decimal form. These units are then added together. The sum of these units is then set equal to the average atomic mass of the element. **LS Visual**

Answers to Practice Problems E
1. (68.926 amu)(0.6000) + (70.925 amu)(0.4000) = 69.73 amu
2. (15.99 amu)(0.9976) + (17.00 amu)(0.00038) + (18.00 amu)(0.0020) = 15.99 amu

Homework — GENERAL

Additional Practice

1. Chlorine exists as chlorine-35, which has a mass of 34.969 amu and makes up 75.8% of chlorine atoms. The rest of naturally occurring chlorine is chlorine-37, with a mass of 36.996 amu. What is the average atomic mass of chlorine? Ans. 35.5 amu

2. U-234 makes up 0.00500% of uranium atoms and has a mass of 234.041 amu. U-235 makes up 0.720% and has a mass of 235.044 amu. U-238 has a mass of 238.051 amu and makes up 99.275%. What is the average atomic mass of uranium? Ans. 238.03 amu

3. Carbon-12 makes up 98.90% of existing carbon. Carbon-13, with a mass of 13.003, makes up 1.10%. Traces of carbon-14 also exist. What is the average atomic mass of carbon? Ans. 12.01 amu
LS Logical

did you know? — GENERAL

Pennies vary in composition and mass depending on when they were made. Pennies made between 1864 and 1962 are 95% copper and 5% tin and zinc. An exception are pennies made in 1943, which are made of steel coated with zinc. Pennies made between 1962 and 1982 are 95% copper and 5% zinc. Pennies made after 1982 are 97.5% zinc and 2.5% copper. Have interested students determine the masses and compositions of various coins and write a report about the history of U.S. coin composition. **LS Verbal**

Chapter Resource File
- Lesson Plan
- Problem Solving

1 Calculate the average atomic mass for gallium if 60.00% of its atoms have a mass of 68.926 amu and 40.00% have a mass of 70.925 amu.

2 Calculate the average atomic mass of oxygen. Its composition is 99.76% of atoms with a mass of 15.99 amu, 0.038% with a mass of 17.00 amu, and 0.20% with a mass 18.00 amu.

Chemical Formulas and Moles

Until now, when you needed to perform molar conversions, you were given the molar mass of compounds in a sample. Where does this molar mass of compounds come from? You can determine the molar mass of compounds the same way that you find the molar mass of individual elements—by using the periodic table.

Formulas Express Composition

The first step to finding a compound's molar mass is understanding what a chemical formula tells you. It tells you which elements, as well as how much of each, are present in a compound. The formula KBr shows that the compound is made up of potassium and bromide ions in a 1:1 ratio. The formula H_2O shows that water is made up of hydrogen and oxygen atoms in a 2:1 ratio. These ratios are shown in **Figure 6.**

You have learned that covalent compounds, such as water and hexachloroethane, consist of molecules as units. Formulas for covalent compounds show both the elements and the number of atoms of each element in a molecule. Hexachloroethane has the formula C_2Cl_6. Each molecule has 8 atoms covalently bonded to each other. Ionic compounds aren't found as molecules, so their formulas do not show numbers of atoms. Instead, the formula shows the simplest ratio of cations and anions.

Figure 6
Although any sample of a compound has many atoms and ions, the chemical formula gives a ratio of those atoms or ions.

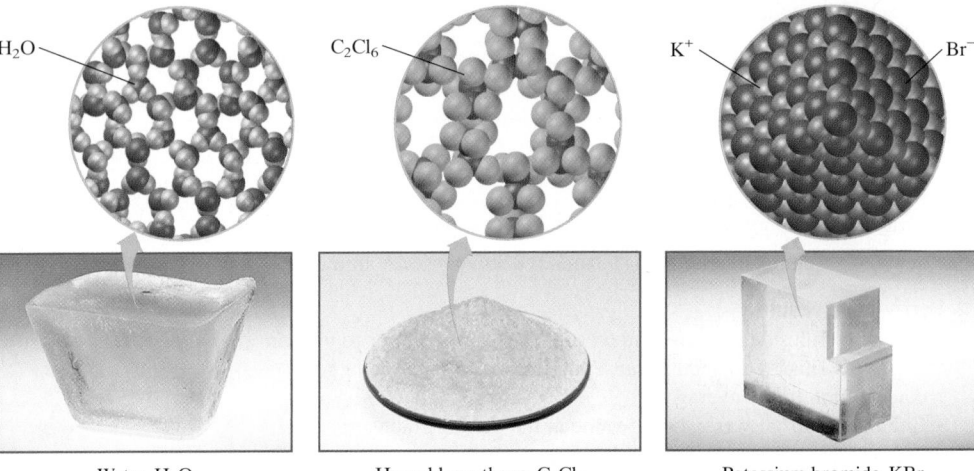

Water, H_2O Hexachloroethane, C_2Cl_6 Potassium bromide, KBr

Using the Figure — ADVANCED

Refer students to **Figure 6.** Emphasize that ionic compounds form an infinite crystal lattice composed of ions held together by ionic bonds. A covalent solid may also form a crystal lattice, but intermolecular forces hold the molecules together, while covalent bonds hold the atoms together within the molecules. Ask them to describe how a microscopic view of the ionic compound calcium chloride, $CaCl_2$, would look. **Ans.** It would show packed ions, as in KBr, except there would be two chloride ions for every calcium ion. **LS** Verbal

Homework — ADVANCED

Avogadro's Number Students can derive the value of Avogadro's number from the mass of the nucleons in an atom and the molar mass of the element. Have students use the mass of 1.672×10^{-24} g for a proton and 1.008 g as the molar mass of hydrogen to determine the value of Avogadro's number. **LS** Logical

Teaching Tip

Emphasize that the subscripts in formulas describe ratios of moles and number of particles but *not* mass ratios.

Figure 7
The formula for a polyatomic ionic compound is the simplest ratio of cations to anions.

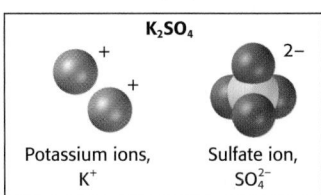
K₂SO₄

Potassium ions, K⁺

Sulfate ion, SO₄²⁻

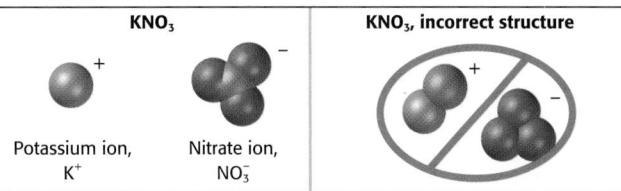
KNO₃

Potassium ion, K⁺

Nitrate ion, NO₃⁻

KNO₃, incorrect structure

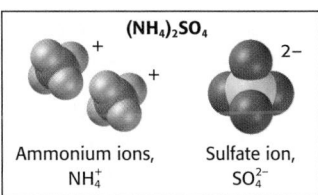
(NH₄)₂SO₄

Ammonium ions, NH₄⁺

Sulfate ion, SO₄²⁻

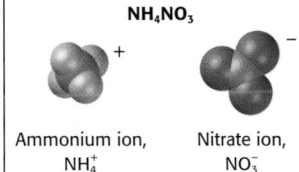
NH₄NO₃

Ammonium ion, NH₄⁺

Nitrate ion, NO₃⁻

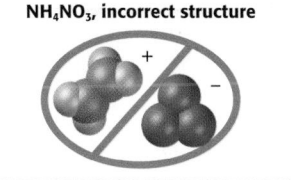
NH₄NO₃, incorrect structure

a Elements in polyatomic ions are bound together in a group and carry a characteristic charge.

b The formula for a compound with polyatomic ions shows how the atoms in each ion are bonded together.

c You cannot move atoms from one polyatomic ion to the next.

Formulas Give Ratios of Polyatomic Ions

The meaning of a formula does not change when polyatomic ions are involved. Potassium nitrate has the formula KNO_3. Just as the formula KBr indicates a 1:1 ratio of K^+ cations to Br^- anions, the formula KNO_3 indicates a ratio of one K^+ cation to one NO_3^- anion.

When a compound has polyatomic ions, such as those in **Figure 7,** look for the cations and anions. Formulas can tell you which elements make up polyatomic ions. For example, in the formula KNO_3, NO_3 is a nitrate ion, NO_3^-. KNO_3 does not have a KN^+ and an O_3 ion. Similarly, the formula of ammonium nitrate is written NH_4NO_3, because NH_4 in a formula stands for the ammonium ion, NH_4^+, and NO_3 stands for a nitrate ion, NO_3^-. If it were written as $H_4N_2O_3$, the number of atoms would be correct. However, the formula would no longer clearly show which ions were in the substance and how many there were. The formula NH_4NO_3 shows that ammonium nitrate is made up of ammonium and nitrate ions in a 1:1 ratio.

Formulas Are Used to Calculate Molar Masses

A formula tells you what atoms (or ions) are present in an element or compound. So, from a formula you can find the mass of a mole of the substance, or its molar mass. The simplest formula for most elements is simply that element's symbol. For example, the symbol for silver is Ag. The molar mass of elements whose formulas are this simple equals the atomic mass of the element expressed in g/mol. So, the molar mass of silver is 107.87 g/mol. Diatomic elements have twice the number of atoms in each molecule, so their molecules have molar masses that are twice the molar mass of each atom. For example, the molar mass of Br_2 molecules is two times the molar mass of Br atoms (2×79.90 g/mol = 159.80 g/mol).

internet connect

www.scilinks.org
Topic : Chemical Formulas
SciLinks code: HW4028

SC/LINKS. Maintained by the National Science Teachers Association

237

Transparencies

TT Understanding Formulas for Polyatomic Ionic Compounds

READING SKILL BUILDER — BASIC

Reading Organizer As students read this section, have them outline the content. They should use the existing heads as main heads and list relevant facts and concepts under them. Students can work in pairs or small groups to compare individual outlines and discuss any similarities and differences. **LS** Verbal

Group Activity — GENERAL
Concept Mapping Have students work in groups to develop concept maps that relate chemical formulas, molar masses, and moles. Have them use key terms as connectors in the map. **Ans.** An example might be a map that shows using the atomic mass and frequency of each element in the chemical formula to determine the mass of a mole, which, expressed in g/mol, is the molar mass. **LS** Interpersonal

Making Models Have students draw formula units for two ionic compounds. Drawings should be similar to those in **Table 2.** Have them use **Appendix A** to obtain relative sizes of some atoms and look through the text for colors and sizes used to represent other atoms and ions. For example, the model atom for oxygen is red, and that for hydrogen is blue. One of their compounds can be a binary compound, but at least one should involve a polyatomic ion. **LS** Visual

Homework ——— BASIC

Determining Numbers of Atoms
Some students may need to practice counting atoms in a molecule before attempting to determine molar mass. Write the following formulas on the board, and ask students how many atoms of each element are present: $CaSO_3$, $K_2Cr_2O_7$, $Al(NO_3)_3$, and $Al_2(SO_4)_3$. Ans. $CaSO_3$: 1 atom Ca, 1 atom S, and 3 atoms O; $K_2Cr_2O_7$: 2 atoms K, 2 atoms Cr, and 7 atoms O; $Al(NO_3)_3$: 1 atom Al, 3 atoms N, and 9 atoms O; and $Al_2(SO_4)_3$: 2 atoms Al, 3 atoms S, and 12 atoms O. **LS** Logical

Transparencies

TT Calculating Molar Mass for Ionic Compounds

Let's say you want to determine the molar mass of a molecular compound. You must use the periodic table to find the molar mass of more than one element. The molar mass of a molecular compound is the sum of the masses of all the atoms in it expressed in g/mol. For example, one mole of H_2O molecules will have two moles of H and one mole of O. Thus, the compound's molar mass is equal to two times the molar mass of a H atom plus the molar mass of an O atom, or 18.02 g (2 × 1.01 g + 16.00 g).

Scientists also use the simplest formula to represent one mole of an ionic compound. They often use the term *formula unit* when referring to ionic compounds, because they are not found as single molecules. A formula unit of an ionic compound represents the simplest ratio of cations to anions. A formula unit of KBr is made up of one K^+ ion and one Br^- ion. One mole of an ionic compound has $6.022 × 10^{23}$ of these formula units. As with molecular compounds, the molar mass of an ionic compound is the sum of the masses of all the atoms in the formula expressed in g/mol. **Table 2** compares the formula units and molar masses of three ionic compounds. **Sample Problem F** shows how to calculate the molar mass of barium nitrate.

Table 2   Calculating Molar Mass for Ionic Compounds

Formula	Formula unit	Calculation of molar mass
$ZnCl_2$	Zn^{2+} Cl^- Cl^-	1 Zn = 1 × 65.39 g/mol = 65.39 g/mol + 2 Cl = 2 × 35.45 g/mol = 70.90 g/mol $ZnCl_2$ = 136.29 g/mol
$ZnSO_4$	Zn^{2+} SO_4^{2-}	1 Zn = 1 × 65.39 g/mol = 65.39 g/mol 1 S = 1 × 32.07 g/mol = 32.07 g/mol + 4 O = 4 × 16.00 g/mol = 64.00 g/mol $ZnSO_4$ = 161.46 g/mol
$(NH_4)_2SO_4$	NH_4^+ NH_4^+ SO_4^{2-}	2 N = 2 × 14.01 g/mol = 28.02 g/mol 8 H = 8 × 1.01 g/mol = 8.08 g/mol 1 S = 1 × 32.07 g/mol = 32.07 g/mol + 4 O = 4 × 16.00 g/mol = 64.00 g/mol $(NH_4)_2SO_4$ = 132.17 g/mol

did you know?

Standards for Atomic Masses Hydrogen was the first element used as a standard for determining atomic masses. The second standard chosen for atomic masses was oxygen because it is so commonly involved in chemical reactions. In the early 1960s, the mass of carbon-12 was chosen as the standard for atomic masses. This standard is still used.

Calculating Molar Mass of Compounds

Find the molar mass of barium nitrate.

1 Gather information.
- simplest formula of ionic barium nitrate: $Ba(NO_3)_2$
- molar mass of $Ba(NO_3)_2$ = ? g/mol

2 Plan your work.
- Find the number of moles of each element in 1 mol $Ba(NO_3)_2$. Each mole has:

  1 mol Ba
  2 mol N
  6 mol O

- Use the periodic table to find the molar mass of each element in the formula.

  molar mass of Ba = 137.33 g/mol
  molar mass of N = 14.01 g/mol
  molar mass of O = 16.00 g/mol

3 Calculate.
- Multiply the molar mass of each element by the number of moles of each element. Add these masses to get the total molar mass of $Ba(NO_3)_2$.

$$\begin{aligned} \text{mass of 1 mol Ba} &= 1 \times 137.33 \text{ g/mol} = 137.33 \text{ g/mol} \\ \text{mass of 2 mol N} &= 2 \times 14.01 \text{ g/mol} = 28.02 \text{ g/mol} \\ + \text{ mass of 6 mol O} &= 6 \times 16.00 \text{ g/mol} = 96.00 \text{ g/mol} \\ \hline \text{molar mass of } Ba(NO_3)_2 &= 261.35 \text{ g/mol} \end{aligned}$$

4 Verify your result.
- The answer has the correct units. The sum of the molar masses of elements can be approximated as 140 + 30 + 100 = 270, which is close to the calculated value.

PRACTICE HINT

Use the same methods for molecular compounds, but use the molecular formula in place of a formula unit.

PRACTICE

1 Find the molar mass for each of the following compounds:

a. CsI   **c.** $C_{12}H_{22}O_{11}$ **e.** $HC_2H_3O_2$

b. $CaHPO_4$ **d.** I_2 **f.** $Mg_3(PO_4)_2$

2 Write the formula and then find the molar mass.

a. sodium hydrogen carbonate   **e.** iron(III) hydroxide

b. cerium hexaboride   **f.** tin(II) chloride

c. magnesium perchlorate   **g.** tetraphosphorus decoxide

d. aluminum sulfate   **h.** iodine monochloride

PROBLEM SOLVING SKILL

continued on next page

239

Answers to Practice Problems F

1. a. 259.80 g/mol
   b. 136.06 g/mol
   c. 342.34 g/mol
   d. 253.80 g/mol
   e. 60.06 g/mol
   f. 262.84 g/mol
2. a. $NaHCO_3$, 84.01 g/mol
   b. CeB_6, 204.98 g/mol
   c. $Mg(ClO_4)_2$, 223.20 g/mol
   d. $Al_2(SO_4)_3$, 342.17 g/mol
   e. $Fe(OH)_3$, 106.88 g/mol
   f. $SnCl_2$, 189.61 g/mol
   g. P_4O_{10}, 283.88 g/mol
   h. ICl, 162.35 g/mol
3. a. 92.15 g/mol
   b. 0.0815 mol $C_6H_5CH_3$
4. a. 300.06 g/mol
   b. 2.050 g $PtCl_2(NH_3)_2$

Homework ── GENERAL

Additional Practice Calculate the molar mass of each of the following compounds:

1. KNO_3 **Ans.** 101.11 g/mol
2. Na_2SO_4 **Ans.** 142.05 g/mol
3. $Ca(OH)_2$ **Ans.** 74.10 g/mol
4. $(NH_4)_2SO_3$ **Ans.** 116.17 g/mol
5. $Ca_3(PO_4)_2$ **Ans.** 310.18 g/mol
6. $Al_2(CrO_4)_3$ **Ans.** 401.96 g/mol
LS Logical

Teaching Tip

Another way to determine molar mass is to add the values of the atomic masses to get the value of the molecular mass. Then, apply the units *g/mol* to the final answer. To make calculations simpler, consider having students round atomic masses to the nearest whole number.

REAL-WORLD CONNECTION

Barium Nitrate The compound used in **Sample Problem F,** barium nitrate, is extremely toxic, as are many barium compounds. This toxicity is more significant than it might be in other barium compounds because barium nitrate is much more soluble in water than other barium compounds. It can cause respiratory irritation, muscle spasms, and skin and eye irritation, as well as other symptoms. It is used in fireworks to produce a green flame.

Chapter Resource File

- **Problem Solving**

Close

Quiz — GENERAL

1. a. What is the formula of aluminum sulfate? Ans. $Al_2(SO_4)_3$

   b. How many moles of each type of atom are present in two moles of the compound? Ans. 4 mol Al, 6 mol S, 24 mol O

2. What is the definition of *molar mass*? Ans. the mass of one mole of a substance

3. You are told that the molar mass of a compound is 209.42. What is incorrect about the way this molar mass is written? Ans. It does not include the unit of g/mol.

4. What is the molar mass of xenon tetrafluoride (XeF_4)? Ans. 207.29 g/mol

LS Logical

Reteaching — BASIC

Have students work in pairs. Have each student give the other student a chemical formula for a compound. Then have each student determine the molar mass of the compound. Check formulas for accuracy before molar masses are determined. **LS** Logical

Chapter Resource File

• Concept Review
• Quiz

One-Stop Planner CD-ROM

• **Supplemental Reading Projects**
Guided Reading Worksheet: The Same and Not the Same "In Praise of Synthesis" *Assign this worksheet for cross-curricular connections to language arts.*

PRACTICE

 PROBLEM SOLVING SKILL

3 a. Find the molar mass of toluene, $C_6H_5CH_3$.

   b. Find the number of moles in 7.51 g of toluene.

4 a. Find the molar mass of cisplatin, $PtCl_2(NH_3)_2$, a cancer therapy chemical.

   b. Find the mass of 4.115×10^{21} formula units of cisplatin.

② Section Review

UNDERSTANDING KEY IDEAS

1. What is a weighted average?

2. On the periodic table, the average atomic mass of carbon is 12.01 g. Why is it not exactly 12.00?

3. What is the simplest formula for cesium carbonate?

4. What ions are present in cesium carbonate?

5. What is the ratio of N and H atoms in NH_3?

6. What is the ratio of calcium and chloride ions in $CaCl_2$?

7. Why is the simplest formula used to determine the molar mass for ionic compounds?

PRACTICE PROBLEMS

8. Calculate the average atomic mass of chromium. Its composition is: 83.79% with a mass of 51.94 amu; 9.50% with a mass of 52.94 amu; 4.35% with a mass of 49.95 amu; 2.36% with a mass of 53.94 amu.

9. Element X has two isotopes. One has a mass of 10.0 amu and an abundance of 20.0%. The other has a mass of 11.0 amu and an abundance of 80.0%. Estimate the average atomic mass. What element is it?

10. Find the molar mass.

   a. CsCl   **d.** $(NH_4)_2HPO_4$

   b. $KClO_3$ **e.** $C_2H_5NO_2$

   c. $C_6H_{12}O_6$

11. Determine the formula, the molar mass, and the number of moles in 2.11 g of each of the following compounds.

   a. strontium sulfide

   b. phosphorus trifluoride

   c. zinc acetate

   d. mercury(II) bromate

   e. calcium nitrate

12. Find the molar mass and the mass of 5.0000 mol of each of the following compounds.

   a. calcium acetate, $Ca(C_2H_3O_2)_2$

   b. iron(II) phosphate, $Fe_3(PO_4)_2$

   c. saccharin, $C_7H_5NO_3S$, a sweetener

   d. acetylsalicylic acid, $C_9H_8O_4$, or aspirin

CRITICAL THINKING

13. In the periodic table, the atomic mass of fluorine is given to 9 significant figures, whereas oxygen is given to only 6. Why? (Hint: fluorine has only one isotope.)

14. Figure 6 shows many K^+ and Br^- ions. Why is the formula not written as $K_{20}Br_{20}$?

15. Why don't scientists use HO as the formula for hydrogen peroxide, H_2O_2?

16. a. How many atoms of H are in a formula unit of $(NH_4)_2SO_4$?

   b. How many atoms of H are in 1 mol of $(NH_4)_2SO_4$?

240

Answers to Section Review

1. It is an average that takes into account both the value and the frequency of each number involved. It is calculated by multiplying the value of each number times its frequency, adding these products together, then dividing the sum by the number of items to be averaged.

2. Some atoms of carbon have a mass greater than 12.00, so they raise the average atomic mass.

3. Cs_2CO_3

4. two Cs^+ ions for every CO_3^{2-} ion

5. one nitrogen atom : three hydrogen atoms

6. one calcium ion : two chloride ions

7. An ionic compound is composed of positive and negative ions in an infinite crystal. The simplest formula reflects the specific ratio of positive and negative ions.

8. (49.95 amu)(0.0435) + (51.94 amu)(0.8379) + (52.94 amu)(0.0950) + (53.94 amu)(0.0236) = 51.99 amu

9. (10.0 amu)(0.20) + (11.0 amu)(0.80) = 10.80 amu; boron

Answers continued on p. 257A

SECTION 3

Formulas and Percentage Composition

KEY TERMS

• percentage composition
• empirical formula
• molecular formula

OBJECTIVES

① **Determine** a compound's empirical formula from its percentage composition.

② **Determine** the molecular formula or formula unit of a compound from its empirical formula and its formula mass.

③ **Calculate** percentage composition of a compound from its molecular formula or formula unit.

Using Analytical Data

Scientists synthesize new compounds for many uses. Once they make a new product, they must check its identity. One way is to carry out a chemical analysis that provides a **percentage composition.** For example, in 1962, two chemists made a new compound from xenon and fluorine. Before 1962, scientists thought that xenon did not form compounds. The scientists analyzed their surprising find. They found that it had a percentage composition of 63.3% Xe and 36.7% F, which is the same as that for the formula XeF_4. Percentage composition not only helps verify a substance's identity but also can be used to compare the ratio of masses contributed by the elements in two substances, as in **Figure 8.**

percentage composition

the percentage by mass of each element in a compound

internet connect

www.scilinks.org
Topic : Percentage Composition
SciLinks code: HW4131

SC*LINKS* Maintained by the National Science Teachers Association

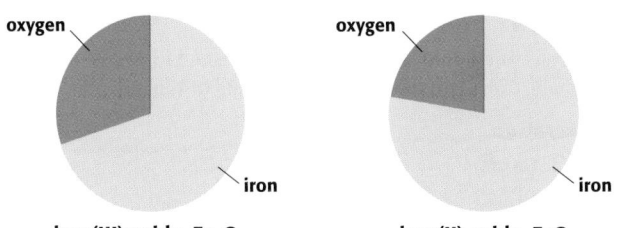

iron(III) oxide, Fe₂O₃

| iron | 69.9% |
| oxygen | 30.1% |

iron(II) oxide, FeO

| iron | 77.7% |
| oxygen | 22.3% |

Figure 8
Iron forms two different compounds with oxygen. The two compounds have different ratios of atoms and therefore have different percentage compositions and different properties.

241

REAL-WORLD CONNECTION

Percentage Composition The percentage composition of any pure substance is always constant, regardless of the source or sample size. Therefore, percentage composition is an identifying characteristic of any substance. For that reason, percentage composition is used by investigators in many different fields—from analytical chemistry to criminal forensics—to determine the identity of unknown materials.

Chapter Resource File

• Lesson Plan

Transparencies

TT Percentage Composition of Iron Oxides

SECTION 3

Focus

Overview

Before beginning this section, review with your students the Objectives listed in the Student Edition. This section investigates the definitions for and the relationships among percentage composition, empirical formulas, and molecular formulas.

Bellringer

Ask students to brainstorm a list of what they know about percentages. **Ans.** Answers should include that a percentage is a type of fraction, that the part divided by the whole multiplied by 100% is used to calculate percentage, and that all the percentages of the whole should total 100%.

Motivate

Demonstration

Modeling Composition Analysis
Everyone should wear safety goggles. Find the mass of a package of unpopped microwave popcorn. Pop the corn, then open the bag and allow the steam to escape. Find the mass of the popped corn and the bag and then the mass of just the bag. From these data, find the mass of the unpopped popcorn and the mass of the water that was contained in the unpopped corn. Find the percentage of water in the unpopped corn by dividing the mass of water by the mass of the unpopped corn, then multiply the results by 100%. Point out to students that this general method of calculation can be used to determine the percentage of each element in a compound.

Teaching Tip

In this text, percentage composition is always given by mass. Some texts may also report percentage composition by mole. When discussing both types of percentage composition, it is important to specify whether a specific value is by mass or by mole.

Determine an Empirical Formula

Magnesium burns in air to form MgO and Mg_3N_2. Water and Mg_3N_2, when heated, produce MgO and NH_3. When only MgO remains in the crucible (NH_3 gas escapes), students can use the data from the demonstration to calculate the empirical formula.

1. Find the mass of a crucible. Then find the mass of the crucible and 65 cm of magnesium ribbon. Calculate the mass of the magnesium.

2. Roll the magnesium loosely, and place it in the crucible. Place the crucible on a clay triangle that is sitting on an iron ring that is attached to a ring stand.

3. Carefully light the magnesium, and allow it to burn completely. **Caution:** Do not look directly at the burning magnesium.

4. Allow the crucible to cool. (Other class work can be done while waiting for the crucible to cool.) After the crucible has cooled, powder the residue. Add several drops of distilled water to the residue to moisten it.

5. Heat the crucible until its contents are dry.

6. After the crucible has cooled, find the mass of the crucible and the magnesium oxide. Calculate the mass of the product by the difference.

7. Use the masses of the magnesium and the magnesium oxide to find the mass of the oxygen used in the reaction.

continued on next page

empirical formula

a chemical formula that shows the composition of a compound in terms of the relative numbers and kinds of atoms in the simplest ratio

Empirical formula NH_2O

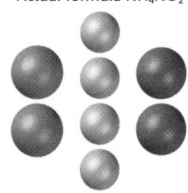

Actual formula NH_4NO_2

Space-filling model

 +

Figure 9
The empirical formula for ammonium nitrite is NH_2O. Its actual formula has 1 ammonium ion, NH_4^+, and 1 nitrite ion, NO_2^-.

Determining Empirical Formulas

Data for percentage composition allow you to calculate the simplest ratio among the atoms found in a compound. The **empirical formula** shows this simplest ratio. For example, ammonium nitrite, shown in **Figure 9**, has the actual formula NH_4NO_2 and is made up of ammonium ions, NH_4^+, and nitrite ions, NO_2^-, in a 1:1 ratio.

But if a chemist does an elemental analysis, she will find the empirical formula to be NH_2O, because it shows the *simplest ratio* of the elements. For some other compounds, the empirical formula and the actual formula are the same.

Let's say that you want to find an empirical formula from the percentage composition. First, convert the mass percentage of each element to grams. Second, convert from grams to moles using the molar mass of each element as a conversion factor. (Keep in mind that a formula for a compound can be read as a number of atoms or as a number of moles.) Third, as shown in **Sample Problem G,** compare these amounts in moles to find the simplest whole-number ratio among the elements in the compound.

To find this ratio, divide each amount by the smallest of all the amounts. This process will give a subscript of 1 for the atoms present in the smallest amount. Finally, you may need to multiply by a number to convert all subscripts to the smallest whole numbers. The final numbers you get are the subscripts in the empirical formula. For example, suppose the subscripts were 1.33, 2, and 1. Multiplication by 3 gives subscripts of 4, 6, and 3.

SAMPLE PROBLEM G

Determining an Empirical Formula from Percentage Composition

Chemical analysis of a liquid shows that it is 60.0% C, 13.4% H, and 26.6% O by mass. Calculate the empirical formula of this substance.

1 **Gather information.**
- percentage C = 60.0%
- percentage H = 13.4%
- percentage O = 26.6%
- empirical formula = $C_?H_?O_?$

2 **Plan your work.**
- Assume that you have a 100.0 g sample of the liquid, and convert the percentages to grams.

for C: $60.0\% \times 100.0 \text{ g} = 60.0 \text{ g C}$

for H: $13.4\% \times 100.0 \text{ g} = 13.4 \text{ g H}$

for O: $26.6\% \times 100.0 \text{ g} = 26.6 \text{ g O}$

242

REAL-WORLD
CONNECTION

Medicinal Compounds Natural compounds that have medicinal applications often exist in minute amounts or are extremely expensive to extract. For that reason, chemists strive to reproduce these substances in a laboratory. Before chemists can synthesize a compound, they must first determine the chemical formula for it. One way to determine the formula of a compound is to find its percentage composition. Once the formula is determined, chemists can then investigate ways of producing it. Synthetic compounds are often as effective as, less expensive than, and more abundant than their natural counterparts.

- To convert the mass of each element into the amount in moles, you must multiply by the proper conversion factor, which is the reciprocal of the molar mass. Find molar mass by using the periodic table.

$$\text{molar mass of C: } 12.01 \text{ g/mol}$$
$$\text{molar mass of H: } 1.01 \text{ g/mol}$$
$$\text{molar mass of O: } 16.00 \text{ g/mol}$$

3 Calculate.
- Calculate the amount in moles of C, H, and O. Round the answers to the correct number of significant figures.

$$60.0 \text{ g C} \times \frac{1 \text{ mol C}}{12.01 \text{ g C}} = 5.00 \text{ mol C}$$

$$13.4 \text{ g H} \times \frac{1 \text{ mol H}}{1.01 \text{ g H}} = 13.3 \text{ mol H}$$

$$26.6 \text{ g O} \times \frac{1 \text{ mol O}}{16.00 \text{ g O}} = 1.66 \text{ mol O}$$

- At this point the formula can be written as $C_5H_{13.3}O_{1.66}$, but you know that subscripts in chemical formulas are usually whole numbers.
- To begin the conversion to whole numbers, divide all subscripts by the smallest subscript, 1.66. This will make at least one of the subscripts a whole number, 1.

$$\frac{5.00 \text{ mol C}}{1.66} = 3.01 \text{ mol C}$$

$$\frac{13.3 \text{ mol H}}{1.66} = 8.01 \text{ mol H}$$

$$\frac{1.66 \text{ mol O}}{1.66} = 1.00 \text{ mol O}$$

- These numbers can be assumed to be the whole numbers 3, 8 and 1. The empirical formula is therefore C_3H_8O.

4 Verify your result.
Verify your answer by calculating the percentage composition of C_3H_8O. If the result agrees with the composition stated in the problem, then the formula is correct.

PRACTICE HINT

When you get fractions for the first calculation of subscripts, think about how you can turn these into whole numbers. For example:
- the subscript 1.33 is roughly $1\frac{1}{3}$, so it will give the whole number 4 when multiplied by 3
- the subscript 0.249 is roughly $\frac{1}{4}$, so it will give the whole number 1 when multiplied by 4
- the subscript 0.74 is roughly $\frac{3}{4}$, so it will give the whole number 3 when multiplied by 4

PRACTICE

Determine the empirical formula for each substance.

1 A dead alkaline battery is found to contain a compound of Mn and O. Its analysis gives 69.6% Mn and 30.4% O.

2 A compound is 38.77% Cl and 61.23% O.

3 Magnetic iron oxide is 72.4% iron and 27.6% oxygen.

4 A liquid compound is 18.0% C, 2.26% H, and 79.7% Cl.

243

HISTORY
CONNECTION

Martin Heinrich Klaproth The German chemist Martin Klaproth is credited with emphasizing the importance of accurately conducting experiments and scrupulously reporting scientific data. Previously, many scientists adjusted the results until percentages of elements in compounds totaled 100%. Klaproth reported actual percentages and predicted that when the total was not 100%, some elements were missing. By using this idea, he discovered uranium, tellurium, zirconium, and titanium.

Transparencies

TT Empirical and Actual Formulas

Chapter Resource File

- **Problem Solving**

continued from previous page

8. For homework, have students use the calculated masses and the molar masses to find the empirical formula for magnesium oxide.

Safety Caution: Wear safety goggles and a lab apron. Use caution when burning the magnesium. Do not look directly at the burning magnesium. Students must be 3 m or more from the demonstration.

Disposal: Place the product in a sealable plastic bag. Contact your local regulatory agency to find out whether or not it is allowable in your area to throw out the product. **LS Logical**

Answers to Practice Problems G
1. Mn_2O_3
2. Cl_2O_7
3. Fe_3O_4
4. $C_2H_3Cl_3$

Homework ——— **GENERAL**

Additional Practice For each of the following examples, find the empirical formula for a compound with the given composition.

1. 63.52% Fe, 36.48% S Ans. FeS
2. 26.58% K, 35.35% Cr, 38.07% O Ans. $K_2Cr_2O_7$
3. 32.37% Na, 22.58% S, 45.05% O Ans. Na_2SO_4
4. 74.51% Pb, 25.49% Cl Ans. $PbCl_2$
5. 52.55% Ba, 10.72% N, 36.73% O Ans. BaN_2O_6 or $Ba(NO_3)_2$
6. 29.15% N, 8.407% H, 12.50% C, 49.94% O Ans. $N_2H_8CO_3$ or $(NH_4)_2CO_3$
LS Logical

Teaching Tip ——— **BASIC**
For students who need math help, provide problems with whole-number answers (given the significant figures). For example, have students determine the empirical formula for the compound oxygen fluoride, which is 30% oxygen and 70% fluorine. Ans. OF_2 **LS Logical**

Vocabulary Have students look up the meaning of the word *empirical* in the dictionary. Have them use the definition to explain where information is obtained about an empirical formula. Their explanations should reflect that empirical formulas are determined from experimental data. **LS** Verbal

READING SKILL BUILDER — BASIC

Discussion Have students get into pairs and take turns reading a paragraph of this section. After each paragraph, have students discuss the content of the paragraph. In this section, the discussion should emphasize the relationships among molecular formulas, empirical formulas, and percentage composition, and the role the mole and molar mass play in determining these three concepts. **LS** Interpersonal

Using the Table — BASIC

Refer students to **Table 3.** Have them examine each of the formulas shown and confirm that the empirical formula for each is CH_2O. Remind students that in the space-filling models, the blue spheres represent H, the red spheres represent O, and the green spheres represent C. **LS** Visual

Transparencies

TT Comparing Empirical and Molecular Formulas

---

molecular formula

a chemical formula that shows the number and kinds of atoms in a molecule, but not the arrangement of the atoms

Figure 10
The formula for glucose, which is found in many sports drinks, is $C_6H_{12}O_6$.

Molecular Formulas Are Multiples of Empirical Formulas

The formula for an ionic compound shows the simplest whole-number ratio of the large numbers of ions in a crystal of the compound. The formula $Ca_3(PO_4)_2$ shows that the ratio of Ca^{2+} ions to PO_4^{3-} ions is 3:2.

Molecular compounds, on the other hand, are made of single molecules. Some molecular compounds have the same molecular and empirical formulas. Examples are water, H_2O, and nitric acid, HNO_3. But for many molecular compounds the **molecular formula** is a whole-number multiple of the empirical formula. Both kinds of formulas are just two different ways of representing the composition of the same molecule.

The molar mass of a compound is equal to the molar mass of the empirical formula times a whole number, *n*. There are several experimental techniques for finding the molar mass of a molecular compound even though the compound's chemical composition and formula are unknown. If you divide the experimental molar mass by the molar mass of the empirical formula, you can figure out the value of *n* needed to scale the empirical formula up to give the molecular formula.

Think about the three compounds in **Table 3**—formaldehyde, acetic acid, and glucose, which is shown in **Figure 10**. Each has the empirical formula CH_2O. However, acetic acid has a molecular formula that is twice the empirical formula. The molecular formula for glucose is six times the empirical formula. The relationship is shown in the following equation.

$$n(\text{empirical formula}) = \text{molecular formula}$$

In general, the molecular formula is a whole-number multiple of the empirical formula. For formaldehyde, *n* = 1, for acetic acid, *n* = 2, and for glucose, *n* = 6. In some cases, *n* may be a very large number.

Table 3   Comparing Empirical and Molecular Formulas

Compound	Empirical formula	Molecular formula	Molar mass (g)	Space-filling model
Formaldehyde	CH_2O	CH_2O • same as empirical formula • *n* = 1	30.03	
Acetic acid	CH_2O	$C_2H_4O_2$ ($HC_2H_3O_2$) • 2 × empirical formula • *n* = 2	60.06	
Glucose	CH_2O	$C_6H_{12}O_6$ • 6 × empirical formula • *n* = 6	180.18	

244

HOME ECONOMICS
CONNECTION — ADVANCED

Hot Peppers Students are familiar with peppers and know that there are many different types and degrees of "hotness" in peppers. Bell peppers are not hot at all, but some peppers, such as the habañero, are so hot that wearing gloves is recommended when handling them. The difference in hotness is determined by the amount of compounds known as *capsaicinoids* in the peppers. The higher the amount of these compounds present, the hotter the pepper. The most common capsaicinoids are capsaicin and dihydrocapsaicin. Have students research and determine the formula and percentage composition of one or more capsaicinoids. **LS** Logical

SAMPLE PROBLEM H

Determining a Molecular Formula from an Empirical Formula

The empirical formula for a compound is P_2O_5. Its experimental molar mass is 284 g/mol. Determine the molecular formula of the compound.

1 Gather information.
- empirical formula = P_2O_5
- molar mass of compound = 284 g/mol
- molecular formula = ?

2 Plan your work.
- Find the molar mass of the empirical formula using the molar masses of the elements from the periodic table.

$$\text{molar mass of P} = 30.97 \text{ g/mol}$$
$$\text{molar mass of O} = 16.00 \text{ g/mol}$$

3 Calculate.
- Find the molar mass of the empirical formula, P_2O_5.

$$
\begin{array}{r}
2 \times \text{molar mass of P} = 61.94 \text{ g/mol} \\
+\ 5 \times \text{molar mass of O} = 80.00 \text{ g/mol} \\
\hline
\text{molar mass of } P_2O_5 = 141.94 \text{ g/mol}
\end{array}
$$

- Solve for n, the factor multiplying the empirical formula to get the molecular formula.

$$n = \frac{\text{experimental molar mass of compound}}{\text{molar mass of empirical formula}}$$

- Substitute the molar masses into this equation, and solve for n.

$$n = \frac{284 \text{ g/mol}}{141.94 \text{ g/mol}} = 2.00 = 2$$

- Multiply the empirical formula by this factor to get the answer.

$$n(\text{empirical formula}) = 2(P_2O_5) = P_4O_{10}$$

4 Verify your result.
- The molar mass of P_4O_{10} is 283.88 g/mol. It is equal to the experimental molar mass.

PRACTICE

1. A compound has an experimental molar mass of 78 g/mol. Its empirical formula is CH. What is its molecular formula?

2. A compound has the empirical formula CH_2O. Its experimental molar mass is 90.0 g/mol. What is its molecular formula?

3. A brown gas has the empirical formula NO_2. Its experimental molar mass is 46 g/mol. What is its molecular formula?

PRACTICE HINT

In some cases, you can figure out the factor n by just looking at the numbers. For example, let's say you noticed that the experimental molar mass was almost exactly twice as much as the molar mass of the empirical formula (as in this problem). That means n must be 2.

Answers to Practice Problems H

1. C_6H_6
2. $C_3H_6O_3$
3. NO_2

Homework ── GENERAL

Additional Practice Determine the molecular formula for each of the following compounds:

1. Molar mass: 232.41 g/mol; empirical formula: OCNCl
   Ans. $O_3C_3N_3Cl_3$

2. Molar mass: 32.06 g/mol; empirical formula: NH_2
   Ans. N_2H_4

3. Molar mass: 120.12 g/mol; empirical formula: CH_2O
   Ans. $C_4H_8O_4$

LS Logical

MISCONCEPTION /// ALERT \\\

Arbitrary Amount of Sample
Students may believe that a 100 g sample of a substance must be used when considering percentage composition. Emphasize that 100.0 g is a matter of convenience because it's so easy to multiply by 100. For example, suppose you want to determine the mass of each element present in a sample of a compound that is 77.9% carbon, 11.7% hydrogen, and 10.4% oxygen. If you choose an arbitrary amount of 100.0 g of sample, it's easy to see that there is 77.9 g carbon in the sample. If you chose a sample that was 23.9 g, the empirical formula would be the same, but the calculations would be more complicated.

Chapter Resource File

- **Problem Solving**

REAL-WORLD
CONNECTION ── GENERAL

Succinic Acid The lichens present on the bark of trees contain a substance called *succinic acid*. This organic acid is also present in fungi. It is extracted from these organisms and is used to make perfumes and dyes. Tell students that the percentage composition of succinic acid is 40.68% carbon, 5.08% hydrogen, and 54.24% oxygen. Its molecular mass is 118.8 g/mol. Ask students to determine the empirical and molecular formulas of succinic acid. Ans. $C_2H_3O_2$, $C_4H_6O_4$

LS Logical

245

Research Skills Have students look through the organic chemistry section of the *CRC Handbook* to find molecular formulas of compounds that have the same empirical formulas. Examples might include acetic acid ($C_2H_4O_2$) and glucose ($C_6H_{12}O_6$). Students might start by skimming the formulas, picking out ones that have subscripts that are not in the lowest ratio. They can keep track of them by writing their empirical formulas and the reference number for the compound. When they find two compounds that have the same empirical formulas, they should list their empirical formulas, molecular formulas, names, and reference numbers. **LS** Visual

SKILL BUILDER — ADVANCED

Graphing Skills Refer students to **Figure 11.** Have them confirm that the percentages shown total up to 100% for each compound. Have them determine and graph the percentage compositions for nitrogen dioxide, NO_2, dinitrogen trioxide, N_2O_3, and nitrous oxide, N_2O. **LS** Visual

Teaching Tip

Parts of the Whole Remind students that if they know the percentage of one element in a binary compound, they can determine the percentage of the other element by subtracting the known percentage from 100%.

Chemical Formulas Can Give Percentage Composition

If you know the chemical formula of any compound, then you can calculate the percentage composition. From the subscripts, you can determine the mass contributed by each element and add these to get the molar mass. Then, divide the mass of each element by the molar mass. Multiply by 100 to find the percentage composition of that element.

Think about the two compounds shown in **Figure 11.** Carbon dioxide, CO_2, is a harmless gas that you exhale, while carbon monoxide, CO, is a poisonous gas present in car exhaust. The percentage composition of carbon dioxide, CO_2, is calculated as follows.

$$1 \text{ mol} \times 12.01 \text{ g C/mol} = 12.01 \text{ g C}$$
$$+ \, 2 \text{ mol} \times 16.00 \text{ g O/mol} = 32.00 \text{ g O}$$
$$\text{mass of 1 mol } CO_2 = 44.01 \text{ g}$$

$$\% \text{ C in } CO_2 = \frac{12.01 \text{ g C}}{44.01 \text{ g } CO_2} \times 100 = 27.29\%$$

$$\% \text{ O in } CO_2 = \frac{32.00 \text{ g O}}{44.01 \text{ g } CO_2} \times 100 = 72.71\%$$

The percentage composition of carbon monoxide, CO, is calculated as follows.

$$1 \text{ mol} \times 12.01 \text{ g C/mol} = 12.01 \text{ g C}$$
$$+ \, 1 \text{ mol} \times 16.00 \text{ g O/mol} = 16.00 \text{ g O}$$
$$\text{mass of 1 mol } CO = 28.01 \text{ g}$$

$$\% \text{ C in } CO = \frac{12.01 \text{ g C}}{28.01 \text{ g } CO} \times 100 = 42.88\%$$

$$\% \text{ O in } CO = \frac{16.00 \text{ g O}}{28.01 \text{ g } CO} \times 100 = 57.12\%$$

Figure 11
Carbon monoxide and carbon dioxide are both made up of the same elements, but they have different percentage compositions.

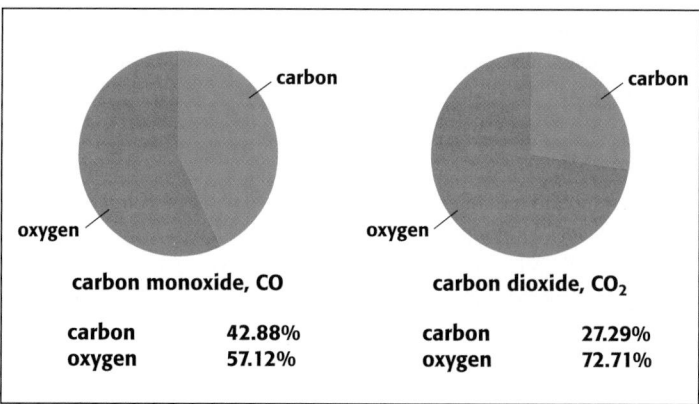

carbon monoxide, CO

| carbon | 42.88% |
| oxygen | 57.12% |

carbon dioxide, CO₂

| carbon | 27.29% |
| oxygen | 72.71% |

246

BIOLOGY
CONNECTION — GENERAL

Serotonin The compound serotonin conducts nerve impulses in the brain. Give students the percentage composition of serotonin, which is 68.2% C, 6.86% H, 15.9% N, and 9.08% O. Have them determine its molecular formula. Tell them that its molecular formula is the same as its empirical formula. Ans. $C_{10}H_{12}N_2O$ **LS** Logical

Using a Chemical Formula to Determine Percentage Composition

Calculate the percentage composition of copper(I) sulfide, a copper ore called chalcocite.

1 Gather information.
- name and formula of the compound: copper(I) sulfide, Cu_2S
- percentage composition: $\%Cu = ?$, $\%S = ?$

2 Plan your work.

To determine the molar mass of copper(I) sulfide, find the molar mass of the elements copper and sulfur using the periodic table.

$$\text{molar mass of } Cu = 63.55 \text{ g/mol}$$
$$\text{molar mass of } S = 32.07 \text{ g/mol}$$

3 Calculate.
- Find the masses of 2 mol Cu and 1 mol S. Use these masses to find the molar mass of Cu_2S.

$$2 \text{ mol} \times 63.55 \text{ g Cu/mol} = 127.10 \text{ g Cu}$$
$$+ 1 \text{ mol} \times 32.07 \text{ g S/mol} = \underline{32.07 \text{ g S}}$$
$$\text{molar mass of } Cu_2S = 159.17 \text{ g/mol}$$

- Calculate the fraction that each element contributes to the total mass. Do this by dividing the total mass contributed by that element by the total mass of the compound. Convert the fraction to a percentage by multiplying by 100.

$$\text{mass \% } Cu = \frac{\text{mass of 2 mol Cu}}{\text{molar mass of } Cu_2S} \times 100$$

$$\text{mass \% } S = \frac{\text{mass of 1 mol S}}{\text{molar mass of } Cu_2S} \times 100$$

- Substitute the masses into the equations above. Round the answers you get on the calculator to the correct number of significant figures.

$$\text{mass \% } Cu = \frac{127.10 \text{ g Cu}}{159.17 \text{ g } Cu_2S} \times 100 = 79.852\% \text{ Cu}$$

$$\text{mass \% } S = \frac{32.07 \text{ g S}}{159.17 \text{ g } Cu_2S} \times 100 = 20.15\% \text{ S}$$

4 Verify your result.
- Add the percentages. The sum should be near 100%.

$$79.852\% + 20.15\% = 100.00\%$$

Practice problems on next page

247

PRACTICE HINT

Sometimes, rounding gives a sum that differs slightly from 100%. This is expected. (However, if you find a sum that differs significantly, such as 112%, you have made an error.)

Answers to Practice Problems I

1. 93.311% Fe, 6.689% C
2. 50.05% S, 49.95% O
3. 35.00% N, 5.05% H, 59.96% O
4. a. 35.41% Sr, 64.59% Br
   b. 29.44% Ca, 23.55% S, 47.01% O
   c. 31.83% Mg, 31.46% C, 36.70% N
   d. 63.70% Pb, 14.77% C, 1.86% H, 19.67% O
5. a. Both are 39.99% C, 6.73% H, and 53.28% O because, if you combine the hydrogen atoms in acetic acid, the empirical formulas are the same.
   b. The percentage composition of the empirical formula is the same as that of the molecular formulas.

Homework ——— GENERAL

Additional Practice Determine the percentage compositions of the following compounds:

1. NaClO **Ans.** 30.88% Na, 47.62% Cl, 21.49% O
2. H_2SO_3 **Ans.** 2.46% H, 39.07% S, 58.47% O
3. C_2H_5COOH **Ans.** 48.63% C, 8.18% H, 43.19% O
LS Logical

Group Activity ——— BASIC

Have groups of four students obtain three index cards and write one of the following terms on each card: *percentage composition*, *molecular formula*, and *empirical formula*. Place the cards face down, and have one student draw a card. The term that is on this card is what is given. Have another student draw another card, and this one is what is unknown. Have the group write a procedure for calculating the unknown information from the known information. They can use any other quantities they need to use, such as molar mass. Repeat the procedure, replacing the cards and drawing again until they have completed each of the six possible combinations.
LS Interpersonal

did you know? ——— ADVANCED

Hydrated Compounds Compounds called *hydrates* contain water in their chemical formulas. Usually, when the compounds are heated, the water of hydration is removed and the anhydrous compound remains. How much water can be removed affects real-life considerations, such as shipping costs or volume for the anhydrous compound versus the hydrate. Have students determine the percentage of water in each of the following hydrates: $ZnSO_4 \cdot 7H_2O$ **Ans.** 43.86%; $CaSO_4 \cdot 2H_2O$ **Ans.** 20.93%; and $MgBr_2 \cdot 6H_2O$ **Ans.** 37.00%.
LS Logical

Chapter Resource File

- **Problem Solving**

Close

PRACTICE

1 Calculate the percentage composition of Fe_3C, a compound in cast iron.

2 Calculate the percentage of both elements in sulfur dioxide.

3 Calculate the percentage composition of ammonium nitrate, NH_4NO_3.

4 Calculate the percentage composition of each of the following:

    a. $SrBr_2$ **c.** $Mg(CN)_2$

    b. $CaSO_4$ **d.** $Pb(CH_3COO)_2$

5 a. Calculate the percentage of each element in acetic acid, $HC_2H_3O_2$, and glucose, $C_6H_{12}O_6$.

    b. These two substances have the same empirical formula. What would you expect the percentage composition of the empirical formula to be?

③ Section Review

UNDERSTANDING KEY IDEAS

1. a. Suppose you know that a compound is 11.2% H and 88.8% O. What information do you need to determine the empirical formula?

    b. What additional information do you need to determine the molecular formula?

2. Isooctane has the molecular formula C_8H_{18}. What is its empirical formula?

3. What information do you need to calculate the percentage composition of CF_4?

PRACTICE PROBLEMS

4. Determine the empirical formula.

    a. The analysis of a compound shows that it is 9.2% B and 90.8% Cl.

    b. An analysis shows that a compound is 50.1% S and 49.9% O.

    c. The analysis of a compound shows that it is 27.0% Na, 16.5% N, and 56.5% O.

5. The experimental molar mass of the compound in item 4b is 64 g/mol. What is the compound's molecular formula?

6. Determine the formula, and then calculate the percentage composition.

    a. calcium sulfate

    b. silicon dioxide

    c. silver nitrate

    d. nitrogen monoxide

7. Calculate the percentage composition.

    a. silver acetate, $AgC_2H_3O_2$

    b. lead(II) chlorate, $Pb(ClO_3)_2$

    c. iron(III) sulfate, $Fe_2(SO_4)_3$

    d. copper(II) sulfate, $CuSO_4$

CRITICAL THINKING

8. When you determine the empirical formula of a compound from analytical data, you seldom get exact whole numbers for the subscripts. Explain why.

9. An amino acid has the molecular formula $C_2H_5NO_2$. What is the empirical formula?

10. A compound has the empirical formula CH_2O. Its experimental molar mass is 45 g/mol. Is it possible to calculate the molecular formula with the information given?

Answers to Section Review

1. a. the molar masses of H and O

    b. the compound's experimental molar mass

2. C_4H_9

3. the molar masses of carbon and fluorine

4. a. BCl_3 **b.** SO_2 **c.** $NaNO_3$

5. SO_2

6. a. $CaSO_4$, 29.44% Ca, 23.55% S, and 47.01% O

    b. SiO_2, 46.75% Si and 53.25% O

    c. $AgNO_3$, 63.50% Ag, 8.247% N, and 28.26% O

    d. NO, 46.68% N and 53.32% O

7. a. 64.62% Ag, 14.39% C, 1.82% H, and 19.17% O

    b. 55.39% Pb, 18.95% Cl, and 25.66% O

    c. 27.93% Fe, 24.06% S, and 48.01% O

    d. 39.81% Cu, 20.09% S, and 40.10% O

8. Experimental error will affect percentage compositions.

9. $C_2H_5NO_2$

10. No. The molar mass of the empirical formula is 30.03 g/mol. The given experimental mass is not a whole-number multiple of this number.

LEAD

82
Pb
Lead
207.2
$[Xe]4f^{14}5d^{10}6s^26p^2$

Element Spotlight

Where Is Pb?
Earth's crust:
< 0.01% by mass

Get the Lead Out

Humans have known for many centuries that lead is toxic, but it is still used in many common materials. High levels of lead were used in white paints until the 1940s. Since then, the lead compounds in paints have gradually been replaced with less toxic titanium dioxide. However, many older buildings still have significant amounts of lead paint, and many also have lead solder in their water pipes.

Lead poisoning is caused by the absorption of lead through the digestive tract, lungs, or skin. Children living in older homes are especially susceptible to lead poisoning. Children eat paint chips that contain lead because the paint has a sweet taste.

The hazards of lead poisoning can be greatly reduced by introducing programs that increase public awareness, removing lead-based paint from old buildings, and screening children for lead exposure.

Industrial Uses

- The largest industrial use of lead is in the manufacture of storage batteries.
- Solder used for joining metals is often an alloy of lead and tin.
- Other lead alloys are used to make bearings for gasoline and diesel engines, type metal for printing, corrosion-resistant cable coverings, and ammunition.
- Lead sheets and lead bricks are used to shield workers and sensitive objects from X rays.

Real-World Connection Lead inside the human body interferes with the production of red blood cells and can cause damage to the kidneys, liver, brain, and other organs.

Posters such as this one are part of public-awareness programs to reduce the hazards of lead poisoning.

A Brief History

3000 BCE   **1000 BCE**   **1 CE**   **1000 CE**

600 BCE: Lead ore deposits are discovered near Athens; they are mined until the second century CE.

1977: The U.S. government restricts lead content in paint.

3000 BCE: Egyptians refine and use lead to make art figurines.

60 BCE: Romans begin making lead pipes, lead sheets for waterproofing roofs, and lead crystal.

Questions

1. How do you perform tests for lead in paint, soil, and water? Present a report that explains how the tests work.

2. Research the laws regarding the recycling of storage batteries that contain lead.

internet connect
www.scilinks.org
Topic : Lead
SciLinks code: HW4074
SCI LINKS Maintained by the National Science Teachers Association

249

Element Spotlight

Lead Until the development of the bright white pigment titanium(IV) oxide, TiO_2, the only acceptable white pigment used in paints was "white lead," whose formula is $Pb(OH)_2 \cdot PbCO_3$. Besides being poisonous, white lead reacts with sulfur compounds in the air to produce black lead(II) sulfide, PbS. In the past, sulfur compounds were prevalent in the air, especially in areas where a great deal of coal was burned by industries. A person who wanted a white house would have to repaint it often because the paint soon turned dirty gray.

Answers to Feature Questions

1. Answers may vary depending on type of home test kit or professional procedures available.

2. Answers may vary depending on local laws.

HISTORY
CONNECTION

Color Trends We see many old houses painted white today, but white houses were relatively uncommon before the 1940s. The original owners of many nineteenth-century houses would be shocked to see their houses painted white today. In their time, houses were painted in colors such as cream, tan, green, blue, and rusty red, and very colorful combinations were often used.

INCLUSION
Strategies

• *Learning Disabled*   • *Developmentally Delayed*

Ask students to make a public awareness poster or other poster about lead poisoning. The students will need to list the types of buildings where lead-based paint or lead solder may have been used. Information about screening children for lead exposure and local poison control phone numbers may be included in the presentation.

7 CHAPTER HIGHLIGHTS

Alternative Assessments

SKILL BUILDER
• Pages 227, 238, 246, 246

Homework
• Pages 227, 236

REAL-WORLD CONNECTION
• Page 227

Reteaching
• Page 233

READING SKILL BUILDER

• Page 235

Group Activity
• Page 237

CHAPTER REVIEW
• Items 88, 89

Portfolio Assessments

SKILL BUILDER
• Pages 229, 230

CHAPTER REVIEW
• Item 90

Chapter Resource File

• **Datasheets for In-Text Labs** Percent Composition of Hydrates

• **Datasheets for In-Text Labs** Gypsum and Plaster of Paris

---

KEY IDEAS

SECTION ONE   Avogadro's Number and Molar Conversions
• Avogadro's number, 6.022×10^{23} units/mol, is the number of units (atoms, ions, molecules, formula units, etc.) in 1 mol of any substance.
• Avogadro's number is used to convert from number of moles to number of particles or vice versa.
• Conversions between moles and mass require the use of molar mass.
• The molar mass of a monatomic element is the number of grams numerically equal to the atomic mass on the periodic table.

SECTION TWO   Relative Atomic Mass and Chemical Formulas
• The average atomic mass of an element is the average mass of the element's isotopes, weighted by the percentage of their natural abundance.
• Chemical formulas reveal composition. The subscripts in the formula give the number of atoms of a given element in a molecule or formula unit of a compound or diatomic element.
• Formulas are used to calculate molar masses of compounds.

SECTION THREE   Formulas and Percentage Composition
• Percentage composition gives the relative contribution of each element to the total mass of one molecule or formula unit.
• An empirical formula shows the elements and the smallest whole-number ratio of atoms or ions that are present in a compound. It can be found by using the percentage composition.
• The molecular formula is determined from the empirical formula and the experimentally determined molar mass.
• Chemical formulas can be used to calculate percentage composition.

KEY TERMS

mole
Avogadro's number
molar mass

average atomic mass

percentage composition
empirical formula
molecular formula

KEY SKILLS

Working Practice Problems
Skills Toolkit 2 p. 227

Converting Between Amount in Moles and Number of Particles
Skills Toolkit 1 p. 226
Sample Problem A p. 228
Sample Problem B p. 229

Converting Between Mass, Amount, and Number of Particles
Skills Toolkit 3 p. 230
Sample Problem C p. 231
Sample Problem D p. 232

Calculating Average Atomic Mass
Sample Problem E p. 235

Calculating Molar Mass of Compounds
Sample Problem F p. 239

Determining an Empirical Formula from Percentage Composition
Sample Problem G p. 242

Determining a Molecular Formula from an Empirical Formula
Sample Problem H p. 245

Using a Chemical Formula to Determine Percentage Composition
Sample Problem I p. 247

250

USING KEY TERMS

1. Distinguish between Avogadro's number and the mole.

2. What term is used to describe the mass in grams of 1 mol of a substance?

3. Why is the ratio between the empirical formula and the molecular formula a whole number?

4. A scientist isolates a compound with the formula CH_2O. She later discovers that its true formula is actually $C_6H_{12}O_6$. What might account for the differences in the two formulas?

5. What do you need to calculate the percentage composition of a substance?

6. Explain the difference between atomic mass and average atomic mass.

UNDERSTANDING KEY IDEAS

Avogadro's Number and Molar Conversions

7. What particular isotope is the basis for defining the atomic mass unit and the mole?

8. How is Avogadro's number related to moles?

9. How would you determine the number of molecules in 3 mol of oxygen, O_2?

10. What conversion factor do you use in converting number of moles into number of formula units?

11. How is molar mass of an element used to convert from number of moles to mass in grams?

12. What result do you get when you multiply the number of moles of a sample by the following conversion factor?

$$\frac{\text{g of element}}{1 \text{ mol element}}$$

13. You convert 10 mol of a substance to grams. Is the number in the answer larger or smaller than 10 g?

Relative Atomic Mass and Chemical Formulas

14. Which has the greater number of molecules: 10 g of N_2 or 10 g of O_2?

15. How is average atomic mass determined from isotopic masses?

16. For an element, what is the relationship between atomic mass and molar mass?

17. How do you determine the molar mass of a compound?

Formulas and Percentage Composition

18. What information does percentage composition reveal about a compound?

19. Compare the empirical formula of a compound with the molecular formula of the compound.

20. Summarize briefly the process of using empirical formula and the value for experimental molar mass to determine the molecular formula.

21. When you calculate the percentage composition of a compound from both the empirical formula and the molecular formula, why are the two results identical?

251

Assignment Guide

Section	Questions
1	1, 2, 7–13, 22–53, 75, 76, 80, 85, 88
2	6, 14–17, 54–61, 74, 77, 79, 81, 82, 89, 96
3	3–5, 18–21, 62–73, 78, 83, 84, 86, 87, 90–95

REVIEW ANSWERS

Using Key Terms

1. Avogadro's number is the number of particles in 1 mol, 6.022×10^{23}/mol, while the mole is the SI unit for measuring the amount of a substance. *Mole* is the name of the unit, while *Avogadro's number* is the quantity of the unit.

2. Molar mass is the mass in grams of 1 mol of a substance.

3. The empirical formula is the simplest ratio of atoms, while the molecular formula shows the actual ratio of atoms. It is not possible to have fractions of atoms in a substance, so there is a whole number factor that compares empirical formulas and molecular formulas.

4. CH_2O is the empirical formula, while $C_6H_{12}O_6$ is the molecular formula.

5. To calculate percentage composition, you need the formula (or the empirical formula) of the substance and the molar masses of the elements.

6. Atomic mass is the mass of an atom of a specific isotope, while average atomic mass is a weighted average of the atomic masses of an element's isotopes.

Understanding Key Ideas

7. carbon-12

8. There are Avogadro's number of particles in a mole.

9. You would multiply 3 mol by Avogadro's number.

10. $\dfrac{6.022 \times 10^{23} \text{ formula units}}{1 \text{ mol}}$

11. You multiply the number of moles of a substance by molar mass to get the mass in grams of the substance.

12. You get the mass in grams of the element.

13. The number in the answer is larger than 10 g.

14. 10 g of N_2

15. Average atomic mass is determined by adding up the products of each of the percentage abundance of the isotopes of an element multiplied by the isotope's atomic mass.

16. Molar mass is the value of the atomic mass with the units g/mol.

17. To determine the molar mass of a compound, you add up the molar masses of each atom in the compound.

18. Percentage composition reveals the mass ratios of the elements that make up a substance. Using percentage composition, you can derive the empirical formula of a substance.

19. The empirical formula shows the simplest ratio of elements in the compound, while the molecular formula shows the actual number of atoms or ions present in the simplest unit of the compound.

20. Find the molar mass of the empirical formula. Solve for n by dividing the experimental molar mass by the molar mass of the empirical formula. Multiply the subscripts in the empirical formula by n to get the subscripts for the molecular formula.

21. Both the empirical formula and molecular formula show equivalent ratios of elements. The percentages of two equivalent ratios are equal.

Practice Problems

22. 1.20×10^{24} ions Na^+
23. 1.20×10^{24} molecules $C_{12}H_{22}O_{11}$
24. 3.01×10^{22} molecules CO_2
25. 7.53×10^{21} atoms Hg
26. 1.56×10^{24} atoms Al

PRACTICE PROBLEMS

Sample Problem A Converting Amount in Moles to Number of Particles

22. How many sodium ions in 2.00 mol of NaCl?

23. How many molecules in 2.00 mol of sucrose, $C_{12}H_{22}O_{11}$?

24. How many molecules of carbon dioxide exit your lungs when you exhale 5.00×10^{-2} mol of carbon dioxide, CO_2?

25. How many atoms are in the 1.25×10^{-2} mol of mercury within the bulb of a thermometer?

26. Assume that you have 2.59 mol of aluminum. How many atoms of aluminum do you have?

27. How many ions are there in a solution with 9.656 mol of Ni^{2+}?

28. **a.** How many formula units are there in a 3.12 mol sample of $MgCl_2$?
    b. How many Cl^- ions are there in the sample?

Sample Problem B Converting Number of Particles to Amount in Moles

29. How many moles of magnesium oxide are there in 2.50×10^{25} formula units of MgO?

30. How many moles of gold are there in 1.00 L of sea water if there are 1.50×10^{17} atoms of gold in the sample?

31. A sample has 7.51×10^{24} molecules of benzene, C_6H_6. How many moles is this?

32. How many moles are in a sample having 9.3541×10^{13} particles?

33. How many moles of sodium ions are there in a sample of salt water that contains 4.11×10^{22} Na^+ ions?

34. How many moles are equal to 3.6×10^{23} molecules of oxygen gas, O_2?

Sample Problem C Converting Number of Particles to Mass

35. How many grams are in the 1.204×10^{20} atoms of phosphorus used to coat your television screen?

36. How many grams are present in 4.336×10^{24} formula units of table salt, NaCl, whose molar mass is 58.44 g/mol?

37. A scientist collects a sample that has 2.00×10^{14} molecules of carbon dioxide gas. How many grams is this, given that the molar mass of CO_2 is 44.01 g/mol?

38. What is the mass in grams of a sample of $Fe_2(SO_4)_3$ that contains 3.59×10^{23} sulfate ions, SO_4^{2-}? The molar mass of $Fe_2(SO_4)_3$ is 399.91 g/mol.

39. Calculate the mass in grams of 2.55 mol of oxygen gas, O_2 (molar mass of O_2 = 32.00 g/mol).

40. How many grams of Ne are in a neon sign that contains 0.0450 mol of neon gas?

41. How many grams are in 2.7 mol of table salt, NaCl (molar mass of NaCl = 58.44 g/mol)?

42. Calculate the mass of each of the following samples:
    a. 0.500 mol I_2 (molar mass of I_2 = 253.80 g/mol)
    b. 2.82 mol PbS (molar mass of PbS = 239.3 g/mol)
    c. 4.00 mol of C_4H_{10} (molar mass of C_4H_{10} = 58.14 g/mol)
    d. 0.300 mol of $Al_2(SO_4)_3$ (molar mass of $Al_2(SO_4)_3$ = 342.17 g/mol)
    e. 0.222 mol of $CuSO_4$ (molar mass of $CuSO_4$ = 159.62 g/mol)

43. How many grams are in each of the following samples?
    a. 1.000 mol NaCl (molar mass of NaCl = 58.44 g/mol)
    b. 2.000 mol H_2O (molar mass of H_2O = 18.02 g/mol)

c. 3.5 mol $Ca(OH)_2$ (molar mass of $Ca(OH)_2 = 74.10$ g/mol)

d. 0.625 mol $Ba(NO_3)_2$ (molar mass of $Ba(NO_3)_2 = 261.35$ g/mol)

Sample Problem D Converting Mass to Number of Particles

44. How many atoms of gold are there in a pure gold ring with a mass of 10.6 g?

45. How many formula units are there in 302.48 g of zinc chloride, $ZnCl_2$? The molar mass of zinc chloride is 136.29 g/mol.

46. Naphthalene, $C_{10}H_8$, an ingredient in mothballs, has a molar mass of 128.18 g/mol. How many molecules of naphthalene are in a mothball that has 2.000 g of naphthalene.

47. How many atoms of aluminum are there in 125 g of aluminum foil?

48. How many moles of water are there if a pipet delivers 99.7 g of water (molar mass of water = 18.02 g/mol)?

49. How many moles of compound are in each of the following samples:

a. 6.60 g $(NH_4)_2SO_4$ (molar mass of $(NH_4)_2SO_4 = 132.17$ g/mol)

b. 4.5 kg of $Ca(OH)_2$ (molar mass of $Ca(OH)_2 = 74.10$ g/mol)

c. 7.35 g of H_2SO_4 (molar mass of $H_2SO_4 = 98.09$ g/mol)

50. Ibuprofen, $C_{13}H_{18}O_2$, an active ingredient in pain relievers has a molar mass of 206.31 g/mol. How many moles of ibuprofen are in a bottle that contains 33 g of ibuprofen?

51. How many moles of $NaNO_2$ are there in a beaker that contains 0.500 kg of $NaNO_2$ (molar mass of $NaNO_2 = 69.00$ g/mol)?

52. How many moles of calcium are in one cup of milk that contains 290.0 mg of calcium?

53. How many moles of propane are in a pressure container that has 2.55 kg of propane, C_3H_8 (molar mass of $C_3H_8 = 44.11$ g/mol)?

Sample Problem E Calculating Average Atomic Mass

54. Naturally occurring silver is composed of two isotopes: Ag-107 is 51.35% with a mass of 106.905092 amu, and the rest is Ag-109 with a mass of 108.9044757 amu. Calculate the average atomic mass of silver.

55. The element bromine is distributed between two isotopes. The first, amounting to 50.69%, has a mass of 78.918 amu. The second, amounting to 49.31%, has a mass of 80.916 amu. Calculate the average atomic mass of bromine.

56. Antimony has two isotopes. The first, amounting to 57.3%, has a mass of 120.9038 amu. The second, amounting to 42.7%, has a mass of 122.9041 amu. What is the average atomic mass of antimony?

57. Calculate the average atomic mass of iron. Its composition is 5.90% with a mass of 53.94 amu, 91.72% with a mass of 55.93 amu, 2.10% with a mass of 56.94 amu, and 0.280% with a mass of 57.93 amu.

Sample Problem F Calculating Molar Mass of Compounds

58. Find the molar mass of the following compounds:

a. lithium chloride

b. sodium sulfate

c. copper(I) cyanide

d. propylene, C_3H_6

e. potassium dichromate

f. magnesium nitrate

g. magnetite, Fe_3O_4

h. uracil, $C_4H_4N_2O_2$

i. tetrasulfur tetranitride

j. cesium tribromide

k. caffeine, $C_8H_{10}N_4O_2$

59. What is the molar mass of the phosphate ion, PO_4^{3-}?

60. Find the molar mass of isopropyl alcohol, C_3H_7OH, used as rubbing alcohol.

27. 5.815×10^{24} ions Ni^{2+}

28. a. 1.88×10^{24} formula units $MgCl_2$

b. 3.76×10^{24} ions Cl^-

29. 41.5 mol MgO

30. 2.49×10^{-7} mol Au

31. 12.5 mol C_6H_6

32. 1.553×10^{-10} mol

33. 6.82×10^{-2} mol Na^+

34. 0.60 mol O_2

35. 0.006192 g P

36. 420.8 g NaCl

37. 1.46×10^{-8} g CO_2

38. 79.5 g $Fe_2(SO_4)_3$

39. 81.6 g O_2

40. 0.908 g Ne

41. 160 g NaCl

42. a. 127 g I_2

b. 675 g PbS

c. 233 g C_4H_{10}

d. 103 g $Al_2(SO_4)_3$

e. 35.4 g $CuSO_4$

43. a. 58.44 g NaCl

b. 36.04 g H_2O

c. 260 g $Ca(OH)_2$

d. 163 g $Ba(NO_3)_2$

44. 3.24×10^{22} atoms Au

45. 1.337×10^{24} formula units $ZnCl_2$

46. 9.396×10^{21} molecules naphthalene

47. 2.79×10^{24} atoms Al

48. 5.53 mol H_2O

49. a. 4.99×10^{-2} mol $(NH_4)_2SO_4$

b. 61 mol $Ca(OH)_2$

c. 7.49×10^{-2} mol H_2SO_4

50. 0.16 mol $C_{13}H_{18}O_2$

51. 7.25 mol $NaNO_2$

52. 7.236×10^{-3} mol Ca

53. 57.8 mol C_3H_8

54. 107.88 amu

55. 79.90 amu

56. 121.8 amu

57. 55.84 amu

253

58. a. LiCl, 42.39 g/mol

  b. Na_2SO_4, 142.05 g/mol

  c. CuCN, 89.57 g/mol

  d. C_3H_6, 42.09 g/mol

  e. $K_2Cr_2O_7$, 294.20 g/mol

  f. $Mg(NO_3)_2$, 148.32 g/mol

  g. 231.55 g/mol

  h. 112.10 g/mol

  i. S_4N_4, 184.32 g/mol

  j. $CsBr_3$, 372.60 g/mol

  k. 194.22 g/mol

59. 94.97 g/mol

60. 60.12 g/mol

61. 75.08 g/mol

62. $AgNO_3$

63. P_2O_3

64. C_2H_6O

65. $C_9H_{18}N_6$

66. C_6H_6

67. $Co_2C_8O_8$

68. $C_{18}H_{34}O_2$

69. $C_4H_4N_2$

70. a. 35.00% N, 5.05% H, and
      59.96% O

  b. 21.23% O and 78.77% Sn

  c. 13.35% Y, 41.23% Ba,
      28.62% Cu, and 16.81% O

71. 37.56% NH_4^+

72. $CaCO_3$, 40.04% Ca, 12.00% C,
    and 47.96% O

73. 22.57% N, 6.51% H, 19.35% C,
    51.56% O

Mixed Review

74. a. 0.050 mol Na_3PO_4

  b. 0.0406 mol $Ca(NO_3)_2$

  c. 0.128 mol SO_2

75. a. 0.00152 mol Na^+

  b. 0.0072 mol Ca^{2+}

61. What is the molar mass of the amino acid
glycine, $C_2H_5NO_2$?

**Sample Problem G Determining an Empirical
Formula from Percentage Composition**

62. A compound of silver has the following
analytical composition: 63.50% Ag, 8.25%
N, and 28.25% O. Calculate the empirical
formula.

63. An oxide of phosphorus is 56.34% phosphorus, and the rest is oxygen. Calculate the
empirical formula for this compound.

64. Determine the empirical formula for a compound made up of 52.11% carbon, 13.14%
hydrogen, and oxygen.

**Sample Problem H Determining a Molecular
Formula from an Empirical Formula**

65. The empirical formula of the anticancer
drug altretamine is $C_3H_6N_2$. The experimental molar mass is 210 g/mol. What is its
molecular formula?

66. Benzene has the empirical formula CH and
an experimental molar mass of 78 g/mol.
What is its molecular formula?

67. Determine the molecular formula for a
compound with the empirical formula
CoC_4O_4 and a molar mass of 341.94 g/mol.

68. Oleic acid has the empirical formula
$C_9H_{17}O$. If the experimental molar mass is
282 g/mol, what is the molecular formula of
oleic acid?

69. Pyrazine has the empirical formula C_2H_2N.
Its experimental formula mass is 80.1 amu.
What is the molecular formula?

**Sample Problem I Using a Chemical Formula
to Determine Percentage Composition**

70. Determine the percentage composition of
the following compounds:

  a. ammonium nitrate, NH_4NO_3, a common
    fertilizer

  b. tin(IV) oxide, SnO_2, an ingredient in
    fingernail polish

  c. $YBa_2Cu_3O_7$, a superconductor

71. What percentage of ammonium carbonate,
$(NH_4)_2CO_3$, an ingredient in smelling salts,
is the ammonium ion, NH_4^+?

72. Some antacids use compounds of calcium, a
mineral that is often lacking in the diet. What
is the percentage composition of calcium
carbonate, a common antacid ingredient?

73. Determine the percentage composition of
ammonium oxalate, $(NH_4)_2C_2O_4$.

MIXED REVIEW

74. Calculate the number of moles in each of
the following samples:

  a. 8.2 g of sodium phosphate

  b. 6.66 g of calcium nitrate

  c. 8.22 g of sulfur dioxide

75. There are exactly 1000 mg in 1 g. A cup of
hot chocolate has 35.0 mg of sodium ions,
Na^+. One cup of milk has 290 mg of calcium
ions, Ca^{2+}.

  a. How many moles of sodium ions are in
    the cup of hot chocolate?

  b. How many moles of calcium ions are in
    the milk?

76. How many atoms are there in a platinum
ring made of 0.0466 mol of platinum?

77. Cyclopentane has the molecular formula
C_5H_{10}. How many moles of hydrogen atoms
are there in 4 moles of cyclopentane?

78. A 1.344 g sample of a compound contains
0.365 g Na, 0.221 g N, and 0.758 g O. What is
its percentage composition? Calculate its
empirical formula.

79. The naturally occurring silicon in sand has
three isotopes; 92.23% is made up of atoms
with a mass of 27.9769 amu, 4.67% is made
up of atoms with a mass of 28.9765 amu, and
3.10% is made up of atoms with a mass of
29.9738 amu. Calculate the average atomic
mass of silicon.

80. How many atoms of Fe are in the formula
Fe_3C? How many moles of Fe are in one
mole of Fe_3C?

81. Shown below are the structures for two sugars, glucose and fructose.

a. What is the molar mass of glucose?

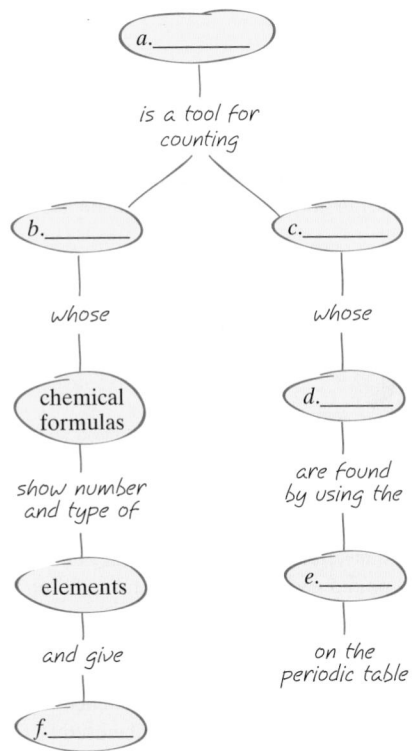

glucose

b. What is the molar mass of fructose?

fructose

82. Chlorine gas is a diatomic molecule, Cl_2. There are 6.00 mol of chlorine atoms in a sample of chlorine gas. How many moles of chlorine gas molecules is this?

83. Which yields a higher percentage of pure aluminum per gram, aluminum phosphate or aluminum chloride?

84. a. Calculate the molar mass of trinitrotoluene, TNT. The chemical formula for TNT is $C_6H_2CH_3(NO_2)_3$.

b. Determine the percentage composition of TNT.

CRITICAL THINKING

85. Use **Skills Toolkit 1** and **Skills Toolkit 3** to make your own graphic model for finding the mass of one atom, reported in grams.

86. Your calculation of the percentage composition of a compound gives 66.9% C and 29.6% H. Is the calculation correct? Explain.

87. Imagine you are a farmer, using NH_3 and NH_4NO_3 as sources of nitrogen. NH_3 costs $0.50 per kg, and NH_4NO_3 costs $0.25 per kg. Use percentage composition to decide which is the best buy for your money.

ALTERNATIVE ASSESSMENT

88. Research methods scientists initially used to find Avogadro's number. Then compare these methods with modern methods.

89. The most accurate method for determining the mass of an element involves a *mass spectrometer*. This instrument is also used to determine the isotopic composition of a natural element. Find out more about how a mass spectrometer works. Draw a model of how it works. Present the model to the class.

CONCEPT MAPPING

90. Use the following terms to complete the concept map below: *atoms, average atomic mass, molecules, mole, percentage composition,* and *molar masses.*

```
            a._____
                |
          is a tool for
            counting
           /          \
    b._____      c._____
        |                |
      whose            whose
        |                |
    chemical          d._____
    formulas             |
        |            are found
  show number        by using the
  and type of            |
        |             e._____
    elements             |
        |             on the
    and give        periodic table
        |
    f._____
```

76. 2.81×10^{22} atoms Pt

77. 40 mol H

78. 27.2% Na, 16.4% N, and 56.4% O; $NaNO_3$

79. 28.09 amu

80. 3 atoms Fe; 3 mol Fe

81. a. 180.18 g/mol

   b. 180.18 g/mol

82. 3.00 mol Cl_2

83. $AlPO_4$ is 22.12% Al, while $AlCl_3$ is 20.23% Al, so aluminum phosphate has more aluminum per gram.

84. a. 227.15 g/mol

   b. 37.01% C, 2.22% H, 18.50% N, and 42.26% O

Critical Thinking

85. Answers may vary. The mass of one atom can be found by dividing the molar mass of an element by Avogadro's number.

86. No; The total does not equal 100% and there is no whole number combination of carbon and oxygen atoms that gives that percentage composition.

87. NH_3 gives 82.22% N, and NH_4NO_3 gives 35.00% N. The cost of the compound divided by the percentage N gives cost per kg N. NH_3 gives $0.61/kg N and NH_4NO_3 gives $0.71/kg N. Thus, NH_3 is the better buy.

Alternative Assessment

88. Historical results might include those of Avogadro, who worked with gases, and those of Jean Baptiste Perrin, who measured the displacement of colloidal particles and first calculated a value of Avogadro's constant that is close to its currently accepted value.

89. Student research should show that in a mass spectrometer, a beam of electrons ionizes a gaseous-element sample. A magnet deflects these ions according to their mass. The results are recorded on a detector, such as a photographic plate.

Concept Mapping

90. a. mole

 b. molecules

 c. atoms

 d. molar masses

 e. average atomic mass

 f. percentage composition

Focus on Graphing

91. The slices represent the mass percentage of each element in the compound.

92. Different compounds that are made up of the same elements have different mass ratios of elements.

93. Iron(III) oxide has a higher percentage of oxygen.

94. She can make more Fe_2O_3.

95. a. 81.68% C and 18.32% H

 b. The chart should have a slice that is 294° to represent carbon and a slice that is 66° to represent hydrogen.

 c. The percentage of hydrogen gets smaller and the percentage for carbon gets larger as you go from methane to ethane to propane.

Technology and Learning

96. a. 233.21 g/mol

 b. 278.1 g/mol

 c. 80.06 g/mol

FOCUS ON GRAPHING

Study the graphs below, and answer the questions that follow.
For help in interpreting graphs, see Appendix B, "Study Skills for Chemistry."

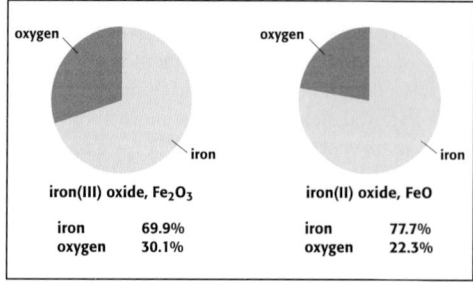

iron(III) oxide, Fe_2O_3	iron(II) oxide, FeO
iron 69.9%	iron 77.7%
oxygen 30.1%	oxygen 22.3%

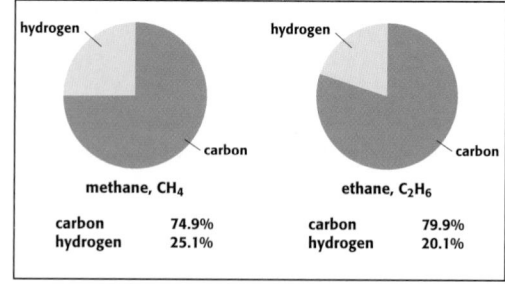

methane, CH_4	ethane, C_2H_6
carbon 74.9%	carbon 79.9%
hydrogen 25.1%	hydrogen 20.1%

91. What do the slices of the pie represent?

92. What do the pie charts show about different compounds that are made up of the same elements?

93. Which has a higher percentage of oxygen, iron(II) oxide or iron(III) oxide?

94. Carlita has 30.0 g of oxygen and 70.0 g of iron. Can she make more FeO or Fe_2O_3 using only the reactants that she has?

95. a. Determine the percentage composition of propane, C_3H_8.

 b. Make a pie chart for propane using a protractor to draw the correct sizes of the pie slices. (Hint: A circle has 360°. To draw the correct angle for each slice, multiply each percentage by 360°.)

 c. Compare the charts for methane, ethane, and propane. How do the slices for carbon and hydrogen differ for each chart?

TECHNOLOGY AND LEARNING

96. Graphing Calculator

Calculating the Molar Mass of a Compound

The graphing calculator can run a program that calculates the molar mass of a compound given the chemical formula for the compound. This program will prompt for the number of elements in the formula, the number of atoms of each element in the formula, and the atomic mass of each element in the formula. It then can be used to find the molar masses of various compounds.

Go to Appendix C. If you are using a TI-83 Plus, you can download the program **MOLMASS** and data sets and run the application as directed. If you are using another calculator, your teacher will provide you with the keystrokes and data sets to use. After you have graphed the data, answer the questions below.

a. What is the molar mass of $BaTiO_3$?

b. What is the molar mass of $PbCl_2$?

c. What is the molar mass of NH_4NO_3?

Answers

1. b
2. c
3. a
4. b
5. d
6. c
7. a
8. b
9. c
10. b
11. c
12. b

1. Element A has two isotopes. One has an atomic mass of 120 amu and constitutes 60%; the other has an atomic mass of 122 amu and constitutes 40%. The average atomic mass is
   a. less than 120.　　c. exactly 121.
   b. less than 121.　　d. more than 121.

2. One mole of NaCl has 6.022×10^{23}
   a. atoms.　　c. formula units.
   b. molecules.　　d. electrons.

3. The molecular formula for acetylene is C_2H_2. The molecular formula for benzene is C_6H_6. The empirical formula for both is
   a. CH.　　c. C_6H_6.
   b. C_2H_2.　　d. $(CH)_2$.

4. If the empirical formula of a compound is known, then
   a. the compound's true formula is also known.
   b. the compound's percentage composition can be calculated.
   c. the arrangement of the compound's atoms is also known.
   d. the molecular mass of the compound can be determined.

5. To calculate the average atomic mass of an element, which of the following must you know?
   a. the atomic number and the atomic mass of each isotope
   b. the atomic mass and the symbol of each isotope
   c. the number of atoms present in each isotope
   d. the atomic mass and relative abundance of each isotope

6. Which of the following compounds has the highest percentage of oxygen?
   a. CH_4O　　c. H_2O
   b. CO_2　　d. Na_2CO_3

7. The units for molar mass are
   a. g/mol.　　c. g/atoms.
   b. atoms/mol.　　d. mol/g.

8. A compound's molar mass is calculated by
   a. adding the number of atoms present in the compound.
   b. adding the atomic masses of all the atoms in the compound and expressing that value in g/mol.
   c. multiplying Avogadro's number times the number of atoms present in the compound.
   d. dividing the number of grams of the sample by Avogadro's number.

9. The empirical formula for a compound that is 1.2% H, 42.0% Cl, and 56.8% O is
   a. HClO.　　c. $HClO_3$.
   b. $HClO_2$.　　d. $HClO_4$.

10. Which of the following shows the percentage composition of H_2SO_4?
    a. 2.5% H, 39.1% S, 58.5% O
    b. 2.1% H, 32.7% S, 65.2% O
    c. 28.6% H, 14.3% S, 57.1% O
    d. 33.3% H, 16.7% S, 50% O

11. You know the empirical formula for a compound. What further data do you need to determine the molecular formula?
    a. Avogadro's number
    b. the formula unit
    c. the molar mass
    d. the number of isotopes

12. To calculate the number of moles of a compound in a sample, which of the following must you know?
    a. atomic mass and symbol of each isotope
    b. mass and molar mass of the compound
    c. atomic mass and Avogadro's number
    d. formula and Avogadro's number

257

Continuation of Answers

Answers to Section 1 Review, *continued*

9. a. 2.86×10^{-7} g He

  b. 15.22 g CH_4

  c. 200.5 g Ca^{2+}

10. a. 4.745×10^{21} ions I^-

  b. 3.3×10^{22} ions Cu^{2+}

  c. 3.97×10^{22} molecules SO_2

11. 206.3 g ibuprofen

12. a. 26.7 g Ca

  b. 50. g boron-11

  c. 7.032×10^{-4} g Na^+

13. a. 1.204×10^{24} molecules H_2

  b. 1.21×10^{23} molecules HF

  c. 2.7×10^{24} molecules $C_6H_{12}O_6$

14. All atoms must have the same mass to be part of the definition. An ordinary sample of carbon would contain carbon atoms of different masses because carbon has two common isotopes.

15. Multiply the number of atoms by 1 mol/ 6.022×10^{23} atoms. Or, divide the number of atoms by 6.022×10^{23} atoms/1 mol.

Answers to Section 2 Review, *continued*

10. a. 168.35 g/mol

  b. 122.55 g/mol

  c. 180.18 g/mol

  d. 132.08 g/mol

  e. 75.08 g/mol

11. a. SrS, 119.69 g/mol, 1.76×10^{-2} mol SrS

  b. PF_3, 87.97 g/mol, 2.40×10^{-2} mol PF_3

  c. $Zn(C_2H_3O_2)_2$, 183.49 g/mol, 1.15×10^{-2} mol $Zn(C_2H_3O_2)_2$

  d. $Hg(BrO_3)_2$, 456.39 g/mol, 4.62×10^{-3} mol $Hg(BrO_3)_2$

  e. $Ca(NO_3)_2$, 164.10 g/mol, 1.29×10^{-2} mol $Ca(NO_3)_2$

12. a. 158.18 g/mol, 790.90 g $Ca(C_2H_3O_2)_2$

  b. 357.49 g/mol, 1787.4 g $Fe_3(PO_4)_2$

  c. 183.20 g/mol, 916.00 g $C_7H_5NO_3S$

  d. 180.17 g/mol, 900.85 g $C_9H_8O_4$

13. Isotopic ratios of oxygen can vary in different parts of the world making it more difficult to determine a precise average atomic mass. Because fluorine has only one isotope, it does not vary throughout the world.

14. Potassium bromide is an ionic compound and forms an infinite crystal, so its formula reflects the lowest ratio of ions, KBr.

15. Hydrogen peroxide is a molecular compound, so the formula shows the numbers of atoms of each element in a molecule.

16. a. 8 H atoms

  b. 4.818×10^{24} H atoms

Chemical Equations and Reactions
Chapter Planning Guide

PACING	CLASSROOM RESOURCES	LABS, ACTIVITIES, AND DEMONSTRATIONS
BLOCK 1 · 45 min pp. 258–259 **Chapter Opener**		**SE Start-Up Activity** Observing a Chemical Reaction, p. 259 ◆
BLOCK 2 · 45 min pp. 260–266 **Section 1** Describing Chemical Reactions	**OSP Career Extension** Book Conservationist GENERAL	**TE Demonstration** Energy: In or Out?, p. 262 ◆ **TE Demonstration** Production of Copper(II) Sulfide, p. 263 ◆ **TE Group Activity,** p. 265 BASIC **TE Demonstration** The Importance of State, p. 265 ◆ GENERAL **CRF Observation Lab** Evidence for a Chemical Change ◆ GENERAL
BLOCKS 3 & 4 · 90 min pp. 267–274 **Section 2** Balancing Chemical Equations		**TE Demonstration** Electrolysis of Water, p. 268 ◆
BLOCKS 5 & 6 · 90 min pp. 275–285 **Section 3** Classifying Chemical Reactions	**OSP Supplemental Reading** *The Pale Horse* ADVANCED **TT Identifying Reactions and Predicting Products** *	**TE Demonstration** Combustion, p. 276 ◆ GENERAL **TE Demonstration** Displacement of Ag with Cu, p. 280 ◆ GENERAL **TE Demonstration** Relative Activity, p. 281 ◆ GENERAL **SE QuickLab** Balancing Equations Using Models, p. 282 ◆ GENERAL **CRF Datasheets for In-Text Labs** * **TE Group Activity,** p. 283 BASIC **CRF Microscale Lab** Simple Qualitative Analysis ◆ ADVANCED
BLOCK 7 · 45 min pp. 286–290 **Section 4** Writing Net Ionic Equations	**TT Writing a Net Ionic Equation** * **SE Consumer Focus** Fire Extinguishers	**TE Group Activity,** p. 287 ◆ ADVANCED **TE Demonstration** Net Ionic Equations and Conservation of Mass, p. 288 ◆ **CRF Microscale Lab** Reacting Ionic Species in Aqueous Solution ◆ ADVANCED

BLOCKS 8 & 9 · 90 min   Chapter Review and Assessment Resources

SE Chapter Review, pp. 292–298
SE Standardized Test Prep, p. 299
CRF Chapter Test *
OSP Test Generator
CRF Test Item Listing *
OSP Scoring Rubrics and Classroom Management Checklists

Holt Chemistry: Online Resources

Visit **go.hrw.com** for a variety of free resources related to this textbook. Enter the keyword **HW4 HOME**.

Holt Online Learning

Students can access interactive problem solving help and active visual concept development with the *Holt Chemistry* Online Edition available at **www.hrw.com**

student CNN News

cnnstudentnews.com

Find the latest chemistry news, lesson plans, and activities related to important scientific events.

KEY

TE	Teacher Edition	**CRF**	Chapter Resource File	*	Also on One-Stop Planner
SE	Student Edition	**TT**	Teaching Transparencies	♦	Requires Advance Prep
OSP	One-Stop Planner				

Compression guide:
To shorten your instruction because of time limitations, omit blocks 1, 7, and 8.

PROBLEM SOLVING AND PRACTICE	SECTION REVIEW AND ASSESSMENT	STANDARDS CORRELATION
		National Science Education Standards
	TE Quiz, p. 266 GENERAL **CRF** Quiz * **TE** Reteaching, p. 266 BASIC **SE** Section Review, p. 266 **CRF** Concept Review *	PS 3a, PS 3b

SE Skills Toolkit 1 Balancing Chemical Equations, p. 268
SE Sample Problem A Balancing an Equation, p. 269
   SE Practice, p. 269 GENERAL   **TE** Homework, p. 269 GENERAL
SE Sample Problem B The Odd-Even Technique, p. 271
   SE Practice, p. 271 GENERAL   **TE** Homework, p. 271 GENERAL
SE Sample Problem C Polyatomic Ions as a Group, p. 273
   SE Practice, p. 273 GENERAL   **TE** Homework, p. 273 GENERAL
	TE Quiz, p. 274 GENERAL
	CRF Quiz *
	TE Reteaching, p. 274 BASIC
	SE Section Review, p. 274
	CRF Concept Review *	

SE Sample Problem D Predicting Products, p. 279
   SE Practice, p. 279 GENERAL   **TE** Homework, p. 279 GENERAL
SE Skills Toolkit 2 Using the Activity Series, p. 281
SE Sample Problem E Determining Products by Using the Activity Series, p. 282
   SE Practice, p. 282 GENERAL   **TE** Homework, p. 282 GENERAL
SE Skills Toolkit 3 Identifying Reactions and Predicting Products, p. 284
	TE Quiz, p. 285 GENERAL
	CRF Quiz *
	TE Reteaching, p. 285 BASIC
	SE Section Review, p. 285
	CRF Concept Review *
		PS 3a, PS 3b

SE Skills Toolkit 4 Writing Net Ionic Equations, p. 288	**TE** Quiz, p. 289 GENERAL **CRF** Quiz * **TE** Reteaching, p. 289 BASIC **SE** Section Review, p. 289 **CRF** Concept Review *	

www.scilinks.org

Topic: Chemical Reactions
SciLinks code: HW4029

Topic: Combustion
SciLinks code: HW4033

Topic: Activity Series
SciLinks code: HW4004

Topic: Reaction Types
SciLinks code: HW4163

Topic: Precipitation Reactions
SciLinks code: HW4160

Topic: Fire Extinguishers
SciLinks code: HW4059

Technology Resources

ChemFile Interactive Tutor CD-Rom

Module 5: Stoichiometry

Topic: Balancing Equations

Overview

This chapter examines different types of evidence of chemical reactions. Students learn to describe chemical reactions by using word equations and unbalanced and balanced formula equations. Students learn how mass is conserved in chemical reactions and how to relate conservation of mass to a balanced equation. Different types of chemical reactions are described, and students learn to predict products for each type. Students also learn to distinguish between and write total and net ionic equations.

Assessing Prior Knowledge

Check for Content Knowledge
Students should be familiar with the following topics:

• reactants and products

• ions

• naming substances

• writing chemical formulas

Using the Figure

The space shuttle is powered by chemical reactions. The main engines on the shuttle are powered by a reaction between oxygen and hydrogen. The boosters on the main fuel tank use a solid fuel comprised of ammonium perchlorate and powdered aluminum that reacts to produce aluminum oxide and ammonium chloride. As these fuels undergo chemical reactions, energy is released in the form of heat and light. When hydrogen burns, gas in the form of water vapor is produced. Both an energy change and production of a gas are signs of a chemical reaction.

CHAPTER

8

CHEMICAL EQUATIONS AND REACTIONS

258

Standards Correlations

National Science Education Standards

PS 3a: Chemical reactions occur all around us, for example in health care, cooking, cosmetics, and automobiles. Complex chemical reactions involving carbon-based molecules take place constantly in every cell in our bodies. (Sections 1, 3)

PS 3b: Chemical reactions may release or consume energy. Some reactions such as the burning of fossil fuels release large amounts of energy by losing heat and by emitting light. Light can initiate many chemical reactions such as photosynthesis and the evolution of urban smog. (Sections 1, 3)

With an eruption of flames and hot gases, a space shuttle leaves the ground on its way into orbit. The brightness and warmth of the flame clearly indicates that a change is occurring. From the jets of the shuttle itself, blue flames emerge. These flames are the result of a reaction between hydrogen and oxygen. The sight is awesome and beautiful. In this chapter, you will learn about chemical reactions, such as the ones that send a space shuttle into space.

START-UP ACTIVITY

Observing a Chemical Reaction

SAFETY PRECAUTIONS

PROCEDURE

1. Place about 5 g (1 tsp) of **baking soda** into a **sealable plastic bag.**

2. Place about 5 mL (1 tsp) of **vinegar** into a **plastic film canister.** Secure the lid.

3. Place the canister into the bag. Squeeze the air out of the bag, and tightly seal the bag.

4. Use a **balance** to determine the total mass of the bag and the bag's contents. Make a note of this value.

5. Open the canister without opening the bag, and allow the vinegar and baking soda to mix.

6. When the reaction has stopped, measure and record the total mass of the bag and the bag's contents.

ANALYSIS

1. What evidence shows that a chemical reaction has taken place?

2. Compare the masses of the bag and its contents before and after the reaction. What does this result demonstrate about chemical reactions?

Pre-Reading Questions

1. What are some signs that a chemical change may be taking place?

2. What are the reactants of a reaction? What are the products of a reaction?

3. Describe the law of conservation of mass.

4. Define the terms *synthesis* and *decomposition,* and describe what you would expect to happen in each of these types of reactions.

CONTENTS

259

Overview

Before beginning this section, review with your students the Objectives listed in the Student Edition. This section describes evidence that suggests that a chemical reaction has occurred and evidence that proves a chemical reaction has taken place. It also shows how to describe a chemical reaction by word and formula equations.

🔊 Bellringer

Have students list observations that they think indicate that a chemical reaction has taken place in the following situations: a cut apple turns brown, an egg changes when it cooks, a log burns, and a car rusts.

Motivate

Discussion ——— GENERAL

Have students look at the photographs of chemical reactions displayed throughout the chapter. Have students write a list of the evidence they see that changes are occurring, and discuss other evidence that is not shown, but can be inferred to happen in other reactions. **Ans.** Students should note bubbling, color change, energy released as heat and as light, and a solid forming. Evidence not readily apparent but that could be inferred includes energy being absorbed and a liquid forming. **LS** Visual

---

SECTION

1 Describing Chemical Reactions

KEY TERMS
• chemical reaction
• chemical equation

OBJECTIVES

① **List** evidence that suggests that a chemical reaction has occurred and evidence that proves that a chemical reaction has occurred.

② **Describe** a chemical reaction by using a word equation and a formula equation.

③ **Interpret** notations in formula equations, such as those relating to states of matter or reaction conditions.

Chemical Change

You witness chemical changes taking place in iron that rusts, in milk that turns sour, and in a car engine that burns gasoline. The processes of digestion and respiration in your body are the result of chemical changes.

A **chemical reaction** is the process by which one or more substances change into one or more new substances whose chemical and physical properties differ from those of the original substances. In any chemical reaction, the original substances, which can be elements or compounds, are known as *reactants*. The substances created are called *products*. A common example of a chemical reaction is shown in **Figure 1**.

chemical reaction

the process by which one or more substances change to produce one or more different substances

Evidence of a Chemical Reaction

It's not always easy to tell that a chemical change is happening, but there are some signs to look for, which are summarized in **Table 1**. For example, certain signs indicate that wood burning in a campfire is undergoing a chemical change. Smoke rises from the wood, and a hissing sound is made. Energy that lights up the campsite and warms the air around the fire is released. The surface of the wood changes color as the wood burns. Eventually, all that remains of the firewood is a grey, powdery ash.

In **Figure 2**, you can see copper reacting with nitric acid. Again, several clues suggest that a chemical reaction is taking place. The color of the solution changes from colorless to blue. The solution bubbles and fizzes as a gas forms. The copper seems to be used up as the reaction continues.

Sometimes, the evidence for a chemical change is indirect. When you place a new battery in a flashlight, you don't see any changes in the battery. However, when you turn the flashlight on, electrical energy causes the filament in the bulb to heat up and emit light. This release of electrical energy is a clue that a chemical reaction is taking place in the battery. Although these signs suggest a change may be chemical, they do not prove that the change is chemical.

Figure 1
Chemical changes occur as wood burns. Two products formed are carbon dioxide and water.

MISCONCEPTION /// ALERT \\\

Chemical Reaction or Not? Students might think that any change of a material involves a chemical reaction. Ask them to brainstorm a list of changes that a material can undergo. Then have them determine which of these changes do not involve permanent changes in the identity of the material. Such changes might include cutting paper or melting ice. Point out that these changes are physical changes and do not involve chemical reactions.

Chapter Resource File

• **Lesson Plan**

🖱 **One-Stop** Planner CD-ROM

• **Career Extension**
Real-World Connections Worksheet 2: Book Conservationist
Assign this worksheet to emphasize relevant applications of text concepts.

Table 1 Evidence of Chemical Change

Changes in energy	Formation of new substances
release of energy as heat	formation of a gas
release of energy as light	formation of a precipitate (an insoluble solid)
production of sound	change in color
reduction or increase of temperature	change in odor
absorption or release of electrical energy	

Chemical Reaction Versus Physical Change

For proof of a chemical change, you need a chemical analysis to show that at least one new substance forms. The properties of the new substance—such as density, melting point, or boiling point—must differ from those of the original substances.

Even when evidence suggests a chemical change, you can't be sure immediately. For example, when paints mix, the color of the resulting paint differs from the color of the original paints. But the change is physical—the substances making up the paints have not changed. When you boil water, the water absorbs energy and a gas forms. But the gas still consists of water molecules, so a new substance has not formed. Even though they demonstrate some of the signs of a chemical change, all changes of state, including evaporation, condensation, melting, and freezing, are physical changes.

Figure 2
When copper reacts with nitric acid, several signs of a reaction are seen. A toxic, brown gas is produced, and the color of the solution changes.

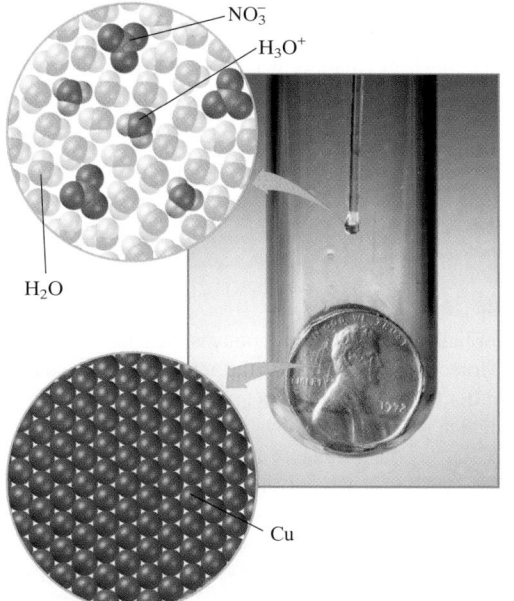
NO$_3^-$
H$_3$O$^+$
H$_2$O
Cu

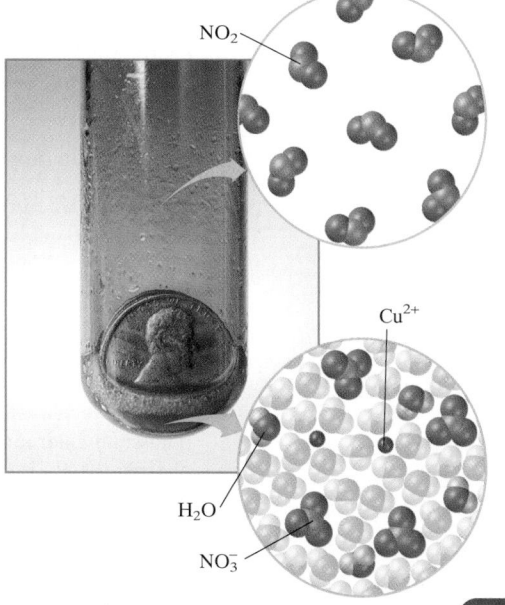
NO$_2$
Cu^{2+}
H$_2$O
NO$_3^-$

261

Assimilating Knowledge Before students read this chapter, have them write a short list of all the things they already know (or think they know) about chemical reactions and chemical equations. Then have students list the things they might want to know about chemical reactions and equations. If students are having trouble, offer the following suggestions to get them on track:

• What do you think happens during a chemical reaction?

• How do you know whether two substances will react with each other or not?

• What characteristics are common to all equations?

Later, when students are reviewing for the chapter test, have them read over their list and correct or expand upon their answers. **LS Logical**

Using the Table

Refer students to **Table 1.** Be sure students know that a precipitate is a solid formed when two solutions are mixed. Also point out that many events that do not involve chemical reactions can exhibit evidence listed in the table. For example, production of a sound when you speak is not evidence of a chemical reaction. Neither is the appearance of a gas that had been dissolved in a liquid and comes out of solution, as when air bubbles form as you heat a pot of water.

Teaching Tip

Water Encourage students to refer to water in the liquid state as *liquid water* instead of just *water.* Point out that ice and steam are also water, and using specific reference terms increases clarity.

MISCONCEPTION ///ALERT

Release of Energy as Heat Students might think that the release or the absorption of energy always indicates a chemical reaction. Point out that many physical changes absorb or release energy. These processes include melting and boiling, which absorb energy, and freezing and condensing, which release energy.

INCLUSION Strategies

• *Gifted and Talented*

Ask students to create a chart that lists an example of a chemical change that exhibits each change in energy in **Table 1.** Then, have students make a second chart that lists an example of a chemical change that exhibits each formation of a new substance in **Table 1.** Students may use the textbook, reference books, and the Internet to complete the charts. The examples may be presented in small groups or to the entire class.

Demonstration

Energy: In or Out?

(Approximate time: 10 minutes)

This demonstration shows energy changes for the solution process.

1. Add 50 mL of water to each of two 125 mL Erlenmeyer flasks. Have two students measure the temperature of the water in the flasks and record each temperature on the board.

2. Add 2 g of solid NaOH to one flask, and add 10 g of solid NH_4Cl to the other flask.

3. Let the solids dissolve for 2 minutes. Stir the solutions with stirring rods to speed up the solution process. Have the students measure the new temperatures and record them on the board. You may want students to carefully feel the outside of each flask as you hold it.

Safety Caution: The teacher and the participating students should wear lab aprons, safety goggles, and safety gloves. The rest of the class must be 3 m from the demonstration. In case of an alkali spill, dilute the spill with water and, while wearing gloves, soak up the spill with cloth or paper towels. Do not stir the solutions with the thermometers.

Disposal: Neutralize the NaOH solution with vinegar, and flush the liquid mixture down the drain. Pour the NH_4Cl solution down the drain with plenty of water. If towels were used to clean up an alkali spill, rinse the towels, and neutralize the rinsings with vinegar.

Figure 3
Energy is released as the elements sodium and chlorine react to form the compound sodium chloride. Breaking down water into hydrogen and oxygen requires the input of electrical energy.

 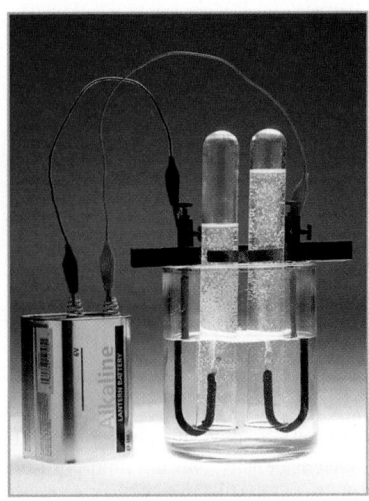

▶ internet connect ▬▬
www.scilinks.org
Topic: Chemical Reactions
SciLinks code: HW4029
SCiLINKS. Maintained by the National Science Teachers Association

Reactions and Energy Changes

Chemical reactions either release energy or absorb energy as they happen, as shown in **Figure 3.** A burning campfire and burning natural gas are examples of reactions that release energy. Natural gas, which is mainly methane, undergoes the following reaction:

$$\text{methane} + \text{oxygen} \longrightarrow \text{carbon dioxide} + \text{water} + \text{energy}$$

Notice that when energy is released, it can be considered a product of the reaction.

If the energy required is not too great, some other reactions that absorb energy will occur because they take energy from their surroundings. An example is the decomposition of dinitrogen tetroxide, which occurs at room temperature.

$$\text{dinitrogen tetroxide} + \text{energy} \longrightarrow \text{nitrogen dioxide}$$

Notice that when energy is absorbed, it can be considered a reactant of the reaction.

Reactants Must Come Together

You cannot kick a soccer ball unless your shoe contacts the ball. Chemical reactions are similar. Molecules and atoms of the reactants must come into contact with each other for a reaction to take place. Think about what happens when a safety match is lighted, as shown in **Figure 4.** One reactant, potassium chlorate ($KClO_3$) is on the match head. The other reactant, phosphorus, P_4, is on the striking surface of the matchbox. The reaction begins when the two substances come together by rubbing the match head across the striking surface. If the reactants are kept apart, the reaction will not happen. Under most conditions, safety matches do not ignite by themselves.

262

REAL-WORLD
CONNECTION ── **ADVANCED**

Bleach Safety Have students examine warning labels on bottles of liquid bleach, especially warnings not to mix the bleach with ammonia. Have students research the active ingredients in these products, what forms from their reaction, and statistics on how many people are injured each year from this reaction. Students could make an informative pamphlet to present their findings. **Ans.** Their results should show that sodium hypochlorite and ammonia gas react to form toxic NCl_3 and toxic chloramines, such as NH_2Cl. **LS Interpersonal**

MISCONCEPTION
ALERT

Bonds and Energy Students often do not recognize that bond breaking always requires energy. The overall process of bond breaking (which requires energy) and bond formation (which releases energy) determines whether a reaction absorbs energy (an endothermic change) or releases energy (an exothermic change).

Figure 4
The reactants KClO₃ (on the match head) and P₄ (on the striking surface) must be brought together for a safety match to ignite.

Constructing a Chemical Equation

You know that symbols represent elements, and formulas represent compounds. In the same way, equations are used to represent chemical reactions. A correctly written **chemical equation** shows the chemical formulas and relative amounts of all reactants and products. Constructing a chemical equation usually begins with writing a word equation. This word equation contains the names of the reactants and of the products separated by an arrow. The arrow means "forms" or "produces." Then, the chemical formulas are substituted for the names. Finally, the equation is balanced so that it obeys the law of conservation of mass. The numbers of atoms of each element must be the same on both sides of the arrow.

chemical equation

a representation of a chemical reaction that uses symbols to show the relationship between the reactants and the products

Writing a Word Equation or a Formula Equation

The first step in writing a chemical equation is to write a word equation. To write the word equation for a reaction, you must write down the names of the reactants and separate the names with plus signs. An arrow is used to separate the reactants from the products. Then, the names of the products are written to the right of the arrow and are separated by plus signs. The word equation for the reaction of methane with oxygen to form carbon dioxide and water is written as follows:

$$\text{methane} + \text{oxygen} \longrightarrow \text{carbon dioxide} + \text{water}$$

To convert this word equation into a formula equation, use the formulas for the reactants and for the products. The formulas for methane, oxygen, carbon dioxide, and water replace the words in the word equation to make a formula equation. The word *methane* carries no quantitative meaning, but the formula CH_4 means a molecule of methane. This change gives the unbalanced formula equation below. The question marks indicate that we do not yet know the number of molecules of each substance.

$$?CH_4 + ?O_2 \longrightarrow ?CO_2 + ?H_2O$$

263

Teaching Tip

Including Energy Although the major focus of this chapter is developing the skills associated with writing and balancing equations, students should recognize that energy is either released or absorbed in the overall process and can therefore be included in the equation for a reaction.

Demonstration

Production of Copper(II) Sulfide

Show students how the identity of a substance changes during a chemical reaction.

1. Cut a piece of copper foil about 1×10 cm.

2. Add flowers of sulfur to a large Pyrex test tube to a depth of about 2 cm.

3. Clamp the test tube vertically onto a ring stand placed in a fume hood. Heat the test tube vigorously.

4. Darken the room so that the reaction will be seen easily.

5. Using crucible tongs, insert the copper strip into the test tube, and wait for the sulfur vapors to reach the copper. A bright light indicates the formation of copper(II) sulfide.

6. Remove the copper strip, and let it cool for inspection. Note how the copper and sulfur have changed to copper(II) sulfide.

Safety Caution: Safety goggles and a lab apron must be worn. Students must be at least 3 m from the reaction. Steps 4 and 5 must be performed in a properly operating fume hood. If the sulfur vapors catch on fire in the test tube, cover the tube with a glass plate until the fire goes out.

Disposal: Clean the copper with steel wool, and save it for reuse. Dispose of the CuS, the steel fragments, and the ruined test tube in the trash can.

HISTORY — CONNECTION

The Statue of Liberty The National Park Service initiated the restoration of the Statue of Liberty in 1981, the hundredth anniversary of the statue's construction. The project was completed in 1986. Many chemical factors involving both the statue's deterioration and its endurance had to be considered. For instance, while formation of iron(III) oxide led to the rusting of the iron ribs, the formation of the patina on the copper skin protected the skin from any further degradation. This protection was the result of a balanced mixture of the compounds brochantite, $CuSO_4 \cdot 3Cu(OH)_2$, and antlerite, $CuSO_4 \cdot 2Cu(OH)_2$. Yet changes observed in the proportions of these two compounds suggest that the patina, and thus the few millimeters of copper skin beneath it, might be endangered. The statue's north side has a higher concentration of antlerite, which is more susceptible to dust erosion. Debate continues on whether or not this degradation is caused by acid precipitation.

Equations and Reaction Information

A chemical equation indicates the amount of each substance in the reaction. But it can also provide other valuable information about the substances or conditions, such as temperature or pressure, that are needed for the reaction.

Equations Are Like Recipes

Imagine that you need to bake brownies for a party. Of course, you would want to follow a recipe closely to be sure that your brownies turn out right. You must know which ingredients to use and how much of each ingredient to use. Special instructions, such as whether the ingredients should be chilled or at room temperature when you mix them, are also provided in the recipe.

Chemical equations have much in common with a recipe. Like a recipe, any instructions shown in an equation can help you or a chemist be sure the reaction turns out the way it should, as shown in **Figure 5.** A balanced equation indicates the relative amounts of reactants and products in the reaction. As discussed below, even more information can be shown by an equation.

Figure 5
The equation for the reaction between baking soda and vinegar provides a lot of information about the reaction.

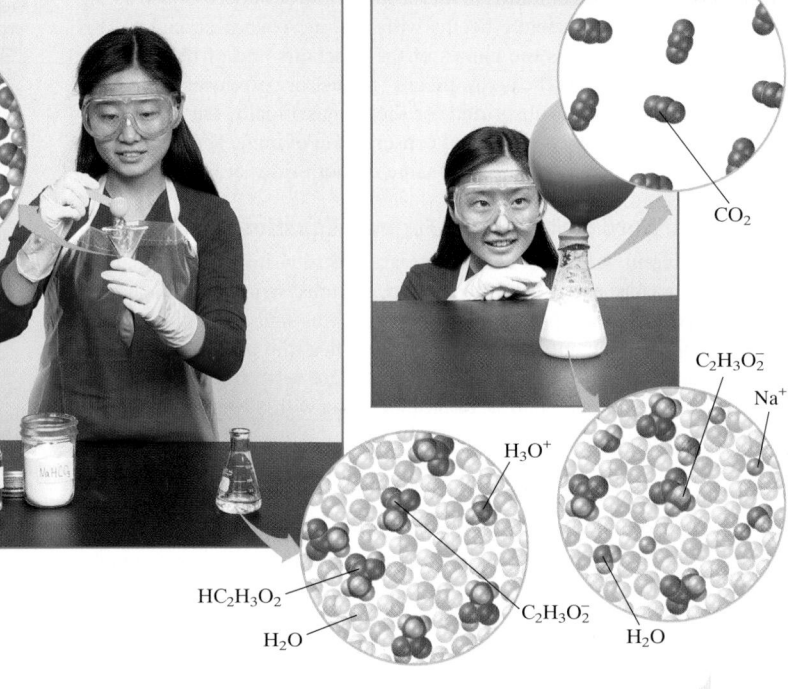

$$NaHCO_3(s) + HC_2H_3O_2(aq) \longrightarrow NaC_2H_3O_2(aq) + CO_2(g) + H_2O(l)$$

Equations Can Show Physical States and Reaction Conditions

The recipe for brownies will specify whether each ingredient should be used in a solid or liquid form. The recipe also may state that the batter should bake at 400°F for 20 min. Additional instructions tell what to do if you are baking at high elevation. Chemical equations are similar. Equations for chemical reactions often list the physical state of each reactant and the conditions under which the reaction takes place.

Look closely at the equation that represents the reaction of baking soda with vinegar.

$$NaHCO_3(s) + HC_2H_3O_2(aq) \longrightarrow NaC_2H_3O_2(aq) + CO_2(g) + H_2O(l)$$

Baking soda, sodium hydrogen carbonate, is a solid, so the formula is followed by the symbol (s). Vinegar, the other reactant, is acetic acid dissolved in water—an aqueous solution. Sodium acetate, one of the products, remains in aqueous solution. So, the formulas for vinegar and sodium acetate are followed by the symbol (aq). Another product, carbon dioxide, is a gas and is marked with the symbol (g). Finally, water is produced in the liquid state, so its formula is followed by the symbol (l).

When information about the conditions of the reaction is desired, the arrow is a good place to show it. Several symbols are used to show the conditions under which a reaction happens. Consider the preparation of ammonia in a commercial plant.

$$N_2(g) + 3H_2(g) \underset{catalyst}{\overset{350°C,\ 25\,000\ kPa}{\rightleftharpoons}} 2NH_3(g)$$

The double arrow indicates that reactions occur in both the forward and reverse directions and that the final result is a mixture of all three substances. The temperature at which the reaction occurs is 350°C. The pressure at which the reaction occurs, 25 000 kPa, is also shown above the arrow. A catalyst is used to speed the reaction, so the catalyst is mentioned, too. Other symbols used in equations are shown in **Table 2**.

Table 2	State Symbols and Reaction Conditions
Symbol	**Meaning**
$(s), (l), (g)$	substance in the solid, liquid, or gaseous state
(aq)	substance in aqueous solution (dissolved in water)
$\longrightarrow$	"produces" or "yields," indicating result of reaction
$\rightleftharpoons$	reversible reaction in which products can reform into reactants; final result is a mixture of products and reactants
$\overset{\Delta}{\longrightarrow}$ or $\overset{heat}{\longrightarrow}$	reactants are heated; temperature is not specified
$\overset{Pd}{\longrightarrow}$	name or chemical formula of a catalyst, added to speed a reaction

Refer to Appendix A to see more symbols used in equations.

did you know?

State Symbols for Solids Occasionally, a symbol more specific than *(s)* is needed for a solid substance. The designations *(cr)* for a crystalline solid and *(amor)* for an amorphous solid are sometimes used when more specific information is needed about the solid.

265

Close

Although chemical equations can be packed with information, most of the ones you will work with will show only the formulas of reactants and products. However, sometimes you need to know the states of the substances. Recognizing and knowing the symbols used will help you understand these equations better. And learning these symbols now will make learning new information that depends on these symbols easier.

Quiz ———— GENERAL

1. Describe how both chemical and physical changes occur in your mouth when you chew food. **Ans.** Physical change occurs when food is broken into tiny pieces by chewing. Chemical change occurs when saliva starts to break down food into simpler substances.

2. How might you use a magnet to show that a completely rusted iron nail has undergone a chemical change? **Ans.** Iron is attracted to a magnet, but rust is not. So, a new substance has formed.

3. When solid aluminum oxide and hydrogen gas are heated, molten aluminum and water vapor form.

   a. Write a word equation for the above reaction. **Ans.** aluminum oxide + hydrogen → aluminum + water

   b. Write an unbalanced formula equation using all appropriate symbols. **Ans.** $Al_2O_3(s) + H_2(g) \xrightarrow{heat} Al(l) + H_2O(g)$

LS Logical

Reteaching ———— BASIC

Have students who are having trouble identifying reactants and products create one list of phrases that are used in sentences to signify reactants, and a second list of phrases that are used to signify products. **Ans.** Reactants are noted by phrases such as *are combined, reacts with,* and *is heated.* Products are noted by phrases such as *are formed, are made,* and *to form.* **LS** Verbal

① Section Review

UNDERSTANDING KEY IDEAS

1. What is a chemical reaction?

2. What is the only way to prove that a chemical reaction has occurred?

3. When water boils on the stove, does a chemical change or a physical change take place?

4. Give four examples of evidence that suggests that a chemical change probably is occurring.

5. When propane gas, C_3H_8, is burned with oxygen, the products are carbon dioxide and water. Write an unbalanced formula equation for the reaction.

6. Assume that liquid water forms in item 5. Write a formula equation for the reaction that shows the physical states of all compounds.

7. What does "Mn" above the arrow in a formula equation mean?

8. What symbol is used in a chemical equation to indicate the phrase "reacts with"?

9. Solid silicon and solid magnesium chloride form when silicon tetrachloride gas reacts with magnesium metal. Write a word equation and an unbalanced formula equation. Include all of the appropriate notations.

10. Magnesium oxide forms from magnesium metal and oxygen gas. Write a word equation and an unbalanced formula equation. Include all of the appropriate notations.

CRITICAL THINKING

11. Describe evidence that burning gasoline in an engine is a chemical reaction.

12. Describe evidence that chemical reactions take place during a fireworks display.

13. The directions on a package of an epoxy glue say to mix small amounts of liquid from two separate tubes. Either liquid alone does not work as a glue. Should the liquids be considered reactants? Explain your answer.

14. When sulfur is heated until it melts and then is allowed to cool, beautiful yellow crystals form. How can you prove that this change is physical?

15. Besides the reactant, what is needed for the electrolysis experiment that breaks down water?

16. Write the word equation for the electrolysis of water, and indicate the physical states and condition(s) of the reaction.

17. For each of the following equations, write a sentence that describes the reaction, including the physical states and reaction conditions.

    a. $Zn(s) + 2HCl(aq) \longrightarrow ZnCl_2(aq) + H_2(g)$

    b. $CaCl_2(aq) + Na_2CO_3(aq) \longrightarrow CaCO_3(s) + 2NaCl(aq)$

    c. $NaOH(aq) + HCl(aq) \longrightarrow NaCl(aq) + H_2O(l)$

    d. $CaCO_3(s) \xrightarrow{\Delta} CaO(s) + CO_2(g)$

Chapter Resource File

• Concept Review

• Quiz

Answers to Section Review

1. the process by which one or more substances are changed into one or more new substances with new and different properties

2. a chemical analysis to determine that at least one new product is formed

3. a physical change

4. Evidence includes the formation of a gas or a precipitate, change in color or odor, release or absorption of energy, and a change in temperature.

5. $C_3H_8 + O_2 \longrightarrow CO_2 + H_2O$

6. $C_3H_8(g) + O_2(g) \longrightarrow CO_2(g) + H_2O(l)$

7. The reaction is carried out using a manganese catalyst.

8. a plus sign

9. silicon tetrachloride + magnesium $\longrightarrow$ silicon + magnesium chloride; $SiCl_4(g) + Mg(s) \longrightarrow Si(s) + MgCl_2(s)$

10. magnesium + oxygen $\longrightarrow$ magnesium oxide; $Mg(s) + O_2(g) \longrightarrow MgO(s)$

11. Answers might include that water and other gases exit the tailpipe or that the gasoline is no longer present after it burns.

Answers continued on p. 299A

SECTION

② Balancing Chemical Equations

KEY TERMS
• coefficient

OBJECTIVES

① **Relate** the conservation of mass to the rearrangement of atoms in a chemical reaction.

② **Write** and interpret a balanced chemical equation for a reaction, and relate conservation of mass to the balanced equation.

Reactions Conserve Mass

A basic law of science is the law of conservation of mass. This law states that in ordinary chemical or physical changes, mass is neither created nor destroyed. If you add baking soda to vinegar, they react to release carbon dioxide gas, which escapes into the air. But if you collect all of the products of the reaction, you find that their total mass is the same as the total mass of the reactants.

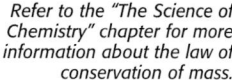
Topic Link

Refer to the "The Science of Chemistry" chapter for more information about the law of conservation of mass.

Reactions Rearrange Atoms

This law is based on the fact that the products and the reactants of a reaction are made up of the same number and kinds of atoms. The atoms are just rearranged and connected differently. Look at the formula equation for the reaction of sodium with water.

$$?Na + ?H_2O \longrightarrow ?NaOH + ?H_2$$

The same types of atoms appear in both the reactants and products. However, **Table 3** shows that the number of each type of atom is not the same on both sides of the equation. To show that a reaction satisfies the law of conservation of mass, its equation must be *balanced*.

Table 3 Counting Atoms in an Equation

	Reactants	Products	Balanced?
Unbalanced formula equation	Na + H$_2$O	NaOH + H$_2$	
Sodium atoms	1	1	yes
Hydrogen atoms	2	3	no
Oxygen atoms	1	1	yes

267

MISCONCEPTION //ALERT\\\\

Coefficients Students might consider coefficients to be numbers of individual particles only. Provide them with the balanced equation $2H_2 + O_2 \longrightarrow 2H_2O$. Ask them how many molecules of hydrogen are present. Provide them with multiples of the equation, showing them that the ratios of reactants and product remain the same. Extend the discussion to convert from molecules to moles, having students recall that a mole is simply 6.022×10^{23} particles. Therefore, the ratio of particles for each component of the reaction can always be considered the same as a mole ratio.

SECTION ②

Focus

Overview
Before beginning this section, review with your students the Objectives listed in the Student Edition. In this section, students will learn to relate the conservation of mass to the rearrangement of atoms during a chemical reaction. They will also learn to balance a formula equation and relate a balanced equation to conservation of mass.

Bellringer
Ask students to write a word equation for baking a cake. Ask them whether a cake has the same properties as the ingredients. **Ans.** Sample answer: sugar + flour + shortening + eggs + baking powder + vanilla + salt → cake. No, it does not.

Motivate

Identifying Preconceptions
Have students describe what it means for an equation to be balanced. Students might think that the sum of the coefficients must be the same on both sides for an equation to be balanced. Have students model what happens during a simple chemical reaction, such as $2H_2 + O_2 \longrightarrow 2H_2O$, to test this idea.

Discussion ——— GENERAL
Ask students if the information given in the word equation for the Bellringer is enough to bake a cake. **Ans.** The amounts of ingredients, the time to bake, and the temperature to use are not given.
LS Logical

Chapter Resource File
• Lesson Plan

Chapter 8 • Chemical Equations and Reactions 267

Teaching Tip

Equation Analogy Write the following items on the board:

1 center + 2 guards + 2 forwards = 1 basketball team

9 girls + 14 boys = 10 lab pairs + 1 group of three

Point out that the same numbers of people are represented on both sides of the equals sign. They are just arranged differently. The same principle applies to atoms represented in a chemical equation.

Coefficients Because half a molecule will not be involved in a reaction, whole-number coefficients reflect the fundamental nature of the materials.

Demonstration

Electrolysis of Water

1. Pour 350 mL of 0.01 M Na_2SO_4 into a 600 mL beaker.

2. Fill two small test tubes with the solution from the beaker. Cover the ends of the tubes, and carefully invert them into the solution. Be sure there are no air bubbles in the test tubes.

3. Place a 9 V battery upright in the center of the beaker.

4. Place the open mouth of each test tube over one terminal of the battery to collect the gases. Students will observe that twice as much H_2 as O_2 is produced. Relate this observation to the balanced chemical equation for the reaction.

Safety Caution: Wear safety goggles, gloves, and a lab apron.

Disposal: Pour the liquid down the drain. Dry the battery, and save it for reuse. Put tape over the terminals before storing the battery.

Balancing Equations

To balance an equation, you need to make the number of atoms for each element the same on the reactants' side and on the products' side. But there is a catch. You cannot change the formulas of any of the substances. For example, you could not change CO_2 to CO_3. You can only place numbers called *coefficients* in front of the formulas. A **coefficient** multiplies the number of atoms of each element in the formula that follows. For example, the formula H_2O represents 2 atoms of hydrogen and 1 atom of oxygen. But $2H_2O$ represents 2 molecules of water, for a total of 4 atoms of hydrogen and 2 atoms of oxygen. The formula $3Ca(NO_3)_2$ represents 3 calcium atoms, 6 nitrogen atoms, and 18 oxygen atoms. Look at **Skills Toolkit 1** as you balance equations.

coefficient

a small whole number that appears as a factor in front of a formula in a chemical equation

1 SKILLS Toolkit

Balancing Chemical Equations

1. Identify reactants and products.
- If no equation is provided, identify the reactants and products and write an unbalanced equation for the reaction. (You may find it helpful to write a word equation first.)
- If not all chemicals are described in the problem, try to predict the missing chemicals based on the type of reaction.

2. Count atoms.
- Count the number of atoms of each element in the reactants and in the products, and record the results in a table.
- Identify elements that appear in only one reactant and in only one product, and balance the atoms of those elements first. Delay the balancing of atoms (often hydrogen and oxygen) that appear in more than one reactant or product.
- If a polyatomic ion appears on both sides of the equation, treat it as a single unit in your counts.

3. Insert coefficients.
- Balance atoms one element at a time by inserting coefficients.
- Count atoms of each element frequently as you try different coefficients. Watch for elements whose atoms become unbalanced as a result of your work.
- Try the odd-even technique (explained later in this section) if you see an even number of a particular atom on one side of an equation and an odd number of that atom on the other side.

4. Verify your results.
- Double-check to be sure that the numbers of atoms of each element are equal on both sides of the equation.

REAL-WORLD CONNECTION

Welding The optimum oxygen-to-acetylene ratio (5:2) of an oxy-acetylene torch, which is used for cutting and welding metals, can be determined by looking at the balanced equation below.

$$2C_2H_2 + 5O_2 \longrightarrow 4CO_2 + 2H_2O$$

ChemFile  CHEMISTRY

- **Module 5: Stoichiometry**
  Topic: Balancing Equations
  This engaging tutorial reviews and reinforces balancing equations through modeling and guided practice.

SAMPLE PROBLEM A

Balancing an Equation

Balance the equation for the reaction of iron(III) oxide with hydrogen to form iron and water.

1 Identify reactants and products.

Iron(III) oxide and hydrogen are the reactants. Iron and water are the products. The unbalanced formula equation is

$$Fe_2O_3 + H_2 \longrightarrow Fe + H_2O$$

2 Count atoms.

	Reactants	Products	Balanced?
Unbalanced formula equation	$Fe_2O_3 + H_2$	$Fe + H_2O$	
Iron atoms	2	1	no
Oxygen atoms	3	1	no
Hydrogen atoms	2	2	yes

3 Insert coefficients.

Add a coefficient of 2 in front of Fe to balance the iron atoms.

$$Fe_2O_3 + H_2 \longrightarrow 2Fe + H_2O$$

Add a coefficient of 3 in front of H_2O to balance the oxygen atoms.

$$Fe_2O_3 + H_2 \longrightarrow 2Fe + 3H_2O$$

Now there are two hydrogen atoms in the reactants and six in the products. Add a coefficient of 3 in front of H_2.

$$Fe_2O_3 + 3H_2 \longrightarrow 2Fe + 3H_2O$$

4 Verify your results.

There are two iron atoms, three oxygen atoms, and six hydrogen atoms on both sides of the equation, so it is balanced.

PRACTICE HINT

One way to know what coefficient to use is to find a lowest common multiple. In this example, there were six hydrogen atoms in the products and two in the reactants. The lowest common multiple of 6 and 2 is 6, so a coefficient of 3 in the reactants balances the atoms.

PRACTICE

Write a balanced equation for each of the following.

1. $P_4 + O_2 \longrightarrow P_2O_5$
2. $C_3H_8 + O_2 \longrightarrow CO_2 + H_2O$
3. $Ca_2Si + Cl_2 \longrightarrow CaCl_2 + SiCl_4$
4. Silicon reacts with carbon dioxide to form silicon carbide, SiC, and silicon dioxide.

269

Answers to Practice Problems A

1. $P_4 + 5O_2 \longrightarrow 2P_2O_5$
2. $C_3H_8 + 5O_2 \longrightarrow 3CO_2 + 4H_2O$
3. $Ca_2Si + 4Cl_2 \longrightarrow 2CaCl_2 + SiCl_4$
4. $2Si + CO_2 \longrightarrow SiC + SiO_2$

Homework ——— GENERAL

Additional Practice Write word equations and balanced chemical equations for each of the following:

1. Magnesium metal and water react to form magnesium hydroxide and hydrogen gas. **Ans.** magnesium + water $\longrightarrow$ magnesium hydroxide + hydrogen; $Mg + 2H_2O \longrightarrow Mg(OH)_2 + H_2$

2. When ethane, C_2H_6, burns, it combines with oxygen to form carbon dioxide and water. **Ans.** ethane + oxygen $\longrightarrow$ carbon dioxide + water; $2C_2H_6 + 7O_2 \longrightarrow 4CO_2 + 6H_2O$

3. Iron metal and aluminum oxide form when aluminum metal reacts with iron(III) oxide. **Ans.** aluminum + iron(III) oxide $\longrightarrow$ iron + aluminum oxide; $2Al + Fe_2O_3 \longrightarrow 2Fe + Al_2O_3$
LS Logical

Teaching Tip ——— GENERAL

Correct Formulas Lead to Correct Equations Stress to students that equations with incorrect formulas are incorrect even if they can be balanced. All of the skill and practice that students have writing formulas will be put to the test in this chapter. Show the balanced equation for the synthesis of sodium chloride.
$2Na + Cl_2 \longrightarrow 2NaCl$

What would happen if the formula for salt were incorrectly written as Na_2Cl? **Ans.** The balanced equation would then be $4Na + Cl_2 \longrightarrow 2Na_2Cl$, which does not represent the reaction. Remind students that it is essential to use charges correctly when writing formulas for ionic compounds. **LS** Logical

---

LANGUAGE ARTS ——
CONNECTION — GENERAL

Natural Resources Provide students with the statement, "Our world is quickly using up its natural resources." Have students write several paragraphs explaining how that statement is true and how it is false. **Ans.** Answers should include that natural resources are limited and are being harvested from Earth, but that they are not being "used up" because the elements are still present somewhere in a product. **LS** Verbal

Chapter 8 • Chemical Equations and Reactions   269

Teaching Tip
Proportions in Equations
Describe to students how the coefficients demonstrate the proportionality of the substances in chemical equations. **Figure 6** clearly shows that coefficients in the equation represent the particles of the substances. Share with students that the coefficients also represent the moles of each substance involved in the reaction. The ratio of moles of substances in an equation is important in many calculations in chemistry.

Using the Figure
Refer students to **Figure 6.** To reinforce that the two products shown are quite different, ask students to compare the properties of water and those of hydrogen peroxide.
Ans. Students might mention the differences in how the compounds react when placed on a wound. Tell them that hydrogen peroxide decomposes in the presence of enzymes, but water does not. You can demonstrate this property by placing samples of a high-enzyme material, such as raw liver or fresh grated carrot or potato, into water and into hydrogen peroxide.

Balanced Equations Show Mass Conservation
The balanced equation for the reaction of sodium with water is

$$2Na + 2H_2O \longrightarrow 2NaOH + H_2$$

Each side of the equation has two atoms of sodium, four atoms of hydrogen, and two atoms of oxygen. The reactants and the products are made up of the same atoms so they must have equal masses. So a balanced equation shows the conservation of mass.

Never Change Subscripts to Balance an Equation
If you needed to write a balanced equation for the reaction of H_2 with O_2 to form H_2O, you might start with this formula equation:

$$H_2 + O_2 \longrightarrow H_2O$$

To balance this equation, some people may want to change the formula of the product to H_2O_2.

$$H_2 + O_2 \longrightarrow H_2O_2$$

Although the equation is balanced, the product is no longer water, but hydrogen peroxide. Look at the models and equations in **Figure 6** to understand the problem. The first equation was balanced correctly by adding coefficients. As expected, the model shows the correct composition of the water molecules formed by the reaction. The second equation was incorrectly balanced by changing a subscript. The model shows that the change of a subscript changes the composition of the substance. As a result, the second equation no longer shows the formation of water, but that of hydrogen peroxide. When balancing equations, never change subscripts. Keep this in mind as you learn about the odd-even technique for balancing equations.

Figure 6
Use coefficients to balance an equation. Never change subscripts.

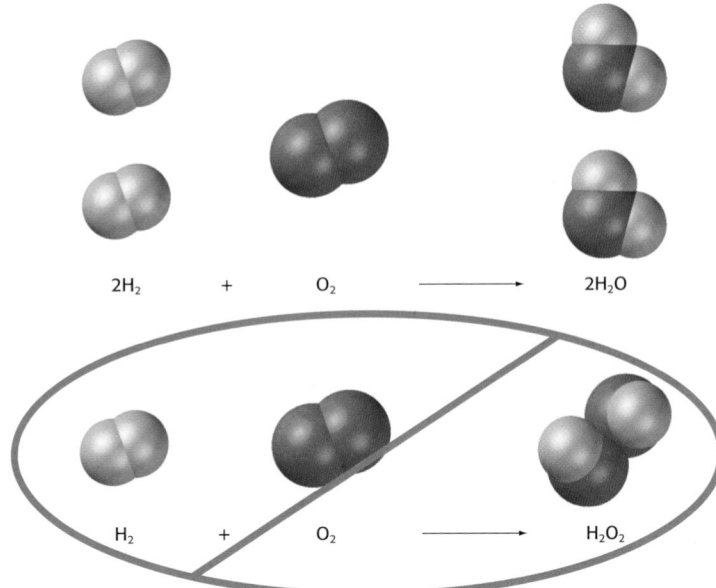

$2H_2$ + O_2 ⟶ $2H_2O$

H_2 + O_2 ⟶ H_2O_2

270

HEALTH CONNECTION
For further emphasis that compounds having different subscripts are different compounds and have different properties, have students compare carbon monoxide, CO, and carbon dioxide, CO_2, in the human body. Carbon dioxide is a waste product of cellular respiration in the human body and helps control the pH in the blood. It is easily eliminated from the blood through the lungs where it is exhaled. On the other hand, carbon monoxide is not naturally found in the human body. If it is inhaled, it reacts with hemoglobin, binding to the hemoglobin so tightly that it cannot efficiently pick up and carry oxygen to cells. If enough carbon monoxide is inhaled, the person suffocates because too little hemoglobin is available to carry oxygen. Only one subscript differs between the two formulas, but CO and CO_2 are totally different compounds with different properties.

SAMPLE PROBLEM B

The Odd-Even Technique

The reaction of ammonia with oxygen produces nitrogen monoxide and water vapor. Write a balanced equation for this reaction.

1 Identify reactants and products.

The unbalanced formula equation is

$$NH_3 + O_2 \longrightarrow NO + H_2O$$

2 Count atoms.

	Reactants	Products	Balanced?
Unbalanced formula equation	$NH_3 + O_2$	$NO + H_2O$	
Nitrogen atoms	1	1	yes
Hydrogen atoms	3	2	no
Oxygen atoms	2	2	yes

The odd-even technique uses the fact that multiplying an odd number by 2 always results in an even number.

3 Insert coefficients.

A 2 in front of NH_3 gives an even number of H atoms. Add coefficients to NO and H_2O to balance the H atoms and N atoms.

$$2NH_3 + O_2 \longrightarrow 2NO + 3H_2O$$

For oxygen, double *all* coefficients to have an even number of O atoms on both sides and keep the other atoms balanced.

$$4NH_3 + 2O_2 \longrightarrow 4NO + 6H_2O$$

Change the coefficient for O_2 to 5 to balance the oxygen atoms.

$$4NH_3 + 5O_2 \longrightarrow 4NO + 6H_2O$$

4 Verify your results.

There are four nitrogen atoms, twelve hydrogen atoms, and ten oxygen atoms on both sides of the equation, so it is balanced.

PRACTICE

Write a balanced chemical equation for each of the following.

1 $C_2H_2 + O_2 \longrightarrow CO_2 + H_2O$

2 $Fe(OH)_2 + H_2O_2 \longrightarrow Fe(OH)_3$

3 $FeS_2 + Cl_2 \longrightarrow FeCl_3 + S_2Cl_2$

PROBLEM SOLVING SKILL

PRACTICE HINT

Watch for cases in which all atoms in an equation are balanced except one, which has an odd number on one side of the equation and an even number on the other side. Multiplying all coefficients by 2 will result in an even number of atoms for the unbalanced atoms while keeping the rest balanced.

did you know?

Hydrogen Peroxide as an Oxygen Supply

Although it is not used as a standard treatment, hydrogen peroxide is sometimes added to the blood through an IV to increase the oxygen content of the blood. In the presence of enzymes in the blood, it breaks down into water and oxygen. It can be used to treat infections involving anaerobic microorganisms and can increase the amount of oxygen in the blood of people who have breathing problems.

Teaching Tip

Charge! To show students that subscripts cannot be changed to balance an equation, remind them that the net charge of a compound must total zero. For example, if the subscripts in NaCl are changed to Na_2Cl, the compound would have a charge of 1+ rather than being neutral, as it should be.

SKILL BUILDER — ADVANCED

Math Skills A slightly different process for using the odd-even technique to balance equations is to use a fractional coefficient to balance an equation. Once the equation is balanced, multiply all coefficients by a number to change the fractions to whole numbers. This is especially helpful for some reactions of hydrocarbons with oxygen (combustion reactions). Have students balance the following equation using a fractional coefficient first, then multiply through to remove the fraction. $C_7H_{14} + O_2 \longrightarrow CO_2 + H_2O$ **Ans.** $C_7H_{14} + \frac{21}{2}O_2 \longrightarrow 7CO_2 + 7H_2O$; Multiply all coefficients by 2 to get: $2C_7H_{14} + 21O_2 \longrightarrow 14CO_2 + 14H_2O$ **LS** Logical

Answers to Practice Problems B

1. $2C_2H_2 + 5O_2 \longrightarrow 4CO_2 + 2H_2O$
2. $2Fe(OH)_2 + H_2O_2 \longrightarrow 2Fe(OH)_3$
3. $2FeS_2 + 5Cl_2 \longrightarrow 2FeCl_3 + 2S_2Cl_2$

Homework — GENERAL

Additional Practice Balance each of the following equations.

1. $Ag_2O \longrightarrow Ag + O_2$
   Ans. $2Ag_2O \longrightarrow 4Ag + O_2$

2. $C_6H_6 + O_2 \longrightarrow CO_2 + H_2O$
   Ans. $2C_6H_6 + 15O_2 \longrightarrow 12CO_2 + 6H_2O$

3. $Al_2O_3 + C \longrightarrow Al + CO_2$
   Ans. $2Al_2O_3 + 3C \longrightarrow 4Al + 3CO_2$

LS Logical

Polyatomic Ions Tell students that some equations have one product that has a polyatomic ion, such as a sulfate ion, that appears in a reactant and a second product that is composed of a key element, such as sulfur, from that ion. In these cases, it is usually best to balance the individual atoms of each element rather than trying to balance the polyatomic ion as a group. As an example, have the students balance the following equation: $Ag + HNO_3 \longrightarrow AgNO_3 + NO + H_2O$. **Ans.** $3Ag + 4HNO_3 \longrightarrow 3AgNO_3 + NO + 2H_2O$ **LS** Logical

READING SKILL BUILDER ── **BASIC**

Discussion As students complete the reading of this section, have them discuss balancing equations. Have each student verbally contribute one fact or hint that will help others balance equations. If possible, have them point out the part of the reading that supports their contribution to the discussion. **LS** Interpersonal

Using the Figure ── **GENERAL**

Refer students to **Figure 7**. Have students write balanced chemical equations for the reaction of aluminum metal with phosphoric acid and with sulfurous acid. For each equation, have students identify the number of the polyatomic ions on each side of the balanced equation. **Ans.** $2Al + 2H_3PO_4 \longrightarrow 2AlPO_4 + 3H_2$, 2 phosphate ions; $2Al + 3H_2SO_3 \longrightarrow Al_2(SO_3)_3 + 3H_2$, 3 sulfite ions **LS** Logical

Polyatomic Ions Can Be Balanced as a Unit

So far, you've balanced equations by balancing individual atoms one at a time. However, balancing some equations is made easier because groups of atoms can be balanced together. This is especially true in the case of polyatomic ions, such as NO_3^-. Often a polyatomic ion appears in both the reactants and the products without changing. The atoms within such ions are not rearranged during the reaction. The polyatomic ion can be counted as a single unit that appears on both sides of the equation. Of course, when you think that you have finished balancing an equation, checking each atom by itself is still helpful.

Look at **Figure 7**. The sulfate ion appears in both the reactant sulfuric acid and in the product aluminum sulfate. You could look at the sulfate ion as a single unit to make balancing the equation easier. Looking at the balanced equation, you can see that there are three sulfate ions on the reactants' side and three on the products' side.

In balancing the equation for the reaction between sodium phosphate and calcium nitrate, you can consider the nitrate ion and the phosphate ion each to be a unit. The resulting balanced equation is

$$2Na_3PO_4 + 3Ca(NO_3)_2 \longrightarrow 6NaNO_3 + Ca_3(PO_4)_2$$

Count the atoms of each element to make sure that the equation is balanced.

Figure 7
In the reaction of aluminum with sulfuric acid, sulfate ions are part of both the reactants and the products.

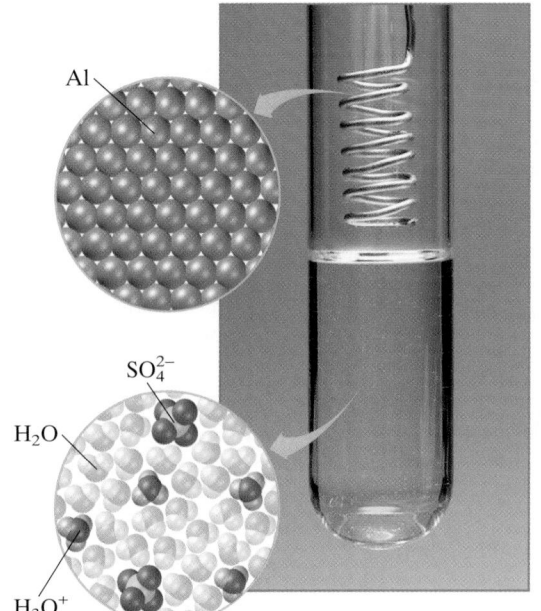 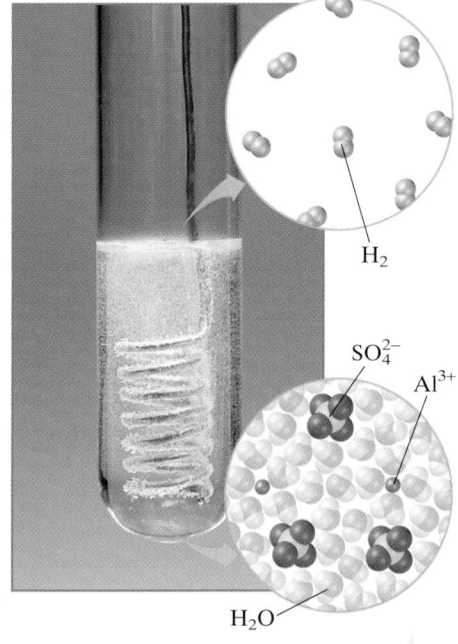

$$2Al(s) + 3H_2SO_4(aq) \longrightarrow Al_2(SO_4)_3(aq) + 3H_2(g)$$

did you know?

Chemical Language Scientists and students all over the world balance equations using the same numbers and symbols. Formulas and other symbols used in equations are standard throughout the world.

MATH CONNECTION ── **ADVANCED**

Have students write a paragraph comparing chemical equations with mathematical equations. **Ans.** Although different symbols are used, the concept of equality is present in both. For example, the arrow in a chemical equation separates equal sides, just as the equals sign does in a mathematical equation. As numerical values are equal on both sides of a mathematical equation, the numbers of each type of atom are equal on both sides of a chemical equation. These equations also have in common that items separated by a plus sign can appear in any order. **LS** Logical

SAMPLE PROBLEM C

Polyatomic Ions as a Group

Aluminum reacts with arsenic acid, $HAsO_3$, to form H_2 and aluminum arsenate. Write a balanced equation for this reaction.

1 Identify reactants and products.

The unbalanced formula equation is

$$Al + HAsO_3 \longrightarrow H_2 + Al(AsO_3)_3$$

2 Count atoms.

	Reactants	Products	Balanced?
Unbalanced formula equation	$Al + HAsO_3$	$H_2 + Al(AsO_3)_3$	
Aluminum atoms	1	1	yes
Hydrogen atoms	1	2	no
Arsenate ions	1	3	no

Because the arsenate ion appears on both sides of the equation, consider it a single unit while balancing.

3 Insert coefficients.

Change the coefficient of $HAsO_3$ to 3 to balance the arsenate ions.

$$Al + 3HAsO_3 \longrightarrow H_2 + Al(AsO_3)_3$$

Double all coefficients to keep the other atoms balanced and to get an even number of hydrogen atoms on each side.

$$2Al + 6HAsO_3 \longrightarrow 2H_2 + 2Al(AsO_3)_3$$

Change the coefficient of H_2 to 3 to balance the hydrogen atoms.

$$2Al + 6HAsO_3 \longrightarrow 3H_2 + 2Al(AsO_3)_3$$

4 Verify your results.

There are 2 aluminum atoms, 6 hydrogen atoms, 6 arsenic atoms, and 18 oxygen atoms on both sides of the equation, so it is balanced.

PRACTICE

Write a balanced equation for each of the following.

1 $HgCl_2 + AgNO_3 \longrightarrow Hg(NO_3)_2 + AgCl$

2 $Al + Hg(CH_3COO)_2 \longrightarrow Al(CH_3COO)_3 + Hg$

3 Calcium phosphate and water are produced when calcium hydroxide reacts with phosphoric acid.

273

PRACTICE HINT

If you consider poly-atomic ions as single units, be sure to count the atoms of each element when you double-check your work.

Homework — GENERAL

Additional Practice Balance each of the following equations.

1. $Pb(NO_3)_2 + Na_2CrO_4 \longrightarrow$ $PbCrO_4 + NaNO_3$
   Ans. $Pb(NO_3)_2 + Na_2CrO_4 \longrightarrow$ $PbCrO_4 + 2NaNO_3$

2. $K_2SO_4 + Ca(NO_3)_2 \longrightarrow CaSO_4 +$ KNO_3
   Ans. $K_2SO_4 + Ca(NO_3)_2 \longrightarrow$ $CaSO_4 + 2KNO_3$

3. $(NH_4)_3PO_4 +$ $Mg(CH_3COO)_2 \longrightarrow Mg_3(PO_4)_2 +$ NH_4CH_3COO **Ans.** $2(NH_4)_3PO_4 +$ $3Mg(CH_3COO)_2 \longrightarrow$ $Mg_3(PO_4)_2 + 6NH_4CH_3COO$

4. $MoO_3 + H_2SO_4 + Zn \longrightarrow$ $Mo_2O_3 + H_2O + ZnSO_4$ **Ans.** $2MoO_3 + 3H_2SO_4 + 3Zn \longrightarrow$ $Mo_2O_3 + 3H_2O + 3ZnSO_4$
   LS Logical

Teaching Tip

Simplest Ratio Remind students that the coefficients in a balanced equation must be in the lowest ratio possible. Even though the equation $4H_2 + 2O_2 \longrightarrow 4H_2O$ has an equal number of each type of atom on each side, its coefficients are not in the simplest ratio. The correctly balanced equation should be $2H_2 + O_2 \longrightarrow 2H_2O$.

REAL-WORLD
CONNECTION — **GENERAL**

Air Purification in Space Throughout the space program, lithium compounds have been used to purify air in space vehicles. Lithium hydroxide reacts with carbon dioxide to form lithium carbonate and water. Lithium peroxide reacts with carbon dioxide to form lithium carbonate and oxygen. Have students write balanced equations for these two reactions. **Ans.** $2LiOH + CO_2 \longrightarrow$ $Li_2CO_3 + H_2O$; $2Li_2O_2 + 2CO_2 \longrightarrow$ $2Li_2CO_3 + O_2$ Ask them why they think lithium compounds were used instead of compounds of more common elements, such as sodium or potassium. **Ans.** Lithium has less mass than sodium or potassium. Therefore a mole of the lithium compounds would weigh less than a mole of the same compound having sodium or potassium. **LS** Logical

---

Close

Quiz — GENERAL

1. Balance the following equation.
$NH_3 + O_2 \longrightarrow NO + H_2O$ **Ans.**
$4NH_3 + 5O_2 \longrightarrow 4NO + 6H_2O$

2. Write a balanced equation for the reaction between chlorine gas and potassium bromide, forming potassium chloride and bromine. **Ans.** $Cl_2 + 2KBr \longrightarrow 2KCl + Br_2$

3. Why should you not balance an equation by changing subscripts? **Ans.** Changing subscripts changes the identity of the substances.
LS Logical

Reteaching — BASIC

Provide students with several simple unbalanced chemical equations and have them work in pairs to explain to each other step-by-step how to balance the equations. Then have the pairs compare their results to those of another group and discuss any differences.
LS Interpersonal

Practice Makes Perfect

You have learned a few techniques that you can use to help you approach balancing equations logically. But don't think that you are done. The more you practice balancing equations, the faster and better you will become. The best way to discover more tips to help you balance equations is to practice a lot! As you learn about the types of reactions in the next section, be aware that these types can provide tips that make balancing equations even easier.

② Section Review

UNDERSTANDING KEY IDEAS

1. What fundamental law is demonstrated in balancing equations?

2. What is meant by a balanced equation?

3. When balancing an equation, should you adjust the subscripts or the coefficients?

PRACTICE PROBLEMS

4. Write each of the following reactions as a word equation, an unbalanced formula equation, and finally as a balanced equation.

a. When heated, potassium chlorate decomposes into potassium chloride and oxygen.

b. Silver sulfide forms when silver and sulfur, S_8, react.

c. Sodium hydrogen carbonate breaks down to form sodium carbonate, carbon dioxide, and water vapor.

5. Balance the following equations.

a. $ZnS + O_2 \longrightarrow ZnO + SO_2$

b. $Fe_2O_3 + CO \longrightarrow Fe + CO_2$

c. $AgNO_3 + AlCl_3 \longrightarrow AgCl + Al(NO_3)_3$

d. $Ni(ClO_3)_2 \longrightarrow NiCl_2 + O_2$

6. Balance the following equations.

a. $(NH_4)_2Cr_2O_7 \longrightarrow Cr_2O_3 + N_2 + H_2O$

b. $NH_3 + CuO \longrightarrow N_2 + Cu + H_2O$

c. $Na_2SiF_6 + Na \longrightarrow Si + NaF$

d. $C_4H_{10} + O_2 \longrightarrow CO_2 + H_2O$

CRITICAL THINKING

7. Use diagrams of particles to explain why four atoms of phosphorus can produce only two molecules of diphosphorus trioxide, even when there is an excess of oxygen atoms.

8. Which numbers in the reactants and products in the following equation are coefficients, and which are subscripts?

$$2Al + 3H_2SO_4 \longrightarrow Al_2(SO_4)_3 + 3H_2$$

9. Write a balanced equation for the formation of water from hydrogen and oxygen. Use the atomic mass of each element to determine the mass of each molecule in the equation. Use these masses to show that the equation demonstrates the law of conservation of mass.

10. A student writes the equation below as the balanced equation for the reaction of iron with chlorine. Is this equation correct? Explain.

$$Fe(s) + Cl_3(g) \longrightarrow FeCl_3(s)$$

Answers to Section Review

1. the law of conservation of mass

2. There are equal numbers of each type of atom on both sides of the arrow.

3. coefficients

4. a. potassium chlorate $\longrightarrow$ potassium chloride + oxygen; $KClO_3 \longrightarrow KCl + O_2$; $2KClO_3 \longrightarrow 2KCl + 3O_2$

b. silver + sulfur $\longrightarrow$ silver sulfide; $Ag + S_8 \longrightarrow Ag_2S$; $16Ag + S_8 \longrightarrow 8Ag_2S$

c. sodium hydrogen carbonate $\longrightarrow$ sodium carbonate + carbon dioxide + water; $NaHCO_3 \longrightarrow Na_2CO_3 + CO_2 + H_2O$; $2NaHCO_3 \longrightarrow Na_2CO_3 + CO_2 + H_2O$

5. a. $2ZnS + 3O_2 \longrightarrow 2ZnO + 2SO_2$

b. $Fe_2O_3 + 3CO \longrightarrow 2Fe + 3CO_2$

c. $3AgNO_3 + AlCl_3 \longrightarrow 3AgCl + Al(NO_3)_3$

d. $Ni(ClO_3)_2 \longrightarrow NiCl_2 + 3O_2$

6. a. $(NH_4)_2Cr_2O_7 \longrightarrow Cr_2O_3 + N_2 + 4H_2O$

b. $2NH_3 + 3CuO \longrightarrow N_2 + 3Cu + 3H_2O$

c. $Na_2SiF_6 + 4Na \longrightarrow Si + 6NaF$

d. $2C_4H_{10} + 13O_2 \longrightarrow 8CO_2 + 10H_2O$

7. Diagrams should show that after making two molecules of diphosphorus trioxide, no more atoms of phosphorus are available to join with additional oxygen atoms.

Answers continued on p. 299A

Classifying Chemical Reactions

KEY TERMS

- combustion reaction
- synthesis reaction
- decomposition reaction
- activity series
- double-displacement reaction

OBJECTIVES

(1) **Identify** combustion reactions, and write chemical equations that predict the products.

(2) **Identify** synthesis reactions, and write chemical equations that predict the products.

(3) **Identify** decomposition reactions, and write chemical equations that predict the products.

(4) **Identify** displacement reactions, and use the activity series to write chemical equations that predict the products.

(5) **Identify** double-displacement reactions, and write chemical equations that predict the products.

Reaction Types

So far in this book, you have learned about a lot of chemical reactions. But they are just a few of the many that take place. To make learning about reations simpler, it is helpful to classify them and to start with a few basic types. Consider a grocery store as an example of how classification makes things simpler. A store may have thousands of items. Even if you have never been to a particular store before, you should be able to find everything you need. Because similar items are grouped together, you know what to expect when you start down an aisle.

Look at the reaction shown in **Figure 8.** The balanced equation for this reaction is

$$2Al + Fe_2O_3 \longrightarrow 2Fe + Al_2O_3$$

By classifying chemical reactions into several types, you can more easily predict what products are likely to form. You will also find that reactions in each type follow certain patterns, which should help you balance the equations more easily.

The five reaction types that you will learn about in this section are not the only ones. Additional types are discussed in other chapters, and there are others beyond the scope of this book. In addition, reactions can belong to more than one type. There are even reactions that do not fit into any type. The value in dividing reactions into categories is not to force each reaction to fit into a single type but to help you see patterns and similarities in reactions.

Figure 8
Knowing which type of reaction occurs between aluminum and iron(III) oxide could help you predict that iron is produced.

275

did you know?

Combustion A combustion reaction is limited in this text to the rapid oxidation of organic materials. Some views of combustion reactions include rapid oxidation of any material. The burning of magnesium or steel wool are combustion reactions under this broader view.

INCLUSION Strategies

- *Learning Disabled*
- *English Language Learners*
- *Attention Deficit Disorder*

Students can create a "Reaction Notebook" by labeling notebook pages with the different types of reactions discussed in the chapter. Students should write on each page specific examples of that reaction type, including the reactants and products of the reaction. Notes about patterns, trends, or similarities with other reactions can be written on each page. The notebook can be used as a study guide for individual study or discussion in small groups.

SECTION 3

Focus

Overview

Before beginning this section, review with your students the Objectives listed in the Student Edition. In this section, students will learn to identify combustion, synthesis, decomposition, displacement, and double-displacement reactions. They will learn to predict the products formed from any one of these reactions and write chemical equations for them.

Bellringer

Ask groups of four students to describe what the terms *synthesis, decomposition,* and *displacement* mean to them. Have groups compare their descriptions and discuss any similarities and any differences. **Ans.** Students should associate synthesis with making something, decomposition with things breaking down, and displacement with movement or with something taking the place of something else.

Motivate

Identifying Preconceptions

Students might be familiar with some types of reactions from previous classes. Ask students what they remember about different ways in which substances change in reactions. Students may only recall generalizations about reactions, such as remembering that sometimes two substances join to form one substance. Encourage students to keep this previously acquired knowledge in mind, and to build on it as they learn about the types of reactions in this section.

Chapter Resource File

- Lesson Plan

Demonstration ——— GENERAL
Combustion

(Approximate time: 5 minutes)

1. Show students several examples of combustion reactions, such as burning butane in a lighter and burning methane in a laboratory burner.

2. Ask students what serves as fuel in each reaction and what else is needed for the reaction to occur. **Ans.** The fuel is whatever material is burning. Other items needed are oxygen and energy to start the reaction.

Safety Caution: Wear safety goggles. Students should be 3 m from the demonstration. Keep all combustible materials away from the flames. Be sure all flames are completely extinguished after the demonstration.

Disposal: Throw solid products in the trash can after they are cool. Save reusable materials for later use. **LS** Logical

Teach

Teaching Tip
Combustion and Energy
Combustion reactions all release energy while they are occurring. However, energy must be added for the reaction to start. This starting energy is known as *activation energy*, and it must be supplied from the environment.

Using the Figure ——— GENERAL
Refer students to **Figure 9.** Tell them that butane, C_4H_{10}, the fuel in a lighter, has a similar flame. Have them write a balanced equation for the combustion of butane.
Ans. $2C_4H_{10} + 13O_2 \longrightarrow 8CO_2 + 10H_2O$ **LS** Logical

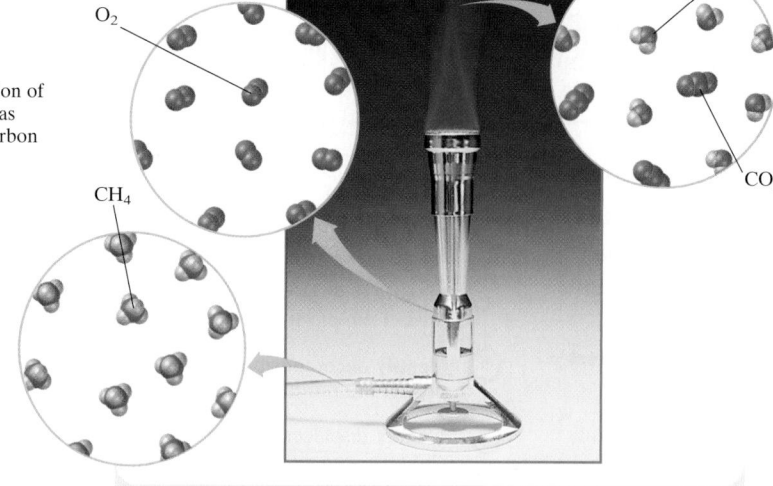

Figure 9
The complete combustion of any hydrocarbon, such as methane, yields only carbon dioxide and water.

O_2 H_2O CH_4 CO_2

$$CH_4(g) + 2O_2(g) \longrightarrow CO_2(g) + 2H_2O(g)$$

Combustion Reactions

Combustion reactions are often used to generate energy. Much of our electrical energy is generated in power plants that work because of the combustion of coal. Combustion of hydrocarbons (as in gasoline) provides energy used in transportation—on the land, in the sea, and in the air. For our purposes, a **combustion reaction** is the reaction of a carbon-based compound with oxygen. The products are carbon dioxide and water vapor. An example of a combustion reaction is shown in **Figure 9.**

Many of the compounds in combustion reactions are called *hydrocarbons* because they are made of only carbon and hydrogen. Propane is a hydrocarbon that is often used as a convenient portable fuel for lanterns and stoves. The balanced equation for the combustion of propane is shown below.

$$C_3H_8 + 5O_2 \longrightarrow 3CO_2 + 4H_2O$$

Some compounds, such as alcohols, are made of carbon, hydrogen, and oxygen. In the combustion of these compounds, carbon dioxide and water are still made. For example, the fuel known as gasohol is a mixture of gasoline and ethanol, an alcohol. The balanced chemical equation for the combustion of ethanol is shown below.

$$CH_3CH_2OH + 3O_2 \longrightarrow 2CO_2 + 3H_2O$$

When enough oxygen is not available, the combustion reaction is incomplete. Carbon monoxide and unburned carbon (soot), as well as carbon dioxide and water vapor, are made.

combustion reaction

the oxidation reaction of an organic compound, in which heat is released

internet connect
www.scilinks.org
Topic: Combustion
SciLinks code: HW4033
SCiLINKS Maintained by the National Science Teachers Association

276

REAL-WORLD ——— CONNECTION

Burning Hydrogen The burning of hydrogen is a potentially important source of energy. Two definite advantages of hydrogen as a fuel are that there is an enormous supply of it and that it can produce no pollution when it burns. The only product of its combustion is water. Hydrogen combustion is a common energy source for spacecraft, and research is currently being conducted on using hydrogen in fuel cells.

One-Stop Planner CD-ROM

• **Supplemental Reading Projects**
Guided Reading Worksheet:
The Pale Horse
Assign this worksheet for cross-curricular connections to language arts.

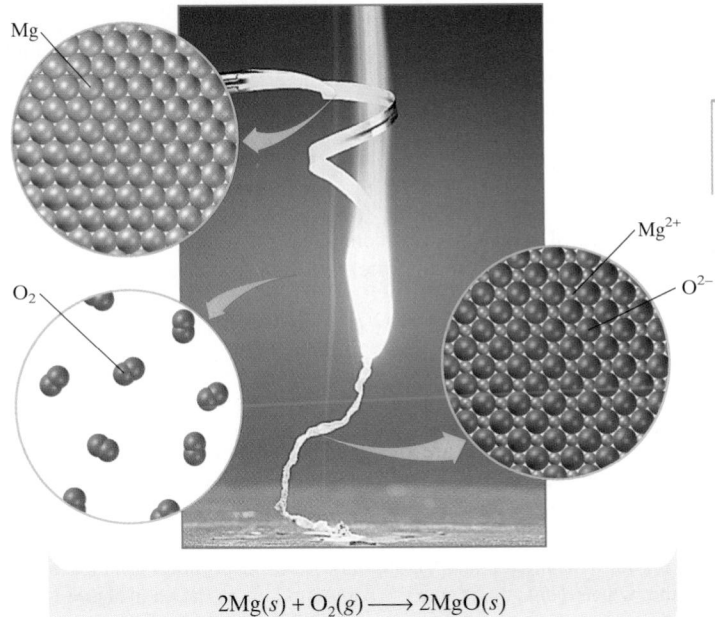

Figure 10
When the elements magnesium and oxygen react, they combine to form the binary compound magnesium oxide.

$$2Mg(s) + O_2(g) \longrightarrow 2MgO(s)$$

Synthesis Reactions

The word *synthesis* comes from a Greek word that means "to put together." In the case of a **synthesis reaction,** a single compound forms from two or more reactants. If you see a chemical equation that has only one product, the reaction is a synthesis reaction. The reactants in many of these reactions are two elements or two small compounds.

synthesis reaction

a reaction in which two or more substances combine to form a new compound

Two Elements Form a Binary Compound

If the reactants in an equation are two elements, the only way in which they can react is to form a binary compound, which is composed of two elements. Often, when a metal reacts with a nonmetal, electrons are transferred and an ionic compound is formed. You can use the charges of the ions to predict the formula of the compound formed. Metals in Groups 1 and 2 lose one electron and two electrons, respectively. Nonmetals in Groups 16 and 17 gain two electrons and one electron, respectively. Using the charges on the ions, you can predict the formula of the product of a synthesis reaction, such as the one in **Figure 10.**

Nonmetals on the far right of the periodic table can react with one another to form binary compounds. Often, more than one compound could form, however, so predicting the product of these reactions is not always easy. For example, carbon and oxygen can combine to form carbon dioxide or carbon monoxide, as shown below.

$$C + O_2 \longrightarrow CO_2 \qquad\qquad 2C + O_2 \longrightarrow 2CO$$

277

HISTORY
CONNECTION

Polymers to Replace Ivory Many scientists consider addition polymerization to be a type of synthesis reaction. Addition polymerization occurs when small units, called *monomers,* join to form a larger unit, called a *polymer.* Since they were first developed, polymers have been very useful. In 1869, John Wesley Hyatt (1837–1920), in an attempt to find a substitute for the fragile ivory in billiard balls, discovered the first synthetic polymer. Called *celluloid,* it is a material that can be molded into shape. In 1884, Louis M. H. Bernigaud pushed the same substance through tiny holes in a nozzle, making small threads that he called *rayon.*

Teaching Tip

Nitrogen Oxides If air is used instead of pure oxygen in the combustion process, nitrogen-oxygen compounds—many of which are poisonous—are produced. Some nitrogen-oxygen compounds are also light sensitive and turn brown when exposed to sunlight, causing the brown smog that appears in some cities.

SKILL BUILDER

Vocabulary Ask students what they think an *anhydride* is. They might correctly assume that it is something that contains no water because *an-* means "without" and *hydride* refers to water. They will learn more about anhydrides when they study acids and bases. Metallic oxides are called basic anhydrides because they combine with water in a synthesis reaction to form a base. For example, $MgO + H_2O \longrightarrow Mg(OH)_2$. Nonmetallic oxides are acidic anhydrides because they combine with water in a synthesis reaction to form an acid. For example, $SO_2 + H_2O \longrightarrow H_2SO_3$.

Teaching Tip

Acid Precipitation Normal precipitation is slightly acidic because carbon dioxide reacts with water in the air in a synthesis reaction to form weak carbonic acid. The small amount of this weak acid present in air does not negatively affect the environment. However, several oxides of nitrogen and sulfur react with water in the air, forming strong acids. For example, sulfur trioxide reacts with water to form sulfuric acid, $SO_3 + H_2O \longrightarrow H_2SO_4$. These acids are harmful to the environment.

Teaching Tip — BASIC

Designing a Model Have students use foam balls and toothpicks to model a simple synthesis reaction. Then have them model the corresponding decomposition reaction. **LS Kinesthetic**

SKILL BUILDER — ADVANCED

Writing Skills Each of the following types of compounds decomposes in a predictable pattern. For each of the following decomposition reactions, the type of reactant and one product is given. Have students predict what the other product will be and give a balanced equation for an example of that type of reaction. Then have students write a mnemonic device to help them remember each of these reaction patterns.

metal carbonate $\longrightarrow$ metal oxide + ?
Ans. carbon dioxide; $CaCO_3 \longrightarrow CaO + CO_2$

metal chlorate $\longrightarrow$ metal chloride + ?
Ans. oxygen; $2KClO_3 \longrightarrow 2KCl + 3O_2$

metal hydroxide $\longrightarrow$ metal oxide + ?
Ans. water; $Mg(OH)_2 \longrightarrow MgO + H_2O$ **LS Logical**

SKILL BUILDER — GENERAL

Writing Skills Although more recycling is being done than has been done in the past, disposal of solid waste is still a problem. Have students write several paragraphs relating decomposition reactions to the possible solution of the disposal of some solid wastes. Paragraphs might discuss addition of materials that increase the rate of decomposition of organic substances, creating compost from organic materials, and increasing exposure of organic materials to air to increase activity of aerobic bacteria that decompose organics. **LS Verbal**

STUDY TIP

WORKING WITH A PARTNER

If you can explain difficult concepts to a study partner, then you know that you understand them yourself.

• Make flashcards that contain examples of chemical reactions. Quiz each other on reaction types by using the flashcards. Explain how you identified each type.

Refer to Appendix B for other studying strategies.

decomposition reaction

a reaction in which a single compound breaks down to form two or more simpler substances

Figure 11
Nitrogen triiodide is a binary compound that decomposes into the elements nitrogen and iodine.

Two Compounds Form a Ternary Compound

Two compounds can combine to form a ternary compound, a compound composed of three elements. One example is the reaction of water and a Group 1 or Group 2 metal oxide to form a metal hydroxide. An example is the formation of "slaked lime," or calcium hydroxide.

$$CaO(s) + H_2O(l) \longrightarrow Ca(OH)_2(s)$$

Some oxides of nonmetals can combine with water to produce acids. Carbon dioxide combines with water to form carbonic acid.

$$CO_2(g) + H_2O(l) \longrightarrow H_2CO_3(aq)$$

Decomposition Reactions

Decomposition reactions are the opposite of synthesis reactions—they have only one reactant. In a **decomposition reaction,** a single compound breaks down, often with the input of energy, into two or more elements or simpler compounds.

If your reactant is a binary compound, then the products will most likely be the two elements that make the compound up, as shown in **Figure 11**. In another example, water can be decomposed into the elements hydrogen and oxygen through the use of electrical energy.

$$2H_2O(l) \xrightarrow{\text{electricity}} 2H_2(g) + O_2(g)$$

The gases produced are very pure and are used for special purposes, such as in hospitals. But these gases are very expensive because of the energy needed to make them. Experiments are underway to make special solar cells in which sunlight is used to decompose water.

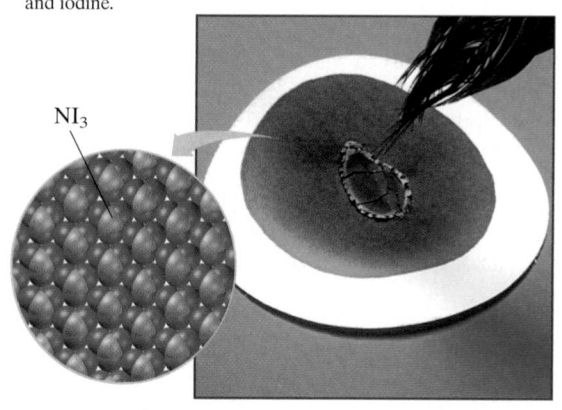
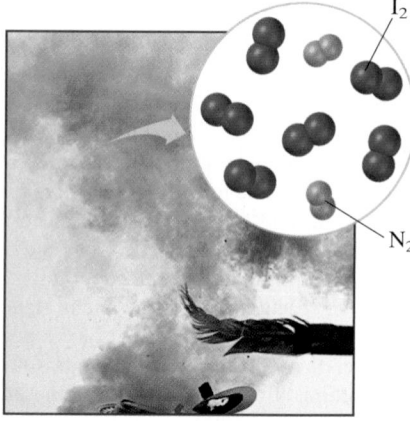

$$2NI_3(s) \longrightarrow N_2(g) + 3I_2(g)$$

---

did you know? — BASIC

Air Bags Air bags in automobiles are inflated by the decomposition of the compound sodium azide, NaN_3. Have students write a balanced equation for this reaction. **Ans.** $2NaN_3(s) \longrightarrow 2Na(s) + 3N_2(g)$ A device that is found in the air bag starts the decomposition with an electric spark. **LS Logical**

Compounds made up of three or more elements usually do not decompose into those elements. Instead, each compound that consists of a given polyatomic ion will break down in the same way. For example, a metal carbonate, such as $CaCO_3$ in limestone, decomposes to form a metal oxide and carbon dioxide.

$$CaCO_3(s) \xrightarrow{\text{heat}} CaO(s) + CO_2(g)$$

Many of the synthesis reactions that form metal hydroxides and acids can be reversed to become decomposition reactions.

SAMPLE PROBLEM D

Predicting Products

Predict the product(s) and write a balanced equation for the reaction of potassium with chlorine.

1 Gather information.

Because the reactants are two elements, the reaction is most likely a synthesis. The product will be a binary compound.

2 Plan your work.

Potassium, a Group 1 metal, will lose one electron to become a 1+ ion. Chlorine, a Group 17 nonmetal, gains one electron to form a 1– ion. The formula for the product will be KCl. The unbalanced formula equation is

$$K + Cl_2 \longrightarrow KCl$$

3 Calculate.

Place a coefficient of 2 in front of KCl and also K.

$$2K + Cl_2 \longrightarrow 2KCl$$

4 Verify your results.

The final equation has two atoms of each element on each side, so it is balanced.

PRACTICE

Predict the product(s) and write a balanced equation for each of the following reactions.

1 the reaction of butane, C_4H_{10}, with oxygen

2 the reaction of water with calcium oxide

3 the reaction of lithium with oxygen

4 the decomposition of carbonic acid

PRACTICE HINT

Look for hints about the type of reaction. If the reactants are two elements or simple compounds, the reaction is probably a synthesis reaction. The reaction of oxygen with a hydrocarbon is a combustion reaction. If there is only one reactant, it is a decomposition reaction.

(279)

did you know?

Equilibrium Some synthesis reactions and their corresponding decomposition reactions are reversible. For example, in the Haber process, ammonia gas is made from hydrogen and nitrogen gases, $3H_2 + N_2 \longrightarrow 2NH_3$. In a closed system, the ammonia will begin to decompose, $2NH_3 \longrightarrow 3H_2 + N_2$. Eventually, the system will reach a stage where both reactions are occurring at the same rate. This stage is known as equilibrium.

Homework ——— GENERAL

Additional Practice Predict the product(s) and write a balanced equation for each of the following reactions.

1. pentane, C_5H_{12}, reacting with oxygen **Ans.** $C_5H_{12} + 8O_2 \longrightarrow 6H_2O + 5CO_2$

2. water reacting with potassium oxide **Ans.** $H_2O + K_2O \longrightarrow 2KOH$

3. calcium reacting with oxygen **Ans.** $2Ca + O_2 \longrightarrow 2CaO$

4. the decomposition of sodium chloride **Ans.** $2NaCl \longrightarrow 2Na + Cl_2$

5. the combustion of octane, C_8H_{18} **Ans.** $2C_8H_{18} + 25O_2 \longrightarrow 16CO_2 + 18H_2O$

LS Logical

READING SKILL BUILDER
BASIC

Reading Organizer As students continue this section, have them create a graphic organizer that shows the types of reactants and products for each type of reaction. **Ans.** The organizers should show that synthesis reactions have two or more reactants, which can be either elements or compounds, and one product, a compound. Decomposition reactions have one reactant, which is a compound, and two or more products, which can be elements or compounds. Combustion reactions have a carbon-based compound and oxygen as reactants and water and carbon dioxide as products. Displacement reactions have an element and a compound as reactants and a different element and compound as products. Double-displacement reactions have two compounds as reactants and two different compounds as products.

LS Visual

280

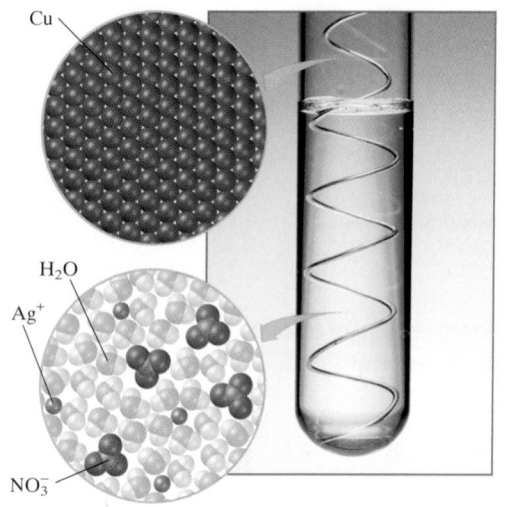
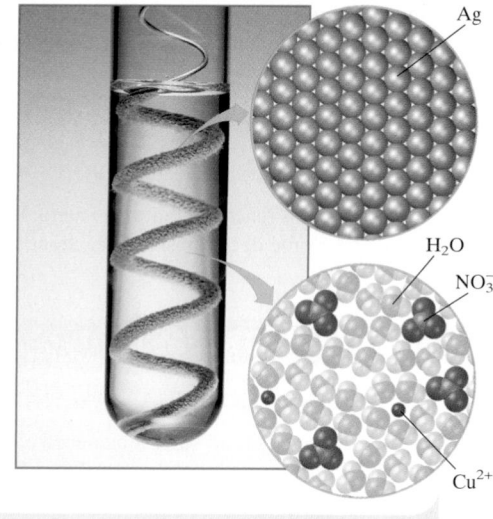

$$Cu(s) + 2AgNO_3(aq) \longrightarrow 2Ag(s) + Cu(NO_3)_2(aq)$$

Demonstration —— GENERAL

Displacement of Ag with Cu

(Approximate time: 10 minutes)

This demonstration provides a vivid example of a displacement reaction.

1. Place a Petri dish on the overhead projector. Pour a 0.1 M AgNO₃ solution into the dish until it is about half full.

2. Cut six 1 cm² pieces of copper foil. Add the copper foil to the silver nitrate solution. Have students describe what happens. **Ans.** The solution becomes blue as the reaction continues, and crystals form on the pieces of copper.

Safety Caution: Safety goggles and a lab apron must be worn. Students must be at least 3 m from the demonstration. Skin should not come in contact with the silver nitrate solution for even a short time, or skin will be stained.

Disposal: Filter the copper and silver out of the solution. Rinse the solids and the filter paper well. Store the metals for later use. Throw away the filter paper. Let the rinse water and the silver nitrate solution evaporate completely. Store and label the residue for future use. **LS** Visual

Teaching Tip —— ADVANCED

The Activity Series Students should not have to memorize the activity series. However, it is useful to challenge students to develop a mnemonic device to help them remember the order of the elements. Have students work individually or in small groups to develop such a device, and then share it with the rest of the class. **LS** Intrapersonal

Figure 12
Copper is the more active metal and displaces silver from the silver nitrate solution. So copper is higher on the activity series than silver is. The Cu²⁺ formed gives the solution a blue color.

activity series

a series of elements that have similar properties and that are arranged in descending order of chemical activity

Displacement Reactions

When aluminum foil is dipped into a solution of copper(II) chloride, reddish copper metal forms on the aluminum and the solution loses its blue color. It is as if aluminum atoms and copper ions have switched places to form aluminum ions and copper atoms.

$$2Al(s) + 3CuCl_2(aq) \longrightarrow 2AlCl_3(aq) + 3Cu(s)$$

In this *displacement reaction,* a single element reacts with a compound and displaces another element from the compound. The products are a different element and a different compound than the reactants are. In general, a metal may displace another metal (or hydrogen), while a nonmetal may displace only another nonmetal.

The Activity Series Ranks Reactivity

Results of experiments, such as the one in **Figure 12,** in which displacement reactions take place are summarized in the **activity series,** a portion of which is shown in **Table 4.** In the activity series, elements are arranged in order of activity with the most active one at the top. In general, an element can displace those listed below it from compounds in solution, but not those listed above it. Thus, you can use the activity series to make predictions about displacement reactions. You could also predict that no reaction would happen, such as when silver is put into a copper(II) nitrate solution.

When a metal is placed in water, the reactivity information in the activity series helps you tell if hydrogen is displaced. If the metal is active enough for this to happen, a metal hydroxide and hydrogen gas form.

REAL-WORLD
CONNECTION —— ADVANCED

Plating Most processes that deposit one metal on another happen through the use of electrical energy in the process known as *electroplating.* However, when a more active metal is placed in a solution containing ions of a less active metal, the less active metal will plate onto the more active metal. This process is shown in **Figure 12.** Ask students

why this procedure would not be useful for plating silver onto pieces of jewelry or flatware. **Ans.** The metal is not plated in a thin, even layer as you would need to plate jewelry. It appears fairly thick and "fuzzy" and does not appear to be strongly attached to the underlying metal. **LS** Visual

Table 4   Activity Series

Element	Reactivity
K Ca Na	react with cold water and acids to replace hydrogen; react with oxygen to form oxides
Mg Al Zn Fe	react with steam (but not with cold water) and acids to replace hydrogen; react with oxygen to form oxides
Ni Pb	do not react with water; react with acids to replace hydrogen; react with oxygen to form oxides
H_2 Cu	react with oxygen to form oxides
Ag Au	fairly unreactive; form oxides only indirectly

Refer to Appendix A for a more complete activity series of metals and of halogens.

SKILLS Toolkit 2

Using the Activity Series

1. Identify the reactants.
- Determine whether the single element is a metal or a halogen.
- Determine the element that might be displaced from the compound if a displacement reaction occurs.

2. Check the activity series.
- Determine whether the single element or the element that might be displaced from the compound is more active. The more active element is higher on the activity series.
- For a metal reacting with water, determine whether the metal can replace hydrogen from water in that state.

3. Write the products, and balance the equation.
- If the more active element is already part of the compound, then no reaction will occur.
- Otherwise, the more active element will displace the less active element.

4. Verify your results.
- Double-check to be sure that the equation is balanced.

internet connect
www.scilinks.org
Topic: Activity Series
SciLinks code: HW4004

SCiLINKS Maintained by the National Science Teachers Association

281

did you know?

Activity of Nonmetals The activity series in **Table 4** involves only metals. Nonmetals, particularly halogens, have their own activity series. Fluorine, being the most electronegative halogen, is the most active, followed by chlorine, bromine, and iodine, in that order. As with metals, a more active halogen will replace a less active halogen from a compound.

Demonstration ── GENERAL
Relative Activity

(Approximate time: 15 minutes)

Show students the relative activity of two different metals.

1. Place a galvanized (zinc-coated) iron nail in a beaker half-full of 1 M HCl. Have students describe what happens. **Ans.** Bubbles form on the nail and the reaction appears to happen at a fast rate, showing that zinc reacts quickly with the acid.

2. After the zinc has reacted, bubbles of hydrogen are still produced, but the reaction is much slower. Have students explain why the reaction slows. **Ans.** As the zinc is removed, the iron is exposed and reacts with the acid. However, iron is less active than zinc, so the reaction is slower. Point out to students how these findings match the positions of the metals on the activity series.

Safety Caution: Wear safety goggles and lab aprons. Students should be at least 3 m from the demonstration. Adequate ventilation should be present.

Disposal: Decant the liquid, and wash any remaining metal. Combine the washings with the decanted liquid. Test the pH of the solution. If the pH is 3 or below, neutralize the liquid with 1 M NaOH, and pour the mixture down the drain with plenty of water. Throw the remaining metal in the trash can. **LS** Visual

Answers to Practice Problems E

1. $2Al + 3Zn(NO_3)_2 \longrightarrow$
   $2Al(NO_3)_3 + 3Zn$
2. $2Na + 2H_2O \longrightarrow 2NaOH + H_2$
3. no reaction

Homework ——— GENERAL

Additional Practice Write a balanced equation if a reaction happens. Otherwise, write "no reaction."

1. Magnesium is dipped into a nickel(II) chloride solution. **Ans.** $Mg + NiCl_2 \longrightarrow MgCl_2 + Ni$

2. Lead is placed into an iron(III) nitrate solution. **Ans.** no reaction

3. Zinc is added to a solution of copper(II) sulfate. **Ans.** $Zn + CuSO_4 \longrightarrow ZnSO_4 + Cu$

 Logical

Quick LAB ———— GENERAL

Skills Acquired:
- Classifying
- Constructing models
- Identifying/Recognizing patterns
- Predicting

Materials:

For each group of 2–3 students:
- gumdrops of at least four different colors
- toothpicks

Teacher's Notes: Students can use pieces of toothpicks rather than whole toothpicks. Caution students that toothpicks are sharp. Remind students to never eat anything in a laboratory setting.

LS Kinesthetic

 **PRACTICE HINT**

You can sometimes use your knowledge of the periodic table to verify how you apply the activity series. In general, Group 1 metals are rarely in atomic form at the end of most reactions. Group 2 metals are less likely than Group 1 metals but more likely than transition metals to be in atomic form after a reaction.

SAMPLE PROBLEM E

Determining Products by Using the Activity Series

Magnesium is added to a solution of lead(II) nitrate. Will a reaction happen? If so, write the equation and balance it.

1 Identify the reactants.

Magnesium will attempt to displace lead from lead(II) nitrate.

2 Check the activity series.

Magnesium is more active than lead and displaces it.

3 Write the products, and balance the equation.

A reaction will occur. Lead is displaced by magnesium.

$$Mg + Pb(NO_3)_2 \longrightarrow Pb + Mg(NO_3)_2$$

4 Verify your results.

The equation is balanced.

PRACTICE

PROBLEM SOLVING SKILL

For the following situations, write a balanced equation if a reaction happens. Otherwise write "no reaction."

1. Aluminum is dipped into a zinc nitrate solution.

2. Sodium is placed in cold water.

3. Gold is added to a solution of calcium chloride.

Quick LAB

SAFETY PRECAUTIONS

Balancing Equations by Using Models

PROCEDURE

1. Use **toothpicks** and **gumdrops of at least four different colors** (representing atoms of different elements) to make models of the substances in each equation below.

2. For each reaction below, use your models to determine the products, if needed, and then balance the equation.

ANALYSIS

Use your models to classify each reaction by type.

a. $H_2 + Cl_2 \longrightarrow HCl$
b. $Mg + O_2 \longrightarrow MgO$
c. $C_2H_6 + O_2 \longrightarrow CO_2 + H_2O$
d. $KI + Br_2 \longrightarrow KBr + I_2$

e. $H_2CO_3 \longrightarrow CO_2 + H_2O$
f. $Ca + H_2O \longrightarrow Ca(OH)_2 + H_2$
g. $KClO_3 \longrightarrow KCl + O_2$
h. $CH_4 + O_2 \longrightarrow CO_2 + H_2O$

i. $Zn + HCl \longrightarrow \underline{\quad}$
j. $H_2O \xrightarrow{\text{electricity}} \underline{\quad}$
k. $C_3H_8 + O_2 \longrightarrow \underline{\quad}$
l. $BaO + H_2O \longrightarrow \underline{\quad}$

282

REAL-WORLD ———
CONNECTION — GENERAL

Sacrificial Metals Sometimes metals that need to be protected from reacting with their environment are placed close to a more active metal. This more active metal, called a *sacrificial metal*, reacts first, protecting the less active metal. Ask students to solve the following problem: Suppose an oceangoing ship has an iron hull that must be protected from rusting. In addition to painting it, a sacrificial metal is attached to the hull below the water line. This metal must be one that will corrode more readily than iron. According to **Table 4**, what metals would be suitable? **Ans.** Any metal more active than Fe but not one that will react with cold water. Acceptable metals are Mg, Al, or Zn.

LS Logical

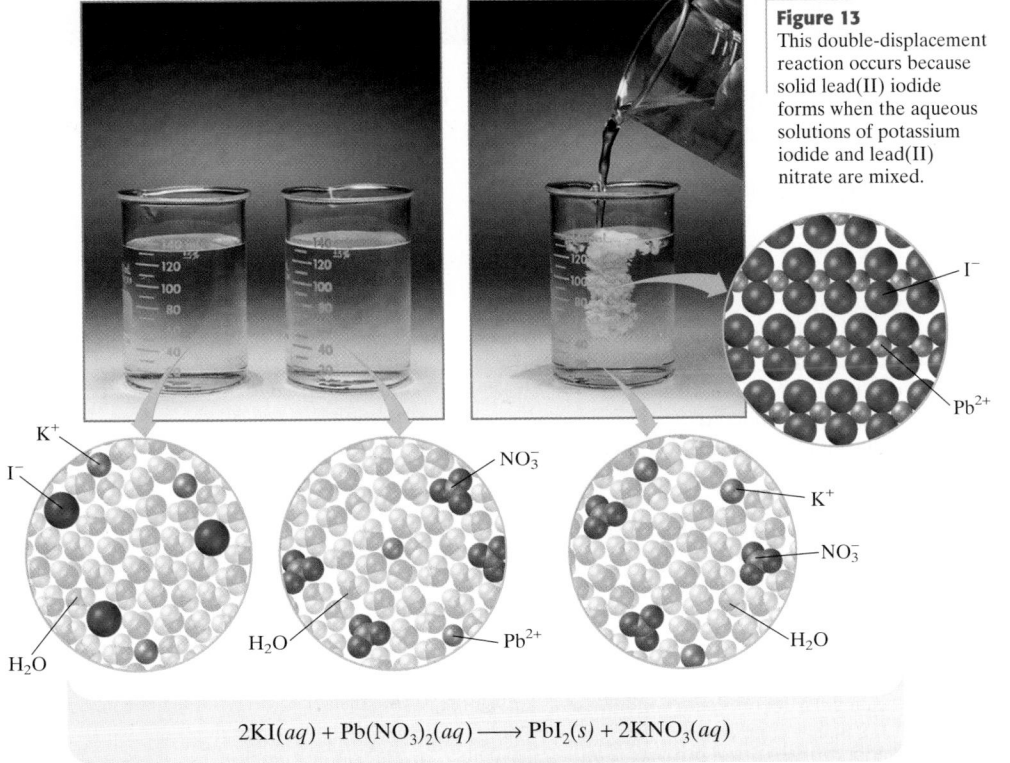

$$2KI(aq) + Pb(NO_3)_2(aq) \longrightarrow PbI_2(s) + 2KNO_3(aq)$$

Double-Displacement Reactions

Figure 13 shows the result of the reaction between KI and $Pb(NO_3)_2$. The products are a yellow precipitate of PbI_2 and a colorless solution of KNO_3. From the equation, it appears as though the parts of the compounds just change places. Early chemists called this a **double-displacement reaction.** It occurs when two compounds in aqueous solution appear to exchange ions and form two new compounds. For this to happen, one of the products must be a solid precipitate, a gas, or a molecular compound, such as water. Water is often written as HOH in these equations.

For example, when dilute hydrochloric acid and sodium hydroxide are mixed, little change appears to happen. However, by looking at the equation for the reaction, you can see that liquid water, a molecular compound, forms.

$$HCl(aq) + NaOH(aq) \longrightarrow HOH(l) + NaCl(aq)$$

Although this type of formula equation is not the best description, the term *double-displacement reaction* is still in use. A better way to represent these reactions is to use a net ionic equation, which will be covered in the next section.

double-displacement reaction

a reaction in which a gas, a solid precipitate, or a molecular compound forms from the apparent exchange of atoms or ions between two compounds

283

Answers to QuickLab

a. $H_2 + Cl_2 \longrightarrow 2HCl$; synthesis

b. $2Mg + O_2 \longrightarrow 2MgO$; synthesis

c. $2C_2H_6 + 7O_2 \longrightarrow 4CO_2 + 6H_2O$; combustion

d. $2KI + Br_2 \longrightarrow 2KBr + I_2$; displacement

e. $H_2CO_3 \longrightarrow CO_2 + H_2O$; decomposition

f. $Ca + 2H_2O \longrightarrow Ca(OH)_2 + H_2$; displacement

g. $2KClO_3 \longrightarrow 2KCl + 3O_2$; decomposition

h. $CH_4 + 2O_2 \longrightarrow CO_2 + 2H_2O$; combustion

i. $Zn + 2HCl \longrightarrow ZnCl_2 + H_2$; displacement

j. $2H_2O \xrightarrow{electricity} 2H_2 + O_2$; decomposition

k. $C_3H_8 + 5O_2 \longrightarrow 3CO_2 + 4H_2O$; combustion

l. $BaO + H_2O \longrightarrow Ba(OH)_2$; synthesis

Chapter Resource File

• Datasheets for In-Text Labs

Group Activity — BASIC

Have pairs of students play a game of concentration. Have them take 10 index cards and write an example equation on each card. There should be two cards for each of the five types of reactions. Then have them shuffle and deal the cards upside down in two rows of five cards each. Taking turns, each player then turns over any two cards, trying to match the two equations that are of the same reaction type. **LS** Visual

---

REAL-WORLD CONNECTION — ADVANCED

Environmental Destruction Chemical reactions are slowly destroying marble statues, monuments, and metal bridges. Have students determine where any old statues, monuments, or bridges are located in your area, and have them gather enough information to perform the following tasks:

1. Learn what materials were used in building or sculpting the structures. Find out when they were built.

2. Use the information in item 1 along with library research to suggest causes of the deterioration.

3. Write equations for the chemical reactions that are taking place during the deterioration.

4. Perform additional research to discover what measures are being taken to slow these reactions and what steps are being used to restore and repair the decaying or broken items. **LS** Verbal

---

SKILL BUILDER — ADVANCED

Vocabulary One common example of a double-displacement reaction is an acid-base reaction. Acid-base reactions are also classified as neutralization reactions because the two components neutralize each other's properties, producing a salt and water.

Have students find out what a *metathesis* reaction is. **Ans.** It is the same thing as a double-displacement reaction. **LS** Verbal

Teaching Tip

May I Have This Dance? To help students remember the types of reactants and products in types of reactions, use an analogy to dancing. Synthesis is like two individual dancers forming a pair. After the dance is over, the pair parts, becoming two individuals again, as in decomposition. Like displacement, when one person cuts in on a dancing pair, a different individual and pair are formed. Double-displacement is like two pairs of dancers exchanging partners.

Transparencies

TT Identifying Reactions and Predicting Products

Identifying Reactions and Predicting Products

1. Is there only one reactant?

If the answer is no, go to step 2.
If the answer is yes, you have a decomposition reaction.
- A binary compound generally breaks into its elements.
- A ternary compound breaks according to the guidelines given earlier in this section.

2. Are the reactants two elements or two simple compounds?

If the answer is no, go to step 3.
If the answer is yes, you probably have a synthesis reaction.
- If both reactants are elements, the product is a binary compound. For a metal reacting with a nonmetal, use the expected charges to predict the formula of the compound.
- If the reactants are compounds, the product will be a single ternary compound according to the guidelines given earlier in this section.

3. Are the reactants oxygen and a hydrocarbon?

If the answer is no, go to step 4.
If the answer is yes, you have a combustion reaction.
- The products of a combustion reaction are carbon dioxide and water.

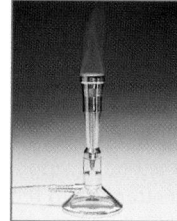

4. Are the reactants an element and a compound other than a hydrocarbon?

If the answer is no, go to step 5.
If the answer is yes, you probably have a displacement reaction.
- Use the activity series to determine the activities of the elements.
- If the more active element is already part of the compound, no reaction will occur. Otherwise, the more active element will displace the less active element from the compound.

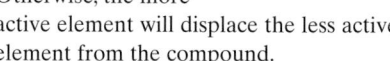

5. Are the reactants two compounds composed of ions?

If the answer is no, go back to step 1 because you might have missed the proper category.
If the answer is yes, you probably have a double-displacement reaction.
- Write formulas for the possible products by forming two new compounds from the ions available.
- Determine if one of the possible products is a solid precipitate, a gas, or a molecular compound, such as water. If neither product qualifies in the above categories, no reaction occurs. Use the rules below to determine whether a substance will be an insoluble solid.

All compounds of Group 1 and NH_4^+ are soluble.
All nitrates are soluble.
All halides, except those of Ag^+ and Pb^{2+}, are soluble.
All sulfates, except Ca^{2+}, Sr^{2+}, Ba^{2+}, Hg_2^{2+}, and Pb^{2+}, are soluble.
All carbonates, except those of Group 1 and NH_4^+, are insoluble.

284

REAL-WORLD CONNECTION

Baking Most baked goods rise as the result of carbon dioxide gas being released and trapped in the batter or dough. Some of this gas is released by the action of yeast. However, many baked goods rise because of a double-displacement reaction, followed by a decomposition reaction. Baking powder consists of baking soda, $NaHCO_3$, and a dry acid, such as tartaric acid, $H_2C_4H_4O_6$. When water is added to baking powder, a reaction takes place between the baking soda and the acid.

$$2NaHCO_3 + H_2C_4H_4O_6 \longrightarrow Na_2C_4H_4O_6 + 2H_2CO_3$$

The carbonic acid formed then decomposes into carbon dioxide gas and water.

$$H_2CO_3 \longrightarrow H_2O + CO_2$$

More Types to Come

This section has been a short introduction to the classification of chemical reactions. Even so, you now have the tools, summarized in **Skills Toolkit 3**, to predict the products of hundreds of reactions. Keep the reaction types in mind as you continue your study of chemistry. And as you learn about other reaction types, think about how they relate to the five types described here.

Section Review

UNDERSTANDING KEY IDEAS

1. Why is the formation of a ternary compound also a synthesis reaction?

2. When a binary compound is the only reactant, what are the products most likely to be?

3. Explain how synthesis and decomposition reactions can be the reverse of one another.

4. What two compounds form when hydrocarbons burn completely?

5. Explain how to use the activity series to predict chemical behavior.

6. In which part of the periodic table are the elements at the top of the activity series?

7. What must be produced for a double-displacement reaction to occur?

PRACTICE PROBLEMS

8. Balance each of the equations below, and indicate the type of reaction for each equation.

a. $Cl_2(g) + NaBr(aq) \longrightarrow NaCl(aq) + Br_2(l)$

b. $CaO(s) + H_2O(l) \longrightarrow Ca(OH)_2(aq)$

c. $Ca(ClO_3)_2(s) \longrightarrow CaCl_2(s) + O_2(g)$

d. $AgNO_3(aq) + K_2SO_4(aq) \longrightarrow$ $Ag_2SO_4(s) + KNO_3(aq)$

e. $Zn(s) + CuBr_2(aq) \longrightarrow ZnBr_2(aq) + Cu(s)$

f. $C_8H_{18}(l) + O_2(g) \longrightarrow CO_2(g) + H_2O(g)$

9. Predict whether a reaction would occur when the materials indicated are brought together. For each reaction that would occur, complete and balance the equation.

a. $Ag(s) + H_2O(l)$

b. $Mg(s) + Cu(NO_3)_2(aq)$

c. $Al(s) + O_2(g)$

d. $H_2SO_4(aq) + KOH(aq)$

10. Predict the products, write a balanced equation, and identify the type of reaction for each of the following reactions.

a. $HgO \longrightarrow$

b. $C_3H_7OH + O_2 \longrightarrow$

c. $Zn + CuSO_4 \longrightarrow$

d. $BaCl_2 + Na_2SO_4 \longrightarrow$

e. $Zn + F_2 \longrightarrow$

f. $C_5H_{10} + O_2 \longrightarrow$

CRITICAL THINKING

11. When will a displacement reaction *not* occur?

12. Explain why the terms *synthesis* and *decomposition* are appropriate names for their respective reaction types.

13. Platinum is used for jewelry because it does not corrode. Where would you expect to find platinum on the activity series?

14. Will a reaction occur when copper metal is dipped into a solution of silver nitrate? Explain.

Answers to Section Review

1. Two compounds combine to form a new compound. When two substances react to form one new substance, it is a synthesis reaction.

2. the elements that make up the compound

3. A synthesis reaction forms a compound from more than one reactant. If that compound is broken down in a decomposition reaction, the products that form might be the same substances that were the reactants in the synthesis reaction. This would result in the two reactions being the reverse of one another.

4. water and carbon dioxide

5. Any metal will replace any other metal below it from a compound.

6. on the left side

7. a solid precipitate, a gas, or a molecular compound, such as water

8. a. $Cl_2(g) + 2NaBr(aq) \longrightarrow 2NaCl(aq) + Br_2(l)$; displacement

b. $CaO(s) + H_2O(l) \longrightarrow Ca(OH)_2(aq)$; synthesis

c. $Ca(ClO_3)_2(s) \longrightarrow CaCl_2(s) + 3O_2(g)$; decomposition

Answers continued on p. 299A

Focus

Overview

Before beginning this section, review with your students the Objectives listed in the Student Edition. In this section, students will define and learn to write total ionic and net ionic equations for double-displacement reactions.

Bellringer

Have students write a definition of the word *spectator* in terms of the part spectators play in a sporting event. **Ans.** Definitions might include that spectators are present at the game but do not actually participate in it.

Motivate

Activity ——— BASIC

Have students outline the opening page of this section. Ask them what is accomplished by outlining. **Ans.** Responses should include that the main topics and the facts that support them are singled out and other words are eliminated. Tell students that outlining is similar to writing a net ionic equation. **LS** Verbal

Teach

READING SKILL BUILDER ——— BASIC

Anticipation Guide As students read the section, have them look for clues as to why net ionic equations tell more about what's happening during chemical reactions in solution than complete chemical equations tell. Have them summarize their clues in several sentences. **LS** Verbal

KEY TERMS

• spectator ions

www.scilinks.org
Topic: Precipitation Reactions
SciLinks code: HW4160

OBJECTIVES

1. **Write** total ionic equations for reactions in aqueous solutions.

2. **Identify** spectator ions and write net ionic equations for reactions in aqueous solutions.

Ionic Equations

When ionic compounds dissolve in water, the ions separate from each other and spread throughout the solution. Thus, the formulas $KI(aq)$ and $Pb(NO_3)_2(aq)$ are actually aqueous ions, as shown below.

$$KI(aq) = K^+(aq) + I^-(aq)$$

$$Pb(NO_3)_2(aq) = Pb^{2+}(aq) + 2NO_3^-(aq)$$

Notice that when lead(II) nitrate dissolves, there are two nitrate ions for every lead ion, so a coefficient of 2 is used for NO_3^-. The reaction between KI and $Pb(NO_3)_2$ can be described by the chemical equation below.

$$2KI(aq) + Pb(NO_3)_2(aq) \longrightarrow PbI_2(s) + 2KNO_3(aq)$$

However, it is more correct to describe the reaction by using a *total ionic equation* as shown below. When you write a total ionic equation, make sure that both the mass and the electric charge are conserved.

$$2K^+(aq) + 2I^-(aq) + Pb^{2+}(aq) + 2NO_3^-(aq) \longrightarrow$$
$$PbI_2(s) + 2K^+(aq) + 2NO_3^-(aq)$$

But even this equation is not the best way to view the reaction.

Identifying Spectator Ions

When two solutions are mixed, all of the ions are present in the combined solution. In many cases, some of the ions will react with each other. However, some ions do not react. These **spectator ions** remain unchanged in the solution as aqueous ions. In the equation above, the K^+ and NO_3^- ions appear as aqueous ions both on the reactants' side and on the products' side. Because K^+ and NO_3^- ions are spectator ions in the above reaction, they can be removed from the total ionic equation. What remains are the substances that do change during the reaction.

$$2K^+(aq) + 2I^-(aq) + Pb^{2+}(aq) + 2NO_3^-(aq) \longrightarrow$$
$$PbI_2(s) + 2K^+(aq) + 2NO_3^-(aq)$$

spectator ions

ions that are present in a solution in which a reaction is taking place but that do not participate in the reaction

286

MISCONCEPTION ALERT

Forming Ions Students might think that all ions in solution come from ionization of compounds. *Ionization* is the process by which molecules of covalent compounds form ions when they are pulled apart by polar water molecules when placed in solution. In ionic compounds, ions already exist. They just change from an organized lattice in a solid to individual ions in solution. The process by which existing ions move apart from each other in solution is called *dissociation*.

Chapter Resource File

• Lesson Plan

Chemical equation:

$K_2SO_4(aq) + Ba(NO_3)_2(aq) \longrightarrow 2KNO_3(aq) + BaSO_4(s)$

Total ionic equation:

$2K^+(aq) + SO_4^{2-}(aq) + Ba^{2+}(aq) + 2NO_3^-(aq) \longrightarrow 2K^+(aq) + 2NO_3^-(aq) + BaSO_4(s)$

Net ionic equation:

$SO_4^{2-}(aq) + Ba^{2+}(aq) \longrightarrow BaSO_4(s)$

Writing Net Ionic Equations

The substances that remain once the spectator ions are removed from the chemical equation make an equation that shows only the net change. This is called a *net ionic equation.* The one for the reaction of KI with $Pb(NO_3)_2$ is shown below.

$$2I^-(aq) + Pb^{2+}(aq) \longrightarrow PbI_2(s)$$

Figure 14 shows the process of determining the net ionic equation for another reaction.

The net ionic equation above is the same as the one for the reaction between NaI and $Pb(ClO_3)_2$. Both compounds are soluble, and their aqueous solutions contain iodide and lead(II) ions, which would form lead(II) iodide. So, the net change is the same.

Net ionic equations can also be used to describe displacement reactions. For example, Zn reacts with a solution of $CuSO_4$ and displaces the copper ion, Cu^{2+}, as shown in the total ionic equation

$$Zn(s) + Cu^{2+}(aq) + SO_4^{2-}(aq) \longrightarrow Cu(s) + Zn^{2+}(aq) + SO_4^{2-}(aq)$$

Only the sulfate ion remains unchanged and is a spectator ion. Thus, the net ionic equation is as follows:

$$Zn(s) + Cu^{2+}(aq) \longrightarrow Cu(s) + Zn^{2+}(aq)$$

Figure 14
For the reaction of potassium sulfate with barium nitrate, the net ionic equation shows that aqueous barium and sulfate ions join to form solid, insoluble barium sulfate.

287

MATH

CONNECTION

Simplifying Equations Relate solving an algebraic equation to removing spectator ions from a reaction. For example, if a mathematical equation is $2x + 4y - 2 = 5x + 4y + 12$, the first step in solving it is to remove common items from both sides. $2x$, $4y$, and -2 should be subtracted from both sides of the equation to simplify it. This process follows the same idea as removing ions that are common to both sides of a chemical equation to simplify it and find the net ionic equation.

 Transparencies

TT Writing a Net Ionic Equation

Teaching Tip

Writing in Rhythm Students who find that they leave out one or two details as they write total and net ionic equations might find establishing a rhythm or pattern helpful. For example, as they write out the total ionic equation, students can write each species in this order: coefficient, symbol, charge, and state. By repeating this pattern on every species in every equation, students will establish a habit and will be less likely to omit any information.

Demonstration

Net Ionic Equations and Conservation of Mass Before class, collect an assortment of small objects. Divide them into two groups of equal mass, and be sure some identical objects are in each group. Be sure that the identical objects are equal in mass as well as in appearance. Place each group on a pan on a balance. Show students that the masses are the same. Then remove equal numbers of objects that are the same on each side of the balance. Show students that the masses of the items on the balance are still the same.

Homework — GENERAL

Writing Ionic Equations Have students choose 5 equations that describe displacement or double-displacement reactions. Have them write total and net ionic equations for each of the equations. Check students' work for accuracy.

LS Logical

Writing Net Ionic Equations

1. List what you know.
- Identify each chemical described as a reactant or product.
- Identify the type of reaction taking place.

2. Write a balanced equation.
- Use the type of reaction to predict products, if necessary.
- Write a formula equation, and balance it. Include the physical state for each substance. Use the rules below with double-displacement reactions to determine whether a substance is an insoluble solid.

> All compounds of Group 1 and NH_4^+ are soluble.
> All nitrates are soluble.
> All halides, except those of Ag^+ and Pb^{2+}, are soluble.
> All sulfates, except Ca^{2+}, Sr^{2+}, Ba^{2+}, Hg_2^{2+}, and Pb^{2+}, are soluble.
> All carbonates, except those of Group 1 and NH_4^+, are insoluble.

3. Write the total ionic equation.
- Write separated aqueous ions for each aqueous ionic substance in the chemical equation.
- Do not split up any substance that is a solid, liquid, or gas.

4. Find the net ionic equation.
- Cancel out spectator ions, and write whatever remains as the net ionic equation.
- Double-check that the equation is balanced with respect to atoms and electric charge.

Check Atoms and Charge

Balanced net ionic equations are no different than other equations in that the numbers and kinds of atoms must be the same on each side of the equation. However, you also need to check that the sum of the charges for the reactants equals the sum of the charges for the products. As an example, recall the net ionic equation from **Figure 14.**

$$SO_4^{2-}(aq) + Ba^{2+}(aq) \longrightarrow BaSO_4(s)$$

One barium atom is on both sides of the equation, and one sulfate ion is on both sides of the equation. The sum of the charges is zero both in the reactants and in the products. Each side of a net ionic equation can have a net charge that is not zero. For example, the net ionic equation below has a net charge of 2+ on each side and is balanced.

$$Zn(s) + Cu^{2+}(aq) \longrightarrow Zn^{2+}(aq) + Cu(s)$$

288

HEALTH CONNECTION

The concentration of various ions in the blood must be maintained within a narrow range. For example, if the concentration of sodium ions is too high, it can have a negative effect on health. However, if the concentration of sodium ions is too low, different health problems might result.

Section Review

UNDERSTANDING KEY IDEAS

1. Explain why the term *spectator ions* is used.

2. What chemicals are present in a net ionic equation?

3. Is the following a correct net ionic equation? Explain.

$$Na^+(aq) + Cl^-(aq) \longrightarrow NaCl(aq)$$

4. Identify the spectator ion(s) in the following reaction:

$$MgSO_4(aq) + 2AgNO_3(aq) \longrightarrow$$
$$Ag_2SO_4(s) + Mg(NO_3)_2(aq)$$

5. Use the rules from **Skills Toolkit 4** to explain how to determine the physical states of the products in item 4.

PRACTICE PROBLEMS

6. Write a total ionic equation for each of the following unbalanced formula equations:

   a. $Br_2(l) + NaI(aq) \longrightarrow NaBr(aq) + I_2(s)$

   b. $Ca(OH)_2(aq) + HCl(aq) \longrightarrow$
   $CaCl_2(aq) + H_2O(l)$

   c. $Mg(s) + AgNO_3(aq) \longrightarrow$
   $Ag(s) + Mg(NO_3)_2(aq)$

   d. $AgNO_3(aq) + KBr(aq) \longrightarrow$
   $AgBr(s) + KNO_3(aq)$

   e. $Ni(s) + Pb(NO_3)_2(aq) \longrightarrow$
   $Ni(NO_3)_2(aq) + Pb(s)$

   f. $Ca(s) + H_2O(l) \longrightarrow Ca(OH)_2(aq) + H_2(g)$

7. Identify the spectator ions, and write a net ionic equation for each reaction in item 6.

8. Predict the products for each of the following reactions. If no reaction happens, write "no reaction." Write a total ionic equation for each reaction that does happen.

   a. $AuCl_3(aq) + Ag(s) \longrightarrow$

   b. $AgNO_3(aq) + CaCl_2(aq) \longrightarrow$

   c. $Al(s) + NiSO_4(aq) \longrightarrow$

   d. $Na(s) + H_2O(l) \longrightarrow$

   e. $AgNO_3(aq) + NaCl(aq) \longrightarrow$

9. Identify the spectator ions, and write a net ionic equation for each reaction that happens in item 8.

10. Write a total ionic equation for each of the following reactions:

    a. silver nitrate + sodium sulfate

    b. aluminum + nickel(II) iodide

    c. potassium sulfate + calcium chloride

    d. magnesium + copper(II) bromide

    e. lead(II) nitrate + sodium chloride

11. Identify the spectator ions, and write a net ionic equation for each reaction in item 10.

CRITICAL THINKING

12. Why is K^+ always a spectator ion?

13. Do net ionic equations always obey the rule of conservation of charge? Explain.

14. Suppose a drinking-water supply contains Ba^{2+}. Using solubility rules, write a net ionic equation for a double-displacement reaction that indicates how Ba^{2+} might be removed.

15. Explain why no reaction occurs if a double-displacement reaction has four spectator ions.

16. Explain why more than one reaction can have the same net ionic equation. Provide at least two reactions that have the same net ionic equation.

289

Answers to Section Review

1. Spectator ions are present but do not participate in the reaction, so they seem to "watch" the change happening.

2. the chemicals that actually react or are formed

3. No, NaCl(*aq*) is ionic and should be written as separated aqueous ions, so both ions shown are spectator ions.

4. Mg^{2+} and NO_3^-

5. Although most sulfates are soluble, silver sulfate is not soluble, so Ag_2SO_4 is a solid precipitate. All nitrates are soluble, so $Mg(NO_3)_2$ is aqueous.

6. a. $Br_2(l) + 2Na^+(aq) + 2I^-(aq) \longrightarrow 2Na^+(aq) + 2Br^-(aq) + I_2(s)$

   b. $Ca^{2+}(aq) + 2OH^-(aq) + 2H^+(aq) + 2Cl^-(aq) \longrightarrow Ca^{2+}(aq) + 2Cl^-(aq) + 2H_2O(l)$

   c. $Mg(s) + 2Ag^+(aq) + 2NO_3^-(aq) \longrightarrow 2Ag(s) + Mg^{2+}(aq) + 2NO_3^-(aq)$

   d. $Ag^+(aq) + NO_3^-(aq) + K^+(aq) + Br^-(aq) \longrightarrow AgBr(s) + K^+(aq) + NO_3^-(aq)$

   e. $Ni(s) + Pb^{2+}(aq) + 2NO_3^-(aq) \longrightarrow Ni^{2+}(aq) + 2NO_3^-(aq) + Pb(s)$

   f. $Ca(s) + 2H_2O(l) \longrightarrow Ca^{2+}(aq) + 2OH^-(aq) + H_2(g)$

Answers continued on p. 299A

CONSUMER FOCUS

Fire Extinguishers

A fire is a combustion reaction. Three things are needed for a combustion reaction: a fuel, oxygen, and an ignition source. If any one of these three is absent, combustion cannot occur. One goal in fighting a fire is to remove one or more of these parts. Many extinguishers are designed to cool the burning material (to hinder ignition) or to prevent air and oxygen from reaching it.

Types of Fires

Each type of fire requires different firefighting methods. Class A fires involve solid fuels, such as wood. Class B fires involve a liquid or a gas, such as gasoline or natural gas. Class C fires involve the presence of a "live" electric circuit. Class D fires involve burning metals.

The type of extinguisher is keyed to the type of fire. Extinguishers for Class A fires often use water. The water cools the fuel so that it does not react as readily. The steam that is produced helps displace the oxygen-containing air around the fire. Carbon dioxide extinguishers can also be used. Because carbon dioxide is denser than air, it forms a layer underneath the air and cuts off the O_2 supply. Water cannot be used on Class B fires.

Because water is usually denser than the fuel, it sinks below the fuel. Carbon dioxide is preferred for Class B fires.

Dry Chemical Extinguishers

Class C fires involving a "live" electric circuit can also be extinguished by CO_2. Water cannot be used because of the danger of electric shock. Some Class C fire extinguishers contain a dry chemical that smothers the fire by interrupting the chain reaction that is occurring. For example, a competing reaction may take place with the contents of the fire extinguisher and the intermediates of the reaction. Class C fire extinguishers usually contain compounds such as ammonium dihydrogen phosphate, $NH_4H_2PO_4$, or sodium hydrogen carbonate, $NaHCO_3$.

Finally, Class D fires involve burning metals. These fires cannot be extinguished with CO_2 or water because these compounds may react with some hot metals. For these fires, nonreactive dry powders are used to cover the metal and to keep it separate from oxygen. One kind of powder contains finely ground sodium chloride crystals mixed with a special polymer that allows the crystals to adhere to any surface, even a vertical one.

Questions

1. Identify the type of fire extinguisher available in your laboratory. On what classes of fires should it be used? Record the steps needed to use the fire extinguisher.

2. Explain why a person whose clothing has caught fire is likely to make the situation worse by running. Explain why wrapping a person in a fire blanket can help extinguish the flames.

internet connect

www.scilinks.org
Topic: Fire Extinguishers
SciLinks code: HW4059

SCiLINKS Maintained by the National Science Teachers Association

290

CHAPTER HIGHLIGHTS 8

CHAPTER HIGHLIGHTS 8

KEY TERMS

chemical reaction
chemical equation

coefficient

combustion reaction
synthesis reaction
decomposition reaction
activity series
double-displacement
   reaction

spectator ions

KEY IDEAS

SECTION ONE   Describing Chemical Reactions

- In a chemical reaction, atoms rearrange to form new substances.
- A chemical analysis is the only way to prove that a reaction has occurred.
- Symbols are used in chemical equations to identify the physical states of substances and the physical conditions during a chemical reaction.

SECTION TWO   Balancing Chemical Equations

- A word equation is translated into a formula equation to describe the change of reactants into products.
- The masses, numbers, and types of atoms are the same on both sides of a balanced equation.
- Coefficients in front of the formulas of reactants and products are used to balance an equation. Subscripts cannot be changed.

SECTION THREE   Classifying Chemical Reactions

- In a combustion reaction, a carbon-based compound reacts with oxygen to form carbon dioxide and water.
- In a synthesis reaction, two reactants form a single product.
- In a decomposition reaction, a single reactant forms two or more products.
- In a displacement reaction, an element displaces an element from a compound. The activity series is used to determine if a reaction will happen.
- In a double-displacement reaction, the ions of two compounds switch places such that two new compounds form. One of the products must be a solid, a gas, or a molecular compound, such as water, for a reaction to occur.

SECTION FOUR   Writing Net Ionic Equations

- A total ionic equation shows all aqueous ions for a reaction.
- Spectator ions do not change during a reaction and can be removed from the total ionic equation.
- Net ionic equations show only the net change of a reaction and are the best way to describe displacement and double-displacement reactions.

KEY SKILLS

Balancing an Equation
Skills Toolkit 1 p. 268
Sample Problem A p. 269

The Odd-Even Technique
Sample Problem B p. 271

Polyatomic Ions as a Group
Sample Problem C p. 273

Predicting Products
Sample Problem D p. 279
Skills Toolkit 3 p. 284

Determining Products by Using the Activity Series
Skills Toolkit 2 p. 281
Sample Problem E p. 282

Writing Net Ionic Equations
Skills Toolkit 4 p. 288

291

Alternative Assessments

SKILL BUILDER
- Page 278

Group Activity
- Page 290

CHAPTER REVIEW
- Items 67, 68, 71

Portfolio Assessments

SKILL BUILDER
- Page 278

CHAPTER REVIEW
- Items 69, 70

Chapter Resource File

- **Observation Lab** Evidence for a Chemical Change
- **Microscale Lab** Simple Qualitative Analysis
- **Microscale Lab** Reacting Ionic Species in Aqueous Solution

Assignment Guide

Section	Questions
1	6–13, 31–36, 50, 63, 73–76
2	2, 5, 14–18, 37–45, 51, 53, 56, 60, 77
3	1, 19–25, 46–49, 52, 54–55, 57, 61–62, 65–72
4	3, 4, 26–30, 58–59, 64

8 CHAPTER REVIEW
REVIEW ANSWERS

Using Key Terms

1. A synthesis reaction is often the reverse of a decomposition reaction.

2. A coefficient multiplies the number of each type of atom in the formula that follows it.

3. a. a reaction in which a single compound breaks down to form two or more simpler substances

b. a reaction in which a gas, a solid precipitate, or a molecular compound forms from the apparent exchange of atoms or ions between two compounds

c. ions that are present in a solution in which a reaction is taking place but that do not participate in the reaction

d. a series of elements that have similar properties and that are arranged in descending order of chemical activity

4. Spectator ions remain unchanged during a reaction. It is as though they are watching the reaction take place, like spectators watching a game.

5. A coefficient is placed before a formula and multiplies each atom in the substance following it. A subscript is placed to the lower right of a chemical symbol and multiplies only the element preceding it.

6. Answers may vary. Accept all reasonable responses.

USING KEY TERMS

1. Describe the relationship between a synthesis reaction and a decomposition reaction.

2. How does a coefficient in front of a formula affect the number of each type of atom in the formula?

3. Define each of the following terms:
a. *decomposition reaction*
b. *double-displacement reaction*
c. *spectator ions*
d. *activity series*

4. How does the term *spectator* apply to spectator ions?

5. How does a coefficient differ from a subscript?

6. Give an example of a word equation, a formula equation, and a chemical equation.

UNDERSTANDING KEY IDEAS

Describing Chemical Reactions

7. A student writes the following statement in a lab report: "During the reaction, the particles of the reactants are lost. The reaction creates energy and particles of the products."
a. Explain the scientific inaccuracies in the student's statement.
b. How could the student correct the inaccurate statement?

8. Write an unbalanced chemical equation for each of the following.
a. Aluminum reacts with oxygen to produce aluminum oxide.

b. Phosphoric acid, H_3PO_4, is produced through the reaction between tetraphosphorus decoxide and water.
c. Iron(III) oxide reacts with carbon monoxide to produce iron and carbon dioxide.

9. Write the symbol used in a chemical equation to represent each of the following:
a. an aqueous solution
b. heated
c. a reversible reaction
d. a solid
e. at a temperature of 25°C

10. Write an unbalanced formula equation for each of the following. Include symbols for physical states in the equation.
a. solid zinc sulfide + oxygen gas $\longrightarrow$ solid zinc oxide + sulfur dioxide gas
b. aqueous hydrochloric acid + solid magnesium hydroxide $\longrightarrow$ aqueous magnesium chloride + liquid water
c. aqueous nitric acid + aqueous calcium hydroxide $\longrightarrow$ aqueous calcium nitrate + liquid water

11. Calcium oxide, CaO, is an ingredient in cement mixes. When water is added, the mixture warms up and calcium hydroxide, $Ca(OH)_2$, forms.
a. Is there any evidence of a chemical reaction?
b. In the reaction above, how can you prove that a chemical reaction has taken place?

12. Evaporating ocean water leaves a mixture of salts. Is this a chemical change? Explain.

13. Translate the following chemical equation into a sentence:

$$CH_4(g) + 2O_2(g) \longrightarrow CO_2(g) + 2H_2O(g)$$

Balancing Chemical Equations

14. Explain what is meant when one says that an equation is balanced.

15. How does the process of balancing an equation illustrate the law of conservation of mass?

16. In balancing a chemical equation, why can you change coefficients, but not subscripts?

17. Differentiate between formula equations and balanced chemical equations.

18. The white paste that lifeguards rub on their nose to prevent sunburn contains zinc oxide, $ZnO(s)$, as an active ingredient. Zinc oxide is produced by burning zinc sulfide.

$$2ZnS(s) + 3O_2(g) \longrightarrow 2ZnO(s) + 2SO_2(g)$$

a. What is the coefficient for sulfur dioxide?
b. What is the subscript for oxygen gas?
c. How many atoms of oxygen react?
d. How many atoms of oxygen appear in the total number of sulfur dioxide molecules?

Classifying Chemical Reactions

19. What are some of the characteristics of each of these five common chemical reactions?
a. combustion
b. synthesis
c. decomposition
d. displacement
e. double-displacement

20. Write a general equation for each of the five types of reaction described in item 19.

21. Explain the difference between displacement and double-displacement reactions.

22. What is an activity series?

23. When would a displacement reaction cause no reaction?

24. What must form in order for a double-displacement reaction to occur?

25. What are the products of the complete combustion of a hydrocarbon?

Writing Net Ionic Equations

26. How do total and net ionic equations differ?

27. Which ions in a total ionic equation are called *spectator ions*? Why?

28. Explain why a net ionic equation is the best way to represent a double-displacement reaction.

29. The saline solution used to soak contact lenses is primarily NaCl dissolved in water. Which of the following ways to represent the solution is *not* correct?
a. $NaCl(aq)$
b. $NaCl(s)$
c. $Na^+(aq) + Cl^-(aq)$

30. How should each of the following substances be represented in a total ionic equation?
a. $KCl(aq)$
b. $H_2O(l)$
c. $Cu(NO_3)_2(aq)$
d. $AgCl(s)$
e. $NH_4ClO_3(aq)$

PROBLEM SOLVING SKILL

PRACTICE PROBLEMS

Sample Problem A Balancing an Equation

31. Balance each of the following:
a. $H_2 + Cl_2 \longrightarrow HCl$
b. $Al + Fe_2O_3 \longrightarrow Al_2O_3 + Fe$
c. $Ba(ClO_3)_2 \longrightarrow BaCl_2 + O_2$
d. $Cu + HNO_3 \longrightarrow Cu(NO_3)_2 + NO + H_2O$

32. Write a balanced equation for each of the following:
a. iron(III) oxide + magnesium $\longrightarrow$ magnesium oxide + iron
b. nitrogen dioxide + water $\longrightarrow$ nitric acid + nitrogen monoxide
c. silicon tetrachloride + water $\longrightarrow$ silicon dioxide + hydrochloric acid
d. ammonium dichromate $\longrightarrow$ nitrogen + chromium(III) oxide + water

293

Understanding Key Ideas

7. a. Chemical reactions do not create new particles, nor are particles lost, they are rearranged. Energy is not created, only changed from one kind to another.

b. During the reaction, the elements are rearranged to make new substances and energy is released.

8. a. $Al + O_2 \longrightarrow Al_2O_3$
b. $P_4O_{10} + H_2O \longrightarrow H_3PO_4$
c. $Fe_2O_3 + CO \longrightarrow Fe + CO_2$

9. a. (aq)
b. $\xrightarrow{\text{heat}}$ or $\xrightarrow{\Delta}$
c. $\rightleftarrows$
d. (s)
e. $\xrightarrow{25°C}$

10. a. $ZnS(s) + O_2(g) \longrightarrow ZnO(s) + SO_2(g)$
b. $HCl(aq) + Mg(OH)_2(s) \longrightarrow MgCl_2(aq) + H_2O(l)$
c. $HNO_3(aq) + Ca(OH)_2(aq) \longrightarrow Ca(NO_3)_2(aq) + H_2O(l)$

11. a. energy is released
b. do a chemical analysis to show that the material formed is a substance other than CaO

12. This is probably not a chemical change. The salts that remained were originally dissolved in the ocean water and simply formed a solid as the water evaporated.

13. Gaseous methane reacts with oxygen gas to form carbon dioxide gas and water vapor.

14. The number of atoms of each element in the reactants is the same as the number of atoms of the element in the products.

15. By balancing an equation, you show that the same atoms that make up the reactants are used to make the products, so that the mass does not change during the reaction.

16. Changing coefficients changes the number of particles of substances involved in the reaction but does not change the identities of the substances. Changing subscripts changes the chemical identity of the substances.

17. Formula equations give the identity of the reactants and the products, but a balanced equation shows equal numbers of atoms of each element on both sides.

18. a. 2
   b. 2
   c. 6
   d. 4

19. a. A compound composed of mainly carbon and hydrogen reacts with oxygen and forms carbon dioxide and water.
   b. There is only one product.
   c. There is only one reactant.
   d. The reactants are an element and a compound, and the products are a different element and a different compound.
   e. Two compounds react to form two new compounds by apparently switching ions.

20. a. $C_xH_y + O_2 \longrightarrow CO_2 + H_2O$
   b. $A + B \longrightarrow AB$
   c. $AB \longrightarrow A + B$
   d. $AB + C \longrightarrow CB + A$
   e. $AB + CD \longrightarrow AD + CB$

21. In a displacement reaction, one element replaces another from a compound. In a double-displacement reaction, two compounds exchange "partners" to form two new compounds.

22. a series of elements that have similar properties and that are arranged in descending order of chemical activity

23. when the single element is lower on the activity series than the element it is trying to replace

24. a solid precipitate, a gas, or a molecular compound, such as water

---

Sample Problem B The Odd-Even Technique

33 Balance each of the following:
   a. $Fe + O_2 \longrightarrow Fe_2O_3$
   b. $H_2O_2 \longrightarrow H_2O + O_2$
   c. $C_8H_{18} + O_2 \longrightarrow CO_2 + H_2O$
   d. $Al + F_2 \longrightarrow AlF_3$

34. Write a balanced equation for each of the following:
   a. propanol (C_3H_7OH) + oxygen $\longrightarrow$
                              carbon dioxide + water
   b. aluminum + iron(II) nitrate $\longrightarrow$
                              aluminum nitrate + iron
   c. iron(III) hydroxide $\longrightarrow$
                              iron(III) oxide + water
   d. lead(IV) oxide $\longrightarrow$ lead(II) oxide + oxygen

Sample Problem C Polyatomic Ions as a Group

35. Balance each of the following:
   a. $Zn + Pb(NO_3)_2 \longrightarrow Pb + Zn(NO_3)_2$
   b. $NH_4CH_3COO + AgNO_3 \longrightarrow$
                              $NH_4NO_3 + AgCH_3COO$
   c. $H_2C_2O_4 + NaOH \longrightarrow Na_2C_2O_4 + H_2O$
   d. $Al + CuSO_4 \longrightarrow Al_2(SO_4)_3 + Cu$

36. Write a balanced equation for each of the following:
   a. copper(II) sulfate + ammonium sulfide $\longrightarrow$
                              copper(II) sulfide + ammonium sulfate
   b. nitric acid + barium hydroxide $\longrightarrow$
                              water + barium nitrate
   c. iron(III) nitrate + lithium hydroxide $\longrightarrow$
                              lithium nitrate + iron(III) hydroxide
   d. barium chloride + phosphoric acid $\longrightarrow$
                              barium phosphate + hydrochloric acid

Sample Problem D Predicting Products

37. Complete and balance the equation for each of the following synthesis reactions.
   a. $Zn + O_2 \longrightarrow$ **c.** $Cl_2 + K \longrightarrow$
   b. $F_2 + Mg \longrightarrow$ **d.** $H_2 + I_2 \longrightarrow$

38. Complete and balance the equation for the decomposition of each of the following.
   a. $HgO \longrightarrow$ **c.** $AgCl \longrightarrow$
   b. $H_2O \longrightarrow$ **d.** $KOH \longrightarrow$

---

39. Complete and balance the equation for the complete combustion of each of the following.
   a. C_3H_6 **c.** CH_3OH
   b. C_5H_{12} **d.** $C_{12}H_{22}O_{11}$

40. Each of the following reactions is a synthesis, decomposition, or combustion reaction. For each reaction, determine the type of reaction and complete and balance the equation.
   a. $C_3H_8 + O_2 \longrightarrow$
   b. $CuCl_2 \longrightarrow$
   c. $Mg + O_2 \longrightarrow$
   d. $Na_2CO_3 \longrightarrow$
   e. $Ba(OH)_2 \longrightarrow$
   f. $C_2H_5OH + O_2 \longrightarrow$

Sample Problem E Determining Products by Using the Activity Series

41. Using the activity series in **Appendix A,** predict whether each of the possible reactions listed below will occur. For the reactions that will occur, write the products and balance the equation.
   a. $Mg(s) + CuCl_2(aq) \longrightarrow$
   b. $Pb(NO_3)_2(aq) + Zn(s) \longrightarrow$
   c. $KI(aq) + Cl_2(g) \longrightarrow$
   d. $Cu(s) + FeSO_4(aq) \longrightarrow$

42. Using the activity series in **Appendix A,** predict whether each of the possible reactions listed below will occur. For the reactions that will occur, write the products and balance the equation.
   a. $H_2O(l) + Ba(s) \longrightarrow$
   b. $Ca(s) + O_2(g) \longrightarrow$
   c. $O_2(g) + Au(s) \longrightarrow$
   d. $Al(s) + O_2(g) \longrightarrow$

Skills Toolkit 3 Identifying Reactions and Predicting Products

43. Identify the type of reaction for each of the following. Then, predict products for the reaction and balance the equation. If no reaction occurs, write "no reaction."
   a. $C_2H_6 + O_2 \longrightarrow$
   b. $H_2SO_4 + Al \longrightarrow$
   c. $N_2 + Mg \longrightarrow$

294

d. $Na_2CO_3 \longrightarrow$

e. $Mg(NO_3)_2 + Na_2SO_4 \longrightarrow$

44. Identify the type of reaction for each of the following. Then, predict products for the reaction, and balance the equation. If no reaction occurs, write "no reaction."
a. water + lithium $\longrightarrow$
b. calcium + bromine $\longrightarrow$
c. silver nitrate + hydrochloric acid $\longrightarrow$
d. hydrogen iodide $\longrightarrow$

45. Identify the type of reaction for each of the following. Then, predict products for the reaction, and balance the equation. If no reaction occurs, write "no reaction."
a. ethanol (C_2H_5OH) + oxygen $\longrightarrow$
b. nitric acid + lithium hydroxide $\longrightarrow$
c. sodium nitrate + calcium chloride $\longrightarrow$
d. lead(II) nitrate + sodium carbonate $\longrightarrow$

Skills Toolkit 4 Writing Net Ionic Equations

46. Write a total ionic equation and a net ionic equation for each of the following reactions.
a. $HCl(aq) + NaOH(aq) \longrightarrow$
$$NaCl(aq) + H_2O(l)$$
b. $Mg(s) + 2HCl(aq) \longrightarrow MgCl_2(aq) + H_2(g)$
c. $CdCl_2(aq) + Na_2CO_3(aq) \longrightarrow$
$$2NaCl(aq) + CdCO_3(s)$$
d. $Mg(s) + Zn(NO_3)_2(aq) \longrightarrow$
$$Zn(s) + Mg(NO_3)_2(aq)$$

47. Identify the spectator ions in each reaction in item 46.

48. Predict the products and write a net ionic equation for each of the following reactions. If no reaction occurs, write "no reaction."
a. $K_2CO_3(aq) + CaCl_2(aq) \longrightarrow$
b. $Na_2SO_4(aq) + AgNO_3(aq) \longrightarrow$
c. $NH_4Cl(aq) + AgNO_3(aq) \longrightarrow$
d. $Pb(s) + ZnCl_2(aq) \longrightarrow$

49. Identify the spectator ions in each reaction in item 48.

MIXED REVIEW

50. For each photograph below, identify evidence of a chemical change, then identify the type of reaction shown. Support your answer.

a.

b.

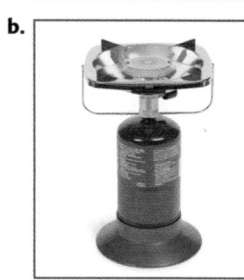

51. Balance the following equations.
a. $CaH_2(s) + H_2O(l) \longrightarrow$
$$Ca(OH)_2(aq) + H_2(g)$$
b. $CH_3CH_2CCH(g) + Br_2(l) \longrightarrow$
$$CH_3CH_2CBr_2CHBr_2(l)$$
c. $Pb^{2+}(aq) + OH^-(aq) \longrightarrow Pb(OH)_2(s)$
d. $NO_2(g) + H_2O(l) \longrightarrow HNO_3(aq) + NO(g)$

52. Write and balance each of the following equations, and then identify each equation by type.
a. hydrogen + iodine $\longrightarrow$ hydrogen iodide
b. lithium + water $\longrightarrow$
$$\text{lithium hydroxide + hydrogen}$$
c. mercury(II) oxide $\longrightarrow$ mercury + oxygen
d. copper + chlorine $\longrightarrow$ copper(II) chloride

53. Write a balanced equation, including all of the appropriate notations, for each of the following reactions.
a. Steam reacts with solid carbon to form the gases carbon monoxide and hydrogen.

25. carbon dioxide and water

26. A total ionic equation contains all ions present in the reaction, including spectator ions. The net ionic equation does not contain the spectator ions and only shows those ions that actually take part in the reaction.

27. Spectator ions are the ions that appear the same on both sides of the total ionic equation. They are unchanged by the reaction and appear in both the reactants and the products.

28. Because some ions in a double-displacement reaction are not changed by the reaction, the net ionic equation is the best description for the change that does happen.

29. Choice **b** is not correct.

30. a. $K^+(aq) + Cl^-(aq)$
b. $H_2O(l)$
c. $Cu^{2+}(aq) + 2NO_3^-(aq)$
d. $AgCl(s)$
e. $NH_4^+(aq) + ClO_3^-(aq)$

Practice Problems

31. a. $H_2 + Cl_2 \longrightarrow 2HCl$
b. $2Al + Fe_2O_3 \longrightarrow 2Fe + Al_2O_3$
c. $Ba(ClO_3)_2 \longrightarrow BaCl_2 + 3O_2$
d. $3Cu + 8HNO_3 \longrightarrow$
$$3Cu(NO_3)_2 + 2NO + 4H_2O$$

32. a. $Fe_2O_3 + 3Mg \longrightarrow$
$$3MgO + 2Fe$$
b. $3NO_2 + H_2O \longrightarrow$
$$2HNO_3 + NO$$
c. $SiCl_4 + 2H_2O \longrightarrow$
$$SiO_2 + 4HCl$$
d. $(NH_4)_2Cr_2O_7 \longrightarrow$
$$N_2 + Cr_2O_3 + 4H_2O$$

33. a. $4Fe + 3O_2 \longrightarrow 2Fe_2O_3$
b. $2H_2O_2 \longrightarrow 2H_2O + O_2$
c. $2C_8H_{18} + 25O_2 \longrightarrow$
$$16CO_2 + 18H_2O$$
d. $2Al + 3F_2 \longrightarrow 2AlF_3$

34. a. $2C_3H_7OH + 9O_2 \longrightarrow$
$$6CO_2 + 8H_2O$$
b. $2Al + 3Fe(NO_3)_2 \longrightarrow$
$$2Al(NO_3)_3 + 3Fe$$
c. $2Fe(OH)_3 \longrightarrow Fe_2O_3 + 3H_2O$
d. $2PbO_2 \longrightarrow 2PbO + O_2$

35. a. $Zn + Pb(NO_3)_2 \longrightarrow$
$Pb + Zn(NO_3)_2$

b. $NH_4CH_3COO + AgNO_3 \longrightarrow$
$NH_4NO_3 + AgCH_3COO$

c. $H_2C_2O_4 + 2NaOH \longrightarrow$
$Na_2C_2O_4 + 2H_2O$

d. $2Al + 3CuSO_4 \longrightarrow$
$Al_2(SO_4)_3 + 3Cu$

36. a. $CuSO_4 + (NH_4)_2S \longrightarrow$
$CuS + (NH_4)_2SO_4$

b. $2HNO_3 + Ba(OH)_2 \longrightarrow$
$2H_2O + Ba(NO_3)_2$

c. $Fe(NO_3)_3 + 3LiOH \longrightarrow$
$3LiNO_3 + Fe(OH)_3$

d. $3BaCl_2 + 2H_3PO_4 \longrightarrow$
$Ba_3(PO_4)_2 + 6HCl$

37. a. $2Zn + O_2 \longrightarrow 2ZnO$

b. $F_2 + Mg \longrightarrow MgF_2$

c. $Cl_2 + 2K \longrightarrow 2KCl$

d. $H_2 + I_2 \longrightarrow 2HI$

38. a. $2HgO \longrightarrow 2Hg + O_2$

b. $2H_2O \longrightarrow 2H_2 + O_2$

c. $2AgCl \longrightarrow 2Ag + Cl_2$

d. $2KOH \longrightarrow K_2O + H_2O$

39. a. $2C_3H_6 + 9O_2 \longrightarrow 6CO_2 +$
$6H_2O$

b. $C_5H_{12} + 8O_2 \longrightarrow 5CO_2 +$
$6H_2O$

c. $2CH_3OH + 3O_2 \longrightarrow 2CO_2 +$
$4H_2O$

d. $C_{12}H_{22}O_{11} + 12O_2 \longrightarrow$
$12CO_2 + 11H_2O$

40. a. combustion; $C_3H_8 + 5O_2 \longrightarrow$
$3CO_2 + 4H_2O$

b. decomposition; $CuCl_2 \longrightarrow$
$Cu + Cl_2$

c. synthesis; $2Mg + O_2 \longrightarrow$
$2MgO$

d. decomposition; $Na_2CO_3 \longrightarrow$
$Na_2O + CO_2$

e. decomposition; $Ba(OH)_2 \longrightarrow$
$BaO + H_2O$

f. combustion; $C_2H_5OH +$
$3O_2 \longrightarrow 2CO_2 + 3H_2O$

b. Heating ammonium nitrate in aqueous solution forms dinitrogen monoxide gas and liquid water.

c. Nitrogen dioxide gas forms from the reaction of nitrogen monoxide gas and oxygen gas.

54. Methanol, CH_3OH, is a clean-burning fuel.
a. Write a balanced chemical equation for the synthesis of methanol from carbon monoxide and hydrogen gas.
b. Write a balanced chemical equation for the complete combustion of methanol.

55. Use the activity series to predict whether the following reactions are possible. Explain your answers.
a. $Ni(s) + MgSO_4(aq) \longrightarrow$
$NiSO_4(aq) + Mg(s)$
b. $3Mg(s) + Al_2(SO_4)_3(aq) \longrightarrow$
$3MgSO_4(aq) + 2Al(s)$
c. $Pb(s) + 2H_2O(l) \longrightarrow$
$Pb(OH)_2(aq) + H_2(g)$

56. Iron(III) chloride, $FeCl_3$, is a chemical used in photography. It can be produced by reacting iron and chlorine. Identify the correct chemical equation from the choices below, and then explain what is wrong with the other two choices.
a. $Fe(s) + Cl_3(g) \longrightarrow FeCl_3(s)$
b. $2Fe(s) + 3Cl_2(g) \longrightarrow 2FeCl_3(s)$
c. $Fe(s) + 3Cl_2(g) \longrightarrow Fe2Cl_3(s)$

57. Write the balanced equation for each of the following:
a. the complete combustion of propane gas, C_3H_8
b. the decomposition of magnesium carbonate
c. the synthesis of platinum(IV) fluoride from platinum and fluorine gas
d. the reaction of zinc with lead(II) nitrate

58. Predict the products for each of the following reactions. Write a total ionic equation and a net ionic equation for each reaction. If no reaction occurs, write "no reaction."
a. $Li_2CO_3(aq) + BaBr_2(aq) \longrightarrow$

b. $Na_2SO_4(aq) + Sr(NO_3)_2(aq) \longrightarrow$
c. $Al(s) + NiCl_2(aq) \longrightarrow$
d. $K_2CO_3(aq) + FeCl_3(aq) \longrightarrow$

59. Identify the spectator ions in each reaction in item 58.

CRITICAL THINKING

60. The following equations are incorrect in some way. Identify and correct each error, and then balance each equation.
a. $Li + O_2 \longrightarrow LiO_2$
b. $MgCO_3 \longrightarrow Mg + C + 3O_2$
c. $NaI + Cl_2 \longrightarrow NaCl + I$
d. $AgNO_3 + CaCl_2 \longrightarrow Ca(NO_3) + AgCl_2$
e. $3Mg + 2FeBr_3 \longrightarrow Fe_2Mg_3 + 3Br_2$

61. Although cesium is not listed in the activity series in this chapter, predict where cesium would appear based on its position in the periodic table.

62. Create an activity series for the hypothetical elements A, J, Q, and Z by using the reaction information provided below.

$$A + ZX \longrightarrow AX + Z$$
$$J + ZX \longrightarrow \text{no reaction}$$
$$Q + AX \longrightarrow QX + A$$

63. When wood burns, the ash weighs much less than the original wood did. Explain why the law of conservation of mass is not violated in this situation.

64. Write the total and net ionic equations for the reaction in which the antacid $Al(OH)_3$ neutralizes the stomach acid HCl. Identify the type of reaction.
a. Identify the spectator ions in this reaction.
b. What would be the advantages of using $Al(OH)_3$ as an antacid rather than $NaHCO_3$, which undergoes the following reaction with stomach acid?

$$NaHCO_3(aq) + HCl(aq) \longrightarrow$$
$$NaCl(aq) + H_2O(l) + CO_2(g)$$

296

65. The images below represent the reactants of a chemical reaction. Study the images, then answer the items that follow.

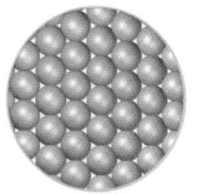 

sodium   water

a. Write a balanced chemical equation for the reaction that shows the states of all substances.
b. What type of reaction is this?

ALTERNATIVE ASSESSMENT

66. Using the materials listed below, describe a procedure that would enable you to organize the metals in order of reactivity. The materials are pieces of aluminum, chromium, and magnesium and solutions of aluminum chloride, chromium(III) chloride, and magnesium chloride.

67. Design an experiment for judging the value and efficacy of different antacids. Include $NaHCO_3$, $Mg(OH)_2$, $CaCO_3$, and $Al(OH)_3$ in your tests. Discover which one neutralizes the most acid and what byproducts form. Show your experiment to your teacher. If your experiment is approved, obtain the necessary chemicals from your teacher and test your procedure.

68. For one day, record situations that suggest that a chemical change has occurred. Identify the reactants and the products, and state whether there is proof of a chemical reaction. Classify each of the chemical reactions according to the five common reaction types discussed in the chapter.

69. Research safety tips for dealing with fires. Create a poster or brochure about fire safety in which you explain both these tips and their basis in science.

70. Many products are labeled "biodegradable." Choose several biodegradable items on the market, and research the decomposition reactions that occur. Take into account any special conditions that must occur for the substance to biodegrade. Present your information to the class to help inform the students about which products are best for the environment.

71. Develop an analogy to help illustrate each of the five types of reactions described in this chapter. Write the analogy as though you were explaining what happens in the reaction to a classmate who was having trouble understanding. You might consider starting your analogy with the phrase "A _____ reaction is like . . ."

CONCEPT MAPPING

72. Use the following terms to complete the concept map below: *a synthesis reaction, a decomposition reaction, coefficients, a chemical reaction,* and *a chemical equation.*

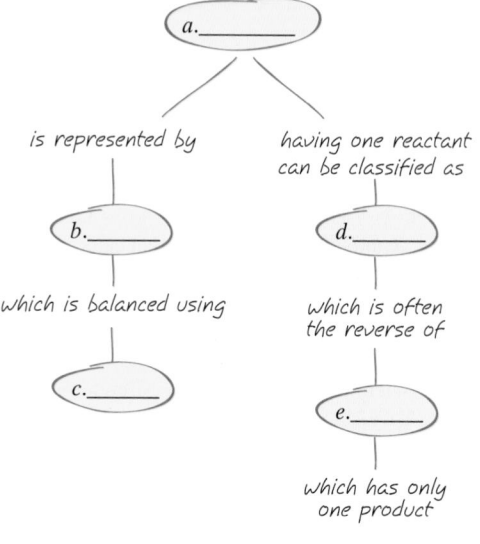

297

41. a. $Mg(s) + CuCl_2(aq) \longrightarrow$
$MgCl_2(aq) + Cu(s)$
b. $Pb(NO_3)_2(aq) + Zn(s) \longrightarrow$
$Zn(NO_3)_2(aq) + Pb(s)$
c. $2KI(aq) + Cl_2(g) \longrightarrow$
$2KCl(aq) + I_2(s)$
d. no reaction

42. a. $2H_2O(l) + Ba(s) \longrightarrow$
$Ba(OH)_2(s) + H_2(g)$
b. $4Ca(s) + O_2(g) \longrightarrow 2CaO(s)$
c. no reaction
d. $4Al(s) + 3O_2(g) \longrightarrow$
$2Al_2O_3(s)$

43. a. combustion; $2C_2H_6 +$
$7O_2 \longrightarrow 4CO_2 + 6H_2O$
b. displacement; $3H_2SO_4 +$
$2Al \longrightarrow Al_2(SO_4)_3 + 3H_2$
c. synthesis; $N_2 + 3Mg \longrightarrow$
Mg_3N_2
d. decomposition; $Na_2CO_3 \longrightarrow$
$Na_2O + CO_2$
e. double-displacement; no reaction

44. a. displacement; $2H_2O +$
$2Li \longrightarrow 2LiOH + H_2$
b. synthesis; $Ca + Br_2 \longrightarrow$
$CaBr_2$
c. double-displacement;
$AgNO_3 + HCl \longrightarrow AgCl(s) +$
HNO_3
d. decomposition; $2HI \longrightarrow H_2 +$
I_2

45. a. combustion; $C_2H_5OH +$
$3O_2 \longrightarrow 2CO_2 + 3H_2O$
b. double-displacement; $HNO_3 +$
$LiOH \longrightarrow LiNO_3 + HOH(l)$
c. double-displacement; no reaction
d. double-displacement;
$Pb(NO_3)_2 + Na_2CO_3 \longrightarrow$
$PbCO_3(s) + 2NaNO_3$

46. a. total: $H^+(aq) + Cl^-(aq) +$
$Na^+(aq) + OH^-(aq) \longrightarrow$
$Na^+(aq) + Cl^-(aq) + H_2O(l)$;
net: $H^+(aq) + OH^-(aq) \longrightarrow$
$H_2O(l)$
b. total: $Mg(s) + 2H^+(aq) +$
$2Cl^-(aq) \longrightarrow Mg^{2+}(aq) +$
$2Cl^-(aq) + H_2(g)$;
net: $Mg(s) + 2H^+(aq) \longrightarrow$
$Mg^{2+}(aq) + H_2(g)$

c. total: $Cd^{2+}(aq) + 2Cl^-(aq) + 2Na^+(aq) + CO_3^{2-}(aq) \longrightarrow 2Na^+(aq) + 2Cl^-(aq) + CdCO_3(s)$; net: $Cd^{2+}(aq) + CO_3^{2-}(aq) \longrightarrow CdCO_3(s)$

d. total: $Mg(s) + Zn^{2+}(aq) + 2NO_3^-(aq) \longrightarrow Zn(s) + Mg^{2+}(aq) + 2NO_3^-(aq)$; net: $Mg(s) + Zn^{2+}(aq) \longrightarrow Zn(s) + Mg^{2+}(aq)$

47. a. Na^+ and Cl^-

b. Cl^-

c. Na^+ and Cl^-

d. NO_3^-

48. a. products: $KCl(aq)$ and $CaCO_3(s)$; $CO_3^{2-}(aq) + Ca^{2+}(aq) \longrightarrow CaCO_3(s)$

b. products: $NaNO_3(aq)$ and $Ag_2SO_4(s)$; $SO_4^{2-}(aq) + 2Ag^+(aq) \longrightarrow Ag_2SO_4(s)$

c. products: $NH_4NO_3(aq)$ and $AgCl(s)$; $Cl^-(aq) + Ag^+(aq) \longrightarrow AgCl(s)$

d. no reaction

49. a. K^+ and Cl^-

b. Na^+ and NO_3^-

c. NH_4^+ and NO_3^-

d. no reaction

Chapter Resource File

• Chapter Test

Answers continued on p. 299B

Study the graph below, and answer the questions that follow.
For help in interpreting graphs, see Appendix B, "Study Skills for Chemistry."

73. Which halogen has the shortest single bond with hydrogen?

74. What is the difference in length between an H–Br bond and an H–I bond?

75. Describe the trend in bond length as you move down the elements in Group 17 on the periodic table.

76. Based on this graph, what conclusion can be drawn about the relative sizes of halogen atoms? Could you draw the same conclusion if an atom of an element other than hydrogen was bonded to an atom of each halogen?

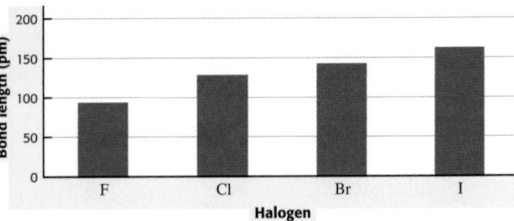

Length of Hydrogen-Halogen Single Bond

TECHNOLOGY AND LEARNING

77. Graphing Calculator

Least Common Multiples When writing chemical formulas or balancing a chemical equation, being able to identify the least common multiple of a set of numbers can often help. Your graphing calculator has a least common multiple function that can compare two numbers. On a TI-83 Plus or similar graphing calculator, press **MATH** ➤ **8**. The screen should read "lcm(." Next, enter one number and then a comma followed by the other number and a closing parenthesis. Press **ENTER**, and the calculator will show the least common multiple of the pair you entered.

Use this function as needed to find the answers to the following questions.

a. Tin(IV) sulfate contains Sn^{4+} and SO_4^{2-} ions. Use the least common multiple of 2 and 4 to determine the empirical formula for this compound.

b. Aluminum ferrocyanide contains Al^{3+} ions and $Fe(CN)_6^{4-}$ ions. Use the least common multiple of 3 and 4 to determine the empirical formula for this compound.

c. Balance the following **unbalanced** equation.

$$\_\_P_4O_{10}(s) + \_\_H_2O(g) \longrightarrow \_\_H_3PO_4(aq)$$

d. Balance the following **unbalanced** equation.

$$\_\_KMnO_4(aq) + \_\_MnCl_2(aq) + 2H_2O(l) \longrightarrow \_\_MnO_2(s) + 4HCl(aq) + 2KCl(aq)$$

e. The combustion of octane, C_8H_{18}, and oxygen, O_2, is one of many reactions that occur in a car's engine. The products are CO_2 and H_2O. Balance the equation for the combustion reaction. (Hint: Balance oxygen last, and use the least common multiple of the number of oxygen atoms on the products' side and on the reactants' side to help balance the equation.)

1. According to the law of conservation of mass, the total mass of the reacting substances is
   a. always more than the total mass of the products.
   b. always less than the total mass of the products.
   c. sometimes more and sometimes less than the total mass of the products.
   d. always equal to the total mass of the products.

2. In a chemical equation, the symbol (*aq*) indicates that the substance is
   a. water.
   b. dissolved in water.
   c. an acid.
   d. insoluble.

3. The ratio of chlorine to hydrogen chloride in the reaction $H_2(g) + Cl_2(g) \longrightarrow 2HCl(g)$ is
   a. 1:1.   **c.** 2:1.
   b. 1:2.   **d.** 2:2.

4. A common interpretation is that the coefficients in a chemical equation
   a. show the number of grams of each substance that would react.
   b. indicate the number of molecules of each substance.
   c. are the molar masses of the substances.
   d. show the valence electrons for each atom.

5. A reaction in which the ions of two compounds exchange places in aqueous solution to form two new compounds is called a
   a. synthesis reaction.
   b. double-displacement reaction.
   c. decomposition reaction.
   d. combustion reaction.

6. To balance a chemical equation, you may adjust the
   a. coefficients.
   b. subscripts.
   c. formulas of the products.
   d. number of products.

7. The use of a double arrow in a chemical equation indicates that the reaction
   a. is reversible.
   b. requires heat.
   c. is written backward.
   d. has not been confirmed in the laboratory.

8. In a word equation, the plus sign, +, represents the word
   a. *and.*   **c.** *produce.*
   b. *heat.*   **d.** *yield.*

9. The ions that do not participate in a reaction and that do not appear in a net ionic equation are called
   a. reactants.   **c.** total ions.
   b. spectator ions.   **d.** products.

10. The reaction type that has only one reactant is
   a. decomposition.   **c.** synthesis.
   b. displacement.   **d.** oxidation.

11. A substance followed by which symbol should be written as separated ions in a total ionic equation?
   a. (*s*)   **c.** (*g*)
   b. (*l*)   **d.** (*aq*)

12. Elements in the activity series are arranged by
   a. atomic number.
   b. position in the family.
   c. ionization energy.
   d. experimentally determined order of reactivity.

299

Continuation of Answers

Answers to Section 1 Review, *continued*

12. loud sound produced, bright light given off, changes in color, production of smoke and gases

13. The liquids could be considered reactants. After mixing, the resulting material has properties that differ from the properties of either liquid, which could indicate that a reaction has happened.

14. A chemical analysis would show that both forms are elemental sulfur; there is no difference in chemical identity.

15. the addition of electrical energy

16. water $(l) \xrightarrow{\text{electricity}}$ hydrogen (g) + oxygen (g)

17. a. Solid zinc reacts with aqueous hydrochloric acid to form aqueous zinc chloride and hydrogen gas.

    b. Aqueous calcium chloride reacts with aqueous sodium carbonate to form solid calcium carbonate and aqueous sodium chloride.

    c. Aqueous sodium hydroxide and aqueous hydrochloric acid react to form aqueous sodium chloride and liquid water.

    d. Heating solid calcium carbonate produces solid calcium oxide plus carbon dioxide gas.

Answers to Section 2 Review, *continued*

8. Coefficients in the reactants are 2 and 3. Coefficients in the products are 1 and 3. Subscripts in the reactants are 1, 2, 1, and 4. Subscripts in the products are 2, 1, 4, 3, and 2.

9. $2H_2 + O_2 \longrightarrow 2H_2O$; Reactants: $2(2.02\ \text{amu}) + 1(32.00\ \text{amu}) = 36.04\ \text{amu}$; Products: $2(18.02\ \text{amu}) = 36.04\ \text{amu}$; So the mass of the reactants is equal to the mass of the products, which demonstrates that mass is conserved.

10. It is incorrect. The formula for chlorine gas should be Cl_2, not Cl_3. The correct equation is $2Fe(s) + 3Cl_2(g) \longrightarrow 2FeCl_3(s)$.

Answers to Section 3 Review, *continued*

    d. $2AgNO_3(aq) + K_2SO_4(aq) \longrightarrow Ag_2SO_4(s) + 2KNO_3(aq)$; double-displacement

    e. $Zn(s) + CuBr_2(aq) \longrightarrow ZnBr_2(aq) + Cu(s)$; displacement

    f. $2C_8H_{18}(l) + 25O_2(g) \longrightarrow 16CO_2(g) + 18H_2O(g)$; combustion

9. a. no reaction

    b. $Mg(s) + Cu(NO_3)_2(aq) \longrightarrow Cu(s) + Mg(NO_3)_2(aq)$

    c. $4Al(s) + 3O_2(g) \longrightarrow 2Al_2O_3(s)$

    d. $H_2SO_4(aq) + 2KOH(aq) \longrightarrow K_2SO_4(aq) + 2H_2O(l)$

10. a. $2HgO \longrightarrow 2Hg + O_2$; decomposition

    b. $2C_3H_7OH + 9O_2 \longrightarrow 6CO_2 + 8H_2O$; combustion

    c. $Zn + CuSO_4 \longrightarrow Cu + ZnSO_4$; displacement

    d. $BaCl_2 + Na_2SO_4 \longrightarrow 2NaCl + BaSO_4$; double-displacement

    e. $Zn + F_2 \longrightarrow ZnF_2$; synthesis

    f. $2C_5H_{10} + 15O_2 \longrightarrow 10CO_2 + 10H_2O$; combustion

11. when the single element is below the element it is trying to replace on the activity series

12. Synthesis implies the making of something, and a synthesis reaction makes a compound. Decomposition implies breaking down something, and a decomposition reaction breaks apart a compound.

13. at or near the bottom, probably in the same category as silver and gold

14. Yes; Copper is higher than silver on the activity series and will replace the silver in silver nitrate.

Answers to Section 4 Review, *continued*

7. a. spectator ion: Na^+; $Br_2(l) + 2I^-(aq) \longrightarrow 2Br^-(aq) + I_2(s)$

    b. spectator ions: Ca^{2+} and Cl^-; $OH^-(aq) + H^+(aq) \longrightarrow H_2O(l)$

    c. spectator ion: NO_3^-; $Mg(s) + 2Ag^+(aq) \longrightarrow 2Ag(s) + Mg^{2+}(aq)$

    d. spectator ions: K^+ and NO_3^-; $Ag^+(aq) + Br^-(aq) \longrightarrow AgBr(s)$

    e. spectator ion: NO_3^-; $Ni(s) + Pb^{2+}(aq) \longrightarrow Ni^{2+}(aq) + Pb(s)$

    f. no spectator ions; $Ca(s) + 2H_2O(l) \longrightarrow Ca^{2+}(aq) + 2OH^-(aq) + H_2(g)$

8. a. products: $Au(s)$ and $AgCl(s)$; $Au^{3+}(aq) + 3Cl^-(aq) + 3Ag(s) \longrightarrow Au(s) + 3AgCl(s)$

    b. products: $AgCl(s)$ and $Ca(NO_3)_2(aq)$; $2Ag^+(aq) + 2NO_3^-(aq) + Ca^{2+}(aq) + 2Cl^-(aq) \longrightarrow 2AgCl(s) + Ca^{2+}(aq) + 2NO_3^-(aq)$

    c. products: $Ni(s)$ and $Al_2(SO_4)_3(aq)$; $2Al(s) + 3Ni^{2+}(aq) + 3SO_4^{2-}(aq) \longrightarrow 3Ni(s) + 2Al^{3+}(aq) + 3SO_4^{2-}(aq)$

    d. products: $NaOH(aq)$ and $H_2(g)$; $2Na(s) + 2H_2O(l) \longrightarrow 2Na^+(aq) + 2OH^-(aq) + H_2(g)$

    e. products: $AgCl(s)$ and $NaNO_3(aq)$; $Ag^+(aq) + NO_3^-(aq) + Na^+(aq) + Cl^-(aq) \longrightarrow AgCl(s) + Na^+(aq) + NO_3^-(aq)$

9. a. no spectator ions; $Au^{3+}(aq) + 3Cl^-(aq) + 3Ag(s) \longrightarrow Au(s) + 3AgCl(s)$

    b. spectator ions: NO_3^- and Ca^{2+}; $Ag^+(aq) + Cl^-(aq) \longrightarrow AgCl(s)$

    c. spectator ion: SO_4^{2-}; $2Al(s) + 3Ni^{2+}(aq) \longrightarrow 3Ni(s) + 2Al^{3+}(aq)$

    d. no spectator ions; $2Na(s) + 2H_2O(l) \longrightarrow 2Na^+(aq) + 2OH^-(aq) + H_2(g)$

    e. spectator ions: NO_3^- and Na^+; $Ag^+(aq) + Cl^-(aq) \longrightarrow AgCl(s)$

10. a. $2Ag^+(aq) + 2NO_3^-(aq) + 2Na^+(aq) + SO_4^{2-}(aq) \longrightarrow 2Na^+(aq) + 2NO_3^-(aq) + Ag_2SO_4(s)$

    b. $2Al(s) + 3Ni^{2+}(aq) + 6I^-(aq) \longrightarrow 3Ni(s) + 2Al^{3+}(aq) + 6I^-(aq)$

    c. $2K^+(aq) + SO_4^{2-}(aq) + Ca^{2+}(aq) + 2Cl^-(aq) \longrightarrow CaSO_4(s) + 2K^+(aq) + 2Cl^-(aq)$

    d. $Mg(s) + Cu^{2+}(aq) + 2Br^-(aq) \longrightarrow Cu(s) + Mg^{2+}(aq) + 2Br^-(aq)$

    e. $Pb^{2+}(aq) + 2NO_3^-(aq) + 2Na^+(aq) + 2Cl^-(aq) \longrightarrow PbCl_2(s) + 2Na^+(aq) + 2NO_3^-(aq)$

11. a. spectator ions: NO_3^- and Na^+; $2Ag^+(aq) + SO_4^{2-}(aq) \longrightarrow Ag_2SO_4(s)$

  b. spectator ion: I^-; $2Al(s) + 3Ni^{2+}(aq) \longrightarrow 3Ni(s) + 2Al^{3+}(aq)$

  c. spectator ions: K^+ and Cl^-; $SO_4^{2-}(aq) + Ca^{2+}(aq) \longrightarrow CaSO_4(s)$

  d. spectator ion: Br^-; $Mg(s) + Cu^{2+}(aq) \longrightarrow Cu(s) + Mg^{2+}(aq)$

  e. spectator ions: NO_3^- and Na^+; $Pb^{2+}(aq) + 2Cl^-(aq) \longrightarrow PbCl_2(s)$

12. All potassium compounds are soluble.

13. Yes; Matter, including electrons, is not created or destroyed during a chemical reaction.

14. $Ba^{2+}(aq) + SO_4^{2-}(aq) \longrightarrow BaSO_4(s)$ or $Ba^{2+}(aq) + CO_3^{2-}(aq) \longrightarrow BaCO_3(s)$

15. If a double-displacement reaction has four spectator ions, then both reactants and both products are soluble. All four ions are spectator ions and no reaction will happen.

16. More than one reaction can have the same net ionic equation as long as they have the same overall reaction. Examples may vary. One set of reactions is $NaCl + AgNO_3 \longrightarrow AgCl + NaNO_3$ and $KCl + AgNO_3 \longrightarrow AgCl + KNO_3$ which have the same net ionic equation of $Ag^+(aq) + Cl^-(aq) \longrightarrow AgCl(s)$.

Answers to Chapter Review, *continued*
Mixed Review

50. a. A cloudy solid forms when the two clear liquids are mixed. The formation of a solid precipitate from two solutions happens in a double-displacement reaction.

  b. The reaction releases energy as heat and light. There is a blue flame, which is similar to the flame of a Bunsen burner. The tank contains propane, a hydrocarbon, which is burning, so this is a combustion reaction.

51. a. $CaH_2(s) + 2H_2O(l) \longrightarrow Ca(OH)_2(aq) + 2H_2(g)$

  b. $CH_3CH_2CCH(g) + 2Br_2(l) \longrightarrow CH_3CH_2CBr_2CHBr_2(l)$

  c. $Pb^{2+}(aq) + 2OH^-(aq) \longrightarrow Pb(OH)_2(s)$

  d. $3NO_2(g) + H_2O(l) \longrightarrow 2HNO_3(aq) + NO(g)$

52. a. $H_2 + I_2 \longrightarrow 2HI$; synthesis

  b. $2Li + 2H_2O \longrightarrow 2LiOH + H_2$; displacement

  c. $2HgO \longrightarrow 2Hg + O_2$; decomposition

  d. $Cu + Cl_2 \longrightarrow CuCl_2$; synthesis

53. a. $H_2O(g) + C(s) \longrightarrow CO(g) + H_2(g)$

  b. $NH_4NO_3(aq) \xrightarrow{heat} N_2O(g) + 2H_2O(l)$

  c. $2NO(g) + O_2(g) \longrightarrow 2NO_2(g)$

54. a. $CO(g) + 2H_2(g) \longrightarrow CH_3OH(l)$

  b. $2CH_3OH + 3O_2 \longrightarrow 2CO_2 + 4H_2O$

55. a. not possible; Ni is below Mg in the activity series

  b. possible; Mg is above Al in the activity series

  c. not possible; Pb does not react with water

56. The correct equation is choice **b**. In choice **a**, the formula for chlorine is incorrect, and in choice **c**, the formula for iron(III) chloride is wrong.

57. a. $C_3H_8 + 5O_2 \longrightarrow 3CO_2 + 4H_2O$

  b. $MgCO_3 \longrightarrow MgO + CO_2$

  c. $Pt + 2F_2 \longrightarrow PtF_4$

  d. $Zn + Pb(NO_3)_2 \longrightarrow Pb + Zn(NO_3)_2$

58. a. $Li_2CO_3(aq) + BaBr_2(aq) \longrightarrow BaCO_3(s) + 2LiBr(aq)$; total: $2Li^+(aq) + CO_3^{2-}(aq) + Ba^{2+}(aq) + 2Br^-(aq) \longrightarrow BaCO_3(s) + 2Li^+(aq) + 2Br^-(aq)$; net: $CO_3^{2-}(aq) + Ba^{2+}(aq) \longrightarrow BaCO_3(s)$

  b. $Na_2SO_4(aq) + Sr(NO_3)_2(aq) \longrightarrow SrSO_4(s) + 2NaNO_3(aq)$; total: $2Na^+(aq) + SO_4^{2-}(aq) + Sr^{2+}(aq) + 2NO_3^-(aq) \longrightarrow SrSO_4(s) + 2Na^+(aq) + 2NO_3^-(aq)$; net: $SO_4^{2-}(aq) + Sr^{2+}(aq) \longrightarrow SrSO_4(s)$

  c. $2Al(s) + 3NiCl_2(aq) \longrightarrow 3Ni(s) + 2AlCl_3(aq)$; total: $2Al(s) + 3Ni^{2+}(aq) + 6Cl^-(aq) \longrightarrow 3Ni(s) + 2Al^{3+}(aq) + 6Cl^-(aq)$; net: $2Al(s) + 3Ni^{2+}(aq) \longrightarrow 3Ni(s) + 2Al^{3+}(aq)$

  d. $3K_2CO_3(aq) + 2FeCl_3(aq) \longrightarrow 6KCl(aq) + Fe_2(CO_3)_3(s)$; total: $6K^+(aq) + 3CO_3^{2-}(aq) + 2Fe^{3+}(aq) + 6Cl^-(aq) \longrightarrow 6K^+(aq) + 6Cl^-(aq) + Fe_2(CO_3)_3(s)$; net: $3CO_3^{2-}(aq) + 2Fe^{3+}(aq) \longrightarrow Fe_2(CO_3)_3(s)$

59. a. Li^+ and Br^-

  b. Na^+ and NO_3^-

  c. Cl^-

  d. K^+ and Cl^-

Critical Thinking

60. a. The formula of the product is incorrect. $4Li + O_2 \longrightarrow 2Li_2O$

  b. A ternary compound does not break into its elements. The reactant is a metal carbonate, so it should break down to form a metal oxide and carbon dioxide. $MgCO_3 \longrightarrow MgO + CO_2$

  c. Chlorine atoms are not balanced, and iodine should be a diatomic element. $2NaI + Cl_2 \longrightarrow 2NaCl + I_2$

  d. The formulas for the products are incorrect. The ions were simply exchanged without using the ion's charges to write the correct formulas. $2AgNO_3 + CaCl_2 \longrightarrow Ca(NO_3)_2 + 2AgCl$

  e. Magnesium should take the place of iron in the compound and bond with bromine. $3Mg + 2FeBr_3 \longrightarrow 2Fe + 3MgBr_2$

61. Cesium is in Group 1, so it would be in the group at the top of the series where potassium and sodium, other Group 1 metals, are listed.

62. Q is the most reactive, followed by A, then Z, then J, which is the least reactive.

63. The major constituent of wood is cellulose, which is composed of carbon, hydrogen, and oxygen. These elements go into the air as CO_2 and H_2O, so their masses are not measured with the mass of the ash.

64. total: $Al^{3+}(aq) + 3OH^-(aq) + 3H^+(aq) + 3Cl^-(aq) \longrightarrow Al^{3+}(aq) + 3Cl^-(aq) + 3HOH(l)$; net: $OH^-(aq) + H^+(aq) \longrightarrow HOH(l)$

  a. Al^{3+} and Cl^-

  b. When $NaHCO_3$ is used, CO_2 gas is formed in the stomach, which could cause discomfort.

Continuation of Answers

65. a. $2Na(s) + 2H_2O(l) \longrightarrow 2NaOH(aq) + H_2(g)$

b. displacement

Alternative Assessment

66. Students' procedures should lead reasonably to a definite order of activity. They should include data tables. If the procedures are sound and safe, you may want to allow students to test them in the lab.

67. Answers will vary. Be sure that students' testing schemes are logical and that proper laboratory procedures are followed.

68. Answers will vary. Some examples are fires or car engines combusting gasoline.

69. Answers will vary. Evaluate students' responses with respect to scientific accuracy and the feasibility of the safety measure.

70. Answers will vary. Be sure students consider reaction conditions such as light, heat, and moisture.

71. Answers will vary. Evaluate students' responses with respect to scientific accuracy and the ability of the analogy to convey the material.

Concept Mapping

72. a. a chemical reaction

b. a chemical equation

c. coefficients

d. a decomposition reaction

e. a synthesis reaction

Focus on Graphing

73. fluorine

74. 20 pm

75. The bond length of a single bond with hydrogen increases as you move down the elements in Group 17.

76. The size of a halogen atom increases as you move down Group 17. So a fluorine atom is the smallest, followed by a chlorine atom, then a bromine atom, then an iodine atom, which is the largest of the four. The same conclusion could be drawn using an atom of an element other than hydrogen as long as an atom of the same element is bonded to an atom of each halogen. In this way, any difference in bond length is a result of the relative sizes of the halogen atoms.

Technology and Learning

77. a. $Sn(SO_4)_2$

b. $Al_4[Fe(CN)_6]_3$

c. $P_4O_{10}(s) + 6H_2O(g) \longrightarrow 4H_3PO_4(aq)$

d. $2KMnO_4(aq) + 3MnCl_2(aq) + 2H_2O(l) \longrightarrow 5MnO_2(s) + 4HCl(aq) + 2KCl(aq)$

e. $2C_8H_{18} + 25O_2 \longrightarrow 16CO_2 + 18H_2O$

Stoichiometry
Chapter Planning Guide

PACING	CLASSROOM RESOURCES	LABS, ACTIVITIES, AND DEMONSTRATIONS
BLOCK 1 · 45 min pp. 300–301 **Chapter Opener**	**OSP** **Supplemental Reading** Consumer's Good Chemical Guide: A Jargon-Free Guide to Controversial Chemicals **ADVANCED**	**SE** **Start-Up Activity** All Used Up, p. 301 ◆
BLOCKS 2 & 3 · 90 min pp. 302–311 **Section 1** Calculating Quantities in Reactions	**TT** **Converting Between Amounts in Moles** * **TT** **Solving Mass-Mass Problems** * **TT** **Solving Volume-Volume Problems** * **TT** **Solving Particle Problems** *	**TE** **Demonstration** Stoichiometry of a Reaction, p. 306 ◆ **SE** **Skills Practice Lab** Stoichiometry and Gravimetric Analysis, p. 786 ◆ **GENERAL** **CRF** **Datasheets for In-Text Labs** * **SE** **Inquiry Lab** Gravimetric Analysis—Hard Water Testing, p. 790 ◆ **GENERAL** **CRF** **Datasheets for In-Text Labs** *
BLOCKS 4 & 5 · 90 min pp. 312–319 **Section 2** Limiting Reactants and Percentage Yield		**TE** **Demonstration** Limiting Reactants, p. 313 ◆ **TE** **Demonstration** Acid and Marble Chips, p. 315 ◆ **ADVANCED** **TE** **Group Activity**, p. 316 **GENERAL**
BLOCK 6 · 45 min pp. 320–327 **Section 3** Stoichiometry and Cars	**TT** **Fuel-Oxygen Ratio** *	

BLOCKS 7 & 8 · 90 min **Chapter Review and Assessment Resources**

Holt Chemistry: Online Resources

go.hrw.com

Visit **go.hrw.com** for a variety of free resources related to this textbook. Enter the keyword **HW4 HOME**.

Holt Online Learning

Students can access interactive problem solving help and active visual concept development with the *Holt Chemistry* Online Edition available at **www.hrw.com**

CNN student News

cnnstudentnews.com

Find the latest chemistry news, lesson plans, and activities related to important scientific events.

PROBLEM SOLVING AND PRACTICE	SECTION REVIEW AND ASSESSMENT	STANDARDS CORRELATION
		National Science Education Standards

SE Skills Toolkit 1 Converting Between Amounts in Moles, p. 303 **SE Sample Problem A** Using Mole Ratios, p. 304 　**SE Practice,** p. 304 `GENERAL`　**TE Homework,** p. 304 `GENERAL` **SE Skills Toolkit 2** Solving Stoichiometry Problems, p. 305 **SE Skills Toolkit 3** Solving Mass-Mass Problems, p. 306 **SE Sample Problem B** Problems Involving Mass, p. 307 　**SE Practice,** p. 307 `GENERAL`　**TE Homework,** p. 307 `GENERAL` **SE Skills Toolkit 4** Solving Volume-Volume Problems, p. 308 **SE Sample Problem C** Problems Involving Volume, p. 309 　**SE Practice,** p. 309 `GENERAL`　**TE Homework,** p. 309 `GENERAL` **SE Skills Toolkit 5** Solving Particle Problems, p. 310 **SE Sample Problem D** Problems Involving Particles, p. 310 　**SE Practice,** p. 311 `GENERAL`　**TE Homework,** p. 310 `GENERAL` **CRF Problem Solving** * `ADVANCED` **SE Problem Bank,** p. 858 `GENERAL`	**TE Quiz,** p. 311 `GENERAL` **CRF Quiz** * **TE Reteaching,** p. 311 `BASIC` **SE Section Review,** p. 311 **CRF Concept Review** *	
SE Sample Problem E Limiting Reactants and Theoretical Yield, p. 314 　**SE Practice,** p. 314 `GENERAL`　**TE Homework,** p. 314 `GENERAL` **SE Sample Problem F** Calculating Percentage Yield, p. 317 　**SE Practice,** p. 317 `GENERAL`　**TE Homework,** p. 317 `GENERAL` **SE Sample Problem G** Calculating Actual Yield, p. 318 　**SE Practice,** p. 318 `GENERAL`　**TE Homework,** p. 318 `GENERAL` **CRF Problem Solving** * `ADVANCED` **SE Problem Bank,** p. 858 `GENERAL`	**TE Quiz,** p. 319 `GENERAL` **CRF Quiz** * **TE Reteaching,** p. 319 `BASIC` **SE Section Review,** p. 319 **CRF Concept Review** *	
SE Sample Problem H Air-Bag Stoichiometry, p. 322 　**SE Practice,** p. 322 `GENERAL`　**TE Homework,** p. 322 `GENERAL` **SE Sample Problem I** Air-Fuel Ratio, p. 324 　**SE Practice,** p. 324 `GENERAL`　**TE Homework,** p. 324 `GENERAL` **SE Sample Problem J** Calculating Yields: Pollution, p. 327 　**SE Practice,** p. 327 `GENERAL`　**TE Homework,** p. 326 `GENERAL` **CRF Problem Solving** * `ADVANCED` **SE Problem Bank,** p. 858 `GENERAL`	**TE Homework,** p. 325 `ADVANCED` **TE Quiz,** p. 327 `GENERAL` **CRF Quiz** * **TE Reteaching,** p. 327 `BASIC` **SE Section Review,** p. 327 **CRF Concept Review** *	PS 3a

SCLINKS.
www.scilinks.org

Topic: Chemical Equations
SciLinks code: HW4141

Topic: Air Bags
SciLinks code: HW4005

Topic: Air Pollution
SciLinks code: HW4133

Topic: Catalytic Converters
SciLinks code: HW4026

Technology Resources

ChemFile Interactive Tutor CD-Rom

Module 5: Equations and Stoichiometry
Topic: Balancing Equations

Module 5: Equations and Stoichiometry
Topic: Stoichiometry

Module 5: Equations and Stoichiometry
Topic: Limiting Reagents

Overview

This chapter discusses the use of a balanced chemical equation to determine quantities of one or more substances involved in a reaction. Mole ratios from the balanced equation are the basis for all of these stoichiometry problems. Stoichiometry problems can involve amount in moles, number of particles, volume, and mass. Other concepts presented include finding limiting reactants, theoretical yield, and percentage yield. The last section of the chapter applies stoichiometric concepts to cars, especially in air bags, engine efficiency, and pollution generation and control.

Assessing Prior Knowledge

Check for Content Knowledge

Students should be familiar with the following topics:

- balancing chemical equations
- conversion factors
- significant figures
- molar masses
- density

Using the Figure

Traditional chess is one of many games that can be played only with a complete set of components. However, some games can take place in a modified form if not all components are available. For example, students might play a game of football or basketball even if there are not the necessary number of players. Often the conditions of the game are modified to accommodate the limited number of players. For example, only half the court will be used. For a chemical reaction to occur, all materials needed for the reaction must be present. However, stoichiometry allows for the calculation of quantities of reactants and products in situations where a reactant is used up before other reactants are used up.

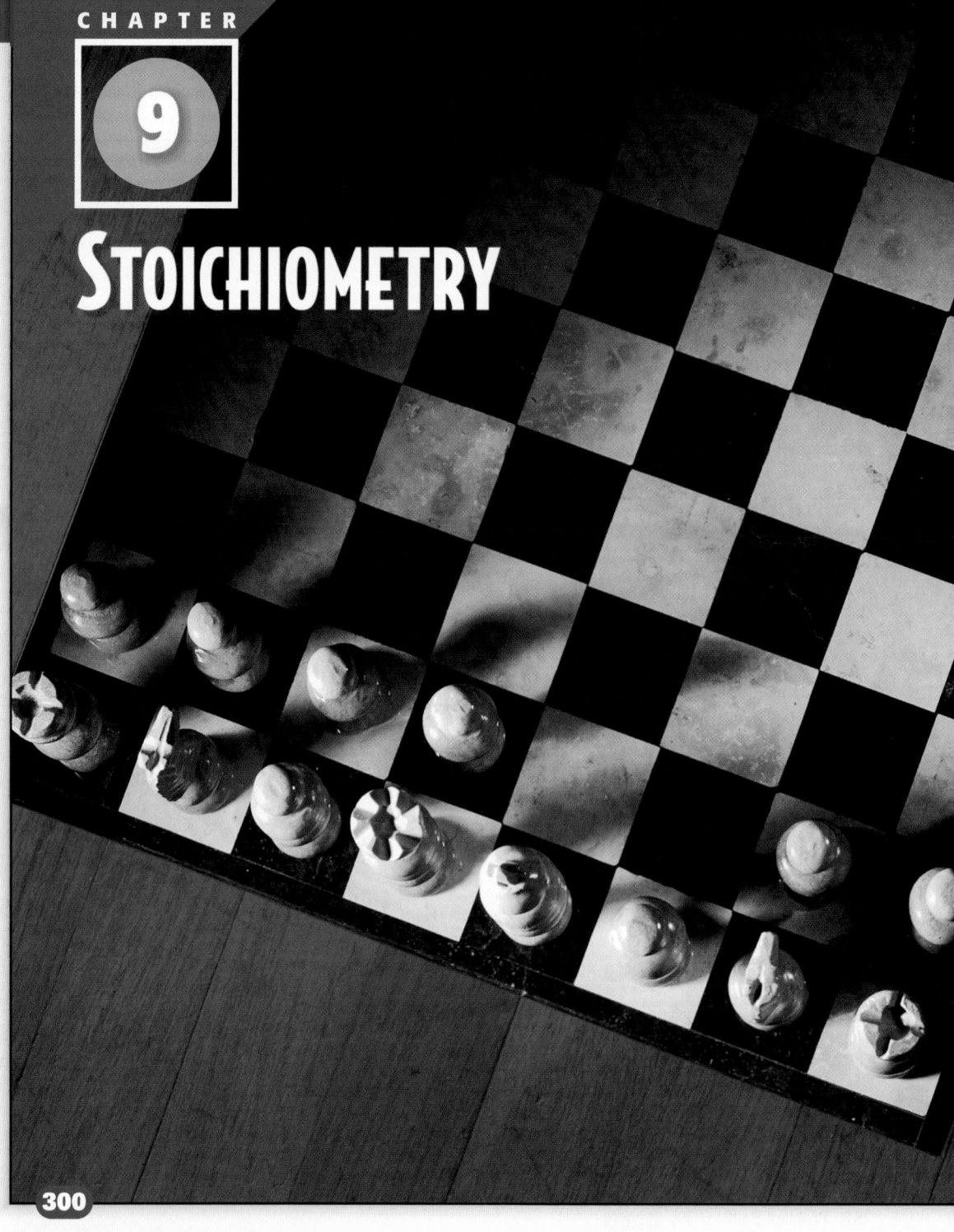

STOICHIOMETRY

300

Standards Correlations

National Science Education Standards

PS 3a: Chemical reactions occur all around us, for example in health care, cooking, cosmetics, and automobiles. Complex chemical reactions involving carbon-based molecules take place constantly in every cell in our bodies. (Section 3)

To play a standard game of chess, each side needs the proper number of pieces and pawns. Unless you find all of them—a king, a queen, two bishops, two knights, two rooks, and eight pawns—you cannot start the game. In chemical reactions, if you do not have every reactant, you will not be able to start the reaction. In this chapter you will look at amounts of reactants present and calculate the amounts of other reactants or products that are involved in the reaction.

START-UP ACTIVITY

All Used Up

PROCEDURE

1. Use a **balance** to find the mass of **8 nuts** and the mass of **5 bolts**.

2. Attach 1 nut (N) to 1 bolt (B) to assemble a nut-bolt (NB) model. Make as many NB models as you can. Record the number of models formed, and record which material was used up. Take the models apart.

3. Attach 2 nuts to 1 bolt to assemble a nut-nut-bolt (N_2B) model. Make as many N_2B models as you can. Record the number of models formed, and record which material was used up. Take the models apart.

ANALYSIS

1. Using the masses of the starting materials (the nuts and the bolts), could you predict which material would be used up first? Explain.

2. Write a balanced equation for the "reaction" that forms NB. How can this equation help you predict which component runs out?

3. Write a balanced equation for the "reaction" that forms N_2B. How can this equation help you predict which component runs out?

4. If you have 18 bolts and 26 nuts, how many models of NB could you make? of N_2B?

Pre-Reading Questions

1. A recipe calls for one cup of milk and three eggs per serving. You quadruple the recipe because you're expecting guests. How much milk and eggs do you need?

2. A bicycle mechanic has 10 frames and 16 wheels in the shop. How many complete bicycles can he assemble using these parts?

3. List at least two conversion factors that relate to the mole.

Answers to Pre-Reading Questions
1. 4 cups milk and 12 eggs
2. Eight; two frames will be left over.
3. molar mass, Avogadro's number, and molar volume of a gas at STP

301

START-UP ACTIVITY

Skills Acquired:
• Experimenting
• Collecting Data
• Interpreting
• Predicting

Materials:
For each group of 2–3 students:
• balance
• bolts (5)
• nuts (8)

Teacher's Notes: This activity leads students to understand that unequal amounts of reactants can be, and usually are, involved in chemical reactions. The reactant that runs out to the point that no more product can be formed limits the reaction.

Students will also see that mass cannot be used to directly determine which material will be used up first. The coefficients of a balanced equation are important to consider.

Answers

1. No, the masses alone do not allow an accurate prediction of which material will be used up first. Each material was the first to run out under different conditions.

2. $N + B \rightarrow NB$; The materials are used in a 1:1 ratio, so whichever component is present in the least number will run out.

3. $2N + B \rightarrow N_2B$; Twice as many nuts as bolts are used, so the number of bolts must be compared with half the number of nuts. Whichever of these is the least will run out.

4. 18 NB; 13 N_2B

Focus

Overview

Before beginning this section, review with your students the Objectives listed in the Student Edition. This section gives students an overview of stoichiometry and emphasizes the importance of the mole ratios determined from the coefficients in a balanced chemical equation. Students will learn to solve problems involving mass, volume, and number of particles.

Bellringer

Have students write the quantities of ingredients they would use to make a sandwich. Then have students determine how many sandwiches they could make from 24 slices of bread. Have students calculate how much of each other ingredient is needed. Tell students that this process models the calculations in this chapter.

Motivate

Identifying Preconceptions — GENERAL

Students might relate coefficients in a balanced chemical equation to mass ratios instead of to mole ratios. Give students the following:

$$2H_2(g) + O_2(g) \rightarrow 2H_2O(l)$$

Ask students to explain why a mass relationship from the coefficients in this equation is not valid.
Ans. Use of coefficients to directly compare masses does not always correctly reflect the law of conservation of mass. For example, 20 g H_2 reacting with 10 g O_2 will not produce 20 g H_2O. **LS Logical**

Calculating Quantities in Reactions

KEY TERMS

• stoichiometry

OBJECTIVES

1. **Use** proportional reasoning to determine mole ratios from a balanced chemical equation.

2. **Explain** why mole ratios are central to solving stoichiometry problems.

3. **Solve** stoichiometry problems involving mass by using molar mass.

4. **Solve** stoichiometry problems involving the volume of a substance by using density.

5. **Solve** stoichiometry problems involving the number of particles of a substance by using Avogadro's number.

Balanced Equations Show Proportions

If you wanted homemade muffins, like the ones in **Figure 1,** you could make them yourself—if you had the right things. A recipe for muffins shows how much of each ingredient you need to make 12 muffins. It also shows the proportions of those ingredients. If you had just a little flour on hand, you could determine how much of the other things you should use to make great muffins. The proportions also let you adjust the amounts to make enough muffins for all your classmates.

A balanced chemical equation is very similar to a recipe in that the coefficients in the balanced equation show the proportions of the reactants and products involved in the reaction. For example, consider the reaction for the synthesis of water.

$$2H_2 + O_2 \rightarrow 2H_2O$$

On a very small scale, the coefficients in a balanced equation represent the numbers of particles for each substance in the reaction. For the equation above, the coefficients show that two molecules of hydrogen react with one molecule of oxygen and form two molecules of water.

Calculations that involve chemical reactions use the proportions from balanced chemical equations to find the quantity of each reactant and product involved. As you learn how to do these calculations in this section, you will assume that each reaction goes to completion. In other words, all of the given reactant changes into product. For each problem in this section, assume that there is more than enough of all other reactants to completely react with the reactant given. Also assume that every reaction happens perfectly, so that no product is lost during collection. As you will learn in the next section, this usually is not the case.

Figure 1
In using a recipe to make muffins, you are using proportions to determine how much of each ingredient is needed.

INCLUSION Strategies

• *Learning Disabled*
• *Attention Deficit Disorder*
• *English Language Learners*

Provide the students with a copy of a muffin recipe. Ask the students to determine how much of each ingredient would be needed if the recipe were doubled. Additionally, have the students determine how much of the ingredients would be needed in order for each student in the class to have one muffin. This activity can be repeated for several recipes to provide more practice on this skill.

Chapter Resource File

• Lesson Plan

Relative Amounts in Equations Can Be Expressed in Moles

Just as you can interpret equations in terms of particles, you can interpret them in terms of moles. The coefficients in a balanced equation also represent the moles of each substance. For example, the equation for the synthesis of water shows that 2 mol H_2 react with 1 mol O_2 to form 2 mol H_2O. Look at the equation below.

$$2C_8H_{18} + 25O_2 \rightarrow 16CO_2 + 18H_2O$$

This equation shows that 2 molecules C_8H_{18} react with 25 molecules O_2 to form 16 molecules CO_2 and 18 molecules H_2O. And because Avogadro's number links molecules to moles, the equation also shows that 2 mol C_8H_{18} react with 25 mol O_2 to form 16 mol CO_2 and 18 mol H_2O.

In this chapter you will learn to determine how much of a reactant is needed to produce a given quantity of product, or how much of a product is formed from a given quantity of reactant. The branch of chemistry that deals with quantities of substances in chemical reactions is known as **stoichiometry**.

internet connect
www.scilinks.org
Topic: Chemical Equations
SciLinks code: HW4141
SCiLINKS. Maintained by the National Science Teachers Association

stoichiometry
the proportional relationship between two or more substances during a chemical reaction

The Mole Ratio Is the Key

If you normally buy a lunch at school each day, how many times would you need to "brown bag" it if you wanted to save enough money to buy a CD player? To determine the answer, you would use the units of dollars to bridge the gap between a CD player and school lunches. In stoichiometry problems involving equations, the unit that bridges the gap between one substance and another is the mole.

The coefficients in a balanced chemical equation show the relative numbers of moles of the substances in the reaction. As a result, you can use the coefficients in conversion factors called *mole ratios*. Mole ratios bridge the gap and can convert from moles of one substance to moles of another, as shown in **Skills Toolkit 1**.

SKILLS Toolkit 1

Converting Between Amounts in Moles

1. Identify the amount in moles that you know from the problem.

2. Using coefficients from the balanced equation, set up the mole ratio with the known substance on bottom and the unknown substance on top.

3. Multiply the original amount by the mole ratio.

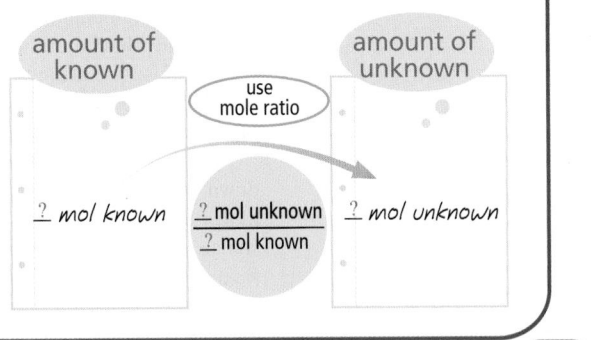

amount of known → use mole ratio → amount of unknown

? mol known   ? mol unknown / ? mol known   ? mol unknown

REAL-WORLD
CONNECTION

Flameless Ration Heater The Flameless Ration Heater (FRH) was developed jointly by the U.S. Army and Zesto Therm, a company based in Cincinnati, Ohio, as an environmentally safe method of preparing hot meals outdoors. Soldiers were able to use the FRH to heat their meals during the 1990–1991 Desert Storm conflict. Because the FRH must be as lightweight as possible, it's important that no materials are used beyond what is necessary.

The exothermic reaction needed to warm the food is magnesium undergoing a supercorrosion process, shown by the following equation:

$$Mg(s) + 2H_2O(l) \rightarrow Mg(OH)_2(s) + H_2(g) + 353 \text{ kJ}$$

NaCl is used in the FRH to enhance the corrosion of magnesium, which is naturally protected by an oxide coating. The Cl^- ion in NaCl reacts with $Mg(OH)_2$ to form MgOHCl. The magnesium oxide coating is soluble in the MgOHCl, so it dissolves. Once the coating is gone, water can continue to react with the magnesium.

Discussion ——— GENERAL

Conversion Factors Remind students that all conversion factors are equal to 1. To determine which conversion factor to use in a problem, students should assign the given unit as the denominator so that the unit will cancel. Give students several units and ask them for conversion factors that could be used to find the equivalent quantity in a different unit. **LS** Logical

Teach

READING SKILL BUILDER — BASIC

Assimilating Knowledge Before reading this chapter, have students write a short list of all the things they know about how the mass of the reactants affects the mass of the products. Then have students list the things they want to know about how the mass of the reactants affects the mass of the products. If students are having trouble, offer the following suggestions to get them on track:

- What do you know about the total mass of reactants compared with the total mass of products?

- What does a balanced chemical equation tell you about the ratio of the amount of reactants to the amount of products?

Later, when students are reviewing for the chapter test, have them read over their list and correct or expand upon their answers.
LS Logical

SKILL BUILDER — BASIC

Using the Figure Have students find several balanced chemical equations in the book and use the equations and **Skills Toolkit 1** to practice using mole ratios to find the amount in moles of a product given 6.0 mol of a reactant.
LS Verbal

Transparencies

TT Converting Between Amounts in Moles

Chapter 9 • Stoichiometry   303

Answers to Practice Problems A

1. a. 0.670 mol O_2
   b. 1.34 mol H_2O
2. a. 6.60 mol Al
   b. 6.60 mol Fe
   c. 3.30 mol Al_2O_3

Homework ——— **GENERAL**

Additional Practice

1. How many moles of H_2CO_3 can form when 2.57 mol CO_2 reacts with excess H_2O?

$$H_2O(l) + CO_2(g) \rightarrow H_2CO_3(aq)$$

Ans. 2.57 mol H_2CO_3

2. How many moles of O_2 are necessary to completely burn 4.33 mol C_3H_8? How many moles of CO_2 form? How many moles of H_2O form?

$$C_3H_8(g) + 5O_2(g) \rightarrow 3CO_2(g) + 4H_2O(l)$$

Ans. 21.6 mol O_2; 13.0 mol CO_2; 17.3 mol H_2O

3. In the combustion of propane, how many moles of C_3H_8 are needed to combine completely with 2.96 mol O_2? How many moles of CO_2 form? How many moles of H_2O form? Ans. 0.592 mol C_3H_8; 1.78 mol CO_2; 2.37 mol H_2O

LS Logical

ChemFile   **CHEMISTRY** INTERACTIVE TUTOR

- **Module 5: Equations and Stoichiometry**
  Topic: Balancing Equations
  This engaging tutorial reviews and reinforces the rules for balancing equations through modeling and guided practice.

SAMPLE PROBLEM A

Using Mole Ratios

Consider the reaction for the commercial preparation of ammonia.

$$N_2 + 3H_2 \longrightarrow 2NH_3$$

How many moles of hydrogen are needed to prepare 312 moles of ammonia?

1 Gather information.
- amount of NH_3 = 312 mol
- amount of H_2 = ? mol
- From the equation: 3 mol H_2 = 2 mol NH_3.

PRACTICE

PROBLEM SOLVING SKILL

1. Calculate the amounts requested if 1.34 mol H_2O_2 completely react according to the following equation.

$$2H_2O_2 \rightarrow 2H_2O + O_2$$

a. moles of oxygen formed
b. moles of water formed

2. Calculate the amounts requested if 3.30 mol Fe_2O_3 completely react according to the following equation.

$$Fe_2O_3 + 2Al \rightarrow 2Fe + Al_2O_3$$

a. moles of aluminum needed
b. moles of iron formed
c. moles of aluminum oxide formed

2 Plan your work.

The mole ratio must cancel out the units of mol NH_3 given in the problem and leave the units of mol H_2. Therefore, the mole ratio is

$$\frac{3 \text{ mol } H_2}{2 \text{ mol } NH_3}$$

3 Calculate.

$$? \text{ mol } H_2 = 312 \text{ mol } NH_3 \times \frac{3 \text{ mol } H_2}{2 \text{ mol } NH_3} = 468 \text{ mol } H_2$$

4 Verify your result.
- The answer is larger than the initial number of moles of ammonia. This is expected, because the conversion factor is greater than one.
- The number of significant figures is correct because the coefficients 3 and 2 are considered to be exact numbers.

PRACTICE HINT

The mole ratio must always have the unknown substance on top and the substance given in the problem on bottom for units to cancel correctly.

Getting into Moles and Getting out of Moles

Substances are usually measured by mass or volume. As a result, before using the mole ratio you will often need to convert between the units for mass and volume and the unit *mol*. Yet each stoichiometry problem has the step in which moles of one substance are converted into moles of a second substance using the mole ratio from the balanced chemical equation. Follow the steps in **Skills Toolkit 2** to understand the process of solving stoichiometry problems.

The thought process in solving stoichiometry problems can be broken down into three basic steps. First, change the units you are given into moles. Second, use the mole ratio to determine moles of the desired substance. Third, change out of moles to whatever unit you need for your final answer. And if you are given moles in the problem or need moles as an answer, just skip the first step or the last step! As you continue reading, you will be reminded of the conversion factors that involve moles.

304

HISTORY CONNECTION

Stoichiometry The word *stoichiometry* is from the Greek word *stoicheion*, which means "to measure the elements." Joseph Black (1728–1799) and his student Daniel Rutherford quantitatively studied the following reaction in both the forward and reverse direction to measure the redistribution of mass during the course of a reaction:

$$CaCO_3(s) \rightleftarrows CaO(s) + CO_2(g)$$

Their findings laid the groundwork for modern stoichiometry.

Chapter Resource File

- Problem Solving

Solving Stoichiometry Problems

You can solve all types of stoichiometry problems by following the steps outlined below.

1. Gather information.

- If an equation is given, make sure the equation is balanced. If no equation is given, write a balanced equation for the reaction described.
- Write the information provided for the given substance. If you are not given an amount in moles, determine the information you need to change the given units into moles and write it down.
- Write the units you are asked to find for the unknown substance. If you are not asked to find an amount in moles, determine the information you need to change moles into the desired units, and write it down.
- Write an equality using substances and their coefficients that shows the relative amounts of the substances from the balanced equation.

2. Plan your work.

- Think through the three basic steps used to solve stoichiometry problems: change to moles, use the mole ratio, and change out of moles. Know which conversion factors you will use in each step.
- Write the mole ratio you will use in the form:

$$\frac{\text{moles of unknown substance}}{\text{moles of given substance}}$$

3. Calculate.

- Write a question mark with the units of the answer followed by an equals sign and the quantity of the given substance.
- Write the conversion factors—including the mole ratio—in order so that you change the units of the given substance to the units needed for the answer.
- Cancel units and check that the remaining units are the required units of the unknown substance.
- When you have finished your calculations, round off the answer to the correct number of significant figures. In the examples in this book, only the final answer is rounded off.
- Report your answer with correct units and with the name or formula of the substance.

4. Verify your result.

- Verify your answer by estimating. You could round off the numbers in the setup in step 3 and make a quick calculation. Or you could compare conversion factors in the setup and decide whether the answer should be bigger or smaller than the initial value.
- Make sure your answer is reasonable. For example, imagine that you calculate that 725 g of a reactant is needed to form 5.3 mg (0.0053 g) of a product. The large difference in these quantities should alert you that there may be an error and that you should double-check your work.

305

Teaching Tips
Understanding Mole Ratios
Stress that when students are solving stoichiometry problems, moles are always involved. The mole ratio is the most critical factor in setting up a correct solution, yet this component presents the most confusion to students. To ensure understanding, relate stoichiometry to a recipe to lead students to think in terms of proportions of items.

Problem Solving Emphasize to students that an important part of problem solving is setting up the problem correctly. To reinforce this concept, have students set up several problems. Have students cancel units to show that the units of the answer will be correct but do not have them calculate the answers. Check the setups for accuracy.

ChemFile | **CHEMISTRY** INTERACTIVE TUTOR

- **Module 5: Equations and Stoichiometry Topic: Stoichiometry**
This engaging tutorial reviews and reinforces stoichiometry concepts through modeling and guided practice.

VISUAL ARTS CONNECTION — GENERAL

Have students choose a step from **Skills Toolkit 2** and illustrate it on a poster or bulletin board. For example, for Step 1, part 1, students could show a balanced chemical equation by drawing a balance and showing that equal numbers of each type of atom are present on each side. Or, when showing conversion factors, they could provide moveable units so students can form their own given, unknown, and conversion factor information.
LS Visual

Demonstration
Stoichiometry of a Reaction

(Approximate time: 20 minutes)

1. Find the mass of a Pyrex evaporating dish to the thousandth of a gram. Record the mass on the board.

2. Add 1 g of anhydrous Na_2CO_3 to the dish. Determine and record the total mass of the dish and the salt.

3. Add 1 M HCl one drop at a time until the bubbling stops. Then add about two more drops of HCl to make sure the reaction goes to completion.

4. Clamp a ring to a ring stand. Make sure you perform this part of the demonstration underneath a fume hood. Place a wire gauze and the evaporating dish on the ring. Heat the dish gently with a lab burner while avoiding splattering until none of the liquid remains.

5. When the dish has cooled, weigh and record the mass of the dish and the dried NaCl.

6. Have students compare the mass of the Na_2CO_3 with the mass of the NaCl. The mass of NaCl will be greater. Ask students to explain this result and what happens to the excess HCl.

7. Save the data for Section 2, when theoretical yield, actual yield, and percentage yield are discussed.

Safety Caution: Wear safety goggles and a lab apron at all times. Use tongs when handling hot glassware. Do not breathe HCl fumes. Students should stand 3 m from the demonstration.

Disposal: Wrap the NaCl in newspaper, and put it in the trash.

Figure 2
These tanks store ammonia for use as fertilizer. Stoichiometry is used to determine the amount of ammonia that can be made from given amounts of H_2 and N_2.

Problems Involving Mass, Volume, or Particles

Figure 2 shows a few of the tanks used to store the millons of metric tons of ammonia made each year in the United States. Stoichiometric calculations are used to determine how much of the reactants are needed and how much product is expected. However, the calculations do not start and end with moles. Instead, other units, such as liters or grams, are used. Mass, volume, or number of particles can all be used as the starting and ending quantities of stoichiometry problems. Of course, the key to each of these problems is the mole ratio.

For Mass Calculations, Use Molar Mass

The conversion factor for converting between mass and amount in moles is the molar mass of the substance. The molar mass is the sum of atomic masses from the periodic table for the atoms in a substance. **Skills Toolkit 3** shows how to use the molar mass of each substance involved in a stoichiometry problem. Notice that the problem is a three-step process. The mass in grams of the given substance is converted into moles. Next, the mole ratio is used to convert into moles of the desired substance. Finally, this amount in moles is converted into grams.

3 SKILLS Toolkit

Solving Mass-Mass Problems

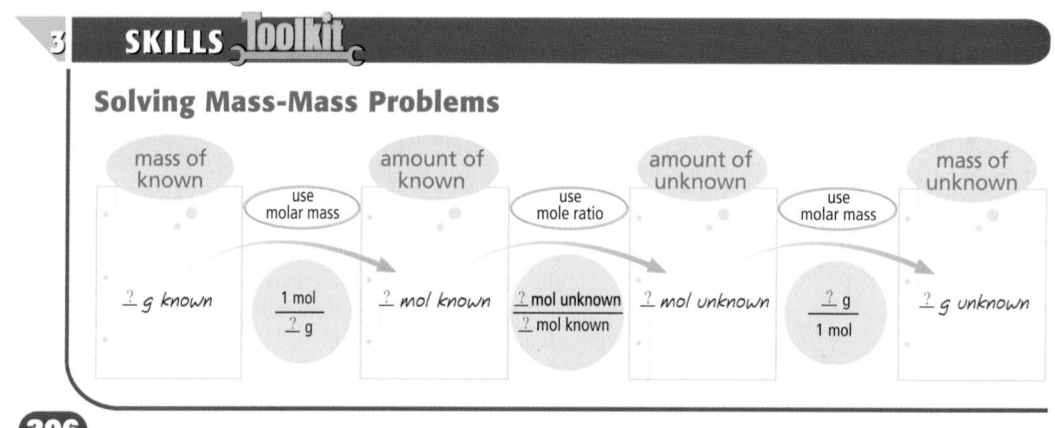

MISCONCEPTION ALERT

Molar Mass and Coefficients Students might think that the molar mass represents the mass of a substance involved in a chemical reaction. Remind them that molar mass is the mass of 1 mol of the substance. To work stoichiometric problems in which molar masses are used, the ratio of coefficients and the number of moles actually present must be considered.

Transparencies

TT Solving Mass-Mass Problems

SAMPLE PROBLEM B

Problems Involving Mass

What mass of NH_3 can be made from 1221 g H_2 and excess N_2?

$$N_2 + 3H_2 \longrightarrow 2NH_3$$

1 **Gather information.**

- mass of H_2 = 1221 g H_2
- molar mass of H_2 = 2.02 g/mol
- mass of NH_3 = ? g NH_3
- molar mass of NH_3 = 17.04 g/mol
- From the balanced equation: 3 mol H_2 = 2 mol NH_3.

2 **Plan your work.**

- To change grams of H_2 to moles, use the molar mass of H_2.
- The mole ratio must cancel out the units of mol H_2 given in the problem and leave the units of mol NH_3. Therefore, the mole ratio is

$$\frac{2 \text{ mol } NH_3}{3 \text{ mol } H_2}$$

- To change moles of NH_3 to grams, use the molar mass of NH_3.

3 **Calculate.**

$$? \text{ g } NH_3 = 1221 \text{ g } H_2 \times \frac{1 \text{ mol } H_2}{2.02 \text{ g } H_2} \times \frac{2 \text{ mol } NH_3}{3 \text{ mol } H_2} \times \frac{17.04 \text{ g } NH_3}{1 \text{ mol } NH_3} =$$

$$6867 \text{ g } NH_3$$

4 **Verify your result.**

The units cancel to give the correct units for the answer. Estimating shows the answer should be about 6 times the original mass.

PRACTICE

Use the equation below to answer the questions that follow.

$$Fe_2O_3 + 2Al \longrightarrow 2Fe + Al_2O_3$$

1 How many grams of Al are needed to completely react with 135 g Fe_2O_3?

2 How many grams of Al_2O_3 can form when 23.6 g Al react with excess Fe_2O_3?

3 How many grams of Fe_2O_3 react with excess Al to make 475 g Fe?

4 How many grams of Fe will form when 97.6 g Al_2O_3 form?

PRACTICE HINT

Remember to check both the units and the substance when canceling. For example, 1221 g H_2 cannot be converted to moles by multiplying by 1 mol NH_3/17.04 g NH_3. The units of grams in each one cannot cancel because they involve different substances.

Teaching Tip

Conserving Mass If the mass for each substance in a reaction is calculated from a given substance, the law of conservation of mass can be used to check the results. The total mass of the reactants must equal the total mass of the products.

Answers to Practice Problems B

1. 45.6 g Al
2. 44.6 g Al_2O_3
3. 679 g Fe_2O_3
4. 107 g Fe

Homework ——— GENERAL

Additional Practice

1. What mass of H_2O is produced if 65.2 g $CaCO_3$ reacts with excess H_3PO_4, to form $Ca_3(PO_4)_2$, H_2O, and CO_2? Ans. $3CaCO_3(s) + 2H_3PO_4(aq) \rightarrow Ca_3(PO_4)_2(s) + 3H_2O(l) + 3CO_2(g)$; 11.7 g H_2O

2. What mass of O_2 forms when 49.89 g $KClO_3$ decomposes? (KCl also forms.) Ans. $2KClO_3(s) \rightarrow 2KCl(s) + 3O_2(g)$; 19.54 g O_2

3. What mass of ammonia is formed when 7.50 g N_2 reacts with excess H_2? Ans. $N_2(g) + 3H_2(g) \rightarrow 2NH_3(g)$; 9.12 g NH_3

LS Logical

Chapter Resource File

- **Problem Solving**

MISCONCEPTION ALERT

Making Calculations Many students use their calculators incorrectly with numbers that are multiplied in the denominator of fractions. Many students will work out problems such as

$$\frac{67 \times 70}{15 \times 35}$$

by pressing $67 \times 70 \div 15 \times 35$. This results in the incorrect answer of 11 000 because the calculator is programmed to follow the order of operations. The result is the calculator finds the product of 67 and 70, divides the result by 15, and then multiplies the answer by 35. Tell students that if the number is in the numerator, it is multiplied, and if it is in the denominator, it is divided. The problem can be worked on a calculator by pressing $67 \times 70 \div 15 \div 35$ or by using parentheses to group the denominator by pressing $67 \times 70 \div (15 \times 35)$. Either method results in the correct answer of 8.9.

Teaching Tip

Calculator Skills Point out to students that a mass-mass problem uses only multiplication and division. Therefore, a calculator is helpful because it allows the student to perform the entire calculation before rounding. This helps to prevent errors caused by calculating and rounding the numerators and the denominators separately. However, point out that using a calculator cannot help if the problem is set up incorrectly or if students do not understand how to enter the data correctly.

SKILL BUILDER — GENERAL

Interpreting Graphics Refer students to **Skills Toolkit 4.** Make sure students understand how to follow the steps to solve problems. Also have students identify how the steps of a mass-mass problem are part of the overall process of solving a volume-volume problem to help reinforce the idea that the problems are building on one another and following the same basic pattern. **LS** Visual

Transparencies

TT Solving Volume-Volume Problems

Solving Volume-Volume Problems

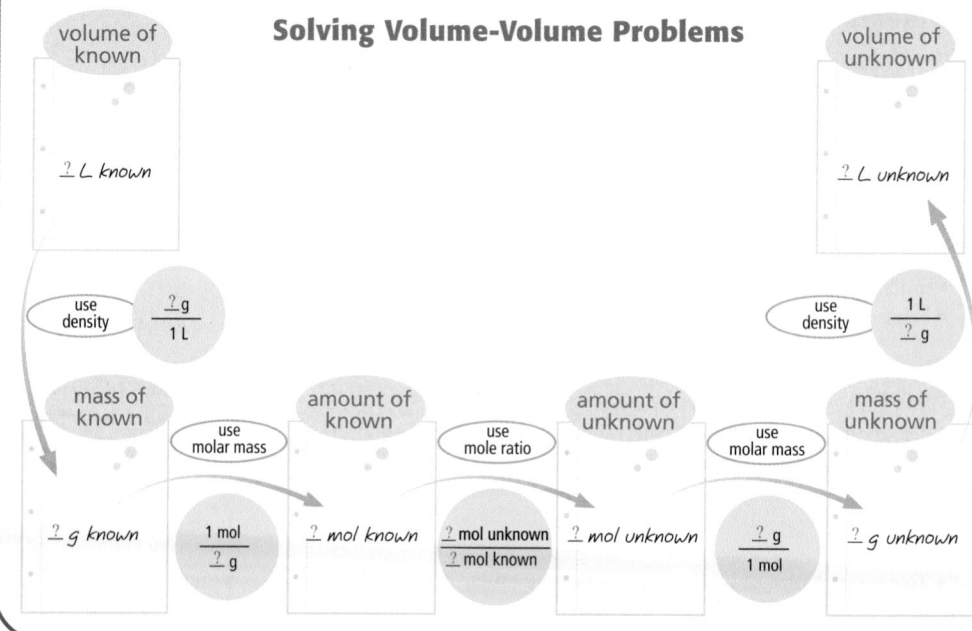

For Volume, You Might Use Density and Molar Mass

When reactants are liquids, they are almost always measured by volume. So, to do calculations involving liquids, you add two more steps to the sequence of mass-mass problems—the conversions of volume to mass and of mass to volume. Five conversion factors—two densities, two molar masses, and a mole ratio—are needed for this type of calculation, as shown in **Skills Toolkit 4.**

To convert from volume to mass or from mass to volume of a substance, use the density of the substance as the conversion factor. Keep in mind that the units you want to cancel should be on the bottom of your conversion factor.

There are ways other than density to include volume in stoichiometry problems. For example, if a substance in the problem is a gas at standard temperature and pressure (STP), use the molar volume of a gas to change directly between volume of the gas and moles. The molar volume of a gas is 22.41 L/mol for any gas at STP. Also, if a substance in the problem is in aqueous solution, then use the concentration of the solution to convert the volume of the solution to the moles of the substance dissolved. This procedure is especially useful when you perform calculations involving the reaction between an acid and a base. Of course, even in these problems, the basic process remains the same: change to moles, use the mole ratio, and change to the desired units.

MISCONCEPTION ALERT

Density and Temperature Students might think that the densities given in this section are constant numbers under all conditions. Density is temperature dependent. In general, as temperature increases, density decreases. For gases, density is also pressure dependent. As pressure increases, so does density. The densities given in this section are standard values measured at temperatures and pressures that approximate laboratory conditions.

Problems Involving Volume

What volume of H_3PO_4 forms when 56 mL $POCl_3$ completely react? (density of $POCl_3$ = 1.67 g/mL; density of H_3PO_4 = 1.83 g/mL)

$$POCl_3(l) + 3H_2O(l) \longrightarrow H_3PO_4(l) + 3HCl(g)$$

1 Gather information.

- volume $POCl_3$ = 56 mL $POCl_3$
- density $POCl_3$ = 1.67 g/mL   • molar mass $POCl_3$ = 153.32 g/mol
- volume H_3PO_4 = ?
- density H_3PO_4 = 1.83 g/mL   • molar mass H_3PO_4 = 98.00 g/mol
- From the equation: 1 mol $POCl_3$ = 1 mol H_3PO_4.

2 Plan your work.

- To change milliliters of $POCl_3$ to moles, use the density of $POCl_3$ followed by its molar mass.
- The mole ratio must cancel out the units of mol $POCl_3$ given in the problem and leave the units of mol H_3PO_4. Therefore, the mole ratio is

$$\frac{1 \text{ mol } H_3PO_4}{1 \text{ mol } POCl_3}$$

- To change out of moles of H_3PO_4 into milliliters, use the molar mass of H_3PO_4 followed by its density.

3 Calculate.

$$? \text{ mL } H_3PO_4 = 56 \text{ mL } POCl_3 \times \frac{1.67 \text{ g } POCl_3}{1 \text{ mL } POCl_3} \times \frac{1 \text{ mol } POCl_3}{153.32 \text{ g } POCl_3} \times$$

$$\frac{1 \text{ mol } H_3PO_4}{1 \text{ mol } POCl_3} \times \frac{98.00 \text{ g } H_3PO_4}{1 \text{ mol } H_3PO_4} \times \frac{1 \text{ mL } H_3PO_4}{1.83 \text{ g } H_3PO_4} = 33 \text{ mL } H_3PO_4$$

4 Verify your result.

The units of the answer are correct. Estimating shows the answer should be about two-thirds of the original volume.

Use the densities and balanced equation provided to answer the questions that follow. (density of C_5H_{12} = 0.620 g/mL; density of C_5H_8 = 0.681 g/mL; density of H_2 = 0.0899 g/L)

$$C_5H_{12}(l) \longrightarrow C_5H_8(l) + 2H_2(g)$$

1 How many milliliters of C_5H_8 can be made from 366 mL C_5H_{12}?

2 How many liters of H_2 can form when 4.53×10^3 mL C_5H_8 form?

3 How many milliliters of C_5H_{12} are needed to make 97.3 mL C_5H_8?

4 How many milliliters of H_2 can be made from 1.98×10^3 mL C_5H_{12}?

PROBLEM SOLVING SKILL

PRACTICE HINT

Do not try to memorize the exact steps of every type of problem. For long problems like these, you might find it easier to break the problem into three steps rather than solving all at once. Remember that whatever you are given, you need to change to moles, then use the mole ratio, then change out of moles to the desired units.

Answers to Practice Problems C

1. 315 mL C_5H_8
2. 2.03×10^3 L H_2
3. 113 mL C_5H_{12}
4. 7.64×10^5 mL H_2

Homework ——— GENERAL

Additional Practice Write balanced chemical equations for each of the following problems, and then solve.

1. When pentane, C_5H_{12}, burns in oxygen, it produces carbon dioxide and water. If 85.5 g of pentane is completely burned, what volume of carbon dioxide is produced? Assume the CO_2 cools to room temperature, where its density is 1.997 g/L. Ans. $C_5H_{12}(g) + 8O_2(g) \rightarrow 5CO_2(g) + 6H_2O(g)$; 131 L CO_2

2. Magnesium burns in oxygen to produce magnesium oxide. What mass of magnesium will burn in the presence of 189 mL of oxygen? The density of oxygen is 1.429 g/L. Ans. $2Mg(s) + O_2(g) \rightarrow 2MgO(s)$; 0.410 g Mg

LS Logical

Chapter Resource File

- **Problem Solving**

309

---

did you know?

Standard Temperature and Pressure

Because the volume of a gas is dependent on temperature and pressure, 0°C (273 K) and 1.00 atm have been chosen as standard temperature and pressure (STP) for gases.

Answers to Practice Problems D

1. 2.89×10^{24} molecules BrF_5
2. 2.22×10^{19} molecules Br_2

Homework ——— GENERAL

Additional Practice

1. If 2.46×10^{25} molecules of chlorine react completely, how many grams of NaCl will form?

$2Na(s) + Cl_2(g) \rightarrow 2NaCl(s)$

Ans. 4.77×10^3 g NaCl

2. How many molecules of carbon dioxide are produced when 79.5 g of K_2CO_3 decompose?

$K_2CO_3(s) \rightarrow K_2O(s) + CO_2(g)$

Ans. 3.46×10^{23} molecules CO_2

3. How many water molecules form from the complete combustion of 1.129×10^{24} molecules C_4H_{10}?

$2C_4H_{10}(g) + 13O_2(g) \rightarrow$ $8CO_2(g) + 10H_2O(l)$

Ans. 5.645×10^{24} molecules H_2O

LS Logical

Teaching Tip

Particle Ratios Remind students that the coefficients in a balanced chemical equation represent particles as well as moles. If the number of particles of a substance is given and the number of particles of a different substance is asked for, the problem can be solved in a single step. The coefficients of the two substances can be used in a particle ratio. It is unnecessary, but not incorrect, to change particles to moles and then back to particles.

SKILLS Toolkit 5

Solving Particle Problems

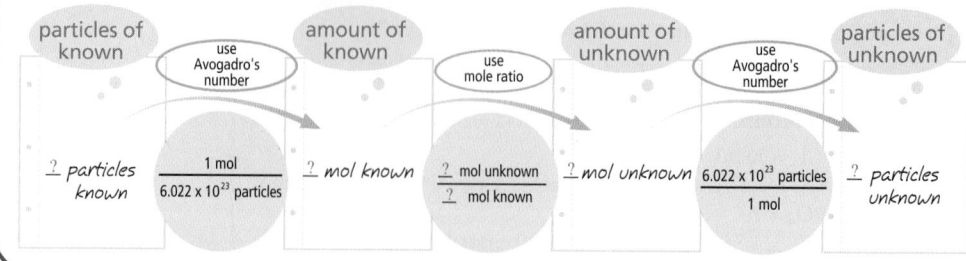

Topic Link

Refer to the chapter "The Mole and Chemical Composition" for more information about Avogadro's number and molar mass.

For Number of Particles, Use Avogadro's Number

Skills Toolkit 5 shows how to use Avogadro's number, 6.022×10^{23} particles/mol, in stoichiometry problems. If you are given particles and asked to find particles, Avogadro's number cancels out! For this calculation you use only the coefficients from the balanced equation. In effect, you are interpreting the equation in terms of the number of particles again.

SAMPLE PROBLEM D

Problems Involving Particles

How many grams of C_5H_8 form from 1.89×10^{24} molecules C_5H_{12}?

$$C_5H_{12}(l) \longrightarrow C_5H_8(l) + 2H_2(g)$$

PRACTICE HINT

Expect more problems like this one that do not exactly follow any single **Skills Toolkit** in this chapter. These problems will combine steps from one or more problems, but all will still use the mole ratio as the key step.

1 **Gather information.**
- quantity of $C_5H_{12} = 1.89 \times 10^{24}$ molecules
- Avogadro's number $= 6.022 \times 10^{23}$ molecules/mol
- mass of $C_5H_8 = ?$ g C_5H_8
- molar mass of $C_5H_8 = 68.13$ g/mol
- From the balanced equation: 1 mol $C_5H_{12} = 1$ mol C_5H_8.

2 **Plan your work.**
Set up the problem using Avogadro's number to change to moles, then use the mole ratio, and finally use the molar mass of C_5H_8 to change to grams.

3 **Calculate.**

$? \text{ g } C_5H_8 = 1.89 \times 10^{24} \text{ molecules } C_5H_{12} \times \dfrac{1 \text{ mol } C_5H_{12}}{6.022 \times 10^{23} \text{ molecules } C_5H_{12}} \times$

$\dfrac{1 \text{ mol } C_5H_8}{1 \text{ mol } C_5H_{12}} \times \dfrac{68.13 \text{ g } C_5H_8}{1 \text{ mol } C_5H_8} = 214 \text{ g } C_5H_8$

4 **Verify your result.**
The units cancel correctly, and estimating gives 210.

310

MISCONCEPTION ALERT

Solution Methods Students might try to memorize the steps for each type of stoichiometric problem. Emphasize to students that converting to and from moles is necessary to solve most stoichiometry problems, and the logical steps to and from the mole involve cancellation of units and multiplying and dividing the numbers that go with the units.

Chapter Resource File

- **Problem Solving**

Transparencies

TT Solving Particle Problems

PRACTICE

Use the equation provided to answer the questions that follow.

$$Br_2(l) + 5F_2(g) \longrightarrow 2BrF_5(l)$$

1 How many molecules of BrF_5 form when 384 g Br_2 react with excess F_2?

2 How many molecules of Br_2 react with 1.11×10^{20} molecules F_2?

Many Problems, Just One Solution

Although you could be given many different problems, the solution boils down to just three steps. Take whatever you are given, and find a way to change it into moles. Then, use a mole ratio from the balanced equation to get moles of the second substance. Finally, find a way to convert the moles into the units that you need for your final answer.

Section Review

UNDERSTANDING KEY IDEAS

1. What conversion factor is present in almost all stoichiometry calculations?

2. For a given substance, what information links mass to moles? number of particles to moles?

3. What conversion factor will change moles CO_2 to grams CO_2? moles H_2O to molecules H_2O?

PRACTICE PROBLEMS

4. Use the equation below to answer the questions that follow.

$$Br_2 + Cl_2 \longrightarrow 2BrCl$$

a. How many moles of BrCl form when 2.74 mol Cl_2 react with excess Br_2?

b. How many grams of BrCl form when 239.7 g Cl_2 react with excess Br_2?

c. How many grams of Br_2 are needed to react with 4.53×10^{25} molecules Cl_2?

5. The equation for burning C_2H_2 is

$$2C_2H_2(g) + 5O_2(g) \longrightarrow 4CO_2(g) + 2H_2O(g)$$

a. If 15.9 L C_2H_2 react at STP, how many moles of CO_2 are produced? (Hint: At STP, 1 mol = 22.41 L for any gas.)

b. How many milliliters of CO_2 (density = 1.977 g/L) can be made when 59.3 mL O_2 (density = 1.429 g/L) react?

CRITICAL THINKING

6. Why do you need to use amount in moles to solve stoichiometry problems? Why can't you just convert from mass to mass?

7. LiOH and NaOH can each react with CO_2 to form the metal carbonate and H_2O. These reactions can be used to remove CO_2 from the air in a spacecraft.

a. Write a balanced equation for each reaction.

b. Calculate the grams of NaOH and of LiOH that remove 288 g CO_2 from the air.

c. NaOH is less expensive per mole than LiOH. Based on your calculations, explain why LiOH is used during shuttle missions rather than NaOH.

311

Answers to Section Review

1. the mole ratio

2. the molar mass of the substance; Avogadro's number, $\dfrac{6.022 \times 10^{23} \text{ particles}}{1 \text{ mole}}$

3. $\dfrac{44.01 \text{ g } CO_2}{1 \text{ mol } CO_2}$; $\dfrac{6.022 \times 10^{23} \text{ molecules } H_2O}{1 \text{ mol } H_2O}$

4. a. 5.48 mol BrCl
  b. 780.0 g BrCl
  c. 1.20×10^4 g Br_2

5. a. 1.42 mol CO_2
  b. 47.2 mL CO_2

6. Coefficients in the balanced chemical equation give mole ratios, not mass ratios.

7. a. $2LiOH + CO_2 \longrightarrow Li_2CO_3 + H_2O$
   $2NaOH + CO_2 \longrightarrow Na_2CO_3 + H_2O$
  b. 524 g NaOH; 313 g LiOH
  c. Less mass of LiOH is needed to remove a given amount of CO_2, so the overall mass of the shuttle and its cargo decreases.

Close

Reteaching ——————— BASIC

Have students make a model that demonstrates all of the steps discussed in solving stoichiometry problems. The model can be a flowchart or a three-dimensional construction, such as a network of roads. **LS** Visual

Quiz ——————————— GENERAL

1. Write the six mole ratios from the following equation:

$$2KClO_3(s) \rightarrow 2KCl(s) + 3O_2(g)$$

Ans. $\dfrac{2 \text{ mol } KClO_3}{2 \text{ mol } KCl}$,

$\dfrac{2 \text{ mol } KClO_3}{3 \text{ mol } O_2}$, $\dfrac{2 \text{ mol } KCl}{2 \text{ mol } KClO_3}$,

$\dfrac{2 \text{ mol } KCl}{3 \text{ mol } O_2}$, $\dfrac{3 \text{ mol } O_2}{2 \text{ mol } KClO_3}$, and

$\dfrac{3 \text{ mol } O_2}{2 \text{ mol } KCl}$

2. How many moles of oxygen can be formed from 2.6 mol $KClO_3$ according to the equation in item 1? Ans. 3.9 mol O_2

3. What do you need to know to change each of the following units to moles?

  a. number of particles
    Ans. Avogadro's number

  b. grams Ans. molar mass

  c. volume Ans. density and molar mass

LS Logical

Chapter Resource File

• Concept Review
• Quiz

Overview

Before beginning this section, review with your students the Objectives listed in the Student Edition. In this section, students will learn about limiting reactants. They will also learn about theoretical yield, actual yield, and percentage yield.

📢 Bellringer

Have students discuss in pairs how the amount of product calculated from a given amount of reactant is an ideal amount. Have students discuss how the amount of product formed from a real reaction might differ from the amount calculated and offer reasons for the difference.

Motivate

Demonstration

Provide students with items they can use to model everyday examples of limiting and excess materials, such as graham crackers, marshmallows, and chocolate squares for "s'mores". Be sure to have one item that is limiting. For example, if a single "s'more" is made from 2 graham crackers, 1 marshmallow, and 1 chocolate square, then 14 graham crackers, 5 marshmallows, and 6 chocolate squares will make 5 "s'mores." The marshmallows are limiting, and there are 4 excess graham crackers and 1 excess chocolate square. This modeling can be done either by students or as a demonstration.

SECTION

Limiting Reactants and Percentage Yield

KEY TERMS
- limiting reactant
- excess reactant
- actual yield

OBJECTIVES

① **Identify** the limiting reactant for a reaction and use it to calculate theoretical yield.

② **Perform** calculations involving percentage yield.

Limiting Reactants and Theoretical Yield

To drive a car, you need gasoline in the tank and oxygen from the air. When the gasoline runs out, you can't go any farther even though there is still plenty of oxygen. In other words, the gasoline limits the distance you can travel because it runs out and the reaction in the engine stops.

In the previous section, you assumed that 100% of the reactants changed into products. And that is what should happen theoretically. But in the real world, other factors, such as the amounts of all reactants, the completeness of the reaction, and product lost in the process, can limit the yield of a reaction. The analogy of assembling homecoming mums for a fund raiser, as shown in **Figure 3,** will help you understand that whatever is in short supply will limit the quantity of product made.

Figure 3
The number of mums these students can assemble will be limited by the component that runs out first.

REAL-WORLD
CONNECTION

Fire Extinguishers In order for a fire to start and continue it must have fuel, oxygen, and energy. If one of these conditions is limited, the fire will not continue, even if the other items are present in large amounts. Spraying a fire with water both lowers the temperature and limits the amount of oxygen that can reach the fuel. Fire extinguishers may also lower temperature, but they commonly lay a "blanket" of heavier-than-air gas that keeps oxygen away from the fire.

Chapter Resource File
• Lesson Plan

The Limiting Reactant Forms the Least Product

The students assembling mums use one helmet, one flower, eight blue ribbons, six white ribbons, and two bells to make each mum. As a result, the students cannot make any more mums once any one of these items is used up. Likewise, the reactants of a reaction are seldom present in ratios equal to the mole ratio in the balanced equation. So one of the reactants is used up first. For example, one way to to make H_2 is

$$Zn + 2HCl \longrightarrow ZnCl_2 + H_2$$

If you combine 0.23 mol Zn and 0.60 mol HCl, would they react completely? Using the coefficients from the balanced equation, you can predict that 0.23 mol Zn can form 0.23 mol H_2, and 0.60 mol HCl can form 0.30 mol H_2. Zinc is called the **limiting reactant** because the zinc limits the amount of product that can form. The zinc is used up first by the reaction. The HCl is the **excess reactant** because there is more than enough HCl present to react with all of the Zn. There will be some HCl left over after the reaction stops.

Again, think of the mums, and look at **Figure 4.** The supplies at left are the available reactants. The products formed are the finished mums. The limiting reactant is the flowers because they are completely used up first. The ribbons, helmets, and bells are excess reactants because there are some of each of these items left over, at right. You can determine the limiting reactant by calculating the amount of product that each reactant could form. Whichever reactant would produce the least amount of product is the limiting reactant.

limiting reactant

the substance that controls the quantity of product that can form in a chemical reaction

excess reactant

the substance that is not used up completely in a reaction

Figure 4
The flowers are in short supply. They are the limiting reactant for assembling these homecoming mums.

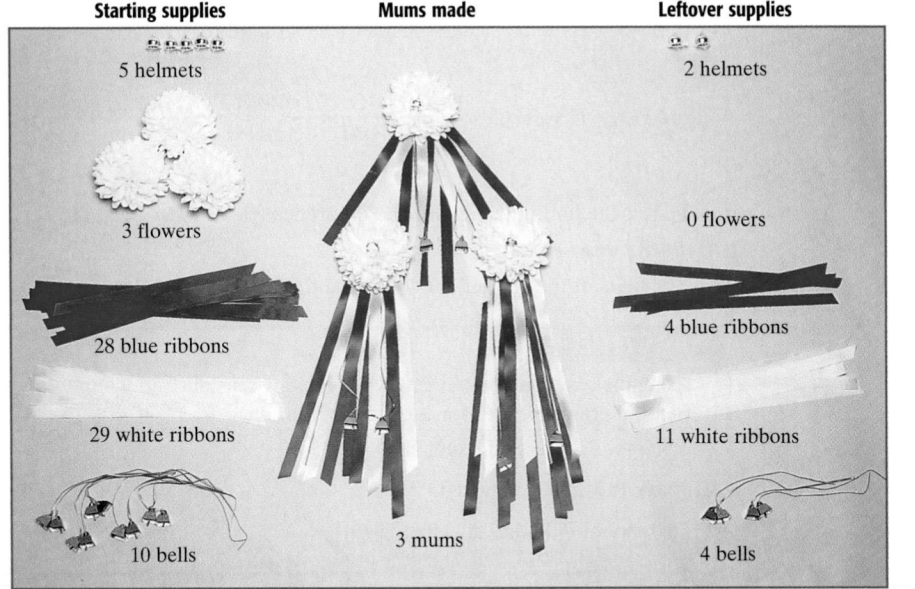

Starting supplies — Mums made — Leftover supplies

5 helmets / 3 flowers / 28 blue ribbons / 29 white ribbons / 10 bells

3 mums

2 helmets / 0 flowers / 4 blue ribbons / 11 white ribbons / 4 bells

Demonstration
Limiting Reactants

(Approximate time: 10 minutes)

Note: Before class, determine if the mass specified will work with the size of balloons you use. Change amounts accordingly.

1. Before class, obtain three 2 L plastic bottles. Remove the labels, and number the bottles *1*, *2*, and *3*. Stretch three 9 to 12 in. balloons by inflating them and deflating them several times. Measure out 4 g, 7.5 g, and 10 g samples of baking soda.

2. Pour 100 mL of vinegar into each bottle.

3. Add a baking soda sample to each of the three balloons. With a permanent marker, label each balloon with the mass of the sample size it contains. Twist each balloon to seal in the sample. In the order the samples were listed in step 1, attach the balloons to the tops of bottles 1–3, respectively.

4. Secure the balloons to the bottle tops with tape or a twist tie. Untwist the balloons, and shake the baking soda into the vinegar.

Teacher's Notes: The reaction in Bottle 1 should produce the least gas. Bottles 2 and 3 should produce about the same amount. Bottle 1 has limited baking soda. Bottle 2 has the correct stoichiometric mass to completely react. Even though bottle 3 has more baking soda, there is not enough vinegar to react with the excess.

Safety Caution: Wear safety goggles and a lab apron. Students should remain at least 3 m from the demonstration. Stay back from the balloons as they fill with gas in case one bursts.

Disposal: Pour the solutions down the drain with plenty of water. Throw the balloons in the trash. Rinse the bottles, and save them for later use.

 **INCLUSION Strategies**

• *Gifted and Talented*

Have the student construct a triangle with mini marshmallows and toothpicks as a model. Provide one bag of marshmallows and one box of toothpicks and have the student determine how many triangles can be constructed from the materials present. The student will need to use a formula to determine how many triangles can be constructed and which ingredient will be the limiting factor and will run out first.

BIOLOGY
CONNECTION

If vitamins and minerals are not present in sufficient quantities in the human body, they become limiting reactants in reactions they are involved in. For example, if the body does not have enough vitamin B-12, the bone marrow does not produce enough mature red blood cells and the person becomes anemic. People with pernicious anemia have lost their ability to make the substance secreted by the lining of the stomach that enables absorption of vitamin B-12 in the intestines.

1. PCl_3 is excess, H_2O is limiting, theoretical yield is 109 g HCl

2. H_2O is excess, PCl_3 is limiting, theoretical yield is 59.7 g HCl

3. PCl_3 is excess, H_2O is limiting, theoretical yield is 101 g HCl

Homework ——— GENERAL

Additional Practice Write a balanced chemical equation for each of the following problems, and then determine the excess reactant, the limiting reactant, and the theoretical yield (in grams) of the first product mentioned.

1. Zinc citrate, $Zn_3(C_6H_5O_7)_2$, an ingredient in toothpaste, is made by reacting zinc carbonate and citric acid, $C_6H_8O_7$. The other products are H_2O and CO_2. There are 6.00 mol $ZnCO_3$ and 10.0 mol $C_6H_8O_7$. **Ans.** $3ZnCO_3 + 2C_6H_8O_7 \rightarrow Zn_3(C_6H_5O_7)_2 + 3H_2O + 3CO_2$; $C_6H_8O_7$ is in excess, $ZnCO_3$ is limiting, and the theoretical yield is 1.15×10^3 g $Zn_3(C_6H_5O_7)_2$.

2. Hydrogen sulfide gas is formed when HCl reacts with FeS. $FeCl_2$ is the other product. 130.5 g of FeS is mixed with 70.4 g of HCl in solution. **Ans.** $FeS + 2HCl \rightarrow H_2S + FeCl_2$; FeS is in excess, HCl is limiting, and the theoretical yield is 32.91 g H_2S.

LS Logical

• **Problem Solving**

Determine Theoretical Yield from the Limiting Reactant

So far you have done only calculations that assume reactions happen perfectly. The maximum quantity of product that a reaction could theoretically make if everything about the reaction works perfectly is called the *theoretical yield.* The theoretical yield of a reaction should always be calculated based on the limiting reactant.

In the reaction of Zn with HCl, the theoretical yield is 0.23 mol H_2 even though the HCl could make 0.30 mol H_2.

SAMPLE PROBLEM E

Limiting Reactants and Theoretical Yield

Identify the limiting reactant and the theoretical yield of phosphorous acid, H_3PO_3, if 225 g of PCl_3 is mixed with 123 g of H_2O.

$$PCl_3 + 3H_2O \longrightarrow H_3PO_3 + 3HCl$$

1 **Gather information.**

• mass PCl_3 = 225 g PCl_3 • molar mass PCl_3 = 137.32 g/mol
• mass H_2O = 123 g H_2O • molar mass H_2O = 18.02 g/mol
• mass H_3PO_3 = ? g H_3PO_3 • molar mass H_3PO_3 = 82.00 g/mol
• From the balanced equation:
   1 mol PCl_3 = 1 mol H_3PO_3 and 3 mol H_2O = 1 mol H_3PO_3.

2 **Plan your work.**

Set up problems that will calculate the mass of H_3PO_3 you would expect to form from each reactant.

3 **Calculate.**

$$? \text{ g } H_3PO_3 = 225 \text{ g } PCl_3 \times \frac{1 \text{ mol } PCl_3}{137.32 \text{ g } PCl_3} \times \frac{1 \text{ mol } H_3PO_3}{1 \text{ mol } PCl_3} \times \frac{82.00 \text{ g } H_3PO_3}{1 \text{ mol } H_3PO_3} =$$
$$134 \text{ g } H_3PO_3$$

$$? \text{ g } H_3PO_3 = 123 \text{ g } H_2O \times \frac{1 \text{ mol } H_2O}{18.02 \text{ g } H_2O} \times \frac{1 \text{ mol } H_3PO_3}{3 \text{ mol } H_2O} \times \frac{82.00 \text{ g } H_3PO_3}{1 \text{ mol } H_3PO_3} =$$
$$187 \text{ g } H_3PO_3$$

PCl_3 is the limiting reactant. The theoretical yield is 134 g H_3PO_3.

4 **Verify your result.**

The units of the answer are correct, and estimating gives 128.

PRACTICE

Using the reaction above, identify the limiting reactant and the theoretical yield (in grams) of HCl for each pair of reactants.

1 3.00 mol PCl_3 and 3.00 mol H_2O

2 75.0 g PCl_3 and 75.0 g H_2O

3 1.00 mol of PCl_3 and 50.0 g of H_2O

PRACTICE HINT

Whenever a problem gives you quantities of two or more reactants, you must determine the limiting reactant and use it to determine the theoretical yield.

314

MISCONCEPTION ///ALERT\\\

Working Stoichiometry Problems Students frequently think there is no flexibility in solving stoichiometry problems. Although this text calculates the product that could form using both reactants to determine the limiting reactant, a single calculation could also be done. In fact, many texts show this other method in which the amount of one reactant needed to completely react with the second reactant is calculated. By comparing the amount needed to the amount on hand, the limiting reactant can be determined. Because this method requires more thought to determine the limiting reactant, it is more difficult to do correctly and students are more likely to make an error.

Limiting Reactants and the Food You Eat

In industry, the cheapest reactant is often used as the excess reactant. In this way, the expensive reactant is more completely used up. In addition to being cost-effective, this practice can be used to control which reactions happen. In the production of cider vinegar from apple juice, the apple juice is first kept where there is no oxygen so that the microorganisms in the juice break down the sugar, glucose, into ethanol and carbon dioxide. The resulting solution is hard cider.

Having excess oxygen in the next step allows the organisms to change ethanol into acetic acid, resulting in cider vinegar. Because the oxygen in the air is free and is easy to get, the makers of cider vinegar constantly pump air through hard cider as they make it into vinegar. Ethanol, which is not free, is the limiting reactant and is used up in the reaction.

Cost is also used to choose the excess reactant when making banana flavoring, isopentyl acetate. Acetic acid is the excess reactant because it costs much less than isopentyl alcohol.

$$CH_3COOH + C_5H_{11}OH \longrightarrow CH_3COOC_5H_{11} + H_2O$$
$$\text{acetic acid} + \text{isopentyl alcohol} \longrightarrow \text{isopentyl acetate} + \text{water}$$

As shown in **Figure 5,** when compared mole for mole, isopentyl alcohol is more than twice as expensive as acetic acid. When a large excess of acetic acid is present, almost all of the isopentyl alcohol reacts.

Choosing the excess and limiting reactants based on cost is also helpful in areas outside of chemistry. In making the homecoming mums, the flower itself is more expensive than any of the other materials, so it makes sense to have an excess of ribbons and charms. The expensive flowers are the limiting reactant.

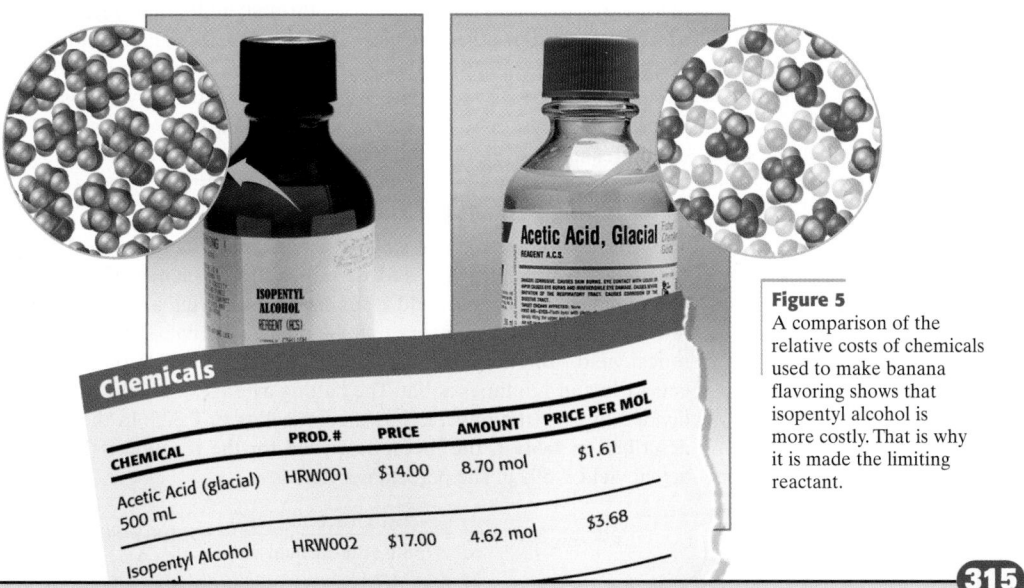

Chemicals

CHEMICAL	PROD.#	PRICE	AMOUNT	PRICE PER MOL
Acetic Acid (glacial) 500 mL	HRW001	$14.00	8.70 mol	$1.61
Isopentyl Alcohol	HRW002	$17.00	4.62 mol	$3.68

Figure 5
A comparison of the relative costs of chemicals used to make banana flavoring shows that isopentyl alcohol is more costly. That is why it is made the limiting reactant.

315

did you know?

Oxygen and Laboratory Burners The efficiency of a Bunsen burner depends on the correct mix of fuel and oxygen. If oxygen is limited, the flame burns bright, yellow, and relatively cool. If you place glassware above this flame, unburned carbon in the form of soot from the natural gas deposits on it. If adequate oxygen is present, the flame burns blue and hot. No carbon would be deposited from this flame because combustion is relatively complete.

ChemFile   CHEMISTRY INTERACTIVE TUTOR

- **Module 5: Equations and Stoichiometry Topic: Limiting Reagents**
This engaging tutorial reviews and reinforces the concept of limiting reagents through modeling and guided practice.

Demonstration —— ADVANCED
Acid and Marble Chips

(Approximate time: 10 minutes)

1. Determine and record the mass of 5–10 marble chips to the nearest thousandth of a gram. Place the marble chips in a Petri dish.

2. Place the Petri dish on an overhead projector, and add 1 M HCl until the dish is half-full.

3. Have students observe the reaction until it nearly stops.

4. Have students answer the following questions:

- What gas forms? **Ans.** CO_2

- How do you know when the reaction has stopped? **Ans.** No more bubbles are given off.

- Look at what remains in the Petri dish. What was the limiting reactant? **Ans.** whatever is used up, probably HCl **What was the excess reactant? Ans.** whatever is left over, probably the marble chips

- How could you find out the theoretical yield of CO_2? **Ans.** Determine the mass of marble that reacted, and use it to calculate the theoretical yield of CO_2.

- What would you need to do to determine the actual yield? **Ans.** Collect the gas given off and find its mass. **the percentage yield? Ans.** Divide the actual yield by the theoretical yield, and multiply by 100%.

Safety Caution: Wear safety goggles and a lab apron. Students should remain at least 3 m from the reaction. Avoid getting acid on your skin or breathing the acid fumes.

Disposal: Decant the liquid, and wash any remaining chips. Combine the washings with the decanted liquid. Test the pH of the solution. If the pH is 3 or below, neutralize the liquid with 1 M NaOH, and pour the mixture down the drain with plenty of water. Save leftover chips.
LS Logical

Equilibrium and Yield When students study chemical equilibrium, they will learn more about reactions that proceed in the opposite direction, thus reducing the amount of actual product and the percentage yield.

Group Activity — GENERAL

Have students work in groups to brainstorm about situations that are analogous to calculating percentage yield, where they compare actual results with theoretical results. Examples might include finding a batting average (the number of hits compared with the number of times at bat) or calculating a percentage average for a test (the number of correct answers compared with the number of possible answers, multiplied by 100%). Have students then make up problems for their own situations, trade the problems, and discuss how the situations are similar to calculating percentage yield. LS Interpersonal

READING SKILL BUILDER — BASIC

Reading Organizer Have students use the content of this section to develop a graphic organizer that defines the different types of yields—theoretical, actual, and percentage—and tells how they are determined. LS Visual

Table 1   Predictions and Results for Isopentyl Acetate Synthesis

Reactants	Formula	Mass present	Amount present	Amount left over
Isopentyl alcohol	$C_5H_{11}OH$	500.0 g	5.67 mol (limiting reactant)	0.0 mol
Acetic acid	CH_3COOH	1.25×10^3 g	20.8 mol	15.1 mol
Products	**Formula**	**Amount expected**	**Theoretical yield (mass expected)**	**Actual yield (mass produced)**
Isopentyl acetate	$CH_3COOC_5H_{11}$	5.67 mol	738 g	591 g
Water	H_2O	5.67 mol	102 g	81.6 g

actual yield

the measured amount of a product of a reaction

Actual Yield and Percentage Yield

Although equations tell you what *should* happen in a reaction, they cannot always tell you what *will* happen. For example, sometimes reactions do not make all of the product predicted by stoichiometric calculations, or the theoretical yield. In most cases, the **actual yield,** the mass of product actually formed, is less than expected. Imagine that a worker at the flavoring factory mixes 500.0 g isopentyl alcohol with 1.25×10^3 g acetic acid. The actual and theoretical yields are summarized in **Table 1.** Notice that the actual yield is less than the mass that was expected.

There are several reasons why the actual yield is usually less than the theoretical yield in chemical reactions. Many reactions do not completely use up the limiting reactant. Instead, some of the products turn back into reactants so that the final result is a mixture of reactants and products. In many cases the main product must go through additional steps to purify or separate it from other chemicals. For example, banana flavoring must be *distilled,* or isolated based on its boiling point. Solid compounds, such as sugar, must be recrystallized. Some of the product may be lost in the process. There also may be other reactions, called *side reactions,* that can use up reactants without making the desired product.

Determining Percentage Yield

The ratio relating the actual yield of a reaction to its theoretical yield is called the *percentage yield* and describes the efficiency of a reaction. Calculating a percentage yield is similar to calculating a batting average. A batter might get a hit every time he or she is at bat. This is the "theoretical yield." But no player has gotten a hit every time. Suppose a batter gets 41 hits in 126 times at bat. The batting average is 41 (the actual hits) divided by 126 (the possible hits theoretically), or 0.325. In the example described in **Table 1,** the theoretical yield for the reaction is 738 g. The actual yield is 591 g. The percentage yield is

$$\text{percentage yield} = \frac{591 \text{ g (actual yield)}}{738 \text{ g (theoretical yield)}} \times 100 = 80.1\%$$

REAL-WORLD CONNECTION

Industry and Percentage Yield Most industrial processes that involve chemical reactions rely heavily on the concepts presented in this chapter. For example, factory owners must know how much of each reactant to order to produce the required amount of product. Percentage yield is constantly determined, and new methods are developed to increase it. Students will learn more about the importance of limiting and excess reactants when they study equilibrium and factors that affect it.

Calculating Percentage Yield

Determine the limiting reactant, the theoretical yield, and the percentage yield if 14.0 g N_2 are mixed with 9.0 g H_2, and 16.1 g NH_3 form.

$$N_2 + 3H_2 \longrightarrow 2NH_3$$

1 Gather information.

- mass N_2 = 14.0 g N_2
- mass H_2 = 9.0 g H_2
- theoretical yield of NH_3 = ? g NH_3
- actual yield of NH_3 = 16.1 g NH_3
- molar mass N_2 = 28.02 g/mol
- molar mass H_2 = 2.02 g/mol
- molar mass NH_3 = 17.04 g/mol

- From the balanced equation:
  1 mol N_2 = 2 mol NH_3 and 3 mol H_2 = 2 mol NH_3.

2 Plan your work.

Set up problems that will calculate the mass of NH_3 you would expect to form from each reactant.

3 Calculate.

$$? \text{ g } NH_3 = 14.0 \text{ g } N_2 \times \frac{1 \text{ mol } N_2}{28.02 \text{ g } N_2} \times \frac{2 \text{ mol } NH_3}{1 \text{ mol } N_2} \times \frac{17.04 \text{ g } NH_3}{1 \text{ mol } NH_3} = 17.0 \text{ g } NH_3$$

$$? \text{ g } NH_3 = 9.0 \text{ g } H_2 \times \frac{1 \text{ mol } H_2}{2.02 \text{ g } H_2} \times \frac{2 \text{ mol } NH_3}{3 \text{ mol } H_2} \times \frac{17.04 \text{ g } NH_3}{1 \text{ mol } NH_3} = 51 \text{ g } NH_3$$

- The smaller quantity made, 17.0 g NH_3, is the theoretical yield so the limiting reactant is N_2.
- The percentage yield is calculated:

$$\text{percentage yield} = \frac{16.1 \text{ g (actual yield)}}{17.0 \text{ g (theoretical yield)}} \times 100 = 94.7\%$$

4 Verify your result.

The units of the answer are correct. The percentage yield is less than 100%, so the final calculation is probably set up correctly.

PRACTICE HINT

If an amount of product actually formed is given in a problem, this is the reaction's actual yield.

PRACTICE

Determine the limiting reactant and the percentage yield for each of the following.

1. 14.0 g N_2 react with 3.15 g H_2 to give an actual yield of 14.5 g NH_3.

2. In a reaction to make ethyl acetate, 25.5 g CH_3COOH react with 11.5 g C_2H_5OH to give a yield of 17.6 g $CH_3COOC_2H_5$.

$$CH_3COOH + C_2H_5OH \longrightarrow CH_3COOC_2H_5 + H_2O$$

3. 16.1 g of bromine are mixed with 8.42 g of chlorine to give an actual yield of 21.1 g of bromine monochloride.

PROBLEM SOLVING SKILL

317

Answers to Practice Problems F

1. N_2 is limiting, 85.3%
2. C_2H_5OH is limiting, 80.0%
3. Br_2 is limiting, 90.9%

Homework — GENERAL

Additional Practice

1. When 4.00×10^5 kg of H_2 is added to an excess of N_2, 1.04×10^6 kg of NH_3 is produced. What is the percentage yield of the reaction? Ans. 46.2%

2. A standard laboratory preparation of iodine is the following reaction.

   $2NaI(aq) + MnO_2(s) + 2H_2SO_4(aq) \rightarrow Na_2SO_4(aq) + MnSO_4(aq) + 2H_2O(l) + I_2(s)$

   Balance the equation, then find the percentage yield of I_2 if the actual yield of I_2 was 39.8 g when the amount of NaI used was 62.6 g. Ans. 75.1%

3. A 15.0 g sample of magnesium reacts with hydrochloric acid to form magnesium chloride and hydrogen. During the reaction, 46.6 g of magnesium chloride was formed. What was the percentage yield? Ans. 79.3%

LS Logical

SKILL BUILDER — BASIC

Math Skills Students might have trouble remembering which number is the numerator and which is the denominator in the equation for determining percentage yield. Remind them that the percentage yield is always less than 100%, so the smaller number (actual yield) must be in the numerator and the larger number (theoretical yield) must be in the denominator. Have students write different masses on slips of paper. Then have them draw masses two at a time and state which of the masses represents the theoretical yield and which one is the actual yield.

LS Logical

REAL-WORLD — CONNECTION

Percentage Yield of Sulfuric Acid The most important industrial chemical produced in the United States is sulfuric acid. It is used extensively in the manufacture of other chemicals and products. Have students do research to compare and contrast the two primary methods used to manufacture sulfuric acid. Research should show that the older lead-chamber process is relatively inexpensive, but its percentage yield is only about 70% and the sulfuric acid produced is relatively impure. The contact process is more commonly used. It is more expensive, but the percentage yield is about 98% and the sulfuric acid is high quality.

Chapter Resource File

• Problem Solving

Answers to Practice Problems G

1. 1.04×10^3 g NH_3
2. 5.9×10^3 g CH_3OH
3. 439 g BrCl

Homework ── **GENERAL**

Additional Practice

1. Huge quantities of sulfur dioxide are produced from zinc sulfide when it reacts with oxygen. The other product is zinc oxide. If the typical yield is 86.78%, what mass of SO_2 should be expected if 4897 g of ZnS is used? *Ans.* 2793 g SO_2

2. Methanol, CH_3OH, can be produced through the reaction of CO and H_2 in the presence of a catalyst. If 75.0 g of CO is used and the typical yield is 79.8%, what mass of methanol is likely to be produced? *Ans.* 68.5 g CH_3OH

LS Logical

Chapter Resource File

• Problem Solving

Determining Actual Yield

Although the actual yield can only be determined experimentally, a close estimate can be calculated if the percentage yield for a reaction is known. The percentage yield in a particular reaction is usually fairly consistent. For example, suppose an industrial chemist determined the percentage yield for six tries at making banana flavoring and found the results were 80.0%, 82.1%, 79.5%, 78.8%, 80.5%, and 81.9%. In the future, the chemist can expect a yield of around 80.5%, or the average of these results.

If the chemist has enough isopentyl alcohol to make 594 g of the banana flavoring theoretically, then an actual yield of around 80.5% of that, or 478 g, can be expected.

SAMPLE PROBLEM G

Calculating Actual Yield

How many grams of $CH_3COOC_5H_{11}$ should form if 4808 g are theoretically possible and the percentage yield for the reaction is 80.5%?

1 Gather information.
- theoretical yield of $CH_3COOC_5H_{11}$ = 4808 g $CH_3COOC_5H_{11}$
- actual yield of $CH_3COOC_5H_{11}$ = ? g $CH_3COOC_5H_{11}$
- percentage yield = 80.5%

2 Plan your work.

Use the percentage yield and the theoretical yield to calculate the actual yield expected.

3 Calculate.

$$80.5\% = \frac{\text{actual yield}}{4808 \text{ g}} \times 100$$

actual yield = 4808 g $\times$ 0.805 = 3.87×10^3 g $CH_3COOC_5H_{11}$

4 Verify your result.

The units of the answer are correct. The actual yield is less than the theoretical yield, as it should be.

PRACTICE HINT

The actual yield should always be less than the theoretical yield. A wrong answer that is greater than the theoretical yield can result if you accidentally reverse the actual and theoretical yields.

PRACTICE

1 The percentage yield of NH_3 from the following reaction is 85.0%. What actual yield is expected from the reaction of 1.00 kg N_2 with 225 g H_2?

$$N_2 + 3H_2 \longrightarrow 2NH_3$$

2 If the percentage yield is 92.0%, how many grams of CH_3OH can be made by the reaction of 5.6×10^3 g CO with 1.0×10^3 g H_2?

$$CO + 2H_2 \longrightarrow CH_3OH$$

3 Suppose that the percentage yield of BrCl is 90.0%. How much BrCl can be made by reacting 338 g Br_2 with 177 g Cl_2?

Section Review

UNDERSTANDING KEY IDEAS

1. Distinguish between limiting reactant and excess reactant in a chemical reaction.

2. How do manufacturers decide which reactant to use in excess in a chemical reaction?

3. How do you calculate the percentage yield of a chemical reaction?

4. Give two reasons why a 100% yield is not obtained in actual chemical manufacturing processes.

5. How do the values of the theoretical and actual yields generally compare?

PRACTICE PROBLEMS

6. A chemist reacts 8.85 g of iron with an excess of hydrogen chloride to form hydrogen gas and iron(II) chloride. Calculate the theoretical yield and the percentage yield of hydrogen if 0.27 g H_2 are collected.

7. Use the chemical reaction below to answer the questions that follow.

$$P_4O_{10} + H_2O \longrightarrow H_3PO_4$$

   a. Balance the equation.

   b. Calculate the theoretical yield if 100.0 g P_4O_{10} react with 200.0 g H_2O.

   c. If the actual mass recovered is 126.2 g H_3PO_4, what is the percentage yield?

8. Titanium dioxide is used as a white pigment in paints. If 3.5 mol $TiCl_4$ reacts with 4.5 mol O_2, which is the limiting reactant? How many moles of each product are produced? How many moles of the excess reactant remain?

$$TiCl_4 + O_2 \longrightarrow TiO_2 + 2Cl_2$$

9. If 1.85 g Al reacts with an excess of copper(II) sulfate and the percentage yield of Cu is 56.6%, what mass of Cu is produced?

10. Quicklime, CaO, can be prepared by roasting limestone, $CaCO_3$, according to the chemical equation below. When 2.00×10^3 g of $CaCO_3$ are heated, the actual yield of CaO is 1.05×10^3 g. What is the percentage yield?

$$CaCO_3(s) \longrightarrow CaO(s) + CO_2(g)$$

11. Magnesium powder reacts with steam to form magnesium hydroxide and hydrogen gas.

   a. Write a balanced equation for this reaction.

   b. What is the percentage yield if 10.1 g Mg reacts with an excess of water and 21.0 g $Mg(OH)_2$ is recovered?

   c. If 24 g Mg is used and the percentage yield is 95%, how many grams of magnesium hydroxide should be recovered?

12. Use the chemical reaction below to answer the questions that follow.

$$CuO(s) + H_2(g) \longrightarrow Cu(s) + H_2O(g)$$

   a. What is the limiting reactant when 19.9 g CuO react with 2.02 g H_2?

   b. The actual yield of copper was 15.0 g. What is the percentage yield?

   c. How many grams of Cu can be collected if 20.6 g CuO react with an excess of hydrogen with a yield of 91.0%?

CRITICAL THINKING

13. A chemist reacts 20 mol H_2 with 20 mol O_2 to produce water. Assuming all of the limiting reactant is converted to water in the reaction, calculate the amount of each substance present after the reaction.

14. A pair of students performs an experiment in which they collect 27 g CaO from the decomposition of 41 g $CaCO_3$. Are these results reasonable? Explain your answer using percentage yield.

319

Answers to Section Review

1. All the limiting reactant is used up in the reaction; some excess reactant is left over.

2. Usually, the least expensive reactant is used in excess.

3. Divide the actual yield by the theoretical yield, then multiply the quotient by 100%.

4. Answers will vary but might include the reaction not going to completion, the reverse reaction or side reactions occurring, or product being lost during purification.

5. Theoretical yield is always greater than actual yield.

6. Theoretical yield is 0.320 g H_2. Percentage yield is 84.4%.

7. a. $P_4O_{10} + 6H_2O \rightarrow 4H_3PO_4$

   b. 138.1 g H_3PO_4

   c. 91.4%

8. $TiCl_4$ is limiting; 3.5 mol TiO_2 and 7.0 mol Cl_2 are produced; 1.0 mol O_2 left over

9. 3.70 g Cu

10. 93.8%

Answers continued on p. 335A

Close

Reteaching — BASIC

Have students who have difficulty with the concept of limiting reactant use a model kit to model reactions in which one reactant is limiting and one is in excess. For example, they might model a reaction as simple as carbon reacting with oxygen to form carbon dioxide. Given equal numbers of each type of atom, students should see that the oxygen is used up and some carbon remains. As students achieve understanding of the concept, the reactions could become more complicated. **LS** Kinesthetic

Quiz — GENERAL

1. If you have equal masses of magnesium and oxygen that will react with each other, which reactant is limiting? Ans. Mg

2. If a question states that a reaction produced 19.0 g of product, is this amount of product the actual yield, the percentage yield, or the theoretical yield? Ans. actual yield

3. Ammonia reacts with oxygen according to the equation below. For each of the following amounts of reactants, identify the excess reactant, the limiting reactant, and the theoretical yield (in grams) of nitrogen.

   $4NH_3 + 3O_2 \rightarrow 2N_2 + 6H_2O$

   a. 0.23 mol NH_3 and 0.19 mol O_2 Ans. O_2 is excess; NH_3 is limiting; theoretical yield is 3.2 g N_2

   b. 47.1 g NH_3 and 55.2 g O_2 Ans. NH_3 is excess; O_2 is limiting; theoretical yield is 32.2 g N_2

LS Logical

Chapter Resource File

- Concept Review
- Quiz

Focus

Overview

Before beginning this section, review with your students the Objectives listed in the Student Edition. This section relates the principles of stoichiometry to cars. Specifically included are the reactions and volumes of gases involved in air bags, limiting reactants and fuel-air ratios for internal combustion engines, and reactions involved in pollution-control devices.

 **Bellringer**

Have students brainstorm reactions that are involved in operating a car. Then have students develop ideas of how stoichiometry is involved in these reactions.

Motivate

Identifying Preconceptions —— GENERAL

Students often think that chemical concepts have few practical applications to everyday life. As students read through this section, have them keep a running list of the everyday applications of stoichiometry concepts. **LS** Intrapersonal

Discussion

Find out how familiar students are with air bags. Ask the following questions to facilitate the discussion.

• How does an air bag work?

• Where does the gas that inflates the bag come from?

Refer to these responses as you cover the stoichiometry of the air bag in this section.

Stoichiometry and Cars

OBJECTIVES

(1) **Relate** volume calculations in stoichiometry to the inflation of automobile safety air bags.

(2) **Use** the concept of limiting reactants to explain why fuel-air ratios affect engine performance.

(3) **Compare** the efficiency of pollution-control mechanisms in cars using percentage yield.

Stoichiometry and Safety Air Bags

So far you have examined stoichiometry in a number of chemical reactions, including making banana flavoring and ammonia. Now it is time to look at stoichiometry in terms of something a little more familiar— a car. Stoichiometry is important in many aspects of automobile operation and safety. First, let's look at how stoichiometry can help keep you safe should you ever be in an accident. Air bags have saved the lives of many people involved in accidents. And the design of air bags requires an understanding of stoichiometry.

An Air Bag Could Save Your Life

Air bags are designed to protect people in a car from being hurt during a high-speed collision. When inflated, air bags slow the motion of a person so that he or she does not strike the steering wheel, windshield, or dashboard with as much force.

Stoichiometry is used by air-bag designers to ensure that air bags do not underinflate or overinflate. Bags that underinflate do not provide enough protection, and bags that overinflate can cause injury by bouncing the person back with too much force. Therefore, the chemicals must be present in just the right proportions. To protect riders, air bags must inflate within one-tenth of a second after impact. The basic components of most systems that make an air bag work are shown in **Figure 6.** A front-end collision transfers energy to a crash sensor that causes an igniter to fire. The igniter provides the energy needed to start a very fast reaction that produces gas in a mixture called the *gas generant*. The igniter also raises the temperature and pressure within the inflator (a metal vessel) so that the reaction happens fast enough to fill the bag before the rider strikes it. A high-efficiency filter keeps the hot reactants and the solid products away from the rider, and additional chemicals are used to make the products safer.

internet connect
www.scilinks.org
Topic: Air Bags
SciLinks code: HW4005
SCINKS Maintained by the National Science Teachers Association

320

HISTORY
CONNECTION

Air Bags Air bags were invented in 1952 by John Hetrick. His design used compressed air to inflate the bag. However, air bags were not widely offered in cars until the late 1980s.

Chapter Resource File
• Lesson Plan

Air-Bag Design Depends on Stoichiometric Precision

The materials used in air bags are constantly being improved to make air bags safer and more effective. Many different materials are used. One of the first gas generants used in air bags is still in use in some systems. It is a solid mixture of sodium azide, NaN_3, and an oxidizer. The gas that inflates the bag is almost pure nitrogen gas, N_2, which is produced in the following decomposition reaction.

$$2NaN_3(s) \longrightarrow 2Na(s) + 3N_2(g)$$

However, this reaction does not inflate the bag enough, and the sodium metal is dangerously reactive. Oxidizers such as ferric oxide, Fe_2O_3, are included, which react rapidly with the sodium. Energy is released, which heats the gas and causes the gas to expand and fill the bag.

$$6Na(s) + Fe_2O_3(s) \longrightarrow 3Na_2O(s) + 2Fe(s) + energy$$

One product, sodium oxide, Na_2O, is extremely corrosive. Water vapor and CO_2 from the air react with it to form less harmful $NaHCO_3$.

$$Na_2O(s) + 2CO_2(g) + H_2O(g) \longrightarrow 2NaHCO_3(s)$$

The mass of gas needed to fill an air bag depends on the density of the gas. Gas density depends on temperature. To find the amount of gas generant to put into each system, designers must know the stoichiometry of the reactions and account for changes in temperature and thus the density of the gas.

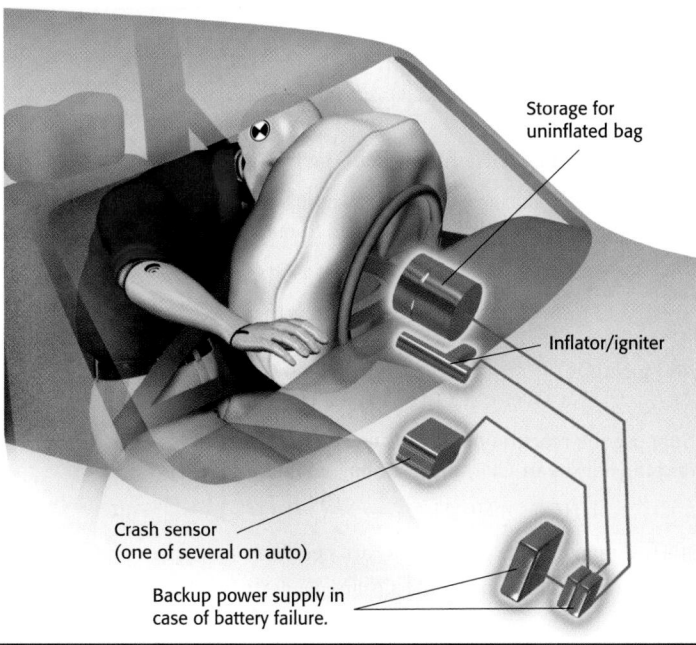

Storage for uninflated bag

Inflator/igniter

Crash sensor (one of several on auto)

Backup power supply in case of battery failure.

Figure 6
Inflating an air bag requires a rapid series of events, eventually producing nitrogen gas to inflate the air bag.

321

did you know?

Air Bags and Injuries Although injured ribs, scrapes, and bruises are common injuries that result from deployment of air bags, more serious injuries probably would have occurred without the presence of the air bag. However, small children and even small adults have been seriously injured or even killed by the force of a deploying air bag. For safety, small people of any age should sit in the back seat, if possible. As an added safety feature, a control switch can be installed that will disengage the passenger-side air bag. A request for installation of a switch must be approved by the National Highway Traffic Safety Administration.

Teach

Using the Figure

Refer students to **Figure 6.** Make students aware that cars—including air bags—are made to absorb some of the energy of a crash. The time it takes for the car to stop increases which means a smaller force is applied to the body of an occupant to bring it to a stop. Unfortunately, the cars sustain more damage. Fortunately, people suffer fewer injuries as a result of crumple zones and air bags.

Teaching Tip

Deflation of Air Bags By the time it comes into contact with a person, a deployed air bag has begun to deflate through vents in its sides. By deflating, the air bag provides more of a cushioning effect for the person.

SKILL BUILDER

Writing **Research Skills** Have students research statistics related to how many lives have been saved or lost as the result of air bags. Have them design a public safety message in the form of a pamphlet or a radio or television spot. Results might include the following statistics, which were accurate as of May, 2001. Since 1990, more than 3.3 million air bags have deployed. It is estimated that more than 6000 lives have been saved. The number of people who have died as a result of air bag use is 175. Of these deaths, 104 were children, almost all of whom were unrestrained or improperly restrained. Of the adults killed, 46 were either unrestrained or improperly restrained. These statistics further emphasize the importance of the use of both seat belts and air bags. **LS** Auditory

Answers to Practice Problems H

1. 33 g Na
2. 40.9 g Fe_2O_3
3. 121 g $NaHCO_3$
4. a. 168 g $NaHCO_3$
   b. 1.20×10^2 g $HC_2H_3O_2$

Homework ———— **GENERAL**

Additional Practice

1. How many grams of NaN_3 are needed to fill an air bag with 23.6 L of N_2 (density = 0.92 g/L)?
   Ans. 33.6 g NaN_3

2. How many grams of $NaHCO_3$ and acetic acid, CH_3COOH, would be needed to inflate an air bag to a volume of 65.1 L when CO_2 gas has a density of 2.68 g/L? The other products are sodium acetate and water.
   Ans. 333 g $NaHCO_3$, 238 g CH_3COOH

LS Logical

Teaching Tip

Different Composition Tell students that in some air bags, the sodium produced in the decomposition of NaN_3 reacts with potassium nitrate instead of with iron(III) oxide. As KNO_3 reacts with the sodium, additional nitrogen gas forms. The sodium oxide and potassium oxide formed in this reaction combine with the silicon dioxide in this air bag to form glass.

Chapter Resource File

• Problem Solving

SAMPLE PROBLEM H

Air-Bag Stoichiometry

Assume that 65.1 L N_2 inflates an air bag to the proper size. What mass of NaN_3 must be used? (density of N_2 = 0.92 g/L)

1 Gather information.

• Write a balanced chemical equation

$$2NaN_3(s) \longrightarrow 2Na(s) + 3N_2(g)$$

• volume of N_2 = 65.1 L N_2
• density of N_2 = 0.92 g/L
• molar mass of N_2 = 28.02 g/mol
• mass of reactant = ? g NaN_3
• molar mass of NaN_3 = 65.02 g/mol
• From the balanced equation: 2 mol NaN_3 = 3 mol N_2.

2 Plan your work.

Start with the volume of N_2, and change it to moles using density and molar mass. Then use the mole ratio followed by the molar mass of NaN_3.

3 Calculate.

$$? \text{ g } NaN_3 = 65.1 \text{ L } N_2 \times \frac{0.92 \text{ g } N_2}{1 \text{ L } N_2} \times \frac{1 \text{ mol } N_2}{28.02 \text{ g } N_2} \times$$

$$\frac{2 \text{ mol } NaN_3}{3 \text{ mol } N_2} \times \frac{65.02 \text{ g } NaN_3}{1 \text{ mol } NaN_3} = 93 \text{ g } NaN_3$$

4 Verify your result.

The number of significant figures is correct. Estimating gives 90.

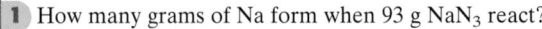

PRACTICE HINT

Gases are measured by volume, just as liquids are. In problems with volume, you can use the density to convert to mass and the molar mass to convert to moles. Then use the mole ratio, just as in any other stoichiometry problem.

PRACTICE

PROBLEM SOLVING SKILL

1 How many grams of Na form when 93 g NaN_3 react?

2 The Na formed during the breakdown of NaN_3 reacts with Fe_2O_3. How many grams of Fe_2O_3 are needed to react with 35.3 g Na?

$$6Na(s) + Fe_2O_3(s) \longrightarrow 3Na_2O(s) + 2Fe(s)$$

3 The Na_2O formed in the above reaction is made less harmful by the reaction below. How many grams of $NaHCO_3$ are made from 44.7 g Na_2O?

$$Na_2O(s) + 2CO_2(g) + H_2O(g) \longrightarrow 2NaHCO_3(s)$$

4 Suppose the reaction below was used to fill a 65.1 L air bag with CO_2 and the density of CO_2 at the air bag temperature is 1.35 g/L.

$$NaHCO_3 + HC_2H_3O_2 \longrightarrow NaC_2H_3O_2 + CO_2 + H_2O$$

a. How many grams of $NaHCO_3$ are needed?
b. How many grams of $HC_2H_3O_2$ are needed?

322

REAL-WORLD CONNECTION — GENERAL

Writing **Air Bags and Car Manufacturers**

Have students research different car manufacturers and gather information about their use of air bags. Information to look for might include when air bags were first used by the manufacturer, where in the current models are they used, and any other interesting and informative information they can find. You might want to assign small groups of students different manufacturers to investigate and then create a class comparison of manufacturers. **LS** Logical

Stoichiometry and Engine Efficiency

The efficiency of a car's engine depends on having the correct stoichiometric ratio of gasoline and oxygen. Although gasoline used in automobiles is a mixture, it can be treated as if it were pure isooctane, one of the many compounds whose formula is C_8H_{18}. (This compound has a molar mass that is about the same as the weighted average of the compounds in actual gasoline.) The other reactant in gasoline combustion is oxygen, which is about 21% of air by volume. The reaction for gasoline combustion can be written as follows.

$$2C_8H_{18}(g) + 25O_2(g) \longrightarrow 16CO_2(g) + 18H_2O(g)$$

Engine Efficiency Depends on Reactant Proportions

For efficient combustion, the above two reactants must be mixed in a mole ratio that is close to the one shown in the balanced chemical equation, that is 2:25, or 1:12.5. If there is not enough of either reactant, the engine might stall. For example, if you pump the gas pedal too many times before starting, the mixture of reactants in the engine will contain an excess of gasoline, and the lack of oxygen may prevent the mixture from igniting. This is referred to as "flooding the engine." On the other hand, if there is too much oxygen and not enough gasoline, the engine will stall just as if the car were out of gas.

Although the best stoichiometric mixture of fuel and oxygen is 1:12.5 in terms of moles, this is not the best mixture to use all the time. **Figure 7** shows a model of a carburetor controlling the fuel-oxygen ratio in an engine that is starting, idling, and running at normal speeds. Carburetors are often used in smaller engines, such as those in lawn mowers. Computer-controlled fuel injectors have taken the place of carburetors in car engines.

Figure 7
The fuel-oxygen ratio changes depending on what the engine is doing.

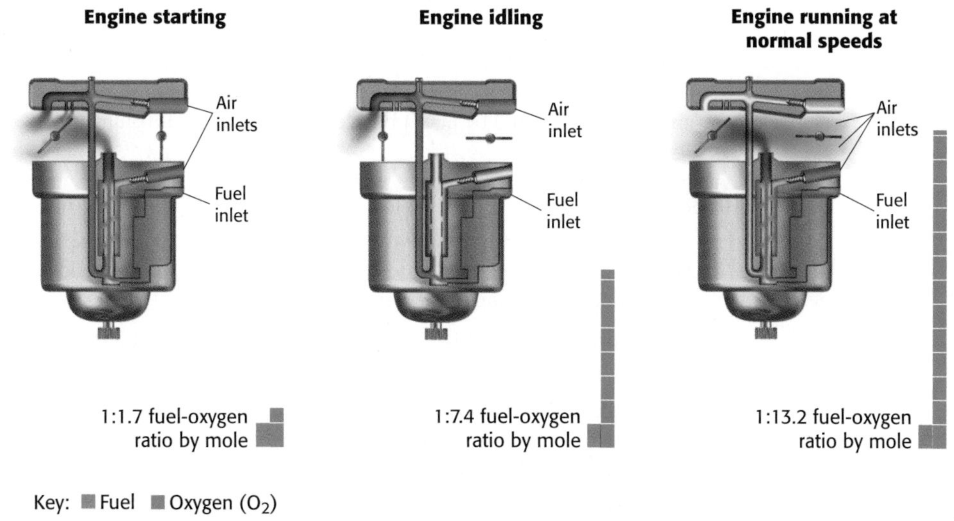

Engine starting
1:1.7 fuel-oxygen ratio by mole

Engine idling
1:7.4 fuel-oxygen ratio by mole

Engine running at normal speeds
1:13.2 fuel-oxygen ratio by mole

Air inlets
Fuel inlet

Key: ■ Fuel ■ Oxygen (O_2)

323

1. 2.17 cycles; after 3 full cycles all of the 1.00 mL of isooctane will have reacted

2. 8.65 mL isooctane

3. $2CH_3OH + 3O_2 \rightarrow 2CO_2 + 4H_2O$; 2.2×10^2 L air

Homework — GENERAL

Additional Practice

1. How many liters of air are needed to completely burn 2.00 mL C_8H_{18}? The density of C_8H_{18} is 0.692 g/mL, and the density of O_2 is 1.33 g/L. Air is 21% oxygen by volume. **Ans.** 17.3 L of air

2. How many grams of O_2 are needed to completely burn 528.7 g C_8H_{18}? **Ans.** 1851 g O_2

LS Logical

SKILL BUILDER — ADVANCED

Writing Skills Have students write a paragraph that relates automobile ratings of miles per gallon to the percentage yield of the combustion reaction. Students can find information on fuel efficiency from automobile manufacturers and dealers. Student information should indicate that factors such as the weight and design of the car are also factors. For example, a heavy car that is not streamlined will have lower miles per gallon than another car with the same percent yield for its combustion reaction. LS Logical

---

SAMPLE PROBLEM I

Air-Fuel Ratio

A cylinder in a car's engine draws in 0.500 L of air. How many milliliters of liquid isooctane should be injected into the cylinder to completely react with the oxygen present? The density of isooctane is 0.692 g/mL, and the density of oxygen is 1.33 g/L. Air is 21% oxygen by volume.

1 Gather information.

- Write a balanced equation for the chemical reaction.

$$2C_8H_{18} + 25O_2 \longrightarrow 16CO_2 + 18H_2O$$

- volume of air = 0.500 L air
- percentage of oxygen in air: 21% by volume
- Organize the data in a table.

Reactant	Formula	Molar mass	Density	Volume
Oxygen	O_2	32.00 g/mol	1.33 g/L	? L
Isooctane	C_8H_{18}	114.26 g/mol	0.692 g/mL	? mL

- From the balanced equation: 2 mol C_8H_{18} = 25 mol O_2.

2 Plan your work.

Use the percentage by volume of O_2 in air to find the volume of O_2. Then set up a volume-volume problem.

3 Calculate.

$$? \text{ mL } C_8H_{18} = 0.500 \text{ L air} \times \frac{21 \text{ L } O_2}{100 \text{ L air}} \times \frac{1.33 \text{ g } O_2}{1 \text{ L } O_2} \times \frac{1 \text{ mol } O_2}{32.00 \text{ g } O_2} \times$$

$$\frac{2 \text{ mol } C_8H_{18}}{25 \text{ mol } O_2} \times \frac{114.26 \text{ g } C_8H_{18}}{1 \text{ mol } C_8H_{18}} \times \frac{1 \text{ mL } C_8H_{18}}{0.692 \text{ g } C_8H_{18}} = 5.76 \times 10^{-2} \text{ mL } C_8H_{18}$$

4 Verify your result.

The denominator is about 10 times larger than the numerator, so the answer in mL should be about one-tenth of the original volume in L.

PRACTICE HINT

Remember that in problems with volumes, you must be sure that the volume unit in the density matches the volume unit given or wanted.

PRACTICE

1. A V-8 engine has eight cylinders each having a 5.00×10^2 cm^3 capacity. How many cycles are needed to completely burn 1.00 mL of isooctane? (One cycle is the firing of all eight cylinders.)

2. How many milliliters of isooctane are burned during 25.0 cycles of a V-6 engine having six cylinders each having a 5.00×10^2 cm^3 capacity?

3. Methyl alcohol, CH_3OH, with a density of 0.79 g/mL, can be used as fuel in race cars. Calculate the volume of air needed for the complete combustion of 51.0 mL CH_3OH.

did you know?

Thermal Inversions Los Angeles often experiences thermal inversions that trap photochemical smog above the city. When this happens, the smog builds up until it is eventually blown eastward into the mountains. Thermal inversions occur when a layer of warm air forms over a layer of cooler air.

Chapter Resource File

- **Problem Solving**

Table 2   Clean Air Act Standards for 1996 Air Pollution

Pollutant	Cars	Light trucks	Motorcycles
Hydrocarbons	0.25 g/km	0.50 g/km	5.0 g/km
Carbon monoxide	2.1 g/km	2.1–3.1 g/km, depending on truck size	12 g/km
Oxides of nitrogen (NO, NO_2)	0.25 g/km	0.25–0.68 g/km, depending on truck size	not regulated

Stoichioimetry and Pollution Control

Automobiles are the primary source of air pollution in many parts of the world. The Clean Air Act was enacted in 1968 to address the issue of smog and other forms of pollution caused by automobile exhaust. This act has been amended to set new, more restrictive emission-control standards for automobiles driven in the United States. **Table 2** lists the standards for pollutants in exhaust set in 1996 by the U.S. Environmental Protection Agency.

The Fuel-Air Ratio Influences the Pollutants Formed

The equation for the combustion of "isooctane" shows most of what happens when gasoline burns, but it does not tell the whole story. For example, if the fuel-air mixture does not have enough oxygen, some carbon monoxide will be produced instead of carbon dioxide. When a car is started, there is less air, so fairly large amounts of carbon monoxide are formed, and some unburned fuel (hydrocarbons) also comes out in the exhaust. In cold weather, an engine needs more fuel to start, so larger amounts of unburned hydrocarbons and carbon monoxide come out as exhaust. These hydrocarbons are involved in forming smog. So the fuel-air ratio is a key factor in determining how much pollution forms.

Another factor in auto pollution is the reaction of nitrogen and oxygen at the high temperatures inside the engine to form small amounts of highly reactive nitrogen oxides, including NO and NO_2.

$$N_2(g) + O_2(g) \longrightarrow 2NO(g)$$

$$2NO(g) + O_2(g) \longrightarrow 2NO_2(g)$$

One of the Clean Air Act standards limits the amount of nitrogen oxides that a car can emit. These compounds react with oxygen to form another harmful chemical, ozone, O_3.

$$NO_2(g) + O_2(g) \longrightarrow 2NO(g) + O_3(g)$$

Because these reactions are started by energy from the sun's ultraviolet light, they form what is referred to as *photochemical smog*. The harmful effects of photochemical smog are caused by very small concentrations of pollutants, including unburned hydrocarbon fuel.

internet connect

www.scilinks.org
Topic: Air Pollution
SciLinks code: HW4133

SCiLINKS. Maintained by the National Science Teachers Association

Homework ───── **ADVANCED**

Designing a Spreadsheet Have students design a spreadsheet that will calculate the theoretical yield of a specific product when the mass of a reactant is entered for a specific balanced chemical equation. You might also want students to include a function that will determine the percentage yield when an actual yield is entered or be able to determine the theoretical yield when quantities of two reactants are given. **LS** Logical

ENVIRONMENTAL SCIENCE
CONNECTION

Horticulture Ask students how people can minimize pollution from lawn-care machinery. Often, lawnmowers, trimmers, and other pieces of equipment that use internal combustion engines do not contain the pollution controls required in automobiles. Per gallon of fuel, they often emit more pollution than cars do.

325

Answers to Practice Problem J
1. 4.01 g CO_2

Homework —— GENERAL

Additional Practice Racecars often burn ethanol, C_2H_5OH, for added performance. The products are carbon dioxide and water. The density of ethanol is 0.816 g/mL, the density of CO_2 is 1.997 g/L, and the density of O_2 is 1.331 g/L.

1. If the car holds 1.00×10^5 mL of ethanol and all of the carbon in it forms carbon dioxide, what volume of carbon dioxide is added to the air? Ans. 7.81×10^4 L CO_2

2. What volume of oxygen is needed to react with 78.3 L of ethanol? Ans. 1.00×10^5 L O_2
LS Logical

Reading Organizer Have students read about how catalytic converters work and draw a graphic organizer that shows how catalytic converters change pollutants into other substances. Organizers might show NO_2, CO, and unburned hydrocarbons undergoing the reactions $NO_2(g) \rightarrow N_2(g) + O_2(g)$, $2CO(g) + O_2(g) \rightarrow 2CO_2(g)$ and the unbalanced equation (unburned hydrocarbons)$(g) + O_2(g) \rightarrow CO_2(g) + H_2O(g)$. **LS** Visual

Chapter Resource File
• **Problem Solving**

internet connect
www.scilinks.org
Topic: Catalytic Converters
SciLinks code: HW4026

SciLINKS Maintained by the National Science Teachers Association

Meeting the Legal Limits Using Stoichiometry

Automobile manufacturers use stoichiometry to predict when adjustments will be necessary to keep exhaust emissions within legal limits. Because the units in **Table 2** are *grams per kilometer*, auto manufacturers must consider how much fuel the vehicle will burn to move a certain distance. Automobiles with better gas mileage will use less fuel per kilometer, resulting in lower emissions per kilometer.

Catalytic Converters Can Help

All cars that are currently manufactured in the United States are built with catalytic converters, like the one shown in **Figure 8,** to treat the exhaust gases before they are released into the air. Platinum, palladium, or rhodium in these converters act as catalysts and increase the rate of the decomposition of NO and of NO_2 into N_2 and O_2, harmless gases already found in the atmosphere. Catalytic converters also speed the change of CO into CO_2 and the change of unburned hydrocarbons into CO_2 and H_2O. These hydrocarbons are involved in the formation of ozone and smog, so it is important that unburned fuel does not come out in the exhaust.

Catalytic converters perform at their best when the exhaust gases are hot and when the ratio of fuel to air in the engine is very close to the proper stoichiometric ratio. Newer cars include on-board computers and oxygen sensors to make sure the proper fuel-air ratio is automatically maintained, so that the engine and the catalytic converter work at top efficiency.

Figure 8
Catalytic converters are used to decrease nitrogen oxides, carbon monoxide, and hydrocarbons in exhaust. Leaded gasoline and extreme temperatures decrease their effectiveness.

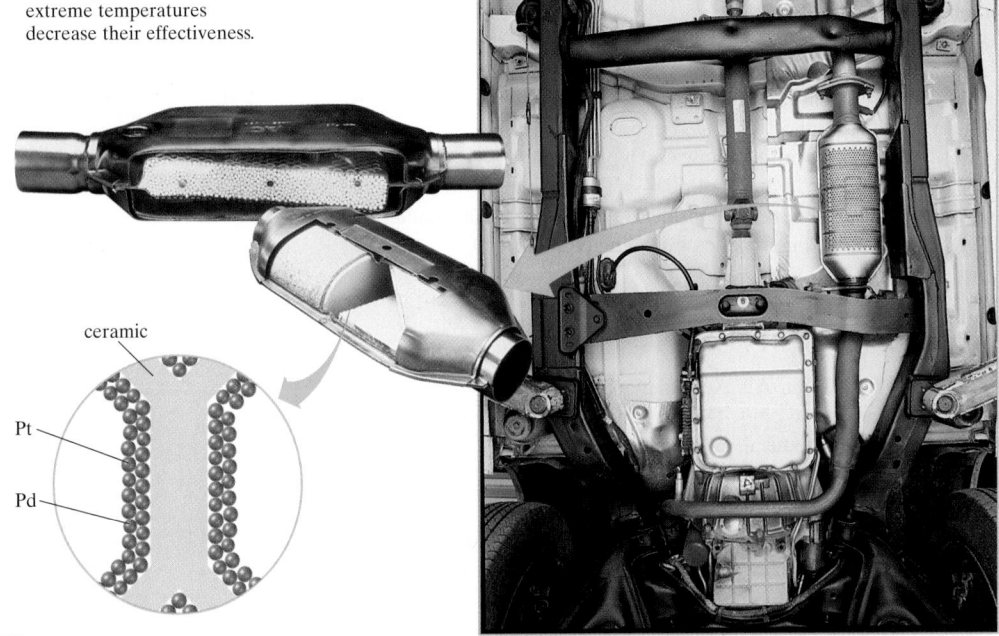

ceramic

Pt

Pd

did you know?

A Place to Gather The catalysts in a catalytic converter do speed up chemical reactions. However, the catalysts are not the only factor that helps in this process. The converter consists of a honeycomb interior, providing increased surface area where the chemical reactions occur.

Calculating Yields: Pollution

What mass of ozone, O_3, can be produced from 3.50 g of NO_2 contained in a car's exhaust? The equation is as follows.

$$NO_2(g) + O_2(g) \longrightarrow NO(g) + O_3(g)$$

1 Gather information.

- mass of NO_2 = 3.50 g NO_2
- molar mass of NO_2 = 46.01 g/mol
- mass of O_3 = ? g O_3
- molar mass of O_3 = 48.00 g/mol
- From the balanced equation: 1 mol NO_2 = 1 mol O_3.

2 Plan your work.

This is a mass-mass problem.

3 Calculate.

$$? \text{ g } O_3 = 3.50 \text{ g } NO_2 \times \frac{1 \text{ mol } NO_2}{46.01 \text{ g } NO_2} \times \frac{1 \text{ mol } O_3}{1 \text{ mol } NO_2} \times \frac{48.00 \text{ g } O_3}{1 \text{ mol } O_3} = 3.65 \text{ g } O_3$$

4 Verify your result.

The denominator and numerator are almost equal, so the mass of product is almost the same as the mass of reactant.

PRACTICE HINT

This is a review of the first type of stoichiometric calculation that you learned.

PRACTICE

1 A catalytic converter combines 2.55 g CO with excess O_2. What mass of CO_2 forms?

PROBLEM SOLVING SKILL

③ Section Review

UNDERSTANDING KEY IDEAS

1. What is the main gas in an air bag that is inflated using the NaN_3 reaction?
2. How do you know that the correct mole ratio of isooctane to oxygen is 1:12.5?
3. What do the catalysts in the catalytic converters accomplish?
4. Give at least two results of too little air being in a running engine.

PRACTICE PROBLEMS

5. Assume that 22.4 g of NaN_3 react completely in an air bag. What mass of Na_2O is produced after complete reaction of the Na with Fe_2O_3?

6. Na_2O eventually reacts with CO_2 and H_2O to form $NaHCO_3$. What mass of $NaHCO_3$ is formed when 44.4 g Na_2O completely react?

CRITICAL THINKING

7. Why are nitrogen oxides in car exhaust, even though there is no nitrogen in the fuel?

8. Why not use the following reaction to produce N_2 in an air bag?

$$NH_3(g) + O_2(g) \longrightarrow N_2(g) + H_2O(g)$$

9. Just after an automobile is started, you see water dripping off the end of the tail pipe. Is this normal? Why or why not?

327

Answers to Section Review

1. nitrogen

2. The ratio of coefficients of fuel to oxygen in the balanced chemical equation is 2:25, which will simplify to 1:12.5.

3. They speed up the conversion of possible pollutants to relatively harmless products.

4. Answers may include the engine stalling, more CO being produced, and more unburned hydrocarbons in the exhaust.

5. 10.7 g Na_2O

6. 1.20×10^2 g $NaHCO_3$

7. Nitrogen is the main component of air. Some of this nitrogen combines with oxygen at the high temperature in the engine to produce nitrogen oxides.

8. The reactants are gases, which would be more difficult to store than the solid sodium azide.

9. It is normal because water is a product of the combustion of the fuel.

9 CHAPTER HIGHLIGHTS

Alternative Assessments

READING SKILL BUILDER

- Page 316

SKILL BUILDER

- Pages 323, 324

REAL-WORLD CONNECTION

- Page 322

Group Activity

- Page 316

CHAPTER REVIEW

- Items 65, 66

Portfolio Assessments

Homework

- Page 325

VISUAL ARTS CONNECTION

- Page 305

SKILL BUILDER

- Page 321

CHAPTER REVIEW

- Item 68

KEY IDEAS

SECTION ONE Calculating Quantities in Reactions

- Reaction stoichiometry compares the amounts of substances in a chemical reaction.
- Stoichiometry problems involving reactions can always be solved using mole ratios.
- Stoichiometry problems can be solved using three basic steps. First, change what you are given into moles. Second, use a mole ratio based on a balanced chemical equation. Third, change to the units needed for the answer.

SECTION TWO Limiting Reactants and Percentage Yield

- The limiting reactant is a reactant that is consumed completely in a reaction.
- The theoretical yield is the amount of product that can be formed from a given amount of limiting reactant.
- The actual yield is the amount of product collected from a real reaction.
- Percentage yield is the actual yield divided by the theoretical yield multiplied by 100. It is a measure of the efficiency of a reaction.

SECTION THREE Stoichiometry and Cars

- Stoichiometry is used in designing air bags for passenger safety.
- Stoichiometry is used to maximize a car's fuel efficiency.
- Stoichiometry is used to minimize the pollution coming from the exhaust of an auto.

KEY TERMS

stoichiometry

limiting reactant
excess reactant
actual yield

KEY SKILLS

Using Mole Ratios
Skills Toolkit 1 p. 303
Sample Problem A p. 304

Solving Stoichiometry Problems
Skills Toolkit 2 p. 305

Problems Involving Mass
Skills Toolkit 3 p. 306
Sample Problem B p. 307

Problems Involving Volume
Skills Toolkit 4 p. 308
Sample Problem C p. 309

Problems Involving Particles
Skills Toolkit 5 p. 310
Sample Problem D p. 310

Limiting Reactants and Theoretical Yield
Sample Problem E p. 314

Calculating Percentage Yield
Sample Problem F p. 317

Calculating Actual Yield
Sample Problem G p. 318

Air-Bag Stoichiometry
Sample Problem H p. 322

Air-Fuel Ratio
Sample Problem I p. 324

Calculating Yields: Pollution
Sample Problem J p. 327

328

Chapter Resource File

- **Datasheets for In-Text Labs**
  Stoichiometry and Gravimetric Analysis

- **Datasheets for In-Text Labs**
  Gravimetric Analysis—Hard Water Testing

USING KEY TERMS

1. Define *stoichiometry*.

2. Compare the limiting reactant and the excess reactant for a reaction.

3. Compare the actual yield and the theoretical yield from a reaction.

4. How is percentage yield calculated?

5. Why is the term *limiting* used to describe the limiting reactant?

UNDERSTANDING KEY IDEAS

Calculating Quantities in Reactions

6. Why is it necessary to use mole ratios in solving stoichiometry problems?

7. What is the key conversion factor needed to solve all stoichiometry problems?

8. When you multiply by a mole ratio, the unit of mol does not change. What is the purpose of using the mole ratio?

9. Why is a balanced chemical equation needed to solve stoichiometry problems?

10. Use the balanced equation below to write mole ratios for the situations that follow.

$$2H_2(g) + O_2(g) \longrightarrow 2H_2O(g)$$

   a. calculating mol H_2O given mol H_2
   b. calculating mol O_2 given mol H_2O
   c. calculating mol H_2 given mol O_2

11. What are two conversions in stoichiometry problems for which you use the molar mass of a substance?

12. Why can density be used in solving stoichiometry problems involving volume?

13. Write the conversion factor needed to convert from g O_2 to L O_2 if the density of O_2 is 1.429 g/L.

14. What conversion factor is used to convert from volume of a gas directly to moles at STP?

15. How is Avogadro's number used as a conversion factor in solving problems involving number of particles?

16. A student uses the equation as written below to calculate the amount of O_2 that reacts with a given amount of CO. Is it likely that the student would get the correct answer? Explain.

$$CO(g) + O_2(g) \longrightarrow CO_2(g)$$

17. Describe a general plan for solving all stoichiometry problems in three steps.

Limiting Reactants and Percentage Yield

18. Explain why cost is often a major factor in choosing a limiting reactant.

19. Give two reasons why the actual yield from chemical reactions is less than 100%.

20. Describe the relationship between the limiting reactant and the theoretical yield.

21. Identify the limiting reactant and the excess reactant in the following situations:
   a. firewood burning in a campfire
   b. stomach acid reacting with a tablet of $Mg(OH)_2$
   c. iron in a nail rusting in the presence of oxygen and water
   d. NO_2 gas reacting with oxygen and water vapor in air to produce acid precipitation

329

Assignment Guide	
Section	**Questions**
1	1, 6–17, 27–37, 54–55, 70–73
2	2–5, 18–21, 38–46, 56, 58, 61, 65–66, 69, 74
3	22–26, 47–53, 57, 59–60, 62–64, 67–68

REVIEW ANSWERS

Using Key Terms

1. the proportional relationship between two or more substances during a chemical reaction

2. A limiting reactant is used up before the excess reactant and thus limits a reaction.

3. The actual yield is the measured quantity of product actually made in a reaction, while the theoretical yield is the maximum calculated quantity of product that could be formed.

4. To find the percentage yield, divide the actual yield by the theoretical yield, and then multiply by 100.

5. Because a limiting reactant runs out during a reaction, it will cause the reaction to stop. Therefore it limits the amount of product that can form.

Understanding Key Ideas

6. The balanced chemical equation relates the relative amounts of chemicals involved in the reaction in terms of moles, so converting from one substance to another requires using a ratio of the moles from the balanced equation.

7. a mole ratio

8. changing from the quantity of one substance in the reaction to the quantity of a different substance

9. The coefficients in a balanced chemical equation give the mole ratios needed for stoichiometry calculations.

10. **a.** $\dfrac{2 \text{ mol } H_2O}{2 \text{mol } H_2}$

   b. $\dfrac{1 \text{ mol } O_2}{2 \text{ mol } H_2O}$

   c. $\dfrac{2 \text{ mol } H_2}{1 \text{ mol } O_2}$

11. changing mass to amount in moles and changing amount in moles to mass

12. Density can be used to convert volume to mass, which can then be converted to moles.

13. $\dfrac{1 \text{ L } O_2}{1.429 \text{ g } O_2}$

14. $\dfrac{1 \text{ mol}}{22.41 \text{ L}}$

15. Avogadro's number can be used to convert between amount in moles and number of particles for any substance.

16. Because the equation is not balanced, the student will most likely get an incorrect answer.

17. Take whatever you are given, and change it into moles. Then, use the mole ratio to get moles of the desired substance. Finally, convert these moles into the units that you need for your final answer.

18. By choosing a limiting reactant that costs less than the other reactants, a company can keep the costs of production low.

19. Answers will vary but might include the reaction not going to completion, the reverse reaction or side reactions occurring, or product being lost during purification.

20. The limiting reactant is used to calculate the theoretical yield of a reaction.

21. a. limiting reactant, firewood; excess reactant, oxygen

    b. limiting reactant, $Mg(OH)_2$; excess reactant, stomach acid

    c. limiting reactant, iron; excess reactants, oxygen and water

    d. limiting reactant, NO_2; excess reactants, oxygen and water

22. air bags, engine efficiency, and pollution control

Stoichiometry and Cars

22. What are three areas of a car's operation or design that depend on stoichiometry?

23. Describe what might happen if too much or too little gas generant is used in an air bag.

24. Why is the ratio of fuel to air in a car's engine important in controlling pollution?

25. Under what conditions will exhaust from a car's engine contain high levels of carbon monoxide?

26. What is the function of the catalytic converter in the exhaust system?

PRACTICE PROBLEMS

Sample Problem A Using Mole Ratios

27. The chemical equation for the formation of water is
$$2H_2 + O_2 \longrightarrow 2H_2O$$

    a. If 3.3 mol O_2 are used, how many moles of H_2 are needed?

    b. How many moles O_2 must react with excess H_2 to form 6.72 mol H_2O?

    c. If you wanted to make 8.12 mol H_2O, how many moles of H_2 would you need?

28. The reaction between hydrazine, N_2H_4, and dinitrogen tetroxide is sometimes used in rocket propulsion. Balance the equation below, then use it to answer the following questions.

$$N_2H_4(l) + N_2O_4(l) \longrightarrow N_2(g) + H_2O(g)$$

    a. How many moles H_2O are produced as 1.22×10^3 mol N_2 are formed?

    b. How many moles N_2H_4 must react with 1.45×10^3 mol N_2O_4?

    c. If 2.13×10^3 mol N_2O_4 completely react, how many moles of N_2 form?

29. Aluminum reacts with oxygen to form aluminum oxide.

    a. How many moles of O_2 are needed to react with 1.44 mol of aluminum?

    b. How many moles of aluminum oxide can be made if 5.23 mol Al completely react?

    c. If 2.98 mol O_2 react completely, how many moles of Al_2O_3 can be made?

Sample Problem B Problems Involving Mass

30. Calcium carbide, CaC_2, reacts with water to form acetylene.

$$CaC_2(s) + 2H_2O(l) \longrightarrow C_2H_2(g) + Ca(OH)_2(s)$$

    a. How many grams of water are needed to react with 485 g of calcium carbide?

    b. How many grams of CaC_2 could make 23.6 g C_2H_2?

    c. If 55.3 g $Ca(OH)_2$ are formed, how many grams of water reacted?

31. Oxygen can be prepared by heating potassium chlorate.

$$2KClO_3(s) \longrightarrow 2KCl(s) + 3O_2(g)$$

    a. What mass of O_2 can be made from heating 125 g of $KClO_3$?

    b. How many grams of $KClO_3$ are needed to make 293 g O_2?

    c. How many grams of KCl could form from 20.8 g $KClO_3$?

32. How many grams of aluminum oxide can be formed by the reaction of 38.8 g of aluminum with oxygen?

33. Ozone, O_3, changes into oxygen. Write the balanced equation. How many grams of oxygen can be obtained from 2.22 mol O_3?

Sample Problem C Problems Involving Volume

34. Use the equation provided to answer the questions that follow. The density of oxygen gas is 1.428 g/L.

$$2KClO_3(s) \longrightarrow 2KCl(s) + 3O_2(g)$$

    a. What volume of oxygen can be made from 5.00×10^{-2} mol of $KClO_3$?

    b. How many grams $KClO_3$ must react to form 42.0 mL O_2?

    c. How many milliliters of O_2 will form at STP from 55.2 g $KClO_3$?

35. Hydrogen peroxide, H_2O_2, decomposes to form water and oxygen.
   a. How many liters of O_2 can be made from 342 g H_2O_2 if the density of O_2 is 1.428 g/L?
   b. The density of H_2O_2 is 1.407 g/mL, and the density of O_2 is 1.428 g/L. How many liters of O_2 can be made from 55 mL H_2O_2?
   c. How many liters of O_2 will form at STP if 22.5 g H_2O_2 react?

Sample Problem D Problems Involving Particles

36. Use the equation provided to answer the questions that follow.

$$2NO + O_2 \longrightarrow 2NO_2$$

   a. How many molecules of NO_2 can form from 1.11 mol O_2 and excess NO?
   b. How many molecules of NO will react with 25.7 g O_2?
   c. How many molecules of O_2 are needed to make 3.76×10^{22} molecules NO_2?

37. Use the equation provided to answer the questions that follow.

$$2Na + 2H_2O \longrightarrow 2NaOH + H_2$$

   a. How many molecules of H_2 could be made from 27.6 g H_2O?
   b. How many atoms of Na will completely react with 12.9 g H_2O?
   c. How many molecules of H_2 could form when 6.59×10^{20} atoms Na react?

Sample Problem E Limiting Reactants and Theoretical Yield

38. In the reaction shown below, 4.0 mol of NO is reacted with 4.0 mol O_2.

$$2NO + O_2 \longrightarrow 2NO_2$$

   a. Which is the excess reactant, and which is the limiting reactant?
   b. What is the theoretical yield, in units of mol, of NO_2?

39. In the reaction shown below, 64 g CaC_2 is reacted with 64 g H_2O.

$$CaC_2(s) + 2H_2O(l) \longrightarrow C_2H_2(g) + Ca(OH)_2(s)$$

   a. Which is the excess reactant, and which is the limiting reactant?
   b. What is the theoretical yield of C_2H_2?
   c. What is the theoretical yield of $Ca(OH)_2$?

40. In the reaction shown below, 28 g of nitrogen are reacted with 28 g of hydrogen.

$$N_2(g) + 3H_2(g) \longrightarrow 2NH_3(g)$$

   a. Which is the excess reactant, and which is the limiting reactant?
   b. What is the theoretical yield of ammonia?
   c. How many grams of the excess reactant remain?

Sample Problem F Calculating Percentage Yield

41. Reacting 991 mol of SiO_2 with excess carbon yields 30.0 kg of SiC. What is the percentage yield?

$$SiO_2 + 3C \longrightarrow SiC + 2CO$$

42. If 156 g of sodium nitrate react, and 112 g of sodium nitrite are recovered, what is the percentage yield?

$$2NaNO_3(s) \longrightarrow 2NaNO_2(s) + O_2(g)$$

43. If 185 g of magnesium are recovered from the decomposition of 1000.0 g of magnesium chloride, what is the percentage yield?

Sample Problem G Calculating Actual Yield

44. How many grams of $NaNO_2$ form when 256 g $NaNO_3$ react? The yield is 91%.

$$2NaNO_3(s) \longrightarrow 2NaNO_2(s) + O_2(g)$$

45. How many grams of Al form from 9.73 g of aluminum oxide if the yield is 91%?

$$Al_2O_3 + 3C \longrightarrow 2Al + 3CO$$

46. Iron and CO are made by heating 4.56 kg of iron ore, Fe_2O_3, and carbon. The yield of iron is 88%. How many kilograms of iron are made?

331

23. Too much gas generant could overinflate the air bag and increase the force of the air bag on an occupant during inflation. Too little gas generant might not inflate the bag sufficiently or quickly enough. Either condition could result in an increased likelihood of injury.

24. Carefully controlling the ratio of fuel to air helps to decrease the amounts of CO and unburned hydrocarbons in exhaust, thus cutting down pollution emissions.

25. when there is too little air in the fuel-air mixture and when starting the engine

26. to decrease the emissions of some pollutants by changing them into less harmful chemicals

Practice Problems

27. a. 6.6 mol H_2
   b. 3.36 mol O_2
   c. 8.12 mol H_2

28. a. 1.63×10^3 mol H_2O
   b. 2.90×10^3 mol N_2H_4
   c. 6.39×10^3 mol N_2

29. a. 1.08 mol O_2
   b. 2.62 mol Al_2O_3
   c. 1.99 mol Al_2O_3

30. a. 273 g H_2O
   b. 58.1 g CaC_2
   c. 26.9 g H_2O

31. a. 49.0 g O_2
   b. 748 g $KClO_3$
   c. 12.7 g KCl

32. 73.3 g Al_2O_3

33. 107 g O_2

34. a. 1.68 L O_2
   b. 0.153 g $KClO_3$
   c. 1.51×10^4 mL O_2

35. a. 113 L O_2
   b. 25 L O_2
   c. 7.41 L O_2

36. a. 1.34×10^{24} molecules NO_2
   b. 9.67×10^{23} molecules NO
   c. 1.88×10^{22} molecules O_2

37. a. 4.61×10^{23} molecules H_2

   b. 4.31×10^{23} atoms Na

   c. 3.30×10^{20} molecules H_2

38. a. excess, O_2; limiting, NO

   b. 4.0 mol NO_2

39. a. excess, H_2O; limiting, CaC_2

   b. 26 g C_2H_2

   c. 74 g $Ca(OH)_2$

40. a. excess, H_2; limiting, N_2

   b. 34 g NH_3

   c. 22 g H_2

41. 75.6%

42. 88.2%

43. 72.5%

44. 1.9×10^2 g $NaNO_2$

45. 4.7 g Al

46. 2.8 kg Fe

47. 46.6 L CO_2

48. a. 84.7 g NaN_3

   b. 43 L N_2

   c. 9.0×10^1 g NaN_3

49. $\dfrac{25 \text{ mol } O_2}{2 \text{ mol } C_8H_{18}}$, or 25:2

50. 2.41×10^3 g O_2

51. 1.71×10^3 L O_2

52. 4.75 g O_3; 96.4%

53. 2.16×10^3 g CO_2; 88.0%

Mixed Review

54. a. 1.2×10^2 g CO_2

   b. 9.70 mL H_2O

   c. 4.49×10^{22} molecules H_2O

55. a. $\dfrac{2 \text{ mol } NaN_3}{2 \text{ mol } Na}$, $\dfrac{2 \text{ mol } NaN_3}{3 \text{ mol } N_2}$,

   $\dfrac{2 \text{ mol } Na}{2 \text{ mol } NaN_3}$, $\dfrac{2 \text{ mol } Na}{3 \text{ mol } N_2}$,

   $\dfrac{3 \text{ mol } N_2}{2 \text{ mol } NaN_3}$, $\dfrac{3 \text{ mol } N_2}{2 \text{ mol } Na}$

   b. 4.0 mol N_2

   c. 11.0 g Na

   d. 4.86×10^{23} molecules N_2

Sample Problem H Air-Bag Stoichiometry

47. Assume that 44.3 g Na_2O are formed during the inflation of an air bag. How many liters of CO_2 (density = 1.35 g/L) are needed to completely react with the Na_2O?

$$Na_2O(s) + 2CO_2(g) + H_2O(g) \longrightarrow 2NaHCO_3(s)$$

48. Assume that 59.5 L N_2 with a density of 0.92 g/L are needed to fill an air bag.

$$2NaN_3(s) \longrightarrow 2Na(s) + 3N_2(g)$$

   a. What mass of NaN_3 is needed to form this volume of nitrogen?

   b. How many liters of N_2 are actually made from 65.7 g NaN_3 if the yield is 94%?

   c. What mass of NaN_3 is actually needed to form 59.5 L N_2?

Sample Problem I Air-Fuel Ratio

49. Write a balanced equation for the combustion of octane, C_8H_{18}, with oxygen to obtain carbon dioxide and water. What is the mole ratio of oxygen to octane?

50. What mass of oxygen is required to burn 688 g of octane, C_8H_{18}, completely?

51. How many liters of O_2, density 1.43 g/L, are needed for the complete combustion of 1.00 L C_8H_{18}, density 0.700 g/mL?

Sample Problem J Calculating Yields: Pollution

52. Nitrogen dioxide from exhaust reacts with oxygen to form ozone. What mass of ozone could be formed from 4.55 g NO_2? If only 4.58 g O_3 formed, what is the percentage yield?

$$NO_2(g) + O_2(g) \longrightarrow NO(g) + O_3(g)$$

53. How many grams CO_2 form from the complete combustion of 1.00 L C_8H_{18}, density 0.700 g/mL? If only 1.90×10^3 g CO_2 form, what is the percentage yield?

MIXED REVIEW

54. The following reaction can be used to remove CO_2 breathed out by astronauts in a spacecraft.

$$2LiOH(s) + CO_2(g) \longrightarrow Li_2CO_3(s) + H_2O(l)$$

   a. How many grams of carbon dioxide can be removed by 5.5 mol LiOH?

   b. How many milliliters H_2O (density = 0.997 g/mL) could form from 25.7 g LiOH?

   c. How many molecules H_2O could be made when 3.28 g CO_2 react?

55. Use the equation below to answer the questions that follow.

$$2NaN_3(s) \longrightarrow 2Na(s) + 3N_2(g)$$

   a. Write the six mole ratios that can be obtained from this equation.

   b. How many moles N_2 can be formed from 2.7 mol NaN_3?

   c. How many grams Na can be made from 31.1 g NaN_3?

   d. How many molecules N_2 are made when 3.24×10^{23} atoms Na are formed?

56. When 55 g NO is mixed with 35 g O_2, 31 g NO_2 are recovered. The density of NO is 1.3388 g/L, and that of NO_2 is 2.053 g/L.

$$2NO(g) + O_2(g) \longrightarrow 2NO_2(g)$$

   a. Which is the limiting reactant?

   b. What is the theoretical yield from this reaction in grams NO_2?

   c. What is the percentage yield?

57. How many liters N_2, density 0.92 g/L, can be made by the decomposition of 2.05 g NaN_3?

$$2NaN_3(s) \longrightarrow 2Na(s) + 3N_2(g)$$

58. The percentage yield of nitric acid is 95%. If 9.88 kg of nitrogen dioxide react, what mass of nitric acid is isolated?

$$3NO_2(g) + H_2O(g) \longrightarrow 2HNO_3(aq) + NO(g)$$

59. If you get 25.3 mi/gal, what mass of carbon dioxide is produced by the complete combustion of C_8H_{18} if you drive 5.40 mi? (Hint: 1 gal = 3.79 L; density of octane = 0.700 g/mL)

CRITICAL THINKING

60. Nitrogen monoxide, NO, reacts with oxygen to form nitrogen dioxide. Then the nitrogen dioxide reacts with oxygen to form nitrogen monoxide and ozone. Write the balanced equations. What is the theoretical yield in grams of ozone from 4.55 g of nitrogen monoxide with excess O_2? (Hint: First calculate the theoretical yield for NO_2, then use that value to calculate the yield for ozone.)

61. Explain the stoichiometry involved in blowing air on the base of a dwindling campfire to keep the coals burning.

62. Why would it be unreasonable for an amendment to the Clean Air Act to call for 0% pollution emissions from cars with combustion engines?

63. Use stoichiometry to explain the following problems that a lawn mower may have.
  a. A lawn mower fails to start because the engine floods.
  b. A lawn mower stalls after starting cold and idling.

64. Air is inexpensive and easily available. Give some reasons why air would not be good to use to fill an air bag.

ALTERNATIVE ASSESSMENT

65. Design an experiment to measure the percentage yields for the reactions listed below. If your teacher approves your design, get the necessary materials, and carry out your plan.
  a. $Zn(s) + 2HCl(aq) \longrightarrow ZnCl_2(aq) + H_2(g)$
  b. $2NaHCO_3(s) \longrightarrow$
    $Na_2CO_3(s) + H_2O(g) + CO_2(g)$
  c. $CaCl_2(aq) + Na_2CO_3(aq) \longrightarrow$
    $CaCO_3(s) + 2NaCl(aq)$
  d. $NaOH(aq) + HCl(aq) \longrightarrow$
    $NaCl(aq) + H_2O(l)$

  (Note: use only dilute NaOH and HCl, less concentrated than 1.0 mol/L.)

66. Your teacher will give you an index card specifying a volume of a gas. Reactants to make the gas will also be listed. Describe exactly how you would make the gas from the reactants. Include a method of collecting the gas without allowing it to mix with the air. Then specify how much of each reactant you need.

67. Calculate the theoretical yield (in kg) of carbon dioxide emitted by a car in one year, assuming 1.20×10^4 mi/y, 25 mi/gal, and octane, C_8H_{18}, as the fuel, 0.700 g/mL. (1 gal = 3.79 L)

68. Ozone is helpful in the upper atmosphere, but we worry about ozone formed from auto exhaust. Research the role of auto exhaust in the formation of ozone near the surface of Earth. Determine the hazards posed by ozone at Earth's surface.

CONCEPT MAPPING

69. Use the following terms to complete the concept map below: *stoichiometry*, *excess reactant*, *theoretical yield*, and *mole ratio*.

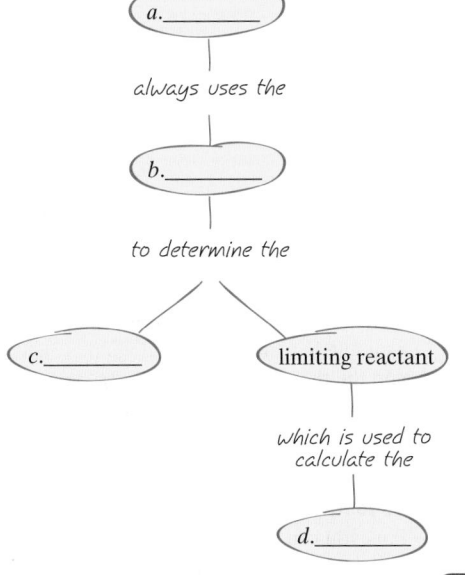

56. a. NO
  b. 84 g NO_2
  c. 37%
57. 1.44 L N_2
58. 8.6×10^3 g HNO_3
59. 1.74×10^3 g CO_2

Critical Thinking

60. 7.28 g O_3

61. Near the center of the fire, oxygen is the limiting reactant. The added oxygen from your breath reacts with the wood and releases more energy.

62. 0% pollution emissions from cars with combustion engines would require 100% yield for the combustion reaction of gasoline. This requirement does not take into account the actual yield or the stoichiometry of nitrogen from the air reacting in the hot engine.

63. a. The ratio of gasoline to air is too high. The combustion reaction will not take place.
  b. The ratio of air to gasoline is too high. The combustion reaction will not take place.

64. Answers may include that the air would need to be pressurized and stored in canisters to be used in air bags. These canisters would likely require more space than using a solid gas generant.

65. For all parts, student procedures should include using the mass of the limiting reactant to determine the theoretical yield of the easiest product to isolate and measure. The procedure should also include finding the actual yield by rinsing, drying, and measuring the mass of the product. The actual yield divided by the theoretical yield times 100 is the percentage yield. When doing the experiments, be sure sufficient excess reactants are used for the reaction to go to completion. Below are some specific hints for each part.

a. Let Zn be the limiting reactant. $ZnCl_2$ should be the product. Be sure students do the drying process in a fume hood, as excess HCl will be driven off also.

b. By drying the product solution, the Na_2CO_3 can be obtained.

c. Choose one reactant to be limiting. Use the mass of a sample of it to determine how much of the other reactant will react with it. Use more than that amount so that it will be in excess. Filter out and rinse the solid $CaCO_3$.

d. Let NaOH be the limiting reactant. Heat the product solution to drive off the water and leave the NaCl behind. Be sure the students do the drying process in a fume hood, as excess HCl will be driven off also.

66. Gases to be collected should differ in density from air so that they can be collected easily. Hydrogen (less dense) and carbon dioxide (more dense) are ideal. Oxygen and nitrogen can be collected, but they are more difficult to collect because they are similar in density to air. Be sure students do not use gases such as H_2S or SO_2. These gases are unsafe and are a hazard to students. Hydrogen can be prepared by reacting a safe, active metal, such as magnesium, with dilute HCl. CO_2 can be prepared by reacting a carbonate with dilute acid. A gas of low density should be collected with an inverted container placed over the reaction. Allowing the gas to fall into the collection container can be used to collect a gas of high density.

Answers continued on p. 335A

FOCUS ON GRAPHING

Study the graph below, and answer the questions that follow.
For help in interpreting graphs, see Appendix B, "Study Skills for Chemistry."

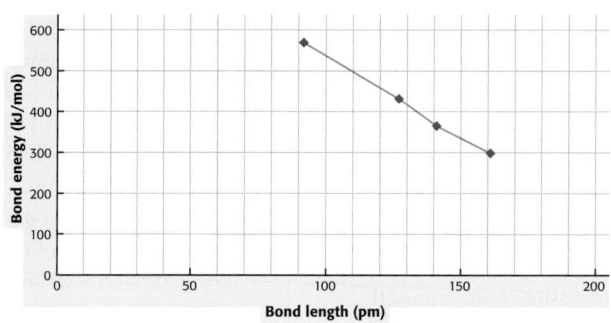

Bond Energy Versus Bond Length

70. Describe the relationship between bond length and bond energy.

71. Estimate the bond energy of a bond of length 100 pm.

72. If the trend of the graph continues, what bond length will have an energy of 200 kJ/mol?

73. The title of the graph does not provide much information about the contents of the graph. What additional information would be useful to better understand and use this graph?

TECHNOLOGY AND LEARNING

74. Graphing Calculator

Calculating Percentage Yield of a Chemical Reaction

The graphing calculator can run a program that calculates the percentage yield of a chemical reaction when you enter the actual yield and the theoretical yield. Using an example in which the actual yield is 38.8 g and the theoretical yield is 53.2 g, you will calculate the percentage yield. First, the program will carry out the calculation. Then you can use it to make other calculations.

Go to Appendix C. If you are using a TI-83 Plus, you can download the program **YIELD** and data and run the application as directed.

If you are using another calculator, your teacher will provide you with keystrokes and data sets to use. After you have run the program, answer the questions.

Note: all answers are written with three significant figures.

a. What is the percentage yield when the actual yield is 27.3 g and the theoretical yield is 44.6 g?

b. What is the percentage yield when the actual yield is 5.40 g and the theoretical yield is 9.20 g?

c. What actual yield/theoretical yield pair produced the largest percentage yield?

334

Chapter Resource File

• Chapter Test
• Problem Solving

Answers
1. d
2. a
3. a
4. a
5. c
6. d
7. a
8. b
9. c
10. c
11. c

1. Stoichiometry problems may require the use of a
   a. table of bond energies.
   b. Lewis structure.
   c. chart of electron configurations.
   d. mole ratio.

2. In the chemical equation $A + B \longrightarrow C + D$, if you know the mass of A, you can determine
   a. the mass of any of the other reactants and products.
   b. only the mass of C and D combined.
   c. only the mass of B.
   d. only the mass of A and B combined.

3. To solve a mass-mass stoichiometry problem, you must know the
   a. coefficients of the balanced equation.
   b. phases of the reactants and products.
   c. rate at which the reaction occurs.
   d. chemical names of the reactants and products.

4. For the reaction below, how many moles of N_2 are required to produce 18 mol NH_3?

   $$N_2 + 3H_2 \longrightarrow 2NH_3$$

   a. 9
   b. 27
   c. 18
   d. 36

5. If a chemical reaction involving substances A and B stops when B is completely used up, then B is referred to as the
   a. excess reactant.
   b. primary reactant.
   c. limiting reactant.
   d. primary product.

6. How much product is collected during a chemical reaction is called the
   a. mole ratio.
   b. theoretical yield.
   c. percentage yield.
   d. actual yield.

7. If a chemist calculates the maximum amount of product that might be obtained in a chemical reaction, he or she is calculating the
   a. theoretical yield.
   b. mole ratio.
   c. percentage yield.
   d. actual yield.

8. Knowing the mole ratio of a reactant and product in a chemical reaction allows one to determine
   a. the energy released in the reaction.
   b. the mass of the product produced from a known mass of reactant.
   c. the speed of the reaction.
   d. whether the reaction was reversible.

9. In stoichiometry, chemists are mainly concerned with
   a. the types of bonds found in compounds.
   b. energy changes occurring in chemical reactions.
   c. mass relationships in chemical reactions.
   d. speed with which chemical reactions occur.

10. What is the mole ratio of CO_2 to $C_6H_{12}O_6$ in the reaction below?

    $$6CO_2 + 6H_2O \longrightarrow C_6H_{12}O_6 + 6O_2$$

    a. 1:2
    b. 1:1
    c. 6:1
    d. 1:4

11. In a chemical reaction, the reactant remaining after all of the limiting reactant is completely used up is referred to as the
    a. product.
    b. controlling reactant.
    c. excess reactant.
    d. catalyst.

335

Continuation of Answers

Answers to Section 2 Review, *continued*

11. a. $Mg + 2H_2O \rightarrow Mg(OH)_2 + H_2$

b. 86.8%

c. 55 g $Mg(OH)_2$

12. a. CuO is limiting

b. 94.3%

c. 15.0 g Cu

13. 0 mol H_2, 10 mol O_2, and 20 mol H_2O

14. The theoretical yield is 23 g CaO. The actual yield reported is 27 g, which gives a percentage yield of 120%. The results are not reasonable because you cannot have a percentage yield greater than 100%.

Answers to Chapter Review, *continued*

67. 3.92×10^3 kg CO_2

68. Student findings might include that ozone forms from the reaction of nitrogen dioxide (from auto exhaust) and oxygen in the air brought about by sunlight (specifically ultraviolet light). Hazards posed by ozone include lung and respiratory disorders and deterioration of rubber, nylon, plastic, and paint.

69. a. stoichiometry

b. mole ratio

c. excess reactant

d. theoretical yield

70. Bond energy decreases as bond length increases.

71. about 540 kJ/mol

72. about 190 pm

73. Answers may include knowing the atoms joined together in each of the bonds plotted, knowing whether one atom was common to all bonds plotted, and knowing what types of bonds are plotted.

74. a. 61.2%

b. 58.7%

c. 27.3 g, 44.6 g

Causes of Change
Chapter Planning Guide

PACING	CLASSROOM RESOURCES	LABS, ACTIVITIES, AND DEMONSTRATIONS
BLOCK 1 · 45 min pp. 336–337 **Chapter Opener**		**SE Start-Up Activity** Heat Exchange, p. 337 ◆
BLOCK 2 · 45 min pp. 338–344 **Section 1** Energy Transfer		**TE Demonstration**, p. 338 **TE Demonstration** Inexpensive Hand Warmer, p. 340 ◆ **ADVANCED** **TE Demonstration** Molar Heat Capacity, p. 341 ◆ **ADVANCED**
BLOCK 3 · 45 min pp. 345–349 **Section 2** Using Enthalpy		**TE Demonstration**, p. 345 ◆
BLOCKS 4 & 5 · 90 min pp. 350–357 **Section 3** Changes in Enthalpy During Chemical Reactions	**TT Bomb Calorimeter** * **OSP Career Extension** Firefighter **GENERAL**	**TE Group Activity**, p. 352 **GENERAL** **TE Demonstration**, p. 353 ◆ **ADVANCED** **TE Group Activity**, p. 354 **BASIC** **CRF CBL™ Probeware Lab** Energy Content in Foods ◆ **ADVANCED** **SE Skills Practice Lab** Calorimetry and Hess's Law, p. 792 ◆ **GENERAL** **CRF Datasheets for In-Text Labs** *
BLOCK 6 · 45 min pp. 358–368 **Section 4** Order and Spontaneity	**SE Science and Technology** Hydrogen-Powered Cars	**TE Activity**, p. 362 **BASIC**

BLOCKS 7 & 8 · 90 min **Chapter Review and Assessment Resources**

SE Chapter Review, pp. 370–374
SE Standardized Test Prep, p. 375
CRF Chapter Test *
OSP Test Generator
CRF Test Item Listing *
OSP Scoring Rubrics and Classroom Management Checklists

Holt Chemistry: Online Resources

Visit **go.hrw.com** for a variety of free resources related to this textbook. Enter the keyword **HW4 HOME**.

Holt Online Learning

Students can access interactive problem solving help and active visual concept development with the *Holt Chemistry* Online Edition available at **www.hrw.com**

cnnstudentnews.com

Find the latest chemistry news, lesson plans, and activities related to important scientific events.

KEY

TE Teacher Edition
SE Student Edition
OSP One-Stop Planner

CRF Chapter Resource File
TT Teaching Transparencies

* Also on One-Stop Planner
◆ Requires Advance Prep

Compression guide:
To shorten your instruction because of time limitations, omit blocks 1, 6, and 7.

PROBLEM SOLVING AND PRACTICE	SECTION REVIEW AND ASSESSMENT	STANDARDS CORRELATION
		National Science Education Standards
SE Sample Problem A Calculating the Molar Heat Capacity of a Sample, p. 342 **SE** Practice, p. 342 GENERAL **TE** Homework, p. 342 GENERAL **CRF** Problem Solving * ADVANCED **SE** Problem Bank, p. 858 GENERAL	**TE** Quiz, p. 344 GENERAL **CRF** Quiz * **TE** Reteaching, p. 344 BASIC **SE** Section Review, p. 344 **CRF** Concept Review *	PS 5b PS 5c
SE Sample Problem B Calculating Molar Enthalpy Change for Heating, p. 346 **SE** Practice, p. 346 GENERAL **TE** Homework, p. 346 GENERAL **SE** Sample Problem C Calculating Molar Enthalpy Change for Cooling, p. 347 **SE** Practice, p. 347 GENERAL **TE** Homework, p. 347 GENERAL **CRF** Problem Solving * ADVANCED **SE** Problem Bank, p. 858 GENERAL	**TE** Quiz, p. 349 GENERAL **CRF** Quiz * **TE** Reteaching, p. 349 BASIC **SE** Section Review, p. 349 **CRF** Concept Review *	PS 3b
SE Sample Problem D Calculating a Standard Enthalpy of Formation, p. 356 **SE** Practice, p. 356 GENERAL **TE** Homework, p. 356 GENERAL **SE** Sample Problem E Calculating a Reaction's Change in Enthalpy, p. 356 **SE** Practice, p. 357 GENERAL **TE** Homework, p. 356 GENERAL **CRF** Problem Solving * ADVANCED **SE** Problem Bank, p. 858 GENERAL	**TE** Quiz, p. 357 GENERAL **CRF** Quiz * **TE** Reteaching, p. 357 BASIC **SE** Section Review, p. 357 **CRF** Concept Review *	PS 3b
SE Sample Problem F Hess's Law and Entropy, p. 361 **SE** Practice, p. 361 GENERAL **TE** Homework, p. 361 GENERAL **SE** Sample Problem G Calculating a Change in Gibbs Energy, p. 364 **SE** Practice, p. 364 GENERAL **TE** Homework, p. 364 GENERAL **SE** Sample Problem H Calculating a Gibbs Energy Change Using ΔG_f^0 Values, p. 365 **SE** Practice, p. 365 GENERAL **TE** Homework, p. 365 GENERAL **CRF** Problem Solving * ADVANCED **SE** Problem Bank, p. 858 GENERAL	**TE** Quiz, p. 367 GENERAL **CRF** Quiz * **TE** Reteaching, p. 367 BASIC **SE** Section Review, p. 367 **CRF** Concept Review *	PS 5d

SCiLINKS
www.scilinks.org

Topic: U.S. National Parks
SciLinks code: HW4126

Topic: Heat and Temperature
SciLinks code: HW4066

Topic: Heat Transfer
SciLinks code: HW4068

Topic: Enthalpy
SciLinks code: HW4052

Topic: Endothermic and Exothermic Reactions
SciLinks code: HW4056

Topic: Nutrition
SciLinks code: HW4090

Topic: Entropy
SciLinks code: HW4053

Topic: Hydrogen
SciLinks code: HW4155

Technology Resources

 Science in the NEWS

Each video segment is accompanied by a Critical Thinking Worksheet.

Segment 20
What is a Calorie?

CHAPTER 10

Overview

In this chapter, students are introduced to thermodynamics. In Section 1, the differences between heat and temperature are explored and enthalpy is defined. In Section 2, enthalpy change is described and thermodynamics is defined. Section 3 describes how enthalpy changes during a reaction and Hess's law. Section 4 describes entropy and Gibbs energy.

Assessing Prior Knowledge

Check for Content Knowledge

Students should be familiar with the following topics:

- kinetic energy
- heat
- temperature
- chemical equations

Using the Figure

A forest fire is a dramatic example of a spontaneous, exothermic reaction. The reaction is the combustion of fuel (grasses, dead leaves, or trees) to produce water, carbon dioxide, and energy as heat and light.

The National Forest Service classifies forest fires as prescribed burns or wildfires. Prescribed burns are controlled burns and reduce the amount of fuel in a forest. Wildfires are uncontrolled burns started naturally or by humans. Although firefighters do use water and other fire retardant chemicals to help quench fires, the primary method of containing wildfires is a *firebreak*. A firebreak is a strip of ground from which the fuel has been removed. When the fire reaches the firebreak, it cannot advance because of the lack of fuel.

CHAPTER 10
CAUSES OF CHANGE

336

Standards Correlations

National Science Education Standards

PS 3b: Chemical reactions may release or consume energy. Some reactions such as the burning of fossil fuels release large amounts of energy by losing heat and by emitting light. Light can initiate many chemical reactions such as photosynthesis and the evolution of urban smog. (Sections 2, 3)

PS 5b: All energy can be considered to be either kinetic energy, which is the energy of motion; potential energy, which depends on relative position; or energy contained by a field, such as electromagnetic waves. (Section 1)

PS 5c: Heat consists of random motion and the vibrations of atoms, molecules, and ions. The higher the temperature, the greater the atomic or molecular motion. (Section 1)

PS 5d: Everything tends to become less organized and less orderly over time. Thus, in all energy transfers, the overall effect is that the energy is spread out uniformly. Examples are the transfer of energy from hotter to cooler objects by conduction, radiation, or convection and the warming of our surroundings when we burn fuels. (Section 4)

A chemical reaction can release or absorb energy and can increase or decrease disorder. The forest fire is a chemical reaction in which cellulose and oxygen form carbon dioxide, water, and other chemicals. This reaction also releases energy and increases disorder because the reaction generates energy as heat and breaks down the long molecules found in living trees into smaller and simpler molecules, such as carbon dioxide, CO_2, and water, H_2O.

START-UP *ACTIVITY*

Heat Exchange

SAFETY PRECAUTIONS

PROCEDURE

1. Fill a **film canister** three-fourths full of **hot water.** Insert the **thermometer apparatus** prepared by your teacher in the hot water.

2. Fill a **250 ml beaker** one-third full of **cool water.** Insert another **thermometer apparatus** in the cool water, and record the water's temperature.

3. Record the temperature of the water in the film canister. Place the film canister in the cool water. Record the temperature measured by each thermometer every 30 s.

4. When the two temperatures are nearly the same, stop and graph your data. Plot temperature versus time on the graph. Remember to write "Time" on the *x*-axis and "Temperature" on the *y*-axis.

ANALYSIS

1. How can you tell that energy is transferred? Is energy transferred to or from the hot water?

2. Predict what the final temperatures would become after a long time.

Pre-Reading Questions

① Can a chemical reaction generate energy as heat?

② Name two types of energy.

③ What is specific heat?

④ Does a thermometer measure temperature or heat?

CONTENTS 10

■ internet connect
www.scilinks.org
Topic: U.S. National Parks
SciLinks code: HW4126
SCiLINKS. Maintained by the National Science Teachers Association

---

START-UP *ACTIVITY*

Skills Acquired:
• Collecting data
• Interpreting
• Identifying/ Recognizing patterns

Materials:
For each group of 2–3 students:
• film canister and lid
• hot water
• thermometer (2)
• 250 ml beaker
• cool water
• stop watch
• graph paper

Teacher's Notes: Prepare the film canister lids so that a thermometer can be held by each lid. Make a hole in each canister lid with a cork borer, drill, or paper punch. A thermometer should fit tightly enough in a lid so that water will not drip out of the lid when the assembly is turned upside down. One hole stoppers can be used.

Safety Caution: Use caution when making a hole in the lids. Be sure the temperature of the hot water is below 49°C. Remind students to handle thermometers carefully.

Answers

1. You can tell that energy is transferred by temperature changes. Energy is transferred from the hot water.

2. The final temperatures should be the same.

LS Kinesthetic

Answers to Pre-Reading Questions

1. yes

2. Student's answers may vary. Sample answer: light and energy as heat

3. the quantity of heat required to raise a unit mass of homogenous material 1K or 1°C in a specified way given constant pressure and volume

4. temperature

Before beginning this section, review with your students the Objectives listed in the Student Edition. In this section, students learn the difference between heat and temperature. They also learn about enthalpy. Molar heat capacity is defined and its relationship to specific heat is described.

⏰ Bellringer

Write the words *heat* and *temperature* on the board. Ask students to write their own definitions for those words and to describe how they are related.

Motivate

Demonstration

Place two beakers of water on an overhead projector. One should be filled with very hot water and the other should be filled with very cold water. As the students watch the screen, turn on the projector and add a drop of food coloring to each beaker at the same time. Ask the students to describe what is happening. **Ans.** The food coloring in one beaker disperses faster than the other. Ask students why there is a difference between the two beakers. **Ans.** The water molecules in the hot water are moving faster than the molecules in the cold water. As a result, the food coloring disperses faster in the hot water because the coloring is pushed more often and with more force by the molecules in the hot water.

KEY TERMS
- heat
- enthalpy
- temperature

OBJECTIVES

① **Define** *enthalpy.*

② **Distinguish** between heat and temperature.

③ **Perform** calculations using molar heat capacity.

Energy as Heat

A sample can transfer energy to another sample. Some examples of energy transfer are the electric current in a wire, a beam of light, a moving piston, and a flame used by a welder as shown in **Figure 1.** One of the simplest ways energy is transferred is as **heat.**

Though energy has many different forms, all energy is measured in units called *joules* (J). So, the amount of energy that one sample transfers to another sample as heat is measured in joules. Energy is never created or destroyed. The amount of energy transferred from one sample must be equal to the amount of energy received by a second sample. Therefore, the total energy of the two samples remains exactly the same.

heat

> the energy transferred between objects that are at different temperatures

Figure 1
A welder uses an exothermic combustion reaction to create a high-temperature flame. The iron piece then absorbs energy from the flame.

MISCONCEPTION
/// ALERT \\\

In everyday usage, the word *heat* is used in a variety of ways, such as warmth or the energy contained in a hot object. However, a much narrower definition of heat is used in science. Students should understand that heat is the energy transferred between samples that are at different temperatures.

Chapter Resource File

- Lesson Plan

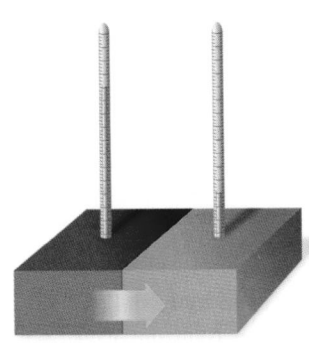

Figure 2
a Energy is always transferred from a warmer sample to a cooler sample, as the thermometers show.

b Even though both beakers receive the same amount of energy, the beakers do not have the same amount of liquid. So, the beaker on the left has a temperature of 30°C, and the beaker on the right has a temperature of 50°C.

Temperature

When samples of different temperatures are in contact, energy is transferred from the sample that has the higher **temperature** to the sample that has the lower temperature. **Figure 1** shows a welder at work; he is placing a high-temperature flame very close to a low-temperature piece of metal. The flame transfers energy as heat to the metal. The welder wants to increase the temperature of the metal so that it will begin to melt. Then, he can fuse this piece of metal with another piece of metal.

If no other process occurs, the temperature of a sample increases as the sample absorbs energy, as shown in **Figure 2a.** The temperature of a sample depends on the average kinetic energy of the sample's particles. The higher the temperature of a sample is, the faster the sample's particles move.

The temperature increase of a sample also depends on the mass of the sample. For example, the liquids in both the beakers in **Figure 2b** were initially 10.0°C, and equal quantities of energy were transferred to each beaker. The temperature increase in the beaker on the left is only about one-half of the temperature increase in the beaker on the right, because the beaker on the left has twice as much liquid in it.

Heat and Temperature are Different

You know that heat and temperature are different because you know that when two samples at different temperatures are in contact, energy can be transferred as heat. Heat and temperature differ in other ways. Temperature is an *intensive property,* which means that the temperature of a sample does not depend on the amount of the sample. However, heat is an *extensive property* which means that the amount of energy transferred as heat by a sample depends on the amount of the sample. So, water in a glass and water in a pitcher can have the same temperature. But the water in the pitcher can transfer more energy as heat to another sample because the water in the pitcher has more particles than the water in the glass.

temperature

a measure of how hot (or cold) something is; specifically, a measure of the average kinetic energy of the particles in an object

Topic Link

Refer to the "Matter and Energy" chapter for a discussion of heat, temperature, the Celsius scale, and the Kelvin scale.

339

did you know?

With the help of a cricket and a watch, you can estimate the temperature of your surroundings without a thermometer. Count the number of chirps a cricket makes in one minute. Divide the number of chirps by four and add 40. The resulting number should be close to the temperature in Fahrenheit. Of course, that's the temperature where the cricket is, which may not be the same temperature as where you are, especially if the cricket is outside and you are inside!

Demonstration

Inexpensive Hand Warmer

1. Place 25 g of iron powder in a small resealable plastic bag, add 1 g NaCl, seal the bag, and shake to mix.

2. Add 1 tablespoon of expanded vermiculite, reseal the bag, and shake it again.

3. To activate the hand warmer, add 5 mL of water to the bag. Seal the bag, and squeeze or shake it thoroughly to mix the contents. The energy as heat given off in the reaction will produce a noticeable temperature increase in about 1 minute.

4. Pass the bag around the class. If it begins to cool too soon, open the bag to let in fresh air, and shake or squeeze it again.

Teacher's Notes:

- The iron must be in powdered form. Iron filings will probably corrode too slowly to increase the temperature noticeably.

- Vermiculite is a mineral that consists of hydrated magnesium-aluminum-iron silicate. Vermiculite also disperses the iron-salt mixture over a large surface area and entraps air, and it acts as an insulator preventing the mixture from cooling too quickly.

- Tell students that the reaction in the bag is the rusting of iron. The reaction is an exothermic reaction and is written as follows:

$$4Fe(s) + 3O_2(g) \longrightarrow 2Fe_2O_3(s) + \text{energy}$$

Tell students that the bag feels warm because energy is transferred from the bag to their hands.

Safety Caution: Wear and have students wear goggles and gloves.

Figure 3
The boiling in a kettle on a stove shows several physical and chemical processes: a combustion reaction, conduction, and a change of state.

enthalpy

the sum of the internal energy of a system plus the product of the system's volume multiplied by the pressure that the system exerts on its surroundings

340

Figure 3 shows a good example of the relationship between heat and temperature. The controlled combustion in the burner of a gas stove transfers energy as heat to the metal walls of the kettle. The temperature of the kettle walls increases. As a result, the hot walls of the kettle transfer energy to the cool water in the kettle. This energy transferred as heat raises the water's temperature to 100°C. The water boils, and steam exits from the kettle's spout. If the burner on the stove was turned off, the burner would no longer transfer energy to the kettle. Eventually, the kettle and the water would have equal temperatures, and the kettle would not transfer energy as heat to the water.

A Substance's Energy Can Be Measured by Enthalpy

All matter contains energy. Measuring the total amount of energy present in a sample of matter is impossible, but changes in energy content can be determined. These changes are determined by measuring the energy that enters or leaves the sample of matter. If 73 J of energy enter a piece of silver and no change in pressure occurs, we know that the enthalpy of the silver has increased by 73 J. **Enthalpy,** which is represented by the symbol H, is the total energy content of a sample. If pressure remains constant, the enthalpy increase of a sample of matter equals the energy as heat that is received. This relationship remains true even when a chemical reaction or a change of state occurs.

A Sample's Enthalpy Includes the Kinetic Energy of Its Particles

The particles in a sample are in constant motion. In other words, these particles have kinetic energy. You know that the enthalpy of a sample is the energy that a sample has. So, the enthalpy of a sample also includes the total kinetic energy of its particles.

Imagine a gold ring being cooled. As the ring transfers energy as heat to its surroundings, there is a decrease in the motions of the atoms that make up the gold ring. The kinetic energies of the atoms decrease. As the total kinetic energy decreases, the enthalpy of the ring decreases. This decrease in the kinetic energy is observed as a decrease in temperature.

You may think that all the atoms in the ring have the same kinetic energy. However, some of the atoms of the gold ring move faster than other atoms in the ring. Therefore, both the total and average kinetic energies of a substance's particles are important to chemistry, because these quantities account for every particle's kinetic energy.

What happens to the motions of the gold atoms if the ring is cooled to absolute zero ($T = 0.00$ K)? The atoms still move! However, the average and total kinetic energies of the atoms at 0.00 K are the *minimum* average and total kinetic energies these atoms can have. This idea is true of any substance and its particles. The minimum average and total kinetic energies of particles that make up a substance occur at 0.00 K.

How can the enthalpy change of a sample be calculated? Enthalpy changes can be calculated by using several different methods. The next section discusses molar heat capacity, which will be used to determine the enthalpy change of a sample.

did you know?

Special ceramic tiles were created for use on the underside of space shuttles. These tiles transfer so little energy that one side of a tile can be hot because of a welder's torch while the other side remains cool to the touch.

MISCONCEPTION ALERT

If a person touches an object that is at a lower temperature than his/her hand is, the "coldness" of the object is not flowing from the object to his/her hand. Instead, energy as heat is flowing from the person's hand to the object, which gives the person a sensation of cold.

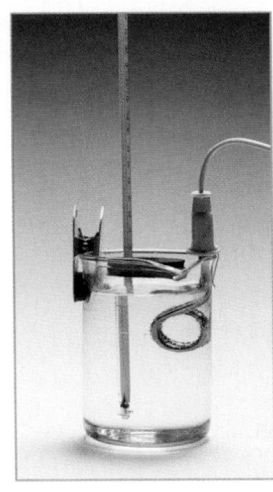

a

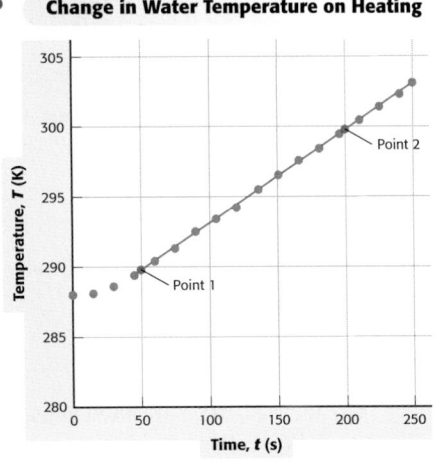

b

Change in Water Temperature on Heating

Temperature, T (K) vs *Time, t (s)*

Point 2

Point 1

Figure 4
a This figure shows apparatus used for determining the molar heat capacity of water by supplying energy at a known constant rate and recording the temperature rise.
b The graph shows the data points from the experiment. The red points are not data points; they were used in the calculation of the line's slope.

Molar Heat Capacity

The *molar heat capacity* of a pure substance is the energy as heat needed to increase the temperature of 1 mol of the substance by 1 K. Molar heat capacity has the symbol C and the unit J/K•mol. Molar heat capacity is accurately measured only if no other process, such as a chemical reaction, occurs.

The following equation shows the relationship between heat and molar heat capacity, where q is the heat needed to increase the temperature of n moles of a substance by ΔT.

$$q = nC\Delta T$$

heat = (amount in moles)(molar heat capacity)(change in temperature)

Experiments and analyses that are similar to **Figure 4** determine molar heat capacity. **Figure 4a** shows 20.0 mol of water, a thermometer, and a 100 W heater in a beaker. The temperature of the water is recorded every 15 s for 250 s. The data are graphed in **Figure 4b**.

The slope of the straight line that is drawn to closely match the data points can be used to determine water's molar heat capacity. During 150 s, the interval between $t = 50$ s and $t = 200$ s, the temperature of the water increased by 9.9 K. The value of the slope is calculated below.

$$\text{slope} = \frac{y_2 - y_1}{x_2 - x_1} = \frac{\Delta T}{\Delta t} = \frac{9.9 \text{ K}}{150 \text{ s}} = 0.066 \text{ K/s}$$

To calculate the molar heat capacity of water, you need to know the heater's power rating multiplied by the amount of time the heater warmed the water. This is because watts are equal to joules per second. So, C for H_2O can be determined by using the following equation. Also notice that Δt divided by ΔT is the inverse of the slope calculated above.

$$C = \frac{q}{n\Delta T} = \frac{1.00 \times 10^2 \text{ J/s}}{n(\text{slope})} = \frac{1.00 \times 10^2 \text{ J/s}}{(20.0 \text{ mol})(0.066 \text{ K/s})} = 76 \text{ J/K•mol}$$

341

METEOROLOGY
CONNECTION

Water has a larger molar heat capacity than land. Because of water's large molar heat capacity, the ocean retains a lot of energy. So, coastal areas stay moderately warm even in the winter. Because water does not heat up as easily as land does, oceans can help keep coastal areas cool during the summer.

did you know?

Heat capacities can apply in baking, too. The filling of a fruit pie generally has a larger heat capacity than the crust. As a result, the crust of the pie cools much quicker than the filling and burns can result from eating the hot filling.

Demonstration
Molar Heat Capacity

1. Obtain samples of three different metals. If possible, use uniform 2.5 cm blocks. Measure the mass of each sample, and write the mass on the board. Have students determine the amount in moles of each metal present while you continue the setup.

2. Pour equal volumes of room-temperature water into each of three 250 mL beakers. Use enough water to cover the samples by at least 1 cm. Each beaker should have the same volume of water. Add a thermometer to each beaker, and place the beakers on folded towels to minimize energy transfer through the bottom. Have students record the initial temperature of the water in each beaker.

3. Use tongs to remove the samples from the beaker and lower them gently into a metal pan of boiling water on a hot plate. After the water returns to a boil, allow the samples to remain in the water for 1 minute.

4. Using tongs, quickly remove each metal sample, and gently return them to their beakers.

5. Have student volunteers watch the thermometers as they swirl the beakers very gently, and record the highest temperature reached by the water in each beaker.

6. Have students use their results to order the metals from highest to lowest molar heat capacity.

7. Because the samples were all at the same temperature, and the quantity of water in each beaker is also the same, the heat capacities are proportional to ΔT/mol of metal. This ratio should be approximately equal for the metals. Explain the different temperature increases they observed in the beakers.

Safety Caution: Wear goggles and a lab apron. Have students wear goggles and student volunteers wear goggles and aprons.

Homework ——— GENERAL

Additional Practice

1. The molar heat capacity for sodium chloride is 50.5 J/K•mol. Calculate the amount of energy needed to raise the temperature of 1.50 mol of sodium chloride from 15.0°C to 30.0°C. Ans. 1140 J or 1.14 kJ

2. The molar heat capacity for nitrobenzene, $C_6H_5NO_2(l)$, which is sometimes used as a solvent in shoe polish, is 185.9 J/K•mol. What will be the temperature of a 50.00 g sample of nitrobenzene at 18.0°C after it absorbs 1.00×10^3 J of heat? Ans. 31.2°C

3. Energy, 4.72 J, is needed to raise the temperature of 4.00 g of gold from 20.0°C to 30.0°C. What is the molar heat capacity for gold? Ans. 23.2 J/K•mol

LS Logical

Answers to Practice Problems A

1. 96.8 J or 97 J
2. 1.3 mol
3. 221.16 J or 220 J
4. 196.85 K or 200 K

Chapter Resource File

• Problem Solving

SAMPLE PROBLEM A

Calculating the Molar Heat Capacity of a Sample

Determine the energy as heat needed to increase the temperature of 10.0 mol of mercury by 7.5 K. The value of *C* for mercury is 27.8 J/K•mol.

PRACTICE HINT

Always convert temperatures to the Kelvin scale before carrying out calculations in this chapter. Notice that in molar heat capacity problems, you will never multiply heat by molar heat capacity. If you did multiply, the joules would not cancel.

1 Gather information.

The amount of mercury is 10.0 mol.
C for Hg = 27.8 J/K•mol
ΔT = 7.5 K

2 Plan your work.

Use the values that are given in the problem and the equation $q = nC\Delta T$ to determine *q*.

3 Calculate.

$q = nC\Delta T$
$q = (10.0 \text{ mol})(27.8 \text{ J/K•mol})(7.5 \text{ K})$
$q = 2085$ J
The answer should only have two significant figures, so it is reported as 2100 J or 2.1×10^3 J.

4 Verify your results.

The calculation yields an energy as heat with the correct unit, *joules*. This result supports the idea that the answer is the energy as heat needed to raise 10.0 mol Hg 7.5 K.

PRACTICE

1 The molar heat capacity of tungsten is 24.2 J/K•mol. Calculate the energy as heat needed to increase the temperature of 0.40 mol of tungsten by 10.0 K.

2 Suppose a sample of NaCl increased in temperature by 2.5 K when the sample absorbed 1.7×10^2 J of energy as heat. Calculate the number of moles of NaCl if the molar heat capacity is 50.5 J/K•mol.

3 Calculate the energy as heat needed to increase the temperature of 0.80 mol of nitrogen, N_2, by 9.5 K. The molar heat capacity of nitrogen is 29.1 J/K•mol.

4 A 0.07 mol sample of octane, C_8H_{18}, absorbed 3.5×10^3 J of energy. Calculate the temperature increase of octane if the molar heat capacity of octane is 254.0 J/K•mol.

342

REAL-WORLD ———
CONNECTION

Snuffing Candles People who put out candles by wetting their fingers and quickly pinching the flame are taking advantage of the heat capacity of water. The gas in the flame is at a high temperature, but there are not very many gas molecules present and the gas has a very small heat capacity. The water on the person's fingers and the water in their skin tissue have very large heat capacities. The water absorbs the energy from the flame and the person's fingers don't get burned.

Table 1 Molar Heat Capacities of Elements and Compounds

Element	C (J/K•mol)	Compound	C (J/K•mol)
Aluminum, Al(s)	24.2	Aluminum chloride, $AlCl_3(s)$	92.0
Argon, Ar(g)	20.8	Barium chloride, $BaCl_2(s)$	75.1
Helium, He(g)	20.8	Cesium iodide, CsI(s)	51.8
Iron, Fe(s)	25.1	Octane, $C_8H_{18}(l)$	254.0
Mercury, Hg(l)	27.8	Sodium chloride, NaCl(s)	50.5
Nitrogen, $N_2(g)$	29.1	Water, $H_2O(g)$	36.8
Silver, Ag(s)	25.3	Water, $H_2O(l)$	75.3
Tungsten W(s)	24.2	Water, $H_2O(s)$	37.4

Molar Heat Capacity Depends on the Number of Atoms

The molar heat capacities of a variety of substances are listed in **Table 1.** One mole of tungsten has a mass of 184 g, while one mole of aluminum has a mass of only about 27 g. So, you might expect that much more heat is needed to change the temperature of 1 mol W than is needed to change the temperature of 1 mol Al. This is not true, however. Notice that the molar heat capacities of all of the metals are nearly the same. The temperature of 1 mol of *any* solid metal is raised 1 K when the metal absorbs about 25 J of heat. The reason the temperature is raised is that the energy is absorbed by increasing the kinetic energy of the atoms in the metal, and every metal has exactly the same number of atoms in one mole.

Notice in **Table 1** that the same "about 25 joule" rule also applies to the molar heat capacities of solid ionic compounds. One mole barium chloride has three times as many ions as atoms in 1 mol of metal. So, you expect the molar heat capacity for $BaCl_2$ to be $C = 3 \times 25$ J/K•mol. The value in **Table 1,** 75.1 J/K•mol, is similar to this prediction.

Molar Heat Capacity Is Related to Specific Heat

The specific heat of a substance is represented by c_p and is the energy as heat needed to raise the temperature of one gram of substance by one kelvin. Remember that molar heat capacity of a substance, C, has a similar definition except that molar heat capacity is related to moles of a substance not to the mass of a substance. Because the molar mass is the mass of 1 mol of a substance, the following equation is true.

$$M \text{ (g/mol)} \times c_p \text{ (J/K•g)} = C \text{ (J/K•mol)}$$

(molar mass)(specific heat) = (molar heat capacity)

Topic Link

Refer to the "Matter and Energy" chapter for a discussion of specific heat.

343

Reteaching —— **BASIC**

To help students understand the difference between heat and temperature, have them derive analogies to situations in everyday life. For example, have students examine **Figure 3.** **LS** Logical

Quiz —————— **GENERAL**

1. What is heat and what are its units? **Ans.** Heat is the energy transferred between two objects that are at different temperatures and is measured in joules.

2. Describe the energy transferred as heat between two objects at different temperatures. **Ans.** Energy is transferred from the object with the higher temperature to the object with the lower temperature.

3. What is the molar heat capacity of a substance? **Ans.** The amount of heat required to raise the temperature of one mole of the substance by 1 K.

4. What equation is used to find the molar heat capacity of a substance? **Ans.** $C = \frac{q}{n\Delta T}$

LS Intrapersonal

Heat Results in Disorderly Particle Motion

When a substance receives energy in the form of heat, its enthalpy increases and the kinetic energy of the particles that make up the substance increases. The direction in which any particle moves is not related to the direction in which its neighboring particles move. The motions of these particles are *random*.

Suppose the substance was a rubber ball and you kicked the ball across a field. The energy that you gave the ball produces a different result than heat because the energy caused the particles in the ball to move together and in the same direction. The kinetic energy that you gave the particles in the ball is not random but is *concerted*.

Do you notice any relationships between energy and motion? Heat often produces disorderly particle motion. Other types of energy can produce orderly motion or orderly positioning of particles.

① Section Review

UNDERSTANDING KEY IDEAS

1. What is heat?

2. What is temperature?

3. How does temperature differ from heat?

4. What is the enthalpy of a substance?

5. Define *molar heat capacity*.

6. How does molar heat capacity differ from specific heat?

7. How is the Kelvin temperature scale different from the Celsius and Fahrenheit scales?

PRACTICE PROBLEMS

8. Calculate the molar heat capacity of diamond, given that 63 J were needed to heat a 1.2 g of diamond by 1.0×10^2 K.

9. Use the molar heat capacity for aluminum from **Table 1** to calculate the amount of energy needed to raise the temperature of 260.5 g of aluminum from 0°C to 125°C.

10. Use the molar heat capacity for iron from **Table 1** to calculate the amount of energy needed to raise the temperature of 260.5 g of iron from 0°C to 125°C.

11. A sample of aluminum chloride increased in temperature by 3.5 K when the sample absorbed 1.67×10^2 J of energy. Calculate the number of moles of aluminum chloride in this sample. Use **Table 1.**

12. Use **Table 1** to determine the final temperature when 2.5×10^2 J of energy as heat is transferred to 0.20 mol of helium at 298 K.

13. Predict the final temperature when 1.2 kJ of energy as heat is transferred from 1.0×10^2 mL of water at 298 K.

14. Use **Table 1** to determine the specific heat of silver.

15. Use **Table 1** to determine the specific heat of sodium chloride.

CRITICAL THINKING

16. Why is a temperature difference the same in Celsius and Kelvin?

17. Predict the molar heat capacities of $PbS(s)$ and $Ag_2S(s)$.

18. Use **Table 1** to predict the molar heat capacity of $FeCl_3(s)$.

19. Use your answer from item 18 to predict the specific heat of $FeCl_3(s)$.

344

Answers to Section Review

1. Heat is the energy transferred between two objects that are at different temperatures.

2. Temperature is a measure of the average kinetic energy of the particles in an object.

3. When two samples at different temperatures are in contact, energy can be transferred as heat.

4. The enthalpy of a substance is the sum of the internal energy of a system plus the product of the system's volume multiplied by the pressure that the system exerts on its surroundings.

Answers continued on p. 375A

Chapter Resource File

• Concept Review

• Quiz

SECTION 2

Using Enthalpy

KEY TERMS
• thermodynamics

OBJECTIVES

① **Define** thermodynamics.

② **Calculate** the enthalpy change for a given amount of substance for a given change in temperature.

Molar Enthalpy Change

Because enthalpy is the total energy of a system, it is an important quantity. However, the only way to measure energy is through a change. In fact, there's no way to determine the true value of H. But ΔH can be measured as a change occurs. The enthalpy change for one mole of a pure substance is called *molar enthalpy change*. The blacksmith in **Figure 5** is causing a molar enthalpy change by heating the iron horseshoe. Though describing a physical change by a chemical equation is unusual, the blacksmith's work could be described as follows.

$$\text{Fe}(s,\ 300\ \text{K}) \longrightarrow \text{Fe}(s,\ 1100\ \text{K}) \quad \Delta H = 20.1\ \text{kJ/mol}$$

This equation indicates that when 1 mol of solid iron is heated from 27°C to 827°C, its molar enthalpy increases by 20 100 joules.

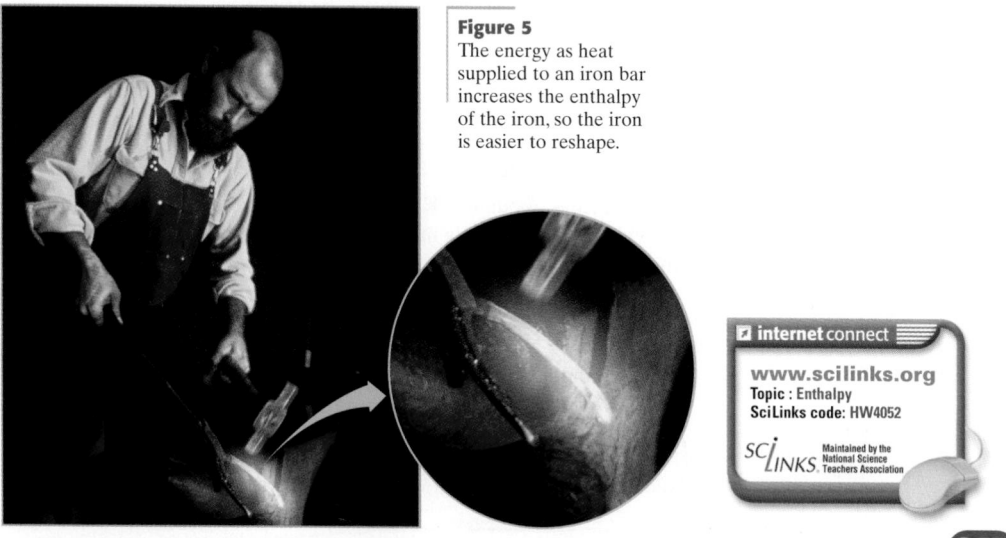

Figure 5
The energy as heat supplied to an iron bar increases the enthalpy of the iron, so the iron is easier to reshape.

⎐ internet connect

www.scilinks.org
Topic : Enthalpy
SciLinks code: HW4052

SCi/LINKS. Maintained by the National Science Teachers Association

345

Chapter Resource File

• Lesson Plan

SECTION 2

Focus

Overview
Before beginning this section, review with your students the Objectives listed in the Student Edition. In this section, students are introduced to molar enthalpy change. In addition, this section introduces the concept of enthalpy change during a reaction. The definition for thermodynamics is also given.

🔔 Bellringer
Ask students to list examples of objects that have absorbed energy and evidence that the objects have gained energy. For example, a burner on an electric stove gains energy when it is turned on. Evidence of this is that the burner glows red and it is hot.

Motivate

Demonstration

1. Place a small paper cup (such as those used in restaurants to serve ketchup) on a wire gauze above a lit Bunsen burner, and watch the cup catch fire and burn.

2. Fill a second paper cup half full with water, and place it on the wire gauze. Place the lit burner under the wire gauze. The water will come to a boil.

3. Explain to students that the water absorbs the energy from the flame and keeps the temperature of the cup below the kindling temperature of paper (233°C). The cup will not burn until the water has boiled away.

4. Ask students why a polystyrene foam cup that has water would melt or catch fire although paper that has water does not. **Ans.** Foam is an insulator, and energy transfer to the water does not take place rapidly enough to keep the foam from melting.

Safety Caution: Wear and have students wear goggles. Also, have students stand at least 3 m away from the demonstration. Have a fire extinguisher near the demonstration.

READING SKILL BUILDER ——— **BASIC**

Assimilating Knowledge Pair each student with a partner. Have each student read Section 2 silently. After reading, both readers should help each other with any part that either (or both) did not understand. Have them create a list of questions to ask the class. **LS** Interpersonal

Teaching Tip

To help students understand why absolute enthalpy of a system cannot be known, ask students to imagine a closed box that could contain any sort of matter imaginable. The only information available to an outside observer is the temperature of the interior of the box. The observer could never know the amount of energy contained within the box. If the thermometer reads 45°C and the box is filled with water, the box contains much more energy than it would if it contained nitrogen. The only way the observer can learn anything about the thermodynamic properties of the box is to alter the conditions, such as the outside temperature, which would cause the box to lose or gain energy. The change in the amount of energy of the box can be known by determining the amount of energy the box absorbed or gave off.

Answers to Practice Problems B
1. 2.60×10^3 J/mol
2. 5.05×10^3 J/mol
3. 3.6×10^2 J/mol

Molar Heat Capacity Governs the Changes

The iron that the blacksmith uses does not change state and is not involved in a chemical reaction. So, the change in enthalpy of the iron horseshoe represents only a change in the kinetic energy of the iron atoms. When a pure substance is only heated or cooled, the amount of heat involved is the same as the enthalpy change. In other words, $\Delta H = q$ for the heating or cooling of substances. So the molar enthalpy change is related to the molar heat capacity by the following equation.

$$\text{molar enthalpy change} = C\Delta T$$
$$\text{molar enthalpy change} = (\text{molar heat capacity})(\text{temperature change})$$

Note that this equation does not apply to chemical reactions or changes of state.

SAMPLE PROBLEM B

Calculating Molar Enthalpy Change for Heating

How much does the molar enthalpy change when ice warms from $-5.4°C$ to $-0.2°C$?

1 Gather information.

$T_{\text{initial}} = -5.4°C = 267.8$ K and $T_{\text{final}} = -0.2°C = 273.0$ K
For $H_2O(s)$, $C = 37.4$ J/K•mol.

2 Plan your work.

The change in temperature is $\Delta T = T_{\text{final}} - T_{\text{initial}} = 5.2$ K.
Because there is no reaction and the ice does not melt, you can use the equation below to determine the molar enthalpy change.

$$\Delta H = C\Delta T$$

PRACTICE HINT

Remember that molar enthalpy change has units of kJ/mol.

3 Calculate.

$$\Delta H = C(\Delta T) = \left(37.4 \ \frac{J}{K•mol}\right)(5.2 \text{ K}) = 1.9 \times 10^2 \ \frac{J}{mol}$$

The molar enthalpy change is 0.19 kJ/mol.

4 Verify your results.

The C of ice is about 40 J/K•mol and its temperature change is about 5 K, so you should expect a molar enthalpy increase of about 200 J/mol, which is close to the calculated answer.

PRACTICE

1. Calculate the molar enthalpy change of $H_2O(l)$ when liquid water is heated from 41.7°C to 76.2°C.

2. Calculate the ΔH of NaCl when it is heated from 0.0°C to 100.0°C.

3. Calculate the molar enthalpy change when tungsten is heated by 15 K.

346

Homework ——————— **GENERAL**

Additional Practice

1. The molar heat capacity of sodium fluoride, NaF(s), is 46.9 J/K•mol. What is the enthalpy change when 2.00 mol of NaF is heated from 5°C to 131°C? Ans. 11800 J or 11.8 kJ

2. The molar heat capacity of carbon tetra-chloride, $CCl_4(l)$, is 130.7 J/K•mol. What is the enthalpy change when 125 g of carbon tetrachloride is heated from 10.0°C to 32.0°C? Ans. 2340 J or 2.34 kJ

LS Logical

Chapter Resource File

• Problem Solving

Calculating the Molar Enthalpy Change for Cooling

Calculate the molar enthalpy change when an aluminum can that has a temperature of 19.2°C is cooled to a temperature of 4.00°C.

1 Gather information.

For Al, $C = 24.2$ J/K•mol.

$T_{initial} = 19.2°C = 292$ K

$T_{final} = 4.00°C = 277$ K

2 Plan your work.

The change in temperature is calculated by using the following equation.

$$\Delta T = T_{final} - T_{initial} =$$
$$277 \text{ K} - 292 \text{ K} = -15 \text{ K}$$

To determine the molar enthalpy change, use the equation $\Delta H = C\Delta T$.

3 Calculate.

$$\Delta H = C\Delta T$$
$$\Delta H = (24.2 \text{ J/K•mol})(-15 \text{ K}) = -360 \text{ J/mol}$$

4 Verify your results.

The calculation shows the molar enthalpy change has units of joules per mole. The enthalpy value is negative, which indicates a cooling process.

PRACTICE HINT

Remember that the Δ notation always represents initial value subtracted from the final value, even if the initial value is larger than the final value.

PRACTICE

1 The molar heat capacity of Al(s) is 24.2 J/K•mol. Calculate the molar enthalpy change when Al(s) is cooled from 128.5°C to 22.6°C.

2 Lead has a molar heat capacity of 26.4 J/K•mol. What molar enthalpy change occurs when lead is cooled from 302°C to 275°C?

3 Calculate the molar enthalpy change when mercury is cooled 10 K. The molar heat capacity of mercury is 27.8 J/K•mol.

Enthalpy Changes of Endothermic or Exothermic Processes

Notice the molar enthalpy change for **Sample Problem B.** This enthalpy change is positive, which means that the heating of a sample requires energy. So, the heating of a sample is an endothermic process. In contrast, the cooling of a sample releases energy or has a negative enthalpy change and is an exothermic process, such as the process in **Sample Problem C.** In fact, you can use enthalpy changes to determine if a process is endothermic or exothermic. Processes that have positive enthalpy changes are endothermic and processes that have negative enthalpy changes are exothermic.

347

HISTORY
CONNECTION

In 1697, German chemist and physician Georg Ernts Stahl suggested the existence of *phlogiston*, a so-called "elastic fluid" that he said was the agent for burning and rusting. Flammable objects, such as wood, contained phlogiston and the object would burn when the phlogiston was released. His theory was generally accepted until 1783 when French chemist Antoine Lavoisier proved that phlogiston could not exist.

Homework ———— **GENERAL**
Additional Practice

1. One mole of a clear liquid loses 6.35 kJ of energy when it is cooled from 28.0°C to 3.0°C? What is the molar heat capacity of the liquid? Using **Table 1,** determine the identity of the clear liquid. Ans. 254 J/K•mol, the liquid is octane, C_8H_{18}.

2. The molar heat capacity of magnesium, Mg(s) is 24.9 J/K•mol. If 5.00 mol of magnesium at 42.0°C loses 4.48 kJ of energy, what is its new temperature? Ans. 6.0°C

3. The molar heat capacity of ethyl alcohol, $C_2H_5OH(l)$, is 112.3 J/K•mol. What is the enthalpy change when 185 g of ethyl alcohol is cooled from 25.0°C to −20.0°C? Ans. −20300 J or −20.3 kJ

4. The molar heat capacity of cyclohexane, $C_6H_{12}(l)$, is 154.9 J/K•mol. What is the enthalpy change when 5.00×10^2 g of cyclohexane is cooled from 28.5°C to −18.8°C? Ans. −43500 J, or −43.5 kJ

5. The molar heat capacity of sulfuric acid, $H_2SO_4(l)$, is 138.9 J/K•mol. What is the enthalpy change when 50.0 g of sulfuric acid is cooled from 25.0°C to 4.0°C? Ans. −1490 J or −1.49 kJ

LS Logical

Answers to Practice Problems C

1. -2.56×10^3 J/mol
2. -7.13×10^2 J/mol
3. -2.8×10^2 J/mol

Chapter Resource File

• **Problem Solving**

Vocabulary The term thermo-dynamics was introduced in 1849 by William Thomson, Lord Kelvin (1824–1907), in a discussion of the theory of heat. The word thermo-dynamics can be broken down into two root words: *thermo-* which comes from the Greek word *therme* meaning heat and *-dynamic* from the Greek word *dunamikos* meaning powerful or force. Have students find other words with the root *thermo-*. **LS** Verbal

Enthalpy of a System of Several Substances

You have read about how a substance's enthalpy changes when the substance receives energy as heat. Enthalpy changes can be found for a system of substances, such as the reaction shown in **Figure 6.** In this figure, hydrogen gas reacts with bromine liquid to form the gas hydrogen bromide, HBr, and to generate energy as heat. Energy transfers out of this system in the form of heat because the enthalpy of the product 2HBr is less than the enthalpy of the reactants H_2 and Br_2. Or, the enthalpy of 2HBr is less than the enthalpy of H_2 and Br_2, so the enthalpy change is negative for this reaction. This negative enthalpy change reveals that the reaction is exothermic.

Enthalpy is the first of three thermodynamic properties that you will encounter in this chapter. **Thermodynamics** is a science that examines various processes and the energy changes that accompany the processes. By studying and measuring thermodynamic properties, chemists have learned to predict whether a chemical reaction can occur and what kind of energy change it will have.

thermodynamics

the branch of science concerned with the energy changes that accompany chemical and physical changes

Figure 6
When hydrogen gas and bromine liquid react, hydrogen bromide gas is formed and energy is released.

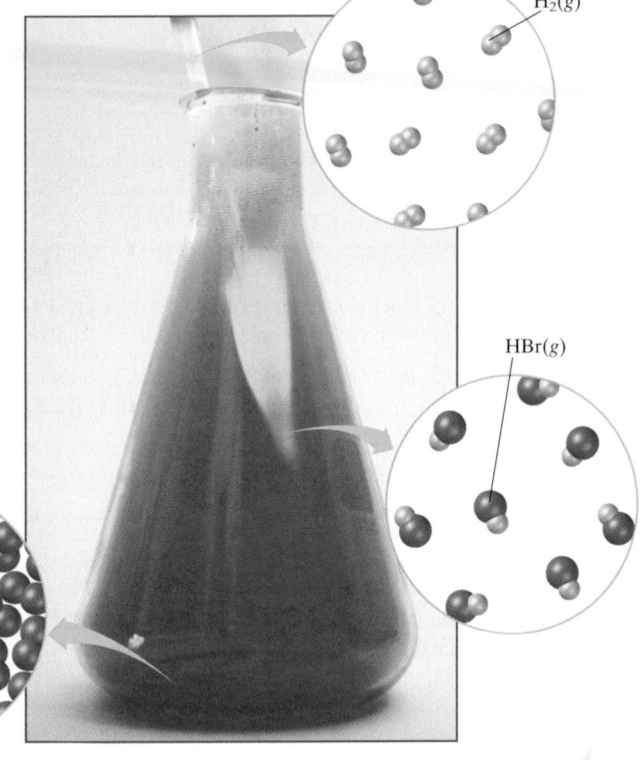

$Br_2(l)$ $H_2(g)$ $HBr(g)$

$$H_2(g) + Br_2(l) \longrightarrow 2HBr(g)$$

348

HISTORY
CONNECTION

In 1702, Guillaume Amontons suggested that a lower limit for temperature exists. This is the first time the idea of an "absolute zero" temperature was put forth. Amontons reached this conclusion because temperature and pressure are known to be directly proportional and pressure can never have a negative value. Thus, when pressure is at its minimum, temperature must be also. William Thomson (Lord Kelvin) independently reached this same conclusion in 1851 using Charles's gas law. Thomson also correctly estimated absolute zero to be at $-273\,°C$.

Writing Equations for Enthalpy Changes

Do you remember the equation that represents the molar enthalpy change when the iron horseshoe is heated?

$$Fe(s, 300 \text{ K}) \longrightarrow Fe(s, 1100 \text{ K}) \quad \Delta H = 20.1 \text{ kJ/mol}$$

Just as an equation can be written for the enthalpy change in the blacksmith's iron, an equation can be written for the enthalpy change that occurs during a change of state or a chemical reaction. The thermodynamics of changes of state are discussed in the chapter entitled "States and Intermolecular Forces." An example of an equation for a chemical reaction is the following equation for the hydrogen and bromine reaction.

$$H_2(g, 298 \text{ K}) + Br_2(l, 298 \text{ K}) \longrightarrow 2HBr(g, 298 \text{ K}) \quad \Delta H = -72.8 \text{ kJ}$$

Notice that the enthalpy change for this reaction and other chemical reactions are written using the symbol ΔH. Also, notice that the negative enthalpy change indicates the reaction is exothermic. Enthalpy changes that are involved in chemical reactions are the subject of section three of this chapter.

 Section Review

UNDERSTANDING KEY IDEAS

1. Name and define the quantity represented by H.

2. During a heating or cooling process, how are changes in enthalpy and temperature related?

3. What is thermodynamics?

PRACTICE PROBLEMS

4. A block of ice is cooled from −0.5°C to −10.1°C. Calculate the temperature change, ΔT, in degrees Celsius and in kelvins.

5. Calculate the molar enthalpy change when a block of ice is heated from −8.4°C to −5.2°C.

6. Calculate the molar enthalpy change when $H_2O(l)$ is cooled from 48.3°C to 25.2°C.

7. The molar heat capacity of benzene, $C_6H_6(l)$, is 136 J/K•mol. Calculate the molar enthalpy change when the temperature of $C_6H_6(l)$ changes from 19.7°C to 46.8°C.

8. The molar heat capacity of diethyl ether, $(C_2H_5)_2O(l)$, is 172 J/K•mol. What is the temperature change if the molar enthalpy change equals −186.9 J/mol?

9. If the enthalpy of 1 mol of a compound decreases by 428 J when the temperature decreases by 10.0 K, what is the compound's molar heat capacity?

CRITICAL THINKING

10. Under what circumstances could the enthalpy of a system be increased without the temperature rising?

11. What approximate enthalpy increase would you expect if you heated one mole of a solid metal by 40 K?

Answers to Section Review

1. Enthalpy; Enthalpy is the sum of the internal energy of a system plus the product of the system's volume multiplied by the pressure that the system exerts on its surroundings.

2. The enthalpy change ΔH and the temperature change ΔT are proportional to each other.

3. The branch of science concerned with the energy changes that accompany chemical and physical processes.

4. $\Delta T = -9.6°C = -9.6 \text{ K}$

5. $\Delta T = (-5.2°C) - (-8.4°C) = 3.2°C = 3.2 \text{ K}$
   $\Delta H = C\Delta T = (37.4 \text{ J/K•mol}) (3.2 \text{ K}) = 120 \text{ J/mol}$

6. $\Delta T = 25.2°C - 48.3°C = -23.1°C = -23.1 \text{ K}$
   $\Delta H = C\Delta T = (75.3 \text{ J/K•mol}) (-23.1 \text{ K}) = -1740 \text{ J} = -1.74 \text{ kJ/mol}$

7. $\Delta T = 46.8°C - 19.7°C = 27.1°C = 27.1 \text{ K}$
   $\Delta H = C\Delta T = (136 \text{ J/K•mol}) (27.1 \text{ K}) = 3690 \text{ J/mol}$

Answers continued on p. 375A

Before beginning this section, review with your students the Objectives listed in the Student Edition. This section describes enthalpy changes during reactions and how the enthalpy change determines whether a reaction is exothermic or endothermic. Students also learn to use Hess's law to determine the ΔH of a chemical reaction.

Bellringer

Have students write a short paragraph describing what they know about food calories and how they relate to nutrition, metabolism, weight gain, and weight loss.

Motivate

Discussion ——— GENERAL

Ask students what *endothermic* and *exothermic* mean. Make sure that they understand that endothermic processes absorb energy while exothermic processes release energy. Have students give examples during the discussion.
LS Intrapersonal

SECTION 3
Changes in Enthalpy During Chemical Reactions

KEY TERMS
- calorimetry
- calorimeter
- Hess's law

OBJECTIVES

① **Explain** the principles of calorimetry.

② **Use** Hess's law and standard enthalpies of formation to calculate ΔH.

Changes in Enthalpy Accompany Reactions

Changes in enthalpy occur during reactions. A change in enthalpy during a reaction depends on many variables, but temperature is one of the most important variables. To standardize the enthalpies of reactions, data are often presented for reactions in which both reactants and products have the *standard thermodynamic temperature* of 25.00°C or 298.15 K.

Chemists usually present a thermodynamic value for a chemical reaction by using the chemical equation, as in the example below.

$$\tfrac{1}{2}H_2(g) + \tfrac{1}{2}Br_2(l) \longrightarrow HBr(g) \quad \Delta H = -36.4 \text{ kJ}$$

This equation shows that when 0.5 mol of H_2 reacts with 0.5 mol of Br_2 to produce 1 mol HBr and all have a temperature of 298.15 K, the enthalpy decreases by 36.4 kJ.

Remember that reactions that have negative enthalpy changes are exothermic, and reactions that have positive enthalpy changes are endothermic.

Topic Link

Refer to the "Science of Chemistry" chapter for a discussion of endothermic and exothermic reactions.

Figure 7
The combustion of wood generates energy as heat and cooks the food in the pan.

internet connect
www.scilinks.org
Topic : Endothermic and Exothermic Reactions
SciLinks code: HW4056
SCiLINKS. Maintained by the National Science Teachers Association

PHYSICS
CONNECTION

The First Law of Thermodynamics is the conservation of energy. (Nuclear reactions are an exception.) Therefore, in a chemical reaction, energy is conserved. Be sure that students understand that although enthalpy can change during a chemical reaction, the total energy does not change. If enthalpy decreases, the reaction is exothermic and the excess energy is usually released as heat or light. If enthalpy increases, the reaction is endothermic and energy is absorbed from the surroundings.

Chapter Resource File

- **Lesson Plan**

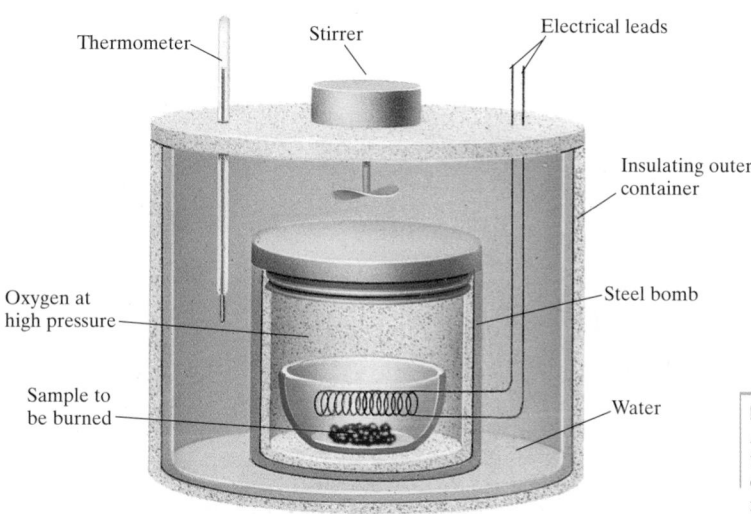

Thermometer Stirrer Electrical leads

Insulating outer
container

Oxygen at
high pressure

Steel bomb

Sample to
be burned

Water

Figure 8
A bomb calorimeter is used
to measure enthalpy changes
caused by combustion
reactions.

Chemical Calorimetry

For the H_2 and Br_2 reaction, in which ΔH is negative, the total energy of the reaction decreases. Energy cannot disappear, so what happens to the energy? The energy is released as heat by the system. If the reaction was endothermic, energy in the form of heat would be absorbed by the system and the enthalpy would increase.

The experimental measurement of an enthalpy change for a reaction is called **calorimetry.** Combustion reactions, such as the reaction in **Figure 7,** are always exothermic. The enthalpy changes of combustion reactions are determined using a bomb **calorimeter,** such as the one shown in **Figure 8.** This instrument is a sturdy, steel vessel in which the sample is ignited electrically in the presence of high-pressure oxygen. The energy from the combustion is absorbed by a surrounding water bath and by other parts of the calorimeter. The water and the other parts of the calorimeter have known specific heats. So, a measured temperature increase can be used to calculate the energy released in the combustion reaction and then the enthalpy change. In **Figure 7,** the combustion of 1.00 mol of carbon yields 393.5 kJ of energy.

$$C(s) + O_2(g) \longrightarrow CO_2(g) \quad \Delta H = -393.5 \text{ kJ}$$

calorimetry

the measurement of heat-related
constants, such as specific heat or
latent heat

calorimeter

a device used to measure the
heat absorbed or released in a
chemical or physical change

Nutritionists Use Bomb Calorimetry

Inside the pressurized oxygen atmosphere of a bomb calorimeter, most organic matter, including food, fabrics, and plastics, will ignite easily and burn rapidly. Some samples of matter may even explode, but the strong walls of the calorimeter contain the explosions. Sample sizes are chosen so that there is excess oxygen during the combustion reactions. Under these conditions, the reactions go to completion and produce carbon dioxide, water, and possibly other compounds.

351

Transparencies

TT Bomb Calorimeter

Group Activity —— GENERAL

Have students work in groups of four to calculate the caloric intake and expenditure for one day for two or more people in the group. Students should carefully record all the types and amounts of food they consumed in one day. If possible, they should save the containers in which the food came. Students should also record the type, duration, and intensity of all the exercise they did during the same day.

Once students have gathered these data, they can start adding up the calories consumed. Students can use the food containers that they saved, calorie-counting books, or websites that contain similar information. If students ate at school, the cafeteria staff and school nutritionist should be able to determine the number of calories consumed. Also, some restaurants have nutritional information available upon request.

To calculate the number of calories expended, students can use "exercise calculators" that can be found on various websites. Do not let students forget to include the calories expended so that their bodies can operate. Students should then compare the amounts of calories consumed and expended in a bar graph and write a paragraph summarizing their findings. **LS** Interpersonal

Figure 9
Nutritionists work with bomb-calorimeter data for a recipe's ingredients to determine the food-energy content of meals.

Nutritionists, such as the nutritionist shown in **Figure 9,** use bomb calorimetry to measure the energy content of foods. To measure the energy, nutritionists assume that all the combustion energy is available to the body as we digest food. For example, consider table sugar, $C_{12}H_{22}O_{11}$, also known as sucrose. Its molar mass is 342.3 g/mol. When 3.423 grams of sugar are burned in a bomb calorimeter, the 1.505 kg of the calorimeter's water bath increased in temperature by 3.524°C. The enthalpy change can be calculated and is shown below.

$$C_{12}H_{22}O_{11}(s) + 12O_2(g) \longrightarrow 12CO_2(g) + 11H_2O(l) \quad \Delta H = -2226 \text{ kJ}$$

When enthalpy changes are reported in this way, a coefficient in the chemical equation indicates the number of moles of a substance. So, the equation above describes the enthalpy change when 1 mol of sucrose reacts with 12 mol of oxygen to produce 12 mol of carbon dioxide and 11 mol of liquid water, at 298.15 K.

Calorimetric measurements can be made with very high precision. In fact, most thermodynamic quantities are known to many significant figures.

Adiabatic Calorimetry Is Another Strategy

Instead of using a water bath to absorb the energy generated by a chemical reaction, _adiabatic calorimetry_ uses an insulating vessel. The word _adiabatic_ means "not allowing energy to pass through." So, no energy can enter or escape this type of vessel. As a result, the reaction mixture increases in temperature if the reaction is exothermic or decreases in temperature if the reaction is endothermic. If the system's specific heat is known, the reaction enthalpy can be calculated. Adiabatic calorimetry is used for reactions that are not ignited, such as for reactions in aqueous solution.

internet connect

www.scilinks.org
Topic: Nutrition
SciLinks code: HW4090

SciLINKS. Maintained by the National Science Teachers Association

MISCONCEPTION ALERT

Inform students that the word Calorie, when spelled with a capital C represents one kilocalorie of energy. These "big C" Calories are always referred to in nutrition labeling of foods. Students should be especially careful when using the abbreviations for "small c" calories, "big C" Calories, and kilocalories. The relationship is 1000 cal = 1 Cal = 1 kcal. "Small c" calories were once used in the physical sciences routinely, but this unit has been supplanted almost entirely by the joule (1 cal = 4.184 J).

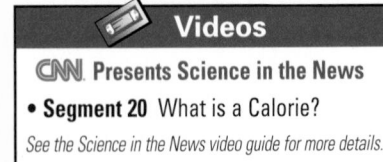

Videos

CNN. Presents Science in the News

• **Segment 20** What is a Calorie?

See the Science in the News video guide for more details.

Hess's Law

Any two processes that both start with the same reactants in the same state and finish with the same products in the same state will have the same enthalpy change. This statement is the basis for **Hess's law,** which states that the overall enthalpy change in a reaction is equal to the sum of the enthalpy changes for the individual steps in the process.

Consider the following reaction, the synthesis of 4 mol of phosphorus pentachloride, PCl_5, when phosphorus is burned in excess chlorine.

$$P_4(s) + 10Cl_2(g) \longrightarrow 4PCl_5(g) \quad \Delta H = -1596 \text{ kJ}$$

Phosphorus pentachloride may also be prepared in a two-step process.

$$\text{Step 1: } P_4(s) + 6Cl_2(g) \longrightarrow 4PCl_3(g) \quad \Delta H = -1224 \text{ kJ}$$
$$\text{Step 2: } PCl_3(g) + Cl_2(g) \longrightarrow PCl_5(g) \quad \Delta H = -93 \text{ kJ}$$

However, the second reaction must take place four times for each occurrence of the first reaction in the two-step process. This two-step process is more accurately described by the following equations.

$$P_4(s) + 6Cl_2(g) \longrightarrow 4PCl_3(g) \quad \Delta H = -1224 \text{ kJ}$$
$$4PCl_3(g) + 4Cl_2(g) \longrightarrow 4PCl_5(g) \quad \Delta H = 4(-93 \text{ kJ}) = -372 \text{ kJ}$$

So, the total change in enthalpy by the two-step process is as follows:

$$(-1224 \text{ kJ}) + (-372 \text{ kJ}) = -1596 \text{ kJ}$$

This enthalpy change, ΔH, for the two-step process is the same as the enthalpy change for the direct route of the formation of PCl_5. This example is in agreement with Hess's law.

Hess's law

the law that states that the amount of heat released or absorbed in a chemical reaction does not depend on the number of steps in the reaction

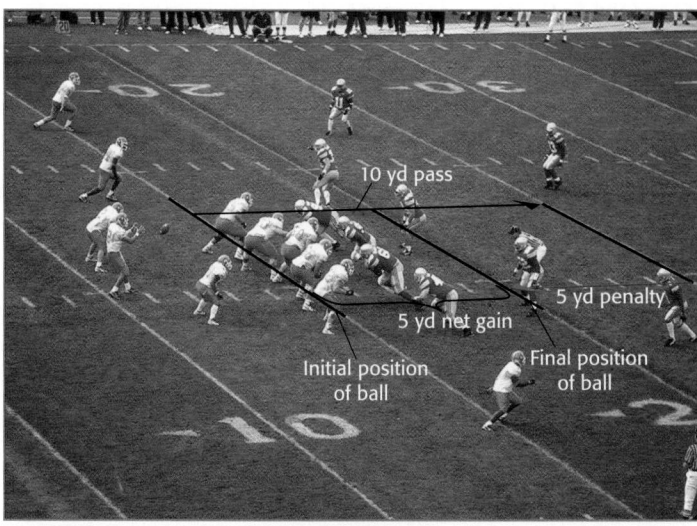

Figure 10
In football, as in Hess's law, only the initial and final conditions matter. A team that gains 10 yards on a pass play but has a five-yard penalty, has the same net gain as the team that gained only 5 yards.

10 yd pass
5 yd penalty
5 yd net gain
Initial position of ball
Final position of ball

353

BIOLOGY
CONNECTION

Combustion vs. Metabolism Ask students to explain how the energy available from the combustion of glucose and the energy available from the metabolism of glucose provide an example of Hess's law. In the oxidation of glucose by cellular respiration, energy is released in small quantities suitable for supplying energy to cell activities. However, the total energy available is the same as the energy obtained from the combustion of glucose in oxygen.

Demonstration

The water and the solutions should all be prepared ahead of time and allowed to stand so that they will be at room temperature.

1. Write the following three equations on the board.

   $NaOH(s) + H_2O(l) \longrightarrow Na^+(aq) + OH^-(aq) + H_2O(l) + \Delta H_1$

   $Na^+(aq) + OH^-(aq) + H_3O^+(aq) + Cl^-(aq) \longrightarrow 2H_2O(aq) + Na^+(aq) + Cl^-(aq) + \Delta H_2$

   $NaOH(s) + H_3O^+(aq) + Cl^-(aq) \longrightarrow 2H_2O(l) + Na^+(aq) + Cl^-(aq) + \Delta H_3$

   Students should note that the first two equations can be added to give the third, so the enthalpy changes of the first two equations add to give the enthalpy change of the third. By keeping the mass of each substance the same in all three reactions, the temperature changes should be proportional to the enthalpy changes, and $\Delta T_1 + \Delta T_2 = \Delta T_3$.

2. Reaction 1: Pour 50 mL of distilled water into a plastic-foam cup, and record the temperature. Add 1 g of NaOH, stir until dissolved. Record the new temperature and calculate ΔT_1.

3. Reaction 2: Pour 25 mL of 1 M HCl into a plastic-foam cup and record the temperature. Stir in 25 mL of 1 M NaOH, record the temperature, and calculate ΔT_2.

4. Reaction 3: Pour 25 mL of 1 M HCl into a plastic foam cup, add 25 mL of distilled water, and record the temperature. Add 1 g of NaOH. Stir until dissolved, and record the temperature, and calculate ΔT_3.

5. Compare $\Delta T_1 + \Delta T_2$ with ΔT_3.

Safety Caution: Wear safety goggles, gloves, and a lab apron. Students should be 3 m from the demonstration. Use a non-mercury thermometer. Follow first-aid instructions on labels and in the MSDS if acid or base gets in the eyes, on skin, or clothing.

Disposal: Rinse out the cups, and put them in the trash. Combine all liquids, neutralize them with 1 M NaOH or 1 M HCl as appropriate, and pour down the drain.

Have students work in small groups to derive other analogies to model Hess's law. For example, students can use different flight paths or different roads on a map to travel from one place to another. Sports that use goals, such as hockey, soccer, and basketball, also provide good analogies. Have students make a poster to explain their analogy. **LS Interpersonal**

Using Hess's Law and Algebra

Chemical equations can be manipulated using rules of algebra to get a desired equation. When equations are added or subtracted, enthalpy changes must be added or subtracted. And when equations are multiplied by a constant, the enthalpy changes must also be multiplied by that constant. For example, the enthalpy of the formation of CO, when CO_2 and solid carbon are reactants, is found using the equations below.

$$2C(s) + O_2(g) \longrightarrow 2CO(g) \quad \Delta H = -221 \text{ kJ}$$
$$C(s) + O_2(g) \longrightarrow CO_2(g) \quad \Delta H = -393 \text{ kJ}$$

You cannot simply add these equations because CO_2 would not be a reactant. But if you subtract or reverse the second equation, carbon dioxide will be on the correct side of the equation. This process is shown below.

$$-C(s) - O_2(g) \longrightarrow -CO_2(g) \quad \Delta H = -(-393 \text{ kJ})$$
$$CO_2(g) \longrightarrow C(s) + O_2(g) \quad \Delta H = 393 \text{ kJ}$$

So, reversing an equation causes the enthalpy of the new reaction to be the negative of the enthalpy of the original reaction. Now add the two equations to get the equation for the formation of CO by using CO_2 and C.

$$2C(s) + O_2(g) \longrightarrow 2CO(g) \quad \Delta H = -221 \text{ kJ}$$
$$CO_2(g) \longrightarrow C(s) + O_2(g) \quad \Delta H = 393 \text{ kJ}$$
$$\overline{2C(s) + O_2(g) + CO_2(g) \longrightarrow 2CO(g) + C(s) + O_2(g) \quad \Delta H = 172 \text{ kJ}}$$

Oxygen and carbon that appear on both sides of the equation can be canceled. So, the final equation is as shown below.

$$C(s) + CO_2(g) \longrightarrow 2CO(g) \quad \Delta H = 172 \text{ kJ}$$

Standard Enthalpies of Formation

The enthalpy change in forming 1 mol of a substance from elements in their standard states is called the *standard enthalpy of formation* of the substance, ΔH_f^0. Many values of ΔH_f^0 are listed in **Table 2**. Note that the values of the standard enthalpies of formation for elements are 0. From a list of standard enthalpies of formation, the enthalpy change of any reaction for which data is available can be calculated. For example, the following reaction can be considered to take place in four steps.

$$SO_2(g) + NO_2(g) \longrightarrow SO_3(g) + NO(g) \quad \Delta H = ?$$

Two of these steps convert the reactants into their elements. Notice that the reverse reactions for the formations of SO_2 and NO_2 are used. So, the standard enthalpies of formation for these reverse reactions are the negative of the standard enthalpies of formation for SO_2 and NO_2.

$$SO_2(g) \longrightarrow \tfrac{1}{8}S_8(s) + O_2(g) \quad \Delta H = -\Delta H_f^0 = -(-296.8 \text{ kJ/mol})$$
$$NO_2(g) \longrightarrow \tfrac{1}{2}N_2(g) + O_2(g) \quad \Delta H = -\Delta H_f^0 = -(33.1 \text{ kJ/mol})$$

354

HISTORY
CONNECTION

Germain Henri Hess (1802–1850) was a Russian chemist and professor of chemistry at the Technological Institute of St. Petersburg. He published the law that bears his name in 1840 after studying the energy changes in a great number of chemical reactions. The law may seem logical today and is actually a statement of the law of conservation of energy to account for multi-step processes.

One-Stop Planner CD-ROM

• **Career Extension**
Real-World Connections Worksheet 8: Firefighter
Assign this worksheet to emphasize relevant applications of text concepts.

Table 2 Standard Enthalpies of Formation

Substance	ΔH_f^0 (kJ/mol)	Substance	ΔH_f^0 (kJ/mol)
$Al_2O_3(s)$	−1676.0	$H_2O(g)$	−241.8
$CaCO_3(s)$	−1206.9	$H_2O(l)$	−285.8
$CaO(s)$	−634.9	$Na^+(g)$	609.4
$Ca(OH)_2(s)$	−985.2	$NaBr(s)$	−361.1
$C_2H_6(g)$	−83.8	$Na_2CO_3(s)$	−1130.7
$CH_4(g)$	−74.9	$NO(g)$	90.3
$CO(g)$	−110.5	$NO_2(g)$	33.1
$CO_2(g)$	−393.5	$Pb(s)$	0
$Fe_2O_3(s)$	−825.5	$SO_2(g)$	−296.8
$H_2(g)$	0	$SO_3(g)$	−395.8
$Hg(l)$	0	$ZnO(s)$	−348.3

Refer to Appendix A for more standard enthalpies of formation.

The two other steps, which are listed below reform those elements into the products.

$$\tfrac{1}{8}S_8(s) + \tfrac{3}{2}O_2(g) \longrightarrow SO_3(g) \quad \Delta H_f^0 = -395.8 \text{ kJ/mol}$$
$$\tfrac{1}{2}N_2(g) + \tfrac{1}{2}O_2(g) \longrightarrow NO(g) \quad \Delta H_f^0 = 90.3 \text{ kJ/mol}$$

In fact, the enthalpy change of any reaction can be determined in the same way—the reactants can be converted to their elements, and the elements can be recombined into the products. Why? Hess's law states that the overall enthalpy change of a reaction is the same, whether for a single-step process or a multiple step one. If you apply this rule, the exothermic reaction that forms sulfur trioxide and nitrogen oxide has the enthalpy change listed below.

$$SO_2(g) + NO_2(g) \longrightarrow SO_3(g) + NO(g)$$
$$\Delta H = (\Delta H_{f,\,NO}^0 + \Delta H_{f,\,SO_3}^0) + (-\Delta H_{f,\,NO_2}^0 - \Delta H_{f,\,SO_2}^0)$$
$$\Delta H = (90.3 \text{ kJ/mol} - 395.8 \text{ kJ/mol}) + (-33.1 \text{ kJ/mol} + 296.8 \text{ kJ/mol}) =$$
$$-41.8 \text{ kJ/mol}$$

When using standard enthalpies of formation to determine the enthalpy change of a chemical reaction, remember the following equation.

$$\Delta H_{reaction} = \Delta H_{products} - \Delta H_{reactants}$$

BIOLOGY CONNECTION

One of the most important endothermic reactions is the one performed by plants and some protists—photosynthesis. The photosynthetic capability of plants is well known. In the uppermost layer of the ocean (the top 100 m or so), tiny geometrically shaped organisms called *phytoplankton* also make food by using photosynthesis. Many fish and other sea animals depend on phytoplankton as their food source.

Homework ── GENERAL

Additional Practice Have students determine the enthalpy change for each of the following reactions.

1. $2H_2O_2(l) \longrightarrow 2H_2O(l) + O_2(g)$
Ans. −196.0 kJ

2. $HCl(g) + NH_3(g) \longrightarrow NH_4Cl(s)$
Ans. −176.2 kJ

3. $CH_4(g) + 2O_2(g) \longrightarrow CO_2(g) + 2H_2O(g)$ Ans. −802.2 kJ

Note that H_2O is a gas.
LS Logical

Answers to Practice Problems D
1. −57.2 kJ
2. −890.2 kJ

Homework ── GENERAL

Additional Practice Have students determine the enthalpy change for the following reactions and determine if the reactions are exothermic or endothermic.

1. $N_2(g) + 3H_2(g) \longrightarrow 2NH_3(g)$
Ans. −91.8 kJ, exothermic

2. $2H_2O(l) \longrightarrow 2H_2(g) + O_2(g)$
Ans. 571.6 kJ, endothermic

3. $C_3H_8(g) + 5O_2(g) \longrightarrow 3CO_2(g) + 4H_2O(g)$ Ans. −2043.0 kJ, exothermic

4. $2H_2O(l) + O_2(g) \longrightarrow 2H_2O_2(l)$
Ans. 196 kJ, endothermic

5. $3C(s, \text{graphite}) + 4H_2(g) \longrightarrow C_3H_8(g)$ Ans. −104.7 kJ, exothermic
LS Logical

Answers to Practice Problems E
1. −1428.6 kJ; exothermic
2. −64.5 kJ; exothermic

SAMPLE PROBLEM D

Calculating a Standard Enthalpy of Formation

Calculate the standard enthalpy of formation of pentane, C_5H_{12}, using the given information.

(1) $C(s) + O_2(g) \longrightarrow CO_2(g)$ $\Delta H_f^0 = -393.5$ kJ/mol

(2) $H_2(g) + \frac{1}{2}O_2(g) \longrightarrow H_2O(l)$ $\Delta H_f^0 = -285.8$ kJ/mol

(3) $C_5H_{12}(g) + 8O_2(g) \longrightarrow 5CO_2(g) + 6H_2O(l)$ $\Delta H = -3535.6$ kJ/mol

1 Gather information.

The equation for the standard enthalpy of formation is

$$5C(s) + 6H_2(g) \longrightarrow C_5H_{12}(g) \quad \Delta H_f^0 = ?$$

2 Plan your work.

C_5H_{12} is a product, so reverse the equation (3) and the sign of ΔH. Multiply equation (1) by 5 to give 5C as a reactant. Multiply equation (2) by 6 to give $6H_2$ as a reactant.

3 Calculate.

(1) $5C(s) + 5O_2(g) \longrightarrow 5CO_2(g)$ $\Delta H = 5(-393.5$ kJ/mol)

(2) $6H_2(g) + 3O_2(g) \longrightarrow 6H_2O(l)$ $\Delta H = 6(-285.8$ kJ/mol)

(3) $5CO_2(g) + 6H_2O(l) \longrightarrow C_5H_{12}(g) + 8O_2(g)$ $\Delta H = 3536.6$ kJ/mol

$$5C(s) + 6H_2(g) \longrightarrow C_5H_{12}(g) \quad \Delta H_f^0 = -145.7 \text{ kJ/mol}$$

4 Verify your results.

The unnecessary reactants and products cancel to give the correct equation.

> **PRACTICE HINT**
>
> A positive ΔH means that the reaction has absorbed energy or that the reaction is endothermic. A negative ΔH means that the reaction has released energy or that the reaction is exothermic.

PRACTICE

1 Calculate the enthalpy change for the following reaction.

$$NO(g) + \frac{1}{2}O_2(g) \longrightarrow NO_2(g)$$

2 Calculate the enthalpy change for the combustion of methane gas, CH_4, to form $CO_2(g)$ and $H_2O(l)$.

SAMPLE PROBLEM E

Calculating a Reaction's Change in Enthalpy

Calculate the change in enthalpy for the reaction below by using data from **Table 2**.

$$2H_2(g) + 2CO_2(g) \longrightarrow 2H_2O(g) + 2CO(g)$$

Then, state whether the reaction is exothermic or endothermic.

356

> **Chapter Resource File**
> • Problem Solving

1 Gather information.

Standard enthalpies of formation for the products are as follows:
For $H_2O(g)$, $\Delta H_f^0 = -241.8$ kJ/mol. For $CO(g)$, $\Delta H_f^0 = -110.5$ kJ/mol.
Standard enthalpies of formation for the reactants are as follows:
For $H_2(g)$, $\Delta H_f^0 = 0$ kJ/mol. For $CO_2(g)$, $\Delta H_f^0 = -393.5$ kJ/mol.

2 Plan your work.

The general rule is $\Delta H = \Delta H$(products) $- \Delta H$(reactants). So,
$\Delta H = $ (mol $H_2O(g)$) ΔH_f^0 (for $H_2O(g)$) + (mol $CO(g)$) ΔH_f^0 (for $CO(g)$) $-$
(mol $H_2(g)$) ΔH_f^0 (for $H_2(g)$) $-$ (mol $CO_2(g)$) ΔH_f^0 (for $CO_2(g)$).

3 Calculate.

$\Delta H = $ (2 mol)(-241.8 kJ/mol) + (2 mol)(-110.5 kJ/mol) $-$
(2 mol)(0 kJ/mol) $-$ (2 mol)(-393.5 kJ/mol) = 82.4 kJ
Because the enthalpy change is positive, the reaction is endothermic.

4 Verify your results.

The enthalpy of the reactants, -787 kJ, is more negative than that of the
products, -704.6 kJ, and shows that the total energy of the reaction
increases by 82.4 kJ.

PRACTICE HINT

Always be sure to check
the states of matter
when you use standard
enthalpy of formation
data. $H_2O(g)$ and
$H_2O(l)$ have different
values.

PRACTICE

1. Use data from **Table 2** to calculate ΔH for the following reaction.

$$C_2H_6(g) + \tfrac{7}{2}O_2(g) \longrightarrow 2CO_2(g) + 3H_2O(g)$$

2. The exothermic reaction known as lime slaking is $CaO(s) + H_2O(l) \longrightarrow$
$Ca(OH)_2(s)$. Calculate ΔH from the data in **Table 2**.

Section Review

UNDERSTANDING KEY IDEAS

1. What is the standard thermodynamic temperature?

2. Why does $\Delta H_f^0 = 0$ for elements, as listed in **Table 2**?

3. How do bomb calorimetry and adiabatic calorimetry differ?

PRACTICE PROBLEMS

4. Use **Table 2** to calculate ΔH for the decomposition of calcium carbonate into calcium oxide and carbon dioxide.

5. What enthalpy change accompanies the reaction $2Al(s) + 3H_2O(l) \longrightarrow$
$Al_2O_3(s) + 3H_2(g)$?

CRITICAL THINKING

6. **Table 2** includes two entries for water. What does the difference between the two values represent?

7. What general conclusion can you draw from observing that most standard enthalpies of formation are negative?

357

Answers to Section Review

1. 298.15 K

2. because their enthalpies of formation are zero, by definition

3. In bomb calorimetry, the enthalpy lost by the reacting system appears as heat transferred to the calorimeter. In adiabatic calorimetry, the enthalpy lost by the reacting system serves to increase the temperature of the reaction products.

4. $CaCO_3(s) \longrightarrow CaO(s) + CO_2(g)$
   $\Delta H = 178.5$ kJ

Answers continued on p. 375A

Chapter Resource File

• Concept Review

• Quiz

Close

Reteaching — BASIC

Have students make a concept map of the main ideas in this section. The following terms should be included in their concept maps: enthalpy, molar enthalpy change, calorimetry, Hess's law, enthalpy of formation, endothermic, exothermic, and nutrition. **LS Intrapersonal**

Quiz — GENERAL

1. How is Hess's law useful when calculating the enthalpy change of a multi-step chemical reaction? **Ans.** Hess's law states that you can find the enthalpy change for the overall reaction rather than finding the enthalpy change for every step.

2. Why is the reaction vessel of a bomb calorimeter surrounded by water? **Ans.** The water will absorb the energy from the combustion reaction. The enthalpy change of the water can be measured and the change is equal to the amount of energy released by the reaction.

3. What is the difference between an exothermic reaction and an endothermic reaction? **Ans.** An exothermic reaction releases energy, but any endothermic reaction absorbs energy.

4. The enthalpy change of a reaction is 395 kJ. Is the reaction exothermic or endothermic? How do you know? **Ans.** Endothermic because the quantity of energy increased.

5. What does adiabatic mean? **Ans.** Adiabatic means that no energy is allowed to leave or enter the system.
LS Intrapersonal

Focus

Overview

Before beginning this section, review with your students the Objectives listed in the Student Edition. In this section, students learn what entropy is and how to determine entropy change during a reaction by using Hess's law. Gibbs energy is defined and students learn how to calculate the Gibbs energy for a reaction using the changes in enthalpy and entropy. Students also learn to predict the spontaneity of chemical reactions.

 **Bellringer**

Have students work in small groups to make lists of examples of both increasing and decreasing disorder. For example, for increasing order students may list making their beds or baking a cake. For decreasing order they may list breaking a dish or shredding paper. The students in each group should evaluate and critique the lists for incorrect examples.

Motivate

Discussion

Give students an introduction to entropy. Tell them that entropy is a measure of disorder in a system of particles and the entropy tends to increase over time. Next have students read items from their lists that they made in the Bellringer without identifying which list the items came from. Ask the rest of the class to decide if the item is an example of increasing disorder or decreasing disorder. Finally, have students evaluate the physics catch phrase, "Entropy isn't what it used to be."

Order and Spontaneity

KEY TERMS
• entropy
• Gibbs energy

OBJECTIVES

① **Define** *entropy*, and discuss the factors that influence the sign and magnitude of ΔS for a chemical reaction.

② **Describe** *Gibbs energy*, and discuss the factors that influence the sign and magnitude of ΔG.

③ **Indicate** whether ΔG values describe spontaneous or nonspontaneous reactions.

Entropy

Some reactions happen easily, but others do not. For example, sodium and chlorine react when they are brought together. However, nitrogen and oxygen coexist in the air you breathe without forming poisonous nitrogen monoxide, NO. One factor you can use to predict whether reactions will occur is enthalpy. A reaction is more likely to occur if it is accompanied by a *decrease in enthalpy* or if ΔH is negative.

But a few processes that are endothermic can occur easily. Why? Another factor known as entropy can determine if a process will occur. **Entropy,** *S,* is a measure of the disorder in a system and is a thermodynamic property. Entropy is not a form of energy and has the units joules per kelvin, J/K. A process is more likely to occur if it is accompanied by an *increase in entropy*; that is, ΔS is positive.

entropy

a measure of the randomness or disorder of a system

Figure 11
a Crystals of potassium permanganate, $KMnO_4$, are dropped into a beaker of water and dissolve to produce the $K^+(aq)$ and $MnO_4^-(aq)$ ions.

b Diffusion causes entropy to increase and leads to a uniform solution.

PHYSICS
CONNECTION

The Second Law of Thermodynamics predicts which processes will occur spontaneously and which will not. As with all three laws of thermodynamics, the Second Law can be stated in a variety of ways. It is usually stated in terms of entropy. According to the Second Law, the total entropy of a system and its surroundings always increases in any natural (spontaneous) process.

Chapter Resource File

• Lesson Plan

Table 3 Standard Entropy Changes for Some Reactions

Reaction	Entropy change, ΔS (J/K)
$CaCO_3(s) + 2H_3O^+(aq) \longrightarrow Ca^{2+}(aq) + CO_2(g) + 3H_2O(l)$	138
$NaCl(s) \longrightarrow Na^+(aq) + Cl^-(aq)$	43
$N_2(g) + O_2(g) \longrightarrow 2NO(g)$	25
$CH_4(g) + 2O_2(g) \longrightarrow CO_2(g) + 2H_2O(g)$	−5
$2Na(s) + Cl_2(g) \longrightarrow 2NaCl(s)$	−181
$2NO_2(g) \longrightarrow N_2O_4(g)$	−176

Factors That Affect Entropy

If you scatter a handful of seeds, you have dispersed them. You have created a more disordered arrangement. In the same way, as molecules or ions become dispersed, their disorder increases and their entropy increases. In **Figure 11,** the intensely violet permanganate ions, $MnO_4^-(aq)$ are initially found only in a small volume of solution. But they gradually spread until they occupy the whole beaker. You can't see the potassium $K^+(aq)$ ions because these ions are colorless, but they too have dispersed. This process of dispersion is called *diffusion* and causes the increase in entropy.

Entropy also increases as solutions become more dilute or when the pressure of a gas is reduced. In both cases, the molecules fill larger spaces and so become more disordered. Entropies also increase with temperature, but this effect is not great unless a phase change occurs.

The entropy can change during a reaction. The entropy of a system can increase when the total number of moles of product is greater than the total number of moles of reactant. Entropy can increase in a system when the total number of particles in the system increases. Entropy also increases when a reaction produces more gas particles, because gases are more disordered than liquids or solids.

Table 3 lists the entropy changes of some familiar chemical reactions. Notice that entropy decreases as sodium chloride forms: 2 mol of sodium combine with 1 mol of chlorine to form 2 mol of sodium chloride.

$$2Na(s) + Cl_2(g) \longrightarrow 2NaCl(s) \quad \Delta S = -181 \text{ J/K}$$

This decrease in entropy is because of the order present in crystalline sodium chloride.

Also notice that the entropy increases when 1 mol of sodium chloride dissolves in water to form 1 mol of aqueous sodium ions and 1 mol of aqueous chlorine ions.

$$NaCl(s) \longrightarrow Na^+(aq) + Cl^-(aq) \quad \Delta S = 43 \text{ J/K}$$

This increase in entropy is because of the order lost when a crystalline solid dissociates to form ions.

359

Using the Figure

Figure 12 Have students study the figure. Ask them to cite evidence they see that a chemical reaction has occurred. Refer them to the "Chemical Equations and Reactions" chapter for the different types of evidence that indicate a chemical change. Next ask them to cite evidence they see that the entropy of the system has increased.

$I_2(g)$ $N_2(g)$

Figure 12
The decomposition of nitrogen triiodide to form nitrogen and iodine has a large entropy increase.

Feather, which starts reaction

Hess's Law Also Applies to Entropy

The decomposition of nitrogen triiodide to form nitrogen and iodine in **Figure 12** creates 4 mol of gas from 2 mol of a solid.

$$2NI_3(s) \longrightarrow N_2(g) + 3I_2(g)$$

This reaction has such a large entropy increase that the reaction proceeds once the reaction is initiated by a mechanical shock.

Molar entropy has the same unit, J/K•mol, as molar heat capacity. In fact, molar entropies can be calculated from molar heat capacity data.

Entropies can also be calculated by using Hess's law and entropy data for other reactions. This statement means that you can manipulate chemical equations using rules of algebra to get a desired equation. But remember that when equations are added or subtracted, entropy changes must be added or subtracted. And when equations are multiplied by a constant, the entropy changes must also be multiplied by that constant. Finally, atoms and molecules that appear on both sides of the equation can be canceled.

The *standard entropy* is represented by the symbol S^0 and some standard entropies are listed in **Table 4.** The standard entropy of the substance is the entropy of 1 mol of a substance at a standard temperature, 298.15 K. Unlike having standard enthalpies of formation equal to 0, elements can have standard entropies that have values other than zero. You should also know that most standard entropies are positive; this is not true of standard enthalpies of formation.

The entropy change of a reaction can be calculated by using the following equation.

$$\Delta S_{\text{reaction}} = S_{\text{products}} - S_{\text{reactants}}$$

Table 4 Standard Entropies of Some Substances

Substance	S^0 (J/K•mol)
C(s) (graphite)	5.7
CO(g)	197.6
$CO_2(g)$	213.8
$H_2(g)$	130.7
$H_2O(g)$	188.7
$H_2O(l)$	70.0
$Na_2CO_3(s)$	135.0
$O_2(g)$	205.1

Refer to Appendix A for more standard entropies.

360

MISCONCEPTION ALERT

Many people have the misconception that absolutely all motion ceases at absolute zero. However, the non-thermal zero-point vibrations remain because of quantum effects. These vibrations are unavailable to do work or release energy because they represent the vibrations and configurations of the particles in their lowest possible ground states. For this fundamental energy to be released, an energy state lower than the ground state would have to be available.

SAMPLE PROBLEM F

Hess's Law and Entropy

Use **Table 4** to calculate the entropy change that accompanies the following reaction.

$$\tfrac{1}{2}H_2(g) + \tfrac{1}{2}CO_2(g) \longrightarrow \tfrac{1}{2}H_2O(g) + \tfrac{1}{2}CO(g)$$

1 Gather information.

Products: $H_2O(g)$ + $CO(g)$

Reactants: $H_2(g)$ + $CO_2(g)$

2 Plan your work.

The general rule is $\Delta S = \Delta S(\text{products}) - \Delta S(\text{reactants})$.

So, $\Delta S = (\text{mol } H_2O(g)) \, S^0 \, (\text{for } H_2O(g)) + (\text{mol } CO(g)) \, S^0$
(for CO(g)) – (mol $H_2(g)$) S^0 (for $H_2(g)$) – (mol $CO_2(g)$) S^0 (for $CO_2(g)$).

The standard entropies from **Table 4** are as follows:
For H_2O, $S^0 = 188.7$ J/K•mol. For CO, $S^0 = 197.6$ J/K•mol.
For H_2, $S^0 = 130.7$ J/K•mol. For CO_2, $S^0 = 213.8$ J/K•mol.

3 Calculate.

Substitute the values into the equation for ΔS.

$\Delta S = \left[\left(\tfrac{1}{2}\,\text{mol}\right)(188.7\ \text{J/K•mol}) + \left(\tfrac{1}{2}\,\text{mol}\right)(197.6\ \text{J/K•mol}) - \right.$
$\left.\left(\tfrac{1}{2}\,\text{mol}\right)(130.7\ \text{J/K•mol}) - \left(\tfrac{1}{2}\,\text{mol}\right)(213.8\ \text{J/K•mol})\right] = 94.35\ \text{J/K} +$
$98.8\ \text{J/K} - 65.35\ \text{J/K} - 106.9\ \text{J/K} = 193.1\ \text{J/K} - 172.2\ \text{J/K} = 20.9\ \text{J/K}$

4 Verify your results.

The sum of the standard entropies of gaseous water and carbon monoxide is larger than the sum of the standard entropies of gaseous hydrogen and carbon dioxide. So, the ΔS for this reaction should be positive.

> **PRACTICE HINT**
>
> Always check the signs of entropy values. Standard entropies are almost always positive, while standard entropies of formation are positive and negative.

PRACTICE

1 Find the change in entropy for the reaction below by using **Table 4** and that S^0 for $CH_3OH(l)$ is 126.8 J/K•mol.

$$CO(g) + 2H_2(g) \longrightarrow CH_3OH(l)$$

2 What is the entropy change for

$$\tfrac{1}{2}CO(g) + H_2(g) \longrightarrow \tfrac{1}{2}CH_3OH(l)?$$

3 Use data from **Table 3** to calculate the entropy change for the following reaction:

$$2Na(s) + Cl_2(g) \longrightarrow 2Na^+(aq) + 2Cl^-(aq)$$

361

Homework ————— **GENERAL**

Additional Practice Have students determine the changes in entropy for the following chemical reactions. Remind students that they must multiply the molar entropy by the number of moles of that substance in the reaction. Assume that the coefficients represent the number of moles involved in the reaction.

1. $2Na(s) + 2HCl(g) \longrightarrow 2NaCl(s) + H_2(g)$ Ans. –201.7 J/K

2. $2C_6H_6(l) + 15O_2(g) \longrightarrow 12CO_2(g) + 6H_2O(l)$ Ans. –437.7 J/K

3. $2Na(s) + 2H_2O(l) \longrightarrow 2NaOH(s) + H_2(g)$ Ans. 16.5 J/K

LS Logical

Answers to Practice Problems F

1. –332.2 J/K

2. –166.1 J/K

3. –95 J/K

Chapter Resource File

• **Problem Solving**

REAL-WORLD CONNECTION

Some packaged meals that are used by the military, campers, hunters, and others are called Meals Ready to Eat, or MREs. MREs come in plastic containers and are fully cooked. The meals can be eaten cold but can also be heated by a flameless ration heater that uses an exothermic reaction. In about 12 minutes, the reaction in the ration heater releases enough energy to warm the MRE to about 60°C.

Discussion

Students may better understand the connection between Gibbs energy and spontaneity by discussing the following idea.

Processes, including physical and chemical changes, are likely to be spontaneous if they increase the entropy of the universe. This is one result of the second law of thermodynamics. The universe consists of a system in which the process is taking place and the system's surroundings. The universe increases in entropy whenever work is done on it. To do work on the universe, a process must be able to supply energy to its surroundings. Gibbs energy is the usable energy available from a process. If a process releases energy, the sign of Gibbs energy is negative and the process can do work on the surroundings. Therefore, the process is spontaneous.

Activity ———— BASIC

Graphic Organizer Have students create a graphic organizer to summarize the thermodynamic quantities that they have learned about in this chapter. The graphic organizer should be in the form of a table and the headings for the columns should be quantity, symbol, units, meaning, and description. The quantities they should include in the table are molar heat capacity, temperature, enthalpy, entropy, and Gibbs energy. **LS** Visual

Gibbs energy

the energy in a system that is available for work

Figure 13
An avalanche is a spontaneous process driven by an increase in disorder and a decrease in energy.

Gibbs Energy

You have learned that the tendency for a reaction to occur depends on both ΔH and ΔS. If ΔH is negative and ΔS is positive for a reaction, the reaction will likely occur. If ΔH is positive and ΔS is negative for a reaction, the reaction will *not* occur. How can you predict what will happen if ΔH and ΔS are both positive or both negative?

Josiah Willard Gibbs, a professor at Yale University, answered that question by proposing another thermodynamic quantity, which now bears his name. **Gibbs energy** is represented by the symbol G and is defined by the following equation.

$$G = H - TS$$

Another name for Gibbs energy is *free energy*.

Gibbs Energy Determines Spontaneity

When the term *spontaneous* is used to describe reactions, it has a different meaning than the meaning that we use to describe other events. A spontaneous reaction is one that does occur or is likely to occur without continuous outside assistance, such as input of energy. A nonspontaneous reaction will never occur without assistance. The avalanche shown in **Figure 13** is a good example of a spontaneous process. On mountains during the winter, an avalanche may or may not occur, but it always *can* occur. The return of the snow from the bottom of the mountain to the mountaintop is a nonspontaneous event, because this event will not happen without aid.

A reaction is spontaneous if the Gibbs energy change is negative. If a reaction has a ΔG greater than 0, the reaction is nonspontaneous. If a reaction has a ΔG of exactly zero, the reaction is at equilibrium.

362

Engineers in the 19th century were interested in developing more efficient engines, usually steam engines, to power the mills that turned out mass-produced goods. They sought to maximize the useful (free) energy available from a process so that the machine could do work most efficiently. Initially, thermodynamics served only industrial applications. But as the understanding of matter developed rapidly during the century, scientists found increasing applications for thermodynamics in explaining the energy changes of any chemical or physical processes.

$$2K(s) + 2H_2O(l) \longrightarrow 2K^+(aq) + 2OH^-(aq) + H_2(g)$$

Figure 14
The reaction of potassium metal with water is spontaneous because a negative ΔH and a positive ΔS both contribute to a negative Gibbs energy change.

Entropy and Enthalpy Determine Gibbs Energy

Reactions that have large negative ΔG values often release energy and increase disorder. The vigorous reaction of potassium metal and water shown in **Figure 14** is an example of this type of reaction. The reaction is described by the following equation.

$$2K(s) + 2H_2O(l) \longrightarrow 2K^+(aq) + 2OH^-(aq) + H_2(g)$$
$$\Delta H = -392 \text{ kJ} \quad \Delta S = 0.047 \text{ kJ/K}$$

The change in Gibbs energy for the reaction above is calculated below.

$$\Delta G = \Delta H - T\Delta S =$$
$$-392 \text{ kJ} - (298.15 \text{ K})(0.047 \text{ kJ/K}) = -406 \text{ kJ}$$

Notice that the reaction of potassium and water releases energy and increases disorder. This example and **Sample Problem G** show how to determine ΔG values at 25°C by using ΔH and ΔS data. However, you can calculate ΔG in another way because lists of standard Gibbs energies of formation exist, such as **Table 5**.

The standard Gibbs energy of formation, ΔG_f^0, of a substance is the change in energy that accompanies the formation of 1 mol of the substance from its elements at 298.15 K. These standard Gibbs energies of formation can be used to find the ΔG for any reaction in exactly the same way that ΔH_f^0 data were used to calculate the enthalpy change for any reaction. Hess's law also applies when calculating ΔG.

$$\Delta G_{\text{reaction}} = \Delta G_{\text{products}} - \Delta G_{\text{reactants}}$$

Table 5   Standard Gibbs Energies of Formation

Substance	ΔG_f^o (kJ/mol)
$Ca(s)$	0
$CaCO_3(s)$	−1128.8
$CaO(s)$	−604.0
$CaCl_2(s)$	−748.1
$CH_4(g)$	−50.7
$CO_2(g)$	−394.4
$CO(g)$	−137.2
$H_2(g)$	0
$H_2O(g)$	−228.6
$H_2O(l)$	−237.2

Refer to Appendix A for more standard Gibbs energies of formation.

363

Using the Figure

Figure 14 shows potassium and hydroxide ions in water. Explain to students that potassium hydroxide is a salt that completely dissociates in water so the potassium ions and hydroxide ions are not really bonded together. Make it clear that the potassium metal has reacted with water, and that potassium ions have displaced the hydrogen in water. The hydrogen that is displaced is evolved as a gas.

Teaching Tip

Gibbs Free Energy Josiah Willard Gibbs (1839–1903) was little known during his lifetime of work at Yale University in Connecticut. Point out to students that Gibbs energy, sometimes called Gibbs free energy, can be viewed as the available, useful energy from a process. When Gibbs energy has a negative value, a system is giving off usable energy and the process taking place is spontaneous.

Teaching Tip

Spontaneous Reactions Emphasize that spontaneous reactions only may occur and often do not. For example, the conversion of diamond to graphite is spontaneous but, for all practical purposes, does not ever occur.

did you know?

Scottish physicist and mathematician William Thomson (a.k.a. Lord Kelvin) contributed many ideas to the study of thermodynamics. However, he was not very good at making predictions about the future. For example, in 1895, Thomson told the Australian Institute of Physics that machines that are heavier than air and fly are impossible.

Teach, continued

Homework —

Additional Practice Have students determine the change in Gibbs energy for the following chemical reactions using the changes in entropy and enthalpy values. Remind students that they must multiply a molar entropy and a molar enthalpy by the number of moles of that substance in the reaction. Assume that the coefficients represent the number of moles involved in the reaction.

1. $2HgO(s) \longrightarrow 2Hg(l) + O_2(g)$ at 25°C Ans. $\Delta G = 111.2$ kJ

2. $CO(g) + H_2O(g) \longrightarrow HCOOH(l)$ at 500°C Ans. $\Delta G = -1013.2$ kJ

3. $4NH_3(g) + 5O_2(g) \longrightarrow 4NO(g) + 6H_2O(g)$ at 25°C Ans. $\Delta G = -1008.7$ kJ

LS Logical

Answers to Practice Problems G

1. $\Delta G = \Delta H - T\Delta S = -76$ kJ $- (298.15 \text{ K})(-0.117 \text{ kJ/K}) = -41$ kJ
Yes, the reaction is spontaneous.

2. $\Delta G = \Delta H - T\Delta S = 11$ kJ $- (298.15 \text{ K})(0.049 \text{ kJ/K}) = -3.6$ kJ
Yes, the reaction is spontaneous.

3. $\Delta G = \Delta H - T\Delta S = 11$ kJ $- (298.15 \text{ K})(0.041 \text{ kJ/K}) = -1.2$ kJ
The reaction is spontaneous.

Chapter Resource File

• Problem Solving

SAMPLE PROBLEM G

Calculating a Change in Gibbs Energy from ΔH and ΔS

Given that the changes in enthalpy and entropy are –139 kJ and 277 J/K respectively for the reaction given below, calculate the change in Gibbs energy. Then, state whether the reaction is spontaneous at 25°C.

$$C_6H_{12}O_6(aq) \longrightarrow 2C_2H_5OH(aq) + 2CO_2(g)$$

This reaction represents the fermentation of glucose into ethanol and carbon dioxide, which occurs in the presence of enzymes provided by yeast cells. This reaction is used in baking.

1 Gather information.

$\Delta H = -139$ kJ
$\Delta S = 277$ J/K
$T = 25°C = (25 + 273.15) \text{ K} = 298 \text{ K}$
$\Delta G = ?$

2 Plan your work.

The equation $\Delta G = \Delta H - T\Delta S$ may be used to find ΔG. If ΔG is positive, the reaction is nonspontaneous. If ΔG is negative, the reaction is spontaneous.

3 Calculate.

$\Delta G = \Delta H - T\Delta S = (-139 \text{ kJ}) - (298 \text{ K})(277 \text{ J/K})$

$= (-139 \text{ kJ}) - (298 \text{ K})(0.277 \text{ kJ/K})$

$= (-139 \text{ kJ}) - (83 \text{ kJ}) = -222 \text{ kJ}$

The negative sign of ΔG shows that the reaction is spontaneous.

4 Verify your results.

The calculation was not necessary to prove the reaction is spontaneous, because each requirement for spontaneity—a negative ΔH and a positive ΔS—was met. In addition, the reaction occurs in nature without a source of energy, so the reaction must be spontaneous.

> **PRACTICE HINT**
>
> Enthalpies and Gibbs energies are generally expressed in kilojoules, but entropies are usually stated in joules (not kilojoules) per kelvin. Remember to divide all entropy values expressed in joules by 1000.

PRACTICE

1 A reaction has a ΔH of –76 kJ and a ΔS of –117 J/K. Is the reaction spontaneous at 298.15 K?

2 A reaction has a ΔH of 11 kJ and a ΔS of 49 J/K. Calculate ΔG at 298.15 K. Is the reaction spontaneous?

3 The gas-phase reaction of H_2 with CO_2 to produce H_2O and CO has a $\Delta H = 11$ kJ and a $\Delta S = 41$ J/K. Is the reaction spontaneous at 298.15 K? What is ΔG?

364

PHYSICS — CONNECTION

The Third Law of Thermodynamics states that it is impossible to reach a temperature of absolute zero. The Third Law was established in 1906 by German physicist and chemist Walther Hermann Nernst. In 1920, Nernst won the Nobel Prize in Chemistry for his work developing this law.

did you know?

Many people have exhibited "perpetual motion" machines and have even attempted to obtain patents for them. In every case, however, examination of the machine has revealed some mechanism for obtaining energy from outside the system.

Calculating a Gibbs Energy Change Using ΔG_f^0 Values

Use **Table 5** to calculate ΔG for the following water-gas reaction.

$$C(s) + H_2O(g) \longrightarrow CO(g) + H_2(g)$$

Is this reaction spontaneous?

1 Gather information.

For $H_2O(g)$, $\Delta G_f^0 = -228.6$ kJ/mol.
For $CO(g)$, $\Delta G_f^0 = -137.2$ kJ/mol.
For $H_2(g)$, $\Delta G_f^0 = 0$ kJ/mol.
For $C(s)$ (graphite), $\Delta G_f^0 = 0$ kJ/mol.

2 Plan your work.

The following simple relation may be used to find the total change in Gibbs energy.

$$\Delta G = \Delta G(\text{products}) - \Delta G(\text{reactants})$$

If ΔG is positive, the reaction is nonspontaneous. If ΔG is negative, the reaction is spontaneous.

3 Calculate.

$\Delta G = \Delta G(\text{products}) - \Delta G(\text{reactants}) =$
$[(\text{mol } CO(g))(\Delta G_f^0 \text{ for } CO(g)) + (\text{mol } H_2(g))(\Delta G_f^0 \text{ for } H_2(g))] -$
$[(\text{mol } C(s))(\Delta G_f^0 \text{ for } C(s)) + (\text{mol } H_2O(g))(\Delta G_f^0 \text{ for } H_2O(g))] =$
$[(1 \text{ mol})(-137.2 \text{ kJ/mol}) + (1 \text{ mol})(0 \text{ kJ/mol})] - [(1 \text{ mol})(0 \text{ kJ/mol}) -$
$(1 \text{ mol})(-228.6 \text{ kJ/mol})] = (-137.2 + 228.6) \text{ kJ} = 91.4 \text{ kJ}$
The reaction is nonspontaneous under standard conditions.

4 Verify your results.

The ΔG_f^0 values in this problem show that water has a Gibbs energy that is 91.4 kJ lower than the Gibbs energy of carbon monoxide. Therefore, the reaction would increase the Gibbs energy by 91.4 kJ. Processes that lead to an increase in Gibbs energy never occur spontaneously.

PRACTICE

1 Use **Table 5** to calculate the Gibbs energy change that accompanies the following reaction.

$$C(s) + O_2(g) \longrightarrow CO_2(g)$$

Is the reaction spontaneous?

2 Use **Table 5** to calculate the Gibbs energy change that accompanies the following reaction.

$$CaCO_3(s) \longrightarrow CaO(s) + CO_2(g)$$

Is the reaction spontaneous?

did you know?

The first person to define temperature as an average of the kinetic energy of particles in a substance was a schoolteacher from Bombay, India by the name of John James Waterston. He submitted a paper outlining his ideas to the Royal Society in 1845, but it was rejected as being "nothing but nonsense." He attempted to publish his work again in 1846 and in 1851, but his work was ignored until 1892 after Joule and Clausius presented similar ideas.

Homework — **GENERAL**

Additional Practice Have students determine the change in Gibbs energy for the following chemical reactions. Remind students that they must multiply a molar Gibbs energy by the number of moles of that substance in the reaction. Assume that the coefficients represent the number of moles involved in the reaction.

1. $2KClO_3(s) \longrightarrow 2KCl(s) + 3O_2(g)$, ΔG_f^0 for $KClO_3(s) = -303.1$ kJ/mol Ans. -212.2 kJ

2. $2AgNO_3(s) + MgCl_2(s) \longrightarrow 2AgCl(s) + Mg(NO_3)_2(s)$, ΔG_f^0 for $Mg(NO_3)_2 = -589.4$ kJ/mol Ans. -150.2 kJ

3. $CH_4(g) + 2O_2(g) \longrightarrow CO_2(g) + 2H_2O(g)$ Ans. -800.8 kJ

LS Logical

Answers to Practice Problems H

1. $\Delta G = [(1 \text{ mol}) (-394.4 \text{ kJ/mol})] - [(0) + (0)] = -394.4$ kJ
Yes, the reaction is spontaneous.

2. $\Delta G = [(1 \text{ mol})(-604.0 \text{ kJ/mol}) + (1 \text{ mol}) (-394.4 \text{ kJ/mol})] - [(1 \text{ mol}) (-1128.8 \text{ kJ/mol})] = 130.4$ kJ
No, the reaction is not spontaneous.

Chapter Resource File

• **Problem Solving**

SKILL BUILDER — ADVANCED

Writing Skills Have interested students research the three laws of thermodynamics. They should learn how the laws were developed, who was involved in their development, and how their development affected chemistry, physics, and engineering. They can also learn how the three laws can be used to prove that a perpetual motion machine cannot exist. Students can use physics textbooks or the Internet to help with their research. The results of their research should be presented in a written report. **LS** Intrapersonal

Teaching Tip

Photosynthesis Process Remind students that the process of photosynthesis is much more complex than its net equation indicates. Students may recall that when light interacts with matter, it usually does so by exciting electrons to higher energy states. This is the role of light in the photo part of photosynthesis. Excited electrons emitted when light strikes chlorophyll molecules are used in one case to reduce water to O_2 gas and H^+ ions and in another case to supply energy for the formation of ATP. These materials are then used in reactions in which CO_2 forms molecules that are assembled into glucose.

SKILL BUILDER — ADVANCED

Writing Skills The study of thermodynamics started during the Industrial Revolution. Much time and money was spent trying to develop a perfectly efficient engine. This work led to the development of the First and Second Laws of Thermodynamics. Encourage interested students to research different types of heat engines, such as external and internal heat engines, the Carnot engine, and Hero's engine, and write a report. Encourage them to include information on the laws of thermodynamics, perpetual motion machines, and entropy.

LS Intrapersonal

Table 6   Relating Enthalpy and Entropy Changes to Spontaneity

ΔH	ΔS	ΔG	Is the reaction spontaneous?
Negative	positive	negative	yes, at all temperatures
Negative	negative	either positive or negative	only if $T < \Delta H/\Delta S$
Positive	positive	either positive or negative	only if $T > \Delta H/\Delta S$
Positive	negative	positive	never

Predicting Spontaneity

Does temperature affect spontaneity? Consider the equation for ΔG.

$$\Delta G = \Delta H - T\Delta S$$

The terms ΔH and ΔS change very little as temperature changes, but the presence of T in the equation for ΔG indicates that temperature may greatly affect ΔG. **Table 6** summarizes the four possible combinations of enthalpy and entropy changes for any chemical reaction. Suppose a reaction has both a positive ΔH value and a positive ΔS value. If the reaction occurs at a low temperature, the value for $T\Delta S$ will be small and will have little impact on the value of ΔG. The value of ΔG will be similar to the value of ΔH and will have a positive value. But when the same reaction proceeds at a high enough temperature, $T\Delta S$ will be larger than ΔH and ΔG will be negative. So, increasing the temperature of a reaction can make a nonspontaneous reaction spontaneous.

Figure 15
Photosynthesis, the nonspontaneous conversion of carbon dioxide and water into carbohydrate and oxygen, is made possible by light energy.

$CO_2(g)$

$H_2O(l)$

$C_6H_{12}O_6(s)$

$O_2(g)$

$$6CO_2(g) + 6H_2O(l) \longrightarrow C_6H_{12}O_6(s) + 6O_2(g)$$

did you know?

The process in which glucose is oxidized to release energy is called *cellular respiration* and requires oxygen.

MISCONCEPTION ALERT

The idea that green plants use energy directly from the sun to power life processes persists. Stress that once green plants manufacture glucose and starch, they carry on cellular respiration just as animals do. Reinforce this idea by pointing out the function of mitochondria in the cell.

Can a nonspontaneous reaction ever occur? A nonspontaneous reaction cannot occur unless some form of energy is added to the system. **Figure 15** shows that the nonspontaneous reaction of photosynthesis occurs with outside assistance. During photosynthesis, light energy from the sun is used to drive the nonspontaneous process. This reaction is described by the equation below.

$$6CO_2(g) + 6H_2O(l) \longrightarrow C_6H_{12}O_6(s) + 6O_2(g) \quad \Delta H = 2870 \text{ kJ/mol}$$

 # Section Review

UNDERSTANDING KEY IDEAS

1. What aspect of a substance contributes to a high or a low entropy?

2. What is diffusion? Give an example.

3. Name three thermodynamic properties, and give the relationship between them.

4. What signs of ΔH, ΔS, and ΔG favor spontaneity?

5. What signs of ΔH, ΔS, and ΔG favor nonspontaneity?

6. How can the Gibbs energy change of a reaction can be calculated?

PRACTICE PROBLEMS

7. The standard entropies for the following substances are 210.8 J/K•mol for $NO(g)$, 240.1 J/K•mol for $NO_2(g)$, and 205.1 J/K•mol for $O_2(g)$. Determine the entropy for the reaction below.

$$2NO(g) + O_2(g) \longrightarrow 2NO_2(g)$$

8. Suppose $X(s) + 2Y_2(g) \longrightarrow XY_4(g)$ has a $\Delta H = -74.8$ kJ and a $\Delta S = -80.8$ J/K. Calculate ΔG for this reaction at 298.15 K.

9. Use **Table 5** to determine whether the reaction below is spontaneous.

$$CaCl_2(s) + H_2O(g) \longrightarrow CaO(s) + 2HCl(g)$$

The standard Gibbs energy of formation for $HCl(g)$ is −95.3 kJ/mol.

10. Calculate the Gibbs energy change for the reaction $2CO(g) \longrightarrow C(s) + CO_2(g)$. Is the reaction spontaneous?

11. Calculate the Gibbs energy change for the reaction $CO(g) \longrightarrow \frac{1}{2}C(s) + \frac{1}{2}CO_2(g)$? How does this result differ from the result in item 10?

CRITICAL THINKING

12. A reaction is endothermic and has a $\Delta H = 8$ kJ. This reaction occurs spontaneously at 25°C. What must be true about the entropy change?

13. You are looking for a method of making chloroform, $CHCl_3(l)$. The standard Gibbs energy of formation for $HCl(g)$ is −95.3 kJ/mol and the standard Gibbs energy of formation for $CHCl_3(l)$ is −73.66 kJ/mol. Use **Table 5** to decide which of the following reactions should be investigated.

$$2C(s) + H_2(g) + 3Cl_2(g) \longrightarrow 2CHCl_3(l)$$

$$C(s) + HCl(g) + Cl_2(g) \longrightarrow CHCl_3(l)$$

$$CH_4(g) + 3Cl_2(g) \longrightarrow CHCl_3(l) + 3HCl(g)$$

$$CO(g) + 3HCl(g) \longrightarrow CHCl_3(l) + H_2O(l)$$

14. If the reaction $X \longrightarrow Y$ is spontaneous, what can be said about the reaction $Y \longrightarrow X$?

15. At equilibrium, what is the relationship between ΔH and ΔS?

16. If both ΔH and ΔS are negative, how does temperature affect spontaneity?

Answers to Section Review

1. disorder

2. the process of dispersion; the gradual, random mixing of the particles of two or more substances; students' answers may vary

3. enthalpy H, entropy S, and Gibbs energy G, $\Delta G = \Delta H - T\Delta S$

4. negative, positive, negative

5. positive, negative, positive

6. either from ΔH and ΔS, or from ΔG_f^0 values of products and reactants

7. −146.5 J/K

8. −50.7 kJ

Answers continued on p. 375A

Reteaching ———— BASIC

Encourage students having difficulty to make an outline of this section or of the entire chapter. This will help organize their thoughts and help them recognize the important topics presented in the section. **LS Logical**

Quiz ———— GENERAL

1. What is entropy? **Ans.** Entropy is a measure of disorder in a system.

2. How is entropy related to temperature? **Ans.** Entropy is directly proportional to temperature. As temperature increases, entropy increases.

3. How can you calculate the entropy change during a reaction? **Ans.** You can use Hess's Law and subtract the entropy of the reactants from the entropy of the products.

4. Which equation should you use to determine if a reaction is spontaneous or not? **Ans.** Gibbs energy equation: $\Delta G = \Delta H - T\Delta S$

5. Describe the two ways you can calculate the change in Gibbs energy during a reaction. **Ans.** You can subtract the Gibbs energy of the reactants from the Gibbs energy of the products or you can use the Gibbs energy equation.

6. A reaction has an enthalpy change that is a positive number and an entropy change that is a negative number. Is the reaction spontaneous or not? Explain. **Ans.** The reaction is nonspontaneous because a positive change in enthalpy and a negative change in entropy will yield a positive value for the change in Gibbs energy.

LS Intrapersonal

Chapter Resource File

• Concept Review

• Quiz

Hydrogen Powered Cars During the past few years several automakers and engineering companies around the world have been attempting to develop affordable hydrogen powered cars and vehicles. Hydrogen-powered vehicles contain hydrogen fuel cells in which hydrogen is combined with oxygen to make water and produce electrical energy. The reaction is essentially the reverse of the electrolysis reaction. NASA has been using hydrogen fuel cells in all of their spacecraft since 1965. Currently, one of the biggest obstacles to overcome before the widespread use of fuel cells can be achieved is the cost of the materials needed to build the cells. Expensive metals such as platinum, niobium, and gold are used to catalyze the reaction. Still, research on hydrogen-powered cars continues.

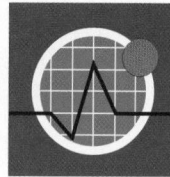

SCIENCE AND TECHNOLOGY

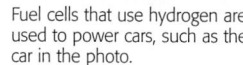

Hydrogen-Powered Cars

Fuel cells that use hydrogen are used to power cars, such as the car in the photo.

Hydrogen As Fuel

When you think of fuel, you probably think of gasoline or nuclear fuel. But did you know that scientists have been studying ways to use the energy generated by the following reaction?

$$H_2(g) + \tfrac{1}{2}O_2(g) \longrightarrow H_2O(l) \quad \Delta H_f^0 = -285.8 \text{ kJ/mol}$$

Engineers use fuel cells that drive an electrochemical reaction, which converts hydrogen or hydrogen-containing materials and oxygen into water, electrical energy, and energy as heat. Fuel cells have already been used by NASA to provide space crews with electrical energy and drinking water. In the future, electrical energy for buildings, ships, submarines, and vehicles may be obtained using the reaction of hydrogen and oxygen to form water.

Cars That Are Powered by Hydrogen Fuel Cells

Many car manufacturers are researching ways to mass produce vehicles that are powered by hydrogen fuel cells. Some hydrogen-powered cars that manufacturers have already developed can reach speeds of over 150 km/h (90 mi/h). These types of cars will also travel 400 to 640 km (250 to 400 mi) before refueling.

These cars have many benefits. Fuel cells have an efficiency of 50 to 60%, which is about twice as efficient as internal combustion engines. These cells are also safe for the environment because they can produce only water as a by product. Unfortunately, fuel cells are expensive because they contain expensive materials, such as platinum.

Questions

1. Research electrical energy and its sources. Which source is the most environmentally safe? Which source is the cheapest? Which source is the most efficient?
2. Research your favorite type of car. How does this car run? How far can this car travel before refueling? What pollutants does this car produce?

CAREER APPLICATION

Engineer

Engineers design, construct, or maintain equipment, buildings, other structures, and transportation. In fact, engineers helped to develop hydrogen-powered cars and the fuel cells. Engineers have also designed and built transportation, such as space shuttles and space stations. And engineers have built structures that you encounter every day, such as your school, your home, and the bridge you cross to get home from school. Most engineers study chemistry in college. There is even a branch of engineering called *chemical engineering*. Some engineers only use computers and paper to create or improve things. Other engineers actually build and maintain equipment or structures. However, the goal of all engineers is to produce items that people use.

internet connect

www.scilinks.org
Topic : Hydrogen
SciLinks code: HW4155

SCILINKS. Maintained by the National Science Teachers Association

368

Answers to Questions

1. Students' answers may vary.

2. Students' answers may vary.

CHAPTER HIGHLIGHTS 10

CHAPTER HIGHLIGHTS 10

KEY TERMS

heat
enthalpy
temperature

thermodynamics

calorimetry
calorimeter
Hess's law

entropy
Gibbs energy

KEY IDEAS

SECTION ONE   Energy Transfer

- Heat is energy transferred from a region at one temperature to a region at a lower temperature.
- Temperature depends on the average kinetic energy of the atoms.
- The molar heat capacity of an element or compound is the energy as heat needed to increase the temperature of 1 mol by 1 K.

SECTION TWO   Using Enthalpy

- The enthalpy of a system can be its total energy.
- When *only* temperature changes, the change in molar enthalpy is represented by $\Delta H = C\Delta T$.

SECTION THREE   Changes in Enthalpy During Reactions

- Calorimetry measures the enthalpy change, which is represented by ΔH, during a chemical reaction.
- Reactions that have positive ΔH are endothermic; reactions that have negative ΔH are exothermic.
- Hess's law indicates that the thermodynamic changes for any particular process are the same, whether the changes are treated as a single reaction or a series of steps.

$$\Delta H_{reaction} = \Delta H_{products} - \Delta H_{reactants}$$

SECTION FOUR   Order and Spontaneity

- The entropy of a system reflects the system's disorder.

$$\Delta S_{reaction} = S_{products} - S_{reactants}$$

- Gibbs energy is defined by $G = H - TS$.
- The sign of ΔG determines spontaneity.

$$\Delta G_{reaction} = \Delta G_{products} - \Delta G_{reactants}$$

KEY SKILLS

Calculating the Molar Heat Capacity of a Sample
Sample Problem A p. 342

Calculating Molar Enthalpy Change for Heating
Sample Problem B p. 346

Calculating the Molar Enthalpy Change for Cooling
Sample Problem C p. 347

Calculating a Standard Enthalpy of Formation
Sample Problem D p. 356

Calculating a Reaction's Change in Enthalpy
Sample Problem E p. 356

Hess's Law and Entropy
Sample Problem F p. 361

Calculating Changes in Gibbs Energy
Sample Problem G p. 364
Sample Problem H p. 365

Alternative Assessments

HISTORY CONNECTION
- Page 343

Group Activity
- Pages 352, 354

Portfolio Assessments

SKILL BUILDER
- Pages 351, 365, 366

Chapter Resource File

- **Datasheets for Intext Labs** Calorimetry and Hess's Law
- **CBL™ Probeware** Energy Content in Foods

REVIEW ANSWERS

1. temperature
2. Heat is the energy transferred between two objects that are at different temperatures.
3. calorimeter
4. negative
5. entropy and enthalpy
6. A spontaneous reaction is one that does occur or is likely to occur without continuous assistance, such as energy.
7. Entropy is a measure of the randomness or disorder of a system.
8. Gibbs energy is a thermodynamic property, which indicates the spontaneity of a chemical reaction.
9. When two samples at different temperatures are in contact, energy can be transferred as heat.
10. Lacking negative values, it closely corresponds to the kinetic energies of the atoms.
11. Energy transfer is always from higher temperature to lower temperature.
12. Molar heat capacity is the energy as heat needed to increase the temperature of 1 mol of pure substance by 1 K. By measuring the energy input and the temperature change during a process.
13. the enthalpy change of one mole of an element or compound
14. molar heat capacity and ΔT
15. heating, cooling, change of state, chemical reaction
16. the enthalpy change when one mole of a compound is formed from its elements in their standard conditions at 298.15 K

USING KEY TERMS

1. What is dependent on the average kinetic energy of the atoms in a substance?
2. Define *heat*.
3. Name a device used for measuring enthalpy changes.
4. Do most exothermic reactions have a negative or positive change in Gibbs energy?
5. What quantities do the Gibbs energy of a reaction depend on?
6. What is a spontaneous reaction?
7. What is entropy?
8. Define *Gibbs energy*, and explain its usefulness.

UNDERSTANDING KEY IDEAS

Energy Transfer

9. Distinguish between heat and temperature.
10. What advantage does the Kelvin temperature scale have?
11. How can you tell which one of two samples will release energy in the form of heat when the two samples are in contact?
12. What is molar heat capacity, and how can it be measured?

Using Enthalpy

13. What is molar enthalpy change?
14. What influences the changes in molar enthalpy?

15. Name two processes for which you could determine an enthalpy change.

Changes in Enthalpy During Reactions

16. Define *standard molar enthalpy of formation*.
17. Explain the meanings of H, ΔH, and ΔH_f^0.
18. State Hess's law. How is it used?
19. Describe a bomb calorimeter.
20. Which thermodynamic property of a food is of interest to nutritionists? Why?
21. What is adiabatic calorimetry?
22. What is the value for the standard molar enthalpy of formation for an element?

Order and Spontaneity

23. Why is entropy described as an extensive property?
24. Explain why Josiah Willard Gibbs invented a new thermodynamic quantity.
25. Explain how a comprehensive table of standard Gibbs energies of formation can be used to determine the spontaneity of any chemical reaction.
26. What information is needed to be certain that a chemical reaction is nonspontaneous?

PRACTICE PROBLEMS

Sample Problem A Calculating the Molar Heat Capacity of a Substance

27. You need 70.2 J to raise the temperature of 34.0 g of ammonia, $NH_3(g)$, from 23.0°C to 24.0°C. Calculate the molar heat capacity of ammonia.

370

Assignment Guide

Section	Questions
1	1, 2, 9–12, 27, 28, 48, 56, 58, 64
2	3, 13–15, 29–32, 46
3	16–22, 33–36, 43–45, 50, 59, 60
4	4–8, 23–26, 37–42, 47, 49, 51–55, 57, 61, 62, 63

28. Calculate C for indium metal given that 1.0 mol In absorbs 53 J during the following process.

$$In(s, 297.5\ K) \longrightarrow In(s, 299.5\ K)$$

Sample Problem B Calculating the Molar Enthalpy Change for Heating

29. Calculate ΔH when 1.0 mol of nitrogen is heated from 233 K to 475 K.

30. What is the change in enthalpy when 11.0 g of liquid mercury is heated by 15°C?

Sample Problem C Calculating the Molar Enthalpy Change for Cooling

31. Calculate ΔH when 1.0 mol of argon is cooled from 475 K to 233 K.

32. What enthalpy change occurs when 112.0 g of barium chloride experiences a change of temperature from 15°C to −30°C?

Sample Problem D Calculating a Standard Enthalpy of Formation

33. The diagram below represents an interpretation of Hess's law for the following reaction.

$$Sn(s) + 2Cl_2(g) \longrightarrow SnCl_4(l)$$

Use the diagram to determine ΔH for each step and the net reaction.

$$Sn(s) + Cl_2(g) \longrightarrow SnCl_2(l) \qquad \Delta H = ?$$

$$SnCl_2(s) + Cl_2(g) \longrightarrow SnCl_4(l) \qquad \Delta H = ?$$

$$Sn(s) + 2Cl_2(g) \longrightarrow SnCl_4(l) \qquad \Delta H = ?$$

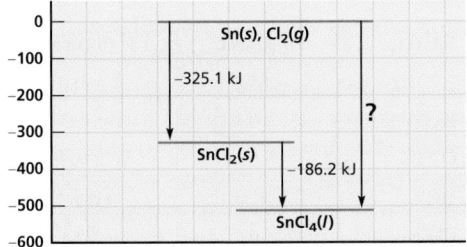

Sample Problem E Calculating a Reaction's Change in Enthalpy

34. Use tabulated values of standard enthalpies of formation to calculate the enthalpy change accompanying the reaction $4Al(s) + 6H_2O(l) \longrightarrow 2Al_2O_3(s) + 6H_2(g)$. Is the reaction exothermic?

35. The reaction $2Fe_2O_3(s) + 3C(s) \longrightarrow 4Fe(s) + 3CO_2(g)$ is involved in the smelting of iron. Use ΔH_f^0 values to calculate the enthalpy change during the production of 1 mol of iron.

36. For glucose, $\Delta H_f^0 = -1263$ kJ/mol. Calculate the enthalpy change when 1 mol of $C_6H_{12}O_6(s)$ combusts to form $CO_2(g)$ and $H_2O(l)$.

Sample Problem F Hess's Law and Entropy

37. Given the entropy change for the first two reactions below, calculate the entropy change for the third reaction below.

$$S_8(s) + 8O_2(g) \longrightarrow 8SO_2(g)\ \Delta S = 89\ J/K$$

$$2SO_2(s) + O_2(g) \longrightarrow 2SO_3(g)\ \Delta S = -188\ J/K$$

$$S_8(s) + 12O_2(g) \longrightarrow 8SO_3(g)\ \Delta S = ?$$

38. The standard entropies for the following substances are 26.9 J/K•mol for MgO(s), 213.8 J/K•mol for $CO_2(g)$, and 65.7 J/K•mol for $MgCO_3(s)$. Determine the entropy for the reaction below.

$$MgCO_3(s) \longrightarrow MgO(s) + CO_2(g)$$

Sample Problem G, Sample Problem H Calculating Changes in Gibbs Energy

39. A reaction has $\Delta H = -356$ kJ and $\Delta S = -36$ J/K. Calculate ΔG at 25°C to confirm that the reaction is spontaneous.

40. A reaction has $\Delta H = 98$ kJ and $\Delta S = 292$ J/K. Investigate the spontaneity of the reaction at room temperature. Would increasing the temperature have any effect on the spontaneity of the reaction?

17. H is the enthalpy. ΔH is the change in enthalpy. ΔH_f^0 is the enthalpy change in forming one mole of a compound from its elements at 298.15 K.

18. The change in enthalpy (or entropy or Gibbs energy) during a process is independent of the route taken. It is useful, for example, in calculating the enthalpy change during a reaction from the molar enthalpies of formation of the reactants and products.

19. It is a sturdy vessel inside which a substance can react with oxygen and the energy as heat generated passes to a surrounding water jacket.

20. its enthalpy of combustion; because that determines the caloric energy of the food

21. Carried out in an insulated vessel, this calorimetric technique allows an exothermic reaction to increase the temperature of the products.

22. 0

23. Entropy of a substance depends on the amount of substance.

24. Prior to the invention, there were two (possibly competing) factors that were responsible for determining the spontaneity of a process.

25. through the equation $\Delta G = \Delta G$ (products) $- \Delta G$ (reactants), which has its origin in Hess's law; If ΔG is negative, the reaction is spontaneous.

26. that ΔG be positive

27. $n = (34.0\ g)/(17.03\ g/mol) = 2.00$ mol
$C = q/n\Delta T = (70.2\ J)/[(2.00\ mol)(1.0\ K)] = 35.2\ J/K$•mol

28. $C = q/n\Delta T = (53\ J)/[(1\ mol)(2.0\ K)] = 26\ J/K$•mol

29. $\Delta H = nC\Delta T = (1.0\ mol)(29.1\ J/K$•mol$)(242\ K) = 7040\ J$

30. $n = (11g)/(200.59\ g/mol) = 0.055$ mol
$\Delta H = nC\Delta T = (0.055\ mol)(27.8\ J/K$•mol$)(15\ K) = 23\ J$

31. $\Delta H = nC\Delta T = (1.0\ mol)(20.8\ J/K$•mol$)(-242\ K) = -5030\ J$

32. $n = (112.0\ g)/(208.32\ g/mol) = 0.54$ mol

$\Delta H = nC\Delta T = (0.54 \text{ mol})$
$(75.1 \text{ J/K} \cdot \text{mol})(-45 \text{ K}) = -1800 \text{ J}$

33. $\Delta H = -325.1 \text{ kJ}, \Delta H = -186.2 \text{ kJ},$
$\Delta H = -511.3 \text{ kJ}$

34. $[(2 \text{ mol})(-1676.0 \text{ kJ/mol}) +$
$(6 \text{ mol})(0)] - [(4 \text{ mol})(0) +$
$(6 \text{ mol})(-285.8 \text{ kJ/mol})] =$
-1637 kJ

Yes, this reaction is exothermic.

35. $[4 \text{ mol}(0) + 3 \text{ mol}(-393.5 \text{ kJ/mol})] -$
$[2 \text{ mol}(-825.5 \text{ kJ/mol}) +$
$3 \text{ mol}(0)] = 470.5 \text{ kJ}$

This is the enthalpy change in manufacturing 4 moles of iron. For each mole produced, $\Delta H = 117.6 \text{ J}$.

36. $C_6H_{12}O_6(s) + 6O_2(g) \longrightarrow$
$6CO_2(g) + 6H_2O(l) =$
$[6(-393.5 \text{ kJ}) + 6(-285.8 \text{ kJ})] -$
$[(-1263 \text{ kJ}) + 6(0)] =$
$-2812.8 \text{ kJ} = -2813 \text{ kJ}$

37. $\Delta S = (89 \text{ J/K}) + 4(-188 \text{ J/K}) =$
-663 J/K

38. $26.9 \text{ J/K} + 213.8 \text{ J/K} - 65.7 \text{ J/K} =$
175 J/K

39. $\Delta G = \Delta H - T\Delta S = (-356 \text{ kJ}) -$
$(298 \text{ K})(-0.036 \text{ kJ/K}) = -345 \text{ kJ}$
The reaction is spontaneous.

40. $\Delta G = \Delta H - T\Delta S = (98 \text{ kJ}) -$
$(298 \text{ K})(-0.292 \text{ kJ/K}) = 11 \text{ kJ}$
The reaction is not spontaneous at 25°C but it will become so at higher temperature (greater than about 63°C)

41. $[(-50.7 \text{ kJ}) + 3(-394.4 \text{ kJ})] -$
$[4(-137.2 \text{ kJ}) + 2(-228.6 \text{ kJ})] =$
-227.9 kJ
The reaction is spontaneous.

42. $[2(-915 \text{ kJ})] - [(-1551 \text{ kJ}) +$
$(-237 \text{ kJ})] = -42 \text{ kJ}$
The reaction is likely to occur.

43. Each term used in the right-hand side of the equation $\Delta H = \Delta H(\text{products}) - \Delta H(\text{reactants})$ must be multiplied by the corresponding coefficient.

41. Use tabulated values of standard Gibbs energies to calculate ΔG for the following reaction.

$$4CO(g) + 2H_2O(g) \longrightarrow CH_4(g) + 3CO_2(g)$$

Is the reaction spontaneous under standard conditions?

42. The sugars glucose, $C_6H_{12}O_6(aq)$, and sucrose, $C_{12}H_{22}O_{11}(aq)$, have ΔG_f^0 values of -915 kJ and -1551 kJ respectively. Is the hydrolysis reaction, $C_{12}H_{22}O_{11}(aq) + H_2O(l) \longrightarrow 2C_6H_{12}O_6(aq)$, likely to occur?

MIXED REVIEW

43. How are the coefficients in a chemical equation used to determine the change in a thermodynamic property during a chemical reaction?

44. Is the following reaction exothermic? The standard enthalpy of formation for $CH_2O(g)$ is approximately -109 kJ/mol.

$$CH_2O(g) + CO_2(g) \longrightarrow H_2O(g) + 2CO(g)$$

45. Nitrogen dioxide, $NO_2(g)$, has a standard enthalpy of formation of 33.1 kJ/mol. Dinitrogen tetroxide, $N_2O_4(g)$, has a standard enthalpy of formation of 9.1 kJ/mol. What is the enthalpy change for synthesizing 1.0 mol of N_2O_4 from NO_2?

46. If the molar heat capacity and the temperature change of a substance are known, why is the mass of a substance not needed to calculate its molar enthalpy change?

47. Predict whether ΔS is positive or negative for the following reaction.

$$Ag^+(aq) + Cl^-(aq) \longrightarrow AgCl(s)$$

48. Explain why $AlCl_3$ has a molar heat capacity that is approximately four times the molar heat capacity of a metallic crystal.

49. At high temperatures, does enthalpy or entropy have a greater effect on a reaction's Gibbs energy?

50. Calculate the enthalpy of formation for sulfur dioxide, SO_2, from its elements, sulfur and oxygen. Use the balanced chemical equation and the following information.

$S(s) + \frac{3}{2}O_2(g) \longrightarrow SO_3(g)$ $\Delta H = -395.8 \text{ kJ/mol}$
$2SO_2(g) + O_2(g) \longrightarrow 2SO_3(g)$ $\Delta H = -198.2 \text{ kJ/mol}$

51. For 1 mol of $HCl(g)$, the $\Delta G_f^0 = -92.307$ kJ/mol. Use this information to calculate ΔG for the following reaction.

$$H_2(g) + Cl_2(g) \longrightarrow 2HCl(g)$$

Is the reaction spontaneous?

52. The photosynthesis reaction is shown below.

$$6CO_2(g) + 6H_2O(l) \longrightarrow C_6H_{12}O_6(s) + 6O_2(g)$$

Calculate ΔG using the following information for the reaction.
$\Delta H = 2870 \text{ kJ}$ $\Delta S = 259 \text{ J/K}$ $T = 297 \text{ K}$

53. Using the following values, compute the ΔG value for each reaction and predict whether they will occur spontaneously.

Reaction	ΔH (kJ)	Temperature	ΔS (J/K)
1	+125	293 K	+35
2	−85.2	127 K	+125
3	−275	500°C	+45

54. Hydrogen gas can be prepared for use in cars in several ways, such as by the decomposition of water or hydrogen chloride.

$$2H_2O(l) \longrightarrow 2H_2(g) + O_2(g)$$
$$2HCl(g) \longrightarrow H_2(g) + Cl_2(g)$$

Use the following data to determine whether these reactions can occur spontaneously at 25°C. Assume that ΔH and ΔS are constant.

Substance	H_f^0 (kJ/mol)	S^0 (J/K·mol)
$H_2O(l)$	−285.8	+70.0
$H_2(g)$	0	+130.7
$O_2(g)$	0	+205.1
$HCl(g)$	−95.3	+186.9
$Cl_2(g)$	0	+223.1

372

55. Find the change in entropy for the reaction below.

$$NH_4NO_3(s) \longrightarrow NH_4NO_3(aq, 1m)$$

The standard entropy for $NH_4NO_3(s)$ is 151.1 J/K•mol and the standard entropy for $NH_4NO_3(aq, 1m)$ is 259.8 J/K•mol.

CRITICAL THINKING

56. Why are the specific heats of $F_2(g)$ and $Br_2(g)$ very different, whereas their molar heat capacities are very similar?

57. Look at the two pictures below this question. Which picture appears to have more order? Why? Are there any similarities between the order of marbles and the entropy of particles?

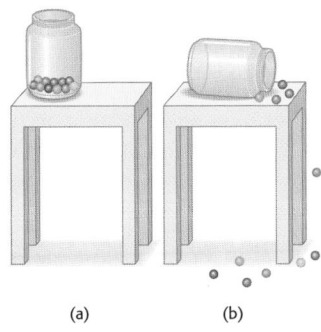

(a)   (b)

58. In what way is specific heat a more useful quantity than molar heat capacity? In what way is molar heat capacity a more useful quantity than specific heat?

59. Why must nutritionists make corrections to bomb calorimetric data if a food contains cellulose or other indigestible fibers?

60. Why are most entries in **Table 2** negative?

61. Why are most entries in **Table 4** negative?

62. Give examples of situations in which **(a)** the entropy is low; **(b)** the entropy is high.

63. Under what circumstances might a nonspontaneous reaction occur?

64. What does ΔT represent?

ALTERNATIVE ASSESSMENT

65. Design an experiment to measure the molar heat capacities of zinc and copper. If your teacher approves the design, obtain the materials needed and conduct the experiment. When you are finished, compare your experimental values with those from a chemical handbook or other reference source.

66. Develop a procedure to measure the ΔH of the following reaction. If your teacher approves your procedure, test your procedure by measuring the ΔH value of the reaction. Determine the accuracy of your method by comparing your ΔH with the accepted ΔH value.

$$CH_3COONa(s) \longrightarrow Na^+(aq) + CH_3COO^-(aq)$$

CONCEPT MAPPING

67. Use the following terms to complete the concept map below: *calorimeter*, *enthalpy*, *entropy*, *Gibbs energy*, and *Hess's Law*

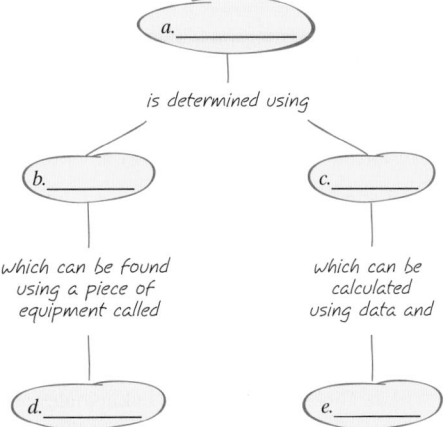

44. [(−242 kJ) + 2(−110 kJ)] − [(−109 kJ) + (−393 kJ)] = 40 kJ No, the reaction is not exothermic.

45. $2NO_2(g) \longrightarrow N_2O_4(g)$ [9.1 kJ] − [2(33.1 kJ)] = −57 kJ

46. because the calculation is being performed in molar, not mass, units

47. negative

48. $AlCl_3$ has a molar heat capacity approximately four times the molar heat capacity of a metal because it has 4 times the number of atoms as atoms in 1 mol of metal.

49. As T increases, the final term in $\Delta G = \Delta H - T\Delta S$ becomes dominant.

50. −296.7 kJ/mol

51. −184.614 kJ; yes

52. 2793 kJ/ mol

53. Reaction 1: ΔG = 115 kJ, not spontaneous; Reaction 2: ΔG = −101 kJ, spontaneous; Reaction 3: ΔG = −310 kJ, spontaneous

54. The reaction using H_2O is not spontaneous at 25°C because ΔG = 474 kJ. The reaction using HCl is likewise not spontaneous at 25°C because ΔG = 196 kJ. It would be spontaneous at temperatures above 830°C.

55. 259.8 J/K − 151.1 J/K = 108.7 J/K

56. One mole each of F_2 and Br_2 contain the same number of atoms. One-gram samples of F_2 and Br_2 contain different numbers of atoms.

57. a, because the marbles are ordered in the jar

The disorder of the marbles increases when the marbles become more random, the entropy of particles increases when the particles become more random.

58. c is useful because it can be applied to mixtures (air, cement). C is useful because it relates more closely to chemistry.

59. Such fibers will combust in a bomb calorimeter, but are indigestible by humans.

60. because most synthesis reactions are exothermic

61. because most synthesis reactions are spontaneous

62. **a.** in a solid at low temperature

    b. in a gas at high temperature and low pressure

63. when energy (not as heat) is supplied from outside, as in photosynthesis or electrolysis

64. The change in temperature, $T_{final} - T_{initial}$

65. Students' answers will vary but should include measuring the mass of samples, heating them, and determining how much they increase the temperature of a known mass of water in a calorimeter. Be sure students use safe procedures and available materials. A calorimeter can easily be made from a foam cup.

66. Answers will vary but should include dissolving a known mass of sodium acetate in a known mass of water and measuring the resulting temperature change. Be sure students follow proper safety procedures, wear proper personal protective equipment, and follow chemical safety guidelines.

67. **a.** Gibbs energy

    b. enthalpy

    c. entropy

    d. calorimeter

    e. Hess's law

68. the slope would be the negative of the slope of the graph regarding the change in water temperature on heating

69. no change in the temperature

Answers continued on p. 375B

FOCUS ON GRAPHING

Study the graph below, and answer the questions that follow.
For help in interpreting graphs, see Appendix B, "Study Skills for Chemistry."

68. How would the slope differ if you were to cool the water at the same rate that graph shows the water was heated?

69. What would a slope of zero indicate about the temperature of water during heating?

70. Calculate the slope given the following data.

    $y_2 = 3.3$ K $x_2 = 50$ s

    $y_1 = 5.6$ K $x_1 = 30$ s

71. Calculate the slope given the following data.

    $y_2 = 63.7$ mL $x_2 = 5$ s

    $y_1 = 43.5$ mL $x_1 = 2$ s

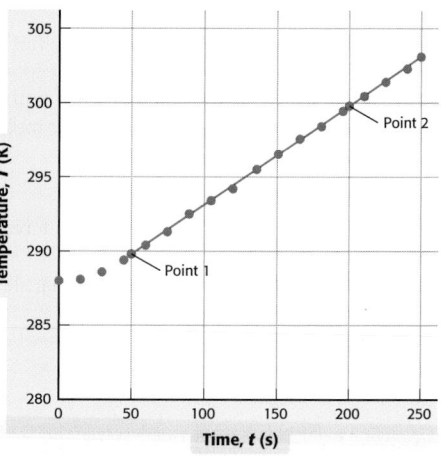

Change in Water Temperature on Heating

TECHNOLOGY AND LEARNING

72. Graphing Calculator

Calculating the Gibbs-Energy Change

The graphing calculator can run a program that calculates the Gibbs-energy change, given the temperature, T, change in enthalpy, ΔH, and change in entropy, ΔS. Given that the temperature is 298 K, the change in enthalpy is 131.3 kJ/mol, and the change in entropy is 0.134 kJ/(mol·K), you can calculate Gibbs-energy change in kilojoules per mole. Then use the program to make calculations.

Go to Appendix C. If you are using a TI-83 Plus, you can download the program **ENERGY** data and run the application as directed. If you are using another calculator, your teacher will provide you with keystrokes and data sets to use. After you have run the program, answer the following questions.

a. What is the Gibbs-energy change given a temperature of 300 K, a change in enthalpy of 132 kJ/mol, and a change in entropy of 0.086 kJ/(mol·K)?

b. What is the Gibbs-energy change given a temperature of 288 K, a change in enthalpy of 115 kJ/mol, and a change in entropy of 0.113 kJ/(mol·K)?

c. What is the Gibbs-energy change given a temperature of 298 K, a change in enthalpy of 181 kJ/mol, and a change in entropy of 0.135 kJ/(mol·K)?

Chapter Resource File

• Chapter Tests

Answers

1. c
2. c
3. d
4. c
5. b
6. b
7. c
8. a
9. d
10. b
11. b
12. d

1. An adiabatic calorimeter can be used to determine the _____ of a reaction.
   a. ΔG
   b. ΔS
   c. ΔH
   d. molar Gibbs energy

2. The temperature of a substance is a measurement of the
   a. total kinetic energy of its atoms.
   b. entropy of the substance.
   c. average kinetic energy of its atoms.
   d. molar heat capacity of the substance.

3. A reaction has a Gibbs energy of −92 kJ. The reaction
   a. is nonspontaneous.
   b. does not have entropy.
   c. does not have free energy.
   d. is spontaneous.

4. Which of the following is not a form of energy?
   a. heat
   b. enthalpy
   c. entropy
   d. kinetic energy

5. Which of the following terms best describes the process of heating a substance?
   a. enthalpy
   b. endothermic
   c. exothermic
   d. standard entropy

6. When carrying out thermodynamic calculations, always
   a. convert joules to calories.
   b. convert temperatures that are in Celsius to temperatures that are in kelvins.
   c. determine the ΔH value first.
   d. convert moles to grams.

7. The following equation
   $2H_2(g) + O_2(g) \longrightarrow 2H_2O(s)\ \Delta H = -560\ kJ$
   reveals that the standard molar enthalpy of formation of solid water is
   a. −560 kJ
   b. 560 kJ
   c. −280 kJ
   d. 560 J/K

8. Which of the following two conditions will favor a spontaneous reaction?
   a. an increase in entropy and a decrease in enthalpy
   b. an increase in entropy and an increase in enthalpy
   c. a decrease in entropy and a decrease in enthalpy
   d. a decrease in entropy and an increase in enthalpy

9. Photosynthesis is a nonspontaneous reaction, as reflected by
   a. $\Delta H > 0$.
   b. $\Delta S < 0$.
   c. $\Delta G < 0$.
   d. $\Delta G > 0$.

10. The gasification of coal is a method of producing methane by the following reaction.
    $C(s) + 2H_2(g) \longrightarrow CH_4(g)\quad \Delta H = ?$
    Find ΔH by using the enthalpy changes in the following combustion reactions
    $C(s) + O_2(g) \longrightarrow CO_2(g)\quad \Delta H = -393\ kJ$
    $2H_2(g) + O_2(g) \longrightarrow 2H_2O(l)\quad \Delta H = -572\ kJ$
    $CH_4(g) + 2O_2(g) \longrightarrow$
    $\qquad\qquad CO_2(g) + 2H_2O(l)\quad \Delta H = -891\ kJ$
    a. 74 kJ
    b. −74 kJ
    c. 1856 kJ
    d. −1856 kJ

11. Which of the following is negative in an exothermic reaction?
    a. ΔT
    b. ΔH
    c. ΔS
    d. ΔG

12. Decreasing the temperature of a nonspontaneous reaction may turn the nonspontaneous reaction into a spontaneous reaction if
    a. $\Delta H > 0$ and $\Delta S > 0$.
    b. $\Delta H > 0$ and $\Delta S < 0$.
    c. $\Delta H < 0$ and $\Delta S > 0$.
    d. $\Delta H < 0$ and $\Delta S < 0$.

375

Continuation of Answers

Answers to Section 1 Review, *continued*

5. Molar heat capacity is the heat required to increase the temperature of 1 mol of a substance by 1 K.

6. Specific heat is the heat required to raise the temperature of one gram of a substance by one kelvin.

7. There are no negative temperatures on the kelvin scale. (Unlike temperatures on the other two scales, kelvin temperatures directly reflect the average kinetic energies of the atoms.)

8. $C = q/(n\Delta T) = (63 \text{ J})/[(0.10 \text{ mol})(100 \text{ K})] = 6.3 \text{ J/K} \cdot \text{mol}$

9. $q = nC\Delta T = [(260.5 \text{ g})/(26.98 \text{ g/mol})](24.2 \text{ J/K} \cdot \text{mol})(125 \text{ K}) = 29207 \text{ J} = 29.2 \text{ kJ}$

10. $q = nC\Delta T = [(260.5 \text{ g})/(55.85 \text{ g/mol})](25.1 \text{ J/K} \cdot \text{mol})(125 \text{ K}) = 14636 \text{ J} = 14.6 \text{ kJ}$

11. 0.52 mol

12. 358 K

13. 295 K

14. $c = C/M = (25.3 \text{ J/K} \cdot \text{mol})/(107.87 \text{ g/mol}) = 0.234 \text{ J/K} \cdot \text{g}$

15. 0.864 J/K·g

16. The temperature *interval* is the same on both scales.

17. Using the "about 25 J/K·mol for each atom in a solid" rule, C values of 50 J/K·mol and 75 J/K·mol, respectively, would be predicted.

18. 100 J/K·mol

19. 0.6 J/K·g

Answers to Section 2 Review, *continued*

8. $\Delta H = -186.9 \text{ J}$

   $\Delta T = \Delta H/C = (-186.9 \text{ J/mol})/(172 \text{ J/K} \cdot \text{mol}) = -1.09 \text{ K}$

9. $C = \Delta H/n\Delta T = (-428 \text{ J})/[(1 \text{ mol})(-10.0 \text{ K})] = 42.8 \text{ J/K} \cdot \text{mol}$

10. if a phase change, such as melting, occurs

11. one kilojoule

Answers to Section 3 Review, *continued*

5. $\Delta H = \Delta H_f^0(\text{products}) - \Delta H_f^0(\text{reactants}) =$
   $[(-1676.0 \text{ kJ}) + 3(0)] - [2(0) + 3(-285.8 \text{ kJ})] = -818.6 \text{ kJ}$

6. the molar enthalpy of vaporization of water

7. that exothermic reactions are more likely to occur than endothermic reactions

Answers to Section 4 Review, *continued*

9. 182.1 kJ; No, the reaction is nonspontaneous.

10. −120 kJ; Yes, the reaction is spontaneous.

11. −60 kJ; The result is half the result of problem 10.

12. It must be positive, and of a value exceeding about 30 J/K.

13. $2C(s) + H_2(g) + 3Cl_2(g) \longrightarrow 2CHCl_3(l)$
    $\Delta G = \Delta G(\text{products}) - \Delta G(\text{reactants}) =$
    $[2(-73.66 \text{ kJ})] - [2(0) + 0 + 3(0)] = -147.32 \text{ kJ}$
    $C(s) + HCl(g) + Cl_2(g) \longrightarrow CHCl_3(l)$
    $\Delta G = \Delta G(\text{products}) - \Delta G(\text{reactants}) =$
    $[1(-73.66 \text{ kJ})] - [0 + (-95.3 \text{ kJ}) + (0)] = 21.6 \text{ kJ}$
    $CH_4(g) + 3Cl_2(g) \longrightarrow CHCl_3(l) + 3HCl(g)$
    $\Delta G = \Delta G(\text{products}) - \Delta G(\text{reactants}) =$
    $[(-73.66 \text{ kJ}) + 3(-95.3 \text{ kJ})] - [(-50.7 \text{ kJ}) + 3(0)] = -308.9 \text{ kJ}$
    $CO(g) + 3HCl(g) \longrightarrow CHCl_3(l) + H_2O(l)$
    $\Delta G = \Delta G(\text{products}) - \Delta G(\text{reactants}) =$
    $[(-73.66 \text{ kJ}) + (-237.2 \text{ kJ})] - [(-137.2 \text{ kJ}) + 3(-95.3 \text{ kJ})] = 112.2 \text{ kJ}$
    Only the first and the third are spontaneous; the other two cannot possibly occur under standard conditions.

14. It is nonspontaneous.

15. $\Delta H = T\Delta S$

16. The reaction will be spontaneous at a sufficiently low temperature; nonspontaneous at a sufficiently high temperature.

Answers to Chapter Review, *continued*

70. −0.115 K/s = −0.11 K/s

71. 6.73 mL/s = 7 mL/s

72. a. 106.200 kJ/mol
    b. 82.456 kJ/mol
    c. 140.770 kJ/mol

States of Matter and Intermolecular Forces
Chapter Planning Guide

PACING	CLASSROOM RESOURCES	LABS, ACTIVITIES, AND DEMONSTRATIONS
BLOCK 1 · 45 min pp. 376–377 **Chapter Opener**		**SE Start-Up Activity** Heating Curve for Water, p. 377 ◆
BLOCK 2 · 45 min pp. 378–384 **Section 1** States and State Changes	**TT Mercury in Three States** * **TT Changes of State** *	**TE Demonstration** A Homemade Compass, p. 378 ◆ **TE Group Activity** Crystal Systems, p. 379 `ADVANCED` **TE Activity** Application of Capillary Action, p. 379 `BASIC` **SE QuickLab** Wetting a Surface, p. 380 ◆ `GENERAL` **CRF Datasheets for In-Text Labs** * **TE Activity**, p. 380 `BASIC` **TE Demonstration** Comparing the Meniscus in H_2O and Hg, p. 381 ◆ **CRF Observation Lab** Viscosity of Liquids ◆ `GENERAL`
BLOCKS 3 & 4 · 90 min pp. 385–392 **Section 2** Intermolecular Forces	**TT Ice and Water** * **TT Temporary Dipoles** *	**TE Demonstration** Polar Versus Nonpolar, p. 386 ◆ `ADVANCED` **TE Group Activity**, p. 389 `GENERAL` **TE Demonstration** Temporary Dipole Model, p. 390 **CRF CBL™ Probeware Lab** Evaporation and Intermolecular Attractions `ADVANCED`
BLOCKS 5 & 6 · 90 min pp. 393–398 **Section 3** Energy of State Changes		**TE Demonstration**, p. 393 **TE Demonstration** Melting Metal, p. 394 ◆ `ADVANCED` **TE Demonstration** Paper That Won't Burn, p. 395 ◆ **TE Demonstration** Synthesis of CuS, p. 396 ◆ **CRF Observation Lab** Constructing a Heating/Cooling Curve `GENERAL`
BLOCK 7 · 45 min pp. 399–406 **Section 4** Phase Equilibrium	**OSP Career Extension** Occupational Applications Worksheet 7: Pest-Control Technician `GENERAL` **TT Energy Distribution of Gas Molecules at Different Temperatures** * **TT Phase Diagram for H_2O** * **TT Phase Diagram for CO_2** * **OSP Career Extension** Occupational Applications Worksheet 9: Air-Quality Specialist `GENERAL`	**TE Demonstration** Boiling Water in a Vacuum, p. 401 ◆ `ADVANCED` **TE Group Activity**, p. 403 `BASIC` **TE Activity**, p. 403 `GENERAL`

BLOCKS 8 & 9 · 90 min **Chapter Review and Assessment Resources**

SE Chapter Review, pp. 408–412
SE Standardized Test Prep, p. 413
CRF Chapter Test *
OSP Test Generator
CRF Test Item Listing *
OSP Scoring Rubrics and Classroom Management Checklists

Holt Chemistry: Online Resources

Visit **go.hrw.com** for a variety of free resources related to this textbook. Enter the keyword **HW4 HOME**.

Holt Online Learning

Students can access interactive problem solving help and active visual concept development with the *Holt Chemistry* Online Edition available at **www.hrw.com**

student CNN News

cnnstudentnews.com

Find the latest chemistry news, lesson plans, and activities related to important scientific events.

Compression guide:
To shorten your instruction because of time limitations, omit blocks 1, 4, 6, 7, and 8.

KEY

TE Teacher Edition	**CRF** Chapter Resource File
SE Student Edition	**TT** Teaching Transparencies
OSP One-Stop Planner	

* Also on One-Stop Planner
◆ Requires Advance Prep

PROBLEM SOLVING AND PRACTICE	SECTION REVIEW AND ASSESSMENT	STANDARDS CORRELATION
		National Science Education Standards
	TE Homework, p. 382 BASIC **TE** Quiz, p. 384 GENERAL **CRF** Quiz * **TE** Reteaching, p. 384 BASIC **SE** Section Review, p. 384 **CRF** Concept Review *	PS 2e
	TE Homework, p. 391 BASIC **TE** Quiz, p. 392 GENERAL **CRF** Quiz * **TE** Reteaching, p. 392 BASIC **SE** Section Review, p. 392 **CRF** Concept Review *	PS 2d, PS 2e
SE **Sample Problem A** Calculating Melting and Boiling Points, p. 397 **SE** Practice, p. 397 GENERAL **TE** Homework, p. 397 GENERAL **SE** Problem Bank, p. 858 GENERAL	**TE** Quiz, p. 398 GENERAL **CRF** Quiz * **TE** Reteaching, p. 398 BASIC **SE** Section Review, p. 398 **CRF** Concept Review *	
SE **Sample Problem B** How to Draw a Phase Diagram, p. 404 **SE** Practice, p. 404 GENERAL **TE** Homework, p. 404 GENERAL	**TE** Quiz, p. 405 GENERAL **CRF** Quiz * **TE** Reteaching, p. 405 BASIC **SE** Section Review, p. 405 **CRF** Concept Review *	

www.scilinks.org

Topic: Snowflakes
SciLinks code: HW4129

Topic: States of Matter
SciLinks code: HW4120

Topic: Hydrogen Bonding
SciLinks code: HW4069

Topic: Changes of State
SciLinks code: HW4393

Topic: Phase Diagrams
SciLinks code: HW4130

Topic: Supercritical Fluids
SciLinks code: HW4122

Technology Resources

 **Science in the NEWS**
Each video segment is accompanied by a Critical Thinking Worksheet.

Segment 15
Coal-Oil Technology

Segment 19
Icebergs

Segment 22
Belching Cows

 ChemFile Interactive Tutor CD-Rom

Module 1: States of Matter
Topic: Solids, Liquids, Gases and Changes of State

Overview

This chapter describes the states of solid, liquid, and gas and the changes from one state to another. It relates the properties of a state to its energy content and particle arrangement. The forces and energy changes involved in change of state are also examined. Intermolecular forces, such as hydrogen bonds and London forces, are studied. Students will learn about energy that is involved in state changes. Topics such as enthalpy and entropy are covered, as well as their application in calculating melting and boiling points and how they are affected by pressure. Students will learn about phase equilibrium, including interpreting phase diagrams.

Assessing Prior Knowledge

Check for Content Knowledge

Before beginning this chapter, be sure your students have had exposure to the following concepts:

- states of matter
- potential and kinetic energy
- heat and temperature
- covalent compounds
- polarity
- energy and change

Using the Figure

The symmetry of snowflakes is due to hydrogen bonding. During crystallization, the water molecules arrange themselves in a state of lowest energy, such that the number of hydrogen bonds formed is maximized for each water molecule. Having symmetry increases the stability of the structure.

CHAPTER 11

STATES OF MATTER AND INTERMOLECULAR FORCES

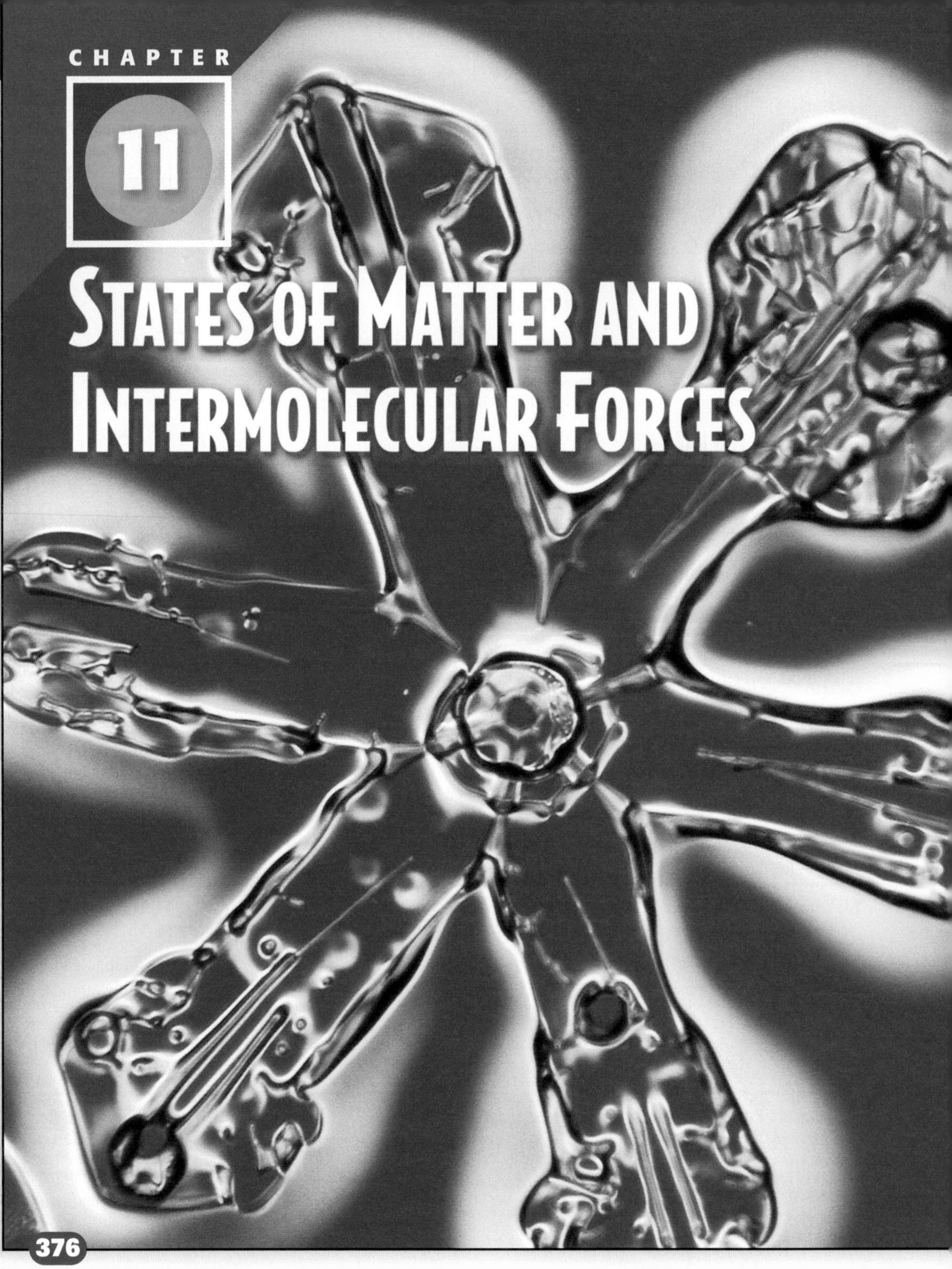

376

Standards Correlations

National Science Education Standards

PS 2d: The physical properties of compounds reflect the nature of the interactions among its molecules. These interactions are determined by the structure of the molecule, including the constituent atoms and the distances and angles between them. (Section 2)

PS 2e: Solids, liquids, and gases differ in the distances and angles between molecules or atoms and therefore the energy that binds them together. In solids the structure is nearly rigid; in liquids molecules or atoms move around each other but do not move apart; and in gases molecules or atoms move almost independently of each other and are mostly far apart. (Sections 1, 2)

Where does a snowflake's elegant structure come from? Snowflakes may not look exactly alike, but all are made up of ice crystals that have a hexagonal arrangement of molecules. This arrangement is due to the shape of water molecules and the attractive forces between them. Water molecules are very polar and form a special kind of attraction, called a *hydrogen bond,* with other water molecules. Hydrogen bonding is just one of the *intermolecular forces* that you will learn about in this chapter.

START-UP ACTIVITY

Heating Curve for Water

SAFETY PRECAUTIONS

PROCEDURE

1. Place several **ice cubes** in a **250 mL beaker.** Fill the beaker halfway with **water.** Place the beaker on a **hot plate.** Using a **ring stand,** clamp a **thermometer** so that it is immersed in the ice water but not touching the bottom or sides of the beaker. Record the temperature of the ice water after the temperature has stopped changing.

2. Turn on the hot plate and heat the ice water. Using a **stirring rod,** carefully stir the water as the ice melts.

3. Observe the water as it is heated. Continue stirring. Record the temperature of the water every 30 s. Note the time at which the ice is completely melted. Also note when the water begins to boil.

4. Allow the water to boil for several minutes, and continue to record the temperature every 30 s. Turn off the hot plate, and allow the beaker of water to cool.

5. Is your graph a straight line? If not, where does the slope change?

ANALYSIS

1. Make a graph of temperature as a function of time.

2. What happened to the temperature of the ice water as you heated the beaker?

3. What happened to the temperature of the water after the water started boiling?

Pre-Reading Questions

① **Name two examples each of solids, liquids, and gases.**

② **What happens when you heat an ice cube?**

③ **What force is there between oppositely charged objects?**

Answers to Pre-Reading Questions

1. Answers may include ice, steel, water, rubbing alcohol, carbon dioxide, and water vapor.
2. The solid ice cube melts to form a puddle of liquid water.
3. Oppositely charged objects are attracted to each other because of coulombic force.

CONTENTS 11

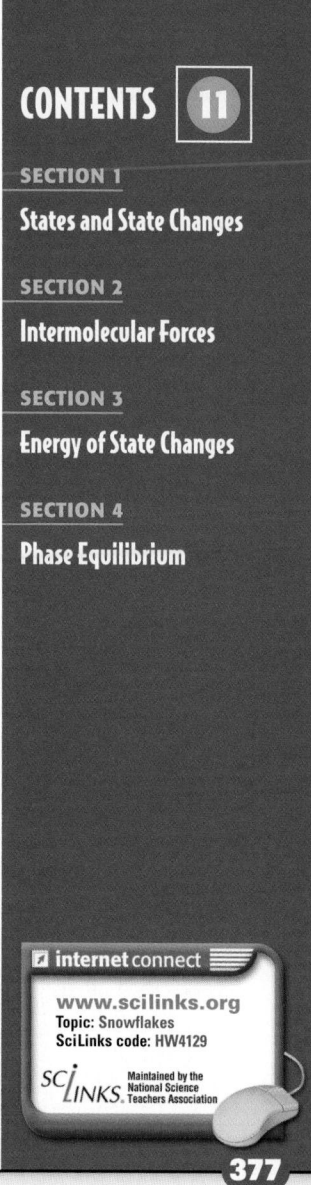

internet connect

www.scilinks.org
Topic: Snowflakes
SciLinks code: HW4129

SCI LINKS. Maintained by the National Science Teachers Association

377

START-UP ACTIVITY

Skills Acquired:
- Measuring
- Collecting data
- Organizing & Analyzing data

Materials:
For each group of 2–3 students:
- beaker, 250-mL
- hot plate
- ice cubes
- ring stand
- stirring rod
- thermometer
- water

Teacher's Notes:
- Be sure students do not stir with the thermometers, but use stirring rods.
- The thermometer should be clamped without touching the sides or bottom of the beaker.
- Caution students to use care with the hot plate and water.
- Have students construct a data table before starting.
- When students study Section 3, they might question the differences between these results and the section content. If solid ice at a temperature below the freezing point had been used instead of ice water, the temperature would have risen until the ice reached the melting point.

Answers

1. The graph should have a line with a slope that is flat until the ice melts, then positive until the water starts to boil, and finally flat again when the water is boiling.

2. The temperature of the ice water remained constant while the ice was melting and then increased after all the ice had melted.

3. The temperature remained constant as the water boiled.

Before beginning this section, review with your students the Objectives listed in the Student Edition. This section discusses three common states of matter (solid, liquid, and gas) and relates the properties of each to its energy content and particle arrangement. Changes of state are investigated.

 **Bellringer**

Asks students to define the term *surface tension* in their own words. What state of matter is associated with surface tension? Ans. liquid

Motivate

Demonstration

A Homemade Compass Practice this demonstration ahead of class. Fill a Petri dish with water. Magnetize a needle by rubbing it on a magnet. Using tweezers, carefully place the needle on the surface of water. (Results will vary depending on the type of needle.) The needle stays on top of the water instead of sinking because of surface tension. If the water's surface is broken when placing the needle, it will sink.

Prove to students that the needle does not float by cleaning the needle with dilute detergent and rinsing it thoroughly. The clean needle will sink, because the water can adhere to the metal surface. The needle will stay at the water's surface when it has a small amount of oily residue on it. The magnetizing step can be omitted.

KEY TERMS

- surface tension
- evaporation
- boiling point
- condensation
- melting
- melting point
- freezing
- freezing point
- sublimation

Topic Link

Refer to the chapter "The Science of Chemistry" for a discussion of states of matter.

OBJECTIVES

(1) **Relate** the properties of a state to the energy content and particle arrangement of that state of matter.

(2) **Explain** forces and energy changes involved in changes of state.

States of Matter

Have you ever had candy apples like those shown in **Figure 1**? Or have you had strawberries dipped in chocolate? When you make these treats, you can see a substance in two states. The fruit is dipped into the liquid candy or chocolate to coat it. But the liquid becomes solid when cooled. However, the substance has the same identity—and delicious taste—in both states. Most substances, such as the mercury shown in **Figure 2**, can be in three states: solid, liquid, and gas. The physical properties of each state come from the arrangement of particles.

Solid Particles Have Fixed Positions

The particles in a solid are very close together and have an orderly, fixed arrangement. They are held in place by the attractive forces that are between all particles. Because solid particles can vibrate only in place and do not break away from their fixed positions, solids have fixed volumes and shapes. That is, no matter what container you put a solid in, the solid takes up the same amount of space. Solids usually exist in crystalline form. Solid crystals can be very hard and brittle, like salt, or they can be very soft, like lead. Another example of a solid is ice, the solid state of water.

Figure 1
When you make candy apples, you see a substance in two states. The warm liquid candy becomes a solid when cooled.

378

Chapter Resource File
- Lesson Plan

INCLUSION Strategies

- *Learning Disabled*
- *Attention Deficit Disorder*
- *English as a Second Language*

Have students use the Key Terms for this section to make a vocabulary file on index cards. Students will title each card with one term and then state the term's definition. They can draw a diagram of each term using the states of water as examples. Have them list boiling, melting, and freezing points for water and isopropyl alcohol. The index cards may be used as a study guide or notes for an oral presentation to the class.

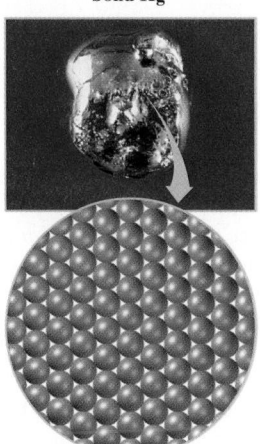
Solid Hg

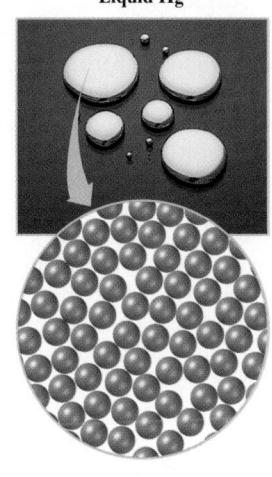
Liquid Hg

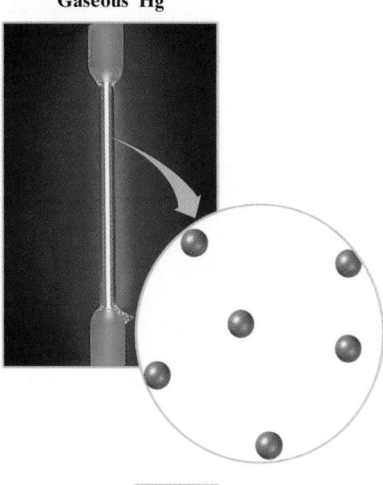
Gaseous Hg

Liquid Particles Can Move Easily Past One Another

If you add energy as heat to ice, the ice will melt and become liquid water. In other words, the highly ordered crystals of ice will break apart to form the random arrangement of liquid particles. Liquid particles are also held close together by attractive forces. Thus, the density of a liquid substance is similar to that of the solid substance. However, liquid particles have enough energy to be able to move past each other readily, which allows liquids to flow. That is, liquids are fluids. Some liquids can flow very readily, such as water or gasoline. Other liquids, such as molasses, are thicker and very *viscous* and flow very slowly. Like solids, liquids have fixed volumes. However, while solids keep the same shape no matter the container, liquids flow to take the shape of the lower part of a container. Because liquid particles can move past each other, they are noticeably affected by forces between particles, which gives them special properties.

Liquid Forces Lead to Surface Wetting and Capillary Action

Why does water bead up on a freshly waxed car? Liquid particles can have *cohesion,* attraction for each other. They can also have *adhesion,* attraction for particles of solid surfaces. The balance of these forces determines whether a liquid will wet a solid surface. For example, water molecules have a high cohesion for each other and a low adhesion to particles in car wax. Thus, water drops tend to stick together rather than stick to the car wax.

Water has a greater adhesion to glass than to car wax. The forces of adhesion and cohesion will pull water up a narrow glass tube, called a *capillary tube,* shown in **Figure 3.** The adhesion of the water molecules to the molecules that make up the glass tube pulls water molecules up the sides of the tube. The molecules that are pulled up the glass pull other water molecules with them because of cohesion. The water rises up the tube until the weight of the water above the surface level balances the upward force caused by adhesion and cohesion.

Figure 2
Mercury is the only metal that is a liquid at room temperature, but when cooled below −40°C, it freezes to a solid. At 357°C, it boils and becomes a gas.

Topic Link

Refer to the "Ions and Ionic Compounds" chapter for a discussion of crystal structure.

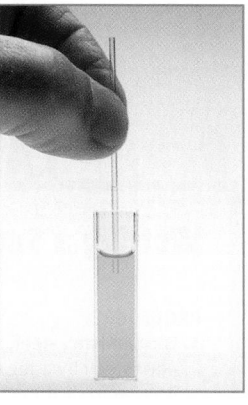

Figure 3
Capillary action, which moves water up through a narrow glass tube, also allows water to move up the roots and stems of plants.

379

REAL-WORLD
CONNECTION
Viscosity of Liquids Viscosity is a function of temperature and intermolecular forces. For example, honey pours slowly. However, if you heat it, it pours more readily; its viscosity has decreased. Viscosity is an important factor in choosing a motor oil. In the winter, oil's viscosity increases because of the cold temperature. Therefore, a less viscous motor oil is needed, one that contains molecules that have less hydrocarbon branching and therefore will "tangle" with each other less. In the summer, higher temperatures decrease viscosity, so a higher-viscosity motor oil is preferable. Many oils are suitable for all seasons unless temperatures are extreme.

Quick LAB

Skills Acquired:
- Collecting data
- Inferring

Materials:

For each group of 2–3 students:
- detergent, diluted with water
- glass plate
- plastic plate
- steel plate
- toothpick
- water

Teacher's Notes:
- To save time consider washing the plates before class and handing them out, being careful not to get fingerprints on them.
- For disposal, pour all liquids down the drain. Dry the plates and save them for reuse.

Answers

1. glass
2. plastic
3. Water has a greater adhesion for glass than steel and for steel than plastic.
4. Water wets glass more than it wets steel or plastic.

Activity ——— BASIC

Hold a contest to determine how many water drops from an eye-dropper can be stacked on a clean, dry penny. Have students discuss their results in terms of surface tension. **LS** Visual

Chapter Resource File

- Datasheets for In-Text Labs

Figure 4
a Water does not wet the feather, because the water's particles are not attracted to the oily film on the feather's surface.

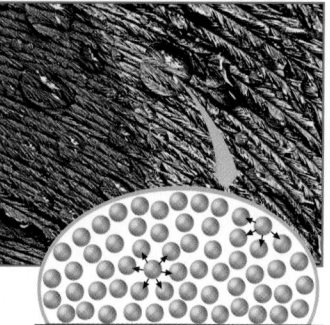

b A drop of water on a surface has particles that are attracted to each other.

surface tension

the force that acts on the surface of a liquid and that tends to minimize the area of the surface

Liquids Have Surface Tension

Why are water drops rounded? Substances are liquids instead of gases because the cohesive forces between the particles are strong enough to pull the particles together so that they are in contact. Below the surface of the liquid, the particles are pulled equally in all directions by these forces. However, particles at the surface are pulled only sideways and downward by neighboring particles, as shown in the model of a water drop in **Figure 4.**

The particles on the surface have a net force pulling them down into the liquid. It takes energy to oppose this net force and increase the surface area. Energy must be added to increase the number of particles at the surface. Liquids tend to decrease energy by decreasing surface area. The tendency of liquids to decrease their surface area to the smallest size possible is called **surface tension.** Surface tension accounts for many liquid properties. Liquids tend to form spherical shapes, because a sphere has the smallest surface area for a given volume. For example, rain and fog droplets are spherical.

Gas Particles Are Essentially Independent

Gas particles are much farther apart than the particles in solids and liquids. They must go far before colliding with each other or with the walls of a container. Because gas particles are so far apart, the attractive forces between them do not have a great effect. They move almost independently of one another. So, unlike solids and liquids, gases fill whatever container they are in. Thus, the shape, volume, and density of an amount of gas depend on the size and shape of the container.

Because gas particles can move around freely, gases are fluids and can flow easily. When you breathe, you can feel how easily the gases that make up air can flow to fill your lungs. Examples of gases include carbon dioxide, a gas that you exhale, and helium, a gas that is used to fill balloons. You will learn more about gases in the "Gases" chapter.

Quick LAB

Wetting a Surface

SAFETY PRECAUTIONS

PROCEDURE
1. Wash **plastic, steel,** and **glass plates** well by using **dilute detergent,** and rinse them completely. Do not touch the clean surfaces.
2. Using a **toothpick,** put a small drop of **water** on each plate. Observe the shape of the drops from the side.

ANALYSIS
1. On which surface does the water spread the most?
2. On which surface does the water spread the least?
3. What can you conclude about the adhesion of water for plastic, steel, and glass?
4. Explain your observations in terms of wetting.

REAL-WORLD CONNECTION ADVANCED

Surface Tension and Laundry Ask students how they think surface tension relates to doing laundry. Why might the surface tension of water keep your clothes from becoming clean? **Ans.** Dirt particles cannot penetrate water drops because of water's high surface tension. Have students find out what a surfactant is, how surfactants relate to surface tension, and a common use of a surfactant. **Ans.** A surfactant, such as soap or detergent, decreases the surface tension of water by disrupting its hydrogen bonds. They enable water to carry away dirt and grease from clothing. **LS** Verbal

Changing States

The hardening of melted candy on an apple is just one example of how matter changes states. *Freezing* is the change of state in which a liquid becomes a solid. You can observe freezing when you make ice cubes in the freezer. *Melting* is the change of state in which a solid becomes a liquid. For example, a solid wax candle melts when it is lit.

Evaporation—the change of state in which a liquid becomes a gas—takes place when water boils in a pot or evaporates from damp clothing. Gases can become liquids. *Condensation* is the change of state in which a gas becomes a liquid. For example, water vapor in the air can condense onto a cold glass or onto grass as dew in the morning.

But solids can evaporate, too. A thin film of ice on the edges of a windshield can become a gas by *sublimation* as the car moves through the air. Gases become solids by a process sometimes called *deposition*. For example, frost can form on a cold, clear night from water vapor in the air. **Figure 5** shows these six state changes. All state changes are physical changes, because the identity of the substance does not change, while the physical form of the substance does change.

Temperature, Energy, and State

All matter has energy related to the energy of the rapid, random motion of atom-sized particles. This energy of random motion increases as temperature increases. The higher the temperature is, the greater the average kinetic energy of the particles is. As temperature increases, the particles in solids vibrate more rapidly in their fixed positions. Like solid particles, liquid particles vibrate more rapidly as temperature increases, but they can also move past each other more quickly. Increasing the temperature of a gas causes the free-moving particles to move more rapidly and to collide more often with one another.

Generally, adding energy to a substance will increase the substance's temperature. But after a certain point, adding more energy will cause a substance to experience a change of state instead of a temperature increase.

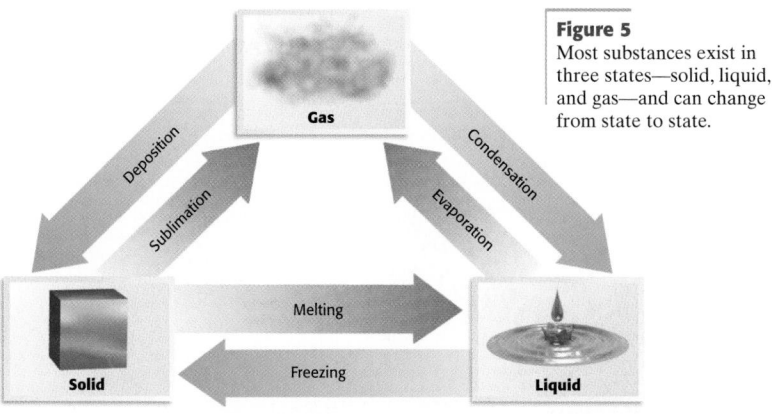

Figure 5
Most substances exist in three states—solid, liquid, and gas—and can change from state to state.

381

Writing The water cycle is essential to life on Earth. The movement of water from one location to another involves changes in state. Have students write a one-page report that explains the water cycle. Students can use terms from **Figure 5** and sketches in their explanations. **Ans.** Liquid water evaporates,

then it condenses in small droplets in clouds. If the temperature is low enough, the droplets freeze. When the droplets become heavy enough, they fall as rain or snow. If the water is frozen, it eventually melts, or it might undergo some sublimation. **LS Verbal**

Demonstration —— GENERAL

Comparing the Meniscus in H$_2$O and Hg A meniscus is the curved surface of a liquid in a container. Students have used the bottom of the meniscus of water to read volumes in graduated cylinders. Show students that water has a concave meniscus, and mercury (in a closed container) has a convex meniscus. Because these meniscuses are small and not easily observed by an entire class, you might want to sketch them on the board.

Capillary action is a measure of the attraction of a liquid for its container. Ask students to explain any differences in shape of the meniscuses as they relate to capillary action. **Ans.** Water has a concave meniscus because the attraction of the water for the container (adhesion) is greater than the attraction of the water molecules for each other (cohesion). Mercury has a convex meniscus because in mercury, cohesion is greater than adhesion).

Safety: Wear safety goggles. Students should be at least 3 m from the demonstration. Mercury and its vapors are extremely toxic. Do not directly observe mercury unless it is in a sealed container. You might want to use a photo of the meniscus of mercury instead of using actual mercury.

Disposal: Save any mercury for later use. Pour water down the drain. **LS Visual**

Melting or Solution? Students might become confused about the difference between a solid's melting and its going into solution. In both cases, what was solid is now in the liquid state. However, for a solution, two substances must be involved, the solute and solvent. Another difference is that a solid going into solution does not occur at a specific temperature.

SKILL BUILDER — **ADVANCED**

Writing Skills As Earth's tectonic plates move, places exist where hot materials from within Earth's mantle come to the surface of the crust. Deep beneath the ocean's surface, hot magma sometimes erupts from these vents and boils the water surrounding them. Have students write a written explanation of why bubbles of steam are not observed rising to the surface of the ocean.
Ans. As the steam rises, it cools and condenses. **LS Verbal**

Homework —— **BASIC**

Have students find out information about a biochemical or geological cycle, sketch a diagram of it, and indicate places in the cycle where changes in state occur. Examples of cycles that could be used are the rock cycle, the water cycle, the carbon cycle, or the nitrogen cycle. For example, in the rock cycle, rock melts to become magma and freezes to become igneous rock. Point out that many of the changes that occur in these cycles do not involve change of state. **LS Visual**

Figure 6
A runner sweats when the body heats as a result of exertion. As sweat evaporates, the body is cooled.

evaporation

the change of a substance from a liquid to a gas

boiling point

the temperature and pressure at which a liquid and a gas are in equilibrium

condensation

the change of state from a gas to a liquid

Figure 7
On a cool night, when humidity is high, water vapor condenses to the liquid state.

Liquid Evaporates to Gas

If you leave an uncovered pan of water standing for a day or two, some of the water disappears. Some of the molecules have left the liquid and gone into the gaseous state. Because even neutral particles are attracted to each other, energy is required to separate them. If the liquid particles gain enough energy of movement, they can escape from the liquid. But where does the energy come from? The liquid particles gain energy when they collide with each other. Sometimes, a particle is struck by several particles at once and gains a large amount of energy. This particle can then leave the liquid's surface through **evaporation.** Because energy must be added to the water, evaporation is an endothermic process. This is why people sweat when they are hot and when they exercise, as shown in **Figure 6.** The evaporation of sweat cools the body.

You may have noticed that a puddle of water on the sidewalk evaporates more quickly on a hot day than on a cooler day. The reason is that the hotter liquid has more high-energy molecules. These high-energy molecules are more likely to gain the extra energy needed to become gas particles more rapidly.

Think about what happens when you place a pan of water on a hot stove. As the liquid is heated, its temperature rises and it evaporates more rapidly. Eventually, it reaches a temperature at which bubbles of vapor rise to the surface, and the temperature of the liquid remains constant. This temperature is the **boiling point.** Why doesn't all of the liquid evaporate at once at the boiling point? The answer is that it takes a large amount of energy to move a molecule from the liquid state to the gaseous state.

Gas Condenses to Liquid

Now, think about what happens if you place a glass lid over a pan of boiling water. You will see liquid form on the underside of the lid. Instead of escaping from the closed pan, the water vapor formed from boiling hits the cooler lid and forms liquid drops through **condensation.** Energy is transferred as heat from the gas particles to the lid. The gas particles no longer have enough energy to overcome the attractive forces between them, so they go into the liquid state. Condensation takes place on a cool night and forms dew on plants, as shown in **Figure 7.** Because energy is released from the water, condensation is an exothermic process.

HISTORY
CONNECTION

Liquefying Chlorine The English chemist Michael Faraday first liquefied chlorine, which is a gas at room temperature. He sealed crystals of a chlorine compound in a bent glass tube, with one leg of the tube cooled. When the crystals were heated, chlorine gas was produced. The increased amount of gas increased pressure in the tube. Between the increased pressure and decreased temperature, the chlorine condensed in the cooled leg of the tube.

Figure 8
When the temperature drops below freezing, farmers spray water on the orange trees. Energy is released by water as the water freezes, which warms the oranges and keeps the crop from freezing.

Solid Melts to Liquid

As a solid is heated, the particles vibrate faster and faster in their fixed positions. Their energy of random motion increases. Eventually, a temperature is reached such that some of the molecules have enough energy to break out of their fixed positions and move around. At this point, the solid is **melting.** That is, the solid is becoming a liquid. As long as both the newly formed liquid and the remaining solid are in contact, the temperature will not change. This temperature is the **melting point** of the solid. The energy of random motion is the same for both states. Energy must be absorbed for melting to happen, so melting is endothermic.

Liquid Freezes to Solid

The opposite process takes place when **freezing,** shown in **Figure 8,** takes place. As a liquid is cooled, the movement of particles becomes slower and slower. The particles' energy of random motion decreases. Eventually, a temperature is reached such that the particles are attracted to each other and pulled together into the fixed positions of the solid state, and the liquid crystallizes. This exothermic process releases energy—an amount equal to what is added in melting. As long as both states are present, the temperature will not change. This temperature is the **freezing point** of the liquid. Note that the melting point and freezing point are the same for pure substances.

Solid Sublimes to Gas

The particles in a solid are constantly vibrating. Some particles have higher energy than others. Particles with high enough energy can escape from the solid. This endothermic process is called **sublimation.** Sublimation is similar to evaporation. One difference is that it takes more energy to move a particle from a solid into a gaseous state than to move a particle from a liquid into a gaseous state. A common example of sublimation takes place when mothballs are placed in a chest, as shown on the next page in **Figure 9.** The solid naphthalene crystals in mothballs sublime to form naphthalene gas, which surrounds the clothing and keeps moths away.

melting

the change of state in which a solid becomes a liquid by adding heat or changing pressure

melting point

the temperature and pressure at which a solid becomes a liquid

freezing

the change of state in which a liquid becomes a solid as heat is removed

freezing point

the temperature at which a liquid substance freezes

sublimation

the process in which a solid changes directly into a gas (the term is sometimes also used for the reverse process)

383

did you know?

Sublime Materials Good examples of materials that sublime are iodine crystals, mothballs, and moth flakes. Moth flakes sublime slowly in a closet or clothes bag and kill the larvae of moths that might be in wool or fur. They are also used in urinals to mask odors. They do not dissolve because they are nonpolar. Because they sublime slowly, they mask the odor in the air for a longer period of time.

Using the Figure

Have students discuss **Figure 8** in terms of exothermic processes. Have them explain to a partner how spraying orange trees with water during freezing weather protects the orange crop. Point out to students that ice also insulates the oranges from the cold air.

MISCONCEPTION ALERT

Chemical vs. State Change
Remind students that a change of state is a physical change and does not change the identity of the substance. Students might confuse the bubbles formed from a chemical reaction with the bubbles formed from boiling. Remind them that boiling is evaporation—the substance changes to a vapor, remains the same substance, and does not undergo a chemical change.

SKILL BUILDER — BASIC

Vocabulary Have students find the meaning of the word *volatile.* Have them identify two volatile substances. Volatile substances are those that easily change from a liquid or a solid to a gas at room temperature. Alcohol and ether are two liquids that are volatile. Naphthalene is a volatile solid. Volatile substances with odors can easily be detected because of the volume of vapor in the surrounding air. If a volatile substance, such as gasoline, is flammable its volatility can be a potential hazard. However *volatile* and *hazardous* are not synonymous terms. **LS** Verbal

Teaching Tip

Identification by Change of State The characteristic temperatures at which a substance melts and evaporates are unique to that substance at a given atmospheric pressure. Others might be similar, but they are not exactly the same. As a result, melting and boiling points can be used to identify substances.

Quiz ──────── GENERAL

1. A sample of water has a fixed volume and shape. What state is it in? Ans. solid

2. What property of a liquid enables you to pour a liquid into a cup until the top of the liquid is slightly higher than the top of the cup? Ans. surface tension

3. During what process does a liquid change to a solid? Ans. freezing

4. What is the main difference between boiling and evaporation? Ans. Boiling is a type of evaporation. Evaporation can occur below the boiling point.
LS Logical

Reteaching ──────── BASIC

Have students describe in their own words the changes of state that occur on a day that starts off colder than the freezing point of water, warms up in the afternoon, and again is cold that night. Have them incorporate as many of the terms that indicate a change in state as they can, such as melting, evaporation, and so on. **LS** Verbal

Chapter Resource File

• Concept Review

• Section Quiz

Figure 9
Molecules of naphthalene sublime from the surface of the crystals in the mothball.

Gas Deposits to Solid

The reverse of sublimation is often called *deposition*. Molecules in the gaseous state become part of the surface of a crystal. Energy is released in the exothermic process. The energy released in deposition is equal to the energy required for sublimation. A common example of deposition is the formation of frost on exposed surfaces during a cold night when the temperature is below freezing. In a laboratory, you may see iodine gas deposit as solid crystals onto the surface of a sealed container.

 Section Review

UNDERSTANDING KEY IDEAS

1. Describe what happens to the shape and volume of a solid, a liquid, and a gas when you place each into separate, closed containers.

2. What is surface tension?

3. You heat a piece of iron from 200 to 400 K. What happens to the atoms' energy of random motion?

4. When water boils, bubbles form at the base of the container. What gas has formed?

5. What two terms are used to describe the temperature at which solids and liquids of the same substance exist at the same time?

6. How are sublimation and evaporation similar?

7. Describe an example of deposition.

CRITICAL THINKING

8. The densities of the liquid and solid states of a substance are often similar. Explain.

9. How could you demonstrate evaporation?

10. How could you demonstrate boiling point?

11. You are boiling potatoes on a gas stove, and your friend suggests turning up the heat to cook them faster. Will this idea work?

12. A dehumidifier takes water vapor from the air by passing the moist air over a set of cold coils to perform a state change. How does a dehumidifier work?

13. Water at 50°C is cooled to −10°C. Describe what will happen.

14. How could you demonstrate melting point?

15. Explain why changes of state are considered physical transitions and not chemical processes.

384

Answers to Section Review

1. The solid will keep its shape and volume. The liquid will keep the same volume but will change to form the shape of the lower part of the container. The gas will change in volume and shape by filling the entire container.

2. Surface tension is the force that acts on the surface of a liquid and that tends to minimize the area of the surface. Liquids will thus assume the shape with the lowest possible surface area.

3. The energy of random motion doubles.

4. water vapor

5. melting point and freezing point

6. In both processes, molecules leave a surface and go into the gaseous state.

7. Sample answer: Deposition takes place when frost forms on a surface on a cold night.

8. In both liquids and solids, the particles are nearly in contact with one another.

9. Answers may include letting a pan of water stand or hanging a wet cloth up to dry.

10. Answers may include putting water in a kettle and heating the water until bubbles form steadily and the temperature remains constant.

Answers continued on p. 413A

Intermolecular Forces

KEY TERMS

- intermolecular forces
- dipole-dipole forces
- hydrogen bond
- London dispersion force

OBJECTIVES

 Contrast ionic and molecular substances in terms of their physical characteristics and the types of forces that govern their behavior.

 Describe dipole-dipole forces.

③ **Explain** how a hydrogen bond is different from other dipole-dipole forces and how it is responsible for many of water's properties.

④ **Describe** London dispersion forces, and relate their strength to other types of attractions.

Comparing Ionic and Covalent Compounds

Particles attract each other, so it takes energy to overcome the forces holding them together. If it takes high energy to separate the particles of a substance, then it takes high energy to cause that substance to go from the liquid to the gaseous state. The boiling point of a substance is a good measure of the strength of the forces that hold the particles together. Melting point also relates to attractive forces between particles. Most covalent compounds melt at lower temperatures than ionic compounds do. As shown in **Table 1,** ionic substances with small ions tend to be solids that have high melting points, and covalent substances tend to be gases and liquids or solids that have low melting points.

Table 1   Comparing Ionic and Molecular Substances

Type of substance	Common use	State at room temperature	Melting point (°C)	Boiling point (°C)
Ionic substances				
Potassium chloride, KCl	salt substitute	solid	770	sublimes at 1500
Sodium chloride, NaCl	table salt	solid	801	1413
Calcium fluoride, CaF_2	water fluoridation	solid	1423	2500
Covalent substances				
Methane, CH_4	natural gas	gas	−182	−164
Ethyl acetate, $CH_3COOCH_2CH_3$	fingernail polish	liquid	−84	77
Water, H_2O	(many)	liquid	0	100
Heptadecane, $C_{17}H_{36}$	wax candles	solid	22	302

385

Chapter Resource File

- Lesson Plan

Focus

Overview

Before beginning this section, review with your students the Objectives listed in the Student Edition. In this section, students will compare ionic and covalent substances in terms of their physical properties and the types of intermolecular forces that determine their behavior. These forces are either dipole-dipole forces, including hydrogen bonding, or London forces.

Bellringer

Ask students to list any terms they know that use the prefixes *inter-* and *intra-*. **Ans.** Answers may include *inter-* : international, interactive software, intercollegiate, or interdenominational; *intra-* : intranuclear, intrapersonal, intrastate, and intravenous.

Motivate

Discussion

Refer students to the lists they compiled in the Bellringer activity. From their lists, ask students what is meant by the prefixes *inter-* and *intra-*. *Inter-* means "between" or "among," and *intra-* means "within." All the *inter-* terms show interaction between two or more things, such as an intercollegiate gathering involving two local colleges. *Intra-* is more often implied than used. For example, international events involve two or more nations. Intranational events are done within one nation, but the prefix is usually not used. Ask students to use their knowledge of these prefixes to define the term *intermolecular forces.*

Demonstration

Polar Versus Nonpolar Show students a candle, and tell them that it is made up of compounds composed of mostly carbon and hydrogen. Tell them that completely burning such compounds yields the products carbon dioxide and water. Light the candle. Turn a 400-mL beaker upside down, and place ice on top of it. Use beaker tongs to pick up the inverted beaker and hold it over the burning candle. After a few minutes, move the beaker from the flame, and remove the ice. Extinguish the flame. Hold the beaker upright. Students should observe drops of a liquid inside of the beaker. Test any liquid with cobalt chloride test paper to determine whether it is water. Ask students to explain the results in terms of boiling points and the intermolecular forces.

Safety: Use caution with open flames and while handling hot glassware. Avoid dripping hot wax when extinguishing the candle.

Disposal: Keep the candle for later use. Throw the cobalt chloride test paper in the trash.

MISCONCEPTION
ALERT

Intermolecular Versus Intramolecular Many students use the terms *intermolecular* and *intramolecular* incorrectly. Intermolecular forces occur between or among molecules. They include the attraction of one polar molecule to another. Intramolecular forces include those within a unit of a compound. Covalent bonds are intramolecular forces.

Topic Link

Refer to the "Ions and Ionic Compounds" chapter for a discussion of crystal lattices.

intermolecular forces

the forces of attraction between molecules

dipole-dipole forces

interactions between polar molecules

Topic Link

Refer to the "Covalent Compounds" chapter for a discussion of dipoles.

386

Oppositely Charged Ions Attract Each Other

Ionic substances generally have much higher forces of attraction than covalent substances. Recall that ionic substances are made up of separate ions. Each ion is attracted to all ions of opposite charge. For small ions, these attractions hold the ions tightly in a crystal lattice that can be disrupted only by heating the crystal to very high temperatures.

The strength of ionic forces depends on the size of the ions and the amount of charge. Ionic compounds with small ions have high melting points. If the ions are larger, then the distances between them are larger and the forces are weaker. This effect helps explain why potassium chloride, KCl, melts at a lower temperature than sodium chloride, NaCl, does. Now compare ions that differ by the amount of charge they have. If the ions have larger charges, then the ionic force is larger than the ionic forces of ions with smaller charges. This effect explains why calcium fluoride, CaF_2 melts at a higher temperature than NaCl does.

Intermolecular Forces Attract Molecules to Each Other

For covalent substances, forces that act between molecules are called **intermolecular forces.** They can be dipole-dipole forces or London dispersion forces. Both forces are short-range and decrease rapidly as molecules get farther apart. Because the forces are effective only when molecules are near each other, they do not have much of an impact on gases. A substance with weak attractive forces will be a gas because there is not enough attractive force to hold molecules together as a liquid or a solid.

The forces that hold the molecules together act only between neighboring molecules. The forces may be weak; some molecular substances boil near absolute zero. For example, hydrogen gas, H_2, boils at −252.8°C. The forces may be strong; some molecular substances have very high boiling points. For example, coronene, $C_{24}H_{12}$, boils at 525°C.

Dipole-Dipole Forces

In **dipole-dipole forces,** the positive end of one molecule attracts the negative end of a neighboring molecule. Bonds are polar because atoms of differing electronegativity are bonded together. The greater the difference in electronegativity in a diatomic molecule, the greater the polarity is.

Dipole-Dipole Forces Affect Melting and Boiling Points

When polar molecules get close and attract each other, the force is significant if the degree of polarity is fairly high. When molecules are very polar, the dipole-dipole forces are very significant. Remember that the boiling point of a substance tells you something about the forces between the molecules. For example, **Table 2** shows that the polar compound 1-propanol, C_3H_7OH, boils at 97.4°C. The less polar compound of similar size, 1-propanethiol, C_3H_7SH, boils at 67.8°C. However, the nonpolar compound butane, C_4H_{10}, also of similar size, boils at −0.5°C. The more polar the molecules are, the stronger the dipole-dipole forces between them, and thus, the higher the boiling point.

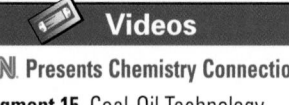
Videos

CNN. Presents Chemistry Connections

• **Segment 15** Coal-Oil Technology

See the Science in the News video guide for more details.

Table 2   Comparing Dipole-Dipole Forces

Substance	Boiling point (°C)	Polarity	State at room temperature	Structure
1-propanol, C_3H_7OH	97.4	polar	liquid	H H H H—C—C—C—O—H H H H
1-propanethiol, C_3H_7SH	67.8	less polar	liquid	H H H H—C—C—C—S—H H H H
Butane, C_4H_{10}	−0.5	nonpolar	gas	H H H H H—C—C—C—C—H H H H H
Water, H_2O	100.0	polar	liquid	O H H
Hydrogen sulfide, H_2S	−60.7	less polar	gas	S H H
Ammonia, NH_3	−33.35	polar	gas	H H—N—H
Phosphine, PH_3	−87.7	less polar	gas	H H—P—H

Hydrogen Bonds

Compare the boiling points of H_2O and H_2S, shown in **Table 2**. These molecules have similar sizes and shapes. However, the boiling point of H_2O is much higher than that of H_2S. A similar comparison of NH_3 with PH_3 can be made. The greater the polarity of a molecule, the higher the boiling point is. However, when hydrogen atoms are bonded to very electronegative atoms, the effect is even more noticeable.

Compare the boiling points and electronegativity differences of the hydrogen halides, shown in **Table 3**. As the electronegativity difference increases, the boiling point increases. The boiling points increase somewhat from HCl to HBr to HI but increase a lot more for HF. What accounts for this jump? The answer has to do with a special form of dipole-dipole forces, called a **hydrogen bond.**

hydrogen bond

the intermolecular force occurring when a hydrogen atom that is bonded to a highly electronegative atom of one molecule is attracted to two unshared electrons of another molecule

internet connect
www.scilinks.org
Topic: Hydrogen Bonding
SciLinks code: HW4069

SCILINKS. Maintained by the National Science Teachers Association

Table 3   Boiling Points of the Hydrogen Halides

Substance	HF	HCl	HBr	HI
Boiling point (°C)	20	−85	−67	−35
Electronegativity difference	1.8	1.0	0.8	0.5

387

EARTH SCIENCE ─
CONNECTION

Weathering Weathering of rocks can be either chemical or physical. Chemical weathering occurs when substances in the environment, such as acidic rainwater, react with the rocks. Physical weathering occurs when physical processes break up rock. Physical processes include ice wedging. When liquid water flows into cracks in rocks and then freezes, its volume increases. The pressure caused by the expanding water breaks up the surrounding rock.

Hydrogen Bonding Because the term *hydrogen bonding* references the element hydrogen, students might think that it is a type of chemical bond. It does not involve the loss, gain, or sharing of electrons. It involves attraction between polar molecules that contain hydrogen and is approximately one one-hundredth the strength of a covalent bond.

Using the Figure

Refer students to **Figure 10.** Remind students that compounds that contain ionic or covalent bonding are also affected by one or more intermolecular forces. Ask students which of the following substances would be more easily broken apart—DNA strands that are held together by hydrogen bonds or DNA strands that are held together by ionic bonds. **Ans.** It would be much easier to "unzip" the DNA strands that are held together by hydrogen bonds, which are stronger than other intermolecular forces, but not as strong as an ionic or a covalent bonds.

SKILL BUILDER — ADVANCED

Writing Skills With water's low molecular mass, it would exist on Earth as a gas only if it did not contain hydrogen bonds. Have students write a story about what they think Earth would be like if hydrogen bonds did not exist in water. Encourage both creativity and scientific accuracy.
LS Verbal

Hydrogen Bonds Form with Electronegative Atoms

Strong hydrogen bonds can form with a hydrogen atom that is covalently bonded to very electronegative atoms in the upper-right part of the periodic table: nitrogen, oxygen, and fluorine. When a hydrogen atom bonds to an atom of N, O, or F, the hydrogen atom has a large, partially positive charge. The partially positive hydrogen atom of polar molecules can be attracted to the unshared pairs of electrons of neighboring molecules. For example, the hydrogen bonds shown in **Figure 10** result from the attraction of the hydrogen atoms in the H—N and H—O bonds of one DNA strand to the unshared pairs of electrons in the complementary DNA strand. These hydrogen bonds hold together the complementary strands of DNA, which contain the body's genetic information.

Hydrogen Bonds Are Strong Dipole-Dipole Forces

It is not just electronegativity difference that accounts for the strength of hydrogen bonds. One reason that hydrogen bonds are such strong dipole-dipole forces is because the hydrogen atom is small and has only one electron. When that electron is pulled away by a highly electronegative atom, there are no more electrons under it. Thus, the single proton of the hydrogen nucleus is partially exposed. As a result, hydrogen's proton is strongly attracted to the unbonded pair of electrons of other molecules. The combination of the large electronegativity difference (high polarity) and hydrogen's small size accounts for the strength of the hydrogen bond.

Figure 10
Hydrogen bonding between base pairs on adjacent molecules of DNA holds the two strands together. Yet the force is not so strong that the strands cannot be separated.

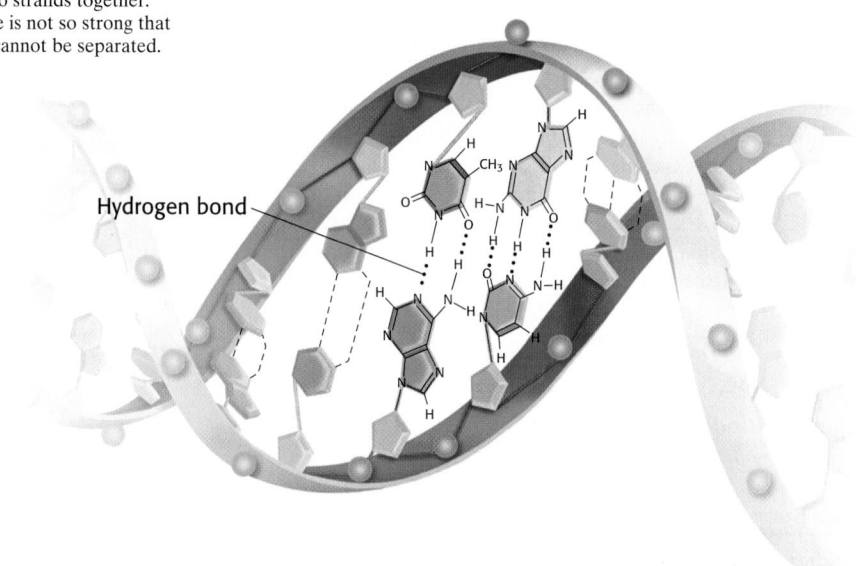

Hydrogen bond

388

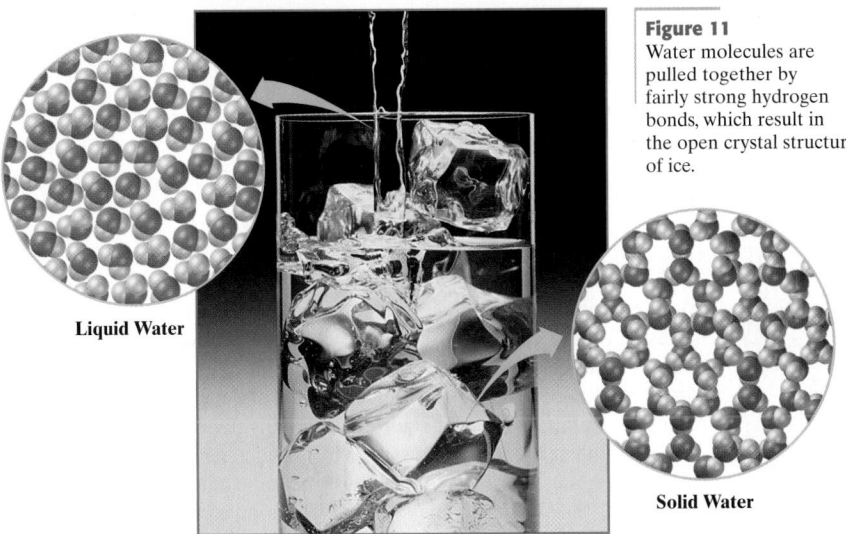

Liquid Water

Solid Water

Figure 11
Water molecules are pulled together by fairly strong hydrogen bonds, which result in the open crystal structure of ice.

Hydrogen Bonding Explains Water's Unique Properties

The energy of hydrogen bonds is lower than that of normal chemical bonds but can be stronger than that of other intermolecular forces. Hydrogen bonding can account for many properties. **Figure 11** shows an example of hydrogen bonding that involves oxygen. Water has unique properties. These unique properties are the result of hydrogen bonding.

Water is different from most other covalent compounds because of how much it can participate in strong hydrogen bonding. In water, two hydrogen atoms are bonded to oxygen by polar covalent bonds. Each hydrogen atom can form hydrogen bonds with neighboring molecules. Because of the water molecule's ability to form multiple hydrogen bonds at once, the intermolecular forces in water are strong.

Another different characteristic of water results from hydrogen bonding and the shape of a water molecule. Unlike most solids, which are denser than their liquids, solid water is less dense than liquid water and floats in liquid water. The angle between the two H atoms in water is 104.5°. This angle is very close to the tetrahedral angle of 109.5°. When water forms solid ice, the angle in the molecules causes the special geometry of molecules in the crystal shown in **Figure 11.** Ice crystals have large amounts of open space, which causes ice to have a low density.

The unusual density difference between liquid and solid water explains many important phenomena in the natural world. For example, because ice floats on water, ponds freeze from the top down and not from the bottom up. Thus, fish can survive the winter in water under an insulating layer of ice. Because water expands when it freezes, water seeping into the cracks of rock or concrete can cause considerable damage due to fracturing. You should never freeze water-containing foods in glass containers, which may break when the water freezes and expands.

389

SOCIAL STUDIES
CONNECTION

Because of its strong hydrogen bonding, water can absorb great amounts of energy before it boils. It often absorbs this energy from objects that come into contact with it. Native Americans made food containers from birch bark. If these containers were filled with water, they could be placed on a fire. As long as the flames did not rise past the water level, the water absorbed energy from the bark, and the bark did not burn.

Transparencies

TT Ice and Water

Teaching Tip

Ice and Aquatic Life The hydrogen bonding that causes ice to float is extremely important to organisms that live in bodies of water. Because ice stays on top of the water, underwater plants and animals can live near the bottom of the water and not be subjected to the temperature fluctuations that occur near the surface.

READING SKILL BUILDER — BASIC

Paired Reading Have student pairs read about hydrogen bonding. As they read, have them write down questions they have that are not answered by the section content. After the reading is completed, have them trade questions with another pair to see which of the questions can be answered by their peers. Any remaining questions can be compiled on the board, and students can check other sources to answer the questions. **LS** Verbal

Group Activity — GENERAL

Have groups of four students compare the density of ice and liquid water. Have them find the mass of a sealable plastic bag. Then, have them pour 50.0 mL of water into the bag, seal the bag, and find its mass. Each group should fill a large beaker with water and mark the water's surface. Have students determine the volume of the water in the bag by displacement of water in the large beaker. Place the sealed bag into a freezer overnight. After the water freezes, determine the volume (by displacement) and the mass of the ice formed. Students can use this mass and volume information to calculate density of both the liquid water and the ice. **LS** Visual

Demonstration

Temporary Dipole Model Refer students to **Figure 13.** Use the following demonstration to clarify what is meant by an induced dipole. Demonstrate that you cannot use an ordinary paperclip to pick up other paperclips. Elicit from students that the paperclip won't pick up the objects because it is not a magnet. Touch one paperclip with a magnet. Now show that you can pull up a long string of paperclips. Each paperclip attracts the next by the magnetism that is induced in each by the magnet. Point out that the magnet has made the paperclips magnetic by creating temporary dipoles in the paperclips. Now remove the magnet and show how the string of paperclips falls apart. Once the magnet is removed, the induced magnetism in the paperclips is also removed and there is no longer any attractive forces between the paperclips.

SKILL BUILDER — BASIC

Graphing Skills Refer students to **Table 4.** Have them use a periodic table to find the atomic mass for each noble gas. Then, have them graph the temperature on the *y*-axis and the atomic mass on the *x*-axis for each noble gas. The resulting graph shows a steady increase in boiling point as atomic mass increases, but the relationship is not linear. **Visual**

Figure 12
The nonpolar molecules in gasoline are held together by London dispersion forces, so it is not a gas at room temperature.

London dispersion force

the intermolecular attraction resulting from the uneven distribution of electrons and the creation of temporary dipoles

Figure 13
Temporary dipoles in molecules cause forces of attraction between the molecules.

Transparencies

TT Temporary Dipoles

Table 4	**Boiling Points of the Noble Gases**					
Substance	**He**	**Ne**	**Ar**	**Kr**	**Xe**	**Rn**
Boiling point (°C)	−269	−246	−186	−152	−107	−62
Number of electrons	2	10	18	36	54	86

London Dispersion Forces

Some compounds are ionic, and forces of attraction between ions of opposite charge cause the ions to stick together. Some molecules are polar, and dipole-dipole forces hold polar molecules together. But what forces of attraction hold together nonpolar molecules and atoms? For example, gasoline, shown in **Figure 12,** contains nonpolar octane, C_8H_{18}, and is a liquid at room temperature. Why isn't octane a gas? Clearly, some sort of intermolecular force allows gasoline to be a liquid.

In 1930, the German chemist Fritz W. London came up with an explanation. Nonpolar molecules experience a special form of dipole-dipole force called **London dispersion force.** In dipole-dipole forces, the negative part of one molecule attracts the positive region of a neighboring molecule. However, in London dispersion forces, there is no special part of the molecule that is always positive or negative.

London Dispersion Forces Exist Between Nonpolar Molecules

In general, the strength of London dispersion forces between nonpolar particles increases as the molar mass of the particles increases. This is because generally, as molar mass increases, so does the number of electrons in a molecule. Consider the boiling point of the noble gases, as shown in **Table 4.** Generally, as boiling point increases, so does the number of electrons in the atoms. For groups of similar atoms and molecules, such as the noble gases or hydrogen halides, London dispersion forces are roughly proportional to the number of electrons present.

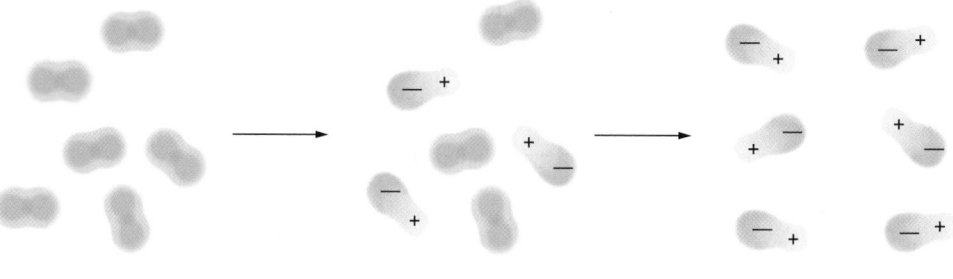

a Nonpolar molecules can become momentarily polar.

b The instantaneous dipoles that form cause adjacent molecules to polarize.

c These London dispersion forces cause the molecules to attract each other.

London Dispersion Forces Result from Temporary Dipoles

How do electrons play a role in London dispersion forces? The answer lies in the way that electrons move and do not stay still. The electrons in atoms and molecules can move. They not only move about in orbitals but also can move from one side of an atom to the other. When the electrons move toward one side of an atom or molecule, that side becomes momentarily negative and the other side becomes momentarily positive. If the positive side of a momentarily charged molecule moves near another molecule, the positive side can attract the electrons in the other molecule. Or the negative side of the momentarily charged molecule can push the electrons of the other molecule away. The temporary dipoles that form attract each other, as shown in **Figure 13,** and make temporary dipoles form in other molecules. When molecules are near each other, they always exert an attractive force because electrons can move.

Properties Depend on Types of Intermolecular Force

Compare the properties of an ionic substance, NaCl, with those of a non-polar substance, I_2, as shown in **Figure 14.** The differences in the properties of the substances are related to the differences in the types of forces that act within each substance. Because ionic, polar covalent, and nonpolar covalent substances are different in electron distribution, they are different in the types of attractive forces that they experience.

While nonpolar molecules can experience only London dispersion forces, polar molecules experience both dipole-dipole forces and London dispersion forces. Determining how much each force adds to the overall force of attraction between polar molecules is not easy. London dispersion forces also exist between ions in ionic compounds, but they are quite small relative to ionic forces and can almost always be overlooked.

Figure 14
Forces between ions are generally much stronger than the forces between molecules, so the melting points of ionic substances tend to be higher.

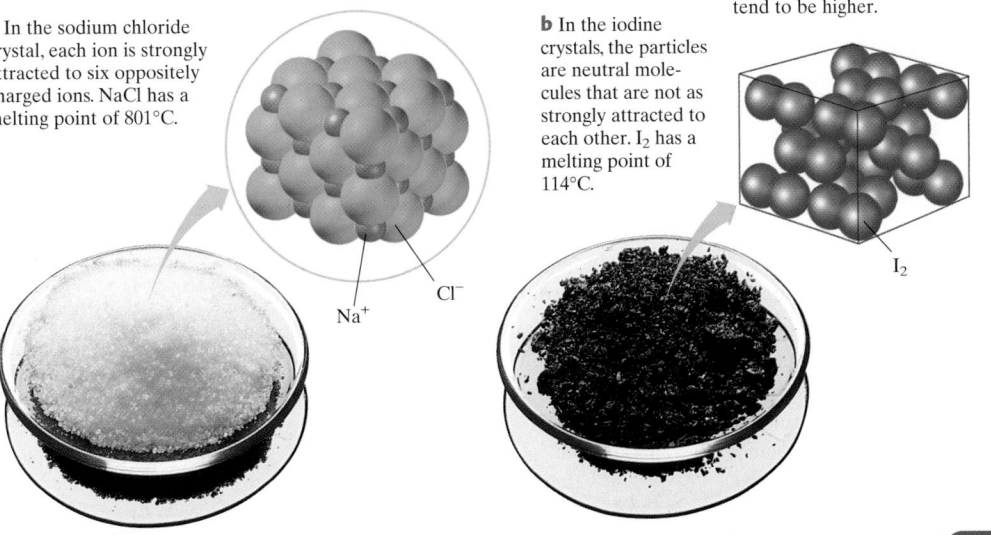

a In the sodium chloride crystal, each ion is strongly attracted to six oppositely charged ions. NaCl has a melting point of 801°C.

Na⁺   Cl⁻

b In the iodine crystals, the particles are neutral molecules that are not as strongly attracted to each other. I_2 has a melting point of 114°C.

I_2

391

Close

Quiz ——————— GENERAL

1. What is an intermolecular force?
Ans. a force that pulls molecules together

2. Which is a type of dipole-dipole force—hydrogen bond or London force? **Ans.** hydrogen bond

3. Which is stronger, a hydrogen bond or a London force? **Ans.** hydrogen bond

4. What is a temporary dipole? **Ans.** an atom or molecule that has a positive side and a negative side because of the electrons being more on one side of the particle than the other

LS Verbal

Reteaching ——————— BASIC

Have students write an outline that includes information about intermolecular forces. The main heads of the outline should be "Dipole-dipole forces" and "London forces." **LS** Verbal

```
┌─────────────────────────────────────┐
│      Chapter Resource File           │
├─────────────────────────────────────┤
│  • Concept Review                    │
│  • Section Quiz                      │
└─────────────────────────────────────┘
```

Figure 15
a The polyatomic ionic compound 1-butylpyridinium nitrate is a liquid solvent at room temperature. The large size of the cations keeps the ionic forces from having a great effect.

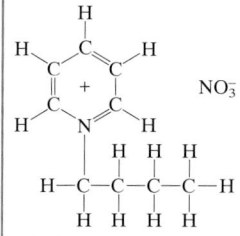

b A molecule of coronene, $C_{24}H_{12}$, is very large, yet its flat shape allows it to have relatively strong London dispersion forces.

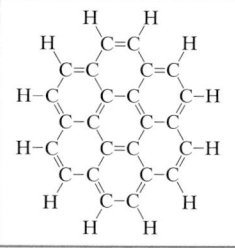

Particle Size and Shape Also Play a Role

Dipole-dipole forces are generally stronger than London dispersion forces. However, both of these forces between molecules are usually much weaker than ionic forces in crystals. There are exceptions. One major factor is the size of the atoms, ions, or molecules. The larger the particles are, the farther apart they are and the smaller the effects of the attraction are. If an ionic substance has very large ions—especially if the ions are not symmetrical—the ionic substance's melting point can be very low. A few ionic compounds are even liquid at room temperature, such as 1-butylpyridinium nitrate, shown in **Figure 15**.

The shape of the particles can also play a role in determining the strength of attraction. For example, coronene molecules, $C_{24}H_{12}$, are very large. However, they are flat, so they can come close together and the attractive forces have a greater effect. Thus, the boiling point of nonpolar coronene is almost as high as that of some ionic compounds.

② Section Review

UNDERSTANDING KEY IDEAS

1. What force holds NaCl units together?

2. Describe dipole-dipole forces.

3. What force gives water unique properties?

4. Why does ice have a lower density than liquid water does?

5. Explain why oxygen, nitrogen, and fluorine are elements in molecules that form strong hydrogen bonds.

6. How is the strength of London dispersion forces related to the number of electrons?

7. How do intermolecular forces affect whether a substance is a solid at room temperature?

CRITICAL THINKING

8. a. Which is nonpolar: CF_4 or CH_2F_2?

   b. Which substance likely has a higher boiling point? Explain your answer.

9. Are the London dispersion forces between water molecules weaker or stronger than the London dispersion forces between molecules of hydrogen sulfide, H_2S?

10. NH_3 has a much higher boiling point than PH_3 does. Explain.

11. Why does argon boil at a higher temperature than neon does?

12. Which will have the higher melting point, KF or KNO_3? Explain your answer.

Answers to Section Review

1. NaCl has strong ionic forces between small ions.

2. Dipole-dipole forces are forces that exist between the positive and negative regions of polar molecules.

3. Hydrogen bonding, a special type of dipole-dipole force, is the most significant of water's intermolecular forces.

4. The ice crystal has an open structure with many spaces. Water forms hydrogen bonds that are oriented in different directions to create this structure.

5. These elements have very high electronegativities, so they form very polar bonds with hydrogen.

6. London dispersion forces result from temporary dipoles formed in electron clouds. The more electrons an atom or molecule has, the greater the likelihood of forming a strong temporary dipole.

7. Substances that have strong intermolecular forces require more energy to pull the molecules apart and therefore melt at higher temperatures.

Answers continued on p. 413A

Energy of State Changes

OBJECTIVES

① **Define** the molar enthalpy of fusion and the molar enthalpy of vaporization, and **identify** them for a substance by using a heating curve.

② **Describe** how enthalpy and entropy of a substance relate to state.

③ **Predict** whether a state change will take place by using Gibbs energy.

④ **Calculate** melting and boiling points by using enthalpy and entropy.

⑤ **Explain** how pressure affects the entropy of a gas and affects changes between the liquid and vapor states.

Enthalpy, Entropy, and Changes of State

Adding enough energy to boil a pan of water takes a certain amount of time. Removing enough energy to freeze a tray of ice cubes also takes a certain amount of time. At that rate, you could imagine that freezing the water that makes up the iceberg in **Figure 16** would take a *very* long time.

Enthalpy is the total energy of a system. *Entropy* measures a system's disorder. The energy added during melting or removed during freezing is called the *enthalpy of fusion*. (*Fusion* means *melting*.) Particle motion is more random in the liquid state, so as a solid melts, the entropy of its particles increases. This increase is the *entropy of fusion*. As a liquid evaporates, a lot of energy is needed to separate the particles. This energy is the *enthalpy of vaporization*. (*Vaporization* means *evaporation*.) Particle motion is much more random in a gas than in a liquid. A substance's *entropy of vaporization* is much larger than its entropy of fusion.

internet connect

www.scilinks.org
Topic: Changes of State
SciLinks code: HW4027

SCi LINKS Maintained by the
National Science
Teachers Association

Topic Link

Refer to the "Causes of Change" chapter for a discussion of enthalpy and entropy.

Figure 16
Melting an iceberg would take a great amount of enthalpy of fusion.

393

PHYSICS
CONNECTION

Enthalpy is a term that refers to energy. In addition to the enthalpy of fusion and enthalpy of vaporization covered in this section, chemists and physicists are concerned with enthalpy of formation, enthalpy of solution, and enthalpy of reaction. All enthalpies involve transfer of energy in the form of heat.

SECTION

3

Focus

Overview

Before beginning this section, review with your students the Objectives listed in the Student Edition. This section examines the concepts of enthalpy and entropy and relates them to state, boiling point, and melting point. Students will learn to identify the enthalpies of fusion and vaporization for a particular substance from a heating curve. They will be able to describe how pressure affects the entropy of a gas and the changes made between the liquid and vapor states.

🔔 Bellringer

Provide each student with a small box and a collection of small objects, such as pennies, dry beans, or gumdrops. Have students shake the box so that the objects are in a random order (high entropy) and then arrange the objects in an orderly pattern (low entropy). Ask students to brainstorm a list of the ways in which the activity relates to different states of matter.

Motivate

Demonstration

Demonstrate an example of a state change to students, such as boiling. Ask students to describe what is taking place at the molecular level. Ask them also to characterize the state change as an increase or decrease in order.

Chapter Resource File

• **Lesson Plan**

Videos

CNN Presents Chemistry Connections

• **Segment 19** Icebergs

See the Science in the News video guide for more details.

Using the Figure —GENERAL

Refer students back to **Figure 2.** Use the diagrams of the three states to discuss entropy changes. Point out that water in the solid state increases in disorder as it changes to the liquid state. The change to a gas represents another increase in disorder because the particles are now farther apart.

Point out the vertical segments of the graph in **Figure 17.** Ask students to speculate on what that energy is doing if it is not increasing the temperature. Ans. The energy is either used to do work on the molecules, moving them out of the crystal lattice at the melting point, or moving them farther from each other into the gaseous state at the boiling point. **LS** Verbal

Demonstration

Melting Metal Obtain a piece of Onion's fusible metal (available from Flinn Scientific and other scientific supply houses). While wearing gloves, show students your attempts to bend the solid metal. Place the metal in a 150-mL beaker of boiling water. Show that the metal has liquefied by moving it with a glass rod. Pour most of the water into a waste beaker. Pour the metal slowly into cold water. While wearing gloves, show the solidified metal to students.

Safety: Observe all precautions described on the label and in the MSDS for Onion's fusible metal. Wear gloves.

Disposal: Save the Onion's fusible metal for another demonstration.

394

Figure 17
Energy is added to 1 mol of ice. At the melting point and boiling point, the temperatures remain constant and large changes in molar enthalpy take place.

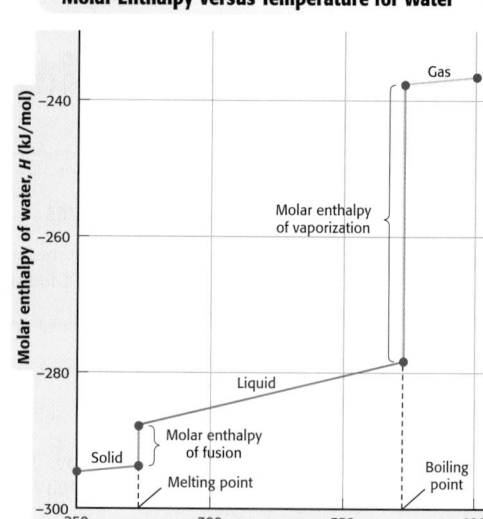

Molar Enthalpy Versus Temperature for Water

Enthalpy and Entropy Changes for Melting and Evaporation

Enthalpy and entropy change as energy in the form of heat is added to a substance, as shown with water in **Figure 17.** The graph starts with 1 mol of solid ice at 250 K (−23°C). The ice warms to 273.15 K. The enthalpy, H, increases slightly during this process. At 273.15 K, the ice begins to melt. As long as both ice and liquid water are present, the temperature remains at 273.15 K. The energy added is the *molar enthalpy of fusion* (ΔH_{fus}), which is 6.009 kJ/mol for ice. ΔH_{fus} is the difference in enthalpy between solid and liquid water at 273.15 K as shown in the following equation:

$$\Delta H_{fus} = H_{(\text{liquid at melting point})} - H_{(\text{solid at melting point})}$$

After the ice melts, the temperature of the liquid water increases as energy is added until the temperature reaches 373.15 K.

At 373.15 K, the water boils. If the pressure remains constant, so does the temperature as long as the two states (liquid and gas) are present. The energy added is the *molar enthalpy of vaporization* (ΔH_{vap}), 40.67 kJ/mol. ΔH_{vap} is the difference in enthalpy between liquid and gaseous water at 373.15 K and is defined in the following equation:

$$\Delta H_{vap} = H_{(\text{vapor at boiling point})} - H_{(\text{liquid at boiling point})}$$

After all of the liquid water has evaporated, the energy added increases the temperature of the water vapor.

Like water, almost all substances can be in the three common states of matter. **Table 5** lists the molar enthalpies and entropies of fusion and vaporization for some elements and compounds. Because intermolecular forces are not significant in the gaseous state, most substances have similar values for *molar entropy of vaporization*, ΔS_{vap}.

MISCONCEPTION ///ALERT\\\

Boiling Point of Water Students may find it strange that the boiling point of water is often given as 372.78 K instead of the expected normal boiling point of 373 K (100°C). Remind them that standard pressure for measuring thermodynamic properties is 100 kPa, a little less than 101.325 kPa (1 atm). Water boils at 99.63°C (372.78 K) at this pressure.

Table 5 **Molar Enthalpies and Entropies of Fusion and Vaporization**

Substance	T_{mp} (K)	ΔH_{fus} (kJ/mol)	ΔS_{fus} (J/mol•K)	T_{bp} (K)	ΔH_{vap} (kJ/mol)	ΔS_{vap} (J/mol•K)
Nitrogen, N_2	63	0.71	11.3	77	5.6	72.2
Hydrogen sulfide, H_2S	188	23.8	126.6	214	18.7	87.4
Bromine, Br_2	266	10.57	39.8	332	30.0	90.4
Water, H_2O	273	6.01	22.0	373	40.7	108.8
Benzene, C_6H_6	279	9.95	35.7	353	30.7	87.0
Lead, Pb	601	4.77	7.9	2022	179.5	88.8

Gibbs Energy and State Changes

As you have learned, the relative values of H and S determine whether any process, including a state change, will take place. The following equation describes a change in Gibbs energy.

$$\Delta G = \Delta H - T\Delta S$$

You may recall that a process is spontaneous if ΔG is negative. That is, the process can take place with a decrease in Gibbs energy. If ΔG is positive, then a process will not take place unless an outside source of energy drives the process. If ΔG is zero, then the system is said to be in a state of *equilibrium*. At equilibrium, the forward and reverse processes are happening at the same rate. For example, when solid ice and liquid water are at equilibrium, ice melts at the same rate that water freezes. You will learn more about equilibrium in the next section.

Enthalpy and Entropy Determine State

At normal atmospheric pressure, water freezes at 273.15 K (0.00°C). At this pressure, pure water will not freeze at any temperature above 273.15 K. Likewise, pure ice will not melt at any temperature below 273.15 K. This point may be proven by looking at ΔG just above and just below the normal freezing point of water. At the normal freezing point, the enthalpy of fusion of ice is 6.009 kJ/mol, or 6009 J/mol. For changes in state that take place at constant temperature, the entropy change, ΔS, is $\Delta H/T$. Thus, ΔS is (6009 J/mol)/(273.15 K) = 22.00 J/mol•K for the melting of ice.

Now let us calculate the Gibbs energy change for the melting of ice at 273.00 K. For this change, ΔH is positive (energy is absorbed), and ΔS is also positive (greater degree of disorder).

$$\Delta G = \Delta H - T\Delta S = +6009 \text{ J/mol} - (273.00 \text{ K} \times +22.00 \text{ J/mol•K})$$
$$= +6009 \text{ J/mol} - 6006 \text{ J/mol} = +3 \text{ J/mol}$$

Because ΔG is positive, the change will not take place on its own. The ordered state of ice is preferred at this temperature, which is below the normal freezing point.

MISCONCEPTION ALERT

Superheated Steam It is often assumed that steam is always at 100°C. Tell students that if energy is added to the steam, its temperature will rise past this point. In a closed system, the pressure will increase as the temperature of the steam increases. Many industrial processes rely on the use of superheated steam. Superheated steam is potentially quite dangerous because it is not visible and is hot enough to cause severe burns to human tissue.

Teaching Tip ——— BASIC

Have students review the symbols used in **Table 5** and make a chart that describes what each symbol means. **LS** Visual

Using the Table

Refer students to **Table 5.** Ask students to scan the list of enthalpies and entropies of fusion and to compare the values with the values of the enthalpies and entropies of vaporization. Ask them what the comparison tells us about state change. **Ans.** Significantly less energy is required to melt a collection of molecules than the energy required to evaporate the same set of molecules. The change from the liquid state to the gaseous state requires a greater input of energy than the change from the solid state to the liquid state.

Demonstration

Paper That Won't Burn Mix approximately 10 mL each of water and ethanol. Fold a piece of notebook or copier paper into a small square and soak it in the solution. Holding the paper with tongs, light it. You might want to turn out the classroom lights to better see the flame. After the flame goes out, examine the paper. Ask students to explain why the paper didn't burn. (The paper may char.) **Ans.** The alcohol vapors were burning, and the cooling effect of the evaporation kept the paper from reaching its kindling temperature.

Safety: Wear safety goggles. Students should be at least 3 m from the demonstration. Be sure there is adequate ventilation. There should be no open flame in the classroom while the ethanol bottle is open.

Disposal: Wet the paper with water and throw it in the trashcan. Save any leftover solution for later use, or pour it down the drain.

Demonstration

Synthesis of CuS A darkened room will help students see the red glow of the exothermic reaction between copper and sulfur that forms copper(II) sulfide, CuS. In this reaction, entropy increases and ΔG is negative, so the reaction is spontaneous. Cut a 1 cm × 10 cm piece of copper foil. Place flowers of sulfur (the powdery form of sulfur) in a heat-resistant test tube to a depth of 2 cm. Clamp the test tube vertically to a ring stand, and heat the bottom strongly with a burner. As sulfur vapors escape from the tube, use crucible tongs to hold the copper strip over the tube so that it can be hit by the sulfur vapors. Point out that no reaction takes place. Insert the strip into the hot, viscous sulfur vapors inside the test tube. The red glow shows that CuS is forming. Remove the copper strip, and let it cool for inspection. Bending the strip should cause flakes of CuS to break off from the surface.

Safety: Ensure adequate ventilation. Wear safety goggles and a lab apron. Students should stand 3 m away from the demonstration. If the sulfur vapors catch fire at the top of the test tube, cover the tube with a glass plate to smother the flames.

Disposal: Put the damaged test tube and used copper foil in the trash.

Similarly, think about the possibility of water freezing at 273.30 K. The ΔH is now negative (energy is released). The ΔS is also negative (greater degree of order in the crystal).

$$\Delta G = \Delta H - T\Delta S = -6009 \text{ J/mol} - (273.30 \text{ K} \times -22.00 \text{ J/mol}\cdot\text{K})$$
$$= -6009 \text{ J/mol} - 6013 \text{ J/mol} = +4 \text{ J/mol}$$

ΔG is positive, so the water will not freeze. The disordered state of liquid water is preferred at 273.30 K, which is above the melting point.

Determining Melting and Boiling Points

For a system at the melting point, a solid and a liquid are in equilibrium, so ΔG is zero. Thus, $\Delta H = T\Delta S$. Rearranging the equation, you get the following relationship, in which *mp* means melting point and *fus* means fusion.

$$T_{mp} = \frac{\Delta H_{fus}}{\Delta S_{fus}}$$

In other words, the melting point of a solid, T_{mp}, is equal to molar enthalpy of fusion, ΔH_{fus}, divided by molar entropy of fusion, ΔS_{fus}. Boiling takes place when the drive toward disorder overcomes the tendency to lose energy. Condensation, shown in **Figure 18,** takes place when the tendency to lose energy overcomes the drive to increase disorder. In other words, when $\Delta H_{vap} > T\Delta S_{vap}$, the liquid state is favored. The gas state is preferred when $\Delta H_{vap} < T\Delta S_{vap}$.

The same situation happens at the boiling point. ΔG is zero when liquid and gas are in equilibrium, so $\Delta H_{vap} = T\Delta S_{vap}$. Thus, given that *bp* stands for boiling point and *vap* stands for vaporization, the following equation is true.

$$T_{bp} = \frac{\Delta H_{vap}}{\Delta S_{vap}}$$

In other words, the boiling point of a liquid, T_{bp}, is equal to molar enthalpy of vaporization, ΔH_{vap}, divided by molar entropy of vaporization, ΔS_{vap}.

Figure 18
a Water condenses on the wings of the dragonfly when $\Delta H_{vap} > T\Delta S_{vap}$.
b Water freezes on the flower when $\Delta H_{fus} > T\Delta S_{fus}$.

396

HISTORY
CONNECTION

Gibbs Energy Josiah Willard Gibbs (1839–1903) was little known during his lifetime of work at Yale University in Connecticut. Point out to students that Gibbs energy, sometimes called Gibbs free energy, can be viewed as the available, useful energy from a process. When Gibbs energy has a negative value, a system is releasing usable energy and the process taking place is spontaneous. Engineers in the 19th century were interested in developing more efficient engines, usually steam engines, to power the mills that turned out mass-produced goods. They sought to maximize the useful (free) energy available from a process so that the machine could do work, such as pushing a piston to turn a wheel, most efficiently. Initially, the heat measurements and calculations of thermodynamics served only practical applications. But as the understanding of matter developed rapidly during the century, scientists found increasing applications for thermodynamics in explaining the energy changes of any chemical or physical processes.

Calculating Melting and Boiling Points

The enthalpy of fusion of mercury is 11.42 J/g, and the molar entropy of fusion is 9.79 J/mol•K. The enthalpy of vaporization at the boiling point is 294.7 J/g, and the molar entropy of vaporization is 93.8 J/mol•K. Calculate the melting point and the boiling point.

1 Gather information.

- molar mass of Hg = 200.59 g/mol
- enthalpy of fusion = 11.42 J/g
- molar entropy of fusion = 9.79 J/mol•K
- enthalpy of vaporization = 294.7 J/g
- molar entropy of vaporization = 93.8 J/mol•K
- melting point, T_{mp} = ?
- boiling point, T_{bp} = ?

2 Plan your work.

First calculate the molar enthalpy of fusion and molar enthalpy of vaporization, which have units of J/mol. Use the molar mass of mercury to convert from J/g to J/mol.

$$\Delta H_{fus} = 11.42 \text{ J/g} \times 200.59 \text{ g/mol} = 2291 \text{ J/mol}$$
$$\Delta H_{vap} = 294.7 \text{ J/g} \times 200.59 \text{ g/mol} = 59\,110 \text{ J/mol}$$

Set up the equations for determining T_{mp} and T_{bp}.

3 Calculate.

$$T_{mp} = \Delta H_{fus}/\Delta S_{fus} = \frac{2291 \text{ J/mol}}{9.79 \text{ J/mol}\bullet\text{K}} = 234 \text{ K}$$

$$T_{bp} = \Delta H_{vap}/\Delta S_{vap} = \frac{59\,110 \text{ J/mol}}{93.8 \text{ J/mol}\bullet\text{K}} = 630 \text{ K}$$

4 Verify your result.

Mercury is a liquid at room temperature, so the melting point must be below 298 K (25°C). Mercury boils well above room temperature, so the boiling point must be well above 298 K. These facts fit the calculation.

PRACTICE

Calculate the freezing and boiling points for each substance.

1. For ethyl alcohol, C_2H_5OH, the enthalpy of fusion is 108.9 J/g, and the entropy of fusion is 31.6 J/mol•K. The enthalpy of vaporization at the boiling point is 837 J/g, and the molar entropy of vaporization is 109.9 J/mol•K.

2. For sulfur dioxide, the molar enthalpy of fusion is 8.62 kJ/mol, and the molar entropy of fusion is 43.1 J/mol•K. ΔH_{vap} is 24.9 kJ/mol, and the molar entropy of vaporization at the boiling point is 94.5 J/mol•K.

3. For ammonia, ΔH_{fus} is 5.66 kJ/mol, and ΔS_{fus} is 29.0 J/mol•K. ΔH_{vap} is 23.33 kJ/mol, and ΔS_{vap} is 97.2 J/mol•K.

PRACTICE HINT

When setting up your equations, use the correct conversion factors so that you get the desired units when canceling. But be careful to keep values for vaporization together and separate from those for fusion. Keep track of your units! You may have to convert from joules to kilojoules or vice versa.

Answers to Practice Problems A

1. T_{mp} = 159 K, T_{bp} = 351 K
2. T_{mp} = 200. K, T_{bp} = 263 K
3. T_{mp} = 195 K, T_{bp} = 240 K

Homework — GENERAL

Additional Practice

1. For cadmium, the molar enthalpy of fusion is 6.19 kJ/mol, and the molar entropy of fusion is 10.42 J/mol•K. The molar enthalpy of vaporization is 99.9 kJ/mol, and the molar entropy of vaporization is 96.1 J/mol•K. Calculate the melting and boiling points. **Ans.** T_{mp} = 594 K, T_{bp} = 1040 K

2. For iodine, I_2, the molar enthalpy of fusion is 15.52 kJ/mol, and the molar entropy of fusion is 40.1 J/mol•K. The molar enthalpy of vaporization is 41.57 kJ/mol, and the molar entropy of vaporization is 90.8 J/mol•K. Calculate the melting and boiling points. **Ans.** T_{mp} = 387 K, T_{bp} = 458 K

3. For methane, CH_4, the molar enthalpy of fusion is 0.94 kJ/mol, and the molar entropy of fusion is 10.35 J/mol•K. The molar enthalpy of vaporization is 8.19 kJ/mol, and the molar entropy of vaporization is 73.1 J/mol•K. Calculate the melting and boiling points. **Ans.** T_{mp} = 91 K, T_{bp} = 112 K

LS Logical

READING SKILL BUILDER — BASIC

Group Activity The many numbers in the "Enthalpy and Entropy Determine State" section might overwhelm some students. Have them read this section in small groups. One student should act as a recorder, as the group develops summaries of each paragraph. **LS** Verbal

did you know?

Skating on Ice For decades, the explanation for why we can skate on ice has been that the pressure exerted by the skate blades melted the ice at the surface. Water is one of very few substances that expand upon freezing. Therefore, an increase in pressure is a stress that favors melting to the denser (lower-volume) liquid state. The assumption was that the film of liquid water caused by the melting greatly reduced the friction between the skate and the ice. However, calculations published in 1995 show that it is practically impossible for a skate to exert enough pressure to melt ice significantly, regardless of the weight of the person wearing the skate. Scientists have reported that water molecules at the surface of ice are mobile and therefore behave in a liquid-like way, even at temperatures of –100°C. Other scientists theorize that the heat generated by friction from the skate blade is enough to melt a thin layer of ice, which refreezes immediately after the skate passes.

1. Which state has the greatest entropy: solid, liquid, or gas? Ans. gas

2. What values does Gibbs energy relate? Ans. enthalpy, entropy, and temperature

3. If you are given the enthalpy of vaporization and entropy of vaporization, which can you calculate—melting point or boiling point? Ans. boiling point

4. Which state is most affected by pressure changes? Ans. gas

LS Verbal

Reteaching ━━━━━━━━ BASIC

Demonstration Wear safety goggles and a lab apron. Have students stand at least 3 m from the demonstration area. Place a small paper cup on a wire gauze above a Bunsen burner, and watch the cup catch fire and burn. Fill a second paper cup half full of water, and place it on the wire gauze. Place the lit burner under it. Ask students to explain what they observe. The water will come to a boil, and the absorption of heat will keep the temperature of the cup below the kindling temperature of paper (233°C). The cup will not burn until the water has boiled away.

LS Visual

━━━━━━━━━━━━━━━━━━━━
Chapter Resource File
━━━━━━━━━━━━━━━━━━━━

• Concept Review

• Section Quiz

Pressure Can Affect Change-of-State Processes

Boiling points are pressure dependent because pressure has a large effect on the entropy of a gas. When a gas is expanded (pressure is decreased), its entropy increases because the degree of disorder of the molecules increases. At sea level, water boils at 100°C. In Denver, Colorado, where the elevation is 1.6 km, atmospheric pressure is about 0.84 times the pressure at sea level. At that elevation, water boils at about 95°C. On Pike's Peak, where the elevation is 4.3 km, water boils at about 85°C. People often use pressure cookers at that altitude to increase the boiling point of water.

Liquids and solids are almost incompressible. Therefore, changes of atmospheric pressure have little effect on the entropy of substances in liquid or solid states. Ordinary changes in pressure have essentially no effect on melting and freezing. Although the elevation is high and atmospheric pressure is very low, water on Pike's Peak still freezes at 273.15 K. You will learn more about pressure effects on state changes in the next section.

③ Section Review

UNDERSTANDING KEY IDEAS

1. What is the molar enthalpy of fusion?

2. What is the molar enthalpy of vaporization?

3. Compare the sizes of the entropy of fusion and entropy of vaporization of a substance.

4. Explain why liquid water at 273.3 K will not freeze in terms of Gibbs energy.

5. The following process has a ΔG equal to zero at 77 K and standard pressure. In how many states can nitrogen be present at this temperature and pressure?

$$N_2(l) \longrightarrow N_2(g)$$

6. a. How does atmospheric pressure affect the boiling point of a liquid?

   b. How does atmospheric pressure affect the melting point of a liquid?

PRACTICE PROBLEMS

7. The enthalpy of fusion of bromine is 10.57 kJ/mol. The entropy of fusion is 39.8 J/mol•K. Calculate the freezing point.

8. Calculate the boiling point of bromine given the following information:

   $$Br_2(l) \longrightarrow Br_2(g)$$
   $\Delta H_{vap} = 30.0$ kJ/mol
   $\Delta S_{vap} = 90.4$ J/mol•K

9. The enthalpy of fusion of nitric acid, HNO_3, is 167 J/g. The entropy of fusion is 45.3 J/mol•K. Calculate the melting point.

CRITICAL THINKING

10. In terms of enthalpy and entropy, when does melting take place?

11. Why is the enthalpy of vaporization of a substance always much greater than the enthalpy of fusion?

12. Why is the gas state favored when

    $$T > \frac{\Delta H_{vap}}{\Delta S_{vap}} ?$$

13. Determine the change-of-state process described by each of the following:

    a. $\Delta H_{vap} > T\Delta S_{vap}$ c. $\Delta H_{vap} < T\Delta S_{vap}$

    b. $\Delta H_{fus} < T\Delta S_{fus}$ d. $\Delta H_{fus} > T\Delta S_{fus}$

Answers to Section Review

1. the amount of energy needed to melt 1 mol of a substance at the melting point

2. the energy needed to turn 1 mol of a substance into a gas

3. The entropy of vaporization is always considerably larger than the entropy of fusion.

4. ΔG is positive, so $\Delta H > T\Delta S$.

5. The liquid and gas states are at equilibrium at 77 K.

6. a. Boiling point decreases as pressure decreases.

   b. Melting point is affected very little by pressure differences.

7. 266 K

8. 332 K

9. 233 K

10. Melting takes place at the temperature at which $\Delta H_{fus} = T\Delta S_{fus}$. If energy is available from the surroundings, the entropy increase will be favored and the solid will spontaneously absorb energy and melt.

11. During evaporation, the particles must be separated completely from each other, which takes more energy.

Answers continued on p. 413A

4 Phase Equilibrium

KEY TERMS
- phase
- equilibrium
- vapor pressure
- phase diagram
- triple point
- critical point

OBJECTIVES

① **Identify** systems that have multiple phases, and determine whether they are at equilibrium.

② **Understand** the role of vapor pressure in changes of state between a liquid and a gas.

③ **Interpret** a phase diagram to identify melting points and boiling points.

Two-Phase Systems

A *system* is a set of components that are being studied. Within a system, a **phase** is a region that has the same composition and properties throughout. The lava lamp in **Figure 19** is a system that has two phases, each of which is liquid. The two phases in a lava lamp are different from each other because their chemical compositions are different. A glass of water and ice cubes is also a system that has two phases. This system has a solid phase and a liquid phase. However, the two phases have the same chemical composition. What makes the two phases in ice water different from each other is that they are different states of the same substance, water.

Phases do not need to be pure substances. If some salt is dissolved in the glass of water with ice cubes, there are still two phases: a liquid phase (the solution) and a solid phase (the pure ice). In this chapter, we will consider only systems like the ice water, that is, systems that contain one pure substance whose phases are different only by state.

phase
| a part of matter that is uniform

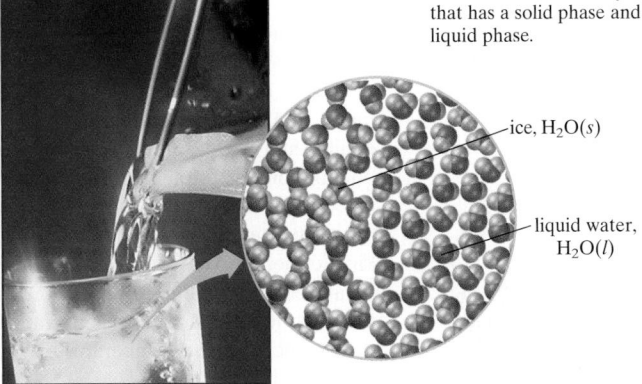

ice, $H_2O(s)$

liquid water, $H_2O(l)$

Figure 19
The lava lamp is a system that has two liquid phases, and the ice water is a system that has a solid phase and a liquid phase.

399

did you know?

Phases The term *phase* has several different meanings, but all of them refer to the term as a part of a whole. Examples include phases of the moon, phases that occur during mitosis or meiosis, and phases that occur during oscillation of light. In this section, phases refer to different states of the same substance.

Chapter Resource File

• Lesson Plan

Focus

Overview

Before beginning this section, review with your students the Objectives listed in the Student Edition. This section identifies systems with multiple phases and determines whether they are at equilibrium or not. Students learn the role of vapor pressure in condensation and evaporation processes. They also will be able to interpret a phase diagram to identify melting and boiling points.

🔊 Bellringer

Set a large block of ice (the size you would have from freezing water in a half-gallon milk container) in a pan in the front of the room. Run a wire across the top of the ice, and attach relatively heavy weights to each end of the wire. Throughout class, have students observe what happens. **Ans.** The wire passes through the ice without cutting it into two pieces.

Motivate

Discussion

You might want to have this discussion at the end of the first class session for this section to allow more observation time. Refer students to the Bellringer activity. Ask them to try to explain why the wire passed through the ice without cutting it in half. **Ans.** At the melting point temperature, ice and liquid water are at equilibrium, which means that the wire can pass through the ice; as water melts and refreezes, the wire will sink into the ice block.

Assimilating Knowledge Give students self-stick notes, and have them mark areas of difficulty as they read this section outside of class. Remind them to study the diagrams and graphs, particularly **Figure 22.** Tell students that as they read they should be prepared to answer the following questions.

- What is the kinetic-molecular explanation of evaporation and condensation?

- How is the vapor pressure of a substance related to temperature? What is the kinetic-molecular explanation for this relationship?

- What are the conditions under which a substance boils?

- What happens during sublimation?

LS Verbal

One-Stop Planner CD-ROM

- **Career Extension**
  Occupational Applications Worksheet 7:
  Pest-Control Technician
  Assign this worksheet to emphasize relevant applications of text concepts.

Figure 20
Iodine sublimes even at room temperature. Molecules escape from the solid and go into the gas phase, which is in equilibrium with the solid.

$I_2(g)$

$I_2(s)$

equilibrium

the state in which a chemical process and the reverse chemical process occur at the same rate such that the concentrations of reactants and products do not change

vapor pressure

the partial pressure exerted by a vapor that is in equilibrium with its liquid state at a given temperature

Figure 21
A few molecules of a liquid have enough energy to overcome intermolecular forces and escape from the surface into the gas phase, which is in equilibrium with the liquid.

400

Equilibrium Involves Constant Interchange of Particles

If you open a bottle of rubbing alcohol, you can smell the alcohol. Some molecules of alcohol have escaped into the gas phase. When you put the cap back on, an equilibrium is quickly reached. A dynamic **equilibrium** exists when particles are constantly moving between two or more phases yet no net change in the amount of substance in either phase takes place. Molecules are escaping from the liquid phase into a gas at the same rate that other molecules are returning to the liquid from the gas phase. That is, the rate of evaporation equals the rate of condensation.

Similarly, if you keep a glass of ice water outside on a 0°C day, a constant interchange of water molecules between the solid ice and the liquid water will take place. The system is in a state of equilibrium. **Figure 20** shows a system that has a solid and a gas at equilibrium.

Vapor Pressure Increases with Temperature

A closed container of water is a two-phase system in which molecules of water are in a gas phase in the space above the liquid phase. Moving randomly above the liquid, some of these molecules strike the walls and some go back into the liquid, as shown in **Figure 21.** An equilibrium, in which the rate of evaporation equals the rate of condensation, is soon created. The molecules in the gas exert pressure when they strike the walls of the container. The pressure exerted by the molecules of a gas, or vapor, phase in equilibrium with a liquid is called the **vapor pressure.** You can define boiling point as the temperature at which the vapor pressure equals the external pressure.

As the temperature of the water increases, the molecules have more kinetic energy, so more of them can escape into the gas phase. Thus, as temperature increases, the vapor pressure increases. This relationship is shown for water in **Table 6.** At 40°C, the vapor pressure of water is 55.3 mm Hg. If you increase the temperature to 80°C, the vapor pressure will be 355.1 mm Hg.

REAL-WORLD
CONNECTION

Pressure Cookers Pressure cookers are a necessity for cooking at high altitudes. Often the boiling point of water at high altitudes is not high enough to cook firm vegetables, pasta, eggs, and similar items in a reasonable amount of time. When water is heated in a sealed pressure cooker, the vapor pressure increases, which in turn increases the pressure inside the pot. The water must become hotter to boil under this pressure. The pressure is regulated by either a weighted seal or a mechanical valve. Many high-altitude restaurants also use pressure ovens to speed the cooking of foods such as roasts and stews.

Table 6	Water-Vapor Pressure	
Temp. (°C)	Pressure (mm Hg)	Pressure (kPa)
0.0	4.6	0.61
10.0	9.2	1.23
20.0	17.5	2.34
30.0	31.8	4.25
40.0	55.3	7.38
50.0	92.5	12.34
60.0	149.5	19.93
70.0	233.7	31.18
80.0	355.1	47.37
90.0	525.8	70.12
100.0	760.0	101.32

Refer to Appendix A to find more values for water-vapor pressure.

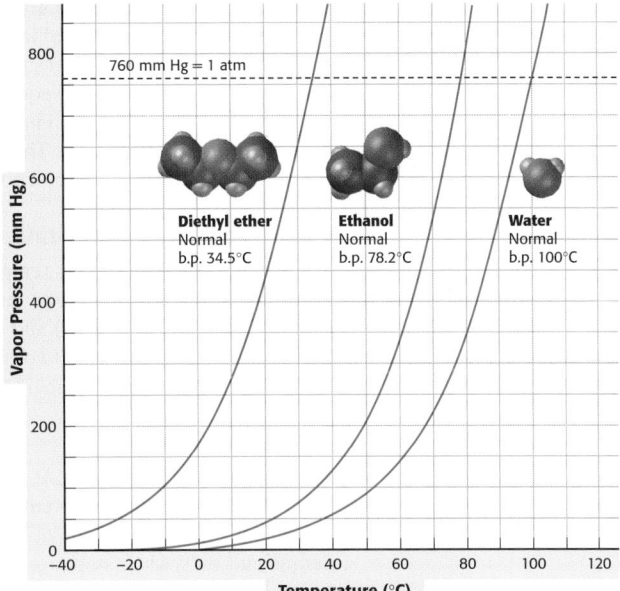

Vapor Pressures of Three Substances at Various Temperatures

760 mm Hg = 1 atm

Diethyl ether Normal b.p. 34.5°C

Ethanol Normal b.p. 78.2°C

Water Normal b.p. 100°C

At 100°C, the vapor pressure has risen to 760.0 mm Hg, which is standard atmospheric pressure, 1 atm (101.32 kPa). The vapor pressure equals the external pressure, and water boils at 100°C. When you increase the temperature of a system to the point at which the vapor pressure of a substance is equal to standard atmospheric pressure—shown as a dotted line in **Figure 22**—you have reached the substance's *normal boiling point.*

The average kinetic energy of molecules increases about 3% for a 10°C increase in temperature, yet the vapor pressure about doubles or triples. The reason is that the fraction of very energetic molecules that can escape about doubles or triples for a 10°C increase in temperature. You can see this relationship at the high energy part of the curves in **Figure 23.**

Figure 22
The dotted line shows standard atmospheric pressure. The point at which the red line crosses the dotted line is the normal boiling point for each substance.

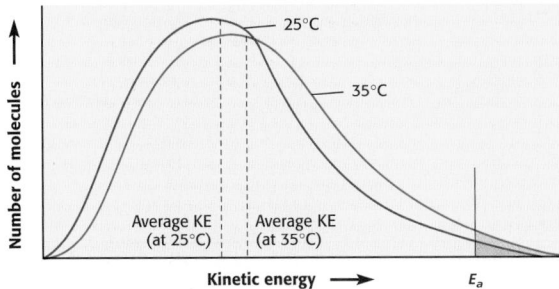

Energy Distribution of Gas Molecules at Different Temperatures

25°C

35°C

Average KE (at 25°C)

Average KE (at 35°C)

Kinetic energy →

E_a

Figure 23
For a 10°C rise in temperature, the average random kinetic energy of molecules increases slightly, but the fraction of molecules that have very high energy (>E_a) increases greatly, as shown by the shaded areas to the right.

401

MISCONCEPTION ALERT

Different Vapor Pressures Students might think that vapor pressure is the same for all liquids. To show that each liquid has its own vapor pressure, pour three different liquids, such as water, acetone, and ethanol, into squirt or spray bottles. Simultaneously squirt or spray equal amounts of each liquid on the board. The liquid with the highest vapor pressure will evaporate first because the intermolecular forces in the liquid are the weakest.

Safety Caution: *Students should wear safety goggles and not breathe vapors.* Check for acetone allergies prior to demonstration. Use a fan for adequate ventilation. Do not perform demonstration in a room with an open flame or spark.

Refer students to **Figure 24.** Ask students to interpret the phase diagram and answer the following questions:

1. What is the relationship between line AD and the boiling point? Ans. The points on this line represent all possible boiling points of water.

2. What can you say about the state of water at any point on the graph that does not lie on a line? Ans. Water will be in the state indicated wherever the point lies.

3. Which point would represent the conditions present in a glass of liquid water and ice cubes held at 0°C at standard atmospheric pressure? Ans. point F

Give students various temperature and pressure combinations and have them determine whether water would be a solid, a liquid, or a gas under those conditions. For example, at 50°C and 90 kPa, water would be a liquid. Because the diagram is not to scale, do not choose temperature and pressure values that intersect close to a line. If possible, supply students with a phase diagram for water that is to scale and have them determine more specific information. **LS Visual**

Teaching Tip

Above the critical point, a substance is a supercritical fluid unless the temperature is very low.

Transparencies

TT Phase Diagram for H_2O
TT Phase Diagram for CO_2

---

phase diagram

a graph of the relationship between the physical state of a substance and the temperature and pressure of the substance

triple point

the temperature and pressure conditions at which the solid, liquid, and gaseous phases of a substance coexist at equilibrium

critical point

the temperature and pressure at which the gas and liquid states of a substance become identical and form one phase

Figure 24
The phase diagram for water shows the physical states of water at different temperatures and pressures. (Note that the diagram is not drawn to scale.)

internet connect

www.scilinks.org
Topic: Phase Diagrams
SciLinks code: HW4130

SCI LINKS Maintained by the National Science Teachers Association

---

Phase Diagrams

You know that a substance's state depends on temperature and that pressure affects state changes. To get a complete picture of how temperature, pressure, and states are related for a particular substance, you can look at a **phase diagram.** A phase diagram has three lines. One line is a vapor pressure curve for the liquid-gas equilibrium. A second line is for the liquid-solid equilibrium, and a third line is for the solid-gas equilibrium. All three lines meet at the **triple point.** The triple point is the only temperature and pressure at which three states of a substance can be in equilibrium.

Phase Diagrams Relate State, Temperature, and Pressure

The *x*-axis of **Figure 24** shows temperature, and the *y*-axis shows pressure. For any given point (x, y), you can see in what state water will be. For example, at 363 K ($x = 90°C$) and standard pressure ($y = 101.3$ kPa), you know that water is a liquid. If you look at the point for these coordinates, it in fact falls in the region labeled "Liquid."

Gas-Liquid Equilibrium

Look at point E on the line AD, where gas and liquid are in equilibrium at 101.3 kPa. If you increase the temperature slightly, liquid will evaporate and only vapor will remain. If you decrease the temperature slightly, vapor condenses and only water remains. Liquid exists to the left of line AD, and vapor exists to the right of AD. Along line AD, the vapor pressure is increasing, so the density of the vapor increases. The liquid decreases in density. At a temperature and pressure called the **critical point,** the liquid and vapor phases of a substance are identical. Above this point, the substance is called a *supercritical fluid*. A supercritical fluid is the state that a substance is in when the liquid and vapor phases are indistinguishable.

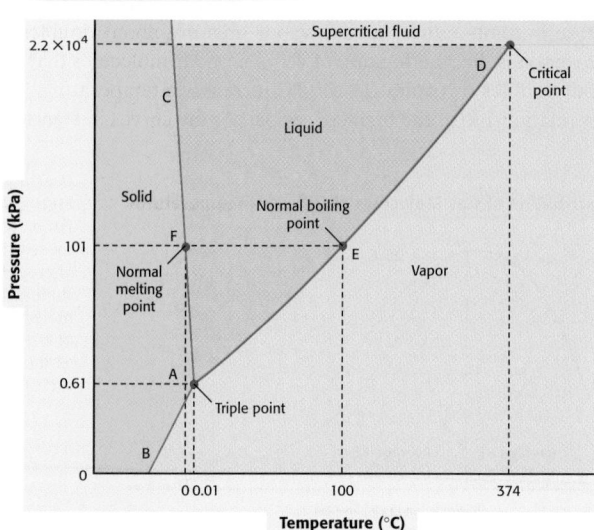

Phase Diagram for H_2O

MISCONCEPTION
///**ALERT**\\\

Triple Point Many students will have trouble understanding that all three states of matter can exist in equilibrium at the same time at the triple point. Liken the triple point to the apex of a pyramid, where a step in any direction places one on a different wall of the structure. At the triple point, a shift in any direction will result in a different physical state of the substance.

REAL-WORLD
CONNECTION

Seasonal Gasoline Because vapor pressure decreases at lower temperatures, the vapor pressure of the liquids that make up gasoline is less on a cold day. Therefore, some gas companies have specific gasoline blends made to be used in cold temperatures because they have higher vapor pressures. Also, the higher vapor pressure of gasoline in hot weather can cause vapor lock. Summer blends of gasoline have lower vapor pressure to avoid this problem.

Solid-Liquid Equilibrium

If you move to the left (at constant pressure) along the line *EF*, you will find a temperature at which the liquid freezes. The line *AC* shows the temperatures and pressures along which solid and liquid are in equilibrium but no vapor is present. If the temperature is decreased further, all of the liquid freezes. Therefore, only solid is present to the left of *AC*. Water is an unusual substance: the solid is less dense than the liquid. If the pressure is increased at point *F*, at constant temperature, water will melt. The line *AC* has a slightly negative slope, which is very rare in phase diagrams of other substances. If the pressure on this system is increased and you move up the line *AC*, you can see that pressure has very little effect on the melting point, so the decrease in temperature is very small.

Solid-Gas Equilibrium

Along the line *AB*, solid is in equilibrium with vapor. If the pressure is decreased below the line *AB*, the solid will sublime. This relationship is the basis of freeze-drying foods, such as those shown in **Figure 25**. The food is frozen, and then a vacuum is applied. Water sublimes, which dehydrates the food very quickly. The food breaks down less when water is removed at the low temperature than when water evaporates at normal temperatures.

Phase Diagrams Are Unique to a Particular Substance

Each pure substance has a unique phase diagram, although the general structure is the same. Each phase diagram has three lines and shows the liquid-solid, liquid-gas, and solid-gas equilibria. These three lines will intersect at the triple point. The triple point is characteristic for each substance and serves to distinguish the substance from other substances.

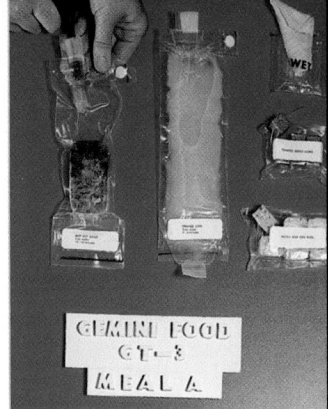

Figure 25
Freeze-drying uses the process of sublimation of ice below the freezing point to dry foods. Many meals prepared for astronauts include freeze-dried foods.

Phase Diagram for CO₂

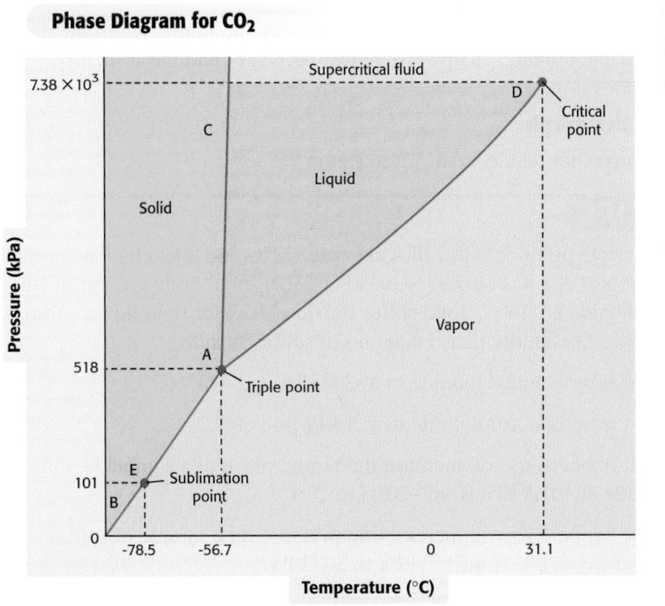

Figure 26
The phase diagram for carbon dioxide shows the physical states of CO_2 at different temperatures and pressures. (Note that the diagram is not drawn to scale.)

403

GEOLOGY
CONNECTION — ADVANCED

Writing — Have students research and write a report that explains the energy source and the mechanisms of geysers. **Ans.** Water is heated by molten rock beneath Earth's surface. If the water has nothing to cause it to stay underground, it forms a hot spring. If it is trapped in an underground chamber, it remains there until it reaches the right combination of temperature and pressure to cause it to instantly evaporate, forcing it out of the ground in a geyser. **LS** Verbal

---

One-Stop Planner CD-ROM

• **Career Extension**
Occupational Applications Worksheet 9: Air-Quality Specialist
Assign this worksheet to emphasize relevant applications of text concepts.

Teaching Tips

Equilibrium Emphasize to students that equilibrium can be either physical or chemical. Equilibrium in this chapter is physical equilibrium because there is no change in identity of the substance involved. They will study chemical equilibrium in detail in another chapter.

States of a Substance Use the phase diagrams to emphasize that any substance can be solid, liquid, or gas, depending on the temperature and pressure of its environment. Substances are often thought of as being whatever state they are in at room temperature.

Using the Figure

Refer students to **Figure 26.** Have them look at **Sample Problem B.** The problem provides information about a phase diagram for carbon dioxide. Have students confirm the values in the sample problem by using the figure.

Group Activity — BASIC

Have small groups of students devise games that involve interpreting phase diagrams. A game could consist of one set of cards with temperatures on them, one set with pressures on them, one set that lists phases, and a phase diagram. Students draw one card from each set. If the substance in the phase diagram is not at that phase when the temperature and pressure are those listed on the cards, the player loses a turn. If the substance is in that phase for the set of conditions, the player is awarded a point. Board games and card games are also possibilities. Have groups exchange games and play them. **LS** Interpersonal

Activity — GENERAL

Have students compare freeze-dried foods and sun-dried foods. Ask them if there is a noticeable difference between the two types of food. Have them discuss why a manufacturer might choose to freeze-dry a food instead of drying it by other methods. **LS** Visual

---

Answers to Practice Problems B

1. a. Phase diagram for sulfur dioxide, SO_2

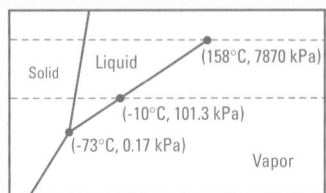

b. solid

c. vapor

d. The sulfur dioxide changes from a liquid to a vapor.

e. The sulfur dioxide remains a liquid.

Homework ── GENERAL

Additional Practice

1. a. The critical point for a substance is at 374°C and 2.2 × 10^4 kPa. The triple point is at 0.01°C and 0.61 kPa. The normal boiling point is at 100°C. Sketch the phase diagram. **Ans.** see Figure 24

b. What phase is the substance in at 200°C and 90 kPa? **Ans.** vapor

c. What phase is the substance in at 0°C and 0.61 kPa? **Ans.** solid

d. What substance is described by the diagram? **Ans.** water

LS Logical

The temperature at which the solid and liquid are in equilibrium—the melting point—is affected little by changes in pressure. Therefore, this line is very nearly vertical when pressure is plotted on the *y*-axis and temperature is plotted on the *x*-axis. Again, water is different in that the solid is less dense than the liquid. Therefore, an increase in pressure decreases the melting point. Most substances, such as carbon dioxide, shown in **Figure 26,** experience a slight increase in melting point when the pressure increases. However, the effect of pressure on boiling point is much greater.

SAMPLE PROBLEM B

How to Draw a Phase Diagram

The triple point of carbon dioxide is at −56.7°C and 518 kPa. The critical point is at 31.1°C and 7.38 × 10^3 kPa. Vapor pressure above solid carbon dioxide is 101.3 kPa at −78.5°C. Solid carbon dioxide is denser than liquid carbon dioxide. Sketch the phase diagram.

1 Gather information.
- triple point of CO_2 = −56.7°C, 518 kPa
- critical point of CO_2 = 31.1°C, 7.38 × 10^3 kPa
- The vapor pressure of the solid is 101.3 kPa (1 atm) at −78.5°C.

2 Plan your work.
- Label the *x*-axis "Temperature" and the *y*-axis "Pressure."
- The vapor pressure curve of the liquid goes from the triple point to the critical point.
- The vapor pressure curve of the solid goes from the triple point through −78.5°C and 101.3 kPa.
- The line for the equilibrium between solid and liquid begins at the triple point, goes upward almost vertically, and has a slightly positive slope.

3 Draw the graph.

The graph that results is shown in **Figure 26.**

PRACTICE HINT

In each phase diagram, the necessary data are triple point, critical point, vapor pressure of the solid or liquid at 1 atm (101.3 kPa), and relative densities of the solid and the liquid. The rough graphs that you draw will be helpful in making predictions but will lack accuracy for most of the data.

PRACTICE

1 a. The triple point of sulfur dioxide is at −73°C and 0.17 kPa. The critical point is at 158°C and 7.87 × 10^3 kPa. The normal boiling point of sulfur dioxide is −10°C. Solid sulfur dioxide is denser than liquid sulfur dioxide. Sketch the phase diagram of sulfur dioxide.

b. What state is sulfur dioxide in at 200 kPa and −100°C?

c. What state is sulfur dioxide in at 1 kPa and 80°C?

d. What happens as you increase the temperature of a sample of sulfur dioxide at 101.3 kPa from −20°C to 20°C?

e. What happens as you increase the pressure on a sample of sulfur dioxide at −11°C from 150 kPa to 300 kPa?

404

MATHEMATICS ─ CONNECTION

A mathematical equation derived by Gibbs is called the phase rule, and it can be used to determine the possible combinations of phases in a system. The equation is $F = C − P + 2$, where *F* is the degrees of freedom, or the number of variables. The number of components in the system is represented by *C*. The number of phases present is given by *P*. For example, if the system has two components and one of the two variables of temperature and pressure remains constant, there will be three phases at equilibrium ($F = 1$ and $C = 2$, so $P = 2 − 1 + 2$, or 3) This rule shows how many phases can exist, but it doesn't reveal which phases they are.

The phase diagram for carbon dioxide is similar to that for water, but there are differences. In the phase diagram for carbon dioxide, the horizontal line at 101.3 kPa does not intersect the solid-liquid line. Thus, carbon dioxide is never a liquid at standard pressure. In fact, if you set dry ice, which is solid carbon dioxide, in a room temperature environment, you can see that it sublimes, or changes directly from a solid to a gas.

The horizontal line at 101.3 kPa intersects the vapor pressure curve for the solid at −78.5°C. Therefore, solid carbon dioxide sublimes at this temperature. This sublimation point is equivalent to the normal boiling point of a liquid such as water. Because dry ice is at equilibrium with carbon dioxide gas at −78.5°C, it is frequently used to provide this low temperature in the laboratory.

 # Section Review

UNDERSTANDING KEY IDEAS

1. A glass of ice water has several ice cubes. Describe the contents of the glass in terms of phase?

2. How is the melting point of a substance defined?

3. What is the connection between vapor pressure and boiling point?

4. What is a supercritical fluid?

5. What happens when dry ice is warmed at 1 atm of pressure?

PRACTICE PROBLEMS

6. Describe what happens if you start with water vapor at 0°C and 0.001 kPa and gradually increase the pressure. Assume constant temperature.

7. **a.** The triple point of benzene is at 5.5°C and 4.8 kPa. The critical point is at 289°C and 4.29×10^3 kPa. Vapor pressure above solid benzene is 101.3 kPa at 80.1°C. Solid benzene is denser than liquid benzene. Sketch the phase diagram of benzene.

   b. In what state is benzene at 200 kPa and 80°C?

   c. In what state is benzene at 10 kPa and 100°C?

   d. What happens as you increase the temperature of a sample of benzene at 101.3 kPa from 0°C to 20°C?

   e. What happens as you decrease the pressure on a sample of benzene at 80°C from 150 kPa to 100 kPa?

CRITICAL THINKING

8. Look at the normal boiling point of diethyl ether in **Figure 22**. What do you think would happen if you warmed a flask of diethyl ether with your hand? (Hint: Normal body temperature is 37°C.)

9. Most rubbing alcohol is isopropyl alcohol, which boils at 82°C. Why does rubbing alcohol have a cooling effect on the skin?

10. **a.** Atmospheric pressure on Mount Everest is 224 mm Hg. What is the boiling point of water there?

    b. What is the freezing point of water on Mount Everest?

11. Why is the triple point near the normal freezing point of a substance?

12. You place an ice cube in a pot of boiling water. The water immediately stops boiling. For a moment, there are three phases of water present: the melting ice cube, the hot liquid water, and the water vapor that formed just before you added the ice. Is this three-phase system in equilibrium? Explain.

Close

Quiz ───────────── GENERAL

1. If sugar is completely dissolved in a glass of water, how many phases are present in the glass? **Ans.** one

2. What is meant by a dynamic equilibrium between a liquid and a gas? **Ans.** Condensation and evaporation occur, but there is no overall (net) change in the amount of liquid or gas present. For every particle that evaporates, a particle condenses.

3. Look at a phase diagram for water. In what phase is water at 100°C and 150 kPa? **Ans.** liquid

4. What three things are related by a phase diagram? **Ans.** state, temperature, and pressure
LS Logical

Reteaching ───────── BASIC

Provide each student with a blank phase diagram for water. Have them fill in the words and numbers that are missing from the diagram. Show a completed diagram on the overhead projector so students can check their work. **LS** Visual

---

Chapter Resource File

• Concept Review
• Quiz

Answers to Section Review

1. A phase is a region of a substance that has identical properties. Each ice cube has the same physical and chemical properties. All of the liquid water in the glass has the same chemical and physical properties.

2. The melting point is the temperature at which the solid and the liquid are in equilibrium.

3. The boiling point is the temperature at which the vapor pressure equals the external pressure.

4. A supercritical fluid is a substance that is above its critical temperature and pressure.

5. The solid turns into a gas, or sublimes.

6. The volume decreases until just before 0.61 kPa, when deposition takes place and ice forms. At very high pressures, the ice melts to liquid.

7. **a.** Phase diagram for benzene, C_6H_6

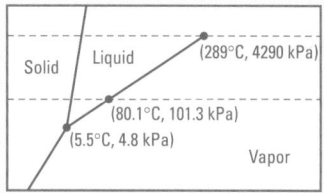

   b. liquid

   c. vapor

Answers continued on p. 413A

Supercritical Fluids

Supercritical CO_2 technology has many commercial applications. It is primarily used in the extraction of essential oils from spices and plants. These essential oils are then used in the production of various foods, flavorings, and fragrances. Supercritical CO_2 is important in the pharmaceutical industry because it extracts integral compounds from plants that are used to synthesize disease-fighting drugs. In the purification of manufactured polymers, supercritical CO_2 removes unreacted monomers from the reaction mix. Also, supercritical CO_2 is useful in the electronics industry because it provides a safe and effective method for cleaning expensive and highly sensitive electrical components.

The commercial application of supercritical CO_2 technology offers several advantages. The solvating power of CO_2 can be adjusted by varying the temperature and pressure of the system, allowing for some widely variable applications of this technology. This technology can be tailored to suit the needs of individual projects. The ability to change the solvating power of supercritical CO_2 ensures an optimum yield of varying products and services. Moreover, applications involving supercritical CO_2 are environmentally friendly, solvent-free processes.

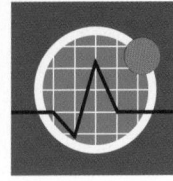

SCIENCE AND TECHNOLOGY

CAREER APPLICATION

Food Scientist

Food science is the study of the chemistry, microbiology, and processing of foods. Food technicians are responsible for testing foods for quality and acceptability in carefully controlled taste tests. Microbiologists in the food industry monitor the safety of food products. Food analysts work in laboratories to monitor the composition of foods and the presence of pesticides. Some food scientists create new food products or food ingredients, such as artificial sweeteners. During their course of study, college students in the field of food science can gain valuable experience by working for food manufacturers and government agencies, such as the U.S. Food and Drug Administration. This experience can help students find jobs after graduation.

internet connect

www.scilinks.org
Topic: Supercritical Fluids
SciLinks code: HW4122

SC*LINKS* Maintained by the National Science Teachers Association

Supercritical Fluids

New Uses for Carbon Dioxide

If the temperature and pressure of a substance are above the critical point, that substance is a *supercritical fluid*. Many supercritical fluids are used for their very effective and selective ability to dissolve other substances. This is very true of carbon dioxide. CO_2 can be made into a supercritical fluid at a relatively low temperature and pressure, so little energy is used in preparing it. Supercritical CO_2 is cheap, nontoxic, and nonflammable and is easy to remove.

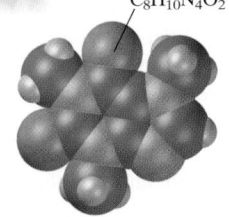

$C_8H_{10}N_4O_2$

Caffeine gives coffee its bitter taste and some people a feeling of restlessness.

Getting a Good Night's Sleep

Supercritical CO_2 is used to remove caffeine from coffee beans. First, the green coffee beans are soaked in water. The beans are then placed in the top of a column that is 70 ft high. Supercritical CO_2 fluid at about 93°C and 250 atm enters at the bottom of the column. The caffeine diffuses out of the beans and into the CO_2. The beans near the bottom of the column mix with almost pure CO_2, which dissolves the last caffeine from the beans. It takes about five hours for fresh beans to move out of the column.

The decaffeinated beans are removed from the bottom, dried, and roasted as usual. The caffeine-rich CO_2 is removed at the top and passed upward through another column. Drops of water fall through the supercritical CO_2 and dissolve the caffeine. The water solution of caffeine is removed and sold to make soft-drinks. The pure CO_2, is recirculated to be used again.

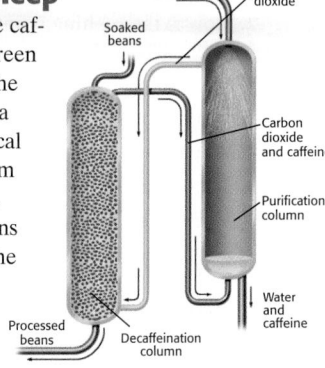

One process used to remove the caffeine from coffee dissolves the caffeine in supercritical CO_2.

Questions

1. Research advantages of using supercritical CO_2 as a solvent.
2. Research and report on other uses of supercritical fluids.

Answers to Feature Questions

1. Answers may vary. Sample answer: Supercritical CO_2 is particularly useful because it is a nonpolar solvent. It cannot be used with products that are damaged when cooled to low temperatures.

2. Answers may vary. Sample answer: Supercritical CO_2 is also used for extracting flavoring oils from various vegetable products, such as ginger oil from ginger root and spearmint oil from mint leaves; determining pesticide and PCB levels in water; and determining organic compounds in soil.

CHAPTER HIGHLIGHTS

CHAPTER HIGHLIGHTS 11

KEY TERMS

surface tension
evaporation
boiling point
condensation
melting
melting point
freezing
freezing point
sublimation

intermolecular forces
dipole-dipole forces
hydrogen bond
London dispersion force

phase
equilibrium
vapor pressure
phase diagram
triple point
critical point

KEY IDEAS

SECTION ONE States and State Changes

- Solid particles vibrate in fixed positions. Thus, solids have a definite shape, volume, and density.
- Liquid particles can move past each other. Thus, liquids change shape and have a definite volume and density.
- Gas particles are far apart from each other. Thus, gases can change shape, volume, and density.
- Solids, liquids, and gases convert from one state to another through freezing, melting, evaporation, condensation, sublimation, and deposition.

SECTION TWO Intermolecular Forces

- The strongest force attracting particles together is the ionic force.
- All ions, atoms, and molecules are attracted by London dispersion forces.
- Polar molecules experience the dipole-dipole force. The dipole-dipole force is usually significant only when the molecules are quite polar.
- Hydrogen bonds are stronger dipole-dipole forces.
- Water's unique properties are due to the combination of the shape of a water molecule and the ability of water to form multiple hydrogen bonds.

SECTION THREE Energy of State Changes

- Energy is needed to change solid to liquid, solid to gas, and liquid to gas. Thus, melting, sublimation, and evaporation are endothermic processes.
- For a given substance, the endothermic state change with the greatest increase in energy is sublimation. Evaporation has a slightly smaller increase in energy, and melting has a much smaller increase in energy.
- The molar enthalpy of fusion of a substance is the energy required to melt 1 mol of the substance at the melting point. The molar enthalpy of vaporization of a substance is the energy required to vaporize 1 mol of the substance.

SECTION FOUR Phase Equilibrium

- A phase diagram shows all of the equilibria between the three states of a substance at various temperatures and pressures.
- On a phase diagram, the triple point is where three phases are in equilibrium. Above the critical point, a substance is a supercritical fluid.
- Water is unique in that increased pressure lowers the freezing point.

KEY SKILLS

Calculating Melting and Boiling Points
Sample Problem A p. 397

How to Draw a Phase Diagram
Sample Problem B p. 404

Alternative Assessments

Homework
- Pages 382, 391

READING SKILL BUILDER
- Page 387

Group Activity
- Page 379

Portfolio Assessments

SKILL BUILDER
- Pages 382, 388

METEOROLOGY CONNECTION
- Page 381

Chapter Resource File

- **Observation Lab** Viscosity of Liquids
- **Observation Lab** Constructing a Heating/Cooling Curve
- **CBL™ Probeware Lab** Evaporation and Intermolecular Attractions

---

407

REVIEW ANSWERS

Using Key Terms

1. solid, liquid, and gas

2. Solid naphthalene undergoes sublimation to a gas, which diffuses through the clothes.

3. the temperature at which solid and liquid are in equilibrium

4. Condensation takes place when water vapor changes to liquid water.

5. The forces of attraction between nonpolar molecules are London dispersion forces.

6. Both involve attractions between polar molecules; hydrogen bonds are a special case of strong dipole-dipole forces.

7. A substance may be a solid, liquid, or gas. These three states are physical conditions of substances. A phase is part of a system and has uniform properties, which may or may not be associated with state.

8. The boiling point is the temperature at which the vapor pressure of a liquid is equal to the external pressure.

9. At the triple point of a substance, solid, liquid, and vapor are at equilibrium.

Understanding Key Ideas

10. Solid particles are close together and vibrate in fixed positions. Liquid particles are close together and can move past each other. Gas particles are independent of one another and move about freely.

CHAPTER REVIEW

USING KEY TERMS

1. Most substances can be in three states. What are they?

2. Explain how solid naphthalene in mothballs is distributed evenly through clothes in a drawer.

3. What is the freezing point of a substance?

4. Clouds form when water vapor in the air collects as liquid water on dust particles in the atmosphere. What physical process takes place?

5. Carbon tetrachloride, CCl_4, is nonpolar. What forces hold the molecules together?

6. Compare dipole-dipole forces and hydrogen bonds.

7. What is the difference between the terms *state* and *phase*?

8. Define *boiling point* in terms of vapor pressure.

9. What is a triple point?

UNDERSTANDING KEY IDEAS

States and State Changes

10. Compare the arrangement and movement of particles in the solid, liquid, and gas states of matter.

11. What is surface tension?

12. A small drop of water assumes an almost spherical form on a Teflon™ surface. Explain why.

13. Compare the temperatures of a substance's melting point and freezing point.

14. What is happening when water is heated from 25°C to 155°C?

15. Give an example of deposition.

16. Give an example of sublimation.

Intermolecular Forces

17. Contrast ionic and molecular substances in terms of the types of attractive forces that govern their behavior.

18. Describe dipole-dipole forces.

19. Why are water molecules polar?

20. Is the melting point of $CaCl_2$ higher than that of NaCl or lower? Explain your answer.

21. A fellow student says, "All substances experience London dispersion forces of attraction between particles." Is this statement true? Explain your answer.

22. Which has larger London dispersion forces between its molecules, CF_4 or CCl_4?

23. Of the three forces, ionic, dipole-dipole, and London dispersion forces, which is the strongest?

24. Hydrogen peroxide, H_2O_2, is shown below. Why is hydrogen peroxide very soluble in water?

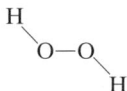

25. Why does ice float in water even though most solids sink in the pure liquid?

26. What accounts for the difference in boiling points between HCl (−85°C) and RbCl (1390°C)?

Assignment Guide	
Section	Questions
1	1–4, 10–16, 68, 73, 74, 78, 80, 81
2	5, 6, 17–28, 70, 71
3	29–40, 52–59, 64, 65, 69, 72
4	7–9, 41–51, 60–63, 66, 67, 75–77, 79, 82–90

27. Why does CBr_4 boil at a higher temperature than CCl_4 does?

28. "All forces between particles are basically polar in nature." Is this statement true? Explain your answer.

Energy of State Changes

29. The molar enthalpy of fusion of water is 6.009 kJ/mol at 0°C. Explain what this statement means.

30. Why is the molar enthalpy of vaporization of a substance much higher than the molar enthalpy of fusion?

31. How do you calculate the entropy change during a change of state at equilibrium?

32. Why is the entropy of a substance higher in the liquid state than in the solid state?

33. During melting, energy is added, the temperature does not change, yet enthalpy increases. How can this be?

34. At 100°C, the enthalpy change for condensation of water vapor to liquid is negative. Is the entropy change positive, or is it negative?

35. ΔH for a process is positive, and ΔS is negative. What can you conclude about the process?

36. What is the enthalpy of fusion for 6 mol of benzene? (Hint: Use **Table 5** to determine the molar enthalpy of fusion of benzene.)

37. Explain why liquid water at 273.0 K will not melt in terms of Gibbs energy.

38. What thermodynamic values do you need to know to calculate a substance's melting point?

39. How does pressure affect the entropy of a gas?

40. How does pressure affect changes between the liquid and vapor states?

Phase Equilibrium

41. You have sweetened iced tea with sugar, and ice cubes are present. How many phases are present?

42. The term *vapor pressure* almost always means the equilibrium vapor pressure. What physical arrangement is needed to measure vapor pressure?

43. What is meant by the statement that a liquid and its vapor in a closed container are in a state of equilibrium?

44. Can solids have vapor pressure?

45. As the temperature of a liquid increases, what happens to the vapor pressure?

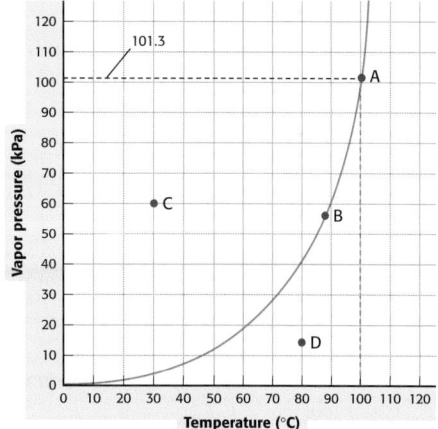

46. Use the above graph of the vapor pressure of water versus temperature to answer the following questions.
  a. At what point(s) does water boil at standard atmospheric pressure?
  b. At what point(s) is water only in the liquid phase?
  c. At what point(s) is water only in the vapor phase?
  d. At what point(s) is liquid water in equilibrium with water vapor?

47. What do the lines of a phase diagram tell you?

11. Surface tension is the force that acts on the surface of a liquid and that tends to minimize the area of the surface. Liquids will thus assume the shape with the lowest possible surface area.

12. The water does not wet the Teflon surface, because the water molecules are not attracted to Teflon. The water forms spherical drops to reduce surface area.

13. For a pure substance, these two points are the same temperature.

14. The molecules gain more kinetic energy of random motion. The water evaporates.

15. Sample answer: Frost forms on a cold night.

16. Sample answer: Molecules of naphthalene escape as a gas from solid crystals in mothballs.

17. Ionic substances experience ionic forces, while molecular substances experience dipole-dipole forces, hydrogen bonds, and London dispersion forces.

18. In dipole-dipole forces, the positive portion of one molecule attracts the negative portion of a neighboring molecule.

19. The difference in electronegativity of hydrogen and oxygen atoms and the bent shape of water molecules cause water to be a polar substance.

20. The melting point of $CaCl_2$ is considerably higher, because the charges on the Ca^{2+} ions are larger than those on the Na^+ ions.

21. Yes, but compared with ionic forces, London dispersion forces are almost negligible.

22. CCl_4 has more electrons and greater London dispersion forces between its molecules even though the molecules are larger and therefore farther apart.

23. Ionic forces are the strongest.

24. Water and hydrogen peroxide form hydrogen bonds with each other.

25. Because of the angle of the covalent bonds in water molecules, the hydrogen bonds in ice are at a particular angle. This arrangement creates a more open crystal structure, which makes ice less dense than water.

26. RbCl has very strong ionic forces (small ions), whereas most of the force between HCl molecules is due to London dispersion forces (and small dipole-dipole forces).

27. CBr_4 has many more electrons in the molecule, so London dispersion forces between molecules are stronger.

28. Yes, even the London dispersion forces are due to temporary dipoles. (However, gravity forces—not discussed here—are not.)

29. There must be an input of 6.009 kJ of energy in the form of heat added to 1 mol of ice to change the ice into liquid water at 0°C.

30. When a substance melts, the particles are still close together. Upon evaporation, the particles are completely separated. More enthalpy is required to separate particles.

31. by dividing the energy of the change by the constant absolute temperature at which the change takes place

32. The particles in the liquid have more-random motion. Randomness gives a higher entropy.

33. The added energy in the form of heat (enthalpy change) is the energy needed to break the particles out of the crystal structure.

48. What two fixed points are on all phase diagrams?

49. Using **Figure 22,** determine the pressure on Pike's Peak, where the boiling point of water is 85°C.

50. Starting at the bottom of **Figure 24,** at a temperature of 100°C, describe what happens as pressure is increased at constant temperature.

51. Starting at the far left of the line *FE* in **Figure 24,** describe what happens as heat is added at constant pressure of 101 kPa.

PRACTICE PROBLEMS

Sample Problem A Calculating Melting and Boiling Points

Calculate the temperatures for the following phase changes. (Liquids are at the normal boiling point.)

52. The enthalpy of fusion of chlorine, Cl_2, is 6.40 kJ/mol, and the entropy of fusion is 37.2 J/mol·K.

53. The enthalpy of fusion of sulfur trioxide is 8.60 kJ/mol, and the entropy of fusion is 29.7 J/mol·K.

54. The enthalpy of fusion of potassium is 2.33 kJ/mol, and the entropy of fusion is 6.91 J/mol·K.

55. The enthalpy of vaporization of butane, C_4H_{10}, is 22.44 kJ/mol, and the entropy of vaporization is 82.2 J/mol·K.

56. ΔH_{vap} for silicon tetrachloride is 28.7 kJ/mol, and ΔS_{vap} is 86.7 J/mol·K.

57. ΔH_{vap} for acetic acid is 23.70 kJ/mol, and ΔS_{vap} is 60.6 J/mol·K.

58. ΔH_{vap} for hydrogen iodide is 19.76 kJ/mol, and ΔS_{vap} is 83.0 J/mol·K.

59. ΔH_{vap} for chloroform, $CHCl_3$, is 29.24 kJ/mol, and ΔS_{vap} is 87.5 J/mol·K.

410

Sample Problem B How to Draw a Phase Diagram

60. The critical point for krypton is at –64°C and a pressure of 5.5×10^3 kPa. The triple point is at –157.4°C and a pressure of 73.2 kPa. At –172°C, the vapor pressure is 13 kPa. The normal boiling point is –152°C. Sketch the phase diagram.

61. The critical point for ammonia is at 132°C and a pressure of 1.14×10^4 kPa. The triple point is at –77.7°C and a pressure of 6 kPa. Normal boiling point is –33.4°C. Sketch the phase diagram.

62. The critical point for carbon tetrachloride is at 283°C and 4.5×10^3 kPa. The triple point is at –87.0°C and 28.9 kPa. The normal boiling point is 76.7°C. Sketch the phase diagram.

MIXED REVIEW

63. The critical point for iodine is at 512°C and 1.13×10^4 kPa. The triple point is at 112.9°C and 11 kPa. The normal boiling point is 183°C. Sketch the phase diagram.

64. Calculate the melting point of acetic acid at standard pressure. The enthalpy of fusion of acetic acid is 11.54 kJ/mol, and the entropy of fusion is 39.8 J/mol·K.

65. ΔH_{vap} for gold is 324 kJ/mol, and ΔS_{vap} is 103.5 J/mol·K. Calculate the boiling point of gold.

66. The critical point for HBr is at 90°C and 8.56×10^3 kPa. The triple point is at –87.0°C and 29 kPa. The normal boiling point is –66.5°C. Sketch the phase diagram.

CRITICAL THINKING

67. How can water be made to evaporate rapidly at room temperature?

68. There is a saying, "Like water off a duck's back." Why does water run off a duck's back?

69. How does a pressure cooker work?

70. If deuterium—an isotope of hydrogen that has one neutron and one proton—were to replace hydrogen in water, would the resulting compound, D_2O, exhibit hydrogen bonds?

71. Which would have the higher boiling point: chloroform, $CHCl_3$, or bromoform, $CHBr_3$?

72. You know that the enthalpy change for vaporizing water is $\Delta H_{vap} = H_{gas} - H_{liq}$. What is the Gibbs energy change for this process?

73. Explain why steam produces much more severe burns than the same amount of boiling water does.

74. Chloroethane ($T_{bp} = -13°C$) has been used as a local anesthetic. When the liquid is sprayed onto the skin, it cools the skin enough to freeze and numb the skin. Explain the cooling effect of this liquid.

75. Is it possible to have *only* liquid water present in a container at 0.00°C? Explain.

76. Consider a system composed of water vapor and liquid at equilibrium at 100°C. Do the molecules of H_2O in the vapor have more kinetic energy than molecules in the liquid do? Explain.

77. Look at the phase diagram for carbon dioxide. How can CO_2 be made to boil?

ALTERNATIVE ASSESSMENT

78. Liquid crystals are substances that have properties of both liquids and crystals. Write a report on these substances and their various uses.

79. Consult reference materials at the library, and prepare a report that discusses freeze-drying, including its history and applications.

80. Some liquids lose all viscosity when cooled to extremely low temperatures—a phenomenon called *superfluidity*. Find out more about the properties of superfluid substances.

81. Many scientists think that more than 99% of the known matter in the universe is made of a fourth state of matter called plasma. Research plasmas, and report your findings to the class.

82. Prepare a report about the adjustments that must be made when cooking and baking at high elevations. Collect instructions for high-elevation adjustments from packages of prepared food mixes. Explain why changes must be made in recipes that will be prepared at high elevations. Check your library for cookbooks that contain information about food preparation at high elevations.

CONCEPT MAPPING

83. Use the following terms to complete the concept map below: *boiling point, liquids, vapor pressure, gases, melting point, states,* and *equilibrium.*

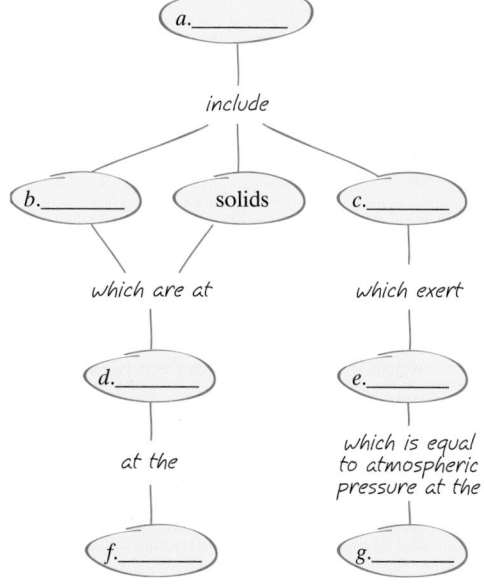

411

34. It is negative. $\Delta S = \Delta H/T$ for a phase change. The sign of the entropy change is the same as the sign of the enthalpy change.

35. ΔG will be positive, so the process will not take place spontaneously.

36. (9.95 kJ/mol)(6 mol) = 59.7 kJ

37. ΔG is positive, so $\Delta H > T\Delta S$.

38. ΔH_{fus} and ΔS_{fus}

39. The entropy of a gas increases as pressure decreases.

40. As pressure decreases, boiling point decreases.

41. two phases: the liquid tea-sugar-water solution and the solid ice

42. The measurement must be in a closed system, in which the rate of evaporation is equal to the rate of condensation.

43. Molecules are constantly leaving the surface of the liquid as vapor and vapor molecules are going into the liquid phase at the same rate.

44. Yes, it is possible for particles to sublime by spontaneously leaving the surface of a solid to become vapor. Such is the case with naphthalene in mothballs.

45. The vapor pressure increases until it reaches the critical pressure.

46. a. point A
b. point C
c. point D
d. points A and B

47. The lines in a phase diagram show the temperatures and pressures at which two phases are in equilibrium: either solid-liquid, liquid-gas, or solid-vapor equilibrium.

48. The critical point and the triple point are each at a definite temperature and pressure.

49. 450 mm Hg

50. Water is a gas at the bottom of the graph. It shrinks as pressure increases until point E is reached. At that point, the vapor condenses to a liquid. Above the line, the liquid increases slightly in density as pressure increases.

51. Water is solid on the far left of the phase diagram. At point *F*, the water melts at 0°C. At point *E*, the water boils at 100°C. To the right of *E*, the water remains gaseous as it is heated further.

Practice Problems

52. 172 K

53. 290 K

54. 337 K

55. 273 K

56. 331 K

57. 391 K

58. 238 K

59. 334 K

60. Phase diagram for krypton, Kr

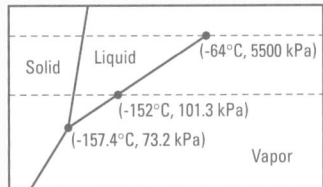

61. Phase diagram for ammonia, NH_3

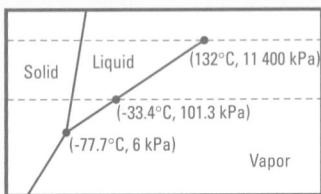

62. Phase diagram for carbon tetrachloride, CCl_4

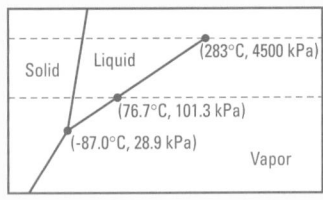

Answers continued on p. 413A

FOCUS ON GRAPHING

Study the graph below, and answer the questions that follow. For help in interpreting graphs, see Appendix B, "Study Skills for Chemistry."

84. What is the normal boiling point of water?

85. Give the coordinates for a point at which only liquid water is present.

86. Give the coordinates for a point at which liquid water is in equilibrium with vapor.

87. Give the coordinates for a point at which only vapor is present.

88. What is the vapor pressure of water at point E?

89. What will happen if the vapor at point D is cooled at constant pressure?

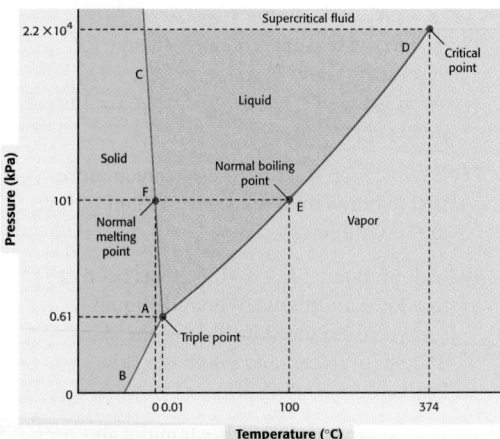

Phase Diagram for H_2O

TECHNOLOGY AND LEARNING

90. Graphing Calculator

Calculating Vapor Pressure by Using a Table

The graphing calculator can run a program that calculates a table for the vapor pressure in atmospheres at different temperatures (K) given the number of moles of a gas and the volume of the gas (V). Given a 0.50 mol gas sample with a volume of 10 L, you can calculate the pressure at 290 K by using a table. Use this program to make the table. Next, use the table to perform the calculations.

Go to Appendix C. If you are using a TI-83 Plus, you can download the program **VAPOR** and data and run the application as directed. If you are using another calculator, your teacher will provide you with keystrokes and data sets to use. After you have run the program, answer the questions.

a. What is the pressure for 1.3 mol of a gas with a volume of 8.0 L and a temperature of 320 K?

b. What is the pressure for 1.5 mol of a gas with a volume of 10.0 L and a temperature of 340 K?

c. Two gases are measured at 300 K. One has an amount of 1.3 mol and a volume of 7.5 L, and the other has an amount of 0.5 mol and a volume of 10.0 L. Which gas has the lesser pressure?

412

Chapter Resource File

• Chapter Test

1. Surface tension is
   a. a skin on the surface of a liquid.
   b. the tendency of the surface of liquids to decrease the area.
   c. a measure of the entropy of a liquid.
   d. the same as vapor pressure.

2. Potatoes cook faster at sea level than at high altitudes because the water
   a. boils more rapidly.
   b. boils at a lower temperature.
   c. increases in temperature while boiling.
   d. boils at a higher temperature.

3. Pure liquids boil at high temperatures under high pressures because
   a. the molecules of liquid are closer together under high pressures.
   b. it takes a high temperature for the vapor pressure to equal the higher external pressure.
   c. the molecules of vapor are farther apart under high pressures.
   d. the entropy change is small at high temperatures and high pressures.

4. In comparing 1 mol of $H_2O(l)$ and $H_2O(g)$ at the same temperature, we find that the
   a. enthalpy of the liquid is higher.
   b. enthalpy of the vapor is higher.
   c. enthalpies are the same and not zero.
   d. enthalpies are both zero.

5. You increase the pressure on a gas. Which of the following would *not* take place?
   a. entropy decrease   **c.** evaporation
   b. enthalpy increase   **d.** sublimation

6. The formation of frost is an example of
   a. condensation.   **c.** deposition.
   b. evaporation.   **d.** melting point.

7. The vapor pressure of alcohol is higher than that of water at room temperature. Thus, the boiling point of alcohol is
   a. higher than that of water.
   b. lower than that of water.
   c. the same as the boiling point of water.
   d. indeterminable.

8. Of solids, liquids, and gases, _____ have the highest degree of disorder.
   a. solids   **c.** gases
   b. liquids   **d.** solids and gases

9. Which of the following substances has the strongest hydrogen bonds?
   a. CH_4 **c.** H_2Se
   b. C_2H_6 **d.** H_2O

10. Which of the following has the highest London dispersion forces between the molecules?
   a. H_2O **c.** H_2Se
   b. H_2S **d.** H_2Te

11. Which of the following has the greatest force between the particles?
   a. NaCl   **c.** Cl_2
   b. HCl   **d.** HOCl

12. Water boils at 100°C. Ethyl alcohol boils at 78.5°C. Which of the following statements are true?
   a. Water has the higher vapor pressure at 78.5°C.
   b. Ethyl alcohol has the higher vapor pressure at 78.5°C.
   c. Both have the same vapor pressure at 78.5°C.
   d. Vapor pressure is not related to boiling point.

413

Continuation of Answers

Answers to Section 1 Review, *continued*

11. No, the water is already at the boiling point, so increasing the heat will not cause the temperature of the water to rise. It will only increase the rate of evaporation.

12. Water vapor is removed from the air when the vapor condenses onto the cold coils.

13. When the water reaches 0°C, ice will begin to form. When all of the water is solid, the ice will cool to −10°C.

14. Answers may include putting some ice cubes and water in a glass. Once some of the cubes have melted, you would determine the temperature.

15. Changes of state do not involve a change in the identity of the substance. Only physical properties change during a state change.

Answers to Section 2 Review, *continued*

8. a. CF_4 is nonpolar because it is symmetrical; CH_2F_2 is polar.

   b. CH_2F_2 most likely has a higher boiling point than CF_4 does because the substances are similar in size but have different polarities. A substance whose polarity is greater generally has a higher boiling point.

9. London dispersion forces are stronger between molecules that have more electrons, so the London dispersion forces in water are much weaker than those in hydrogen sulfide.

10. PH_3 is a nonpolar substance, because the electronegativity difference of the bonds is zero. Thus, PH_3 has only London dispersion forces, whereas molecules of NH_3 can form hydrogen bonds with other NH_3 molecules. Because NH_3 has stronger forces holding its molecules together, it boils at a higher temperature.

11. Both substances are nonpolar and have only London dispersion forces, but argon has more electrons and therefore a stronger London dispersion force.

12. The melting point of KF is higher because fluoride ions are smaller than nitrate ions.

Answers to Section 3 Review, *continued*

12. The temperature exceeds the boiling point, and $T\Delta S_{vap} > \Delta H_{vap}$, so the gas is favored.

13. a. condensation

    b. melting

    c. evaporating

    d. freezing

Answers to Section 4 Review, *continued*

   d. The benzene changes from a solid to a liquid.

   e. The benzene changes from a liquid to a vapor.

8. The ether absorbs energy from your hand and evaporates and cools the skin.

9. The alcohol evaporates readily at body temperature. Heat is absorbed during the endothermic process, and the heat comes from the skin.

10. a. Water boils at about 65°C.

    b. 0.0°C

11. Small changes in pressure have little effect on freezing points.

12. No. The solid → liquid process is not occurring at the same rate as the liquid → solid process. The same can be said for the liquid → vapor process and its reverse process. Also, there is no equilibrium between the solid and vapor phases in this system.

Answers to Chapter Review, *continued*
Mixed Review

63. Phase diagram for iodine, I_2

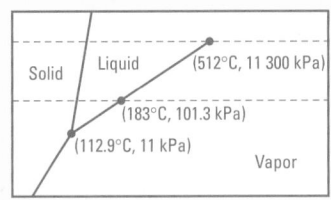

Solid Liquid (512°C, 11 300 kPa)
(183°C, 101.3 kPa)
(112.9°C, 11 kPa)
Vapor

64. 290 K

65. 3130 K

66. Phase diagram for hydrogen bromide, HBr

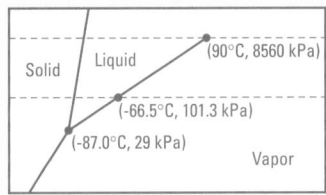

Solid Liquid (90°C, 8560 kPa)
(−66.5°C, 101.3 kPa)
(−87.0°C, 29 kPa)
Vapor

Critical Thinking

67. by applying a vacuum, which will cause the water to boil at a sufficiently low pressure

68. The duck's feathers are covered with a thin film of natural oil. Water forms droplets and runs off the feathers, because the oil prevents the water molecules from adhering to and wetting the feathers.

69. At the higher pressure in the cooker, the boiling point of water is higher. Cooking is faster at this higher temperature.

70. Yes, deuterium is a heavier isotope of hydrogen and is almost identical chemically, so it would still form very polar bonds with oxygen.

71. Bromoform, which has more electrons, would have higher London dispersion forces and therefore a higher boiling point.

72. $\Delta G_{vap} = G_{gas} - G_{liq}$

73. The steam has more thermal energy than does liquid water at its boiling point, 100°C. When the steam comes into contact with the much cooler skin, it releases the heat of condensation. Then, as a liquid, it is still at 100°C, at which temperature additional heat can be transferred to the skin.

74. This liquid boils at a low temperature. Thus, chloroethane evaporates as it hits the skin and absorbs energy of evaporation, thus cooling the skin.

75. Yes, but the slightest addition of heat will raise the temperature, and the slightest removal of heat will produce some ice crystals.

76. No, the random kinetic energy is proportional to the absolute temperature. The average kinetic energy is the same for both.

77. The triple point is 518 kPa (5.11 atm) and –56.7°C. If we put CO_2 under a pressure of 10 atm at about –55°C and then raise the temperature, CO_2 will boil at about –50°C.

Alternative Assessment

78. Answers may vary.

79. Answers may vary.

80. Answers may vary. Sample answer: At about –271°C, helium becomes superfluid. It will flow through extremely small holes that it normally could not flow through. It also forms a thin film on the walls of its container. This film is made up of helium atoms that are flowing up and out of the container.

81. Answers may vary. Sample answer: Plasmas do not have a definite shape or volume and have particles that have broken apart. Plasmas conduct electric current and are affected by electric and magnetic fields. Strong magnetic fields are used to hold plasmas that would destroy a normal container. Natural plasmas are found in such phenomena as lightning, fire, and aurora borealis. Artificial plasmas are made by passing electric charges through gases and are found in fluorescent lights and plasma balls.

82. Answers may vary but should mention that boiling point is affected by external pressure and external pressures are lower at high altitudes.

Concept Mapping

83. a. states

   b. liquids

   c. gases

   d. equilibrium

   e. vapor pressure

   f. melting point

   g. boiling point

84. 100°C

85. Sample answer: (50°C, 100 kPa)

86. Sample answer: (100°C, 101 kPa)

87. Sample answer: (100°C, 0.61 kPa)

88. 101 kPa

89. It will begin to condense.

Technology and Learning

90. a. 4.3 atm

   b. 4.2 atm

   c. the gas with an amount of 0.5 mol and volume 10.0 L

Gases
Chapter Planning Guide

PACING	CLASSROOM RESOURCES	LABS, ACTIVITIES, AND DEMONSTRATIONS
BLOCK 1 · 45 min pp. 414–415 **Chapter Opener**		**SE** **Start-Up Activity** Pressure Relief, p. 415 ◆
BLOCK 2 · 45 min pp. 416–422 **Section 1** Characteristics of Gases	**TT** **Mercury Barometer** *	**TE** **Demonstration** Elastic and Inelastic Collisions, p. 418 **TE** **Group Activity** Representations of Gas Particles, p. 421 GENERAL
BLOCKS 3 & 4 · 90 min pp. 423–432 **Section 2** The Gas Laws	**TT** **Volume Versus Pressure for a Gas at Constant Temperature** * **TT** **Volume Versus Temperature for a Gas at Constant Pressure** * **TT** **Pressure Versus Temperature for a Gas at Constant Volume** * **TT** **Avogadro's Law** *	**TE** **Demonstration** Pressure-Volume Relationships, p. 423 **TE** **Group Activity** Understanding the Gas Laws, p. 430 GENERAL **CRF** **CBL™ Probeware Lab** Pressure-Volume Relationships: Understanding Boyle's Law ◆ ADVANCED
BLOCKS 5 & 6 · 90 min pp. 433–443 **Section 3** Molecular Composition of Gases	**OSP** **Supplemental Reading** The Same and Not the Same: "Static/Dynamic" ADVANCED **TT** **Finding the Volume of an Unknown** * **SE** **Element Spotlight** Nitrogen	**TE** **Demonstration** Amount of Oxygen in Air, p. 434 ◆ **TE** **Demonstration** An Application for Graham's Law, p. 436 ◆ **CRF** **Microscale Lab** Generating the Collecting O_2 ◆ ADVANCED **CRF** **Microscale Lab** Generating the Collecting H_2 ◆ ADVANCED

BLOCKS 7 & 8 · 90 min **Chapter Review and Assessment Resources**

- **SE** Chapter Review, pp. 445–450
- **SE** Standardized Test Prep, p. 451
- **CRF** Chapter Test *
- **OSP** Test Generator
- **CRF** Test Item Listing *
- **OSP** Scoring Rubrics and Classroom Management Checklists

Holt Chemistry: Online Resources

Visit **go.hrw.com** for a variety of free resources related to this textbook. Enter the keyword **HW4 HOME**.

Holt
Online
Learning

Students can access interactive problem solving help and active visual concept development with the *Holt Chemistry* Online Edition available at **www.hrw.com**

student
CNN News

cnnstudentnews.com

Find the latest chemistry news, lesson plans, and activities related to important scientific events.

Compression guide:
To shorten from your instruction because of time limitations, omit blocks 1, 4, 6, and 7.

KEY

TE Teacher Edition	**CRF** Chapter Resource File	***** Also on One-Stop Planner
SE Student Edition	**TT** Teaching Transparencies	**◆** Requires Advance Prep
OSP One-Stop Planner		

PROBLEM SOLVING AND PRACTICE	SECTION REVIEW AND ASSESSMENT	STANDARDS CORRELATION
		National Science Education Standards
SE Sample Problem A Converting Pressure Units, p. 420 **SE Practice**, p. 421 `GENERAL`　　　**TE Homework**, p. 420 `GENERAL` **CRF Problem Solving** * `ADVANCED` **SE Problem Bank**, p. 858 `GENERAL`	**TE Quiz**, p. 422 `GENERAL` **CRF Quiz** * **TE Reteaching**, p. 422 `BASIC` **SE Section Review**, p. 422 **CRF Concept Review** *	PS 2e, PS 5b, PS 5c
SE Sample Problem B Solving Pressure-Volume Problems, p. 425 **SE Practice**, p. 425 `GENERAL`　　　**TE Homework**, p. 425 `GENERAL` **SE Sample Problem C** Solving Volume-Temperature Problems, p. 428 **SE Practice**, p. 428 `GENERAL`　　　**TE Homework**, p. 428 `GENERAL` **SE Sample Problem D** Solving Pressure-Temperature Problems, p. 430 **SE Practice**, p. 431 `GENERAL`　　　**TE Homework**, p. 431 `GENERAL` **CRF Problem Solving** * `ADVANCED` **SE Problem Bank**, p. 858 `GENERAL`	**TE Quiz**, p. 432 `GENERAL` **CRF Quiz** * **TE Reteaching**, p. 432 `BASIC` **SE Section Review**, p. 432 **CRF Concept Review** *	PS 2e, PS 5b, PS 5c
SE Sample Problem E Using the Ideal Gas Law, p. 435 **SE Practice**, p. 435 `GENERAL`　　　**TE Homework**, p. 435 `GENERAL` **SE Sample Problem F** Comparing Molecular Speeds, p. 438 **SE Practice**, p. 438 `GENERAL`　　　**TE Homework**, p. 438 `GENERAL` **SE Skills Toolkit 1** Finding Volume of Unknown, p. 441 **SE Sample Problem G** Using the Ideal Gas Law to Solve Stoichiometry Problems, p. 441 **SE Practice**, p. 442 `GENERAL`　　　**TE Homework**, p. 441 `GENERAL` **CRF Problem Solving** * `ADVANCED` **SE Problem Bank**, p. 858 `GENERAL`	**TE Quiz**, p. 442 `GENERAL` **CRF Quiz** * **TE Reteaching**, p. 442 `BASIC` **SE Section Review**, p. 442 **CRF Concept Review** *	PS 2e, PS 5b, PS 5c

SC/LINKS.
www.scilinks.org

Topic: Gases
SciLinks code: HW4152

Topic: Atmospheric Pressure
SciLinks code: HW4013

Topic: The Gas Laws
SciLinks code: HW4063

Topic: Avogadro's Law
SciLinks code: HW4137

Topic: Effusion/Diffusion
SciLinks code: HW4041

Topic: Nitrogen Cycle
SciLinks code: HW4082

Technology Resources

  **Science in the NEWS**
Each video segment is accompanied by a Critical Thinking Worksheet.

Segment 21
Global Warming

Segment 22
Belching Cows

 ChemFile Interactive Tutor CD-Rom

Module 6: Gas Laws

Topic: Behavior of Gases

Overview

This chapter focuses on one specific state of matter—gases. In studying this chapter, students will become familiar with the characteristics of gases such as compressibility and density. The concept of pressure is explained, as well as standard conditions. The kinetic-molecular theory is introduced. A large portion of the chapter is devoted to developing the mathematical relationships between pressure, volume, temperature and amount of gas by means of the gas laws. The final portion of the chapter develops molar relationships for gases, among them the ideal gas law and stoichiometry.

Assessing Prior Knowledge

Check for Content Knowledge

Students should be familiar with the following topics:

• states of matter
• kinetic energy
• heat
• temperature
• the mole
• chemical equations
• stoichiometry

Using the Figure

A hot-air balloon is first inflated with a giant motorized fan. A propane heater warms the air trapped in the cloth balloon. The warmed air can escape from the bottom, making the balloon less dense than the surrounding air; therefore, it rises. People who pilot balloons need a knowledge of the gas laws in order to safely fly their aircraft.

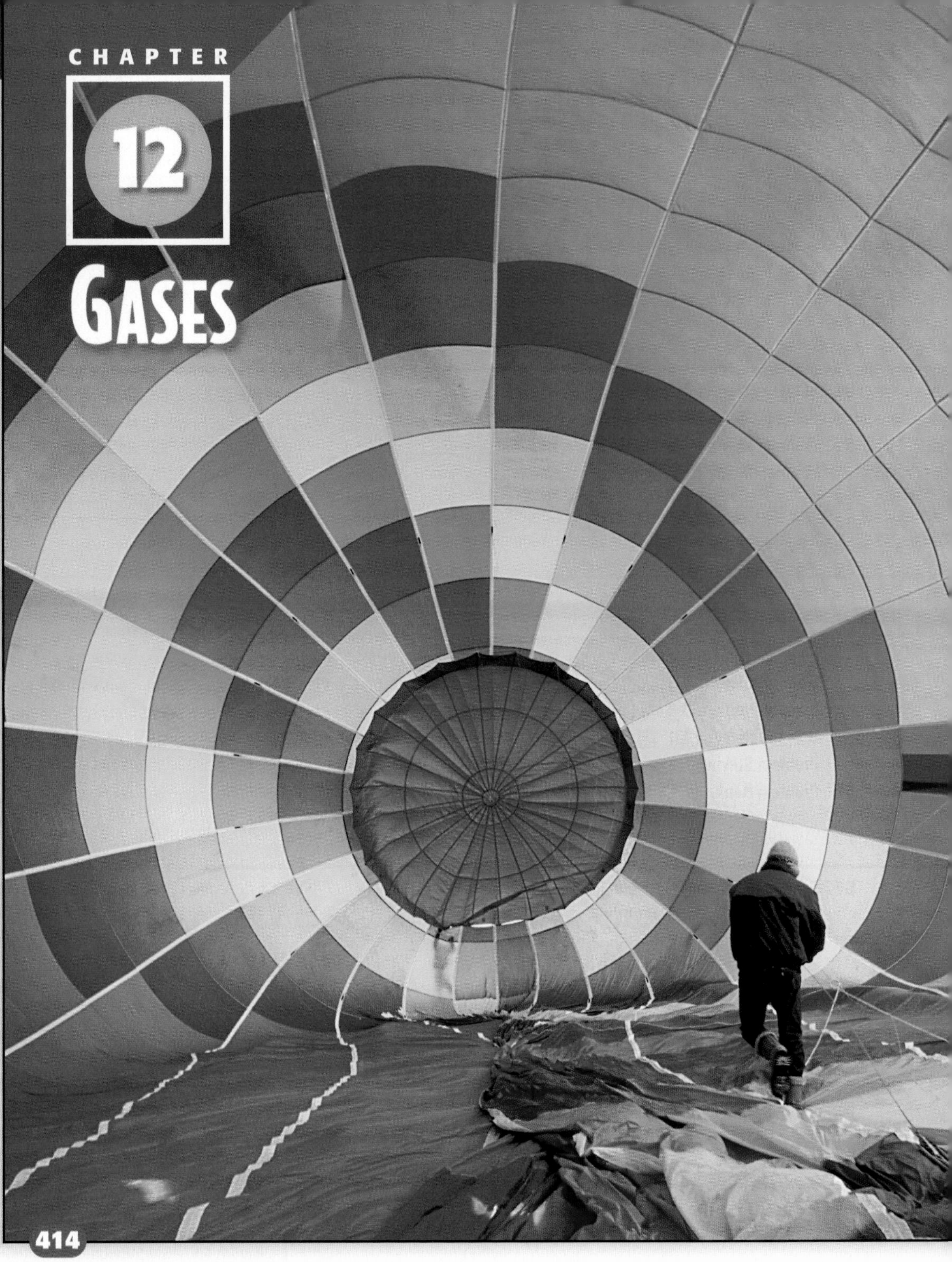

CHAPTER

12
GASES

414

Standards Correlations

National Science Education Standards

PS 2e: Solids, liquids, and gases differ in the distances and angles between molecules or atoms and therefore the energy that binds them together. In solids the structure is nearly rigid; in liquids molecules or atoms move around each other but do not move apart; and in gases molecules or atoms move almost independently of each other and are mostly far apart. (Sections 1, 2, and 3)

PS 5b: All energy can be considered to be either kinetic energy, which is the energy of motion; potential energy, which depends on relative position; or energy contained by a field, such as electromagnetic waves. (Section 1)

PS 5c: Heat consists of random motion and the vibrations of atoms, molecules, and ions. The higher the temperature, the greater the atomic or molecular motion. (Sections 1 and 2)

The hot-air balloon pictured here is being filled with hot air. Hot air expands, so the air in the balloon will be less dense. The balloon will be lifted up by the cooler, denser air outside it. The hot-air balloon demonstrates some important facts about gases: gases expand when heated, they have weight, and they have mass.

START-UP *ACTIVITY*

Pressure Relief

SAFETY PRECAUTIONS

PROCEDURE

1. Blow up a **round latex balloon,** and let the air out.
2. Put the round part of the balloon inside a **PET bottle.** Roll the neck of the balloon over the mouth of the bottle, and secure the neck of the balloon with a **rubber band** so that it dangles into the bottle.
3. Try to blow up the balloon. Record the results.
4. Answer the first two analysis questions below.
5. Design a modification to the balloon-in-a-bottle apparatus that will allow the balloon to inflate. Answer the third analysis question below.
6. If your teacher approves of your design, try it out.

ANALYSIS

1. What causes the balloon to expand?
2. What happens to air that is trapped in the bottle when you blow into the balloon?
3. Describe your design modification, and include a sketch, if applicable. Explain why your design works.

Pre-Reading Questions

1. **Among the states of matter, what is unique about gases?**
2. **Do gases have mass and weight? How can you tell?**
3. **Gases are considered fluids. Why?**

CONTENTS 12

internet connect

www.scilinks.org
Topic: Gases
SciLinks code: HW4152

SC*LINKS*. Maintained by the National Science Teachers Association

START-UP *ACTIVITY*

Skills Acquired:
• Experimenting
• Designing Experiments
• Interpreting
• Inferring

Materials:
For each group of 2–4 students:
• 2-liter plastic soft drink bottle
• Balloon
• Rubber band

Safety Caution: Check prior to the activity if any students are allergic to latex and should avoid contact with the balloon. Make sure only one student blows up each balloon.

Answers

1. The balloon expands because of the pressure of the air being blown into it. After the balloon expands slightly, the pressure of the air inside the bottle becomes large enough to match the person's ability to exert air pressure into the balloon.
2. It becomes compressed in the space between the bottle and the balloon.
3. Designs should incorporate a means for air to leave the bottle as the balloon is inflated. This could include a hole in the side or bottom, or a tube that delivers air back up the neck, underneath the balloon.

Answers to Pre-Reading Questions

1. Particles of a gas are much farther apart from one another than particles of a liquid or a solid are.
2. Gases have mass and therefore weight. The example of the hot-air balloon, for instance, demonstrates that denser air has the capability to buoy up an object containing less-dense air.
3. Gas particles can easily flow and move past each other, so gases are considered fluids.

Before beginning this section, review with your students the Objectives listed in the Student Edition. This section develops concepts introduced in previous discussions of the states of matter such as the fluidity, compressibility and low density of gases. It also introduces pressure and its units as well as the kinetic molecular theory.

Have students make a list of gases. Ask them to separate the list into elements compounds and mixtures. Students can share these lists with the class by putting them on the board.

MISCONCEPTION
///ALERT\\\

Check to see if some students consider "air" to be an element.

Motivate

Identifying Preconceptions

Show students an "empty bottle". Ask them to describe what is in the bottle. Most of them will say "nothing". Help them come to the realization that a gas takes up space (ask them why an inflated balloon resists being compressed) and has mass. Explain that the reason a helium-filled balloon floats in air is because helium is less dense than air.

KEY TERMS
- pressure
- newton
- pascal
- standard temperature and pressure
- kinetic-molecular theory

OBJECTIVES

(1) **Describe** the general properties of gases.

(2) **Define** pressure, give the SI unit for pressure, and convert between standard units of pressure.

(3) **Relate** the kinetic-molecular theory to the properties of an ideal gas.

Properties of Gases

Each state of matter has its own properties. Gases have unique properties because the distance between the particles of a gas is much greater than the distance between the particles of a liquid or a solid. Although liquids and solids seem very different from each other, both have small intermolecular distances. Gas particles, however, are much farther apart from each other than liquid and solid particles are. In some ways, gases behave like liquids; in other ways, they have unique properties.

Gases Are Fluids

Gases are considered *fluids*. People often use the word *fluid* to mean "liquid." However, the word *fluid* actually means "any substance that can flow." Gases are fluids because they are able to flow. Gas particles can flow because they are relatively far apart and therefore are able to move past each other easily. In **Figure 1,** a strip of copper is reacting with nitric acid to form nitrogen dioxide, a brown gas. Like all gases, nitrogen dioxide is a fluid. The gas flows over the sides of the beaker.

Figure 1
The reaction in the beaker in this photo has formed NO_2, a brown gas, which flows out of the container.

Chapter Resource File

- Lesson Plan

 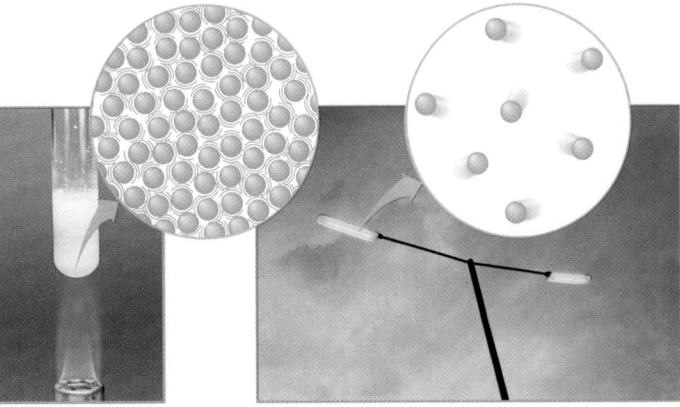

Gases Have Low Density

Gases have much lower densities than liquids and solids do. Because of the relatively large distances between gas particles, most of the volume occupied by a gas is empty space. As shown in **Figure 2,** particles in solids and liquids are almost in contact with each other, but gas particles are much farther apart. This distance between particles shows why a substance in the liquid or solid state always has a much greater density than the same substance in the gaseous state does. The low density of gases also means that gas particles travel relatively long distances before colliding with each other.

Gases Are Highly Compressible

Suppose you completely fill a syringe with liquid and try to push the plunger in when the opening is plugged. You cannot make the space the liquid takes up become smaller. It takes very great pressure to reduce the volume of a liquid or solid. However, if only gas is in the syringe, with a little effort you can move the plunger down and compress the gas. As shown in **Figure 3,** gas particles can be pushed closer together. The space occupied by the gas particles themselves is very small compared with the total volume of the gas. Therefore, applying a small pressure will move the gas particles closer together and will decrease the volume.

Figure 2
Particles of sodium in the solid, liquid, and gas phases are shown. The atoms of gaseous sodium are much farther apart, as in a sodium vapor street lamp.

Figure 3
The volume occupied by a gas can be reduced because gas molecules can move closer together.

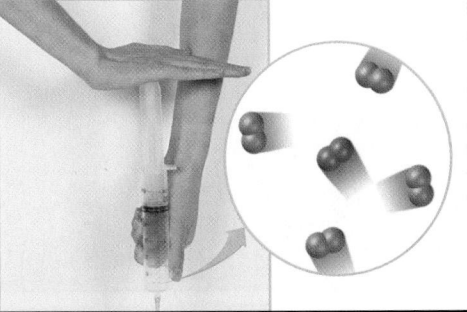 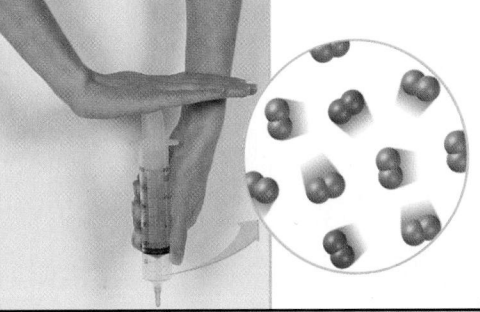

417

did you know?

Distance Between Gas Particles Gas particles are approximately 10 times farther apart than the particles of a liquid or a solid. As a result, the volume of a substance in the gas state is roughly 1000 ($10 \times 10 \times 10$) times the volume of the same amount of the substance in the liquid or solid state. An easy way to show this is to display a graduated cylinder with 18 mL of water and a box or bucket whose volume is roughly equal to 22.4 L to show how much volume the equivalent amount of water vapor would occupy at 0°C and 1 atmosphere pressure.

Teach

READING SKILL BUILDER — BASIC

Assimilating Knowledge Before students read this chapter, have them write a short list of all the things they know (or think they know) about the structure of matter and the changes in matter. Then have students list the things they might want to know about the nature of matter. For example, they might ask the following questions:

• Why must a cork be removed from a test tube before heating it?

• When the weather report talks of air pressure, why do I not feel any pressure?

• Why does a balloon change sizes as you go from a warm room to a cold car or vice versa?

Later, when students are reviewing for the chapter test, have them read over their list and correct or expand upon their answers.
LS Logical

READING SKILL BUILDER — GENERAL

Writing **Paired Summarizing** Have students read the section and then join with a partner. Have each person write a summary of what they have read. Then exchange papers and write an abstract of what the other person wrote. In this step have them note what they both understood, both did not understand and then formulate questions for the teacher and their classmates on the reading. When this step is completed have the entire class share questions and/or abstracts. **LS** Interpersonal

Using the Figure

If you have a plastic syringe, demonstrate **Figure 3,** and allow students to do it themselves to feel the increase in pressure. Ask students to draw a diagram to explain why the compression shown in **Figure 3** is possible. Ask them if they think the compression would be possible if a liquid were in the syringe.

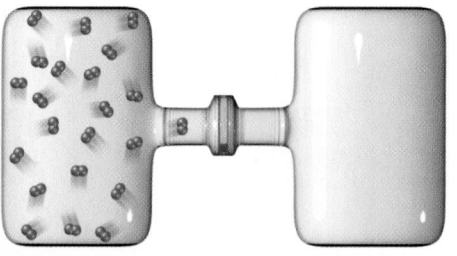
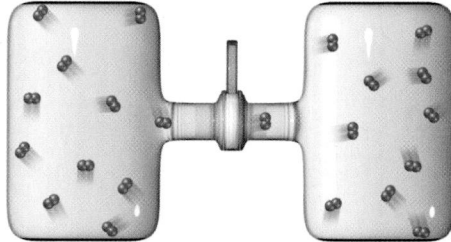

Demonstration

Elastic and Inelastic Collisions

(Approximate time: 5 minutes)

1. Obtain a set of "happy" and "sad" balls from a scientific supply house. (The "happy" ball is made of a very elastic rubber, like that used in "super balls." The "sad" ball is made of a very inelastic rubber, like that used for cushioning collisions and for making the soles of shoes.)

2. Bounce the "happy" ball on the table several times while initiating a discussion on elastic collisions.

3. Ask for two volunteers to help determine the ball's elasticity by measuring how high it bounces.

4. Hand the "sad" ball to a student. Ask the other student to hold a meter-stick while the other drops the ball, and stand back.

5. Explain how the two balls differ, and discuss the types of collisions each underwent. Ask students what happened to the kinetic energy of the inelastic ball when it hit the floor. It was converted to heat by internal friction as the ball became distorted.

Safety: Safety goggles are recommended for all participants. The rest of the class must stand back at least 10 ft.

Disposal: None

Figure 4
When the partition between the gas and the empty container (vacuum) is removed, the gas flows into the empty container and fills the entire volume of both containers.

Figure 5
Gas molecules in the atmosphere collide with Earth's surface, creating atmospheric pressure.

Gases Completely Fill a Container

A solid has a certain shape and volume. A liquid has a certain volume but takes the shape of the lower part of its container. In contrast, a gas completely fills its container. This principle is shown in **Figure 4.** One of the containers contains a gas, and the other container is empty. When the barrier between the two containers is removed, the gas rushes into the empty container until it fills both containers equally. Gas particles are constantly moving at high speeds and are far apart enough that they do not attract each other as much as particles of solids and liquids do. Therefore, a gas expands to fill the entire volume available.

Gas Pressure

Gases are all around you. Earth's atmosphere, commonly known as *air,* is a mixture of gases: mainly nitrogen and oxygen. Because you cannot always feel air, you may have thought of gases as being weightless, but all gases have mass; therefore, they have weight in a gravitational field. As gas molecules are pulled toward the surface of Earth, they collide with each other and with the surface of Earth more often, as shown in **Figure 5.** Collisions of gas molecules are what cause *air pressure.*

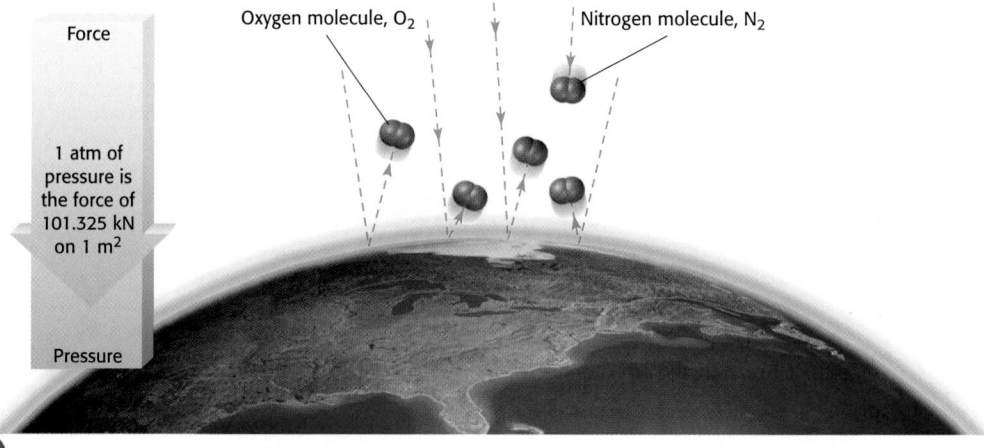

Force

1 atm of pressure is the force of 101.325 kN on 1 m²

Pressure

Oxygen molecule, O_2

Nitrogen molecule, N_2

HISTORY CONNECTION

In the seventeenth century, many scientists believed that liquid suspended in eudiometer tubes was held in place by invisible "ropes." Robert Boyle showed this idea to be untrue by inserting a long wire into a eudiometer tube and wiggling it back and forth. Because the wire did not catch on the rope, the rope did not exist. It is for simple and straightforward experiments such as this that Boyle is considered the father of modern chemistry.

You may have noticed that your ears sometimes "pop" when you ascend to high altitudes or fly in an airplane. This popping happens because the density of the air changes when you change altitudes. The atmosphere is denser as you move closer to Earth's surface because the weight of atmospheric gases at any elevation compresses the gases below. Less-dense air exerts less pressure. Your ears pop when the air inside your ears changes to the same pressure as the outside air.

Measuring Pressure

The scientific definition of **pressure** is "force divided by area." To find pressure, you need to know the force and the area over which that force is exerted.

The unit of force in SI units is the **newton.** One newton is the force that gives an acceleration of 1 m/s^2 to an object whose mass is 1 kg.

$$1 \text{ newton} = 1 \text{ kg} \times 1 \text{ m/s}^2 = 1 \text{ N}$$

The SI unit of pressure is the **pascal,** Pa, which is the force of one newton applied over an area of one square meter.

$$1 \text{ Pa} = 1 \text{ N/1 m}^2$$

One pascal is a small unit of pressure. It is the pressure exerted by a layer of water that is 0.102 mm deep over an area of one square meter.

Atmospheric pressure can be measured by a barometer, as shown in **Figure 6.** The atmosphere exerts pressure on the surface of mercury in the dish. This pressure goes through the fluid and up the column of mercury. The mercury settles at a point where the pressure exerted downward by its weight equals the pressure exerted by the atmosphere.

pressure

the amount of force exerted per unit area of surface

newton

the SI unit for force; the force that will increase the speed of a 1 kg mass by 1 m/s each second that the force is applied (abbreviation, N)

pascal

the SI unit of pressure; equal to the force of 1 N exerted over an area of 1 m^2 (abbreviation, Pa)

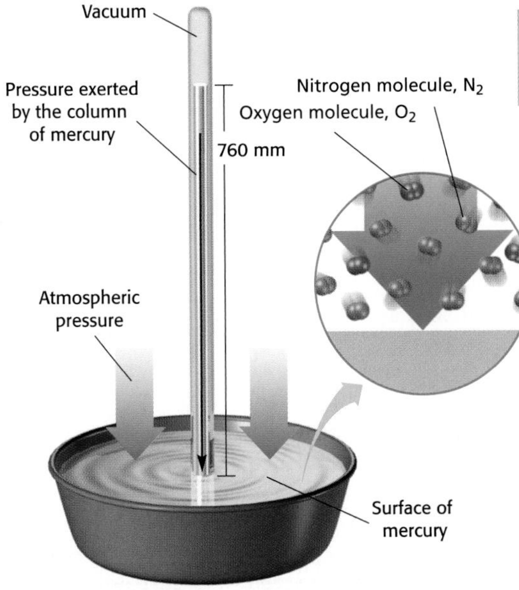

Vacuum

Pressure exerted by the column of mercury

760 mm

Nitrogen molecule, N$_2$
Oxygen molecule, O$_2$

Atmospheric pressure

Surface of mercury

Figure 6
In the sealed column of a barometer, the mercury falls until the pressure exerted by its weight equals the atmospheric pressure.

Teaching Tip

Help students to understand the ideas of force and pressure by discussing how manufacturers test airplane escape slides. Manufacturers have a machine that stretches the fabric of the slide and then subjects it to rapid repeated impacts from a small-diameter metal rod. When passengers are asked to evacuate a plane via one of these slides, they are asked to remove their shoes. Some people do not obey these requests. This impact test can tell if the slide will hold up in a real escape situation.

Discuss why a high heeled shoe worn by a 150 pound person would prove to be more of a hazard than a standard oxford/loafer style shoe worn by a 150 pound person. **Ans.** The same force is being applied over a smaller area, so there is greater pressure.

Transparencies

TT Mercury Barometer

HISTORY
CONNECTION

The barometer was invented by the Italian physicist and mathematician Evangelista Torricelli (1608–1647) in 1643. Torricelli was a student of Galileo. The pressure unit torr was named in his honor.

Table 1   Pressure Units

Unit	Abbreviation	Equivalent number of pascals
Atmosphere	atm	1 atm = 101 325 Pa
Bar	bar	1 bar = 100 025 Pa
Millimeter of mercury	mm Hg	1 mm Hg = 133.322 Pa
Pascal	Pa	1
Pounds per square inch	psi	1 psi = $6.892\ 86 \times 10^3$ Pa
Torr	torr	1 torr = 133.322 Pa

Teach, *continued*

Answers to Practice Problems A

1. 7.37×10^6 Pa
2. 92.48 mm Hg
3. 0.9869 atm
4. 3.08×10^5 Pa

Homework ————— GENERAL

Additional Practice

1. Define a Pascal in terms of Newtons and meters squared. **Ans.** 1 Pascal = 1 Newton/ 1 meter squared.

2. How many Pascals are in a kiloPascal? **Ans.** 1000 Pa

3. What pressure in Pascals is exerted if a force of 10.0 Newtons is applied over an area of 2.5 square meters? **Ans.** 4.0 Pa

4. What pressure in Pascals is exerted if a force of 400.0 Newtons is applied over an area of 4.0 square meters? **Ans.** 100 Pa

5. Express the answer to number 4 in kiloPascals. **Ans.** 0.100 kPa

6. How many kiloPascals equal an atmosphere? **Ans.** 101.3 kPa

7. Express the answer to number 4 in atmospheres. **Ans.** 9.87×10^{-4} atm

LS Logical

standard temperature and pressure

for a gas, the temperature of 0°C and the pressure 1.00 atmosphere

At sea level, the atmosphere keeps the mercury in a barometer at an average height of 760 mm, which is 1 atmosphere. One millimeter of mercury is also called a *torr*, after Evangelista Torricelli, the Italian physicist who invented the barometer. Other units of pressure are listed in **Table 1.**

In studying the effects of changing temperature and pressure on a gas, one will find a standard for comparison useful. Scientists have specified a set of standard conditions called **standard temperature and pressure,** or STP, which is equal to 0°C and 1 atm.

SAMPLE PROBLEM A

Converting Pressure Units

Convert the pressure of 1.000 atm to millimeters of mercury.

1 Gather information.

From **Table 1,** 1 atmosphere = 101 325 Pa, 1 mm Hg = 133.322 Pa

2 Plan your work.

Both units of pressure can be converted to pascals, so use pascals to convert atmospheres to millimeters of mercury.

The conversion factors are $\dfrac{101\ 325\ \text{Pa}}{1\ \text{atm}}$ and $\dfrac{1\ \text{mm Hg}}{133.322\ \text{Pa}}$.

3 Calculate.

$$1.000\ \text{atm} \times \frac{101\ 325\ \text{Pa}}{1\ \text{atm}} \times \frac{1\ \text{mm Hg}}{133.322\ \text{Pa}} = 760.0\ \text{mm Hg}$$

4 Verify your results.

By looking at **Table 1,** you can see that about 100 000 pascals equal one atmosphere, and the number of pascals that equal 1 mm Hg is just over 100. Therefore, the number of millimeters of mercury equivalent to 1 atm is just below 1000 (100 000 divided by 100). The answer is therefore reasonable.

PRACTICE HINT

Remember that millimeters of mercury is actually a unit of pressure, so it cannot be canceled by a length measurement.

420

Chapter Resource File

• Problem Solving

1. The critical pressure of carbon dioxide is 72.7 atm. What is this value in units of pascals?

2. The vapor pressure of water at 50.0°C is 12.33 kPa. What is this value in millimeters of mercury?

3. In thermodynamics, the standard pressure is 100.0 kPa. What is this value in units of atmospheres?

4. A tire is inflated to a gauge pressure of 30.0 psi (Which must be added to the atmospheric pressure of 14.7 psi to find the total pressure in the tire). Calculate the total pressure in the tire in pascals.

PROBLEM SOLVING SKILL

The Kinetic-Molecular Theory

The properties of gases stated earlier are explained on the molecular level in terms of the **kinetic-molecular theory.** The kinetic-molecular theory is a model that is used to predict gas behavior.

The kinetic-molecular theory states that gas particles are in constant rapid, random motion. The theory also states that the particles of a gas are very far apart relative to their size. This idea explains the fluidity and compressibility of gases. Gas particles can easily move past one another or move closer together because they are farther apart than liquid or solid particles.

Gas particles in constant motion collide with each other and with the walls of their container. The kinetic-molecular theory states that the pressure exerted by a gas is a result of collisions of the molecules against the walls of the container, as shown in **Figure 7.** The kinetic-molecular theory considers collisions of gas particles to be perfectly elastic; that is, energy is completely transferred during collisions. The total energy of the system, however, remains constant.

kinetic-molecular theory

a theory that explains that the behavior of physical systems depends on the combined actions of the molecules constituting the system

internet connect

www.scilinks.org
Topic: Kinetic Theory
SciLinks code: HW4157

SCiLINKS. Maintained by the National Science Teachers Association

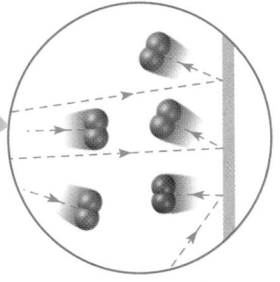

Figure 7
Gas molecules travel far, relative to their size, in straight lines until they collide with other molecules or the walls of the container.

421

Using the Figure

Figure 8 This figure should be carefully explained to students. Use it to stress that temperature is proportional to the *average* kinetic energy of a collection of particles. In that collection, any one particle may have nearly any kinetic energy. This is indicated by the fact that both curves start at zero kinetic energy and extend far beyond the average temperature. As the temperature of the collection increases, the average kinetic energy of the collection increases, as illustrated by the red curve.

Close

Quiz ———— GENERAL

1. What does it mean to be a fluid? Ans. A fluid is capable of flowing.

2. How do densities of gases compare to that of solids and liquids? Ans. they are much smaller

3. Why is it easy to compress a gas? Ans. Particles in a gas are far apart. Exerting pressure on it forces them closer together.

4. What causes atmospheric pressure? Ans. the force of the blanket of air on a surface
LS Verbal

Reteaching ———— BASIC

Have students create a graphic organizer that describes the behavior of gases in terms of the motion of their particles. LS Visual

Chapter Resource File
- **Concept Review**
- **Section Quiz**

Figure 8
Increasing the temperature of a gas shifts the energy distribution in the direction of greater average kinetic energy.

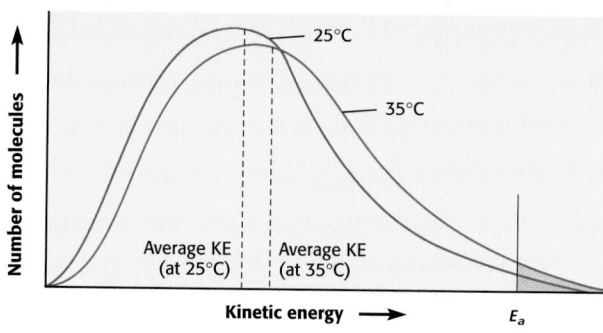

Energy Distribution of Gas Molecules at Different Temperatures

Gas Temperature Is Proportional to Average Kinetic Energy

Molecules are always in random motion. The average kinetic energy of random motion is proportional to the absolute temperature, or temperature in kelvins. Heat increases the energy of random motion of a gas.

But not all molecules are traveling at the same speed. As a result of multiple collisions, the molecules have a range of speeds. **Figure 8** shows the numbers of molecules at various energies. For a $10°C$ rise in temperature from STP, the average energy increases about 3%, while the number of very high-energy molecules approximately doubles or triples.

① Section Review

UNDERSTANDING KEY IDEAS

1. What characteristic of gases makes them different from liquids or solids?

2. Why are gases considered fluids?

3. What happens to gas particles when a gas is compressed?

4. What is the difference between force and pressure?

5. What is the SI unit of pressure, and how is it defined?

6. How does the kinetic-molecular theory explain the pressure exerted by gases?

7. How is a gas's ability to fill a container different from that of a liquid or a solid?

PRACTICE PROBLEMS

8. The atmospheric pressure on top of Mount Everest is 58 kPa. What is this pressure in atmospheres?

9. The vapor pressure of water at $0°C$ is 4.579 mm Hg. What is this pressure in pascals?

10. A laboratory high-vacuum system may operate at 1.0×10^{-5} mm Hg. What is this pressure in pascals?

CRITICAL THINKING

11. How does the kinetic-molecular theory explain why atmospheric pressure is greater at lower altitudes than at higher altitudes?

12. Molecules of hydrogen escape from Earth, but molecules of oxygen and nitrogen are held to the surface. Why?

422

Answers to Section Review

1. Their relatively large intermolecular distances.
2. Gas particles are able to flow past one another easily.
3. They are pushed closer together.
4. A force causes acceleration of an object that is free to move. Pressure is force divided by area on which it impinges.
5. The Pascal: newtons per square meter.
6. Pressure exerted by gases is caused by collisions of molecules with each other and with the walls of the container.

7. A gas assumes the entire shape and volume of its container.
8. 0.57 atm
9. 610.5 Pa
10. 1.3×10^{-3} Pa
11. Gravity pulls the air molecules towards the earth. At lower altitudes, there are more gas molecules and therefore more collisions. This means that the air pressure will be higher.
12. Their lower mass means they are held less strongly by gravity.

② The Gas Laws

KEY TERMS
- Boyle's law
- Charles's law
- Gay-Lussac's law
- Avogadro's law

OBJECTIVES

① **State** Boyle's law, and use it to solve problems involving pressure and volume.

② **State** Charles's law, and use it to solve problems involving volume and temperature.

③ **State** Gay-Lussac's law, and use it to solve problems involving pressure and temperature.

④ **State** Avogadro's law, and explain its importance in determining the formulas of chemical compounds.

Measurable Properties of Gases

In this section, you will study the relationships between the measurable properties of a gas, represented by the variables shown below.

P = pressure exerted by the gas
T = temperature in kelvins of the gas

V = total volume occupied by the gas
n = number of moles of the gas

Pressure-Volume Relationships

As you read in the last section, gases have pressure, and the space that they take up can be made smaller. In 1662, the English scientist Robert Boyle studied the relationship between the volume and the pressure of a gas. He found that as pressure on a gas increases in a closed container, the volume of the gas decreases. In fact, the product of the pressure and volume, PV, remains almost constant if the temperature remains the same. **Table 2** shows data for experiments similar to Boyle's.

internet connect
www.scilinks.org
Topic : The Gas Laws
SciLinks code: HW4063

SCILINKS. Maintained by the National Science Teachers Association

Table 2	Pressure-Volume Data for a Sample of Gas at Constant Temperature	
Pressure (kPa)	**Volume (L)**	**PV (kPa × L)**
150	0.334	50.1
200	0.250	50.0
250	0.200	50.0
300	0.167	50.1

423

Chapter Resource File

• Lesson Plan

Focus

Overview

Before beginning this section, review with your students the Objectives listed in the Student Edition. This section introduces students to the mathematical relationships found in the behavior of gases. In this section students will learn the relationships between the variables of temperature, pressure, volume and number of gas particles both in theory and by applying mathematical calculations.

🔊 Bellringer

Have students review and write down the definitions of pressure, volume, temperature and moles. Have them also write down possible units of measure for each of these quantities.

Motivate

Demonstration

Pressure-Volume Relationship
Repeat the demonstration of the syringe that was illustrated in **Figure 3** to show that when pressure increases, volume decreases. For an even more effective demonstration, obtain an ordinary T-handled cylindrical air pump like the kind used to inflate bicycle tires by hand. Pull the handle about three-fourths of the way up, and clamp the delivery hose shut with a C-clamp or a sturdy laboratory screw clamp. Demonstrate that the upward force on the piston increases as the handle is pushed down. Ask students what this increase indicates. **Ans.** It is evidence of an increase in pressure in the cylinder. Leave the pump set up, and let students feel the change in pressure for themselves during a free moment. Challenge students to explain why the pressure increases when the piston is pushed down. **Ans.** The decrease in volume causes an increase in gas pressure within the cylinder.

K-W-L Have students draw three columns on their papers. Label column 1 K, column 2 W and column 3 L. In the K column students should write down what they already know about the behavior of gas particles (emphasizing pressure and temperature). In the W column they should write down what they want to/will learn about the behavior of gas particles. They should fill in the L column with what they have learned after they read the section. **LS** Logical

Teaching Tip — BASIC

Distribute pieces of paper about 9 inches by 2 inches in size to students. Have them write P at the left end, T in the middle and V at the right end of the paper. Holding the constant variable as the pivot point the student can determine the direction of change in the other variables. Ask them what happens to V when P goes up? Ans. V goes down. What happens to T when P goes up? Ans. T goes up. What happens to V when T goes up? Ans. V goes up. This is a good way for students to get a mental picture of the relationships found in the gas laws. **LS** Visual

SKILL BUILDER — GENERAL

Math Skills and Graphing Have students explain what a direct relationship is. Have them draw a graph that illustrates such a relationship.

Have students explain what an inverse relationship is. Have them draw a graph that illustrates such a relationship. **LS** Logical

---

Figure 9

a Gas molecules in a car-engine cylinder spread apart to fill the cylinder.

b As the cylinder's volume decreases, there are more molecular collisions, and the pressure of the gas increases.

Boyle's Law

Figure 9 shows what happens to the gas molecules in a car cylinder as it compresses. As the volume decreases, the concentration, and therefore pressure, increases. This concept is shown in graphical form in **Figure 10**. The inverse relationship between pressure and volume is known as **Boyle's law**. The third column of **Table 2** shows that at constant temperature, the product of the pressure and volume of a gas is constant.

$$PV = k$$

If the temperature and number of particles are not changed, the PV product remains the same, as shown in the equation below.

$$P_1V_1 = P_2V_2$$

Boyle's law

the law that states that for a fixed amount of gas at a constant temperature, the volume of the gas increases as the pressure of the gas decreases and the volume of the gas decreases as the pressure of the gas increases

Figure 10
This pressure-volume graph shows an inverse relationship: as pressure increases, volume decreases.

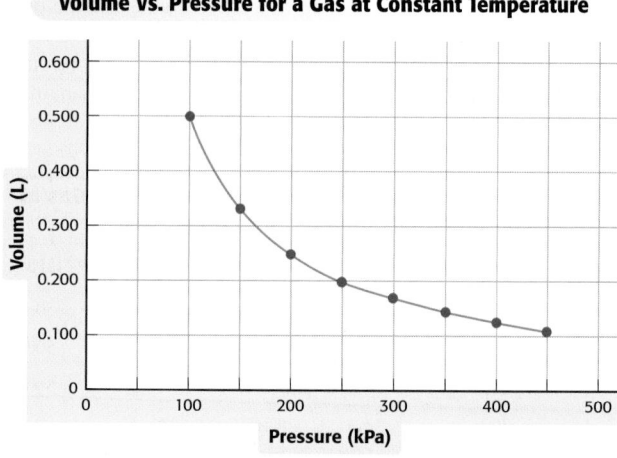

Volume Vs. Pressure for a Gas at Constant Temperature

HISTORY CONNECTION

The English physicist James Clerk Maxwell (1831–1879) formulated the kinetic-molecular theory, which states that gases consist of a large number of randomly moving particles, in the 1860s. He and Ludwig Boltzmann (1844–1906) used the theory to derive Boyle's law mathematically.

Transparencies

TT Volume Versus Pressure for a Gas at Constant Temperature

Solving Pressure-Volume Problems

A given sample of gas occupies 523 mL at 1.00 atm. The pressure is increased to 1.97 atm, while the temperature remains the same. What is the new volume of the gas?

1 Gather information.

The initial volume and pressure and the final pressure are given. Determine the final volume.

$$P_1 = 1.00 \text{ atm} \qquad V_1 = 523 \text{ mL}$$
$$P_2 = 1.97 \text{ atm} \qquad V_2 = ?$$

2 Plan your work.

Place the known quantities into the correct places in the equation relating pressure and volume.

$$P_1 V_1 = P_2 V_2$$
$$(1.00 \text{ atm})(523 \text{ mL}) = (1.97 \text{ atm})V_2$$

3 Calculate.

$$V_2 = \frac{(1.00 \text{ atm})(523 \text{ mL})}{1.97 \text{ atm}} = 265 \text{ mL}$$

4 Verify your results.

The pressure was almost doubled, so the new volume should be about one-half the initial volume. The answer is therefore reasonable.

PRACTICE HINT

It does not matter what units you use for pressure and volume when using Boyle's law as long as they are the same on both sides of the equation.

PRACTICE

1 A sample of oxygen gas has a volume of 150.0 mL at a pressure of 0.947 atm. What will the volume of the gas be at a pressure of 1.000 atm if the temperature remains constant?

2 A sample of gas in a syringe has a volume of 9.66 mL at a pressure of 64.4 kPa. The plunger is depressed until the pressure is 94.6 kPa. What is the new volume, assuming constant temperature?

3 An air mass of volume 6.5×10^5 L starts at sea level, where the pressure is 775 mm Hg. It rises up a mountain where the pressure is 622 mm Hg. Assuming no change in temperature, what is the volume of the air mass?

4 A balloon has a volume of 456 mL at a pressure of 1.0 atm. It is taken under water in a submarine to a depth where the air pressure in the submarine is 3.3 atm. What is the volume of the balloon? Assume constant temperature.

PROBLEM SOLVING SKILL

Answers to Practice Problems B

1. 142 mL
2. 6.58 mL
3. 7.9×10^5 L
4. 1.4×10^2 mL

Homework — GENERAL

Additional Practice

1. If 2.5 L of a gas at 110.0 kPa is expanded to 4.0 L at constant temperature what will happen to the pressure? What will be the new value of pressure? Ans. the pressure will decrease to 69 kPa

2. If 650 mL of hydrogen is stored in a cylinder with a moveable piston at 225 kPa and the pressure is increased to 545 kPa at constant temperature, what is the new volume? Ans. 268 mL

3. If the volume of a gas is tripled at constant temperature, what will happen to the pressure? Ans. It will decrease to 1/3 of its original value.
LS Logical

Teaching Tip

In addition to the decrease in volume due to lower temperature, gases condense into liquid at lower temperature, which also contributes to a decrease in volume.

425

Chapter Resource File

• Problem Solving

Teaching Tip

The Pressure of a Gas Ask students to list the factors that determine the pressure exerted by the randomly moving molecules of a gas. **Ans.** The pressure of an enclosed gas depends on two factors—the frequency of collisions and the average energy of the collisions. The energy of the collisions depends on temperature. At higher temperatures, particles move faster, so both the frequency and the average energy of the collisions are greater. At constant temperature, frequency of collisions is the only variable factor.

SKILL BUILDER — BASIC

Math Skills As a homework assignment, ask students to inflate a round balloon at normal room temperature. Have them measure the circumference of the balloon and then calculate the approximate volume of the balloon. Have them put the balloon into the freezer for about an hour and then remove it and quickly measure its circumference and calculate the new approximate volume of the balloon. How are the volume and temperature related? **LS Logical**

Temperature-Volume Relationships

Heating a gas makes it expand. Cooling a gas makes it contract. This principle is shown in **Figure 11.** As balloons are dipped into liquid nitrogen, the great decrease in temperature makes them shrink. When they are removed from the liquid nitrogen, the gas inside the balloons warms up, and the balloons expand to their original volume.

In 1787, the French physicist Jacques Charles discovered that a gas's volume is directly proportional to the temperature on the Kelvin scale if the pressure remains the same.

Charles's Law

The direct relationship between temperature and volume is known as **Charles's law.** The kinetic-molecular theory states that gas particles move faster on average at higher temperatures, causing them to hit the walls of their container with more force. Repeated strong collisions cause the volume of a flexible container, such as a balloon, to increase. Likewise, gas volume decreases when the gas is cooled, because of the lower average kinetic energy of the gas particles at the lower temperature. If the absolute temperature is reduced by half, then the average kinetic energy is reduced by half, and the particles will strike the walls with half of the energy they had at the higher temperature. In that case, the volume of the gas will be reduced to half of the original volume if the pressure remains the same.

The direct relationship between volume and temperature is shown in **Figure 12,** in which volume-temperature data are graphed using the Kelvin scale. If you read the line in **Figure 12** all the way down to 0 K, it looks as though the gas's volume becomes zero. Does a gas volume really become zero at absolute zero? No. Before this temperature is reached, the gas becomes a liquid and then freezes to a solid, each of which has a certain volume.

Charles's law

the law that states that for a fixed amount of gas at a constant pressure, the volume of the gas increases as the temperature of the gas increases and the volume of the gas decreases as the temperature of the gas decreases

Figure 11

 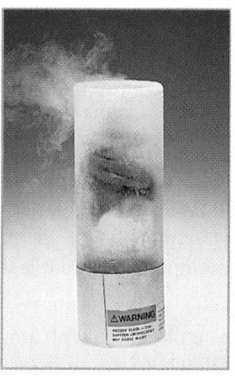 

a Air-filled balloons are dipped into liquid nitrogen.

b The extremely low temperature of the nitrogen causes them to shrink in volume.

c When the balloons are removed from the liquid nitrogen, the air inside them quickly warms, and the balloons expand to their original volume.

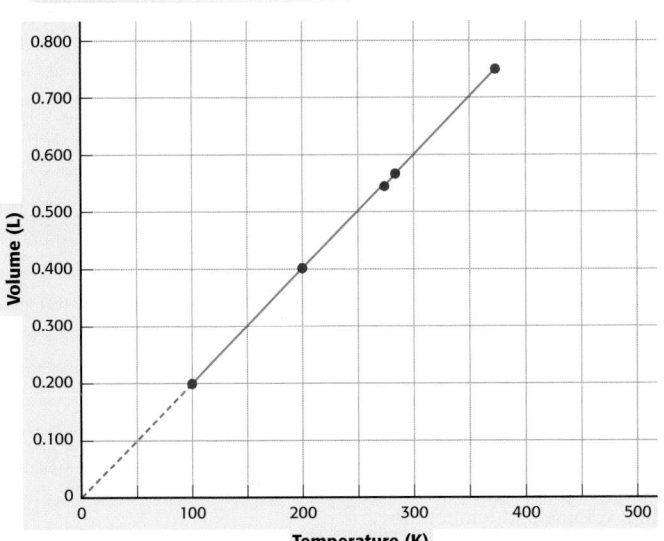

Volume Vs. Temperature for a Gas at Constant Pressure

Data for an experiment of the type carried out by Charles are given in **Table 3.** At constant pressure, the volume of a sample of gas divided by its absolute temperature is a constant, k. Charles's law can be stated as the following equation.

$$\frac{V}{T} = k$$

If all other conditions are kept constant, V/T will remain the same. Therefore, Charles's law can also be expressed as follows.

$$\frac{V_1}{T_1} = \frac{V_2}{T_2}$$

Table 3   Volume-Temperature Data for a Sample of Gas at a Constant Pressure

Volume (mL)	Temperature (K)	V/T (mL/K)
748	373	2.01
567	283	2.00
545	274	1.99
545	273	2.00
546	272	2.01
402	200	2.01
199	100	1.99

Using the Figure

Figure 12 Students will see that if the line on the graph were extended, it would intersect with the (0, 0) point. Ask them to explain the significance of this point. **Ans.** The point represents the volume of the gas at 0 K, or absolute zero. Theoretically, the gas would have zero volume at 0 K, but in reality it would have solidified before reaching that point.

SKILL BUILDER – **GENERAL**

Graphing Discuss the shape of curves for inverse and direct relationships. Have students graph the following data and determine if it illustrates a direct or inverse relationship.

A	B
1	40
2	20
3	13
4	10
5	8

Students can also determine which relationship gives a better constant, A × B (inverse relationship) or A/B (direct relationship). **LS** Visual

Transparencies

TT Volume Versus Temperature for a Gas at Constant Pressure

Answers to Practice Problems C

1. 0.67 L

2. 815 mL

3. –11.0°C

4. 1.64×10^3 L

Homework ──── GENERAL

Additional Practice

1. A balloon with a volume of 15.5 L is inflated in a room at 20.0°C and then taken outside where the temperature is 7.0°C, what will be the new volume of the balloon if the pressure remains constant? Ans. 14.8 L

2. If the original temperature of a 62.2 L sample of a gas is 150°C, what is the final temperature of the gas (in degrees Celsius) if the new volume of the gas is 24.4 L and the pressure remains constant? Ans. 107°C

3. The volume of gas in a syringe is 15.0 mL at 23.5°C. What will the volume of the gas be at 72.5°C if the pressure is held constant? Ans. 17.5 mL

LS Logical

SAMPLE PROBLEM C

Solving Volume-Temperature Problems

A balloon is inflated to 665 mL volume at 27°C. It is immersed in a dry-ice bath at –78.5°C. What is its volume, assuming the pressure remains constant?

1 Gather information.

The initial volume and temperature and the final temperature are given. Determine the final volume.

$$V_1 = 665 \text{ mL} \qquad\qquad T_1 = 27°C$$
$$V_2 = ? \qquad\qquad T_2 = -78.5°C$$

2 Plan your work.

Convert the temperatures from degrees Celsius to kelvins:

$$T_1 = 27°C + 273 = 300 \text{ K} \qquad T_2 = -78.5°C + 273 = 194.5 \text{ K}$$

Place the known quantities into the correct places in the equation relating volume and temperature.

$$\frac{V_1}{T_1} = \frac{V_2}{T_2}$$
$$\frac{665 \text{ mL}}{300 \text{ K}} = \frac{V_2}{194.5 \text{ K}}$$

3 Calculate.

$$V_2 = \frac{(665 \text{ mL})(194.5 \text{ K})}{300 \text{ K}} = 431 \text{ mL}$$

4 Verify your results.

Charles's law tells you that volume decreases as temperature decreases. The temperature decreased by about one-third, and according to the calculation, so did the volume. The answer is therefore reasonable.

PRACTICE HINT

In gas law problems, always convert temperatures to kelvins. The gas law equations do not work for temperatures expressed in the Celsius or Fahrenheit scales.

PRACTICE

1 Helium gas in a balloon occupies 2.5 L at 300.0 K. The balloon is dipped into liquid nitrogen that is at a temperature of 80.0 K. What will the volume of the helium in the balloon at the lower temperature be?

2 A sample of neon gas has a volume of 752 mL at 25.0°C. What will the volume at 50.0°C be if pressure is constant?

3 A helium-filled balloon has a volume of 2.75 L at 20.0°C. The volume of the balloon changes to 2.46 L when placed outside on a cold day. What is the temperature outside in degrees Celsius?

4 When 1.50×10^3 L of air at 5.00°C is injected into a household furnace, it comes out at 30.0°C. Assuming the pressure is constant, what is the volume of the heated air?

428

Chapter Resource File

• Problem Solving

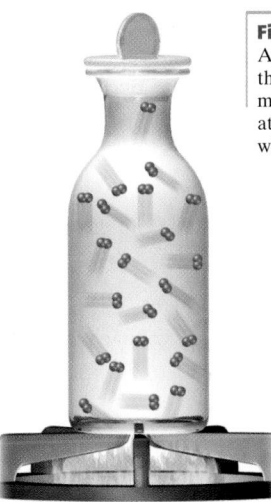

Figure 13
As the temperature of a gas increases, the average kinetic energy of the molecules increases. This means that at constant volume, pressure increases with temperature.

Temperature-Pressure Relationships

As you have learned, pressure is the result of collisions of particles with the walls of the container. You also know that the average kinetic energy of particles is proportional to the sample's average absolute temperature. Therefore, if the absolute temperature of the gas particles is doubled, their average kinetic energy is doubled. If there is a fixed amount of gas in a container of fixed volume, the collisions will have twice the energy, so the pressure will double, as shown in **Figure 13.** The French scientist Joseph Gay-Lussac is given credit for discovering this relationship in 1802. **Figure 14** shows a graph of data for the change of pressure with temperature in a gas of constant volume. The graph is a straight line with a positive slope, which indicates that temperature and pressure have a directly proportional relationship.

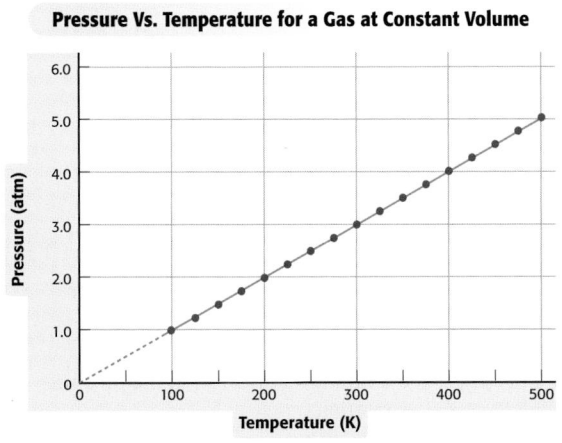

Pressure Vs. Temperature for a Gas at Constant Volume

Figure 14
This graph shows that gas pressure is directly proportional to Kelvin temperature, at constant volume.

429

Teach, continued

Group Activity — GENERAL

Check students' conceptual understanding of the gas laws by asking students to form groups of 3–4 to determine the unknown quantity in each column without using a calculator. Students should determine the degree of change by inspection. Note that all changes are by factors of 2 or 1/2. **LS** Interpersonal

Condition 1	Sample 1	Sample 2
Temperature	300 K	300 K
Pressure	100 kPa	100 kPa
Volume	200 mL	200 mL

Condition 2		
Temperature	300 K	300 K
Pressure	200 kPa	? kPa
Volume	? mL	400 mL

Sample 3	Sample 4	Sample 5	Sample 6
300 K	400 K	300 K	300 K
? kPa	100 kPa	100 kPa	100 kPa
400 mL	600 mL	400 mL	200 mL
no change	200 K	? K	150 K
200 kPa	no change	no change	? kPa
200 mL	? mL	200 mL	no change

Gay-Lussac's law

the law that states that the pressure of a gas at a constant volume is directly proportional to the absolute temperature

Gay-Lussac's Law

The direct relationship between temperature and pressure is known as **Gay-Lussac's law.** Because the pressure of a gas is proportional to its absolute temperature, the following equation is true for a sample of constant volume.

$$P = kT$$

This equation can be rearranged to the following form.

$$\frac{P}{T} = k$$

At constant volume, the following equation applies.

$$\frac{P_1}{T_1} = \frac{P_2}{T_2}$$

If any three of the variables in the above equation are known, then the unknown fourth variable can be calculated.

SAMPLE PROBLEM D

Solving Pressure-Temperature Problems

An aerosol can containing gas at 101 kPa and 22°C is heated to 55°C. Calculate the pressure in the heated can.

1 Gather information.

The initial pressure and temperature and the final temperature are given. Determine the final pressure.

$P_1 = 101 \text{ kPa}$ $\qquad$ $T_1 = 22°C$

$P_2 = ?$ $\qquad$ $T_2 = 55°C$

PRACTICE HINT

When solving gas problems, be sure to identify which variables (P, V, n, T) remain constant and which change. That way you can pick the right equation to use.

2 Plan your work.

Convert the temperatures from degrees Celsius to kelvins:

$T_1 = 22°C + 273 = 295 \text{ K}$ $\qquad$ $T_2 = 55°C + 273 = 328 \text{ K}$

Use the equation relating temperature and pressure.

$$\frac{P_1}{T_1} = \frac{P_2}{T_2}$$

$$\frac{101 \text{ kPa}}{295 \text{ K}} = \frac{P_2}{328 \text{ K}}$$

3 Calculate.

$$P_2 = \frac{(101 \text{ kPa})(328 \text{ K})}{295 \text{ K}} = 112 \text{ kPa}$$

4 Verify your results.

The temperature of the gas increases by about 10%, so the pressure should increase by the same proportion.

430

HISTORY
CONNECTION

Joseph Louis Gay-Lussac (1778-1850) was a French chemist and physicist who was interested in the properties of gases. He was also a pioneer in ballooning, reaching an altitude of 7000 m (4.35 mi).

1 At 122°C the pressure of a sample of nitrogen is 1.07 atm. What will the pressure be at 205°C, assuming constant volume?

2 The same sample of nitrogen as in item 1 starts at 122°C and 1.07 atm. After cooling, the pressure is measured to be 0.880 atm. What is the new temperature?

3 A sample of helium gas is at 122 kPa and 22°C. Assuming constant volume, what will the temperature be when the pressure is 203 kPa?

4 The air in a steel-belted tire is at a gauge pressure of 29.8 psi at a temperature of 20°C. After the tire is driven fast on a hot road, the temperature in the tire is 48°C. What is the tire's new gauge pressure?

PROBLEM SOLVING SKILL

Volume-Molar Relationships

In 1811, the Italian scientist Amadeo Avogadro proposed the idea that equal volumes of all gases, under the same conditions, have the same number of particles. This idea is shown in **Figure 15,** which shows equal numbers of molecules of the gases H_2, O_2, and CO_2, each having the same volume. A result of this relationship is that molecular masses can be easily determined. Unfortunately, Avogadro's insight was not recognized right away, mainly because scientists at the time did not know the difference between atoms and molecules. Later, the Italian chemist Stanislao Cannizzaro used Avogadro's principle to determine the true formulas of several gaseous compounds.

Figure 15
At the same temperature and pressure, balloons of equal volume contain equal numbers of molecules, regardless of which gas they contain.

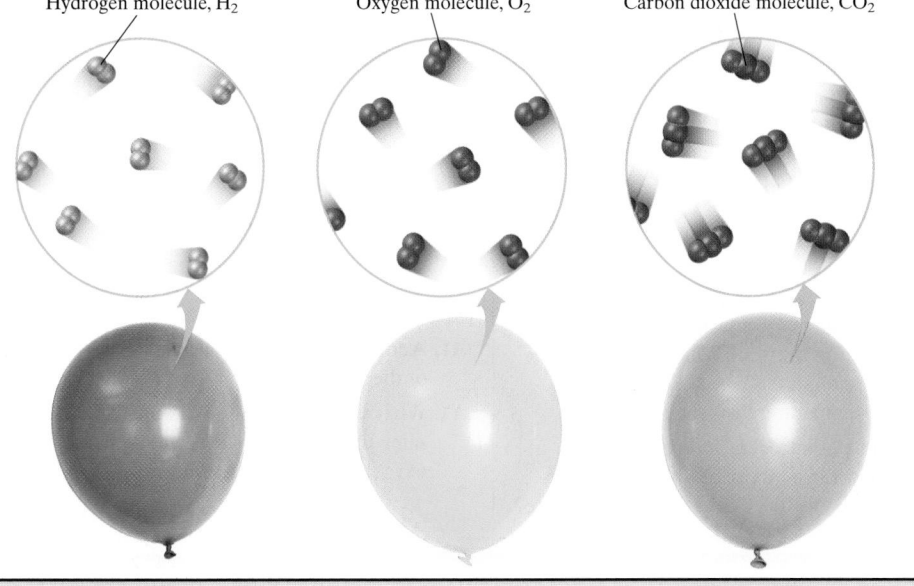

Hydrogen molecule, H_2 Oxygen molecule, O_2 Carbon dioxide molecule, CO_2

431

Answers to Practice Problems D

1. 1.29 atm

2. 325 K, or 52°C

3. 491 K, or 218°C

4. 32.6 psi

Homework ——— GENERAL

Additional Practice

1. The pressure in a tire is 101 kPa at 10.0°C, what will be the pressure of the tire at 45.0°C? (Hint: the volume remains constant.) Ans. 113 kPa

2. The pressure in a bottle of soda pop is 505 kPa at 20.0°C. What is the new pressure if someone warms the sealed bottle to 65.0°C? Ans. 583 kPa

3. If a gas at 600°C exerts a pressure of 1515 kPa, at what temperature (in degrees Celsius) will the gas exert 151.5 kPa of pressure if the volume remains constant? Ans. –186°C

LS Logical

Chapter Resource File

• **Problem Solving**

Transparencies

TT Avogadro's Law

Close

Quiz — GENERAL

1. What is the relationship between volume and pressure at a constant temperature? Ans. inverse

2. Which law shows the mathematical relationship between volume and pressure? Ans. Boyle's law

3. How are volume and Kelvin temperature related at constant pressure? Ans. directly

4. Which law shows the mathematical relationship between volume and Kelvin temperature? Ans. Charles's law

5. What is the relationship between Kelvin temperature and pressure at a constant pressure? Ans. direct

6. Whose law shows the mathematical relationship between Kelvin temperature and pressure? Ans. Gay-Lussac's law

LS Logical

Reteaching — BASIC

Have students create a flowchart, concept map, list or other graphic organizer that shows the relationships between pressure, volume and temperature. **LS** Visual

Chapter Resource File

• Concept Review
• Section Quiz

Avogadro's Law

Avogadro's law

the law that states that equal volumes of gases at the same temperature and pressure contain equal numbers of molecules

internet connect

www.scilinks.org
Topic : Avogadro's Law
SciLinks code: HW4137

SCILINKS. Maintained by the National Science Teachers Association

Avogadro's idea turned out to be correct and is now known as **Avogadro's law.** With this knowledge, chemists gained insight into the formulas of chemical compounds for the first time. In 1858, Cannizzaro used Avogadro's law to deduce that the correct formula for water is H_2O. This important discovery will be discussed in more detail in the next section.

Avogadro's law also means that gas volume is directly proportional to the number of moles of gas at the same temperature and pressure. This relationship is expressed by the equation below, in which k is a proportionality constant.

$$V = kn$$

But volumes of gases change with changes in temperature and pressure. A set of conditions has been defined for measuring volumes of gases. For example, we know that argon exists as single atoms, and that its atomic mass is 39.95 g/mol. It has been determined that 22.41 L of argon at 0°C and 1 atm have a mass of 39.95 g. *Therefore, 22.41 L is the volume of 1 mol of any gas at STP.* The mass of 22.41 L of a gas at 0°C and a pressure of 1 atm will be equal to the gas's molecular mass.

② Section Review

UNDERSTANDING KEY IDEAS

1. What is the name of the gas law relating pressure and volume, and what does it state?

2. What is the name of the gas law relating volume and absolute temperature, and what does it state?

3. What is the name of the gas law relating pressure and absolute temperature, and what does it state?

4. What relationship does Avogadro's law express?

PRACTICE PROBLEMS

5. A sample of gas occupies 1.55 L at 27.0°C and 1.00 atm pressure. What will the volume be if the pressure is increased to 50.0 atm, but the temperature is kept constant?

6. A sample of nitrogen gas occupies 1.55 L at 27°C and 1.00 atm pressure. What will the volume be at −100°C and the same pressure?

7. A 1.0 L volume of gas at 27.0°C exerts a pressure of 85.5 kPa. What will the pressure be at 127°C? Assume constant volume.

8. A sample of nitrogen has a volume of 275 mL at 273 K. The sample is heated and the volume becomes 325 mL. What is the new temperature in kelvins?

9. A small cylinder of oxygen contains 300.0 mL of gas at 15 atm. What will the volume of this gas be when released into the atmosphere at 0.900 atm?

CRITICAL THINKING

10. A student has the following data: $V_1 =$ 822 mL, $T_1 = 75°C$, $T_2 = −25°C$. He calculates V_2 and gets −274 mL. Is this correct? Explain why or why not.

11. Aerosol cans have a warning not to dispose of them in fires. Why?

12. What volume of carbon dioxide contains the same number of molecules as 20.0 mL of oxygen at the same conditions?

432

Answers to Section Review

1. Boyle's law states that pressure and volume are inversely proportional, if the temperature is kept constant.

2. Charles's law states that volume and absolute temperature are directly proportional, if the pressure is kept constant.

3. Gay-Lussac's law states that pressure and absolute temperature are directly proportional, if the volume is kept constant.

4. The direct relationship between amount and volume of a gas.

5. 31.0 mL
6. 0.894 L
7. 114 kPa
8. 323 K
9. 5.00 L
10. No. The volume cannot be negative. The temperatures must be changed to Kelvins.
11. The pressure increases in the closed container, and the can may explode.
12. 20.0 mL

SECTION

3 Molecular Composition of Gases

KEY TERMS
- ideal gas
- ideal gas law
- diffusion
- effusion
- Graham's law of diffusion
- Gay-Lussac's law of combining volumes
- partial pressure
- Dalton's law of partial pressure

OBJECTIVES

1 **Solve** problems using the ideal gas law.

2 **Describe** the relationships between gas behavior and chemical formulas, such as those expressed by Graham's law of diffusion, Gay-Lussac's law of combining volumes, and Dalton's law of partial pressures.

3 **Apply** your knowledge of reaction stoichiometry to solve gas stoichiometry problems.

The Ideal Gas Law

You have studied four different gas laws, which are summarized in **Table 4**. Boyle's law states the relationship between the pressure and the volume of a sample of gas. Charles's law states the relationship between the volume and the absolute temperature of a gas. Gay-Lussac's law states the relationship between the pressure and the temperature of a gas. Avogadro's law relates volume to the number of moles of gas.

No gas perfectly obeys all four of these laws under all conditions. Nevertheless, these assumptions work well for most gases and most conditions. As a result, one way to model a gas's behavior is to assume that the gas is an **ideal gas** that perfectly follows these laws. An ideal gas, unlike a real gas, does not condense to a liquid at low temperatures, does not have forces of attraction or repulsion between the particles, and is composed of particles that have no volume.

ideal gas

an imaginary gas whose particles are infinitely small and do not interact with each other

Table 4 Summary of the Basic Gas Laws

Boyle's law	$P_1V_1 = P_2V_2$
Charles's law	$\dfrac{V_1}{T_1} = \dfrac{V_2}{T_2}$
Gay-Lussac's law	$\dfrac{P_1}{T_1} = \dfrac{P_2}{T_2}$
Avogadro's law	$V = kn$

STUDY ⟩ TIP

ORGANIZING THE GAS LAWS

You can use **Table 4** as a reference to help you learn and understand the gas laws.

- You may want to make a more elaborate table in which you write down each law's name, a brief explanation of its meaning, its mathematical representation, and the variable that is kept constant.

433

SECTION 3

Focus

Overview
Before beginning this section, review with your students the Objectives listed in the Student Edition. This section of the chapter focuses on the concepts related to moles of gas. The ideal gas law relates pressure, temperature, volume and moles. Effusion and diffusion are dependent on the molar masses of the gases involved. Dalton's law is briefly introduced. Finally, Gay-Lussac's law of combining volumes introduces gas stoichiometry which relates volumes and moles.

🔊 Bellringer
Have students write down what they think will happen in the following situations if the number of moles of gas increase.

- Situation 1—With pressure and temperature constant, what will happen to volume?
- Situation 2—With volume and temperature constant, what will happen to pressure?

Motivate

Identifying Preconceptions

1. Continuing with the Bellringer activity, have students discuss why the volume and pressure will increase in terms of molecular motion.

2. Ask students to explain the difference between inverse and direct relationships.

3. Ask students to identify what kind of relationship describes that between pressure and the number of moles of gas and between volume and the number of moles of gas.

Chapter Resource File
- Lesson Plan

Demonstration
Amount of Oxygen in Air

(Approximate time: 5 minutes for setup)

1. Wet a tuft of fine steel wool, and push it into the bottom of a 100 mL or larger graduated cylinder. If the steel wool is oily, wash it with soap and water to remove the oil before putting it in the cylinder.

2. Invert the cylinder, and submerge the mouth in a beaker of water. Make sure that enough water is above the mouth of the cylinder to fill one-fifth of its volume. Clamp the cylinder in place.

3. As the iron oxidizes, oxygen will be removed from the air. The water level will rise about one-fifth of the length of the cylinder. The effect will reach completion by the next day.

4. On the following day, have students note the change in water level. Then test to see if there is any oxygen left in the tube. Place a small glass plate under the mouth of the cylinder while it is still submerged. Withdraw the cylinder, and immediately set it upright on the table with the plate still in place. Light a wooden splint, remove the cover from the cylinder, and quickly thrust the burning splint inside. The splint should go out immediately.

5. After the demonstration, ask students why only one fifth of the air disappeared. What else is left?

Safety: Safety goggles and a lab apron must be worn. Students must be 10 ft or more from the demonstration.

Disposal: Throw rusty iron into the trash.

ideal gas law

the law that states the mathematical relationship of pressure (*P*), volume (*V*), temperature (*T*), the gas constant (*R*), and the number of moles of a gas (*n*); $PV = nRT$

The Ideal Gas Law Relates All Four Gas Variables

In using the basic gas laws, we have made use of four variables: pressure, *P*, volume, *V*, absolute temperature, *T*, and number of moles, *n*. Boyle's law, Charles's law, Gay-Lussac's law, and Avogadro's law can be combined into one equation that gives the relationship between all four variables, *P*, *V*, *T*, and *n*, for any sample of gas. This relationship is called the **ideal gas law.** When any three variables are given, the fourth can be calculated. The ideal gas law is represented mathematically below.

$$PV = nRT$$

R is a proportionality constant. The value for *R* used in any calculation depends on the units used for pressure and volume. In this text, we will normally use units of kilopascals and liters when using the ideal gas law, so the value you will use for *R* is as follows.

$$R = \frac{8.314 \text{ L} \cdot \text{kPa}}{\text{mol} \cdot \text{K}}$$

If the pressure is expressed in atmospheres, then the value of *R* is 0.0821 (L•atm)/(mol•K).

The ideal gas law describes the behavior of real gases quite well at room temperature and atmospheric pressure. Under those conditions, the volume of the particles themselves and the forces of attraction between them can be ignored. The particles behave in ways that are similar enough to an ideal gas that the ideal gas law gives useful results. However, as the volume of a real gas decreases, the particles attract one another more strongly, and the volume is less than the ideal gas law would predict. At extremely high pressures, the volume of the particles themselves is close to the total volume, so the actual volume will be higher than calculated. This relationship is shown in **Figure 16.**

Figure 16
For an ideal gas, the ratio of *PV/nRT* is 1, which is represented by the dashed line. Real gases deviate somewhat from the ideal gas law and more at very high pressures.

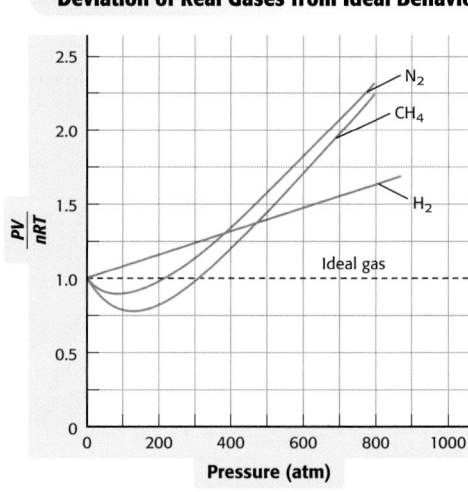

Deviation of Real Gases from Ideal Behavior

SAMPLE PROBLEM E

Using the Ideal Gas Law

How many moles of gas are contained in 22.41 liters at 101.325 kPa and 0°C?

1 Gather information.

Three variables for a gas are given, so you can use the ideal gas law to compute the value of the fourth.

$V = 22.41$ L $\qquad\qquad$ $T = 0°C$

$P = 101.325$ kPa $\qquad\qquad$ $n = ?$

2 Plan your work.

Convert the temperature to kelvins.

$$0°C + 273 = 273 \text{ K}$$

Place the known quantities into the correct places in the equation relating the four gas variables, $PV = nRT$.

$$(101.325 \text{ kPa})(22.41 \text{ L}) = n\left(\frac{8.314 \text{ L}\cdot\text{kPa}}{\text{mol}\cdot\text{K}}\right)(273 \text{ K})$$

3 Calculate.

Solve for the unknown, n.

$$n = \frac{(101.325 \text{ kPa})(22.41 \text{ L})}{\left(\dfrac{8.314 \text{ L}\cdot\text{kPa}}{\text{mol}\cdot\text{K}}\right)(273 \text{ K})} = 1.00 \text{ mol}$$

4 Verify your results.

The product of pressure and volume is a little over 2000, and the product of R and the temperature is a little over 2000, so one divided by the other should be about 1. Also, recall that at STP the volume of 1 mol of gas is 22.41 L. The calculated result agrees with this value.

PRACTICE HINT

When using the ideal gas law, be sure that the units of P and V match the units of the value of R that is used.

PRACTICE

1 How many moles of air molecules are contained in a 2.00 L flask at 98.8 kPa and 25.0°C?

2 How many moles of gases are contained in a can with a volume of 555 mL and a pressure of 600.0 kPa at 20°C?

3 Calculate the pressure exerted by 43 mol of nitrogen in a 65 L cylinder at 5°C.

4 What will be the volume of 111 mol of nitrogen in the stratosphere, where the temperature is −57°C and the pressure is 7.30 kPa?

435

Chapter Resource File

• Problem Solving

Demonstration

Application for Graham's Law

Preparation time: 15 minutes

Time: 10 minutes

Preparation

Select glass tubing with an inner diameter about that of a cotton swab. Cut the tubing to about 24 inches. Fire polish both ends. Mount it on a meter stick with string or rubber bands.

Procedure

1. Dip a cotton swab into $NH_3(aq)$ and another swab into $HCl(aq)$.

2. Insert one swab into each end of the piece of glass tubing, both at the same time.

3. Observe where the white ring of NH_4Cl forms and mark this point with a wax pencil. Record the distance each gas travels.

4. Have students note which molecules traveled further, the $NH_3(aq)$ or the $HCl(aq)$. Relate these distances to the speed of each molecule predicted by Graham's law.

Safety: All students must wear safety goggles. Because of the vapors given off by both of these chemicals do this demonstration in a well ventilated room. You must wear safety goggles and a lab apron. Students must be 10 ft or more from the demonstration.

Disposal: First rinse the swabs with water and then throw materials in the trash. Rinse, wash and store the prepared glass tube for future use.

Gas Behavior and Chemical Formulas

So far we have dealt only with the general behavior of gases. No calculation has depended on knowing which gas was involved. For example, in **Sample Problem E,** we determined the number of moles of a gas. The calculation would have been the same no matter which gas or mixture of gases was involved. Now we will consider situations in which it is necessary to know more about a gas, such as its molar mass.

Diffusion

diffusion

the movement of particles from regions of higher density to regions of lower density

Household ammonia is a solution of ammonia gas, NH_3, in water. When you open a bottle of household ammonia, the odor of ammonia gas doesn't take long to fill the room. Gaseous molecules, including molecules of the compounds responsible for the smell, travel at high speeds in all directions and mix quickly with molecules of gases in the air in a process called **diffusion.** Gases diffuse through each other at a very rapid rate. Even if the air in the room was very still, it would only be a matter of minutes before ammonia molecules were evenly distributed throughout the room. During diffusion, a substance moves from a region of high concentration to a region of lower concentration. Eventually, the mixture becomes homogeneous, as seen in the closed bottle of bromine gas in **Figure 17.** Particles of low mass diffuse faster than particles of higher mass.

The process of diffusion involves an increase in entropy. Entropy can be thought of as a measure of randomness. One way of thinking of randomness is to consider the probability of finding a particular particle at a particular location. This probability is much lower in a mixture of gases than in a pure gas. Diffusion of gases is a way in which entropy is increased.

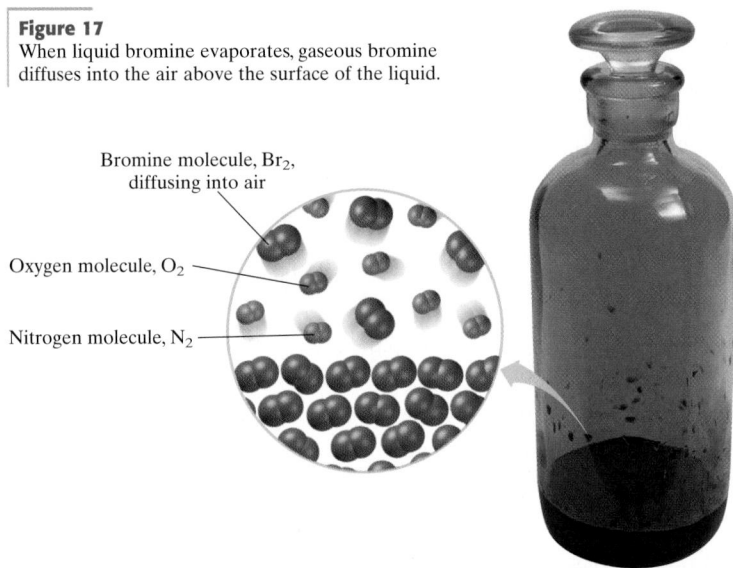

Figure 17

When liquid bromine evaporates, gaseous bromine diffuses into the air above the surface of the liquid.

Bromine molecule, Br_2, diffusing into air

Oxygen molecule, O_2

Nitrogen molecule, N_2

436

did you know?

Strictly speaking, Graham's law of effusion applies only to gases effusing into a vacuum. Experiments that show the diffusion of $HCl(g)$ and $NH_3(g)$ through air to form a cloud of $NH_4Cl(s)$ do not demonstrate Graham's law. However, because both gases are diffusing through the same air, their relative rates of diffusion reflect the relative velocities predicted by the law.

Figure 18
Molecules of oxygen and nitrogen from inside the bicycle tire effuse out through a small nail hole.

Oxygen molecule, O_2

Nitrogen molecule, N_2

Effusion

The passage of gas particles through a small opening is called **effusion.** An example of effusion is shown in **Figure 18.** In a tire with a very small leak, the air in the tire effuses through the hole. The Scottish scientist Thomas Graham studied effusion in detail. He found that at constant temperature and pressure, the rate of effusion of a gas is inversely proportional to the square root of the gas's molar mass, *M*. This law also holds when one compares rates of diffusion, and molecular speeds in general. The molecular speeds, v_A and v_B, of gases A and B can be compared according to **Graham's law of diffusion,** shown below.

$$\frac{v_A}{v_B} = \sqrt{\frac{M_B}{M_A}}$$

For example, compare the speed of effusion of H_2 with that of O_2.

$$\frac{v_{H_2}}{v_{O_2}} = \sqrt{\frac{M_{O_2}}{M_{H_2}}} = \sqrt{\frac{32 \text{ g/mol}}{2 \text{ g/mol}}} = \sqrt{16} = 4$$

Hydrogen gas effuses four times as fast as oxygen. Particles of low molar mass travel faster than heavier particles.

According to the kinetic-molecular theory, gas particles that are at the same temperature have the same average kinetic energy. Therefore, the kinetic energy of two gases that are at the same temperature is

$$\frac{1}{2}M_A v_A{}^2 = \frac{1}{2}M_B v_B{}^2$$

Solving this equation for the ratio of speeds between v_A and v_B gives Graham's law of diffusion.

effusion

the passage of a gas under pressure through a tiny opening

Graham's law of diffusion

the law that states that the rate of diffusion of a gas is inversely proportional to the square root of the gas's density

internet connect

www.scilinks.org
Topic: Effusion/Diffusion
SciLinks code: HW4041

SC*i*LINKS. Maintained by the National Science Teachers Association

437

PHYSICS CONNECTION

The kinetic energy of an object is equal to $\frac{1}{2}mv^2$, where *m* is an object's mass, and *v* is its velocity. Since the kinetic-molecular theory states that gas particles at the same temperature have the same average kinetic energy, differing masses will result in differing velocities in order for the equation given above to hold true.

Answers to Practice Problems F

1. N_2 has higher speed; 1.069 times faster

2. 1.9×10^3 m/s

3. 48.6 g/mol

4. $^{235}UF_6$ diffuses at 1.0043 times the speed of $^{238}UF_6$.

Homework ── **GENERAL**

Additional Practice

1. The average velocity of O_2 molecules at room temperature is 480 m/s. How fast would a hydrogen cyanide, HCN, molecule travel on average under the same conditions? Ans. 522 m/s

2. The average velocity of ammonia, NH_3, molecules at room temperature is 658 m/s. How fast would hydrogen sulfide, H_2S, molecules travel on average under the same conditions? Ans. 465 m/s

3. The average velocity of CO_2 molecules at room temperature is 409 m/s. What is the molar mass of a gas whose molecules have an average velocity of 322 m/s under the same conditions? Ans. 71.0 g/mol

LS Logical

SAMPLE PROBLEM F

Comparing Molecular Speeds

Oxygen molecules have an average speed of about 480 m/s at room temperature. At the same temperature, what is the average speed of molecules of sulfur hexafluoride, SF_6?

1 Gather information.

$$v_{O_2} = 480 \text{ m/s}$$

$$v_{SF_6} = ? \text{ m/s}$$

2 Plan your work.

This problem can be solved using Graham's law of diffusion, which compares molecular speeds.

$$\frac{v_{SF_6}}{v_{O_2}} = \sqrt{\frac{M_{O_2}}{M_{SF_6}}}$$

You need molar masses of O_2 and SF_6.
Molar mass of O_2 = 2(16.00 g/mol) = 32.00 g/mol
Molar mass of SF_6 = 32.07 g/mol + 6(19.00 g/mol) = 146 g/mol

3 Calculate.

Place the known quantities into the correct places in Graham's law of diffusion.

$$\frac{v_{SF_6}}{480 \text{ m/s}} = \sqrt{\frac{32 \text{ g/mol}}{146 \text{ g/mol}}}$$

Solve for the unknown, v_{SF_6}.

$$v_{SF_6} = (480 \text{ m/s})\sqrt{\frac{32 \text{ g/mol}}{146 \text{ g/mol}}} = 480 \text{ m/s} \times 0.47 = 220 \text{ m/s}$$

4 Verify your results.

SF_6 has a mass about 4 times that of O_2. The square root of 4 is 2, and the inverse of 2 is $\frac{1}{2}$. SF_6 should travel about half as fast as O_2. The answer is therefore reasonable.

PRACTICE

1 At the same temperature, which molecule travels faster, O_2 or N_2? How much faster?

2 At room temperature, Xe atoms have an average speed of 240 m/s. At the same temperature, what is the speed of H_2 molecules?

3 What is the molar mass of a gas if it diffuses 0.907 times the speed of argon gas?

4 Uranium isotopes are separated by effusion. What is the relative rate of effusion for $^{235}UF_6$ (M = 349.03 g/mol) and $^{238}UF_6$ (M = 352.04 g/mol)?

438

Chapter Resource File

• **Problem Solving**

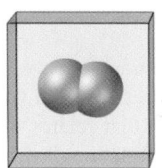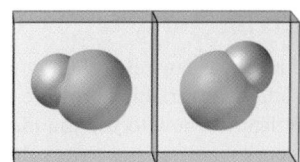

Gas Reactions Allow Chemical Formulas to Be Deduced

In 1808, Joseph Gay-Lussac made an important discovery: if the pressure and temperature are kept constant, gases react in volume proportions that are whole numbers. This is called **Gay-Lussac's law of combining volumes.** Consider the formation of gaseous hydrogen chloride from the reaction of hydrogen gas and chlorine gas. Gay-Lussac showed in an experiment that one volume of hydrogen gas reacts with one volume of chlorine gas to form two volumes of hydrogen chloride gas. **Figure 19** illustrates the volume ratios in the reaction in the form of a model. Let us use Avogadro's law and assume that two molecules of hydrogen chloride are formed, one in each of the two boxes on the right in **Figure 19**. Therefore, we must start with two atoms of hydrogen and two atoms of chlorine. Each box must contain one molecule of hydrogen and one molecule of chlorine, so there must be two atoms in each of these molecules.

Using several other reactions, such as the reaction of gaseous hydrogen and gaseous oxygen to form water vapor, the Italian chemist Stanislao Cannizzaro was able to deduce that oxygen is also diatomic and that the formula for water is H_2O. Dalton had guessed that the formula for water was HO, because this seemed the most likely combination of atoms for such a common compound. Before knowing atomic masses, chemists had only a set of relative weights. For example, it was known that 1 g of hydrogen can react with 8 g of oxygen to form water, so it was assumed that an oxygen atom was eight times as heavy as a hydrogen atom. It was not until the mid 1800s, just before the Civil War, that chemists knew the correct formula for water.

Dalton's Law of Partial Pressure

In 1805, John Dalton showed that in a mixture of gases, each gas exerts a certain pressure as if it were alone with no other gases mixed with it. The pressure of each gas in a mixture is called the **partial pressure.** The total pressure of a mixture of gases is the sum of the partial pressures of the gases. This principle is known as **Dalton's law of partial pressure.**

$$P_{total} = P_A + P_B + P_C$$

P_{total} is the total pressure, and P_A, P_B, and P_C are the partial pressures of each gas.

How is Dalton's law of partial pressure explained by the kinetic-molecular theory? All the gas molecules are moving randomly, and each has an equal chance to collide with the container wall. Each gas exerts a pressure proportional to its number of molecules in the container. The presence of other gas molecules does not change this fact.

Gay-Lussac's law of combining volumes

the law that states that the volumes of gases involved in a chemical change can be represented by the ratio of small whole numbers

partial pressure

the pressure of each gas in a mixture

Dalton's law of partial pressure

the law that states that the total pressure of a mixture of gases is equal to the sum of the partial pressures of the component gases

439

Students may have difficulty seeing where to start on the problems involving stoichiometry and the ideal gas law. Point out that if they are given the mass of a substance in the reaction they should begin with the stoichiometry problem and determine the number of moles of the gaseous substance in question. Using this answer, they can then solve the ideal gas law equation for the unknown variable.

If they are given the volume of a gaseous substance in the equation, they should use the ideal gas law equation to determine the number of moles in the problem. With this answer they can solve for the grams of the unknown through stoichiometry.

Gas Stoichiometry

The ideal gas law relates amount of gaseous substance in moles, *n*, with the other gas variables: pressure, volume, and temperature. Now that you have learned how to use the ideal gas law, an equation that relates the number of moles of gas to its volume, you can use it in calculations involving gases that react.

Gas Volumes Correspond to Mole Ratios

Ratios of gas volumes will be the same as mole ratios of gases in balanced equations. Avogadro's law shows that the mole ratio of two gases at the same temperature and pressure is the same as the volume ratio of the two gases. This greatly simplifies the calculation of the volume of products or reactants in a chemical reaction involving gases. For example, consider the following equation for the production of ammonia.

$$3H_2(g) + N_2(g) \longrightarrow 2NH_3(g)$$

Consequently, 3 L of H_2 react with 1 L of N_2 to form 2 L of NH_3, and no H_2 or N_2 is left over (assuming an ideal situation of 100% yield).

Consider the electrolysis of water, a reaction expressed by the following chemical equation.

$$2H_2O(l) + \text{electricity} \longrightarrow 2H_2(g) + O_2(g)$$

The volume of hydrogen gas produced will be twice the volume of oxygen gas, because there are twice as many moles of hydrogen as there are moles of oxygen. As you can see in **Figure 20,** the volume of hydrogen gas produced is, in fact, twice as large as the volume of oxygen produced.

Furthermore, if we know the number of moles of a gaseous substance, we can use the ideal gas law to calculate the volume of that gas. **Skills Toolkit 1** on the following page shows how to find the volume of product given the mass of one of the reactants.

Topic Link

Refer to the "Stoichiometry" chapter for a discussion of stoichiometry.

Figure 20
In the electrolysis of water, the volume of hydrogen is twice the volume of oxygen, because the mole ratio between the two gases is two to one.

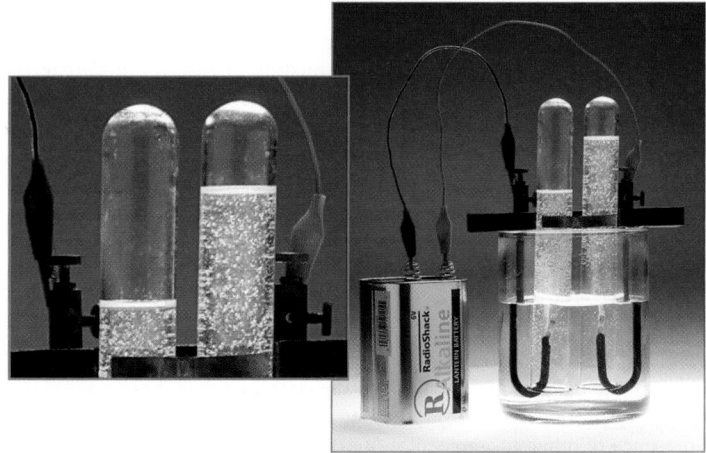

440

Finding Volume of Unknown

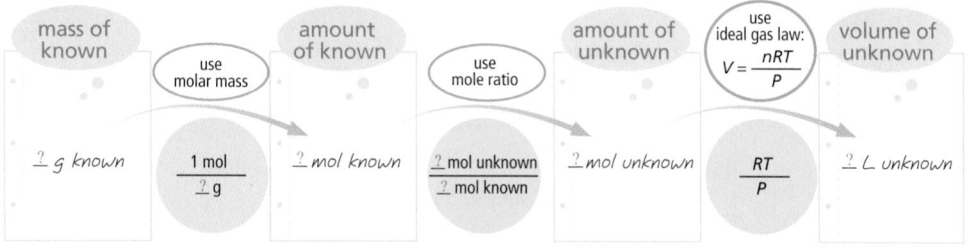

| mass of known | → use molar mass | amount of known | → use mole ratio | amount of unknown | → use ideal gas law: $V = \dfrac{nRT}{P}$ | volume of unknown |

? g known $\dfrac{1 \text{ mol}}{? \text{ g}}$ _? mol known_ $\dfrac{? \text{ mol unknown}}{? \text{ mol known}}$ _? mol unknown_ $\dfrac{RT}{P}$ _? L unknown_

SAMPLE PROBLEM G

Using the Ideal Gas Law to Solve Stoichiometry Problems

How many liters of hydrogen gas will be produced at 280.0 K and 96.0 kPa if 1.74 mol of sodium react with excess water according to the following equation?

$$2Na(s) + 2H_2O(l) \longrightarrow 2NaOH(aq) + H_2(g)$$

1 Gather information.

$T = 280.0$ K $P = 96.0$ kPa $R = 8.314$ L•kPa/mol•K

$n = ?$ mol $V = ?$ L

2 Plan your work.

Use the mole ratio from the equation to compute moles of H_2.

$$1.74 \text{ mol Na} \times \frac{1 \text{ mol } H_2}{2 \text{ mol Na}} = 0.870 \text{ mol } H_2$$

Rearrange the ideal gas law to solve for the volume of hydrogen gas.

$$V = \frac{nRT}{P}$$

3 Calculate.

Substitute the three known values into the rearranged equation.

$$V = \frac{(0.870 \text{ mol } H_2)\left(\dfrac{8.314 \text{ L•kPa}}{\text{mol•K}}\right)(280.0 \text{ K})}{(96.0 \text{ kPa})} = 21.1 \text{ L } H_2$$

4 Verify your results.

Recall that 1 mol of gas at 0°C and 1 atm occupies 22.4 L. The conditions in this problem are close to this, so the volume should be near 22.4 L.

Practice problems on next page

 PRACTICE HINT

Whenever you can relate the given information to moles, you can solve stoichiometry problems. In this case, the ideal gas law is the bridge that you need to get from moles to the answer.

Answers to Practice Problems G

1. 11.4 L

2. 2.08×10^5 L H_2O

3. 3.87 g Na

Homework ── GENERAL

Additional Practice

1. In the combustion reaction of 149 g of propane with excess oxygen, what volume of carbon dioxide is produced at STP? $C_3H_8 + 5O_2 \rightarrow 3CO_2 + 4H_2O$ Ans. 227 L CO_2

2. A student wishes to prepare oxygen by using the thermal decomposition of potassium chlorate, $KClO_3$. Given that the gas will have a temperature of 700°C and a pressure of 98.6 kPa, how much potassium chlorate will be necessary to produce 125 mL of oxygen? Ans. 0.124 g $KClO_3$

LS Logical

Chapter Resource File

• **Problem Solving**

Transparencies

TT Finding Volume of Unknown

ChemFile   **CHEMISTRY** INTERACTIVE TUTOR

• **Module 6: Gas Laws**
  Topic: Behavior of Gases
  This engaging tutorial reviews and reinforces the behavior of gases through modeling and guided practice.

Close

Quiz ——— GENERAL

1. Why can we use an "ideal gas law" for real gases? **Ans.** At conditions near room temperature and pressure, real gases behave as if they were ideal.

2. Why do gases diffuse? **Ans.** Because by diffusing, they move from an area of high concentration to one of lower concentration.

3. Which gas will travel faster, a heavy one or a light one? **Ans.** a light gas

4. When gas escapes rapidly from a small opening this is called _____. **Ans.** effusion

5. What is partial pressure? **Ans.** the amount of pressure exerted by one of the gases in a mixture that is part of the total pressure of gases in a container

6. When do gas volumes correspond to mole ratios? **Ans.** When gases are under the same conditions of temperature and pressure.

7. Why do the gas volumes correspond to mole ratios in balanced equations? **Ans.** According to Avogadro's law, equal volumes of gases at the same temperature and pressure contain equal numbers of particles, and a mole contains a definite number of particles.

LS Logical

Reteaching ——— BASIC

Have students create a flowchart, concept map or other graphic organizer that connects the temperature, pressure and volume of a gas to its number of moles and how that can be used to determine the number of moles or grams of another substance through a balanced equation. **LS** Visual

---

PRACTICE

1 What volume of oxygen, collected at 25°C and 101 kPa, can be prepared by decomposition of 37.9 g of potassium chlorate?

$$2KClO_3(s) \longrightarrow 2KCl(s) + 3O_2(g)$$

2 Liquid hydrogen and oxygen are burned in a rocket. What volume of water vapor, at 555°C and 76.4 kPa, can be produced from 4.67 kg of H_2?

$$2H_2(l) + O_2(l) \longrightarrow 2H_2O(g)$$

3 How many grams of sodium are needed to produce 2.24 L of hydrogen, collected at 23°C and 92.5 kPa?

$$2Na(s) + 2H_2O(l) \longrightarrow 2NaOH(aq) + H_2(g)$$

---

③ Section Review

UNDERSTANDING KEY IDEAS

1. What gas laws are combined in the ideal gas law?

2. State the ideal gas law.

3. What is the relationship between a gas's molecular weight and speed of effusion?

4. How did Gay-Lussac's experiment allow the chemical formulas of hydrogen gas and chlorine gas to be deduced?

5. How does the total pressure of a mixture of gases relate to the partial pressures of the individual gases in a mixture?

6. In gas stoichiometry problems, what is the "bridge" between amount in moles and volume?

PRACTICE PROBLEMS

7. How many moles of argon are there in 20.0 L, at 25°C and 96.8 kPa?

8. A sample of carbon dioxide has a mass of 35.0 g and occupies 2.5 L at 400.0 K. What pressure does the gas exert?

9. How many moles of SO_2 gas are contained in a 4.0 L container at 450 K and 5.0 kPa?

10. Two gases effuse through a hole. Gas A has nine times the molecular mass of gas B. What is the ratio of the two molecular speeds?

11. What volume of ammonia can be produced from the reaction of 22.5 L of hydrogen with nitrogen?

$$N_2(g) + 3H_2(g) \longrightarrow 2NH_3(g)$$

12. What will be the volume, at 115 kPa and 355 K, of the nitrogen from the decomposition of 35.8 g of sodium azide, NaN_3?

$$2NaN_3(s) \longrightarrow 2Na(s) + 3N_2(g)$$

CRITICAL THINKING

13. Explain why helium-filled balloons deflate over time faster than air-filled balloons do.

14. Nitrous oxide is sometimes used as a source of oxygen gas:

$$2N_2O(g) \longrightarrow 2N_2(g) + O_2(g)$$

What volume of each product will be formed from 2.22 L N_2O? At STP, what is the density of the product gases when mixed? (Hint: Keep in mind Avogadro's law.)

442

Answers to Section Review

1. Boyle's law, Charles' law, Gay-Lussac's law and Avogadro's law.

2. $PV = nRT$

3. The speed of the gas particles is inversely related to the square root of the molecular weight.

4. Equal volumes of each gas produced twice the volume of hydrogen chloride gas. Therefore, there must be two atoms in each molecule of hydrogen gas and chlorine gas, meaning that the gases are diatomic.

5. It is the sum of the individual pressures

6. the ideal gas law, and also the fact that at STP, 1 mole of gas = 22.41 L

7. 0.781 mol

8. 1.1×10^3 kPa

9. 5.3×10^{-3} mol SO_2

10. gas B has three times the speed of gas A

11. 15.0 L

12. 21.2 L

13. The average velocity of helium molecules is higher than the molecules of any gas found in air, so they escape through the pores in the balloon more quickly.

14. 2.22 L N_2 and 1.11 L O_2; 1.31 g/L

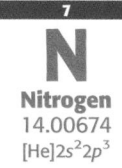

7

N

Nitrogen

14.00674

$[He]2s^22p^3$

Element Spotlight

Element Spotlight

Where Is N?

Earth's crust:
<0.01% by mass

Air: 75% by mass

Sea water:
0.0016% by mass

Nitrogen

Nitrogen gas is the most abundant gas in the atmosphere. Nitrogen is important for making the proteins, nucleic acids, vitamins, enzymes, and hormones needed by plants and animals to live. However, nitrogen gas is too unreactive for plants and animals to use directly. Nitrogen-fixing bacteria convert atmospheric nitrogen into substances that green plants absorb from the soil. Animals then eat these plants or eat other animals that feed on these plants. When the animals and plants die and decay, the nitrogen in the decomposed organic matter returns as nitrogen gas to the atmosphere and as compounds to the soil. The nitrogen cycle then starts all over again.

Industrial Uses

- Nitrogen is used in the synthesis of ammonia.
- Ammonia, NH_3, is used to produce fertilizer, explosives, nitric acid, urea, hydrazine, and amines.
- Liquid nitrogen is used in superconductor research and as a cryogenic supercoolant for storing biological tissues.
- Nitrogen gas is used as an inert atmosphere for storing and processing reactive substances.
- Nitrogen gas, usually mixed with argon, is used for filling incandescent light bulbs.

Real-World Connection The atmosphere contains almost 4×10^{18} kg of N_2. That's more than the mass of 10 000 typical mountains!

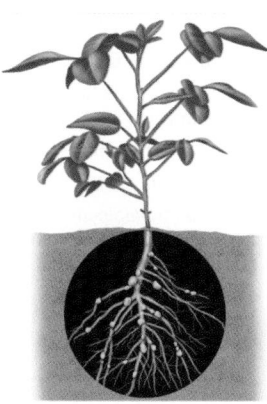

Legumes, such as this soybean plant, have nitrogen-fixing bacteria on their roots. These bacteria can convert atmospheric nitrogen into compounds that can be used by plants in the formation of proteins.

A Brief History

1774–1777: Antoine Lavoisier determines that nitrogen is an element.

1909: Fritz Haber, a German chemist, discovers a method for synthesizing ammonia from hydrogen gas and nitrogen gas. The method is still used today and is called the Haber-Bosch process.

| 1600 | 1700 | 1800 | 1900 |

1772: Nitrogen is discovered by Daniel Rutherford in Scotland, Joseph Priestley and Henry Cavendish in England, and Carl Scheele in Sweden.

Questions

1. Name three nitrogen-containing products you come into contact with regularly.

2. Research the historical impacts of the invention of the Haber-Bosch process, which is mentioned in the timeline above.

internet connect

www.scilinks.org
Topic: Nitrogen Cycle
SciLinks code: HW4082

SCI LINKS. Maintained by the National Science Teachers Association

443

Element Spotlight

Nitrogen Daniel Rutherford discovered nitrogen in 1772. From 1774 to 1777, Antoine Lavoisier researched nitrogen and determined that it is an element. Nitrogen accounts for 78% of the air by volume; an estimated 4000 trillion tons of nitrogen is contained in Earth's atmosphere. Lavoisier called nitrogen *azote*, which means "without life," because it is so inert. The compounds of nitrogen, however, are so active that they are essential in the production of foods, poisons, fertilizers, and explosives.

Elemental nitrogen has several industrial applications. The largest consumer of nitrogen gas is the ammonia industry. Nitrogen gas is used in the electronics industry as an inert blanketing agent during the production of sensitive components as well as in the pharmaceutical industry for the synthesis of many drugs. Liquid nitrogen is used both as a refrigerant in the transportation of food products and as a coolant in superconductor research.

Recently, research was conducted at the University of Washington to produce a liquid nitrogen-propelled automobile. A car powered by liquid nitrogen would have a range of 200 miles or more than twice that of a typical electric car. Refilling its tank would take only 10–15 minutes and its cost would be competitive to that of a gasoline powered car.

Other researchers at the Lawrence Livermore National Laboratory are trying to create nitrogen fullerene molecules (based on the buckyballs of carbon). These nitrogen fullerenes would be useful as propellants, especially for supersonic transport vehicles. Currently these polynitrogens are only hypothetical but research continues.

Answers to Feature Questions

1. Answers may include proteins, ammonia-containing cleaning products, many pharmaceuticals, fertilizers, etc.

2. The Haber-Bosch process for producing ammonia was crucial to Germany's war effort in World War I. It is estimated that Germany would have run out of nitrates, used in explosives, by 1916 if not for the Haber-Bosch process. The process is also responsible for the large-scale production of nitrate fertilizers, without which it is estimated that the world's present population could not be fed.

12 CHAPTER HIGHLIGHTS

Alternative Assessments

SKILL BUILDER
• Page 426

CHAPTER REVIEW
• Items 94–96

Portfolio Assessments

READING SKILL BUILDER
• Pages 417, 424

Group Activity
• Page 421

Reteaching
• Pages 422, 432, 442

KEY IDEAS

SECTION ONE Characteristics of Gases
• Gases are fluids, have low density, and are compressible, because of the relatively large intermolecular distances between gas particles.
• Gases expand to fill their entire container.
• Gases exert pressure in all directions.
• The kinetic-molecular theory states that gas particles are in constant random motion, are relatively far apart, and have volumes that are negligible when compared with the total volume of a gas.
• The average kinetic energy of a gas is proportional to its absolute temperature.

SECTION TWO The Gas Laws
• Pressure and volume of a gas at constant temperature are inversely proportional.
• The volume of a gas at constant pressure is proportional to the absolute temperature.
• The pressure of a gas at constant volume is proportional to the absolute temperature.
• Equal volumes of gas under the same conditions contain an equal number of moles of gas.

SECTION THREE Molecular Composition of Gases
• The ideal gas law is the complete statement of the relations between P, V, T, and n in a quantity of gas.
• Gases diffuse rapidly into each other.
• The rate of diffusion of a gas is inversely proportional to the square root of its molar mass.
• Each gas in a mixture produces a pressure as if it were alone in a container.
• Gay-Lussac's law of combining volumes can be used to deduce the chemical formula of a gas through observation of volume changes in a chemical reaction.

KEY TERMS

pressure
newton
pascal
standard temperature and pressure
kinetic-molecular theory

Boyle's law
Charles's law
Gay-Lussac's law
Avogadro's law

ideal gas
ideal gas law
diffusion
effusion
Graham's law of diffusion
Gay-Lussac's law of combining volumes
partial pressure
Dalton's law of partial pressure

KEY SKILLS

Converting Pressure Units
Sample Problem A p. 420

Solving Pressure-Volume Problems
Sample Problem B p. 425

Solving Volume-Temperature Problems
Sample Problem C p. 428

Solving Pressure-Temperature Problems
Sample Problem D p. 430

Using the Ideal Gas Law
Sample Problem E p. 435

Comparing Molecular Speeds
Sample Problem F p. 438

Using the Ideal Gas Law to Solve Stoichiometry Problems
Skills Toolkit 1 p. 441
Sample Problem G p. 441

444

Chapter Resource File

• **Microscale Lab** Generating and Collecting O_2
• **Microscale Lab** Generating and Collecting H_2
• **CBL™ Probeware Lab** Pressure-Volume Relationship: Understanding Boyle's Law

USING KEY TERMS

1. What is the definition of *pressure?*

2. What is a newton?

3. Write a paragraph that describes how the kinetic-molecular theory explains the following properties of a gas: fluidity, compressibility, and pressure. **WRITING SKILLS**

4. How do pascals relate to newtons?

5. What relationship does Boyle's law express?

6. What relationship does Charles's law express?

7. What relationship does Gay-Lussac's law express?

8. What gas law combines the basic gas laws?

9. What are two characteristics of an ideal gas?

10. Describe in your own words the process of diffusion.

UNDERSTANDING KEY IDEAS

Characteristics of Gases

11. How does wind illustrate that gases are fluids?

12. Using air, water, and a syringe, how can you show the difference in compressibility of liquids and gases?

13. When you drive a car on hot roads, the pressure in the tires increases. Explain.

14. As you put more air in a car tire, the pressure increases. Explain.

15. When you expand your lungs, air flows in. Explain.

16. Even when there is air in bicycle tires, you can still push down on the handle of the pump rather easily. Explain.

17. When you put air in a completely flat bicycle tire, the entire tire expands. Explain.

18. What assumptions does the kinetic-molecular theory make about the nature of a gas?

19. How does the average kinetic energy of a gas relate to its temperature?

The Gas Laws

20. How are the volume and pressure of a gas related, if its temperature is kept constant?

21. Explain why pressure increases as a gas is compressed into a smaller volume.

22. How are the absolute temperature and volume of a gas related, at constant pressure?

23. Explain Charles's law in terms of the kinetic-molecular theory.

24. How is a gas's pressure related to its temperature, at constant volume?

25. Explain Gay-Lussac's law in terms of the kinetic-molecular theory.

26. What does Avogadro's law state about the relationship between gas volumes and amounts in moles?

Molecular Composition of Gases

27. When using the ideal gas law, what is the proportionality constant, and in what units is it usually expressed?

28. Ammonia, NH_3, and alcohol, C_2H_6O, are released together across a room. Which will you smell first?

445

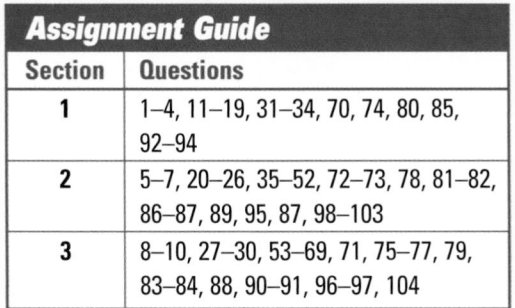

Assignment Guide	
Section	**Questions**
1	1–4, 11–19, 31–34, 70, 74, 80, 85, 92–94
2	5–7, 20–26, 35–52, 72–73, 78, 81–82, 86–87, 89, 95, 87, 98–103
3	8–10, 27–30, 53–69, 71, 75–77, 79, 83–84, 88, 90–91, 96–97, 104

REVIEW ANSWERS

1. Pressure is force divided by area on which it impinges.

2. A newton is that force which gives an acceleration of 1 m/s² to a mass of 1 kg.

3. The fluidity of a gas is due to the fact that gas particles are very widely separated and moving at high speeds, allowing them to move past each other easily. The wide separation of gas particles also explains a gas's compressibility. The constant rapid, random motion of gas particles and the resulting collisions with each other and the walls of their container is what causes the pressure exerted by a gas.

4. A pascal is a unit of pressure: 1 newton exerted over an area of 1 m².

5. Pressure and volume.

6. Volume and temperature

7. Pressure and temperature

8. The ideal gas law

9. It perfectly obeys the ideal gas law and conforms to the kinetic molecular theory.

10. Answers should include the idea that gas molecules initially concentrated in one area will spread evenly over the space available to it.

11. The air flows across the countryside, just as water in a stream.

12. When filled with water, and plugged, the plunger can't be pushed in, but with air, a small pressure decreases the volume in the syringe.

13. The volume of the tire is almost constant; the average velocity of the molecules increases with increase in temperature.

14. There are more molecules, which strike the walls of the tire more often.

15. As the lungs expand, pressure decreases. Air then flows in.

16. gases compress easily

17. the gas fills the entire container

18. The kinetic-molecular theory states that gas particles are in constant random motion, are very far apart, have volumes which are negligible when compared with the total volume of a gas, and have collisions which are perfectly elastic.

19. The average kinetic energy of a gas is directly proportional to temperature.

20. They are inversely related.

21. The molecules are closer together, and strike the walls of the container more often.

22. They are directly proportional.

23. Gas particles at higher temperatures will have a higher average kinetic energy, which means that their molecular speeds will be higher and therefore the volume of the gas will increase in order to maintain the same frequency of collisions and therefore maintain constant pressure.

24. They are directly proportional.

25. Gas particles at higher temperature will have a higher average kinetic energy, which means that at constant volume they will collide with each other and with the walls of their container more often, thus exerting more pressure.

26. Equal volumes of gases at the same temperature and pressure will contain the same number of gas molecules.

27. $R = 8.314$ L•kPa/mol•K

28. Both travel by diffusion. Ammonia, of lower mass, travels faster.

29. Volume changes in chemical reactions, if one of the formulas is known, you can show how what the mole proportions are in the reaction.

29. How can Gay-Lussac's law of combining volumes be used to deduce chemical formulas?

30. Write the equation that expresses Dalton's law of partial pressures.

PRACTICE PROBLEMS

Sample Problem A Converting Pressure Units

31. The standard pressure at sea level is 101 325 pascals. What force is being exerted on each square meter of Earth's surface?

32. The vapor pressure of hydrogen peroxide is 100.0 torr at 97.9°C. What is this pressure in kPa?

33. The gauge pressure in a tire is 28 psi, which adds to atmospheric pressure of 14.0 psi. What is the internal tire pressure in kPa?

34. The weather bureau reports the atmospheric pressure as 925 millibars. What is this pressure in kPa?

Sample Problem B Solving Pressure-Volume Problems

35. A gas sample has a volume of 125 mL at 91.0 kPa. What will its volume be at 101 kPa?

36. A 125 mL sample of gas at 105 kPa has its volume reduced to 75.0 mL. What is the new pressure?

37. A diver at a depth of 1.0×10^2 m, where the pressure is 11.0 atm, releases a bubble with a volume of 100.0 mL. What is the volume of the bubble when it reaches the surface? Assume a pressure of 1.00 atm at the surface.

38. In a deep-sea station 2.0×10^2 m below the surface, the pressure in the module is 20.0 atm. How many liters of air at sea level are needed to fill the module with 2.00×10^7 L of air?

39. The pressure on a 240.0 mL sample of helium gas is increased from 0.428 atm to 1.55 atm. What is the new volume, assuming constant temperature?

40. A sample of air with volume 6.6×10^7 L changes pressure from 99.4 kPa to 88.8 kPa. Assuming constant temperature, what is the new volume?

Sample Problem C Solving Volume-Temperature Problems

41. Use Charles's law to solve for the missing value in the following. $V_1 = 80.0$ mL, $T_1 = 27°C$, $T_2 = 77°C$, $V_2 = ?$

42. A balloon filled with helium has a volume of 2.30 L on a warm day at 311 K. It is brought into an air-conditioned room where the temperature is 295 K. What is its new volume?

43. The balloon in item 42 is dipped into liquid nitrogen at −196°C. What is its new volume?

44. A gas at 65°C occupies 4.22 L. At what Celsius temperature will the volume be 3.87 L, at the same pressure?

45. A person breathes 2.6 L of air at −11°C into her lungs, where it is warmed to 37°. What is its new volume?

46. A scientist warms 26 mL of gas at 0.0°C until its volume is 32 mL. What is its new temperature in degrees Celsius?

Sample Problem D Solving Pressure-Temperature Problems

47. Use Gay-Lussac's law to solve for the unknown. $P_1 = 111$ kPa, $T_1 = 273$ K, $T_2 = 373$ K, $P_2 = ?$

48. A sample of hydrogen exerts a pressure of 0.329 atm at 47°C. What will the pressure be at 77°C, assuming constant volume?

49. A sample of helium exerts a pressure of 101 kPa at 25°C. Assuming constant volume, what will its pressure be at liquid nitrogen temperature, −196°C?

50. The pressure inside a tire is 39 psi at 20°C. What will the pressure be after the tire is driven at high speed on a hot highway, when the temperature in the tire is 48°C?

51. A tank of oxygen for welding is at 31°C and 11 atm. What is the pressure when it is taken to the South Pole, where the temperature is −41°C?

52. A cylinder of gas at 555 kPa and 22°C is heated until the pressure is 655 kPa. What is the new temperature?

Sample Problem E Using the Ideal Gas Law

53. How many moles of argon are there in 20.0 L, at 25°C and 96.8 kPa?

54. How many moles of air are in 1.00 L at −23°C and 101 kPa?

55. A 4.44 L container holds 15.4 g of oxygen at 22.55°C. What is the pressure?

56. A polyethylene plastic weather balloon contains 65 L of helium, which is at 20.0°C and 94.0 kPa. How many moles of helium are in the balloon?

57. What will be the volume of the balloon in item 56 in the stratosphere at −61°C and 1.1×10^3 Pa?

58. A polyethylene weather balloon is inflated with 12.0 g of helium at −23°C and 100.0 kPa. What is its volume?

Sample Problem F Comparing Molecular Speeds

59. An unknown gas effuses at a speed one-quarter of that of helium. What is the molar mass of the unknown gas? It is either sulfur dioxide or sulfur trioxide. Which gas is it?

60. An unknown gas effuses at one half the speed of oxygen. What is the molar mass of the unknown? It is either HBr or HI. Which gas is it?

61. Oxygen molecules have an average speed of 4.80×10^2 m/s at 25°C. What is the average speed of H_2 molecules at the same temperature?

62. Oxygen molecules have an average speed of 480 m/s at 25°C. What is the average speed of HI molecules at that temperature?

Sample Problem G Using the Ideal Gas Law to Solve Stoichiometry Problems

63. How many liters of hydrogen gas can be produced at 300.0 K and 104 kPa if 20.0 g of sodium metal is reacted with water according to the following equation?

$$2Na(s) + 2H_2O(l) \longrightarrow 2NaOH(aq) + H_2(g)$$

64. Magnesium will burn in oxygen to form magnesium oxide as represented by the following equation.

$$2Mg(s) + O_2(g) \longrightarrow 2MgO(s)$$

What mass of magnesium will react with 500.0 mL of oxygen at 150.0°C and 70.0 kPa?

65. Suppose a certain automobile engine has a cylinder with a volume of 500.0 mL that is filled with air (21% oxygen) at a temperature of 55°C and a pressure of 101.0 kPa. What mass of octane must be injected to react with all of the oxygen in the cylinder?

$$2C_8H_{18}(l) + 25O_2(g) \longrightarrow 16CO_2(g) + 18H_2O(g)$$

66. Methanol, CH_3OH, is made by using a catalyst to react carbon monoxide with hydrogen at high temperature and pressure. Assuming that 450.0 mL of CO and 825 mL of H_2 are allowed to react, answer the following questions. (Hint: First write the balanced chemical equation for this reaction.)
  a. Which reactant is in excess?
  b. How much of that reactant remains when the reaction is complete?
  c. What volume of $CH_3OH(g)$ is produced?

67. What volume of oxygen, measured at 27°C and 101.325 kPa, is needed for the combustion of 1.11 kg of coal? (Assume coal is 100% carbon.)

$$C(s) + O_2(g) \longrightarrow CO_2(g)$$

447

30. $P_{total} = P_A + P_B + P_C$
31. 101 325 newtons
32. 13.3 kPa
33. 290 kPa
34. 92.5 kPa
35. 113 mL
36. 175 kPa
37. 1100 mL
38. 4.00×10^8 L
39. 66.3 mL
40. 7.4×10^7 L
41. 93.3 mL
42. 2.18 L
43. 0.570 L
44. 37°C
45. 3.1 L
46. 63°C
47. 152 kPa
48. 0.360 atm
49. 26 kPa
50. 43 psi
51. 8.4 atm
52. 75°C
53. 0.781 mol
54. 0.0486 mol
55. 266 kPa
56. 2.5 mol
57. 4.0×10^3 L
58. 62.4 L
59. M = 64 g/mol. It is SO_2.
60. M = 128 g/mol. It is HI.
61. 1.91×10^3 m/s
62. 240 m/s
63. 10.4 L
64. 0.484 g Mg
65. 3.56×10^{-2} g C_8H_{18}
66. a. CO
  b. 37.5 mL CO
  c. 412.5 mL CH_3OH

67. 2.28×10^3 L

68. 2.64 L

69. 1.80 g CO_2; 1.80 g/L CO_2; 1.31 g/L O_2

70. Mercury is under higher pressure. Its density is greater than water, so it takes a higher pressure for it to be at the same height as water.

71. 8.1×10^3 g LiOH

72. 0.179 g/L

73. 5.71 g/L

74. 10.3 meters

75. 171 kPa

76. 0.659 atm

77. 12.5 g O_2

78. 134 kPa

79. 2.64 L H_2

80. The average kinetic energy of the molecules of a gas is proportional to the absolute temperature. Increased temperature increases the kinetic energy and so also the pressure.

81. 167°C

82. 194°C

83. 44 g He

84. Diagrams should show two molecules of H_2 and one molecule of O_2 reacting to form two molecules of H_2O.

85. Although sealed, there will be small leaks. Air effusing outward hinders microbes from entering.

86. 0.18 atm

87. Doubling the absolute temperature will increase the volume to twice original, doubling the pressure will reduce volume back to original volume.

MIXED REVIEW

68. How many liters of hydrogen are obtained from the reaction of 4.00 g calcium with excess water, at 37°C and 0.962 atm?

69. How many grams of carbon dioxide are contained in 1.000 L of the gas, at 25.0°C and 101.325 kPa? What is the density of the gas at these conditions? What would the density of oxygen be at these conditions?

70. Below is a diagram showing the effects of pressure on a column of mercury and on a column of water. Which system is under a higher pressure? Explain your choice.

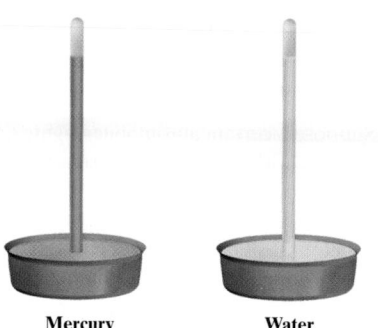

Mercury　　　　Water

71. Solid LiOH can be used in spacecraft to remove CO_2 from the air.

$$2\text{LiOH}(s) + \text{CO}_2(g) \longrightarrow \text{Li}_2\text{CO}_3(s) + \text{H}_2\text{O}(l)$$

What mass of LiOH must be used to absorb the carbon dioxide that exerts a partial pressure of 5.0 kPa at 15°C in a spacecraft with volume of 8.0×10^4 L?

72. What is the density of helium gas at 0°C and 101.325 kPa?

73. What is the density of HI gas under the same conditions as in item 71?

74. The density of mercury is 13.6 times that of water. What would be the height of a water barometer at sea level?

75. A 6.55 L container has oxygen at 25°C and 95.5 kPa. What will the pressure be after 0.200 mol of nitrogen is added?

76. A sample of carbon dioxide occupies 638 mL at 0.893 atm and 12°C. What will the pressure be at a volume of 881 mL and temperature of 18°C?

77. How many grams of oxygen gas must be in a 10.0 L container to exert a pressure of 97.0 kPa at a temperature of 25°C?

78. A 3.00 L sample of air exerts a pressure of 101 kPa at 300.0 K. What pressure will the air exert if the temperature is increased to 400.0 K?

79. How many liters of hydrogen can be obtained from the reaction of 4.00 g Ca with excess water, at 37°C and 0.962 atm?

80. Explain in terms of the kinetic-molecular theory why increasing the temperature of a gas at constant volume increases the pressure of the gas.

81. A sample of air has a pressure of 80.0 kPa at 20.0°C. The temperature is changed, and the pressure becomes 120.0 kPa. What is the new temperature in degrees Celsius? Assume constant volume.

82. A sample of nitrogen has a volume of 421 mL at 27°C. To what temperature must it be heated so that its volume becomes 656 mL, assuming constant pressure?

83. A 10 L tank is filled with helium until its pressure is 2.7×10^3 kPa at 23°C. What mass of helium is contained in the tank?

84. Above 100°C and at constant pressure, two volumes of hydrogen react with one volume of oxygen to form two volumes of gaseous water. Set up a diagram for this reaction that is similar to **Figure 19**, and that shows how the molecular formulas for oxygen and water can be determined.

CRITICAL THINKING

85. Clean rooms, used for sterile biological research, are sealed tightly and operate under high air pressure. Explain why.

86. The partial pressure of oxygen in the air is 0.21 atm at sea level, where the total pressure is 1.00 atm. What is the partial pressure of oxygen when this air rises to where the total pressure is 0.86 atm?

87. A sample of gas has a volume of 100 mL. Suppose the temperature is raised from 200 to 400 K, and the pressure changed from 100 to 200 kPa. What is the final volume?

88. A 1.00 L sample of chlorine gas is at 0°C and 101 kPa. This is liquefied, and the liquid has a density of 1.37 g/mL. What is the volume of the liquid chlorine?

89. You have a 1 L container of a gas at 20°C and 1 atm. Without opening the container, how could you tell whether the gas is chlorine or fluorine?

90. The density of oxygen gas is 1.31 g/L at 25°C and 101.325 kPa. What is the density of SO_2 under the same conditions?

91. If a breath of air is 2.4 L at 28°C and 101.3 kPa, how many grams of oxygen are in it? (Air is 21% oxygen by volume.)

92. Gas companies often store their fuel supplies in liquid form and in large storage tanks. Liquid nitrogen is used to keep the temperature low enough for the fuel to remain condensed in liquid form. Although continuous cooling is expensive, storing a condensed fuel is more economical than storing the fuel as gas. Suggest a reason that storing a liquid is more economical than storing a gas.

93. How would the shape of a curve showing the kinetic-energy distribution of gas molecules at 50°C compare with the blue and red curves in **Figure 8?**

ALTERNATIVE ASSESSMENT

94. The air pressure of car tires should be checked regularly for safety reasons and to prevent uneven tire wear. Find out the units of measurement on a typical tire gauge, and determine how gauge pressure relates to atmospheric pressure.

95. Find a local hot-air balloon group, and discuss with group members how they use the gas laws to fly their balloons. The group may be willing to give a demonstration. Report your experience to the class.

96. Qualitatively compare the molecular masses of various gases by noting how long it takes you to smell them from a fixed distance. Work only with materials that are not dangerous, such as flavor extracts, fruit peels, and onions.

CONCEPT MAPPING

97. Use the following terms to complete the concept map below: *amount in moles, ideal gas law, pressure, temperature,* and *volume.*

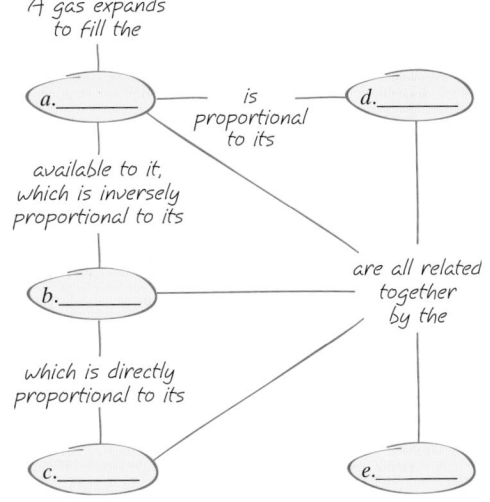

A gas expands to fill the

a._____ — is proportional to its — d._____

available to it, which is inversely proportional to its

b._____

which is directly proportional to its

c._____

are all related together by the

e._____

449

REVIEW ANSWERS
continued

88. 2.31 mL Cl_2

89. Use the ideal gas law to compute moles of gas. Weigh the container. The mass of the contents will match one of the two elements.

90. 2.62 g/L SO_2

91. 0.65 g O_2

92. As a liquid, the fuel occupies about a much smaller volume than it would have as a gas, so only a small number of tanks are needed. This savings more than compensates for the cost of maintaining natural gas in liquid form.

93. The peak of the curve would be broader and lower and would be farther to the right than the red curve, which peaks at 35°C.

94. The air pressure of car tires is usually measured in units of psi, pounds per square inch. Gauge pressure is defined as the difference between the total pressure in the tire and atmospheric pressure.

95. Hot-air ballooning depends on the buoyancy of the balloon, which in turn depends on the density of the gas inside the balloon. The report should include factors such as the temperature of the air in the balloon, the average time it takes to heat the air, and the method used to determine the lifting force of the balloon.

96. Students should be able to roughly inversely correlate the time it takes to smell a certain gas with that gas's molecular weight.

97. a. volume
   b. pressure
   c. temperature
   d. amount in moles
   e. ideal gas law

98. volume

99. about 0.340 L

100. 250 kPa

101. Answers should state that volume of a gas decreases as the pressure increases; or, that the volume of a gas increases as the pressure decreases.

102. Paired values should demonstrate that $PV = k$.

103. The pressure increases as the volume decreases.

104. a. 4.0×10^2 kPa

   b. 8.2×10^2 kPa

   c. at the higher temperature

FOCUS ON GRAPHING

Study the graph below, and answer the questions that follow.
For help in interpreting graphs, see Appendix B, "Study Skills for Chemistry."

98. What variable decreases as you go down the *y*-axis of the graph?

99. What is the volume of the gas at a pressure of 150 kPa?

100. What is the pressure of the gas at a volume of 0.200 L?

101. Describe in your own words the relationship this graph illustrates.

102. List the values of volume and pressure that correspond to any two points on the graph, and show why they demonstrate Boyle's law.

103. Explain why you feel resistance if you try to compress a sample of gas inside a plugged syringe.

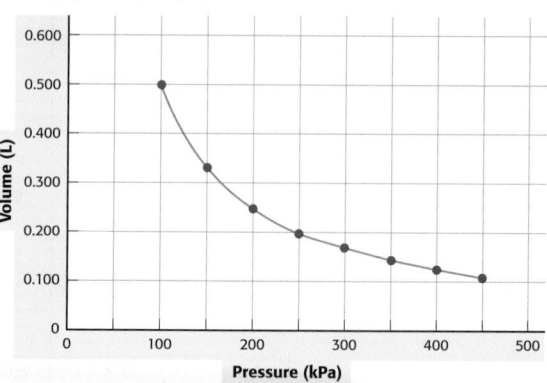

TECHNOLOGY AND LEARNING

104. Graphing Calculator

Calculating Pressure Using the Ideal Gas Law

The graphing calculator can run a program that calculates the pressure in atmospheres, given the number of moles of a gas (*n*), volume (*V*), and temperature (*T*).

Go to Appendix C. If you are using a TI-83 Plus, you can download the program **IDEAL** and run the application as directed. If you are using another calculator, your teacher will provide you with keystrokes and data sets to use. Use the program on your calculator to answer these questions.

a. What is the pressure for a gas with an amount of 1.3 mol, volume of 8.0 L, and temperature of 293 K?

b. What is the pressure for a gas with an amount of 2.7 mol, volume of 8.5 L, and temperature of 310 K?

c. A gas with an amount of 0.75 mol and a volume of 6.0 L is measured at two different temperatures: 300 K and 275 K. At which temperature is the pressure greater?

450

Chapter Resource File

• **Chapter Test**

STANDARDIZED TEST PREP **12**

Answers
1. b
2. c
3. a
4. c
5. b
6. d
7. a
8. b
9. b
10. d
11. c
12. a

1. The kinetic-molecular theory states that ideal gas molecules
   a. have weight and take up space.
   b. are in constant, rapid, random motion.
   c. exert forces of attraction and repulsion on each other.
   d. have high densities compared with liquids and solids.

2. Pressure can be measured in
   a. grams.   c. pascals.
   b. meters.   d. liters.

3. A sample of oxygen gas has a volume of 150 mL when its pressure is 0.947 atm. If the pressure is increased to 0.987 atm and the temperature remains constant, what will the new volume be?
   a. 140 mL   c. 200 mL
   b. 160 mL   d. 240 mL

4. A sample of neon gas occupies a volume of 752 mL at 25.0°C. If the pressure remains constant, what will be the volume of the gas at 50.0°C?
   a. 694 mL   c. 815 mL
   b. 752 mL   d. 955 mL

5. Which gas will effuse through a small opening fastest?
   a. the gas with the greatest molar mass
   b. the gas with the smallest molar mass
   c. the gas with the most polar molecules
   d. all will effuse at the same rate if temperature is constant

6. What is the pressure exerted by a 0.500 mol sample of nitrogen in a 10.0 L container at 25°C?
   a. 1.2 kPa   c. 0.10 kPa
   b. 10 kPa   d. 120 kPa

7. What volume of chlorine gas at 38°C and 1.63 atm is needed to react completely with 12.4 g of sodium to form NaCl?
   a. 4.22 L   c. 7.90 L
   b. 4.45 L   d. 10.2 L

8. A sample of gas in a closed container at a temperature of 100.0°C and 3.0 atm is heated to 300.0°C. What is the pressure of the gas at the higher temperature?
   a. 3.5 atm   c. 5.9 atm
   b. 4.6 atm   d. 9.0 atm

9. At room temperature, oxygen molecules have an average speed near 480 m/s. What is the speed of helium atoms at the same conditions?
   a. 3840 m/s   c. 170 m/s
   b. 1360 m/s   d. 60 m/s

10. An unknown gas effuses twice as fast as CH_4. What is the molar mass of the gas?
    a. 64 g/mol   c. 8 g/mol
    b. 32 g/mol   d. 4 g/mol

11. If 3 L N_2 is reacted completely with 3 L H_2, how many liters of *unreacted* gas remain?

    $$N_2 + 3H_2 \longrightarrow 2NH_3$$

    a. 4 L   c. 2 L
    b. 3 L   d. 1 L

12. Avogadro's law states that
    a. equal amounts of gases at the same conditions occupy equal volumes, regardless of the identity of the gases.
    b. at constant pressure, gas volume is directly proportional to absolute temperature.
    c. the volume of a gas is inversely proportional to its amount in moles.
    d. at constant temperature, gas volume is inversely proportional to pressure.

451

Solutions
Chapter Planning Guide

PACING	CLASSROOM RESOURCES	LABS, ACTIVITIES, AND DEMONSTRATIONS
BLOCK 1 • 45 min pp. 452–453 **Chapter Opener**		SE **Start-Up Activity** Exploring Types of Mixtures, p. 453 ◆
BLOCK 2 • 45 min pp. 454–459 **Section 1** What Is a Solution?	TT **Freshwater Versus Salt Water** *	TE **Group Activity** Examples of Solutions, p. 456 **BASIC** TE **Demonstration** Separation of Substances, p. 457 ◆ SE **QuickLab** Design Your Own Lab: The Colors of Candies, p. 458 ◆ CRF **Datasheets for In-Text Labs** * SE **Skills Practice Lab** Paper Chromatography, p. 800 ◆ **GENERAL** CRF **Datasheets for In-Text Labs** * CRF **Observation Lab** Paper Chromatography ◆ **GENERAL**
BLOCKS 3 & 4 • 90 min pp. 460–467 **Section 2** Concentration and Molarity	TT **Calculating Molarity Given Mass of Solute and Volume of Solution** * TT **Calculating Mass of Solute Given Molarity and Volume of Solution** *	TE **Demonstration** Concentration and Color, p. 460 ◆ TE **Demonstration** Total Volume of a Solution, p. 463 ◆ TE **Demonstration** Volume of Solvent, Solute, and Solution, p. 464 ◆
BLOCKS 5 & 6 • 90 min pp. 468–477 **Section 3** Solubility and the Dissolving Process	TT **Mass of Solute Added Versus Mass of Solute Dissolved** *	TE **Demonstration** Polarity, pp. 468–469 ◆ TE **Group Activity** "Like Dissolves Like", p. 470 **BASIC** TE **Demonstration** Saturation, p. 474 ◆ TE **Demonstration** Supersaturation, p. 475 ◆ TE **Demonstration** Temperature and Solubility of Gases, p. 476 ◆ CRF **CBL™ Probeware Lab** Effect of Temperature on Solubility of a Salt ◆ **ADVANCED**
BLOCKS 7 & 8 • 90 min pp. 478–486 **Section 4** Physical Properties of Solutions	TT **Electrical Conductivity of Solutions** *	TE **Demonstration** Conductance of Solutions, pp. 478–479 ◆ TE **Activity** Edible Example of Freezing Point Depression, p. 482 ◆ **BASIC** CRF **Consumer Lab** A Close Look at Soaps and Detergents ◆ **GENERAL**

BLOCKS 8 & 9 • 90 min **Chapter Review and Assessment Resources**

SE **Chapter Review,** pp. 488–492
SE **Standardized Test Prep,** p. 493
CRF **Chapter Test** *
OSP **Test Generator**
CRF **Test Item Listing** *
OSP **Scoring Rubrics and Classroom Management Checklists**

Holt Chemistry: Online Resources

Visit **go.hrw.com** for a variety of free resources related to this textbook. Enter the keyword **HW4 HOME**.

Holt Online Learning

Students can access interactive problem solving help and active visual concept development with the *Holt Chemistry* Online Edition available at **www.hrw.com**.

student CNN News

cnnstudentnews.com

Find the latest chemistry news, lesson plans, and activities related to important scientific events.

KEY

TE	Teacher Edition	CRF	Chapter Resource File	*	Also on One-Stop Planner
SE	Student Edition	TT	Teaching Transparencies	◆	Requires Advance Prep
OSP	One-Stop Planner				

PROBLEM SOLVING AND PRACTICE	SECTION REVIEW AND ASSESSMENT	STANDARDS CORRELATION
		National Science Education Standards
	TE **Quiz**, p. 459 GENERAL CRF **Quiz** * TE **Reteaching**, p. 459 BASIC SE **Section Review**, p. 459 CRF **Concept Review** *	PS 2d
SE **Sample Problem A** Calculating Parts per Million, p. 461 SE **Practice**, p. 461 GENERAL TE **Homework**, p. 461 GENERAL SE **Skills Toolkit 1** Preparing 1.000 L of a 0.5000 M Solution, p. 463 SE **Skills Toolkit 2** Calculating with Molarity, p. 464 SE **Sample Problem B** Calculating Molarity, p. 465 SE **Practice**, p. 465 GENERAL TE **Homework**, p. 465 GENERAL SE **Sample Problem C** Solution Stoichiometry, p. 466 SE **Practice**, p. 466 GENERAL TE **Homework**, p. 466 GENERAL CRF **Problem Solving** * ADVANCED SE **Problem Bank**, p. 858 GENERAL	TE **Quiz**, p. 467 GENERAL CRF **Quiz** * TE **Reteaching**, p. 467 BASIC SE **Section Review**, p. 467 CRF **Concept Review** *	
	TE **Quiz**, p. 477 GENERAL CRF **Quiz** * TE **Reteaching**, p. 477 BASIC SE **Section Review**, p. 477 CRF **Concept Review** *	PS 2d, PS 2e
SE **Problem Bank**, p. 858 GENERAL	TE **Quiz**, p. 486 GENERAL CRF **Quiz** * TE **Reteaching**, p. 486 BASIC SE **Section Review**, p. 486 CRF **Concept Review** *	PS 2d, PS 2e

SC/LINKS.

www.scilinks.org

Topic: Solution
SciLinks code: HW4118

Topic: Colloids
SciLinks code: HW4032

Topic: Solubility
SciLinks code: HW4117

Topic: Vitamins
SciLinks code: HW4127

Topic: Electrolytes and Nonelectrolytes
SciLinks code: HW4047

Topic: Emulsions
SciLinks code: HW4050

Topic: Hard Water
SciLinks code: HW4154

Overview

This chapter introduces students to mixtures and solutions. Molarity is defined and students are told that it is the most commonly used measure of concentration in chemistry. Students learn how to calculate molarity and learn how to use molarity when solving stoichiometry problems. Factors that affect solubility and the dissolving process are described, and the concept of saturation is explained. Finally, properties of solutions such as electrical conductivity and colligative properties are described and explained.

Assessing Prior Knowledge

Check for Content Knowledge

Students should be familiar with the following topics:

• Ions and ionic compounds

• Molecular structure and properties of water

• Measuring amounts in moles

• Boiling and freezing points

Using the Figure

The seawater in the oceans of the world is a very complex solution. Chemists have studied seawater for over a hundred years and have discovered that at least 72 elements are dissolved in it. They speculate that all naturally occurring elements can be found in seawater. The average salinity of the ocean is approximately 35 parts per thousand. The most abundant ions in seawater include sodium, chloride, sulfate, magnesium, calcium, and potassium.

CHAPTER

13

SOLUTIONS

452

Standards Correlations

National Science Education Standards

PS 2d: The physical properties of compounds reflect the nature of the interactions among its molecules. These interactions are determined by the structure of the molecule, including the constituent atoms and the distances and angles between them. (Sections 1, 3, and 4)

PS 2e: Solids, liquids, and gases differ in the distances and angles between molecules or atoms and therefore the energy that binds them together. In solids the structure is nearly rigid; in liquids molecules or atoms move around each other but do not move apart; and in gases molecules or atoms move almost independently of each other and are mostly far apart. (Sections 3 and 4)

Ocean water is an excellent example of a solution. A solution is a homogeneous mixture of two or more substances. *Homogeneous* means that the solution looks the same throughout, even under a microscope. In this case, water is the solvent—the substance in excess. Many substances are dissolved in the water. The most abundant substance is ordinary table salt, NaCl, which is present as the ions Na^+ and Cl^-. Other ions present include Mg^{2+}, Ca^{2+}, and Br^-. Oxygen gas is dissolved in the water. Without oxygen, the fish could not live. Because humans do not have gills like fish do, the diver needs scuba gear.

START-UP ACTIVITY

Exploring Types of Mixtures

SAFETY PRECAUTIONS

PROCEDURE

1. Prepare five mixtures in **five different 250 mL beakers.** Each should contain about 200 mL of water and one of the following substances: **12 g of sucrose, 3 g of soluble starch, 5 g of clay, 2 mL of food coloring,** and **20 mL of cooking oil.**

2. Observe the five mixtures and their characteristics. Record the appearance of each mixture after stirring.

3. Note which mixtures do not separate after standing. Transfer 10 mL of each of these mixtures to an individual **test tube.** Shine a **flashlight** through each mixture in a darkened room. Make a note of the mixture(s) in which the path of the light beam is visible.

ANALYSIS

1. What characteristic did the mixture(s) that separated after stirring share?

2. How do you think the mixture(s) in which the light's path was visible differed from those in which it was not?

Pre-Reading Questions

1. Give three examples of solutions you find in everyday life.

2. What main components do these solutions consist of?

3. How do you know that each of these examples is actually a solution?

CONTENTS 13

internet connect

www.scilinks.org
Topic : Solutions
SciLinks code: HW4118

SC LINKS. Maintained by the National Science Teachers Association

453

START-UP ACTIVITY

Skills Acquired:
- Experimenting
- Collecting data
- Interpreting
- Inferring

Materials:
For each group of 2–4 students:
- 250 mL beakers (5)
- test tubes (5)
- 12 g sucrose
- 3 g soluble starch
- 5 g clay
- 2 mL food coloring
- 20 mL cooking oil
- distilled water, about 1 L
- flashlight

Teacher's Notes: This activity will introduce students to the concrete distinctions between *suspensions, colloids,* and *solutions.* Through a method of progressive elimination, suspensions will be identified, then colloids based on the Tyndall effect, and the remaining mixtures, the most homogeneous ones, are the solutions.

Answers

1. The particles of the substances were temporarily suspended in the water.

2. Answers should include the idea that those mixtures are not in a single continuous phase, as the solutions were.

Answers to Pre-Reading Questions

1. Examples may include tap water, soft drinks, fruit juices, etc.

2. Accept all reasonable answers.

3. Answers should include the idea that the solutions' components are in a single phase, completely mixed at the molecular level.

Focus

Overview

Before beginning this section, review with your students the Objectives listed in the Student Edition. In this section students will learn to distinguish between the different types of mixtures. Several examples of solutions are described and definitions of solute and solvent are given. Students will also learn about techniques that can be used to separate mixtures, such as chromatography and distillation.

 **Bellringer** ━━━ **BASIC**

Write the definitions for suspension, solution, and colloid on the board. Ask students to work in small groups to brainstorm mixtures that can fit into each category. To help them get started, suggest thinking about different foods and drinks. After approximately five minutes, have each group share their lists with the rest of the class.

LS Interpersonal

Motivate

Discussion

As students come into class, call their attention to a beaker containing water and a bit of potter's clay as shown in **Figure 1,** on a stir plate. Ask students to predict what will happen when you stop stirring the mixture. Introduce the word *suspension*, and ask students to name things that are suspended. Stop stirring the mixture, and have students watch what happens. Ask them if the solid material formed a suspension or a solution.

What Is a Solution?

KEY TERMS

- solution
- suspension
- solvent
- solute
- colloid

solution

a homogeneous mixture of two or more substances uniformly dispersed throughout a single phase

suspension

a mixture in which particles of a material are more or less evenly dispersed throughout a liquid or gas

Topic Link

Refer to the "The Science of Chemistry" chapter for a discussion of the classification of mixtures.

OBJECTIVES

1. **Distinguish** between solutions, suspensions, and colloids.
2. **Describe** some techniques chemists use to separate mixtures.

Mixtures

You have learned about the difference between pure substances and mixtures. Mixtures can either be *heterogeneous* or *homogeneous.* The particles of a heterogeneous mixture are large enough to see under a microscope. In a homogeneous mixture, however, the particles are molecule-sized, so the mixture appears uniform, even under a microscope. A homogenous mixture is also known as a **solution.**

Suspensions Are Temporary Heterogeneous Mixtures

The potter shown in **Figure 1** uses water to help sculpt the sides of a clay pot. As he dips his clay-covered fingers into a container of water, the water turns brown. The clay-water mixture appears uniform. However, if the container sits overnight, the potter will see a layer of mud on the bottom and clear water on top. The clay does not dissolve in water. The clay breaks up into small pieces that are of such low mass that, for a while, they remain suspended in the water. This type of mixture, in which the different parts spontaneously separate over time, is called a **suspension.** In a suspension, the particles may remain mixed with the liquid while the liquid is being stirred, but later they settle to the bottom.

Figure 1
Clay particles suspended in water form a suspension.

Before settling   **After settling**

454

REAL-WORLD
CONNECTION

Salad Dressing Suspensions Oil and vinegar are immiscible liquids that are often combined to make salad dressing. Before pouring oil and vinegar dressings on their salads, people usually stir or shake the two liquids together to form a suspension. However, because it is a suspension, the two layers will separate over time. Some kitchenware companies have developed special salad dressing shakers that cause the liquids to form smaller droplets, thereby forming a more-mixed suspension.

Chapter Resource File

- Lesson Plan

Water molecule, H$_2$O

Fresh water

Water molecule, H$_2$O

Chloride ion, Cl$^-$

Sodium ion, Na$^+$

Salt water solution

Solutions Are Stable Homogeneous Mixtures

A student working in a pet shop is asked to prepare some water for a salt water aquarium. She prepares a bucket of fresh water. Then she adds a carefully measured quantity of salt crystals to the water, as shown in **Figure 2,** and stirs the water. These crystals consist of a mixture of salts which, when dissolved in water, will produce a solution with the same composition as sea water. After stirring, the student can no longer see any grains of salt in the water. No matter how long she waits, the salt will not separate from the water. The salt has dissolved in the water to form a stable homogeneous mixture. The particles are evenly distributed throughout the mixture, making it a true solution. The dissolved particles, which are ions in this case, are close to the size of the water molecules and are not clustered together.

Solution Is a Broad Term

Any mixture that is homogeneous on a microscopic level is a solution. According to that definition, air is a gaseous solution. However, when most people use the word *solution,* they are usually referring to a homogeneous *liquid* mixture. A homogeneous liquid mixture has one main component—a liquid—as well as one or more additional ingredients that are usually present in smaller amounts. The primary ingredient in a solution is called the **solvent,** and the other ingredients are the **solutes** and are said to be dissolved in the solvent. Water is the most common solvent. Although it is a very common substance, water is a unique solvent because so many substances can dissolve in it. Solutions in which water is the solvent are called *aqueous* solutions.

Figure 2
Fresh water is stable and homogeneous. The saltwater mixture is also stable and homogeneous because mixing occurs between molecules and ions.

solvent

in a solution, the substance in which the solute is dissolved

solute

in a solution, the substance that is dissolved in the solvent

(455)

HISTORY

CONNECTION

The Sinking of the *Titanic* The RMS *Titanic* sank in the morning of April 15, 1912 after colliding with an iceberg. The steel used to make the hull is one suggested reason why the *Titanic* sank so quickly. Steel is an alloy made from a variety of elements including carbon, iron, sulfur, and chromium. An analysis of a recovered piece of steel from the hull showed that it contained unusually large amounts of sulfur. The excess sulfur may have made steel brittle. As a result, the brittle steel may have cracked when the *Titanic* collided with the iceberg.

 Transparencies

TT Freshwater Versus Salt Water

Have students work in small groups to find everyday examples of solutions in all three states of matter. Each group should find at least one example of solutes and solvents in each state of matter, for a total of nine different types of solutions. **LS** Interpersonal

Teaching Tip

Characteristics of Colloids
Point out that colloids do not settle upon standing. Also, some colloids can appear to be clear. An example is gelatin that has set. These colloids can always be identified by the Tyndall effect, the scattering of a beam of light passing through them.

To demonstrate the Tyndall effect, stir one or two drops of milk into an aquarium or other large, transparent container filled with clear water. In a darkened room, shine a focused beam from a flashlight or laser pointer through the tank. The colloidal particles from the milk will make the beam visible. Place a small mirror on the bottom of the tank, and shine the beam at an angle toward the mirror. Students will be able to see both the incident beam and the reflected beam.

Teaching Tip

Examples of Colloids Students may be unfamiliar with the term *colloid*, but they have probably encountered many examples of colloids in their everyday lives. Tell students that gelatin desserts, whipped cream, milk, mayonnaise, and fog are all examples of colloids.

Copper atom

Zinc atom

Brass

Figure 3
Brass is a solid solution of zinc atoms in copper atoms.

Another type of solution involves one solid mixed with another solid. Examples include solid alloys, such as brass, bronze, and steel. Brass, shown in **Figure 3,** is a mixture of copper and zinc. Brass is widely used in musical instruments because it is harder and more resistant to corrosion than pure copper.

Colloids Are Stable Heterogeneous Mixtures

Milk appears to be homogeneous. But as **Figure 4** shows, under a microscope you see that milk contains globules of fat and small lumps of the protein casein dispersed in a liquid called *whey*. Milk is actually a **colloid,** and not a solution. The particles of casein do not settle out after standing.

The particles in a colloid usually have an electric charge. These likecharged particles repel each other, so they do not collect into larger particles that would settle out. If you added acid to milk, the acid would neutralize the charge, and the particles would coagulate and settle to the bottom of the container.

colloid

a mixture consisting of tiny particles that are intermediate in size between those in solutions and those in suspensions and that are suspended in a liquid, solid, or gas

Figure 4
Milk is a colloidal suspension of proteins and fat globules in whey.

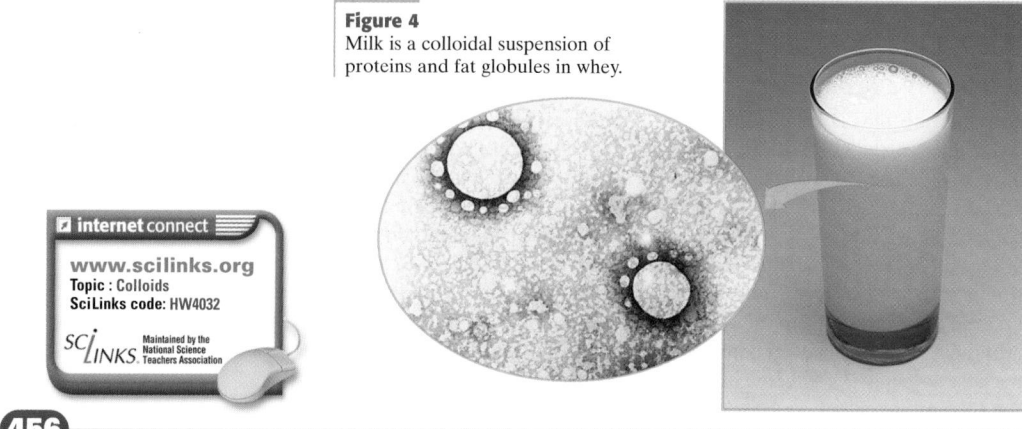

internet connect
www.scilinks.org
Topic : Colloids
SciLinks code: HW4032

SCiLINKS. Maintained by the National Science Teachers Association

456

did you know?

Some common colloids, such as milk and mayonnaise, appear white primarily because of the complete scattering of light striking them. On the microscopic level, they generally appear to be clear droplets suspended in another clear liquid.

Separating Mixtures

There are many ways to separate mixtures into their components. The best method to use in a particular case depends on the kind of mixture and on the properties of the ingredients. The methods shown in **Figure 5** rely on the physical properties of the ingredients to separate them. Notice that some of these methods can be used outside a laboratory. In fact, you may use one or more of these methods in your own home. For example, you may use filtration to make coffee or evaporation when you cook. Centrifuges are used in dairies to separate the cream from the milk to make fat-free milk.

Figure 5

a Decanting separates a liquid from solids that have settled. To decant a mixture, carefully pour off the liquid to leave the solids behind.

b A centrifuge is used to separate substances of different densities. The centrifuge spins rapidly, and the denser substances collect at the bottom of the tube.

c Ground coffee is separated from liquid coffee by filtration. The filtrate—the liquid and whatever passes through the filter—collects in the coffeepot. The solid grounds stay on the filter.

d In saltwater ponds such as this one, sea water evaporates, and salts, mainly sodium chloride, are left behind.

457

REAL-WORLD CONNECTION

The property of density is used to separate a mixture of ripe and unripe cranberries. Ripe cranberries float in water, but unripe cranberries sink. During harvesting, cranberry bogs are flooded, and the floating cranberries are skimmed from the water.

Using the Figure

Figure 5 Stress to students that all separation techniques take advantage of the properties of the materials involved. The separation of the coffee grounds is based on particle size. Separation by centrifugation is based on density.

Demonstration
Separation of Substances

(Approximate time: 15 minutes)

1. Place 4 g of NaCl and 5 g of white sand (mostly SiO_2) into 20 mL of water in a large heat-resistant test tube.

2. Ask students how they might separate the mixture based on the properties of the three substances. Ask them if filtering the sand will also separate the salt.

3. Filter the mixture, and then ask for methods to separate the salt and water. An easy method is to place the solution in an evaporating dish and to heat it gently until the water evaporates, leaving dry salt crystals. Heat very gently as the mixture approaches dryness.

4. Ask students how they could also recover the water that evaporated.

Safety: Wear safety goggles and a lab apron. Tie back loose clothing and long hair.

Disposal: Save the sand for reuse at another time. Dissolve the salt in water, and pour the mixture down the drain.

Group Activity ── BASIC
Depicting Molecules in Mixtures
Ask students to work in small groups to make drawings to illustrate the following mixtures at the molecular level:

- a solution of glucose in water

- a mercury-silver dental amalgam (alloy)

- a suspension of $Al(OH)_3$ (insoluble) in water

LS Visual

Quick LAB

Skills Acquired:
- Designing Experiments
- Experimenting
- Interpreting

Teacher's Notes: M&Ms® or other candies that have deeply colored sugar coatings can be used. Students should use no more solution than is necessary to dissolve the color coating. Check that they do not dissolve too much of the underlying sugar. It may interfere with the separation.

Safety: Remind students never to eat in laboratory.

Disposal: Dispose of all solids in the garbage. Liquids can be poured down the drain.

Answers
1. Students should report each of the original colors tested and the color components that each color contains.
2. Students will find that most colors are mixtures.

Chapter Resource File
- **Datasheets for In-Text Labs**

Figure 6
By using paper chromatography, you can separate different dyes mixed together in a product.

Chromatography Separates by Differences in Attraction

You may have noticed that stains stick to some fabrics more than other fabrics. Also, different processes are used to remove different stains. This illustrates the principles used in *chromatography*. Chromatography separates components of a mixture based on how quickly different molecules dissolved in a *mobile phase* solvent move along a *solid phase*.

Paper chromatography is a powerful technique for separating components of a solution. For example, it can be used to separate the dyes in ink. The dyes are blotted onto the paper (solid phase), and a solvent such as water (mobile phase) travels up the paper. The solvent dissolves the ink as it travels up the paper. Dyes that are attracted more strongly to the paper than other dyes travel more slowly along the paper. **Figure 6** shows the separation of dyes that make up different colors of ink.

Quick LAB

SAFETY PRECAUTIONS

Design Your Own Lab: The Colors of Candies

In this activity you will investigate whether the colors of candy-coated chocolates are single dyes or mixtures of dyes.

PROCEDURE
Design a paper chromatography experiment that uses **candy-coated chocolates,** **chromatography paper, small beakers,** and **0.1% NaCl developing solution** to determine whether the dyes in the candies are mixtures of dyes or are single dyes. For example, is the green color a result of mixing two primary colors? Are the other colors mixtures?

ANALYSIS
1. Prepare a report that includes your experimental procedure, a data table that summarizes your results, and the experimental evidence.
2. Be sure to answer the questions posed in the procedure.

458

TECHNOLOGY
CONNECTION — ADVANCED

Crude oil is separated into its components through a process called fractional distillation. Have interested students study fractional distillation in detail. Students should summarize their findings in a written report or a poster. **LS Logical**

did you know?

Air can be separated into its component gases by fractional distillation. To do this, the air is cooled and is condensed into a liquid. As the air warms up the different gases boil off at different temperatures and can be collected.

Distillation Separates by Differences in Boiling Point

Sometimes mixtures of liquids need to be purified or have their components separated. If the boiling points of the components are different, *distillation* can separate them based on their boiling points. As one component reaches its boiling point, it evaporates from the mixture and is allowed to cool and condense. This process continues until all the desired components have been separated from the mixture. For example, fermentation produces a solution of alcohol in water. If this is placed in a pot and boiled, the alcohol boils first. This alcohol-rich *distillate* can be collected by a distilling column. The distilling column is a cooler surface upon which the distillate recondenses, and can be collected as a liquid.

In some places where fresh water is scarce, distillation is used to obtain drinking water from sea water. However, because distillation requires a lot of energy, the process is expensive. Distillation is also used in the petroleum industry to separate crude oil into fractions according to their boiling points. The first fractions to distill are fluids with low boiling points, used as raw material in the plastics industry. Next comes gasoline, then at higher temperatures diesel fuel, then heating oil, then kerosene distill. What remains is the basis for lubricating greases.

Section Review

UNDERSTANDING KEY IDEAS

1. Explain why a suspension is considered a heterogeneous mixture.

2. Classify the following mixtures as homogeneous or heterogeneous:
   a. lemon juice   c. blood
   b. tap water   d. house paint

3. In a solution, which component is considered the solvent? Which is the solute?

4. Name the solvent and solute(s) in the following solutions:
   a. carbonated water   c. coffee
   b. apple juice   d. salt water

5. Does a solution have to involve a liquid? Explain your answer.

6. How is a colloid distinguished from a solution or a suspension?

7. What is the basic physical principle that chromatography is based upon?

8. How can distillation be used to prepare pure water from tap water?

CRITICAL THINKING

9. Explain how you could determine that brass is a solution, and not a colloid or suspension.

10. Explain why fog is a colloid.

11. You get a stain on a table cloth. Soapy water will not take the stain out, but rubbing alcohol will. How does this relate to chromatography?

12. If you allow a container of sea water to sit in the sun, the liquid level gets lower and lower, and finally crystals appear. What is happening?

459

Answers to Section Review

1. A suspension consists of particles which are large enough to be viewed under a microscope.

2. a. heterogeneous
   b. homogeneous
   c. heterogeneous
   d. heterogeneous

3. The solvent is the substance in excess, and the solute is the substance dissolved in the solvent.

4. a. solvent: water; solute: carbon dioxide
   b. solvent: water; solute: apple extract
   c. solvent: water; solute: coffee filtrate
   d. solvent: water; solute: salt

5. No; a solution is any homogeneous mixture and can be, for instance, gas/gas, or solid/solid.

6. A colloid is different from a suspension in that a colloid is a permanent mixture, while a suspension will settle out if left alone. A colloid is different from a solution in that a colloid is heterogeneous at the microscopic level, while a solution is homogeneous all the way down to the molecular level.

7. Chromatography is based on differences of adhesion of substances to a solid surface.

Answers continued on p. 493A

Close

Reteaching ———— BASIC

Have students create a concept map to help them remember the differences between suspensions, colloids, and solutions. The concept map should show how mixtures are divided into heterogeneous and homogeneous mixtures and how suspensions, colloids, and solutions fit into those two groups. The map should also show the properties of suspensions, colloids, and solutions that distinguish each type of mixture. **LS** Visual

Quiz ———— GENERAL

1. How is a suspension different from a colloid? **Ans.** Suspensions will settle out if allowed to rest but colloids will not. Suspensions also contain larger particles than colloids.

2. Describe how one can separate a mixture of sugar in water. **Ans.** Boil off the water or allow the water to evaporate.

3. Name one way to distinguish between a colloid and a solution. **Ans.** Accept all reasonable answers. Possible answers include: examine mixture with a microscope, if particles can be seen then it is a colloid; shine light through the mixture, if it scatters light then it is a colloid.

4. Blood is a suspension of different liquids and cells that have different densities. What is the best way to separate blood into its different components? **Ans.** Use a centrifuge to separate blood into its components.

Chapter Resource File

- Concept Review
- Quiz

Focus

Overview

Before beginning this section, review with your students the Objectives listed in the Student Edition. This section introduces students to the concept of concentration. Students learn that concentration can be measured by a variety of units, and that molarity is the most commonly used unit in chemistry. Students also learn how to solve problems that find the molarity of a solution and problems that use molarity in stoichiometry. In addition, the preparation of a solution with a known concentration is illustrated.

 Bellringer

Have students write a brief paragraph about the preparation of lemonade. Ask them to address the question of why the amount of lemon juice and sugar is important to making a good glass of lemonade.

Motivate

Demonstration

Concentration and Color
Display several identical flasks or beakers containing solutions of potassium permanganate of various concentrations (made by diluting a stock solution of $KMnO_4$ with varying volumes of water). Ask students which container has the solution of greatest concentration and how they made that judgment. Elicit the idea that the solutions have varying numbers of ions per unit volume, which accounts for the varying depths of color.

Concentration and Molarity

KEY TERMS
• concentration
• molarity

OBJECTIVES

① **Calculate** concentration using common units.

② **Define** molarity, and calculate the molarity of a solution.

③ **Describe** the procedure for preparing a solution of a certain molarity.

④ **Use** molarity in stoichiometric calculations.

Concentration

concentration

the amount of a particular substance in a given quantity of a solution

In a solution, the solute is distributed evenly throughout the solvent. This means that any part of a solution has the same ratio of solute to solvent as any other part of the solution. This ratio is the **concentration** of the solution. Some common ways of expressing concentration are given in **Table 1**.

Calculating Concentration

Concentrations can be expressed in many forms. One unit of concentration used in pollution measurements that involve very low concentrations is *parts per million*, or ppm. Parts per million is the number of grams of solute in 1 million grams of solution. For example, the concentration of lead in drinking water may be given in parts per million.

When you want to express concentration, you will begin with analytical data which may be expressed in units other than the units you want to use. In that case, each value must be converted into the appropriate units. Then, you must be sure to express the concentration in the correct ratio. **Sample Problem A** shows a typical calculation.

Table 1 Some Measures of Concentration

Name	Abbreviation or symbol	Units	Areas of application
Molarity	M	$\dfrac{\text{mol solute}}{\text{L solution}}$	in solution stoichiometry calculations
Molality	m	$\dfrac{\text{mol solute}}{\text{kg solvent}}$	with calculation of properties such as boiling-point elevation and freezing-point depression
Parts per million	ppm	$\dfrac{\text{g solute}}{1000\ 000\ \text{g solution}}$	to express small concentrations

See Appendix A for additional units of concentration.

REAL-WORLD CONNECTION

Espresso and Drip Coffee Although the main ingredients in espresso and coffee are the same, the two beverages have several differences. One of the main differences is that espresso has a stronger flavor than coffee prepared by a drip coffee maker. This is because it has a higher concentration than drip coffee. The difference in concentration is because espresso is prepared by quickly forcing boiling water through finely ground coffee beans, while drip coffee is prepared by dripping boiling water over more coarsely ground beans.

Chapter Resource File

• Lesson Plan

Calculating Parts per Million

A chemical analysis shows that there are 2.2 mg of lead in exactly 500 g of a water sample. Convert this measurement to parts per million.

1 Gather information.

mass of solute: 2.2 mg
mass of solution: 500 g
parts per million = ?

2 Plan your work.

First, change 2.2 mg to grams:

$$2.2 \text{ mg} \times \frac{1 \text{ g}}{1000 \text{ mg}} = 2.2 \times 10^{-3} \text{ g}.$$

Divide this by 500 g to get the amount of lead in 1 g water, then multiply by 1 000 000 to get the amount of lead in 1 000 000 g water.

3 Calculate.

$$\frac{0.0022 \text{ g Pb}}{500 \text{ g H}_2\text{O}} \times \frac{1\,000\,000 \text{ parts}}{1 \text{ million}} =$$
$$4.4 \text{ ppm (parts Pb per million parts sample)}$$

4 Verify your results.

Work backwards. If you divide 4.4 by 1 000 000 you get 4.4×10^{-6}. This result is the mass in grams of lead found per gram of the water sample. Multiply by 500 g to find the total amount of lead in the sample. The result is 2.2×10^{-3}, which is the given number of grams of lead in the sample.

PRACTICE HINT

Be sure to keep track of units in all concentration calculations. In the case of mass-to-mass ratios, such as parts per million, the masses of solute and solution must be expressed in the same units to obtain a correct ratio.

PRACTICE

1. Helium gas, 3.0×10^{-4} g, is dissolved in 200.0 g of water. Express this concentration in parts per million.

2. A sample of 300.0 g of drinking water is found to contain 38 mg Pb. What is this concentration in parts per million?

3. A solution of lead sulfate contains 0.425 g of lead sulfate in 100.0 g of water. What is this concentration in parts per million?

4. A 900.0 g sample of sea water is found to contain 6.7×10^{-3} g Zn. Express this concentration in parts per million.

5. A 365.0 g sample of water, contains 23 mg Au. How much gold is present in the sample in parts per million?

6. A 650.0 g hard-water sample contains 101 mg Ca. What is this concentration in parts per million?

7. An 870.0 g river water sample contains 2 mg of cadmium. Express the concentration of cadmium in parts per million.

461

ENVIRONMENTAL SCIENCE
CONNECTION — GENERAL

The concentrations of air and water pollutants are often measured in either parts per million (ppm) or parts per billion (ppb). Have interested students research changes in air pollution in major cities around the world. They should find historical smog reports, information about what measures were taken to reduce air pollution, and evidence showing the effectiveness of the changes made. **LS Logical**

Chapter Resource File

• Problem Solving

Teach

READING SKILL BUILDER — GENERAL

Paired Summarizing Pair each student with a partner and have them read silently about concentration and molarity. Then have one student summarize the steps to calculate concentration. The other student should listen to the retelling and point out any inaccuracies or steps that were left out. The students should repeat this process for calculating with molarity and using molarity in stoichiometric calculations and should take turns summarizing each part.
LS Interpersonal

Answers to Practice Problems A

1. 1.5 ppm
2. 130 ppm
3. 4250 ppm
4. 7.4 ppm
5. 63 ppm
6. 155 ppm
7. 2.3 ppm

Homework

Additional Practice Have students calculate the concentrations in the following problems.

1. How many parts per million of mercury are there in a sample of tap water with a mass of 750 g containing 2.2 mg of Hg? **Ans.** 2.9 ppm mercury

2. What is the concentration in parts per million of cobalt in a water sample of 300 g which contains 1 mg of Co? **Ans.** 3 ppm cobalt

3. A sample of water from a stream has a mass of 625 g. The sample is found to contain 0.35 mg of arsenic. What is the concentration of arsenic in parts per million? **Ans.** 0.56 ppm arsenic

Some students have a difficult time with the idea that the concentration of a specific solution is the same regardless of the quantity present. Make 1 L of a $CuSO_4$ solution, as shown in the **Skills Toolkit** feature on the next page. Pour samples of various volumes into several beakers. Ask students which sample contains the most solute (the largest sample). Then ask which sample has the highest concentration. Students should realize that all are the same concentration. Refer students to **Figure 7** to reinforce this concept. To help students learn that the concentration of a solution does not depend on the amount of solution present, tell students that concentration is an intensive property of a solution. Have students think of other intensive properties such as density and color to help them remember what it means when a property is intensive.

Figure 7
Solutions of the same molarity of a solute, regardless of the volume, all contain the same ratio of solute to solvent. In this case, various samples of 0.75 M KBr are shown.

Potassium ion, K^+
Water molecule, H_2O
Bromine ion, Br^-

molarity

a concentration unit of a solution expressed as moles of solute dissolved per liter of solution

Topic Link

Refer to the "The Mole and Chemical Composition" chapter for a discussion of the use of the mole to express chemical amounts.

Molarity

It is often convenient for chemists to discuss concentrations in terms of the number of solute particles in solution rather than the mass of particles in solution. Since the mole is the unit chemists use to measure the number of particles, they often specify concentrations using **molarity.** Molarity describes how many moles of solute are in each liter of solution.

Suppose that 0.30 moles of KBr are present in 0.40 L of solution. The molarity of the solution is calculated as follows:

$$\frac{0.30 \text{ mol KBr}}{0.40 \text{ L solution}} = 0.75 \text{ M KBr}$$

The symbol M is read as "molar" or as "moles per liter." Any amount of this solution has the same ratio of solute to solution, as shown in **Figure 7.**

Chemists often refer to the molarity of a solution by placing the formula for the solute in brackets. For example, $[CuSO_4]$ would be read as "the molarity of copper sulfate."

Preparing a Solution of Specified Molarity

Note that molarity describes concentration in terms of *volume of solution,* not volume of solvent. If you simply added 1.000 mol solute to 1.000 L solvent, the solution would not be 1.000 M. The added solute will change the volume, so the solution would not have a concentration of 1.000 M. The solution must be made to have exactly the specified volume of solution. The process of preparing a solution of a certain molarity is described in **Skills Toolkit 1.**

462

did you know?

The term *karat* is a measure of the concentration of gold in an alloy. Twenty-four karat (24k) gold is pure gold and 12k gold is 50 percent gold and 50 percent other metals. Because 24k gold is very malleable, gold used in jewelry is usually alloyed with other metals such as copper and silver to make it more durable. Most jewelry in the United States is made from either 10k, 14k, or 18k gold. However, people in other countries such as China and India prefer jewelry made with a higher concentration of gold. Jewelry in those countries is often made from 22k or 24k gold.

Preparing 1.000 L of a 0.5000 M Solution

Copper(II) sulfate, $CuSO_4$, is one of the compounds used to produce the chemiluminescence in light sticks. To make a 0.5000 M $CuSO_4$ solution, you need 0.5000 mol of the hydrate, $CuSO_4 \cdot 5H_2O$, for each liter of solution. To convert this amount of $CuSO_4 \cdot 5H_2O$ to a mass, multiply by the molar mass of $CuSO_4 \cdot 5H_2O$ (mass of $CuSO_4 \cdot 5H_2O$ = 0.5000 mol × 249.68 g/mol = 124.8 g).

Add some solvent (water) to the calculated mass of solute in the beaker to dissolve it, and then pour the solution into a 1.000 L volumetric flask.

Rinse the beaker with more water several times, and each time pour the rinse water into the flask until the solution almost reaches the neck of the flask.

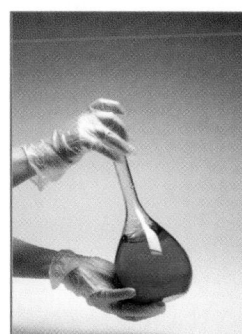

Stopper the flask, and swirl thoroughly until all of the solid is dissolved.

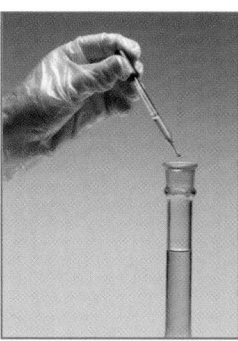

Carefully fill the flask with water to the 1.000 L mark.

Restopper the flask, and invert the flask at least 10 more times to ensure complete mixing.

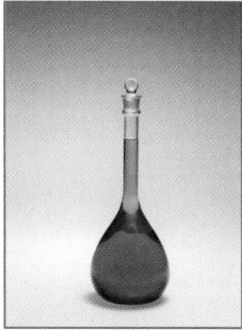

The solution that results has 0.5000 mol of $CuSO_4$ dissolved in 1.000 L of solution—a 0.5000 M concentration.

463

Teaching Tip

Water of Hydration Call students' attention to the fact that when solutions of hydrates are being prepared, the water of hydration must be added to the molar mass.

Demonstration

Students may not realize that the total volume of a solution is not equal to the total volume of solvent used to dissolve the solute. To illustrate this, place 50 mL of water in a 100 mL graduated cylinder and allow students to observe the volume. Pour the water into a dry beaker. Add several grams of sodium chloride (NaCl) to the water and stir to dissolve. Pour the newly made solution into the graduated cylinder and allow students to observe the new volume. The volume of the solution will be greater than the original volume of the water. Explain to students that this is why one does not fill a volumetric flask up to the line with the solvent before dissolving all of the solute when preparing a solution of known concentration.

did you know?

Glue is a solution. Inside a container of glue, adhesives are dissolved in a liquid solvent. Once the glue is applied to a surface, the solvent evaporates, leaving the adhesives on the surface. The adhesives link together and can bind surfaces together. Some glues, such as school glue, are water soluble, but others are only soluble in organic solvents such as acetone.

Teach, *continued*

Demonstration

Volume of Solvent, Solute, and Solution

(Approximate time: 10 minutes)

Use this demonstration of a discrepant event to generate interest in the mechanism of solution formation.

1. Fill a 100 mL graduated cylinder to the 100.0 mL mark with H_2O. Adjust the volume with an eyedropper.

2. Fill another 100 mL cylinder to the 100.0 mL mark with 95% ethanol (available as denatured alcohol at most pharmacies).

3. Have a student verify the cylinder readings and write them on the board in the following form.

100.0 mL H_2O + 100.0 mL ethanol = _____ mL solution

4. Pour the contents of both cylinders into a single 250 mL cylinder. Mix the contents thoroughly.

5. Have the same student record the combined volume and write it on the board. The volume should be around 190 to 192 mL. Challenge students to account for the missing volume. Strong hydrogen bonding occurs among water molecules and ethanol molecules. When these molecules are combined, this bonding pulls them closer together than they were before mixing. The formation of this tighter arrangement is exothermic, as is usually evident to the touch.

Safety: Wear safety goggles and an apron. Avoid flames. Students should remain 10 ft from the demonstration.

Disposal: Mix all liquids together. Add 500 mL of water, and pour the mixture down the drain.

Calculating with Molarity

In working with solutions in chemistry, you will find that numerical calculations often involve molarity. The key to all such calculations is the definition of molarity, which is stated as an equation below.

$$\text{molarity} = \frac{\text{moles of solute}}{\text{liters of solution}}$$

Skills Toolkit 2, below, shows how to use this equation in two common types of problems.

2 SKILLS Toolkit

1. Calculating the molarity of a solution when given the mass of solute and volume of solution

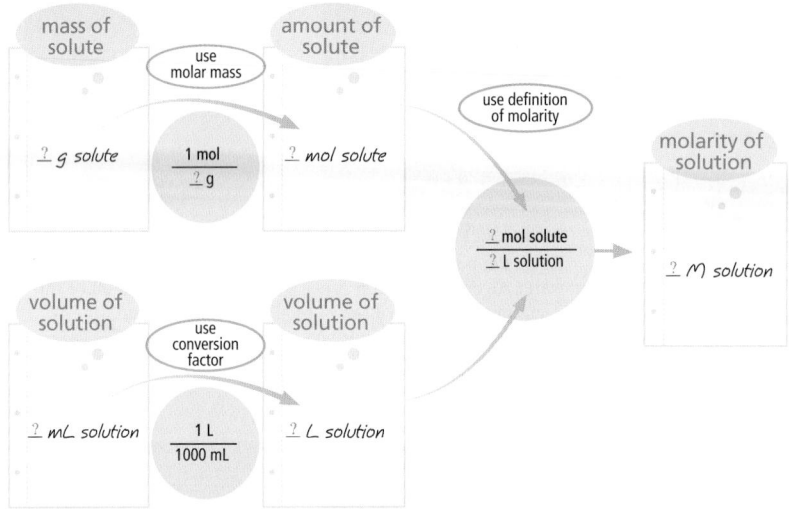

2. Calculating the mass of solute when given the molarity and volume of solution

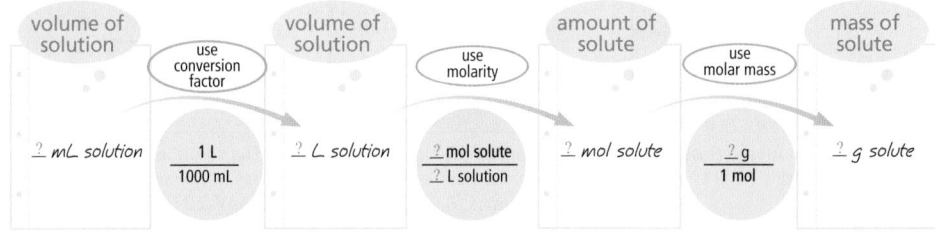

Transparencies

TT Calculating Molarity Given Mass of Solute and Volume of Solution

TT Calculating Mass of Solute Given Molarity and Volume of Solution

SAMPLE PROBLEM B

Calculating Molarity

What is the molarity of a potassium chloride solution that has a volume of 400.0 mL and contains 85.0 g KCl?

1 Gather information.

volume of solution = 400.0 mL
mass of solute = 85.0 g KCl
molarity of KCl solution = ?

2 Plan your work.

Convert the mass of KCl into moles of KCl by using the molar mass:

$$85.0 \text{ g KCl} \times \frac{1 \text{ mol}}{74.55 \text{ g KCl}} = 1.14 \text{ mol KCl}$$

Convert the volume in milliliters into volume in liters:

$$400.0 \text{ mL} \times \frac{1 \text{ L}}{1000 \text{ mL}} = 0.4000 \text{ L}$$

3 Calculate.

Molarity is moles of solute divided by volume of solution:

$$\frac{1.14 \text{ mol KCl}}{0.4000 \text{ L}} = 2.85 \text{ mol/L} = 2.85 \text{ M KCl}$$

4 Verify your results.

As a rough estimate, 85 g divided by 75 g/mol is about 1 mol. If you divide 1 mol by 0.4 L you get 2.5 M. This approximation agrees with the answer of 2.85 M.

PRACTICE HINT

Remember to check that any solution volumes are converted to liters before you begin calculations that involve molarity.

PRACTICE

1. Vinegar contains 5.0 g of acetic acid, CH_3COOH, in 100.0 mL of solution. Calculate the molarity of acetic acid in vinegar.

2. If 18.25 g HCl is dissolved in enough water to make 500.0 mL of solution, what is the molarity of the HCl solution?

3. If 20.0 g H_2SO_4 is dissolved in enough water to make 250.0 mL of solution, what is the molarity of the sulfuric acid solution?

4. A solution of $AgNO_3$ contains 29.66 g of solute in 100.0 mL of solution. What is the molarity of the solution?

5. A solution of barium hydroxide, $Ba(OH)_2$, contains 4.285 g of barium hydroxide in 100.0 mL of solution. What is the molarity of the solution?

6. What mass of KBr is present in 25 mL of a 0.85 M solution of potassium chloride?

7. If all the water in 430.0 mL of a 0.45 M NaCl solution evaporates, what mass of NaCl will remain?

PROBLEM SOLVING SKILL

465

Answers to Practice Problems B

1. 0.83 M acetic acid
2. 1.001 M HCl
3. 0.816 M sulfuric acid
4. 1.75 M $AgNO_3$
5. 0.2501 M $Ba(OH)_2$
6. 2.5 g KBr
7. 11 g NaCl

Homework

Additional Practice Have students solve the following molarity problems.

1. Determine the molarity of a solution prepared by dissolving 16.9 g of NaOH in enough water to make 250.0 mL of solution. Ans. 1.69 M

2. A solution is prepared by dissolving 30.05 g of ammonium dichromate, $(NH_4)_2Cr_2O_7$, in water and diluting it to 500.0 mL in a volumetric flask. What is the molarity of the solution? Ans. 0.2384 M

3. A mass of 158.0 g of calcium nitrate tetrahydrate is dissolved in enough water to make 1.500 L of solution. Calculate the molarity of this solution. Ans. 0.4460 M

SKILL BUILDER — ADVANCED

Research Skills Encourage interested students to research various oscillating reactions such as the Belousov-Zhabotinsky reaction or the Briggs-Rauscher reaction. Explain to them that the changing colors of the reactions are a result of the changing concentrations of the products and reactants. Also explain that the colors oscillate because the equilibrium of the reactions shifts back and forth between the product side and the reactant side. Have the students find recipes for the reactions (they can be found on the Internet and in chemistry teaching journals) and have them demonstrate the reactions for the rest of the class. The students should also write a report about the reaction mechanism for one of the reactions.
LS Verbal

HISTORY
CONNECTION

Alchemy Chemistry is thought to have grown out of the ancient practice of alchemy. One goal of alchemists was to develop a universal solvent. Of course, the problem with a universal solvent is that it cannot be held in any physical container! Water is sometimes called the universal solvent because many things can dissolve in it. However, water is not as "universal" as the alchemists wished because not all materials dissolve well in it.

Chapter Resource File

• Problem Solving

Answers to Practice Problems C

1. 109 g HCl
2. 0.852 g ZnCl$_2$
3. 451 g CdS

Homework

Additional Practice Have students solve the following stoichiometry problems.

1. What volume in milliliters of a 1.50 M HCl solution would be needed to react completely with 28.4 g of Na$_2$CO$_3$ to produce water, CO$_2$, and NaCl? Ans. 357 mL

2. A zinc bar is placed in 435 mL of a 0.770 M solution of CuCl$_2$. What mass of zinc would be replaced by copper if all of the copper ions were used up? Ans. 21.9 g

3. What volume of a 0.232 M solution of barium nitrate would be needed to precipitate all of the sulfate ions in 150.0 mL of a 0.086 M solution of sodium sulfate, Na$_2$SO$_4$? Ans. 56 mL

4. Zinc combines with hydrochloric acid to form aqueous zinc chloride and hydrogen gas. If 300.0 mL of a 0.150 M solution of HCl is allowed to react with excess zinc, how many moles of hydrogen gas would be produced? If this reaction takes place at standard temperature and pressure, what volume of hydrogen gas is produced? Ans. 0.0225 mol H$_2$; 0.501 L H$_2$

Topic Link

Refer to the "Stoichiometry" chapter for a discussion of stoichiometric calculations.

PRACTICE HINT

As in all stoichiometry problems, the mole ratio is the key. In solution stoichiometry, molarity provides the bridge between volume of solution and amount of solute.

Using Molarity in Stoichiometric Calculations

There are many instances in which solutions of known molarity are used in chemical reactions in the laboratory. Instead of starting with a known mass of reactants or with a desired mass of product, the process involves a solution of known molarity. The substances are measured out by volume, instead of being weighed on a balance. An example of such an application in stoichiometry is shown in **Sample Problem C.**

SAMPLE PROBLEM C

Solution Stoichiometry

What volume (in milliliters) of a 0.500 M solution of copper(II) sulfate, CuSO$_4$, is needed to react with an excess of aluminum to provide 11.0 g of copper?

1 Gather information.

[CuSO$_4$] = 0.500 M
mass of product = 11.0 g Cu
solution volume = ? L

2 Plan your work.

Write the balanced chemical equation for the reaction:

$$3CuSO_4(aq) + 2Al(s) \longrightarrow 3Cu(s) + Al_2(SO_4)_3(aq)$$

Look up the molar mass of Cu:

$$\text{molar mass of Cu} = 63.55 \text{ g/mol}$$

Convert the mass of Cu to moles, and then use the mole ratio of CuSO$_4$:Cu from the balanced chemical equation to determine the number of moles of CuSO$_4$ needed. The moles of CuSO$_4$ can be converted into volume of solution using the reciprocal of molarity.

$$\text{g Cu} \times \frac{1 \text{ mol Cu}}{\text{g Cu}} \times \frac{\text{mol CuSO}_4}{\text{mol Cu}} \times \frac{\text{L solution}}{\text{mol CuSO}_4} = \text{L CuSO}_4$$

3 Calculate.

Substitute the values given:

$$11.0 \text{ g Cu} \times \frac{1 \text{ mol Cu}}{63.55 \text{ g Cu}} \times \frac{3 \text{ mol CuSO}_4}{3 \text{ mol Cu}} \times \frac{1 \text{ L solution}}{0.500 \text{ mol CuSO}_4} \times$$

$$\frac{1000 \text{ mL solution}}{1 \text{ L solution}} = 346 \text{ mL CuSO}_4 \text{ solution}$$

4 Verify your results.

Work backwards. A volume of 0.346 L of a 0.500 M CuSO$_4$ solution contains 0.173 mol CuSO$_4$ (0.346 L × 0.500 M = 0.173 mol). A 0.173 mol sample of CuSO$_4$ contains 0.173 mol Cu, which has a mass of 11.0 g, so the answer is correct.

466

Chapter Resource File

• **Problem Solving**

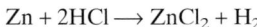

1. Commercial hydrochloric acid, HCl, is 12.0 molar. Calculate the mass of HCl in 250.0 mL of the solution.

2. An excess of zinc is added to 125 mL of 0.100 M HCl solution. What mass of zinc chloride is formed?

$$Zn + 2HCl \longrightarrow ZnCl_2 + H_2$$

PROBLEM SOLVING SKILL

3. Yellow CdS pigment is prepared by reacting ammonium sulfide with cadmium nitrate. What mass of CdS can be prepared by mixing 2.50 L of a 1.25 M $Cd(NO_3)_2$ solution with an excess of $(NH_4)_2S$?

$$Cd(NO_3)_2(aq) + (NH_4)_2S(aq) \longrightarrow CdS(s) + 2NH_4NO_3(aq)$$

② Section Review

UNDERSTANDING KEY IDEAS

1. Why did chemists develop the concept of molarity?

2. In what units is molarity expressed?

3. Describe in your own words how to prepare 100.0 mL of a 0.85 M solution of sodium chloride.

4. If you dissolve 2.00 mol KI in 1.00 L of water, will you get a 2.00 M solution? Explain your answer.

PRACTICE PROBLEMS

5. A sample of 400.0 g of water is found to contain 175 mg Cd. What is this concentration in parts per million?

6. If 1.63×10^{-4} g of helium dissolves in 100.0 g of water, what is the concentration in parts per million?

7. A standard solution of NaOH is 1.000 M. What mass of NaOH is present in 100.0 mL of the solution?

8. A 32 g sample of LiCl is dissolved in water to form 655 mL of solution. What is the molarity of the solution?

9. Most household bleach contains sodium hypochlorite, NaOCl. A 2.84 L bottle contains 177 g NaOCl. What is the molarity of the solution?

10. What mass of $AgNO_3$ is needed to prepare 250.0 mL of a 0.125 M solution?

11. Calcium phosphate used in fertilizers can be made in the reaction described by the following equation:

$$2H_3PO_4(aq) + 3Ca(OH)_2(aq) \longrightarrow$$
$$Ca_3(PO_4)_2(s) + 6H_2O(aq)$$

What mass in grams of each product would be formed if 7.5 L of 5.00 M phosphoric acid reacted with an excess of calcium hydroxide?

CRITICAL THINKING

12. You have 1 L of 1 M NaCl, and 1 L of 1 M KCl. Which solution has the greater mass of solute?

13. Under what circumstances might it be easier to express solution concentrations in terms of molarity? in terms of parts per million?

14. One solution contains 55 g NaCl per liter, and another contains 55 g KCl per liter. Which solution has the higher molarity? How can you tell?

467

Answers to Section Review

1. To easily connect the amount in moles of a substance with its concentration in a solution.

2. mol/L

3. Carefully weigh out 5.0 g NaCl, dissolve it in some water, pour the water into a 100 mL volumetric flask, rinse out the original container several times into the volumetric flask, swirl the flask to ensure complete dissolving, and top off the flask with water to the volumetric line.

4. No; the total volume may be different by the time the solute is dissolved.

5. 438 ppm Cd

6. 1.63 ppm He

7. 4.00 g NaOH

8. 1.1 M LiCl

9. 0.838 M NaOCl

10. 5.30 g $AgNO_3$

11. 5.8×10^3 g $Ca_3(PO_4)_2$ and 2.0×10^3 g H_2O

12. KCl; there is the same amount of moles of each substance, but KCl has the higher molecular weight.

13. Molarity can be used for normal lab concentrations, and ppm for very dilute solutions, such as pollutants in water.

Answers continued on p. 493A

Close

Reteaching ——— BASIC

If students are having difficulty remembering how to solve molarity problems, have them make a chart that shows the generalized steps on how to work molarity problems and shows an example problem worked beside these steps. To do this, have students divide a sheet of paper into two columns. In the left column, the students should copy the steps in **Skills Toolkit 2**. In the right column, students should work out example problems following the steps in the left column. Students may use their chart to help them solve practice problems and homework problems.
LS Logical

Quiz ——— GENERAL

1. What is the molarity of a solution containing 0.15 mol of KCl in 350 mL of solution? Ans. 0.43 M

2. What is the molarity of a solution containing 35.0 g of $CuCl_2$ in enough H_2O to make 300.0 mL of solution? Ans. 0.868 M

3. How many moles of NaOH would be required to prepare 2.0 liters of a 3.5 M solution? Ans. 7.0 mol NaOH

4. Calcium hydroxide reacts with phosphoric acid by the following equation:

$$2H_3PO_4(aq) + 3Ca(OH)_2(aq) \longrightarrow$$
$$Ca_3(PO_4)_2(s) + 6H_2O(l)$$

What mass of calcium phosphate is produced when 400.0 mL of 3.00 M phosphoric acid is allowed to react with an excess of calcium hydroxide? Ans. 186 g $Ca_3(PO_4)_2$

Chapter Resource File

- **Concept Review**
- **Quiz**

Solubility and the Dissolving Process

Focus

Overview

Before beginning this section, review with your students the Objectives listed in the Student Edition. In this section, students learn about solubility and the factors that affect it. The effect of polarity on solubility is described and the phrase "like dissolves like" is explained in terms of polarity. Students also learn how the surface area of a solid and the temperature of the solution influence the solubility of a solid. The terms *unsaturated*, *saturated*, and *supersaturated* are defined. Finally, students learn about the solubility of gases in liquids.

 **Bellringer**

Have students work in groups of 3–4 to brainstorm ideas on how to make a solid such as salt or sugar dissolve more quickly in a given amount of water.

Motivate

Demonstration

Polarity

(Approximate time: 15 minutes)

This demonstration should provoke discussion about solvent-solute interactions.

1. Place 10 mL of H_2O and 10 mL of ethanol in separate test tubes. Use forceps to add two crystals of I_2 to each tube.

2. Stopper and shake each tube. The nonpolar iodine should give a brown color to the less polar ethanol but not to the water.

Continued on next page

KEY TERMS

- **solubility**
- **miscible**
- **immiscible**
- **dissociation**
- **hydration**
- **saturated solution**
- **unsaturated solution**
- **supersaturated solution**
- **solubility equilibrium**
- **Henry's law**

solubility

the ability of one substance to dissolve into another at a given temperature and pressure; expressed in terms of the amount of solute that will dissolve in a given amount of solvent to produce a saturated solution

internet connect

www.scilinks.org
Topic : Solubility
SciLinks code: HW4117

SCiLINKS Maintained by the National Science Teachers Association

OBJECTIVES

1. **Identify** applications of solubility principles, and relate them to polarity and intermolecular forces.

2. **Explain** what happens at the particle level when a solid compound dissolves in a liquid.

3. **Predict** the solubility of an ionic compound by using a solubility table.

4. **Describe** solutions in terms of their degree of saturation.

5. **Describe** factors involved in the solubility of gases in liquids.

Solubility and Polarity

Some pairs of liquids form a solution when they are mixed. For example, any amount of ethylene glycol, a common antifreeze, mixes with any amount of water to form antifreeze solutions in radiators. These two compounds are both very polar and have 100% **solubility** with each other.

Oils, such as cooking oil, do not mix with water. An oil is nonpolar, and water is polar. However, paint thinner is soluble with the oil in oil-based paints. Both the paint and paint thinner are nonpolar. Polar compounds tend to dissolve in other polar compounds, and nonpolar compounds tend to dissolve in other nonpolar compounds.

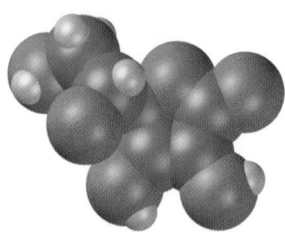 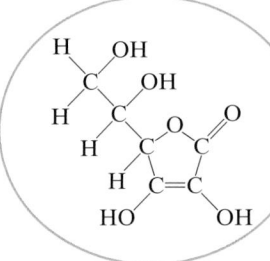 

Figure 8
a The most common form of vitamin C is ascorbic acid, which is shown here.

b Because ascorbic acid is polar, it is very soluble in water but insoluble in fats and oils.

c Lemons, oranges, grapefruits, and limes are good sources of vitamin C.

---

did you know?

Many vitamins, especially the water-soluble vitamins, act as coenzymes or are precursors to coenzymes. These molecules must be present and bind with an enzyme and substrate for the enzyme-catalyzed reaction to take place. Usually, the coenzymes are changed in the reaction and then enter a biochemical cycle that returns the coenzyme to its original form. Coenzymes might better be called co-substrates.

Chapter Resource File

- Lesson Plan

Vitamin C Is a Water-Soluble Vitamin

The human body cannot make its own vitamin C; it must be obtained from external sources. Vitamin C also cannot be stored in the body. The disease scurvy, caused by a lack of vitamin C, has always been a threat to people with a limited diet. In 1747, Dr. James Lind studied the effect of diet on sailors who had scurvy. Those whose diet included citrus fruits recovered. In 1795, long before people knew that citrus fruits were rich in vitamin C, the British navy began to distribute lime juice during long sea voyages. For this reason, British sailors were often called "limeys."

Vitamin C was isolated and identified in the early 1930s by American chemists. The most important function of vitamin C is in the synthesis of collagen, a protein that makes up tendons and that enables muscle movements. **Figure 8** on the previous page shows that vitamin C has several –OH groups. These –OH groups form strong hydrogen bonds with the –OH groups in water, so vitamin C is very soluble in water. At room temperature, 33 g of vitamin C will dissolve in 100 g of water. Any excess in the diet is quickly eliminated by the kidneys. It is almost impossible to overdose on vitamin C.

Vitamin A Is a Fat-Soluble Vitamin

Vitamin A also must be obtained in food, especially yellow vegetables. It has many functions in the body, and is essential for good vision. Vitamin A is also needed for the respiratory tract, skin, and for normal growth of bones. Fortunately, vitamin A is fairly abundant in foods.

Vitamin A has a long, nonpolar carbon-hydrogen chain, as shown in **Figure 9.** Consequently, it has very low solubility in water. Its nonpolarity makes it very soluble in fats and oils, which are also nonpolar. Any excess of vitamin A in the diet builds up in body fat and is not easily eliminated from the body. So much can accumulate in fat that the amount of vitamin A may become toxic. So, as with other fat-soluble vitamins, it is possible to take too much vitamin A.

internet connect

www.scilinks.org
Topic : Vitamins
SciLinks code: HW4127

SCI LINKS Maintained by the National Science Teachers Association

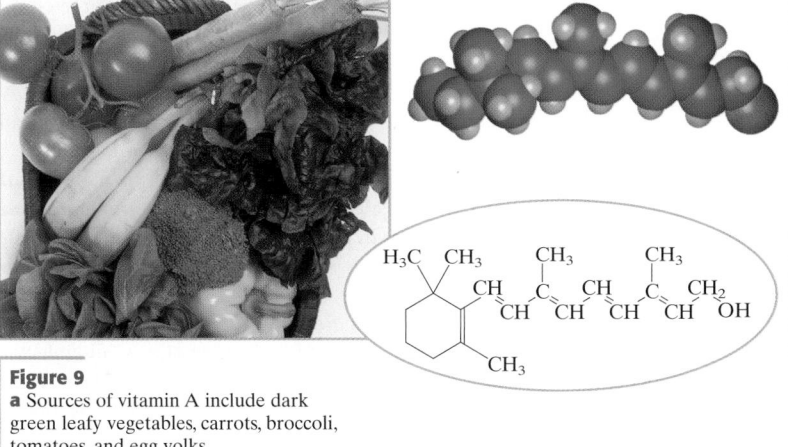

b Vitamin A is also known as retinol because it plays a vital role in helping the retina of your eye detect light.

c The vitamin A molecule is composed of mostly carbon and hydrogen, which makes the molecule nonpolar.

Figure 9
a Sources of vitamin A include dark green leafy vegetables, carrots, broccoli, tomatoes, and egg yolks.

did you know?

Some substances that cannot be eliminated by the kidneys, such as excess vitamin A, are usually processed by liver cells into materials that can be eliminated. For this reason, toxic substances, such as excess ethanol from alcoholic beverages, often damage the liver.

Group Activity ─── BASIC

Divide the class into groups of 3–4, and on the board write the expression, "like dissolves like." Ask each group to formulate an idea about the meaning of the expression and about what is meant by "like" substances. Invite them to draw on experience with materials that dissolve in water. Ask them to address, for example, what is "like" about sugar, salt, and water. Suggest that they examine **Figure 8** and **Figure 9** and speculate on the reasons for the differences in solubility. Have one student from each group report the conclusions of the group to the class. Students should suggest reasons related to attractions between solvent and solute particles.

LS Interpersonal

MISCONCEPTION /ALERT\

Some students may still look at the solution process as a chemical reaction and even view solutions as compounds. This idea may stem partly from our habit of saying, for example, that metals dissolve in acids and that fats dissolve in bases when we refer to these chemical reactions. Emphasize that solutions are mixtures and that they are not formed through chemical changes but through a process of very thorough mixing at the particle level. For example, the dissociation of NaCl into Na^+ and Cl^- is not an ionization reaction; the sodium and chlorine are already ions in solid NaCl.

Figure 10

a The nonpolar hydrocarbons in paint dissolve in the nonpolar oil in the paint thinner.

b Oil and water do not mix because oil is nonpolar and water is very polar.

miscible

describes two or more liquids that are able to dissolve into each other in various proportions

Topic Link

Refer to the "States of Matter and Intermolecular Forces" chapter for a discussion of intermolecular forces.

immiscible

describes two or more liquids that do not mix with each other

The Rule Is "Like Dissolves Like"

In nonpolar molecules, such as vitamin A, London forces are the only forces of attraction between molecules. When nonpolar molecules are mixed with other nonpolar molecule, the intermolecular forces of the molecules easily match. Thus, nonpolar molecules are generally soluble with each other, as shown in **Figure 10a.** This is one part of the rule "like dissolves like": liquids that are completely soluble with each other are described as being **miscible** in each other.

If molecules are sufficiently polar, there is an additional electrical force pulling them toward each other. The negative partial charge on one side of a polar molecule attracts the positive partial charge on the other side of the next polar molecule. If you add polar molecules to other polar molecules, such as water, the attraction between the two is strong. An example is vitamin C dissolving in water. This is another part of the rule "like dissolves like": polar molecules dissolve other polar molecules.

However, if you try to mix oil and water, the nonpolar oil molecules do not mix with the polar water molecules. The two liquids are **immiscible.** They form two layers, as shown in **Figure 10b.** The polar water molecules attract each other, so they cannot be pushed apart by the nonpolar oil molecules to form a solution.

Miscibility can be difficult to determine for some substances. Ethyl alcohol, CH_3CH_2OH, is sufficiently polar to be completely miscible with water. But the alcohol octanol, $CH_3CH_2CH_2CH_2CH_2CH_2CH_2CH_2OH$, has a long nonpolar tail which causes it to be only slightly soluble in water.

470

INCLUSION Strategies

• *Learning Disabled*   • *English Language Learners*

Ask students to fold a piece of paper in half vertically. Label one half of the paper "Polar Compounds" and the other half of the paper "Nonpolar Compounds." Students can list various compounds under each heading. Compounds and their polarity can be found in the textbook or in other reference books. Discuss whether combinations of various compounds would be miscible or immiscible in each other to reinforce the concept further.

Solubilities of Solid Compounds

Even two polar liquids placed in the same container may not dissolve in each other rapidly. Their strong intermolecular forces can only act on nearby molecules—not between molecules at the top of the container and those at the bottom.

The speed of the process can be increased by shaking the mixture. This action breaks the two liquids into small droplets and thereby increases the amount of contact between the surfaces of the liquids. This process works because the only place that dissolving can occur is at the surface between the two liquids, where the different molecules are near each other.

Similarly, in considering the solubility of solids in liquids, the only place where dissolution can occur is at the surface of the solid particles. The solid must be broken into smaller particles and then into molecules or ions, which can form a solution with the solvent molecules.

Greater Surface Area Speeds Up the Dissolving Process

As the discussion above indicated, the only place where dissolving can take place is at the surfaces where solute and solvent molecules are in contact. So if a solid has been broken into small particles, the surface area is much greater and the rate of the dissolving process is increased.

This is illustrated in **Figure 11.** The sugar granules dissolve more quickly than the sugar cubes. Because the sugar granules have more surface area, more of their molecules are directly exposed to the solvent and the dissolving process takes place faster. In the case of sugar cubes, most of the sugar molecules are inside the cubes and cannot dissolve until after the molecules at the outside of the cubes are dissolved.

Figure 11
Sugar granules dissolve in water more quickly than sugar cubes, because sugar granules have more surface area than sugar cubes.

Undissolved sugar molecules

Dissolved sugar molecule

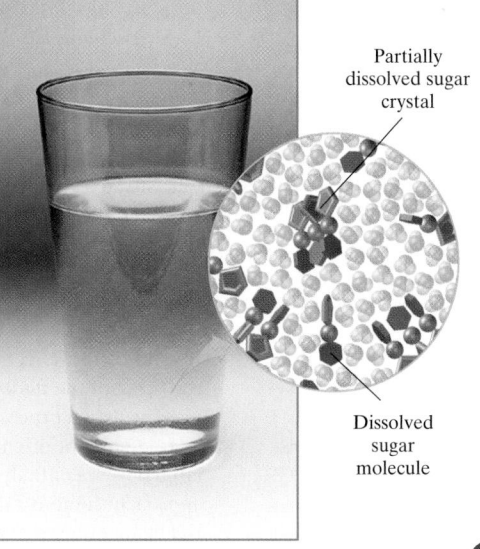
Partially dissolved sugar crystal

Dissolved sugar molecule

did you know?

Table sugar is manufactured with different sized granules. Students will be most familiar with white granulated sugar that is often used in cooking and baking. The crystal size of granulated sugar is considered to be large. Superfine sugar has smaller crystals and is often used to sweeten beverages such as iced tea and lemonade. The smaller crystals dissolve faster in the drinks. Powdered sugar is finely pulverized sugar that is used to make icings for desserts. Powdered sugar dissolves even faster than superfine sugar.

Figure 13 Have students study the graph on this page. Ask them to make a generalized statement about the effect of temperature on the solubility of ionic compounds. (Students should note that, in general, the solubility of ionic compounds increases with temperature.) Be sure students notice that this generalized statement does not apply to all compounds. The solubilities of some compounds such las cerium sulfate ($Ce_2(SO_4)_3$) and ytterbium sulfate ($Yb_2(SO_4)_3$) decrease with increasing temperature. Students should also notice that the effect of temperature varies from compound to compound. The solubilities of some compound increase very rapidly with increasing temperature, while the solubilities of other compound change very little as temperature increases. **LS** Visual

Figure 13 Astute students may note the discontinuity in the slope of the solubility curve for sodium acetate. Point it out, and explain that this is because below 60°C the solute in equilibrium with the solvent is the hydrated form $NaOOCH_3\cdot3H_2O$, but above 60°C the solute is in its anhydrous form, $NaOOCH_3$. These two forms of sodium acetate have different solute properties, and therefore their solubilities are affected differently by temperature. **LS** Visual

Figure 12
The effect of temperature on the solubility of some ionic solids is graphed here.

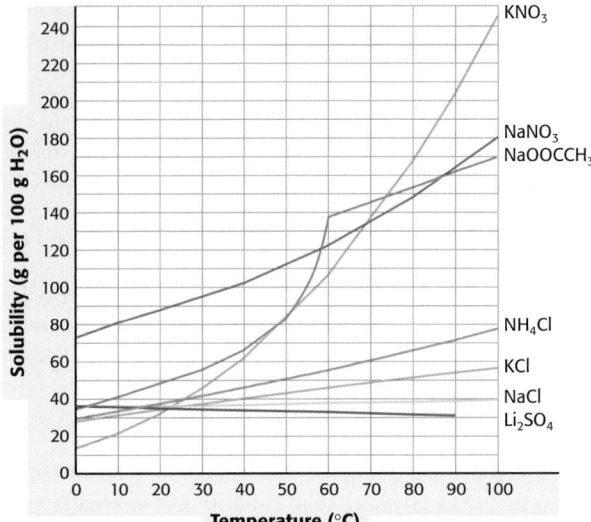

Solubility Vs. Temperature for Some Solid Solutes

Solubilities of Solids Generally Increase with Temperature

Another way to make most solids dissolve more and faster is to increase the temperature. Increasing the temperature is effective because, in general, solvent molecules with greater kinetic energy can dissolve more solute particles. **Figure 12** shows how an increase in temperature affects the solubility of several ionic compounds. In most cases, such as in the case of KNO_3, the solubility increases with temperature. However, temperature has little effect on the solubility of NaCl. The solubility of Li_2SO_4 actually decreases slightly as temperature increases.

Both Enthalpy and Entropy Affect the Solubility of Salts

Until now, we have not made a distinction between the dissolving process of a covalent solid, such as sugar, and that of an ionic solid, such as table salt. Surface area and temperature affect both covalent and ionic solids. However, the dissolving of an ionic compound involves a unique factor: the separation of ions from the lattice into individual dissolved ions. This process, called **dissociation,** can be represented as an equation.

$$NaCl(s) \longrightarrow Na^+(aq) + Cl^-(aq)$$

If water is the solvent, as above, dissociation involves **hydration,** the surrounding of the dissociated ions by water molecules.

The actual process of dissociation, however, is more complex. It takes a large amount of energy to separate the ions. The separation requires a large positive enthalpy change, ΔH. The polar ends of the water molecules approach the ions and release energy, and this ΔH is very negative. The ΔH changes nearly cancel.

Topic Link

Refer to the "Causes of Change" chapter for a discussion of enthalpy and entropy.

dissociation

the separating of a molecule into simpler molecules, atoms, radicals, or ions

hydration

the strong affinity of water molecules for particles of dissolved or suspended substances that causes electrolytic dissociation

472

Review the terms ionization and dissociation to help clear up any confusion students may have with the meanings. Dissociation refers to the separation of already existing ions when an ionic substance dissolves. Ionization refers to the formation of ions by a molecular compound when it dissolves. Ionization can also refer to many other ways that ions form.

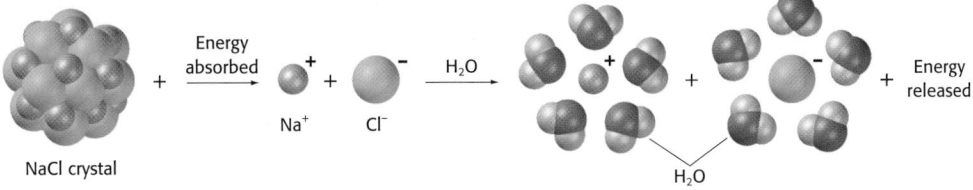

NaCl crystal

Figure 13

Individual ions are separated from the solid lattice by absorbed energy before they are hydrated by water molecules. Energy is released when hydrated ions are removed from the lattice.

Entropy increases as the ions are scattered throughout the solution. On the other hand, there is a large decrease in entropy as the water molecules are structured around the ions. Smaller ions, such as the Na^+ ions in **Figure 13**, have a greater decrease in entropy than larger ions. The net result of all of the enthalpy and entropy changes that accompany the dissolving process determines the solubility of an ionic solid.

Solubilities of Ionic Compounds

Solubilities are difficult to predict because of the many factors involved, so they must be measured experimentally. From experimental results of ionic solubilities in water, some patterns emerge, as shown in **Table 2**. Categories such as *soluble* and *insoluble* can be useful in many cases. The solubility of $Ba(OH)_2$ is 3.5 g per 100 g of water. It is described as slightly soluble. However, most substances are, at least to some extent, soluble in everything else. Even glass is very slightly soluble in water. In some delicate measurements, glass cannot be used as a container.

Table 2   Solubility Rules for Some Common Ionic Compounds

Compounds containing these ions are soluble in water:
Acetates, $CH_3CO_2^-$, except that of Fe^{3+}
Alkali metals (Group 1), except LiF
Ammonium, NH_4^+
Bromides, Br^-, except those of Ag^+, Pb^{2+}, and Hg_2^{2+}
Chlorides, Cl^-, except those of Ag^+, Pb^{2+}, and Hg_2^{2+}
Nitrates, NO_3^-
Sulfates, SO_4^{2-}, except those of Ca^{2+}, Sr^{2+}, Ba^{2+}, Pb^{2+}, and Hg_2^{2+}
Compounds containing these ions are insoluble in water:
Carbonates, CO_3^{2-}, except those of Group 1 and NH_4^+
Chromates, CrO_4^{2-}, except those of Group 1 and NH_4^+
Hydroxides, OH^-, except those of Group 1
Oxides, O^{2-}, except those of Group 1, Ca^{2+}, Sr^{2+}, and Ba^{2+} (which form hydroxides)
Phosphates, PO_4^{3-}, except those of Group 1 and NH_4^+
Sulfides, S^{2-}, except those of Group 1, Mg^{2+}, Ca^{2+}, Ba^{2+}, and NH_4^+

For a more detailed solubility table, see Appendix A.

HISTORY
CONNECTION

The Swedish chemist Svante Arrhenius was working toward his Ph.D. degree in 1884 when he proposed that the salt $CuCl_2$ was pulled apart into ions when it dissolved in water. His idea was thought to be so weak that he was awarded his degree with the lowest possible grade. In 1903, Arrhenius won the Nobel Prize for this theory of ionic dissociation in water.

Using the Figure

Figure 14 Have students review the energetics of crystal formation. Point out that the lattice energy of the crystal must be overcome when ions are hydrated and pulled away from the crystal. The overall enthalpy change depends on whether the enthalpy of hydration (exothermic) is greater than the input of energy needed to dislodge ions (endothermic).

Teaching Tip

Heats of Solution Chemists and technicians who prepare solutions know that certain substances dissolve in a highly exothermic way. Sometimes this release of heat is helpful, as in the dissolving of crystalline NaOH. Solutions of NaOH are often required in laboratory work. Usually, the NaOH to be dissolved is placed in a relatively small amount of water. The solution becomes hot as the NaOH dissolves, which helps the remaining NaOH dissolve rapidly. The small volume of solution produced is easy to cool, after which it can be diluted to the desired final concentration.

In other cases, the release of heat is hazardous. The ionization and hydration of H_2SO_4 presents a hazard because the heat released can cause the solution to boil and spatter, especially when water is poured into the concentrated acid. For this reason, H_2SO_4 solutions are always made by adding the acid slowly to water while stirring. In this way the whole volume of water is already present and acts as a heat sink. The solution may get very hot but usually does not reach its boiling temperature if monitored carefully.

Demonstration

Saturation

(Approximate time: 40 minutes)

This demonstration can be set up at the beginning of class and allowed to run during class.

1. To four separate test tubes, add 10 mL of distilled H_2O and one of the following: 3 g, 4 g, 5 g, and 6 g of NH_4Cl. Mark each test tube with the amount of ammonium chloride it contains.

2. Place the test tubes in a water bath using a 600 mL beaker. The water should come just above the solution level.

3. Heat the beaker until the NH_4Cl in the four tubes completely dissolves.

4. Remove the heat source, and place a nonmercury thermometer in each test tube. After one minute, remove the tubes and their thermometers from the bath and place them side-by-side in a test-tube rack.

5. Have students watch the tubes. When they notice crystallization occurring in a tube, have them read the temperature of the crystallizing solution. Record the tube contents and temperature on the board. You may wish to graph the solubility curve at the end of class.

Safety: Wear goggles and a lab apron; tie back hair and loose clothing. Students should remain 10 ft from the demonstration.

Disposal: Combine all liquids and solids from the tubes in an Erlenmeyer flask. Allow the water to evaporate, and save the NH_4Cl for future use.

Figure 14
a When a solution is saturated, additional solute added to the solvent will remain undissolved.

b As long as a solution is unsaturated, more solute can be added to the solvent and be dissolved.

Saturation

If you look in a chemistry handbook, or check the table in Appendix A, you will find that the solubility of potassium chloride, KCl, is 23.8 grams per 100 grams of water, at 20°C. This suggests some sort of limit.

When the maximum amount of solute, such as the 23.8 g of KCl mentioned above, is dissolved in a solution, the solution is said to be a **saturated solution.** As shown in **Figure 14a,** if a solution is saturated, any additional solute that is added collects at the bottom of the container. If more solute can be added to a solution and dissolve, as in **Figure 14b,** the solution is considered to be an **unsaturated solution. Figure 15** illustrates the relationship between solute added and solute dissolved.

The amount of solute that can dissolve depends on the forces between the solute particles and on forces between the solute particles and the solvent particles. When a solute is placed in contact with a solvent, molecules or ions of solute dissolve into the solvent. As soon as this happens, these same dissolved ions or molecules are capable of rejoining the undissolved solute. As the concentration of solute increases, the rate of return to the solute increases.

saturated solution

a solution that cannot dissolve any more solute under the given conditions

unsaturated solution

a solution that contains less solute than a saturated solution and that is able to dissolve additional solute

Figure 15
This graph shows the masses of solute that can be dissolved before saturation is reached. Any additional solute added beyond this point will remain undissolved.

Mass of Solute Added Vs. Mass of Solute Dissolved

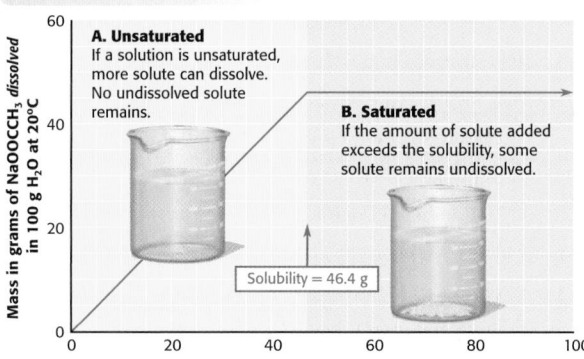

A. Unsaturated
If a solution is unsaturated, more solute can dissolve. No undissolved solute remains.

B. Saturated
If the amount of solute added exceeds the solubility, some solute remains undissolved.

Solubility = 46.4 g

Mass in grams of $NaOOCCH_3$ dissolved in 100 g H_2O at 20°C

Mass in grams of $NaOOCCH_3$ *added* to 100 g H_2O at 20°C

474

GEOLOGY CONNECTION

Many caves, such as Carlsbad Caverns in New Mexico and Mammoth Cave in Kentucky, were formed by calcium carbonate alternately dissolving in water and being deposited when the water evaporated. This process forms the cave formations called stalactites and stalagmites. Calcium carbonate is a white solid, but sometimes many colors can be seen in stalactites and stalagmites. These colors are a result of other minerals that have deposited with the calcium carbonate.

Transparencies

TT Mass of Solute Added Versus Mass of Solute Dissolved

Solubility Can Be Exceeded

In a saturated solution, some excess solute remains undissolved, and the mass that dissolves is equal to the solubility value for that temperature. Under special conditions, **supersaturated solutions** can also exist. Supersaturated solutions have more solute dissolved than the solubility indicates would normally be possible, but only as long as there is no excess undissolved solute remaining.

Supersaturation is the reason why hand warmers, such as those shown in **Figure 16,** work. Inside the plastic pack, 60 g of sodium acetate, $NaOOCCH_3$, has been combined with 100 mL of water. This amount of sodium acetate is more than the amount that can dissolve in 100 mL of water at 20°C. As shown in **Figure 12,** only about 48 g $NaOOCCH_3$ can dissolve in 100 mL of water. When the solution is heated to 100°C, all of it dissolves, as shown in **Figure 16a.** When the solution is allowed to cool to 20°C, crystals of solute should form. **Figure 16b** shows that crystals do not form. Instead, the solution becomes supersaturated. However, if you disturb the cooled solution by clicking the disk in the center of the pack, crystallization immediately occurs, as shown in **Figure 16c.** The recrystallization of sodium acetate is exothermic, so its reappearance in **Figure 16c** releases heat.

supersaturated solution

a solution holding more dissolved solute than what is required to reach equilibrium at a given temperature

Figure 16
On a cold day it can be comforting to use a hand warmer. Hand warmers use the principle of supersaturation to provide heat.

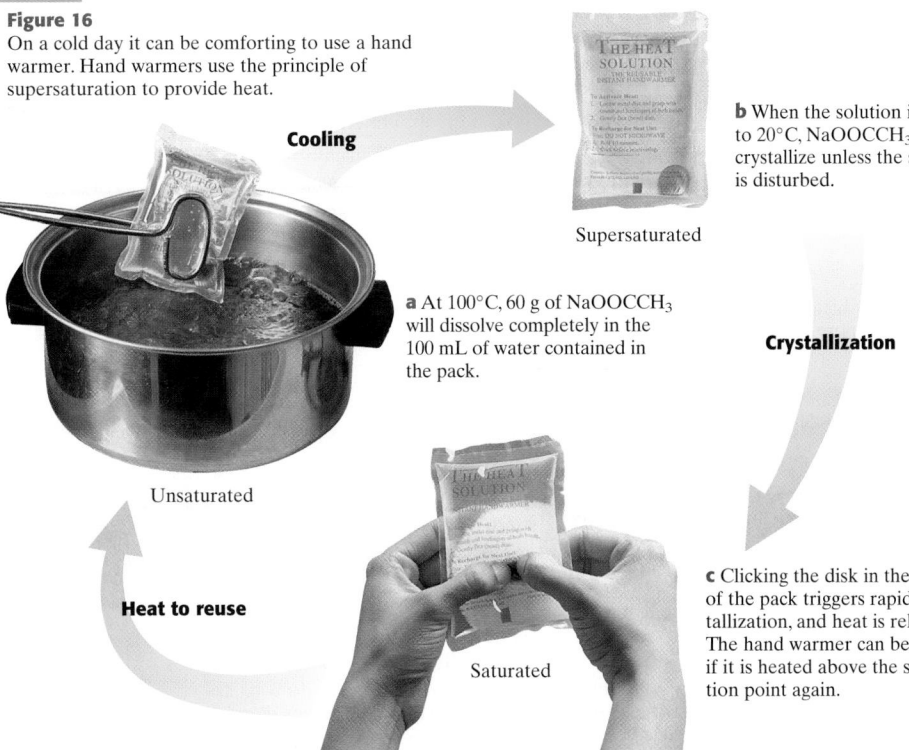

Cooling

b When the solution is cooled to 20°C, $NaOOCCH_3$ does not crystallize unless the solution is disturbed.

Supersaturated

a At 100°C, 60 g of $NaOOCCH_3$ will dissolve completely in the 100 mL of water contained in the pack.

Crystallization

Unsaturated

Heat to reuse

c Clicking the disk in the center of the pack triggers rapid recrystallization, and heat is released. The hand warmer can be reused if it is heated above the saturation point again.

Saturated

did you know?

Cold, aerated water typically forms white ice, which contains many small bubbles of gas that come out of solution as the ice freezes. This type of ice catalyzes the release of CO_2 from soda that is poured over it, causing the soda to foam and go flat. Fast-food restaurant icemakers are sometimes connected to hot-water supplies in order to make ice that is clear. Hot water contains much less dissolved gas than cold water. The clear ice allows servers to fill drink containers more rapidly because less foaming occurs.

Using the Figure

Figure 19 Ask students to describe what is taking place in the bottle before the cap is removed. They should understand that a stable system exists in which $CO_2(g)$ is in a dynamic equilibrium with $CO_2(aq)$.

Demonstration

Temperature and Solubility of Gases

(Approximate time: 10 minutes)

1. At the beginning of class, submerge a bottle of soda in an ice bath. Place an identical bottle in a warm-water bath. Allow a third identical bottle to stand at room temperature. Monitor the warm-water bath, and add ice to the ice bath as needed.

2. Near the end of class, turn the cap of each bottle so that the gas just begins to escape. Have student volunteers record the time it takes for each bottle to stop hissing. Have students draw conclusions about the relationship between temperature and the solubility of CO_2.

Safety: Goggles and a lab apron should be worn by the teacher and the volunteers. Other students should remain 10 ft from the demonstration. Do not allow anyone to drink the sodas.

Disposal: Pour the sodas and the water down the drain.

Teaching Tip

Dissolved Air in Water Ask students to suggest a reason why boiled water tastes "flat" and why lake fish sometimes die in large numbers during a period of hot weather.

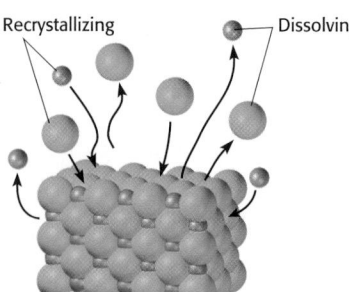
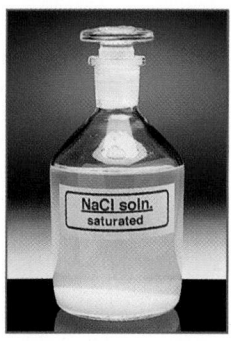

Figure 17
In a saturated solution, the solute is recrystallizing at the same rate that it is dissolving.

Recrystallizing   Dissolving

Saturation Occurs at a Point of Solubility equilibrium

In a saturated solution, solute particles are dissolving and recrystallizing at the same rate. It is a state of *dynamic equilibrium*. There is constant exchange, yet there is no net change.

When the amount of solute added to a solvent has reached its solubility limit, it is understood that the particles of solute in solution are in dynamic equilibrium with excess solute. This is illustrated in **Figure 17.** Na^+ ions and Cl^- ions are leaving the solid surface at the same rate as ions are also returning to the pile of excess solute at the bottom. These ions are considered to be in **solubility equilibrium.**

solubility equilibrium

the physical state in which the opposing processes of dissolution and crystallization of a solute occur at equal rates

Gases Can Dissolve in Liquids

When you first look at an unopened bottle of soda, you see very few bubbles. The liquid is homogeneous, as shown in **Figure 18a.** But when you open the bottle, you can hear gas escaping. Then you see many bubbles rising in the liquid, as pictured in **Figure 18b.** You probably know that the bubbles from soda are carbon dioxide, CO_2.

CO_2 under high pressure above solvent

Dissolved CO_2 molecules

Figure 18
a There are no bubbles in an unopened bottle of soda because carbon dioxide is dissolved in the liquid.

Air at atmospheric pressure

CO_2 gas bubble

Dissolved CO_2 molecules

b When the bottle cap is removed, the pressure inside the bottle decreases rapidly. Carbon dioxide escapes due to its lowered solubility.

did you know?

Beverages in which plant materials are infused into boiling water have been made for thousands of years. For much of history, water near settled communities was dangerous to drink without being boiled first. Drinks such as tea and coffee arose from people's desire to add flavor to the flat boiled water.

Gas Solubility Depends on Pressure and Temperature

Because a gas escapes when you open the bottle, you know that there is gaseous carbon dioxide present above the solution. Eventually, almost all the gas escapes when the bottle is opened to the atmosphere. Why?

In a gas, there is low attraction between the molecules. Likewise, there is usually little attraction between molecules of a gas and molecules of a liquid solvent. **Henry's law** states that the solubility of a gas increases as the partial pressure of the gas on the surface of the liquid increases.

At the high pressure of CO_2 in the unopened can, the gaseous CO_2 is in equilibrium with the dissolved gas. Therefore, the solution is saturated. When the bottle is opened and the pressure is released, the solubility decreases, and CO_2 bubbles escape. Finally, the dissolved CO_2 comes to equilibrium with the carbon dioxide in the air. The solution is saturated, but at this lower pressure, it is at a much lower concentration.

Temperature also affects gas solubility. After the soda bottle is open and becomes warm, the soda forms fewer bubbles and tastes flat. Even if you compare the taste of a newly opened warm soda with that of a newly opened cold soda, you will find that the warm soda will taste somewhat flat. Warm soda tastes flat because there is less CO_2 dissolved in it. Gases are less soluble in a liquid of higher temperature because the increased molecular motion in the solution allows gas molecules to escape their loose association with the solvent molecules.

Henry's law
the law that states that at constant temperature, the solubility of a gas in a liquid is directly proportional to the partial pressure of the gas on the surface of the liquid

Topic Link

Refer to the "Gases" chapter for a definition and discussion of partial pressure.

Section Review

UNDERSTANDING KEY IDEAS

1. Why is it possible to overdose on Vitamin A, but not on Vitamin C?

2. Why is ethanol miscible in water?

3. Why do sugar cubes dissolve more slowly than granulated sugar?

4. What factors are involved in determining the solubility of an ionic salt?

5. Would the compound $MgSO_4$ be considered soluble in water?

6. Would the compound PbS be considered soluble in water?

7. You keep adding sugar to a cold cup of coffee and stirring the coffee. Finally, solid sugar remains on the bottom of the cup. Explain why no more sugar dissolves.

8. What is the relation between supersaturation and the hand warmer in **Figure 16?**

CRITICAL THINKING

9. Ethylene glycol is represented by the formula $HOCH_2CH_2OH$. Is it likely to be soluble in water? in paint thinner? Explain your answer.

10. A solution of $BaCl_2$ is added to a solution of $AgNO_3$. Use **Table 2** to decide what reaction happens, and write a balanced equation.

11. Suppose a salt, when dissolving, has a positive ΔH and a negative ΔS. Is the solubility of the salt low or high?

12. A commercial "fizz saver" pumps helium under pressure into a soda bottle to keep gas from escaping. Will this keep CO_2 in the soda bottle? Explain your answer.

Answers to Section Review

1. Excess vitamin A collects in body fat, while excess vitamin C is eliminated in the urine.

2. It is polar, like water. Also, ethanol and water molecules are able to hydrogen bond with each other.

3. Sugar cubes have less surface area available to the solvent to dissolve solute particles.

4. Entropy and enthalpy.

5. Yes.

6. No.

7. The sugar in the coffee has reached its saturation point, and therefore no more sugar will dissolve in it.

8. Supersaturation enables a greater amount of sodium acetate than the solubility level to be dissolved, which then releases heat upon recrystallization.

9. It is likely to be soluble in water because of its polarity, and because of the –OH groups, which will enable ethylene glycol molecules to hydrogen bond with water molecules. It is not likely to be soluble in paint thinner, because paint thinner is nonpolar.

10. $BaCl_2(aq) + 2AgNO_3(aq) \longrightarrow 2AgCl(s) + Ba(NO_3)_2(aq)$; it will form the precipitate AgCl.

Answers continued on p. 493A

Focus

Overview

Before beginning this section, review with your students the Objectives listed in the Student Edition. This section describes some of the physical properties of solutions. Students learn why some solutions conduct electricity, while others do not. They also learn about two colligative properties of solutions—freezing point depression and boiling point elevation. Finally, students learn about how surfactants work and how emulsions are formed.

 Bellringer

Bring a hair dryer with a shock-warning label on its cord to class and allow students to examine and read the label. Ask students to write a brief paragraph on why such a label is necessary.

Motivate

Demonstration

Conductance of Solutions

(Approximate time: 10 minutes)

This demonstration will help clarify the conditions that produce conductance.

1. Write the formulas CH_3COOH, HCl, NaCl, and H_2O on the board. Ask students what kind of bonding exists in each of the compounds, and ask them to predict the conductivity of each of these substances in solution.

2. Half-fill four equal-size beakers with 1 M acetic acid, 1 M hydrochloric acid, 0.5 M sodium chloride solution, and distilled water, and label the beakers.

Continued on next page

Physical Properties of Solutions

KEY TERMS

- **conductivity**
- **electrolyte**
- **nonelectrolyte**
- **hydronium ion**
- **colligative property**
- **surfactant**
- **detergent**
- **soap**
- **emulsion**

conductivity

the ability to conduct an electric current

electrolyte

a substance that dissolves in water to give a solution that conducts an electric current

OBJECTIVES

① **Distinguish** between nonelectrolytes, weak electrolytes, and strong electrolytes.

② **Describe** how a solute affects the freezing point and boiling point of a solution.

③ **Explain** how a surfactant stabilizes oil-in-water emulsions.

Electrical Conductivity in Solutions

Some substances conduct electricity and some cannot. The **conductivity** of a substance depends on whether it contains charged particles, and these particles must be able to move. Electrons move freely within a metal, thus allowing it to conduct electricity. Solid NaCl contains ions, but they cannot move, so solid NaCl is a nonconductor by itself. But an aqueous solution of ionic compounds such as NaCl contains charged ions, which can move about. Solutions of ionic compounds conduct electricity. Pure water does not conduct electricity.

Electrolytes Provide Ions in Solution

An **electrolyte** is a substance that dissolves in a liquid solvent and provides ions that conduct electricity. For example, sports drinks such as the one pictured in **Figure 19** contain electrolytes that your body needs replenished after strenuous physical activity. Electrolytes are considered to belong to one of two classes depending on their tendency to dissociate.

Figure 19
A sports drink not only supplies water but also supplies electrolytes.

MISCONCEPTION ///ALERT\\\

Electrolyte Strength Point out that the strength of a molecular electrolyte, such as HCl and CH_3COOH, depends on the extent to which it dissociates into ions in aqueous solution, not on its concentration. Also note that electrolyte strength is independent of molecular complexity. Compounds with polyatomic ions, such as HNO_3, H_2SO_4, and H_3PO_4, can be either weak or strong electrolytes, depending on the degree to which they dissociate in water.

Chapter Resource File

- **Lesson Plan**

Strong electrolytes completely dissociate into ions and conduct electricity well. *Weak electrolytes* provide few ions in solution. Therefore, even in high concentrations, solutions of weak electrolytes conduct electricity weakly. Ionic compounds are usually strong electrolytes. Covalent compounds may be strong electrolytes, weak electrolytes, or nonconductors.

internet connect

www.scilinks.org
Topic : Electrolytes and Nonelectrolytes
SciLinks code: HW4047

SC*LINKS* Maintained by the National Science Teachers Association

Electrical Conductivities Span a Wide Range

As shown in **Figure 20,** the extent to which electrolytes dissociate into ions is indicated by the conductivity of their solutions. The apparatus shown has a light bulb attached to a battery, and there is a gap in the circuit between two electrodes. The electrodes are dipped in a solution. If the solution conducts electricity, the circuit is completed and the bulb lights. The amount of current that can be carried depends on the concentration of ions in the solution. A solution of a strong electrolyte has a high concentration of ions, so the bulb lights up brightly. A solution of a weak electrolyte has a low concentration of ions, so the bulb lights up dimly.

Virtually all of a strong electrolyte dissociates as it dissolves in a solvent. Sodium chloride, for example, ionizes completely in solution:

$$NaCl(s) \longrightarrow Na^+(aq) + Cl^-(aq)$$

The ions in solution can move about. NaCl is a strong electrolyte, and a solution of NaCl can conduct electricity. The sugar sucrose, on the other hand, does not ionize at all in solution. It is a **nonelectrolyte,** and does not conduct electricity.

nonelectrolyte

a liquid or solid substance that does not allow the flow of an electric current, either in solution or in its pure state, such as water or sucrose

Figure 20
All four solutions have the same concentration. The brightness of the bulb indicates the degree of conduction and the degree of dissociation (ionization).

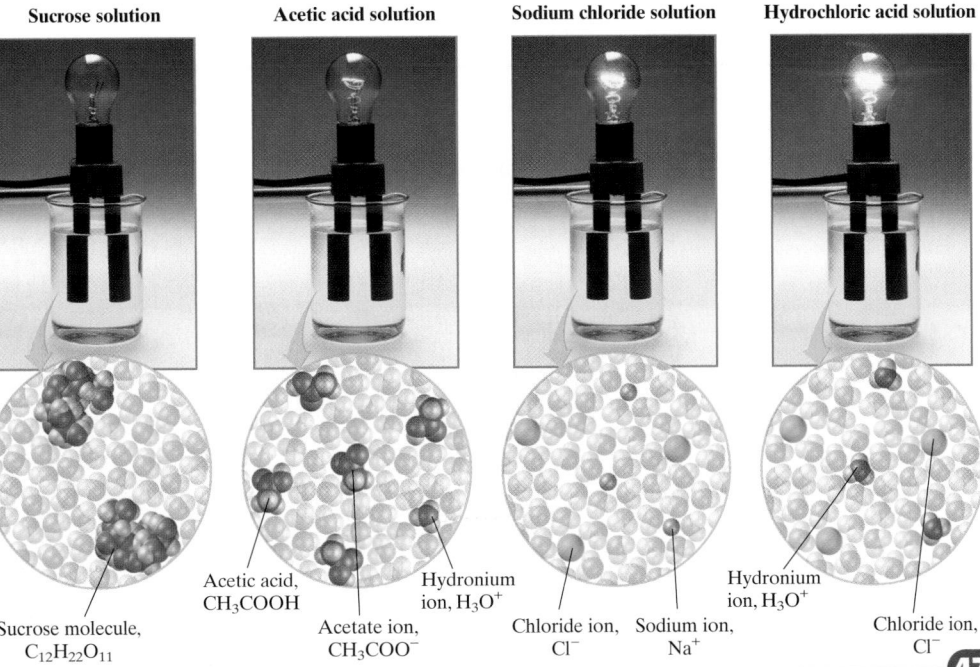

Sucrose solution **Acetic acid solution** **Sodium chloride solution** **Hydrochloric acid solution**

Sucrose molecule, $C_{12}H_{22}O_{11}$

Acetic acid, CH_3COOH
Acetate ion, CH_3COO^-
Hydronium ion, H_3O^+

Chloride ion, Cl^- Sodium ion, Na^+

Hydronium ion, H_3O^+
Chloride ion, Cl^-

479

did you know?

The H^+ ion from the ionization of an acid is hydrated by water molecules. The formula H_3O^+ is used to represent the hydrated state. However, experimental evidence indicates that the proton is more likely to be hydrated by three or four water molecules, giving it a formula of $H_7O_3^+$ or $H_9O_4^+$. In all but advanced applications, the actual formula is unimportant as long as students recognize that H^+ does not exist in an unhydrated form in water. Writing the hydronium ion, H_3O^+, is convenient to remind students how the acidic proton ionizes.

Transparencies

TT Electrical Conductivity of Solutions

Motivate, *continued*

Continued from previous page

3. Use a battery-powered conductivity tester to examine the conductivity of each solution. Compare the results with students' predictions and with **Figure 20** on the facing page.

4. Ask what happened to the covalent compounds to produce a solution that conducts electricity.

Safety: Safety goggles and a lab apron must be worn. Students must be 10 ft from the demonstration.

Disposal: Combine all liquids, neutralize with 1 M NaOH, and pour the mixture down the drain.

Teach

Discussion

Conduct a short review of ideas about ionic compounds. Ask the following questions:

• Describe the composition of an ionic substance in terms of its particles and charge.

• Explain why a solid ionic compound does not conduct electricity, even though it is composed of charged particles.

• Under what conditions will an ionic compound conduct electricity?

Using the Figure

Figure 20 Trace the path of the electric current in the figure so students understand that there is a current source and that the solution must complete the circuit in order to light the bulb. Be sure students do not think that the solution and electrodes generate electricity.

Ask students to interpret the diagrams above each solution, telling what is present in each and why such a mixture conducts (or does not conduct) electricity. Also, make students aware that solution strength (concentration) is independent of whether the solute is a strong or weak electrolyte.

Testing Conductivity It is also possible to use an ohmmeter or multimeter as a conductivity tester. The solutions being tested may corrode the tips of the meter probes. If possible, use leads with alligator clips and attach the clips to 8 cm pieces of bare copper wire that are then used as probes. For convenience, push the wires through a cork and use the cork as a grip.

Paired Reading Pair each student with a partner. Have each student read the section about the electrical conductivity of solutions silently while making a question mark on a sticky note next to the passages they found confusing. After finishing reading, ask one student to summarize and the second to add anything omitted. Both readers should then help each other with any parts that either, or both, did not understand. Have them create a list of questions to ask the class. Students should repeat this process for the section on colligative properties and for the section on surfactants. **LS** Interpersonal

Figure 21
When acetic acid dissolves in water, very little of it is changed into ions.

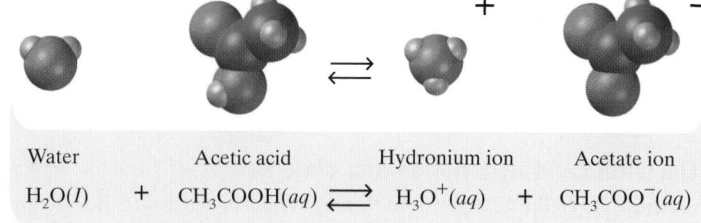

Water		Acetic acid		Hydronium ion		Acetate ion
$H_2O(l)$	$+$	$CH_3COOH(aq)$	$\rightleftharpoons$	$H_3O^+(aq)$	$+$	$CH_3COO^-(aq)$

hydronium ion

an ion consisting of a proton combined with a molecule of water; H_3O^+

Acids react with water to form the **hydronium ion,** H_3O^+. The reaction of acetic acid with water is shown in **Figure 21.** Acetic acid is a weak electrolyte. In water, only about 1% of acetic acid molecules ionize.

Hydrogen chloride dissolves in water to form a strongly conducting solution called *hydrochloric acid.* Hydrogen chloride is a strong electrolyte because it ionizes completely, as shown by the following equation:

$$HCl(g) + H_2O(l) \longrightarrow H_3O^+(aq) + Cl^-(aq)$$

Keep in mind that the use of the terms *strong* and *weak* have nothing to do with concentration. The term *strong* means that the substance provides a high *proportion* of ions in solution. Hydrochloric acid is a strong electrolyte at any concentration.

Tap Water Conducts Electricity

Have you wondered why you are warned not to use electrical appliances when you are near water? Have you wondered why you are also warned not to go swimming when a thunderstorm is near? The reason is because of the electrolytes in the water. Sea water, as shown in **Figure 22,** also conducts electricity.

Figure 22
The electricity from lightning is conducted through groundwater, ponds, or ocean water, which all contain electrolytes.

MISCONCEPTION ALERT

Many manufacturers of bottled water put phrases such as "pure water" or "100% water" on their labels. Students may believe that all bottled water is similar to distilled water and that they do not contain any solutes. However, most bottled water is not distilled water and usually contains some amount of dissolved minerals and salts. You can use a conductivity tester to demonstrate the differences in conductivity between distilled water and bottled water.

If you collect rainwater in a relatively unpolluted area, you will discover that the rainwater is essentially a nonconductor of electricity. A small concentration of carbonic acid from the carbon dioxide in the air added to the rainwater causes the rain water to be a weak conductor. Pure rainwater conducts almost as poorly as distilled water. However, most of the water we use comes from wells, lakes, or rivers. This water has been in contact with soil and rocks, which contain ionic compounds that dissolve in the water. Consequently, tap water conducts electricity. The conduction is not high, but the water can conduct enough current to stop a person's heart. So, for example, a person should not use an electrical appliance when in the bathtub or shower.

You should also not seek shelter under a tree during a thunderstorm. The tree not only sticks up like a lightning rod, but the sap in the tree also contains electrolytes, and conducts the electricity. Lightning finds a path to the ground through the trunk of the tree.

Colligative Properties

A solution made by dissolving a solute in a liquid, such as adding sulfuric acid to water, has particular chemical properties that the solvent alone did not have. The *physical properties* of water, such as how well the water mixes with other compounds, are also changed when substances dissolve in it.

As shown in **Figure 23,** salt can be added to icy sidewalks to melt the ice. The salt actually lowers the freezing point of water. Therefore, ice is able to melt at a lower temperature than it normally would. This change is called *freezing-point depression.* Nonvolatile solutes such as salt also increase the boiling point of a solvent. This change is called *boiling-point elevation.* For example, glycol in a car's radiator increases the boiling point of water in the radiator, which prevents overheating. It also lowers the freezing point, preventing freezing in cold weather.

Figure 23
Dissolved salt lowers the freezing point of water, causing it to melt at a lower temperature than it normally would.

481

Discussion
Ask students why salt is spread on icy roads and used in making homemade ice cream. Most will say that it melts the ice. Respond by asking how it melts the ice. Ask them why calcium chloride is better than sodium chloride for melting ice. Return to these questions after students have studied the next three pages.

Group Activity ─── BASIC
Sports Drinks and Electrolytes
Sports drinks such as Gatorade and Powerade are marketed to athletes as being necessary to replenish electrolytes that were lost by the body during exercise. Have students work in small groups to research and create a poster about sports drinks and electrolytes. They should describe why the human body needs electrolytes, what types of electrolytes are contained in sports drinks, and whether or not a person truly needs to consume sports drinks in order to main the correct electrolyte balance in his or her body. **LS Verbal**

REAL-WORLD
CONNECTION
Ground Fault Interrupters In newer buildings, electrical outlets near water supplies, such as in bathrooms, kitchens, and laundry areas, are wired with ground-fault interrupters. These devices minimize the hazard of electrocution. A ground-fault interrupter constantly checks to see if current is flowing to ground and breaks the circuit within a few thousandths of a second when this occurs—in time to prevent electrocution. Most older buildings are being fitted with these devices whenever the wiring is upgraded.

INCLUSION *Strategies*
- *Learning Disabled*
- *Attention Deficit Disorder*
- *English Language Learners*

Have students draw models of any of the compound formulas listed in the chapter. Students can use colored markers or pencils to draw models of the compounds similar to those depicted in **Figure 20.** The models should be clearly labeled to show understanding of the composition of the formula. Students could also use colored clay to make three-dimensional models of the compound formulas.

Activity ———————— BASIC

Edible Example of Freezing Point Depression Students can experience a fun example of freezing point depression by making their own ice cream. Pair each student with a partner. Each pair of students should have 1 quart-size ziplock bag, 1 gallon-size ziplock bag, 1 cup half & half (not fat free), a teaspoon of vanilla extract, 4 packets of sugar (similar to those found in restaurants), ice, rock salt, and spoons. Combine the half & half, vanilla, and sugar in the quart-size bag. Seal and cover the top of the bag with duct tape. Place the small bag inside the large bag, and pack ice around the small bag. Add 1 cup of rock salt. Knead the two bags together. It takes 15–20 minutes to produce an ice cream of a good consistency. Students must knead the bags continuously to break up ice crystals that may form in the ice cream. When the ice cream is finished, remove the small bag and open it with scissors. Eat it with spoons. During this activity, students directly observe freezing point depression by taking the temperature of the bags at regular intervals. The easiest (least messy) way to take the temperature is to wrap the bags around the thermometer. **LS** **Kinesthetic**

colligative property

a property of a substance or system that is determined by the number of particles present in the system but independent of the properties of the particles themselves.

Figure 24
Colligative properties depend on the concentration of solute particles. Equal amounts in moles of sugar, table salt, and calcium chloride affect the solvent in different degrees because of the different numbers of particles they form when dissolved.

Only the Concentration of Dissolved Particles Is Important

Any physical effect of the solute on the solvent is a **colligative property.** The lowering of the freezing point and the raising of the boiling point are examples of colligative properties. Only nonvolatile solutes have predictable effects on boiling point, but besides that requirement, the identity of the solute is relatively unimportant.

Any solute, whether an electrolyte or a nonelectrolyte, contributes to the colligative properties of the solvent. The degree of the effect depends on the concentration of solute particles (either molecules or ions) in a certain mass of solvent. The greater the particle concentration is, the greater the boiling-point elevation or the freezing-point depression is. For example, based on the number of moles of solute particles, 1 mol of sodium chloride, NaCl, is expected to give twice the amount of change as 1 mol of sucrose, $C_{12}H_{22}O_{11}$. This result occurs because NaCl dissolves to give two moles of particles per mole, and sucrose dissolves to give only one. Likewise, 1 mol of calcium chloride, $CaCl_2$, has about three times the effect as 1 mol of sucrose because $CaCl_2$ dissolves to give three dissolved particles per mole. The following equations illustrate the logic.

$$C_{12}H_{22}O_{11}(s) \longrightarrow C_{12}H_{22}O_{11}(aq) \qquad \text{(1 dissolved particle)}$$

$$NaCl(s) \longrightarrow Na^+(aq) + Cl^-(aq) \qquad \text{(2 dissolved particles)}$$

$$CaCl_2(s) \longrightarrow Ca^{2+}(aq) + 2Cl^-(aq) \qquad \text{(3 dissolved particles)}$$

Figure 24, below, illustrates the differing numbers of solute particles generated by equal concentrations of each compound in solution.

Sucrose molecule, $C_{12}H_{22}O_{11}$

Sodium ion, Na^+ Chloride ion, Cl^-

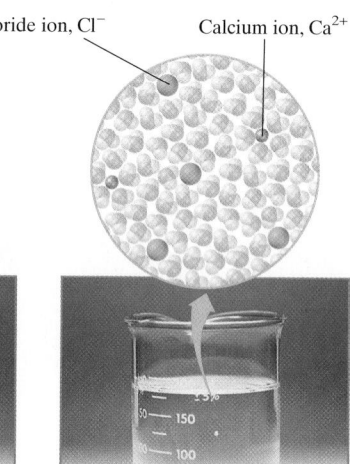
Calcium ion, Ca^{2+}

1.3 M Sucrose solution   **1.3 M Sodium chloride solution**   **1.3 M Calcium chloride solution**

MISCONCEPTION ALERT

Be sure students understand that the amount of salt put in water to cook pasta or vegetables raises the boiling point only very slightly. Raising the boiling point of 2 L of water just 2°C requires about 240 g, or more than 0.5 lb, of salt.

Dissolved Solutes Lower the Vapor Pressure of the Solvent

Colligative properties are all caused by a decrease in the vapor pressure of the solvent. Recall that all gases exert pressure, and vapor pressure is the pressure exerted by the vapor in equilibrium with its liquid state at a given temperature. The effect of a dissolved solute on the vapor pressure of a solvent can be understood when you consider the number of solvent particles in a solution. A solution has fewer solvent particles per volume than the pure solvent has, so fewer solvent particles are available to vaporize. Vapor pressure will therefore be decreased in proportion to the number of solute particles.

Figure 25 illustrates the difference between water's boiling point and freezing point (red lines), and the boiling point and freezing point of an aqueous solution (blue lines). Recall that the boiling point of a liquid is the temperature at which the liquid's vapor pressure is equal to the atmospheric pressure above the liquid. Because the vapor pressure of the water is lowered by the addition of a nonvolatile solute, the solution must be heated to a higher temperature for its vapor pressure to reach atmospheric pressure, at which point the solution boils.

The freezing point is the temperature at which water and ice are in equilibrium. Ice has a vapor pressure that is indicated by the line down to the left in **Figure 25.** The freezing point of water is the temperature at which the vapor pressure of pure water and ice are equal. Because the vapor pressure of the solution is lower, the vapor pressure of the solution intersects the line for the vapor pressure of ice at a lower temperature. Ice and water in the solution are in equilibrium at a lower temperature. The freezing point of the solution is therefore lower than that of pure water.

Topic Link

Refer to the "States of Matter and Intermolecular Forces" chapter for a discussion of vapor pressure.

Solute Effects on the Vapor Pressure of a Pure Solvent

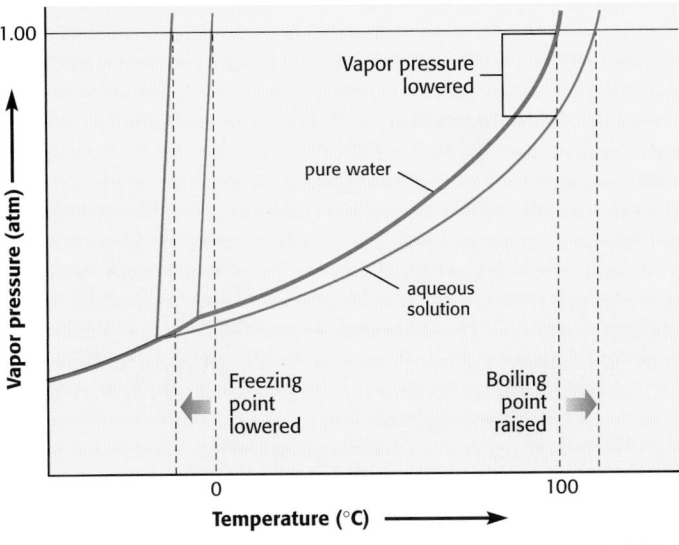

Figure 25
This is a modified phase diagram for pure water (red lines) and for an aqueous solution (blue lines). The addition of a solute has the effect of extending the range of the liquid phase.

Demonstration

Insta-Freeze! This demonstration is a good example to link the ideas of freezing point depression and gas solubility. You will need a small (pint size), unopened bottle of soda water in a glass bottle. Show the students the bottle and remove the label. Pack the bottle in a large beaker of ice and sprinkle rock salt on top of the ice. Approximately ten minutes later, remove the bottle from the ice—the soda water should still be a liquid. With students watching, open the bottle. As the gas effervesces from the water, the water in the bottle will instantly freeze. Ask students the following questions: How did the temperature of the ice bath compare to the normal freezing point of water? **Ans.** The temperature was lower than 0°C because of freezing point depression. Why did the soda water remain a liquid while in the ice bath? **Ans.** The CO_2 gas was dissolved in the water and lowered its freezing point. What happened to the CO_2 when the bottle was opened? Why? **Ans.** The CO_2 came out of solution because the pressure was lowered when the bottle was opened. Why did the water freeze when the bottle was opened? **Ans.** When the CO_2 came out of solution, the freezing point of the water was raised back to 0°C, and the temperature of the water was already below 0°C.

MISCONCEPTION ALERT

Be sure that students do not confuse boiling point elevation with the change in boiling point due to geographical elevation. In general, the boiling point of water (and other liquids) decreases as one travels to higher geographical elevations. This is because the atmospheric pressure is usually lower at high elevations than at sea level and has nothing to do with colligative properties.

Vocabulary Surfactant is short for "surface-active agent." Many household cleaning solutions contain nonsoap surfactants to help them "mobilize" grease, enabling it to be suspended in water.

Using the Figure

Figure 26 Make sure that students understand that soaps and detergents do not exactly dissolve oils; rather, they cause the formation of suspended droplets of oil that can be washed away.

Teaching Tip

Surfactants are usually classified by their ionic charge. Anionic surfactants have a negative charge, are high sudsing, and are used as laundry detergents and personal cleaning products such as shampoo. Cationic surfactants have a positive charge and are often used as fabric softeners or as disinfectants in household cleaning supplies. Nonionic surfactants have no charge, are low sudsing, and are used as laundry or dishwasher detergents. Amphoteric surfactants can become either positively charged or negatively charged, or can remain uncharged when in solution. The charge they take depends on the pH of the solution. Amphoteric surfactants are used in household and personal cleaning products.

surfactant

a compound that concentrates at the boundary surface between two immiscible phases, solid-liquid, liquid-liquid, or liquid-gas

detergent

a water-soluble cleaner that can emulsify dirt and oil

soap

a substance that is used as a cleaner and dissolves in water

emulsion

any mixture of two or more immiscible liquids in which one liquid is dispersed in the other

Figure 26
When you wash with soap, you create an emulsion of oil droplets dispersed in water, and stabilized by the soap.

☑ **internet** connect

www.scilinks.org
Topic: Emulsions
SciLinks code: HW4050

*SCi*LINKS Maintained by the National Science Teachers Association

Surfactants

Have you ever tried to wash your hands when they were very dirty without using soap? You were probably not very successful. Perspiration contains water and oils. The water evaporates, but the oil remains behind and coats the dirt particles. Oil and water do not mix, so washing without soap does not clean very well. However, if you use soap, the cleaning process is much more successful. Why?

The action of scrubbing your skin breaks the oil into tiny droplets. Soap molecules contain long nonpolar hydrocarbon chains, which are soluble in the nonpolar oil. As shown in **Figure 26,** soap molecules also have negatively charged ends, which are soluble in the water just outside the oil droplet. The negatively charged droplets repel each other and are carried away from the skin, along with any dirt that was on your skin.

Soap belongs to a general class of substances called **surfactants.** A surfactant is a substance that concentrates at the interface between two phases, either the solid-liquid, liquid-liquid, or gas-liquid phase. A **detergent** is a surfactant that is used for cleaning purposes. Usually, when we speak of detergents, we are talking about *synthetic detergents,* substances that are not natural products. A **soap** is a particular type of detergent and one that is a natural product. Soaps are sodium or potassium salts of fatty acids with long hydrocarbon chains. The formula for a typical soap, sodium palmitate, is shown below.

$$CH_3CH_2CH_2CH_2CH_2CH_2CH_2CH_2CH_2CH_2CH_2CH_2COO^-Na^+$$

Soap is an emulsifying agent. An **emulsion** is made of colloid-sized droplets suspended in a liquid in which they would ordinarily be insoluble, unless stabilized by an *emulsifying agent,* such as a soap. Without an emulsifying agent, polar and nonpolar molecules remain separate, as pictured in **Figure 27** on the next page.

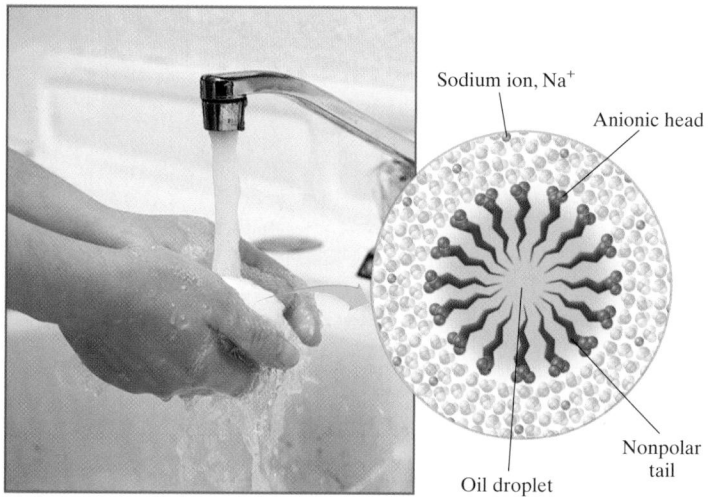

Sodium ion, Na⁺

Anionic head

Nonpolar tail

Oil droplet

484

Soap was used in ancient Sumer (now Iran and Iraq) in 2500 B.C. Later the Greek physician Galen, who lived from about A.D. 130 to about A.D. 200, made soap from the reaction of fat with an alkali.

Cincinnati, Ohio had the nickname "Porkopolis" in the 1830s and 1840s because it had many slaughterhouses for the processing of hogs. By 1850, Cincinnati had become the nation's chief pork-packing center. Because of the abundance of animal fat and the city's location on a major transportation artery, the Ohio River, Cincinnati also became the soap-manufacturing center of the United States.

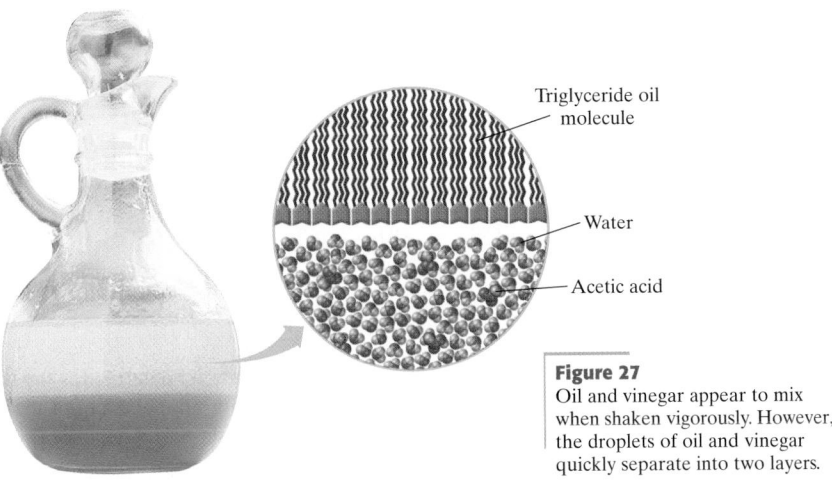

Triglyceride oil molecule

Water

Acetic acid

Figure 27
Oil and vinegar appear to mix when shaken vigorously. However, the droplets of oil and vinegar quickly separate into two layers.

Hard Water Limits Soap's Detergent Ability

Soaps are actually salts, although their physical properties are quite different from the salts you have studied so far. Like other salts, when dissolved, soaps form ions. Unlike other salts, the polyatomic anion of soap contains a long nonpolar part. It is this nonpolar hydrocarbon chain that is soluble with oils and dirt.

Soaps are not ideal cleansing agents because the salts of some of their anions are insoluble in water, especially salts of calcium, magnesium, and iron(II). Hard water has high concentrations of these cations, which react with anions such as the palmitate anion to form insoluble salts, such as the one shown in the following equation.

$$2C_{12}H_{25}COO^-(aq) + Ca^{2+}(aq) \longrightarrow (C_{12}H_{25}COO)_2Ca(s)$$

This is the type of substance responsible for bathtub rings. Before synthetic detergents were introduced for shampoos, some of the scum left over from soap would remain in people's hair after they washed it with soap. People would have to rinse their hair with vinegar to wash out the solid salts left over from the soap.

Synthetic Detergents Outperform Soaps in Hard Water

As noted above, soaps form precipitates when used in hard water. In the 1930s, chemists developed synthetic detergents as a substitute for soap to avoid this problem. Synthetic detergents can be used in hard water without forming precipitates. Today, almost all laundry products and shampoos contain synthetic detergents.

The early synthetic detergents were not biodegradable, gradually collected in the groundwater, and caused a serious pollution problem in some regions. Streams going over waterfalls developed huge mounds of soap suds. Most synthetic detergents are now made to biodegrade when they are disposed of.

internet connect

www.scilinks.org
Topic : Hard Water
SciLinks code: HW4154

SCILINKS. Maintained by the National Science Teachers Association

485

did you know?

Commercial lecithin, an emulsifying agent, is obtained as a byproduct of the production of soybean oil. It is used in margarine, mayonnaise, chocolate candy, lubricant sprays for cooking pans, baked goods, animal feeds, printing inks, paints, soaps, and cosmetics. Its structure is similar to the triglyceride fat shown in **Figure 27.** Instead of being bonded to the third fatty acid, the third carbon is bonded to the group $^-OPO_3$–$CH_2CH_2N^+(CH_3)_3$, which is ionic and soluble in water.

Reteaching   BASIC

Have students work in pairs to review the material in this section. Each pair should then write five quiz questions based on the material in the section. The pairs can swap quizzes with other groups and try to answer each other's questions. **LS** Interpersonal

Quiz   GENERAL

1. How are electrolytes related to the conductivity of a solution?
Ans. Electrolytes form ions in solution and the ions increase the conductivity of the solution.

2. Suppose you add 100 g of salt to a certain amount of water and manage to raise the boiling point by a few degrees. Predict what would happen to the boiling point if you added an additional 100 g of salt to the water.
Ans. The boiling point would be raised to a higher temperature.

3. Why do you need to wash your hands with both soap and water after eating buttered popcorn?
Ans. Butter is a nonpolar fat. Therefore, you need soap to form micelles around the butter that can be washed away with water.

Chapter Resource File

- **Concept Review**
- **Quiz**

Figure 28
The structure of sodium laurate (top), a typical soap, is shown here. Sodium dodecylbenzene sulfonate (bottom), is a synthetic detergent. It does not form an insoluble precipitate in hard water.

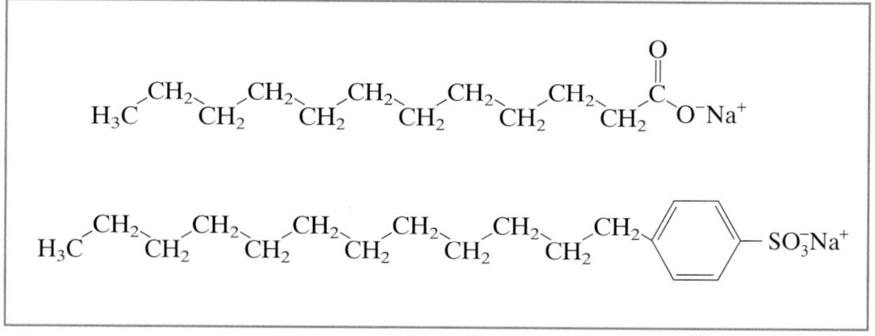

The basic structure of a synthetic detergent is the same as the structure of a soap. However, the long nonpolar tail of the detergent is connected to the salt of a sulfonic acid, –SOOOH, instead of the organic acid, –COOH. The difference between the structures of a synthetic detergent and soap is shown in **Figure 28.** The hexagon represents a six-carbon benzene ring, which is also nonpolar. The sulfonate group is negatively charged, just as the anionic group on a typical soap is. However, the sulfonate group does not form precipitates with magnesium, calcium, and iron(II) ions, which are found in hard water. The carboxylate anion, –COO⁻, reacts with hard water ions to form salts that come out of solution, but the sulfonate anion, –SOOO⁻, does not.

④ Section Review

UNDERSTANDING KEY IDEAS

1. What carries an electric current through a solution?

2. Is sugar an electrolyte? Why or why not?

3. How is a *weak electrolyte* different from a *strong electrolyte*?

4. Why does tap water conduct electricity, whereas distilled water does not?

5. What effect does a solute have on the boiling point of a solvent?

6. Why does spreading salt on an icy sidewalk cause the ice to melt?

7. What is the difference between the meaning of the terms *detergent* and *soap*?

8. What is hard water?

9. What is an emulsion?

10. What is an emulsifying agent?

CRITICAL THINKING

11. Will 1 mol of sugar have the same effect as 1 mol of table salt in lowering the freezing point of water? Explain.

12. Suppose you were taking a bath in distilled water but were using soap. Should you still worry about electric shock?

13. Are soap and synthetic detergents equally good as emulsifiers? Explain your answer.

14. A water softener removes calcium and magnesium ions from water. Why does a softening agent improve the cleansing ability of soap?

15. Why is soap described as a detergent? Why is it described as a surfactant?

Answers to Section Review

1. mobile ions

2. No; it is a covalent compound that does not ionize in solution.

3. Only a small proportion of the molecules of a weak electrolyte ionize in solution, while a strong electrolyte ionizes essentially completely.

4. Tap water contains dissolved ions.

5. A dissolved solute raises the boiling point of a solvent.

6. It lowers the vapor pressure of ice, causing it to melt at a lower temperature.

7. A *detergent* is a surfactant used for cleansing; a *soap* is a natural detergent.

8. "Hard water" contains ions, such as Ca²⁺ and Mg²⁺, which form insoluble compounds with soap.

9. A colloidal suspension of one liquid in another.

10. An emulsifying agent can form an emulsion, such as oil droplets in water.

Answers continued on p. 493A

CHAPTER HIGHLIGHTS 13

KEY TERMS

solution
suspension
solvent
solute
colloid

concentration
molarity

solubility
miscible
immiscible
dissociation
hydration
saturated solution
unsaturated solution
supersaturated solution
solubility equilibrium
Henry's law

conductivity
electrolyte
nonelectrolyte
hydronium ion
colligative property
surfactant
detergent
soap
emulsion

KEY IDEAS

SECTION ONE What Is a Solution?

- A solution is a homogeneous mixture of a solute dissolved in a solvent.
- Several methods can be used to separate the components in a mixture.

SECTION TWO Concentration and Molarity

- Units of concentration express the ratio of solute to solution, or solute to solvent, that is present throughout a solution.
- Molarity is moles of solute per liter of solution.

SECTION THREE Solubility and the Dissolving Process

- Whether substances dissolve in each other depends on their chemical nature, on temperature, and on their ability to form hydrogen bonds.
- In general, polar dissolves in polar, and nonpolar dissolves in nonpolar.
- Ionic solubility can be roughly predicted using a table of ionic solubilities.
- A saturated solution has the solute in equilibrium with excess solute. A supersaturated solution has more dissolved solute than the equilibrium amount.
- Pressure and temperature affect the solubility of gases.

SECTION FOUR Physical Properties of Solutions

- Ions are mobile in solution, so ionic solutions conduct electricity.
- Colligative properties involve the number of solute particles in solution.
- Surfactants make oil and water miscible.

KEY SKILLS

Calculating Parts per Million	Preparing 1.000 L of a 0.5000 M Solution	Calculating Molarity	Solution Stoichiometry
Sample Problem A p. 461	Skills Toolkit 1 p. 463	Skills Toolkit 2 p. 464	Sample Problem C p. 466
		Sample Problem B p. 465	

487

Alternative Assessments

Group Activity
Pages 456, 481

SKILL BUILDER
Pages 465, 469, 474

Reteaching
Page 486

Chapter Review
Items 85–87

Portfolio Assessments

Group Activity
Pages 457, 467, 477

TECHNOLOGY CONNECTION
Page 458

Reteaching
Pages 459, 477

ENVIRONMENTAL SCIENCE CONNECTION
Page 461

Chapter Resource File

- **Datasheets for In-Text Lab** Paper Chromatography
- **Observation Lab** Paper Chromatography
- **Consumer Lab** A Close Look at Soaps and Detergents
- **CBL™ Probeware Lab** Effect of Temperature on Solubility of a Salt

REVIEW ANSWERS

1. Its two components settle out in separate layers.

2. Sugar is the solute, and water is the solvent.

3. Milk contains suspended microscopic globules of casein, which are not interspersed with the solvent at the molecular level.

4. moles of solute per liter of solution

5. water and ethanol

6. The dissolution of NaCl involves dissociation.

7. about 38 g

8. unsaturated

9. When an ionic salt is dissolved in water, its ions dissociate and become interspersed with the water molecules. As more of it than its solubility amount is added to the solution, ions begin to recrystallize into salt particles. As the solution is heated, the crystallized solute dissolves. When the solution cools, the excess solute remains dissolved, and the solution is supersaturated. When more solute is added to the solution, the excess solute recrystallizes.

10. Answers should state that the solubility of a gas in a liquid depends on the partial pressure of that gas above the liquid.

11. It dissociates completely in aqueous solution.

12. yes; The definition of a surfactant is a substance that concentrates at the boundary of two immiscible phases. A detergent does this when it forms a water-oil emulsion.

13. Answers can include any covalent compound that does not dissociate in aqueous solution.

13 CHAPTER REVIEW

USING KEY TERMS

1. What happens to a suspension when it is allowed to stand over a period of time?

2. If sugar is dissolved in water, which component is the solute, and which component is the solvent?

3. Explain why milk is a colloid.

4. What ratio does molarity express?

5. Of the following three substances, which two are miscible with one another: oil, water, and ethanol?

6. One solution is made by dissolving sucrose in water. Another solution is made by dissolving NaCl in water. Which of these dissolving processes involves dissociation?

7. What mass of ammonium chloride can be added to 100 g of water at 20°C before the solution becomes saturated? (See **Figure 12.**)

8. If 20 g KCl is dissolved in 100 g of water at 20°C, is the solution unsaturated, saturated, or supersaturated? (See **Figure 12.**)

9. Write a paragraph explaining what happens to an ionic salt in the following steps: it is dissolved in water, more of it than its solubility amount is added to the solution, the solution is heated, the solution is cooled to room temperature, and the solution is disturbed by adding more solute.

WRITING SKILLS

10. State Henry's law in your own words.

11. Why is NaCl described as a strong electrolyte?

12. Are all detergents surfactants? Explain.

488

13. Give an example of a nonelectrolyte.

14. What kind of mixture is soap able to form in order to make oil and water soluble?

15. Name two colligative properties.

16. What is the formula of a hydronium ion?

UNDERSTANDING KEY IDEAS

Solutions, Suspensions, and Colloids

17. What is a solution? How does it differ from a colloid?

18. What are the two components of a solution, and how do they relate to each other?

19. Explain how distillation can be used to obtain drinking water from sea water.

20. Explain how paper chromatography separates the components in a solution.

21. List these mixtures in order of increasing particle size: muddy water, sugar water, sand in water, and milk.

22. A few drops of milk are added to a glass of water, producing a cloudy mixture. The water is still cloudy after standing in the refrigerator for a week. What is this mixture called?

Concentration and Molarity

23. Name a unit of concentration commonly used to express small concentrations.

24. State the following expression in words: [K₃PO₄]

25. A solution of NaCl is 1 M. Why is the concentration of particles 2 M?

Assignment Guide

Section	Questions
1	1–3, 17–22
2	4–5, 23–26, 42–74, 76, 86
3	6–12, 27–35, 77–78, 80–85, 88–94
4	13–18, 36–41, 75, 79, 87

26. Describe how you would prepare 250.0 mL of a 0.500 M solution of NaCl by using apparatus found in a chemistry lab.

Solubility and the Dissolving Process

27. Explain why vitamin C is soluble in water.

28. Explain why gasoline is insoluble in water.

29. Why do small solid crystals dissolve in liquid more quickly than large crystals?

30. Would ammonium chloride be considered soluble in water?

31. Would the compound $BaSO_4$ be considered soluble in water?

32. Would the compound K_2O be considered soluble in water?

33. If a solution is saturated, how does the rate of dissolution of a solute compare with its rate of recrystallization?

34. Explain the effect of pressure on the solubility of a gas in a liquid.

35. Why does warmer liquid dissolve less gas than colder liquid?

Physical Properties of Solutions

36. A solution of salt in water conducts electricity, but a solution of sugar does not. Explain why.

37. A 1 M solution of NaCl in water has a freezing point that is 3.7°C lower than pure water. Estimate what the freezing point would be for a 1 M solution of $CaCl_2$.

38. Explain why soap is a surfactant, a detergent, and an emulsifying agent.

39. Explain why acetic acid is considered a weak electrolyte and why HCl is considered a strong electrolyte.

40. Draw a diagram of an oil droplet suspended in soapy water.

41. What is a colligative property?

PRACTICE PROBLEMS

Sample Problem A Calculating Parts per Million

42. A saturated solution of $PbCO_3$ contains 0.00011 g $PbCO_3$ in 100 g of water. What is this concentration in parts per million?

43. A 150 g sample of water contains 0.26 mg of cadmium ions. What is this concentration in parts per million?

44. Community water supplies usually contain 1.0 ppm of sodium fluoride. A particular water supply contains 0.0016 g of NaF in 1.60 L of water. Does it have enough NaF?

45. Most community water supplies have 0.5 ppm of chlorine added for purification. What mass of chlorine must be added to 100.0 L of water to achieve this level?

46. A 12.5 kg sample of shark meat contained 22 mg of methyl mercury, CH_3Hg^+. Is this amount within the legal limit of 1.00 ppm of methyl mercury in meat?

Sample Problem B Calculating Molarity

47. If 15.55 g NaOH are dissolved in enough water to make a 500.0 mL solution, what is the molarity of the solution?

48. A solution contains 32.7 g H_3PO_4 in 455 mL of solution. Calculate its molarity.

49. How many moles of $AgNO_3$ are needed to prepare 0.50 L of a 4.0 M solution?

50. What is the molarity of a solution that contains 20.0 g NaOH in 2.00 L of solution?

51. Calculate the molarity of a solution that contains 65.0 g $CuCl_2$ per 300.0 mL.

52. Calculate the molarity of a solution that contains 5.85 g KI in 0.125 L of solution.

53. Calculate the molarity of a H_3PO_4 solution of 6.66 g in 555 mL of solution.

54. What is the molarity of a solution that contains 0.0049 g NaCN in 1.55 L of solution?

489

REVIEW ANSWERS
continued

14. an emulsion

15. Answers can include freezing point depression, boiling point elevation, osmotic pressure, or vapor pressure lowering.

16. H_3O^+

17. A solution is homogeneous on a microscopic scale, while particles of a colloid can be seen in a microscope.

18. A solution consists of a solute and a solvent. The solute is *dissolved in* the solvent (the substance in excess), meaning that the particles of the solute are interspersed homogeneously with the particles of the solvent.

19. Water will boil, leaving behind the solutes. The recondensed distillate is pure water.

20. Components separate by differences in attraction to the chromatography paper.

21. sugar water, milk, muddy water, sand in water

22. a colloid

23. parts per million

24. the concentration of tripotassium phosphate

25. NaCl dissolves to give two particles, Na^+ and Cl^-, per mole of NaCl.

26. 250.0 mL of a 0.500 M solution requires 0.125 mol solute, which equals 7.31 g NaCl. Carefully weigh out this amount in a beaker. Dissolve it in some distilled water, and pour it into a 250 mL volumetric flask. Rinse out the beaker several times into the volumetric flask, until the level of the solution almost reaches the volumetric mark of the flask. Stopper the flask, and swirl the flask until all the crystals are dissolved. Then, top off the flask to the volumetric line with distilled water.

27. Vitamin C is a polar molecule, and its molecules are therefore able to match intermolecular forces with water.

28. Gasoline is nonpolar, and therefore its molecules are not able to come between the strongly attracted polar water molecules.

29. Smaller crystals have more surface area between solvent and solute particles than larger crystals do, and so the dissolving process is able to take place more rapidly.

30. yes

31. no

32. yes

33. They are the same.

34. Increased partial pressure of a gas on the surface of a liquid increases the solubility of that gas in the liquid.

35. Particles of a warmer liquid are in more rapid motion, which makes it easier for dissolved gases to come out of solution.

36. Salt dissociates to form ions, which are able to move in solution and thus conduct electricity. Sugar is a nonelectrolyte.

37. −5.6°C

38. Soap is a surfactant because it concentrates at the boundary between two phases. It is a detergent because it is used for cleaning. It is an emulsifying agent because it allows oil and water to form an emulsion.

39. Acetic acid is a weak electrolyte because only a small proportion of its molecules ionize in aqueous solution. Hydrogen chloride is a strong electrolyte because it dissociates completely in aqueous solution.

40. Answers should resemble the diagram of an oil droplet surrounded by soap molecules in water in Figure 26.

41. A colligative property is a property of a solution which is dependent only on the concentration of dissolved particles, and not on the identities of the particles.

55. Calculate the mass of NaOH in 65.0 mL of 2.25 M solution.

56. What mass of HCl is contained in 645 mL of 0.266 M solution?

57. What is the molarity of a hydrochloric acid solution that contains 18.3 g HCl in 100.0 mL of solution?

58. A saturated solution of NaCl contains 36 g NaCl in 114 mL of solution. What is the molarity of the solution?

59. Calculate the mass of LiF in 100.0 mL of 0.100 M solution.

60. How many grams of glucose, $C_6H_{12}O_6$, are in 255 mL of a 3.55 M solution?

61. Calculate the molarity of a solution of glucose, $C_6H_{12}O_6$, that contains 66 g in 500.0 mL of solution.

Sample Problem C Solution Stoichiometry

62. If 0.125 L of a 0.100 M solution of $Ba(OH)_2$ are mixed with 0.200 L of a 0.0750 M solution of H_3PO_4, the following reaction takes place.

$$3Ba(OH)_2 + 2H_3PO_4 \longrightarrow Ba_3(PO_4)_2 + 6H_2O$$

What mass of barium phosphate is formed?

63. You mix 1.00 L of 2.00 M $BaCl_2$ with 1.00 L of 2.00 M $AgNO_3$. What compounds remain in solution, and what are their concentrations?

$$BaCl_2(aq) + 2AgNO_3(aq) \longrightarrow$$
$$2AgCl(s) + Ba(NO_3)_2(aq)$$

64. How many milliliters of 18.0 M H_2SO_4 are required to react with 250 mL of 2.50 M $Al(OH)_3$ if the products are aluminum sulfate and water?

65. If 75.0 mL of an $AgNO_3$ solution reacts with enough Cu to produce 0.250 g Ag by single displacement, what is the molarity of the initial $AgNO_3$ solution if $Cu(NO_3)_2$ is the other product?

MIXED REVIEW

66. A 250 g sample of river water contained 0.0013 mg of selenium. What is this concentration in parts per million?

67. Commercial concentrated sulfuric acid solution is 18 M. How many grams of sulfuric acid are in 250 mL of the solution?

68. How many milliliters of 1.0 M $AgNO_3$ are needed to provide 168.88 g of pure $AgNO_3$?

69. What is the mass of potassium chromate, K_2CrO_4, in 20.0 mL of 6.0 M solution?

70. Sodium ions in blood serum normally are 0.145 M. How many grams of sodium ions are in 10.0 mL of serum?

71. A package of compounds used to achieve rehydration in sick patients contains 20.0 g of glucose, $C_6H_{12}O_6$. When this material is diluted to 1.00 L, what is the molarity of glucose?

72. The concentration of the solution in item 70 is also 0.020 M in KCl. How many grams of KCl are needed to prepare the 1.00 L of 0.020 M KCl?

73. A saturated solution contains 0.844 g of lead(II) bromide in 100.0 mL. What is the molarity of this solution?

74. Reagent nitric acid in the laboratory is usually 3.0 M. What mass of nitric acid is contained in 50.0 mL of this solution?

75. You are determining the concentration of a salt water solution by evaporating different samples and measuring the mass of salt that remains. Your data are shown below. Calculate the molar concentration of the original solution.

Sample volume (mL)	Mass of NaCl (g)
25.0	2.9
50.0	5.6
75.0	8.5
100.0	11.3

CRITICAL THINKING

76. Which will have the lower freezing point, 2.0 M NaCl or 1.0 M AlCl$_3$?

77. You evaporate 100.0 mL NaCl solution and obtain 11.3 g NaCl. What was the molarity of the original solution?

78. Calcium phosphate, Ca$_3$(PO$_4$)$_2$, is quite cheap and causes few pollution problems. Why is it not used to de-ice sidewalks? (Hint: See **Table 2** in Section 3.)

79. A calculation shows that a salt will have a negative ΔH and a positive ΔS when it dissolves. Is it actually soluble?

80. Imagine you are a sailor who must wash in sea water. Which is better to use, soap or synthetic detergent? Why?

81. You get a small amount of lubricating oil on your clothing. Which would work better to remove the oil, water or paint thinner? Explain your answer.

82. Air pressure in an airplane cabin while in flight is significantly lower than at sea level. Explain in terms of Henry's law how this affects the speed at which a carbonated beverage, after opening, loses its fizz.

83. Explain why oil and water do not mix.

84. Why would a substance that contains only ionic bonds not work as an emulsifying agent?

ALTERNATIVE ASSESSMENT

85. Design a solubility experiment that would identify an unknown substance that is either CsCl, RbCl, LiCl, NH$_4$Cl, KCl, or NaCl. (Hint: You will need a solubility versus temperature graph for each of the salts.) If your instructor approves your design, get a sample from the instructor, and perform your experiment.

86. Emergency-response teams working with chemical spills use solubility principles to control and clean up spills. Research the techniques they use, and explain why they work. Present your findings to the class.

87. Many reagent chemicals used in the lab are sold in the form of concentrated aqueous solutions, as shown in the table below. Different volumes are diluted to 1.00 L to make less-concentrated solutions. Create a computer spreadsheet that will calculate the volume of concentrated reagent needed to make 1.00 L solutions of any molar concentration that you enter.

Reagent	Concentration (M)
H$_2$SO$_4$	18.0
HCl	12.1
HNO$_3$	16.0
H$_3$PO$_4$	14.8
CH$_3$COOH	17.4
NH$_3$	15.0

CONCEPT MAPPING

88. Use the following terms to complete the concept map below: *concentration, dissociates, electrical conductivity, solute,* and *solvent.*

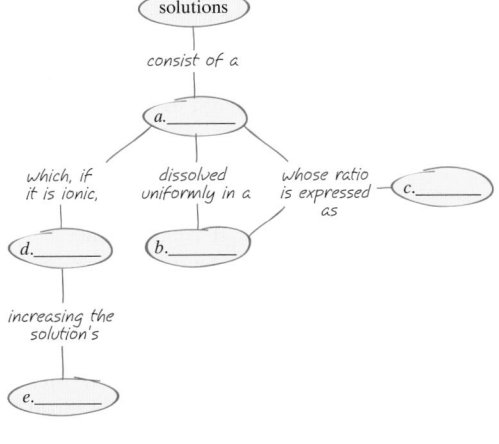

42. 1.1 ppm

43. 1.7 ppm

44. yes; It has a concentration of 1.0 ppm NaF.

45. 5×10^{-2} g Cl$_2$

46. This is a concentration of 1.76 ppm, which is above the legal limit.

47. 0.7776 M NaOH

48. 0.734 M H$_3$PO$_4$

49. 2.0 mol AgNO$_3$

50. 0.250 M NaOH

51. 1.61 M CuCl$_2$

52. 0.282 M KI

53. 0.123 M H$_3$PO$_4$

54. 6.5×10^{-5} M NaCN

55. 5.85 g NaOH

56. 6.27 g HCl

57. 5.02 M HCl

58. 5.4 M NaCl

59. 0.259 g LiF

60. 163 g C$_6$H$_{12}$O$_6$

61. 0.74 M C$_6$H$_{12}$O$_6$

62. 2.51 g Ba$_3$(PO$_4$)$_2$

63. 0.500 M Ba(NO$_3$)$_2$ and 0.500 M BaCl$_2$

64. 52.1 mL

65. 0.0309 M AgNO$_3$

66. 0.0052 ppm Se

67. 440 g H$_2$SO$_4$

68. 994 mL

69. 23 g K$_2$CrO$_4$

70. 0.033 g Na$^+$

71. 0.111 M C$_6$H$_{12}$O$_6$

72. 1.5 g KCl

73. 0.0230 M PbBr$_2$

74. 9.5 HNO$_3$

75. 1.9 M NaCl

76. The two solutions should have about the same freezing point.

77. 1.93 M NaCl

78. Calcium phosphate is insoluble. It must be dissolved to lower the freezing point.

79. yes; The two terms give a negative ΔG.

80. Detergent will work better, because soap will form insoluble compounds with the salts in sea water.

81. Paint thinner; because both oil and paint thinner are nonpolar, they will dissolve in one another.

82. Because the air pressure is relatively low, the partial pressure of carbon dioxide in the air will be accordingly lower. Therefore, carbon dioxide dissolved in the beverage will come out of solution more quickly than at sea level.

83. Water is polar, and oil is nonpolar. When the two are combined, each type of molecule preferentially attracts other molecules that have the same type of intermolecular forces. Therefore, they remain separate.

84. Such a compound will not be soluble in fats and oils.

85. Experimental designs may vary but should include a method for determining the mass of solute needed to form a saturated solution at a certain temperature.

86. Research should include both biological and mechanical means of cleaning up spills as well as chemical methods.

87. If all solutions are to be 1.00 L in volume, the program should calculate the volume of stock solution to use in liters as $V_{stock} = M_{solution}/M_{stock}$. To calculate the volume in milliliters, the expression is $V_{stock} = 1000(M_{solution}/M_{stock})$. The volume to use cannot have more significant figures than the given molarity of the concentration reagents. Students cannot expect to know what volume to use to make, for example, a 0.8150 M solution. Solutions of such accurate concentrations would have to be made roughly and then titrated against a standard.

Answers continued on p. 493B

FOCUS ON GRAPHING

*Study the graph below, and answer the questions that follow.
For help in interpreting graphs, see Appendix B, "Study Skills for Chemistry."*

89. What do the numbers on the *y*-axis represent?

90. What does each curve on the graph represent?

91. Are most of the substances represented on the graph more or less soluble at higher temperatures?

92. Which salt is most soluble at 10°C? at 60°C? at 80°C?

93. If you heat water to 80°C, what amount of NaCl could you dissolve in it as compared to water that is at 20°C?

94. Which salt's solubility is most strongly affected by changes in temperature?

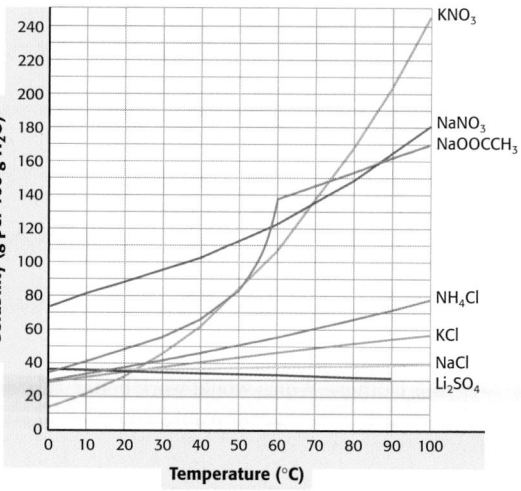

Solubility Vs. Temperature for Some Solid Solutes

TECHNOLOGY AND LEARNING

95. Graphing Calculator

Predicting Solubility from Tabular Data

The graphing calculator can run a program that graphs solubility data. Given solubility measurements for KCl, you will use the data to predict its solubility at various temperatures.

Go to Appendix C. If you are using a TI-83 Plus, you can download the program and data sets and run the application as directed. Press the **APPS** key on your calculator, then choose the application **CHEMAPPS**. Press **3**, then highlight **ALL** on the screen, press **1**, then highlight **LOAD** and press **2** to load the data into your calculator. Press the keys **2nd** and then **QUIT**, and then run the program **SOLUBIL**. For L₁, press **2nd** and **LIST**, and choose **TMP21**. For L₂, press **2nd** and **LIST** and choose **SOL21**.

If you are using another calculator, your teacher will provide you with keystrokes and data sets to use.

a. At what temperature would you expect the solubility to be 48.9 g per 100 g H₂O?

b. At what temperature would you expect the solubility to be 35 g per 100 g H₂O?

c. What would you expect the solubility to be at a temperature of 100°C?

492

Chapter Resource File

• Chapter Test

Answers
1. b
2. b
3. c
4. c
5. d
6. d
7. b
8. d
9. d
10. d
11. b
12. a

1. Two immiscible liquids can form a stable colloidal suspension by the addition of a(n)
   a. weak electrolyte.
   b. emulsifying agent.
   c. colloid.
   d. hydronium ion.

2. What mass of NaOH is contained in 2.5 L of a 0.010 M solution?
   a. 0.010 g
   b. 1.0 g
   c. 2.5 g
   d. 0.40 g

3. Water is an excellent solvent because
   a. it is a covalent compound.
   b. it is a nonconductor of electricity.
   c. of its high polarity.
   d. of its colligative properties.

4. Two liquids are likely to be immiscible if
   a. both have polar molecules.
   b. both have nonpolar molecules.
   c. one is polar and the other is nonpolar.
   d. one is water and the other is methyl alcohol, CH_3OH.

5. An _____ would increase the solubility of a gas in a liquid.
   a. addition of an electrolyte
   b. addition of an emulsifier
   c. agitation of the solution
   d. increase in its partial pressure

6. Acetic acid is considered a weak electrolyte because it
   a. is miscible with water.
   b. forms hydronium and hydroxide ions in aqueous solution.
   c. lowers the freezing point of water.
   d. ionizes only slightly in aqueous solution.

7. Because vitamin A is fat-soluble, it
   a. does not mix well with nonpolar substances.
   b. is stored in body fat.
   c. passes out of the body in the urine.
   d. cannot accumulate in the body.

8. Which of the following types of compounds is most likely to be a strong electrolyte?
   a. a polar compound
   b. a nonpolar compound
   c. a covalent compound
   d. an ionic compound

9. Which of the following will have the lowest freezing point?
   a. 1 M sugar solution
   b. 1 M NaCl solution
   c. 1 M $MgSO_4$ solution
   d. 1 M $CaCl_2$ solution

10. Soap, such as $CH_3(CH_2)_{10}COO^-Na^+$, is considered a detergent because it
    a. is an electrolyte.
    b. is nonpolar.
    c. is soluble in water.
    d. makes oil and water miscible.

11. A saturated solution can become supersaturated under which of the following conditions?
    a. it contains electrolytes
    b. the solution is heated and then allowed to cool
    c. more solvent is added
    d. more solute is added

12. Molarity is expressed in units of
    a. moles of solute per liter of solution.
    b. liters of solution per mole of solute.
    c. moles of solute per liter of solvent.
    d. liters of solvent per mole of solute.

493

Continuation of Answers

Answers to Section 1 Review, *continued*

8. The tap water can be boiled, and the resulting water vapor, when recondensed, will be pure water. Almost all solutes will be left behind.

9. The components of brass do not settle out over time, so it cannot be a suspension. You can look at the surface of brass under a microscope and see that it is homogeneous, so it is not a colloid.

10. Fog consists of water droplets suspended in the air. If you looked at it through a microscope, you would see the individual water droplets with air between them.

11. The example pertains to relative adhesion of a substance to a surface versus solubility in a solvent.

12. Water is evaporating, leaving behind the salt solutes.

Answers to Section 2 Review, *continued*

14. The NaCl solution has the higher molarity: it has the lower molecular mass, so there are more moles of NaCl in 55 g.

Answers to Section 3 Review, *continued*

11. Its solubility would be low: a positive ΔH and a negative ΔS, when substituted into the Gibbs free energy equation $\Delta G = \Delta H - T\Delta S$ would give a positive number, signifying nonspontaneity.

12. It will not keep CO_2 from escaping from solution: Henry's law states that it is the partial pressure of the same gas that is in solution which keeps the gas in solution. Increased pressure of only helium will have no effect on CO_2's solubility.

Answers to Section 4 Review, *continued*

11. 1 mole of table salt, NaCl, will have approximately twice the effect in lowering the freezing point of water as 1 mole sugar, because 1 mole of NaCl produces 2 moles of solute particles in solution, and 1 mole of sugar produces only 1 mole of solute particles.

12. Yes; soap is a salt and therefore an electrolyte.

13. Yes, but not in hard water, where soap will form soap scum with metal ions, whereas a synthetic detergent will not.

14. The ions the water softener removes would form insoluble compounds with soap.

15. Soap is a surfactant because it concentrates at the interface between two phases (water and oil); it is a detergent because it is a cleansing agent.

Answers to Chapter Review, *continued*

88. **a.** solute

b. solvent

c. concentration

d. dissociates

e. electrical conductivity

89. the mass of solute that can dissolve in 100 g of water at a given temperature

90. a certain solute

91. more soluble

92. $NaNO_3$ is most soluble at 10°C. $NaOOCCH_3$ is most soluble at 60°C. KNO_3 is most soluble at 80°C.

93. Only slightly more.

94. KNO_3, since its curve shows the greatest change between the temperature range shown in the graph.

95. **a.** about 70°C

b. about 20°C

c. about 55 g/100 g H_2O

Note: additional questions can be given if "TMP22" and "SOL22" are loaded and chosen as datasets.

CHAPTER 14

Chemical Equilibrium
Chapter Planning Guide

PACING	CLASSROOM RESOURCES	LABS, ACTIVITIES, AND DEMONSTRATIONS
BLOCK 1 · 45 min pp. 494–495 **Chapter Opener**		**SE** **Start-Up Activity** Finding Equilibrium, p. 495 ◆
BLOCK 2 · 90 min pp. 496–501 **Section 1** Reversible Reactions and Equilibrium	**TT** **Rate Comparison for** $H_2(g) + I_2(g) \rightleftharpoons 2HI(g)$ *	**TE** **Demonstration**, p. 497 ◆ **ADVANCED** **TE** **Group Activity**, p. 499 **GENERAL** **TE** **Demonstration** Complex Ion Equilibria, p. 500 ◆ **ADVANCED** **TE** **Demonstration** Weather Indicators, p. 501 ◆ **CRF** **Microscale Lab** Equilibrium ◆ **ADVANCED**
BLOCKS 3 & 4 · 90 min pp. 502–511 **Section 2** Systems at Equilibrium	**OSP** **Supplemental Reading** Guided Reading Worksheet: The Same and Not the Same "Fritz Haber" **ADVANCED**	**TE** **Demonstration** Kinetics Simulation, p. 505 ◆ **TE** **Group Activity**, p. 511 **BASIC** **CRF** **Observation Lab** Solubility Product Constant ◆ **GENERAL** **CRF** **Observation Lab** Solubility Product Constant— Algae Blooms ◆ **GENERAL**
BLOCK 5 · 45 min pp. 512–519 **Section 3** Equilibrium Systems and Stress	**TT** **NO_2-N_2O_4 Equilibrium System** * **SE** **Element Spotlight** Chlorine Gives Us Clean Drinking Water	**TE** **Demonstration**, p. 512 **TE** **Demonstration** Temperature Effects, p. 515 ◆ **ADVANCED** **TE** **Group Activity**, p. 516 **GENERAL** **TE** **Demonstration** Common Ion Effect, p. 517 ◆

BLOCKS 6 & 7 · 90 min **Chapter Review and Assessment Resources**

SE Chapter Review, pp. 521–526
SE Standardized Test Prep, p. 527
CRF Chapter Test *
OSP Test Generator
CRF Test Item Listing *
OSP Scoring Rubrics and Classroom Management Checklists

Holt Chemistry: Online Resources

Visit **go.hrw.com** for a variety of free resources related to this textbook. Enter the keyword **HW4 HOME**.

Holt Online Learning

Students can access interactive problem solving help and active visual concept development with the *Holt Chemistry* Online Edition available at **www.hrw.com**

student CNN News

cnnstudentnews.com

Find the latest chemistry news, lesson plans, and activities related to important scientific events.

KEY

TE	Teacher Edition	**CRF**	Chapter Resource File	*****	Also on One-Stop Planner
SE	Student Edition	**TT**	Teaching Transparencies	**◆**	Requires Advance Prep
OSP	One-Stop Planner				

Compression guide:
To shorten your instruction because of time limitations, omit blocks 1, 3, 4, and 6.

PROBLEM SOLVING AND PRACTICE	SECTION REVIEW AND ASSESSMENT	STANDARDS CORRELATION
		National Science Education Standards
	TE **Homework,** p. 499 `ADVANCED` TE **Quiz,** p. 501 `GENERAL` CRF **Quiz** * TE **Reteaching,** p. 501 `BASIC` SE **Section Review,** p. 501 CRF **Concept Review** *	PS 3a
SE **Skills Toolkit 1** Determining K_{eq} for Reactions at Chemical Equilibrium, p. 503 SE **Sample Problem A** Calculating K_{eq} from Concentrations of Reactants and Products, p. 504 　SE **Practice,** p. 504 `GENERAL`　　　TE **Homework,** p. 504 `GENERAL` SE **Sample Problem B** Calculating Concentrations of Products from Keq and Concentrations of Reactants, p. 506 　SE **Practice,** p. 506 `GENERAL`　　　TE **Homework,** p. 506 `GENERAL` SE **Skills Toolkit 2** Determining K_{sp} for Reactions at Chemical Equilibrium, p. 508 SE **Sample Problem C** Calculating K_{sp} from Solubility, p. 509 　SE **Practice,** p. 509 `GENERAL`　　　TE **Homework,** p. 509 `GENERAL` SE **Sample Problem D** Calculating Ionic Concentrations Using K_{eq}, p. 510 　SE **Practice,** p. 510 `GENERAL`　　　TE **Homework,** p. 510 `GENERAL` CRF **Problem Solving** * `ADVANCED` SE **Problem Bank,** p. 858 `GENERAL`	TE **Quiz,** p. 511 `GENERAL` CRF **Quiz** * TE **Reteaching,** p. 511 `BASIC` SE **Section Review,** p. 511 CRF **Concept Review** *	PS 3a
	TE **Homework,** p. 516 `GENERAL` TE **Quiz,** p. 518 `GENERAL` CRF **Quiz** * TE **Reteaching,** p. 518 `BASIC` SE **Section Review,** p. 518 CRF **Concept Review** *	PS 3a

www.scilinks.org

Topic: Chemical Reactions
SciLinks code: HW4029

Topic: Factors Affecting Equilibrium
SciLinks code: HW4057

Topic: Common Ion Effect
SciLinks code: HW4034

Topic: Haber Process
SciLinks code: HW4153

Topic: Chlorine
SciLinks code: HW4031

Technology Resources

ChemFile Interactive Tutor CD-Rom

Module 7: Equilibrium
Topic: Equilibrium

Module 7: Equilibrium
Topic: Shifting Equilibrium

Overview

This chapter introduces students to chemical equilibrium. Students will learn that chemical equilibrium is dynamic and involves processes that proceed in opposite directions at the same rate. They will also learn how stress on a system in equilibrium causes the system to adjust to establish a new equilibrium that relieves the stress.

Assessing Prior Knowledge

Check for Content Knowledge

Students should be familiar with the following topics:

• ions

• ionic compounds

• the mole

• chemical equations

• endothermic and exothermic reactions

• the effects of temperature, pressure, volume, and amount on gases

• solutions

Using the Figure

The shell on the beach provides an example of a system that is not in equilibrium. The ocean (as well as the wind and sand) is slowly "weathering" the shell by slowly dissolving it. Because the marine animal that inhabited the shell no longer is depositing new calcium carbonate on the shell, the weathering has no reverse process. In addition, the dissolving and solidifying of insoluble salts such as calcium carbonate can be discussed as an introduction to the solubility product constant.

ChemFile | **CHEMISTRY** INTERACTIVE TUTOR

• Module 7: Equilibrium
  Topic: Equilibrium

CHAPTER

14

CHEMICAL EQUILIBRIUM

494

Standards Correlations

National Science Education Standards

PS 3a: Chemical reactions occur all around us, for example in health care, cooking, cosmetics, and automobiles. Complex chemical reactions involving carbon-based molecules takes place constantly in every cell in our bodies. (Sections 1, 2, 3)

Some marine animals use a salt called *calcium carbonate* to make their shells, such as the shell in the photo. You may think that these animals must constantly build up their shells so that their shells will not dissolve. Calcium carbonate, however, is not very soluble in water. So, even a shell that does not have an animal living in it will take a long time to dissolve. In this chapter, you will learn how to calculate the concentration of slightly soluble salts in water.

START-UP ACTIVITY

Finding Equilibrium

SAFETY PRECAUTIONS

PROCEDURE

1. Label one **rectangular plastic bucket** as "Reactants," and fill the bucket with **water.**

2. Label the other **rectangular plastic bucket** as "Products." Do not put any water in this bucket.

3. One student should be in charge of the "Reactants," and another student should be in charge of the "Products." Each student should use a **250 mL plastic beaker** to dip into the other student's bucket and get as much water as possible in the beaker. Then the students should pour the water from their beakers into their own bucket.

4. The students should dip and pour at equal rates.

ANALYSIS

1. What happens to the amount of "Reactants"? What happens to the amount of "Products"?

2. Does the system reach a state in which the system does not change?

3. How would the outcome of the experiment be different if the beakers had been of different sizes?

Pre-Reading Questions

(1) How do you know that a reaction has taken place?

(2) How do reactants differ from products?

(3) What is a solution?

internet connect

www.scilinks.org
Topic: Chemical Reactions
SciLinks code: HW4029

SCiLINKS Maintained by the National Science Teachers Association

495

Answers to Pre-Reading Questions

1. Observations that suggest a reaction has taken place are the evolution of energy as heat, light, or sound; the production of a gas; the formation of a precipitate; or a change in color.

2. The reactants form products during a reaction.

3. A solution is a mixture whose components form a single homogenous phase and does not separate.

START-UP ACTIVITY

Skills Acquired:
- Experimenting
- Collecting data
- Interpreting

Materials:
For each group of 2–3 students:
- plastic beaker, 250 mL (2)
- rectangular plastic bucket (2)
- water

Teachers Notes: You may want to tint the water with food coloring to make the water levels more visible. Use clear plastic buckets so that students may better see the water levels. For a more accurate simulation of a reaction coming to equilibrium, have the "Products" student start by dipping the beaker once for every fourth dip of the "Reactants" student. After some time, increase the frequency to one for every three, then one for every two, and finally one for one, which represents equal rates. The time intervals between increases will determine whether the equilibrium favors products or reactants.

Safety Caution: Students should wear lab aprons and safety goggles during this activity.

Answers

1. The amount of "Reactants" decreases initially and then does not change. The amount of "Products" increases initially and then does not change.

2. Yes, the amount of "Products" and the amount of "Reactants" eventually do not change.

3. The student that has the larger beaker would eventually have all of the water in her or his bucket and this system would not reach equilibrium.

Focus

Overview

Before beginning this section, review with your students the Objectives listed in the Student Edition. This section explains chemical reactions that go to completion and chemical reactions that are reversible. Students learn to define and describe chemical equilibrium in terms of reversible chemical reactions. Students also learn that the formations of complex ions are examples of chemical equilibria.

 Bellringer

Have students describe what *reversible* means. Ask them if they can find a synonym for reversible.

Motivate

Identifying Preconceptions

Discuss students' answers to the Bellringer activity. When students discuss the term reversible, they are likely to discuss reversibility in terms of physical processes. Tell them that nearly all of the chemical reactions of life processes are reversible. Make sure that they discuss reactions that go to completion.

Chapter Resource File
• Lesson Plan

Reversible Reactions and Equilibrium

KEY TERMS
• reversible reaction
• chemical equilibrium

OBJECTIVES

(1) **Contrast** reactions that go to completion with reversible ones.

(2) **Describe** chemical equilibrium.

(3) **Give** examples of chemical equilibria that involve complex ions.

Completion Reactions and Reversible Reactions

Do all reactants change into products during a reaction? Sometimes only a trace of reactants remains after the reaction is over. **Figure 1** shows an example of such a reaction. Oxygen gas reacts with sulfur to form sulfur dioxide, as shown in the following chemical equation:

$$S_8(s) + 8O_2(g) \longrightarrow 8SO_2(g)$$

If there is enough oxygen, almost all of the sulfur will react with it. In addition, the sulfur dioxide does not significantly break down into sulfur and oxygen. Reactions such as this one are called *completion reactions*. But in other reactions, the products can re-form reactants. Such reactions are called reversible reactions.

> **Topic Link**
>
> *Refer to the "Chemical Equations and Reactions" chapter for a discussion of chemical reactions.*

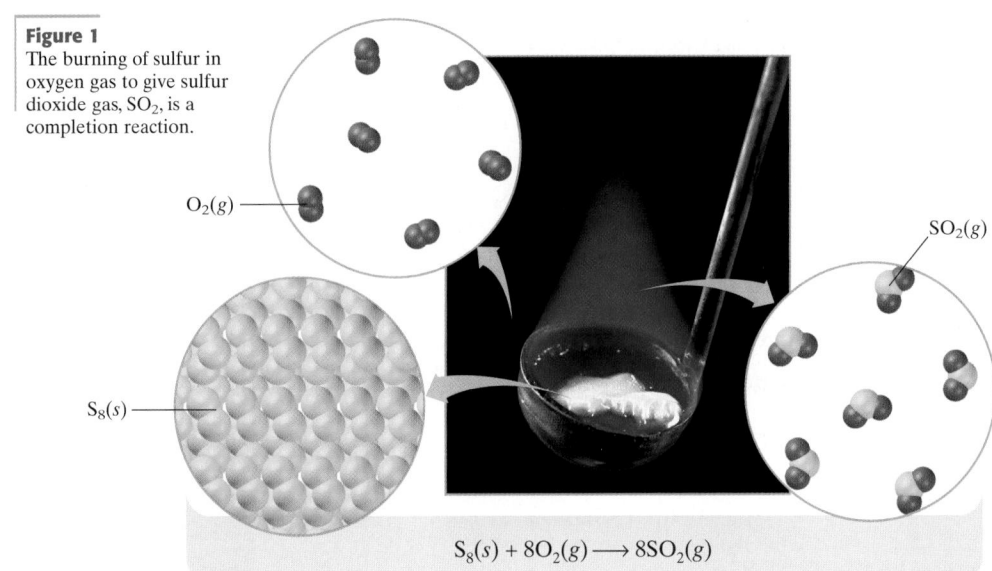

Figure 1
The burning of sulfur in oxygen gas to give sulfur dioxide gas, SO_2, is a completion reaction.

$O_2(g)$

$S_8(s)$

$SO_2(g)$

$$S_8(s) + 8O_2(g) \longrightarrow 8SO_2(g)$$

496

REAL-WORLD
CONNECTION

Sunglasses One common example of a reversible reaction is observed when eyeglasses darken when exposed to the sun. When light-sensitive silver chloride in the glasses is exposed to sunlight, the silver chloride dissociates and elemental silver forms, which darkens the glasses. This reaction is endothermic. When the source of light is no longer present, the light-colored silver chloride re-forms. This reverse reaction is an exothermic reaction.

Figure 2
The precipitation of calcium sulfate, $CaSO_4$, is an example of equilibrium.

$Ca^{2+}(aq)$

$Cl^-(aq)$

$Na^+(aq)$

$SO_4^{2-}(aq)$

$CaSO_4(s)$

$$CaCl_2(aq) + Na_2SO_4(aq) \longrightarrow CaSO_4(s) + 2NaCl(aq)$$

Reversible Reactions Reach Equilibrium

An example of a **reversible reaction** is the formation of solid calcium sulfate by mixing solutions of calcium chloride and sodium sulfate, as shown in **Figure 2** and written below.

$$CaCl_2(aq) + Na_2SO_4(aq) \longrightarrow CaSO_4(s) + 2NaCl(aq)$$

Because the chloride and sodium ions are spectator ions, the net ionic equation better describes what is happening.

$$Ca^{2+}(aq) + SO_4^{2-}(aq) \longrightarrow CaSO_4(s)$$

Although solid calcium sulfate is the product, it can break down to make calcium ions and sulfate ions, as described below:

$$CaSO_4(s) \longrightarrow Ca^{2+}(aq) + SO_4^{2-}(aq)$$

This chemical equation is the reverse of the previous one. Another way to think about reversible reactions is that they form both products and reactants. Use arrows that point in opposite directions when writing a chemical equation for a reversible reaction.

$$Ca^{2+}(aq) + SO_4^{2-}(aq) \rightleftharpoons CaSO_4(s)$$

The forward and reverse reactions take place at the same time. Some Ca^{2+} and SO_4^{2-} ions form $CaSO_4$, while some $CaSO_4$ is dissolving in water to form Ca^{2+} and SO_4^{2-} ions. The reactions occur at the same rate after the initial mixing of $CaCl_2$ and Na_2SO_4. As one unit of $CaSO_4$ forms, another unit of $CaSO_4$ dissolves. Therefore, the amounts of the products and reactants do not change. Reactions in which the forward and reverse reaction rates are equal are at **chemical equilibrium.**

reversible reaction

a chemical reaction in which the products re-form the original reactants

chemical equilibrium

a state of balance in which the rate of a forward reaction equals the rate of the reverse reaction and the concentrations of products and reactants remain unchanged

497

MATH

CONNECTION — **GENERAL**

Be sure students understand what *rate* means. Explain that rate is the speed at which a process occurs and is measured as a quantity per unit of time. Have them list different rates with which they are familiar. Examples might include speed limits in km/h or mi/h, the number of words/min typed, or the number of parts/min on an assembly line. For a chemical reaction, the rate is the amount of reactant lost or product formed per unit of time. **LS** Intrapersonal

---

Teach

READING SKILL BUILDER — BASIC

Assimilating Knowledge Before students read this chapter, have them write a short list of all the things they know about the word *equilibrium* and about chemical reactions that do not go to completion. Then, ask students to compile their lists into a group list on the chalkboard. Explain any misconceptions students may have. Then have students list the things they might want to know about chemical equilibrium. Offer some suggestions if students have trouble creating questions. **LS** Verbal

Demonstration

This demonstration shows students an example of a completion reaction. Place 15 mL of 0.1 M $AgNO_3$ solution in a test tube. Add 15 mL of 0.2 M NaCl solution. A white AgCl precipitate forms. Note that NaCl is in excess. Wet a filter paper, and use a small funnel to filter the solution into another test tube. Add 5 mL more of 0.2 M NaCl solution to show that the reaction is complete and that almost no Ag^+ is present (NO_3^- still exists in the solution, it is a spectator ion). Add 1 drop of $AgNO_3$ to show what happens when Ag^+ is added.

Safety: Wear safety goggles, gloves, and a lab apron. Have students wear safety goggles. Students must be 3 m or more from the demonstration.

Disposal: Add a few drops of NaCl solution to any leftover $AgNO_3$ solution until all of the silver ions precipitate. Decant the liquid from the AgCl crystals, and evaporate the liquid. Put the AgCl in a bottle labeled "AgCl(s)" and keep it for future display. Do not dispose of AgCl in a landfill.

Teach, continued

Discussion ————— GENERAL

Designing a Model Have students describe or demonstrate models of dynamic equilibrium. Examples could include bailing water from a leaky boat at the same rate that water enters the boat. **LS** Logical

Teaching Tip

Have students examine **Figure 3.** Help students understand that the straight horizontal line means only that the rates of the forward and reverse reactions are equal. The line reveals nothing about the amount or concentration of reactants and products at equilibrium. It also reveals nothing about the extent to which the reaction formed products.

Paired Reading Have each student read the section on reversible and completion reactions and study the figures. After reading, students should pair up and take turns asking and answering the following questions:

- Under what conditions do reactions go to completion?

- Contrast reversible reactions and completion reactions.

- What is taking place when a reaction reaches chemical equilibrium?

- What happens to the rate of the forward reaction for a system in equilibrium after reactants are combined? Why does the rate not drop to zero?

LS Visual

Opposing Reaction Rates Are Equal at Equilibrium

The reaction of hydrogen, H_2, and iodine, I_2, to form hydrogen iodide, HI, also reaches chemical equilibrium.

$$H_2(g) + I_2(g) \rightleftharpoons 2HI(g)$$

When equal amounts of H_2 and I_2 gases are mixed, only a very small fraction of the collisions between H_2 and I_2 result in the formation of HI.

$$H_2(g) + I_2(g) \longrightarrow 2HI(g)$$

After some time, the concentration of HI in the gas mixture goes up. As a result, fewer collisions occur between H_2 and I_2 molecules, and the rate of the forward reaction drops. This decline in the forward reaction rate is shown in the upper curve of **Figure 3.** Similarly, in the beginning, few HI molecules exist in the system, so they rarely collide with each other. As more HI molecules are made, they collide more often. A few of these collisions form H_2 and I_2 by the reverse reaction.

$$2HI(g) \longrightarrow H_2(g) + I_2(g)$$

The greater the number of HI molecules that form, the more often the reverse reaction occurs, as the lower curve in the figure shows. When the forward rate and the reverse rate are equal, the system is at chemical equilibrium.

If you repeated this experiment at the same temperature, starting with a similar amount of pure HI instead of the H_2 and I_2, the reaction would reach chemical equilibrium again. This equilibrium mixture would have exactly the same concentrations of each substance whether you started with HI or with a mixture of H_2 and I_2!

Figure 3
This graph shows how the forward and reverse rates of a reaction become equal at equilibrium.

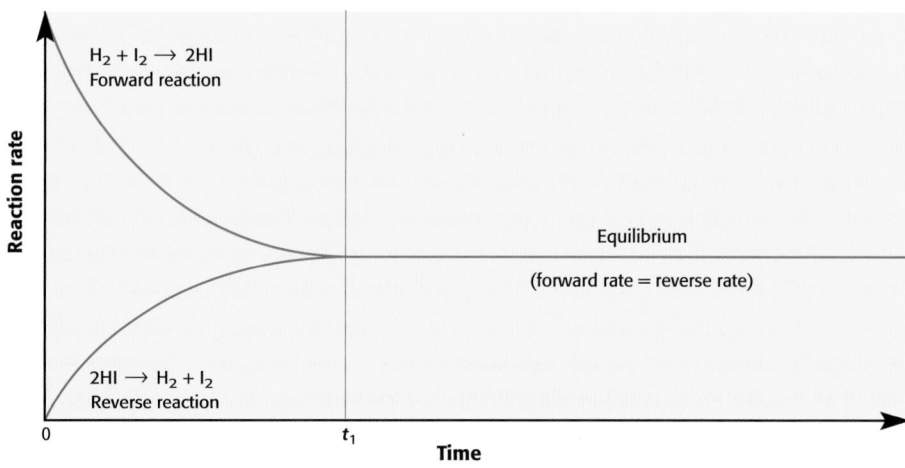

Rate Comparison for $H_2(g) + I_2(g) \rightleftharpoons 2HI(g)$

498

PHYSICS
CONNECTION

Remind students that objects don't move if they are subject to balanced forces. For example, if two students are pushing on a wagon with forces that are equal in magnitude but opposite in direction, the wagon does not move. Forces act on the wagon, but the forces are equal and opposite. Show how this situation is analogous to equilibrium. No apparent reaction takes place because both the forward and reverse reactions occur at the same rate.

Transparencies

TT Rate Comparison for $H_2(g) + I_2(g)$ $\rightleftharpoons$ 2HI(g)

Chemical Equilibria Are Dynamic

If you drop a small ball into a bowl, the ball will bounce around and then come to rest in the center of the bowl. The ball has reached static equilibrium. *Static equilibrium* is a state in which nothing changes.

Chemical equilibrium is different from static equilibrium because it is dynamic. In a *dynamic equilibrium,* there is no net change in the system. Two opposite changes occur at the same time in a dynamic equilibrium. A moment during a hockey game illustrates dynamic equilibrium shown in **Figure 4.** Even though 12 players are on the ice, the players are changing as some players enter the game and others return to the bench. In a chemical equilibrium, an atom may change from being a part of the products to being a part of the reactants many times. But the overall concentrations of products and reactants stay the same. For chemical equilibrium to be maintained, the rates of the forward and reverse reactions must be equal. Arrows of equal length also show equilibrium.

$$\text{reactants} \rightleftharpoons \text{products}$$

In some cases, the forward reaction is nearly done before the rate of the reverse reaction becomes large enough to create equilibrium. So, this equilibrium has a higher concentration of products than reactants. This type of equilibrium, also shown by using two arrows, is shown below. Notice that the forward reaction has a longer arrow to show that the products are favored.

$$\text{reactants} \xrightleftharpoons{} \text{products}$$

499

Teach, continued

Demonstration

Complex Ion Equilibria Fill a test tube half-way with a 1 M $CuSO_4$ solution. The blue color is characteristic of the $[Cu(H_2O)_4]^{2+}$ ions. Add NaCl crystals to the test tube, and stir the solution. Continue to add salt until the solution turns deep green. The new color indicates the replacement of the water ligands with chloride ions to form $[CuCl_4]^{2-}$ ions. Add more water to reduce the Cl^- ion concentration. You should see the solution become bluer in color. Note that the water concentration changes very little. Write a list of the species involved in this reaction, and ask students to write an equilibrium equation.

$[Cu(H_2O)_4]^{2+}(aq) + 4Cl^-(aq) \rightleftharpoons$
$[CuCl_4]^{2-}(aq) + 4H_2O(l)$

Safety Caution: Safety goggles, gloves, and a lab apron must be worn. Have students wear safety goggles. Students must be 3 m or more from the demonstration.

Disposal: Add an excess of 1 M NaOH to precipitate the copper as $Cu(OH)_2$. Filter the precipitate and wash it with water. Neutralize the combined filtrate and washings with 1 M HCl, and pour down the drain. Wrap the wet $Cu(OH)_2$ crystals in newspaper, and place them in the trash.

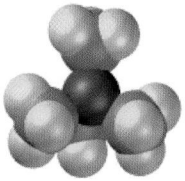

Figure 5
In this complex ion, $[Cu(NH_3)_4]^{2+}$, ammonia molecules bond to the central copper(II) ion.

Topic Link

Refer to the "Periodic Table" chapter for a discussion of transition metals.

More Examples of Equilibria

Our environment has many examples of chemical equilibria. Even when systems are not in equilibrium, they are continuously changing in an effort to reach equilibrium. Equilibria control the composition of Earth's atmosphere, oceans, and lakes, as well as the composition of our body fluids. Even soot provides an example of an equilibrium.

Have you ever thought about how soot forms? Soot is the powdery form of carbon found in chimneys, around smoky candles, and often when combustion occurs. As you know, combustion produces carbon dioxide, CO_2, and poisonous carbon monoxide, CO. As carbon monoxide and carbon dioxide cool after combustion, a reversible reaction that produces soot occurs.

$$2CO(g) \rightleftharpoons C(s) + CO_2(g)$$

This reaction of gases and a solid will reach chemical equilibrium. Equilibria can involve any state of matter, including aqueous solutions.

Equilibria Involving Complex Ions

Complex ion, or *coordination compound,* is the name given to any metal atom or ion that is bonded to more than one atom or molecule. Some of the most interesting ions have a metal ion surrounded by a number of ligands. *Ligands* are molecules, such as ammonia, NH_3, or anions, such as cyanide, CN^-, that readily bond to metal ions. **Figure 5** shows a model of one complex ion, $[Cu(NH_3)_4]^{2+}$. Complex ions may be positively charged cations or negatively charged anions. Those formed from transition metals are often deeply colored, as **Table 1** shows. You can also see that the charge on the complex ion is the sum of the charges on the species from which the complex ion forms. For example, when the cobalt ion, Co^{2+}, bonds with four Cl^- ligands, the total charge is $(+2) + 4(-1) = -2$. Note that metal ions and ligands can form complexes that have no charge, such as $[Cu(NH_3)_2Cl_2]$, but are not complex ions because they are not charged.

Complex ions often form in systems that reach equilibrium. The following equation represents zinc nitrate dissolving in water.

$$Zn(NO_3)_2(s) \longrightarrow Zn^{2+}(aq) + 2NO_3^-(aq)$$

Table 1   Examples of Complex Ions

Complex cation	Color	Complex anion	Color
$[Co(NH_3)_6]^{3+}$	yellow-orange	$[Co(CN)_6]^{3-}$	pale yellow
$[Co(NH_3)_5Cl]^{2+}$	violet	$[CoCl_4]^{2-}$	blue
$[Fe(H_2O)_5SCN]^{2+}$	deep red	$[Fe(CN)_6]^{3-}$	red
$[Cu(NH_3)_4]^{2+}$	blue-purple	$[Fe(CN)_6]^{4-}$	yellow

Refer to Appendix A for more examples of complex ions.

500

INCLUSION Strategies

• Gifted and Talented

Have the students interview a local chimney sweep or research chimney safety on the internet, and report back to the class with a presentation of some sort. Included in the presentation should be knowledge of the periodic care of a chimney, information about the reversible reaction that causes soot, and the procedures used by a chimney sweep in order to keep a chimney functioning well.

Chapter Resource File

- Concept Review
- Quiz

Notice from the table that water molecules can be ligands. In the absence of other ligands, water molecules bond with zinc ions. So, a more accurate description of this reaction is

$$Zn(NO_3)_2(s) + 4H_2O(l) \rightleftharpoons [Zn(H_2O)_4]^{2+}(aq) + 2NO_3^-(aq)$$

If another ligand, such as CN^-, is added, the new system will again reach chemical equilibrium. Both water molecules and cyanide ions "compete" to bond with zinc ions, as shown in the equation below.

$$[Zn(H_2O)_4]^{2+}(aq) + 4CN^-(aq) \rightleftharpoons [Zn(CN)_4]^{2-}(aq) + 4H_2O(l)$$

All of the ions in the previous equation are colorless, so you cannot see which complex ion has the greater concentration. However, in the following chemical equilibrium of nickel ions, ammonia, and water, the complex ions have different colors. Therefore, you can see which complex ion has the greater concentration.

$$[Ni(H_2O)_6]^{2+}(aq) + 6NH_3(aq) \rightleftharpoons [Ni(NH_3)_6]^{2+}(aq) + 6H_2O(l)$$
    green   blue-violet

The starting concentration of NH_3 will determine which one will have the greater concentration.

Section Review

UNDERSTANDING KEY IDEAS

1. Which of the following equations best represents a reaction that goes to completion?
   a. reactants $\xleftarrow{\longrightarrow}$ products
   b. reactants $\rightleftharpoons$ products
   c. reactants $\longrightarrow$ products

2. At equilibrium, what is the relationship between the rates of the forward and reverse reactions?

3. Explain what each reaction below shows. Describe the relationship between the amounts of products and reactants for each case.
   a. reactants $\rightleftharpoons$ products
   b. reactants $\xrightarrow{\longleftarrow}$ products

4. What is the difference between dynamic and static equilibria?

5. Write the formula and charge of the complex ion that forms when a copper ion, Cu^{2+}, bonds with four chloride ligands.

CRITICAL THINKING

6. What evidence might lead you to believe that a chemical reaction was *not* at equilibrium?

7. In what way are the number of players on the ice in a hockey game *not* like chemical equilibrium?

8. The final equation on this page describes a reaction in which all six water ligands are replaced by six ammonia molecules. Write the formula of a complex ion that would form if only two ligands were exchanged.

9. Which of the following nitrogen compounds or ions, NH_4^+, NH_3, or NH_2^-, cannot be a ligand? Why?

501

Overview

Before beginning this section, review with your students the Objectives listed in the Student Edition. Students will use the equilibrium constant, K_{eq}, to write equilibrium expressions and calculate the concentrations of reactants and products as well as K_{eq}. They will also learn and apply the solubility product constant, K_{sp}, for slightly soluble salts.

Bellringer

Ask students to make a list of numbers that are "constants" under constant conditions. An example is the speed of light. Ask them what these constants have in common. **Ans.** Each constant is always the same number for a certain and constant set of conditions.

Motivate

Identifying Preconceptions

Students might think that a constant has the same value for all conditions. Remind students that the volume of one mole of gas at STP is a constant value of 22.4 L. Ask them if that volume stays the same when temperature or pressure changes.

Discussion ──── BASIC

Write the equilibrium equation $CaCO_3(s) + CO_2(aq) + H_2O(l) \rightleftharpoons Ca^{2+}(aq) + 2HCO_3^-(aq)$ on the chalkboard, and ask students to explain the formation of stalactites and stalagmites as you describe the changes in CO_2 concentration.
LS Interpersonal

Systems at Equilibrium

KEY TERMS
- equilibrium constant, K_{eq}
- solubility product constant, K_{sp}

OBJECTIVES

(1) **Write K_{eq}** expressions for reactions in equilibrium, and perform calculations with them.

(2) **Write K_{sp}** expressions for the solubility of slightly soluble salts, and perform calculations with them.

The Equilibrium Constant, K_{eq}

Limestone caverns, shown in **Figure 6,** form over millions of years. They are made as rainwater, slightly acidified by H_3O^+, gradually dissolves rocks made of calcium carbonate. This reaction is still going today and is slowly enlarging caverns. The reverse reaction also takes place, and solid calcium carbonate is deposited as beautiful stalactites and stalagmites.

$$CaCO_3(s) + 2H_3O^+(aq) \rightleftharpoons Ca^{2+}(aq) + CO_2(g) + 3H_2O(l)$$

Figure 6
Stalactites and stalagmites in limestone caverns form because of a slight displacement of a reaction from equilibrium.

502

HISTORY
CONNECTION

Early Equilibrium Constants Norwegian chemists Cato Maximilian Guldberg (l836–1902) and Peter Waage (1833–1900) published the idea of equilibrium constants in 1863. Unfortunately, their work was published in Norwegian, and it went unnoticed until 1879, when it was translated into German.

Chapter Resource File
- Lesson Plan

When some $CaCO_3$ first reacts with H_3O^+, the rate of the forward reaction is large. The rate of the reverse reaction is zero until some products form. As the reaction proceeds, the forward reaction rate slows as the reactant concentrations decrease. At the same time, the reverse rate increases as more products of the forward reaction form. When the two rates become equal, the reaction reaches chemical equilibrium. Because reaction rates depend on concentrations, there is a mathematical relationship between product and reactant concentrations at equilibrium. For the reaction of limestone and acidified water, the relationship is

$$K_{eq} = \frac{[Ca^{2+}][CO_2]}{[H_3O^+]^2}$$

where K_{eq} is a number called the **equilibrium constant** of the reaction. This expression has a specific numerical value that depends on temperature and must be found experimentally or from tables. The value of K_{eq} for this reaction is 1.4×10^{-9} at 25°C.

Notice that equilibrium constants are unitless. In addition, equilibrium constants apply only to systems in equilibrium. **Skills Toolkit 1** contains the rules for writing the equilibrium constant expression for *any* reaction, including the one above.

equilibrium constant, K_{eq}

a number that relates the concentrations of starting materials and products of a reversible chemical reaction to one another at a given temperature

---

SKILLS Toolkit 1

Determining K_{eq} for Reactions at Chemical Equilibrium

1. Write a balanced chemical equation.

- Make sure that the reaction is at equilibrium before you write a chemical equation.

2. Write an equilibrium expression.

- To write the expression, place the product concentrations in the numerator and the reactant concentrations in the denominator.

- The concentration of any solid or a pure liquid that takes part in the reaction is left out because these concentrations never change.

- For a reaction occurring in aqueous solution, water is omitted because the concentration of water is almost constant during the reaction.

3. Complete the equilibrium expression.

- To complete the expression, raise each substance's concentration to the power equal to the substance's coefficient in the balanced chemical equation.

503

Teach

Teaching Tip

K_{eq} and Units of Concentration
In this chapter, concentrations expressed in the notation of the bracketed solute will be expressed without units, the bracketed notation being considered to have the unit of molarity already implicit. In this way, K_{eq} can also be accurately expressed as a dimensionless constant without having to worry about units.

Temperature Dependence of K_{eq}

The distribution of products and reactants of a reaction at equilibrium is quantified in the equilibrium constant, K_{eq}, for that reaction. Students should be made aware that an equilibrium constant is specific to only one reaction at a certain temperature. If the temperature changes, the reaction will have a different K_{eq}.

Using the Figure

Draw students' attention to **Figure 6.** Caves that have stalactites and stalagmites occur in regions where there are large deposits of limestone, or calcium carbonate, $CaCO_3$. Rainwater that runs down into limestone formations contains dissolved CO_2 from the air and can dissolve limestone by the equilibrium process described on the facing page. When the resulting solution seeps through the roof of a cave, the air in the cave often has a lower CO_2 concentration than the outside air. As a result, the solution loses CO_2 as it drips from the ceiling. This loss shifts the equilibrium to the left, and a very small quantity of insoluble $CaCO_3$ precipitates.

In time, a stalactite made of crystalline $CaCO_3$ develops. Then, the solution can travel down the stalactite to the floor, where more CO_2 is lost and more $CaCO_3$ is deposited, leading to a stalagmite. After many thousands of years, the two deposits meet, forming a solid column.

SKILL BUILDER

Vocabulary — BASIC

Have students define the word *concentration* as it applies to equilibrium constants. The word *concentration* is taken to apply only to substances in dispersed phases. *Dispersed* refers to matter that is spread out throughout a volume of space, such as a gas or a solute in a solution. Concentrations are usually measured in moles per liter (of space). **LS** Logical

Homework — GENERAL

Additional Practice Calculate K_{eq} for the following reactions from the given data.

a. $COCl_2(g) \rightleftarrows CO(g) + Cl_2(g)$
At equilibrium $[CO] = [Cl_2] = 0.0178$, $[COCl_2] = 0.00740$
Ans. $K_{eq} = 0.043$

b. $Br_2 \rightleftarrows 2Br(g)$
At equilibrium $[Br_2] = 0.99$, $[Br] = 0.020$ Ans. $K_{eq} = 4.0 \times 10^{-4}$

LS Intrapersonal

Answers to Practice Problems A

1. $K_{eq} = \dfrac{[N_2O_4]}{[NO_2]^2} = \dfrac{(4.0 \times 10^{-2})}{(1.4 \times 10^{-1})^2} = 2.0$

2. $K_{eq} = \dfrac{[SO_3]^2}{([SO_2]^2[O_2])} =$

   $\dfrac{(1.01 \times 10^{-2})^2}{(3.61 \times 10^{-3})^2(6.11 \times 10^{-4})} = 1.28 \times 10^4$

SAMPLE PROBLEM A

Calculating K_{eq} from Concentrations of Reactants and Products

An aqueous solution of carbonic acid reacts to reach equilibrium as described below.

$$H_2CO_3(aq) + H_2O(l) \rightleftarrows HCO_3^-(aq) + H_3O^+(aq)$$

The solution contains the following solute concentrations: carbonic acid, 3.3×10^{-2} mol/L; bicarbonate ion, 1.19×10^{-4} mol/L; and hydronium ion, 1.19×10^{-4} mol/L. Determine the K_{eq}.

1 Gather information.

$[H_2CO_3] = 3.3 \times 10^{-2}$, $[HCO_3^-] = [H_3O^+] = 1.19 \times 10^{-4}$

2 Plan your work.

Write the equilibrium constant expression. For this reaction, the equilibrium constant expression is

$$K_{eq} = \frac{[HCO_3^-][H_3O^+]}{[H_2CO_3]}$$

3 Calculate.

Substitute the concentrations into the expression.

$$K_{eq} = \frac{[HCO_3^-][H_3O^+]}{[H_2CO_3]} = \frac{(1.19 \times 10^{-4}) \times (1.19 \times 10^{-4})}{(3.3 \times 10^{-2})} = 4.3 \times 10^{-7}$$

4 Verify your results.

The numerator of the expression is approximately 1×10^{-8}. The denominator is approximately 3×10^{-2}. The rough calculation below supports the value 4.3×10^{-7} for the K_{eq}.

$$K_{eq} \approx \frac{1 \times 10^{-8}}{3 \times 10^{-2}} = 3 \times 10^{-7}$$

> **PRACTICE HINT**
>
> Remember that $[H_2CO_3]$ means "the concentration of H_2CO_3 in mol/L." So, these quantities do not carry units because the use of square brackets around the formula implies a concentration in moles per liter.
>
> Note that the chemical equation has coefficients of 1, so no concentration used in this expression is raised to a power.

PRACTICE

1 For the system involving N_2O_4 and NO_2 at equilibrium at a temperature of 100°C, the product concentration of N_2O_4 is 4.0×10^{-2} mol/L and the reactant concentration of NO_2 is 1.4×10^{-1} mol/L. What is the K_{eq} value for this reaction?

2 An equilibrium mixture at 852 K is found to contain 3.61×10^{-3} mol/L of SO_2, 6.11×10^{-4} mol/L of O_2, and 1.01×10^{-2} mol/L of SO_3. Calculate the equilibrium constant, K_{eq}, for the reaction where SO_2 and O_2 are reactants and SO_3 is the product.

504

MISCONCEPTION ALERT

Equilibrium Constants Students may be confused by the use of K to represent several different constants. Explain that K is used to represent the constant value of an equilibrium expression. The subscript used specifies the type of equilibrium. K_{eq} is the general term for the equilibrium constant. K_{sp}, which students will encounter later in this section, is a more specific type of equilibrium constant. When students study acids and bases, they will again encounter equilibrium constants in the form of K_w for water, K_a for an acid, and K_b for a base.

Chapter Resource File

• Problem Solving

Table 2 Equilibrium Constants at 25°C

Equation	K_{eq} expression and value
$N_2(g) + 3H_2(g) \rightleftharpoons 2NH_3(g)$	$\dfrac{[NH_3]^2}{[N_2][H_2]^3} = 3.3 \times 10^8$
$2NO_2(g) \rightleftharpoons N_2O_4(g)$	$\dfrac{[N_2O_4]}{[NO_2]^2} = 165$
$Hg^{2+}(aq) + Hg(l) \rightleftharpoons Hg_2^{2+}(aq)$	$\dfrac{[Hg_2^{2+}]}{[Hg^{2+}]} = 81$
$HCO_3^-(aq) + H_2O(l) \rightleftharpoons CO_3^{2-}(aq) + H_3O^+(aq)$	$\dfrac{[CO_3^{2-}][H_3O^+]}{[HCO_3^-]} = 4.7 \times 10^{-11}$
$N_2(g) + O_2(g) \rightleftharpoons 2NO(g)$	$\dfrac{[NO]^2}{[N_2][O_2]} = 4.5 \times 10^{-31}$

Refer to Appendix A for more K_{eq} values.

K_{eq} Shows If the Reaction Is Favorable

When K_{eq} is large, the numerator of the equilibrium constant expression is larger than the denominator. Thus, the concentrations of the products will usually be greater than those of the reactants. In other words, when a reaction that has a large K_{eq} reaches equilibrium, the system's contents may be mostly products. Reactions in which more products form than reactants form are said to be "favorable." Look at the first entry in **Table 2**, where the reaction has a large K_{eq} value. The synthesis of ammonia is very favorable at 25°C.

When K_{eq} is small, the denominator of the equilibrium constant expression is larger than the numerator. The larger denominator shows that the concentrations of reactants at chemical equilibrium may be greater than those of products. A reaction that has larger concentrations of reactants than concentrations of products is an "unfavorable" reaction. Our air would be unbreathable if the reaction of oxygen and nitrogen to give nitrogen monoxide was favorable at 25°C! (See **Table 2** to know the value for this reaction.)

Figure 7 shows what the composition of a reaction mixture ($R \rightleftharpoons P$) would be at equilibrium for three different K_{eq} values. Notice that for $K_{eq} = 1$, there would be a 50:50 mixture of reactants (R) and products (P).

Figure 7
These pie charts show the relative amounts of reactants and products for three K_{eq} values of a reaction.

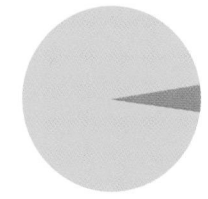

$K_{eq} = 0.02$

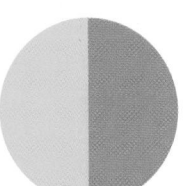

$K_{eq} = 1$

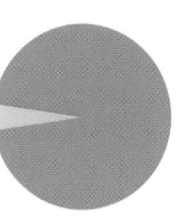

$K_{eq} = 50$

Reactants ▢ Products ▢

Teach, *continued*

Teaching Tip

A Unitless Quantity Units are not used because the unit of K_{eq} would vary depending on the powers to which the concentrations are raised. For consistency, bracketed concentration notations in this chapter are considered to already imply the unit of molarity, so their concentration values will be expressed as unitless. In this way, K_{eq} can also be expressed as unitless.

Homework ——— GENERAL

Additional Practice

a. Determine the equilibrium concentrations of H_3O^+ and HCO_3^- in a solution of carbonic acid at 25°C in which $[H_2CO_3] = 0.027$, $K_{eq} = 4.3 \times 10^{-7}$, and $H_2CO_3(aq) + H_2O(l) \rightleftharpoons H_3O^+(aq) + HCO_3^-(aq)$ Ans. 1.1×10^{-4} M ($[H_3O^+] = [HCO_3^-]$)

b. Determine $[H_2]$ equilibrium at 700 K when $[CH_3OH] = 0.25$, $[CO] = 0.0098$, $K_{eq} = 290$, and $CO(g) + 2H_2(g) \rightleftharpoons CH_3OH(g)$ Ans. $[H_2] = 0.30$

LS Intrapersonal

Answers to Practice Problems B

1. $[NO]^2 = K_{eq} [N_2] [O_2]$; $[NO]^2 = (1.65 \times 10^{-3})(1.8 \times 10^{-3})^2(4.2 \times 10^{-4})$; $[NO] = 3.5 \times 10^{-5}$

2. $[SO_2]^2 = \dfrac{[SO_3]^2}{K_{eq}[O_2]} =$

$= (0.260)^2/(4.32)(0.045) = 0.59$

Chapter Resource File
• Problem Solving

506

Calculating Concentrations of Products from K_{eq} and Concentrations of Reactants

K_{eq} for the equilibrium below is 1.8×10^{-5} at a temperature of 25°C. Calculate $[NH_4^+]$ when $[NH_3] = 6.82 \times 10^{-3}$.

$$NH_3(aq) + H_2O(l) \rightleftharpoons NH_4^+(aq) + OH^-(aq)$$

1 Gather information.

Determine the equilibrium expression first.

$$K_{eq} = \frac{[NH_4^+][OH^-]}{[NH_3]}$$

NH_4^+ and OH^- ions are produced in equal numbers, so $[OH^-] = [NH_4^+]$.

> **PRACTICE HINT**
>
> Remember to keep track of significant figures and to give no more significant figures than are needed.

2 Plan your work.

The numerator can be written as x^2 because $[OH^-] = [NH_4^+]$. Both K_{eq} and $[NH_3]$ are known and can be put into the expression.

3 Calculate.

Substitute known quantities.

$$1.8 \times 10^{-5} = K_{eq} = \frac{[NH_4^+][OH^-]}{[NH_3]} = \frac{x^2}{6.82 \times 10^{-3}}$$

Rearrange the equation.

$$x^2 = (1.8 \times 10^{-5}) \times (6.82 \times 10^{-3}) = 1.2 \times 10^{-7}$$

Take the square root of x^2.

$$[NH_4^+] = \sqrt{1.2 \times 10^{-7}} = 3.5 \times 10^{-4}$$

4 Verify your results.

Test to see whether the answer gives the equilibrium constant.

$$K_{eq} = \frac{[NH_4^+][OH^-]}{[NH_3]} = \frac{(3.5 \times 10^{-4})(3.5 \times 10^{-4})}{6.82 \times 10^{-3}} = 1.8 \times 10^{-5}$$

PRACTICE

1 If K_{eq} is 1.65×10^{-3} at 2027°C for the reaction below, what is the equilibrium concentration of NO when $[N_2] = 1.8 \times 10^{-3}$ and $[O_2] = 4.2 \times 10^{-4}$?

$$N_2(g) + O_2(g) \rightleftharpoons 2NO(g)$$

2 At 600°C, the K_{eq} for the reaction below is 4.32 when $[SO_3] = 0.260$ and $[O_2] = 0.045$. Calculate the equilibrium concentration for SO_2.

$$2SO_2(g) + O_2(g) \rightleftharpoons 2SO_3(g)$$

PROBLEM SOLVING SKILL

506

---

did you know?

Writing Equilibrium Equations The equation for a reversible reaction can legitimately be written backwards. For example, the equation $2NO_2(g) \rightleftharpoons N_2O_4(g)$ can also be written as $N_2O_4(g) \rightleftharpoons 2NO_2(g)$. The K_{eq} for the second reaction is the inverse of the K_{eq} for the first reaction. Thus, a reaction that has a large K_{eq} when written in one direction will have a small K_{eq} when written in reverse, but both constants represent the same equilibrium.

The Solubility Product Constant, K_{sp}

The maximum concentration of a salt in an aqueous solution is called the solubility of the salt in water. Solubilities can be expressed in moles of solute per liter of solution (mol/L or M). For example, the solubility of calcium fluoride in water is 3.4×10^{-4} mol/L. So, 0.00034 mol (or less than 0.03 g) of CaF_2 will dissolve in 1 L of water to give a saturated solution. If you try to dissolve 0.00100 mol of CaF_2 in 1 L of water, 0.00066 mol of CaF_2 will remain undissolved.

Like most salts, calcium fluoride is an ionic compound that dissociates into ions when it dissolves in water. Calcium fluoride is also one of a large class of salts that are said to be slightly soluble in water. The ions in solution and any solid salt are at equilibrium, as the following equation for CaF_2 and water shows.

$$CaF_2(s) \rightleftarrows Ca^{2+}(aq) + 2F^-(aq)$$

Recall that solids are not a part of equilibrium constant expressions, so K_{eq} for this reaction is the product of $[Ca^{2+}]$ and $[F^-]^2$, which is equal to a constant. Equilibrium constants for the dissolution of slightly soluble salts are given a special name and symbol. They are called **solubility product constants, K_{sp}**, and have no units. The K_{sp} value for calcium fluoride at 25°C is 1.6×10^{-10}.

$$K_{sp} = [Ca^{2+}][F^-]^2 = 1.6 \times 10^{-10}$$

This relationship is true whenever calcium ions and fluoride ions are in equilibrium with calcium fluoride, not just when the salt dissolves. For example, if you mix solutions of calcium nitrate and sodium fluoride, calcium fluoride precipitates. The net ionic equation for this precipitation is the reverse of the dissolution.

$$Ca^{2+}(aq) + 2F^-(aq) \rightleftarrows CaF_2(s)$$

This equation is the same equilibrium. So, the K_{sp} for the dissolution of CaF_2 in this system is the same and is 1.6×10^{-10}.

Refer to the "Solutions" chapter for a discussion of solubility and concentrations.

solubility product constant, K_{sp}

the equilibrium constant for a solid that is in equilibrium with the solid's dissolved ions

Figure 8
Seashells are made mostly of calcium carbonate, which is only slightly soluble in water. The calcium carbonate is in equilibrium with ions present in sea water.

ENVIRONMENTAL SCIENCE
CONNECTION

As water flows through rivers and streams, it carries dissolved salts, including a few ions of materials that are very slightly soluble. When the water reaches the ocean, it carries these ions with it. Over millions of years, water with ions has entered the ocean, where the water evaporates through the water cycle, leaving the salts behind. The salts stay dissolved in the ocean until enough are present so that their concentration exceeds that of a saturated solution. The ions crystallize and precipitate, forming sediments.

Teaching Tip

K_{sp} Students should understand that a solubility product constant is simply an equilibrium constant that represents the reversible process which takes place when a saturated solution is in contact with undissolved solute.

Solubility Equilibrium Equations
Although most equilibrium equations can be written using reactants and products on either side of the equation, equations for solubility equilibrium are different. Although a reaction in which a solubility product constant applies is a reversible process, the formula is usually written with the solid on the left side of the equation, and the ions are written on the right side.

READING SKILL BUILDER — BASIC

Assimilating Knowledge Have each student read the section on the solubility product constant and look over the problem setups in Sample Problems C and D. After reading, students should pair up and take turns asking and answering the following questions as a summary of the section:

• Why is the word *insoluble* an inexact term when applied to real solutions? **Ans.** Each solid present has at least a few ions in solution.

• Describe the dynamic equilibrium that occurs in a saturated solution. **Ans.** Ions crystallize at the same rate as ions enter solution.

• Why is the concentration of undissolved substance not included in the equilibrium expression? **Ans.** The concentration of a solid is a constant.

• Why do K_{sp} values depend on temperature? **Ans.** Changing temperature changes molecular motion. When particles speed up or slow down, their ability to enter or stay in solution changes.

LS Verbal

SKILL BUILDER — ADVANCED

Math Skills Ask students to study **Table 3** and determine which substance has the lowest solubility. Many will choose Ag_2S because of its very small K_{sp}. Explain that this compound is not the least soluble.

Direct students' attention to the compounds $Ca_3(PO_4)_2$ and $Fe(OH)_3$, and ask them to write K_{sp} expressions for these substances as well as for Ag_2S. Have them explain why you cannot judge solubility from K_{sp} alone. The powers of 2 and 3 to which the concentrations are raised contribute to the very small K_{sp}.

The least soluble compound on the list is CuS because it has Cu^{2+} and S^{2-} concentrations of 1.14×10^{-18} mol/L. By contrast, the concentration of Ag^+ ions in a saturated solution of Ag_2S is 6.0×10^{-17} mol/L, roughly five times that of Cu^{2+} ions in a solution of CuS. **LS** Logical

MISCONCEPTION ///ALERT\\\

Determining Concentrations

Students might master using coefficients from a chemical equation as exponents but not remember that the concentrations must reflect the number of ions formed by dissociation of one unit of the compound. Consider the compound Ag_2S. The solubility product constant equation for this compound is $K_{sp} = [Ag^+]^2[S^{2-}]$. But students must remember that $[Ag^+]$ is two times $[S^{2-}]$ because two silver ions form for every sulfide ion produced.

Table 3   Solubility Product Constants at 25°C

Salt	K_{sp}	Salt	K_{sp}
Ag_2CO_3	8.4×10^{-12}	CuS	1.3×10^{-36}
Ag_2CrO_4	1.1×10^{-12}	FeS	1.6×10^{-19}
Ag_2S	1.1×10^{-49}	$MgCO_3$	6.8×10^{-6}
AgBr	5.4×10^{-13}	$MnCO_3$	2.2×10^{-11}
AgCl	1.8×10^{-10}	PbS	9.0×10^{-29}
AgI	8.5×10^{-17}	$PbSO_4$	1.8×10^{-8}
$BaSO_4$	1.1×10^{-10}	$SrSO_4$	3.4×10^{-7}
$Ca_3(PO_4)_2$	2.1×10^{-33}	$ZnCO_3$	1.2×10^{-10}
$CaSO_4$	7.1×10^{-5}	ZnS	2.9×10^{-25}

Refer to Appendix A for more K_{sp} values.

2 SKILLS Toolkit

Determining K_{sp} for Reactions at Chemical Equilibrium

1. Write a balanced chemical equation.

- Remember that the solubility product is only for those salts that have low solubility. Soluble salts such as sodium chloride and ammonium nitrate do not have K_{sp} values.
- Make sure that the reaction is at equilibrium.
- Chemical equations should always be written so that the solid salt is the reactant and the ions are products.

2. Write a solubility product expression.

- To write the expression, write the product of the ion concentrations.
- Concentrations of any solid or pure liquid that take part in the reaction are left out because they never change.

3. Complete the solubility product expression.

- To complete the expression, raise each concentration to a power equal to the substance's coefficient in the balanced chemical equation.
- Remember that K_{sp} values depend on temperature.

508

did you know?

Interpreting K_{sp} Note that the K_{sp} of CuS is 1.3×10^{-36}. This value means that each ion in a saturated solution would have a concentration of $(1.3 \times 10^{-36})^{1/2}$ M, 1.14×10^{-18} M, or only about 690000 ions per liter.

SAMPLE PROBLEM C

Calculating K_{sp} from Solubility

Most parts of the oceans are nearly saturated with CaF_2. The mineral fluorite, CaF_2, may precipitate when ocean water evaporates. A saturated solution of CaF_2 at 25°C has a solubility of 3.4×10^{-4} M. Calculate the solubility product constant for CaF_2.

1 **Gather information.**

$$CaF_2(s) \rightleftharpoons Ca^{2+}(aq) + 2F^-(aq)$$

$$[CaF_2] = 3.4 \times 10^{-4}, [F^-] = 2[Ca^+]$$

$$K_{sp} = [Ca^{2+}][F^-]^2$$

2 **Plan your work.**

Because 3.4×10^{-4} mol CaF_2 dissolves in each liter of solution, you know from the balanced equation that every liter of solution will contain 3.4×10^{-4} mol Ca^{2+} and 6.8×10^{-4} mol F^-.

3 **Calculate.**

$$K_{sp} = [Ca^{2+}][F^-]^2 = (3.4 \times 10^{-4})(6.8 \times 10^{-4})^2 = 1.6 \times 10^{-10}$$

4 **Verify your result.**

To verify that the K_{sp} value is correct, determine the concentration of the F^- ion using $K_{sp} = 1.6 \times 10^{-10}$ and $[Ca^{2+}] = 3.4 \times 10^{-4}$.

$$[F^-]^2 = \frac{K_{sp}}{[Ca^{2+}]} = \frac{1.6 \times 10^{-10}}{3.4 \times 10^{-4}} = 4.7 \times 10^{-7}$$

$$[F^-] = 6.9 \times 10^{-4}$$

This calculation provides the concentration of the fluorine ions and confirms that the solubility product constant is 1.6×10^{-10}.

PRACTICE HINT

Do not forget to balance the chemical equation before using it to solve solubility problems.

PRACTICE

1 Copper(I) bromide is dissolved in water to saturation at 25°C. The concentration of Cu^+ ions in solution is 7.9×10^{-5} mol/L. Calculate the K_{sp} for copper(I) bromide at this temperature.

2 What is the K_{sp} value for $Ca_3(PO_4)_2$ at 298 K if the concentrations in a solution at equilibrium with excess solid are 3.42×10^{-7} M for Ca^{2+} ions and 2.28×10^{-7} M for PO_4^{3-} ions?

3 If a saturated solution of silver chloride contains an AgCl concentration of 1.34×10^{-5} M, confirm that the solubility product constant of this salt has the value shown in **Table 3**.

509

Teaching Tip

Predicting Precipitation Solubility product constants can be used to determine whether a precipitate will form when two solutions containing ions that will form a slightly soluble solid are mixed. Substitute the concentrations of the ions in solution into the solubility product constant expression. If the product is greater than the K_{sp} for the compound, a precipitate forms. If it is less, no precipitate forms.

Homework ——— GENERAL

Additional Practice

a. Calculate the solubility product constant of HgI_2 if the Hg^{2+} concentration in a saturated solution is 1.9×10^{-10} M. Ans. $K_{sp} = 2.7 \times 10^{-29}$

b. Calculate the solubility product constant of $Fe(OH)_2$ if the OH^- concentration in a saturated solution is 4.6×10^{-6} M. Ans. $K_{sp} = 4.9 \times 10^{-17}$

c. The K_{sp} of CdF_2 is 6.4×10^{-3}. Calculate the concentration of the ions in a saturated solution of CdF_2. Ans. $[Cd^{2+}] = 0.12$, $[F^-] = 0.24$

LS Logical

Answers to Practice Problems C

1. $K_{sp} = [Cu^+][Br^-] = (7.9 \times 10^{-5})^2 = 6.2 \times 10^{-9}$

2. $K_{sp} = [Ca^{2+}]^3 [PO_4^{3-}]^2 = (3.42 \times 10^{-7})^3(2.28 \times 10^{-7})^2 = 2.08 \times 10^{-33}$

3. $K_{sp} = [Ag^+][Cl^-] = (1.34 \times 10^{-5})^2 = 1.80 \times 10^{-10}$

Chapter Resource File

• Problem Solving

Homework ——— GENERAL

Additional Practice

a. Calculate the concentration of Ba^{2+} ion in a saturated solution of $BaSO_4$ both before and after the SO_4^{2-} concentration has been boosted to 0.010 M by the addition of Na_2SO_4. The K_{sp} of $BaSO_4$ is 1.1×10^{-10}. By what factor is the Ba^{2+} concentration decreased? Ans. 1.0×10^{-5} M; after: 1.1×10^{-8} M. The Ba^{2+} concentration is reduced to approximately 0.001 of its original concentration.

b. A chemist wishes to reduce the silver ion concentration in saturated AgCl solution to 2.0×10^{-6} M. What concentration of Cl^- would achieve this goal? Ans. $[Cl^-] = 9.0 \times 10^{-5}$

c. The K_{sp} of $MgCO_3$ is 6.8×10^{-6}. The concentration of CO_3^{2-} ions in a solution containing both $MgCO_3$ and Na_2CO_3 is 4.0×10^{-2} M. What is the concentration of magnesium ions if the solution is saturated with respect to $MgCO_3$? Ans. $[Mg^{2+}] = 1.7 \times 10^{-4}$

LS Logical

Answers to Practice Problems D

1. $K_{sp} = 8.4 \times 10^{-12} = [Ag^+]^2(1.28 \times 10^{-4})$; $[Ag] = 2.6 \times 10^{-4}$

2. $[Pb]^{2+} = K_{sp}/[SO_4^{2-}] = 1.8 \times 10^{-8}/1.0 = 1.8 \times 10^{-8}$

3. $K_{sp} = 1.17 \times 10^{-5} = [Pb^{2+}](2.86 \times 10^{-2})^2$; $[Pb^{2+}] = 1.43 \times 10^{-2}$

4. $K_{sp} = 1.72 \times 10^{-7} = [Cu^+][Cl^-]$; $[Cu^+] = 4.15 \times 10^{-4}$

SAMPLE PROBLEM D

Calculating Ionic Concentrations Using K_{sp}

Copper(I) chloride has a solubility product constant of 1.2×10^{-6} and dissolves according to the equation below. Calculate the solubility of this salt in ocean water in which the $[Cl^-] = 0.55$.

$$CuCl(s) \rightleftarrows Cu^+(aq) + Cl^-(aq)$$

1 Gather information.

The given data and the chemical equation show that $[Cu^+][Cl^-] = K_{sp} = 1.2 \times 10^{-6}$. Additional information reveals that $[Cl^-] = 0.55$.

2 Plan your work.

The product of $[Cu^+][Cl^-]$ must equal $K_{sp} = 1.2 \times 10^{-6}$.

$$K_{sp} = [Cu^+][Cl^-] = 1.2 \times 10^{-6}$$

3 Calculate.

Using the following equation, you can determine the concentration of copper ions.

$$[Cu^+] = \frac{K_{sp}}{[Cl^-]} = \frac{1.2 \times 10^{-6}}{0.55} = 2.2 \times 10^{-6}$$

The quantity is also the solubility of copper(I) chloride in ocean water because the dissolution of 1 mol of CuCl produces 1 mol of Cu^+. Therefore, the solubility of CuCl is 2.2×10^{-6} mol/L.

4 Verify your results.

You can recalculate the solubility product constant to check the answers.

$$K_{sp} = [Cu^+][Cl^-] = (0.55)(2.2 \times 10^{-6}) = 1.2 \times 10^{-6}$$

> **PRACTICE HINT**
>
> Recall that the K_{sp} expression holds even if the ions are from a source other than the salt.

PRACTICE

1. The K_{sp} for silver carbonate is 8.4×10^{-12} at 298 K. The concentration of carbonate ions in a saturated solution is 1.28×10^{-4} M. What is the concentration of silver ions?

2. Lead-acid batteries employ lead(II) sulfate plates in a solution of sulfuric acid. Use data from **Table 3** to calculate the solubility of $PbSO_4$ in a battery acid that has an SO_4^{2-} concentration of 1.0 M.

3. Calculate the concentration of Pb^{2+} ions in solution when $PbCl_2$ is dissolved in water. The concentration of Cl^- ions in this solution is found to be 2.86×10^{-2} mol/L. At 25°C, the K_{sp} of $PbCl_2$ is 1.17×10^{-5}.

4. What is the concentration of Cu^+ ions in a saturated solution of copper(I) chloride given that the K_{sp} of CuCl is 1.72×10^{-7} at 25°C?

REAL-WORLD — CONNECTION

Usefulness of a Low K_{sp} Some compounds are useful because they are quite soluble, but certain compounds are useful because they are only very slightly soluble. For example, window glass would not be useful if it dissolved in rain. Glass jars and tumblers would be useless if they dissolved in liquids placed in them. Most items that we use every day would not be as useful if they were more soluble in liquids such as water.

Chapter Resource File

• Problem Solving

Using K_{sp} to Make Magnesium

Though slightly soluble hydroxides are not salts, they have solubility product constants. Magnesium hydroxide is an example.

$$Mg(OH)_2(s) \rightleftharpoons Mg^{2+}(aq) + 2OH^-(aq)$$

$$[Mg^{2+}][OH^-]^2 = K_{sp} = 1.8 \times 10^{-11}$$

This equilibrium is the basis for obtaining magnesium.

Table 4 lists the most abundant ions in ocean water and their concentrations. Notice that Mg^{2+} is the third most abundant ion in the ocean. Magnesium is so abundant that ocean water is used as the raw material from which magnesium is gotten. To get magnesium, calcium hydroxide is added to sea water. This raises the hydroxide ion concentration to a large value so that $[Mg^{2+}][OH^-]^2$ would be greater than 1.8×10^{-11}. As a result, magnesium hydroxide precipitates and can be collected. The next step in the process is to treat magnesium hydroxide with hydrochloric acid to make magnesium chloride. Finally, magnesium is obtained by the electrolysis of $MgCl_2$ in the molten state. One cubic meter of sea water yields 1 kg of magnesium metal.

Magnesium has a density of 1.7 g/cm^3, so it is one of the lightest metals. Because of magnesium's low density and rigidity, alloys of magnesium are used when light weight and strength are needed. Magnesium is found in ladders, cameras, cars, and airplanes.

Table 4   Ions in the Ocean

Ions	Concentration in the ocean (mol/L)
Cl^-	0.554
Na^+	0.470
Mg^{2+}	0.047
SO_4^{2-}	0.015
K^+	0.010
Ca^{2+}	0.009

Refer to Appendix A for more information on ocean water.

② Section Review

UNDERSTANDING KEY IDEAS

1. Giving an example, explain how to write an expression for K_{eq} from a chemical equation.

2. Which species are left out from the K_{eq} expression, and why?

3. To which chemical systems can a K_{sp} be assigned?

4. When does K_{sp} not apply?

PRACTICE PROBLEMS

5. For the reaction in which hydrogen iodide is made at 425°C in the gas phase from its elements, calculate [HI], given that $[H_2] = [I_2] = 4.79 \times 10^{-4}$ and $K_{eq} = 54.3$.

6. Given that the K_{sp} value of CuS is 1.3×10^{-36}, what is $[Cu^{2+}]$ in a saturated solution?

7. Write the equation for the reaction in which solid carbon reacts with gaseous carbon dioxide to form gaseous carbon monoxide. At equilibrium, a 2.0 L reaction vessel is found to contain 0.40 mol of C, 0.20 mol of CO_2, and 0.10 mol of CO. Find the K_{eq}.

CRITICAL THINKING

8. A reaction has a single product and a single reactant, and both the product and the reactant are gases. If the concentration of each one was 0.1 M at equilibrium, what would the equilibrium constant be?

9. Write the solubility product expression for the slightly soluble compound aluminum hydroxide, $Al(OH)_3$.

Answers to Section Review

1. The equilibrium constant is equal to a ratio. The numerator is the product concentration(s), each raised to the power of its coefficient in the equation. The reactant concentrations make up the denominator, similarly exponentiated. For example, for the reaction $N_2(g) + 3H_2(g) \rightleftharpoons 2NH_3(g)$, the equilibrium constant expression is

$$K_{eq} = \frac{[NH_3]^2}{[N_2][H_2]^3}$$

where a formula in brackets implies the concentration of that species ions in moles per liter.

Answers continued on p. 527A

Group Activity — BASIC

Divide students into groups of four. Provide two of the four students with a list of chemical equations for reversible chemical reactions. Have them write the equilibrium expressions for half the reactions. Then have the other two students write the equations from the equilibrium expressions. Have students reverse roles for the other half of the equations. **LS Interpersonal**

Close

Reteaching — BASIC

Have students write the chemical equation for the equilibrium that has the following K_{eq}.

$$K_{eq} = \frac{[C]^2[D]^3}{[A][B]^4}$$

$(A + 4B \rightleftharpoons 2C + 3D)$
LS Logical

Quiz — GENERAL

1. What is an equilibrium constant? **Ans.** for a reversible reaction in equilibrium, an expression that relates the concentrations of the reactants and products at a specific temperature to a constant

2. Write the equilibrium expression for the following equilibrium equation.
$N_2(g) + 3H_2(g) \rightleftharpoons 2NH_3(g)$
Ans. $K_{eq} = \dfrac{[NH_3]^2}{[N_2][H_2]^3}$

3. For a certain reaction, $A \rightleftharpoons B$, $K_{eq} = 1.2$. Does the reaction probably favor formation of reactants or formation of products? **Ans.** It probably favors formation of products because $K_{eq} > 1$.

4. Which of the following compounds is more soluble, compound A or compound B? For compound A, $K_{sp} = 1.2 \times 10^{-10}$. For compound B, $K_{sp} = 3.4 \times 10^{-25}$. **Ans.** Compound A is probably more soluble because it has a larger K_{sp}.

5. What happens to K_{eq} for a particular reaction when the reaction's temperature changes? **Ans.** K_{eq} changes also.
LS Logical

511

Overview

Before beginning this section, review with your students the Objectives listed in the Student Edition. This section defines and applies Le Châtelier's principle. Students learn that if a stress is placed on an equilibrium, the equilibrium will shift to relieve the stress. One specific stress, the increase in concentration of a specific ion in solution, known as the common ion effect, is investigated in detail.

Bellringer

Have students list examples of everyday adjustments that are made to relieve stress on a system. Answers might include putting on a coat when it's cold outside or breathing more deeply or quickly when exercising and your muscles need more oxygen.

Motivate

Demonstration

In a closed soft-drink bottle, the gases above the liquid are in equilibrium with the gases in solution. Show students an unopened bottle of soft drink, then open it. Ask students to explain what effect the change in pressure had on the equilibrium. **Ans.** As pressure decreased, the equilibrium shifted and more gas came out of solution.

Chapter Resource File

• Lesson Plan

Equilibrium Systems and Stress

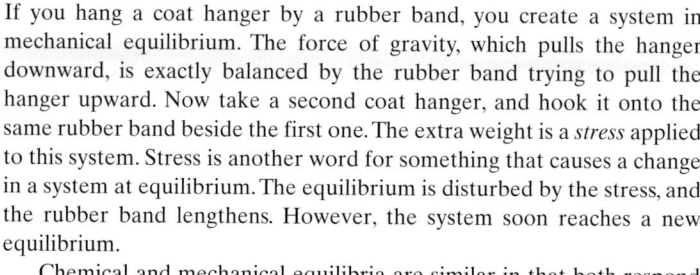

KEY TERMS
• Le Châtelier's principle
• common-ion effect

OBJECTIVES

(1) **State** Le Châtelier's principle.

(2) **Apply** Le Châtelier's principle to determine whether the forward or reverse reaction is favored when a stress such as concentration, temperature, or pressure is applied to an equilibrium system.

(3) **Discuss** the common-ion effect in the context of Le Châtelier's principle.

(4) **Discuss** the practical uses of Le Châtelier's principle.

Le Châtelier's Principle

Figure 9
Henri Louis Le Châtelier (1850–1936) was particularly interested in applying science to industry and in getting maximum yield from chemical reactions.

Le Châtelier's principle

the principle that states that a system in equilibrium will oppose a change in a way that helps eliminate the change

If you hang a coat hanger by a rubber band, you create a system in mechanical equilibrium. The force of gravity, which pulls the hanger downward, is exactly balanced by the rubber band trying to pull the hanger upward. Now take a second coat hanger, and hook it onto the same rubber band beside the first one. The extra weight is a *stress* applied to this system. Stress is another word for something that causes a change in a system at equilibrium. The equilibrium is disturbed by the stress, and the rubber band lengthens. However, the system soon reaches a new equilibrium.

Chemical and mechanical equilibria are similar in that both respond to stresses by adjusting until new equilibria are reached. Henri Le Châtelier, the chemist pictured in **Figure 9,** studied the way in which chemical equilibria respond to changes. His findings are known as **Le Châtelier's principle.** This principle states that *when a system at equilibrium is disturbed, the system adjusts in a way to reduce the change.* Some situations in our lives are analogies for Le Châtelier's principle. For example, if you are disturbed by a loud noise, you may move to a quieter location.

Chemical equilibria respond to three kinds of stress: changes in the concentrations of reactants or products, changes in temperature, and changes in pressure. When a stress is first applied to a system, any chemical equilibrium is disturbed. As a result, the rates of the forward and backward reactions in the system are no longer equal. The system responds to the stress by forming more products or by forming more reactants. A new chemical equilibrium is reached when enough reactants or products form. At this point, the rates of the forward and backward reactions are equal again.

512

Teaching Tip

Pressure and Concentration Point out to students how for gases, changes in pressure are actually changes in concentration. For increased pressure, there are more gas particles per volume, and there are fewer gas particles per volume for decreased pressure.

Changes in Concentration Alter Equilibrium Composition

What happens if you add more reactant to a system in chemical equilibrium? This increase in the reactant's concentration is a stress on the system. The system will respond to decrease the concentration of the reactant by changing some of the reactant into product. Therefore, the rate of the forward reaction must be greater than the rate of the reverse reaction. Because the forward reaction is increasing, the equilibrium is said to *shift right*. The reactant concentration will continue to drop until the reaction reaches equilibrium. Then, the forward and reverse reaction rates will be equal. Remember that changes in the amounts of solids and pure liquids do not affect K_{eq} values.

A reaction of two colored complex ions is described by the equation

$$[Cu(H_2O)_4]^{2+}(aq) + 4NH_3(aq) \rightleftharpoons [Cu(NH_3)_4]^{2+}(aq) + 4H_2O(l)$$
pale blue · · · · · · · · · · · · · · · · · · · blue-purple

The beaker on the left in **Figure 10** contains this copper complex ion reaction in a chemical equilibrium that favors the formation of reactants. We know that the reverse reaction is favored, because the reaction mixture in the beaker is pale blue. But if additional ammonia is added to this beaker, the system responds to offset the increase by forming more of the product. This increase in the presence of product can be seen in the beaker to the right in **Figure 10,** which contains a blue-purple solution.

The equilibrium below occurs in a closed bottle of a carbonated liquid.

$$H_3O^+(aq) + HCO_3^-(aq) \rightleftharpoons 2H_2O(l) + CO_2(aq)$$

After you uncap the bottle, the dissolved carbon dioxide leaves the solution and enters the air. The forward reaction rate of this system will increase to produce more CO_2. This increase in the rate of the forward reaction decreases the concentration of H_3O^+ (H_3O^+ ions make soda taste sharp). As a result, the drink gets "flat." What would happen if you could increase the concentration of CO_2 in the bottle? The reverse reaction rate would increase, and $[H_3O^+]$ and $[HCO_3^-]$ would increase.

www.scilinks.org
Topic: Factors Affecting Equilibrium
SciLinks code: HW4057

SCILINKS Maintained by the National Science Teachers Association

Figure 10
The color of the solution makes it easy to see which complex ion is dominant. Adding extra ammonia shifts the equilibrium in favor of the ammoniated complex ion.

$NH_3(aq)$

$[Cu(H_2O)_4]^{2+}(aq)$

$H_2O(l)$

$[Cu(NH_3)_4]^{2+}(aq)$

$$[Cu(H_2O)_4]^{2+}(aq) + 4NH_3(aq) \rightleftharpoons [Cu(NH_3)_4]^{2+}(aq) + 4H_2O(l)$$

513

MISCONCEPTION ALERT

Le Châtelier's Principle Students sometimes mistakenly interpret Le Châtelier's principle to mean that an equilibrium system somehow "knows" what to do when a stress is placed upon it. Emphasize that this principle, like all scientific laws, simply describes a phenomenon that occurs unfailingly in nature. Nothing in nature happens because of a law.

Teach

READING SKILL BUILDER — BASIC

Discussion

Summarizing As students read through each part of Section 3, have them write two or three sentences that summarize the topic. Have them trade sentences with other students and discuss any differences. **LS** Interpersonal

Teaching Tip

Reaction Spotlight Explain the demonstration on the previous page by telling students that the dissolving of carbon dioxide in water demonstrates Le Châtelier's principle. An equilibrium exists between gaseous and dissolved CO_2.

$$CO_2(g) \rightleftharpoons CO_2(aq)$$

A chemical equilibrium exists between dissolved CO_2 and H_2CO_3.

$$CO_2(aq) + H_2O(l) \rightleftharpoons H_2CO_3(aq)$$

Students should recall the relationship between pressure and gas solubility. When CO_2 is added to water under higher pressure, the concentration of $CO_2(aq)$ increases. When a bottle of soda is opened, the CO_2 pressure drops, and CO_2 bubbles out of solution. Write the equilibrium equations on the chalkboard, and ask students what happens when carbonated water is made.

SKILL BUILDER — ADVANCED

Writing Skills Have students find reactions and reaction conditions that would exhibit the stresses. Also ask students to describe the stresses and what reactions would do to adjust to the stresses. **LS** Intrapersonal

Using the Figure

Refer students to **Figure 11.**
Encourage them to think of when
energy is released and absorbed
during this reversible reaction.
Then ask your students what the
effect temperature has on this reaction and the reasons for this effect.
Have students also consider an
additional reaction as shown by
the following equation during this
discussion. $2H_2(g) + O_2(g) \longrightarrow 2H_2O(g) + 483.6$ kJ

READING SKILL BUILDER — BASIC

Reading Organizer Have students
label two columns on a piece of
paper "Cause" and "Effect." Have
students list causes of shifts in equilibrium in the "Cause" column;
as they read through this section.
Then have them list the corresponding effect in the "Effect" column.
Be sure causes are specific enough
that students can predict their
effects. For example, "temperature
change" is not specific enough to
predict what the effect will be. The
cause should indicate an increase or
a decrease in temperature. **LS** Visual

Teaching Tip

The Haber Process The formation of ammonia from nitrogen
and hydrogen is an exothermic
reaction that reaches equilibrium.
The Haber process uses high pressure and low temperature to favor
the formation of the product. Most
of the ammonia used in making
fertilizers and other products is
produced by this method.

Temperature Affects Equilibrium Systems

The effect of temperature on the gas-phase equilibrium of nitrogen
dioxide, NO_2, and dinitrogen tetroxide, N_2O_4, can be seen because of
the difference in color of NO_2 and N_2O_4. The intense brown NO_2 gas is
the pollution that is responsible for the colored haze that you sometimes
see on smoggy days.

$$2NO_2(g) \rightleftharpoons N_2O_4(g)$$
$$\text{brown} \qquad \text{colorless}$$

To understand how the nitrogen dioxide–dinitrogen tetroxide equilibrium
is affected by temperature, we need to review endothermic and exothermic reactions. Recall that endothermic reactions absorb energy and have
positive ΔH values. Exothermic reactions release energy and have negative
ΔH values. The forward reaction is an exothermic process, as the equation
below shows.

$$2NO_2(g) \longrightarrow N_2O_4(g) \qquad\qquad \Delta H = -55.3 \text{ kJ}$$

Consider that you are given the flask shown in the middle in **Figure 11.**
The flask contains an equilibrium mixture of NO_2 and N_2O_4 and has a
temperature of 25°C. Now suppose that you heat the flask to 100°C.
The heated flask will look like the flask at the far right. The mixture
becomes dark brown because the reverse reaction rate increased to
remove some of the energy that you added to the system. The equilibrium
shifts to the left, toward the formation of NO_2. Because this reaction is
endothermic, the temperature of the flask drops as energy is absorbed.
This equilibrium shift is true for all exothermic forward reactions:
*Increasing the temperature of an equilibrium mixture usually leads to a
shift in favor of the reactants.*

The opposite statement is true for endothermic forward reactions:
*Increasing the temperature of an equilibrium mixture usually leads to a
shift in favor of the products.*

Figure 11
The NO_2-N_2O_4 equilibrium
system is shown at three
different temperatures.
Temperature changes put
stress on equilibrium
systems and cause either
the forward or reverse
reaction to be favored.

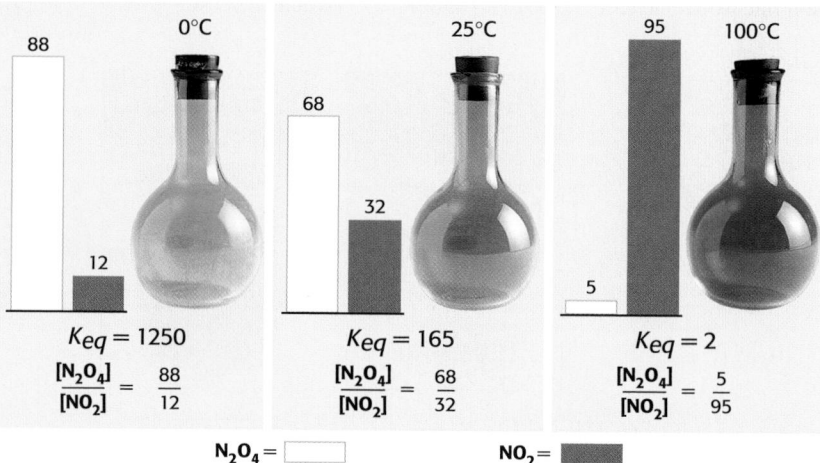

HISTORY CONNECTION

Gibbs and Le Châtelier When French
chemist Henri Louis Le Châtelier (1850–1936)
proposed his principle in 1888, he was trans-
lating a work by American physicist Josiah
Willard Gibbs (1839–1903). As it turned
out, Gibbs's theories of thermodynamics explained Le Châtelier's principle quite well.

Transparencies

TT NO_2-N_2O_4 Equilibrium System

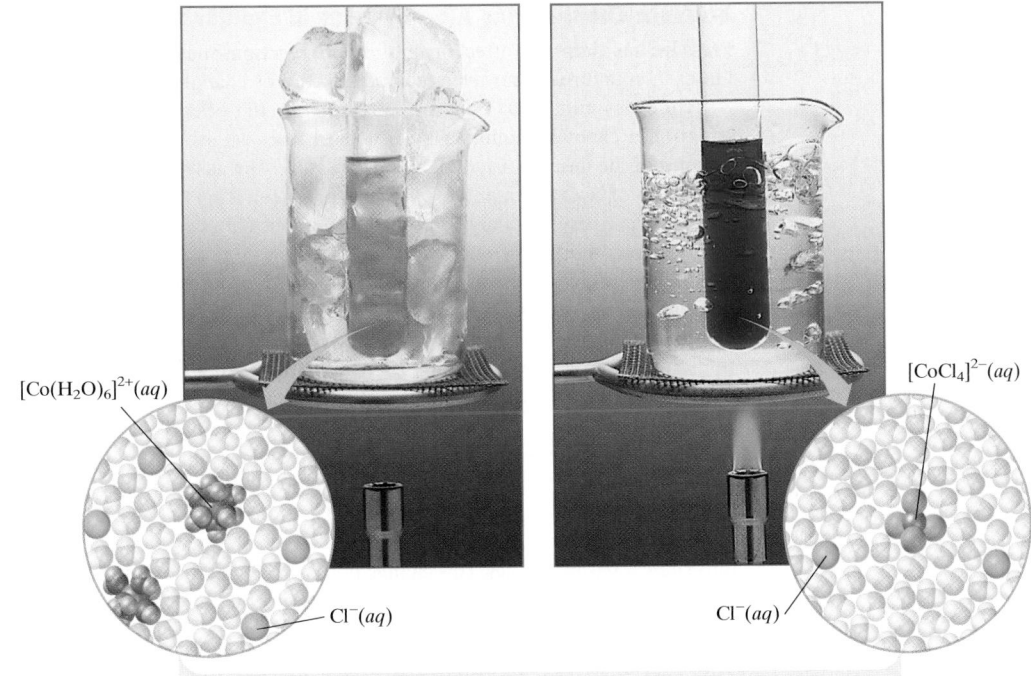

$[Co(H_2O)_6]^{2+}(aq)$

$Cl^-(aq)$

$[CoCl_4]^{2-}(aq)$

$Cl^-(aq)$

$$[Co(H_2O)_6]^{2+}(aq) + 4Cl^-(aq) \rightleftharpoons [CoCl_4]^{2-}(aq) + 6H_2O(l)$$

An example of the effect of temperature on an endothermic reaction is illustrated in **Figure 12**. The following equation describes an equilibrium that involves the two colored cobalt complex ions.

$$[Co(H_2O)_6]^{2+}(aq) + 4Cl^-(aq) \rightleftharpoons [CoCl_4]^{2-}(aq) + 6H_2O(l)$$
pink   blue

This particular equilibrium has an endothermic forward reaction. Therefore, this forward reaction is favored as the temperature of the reaction rises. The reverse reaction is favored at lower temperatures. The experiment illustrated in **Figure 12** confirms these predictions. The solution is pink at 0°C, which shows that the reactants are favored at low temperatures. At 100°C, the solution is blue, which shows that the products are favored at high temperatures.

Temperature changes affect not only systems at equilibrium but also the value of equilibrium constants. In fact, equilibrium constants changing with temperature is the reason that equilibria change with temperature. For example, consider K_{eq} for the ammonia synthesis equilibrium.

$$N_2(g) + 3H_2(g) \rightleftharpoons 2NH_3(g)$$

The forward reaction is exothermic ($\Delta H = -91.8$ kJ), so the equilibrium constant decreases a lot as temperature increases.

Figure 12
Depending on whether energy is removed or added, the reverse or forward reaction will be favored and a pink solution or a blue solution, respectively, will form.

Topic Link

Refer to the "Causes of Change" chapter for a discussion of enthalpy.

515

REAL-WORLD CONNECTION

Tooth Enamel Tooth enamel is very slightly soluble in water. Tooth enamel also reacts with acids present in saliva, which depletes its presence on teeth. This reaction would eventually destroy the enamel, and teeth would decay. However, the addition of fluorides to water supplies and tooth products has eliminated some of this problem. The dissolved tooth enamel reacts with the fluoride ions, forming a compound similar to tooth enamel. Because of its low K_{sp}, this compound precipitates on teeth, replacing dissolved tooth enamel with a more durable compound.

Demonstration
Temperature Effects

This demonstration illustrates how temperature affects an equilibrium system.

Set up the demonstration before class in time for the flasks to reach the desired temperatures. To set up, inflate and stretch three identical balloons several times. Check to make sure the balloons will fit over the mouth of a 250-mL Erlenmeyer flask. Place 10 g of sodium carbonate, Na_2CO_3, in each uninflated balloon. Place 50 mL of 1 M HCl and a thermometer in each flask. Place one flask in the refrigerator, the second such that it is at room temperature, and the third in a warm place at about 50°C.

When performing the demonstration, twist each balloon so that no sodium carbonate can fall into a flask. Working rapidly to minimize temperature change, place the balloons over the mouths of the three flasks. Have three student volunteers release the sodium carbonate into the flasks at the same time. The balloons should inflate at different rates, based on the temperature.

Safety Caution: Safety goggles, gloves, and lab aprons must be worn by you and your volunteers. Have other students wear safety goggles. Students must be 3 m or more from the demonstration. In case of an acid spill, dilute the spill with water and, while wearing gloves, soak up the spill with cloth or paper towels. Rinse the towels, neutralize the rinses with 1 M NaOH, and flush down the drain.

Disposal: If the resulting solution is alkaline, pour it down the drain. If the solution is acidic, neutralize it with 1 M NaOH and then pour it down the drain.

ChemFile   CHEMISTRY

• Module 7: Equilibrium
  Topic: Shifting Equilibrium

Using the Figure

Refer students to **Figure 13.** Ask students whether the reaction stops when the color becomes constant. Student responses should include that when no further color change occurs, the reactions don't stop, but both the forward and reverse reactions occur at the same rate.

Homework ——— GENERAL

Concept Map Have students draw a concept map that shows how temperature, pressure, and concentration changes affect equilibria. LS **Visual**

Group Activity ——— GENERAL

Divide the class into groups and each group into two teams. Write the following equation on the chalkboard.

$A(g) + 3X(g) \rightleftharpoons Z(g) + \text{energy}$

Have one team ask the other team how a change in temperature, pressure, or concentration affects the reaction, and have the other team indicate whether the reaction will shift to the right or to the left. Have the teams alternate giving conditions and indicating the shift of the reaction. An example might be increasing the concentration of X. The result is a shift to the right to relieve the stress. Usually, changes in temperature and concentration are easy for students to predict the result. Be sure they understand that increasing pressure favors the direction that produces fewer gas molecules because fewer molecules produce less pressure. LS **Interpersonal**

Pressure Changes May Alter Systems in Equilibrium

Pressure has almost no effect on equilibrium reactions that are in solution. Gases in equilibrium, however, may be affected by changes in pressure.

The NO_2 and N_2O_4 equilibrium can show the effect of a pressure stress on a chemical equilibrium. In **Figure 13a,** you see an equilibrium mixture of the gases in a syringe at low pressure. The gas has a larger concentration of N_2O_4 than of NO_2, which can be seen from the pale color of the gas. The gas is suddenly compressed to about half its former volume, which, as a consequence of Boyle's law, doubles the pressure. Before the system has time to adjust to the pressure stress, the concentration of each gas doubles. This effect is seen in the darker color of the gas in **Figure 13b.** Le Châtelier's principle predicts that the system will adjust in an attempt to reduce the pressure. According to the equation, 2 mol of NO_2 produce 1 mol of N_2O_4. The gas laws tell us that at constant volume and temperature, pressure is proportional to the number of moles. So the pressure reduces when there are fewer moles of gas. Thus, the equilibrium shifts to the right, and more N_2O_4 is produced. In **Figure 13c,** the color of the gas shows that the NO_2 concentration has fallen almost to its original level.

In an equilibrium, a pressure increase favors the reaction that produces fewer gas molecules. This change does not always favor the forward direction. The reverse reaction is favored for the equilibrium

$$2NOCl(g) \rightleftharpoons 2NO(g) + Cl_2(g)$$

There is no change in the number of molecules in the following reaction:

$$H_2O(g) + CO(g) \rightleftharpoons H_2(g) + CO_2(g)$$

In such cases, a change in pressure will not affect equilibrium.

Topic Link

Refer to the "Gases" chapter for a discussion of gas laws.

Figure 13
A sudden decrease in the volume of an NO_2-N_2O_4 mixture increases the pressure and makes the mixture darker at first. The color fades as more N_2O_4 forms to reduce the pressure and reach a new equilibrium.

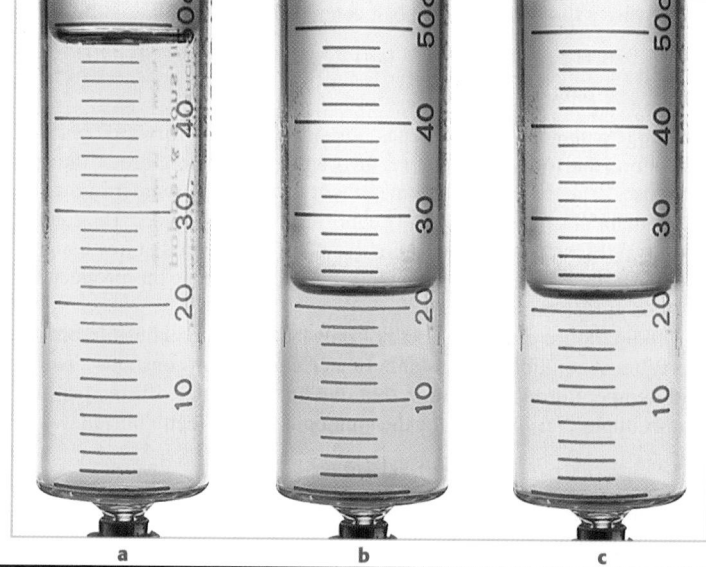

a   b   c

516

HISTORY
CONNECTION

During World War I, explosives made from nitrates were used extensively. The allied forces blockaded the supplies of these nitrates from Chile to Germany so that Germany could not produce explosives. However, in 1913, Carl Bosch, a German chemist, refined a procedure by Fritz Haber in which ammonia could be produced from hydrogen and nitrogen gases. This ammonia could be used to make nitric acid, which contains the essential nitrate ion. Bosch more closely controlled the temperature and pressure used by Haber, and nitrates were produced in large quantities.

The Common-Ion Effect

In pure water, the solubility of CuCl is 1.1×10^{-3} mol/L. Remember that you calculated the solubility of CuCl in ocean water to be 2.2×10^{-6} mol/L in **Sample Problem D**. So, CuCl is 500 times less soluble in sea water than it is in pure water. This dramatic reduction in solubility demonstrates Le Châtelier's principle and the dissolution of CuCl is shown below.

$$CuCl(s) \rightleftharpoons Cu^+(aq) + Cl^-(aq) \qquad [Cu^+][Cl^-] = K_{sp} = 1.2 \times 10^{-6}$$

If you have a saturated solution of copper(I) chloride in water and then add chloride-rich ocean water, you will increase $[Cl^-]$. However, the K_{sp}, which is the mathematical product of $[Cl^-]$ and $[Cu^+]$, must remain constant. Hence $[Cu^+]$ must decrease. This decrease can occur only by the precipitation of the CuCl salt. The ion Cl^- is the *common-ion* in this case because the Cl^- ion comes from two sources. The reduction of the solubility of a salt in the solution due to the addition of a common ion is called the **common-ion effect.**

Doctors use barium sulfate solutions to diagnose problems in the digestive tract. A patient swallows it, $BaSO_4$, so that the target organ becomes "visible" in X-ray film, as shown in **Figure 14.** $BaSO_4$ is a not a very soluble salt and is not absorbed by the body. But $BaSO_4$ is in equilibrium with a small concentration of $Ba^{2+}(aq)$, a poison. To reduce the Ba^{2+} concentration to a safe level, a common ion is added by mixing sodium sulfate, Na_2SO_4, into the $BaSO_4$ solution that a patient must swallow.

Another example of the common-ion effect emerges when solutions of potassium sulfate and calcium sulfate are mixed and either solution is saturated. Immediately after the two solutions are mixed, the product of the $[Ca^{2+}]$ and $[SO_4^{2-}]$ is greater than the K_{sp} of calcium sulfate. So, $CaSO_4$ precipitates to establish the equilibrium shown below.

$$CaSO_4(s) \rightleftharpoons Ca^{2+}(aq) + SO_4^{2-}(aq)$$

common-ion effect

the phenomenon in which the addition of an ion common to two solutes brings about precipitation or reduces ionization

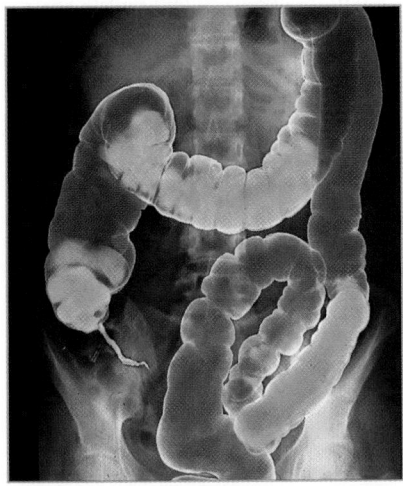

Figure 14
The light areas on this X ray of the digestive tract show insoluble barium sulfate.

internet connect
www.scilinks.org
Topic: Common-Ion Effect
SciLinks code: HW4034
SC LINKS Maintained by the National Science Teachers Association

REAL-WORLD CONNECTION

Making More Products In industry, marketable products are produced. It is obvious that any product-producing chemical reaction that reaches equilibrium must be shifted to favor production of the desired product. This shift can be done in several ways. Reactants are constantly added, and temperature and pressure conditions that favor formation of product are used. Commonly, the product will be removed as it is formed, shifting the reaction in the direction that produces more product.

Demonstration
Common Ion Effect

This demonstration shows how an excess of ammonium ions, NH_4^+, reduces the concentration of hydroxide ions, OH^- for the following reaction.

$$NH_3(aq) + H_2O(l) \rightleftharpoons NH_4^+(aq) + OH^-(aq)$$

1. Place half a Petri dish on an overhead projector and fill it half-way with 1 M $NH_3(aq)$ (sometimes labeled as NH_4OH). Add 2 drops of phenolphthalein. The solution should be magenta. If it is too pale, add more indicator. Write the equilibrium equation for the ionization of NH_3 on the chalkboard. Point out that the presence of OH^- ions causes the indicator to be magenta. Ask students to predict what will happen when the NH_4^+ ion concentration is increased.

2. Stir a small amount of solid NH_4Cl into the solution. The magenta color should disappear. The added NH_4^+ ions suppress the ionization of the aqueous ammonia.

Safety Caution: Safety goggles, a lab apron, and gloves must be worn. NH_4Cl is toxic. Have students wear safety goggles. Students must be 3 m or more from the demonstration.

Disposal: Rinse all of the liquid from the Petri dish into a beaker, and add an excess of 1 M NaOH to convert the NH_4Cl into $NH_3(aq)$. In an operating fume hood, warm the solution to expel the NH_3. Neutralize the remaining liquid with 1 M HCl, and pour it down the drain.

518   **Chapter 14 · Chemical Equilibrium**

Close

Quiz ———————— GENERAL

1. Define Le Châtelier's principle. **Ans.** When a system at equilibrium is disturbed, the system adjusts in a way to reduce the change.

2. If the concentration of a reactant is decreased, what will be the resulting effect on the reaction at equilibrium? **Ans.** The equilibrium will shift to produce more reactants.

3. If the pressure on the following reaction is increased, what will be the resulting effect on the equilibrium? Explain. $2AB(g) + C_2(g) \rightleftarrows 2AC(g) + B_2(g)$ **Ans.** A change in pressure will have no effect because there are equal numbers of gas molecules on both sides of the equation.

4. What effect does an increase in temperature have on an equilibrium in which the forward reaction is exothermic? **Ans.** The equilibrium will shift it to the left.

5. You have a solution of $NaNO_3$. Which of the following could be added to produce the common ion effect—KNO_3, $AgNO_3$, or KNO_2? **Ans.** $AgNO_3$ **LS** Logical

Reteaching ———————— BASIC

Ask students to list the factors that can cause stress on an equilibrium and give an example of the effect of each. Accept answers that are not specific. For example, a student may say that a chemical equilibrium that involves an exothermic reaction is used to produce a product. Decreasing the temperature (removing energy) increases the amount of product produced. **LS** Logical

Figure 15
Today ammonia is produced on a large scale in industrial plants such as this one.

internet connect

www.scilinks.org
Topic: Haber Process
SciLinks code: HW4153

SCi LINKS Maintained by the National Science Teachers Association

Practical Uses of Le Châtelier's Principle

The chemical industry, such as the factory in **Figure 15,** makes use of Le Châtelier's principle in many ways. One way is in the synthesis of ammonia by the Haber Process. High pressure is used to drive the following equilibrium to the right.

$$N_2(g) + 3H_2(g) \rightleftarrows 2NH_3(g)$$

Notice that the forward reaction converts 4 mol of gas into 2 mol of another gas. Therefore, the forward reaction is favored during an increase in pressure. The manufacturing process employs pressures as high as 400 times atmospheric pressure.

The ammonia synthesis is an exothermic reaction. Therefore, the forward reaction is favored at low temperatures.

0°C	$K_{eq} = 6.5 \times 10^8$
250°C	$K_{eq} = 52$
500°C	$K_{eq} = 5.8 \times 10^{-2}$

The Haber Process is operated at temperatures of 500°C, which may seem odd because the K_{eq} is so small at that temperature. The reason that this process must be run at such a high temperature is that the reaction proceeds too slowly at lower temperatures.

③ Section Review

UNDERSTANDING KEY IDEAS

1. State Le Châtelier's principle.

2. To what stresses does a chemical equilibrium respond?

3. How does an equilibrium reaction respond to the addition of extra reactant? extra product?

4. What role does ΔH play in determining how temperature influences K_{eq}?

5. Which stress affects the value of an equilibrium constant?

6. Which equilibria are affected by pressure change?

7. Describe the common-ion effect, and give an example.

CRITICAL THINKING

8. If a reaction could exist that has $\Delta H = 0$, what effect would temperature have on this reaction at equilibrium?

9. For a purely gaseous equilibrium, does doubling the pressure have the same effect on equilibrium as doubling the concentration?

10. Do "spectator ions" affect K_{sp}?

11. Which of the two common ions, Ca^{2+} or F^-, is more effective in precipitating CaF_2?

12. In the Haber Process, how would removing NH_3 as it forms affect the equilibrium?

13. How does adding extra reactant affect K_{eq}?

518

Answers to Section Review

1. The principle states that when a system at equilibrium is disturbed, the system adjusts in a way to reduce the change.

2. changes in pressure, temperature, and the concentrations of reactants or products

3. The addition of extra reactant causes the equilibrium to shift to the right, forming more product. The addition of extra product causes the equilibrium to shift to the left, producing

more reactant. (Unless the reactant or product is a pure liquid or solid.)

4. If ΔH is negative, increasing temperature (adding energy) usually causes the equilibrium to shift to the left, producing more reactant. If ΔH is positive, increasing temperature usually causes the equilibrium to shift to the right, forming more product.

5. temperature changes

Answers continued on p. 527A

Cl
Chlorine
35.4527
$[Ne]3s^23p^5$

Element Spotlight

Where Is Cl?

Earth's crust
0.045% by mass

Sea water
1.9% by mass
30.61% of dissolved materials

Chlorine Gives Us Clean Drinking Water

The practice of chlorinating drinking water began in the early 20th century. Chlorinating water greatly reduced the number of people who are affected by infectious diseases. No other method of disinfecting drinking water is as inexpensive and reliable as chlorination. Today, chlorine is used as a disinfectant in almost every drinking-water treatment plant in the United States and Canada. This method will probably continue to be used well into the 21st century.

Industrial Uses

- Chlorine is a strong oxidizing agent and is used as a bleaching agent for paper.
- Chlorine gas is used to produce bromine compounds for use as a fire retardant.
- Table salt (sodium chloride) is one of the most important chlorine-containing compounds.
- Plastics, such as the vinyl used in sporting goods, contain chlorine.
- Many pharmaceuticals, including antibiotics and allergy medications, require chlorine-containing compounds for their manufacture.

Real-World Connection Many Third World regions have outbreaks of deadly diseases, such as cholera and typhoid fever, that could easily be prevented by disinfecting water supplies with chlorine.

Chlorine gas and chlorine compounds are used to purify swimming pools.

A Brief History

1801: W. Cruickshank recommends the use of Cl_2 as a disinfectant.

1902: J. C. Downs patents the Downs cell for the production of $Cl_2(g)$ and $Na(s)$.

1600	1700	1800	1900

1648: J. R. Glauber prepares concentrated hydrochloric acid.

1774: C. W. Scheele isolates chlorine gas, which was called *dephlogisticated marine acid air* at the time.

1810: H. Davy proves that chlorine is an element and names it.

1908: P. Sommerfeld shows that HCl is present in the gastric juices of animals.

Questions

1. Research environmentally hazardous materials that contain chlorine. Also research ways that people are trying to reduce these hazards.

2. Research how chlorine was discovered.

internet connect

www.scilinks.org
Topic: Chlorine
SciLinks code: HW4031

SCiLINKS. Maintained by the National Science Teachers Association

519

Element Spotlight

All halogens kill bacteria and other microorganisms. Chlorine is the only halogen that is safe enough and readily available for large-scale treatment of public water supplies.

When chlorine is added to water, the following reaction produces HCl and hypochlorous acid.

$Cl_2(g) + H_2O(l) \longrightarrow HCl(aq) + HOCl(aq)$

Hypochlorous acid is a weak acid that ionizes to give hydronium ions and hypochlorite ions, OCl^-.

$HOCl(aq) + H_2O(l) \longrightarrow H_3O^+(aq) + OCl^-(aq)$

The OCl^- ions are strong oxidizing agents that can destroy microorganisms.

Answers to Feature Questions

1. Students' answers may vary.
2. Students' answers may vary.

Chapter Resource File

- **Concept Review**
- **Quiz**

14 CHAPTER HIGHLIGHTS

Alternative Assessments

Group Activity
• Pages 499, 516

MATH CONNECTION
• Page 505

READING SKILL BUILDER
• Pages 513, 514

Portfolio Assessments

SKILL BUILDER
• Page 513

Homework
• Page 516

Chapter Resource File

• **Microscale Lab** Equilibrium
• **Observation Lab** Solubility Product Constant
• **Observation Lab** Solubility Product Constant—Algae Blooms

---

14 CHAPTER HIGHLIGHTS

KEY IDEAS

SECTION ONE Reversible Reactions and Equilibrium
• During completion reactions, products do not significantly re-form reactants.
• During reversible reactions, products re-form the original reactants.
• Reversible reactions can reach equilibrium.
• The forward and reverse reaction rates are equal at chemical equilibrium.
• At chemical equilibrium, reactant and product concentrations remain unchanged.
• Chemical equilibria are dynamic equilibria.
• Complex ions are metal ions or atoms that are bonded to more than one atom or molecule.
• Complex ion formations are often examples of equilibria.

SECTION TWO Systems at Equilibrium
• The constant K_{eq} is equal to an expression that has the concentrations of species in equilibrium.
• The equilibrium constant for the dissolution of a slightly soluble salt is the solubility product constant K_{sp}.
• Equilibrium constants and solubility product constants have no units.

SECTION THREE Equilibrium Systems and Stress
• Le Châtelier's principle states that chemical equilibria adjust to relieve applied stresses.
• Stresses due to changes in concentration, temperature, and pressure are subject to Le Châtelier's principle.
• Temperature changes affect the values of equilibrium constants.
• Pressure changes have almost no affect on equilibrium reactions in solution. Pressure changes can affect equilibrium reactions in the gas phase.
• The common-ion effect reduces the solubility of slightly soluble salts.

KEY TERMS

reversible reaction
chemical equilibrium

equilibrium constant, K_{eq}
solubility product constant, K_{sp}

Le Châtelier's principle
common-ion effect

KEY SKILLS

Determining K_{eq} for Reactions at Chemical Equilibrium
Skills Toolkit 1, p. 503
Sample Problem A, p. 504

Calculating Concentrations from K_{eq}
Sample Problem B, p. 506

Determining K_{sp} for Reactions at Chemical Equilibrium
Skills Toolkit 2, p. 508
Sample Problem C, p. 509

Calculating Concentrations from K_{sp}
Sample Problem D, p. 510

520

520   Chapter 14 • Chemical Equilibrium

CHAPTER REVIEW 14

USING KEY TERMS

1. Explain the meaning of the terms *reversible reaction*, *completion reaction*, and *reaction at equilibrium*.

2. What is the distinction between a static equilibrium and a dynamic equilibrium? In which class do chemical equilibria fit?

3. Define *complex ion* and give two examples.

4. What are ligands?

5. What is an equilibrium constant?

6. Distinguish between solubility and solubility product constant. Explain how one may be calculated from the other.

7. In the context of Le Châtelier's principle, what is a "stress"? List the stresses that affect chemical equilibria.

8. Give one example of a stress on a reaction in aqueous solution and at equilibrium.

9. In the precipitation of cadmium iodate, $Cd(IO_3)_2$, which of the following salts would exert a common-ion effect: $NaIO_3$, KI, $Cd(NO_3)_2$, NH_4IO_4, $CdCl_2$, or $CsNO_3$?

UNDERSTANDING KEY IDEAS

Reversible Reactions and Equilibrium

10. Use **Figure 3** to answer the following questions:
    a. What is happening to the rate of formation of $HI(g)$ before the system reaches equilibrium?
    b. When is the rate of the forward reaction the greatest?

11. Give two examples of static equilibrium and two examples of dynamic equilibrium. Your examples do not have to be chemical examples.

12. Identify the ligands in the following complex ions.

    a. 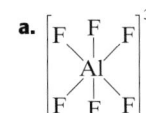   b.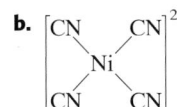

Systems at Equilibrium

13. Why must a balanced chemical equation be used when determining K_{eq}?

14. Describe and explain how the concentrations of A, B, C, and D change from the time when A and B first combine to the point at which equilibrium is established for the reaction

    $$A + B \rightleftarrows C + D.$$

15. In general, which reaction (forward, reverse, or neither) is favored if the value of K at a specified temperature is
    a. equal to 1?
    b. very small?
    c. very large?

16. When nitrogen monoxide reacts with oxygen to produce nitrogen dioxide, an equilibrium is established.
    a. Write the balanced equation.
    b. Write the equilibrium constant expression.

17. Write equilibrium constant expressions for the following reactions:
    a. $2NO_2(g) \rightleftarrows N_2O_4(g)$

521

Assignment Guide	
Section	Questions
1	1–4, 10–12, 58
2	5, 6, 13–19, 30–47, 48, 49, 52, 54, 55–57, 59, 60, 62, 63
3	7–9, 20–29, 50, 51, 53, 61

REVIEW ANSWERS

Using Key Terms

1. A reversible reaction can proceed in either direction. A completion reaction is one that proceeds to such an extent that the reactant (or one of the reactants if there are several) is undetectable when the reaction is complete. When a reversible reaction is occurring, but there is no longer any change in concentrations (the forward rate equals the reverse rate), the reaction is at equilibrium.

2. Static equilibrium is a state in which nothing changes. Dynamic equilibrium is a state in which there is no net change. However, opposite changes occur at the same rate in a dynamic equilibrium. Chemical equilibria are dynamic.

3. A complex ion is an ion that has a metal atom or ion that is bonded to more than one ligand. Examples will vary.

4. an atom, molecule, or ion that is bonded to a central atom of a complex ion

5. a number that relates the concentrations of reactants and products of a reversible chemical reaction to one another at a given temperature

6. The solubility of a compound M_xX_m is its concentration s in a saturated solution, equal to $[M^{m+}]/x$ or $[X^{x-}]/m$. The solubility product constant is the numerical value K_{sp} of the quantity $[M^{m+}]^x[X^{x-}]^m$. The relationship that permits one to be calculated from the other is $K_{sp} = (xs)^x(ms)^m$

7. A stress is a change in conditions; for example, changes in concentration, pressure, or temperature.

8. temperature or concentration

9. $NaIO_3$, $Cd(NO_3)_2$, $CdCl_2$

10. a. The rate of formation of HI is decreasing.
    b. The forward rate is greatest when the reaction begins.

11. Students' answers may vary.

12. a. fluoride ions, F^-

  b. cyanide ions, CN^-

13. A balanced equation must be used when determining K_{eq} because the coefficients of the balanced equation are used in the equilibrium constant expression.

14. At the start, A and B are at their maximum concentrations and there is no C or D. As A and B react, their concentrations decrease and those of C and D increase. The rate at which C and D combine increases while the rate at which A and B combine decreases. Eventually, the two rates become equal and equilibrium is established. The concentration of A, B, C, and D then remain constant as long as conditions remain the same.

15. a. neither

  b. reverse

  c. forward

16. a. $2NO(g) + O_2(g) \rightleftarrows 2NO_2(g)$

  b. $K_{eq} = \dfrac{[NO_2]^2}{[NO]^2[O_2]}$

17. a. $K_{eq} = \dfrac{[N_2O_4]}{[NO_2]^2}$

  b. $K_{eq} = \dfrac{[COCl_2]}{[CO][Cl_2]}$

  c. $K_{sp} = [Ag^+][Cl^-]$

  d. $K_{eq} = \dfrac{[H_3O^+][CH_3COO^-]}{[CH_3COOH]}$

18. a. $\dfrac{[NH_4^+][OH^-]}{[NH_3]} = 1.8 \times 10^{-5}$

  b. 5.6×10^4

19. $[Ag^+][I^-]$, $[Sr^{2+}][SO_4^{2-}]$, $[Ag^+]^2[CO_3^{2-}]$, $[Ag^+]^2[S^{2-}]$, $[Pb^{2+}][I^-]^2$, $[Ag^+][IO_3^-]$, $[Mg^{2+}]^3[PO_4^{3-}]^2$, $[Hg_2^{2+}][Cl^-]^2$

---

  b. $CO(g) + Cl_2(g) \rightleftarrows COCl_2(g)$

  c. $AgCl(s) \rightleftarrows Ag^+(aq) + Cl^-(aq)$

  d. $CH_3COOH(aq) + H_2O(l) \rightleftarrows$
  $H_3O^+(aq) + CH_3COO^-(aq)$

18. The equilibrium constant for the reaction below is 1.8×10^{-5}.

  $NH_3(aq) + H_2O(l) \rightleftarrows NH_4^+(aq) + OH^-(aq)$

  a. Write the equilibrium constant expression.

  b. What is the numerical value of equilibrium constant for the following reaction?

  $NH_4^+(aq) + OH^-(aq) \rightleftarrows NH_3(aq) + H_2O(l)$

19. Write the solubility product expressions for the following slightly soluble salts: AgI, $SrSO_4$, Ag_2CO_3, Ag_2S, PbI_2, $AgIO_3$, $Mg_3(PO_4)_2$, and Hg_2Cl_2.

Equilibrum Systems and Stress

20. How would you explain Le Châtelier's principle in your own words to someone who finds the concept difficult to understand?

21. Predict whether each of the following pressure changes would favor the forward reaction or the reverse reaction.

  $2NO(g) + O_2(g) \rightleftarrows 2NO_2(g)$

  a. increased pressure
  b. decreased pressure

22. Predict the effect of each of the following on the indicated equilibrium system in terms of which reaction (forward, reverse, or neither) will be favored.

  $H_2(g) + Cl_2(g) \rightleftarrows 2HCl(g) + 184 \text{ kJ}$

  a. addition of Cl_2
  b. removal of HCl
  c. increased pressure
  d. decreased temperature
  e. removal of H_2
  f. decreased pressure

---

  g. increased temperature
  h. decreased system volume

23. What relative pressure (high or low) would result in the production of the maximum level of CO_2 according to the following equation? Explain.

  $2CO(g) + O_2(g) \rightleftarrows 2CO_2(g)$

24. What relative conditions (reactant concentrations, pressure, and temperature) would favor a high equilibrium concentration of the substance in bold in each of the following equilibrium systems?

  a. $2CO(g) + O_2(g) \rightleftarrows \mathbf{2CO_2(g)} + 167 \text{ kJ}$

  b. $Cu^{2+}(aq) + 4NH_3(aq) \rightleftarrows$
  $\mathbf{[Cu(NH_3)_4]^{2+}(aq)} + 42 \text{ kJ}$

  c. $2HI(g) + 12.6 \text{ kJ} \rightleftarrows \mathbf{H_2(g)} + \mathbf{I_2(g)}$

  d. $4HCl(g) + O_2(g) \rightleftarrows$
  $2H_2O(g) + \mathbf{2Cl_2(g)} + 113 \text{ kJ}$

25. What changes in conditions would favor the reactants in the following equilibrium?

  $2SO_2(g) + O_2(g) \rightleftarrows 2SO_3(g) \qquad \Delta H = -198 \text{ kJ}$

26. Write the equilibrium constant expression for the reaction in problem 25.

27. What changes in conditions would favor the products in the following equilibrium?

  $PCl_5(g) \rightleftarrows Cl_2(g) + PCl_3(g) \qquad \Delta H = 88 \text{ kJ}$

28. Write the equilibrium constant expression for the reaction in problem 27.

29. Relate Le Châtelier's principle to the common-ion effect.

PRACTICE PROBLEMS

Sample Problem A Calculating K_{eq} from Concentrations of Reactants and Products

30. Vinegar—a solution of acetic acid, CH_3COOH, in water—is used in varying concentrations for different household tasks.

The following equilibrum exists in vinegar.

$$CH_3COOH(aq) + H_2O(l) \rightleftarrows$$
$$H_3O^+(aq) + CH_3COO^-(aq)$$

If the concentration of the acetic acid solution at equilibrium is 3.00 M and the $[H_3O^+] = [CH_3COO^-] = 7.22 \times 10^{-3}$, what is the K_{eq} value for acetic acid?

31. Write the K_{eq} expressions for the following reactions:

a. $4H_3O^+(aq) + 2Cl^-(aq) + MnO_2(s) \rightleftarrows$
$$Mn^{2+}(aq) + 6H_2O(l) + Cl_2(g)$$

b. $As_4O_6(s) + 6H_2O(l) \rightleftarrows 4H_3AsO_3(aq)$

32. The following equation shows an equilibrium reaction between hydrogen and carbon dioxide.

$$H_2(g) + CO_2(g) \rightleftarrows H_2O(g) + CO(g)$$

At 986°C, the following data were obtained at equilibrium in a 2.0 L reaction vessel.

Substance	Amount (mol)
H_2	0.4693
CO_2	0.0715
H_2O	0.2296
CO	0.2296

Calculate the K_{eq} for this reaction.

33. Determine the value of the equilibrium constant for each reaction below assuming that the equilibrium concentrations are as specified.
a. $A + B \rightleftarrows C$; $[A] = 2.0$; $[B] = 3.0$; $[C] = 4.0$
b. $D + 2E \rightleftarrows F + 3G$; $[D] = 1.5$; $[E] = 2.0$; $[F] = 1.8$; $[G] = 1.2$
c. $N_2(g) + 3H_2(g) \rightleftarrows 2NH_3(g)$; $[N_2] = 0.45$; $[H_2] = 0.14$; $[NH_3] = 0.62$

34. An equilibrium mixture at a specific temperature is found to consist of 1.2×10^{-3} mol/L HCl, 3.8×10^{-4} mol/L O_2, 5.8×10^{-2} mol/L H_2O, and 5.8×10^{-2} mol/L Cl_2 according to the following:

$$4HCl(g) + O_2(g) \rightleftarrows 2H_2O(g) + 2Cl_2(g).$$

Determine the value of the equilibrium constant for this system.

Sample Problem B Calculating Concentrations of Products from K_{eq} and Concentrations of Reactants

35. Analysis of an equilibrium mixture in which the following equilibrium exists gave $[OH^-] = [HCO_3^-] = 3.2 \times 10^{-3}$.

$$HCO_3^-(aq) + OH^-(aq) \rightleftarrows CO_3^{2-}(aq) + H_2O(l)$$

The equilibrium constant is 4.7×10^3. What is the concentration of the carbonate ion?

36. When a sample of NO_2 gas equilibrated in a closed container at 25°C, a concentration of 0.0187 mol/L of N_2O_4 was found to be present. Use the data in **Table 2** to calculate the NO_2 concentration.

37. Methanol, CH_3OH, can be prepared in the presence of a catalyst by the reaction of H_2 and CO at high temperatures according to the following equation:

$$CO(g) + 2H_2(g) \rightleftarrows CH_3OH(g)$$

What is the concentration of $CH_3OH(g)$ in moles per liter if the concentration of $H_2 = 0.080$ mol/L, the concentration of $CO = 0.025$ mol/L, and $K_{eq} = 290$ at 700 K?

38. At 450°C, the value of the equilibrium constant for the following system is 6.59×10^{-3}. If $[NH_3] = 1.23 \times 10^{-4}$ and $[H_2] = 2.75 \times 10^{-3}$ at equilibrium, determine the concentration of N_2 at that point.

$$N_2(g) + 3H_2(g) \rightleftarrows 2NH_3(g)$$

Sample Problem C Calculating K_{sp} from Solubility

39. The solubility of cobalt(II) sulfide, CoS, is 1.7×10^{-13} M. Calculate the solubility product constant for CoS.

40. What is the solubility product for copper(I) sulfide, Cu_2S, given that the solubility of Cu_2S is 8.5×10^{-17} M?

523

20. By using an analogy. For example, soda enters a straw because air pressure on the surface of the soda outside the straw is greater than the air pressure inside the straw. Soda goes up the straw by the air pressure outside the straw. Equilibrium is reached when the weight of the raised soda is equal to the force attempting to push the soda down.

21. a. forward
b. reverse

22. a. forward
b. forward
c. neither
d. forward
e. reverse
f. neither
g. reverse
h. neither

23. high pressure, because the forward reaction converts three molecules into two, relieving the stress imposed by the pressure increase

24. a. high reactant concentrations, high pressure, low temperature
b. high reactant concentrations, pressure not relevant, low temperature
c. high reactant concentrations, pressure not relevant, high temperature
d. high reactant concentrations, high pressure, low temperature

25. increasing the sulfur trioxide concentration, decreasing the concentration of either reactant, decreasing the pressure, or increasing the temperature

26. $K_{eq} = \dfrac{[SO_3]^2}{[SO_2]^2[O_2]}$

27. increasing the temperature or the phosphorus pentachloride concentration; decreasing the pressure or either of the product concentrations

28. $K_{eq} = \dfrac{[Cl_2][PCl_3]}{[PCl_5]}$

Chapter 14 • Chemical Equilibrium   523

29. When added to a solution, a common ion becomes more concentrated, placing a stress on the equilibrium that exists between a solid substance and its dissolved ions in a saturated solution. The equilibrium shifts, relieving the stress by removing some of the common ion by precipitation. Thus, the concentration of the other ion in the equilibrium is also reduced.

30. 1.74×10^{-5}

31. a. $K_{eq} = \dfrac{[Mn^{2+}][Cl_2]}{[H_3O^+]^4[Cl^-]^2}$

   b. $K_{eq} = [H_3AsO_4]^4$

32. 1.57

33. a. 0.67

   b. 0.52

   c. 311

34. 1.4×10^{10}

35. 0.048 mol/L

36. 0.0106 mol/L

37. 0.046 mol/L

38. 110 M

39. 2.9×10^{-26}

40. 2.5×10^{-48}

41. 8.22×10^{-96}

42. 7.14×10^{-11}

43. 1.7×10^{-14}

   1.3×10^{-7}

   2.7×10^{-7}

   1.6×10^{-4}

44. For these salts, the concentration s of a saturated solution (in moles per liter) is related to the solubility product constant by $K_{sp} = 4s^3$, so that cube roots (difficult on some calculators) are needed to find s.

45. 4.0×10^{-5} M

46. 7.3×10^{-7} M

---

41. The ionic substance T_3U_2 ionizes to form T^{2+} and U^{3-} ions. The solubility of T_3U_2 is 3.77×10^{-20} mol/L. What is the value of the solubility-product constant?

42. The ionic substance EJ dissociates to form E^{2+} and J^{2-} ions. The solubility of EJ is 8.45×10^{-6} mol/L. What is the value of the solubility-product constant?

Sample Problem D Calculating Ionic Concentrations Using K_{sp}

43. Four lead salts and their solubility products are as follows: PbS, 3.0×10^{-28}; PbCrO$_4$, 1.8×10^{-14}; PbCO$_3$, 7.4×10^{-14}; and PbSO$_4$, 2.5×10^{-8}. Calculate the Pb^{2+} concentration in a saturated solution of each of these salts.

44. Review problem 43, and note that all of the lead salts listed have anions whose charges are 2–. Why is the calculation more difficult for salts such as PbCl$_2$ or PbI$_2$?

45. What is the concentration of F$^-$ ions in a saturated solution that is 0.10 M in Ca^{2+}? The K_{sp} of CaF$_2$ is 1.6×10^{-10}.

46. Silver bromide, AgBr, is used to make photographic black-and-white film. Calculate the concentration of Ag$^+$ and Br$^-$ ions in a saturated solution at 25°C using **Table 3**.

47. The figure below shows the results of adding three different chemicals to distilled water and stirring well.

   a. Which substance(s) are completely soluble in water?

   b. Is it correct to say that AgCl is completely insoluble? Explain your answer. Is Ba(OH)$_2$ completely insoluble?

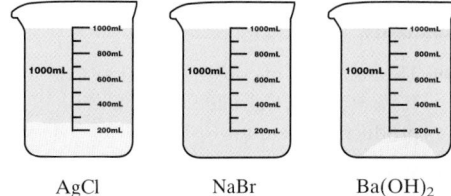

AgCl   NaBr   Ba(OH)$_2$

MIXED REVIEW

48. At 1400°C, the following reaction attained equilibrium in a 1.0 L container.

$$H_2(g) + CO_2(g) \rightleftharpoons H_2O(g) + CO(g)$$

An analysis of the equilibrium mixture found the following amounts: hydrogen, 0.060 mol; carbon dioxide, 0.060 mol; water vapor, 0.040 mol; and carbon monoxide, 0.040 mol. Calculate K_{eq} for the reaction.

49. Was it necessary to know the volume of the container in problem 48? Why or why not?

50. Why might an industrial process be operated at high temperature even though the reaction is more favorable at lower temperatures?

51. How does temperature affect equilibrium constants?

52. The reaction below has an equilibrium constant of 4.9×10^{11}.

$$Fe(OH)_2(s) + 2H_3O^+(aq) \rightleftharpoons Fe^{2+}(aq) + 4H_2O(l)$$

Write the equilibrium constant expression, and determine the concentration of Fe^{2+} ions in equilibrium when the hydronium ion concentration is 1.0×10^{-7} mol/L.

53. When does a pressure change affect a chemical equilibrium?

54. What information may be conveyed by the knowledge that the equilibrium constant of a reaction is very small?

55. The K_{sp} of potassium periodate, KIO$_4$, is 3.7×10^{-4}. Determine whether a precipitate will form when 2.00 g of KCl and 2.00 g of NaIO$_4$ are dissolved in 1.00 L of water.

56. What is the concentration of Cu$^+$ ions in a saturated solution of copper(I) chloride given that the K_{sp} of CuCl is 1.72×10^{-7} at 25°C?

57. Calculate the solubility of a substance MN that ionizes to form M^{2+} and N^{2-} ions given that $K_{sp} = 8.1 \times 10^{-6}$.

CRITICAL THINKING

58. A student wrote, "The larger the equilibrium constant, the greater the rate at which reactants convert to products." How was he wrong?

59. Write the balanced equation for the reaction to which the following equilibrium constant applies.

$$\frac{[CO]^2}{[CO_2]} = K_{eq}$$

60. Write the equilibrium constant expression for the reaction below.

$$CaCO_3(s) \rightleftarrows CaO(s) + CO_2(g)$$

Demonstrate that at a given temperature, the pressure of carbon dioxide in equilibrium with a mixture of calcium carbonate and calcium oxide cannot vary.

61. Imagine the following hypothetical reaction, taking place in a sealed, rigid container, to be neither exothermic nor endothermic.

$$A(g) + B(g) \rightleftarrows C(g) \qquad \Delta H = 0$$

Would an increase in temperature favor the forward reaction or the reverse reaction? (Hint: Recall the gas laws.)

62. A very dilute solution of silver nitrate is added dropwise to a solution that contains equal concentrations of sodium chloride and potassium bromide. What salt will precipitate first?

63. Changes in the concentrations of the reactants and products at equilibrium have no effect on the value of the equilibrium constant. Explain this statement.

ALTERNATIVE ASSESSMENT

64. Your instructor will give you an index card with a specific equilibrium reaction on it. Describe how you would alter the reaction to produce either more of the products or more of the reactants. Show your method to your instructor. If your method is approved, obtain the necessary materials from your instructor and perform the experiment.

65. Research the practical uses of Le Châtelier's principle. Present your results to your class.

66. Develop a model that shows the concept of equilibrium. Be sure that your model includes the impact of Le Châtelier's principle on equilibrium.

CONCEPT MAPPING

67. Use the following terms to complete the concept map below: *chemical equilibrium, equilibrium constant, solubility product constant, reversible reactions,* and *Le Châtelier's principle.*

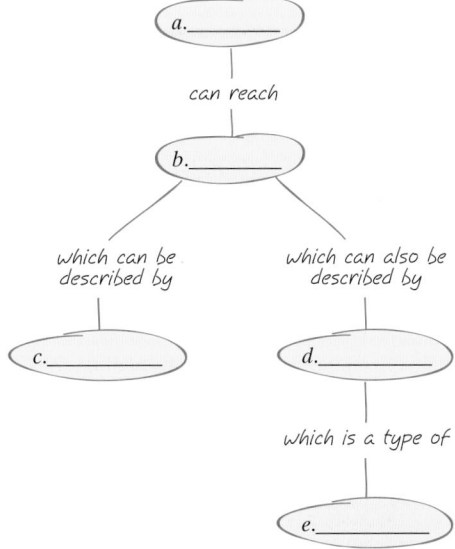

47. a. NaBr is completely soluble in water.
   b. No. A very small amount of AgCl dissolves as equilibrium is established. $Ba(OH)_2$ is likewise slightly soluble in water and establishes an equilibrium between dissolved ions and solid $Ba(OH)_2$.

48. 0.44

49. No, because the volume term cancels.

50. The reaction proceeds too slowly at lower temperatures.

51. increasing T usually increases K_{eq} for an endothermic reaction and decreases K_{eq} for an exothermic reaction

52. 4.9×10^{-3}

53. Pressure affects equilibria in the gas phase, but not when there are equal numbers of gas molecules on each side of the chemical equation.

54. that the reaction may proceed only to a small extent

55. almost no precipitate is formed

56. 4.15×10^{-4} mol/L

57. 2.8×10^{-3} mol/L

58. Equilibrium constants have no bearing on reaction rates, only on conditions when the reaction is at equilibrium.

59. probably $C(s) + CO_2(g) \rightleftarrows 2CO(g)$ (But not necessarily. It could be $SiC(s) + CO_2(g) \rightleftarrows Si(s) + 2CO(g)$ for example).

60. $K_{eq} = [CO_2]$. Because equilibrium constants do not vary at a constant temperature, the concentration of carbon dioxide must be constant. Therefore, its pressure must also be constant.

61. Because $\Delta H = 0$, the equilibrium constant is unaffected by temperature. However, the temperature increase will increase pressure, favoring the forward reaction.

62. Because AgBr has the smaller K_{sp}, its value will be reached first, so the initial precipitate will be AgBr.

63. Changes in the concentrations of products or reactants do not affect equilibrium constants because these changes cause systems to no longer be in equilibrium. Therefore, the equilibrium constants do not apply.

64. Methods chosen by students may vary. Many of the equilibrium reactions cited in text cannot be used because of the temperature and pressure conditions needed. In other reactions, it is difficult to demonstrate changes in equilibrium. Consider using the colored complex ions and extending the assessment into the Acids and Bases chapter using indicators, pH meters, and buffers to demonstrate changes in ionization equilibria.

65. Students' research may vary.

66. Students' answers may vary.

67. a. reversible reactions

b. chemical equilibrium

c. Le Châtelier's principle

d. solubility product constant

e. equilibrium constant

68. 0

69. According to the graph, the reverse reaction rate does equal 0 at $t = 0$. This rate does equal zero at the very beginning of the reaction because no products have formed.

70. According to the graph, the forward reaction rate never is 0. This rate is never 0 because reactants are always present in the system.

71. a. 54.000205

b. 54.01344

c. 53.99955

Study the graph below, and answer the questions that follow.
For help in interpreting graphs, see Appendix B, "Study Skills for Chemistry."

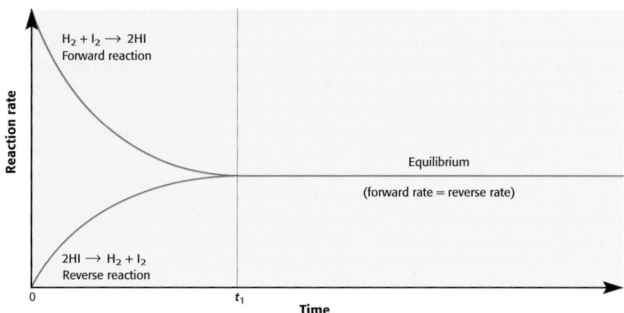

Rate Comparison for $H_2(g) + I_2(g) \rightleftharpoons 2HI(g)$

68. What is the change in the rate of the forward reaction after the reaction is at equilibrium?

69. Does the reverse reaction rate ever equal zero? Why or why not?

70. Does the forward reaction rate ever equal zero? Why or why not?

TECHNOLOGY AND LEARNING

71. Graphing Calculator

Calculating the Equilibrium Constant, K, for a System

The graphing calculator can run a program that calculates K for a system, given the concentrations of the products and the concentrations of the reactants.

Given the balanced chemical equation

$$H_2(g) + I_2(g) \longrightarrow 2HI(g)$$

and the equilibrium mixture at 425°C, you can calculate the equilibrium constant for the system. Then you can use the program to make calculations.

Go to Appendix C. If you are using a TI-83 Plus, you can download the program

CONSTANT and data and run the application as directed. If you are using another calculator, your teacher will provide you with keystrokes and data sets to use. After you have graphed the data, answer the questions below.

a. What is the equilibrium constant given the following equilibrium concentrations: 0.012840 mol/L of H_2, 0.006437 mol/L of I_2, and 0.066807 mol/L of HI?

b. What is the equilibrium constant given the following equilibrium concentrations: 0.000105 mol/L of H_2, 0.000107 mol/L of I_2, and 0.000779 mol/L of HI?

c. What is the equilibrium constant given the following equilibrium concentrations: 0.000527 mol/L of H_2, 0.000496 mol/L of I_2, and 0.003757 mol/L of HI?

526

Chapter Resource File

• **Chapter Test**

Answers
1. c
2. c
3. a
4. c
5. d
6. d
7. b
8. a

1. A chemical reaction is in equilibrium when
  a. forward and reverse reactions have ceased.
  b. the equilibrium constant equals 1.
  c. forward and reverse reaction rates are equal.
  d. no reactants remain.

2. A ligand is
  a. a component of wood.
  b. the metal atom in the center of a complex ion.
  c. a molecule or ion that bonds to the central metal ion in a complex ion.
  d. a colored transition metal ion.

3. Consider the following reaction:

$$2C(s) + O_2(g) \rightleftarrows 2CO(g)$$

The equilibrium constant expression for this reaction is

  a. $\dfrac{[CO]^2}{[O_2]}$.
  c. $\dfrac{2[CO]}{[O_2][2C]}$.

  b. $\dfrac{[CO]^2}{[O_2][C]^2}$.
  d. $\dfrac{[CO]}{[O_2]^2}$.

4. The solubility product of cadmium carbonate, $CdCO_3$, is 1.0×10^{-12}. In a saturated solution of this salt, the concentration of $Cd^{2+}(aq)$ ions is

  a. 5.0×10^{-13} mol/L.
  c. 1.0×10^{-6} mol/L.

  b. 1.0×10^{-12} mol/L.
  d. 5.0×10^{-7} mol/L.

5. Consider the following equation for an equilibrium system:

$$2PbS(s) + 3O_2(g) + C(s) \rightleftarrows$$
$$2Pb(l) + CO_2(g) + 2SO_2(g)$$

The concentration(s) that would be included in the denominator of the equilibrium constant expression are

  a. $Pb(l)$, $CO_2(g)$, and $SO_2(g)$.

  b. $PbS(s)$, $O_2(g)$, and $C(s)$.

  c. $O_2(g)$, $Pb(l)$, $CO_2(g)$, and $SO_2(g)$.

  d. $O_2(g)$.

6. Le Châtelier's principle states that
  a. at equilibrium, the forward and reverse reaction rates are equal.
  b. stresses include changes in concentrations, pressure, and temperature.
  c. to relieve stress, solids and solvents are omitted from equilibrium constant expressions.
  d. chemical equilibria respond to reduce applied stress.

7. If an exothermic reaction has reached equilibrium, then increasing the temperature will
  a. favor the forward reaction.
  b. favor the reverse reaction.
  c. favor both the forward and reverse reactions.
  d. have no effect on the equilibrium.

8. For the reaction _____, products are favored by increasing the pressure.

  a. $2NO(g) + O_2(g) \longrightarrow 2NO_2(g)$

  b. $CO_2(aq) + 2H_2O(l) \longrightarrow$
  $$H_3O^+(aq) + HCO_3^-(aq)$$

  c. $C(s) + O_2(g) \longrightarrow CO_2(g)$

  d. $2O_3(g) \longrightarrow 3O_2(g)$

527

Continuation of Answers

Answers to Section 2 Review, *continued*

2. solids, pure liquids, and solvents; because the concentrations of such species do not change their during a reaction

3. K_{sp} can be assigned to salts that are slightly soluble in water and to similar hydroxides.

4. K_{sp} does not apply to salts that are very soluble in water.

5. $K_{eq} = 54.3 = [HI]^2/([H_2][I_2]); [HI]^2 = (54.3)(4.79 \times 10^{-4})^2;$ $[HI] = 3.53 \times 10^{-3}$

6. $K_{sp} = 1.3 \times 10^{-36} = [Cu^{2+}][S^{2-}]; [Cu^{2+}] = 1.1 \times 10^{-18}$

7. $C(s) + CO_2(g) \rightleftarrows 2CO(g); K_{eq} = [CO]^2/[CO_2] =$ $(0.10/2.0)^2/(0.20/2.0) = 2.5 \times 10^{-2}$

8. 1

9. $K_{sp} = [Al^{3+}][OH^-]^3$

Answers to Section 3 Review, *continued*

6. Those equilibria in which the number of moles of gases differs between the products and reactants.

7. The solubility of a salt is decreased by the addition of a soluble compound that provides an ion that is also produced by the salt. For example, the solubility of CuCl is reduced by the presence of either Cu^+ or Cl^-.

8. None

9. Yes, immediately after the pressure is doubled, all concentrations are doubled.

10. Not significantly

11. Because its concentration is squared in the expression for K_{sp}, the F^- ion is the more effective common ion.

12. It causes more ammonia to form, improving the yield.

13. Not at all, K_{eq} is a constant at a fixed temperature.

Acids and Bases
Chapter Planning Guide

PACING	CLASSROOM RESOURCES	LABS, ACTIVITIES, AND DEMONSTRATIONS
BLOCK 1 • 45 min pp. 528–529 **Chapter Opener**		**SE Start-Up Activity** What Does an Antacid Do? p. 529 ◆
BLOCKS 2 & 3 • 90 min pp. 530–538 **Section 1** What Are Acids and Bases?		**TE Demonstration** Zinc in an HCl solution, p. 532 ◆ **TE Demonstration** Ionization in Aqueous Acids, p. 533 ◆ **SE QuickLab** Acids and Bases in the Home, p. 535 **GENERAL** **TE Group Activity** Identifying Conjugate Acid and Base Pairs, p. 537 **GENERAL**
BLOCKS 4 & 5 • 90 min pp. 539–547 **Section 2** Acidity, Basicity, and pH	**TT** Relationship of $[H_3O^+]$ to $[OH^-]$ * **TT** pH Values of Some Common Materials *	**TE Demonstration** pH of Various Materials, p. 542 ◆ **TE Activity** Working with Logarithms, p. 544 **BASIC** **TE Demonstration** Effects of pH on Chlorophyll Stability, p. 546 ◆ **ADVANCED**
BLOCKS 6 & 7 • 90 min pp. 548–556 **Section 3** Neutralization and Titrations	**TT** How to Perform a Titration, Part 1 * **TT** How to Perform a Titration, Part 2 *	**TE Activity** Understanding Neutralizations, p. 549 **GENERAL** **TE Activity** Calculating Molarity from Titrations, p. 550 **GENERAL** **TE Activity** Titration Curve of a Weak Acid with a Strong Base, p. 550 **ADVANCED** **TE Demonstration** Indicators, p. 554 ◆ **SE Skills Practice Lab** Drip-Drop Acid-Base Experiment, p. 801 ◆ **GENERAL** **SE Skills Practice Lab** Acid-Base Titration of an Eggshell, p. 804 ◆ **GENERAL** **SE Inquiry Lab** Acid-Base Titration of an Industrial Spill, p. 809 ◆ **ADVANCED** **CRF Datasheets for In-Text Labs** * **CRF Consumer Lab** How Effective is an Antacid? ◆ **GENERAL** **CRF Consumer Lab** Titration of an Antacid ◆ **GENERAL** **CRF CBL™ Probeware Lab** Acid-Base Titration ◆ **ADVANCED**
BLOCK 8 • 45 min pp. 557–564 **Section 4** Equilibria of Weak Acids and Bases	**OSP Supplemental Reading** Chemistry and Physics in the Kitchen **ADVANCED** **SE Consumer Focus** Antacids	**TE Activity** Ionization of Weak Acids, p. 558 **GENERAL** **TE Activity** Hydrolysis of Weak Bases, p. 559 ◆ **GENERAL** **TE Demonstration** Buffers in Solution, p. 561 **CRF CBL™ Probeware Lab** Buffer Capacity in Commercial Beverages ◆ **ADVANCED**

BLOCKS 9 & 10 • 90 min **Chapter Review and Assessment Resources**

- **SE Chapter Review,** pp. 566–572
- **SE Standardized Test Prep,** p. 573
- **CRF Chapter Test** *
- **OSP Test Generator**
- **CRF Test Item Listing** *
- **OSP Scoring Rubrics and Classroom Management Checklists**

Holt Chemistry: Online Resources

go.hrw.com

Visit **go.hrw.com** for a variety of free resources related to this textbook. Enter the keyword **HW4 HOME.**

Holt Online Learning

Students can access interactive problem solving help and active visual concept development with the *Holt Chemistry* Online Edition available at **www.hrw.com**

student CNN News

cnnstudentnews.com

Find the latest chemistry news, lesson plans, and activities related to important scientific events.

PROBLEM SOLVING AND PRACTICE	SECTION REVIEW AND ASSESSMENT	STANDARDS CORRELATION
		National Science Education Standards
TE Homework, p. 534 [BASIC]	**TE** Quiz, p. 538 [GENERAL] **CRF** Quiz * **TE** Reteaching, p. 538 [BASIC] **SE** Section Review, p. 538 **CRF** Concept Review *	PS 3c
SE Sample Problem A Determining [H_3O^+] or [OH^-] using K_w, p. 541 **SE** Practice, p. 541 [GENERAL] **TE** Homework, p. 541 [GENERAL] **TE** Homework, p. 542 [BASIC] **SE** Skills Toolkit 1 Using Logarithms in pH Calculations, p. 543 **SE** Sample Problem B Calculating pH for an Acidic or Basic Solution p. 544 **SE** Practice, p. 544 [GENERAL] **TE** Homework, p. 544 [GENERAL] **SE** Sample Problem C Calculating [H_3O^+] and [OH^-] from pH, p. 545 **SE** Practice, p. 545 [GENERAL] **TE** Homework, p. 545 [GENERAL] **CRF** Problem Solving * [ADVANCED]	**TE** Quiz, p. 547 [GENERAL] **CRF** Quiz * **TE** Reteaching, p. 547 [BASIC] **SE** Section Review, p. 547 **CRF** Concept Review *	PS 3c
SE Skills Toolkit 2 Performing a Titration, pp. 552–553 **TE** Homework, p. 553 [BASIC] **SE** Sample Problem D Calculating Concentration from Titration Data, p. 555 **SE** Practice, p. 556 [GENERAL] **TE** Homework, p. 555 [GENERAL] **CRF** Problem Solving * [ADVANCED] **SE** Problem Bank, p. 858 [GENERAL]	**TE** Quiz, p. 556 [GENERAL] **CRF** Quiz * **TE** Reteaching, p. 556 [BASIC] **SE** Section Review, p. 556 **CRF** Concept Review *	PS 3c
SE Sample Problem E Calculating K_a of a Weak Acid, p. 560 **SE** Practice, p. 560 [GENERAL] **TE** Homework, p. 560 [GENERAL] **CRF** Problem Solving * [ADVANCED] **SE** Problem Bank, p. 858 [GENERAL]	**TE** Quiz, p. 563 [GENERAL] **CRF** Quiz * **TE** Reteaching, p. 563 [BASIC] **SE** Section Review, p. 563 **CRF** Concept Review *	PS 3c

SCiLINKS®

www.scilinks.org

Topic: Acids and Bases
SciLinks code: HW4002

Topic: Acids
SciLinks code: HW4001

Topic: Bases
SciLinks code: HW4020

Topic: pH
SciLinks code: HW4095

Topic: Titrations and Indicators
SciLinks code: HW4125

Topic: Buffers
SciLinks code: HW4023

Topic: Antacids
SciLinks code: HW4010

Technology Resources

  **Science in the NEWS**

Each video segment is accompanied by a Critical Thinking Worksheet.

Segment 26
Acids in the Environment

 ChemFile Interactive Tutor CD-Rom

Module 8: Strong and Weakly Ionized Species, pH and Titrations
Topic: pH

Module 8: Strong and Weakly Ionized Species, pH and Titrations
Topic: Titrations

Overview

The chapter begins by defining acids and bases according to the Arrhenius and Brønsted-Lowry definitions. Students will learn the significance of the self-ionization of water and how pH is used to measure the acidity or basicity of solutions. They will understand the meaning of neutralization, learn how to carry out a titration, and make calculations based on titration data. The acid-ionization constant and buffer solutions will be introduced.

Assessing Prior Knowledge

Check for Content Knowledge

Students should be familiar with the following topics:

- the mole
- balanced chemical equations
- stoichiometry
- solutions
- ionization
- molarity

Using the Figure

pH affects every other chemical balance in pool water. A slightly alkaline pH of 7.4 to 7.6 is most desirable, because this range is most comfortable to the human eye and provides for optimum use of free chlorine while maintaining water that is not corrosive or scale-forming. The pH test kit shown here has a receptacle containing a sample of pool water plus an indicator. The colored capsules along the sides of the test kit contain water samples of different pH, plus an indicator, so that the pH of the pool water can be determined by color comparison.

CHAPTER

15

ACIDS AND BASES

528

Standards Correlations

National Science Education Standards

PS 3c: A large number of important reactions involve the transfer of either electrons (oxidation/reduction reactions) or hydrogen ions (acid/base reactions) between reacting ions, molecules, or atoms. In other reactions, chemical bonds are broken by heat or light to form very reactive radicals with electrons ready to form new bonds. Radical reactions control many processes such as the presence of ozone and greenhouse gases in the atmosphere, burning and processing of fossil fuels, the formation of polymers, and explosions. (Sections 1, 2, 3, 4)

For health reasons, swimming pools are regularly treated with chemicals that are usually called "chlorine." However, the chemicals are more likely to be hypochlorites or similar compounds. The effectiveness of these disinfectants depends on how acidic or basic the water is. Pool owners must therefore measure and, if necessary, adjust, the acidity of their pool. A pH test kit, such as the one shown at left, can measure the balance of acids and bases dissolved in the pool water to ensure that it is in a healthy range.

START-UP ACTIVITY

What Does an Antacid Do?

SAFETY PRECAUTIONS

PROCEDURE

1. Pour **100 mL of water** into a **150 mL beaker**. Use a **plastic pipet** to add **vinegar** one drop at a time while stirring with a **glass stirrer**. Keep **a piece of blue litmus paper** dipped in the solution, and record the number of drops needed to make the solution turn the litmus paper bright red.

2. Use a **mortar and pestle** to crush an **antacid tablet** until the tablet is powdered. Pour **100 mL of water** into another **150 mL beaker**, add the powdered tablet, and stir until the tablet is at least mostly dissolved.

3. Add vinegar dropwise to the antacid solution and monitor the solution by using **another piece of blue litmus paper.** Record the number of drops needed to turn the litmus paper bright red.

ANALYSIS

1. Which required more acid to turn the blue litmus paper red: the water or the antacid solution?

2. How does an antacid work to counteract excess stomach acid?

Pre-Reading Questions

① Give examples of acids and bases that you encounter in your everyday life.

② What polyatomic ion does the hydrogen ion, H^+, form in aqueous solution?

③ Acids are said to "neutralize" bases, and vice versa. How would you define the term *neutralize*?

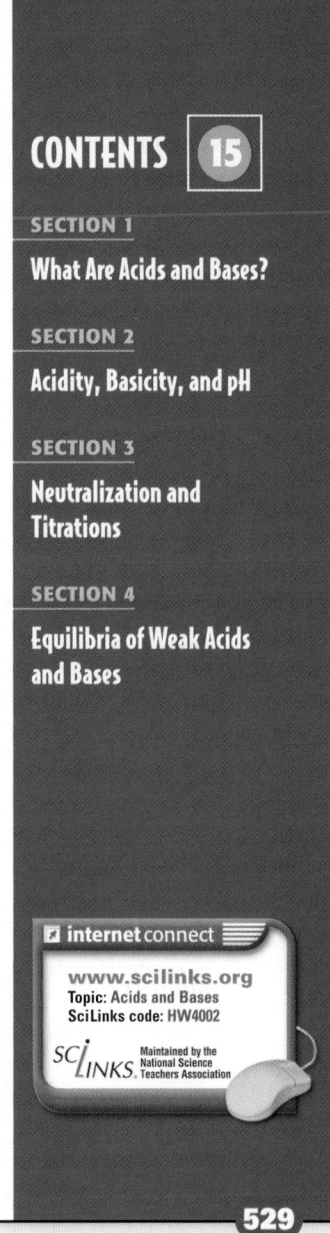

CONTENTS 15

internet connect

www.scilinks.org
Topic: Acids and Bases
SciLinks code: HW4002

SCI LINKS. Maintained by the National Science Teachers Association

529

START-UP ACTIVITY

Skills Acquired:
• Experimenting
• Collecting data
• Interpreting

Materials:
For each group of 2–4 students:
• antacid tablet
• beaker, 150 mL (2)
• blue litmus paper
• mortar and pestle
• pipets, plastic
• stirrer, glass
• vinegar, approx. 100 mL
• water, distilled, 200 mL

Teacher Notes:
• This activity should demonstrate in an observable way that bases neutralize acids. It will also introduce students to the use of indicators (in the form of litmus paper, in this case).
• Remind students that the tablet needs to be powdered so that it has the maximum surface area, which will allow the tablet to dissolve quickly and as fully as possible.
• The antacid tablets may or may not fully dissolve. Inform students that the experiment will work even if a small quantity of the tablet remains undissolved.

Answers

1. The antacid solution should have required more acid to turn the blue litmus paper red.

2. The antacid neutralizes some of the acid.

Answers to Pre-Reading Questions

1. Answers may include citric acid or vinegar as acids, ammonia or lye as bases, etc.

2. the hydronium ion, H_3O^+

3. Answers should include the idea that one substance nullifies the properties of the other.

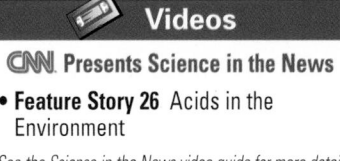

Videos

CNN. Presents Science in the News

• **Feature Story 26** Acids in the Environment

See the Science in the News video guide for more details.

Focus

Overview

Before beginning this section, review with your students the Objectives listed in the Student Edition. This section will teach students to distinguish between acids and bases by means of physical and chemical properties. Strong and weak acids and bases will be defined in terms of electrolyte properties. Students will learn the difference between the Arrhenius and Brønsted-Lowry definitions of acids and bases. The role of conjugate acid-base pairs will become clear as students write chemical equations. Students will see how an amphoteric substance can act either as an acid or as a base.

 Bellringer

Ask students to think about substances that they encounter in a typical day and to make two lists. One list should contain substances that might be acids and the other should contain substances that might be bases. Have students retain their lists and review them upon finishing the chapter to see if they would make any changes.

Motivate

Identifying Preconceptions

Ask students in what physical state acids and bases are found. Having used vinegar and dilute solutions of other acids and bases in previous science classes, some students may think that acids and bases are liquids. This preconception will allow you to discuss that many acids and bases are solids and some are gases dissolved in water.

SECTION 1
What Are Acids and Bases?

KEY TERMS
- strong acid
- weak acid
- strong base
- weak base
- Brønsted-Lowry acid
- Brønsted-Lowry base
- conjugate acid
- conjugate base
- amphoteric

OBJECTIVES

(1) Describe the distinctive properties of strong and weak acids, and relate their properties to the Arrhenius definition of an acid.

(2) Describe the distinctive properties of strong and weak bases, and relate their properties to the Arrhenius definition of a base.

(3) Compare the Brønsted-Lowry definitions of acids and bases with the Arrhenius definitions of acids and bases.

(4) Identify conjugate acid-base pairs.

(5) Write chemical equations that show how an amphoteric species can behave as either an acid or a base.

Acids

Vinegar is acidic. So are the juices of the fruits that you see in **Figure 1.** Colas and some other soft drinks are also acidic. You can recognize these liquids as acidic by their tart, sour, or sharp taste. What they have in common is that they contain dissolved compounds that chemists describe as *acids*. Many other acids are highly caustic and should not be put to the taste test. One example of a hazardous acid is sulfuric acid, H_2SO_4, which is important in car batteries. Another example is hydrochloric acid, HCl, which is used to treat the water in swimming pools.

Figure 1
Fruits and fruit juices contain a variety of acids, including those shown here. Carbonic acid is found in cola. Vinegar contains acetic acid.

Carbonic acid

Citric acid

Acetic acid

Ascorbic acid

530

did you know?

Muriatic Acid Hydrochloric acid is sold at hardware and building-supply stores as muriatic acid. It is used for cleaning mortar stains from bricks and tile and lime deposits (resulting from hard water) from tile, porcelain, and some metals. Limitations on the use of muriatic acid include cleaning marble (limestone), because the acid reacts with this rock. The name *muriatic acid* is seldom used by chemists.

Chapter Resource File
- Lesson Plan

Acid Solutions Conduct Electricity Well

Acids are electrolytes, so their solutions in water are conductors of electric current, as **Figure 2** demonstrates. To understand why, consider what happens as hydrogen chloride, HCl, dissolves in water. Like other electrolytes, hydrogen chloride dissociates to produce ions. The hydrogen ion immediately reacts with a water molecule to form a hydronium ion, as shown in the equation below.

$$HCl(g) + H_2O(l) \longrightarrow H_3O^+(aq) + Cl^-(aq)$$

The resulting solution is called *hydrochloric acid.*

The hydronium ion, H_3O^+, is able to transfer charge through aqueous solutions much faster than other ions do. The positive charge is simply passed from water molecule to water molecule. The result is that acid solutions are excellent conductors of electricity.

Acids React with Many Metals

Another property shared by aqueous solutions of acids is that they react with many metals. All metals that are above hydrogen in the activity series react with acids to produce hydrogen gas. The reaction is caused by the hydronium ion present in the solution. The presence of the hydronium ion explains why all acids behave in this way. An example is the reaction of hydrochloric acid with zinc, which is shown in **Figure 3** and is represented by the following net ionic equation.

$$2H_3O^+(aq) + Zn(s) \longrightarrow 2H_2O(l) + H_2(g) + Zn^{2+}(aq)$$

Notice that even though hydrochloric acid is often represented by the formula HCl(aq), in aqueous solution hydrochloric acid actually consists of dissolved $H_3O^+(aq)$ and $Cl^-(aq)$ ions. The chloride ions do not play a part in the reaction and so do not appear in the net ionic equation.

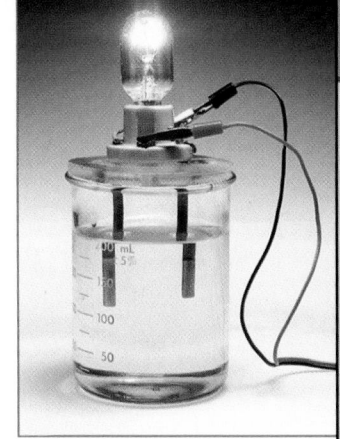

Figure 2
The beaker of hydrochloric acid solution contains $H_3O^+(aq)$ and $Cl^-(aq)$ ions. Because the solution is a good conductor, electricity can pass through the solution to cause the bulb to be brightly lit.

Topic Link

Refer to the "Solutions" chapter for a discussion of strong and weak electrolytes.

Chloride ion, Cl⁻
Hydronium ion, H_3O^+
Zinc metal, Zn
Hydrogen gas, H_2
Zinc ion, Zn^{2+}
Chloride ion, Cl⁻

Figure 3
The hydronium ions present in a hydrochloric acid solution react with zinc to form hydrogen gas bubbles and dissolved zinc ions.

531

MISCONCEPTION ALERT

Having seen that a vigorous reaction occurs when zinc or magnesium reacts with HCl and that the metals disappear, students may be strengthened in the misconception that acids are substances that "eat away" other substances. Point out that acids do not react with all metals, only those that are above hydrogen on the activity series. For instance, if you dropped a piece of copper into hydrochloric acid, no reaction would occur.

Discussion

Pour equal volumes of vinegar (5% CH_3COOH), household ammonia (NH_3), 1 M HCl, and 1 M NaOH into small, labeled beakers. Ask students which solutions will conduct electricity and how well, what ions are in each solution, and how each solution will react with other substances. Test the conductivity of each solution with a conductivity tester. Determine pH with pH paper. Drop a 1 cm strip of magnesium ribbon into each beaker, and observe what happens. Summarize the results in a table, and have students identify each solution as a strong or weak acid or base.

Teach

READING SKILL BUILDER — **BASIC**

Assimilating Knowledge Before reading this chapter, students should write a short list of the things they already know or think they know about acids and bases in aqueous solution, strengths of acids and bases, pH, and titrations. Then, have students list what they want to know about the chemistry of acids and bases. If students have trouble coming up with questions, offer the following questions:

• How do weak acids and strong acids differ?

• What are typical pH values of acids and bases?

• What forms when an acid reacts with a base?

• What are some properties that acids and bases have in common?

• What ions exist in acid solution?

LS Intrapersonal

Using the Figure — **BASIC**

Ask students to suppose that the acid being tested in **Figure 2** were only 10% ionized. How would the brightness of the bulb in that case compare with the brightness produced by hydrochloric acid? The brightness depends on the concentration of ions in solution. **LS** Visual

Teach, continued

Demonstration

Reproduce the reaction shown in **Figure 3** by dropping a small piece of zinc into a small beaker of 1 M HCl. Write the equation for the reaction on the board. Then, drop a small piece of magnesium ribbon into a similar beaker of HCl. Ask students to write the equation for this reaction.

Safety: Wear safety goggles, gloves, and a lab apron while handling chemicals. Make sure all students are at least 3 m away from the reaction beaker.

Disposal: Once the reaction is complete, carefully remove the remaining magnesium from the solution with tongs, and dispose of the $MgCl_2$ solution in a chemical waste container.

Homework ——— BASIC

Have students recall Lewis structures from the chapter "Covalent Compounds." Ask them to draw a Lewis structure for H_2O. Then, have them show how a hydrogen ion can be added to form H_3O^+. Ask students to draw a Lewis diagram for the OH^- and to add a hydrogen ion to form H_2O.
LS Visual

Discussion

Hydrogen ions are protons, and both names are often used to refer to these particles. Acids are called *proton donors,* and bases are called *proton acceptors.* Ask students to justify the use of the word *proton* for hydrogen ion.

strong acid
an acid that ionizes completely in a solvent

weak acid
an acid that releases few hydrogen ions in aqueous solution

Figure 4
According to Arrhenius, an acid in aqueous solution ionizes to form hydronium ions, H_3O^+, as shown here for nitric acid, HNO_3.

HNO₃ + H₂O ⟶

NO₃⁻ + H₃O⁺

532

Table 1	Some Strong Acids and Some Weak Acids
Strong acids	**Weak acids**
hydrochloric acid, HCl	acetic acid, CH_3COOH
hydrobromic acid, HBr	hydrocyanic acid, HCN
hydriodic acid, HI	hydrofluoric acid, HF
nitric acid, HNO_3	nitrous acid, HNO_2
sulfuric acid, H_2SO_4	sulfurous acid, H_2SO_3
perchloric acid, $HClO_4$	hypochlorous acid, HOCl
periodic acid, HIO_4	phosphoric acid, H_3PO_4

Acids Generate Hydronium Ions

Recall that some electrolytes are *strong* and others are *weak,* depending on whether they dissociate completely or partially. Because acids are also electrolytes, they can be classified as strong or weak. **Table 1** lists the names and formulas of several strong acids and several weak acids.

Nitric acid, HNO_3, is an example of a **strong acid.** Its reaction with water is shown in **Figure 4** and is represented by the following equation:

$$HNO_3(l) + H_2O(l) \longrightarrow H_3O^+(aq) + NO_3^-(aq)$$

No HNO_3 molecules are present in a solution of nitric acid.

When a **weak acid** is dissolved in water, only a small fraction of its molecules are ionized at any given time. Hypochlorous acid, HOCl, is a weak acid. Its reaction with water is described by the equation below.

$$HOCl(l) + H_2O(l) \rightleftharpoons H_3O^+(aq) + ClO^-(aq)$$

Recall that the opposed arrows in this equation indicate equilibrium. Hypochlorite ions, ClO^-, react with hydronium ions to form HOCl at exactly the same rate that HOCl molecules react with water to form ions.

The presence of a considerable number of hydronium ions identifies an aqueous solution as acidic. A Swedish chemist, Svante Arrhenius, was among the first to recognize this fact. In 1890, he proposed that an acid be defined as any substance that, when added to water, increases the hydronium ion concentration. Later, you will learn about another way to define acids that goes beyond aqueous solutions.

In some acids, a single molecule can react to form more than one hydronium ion. This happens when sulfuric acid dissolves in water, as described by the following equations:

$$H_2SO_4(l) + H_2O(l) \longrightarrow H_3O^+(aq) + HSO_4^-(aq)$$

$$HSO_4^-(aq) + H_2O(l) \rightleftharpoons H_3O^+(aq) + SO_4^{2-}(aq)$$

As shown above, sulfuric acid has two ionizable hydrogens. One of them ionizes completely, after which the other ionizes partially as a weak acid.

did you know?

The Hydrated Hydrogen Ion When an acid ionizes in solution, a hydrogen ion, H^+, is released. But because the hydrogen ion is small and its charge-to-mass ratio is large, the ion immediately attracts and becomes surrounded by several water molecules. One of the water molecules bonds covalently with the hydrogen ion by sharing a pair of electrons on its oxygen molecule, which produces the hydronium ion, H_3O^+. In addition, any number of water molecules may attach themselves to the hydronium ion by means of hydrogen bonds. Thus, the nature of the hydrogen ion in solution can be represented by the formula $H(H_2O)_n^+$. For simplicity, however, chemists let $n = 1$ and write the ion as H_3O^+.

| Table 2 | Some Strong Bases and Some Weak Bases |

Table 2 Some Strong Bases and Some Weak Bases

Strong bases	Weak bases
sodium hydroxide, NaOH	ammonia, NH_3
potassium hydroxide, KOH	sodium carbonate, Na_2CO_3
calcium hydroxide, $Ca(OH)_2$	potassium carbonate, K_2CO_3
barium hydroxide, $Ba(OH)_2$	aniline, $C_6H_5NH_2$
sodium phosphate, Na_3PO_4	trimethylamine, $(CH_3)_3N$

internet connect
www.scilinks.org
Topic: Bases
SciLinks code: HW4020

SCiLINKS Maintained by the National Science Teachers Association

Bases

Bases are another class of electrolytes. Unlike acids, which are usually liquids or gases, many common bases are solids. Solutions of bases are slippery to the touch, but touching bases is an unsafe way to identify them. The slippery feel comes about because bases react with oils in your skin, converting them into soaps. This property of attacking oils and greases makes bases useful in cleaning agents, such as those in **Figure 5.**

Some bases, such as magnesium hydroxide, $Mg(OH)_2$, are almost insoluble in water. Other bases, such as potassium hydroxide, are so soluble that they will absorb water vapor from the air and dissolve in the water. A base that is very soluble in water is called an *alkali*, a term that you also know describes the Group 1 metals of the periodic table. These metals react with water to form hydroxides that are water-soluble alkalis. The hydroxide-rich solutions that form when bases dissolve in water are said to be *basic* or *alkaline.*

Just as acids may be strong or weak depending on whether they ionize completely or reach an equilibrium between ionized and un-ionized forms, bases are also classified as strong or weak. **Table 2** lists several bases of each class.

Figure 5
A variety of bases, including ammonia, sodium hydroxide, and sodium bicarbonate, can be found in products used around the home.

Demonstration

Ionization in Aqueous Acids

(Approximate time: 10 min)

1. Pour 150 mL of distilled water into each of four beakers. Because totally pure water is very difficult to obtain, you may find a small conductivity in distilled water. Explain to students that any conduction shown by water is due to dissolved solutes.

2. Write the formulas $C_{12}H_{22}O_{11}$ (sucrose), CH_3COOH (acetic acid), HCl (hydrochloric acid), and NaCl (sodium chloride) on the board. Ask students to describe the bonding in each substance.

3. Add 2 g of sucrose to beaker 1, 10 mL of 1 M CH_3COOH to beaker 2, 10 mL of 1 M HCl to beaker 3, and 2 g NaCl to beaker 4. Stir the solutions.

4. Test each solution by using a battery-powered conductivity tester. Ask students why acetic acid and HCl are conductors even though they are covalent compounds.

Challenge students to account for the difference in conductivity between acetic acid and HCl. Have them recall what they learned about molecular electrolytes in the chapter "Solutions."

Safety: Wear safety goggles and a lab apron. Students should remain 3 m from the demonstration.

Disposal: Combine the acid solutions, and neutralize them with 1 M NaOH. Pour all the solutions down the drain.

533

MISCONCEPTION ALERT

Studies show that more that 54% of entering university students cannot name two common bases. It may be that bases are less easy to identify because their names do not have the word *base* in them, whereas the word *acid* appears in the names of acids. Also, bases are often discussed in terms of their ion formulas, such as OH^-, CO_3^{2-}, SO_4^{2-}, and CH_3COO^-. Students may not always think of these bases as parts of ionic compounds, such as NaOH, $CaCO_3$, Na_2SO_4, or CH_3COONa. As your discussion of bases continues throughout this chapter, emphasize that the negative ions are the part of the ionic compound that can accept a hydrogen ion, but that the entire ionic compound is dissolved in a basic solution.

1. What ion is characteristic of aqueous solutions of all acids? all bases? **Ans.** The hydronium ion, H_3O^+, is characteristric of all acids; the hydroxide ion, OH^-, is characteristic of all bases.

2. What is the difference between a strong acid and a weak acid? **Ans.** A strong acid is completely ionized; a weak acid is only partially ionized.

3. List three characteristics of an acid solution. **Ans.** It tastes sour, conducts electricity, and reacts with metals above hydrogen in the activity series.

4. Write the equation for the reaction of the strong acid, HBr, with water. **Ans.** $HBr(g) + H_2O(l) \longrightarrow H_3O^+(aq) + Br^-(aq)$

LS Logical

Teaching Tip
Breaking the Covalent Bond
You may wish to refer students to the formulas for some organic acids in **Figure 1** and to explain why the carboxyl group can ionize in water and produce an acidic solution. Because the oxygen atoms in the carboxyl group are strongly electronegative, they attract electrons to themselves, including the electron pair that bonds the hydrogen atom. The hydrogen atom is thereby "deprived" of its electron, and the proton alone can easily transfer to another atom (base) that offers an unshared pair of electrons.

Figure 6
a Sodium hydroxide is a strong base. It dissociates completely in aqueous solution.

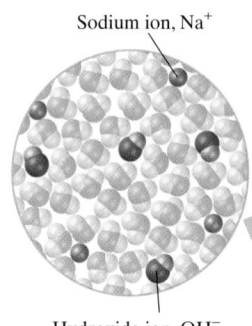

Sodium ion, Na^+

Hydroxide ion, OH^-

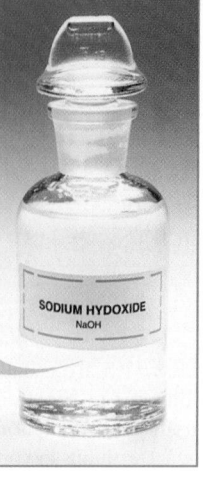

b Only a small portion of the dissolved molecules of ammonia, a weak base, react with water. Most of the ammonia remains as neutral molecules.

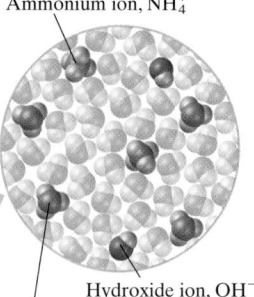

Ammonium ion, NH_4^+

Hydroxide ion, OH^-
Ammonia molecule, NH_3

strong base

a base that ionizes completely in a solvent

weak base

a base that releases few hydroxide ions in aqueous solution

Bases Generate Hydroxide Ions

The following equation can be used to describe the dissolving of sodium hydroxide, a **strong base,** in water.

$$NaOH(s) \longrightarrow Na^+(aq) + OH^-(aq)$$

The ions dissociate in solution as they become surrounded by water molecules and float away independently.

Ammonia is a typical **weak base.** At room temperature, ammonia is a gas, $NH_3(g)$, but it is very soluble in water, forming $NH_3(aq)$. A few of the ammonia molecules in solution react with water to reach the equilibrium system described by the equation below.

$$NH_3(aq) + H_2O(l) \rightleftharpoons NH_4^+(aq) + OH^-(aq)$$

$NH_4^+(aq)$ is the ammonium ion. The vast majority of ammonia molecules, however, remain un-ionized at any given time.

Both strong and weak bases generate hydroxide ions when they dissolve in water, as **Figure 6** shows. This property is the basis of the Arrhenius definitions of a base.

Many oxides, carbonates, and phosphates are bases, too. Potassium oxide is a strong base. It reacts with water as shown below:

$$K_2O(s) + H_2O(l) \longrightarrow 2K^+(aq) + 2OH^-(aq)$$

Soluble carbonates are weak bases, however, because the dissolved carbonate ion establishes the following equilibrium in water:

$$CO_3^{2-}(aq) + H_2O(l) \rightleftharpoons HCO_3^-(aq) + OH^-(aq)$$

The equilibrium system contains a low concentration of hydroxide ions, a low concentration of hydrogen carbonate ions, $HCO_3^-(aq)$, and a greater concentration of unreacted carbonate ions, $CO_3^{2-}(aq)$.

MISCONCEPTION
**//// ALERT **

Bottles of aqueous ammonia, $NH_3(aq)$, are sometimes labeled in a misleading way. For example, a solution in which 1 mol of ammonia gas is dissolved in a liter of water might be labeled 1 M NH_4OH. Because ammonium hydroxide is an ionic compound, this label implies that the concentration of OH^- is 1 M. This is not the case because ammonia is a weak base and 1 mol of NH_3 does not react completely to produce 1 mol of hydroxide ion. The proper labeling should be 1 M $NH_3(aq)$. The actual concentration of the hydroxide ion can be obtained through calculation using the equilibrium constant K_b.

Quick LAB

Acids and Bases in the Home

Because taste and feel are not safe ways to determine whether a substance is an acid or a base, you should use an indicator to recognize acids and bases. As its name suggests, an *indicator* indicates whether a solution is acidic or basic. It does this by changing color.

PROCEDURE

1. Using a **blender**, grind **red cabbage** with **water**.

2. Strain the liquid into a **large beaker**, and dilute it with water. You now have an indicator.

3. Using your indicator, test **various household products** by adding each product to a separate **small beaker** containing a sample of the indicator. Start with an item that you know is an acid and an item that you know is a base.

ANALYSIS

1. What color is the indicator in acidic solution? in basic solution?

2. Are cleaning products more likely to be acidic or basic?

3. Are food products more likely to be acidic or basic?

Brønsted-Lowry Classification

The definitions of *Arrhenius acid* and *Arrhenius base* given earlier in this book are variants of the definitions of *acid* and *base* originally proposed by Arrhenius in the late 19th century. One drawback that the Arrhenius definitions have is that they are limited to aqueous solutions: HCl, for instance, should be considered an acid whether it is in the form of a pure gas or in aqueous solution. Another limitation is that the Arrhenius definition cannot classify substances that sometimes act as acids and sometimes act as bases.

Brønsted-Lowry Acids Donate Protons

In 1923, the Danish chemist Johannes Brønsted proposed a broader definition of *acid.* Surprisingly, the same year, the British scientist Thomas Lowry happened to make exactly the same proposal independently. Their idea was to apply the name *acid* to any species that can donate a proton. Recall that a proton is a hydrogen atom that has lost its electron; it is a hydrogen ion and can be represented as H^+. Such molecules or ions are now called **Brønsted-Lowry acids.** A reaction showing hydrochloric acid, a representative Brønsted-Lowry acid, is depicted in **Figure 7.**

Brønsted-Lowry acid

a substance that donates a proton to another substance

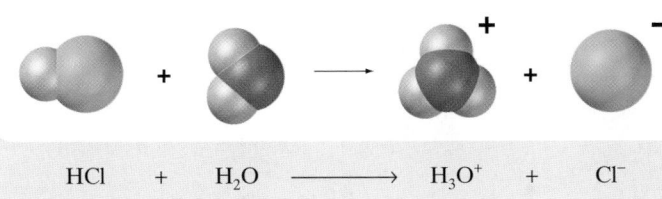

$$HCl \quad + \quad H_2O \quad \longrightarrow \quad H_3O^+ \quad + \quad Cl^-$$

Figure 7
When hydrogen chloride dissolves in water, a proton is transferred from the HCl molecule (leaving behind Cl^-) to an H_2O molecule (forming H_3O^+).

did you know?

A Third Acid-Base Definition G. N. Lewis, the American chemist who introduced the electron dot diagram as a way of modeling covalent bonds, also developed a broader way of viewing acids and bases. His definition includes not only all Brønsted-Lowry acid-base reactions but also other reactions that do not involve protons. Lewis defined an acid as an electron pair acceptor and a base as an electron pair donor. For example, the reaction of boron trifluoride, BF_3, with ammonia, NH_3, is a Lewis acid-base reaction. Boron has an empty orbital. Nitrogen has an unshared pair of electrons. When boron accepts a pair of electrons from nitrogen, the result is F_3B-NH_3. Metal ions also act as Lewis acids. Most have empty inner orbitals. When metal ions are dissolved in water, they accept electron pairs from water molecules to form complex ions such as $Ni(H_2O)_6^{2+}$.

Quick LAB — GENERAL

Acids and Bases in the Home
Materials:
For each group of 2–4 students:
- assorted household cleaners and food products
- beaker, 500 mL
- beakers, small (one for each item to be tested)
- blender
- red cabbage, cut up
- spatula
- strainer
- water

Teacher Notes: This activity should test a variety of acids and bases. It will give students an introduction to indicators, which will be covered in more detail in Section 2.

Safety: Ensure that students use gloves, an apron, and goggles.

Disposal: Neutralize all solutions and dispose of in the drain.

Answers

1. Red cabbage juice varies from light red at pH 2 to purple at pH 7, to green and yellow at about pH 9 through 14.

2. In general, cleaning products are basic (alkaline solutions are good at removing grease).

3. Food products are likely to be somewhat acidic. Plants and fruits tend to be naturally acidic, and processed products are generally acidic, because keeping the pH low retards growth of bacteria. Also, the taste of basic materials is disagreeable to most people.
LS Visual

Using the Figure — GENERAL

Refer students to **Figure 7,** and ask them to describe what is happening. They should recognize that a hydrogen ion is being transferred from HCl to H_2O, creating a hydronium ion and a chloride ion, and that HCl is an acid according to the Brønsted-Lowry definition. **LS** Visual

Chapter Resource File

- Datasheets for In-Text Labs

Using the Figure

Figure 8 illustrates the important concept that an acid-base reaction can occur without water. However, students may look at the photograph and think that acid-base reactions such as the one pictured are benign. Emphasize to students that acid-base reactions give off a great deal of heat such that acids and bases should not be kept in the same storage cabinet. Make sure students know that opening a bottle of acid and a bottle of base together in such close proximity as is shown in **Figure 8** is dangerous and should not be attempted.

Teaching Tip

Organic Bases Draw a Lewis dot diagram of ammonia on the board. One by one, replace the hydrogens with small hydrocarbon groups, such as the methyl group, $-CH_3$. One methyl group substitution produces methylamine, two methyl groups result in dimethylamine, and three methyl groups produce trimethylamine. Point out that regardless of the hydrocarbon substitutions, the nitrogen atom retains its nonbonding pair of electrons which gives it the ability to act as a base by accepting a proton. Amines such as these are organic bases.

We might also think of the reaction occurring when HCl dissolves in water as proceeding in two steps:

$$HCl \longrightarrow H^+ + Cl^-$$

followed by $H^+ + H_2O \longrightarrow H_3O^+$

The reaction shows HCl acting as an Arrhenius acid, forming a hydronium ion. Because the reaction also involves a proton transfer, hydrochloric acid is also a Brønsted-Lowry acid. All Arrhenius acids are, by definition, also Brønsted-Lowry acids.

Brønsted-Lowry Bases Accept Protons

As you might expect based on their definition of an acid, Brønsted and Lowry defined a base as a proton acceptor. In the reaction taking place in **Figure 8,** ammonia, NH_3, serves as the proton acceptor and is therefore a **Brønsted-Lowry base.** Ammonia also functions as a proton acceptor when it dissolves in water, as in the equation below.

$$NH_3(aq) + H_2O(l) \rightleftharpoons NH_4^+(aq) + OH^-(aq)$$

Again, notice that ammonia, a Brønsted-Lowry base, is also an Arrhenius base. All Arrhenius bases are also Brønsted-Lowry bases. Ammonia does not have to react in aqueous solution to be considered a Brønsted-Lowry base. Even as a gas, ammonia accepts a proton from hydrogen chloride, as **Figure 8** shows.

Brønsted-Lowry base

a substance that accepts a proton

Figure 8
$HCl(g)$ and $NH_3(g)$ that have each escaped from aqueous solution combine to form a cloud of solid ammonium chloride, $NH_4Cl(s)$

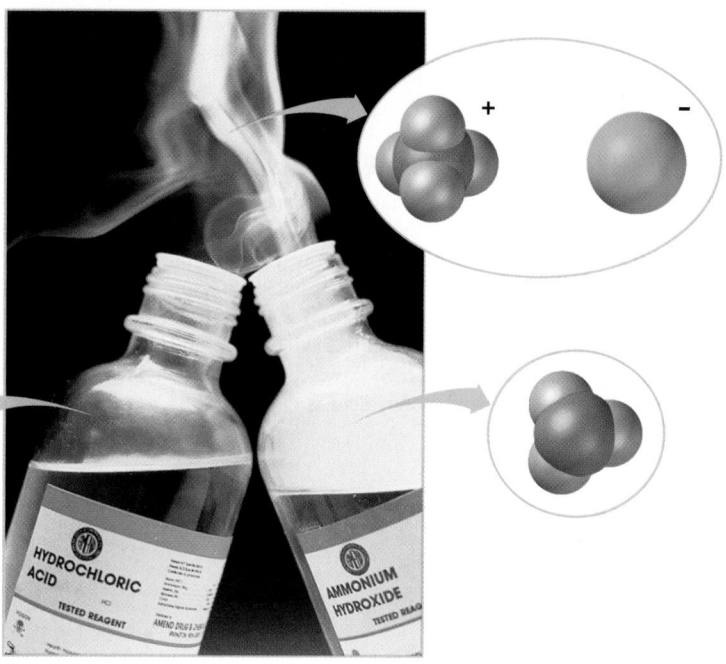

536

REAL-LIFE CONNECTION — ADVANCED

The Importance of Sulfuric Acid

Writing Every year, the United States produces more than 1 billion tons of sulfuric acid. Of this enormous production, a very small portion is purchased directly by consumers. The bulk of the production is used by the chemical industry as a raw material in the manufacture of myriad ordinary goods used by people every day. The largest fraction, over 60 percent, is used in making fertilizers. Smaller amounts are used to make detergents, dyes, gasoline, inks, paints, explosives, fabrics, synthetic rubber, iron, steel, rayon, film, and many more products. Sulfuric acid is made by the contact process in which sulfur is oxidized to SO_3 and the gas reacts with water to produce an 18 M solution of concentrated acid. Have interested students investigate the contact process and write a short paper or oral report for the class. **LS Verbal**

Conjugate Acids and Bases

In the language of Brønsted and Lowry, an acid-base reaction is very simple: one molecule or ion passes a proton to another molecule or ion. Whatever loses the proton is an acid, and whatever accepts the proton is a base.

Look again at the equation for the reversible reaction of ammonia, NH_3, with water:

$$\underset{\text{base}}{NH_3(aq)} + \underset{\text{acid}}{H_2O(l)} \rightleftharpoons NH_4^+(aq) + OH^-(aq)$$

Water donates a proton to ammonia, so it is an acid. Ammonia accepts the proton, so it is a base. The ammonium ion forms when NH_3 receives the proton, and the hydroxide ion forms when H_2O has lost the proton.

Notice that the reaction is reversible, so it can be written with the products as reactants. When the reaction is written that way, we can identify another acid and another base.

$$\underset{\text{acid}}{NH_4^+(aq)} + \underset{\text{base}}{OH^-(aq)} \rightleftharpoons NH_3(aq) + H_2O(l)$$

In this reaction, the ammonium ion donates a proton to the hydroxide ion. NH_4^+ is the acid, and OH^- is the base. The ammonium ion is called the **conjugate acid** of the base, ammonia. The hydroxide ion is called the **conjugate base** of the acid, water.

$$\underset{\text{base}}{NH_3(aq)} + \underset{\text{acid}}{H_2O(l)} \rightleftharpoons \underset{\text{conjugate acid}}{NH_4^+(aq)} + \underset{\text{conjugate base}}{OH^-(aq)}$$

Every Brønsted-Lowry acid has a conjugate base, and every Brønsted-Lowry base has a conjugate acid. **Table 3** lists many such acid-base pairs.

conjugate acid

an acid that forms when a base gains a proton

conjugate base

a base that forms when an acid loses a proton

Table 3 Conjugate Acid-Base Pairs

Acid	Conjugate base
hydrochloric acid, HCl	chloride ion, Cl^-
sulfuric acid, H_2SO_4	hydrogen sulfate ion, HSO_4^-
hydronium ion, H_3O^+	water, H_2O
hydrogen sulfate ion, HSO_4^-	sulfate ion, SO_4^{2-}
hypochlorous acid, HOCl	hypochlorite ion, ClO^-
dihydrogen phosphate ion, $H_2PO_4^-$	monohydrogen phosphate ion, HPO_4^{2-}
ammonium ion, NH_4^+	ammonia, NH_3
hydrogen carbonate ion, HCO_3^-	carbonate ion, CO_3^{2-}
water, H_2O	hydroxide ion, OH^-
Conjugate acid	**Base**

537

INCLUSION Strategies

- *Learning Disabled*
- *English Language Learners*
- *Attention Deficit Disorder*

Ask students to use product labels from vinegar, citrus, and other household products to identify acids used in the ingredients. Using **Table 1,** students can classify those bases as strong or weak. Ask students to use product labels from soaps, cleaning agents, and other household products to identify bases used in the ingredients. Using Table 2, students can classify those bases as strong or weak. Students may create two charts showing the products containing acids and bases along with their classification as strong or weak.

Reading Organizer Have students make a table with two columns. One column should be labeled "Acids," the other "Bases." Have students organize the following information in the table: Arrhenius definition, Brønsted-Lowry definition, characteristics (taste, conductivity, color of litmus, reactivity with metals, etc.), and characteristic ion in solution. Have them keep their organizer for future reference. **LS Logical**

Group Activity —— GENERAL

To provide students with practice in identifying conjugate acid and base pairs, divide your class into groups of three. Assign each group two or three of the weak acids listed in **Table 1.** Have the members of the groups collaborate on writing the equation for the reaction of each acid with water. Have them use double arrows and include the physical states of all reactants and products. Then, ask them to identify the reacting acid and its conjugate base and the reacting base and its conjugate acid. Have them connect the pairs by using lines below the equation. When they have finished, groups may exchange papers and check each other's work. If there is any disagreement, urge the groups to resolve the difficulty. **LS Interpersonal**

SKILL BUILDER — ADVANCED

Writing Skill Nonmetal elements form covalent oxides that act like acids. Most metal elements form ionic oxides that act as bases. Some metals and metalloids form oxides that are amphoteric. Have interested students use a college text or other reference to explore the acid-base reactivity of the elements and write a short paper describing their results. Have them include representative equations for the reactions they uncover. **LS Verbal**

Chapter 15 • Acids and Bases   537

Reteaching —————— BASIC

Refer students to the acids and bases in **Tables 1** and **2**. Tell them that the essential ideas in this chapter are incorporated in the reactions of each of these compounds with water. Lead students through some of these reactions, and have them identify the conjugate acid and base in each reaction.

LS Verbal

Quiz —————— GENERAL

1. State whether the following observations indicate an acid or a base.

   a. A solution conducts electricity and feels slippery. Ans. base

   b. A solution conducts electricity and reacts with metals. Ans. acid

   c. A solution is sour and turns blue litmus paper red. Ans. acid

2. What is the definition of a Brønsted-Lowry acid? Ans. any species that can donate a proton, H^+

3. What is the definition of a Brønsted-Lowry base? Ans. any species that can accept a proton, H^+

4. Identify the conjugate acid-base pairs in the following equation: $HS^- + H_2O \rightleftarrows S^{2-} + H_3O^+$ Ans. HS^-, acid; S^{2-}, conjugate base; H_2O, base; H_3O^+, conjugate acid.

Chapter Resource File

• Concept Review
• Quiz

Amphoteric Species Are Both Acids and Bases

amphoteric

describes a substance, such as water, that has the properties of an acid and the properties of a base

Did you notice that some species appear in both the "Acid" and the "Base" columns in **Table 3**? Several species can both donate and accept protons. Such species are described as **amphoteric.** The hydrogen carbonate ion, HCO_3^-, is amphoteric in aqueous solution, for example. It can act as an acid by donating a proton to an ammonia molecule, as represented by the following equation:

$$HCO_3^-(aq) + NH_3(aq) \rightleftarrows CO_3^{2-}(aq) + NH_4^+(aq)$$
$$\quad\quad acid \quad\quad\quad\quad base$$

Alternatively, the hydrogen carbonate ion can act as a base when, for example, it accepts a proton from a hydronium ion in the reaction described by the equation below.

$$HCO_3^-(aq) + H_3O^+(aq) \rightleftarrows H_2CO_3(aq) + H_2O(l)$$
$$\quad base \quad\quad\quad acid$$

Water itself is amphoteric. A water molecule can donate a proton and become a hydroxide ion in the process. Or it can accept a proton to become a hydronium ion. The consequences of this fact are described in the next section.

① Section Review

UNDERSTANDING KEY IDEAS

1. List the observable properties of an acid.

2. How did Arrhenius define a base?

3. Giving examples, explain how strong acids and weak acids differ.

4. How does the Brønsted-Lowry definition of an acid differ from the Arrhenius definition of an acid?

5. Write an equation that demonstrates the properties of acids and bases, as defined by Brønsted and Lowry.

6. Define a conjugate acid-base pair, and give an example.

7. Show chemical equations for the reaction of water with (a) an acid of your choosing and (b) a base of your choosing.

CRITICAL THINKING

8. Could a Brønsted-Lowry acid *not* be an Arrhenius acid? Explain.

9. How would $[OH^-]$ in an ammonia solution compare with $[OH^-]$ in a sodium hydroxide solution of similar concentration?

10. Write the formulas of the conjugate acid and the conjugate base of the $H_2PO_4^-$ ion.

11. Write an equation describing a proton transfer between $H_2SO_4(aq)$ and $SO_4^{2-}(aq)$.

12. Why can magnesium hydroxide be described as a strong base even though it is only slightly soluble in water?

13. Identify two acids and their conjugate bases in the following reaction.

$$H_2SO_4(aq) + SO_3^{2-}(aq) \rightleftarrows$$
$$\quad\quad\quad\quad\quad HSO_4^-(aq) + HSO_3^-(aq)$$

Answers to Section Review

1. Acids have a sharp taste, and they dissolve many metals, producing hydrogen gas. Their solutions in water are good conductors of electricity. They react with bases to form salts.

2. as a substance that increases the concentration of hydroxide ions in aqueous solution

3. When it dissolves in water, a strong acid produces hydronium ions in stoichiometric quantity, such as hydrogen chloride in the following reaction.
$HCl(g) + H_2O(l) \longrightarrow H_3O^+(aq) + Cl^-(aq)$
When a weak acid dissolves in water, an equilibrium is set up and one of the products is the

hydronium ion. An example is provided by hypochlorous acid:
$HOCl(aq) + H_2O(l) \rightleftarrows ClO^- + H_3O^+(aq)$

4. The Brønsted-Lowry definition recognizes an acid as any species that can donate a proton, whereas an Arrhenius acid must generate hydronium ions in aqueous solution.

5. An acceptable answer could be either a generic equation, such as: base + acid $\longrightarrow$ conjugate acid + conjugate base, or an example, such as:
$NH_3(aq) + H_2O(l) \rightleftarrows NH_4^+(aq) + OH^-(aq)$

Answers continued on p. 573A

Acidity, Basicity, and pH

KEY TERMS

- self-ionization constant of water, K_w
- neutral
- pH
- indicator

OBJECTIVES

① **Use** K_w in calculations.

② **Explain** the relationship between pH and H_3O^+ concentration.

③ **Perform** calculations using pH, $[H_3O^+]$, $[OH^-]$, and K_w.

④ **Describe** two methods of measuring pH.

The Self-Ionization of Water

You have just learned that water is both an acid and a base. This means that a water molecule can either give or receive a proton. So what happens when one molecule of water donates a proton to another molecule of water? The reaction is described by the equation below.

$$H_2O(l) + H_2O(l) \rightleftharpoons H_3O^+(aq) + OH^-(aq)$$
$$\text{base} \qquad \text{acid}$$

As also shown in **Figure 9,** a pair of water molecules are in equilibrium with two ions—a hydronium ion and a hydroxide ion—in a reaction known as the *self-ionization of water.*

Thus, even pure water contains ions. The chemical equation shows that the two ions are produced in equal numbers. Therefore, in pure water, the two ions must share the same concentration. Experiments show that this concentration is 1.00×10^{-7} M at 25°C.

$$[H_3O^+] = [OH^-] = 1.00 \times 10^{-7} \text{ M}$$

Figure 9
In water and aqueous solutions, an equilibrium exists between H_2O and the two ions shown.

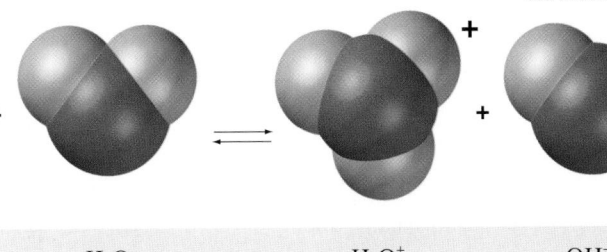

| H_2O | + | H_2O | ⇌ | H_3O^+ | + | OH^- |
| Water molecule | | Water molecule | | Hydronium ion | | Hydroxide ion |

539

did you know? ——— ADVANCED

K_w **and Temperature** Like all equilibrium constants, the self-ionization constant for water is temperature dependent. K_w equals 1.00×10^{-14} only at 25°C. For example, at 90°C, $K_w = 3.73 \times 10^{-13}$. Interested students could obtain data from the Handbook of Chemistry and Physics for temperatures ranging from 0°C to 100°C and could plot a graph to find out how the values change with temperature. **LS Logical**

Chapter Resource File

- **Lesson Plan**

Focus

Overview

Before beginning this section, review with your students the Objectives listed in the Student Edition. Students will be introduced to water's self-ionization constant, K_w, and will use it to calculate the concentration of either H_3O^+ or OH^-. They will learn how pH is measured, how it relates to ion concentrations, and how it is used in calculations.

🔊 Bellringer

Refer students to the chemical equation on this page. Ask them to write a short paragraph explaining what they think is happening in the equation. Use their explanations as a springboard for beginning the lesson.

Motivate

Identifying Preconceptions

Some students may think that K_w applies only to pure water. They may not recognize that the equilibrium defined by K_w exists in all water solutions no matter how many substances may be dissolved in them. Tell students that it does not matter if a solute increases the concentration of H_3O^+ or OH^-; the product of the two concentrations will always equal 1×10^{-14}.

Activity ——————— BASIC

Have students use whatever kinds of models you have available to model the equation for the self-ionization of water shown on this page. Have them build models of two molecules of water. Then, have them transfer a proton from one molecule to the other to obtain the products. Toothpicks and mini marshmallows or gumdrops can be used, or students can draw Lewis diagrams. Ask students to think of other innovative ways to model the essential ideas in the equation. **LS Kinesthetic**

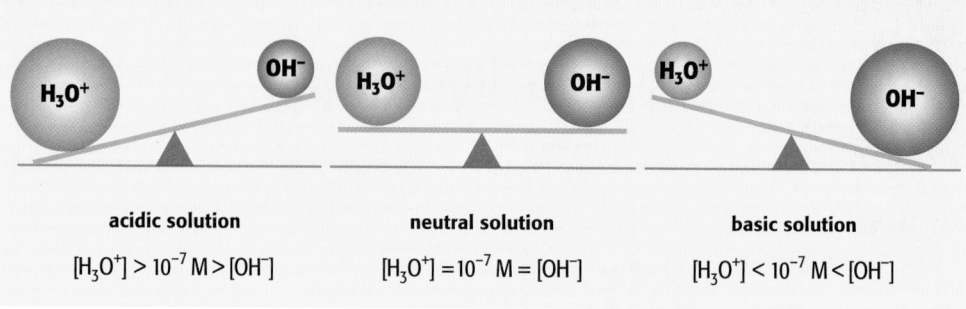

acidic solution	neutral solution	basic solution
$[H_3O^+] > 10^{-7}\,M > [OH^-]$	$[H_3O^+] = 10^{-7}\,M = [OH^-]$	$[H_3O^+] < 10^{-7}\,M < [OH^-]$

Teach

Using the Figure

Ask students to look at **Figure 10** and to find where the concentration of H_3O^+ is the greatest. Then have them tell you where the concentration of OH^- is the lowest. Ask where the concentrations of the two ions are equal. Have students describe in their own words how the two ions vary with respect to each other. Now, have them refer to **Table 4**, and ask if their conclusions are verified.

Discussion

In your discussion of K_w, you may want to emphasize that the equation for the self-ionization of water represents a reaction at equilibrium. Ask students how this is shown in the equation. Elicit from them that both the forward and backward reactions are proceeding at equal rates. If necessary, review Le Châtelier's principle. Be sure students recognize that the K_w expression is an equilibrium-constant expression in which the concentration of the reactant (H_2O) in the denominator is constant and therefore is omitted from the expression. Ask students to assume that they have a beaker of pure water. Ask how the equilibrium would shift if a small amount of an acid were added to the beaker. Ask what would happen to the concentration of the OH^- ion. Ask what would happen if a small amount of a base were added and what would happen to the concentration of the H_3O^+ ion. Finally, help them conclude that, as with all equilibria, the equilibrium constant, K_w, does not change even when the concentrations of the products change.

Figure 10
When the concentration of H_3O^+ goes up, the concentration of OH^- goes down, and vice versa.

the self-ionization constant of water, K_w

the product of the concentrations of the two ions that are in equilibrium with water; $[H_3O^+][OH^-]$

Topic Link

Refer to the "Chemical Equilibrium" chapter for a discussion of equilibrium constants.

The Self-Ionization Constant of Water

The equilibrium between water and the ions it forms is described by the following equation:

$$2H_2O(l) \rightleftharpoons H_3O^+(aq) + OH^-(aq)$$

Recall that an equilibrium-constant expression relates the concentrations of species involved in an equilibrium. The relationship for the water equilibrium is simply $[H_3O^+][OH^-] = K_{eq}$. This equilibrium constant, called the **self-ionization constant of water,** is so important that it has a special symbol, K_w. Its value can be found from the known concentrations of the hydronium and hydroxide ions in pure water, as follows:

$$[H_3O^+][OH^-] = K_w = (1.00 \times 10^{-7})(1.00 \times 10^{-7}) = 1.00 \times 10^{-14}$$

The product of these two ion concentrations is always a constant. Thus, anything that increases one of the ion concentrations decreases the other, as elaborated in **Table 4** and illustrated in **Figure 10**. Likewise, if you know one of the ion concentrations, you can calculate the other. The concentration of hydronium ions in a solution expresses its *acidity*. The concentration of hydroxide ions in a solution expresses its *basicity*.

Table 4   Concentrations and K_w

Solution	$[H_3O^+]$ (M)	$[OH^-]$ (M)	$K_w = [H_3O^+][OH^-]$
Pure water	1.0×10^{-7}	1.0×10^{-7}	1.0×10^{-14}
0.10 M strong acid	1.0×10^{-1}	1.0×10^{-13}	1.0×10^{-14}
0.010 M strong acid	1.0×10^{-2}	1.0×10^{-12}	1.0×10^{-14}
0.10 M strong base	1.0×10^{-13}	1.0×10^{-1}	1.0×10^{-14}
0.010 M strong base	1.0×10^{-12}	1.0×10^{-2}	1.0×10^{-14}
0.025 M strong acid	2.5×10^{-2}	4.0×10^{-13}	1.0×10^{-14}
0.025 M strong base	4.0×10^{-13}	2.5×10^{-2}	1.0×10^{-14}

MISCONCEPTION
ALERT

Students may begin to think that because water produces ions, it is an ionic compound. Remind them that the bonds in the water molecule are covalent and that the extent of ionization of the molecules is extremely small. A concentration of 1×10^{-7} M for the two ions in pure water indicates that only one molecule in 10,000,000 (10^7) undergoes ionization. In addition, students do not easily grasp negative exponents. As a brief math reminder, tell students that, for example, 10^{-7} M is 10 times *less* than 10^{-6} M.

Transparencies

TT Relationship of $[H_3O^+]$ to $[OH^-]$

Determining [OH⁻] or [H₃O⁺] Using K_w

What is [OH⁻] in a 3.00×10^{-5} M solution of HCl?

1 Gather information.

Because HCl is a strong acid, all HCl in an aqueous solution ionizes according to the equation below.

$$HCl(g) + H_2O(l) \longrightarrow H_3O^+(aq) + Cl^-(aq)$$

Therefore, a 3.00×10^{-5} M solution of HCl has

$$[H_3O^+] = 3.00 \times 10^{-5} \text{ M}.$$

The self-ionization constant of water is

$$K_w = [H_3O^+][OH^-] = 1.00 \times 10^{-14}.$$

2 Plan your work.

$$K_w = 1.00 \times 10^{-14} = [H_3O^+][OH^-] = (3.00 \times 10^{-5})[OH^-]$$

Values of [H₃O⁺] and of [H₃O⁺][OH⁻], K_w, are known. Therefore, [OH⁻] can be found by division.

3 Calculate.

$$[OH^-] = \frac{K_w}{[H_3O^+]} = \frac{1.00 \times 10^{-14}}{3.00 \times 10^{-5}} = 3.33 \times 10^{-10} \text{ M}$$

4 Verify your results.

Multiplying the values for [H₃O⁺] and [OH⁻] gives the known K_w and confirms that the concentration of hydroxide ion in the solution is 3.33×10^{-10} M.

PRACTICE

1. Calculate the hydronium ion concentration in an aqueous solution that has a hydroxide ion concentration of 7.24×10^{-4} M.

2. What is [OH⁻] in a 0.450 M solution of HNO₃?

3. What is [H₃O⁺] in a solution of NaOH whose concentration is 3.75×10^{-2} M?

4. Calculate the hydroxide ion concentration of a 0.200 M solution of HClO₄.

5. If 1.2×10^{-4} moles of magnesium hydroxide, Mg(OH)₂, are dissolved in 1.0 L of aqueous solution, what are [OH⁻] and [H₃O⁺]?

PRACTICE HINT

Remember that [H₃O⁺] is equivalent to the concentration of the acid itself only if the acid is a strong acid. The same is true for [OH⁻] and strong bases.

PROBLEM SOLVING SKILL

Answers to Practice Problems A

1. $[H_3O^+] = 1.38 \times 10^{-11}$ M
2. $[OH^-] = 2.22 \times 10^{-14}$ M
3. $[H_3O^+] = 2.67 \times 10^{-13}$ M
4. $[OH^-] = 5.00 \times 10^{-14}$ M
5. $[OH^-] = 2.4 \times 10^{-4}$ M; $[H_3O^+] = 4.2 \times 10^{-11}$ M

Homework ——— BASIC

Additional Practice

1. Calculate the hydroxide ion concentration in an aqueous solution that has a concentration of hydronium ion equal to 1.55×10^{-2} M. Ans. 6.45×10^{-13} M

2. What is [OH⁻] for a 0.125 M solution of HCl? Ans. 8.00×10^{-14} M

3. What is [H₃O⁺] in a solution of 0.000500 M NaOH? Ans. 2.00×10^{-11} M

4. What is [OH⁻] in a 0.00240 M solution of the strong acid HBr? Ans. 4.17×10^{-12} M

5. Based on the definition of K_w, show mathematically that in pure water, $[H_3O^+] = [OH^-] = 1 \times 10^{-7}$ M. Ans. $[H_3O^+] = [OH^-] = x$; $x^2 = 1.00 \times 10^{-14}$; $x = 1 \times 10^{-7}$ M

LS Logical

541

did you know?

Pickling Steel Pickling is one of the finishing processes in the production of steel from iron ore. It consists of submerging the steel in a bath of sulfuric or hydrochloric acid. The acid slowly removes a coating of scale that developed during the high-temperature processes of drawing and rolling. The scale dissolves in the acid, leaving behind a clean steel surface. The dissolved compounds are reclaimed by conversion to iron oxide, which is fed back into the blast furnace for reprocessing. Relieved of its solute, the acid is recycled back into the pickling bath.

Chapter Resource File

• Problem Solving

Discussion

Many of your students will have heard of pH and will have encountered it in previous science courses. Before starting the section, conduct a short inquiry to gauge students' understanding of the pH concept by asking the following questions: "What pH values are typical of acids and bases? What pH is considered neutral? What is the pH of pure water?"

Demonstration

pH of Various Materials

Approximate time: 10 min

Prepare small samples of the following for pH testing with pH paper: 1 M NaOH in a dropper bottle, household ammonia, milk of magnesia, lake water or sea water, 1 M sodium hydrogen carbonate solution, distilled water, milk, rainwater, apple juice, orange juice, vinegar, 1 M HCl in a dropper bottle, and lemon juice. Have students note the range of pH represented by these materials.

Safety: Wear safety goggles and a lab apron.

Disposal: Combine all liquids, and neutralize them with 1 M acid or base as appropriate. Pour the neutralized solution down the drain.

Homework ———— BASIC

Ask students to find products at home that make pH claims and to bring these products to class the next day. Make a display of the contributions from all classes.
LS Visual

internet connect
www.scilinks.org
Topic: pH
SciLinks code: HW4095
SCILINKS Maintained by the
National Science
Teachers Association

neutral

describes an aqueous solution that contains equal concentrations of hydronium ions and hydroxide ions

pH

a value used to express the acidity or alkalinity of a solution; it is defined as the logarithm of the reciprocal of the concentration of hydronium ions; a pH of 7 is neutral, a pH of less than 7 is acidic, and a pH of greater than 7 is basic

The Meaning of pH

You have probably seen commercials in which products, such as that pictured in **Figure 11** on the next page, are described as "pH balanced." Perhaps you know that pH has to do with how basic or acidic something is. You may have learned that the pH of pure water is 7 and that acid rain has a lower pH. But what does pH actually mean?

pH and Acidity

When acidity and basicity are exactly balanced such that the numbers of H_3O^+ and OH^- ions are equal, we say that the solution is **neutral.** Pure water is neutral because it contains equal amounts of the two ions.

Two of the solutions listed in **Table 5** are neutral: both have a hydronium ion concentration of 1.00×10^{-7} M. The other solutions are either acidic or basic, depending on whether a strong acid or a strong base was dissolved in water. The solution listed last in **Table 5** was made by dissolving 0.100 mol of NaOH in 1.00 L of water, so it has a hydroxide ion concentration of 0.100 M. Its hydronium ion concentration can be calculated using K_w, as shown below.

$$[H_3O^+] = \frac{[H_3O^+][OH^-]}{[OH^-]} = \frac{1.00 \times 10^{-14}}{0.100} = 1.00 \times 10^{-13}$$

Notice that the hydronium ion concentrations in the listed solutions span a very wide range—in fact a trillionfold range. You can see that working with $[H_3O^+]$ can involve awkward negative exponents. In part to avoid this inconvenience, scientists adopted the suggestion, made by the Danish chemist Søren Sørensen in 1909, to focus not on the value of $[H_3O^+]$ but on the power of 10 that arises when $[H_3O^+]$ is expressed in scientific notation. Sørensen proposed using the negative of the power of 10 (that is, the negative logarithm) of $[H_3O^+]$ as the index of basicity and acidity. He called this measure the **pH.** The letters *p* and *H* represent **p**ower of **h**ydrogen. Keep in mind that because pH is a negative logarithmic scale, a *lower* pH reflects a *higher* hydronium ion concentration.

Table 5 **pH Values at Specified $[H_3O^+]$**		
Solution	**$[H_3O^+]$ (M)**	**pH**
1.00 L of H_2O	1.00×10^{-7}	7.00
0.100 mol HCl in 1.00 L of H_2O	1.00×10^{-1}	1.00
0.0100 mol HCl in 1.00 L of H_2O	1.00×10^{-2}	2.00
0.100 mol NaCl in 1.00 L of H_2O	1.00×10^{-7}	7.00
0.0100 mol NaOH in 1.00 L of H_2O	1.00×10^{-12}	12.00
0.100 mol NaOH in 1.00 L of H_2O	1.00×10^{-13}	13.00

BIOLOGY CONNECTION

Most garden plants perform well when the pH of soil is neutral or slightly acidic (6.0 to 7.0). However, woodland plants, such as blueberries, rhododendrons, and potatoes, prefer more acidic soil (pH of 5.5 to 6.5). Most plants can absorb mineral nutrients when the soil pH is between 6.0 and 7.5. If the pH is below 6.0, the important nutrients nitrogen, phosphorus, and potassium are less easily absorbed. If the pH is above 7.5, iron, phosphorus, and manganese are less available. In addition, acidic soils release heavy metals from the soil in concentrations high enough to adversely affect beneficial organisms. Alkaline soils tend to be saline, which is toxic to plants. Overly acidic soils can be neutralized by adding ground limestone, $CaCO_3$, or wood ashes. Excessive alkalinity can be overcome by treating the soil with sulfur.

Calculating pH from $[H_3O^+]$

Based on Sørensen's definition, pH can be calculated by the following mathematical equation:

$$pH = -\log [H_3O^+]$$

Because of the negative sign, as the hydronium ion concentration increases, the pH will *decrease*. A solution of pH 0 is very acidic. A solution of pH 14 is very alkaline. A solution of pH 7 is neutral.

The equation above may be rearranged to calculate the hydronium ion concentration from the pH. In that form, the equation is as follows:

$$[H_3O^+] = 10^{-pH}$$

When the pH is a whole number, you can do this calculation in your head. For example, if a solution has a pH of 3, its $[H_3O^+]$ is 10^{-3} M, or 0.001 M.

Because pH is related to powers of 10, a change in one pH unit corresponds to a tenfold change in the concentrations of the hydroxide and hydronium ions. Therefore, a solution whose pH is 2.0 has a $[H_3O^+]$ that is ten times greater than a solution whose pH is 3.0 and 100 times greater than a solution whose pH is 4.0.

Figure 11
Soap manufacturers sometimes make claims about the pH of their products.

SKILLS Toolkit

Using Logarithms in pH Calculations

It is easy to find the pH or the $[H_3O^+]$ of a solution by using a scientific calculator. Because calculators differ, check your manual to find out which keys are used for log and antilog functions and how to use these functions.

1. Calculating pH from $[H_3O^+]$
(see Sample Problem B for an example)
Use the definition of pH:

$$pH = -\log [H_3O^+]$$

- Take the logarithm of the hydronium ion concentration.
- Change the sign (+/−).
- The result is the pH.

2. Calculating $[H_3O^+]$ from pH
(see Sample Problem C for an example)
If you rearrange $pH = -\log [H_3O^+]$ to solve for $[H_3O^+]$, the equation becomes

$$[H_3O^+] = 10^{-pH}$$

- Change the sign of the pH (+/−).
- Raise 10 to the negative pH power (take the antilog).
- The result is $[H_3O^+]$.

543

HISTORY
CONNECTION

Søren Sørensen, a Danish biochemist, introduced the concept of pH in 1909 as a tool to simplify his studies of the effects of hydronium concentration on the activity of enzymes. He refined the idea and presented it in 1923 in his research on dyes that can act as acid-base indicators. Chemists did not generally accept the use of the pH scale until the mid-1930s, partly because many early physical, analytical, and inorganic chemists looked with suspicion and disdain on any idea that originated in the realm of biological science or biochemistry.

Activity ——— **BASIC**

To promote student understanding of the logarithmic relationship of pH before they begin to use a calculator, have students work in groups of three or four to estimate the magnitude of each of the following examples:

1. the logarithm of 3.5×10^8
Ans. 8.54

2. the logarithm of 0.000644
Ans. −3.191

3. the antilog of 10.415
Ans. 2.600×10^{10}

4. the antilog of 0.499. Ans. 3.155
LS Interpersonal

Answers to Practice Problems B

1. 2.3

2. 0.70

3. 11.30

4. 13.54

Homework ——— **BASIC**

Additional Practice

1. Calculate pH if $[H_3O^+] = 3.5 \times 10^{-4}$ M. Ans. 3.5

2. What is the pH of a 3.0 M solution of a strong acid? Ans. −0.48

3. Calculate pH if $[OH^-]$ is 5.0×10^{-5} M. Ans. 9.7

4. What is the pH of a solution of the strong acid HCl that contains 0.15 moles per liter? Ans. 0.82
LS Logical

SAMPLE PROBLEM B

Calculating pH for an Acidic or Basic Solution

What is the pH of (a) a 0.000 10 M solution of HNO_3, a strong acid, and (b) a 0.0136 M solution of KOH, a strong base?

1 Gather information.

(a) Concentration of HNO_3 solution = 0.000 10 M = 1.0×10^{-4} M; pH = ?
(b) Concentration of KOH solution = 0.0136 M = 1.36×10^{-2} M; pH = ?
$K_w = 1.00 \times 10^{-14}$

2 Plan your work.

Because HNO_3 and KOH are strong electrolytes, their aqueous solutions are completely ionized.
(a) Therefore, for HNO_3 solution, $[H_3O^+] = 1.0 \times 10^{-4}$ M.
(b) Therefore, for KOH solution, $[OH^-] = 1.36 \times 10^{-2}$ M.
The equation relating pH to $[H_3O^+]$ is pH = −log $[H_3O^+]$. This equation alone is adequate for (a). For (b), you must first use K_w to calculate $[H_3O^+]$ from $[OH^-]$.

3 Plan your work.

Using a scientific calculator and following the instructions under item 1 of **Skills Toolkit 1**, one calculates for (a)

$$pH = -\log [H_3O^+] = -\log (1.0 \times 10^{-4}) = -(-4.00) = 4.00$$

For (b), $[H_3O^+] = \dfrac{K_w}{[OH^-]} = \dfrac{1.00 \times 10^{-14}}{1.36 \times 10^{-2}} = 7.35 \times 10^{-13}$ and then

$$pH = -\log [H_3O^+] = -\log (7.35 \times 10^{-13}) = -(-12.13) = 12.13.$$

4 Verify your results.

Because the solution in (a) is acidic, a pH between 0 and 7 is expected, so the calculated value of 4.00 is reasonable. The solution in (b) will be basic; therefore, a pH between 7 and 14 is expected. Therefore, the answer pH = 12.13 is reasonable.

PRACTICE

1 Calculate pH if $[H_3O^+] = 5.0 \times 10^{-3}$ M.

2 What is the pH of a 0.2 M solution of a strong acid?

3 Calculate pH if $[OH^-] = 2.0 \times 10^{-3}$ M.

4 What is the pH of a solution that contains 0.35 mol/L of the strong base NaOH?

544

---

did you know?

pOH Sometimes it's more convenient to use another measure of the relative acidity or basicity of solutions based on $[OH^-]$ rather than $[H_3O^+]$. This measure is pOH, and it is defined in the same way as pH; namely, pOH = −log $[OH^-]$. In this system, basic solutions have pOH values of less than 7, and acid solutions have pOH values of greater than 7. The pH of a solution plus pOH equals 14. For example, if pOH = 6.6, then pH = 14.0 − 6.6 = 7.4.

Chapter Resource File

• **Problem Solving**

Calculating [H₃O⁺] and [OH⁻] from pH

What are the concentrations of the hydronium and hydroxide ions in a sample of rain that has a pH of 5.05?

1 Gather information.

$pH = 5.05$
$K_w = 1.00 \times 10^{-14}$
$[H_3O^+] = ?$
$[OH^-] = ?$

2 Plan your work.

Because you know the rain's pH, the equation $[H_3O^+] = 10^{-pH}$ can be used to find the hydronium ion concentration.

Then the equation $[H_3O^+][OH^-] = K_w$ can be rearranged and used to find $[OH^-]$.

3 Calculate.

Using a scientific calculator and following the instruction under item 2 of **Skills Toolkit 1**, one finds

$$[H_3O^+] = 10^{-pH} = 10^{-5.05} = 8.9 \times 10^{-6} \text{ M.}$$

$$\text{Next, } [OH^-] = \frac{K_w}{[H_3O^+]} = \frac{1.00 \times 10^{-14}}{8.9 \times 10^{-6}} = 1.1 \times 10^{-9} \text{ M.}$$

The concentrations of the hydronium and hydroxide ions are 8.9×10^{-6} M and 1.1×10^{-9} M, respectively.

4 Verify your results.

The rain's pH is mildly acidic, so the hydronium ion concentration should be more than 1.0×10^{-7} M, and the hydroxide concentration should be less than 1.0×10^{-7} M. Thus, the answers are reasonable.

PRACTICE HINT

For further verification, you could recalculate the pH from the found concentrations by using the methods in Sample Problems B and C.

PRACTICE

1 What is the hydronium ion concentration in a fruit juice that has a pH of 3.3?

2 A commercial window-cleaning liquid has a pH of 11.7. What is the hydroxide ion concentration?

3 If the pH of a solution is 8.1, what is $[H_3O^+]$ in the solution? What is $[OH^-]$ in the solution?

4 Normal human blood has a hydroxide ion concentration that ranges from 1.7×10^{-7} M to 3.5×10^{-7} M, but diabetics often have readings outside this range. A patient's blood has a pH of 7.67. Is there cause for concern?

545

REAL-LIFE
CONNECTION

The pH of stomach acid, HCl, is about 2, which means that $[H_3O^+]$ is 0.01 M. The stomach is protected from the acid by a mucous membrane. This protection does not extend into the esophagus, so when stomach acid rises in the esophagus as a result of indigestion, it can cause a painful burning sensation. This discomfort is called *heartburn* because it seems to come from the location of the heart. The action of stomach acid as a result of continued indigestion or bulimia can permanently damage the tongue, throat, and esophagus. In addition, the acid can damage teeth by dissolving the enamel.

Teaching Tip — **GENERAL**

Emphasize that the pH scale is a logarithmic scale, so an increase in $[OH^-]$ by a factor of 10 results in a pH increase of 1 unit. An increase in $[OH^-]$ by a factor of 100 results in a pH increase of 2 units. Similarly, an increase in $[H_3O^+]$ by factors of 10 and 100 results in decreases in pH of 1 and 2 units, respectively. Ask students by how many units the pH would change if one of the ions increased a thousandfold. Ans. 3 Have students attempt to draw a scale of $[H_3O^+]$ that goes from 10^{-14} M to 1.0 M by using a linear axis, and then by using a logarithmic axis. This exercise should demonstrate the utility of using the pH scale. **LS Logical**

Answers to Practice Problems C

1. $[H_3O^+] = 5.0 \times 10^{-4}$ M
2. $[OH^-] = 5.0 \times 10^{-3}$ M
3. $[H_3O^+] = 7.9 \times 10^{-9}$ M; $[OH^-] = 1.3 \times 10^{-6}$ M
4. $[H_3O^+] = 2.14 \times 10^{-8}$ M; $[OH^-] = 4.7 \times 10^{-7}$ M. The hydroxide ion concentration of 4.7×10^{-7} M does lie somewhat outside the normal range. The patient has mild alkalosis and should be concerned.

Homework — **GENERAL**

Additional Practice

1. What is the hydronium ion concentration of lemon juice, which has a pH of 2.25? Ans. 5.6×10^{-3} M

2. Milk of magnesia has a pH of 10.65. What is $[H_3O^+]$? What is $[OH^-]$? Ans. 2.2×10^{-11} M; 4.5×10^{-4} M

3. Determine the concentrations of hydronium ion and hydroxide ion in stomach acid, which has a pH of 2.0. Ans. $[H_3O^+] = 0.01$ M; $[OH^-] = 1 \times 10^{-12}$ M **LS Logical**

Chapter Resource File

• Problem Solving

Demonstration

Effects of pH on Chlorophyll Stability

(Approximate time: 20 min)

Add approximately 300 mL of tap water to two 600 mL beakers labeled "A" and "B," and heat the beakers with Bunsen burners until the water boils. Check the pH of the boiling water in each beaker.

Slowly add baking soda to the boiling water in beaker B. Evolution of gas may occur rapidly upon the addition of baking soda. Make sure that all baking soda has dissolved completely. Check the pH of the solution in beaker B. It should be approximately 8.

Slowly add frozen green peas to each beaker. After 10 min, check the pH of each beaker's water. The tap water in beaker A will have a pH of about 6 because of the acids that are liberated from the frozen green peas upon heating. The pH of the solution in beaker B should remain at about 8 because an excess of baking soda was added.

Carefully place the cooked peas on two white paper plates that have been labeled "A" and "B." The peas from beaker B should be greener than those from beaker A because chlorophyll, the pigment in green vegetables, decomposes readily upon heating in an acidic environment.

Safety: Wear safety goggles and a lab apron. Students should remain back at least 3 m.

Disposal: Combine all solutions, and neutralize the mixture with 1 M HCl or 1 M NaOH as appropriate. Pour the neutralized solution down the drain.

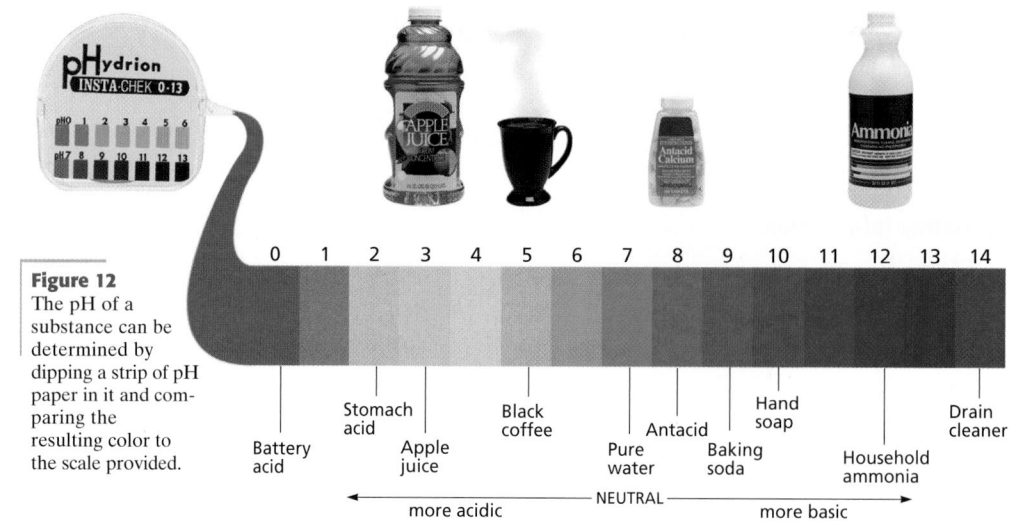

Figure 12
The pH of a substance can be determined by dipping a strip of pH paper in it and comparing the resulting color to the scale provided.

Battery acid • Stomach acid • Apple juice • Black coffee • Pure water • Antacid • Baking soda • Hand soap • Household ammonia • Drain cleaner

← more acidic — NEUTRAL — more basic →

Measuring pH

Measuring pH is an operation that is carried out frequently, for a variety of reasons, in chemical laboratories. There are two ways to measure pH. The first method, which uses indicators, is quick and convenient but does not give very precise results. The second method, which uses a pH meter, is very precise but is more complicated and expensive.

Indicators

indicator

a compound that can reversibly change color depending on the pH of the solution or other chemical change

Certain dyes, known as **indicators,** turn different colors in solutions of different pH. The pH paper pictured in **Figure 12** contains a variety of indicators and can develop a rainbow of colors, each of which corresponds to a particular pH value.

Thymol blue is an example of an indicator. It is yellow in solutions whose pH is between 3 and 8 but blue in solutions whose pH is 10 or higher. **Figure 13** shows the structure of the organic ion responsible for the yellow color—it is a weak acid. The blue form is the conjugate base.

Figure 13
The indicator thymol blue is yellow in neutral and acidic solutions. As [OH⁻] in the solution increases, the indicator turns blue.

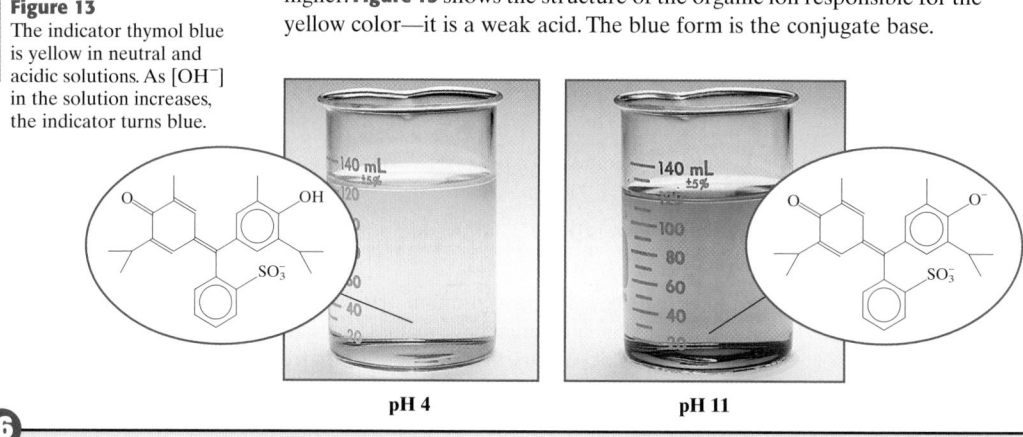

pH 4   pH 11

546

Transparencies

TT pH Values of Some Common Materials

Dozens of indicator dyes are available. Some indicators, such as litmus, are natural products, but most are synthetic. Each indicator has its own colors and its individual range of pH over which it changes shade. By suitably blending several indicators, chemists have prepared "universal indicators," which turn different colors throughout the entire pH range. One such universal indicator is incorporated into the "pH paper" shown in **Figure 12.** By matching the color the paper develops to a standard chart, one can easily measure the solution's approximate pH.

pH Meters

The pH of a solution is being measured by a pH meter in **Figure 14.** A pH meter is an electronic instrument equipped with a probe that can be dipped into a solution of unknown pH. The probe has two electrodes, one of which is sensitive to the hydronium ion. An electrical voltage develops between the two electrodes, and the circuitry measures this voltage very precisely. The instrument converts the measurement into a pH reading, which is displayed on the meter's screen.

After calibration with standard solutions of known pH, a pH meter can measure pH with a precision of 0.01 pH units, which is much greater that the precision of measurements with indicators.

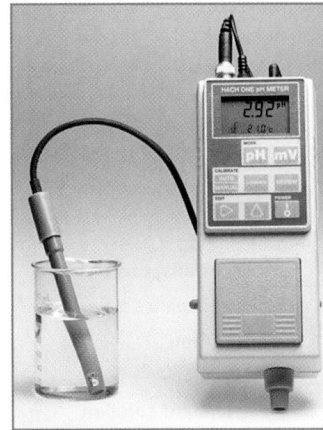

Figure 14
A pH meter is an electrochemical instrument that can measure pH accurately.

② Section Review

UNDERSTANDING KEY IDEAS

1. Describe the relationship between hydronium and hydroxide ion concentrations in an aqueous solution.

2. What does pH measure? How is pH defined?

3. What is a *neutral* solution? What is its pH?

4. Write equations linking the terms K_w, pH, $[H_3O^+]$, and $[OH^-]$.

5. The pH of pancreatic juice is 7.9. Is pancreatic juice acidic or basic?

6. What methods are used to measure pH? Briefly describe how each method works.

PRACTICE PROBLEMS

7. The hydronium ion concentration in a solution is 3.16×10^{-3} mol/L. What is $[OH^-]$? What is the pH?

8. The pH of vinegar is 2.9. Calculate the concentrations of H_3O^+ and OH.

9. If 5.3 g of the strong base NaOH is dissolved in water to form 1500 mL of solution, what are the pH, $[H_3O^+]$, and $[OH^-]$?

10. The $[OH^-]$ of a fruit juice is 3.2×10^{-11} M. What is the pH?

11. What amount in moles of a strong acid such as HBr must be dissolved in 1.00 L of water to prepare a solution whose pH is 2.00?

12. What volume of solution is needed to dissolve 1.0 mol of a strong base such as KOH to make a solution whose pH is 12.5?

CRITICAL THINKING

13. Why is "deionized water" *not* an entirely accurate description of pure water?

14. Can pH be negative? Why or why not?

15. Why would pH paper be unsuitable for measuring blood pH?

547

Answers to Section Review

1. The product of $[H_3O^+]$ and $[OH^-]$ is a constant, K_w, equal to 1.00×10^{-14} at 25°C. Therefore, as one increases, the other decreases.

2. pH measures the acidity of an aqueous solution and is defined as $-\log[H_3O^+]$.

3. a solution that contains equal concentrations of H_3O^+ and OH^-; 7.00 at 25°C

4. pH = $-\log [H_3O^+]$
   $K_w = [H_3O^+][OH^-]$

5. pH values of greater than 7 correspond to basic solutions, so pancreatic juice is basic.

6. Indicators rely on color changes that certain dyes undergo as pH changes. Indicators can be used to roughly measure pH within ranges specific to the indicator. pH meters measure the voltage between two electrodes and convert it to a precise pH reading.

7. $[OH^-] = 3.16 \times 10^{-12}$ M; pH = 2.50

8. $[H_3O^+] = 1.3 \times 10^{-3}$ M; $[OH^-] = 7.7 \times 10^{-12}$ M

9. $[OH^-] = 0.088$ M; $[H_3O^+] = 1.1 \times 10^{-13}$ M; pH = 12.95

10. $[H_3O^+] = 3.1 \times 10^{-4}$ M; pH = 3.51

Answers continued on p. 573A

Close

Reteaching —————— BASIC

Have students make a table that has 15 columns. Tell them to number the columns from 0 to 14 to represent pH values. In the first row, have students write the concentration of hydronium ion that corresponds to each pH value. In the second row, have them write the concentration of the hydroxide ion. Ask them to devise a way to show in which column a solution is neutral. Then, ask them to indicate the direction along the rows in which acidity increases and the direction in which basicity increases. **LS** Logical

Quiz —————— GENERAL

1. Write a short paragraph in which you tell everything that you know about the following information.

$$2H_2O(l) \rightleftharpoons H_3O^+(aq) + OH^-(aq)$$
$$K_w = [H_3O^+][OH^-] = 1.00 \times 10^{-14}$$

   Ans. The equation shows the self-ionization of water. It is an equilibrium reaction between water molecules and water's dissociated ions, H_3O^+ and OH^-, which exist in any aqueous solution. K_w is the equilibrium constant for the reaction, and the equilibrium-constant expression is the product of the concentrations of the hydronium and hydroxide ions. The size of the equilibrium constant reveals that the forward reaction is not favored.

2. Explain how $[H_3O^+]$ is related to pH. **Ans.** pH is the negative log of $[H_3O^+]$.

3. Determine $[OH^-]$ in a 0.050 M solution of the strong acid HI. **Ans.** 2.0×10^{-13} M

4. If the pH of a solution is 8.50, what is $[OH^-]$? **Ans.** 3.2×10^{-6} M
   LS Logical

Chapter Resource File

- **Concept Review**
- **Quiz**

Overview

Before beginning this section, review with your students the Objectives listed in the Student Edition. Students will learn how to write equations for neutralization reactions and how to carry out a neutralization titration. They will learn that the equivalence point is not at the same pH for all titrations and will choose appropriate indicators. Students will use titration data to calculate molarity.

🔊 Bellringer

Ask students to read the first paragraph of this section and to answer the question posed in it. Use their answers to start your discussion about neutralization.

Motivate

Discussion

To help students analyze the question posed in the first paragraph of this page, ask them what is in the acid solution that reacts with metals. Ans. H_3O^+ Ask what is in the base solution that reacts with grease. Ans. OH^- Finally, ask them what happens to these ions when the solutions are combined. Ans. They react to form water.

Identifying Preconceptions

Ask students if the pH of a neutralized solution is 7. Many will agree that it is. Tell them that they are correct if they are considering only solutions of strong acids neutralized by strong bases. Explain the situation differs for neutralizations involving weak acids and bases.

KEY TERMS

- **neutralization reaction**
- **equivalence point**
- **titration**
- **titrant**
- **standard solution**
- **transition range**
- **end point**

neutralization reaction

the reaction of the ions that characterize acids (hydronium ions) and the ions that characterize bases (hydroxide ions) to form water molecules and a salt

OBJECTIVES

1. **Predict** the product of an acid-base reaction.

2. **Describe** the conditions at the equivalence point in a titration.

3. **Explain** how you would select an indicator for an acid-base titration.

4. **Describe** the procedure for carrying out a titration to determine the concentration of an acid or base solution.

Neutralization

The solution of strong acid in the beaker on the left in **Figure 15** contains a high H_3O^+ concentration: high enough to react with and dissolve metals. The solution of strong base on the right is concentrated enough in OH^- to free a grease-clogged drain. Yet when these acidic and basic solutions are mixed in equal amounts, the solution formed has little effect on metal or grease. What has occurred? Because the relationship $[H_3O^+][OH^-] = 1.0 \times 10^{-14}$ must always be true, high concentrations of $H_3O^+(aq)$ and $OH^-(aq)$ cannot coexist. Most of these ions have reacted with each other in a process known as a **neutralization reaction.**

Figure 15
a This beaker contains a solution of nitric acid, a strong acid. This solution turns pH paper red.

b This beaker contains a solution of sodium hydroxide, a strong base. This solution turns pH paper blue.

c The neutralization reaction produces a sodium nitrate solution, which has a neutral pH of 7.

548

Chapter Resource File

• Lesson Plan

All Neutralizations Are the Same Reaction

When solutions of a strong acid and a strong base, having exactly equal amounts of $H_3O^+(aq)$ and $OH^-(aq)$ ions, are mixed, almost all of the hydronium and hydroxide ions react to form water. The reaction is described by the equation below.

$$H_3O^+(aq) + OH^-(aq) \longrightarrow 2H_2O(l)$$

This same reaction happens regardless of the identities of the strong acid and strong base.

Suppose, as in **Figure 16,** that the acid was hydrochloric acid, HCl, and the base was sodium hydroxide, NaOH. When these solutions are mixed, the result will be a solution of only water and the spectator ions sodium and chloride. This is just a solution of sodium chloride. You can prepare the same solution by dissolving common salt, NaCl(s), in water.

You may sometimes see this reaction described as follows:

$$HCl + NaOH \longrightarrow NaCl + H_2O$$

Arrhenius might have said "an acid plus a base produces a salt plus water." This representation can be misleading because the only reactants are $H_3O^+(aq)$ and $OH^-(aq)$ ions and the only product is H_2O.

Figure 16
After hydrochloric acid neutralizes a solution of sodium hydroxide, the only solutes remaining are $Na^+(aq)$ and $Cl^-(aq)$. When the water is evaporated, a small amount of sodium chloride crystals, which will be just like the ones shown, will be left.

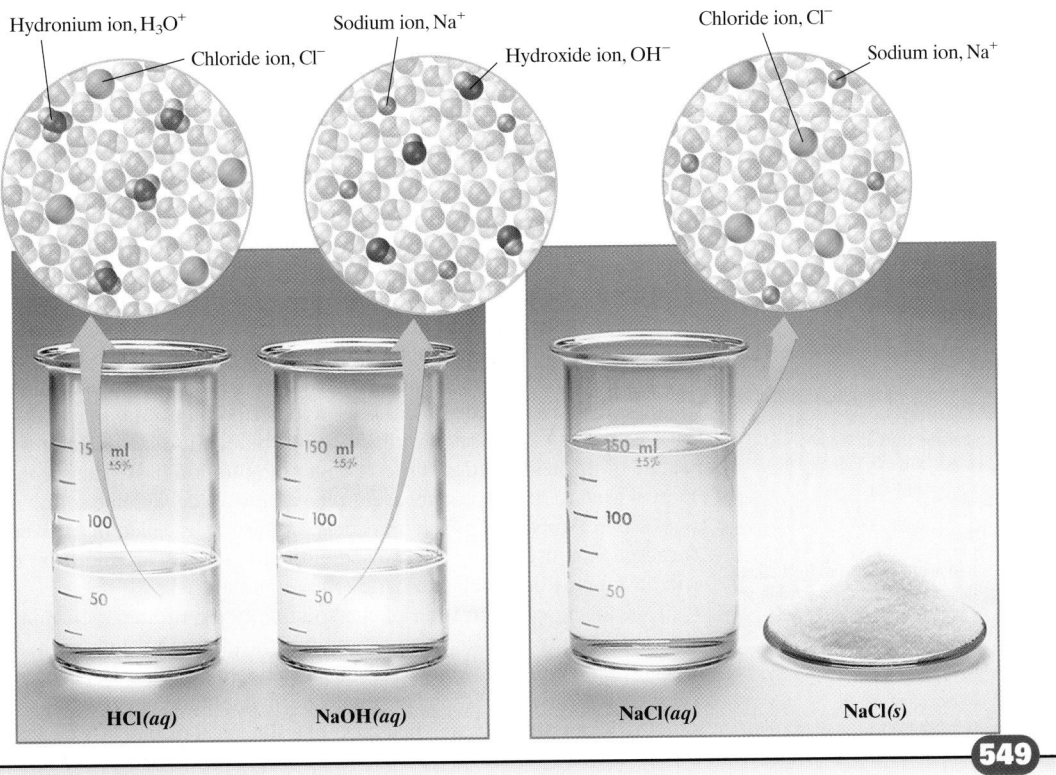

Hydronium ion, H_3O^+ — Chloride ion, Cl^- — Sodium ion, Na^+ — Hydroxide ion, OH^- — Chloride ion, Cl^- — Sodium ion, Na^+

HCl(aq)   NaOH(aq)   NaCl(aq)   NaCl(s)

549

did you know?

Each of the 22 amino acids that make up the proteins of common plants and animals is both an acid and a base. Each contains a carboxylic acid group and an amine group. The amine is located on the carbon atom adjacent to the carboxyl carbon and can be considered a weak base much like a substituted ammonia molecule. When an amino acid is in alkaline solution, it acts as an acid. When an amino acid is in an acid solution, it acts as a base. At some pH, characteristic of the individual amino acid, the molecule is ionized but neutral. That is, the carboxyl group loses its proton and is negatively charged; the amine group gains a proton and is positively charged. This form of an amino acid is called a *zwitterion*.

Teach

Using the Figure ——— BASIC
Refer students to **Figure 16.** Ask students to list the contents of the half-filled beakers. Ask what happens when the two solutions are combined. What has happened to the hydronium and hydroxide ions in the third beaker? Point out that although no hydronium or hydroxide ions are shown in the NaCl solution, they are still present in solution at a concentration of 1×10^{-7} M, which is very small compared with the concentration of Na^+ and Cl^-. Also remind students that mixing a strong acid with a strong base causes a violent reaction and should not be attempted in the laboratory. **LS** Visual

READING SKILL BUILDER ——— BASIC

Brainstorming Have students look up the definitions of *neutral* and *neutralization* in the dictionary. Then, have the class brainstorm areas other than acids and bases in which situations, people, or things are metaphorically neutralized. **LS** Verbal

Activity ——— GENERAL
To make it clear that the reaction of all strong acids with all strong bases results in the same simple ionic neutralization equation, assign an acid and a base to each student in class. Use only the strong acids and bases in **Table 1** and **Table 2.** Have students write the complete ionic equation for their reactions. Then, ask them to eliminate the spectator ions. Each student should end up with the same equation. If not, you can correct any misunderstanding. **LS** Logical

Activity ——————— GENERAL

Tell students that solving titration problems provides an opportunity to review stoichiometry. As always, have students include all units in their calculations and cancel the units to check that their answer has the correct unit. Perform a titration demonstration for the class before students attempt a titration. Have a student write down initial and final readings on the board. Use the data collected in the demonstration, and have students work through the steps with you as you calculate molarity.
LS Logical

Activity ——————— ADVANCED

Ask interested students to use other textbooks or references to find the titration curve for titration of a weak acid with a strong base.
LS Logical

ChemFile   *CHEMISTRY*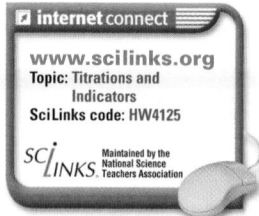

- **Module 8: Strong and Weakly Ionized Species, pH, and Titrations**
  Topic: Titrations

Figure 17
a A titration is done by using a buret, as shown here, to deliver a measured volume of titrant into a solution of unknown concentration.

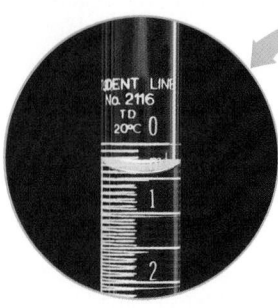

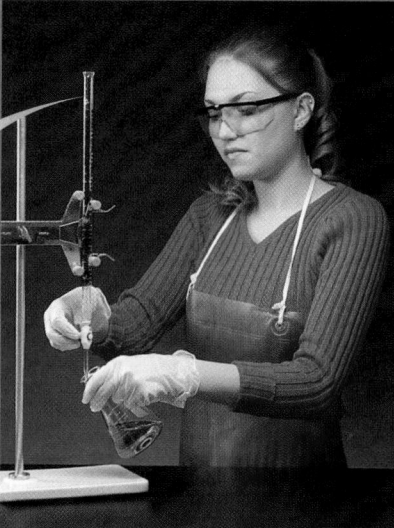

b When reading the liquid level in a buret, you must read the level at the bottom of the meniscus. Here, the reading is 0.42 mL.

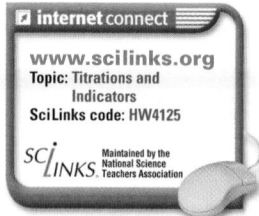

internet connect
www.scilinks.org
Topic: Titrations and Indicators
SciLinks code: HW4125

SC/LINKS Maintained by the National Science Teachers Association

equivalence point

the point at which the two solutions used in a titration are present in chemically equivalent amounts

titration

a method to determine the concentration of a substance in solution by adding a solution of known volume and concentration until the reaction is completed, which is usually indicated by a change in color

titrant

a solution of known concentration that is used to titrate a solution of unknown concentration

standard solution

a solution of known concentration

Titrations

If an acidic solution is added gradually to a basic solution, at some point the neutralization reaction ends because the hydroxide ions become used up. Likewise, if a basic solution is added to an acid, eventually all of the hydronium ions will be used up. The point at which a neutralization reaction is complete is known as the **equivalence point.**

When a solution of a strong base is added to a solution of a strong acid, the equivalence point occurs when the amount of added hydroxide ions equals the amount of hydronium ions originally present. As you have learned, at 25°C this is the point at which both $H_3O^+(aq)$ and $OH^-(aq)$ ions have concentrations of 1.0×10^{-7} M, and the pH is 7.

The gradual addition of one solution to another to reach an equivalence point is called a **titration.** The purpose of a titration is to determine the concentration of an acid or a base. In addition to the two solutions, the equipment needed to carry out a titration usually includes two burets, a titration flask, and a suitable indicator. **Skills Toolkit 2,** later in this section, will describe how this equipment is used to perform a titration.

If an acid is to be titrated with a base, one buret is used to measure the volume of the acid solution dispensed into the titration flask. The second buret is used to deliver and measure the volume of the alkaline solution, as shown in **Figure 17.** The solution added in this way is called the **titrant.** Titrations can just as easily be carried out the other way around. That is, acid titrant may be added to a basic solution in the flask.

To find the concentration of the solution being titrated, you must, of course, already know the concentration of the titrant. A solution whose concentration is already known is called a **standard solution.** The concentration of a standard solution has usually been determined by reacting the solution with a precisely weighed mass of a solid acid or base.

550

HISTORY
CONNECTION

A German chemist named Jeremias Benjamin Richter (1762–1807) was the first to study neutralization reactions quantitatively. In 1792, he published his work, which stated that compounds react in fixed proportions by mass. Richter introduced the term *stoichiometry* to chemistry.

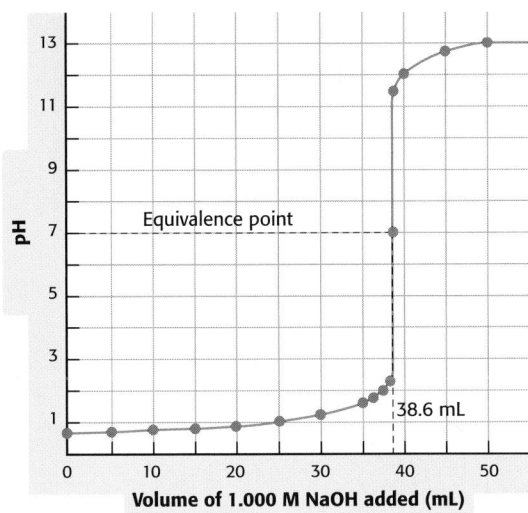

Strong Acid Titrated with Strong Base

Y-axis: pH (13, 11, 9, 7, 5, 3, 1)

Equivalence point

38.6 mL

X-axis: Volume of 1.000 M NaOH added (mL) (0, 10, 20, 30, 40, 50)

Figure 18
This graph of pH versus the volume of 1.000 M NaOH added to an HCl solution indicates that the equivalence point occurred after 38.6 mL of titrant was added.

The Equivalence Point

As titrant is added to the flask containing the solution of unknown concentration, pH is measured. A distinctively shaped graph, called a *titration curve,* results when pH is plotted against titrant volume. **Figure 18** shows a typical example. Because the curve is steep at the equivalence point, it is easy to locate the exact volume that corresponds to a pH of 7.00.

A titration is exact only if the equivalence point can be accurately detected. A pH meter can be used to monitor the pH during the titration, and indicators are also commonly used to detect the equivalence point.

Carrying Out a Titration

Skills Toolkit 2, on the next two pages, has step-by-step instructions to help you carry out an acid-base titration. Study and understand all of the steps *before* you start to perform a titration experiment. If your attention alternates between book and buret, you're likely to make mistakes. Experience helps, and your second titration should be much better than your first.

With each addition of titrant, the indicator will begin to change color but then will go back to its original color as you swirl the flask. The color will fade ever more slowly as the end point gets near. Immediately slow down to a drop-by-drop flow rate. Otherwise, you may miss the end point.

If you do miss the end point by adding too much titrant, however, you do not have to start all over. You can "back-titrate" by adding more unknown solution to the flask until the indicator turns back to its original color. Measure the volume of unknown solution that you added, then slowly add titrant again until the equivalence point is reached. You can then use the total volumes of unknown and titrant in your calculations.

551

Using the Figure — GENERAL

Remind students that the graph in **Figure 18** represents a titration of a strong acid with a strong base. Challenge them to explain why the pH rises so rapidly at the equivalence point. Students should recognize that when the solution has been neutralized, a single drop of NaOH causes a large change in pH. **LS** Visual

Discussion

Students may wonder why the equivalence point of all titrations is not at pH 7. For titrations of strong acids and strong bases, the equivalence point is at pH 7 because there is little or no tendency for the conjugate acid or conjugate base to react with water in a backward reaction. The same is not true for weak acids and weak bases, which can produce conjugate acids and bases that do hydrolyze or react with water in the reverse direction. For example, in the titration of acetic acid, CH_3COOH, the conjugate base CH_3COO^- forms. The acetate ion undergoes the following hydrolysis reaction:

$$CH_3COO^- + H_2O \longrightarrow CH_3COOH + OH^-$$

The hydroxide ion formed causes the pH of the titration mixture at the equivalence point to be greater than 7. Similarly, when the weak base ammonia, NH_3, is titrated, the pH at the equivalence point is less than 7 because the hydrolysis of its conjugate acid, NH_4^+ produces hydronium ions according to the below equation.

$$NH_4^+ + H_2O \longrightarrow NH_3 + H_3O^+$$

Discussion — GENERAL

Ask students to describe the composition of the titration mixture represented by the graph in **Figure 18** at pH 1 and at pH 7. **LS** Visual

Performing a Titration

The following procedure is used to determine the unknown concentration of an acid solution by titrating the solution with a standardized base solution.

Decide which buret will be used for the acid and which will be used for the base. Label each buret to avoid confusion. Rinse the acid buret three times with the acid to be used in the titration. Use the base solution to rinse the other buret in a similar manner.

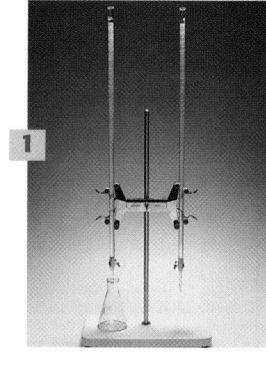

2 Fill the acid buret with the acid solution to a point above the 0 mL mark.

3 Release some acid into a waste flask to lower the volume into the calibrated portion of the buret.

4 Record the volume of the acid in the buret to the nearest 0.01 mL as your starting point.

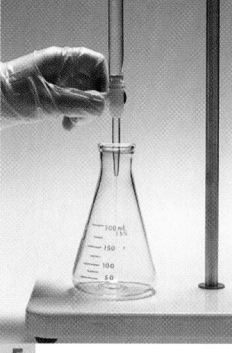

5 Release a volume of acid (determined by your lab procedure) into a clean Erlenmeyer flask.

6 Record the new volume reading, and subtract the starting volume to find the volume of acid added.

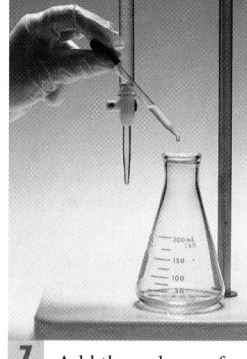

7 Add three drops of an appropriate indicator (phenolphthalein in this case) to the flask.

552

MISCONCEPTION ALERT

Students are advised to titrate until the light pink color of the phenolphthalein indicator remains after 30 s of swirling. Sometimes students are surprised and concerned when they discover that the color has disappeared a few minutes later. They begin to wonder whether they reached the true end point and whether they should add more base. Tell students that the color disappeared because carbon dioxide from the air dissolved in the neutral solution to form carbonic acid, H_2CO_3. The carbonic acid donates a proton to water, and the solution becomes slightly more acidic.

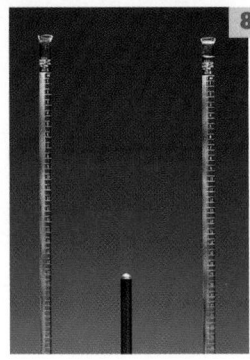

8 Fill the other buret with standardized base solution to a point above the 0 mL mark. Record the concentration of the standardized solution.

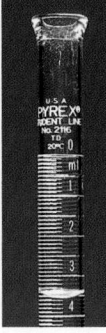

9 Release some base from the buret into a waste flask so that the top of the liquid is in the calibrated portion of the buret.

10 Record the volume of the base to the nearest 0.01 mL as your starting point.

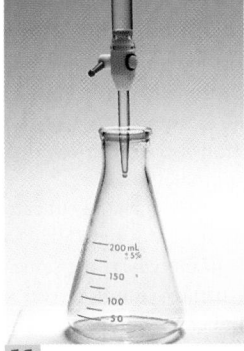

11 Place the flask containing the acid under the base buret. Notice that the tip of the buret extends into the mouth of the flask.

12 Slowly release base from the buret into the flask while constantly swirling the flask. The pink color should fade with swirling.

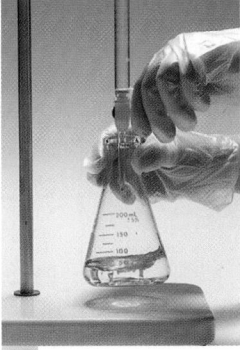

13 Near the end point, add base drop by drop.

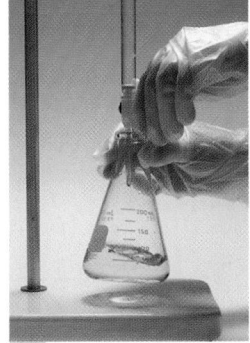

14 The end point is reached when a very light pink color remains after 30 s of swirling.

15 Record the new volume, and determine the volume of base added.

553

Transparencies

TT How to Perform a Titration, Part 1
TT How to Perform a Titration, Part 2

Teaching Tip

Accurate Buret Reading Remind students that a buret should be read to the nearest 0.01 mL by visualizing 10 equal spaces between the 0.1 mL markings.

Homework ——— **BASIC**

1. In preparing for a titration, explain why cleaning burets and eliminating air bubbles are important. **Ans.** When solutions are measured to hundreths of a milliliter, a bit of dirt or a bubble can cause inaccurate results.

2. Why is recording the initial volume necessary? **Ans.** The initial volume is not always 0.00 mL, so it must be recorded and subtracted from the final volume.

3. Why is swirling the flask necessary? **Ans.** Swirling thoroughly mixes the two solutions so that the reaction happens quickly and any local color of the indicator dissipates.

4. What conclusion can you draw if the indicator phenolphthalein is bright red at the end of the titration? **Ans.** You have added too much base.

5. How does it help to have a white background for the flask when performing a titration? **Ans.** It helps you see the faint pink color and not overrun the end point.

6. Is the end point of the titration the same as the equivalence point? **Ans.** No, the end point is the point at which the indicator changes color. The equivalence point is the point at which moles of acid equal moles of base. In a well-planned titration, the end point and the equivalence point should be close.

LS Logical

Demonstration
Indicators

(Approximate time: 5 min)

It is advisable to do a trial run of this demonstration to see how deep the resulting color is. If it is too pale, increase the phenolphthalein concentration.

1. The day before class, mix 10 drops of phenolphthalein indicator with 20 mL of distilled water.

2. Using the indicator-water mixture, make a sign with a saying of your choice on a large piece of white paper, and allow the paper to dry.

3. Hang the paper on the wall or board, and have a bottle of ammonia-based window cleaner on the desktop.

4. Pick up the cleaner, and spray the sign. The message should appear in pink letters. Ask students to explain what happened.

Safety: Wear safety goggles and a lab apron.

Disposal: Pour all solutions down the drain.

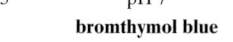

pH 5   pH 7   pH 9
bromthymol blue

pH 7   pH 9   pH 11
phenolphthalein

Figure 19
Bromthymol blue changes color between a pH of 6.0 and 7.6, as in the neutralization of a strong acid and a strong base. Phenolphthalein changes color between a pH of 8.0 and 9.6, as in the neutralization of a weak acid with a strong base.

transition range

the pH range through which an indicator changes color

end point

the point in a titration at which a marked color change takes place

Selecting a Suitable Indicator

All indicators have a **transition range.** In this range, the indicator is partly in its acidic form and partly in its basic form. Thus, the indicator's color is intermediate between those of its acid and base colors. **Figure 19** illustrates the transition range for two typical indicators, bromthymol blue and phenolphthalein.

The instant at which the indicator changes color is the **end point** of the titration. If an appropriate indicator is chosen, the end point and the equivalence point will be the same. In order to determine the concentration of the titrated solution, you must determine the titrant volume at which the indicator changes color.

In titrations of a strong acid by a strong base, the equivalence point occurs at pH 7, and bromthymol blue would be an appropriate indicator, as **Table 6** confirms. However, when a weak acid is titrated by a strong base, the equivalence point is at a pH greater than 7 and thymol blue or phenolphthalein would be a better choice. On the other hand, methyl orange could be the best choice if your titration uses a weak base and a strong acid, because the equivalence point might be at pH 4.

Table 6   Transition Ranges of Some Indicators

Indicator name	Acid color	Transition range (pH)	Base color
Thymol blue	red	1.2–2.8	yellow
Methyl orange	red	3.1–4.4	orange
Litmus	red	5.0–8.0	blue
Bromthymol blue	yellow	6.0–7.6	blue
Thymol blue	yellow	8.0–9.6	blue
Phenolphthalein	colorless	8.0–9.6	red
Alizarin yellow	yellow	10.1–12.0	red

554

did you know?

Indicators Because of the large pH change at the equivalence point of a strong acid–strong base titration, the choice of indicator is not as critical as it is in titrations involving weak acids or bases. Bromthymol blue, thymol blue, and phenolphthalein can be used for strong acid–strong base titrations.

Titration Calculations: From Volume to Amount in Moles

The goal of a titration is to determine either the original concentration of the solution in the titration flask or the original amount of acid or base.

Recall the simple equation, given below, that relates the amount n (in moles) of a solute to the concentration and volume.

$$n = cV$$

Here c is the concentration (in moles per liter) and V is the volume (in liters) of the solution. At the equivalence point in a titration of a strong acid by a strong base, the amount of hydroxide ion added equals the initial amount of hydronium ion. This relationship may be represented as $n_{H_3O^+} = n_{OH^-}$. If each of these amounts is replaced by the corresponding product of concentration and volume, the following equation is the result.

$$(c_{H_3O^+})(V_{H_3O^+}) = (c_{OH^-})(V_{OH^-})$$

This relationship is the one that you will need for most titration calculations. The equation applies whether the titrant is an acid or a base.

SAMPLE PROBLEM D

Calculating Concentration from Titration Data

A student titrates 40.00 mL of an HCl solution of unknown concentration with a 0.5500 M NaOH solution. The volume of base solution needed to reach the equivalence point is 24.64 mL. What is the concentration of the HCl solution in moles per liter?

1 Gather information.

$V_{H_3O^+} = 40.00$ mL $= 0.040\ 00$ L $V_{OH^-} = 24.64$ mL $= 0.024\ 64$ L

$c_{OH^-} = 0.5500$ mol/L $c_{H_3O^+} = ?$

2 Plan your work.

The general equation $(c_{H_3O^+})(V_{H_3O^+}) = (c_{OH^-})(V_{OH^-})$ can be rearranged into the following equation:

$$c_{H_3O^+} = \frac{(c_{OH^-})(V_{OH^-})}{V_{H_3O^+}}$$

3 Calculate.

$$c_{H_3O^+} = \frac{(c_{OH^-})(V_{OH^-})}{V_{H_3O^+}} = \frac{(0.5500\ \text{mol/L})(0.024\ 64\ \text{L})}{0.040\ 00\ \text{L}} = 0.3388\ \text{mol/L}$$

4 Verify your results.

Amounts of hydronium and hydroxide ions should be the same.

$n_{H_3O^+} = c_{H_3O^+}V_{H_3O^+} = (0.3388\ \text{mol/L})(0.040\ 00\ \text{L}) = 0.013\ 55$ mol

$n_{OH^-} = c_{OH^-}V_{OH^-} = (0.5500\ \text{mol/L})(0.024\ 64\ \text{L}) = 0.013\ 55$ mol

Practice problems on next page

555

PRACTICE HINT

If you get confused, remember to keep track of amounts of acid and base. The rest is just a matter of converting molarity to moles by multiplying molarity by volume.

Homework ——— GENERAL

Additional Practice

1. A volume of 20.00 mL of a solution of HNO_3 that has an unknown concentration is titrated with 34.37 mL of a 0.8220 M solution of NaOH. What is the concentration of the HNO_3 solution? Ans. 1.413 M HNO_3

2. A lab worker makes up 1000.00 mL of a KOH solution but forgets to record the mass of dissolved KOH. When a 42.82 mL sample of the solution is titrated with a 1.209 M solution of HCl, 28.35 mL of the acid solution are required to reach the equivalence point. What is the concentration of the KOH solution, and what mass of KOH was dissolved? Ans. 0.8004 M KOH; 44.91 g KOH

3. What volume of a 1.366 M solution of NaOH would be required to titrate 47.22 mL of a 2.075 M solution of H_2SO_4? Ans. 143.5 mL (note that H_2SO_4 has two ionizable protons)

4. A solution of sodium hydroxide was made by dissolving 4.500 g NaOH in water to form 1.000 L of solution. Then, 25.00 mL of the solution were titrated with 0.1020 M HCl. How many milliliters of HCl were required? Ans. 27.57 mL

LS Logical

Chapter Resource File

• Problem Solving

VISUAL-ARTS CONNECTION — GENERAL

Acid-base reactions are destroying the art and architecture of ancient civilizations. Many of the world's most beautiful statues and buildings have stood in the open for centuries without significant change. However, the Industrial Revolution marked the beginning of the widespread burning of fossil fuels, which has escalated during the past two centuries and contributes to much higher concentrations of corrosive pollutants in the atmosphere. Ask interested students to investigate the reactions that are causing these changes and the methods that scientists are using to help protect the treasures of Greece, Rome, and other ancient sites. LS Verbal

Answers to Practice Problems D

1. 6.9×10^{-3} M
2. 0.585 M
3. 4.674×10^{-3} moles
4. 2.31×10^{-5} M

Close

Reteaching ——————— BASIC

Have each student make up a titration problem that requires the calculation of either a volume of solution or a molarity. Have students choose an acid from among the strong acids in **Table 1** and a base from among the strong bases in **Table 2**. Students should solve their problems and then exchange problems with another student.
LS Interpersonal

Quiz ——————— GENERAL

1. Write a stepwise description of the process of titration. **Ans.** Students should include the steps in **Skills Toolkit 2.**

2. How is an indicator choosen for a titration? **Ans.** The indicator should change color at the expected pH of the equivalence point.

3. How many milliliters of 0.1000 M HCl would be required to completely titrate 25.00 mL of 0.2500 M NaOH? **Ans.** 62.50 mL
LS Logical

Chapter Resource File
- Concept Review
- Quiz

PRACTICE

1. If 20.6 mL of 0.010 M aqueous HCl is required to titrate 30.0 mL of an aqueous solution of NaOH to the equivalence point, what is the molarity of the NaOH solution?

2. In the titration of 35.0 mL of drain cleaner that contains NaOH, 50.08 mL of 0.409 M HCl must be added to reach the equivalence point. What is the concentration of the base in the cleaner?

3. Titrating a sludge sample of unknown origin required 41.55 mL of 0.1125 M NaOH. How many moles of H_3O^+ did the sample contain?

4. Neutralizing 5.00 L of an acid rain sample required 11.3 mL of 0.0102 M KOH. Calculate the hydronium ion concentration in the rain sample.

3 Section Review

UNDERSTANDING KEY IDEAS

1. What are the reactants and the product common to all neutralization reactions?

2. Define *equivalence point*. How does the equivalence point differ from the *end point* of a titration?

3. What are standard solutions, and how are they standardized?

4. How would you choose an indicator for titrating a strong acid with a strong base?

5. What titration data are needed to calculate an unknown acid concentration?

6. What are the roles of the two burets in a titration experiment?

7. At the equivalence point of a titration, what is present in the solution?

PRACTICE PROBLEMS

8. If 29.5 mL of 0.150 M HCl neutralizes 25.0 mL of a basic solution, what was $[OH^-]$ in the basic solution?

9. What volume of 0.250 M nitric acid is needed to neutralize 17.35 mL of 0.195 M KOH solution?

10. In a titration of 30.00 mL of 0.0987 M HBr solution with a strong base of unknown concentration, the pH reached 7 after the addition of 37.43 mL of titrant. What was the concentration of the base?

11. If it took 72 mL of 0.55 NaOH titrant to neutralize 220 mL of an acidic solution, what was the hydronium ion concentration in the acidic solution?

12. In a titration of a sample of 0.31 M HNO_3, it took 75 mL of a 0.24 M KOH solution to reach a pH of 7. What was the volume of the sample?

CRITICAL THINKING

13. What indicator would you choose for the titration of acetic acid with potassium hydroxide?

14. Why is the steepness of a titration curve helpful in locating the equivalence point?

15. Explain why the titration of a strong acid with a weak base ends at a pH lower than 7.

Answers to Section Review

1. All neutralization reactions involve the reaction of hydronium and hydroxide ions to form water.

2. The equivalence point is the point in a titration at which the two solutions are present in chemically equivalent amounts. The end point is the point in a titration at which the indicator changes color.

3. solutions of accurately known concentration, by titrating a known amount of reactant, usually a solid sample

4. Choose an indicator which changes color close to pH 7.

5. the concentration and volume of the base solution and the volume of the acid solution

6. One buret is used to dispense and accurately measure the volume of the solution to be titrated, the other buret is used to dispense and accurately measure the volume of titrant.

7. only equilibrium amounts of hydronium and hydroxide ions, plus whatever spectator ions are left

8. 0.177 M

9. 13.5 mL

10. 0.0791 M

Answers continued on p. 573A

Equilibria of Weak Acids and Bases

KEY TERMS

- acid-ionization constant, K_a
- buffer solution

OBJECTIVES

(1) **Write** an equilibrium equation that shows how a weak acid is in equilibrium with its conjugate base.

(2) **Calculate** K_a from the hydronium ion concentration of a weak acid solution.

(3) **Describe** the components of a buffer solution, and explain how a buffer solution resists changes in pH.

Weak Acids and Bases

Consider the reaction represented by the following equation, in which one arrow is longer than the other:

$$A(aq) + B(aq) \xrightleftharpoons{} C(aq) + D(aq)$$

Chemists use this notation to indicate that the forward reaction is *favored*. In other words, when the reaction has reached equilibrium, there will be more products than reactants.

Some Acids are Better Proton Donors Than Others

Some aspects of formic acid, HCOOH, are illustrated in **Figure 20.** Formic acid is a typical Brønsted-Lowry acid, able to donate a proton to a base, such as the acetate ion, CH_3COO^-. Thus, in a solution prepared by dissolving formic acid and sodium acetate in water, a reaction will occur.

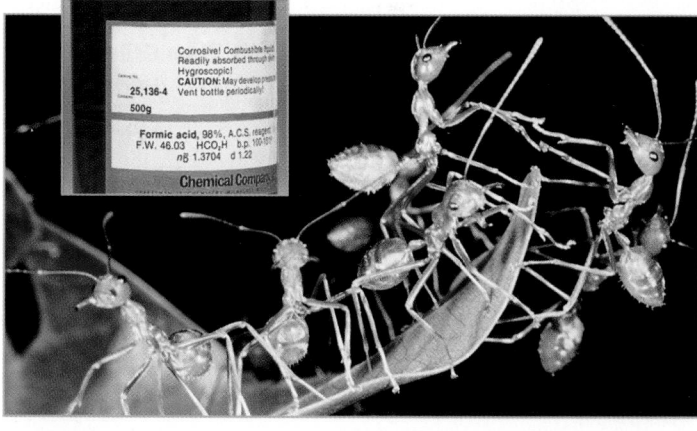

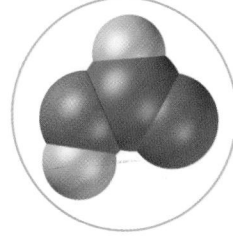

Figure 20
The name *formic acid* comes from *formica*, the Latin word for "ant." Formic acid was first isolated by distillation from ants in 1670. A molecular model of formic acid is shown.

557

BIOLOGY
CONNECTION

Students have heard about DNA in their previous science classes, but they may not have focused on the significance of the name. Deoxyribonucleic acid is a polyprotic acid. The two spines of the helical molecule are made up of phosphoric acid groups alternating with ribose sugar groups, so one molecule of DNA contains thousands of proton donors. In addition, pairs of bases occupy the center of the DNA helix; each pair is held together by hydrogen bonds.

Focus

Overview

Before beginning this section, review with your students the Objectives listed in the Student Edition. In this section, students will write equations for the equilibrium between a weak acid and its conjugate base. They will learn that acids and bases have different abilities to donate or accept a proton and that K_a is a measure of acid strength. Students will find out how buffers control pH.

🔔 Bellringer

Refer students to **Figure 20,** and ask them to write the equation for the reaction of formic acid with water. You can use their equations to begin the discussion of weak acids.

Motivate

Identifying Preconceptions

The concept of conjugate acids and bases is often confusing to students. Have students note that each acid-base pair in **Table 3** differs from one another by one proton. Help students understand that if a molecule or ion donates a proton to another species, the resulting molecule or ion can then accept a proton from another species to re-form the original molecule or ion. As an example, write the equation for the ionization equilibrium of HCN on the board. HCN ionizes weakly in water to form hydronium ions and CN^- ions.

$$HCN + H_2O \rightleftharpoons H_3O^+ + CN^-$$

Have students look at the left side of the equation. Point out that HCN acts as an acid by donating a proton to H_2O to form H_3O^+. However, the backward reaction occurs readily. In the backward reaction, students should see that CN^- acts as a base by accepting a proton from H_3O^+ to re-form HCN. Ask students to identify the acid and its conjugate base as well as the base and its conjugate acid.

Chapter Resource File

- Lesson Plan

Activity ——— BASIC

Assign one of the acids in the second column of **Table 7** to each student, and ask students to write the equation for the ionization of their acid on the board. Check for any mistakes so that you can address any misconceptions. **LS** Logical

Teach

Discussion

Help students understand the italicized statement at the bottom of the page by providing extra examples. Discuss the ionization of HNO_3, as shown below.

$$HNO_3 + H_2O \longrightarrow$$
strong acid   weak acid/base

$$H_3O^+ + NO_3^-$$
strong acid   weak base

The conjugate base of HNO_3 is NO_3^-. Nitric acid is strong because it loses a proton readily, even to a weak base like H_2O. It follows, then, that the conjugate base of nitric acid, NO_3^-, would have little tendency to regain (accept) a proton and become HNO_3 again. Thus, NO_3^- is a weak base.

What happens when this nitric acid solution is combined with a KOH solution that consists of K^+, OH^-, and H_2O?

$$H_3O^+ + NO_3^- + K^+ + OH^- \longrightarrow$$
strong acid   weak base   strong base

$$2H_2O + K^+ + NO_3^-$$

In this mixture, OH^- is a much better proton acceptor (stronger base) than NO_3^- is, so it reacts with H_3O^+ to leave NO_3^- as a spectator ion.

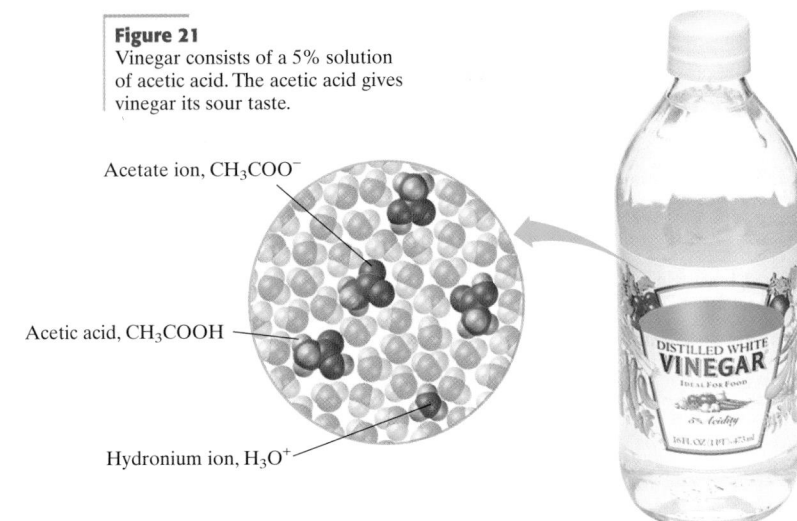

Figure 21
Vinegar consists of a 5% solution of acetic acid. The acetic acid gives vinegar its sour taste.

Acetate ion, CH_3COO^-

Acetic acid, CH_3COOH

Hydronium ion, H_3O^+

The reaction will produce the conjugate base of the formic acid, the formate ion, $HCOO^-$, and the conjugate acid of the acetate ion, acetic acid, CH_3COOH, as shown below.

$$HCOOH(aq) + CH_3COO^-(aq) \underset{}{\overset{}{\rightleftharpoons}} HCOO^-(aq) + CH_3COOH(aq)$$
acid   acid

Acetic acid is the active ingredient in vinegar, as shown in **Figure 21**. The unequal arrows in the equation above indicate that if you dissolved equal amounts of all four substances in water, the concentrations of $HCOO^-$ and CH_3COOH would be greater than the concentrations of $HCOOH$ and CH_3COO^- at equilibrium. If you think of this as a contest between the two acids to see which is better able to donate protons, formic acid would be the winner. $HCOOH$ is more willing to lose a proton than CH_3COOH is. Therefore, formic acid is considered a *stronger* acid than acetic acid is.

Some Bases Accept Protons More Readily Than Others

Now look at the same reaction from the standpoint of the two bases:

$$HCOOH(aq) + CH_3COO^-(aq) \underset{}{\overset{}{\rightleftharpoons}} HCOO^-(aq) + CH_3COOH(aq)$$
base   base

Both bases can accept protons, but the acetate ion, CH_3COO^-, has been more successful in accepting protons than the formate ion, $HCOO^-$, has. The formate ion is a weaker base than the acetate ion is. At equilibrium, there is more formate ion than acetate ion in solution.

In this example, formic acid is the stronger acid, but its conjugate base, the formate ion, is the weaker base. This example illustrates a general principle: *In an acid-base reaction, the conjugate base of the stronger acid is the weaker base, and the conjugate acid of the stronger base is the weaker acid.*

EARTH SCIENCE
CONNECTION ——— GENERAL

Limestone caves are some of the most impressive geological features of the United States and other parts of the world. Gigantic rooms and passages decorated with formations called *stalagmites* and *stalactites* sometimes extend for miles. These caves were hollowed out by the action of acids and bases. Some students may be interested in learning more about the chemistry of cave formation. Suggest that they investigate a nearby cave or a cave that they have visited or read about. Have them make posters entitled "Cave Formation" that shows the pertinent equations. Have them present their posters to the class. **LS** Verbal

Table 7 Relative Strengths of Acids and Bases

Acid	Formula	K_a of acid	Conjugate base	Formula
Hydronium ion	H_3O^+	5.53×10^1	water	H_2O
Hydrogen sulfate ion	HSO_4^-	1.23×10^{-2}	sulfate ion	SO_4^{2-}
Phosphoric acid	H_3PO_4	7.52×10^{-3}	dihydrogen phosphate ion	$H_2PO_4^-$
Formic acid	HCOOH	1.82×10^{-4}	formate ion	$HCOO^-$
Benzoic acid	C_6H_5COOH	6.46×10^{-5}	benzoate ion	$C_6H_5COO^-$
Acetic acid	CH_3COOH	1.75×10^{-5}	acetate ion	CH_3COO^-
Carbonic acid	H_2CO_3	4.30×10^{-7}	hydrogen carbonate ion	HCO_3^-
Dihydrogen phosphate ion	$H_2PO_4^-$	6.31×10^{-8}	monohydrogen phosphate ion	HPO_4^{2-}
Hypochlorous acid	HOCl	2.95×10^{-9}	hypochlorite ion	ClO^-
Ammonium ion	NH_4^+	5.75×10^{-10}	ammonia	NH_3
Hydrogen carbonate ion	HCO_3^-	4.68×10^{-11}	carbonate ion	CO_3^{2-}
Monohydrogen phosphate ion	HPO_4^{2-}	4.47×10^{-13}	phosphate ion	PO_4^{3-}
Water	H_2O	1.81×10^{-16}	hydroxide ion	OH^-
Conjugate acid	**Formula**	**K_a of acid**	**Base**	**Formula**

Increasing acid strength ↑ (left axis) *Increasing base strength* ↓ (right axis)

The Acid-Ionization Constant

The strengths of acids may be described in relative terms of *stronger* or *weaker,* but the strength of an acid may also be expressed quantitatively by its **acid-ionization constant, K_a.** This is just the equilibrium constant, K_{eq}, that describes the ionization of an acid in water.

Consider the following equation, which describes the equilibrium established when acetic acid dissolves in water.

$$CH_3COOH(aq) + H_2O(l) \rightleftharpoons H_3O^+(aq) + CH_3COO^-(aq)$$

The equilibrium expression for this reaction is written as follows:

$$\frac{[H_3O^+][CH_3COO^-]}{[CH_3COOH]} = K_a = 1.75 \times 10^{-5}$$

Recall that only solutes appear in equilibrium expressions. When water is a solvent, it is omitted. Remember, too, that K_a is unitless.

Table 7 lists many acid-ionization constants. Note that the stronger the acid is, the weaker its conjugate base is. Accordingly, the stronger the base is, the weaker its conjugate acid is.

acid-ionization constant, K_a

the equilibrium constant for a reaction in which an acid donates a proton to water

Topic Link

Refer to the "Chemical Equilibrium" chapter for a discussion of equilibrium constants.

559

Activity

Students will probably recognize the weak acids in the acid list in **Table 7.** However, with the exception of ammonia and the hydroxide ion, students may not recognize as bases many of the ions in the base list. To reinforce that these ions are bases, dissolve a small amount of sodium carbonate, Na_2CO_3, and sodium phosphate, Na_3PO_4 (sold in paint-supply stores as TSP, trisodium phosphate), in 200 mL of water in two labeled 400 mL beakers. Use litmus paper, pH paper, or a pH meter to demonstrate the alkalinity of these solutions. Write on the board the corresponding equations for the hydrolysis of the ions.

$$CO_3^{2-} + H_2O \longrightarrow HCO_3^- + OH^-$$
$$PO_4^{3-} + H_2O \longrightarrow HPO_4^{2-} + OH^-$$
$$HPO_4^{2-} + H_2O \longrightarrow H_2PO_4^- + OH^-$$

These hydrolysis reactions produce fairly large concentrations of hydroxide ions, which result in solutions of high pH.

One-Stop Planner CD-ROM

- **Supplemental Reading Projects** Guided Reading Worksheet: "Chemistry and Physics in the Kitchen"

did you know?

Ranking Strong Acids K_a values provide a way of distinguishing the relative strengths of weak acids, but in aqueous solution, it is impossible to measure the relative strengths of strong acids. All strong acids are equally strong when measured against the base H_2O because all are completely ionized in water. To determine a ranking of acid strength for strong acids, one would have to test their proton-donating capability against a base that is weaker than H_2O, such as ethanol, C_2H_5OH, a compound that is usually not considered a base. In such an investigation, ethanol would take the place of water as the solvent for the reaction.

Answers to Practice Problem E

1. $[H_3O^+] = 1.62 \times 10^{-3}$ M

2. $K_a = 3.4 \times 10^{-8}$

3. $K_a = 6.4 \times 10^{-5}$

4. $[HCOO^-] = 3.9 \times 10^{-3}$ M

Homework ——— GENERAL

1. What is $[H_3O^+]$ in a 0.250 M solution of benzoic acid, C_6H_5COOH? Ans. 4.02×10^{-3} M

2. In a 0.025 M solution of formic acid, the hydronium ion concentration is 2.03×10^{-3} M. Calculate the K_a for HCOOH. Ans. $K_a = 1.65 \times 10^{-4}$

3. $[H_3O^+]$ in a 1.20 M solution of dibromoacetic acid is 0.182 M. Calculate K_a for this acid. Ans. $K_a = 2.76 \times 10^{-2}$

LS Logical

Teaching Tip

A Simplifying Assumption In actuality, the concentration of un-ionized acid, which appears in the denominator of the acid-ionization constant expression, is the initial concentration minus the concentration of the hydronium ion. The sample problem makes the simplifying assumption that when K_a is small and the initial concentration of acid is reasonably large, very little hydronium ion is formed compared to the concentration of the acid. Solving the problem without this simplifying assumption requires solving a quadratic equation, which is unnecessarily tedious for most purposes.

SAMPLE PROBLEM E

Calculating K_a of a Weak Acid

A vinegar sample is found to have 0.837 M CH_3COOH. Its hydronium ion concentration is found to be 3.86×10^{-3} mol/L. Calculate K_a for acetic acid.

1 Gather information.

$[CH_3COOH] = 0.837$ M
$[H_3O^+] = 3.86 \times 10^{-3}$ M
$K_a = ?$

2 Plan your work.

The equation for the equilibrium is

$$CH_3COOH(aq) + H_2O(l) \rightleftharpoons H_3O^+(aq) + CH_3COO^-(aq)$$

which establishes that the expression for K_a is

$$K_a = \frac{[H_3O^+][CH_3COO^-]}{[CH_3COOH]}$$

The equation also shows that hydronium and acetate ions are produced in equal amounts, so $[CH_3COO^-] = [H_3O^+]$. Hence, all of the necessary concentration data are known.

3 Calculate.

$$K_a = \frac{[H_3O^+][CH_3COO^-]}{[CH_3COOH]} = \frac{[H_3O^+][H_3O^+]}{[CH_3COOH]} =$$

$$\frac{(3.86 \times 10^{-3})(3.86 \times 10^{-3})}{0.837} = 1.78 \times 10^{-5}$$

4 Verify your results.

The calculated acid-ionization constant is very close to the value listed in **Table 7,** so the answer seems reasonable.

> **PRACTICE HINT**
>
> Earlier in this chapter, a sample problem demonstrated how to calculate $[H_3O^+]$ from pH. In some K_a problems, you may need to perform this step first.

PRACTICE

1 Calculate $[H_3O^+]$ of a 0.150 M acetic acid solution.

2 Find K_a if a 0.50 M solution of a weak acid has a hydronium ion concentration of 1.3×10^{-4} M.

3 A solution prepared by dissolving 1.0 mol of benzoic acid in water to form 1.0 L of solution has a pH of 2.1. Calculate the acid-ionization constant.

4 Use **Table 7** to calculate the concentration of formate ion in 0.085 M formic acid.

560

TECHNOLOGY — CONNECTION

Acetic acid is the most important organic hydrocarbon acid. The solution called *vinegar,* which is 5% acetic acid, is made by the fermentation and oxidation of carbohydrates, such as those found in apple juice. However, large amounts of acetic acid are produced synthetically for many industrial uses. One process starts with acetylene, C_2H_2, which is hydrated to produce acetaldehyde, CH_3CHO, and then oxidized to acetic acid. In another process, ethyl alcohol is oxidized directly to acetic acid. Large amounts of acetic acid are converted to both metal and organic acetates. The latter are used as solvents for resins, paints, and lacquers. Acetic acid is also used in film coatings, plastics, fabrics, and many other products.

Table 8 **Typical pH Values of Human Body Fluids**

Solution	pH	Solution	pH
Gastric juice	1.5	Blood	7.4
Urine	6.0	Tears	7.4
Saliva	6.5	Pancreatic juice	7.9
Milk	6.6	Bile	8.2

www.scilinks.org
Topic: Buffers
SciLinks code: HW4023
*SCI*LINKS. Maintained by the National Science Teachers Association

Buffer Solutions

You can see in **Table 8** that the pH of blood is 7.4. Keeping your blood pH between 7.35 and 7.45 is vital to your health. If your blood's pH goes outside this very narrow range, you will become ill. If your blood pH is lower than 7.35, you suffer *acidosis.* If your blood's pH rises above 7.45, symptoms of *alkalosis* appear. How does your body control the pH of blood within such narrow bounds? Your body relies on the properties of **buffer solutions**—solutions that resist changes in pH that would otherwise be caused by the addition of acids or bases. These solutions are said to be "buffered" against pH changes.

A Buffer Has Two Ingredients

A buffer solution, often simply called a *buffer,* is a solution that contains approximately equal amounts of a weak acid and its conjugate base.

Imagine preparing two solutions. In the first, you dissolve one mole of sodium acetate in one liter of water. Sodium acetate is a strong electrolyte and ionizes completely in solution.

$$CH_3COONa(s) \longrightarrow CH_3COO^-(aq) + Na^+(aq)$$

For the second solution, you prepare one mole of acetic acid in one liter of water. Acetic acid is a weak acid that ionizes very little in water. The following equilibrium equation describes the solution:

$$CH_3COOH(aq) + H_2O(l) \underset{\longrightarrow}{\longleftarrow} H_3O^+(aq) + CH_3COO^-(aq)$$

As the unequal arrows suggest, this equilibrium favors the reactants on the left side. About 99.6% of the acetic acid is un-ionized. Its pH is 2.4.

Now mix the two solutions. Both contain the acetate ion, so the common ion effect comes into play. Recall that Le Châtelier's principle predicts that the equilibrium will adjust to reduce the stress imposed by the increase in the $CH_3COO^-(aq)$ concentration. It does this by shifting even more heavily toward the left. In fact, now 99.996% of the acetic acid is un-ionized. The pH has doubled to 4.8.

The mixture is a buffer solution. It contains nearly equal amounts of the weak acid acetic acid, $CH_3COOH(aq)$, and its conjugate base, the acetate ion, $CH_3COO^-(aq)$. It is not necessary that the acid and its conjugate base be present in equal amounts to act as a buffer, but there must be a substantial concentration of each.

buffer solution

a solution made from a weak acid and its conjugate base that neutralizes small amounts of acids or bases added to it

Topic Link

Refer to the "Chemical Equilibrium" chapter for a discussion of Le Châtelier's principle.

561

did you know?

Acidosis and Alkalosis Acidosis is a serious condition that occurs when the blood is overwhelmed by excess acid and the buffer system is insufficient to cope with it. The condition can lead to death. Most students will have experienced a symptom of mild acidosis when they have suffered from leg cramps after exhausting exercise. The cramps are caused by excess lactic acid deposited in muscle tissue. Lactic acid, $CH_3CHOHCOOH$, is a carboxylic acid found in sour milk. Alkalosis can be caused by hyperventilation, in which the body loses CO_2 too quickly. Symptoms include confusion, lightheadedness, and muscle tremors.

Demonstration
Buffers in Solution
(Approximate time: 10 min)

1. Make a pH 7 buffer solution by dissolving 6.8 g of KH_2PO_4 in water. Add 295 mL of 0.1 M KOH solution, and dilute to 1 L.

2. Pour 50 mL of distilled water into a 250 mL beaker, and pour 50 mL of the buffer solution into another beaker. Place the beakers on the overhead projector.

3. Add three drops of phenolphthalein indicator to each beaker, and mix well. Using a dropper, add 0.1 M KOH dropwise to both solutions, and swirl the beaker after adding each drop. The water should change after a few drops, but the buffer solution will require more than 7 mL of the base before a color change occurs.

4. Repeat the demonstration by using methyl orange or bromcresol green as the indicator, and add 0.1 M HCl dropwise. Again, only a few drops are needed in the water, but several milliliters are needed to change the color in the buffer solution.

Safety: Wear safety goggles and a lab apron. Students should remain at least 3 m from the demonstration.

Disposal: Combine all solutions, and neutralize the mixture with 1 M HCl or 1 M NaOH as appropriate. Pour the neutralized solution down the drain.

Discussion — GENERAL

Perform the demonstration, and challenge students to explain the results in terms of the $H_2PO_4^- + H_2O \rightleftharpoons HPO_4^{2-} + H_3O^+$ equilibrium. Have students identify the weak acid and its conjugate base.
LS Logical

Vocabulary Ask students to look up the word *buffer* in the dictionary, or read all of the definitions aloud. Have students think of situations other than in chemistry in which the word would be used. Have them create sentences that would be used in those contexts. Finally, ask if the word seems appropriate as it is used in acid-base chemistry. **LS** Verbal

Activity ——— ADVANCED

Have students design an activity that would show how buffered aspirin differs from ordinary aspirin. If you approve their design, you could let them perform the activity. **LS** Logical

Teaching Tip

Buffering Blood Although several buffer systems operate to keep the pH of blood within healthy limits, the principal system is described by the following equation:

$$CO_2(g) + H_2O(l) \rightleftharpoons H_2CO_3(aq)$$
$$\rightleftharpoons H^+(aq) + HCO_3^-(aq)$$

If excess acid invades the blood, the equilibrium shifts to the left to produce more CO_2, which is expelled by an automatic increase in breathing rate. If the blood must accommodate excess base, the reaction shifts to the right as the hydrogen ion reacts with the base. Breathing rate automatically decreases to again increase the concentration of CO_2 and shift the equilibrium back to the left.

Figure 22
The left-hand beaker in each photo contains a neutral solution. The right-hand beaker in each photo contains the same solution plus hydrochloric acid.

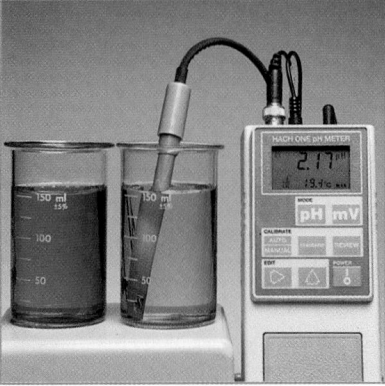
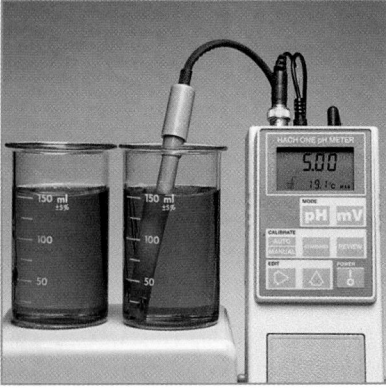

a When a small amount of HCl is added to an unbuffered solution, the solution's pH drops significantly.

b When the same amount of HCl is added to a buffered solution, the pH of the solution does not change very much.

Buffer Solutions Stabilize pH

How do buffer solutions prevent large changes in pH when small amounts of acid or base are added, as demonstrated in **Figure 22**? Le Châtelier's principle can help us understand the effect. If HX is a weak acid and X^- is its conjugate base, then in a buffer solution composed of the two, the following equilibrium is established:

$$HX(aq) + H_2O(l) \rightleftharpoons H_3O^+(aq) + X^-(aq)$$

If a base is added to the buffer solution, the base will react with the H_3O^+ and remove some of this ion from solution. According to Le Châtelier's principle, the equilibrium will adjust by shifting to the right to make more H_3O^+, preventing too great a pH change. It is a similar story if an acid is added. The tendency for $[H_3O^+]$ to increase is countered by a shift of the equilibrium to the left and the formation of more HX molecules.

The greater the concentrations of the two buffer components, the greater the ability of the buffer to resist changes in pH. The efficiency of the buffer is greatest when the concentrations of the two components are equal, but this condition is not necessary for the buffer to work.

The equilibrium-constant expression for the reaction above is simply $[H_3O^+][X^-]/[HX] = K_a$. From this expression, it is easy to see that when the concentration of each member of the conjugate pair is equal, $[H_3O^+] = K_a$. Thus the pH of such a buffer solution is $-\log(K_a)$.

Buffers Are All Around

Now you understand what manufacturers of shampoos and antacids mean when they say that their products are buffered: the products have ingredients that resist pH changes. The pH of foods affects their taste and texture, so many packaged foods are buffered, too. Check ingredient labels for phosphates. The presence of phosphates probably means that the product contains the acid-base pair $H_2PO_4^-/HPO_4^{2-}$ to control the pH.

562

INCLUSION Strategies

- *Learning Disabled*
- *English Language Learners*
- *Attention Deficit Disorder*

Using household product lables from shampoos, antacids, and packaged foods, find those containing phosphates in their ingredients. Discuss in small groups the purposes for manufacturers to use phosphates in the products identified. Be sure to include some that are used for controlling pH, altering taste, and texture. List those products using phosphates for the same purpose in categories and discuss their similarities and differences.

The liquid portion of blood is an example of a buffer solution. To keep the blood's pH very close to 7.40, the body uses a buffer in which the weak acid H_2CO_3, carbonic acid, is paired with its conjugate base, the hydrogen carbonate ion HCO_3^-. The equation below describes the equilibrium that is established.

$$H_2CO_3(aq) + H_2O(l) \rightleftharpoons H_3O^+(aq) + HCO_3^-(aq)$$

There are many medical conditions that can disrupt the equilibrium of this system. Uncontrolled diabetes can cause acidosis, in which the equilibrium is displaced too far to the right. Alcoholic intoxication causes alkalosis, in which the equilibrium lies too far to the left. Hyperventilation removes CO_2, which is also in equilibrium with H_2CO_3. The equilibrium will shift to the left, causing alkalosis.

 Section Review

UNDERSTANDING KEY IDEAS

1. Identify the stronger acid and the stronger base in the reaction described by the following equation:

$$HOCl(aq) + NH_3(aq) \rightleftharpoons NH_4^+(aq) + ClO^-(aq)$$

2. Write the acid-ionization constant expression for the weak acid H_2SO_3.

3. The hydrogen sulfite ion, HSO_3^-, is a weak acid in aqueous solution. Write an equation showing the equilibrium established when hydrogen sulfite is dissolved in aqueous solution, using unequal arrows to show the equilibrium.

4. What is a buffer solution?

5. Give two examples of the practical uses of buffers.

PRACTICE PROBLEMS

6. Use **Table 7** to determine which direction is favored in the following reaction. Explain your answer.

$$H_2CO_3 + H_2O \rightleftharpoons HCO_3^- + H_3O^+$$

7. A 0.105 M solution of HOCl has a pH of 4.19. What is the acid-ionization constant?

8. A buffer solution, prepared from equal amounts of an acid and its conjugate base, has a pH of 10.1. What is the K_a of the acid?

9. Calculate the K_a of nitrous acid, given that a 1.00 M solution of the acid contains 0.026 mol of NO_2^- per liter of solution.

CRITICAL THINKING

10. Ammonia is a weak base. A 0.0123 M solution of ammonia has a hydroxide ion concentration of 4.63×10^{-4} M. Calculate the K_a of NH_4^+.

11. What would be the value of the acid-ionization constant for an acid that was so strong that not a single molecule remained un-ionized?

12. What would be a good acid-base pair from which to prepare a buffer solution whose pH is 10.3?

13. If 99.0% of the weak acid HX stays un-ionized in 1.0 M aqueous solution, what is the K_a?

14. Write all three K_a expressions for H_3PO_4. Which will have the smallest value?

15. Calculate K_{eq} for the following reaction:

$$H_2CO_3(aq) + CO_3^{2-}(aq) \rightleftharpoons 2HCO_3^-(aq)$$

Answers to Section Review

1. HOCl is the stronger acid, and NH_3 is the stronger base.

2. $K_a = \dfrac{[H_3O^+][HSO_3^-]}{[H_2SO_3]}$

3. $HSO_3^-(aq) + H_2O(l) \rightleftharpoons SO_3^{2-}(aq) + H_3O^+(aq)$

4. a solution that contains a weak acid and its conjugate base and that resists changes in pH

5. Accept all reasonable answers, which may include the following: in foods, antacids, shampoos, and the blood.

6. The reverse reaction is favored because H_3O^+ is a much stronger acid than H_2CO_3 is.

7. $K_a = 3.97 \times 10^{-8}$

8. $K_a = 7.94 \times 10^{-11}$

9. $K_a = 6.8 \times 10^{-4}$

10. The concentrations of NH_4^+ and OH^- will be almost equal, so $K_a = 3.74 \times 10^{-20}$.

11. undefined (division by zero)

12. HCO_3^-/CO_3^{2-}

13. 1.0×10^{-4}

Answers continued on p. 573A

CONSUMER
FOCUS

Antacids

The medical term for "heartburn" is *gastroesophageal reflux*. The cause of heartburn is momentary relaxation of the lower esophageal sphincter muscle, which lies at the junction of the esophagus and the stomach. Normally, this muscle keeps the esophagus closed off from the stomach except during swallowing. Occasional heartburn is not unusual, but if the problem occurs frequently, it should not be ignored because it can be a symptom of a hiatal hernia, an ulcer, stomach or esophageal cancer, or a neurological disorder.

CONSUMER FOCUS

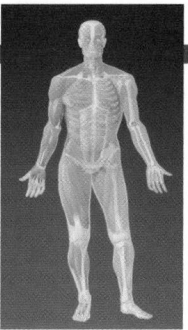

Antacids

The pH of gastric juice in the human stomach is 1.5. This strongly acidic environment activates digestive enzymes, such as pepsin, that work in the stomach.

Stomach acids and antacids

Acidity in the stomach is provided by 0.03 M hydrochloric acid, HCl(*aq*). Sometimes, a person's stomach generates too much acid. The discomfort known as *heartburn* results when the acid solution is forced into the esophagus. Heartburn can be temporarily relieved by taking an antacid to neutralize the excess stomach acid.

Although antacids contain other ingredients, all antacids contain a base that counteracts stomach acid. The base is either sodium hydrogen carbonate, $NaHCO_3$, calcium carbonate, $CaCO_3$, aluminum hydroxide, $Al(OH)_3$, or magnesium hydroxide, $Mg(OH)_2$.

Dangers of excess metals from antacids

In any antacid, the anion is the base that neutralizes the stomach acid. However, the cation in the antacid is also important. Antacids containing $NaHCO_3$ work fastest because $NaHCO_3$ is much more soluble than other antacid substances are. Overusing these antacids, however, can raise the level of positive ions in the body, just as salt does. Overuse can also seriously disrupt the acid-base balance in your blood.

Because of the risks associated with an excess of sodium, some antacid manufacturers have substituted calcium carbonate, $CaCO_3$, for $NaHCO_3$. But if calcium is taken in large amounts, it can promote kidney stones. Ingesting too much aluminum from antacid products, such as $Al(OH)_3$, can interfere with the body's absorption of phosphorus, which is needed for healthy bones. Excess magnesium from antacids that contains $Mg(OH)_2$ may pose problems for people who have kidney disorders.

You should know the active ingredient in any antacid product before you ingest the product, and you should never use an antacid product for more than a few days without consulting a doctor. It is best to avoid the need for an antacid in the first place. You can minimize the production of excess stomach acid by following a healthy diet, avoiding stress, and limiting your consumption of coffee, fatty foods, and chocolate.

Questions

1. What class of compound is common to all antacids?
2. Why should you pay attention to which ions an antacid contains?

☑ **internet** connect

www.scilinks.org
Topic: Antacids
SciLinks code: HW4010

SC*i*LINKS. Maintained by the National Science Teachers Association

564

Answers to the Feature Questions

1. a base that neutralizes stomach acid
2. Some metal ions may be unhealthful if they are taken in excess.

CHAPTER HIGHLIGHTS 15

KEY TERMS

strong acid
weak acid
strong base
weak base
Brønsted-Lowry acid
Brønsted-Lowry base
conjugate acid
conjugate base
amphoteric

self-ionization
  constant of water, K_w
neutral
pH
indicator

neutralization
  reaction
equivalence point
titration
titrant
standard solution
transition range
end point

acid-ionization
  constant, K_a
buffer solution

KEY IDEAS

SECTION ONE   What Are Acids and Bases?

• Acid solutions have distinctive properties attributable to the H_3O^+ ion.
• Bases have distinctive properties attributable to the OH^- ion.
• Brønsted and Lowry defined an acid as donating a proton, and a base as accepting a proton.
• Every acid has a conjugate base, and every base has a conjugate acid.
• An amphoteric species, such as water, can behave as an acid or a base.

SECTION TWO   Acidity, Basicity, and pH

• In aqueous solutions, $[H_3O^+]$ and $[OH^-]$ are interrelated by K_w.
• pH, which is a quantitative measure of acidity and basicity, is the negative logarithm of $[H_3O^+]$.

SECTION THREE   Neutralization and Titrations

• A neutralization reaction between an acid and a base produces water.
• In a titration, a solution of unknown concentration is neutralized by a standard solution of known concentration.
• An indicator has a transition range of pH, within which lies its end point pH.

SECTION FOUR   Equilibria of Weak Acids and Bases

• The acid-ionization constant reflects the strength of a weak acid and the strength of the acid's conjugate base.
• K_a can be used to calculate $[H_3O^+]$ in a solution of a weak acid.
• Buffer solutions are mixtures of a weak acid and its conjugate base, and resist pH changes.

KEY SKILLS

Determining [OH⁻] using K_w
Sample Problem A p. 541

Calculating pH for an Acidic or Basic Solution
Skills Toolkit 1 p. 543

Sample Problem B p. 544

Calculating [H₃O⁺] and [OH⁻] from pH
Skills Toolkit 1 p. 543
Sample Problem C p. 545

Performing a Titration
Skills Toolkit 2 p. 552

Calculating Concentration from Titration Data
Sample Problem D p. 555

Calculating K_a of a Weak Acid
Sample Problem E p. 560

565

Alternative Assessments

Group Activity
• Page 537

Activity
• Pages 539, 549, 559

Reteaching
• Page 547

Chapter Review
• Items 114–118

Portfolio Assessments

• Page 537

Teaching Tip
• Page 537

Reteaching
• Page 556

EARTH SCIENCE CONNECTION
• Page 558

Chapter Resource File

• **Consumer Lab** How Effective Is an Antacid?
• **Consumer Lab** Titration of Aspirin
• **CBL™ Probeware Lab** Acid-Base Titration
• **CBL™ Probeware Lab** Buffer Capacity in Commercial Beverages
• **Datasheets for In-Text Lab** Drip-Drop Acid-Base Experiment
• **Datasheets for In-Text Lab** Acid-Base Titration of an Eggshell
• **Datasheets for In-Text Lab** Acid-Base Titration–Industrial Spill

CHAPTER REVIEW

REVIEW ANSWERS

Using Key Terms

1. A strong acid dissociates completely in solution. A weak acid dissociates only to a small extent in solution.

2. A weak base is a weak electrolyte.

3. Arrhenius defined an acid as a substance that increases the hydronium ion concentration of a solution. Brønsted and Lowry defined an acid as a substance that donates a proton to another substance. Because an Arrhenius acid is also a Brønsted-Lowry acid, the Brønsted-Lowry definition is broader.

4. the ammonium ion, NH_4^+

5. It can act either as an acid, donating a proton, or as a base, accepting a proton.

6. $[OH^-] = 1.00 \times 10^{-7}$ M

7. $K_w = 1.0 \times 10^{-14}$

8. 7

9. pH = $-$log $[H_3O^+]$

10. The strength of an acid does not relate to its concentration. The greater the strength of the acid, the lower the pH of a solution of a given concentration of the acid. The greater the concentration of an acid of given strength, the lower the pH.

11. water

12. the equivalence point

13. The end point of a titration occurs when the pH of the indicator passes through the indicator's transition range. The titrant used in a titration is normally a standard solution.

14. K_a

15. It does not change the pH.

15 CHAPTER REVIEW

USING KEY TERMS

1. How does a strong acid differ from a weak acid?

2. What kind of an electrolyte is a weak base?

3. How does Brønsted and Lowry's definition of an acid differ from Arrhenius's definition of an acid? Explain which definition is broader.

4. What is the conjugate acid of the base ammonia, NH_3?

5. Why is water considered amphoteric?

6. What is the concentration of hydroxide ions in pure water?

7. What is the value of K_w at 25°C?

8. What is the pH of a neutral solution?

9. Give the equation that relates pH to hydronium ion concentration.

10. How does the strength of an acid **WRITING SKILLS** relate to the concentration of the acid? How does the strength of an acid relate to the pH of an aqueous solution of the acid? How does the concentration of an acid solution relate to the solution's pH?

11. What product do all neutralization reactions have in common?

12. At what point in a titration are the amounts of hydronium ions and hydroxide ions equal?

13. Group the following four terms into two pairs according to how the terms are related, and explain how they are related: *end point, standard solution, titrant,* and *transition range.*

14. What is the equilibrium constant that is applied to a weak acid?

15. How does the addition of a small amount of acid or base affect a buffered solution?

UNDERSTANDING KEY IDEAS

What Are Acids and Bases?

16. Compare the properties of an acid with those of a base.

17. What is a base according to Arrhenius? according to Brønsted and Lowry?

18. Why are weak acids and weak bases poor electrical conductors?

19. What is the difference between the strength and the concentration of an acid?

20. Identify each of the following compounds as an acid or a base according to the Brønsted-Lowry classification. For each species, write the formula and the name of its conjugate.
    a. CH_3COO^-
    b. HCN
    c. HOOCCOOH
    d. $C_6H_5NH_3^+$

21. Write an equation for the reaction between hydrocyanic acid, HCN, and water. Label the acid, base, conjugate acid, and conjugate base.

22. Write chemical equations that show how the hydrogen carbonate ion, HCO_3^-, acts as an amphoteric ion.

Acidity, Basicity, and pH

23. Explain the relationship between the self-ionization of water and K_w.

566

Assignment Guide

Section	Questions
1	1–5, 16–22, 113, 116, 118
2	6–10, 23–28, 40–71, 91–92, 97, 108, 114–115
3	11–13, 29–34, 72–82, 90, 94, 99, 102–105, 107–112, 119–127
4	14–15, 35, 39, 83–89, 93, 95, 98, 100–101, 106–117

24. Write an equation that shows the self-ionization of water.

25. Three solutions have pHs of 3, 7, and 11. Which solution is basic? Which is acidic? Which is neutral?

26. By what factor does $[OH^-]$ change when the pH increases by 3? by 2? by 1? by 0.5?

27. Explain how you can calculate pH from $[H_3O^+]$ by using your calculator.

28. Describe two methods of measuring pH, and explain the advantages and disadvantages of each method.

Neutralization and Titrations

29. What is a neutralization reaction?

30. Describe two precautions that should be taken to ensure an accurate titration.

31. Explain what a titration curve is, and sketch its shape.

32. How would you select an indicator for a particular acid-base titration?

33. Would the pH at the equivalence point of a titration of a weak acid with a strong base be less than, equal to, or greater than 7.0?

34. Name an indicator you might use to titrate ammonia with hydrochloric acid.

Equilibria of Weak Acids and Bases

35. The K_a of nitrous acid, HNO_2, is 6.76×10^{-4}. Write the equation describing the equilibrium established when HNO_2 reacts with NH_3. Use unequal arrows to indicate whether reactants or products are favored.

36. **a.** What is the relationship between the strength of an acid and the strength of its conjugate base?

   b. What is the realtionship between the strength of a base and the strenght of its conjugate acid?

37. Propanoic acid, C_2H_5COOH, is a weak acid. Write the expression defining its acid-ionization constant.

38. Place the following acids in order of increasing strength:
   a. valeric acid, $K_a = 1.5 \times 10^{-5}$
   b. glutaric acid, $K_a = 3.4 \times 10^{-4}$
   c. hypobromous acid, $K_a = 2.5 \times 10^{-9}$
   d. acetylsalicylic acid (aspirin),
   $$K_a = 3.3 \times 10^{-4}$$

39. What are the components of a buffer solution? Give an example.

PRACTICE PROBLEMS

Sample Problem A Determining $[OH^-]$ or $[H_3O^+]$ Using K_w

40. If the hydronium ion concentration of a solution is 1.63×10^{-8} M, what is the hydroxide ion concentration?

41. Calculate the hydronium ion concentration in a solution of 0.365 mol/L of NaOH.

42. How much HCl would you need to dissolve in 1.0 L of water so that $[OH^-] = 6.0 \times 10^{-12}$ M?

43. The hydronium ion concentration in a solution is 1.87×10^{-3} mol/L. What is $[OH^-]$?

44. If 0.150 mol of KOH is dissolved in 500 mL of water, what are $[OH^-]$ and $[H_3O^+]$?

45. If a solution contains twice the concentration of hydronium ions as hydroxide ions, what is the hydronium ion concentration?

Sample Problem B Calculating pH for an Acidic or Basic Solution

46. Stomach acid contains HCl, whose concentration is about 0.03 mol/L. What is the pH of stomach acid?

47. If $[OH^-]$ of an aqueous solution is 0.0134 mol/L, what is the pH?

567

Understanding Key Ideas

16. Acids are electrolytes that have a sour taste and react with many metals to produce hydrogen gas. Bases are electrolytes that have a slippery feel and react with acids.

17. a producer of hydroxide ions; a proton acceptor

18. Because they ionize incompletely, they produce few ions to conduct the electricity.

19. Strength refers to the extent to which the acid ionizes. Concentration refers to the amount of acid per unit volume.

20. **a.** base; CH_3COOH
    b. acid; CN^-
    c. acid; $^-OOCCOO^-$
    d. acid; $C_6H_5NH_2$

21. $HCN + H_2O \longrightarrow CN^- + H_3O^+$
    acid   base   conjugate conjugate
                     base   acid

22. $HCO_3^- + H_2O \rightleftarrows$
    $$H_2CO_3 + OH^-$$
    $HCO_3^- + H_2O \rightleftarrows CO_3^{2-} + H_3O^+$

23. K_w is the equilibrium constant, equal to $[H_3O^+][OH^-]$, for the self-ionization reaction of water.

24. $2H_2O(l) \longrightarrow$
    $$H_3O^+(aq) + OH^-(aq)$$

25. pH = 11 is basic, pH = 3 is acidic, and pH = 7 is neutral.

26. 1000; 100; 10; 3.16

27. by taking the logarithm of $[H_3O^+]$ and changing the sign of the answer

28. pH can be measured by an indicator or by a pH meter. Indicators are inexpensive and convenient but are not very precise. pH meters are more expensive and complicated to use, but they give very precise results.

29. the reaction of hydronium and hydroxide ions to produce water

30. Accept all reasonable answers. Answers may include adding the titrant drop by drop as the end point nears, viewing the color of the flask against a white background, or reading the buret with the meniscus at eye level.

31. A graph that shows how pH changes as titrant is added; its shape should resemble the graph in **Figure 18.**

32. by choosing an indicator with a transition range that includes the pH at the equivalence point

33. greater than 7.0

34. Methyl orange or thymol blue are acceptable answers.

35. $HNO_2(aq) + NH_3(aq) \rightleftharpoons NO_2^-(aq) + NH_4^+(aq)$

36. a. If an acid is strong, its conjugate base is weak.

   b. If a base is strong, its conjugate acid is weak.

37. $K_a =$

$$\frac{[H_3O^+][C_2H_5COO^-]}{[C_2H_5COOH]}$$

38. c, a, d, b

39. a weak acid and a salt of its conjugate base; for example, acetic acid and sodium acetate

Practice Problems

40. 6.13×10^{-7} M

41. 2.74×10^{-14} M

42. 1.7×10^{-3} mol

43. 5.35×10^{-12} M

44. $[OH^-] = 0.300$ M; $[H_3O^+] = 3.33 \times 10^{-14}$ M

45. 1.41×10^{-7} M

46. 1.52

47. 12.13

48. 0.82

49. 12.91

50. 0.54

51. a. 2.3

   b. 1.3

   c. 0.3

   d. −0.7

52. 13.81

53. 13.48

54. 9.00

55. 0.17

48. What is the pH of a 0.15 M solution of $HClO_4$, a strong acid?

49. LiOH is a strong base. What is the pH of a 0.082 M LiOH solution?

50. Find the pH of a solution consisting of 0.29 mol of HBr in 1.0 L of water.

51. What is the pH of aqueous solutions of the strong acid HNO_3, nitric acid, if the concentrations of the solutions are as follows: (a) 0.005 M, (b) 0.05 M, (c) 0.5 M, (d) 5 M?

52. Find the pH of a solution prepared by dissolving 0.65 mol of the strong base NaOH in 1.0 L of water.

53. What is the pH of a solution prepared by dissolving 0.15 mol of the strong base $Ba(OH)_2$ in one liter of water? (Hint: How much hydroxide ion does barium hydroxide generate per mole in solution?)

54. A solution has a hydronium ion concentration of 1.0×10^{-9} M. What is its pH?

55. If a solution has a hydronium ion concentration of 6.7×10^{-1} M, what is its pH?

56. What is the pH of a solution whose hydronium ion concentration is 2.2×10^{-12} M?

57. What is the pH of a solution whose H_3O^+ concentration is 1.9×10^{-6} M?

58. Calculate the pH of a 0.0316 M solution of the strong base RbOH.

Sample Problem C Calculating $[H_3O^+]$ and $[OH^-]$ from pH

59. The pH of a solution is 9.5. What is $[H_3O^+]$? What is $[OH^-]$?

60. A solution of a weak acid has a pH of 4.7. What is the hydronium ion concentration?

61. A 50 mL sample of apple juice has a pH of 3.2. What amount, in moles, of H_3O^+ is present?

62. Find $[H_3O^+]$ in a solution of pH 4.

63. What is the hydroxide ion concentration in a solution of pH 8.72?

64. Calculate the concentration of the H_3O^+ and OH^- ions in an aqueous solution of pH 5.0.

65. A solution has a pH of 10.1. Calculate the hydronium ion concentration and the hydroxide ion concentration.

66. What is the hydronium ion concentration in a solution of pH 5.5?

67. If the pH of a solution is 4.3, what is the hydroxide ion concentration?

68. What is the hydronium ion concentration in a solution whose pH is 10.0?

69. The pH of a solution is 3.0. What is $[H_3O^+]$?

70. What is $[H_3O^+]$ in a solution whose pH is 1.9?

71. If a solution has a pH of 13.3, what is its hydronium ion concentration?

Skills Toolkit Performing a Titration

72. To what volumetric mark should a buret be filled?

73. Why is it important to slow down the drop rate of the buret near the end of a titration?

74. What two buret readings need to be recorded in order to determine the volume of solution dispensed by the buret?

Sample Problem D Calculating Concentration from Titration Data

75. What volume of 0.100 M NaOH is required to neutralize 25.00 mL of 0.110 M H_2SO_4?

76. What volume of 0.100 M NaOH is required to neutralize 25.00 mL of 0.150 M HCl?

77. If 35.40 mL of 1.000 M HCl is neutralized by 67.30 mL of NaOH, what is the molarity of the NaOH solution?

78. If 50.00 mL of 1.000 M HI is neutralized by 35.41 mL of KOH, what is the molarity of the KOH solution?

568

79. If 133.73 mL of a standard solution of KOH, of concentration 0.298 M, exactly neutralized 50.0 mL of an acidic solution, what was the acid concentration?

80. To standardize a hydrochloric acid solution, it was used as titrant with a solid sample of sodium hydrogen carbonate, $NaHCO_3$. The sample had a mass of 0.3967 g, and 41.77 mL of acid was required to reach the equivalence point. Calculate the concentration of the standard solution.

Strong Acid Titrated with Strong Base

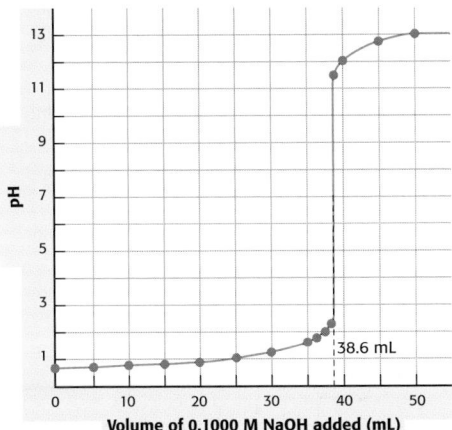

81. The graph above shows a titration curve obtained during the titration of a 25.00 mL sample of an acid with 0.1000 M NaOH. Calculate the concentration of the acid.

82. An HNO_3 solution has a pH of 3.06. What volume of 0.015 M LiOH will be required to titrate 65.0 mL of the HNO_3 solution to reach the equivalence point?

Sample Problem E Calculating K_a of a Weak Acid

83. The hydronium ion concentration in a 0.100 M solution of formic acid is 0.0043 M. Calculate K_a for formic acid.

84. $[NO_2^-] = 9.1 \times 10^{-3}$ mol/L in a nitrous acid solution of concentration 0.123 mol/L. What is K_a for HNO_2?

85. A solution of acetic acid had the following solute concentrations: $[CH_3COOH] = 0.035$ M, $[H_3O^+] = 7.4 \times 10^{-4}$ M, and $[CH_3COO^-] = 7.4 \times 10^{-4}$ M. Calculate the K_a of acetic acid based on these data.

86. Hydrazoic acid, HN_3, is a weak acid. A 0.01 M solution of hydrazoic acid contained a concentration of 0.0005 M of the N_3^- ion. Find the acid-ionization constant of hydrazoic acid.

87. Find K_a for a weak acid that contains 0.050 M H_3O^+ in a solution of 1.00 M concentration.

88. The pH of a solution prepared by dissolving 1.0 mol of a weak acid in water to form 1.0 L of solution was 3.1. What is the acid-ionization constant of the weak acid?

89. An aqueous solution of periodic acid had the following concentrations: $[HIO_4] = 1.1 \times 10^{-4}$ M and $[IO_4^-] = [H_3O^+] = 1.2 \times 10^{-2}$ M. What is the K_a of HIO_4?

MIXED REVIEW

90. If 25 mL of 1.00 M HCl is mixed with 75 mL of 1.00 M NaOH, what are the final amounts and concentrations of all ions present?

91. If 0.30 mol of HI is dissolved in 750 mL of water, what are $[H_3O^+]$, pH, and $[OH^-]$?

92. An aqueous solution is prepared by dissolving 0.321 g of $Ca(OH)_2$ in 555 mL of water. Find $[H_3O^+]$ and $[OH^-]$.

93. When 1.0 mol of a weak acid was dissolved in 10.0 L of water, the pH was found to be 3.90. What is K_a for the acid?

94. At the end point of a titration of 25 mL of 0.300 M NaOH with 0.200 M HNO_3, what would the concentration of sodium nitrate in the titration flask be?

93. 1.7×10^{-7}

94. 0.12 M

95. $[H_3O^+] = 1.1 \times 10^{-3}$ M;
$[C_6H_5COOH] = 0.020$ M;
$[C_6H_5COO^-] = 1.2 \times 10^{-3}$ M

96. 1.75×10^{-5}

97.

pH	$[H_3O^+]$	$[OH^-]$
14.25	5.6×10^{-15}	1.8
14.00	1.0×10^{-14}	1.0
13.75	1.8×10^{-14}	0.56
13.25	5.6×10^{-14}	0.18
13.00	1.0×10^{-13}	0.10
7.25	5.6×10^{-8}	1.8×10^{-7}
7.00	1.0×10^{-7}	1.0×10^{-7}
6.75	1.8×10^{-7}	5.6×10^{-8}
1.00	0.10	1.0×10^{-13}
0.75	0.18	5.6×10^{-14}
0.50	0.32	3.2×10^{-14}
0.25	0.56	1.8×10^{-14}
0.00	1.0	1.0×10^{-14}
-0.25	1.8	5.6×10^{-15}

98. $NH_4^+(aq) + H_2O(l) \underset{\longleftarrow}{\longrightarrow}$
$\qquad\qquad NH_3(aq) + H_3O^+(aq)$

$$K_a = \frac{[H_3O^+][NH_3]}{[NH_4^+]}$$

99. 2.50 g

100. 7.20

101. 2.42

Critical Thinking

102. Enough H_3O^+ ions have reacted with OH^- ions to reduce the H_3O^+ concentration from 0.1 M to 0.066 M. Approximately 0.034 mol/L of OH^- have been added. We cannot know the precise amounts because we do not know the volumes.

95. What are the concentrations of all of the components of a benzoic acid solution if K_a is 6.5×10^{-5}, pH is 2.96, and C_6H_5COOH has a concentration of 0.020 M?

96. What is $[H_3O^+]$ in a buffer solution containing equal concentrations of acetic acid and sodium acetate?

97. Make a table listing the ionic concentrations in solutions of the following pH values: 14.25, 14.00, 13.75, 13.25, 13.00, 7.25, 7.00, 6.75, 1.00, 0.75, 0.50, 0.25, 0.00, and −0.25.

98. Write the equilibrium equation and the equilibrium constant expression for an ammonia–ammonium ion buffer solution.

99. If 18.5 mL of a 0.0350 M H_2SO_4 solution neutralizes 12.5 mL of aqueous LiOH, what mass of LiOH was used to make 1.00 L of the LiOH solution?

100. Use **Table 7** to calculate the pH of a buffer solution made from equal amounts of sodium monohydrogen phosphate and potassium dihydrogen phosphate.

101. A solution of 5% acetic acid ($K_a = 1.76 \times 10^{-5}$) has a concentration of 0.83 M. What is the pH of the solution?

CRITICAL THINKING

102. A solution of HCl has a pH of 1.00. You add NaOH to the solution, and the pH rises to 1.18. What happened? What does this outcome have to do with the relationship of acids to bases along the pH scale? Can you estimate what amounts of HCl and NaOH were used?

103. Why is a buret, rather than a graduated cylinder, used in titrations?

104. A small volume of indicator solution is usually added to the titration flask right before the titration. As a result, the sample is diluted slightly. Does this matter? Why or why not?

105. A student passes an end point in a titration. Is it possible to add an additional measured amount of the unknown and continue the titration? Explain how this process might work. How would the answer for the calculation of the molar concentration of the unknown differ from the answer the student would have gotten if the titration had been performed properly?

106. Design a buffer solution suitable for controlling the pH of a solution at a value close to 4.0.

Indicator	Acid color	pH transition range	Base color
Thymol blue	red	1.2–2.8	yellow
Bromphenol blue	yellow	3.0–4.6	blue
Bromcresol green	yellow	2.0–5.6	blue
Bromthymol blue	yellow	6.0–7.6	blue
Phenol red	yellow	6.6–8.0	red
Alizarin yellow	yellow	10.1–12.0	red

107. Refer to the table above to answer the following questions:
   a. Which indicator would be the best choice for a titration with an end point at a pH of 4.0?
   b. Which indicators would work best for a titration of a weak base with a strong acid?

108. Why does an indicator need to be a weak acid or a weak base?

109. A 0.0200 M solution of NaOH is being used to titrate 20.00 mL of 0.0400 M HCl using bromthymol blue as indicator. The end point, corresponding to a pH of 7.0, occurs when 40.00 mL of base has been added. Calculate the volumes of titrant that correspond to the beginning and to the end of bromthymol blue's transition range, 6.0–7.6.

110. What would be a suitable titrant (compound and concentration) with which to titrate 20.00 mL of a strong acid that has a concentration of about 0.015 M?

111. Explain the difference between *end point* and *equivalence point*. Why is it important that both occur at approximately the same pH in a titration?

112. Can you neutralize a strong acid solution by adding an equal volume of a weak base having the same molarity as the acid? Support your position.

113. In the 18th century, Antoine Lavoisier experimented with oxides such as CO_2 and SO_2. He observed that they formed acidic solutions. His observations led him to infer that for a substance to exhibit acidic behavior, it must contain oxygen. However, today that is known to be incorrect. Provide evidence to refute Lavoisier's conclusion.

ALTERNATIVE ASSESSMENT

114. Design an experiment to measure the pH of four types of hair shampoo: baby shampoo, shampoo for extra body, shampoo for oily hair, and shampoo with conditioner. Also compare two brands of "pH-balanced" shampoo. If your teacher approves your plan, carry it out.

115. Describe exactly how you would test the pH of an unknown solution. If your teacher approves your plan and has a solution available, complete the test.

116. Design an experiment to test the neutralization effectiveness of various brands of antacid. Show your procedure, including all safety procedures and cautions, to your teacher for approval. If your teacher approves your plan, carry it out. After experimenting, write an advertisement for the antacid you judge to be the most effective. Cite data from your experiments as part of your advertising claims.

117. Describe how you would prepare one or more buffer solutions, including which compounds to use. Predict the pH of each solution. If your teacher provides the needed materials, measure the pH to test your prediction.

118. Design an experiment to extract possible acid-base indicators from sources such as red cabbage, berries, and flower petals, using known acidic, basic, and neutral solutions to test the action of each indicator that you are able to isolate. If your teacher approves your design, carry it out.

CONCEPT MAPPING

119. Use the following terms to complete the concept map below: *hydronium ions (H₃O)*, *hydroxide ions (OH⁻)*, *neutralization reaction*, *pH,* and *titration.*

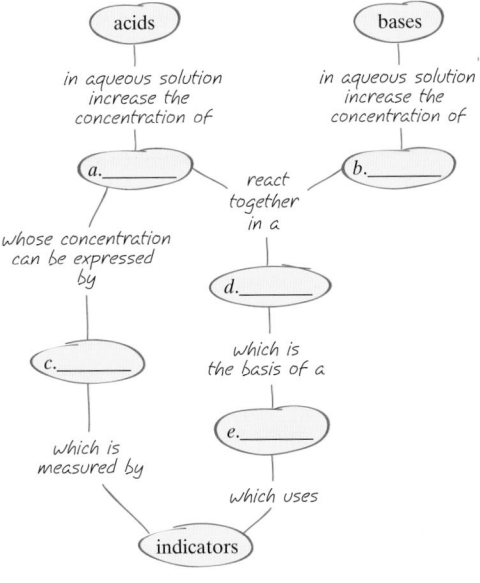

103. The buret has a mechanism for delivering solution and adjusting the rate of its delivery. Also, a buret can be read to greater precision than a graduated cylinder can.

104. No, the added volume does not matter. The volume of titrant reflects the amount $n = cV$ in the flask, not the individual values of c or V. Dilution reduces c but increases V to compensate.

105. An additional amount of the unknown can be added as long as the volume is known with the same precision as in the original sample. The student would continue titrating to the new end point. The concentration of unknown is not affected because the second sample is of the same concentration as the first.

106. Answers may vary. One possibility is a mixture of sodium formate and formic acid in the mole ratio 1.82:1.

107. a. bromphenol blue
    b. Thymol blue, bromphenol blue, or bromcresol green would be chosen depending on the pH of the equivalence point, which would be in the acidic pH range.

108. The indicator must respond to increasing or decreasing acidity by being titrated itself.

109. beginning: 39.997 mL; end: 40.002 mL

110. 0.0100 M concentration of a strong base, such as NaOH or KOH

111. The end point is the point at which an indicator changes color. An equivalence point (in a strong acid–strong base titration) is the point at which stoichiometrically equal amounts of H_3O^+ and OH^- have been combined. The two should occur at about the same pH so that the titration will be concluded at the equivalence point.

112. yes; The ionization of the weak base will continue to produce OH^- ions as others are used up. This will continue until practically all of the OH^- has reacted with H_3O^+.

Chapter 15 · Acids and Bases   **571**

REVIEW ANSWERS
continued

113. Answers may vary. Students should provide examples of acids that do not contain oxygen, such as HCl or HBr. They should discuss the different ways that an acid can be defined that can either narrow or broaden the definition and that do not in any case require the presence of oxygen in the acid.

Alternative Assessment

114. Answers may vary depending on the shampoos selected. Shampoos may have to be diluted with distilled water (the same volume in each sample) to get a reliable pH reading.

115. Plans will vary but should include the use of a pH meter or pH test paper. Be sure students follow all lab safety precautions.

116. Answers should involve titrations with standardized acid solution. Students should consider what to do with insoluble antacids, such as $Mg(OH)_2$. One technique is to react them completely with excess standard acid to form soluble products and then back-titrate with a base to the equivalence point.

Answers continued on p. 573B

FOCUS ON GRAPHING

Study the graph below, and answer the questions that follow.
For help in interpreting graphs, see Appendix B, "Study Skills for Chemistry."

120. What variable is being measured along the *x*-axis?

121. What is the pH at the beginning of the titration?

122. What was the pH after 25 mL of titrant had been added?

123. What volume of titrant was needed to reach a pH of 2.0?

124. Where on the graph do you find the single most important data point?

125. If the titration continued beyond what the graph shows, how would you expect the pH to change past the end of the graph?

126. Roughly sketch the titration curve (pH versus volume) that you would expect if you titrated a weak base with a strong acid. Mark the equivalence point.

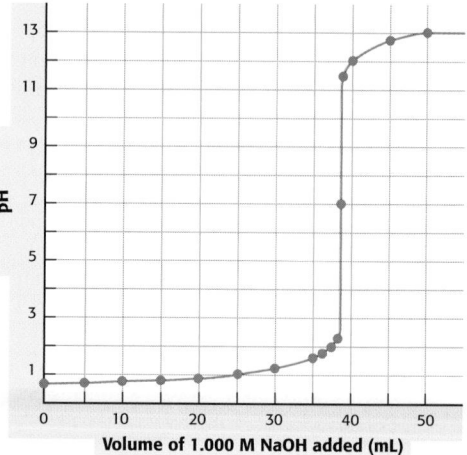

Strong Acid Titrated with Strong Base

TECHNOLOGY AND LEARNING

127. Graphing Calculator

Graphing Titration Data

The graphing calculator can run a program that graphs data such as pH versus volume of base. Graphing the titration data will allow you to determine which combination of acid and base is represented by the shape of the graph.

Go to Appendix C.. If you are using a TI-83 Plus, you can download the program and data and run the application as directed. Press the **APPS** key on your calculator, then choose the application **CHEMAPPS**. Press 5,

then highlight **ALL** on the screen, press **1**, then highlight **LOAD** and press **2** to load the data into your calculator. Quit the application, and then run the program **TITRATN**. For L_1, press **2nd** and **LIST**, and choose **VOL1**. For L_2, press **2nd** and **LIST** and choose **PH1**. If you are using another calculator, your teacher will provide you with keystrokes and data sets to use.

a. At what approximate volume does the pH change from acidic to basic?

b. If the titrant was 0.24 M NaOH, and the volume of unknown was 230 mL, what was $[H_3O^+]$ in the unknown solution?

572

Chapter Resource File

• Chapter Test

STANDARDIZED TEST PREP 15

Answers
1. a
2. c
3. b
4. b
5. c
6. b
7. a
8. d
9. c
10. a
11. c

1. An aqueous solution of which of the following would have the highest concentration of hydronium ions?
   a. formic acid
   b. acetic acid
   c. potassium carbonate
   d. sodium formate

2. When an acid reacts with a metal,
   a. the hydronium ion concentration increases.
   b. the metal forms anions.
   c. hydrogen gas is produced.
   d. the K_w value changes.

3. The K_w value can be affected by
   a. dissolving a salt in the solution.
   b. changes in temperature.
   c. changes in the hydroxide ion concentration.
   d. the presence of a strong acid.

4. A solution of an Arrhenius acid would have
   a. no measurable pH.
   b. a pH of less than 7.
   c. a pH of 7.
   d. a pH of greater than 7.

5. The stronger an acid is,
   a. the smaller its K_a value.
   b. the stronger its conjugate base.
   c. the more hydronium ions produced in aqueous solution.
   d. the more hydroxide ions produced in aqueous solution.

6. Which of the following species is the conjugate acid of another species in the list?
   a. PO_4^{3-}
   b. H_3PO_4
   c. H_2O
   d. $H_2PO_4^-$

7. Which of the following solutions would have a pH value greater than 7?
   a. $[OH^-] = 2.4 \times 10^{-2}$ M
   b. $[H_3O^+] = 1.53 \times 10^{-2}$ M
   c. 0.0001 M HCl
   d. $[OH^-] = 4.4 \times 10^{-9}$ M

8. If the pH of a solution of a strong base is known, the _____ of the solution can be calculated.
   a. molar concentration
   b. $[OH^-]$
   c. $[H_3O^+]$
   d. All of the above

9. A neutral aqueous solution
   a. has a 7 M H_3O^+ concentration.
   b. contains neither hydronium ions nor hydroxide ions.
   c. has an equal number of hydronium ions and hydroxide ions.
   d. contains buffers that resist changes in pH.

10. Identify the salt that remains when a solution of H_2SO_4 is titrated with a solution of $Ca(OH)_2$.
    a. calcium sulfate
    b. calcium hydroxide
    c. calcium oxide
    d. calcium phosphate

11. Which of the following could not be titrated by 0.100 M HCl solution no matter which indicator you used?
    a. $OH^-(aq)$
    b. $NaHCO_3(s)$
    c. $H_2SO_4(aq)$
    d. $Na_3PO_4(s)$

573

Answers to Section 1 Review, *continued*

6. A conjugate acid-base pair is a pair of species such that the acid member is formed by adding of a proton to the base member. One of many examples is NH_4^+/NH_3.

7. **a.** with any acid: $H_2O + acid \longrightarrow H_3O^+ + conjugate\ base^-$

   b. with any base: $H_2O + base \longrightarrow OH^- + conjugate\ acid^+$

8. yes; For instance, in the gaseous phase, HCl donates a proton to NH_3 to form NH_4Cl. This reaction does not involve the production of H_3O^+ ions, so HCl is not acting as an Arrhenius acid in this case.

9. $[OH^-]$ would be lower in an ammonia solution than in a sodium hydroxide solution of similar concentration.

10. conjugate acid: H_3PO_4; conjugate base: HPO_4^{2-}

11. $H_2SO_4(aq) + SO_4^{2-}(aq) \longrightarrow 2HSO_4^-(aq)$

12. because what little $Mg(OH)_2$ does dissolve is entirely dissociated as $Mg^{2+}(aq)$ and $OH^-(aq)$ ions

13. H_2SO_4 is an acid, and HSO_4^- is its conjugate base. HSO_3^- is an acid, and SO_3^- is its conjugate base.

Answers to Section 2 Review, *continued*

11. 1.00×10^{-2} mol HBr

12. 32 L

13. Even the purest water contains equilibrium amounts of hydronium and hydroxide ion.

14. Yes, the pH is negative whenever the hydronium concentration exceeds 1.0 M.

15. The red color of the blood would obscure the color of the pH paper. Also, to be useful, the blood pH must be read to higher precision than is possible by indicator methods.

Answers to Section 3 Review, *continued*

11. 0.18 M

12. 58 mL

13. thymol blue or phenolphthalein

14. because a steep titration curve implies that the pH changes very rapidly near the end point, so a precise volume can be measured with little uncertainty

15. The titration of hydrochloric acid by ammonia is a case in point. At the equivalence point, the solution consists only of ammonium chloride dissolved in water. This salt is fully ionized, but the ammonium ion is involved in the following equilibrium: $NH_4^+ (aq) + H_2O(l) \rightleftarrows NH_3(aq) + H_3O^+(aq)$. It is the hydronium ion produced by this reaction that lowers the pH to less than 7 at the equivalence point.

Answers to Section 4 Review, *continued*

14. $K_a = \dfrac{[H_3O^+][H_2PO_4^-]}{[H_3PO_4]}$

$K_a = \dfrac{[H_3O^+][HPO_4^{2-}]}{[H_2PO_4^-]}$

$K_a = \dfrac{[H_3O^+][PO_4^{3-}]}{[HPO_4^{2-}]}$

The third expression will have the smallest value.

15. $K_{eq} = 9.19 \times 10^3$

Answers to Chapter Review, continued

Alternative Assessment

117. Provide samples of either sodium hydrogen carbonate and sodium carbonate or disodium hydrogen phosphate and potassium dihydrogen phosphate. If 250 mL flasks are used to prepare the solutions, one-hundredth of a mole of each salt will be needed. As stated in the text, if the amounts of the weak acid and its conjugate base in a buffer solution are equal, the pH will equal –log (K_a). The predicted pH values will be as follows:

carbonate buffer: $pH = -\log (K_a) = -\log (4.68 \times 10^{-11}) = 10.33$

phosphate buffer: $pH = -\log (K_a) = -\log (6.31 \times 10^{-8}) = 7.12$

118. Experimental designs should contain appropriate safety cautions and should have the necessary controls to ensure that the indicators tested are responsible for any changes in color upon addition of acid or base.

Concept Mapping

119. a. hydronium ions (H_3O^+)

b. hydroxide ions (OH^-)

c. pH

d. neutralization reaction

e. titration

Focus on Graphing

120. volume of titrant added

121. about 0.70

122. about 1.00

123. about 37.5 mL

124. at the straight vertical line, which centers on pH = 7

125. the pH would change very little

126. The shape of the curve should be an inverted form of the graph used to answer questions 120–125, with the starting pH between 8–11, the equivalence point between pH = 4–6, and the pH approaching 0–2 towards the end of the graph.

Technology and Learning

127. a. About 45 mL

b. 0.047 M

Note: additional questions can be given if VOL2 and PH2 are loaded from CHEMAPPS and chosen as datasets.

Reaction Rates
Chapter Planning Guide

PACING	CLASSROOM RESOURCES	LABS, ACTIVITIES, AND DEMONSTRATIONS
BLOCK 1 · 45 min pp. 574–575 **Chapter Opener**		SE **Start-Up Activity** Temperature and Reaction Rates, p. 575
BLOCKS 2 & 3 · 90 min pp. 576–585 **Section 1** **What Affects the Rate of a Reaction?**	OSP **Supplemental Reading** George Bear's "Heads" ADVANCED	TE **Demonstration** Calculating the Rate of Reaction, p. 579 ◆ TE **Activity** Calculus and Reaction Rates, p. 581 ADVANCED TE **Demonstration** Oxidizing Iron, p. 582 TE **Demonstration** Change in Rate Over Time, p. 583 ◆ SE **QuickLab** Concentration Affects Reaction Rate, p. 577 GENERAL CRF **Datasheets for In-Text Labs** *
BLOCKS 4 & 5 · 90 min pp. 586–595 **Section 2** **How Can Reaction Rates Be Explained?**	TT Activation Energies * TT Particle Collisions *	TE **Demonstration** Effect of Particle Size, p. 588 ◆ SE **QuickLab** Modeling a Rate-Determining Step, p. 589 GENERAL CRF **Datasheets for In-Text Labs** * TE **Group Activity** Effect of Temperature on Reaction Rate, p. 590 BASIC TE **Demonstration** Three Types of Catalysts, p. 594 ◆ SE **Observation Lab** Reaction Rates, p. 814 ◆ GENERAL CRF **Datasheets for In-Text Labs** * CRF **CBL™ Probeware Lab** A Leaky Reaction ◆ ADVANCED

BLOCKS 6 & 7 · 90 min **Chapter Review and Assessment Resources**

SE **Chapter Review,** pp. 596–600
SE **Standardized Test Prep,** p. 601
CRF **Chapter Test** *
OSP **Test Generator**
CRF **Test Item Listing** *
OSP **Scoring Rubrics and Classroom Management Checklists**

Holt Chemistry: Online Resources

Visit **go.hrw.com** for a variety of free resources related to this textbook. Enter the keyword **HW4 HOME.**

Holt Online Learning

Students can access interactive problem solving help and active visual concept development with the *Holt Chemistry* Online Edition available at **www.hrw.com**.

student cnn News

cnnstudentnews.com

Find the latest chemistry news, lesson plans, and activities related to important scientific events.

PROBLEM SOLVING AND PRACTICE	SECTION REVIEW AND ASSESSMENT	STANDARDS CORRELATION
		National Science Education Standards
SE Sample Problem A Calculating a Reaction Rate, p. 581 **SE Practice**, p. 581 `GENERAL` **TE Homework**, p. 581 `GENERAL` **CRF Problem Solving** * `ADVANCED` **SE Problem Bank**, p. 858 `GENERAL`	**TE Quiz**, p. 585 `GENERAL` **CRF Quiz** * **TE Reteaching**, p. 585 `BASIC` **SE Section Review**, p. 585 **CRF Concept Review** *	PS 3d
SE Sample Problem B Determining a Rate Law, p. 587 **SE Practice**, p. 587 `GENERAL` **TE Homework**, p. 587 `GENERAL` **CRF Problem Solving** * `ADVANCED` **SE Problem Bank**, p. 858 `GENERAL`	**TE Quiz**, p. 595 `GENERAL` **CRF Quiz** * **TE Reteaching**, p. 595 `BASIC` **SE Section Review**, p. 595 **CRF Concept Review** *	PS 3d, PS 3e

www.scilinks.org

Topic: Inhibitors
SciLinks code: HW4172

Topic: Factors Affecting Equilibrium
SciLinks code: HW4057

Topic: Factors Affecting Reaction Rate
SciLinks code: HW4058

Topic: Rate Laws
SciLinks code: HW4161

Topic: Reaction Mechanisms
SciLinks code: HW4162

Topic: Catalysts
SciLinks code: HW4025

Topic: Enzymes
SciLinks code: HW4054

Technology Resources

   Science in the NEWS

Each video segment is accompanied by a Critical Thinking Worksheet.

Segment 27
Synthetic Soil

Segment 28
Ozone-Eater Radiator

Overview

In this chapter students will define the rate of a reaction and learn what factors affect the rate. They will learn that activation energy determines the speed of a reaction and how catalysts, including enzymes, operate to lower the activation energy.

Assessing Prior Knowledge

Check for Content Knowledge

Students should be familiar with the following topics:

• Stoichiometry

• Energy changes in chemical reactions

• Concentration of solutions

Using the Figure —— BASIC

Have students discuss how the fire-retardant chemical in the figure works to slow the spread of a forest fire. Ask them why a mixture that contains other chemicals in addition to water may be more effective than water alone. Students may suggest that if the intense heat generated by a fire is enough to evaporate all of the water dropped on it, the barrier between the fuel and the oxygen will be removed completely. **LS** Logical

CHAPTER

16

REACTION RATES

574

Standards Correlations

National Science Education Standards

PS 3d: Chemical reactions can take place in time periods ranging from the few femtoseconds (10^{-15} seconds) required for an atom to move a fraction of a chemical bond distance to geologic time scales of billions of years. Reaction rates depend on how often the reacting atoms and molecules encounter one another, on the temperature, and on the properties—including shape—of the reacting species. (Sections 1, 2)

PS 3e: Catalysts, such as metal surfaces, accelerate chemical reactions. Chemical reactions in living systems are catalyzed by protein molecules called enzymes. (Section 2)

A forest fire is an enormous combustion reaction that can go on as long as it has fuel, oxygen, and heat. The air tanker in the photograph is dropping a fire-retardant mixture to slow the spread of one of these fires. Fire retardants, which usually contain chemicals such as water, ammonium sulfate, and ammonium phosphate, work by forming a barrier between the fuel (brush and trees) and the oxygen. These chemicals help firefighters slow and eventually stop the combustion reaction. In this chapter, you will learn about the many factors that affect how fast a chemical reaction takes place.

START-UP **ACTIVITY**

SAFETY PRECAUTIONS

Temperature and Reaction Rates

PROCEDURE

1. Submerge one **light stick** in a **bath of cold water** (about 10°C).

2. Submerge a second **light stick** in a **bath of hot water** (about 50°C).

3. Allow each light stick to reach the same temperature as its bath.

4. Remove the light sticks, and activate them.

5. In a dark corner of the room, observe and compare the light intensities of the two sticks.

ANALYSIS

1. Which stick was brighter?

2. Light is emitted from the stick because of a chemical reaction. What can you conclude about how temperature affects this reaction?

Pre-Reading Questions

① Give two examples of units that could be used to measure a car's rate of motion.

② What can you do to slow the rate at which milk spoils?

③ What is a catalytic converter in an automobile?

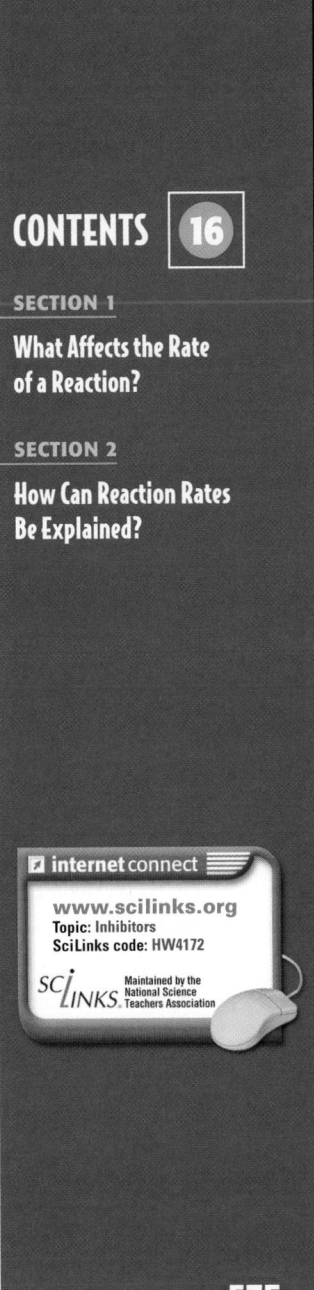

CONTENTS 16

internet connect

www.scilinks.org
Topic: Inhibitors
SciLinks code: HW4172

SCiLINKS Maintained by the National Science Teachers Association

575

START-UP **ACTIVITY**

Skills Acquired:
• Experimenting
• Inferring

Materials:
For each group of 2–4 students:
• two light sticks
• water bath, about 10°C
• water bath, about 50°C

Teacher Notes: If only one light stick is available for each group of students, have students start with the cold water and transfer the light stick from the cold water to the hot water.

Answers

1. The stick that was submerged in the hot water bath should be brighter.

2. Students should infer that higher temperatures enhance the rates of chemical reactions.

Answers to Pre-Reading Questions

1. Answers should be a measure of distance per measure of time.

2. refrigerate the milk

3. A catalytic converter changes any carbon monoxide resulting from fuel combustion to carbon dioxide.

What Affects the Rate of a Reaction?

Overview

Before beginning this section, review with your students the Objectives listed in the Student Edition. In this section, students will learn that the speed of a reaction is a function of the concentration of a reactant or product and time. They will understand how concentration, temperature, pressure, and surface area affect reaction rate.

 **Bellringer**

Ask students to make a list of rates that they have heard about in their daily lives. Besides rate of speed, they may come up with rate of aging, rate of decay, rate of growth, rate of pay, and rate of flow. Compile their lists and start your discussion by eliciting from them that most rates have to do with something happening in a unit of time.

Motivate

Identifying Preconceptions

Because students have learned to identify chemical reactions by obvious changes they may think that if no obvious signs of reaction are perceptible, no reaction is occurring. Tell them that some reactions are so slow that it's difficult, on a macroscopic scale, to verify that they are happening.

Chapter Resource File

• Lesson Plan

KEY TERMS

• chemical kinetics
• reaction rate

OBJECTIVES

① **Define** the rate of a chemical reaction in terms of concentration and time.

② **Calculate** the rate of a reaction from concentration-versus-time data.

③ **Explain** how concentration, pressure, and temperature may affect the rate of a reaction.

④ **Explain** why, for surface reactions, the surface area is an important factor.

Rates of Chemical Change

A *rate* indicates how fast something changes with time. In a savings account, the rate of interest tells how your money is growing over time. Speed is also a rate. From the speed of one of the race cars shown in **Figure 1,** you can tell the distance that the car travels in a certain time. If a car's speed is 67 m/s (150 mi/h), you know that it travels a distance of 67 meters every second. Rates are always measured in a unit of something per time interval. The rate at which the car's wheels turn would be measured in revolutions per second. The rate at which the car burns gasoline could be measured in liters per minute.

The rate of a chemical reaction measures how quickly reactants are changed into products. Some reactions are over in as little as 10^{-15} s; others may take hundreds of years. The study of reaction rates is called **chemical kinetics.**

chemical kinetics

the area of chemistry that is the study of reaction rates and reaction mechanisms

Figure 1
The winner of the race is the car that has the highest rate of travel.

REAL-WORLD CONNECTION

Fast and Slow Reactions Some reactions move imperceptibly, like the rusting of iron. Other are over in an instant, like the explosion of a firecracker. Have students make up a list of reactions or processes they see happening around them or read about in the news. Have them divide the list into fast, ordinary, and slow. Get them thinking ahead by asking them to suggest possible ways to increase the speed of slow reactions. For example, if they list as a slow process the boiling of water, they might suggest that turning up the heat on the stove would increase the rate.

Rate Describes Change over Time

At 500°C, the compound dimethyl ether slowly decomposes according to the equation below to give three products—methane, carbon monoxide, and hydrogen gas.

$$CH_3OCH_3(g) \longrightarrow CH_4(g) + CO(g) + H_2(g)$$

The concentration of dimethyl ether will keep decreasing during the reaction. Recall that the symbol Δ represents a change in some quantity. If the concentration of dimethyl ether changes by $\Delta[CH_3OCH_3]$ during a small time interval Δt, then the rate of the reaction is defined as

$$rate = \frac{-\Delta[CH_3OCH_3]}{\Delta t}$$

The sign is negative because, while $\Delta[CH_3OCH_3]$ is negative, the rate during the reaction must be a positive number.

The chemical equation shows that for every mole of dimethyl ether that decomposes, 1 mol each of methane, carbon monoxide, and hydrogen is produced. Thus, the concentrations of CH_4, CO, and H_2 will *increase* at the same rate that $[CH_3OCH_3]$ *decreases*. This means that the rate for this reaction can be defined in terms of the changes in concentration of any one of the products, as shown below.

$$rate = \frac{-\Delta[CH_3OCH_3]}{\Delta t} = \frac{\Delta[CH_4]}{\Delta t} = \frac{\Delta[CO]}{\Delta t} = \frac{\Delta[H_2]}{\Delta t}$$

The concentrations of the products are all increasing, so the signs of their rate expressions are positive.

Quick LAB

SAFETY PRECAUTIONS

Concentration Affects Reaction Rate

PROCEDURE

1. Prepare two labeled beakers, one containing 0.001 M hydrochloric acid and the second containing 0.1 M hydrochloric acid.

2. Start a **stopwatch** at the moment you drop an **effervescent tablet** into the first beaker.

3. Stop the stopwatch when the tablet has finished dissolving.

4. Repeat steps 2–3 with a second **effervescent tablet,** using the second beaker.

ANALYSIS

1. What evidence is there that a chemical reaction occurred?

2. Were the dissolution times different? Did the tablet dissolve faster or slower in the more concentrated solution?

3. What conclusion can you draw about how the rate of a chemical reaction depends on the concentration of the reactants?

577

Videos

CNN. Presents Chemistry Connections

• **Segment 27** Synthetic Soil

See the Science in the News video guide for more details.

Average, Instantaneous, and Initial Rate Only the average rate of a reaction can be determined by taking one measurement at the beginning and another at the end of a reaction. In order to examine the progress of a reaction in more detail, a chemist must conduct an experiment in which the reaction is sampled at several time intervals. Graphing these concentrations over time produces a rate curve. Chemists can determine the reaction rate at any given instant by calculating the slope of the curve at any given time. This is called the instantaneous rate. Chemists also determine and record initial rates, which are based on the first interval after the initiation of the reaction.

Chapter Resource File

• Datasheets for In-Text Labs

Balanced Coefficients Appear in the Rate Definition

Now consider the following reaction, which is the one illustrated in **Figure 2** below.

$$2N_2O_5(s) \longrightarrow 4NO_2(g) + O_2(g)$$

The stoichiometry is more complicated here because 2 mol of dinitrogen pentoxide produce 4 mol of nitrogen dioxide and 1 mol of oxygen. So, it is no longer true that the rate of decrease of the reactant concentration equals the rates of increase of the product concentrations. However, this difficulty can be overcome if, in order to define the reaction rate, we divide by the coefficients from the balanced equation. For this reaction, we get the following.

$$\text{rate} = \frac{-\Delta[N_2O_5]}{2\Delta t} = \frac{\Delta[NO_2]}{4\Delta t} = \frac{\Delta[O_2]}{\Delta t}$$

The definition of **reaction rate** developed in these two examples may be generalized to cover any reaction.

It is important to realize that a reaction does not have a single, specific rate. Reaction rates depend on conditions such as temperature and pressure. Also, the rate of a reaction changes *during* the reaction. Usually, the rate decreases gradually as the reaction proceeds. The rate becomes zero when the reaction is complete.

reaction rate

the rate at which a chemical reaction takes place; measured by the rate of formation of the product or the rate of disappearance of the reactants

Figure 2
Dinitrogen pentoxide decomposes to form oxygen and the orange-brown gas nitrogen dioxide.

MATH CONNECTION — BASIC

Graphing Give students the following equation and data for the decomposition of NO_2 at 300°C and have them draw a graph with concentration on the y-axis and time on the x-axis. $2NO_2(g) \rightarrow 2NO(g) + O_2(g)$ **LS** Visual

Time	[NO₂] (mol/L)
0	0.0100
100	0.0065
200	0.0048
300	0.0038
400	0.0031

Table 1 Concentration Data and Calculations for the Decomposition of N_2O_5

t (s)	$[NO_2]$ (M)	$\Delta[NO_2]$ (M)	Δt (s)	$\Delta[NO_2]/\Delta t$ (M/s)	Rate (M/s)
0	0				
		4.68×10^{-3}	20.0	2.34×10^{-4}	5.85×10^{-5}
20.0	0.00468				
		4.22×10^{-3}	20.0	2.11×10^{-4}	5.28×10^{-5}
40.0	0.00890				
		3.82×10^{-3}	20.0	1.91×10^{-4}	4.78×10^{-5}
60.0	0.01272				
		3.44×10^{-3}	20.0	1.72×10^{-4}	4.30×10^{-5}
80.0	0.01616				

Reaction Rates Can Be Measured

To measure a reaction rate, you need to be able to keep track of how the concentration of one or more reactants or products changes over time. There are many ways of tracking these changes depending on the reaction you are studying.

For the reaction in **Figure 2**, you could measure how quickly the concentration of one product changes by measuring a change in color. Because nitrogen dioxide is the only gas in the reaction that has a color, you could use the red-brown color of the gas mixture to calculate $[NO_2]$. On the other hand, because the pressure of the system changes during the reaction, you could measure this change and, with help from the gas laws, calculate the concentrations.

internet connect
www.scilinks.org
Topic: Factors Affecting Equilibrium
SciLinks code: HW04057
SCiLINKS
Maintained by the National Science Teachers Association

Concentrations Must Be Measured Often

Remember that the Δt that occurs in the equations defining reaction rate is a *small* time interval. This means that studies of chemical kinetics require that concentrations be measured frequently. **Table 1** shows the results from a study of the following reaction.

$$2N_2O_5(g) \longrightarrow 4NO_2(g) + O_2(g)$$

The NO_2 concentrations were used to calculate the reaction rate in this example, but $[N_2O_5]$ or $[O_2]$ data could also have been used. As expected, the reaction rate decreases with time. It takes about 900 s before the reaction is 99% complete, and at that point, the rate is only 6.2×10^{-7} M/s. Reaction rates are generally expressed, as they are here, in moles per liter-second or M/s.

Notice in the table how the rate is calculated from pairs of data points—two different time readings and two different concentrations of NO_2. For example, the last rate in the table comes from the calculation shown below.

$$\text{rate} = \frac{\Delta[NO_2]}{4\Delta t} = \frac{0.01616\ \text{M} - 0.01272\ \text{M}}{4(80.0\ \text{s} - 60.0\ \text{s})} = 4.30 \times 10^{-5}\ \text{M/s}$$

This result shows the rate of the reaction after it has been going on for about 70 s.

MISCONCEPTION ALERT

Students hear the expression "*the* rate of the reaction" and may think that there is only one rate for a given reaction and that it remains constant throughout the reaction. As you begin to discuss the factors that affect reaction rates, in particular concentration and temperature, it would be well to keep this in mind. Point out to students that a reaction starts with specific concentrations of the reactants which produce a specific initial rate. However, as these concentrations dwindle during the course of the reaction, so does the rate of reaction. Point out also that the temperature at which a reaction takes place also influences the rate. Rate data should always be accompanied by the temperature at which it was obtained.

Demonstration
Calculating the Rate of a Reaction

(Approximate time: 5 minutes)

1. Cut two 4-cm pieces of Mg ribbon.

2. Tie a 10-cm length of thread around each ribbon. Tape the free ends of thread just outside the mouths of two 50 mL eudiometer tubes (closed burets turned upside down can be substituted for the eudiometer tubes).

3. With the Mg pieces outside the tube, pour 15 mL of 1 M HCl solution into each tube.

4. Fill one tube with room-temperature tap water. Pour the water slowly down the side of the tube to prevent too much water from mixing with the HCl. Repeat the procedure with the second eudiometer, but this time use ice water. Record the temperature of the contents of each tube.

5. Fill a 600-mL beaker almost full with tap water, and set the beaker in a pan to catch any overflow.

6. Place the Mg ribbon inside the eudiometer with the ice water, and plug the mouth with a one-hole stopper. Immediately invert the eudiometer tube while submerging the stoppered end in the water in the beaker.

7. Every 15 seconds, record the total volume of gas that has been produced until the reaction stops. Calculate the reaction rate.

8. Repeat steps 6 and 7 for the eudiometer tube filled with room-temperature water.

9. Have students write a sentence that relates reaction rate and temperature.

Safety Caution: Wear safety goggles and a laboratory apron. Use a nonmercury thermometer.

Disposal: Rinse any unreacted Mg, and put it in the trash. Combine the rinsings with the liquid from the eudiometers and beakers, neutralize with 0.1 M NaOH until the pH is between 6 and 8, and flush the rinsings down the drain.

Homework —— GENERAL

Have students create a flow chart for the rate of the reaction A + 2B → 2C + D. Have them define the rate in terms of both the reactants and products and include the four rate expressions. **LS Logical**

Teaching Tip

Units for Rate Point out the advantage of expressing reaction rate as the change in concentration per unit of time. Concentration is expressed as the amount of a substance per unit of volume. Therefore, the concentration term is independent of the total volume of a reaction or the total mass of a reactant or product. The rate calculated for a reaction done on test-tube scale will theoretically be the same as the reaction taking place in 1000 kL as long as other conditions remain the same.

Using the Figure —— GENERAL

Refer students to **Figure 3** and to the text which states that at $t = 70$ s the slope of the line for the production of O_2 is 4.30×10^{-5} M/s. Ask them what the rate of the reaction is. Ask them to predict the slopes of the lines for NO_2 and N_2O_5 at 70 s. They should respond that the slope of the NO_2 curve should be four times that of O_2 or 1.72×10^{-4} M/s, and the absolute value of the slope of the N_2O_5 should be two times that of O_2 or 8.60×10^{-5} M/s. Students can check these values themselves. **LS Logical**

Figure 3
The graph shows the changes in concentration with time during the decomposition of dinitrogen pentoxide. The points represent the data used in **Table 1**.

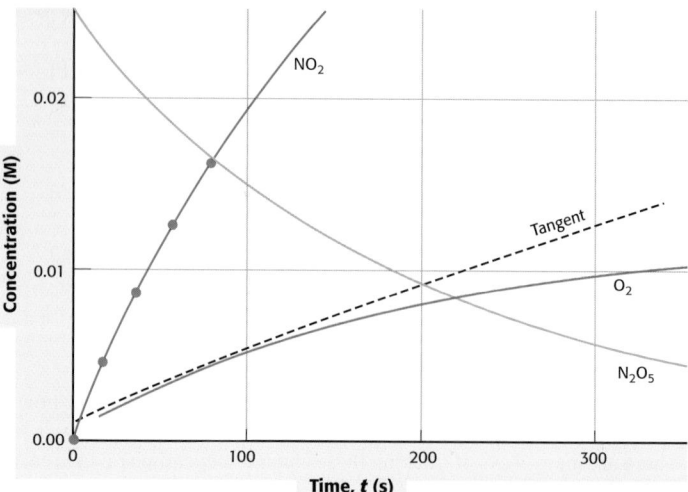

N₂O₅ Decomposition Data

Reaction Rates Can Be Represented Graphically

Chemists often use graphs to help them think about chemical changes. Graphs are especially helpful in the field of chemical kinetics. For one example of how a graph can be useful, we can take another look at the decomposition reaction $2N_2O_5(g) \longrightarrow 4NO_2(g) + O_2(g)$. **Figure 3** is a graph that keeps track of this reaction with three curves, which show how the concentrations of the reactant and the products change with time. Notice that the concentration of dinitrogen pentoxide steadily falls. Also note that the concentration of oxygen and the concentration of nitrogen dioxide steadily increase.

Finally, notice that the graph also shows that the concentration of nitrogen dioxide increases four times faster than the concentration of oxygen increases. This result agrees with the 4:1 ratio of nitrogen dioxide to oxygen in the balanced equation.

Now, when some quantity is plotted versus time, the slope of the line tells you how fast that quantity is changing with time. So the slopes of the three curves in **Figure 3** measure the rates of change of each concentration. The slope of a curve at a particular point is just the slope of a straight line drawn as a tangent to the curve at that point. Because oxygen is a product and its coefficient in the equation is 1, the slope of the O_2 curve is simply the reaction rate.

$$\text{slope of } O_2 \text{ curve} = \frac{\Delta[O_2]}{\Delta t} = \text{rate of the reaction}$$

A line has been drawn as a tangent to the O_2 curve at $t = 70$ s. Its slope was measured in the usual way as rise/run and is 4.30×10^{-5} M/s. This value agrees with the rate calculated in **Table 1** at the same instant.

PHYSICS
CONNECTION

Students may have learned in previous science classes about nuclear decay or transmutations, which proceed at regular, measurable rates. The rate of nuclear decay is called half-life. Half-life is defined as the length of time in which one-half of a given sample decays. This rate of reaction is different from rates of reaction in chemistry. Although it does involve the chemical elements, transmutation is a reaction of the nucleus of the atom not the electrons outside the nucleus as in chemical reactions. In addition, there are no factors that can influence the rate of decay—not temperature or pressure or concentration.

Calculating a Reaction Rate

The data below were collected during a study of the following reaction.

$$2Br^-(aq) + H_2O_2(aq) + 2H_3O^+(aq) \longrightarrow Br_2(aq) + 4H_2O(l)$$

Time t (s)	[H₃O⁺] (M)	[Br₂] (M)
0	0.0500	0
85	0.0298	0.0101
95	0.0280	0.0110
105	0.0263	0.0118

Use two methods to calculate what the reaction rate was after 100 s.

1 Gather information.

During the interval $\Delta t = 10$ s between $t = 95$ s and $t = 105$ s, the changes in the concentrations of hydronium ion and bromine were

$$\Delta[H_3O^+] = (0.0263\ M) - (0.0280\ M) = -0.0017\ M$$

$$\Delta[Br_2] = (0.0118\ M) - (0.0110) = 0.0008\ M$$

2 Plan your work.

For this reaction, two definitions of the reaction rate are as follows.

$$\text{rate} = \frac{-\Delta[H_3O^+]}{2\Delta t} = \frac{\Delta[Br_2]}{\Delta t}$$

3 Calculate.

From the change in hydronium ion concentration,

$$\text{rate} = \frac{-\Delta[H_3O^+]}{2\Delta t} = \frac{-(-0.0017\ M)}{2(10\ s)} = 8.5 \times 10^{-5}\ M/s$$

From the change in bromine concentration,

$$\text{rate} = \frac{\Delta[Br_2]}{\Delta t} = \frac{0.0008\ M}{10\ s} = 8 \times 10^{-5}\ M/s$$

4 Verify your results.

The two ways of solving the problem provide approximately the same answer.

PRACTICE HINT

The coefficient from the chemical equation, unless it is 1, must be included when calculating a reaction rate.

PRACTICE

1 For the reaction in **Sample Problem A,** write the expressions that define the rate in terms of the hydrogen peroxide and bromide ion concentrations.

2 The initial rate of the $N_2O_4(g) \longrightarrow 2NO_2(g)$ reaction is 7.3×10^{-6} M/s. What are the rates of concentration change for the two gases?

3 Use the data from **Sample Problem A** to calculate the reaction rate after 90 s.

581

TECHNOLOGY CONNECTION — ADVANCED

The study of reaction rates is important to chemical engineers, who often must design processes to produce substances on an industrial scale. Their job is to get the best possible yield of a product without producing large amounts of impurities or products of side reactions. After studying the kinetics of a reaction, an engineer designs ways to regulate temperatures, pressures, rates of flow of reactants, rates of removal of products, and other factors to produce the substance in an efficient and economical way. Two such industrial processes are the Haber Process for making ammonia and the Contact Process for making sulfuric acid. Have interested and capable students research either of these processes and relate what they learn to reaction kinetics. **LS** Logical

Activity ——————— ADVANCED

Ask students who are taking calculus to show how calculus could be useful in determining the rate of a reaction. **LS** Logical

Answers to Practice Problems A

1. $\text{rate} = \dfrac{-\Delta[Br^-]}{2\Delta t} = \dfrac{-\Delta[H_2O_2]}{\Delta t}$

2. rate of change of $[N_2O_4] = -7.3 \times 10^{-6}$ M/s; rate of change of $[NO_2] = 1.5 \times 10^{-5}$ M/s

3. 9.0×10^{-5} M/s

Homework ——————— GENERAL

Additional Practice The following reaction between ethylene, C_2H_4, and ozone, O_3, occurs in the atmosphere to produce smog. The table provides data collected in an experiment to determine the rate of the reaction. **LS** Logical

$$C_2H_4(g) + O_3(g) \rightarrow C_2H_4O(g) + O_2(g)$$

Time (s)	[O₃] (mol/L)
0.0	3.20×10^{-5}
10.0	2.42×10^{-5}
20.0	1.95×10^{-5}
30.0	1.63×10^{-5}
40.0	1.40×10^{-5}
50.0	$1.23\ v\ 10^{-5}$
60.0	1.10×10^{-5}

1. Determine the average rate of the reaction. Ans. 3.50×10^{-7} mol/L

2. Determine the initial rate in the first 10.0 s of the reaction. Ans. 7.80×10^{-7} mol/L

3. Determine the rate between 50.0s and 60.0 s. Ans. 1.30×10^{-7} mol/L

4. Describe how the rate of the reaction changes with time. Ans. The rate decreases with time.

5. Describe how you could obtain the instantaneous rate at 35.0 s. Ans. Make a graph of the data and determine the tangent to the line at 35.0 s.

Demonstration

Oxidizing Iron

(Approximate time: 10 minutes)

1. Pour about 25 mL of fresh 3% H_2O_2 into a heavy glass bottle of about 250 mL capacity, and add about a gram of MnO_2 powder. The mixture will bubble as the MnO_2 catalyzes the decomposition of H_2O_2 to H_2O and O_2 gas.

2. Place a glass plate over the mouth of the bottle so that the O_2 will not diffuse out.

3. Hold a small loose tuft of fine steel wool with crucible tongs, and thrust the steel wool into the flame of a laboratory burner. Have students note that it stops burning soon after it is withdrawn from the flame.

4. Light the tuft again, and while it is glowing, thrust it into the bottle of oxygen. It will burn brightly as the iron oxidizes.

5. Ask students to account for the difference in the way the iron burned in air and in nearly pure oxygen. Air is about 20% O_2, whereas the gas in the bottle was nearly 100% O_2.

Safety Caution: Wear a lab apron, gloves, and goggles during this demonstration. Keep students at least 10 ft away from the demonstration.

Disposal: Filter the solids, and rinse them with tap water. Pour the liquid down the drain, provided your school drains are connected to a sanitary sewer system with a treatment plant. Dispose of the solids in the trash.

Factors Affecting Rate

Concentration, pressure, temperature, and surface area are the most important factors on which the rate of a chemical reaction depends. Consider each of these effects for a type of reaction that is already familiar to you—combustion.

You know that the more fuel and oxygen there is, the faster a fire burns. This is an example of the general principle that the rate of a chemical reaction increases as the concentration of a reactant increases.

Many combustion processes, such as those of sulfur or wood, take place at a surface. The larger the surface area, the greater the chances that each particle will be involved in a reaction.

Concentration Affects Reaction Rate

Though there are exceptions, almost all reactions, including the one shown in **Figure 4,** increase in rate when the concentrations of the reactants are increased.

It is easy to understand why reaction rates increase as the concentrations of the reactants increase. Think about the following reaction taking place within a container.

$$NO_2(g) + CO(g) \longrightarrow NO(g) + CO_2(g)$$

Clearly, the reaction can take place only when a nitrogen dioxide molecule collides with a carbon monoxide molecule. If the concentration of NO_2 is doubled, there are twice as many nitrogen dioxide molecules, and so the number of collisions with CO molecules will double. Only a very small fraction of those collisions will actually result in a reaction. Even so, the possibility that each reaction will take place is twice as much when the NO_2 concentration is doubled.

Reaction rates decrease with time because the reaction rate depends on the concentration of the reactants. As the reaction proceeds, the reactant is consumed and its concentration declines. This change in concentration, in turn, decreases the reaction rate.

Topic Link

Refer to the "Chemical Reactions and Equations" chapter for a discussion of collisions between molecules.

Figure 4
Carbon burns faster in pure oxygen **a** than in air **b** because the concentration of the reacting species, O_2, is greater.

a   b

582

did you know?

Exceptions to the Rule Increasing concentration does not always produce the predicted increase in reaction rate. Higher reactant concentrations may make side reactions or competing reactions more of a problem. These reactions may even slow down the rate of the desired reaction. This phenomenon is one problem that chemical engineers face when they must scale up reactions from the volume of a flask in a laboratory to that of a vat in a chemical plant.

Concentration Affects Noncollision Reaction Rates

Not all reactions require a collision. The gas cyclopropane has a molecule in which three bonded carbon atoms form a triangle, with two hydrogen atoms attached to each carbon atom. Above room temperature, cyclopropane slowly changes into propene.

$$(CH_2)_3(g) \longrightarrow CH_2{=}CH{-}CH_3(g)$$

A collision is not necessary for this reaction, but the rate of the reaction still increases as the concentration of cyclopropane increases. In fact, the rate doubles if the $(CH_2)_3$ concentration doubles. This is not surprising. Because there are twice as many molecules, their reaction is twice as likely, and so the reaction rate doubles.

■ internet connect ▤
www.scilinks.org
Topic: Factors Affecting
Reaction Rate
SciLinks code: HW4058

SCiLINKS Maintained by the National Science Teachers Association

Pressure Affects the Rates of Gas Reactions

Pressure has almost no effect on reactions taking place in the liquid or solid states. However, it does change the rate of reactions taking place in the gas phase, such as the reaction shown in **Figure 5.**

As the gas laws confirm, doubling the pressure of a gas doubles its concentration. So changing the pressure of a gas or gas mixture is just another way of changing the concentration.

Temperature Greatly Influences the Reaction Rate

All chemical reactions are affected by temperature. In almost every case, the rate of a chemical reaction *increases* with increasing temperature. The increase in rate is often very large. A temperature rise of only 10%, say from 273 K to 300 K, will frequently increase the reaction rate tenfold. Our bodies work best at around 37°C or 310 K. Even a 1°C change in body temperature affects the rates of the body's chemical reactions enough that we may become ill as a result.

Figure 5
This reaction between two gases, ammonia and hydrogen chloride, forms solid ammonium chloride in a white ring near the center of the glass tube.

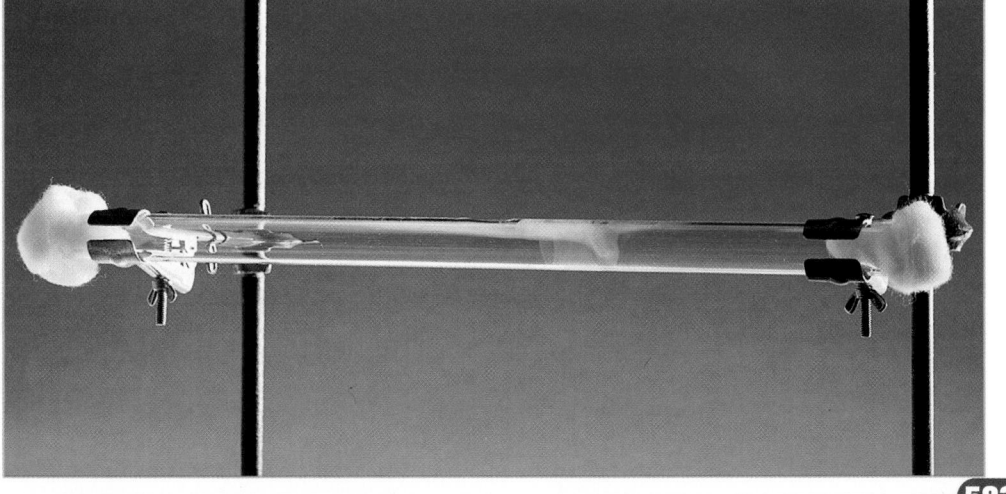

583

BIOLOGY CONNECTION

Reptiles have no internal mechanism for generating heat and so must depend upon absorbing energy from their surroundings. A snake or lizard basking in the sun is renewing its energy supply in order to carry out the metabolic functions of a healthy organism. Each animal has a temperature range in which it can function. Should its temperature fall below the lower limit, the animal becomes sluggish, its metabolic functioning slows, breathing rate declines and it can no longer function successfully in its environment. Thus, temperature is a factor which increases the rates of metabolic reactions in cold-blooded animals.

Demonstration
Change in Rate over Time
(Approximate time: 15 minutes)

1. Cut four 20.0-cm pieces of clean magnesium ribbon. Check masses to be sure the pieces are within 0.002 g of each other. Tie a 15-cm length of thread to each.

2. Write the mass of 20.0 cm of Mg ribbon on the board to the nearest 0.001 g. This is the mass at time 0.

3. Tie the other end of the threads to each of four glass or plastic stirring rods, keeping about 1 cm of thread between the rod and the Mg strip.

4. Place 150 mL of 0.1 M HCl solution in each of four 250 mL beakers. Have a 400 mL beaker of water available.

5. Place one of the Mg pieces in the first beaker. After 1 minute, remove and rinse it in the beaker of water.

6. Dry the piece, remove the thread, and measure the mass of the Mg to the nearest 0.001 g.

7. Repeat steps 5 and 6 with the other three pieces of Mg ribbon, leaving the pieces in the acid for 2, 3, and 4 min, respectively.

8. Plot the four masses on the *y*-axis of a graph, plot the time on the *x*-axis, and draw a smooth curve through the data points.

Safety Caution: Wear safety goggles and a laboratory apron. Keep students at least 10 ft away from the demonstration.

Disposal: Rinse the unreacted Mg, and put it in the trash. Combine the rinsings with the liquid from the beakers, neutralize the liquid with 0.1 M NaOH, and flush it down the drain.

Teaching Tip

Surface Area You can demonstrate that dividing a solid increases its surface area by forming a cube with modeling clay. Measure the cube, and calculate its total surface area. Cut the cube into eight equal smaller cubes. Determine the total surface area of each cube. Multiply this area by 8 to get the total surface area of the cubes. Students can see that cutting the cube in this way has doubled the surface area.

Discussion

Ask students why food is kept in a refrigerator. Ask why sugar dissolves more quickly in hot tea than in iced tea. Ask why unused batteries are sometimes kept in the freezer. Students will see that chemical reactions tend to proceed more quickly at higher temperatures. The rule of thumb is that for many reactions, every increase of 10°C in temperature doubles the rate of the reaction.

Figure 6
The reactions that cause food such as these grapes to spoil occur much more slowly when food is placed in a refrigerator or freezer.

Temperature Affects Reactions in Everyday Life

The fact that reaction rates respond to temperature changes is part of everyday life. In the kitchen, we increase the temperature to speed up the chemical processes of cooking food, and we lower the temperature to slow down the chemical processes of food spoilage. When you put food in a refrigerator, you slow down the chemical reactions that cause food, such as the grapes shown in **Figure 6** to decompose. Most manufacturing operations use either heating or cooling to control their processes for optimal performance.

Why do chemical reactions increase in rate so greatly when the temperature rises? You have seen, in discussing reactions such as $NO_2(g) + CO(g) \longrightarrow NO(g) + CO_2(g)$, that a collision between molecules (or other particles, such as ions or atoms) is necessary for a reaction to occur. A common misconception is that a rise in temperature increases the number of collisions and thereby boosts the reaction rate. It is true that a temperature rise does increase the collision frequency somewhat, but that effect is small. The main reason for the increase in reaction rate is that a temperature rise increases the fraction of molecules that have an energy great enough for collision to lead to reaction. If they are to react, molecules must collide with enough energy to rearrange bonds. A rise in temperature means that many more molecules have the required energy.

Surface Area Can Be an Important Factor

Most of the reactions that we have considered so far happen uniformly in three-dimensional space. However, many important reactions—such as precipitations, corrosions, and many combustions—take place at surfaces. The definition of *rate* given earlier does not apply to surface reactions. Even so, these reactions respond to changes in concentration, pressure, and temperature in much the same way as do other reactions.

A feature of surface reactions is that the amount of matter that reacts is proportional to the surface area. As **Figure 7** shows, you get a bigger blaze with small pieces of wood, because the surface area of many small pieces is greater than that of one larger piece of wood.

584

The Hazard of Dust Disastrous explosions can occur in mines, grain elevators, and flour mills when the air becomes filled with finely divided, combustible material, such as coal dust, bran, or flour. Because of the high surface area of these particles mixed with air, a spark can readily initiate a combustion reaction. The sudden increase in pressure because of the release of gaseous CO_2 and H_2O as the dust burns causes a second violent explosion.

INCLUSION Strategies

- **Attention Deficit Disorder**
- **Learning Disability**
- **English as a Second Language**

Have students test the rates of a seltzer tablet dissolving, using different temperatures of water. Ask students to drop a seltzer tablet in a glass of room-temperature water and record the time it takes for the tablet to completely dissolve. Repeat the procedure with refrigerated water and with boiling water. Have students record the times for the tablets to dissolve in each of these two tests. Ask them to explain why the rates for dissolving may be different for the three temperatures.

Figure 7

a Division of a solid makes the exposed surface of the solid larger.

b More divisions mean more exposed surface.

c Hence, more surface is available for other reactant molecules to come together.

① Section Review

UNDERSTANDING KEY IDEAS

1. What does the word *rate* mean in everyday life, and what do chemists mean by *reaction rate*?

2. What is the name given to the branch of chemistry dealing with reaction rates? Why are such studies important?

3. Why is a collision between molecules necessary in many reactions?

4. How may reaction rates be measured?

5. Explain why reactant concentration influences the rate of a chemical reaction.

6. Give examples of the strong effect that temperature has on chemical reactions.

7. What is unique about surface reactions?

CRITICAL THINKING

8. Why must coefficients be included in the definition of reaction rate?

9. Calculating the reaction rate from a product appeared to give an answer different from that calculated from a reactant. Suggest a possible explanation.

10. The usual unit for reaction rate is M/s. Suggest a different unit that could be used for reaction rate, and explain why this unit would be appropriate.

11. Explain why an increase in the frequency of collisions is not an adequate explanation of the effect of temperature on reaction rate.

12. Would the factors that affect the rate of a chemical reaction influence a physical change in the same way? Explain, and give an example.

13. Why does pressure affect the rates of gas reactions?

585

Before beginning this section, review with your students the Objectives listed in the Student Edition. In this section, the rate law will be introduced. Students will learn the importance of activation energy and collision orientation in determining the dynamics of chemical reactions. The role of catalysts, including enzymes, will be explored.

Bellringer

Have students make an outline of the heads and subheads in this section and write a few sentences about what they think they will learn. If you have time to discuss their ideas and affirm those that are correct, students will be more likely to master those ideas.

Chapter Resource File

• Lesson Plan

KEY TERMS
- rate law
- reaction mechanism
- order
- rate-determining step
- intermediate
- activation energy
- activated complex
- catalyst
- catalysis
- enzyme

rate law

the expression that shows how the rate of formation of product depends on the concentration of all species other than the solvent that take part in a reaction

reaction mechanism

the way in which a chemical reaction takes place; expressed in a series of chemical equations

order

in chemistry, a classification of chemical reactions that depends on the number of molecules that appear to enter into the reaction

OBJECTIVES

① **Write** a rate law using experimental rate-versus-concentration data from a chemical reaction.

② **Explain** the role of activation energy and collision orientation in a chemical reaction.

③ **Describe** the effect that catalysts can have on reaction rate and how this effect occurs.

④ **Describe** the role of enzymes as catalysts in living systems, and give examples.

Rate Laws

You have learned that the rate of a chemical reaction is affected by the concentration of the reactant or reactants. The **rate law** describes the way in which reactant concentration affects reaction rate. A rate law may be simple or very complicated, depending on the reaction.

By studying rate laws, chemists learn *how* a reaction takes place. Researchers in chemical kinetics can often make an informed guess about the **reaction mechanism.** In other words, they can create a model to explain how atoms move in rearranging themselves from reactants into products.

Determining a General Rate Law Equation

For a reaction that involves a single reactant, the rate is often proportional to the concentration of the reactant raised to some power. That is, the rate law takes the following form.

$$\text{rate} = k[\text{reactant}]^n$$

This is a general expression for the rate law. The exponent, n, is called the **order** of the reaction. It is usually a whole number, often 1 or 2, but it could be a fraction. Occasionally, n equals 0, which means that the reaction rate is independent of the reactant concentration. The term k is the *rate constant*, a proportionality constant that varies with temperature.

Reaction orders cannot be determined from a chemical equation. They must be found by experiment. For example, you might guess that $n = 1$ for the following reaction.

$$CH_3CHO(g) \longrightarrow CH_4(g) + CO(g)$$

However, experiments have shown that the reaction order is 1.5.

586

MISCONCEPTION ALERT

Students have learned that for the reaction $X \rightarrow Y$, rate = $-\Delta[X]/\Delta t$ or rate = $\Delta[Y]/\Delta t$. In this section, students will see a different equation for rate, the rate law, which for the reaction above is rate = $k[X]^n$. Students may be confused about how they are related. Tell them that the rate of a specific reaction can be determined experimentally by following the changes in concentration of a reactant or product as a function of time. From data

obtained in this way, a chemist can also determine the rate law for a particular reaction. Multiple experiments varying the concentrations of the reactants one by one must be done to determine the value of k and the order of the reaction n. Once the law is established, it can be used to calculate the rate of the reaction at a specified temperature using any initial concentrations.

Determining a Rate Law

Three experiments were performed to measure the initial rate of the reaction $2HI(g) \longrightarrow H_2(g) + I_2(g)$. Conditions were identical in the three experiments, except that the hydrogen iodide concentrations varied. The results are shown below.

Experiment	[HI] (M)	Rate (M/s)
1	0.015	1.1×10^{-3}
2	0.030	4.4×10^{-3}
3	0.045	9.9×10^{-3}

1 Gather information.

The general rate law for this reaction is as follows: $rate = k[HI]^n$
$n = ?$

2 Plan your work.

Find the ratio of the reactant concentrations between experiments
1 and 2, $\dfrac{[HI]_2}{[HI]_1}$

Then see how this affects the ratio $\dfrac{(rate)_2}{(rate)_1}$ of the reaction rates.

3 Calculate.

$$\dfrac{[HI]_2}{[HI]_1} = \dfrac{0.030 \text{ M}}{0.015 \text{ M}} = 2.0 \qquad \dfrac{(rate)_2}{(rate)_1} = \dfrac{4.4 \times 10^{-3} \text{ M/s}}{1.1 \times 10^{-3} \text{ M/s}} = 4.0$$

Thus, when the concentration changes by a factor of 2, the rate changes by 4, or 2^2. Hence n, the reaction order, is 2.

4 Verify your results.

On inspecting items 1 and 3 in the table, one sees that when the concentration triples, the rate changes by a factor of 9, or 3^2. This confirms that the order is 2.

PRACTICE HINT

To find a reaction order, compare a rate ratio with a concentration ratio.

PRACTICE

1. In a study of the $2NH_3(g) \longrightarrow N_2(g) + 3H_2(g)$ reaction, when the ammonia concentration was changed from 3.57×10^{-3} M to 5.37×10^{-3} M, the rate increased from 2.91×10^{-5} M/s to 4.38×10^{-5} M/s. Find the reaction order.

2. What is the order of a reaction if its rate increases by a factor of 13 when the reactant concentration increases by a factor of 3.6?

3. What concentration increase would cause a tenfold increase in the rate of a reaction of order 2?

4. When the CH_3CHO concentration was doubled in a study of the $CH_3CHO(g) \longrightarrow CH_4(g) + CO(g)$ reaction, the rate changed from 7.9×10^{-5} M/s to 2.2×10^{-4} M/s. Confirm that the order is 3/2.

587

MISCONCEPTION ALERT

When students see the form of the rate law, they may conclude that the exponents designated as n should be the coefficients preceding the species in the chemical equation. Make it clear from the beginning that the exponents have no relationship to the coefficients and must be found experimentally.

Motivate

Discussion —————— GENERAL

Students may wonder how chemists make the measurements that allow them to calculate rate laws. Tell them that there are many ways and that each method must be tailored to the individual reaction. Give them the equations below and challenge them to think of a method for each.

1. $I_2(s) \rightarrow I_2(g)$

2. $2ClO_2(aq) + 2OH^-(aq) \rightarrow ClO_3^-(aq) + ClO_2^-(aq) + H_2O(l)$

3. $2I^-(aq) + S_2O_8^{2-}(aq) \rightarrow I_2(aq) + 2SO_4^{2-}(aq)$

In reaction 1, the pressure of evolving iodine gas could be measured. In reaction 2, measured quantities of the reaction mixture (aliquots) could be taken and the hydroxide ion titrated with acid. In reaction 3, the intensity of the color of I_2 can be monitored either through color standards or by using a spectrometer. **LS Logical**

Teach

Answers to Practice Problems B

1. 1

2. 2

3. a factor of 3.2

4. answer should show that $2^{1.5} = 2.8$

Homework —————— GENERAL

Additional Practice

1. Determine the rate laws of the following reactions from the data given:

a. $A \rightarrow 2B + C$

[A]	Reaction rate
0.30 M	0.19 M/s
0.60 M	0.38 M/s

Ans. rate = $k[A]$

b. $A \rightarrow B + 3C$

[A]	Reaction rate
0.072 M	0.0013 M/s
0.144 M	0.0104 M/s

Ans. rate = $k[A]^3$
LS Logical

Reaction Mixtures for NO + O₃ ⟶ NO₂ + O₂

Start with equal
concentrations
of reactants

0:09

Completion time

Triple the
concentration
of either
reactant

Reaction
rate is
3 times
as fast

0:03

Completion time

Triple the
concentration
of both
reactants

Reaction
rate is
9 times
as fast

0:01

Completion time

Figure 8
Nitrogen monoxide reacts with ozone. Increasing the concentration of either NO or O_3 will increase the reaction rate.

Teach, *continued*

Discussion ———— GENERAL

Reinforce the idea of rate-determining steps by posing the following hypothetical situation to students. Suppose you work in a shop where submarine sandwiches are made on an assembly line, with each person doing a specific task. During the lunchtime rush, the sandwich line always gets backed up. What would you look for to determine the cause of the backup? **Ans.** the slowest step What would you do if you found that the delay was caused by the sandwich bagger, who can bag the subs only about half as fast as they can be made? **Ans.** Add another bagger to double the rate. Suppose that after you add the extra bagger, there is a new backup—the two baggers are waiting for sandwiches. What would you do then? **Ans.** Look for the new rate-determining step.
LS Logical

Demonstration

(Approximate time: 10 minutes)

1. Place a small pile of lycopodium powder in an evaporating dish or watch glass, and play the flame of a laboratory burner over it. Students will note that the powder will scorch.

2. Next, set the burner on the table. Toss a small pinch of the powder into the flame. A dramatic fireball will occur as the separated particles burn rapidly.

Safety Caution: Wear safety goggles and a lab apron. Do not wear long sleeves. Practice this before exhibiting it to the class so that you know how much lycopodium to use. Remove all flammable material within 3 ft of the burner, and keep students at least 10 ft away.

☑ internet connect
www.scilinks.org
Topic: Rate Laws
SciLinks code: HW4161

SC/LINKS. Maintained by the National Science Teachers Association

588

Rate Laws for Several Reactants

When a reaction has more than one reactant, a term in the rate law corresponds to each. There are three concentration terms in the rate law for the following reaction.

$$2Br^-(aq) + H_2O_2(aq) + 2H_3O^+(aq) \longrightarrow Br_2(aq) + 4H_2O(l)$$

There is an order associated with each term:

$$\text{rate} = k[Br^-]^{n_1}[H_2O_2]^{n_2}[H_3O^+]^{n_3}$$

For example, n_1 is the reaction order with respect to Br^-.

To be sure of the orders of reactions that have several reactants, one must perform many experiments. Often the concentration of only a single reactant is varied during a series of experiments. Then a new series is begun and a second reactant is varied, and so on.

Figure 8 shows the results of changing conditions during a study of the reaction represented by the equation below.

$$NO(g) + O_3(g) \longrightarrow NO_2(g) + O_2(g)$$

This is an important reaction because it participates in the destruction of the ozone layer high in the atmosphere. There are two terms in the rate law for this reaction, which is shown below.

$$\text{rate} = k[NO]^{n_1}[O_3]^{n_2}$$

In this case, it turns out that $n_1 = n_2 = 1$. The fact that the orders for each reactant are equal to one suggests that this reaction has a simple one-step mechanism in which an oxygen atom is transferred when the two reactant molecules collide.

did you know?

The How of Explosives In a tiny fraction of a second, the reactions of explosives such as nitroglycerin, trinitrotoluene (TNT), and dynamite are over. These materials are primarily organic substances containing mostly carbon, hydrogen, oxygen, and nitrogen atoms held together by relatively weak bonds. When "set off," explosive materials experience rapid decomposition. The released elements immediately react to form gaseous N_2, CO, CO_2, and NO_2. The bonds in these small molecules are much stronger than those in the original explosive material, and so an enormous amount of energy is released. In addition, the sudden formation of gaseous material causes a tremendous increase in pressure that provides the force to demolish an unwanted building, break rocks for building roads, or propel a bullet on its way.

Rate-Determining Step Controls Reaction Rate

Although a chemical equation can be written for the overall reaction, it does not usually show how the reaction actually takes place. For example, the reaction shown below is believed to take place in four steps, in the mechanism that follows.

internet connect
www.scilinks.org
Topic: Reaction Mechanisms
SciLinks code: HW4162
SCILINKS Maintained by the National Science Teachers Association

$$2Br^-(aq) + H_2O_2(aq) + 2H_3O^+(aq) \longrightarrow Br_2(aq) + 4H_2O(l)$$

The order with respect to each of the three reactants was found to be 1.

(1) $Br^-(aq) + H_3O^+(aq) \rightleftharpoons HBr(aq) + H_2O(l)$ (1)

(2) $HBr(aq) + H_2O_2(aq) \longrightarrow HOBr(aq) + H_2O(l)$

(3) $Br^-(aq) + HOBr(aq) \rightleftharpoons Br_2(aq) + OH^-(aq)$

(4) $OH^-(aq) + H_3O^+(aq) \rightleftharpoons 2H_2O(l)$

These four steps add up to the overall reaction that was shown above. Three of the steps are shown as equilibria; these are fast reactions. Step 2, however, is slow. If one step is slower than the others in a sequence of steps, it will control the overall reaction rate, because a reaction cannot go faster than its slowest step. Such a step is known as the **rate-determining step.** Step 2 is the rate-determining step of the mechanism shown by steps 1–4. Species such as HOBr that form during a reaction but are then consumed are called **intermediates.**

rate-determining step

in a multistep chemical reaction, the step that has the lowest velocity, which determines the rate of the overall reaction

intermediate

a substance that forms in a middle stage of a chemical reaction and is considered a stepping stone between the parent substance and the final product

Quick LAB

Modeling a Rate-Determining Step

SAFETY PRECAUTIONS

PROCEDURE

1. Attach a **large-bore funnel** above a **small-bore funnel** onto a **ring stand.** Set a large **bowl** on the table, directly below the funnels.

2. Pour one cup of **sand** into the top funnel, and start a **stopwatch.**

3. When the last of the sand has fallen into the bowl, stop the stopwatch.

4. Write down the elapsed time.

5. Repeat steps 1 through 4 using the large-bore funnel above a **medium-bore funnel.**

6. Repeat steps 1 through 4 using the medium-bore funnel above the small-bore funnel.

7. Repeat steps 1 through 4 using the small-bore funnel above the large-bore funnel.

ANALYSIS

1. Which combination of funnels made the process go the fastest?

2. Which funnel controlled the rate of the process?

3. Does reversing the order of the two funnels in a trial change the results? Explain.

4. What strengths does this process have as a model for a chemical reaction? What weaknesses does it have?

589

GEOLOGY CONNECTION

Two mountain ranges in the United States provide a picture of the process of weathering, the slow wearing away of rock by the action of wind, rain, ice, and snow. The Rocky Mountains with their high, sharp, craggy profiles are considered young at 100 million years. Yet evidence of weathering can be seen at the base of the mountains where "skirts" of finely divided rock fan out at the bottom of each runoff stream, thus broadening the base of the mountains at the expense of their height. The Appalachian range is 245 million years older than the Rockies. During those years, Appalachian peaks have become lower, smoother, rounder. Today they could be a snapshot of the Rockies of the future as the slow, imperceptible erosion process continues for another 245 million years.

Using the Figure

Have students note that in **Figure 8,** "reaction progress" is plotted along the horizontal axis. Have them imagine that two molecules are moving toward each other on a collision path. Ask what happens as the molecules come close enough to repel each other. Ans. Repulsion slows them down and their kinetic energy decreases. Have students recall that potential energy is energy of position. Lead them to the understanding that an increase in potential energy accompanies the decrease in kinetic energy. Ask what is happening at the peak of the potential energy curve. Students should understand that at this point a high-energy supermolecule has a momentary existence. If the orientation is correct, new bonds may form as old bonds break and a product is formed.

Group Activity ────── BASIC

Divide your class into groups of six, and ask each group to write an explanation of how increasing temperature affects reaction rate. Have a member of each group read the group's explanation. Compare explanations until students have established that increasing temperature causes particles to move faster and collide more often with more energy. Next, ask them to explain the effects of increasing concentration and surface area. Guide students to an understanding that neither of these factors affects the energy of the collisions, but by increasing the overall frequency of collisions, they increase the frequency of those collisions that have the needed activation energy and orientation. **LS Interpersonal**

Topic Link

Refer to the "Gases" chapter for a discussion of the energy distribution of gas molecules.

activation energy

the minimum energy required to start a chemical reaction

activated complex

a molecule in an unstable state intermediate to the reactants and the products in the chemical reaction.

Reaction Pathways and Activation Energy

If two molecules approach each other, the outer electrons of each molecule repel the outer electrons of the other. So, ordinarily, the molecules just bounce off each other. For two molecules to react, they must collide violently enough to overcome the mutual repulsion, so that the electron clouds of the two molecules merge to some extent. This merging may lead to a distortion of the shapes of the colliding molecules and, ultimately, to the creation of new bonds.

Violent collisions happen only when the colliding pair of molecules have an unusually large amount of energy. The kinetic energies of individual gas molecules vary over a wide range. Only the molecules with especially high kinetic energy are likely to react. The other molecules must wait until a succession of "lucky" collisions brings their kinetic energies up to the necessary amount.

The minimum energy that a pair of colliding molecules (or atoms or ions) need to have before a chemical change becomes a possibility is called the **activation energy** of the reaction. It is represented by the symbol E_a. No reaction is possible if the colliding pair has less energy than E_a.

Activation-Energy Diagrams Model Reaction Progress

Imagine rolling a ball toward a speed bump in a parking lot. If you do not give the ball enough kinetic energy, it will roll partway up the bump, stop, reverse its direction, and come back toward you. If you give it enough energy, the ball will make it just to the top of the bump and stay there for a moment. After that, it may go either way. Given plenty of energy, the ball will pass easily over the bump. Then, gaining more kinetic energy as it descends, it will roll away down the far side of the speed bump.

The model of the ball and speed bump provides a good analogy of the reaction between two colliding molecules. Without enough kinetic energy, the two molecules will not change chemically. With a combined kinetic energy equal to the activation energy, the molecules reach a state where there is a 50:50 chance of either returning to the initial state without reacting, or of being rearranged and becoming products. This point, similar to the top of the speed bump, is called the **activated complex** or *transition state* of the reaction.

Figure 9a is a graph of how the energy changes as a pair of hydrogen iodide molecules collide, form an activated complex, and then go on to become hydrogen and iodine molecules. As a chemical equation, the process could be written as follows.

$$2HI \quad \longrightarrow \quad H_2I_2 \quad \longrightarrow \quad H_2 + I_2$$

initial state   activated   final state
(reactant)   complex   **(products)**

In the initial state, the bonds are between the hydrogen and iodine atoms, H–I. In the activated complex, four weak bonds link the four atoms into a deformed square. In the final state the bonds link hydrogen to hydrogen, H–H, and iodine to iodine, I–I.

590

did you know?

Activation-energy diagrams can have more than one peak. The number of peaks depends upon how many elementary steps are in the reaction mechanism. Consider a three-step mechanism. Each step has its own activation energy and activated complex for the production of an intermediate or, in the final step, a product. The energy of the first intermediate is shown by the valley or lowest point between the first and second peaks. The intermediate reacts in the second step by surmounting the second potential energy barrier and forming a second intermediate. The potential energy of the second intermediate is the energy at the lowest point between the second and third peaks. In the final step, the second intermediate forms a third activated complex that results in the final product.

Activation Energies for the Decomposition of HI and HBr

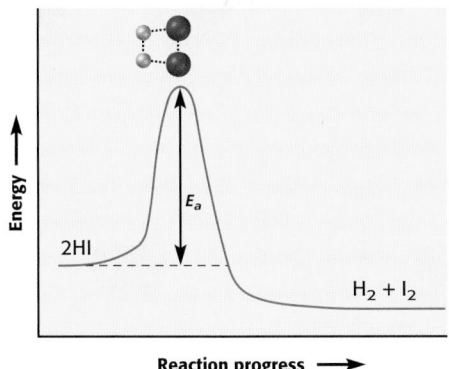

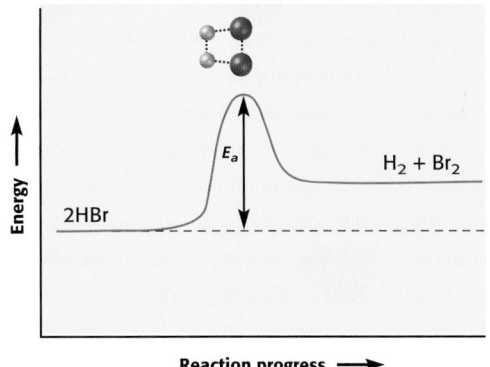

Figure 9
a The difference in energy between the bottom of this curve and the peak is the energy of activation for the decomposition of HI.

b The decomposition of HBr occurs at a faster rate than the decomposition of HI because this reaction has a lower activation energy.

Hydrogen Bromide Requires a Different Diagram

Figure 8b similarly represents how potential energy changes with reaction progress for the reaction below.

$$2HBr \longrightarrow H_2Br_2 \longrightarrow H_2 + Br_2$$

initial state	activated	final state
(reactant)	complex	**(products)**

One difference between the two graphs is that the activation energy is lower in the case of hydrogen bromide. Because the activation energy of HBr is lower than that of HI, a larger fraction of the HBr molecules have enough energy to clear the activation energy barrier than in the HI case. As a result, hydrogen bromide decomposes more quickly than hydrogen iodide does.

Notice in both **Figure 9** graphs that the initial states are not at the same energy as the final states. Note also that the products have a lower energy than the reactants in the case of the HI decomposition reaction in **Figure 9a,** while the opposite is true for hydrogen bromide decomposition in **Figure 9b.** This distinction reflects the fact that hydrogen iodide decomposition is exothermic,

$$2HI(g) \longrightarrow H_2(g) + I_2(g) \qquad \Delta H = -53 \text{ kJ}$$

while the decomposition of hydrogen bromide is endothermic.

$$2HBr(g) \longrightarrow H_2(g) + Br_2(g) \qquad \Delta H = 73 \text{ kJ}$$

Topic Link

Refer to the "Causes of Change" chapter for a discussion of energy changes in chemical reactions.

591

did you know?

Elementary Steps Consider the reaction of NO_2 with CO:

$$NO_2(g) + CO(g) \rightarrow NO(g) + CO_2(g)$$

A proposed two-step mechanism is

$$NO_2(g) + NO_2(g) \rightarrow NO_3(g) + NO(g)$$
$$NO_3(g) + CO(g) \rightarrow NO_2(g) + CO_2(g)$$

The species NO_3, formed in the first step, is an intermediate. Each of the reactions in the mechanism is an elementary step. The rate law for an elementary step can be written on the basis of the number of species that must collide in order to produce the product. In both steps of this mechanism, the number of colliding species is two. Thus, the rate law of the first step is rate = $[NO_2]^2$ and for second step, rate = $[NO_3][CO]$

Teaching Tip ───── GENERAL

Activation Energy Show students a box or book of matches. The matches are exposed to oxygen in the air, which is all that the matches requires to burn. However, the matches do not burst into flames. Now strike a match to produce the expected combustion. Help students understand that the heat created by the friction of the match against the striking surface is sufficient to provide the activation energy for the combustion. The match head is made of readily combustible materials that have a low activation energy. Ask students to sketch a potential energy curve for this reaction. The curve on the right side of the peak should be considerably lower than on the left side and the energy peak should be low. **LS** Visual

Discussion

Have a students draw a ball and stick model of a hydrogen iodide molecule, HI, on the board. Now, have another student add a second model of HI to the first as if the two molecules were colliding. Ask if bonds could form between the two hydrogen atoms and between the two iodine atoms to form the products H_2 and I_2. Depending upon the drawing, the hydrogen and iodine atoms may or may not be in position to form bonds. Ask students if bonding can take place. If not, tell students that the product-forming orientation would have the two molecules colliding so that they form an approximate square with like atoms at adjacent corners. Have them draw this model and elicit from them the fact that this is the activated complex.

Transparencies

TT Activation Energies

Teaching Tip

A Collision Model Model the effect of increasing concentration on reaction rate by using a CD case and metal ball bearings. Remove the plastic insert that holds the CD in the case. This will give you a clear-plastic, shallow, lidded box. Place 25 to 50 ball bearings in the case, and place the case on an overhead projector. Swirl and jerk the case at random so that students can note the frequency of collisions between the ball bearings. Next double the number of ball bearings in the box. This is equivalent to doubling the concentration because the ball bearings are confined to the same space. Swirl the box again; students should be able to see that the frequency of collisions increases. You may want to cover the outer $\frac{3}{4}$ in. of all sides of the box so that students will not be distracted by the ball bearings hitting and clumping against the sides of the box.

Using the Figure

Ask students which reaction in **Figure 9** is likely to have the faster rate. They should recognize that the lower the activation energy, the easier it is for molecules at the same temperature to make it over the potential energy hump and produce products.

Transparencies

TT Particle Collisions

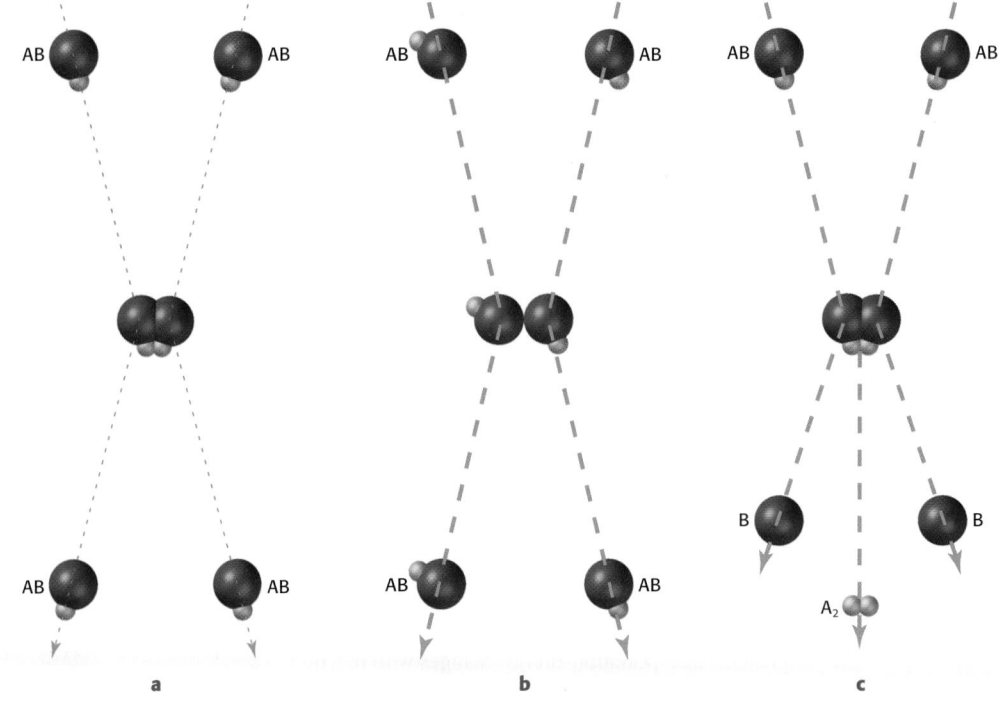

a   b   c

Figure 10
A reaction will not occur if the collision occurs too gently, as in **a,** or with the wrong orientation as in **b.** An effective collision, as in **c,** must deliver sufficient energy and bring together the atoms that bond in the products.

Not All Collisions Result in Reaction

Much of what we know about the collisions of molecules (and atoms) has come from studies of reactions between gases. However, it is believed that collisions happen similarly in solution. The principles of rate laws and activation energies apply in reactions that occur in solutions as well as in gas-phase reactions.

Collision between the reacting molecules is necessary for almost all reactions. Collision is not enough, though. The molecules must collide with enough energy to overcome the activation energy barrier. But another factor is also important. **Figure 10** illustrates the need for adequate energy and correct orientation in a collision.

A chemical reaction produces new bonds, and those bonds are formed between specific atoms in the colliding molecules. Unless the collision brings the correct atoms close together and in the proper orientation, the molecules will not react, no matter how much kinetic energy they have. For example, if a chlorine molecule collides with the oxygen end of the nitrogen monoxide molecule, the following reaction may occur.

$$NO(g) + Cl_2(g) \longrightarrow NOCl(g) + Cl(g)$$

This reaction will not occur if the chlorine molecule strikes the nitrogen end of the molecule.

592

did you know?

Uncooked pineapple cannot be used in gelatin desserts or salads because the enzymes in the pineapple break down the protein chains of the gelatin and keep it in the liquid state. The principal enzyme responsible is bromelain. Cooking the pineapple destroys this enzyme, so cooked pineapple does not prevent the gelling of gelatin. Bromelain is also used to break down proteins in the tenderizing of meats.

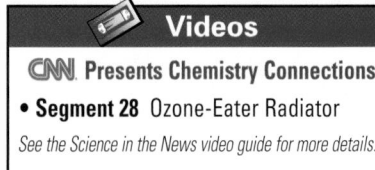
Videos

CNN Presents Chemistry Connections
• **Segment 28** Ozone-Eater Radiator
See the Science in the News video guide for more details.

Catalysts Increase Reaction Rate

Adding more reactant will usually increase the rate of a reaction. Adding extra product will sometimes cause the rate to decrease. Often, adding substances called **catalysts** to a reaction mixture will increase the reaction rate, even though the catalyst is still present and unchanged at the end of the reaction. The process, which is called **catalysis,** is shown in **Figure 11.**

Hydrogen peroxide solution, commonly used as a mild antiseptic and as a bleaching agent, decomposes only very slowly when stored in a bottle, forming oxygen as shown in the following equation.

$$2H_2O_2(aq) \longrightarrow 2H_2O(l) + O_2(g)$$

Adding a drop of potassium iodide solution speeds up the reaction. On the other hand, adding a few crystals of insoluble manganese dioxide, $MnO_2(s)$, causes a violent decomposition to occur. The iodide ion, $I^-(aq)$, and manganese dioxide are two of many catalysts for the decomposition of hydrogen peroxide.

Catalysis is widely used in the chemical industry, particularly in the making of gasoline and other petrochemicals. Catalysts save enormous amounts of energy. As you probably know, carbon monoxide is a poisonous gas that is found in automobile exhaust. The following oxidation reaction could remove the health hazard, but this reaction is very slow.

$$2CO(g) + O_2(g) \longrightarrow 2CO_2(g)$$

It is the job of the catalytic converter, built into the exhaust system of all recent models of cars, to catalyze this reaction.

Catalysis does not change the *overall* reaction at all. The stoichiometry and thermodynamics of the reaction are not changed. The changes affect only the path the reaction takes from reactant to product.

catalyst

a substance that changes the rate of a chemical reaction without being consumed or changed significantly

catalysis

the acceleration of a chemical reaction by a catalyst

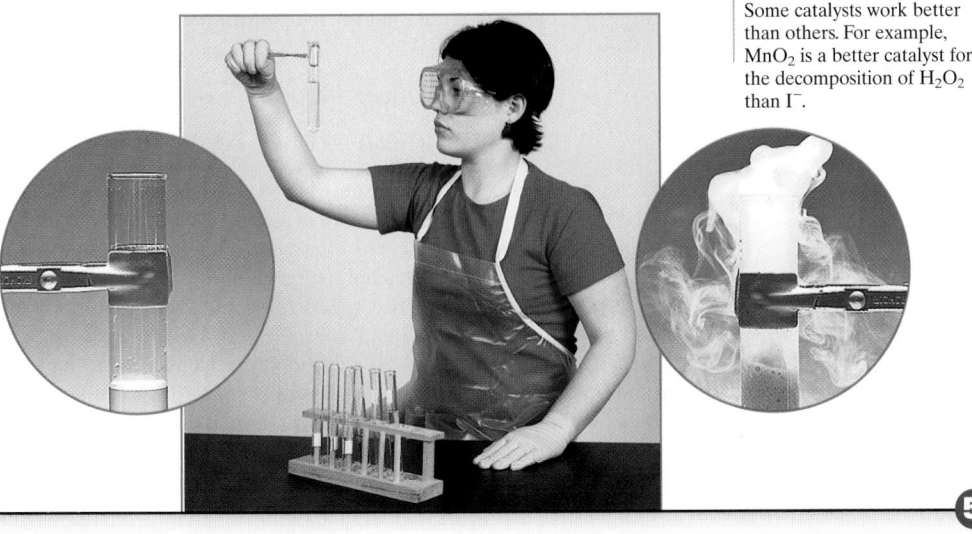

Figure 11
Some catalysts work better than others. For example, MnO_2 is a better catalyst for the decomposition of H_2O_2 than I^-.

(593)

HISTORY
CONNECTION — ADVANCED

The Contributions of Berzelius In 1836, Jöns Jakob Berzelius (1779–1848), a Swedish chemist, first used the terms *catalysis* and *catalyst* in their modern sense. He chose the words because the catalyzed reactions he had seen were all reactions in which a reactant was thought to break down. Have interested students investigate the work of Berzelius and write a report or oral presentation. Berzelius is considered one of the founders of modern chemistry. He discovered several elements, determined the atomic masses of many more, and investigated approximately 2,000 compounds. His many published papers and textbook of chemistry were instrumental in advancing the young science of chemistry worldwide. **LS** Verbal

Demonstration

Three Kinds of Catalysts

(Approximate time: 15 minutes)

In this demonstration students can compare the actions of three kinds of catalysts.

1. Place three Petri dishes on an overhead projector. Hold one back if all three do not fit.

2. Pour 10 mL of 3% H_2O_2 (ordinary hydrogen peroxide) into each Petri dish.

3. To the first dish, add 3 drops of 1 M KI solution (homogeneous catalyst).

4. To the second dish, add juice from a crushed turnip (the enzyme catalase).

5. Allow students to compare the activities in the first two dishes. Next, drop a very small amount of powdered MnO_2 (heterogeneous catalyst) from the end of a small spatula into the third dish. **CAUTION:** The reaction may be vigorous.

6. Ask students to compare their observations with the graph in **Figure 11** and explain any observed differences in activity.

Safety Caution: Wear safety goggles and a laboratory apron.

Disposal: Filter any remaining solids, and rinse them with tap water. Combine solutions. When bubbling stops, flush the liquid down the drain. Dispose of the solids in the trash.

Figure 12
The four curves show that various catalysts reduce the activation energy for the hydrogen peroxide decomposition reaction, but by different amounts. Notice that the enzyme catalase almost cancels the activation energy.

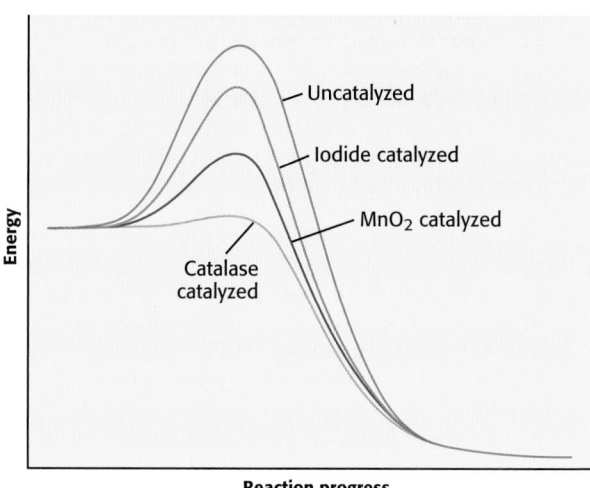

Comparison of Pathways for the Decomposition of H_2O_2

internet connect
www.scilinks.org
Topic: Catalysts
SciLinks code: HW4025
SCiLINKS. Maintained by the National Science Teachers Association

internet connect
www.scilinks.org
Topic: Enzymes
SciLinks code: HW4054
SCiLINKS. Maintained by the National Science Teachers Association

594

Catalysts Lower the Activation Energy Barrier

Catalysis works by making a different pathway available between the reactants and the products. This new pathway has a different mechanism and a different rate law from that of the uncatalyzed reaction. The catalyzed pathway may involve a surface reaction, as in the decomposition of hydrogen peroxide catalyzed by manganese dioxide, and in biological reactions catalyzed by enzymes. Or, the catalytic mechanism may take place in the same phase as the uncatalyzed reaction.

The iodide-catalyzed decomposition of hydrogen peroxide is an example of catalysis that does not involve a surface. It probably works by the following mechanism.

(1) $$I^-(aq) + H_2O_2(aq) \longrightarrow IO^-(aq) + H_2O(l)$$

(2) $$IO^-(aq) + H_2O_2(aq) \longrightarrow I^-(aq) + O_2(g) + H_2O(l)$$

Notice that the iodide ion, I^-, consumed in step 1 is regenerated in step 2, and the hypoiodite ion, IO^-, generated in step 1 is consumed in step 2. In principle, a single iodide ion could break down an unlimited amount of hydrogen peroxide. This is the characteristic of all catalytic pathways—the catalyst is never used up. It is regenerated and so becomes available for use again and again.

Each pathway corresponds to a different mechanism, a different rate law, and a different activation energy. **Figure 12** shows the potential energy profiles for the uncatalyzed reaction and for catalysis by three different catalysts. Because the catalyzed pathways have lower activation energy barriers, the catalysts speed up the rate of the reaction.

HISTORY CONNECTION

Catalytic Cracking The demand for gasoline increased rapidly in the early part of the twentieth century as automobiles and trucks replaced other forms of transportation. Only a certain amount of gasoline (called straight-run gasoline) can be obtained from the fractional distillation of raw petroleum—not nearly enough to meet demand. In 1912, large-scale "cracking" first began in the United States. In cracking, larger molecules, for which there is less commercial demand, are broken down into smaller molecules suitable for blending into gasoline. Initially, distillation at high temperatures was used to crack molecules. Catalytic cracking was developed in 1936, just in time to supply sufficient gasoline to the Allies in World War II.

Enzymes Are Catalysts Found in Nature

The most efficient of the three catalysts compared in **Figure 12** is an **enzyme.** Enzymes are large protein molecules. Their biological role is to catalyze metabolic processes that otherwise would happen too slowly to help the organism. For example, the enzyme lactase catalyzes the reaction of water with the sugar lactose, present in milk. People whose bodies lack the ability to produce lactase have what is known as *lactose intolerance.*

Enzymes are very specific and catalyze only one reaction. This is because the surface of an enzyme molecule has a detailed arrangement of atoms that interacts with the target molecule (lactose, for instance). The enzyme site and the target molecule are often said to have a "lock and key" relationship to each other.

Hydrogen peroxide is a toxic metabolic product in higher animals, and the enzyme catalase is present in their blood and other tissues to destroy H_2O_2. On the other hand, the bombardier beetle stores a supply of hydrogen peroxide for use as a defense mechanism. When threatened by a predator, the beetle injects catalase into its hydrogen peroxide store. The rapidly released oxygen gas provides pressure for a spray of irritating liquid that the beetle can squirt at its enemy, as shown in **Figure 13**.

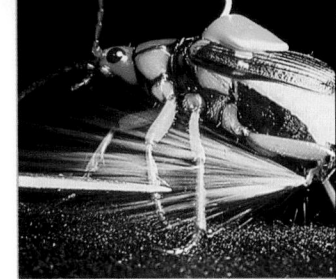

Figure 13
Bombardier beetles can repel predators such as frogs with a chemical defense mechanism powered by the catalytic decomposition of hydrogen peroxide.

enzyme
a type of protein that speeds up metabolic reactions in plants and animals without being permanently changed or destroyed

② Section Review

UNDERSTANDING KEY IDEAS

1. How can reaction orders be measured?

2. What can be learned from reaction orders?

3. Explain why not all collisions between reactant molecules lead to reaction.

4. What are catalysts and how do they function?

5. Give an example of an enzyme-catalyzed reaction.

PRACTICE PROBLEMS

6. What is the order of a reaction if its rate triples when the reactant concentration triples?

7. The reaction $CH_3NC(g) \longrightarrow CH_3CN(g)$ is of order 1, with a rate of 1.3×10^{-4} M/s when the reactant concentration was 0.040 M. Predict the rate when $[CH_3NC] = 0.025$ M.

8. The following data relate to the reaction $A + B \longrightarrow C$. Find the order with respect to each reactant.

[A](M)	[B] (M)	Rate (M/s)
0.08	0.06	0.012
0.08	0.03	0.006
0.04	0.06	0.003

CRITICAL THINKING

9. Which corresponds to the faster rate: a mechanism with a small activation energy or one with a large activation energy?

10. If the reaction $NO_2(g) + CO(g) \longrightarrow NO(g) + CO_2(g)$ proceeds by a one-step mechanism, what is the rate law?

11. What happens if a pair of colliding molecules possesses less energy than E_a?

12. Why is the phrase "lock and key" used to describe enzyme catalysis?

13. How are a catalyst and an intermediate similar? How are they different?

14. Draw a diagram similar to **Figure 10** to show (a) an unsuccessful and (b) a successful collision between $H_2(g)$ and $Br_2(g)$.

Close

Quiz ———— GENERAL

1. Two molecules collide but bounce apart unchanged. What two reasons could account for their failure to react? **Ans.** They had insufficient energy or did not collide in the necessary orientation.

2. Explain the role of a catalyst in a chemical reaction and how it works. **Ans.** A catalyst increases the rate of the reaction by lowering the activation energy required for reaction.

Reteaching ———— BASIC

Have each student make up three questions about the material covered in this section. Tell them the questions should not be just definitions. You might divide up the chapter among groups of students so that all topics are covered. Students should provide answers to the questions they write. Use the questions to quiz the class.
LS Intrapersonal

Chapter Resource File
• Concept Review
• Quiz

Answers to Section Review

1. by changing a concentration by a certain factor and comparing this with the factor by which the reaction rate changes

2. they can provide valuable clues about the mechanism of a reaction

3. It is necessary that the collision be sufficiently violent that energy in excess of the activation energy becomes available. Orientation is also important.

4. Catalysts are substances, other than reactants or products, that increase the rate of a reaction. They operate by providing a path of lower activation energy.

5. answers may vary; acceptable answers may include peroxidase and lactase

6. one

7. Rate changes by a factor of (0.025 M)/(0.040 M) = 0.625. Because the order is 1, rate will change by the same factor. Therefore new rate = $0.625 \times 1.3 \times 10^{-4}$ M = 8.1×10^{-5} M

8. one with respect to B (because rate halves when concentration halves); two with respect to A (because rate changes by a factor of 4 when concentration halves)

9. small activation energy

Answers continued on p. 601A

Chapter 16 • Reaction Rates   595

Alternative Assessments

SKILL BUILDER
• Page 593

Homework
• Page 580

Discussion
• Page 587

Group Activity
• Page 590

Reteaching
• Page 595

CHAPTER REVIEW
• Page 599

Portfolio Assessments

TECHNOLOGY CONNECTION
•Page 581

Reteaching
• Page 585

KEY IDEAS

SECTION ONE   What Affects the Rate of a Reaction?

• The rate of a chemical reaction is calculated from changes in reactant or product concentration during a small time interval.

• Reaction rates generally increase with reactant concentration or, in the case of gases, pressure.

• Rate increases with temperature because at a higher temperature a greater fraction of collisions have enough energy to cause a reaction.

SECTION TWO   How Can Reaction Rates Be Explained?

• Rate laws, which are used to suggest mechanisms, are determined by studying how reaction rate depends on concentration.

• An activated complex occupies the energy high point on the route from reactant to product.

• Catalysts provide a pathway of lower activation energy.

• Enzymes are biological catalysts that increase the rates of reactions important to an organism.

KEY TERMS

chemical kinetics
reaction rate

rate law
reaction mechanism
order
rate-determining step
intermediate
activation energy
activated complex
catalyst
catalysis
enzyme

KEY SKILLS

Calculating a Reaction Rate
Sample Problem A p. 581

Determining a Rate Law
Sample Problem B p. 587

596

Chapter Resource File

• **Datasheets for In-Text Lab**
  Reaction Rates
• **CBL™ Probeware Lab** A Leaky
  Reaction

CHAPTER REVIEW 16

USING KEY TERMS

1. Explain why you must use a negative sign when you write the rate expression for a chemical reaction in terms of the change in concentration of a reactant.

2. Define *reaction rate*.

3. Explain the difference between a reaction rate and a rate law.

4. What is a mechanism, and what is its rate-determining step?

5. Explain why the names *activated complex* and *transition state* are suitable for describing the highest energy point on a reaction's route from reactant to product.

6. Explain the role of an intermediate in a reaction mechanism.

7. What are enzymes, and what common features do they all share?

UNDERSTANDING KEY IDEAS

What Affects the Rate of a Reaction?

8. What unit is most commonly used to express reaction rate?

9. Explain how to calculate a reaction rate from concentration-versus-time data.

10. Mention uses of the term *rate* other than in chemical kinetics.

11. Explain how a graph can be useful in defining and measuring the rate of a chemical reaction.

12. Suggest ways of measuring concentration in a reaction mixture.

13. Why is it necessary to divide by the coefficient in the balanced chemical equation when calculating a reaction rate? When can that step be omitted?

14. What does Δ[A] mean if A is the reactant in a chemical reaction?

Figure 14

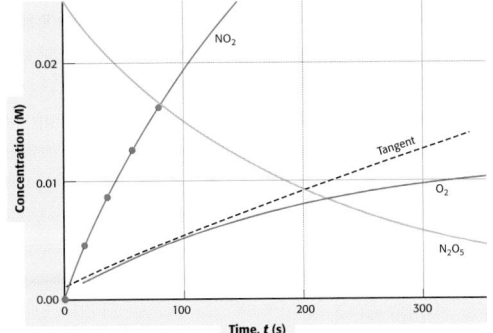

N₂O₅ Decomposition Data

15. In a graph like the one in **Figure 14,** what are the signs of the slopes for reactants and for products?

16. Which would have the greater effect on the rate of a chemical reaction in most cases—doubling the concentration of the reactant(s) or doubling the temperature?

17. Why does pressure affect the rate of the following reaction?

$$(CH_2)_3(g) \longrightarrow CH_2{=}CH{-}CH_3(g)$$

18. Explain the effect that area has on reactions that occur on surfaces.

597

Assignment Guide	
Section	**Questions**
1	1–2, 8–18, 24–28, 34–35, 40–41, 44–45, 48, 51
2	3–7, 19–23, 29–33, 36–39, 42–43, 46–47, 49–50, 52–56

CHAPTER REVIEW 16

REVIEW ANSWERS
Using Key Terms

1. Because the concentration of a reactant decreases with forward reaction progress, it must be expressed in negative form to obtain the rate expression for the reaction.

2. The decrease in the concentration of a reactant (or the increase in the concentration of a product) during a small time interval, divided by the duration of that interval and by the corresponding coefficient in the chemical equation.

3. The reaction rate describes how fast the reaction is proceeding; the rate law describes how the rate of the reaction depends on the concentration(s) of the reactant(s).

4. A mechanism describes the path the reaction takes in going from reactants to products. The rate-determining step is the slowest of the steps in a mechanism.

5. "Activated" is a suitable adjective because of the excess energy present. "Complex" refers to the complicated molecular configuration, to a more complex arrangement. "Transition state" reflects the condition of the atoms: they are in the middle of a transit from the initial state to the final state.

6. An intermediate occurs in a multi-step reaction, being formed in an early step and destroyed in a later step.

7. They are catalysts produced by living organisms to aid their metabolic processes. They are invariably large protein molecules, and often very efficient and specific.

Understanding Key Ideas

8. Mole per (liter second), or M/s.

9. If, in a short time interval Δt, the concentration of a reactant changes by ΔM, then the reaction rate is calculated by dividing –ΔM by Δt and by the coefficient of the reactant in the chemical equation describing the reaction.

10. Answers may vary, but should represent some measure of change over time.

11. The slope of a graph of a concentration c versus time t is $\Delta c/\Delta t$, from which the reaction rate is calculable. The concentration may be that of a reactant or a product.

12. Among the methods that might be suggested are: via a color intensity, by removing samples and titrating, by measuring pressure change, etc.

13. Unless their coefficients are equal, the concentrations of reactants do not change at the same rate; division by the coefficient gives a measure of reaction rate that is unambiguous. The coefficient may be omitted if it is 1.

14. It represents the change in the concentration of A.

15. The slopes are negative for reactants, positive for products.

16. doubling the temperature

17. Increasing the pressure increases the number of cyclopropane molecules present in a given volume. More molecules are present, so more molecules decompose.

18. If the surface available for reaction doubles, the amount of reaction doubles.

19. Reactions often occur via a multi-step mechanism. Reaction orders reflect the coefficients in the rate-determining step of such a mechanism, not in the overall equation.

20. rate $= k[A]^{n_1}[B]^{n_2}[C]^{n_3}$ where A, B and C are the three reactants.

21. A catalyst is a species, other than a reactant or product, that increases the rate of a chemical reaction. It does so by providing a reaction pathway of lower activation energy.

How Can Reaction Rates Be Explained?

19. Why are reaction orders not always equal to the coefficients in a chemical equation?

20. Write the general expression for the rate law of a reaction with three reactants A, B, and C.

21. Explain what a catalyst is and how it works.

22. Sketch a diagram showing how the potential energy changes with the progress of an endothermic reaction. Label the curve "Initial state," "Final state," and "Transition state." Then, draw a second curve to show the change brought about by a catalyst.

23. How do enzymes differ from other catalysts?

PRACTICE PROBLEMS

Sample Problem A   Calculating a Reaction Rate

24. The concentration of a reaction product with a coefficient of 3 in the chemical equation increases by 1.5×10^{-4} M in 11 s. Calculate the reaction rate.

25. What is the rate of the reaction
$$2NO(g) + Br_2(g) \longrightarrow 2NOBr$$
given that the bromine concentration decreased by 5.3×10^{-5} M during an interval of 38 s?

26. During the same 38 s interval cited in problem 25, the nitric oxide concentration decreased by 1.04×10^{-4} M. Recalculate the rate.

27. Calculate the rate of a reaction, knowing that a graph of the concentration of a product versus time had a slope of 3.6×10^{-6} M/s. The product had a coefficient of 2.

28. Exposed to the air, a solution of potassium iodide slowly oxidizes in the following reaction.
$$6I^-(aq) + O_2(aq) + 2H_2O(l) \longrightarrow$$
$$2I_3^-(aq) + 4OH^-(aq)$$
A 250-mL volume of such a solution was found to contain 2.0×10^{-5} mol of I_3^- ions after 15 days. What was the reaction rate?

Sample Problem B   Determining a Rate Law

29. In the reaction
$$2NO(g) + Br_2(g) \longrightarrow 2NOBr(g)$$
doubling the Br_2 concentration doubles the rate, but doubling the NO concentration quadruples the rate. Write the rate law.

30. What is the reaction order if the reaction rate triples when the concentration of a reactant is increased by a factor of 3?

31. The following reaction is first order.
$$(CH_2)_3(g) \longrightarrow CH_2 = CH-CH_3(g)$$
What change in reaction rate would you expect if the pressure of $(CH_2)_3$ doubled?

32. The rate of the reaction
$$2NOCl(g) \longrightarrow 2NO(g) + Cl_2(g)$$
increased by a factor of 3.14 when the concentration of NOCl was changed by a factor of 1.77. What is the order of this reaction?

MIXED REVIEW

33. Explain why, even though a collision may have energy in excess of the activation energy, reaction may not occur.

34. Describe the procedure for measuring the slope of a curve.

35. What is the sign of $\Delta[A]$ if A is the reactant in a chemical reaction?

36. The names of most enzymes, for instance protease and carbonic anhydrase, end with –ase. Which one of these enzymes do you think would be the catalyst for the reaction shown below?
$$H_2CO_3(aq) \longrightarrow H_2O(l) + CO_2(aq)$$
(Hint: The other one breaks proteins into amino acids.)

37. $2SO_2(g) + O_2(g) \longrightarrow 2SO_3(g)$
is a slow reaction, but nitrogen dioxide acts as a catalyst. The first step in the proposed two-step mechanism is shown below.
$$SO_2(g) + NO_2(g) \longrightarrow SO_3(g) + NO(g).$$
Suggest the second step.

38. What is meant by the *rate-determining step* in a reaction mechanism?

39. When hydrogen peroxide solution, used as an antiseptic, is applied to a wound, it often bubbles. Explain why.

40. Using chemical terminology, explain the purpose of food refrigeration.

41. Why is a negative sign present in the equation defining the rate of a reaction in terms of the concentration of a reactant?

42. What is lactose? What is lactase?

43. Discuss whether or not an activated complex is an intermediate in a chemical reaction.

44. Why do reptiles move more sluggishly in cold weather?

45. Explain how to measure a reaction rate from a graph of reactant concentration versus time.

46. Write the equation for the decomposition of gaseous ammonia into its elements. Also write the general rate law. Are you able to assign a reaction order? Explain.

CRITICAL THINKING

47. Sketch a picture of the shape and bonding in the activated complex during the following reaction.
$$NO(g) + Cl_2(g) \longrightarrow NOCl(g) + Cl(g)$$

48. Why is it necessary, in defining the rate of a reaction, to require that Δt be small?

49. Point out any shortcomings of the "ball over a speed bump" analogy of a reaction path.

50. In a 1 L flask filled with NO_2 at 1 atm pressure, the content of dinitrogen tetroxide was 0.00027 mol after 15 s. Calculate the reaction rate. In a second experiment, at 2 atm of pressure, the N_2O_4 content after 15 s was 0.00108 mol. Write the rate law.

51. Explain why, unlike gas-phase reactions, a reaction in solution is hardly affected at all by pressure.

52. Could a catalyzed reaction pathway have an activation energy higher than the uncatalyzed reaction? Explain.

53. Would you expect the concentration of a catalyst to appear in the rate law of a catalyzed reaction? Explain.

ALTERNATIVE ASSESSMENT

54. Boilers are sometimes used to heat large buildings. Deposits of $CaCO_3$, $MgCO_3$, and $FeCO_3$ can hinder the boiler operation. Aqueous solutions of hydrochloric acid are commonly used to remove these deposits. The general equation for the reaction is written below.
$$MCO_3(s) + 2H_3O^+(aq) \longrightarrow$$
$$M^{2+}(aq) + 3H_2O(l) + CO_2(g)$$
In the equation, M stands for Ca, Mg, or Fe. Design an experiment to determine the effect of various HCl concentrations on the rates of this reaction. Present your design to the class.

CONCEPT MAPPING

55. Use the following terms to complete the concept map below: *activation energy, alternative reaction pathway, catalysts, enzymes,* and *reaction rate*.

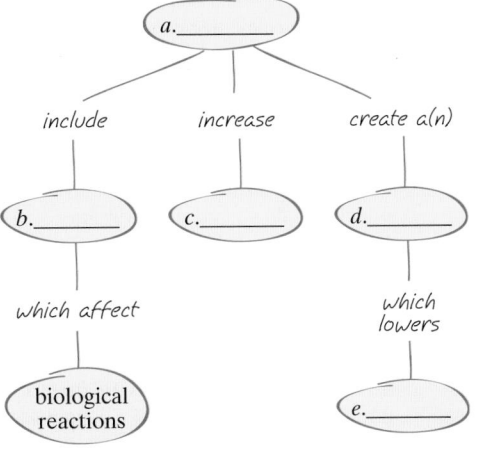

22. The humped curve should have its final plateau higher than its initial plateau. The second curve should share the same plateaus as the first, but have a lower hump.

23. In being proteins of biological origin, and being very substrate-specific

Practice Problems
24. Rate = 4.5×10^{-6} M/s
25. Rate = 1.4×10^{-6} M/s
26. Rate = 1.4×10^{-6} M/s
27. Rate = 1.8×10^{-6} M/s
28. Rate = 3×10^{-11} M/s
29. rate = $k[NO]^2[Br_2]$
30. 1
31. an increase by a factor of 2
32. 2

Mixed Review
33. It is necessary, not only that the colliding species have enough energy to react, but also that they are in the correct orientation to react.

34. Draw a straight line that just touches the curve, without crossing it, at the point of interest. The slope of the curve then equals the slope of the straight line, which can be found by the usual rise/run method.

35. negative

36. carbonic anhydrase

37. $2NO(g) + O_2(g) \rightarrow 2NO_2(g)$

38. It is the slowest step in a sequence of steps, and therefore controls the overall reaction rate.

39. Hydrogen peroxide is toxic to our bodies, and our body fluids contain an enzyme for decomposing H_2O_2. The bubbles are oxygen formed by this decomposition.

40. Foods decompose by slow chemical reactions. The rates of all reactions decrease massively on cooling. Therefore keeping food at a lower temperature delays spoilage.

41. Because the concentration change is negative. The presence of the negative sign ensures that the reaction rate is a positive quantity.

42. Lactose is a sugar, present in milk. Lactase is an enzyme that breaks down lactose into simpler sugars, which are digestible.

43. Intermediates occur as part of a multi-step mechanism, being formed by one step and destroyed by another. An activated complex could be considered as a very simple example of an intermediate.

44. Unlike mammals and birds, reptiles do not generate significant heat internally and hence their temperatures are those of their surroundings. Like all chemical reactions, the metabolic processes of reptiles, including those responsible for movement, are slower at low temperatures.

45. Measure the slope of the curve at the point of interest. It will be negative. Changing its sign and dividing by the reactant's coefficient will give the reaction rate.

46. $2NH_3(g) \rightarrow N_2(g) + 3H_2(g)$. Rate $= k[NH_3]^n$ No, rate laws cannot be found from the balanced chemical equation.

Critical Thinking

47. something like N—O---Cl---Cl

48. Reaction rates change with time and the ratio $(c_2-c_1)/(t_2-t_1) = \Delta c/\Delta t$ actually gives an average value of the rate between the times t_1 and t_2. To make this measurement relate closely to a particular time, it is necessary to minimize the time interval between t_1 and t_2; that is, Δt must be small.

Answers continued on p. 601A

FOCUS ON GRAPHING

Study the graph below, and answer the questions that follow.
For help in interpreting graphs, see Appendix B, "Study Skills for Chemistry."

The graph relates to an experiment in which the concentrations of bromide ion, hydrogen peroxide, and bromine were monitored as the following reaction took place.

$$2Br^-(aq) + H_2O_2(aq) + 2H_3O^+(aq) \longrightarrow$$
$$Br_2(aq) + 4H_2O(l)$$

56. The three curves are lettered **a, b,** and **c.** Which curves have positive slopes and which have negative slopes?

57. Associate each curve with one of the species being monitored.

58. What were the initial concentrations of bromine and hydrogen peroxide?

59. Measure the slope of each of the three curves at $t = 500$ s.

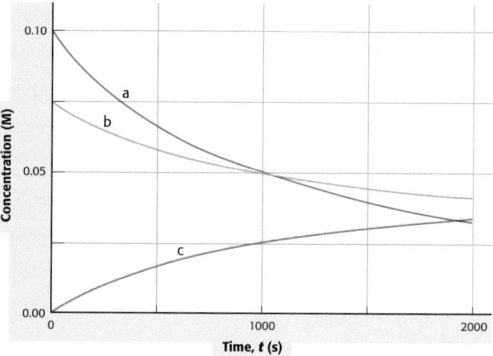

Changes in Concentration During Reaction

60. From each slope calculate a reaction rate. Do your three values agree?

TECHNOLOGY AND LEARNING

61. Graphing Calculator

Reaction Order

The graphing calculator can run a program that can tell you the order of a chemical reaction, provided you indicate the reactant concentrations and reaction rates for two experiments involving the same reaction.

Go to Appendix C. If you are using a TI-83 Plus, you can download the program RXNORDER and run the application as directed. If you are using another calculator, your teacher will provide you with key-strokes and data sets to use. At the prompts, enter the reactant concentrations and reaction rates. Run the program as needed to find the order of the following reactions. (All rates are given in M/s.)

a. $2N_2O_5(g) \longrightarrow 4NO_2(g) + O_2(g)$
N_2O_5: conc. 1 = 0.025 M; conc. 2 = 0.040 M
rate 1 = 8.1×10^{-5}; rate 2 = 1.3×10^{-4}

b. $2NO_2(g) \longrightarrow 2NO(g) + O_2(g)$
NO_2: conc. 1 = 0.040 M; conc. 2 = 0.080 M
rate 1 = 0.0030; rate 2 = 0.012

c. $2H_2O_2(g) \longrightarrow 2H_2O(g) + O_2(g)$
H_2O_2: conc. 1 = 0.522 M; conc. 2 = 0.887 M
rate 1 = 1.90×10^{-4}; rate 2 = 3.23×10^{-4}

d. $2NOBr(g) \longrightarrow 2NO(g) + Br_2(g)$
NOBr: conc. 1 = 1.27×10^{-4} M; conc. 2 = 4.04×10^{-4} M
rate 1 = 6.26×10^{-5}; rate 2 = 6.33×10^{-4}

e. $2HI(g) \longrightarrow H_2(g) + I_2(g)$
HI: conc. 1 = 4.18×10^{-4} M; conc. 2 = 8.36×10^{-4} M
rate 1 = 3.86×10^{-5}; rate 2 = 1.54×10^{-4}

600

STANDARDIZED TEST PREP 16

1. The sequence of steps that occurs in a reaction process is called the
   a. heterogeneous reaction.
   b. rate law.
   c. overall reaction.
   d. reaction mechanism.

2. To be effective, a collision requires
   a. enough energy only.
   b. favorable orientation only.
   c. enough energy and a favorable orientation.
   d. a reaction mechanism.

3. The usual unit of reaction rate is
   a. mol/s.   c. L/s.
   b. mol/L•s   d. mol s/L.

4. How does the potential energy of the activated complex compare with the potential energies of the reactants and products?
   a. It is lower than both the potential energy of the reactants and the potential energy of the products.
   b. It is lower than the potential energy of the reactants but higher than the potential energy of the products.
   c. It is higher than the potential energy of the reactants but lower than the potential energy of the products.
   d. It is higher than both the potential energy of the reactants and the potential energy of the products.

5. If a collision between molecules is very gentle, the molecules are
   a. more likely to be orientated favorably.
   b. less likely to be orientated favorably.
   c. likely to react.
   d. likely to rebound without reacting.

6. A species that changes the rate of a reaction but is neither consumed nor changed is

a. a catalyst.
b. an activated complex.
c. an intermediate.
d. an enzyme.

7. Chemical kinetics is useful because
   a. the rate law enables the equation of the reaction to be derived.
   b. the rate law suggests possible reaction mechanisms.
   c. thermodynamic data can be obtained from activation energies.
   d. catalysts save money.

8. Which of the following is not a catalyst for the decomposition of hydrogen peroxide?
   a. $I^-(aq)$.   c. $MnO_2(s)$.
   b. catalase.   d. lactase.

9. A rate law relates
   a. reaction rate and temperature.
   b. reaction rate and concentration.
   c. temperature and concentration.
   d. energy and concentration.

10. The human body uses which of the following to speed up vital reactions?
    a. reactants   c. collisions
    b. enzymes   d. temperature

11. In a graph of how potential energy changes with reaction progress, the activated complex appears at the
    a. left end of the curve.
    b. right end of the curve.
    c. bottom of the curve.
    d. peak of the curve.

12. The slowest step in a mechanism is called
    a. the rate-determining step.
    b. the uncatalyzed reaction.
    c. the activation step.
    d. None of the above

601

Answers to Section 1 Review, *continued*

8. The rates of concentration change, for the species that participate in the reaction, are proportional to that species' coefficient in the chemical equation. Therefore, in order that the rate of reaction be defined in a way that is independent of the choice of species, the rate of concentration change must be divided by the corresponding coefficient.

9. One possibility is that the product might be involved in some other reaction, a "side reaction". Other valid answers are possible.

10. mole per cubic centimeter per minute; many other valid answers are possible

11. There is only a very mild increase in collision rate with temperature. In contrast, reaction rates increase massively.

12. Generally, yes; for example, an increase in temperature greatly increases the rate of evaporation of a liquid.

13. because it changes concentrations

Answers to Section 2 Review, *continued*

10. Rate = $k[NO_2][CO]$

11. They bounce off each other, unchanged.

12. because the enzyme and the target molecule have complementary shapes in the same way that a lock and key do

13. They are similar in that they both play roles in the mechanism of a reaction. They are different in that an intermediate is produced by the reactants themselves, whereas a catalyst is an outside agent, and in that an intermediate changes into product, and a catalyst remains unchanged during a reaction.

14. **a.** Diagram should show the two molecules colliding in any orientation other than that of both hydrogen atoms colliding with both bromine atoms.

   b. Diagram should show the two molecules colliding in the orientation of each hydrogen atom colliding with one bromine atom.

Answers to Chapter Review, *continued*

49. Unlike the reaction path analogy, the levels before and after a speed bump are the same. (There could be other valid criticisms)

50. The rate for the first experiment = 1.8×10^{-5} M/s; rate law = $k[NO_2]^2$.

51. In the gas phase, a pressure change results in a proportional change in concentrations with a consequential change in reaction rates. In solution, there is no significant effect on volume or concentration and therefore reaction rates are unchanged.

52. no; By definition, a catalyst enhances the reaction rate by lowering the activation energy for the reaction.

53. yes; It logically follows that the greater the amount of catalyst present, more reactant can use the catalyst to react at a lower activation level.

Alternative Assessment

54. Designed experiments should include all appropriate safety cautions and controls.

Concept Mapping

55. **a.** catalysts
   b. enzymes
   c. reaction rate
   d. alternative reaction pathway
   e. activation energy

Focus on Graphing

56. curve **c** has a positive slope; **a** and **b** have negative slopes

57. **a** is Br^-, **b** is H_2O_2, **c** is Br_2.

58. $[Br_2]_{t=0} = 0$, $[H_2O_2]_{t=0} = 0.075$ M

59. The slopes are as follows: **a.** -4.4×10^{-5} M/s; **b.** -2.2×10^{-5} M/s **c.** 2.2×10^{-5} M/s but considerable error in the measurement of these slopes might be expected.

60. all give 2.2×10^{-5} M/s

Technology and Learning

61. **a.** 1
   b. 2
   c. 1
   d. 2
   e. 2

CHAPTER 17
Oxidation, Reduction, and Electrochemistry
Chapter Planning Guide

PACING	CLASSROOM RESOURCES	LABS, ACTIVITIES, AND DEMONSTRATIONS
BLOCK 1 • 45 min pp. 602–603 **Chapter Opener**		SE **Start-Up Activity** Lights On, p. 603 ◆
BLOCKS 2 & 3 • 90 min pp. 604–611 **Section 1** Oxidation-Reduction Reactions		TE **Demonstration** Three Oxidation States of Mn, p. 606 ◆ TE **Group Activity**, p. 608 BASIC TE **Demonstration** A Redox Reaction, p. 608 ◆ SE **Skills Practice Lab** Redox Titration, p. 818 ◆ GENERAL CRF **Datasheets for In-Text Labs** *
BLOCK 4 • 45 min pp. 612–615 **Section 2** Introduction to Electrochemistry		
BLOCKS 5 & 6 • 90 min pp. 616–624 **Section 3** Galvanic Cells	TT **Galvanic Cell** * TT **Dry Cells** * TT **Fuel Cell** * TT **Corrosion Cell** * SE **Science and Technology** Fuel Cells	TE **Activity** Designing a Model, p. 616 BASIC SE **QuickLab** Listen Up, p. 618 ◆ GENERAL CRF **Datasheets for In-Text Labs** * TE **Demonstration** Making a Six-Cell Fruit Battery, p. 619 ◆ TE **Demonstration** Predicting Reactions, p. 621 ◆ TE **Activity**, p. 622 ◆ GENERAL SE **Inquiry Lab** Redox Titration—Mining Feasibility Study, p. 822 ◆ GENERAL CRF **Datasheets for In-Text Labs** * CRF **CBL™ Probeware Lab** Establishing a Table of Reduction Potentials ◆ ADVANCED
BLOCK 7 • 45 min pp. 626–631 **Section 4** Electrolytic Cells	TT **Refining Copper Using an Electrolytic Cell** * TT **Downs Cell** * TT **Hall-Héroult Process** * TT **Electroplating** *	TE **Demonstration** The Electrolysis of KI, p. 628 ◆ TE **Group Activity**, p. 629 ADVANCED CRF **Observation Lab** Electroplating for Corrosion Protection ◆ GENERAL

BLOCKS 8 & 9 • 90 min   **Chapter Review and Assessment Resources**

SE **Chapter Review**, pp. 633–638
SE **Standardized Test Prep**, p. 639
CRF **Chapter Test** *
OSP **Test Generator**
CRF **Test Item Listing** *
OSP **Scoring Rubrics and Classroom Management Checklists**

Holt Chemistry: Online Resources

Visit **go.hrw.com** for a variety of free resources related to this textbook. Enter the keyword **HW4 HOME**.

Holt Online Learning

Students can access interactive problem solving help and active visual concept development with the *Holt Chemistry* Online Edition available at **www.hrw.com**

student CNN News

cnnstudentnews.com

Find the latest chemistry news, lesson plans, and activities related to important scientific events.

Compression guide:
To shorten your instruction because of time limitations, omit blocks 1, 3, 6, and 8.

KEY

TE	Teacher Edition	**CRF** Chapter Resource File	* Also on One-Stop Planner
SE	Student Edition	**TT** Teaching Transparencies	◆ Requires Advance Prep
OSP	One-Stop Planner		

PROBLEM SOLVING AND PRACTICE	SECTION REVIEW AND ASSESSMENT	STANDARDS CORRELATION
		National Science Education Standards
SE **Skills Toolkit 1** Assigning Oxidation Numbers, p. 606 **SE** **Sample Problem A** Determining Oxidation Numbers, p. 607 **SE** **Practice,** p. 607 (GENERAL) **TE** **Homework,** p. 607 (GENERAL) **SE** **Skills Toolkit 2** Balancing Redox Equations Using the Half-Reaction Method, p. 609 **SE** **Sample Problem B** The Half-Reaction Method, p. 610 **SE** **Practice,** p. 610 (GENERAL) **TE** **Homework,** p. 610 (GENERAL) **CRF** **Problem Solving** * (ADVANCED) **SE** **Problem Bank,** p. 858 (GENERAL)	**TE** **Quiz,** p. 611 (GENERAL) **CRF** **Quiz** * **TE** **Reteaching,** p. 611 (BASIC) **SE** **Section Review,** p. 611 **CRF** **Concept Review** *	PS 3c
	TE **Quiz,** p. 615 (GENERAL) **CRF** **Quiz** * **TE** **Reteaching,** p. 615 (BASIC) **SE** **Section Review,** p. 615 **CRF** **Concept Review** *	PS 3a, PS 3b
SE **Sample Problem C** Calculating Cell Voltage, p. 623 **SE** **Practice,** p. 623 (GENERAL) **TE** **Homework,** p. 623 (GENERAL) **CRF** **Problem Solving** * (ADVANCED)	**TE** **Quiz,** p. 624 (GENERAL) **CRF** **Quiz** * **TE** **Reteaching,** p. 624 (BASIC) **SE** **Section Review,** p. 624 **CRF** **Concept Review** *	PS 3a, PS 3b, PS 3c
TE **Homework,** p. 631 (BASIC)	**TE** **Quiz,** p. 631 (GENERAL) **CRF** **Quiz** * **TE** **Reteaching,** p. 631 (BASIC) **SE** **Section Review,** p. 631 **CRF** **Concept Review** *	PS 3a, PS 3b, PS 3c

SCiLINKS

www.scilinks.org
Topic: Redox Reaction
SciLinks code: HW4110

Topic: Electrical Energy
SciLinks code: HW4045

Topic: Electrochemical Cells
SciLinks code: HW4046

Topic: Corrosion
SciLinks code: HW4147

Topic: Electroplating
SciLinks code: HW4049

Technology Resources

 CNN **Science in the NEWS**

Each video segment is accompanied by a Critical Thinking Worksheet.

Segment 29
Electric Car

Segment 30
Battery Technology

 ChemFile Interactive Tutor CD-Rom

Module 10: Electrochemistry
Topic: Electricity

Module 10: Electrochemistry
Topic: Electrochemical Cells

Overview

This chapter introduces students to electrochemistry. Students learn about the processes of oxidation and reduction and learn how to assign oxidation numbers to atoms. Students also learn how to recognize redox reactions and to balance them using half-reactions. Students then learn the basic components of electrochemical cells. Students study several examples of galvanic cells and learn how to calculate the voltage of a cell. Finally, students study several examples of electrolytic cells, including those used in electrolysis and electroplating.

Assessing Prior Knowledge

Students should be familiar with the following topics:

- electronegativity
- the mole
- balancing equations
- the activity series
- the nature of solutions
- ionization

Using the Figure

Robots are used to explore a variety of locations. Robots have been sent to other planets and moons, as well as into hostile environments on Earth. In order to accomplish its mission of data collection, a robot must have a reliable energy source. Some or all of this energy can be supplied using electrochemical cells. Through the processes of oxidation and reduction in the cells, an electric current is generated. Electrochemical cells used in this way are better known to most people as batteries.

CHAPTER

17

OXIDATION, REDUCTION, AND ELECTROCHEMISTRY

602

Standards Correlations

National Science Education Standards

PS 3a: Chemical reactions occur all around us, for example in health care, cooking, cosmetics, and automobiles. Complex chemical reactions involving carbon-based molecules take place constantly in every cell in our bodies. (Sections 2, 3, 4)

PS 3b: Chemical reactions may release or consume energy. Some reactions such as the burning of fossil fuels release large amounts of energy by losing heat and by emitting light. Light can initiate many chemical reactions such as photosynthesis and the evolution of urban smog. (Sections 2, 3, 4)

PS 3c: A large number of important reactions involve the transfer of either electrons (oxidation/reduction reactions) or hydrogen ions (acid/base reactions) between reacting ions, molecules, or atoms. In other reactions, chemical bonds are broken by heat or light to form very reactive radicals with electrons ready to form new bonds. Radical reactions control many processes such as the presence of ozone and greenhouse gases in the atmosphere, burning and processing of fossil fuels, the formation of polymers, and explosions. (Sections 1, 3, 4)

If you run out of gasoline in a car, you might have to walk to the nearest gas station for more fuel. But running out of energy when you are millions of kilometers from Earth is a different story! A robot designed to collect data on other bodies in our solar system needs a reliable, portable source of energy that can work in the absence of an atmosphere to carry out its mission. Many of the power sources for these explorer robots are batteries. In this chapter, you will learn about the processes of oxidation and reduction and how they are used in batteries to provide energy. You will also learn how these processes are used to purify metals and protect objects from corrosion.

START-UP **ACTIVITY**

Lights On

SAFETY PRECAUTIONS

PROCEDURE

1. Assemble **batteries**, a **light-emitting diode**, and **wires** so that the diode lights. Make a diagram of your construction.

2. Remove the batteries, and reconnect them in the opposite direction. Record the results.

ANALYSIS

1. A light-emitting diode allows electrons to move through it in only one direction—into the short leg and out of the longer leg. Based on your results, from which end of the battery must electrons leave?

2. How are the atoms changing in the end of the battery from which electrons leave?

3. What must happen to the electrons as they enter into the other end of the battery?

4. Why does a battery eventually run down?

Pre-Reading Questions

(1) **What type of charge results from losing electrons? from gaining electrons?**

(2) **Name a device that converts chemical energy into electrical energy.**

(3) **Why are batteries marked with positive and negative terminals?**

CONTENTS

603

START-UP **ACTIVITY**

Skills Acquired:
- Constructing Models
- Experimenting
- Interpreting

Materials:
For each group of 2–3 students:
- battery, 1.5 V (2)
- light-emitting diode
- wires, insulated copper with ends stripped (4)

Teacher's Notes: Make sure students do not connect the two terminals of a battery directly to one another. The resulting short circuit could cause the battery to overheat. To save time, you might want to tell students that the batteries must be joined in a series (positive terminal to negative terminal) to light the LED. The short leg of the LED is the leg on the side of the LED that has a flat edge.

Answers
1. the negative end
2. The atoms must be losing electrons and becoming positively charged.
3. The electrons must be absorbed by atoms in the battery.
4. Eventually the battery runs down because the atoms in the battery no longer lose and gain electrons.

Answers to Pre-Reading Questions
1. Losing electrons causes the material to have a positive charge. Gaining electrons causes the material to have a negative charge.
2. a battery
3. to show the end where electrons are lost and the end where electrons are gained

Focus

Overview

Before beginning this section, review with your students the Objectives listed in the Student Edition. In this section students learn about oxidation and reduction. They learn how to assign oxidation numbers to atoms, how to identify redox reactions, and how to balance them using the half-reaction method.

 Bellringer

Have students list the types of reactions they have learned and compare their answers. Tell students that they will learn about oxidation-reduction reactions, which encompass the other types they have learned.

Motivate

Discussion ——— GENERAL

Before class, cut an apple into slices and allow it to turn brown. Display the apple slices, a rusted piece of iron, and a burning candle. Ask students to list evidence of chemical reactions. **Ans.** the apple turned brown, the iron developed a reddish, flaky layer, and the candle emits energy as light and heat Ask students what could have been done to prevent or stop these reactions. **Ans.** don't cut the apple (or wrap it up after it is cut), paint the piece of iron, blow out the candle Ask students what is a common reactant in the three reactions. **Ans.** oxygen Explain that these reactions are oxidation reactions, but not all oxidation reactions involve oxygen.
LS Verbal

Oxidation-Reduction Reactions

KEY TERMS

- oxidation
- reduction
- oxidation-reduction reaction
- oxidation number
- half-reaction
- oxidizing agent
- reducing agent

Topic Link

Refer to the "Ions and Ionic Compounds" and "Covalent Compounds" chapters for more information about chemical bonding.

oxidation

a reaction that removes one or more electrons from a substance such that the substance's valence or oxidation state increases

OBJECTIVES

1. **Identify** atoms that are oxidized or reduced through electron transfer.

2. **Assign** oxidation numbers to atoms in compounds and ions.

3. **Identify** redox reactions by analyzing changes in oxidation numbers for different atoms in the reaction.

4. **Balance** equations for oxidation-reduction reactions through the half-reaction method.

Electron Transfer and Chemical Reactions

You already know that atoms with very different electronegativities bond by an electron transfer. For example, sodium chloride is formed by the transfer of electrons from sodium atoms to chlorine atoms in the reaction shown in **Figure 1** and described by the following equation:

$$2Na(s) + Cl_2(g) \longrightarrow 2NaCl(s)$$

Though NaCl is the way the formula of sodium chloride is usually written, the compound is made up of ions. Therefore, it might be helpful to think of sodium chloride as if its formula were written Na^+Cl^- so that you remember the ions.

When the electronegativity difference between the atoms is smaller, a polar covalent bond can form when the atoms join, as shown below.

$$2C(s) + O_2(g) \longrightarrow 2CO(g)$$

The C–O bond has some ionic character because there is an unequal sharing of electrons between the carbon atom and the oxygen atom. The oxygen atom attracts the shared electrons more strongly than the carbon atom does.

Oxidation Involves a Loss of Electrons

In the examples above, electrons were transferred at least in part from one atom to another. The sodium atom lost an electron to the chlorine atom. The carbon atom lost some of its control over its electrons to the oxygen atom. The loss, wholly or in part, of one or more electrons is called **oxidation.**

Thus, in making NaCl, the sodium atom is oxidized from Na to Na^+. Likewise, the carbon atom is oxidized when CO forms, even though the carbon atom does not become an ion.

604

MISCONCEPTION
/// **ALERT** \\\

Some books define *oxidation* as a reaction in which a substance gains oxygen or loses hydrogen and define *reduction* as a reaction in which a substance loses oxygen or gains hydrogen. The more accurate and generally accepted definitions describe the gain or loss of electrons. Students can associate the consumption of oxygen with oxidation reactions, however be sure they understand that this is only one clue they should look for.

Chapter Resource File

- Lesson Plan

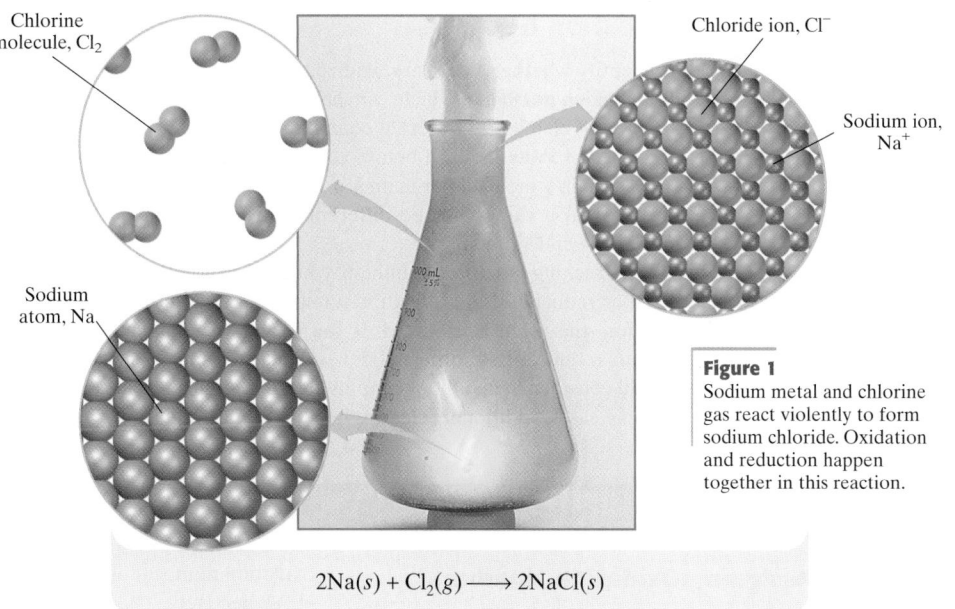

Chlorine molecule, Cl_2

Chloride ion, Cl^-

Sodium ion, Na^+

Sodium atom, Na

Figure 1
Sodium metal and chlorine gas react violently to form sodium chloride. Oxidation and reduction happen together in this reaction.

$$2Na(s) + Cl_2(g) \longrightarrow 2NaCl(s)$$

Reduction Involves a Gain of Electrons

In making NaCl, the electrons lost by the sodium atoms do not just disappear. They are gained by the chlorine atoms. The gain of electrons is described as **reduction.** The chlorine atoms are reduced as they change from Cl_2 to $2Cl^-$.

When joining with carbon atoms to make CO, oxygen atoms do not gain electrons but gain only a partial negative charge. But because the electrons in the C—O bonds spend more time near the oxygen atoms, the change is still a reduction.

More than one electron may be gained in a reduction. In the formation of Li_3N, described by the equation below, three electrons are gained by each nitrogen atom.

$$6Li(s) + N_2(g) \longrightarrow 2Li_3N(s)$$

Oxidation and Reduction Occur Together

When oxidation happens there must also be reduction taking place. You will learn later in this chapter that oxidation and reduction can happen at different places. In most situations, however, oxidation and reduction happen in a single place. Consider HgO being broken down into its elements, as described by the following equation:

$$2HgO(s) \longrightarrow 2Hg(l) + O_2(g)$$

In this reaction, mercury atoms are reduced, while oxygen atoms are oxidized. A single reaction in which an oxidation and a reduction happen is called an **oxidation-reduction reaction** or *redox reaction*.

reduction

a chemical change in which electrons are gained, either by the removal of oxygen, the addition of hydrogen, or the addition of electrons

internet connect

www.scilinks.org
Topic: Redox Reactions
SciLinks code: HW4110

SCI LINKS Maintained by the National Science Teachers Association

oxidation-reduction reaction

any chemical change in which one species is oxidized (loses electrons) and another species is reduced (gains electrons); also called *redox reaction*

605

Teach

READING SKILL BUILDER — BASIC

Assimilating Knowledge Before reading this chapter, have students write a short list of all the things they already know or think they know about electrochemistry. When students have finished their list, ask them to contribute their entries to a group list on the board or on an overhead projector. Then have students list the things they might want to know about electrochemistry. If students have trouble coming up with questions, offer the following:

• What are the components of various types of batteries?

• How is electrical energy used to electroplate metals onto jewelry and flatware?

• What happens when iron corrodes?

Later, when students are reviewing for the chapter test, have them read over their list and correct or expand upon their answers.
LS Logical

Teaching Tips

Mnemonic Devices An easy mnemonic device for associating oxidation and reduction with electron transfer is **OIL RIG**: **O**xidation **I**s the **L**oss of electrons, and **R**eduction **I**s the **G**ain of electrons. Students may prefer **LEO** the lion says "**GER**": **L**oss of **E**lectrons is **O**xidation, and **G**ain of **E**lectrons is **R**eduction.

Vocabulary The idea of reduction relates to the fact that when an ion or atom gains electrons, its oxidation state is reduced to a lower value. The opposite process, oxidation, relates to the fact that elements lose electrons in the same way they do when combining with oxygen, taking on higher oxidation states. This terminology originated before Mendeleev's time, when the combining abilities of elements were studied by examining and comparing the proportions in their various oxides.

BIOLOGY CONNECTION

Photosynthesis is the process in which organisms, such as green plants and blue-green algae, convert light energy into chemical energy stored in sugars (glucose). The equation for photosynthesis is usually written as:

$$6CO_2 + 6H_2O \rightarrow C_6H_{12}O_6 + 6O_2$$

However, this equation is an overall equation that does not show all of the chemical reactions involved. Photosynthesis is a very complex series of oxidations and reductions that involve chemical compounds not shown in the overall equation. The end products of photosynthesis, glucose and oxygen, are used by plants and animals in a complimentary process called *respiration*. Respiration can be thought of as the reverse of photosynthesis because the overall reaction shows the conversion of glucose and oxygen to carbon dioxide and water. Encourage interested students to study the reactions in photosynthesis or respiration and determine which atoms are being oxidized and which are being reduced.

Demonstration

Three Oxidation States of Mn

1. Pour 0.05 M $KMnO_4$ solution into a Petri dish until the dish is one-fourth full. Point out to students that manganese is in the +7 oxidation state.

2. Place the dish on the overhead projector.

3. Add 1 mL of 0.1 M NaOH to the solution.

4. Gently add one drop of acetaldehyde, which acts as a reducing agent.

5. Let the acetaldehyde remain undisturbed on the surface. Three colored rings should appear, corresponding to three oxidation states of Mn: brown, +4; green, +6; and purple, +7.

Safety Caution: Wear safety goggles and a lab apron. Keep students at least 3 m away from the demonstration.

Disposal: While stirring, add 1.0 M $Na_2S_2O_3$ until the mixture is brown. Adjust the pH to between 5 and 9, dilute tenfold, and pour down the drain.

Teaching Tip

Oxidation Numbers and Bonds

In ionic bonding, the oxidation number of a monatomic ion is the same as its charge. In a covalent bond, the higher oxidation number is assigned to the element that has the lower electronegativity because it "loses" electrons to the other element. However, students should remember that electrons are not transferred in the bond and the oxidation numbers do not imply charges.

oxidation number

the number of electrons that must be added to or removed from an atom in a combined state to convert the atom into the elemental form

Oxidation Numbers

To identify whether atoms are oxidized or reduced, chemists use a model of **oxidation numbers,** which can help them identify differences in an atom of an element in different compounds. By following the set of rules described in **Skills Toolkit 1** below, you can assign an oxidation number to each atom in a molecule or in an ion. **Sample Problem A** shows how to use the rules. You can see three different oxidation numbers for atoms of manganese in **Figure 2.**

By tracking oxidation numbers, you can tell whether an atom is oxidized or reduced. If the oxidation number of an atom increases during a reaction, the atom is oxidized. If the oxidation number decreases, the atom is reduced. Like other models, oxidation numbers have limits. You should consider them a bookkeeping tool to help keep track of electrons. In some cases, additional rules are needed to find values that make sense.

---

1 SKILLS Toolkit

Assigning Oxidation Numbers

1. Identify the formula.
  • If no formula is provided, write the formula of the molecule or ion.

2. Assign known oxidation numbers.
  • Place an oxidation number above each element's symbol according to the following rules.
  a. The oxidation number of an atom of any free (uncombined) element in atomic or molecular form is zero.
  b. The oxidation number of a monatomic ion is equal to the charge on the ion.
  c. The oxidation number of an atom of fluorine in a compound is always –1 because it is the most electronegative element.
  d. An atom of the more electronegative element in a binary compound is assigned the number equal to the charge it would have if it were an ion.
  e. In compounds, atoms of the elements of Group 1, Group 2, and aluminum have positive

oxidation numbers of +1, +2, and +3, respectively.
  f. The oxidation number of each hydrogen atom in a compound is +1, unless it is combined with a metal atom; then it is –1.
  g. The oxidation number of each oxygen atom in compounds is usually –2. When combined with fluorine atoms, oxygen becomes +2. In peroxides, such as H_2O_2, an oxygen atom has an oxidation number of –1.

3. Calculate remaining oxidation numbers, and verify the results.
  • Use the total oxidation number of each element's atoms (the oxidation number for an atom of the element multiplied by the subscript for the element) and the following rules to calculate missing oxidation numbers.
  h. The sum of the oxidation numbers for all the atoms in a molecule is zero.
  i. The sum of the oxidation numbers for all atoms in a polyatomic ion is equal to the charge on that ion.

606

REAL-WORLD
CONNECTION

Potters use a type of coating called a *glaze* to give their pottery color and shine. Glazes produce different colors because they contain different elements—usually transition metals. The glazes are brushed on a piece of pottery and then the pottery is placed in a high temperature oven called a *kiln* to be baked or "fired." Before the glazes are fired they are often a dull pastel color, but as they are fired, the metals are oxidized and the glaze changes into different, often bright colors. Sometimes the amount of oxygen in the kiln determines the color of the glaze. For example, in a kiln that has a high oxygen concentration, a glaze containing iron may form iron(III) oxide and have a red color. In a kiln that has a lower oxygen concentration, the same glaze may turn black if it forms iron(II) oxide.

$$\overset{0}{Mn} \quad \overset{+4 \; -2}{MnO_2} \quad \overset{+1}{K}^+ \quad \overset{+7 \; -2}{MnO_4^-}$$

Figure 2
An atom of manganese in its elemental form has an oxidation number of 0. In MnO_2, the oxidation number of the manganese atom is +4. In the permanganate ion, MnO_4^-, the oxidation number of the manganese atom is +7.

SAMPLE PROBLEM A

Determining Oxidation Numbers

Assign oxidation numbers to the sulfur and oxygen atoms in the pyrosulfate ion, $S_2O_7^{2-}$.

1 Identify the formula.

The pyrosulfate ion has the formula $S_2O_7^{2-}$.

2 Assign known oxidation numbers.

According to Rule **g,** the oxidation number of the O atoms is –2, so this number is written above the O symbol in the formula. Because the oxidation number of the sulfur atoms is unknown, x is written above the S symbol. Thus the formula is as follows:

$$\overset{x \quad -2}{S_2O_7^{2-}}$$

3 Calculate remaining oxidation numbers, and verify the results.

- Multiplying the oxidation numbers by the subscripts, we see that the S atoms contribute $2x$ and the O atoms contribute $7(-2) = -14$ to the total oxidation number. To come up with the correct total charge, (Rule **i**), $2x + (-14) = -2$. Solve this equation to find $x = +6$.

- In $S_2O_7^{2-}$, the oxidation number of the S atoms is +6, and the oxidation number of the O atoms is –2. The sum of the total oxidation numbers for each element is $2(+6) + 7(-2) = -2$, which is the charge on the ion.

> **PRACTICE HINT**
>
> In this book, the oxidation number for a single atom is written above its chemical symbol. However, be sure to use the total number of atoms for each element when finding the sum of the oxidation numbers for all atoms in the molecule or ion.

PRACTICE

Determine the oxidation number for each atom in each of the following.

1
- **a.** NH_4^+
- **b.** Al
- **c.** H_2O
- **d.** Pb^{2+}
- **e.** H_2
- **f.** $PbSO_4$
- **g.** $KClO_3$
- **h.** BF_3
- **i.** $Ca(OH)_2$
- **j.** $Fe_2(CO_3)_3$
- **k.** $H_2PO_4^-$
- **l.** NH_4NO_3

607

BIOLOGY

CONNECTION — GENERAL

Nitrogen Fixation Nitrogen is an essential element in organic compounds used by all living things. Although nitrogen (N_2) is the most abundant gas in air, plants and animals cannot readily use nitrogen gas. Nitrogen must be reduced and combined with hydrogen to form ammonium ions or combined with oxygen to form nitrate or nitrite ions before it can be used by living organisms.

The different processes through which nitrogen is reduced are collectively called *nitrogen fixation*. Have interested students study the different ways nitrogen can be fixed, and create a poster with the information they gather. They should include the different chemical equations showing the reduction of nitrogen on their posters. **LS Visual**

Answers to Practice Problems A

1. a. N = –3; H = +1
   b. Al = 0
   c. H = +1; O = –2
   d. Pb = +2
   e. H = 0
   f. Pb = +2; S = +6; O = –2
   g. K = +1; Cl = +5; O = –2
   h. B = +3; F = –1
   i. Ca = +2; O = –2; H = +1
   j. Fe = +3; C = +4; O = –2
   k. H = +1; P = +5; O = –2
   l. N = –3; H = +1; N = +5; O = –2

Homework ——— GENERAL

Additional Practice Determine the oxidation number for each atom in each of the following.

1. KOH Ans. K = +1; O = –2; H = +1

2. Na_3AsO_4 Ans. Na = +1; As = +5; O = –2

3. NO_3^- Ans. N = +5; O = –2

4. NO_2^- Ans. N = +3; O = –2

5. $Ca(ClO_3)_2$ Ans. Ca = +2; Cl = +5; O = –2

6. $NaClO_4$ Ans. Na = +1; Cl = +7; O = –2

LS Logical

> **Chapter Resource File**
>
> • Problem Solving

Figure 3
Zinc metal reacts with hydrochloric acid, making bubbles of hydrogen gas.

Group Activity ——— BASIC

Oxidation-reduction reactions may initially seem uncommon to students, yet they occur all around us. Have students work in small groups to research four naturally occurring redox reactions or redox reactions that are useful for humans. Have them write the equations for these reactions on poster boards and have them use the half-reaction method to balance them. Finally, ask the students to illustrate their reactions with photographs and drawings. Display the posters around the room. **LS** Interpersonal

Demonstration

A Redox Reaction Be sure students understand that this simple redox demonstration is not a Cu/Zn galvanic cell.

1. Half-fill a Petri dish with a 0.5 M $CuSO_4$ solution.

2. Place the dish on an overhead projector.

3. Add to the Petri dish a few pieces of mossy zinc or some 1 cm^2 pieces of zinc sheet.

4. Copper metal should form on the zinc as the blue color of the solution fades, indicating the removal of Cu^{2+} ions from solution.

5. Allow students to view the dish under room light so that they can see the reddish color of the copper metal formed. Discuss the process in terms of electron transfer, oxidation, reduction, and the ions present in solution both before and after the reaction.

continued on next page

Identifying Redox Reactions

Figure 3 shows the reaction of Zn with HCl. Is this a redox reaction? Hydrochloric acid is a solution in water of Cl^-, which plays no part in the reaction, and H_3O^+. The net change in this reaction is

$$2H_3O^+(aq) + Zn(s) \longrightarrow H_2(g) + 2H_2O(l) + Zn^{2+}(aq)$$

Using rules **a, b, f, g, h,** and **i** from **Skills Toolkit 1,** you can give oxidation numbers to all atoms as follows:

$$2\overset{+1\ -2}{H_3O}{}^+(aq) + \overset{0}{Zn}(s) \longrightarrow \overset{0}{H_2}(g) + 2\overset{+1\ -2}{H_2O}(l) + \overset{+2}{Zn}{}^{2+}(aq)$$

Comparing oxidation numbers, you see that the zinc atom changes from 0 to +2 and that two hydrogen atoms change from +1 to 0. So, this is a redox reaction. In a redox reaction, the oxidation numbers of atoms that are oxidized increase, and those of atoms that are reduced decrease.

Half-Reactions

In the reaction shown in **Figure 3,** each zinc atom loses two electrons and is oxidized. One way to show only this half of the overall redox reaction is by writing a **half-reaction** for the change.

$$Zn(s) \longrightarrow Zn^{2+}(aq) + 2e^-$$

Note that electrons are a product. Of course, there is also a half-reaction for reduction in which electrons are a reactant.

$$2e^- + 2H_3O^+(aq) \longrightarrow H_2(g) + 2H_2O(l)$$

By adding the two half-reactions together, you get the overall redox reaction shown earlier. Notice that enough electrons are in each half-reaction to keep the charges balanced. Keep in mind that free electrons do not actually leave the zinc atoms and float around before being picked up by the hydronium ions. Instead, they are "handed off" directly from one to the other.

half-reaction

the part of a reaction that involves only oxidation or reduction

did you know?

Redox reactions save lives! Airbags in automobiles are inflated with nitrogen gas produced by two redox reactions. The gas generator in some airbags contains sodium azide (NaN_3) and iron(III) oxide. The mixture is automatically ignited during a head-on collision. When this happens, the sodium azide decomposes in a redox reaction to form sodium and nitrogen:

$$2NaN_3 \rightarrow 2Na + 3N_2(g)$$

The sodium produced by this reaction is oxidized by iron(III) oxide:

$$6Na + Fe_2O_3 \rightarrow 3Na_2O + 2Fe$$

Balancing Oxidation-Reduction Equations

Equations for redox reactions are sometimes difficult to balance. Use the steps in **Skills Toolkit 2** below to balance redox equations for reactions in acidic aqueous solution. An important step is to identify the key ions or molecules that contain atoms whose oxidation numbers change. These atoms are the starting points of the unbalanced half-reactions. For the reaction of zinc and hydrochloric acid, the unbalanced oxidation and reduction half-reactions would be as follows:

$$Zn(s) \longrightarrow Zn^{2+}(aq) \quad \text{and} \quad H_3O^+(aq) \longrightarrow H_2(g)$$

These reactions are then separately balanced. Finally, the balanced equations of the two half-reactions are added together to cancel the electrons.

SKILLS Toolkit 2

Balancing Redox Equations Using the Half-Reaction Method

1. Identify reactants and products.
- Write the unbalanced equation in ionic form, excluding any spectator ions.
- Assign oxidation numbers, and identify the atoms that change their oxidation numbers. Ignore all species whose atoms do not change their oxidation number.

2. Write and balance the half-reactions.
- Separate the equation into its two half-reactions.
- For each half-reaction, do the following:
  a. Balance atoms other than hydrogen and oxygen.
  b. Balance oxygen atoms by adding water molecules as needed.
  c. Balance hydrogen atoms by adding one hydronium ion for each hydrogen atom needed and then by adding the same number of water molecules to the other side of the equation.
  d. Balance the overall charge by adding electrons as needed.

3. Make the electrons equal, and combine half-reactions.
- Multiply each half-reaction by an appropriate number so that both half-reactions have the same number of electrons. Now the electrons lost equal the electrons gained, so charge is conserved.
- Combine the half-reactions, and cancel anything that is common to both sides of the equation.

4. Verify your results.
- Double-check that all atoms and charge are balanced.

609

MISCONCEPTION ALERT

Half-Reaction Method Versus Balancing by Inspection Students who have had good success balancing chemical equations by inspection might be reluctant to learn the half-reaction method. To overcome resistance, tell students that redox reactions are sometimes difficult or impossible to balance by inspection because some species (such as hydronium ions and water) must be added to an equation before it can be properly balanced.

BIOLOGY CONNECTION

Fireflies produce light through a process called *bioluminescence*. Bioluminescence occurs through a two-step oxidation reaction mechanism in which a chemical called *luciferin* combines with adenosine triphosphate and oxygen to form oxyluciferin, adenosine monophosphate, and light.

Demonstration, continued

Safety Caution: Wear safety goggles and a lab apron. Keep students at least 3 m away from the demonstration.

Disposal: Separate the solids from the supernatant liquid, and rinse the solids with water. Throw the solids in the trash. Combine the rinsings with the supernatant. Slowly with stirring add 1 M sulfuric acid solution until the pH is between 5 and 6. Scour six 6d iron nails with steel wool until they are bright and shiny. Immerse the nails in the acidified solution. Let the nails remain immersed overnight until all the copper has precipitated. Remove the nails, and filter the solution. Heat the nails and any precipitate obtained from filtration sufficiently to convert the copper precipitate and the copper on the nails to copper oxide. Let them cool, and put the copper oxide and the nails in the trash. Treat the filtrate with sufficient 1 M NaOH to bring the pH to between 8 and 10. Filter. Let the precipitate dry, and put it in the trash. Pour the filtrate down the drain.

Teaching Tip ——— ADVANCED

The Half-Reaction Method and Basic Solutions For simplicity, the half-reaction method in this book shows how to balance equations only in acidic solutions. One way to balance a reaction in basic solution is to first balance each half-reaction in acidic solution. Then add one OH^- ion per hydronium ion shown to each side of the equation. Combine the H_3O^+ and OH^- to form $2H_2O$. Then simplify each half-reaction by canceling water molecules before combining the half-reactions. Have students balance the reaction of permanganate ions with sulfite ions forming manganese dioxide and sulfate ions in basic solution. **Ans.** $2MnO_4^- + 3SO_3^{2-} + H_2O \rightarrow 2MnO_2 + 3SO_4^{2-} + 2OH^-$ **LS** **Logical**

Answers to Practice Problems B

1. half-reactions: $4Fe(s) \rightarrow 4Fe^{3+}(aq) + 12e^-$ and $3O_2(aq) + 12H_3O^+(aq) + 12e^- \rightarrow 18H_2O(l)$; overall: $4Fe(s) + 3O_2(aq) + 12H_3O^+(aq) \rightarrow 4Fe^{3+}(aq) + 18H_2O(l)$

2. half-reactions: $2Al(s) \rightarrow 2Al^{3+}(aq) + 6e^-$ and $6H_3O^+(aq) + 6e^- \rightarrow 3H_2(g) + 6H_2O(l)$; overall: $2Al(s) + 6H_3O^+(aq) \rightarrow 2Al^{3+}(aq) + 3H_2(g) + 6H_2O(l)$

3. half-reactions: $2Br^-(aq) \rightarrow Br_2(aq) + 2e^-$ and $H_2O_2(aq) + 2H_3O^+(aq) + 2e^- \rightarrow 4H_2O(l)$; overall: $2Br^-(aq) + H_2O_2(aq) + 2H_3O^+(aq) \rightarrow Br_2(aq) + 4H_2O(l)$

4. half-reactions: $MnO_2(s) + 4H_3O^+(aq) + 2e^- \rightarrow Mn^{2+}(aq) + 6H_2O(l)$ and $2Cu^+(aq) \rightarrow 2Cu^{2+}(aq) + 2e^-$; overall: $MnO_2(s) + 2Cu^+(aq) + 4H_3O^+(aq) \rightarrow Mn^{2+}(aq) + 2Cu^{2+}(aq) + 6H_2O(l)$

Homework ── GENERAL

Additional Practice

1. $Cr_2O_7^{2-}(aq) + Br^-(aq) \rightarrow Cr^{3+}(aq) + Br_2(l)$ Ans. half-reactions: $Cr_2O_7^{2-}(aq) + 14H_3O^+(aq) + 6e^- \rightarrow 2Cr^{3+}(aq) + 21H_2O(l)$ and $6Br^-(aq) \rightarrow 3Br_2(l) + 6e^-$; overall: $Cr_2O_7^{2-}(aq) + 6Br^-(aq) + 14H_3O^+(aq) \rightarrow 2Cr^{3+}(aq) + 3Br_2(l) + 21H_2O(l)$

2. Copper reacts with sulfate ions to form copper(II) ions and sulfur dioxide gas. Ans. half-reactions: $Cu(s) \rightarrow Cu^{2+}(aq) + 2e^-$ and $SO_4^{2-}(aq) + 4H_3O^+(aq) + 2e^- \rightarrow SO_2(g) + 6H_2O(l)$; overall: $Cu(s) + SO_4^{2-}(aq) + 4H_3O^+(aq) \rightarrow Cu^{2+}(aq) + SO_2(g) + 6H_2O(l)$

LS Logical

The Half-Reaction Method

Write and balance the equation for the reaction when an acidic solution of MnO_4^- reacts with a solution of Fe^{2+} to form a solution containing Mn^{2+} and Fe^{3+} ions.

1 Identify reactants and products.

- The unbalanced equation in ionic form is as follows:

$$H_3O^+(aq) + MnO_4^-(aq) + Fe^{2+}(aq) \longrightarrow Mn^{2+}(aq) + Fe^{3+}(aq)$$

- Oxidation numbers for the atoms are as follows:

$$\overset{+1\ -2}{H_3O^+} + \overset{+7\ -2}{MnO_4^-} + \overset{+2}{Fe^{2+}} \longrightarrow \overset{+2}{Mn^{2+}} + \overset{+3}{Fe^{3+}}$$

- Atoms of Mn and Fe change oxidation numbers.

2 Write and balance the half-reactions.

Unbalanced:	$Fe^{2+} \longrightarrow Fe^{3+}$	$MnO_4^- \longrightarrow Mn^{2+}$
Balance O:	$Fe^{2+} \longrightarrow Fe^{3+}$	$MnO_4^- \longrightarrow Mn^{2+} + 4H_2O$
Balance H:	$Fe^{2+} \longrightarrow Fe^{3+}$	$8H_3O^+ + MnO_4^- \longrightarrow Mn^{2+} + 12H_2O$
Balance e^-:	$Fe^{2+} \longrightarrow Fe^{3+} + e^-$	$5e^- + 8H_3O^+ + MnO_4^- \longrightarrow Mn^{2+} + 12H_2O$

3 Make the electrons equal, and combine half-reactions.

Multiply the half-reaction that involves iron by 5 to make the numbers of electrons the same in each half-reaction. Add the half-reactions, and cancel the electrons to get the final balanced equation:

$$8H_3O^+(aq) + MnO_4^-(aq) + 5Fe^{2+}(aq) \longrightarrow Mn^{2+}(aq) + 5Fe^{3+}(aq) + 12H_2O(l)$$

4 Verify your results.

Note that there are equal numbers of all atoms and a net charge of +17 on each side of the equation.

> **PRACTICE HINT**
>
> To help avoid confusion between charges and oxidation numbers, the sign of a charge is written last and the sign of an oxidation number is written first. Thus the Fe^{2+} ion has a 2+ charge and a +2 oxidation number.

PRACTICE

Use the half-reaction method to write a balanced equation for each of the following reactions in acidic, aqueous solution.

1 The reactants are $Fe(s)$ and $O_2(aq)$, and the products are $Fe^{3+}(aq)$ and $H_2O(l)$.

2 $Al(s)$ is placed in the acidic solution and forms $H_2(g)$ and $Al^{3+}(aq)$.

3 The reactants are sodium bromide and hydrogen peroxide, and the products are bromine and water.

4 The reactants are manganese dioxide and a soluble copper(I) salt, and the products are soluble manganese(II) and copper(II) salts.

610

Identifying Agents in Redox Reactions

In **Sample Problem B,** the permanganate ion caused the oxidation of the iron(II) ion. Substances that cause the oxidation of other substances are called **oxidizing agents.** They accept electrons easily and so are reduced. Common oxidizing agents are oxygen, hydrogen peroxide, and halogens.

   Reducing agents cause reduction to happen and are themselves oxidized. The iron(II) ion caused the reduction of the permanganate ion and was the reducing agent. Common ones are metals, hydrogen, and carbon.

oxidizing agent

the substance that gains electrons in an oxidation-reduction reaction and is reduced

reducing agent

a substance that has the potential to reduce another substance

---

 # Section Review

UNDERSTANDING KEY IDEAS

1. Explain oxidation and reduction in terms of electron transfer.

2. How can you identify a reaction as a redox reaction?

3. Describe how an oxidation-reduction reaction may be broken down into two half-reactions, and explain why the latter are useful in balancing redox equations.

4. Compare the number of electrons lost in an oxidation half-reaction with the number of electrons gained in the corresponding reduction half-reaction.

5. Describe what an oxidizing agent and a reducing agent are.

PRACTICE PROBLEMS

6. Assign oxidation numbers to the atoms in each of the following:
   a. H_2SO_3 c. SF_6
   b. Cl_2 d. NO_3^-

7. Assign oxidation numbers to the atoms in each of the following:
   a. CH_4 c. $NaHCO_3$
   b. HSO_3^- d. $NaBiO_3$

8. Identify the oxidation number of a Cr atom in each of the following: CrO_3, CrO, $Cr(s)$, CrO_2, Cr_2O_3, $Cr_2O_7^{2-}$, and CrO_4^{2-}.

9. Which of the following equations represent redox reactions? For each redox reaction, determine which atom is oxidized and which is reduced, and identify the oxidizing agent and the reducing agent.
   a. $MgO(s) + H_2CO_3(aq) \longrightarrow$
      $$MgCO_3(s) + H_2O(l)$$
   b. $2KNO_3(s) \longrightarrow 2KNO_2(s) + O_2(g)$
   c. $H_2(g) + CuO(s) \longrightarrow Cu(s) + H_2O(l)$
   d. $NaOH(aq) + HCl(aq) \longrightarrow$
      $$NaCl(aq) + H_2O(l)$$
   e. $H_2(g) + Cl_2(g) \longrightarrow 2HCl(g)$
   f. $SO_3(g) + H_2O(l) \longrightarrow H_2SO_4(aq)$

10. Use the half-reaction method in acidic, aqueous solution to balance each of the following redox reactions:
   a. $Cl^-(aq) + Cr_2O_7^{2-}(aq) \longrightarrow$
      $$Cl_2(g) + Cr^{3+}(aq)$$
   b. $Cu(s) + Ag^+(aq) \longrightarrow Cu^{2+}(aq) + Ag(s)$
   c. $Br_2(l) + I^-(aq) \longrightarrow I_2(s) + Br^-(aq)$
   d. $I^-(aq) + NO_2^-(aq) \longrightarrow NO(g) + I_2(s)$

11. Determine which atom is oxidized and which is reduced, and identify the oxidizing agent and the reducing agent for each reaction in item 10.

CRITICAL THINKING

12. How is it possible for hydrogen peroxide to be both an oxidizing agent and a reducing agent?

611

Close

Chapter Resource File

• Concept Review

• Quiz

Introduction to Electrochemistry

Overview

Before beginning this section, review with your students the Objectives listed in the Student Edition. This section gives students a brief introduction to electricity that includes a definition for voltage. Students also learn about the components of electrochemical cells and how they are related to oxidation and reduction.

🔔 Bellringer

Have students write a brief paragraph about the requirements of an electric circuit. Have groups of 2 or 3 students share their paragraphs and discuss any differences.

Motivate

Activity ———— BASIC

Batteries Provide students a large variety of batteries. Have students examine the batteries and locate the voltage rating and the positive and negative terminals. Inform students that electrons flow from the negative terminal into the external circuit and re-enter the battery at the positive terminal. Have students decide whether oxidation or reduction should occur at the negative terminal and explain their reasoning. **Ans.** Electrons leave the battery from the negative terminal, so oxidation must happen there to provide the electrons. **LS Visual**

Chapter Resource File

• Lesson Plan

KEY TERMS

• electrochemistry
• voltage
• electrode
• electrochemical cell
• cathode
• anode

---

electrochemistry

the branch of chemistry that is the study of the relationship between electric forces and chemical reactions

---

☑ internet connect

www.scilinks.org
Topic: Electrical Energy
SciLinks code: HW4045

SCILINKS. Maintained by the National Science Teachers Association

---

Figure 4
When the switch of a flashlight is closed, electrons "pushed" by the battery are forced through the thin tungsten filament of the bulb. This flow of electrons makes the filament hot enough to emit light.

OBJECTIVES

① **Describe** the relationship between voltage and the movement of electrons.

② **Identify** the parts of an electrochemical cell and their functions.

③ **Write** electrode reactions for cathodes and anodes.

Chemistry Meets Electricity

The science of **electrochemistry** deals with the connections between chemistry and electricity. It is an important subject because it is involved in many of the things you use every day. Electrochemical devices change electrical energy into chemical energy and vice versa. A simple flashlight is an everyday example of something that converts chemical energy into electrical energy, which is then converted into light energy.

Figure 4 shows the components of a typical flashlight. The power source is two batteries or cells. Each cell has a metal terminal at each end. If you examine a battery, you will find a $\oplus$ or + symbol near the top identifying the positive terminal. The bottom is the negative terminal, and is possibly unmarked or marked with a $\ominus$ or – symbol. By closing the switch, you turn the flashlight on. Electrons move from the negative terminal of the lower battery, through a metal circuit that includes the light bulb, and continue to the positive terminal of the upper battery. The circuit is completed by electrons and ions moving charge through each of the batteries and electrons moving charge from the upper battery to the lower one. When you turn off the flashlight, you break the pathway, and the movement of electrons and ions stops.

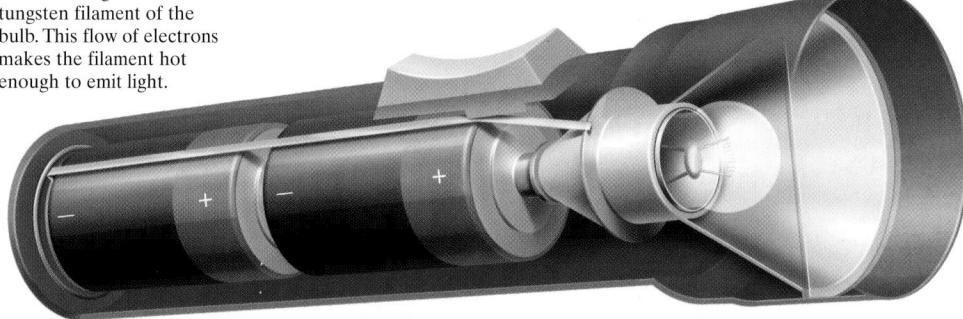

612

MISCONCEPTION
⁄⁄ALERT ⁄⁄⁄

Surprisingly, many students do not understand that a completed circuit is required for an electric current to exist. Some students may think that just touching a bulb to a battery terminal will cause the bulb to light. Use a battery, bulb, and two 20 cm lengths of insulated wire to demonstrate that a complete circuit is necessary to light the bulb.

"Electrical Pressure" Is Expressed in Volts

Electrochemical reactions in a battery cause a greater electron density in the negative terminal than in the positive terminal. Electrons repel each other, so there is a higher "pressure" on the electrons in the negative terminal, which drives the electrons out of the battery and through the flashlight.

Electrical "pressure," often called *electric potential,* or **voltage,** is expressed in units of *volts.* The voltage of an ordinary flashlight cell is 1.5 volts or 1.5 V. When two cells are placed end-to-end, as in **Figure 4,** the voltages add together, so the overall voltage driving electrons is 3.0 V. The movement of electrons or other charged particles is described as *electric current* and is expressed in units of *amperes.*

Components of Electrochemical Cells

A flashlight battery is an electrochemical cell. An **electrochemical cell** consists of two electrodes separated by an electrolyte. An **electrode** is a conductor that connects with a nonmetallic part of a circuit. You have learned about two kinds of conductors. One kind includes metals, which conduct electric current through moving electrons. The second kind includes electrolyte solutions, which conduct through moving ions.

The Cathode Is Where Reduction Occurs

Electrode reactions happen on the surfaces of electrodes. **Figure 5** shows half of an electrochemical cell. The copper strip is an electrode because it is a conductor in contact with the electrolyte solution. The reaction on this electrode is the reduction described by the following equation:

$$Cu^{2+}(aq) + 2e^- \longrightarrow Cu(s)$$

The copper electrode is a **cathode** because reduction happens on it.

voltage

the potential difference or electromotive force, measured in volts; it represents the amount of work that moving an electric charge between two points would take

electrochemical cell

a system that contains two electrodes separated by an electrolyte phase

electrode

a conductor used to establish electrical contact with a non-metallic part of a circuit, such as an electrolyte

cathode

the electrode on whose surface reduction takes place

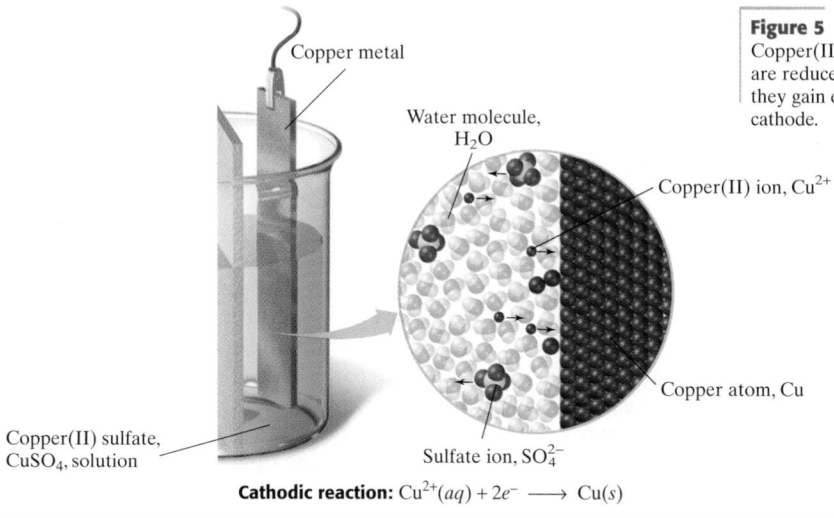

Figure 5
Copper(II) ions, $Cu^{2+}(aq)$, are reduced to atoms as they gain electrons on the cathode.

Copper metal

Water molecule, H_2O

Copper(II) ion, Cu^{2+}

Copper atom, Cu

Copper(II) sulfate, $CuSO_4$, solution

Sulfate ion, SO_4^{2-}

Cathodic reaction: $Cu^{2+}(aq) + 2e^- \longrightarrow Cu(s)$

613

did you know?

Volta was also the first to derive the electromotive series—a list that ranks metals in order of their ability to generate electric current. He tested this ability by touching two metals to his tongue and rating the strength of the electric shock he received.

Using the Figure

Figures 5 and 6 Be sure students realize that each of these figures represents half of an electrochemical cell. In **Figure 5,** electrons for the reduction of Cu^{2+} to Cu come from the oxidation of Zn to Zn^{2+} in **Figure 6.** Use these figures to reinforce the idea that an oxidation process never occurs without a simultaneous reduction process.

Discussion ———— GENERAL

Lead a discussion on the function of the porous barrier and why it is needed in order to get usable energy from a galvanic cell. Help students by asking them to discuss what would happen if the barrier were removed. **Ans.** Copper ions would be able to reach the zinc strip and would be reduced in a direct exchange of electrons with zinc rather than causing the electrons to pass through the external circuit. Students should understand that a cell can generate an electric current only if the oxidation and reduction halves of an overall redox reaction are isolated from each other. This way the electrons being lost at the anode pass through an external wire, reaching the cathode, where reduction takes place. **LS** Verbal

---

ChemFile **CHEMISTRY** INTERACTIVE TUTOR

- **Module 10: Electrochemistry Topic: Electrochemical Cells**

  This engaging tutorial reviews and reinforces the components and reactions of electrochemical cells.

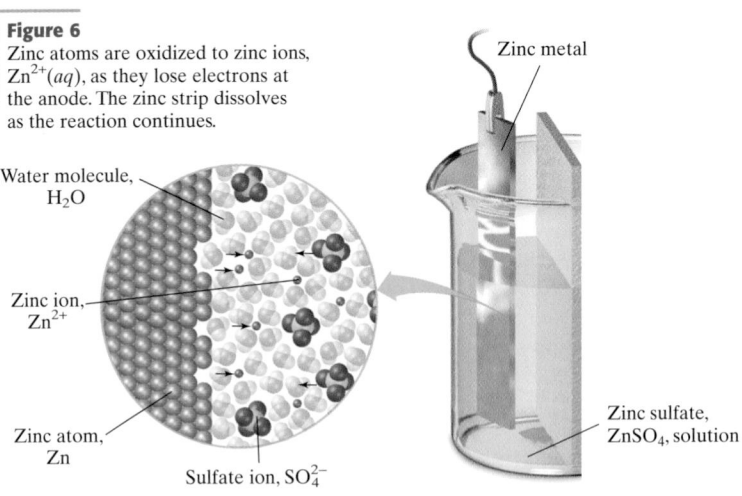

Figure 6
Zinc atoms are oxidized to zinc ions, $Zn^{2+}(aq)$, as they lose electrons at the anode. The zinc strip dissolves as the reaction continues.

Zinc metal

Water molecule, H_2O

Zinc ion, Zn^{2+}

Zinc atom, Zn

Sulfate ion, SO_4^{2-}

Zinc sulfate, $ZnSO_4$, solution

Anodic reaction: $Zn(s) \longrightarrow Zn^{2+}(aq) + 2e^-$

---

anode

the electrode on whose surface oxidation takes place; anions migrate toward the anode, and electrons leave the system from the anode

---

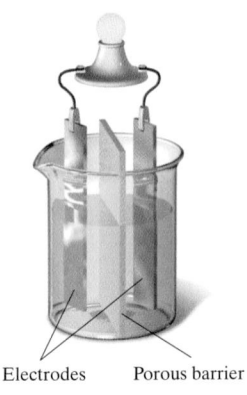

Electrodes   Porous barrier

Figure 7
The light bulb is powered by the reaction in this cell.

The Anode Is Where Oxidation Occurs

The electrons that cause reduction at the cathode are pushed there from a reaction at the second electrode of a cell. The **anode** is the electrode on which oxidation occurs. **Figure 6** shows the second half of the electrochemical cell. The zinc strip is an electrode because it is a conductor in contact with the solution. The reaction on this electrode is the oxidation described by the following equation:

$$Zn(s) \longrightarrow Zn^{2+}(aq) + 2e^-$$

The zinc electrode is an anode because oxidation happens on it.

The electrode reactions described here would have been called half-reactions in the last section. The difference is that a half-reaction is a helpful model, but the electrons shown are never really free. An electrode reaction describes reality because the electrons actually move from one electrode to another to continue the reaction.

Pathways for Moving Charges

The electrode reactions cannot happen unless the electrodes are part of a complete circuit. So, a cell must have pathways to move charges. Wires are often used to connect the electrodes through a meter or a light bulb. Electrons carry charges in the wires and electrodes.

Ions in solution carry charges between the electrolytes to complete the circuit. A porous barrier, as shown in **Figure 7,** or a salt bridge keeps the solutions from mixing, but lets the ions move. In the cell shown, charge is carried through the barrier by a combination of $Zn^{2+}(aq)$ ions moving to the right and $SO_4^{2-}(aq)$ ions moving to the left. Understand that positive charge may be carried from left to right through this cell either by negative particles (electrons or anions) moving from right to left or by cations moving from left to right.

MISCONCEPTION ALERT

The words *anode* and *cathode* are often incorrectly interchanged with *positive* and *negative* electrodes. An anode may be the positive *or* the negative electrode of a cell; the anode is wherever oxidation occurs. The cathode may be the positive *or* the negative electrode; it is wherever reduction occurs.

The Complete Cell

Figure 7 shows a complete cell composed of the electrodes shown separately in earlier figures. The overall process when the anode reaction is added to the cathode reaction for this cell is the same as the following redox reaction:

$$Zn(s) + Cu^{2+}(aq) \longrightarrow Zn^{2+}(aq) + Cu(s)$$

Although the two electrode reactions occur at the same time, they occur at different places in the cell. This is an important distinction from the redox reactions discussed in the last section.

internet connect
www.scilinks.org
Topic: Electrochemical Cells
SciLinks code: HW4046
SCiLINKS Maintained by the National Science Teachers Association

 # Section Review

UNDERSTANDING KEY IDEAS

1. What is voltage?

2. How does voltage relate to the movement of electrons?

3. List the components of an electrochemical cell, and describe the function of each.

4. What are the names of the electrodes in an electrochemical cell? What type of reaction happens on each?

5. Describe the difference in how charge flows in wires and in electrolyte solutions.

6. A 12 V car battery has six cells connected end-to-end. What is the voltage of a single cell?

7. Will the reaction below happen at an anode or a cathode? Explain.

$$Fe(CN)_6^{3-}(aq) + e^- \longrightarrow Fe(CN)_6^{4-}(aq)$$

8. Describe the changes in oxidation number that happen in an anode reaction and in a cathode reaction.

CRITICAL THINKING

9. What would happen if you put one of the batteries in backward in a two-cell flashlight?

10. What would happen if you put both batteries in backward in a two-cell flashlight?

11. If an electrode reaction has dissolved oxygen, $O_2(aq)$, as a reactant, is the electrode an anode or a cathode? Explain.

12. Write an electrode reaction in which you change $Br^-(aq)$ to $Br_2(aq)$. Would this reaction happen at an anode or a cathode?

13. Write an electrode reaction in which $Sn^{4+}(aq)$ is changed to $Sn^{2+}(aq)$. Would this reaction happen at an anode or a cathode?

14. If you wanted to use an electrochemical cell to deposit a thin layer of silver metal onto a bracelet, which electrode would you make the bracelet? Explain using the equation for the electrode reaction that would occur.

15. What would happen at each electrode if batteries were connected to the cell in **Figure 7** so that electrons flowed in the opposite direction of the direction described in the text?

16. Compare the equations for electrode reactions with the equations for half-reactions.

17. Write the electrode reactions for a cell that involves only $Cu(s)$ and $Cu^{2+}(aq)$ in which the anode reaction is the reverse of the cathode reaction. What is the net result of operating this cell?

18. Is it correct to say that the net chemical result of an electrochemical cell is a redox reaction? Explain.

Answers to Section Review

1. the potential difference or electromotive force, measured in volts

2. A voltage is needed for electrons to move from one place to another.

3. anode: electrode where oxidation occurs; cathode: electrode where reduction occurs; wires: pathway for electrons to move from one electrode to the other; electrolyte: pathway for ions to move to complete the circuit; porous barrier or salt bridge: separate the electrolytes to prevent them from mixing

4. The electrodes are the anode and the cathode. Oxidation occurs at the anode, and reduction occurs at the cathode.

5. Charge flows in wires by electron movement. Charge flows in electrolyte solutions by ion movement.

6. 2 V

7. The reaction will happen at a cathode because reduction is happening. An electron is gained in the reaction.

Answers continued on p. 639A

Focus

Overview

Before beginning this section, review with your students the Objectives listed in the Student Edition. In this section, students learn about galvanic cells, including dry cells, lead-acid batteries, fuel cells, and corrosion cells. Methods of preventing corrosion are described. Finally, students learn how the voltage of an electrochemical cell is determined using a standard hydrogen electrode.

🔊 Bellringer

Have students write a brief paragraph about batteries and their usage. Students may write about the types of devices that use batteries, the importance of orientation when placing batteries in a device, the differences between rechargeable and single-use batteries, and the different types and sizes of batteries. After the students complete their paragraphs, invite four or five students to share their paragraphs with the rest of the class.

Motivate

Activity ——————— **BASIC**

Designing a Model Have students model the electrochemical cell in **Figure 8** using objects that can be manipulated to show electron and ion movement. For example, students could cut electron and ion models out of poster board and attach magnets to the back of them. The models could be moved around on a magnetic board on which the basic cell is drawn.
LS Visual

Key Terms
• corrosion
• standard electrode potential

OBJECTIVES

① **Describe** the operation of galvanic cells, including dry cells, lead-acid batteries, and fuel cells.

② **Identify** conditions that lead to corrosion and ways to prevent it.

③ **Calculate** cell voltage from a table of standard electrode potentials.

Types of Galvanic Cells

A battery is one kind of *galvanic cell*, a device that can change chemical energy into electrical energy. In these cells, a spontaneous reaction happens that causes electrons to move. **Figure 8** shows a kind of galvanic cell known as a Daniell cell. Daniell cells were used as energy sources in the early days of electrical research. Of course, Daniell cells would be impractical to use in a radio or portable computer today. There are many other kinds of galvanic cells. These include dry cells, lead-acid batteries, and fuel cells.

Figure 8
As a result of the reaction in a galvanic cell, the bulb lights up as electrons move in the wires from the anode to the cathode.

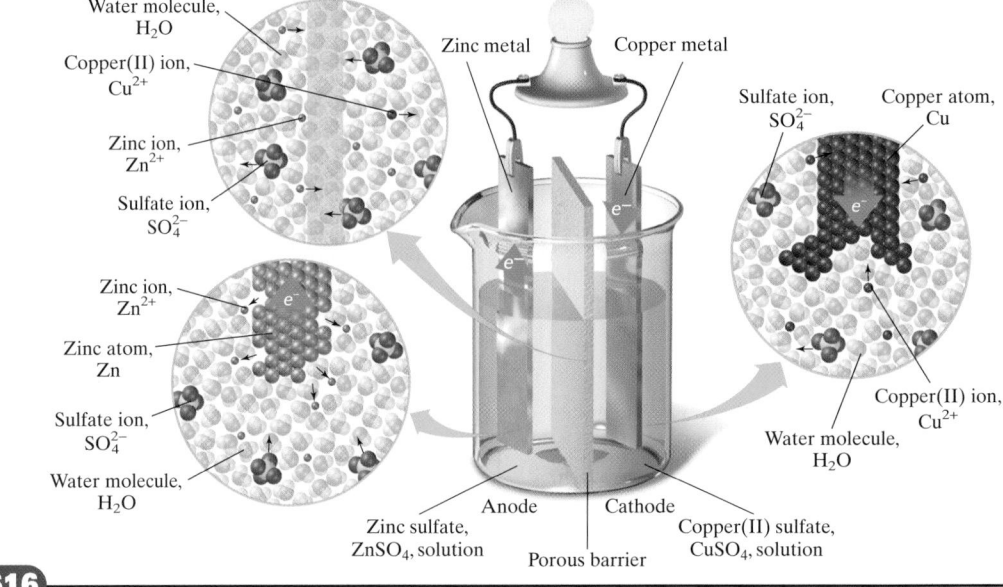

Water molecule, H_2O
Copper(II) ion, Cu^{2+}
Zinc ion, Zn^{2+}
Sulfate ion, SO_4^{2-}
Zinc metal
Copper metal
Sulfate ion, SO_4^{2-}
Copper atom, Cu
Zinc ion, Zn^{2+}
Zinc atom, Zn
Sulfate ion, SO_4^{2-}
Water molecule, H_2O
Copper(II) ion, Cu^{2+}
Water molecule, H_2O
Zinc sulfate, $ZnSO_4$, solution
Anode
Cathode
Porous barrier
Copper(II) sulfate, $CuSO_4$, solution

616

did you know?

Some textbooks distinguish between a battery and a single cell. The original use of the word *battery* as a chemical device to generate electrical energy was to describe a group of connected cells, a battery of galvanic cells. Some batteries, such as standard 9 V batteries and 12 V automobile batteries, do consist of several cells. Through common usage, the term is now accepted as referring to a single cell, such as a D cell.

Chapter Resource File
• Lesson Plan

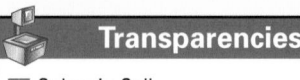

Transparencies
TT Galvanic Cell

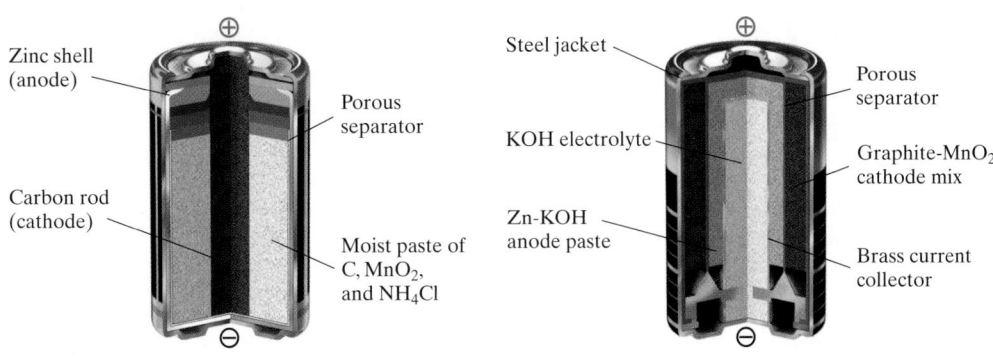

Figure 9
Two familiar kinds of dry cells use different electrolytes. The electrolytes make the cell on the left acidic and the cell on the right alkaline.

Dry Cells

Although the Daniell cell was useful for supplying energy in the lab, it wasn't very portable because it held solutions. The energy source that you know as a *battery* and use in radios and remote controls is a dry cell. In a dry cell, moist electrolyte pastes are used instead of solutions. This cell was invented over a century ago by Georges Leclanché, a French chemist. His original design was close to that shown on the left in **Figure 9.** A carbon rod, the battery's positive terminal, connects with a wet paste of carbon; ammonium chloride, NH_4Cl; manganese(IV) oxide, MnO_2; starch; and water. When the cell is used, the carbon rod is the cathode, and the following electrode reaction happens:

Cathode: $2MnO_2(s) + 2NH_4^+(aq) + 2e^- \longrightarrow Mn_2O_3(s) + 2NH_3(aq) + H_2O(l)$

The zinc case serves as the negative terminal of the battery. When the cell is used, zinc dissolves in the following electrode reaction:

Anode: $Zn(s) + 4NH_3(aq) \longrightarrow 2e^- + Zn(NH_3)_4^{2+}(aq)$

The white powder that you see on old corroded batteries is the chloride salt of this $Zn(NH_3)_4^{2+}(aq)$ complex ion.

The $NH_4^+(aq)$ ion is a weak acid, and for this reason the Leclanché cell is called the *acidic* version of the dry cell. The *alkaline cell,* shown in **Figure 9,** is a newer, better version. The ingredients of the alkaline cell are similar to the acidic version, but the carbon cathode is replaced by a piece of brass, and ammonium chloride is replaced by potassium hydroxide. The presence of this strong base gives the alkaline cell its name. The electrode reactions that occur when the cell is used are described by the equations below.

Cathode: $2MnO_2(s) + H_2O(l) + 2e^- \longrightarrow Mn_2O_3(s) + 2OH^-(aq)$

Anode: $Zn(s) + 2OH^-(aq) \longrightarrow 2e^- + Zn(OH)_2(s)$

A sturdy steel shell is needed to prevent the caustic contents from leaking out of the battery. Because of this extra packaging, alkaline cells are more expensive than cells of the older, acidic version.

617

Skills Acquired:
• Experimenting
• Inferring

Materials:
For each group of 2–4 students:
• alligator clips (2)
• copper strip
• earphone
• potato, raw
• zinc strip

Teacher's Notes:
• The metal strips should be about 1 cm wide and 5 cm long.
• If you have a very small speaker, like those in very small portable radios, try it as a demonstration of the lab.

Safety Caution: Do not eat the potato. Wear goggles and a lab apron.

Disposal: Mark the potato "Unfit for food use," and throw it in the trash. Save the metal strips for the next time.

Answers

1. When the wires from the earphone are touched to both strips, the earphone makes a crackling noise. Touching the wire to only one strip produces no sound. A complete circuit exists only when the wires are connected to separate strips.

2. The potato "battery" is producing an electric current in the earphone, causing it to produce a crackling sound.

LS Kinesthetic

Quick **LAB**

Listen Up

PROCEDURE

1. Press a **zinc strip** and a **copper strip** into a **raw potato.** The strips should be about 0.5 cm apart but should not touch one another.

2. While listening to the **earphone,** touch one wire from the earphone to one of the metal strips and the other wire from the earphone to the second metal strip using **alligator clips.** Record your observations.

3. While listening to the earphone, touch both wires to a single metal strip. Record your observations.

ANALYSIS

1. Compare your results from step 2 with your results from step 3. Suggest an explanation for any similarities or differences.

2. Suggest an explanation for the sound.

Lead-Acid Batteries

The batteries just discussed are called *dry cells* because the water is not free, but absorbed in pastes. In contrast, the Daniell cell and the lead-acid battery use aqueous solutions of electrolytes, so they should be used in an upright position.

Most car batteries are lead-acid storage batteries. Usually they have six cells mounted side-by-side in a single case, as shown in **Figure 10.** Though many attempts have been made to replace the heavy lead-acid battery by lighter alternatives, no other material has been found that can reliably and economically give the large surges of electrical energy needed to start a cold engine.

A fully charged lead-acid cell is made up of a stack of alternating lead and lead(IV) oxide plates isolated from each other by thin porous separators. All these components are in a concentrated solution of sulfuric acid. The positive terminal of one cell is linked to the negative terminal of the next cell in the same way that the batteries in the flashlight were connected. This arrangement of cells causes the outermost terminals of the lead-acid battery to have a voltage of 12.0 V. When the cell is used, it acts as a galvanic cell with the following reactions:

Cathode: $PbO_2(s) + HSO_4^-(aq) + 3H_3O^+(aq) + 2e^- \longrightarrow PbSO_4(s) + 5H_2O(l)$

Anode: $Pb(s) + HSO_4^-(aq) + H_2O(l) \longrightarrow 2e^- + PbSO_4(s) + H_3O^+(aq)$

Notice that $PbSO_4$ is produced at both electrodes.

Unlike the Daniell and Leclanché cells, the lead-acid cell is rechargeable. So, when the battery runs down, you do not need to replace it. Instead, an electric current is applied in a direction opposite to that discussed above. As a result of the input of energy, the reactions are reversed. The cell is eventually restored to its charged state. During recharge, the cell functions as an *electrolytic cell,* which you will learn about in the next section.

618

Chapter Resource File

• Datasheets for In-Text Labs

HISTORY
CONNECTION

Fifty years ago, a large amount of lead was used to manufacture paint pigments and tetraethyl lead, $Pb(C_2H_5)_4$, which was used as an antiknock agent in gasoline. Today lead compounds in gasoline have been replaced by other substances in most developed countries. The use of lead-based pigments in paints has been reduced significantly in recent years because exposure to lead-based paints has been found to be a health hazard. Today, the main use for lead is in storage batteries.

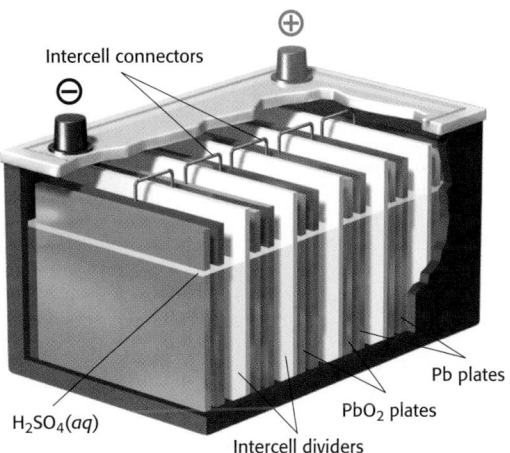

Intercell connectors

⊕

⊖

Pb plates

PbO₂ plates

Intercell dividers

$H_2SO_4(aq)$

Figure 10
The lead-acid battery is used to store energy in almost all vehicles. Although the cutaway view shows a single PbO_2 plate and a single Pb plate in each cell, there are actually several of each.

Fuel Cells

In a *fuel cell,* the oxidizing and reducing agents are brought in, often as gases, from outside of the cell, rather than being part of it. Unlike a dry cell, a fuel cell can work forever, in principle, changing chemical energy into electrical energy. **Figure 11** models a fuel cell that uses the reactions below.

$$\text{Cathode: } O_2(g) + 2H_2O(l) + 4e^- \longrightarrow 4OH^-(aq)$$

$$\text{Anode: } 2H_2(g) + 4OH^-(aq) \longrightarrow 4e^- + 4H_2O(l)$$

Because fuel cells directly change chemical energy into electrical energy, they are very efficient and are cleaner than the burning of fuels in power plants to generate electrical energy. Research into fuel cells continues, and fuel cells are used in a few experimental power plants.

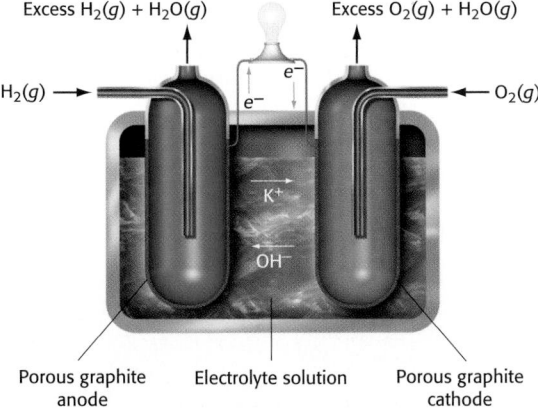

Excess $H_2(g) + H_2O(g)$

Excess $O_2(g) + H_2O(g)$

$H_2(g)$ →

e^-

e^-

← $O_2(g)$

K^+

OH^-

Porous graphite anode

Electrolyte solution

Porous graphite cathode

Figure 11
The reactions in this fuel cell take place at carbon electrodes that contain metal catalysts. The water formed is removed as a gas.

619

Transparencies

TT Fuel Cell

Teaching Tips

Disposable Heat Packs Single-use heat packs generate thermal energy through the rapid oxidation or corrosion of iron. These heat packs contain iron powder, water, saw dust, vermiculite, sodium chloride, and activated carbon. The reactions involved are believed to be complex and are not completely understood. These heat packs are different from the reusable type that contains supersaturated sodium acetate solutions.

Passive Films Some metals that have a tendency to corrode can become resistant to corrosion when a thin film of metal oxide forms on their surface. This film is called a *passive film* and is chemically very stable. For example, stainless steel—an alloy containing several metals including iron, nickel, and chromium—forms a thin chromium oxide layer that is highly stable. Passive films can be broken down by reactive anions such as halide ions. Chloride ions are particularly destructive which is why corrosion tends to be more problematic in areas exposed to seawater.

Transparencies
TT Corrosion Cell

Corrosion Cells

Oxygen is so reactive that many metals spontaneously oxidize in air. Fortunately, many of these reactions are slow. The disintegration of metals is called **corrosion.** Usually $O_2(g)$ is the oxidizing agent, but the direct reaction of O_2 with the metal is not usually how corrosion happens.

Water is usually involved in corrosion. Consider the corrosion of iron. Hydrated iron(III) oxide, or rust, forms by the following overall reaction:

$$4Fe(s) + 3O_2(aq) + 4H_2O(l) \longrightarrow 2Fe_2O_3 \cdot 2H_2O(s)$$

However, the reaction mechanism is more complicated than this equation suggests. An even simpler version of what happens is shown in **Figure 12.** The iron dissolves by the oxidation half-reaction

$$Fe(s) \longrightarrow Fe^{3+}(aq) + 3e^-$$

Any oxidation must be accompanied by a reduction taking place at the same time, but not necessarily at the same location. In fact, the electrons produced by the oxidation are consumed by the reduction of oxygen at a cathodic site elsewhere on the iron's surface. The reduction half-reaction is

$$O_2(aq) + 2H_2O(l) + 4e^- \longrightarrow 4OH^-(aq)$$

Electrons move in the metal and ions move in the water layer between the two reaction sites, as in an electrochemical cell. In fact, a corrosion cell is an unwanted galvanic cell. Chemical energy is converted into electrical energy, which heats the metal.

The three ingredients—oxygen, water, and ions—needed for the corrosion of metals are present almost everywhere on Earth. Even pure rainwater contains a few H_3O^+ and HCO_3^- ions from dissolved carbon dioxide. Higher ion concentrations—from airborne salt near the ocean, from acidic air pollutants, or from salts spread on icy roads—make corrosion worse in certain areas.

corrosion

the gradual destruction of a metal or alloy as a result of chemical processes such as oxidation or the action of a chemical agent

Figure 12
The cathodic reaction happens where the O_2 concentration is high. The anodic reaction happens in a region where the O_2 concentration is low, such as in a pit in the metal.

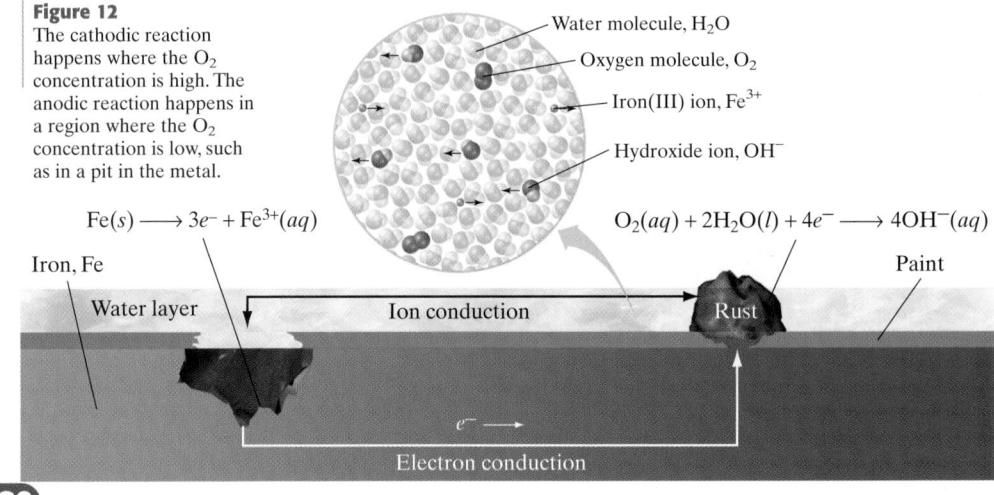

Water molecule, H_2O
Oxygen molecule, O_2
Iron(III) ion, Fe^{3+}
Hydroxide ion, OH^-

$Fe(s) \longrightarrow 3e^- + Fe^{3+}(aq)$
$O_2(aq) + 2H_2O(l) + 4e^- \longrightarrow 4OH^-(aq)$

Iron, Fe
Paint
Water layer
Ion conduction
Rust
$e^- \longrightarrow$
Electron conduction

BIOLOGY
CONNECTION **ADVANCED**

Writing **Bacteria and Corrosion** Certain bacteria have been known to facilitate the process of corrosion in a variety of ways. For example, some bacteria produce organic acids that help dissolve metals. Other bacteria oxidize metals to generate energy. Corrosion cells can form under colonies of such bacteria and pitting can occur. Have interested students study the types of bacteria that contribute to corrosion problems and have them propose ways to minimize the effects of these bacteria.
LS Logical

Figure 13
The Alaskan oil pipeline is cathodically protected by a parallel zinc cable.

Methods to Prevent Corrosion

Corrosion is a major economic problem. About 20% of all the iron and steel produced is used to repair or replace corroded structures. That is why the prevention of corrosion is a major focus of research in materials science and electrochemistry. An obvious response to corrosion is to paint the metal or coat it with some other material that does not corrode. However, once a crack or scrape occurs in the coating, corrosion can begin and often spread even faster than on an uncoated surface.

Some metals corrode more easily than others do. Electronegativity is one factor. Gold, the metal with the highest electronegativity, is the most corrosion resistant. The alkali metals, with the lowest electronegativities, easily corrode. The properties of the metal oxides that form are also important. Despite having low electronegativities, aluminum, chromium, and titanium are corrosion-resistant metals. This is because the oxides of these metals form layers that cover the underlying metal, stopping oxidation. In contrast, rust is a porous powder that flakes off, so it does not protect the iron surface.

Surprisingly, it is better to coat steel with another metal that *does* corrode. Trash cans, for example, are made of zinc-coated steel. This coating does not stop corrosion. But the zinc corrodes first, making the steel underneath last much longer than it would without the zinc layer.

Whenever two metals are in electrical contact, a corrosion cell is likely to form. In fact, the metal that is the anode in the cell corrodes faster than it would if it were not connected to another metal. This idea explains the use of sacrificial anodes on ships and pipelines, as shown in **Figure 13**. As the anode corrodes, it gives electrons to the cathode. The corrosion of the anode slows or stops the corrosion of the important structural metal in a process called *cathodic protection*.

internet connect

www.scilinks.org
Topic: Corrosion
SciLinks code: HW4147

SCLINKS Maintained by the National Science Teachers Association

621

HISTORY
CONNECTION

British chemist Sir Humphry Davy (1778–1829) developed cathodic protection in 1824. The method was first used to control corrosion on British naval ships.

Using the Table

Have students look ahead to **Table 1.** Theoretically, any metal near the top of the list should be able to act as a sacrificial anode and protect any metal lower on the table. However, Li, K, Ca, and Na are too reactive to use under most conditions. Magnesium and zinc are less reactive and make excellent sacrificial materials.

Demonstration
Predicting Reactions

1. Half-fill one Petri dish with 0.1 M $AgNO_3$ solution and another with 0.1 M $Zn(NO_3)_2$ solution. Place both dishes on an overhead projector and note their contents on the board.

2. Display two polished copper strips about 0.5 cm × 2 cm in size, and ask students to predict what will happen when the strips are placed in the dishes.

3. Place one strip in each dish, and observe the results. Silver forms on the copper strip placed in the $AgNO_3$ solution. No reaction will take place in the other dish.

4. Allow students to observe the dishes in room light to see the silver crystals that form.

5. Ask students to explain the difference in the results.

Safety Caution: Wear safety goggles and a lab apron. Keep students at least 3 m away from the demonstration.

Disposal: Remove the Cu strip coated with Ag, and scour off the Ag with emery paper. Put the emery paper in a labeled container for future use; do not discard. Remove the other Cu strip. Save both Cu strips for future use. Save left-over $AgNO_3$ and $Zn(NO_3)_2$ solutions for reuse, or treat with an excess of 1.0 M NaOH while stirring to precipitate all of the silver and zinc as oxides. Filter. Adjust the pH of the solutions to be between 8 to 10 with 0.5 M HCl, and pour down the drain. After the solids have dried, wrap them in newspaper, and put them in the trash.

Chapter 17 • Oxidation, Reduction and Electrochemistry   **621**

The Statue of Liberty In 1984, a two year restoration project was started at the Statue of Liberty to prepare her for the 100th anniversary of her dedication. Most of the copper "skin" of the statue did not need to be replaced. This was because a patina of copper compounds covered the surface of the statue and protected the copper. This patina is what gives the statue its green color. During the restoration, the rivets that hold the skin to the skeleton inside were replaced with high-alloy copper fastenings. These fastenings help guard against galvanic reaction problems.

Activity ——————— GENERAL

Use a voltmeter to measure the voltage of several batteries with different voltage ratings. Review the previous sections by asking students to explain how a battery generates current. Challenge them to explain why different types of galvanic cells produce different voltages. **LS Logical**

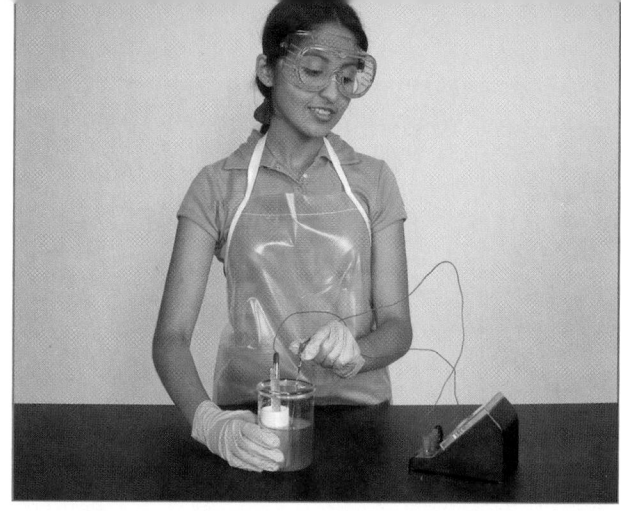

Figure 14
It would be difficult to measure every possible combination of electrodes using a technique like the one shown here. This is why the SHE is used as a reference for all other electrodes.

Determining the Voltage of a Cell

A voltmeter is an electronic instrument that measures the voltage between its two leads. The student in **Figure 14** is using the meter to measure the difference in potentials between the electrodes in a Daniell cell. With such a meter, measuring the voltage of a cell is easy. However, the voltage of a cell depends on such factors as temperature and concentration. And because there are so many combinations of electrode reactions, it would be very difficult to measure the voltage for each combination.

Standard Electrode Potentials

Picking one electrode as a standard and determining electrode potentials in reference to that standard is much easier than measuring the potential between every combination of electrodes is. The electrode that has been chosen as a standard is the *standard hydrogen electrode* (SHE). It consists of a platinum electrode in a 1.00 M H_3O^+ solution in the presence of H_2 gas at 1 atm pressure and 25°C. The SHE is assigned a potential of 0.0000 V and its reaction is

$$2H_3O^+(aq) + 2e^- \rightleftharpoons H_2(g) + 2H_2O(l)$$

standard electrode potential

the potential developed by a metal or other material immersed in an electrolyte solution relative to the potential of the hydrogen electrode, which is set at zero

When measuring potentials, a salt bridge, a narrow tube filled with a concentrated solution of a salt, must be used to link the compartments. When a SHE is joined to another electrode by a salt bridge, a voltmeter can be used to determine the **standard electrode potential,** $E°$, for the electrode. Some standard electrode potentials are shown in **Table 1,** on the last page of this section.

The standard electrode potential is sometimes called the standard reduction potential because it is listed by the reduction half-reactions. However, a voltmeter allows no current in the cell during the measurement. Therefore, the conditions are neither galvanic nor electrolytic—the cell is at equilibrium. As a result, the half-reactions listed in the table are shown as reversible. If the reaction occurs in the opposite direction, as an oxidation half-reaction, $E°$ will have the opposite sign.

622

did you know?

In addition to 1.000 M H_3O^+ concentration at 1 atm pressure of H_2, the standard hydrogen electrode must also be at 25°C. The standard potentials of other electrodes also require conditions of 25°C and a 1.000 M concentration of the ions involved in the electrode reaction.

Calculating the Voltage of a Cell

Think of $E°$ as a measure of the ability of an electrode to gain electrons. A more positive value means the electrode is more likely to be a cathode. The standard cell voltage—the voltage of a cell under standard conditions—can be found by subtracting the standard potentials of the two electrodes, as follows:

$$E°_{cell} = E°_{cathode} - E°_{anode}$$

To determine the reaction that will happen naturally, use the electrode with the most positive $E°$ value as the cathode, as shown in **Sample Problem C.** Otherwise, a given reaction happens naturally if $E°_{cell}$ is positive. If $E°_{cell}$ is negative, the reaction could be made to happen if energy is added.

SAMPLE PROBLEM C

Calculating Cell Voltage

Calculate the voltage of a cell for the naturally occurring reaction between a liquid mercury electrode in a solution of mercury(I) nitrate and a cadmium metal electrode in a solution of cadmium nitrate.

1 Gather information.

From **Table 1** the standard electrode potentials are

$$Hg_2^{2+}(aq) + 2e^- \longrightarrow 2Hg(l) \qquad E° = +0.7973 \text{ V}$$
$$Cd^{2+}(aq) + 2e^- \longrightarrow Cd(s) \qquad E° = -0.4030 \text{ V}$$

2 Plan your work.

The mercury electrode has the more positive $E°$, so it is the cathode.

3 Calculate.

$$E°_{cell} = E°_{Hg} - E°_{Cd} = (+0.7973 \text{ V}) - (-0.4030 \text{ V}) = +1.2003 \text{ V}$$

4 Verify your results.

The positive value of $E°_{cell}$ shows that the reaction is spontaneous and will occur naturally with the mercury electrode as the cathode.

PRACTICE HINT

When asked to calculate the voltage of a cell for a particular chemical equation, you must determine which atom is oxidized and which is reduced based on the change of oxidation numbers.

PRACTICE

Use data from **Table 1** to answer the following:

1 Calculate the voltage of a cell if the reactions are as follows:
$$Fe(s) \longrightarrow Fe^{3+}(aq) + 3e^-$$
$$O_2(g) + 2H_2O(l) + 4e^- \longrightarrow 4OH^-(aq)$$

2 Calculate the voltage of a cell for the naturally occurring reaction between a copper electrode in a copper(II) solution and a zinc electrode in a solution containing zinc ions.

3 Calculate the voltage of a cell for the naturally occurring reaction between a silver electrode in a solution containing silver ions and a copper electrode in a copper(II) solution.

PROBLEM SOLVING SKILL

623

Using the Table —— ADVANCED

Table 1 Refer students to the activity series shown in **Appendix A** and ask them to compare its arrangement with that of **Table 1.** The two do not contain the exact same set of elements, but students can see that the elements common to the two tables are in the same order. Help students reinterpret the activity series in light of what they have learned in this chapter. Ans. A displacement reaction is also an oxidation-reduction reaction. The higher an element is in the series, the more likely it is to be oxidized by losing electrons to the cation of an element lower in the table. This includes the replacement of hydrogen from acids, which is the reduction reaction of the SHE. Reaction pairs that are farther apart on the table of electrode potentials have greater voltages in the galvanic cell involving that pair. This correlates to the observation that reaction pairs farther apart on the activity series react more vigorously. **LS** Logical

Answers to Practice Problems C

1. $(+0.401 \text{ V}) - (-0.037 \text{ V}) = +0.438 \text{ V}$

2. $(+0.3419 \text{ V}) - (-0.7618 \text{ V}) = +1.1037 \text{ V}$

3. $(+0.7996 \text{ V}) - (+0.3419 \text{ V}) = +0.4577 \text{ V}$

Chapter Resource File

• **Problem Solving**

Homework —— GENERAL

Additional Practice Calculate the voltage of each of the following cells for the naturally occurring reaction between the given electrodes. Use the standard electrode potentials in **Table 1.**

1. I^-/I_2 and Pb^{2+}/Pb
Ans. $(+0.5355 \text{ V}) - (-0.1262 \text{ V}) = +0.6617 \text{ V}$

2. Ag^+/Ag and Zn^{2+}/Zn
Ans. $(+0.7996 \text{ V}) - (-0.7618 \text{ V}) = +1.5614 \text{ V}$
LS Logical

Quiz — GENERAL

1. What is a galvanic cell? Ans. A galvanic cell is an electrochemical cell in which the electrode reactions occur spontaneously and chemical energy is converted into electrical energy.

2. What type of galvanic cell is used in automobiles? Ans. lead-acid batteries

3. Describe one difference between fuel cells and lead-acid batteries. Ans. Answers may vary. Sample answer: The reactants used in fuel cells can be replenished as the cell provides energy, but while a lead-acid battery is recharging, it cannot also provide energy.

4. How is the standard electrode potential of an electrode measured? Ans. The electrode is paired with a standard hydrogen electrode in an electrochemical cell and the cell potential is measured with a voltmeter.

LS Logical

Reteaching — BASIC

Have students make a concept map of the ideas in this section. Their map should include all of the Key Terms. **LS** Visual

Chapter Resource File

• Concept Review

• Quiz

Table 1   Standard Electrode Potentials

Electrode reaction	$E°$ (V)	Electrode reaction	$E°$ (V)
$Li^+(aq) + e^- \rightleftarrows Li(s)$	−3.0401	$Cu^{2+}(aq) + 2e^- \rightleftarrows Cu(s)$	+0.3419
$K^+(aq) + e^- \rightleftarrows K(s)$	−2.931	$O_2(g) + 2H_2O(l) + 4e^- \rightleftarrows 4OH^-(aq)$	+0.401
$Na^+(aq) + e^- \rightleftarrows Na(s)$	−2.71	$I_2(s) + 2e^- \rightleftarrows 2I^-(aq)$	+0.5355
$2H_2O(l) + 2e^- \rightleftarrows H_2(g) + 2OH^-(aq)$	−0.828	$Fe^{3+}(aq) + e^- \rightleftarrows Fe^{2+}(aq)$	+0.771
$Zn^{2+}(aq) + 2e^- \rightleftarrows Zn(s)$	−0.7618	$Hg_2^{2+}(aq) + 2e^- \rightleftarrows 2Hg(l)$	+0.7973
$Fe^{2+}(aq) + 2e^- \rightleftarrows Fe(s)$	−0.447	$Ag^+(aq) + e^- \rightleftarrows Ag(s)$	+0.7996
$PbSO_4(s) + H_3O^+(aq) + 2e^- \rightleftarrows Pb(s) + HSO_4^-(aq) + H_2O(l)$	−0.42	$Br_2(l) + 2e^- \rightleftarrows 2Br^-(aq)$	+1.066
$Cd^{2+}(aq) + 2e^- \rightleftarrows Cd(s)$	−0.4030	$Cl_2(g) + 2e^- \rightleftarrows 2Cl^-(aq)$	+1.358
$Pb^{2+}(aq) + 2e^- \rightleftarrows Pb(s)$	−0.1262	$PbO_2(s) + 4H_3O^+(aq) + 2e^- \rightleftarrows Pb^{2+}(aq) + 6H_2O(l)$	+1.455
$Fe^{3+}(aq) + 3e^- \rightleftarrows Fe(s)$	−0.037	$PbO_2(s) + HSO_4^-(aq) + 3H_3O^+(aq) + 2e^- \rightleftarrows PbSO_4(s) + 5H_2O(l)$	+1.691
$2H_3O^+(aq) + 2e^- \rightleftarrows H_2(g) + 2H_2O(l)$	0.0000	$Ce^{4+}(aq) + e^- \rightleftarrows Ce^{3+}(aq)$	+1.72
$AgCl(s) + e^- \rightleftarrows Ag(s) + Cl^-(aq)$	+0.222	$F_2(g) + 2e^- \rightleftarrows 2F^-(aq)$	+2.866

Refer to Appendix A for additional standard electrode potentials.

 # Section Review

UNDERSTANDING KEY IDEAS

1. How does a fuel cell differ from a battery?

2. How do an acidic dry cell and an alkaline dry cell differ?

3. Of the metals Zn, Fe, and Ag, which will corrode the easiest? Explain.

4. Describe how a standard electrode potential is measured.

PRACTICE PROBLEMS

5. Calculate the voltage and identify the anode for a cell in which the following electrode reactions take place:

$$Ag(s) + Cl^-(aq) \longrightarrow AgCl(s) + e^-$$
$$Cl_2(g) + 2e^- \longrightarrow 2Cl^-(aq)$$

6. Calculate the voltage and identify the cathode for a cell in which the natural reaction between the following electrodes happens:

$$Zn^{2+}(aq) + 2e^- \rightleftarrows Zn(s)$$
$$2H_2O(l) + 2e^- \rightleftarrows H_2(g) + 2OH^-(aq)$$

CRITICAL THINKING

7. Write the overall equation for the reaction occurring in a lead-acid cell during discharge. What happens to the sulfuric acid concentration during this process? Why is it possible to use a hydrometer, which measures the density of a liquid, to determine if a lead-acid cell is fully charged?

8. A sacrificial anode is allowed to corrode. Why is use of a sacrificial anode considered to be a way to prevent corrosion?

624

Answers to Section Review

1. In a fuel cell, the oxidizing and reducing agents are supplied from outside the cell. In a battery, the materials must be a part of the cell.

2. The materials in the dry cells are different. In an acidic dry cell, a carbon rod and zinc case are used along with NH_4Cl, which is acidic. In an alkaline dry cell, a brass rod and zinc case are used along with KOH, which is basic. Alkaline dry cells tend to last longer than acidic dry cells.

3. Zn will corrode the easiest. It is the most likely to be oxidized because it has the most negative electrode potential.

4. Connect a standard hydrogen electrode to a half-cell in which the reaction of interest happens using a salt bridge. The voltage of the electrode with respect to the standard hydrogen electrode is the standard electrode potential.

5. The cell voltage is (+1.358 V) − (+0.222 V) = +1.136 V. The silver electrode is the anode.

6. The cell voltage is (−0.7618 V) − (−0.828 V) = +0.066 V. The zinc electrode is the cathode.

Answers continued on p. 639A

SCIENCE AND TECHNOLOGY

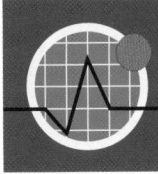

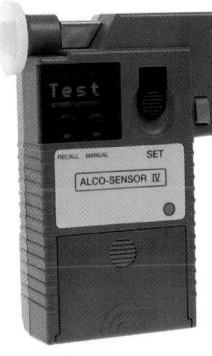

A small pump sends a precise sample size into the fuel cell inside this device.

Fuel Cells

Historical Perspective

In 1839, Sir William Robert Grove, a British lawyer and physicist, built the first fuel cell. More than 100 years later, fuel cells finally found a practical application—in space exploration. During short space missions, batteries can provide enough energy to keep the astronauts warm and to run electrical systems. But longer missions need energy for much longer periods of time, and fuel cells are better suited for this than batteries are. Today, fuel cells are critical to the space shuttle missions and to future missions on the international space station.

Blood Alcohol Testing

A more down-to-Earth use of fuel cells is found in traffic-law enforcement. Police officers need quick and simple ways to determine a person's blood alcohol level in the field. In the time it takes to bring a person to the station or to a hospital for a blood or urine test, the person's blood alcohol content (BAC) might change. Fuel cells, such as the one in the device shown above, provide a quick and accurate way to measure BAC from a breath sample. The alcohol ethanol from the person's breath is oxidized to acetic acid at the anode. At the cathode, gaseous oxygen is reduced and combined with hydronium ions (released from the anode) to form water. The reactions generate an electric current. The size of this current is related to the BAC.

Questions

1. Research at least two more uses for fuel cells. Identify the reactants and products for the cell in each use.
2. Research careers open to chemical engineers. Determine the level of education, the approximate salaries, and the areas of study needed to become a chemical engineer.

Chemical Engineer

Chemical engineers do much of the ongoing fuel-cell research. There are many careers open to chemical engineers. They can work to find alternative, renewable fuel sources, to design new recyclable materials, and to devise new recycling methods. These scientists combine knowledge of chemistry, physics, and mathematics to link laboratory chemistry with its industrial applications. As with any scientist, they also must be good problem solvers.

Chemical engineers use lab techniques that you may know, including distillation, separation, and mixing. The chemical engineer in the photo is part of a team that is developing a compound to help locate explosives, such as those used in land mines. The compound detects small amounts of nitrogen-containing compounds in the air that are often found with explosives. The bright fluorescence of the compound dims when it contacts these compounds.

internet connect

www.scilinks.org
Topic: Fuel Cells
SciLinks code: HW4061

SCiLINKS. Maintained by the National Science Teachers Association

625

SCIENCE AND TECHNOLOGY

Fuel Cells

Research in fuel-cell technology is progressing at a rapid pace as automobile manufacturers compete to develop cars powered by fuel cells. Several automakers have developed fuel cell prototype cars, but the cost of manufacturing these cars is currently too high to make them marketable to the public.

Blood Alcohol Monitors There are several different types of monitors that can test blood alcohol concentration using a person's breath. Some use chemical reactions, and others use infrared spectroscopy to measure the alcohol concentration.

One-Stop Planner CD-ROM

- **Career Extension**
  Real-World Connections Worksheet 11: Photo-Lab Technician
  Assign this worksheet to emphasize relevant applications of text concepts.

Answers to Feature Questions

1. Additional uses of fuel cells include powering cars, buses, and bicycles. They are used to generate electrical energy for individual homes and at large power plants. Some fuel cells are small enough to power sensors used in the military. Many of these fuel cells use hydrogen as a fuel and oxygen from the air and produce water.

2. Students will likely find that chemical engineers work to develop ways to make large amounts of chemicals and other products, such as detergents, plastics, and medicine. These engineers are also involved in designing and operating equipment and chemical plants. A bachelor's degree in chemical engineering is the minimum level of education needed, but many areas require a master's or a doctorate. Salaries depend on level of education and experience, as well as whether the engineer works for the government or in private industry. Areas of study in high school include math, including algebra and geometry, and science, including chemistry and physics. Additional subjects, including calculus, biology, and earth science, are also recommended.

Overview

Before beginning this section, review with your students the Objectives listed in the Student Edition. In this section, students learn that electrolytic cells are electrochemical cells that consume electrical energy. Students further learn that electrolytic cells can perform electrolysis and that electrolysis can be used to decompose water into hydrogen and oxygen gas, and can be used to isolate sodium and aluminum. Finally, students are introduced to the process of electroplating and its benefits and concerns.

 **Bellringer**

Have pairs of students review what they have learned about electrochemical cells, their components, and the reactions that occur at each electrode.

Motivate

Discussion ——— BASIC

Have students study **Figure 15** and answer the following questions:

1. What is the source of electrical energy for this cell? **Ans.** a battery or other power source

2. What are the two electrodes made from? **Ans.** pure copper and impure copper

3. What is the purpose of this cell? **Ans.** to purify the copper

4. What happens to the impurities from the "dirty" electrode? **Ans.** they fall to the bottom of the beaker

LS Visual

KEY TERMS

- electrolytic cell
- electrolysis
- electroplating

OBJECTIVES

① **Describe** how electrolytic cells work.

② **Describe** the process of electrolysis in the decomposition of water and in the production of metals.

③ **Describe** the process of electroplating.

Cells Requiring Energy

Galvanic cells generate electrical energy, but another kind of cell *consumes* electrical energy. This energy is used to drive a chemical reaction. The cell shown in **Figure 15** is a laboratory-scale version of the industrial process used to refine copper. The anode is impure copper, which includes such metals as zinc, silver, and gold. The oxidation reaction at the anode changes Cu atoms in the impure sample to $Cu^{2+}(aq)$. The opposite reaction happens at the cathode. The $Cu^{2+}(aq)$ ions are reduced to Cu atoms. Pure copper is formed, adding to the pure copper cathode.

Impurities such as Zn and other active metals also dissolve as cations. But they are not reduced at the cathode. Inactive metals, such as Au and Ag, fall to the bottom as an *anode sludge*. This sludge is a valuable source of these more-expensive metals in the industrial process.

Figure 15
The experiment shown here mirrors the industrial refining of copper.

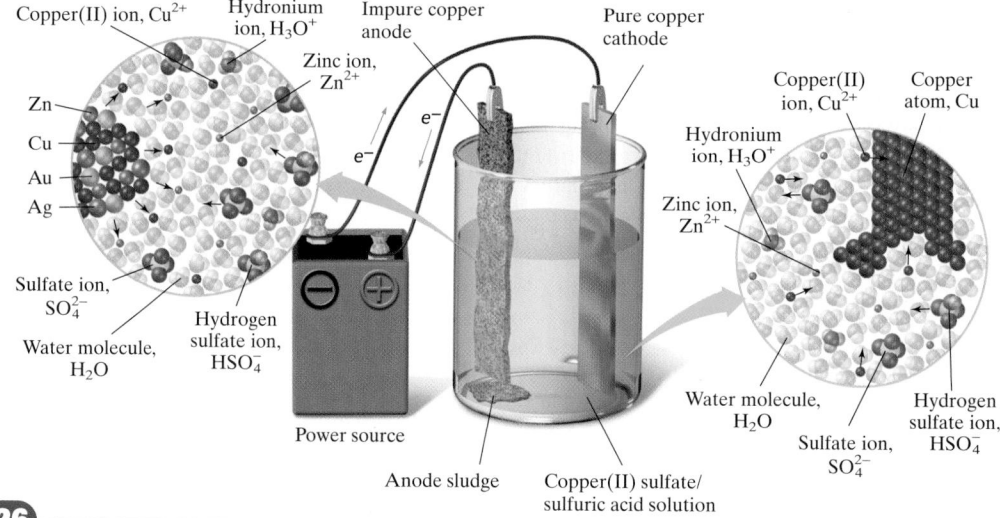

Copper(II) ion, Cu^{2+} Hydronium ion, H_3O^+ Impure copper anode   Pure copper cathode

Copper(II) ion, Cu^{2+} Copper atom, Cu

Zinc ion, Zn^{2+}

Zn

Cu

Au

Ag

Hydronium ion, H_3O^+

Zinc ion, Zn^{2+}

Sulfate ion, SO_4^{2-}

Water molecule, H_2O

Hydrogen sulfate ion, HSO_4^-

Power source

Anode sludge

Copper(II) sulfate/ sulfuric acid solution

Water molecule, H_2O

Sulfate ion, SO_4^{2-}

Hydrogen sulfate ion, HSO_4^-

626

Chapter Resource File

- Lesson Plan

Transparencies

TT Refining Copper Using an Electrolytic Cell

Electrolysis

In an **electrolytic cell,** chemical changes are brought about by driving electrical energy through an electrochemical cell. In fact, the words *electrolysis* and *electrolytic* mean "splitting by electricity" and **electrolysis** does refer to the decomposition of a compound, usually into its elements. However, many changes other than decomposition can happen when you use an electrolytic cell. For example, adiponitrile, one of the raw materials used to make nylon, is synthesized in an electrolytic cell. What makes an electrolytic cell useful is that the overall reaction is a *nonspontaneous* process that is forced to happen by an input of energy.

Electrolysis of Water

The electrolysis of water, shown in **Figure 16,** leads to the overall reaction in which H_2O is broken down into its elements, H_2 and O_2. Pure water does not have enough ions in it and is not conductive enough for electrolysis. An electrolyte, such as sodium sulfate, must be added. The $Na^+(aq)$ and $SO_4^{2-}(aq)$ ions play no part in the electrode reactions, which are as follows:

$$Anode: 6H_2O(l) \longrightarrow 4e^- + O_2(g) + 4H_3O^+(aq)$$

$$Cathode: 4H_2O(l) + 4e^- \longrightarrow 2H_2(g) + 4OH^-(aq)$$

As always, oxidation happens on the anode, while reduction happens on the cathode. Note that hydronium ions form at the anode. Thus, the solution near the anode becomes acidic. But hydroxide ions form at the cathode. So, the solution near the cathode becomes basic.

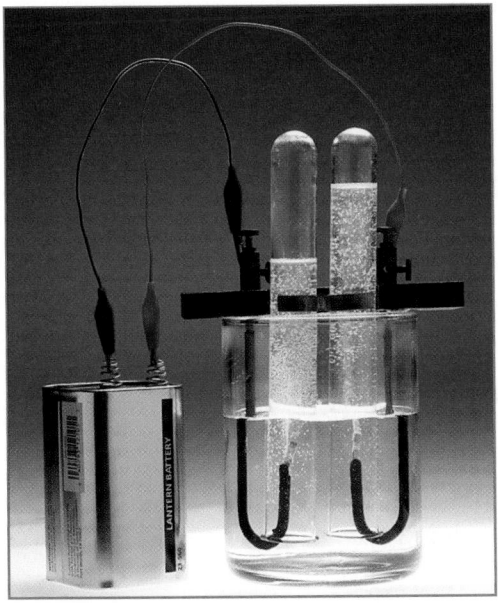

Figure 16
Electrical energy from the battery is used to break down water. Hydrogen forms at the cathode, and oxygen forms at the anode.

electrolytic cell

an electrochemical device in which electrolysis takes place when an electric current is in the device

electrolysis

the process in which an electric current is used to produce a chemical reaction, such as the decomposition of water

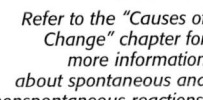

Refer to the "Causes of Change" chapter for more information about spontaneous and nonspontaneous reactions.

627

HISTORY
CONNECTION

Sir Humphry Davy was a pioneer in the field of electrochemistry. His work with electrolysis led him to discover sodium, calcium, potassium, and other metals.

Teach

READING SKILL BUILDER — BASIC

Paired Summarizing Group students into pairs and have them read silently about electrolysis. Then have one student summarize the ideas and the steps in the process of electrolysis. The other student should listen to the retelling and point out any inaccuracies or ideas that were left out. Have students work with partners during this clarification process and refer to the text as needed. Students should repeat this process for the section on electroplating.
LS Interpersonal

Using the Figure — ADVANCED

Figure 16 Students should note that H_2O is both reduced at the cathode and oxidized at the anode during electrolysis. Have them take note of the products at each electrode. Ask students to determine the overall reaction and use the equation to determine which test tube is filling with oxygen and which is filling with hydrogen. **Ans.** The test tube on the left is filling with hydrogen. The test tube on the right is filling with oxygen. The balanced equation shows that twice as much hydrogen as oxygen is made.
LS Logical

Teaching Tip

Uses for Sodium Sodium is manufactured by the electrolysis of molten sodium chloride. Much metallic sodium was once used in the production of tetraethyl lead, used as an antiknock agent in gasoline. Today, not much sodium is produced for that purpose because of environmental concerns about lead pollution. As a result, the production of sodium has declined in recent years. Metallic sodium is used in the industrial synthesis of some organic compounds. Sodium is also used in the molten state as a coolant in nuclear reactors.

Demonstration
The Electrolysis of KI

1. Place a Petri dish on an overhead projector. Add 20 mL of 0.1 M KI solution.

2. Add a few drops of phenolphthalein and a few drops of starch solution to the KI solution.

3. Using insulated leads, attach two electrodes to a 9 V transistor-radio battery.

4. Hold the electrodes in the KI solution. I_2 can be detected at the anode by the characteristic blue-black color of starch in the presence of I_2. H_2 bubbles can be detected at the cathode. OH^- ions can be detected near the cathode by the pink color of phenolphthalein.

5. Have students write the electrode reactions. **Ans.**

Anode: $2I^- \rightarrow I_2 + 2e^-$

Cathode: $2H_2O + 2e^- \rightarrow H_2 + 2OH^-$

6. Discuss the fact that the reaction did not proceed until the electrodes and battery were added. When students realize that this is a nonspontaneous reaction, have them trace the path of the electrons. **Ans.** Electrons are pumped from the anode, where I^- ions are oxidized, and forced through the external circuit to the cathode, where H^+ ions from the water are reduced to H_2.

Safety Caution: Wear safety goggles and a lab apron.

Disposal: Add sufficient 1.0 M $Na_2S_2O_3$ while stirring to convert all iodine to iodide. Adjust the pH to between 5 and 9, and pour down the drain.

Sodium Production by Electrolysis

Sodium is such a reactive metal that preparing it through a chemical process can be dangerous. Sir Humphry Davy first isolated it in 1807 by the electrolysis of molten sodium hydroxide. Today, sodium is made by the electrolysis of molten sodium chloride in a Downs cell, as shown in **Figure 17.**

Pure sodium chloride melts at 801°C. The addition of calcium chloride, $CaCl_2$, to the NaCl lowers the melting point. The Downs cell can then work at 590°C, and less energy is needed to run the cell. The equations below describe the major reactions that occur.

$$\text{Anode: } 2Cl^-(l) \longrightarrow 2e^- + Cl_2(g)$$

$$\text{Cathode: } 2Na^+(l) + 2e^- \longrightarrow 2Na(s)$$

Because chlorine reacts with most metals, the anode is made of graphite. The cathode is steel. Because sodium melts at 98°C and is less dense than the molten salts, it floats to the top and can be removed.

In addition to sodium, a small amount of calcium also forms at the cathode by the reaction:

$$\text{Cathode: } Ca^{2+}(l) + 2e^- \longrightarrow Ca(s)$$

The calcium is more dense than the molten salts, so it falls to the bottom of the cell, where it slowly changes back to Ca^{2+} by the following reaction:

$$Ca(l) + 2Na^+(l) \longrightarrow 2Na(l) + Ca^{2+}(l)$$

Figure 17
In a Downs cell the electrolysis of molten NaCl forms the elements sodium and chlorine.

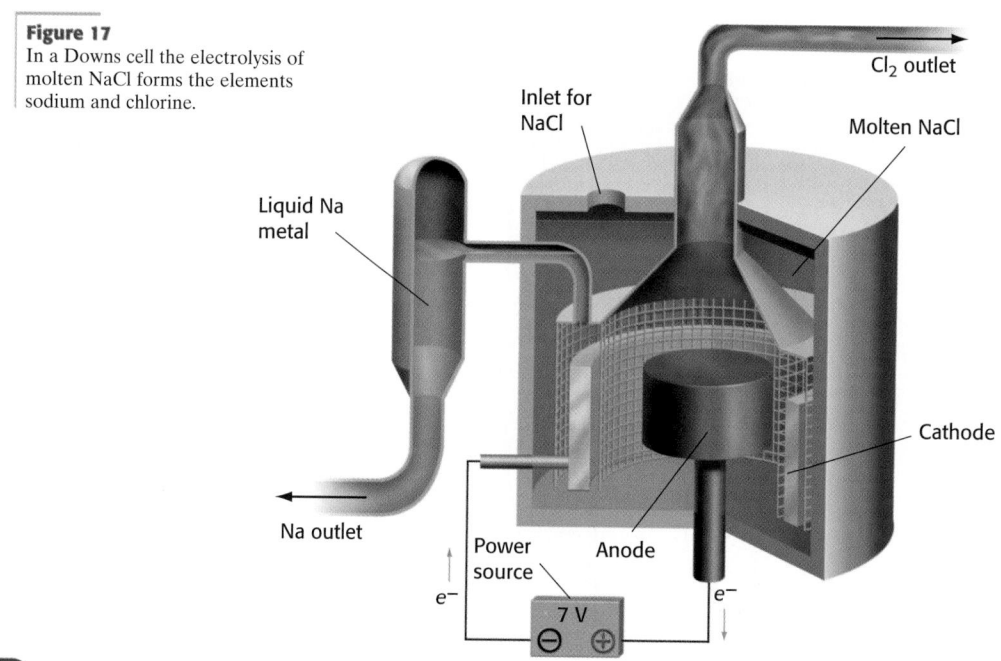

Cl₂ outlet

Inlet for NaCl

Molten NaCl

Liquid Na metal

Cathode

Na outlet

Power source

e^-

Anode

e^-

7 V

HISTORY
CONNECTION

The Downs cell was first used to produce sodium by the DuPont Company in 1921. Because of its great consumption of electrical energy, the cell was located at Niagara Falls to take advantage of the hydroelectric station located there.

Transparencies

TT Downs Cell

Aluminum Production by Electrolysis

Aluminum is the most abundant metal in Earth's crust. Aluminum is light, weather resistant, and easily worked. You have seen aluminum in use in drink cans, in food packaging, and even in airplanes. However, aluminum is never found in nature as a pure metal. Instead, it is isolated from its ore through electrolysis.

The process used to get aluminum from its ore, bauxite, is the electrochemical Hall-Héroult process. This process is the largest single user of electrical energy in the United States—nearly 5% of the national total. This need for energy makes the manufacture of aluminum expensive. Recycling aluminum saves almost 95% of the cost! Aluminum recycling is one of the most economically worthwhile recycling programs that has been developed.

The bauxite is processed to extract and purify hydrated alumina, Al_2O_3. The alumina is fed into huge carbon-lined tanks, like the one in **Figure 18.** There the alumina dissolves in molten cryolite, Na_3AlF_6, at 970°C. Liquid aluminum forms at the cathode. Being more dense than the molten cryolite, aluminum sinks to the floor of the tank. As reduction continues, the level of aluminum rises. As needed, the liquid aluminum is drained and allowed to cool.

Carbon rods serve as the anode. The carbon is oxidized during the anodic reaction, forming CO_2. The rods are eaten away by this oxidation and must be replaced from time to time.

The Hall-Héroult process has been in use for more than a century. But scientists do not completely understand how alumina dissolves and what exactly the species are that participate in the electrode reactions. Although scientists still debate how the process works, they agree that the overall reaction is

$$2Al_2O_3(l) + 3C(s) \longrightarrow 4Al(l) + 3CO_2(g)$$

Figure 18
The Hall-Héroult process is used to make aluminum by the electrolysis of dissolved alumina, Al_2O_3.

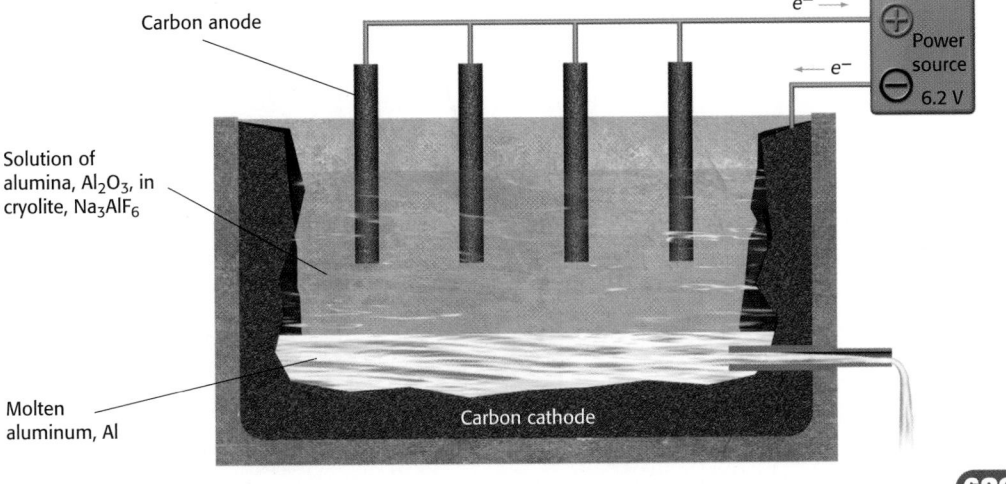

Carbon anode

Solution of alumina, Al_2O_3, in cryolite, Na_3AlF_6

Molten aluminum, Al

Carbon cathode

e^-

Power source 6.2 V

(629)

HISTORY
CONNECTION

An economical process for making aluminum by electrolysis was discovered by Charles Martin Hall in the United States and Paul-Louis-Toussaint Héroult in France. Neither man knew of the other's work, but they discovered the same process within a few months of each other in 1886. After a long patent litigation, the two men reached an agreement to share their invention. Hall and Héroult were born and died in the same years, 1863–1914. They were about 23 years old when they discovered the process named for them.

Teaching Tip

Cryolite is a naturally occurring mineral, but it is not abundant enough to supply the needs of the aluminum industry. Therefore, cryolite must be produced synthetically by the action of HF on sodium aluminate, $NaAlO_2$, a product of the aluminum ore refining process.

Group Activity ——— ADVANCED

Aluminum and its alloys have many uses. Consider having groups of students do visual presentations on the uses of aluminum and how those uses relate to aluminum's chemical and physical properties. Each group should focus on a particular application. One group could prepare a video presentation demonstrating the overall properties of aluminum. Consider some of the following lines of inquiry:

- Why does aluminum make up 90% of overhead power lines though it is only about 60% as effective a conductor as copper?

- Aluminum is called a self-protecting metal. What does this mean, and what applications of aluminum are related to this property?

- What other elements are alloyed with aluminum, and what properties does each confer on the alloy?

- Why is aluminum used in the aerospace industry, and what alloys are suitable for such uses?

LS Visual

Teaching Tip

Recycling Inform students that the cost of refining aluminum from recycled materials is between 5% and 10% of the cost of producing the same amount of aluminum from ore. As a result, aluminum scrap has considerable value, and aluminum recycling has been highly successful.

Transparencies

TT Hall-Héroult Process

Chapter 17 · Oxidation, Reduction, and Electrochemistry **629**

Writing **Research Skills** Some students may wonder how nonconductors such as organic materials and plastics are plated with metals. Often the surfaces of such objects are made conductive by coating them with paint containing metal particles so that they may be electroplated. Plastic objects are plated using a process called *electroless plating*. To do this, the plastic object is etched in the places where plating is desired. Then it is dipped in solutions of tin(II) chloride and palladium chloride. Copper or nickel can then be plated onto the plastic. Have interested students research the electroless plating process. Have them report their findings to the class in a report that compares electroplating with electroless plating.
LS Logical

MISCONCEPTION ALERT

Some students may think that any process that coats an object with a metal is an example of electroplating. Explain to students that electroplating is just one of the ways an object can be coated with a metal. For example, gold, silver, and copper foils are often rubbed on objects made from ceramics and wood for decorative purposes. In fact, gold foil is sometimes used as decorations on cakes and other foods—and chefs don't use electric currents to apply the gold to the food!

---

electroplating

the electrolytic process of plating or coating an object with a metal

internet connect

www.scilinks.org
Topic: Electroplating
SciLinks code: HW4049

*SCi*LINKS Maintained by the National Science Teachers Association

---

Electroplating

Many of the metal things that you use every day—forks and spoons, cans for food and drinks, plumbing fixtures, jewelry and decorative ornaments, automobile and appliance parts, nails, nuts, and bolts—have been treated to change their surfaces. Often this involves putting a layer of another metal on top of the main metal. **Electroplating** is one way of applying these finishes. Forks, spoons, and jewelry are often electroplated to give the objects the appearance of silver or gold while still keeping the cost of the objects low. The chrome parts on automobiles have been electroplated to improve the parts' appearance and protect them from corrosion. Electroplating is also used for many electronic and computer parts to give them a certain physical property or to make them last longer by protecting them against corrosion.

To electroplate a bracelet with silver, as shown in the simplified model in **Figure 19,** the bracelet is made the cathode of an electrolytic cell. The anode is a strip of pure silver metal. Both electrodes are placed in a solution of silver ions. The net reactions that happen are very simple. The anode slowly dissolves by the following oxidation reaction:

$$\text{Anode: Ag}(s) \longrightarrow e^- + \text{Ag}^+(aq)$$

The cathode reaction on the surface of the bracelet is the reverse of the anode reaction.

$$\text{Cathode: Ag}^+(aq) + e^- \longrightarrow \text{Ag}(s)$$

The result is that a thin coating of silver forms on the bracelet. The longer the plating is continued, the thicker the silver layer becomes.

Figure 19
The bracelet in this cell is being coated with a thin layer of silver. Silver ions are replaced in the solution as the pure silver anode dissolves.

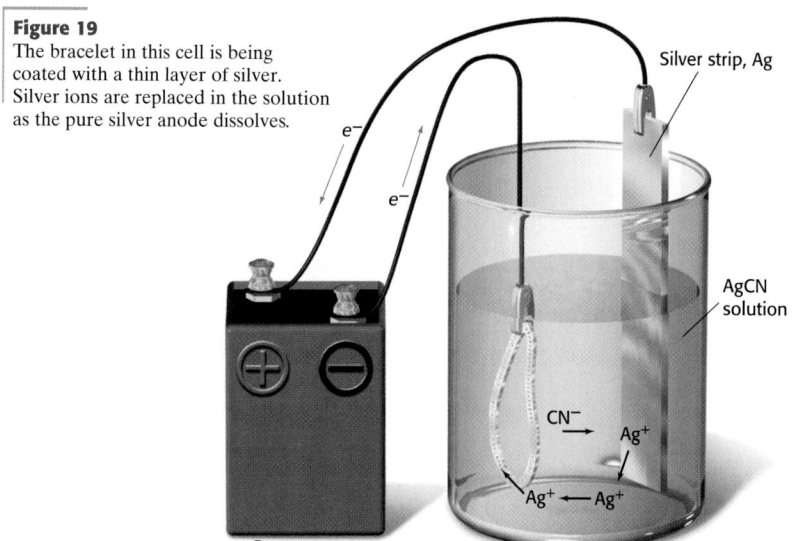

Silver strip, Ag

AgCN solution

e^-

e^-

CN^- → Ag$^+$

Ag$^+$ ← Ag$^+$

Power source

Cathode

Anode

---

did you know?

Vermeil jewelry may look like solid gold jewelry but it is actually made from a metal (usually silver) that is electroplated with gold. In fact, sometimes vermeil is plated with 24 karat gold to give the jewelry a richer look. Stores that sell "gold by the inch" at low prices are often selling vermeil chains rather than solid gold chains.

Transparencies

TT Electroplating

Benefits and Concerns About Electroplating

A major benefit of electroplating a metal is that it becomes more resistant to corrosion. Chrome-plated car parts and zinc-plated food cans are two uses of electroplating to reduce corrosion. Another benefit of electroplating is that it improves the appearance of an object. An item that is gold-plated or silver-plated is much cheaper than the same item in solid gold or silver but looks the same.

However, there are also drawbacks with the electroplating process. In practice, electroplating is not as simple as the description above suggests. It is difficult to get metal to deposit in depressions, so the metal layer often is not uniform. Sometimes, deposits are loose and powdery. Additives are used and conditions such as temperature and pH are carefully controlled to overcome these problems. Over time, impurities build up in the solutions used for electroplating. Eventually, the spent solutions must be discarded. They can contain high concentrations of such toxic metals as cadmium or chromium and require careful disposal to protect the environment.

 Section Review

UNDERSTANDING KEY IDEAS

1. How does an electrolytic cell differ from a galvanic cell?

2. What chemical process occurs at the anode of an electrolytic cell?

3. What chemical process occurs at the cathode of an electrolytic cell?

4. What form of energy is used to drive an electrolytic cell?

5. List three commercial products made by using electrolytic cells.

6. What does *electrolysis* mean?

7. Write the equation for the cathodic reaction that occurs in the Downs cell used to make sodium.

8. How is aluminum manufactured?

9. Describe the electroplating process.

CRITICAL THINKING

10. In the copper refining process, why does zinc not also deposit on the cathode?

11. Elemental aluminum was first prepared in 1827 by the reaction of aluminum chloride with potassium.
    a. Write the balanced equation for this reaction.
    b. Determine if the reaction is a redox reaction.
    c. Would this reaction need to happen in an electrolytic cell? Explain.

12. Cryolite, Na_3AlF_6, is an ionic mineral used in the preparation of aluminum. Explain why the sodium ions are not reduced during the electrolytic process that produces aluminum.

13. The following reaction happens naturally:

    $$Zn(s) + Cu^{2+}(aq) \longrightarrow Cu(s) + Zn^{2+}(aq)$$

    Explain why it is necessary to use an electric current to deposit a layer of zinc on a copper bracelet.

14. Explain why a galvanic cell is often used in an electrolytic cell. What function does the galvanic cell serve?

15. Why is it so important to recycle rather than discard aluminum products?

631

Answers to Section Review

1. A galvanic cell generates an electric current spontaneously from a chemical reaction. An electrolytic cell uses electrical energy to cause a reaction to happen.

2. oxidation

3. reduction

4. electrical energy

5. Answers may vary and may include sodium metal, chlorine gas, aluminum, and hydrogen gas.

6. Literally "splitting by electricity"; chemical changes brought about by an electric current.

7. $Na^+(l) + e^- \rightarrow Na(l)$

8. By the electrolysis of alumina dissolved in molten cryolite.

9. An object to be plated is made the cathode in an electrolytic cell. As electrical energy is provided to the cell, metal ions in solution are reduced to atoms on the cathode, and the object becomes plated with metal.

10. Zinc is more active than copper is and is much more difficult to reduce than copper is.

Answers continued on p. 639A

17 CHAPTER HIGHLIGHTS

Alternative Assessments

Homework
• Page 631

|SKILL BUILDER
• Page 630

CHAPTER REVIEW
• Items 90, 92, 94, 96

Portfolio Assessments

REAL-WORLD ─CONNECTION─
• Page 619

Group Activity
• Pages 608, 629

CHAPTER REVIEW
• Items 91, 93, 97

Chapter Resource File

• **Observation Lab** Electroplating for Corrosion Protection
• **CBL™ Probeware Lab** Establishing a Table of Reduction Potentials
• **Datasheets for In-Text Lab** Redox Titration
• **Datasheets for In-Text Lab** Redox Titration—Mining Feasibility Study

KEY IDEAS

SECTION ONE Oxidation-Reduction Reactions
• The loss or gain of electrons in a chemical reaction is called *oxidation* or *reduction,* respectively.
• In a redox reaction, oxidation and reduction occur at the same time.
• An oxidation number may be assigned to each atom in a molecule or ion.
• Half-reactions, in which only the oxidation or the reduction is described, are useful in balancing redox equations.
• Reducing agents readily donate electrons; oxidizing agents readily accept electrons.

SECTION TWO Introduction to Electrochemistry
• An electrochemical cell is made up of two electrodes linked by one or more ionic conductors.
• When electric current is in a cell, electrode reactions take place.
• Oxidation happens at the anode. Reduction happens at the cathode.

SECTION THREE Galvanic Cells
• Many examples of galvanic cells are power sources that generate electrical energy from chemical energy.
• Fuel cells differ from batteries in that their oxidizing and reducing agents are gases introduced to the cell from outside.
• The corrosion of metals generally happens in a galvanic cell.
• Whether there is electric current or not, the electrodes of a cell have different potentials. The difference between these is the voltage of the cell.

SECTION FOUR Electrolytic Cells
• Electrical energy is used to power an electrolytic cell.
• In electrolysis, a compound is decomposed, usually to its elements; H_2, O_2, Al, Na, and Cl_2 are among the elements that can be isolated by electrolysis.
• A layer of a second metal is deposited cathodically in electroplating.

KEY TERMS

oxidation
reduction
oxidation-reduction reaction
oxidation number
half-reaction
oxidizing agent
reducing agent

electrochemistry
voltage
electrode
electrochemical cell
cathode
anode

corrosion
standard electrode potential

electrolytic cell
electrolysis
electroplating

KEY SKILLS

Assigning Oxidation Numbers
Skills Toolkit 1 p. 606
Sample Problem A p. 607

Balancing Redox Equations Using the Half-Reaction Method
Skills Toolkit 2 p. 609
Sample Problem B p. 610

Calculating Cell Voltage
Sample Problem C p. 623

632

USING KEY TERMS

1. What is a redox reaction?

2. Explain the terms *oxidation* and *reduction* in terms of electrons.

3. Explain how oxidation numbers are used to identify redox reactions?

4. Explain what half-reactions are and why they are useful.

5. Define *electrochemistry*.

6. Define *electrode, anode,* and *cathode*.

7. Explain voltage and current in terms of electrons.

8. Distinguish between galvanic and electrolytic cells.

9. What is a fuel cell?

10. What is corrosion, and what is a corrosion cell?

11. Why does a sacrificial anode provide cathodic protection?

12. What is electroplating?

UNDERSTANDING KEY IDEAS

Oxidation-Reduction Reactions

13. Assign oxidation numbers to the atoms in the ionic compound $MgBr_2(s)$.

14. Assign oxidation numbers to the atoms in the ionic compound $NH_4NO_3(s)$.

15. Assign oxidation numbers to the atoms in the ion $PF_6^-(aq)$.

16. Identify each of the following half-reactions as oxidation or reduction reactions.
a. $K(s) \longrightarrow e^- + K^+(aq)$
b. $Cu^{2+}(aq) + e^- \longrightarrow Cu^+(aq)$
c. $Br_2(l) + 2e^- \longrightarrow 2Br^-(aq)$

17. Is the reaction below a redox reaction? Explain your answer.
$$Ca(s) + Cl_2(g) \longrightarrow CaCl_2(s)$$

18. Describe how to identify the oxidizing agent and the reducing agent in a reaction.

19. Nitrogen monoxide, $NO(g)$, reacts with phosphorus, $P_4(s)$, to produce nitrogen, $N_2(g)$, and diphosphorus pentoxide, P_2O_5. Write the balanced equation for this reaction. Identify the atoms that have been oxidized and reduced, and identify the oxidizing and reducing agents.

Introduction to Electrochemistry

20. What is the distinction between a half-reaction and an electrode reaction?

21. Describe the components of an electro-chemical cell.

22. Identify which of the following reactions (as written) is an anodic reaction and which is a cathodic reaction. Write the balanced over-all ionic equation for the redox reaction of the cell.
$$Cd(s) \longrightarrow Cd^{2+}(aq) + 2e^-$$
$$Ag^+(aq) + e^- \longrightarrow Ag(s)$$

23. What is the significance of the $\oplus$ symbol on a dry cell or battery?

24. What reaction happens at the cathode of an electrochemical cell?

633

Assignment Guide

Section	Questions
1	1–4, 13–19, 39–60, 68, 69, 75–77, 81, 82, 89
2	5–7, 20–26, 61–67, 70, 71, 78–80, 83, 84, 86, 100
3	8–11, 27–33, 72, 85, 87, 90–94, 96–98, 105
4	12, 34–38, 73, 74, 88, 95, 99, 101–104

REVIEW ANSWERS

Using Key Terms

1. a reaction in which one species is oxidized (loses electrons) while another species is reduced (accepts electrons)

2. Oxidation involves the loss of electrons. Reduction involves the gain of electrons.

3. After assigning an oxidation number to each atom involved in a reaction, look for an atom that has a different oxidation number in the reactants than it has in the products. A reaction in which atoms change their oxidation numbers is a redox reaction.

4. Half-reactions are shown as equations that involve electrons and represent the reduction and oxidation taking place in a redox reaction. They are useful in balancing equations for redox reactions.

5. the branch of chemistry that is the study of the relationship between electric forces and chemical reactions

6. An electrode is a conductor (usually a metal) used to establish electrical contact with a non-metallic part of a circuit, such as an electrolyte. An anode is an electrode at which oxidation occurs. A cathode is an electrode at which reduction occurs.

7. Voltage is the potential difference between two points as a result of differences in electron density. As a result of a voltage, electrons can move from the point of high electron density to the point of low electron density. This movement of electrons is a current.

8. Galvanic cells generate electrical energy as a result of a naturally occurring chemical reaction. Electrolytic cells use electrical energy to cause a reaction that would not occur naturally.

9. a galvanic cell in which the oxidizing and reducing agents are supplied from outside the cell

10. Corrosion is the gradual destruction of a metal or alloy as a result of chemical processes such as oxidation or the action of a chemical agent. A corrosion cell is a galvanic cell in which oxidation of the metal occurs at one site, with reduction (usually of oxygen) taking place elsewhere.

11. The sacrificial anode corrodes more easily than the metal it protects. As a result, the protected metal is a cathode and receives electrons from the sacrificial anode, which helps protect the cathode from corrosion.

12. the electrolytic process of plating or coating an object with a metal

Understanding Key Ideas

13. Mg = +2; Br = −1

14. N = −3; H = +1; N = +5; O = −2

15. P = +5; F = −1

16. a. oxidation

    b. reduction

    c. reduction

17. The reaction is a redox reaction. Calcium is oxidized (0 to +2), and chlorine is reduced (0 to −1).

18. Identify the atom that is oxidized and the atom that is reduced. The reactant ion or molecule that contains the atom that is oxidized is the reducing agent. The reactant ion or molecule that contains the atom that is reduced is the oxidizing agent.

19. $P_4 + 10NO \rightarrow 2P_2O_5 + 5N_2$; P is oxidized, and N is reduced. P_4 is the reducing agent, and NO is the oxidizing agent.

25. In an electrochemical cell, what role does the porous barrier play? What would happen without it?

26. Explain why the combination of the two electrode reactions of an electrochemical cell always gives the equation of a redox reaction.

Galvanic Cells

27. Describe a galvanic cell, and give an example.

28. Write the equations of the two electrode reactions that occur when a Daniell cell is in use. Identify the anode reaction and the cathode reaction.

29. What is the essential advantage of a fuel cell over other types of galvanic cells that are used to generate electrical energy?

30. Explain why a corrosion cell is a galvanic cell.

31. Discuss methods of reducing corrosion.

32. The standard electrode potential for the reduction of $Zn^{2+}(aq)$ to $Zn(s)$ is −0.762 V. What does this value indicate?

33. Which half-reaction would be more likely to be an oxidation: one with a standard electrode potential of −0.42 V, or one with a standard electrode potential of +0.42 V? Explain your answer.

Electrolytic Cells

34. Define *electrolytic cell,* and give an example.

35. Describe the apparatus used in the electrolysis of water.

36. Explain why sodium can be prepared by electrolysis.

37. Describe some benefits of electroplating.

38. What are some problems in the electroplating industry?

PRACTICE PROBLEMS

Determining Oxidation Numbers

39. Determine the oxidation number of each atom in CO_2.

40. Determine the oxidation number of each atom in CoO.

41. Determine the oxidation number of each atom in $BaCl_2$.

42. Determine the oxidation number of each atom in K_2SO_4.

43. Determine the oxidation number of the sulfur atom in S^{2-}.

44. Determine the oxidation number of the lanthanum atom in La^{3+}.

45. Determine the oxidation number of each atom in CH_4.

46. Determine the oxidation number of each atom in NH_4^+.

47. Determine the oxidation number of each atom in $CaCO_3$.

48. Determine the oxidation number of each atom in $PtCl_6^{2-}$.

49. Determine the oxidation number of each atom in $COCl_2$.

50. Determine the oxidation number of each atom in PO_4^{3-}.

The Half-Reaction Method

51. Write the balanced half-reaction for the conversion of $Fe(s)$ to $Fe^{2+}(aq)$.

52. Write the balanced half-reaction for the conversion of $Cl_2(g)$ to $Cl^-(aq)$.

53. Combine the half-reactions from items 51 and 52 into a single reaction.

54. Write the balanced half-reaction for the conversion of $HOBr(aq)$ to $Br_2(aq)$ in acidic solution.

55. Write the balanced half-reaction for the conversion of $H_2O(l)$ to $O_2(aq)$ in acidic solution.

56. Combine the half-reactions from items 54 and 55 into a single reaction.

57. Write the balanced half-reaction for the change of $O_2(aq)$ to $H_2O(l)$ in acidic solution.

58. Write the balanced half-reaction for the change of $SO_2(aq)$ to $HSO_4^-(aq)$ in acidic solution.

59. Combine the reactions from items 57 and 58 into a single reaction.

60. Using half-reactions, balance the redox equation of $Zn(s)$ and $Fe^{3+}(aq)$ reacting to form $Zn^{2+}(aq)$ and $Fe^{2+}(aq)$.

Calculating Cell Voltage

61. The standard electrode potentials of two electrodes in a cell are 1.30 V and 0.45 V. What is the voltage of the cell?

Use the information from Table 1 to answer the following items.

62. Calculate the voltage of a cell for the naturally occurring reaction between the following electrodes:

$$AgCl(s) + e^- \rightleftarrows Ag(s) + Cl^-(aq)$$

$$2H_3O^+(aq) + 2e^- \rightleftarrows 2H_2O(l) + H_2(g)$$

63. Calculate the voltage of a cell that has the following electrode reactions:

$$2H_3O^+(aq) + 2e^- \longrightarrow 2H_2O(l) + H_2(g)$$

$$Fe^{2+}(aq) \longrightarrow Fe^{3+}(aq) + e^-$$

64. Calculate the voltage of a cell in which the overall reaction is

$$2Fe^{3+}(aq) + Cd(s) \longrightarrow Cd^{2+}(aq) + 2Fe^{2+}(aq)$$

65. Calculate the voltage of a cell that has the following electrode reactions:

$$O_2(g) + 2H_2O(l) + 4e^- \longrightarrow 4OH^-(aq)$$

$$H_2(g) + 2OH^-(aq) \longrightarrow 2H_2O(l) + 2e^-$$

66. Calculate the voltage of a cell in which the two electrode reactions are the reduction of chlorine gas to chloride ions and the oxidation of copper metal to copper(II) ions.

67. Calculate the voltage of a cell in which the two electrode reactions are those of the lead-acid battery.

MIXED REVIEW

68. Assign an oxidation number to the N atom in each of the following oxides of nitrogen: N_2O, NO, NO_2, N_2O_3, N_2O_4, and N_2O_5.

69. Refer to the figure below to answer the following questions:

a. What observations suggest that a chemical reaction has taken place?

b. Write a balanced equation for the reaction taking place, and explain how you can identify it as a redox reaction.

c. Identify what is oxidized and what is reduced in the reaction.

d. Describe how the quantity of zinc metal and of copper metal will change as the reaction continues.

e. Describe the role of the chloride ions in the reaction.

70. What name is given to an electrode at which oxidation occurs? at which reduction occurs?

71. Explain how electric charges travel through the various parts of an electrochemical cell, including how they cross the electrode surfaces.

20. A half-reaction is a model used to help balance redox equations. The electrons shown in a half-reaction are not free, but transferred directly between the reacting species. An electrode reaction is an actual reaction taking place at an electrode.

21. An electrochemical cell is made up of two electrodes that are connected by wires that allow electrons to move and an electrolyte which provides a pathway for ions to move.

22. The reaction involving cadmium is anodic. The reaction involving silver is cathodic. The overall reaction is $Cd(s) + 2Ag^+(aq) \rightarrow Cd^{2+}(aq) + 2Ag(s)$.

23. It represents the positive terminal, or cathode, of the battery. This end of the battery is where reduction occurs.

24. reduction

25. The porous barrier allows the movement of charge, by ion transfer, from one solution to the other. Without the barrier, the solutions would mix and the electric current would stop.

26. The anodic reaction is an oxidation. The cathodic reaction is a reduction. When these reactions are combined, the result is a redox reaction because it includes both an oxidation and a reduction.

27. A galvanic cell is an electrochemical cell in which a reaction happens that can often be used to generate electrical energy. Examples include dry cells, lead-acid batteries, fuel cells, and corrosion cells.

28. anode reaction: $Zn(s) \rightarrow 2e^- + Zn^{2+}(aq)$; cathode reaction: $Cu^{2+}(aq) + 2e^- \rightarrow Cu(s)$

29. A fuel cell can, in principle, function indefinitely without stopping to recharge.

30. The oxidation and reduction half-reactions occur naturally at different sites. The two sites are connected by both a metal conductor and an electrolyte.

31. Answers may vary. Answers might include painting, using a sacrificial anode, or coating with another metal.

32. This is the voltage produced by a zinc electrode in a solution of zinc ions when measured against a SHE. The negative value indicates that zinc will undergo oxidation to Zn^{2+} when coupled with a SHE.

33. The half-reaction with a standard electrode potential of -0.42 V would be more likely to be an oxidation. The negative value indicates that the electrode will be an anode when coupled with a SHE.

34. An electrolytic cell is a cell in which electrode reactions occur as a result of electrical energy being added to the cell. Examples include cells in which sodium or aluminum are made or copper is purified and a cell used for electroplating.

35. An electrolytic cell is made. A battery can be used as the power source. An electrolyte, such as sulfuric acid or sodium chloride, is added to the water to increase conductivity.

36. By supplying electrons to sodium ions, the ions can be made to gain electrons and become sodium atoms. This process occurs in electrolysis.

37. Electroplating can help a metal resist corrosion. It can improve the appearance of objects using small amounts of precious metals.

72. Why are dry cells so called, even though their chemistry involves water?

73. Explain how copper is refined.

74. Commercial aluminum smelters are located, not near bauxite mines, but near electrical power plants. Why?

75. Determine the oxidation number of each phosphorus atom in P_4.

76. Determine the oxidation number of each atom in H_2SO_3.

77. Write the half-reaction for the conversion of $Fe^{3+}(aq)$ to $Fe^{2+}(aq)$. Is this change an oxidation or a reduction?

78. Using half-reactions, balance the equation for the redox reaction when $Cr_2O_7^{2-}(aq)$ and $Fe^{2+}(aq)$ react to form $Cr^{3+}(aq)$ and $Fe^{3+}(aq)$ in acidic solution.

79. Calculate the voltage of a cell in which the overall reaction is the electrolysis of aqueous cadmium chloride into its elements.

80. Calculate the voltage of a cell in which the overall reaction is the electrolysis of solid AgCl into its elements.

CRITICAL THINKING

81. Suggest how the word *oxidation* might have come to be the general term for a loss of electrons.

82. Think of ammonium nitrite as $(NH_4^+)(NO_2^-)$, and assign oxidation numbers within each ion. Now think of ammonium nitrate as $N_2H_4O_2$, and assign oxidation numbers. Which assignment makes more sense?

83. Why is the negative battery terminal the one with the greater "electron pressure"?

84. What is different about electric current in metals compared with current in electrolyte solutions?

85. The activity of the halogens decreases as you move down the group on the periodic table. Use the information in **Table 1** to explain this trend.

86. A fuel cell uses methanol, CH_3OH, as fuel. Assuming complete oxidation of the fuel to CO_2 and H_2O in acidic aqueous solution, how many electrons are transferred per CH_3OH molecule?

87. Using **Table 1,** calculate the voltage of a cell for the naturally occurring reaction between the electrodes below. If a wire were connected between the Ag and Cu, which way would electrons travel? Which electrode would be the anode?

$$Ag^+(aq) + e^- \rightleftharpoons Ag(s)$$
$$Cu^{2+}(aq) + 2e^- \rightleftharpoons Cu(s)$$

88. Electrolytic cells can be considered the opposite of galvanic cells. Discuss whether they are closer to being the opposite of batteries or of fuel cells.

89. How does the oxidation number of an atom of Mn change in the reaction below?

$$4Mn^{2+}(aq) + MnO_4^-(aq) + 8H_3O^+(aq) + 15H_2P_4O_7^{2-}(aq) \longrightarrow 12H_2O(l) + 5Mn(H_2P_4O_7)_3^{3-}(aq)$$

ALTERNATIVE ASSESSMENT

90. Your teacher will assign you a known metal with an unknown reduction potential. Devise a method to determine the $E°$ value of the metal from a list of metals with known $E°$ values. Present your method to your teacher.

91. Investigate the development and operation of the sodium-sulfur battery that has been proposed to run electric cars. Choose a stand for or against its use, and present your findings in a persuasive speech to your classmates.

636

92. Consumer use of rechargeable batteries is growing. Nickel-cadmium batteries, a common type of rechargeable battery, are used in cellular phones, electric shavers, and portable video-game systems. Make a list of the items with which you come into contact that use nickel-cadmium batteries or other rechargeable batteries. Write a short essay about technology that was not and could not have been available before the development of the nickel-cadmium battery.

93. For one week, keep a record of how many times you use devices powered by batteries. Record what kind of device you used and the number and type of batteries it contained. Your teacher will provide you with various batteries and a balance. Record the mass of each type of battery you used during the week. Assuming that your battery usage is typical of everyone in the country, estimate the mass of waste material produced by battery usage in one year. Write a short report offering ways to reduce the amount of waste.

94. Read the book *Apollo 13* on which the movie of the same title was based, and report information about the power sources in the spacecraft.

95. Research metals other than copper, sodium, and aluminum that are manufactured or refined electrochemically.

96. Using a voltmeter from home or one borrowed from your teacher, devise a method of measuring the voltage of a flashlight battery *while it is delivering power to the bulb.* Compare that voltage with that of a new battery. Also record the voltage as the battery "runs down." Write a report on your results.

97. Manufacturers often claim that their batteries are "heavy duty" or "long lasting." Design an experiment to test the value and efficacy of AA batteries. If your teacher approves, carry out your procedure.

98. Carpentry nails are steel, but some are plated with a second metal. Obtain a variety of nails and evaluate their tendency to rust by laying them on a piece of cloth moistened by water containing salt and vinegar, so they are exposed both to the liquid and to air. After two weeks, report your findings.

99. Find at least five items that have been electroplated. Describe why each object was electroplated.

CONCEPT MAPPING

100. Use the following terms to complete the concept map below: *cathode, electrodes, electrochemical cell, anode, oxidation,* and *reduction.*

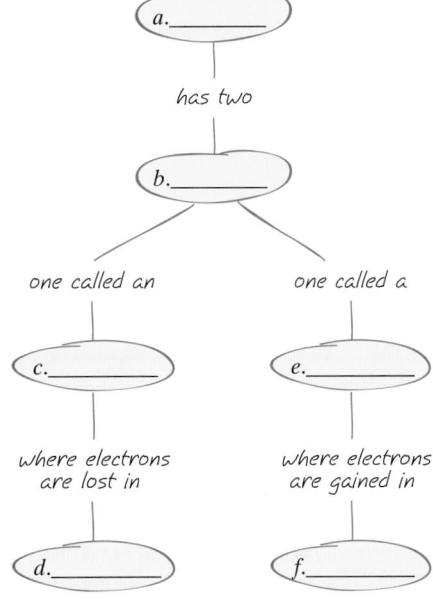

38. The metal layer is often not uniform due to surface conditions on the underlying metal. The used solutions often contain high concentrations of toxic metals and require careful disposal.

Practice Problems

39. C = +4; O = −2

40. Co = +2; O = −2

41. Ba = +2; Cl = −1

42. K = +1; S = +6; O = −2

43. S = −2

44. La = +3

45. C = −4; H = +1

46. N = −3; H = +1

47. Ca = +2; C = +4; O = −2

48. Pt = +4; Cl = −1

49. C = +4; O = −2; Cl = −1

50. P = +5; O = −2

51. $Fe(s) \rightarrow Fe^{2+}(aq) + 2e^-$

52. $Cl_2(g) + 2e^- \rightarrow 2Cl^-(aq)$

53. $Fe(s) + Cl_2(g) \rightarrow Fe^{2+}(aq) + 2Cl^-(aq)$

54. $2HOBr(aq) + 2H_3O^+(aq) + 2e^- \rightarrow Br_2(l) + 4H_2O(l)$

55. $6H_2O(l) \rightarrow O_2(g) + 4H_3O^+(aq) + 4e^-$

56. $4HOBr(aq) \rightarrow O_2(g) + 2Br_2(l) + 2H_2O(l)$

57. $O_2(g) + 4H_3O^+(aq) + 4e^- \rightarrow 6H_2O(l)$

58. $5H_2O(l) + SO_2(aq) \rightarrow HSO_4^-(aq) + 3H_3O^+(aq) + 2e^-$

59. $4H_2O(l) + 2SO_2(aq) + O_2(g) \rightarrow 2HSO_4^-(aq) + 2H_3O^+(aq)$

60. $Zn(s) + 2Fe^{3+}(aq) \rightarrow Zn^{2+}(aq) + 2Fe^{2+}(aq)$

61. (+1.30 V) − (+0.45 V) = +0.85 V

62. (+0.222 V) − (0.0000 V) = +0.222 V

63. (0.0000 V) − (+0.771 V) = −0.771 V

64. (+0.771 V) − (−0.4030 V) = +1.174 V

65. (+0.401 V) − (−0.828 V) = +1.229 V

66. (+1.358 V) − (+0.3419 V) = +1.016 V

67. (+1.691 V) − (−0.42 V) = +2.11 V

637

Mixed Review

68. N_2O, +1; NO, +2; NO_2, +4; N_2O_3, +3; N_2O_4, +4; N_2O_5, +5

69. a. The zinc strip appears to be eaten away, and copper metal has collected at the bottom of the beaker.

b. $Zn(s) + Cu^{2+}(aq) \rightarrow Zn^{2+}(aq) + Cu(s)$; The oxidation number of the atoms of zinc and of copper change, so the reaction is a redox reaction.

c. $Zn(s)$ is oxidized, and $Cu^{2+}(aq)$ is reduced.

d. As the reaction continues, the quantity of zinc metal present will decrease and the quantity of copper metal will increase.

e. The chloride ions are spectator ions. They do not participate in the reaction, but they do keep the overall charge of the contents of the beaker electrically neutral.

70. anode; cathode

71. Electric charges move through the metal wires as a result of moving electrons and through the electrolyte solutions as a result of moving ions. The charges cross the surfaces of the electrodes as a result of the reactions taking place there.

72. Dry cells use a moist paste as an electrolyte, so they are much drier than cells that use solutions.

73. Copper is refined electrolytically, using an anode of impure copper and a cathode of pure copper. The cathode becomes plated with copper as the cell operates.

74. The smelters use so much electrical energy that it is cheaper to carry the bauxite to the electrical power plant than it is to transmit the electrical energy to the mine.

Answers continued on p. 639B

FOCUS ON GRAPHING

Study the graph below, and answer the questions that follow.
For help in interpreting graphs, see Appendix B, "Study Skills for Chemistry."

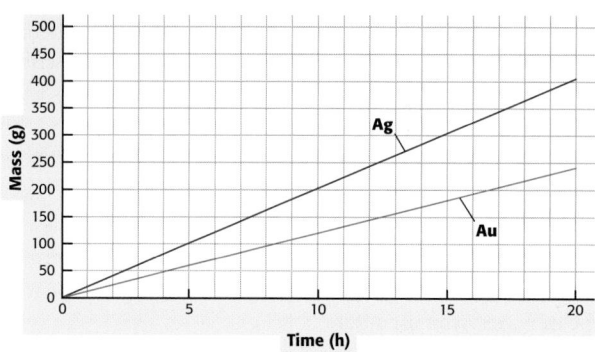

Mass of Metal Deposited by Electroplating Using a 5 A Current

101. What is the rate of production, in g/h, for each metal?

102. What is the rate of production, in mol/h, for each metal?

103. How much time is needed for 1.00 mol of electrons to be absorbed?

104. Write the half-reaction for the production of silver metal from silver ions. Write the half-reaction for the production of gold metal from gold(III) ions. Using these half-reactions, provide an explanation for the difference in the rates of production of the metals.

TECHNOLOGY AND LEARNING

105. Graphing Calculator

Calculate the Equilibrium Constant, using the Standard Cell Voltage

The graphing calculator can run a program that calculates the equilibrium constant for an electrochemical cell using an equation called the Nernst equation, given the standard potential and the number of electrons transferred. Given that the standard potential is 2.041 V and that two electrons are transferred, you will calculate the equilibrium constant. The program will be used to make the calculations.

Go to Appendix C. If you are using a TI-83 Plus, you can download the program NERNST and data and run the application as directed. If you are using another calculator, your teacher will provide you with key-strokes and data sets to use. After you have run the program, answer the following questions.

a. What is the equilibrium constant when the standard potential is 0.099?

b. What is the equilibrium constant when the standard potential is 1.125?

c. What is the equilibrium constant when the standard potential is 2.500?

638

Chapter Resource File

• Chapter Test

STANDARDIZED TEST PREP **17**

STANDARDIZED
TEST PREP **17**

Answers

1. b
2. b
3. d
4. d
5. a
6. c
7. c
8. b
9. d
10. c
11. d

1. In the following reaction, which species is reduced?

$$2K + Br_2 \longrightarrow 2K^+ + 2Br^-$$

a. K only
b. Br_2 only
c. both K and Br_2
d. neither K nor Br_2

2. The electrode at which reduction occurs is
a. always the anode.
b. always the cathode.
c. either the anode or the cathode.
d. always the half-cell.

3. In a galvanic cell, the anode
a. can be either positive or negative.
b. is positive.
c. is not charged.
d. is negative.

4. A cell contains two electrodes. The first electrode is a strip of zinc metal in a solution containing zinc ions. The second is a strip of copper metal in a solution containing copper ions. When this cell operates as a galvanic cell,
a. Cu is oxidized and Zn^{2+} is reduced.
b. Cu is reduced and Zn^{2+} is oxidized.
c. Cu^{2+} is oxidized and Zn is reduced.
d. Cu^{2+} is reduced and Zn is oxidized.

5. When water is electrolyzed, oxygen gas is formed at
a. the anode only.
b. the cathode only.
c. the midpoint between the anode and the cathode.
d. both the anode and the cathode.

6. When sulfuric acid, H_2SO_4, is used in the electrolysis of water, its purpose is to
a. react with the water.
b. keep the electrode clean.
c. provide adequate conductivity.
d. supply energy.

7. During the electrolysis of molten $CaBr_2$, which of the following would you expect to be made at the anode?
a. H_2
b. Ca
c. Br_2
d. O_2

8. When a rechargeable cell is being recharged, the cell acts as a(n)
a. fuel cell.
b. electrolytic cell.
c. galvanic cell.
d. Leclanché cell.

9. When silver is electroplated onto another metal, Ag^+ is
a. oxidized at the anode.
b. reduced at the anode.
c. oxidized at the cathode.
d. reduced at the cathode.

10. The oxidation number of the sulfur atom in the SO_4^{2-} ion is
a. +2.
b. −2.
c. +6.
d. +4.

11. A half-reaction
a. involves electrons.
b. is useful in balancing equations.
c. could occur at an electrode.
d. All of the above

639

Answers to Section 1 Review, *continued*

  c. Na = +1; H = +1; C = +4; O = –2

  d. Na = +1; Bi = +5; O = –2

8. CrO_3, +6; CrO, +2; Cr(s), 0; CrO_2, +4; Cr_2O_3, +3; $Cr_2O_7^{2-}$, +6; CrO_4^{2-}, +6

9. a. not redox

  b. redox; O is oxidized (–2 to 0), so NO_3^- is the reducing agent. N is reduced (+5 to +3), so NO_3^- is the oxidizing agent.

  c. redox; H is oxidized (0 to +1), so H_2 is the reducing agent. Cu is reduced (+2 to 0), so Cu^{2+} is the oxidizing agent.

  d. not redox

  e. redox; H is oxidized (0 to +1), so H_2 is the reducing agent. Cl is reduced (0 to –1), so Cl_2 is the oxidizing agent.

  f. not redox

10. a. half-reactions: $6Cl^-(aq) \rightarrow 3Cl_2(g) + 6e^-$ and $Cr_2O_7^{2-}(aq) + 14H_3O^+(aq) + 6e^- \rightarrow 2Cr^{3+}(aq) + 21H_2O(l)$; overall: $6Cl^-(aq) + Cr_2O_7^{2-}(aq) + 14H_3O^+(aq) \rightarrow 3Cl_2(g) + 2Cr^{3+}(aq) + 21H_2O(l)$

  b. half-reactions: $Cu(s) \rightarrow Cu^{2+}(aq) + 2e^-$ and $2Ag^+(aq) + 2e^- \rightarrow 2Ag(s)$; overall: $Cu(s) + 2Ag^+(aq) \rightarrow Cu^{2+}(aq) + 2Ag(s)$

  c. half-reactions: $Br_2(l) + 2e^- \rightarrow 2Br^-(aq)$ and $2I^-(aq) \rightarrow I_2(s) + 2e^-$; overall: $Br_2(l) + 2I^-(aq) \rightarrow 2Br^-(aq) + I_2(s)$

  d. half-reactions: $2I^-(aq) \rightarrow I_2(s) + 2e^-$ and $4H_3O^+(aq) + 2NO_2^-(aq) + 2e^- \rightarrow 2NO(g) + 6H_2O(l)$; overall: $2I^-(aq) + 2NO_2^-(aq) + 4H_3O^+(aq) \rightarrow I_2(s) + 2NO(g) + 6H_2O(l)$

11. a. Cl is oxidized, so Cl^- is the reducing agent. Cr is reduced, so $Cr_2O_7^{2-}$ is the oxidizing agent.

  b. Cu is oxidized, so Cu is the reducing agent. Ag is reduced, so Ag^+ is the oxidizing agent.

  c. I is oxidized, so I^- is the reducing agent. Br is reduced, so Br_2 is the oxidizing agent.

  d. I is oxidized, so I^- is the reducing agent. N is reduced, so NO_2^- is the oxidizing agent.

12. The oxidation number of O in H_2O_2 is –1. Oxygen atoms may be reduced to –2 and oxidized to 0 during a single reaction.

Answers to Section 2 Review, *continued*

8. In an anode reaction, at least one atom must increase its oxidation number so that it becomes more positive. In a cathode reaction, at least one atom must decrease its oxidation number so that it becomes more negative.

9. The cell voltages would oppose each other, and the flashlight would not work.

10. The flashlight would probably work. The direction of the moving electrons would be reversed.

11. The electrode is a cathode. An oxygen atom starting with an oxidation number of 0 is likely to change to an oxidation number of –2. This requires a gain of electrons and would be a reduction, which occurs at a cathode.

12. $2Br^-(aq) \rightarrow Br_2(l) + 2e^-$; at an anode

13. $Sn^{4+}(aq) + 2e^- \rightarrow Sn^{2+}(aq)$; at a cathode

14. cathode; Depositing silver onto a bracelet requires that the reaction $Ag^+(aq) + e^- \rightarrow Ag(s)$ takes place on the bracelet. This is a reduction reaction which must happen at a cathode.

15. The reactions would be the opposites of those discussed in the text: $Zn^{2+}(aq) + 2e^- \rightarrow Zn(s)$ and $Cu(s) \rightarrow 2e^- + Cu^{2+}(aq)$.

16. Both equations represent either an oxidation or a reduction and include electrons.

17. The reactions would be $Cu(s) \rightarrow 2e^- + Cu^{2+}(aq)$ at the anode and $Cu^{2+}(aq) + 2e^- \rightarrow Cu(s)$ at the cathode. The result would be the transfer of copper from the anode to the cathode.

18. yes; The electrode reactions correspond to two half-reactions that combine to form an overall redox reaction.

Answers to Section 3 Review, *continued*

7. $Pb(s) + PbO_2(s) + 2H_3O^+(aq) + 2HSO_4^-(aq) \rightarrow 2PbSO_4(s) + 4H_2O(l)$. Sulfuric acid is consumed and water is produced, so the sulfuric acid concentration decreases as the battery is discharged. As the concentration decreases, the density of the solution also decreases. Thus, you can use density to determine whether the battery is fully charged.

8. Using a sacrificial anode is a way to prevent corrosion because the anode is connected to another metal which is usually an important structural component. This metal is protected from oxidation as the anode gives up electrons to the structural metal because the sacrificial metal oxidizes more easily.

Answers to Section 4 Review, *continued*

11. a. $AlCl_3 + 3K \rightarrow Al + 3KCl$

  b. The reaction is a redox reaction because atoms of Al and of K change oxidation numbers.

  c. The reaction would not need to happen in an electrolytic cell because potassium is more active than aluminum and should replace it.

12. Sodium is more active than aluminum is and is much more difficult to reduce than aluminum is.

13. To deposit zinc on a bracelet, the zinc must be dissolved in solution and the copper must be in solid form. So, the reverse of the reaction shown is needed. If the reaction happens naturally in the direction shown, then an electric current is needed to cause the reverse reaction to happen.

14. A galvanic cell generates electrical energy, so one can be used in an electrolytic cell as the power source driving the reaction.

15. The production of aluminum from ore requires much more energy than is needed to process recycled aluminum. Aluminum production also depletes natural ore resources.

75. 0

76. H = +1; S = +4; O = –2

77. $Fe^{3+}(aq) + e^- \rightarrow Fe^{2+}(aq)$; reduction

78. half-reactions: $Cr_2O_7^{2-}(aq) + 14H_3O^+(aq) + 6e^- \rightarrow 2Cr^{3+}(aq) + 21H_2O(l)$ and $6Fe^{2+}(aq) + 6e^- \rightarrow 6Fe^{3+}(aq)$; overall: $Cr_2O_7^{2-}(aq) + 6Fe^{2+}(aq) + 14H_3O^+(aq) \rightarrow 2Cr^{3+}(aq) + 6Fe^{3+}(aq) + 21H_2O(l)$

79. (–0.4030 V) – (+1.358 V) = –1.761 V

80. (+0.222 V) – (+1.358 V) = –1.136 V

Critical Thinking

81. When something reacts with oxygen gas, the oxygen atoms generally gain electrons to become reduced. The substance that reacted with oxygen is oxidized. Because reactions with oxygen were known long before the ideas of electrochemistry, the term *oxidation* was already in use and was expanded to include any change in which electrons were lost.

82. For $(NH_4)(NO_2)$, N = –3 and H = +1 in NH_4^+ and N = +3 and O = –2 in NO_2^-. For $N_2H_4O_2$, N = 0, H = +1, and O = –2. The first assignment makes more sense because it uses the fact that the nitrogen atoms belong to different polyatomic ions and therefore must satisfy the rule which states that the sum of the oxidation numbers of atoms in an ion must equal the charge of the ion.

83. The negative terminal is the anode. Oxidation occurs there, so free electrons are collected at the negative terminal and cause a higher "electron pressure."

84. Electric current in metals is a result of moving electrons. Electric current in electrolyte solutions is a result of moving ions.

85. Fluorine has the most positive value for the standard electrode potential This means that fluorine gains electrons and is reduced more easily than other substances. The standard electrode potentials for other halogens become less positive in the order of chlorine, bromine, and iodine. So as you move down the group on the periodic table, the halogens become less likely to be reduced and more likely to be oxidized to their elemental forms.

86. Six electrons are transferred per CH_3OH molecule in the anode reaction $CH_3OH(aq) + 7H_2O(l) \rightarrow 6e^- + CO_2(g) + 6H_3O^+(aq)$.

87. (+0.7996 V) – (+0.3419 V) = +0.4577 V; Electrons would travel from Cu to Ag. Copper is the anode.

88. Electrolytic cells are closer to being the opposite of batteries because the materials for the cell are contained within the cell and can eventually be used up.

89. In the reactants, Mn = +2 in Mn^{2+} and Mn = +7 in MnO_4^-. In the products, Mn = +3 in $Mn(H_2P_4O_7)_3^{3-}$.

Alternative Assessment

90. Answers will vary. Plans should involve making the metal an electrode in a cell along with a second electrode of known $\Delta E°$. The resulting cell voltage and the direction of current are the needed data. Students should note that electrode potentials must be compared at 25°C and at a 1.000 M concentration of the respective ions.

91. Student presentations should clearly show the electrode reactions, the products formed, and the nature of the starting materials. The sodium must be in the liquid state, so operating temperatures are around 200°C. Students should explain why such a cell cannot work in an aqueous medium.

92. Answers will vary. Students should note that many devices that use rechargeable batteries would be too expensive to use with nonrechargeable batteries or that the user would have to stay near an electrical outlet. Video camcorders are a good example of devices dependent on rechargeable batteries.

93. Answers will vary. Students should consider several factors, including the various types of batteries used and their typical lifetime in each application.

94. Answers will vary.

95. Answers will vary.

96. Answers will vary.

97. Answers will vary. Check the design of student investigations before they begin. Students should realize that the device used to test the batteries must be the same for each trial and that there is a possibility that different batteries might have been manufactured at different times.

98. Answers will vary depending on the types of nails used.

99. Answers will vary. Students should be able to find items that are coated with precious metals to improve their appearance and items that have been coated to prevent corrosion.

Concept Mapping

100. a. electrochemical cell

    b. electrodes

    c. anode

    d. oxidation

    e. cathode

    f. reduction

Focus on Graphing

101. silver: about 20.1 g/h; gold: about 12.2 g/h

102. silver: about 0.19 mol/h; gold: about 0.062 mol/h

103. about 5.3 h

104. $Ag^+ + e^- \rightarrow Ag$; $Au^{3+} + 3e^- \rightarrow Au$; Producing 1 mol of gold requires three times more electrons as producing 1 mol of silver does. Therefore, the rate of production of gold (in mol/h) is one-third the rate of production of silver (in mol/h).

Technology and Learning

112. a. 2211.0298

    b. 1.0157×10^{38}

    c. 2.8804×10^{84}

Nuclear Chemistry
Chapter Planning Guide

PACING	CLASSROOM RESOURCES	LABS, ACTIVITIES, AND DEMONSTRATIONS
BLOCK 1 · 45 min pp. 640–641 **Chapter Opener**		**SE** **Start-Up Activity** Half-Lives and Pennies, p. 641 ♦
BLOCK 2 · 45 min pp. 642–647 **Section 1** Atomic Nuclei and Nuclear Stability	**OSP** **Supplemental Reading** Chapter 7 of Linus Pauling's *In His Own Words,* "Atomic Politics" **ADVANCED**	**TE** **Demonstration,** p. 643 **TE** **Group Activity,** p. 643 **ADVANCED**
BLOCKS 3 & 4 · 90 min pp. 648–657 **Section 2** Nuclear Change	**TT** **Model of a Pressurized, Light Water Nuclear Reactor** *	**TE** **Demonstration,** p. 649 ♦ **TE** **Group Activity,** p. 650 **BASIC** **TE** **Demonstration,** p. 654
BLOCKS 5 & 6 · 90 min pp. 658–667 **Section 3** Uses of Nuclear Chemistry	**OSP** **Career Extension** Archaeologist **GENERAL** **TT** **Rate of Decay of Potassium-40** * **SE** **Element Spotlight** Hydrogen is an Element unto Itself	**TE** **Demonstration,** p. 658 ♦ **TE** **Demonstration** Ionizing Smoke Detectors, p. 663 ♦ **CRF** **Observation Lab** Radioactivity ♦ **GENERAL** **CRF** **Observation Lab** Detecting Radioactivity ♦ **GENERAL**

BLOCKS 7 & 8 · 90 min **Chapter Review and Assessment Resources**

- **SE** **Chapter Review,** pp. 669–674
- **SE** **Standardized Test Prep,** p. 675
- **CRF** **Chapter Test** *
- **OSP** **Test Generator**
- **CRF** **Test Item Listing** *
- **OSP** **Scoring Rubrics and Classroom Management Checklists**

Holt Chemistry: Online Resources

Visit **go.hrw.com** for a variety of free resources related to this textbook. Enter the keyword **HW4 HOME**.

Holt Online Learning

Students can access interactive problem solving help and active visual concept development with the *Holt Chemistry* Online Edition available at **www.hrw.com**

student CNN News

cnnstudentnews.com

Find the latest chemistry news, lesson plans, and activities related to important scientific events.

Compression guide:
To shorten your instruction
because of time limitations,
omit blocks 1, 5, 6, and 7.

KEY		
TE	Teacher Edition	CRF Chapter Resource File	* Also on One-Stop Planner
SE	Student Edition	TT Teaching Transparencies	◆ Requires Advance Prep
OSP	One-Stop Planner		

PROBLEM SOLVING AND PRACTICE	SECTION REVIEW AND ASSESSMENT	STANDARDS CORRELATION
		National Science Education Standards
	TE **Homework,** p. 644 ADVANCED TE **Quiz,** p. 647 GENERAL CRF **Quiz** * TE **Reteaching,** p. 647 BASIC SE **Section Review,** p. 647 CRF **Concept Review** *	PS 1c PS 1d PS 5a
SE **Skills Toolkit 1** Balancing Nuclear Equations, p. 652 SE **Sample Problem A** Balancing a Nuclear Equation, p. 653 SE **Practice,** p. 653 GENERAL TE **Homework,** p. 653 GENERAL SE **Problem Bank,** p. 858 GENERAL	TE **Homework,** p. 650 BASIC TE **Quiz,** p. 657 GENERAL CRF **Quiz** * TE **Reteaching,** p. 657 BASIC SE **Section Review,** p. 657 CRF **Concept Review** *	PS 1c PS 1d PS 5a
SE **Sample Problem B** Determining the Age of an Artifact or Sample, p. 660 SE **Practice,** p. 660 GENERAL TE **Homework,** p. 660 GENERAL SE **Sample Problem C** Determining the Original Mass of a Sample, p. 662 SE **Practice,** p. 662 GENERAL TE **Homework,** p. 662 GENERAL SE **Problem Bank,** p. 858 GENERAL	TE **Quiz,** p. 666 GENERAL CRF **Quiz** * TE **Reteaching,** p. 666 BASIC SE **Section Review,** p. 666 CRF **Concept Review** *	PS 1d PS 5a

www.scilinks.org

Topic: Nuclear Power
SciLinks code: HW4087

Topic: Nuclear Reactors
SciLinks code: HW4089

Topic: Radioactive Decay
SciLinks code: HW4106

Topic: Radioactive Emissions
SciLinks code: HW4107

Topic: Nuclear Reactions
SciLinks code: HW4088

Topic: Fission
SciLinks code: HW4085

Topic: Nuclear Fusion
SciLinks code: HW4086

Topic: Nuclear Energy
SciLinks code: HW4084

Topic: Radioactive Dating
SciLinks code: HW4105

Topic: Discovering Radioactivity
SciLinks code: HW4150

Topic: Hydrogen
SciLinks code: 4155

Technology Resources

 **Science in the NEWS**

Each video segment is accompanied by a Critical Thinking Worksheet.

Segment 31
Nuclear Waste

Segment 32
Radioisotopes in Medicine

Overview

This chapter explores the area of nuclear chemistry. In Section 1, students learn about the binding energy of nuclei and how it is related to nuclear stability. Rules for predicting nuclear stability are also given. In Section 2, students learn about different types of nuclear change. Students also learn about the particles and electromagnetic waves that are produced by radioactive decay and how to balance nuclear equations. Section 3 describes the applications of nuclear chemistry, such as radioactive dating and nuclear medicine.

Assessing Prior Knowledge

Check for Content Knowledge

Students should be familiar with the following topics:

• proton

• neutron

• atomic number

• mass number

• isotope

Using the Figure

The photograph on this page shows a nuclear submarine. Nuclear submarines contain compact nuclear reactors that act as heat engines. Inside the reactor, uranium is split into smaller atoms through nuclear fission. The fission reactions generate a large amount of energy. This energy is used to generate steam that turns turbines. Propulsion turbines turn the submarine's propellers to move the vessel through water. Turbine generators generate electrical energy that powers the auxiliary systems, such as oxygen generation and climate control, on the submarine.

CHAPTER
18
NUCLEAR CHEMISTRY

640

Standards Correlations

National Science Education Standards

PS 1c: The nuclear forces that hold the nucleus of an atom together, at nuclear distances, are usually stronger than the electric forces that would make it fly apart. Nuclear reactions convert a fraction of the mass of interacting particles into energy, and they can release much greater amounts of energy than atomic interactions. Fission is the splitting of a large nucleus into smaller pieces. Fusion is the joining of two nuclei at extremely high temperature and pressure, and is the process responsible for the energy of the sun and other stars. (Sections 1 and 2)

PS 1d: Radioactive isotopes are unstable and undergo spontaneous nuclear reactions, emitting particles and/or wavelike radiation. The decay of any one nucleus cannot be predicted, but a large group of identical nuclei decay at a predictable rate. This predictability can be used to estimate the age of materials that contain radioactive isotopes. (Sections 1, 2, 3)

PS 5a: The total energy of the universe is constant. Energy can be transferred by collisions in chemical and nuclear reactions, by light waves and other radiations, and in many other ways. However, it can never be destroyed. As these transfers occur, the matter involved becomes steadily less ordered. (Sections 1, 2, 3)

Before nuclear power was used, submarines could stay submerged for only a brief period of time. A diesel-powered submarine had to surface regularly to recharge its batteries and refuel. But with a lump of nuclear fuel about the size of a golf ball, the first nuclear-powered submarine could remain underwater for months and travel about 97 000 km (about 60 000 mi). Today, nuclear power enables submarines to refuel only once every nine years.

START-UP ACTIVITY

Half-Lives and Pennies

SAFETY PRECAUTIONS

PROCEDURE

1. Make a data table with two columns. Label the first column "Trials." Label the second column "Number of pennies." Count the **pennies** your teacher has given you, and record this number in the table. Also, record "0" in the column labeled "Trials."

2. Place the pennies in a **plastic cup.** Cover the cup with one hand, and gently shake it for several seconds.

3. Pour the pennies on your desk or laboratory table. Remove all the pennies that are heads up. Count the remaining pennies, and record this number in column two. In the first column, record the number of times you performed step 2.

4. Repeat steps 2 and 3 until you have no pennies to place in your cup.

5. Plot your data on **graph paper.** Label the x-axis "Trial," and label the y-axis "Number of pennies."

ANALYSIS

1. What does your graph look like?

2. Describe any trend that your data display.

Pre-Reading Questions

① **What particles make up an atom?**

② **Name some types of radiation that compose the electromagnetic spectrum.**

③ **Can energy be created? Explain.**

④ **What quantities are conserved in a chemical reaction?**

CONTENTS 18

internet connect

www.scilinks.org
Topic: Nuclear Power
SciLinks code: HW4087

SCI LINKS. Maintained by the National Science Teachers Association

internet connect

www.scilinks.org
Topic : Nuclear Reactors
SciLinks code: HW4089

SCI LINKS. Maintained by the National Science Teachers Association

START-UP ACTIVITY

Skills Acquired:
• Experimenting
• Collecting Data
• Interpreting

Materials:
For each group of 2–3 students:
• pennies (at least 20 pennies)
• plastic cup
• graph paper

Safety Caution: Students should wear safety goggles during this activity. You may use candy, buttons, or any other small safe items that have two distinct sides instead of pennies.

Answers

1. Students' answers may vary. Students' graphs should reflect that the amount of pennies decreases by about half each time a trial is performed.

2. Students' answers may vary. Students should observe a graph with an inverse relationship—as the trial number increases, the number of pennies decreases.

Answers to Pre-Reading Questions

1. protons, neutrons, and electrons

2. Students' answers may vary. Sample answer: gamma, X-ray, ultraviolet, visible, infrared, microwave, and radiowave

3. No (However, students will learn in this chapter that mass can be transformed into energy during the formation of nuclei.) The law of conservation of energy states that energy cannot be created or destroyed but can be changed from one form to another.

4. mass and energy (However, students will learn in this chapter that not all reactions conserve mass and energy.)

Focus

Overview

Before beginning this section, review with your students the Objectives listed in the Student Edition. Section 1 includes a review of subatomic particles and the definitions of the terms *nucleon* and *nuclide*. Students also learn about the relationship between the strong force and electrostatic repulsion within the nucleus and how the nucleus is held together. Binding energy is defined, and its relationship to nuclear stability is explained. Finally, students learn about the band of stability for nuclei and learn some rules to predict nuclear stability.

🔊 Bellringer

Have students draw a picture of a helium atom and label the particles that make it up. Ask them to give the mass number and atomic number for their atoms. **Ans.** The drawings should show two protons and two neutrons forming the nucleus and an electron cloud appearing around the nucleus. The mass number for helium is four, and the atomic number is two.

Motivate

Identifying Preconceptions

Ask students what particles make up an atom. Most students will answer protons, neutrons, and electrons. Then ask if these particles can be broken down into smaller particles. Most students will answer no. Explain to students that protons and neutrons are made of smaller particles called *quarks*.

Atomic Nuclei and Nuclear Stability

KEY TERMS
- **nucleons**
- **nuclide**
- **strong force**
- **mass defect**

OBJECTIVES

1. **Describe** how the strong force attracts nucleons.
2. **Relate** binding energy and mass defect.
3. **Predict** the stability of a nucleus by considering factors such as nuclear size, binding energy, and the ratio of neutrons to protons in the nucleus.

Nuclear Forces

In 1911, Ernest Rutherford's famous gold-foil experiment determined the distribution of charge and mass in an atom. Rutherford's results showed that all of an atom's positive charge and almost all of its mass are contained in an extremely small nucleus.

Other scientists later determined more details about the nuclei of atoms. Atomic nuclei are composed of protons. The nuclei of all atoms except hydrogen also are composed of neutrons. The number of protons is the atomic number, Z, and the total number of protons and neutrons is the mass number, A. The general symbol for the nucleus of an atom of element X is shown in **Figure 1.**

The protons and neutrons of a nucleus are called **nucleons.** A **nuclide** is a general term applied to a specific nucleus with a given number of protons and neutrons. Nuclides can be represented in two ways. One way, shown in **Figure 1,** shows an element's symbol with its atomic number and mass number. A second way is to represent the nuclide by writing the element's name followed by its mass number, such as radium-228 or einsteinium-253. It is not essential to include the atomic number when showing a nuclide because all nuclides of an element have the same atomic number.

Recall that isotopes are atoms that have the same atomic number but different mass numbers. So, isotopes are nuclides that have the same number of protons but different numbers of neutrons. The following symbols represent nuclei of isotopes of tellurium.

$$^{122}_{52}\text{Te} \quad ^{124}_{52}\text{Te} \quad ^{128}_{52}\text{Te}$$

These three isotopes of tellurium are stable. So, their nuclei do not break down spontaneously. Yet, each of these nuclei are composed of 52 protons. How can these positive charges exist so close together? Protons repel each other because of their like charges. So, why don't nuclei fall apart? There must be some attraction in the nucleus that is stronger than the repulsion due to the positive charges on protons.

Topic Link

Refer to the "Atoms and Moles" chapter for a discussion of Rutherford's experiment.

nucleon
| a proton or a neutron

nuclide
| an atom that is identified by the number of protons and neutrons in its nucleus

Figure 1
In this figure, X represents the element, Z represents the atom's atomic number, and A represents the element's mass number.

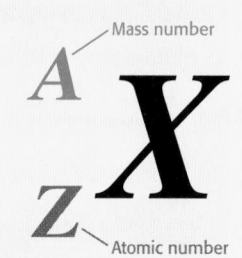

Mass number — A

X

Z — Atomic number

642

did you know?

The term *isobar* is applied to two or more nuclei that have the same mass number, A, but different atomic numbers, Z. The following nuclei are isobars:

$$^{124}_{50}\text{Sn}, \ ^{124}_{51}\text{Sb}, \ ^{124}_{52}\text{Te}, \ ^{124}_{53}\text{I}, \ ^{124}_{54}\text{Xe}, \ ^{124}_{55}\text{Cs}$$

Point out that at least one of any two neighboring isobars is unstable. For example, in the group above, Sb-124, I-124, and Cs-124 are radioactive, and the other three are stable. The term *isotone* is applied to two or more nuclei that have the same number of neutrons but different mass numbers, such as $^{30}_{14}\text{Si}$ and $^{31}_{15}\text{P}$.

The Strong Force Holds the Nucleus Together

In 1935, the Japanese physicist Hideki Yukawa proposed that a force between protons that is stronger than the electrostatic repulsion can exist between protons. Later research showed a similar attraction between two neutrons and between a proton and a neutron. This force is called the **strong force** and is exerted by nucleons only when they are very close to each other. All the protons and neutrons of a stable nucleus are held together by this strong force.

Although the strong force is much stronger than electrostatic repulsion, the strong force acts only over very short distances. Examine the nuclei shown in **Figure 2**. The nucleons are close enough for each nucleon to attract all the others by the strong force. In larger nuclei, some nucleons are too far apart to attract each other by the strong force. Although forces due to charges are weaker, they can act over greater distances. If the repulsion due to charges is not balanced by the strong force in a nucleus, the nucleus will break apart.

Protons and Neutrons Are Made of Quarks

In the early 1800s, John Dalton suggested that atoms could not be broken down. However, the discovery of electrons, protons, and neutrons showed that this part of his atomic theory is not correct. So, scientists changed the atomic theory to state that these subatomic particles were indivisible and were the basic building blocks of all matter. However, the atomic theory had to change again when scientists discovered in the 1960s that protons and neutrons are made of even smaller particles called *quarks,* as shown in **Figure 3**.

Quarks were first identified by observing the products formed in high-energy nuclear collisions. Six types of quarks are recognized. Each quark type is known as a flavor. The six flavors are up, down, top, bottom, strange, and charm. Only two of these—the up and down quarks—compose protons and neutrons. A proton is made up of two up quarks and one down quark, while a neutron consists of one up quark and two down quarks. The other four types of quarks exist only in unstable particles that spontaneously break down during a fraction of a second.

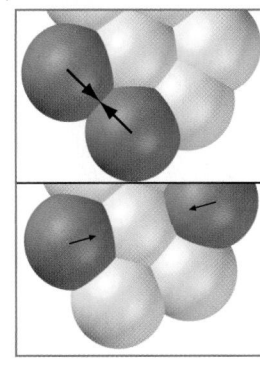

Figure 2
In the nucleus, the nuclear force acts only over a distance of a few nucleon diameters. Arrows describe magnitudes of the strong force acting on the protons.

strong force

the interaction that binds nucleons together in a nucleus

Topic Link

Refer to the "Atoms and Moles" chapter for a discussion of protons, neutrons, and Dalton's theory.

Figure 3
Protons and neutrons, which are made of quarks, make up nuclei.

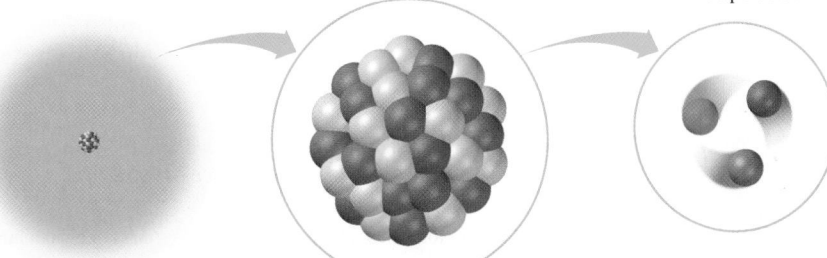

Atom, which has a nucleus

Nucleus, which is made of protons and neutrons

Quarks, which make up protons and neutrons

643

Demonstration

Use two bar magnets to model the difference between electrostatic attraction and electrostatic repulsion. Tell students to pretend that the north poles of the magnets are like positive charges and the south poles are like negative charges. Show students that positive charges and negative charges attract each other and explain how this attraction holds the electrons around the nucleus of an atom. Then, show students that two positive charges repel each other. Explain that this repulsion exists between the protons of a nucleus, but that the protons do not fly apart because of another force called the *strong force.*

Teach

READING SKILL BUILDER — BASIC

Assimilating Knowledge Have students write short lists of all the things they know about the forces that hold an atom's nucleus together, the role of neutrons in nuclear stability, radioactivity, nuclear fission and nuclear fusion, and applications of nuclear changes before students read this chapter. Then, have students contribute their entries to a group list on the board. Finally, have students list what they want to know about nuclear changes.

If students have trouble creating questions, use these suggestions:

• How do nuclear changes differ from chemical changes?

• What effect must protons of a nucleus have on one another?

• What keeps a nucleus from flying apart due to electrostatic repulsion? **LS** Interpersonal

Group Activity — ADVANCED

Have students work in small groups to research quarks and the history of quark theory. They should learn the properties of the six different quarks and where they can be found. The students should prepare a poster and present their findings to the class. **LS** Interpersonal

LANGUAGE-ARTS
CONNECTION

American physicist Murray Gell-Mann was the first to predict the existence of quarks and was the one who choose the name *quark.* He found the name in a nonsensical poem in James Joyce's novel *Finnegans Wake.* The poem begins with the line, "Three quarks for Muster Mark!" At the time, Gell-Mann thought there were only three particles so the name quark seemed appropriate. Furthermore, the idea of quarks seemed strange at the time so taking the name quark from a strange poem also seemed to fit.

Paired Reading Have each student read Section 1. Divide the class into pairs, and have students alternate with their partners in asking and in answering the following questions:

- How do isotopes of the same element differ from one another?
- Why do both repulsive and attractive forces act in a nucleus?
- How and why does the mass of a nucleus differ from the sum of the masses of its individual particles?
- Describe the trends in nuclear stability shown by the graph in **Figure 5.** **LS** Interpersonal

Homework ━━ **ADVANCED**

Calculating Binding Energy
Have students calculate the binding energy for carbon-14, phosphorus-33, and iodine-126. They should first calculate the mass defect for each isotope in amu, then convert the amu to kilograms, and use Einstein's mass-energy equation to find the binding energy in Joules.
LS Intrapersonal

Using the Figure ━━ **GENERAL**

Figure 4 Inform students that mass defect takes into account the total number of protons and neutrons. (In more advanced calculations, electrons are considered.) Ask students which isotope would have the greater mass defect, $^{122}_{52}$Te or $^{128}_{52}$Te. Ans. Te-122 has six fewer neutrons than Te-128. Therefore, Te-128 has the greater mass defect.
LS Logical

Binding Energy and Nuclear Stability

When protons and neutrons that are far apart come together and form a nucleus, energy is released. As a result, a nucleus is at a lower energy state than the separate nucleons were. A system is always more stable when it reaches a lower energy state. One way to describe this reaction is as follows:

$$\text{separate nucleons} \longrightarrow \text{nucleus} + \text{energy}$$

The energy released in this reaction is enormous compared with the energy changes that take place during chemical reactions. The energy released when nucleons come together is called *nuclear binding energy*. Where does this enormous quantity of energy come from? The answer can be found by comparing the total mass of the nucleons with the nucleus they form.

The mass of any atom is less than the combined masses of its separated parts. This difference in mass is known as the **mass defect**, also called *mass loss*. Electrons have masses so small that they can be left out of mass defect calculations. For helium, $^4_2$He, the mass of the nucleus is about 99.25% of the total mass of two protons and two neutrons. According to the equation $E = mc^2$, energy can be converted into mass, and mass can be converted into energy. So, a small quantity of mass is converted into an enormous quantity of energy when a nucleus forms.

mass defect

the difference between the mass of an atom and the sum of the masses of the atom's protons, neutrons, and electrons

Binding Energy Can Be Calculated for Each Nucleus

As **Figure 4** shows, the mass defect for one $^4_2$He nucleus is 0.0304 amu. The equation $E = mc^2$ can be used to calculate the binding energy for this nucleus. Remember to first convert the mass defect, which has units of amu to kilograms, to match the unit for energy which is joules (kg·m²/s²).

$$0.0304 \text{ amu} \times \frac{1.6605 \times 10^{-27} \text{ kg}}{1 \text{ amu}} = 5.05 \times 10^{-29} \text{ kg}$$

The binding energy for one $^4_2$He nucleus can now be calculated.

$$E = (5.05 \times 10^{-29} \text{ kg})(3.00 \times 10^8 \text{ m/s})^2 = 4.54 \times 10^{-12} \text{ J}$$

This quantity of energy may seem rather small, but remember that 4.54×10^{-12} J is released for every $^4_2$He nucleus that forms. The binding energy for 1 mol of $^4_2$He nuclei is much more significant.

$$4.54 \times 10^{-12} \frac{\text{J}}{\text{He nucleus}} \times 6.022 \times 10^{23} \frac{\text{He nuclei}}{\text{mol}} = 2.73 \times 10^{12} \text{ J/mol}$$

Figure 4
The mass defect represents the difference in mass between the helium nucleus and the total mass of the separated nucleons.

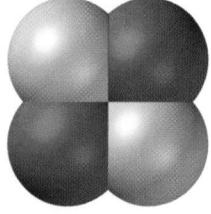

Helium nucleus

$^4_2$He nucleus = 2(mass of proton) + 2(mass of neutron)

= 2(1.007 276 47 amu) + 2(1.008 664 90 amu)

= 4.031 882 74 amu

mass defect = (total mass of separate nucleons) − (mass of helium nucleus)

= 4.031 882 74 amu − 4.001 474 92 amu

= 0.030 407 82 amu per nucleus of $^4_2$He

MISCONCEPTION /// **ALERT** \\\

The term *mass defect* sometimes is misinterpreted as an unexplained mistake in mass. In reality, it refers to mass that is unaccounted for when the mass of the nucleus is compared with the total mass of the individual nucleons. The fact that this mass is converted into energy is an indication that the nucleus is at a lower energy state than the collection of individual nucleons and is therefore more stable.

Binding Energy Is One Indicator of Nuclear Stability

A system's stability depends on the amount of energy released as the system is established. When 16 g of oxygen nuclei is formed, 1.23×10^{13} J of binding energy is released. This amount of energy is about equal to the energy needed to heat 4.6×10^6 L of liquid water from 0°C to 100°C and to boil the water away completely.

The binding energy of a selenium nucleus, $^{80}_{34}\text{Se}$, is much greater than that of an $^{16}_{8}\text{O}$ nucleus. Does this difference in energy mean that the $^{80}_{34}\text{Se}$ nucleus is more stable than the $^{16}_{8}\text{O}$ nucleus? Not necessarily. After all, $^{80}_{34}\text{Se}$ contains 64 more nucleons than $^{16}_{8}\text{O}$ does. To make a good comparison of these nuclei, you must look at the binding energy per nucleon. Examine the graph in **Figure 5.** Notice that the binding energy per nucleon rises rapidly among the lighter nuclei. The greater the binding energy per nucleon is, the more stable the nucleus is.

In the graph, the binding energy per nucleon levels off when the mass number is approximately 60. The curve reaches a maximum when the mass number is around 55. Therefore, the most stable nuclei are $^{56}_{26}\text{Fe}$ and $^{58}_{28}\text{Ni}$. These isotopes are relatively abundant in the universe in comparison to other heavy metals, and they are the major components of Earth's core.

Atoms that have larger mass numbers than $^{56}_{26}\text{Fe}$ and $^{58}_{28}\text{Ni}$ have nuclei too large to have larger binding energies per nucleon than these iron and nickel isotopes. In these cases, the net attractive force on a proton is reduced because of the increased distance of the neighboring protons. So, the repulsion between protons results in a decrease in the binding energy per nucleon. Nuclei that have mass numbers greater than 209 and atomic numbers greater than 83 are never stable.

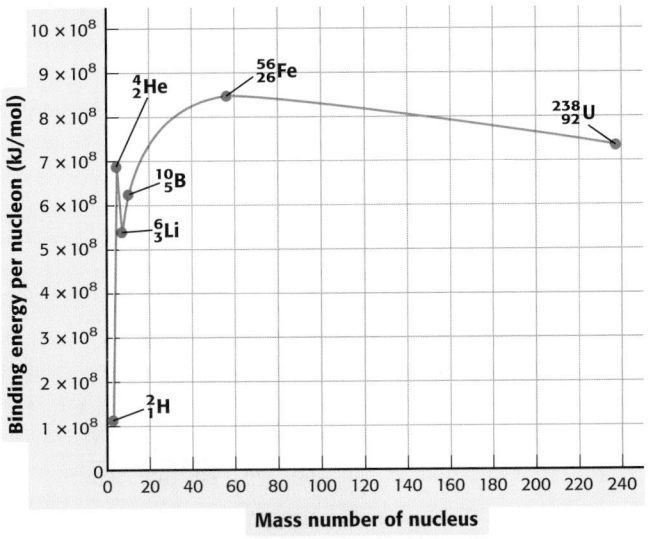

Relative Stability of Nuclei

Figure 5
This graph indicates the relative stability of nuclei. Isotopes that have a high binding energy per nucleon are more stable. The most stable nucleus is $^{56}_{26}\text{Fe}$.

Teaching Tips
Tell students to pay attention to the significant figures when doing mass defect calculations because the differences between two numbers may be very small. Remind students that exact values—such as the exact number of protons and neutrons of a nucleus—have an infinite number of significant figures. This rule explains why the numbers in **Figure 4** were not rounded to one significant figure.

Using the Figure
Graphing Emphasize that what is being shown in **Figure 5** is binding energy *per nucleon*, not total binding energy. The binding energies of elements that have mass numbers larger than Fe-56 increase but at a slower rate than the rate that the numbers of nucleons of those elements increase. Therefore, binding energy per nucleon for these elements decreases.

MISCONCEPTION ALERT

Some students may already realize that isotopes, whether naturally occurring or artificially produced, are classified simply as nonradioactive (stable) or radioactive (unstable). Students may not realize that nuclear stability can be affected by extreme temperatures and pressures. Inform students that atomic nuclei that are stable under the conditions of temperature and pressure found on Earth might not be stable under the extreme conditions found in stars, novae, and supernovae.

Teaching Tip

Isobar Stability Caution students that the rule regarding the stability of isobars should not be construed to mean that for any two neighboring isobars, one will be stable and the other will be unstable. They are likely to follow this rule if both are in the range of stability for nuclei of their size. However, both isobars will be unstable if both are outside the band of stability. For example, of the isobars O-16 and F-16, the oxygen nucleus is stable and the fluorine nucleus is unstable. On the other hand, of the isobars F-16 and Ne-16, both are unstable.

Predicting Nuclear Stability

Finding the binding energy per nucleon for an atom is one way to predict a nucleus's stability. Another way is to compare the number of neutrons with the number of protons of a nucleus. Examine the graph in **Figure 6.** The number of neutrons, N, is plotted against the number of protons, Z, of each stable nucleus. All known stable nuclei are shown as red dots.

The maroon line shows where the data lie for $N/Z = 1$. For elements that have small atomic numbers, the most stable nuclei are those for which $N/Z = 1$. Notice in **Figure 6** that the dots that represent elements that have small atomic numbers are clustered near the line that represents $N/Z = 1$. The green line shows where the data would lie for $N/Z = 1.5$. For elements that have large atomic numbers, the most stable nuclei are those where $N/Z = 1.5$. The reason for the larger N/Z number is that neutrons are needed to stabilize the nuclei of heavier atoms. Notice in **Figure 6** that the dots that represent elements with large atomic numbers are clustered near the line $N/Z = 1.5$.

The dots representing 256 known stable nuclei cluster over a range of neutron-proton ratios, which are referred to as a *band of stability*. This band of stability is shown in yellow in **Figure 6.**

Figure 6
The graph shows the ratio of protons to neutrons for 256 of the known stable nuclei.

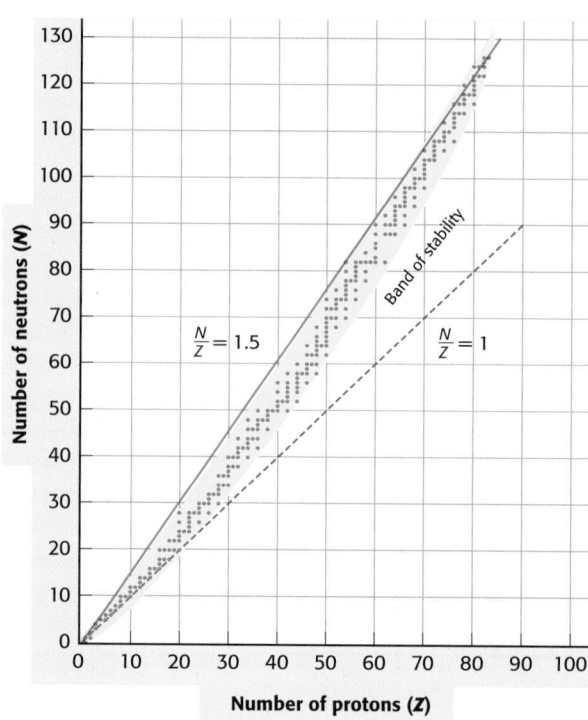

Neutron-Proton Ratios of Stable Nuclei

Some Rules to Help You Predict Nuclear Stability

You probably see that the graph in **Figure 6** shows several trends. The following rules for predicting nuclear stability are based on this graph.

1. Except for $_1^1$H and $_2^3$He, all stable nuclei have a number of neutrons that is equal to or greater than the number of protons.

2. A nucleus that has an *N/Z* number that is too large or too small is unstable. For small atoms, *N/Z* is very close to 1. As the nuclei get larger, this number increases gradually until the number is near 1.5 for the largest nuclei.

3. Nuclei with even numbers of neutrons and protons are more stable. Almost 60% of all stable nuclei have even numbers of protons and even numbers of neutrons.

4. Nuclei that have so-called magic numbers of protons and neutrons tend to be more stable than others. These numbers—2, 8, 20, 28, 50, 82, and 126—apply to the number of protons or the number of neutrons. Notice in **Figure 5** the large binding energy of $_2^4$He. This nucleus is very small and has two protons and two neutrons. Such "extra stability" also is true of the element calcium, which has six stable isotopes that range from $_{20}^{40}$Ca to $_{20}^{48}$Ca, all of which have 20 protons. Tin, having the magic number of 50 protons, has 10 stable isotopes, the largest number of any element. The heaviest stable element, bismuth, having only one stable isotope, has the magic number of 126 neutrons in $_{83}^{209}$Bi.

5. No atoms that have atomic numbers larger than 83 and mass numbers larger than 209 are stable. The nuclei of these atoms are too large to be stable.

 ## Section Review

UNDERSTANDING KEY IDEAS

1. What are the nucleons of an atom?

2. What role does the strong force play in the structure of an atom?

3. What is the band of stability?

4. What is mass defect?

5. Explain what happens to the mass that is lost when a nucleus forms.

6. How do the nuclides $_8^{16}$O and $_8^{15}$O differ?

7. Why is bismuth, $_{83}^{209}$Bi, stable?

8. Which are more stable, nuclei that have an even number of nucleons or nuclei that have an odd number of nucleons?

CRITICAL THINKING

9. Which is generally more stable, a small nucleus or a large nucleus? Explain.

10. How does nuclear binding energy relate to the stability of an atom?

11. Which is expected to be more stable, $_3^6$Li or $_3^9$Li? Explain.

12. Use **Figure 6** and the rules for predicting nuclear stability to determine which of the following isotopes are stable and which are unstable.

a. $_{15}^{32}$P   **d.** $_{12}^{24}$Mg

b. $_6^{14}$C   **e.** $_{43}^{97}$Tc

c. $_{23}^{51}$V

Answers to Section Review

1. The nucleons are the protons and neutrons that make up the nucleus of an atom.

2. The strong force is greater than the repulsion of the electrostatic force between protons and binds the nucleus together.

3. The band of stability is the area on a graph showing the neutron/proton ratios for stable nuclei. For small nuclei, this ratio is near 1:1. For the largest nuclei, this ratio reaches about 1.5:1.

4. Mass defect is the difference between the sum of all the masses of nucleons and electrons that make up an atom and the actual mass of that atom.

5. The mass that is lost is converted into nuclear binding energy.

6. They are isotopes of the same element, oxygen. They have different mass numbers indicating that $^{16}_8$O has one more neutron than $^{15}_8$O.

7. This isotope contains 126 neutrons, which is one of the magic numbers.

8. Nuclei that have an even number of protons and neutrons are more stable than nuclei that have an odd number of protons and neutrons.

Answers continued on p. 675A

Close

Reteaching ——— BASIC

Two of the four fundamental forces were discussed in this chapter. To help students understand why the strong force has a greater effect on protons in a nucleus than the electrostatic force, have students make a table that compares the strengths and effective distances of the four forces. **LS** Logical

Quiz ——— GENERAL

1. What force keeps nucleons together? **Ans.** the strong force

2. Which equation should you use to calculate the binding energy of an atom? **Ans.** Einstein's mass-energy equation: $E=mc^2$

3. What is the mass defect of an atom? **Ans.** the difference in mass when a nucleus forms from separated nucleons

4. Explain the differences between the nuclei of carbon-13, carbon-14, and nitrogen-14. **Ans.** Carbon-13 and carbon-14 have different numbers of neutrons but the same number of protons. The atomic number of carbon-13 and carbon-14 is different from the atomic number of nitrogen-14.

5. To predict the stability of a nucleus you would look at the ratio between its _____ and its _____. **Ans.** protons, neutrons

LS Intrapersonal

Chapter Resource File

• Concept Review
• Quiz

Focus

Overview

Before beginning this section, review with your students the Objectives listed in the Student Edition. This section describes the different types of nuclear change. One type of nuclear change is radioactive decay. In radioactive decay, an unstable nucleus releases particles or electromagnetic waves in order to increase its stability. Another type of nuclear change is fission, which occurs when large unstable nuclei are split into two smaller nuclei. Still another type of nuclear change is fusion. Fusion occurs when two small nuclei combine to form a larger nucleus. In all nuclear reactions, the total mass, total energy, and total charge of the reactants and products must be balanced.

🎧 Bellringer

Ask students to write a short paragraph about nuclear power plants. They should discuss what they know about the power source, the benefits, safety issues, and concerns.

Motivate

Discussion ——— GENERAL

Have students debate the benefits and disadvantages of various fuel sources. Be sure that students include nuclear power in their discussion. You may want to have students research the fuel sources and formally debate.
LS Interpersonal

2 Nuclear Change

KEY TERMS

- **radioactivity**
- **beta particle**
- **gamma ray**
- **nuclear fission**
- **chain reaction**
- **critical mass**
- **nuclear fusion**

radioactivity

the process by which an unstable nucleus emits one or more particles or energy in the form of electromagnetic radiation

OBJECTIVES

① **Predict** the particles and electromagnetic waves produced by different types of radioactive decay, and write equations for nuclear decays.

② **Identify** examples of nuclear fission, and describe potential benefits and hazards of its use.

③ **Describe** nuclear fusion and its potential as an energy source.

Radioactive Decay

Nuclear changes can be easier to understand than chemical changes because only a few types of nuclear changes occur. One type is the spontaneous change of an unstable nucleus to form a more stable one. This change involves the release of particles, electromagnetic waves, or both and is generally called **radioactivity** or radioactive decay. Specifically, radioactivity is the spontaneous breakdown of unstable nuclei to produce particles or energy. **Table 1** summarizes the properties of both the particles and the energy released by radioactive decay.

Table 1   Characteristics of Nuclear Particles and Rays

Particle	Mass (amu)	Charge	Symbol	Stopped by
Proton	1.007 276 47	+1	$p, p^+, {}_{+1}^{1}p, {}_{1}^{1}\text{H}$	a few sheets of paper
Neutron	1.008 664 90	0	$n, n^0, {}_{0}^{1}n$	a few centimeters of lead
β particle (electron)	0.000 548 580	−1	$\beta, \beta^-, {}_{-1}^{0}e^*$	a few sheets of aluminum foil
Positron†	0.000 548 580	+1	$\beta^+, {}_{+1}^{0}e^*$	same as electron
α particle (He-4 nucleus)	4.001 474 92	+2	$\alpha, \alpha^{2+}, {}_{2}^{4}\text{He}$	skin or one sheet of paper
Gamma ray	0	0	γ	several centimeters of lead

*The superscript zero in the symbols for electron and positron does not mean that they have zero mass. It means their mass number is zero.

†The positron is the antiparticle of the electron. Each particle has an antiparticle, but only the positron is frequently involved in nuclear changes.

Chapter Resource File

- **Lesson Plan**

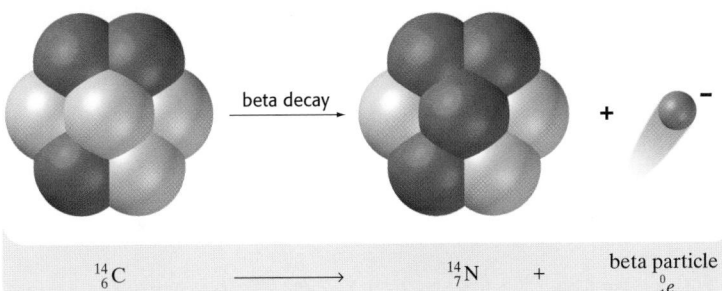

$$^{14}_{6}\text{C} \xrightarrow{} ^{14}_{7}\text{N} \quad + \quad \begin{array}{c}\text{beta particle}\\ ^{0}_{-1}e\end{array}$$

Stabilizing Nuclei by Converting Neutrons into Protons

Recall that the stability of a nucleus depends on the ratio of neutrons to protons, or the N/Z number. If a particular isotope has a large N/Z number or too many neutrons, the nucleus will decay and emit radiation.

A neutron in an unstable nucleus may emit a high-energy electron, called a **beta particle** (β particle), and change to a proton. This process is called *beta decay*. This process often occurs in unstable nuclei that have large N/Z numbers.

$$^{1}_{0}n \xrightarrow{\text{beta decay}} ^{1}_{+1}p + ^{0}_{-1}e$$

Because this process changes a neutron into a proton, the atomic number of the nucleus increases by one, as you can see in **Figure 7**. As a result of beta decay, carbon becomes a different element, nitrogen. However, the mass number does not change because the total number of nucleons does not change as shown by the following equation.

$$^{14}_{6}\text{C} \longrightarrow ^{14}_{7}\text{N} + ^{0}_{-1}e$$

Stabilizing Nuclei by Converting Protons into Neutrons

One way that a nucleus that has too many protons can become more stable is by a process called *electron capture*. In this process, the nucleus merely absorbs one of the atom's electrons, usually from the $1s$ orbital. This process changes a proton into a neutron and decreases the atomic number by one. The mass number stays the same.

$$^{1}_{+1}p + ^{0}_{-1}e \xrightarrow{\text{electron capture}} ^{1}_{0}n$$

A typical nucleus that decays by this process is chromium-51.

$$^{51}_{24}\text{Cr} + ^{0}_{-1}e \xrightarrow{\text{electron capture}} ^{51}_{23}\text{V} + \gamma$$

The final symbol in the equation, γ, indicates the release of **gamma rays.** Many nuclear changes leave a nucleus in an energetic or excited state. When the nucleus stabilizes, it releases energy in the form of gamma rays. **Figure 8** shows a thunderstorm during which gamma rays may also be produced.

beta particle

a charged electron emitted during a certain type of radioactive decay, such as beta decay

gamma ray

the high-energy photon emitted by a nucleus during fisson and radioactive decay

Figure 8
Thunderstorms may produce terrestrial gamma-ray flashes (TGFs).

Teach

Demonstration

Obtain a Geiger-Mueller counter and a safe beta source from a scientific supply company. Point the Geiger-Mueller counter at the beta source so students can hear what the counter does when it detects radiation. Place pieces of paper, aluminum foil, and plastic between the beta source and the Geiger-Mueller counter to show the effects of shielding. Explain that everyday objects may give off radiation, but the levels are low enough to be safe. Finally point the Geiger-Mueller counter to the air to demonstrate that low levels of background radiation are always present.

Teaching Tip

Beta Particles Make students aware that $^{0}_{-1}e$ is the symbol for a beta particle that is a high energy electron. These high-energy electrons result from the decay of a neutron into a proton. In this decay process, the total number of nucleons doesn't change, so the atomic number increases while the mass number stays the same.

Using the Figure

Figure 7 Stress that when a nucleus of an atom of one element changes into a nucleus of an atom of another element by radioactive decay, either a proton in the atomic nucleus changes into a neutron or a neutron in the nucleus changes into a proton. The total number of nucleons remains unchanged.

HISTORY CONNECTION

Marie Sklodowska Curie (1867–1934) and her husband, Pierre, discovered the elements radium and polonium, the latter of which was named for Marie's homeland, Poland (*Polonia* in Latin). She also determined that the only radioactive parts of uranium compounds and thorium compounds were the uranium and thorium atoms.

WORLD LANGUAGE CONNECTION

Ernest Rutherford named the three types of radioactive emission after the first three letters of the Greek alphabet—*alpha, beta,* and *gamma* (α, β, and γ, respectively).

READING SKILL BUILDER ——— BASIC

Paired Reading Pair each student with a partner. Have each student read silently Section 2 while marking confusing passages with a small sticky note. After they have finished reading, students in each pair should help each other with the parts they have difficulty understanding. **LS** Interpersonal

Group Activity ——— BASIC

Have students work in small groups to construct a model that can be used to show radioactive decays. Students can represent protons and neutrons with balls of modeling clay of two different colors. Electrons and positrons can be small, hard balls or beads of contrasting colors. Each group should use examples of nuclei that undergo alpha decay, beta decay, positron emission, and electron capture and then manipulate their models to show the changes. **LS** Interpersonal

Homework ——— BASIC

Have students make a table to organize the information about radioactive decays. The first column should contain the names: *alpha decay, beta decay, positron emission,* and *electron capture.* The second column should list the change in the atomic number of a nucleus. The third column should list the change in the mass number of a nucleus. The fourth column should list the particle or rays that are emitted during the reaction. The last column should show a sample reaction. **LS** Intrapersonal

Figure 9
Nuclei can release positrons to form new nuclei. Matter is then converted into energy when positrons and electrons collide and are converted into gamma rays.

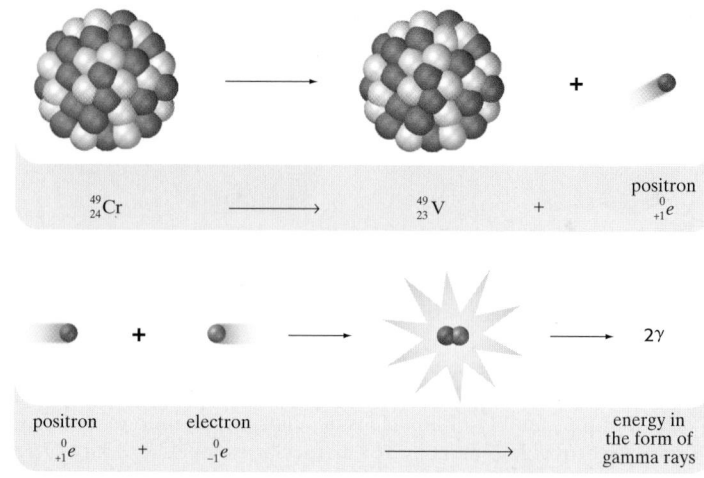

Topic Link

Refer to the "Atoms and Moles" chapter for a discussion of electromagnetic waves.

Gamma Rays Are Also Emitted in Positron Emission

Some nuclei that have too many protons can become stable by emitting positrons, which are the antiparticles of electrons. The process is similar to electron capture in that a proton is changed into a neutron. However, in *positron emission,* a proton emits a positron.

$$_{+1}^{1}p \xrightarrow{\text{positron emission}} _{0}^{1}n + _{+1}^{0}e$$

Notice that when a proton changes into a neutron by emitting a positron, the mass number stays the same, but the atomic number decreases by one. The isotope chromium-49 decays by this process, as shown by the model in **Figure 9**.

$$_{24}^{49}Cr \longrightarrow _{23}^{49}V + _{+1}^{0}e$$

Another example of an unstable nucleus that emits a positron is potassium-38, which changes into argon-38.

$$_{19}^{38}K \longrightarrow _{18}^{38}Ar + _{+1}^{0}e$$

The positron is the opposite of an electron. Unlike a beta particle, a positron seldom makes it into the surroundings. Instead, the positron usually collides with an electron, its antiparticle. Any time a particle collides with its antiparticle, all of the masses of the two particles are converted entirely into electromagnetic energy or gamma rays. This process is called *annihilation of matter,* which is illustrated in **Figure 9**.

$$_{-1}^{0}e + _{+1}^{0}e \xrightarrow{\text{annihilation}} 2\gamma$$

The gamma rays from electron-positron annihilation have a characteristic wavelength; therefore, these rays can be used to identify nuclei that decay by positron emission. Such gamma rays have been detected coming from the center of the Milky Way galaxy.

650

ASTRONOMY ——— CONNECTION

Astronomers use specialized telescopes to observe gamma rays that come from deep space. Gamma ray telescopes detect the production of electrons and positrons (pair production) when gamma rays from space interact with matter. The movement of the electrons and positrons tells astronomers where the gamma rays came from. Astronomers use gamma ray telescopes to study black holes, neutron stars, and supernovae. Gamma ray telescopes must be placed in space because most gamma rays are absorbed by Earth's atmosphere.

Stabilizing Nuclei by Losing Alpha Particles

An unstable nucleus that has an N/Z number that is much larger than 1 can decay by emitting an alpha particle. In addition, none of the elements that have atomic numbers greater than 83 and mass numbers greater than 209 have stable isotopes. So, many of these unstable isotopes decay by emitting alpha particles, as well as by electron capture or beta decay. Uranium-238 is one example.

$$^{238}_{92}\text{U} \xrightarrow{\text{alpha decay}} {}^{234}_{90}\text{Th} + {}^{4}_{2}\text{He}$$

Notice that the atomic number in the equation decreases by two while the mass number decreases by four. Alpha particles have very low penetrating ability because they are large and soon collide with other matter. Exposure to external sources of alpha radiation is usually harmless. However, if substances that undergo alpha decay are ingested or inhaled, the radiation can be quite damaging to the body's internal organs.

Many heavy nuclei go through a series of reactions called a decay series before they reach a stable state. The decay series for uranium-238 is shown in **Figure 10.** After the $^{238}_{92}\text{U}$ nucleus decays to $^{234}_{90}\text{Th}$, the nucleus is still unstable because it has a large N/Z number. This nucleus undergoes beta decay to produce $^{234}_{91}\text{Pa}$. By another beta decay, $^{234}_{91}\text{Pa}$ changes to $^{234}_{92}\text{U}$. After a number of other decays (taking millions of years), the nucleus finally becomes a stable isotope, $^{206}_{82}\text{Pb}$.

Topic Link

Refer to the "Atoms and Moles" chapter for a discussion of alpha particles.

Figure 10
Uranium-238 decays to lead-206 through a decay series.

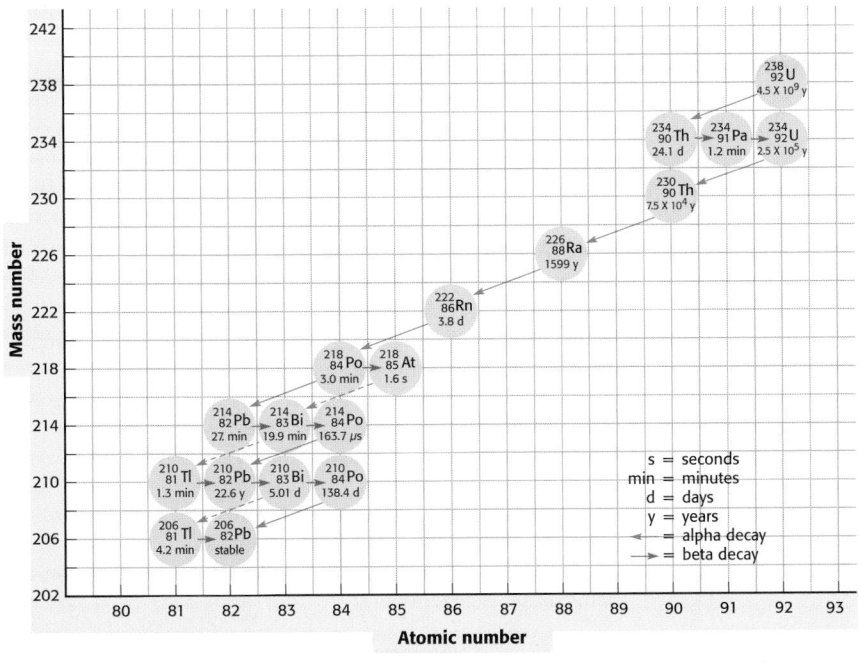

Uranium-238 Decay Series

Teaching Tip

Alpha Decay Point out to students that alpha decay is a common form of decay for elements that are more massive than bismuth. Usually, only the lightest isotopes of elements that have atomic numbers less than the atomic number of bismuth will undergo alpha decay. However, few examples of initial alpha decay exist for elements that have atomic numbers less than $Z = 72$.

Using the Figure

Figure 10 Stress that the U-238 nucleus does not decay directly into Pb-206. Draw students' attention to the alternative pathways of disintegration after the decay series reaches Po-218. Ask students how they could determine the form of decay of each isotope if there were no color-coded arrows. Ans. An alpha decay is indicated by a decrease of four in the mass number and a decrease of two in the atomic number. A beta decay is indicated by an increase of one in the atomic number and no change in the mass number. A decay series proceeds until a stable (nonradioactive) nucleus is achieved.

GEOLOGY CONNECTION

Uranium-238 is a radioactive element that is present in most rocks. Two products of the uranium decay series are radium-226 and radon-222. When radium-226 decays by alpha emission, the resulting radon-222 atom recoils and is pushed in the opposite direction as the alpha particle. The recoil pushes the radon atom through pores in the rock. Radon is a gas, so it can travel through pores and between grains of soil. Because of its ability to travel through the ground, radon can accumulate in basements of buildings.

Teaching Tip

Units of Radioactivity Marie and Pierre Curie, Henri Becquerel, and Wilhelm Conrad Roentgen were pioneers in radioactivity research, and they all have units of radiation measurement named after them.

- A *curie* (Ci) equals 3.7×10^{10} decays per second—approximately the activity of 1 gram of radium.

- The *becquerel* (Bq) is an SI unit and equals one disintegration per second.

- The *roentgen* is used to express exposure to radiation.

Teaching Tip

Natural Exposure to Radiation
After reading this chapter, some students may be concerned about overexposure to radiation. Tell students that humans are exposed to radiation from a variety of natural sources every day. Background radiation comes from uranium and thorium in rocks and minerals, cosmic rays from space, radon gas in the atmosphere, and radioactive elements in food and in the human body. Only 18 percent of the radiation a person receives during a year comes from artificial sources, such as medical X-rays and nuclear power plants. Have interested students calculate their yearly radiation exposure and compare it to the government's radiation standard. Questionnaires can be found on the Internet that will help students with their calculations. The government's standard can also be found on the Internet.

internet connect
www.scilinks.org
Topic: Nuclear Reactions
SciLinks code: HW4088
SCiLINKS Maintained by the National Science Teachers Association

Nuclear Equations Must Be Balanced

Look back at all of the nuclear equations that have appeared so far in this chapter. Notice that the sum of the mass numbers (superscripts) on one side of the equation always equals the sum of the mass numbers on the other side of the equation. Likewise, the sums of the atomic numbers (subscripts) on each side of the equation are equal. Look at the following nuclear equations, and notice that they balance in terms of both mass and nuclear charge.

$$^{238}_{92}\text{U} \longrightarrow {}^{234}_{90}\text{Th} + {}^{4}_{2}\text{He} \quad \begin{matrix}[238 = 234 + 4 \text{ mass balance}]\\ [92 = 90 + 2 \text{ charge balance}]\end{matrix}$$

$$^{234}_{90}\text{Th} \longrightarrow {}^{234}_{91}\text{Pa} + {}^{0}_{-1}e \quad \begin{matrix}[234 = 234 + 0 \text{ mass balance}]\\ [90 = 91 + (-1) \text{ charge balance}]\end{matrix}$$

Remember that whenever the atomic number changes, the identity of the element changes. In the above examples, uranium changes into thorium, and thorium changes into protactinium.

1 SKILLS Toolkit

Balancing Nuclear Equations

The following rules are helpful for balancing a nuclear equation and for identifying a reactant or a product in a nuclear reaction.

1. **Check mass and atomic numbers.**
   - The total of the mass numbers must be the same on both sides of the equation.
   - The total of the atomic numbers must be the same on both sides of the equation. In other words, the nuclear charges must balance.
   - If the atomic number of an element changes, the identity of the element also changes.

2. **Determine how nuclear reactions change mass and atomic numbers.**
   - If a beta particle, ${}^{0}_{-1}e$, is released, the mass number does not change but the atomic number increases by one.
   - If a positron, ${}^{0}_{+1}e$ is released, the mass number does not change but the atomic number decreases by one.
   - If a neutron, ${}^{1}_{0}n$, is released, the mass number decreases by one and the atomic number does not change.
   - Electron capture does not change the mass number but decreases the atomic number by one.
   - Emission of an alpha particle, ${}^{4}_{2}\text{He}$, decreases the mass number by four and decreases the atomic number by two.
   - When a positron and an electron collide, energy in the form of gamma rays is generated.

652

REAL-WORLD CONNECTION

Irradiating Mail After the terrorist attacks on the World Trade Center and the Pentagon on September 11, 2001, several letters containing anthrax spores were sent to U.S. Senators, a Florida newspaper, and two television studios. These events caused many people to worry about the spread of anthrax through the U.S.

Postal Service. As a result, the Postal Service looked into irradiating the mail with beta particles to kill any anthrax spores. However, anthrax is so hardy that an extremely high dose of radiation is needed to kill it. At present, the cost of irradiating all mail in the United States is much too high.

SAMPLE PROBLEM A

Balancing a Nuclear Equation

Identify the product formed when polonium-212 emits an alpha particle.

1 Gather information.
- Check the periodic table to write the symbol for polonium-212: $^{212}_{84}\text{Po}$.
- Write the symbol for an alpha particle: $^{4}_{2}\text{He}$.

2 Plan your work.
- Set up the nuclear equation.

$$^{212}_{84}\text{Po} \longrightarrow\ ^{4}_{2}\text{He} + ?$$

3 Calculate.
- The sums of the mass numbers must be the same on both sides of the equation: $212 = 4 + A; A = 212 - 4 = 208$

$$^{212}_{84}\text{Po} \longrightarrow\ ^{4}_{2}\text{He} +\ ^{208}?$$

- The sums of the atomic numbers must be the same on both sides of the equation: $84 = 2 + Z; Z = 84 - 2 = 82$

$$^{212}_{84}\text{Po} \longrightarrow\ ^{4}_{2}\text{He} +\ ^{208}_{82}?$$

- Check the periodic table to identify the element that has an atomic number of 82, and complete the nuclear equation.

$$^{212}_{84}\text{Po} \longrightarrow\ ^{4}_{2}\text{He} +\ ^{208}_{82}\text{Pb}$$

4 Verify your results.
- Emission of an alpha particle does decrease the atomic number by two (from 84 to 82) and does decrease the mass number by four (from 212 to 208).

PRACTICE HINT

Unlike a chemical equation, the elements are usually different on each side of a balanced nuclear equation.

PRACTICE

Write balanced equations for the following nuclear equations.

1. $^{218}_{84}\text{Po} \longrightarrow\ ^{4}_{2}\text{He} + ?$

2. $^{142}_{61}\text{Pm} + ? \longrightarrow\ ^{142}_{60}\text{Nd}$

3. $^{253}_{99}\text{Es} +\ ^{4}_{2}\text{He} \longrightarrow\ ^{1}_{0}n + ?$

4. Write the balanced nuclear equation that shows how sodium-22 changes into neon-22.

Homework —— **GENERAL**

Additional Practice Have students fill in the blanks in the following nuclear equations:

1. $^{222}_{86}\text{Rn} \longrightarrow\ ^{218}_{84}\text{Po} + \underline{\hspace{1cm}}$
   Ans. $^{4}_{2}\text{He}$

2. $^{87}_{37}\text{Rb} \longrightarrow\ ^{87}_{38}\text{Sr} + \underline{\hspace{1cm}}$ Ans. $^{0}_{-1}e$

3. $^{188}_{79}\text{Au} \longrightarrow\ ^{188}_{78}\text{Pt} + \underline{\hspace{1cm}}$
   Ans. $^{0}_{+1}e$

4. $^{37}_{18}\text{Ar} +\ ^{0}_{-1}e \longrightarrow \underline{\hspace{1cm}}$ Ans. $^{37}_{17}\text{Cl}$

5. $^{28}_{13}\text{Al} \longrightarrow \underline{\hspace{1cm}} +\ ^{0}_{-1}e$ Ans. $^{28}_{14}\text{Si}$

6. $^{57}_{27}\text{Co} + \underline{\hspace{1cm}} \longrightarrow\ ^{57}_{26}\text{Fe}$ Ans. $^{0}_{-1}e$

7. $^{243}_{95}\text{Am} \longrightarrow \underline{\hspace{1cm}} +\ ^{4}_{2}\text{He}$
   Ans. $^{239}_{93}\text{Np}$

8. $^{125}_{56}\text{Ba} \longrightarrow \underline{\hspace{1cm}} +\ ^{0}_{+1}e$
   Ans. $^{125}_{55}\text{Cs}$

9. $\underline{\hspace{1cm}} \longrightarrow\ ^{61}_{28}\text{Ni} +\ ^{0}_{+1}e$ Ans. $^{61}_{29}\text{Cu}$

10. $\underline{\hspace{1cm}} +\ ^{0}_{-1}e \longrightarrow\ ^{82}_{37}\text{Rb}$ Ans. $^{82}_{38}\text{Sr}$

11. $\underline{\hspace{1cm}} \longrightarrow\ ^{226}_{90}\text{Th} +\ ^{4}_{2}\text{He}$
    Ans. $^{230}_{92}\text{U}$

12. $\underline{\hspace{1cm}} \longrightarrow\ ^{198}_{80}\text{Hg} +\ ^{0}_{-1}e$
    Ans. $^{198}_{79}\text{Au}$

LS Logical

Answers to Practice Problems A

1. $^{214}_{82}\text{Pb}$
2. $^{0}_{-1}e$
3. $^{256}_{101}\text{Md}$
4. $^{22}_{11}\text{Na} \longrightarrow\ ^{22}_{10}\text{Ne} +\ ^{0}_{+1}e$

653

MISCONCEPTION
/// **ALERT** \\\

When doing problems like the ones above, the most common mistake students make is that they forget that a change in the atomic number changes the identity of the element. Remind students that they should look up new atomic numbers on the periodic table to determine the identity of the new element.

Demonstration

Use dominoes to model a chain reaction. Set the dominoes on their ends so that the first row has one domino, the second row has two dominoes, the third row has three dominoes, and so on. The arrangement should resemble a pyramid. The dominoes should be close together (less than one domino-length apart) but not touching. While the students are watching push the domino in the first row over so that it hits the dominoes in the second row. Explain to students that the falling dominoes are like a uranium fission chain reaction—each domino causes other dominoes to fall over, and the number of dominoes falling at each step increases.

Anticipation Guide Before students read the sections on nuclear fission and nuclear fusion, have students decide whether the following statements are true or false. Have them check their answers after reading.

• Nuclear fusion occurs in nuclear reactors. Ans. false

• A chain reaction can begin when a uranium-235 atom splits by nuclear fission. Ans. true

• You can perform a nuclear fusion reaction in your high school chemistry lab. Ans. false

LS Auditory

---

nuclear fission

the splitting of the nucleus of a large atom into two or more fragments, a process that produces additional neutrons and a lot of energy

chain reaction

a reaction in which a change in a single molecule makes many molecules change until a stable compound forms

critical mass

the minimum mass of a fissionable isotope that provides the number of neutrons needed to sustain a chain reaction

---

Nuclear Fission

So far, you have learned about one class of nuclear change in which a nucleus decays by adding or losing particles. Another class of nuclear change is called **nuclear fission.** Nuclear fission occurs when a very heavy nucleus splits into two smaller nuclei, each more stable than the original nucleus. Some nuclei undergo fission without added energy. A very small fraction of naturally occurring uranium nuclei is of the isotope $^{235}_{92}U$, which undergoes spontaneous fission. However, most fission reactions happen artificially by bombarding nuclei with neutrons.

Figure 11 shows what happens when an atom of uranium-235 is bombarded with a neutron. The following equation represents the first reaction shown in **Figure 11.**

$$^{235}_{92}U + ^{1}_{0}n \xrightarrow{\text{nuclear fission}} ^{93}_{36}Kr + ^{140}_{56}Ba + 3\,^{1}_{0}n$$

Notice that the products include Kr-93, Ba-140, and three neutrons.

As shown in **Figure 11,** each of the three neutrons emitted by the fission of one nucleus can cause the fission of another uranium-235 nucleus. Again, more neutrons are emitted. These reactions continue one after another as long as enough uranium-235 remains. This process is called a **chain reaction.** One characteristic of a chain reaction is that the particle that starts the reaction, in this case a neutron, is also produced from the reaction. A minimum quantity of radioactive material, called **critical mass,** is needed to keep a chain reaction going.

Figure 11
A neutron strikes a uranium-235 nucleus, which splits into a krypton nucleus and a barium nucleus. Three neutrons are also produced. Each neutron may cause another fission reaction.

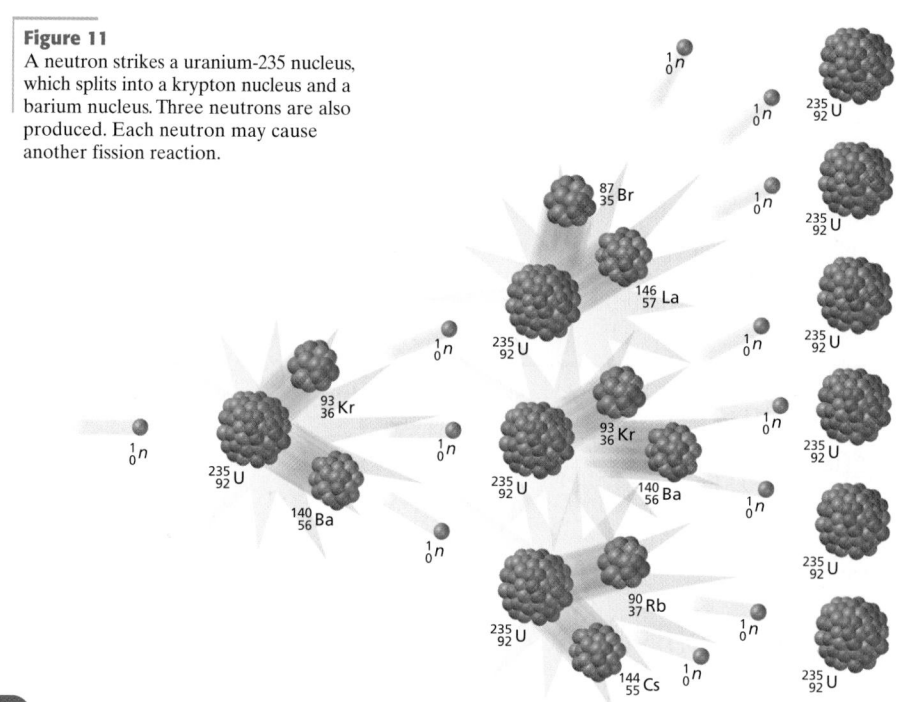

HISTORY
CONNECTION

The German physicist Lise Meitner was the first to use the word *splitting* to describe the breaking of the nucleus. She used the term in 1939 when suggesting a reason for the formation of barium atoms in an experiment in which uranium nuclei were bombarded with neutrons.

Chain Reactions Occur in Nuclear Reactors

Fission reactions can produce a large amount of energy. For example, the fission of 1 g of uranium-235 generates as much energy as the combustion of 2700 kg of coal. Fission reactions are used to generate electrical energy in nuclear power plants. Uranium-235 and plutonium-239 are the main radioactive isotopes used in these reactors.

In a nuclear reactor, represented in **Figure 12,** the fuel rods are surrounded by a moderator. The moderator is a substance that slows down neutrons. Control rods are used to adjust the rate of the chain reactions. These rods absorb some of the free neutrons produced by fission. Moving these rods into and out of the reactor can control the number of neutrons that are available to continue the chain reaction. Chain reactions that occur in reactors can be very dangerous if they are not controlled. An example of the danger that nuclear reactors can create is the accident that happened at the Chernobyl reactor in the Ukraine in 1986. This accident occurred when technicians briefly removed most of the reactor's control rods during a safety test. However, most nuclear reactors have mechanisms that can prevent most accidents.

As shown in **Figure 12,** water is heated by the energy released from the controlled fission of U-235 and changed into steam. The steam drives a turbine to produce electrical energy. The steam then passes into a condenser and is cooled by a river or lake's water. Notice that water heated by the reactor or changed into steam is isolated. Only water used to condense the steam is gotten from and is returned to the environment.

internet connect

www.scilinks.org
Topic: Fission
SciLinks code: HW4085

SCi LINKS Maintained by the National Science Teachers Association

Figure 12
This model shows a pressurized, light-water nuclear reactor, the type most often used to generate electrical energy in the United States. Note that each of the three water systems is isolated from the others for safety reasons.

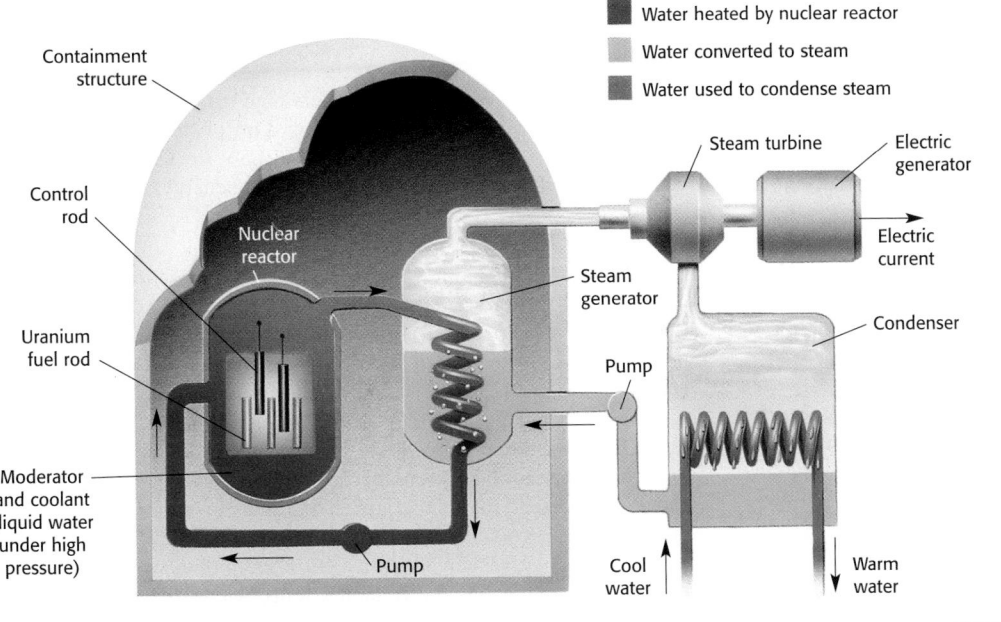

- ■ Water heated by nuclear reactor
- ■ Water converted to steam
- ■ Water used to condense steam

Containment structure

Control rod

Nuclear reactor

Uranium fuel rod

Moderator and coolant (liquid water under high pressure)

Pump

Steam generator

Steam turbine

Electric generator

Electric current

Condenser

Pump

Cool water

Warm water

655

Using the Figure

Figure 11 Make sure students understand that in this figure, the ratio of neutron release during a collision event per neutron release during the initial collision event is ideally 1:3:9:27:81:243, and so on, and that the picture has been simplified for easier study. The escaping neutrons will continue to bombard other nearby atoms unless they encounter absorbing material, such as graphite, europium, or cobalt, or are lost by the system.

Using the Figure

Figure 12 The reactor shown is called a light-water reactor because it uses ordinary water instead of heavy water, which contains increased amounts of D_2O. Students should note that the reactor uses three different water systems, each of which is isolated from the others. The only water that reaches the environment is the final cooling water, which cools the turbine water. The turbine water had been heated by the water that circulates through the reactor. With this system, no water containing radioactive materials should ever reach the outside environment.

INCLUSION Strategies

- *Learning Disabled*
- *Attention Deficit Disorder*
- *Developmentally Delayed*

Have the students create a poster, bulletin board, or computer sketch of the nuclear reactor shown in **Figure 12.** Students can lable each part of the reactor. Students may include information about a nuclear reactor in your state or a nearby state. If possible, include a picture of the local reactor in the display.

Transparencies

TT Model of a Pressurized, Light-Water Nuclear Reactor

MISCONCEPTION ALERT

Many people mistakenly believe that an accident in a nuclear power plant could result in a nuclear explosion like that of a nuclear bomb. During the worst case, a nuclear reactor in a power plant can undergo meltdown from the simultaneous failures of cooling systems and safety systems. Meltdown could result in the release of dangerous amounts of radioactive materials into the environment, not a nuclear explosion.

ENVIRONMENTAL SCIENCE CONNECTION

The Department of Energy (DOE) is hoping to use Yucca Mountain in Nevada as a nuclear spent-fuel repository. The department is currently studying the site to ensure that radioactive spent fuel can be safely stored for many years without harming the surrounding environment. The DOE is also planning an extensive reclamation plan through which they will restore the Yucca Mountain site to its original state after the project is finished.

Chapter 18 • Nuclear Chemistry   655

A Scientific Flop In 1989, B. Stanley Pons and Martin Fleischmann, two chemists at the University of Utah, announced that they had produced nuclear fusion at room temperature. Scientists around the world rushed to test the claim of this so-called "cold-fusion." Results varied, but no one has been able to duplicate what was thought to be a cold-fusion reaction. Many scientists have concluded that the original reports were incorrect and that further research would be a waste of time and money. Have interested students research the history of cold fusion and summarize their findings in a poster to be presented to the class. **LS** Interpersonal

Teaching Tip

Helium Bombs Make students aware that fusion bombs use uncontrolled fusion reactions. The goal of scientific research in this area is to control the fusion reaction to generate electrical energy.

Close

Reteaching ——— BASIC

Have students create a concept map to organize the ideas in this section. Their map should include the terms: radioactive decay, beta particles, alpha particles, positrons, gamma rays, nuclear fission, and nuclear fusion. Students can add any other terms that they feel are important. **LS** Intrapersonal

Figure 13
In the stars of this galaxy, four hydrogen nuclei fuse to form a single $^4_2$He nucleus.

nuclear fusion

the combination of the nuclei of small atoms to form a larger nucleus, a process that releases energy

internet connect

www.scilinks.org
Topic: Nuclear Fusion
SciLinks code: HW4086

SCLINKS Maintained by the National Science Teachers Association

Topic Link

Refer to the "Periodic Table" chapter for a discussion of nuclear fusion.

Nuclear Fusion

Nuclear fusion, which is when small nuclei combine, or fuse, to form a larger, more stable nucleus, is still another type of nuclear change. The new nucleus has a higher binding energy per nucleon than each of the smaller nuclei does, and energy is released as the new nucleus forms. In fact, fusion releases greater amounts of energy than fission for the same mass of starting material. Fusion is the process by which stars, including our sun, generate energy.

In the sun, four hydrogen nuclei fuse to form a single $^4_2$He nucleus

$$4^1_1\text{H} \longrightarrow {}^4_2\text{He} + 2^{\,0}_{+1}e.$$

The reaction above is a net reaction. Very high temperatures are required to bring the nuclei together. The temperature of the sun's core, where some of the fusion reactions occur, is about $1.5 \times 10^7\,°$C. When the hydrogen nuclei are fused, some mass is converted to energy.

Fusion Reactions Are Hard to Maintain

Scientists are investigating ways to control fusion reactions so that they may be used for both energy generation and research. One problem is that starting a fusion reaction takes a lot of energy. So far, researchers need just as much energy to start a fusion reaction as is released by the reaction. As a result, fusion is not a practical source of energy.

Another challenge is finding a suitable place for a fusion reaction. In fusion reactions, the reactants are in the form of a plasma, a random mixture of positive nuclei and electrons. Because no form of solid matter can withstand the tremendous temperatures required for fusion to occur, this plasma is hard to contain. Scientists currently use extremely strong magnetic fields to suspend the charged plasma particles. In this way, the plasma can be kept from contacting the container walls. Scientists have also experimented with high-powered laser light to start the fusion process.

656

ASTRONOMY
CONNECTION

Neutrinos have very little mass and are produced during the fusion reactions that occur in the core of stars. Neutrinos travel at the speed of light. Because they rarely interact with matter, they are very difficult to detect. A neutrino telescope is usually a tank of cleaning fluid (C_2Cl_4). Trillions of neutrinos from the sun travel regularly through Earth and through the cleaning fluid every second. About once every 12 h a neutrino will interact with a Cl atom to produce a radioactive Ar atom. The amount of Ar produced is then measured and an estimate of the number of neutrinos passing through the tank is made.

Nuclear Energy and Waste

The United States depends on nuclear power to generate electrical energy. In fact, about 100 nuclear reactors generate 20% of electrical energy needs in the United States. Nuclear power also generates waste like many other sources of energy, such as fossil fuels. Nuclear waste is "spent fuel" that can no longer be used to create energy. But this material is still radioactive and dangerous and must be disposed of with care.

Nuclear waste is often stored in "spent-fuel pools" that cover the spent fuel with at least 6 m of water. This amount of water prevents radiation from the waste from harming people. Nuclear waste can also be stored in a tightly sealed steel container. These containers have inert gases that surround the waste. These containers can also be surrounded by steel or concrete. Most of the nuclear waste that is put into a container has first been put in a spent-fuel pool to cool for about one year.

Some isotopes from the spent fuel can be extracted and used again as reactor fuel. However, this process is not currently done on a large scale in the United States.

internet connect
www.scilinks.org
Topic: Nuclear Energy
SciLinks code: HW4084

SCiLINKS Maintained by the National Science Teachers Association

② Section Review

UNDERSTANDING KEY IDEAS

1. What is the name of a high-energy electron that is emitted from an unstable nucleus?

2. How are nuclear fission and nuclear fusion similar? How are they different?

3. Describe what happens when a positron and an electron collide.

4. How is critical mass related to a chain reaction?

PRACTICE PROBLEMS

5. Write the balanced equations for the following nuclear reactions.

   a. Uranium-233 undergoes alpha decay.

   b. Copper-66 undergoes beta decay.

   c. Beryllium-9 and an alpha particle combine to form carbon-13. The carbon-13 nucleus then emits a neutron.

   d. Uranium-238 absorbs a neutron. The product then undergoes successive beta emissions to become plutonium-239.

6. A fusion reaction that takes place in the sun is the combination of two helium-3 nuclei to form two hydrogen nuclei and one other nucleus. Write the balanced nuclear equation for this fusion reaction. Be sure to include both products that are formed.

CRITICAL THINKING

7. In electron capture, why is the electron that is absorbed by the nucleus usually taken from the $1s$ orbital?

8. Can annihilation of matter occur between a positron and a neutron? Explain your answer.

9. Why do the nuclear reactions in a decay series eventually stop?

10. Cobalt-59 is bombarded with neutrons to produce cobalt-60, which is then used to treat certain cancers. The nuclear equation for this reaction shows the gamma rays that are released when cobalt-60 is produced.

$$^{59}_{27}\text{Co} + ^{1}_{0}n \longrightarrow ^{60}_{27}\text{Co} + \gamma$$

Is this an example of a nuclear change that involves the creation of a nucleus of another element? Explain your answer.

657

Quiz ——————— GENERAL

1. List the different types of nuclear change. **Ans.** radioactive decay, nuclear fission, and nuclear fusion

2. Which two processes can stabilize a nucleus by converting a proton into a neutron? **Ans.** electron capture and positron emission

3. A high energy electron can also be called a ____. **Ans.** beta particle

4. Which type of nuclear radiation is the most difficult to stop? **Ans.** gamma rays

5. Why don't we observe positrons after positron emission occurs? **Ans.** Positrons usually collide with electrons before they leave an atom. The positron and the electron it meets are destroyed through the annihilation of matter.

6. An alpha particle is the nucleus of what atom? **Ans.** a helium-4 atom

7. What type of nuclear reaction occurs in a nuclear reactor? **Ans.** nuclear fission

8. The smallest amount of radioactive material needed to sustain a chain reaction is called the ____. **Ans.** critical mass

9. Which type of nuclear reaction occurs in the sun and other stars? **Ans.** nuclear fusion

LS Intrapersonal

Videos

CNN Presents Science in the News

• **Segment 31** Nuclear Waste

See the Science in the News video guide for more details.

Chapter Resource File

• **Concept Review**

• **Quiz**

Answers to Section Review

1. a beta particle

2. They are similar in that both are nuclear changes that can occur spontaneously and release energy. They are different in that fission involves splitting a nucleus while fusion involves combining two nuclei into a single nucleus. Fusion also produces much more energy per gram of starting material than fission does.

3. They annihilate each other, which produces two gamma rays. Matter is totally converted into energy.

4. A chain reaction can occur only when a minimal amount of radioactive material, called the critical mass, is present.

5. a. $^{233}_{92}\text{U} \longrightarrow ^{229}_{90}\text{Th} + ^{4}_{2}\text{He}$

   b. $^{66}_{29}\text{Cu} \longrightarrow ^{66}_{30}\text{Zn} + ^{0}_{-1}e$

   c. $^{9}_{4}\text{Be} + ^{4}_{2}\text{He} \longrightarrow ^{13}_{6}\text{C}$
   $^{13}_{6}\text{C} \longrightarrow ^{12}_{6}\text{C} + ^{1}_{0}n$

   d. $^{238}_{92}\text{U} + ^{1}_{0}n \longrightarrow ^{239}_{92}\text{U}$
   $^{239}_{92}\text{U} \longrightarrow ^{239}_{93}\text{Np} + ^{0}_{-1}e$
   $^{239}_{93}\text{Np} \longrightarrow ^{239}_{94}\text{Pu} + ^{0}_{-1}e$

6. $2^{3}_{2}\text{He} \longrightarrow 2^{1}_{1}\text{H} + ^{4}_{2}\text{He}$

7. An electron in this orbital is closest to the nucleus and therefore most likely to be absorbed.

Answers continued on p. 675A

Focus

Overview

Before beginning this section, review with your students the Objectives listed in the Student Edition. This section discusses several uses for nuclear chemistry. First, the half-life of radioactive isotopes is defined and the process of determining the approximate age of artifacts and fossils using radioactive dating is described. Next, some of the uses of radioactivity are described. Finally, the effect of radiation on the body is discussed.

Bellringer

Have students write an opinion paragraph on one of the two following topics: "Radioactivity can be beneficial." or "Radioactivity can be harmful."

Motivate

Demonstration

Show students an X ray and an image from an MRI or a CAT scan. Explain to students that the X ray can easily show doctors bones, but cannot show any detail in soft tissue like the human brain or heart. However, devices such as the MRI or the CAT scan use radioactivity to create images of internal organs. Further explain that although patients are exposed to radiation, the dose is not harmful and the overall result is beneficial. Before such imaging devices were available the only way a doctor could see inside a patient was through exploratory surgery which is much more invasive.

Uses of Nuclear Chemistry

KEY TERMS
• half-life

OBJECTIVES

① **Define** the half-life of a radioactive nuclide, and explain how it can be used to determine an object's age.

② **Describe** some of the uses of nuclear chemistry.

③ **Compare** acute and chronic exposures to radiation.

Half-Life

The start-up activity for this chapter involved shaking pennies and then removing those that landed heads up after they were poured out of the cup. Each time you repeated this step, you should have found that about half the pennies were removed. Therefore, if you started with 100 pennies, about 50 should have been removed after the first shake. After the second shake, about 25 should have been removed, and so on. So, half of the amount of pennies remained after each step. This process is similar to what happens to radioactive materials that undergo nuclear decay. A radioactive sample decays at a constant rate. This rate of decay is measured in terms of its **half-life.**

half-life

the time required for half of a sample of a radioactive substance to disintegrate by radioactive decay or natural processes

Constant Rates of Decay Are Key to Radioactive Dating

The half-life of a radioactive isotope is a constant value and is not influenced by any external conditions, such as temperature and pressure. The use of radioactive isotopes to determine the age of an object, such as the one shown in **Figure 14,** is called *radioactive dating*. The radioactive isotope carbon-14 is often used in radioactive dating.

Nearly all of the carbon on Earth is present as the stable isotope carbon-12. A very small percentage of the carbon in Earth's crust is carbon-14. Carbon-14 undergoes decay to form nitrogen-14. Because carbon-12 and carbon-14 have the same electron configuration, they react chemically in the same way. Both of these carbon isotopes are in carbon dioxide, which is used by plants in photosynthesis.

As a result, all animals that eat plants contain the same ratio of carbon-14 to carbon-12 as the plants do. Other animals eat those animals, and so on up the food chain. So all animals and plants have the same ratio of carbon-14 to carbon-12 throughout their lives. Any carbon-14 that decays while the organism is alive is replaced through photosynthesis or eating. But when a plant or animal dies, it stops taking in carbon-containing substances, so the carbon-14 that decays is not replaced.

Figure 14
Using radioactive-dating techniques, scientists determined this Egyptian cat was made between 950–342 BCE.

658

PHILOSOPHY
CONNECTION

The concept of half-life is similar to Zeno's Paradox. Zeno (c. 490–430 BCE) was a Greek philosopher and was part of the Eleatic school. You can find many variations of his paradox on the Internet. However, unlike the paradox, there is a definite end to a half-life problem because atoms are not infinitely divisible. In radioactive decay, you will eventually reach the point of only one remaining atom. That atom will either decay or not decay in the next half-life. If it does decay, no more radioactive material will be left.

One-Stop Planner CD-ROM

• **Career Extension**
Real-World Connections Worksheet 3: Archaeologist
Assign this worksheet to emphasize relevant applications of text concepts.

Chapter Resource File

• **Lesson Plan**

TABLE 2 Half-Lives of Some Radioactive Isotopes

Isotope	Half-life	Radiation emitted	Isotope formed
Carbon-14	5.715×10^3 y	β^-, γ	nitrogen-14
Iodine-131	8.02 days	β^-, γ	xenon-131
Potassium-40	1.28×10^9 y	β^+, γ	argon-40
Radon-222	3.82 days	α, γ	polonium-218
Radium-226	1.60×10^3 y	α, γ	radon-222
Thorium-230	7.54×10^4 y	α, γ	radium-226
Thorium-234	24.10 days	β^-, γ	protactinium-234
Uranium-235	7.04×10^8 y	α, γ	thorium-231
Uranium-238	4.47×10^9 y	α, γ	thorium-234
Plutonium-239	2.41×10^4 y	α, γ	uranium-235

Table 2 shows that the half-life of carbon-14 is 5715 years. After that interval, only half of the original amount of carbon-14 will remain. In another 5715 years, half of the remaining carbon-14 atoms will have decayed and leave one-fourth of the original amount.

Once amounts of carbon-12 and carbon-14 are measured in an object, the ratio of carbon-14 to carbon-12 is compared with the ratio of these isotopes in a sample of similar material whose age is known. Using radioactive dating, with carbon-14, scientists can estimate the age of the object.

A frozen body that was found in 1991 in the Alps between Austria and Italy was dated using C-14. The body is known as the Iceman. A small copper ax was found with the Iceman's body, which shows that the Iceman lived during the Age of copper (4000 to 2200 BCE). Radioactive dating with C-14 revealed that the Iceman lived between 3500 and 3000 BCE and is the oldest prehistoric human found in Europe.

Generally, the more unstable a nuclide is, the shorter its half-life is and the faster it decays. **Figure 15** shows the radioactive decay of iodine-131, which is a very unstable isotope that has a short half-life.

internet connect

www.scilinks.org
Topic: Radioactive Dating
SciLinks code: HW4105

SC*LINKS* Maintained by the National Science Teachers Association

internet connect

www.scilinks.org
Topic: Discovering Radioactivity
SciLinks code: HW4150

SC*LINKS* Maintained by the National Science Teachers Association

Figure 15
The radioactive isotope $^{131}_{53}$I has a half-life of 8.02 days. In each successive 8.02-day period, half the atoms of $^{131}_{53}$I in the original sample decay to $^{131}_{54}$Xe.

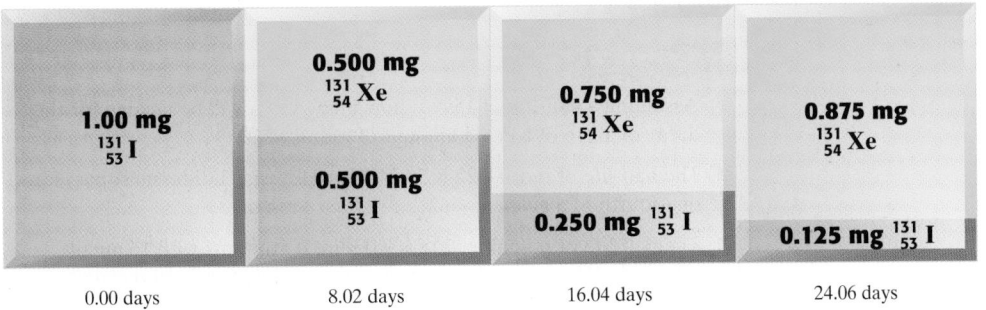

| 1.00 mg $^{131}_{53}$I | 0.500 mg $^{131}_{54}$Xe / 0.500 mg $^{131}_{53}$I | 0.750 mg $^{131}_{54}$Xe / 0.250 mg $^{131}_{53}$I | 0.875 mg $^{131}_{54}$Xe / 0.125 mg $^{131}_{53}$I |

0.00 days   8.02 days   16.04 days   24.06 days

659

HISTORY
CONNECTION

Carbon-14 dating (radiocarbon dating) was developed in 1946 by Willard F. Libby, an American chemist. Libby was awarded the Nobel Prize in chemistry in 1960 for his discovery. He also worked on the Manhattan Project to develop the atomic (fission) bomb. British physicist Ernest Rutherford was the first to suggest using radioactivity to measure geologic time and in 1907, an American radiochemist, B. R. Boltwood, first used the decay of uranium into lead to date rocks.

Teach

READING SKILL BUILDER — BASIC

Reading Organizer Have students survey Section 3, noting the headings and subheadings. Also have students note the boldface words and the use of illustrations. Discuss with students how the section is structured. **LS** Intrapersonal

Teaching Tip — BASIC

Rate of Decay Start the lesson by making an analogy between the rates of chemical reactions and the rates of radioactive decay. Remind students that most reactions depend on collisions between particles of reactants and that the rate is determined by the frequency of collisions. Ask students the following questions:

- What happens to the rate as reactants are used up, assuming the temperature remains the same? **Ans.** The rate decreases.

- Why does the rate decrease? **Ans.** The concentration of reactants is reduced, so collisions are less frequent.

- If 10 minutes are required for half the substances to react, will it take another 10 minutes for the remaining half to react? **Ans.** No. It will take much longer because the rate has decreased.

- Does this principle also apply to nuclear reactions? **Ans.** No, the decay rate does not depend on the concentration.

LS Auditory

Using the Figure

Figure 15 Students who are confused by the concept of half-life may have trouble understanding why the mass of iodine-131 that decays each half-life changes. Some students may find it easier to think of the percentage that decays rather than the mass that decays. For example, in **Figure 15** students should think about multiplying the mass remaining by 50 percent for each half-life.

Homework ——— GENERAL

Additional Practice

1. A fossil of an unknown age is found. Scientists determine that the C-14/C-12 ratio in the fossil is 1/32 of the ratio found in living substances today. Calculate the age of the fossil. Ans. 28 580 yr

2. Another fossil of an unknown age is found. Scientists determine that the C-14/C-12 ratio in the fossil is 1/128 of the ratio found in living substances today. Calculate the age of the fossil. Ans. 40 005 yr

LS Logical

Answers to Practice Problems B

1. 6396 yr
2. 7.648 days
3. 12 min

Teaching Tip

Radioactive Waste and Half-Lives In the late 1980s and the 1990s molecular biology laboratories used radioactive phosphorus-32 and sulfur-35 to study DNA. The disposal of phosphorus-32 waste was much easier than the disposal of sulfur-35 waste because of the difference in their half-lives. The half-life of phosphorus-32 is only 14.3 days while the half-life of sulfur-35 is 87.2 days. After approximately seven half-lives, the radioactive emissions would be close to background levels and the waste could be disposed of as nonradioactive waste. Therefore, laboratories only had to store phosphorus-32 waste for approximately 100 days before disposal, but had to store sulfur-35 waste for 610 days.

SAMPLE PROBLEM B

Determining the Age of an Artifact or Sample

An ancient artifact is found to have a ratio of carbon-14 to carbon-12 that is one-eighth of the ratio of carbon-14 to carbon-12 found in a similar object today. How old is this artifact?

1 **Gather information.**
- The half-life of carbon-14 is 5715 years.
- The artifact has a ratio of carbon-14 to carbon-12 that is one-eighth of the ratio of carbon-14 to carbon-12 found in a modern-day object.

2 **Plan your work.**
- First, determine the number of half-lives that the carbon-14 in the artifact has undergone.
- Next, find the age of the artifact by multiplying the number of half-lives by 5715 y.

3 **Calculate.**
- For an artifact to have one-eighth of the ratio of carbon-14 to carbon-12 found in a modern-day object, three half-lives must have passed.

$$\frac{1}{8} = \frac{1}{2} \times \frac{1}{2} \times \frac{1}{2}$$

- To find the age of the artifact, multiply the half-life of carbon-14 three times for the three half-lives that have elapsed.

$$3 \times 5715 \text{ y} = 17\ 145 \text{ y}$$

4 **Verify your results.**
- Start with your answer, and work backward through the solution to be sure you get the information found in the problem.

$$\frac{17\ 145 \text{ y}}{3} = 5715 \text{ y}$$

> **PRACTICE HINT**
>
> Make a diagram that shows how much of the original sample is left to solve half-life problems.
>
> $1 \longrightarrow 1/2 \longrightarrow 1/4 \longrightarrow 1/8$
> $\longrightarrow 1/16 \longrightarrow 1/32 \longrightarrow$ etc.
>
> Each arrow represents one half-life.

PRACTICE

1 Assuming a half-life of 1599 y, how many years will be needed for the decay of 15/16 of a given amount of radium-226?

2 The half-life of radon-222 is 3.824 days. How much time must pass for one-fourth of a given amount of radon to remain?

3 The half-life of polonium-218 is 3.0 min. If you start with 16 mg of polonium-218, how much time must pass for only 1.0 mg to remain?

660

ARCHEOLOGY ——— CONNECTION

Iceman In 1991, Erika and Helmut Simon discovered a body encased in a glacier in the Alps near the border between Austria and Italy. The body was frozen and was a very well preserved body of a prehistoric man who died between 3500 and 3000 BCE. The man's clothes, tools, and hair were still intact. The Iceman is currently stored in a freezer set at –6°C and 98% humidity at the Institute for Anatomy in Innsbruck, Austria. The freezer conditions keep the body from decomposing by replicating the conditions of the glacier in which the body was found.

Some Isotopes Are Used for Geologic Dating

By analyzing organic materials in the paints, scientists used carbon-14 to date the cave painting shown in **Figure 16**. Two factors limit dating with carbon-14. The first limitation is that C-14 cannot be used to date objects that are completely composed of materials that were never alive, such as rocks or clay. The second limitation is that after four half-lives, the amount of radioactive C-14 remaining in an object is often too small to give reliable data. Consequently, C-14 is not useful for dating specimens that are more than about 50 000 years old. Anything older must be dated on the basis of a radioactive isotope that has a half-life longer than that of carbon-14. One such isotope is potassium-40.

Potassium-40, which has a half-life of 1.28 billion years, represents only about 0.012% of the potassium present in Earth today. Potassium-40 is useful for dating ancient rocks and minerals. Potassium-40 produces two different isotopes in its radioactive decay. About 11% of the potassium-40 in a mineral decays to argon-40 by emitting a positron.

$$^{40}_{19}K \longrightarrow \, ^{40}_{18}Ar + \, ^{0}_{+1}e$$

The argon-40 may remain in the sample. The remaining 89% of the potassium-40 decays to calcium-40 by emitting a beta particle.

$$^{40}_{19}K \longrightarrow \, ^{40}_{20}Ca + \, ^{0}_{-1}e$$

The calcium-40 is not useful for radioactive dating because it cannot be distinguished from other calcium in the rock. The argon-40, however, can be measured. **Figure 17** shows the decay of potassium-40 through four half-lives.

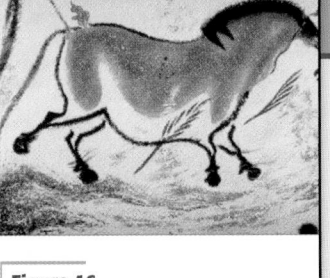

Figure 16
Scientists determined that this cave painting at Lascaux, called *Chinese Horse*, was created approximately 13 000 BCE.

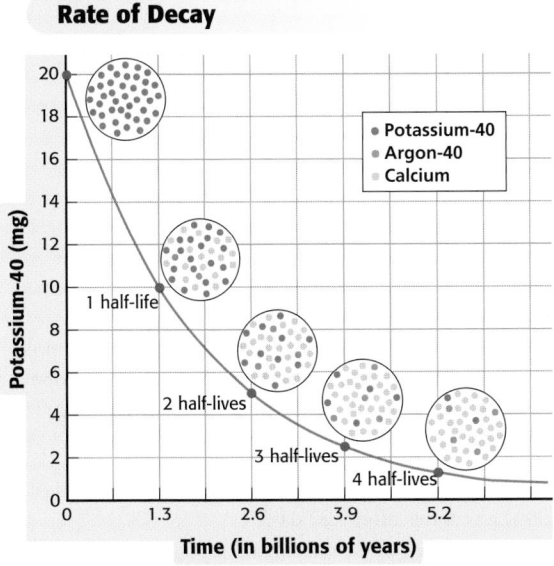

Rate of Decay

Legend:
- Potassium-40
- Argon-40
- Calcium

1 half-life
2 half-lives
3 half-lives
4 half-lives

y-axis: Potassium-40 (mg) — 0, 2, 4, 6, 8, 10, 12, 14, 16, 18, 20
x-axis: Time (in billions of years) — 0, 1.3, 2.6, 3.9, 5.2

Figure 17
Potassium-40 decays to argon-40 and calcium-40, but scientists monitor only the ratio of potassium-40 to argon-40 to determine the age of the object.

661

Transparencies

TT Rate of Decay of Potassium-40

Teaching Tip

Carbon Dating Students should understand that an ancient object that contains carbon and was once living cannot always be carbon dated. Ask students if they think C-14 dating could be used to determine the age of a lump of coal, which is mostly carbon. Lead them to understand that C-14 dating of coal would not be possible because the organisms that formed it lived too long ago, about 300 million to 350 million years ago.

Homework ──── GENERAL

Additional Practice Have students use the equation below to solve the following problems.

$$t = \frac{t_{1/2}}{\ln 2} \times \ln\left(1 + \frac{\text{Ar-40}}{\text{K-40}}\right)$$

t = age of sample

$t_{1/2}$ = half-life of potassium-40 = 1.28 billion years

$\dfrac{\text{Ar-40}}{\text{K-40}}$ = the ratio of argon-40 to potassium-40

1. A sample of rock is found and its ratio of argon-40 to potassium-40 is 0.850. How old is the rock? **Ans.** 1.14 billion years old

2. A moon rock is studied and its ratio of argon-40 to potassium-40 is 7.30. How old is the moon rock? **Ans.** 3.91 billion years old

LS **Logical**

Answers to Practice Problems C

1. 0.25 mg
2. 0.0312 g
3. 0.32 g

> **PRACTICE HINT**
>
> Remember to double the amount of radioactive isotope each time you go back one half-life.

SAMPLE PROBLEM C

Determining the Original Mass of a Sample

A rock is found to contain 4.3 mg of potassium-40. The rock is dated to be 3.84 billion years old. How much potassium-40 was originally present in this rock?

1 Gather information.
- The rock is 3.84 billion years old and contains 4.3 mg of $^{40}_{19}$K.
- The half-life of potassium-40 is 1.28 billion years.

2 Plan your work.
- Find the number of half-lives that the $^{40}_{19}$K in the rock has undergone.
- Next, find the mass of the $^{40}_{19}$K that was originally in the rock. Double the present amount for every half-life that the isotope has undergone.

3 Calculate.
- Divide the age of the rock by the half-life of the isotope to find the number of half-lives.

$$\frac{3.84 \text{ billion y}}{1.28 \text{ billion y}} = 3 \text{ half-lives have elapsed}$$

- The mass of the original potassium-40 sample is calculated by doubling 4.3 mg three times.

  4.3 mg × 2 = 8.6 mg were present in the rock 1 half-life ago

  8.6 mg × 2 = 17 mg were present in the rock 2 half-lives ago

  17 mg × 2 = 34 mg were present in the rock 3 half-lives ago

4 Verify your results.

After three half-lives, one-eighth of the original $^{40}_{19}$K remains. So, 8 × 4.3 = 34 mg.

PRACTICE

1 The half-life of polonium-210 is 138.4 days. How many milligrams of polonium-210 remain after 415.2 days if you start with 2.0 mg of the isotope?

2 After 4797 y, how much of an original 0.250 g sample of radium-226 remains? Its half-life is 1599 y.

3 The half-life of radium-224 is 3.66 days. What was the original mass of radium-224 if 0.0800 g remains after 7.32 days?

ASTRONOMY ─ CONNECTION

Potassium-40 dating was used to date moon rocks. The oldest lunar rocks were found to be around 4.5 billion years old, but some rocks were found to be as young as 3.2 billion years old. Astronomers believe that the younger rock formed when volcanoes on the moon erupted. Very little rock formation occurred after that time. The results of the dating determined that the moon formed at about the same time as Earth formed.

Other Uses of Nuclear Chemistry

Scientists create new elements by using nuclear reactions. But the use of nuclear reactions has extended beyond laboratories. Today, nuclear reactions have become part of our lives. Nuclear reactions that protect your life may be happening in your home.

Smoke Detectors Contain Sources of Alpha Particles

Smoke detectors depend on nuclear reactions to sound an alarm when a fire starts. Many smoke detectors contain a small amount of americium-241, which decays to form neptunium-237 and alpha particles.

$$^{241}_{95}\text{Am} \longrightarrow ^{237}_{93}\text{Np} + ^{4}_{2}\text{He}$$

The alpha particles cannot penetrate the plastic cover and can travel only a short distance. When alpha particles travel through the air, they ionize gas molecules in the air, which change the molecules into ions. These ions conduct an electric current. Smoke particles reduce this current when they mix with the ionized molecules. In response, the smoke detector sets off an alarm.

Detecting Art Forgeries with Neutron Activation Analysis

Nuclear reactions can be used to help museum directors detect whether an artwork, such as the one shown in **Figure 18,** is a fake. The process is called *neutron activation analysis.* A tiny sample from the suspected forgery is placed in a machine. A nuclear reactor in the machine bombards the sample with neutrons. Some of the atoms in the sample absorb neutrons and become radioactive isotopes. These isotopes emit gamma rays as they decay.

Scientists can identify each element in the sample by the characteristic pattern of gamma rays that each element emits.

Figure 18
Neutron activation analysis can be used to determine if this artwork is real.

Teaching Tip
Labeling Smoke Detectors
Emphasize that ionizing smoke detectors usually carry a label explaining where and how to dispose of the unit when it no longer functions. The long half-lives of the isotopes (432 years for americium-241 and 2.14 million years for neptunium-237) will pose long-term environmental problems if the detector is disposed of in a landfill.

Demonstration
Ionizing Smoke Detectors Show students a household ionizing smoke detector. Point out the label that warns consumers about the radioactive contents of the detector. To demonstrate how the smoke detector works, light a wooden splint and blow out the flame. Hold the smoke detector in the smoke produced by the splint and the alarm should sound. To silence the detector, move it away from the smoke and wave it around a little bit to air it out. Explain to students that if they accidentally burn food while cooking, they can prevent their smoke detector from sounding by fanning the air in front of the detector. Of course, they should only do this after removing the burning food from the stove!

If your school is equipped with a sprinkler system you can explain the difference between it and a household smoke detector. Most sprinkler systems are activated by energy as heat while smoke detectors are activated by smoke particles in the air.

663

Making Models Have students build a model or make a poster of the reactions involved in neutron activation analysis. They should demonstrate their model or present their poster to the class.
LS Interpersonal

Teaching Tip

Neutron Activation Analysis
Neutron activation analysis (NAA) is so sensitive that forensic laboratories can use it to study extremely small samples of evidence such as gun shot residues, hair, and paint chips. For example, the bullet lead from President John F. Kennedy's body was analyzed in the investigation of his assassination. NAA has also been used to trace prehistoric trade routes by studying obsidian artifacts and the location of their discovery. Harvard Medical School is currently using NAA to study the effects of selenium supplements on colon, lung, and prostate cancer.

Teaching Tip

PET Scans Positron Emission Tomography (PET) is another application of nuclear decay that is an important diagnostic tool in medicine. It allows medical personnel to observe images of internal organs and their activities without invading the body. In PET, an isotope such as C-11, N-13, O-15, or one of several isotopes of Tc decays by positron emission. Positron emission results in the annihilation of the positron and an electron that produces gamma rays. These rays are detected and measured, and the data are used by a computer to generate an image.

Scientists can then determine the exact proportions of the elements present. This method gives scientists a "fingerprint" of the elements in the sample. If the fingerprint matches materials that were not available when the work was supposedly created, then the artwork is a fake.

Nuclear Reactions Are Used in Medicine

The use of nuclear reactions by doctors has grown to the point where a whole field known as nuclear medicine has developed. Nuclear medicine includes the use of nuclear reactions both to diagnose certain conditions and to treat a variety of diseases, especially certain types of cancer.

For years, doctors have used a variety of devices, such as X-ray imaging, to get a view inside a person's body. Nuclear reactions have enabled them to get a much more detailed view of the body. For example, doctors can take a close look at a person's heart by using a thallium stress test. The person is given an intravenous injection of thallium-201, which acts chemically like calcium and collects in the heart muscle. As the thallium-201 decays, low-energy gamma rays are emitted and are detected by a special camera that produces images, such as the one shown in **Figure 19.**

The radioactive isotope most widely used in nuclear medicine is technetium-99, which has a short half-life and emits low-energy gamma rays. This radioactive isotope is used in bone scans. Bone repairs occur when there is a fracture, infection, arthritis, or an invading cancer. Bones that are repairing themselves take in minerals and absorb the technetium at the same time. If an area of bone has an unusual amount of repair, the technetium will gather there. Cameras detect the gamma rays that result from its decay.

Another medical procedure that uses nuclear reactions is called *positron emission tomography* (PET), which is shown in **Figure 20.** PET uses radioactive isotopes that have short half-lives. An unstable isotope that contains too many protons is injected into the person.

Figure 19
This image reveals the size of the heart, how well the chambers are pumping, and whether there is any scarring of muscle from previous heart attacks.

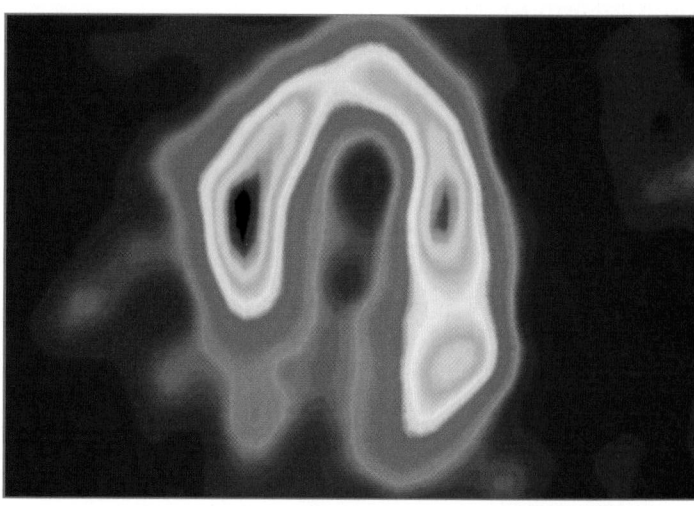

HISTORY — CONNECTION

On July 4, 1850, President Taylor dedicated the cornerstone for the Washington Monument. Five days later, he became seriously ill and died. Some historians believed that Taylor had been poisoned with arsenic. So, forensic scientists opened Taylor's sealed, lead coffin almost 150 years after his death and took small samples of his hair, nails, and body tissues. Neutron activation analysis revealed that arsenic was present but only in the amount that is normally present in the human body. So, President Taylor had not been poisoned with arsenic.

Videos
CNN. **Presents Science in the News**
• **Segment 32** Radioisotopes in Medicine
See the Science in the News video guide for more details.

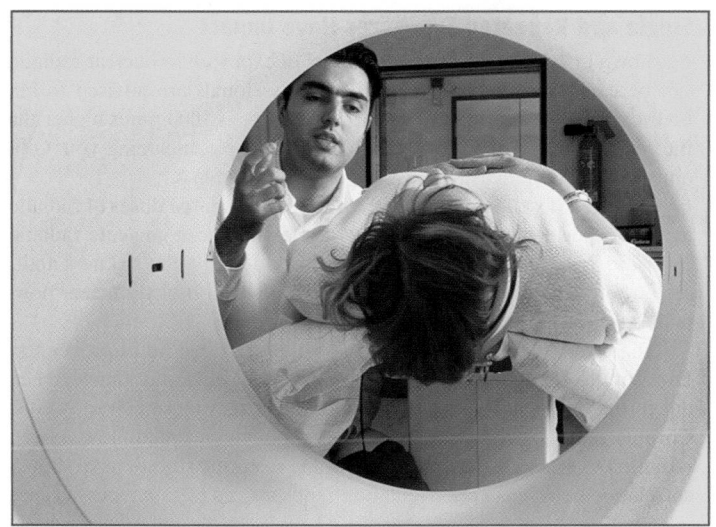

Figure 20
This person is undergoing a PET scan. The scan will provide information about how well oxygen is being used by the person's brain.

As this isotope decays, positrons are emitted. Recall that when a positron collides with an electron, both are annihilated, and two gamma rays are produced. These gamma rays leave the body and are detected by a scanner. A computer converts the images into a detailed three-dimensional picture of the person's organs.

Exposure to Radiation Must Be Checked

Table 3 shows how radiation can affect a person's health using the unit *rem*, which expresses the biological effect of an absorbed dose of radiation in humans. People who work with radioactivity wear a film badge to monitor the amount of radiation to which they are exposed. Radioactivity was discovered when sealed photographic plates exposed to radiation became fogged. A film badge works on the same principle. Any darkening of the film indicates that the badge wearer was exposed to radiation, and the degree of darkening indicates the total exposure.

Table 3	Effect of Whole-Body Exposure to a Single Dose of Radiation	
Dose (rem)	**Probable effect**
0–25	no observable effect
25–50	slight decrease in white blood cell count
50–100	marked decrease in white blood cell count
100–200	nausea, loss of hair
200–500	ulcers, internal bleeding
> 500	death

665

did you know?

In the United States, the average yearly exposure to radiation is approximately 360 millirem. Of that, about 300 millirem comes from natural sources. A person's exposure depends on several things. People who live at higher altitudes usually have a higher exposure because of increased exposure to cosmic rays from space. Furthermore, the amount of radioactive elements in soil and rocks varies from place to place. People living in northeast Washington state have a higher than average yearly exposure rate because of radon in the ground.

Teaching Tip

Origins of Radiology The first medical use of radiation occurred during World War I. During that war, Marie Curie took charge of the Military Radiological Service and taught military doctors about radiology, the branch of medicine that uses ionizing radiation for diagnosis and treatment. Early radiology relied heavily on X rays and they were primarily used to locate fractures and to detect the position of bullets.

SKILL BUILDER — GENERAL

Writing Have students research different types of radiological medical treatments or diagnostic procedures. Each student should concentrate on one treatment or procedures for one disease. In addition to learning how the procedure works, students should find out when the procedure was first used, recent developments, and statistics about its success rate or benefits. Students can present their findings in a written report or poster to share with the rest of the class.
LS Intrapersonal

Teaching Tip

Tracer Studies Doctors can use tracer studies as a noninvasive way to detect functional problems or damaged areas in the human body. During a tracer study, the patient is given a radioisotope that can be detected with an imaging system. Different radioisotopes are used to study different parts of the body because certain elements tend to collect in specific parts of the body. For example, iodine-131 will travel to the thyroid and is used to detect problems with that gland. Another example is technetium-99 which is used to detect tumors because it collects in areas with rapid cell growth.

Reteaching — BASIC

Have students write articles for the school newspaper about one positive use for nuclear chemistry. They should describe the nuclear reactions involved, why radioactivity is essential, and the benefits and disadvantages from the use of radioactivity. LS Intrapersonal

Quiz — GENERAL

1. What is a half-life? **Ans.** the amount of time for half of a sample of radioactive atoms to decay

2. You find an artifact and believe it to be less than 200 years old. Should you use potassium-40 to determine its age? Why or why not? **Ans.** No. The half-life of potassium-40 is 1.28 billion years which is much too long to date your object.

3. What type of radioactive emissions are produced in a household smoke detector? **Ans.** alpha particles

4. Why do the dentist and the dental hygienist leave the room when they take an X-ray image of a patient? **Ans.** They take several X-ray images every day, so they leave the room to limit their exposure to ionizing radiation. LS Intrapersonal

Chapter Resource File
• Concept Review
• Quiz

Table 4   Units Used in Measurements of Radioactivity

Units	Measurements
Curie (C)	radioactive decay
Becquerel (Bq)	radioactive decay
Roentgens (R)	exposure to ionizing radiation
Rad (rad)	energy absorption caused by ionizing radiation
Rem (rem)	biological effect of the absorbed dose in humans

Single and Repeated Exposures Have Impact

As shown in **Table 3**, the biological effect of exposure to nuclear radiation can be expressed in rem. Healthcare professionals are advised to limit their exposure to 5 rem per year. This exposure is 1000 times higher than the recommended exposure level for most people, including you. Other units of radiation measurement can be seen in **Table 4**.

People exposed to a single large dose or a few large doses of radiation in a short period of time are said to have experienced an acute radiation exposure. More than 230 people suffered acute radiation sickness and 28 died when a meltdown occurred in 1986 at the Chernobyl nuclear power plant in the Ukraine.

The effects of nuclear radiation on the body can add up over time. Exposure to small doses of radiation over a long period of time can be as dangerous as a single large dose if the total radiation received is equal. Chronic radiation exposure occurs when people get many low doses of radiation over a long period of time. Some scientific studies have shown a correlation between chronic radiation exposure and certain types of cancer.

3 Section Review

UNDERSTANDING KEY IDEAS

1. What is meant by the *half-life* of a radioactive nuclide?
2. Explain how carbon-14 dating is used to determine the age of an object.
3. Why is potassium-40 used to date objects older than 50 000 years old?
4. Identify three practical applications of nuclear chemistry.

PRACTICE PROBLEMS

5. What fraction of an original sample of a radioactive isotope remains after three half-lives have passed?
6. How many half-lives of radon-222 have passed in 11.46 days? If 5.2×10^{-8} g of radon-222 remain in a sealed box after 11.46 days, how much was present in the box initially? Refer to **Table 2**.

7. The half-life of protactinium-234 in its ground state is 6.69 h. What fraction of a given amount remains after 26.76 h?
8. The half-life of thorium-227 is 18.72 days. How many days are required for three-fourths of a given amount to decay?

CRITICAL THINKING

9. Someone tells you that neutron activation analysis can reveal whether a famous painter or a rival living at the same time created a painting. What is wrong with this reasoning?
10. Why are isotopes that have relatively short half-lives the only ones used in medical diagnostic tests?
11. A practical rule is that a radioactive nuclide is essentially gone after 10 half-lives. What percentage of the original radioactive nuclide is left after 10 half-lives? How long will it take for 10 half-lives to pass for plutonium-239? Refer to **Table 2**.

666

Answers to Section Review

1. Half-life is the time required for half of the nuclei in a radioactive element to decay.
2. The amount of carbon-12 and carbon-14 in a sample are measured. The ratio of C-14 to C-12 compared to that found in a comparable sample today (or in a sample whose age is known) reveals how many half-lives have elapsed since the carbon was incorporated into the object being dated.
3. Potassium-40 has a long enough high-life, 1.28 billion years, which is useful in dating objects that are millions or even billions of years old.

4. Uses include radioactive dating, smoke detectors, generating electrical energy, creating new elements, medical diagnosis and treatment, and detecting art forgeries.
5. After three half-lives, one-eighth, or 0.125, of the original radioactive sample remains.
6. Three half-lives have elapsed. The original sample contained 4.2×10^{-7} g of radioactive material.
7. 1/16
8. 37.44 days

Answers continued on p. 675A

H
Hydrogen
1.007 94
$1s^1$

Element Spotlight

Where Is H?

Earth's crust
0.9 by mass

Universe
approximately 93% of
all atoms

Hydrogen Is an Element unto Itself

Hydrogen is a unique element in many respects. Its scarcity as a free element on Earth is partially due to the low density of hydrogen gas. The low density permits hydrogen molecules to escape Earth's gravitational pull and drift into space.

Hydrogen does not fit precisely anywhere in the periodic table. It could be placed in Group 1 because it has a single valence electron. But it could also be placed with the halogens in Group 17 because it needs only one electron to get a full outer shell.

Industrial Uses

- Hydrogen gas is prepared industrially by the thermal decomposition of hydrocarbons, such as natural gas, oil-refinery gas, gasoline, fuel oil, and crude oil.
- Most of the hydrogen gas produced is used for synthesizing ammonia.
- Hydrogen is used in the hydrogenation of unsaturated vegetable oils to make solid fats.
- Liquid hydrogen is a clear, colorless liquid that has a boiling point of −252.87°C, the lowest boiling point of any known liquid other than liquid helium. Because of its low temperature, liquid hydrogen is used to cool superconducting materials.
- Liquid hydrogen is used to fuel rockets, satellites, and spacecraft.

Liquid hydrogen is used as fuel for some rockets.

Real-World Connection Nuclear fusion, in which hydrogen atoms form helium atoms, occurs in our sun.

A Brief History

| 1600 | 1700 | 1800 | 1900 |

1783: Jacques Charles fills a balloon with hydrogen and flies in a basket over the French countryside.

1931: Harold Urey discovers deuterium, an isotope of hydrogen, in water.

1937: The Hindenburg, a hydrogen-filled dirigible, explodes during a landing in Lakehurst, New Jersey.

1660: Robert Boyle prepares hydrogen from a reaction between iron and sulfuric acid.

1766: Henry Cavendish prepares a pure sample of hydrogen and distinguishes it from other gases. He names it "inflammable air."

1898: James Dewar produces liquid hydrogen and develops a glass vacuum flask to hold it.

1934: Ernest Rutherford, Marcus Oliphant, and Paul Harteck discover tritium.

1996: Scientists at Lawrence Livermore National Laboratory succeed in making solid, metallic hydrogen.

Questions

1. Research how hydrogen is used to fuel rockets and spacecrafts.

2. Write a paragraph about stars and fusion.

internet connect

www.scilinks.org
Topic : Hydrogen
SciLinks code: HW4155

SCI LINKS Maintained by the National Science Teachers Association

667

18 CHAPTER HIGHLIGHTS

KEY IDEAS

SECTION ONE   Atomic Nuclei and Nuclear Stability

• The strong force overcomes the repulsive force between protons to keep a nucleus intact.

• The mass that is converted to energy when nucleons form a nucleus is known as the mass defect.

• If the mass defect is known, the nuclear binding energy can be calculated by using the equation $E = mc^2$.

• The ratio of neutrons to protons defines a band of stability that includes the stable nuclei.

SECTION TWO   Nuclear Change

• Unstable nuclei are radioactive and can emit radiation in the form of alpha particles, beta particles, and gamma rays.

• Unstable nuclei that have large N/Z usually emit beta particles.

• Unstable nuclei that have small N/Z or have too few neutrons can undergo either electron capture or positron emission, emitting gamma rays in the process.

• Large nuclei that have large N/Z frequently emit alpha particles.

• Nuclear equations are balanced in terms of mass and nuclear charge.

• In nuclear fission, a heavy nucleus splits into two smaller nuclei; in nuclear fusion, two or more smaller nuclei combine to form one larger nucleus.

• Nuclear fission reactions that cause other fissions are chain reactions. Chain reactions must be controlled to generate usable energy.

SECTION THREE   Uses of Nuclear Chemistry

• Half-life is the time required for one half of the mass of a radioactive isotope to decay.

• The half-life of the carbon-14 isotope can be used to date organic material that is up to 50 000 years old. Other radioactive isotopes are used to date older rock and mineral formations.

• Radioactive isotopes have a number of practical applications in industry, medicine, and chemical analysis.

KEY TERMS

nucleons
nuclide
strong force
mass defect

radioactivity
beta particle
gamma ray
nuclear fission
chain reaction
critical mass
nuclear fusion

half-life

KEY SKILLS

Balancing a Nuclear Equation	Determining the Age of an Artifact or Sample	Determining the Original Mass of a Sample
Skills Toolkit 1 p. 652	Sample Problem B p. 660	Sample Problem C p. 662
Sample Problem A p. 653		

668

USING KEY TERMS

1. What is the energy emitted when a nucleus forms?

2. What is a nucleon?

3. What is the high-energy electromagnetic radiation produced by decaying nuclei?

4. What nuclear reaction happens when two small nuclei combine?

5. Explain the difference between fission and fusion.

6. Name the process that describes an unstable nucleus that emits particles and energy.

7. Define *critical mass*.

8. Define *half-life*.

9. What is the combination of neutrons and protons in a nucleus known as?

10. Name two types of nuclear changes.

UNDERSTANDING KEY IDEAS

Atomic Nuclei and Nuclear Stability

11. Explain how the strong force holds a nucleus together despite the repulsive forces between protons.

12. Describe what happens to unstable nuclei.

13. **a.** What is the relationship among the number of protons, the number of neutrons, and the stability of the nucleus for small atoms?

    b. What is the relationship among number of protons, the number of neutrons, and the stability of the nucleus for large atoms?

14. What is the relationship between binding energy and the formation of a nucleus from protons and neutrons?

15. What is the relationship between mass defect and binding energy?

16. Why is nuclear stability better indicated by binding energy per nucleon than by total binding energy per nucleus?

17. What is a quark?

Nuclear Change

18. What is the relationship between an alpha particle and a helium nucleus?

19. Compare the penetrating powers of alpha particles, beta particles, and gamma rays.

20. Is the decay of an unstable isotope into a stable isotope always a one-step process? Explain.

21. **a.** What role does a neutron serve in starting a nuclear chain reaction and in keeping it going?

    b. Why must neutrons in a chain reaction be controlled?

    c. Why must there be a minimum mass of material in order to sustain a chain reaction?

22. Under what conditions does fusion occur?

23. Why do positron emission and electron capture have the same effect on a nucleus?

Uses of Nuclear Chemistry

24. Explain why nuclei that emit alpha particles, such as americium-241, are safe to use in smoke detectors.

 669

Assignment Guide

Section	Questions
1	1, 2, 11–17, 45–48, 55, 60, 61, 67, 78
2	3–7, 9, 10, 18–23, 30–36, 49–52, 56, 58, 62, 64–66, 70, 77, 79, 81, 85, 86
3	8, 24–29, 37–44, 53, 54, 57, 59, 63, 68, 69, 71–76, 80, 82, 83, 84, 87

REVIEW ANSWERS

1. nuclear binding energy

2. protons or neutrons that make up a nucleus

3. gamma ray

4. nuclear fusion

5. Large nuclei break apart to form smaller nuclei during fission, and small nuclei fuse to form larger nuclei during fusion.

6. radioactivity

7. Critical mass is the smallest mass of radioactive material needed to sustain a chain reaction.

8. The time required for 50% of a sample of a radioactive isotope to decay is known as half-life.

9. nuclide

10. Students' answers may vary. Sample answer: fission and fusion

11. The strong force is an attractive force that acts only at very close distances. At such distances, it is much stronger than the electrostatic repulsion between protons.

12. Unstable nuclei decay in a variety of ways, releasing radioactivity and eventually forming stable nuclei.

13. **a.** In small, stable nuclei, the numbers of protons and neutrons are approximately equal.

    b. In large, stable nuclei, the numbers of neutrons and protons approach a ratio of 1.5:1.

14. As a nucleus forms from separated protons and neutrons, energy is released. This is the binding energy, which also represents the energy needed to break apart the stable nucleus that has formed into neutrons and protons.

15. The mass defect is the mass equivalent of the binding energy as given by Einstein's equation $E = mc^2$. The mass defect is the difference between the sum of the separate masses of the nucleons and the mass of the formed nucleus.

16. Calculating the binding energy per nucleon allows for a comparison of the stability of all nuclei with one another regardless of size.

17. Quarks are the building blocks of protons and neutrons. Each of these subatomic particles is made up of three quarks.

18. An alpha particle is a helium-4 nucleus.

19. Alpha particles have very low penetrating power because they are massive, charged, and slow. Beta particles have more penetrating power because they have little mass and much higher velocity than alpha particles have. Gamma rays have high penetrating power because they are massless, travel at the velocity of light, and have no charge.

20. No. Often, the nucleus decays into another nucleus that is also unstable. This process can occur a number of times, leading to a decay series.

21. a. A neutron begins the reaction by colliding with a nucleus. Additional neutrons are ejected from the bombarded nucleus and go on to collide with other nuclei, continuing the reaction.

b. Controlling the neutrons controls the rate of the nuclear chain reaction, preventing it from producing too much energy, especially in the form of heat.

c. The critical mass is the minimum mass that provides enough nuclei so that the neutrons from each fission will collide with unsplit nuclei.

22. Fusion occurs only at extremely high temperatures.

23. Both result in the change of a proton to a neutron, thereby reducing the atomic number by one without changing the mass number.

25. How does acute radiation exposure differ from chronic radiation exposure?

26. Why do animals contain the same ratio of carbon-14 to carbon-12 as plants do?

27. What type of radioactive nuclide is injected into a person who is about to undergo a PET scan?

28. Describe how nuclear chemistry can be used to detect an art forgery.

29. What does the unit *rem* describe?

PRACTICE PROBLEMS

Sample Problem A Balancing a Nuclear Equation

30. The decay of uranium-238 results in the spontaneous ejection of an alpha particle. Write the nuclear equation that describes this process.

31. What type of radiation is emitted in the decay described by the following equation?

$$^{43}_{19}K \longrightarrow {}^{43}_{20}Ca + ?$$

32. When a radon-222 nucleus decays, an alpha particle is emitted. Write the nuclear equation to show what happens when a radon-222 nucleus decays. What is the other product that forms?

33. One radioactive decay series that begins with uranium-235 and ends with lead-207 shows the partial sequence of emissions: alpha, beta, alpha, beta, alpha, alpha, alpha, alpha, beta, beta, and alpha. Write an equation for each reaction in the series.

34. Balance the following nuclear reactions.

a. $^{239}_{93}Np \longrightarrow {}^{0}_{-1}e + ?$

b. $^{9}_{4}Be + {}^{4}_{2}He \longrightarrow ?$

c. $^{32}_{15}P + ? \longrightarrow {}^{33}_{15}P$

d. $^{236}_{92}U \longrightarrow {}^{94}_{36}Kr + ? + 3{}^{1}_{0}n$

35. Complete and balance the following nuclear equations:

a. $^{187}_{75}Re + ? \longrightarrow {}^{188}_{75}Re + {}^{1}_{1}H$

b. $^{9}_{4}Be + {}^{4}_{2}He \longrightarrow ? + {}^{1}_{0}n$

c. $^{22}_{11}Na + ? \longrightarrow {}^{22}_{10}Ne$

36. Write the nuclear equation for the release of a positron by $^{117}_{54}Xe$.

Sample Problem B Determining the Age of an Artifact or Sample

37. Copper-64 is used to study brain tumors. Assume that the original mass of a sample of copper-64 is 26.00 g. After 64 hours, all that remains is 0.8125 g of copper-64. What is the half-life of this radioactive isotope?

38. The half-life of thorium-234 is 24.10 days. How many days until only one-sixteenth of a 52.0 g sample of thorium-234 remains?

39. The half-life of carbon-14 is 5715 y. How long will it be until only half of the carbon-14 in a sample remains?

40. The half-life of of element *Y* is 350 s. How many minutes are required for one-half of a given amount of *Y* to decay?

Sample Problem C Determining the Original Mass of a Sample

41. The half-life of one radon isotope is 3.8 days. If a sample of gas contains 4.38 g of radon-222, how much radon will remain in the sample after 15.2 days?

42. Phosphorus-32 is used to treat a certain form of leukemia. Starting with 10.0 mg, what mass of phosphorus-32 would remain after 57 days? The half-life of phosphorus-32 is 14.3 days.

43. After 4797 y, how much of an original 0.450 g of radium-226 remains? The half-life of radium-226 is 1599 y.

44. The half-life of cobalt-60 is 10.47 min. How many milligrams of Co-60 remain after 104.7 min if you start with 10.0 mg of Co-60?

MIXED REVIEW

45. Calculate the neutron-proton ratios for the following nuclides, and determine where they lie in relation to the band of stability.

  a. $^{235}_{92}U$

  b. $^{16}_{8}O$

  c. $^{56}_{26}Fe$

  d. $^{156}_{60}Nd$

46. Calculate the binding energy per nucleon of $^{238}_{92}U$ in joules. The atomic mass of a $^{238}_{92}U$ nucleus is 238.050 784 amu.

47. The energy released by the formation of a nucleus of $^{56}_{26}Fe$ is 7.89×10^{-11} J. Use Einstein's equation, $E = mc^2$, to determine how much mass is lost (in kilograms) in this process.

48. Calculate the binding energy for one mole of deuterium atoms. The measured mass of deuterium is 2.0140 amu.

49. What nuclear process is occuring in the sun shown? Also, write a nuclear reaction that describes this process.

50. The radiation given off by iodine-131 in the form of beta particles is used to treat cancer of the thyroid gland. Write the nuclear equation to describe the decay of an iodine-131 nucleus.

51. The plutonium isotope $^{239}_{94}Pu$ is sometimes detected in nuclear reactors. Consider that nuclear fuel contains a large portion of the common uranium isotope $^{238}_{92}U$ in addition to fissionable $^{235}_{92}U$. Describe a process by which $^{239}_{94}Pu$ might form in a nuclear reactor.

52. The parent nuclide of the thorium decay series is $^{232}_{90}Th$. The first four decays are as follows: alpha emission, beta emission, beta emission, and alpha emission. Write the nuclear equations for this series of emissions.

53. The half-life of radium-224 is 3.66 days. What was the original mass of radium-224 if 0.0500 g remains after 7.32 days?

54. How many milligrams remain of a 15.0 mg sample of radium-226 after 6396 y? The half-life of this isotope is 1599 y.

55. The mass of a $^{7}_{3}Li$ atom is 7.016 00 amu. Calculate its mass defect.

56. Determine whether each of the following nuclear reactions involves alpha decay, beta decay, positron emission, or electron capture.

  a. $^{234}_{90}Th \longrightarrow {}^{0}_{-1}e + {}^{234}_{91}Pa$

  b. $^{238}_{92}U \longrightarrow {}^{4}_{2}He + {}^{234}_{90}Th$

  c. $^{15}_{8}O \longrightarrow {}^{0}_{+1}e + {}^{15}_{7}N$

57. Uranium-238 decays through alpha decay with a half-life of 4.46×10^9 y. How long would it take for seven-eighths of a sample of uranium-238 to decay?

58. Write the nuclear equation for the release of an alpha particle by $^{157}_{70}Yb$.

59. The half-life of iodine-131 is 8.02 days. What percentage of an iodine-131 sample will remain after 40.2 days?

60. The mass of a $^{20}_{10}Ne$ atom is 19.992 44 amu. Calculate its mass defect.

61. Calculate the nuclear binding energy of one lithium-6 atom. The measured atomic mass of lithium-6 is 6.015 amu.

62. Write the nuclear equation for the release of a beta particle by $^{210}_{82}Pb$.

63. The half-life of an element X is 5.25 y. How many days are required for one-fourth of a given amount of X to decay?

671

24. Because of their low penetrating power, alpha particles cannot pass through the cover of the smoke detector.

25. Acute radiation exposure refers to a situation where a person is exposed to one or more large doses of radiation in a short period of time. Chronic radiation exposure refers to a situation where a person is exposed to many small doses of radiation over a long period of time.

26. The carbon-12 which plants take in as carbon dioxide is used to make food during photosynthesis. Animals eat plants or other animals that have eaten plants. Therefore, animals contain the same C-14 to C-12 ratio as plants.

27. The person is injected with a nuclide that has too many protons and that will undergo positron emission to decrease that number.

28. A sample of the object is subjected to neutron activation analysis. Bombardment with neutrons produces unstable nuclei that emit gamma rays. These rays are analyzed to identify the elements present in the sample and the source of those elements. If the source was not available at the time the work was supposedly done, then it is a forgery.

29. The unit *rem* describes the biological effect of an absorbed dose of radiation in humans.

30. $^{238}_{92}U \longrightarrow {}^{234}_{90}Th + {}^{4}_{2}He$

31. beta particle, $^{0}_{-1}e$

32. $^{222}_{86}Rn \longrightarrow {}^{218}_{84}Po + {}^{4}_{2}He$

33. $^{235}_{92}U \longrightarrow ^{231}_{90}Th + ^{4}_{2}He$

$^{231}_{90}Th \longrightarrow ^{231}_{91}Pa + ^{0}_{-1}e$

$^{231}_{91}Pa \longrightarrow ^{227}_{89}Ac + ^{4}_{2}He$

$^{227}_{89}Ac \longrightarrow ^{227}_{90}Th + ^{0}_{-1}e$

$^{227}_{90}Th \longrightarrow ^{223}_{88}Ra + ^{4}_{2}He$

$^{223}_{88}Ra \longrightarrow ^{219}_{86}Rn + ^{4}_{2}He$

$^{219}_{86}Rn \longrightarrow ^{215}_{84}Po + ^{4}_{2}He$

$^{215}_{84}Po \longrightarrow ^{211}_{82}Pb + ^{4}_{2}He$

$^{211}_{82}Pb \longrightarrow ^{211}_{83}Bi + ^{0}_{-1}e$

$^{211}_{83}Bi \longrightarrow ^{211}_{84}Po + ^{0}_{-1}e$

$^{211}_{84}Po \longrightarrow ^{207}_{82}Pb + ^{4}_{2}He$

34. a. $^{239}_{94}Pu$

b. $^{13}_{6}C$

c. $^{1}_{0}n$

d. $^{139}_{56}Ba$

35. a. $^{2}_{1}H$

b. $^{12}_{6}C$

c. $^{0}_{-1}e$

36. $^{117}_{54}Xe \longrightarrow ^{117}_{53}I + ^{0}_{+1}e$

37. The quantity 0.8215 g is 1/32 of the original quantity, 26.00 g, so five half-lives have passed (2^5 = 32). The half-life is 64/5 = 12.8 h.

38. 96.40 days

39. 5715 y

40. 5.83 min

41. The sample has passed through four half-lives after 15.2 days. Therefore, one-sixteenth of the original quantity, or 0.274 g, will remain.

42. 0.625 mg would remain after 4 half-lives (57.2 days)

43. 0.056 g

44. 9.77×10^{-3} mg

45. a. 1.55:1; outside

b. 1:1; within

c. 1.15:1; within

d. 1.6:1; outside

64. Complete the following nuclear reactions.

a. $^{12}_{5}B \longrightarrow ^{12}_{6}C + ?$

b. $^{225}_{89}Ac \longrightarrow ^{221}_{87}Fr + ?$

c. $^{63}_{28}Ni \longrightarrow ? + ^{0}_{-1}e$

d. $^{212}_{83}Bi \longrightarrow ? + ^{4}_{2}He$

65. Actinium-217 decays by releasing an alpha particle. Write an equation for this decay process, and determine what element is formed.

66. Indicate if the following equations represent fission reactions or fusion reactions.

a. $^{1}_{1}H + ^{2}_{1}H \longrightarrow ^{3}_{2}He + \gamma$

b. $^{1}_{0}n + ^{235}_{92}U \longrightarrow ^{146}_{57}La + ^{87}_{35}Br + 3^{1}_{0}n$

c. $^{21}_{10}Ne + ^{4}_{2}He \longrightarrow ^{24}_{12}Mg + ^{1}_{0}n$

d. $^{208}_{82}Pb + ^{58}_{26}Fe \longrightarrow ^{265}_{108}Hs + ^{1}_{0}n$

67. Predict whether the total mass of the 26 protons and neutrons that make up the iron nucleus will be more, less, or equal to 55.847 amu, the mass of an iron atom from the periodic table. If it is not equal, explain why not.

68. A sample of francium-212 will decay to one-sixteenth its original amount after 80 min. What is the half-life of francium-212?

69. A sample of strontium-90 is found to have decayed to one-eighth of its original amount after 87.3 y. What is the half-life of strontium-90?

70. Identify which of the four common types of nuclear radiation (beta, neutron, alpha, or gamma) correspond to the following descriptions:

a. an electron

b. uncharged particle

c. can be stopped by a piece of paper

d. high-energy light

71. Calculate the time required for three-fourths of a sample of cesium-138 to decay given that its half-life is 32.2 min.

72. Calculate that half-life of cesium-135 if seven-eighths of a sample decays in 6×10^6 y.

73. An archaeologist discovers a wooden mask whose carbon-14 to carbon-12 ratio is one-sixteenth the ratio measured in a newly fallen tree. How old does the wooden mask seem to be, given this evidence?

74. The half-life of tritium, $^{3}_{1}H$, is 12.3 y. How long will it take for seven-eighths of the sample to decay?

75. The longest-lived radioactive isotope yet discovered is the beta-emitter tellurium-130. It has been determined that it would take 2.5×10^{21} y for 99.9% of this isotope to decay. Write the equation for this reaction, and identify the isotope into which tellurium-130 decays.

76. It takes about 10^6 y for just half the samarium-149 in nature to decay by alpha-particle emission. Write the decay equation, and find the isotope that is produced by the reaction.

77. Describe some of the similarities and differences between atomic electrons and beta particles.

CRITICAL THINKING

78. Medium-mass nuclei have larger binding energies per nucleon than heavier nuclei do. What can you conclude from this fact?

79. Why are elevated temperatures necessary to initiate fusion reactions but not fission reactions?

80. The release of radioactive strontium into the atmosphere was once a major concern, especially for infants whose main food source was milk. Why were scientists concerned about radioactive strontium? (Hint: Check a periodic table for the members of the group that include strontium.)

81. Why do lighter elements undergo fusion more readily than heavier elements do?

82. Why is the constant rate of decay of radioactive nuclei so important in radioactive dating?

83. Why would someone working around radio-active waste in a landfill use a radiation monitor instead of a watch to determine when the workday is over? At what point would that person decide to stop working?

84. Radioactive isotopes are often used as "tracers" to follow the path of an element through a chemical reaction. For example, the oxygen atoms in O_2 produced by a plant during photosynthesis come from the oxygen in H_2O and not the oxygen in carbon dioxide, CO_2. Explain how you could use a radioactive isotope of oxygen to identify the source of the oxygen atoms in the O_2 produced during photosynthesis.

85. Suppose a nucleus captures two neutrons and decays to produce one neutron. Is this process likely to produce a chain reaction? Explain your reasoning.

86. Explain why charged particles do not pene-trate matter deeply.

87. Many radioactive isotopes have half-lives of several billion years. Other radioactive iso-topes have half-lives of billionths of a sec-ond. Suggest a way in which the half-lives of such isotopes are measured.

ALTERNATIVE ASSESSMENTS

88. Your local grocery store may sell perishable foods that have been irradiated. Find out what stores in your area sell irradiated foods, and determine the shelf life of these foods. What are the shelf lives of the same foods without irradiation? Do research to identify foods that are routinely irradi-ated even though their label may not indicate irradiation. Report your findings to the class.

89. Research some important historical findings that have been validated through radioactive dating. Report your findings to the class.

90. Design an experiment that illustrates the concept of half-life.

91. Research and evaluate environmental issues regarding the storage, containment, and disposal of nuclear wastes.

92. Suppose you are an energy consultant who has been asked to evaluate a proposal to build a power plant in a remote area of the desert. Research the requirements for each of the following types of power plant: nuclear-fission power plant, coal-burning power plant, solar-energy farm. Decide which of these power plants would be best for its surroundings, and write a paragraph supporting your decision.

CONCEPT MAPPING

93. Use the following terms to complete the concept map below: *critical mass, chain reaction, nuclear fission,* and *nucleon.*

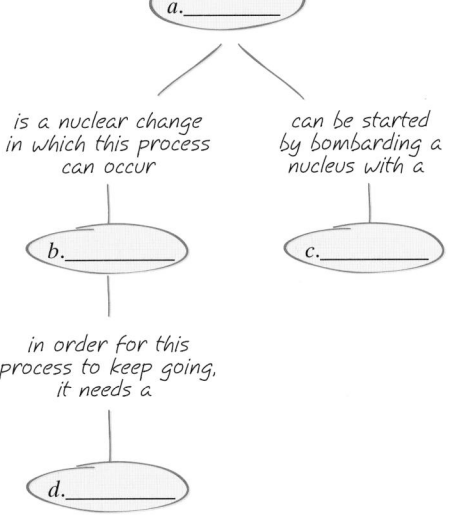

46. 1.18×10^{-12} J/nucleon

47. 8.77×10^{-28} kg

48. 1.75×10^{11} J/mol

49. fusion; one example is
$4{}^{1}_{1}\text{H} \longrightarrow {}^{4}_{2}\text{He} + 2{}^{0}_{+1}e$

50. ${}^{131}_{53}\text{I} \longrightarrow {}^{131}_{54}\text{Xe} + {}^{0}_{-1}e$

51. Students' answers may vary. The most common pathway that occurs is the following:
${}^{238}_{92}\text{U} + {}^{1}_{0}n \longrightarrow {}^{239}_{92}\text{U}$
${}^{239}_{92}\text{U} \longrightarrow {}^{239}_{93}\text{Np} + {}^{0}_{-1}e$
${}^{239}_{93}\text{Np} \longrightarrow {}^{239}_{94}\text{Pu} + {}^{0}_{-1}e$

52. ${}^{232}_{90}\text{Th} \longrightarrow {}^{4}_{2}\text{He} + {}^{228}_{88}\text{Ra}$
${}^{228}_{88}\text{Ra} \longrightarrow {}^{0}_{-1}e + {}^{228}_{89}\text{Ac}$
${}^{228}_{89}\text{Ac} \longrightarrow {}^{0}_{-1}e + {}^{228}_{90}\text{Th}$
${}^{228}_{90}\text{Th} \longrightarrow {}^{4}_{2}\text{He} + {}^{224}_{88}\text{Ra}$

53. 0.200 g

54. 0.937 mg

55. 0.04049 amu per atom

56. a. beta decay

   b. alpha decay

   c. positron emission

57. 3 half-lives or 1.34×10^{10} y

58. ${}^{157}_{70}\text{Yb} \longrightarrow {}^{4}_{2}\text{He} + {}^{153}_{68}\text{Er}$

59. 3%

60. 0.166 97 amu per atom

61. 4.9×10^{-12} J

62. ${}^{210}_{82}\text{Pb} \longrightarrow {}^{0}_{-1}e + {}^{210}_{83}\text{Bi}$

63. 796 days

64. a. ${}^{0}_{-1}e$

   b. ${}^{4}_{2}\text{He}$

   c. ${}^{63}_{29}\text{Cu}$

   d. ${}^{208}_{81}\text{Tl}$

65. ${}^{217}_{89}\text{Ac} \longrightarrow {}^{213}_{87}\text{Fr} + {}^{4}_{2}\text{He}$

66. a. fusion

   b. fission

   c. fusion

   d. fusion

67. The total mass of this nucleus—56 amu—will be greater than 55.847 amu. The mass is not equal to 55.847 amu because that value is an average of the masses of several different isotopes.

68. 20 min

69. 29.1 y

70. a. beta

   b. neutron

   c. alpha particle

   d. gamma

71. 64.4 min

72. 2×10^6 y

73. 22 860 y

74. 36.9 y

75. $^{130}_{52}\text{Te} \longrightarrow {}^{130}_{53}\text{I} + {}^{0}_{-1}e$

76. $^{149}_{62}\text{Sm} \longrightarrow {}^{145}_{60}\text{Nd} + {}^{4}_{2}\text{He}$

77. Atomic electrons and beta particles are electrons. Atomic electrons are found around (in an orbital) or are associated with a nucleus, while beta particles are emitted by a nucleus and cannot be found in an atom.

78. Medium-mass nuclei are more stable than very massive nuclei.

79. In order for fusion to occur, nuclei must collide with a certain amount of energy that can be attained only at extremely high temperatures.

80. Strontium is chemically similar to calcium and can replace calcium in bones and teeth. Milk is rich in calcium, so radioactive strontium could become concentrated in milk by grazing animals that have consumed strontium-containing foliage.

81. When light elements fuse, the nuclei formed are more likely to have a higher binding energy per nucleon and thus be more stable.

82. Radioactive dating will work only if the decay rates are the same today as they were in the past. In other words, the rate of decay must be constant in order to date an object using radioactive dating.

83. A safe dose of radiation exposure will have been established. That level is determined by the total exposure as detected by the film badge and not by the hours spent on the job.

Answers continued on p. 675A

FOCUS ON GRAPHING

Study the graph below, and answer the questions that follow.
For help in interpreting graphs, see Appendix B, "Study Skills for Chemistry."

94. Do stable nuclei that have *N/Z* numbers approximately equal to 1 have small or large atomic numbers?

95. Do stable nuclei that have *N/Z* numbers approximately equal to 1.5 have small or large atomic numbers?

96. Calculate the *N/Z* number for a nucleus A that has 70 neutrons and 50 protons.

97. Calculate the *N/Z* number for a nucleus B that has 90 neutrons and 60 protons.

98. Does nucleus A or nucleus B have an *N/Z* number closer to 1.5?

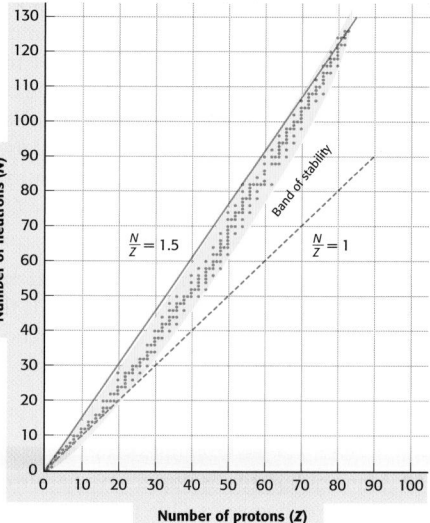

Neutron-Proton Ratios of Stable Nuclei

TECHNOLOGY AND LEARNING

99. Graphing Calculator

Calculating the Amount of Radioactive Material

The graphing calculator can run a program that graphs the relationship between the amount of radioactive material and elapsed time. Given the half-life of the radioactive material and the initial amount of material in grams, you will graph the relationship between the amount of radioactive material and the elapsed time. Then, with the elapsed time, you will trace the graph to calculate the amount of radioactive material.

Go to Appendix C. If you are using a TI-83 Plus, you can download the program **RADIOACT** and run the application as

directed. If you are using another calculator, your teacher will provide you with keystrokes and data sets to use. After you have run the program, answer these questions.

a. Determine the amount of neptunium-235 left after 2.0 years, given the half-life of neptunium-235 is 1.08 years and the initial amount was 8.00 g.

b. Determine the amount of neptunium-235 left after 5.0 years, given the half-life of neptunium-235 is 1.08 years and the initial amount was 8.00 g.

c. Determine the amount of uranium-232 left after 100 years, given the half-life of uranium-232 is 69 years and the initial amount was 10.0 g.

674

Chapter Resource File

• Chapter Test

Answers

1. c
2. c
3. d
4. b
5. a
6. c
7. b
8. d
9. b
10. b
11. a
12. c

1. Which of the following pairs represent a nucleon combination?
   a. protons and electrons
   b. neutrons and electrons
   c. protons and neutrons
   d. alpha particle and beta particle

2. When a nucleus is formed,
   a. some mass is gained.
   b. energy is absorbed.
   c. some mass is converted to energy.
   d. electrons are captured by the nucleus that forms.

3. Which of the following combinations of proton number and neutron number yields a nucleus that is least stable?
   a. even/even **c.** odd/even
   b. even/odd   **d.** odd/odd

4. When several smaller nuclei form a larger nucleus, _____ has occurred.
   a. nuclear fission
   b. nuclear fusion
   c. alpha emission
   d. gamma ray production

5. Which of the following nuclear equations is correctly balanced?
   a. $^{37}_{18}Ar + ^{0}_{-1}e \longrightarrow ^{37}_{17}Cl$
   b. $^{6}_{3}Li + 2^{1}_{0}n \longrightarrow ^{4}_{2}He + ^{3}_{1}H$
   c. $^{254}_{99}Es + ^{4}_{2}He \longrightarrow ^{258}_{101}Md + 2^{1}_{0}n$
   d. $^{14}_{7}N + ^{4}_{2}He \longrightarrow ^{17}_{8}O + ^{2}_{1}H$

6. Gamma rays
   a. have the same energy as beta particles do.
   b. are visible light.
   c. have no charge and no mass.
   d. are not a form of electromagnetic radiation.

7. Combining a nucleus of curium-246 with a nucleus of carbon-12 will produce a nucleus of
   a. einsteinium-254.
   b. nobelium-258.
   c. fermium-254.
   d. mendelevium-256.

8. The half-life of thorium-234 is 24 days. If you start with a 42.0 g sample of thorium-24, how much will remain after 72 days?
   a. 42.0 g   **c.** 10.5 g
   b. 21.0 g   **d.** 5.25 g

9. Alpha particles can safely be used in smoke detectors because they
   a. are not a form of radiation.
   b. cannot even penetrate paper.
   c. are harmless even if swallowed.
   d. combine with beta particles to produce a neutron.

10. Every radioactive isotope
    a. emits an alpha particle.
    b. has a characteristic half-life.
    c. has the same atomic mass.
    d. has an atomic number greater than 92.

11. A nucleus that is stable
    a. does not have a half-life.
    b. always emits an alpha particle.
    c. must have an equal number of protons and neutrons.
    d. has a low binding energy.

12. When a radioactive nuclide emits a beta particle, its
    a. atomic number decreases by one.
    b. mass number decreases by one.
    c. atomic number increases by one.
    d. atomic number remains the same.

675

Continuation of Answers

Answers to Section 1 Review, *continued*

9. Smaller nuclei tend to be more stable because the strong force acts among all or most of the particles. As the nuclei get larger, electrostatic repulsion between protons is greater than the strong force, breaking apart the nucleus.

10. The higher the nuclear binding energy per nucleon, the more stable the nucleus is.

11. With 3 protons and 3 neutrons, lithium-6 has a *N/Z* ratio of 1:1 and is therefore more stable.

12. **a.** unstable

   b. unstable

   c. stable

   d. stable

   e. unstable

Answers to Section 2 Review, *continued*

8. The two are not antiparticles and therefore will not annihilate each other when they collide.

9. The series stops when a stable isotope, which does not decay, is finally obtained.

10. This reaction is not an example of a nuclear change in that cobalt is not changed into an atom of another element. Only a neutron has been added to the nucleus of a cobalt-59 atom.

Answers to Section 3 Review, *continued*

9. The analysis reveals the samples' elemental composition and can tell you about the source materials used. It cannot tell you about who used them.

10. With short half-lives, these elements will decay quickly and not cause long-term radiation exposure to the patient.

11. The percentage remaining is 0.098%. With a half-life of 2.41×10^4 years, Pu-239 would take 2.41×10^5 years to pass through ten half-lives.

Answers to Chapter Review, *continued*

84. Set up an experiment in which one group of plants is supplied with water in which some oxygen-16 atoms have been replaced by oxygen-15 atoms, the radioactive isotope that has the longest half-life, 122 seconds. This group should be supplied with CO_2 in which none of the oxygen atoms are O-15. The second group should be supplied with water whose atoms all contain oxygen-16. However, this second group will be supplied with CO_2 where some of the O atoms are oxygen-15. After both groups have been photosynthesizing for some time, the atmosphere in the growth chambers is tested for oxygen-15. The presence of this isotope in the chamber with the first group of plants indicates that water is the source of the oxygen produced in photosynthesis.

85. A continued chain reaction will not occur. Each step of fission reactions requires more neutrons than the previous fission releases.

86. They ionize the materials that they go through. Each ionization transfers energy from the alpha or beta particle to the ionized particle. Less energy means less penetration.

87. Answers may include that it is calculated mathematically using an equation based on current and original amounts of radioactivity and lapsed time.

88. Students should find that irradiation significantly increases the shelf life of foods such as strawberries, onions, and potatoes.

89. Two examples that students might find interesting to investigate include the Shroud of Turin and the Iceman.

90. A logical choice of an object to use would be something that has two sides or faces, like a penny that was used in the Start-Up Activity.

91. Students' answers may vary.

92. Students' answers may vary.

93. **a.** nuclear fission

   b. chain reaction

   c. nucleon

   d. critical mass

94. small

95. large

96. 1.4

97. 1.5

98. nucleus B

99. **a.** 2.22 g

   b. 0.32 g

   c. 3.7 g

Carbon and Organic Compounds
Chapter Planning Guide

PACING	CLASSROOM RESOURCES	LABS, ACTIVITIES, AND DEMONSTRATIONS
BLOCK 1 • 45 min pp. 676–677 **Chapter Opener**		SE **Start-Up Activity** Testing Plastics, p. 677
BLOCK 2 • 45 min pp. 678–686 **Section 1** Compounds of Carbon	TT **Allotropes of Carbon** * OSP **Career Extension** Perfumer **GENERAL**	TE **Activity** Lewis Diagrams, p. 681 **GENERAL** TE **Activity** Functional Groups p. 683 **BASIC** TE **Group Activity** Isomers, p. 685 **BASIC** TE **Demonstration** Isomers and Properties, p. 685 CRF **Observation Lab** Carbon ◆ **GENERAL**
BLOCKS 3 & 4 • 90 min pp. 687–695 **Section 2** Names and Structures of Organic Compounds	OSP **Supplemental Reading** Cholesterol, Fat, and Fiber **ADVANCED** OSP **Career Extension** Organic Composter **GENERAL** TT **Types of Molecular Models** * OSP **Supplemental Reading** The Same and Not the Same: "Molecular Mimicry" **ADVANCED**	TE **Activity** Naming of Compounds with Functional Groups, p. 691 **GENERAL** TE **Demonstration** Unsaturated Fats, p. 693 CRF **Consumer Lab** Cloth of Many Colors ◆ **GENERAL**
BLOCKS 5 & 6 • 90 min pp. 696–702 **Section 3** Organic Reactions	OSP **Career Extension** Recycling Engineer **GENERAL**	TE **Group Activity** Types of Polymers, p. 698 **GENERAL** TE **Demonstration** What Type of Fiber Is It?, p. 700 SE **Observation Lab** Polymers and Toy Balls, p. 824 ◆ **GENERAL** CRF **Datasheets for In-Text Labs** *

BLOCK 7 • 45 min   **Chapter Review and Assessment Resources**

SE **Chapter Review,** pp. 704–708
SE **Standardized Test Prep,** p. 709
CRF **Chapter Test** *
OSP **Test Generator**
CRF **Test Item Listing** *
OSP **Scoring Rubrics and Classroom Management Checklists**

Holt Chemistry: Online Resources

Visit **go.hrw.com** for a variety of free resources related to this textbook. Enter the keyword **HW4 HOME**.

Holt Online Learning

Students can access interactive problem solving help and active visual concept development with the *Holt Chemistry* Online Edition available at **www.hrw.com**.

student CNN News

cnnstudentnews.com

Find the latest chemistry news, lesson plans, and activities related to important scientific events.

Compression guide:
To shorten your instruction
because of time limitations,
omit blocks 1, 4, and 6.

KEY

TE	Teacher Edition	**CRF**	Chapter Resource File	* Also on One-Stop Planner
SE	Student Edition	**TT**	Teaching Transparencies	◆ Requires Advance Prep
OSP	One-Stop Planner			

PROBLEM SOLVING AND PRACTICE	SECTION REVIEW AND ASSESSMENT	STANDARDS CORRELATION
		National Science Education Standards
	TE Quiz, p. 686 GENERAL **CRF** Quiz * **TE** Reteaching, p. 686 BASIC **SE** Section Review, p. 686 **CRF** Concept Review *	PS 2f
SE Sample Problem A Naming a Branched Hydrocarbon, p. 690 **SE** Practice, p. 690 GENERAL **TE** Homework, p. 690 GENERAL **SE** Sample Problem B Naming a Compound with a Functional Group, p. 692 **SE** Practice, p. 692 GENERAL **TE** Homework, p. 692 GENERAL **SE** Sample Problem C Drawing Structural and Skeletal Formulas, p. 694 **SE** Practice, p. 695 GENERAL **CRF** Problem Solving * ADVANCED **SE** Problem Bank, p. 858 GENERAL	**TE** Quiz, p. 695 GENERAL **CRF** Quiz * **SE** Section Review, p. 695 **CRF** Concept Review *	PS 2f
	TE Quiz, p. 701 GENERAL **CRF** Quiz * **TE** Reteaching, p. 701 BASIC **SE** Section Review, p. 701 **CRF** Concept Review *	PS 3a

www.scilinks.org

Topic: Organic Compounds
SciLinks code: HW4092

Topic: Carbon
SciLinks code: HW4138

Topic: Allotropes
SciLinks code: HW4009

Topic: Diamond/Graphite
SciLinks code:: HW4149

Topic: Alkanes
SciLinks code: HW4134

Topic: Aromatic Compounds
SciLinks code: HW4011

Topic: Polymers
SciLinks code: HW4098

Technology Resources

  Science in the NEWS

Each video segment is accompanied by a Critical Thinking Worksheet.

Segment 16
Carbon Nitride

Segment 17
Advances in Fuel
Technology

Segment 18
Chemical Separation
Techniques for Plastics

Overview

After the special bonding capacity of carbon is explained through its allotropes, students are introduced to straight-chain, cyclic, and aromatic hydrocarbons. They will learn that some hydrocarbons contain functional groups that determine their properties and identify these as belonging to specific groups of organic compounds. Students will write structural formulas and name compounds. They will study some typical organic reactions—substitution, addition, condensation, and elimination.

Assessing Prior Knowledge

Check for Content Knowledge
Students should be familiar with the following topics:

- electron configurations
- covalent and hydrogen bonds, Lewis structures

Using the Figure

Tomatoes like these are not only organic chemical warehouses; they are also chemical factories. Not long before this photo was taken, these luscious fruits were hard, green, and sour. The green color disappeared as chlorophyll broke down and was replaced by two newly synthesized carotene pigments that changed the color to yellow. Then a third pigment, lycopene, was synthesized, creating the red color evident here.

Meanwhile, calcium pectate—a compound that binds the cells together—was converted to pectin, and the fruit softened. Sour turned to sweet as the sugar content rose and the concentration of organic acids dropped. All of these substances, plus proteins, starches, vitamins, hormones, and many other kinds of organic compounds, are contained in a tomato, a kind of chemical soup with a cellulose casing.

CHAPTER

19

CARBON AND ORGANIC COMPOUNDS

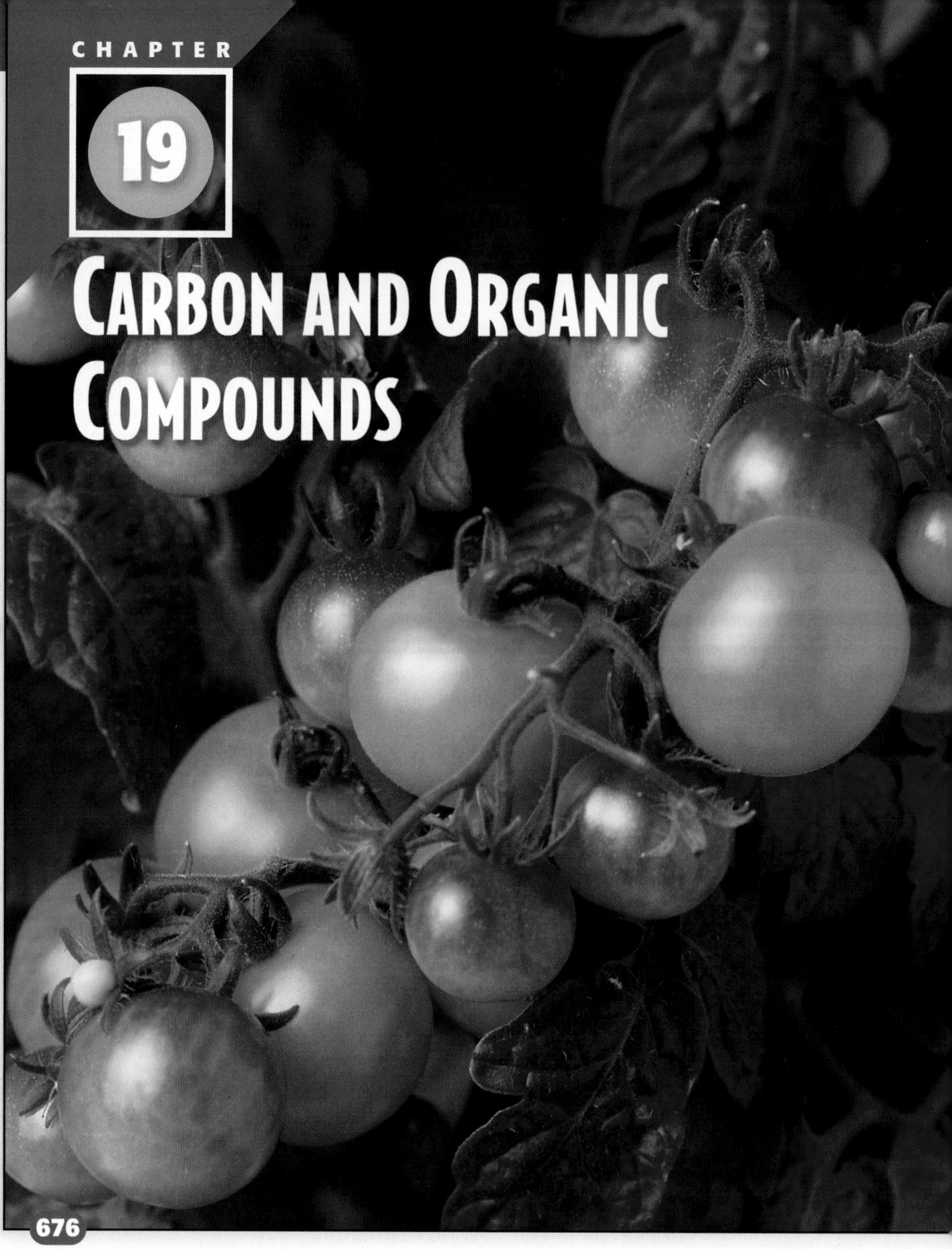

676

Standards Correlations

National Science Education Standards

PS 2f Carbon atoms can bond to one another in chains, rings, and branching networks to form a variety of structures, including synthetic polymers, oils, and the large molecules essential to life. (Sections 1 and 2)

PS 3a Chemical reactions occur all around us, for example in health care, cooking, cosmetics, and automobiles. Complex chemical reactions involving carbon-based molecules take place constantly in every cell in our bodies. (Section 3)

Tomatoes contain many compounds of carbon, including some that have properties that help people stay healthy. Two of these compounds are lycopene and beta-carotene. Lycopene gives tomatoes their red color and is believed to help prevent heart disease and some forms of cancer. In the human body, beta-carotene is converted to vitamin A, an essential nutrient.

Like the vine that supports the tomatoes in this picture, carbon forms the backbone for the chemicals that make up living organisms. In this chapter, you will learn about the nature of carbon and its many compounds.

START-UP ACTIVITY

Testing Plastics

PROCEDURE

1. Examine **two plastic samples** with a **magnifying lens** to look for any structural differences.

2. To test the rigidity of each sample, try to bend both pieces.

3. To test the hardness of each sample, press into each sample with your fingernail and try to make a permanent mark.

4. To test the strength of each sample, try tearing each plastic piece.

ANALYSIS

1. Which plastic sample would you use to hold liquids?

2. What physical differences did you observe between the two samples?

3. Why do you think most communities recycle only one of these plastics?

Pre-Reading Questions

① How many covalent bonds can a carbon atom form?

② How does the structure of a compound affect its chemical reactivity?

③ What are two possible ways to show the structure of CH_4?

CONTENTS 19

internet connect

www.scilinks.org
Topic: Organic Compounds
SciLinks code: HW4092

SCiLINKS. Maintained by the National Science Teachers Association

677

Overview

Before beginning this section, review with your students the Objectives listed in the Student Edition. Students will learn that the bonding characteristics of carbon account not only for the vast number of organic compounds but for the differences in carbon's allotropes. They will understand that in hydrocarbons, carbon forms single, double, and triple bonds and recognize the bonding characteristics of aromatic compounds. They will distinguish functional groups and classify compounds accordingly.

 **Bellringer**

Have students write a list of facts that they already know about carbon including its Lewis diagram. A quickly prepared master list will tell you what background you need to strengthen before going on.

Motivate

Discussion

Hold up a chemistry handbook. Show students the number of pages that apply to carbon compounds and the number that apply to compounds that do not contain carbon. Carbon compounds take up many more pages. Ask students to speculate about what makes carbon compounds special and how they differ from other compounds.

KEY TERMS

- hydrocarbon
- alkane
- alkene
- alkyne
- aromatic hydrocarbon
- functional group
- isomer

OBJECTIVES

1. **Explain** the unique properties of carbon that make the formation of organic molecules possible.

2. **Relate** the structures of diamond, graphite, and other allotropes of carbon to their properties.

3. **Describe** the nature of the bonds formed by carbon in alkanes, alkenes, alkynes, aromatic compounds, and cyclic compounds.

4. **Classify** organic compounds such as alcohols, esters, and ketones by their functional groups.

5. **Explain** how the structural difference between isomers is related to the difference in their properties.

Properties of Carbon

internet connect

www.scilinks.org
Topic: Carbon
SciLinks code: HW4138

SciLINKS Maintained by the National Science Teachers Association

The water bottle shown in **Figure 1** is made of a strong but flexible plastic. These properties result from the bonds formed by the carbon atoms that make up the plastic. Carbon atoms nearly always form covalent bonds. Three factors make the bonds that carbon atoms form with each other unique.

First, even a single covalent bond between two carbon atoms is quite strong. In contrast, the single covalent bond that forms between two oxygen atoms, such as in hydrogen peroxide (HO—OH), is so weak that this compound decomposes at room temperature. Second, carbon compounds are not extremely reactive under ordinary conditions. Butane, C_4H_{10}, is stable in air, but tetrasilane, Si_4H_{10}, catches fire spontaneously in air. Third, because carbon can form up to four single covalent bonds, a wide variety of compounds is possible.

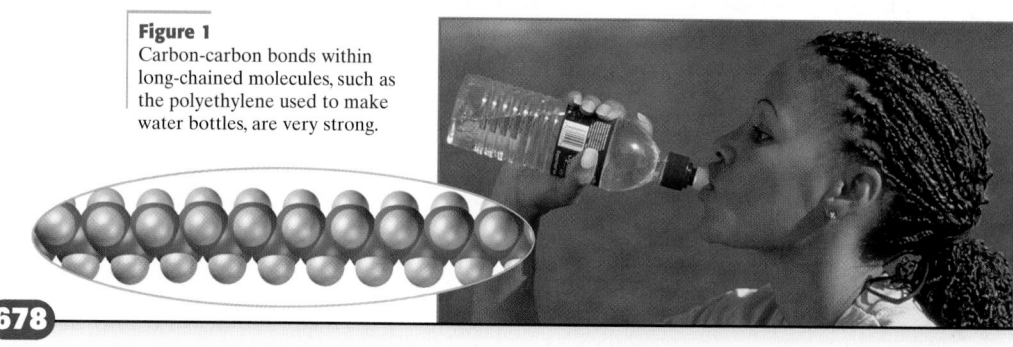

Figure 1
Carbon-carbon bonds within long-chained molecules, such as the polyethylene used to make water bottles, are very strong.

678

Chapter Resource File

- **Lesson Plan**

Videos

CNN **Presents Chemistry Connections**
- **Segment 16** Carbon Nitride
See the Science in the News video guide for more details.

One-Stop **Planner CD-ROM**

- **Supplemental Reading Projects**
  Guided Reading Worksheet:
  The Same and Not the Same, Essay 9
  "Handshakes in the Dark" and Essay 16
  "Representation and Reality"
  Assign this worksheet for cross-curricular connections to language arts.

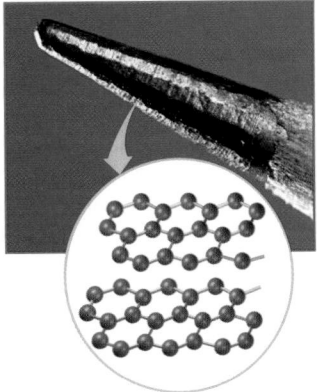
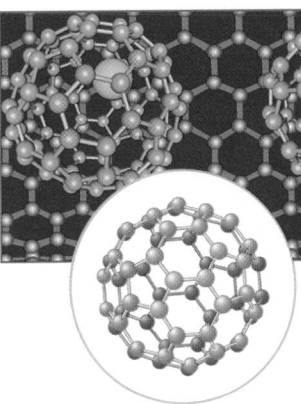

Figure 2

a Diamond is a carbon allotrope in which the atoms are densely packed in a tetrahedral arrangement.

b Graphite is a carbon allotrope in which the atoms form separate layers that can slide past one another.

c Buckminsterfullerene is a carbon allotrope in which 60 carbon atoms form a sphere.

Carbon Exists in Different Allotropes

As an element, carbon atoms can form different bonding arrangements, or *allotropes*. Three carbon allotropes are illustrated in **Figure 2**. As shown in **Figure 2a**, a diamond contains an enormous number of carbon atoms that form an extremely strong, tetrahedral network, which makes diamond the hardest known substance.

In contrast, graphite, another allotrope of carbon, is very soft. As illustrated in **Figure 2b**, the carbon atoms in graphite are bonded in a hexagonal pattern and lie in planes. The covalent bonds in each plane are very strong. However, weaker forces hold the planes together so that the planes can slip past each other. The sliding layers make graphite useful as a lubricant and as pencil lead. As you write with a pencil, the graphite layers slide apart, leaving a trail of graphite on the paper.

Other Carbon Allotropes Include Fullerenes and Nanotubes

In the mid-1980s, another type of carbon allotrope, the fullerene, was discovered. As illustrated in **Figure 2c**, fullerenes consist of near-spherical cages of carbon atoms. The most stable of these structures is C_{60}, which is formed by 60 carbon atoms arranged in interconnecting rings. The discoverers of these allotropes named C_{60} *buckminsterfullerene* in honor of the architect and designer Buckminster Fuller, whose geodesic domes had a similar shape. These allotropes can be found in the soot that forms when carbon-containing materials burn with limited oxygen.

In 1991, yet another carbon allotrope was discovered. Hexagons of carbon atoms were made to form a hollow cylinder known as a *nanotube*. A nanotube has a diameter about 10 000 times smaller than a human hair. Despite its thinness, a single nanotube is between 10 and 100 times stronger than steel by weight. Scientists are currently experimenting to find ways in technology and industry to use the unique properties of nanotubes.

679

REAL-WORLD CONNECTION

Other Carbon Allotropes It's a good idea to have chimneys cleaned regularly to remove deposits of incompletely burned organic materials called carbon black. This form of carbon consists of layers of carbon atoms similar to graphite but lacking the organization of graphite. Carbon black is used in the production of inks and rubber products. Charcoal is another form of carbon that also has a structure more haphazard than graphite. A close look at charcoal reveals that it is porous and therefore, has a large surface area. Some charcoal, called activated charcoal, can have a surface area of 1000 m^2 per gram. This property makes activated charcoal particularly useful for filtering any substances from water that may cause unwanted odors and flavors.

Paired Summarizing Divide your class into pairs and have each pair read the chapter or part of the chapter together and summarize the contents. Check each summary for completeness and have students save them for review.
LS Interpersonal

MISCONCEPTION
///ALERT

Students may think that all carbon compounds are organic compounds. Carbon oxides and sulfides, carbides, cyanides, carbonic acid, and carbonates are considered inorganic compounds.

Teaching Tip

As students review what they already know about alkanes, make a ball and stick model of methane and build up carbon by carbon to hexane or heptane. Tell students you are making straight-chain hydrocarbons, but that they should note that the chains zigzag back and forth as a result of the tetrahedral orientation of the bonds around carbon. As you make each model, write the formula on the board so that students can relate the way the model looks to the formula.

Videos

 Presents Chemistry Connections

• **Segment 17** Advances in Fuel Technology

See the Science in the News video guide for more details.

Organic Compounds

Most compounds of carbon are referred to as *organic compounds*. Organic compounds contain carbon, of course, and most also contain atoms of hydrogen.

In addition to hydrogen, many other elements can bond to carbon. These elements include oxygen, nitrogen, sulfur, phosphorus, and the halogens. These bonded atoms are found in the different types of organic compounds found in living things, including proteins, carbohydrates, lipids (fats), and nucleic acids. In addition, these atoms are used to make a wide variety of synthetic organic compounds including plastics, fabrics, rubber, and pharmaceutical drugs. **Figure 3** shows examples of some natural and synthetic organic compounds.

More than 12 million organic compounds are known, and thousands of new ones are discovered or synthesized each year. There are more known compounds of carbon than compounds of all the other elements combined. To make the study of these many organic compounds easier, chemists group those with similar characteristics. The simplest class of organic compounds are those that contain only carbon and hydrogen and are known as **hydrocarbons.** Hydrocarbons can be classified into three categories based on the type of bonding between the carbon atoms.

hydrocarbon

an organic compound composed only of carbon and hydrogen

Figure 3
a This shirt and the paper are both made of cellulose. Cellulose is made from chains of glucose molecules.

b Your hair is made of proteins that are made from smaller organic compounds called amino acids. Serine is an example of an amino acid.

c Citrus fruits contain citric acid, an organic acid.

d Caffeine is an organic compound that contains nitrogen.

680

---

did you know?

Products from Petroleum Petroleum is the direct source of numerous organic products, and provides the raw materials for uncountable others. Crude oil is a complex mixture of hydrocarbons of varying lengths. The mixture is roughly separated by means of distillation. To further separate these fractions, physical methods such as adsorption, stripping, solvent extraction, and crystallization are used. These methods produce products suitable for use as gasoline, kerosene, and lubricating oils. Other fractions may be subjected to chemical processes such as cracking and reforming. Cracking breaks large molecules into smaller ones. Reforming changes the molecule without breaking it.

Alkanes Are the Simplest Hydrocarbons

The simplest hydrocarbons, **alkanes,** have carbon atoms that are connected only by single bonds. Three examples include methane, ethane, and propane. The structural formulas for each of these alkanes are drawn as follows.

methane, CH_4 ethane, C_2H_6 propane, C_3H_8

If you examine the structural formulas for these three alkanes, you will notice that each member of the series differs from the one before by one carbon atom and two hydrogen atoms. This difference is more obvious when you compare the molecular formulas of each compound. The molecular formulas of the alkanes fit the general formula C_nH_{2n+2}, where n represents the number of carbon atoms. If the alkane contains 30 carbon atoms, then its formula is $C_{30}H_{62}$.

Many Hydrocarbons Have Multiple Bonds

The second class of hydrocarbons is the **alkenes,** which contain at least one double bond between two carbon atoms. The structural formulas for two alkenes are drawn as follows.

ethene   propene

Because alkenes with one double bond have twice as many hydrogen atoms as carbon atoms, their general formula is written C_nH_{2n}.

The third class of hydrocarbons is the **alkynes,** which contain at least one triple bond between two carbon atoms. The simplest alkyne is ethyne, C_2H_2, which is shown in **Figure 4.** The general formula for an alkyne with one triple bond is C_nH_{2n-2}.

alkane

a hydrocarbon characterized by a straight or branched carbon chain that contains only single bonds

internet connect

www.scilinks.org
Topic: Alkanes
SciLinks code: HW4134

*SCi*LINKS. Maintained by the National Science Teachers Association

alkene

a hydrocarbon that contains one or more double bonds

alkyne

a hydrocarbon that contains one or more triple bonds

Figure 4
Ethyne, commonly called *acetylene,* is one of the very few alkynes that are of practical importance. This welder is using an acetylene torch.

$$H-C\equiv C-H$$

Homework ——— BASIC

Graphic Organizer Have students create a table in which they can record the formulas of the alkanes, alkenes, and alkynes from the simplest formula up to ten carbon atoms. The table should have four columns with the headings *Carbon atoms, Alkanes, Alkenes,* and *Alkynes.* The first column, *Carbon atoms,* lists the number of carbon atoms in each compound, from one to 10. The column headed *Alkanes* should begin with CH_4 for one carbon atom and C_2H_6 for two carbon atoms. The column headed *Alkenes* should start in row two with C_2H_4, and the column headed *Alkynes* should begin in row two with C_2H_2. Students can apply the general formulas for the alkanes (C_nH_{2n+2}), the alkenes (C_nH_{2n}), and the alkynes (C_nH_{2n-2}). **LS** Visual

Activity ——— GENERAL

Have each student draw a Lewis diagram for ethane. Then have them draw a Lewis diagram for ethene and one for ethyne. Have them check their diagrams by drawing a circle around each carbon atom enclosing eight electrons and a circle around each hydrogen atom enclosing two electrons. If students need more practice, have them draw diagrams for propane, propene and propyne. **LS** Logical

---

MISCONCEPTION
///ALERT\\\

When students see formulas written CH_3–CH_2–CH_2–CH_3, they may think of the bonds shown as attaching two carbon atoms together along with two or three hydrogen atoms. Be sure that, in the beginning, you write formulas completely showing all the bonds to all the hydrogen atoms.

INCLUSION
Strategies

• *Attention Deficit Disorder*

Ask students to create a poster or bulletin board showing three categories of hydrocarbons. Students should identify each hydrocarbon, describe what makes these compounds hydrocarbons, draw a diagram of each of the compounds, and give examples of the different types of hydrocarbons. Students may also research acetylene as a hydrocarbon and the uses of acetylene.

A Natural Insecticide Cyclopropane is a strained ring structure and therefore, somewhat unstable because the bond angles required for the carbon atoms in forming a three-membered ring are 60°, considerably smaller than the tetrahedral angle, 109°, common for carbon. Nevertheless, compounds known as pyrethrins contain the cyclopropane group. These compounds form the natural insecticides found in chrysanthemums. Scientists are studying these compounds with hopes that they can be used with other types of plants, thus cutting down on the use of synthetic insecticides. Synthetic insecticides are usually more toxic than natural insecticides and may persist in the environment, causing pollution.

Discussion

Draw the two possible resonance structures for benzene on the board. Remind students that neither resonance structure represents the bonding in benzene and that they should not think of the molecule as shifting back and forth from one structure to the other. Refer them to **Figure 5** which shows that all the carbon and hydrogen atoms are in one plane with bonding electron clouds above and below the plane. Chemists use the hexagon with the circle inside to represent this model.

Carbon Atoms Can Form Rings

Carbon atoms that form covalent bonds with one another can be arranged in a straight line or in a ring structure. They can also be branched. For example, 4 carbon atoms and 10 hydrogen atoms can be arranged to form butane, C_4H_{10}, which has a linear structure. Four carbon atoms can also form a compound called cyclobutane, C_4H_8, which has a ring structure.

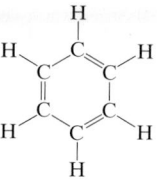

butane   cyclobutane

Notice that the prefix *cyclo-* is added to the name of the alkane to indicate that it has a ring structure.

Benzene Is an Important Ring Compound

A most important organic ring compound is the hydrocarbon benzene, C_6H_6. Benzene is the simplest member of a class of organic compounds known as **aromatic hydrocarbons.** These compounds have a variety of practical uses from insecticides to artificial flavorings. Benzene can be drawn as a six-carbon ring with three double bonds, as shown below.

aromatic hydrocarbon

a member of the class of hydrocarbons (of which benzene is the first member) that consists of assemblages of cyclic conjugated carbon atoms and that is characterized by large resonance energies

However, experiments show that all the carbon-carbon bonds in benzene are the same. In other words, benzene is a molecule with resonance structures. **Figure 5** illustrates how the electron orbitals in benzene overlap to form continuous molecular orbitals known as delocalized clouds. The following structural formula is often used to show the ring structure of benzene.

Topic Link

Refer to the "Covalent Compounds" chapter for a discussion of resonance structures.

The hexagon represents the six carbon atoms, while the circle represents the delocalized electron clouds. The hydrogen atoms are not shown in this simplified structural formula.

🖳 internet connect

www.scilinks.org
Topic: Aromatic Compounds
SciLinks code: HW4011

SC_iLINKS Maintained by the National Science Teachers Association

Figure 5
Electron orbitals in benzene overlap to form continuous orbitals that allow the delocalized electrons to spread uniformly over the entire ring.

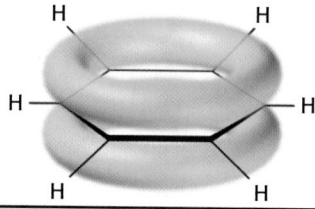

HISTORY
──
|CONNECTION|
──

Benzene Michael Faraday, the famous electrochemist, discovered benzene in 1825 as a component of the oily residue found in London gas mains. The structure of benzene was a mystery to chemists for many years. The German chemist Friedrich Kekulé, who had established that carbon was tetravalent and formed long chains of carbon atoms, was intrigued by the problem. As the story goes, he dreamt one night of a snake that curled around and bit its own tail, which gave Kekulé the idea that benzene is a six-membered ring of carbon atoms with six bonded hydrogen atoms.

Other Organic Compounds

Hydrocarbons are only one class of organic compounds. The other classes of organic compounds include other atoms such as oxygen, nitrogen, sulfur, phosphorus, and the halogens along with carbon (and usually hydrogen).

Less than 200 years ago, scientists believed that organic compounds could be made only by living things. The word *organic* that is used to describe these compounds comes from this belief. Then in 1828 a German chemist named Friedrich Wöhler synthesized urea, an organic compound, from inorganic substances.

Many Compounds Contain Functional Groups

Like most organic compounds, urea contains a group of atoms that is responsible for its chemical properties. Such a group of atoms is known as a **functional group.** Many common organic functional groups can be seen in **Figure 6.** Because single bonds between carbon atoms are rarely involved in most chemical reactions, functional groups, which contain bonds between carbon atoms and atoms of other elements, are often responsible for how an organic compound reacts. Organic compounds are commonly classified by the functional groups they contain. **Table 1** on the next page provides an overview of some common classes of organic compounds and their functional groups.

functional group

the portion of a molecule that is active in a chemical reaction and that determines the properties of many organic compounds

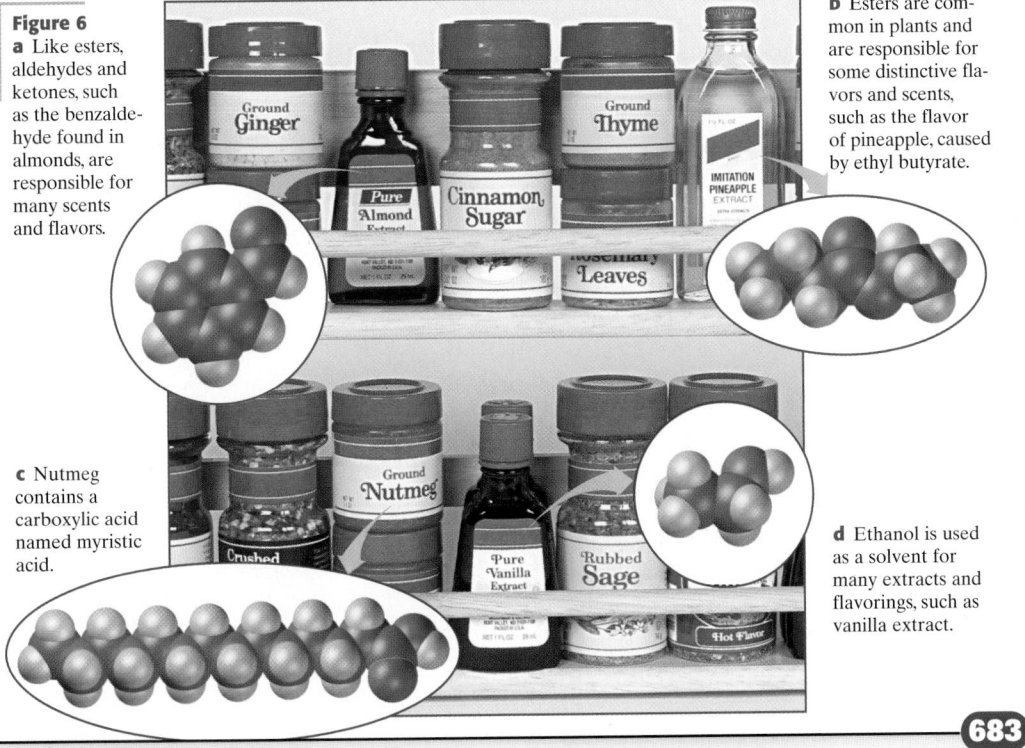

Figure 6
a Like esters, aldehydes and ketones, such as the benzaldehyde found in almonds, are responsible for many scents and flavors.

b Esters are common in plants and are responsible for some distinctive flavors and scents, such as the flavor of pineapple, caused by ethyl butyrate.

c Nutmeg contains a carboxylic acid named myristic acid.

d Ethanol is used as a solvent for many extracts and flavorings, such as vanilla extract.

683

did you know?

The common names of many carboxylic acids come from the Latin names of the organisms or products in which they occur. The following table gives some examples.

Acid	Source	Latin name for source
formic	ants	*formica*
acetic	vinegar	*acetum*
butanoic	butter	*butyrum*
capric	goats	*caper*

INCLUSION Strategies

• Gifted and Talented

Have students use the Internet or reference books to research Friedrich Wöhler and his work with organic compounds. Students should include a description of the belief that scientists had concerning where organic compounds come from, a brief biography of Wöhler, a summary of Wöhler's synthesis of organic compounds from inorganic compounds, and a list of what compounds Wöhler developed from his experiments.

Activity — BASIC

To give students more familiarity with compounds containing functional groups, have them write formulas for the series of alcohols beginning with methanol. Probably they will write the formulas for the primary alcohols. When they have done so, you might encourage them to place the –OH group on a carbon other than the end carbon and form a secondary or even tertiary alcohol. Have them repeat the exercise with the other functional groups. When writing formulas for amines, ask students to write primary ($-NH_2$), secondary ($-NH-$), and tertiary amines. **LS Visual**

Teaching Tip

Fragrances The chemical laboratory has the reputation for being the site of smelly concoctions, but when esters are being synthesized, the odors are often sweet, fruity, and pleasant. Esters are responsible for the fragrance of many fruits, herbs, and flowers. The odor of bananas is due to isopentyl ethanoate, methyl salicylate flavors wintergreen, and ethyl butyrate is responsible for the fragrance of pineapple. Esters are easily synthesized by combining a carboxylic acid with an alcohol in the presence of a strong acid, which acts as a catalyst. Synthetic esters used in foods and perfumes mimic the flavor and aroma of natural flavors, which are often a combination of more than one compound.

SKILL BUILDER — ADVANCED

Research Skills Vanillin is a natural product that contains an alcohol group, an aldehyde, and an ether. Ask interested students to research the structure of vanillin. **LS Logical**

Table 1   Classes of Organic Compounds

Teach, continued

Using the Table —— GENERAL

Refer students to **Table 1**, column 2. Tell them that the compounds that appear in column 3 are natural products, each containing the functional group shown. Ask them how many bonds connect an alcohol group to a molecule. Ans. one How many bonds attach an aldehyde group? Ans. one a ketone group? Ans. two an amine group? Ans. three Tell them that in the case of an amine, one, two, or three bonds can attach an amine group because it also can be $-NH_2$ or $-NH-$. Have students construct molecules containing each of the functional groups in the table by attaching the correct number of hydrocarbon groups to the functional group. **LS Visual**

Homework —— BASIC

Give students a list of simple organic structures, each containing at least one of the eight functional groups listed in **Table 1.** Have them circle the group or groups and name the functional group.
LS Logical

SKILL BUILDER — GENERAL

Writing Skills Tell students about chlorofluorocarbons (CFCs), such as trichlorofluoromethane in **Table 1**. These compounds have been implicated in the destruction of Earth's ozone layer and their use is banned in some countries. Ask students to investigate both the concern about ozone depletion and the ways in which CFCs have been used. Have them write a report about their findings.
LS Verbal

Class	Functional group	Example	Use
Alcohol	—OH	2-propanol	disinfectant
Aldehyde	O⫶ —C—H	benzaldehyde	almond flavor
Halide	—F, Cl, Br, I	trichlorofluoromethane (Freon-11)	refrigerant
Amine	—N⫶	caffeine	beverage ingredient
Carboxylic acid	O⫶ —C—OH	tetradecanoic acid (myristic acid)	soap-making ingredient
Ester	O⫶ —C—O—	ethyl butanoate	perfume ingredient
Ether	—O—	methyl phenyl ether (anisole)	perfume ingredient
Ketone	O⫶ —C—	propanone (acetone)	solvent in nail-polish remover

HISTORY CONNECTION

Synthesizing Medicines Obtaining organic compounds from natural sources can be costly and often yields amounts too small to be of practical use. Many organic compounds used as medicines are synthesized in a laboratory. In 1945, Dorothy Hodgkin determined the structure of penicillin, thus providing a crucial step toward its synthesis. In 1957, John Sheehan and his co-workers synthesized penicillin after a nine-year effort.

One-Stop Planner CD-ROM

• **Career Extension**
Real-World Connections Worksheet 12: Perfumer
Assign this worksheet to emphasize relevant applications of text concepts.

Functional Groups Determine Properties

The presence of a functional group in an organic compound causes the compound to have properties that differ greatly from those of the corresponding hydrocarbon. In fact, while molecules of very different sizes with the same functional group will have similar properties, molecules of similar sizes with different functional groups will have very different properties.

Compare the structural formulas of the molecules shown in **Table 2.** Notice that each of these molecules consists of four carbon atoms joined to one another by a single bond and arranged in a linear fashion. Notice, however, that each molecule, with the exception of butane, has a different functional group attached to one or more of these carbon atoms. As a result, each molecule has properties that differ greatly from butane.

For example, compare the boiling point of butane with those of the other compounds in **Table 2.** Butane is a gas at room temperature. Because of the symmetrical arrangement of the atoms, butane is nonpolar. Because the intermolecular forces between butane molecules are weak, butane has very low boiling and melting points and a lower density than the other four-carbon molecules.

Next compare the structural formulas of butane and 1-butanol in **Table 2.** Notice that the only difference between these two molecules is the presence of the functional group —OH on one of the carbon atoms in 1-butanol. The presence of this functional group causes 1-butanol to exist as a liquid at room temperature with much higher melting and boiling points and a significantly greater density than butane.

Table 2 Comparing Classes of Organic Compounds

Name	Structural formula	Melting point (°C)	Boiling point (°C)	Density (g/mL)
Butane		−138.4	−0.5	0.5788
1-butanol		−89.5	117.2	0.8098
Butanoic acid		−4.5	163.5	0.9577
2-butanone		−86.3	79.6	0.8054
Diethyl ether		−116.2	34.5	0.7138

685

did you know?

Organic Communicators Pheromones are organic compounds produced by an organism and used to communicate with others. For example, bees use pheromones to communicate the location of nectar or the presence of danger. Ants use them to leave a trail other ants can follow to a food source. Some pheromones are used to attract members of the opposite sex. Many pheromones contain ketone, ester, or ether functional groups.

Close

Quiz ———————— GENERAL

1. Write structural formulas for a straight-chain alkane, an alkene, and an alkyne that have four carbons atoms. **Ans.** alkane, no double or triple bond; alkene, one or two double bonds; alkyne, one or two triple bonds

2. Write the formula for any simple molecule containing each of the following functional groups: alcohol, aldehyde, chloride, amine, carboxylic acid, ester, ether, ketone. **Ans.** Alcohol, aldehyde, chloride, and carboxylic acid, each should be attached to one carbon atom. Amine could be attached to one, two, or three carbon atoms. Ester, ether, and ketone should be attached to two carbon atoms.

3. What is meant by an isomer? **Ans.** two or more compounds that have the same molecular formula, but different structures

LS Logical

Reteaching ———————— BASIC

Have students list the key terms for this section and write a clear definition of each. Ask them to give examples of each. **LS** Logical

---

Chapter Resource File

- Concept Review
- Quiz

Figure 7
Both of these molecules are alcohols. They are isomers of each other because they both have the molecular formula $C_4H_{10}O$.

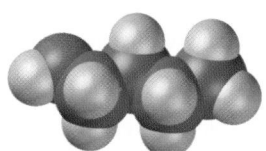
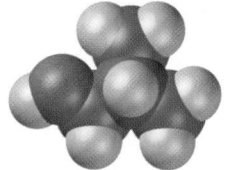

1-butanol   **2-methyl-1-propanol**
                              (isobutyl alcohol)

Different Isomers Have Different Properties

Examine the two molecules shown in **Figure 7**. Both have the same molecular formula: $C_4H_{10}O$. They differ, however, in the way in which their atoms are arranged. These two molecules are known as **isomers.** Isomers are compounds that have the same formula but differ in their chemical and physical properties because of the difference in the arrangement of their atoms. The greater the structural difference between two isomers, the more significant is the difference in their properties. Because the structural difference between the two isomers shown in **Figure 7** is minor, both molecules have similar boiling points and densities.

isomer

one of two or more compounds that have the same chemical composition but different structures

---

① Section Review

UNDERSTANDING KEY IDEAS

1. List the three factors that make the bonding of carbon atoms unique.

2. What are allotropes?

3. How are alkanes, alkenes, and alkynes similar? How are they different from each other?

4. Draw the simplified representation of the resonance structure for benzene.

5. List four elements other than carbon and hydrogen that can bond to carbon in organic compounds.

6. What is an aromatic compound?

7. What is a functional group?

8. What is an isomer? What do two molecules that are isomers of each other have in common?

CRITICAL THINKING

9. Draw a structural formula for the straight-chain hydrocarbon with the molecular formula C_3H_6. Is this an alkane, alkene, or alkyne?

10. Can molecules with molecular formulas C_4H_{10} and $C_4H_{10}O$ be isomers of one another? Why or why not?

11. Draw a structural formula for an alkyne that contains seven carbon atoms.

12. Draw the structural formulas for two isomers of C_4H_{10}.

13. Why is benzene not considered a cycloalkene even though double bonds exist between the carbon atoms that are arranged in a ring structure?

14. Write the molecular formulas for an alkane, alkene, and alkyne with 5 carbon atoms each. Why are these three hydrocarbons not considered isomers?

15. Draw C_4H_6 as a cycloalkene.

686

Answers to Section Review

1. 1) Strong bonds exist between carbon atoms
2) Carbon compounds display low reactivity compared to compounds of other elements
3) By being able to form four single bonds, carbon can form a wide variety of compounds.

2. forms of an element that differ in either structure or in bonding

3. All are hydrocarbons. Alkanes contain only single bonds. Alkenes contain at least one double bond. Alkynes contain one or more triple bonds.

4.

5. Answers should include four of the following: oxygen, nitrogen, phosphorus, sulfur, and the halogens.

6. An aromatic compound contains a ring of carbon atoms that displays resonance because of delocalized electrons that provide great stability.

Answers continued on p. 709A

SECTION 2

Names and Structures of Organic Compounds

KEY TERMS

- **saturated hydrocarbon**
- **unsaturated hydrocarbon**

OBJECTIVES

1. **Name** simple hydrocarbons from their structural formulas.
2. **Name** branched hydrocarbons from their structural formulas.
3. **Identify** functional groups from a structural formula, and assign names to compounds containing functional groups.
4. **Draw** and interpret structural formulas and skeletal structures for common organic compounds.

Naming Straight-Chain Hydrocarbons

Inorganic carbon compounds, such as carbon dioxide, are named by using a system of prefixes and suffixes. Organic compounds have their own naming scheme, which includes prefixes and suffixes that denote the class of organic compound. Learning just a few rules will help you decipher the names of most common organic compounds.

For example, the names of all alkanes end with the suffix *-ane*. The simplest alkane is methane, CH_4, the main component of natural gas. **Table 3** lists the names and formulas for the first 10 straight-chain alkanes. For alkanes that consist of five or more carbon atoms, the prefix comes from a Greek or Latin word that indicates the number of carbon atoms in the chain.

Table 3 Straight-Chain Alkane Nomenclature

Number of carbon atoms	Name	Formula
1	methane	CH_4
2	ethane	$CH_3—CH_3$
3	propane	$CH_3—CH_2—CH_3$
4	butane	$CH_3—CH_2—CH_2—CH_3$
5	pentane	$CH_3—CH_2—CH_2—CH_2—CH_3$
6	hexane	$CH_3—CH_2—CH_2—CH_2—CH_2—CH_3$
7	heptane	$CH_3—CH_2—CH_2—CH_2—CH_2—CH_2—CH_3$
8	octane	$CH_3—CH_2—CH_2—CH_2—CH_2—CH_2—CH_2—CH_3$
9	nonane	$CH_3—CH_2—CH_2—CH_2—CH_2—CH_2—CH_2—CH_2—CH_3$
10	decane	$CH_3—CH_2—CH_2—CH_2—CH_2—CH_2—CH_2—CH_2—CH_2—CH_3$

687

Chapter Resource File

- Lesson Plan

SECTION 2

Focus

Overview

Before beginning this section, review with your students the Objectives listed in the Student Edition. In this section, students will learn to name organic compounds starting with simple and branched hydrocarbons and followed by compounds containing functional groups. Once nomenclature is mastered, students will draw structural formulas for compounds from their formulas.

Bellringer

Have students examine **Table 3** and determine what organizing principles they see that will help them quickly process all the information contained in the table. **Ans.** Each formula beyond ethane contains one more $–CH_2–$ group. Beyond butane, the formulas are designated by their prefixes.

Motivate

Discussion

Ask students to describe a pineapple upside-down cake. Point out that the name indicates that the cake has been turned upside down and contains pineapple. Have students think of names of objects that express part or all of what makes up the objects. Red-headed woodpecker is another example. Emphasize that names of objects can describe the object's makeup, as do the names of organic compounds.

Teaching Tip

Students may not realize that the list of compounds in **Table 3** could continue through compounds with chain length many times the length of decane, and that the properties of the compounds change with length. For example, methane, ethane, propane, and butane are gases at ordinary temperatures. Pentane is a liquid, as are all the remaining compounds listed in the table. Long-chain alkanes are solids and are called *paraffins*.

Reading Organizer Have students go through the section and make an outline using the head and subheads. Tell them to leave at least five lines of space under each head. Then as they read the chapter, have them summarize the content of each segment. **LS** Logical

Teaching Tip

Emphasize to students that *unsaturated* means "able to hold more," just as it did in the chapter "Solutions." An unsaturated sponge can hold more water; an unsaturated saltwater solution can hold more salt. Unsaturated compounds can hold more hydrogen. An unsaturated hydrocarbon contains fewer hydrogen atoms than does a saturated hydrocarbon with the same number of carbon atoms.

Homework — GENERAL

Graphic Organizer Have students organize what they have learned about hydrocarbons in a table. The table should have four columns labeled *Hydrocarbon, Bonding, Suffix,* and *General formula*. Tell them to list alkane, alkene, and alkyne in the first column. In the second column, students should state what is characteristic about the bonding in each class of compounds. In the third column, the suffixes that denote each class of compounds. In the fourth column, the general formula that applies to all in the group. **LS** Logical

saturated hydrocarbon

an organic compound formed only by carbon and hydrogen linked by single bonds

unsaturated hydrocarbon

a hydrocarbon that has available valence bonds, usually from double or triple bonds with carbon

Naming Short-Chain Alkenes and Alkynes

The scheme used to name straight-chain hydrocarbons applies to both saturated and unsaturated compounds. A **saturated hydrocarbon** is a hydrocarbon in which each carbon atom forms four single covalent bonds with other atoms. The alkanes are saturated hydrocarbons. An **unsaturated hydrocarbon** is a hydrocarbon in which not all carbon atoms have four single covalent bonds. The alkenes and alkynes are unsaturated hydrocarbons.

The rules for naming an unsaturated hydrocarbon with fewer than four carbon atoms are similar to those for naming alkanes. A two-carbon alkene is named *ethene*, with the suffix *-ene* indicating that the molecule is an alkene. A three-carbon alkyne is named *propyne*, with the suffix *-yne* indicating that the molecule is an alkyne.

Naming Long-Chain Alkenes and Alkynes

The name for an unsaturated hydrocarbon containing four or more carbon atoms must indicate the position of the double or triple bond within the molecule. First number the carbon atoms in the chain so that the first carbon atom in the double bond has the lowest number. Examine **Figure 8,** which shows structural formulas for two alkenes with five carbon atoms.

The correct name for the alkene shown on the left in **Figure 8** is *1-pentene*. The molecule is correctly numbered from left to right because the first carbon atom with the double bond must have the lowest number. The name *1-pentene* indicates that the double bond is present between the first and second carbon atoms. The alkene shown on the right in **Figure 8** is correctly named *2-pentene*, indicating that the double bond is present between the second and third carbon atoms. Note that 1-pentene and 2-pentene are the only possible pentenes, because 3-pentene would be the same molecule as 2-pentene and the lower numbering is preferred.

If there is more than one multiple bond in a molecule, number the position of each multiple bond, and use a prefix to indicate the number of multiple bonds. For example, the following molecule is called *1,3-pentadiene*. (Note the placement of the prefix *di-*.)

Figure 8
Both the names and structural formulas indicate the position of the double bond in each alkene. Notice that you cannot tell from the space-filling models where the double bond is located.

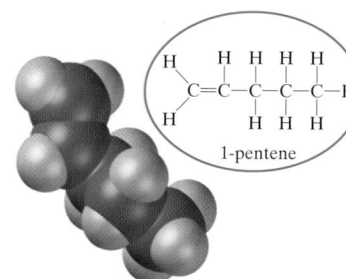

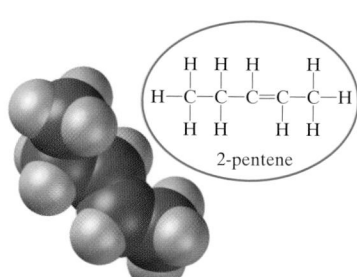

Naming Branched Hydrocarbons

When naming a hydrocarbon that is not a simple straight chain, first determine the number of carbon atoms in the longest chain. It can be named based on the corresponding alkane in **Table 3.** The longest chain may not appear straight in a structural formula, as in the example below.

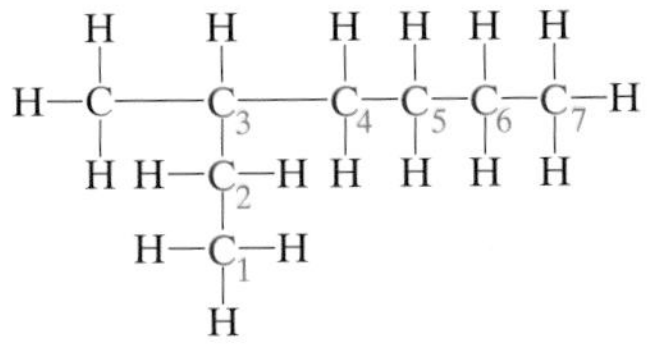

The "parent" chain in the compound shown above contains seven carbon atoms, so it is heptane. Next, number the carbon atoms on the parent chain so that any branches on the chain have the lowest numbers possible.

Name the Attached Groups and Indicate Their Positions

In the structural formula above, all the numbered carbon atoms, with one exception, are bonded to at least two hydrogen atoms. The exception is the third carbon atom, which has a $—CH_3$ group attached. This group is called a *methyl group,* because it is similar to a methane molecule, but with one less hydrogen atom. Because the methyl group is attached to the third carbon, the complete name for this branched alkane is *3-methylheptane*.

You can omit the numbers if there is no possibility of ambiguity. For example, a propane chain can have a methyl group only on its second carbon (if the methyl group were on the first or third carbon of propane, the molecule would be butane). So, what you might want to call *2-methylpropane* would be called *methylpropane.*

With unsaturated hydrocarbons that have attached groups, the longest chain containing the double bond is considered the parent compound. In addition, if more than one group is attached to the longest chain, the position of attachment of each group is given. Prefixes are used if the same group is attached more than once. Examine the following structural formula for a branched alkene.

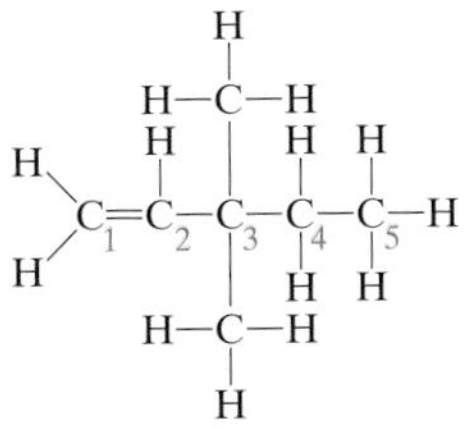

The chain containing the double bond has five carbon atoms. Therefore, the compound is a pentene. Notice that the first carbon atom has a double bond, making the chain 1-pentene. Because two methyl groups are attached to the third carbon atom, the correct name for this branched alkene is *3,3-dimethyl-1-pentene*.

Teaching Tip

Nomenclature Rules Give students the following additional naming rules:

- The parent chain is the longest continuous chain that contains either the most straight chain branches or a multiple bond.
- If more than one type of branch is present, the names are arranged in alphabetical order in front of the name of the parent hydrocarbon.
- If a compound contains both branches and a multiple bond, the carbon atoms are numbered so that the carbon with the multiple bond has the lowest number possible.

TECHNOLOGY CONNECTION

How do chemists figure out the structures of complex organic molecules? They can determine the molar mass by mass spectrometry and use laboratory methods to determine the percent of the elements in a compound, but how the atoms are arranged is a puzzle that can be solved by nuclear magnetic resonance (NMR) spectroscopy. In this technique, hydrogen atoms in different environments within a molecule absorb different amounts of energy when irradiated by radio frequencies in a magnetic field. The absorbed energy causes the spins of the hydrogen atoms to align opposite to the magnetic field. By using a standard for comparison, the number of hydrogen atoms in particular environments can be determined and related to the suspected structure of the molecule.

Answers to Practice Problem A

a. 2,2,4-trimethylpentane

b. 1-pentyne

c. 2,3,4-trimethylnonane

d. 2-methyl-3-hexene

Homework —— GENERAL

Additional Practice Name the following compounds.

1.

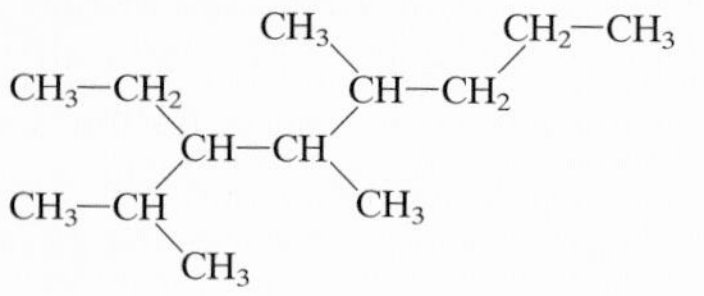

$$CH_3-CH_2 \quad CH_3 \quad CH_2-CH_3$$
$$CH-CH_2$$
$$CH-CH$$
$$CH_3-CH \quad CH_3$$
$$CH_3$$

Ans. 3-ethyl-2, 4, 5-trimethyloctane

2.

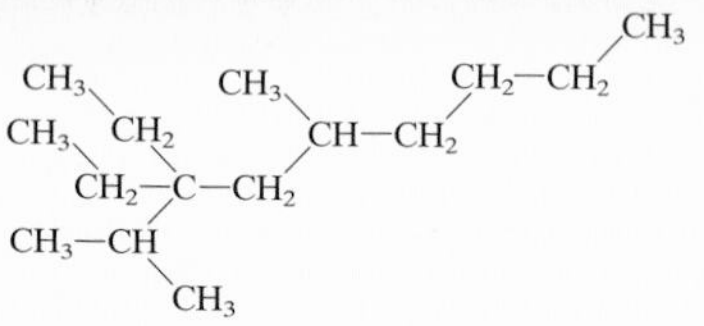

$$CH_3 \quad CH_3 \quad CH_3$$
$$CH_3 \quad CH_2 \quad CH_2-CH_2$$
$$CH_2-C-CH_2 \quad CH-CH_2$$
$$CH_3-CH$$
$$CH_3$$

Ans. 3,3-diethyl-2, 5-dimethylnonane

LS Logical

Naming a Branched Hydrocarbon

Name the following hydrocarbon.

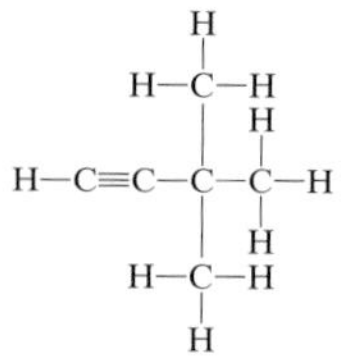

$$\begin{array}{c} H \\ H-C-H \\ | \\ H \\ H-C\equiv C-C-C-H \\ | \\ H \\ H-C-H \\ | \\ H \end{array}$$

1 **Gather information.**

• The triple bond makes the branched hydrocarbon an alkyne.

2 **Plan your work.**

• Identify the longest continuous chain (the "parent" chain), and name it.
• Number the parent chain so that the triple bond is attached to the carbon atom with the lowest possible number.
• Name the groups that make up the branches.
• Identify the positions that the branches occupy on the longest chain.

3 **Name the structure.**

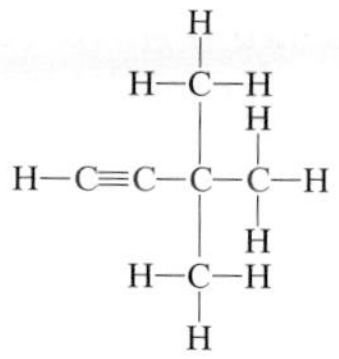

• The longest continuous chain has four carbon atoms.
• The parent chain is butyne.
• The numbering begins with the triple bond.
• Two methyl, —CH₃, groups are present.
• Both methyl groups are attached to the third carbon atom.
• The name of this branched hydrocarbon is *3, 3-dimethyl-1-butyne*.

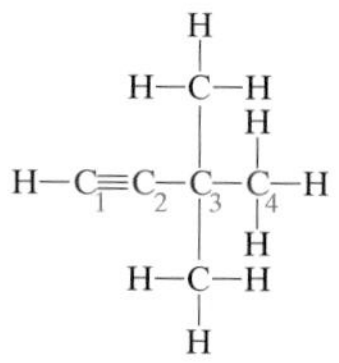

4 **Verify your results.**

• The parent name *butyne* indicates that four carbon atoms are present in the longest chain. The *1-butyne* indicates that the first carbon atom has a triple bond. The *3,3-dimethyl-* indicates that two methyl groups, —CH₃, are attached to the third carbon atom in the longest chain.

PRACTICE

Name the following branched hydrocarbons.

1 a.

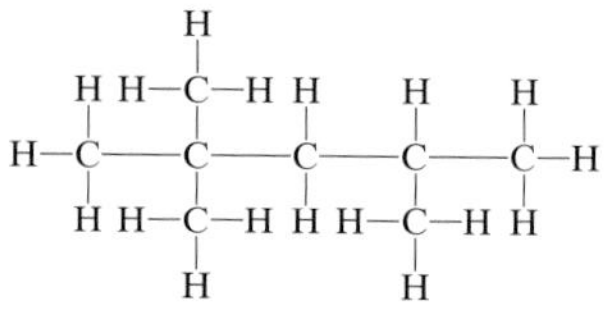

$$\begin{array}{c} H \\ H \; H-C-H \; H \quad H \quad H \\ | \\ H-C \quad C \quad C \quad C \quad C-H \\ | \\ H \; H-C-H \; H \; H \; H-C-H \; H \\ | \qquad\qquad\quad | \\ H \qquad\qquad\quad H \end{array}$$

b.

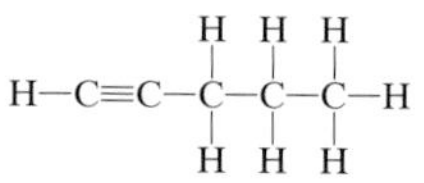

$$\begin{array}{c} H \; H \; H \\ H-C\equiv C-C-C-C-H \\ | \; | \; | \\ H \; H \; H \end{array}$$

690

---

1 c.

```
            H               H
      H H—C—H H H—C—H H H H H
      |   |  | |   |  | | | | |
  H—C—C—C—C—C—C—C—C—C—H
      |   |  | |   |  | | | | |
      H   H H—C—H H H H H H
                |
                H
```

d.

```
            H
      H H—C—H           H H
      |   |             | |
  H—C—C—C=C—C—C—H
      |   |     | | | |
      H   H     H H H H
```

Names of Compounds Reflect Functional Groups

Names for organic compounds with functional groups are based on the same system used for hydrocarbons with branched chains. First, the longest chain is named. Then a prefix or suffix indicating the functional group is added to the hydrocarbon name. **Table 4** lists the prefixes and suffixes for various functional groups. When necessary, the position of the functional group is noted in the same way that the position of hydrocarbon branches is noted. Consider the following structural formula.

```
      H   H   H
      |   |   |
  H—C—C—C—H
      |   |   |
      H   O   H
          |
          H
```

Because the longest chain consists of three carbon atoms, the name for this compound is based on propane. From **Table 1,** you can see that the presence of the —OH functional group classifies this compound as an alcohol. Therefore, as indicated by **Table 4,** the name for this compound is *propanol,* whose suffix *-ol* indicates that this molecule is an alcohol. Because the functional group is attached to the second carbon atom, the correct name for this compound is *2-propanol.* A number of organic compounds are often referred to by their common names, even by chemists. The common name for 2-propanol is *isopropyl alcohol.*

Table 4 Naming Compounds with Functional Groups

Class of compound	Suffix or prefix	Example
Alcohol	*-ol*	propanol
Aldehyde	*-al*	butanal
Amine	*-amine* or *amino-*	methylamine
Carboxylic acid	*-oic acid*	ethanoic acid
Ketone	*-one*	propanone

691

did you know?

Methanol and Ethanol Methanol is the principal ingredient in dry gas, a product that is sometimes added to gasoline to prevent gas lines from freezing up during cold weather. The methanol combines with any water in the gas tank and lowers its freezing point. Ethanol is being used in some areas as an additive to gasoline in a product called gasohol. Gasohol is a blend of approximately 10 percent ethanol and 90 percent gasoline.

Teach, *continued*

Teaching Tip

Multiple Functional Groups Be sure students know that if an alcohol has two –OH groups, it is called a diol; three would make it a triol. Similarly, a compound could be designated diamino- or trichloro-.

Answers to Practice Problem B

a. ethanol

b. 2-pentanone

c. butanoic acid

d. 3-hexanol

Homework ——— BASIC

Additional Practice Name the following compounds.

1.

OH
|
CH   CH₃
| |
CH₃   CH₂

Ans. 2-butanol

2.

NH₂
|
CH   CH₂   CH₃
| |
CH₃   CH₂   CH₂

Ans. 2-aminohexane

3.

                    O
                    ‖
CH₃   CH₂   C   CH₃
    | | |
   CH₂   CH₂   CH₂

Ans. 3-heptanone

4.

         OH
          |
CH₃   CH
    | |
   CH   CH₃
    |
   OH

Ans. 2,3-butanediol

LS Logical

SAMPLE PROBLEM B

Naming a Compound with a Functional Group

Name the following organic compound.

```
        H  H  H  O  H  H
        |  |  |  ‖  |  |
     H—C—C—C—C—C—C—H
        |  |  |     |  |
        H  H  H     H  H
```

1 **Gather information.**

- Notice that the functional group indicates that this compound is a ketone.

2 **Plan your work.**

- Identify the longest continuous chain (the "parent" chain), and name it.
- Number the parent chain so that the functional group is attached to the carbon atom with the lowest possible number.
- Identify the position that the functional group occupies on the longest chain.
- Name the organic compound.

3 **Name the structure.**

- The longest continuous chain has six carbon atoms: the parent chain is hexane.
- The carbon atoms are numbered from right to left to give the ketone functional group the lowest number.

```
        H  H  H  O  H  H
        |  |  |  ‖  |  |
     H—C—C—C—C—C—C—H
        |6 |5 |4 3 |2 |1
        H  H  H     H  H
```

- The name of this organic compound is *3-hexanone*.

4 **Verify your results.**

- The name *3-hexanone* indicates that six carbon atoms are present in the parent chain. The suffix *-one* indicates that this compound is a ketone. The *3-* indicates that the functional group is attached to the third carbon atom in the parent chain.

> **PRACTICE HINT**
>
> The steps to follow for naming organic compounds with functional groups are similar to those for naming branched hydrocarbons.

PRACTICE

Name the following organic compounds.

1 **a.**

```
       H  OH
       |  |
    H—C—C—H
       |  |
       H  H
```

b.

```
       H  H  H  O  H
       |  |  |  ‖  |
    H—C—C—C—C—C—H
       |  |  |     |
       H  H  H     H
```

c.

```
       H  H  H  O
       |  |  |  ‖
    H—C—C—C—C—OH
       |  |  |
       H  H  H
```

d.

```
       H  H  OH H  H  H
       |  |  |  |  |  |
    H—C—C—C—C—C—C—H
       |  |  |  |  |  |
       H  H  H  H  H  H
```

Representing Organic Molecules

Table 5 shows four ways of representing the organic molecule cyclohexane. Each type of model used to represent an organic compound has both advantages and disadvantages. Each one highlights a different feature of the molecule, from the number and kinds of atoms in a chemical formula to the three-dimensional shape of the space-filling model. Keep in mind that a picture or model cannot fully convey the true three-dimensional shape of a molecule or show the motion within a molecule caused by the atoms' constant vibration.

Structural Formulas Can Be Simplified

Structural formulas are sometimes represented by what are called *skeletal structures*, which show bonds, but leave out some or even all of the carbon and hydrogen atoms. You have already seen the skeletal structure for benzene, which is a hexagon with a ring inside it.

A skeletal structure usually shows the carbon framework of a molecule only as lines representing bonds. These lines are often drawn in a zigzag pattern to indicate the tetrahedral arrangement of bonds between a carbon atom and other atoms. Carbon atoms are understood to be at each bond along with enough hydrogen atoms so that each carbon atom has four bonds. Atoms other than carbon and hydrogen are always shown, which highlights any functional groups present.

Table 5   Types of Molecular Models

Type of model	Example	Advantages	Disadvantages
Chemical formula	C_6H_{12}	shows number of atoms in a molecule	does not show bonds, atom sizes, or shape
Structural formula		shows arrangement of all atoms and bonds in a molecule	does not show actual shape of molecule or atom sizes; larger molecules can be too complicated to draw easily
Skeletal structure		shows arrangements of carbon atoms; is simple	does not show actual shape or atom sizes; does not show all atoms or bonds
Space-filling model		shows three-dimensional shape of molecule; shows most of the space taken by electrons	uses false colors to differentiate between elements; bonds are not clearly indicated; parts of large molecules may be hidden

693

Demonstration ——— GENERAL
Unsaturated Fats

(Approximate time: 10 minutes)

1. Under a fume hood, prepare an iodine solution by dissolving 1.5 g of iodine in 50 mL of mineral oil.

2. Pour 10 mL of cooking oil into a 50 mL beaker.

3. Put the beaker on an overhead projector, and project the image onto a screen.

4. Add 3 mL of the iodine solution to the cooking oil, and stir thoroughly. The color will disappear, indicating that iodine has been added to the double bonds of the oil.

5. Ask students to use these results to determine whether mineral oil and cooking oil are saturated or unsaturated. **Ans.** mineral oil, saturated; cooking oil, unsaturated

LS Visual

One-Stop Planner CD-ROM

• **Supplemental Reading Projects**
Guided Reading Worksheet: The Same and Not the Same, Essay 10 "Molecular Mimicry"
Assign this worksheet for cross-curricular connections to language arts.

Transparencies

TT Types of Molecular Models

HEALTH
CONNECTION — ADVANCED

Carcinogens Aromatic compounds, those built around the benzene ring are known to cause cancer. Benzene itself is known to cause cancer, but structures involving fused benzene rings are more seriously implicated. Fused benzene rings share one or more sides of their hexagonal structure with another ring. Naphthalene is a two-ring fused system and the simplest example. Naphthalene is a well-known moth repellent and is used in making a certain class of dyes. More complicated fused-ring systems are found in cigarette smoke, exhaust from automobiles, and from smoke from barbecues. Have interested students find out more about the names and structures of the carcinogens found in cigarette smoke and make a poster to display in class. **LS** Logical

Teach, *continued*

Answers to Practice Problems C

1. structural:

CH₃   CH₂   CH₂   CH₂
    |C| CH₂   CH₂   CH₃
    ‖
    O

skeletal:

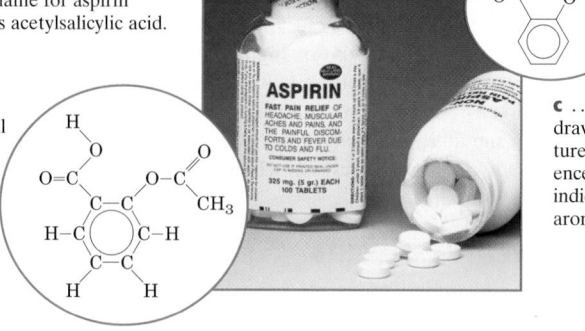 *(skeletal structure drawing with O)*

2. structural:

CH₃   CH₂   OH
    CH₂   C
         ‖
         O

skeletal:

(zigzag skeletal structure with OH and O)

3. structural:

Br
|
Br—C   CH₂
    | | CH₃
    Br CH
        |
        Br

skeletal:

Br
|
Br—
|
Br |
     Br

4. structural:

Cl   Cl
 \   /
F   C
 \ / \
  CH   CH₃
  |
  F

skeletal:

Cl   Cl
 \   /
F
  \
   F

Figure 9

a The chemical name for aspirin is acetylsalicylic acid.

b Because the complete structural formula of acetyl-salicylic acid is complex . . .

H
|
O
‖
O=C   O—C
   \   /   ‖
    C—C   O
   /   \   CH₃
H—C   C—H
   \   /
    C—C
   /   \
  H   H

c . . . chemists usually draw its skeletal structure instead. The presence of a benzene ring indicates that it is an aromatic compound.

SAMPLE PROBLEM C

Drawing Structural and Skeletal Formulas

Draw both the structural formula and the skeletal structure for 1,2,3-propanetriol.

1 Gather information.

- The name *propanetriol* indicates that the molecule is an alcohol that consists of three carbon atoms making up the parent chain.
- The suffix -*triol* indicates that three alcohol groups are present.
- The *1,2,3-* prefix indicates that an alcohol group is attached to the first, second, and third carbon atoms.

2 Plan your work.

- Draw the carbon framework showing the parent chain.
- Add the alcohol groups to the appropriate carbon atoms.
- Add enough hydrogen atoms so that each carbon atom has four bonds.
- Show the carbon framework as a zigzag line.
- Include the functional groups as part of the skeletal structure.

3 Draw the structures.

Structural formula:

C—C—C ⟶ HO—C—C—C—OH ⟶ HO—C—C—C—OH
                    | | | |
                    OH   H H H
                                    (with OH on middle and H's)

Skeletal formula:

(zigzag) ⟶ HO⌒⌒OH with OH on top

4 Verify your results.

- The structural formula should show all bonds and atoms in the compound 1, 2, 3-propanetriol.
- The skeletal formula should show only carbon-carbon bonds plus any functional groups present in the molecule.

> **PRACTICE HINT**
>
> Unless it is a part of a functional group, hydrogen is not shown in a skeletal structure. In the sample, the hydrogens shown are part of the alcohol functional group. The other hydrogen atoms bonded to carbon are not shown.

694

694   Chapter 19 • Carbon and Organic Compounds

 **PRACTICE**

Draw both structural and skeletal formulas for each of the following compounds.

1 2-octanone

2 butanoic acid

3 1,1,1,2-tetrabromobutane
(Hint: *Bromo-* indicates that a Br atom is attached to the parent chain.)

4 2,2-dichloro-1,1-difluoropropane
(Hint: Both Cl and F atoms are attached to the parent chain.)

PROBLEM SOLVING SKILL

 # Section Review

UNDERSTANDING KEY IDEAS

1. How does a saturated hydrocarbon differ from an unsaturated hydrocarbon?

2. What does the prefix *dec-* indicate about the composition of an organic compound?

3. What is the functional group for an aldehyde?

4. How are the carbon atoms in the parent chain numbered in a branched alkene or alkyne?

5. Which class of compounds forms the basis for naming most other carbon compounds?

PRACTICE PROBLEMS

6. Name the following branched hydrocarbon. (Hint: The —CH₂—CH₃ group is an ethyl group.)

CH₃
|
CH₃—CH—C=CH₂
|
CH₂—CH₃

7. Name the following branched hydrocarbon.

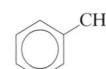

8. Draw the structural formula for dichloromethane.

9. Draw the structural and skeletal formulas for 2-bromo-4-chloroheptane.

10. Write the molecular formula for the compound with the following skeletal structure.

(Hint: Draw a full structural formula, and include all the carbon and hydrogen atoms in your count.)

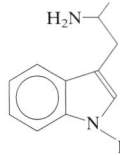

CRITICAL THINKING

11. Why do the names of organic acids not contain any numbers to indicate the position of the functional group?

12. What is incorrect about the name *nonene*?

13. How is methanol different from methanal? How are they similar?

14. How many double bonds are present in 1,3-butadiene? Where are they located in the molecule?

695

Answers to Section Review

1. The carbon atoms in a saturated hydrocarbon form four single covalent bonds with other atoms. Not all the carbon atoms in an unsaturated hydrocarbon have four single covalent bonds.

2. The compound consists of a parent chain with 10 carbon atoms.

3. The functional group is a carbonyl group —CO attached to a carbon atom at the end of a carbon-hydrogen chain.

4. The carbon atoms are numbered from the end closest to the double or triple bond.

5. alkanes

6. 2-ethyl-3-methyl-1-butene

7. methylbenzene

8. Cl—CH₂—Cl

9.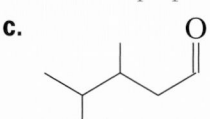

10. $C_{11}H_{12}N_2$

11. The acid group is always at the end of the molecule.

Answers continued on p. 709A

Close

Quiz ——————— GENERAL

1. List eight functional groups and write the structural formula for the simplest possible organic compound with that functional group, and its name.

Ans.

CH₃—OH   alcohol; methanol

aldehyde; ethanal

CH₃—Cl   chloride; chloromethane

CH₃—NH₂   amine; methylamine

carboxylic acid; ethanoic acid

ester; methyl ethanoate

CH₃—O—CH₃   ether; dimethyl ether

ketone; propanone

2. Name the following compounds.

a.

Ans. 4-methyl-2-pentanone

b.

Ans. 2-aminopropanoic acid

c.

Ans. 3,4-dimethylpentanal

d.

Ans. methyl-2-butene

LS Logical

Chapter Resource File

• Concept Review

• Quiz

Chapter 19 • Carbon and Organic Compounds   695

Overview

Before beginning this section, review with your students the Objectives listed in the Student Edition. Students will become familiar with addition and substitution reactions and learn that an enormous number of products can be made by means of these reactions. They will distinguish between condensation and elimination reactions.

 **Bellringer**

Have students read the definitions of the key terms and find a reaction that would illustrate each. Have them keep their predictions and check for correctness as they study this section.

Motivate

Discussion

Ask students for examples of substituting one thing for another. For example, in baking a cake, one-half cup of margarine might be substituted for one-half cup of butter. No butter is found in the cake, but margarine is. Ask how substitutions differs from addition. Continue the analogy by proposing that a cup of raisins could be added to the cake recipe. In addition to all the cake ingredients called for by the recipe, raisins are also present.

KEY TERMS

- **substitution reaction**
- **addition reaction**
- **polymer**
- **condensation reaction**
- **elimination reaction**

---

substitution reaction

a reaction in which one or more atoms replace another atom or group of atoms in a molecule

addition reaction

a reaction in which an atom or molecule is added to an unsaturated molecule

OBJECTIVES

1 **Describe** and distinguish between substitution and addition reactions.

2 **Describe** and distinguish between condensation and elimination reactions.

Substitution and Addition Reactions

The single bonds between carbon and hydrogen atoms in organic compounds are not highly reactive. However, these compounds do participate in a variety of chemical reactions, one of which is called a *substitution reaction*. A **substitution reaction** is a reaction in which one or more atoms replace another atom or group of atoms in a molecule. Another type of reaction involving organic compounds is an **addition reaction** in which an atom or molecule is added to an unsaturated molecule and increases the saturation of the molecule.

Halogens Often Replace Hydrogen Atoms

As saturated hydrocarbons, the alkanes have the lowest chemical reactivity of organic compounds. However, under certain conditions these compounds can undergo substitution reactions, especially with the halogens. An example of such a reaction is that between an alkane, such as methane, and a halogen, such as chlorine. In this substitution reaction, a chlorine atom replaces a hydrogen atom on the methane molecule.

$$
\begin{array}{ccccccc}
& \text{H} & & & \text{H} & & \\
& | & & & | & & \\
\text{H}-\text{C}-\text{H} & + & \text{Cl}-\text{Cl} & \longrightarrow & \text{H}-\text{C}-\text{Cl} & + & \text{H}-\text{Cl} \\
& | & & & | & & \\
& \text{H} & & & \text{H} & & \\
\text{methane} & & \text{chlorine} & & \text{chloromethane} & & \text{hydrogen chloride}
\end{array}
$$

The substitution reactions can continue, replacing the remaining hydrogen atoms in the methane molecule one at a time. The products are dichloromethane, trichloromethane, and tetrachloromethane. Trichloromethane is commonly known as *chloroform*, which was once used as an anesthetic. The common name for tetrachloromethane is carbon *tetrachloride*, which for many years was commonly used as a solvent.

Because the single covalent bonds are hard to break, catalysts are often added to the reaction mixture. For example, trichlorofluoromethane, CCl_3F, commonly known as *Freon-11*, was used as a refrigerant. It was made by a substitution reaction catalyzed by SbF_3.

Chapter Resource File

- Lesson Plan

Hydrogenation Is a Common Addition Reaction

A common type of addition reaction is hydrogenation, in which one or more hydrogen atoms are added to an unsaturated molecule. As a result of hydrogenation, the product of the reaction contains fewer double or triple bonds than the reactant. Hydrogenation is used to convert vegetable oils into fats. Vegetable oils are long chains of carbon atoms with many double bonds. When hydrogen gas is bubbled through an oil, double bonds between carbon atoms in the oil are broken and hydrogen atoms are added. Only a portion of the very long oil and fat molecules are shown in the following hydrogenation reaction.

$$\begin{pmatrix} \overset{\displaystyle H}{\underset{\displaystyle H}{|}} & \overset{\displaystyle H}{|} & \overset{\displaystyle H}{|} & \overset{\displaystyle H}{|} & \overset{\displaystyle H}{|} & \overset{\displaystyle H}{|} & \overset{\displaystyle H}{|} \\ -C-C=C-C-C=C-C- \\ \underset{\displaystyle H}{|} & & \underset{\displaystyle H}{|} & & & \underset{\displaystyle H}{|} \end{pmatrix} + H_2 \xrightarrow{\text{catalyst}} \begin{pmatrix} \overset{\displaystyle H}{\underset{\displaystyle H}{|}} & \overset{\displaystyle H}{|} & \overset{\displaystyle H}{|} & \overset{\displaystyle H}{|} & \overset{\displaystyle H}{|} & \overset{\displaystyle H}{|} & \overset{\displaystyle H}{|} \\ -C-C=C-C-C-C-C- \\ \underset{\displaystyle H}{|} & & \underset{\displaystyle H}{|} & & \underset{\displaystyle H}{|} & \underset{\displaystyle H}{|} & \underset{\displaystyle H}{|} \end{pmatrix}$$

oil   fat

Making Consumer Products by Hydrogenation

The margarine and vegetable shortening shown in **Figure 10** are two products made by the hydrogenation of oil. Although they contain double bonds, oils are still not very reactive. As a result, the hydrogenation of an oil requires the addition of a catalyst and temperatures of about 260°C.

Another application of hydrogenation is the manufacture of cyclohexane from benzene as shown by the following reaction.

benzene $+ 3H_2 \xrightarrow{\text{catalyst}}$ cyclohexane

Over 90% of the cyclohexane that is made is used in the manufacture of nylon. The rest is used mostly as a solvent for paints, varnish, and oils.

Figure 10
Hydrogenation is used to turn vegetable oil into solid margarine and butter.

did you know?

Chloroform and carbon tetrachloride were important chemicals and still are used as solvents when needed in the laboratory. Chloroform was one of the first anesthetics and was welcomed by nineteenth century doctors as a means of performing surgery without the concern of a conscious patient. Chloroform is a liquid that could be carried in a small bottle as part of a country doctor's regular equipment. Carbon tetrachloride was commonly used as a dry cleaning fluid. Neither compound is now recommended for these purposes because they are carcinogens.

Show students different types of polymers. Examples might include plastic wrap, rubber tubing, polyester thread, nylon pantyhose, foam cups, trash bags, a plastic that is firm but flexible (such as a margarine container), and an inflexible plastic (such as the material used to make a telephone). Tell students that all of the examples are made up of the same type of compound. Ask groups of students to write down what they think the items have in common.
LS Interpersonal

SKILL BUILDER —— BASIC

Vocabulary Emphasize that the prefix poly- in the term polygon means "many." Have students think of other examples of terms in which poly- has the same meaning. **Ans.** Examples might include math terms, such as polyhedron or polynomial. Other familiar terms might include polygamy, polygraph, or polysyllabic. Ask students to distinguish between a monomer and a polymer. **LS** Verbal

polymer
a large molecule that is formed by more than five monomers, or small units

internet connect
www.scilinks.org
Topic: Polymers
SciLinks code: HW4098
SciLINKS Maintained by the National Science Teachers Association

Some Addition Reactions Form Polymers

The addition reactions you have examined so far involve adding atoms to a molecule. Some addition reactions involve joining smaller molecules together to make larger ones. The smaller molecules are known as *monomers*. The larger molecule that is made by the addition reaction is called a **polymer.**

Consider how polyethylene is made. Polyethylene is a strong but flexible plastic used to make a variety of consumer products, including the water bottle shown at the beginning of this chapter. The monomer from which polyethylene is made is ethene, C_2H_4. Because ethene is commonly known as *ethylene*, the polymer it forms is often called *polyethylene*. The following equation shows how a portion of the polymer forms. Notice that these are condensed formulas that show all the atoms but not the bonds between the carbon and hydrogen atoms.

$$CH_2{=}CH_2 \; + \; CH_2{=}CH_2 \; \longrightarrow \; {-}CH_2{-}CH_2{-}CH_2{-}CH_2{-}$$

Monomers Can Be Added in Different Ways

Notice the open single bonds at each end of the product in the reaction shown above. An ethene molecule can be added at each end. The process of adding ethene molecules, one at each end, continues until polyethylene is eventually produced. Polyethylene is a very long alkane polymer chain. These chains form a product that is strong yet flexible.

Occasionally, monomers are added so that a chain branches. For example, an ethene monomer is sometimes added to form a side chain. A polymer with many side chains remains flexible. Such polymers are used to manufacture the plastic that wraps a variety of consumer products such as those shown in **Figure 11.**

Figure 11
Plastic wrap is used to protect foods from spoiling. It is flexible because the side chains in the polymer prevent side-by-side molecules from packing together rigidly.

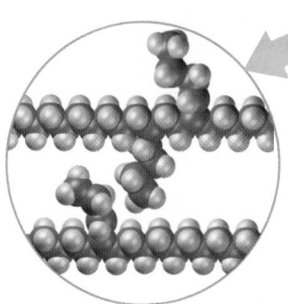
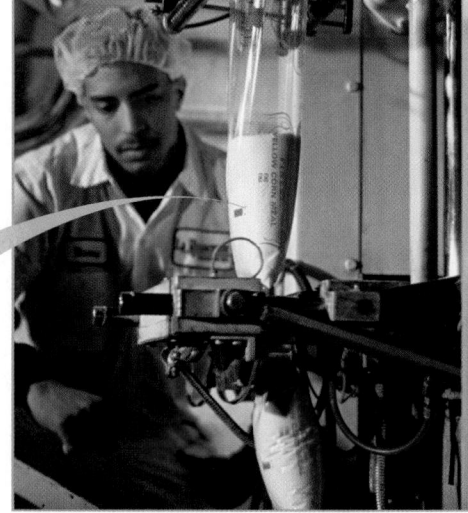

did you know?

Vulcanization An accidental discovery in 1839 helped to develop the field of polymer chemistry. In that year, Charles Goodyear found a way to make rubber, a natural polymer, stronger and more durable. Natural rubber is soft and weak. Goodyear was working with a mixture of natural rubber and sulfur when he accidentally dropped some onto a hot stove. He noticed that the rubber did not melt but instead got much stronger. What Goodyear had discovered is known today as vulcanized rubber. The vulcanization process makes rubber stronger by joining the polymer chains together with sulfur links. Vulcanized rubber is used today to make tires, rain gear, and rubber bands.

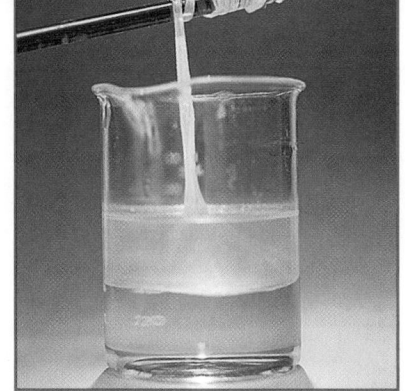

Figure 12
Nylon 66, shown here being wound onto a stirring rod, is one of the most widely used of all synthetic polymers.

Condensation and Elimination

Polymers can also be formed by a **condensation reaction** in which two molecules combine, usually accompanied by the loss of a water molecule. The formation of water as a reaction product is the reason for the name of this type of reaction. In some instances, hydrochloric acid is formed as a byproduct of a condensation reaction.

Another type of reaction that produces water is known as an *elimination reaction*. An **elimination reaction** is a reaction in which a simple molecule is removed from adjacent carbon atoms on the same organic molecule. Another simple molecule that can be a product of an elimination reaction is ammonia.

condensation reaction

a chemical reaction in which two or more molecules combine to produce water or another simple molecule

elimination reaction

a reaction in which a simple molecule, such as water or ammonia, is removed and a new compound is produced

Condensation Reactions Produce Nylon

Figure 12 shows a polymer being formed in a condensation reaction. The bottom layer in the beaker shown in **Figure 12** is hexanediamine, an organic molecule with an amine group at each end. The top layer in the beaker is adipic acid, an organic molecule with a carboxyl group at each end. The condensation reaction takes place between an amine group on hexanediamine and a carboxyl group on adipic acid as shown below.

$$H-\underset{\underset{H}{|}}{N}-CH_2-CH_2-CH_2-CH_2-CH_2-CH_2-N-H \;+\; HO-\overset{\overset{O}{||}}{C}-CH_2-CH_2-CH_2-CH_2-\overset{\overset{O}{||}}{C}-OH \;\longrightarrow$$

hexanediamine   adipic acid

$$H-\underset{\underset{H}{|}}{N}-CH_2-CH_2-CH_2-CH_2-CH_2-CH_2-\underset{\underset{H}{|}}{N}-\overset{\overset{O}{||}}{C}-CH_2-CH_2-CH_2-CH_2-\overset{\overset{O}{||}}{C}-OH \;+\; H_2O$$

water

Notice that a water molecule is eliminated when an H atom from the amine group and an —OH group from the carboxyl group are removed. Another adipic acid molecule is then added to the amine group shown on the left, while another hexanediamine molecule is added to the carboxyl group shown on the right. This process continues, linking hundreds of reactants to form a product called *nylon 66*.

699

did you know?

In the late 1930s, Wallace Carothers, a chemist working for the Dupont Company, mixed a dicarboxylic acid and a di-amino compound together and produced a substance that was sticky and easily fell apart. Because of these physical properties, Carothers thought the substance was of little practical value. But his co-worker, Julian Hill, put a small ball of the substance on the end of a glass rod. He slowly pulled the rod away from the sticky mass and observed that the material could be stretched into long, silky fibers. What Carothers and Hill had discovered was nylon 66, a compound that can be woven into strong cloth to take the place of silk for uses such as parachutes and sheer hosiery. The linkage in nylon is an amide linkage like those in proteins.

SKILL BUILDER — **ADVANCED**

Research Skills Have interested students research the reactions that produce the fibers with the trade names Dacron or Orlon. **LS** Logical

Demonstration

What Type of Fiber Is It?

(Approximate time: 10 minutes)

1. Use tongs to hold a square of a synthetic fabric, such as nylon, in a flame for 2 seconds.

2. Remove the fabric from the flame. Blow out any flame if the fabric is burning.

3. Have students carefully observe the odor and appearance of the results.

4. Repeat steps 1–3 for a square of a natural fabric, such as cotton.

5. Have students compare the results. Students should notice that the synthetic fabric gives off a "chemical" odor and appears to melt. The natural fabric gives off an odor similar to that of burning paper or hair and appears to char.

Homework — **BASIC**

Graphic Organizer Have students create a graphic organizer that pulls together the four types of reactions covered in this section with examples of each type of reaction. **LS** Visual

Figure 13
These colorful threads are made from polyester that can be woven into fabrics to make many types of clothing.

Many Polymers Form by Condensation Reactions

In addition to nylon 66, many other polymers are made by condensation reactions. The polymer shown in **Figure 13** is polyethylene terephthalate, abbreviated PET, which is used to make permanent-press clothing and soda bottles. The following formulas show how two monomers are combined in this condensation reaction.

$$n\text{HO—CH}_2\text{—CH}_2\text{—OH} \ + \ n\text{HO—C} \overset{O}{\Vert} \overset{\bigcirc}{} \text{C—OH} \longrightarrow$$

$$\text{(O—CH}_2\text{—CH}_2\text{—O—C} \overset{\bigcirc}{} \overset{O}{\underset{\Vert}{C}}\text{)}n \ + \ 2n\text{H}_2\text{O}$$

Notice that the first reactant shown is an alcohol because it contains two —OH groups. The second reactant shown is an organic acid because it contains two —COOH groups. When PET is made, water is formed from an —H from the alcohol and an —OH from the acid. The two monomers then bond. The functional group present in the product shown above classifies this molecule as an ester. Therefore, PET is a polyester.

Elimination Reactions Often Form Water

An elimination reaction involves the removal of a small molecule from two adjacent carbon atoms, as shown below.

$$\begin{array}{c} \text{H} \ \ \text{OH} \\ | \ \ \ | \\ \text{H—C—C—H} \\ | \ \ \ | \\ \text{H} \ \ \text{H} \\ \text{ethanol} \end{array} \xrightarrow[\Delta]{\text{conc. H}_2\text{SO}_4} \begin{array}{c} \text{H—C=C—H} \\ | \ \ \ | \\ \text{H} \ \ \text{H} \\ \text{ethene} \end{array} + \ \text{H}_2\text{O} \ \ \ \text{water}$$

The acid catalyzes a reaction that eliminates water from the ethanol molecule, which leaves a double bond.

700

REAL-WORLD CONNECTION

Nylon was the first fiber to result from a deliberate effort to create a synthetic polymer. As women's dresses became shorter during the early 1900s, stockings became more visible and fashionable. However, silk, the main material used to make sheer stockings, was expensive. When stockings made of nylon-66 came on the market, they were in such high demand that they had to be rationed.

Figure 14
An elimination reaction occurs when sucrose and concentrated sulfuric acid are mixed. Water is formed, which leaves a product that is mostly carbon.

Figure 14 shows another example of an elimination reaction; one whose results can be seen easily. When sucrose reacts with concentrated sulfuric acid, water is eliminated, which leaves behind mostly carbon. Carbon is the black substance you can see forming in the photos and rising out of the beaker on the far right.

 # Section Review

UNDERSTANDING KEY IDEAS

1. Explain why an addition reaction increases the saturation of a molecule.

2. What molecule is often a product of both condensation and elimination reactions?

3. What kind of organic reaction can form fluoromethane, CH_3F, from methane?

4. Give an example of a polymer, and tell what monomers it consists of.

5. How does a condensation reaction get its name?

6. Name the type of organic reaction that results in the formation of a double bond.

CRITICAL THINKING

7. Explain why alkanes do not undergo addition reactions.

8. Explain how an elimination reaction can be considered the opposite of an addition reaction.

9. Draw the skeletal structure of part of a polyethylene molecule consisting of eight monomers.

10. Can two different monomers be involved in an addition reaction? Why or why not?

11. Why is a molecule with only one functional group unable to undergo a condensation reaction to form a polymer?

12. Why does a substitution reaction involving an alkane and a halogen not increase the saturation of the organic compound?

701

Answers to Section Review

1. Double bonds are replaced by single bonds between the carbon atoms and atoms are added to the molecule.

2. water

3. substitution

4. A protein is a polymer made of hundreds of amino acid monomers that have been bonded to form a single large molecule.

5. During condensation reactions, two molecules combine to form one molecule and water.

6. elimination

7. Addition can occur only with reactants that are unsaturated.

8. An elimination reaction results in the formation of double bonds, while an addition reaction results in the loss of double bonds.

9. ⌇⌇⌇⌇⌇⌇⌇⌇⌇⌇

10. An addition reaction between them is possible as long as each monomer contained either a double or triple bond.

11. The molecule must contain two functional groups, one on each end. Only in this way can a polymer be formed when the condensation reaction occurs.

Answers continued on p. 709A

CONSUMER FOCUS

Recycling Codes for Plastic Products

More than half the states in the United States have enacted laws that require plastic products to be labeled with numerical codes that identify the type of plastic used in them.

Sorting your plastics

Used plastic products can be sorted by the codes shown in **Table 6** and properly recycled or processed. Only Codes 1 and 2 are widely accepted for recycling. Codes 3 and 6 are rarely recycled. Find out what types of plastics are recycled in your area. If you know what the codes

mean, you will have an idea of how successfully a given plastic product can be recycled. This information may affect your decision to buy or not buy particular items.

Questions

1. What do the recycling codes on plastic products indicate?
2. Why is it important to sort plastics before recycling them?

Table 6	**Recycling Codes for Plastic Products**			
Recycling code	**Type of plastic**	**Physical properties**	**Examples**	**Uses for recycled products**
1	polyethylene terephthalate (PET)	tough, rigid; can be a fiber or a plastic; solvent resistant; sinks in water	soda bottles, clothing, electrical insulation, automobile parts	backpacks, sleeping bags, carpet, new bottles, clothing
2	high density polyethylene (HDPE)	rough surface; stiff plastic; resistant to cracking	milk containers, bleach bottles, toys, grocery bags	furniture, toys, trash cans, picnic tables, park benches, fences
3	polyvinyl chloride (PVC)	elastomer or flexible plastic; tough; poor crystallization; unstable to light or heat; sinks in water	pipe, vinyl siding, automobile parts, clear bottles for cooking oil, bubble wrap	toys, playground equipment
4	low density polyethylene (LDPE)	moderately crystalline, flexible plastic; solvent resistant; floats on water	shrink wrapping, trash bags, dry-cleaning bags, frozen-food packaging, meat packaging	trash cans, trash bags, compost containers
5	polypropylene (PP)	rigid, very strong; fiber or flexible plastic; lightweight; heat-and-stress-resistant	heatproof containers, rope, appliance parts, outdoor carpet, luggage, diapers, automobile parts	brooms, brushes, ice scrapers, battery cable, insulation, rope
6	polystyrene (P/S, PS)	somewhat brittle, rigid plastic; resistant to acids and bases but not organic solvents; sinks in water, unless it is a foam	fast-food containers, toys, videotape reels, electrical insulation, plastic utensils, disposable drinking cups, CD jewel cases	insulated clothing, egg cartons, thermal insulation

702

Answers to Feature Questions

1. the type of plastic
2. A mixture of different types of plastics cannot be properly processed for recycling.

KEY TERMS

hydrocarbon
alkane
alkene
alkyne
aromatic hydrocarbon
functional group
isomer

saturated hydrocarbon
unsaturated
  hydrocarbon

substitution reaction
addition reaction
polymer
condensation reaction
elimination reaction

KEY IDEAS

SECTION ONE Compounds of Carbon

• The properties of carbon allotropes depend on the arrangement of the atoms and how they are bonded to each other.
• The simplest organic compounds are the hydrocarbons, which consist of only carbon and hydrogen atoms.
• Alkanes, alkenes, and alkynes are hydrocarbons. Organic compounds containing one or more rings with delocalized electrons are aromatic hydrocarbons.
• Organic compounds are classified by their functional groups.

SECTION TWO Names and Structures of Organic Compounds

• The names of the alkanes form the basis for naming most other organic compounds.
• When an organic compound is named, the parent chain is identified, and the carbon atoms are numbered so that any branches or multiple bonds have the lowest possible numbers.
• Organic molecules can be represented in various ways, and each model has advantages and disadvantages.

SECTION THREE Organic Reactions

• In a substitution reaction, an atom or group of atoms is replaced.
• In an addition reaction, an atom or group of atoms is added to replace a double or triple bond.
• Polymers are very long organic molecules formed by successive addition of monomers and are used in plastics.
• In a condensation reaction, two molecules or parts of the same molecule combine, which usually forms water.
• In an elimination reaction, a molecule, usually water, is formed by combining atoms from adjacent carbon atoms.

KEY SKILLS

Naming a Branched Hydrocarbon
Sample Problem A p. 690

Naming a Compound with a Functional Group
Sample Problem B p. 692

Drawing Structural and Skeletal Formulas
Sample Problem C p. 694

Alternative Assessments

Teaching Tip
• Page 677

SKILL BUILDER
• Pages 681, 682, 697, 698

CHAPTER REVIEW
• Items 61, 62, 63

Portfolio Assessments

READING SKILL BUILDER
• Page 677

Homework
• Pages 679, 686

SKILL BUILDER
• Page 698

703

Chapter Resource File

• **Datasheets for In-Text Lab**
  Polymers and Toy Balls
• **Observation Lab** Carbon
• **Consumer Lab** Cloth of Many Colors

REVIEW ANSWERS

Using Key Terms

1. aromatic compounds
2. isomer
3. alkanes
4. allotrope
5. substitution reaction
6. functional group
7. condensation and elimination reactions
8. polymer
9. alkyne
10. skeletal structure

Understanding Key Ideas

11. The triple bond in an alkyne is less stable, making the compound more reactive.

12. **a.** The carbon atoms are bonded to one another in three dimensions to form a tetrahedral network.

    b. Although the carbon atoms in each layer are strongly bonded to one another, adjacent layers are held together only by weak, intermolecular forces.

13. Both fullerenes and nanotubes are carbon allotropes. However, the carbon atoms in fullerenes are arranged as near-spherical cages, while those in nanotubes are arranged as hollow cylinders.

14. $C_{14}H_{30}$

15.

16. A single carbon-carbon bond is stronger than most other covalent bonds, so a molecule can contain many carbon-carbon bonds and be relatively stable.

---

19 CHAPTER REVIEW

USING KEY TERMS

1. The benzene ring is the simplest member of what class of organic compounds?

2. Two compounds may have the same molecular formulas but different structural formulas. Each of these compounds is known as a(n) _____.

3. What class of organic compounds includes all saturated hydrocarbons?

4. If an element exists in more than one bonding pattern, what term is used for each of these forms?

5. What type of reaction involves the replacement of a hydrogen atom by a halogen atom?

6. The chemical and physical properties of an organic compound are largely determined by the presence of a(n) _____.

7. Which two types of organic reactions usually form small molecules such as water?

8. What type of molecule results when many smaller units are joined in addition reactions?

9. What type of hydrocarbon molecule contains a triple bond between two of its carbon atoms?

10. The hexagon and circle often used to depict a benzene molecule is an example of what kind of structure?

UNDERSTANDING KEY IDEAS

Compounds of Carbon

11. Explain why alkynes are more reactive than alkanes.

12. **a.** Why is diamond so hard and strong?

    b. Why is graphite so soft and easy to break apart?

13. How are fullerenes and nanotubes alike? How are they different?

14. What is the molecular formula for the alkane that contains 14 carbon atoms?

15. Draw the two possible resonance structures for benzene.

16. Explain the connection between the strength of the carbon-carbon single bond and the ability of carbon to be the basis of large molecules.

17. How does pentane differ from cyclopentane?

18. Explain why isomers have different chemical and physical properties.

19. Explain why the properties of butane differ from those of butanol.

Names and Structures of Organic Compounds

20. Use **Table 4** to identify the functional group from the name for each of the following organic compounds.
    a. propanol
    b. ethanoic acid
    c. propanal
    d. hexanone

21. What functional groups are present in a molecule of adrenaline, whose structural formula is shown below?

Assignment Guide	
Section	**Questions**
1	1–2, 4, 6, 9, 11–19, 48, 65
2	3, 10, 20–27, 39–47, 49–51, 54–55, 58–59, 63
3	5, 7–8, 28–38, 52–53, 56–57, 60

22. What group of organic compounds forms the basis for naming the other organic compounds?

23. What rule must be followed when the carbon atoms in an alkene or alkyne is numbered?

24. How does a skeletal structure differ from a structural formula? How are they the same?

25. Why is the name *pentyne* not completely correct?

26. What information does the name *1-aminobutane* provide about the structure of this organic compound?

27. List the main advantage and disadvantage of using a skeletal structure as a model.

Organic Reactions

28. Identify the functional groups that are the source of the water molecule produced in the condensation reaction between hexanediamine and adipic acid.

29. What are two reactions by which polymers can be formed?

30. Compare substitution and addition reactions.

31. What is the structural requirement for a molecule to be a monomer in an addition reaction?

32. Explain what hydrogenation is.

33. How does adding monomers as branches to a parent chain affect the properties of a polymer?

34. What is the difference between condensation and elimination?

35. Why are catalysts added to substitution reactions involving alkanes?

36. What is the chemical difference between an oil and a fat?

37. How are a nylon and polyethylene similar? How are they different?

PRACTICE PROBLEMS

38. Name the following compounds.

a. 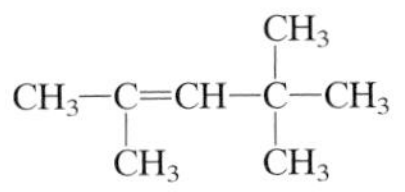
$$CH_3-C=CH-C-CH_3$$
with CH_3 on top carbon and CH_3, CH_3 below

b.
$$CH_3-CH_2-C-C=C-CH_3$$
with CH_3 above and CH_3 below the third carbon

c.
$$CH_3-CH_2-C=CH-CH_2-CH_3$$
with CH_2-CH_3 below

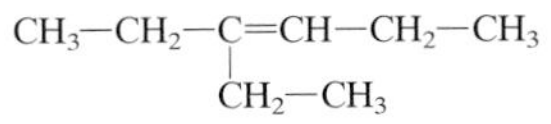

d. 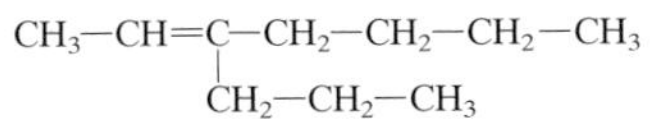
$$CH_3-CH=C-CH_2-CH_2-CH_2-CH_3$$
with $CH_2-CH_2-CH_3$ below

39. Draw the structural formulas for each of the following compounds.
a. 1,4-dichlorohexane
b. 2-bromo-4-chloroheptane

40. Name the following organic compounds, and then write the molecular formula for each compound.

a. 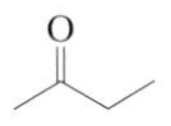 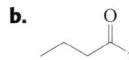   **b.**

41. Draw structural formulas for the following compounds.
a. butanoic acid
b. 1-nonanol
c. 2-pentanone

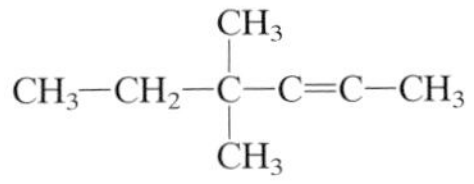

42. The skeletal structure for proline, an amino acid, is shown below. Draw its structural formula.

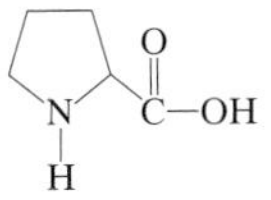

(structure with ring, N–H, and C–OH with O)

43. Name the following hydrocarbons.
a. CH_3-I
b. $Cl-CH_2-CH_2-Cl$

705

---

17. Pentane is a straight-chain hydrocarbon, whereas cyclopentane consists of a ring structure.

18. Properties are determined both by the kinds of atoms making up a compound and the way in which they are arranged. Because the atoms making up isomers are arranged differently, they have different properties.

19. Butane is a hydrocarbon, whereas butanol is an alcohol. The presence of the —OH group causes butanol to have properties different from those of butane.

20. a. alcohol
 b. carboxylic acid
 c. aldehyde
 d. ketone

21. alcohol and amine

22. alkanes

23. The carbon atoms must be numbered so that first atom with the multiple bond must have the lowest number possible.

24. A skeletal structure shows the carbon and hydrogen atom making up the parent chain as a series of zigzag lines. Both show the atoms in any branches, including functional groups.

25. The name must include a number as a prefix to indicate the position of the triple bond in the molecule.

26. The parent chain consists of four carbons, all with single bonds. An amino group is attached to the first carbon atom.

27. The main advantage is its simplicity. The main disadvantage is that this model does not show all the atoms or bonds.

28. The water molecule comes from the hydrogen on the amine group and a hydroxyl group on the carboxyl group.

29. addition and condensation

30. In a substitution reaction, an atom or group of atoms replaces a hydrogen atom in a molecule. In an addition reaction, an atom or group of atoms adds to a double or triple bond, forming single bonds.

Chapter 19 • Carbon and Organic Compounds   705

31. The monomer must contain a double or triple bond.

32. Hydrogenation is an example of an addition reaction in which one or more hydrogen atoms are added to an unsaturated molecule.

33. The presence of branches prevents the polymer from packing as tightly as it otherwise would.

34. A water molecule is removed from two adjacent carbon atoms on the same molecule during elimination, but involves two functional groups on two molecules in condensation.

35. With only single bonds, alkanes have low chemical reactivity. Catalysts are required for these reactions to occur in a reasonable time.

36. An oil contains more multiple bonds than a fat.

37. Both are polymers. Nylon is made by a succession of condensation reactions, and is made of two different kinds of monomers (hexanediamine and adipic acid), while polyethylene is made by a succession of addition reactions, and is made from a single kind of monomer (ethylene).

Practice Problems

38. a. 2,4,4,-trimethyl-2-pentene

b. 4,4-dimethyl-2-hexene

c. 3-ethyl-3-hexene

d. 3-propyl-2-heptene

39. a.

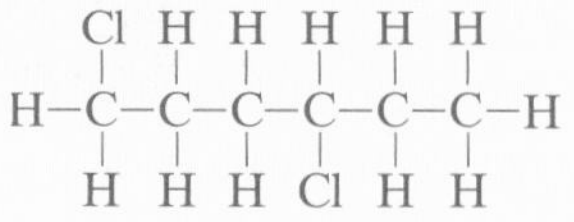

b.

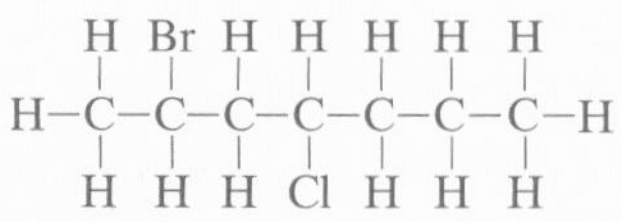

40. a. butanone

b. butanal

c. both formulas are C_4H_8O

---

44. Name the following alcohols.

a. $CH_3\!-\!OH$

b.

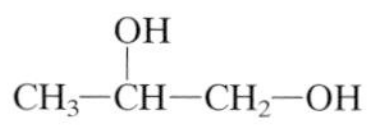

c.

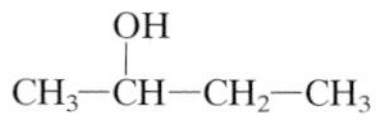

d.

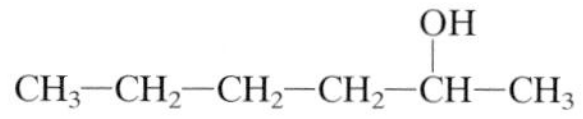

45. The skeletal structure for vitamin A is shown below. Draw its structural formula.

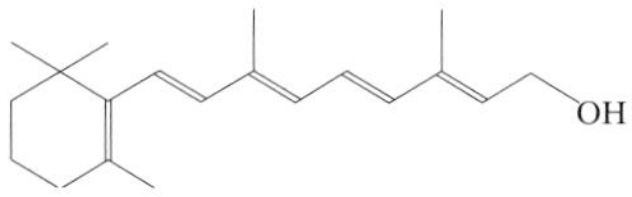

46. Draw skeletal structures for the following organic compounds.
a. 2,3,4-trichloropentane
b. 2,2 dichloro-1,1-difluoropropane

47. Name the following compounds.

a.

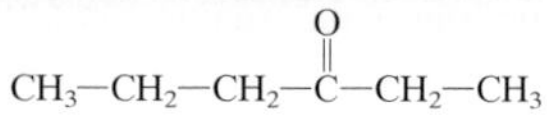

b.

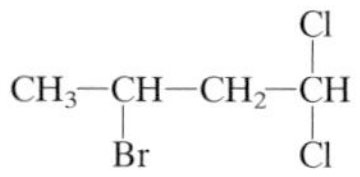

c.

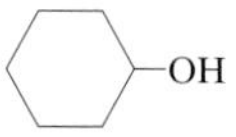

CRITICAL THINKING

48. Explain why some alcohols and organic acids are soluble in water, whereas hydrocarbons are virtually insoluble.

49. Identify all the functional groups in the following molecules.
a. cinnamaldehyde
b. salicylic acid

50. Correctly number the carbon atoms in the following hydrocarbon.

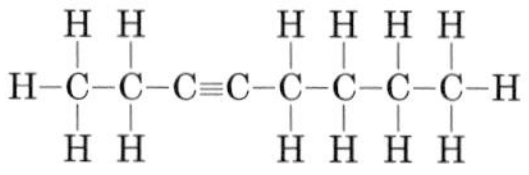

---

51. Draw the structural formula for 2,3, 4-trimethylnonane.

52. Copolymers are made from two different monomers. For example, some plastic food wrap is an addition polymer made from 1,1-dichloroethene and chloroethene. Draw a possible structure for this copolymer showing a structure that is four monomers in length.

53. When propyne reacts with H_2 under the proper conditions, the triple bond is broken and hydrogen atoms are added to the alkyne to form an alkane.
a. Draw the structural formula for the alkane product.
b. What is the name of this alkane?

54. When 2-methylpropene is mixed with HI, 2-iodo-2-methylpropane is produced.
a. Draw the structural formula for the organic reactant.
b. Draw the structural formula of the product.

55. Convert the following structural formula to a skeletal structure.

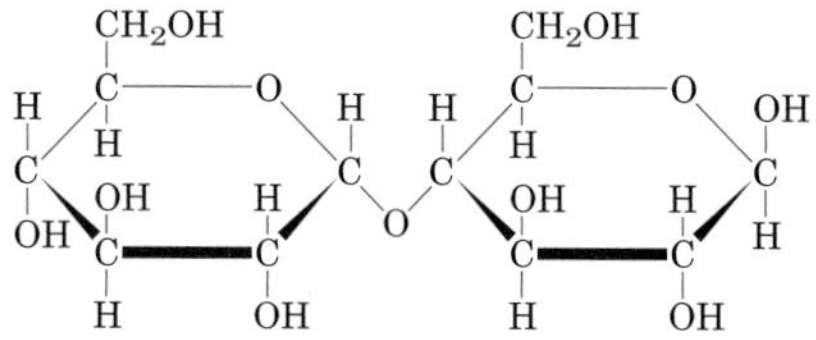

Maltose

56. Polymethylmethacrylate, sold under the trade names Plexiglas® and Lucite®, is an addition polymer made from the monomer methyl methacrylate, whose structural formula is shown below.

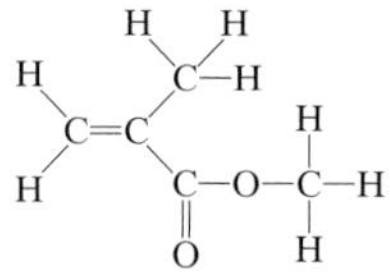

Draw a portion of the polymer showing four monomers that have combined.

---

57. The Kevlar™ that is used in bulletproof vests is a condensation polymer that can be made from the following monomer.

$$H_2N- \bigcirc -N-C- \bigcirc -C-OH$$

Draw a portion of a Kevlar™ polymer showing four molecules of a monomer that have combined.

58. Draw two structural formulas for an alcohol with the molecular formula C_3H_8O.

59. Classify the organic compounds shown below by their functional groups.

a.
$$\begin{array}{c} O \\ \| \\ H-C-OH \end{array}$$

b.
$$\begin{array}{cccccc} H & H & & H & H \\ | & | & & | & | \\ H-C-C-O-C-C-H \\ | & | & & | & | \\ H & H & & H & H \end{array}$$

c.
$$\begin{array}{ccccc} H & O & H & O & H \\ | & \| & | & \| & | \\ H-C-C-C-C-C-H \\ | & & | & & | \\ H & & H & & H \end{array}$$

d.
$$\begin{array}{cc} H & H \\ | & | \\ H-C-N \\ | & | \\ H & H \end{array}$$

e.
$$\begin{array}{c} H \\ | \\ H-C-H \\ | \\ Cl \end{array}$$

60. Ethylene glycol, $HO-CH_2-CH_2-OH$, is a monomer that forms a condensation polymer used as a car wax. Draw a portion of this wax polymer showing four molecules of the monomer that have combined.

ALTERNATIVE ASSESSMENT

61. Dimethyl mercury is an organic compound that poses a serious environmental threat to all living things. Research how this compound affects living things. Include information on whether dimethyl mercury poses a threat to your local environment. If so, determine what is being done to eliminate this problem.

62. Environmental concerns have led to the development of plastics that are labeled "biodegradable." Devise a set of experiments to study how well biodegradable plastics break down. If your teacher approves your plan, carry out your experiments on various consumer products labeled "biodegradable."

63. At one time, a group of compounds known as PCBs were very popular for a number of industrial applications. Find out the general structural formula for these compounds and the properties that made them so popular. Why were PCBs eventually banned for most industrial uses? Present your findings to the class.

CONCEPT MAPPING

64. Use the following terms to complete the concept map below: *organic reactions, substitution, addition, condensation, hydrogen, halogen,* and *water.*

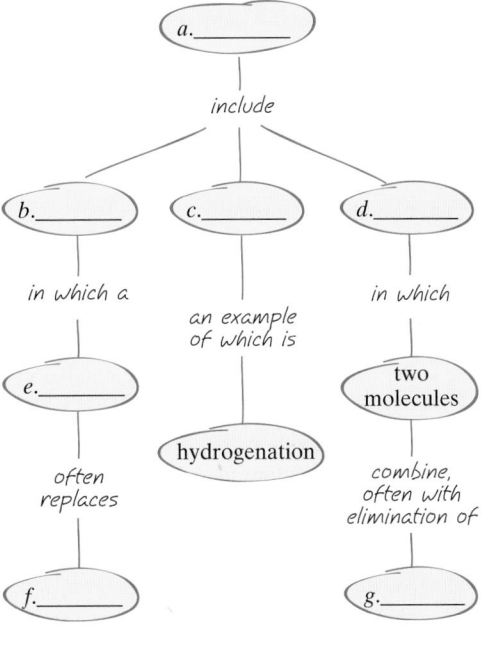

41. a.
$$\begin{array}{cccc} H & H & H & O \\ | & | & | & \| \\ H-C-C-C-C-OH \\ | & | & | \\ H & H & H \end{array}$$

b.
$$\begin{array}{ccccccccc} H & H & H & H & H & H & H & H & H \\ | & | & | & | & | & | & | & | & | \\ H-C-C-C-C-C-C-C-C-C-OH \\ | & | & | & | & | & | & | & | & | \\ H & H & H & H & H & H & H & H & H \end{array}$$

c.
$$\begin{array}{ccccc} H & O & H & H & H \\ | & \| & | & | & | \\ H-C-C-C-C-C-H \\ | & & | & | & | \\ H & & H & H & H \end{array}$$

42.
$$\begin{array}{c} H \ \ H \ H \\ | \ \ | \ | \\ H-C-C \ \ O \\ | \ \ \ \ \ \ \ \ \| \\ H-C \ \ \ \ C-C-OH \\ | \ \ N \ \ H \\ H \ | \\ H \end{array}$$

43. a. iodomethane
b. 1,2-dichloroethane

44. a. methanol
b. 1,2-propanediol
c. 2-butanol
d. 2-hexanol

45.
(structural formula)

46. a.
(structural formula with Cl, Cl, Cl)

b.
(structural formula with F, F, Cl, Cl)

47. a. 3-hexanone
b. 3-bromo-1,1-dichlorobutane
c. cyclohexanol

48. Alcohols and organic acids are polar and form hydrogen bonds with water molecules. In contrast, hydrocarbons are nonpolar and do not form hydrogen bonds with water molecules.

49. a. aldehyde
b. alcohol; carboxylic acid

50.

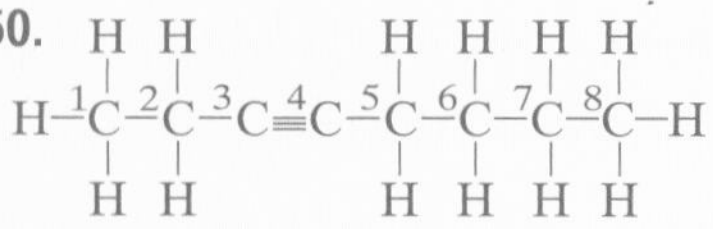

H–$\overset{\overset{\displaystyle H}{|}}{\underset{\underset{\displaystyle H}{|}}{C^1}}$–$\overset{\overset{\displaystyle H}{|}}{\underset{\underset{\displaystyle H}{|}}{C^2}}$–$C^3$≡$C^4$–$\overset{\overset{\displaystyle H}{|}}{\underset{\underset{\displaystyle H}{|}}{C^5}}$–$\overset{\overset{\displaystyle H}{|}}{\underset{\underset{\displaystyle H}{|}}{C^6}}$–$\overset{\overset{\displaystyle H}{|}}{\underset{\underset{\displaystyle H}{|}}{C^7}}$–$\overset{\overset{\displaystyle H}{|}}{\underset{\underset{\displaystyle H}{|}}{C^8}}$–H

51. 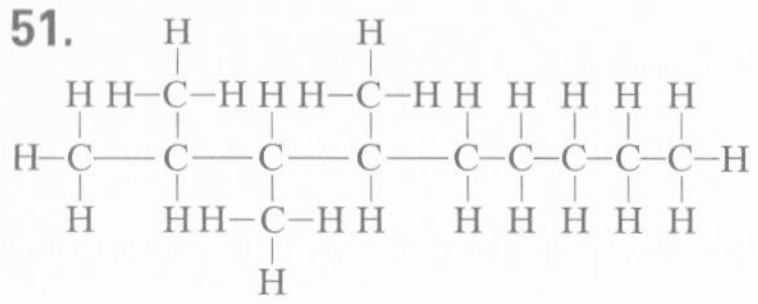 [structure]

52. 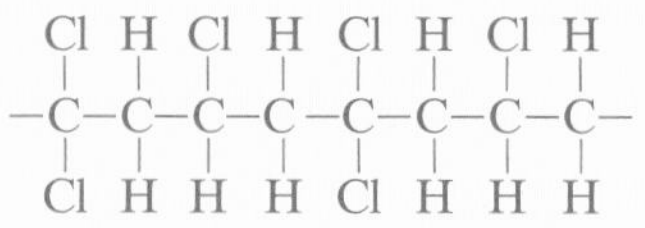 [structure with Cl and H]

53. a.

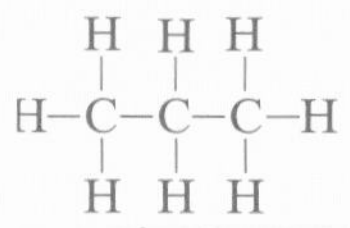

H–$\overset{\overset{\displaystyle H}{|}}{\underset{\underset{\displaystyle H}{|}}{C}}$–$\overset{\overset{\displaystyle H}{|}}{\underset{\underset{\displaystyle H}{|}}{C}}$–$\overset{\overset{\displaystyle H}{|}}{\underset{\underset{\displaystyle H}{|}}{C}}$–H

b. propane

54. a. 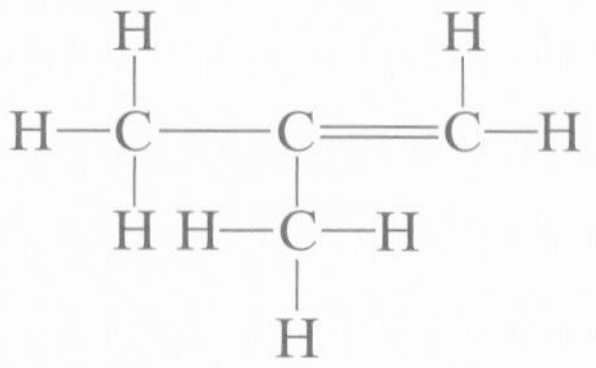 [structure]

b. 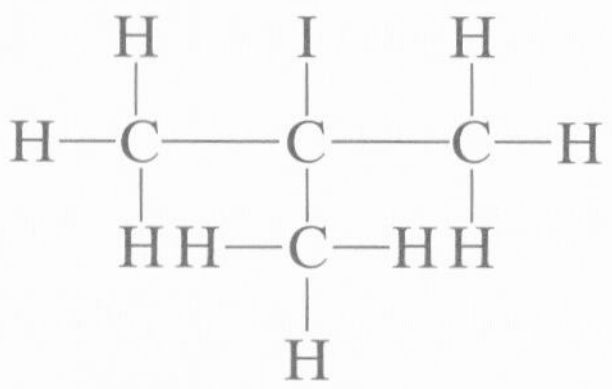 [structure]

55. 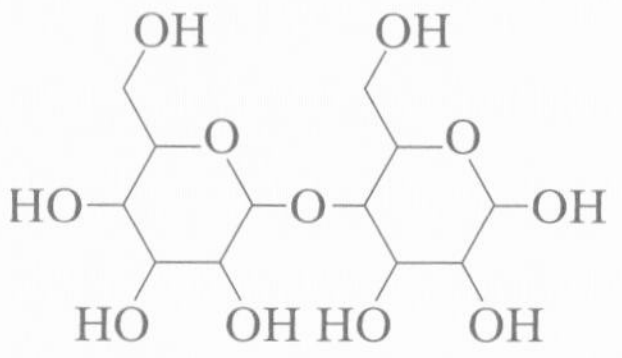 [structures with OH groups]

56. 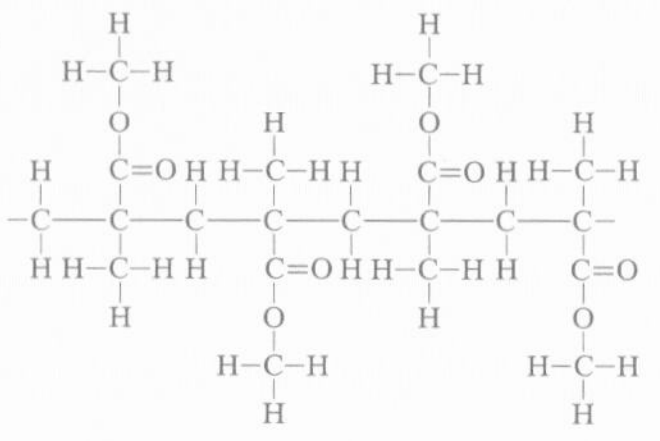 [structure]

Answers continued on p. 709A

Study the graph below, and answer the questions that follow.
For help in interpreting graphs, see Appendix B, "Study Skills for Chemistry."

65. a. Determine the percentage composition by weight of hexane, C_6H_{14}.
b. Using the charts to the right as a model, make a pie chart for hexane using a protractor to draw the correct sizes of the pie slices. (Hint: A circle has 360°. To draw the correct angle for each slice, multiply each percentage by 360°.)

66. a. Compare the charts for methane, ethane, and hexane. In which of these three charts is the slice for carbon the largest?
b. In which of the three charts is the slice for carbon the smallest?

67. Based on your answers to the previous item, complete the following statement:
For saturated hydrocarbons, as the number of carbon atoms in the molecule increases, the percentage of carbon in the molecule will _____.

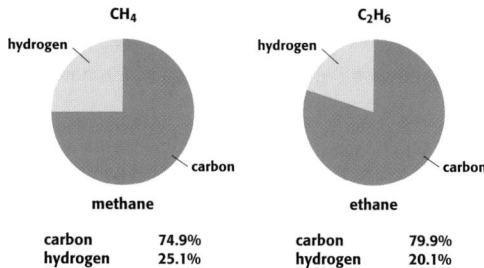

CH_4			C_2H_6	
carbon	74.9%		carbon	79.9%
hydrogen	25.1%		hydrogen	20.1%

68. a. Determine the percentage composition of hexene, C_6H_{12}.
b. Using the charts above as a model, make a pie chart for hexene.
c. Compare the charts for hexane and hexene. Which of these charts shows a larger slice for carbon?

TECHNOLOGY AND LEARNING

69. Graphing Calculator

Hydrocarbon formulas

The graphing calculator can run a program that can tell you the formula of any straight-chain hydrocarbon, provided you indicate the number of carbons and the number of double bonds in the compound.

Go to Appendix C. If you are using a TI-83 Plus, you can download the program **HYDROCAR** and run the application as directed. If you are using another calculator, your teacher will provide you with keystrokes and data sets to use. At the prompts, enter the number of carbon atoms and the

number of double bonds in the molecule. Run the program as needed to answer the following questions.

a. Dodecane is an alkane with 12 carbons and no double bonds. What is its formula?

b. The name 1,5-hexadiene describes a molecule with six carbons (hexa–) and two double bonds (–diene). What is its formula?

c. What is the formula for 1, 3, 5-hexatriene?

d. What is the formula for 3-nonene?

e. What is the formula for 1,3,5,7-octatetraene?

f. What is the formula for 2,4,6-octatriene?

708

Chapter Resource File
• **Chapter Test**

1. Which of the following hydrocarbons must be an alkane?

   a. C_2H_2 **c.** C_7H_{12}

   b. C_5H_{10} **d.** $C_{14}H_{30}$

2. A hydrocarbon with the formula C_8H_{18} is called

   a. octene.

   b. octyne.

   c. octane.

   d. propane.

3. During a condensation polymerization reaction,

   a. single bonds replace all double bonds that are present in the monomer.

   b. water is often produced.

   c. alcohol groups are formed.

   d. an aldehyde group is changed to a ketone group.

4. In naming an organic compound,

   a. remember that the locations of all functional groups are optional.

   b. do not consider the number of carbon atoms in the molecule as a factor.

   c. begin by identifying and naming the longest hydrocarbon chain.

   d. use side chains as the basis for naming the molecule.

5. Which of the following can have the greatest number of isomers?

   a. C_2H_6 **c.** C_6H_6

   b. C_3H_8 **d.** $C_{20}H_{42}$

6. Which of the following compounds cannot have different isomers?

   a. C_7H_{16}

   b. C_5H_{10}

   c. C_3H_8

   d. $C_6H_{12}O_6$

7. Examine the following structural formula.

The correct name for this compound is

   a. 2,2-dimethylbutane.

   b. 1,1,1-trimethylpropane.

   c. 2-ethyl-2-methylpropane.

   d. 3,3-dimethylbutane.

8. Examine the following skeletal structure.

The correct molecular formula for this compound is

   a. $C_2H_4O_2$.   **c.** $C_5H_8O_2$.

   b. $C_5H_4O_2$.   **d.** CHO.

9. Which of the following is an aromatic compound?

   a.

   b.

   c.

   d.

709

Continuation of Answers

Answers to Section 1 Review, *continued*

7. A functional group is an atom or group of atoms that gives characteristic properties to organic compounds.

8. one of two or more compounds that have the same chemical composition but different structures; their chemical composition

9. ; an alkene

10. They cannot be isomers because they contain different types of elements.

11. Any structure that contains seven C atoms, an appropriate number of H atoms, and at least one triple bond is correct.

12. Any two alkane structures that contain four C atoms, 10 H atoms, and only single bonds are correct as long as they have different structural formulas.

13. It is an aromatic ring, and therefore has a completely different character from that of an alkene.

14.

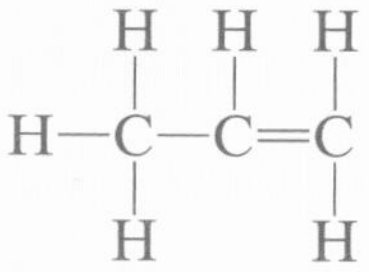

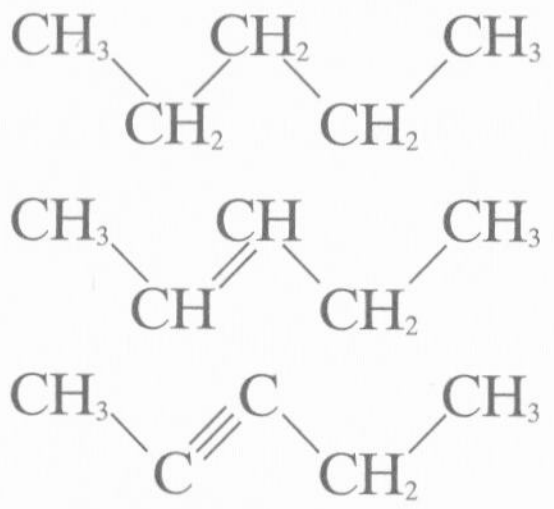

(other examples for the alkene and alkyne are possible.)

They have different chemical formulas and therefore are not isomers.

15. A ring structure consisting of 4 C atoms, an appropriate number of H atoms, and one double bond is correct.

Answers to Section 2 Review, *continued*

12. The correct name must include a number to indicate the position of the double bond, such as 2-nonene.

13. Methanol is an alcohol, whereas methanal is an aldehyde. Both contain only one carbon atom.

14. There are two double bonds, one between the first and second carbon atoms, and the other bonding the third and fourth carbon atoms.

Answers to Section 3 Review, *continued*

12. The alkane in the reaction is already a saturated compound.

Answers to Chapter Review, *continued*

57.

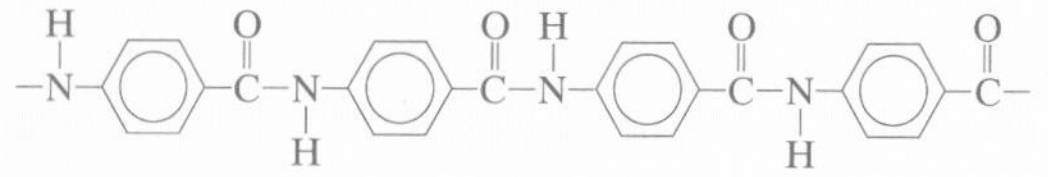

58.

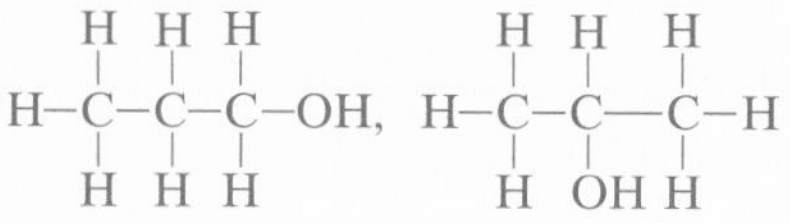

59. a. carboxylic acid
b. ether
c. ketone
d. amine
e. halide

60. $-O-CH_2-CH_2-O-CH_2-CH_2-O-CH_2-CH_2-O-CH_2-CH_2-$

61. Answers may vary.

62. Answers may vary.

63. Answers may vary.

64. a. organic reactions
b. substitution
c. addition
d. condensation
e. halogen
f. hydrogen
g. water

65. a. carbon 83.6%; hydrogen 16.4%
b. The chart should show hydrogen covering about $\frac{1}{6}$ of the area and carbon covering the rest.

66. a. hexane
b. methane

67. increase

68. a. carbon 83.6%; hydrogen 16.4%
b. The chart should show hydrogen covering about $\frac{1}{6}$ of the area and carbon covering the rest.
c. hexene

69. a. $C_{12}H_{26}$
b. C_6H_{10}
c. C_6H_8
d. C_9H_{18}
e. C_8H_{10}
f. C_8H_{12}

Biological Chemistry
Chapter Planning Guide

PACING	CLASSROOM RESOURCES	LABS, ACTIVITIES, AND DEMONSTRATIONS
BLOCK 1 • 45 min pp. 710–711 **Chapter Opener**		SE **Start-Up Activity** Exploring Carbohydrates, p. 711 ◆
BLOCKS 2 & 3 • 45 min pp. 712–716 **Section 1** Carbohydrates and Lipids	TT **Types of Carbohydrates** *	TE **Demonstration** Testing for Simple Sugars, p. 713 ◆ CRF **Consumer Lab** All Fats Are Not Equal ◆ GENERAL
BLOCK 4 • 45 min pp. 717–724 **Section 2** Proteins	TT **Structures and Roles of Several Amino Acids** *	TE **Demonstration** Modeling Amino Acids, p. 718 TE **Demonstration** Pineapple Enzyme, p. 722 ◆ BASIC SE **QuickLab** Denaturing an Enzyme, p. 723 ◆ GENERAL CRF **Datasheets for In-Text Labs** *
BLOCK 5 • 45 min pp. 725–733 **Section 3** Nucleic Acids	TT **Nitrogenous Bases of Nucleic Acids** * SE **Science and Technology** Protease Inhibitors	TE **Group Activity**, p. 726 BASIC SE **QuickLab** Isolation of Onion DNA, p. 727 ◆ GENERAL CRF **Datasheets for In-Text Labs** * TE **Demonstration** DNA Strings, p. 729
BLOCK 6 • 45 min pp. 734–738 **Section 4** Energy in Living Systems	TT **Hydrolysis of ATP** * SE **Element Spotlight** Magnesium: An Unlimited Resource	TE **Activity**, p. 735 ADVANCED TE **Activity**, p. 736 BASIC TE **Group Activity**, p. 736 GENERAL

BLOCKS 7 & 8 • 90 min

Chapter Review and Assessment Resources

SE **Chapter Review**, pp. 741–744
SE **Standardized Test Prep**, p. 745
CRF **Chapter Test** *
OSP **Test Generator**
CRF **Test Item Listing** *
OSP **Scoring Rubrics and Classroom Management Checklists**

Holt Chemistry: Online Resources

go.hrw.com

Visit **go.hrw.com** for a variety of free resources related to this textbook. Enter the keyword **HW4 HOME**.

Holt Online Learning

Students can access interactive problem solving help and active visual concept development with the *Holt Chemistry* Online Edition available at **www.hrw.com**

student CNN News

cnnstudentnews.com

Find the latest chemistry news, lesson plans, and activities related to important scientific events.

KEY

TE	Teacher Edition	**CRF**	Chapter Resource File	*	Also on One-Stop Planner
SE	Student Edition	**TT**	Teaching Transparencies	◆	Requires Advance Prep
OSP	One-Stop Planner				

PROBLEM SOLVING AND PRACTICE	SECTION REVIEW AND ASSESSMENT	STANDARDS CORRELATION
		National Science Education Standards
	TE **Homework,** p. 714 GENERAL TE **Quiz,** p. 716 GENERAL CRF **Quiz** * TE **Reteaching,** p. 716 BASIC SE **Section Review,** p. 716 CRF **Concept Review** *	PS 2f, PS 3a
	TE **Homework,** p. 721 BASIC TE **Quiz,** p. 724 GENERAL CRF **Quiz** * TE **Reteaching,** p. 724 BASIC SE **Section Review,** p. 724 CRF **Concept Review** *	PS 2f, PS 3a, PS 3e
SE **Skills Toolkit 1** Using the Genetic Code, p. 729 TE **Homework,** p. 729 GENERAL	TE **Quiz,** p. 732 GENERAL CRF **Quiz** * TE **Reteaching,** p. 732 BASIC SE **Section Review,** p. 732 CRF **Concept Review** *	PS 2f
	TE **Homework,** p. 737 BASIC TE **Quiz,** p. 738 GENERAL CRF **Quiz** * TE **Reteaching,** p. 738 BASIC SE **Section Review,** p. 738 CRF **Concept Review** *	PS 3a

www.scilinks.org

Topic: Spider Proteins **SciLinks code:** HW4711	**Topic:** DNA **SciLinks code:** HW4042	**Topic:** DNA Fingerprinting **SciLinks code:** HW4043	**Topic:** Photosynthesis **SciLinks code:** HW4096	**Topic:** Magnesium **SciLinks code:** HW4077
Topic: Carbohydrates **SciLinks code:** HW4024	**Topic:** Replication **SciLinks code:** HW4044	**Topic:** Protease Inhibitors **SciLinks code:** HW4102	**Topic:** Respiration **SciLinks code:** HW4111	
Topic: Proteins **SciLinks code:** HW4104	**Topic:** Synthesis **SciLinks code:** HW4103	**Topic:** Biochemical Processes **SciLinks code:** HW4022	**Topic:** ATP **SciLinks code:** HW4018	

Overview

This chapter introduces students to the chemistry of biological systems. Students will learn about the structure of carbohydrates, how carbohydrates combine to form large biological polymers, and how carbohydrates function in living systems. Students will also explore the structure and function of amino acids, proteins, enzymes, and nucleic acids, as well as learn how nucleic acids store genetic coding information. Finally, students will learn how plants and animals produce energy through photosynthesis and cellular respiration.

Assessing Prior Knowledge

Check for Content Knowledge
Students should be familiar with the following topics:

- chemical formulas
- reaction types
- polymerization
- chemical bonding and inter-molecular forces

Using the Figure

Strong Proteins In some ways, human hair and spider silk are very similar. They are both flexible solids made up primarily of proteins, a type of biological polymer made of amino acids. However, fibroin, the protein in spider silk, is considerably stronger than keratin, the protein in human hair. When considered on a strength per weight basis, fibroin is actually six times stronger than steel.

CHAPTER

20

BIOLOGICAL CHEMISTRY

710

Standards Correlations

National Science Education Standards

PS 2f: Carbon atoms can bond to one another in chains, rings, and branching networks to form a variety of structures, including synthetic polymers, oils, and the large molecules essential to life. (Sections 1, 2, 3)

PS 3a: Chemical reactions occur all around us, for example in health care, cooking, cosmetics, and automobiles. Complex chemical reactions involving carbon-based molecules take place constantly in every cell in our bodies. (Sections 1, 2, 4)

PS 3e: Catalysts, such as metal surfaces, accelerate chemical reactions. Chemical reactions in living systems are catalyzed by protein molecules called enzymes. (Section 2)

A spider web can stop an insect that is flying at top speed, and a single thread of spider silk can hold the weight of a spider that is large in size. Scientists have marveled that a material as lightweight as spider silk can be so strong. The silk that spiders use to form their webs is made up of a biological chemical—a protein—called fibroin. Scientists are searching for ways to use fibroin to make building materials that are strong and lightweight, like spider silk. The study of spider silk is just one example of how biological chemists are looking to nature to solve problems in the industrial world.

START-UP ACTIVITY

Exploring Carbohydrates

SAFETY PRECAUTIONS

PROCEDURE

1. Measure out one-half teaspoon of **sugar** into a **small beaker.**

2. Measure out one-half teaspoon of **cornstarch** into a **second small beaker.**

3. Your teacher will provide you with a **slice of apple**, a **slice of potato**, and a **slice of turkey.**

4. Add a drop of **iodine solution** to all five samples.

ANALYSIS

1. In the presence of starch, iodine turns dark blue-black. Note which samples test positive for starch.

2. Explain your observations.

Pre-Reading Questions

① **Describe at least one way that the laws of chemistry apply to living systems.**

② **What biological molecule contains the information that determines your traits?**

③ **In chemical terms, what is the purpose of the food we eat?**

CONTENTS 20

internet connect

www.scilinks.org
Topic: Spider Proteins
SciLinks code: HW4132

SCi LINKS. Maintained by the National Science Teachers Association

711

START-UP ACTIVITY

Skills Acquired:
• Collecting Data
• Interpreting
• Inferring

Materials:
For each group of 2–3 students:
• apple, sliced
• beakers, small (2)
• cornstarch, $\frac{1}{2}$ tsp.
• eyedropper
• iodine solution, 5 drops
• paper plates, small (3)
• potato, sliced
• sugar, $\frac{1}{2}$ tsp.
• teaspoon
• turkey, sliced

Teacher Notes: Iodine solution, which contains potassium iodide and can be purchased at a drug store, tests for the presence of starch.

Safety Caution: Remind students to use goggles, aprons, and gloves for this activity. Students should not ingest the iodine solution or any of the test samples.

Answers

1. The cornstarch, apple, and potato should test positive for starch. The sugar and turkey should not test positive for starch.

2. The sugar and turkey do not contain starch, while the cornstarch, apple, and potato are starchy foods and therefore turn the iodine blue-black.

Answers to Pre-Reading Questions

1. The structures in living systems are made up of many of the same atoms that make up nonliving objects. Thus, the rules of chemistry apply equally to living systems and nonliving objects. For example, a water molecule has the same chemical properties whether it is found in a lake or in a human body.

2. The nucleic acid DNA contains the information that determines your body's traits.

3. The food we eat provides energy and reactants to fuel the chemical reactions that take place in our bodies.

Focus

Overview

Before beginning this section, review with your students the Objectives listed in the Student Edition. This section introduces carbohydrates, their function as a food source, and the biological reactions that form them and break them down. Students will also learn the characteristic structure of mono-, di-, and polysaccharides.

Bellringer

Write the chemical formula for cellulose, $(C_6H_{10}O_5)_n$, on the board and identify it as such. Ask students to write the complete balanced chemical reaction for the combustion of cellulose.

Ans. $(C_6H_{10}O_5)_n(s) + 6nO_2(g) \rightarrow 6nCO_2(g) + 5nH_2O(g)$

Motivate

Identifying Preconceptions — GENERAL

1. Ask students to list three basic sources of Calories in their diets. Ans. fats (lipids), proteins, and carbohydrates

2. Ask students which of the following are carbohydrates: sugar, starch, cellulose (plant stems, leaves, and roots), and chitin (shells of crustaceans). Ans. All are examples of carbohydrates.

3. Ask students if polymers can be formed from organic reactions as well as inorganic reactions. Ans. yes

LS Verbal

Chapter Resource File

• Lesson Plan

Carbohydrates and Lipids

KEY TERMS

• carbohydrate
• monosaccharide
• disaccharide
• polysaccharide
• condensation reaction
• hydrolysis
• lipid

OBJECTIVES

① **Describe** the structure of carbohydrates.

② **Relate** the structure of carbohydrates to their role in biological systems.

③ **Identify** the reactions that lead to the formation and breakdown of carbohydrate polymers.

④ **Describe** a property that all lipids share.

carbohydrate

any organic compound that is made of carbon, hydrogen, and oxygen and that provides nutrients to the cells of living things

monosaccharide

a simple sugar that is the basic subunit of a carbohydrate

disaccharide

a sugar formed from two monosaccharides

polysaccharide

one of the carbohydrates made up of long chains of simple sugars; polysaccharides include starch, cellulose, and glycogen

Carbohydrates in Living Systems

Most of the energy that you get from food comes in the form of **carbohydrates.** For most of us, starch, found in such foods as potatoes, bread, and rice, is our major carbohydrate source. Sugars—in fruit, honey, candy, and many packaged foods—are also carbohydrates. Plants make carbohydrates, such as the starch in potato tubers, shown in **Figure 1.**

Raw potato is difficult to digest because the starch is present in tight granules. Cooking bursts the granules, so that starch can be attacked by our digestive juices. During digestion, the starch is broken down into another carbohydrate called glucose, which—unlike starch—can be carried by the bloodstream.

Carbohydrates are compounds of carbon, hydrogen, and oxygen. They usually have the general formula $C_{6n}H_{10n+2}O_{5n+1}$. When $n = 1$ (6 C atoms), the carbohydrate is a **monosaccharide;** glucose is an example. A **disaccharide** is a carbohydrate with $n = 2$ (12 C atoms). Starch is an example of a **polysaccharide,** in which n can be many thousands.

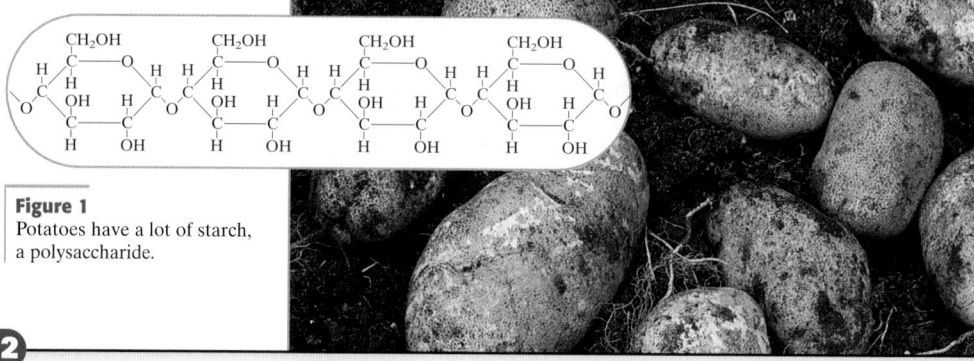

Figure 1
Potatoes have a lot of starch, a polysaccharide.

---

did you know?

How Sweet It Is Not all sugars are equally as sweet. Fructose, a simple sugar that occurs in fruits, is considerably sweeter on a per gram or per mole basis than sucrose, the sugar commonly used in baking. Baked foods containing fructose can be as sweet as baked foods containing sucrose, but actually contain less sugar and fewer Calories. Controlling sugar intake is important in all diets, but especially for diabetics, whose bodies are unable to properly regulate blood sugar levels. The table below shows the sweetness of several sugars relative to sucrose.

Sugar	Relative sweetness
Lactose	0.16
Galactose	0.32
Sucrose	1.00
Fructose	1.73

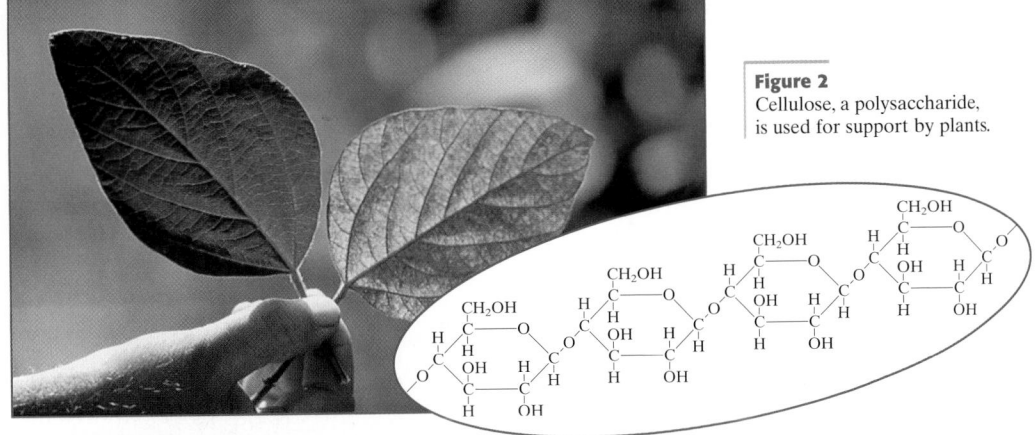

Figure 2
Cellulose, a polysaccharide, is used for support by plants.

Carbohydrates Have Many Functions

Starch is the polysaccharide that plants use for storing energy. Many animals make use of a similar energy-storage carbohydrate called *glycogen*. It is often stored in muscle tissue as an energy source.

Mammals rely on bones and muscles, which are made primarily of proteins, to give their bodies structure and support. However, insects and crustaceans, such as crabs and lobsters, rely on hard shells made of the polysaccharide chitin for structure.

The carbohydrate you come into contact with the most is the one you are looking at right now—cellulose, in paper, which comes from wood fiber. Cellulose is the most abundant organic compound on Earth. It is the polysaccharide that most plants use to give their structures rigidity. The leaves, stems, and roots of these plants are all made of cellulose, shown in **Figure 2.**

Structure of Simple Sugars

To a chemist, *sugar* is the name given to all monosaccharides and disaccharides. To a cook, *sugar* means one particular disaccharide, sucrose. The cyclic sugar glucose is important to the body because it is the chemical that the bloodstream uses to carry energy to every cell in the body. Shown below are the structures for glucose, $C_6H_{12}O_6$, and fructose, another sugar.

glucose   fructose

The glucose molecule has a ring made of six atoms—five carbon atoms and one oxygen atom. A sixth carbon atom is part of a $-CH_2OH$ side chain. Four other hydroxyl, $-OH$, groups connect to the carbons in the ring, as do four H atoms. The fructose molecule has a ring of five atoms, four carbon and one oxygen. Fructose has two $-CH_2OH$ side chains. Fructose and glucose have the same molecular formula, $C_6H_{12}O_6$, even though they have very different structures.

internet connect

www.scilinks.org
Topic: Carbohydrates
SciLinks code: HW4024

SCiLINKS. Maintained by the National Science Teachers Association

713

Teaching Tip

Lactose Intolerance Lactose is a component of a nursing mother's milk. After infancy, however, some people lose their ability to digest lactose. The result is gastrointestinal discomfort whenever milk is consumed. Such "lactose-intolerant" people must either avoid dairy products in their diet or take an enzyme supplement along with their meal. The enzyme catalyzes the digestive reaction needed to break down the lactose.

Homework —— GENERAL

Have students examine nutrition labels to determine the number of grams of carbohydrates per serving. What is the recommended daily carbohydrate intake for a balanced 2,500 Calorie diet?
Ans. 375 g

Have students summarize data for three different foods or beverages in a table. They should include the total carbohydrates per serving (g), sugar per serving (g), and percentage of total Calories from sugar. Inform students that 1 g of sugar, carbohydrate, or protein contains 4 Calories, and 1 g of fat contains 9 Calories. **LS** Logical

SKILL BUILDER —ADVANCED

Math Skills Reinforce the everyday relevance of molar-based calculations by having students determine the volume of sugar (sucrose), in milliliters and in teaspoons, contained in a typical cola beverage. Also have them verify the information of the nutrition label by calculating the number of Calories contained in the cola. **LS** Logical

Figure 3
Three different disaccharides—sucrose, maltose, and lactose—are present in a malted milk shake.

Sugars Combine to Make Disaccharides

Monosaccharides, such as glucose, have one ring. However, two can combine to form a double-ringed disaccharide. Three examples of disaccharides—lactose, maltose, and sucrose—are found in the malted milk shake shown in **Figure 3**. Notice that the disaccharides are each made up of two monosaccharides. Each molecule of maltose, the sugar that adds to the flavor of malted milk shakes, is made up of two glucose units. Each molecule of sucrose, the sugar you use to sweeten food, is made up of a glucose and a fructose unit.

Structure of Polysaccharides

Just as two monosaccharides combine to form a disaccharide, many monosaccharides or disaccharides can combine to form a long chain called a polysaccharide. Polysaccharides may be represented by the general formula below or by structural models such as the ones shown in **Figures 1 and 2**.

$$\cdots O-(C_6H_{10}O_4)-O-(C_6H_{10}O_4)-O-(C_6H_{10}O_4)-O-(C_6H_{10}O_4)\cdots$$

Earlier, you learned about the linking together of small molecular units in a process known as *polymerization*. Polymerization is a series of synthesis reactions that link many monomers together to make a very large, chainlike molecule. The formation of polysaccharides is similar to polymerization. In fact, polysaccharides and other large, chainlike molecules found in living things are called *biological polymers*. Amylose, a biological polymer listed in **Table 1**, is a form of starch.

Topic Link

Refer to the "Carbon and Organic Compounds" chapter for a discussion of polymers.

714

did you know?

Sugar Consumption and Health According to the United States Department of Agriculture (USDA) the average person who eats a healthy diet and consumes about 2000 Calories a day should limit sugar consumption to about 10 teaspoons per day of added sugar. The average American consumes twice this amount—an extra 20 teaspoons of added sugar per day! There are many possible health problems caused by prolonged high sugar consumption, including diabetes, heart disease, obesity, and dental problems.

did you know? —— ADVANCED

Blood Chemistry One milliliter of healthy blood contains approximately 1 mg of glucose. Have students calculate the concentration (molarity) of glucose in healthy blood.
Ans. 0.006 M **LS** Logical

Table 1 Types of Carbohydrates

Type	Example	Role
Monosaccharides	fructose	sweetener found in fruits
	glucose	cell fuel
Disaccharides	sucrose	sweetener (table sugar)
Polysaccharides	chitin	insect exoskeleton, support, protection
	amylose	energy storage (plants)
	glycogen	energy storage (animals)

Carbohydrate Reactions

Photosynthesis and respiration, described later, are the main ways that carbohydrates are made and broken down in living systems. These processes are also the primary ways that living things capture and use energy. Thus, carbohydrate reactions play a major role in the chemistry of life.

Formation of Disaccharides and Polysaccharides

Because glucose and other sugars dissolve easily in water, they are not useful for long-term energy storage. This is why living things change sugars to starch or glycogen, neither of which is soluble in water.

Disaccharides and polysaccharides are formed from sugars during **condensation reactions,** in which water is a byproduct. Though there are many more steps that are not shown here, the net equation below describes the formation of the disaccharide sucrose.

condensation reaction

a chemical reaction in which two or more molecules combine to produce water or another simple molecule

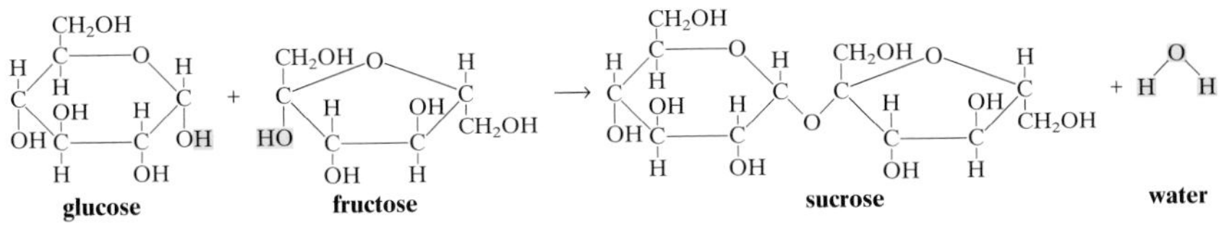

glucose   fructose   →   sucrose   +   water

Breakdown of Carbohydrates

When an organism is ready to use energy that was previously stored as a polysaccharide, a different kind of reaction takes place. Polysaccharides are changed back to sugars during **hydrolysis** reactions. In these reactions, the decomposition of a biological polymer takes place along with the breakdown of a water molecule, as shown in the equation below.

hydrolysis

a chemical reaction between water and another substance to form two or more new substances

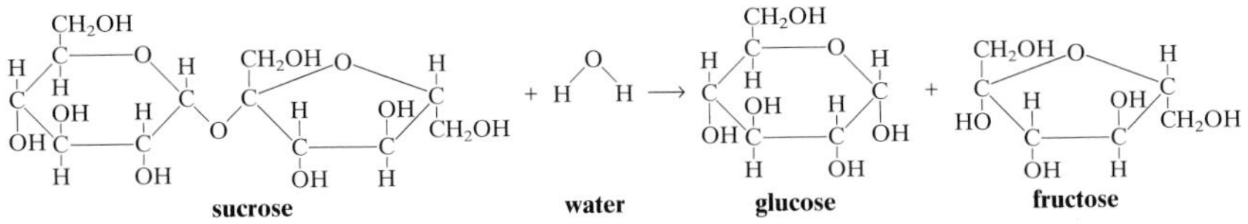

sucrose   +   water   →   glucose   +   fructose

The reaction is the reverse of the condensation reaction by which sucrose formed. In humans, polysaccharides, such as starch and glycogen, and disaccharides, such as sucrose, are broken down in this way to make glucose.

715

did you know?

Anaerobic Processes Most sugars formed by hydrolysis reactions will eventually be converted to glucose and used, with oxygen, to generate energy through cellular respiration. There are, however, two alternative pathways that generate energy without using oxygen. Alcoholic fermentation produces ethanol and carbon dioxide. The reaction is $C_6H_{12}O_6 \rightarrow 2C_2H_5OH + 2CO_2$. Alcoholic fermentation occurs in natural decomposition and is used in baking, winemaking, and the commercial brewing industry. A similar process known as lactic fermentation permits oxygen-starved muscles to continue functioning, though painfully and less efficiently, while producing lactic acid. The reaction is $C_6H_{12}O_6 \rightarrow 2CH_3CH(OH)COOH$. Essentially the same reaction occurs when lactose in milk ferments in the presence of bacterial enzymes to produce sour milk.

Close

lipid

a type of biochemical that does not dissolve in water, including fats and steroids; lipids store energy and make up cell membranes

Figure 4
Like all steroids, cholesterol has a structure with four connected rings.

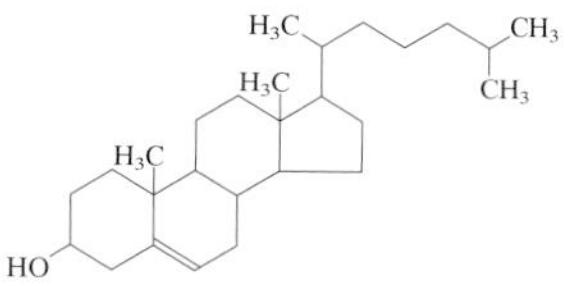

Lipids

Lipids are a class of biological molecules that do not dissolve in water. However, they generally can have a polar, hydrophilic region at one end of the molecule. For example, the lipid shown below is oleic acid, which is found in the fat of some animals.

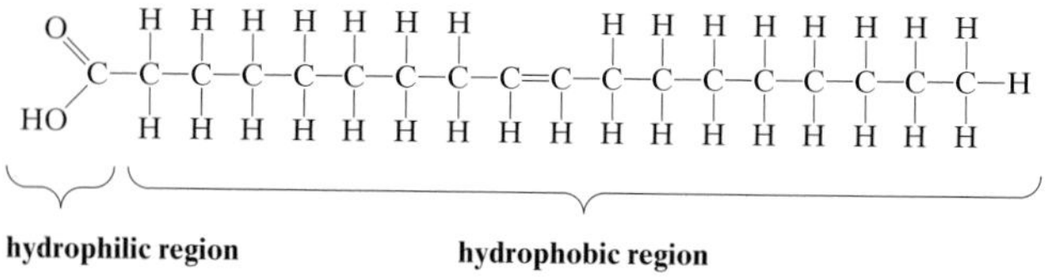

hydrophilic region   hydrophobic region

The hydrophilic region on the right side of the molecule allows it to interact with polar molecules. The hydrophobic region on the left side of the molecule allows it to interact with nonpolar molecules.

Lipids have a variety of roles in living systems. They are used in animals for energy storage as *fats*. Cell membranes are made up of lipids called *phospholipids*. Steroids—such as cholesterol, shown in **Figure 4**—are lipids used for chemical signaling. Waxes, such as those found in candles and beeswax are also lipids.

① Section Review

UNDERSTANDING KEY IDEAS

1. Describe the general chemical formula of carbohydrates.

2. What do chemists mean by a *sugar*, and what are the two principal classes of sugars?

3. What role do carbohydrates play in the survival of animals and plants?

4. Name several polysaccharides, and explain the biological role of each.

5. What is the molecular formula of glucose, and what is the role of this compound in human body systems?

6. What names are given to the reactions by which large carbohydrate molecules are built up and broken down?

7. How does the formation of a biological polymer compare to the formation of most manufactured polymers?

8. What property do all lipids share?

CRITICAL THINKING

9. What is the formula of the compound formed by the condensation of two disaccharides?

10. Why do we cook starchy foods?

11. Classify the following carbohydrates into monosaccharides, disaccharides, or polysaccharides: cellulose, glucose, lactose, starch, maltose, sucrose, chitin, and fructose.

12. Why is glycogen often called *animal starch*?

13. a. What type of reaction does the following equation describe?

b. Name the reactants and the products.

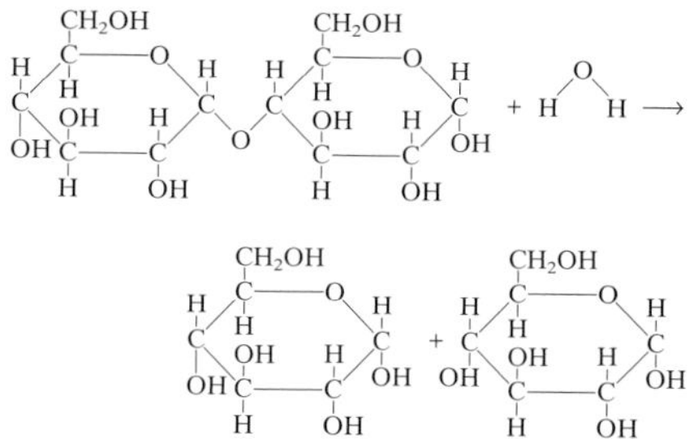

Answers to Section Review

1. Carbohydrates have the general formula $C_{6n}H_{10n+2}O_{5n+1}$.

2. Sugars are small carbohydrate molecules. The two principal classes are monosaccharides and disaccharides.

3. They are responsible for energy storage and delivery, as well as providing structural support.

4. Sample answer: Starch and glycogen serve as energy stores, and cellulose and chitin provide structural support.

5. $C_6H_{12}O_6$; Glucose is present in the bloodstream as a means of distributing energy throughout the body.

6. condensation and hydrolysis, respectively

7. Both are long molecular strands with repeating units and generally form through condensation reactions.

8. All lipids are insoluble in water.

9. $C_{24}H_{42}O_{21}$

10. It breaks up the starch granules, permitting our digestive juice to reach the starch molecules.

11. monosaccharides: glucose, fructose; disaccharides: lactose, maltose, sucrose; polysaccharides: cellulose, starch, chitin

Answers continued on p. 745A

Answers continued on p. 745A

② Proteins

KEY TERMS
- protein
- amino acid
- polypeptide
- peptide bond
- enzyme
- denature

OBJECTIVES

① **Describe** the general amino acid structure.

② **Explain** how amino acids form proteins through condensation reactions.

③ **Explain** the significance of amino-acid side chains to the three-dimensional structure and function of a protein.

④ **Describe** how enzymes work and how the structure and function of an enzyme is affected by changes in temperature and pH.

Amino Acids and Proteins

A **protein** is a biological polymer that is made up of nitrogen, carbon, hydrogen, oxygen, and sometimes other elements. Our bodies are mostly made out of proteins. For example, the most abundant protein in your body is collagen, which is found in skin and bones. Your hair has structural proteins, such as keratin, shown in **Figure 5**. Proteins in muscles allow your muscles to contract, making body movement possible.

Different proteins have different physical properties. Some—such as casein in milk, ovalbumin in egg whites, and hemoglobin in blood—are water-soluble. Others—such as keratin in hair, fibroin in spider silk, and collagen in connective tissue—are flexible solids.

What do all these proteins have in common? They are all made up of **amino acids.** In the same way that sugars are the building blocks of carbohydrates, amino acids are the building blocks of proteins.

protein

an organic compound that is made of one or more chains of amino acids and that is a principal component of all cells

amino acid

any one of 20 different organic molecules that contain a carboxyl and an amino group and that combine to form proteins

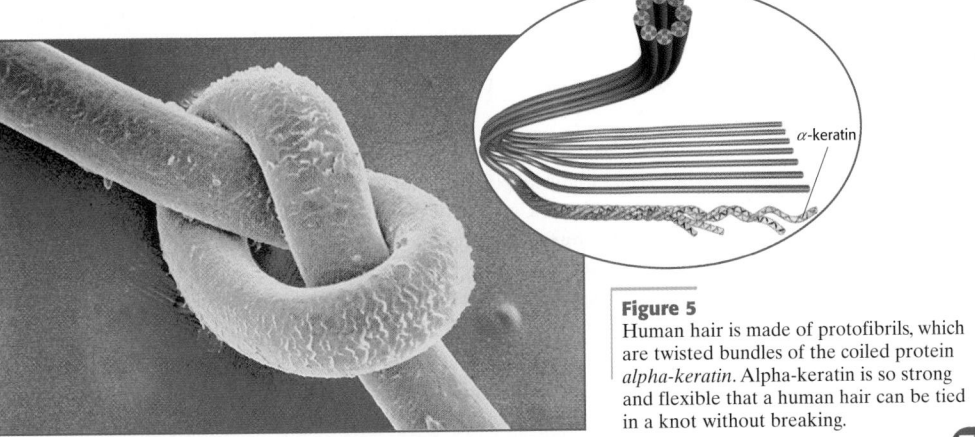

α-keratin

Figure 5
Human hair is made of protofibrils, which are twisted bundles of the coiled protein *alpha-keratin*. Alpha-keratin is so strong and flexible that a human hair can be tied in a knot without breaking.

717

did you know?

Protein Facts Proteins make up about 15% of the mass of our bodies. Typical proteins have molar masses ranging from 6,000 to over 1,000,000 g/mol. As discussed in the section, proteins have a wide range of functions in the body. Some proteins, called *fibrous proteins*, provide strength and structure to many types of tissue, including hair, muscle, and cartilage.

Another class of proteins, known as *globular proteins* because of their sphere-like shapes, carry out much of the work done within the body. Just some of the various roles globular proteins perform include catalyzing countless biological reactions, attacking invading objects and pathogens, and transporting and storing nutrients.

Focus

Overview
Before beginning this section, review with your students the Objectives listed in the Student Edition. In this section, students learn how proteins are formed by the synthesis of amino acids, how a protein's composition and shape influence its function, and how enzymes function in biological systems.

🔊 Bellringer
Write the following question on the board: "If you have 20-piece sets of 20 different objects, how many ways can these objects be arranged into a 20-object long sequence, if each object can be used once, multiple times, or not at all?" Ans. 20^{20} possibilities

Motivate

Identifying Preconceptions —— GENERAL

1. How does a catalyst affect the forward and reverse rates of a chemical reaction? Ans. Both the forward and reverse reaction rates increase by the same amount.

2. Does hydrogen bonding only occur between water molecules? Ans. no (Students may think that hydrogen bonding occurs only in water, when in fact it can occur anywhere that a hydrogen atom that is covalently bonded to a strongly electronegative atom is also attracted to an unshared electron pair of another electronegative atom.)

LS Verbal

Chapter Resource File
- Lesson Plan

Chapter 20 • Biological Chemistry   717

Motivate, *continued*

Demonstration

Modeling Amino Acids Make ball-and-stick models of several amino acids. Identify the amino group and the carboxylic acid group on each model. Because the central carbon of an amino acid is an asymmetric carbon, each amino acid can exist as two optical, or mirror image, isomers (enantiomers). Use a mirror to show that the image of a structure is the same as its enantiomer. Discuss how knowing whether the central carbon in the amino acid is chiral or not determines if the amino acid has mirror image isomers. Also use the ball-and-stick models to show how amino acids join through condensation polymerization.

Teach

Using the Table — GENERAL

Table 2 Draw students' attention to the similarities in the structure of the various amino acids. Identify the location of $-NH_2$ and $-COOH$ groups on each amino acid. Ask students to describe the general structure of an amino acid using R to represent the amino acid's side chain. **Ans.** Each amino acid has a "backbone" consisting of a carboxylic acid (COOH) group and an amino (NH_2) group covalently bonded to a central carbon atom. The central carbon's two remaining bonding sites are occupied by a hydrogen atom and a side chain, R, of varying composition. Then, ask students to write the general formula for an amino acid, using R to represent the amino acid's side chain. **Ans.** $CH(R)(NH_2)COOH$
LS Logical

Table 2 Structures and Roles of Several Amino Acids

Name	Structure	Role	Name	Structure	Role
Cysteine		cross-links to other cysteine units	Valine		contributes to hydrophobicity (nonpolar)
Glutamic acid		gives an acidic side chain	Asparagine		gives hydrogen-bonding sites (polar)
Glycine		acts as a spacer	Histidine		gives a basic side chain

Amino-Acid Structure and Protein Synthesis

Amino refers to the $-NH_2$ group of atoms. Generally, organic acids have the carboxylic acid group, $-COOH$. Thus, *amino acids* are compounds that have both the basic $-NH_2$ and the acidic $-COOH$ groups. There are 20 amino acids from which natural proteins are made. All of them have the same basic structure shown below. The R represents a *side chain*.

A side chain is a chemical group that differs from one amino acid to another. **Table 2** shows the detailed structure of six of these amino acids.

The reaction by which proteins are made from amino acids is similar to the condensation of carbohydrates. A water molecule forms from the $-OH$ of the carboxylic acid group of one amino acid and an $-H$ of the amino group of another. The condensation of amino acids is shown below.

The biological polymer that forms is called a **polypeptide.** The link that joins the N and C atoms of two different amino acids in a protein is called a **peptide bond.** In protein synthesis, hundreds of peptide bonds are formed one after another. This process makes a long polypeptide chain. The chain's backbone has the pattern $-N-C-C-N-C-C-N-C-C-$. Half the C atoms have side chains (R), as shown below.

internet connect

www.scilinks.org
Topic: Proteins
SciLinks code: HW4104

*SCi*LINKS. Maintained by the National Science Teachers Association

polypeptide
a long chain of several amino acids

peptide bond
the chemical bond that forms between the carboxyl group of one amino acid and the amino group of another amino acid

718

HEALTH CONNECTION — GENERAL

Amino Acids and Nutrition Of the greater than 2000 naturally occurring and synthetic amino acids that exist, only 20 occur in the proteins of organisms. Human cells are only able to make about half of these amino acids. The others must be obtained from foods, and are thus known as essential amino acids. The body does not store amino acids effectively, so it is necessary for people to have a regular intake of high-quality protein rich foods. High-quality protein rich foods, such as eggs, fish, meat, and dairy products, contain the essential amino acids in ratios similar to those found in the proteins in our bodies. Other foods, such as beans and grains, do not and are lesser-quality protein sources. Pairing certain lesser-quality protein sources, such as rice and beans, results in a high-quality protein rich meal. For proper health it is important to eat a consistent diet that is high in all of the essential amino acids. Have interested students investigate all 20 amino acids, determine their structure and properties, and research their role in the human diet. **LS** Logical

Properties and Interactions of Side Chains

The properties of a part of a polypeptide chain depend on the properties of the side chains present. For example, the side chain of glutamic acid is acidic. The side chain of histidine is basic. The side chains of asparagine and several other amino acids are strongly polar. On the other hand, amino acids with nonpolar side groups, such as valine, are nonpolar.

Some amino acid side chains can form ionic or covalent bonds with other side chains. Cysteine is a unique amino acid, because the –SH group in cysteine can form a covalent bond with other cysteine units. Two cysteine units, at different points on a protein molecule, can bond to form a *disulfide bridge,* shown in **Figure 6.** Such bonding can form a looped protein or link two separate polypeptides. In fact, curly hair is a result of the presence of disulfide bridges in hair protein. Some amino acid side chains can form ionic bonds with other amino acid side chains. These bonds also link different points on a protein. For example, glutamic acid can give up a proton to histidine. When this happens, an ionic bond will form between the two amino acids.

Also, weaker interactions can affect how segments of proteins interact with one another. You have read about these interactions in earlier chapters. Two are shown in **Figure 6.** One of these weak interactions is between the nonpolar hydrocarbon side chains present on many amino acids. These groups are hydrophobic and do not tend to be found in polar and ionic environments. Instead, nonpolar segments of a protein tend to be found with nonpolar molecules or with other nonpolar segments of the same protein.

The side chains of certain amino acids, such as asparagine, allow for another kind of interaction—hydrogen bonding. The hydrogen atoms on hydroxyl groups, –OH, and amino groups, –NH$_2$, are drawn to places where they can hydrogen bond to oxygen atoms, especially to carboxyl groups, –C=O, in the polypeptide backbone or in the side chains.

Topic Link

Refer to the "States of Matter and Intermolecular Forces" chapter for a discussion of intermolecular forces.

Figure 6
Four different kinds of interaction between side chains on a polypeptide molecule help to make the shape that a protein takes. Three are shown here.

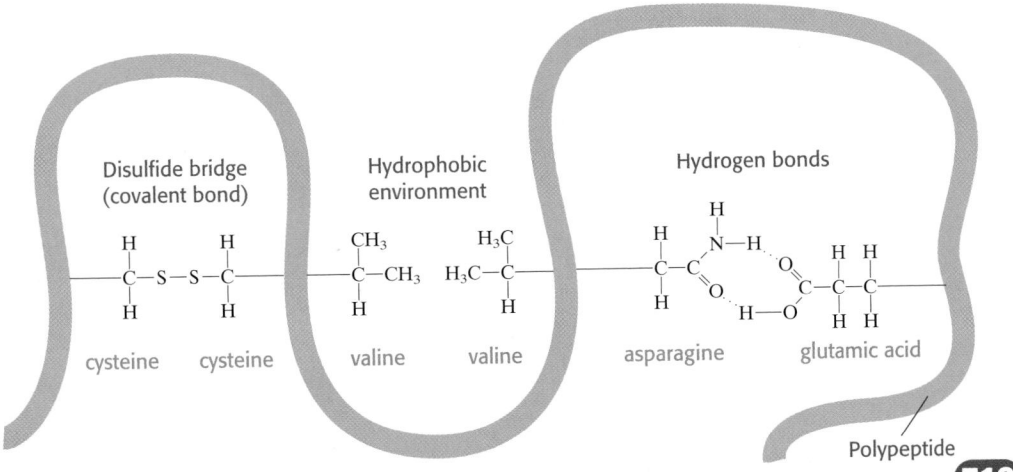

SKILL BUILDER — **ADVANCED**

Writing Skills Have students make a written narrative of the process they use to determine the four levels of structure of an unknown protein. The narrative should include all steps of the thought process used in determining each level of the structure. Have them present their instructions as a brochure for peers who might need help in understanding protein structure. Students may need to research protein structure using sources other than their textbook in order to develop a detailed set of instructions. **LS** Verbal

SKILL BUILDER — **ADVANCED**

Research Skills Have interested students research the α-helix and the β-pleated sheet. What types of proteins have these structures? What factors influence their formation? What kinds of bonds and intermolecular forces contribute to the formation of helices and pleats? Students may wish to draw a detailed diagram or build a model of one or both of these structures to share with the class. **LS** Visual

Transparencies

TT Levels of Protein Structure

Four Levels of Protein Structure

Proteins are not just long polypeptide chains. Because of the interactions of the side chains and other forces, each protein usually folds up into a unique shape. The three-dimensional shape that the chain forms gives characteristic properties to each protein. If a polypeptide chain folds into the wrong shape, it can function differently. It may also be unable to carry out its biological role. The levels of protein structure are shown in **Table 3.**

The amino-acid sequence of the polypeptide chain is said to be the *primary structure* of a protein. Thus, the primary structure of a protein is simply the order in which the amino acids bonded together.

Most proteins have segments in which the polypeptide chain is coiled or folded. These coils and folds are often held in place by hydrogen bonding. They give the protein its *secondary structure*. Two common kinds of secondary structures are the *alpha helix* and the *beta pleated sheet*, both of which are shown in the table. The alpha (α) helix is shaped like a coil with hydrogen bonds that form along a single segment of a polypeptide. The beta (β) pleated sheet is shaped like an accordion with hydrogen bonds that form between adjacent polypeptide segments.

In alpha-keratin, shown in **Figure 5,** the entire length of the protein has an α-helix structure. However, other proteins will have only sections that are α-helixes. Different sections of the same protein may have a pleated sheet secondary structure. These different sections of a protein can fold in different directions. These factors, combined with the intermolecular forces acting between side chains give each protein a distinct three-dimensional shape. This shape is the *tertiary structure* of the protein.

A *quaternary structure* arises when different polypeptide chains that have their own three-dimensional structure come together to form a larger protein. For example, four separate polypeptides make up a single molecule of hemoglobin, the protein that carries O_2 within red blood cells.

Table 3   Levels of Protein Structure

Primary structure	Secondary structure	Tertiary structure	Quaternary structure
valine proline	α-helix β-pleated sheet		

720

Amino-Acid Substitution Can Affect Shape

The sequence of amino acids—the primary structure—helps dictate the protein's final shape. A substitution of just one amino acid in the polypeptide sequence can have major effects on the final shape of the protein.

A hereditary blood cell disease called *sickle cell anemia* gives one example of the importance of amino-acid sequence. As the blood circulates, hemoglobin proteins in red blood cells pick up oxygen in the lungs and deliver it to all regions of the body. Normal red blood cells have the dimpled disk shape shown on the left in **Figure 7**. However, people with sickle cell anemia have blood cells with a crescent, or "sickle," shape. These cells are less efficient at carrying oxygen, which can cause respiration difficulties. Worse, the sickled cells tend to clump together in narrow blood vessels, causing clotting and sometimes death.

The cause of the sickle cell shape lies in the amino-acid sequence of the polypeptide. In sickle cell hemoglobin, the sixth amino acid in one of the polypeptide chains is valine. The sixth amino acid in healthy hemoglobin is glutamic acid. Because of the difference in only one amino acid, the entire shape of the hemoglobin is different in the unhealthy blood cells. This tiny change in the primary structure of the protein is enough to affect the health and life of people who have this disease.

Figure 7

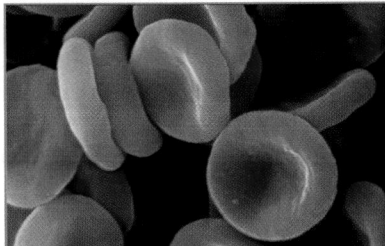

a The round, flat shape of healthy red blood cells shows they have normal hemoglobin molecules.

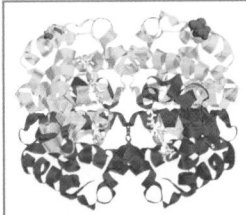

b Hemoglobin consists of four polypeptide chains; a fragment of one chain is shown in green.

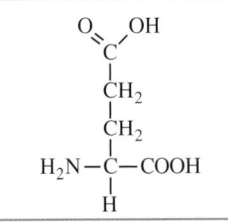

c Each of the chains is a polymer of 141 or 146 amino acid units, such as the glutamic acid monomer shown here.

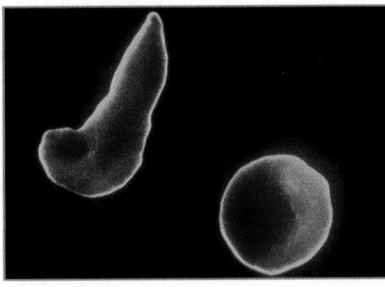

d Because of their shape, sickle cells clog small blood vessels.

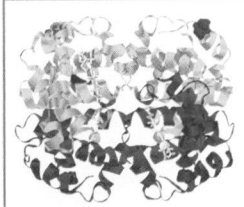

e A genetic mutation causes one glutamic acid to be replaced by valine in the hemoglobin molecules, as shown in red.

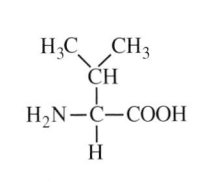

f The sickle shape of the cell comes from the different shape of the hemoglobin caused by the valine substitution.

721

Discussion

Form and Function Ask students to describe what would happen if they tried to unscrew a single-slotted screw with a hexagonal screwdriver. Point out that the problems encountered in this example are similar to how a misshapen protein cannot carry out the biological task it is designed to do. Ask students for other examples of shape-dependent processes. Lead the discussion to the importance of amino acid sequence in protein function. Tell students that there are over 200 known diseases caused by incorrect amino acid sequences, or gene mutations.

Homework ── BASIC

Graphic Organizer Have students create a graphic organizer or concept map that shows the relationship between the following key ideas: *protein structure, protein function, denaturation, primary structure, tertiary structure, amino acid substitution,* and *quaternary structure.* **LS Visual**

SKILL BUILDER — ADVANCED

Math Skills Write the equation for the decomposition of carbonic acid on the board.

$$H_2CO_3(aq) \rightarrow CO_2(aq) + H_2O(l)$$

Have students calculate the time required for 1000 carbonic anhydrase molecules to dehydrate 1.0 mol H_2CO_3. Ans. about 10^{15} s

LS Logical

Demonstration — BASIC
Pineapple Enzyme

1. Add a small volume of fresh or frozen pineapple juice to a test tube. Add a similar volume of canned pineapple juice to a second test tube.

2. Dissolve some powdered gelatin into each sample. Clean the stirring rod between samples.

3. Place the two samples in a place they can rest overnight.

4. The next day, show the two samples to the class. The fresh or frozen juice sample did not gel, whereas the canned juice sample did form a gel.

5. Ask students to offer an explanation for the results. If necessary, point out that the canned pineapple juice is heated during processing. Ans. Pineapple juice contains an enzyme that breaks down gelatin and prevents it from forming a gel. When the canned juice is processed, the heat denatures the enzyme. The denatured enzyme in canned juice is unable to break down the gelatin.

Safety Caution: Wear goggles and an apron.

Disposal: All solutions can go down the drain with excess water.

LS Visual

---

enzyme

a type of protein that speeds up metabolic reactions in plant and animals without being permanently changed or destroyed

Topic Link

Refer to the "Reaction Rates" chapter for a discussion of catalysis.

Enzymes

An **enzyme** is a protein that catalyzes a chemical reaction. Almost all of the chemical reactions in living systems take place with the help of enzymes. In fact, some biochemical processes would not take place at all without enzymes.

Enzymes have remarkable catalytic power. For example, blood cells change carbon dioxide, CO_2, to carbonic acid, H_2CO_3, which is easily carried to the lungs. Once in the lungs, carbonic acid decomposes back into carbon dioxide so that the CO_2 can be exhaled by the lungs. The reaction described by the equation below takes place in our lungs and tissues.

$$CO_2(aq) + H_2O(l) \rightleftharpoons H_2CO_3(aq)$$

The enzyme *carbonic anhydrase* allows this reaction to take place 10 million times faster than it normally would. The forward and reverse processes are accelerated equally. Hence the reaction's equilibrium constant is unaffected by the enzyme's presence. Enzymes are very efficient. A single molecule of carbonic anhydrase can cause 600 000 carbon dioxide molecules to react each second.

How Enzymes Work

In the late 19th century, the German chemist Emil Fischer proposed that enzymes work like a lock and key. That is, only an enzyme of a specific shape can fit the reactants of the reaction that it is catalyzing. A model of an enzyme mechanism is shown in **Figure 8.** Only a small part of the enzyme's surface, known as the *active site*, is believed to make the enzyme active. In reactions that use an enzyme, the reactant is called a *substrate*. The substrate has bumps and dips that fit exactly into the dips and bumps of the active site, much like three-dimensional puzzle pieces. Also, the active site has groups of side chains that form hydrogen bonds and other interactions with parts of the substrate. While the enzyme and the substrate hold this position, the bond breaking (or bond formation) takes place and the products are released. Once the products are released, the enzyme is available for a new substrate.

Figure 8

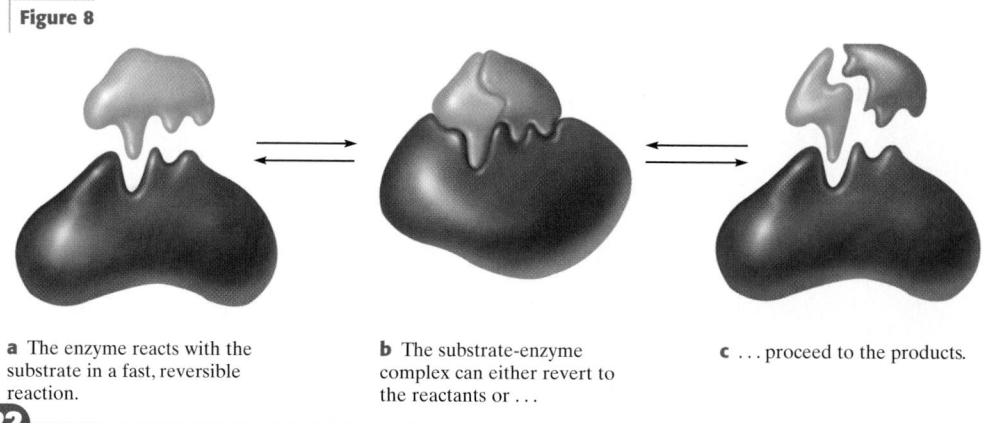

a The enzyme reacts with the substrate in a fast, reversible reaction.

b The substrate-enzyme complex can either revert to the reactants or ...

c ... proceed to the products.

722

MISCONCEPTION ALERT

Many students think that enzymes favor only the forward reaction, resulting in a greater amount of product than would form without the enzyme present. Reinforce the fact that enzymes favor *both* the forward and reverse reactions, and thus the amount of product formed is not altered. Explain that enzymes affect the reaction rate, but not the reaction equilibrium. Also reinforce the point that an enzyme functions by lowering the activation energy required for a reaction.

Transparencies

TT Enzyme Mechanism

Scientists have added to Fischer's idea and suggested that some enzymes are flexible structures. An enzyme might wrap its active site around the substrate as the substrate approaches. Further flexing of the enzyme causes some bonds in the substrate to break and frees the products. Whatever the actual mechanism of an enzyme, its shape is very important to its ability to catalyze a reaction. Because protein function depends so much on the shape of the protein, changing a protein's shape can inactivate a protein.

Denaturing an Enzyme Destroys Its Function

You do not have to change the primary structure of an enzyme to inactivate it. You can **denature** a protein. To denature a protein means to cause it to lose its tertiary and quaternary structures so that the polypeptide becomes a random tangle. Mild changes, such as shifts in solvent, temperature, pH, or salinity, may be enough to denature the enzyme. For example, the enzymatic ability to decompose hydrogen peroxide is lost by plant and animal cells when they are heated.

Of course, many proteins other than enzymes can also be easily denatured. When you prepare protein foods for meals you are usually denaturing proteins. For example, when you cook an egg, the egg white changes from runny and clear to firm and white, because the proteins are denatured by the change in temperature. Denaturing is the reason you can "cook" some foods without heating them. For example, when you make a dish called *ceviche* (suh VEE CHAY), you denature the proteins in raw fish by changing the pH of the protein's environment. By marinating the fish in acidic lime juice, you are denaturing the proteins much in the same way as if you heated the fish. Some recipes for pickled herring work in the same way, using vinegar (acetic acid) to denature the raw fish proteins.

denature

to change irreversibly the structure or shape—and thus the solubility and other properties—of a protein by heating, shaking, or treating the protein with acid, alkali, or other species

Quick LAB

Denaturing an Enzyme

SAFETY PRECAUTIONS

PROCEDURE

1. Get **15 potato cubes** from your teacher. Place one potato cube on a **paper plate.**

2. Using a **dropper,** drop **hydrogen peroxide solution** onto the potato cube. Note the amount of bubbling (the enzymatic activity). Let this

amount of bubbling count as a score of 10.

3. Place the remaining potato cubes in a **beaker of water** at room temperature. Place the beaker on a preheated **hot plate** that remains switched on.

4. Using **tongs,** remove one cube every 30 s, and test its enzymatic activity, assigning

a score between 0 and 10 based on the amount of bubbling.

ANALYSIS

1. Graph the enzymatic activity score versus heating time.

2. What happens to the enzymatic activity of a potato with heating? Explain.

723

Quick LAB GENERAL

Skills Acquired:
• Collecting Data
• Organizing/Analyzing Data

Materials:
For each group of 2–3 students:
• beaker, 100 mL
• dropper
• hot plate
• hydrogen peroxide solution, 3%
• paper plate
• potato cubes, 1 cm^3 (15)
• stopwatch
• tongs
• water

Teacher's Notes: Students may require assistance with coming up with a qualitative scheme for determining enzymatic activity. You may wish to show them examples of scores of 0, 5, and 10.

Safety Caution: Students should wear gloves, goggles, and aprons and use caution with hot plates and the heated materials.

Answers

1. Answers may vary. Graphs should show a decrease in enzymatic activity as heating time increases.

2. Enzymatic activity decreases with heating time. Heat denatures the enzyme so that it no longer breaks down the hydrogen peroxide.

LS Visual

Chapter Resource File

• Datasheets for In-Text Labs

REAL-WORLD
CONNECTION

Familiar Enzymes Students may be aware of enzyme-driven reactions and examples of protein denaturation from their everyday experiences. Ask probing questions to elicit examples from the class. Common examples include the beating of egg whites to form meringue (denaturation), the use of proteases to clean contact lenses and tenderize meat (enzyme activity), and the taking of lactase pills by lactose intolerant people (enzyme activity).

Quiz — GENERAL

1. What type of bond can be formed between two cysteine units? **Ans.** a disulfide bond

2. What type of reaction occurs and what byproduct is always formed when proteins are synthesized from amino acids? **Ans.** condensation reaction; Water is a byproduct.

3. What is the primary factor that determines the properties of a protein? **Ans.** the way in which the polypeptide chain folds on itself into a complicated three-dimensional shape

4. What element is contained in proteins but not in carbohydrates? **Ans.** nitrogen

5. Describe two of the possible shapes that result from the folding of a polypeptide chain in secondary structure? **Ans.** α-helix and β-pleated sheet are two of the most common shapes.

LS Verbal

Reteaching — BASIC

Cut a long inch-wide strip from a sheet of paper and tape a small magnet one-third of the way in from each end. Move the strip around until the magnets attract one another, pulling the strip into a contorted shape. Discuss how this simple demonstration models the intermolecular forces that give proteins their unique shapes. **LS** Visual

Chapter Resource File

- Concept Review
- Quiz

Curbing Enzyme Action

Enzymes can be too strong by themselves. One example of an overly strong enzyme is a *proteolytic* (or protein-splitting) enzyme called *trypsin,* which plays a part in the digestion of protein food. Trypsin is used in the small intestine to help break down proteins into amino acids through hydrolysis. However, the small intestine is itself made of proteins, which can also be broken down by trypsin! Rather than producing trypsin that will destroy its own organs, the body makes an inactive form of trypsin, a protein called *trypsinogen.*

Trypsinogen is stored in the pancreas. It is added to semidigested food as it passes through the small intestine. Small amounts of another protein, *enteropeptidase,* which is enzymatically active, are also added. When an enteropeptidase molecule meets a molecule of trypsinogen, enteropeptidase attacks one of the bonds in trypsinogen. When this bond is broken, one of the products is trypsin. Thus, this strong enzyme is made only at a time and place when it can break down food with the fewest dangerous side effects.

② Section Review

UNDERSTANDING KEY IDEAS

1. Describe the meaning of the two parts of the name *amino acid*.

2. Draw the general structure of an amino acid.

3. What is a peptide bond, and what name is given to enzymes that catalyze its hydrolysis?

4. **a.** Identify three side chains found in amino acids.
   b. Draw the three amino acids that have these side chains.
   c. What property does each of these chains give to a polypeptide chain?

5. What causes sickle cell anemia?

6. Describe the secondary structure of proteins.

7. What is meant by *denaturing* an enzyme, and what changes in conditions might bring it about?

8. Briefly describe how enzymes are believed to work to catalyze a reaction.

CRITICAL THINKING

9. What do condensation of sugars and condensation of amino acids have in common?

10. What different meanings do the words *polypeptide* and *protein* have?

11. List four different ways in which one part of a polypeptide chain may interact with another part. List them in the order that reflects *decreasing* strength of the interaction. (Hint: Apply what you have learned in previous chapters about the strength of different types of bonds and intermolecular forces.)

12. Proteolytic enzymes catalyze the hydrolysis of polypeptides. Predict the products if you carried out the hydrolysis of the following molecule, a dipeptide.

$$\begin{array}{c} H_3C \quad CH_3 \\ \diagdown \quad \diagup \\ CH \qquad\qquad H \\ | \qquad\qquad\quad | \\ H-N-C-C-N-C-C-OH \\ \;\;\;| \;\;\; | \;\;\; \| \;\;\; | \;\;\; | \;\;\; \| \\ \;\;\;H \;\; H \;\; O \;\; H \;\; H \;\; O \end{array}$$

Answers to Section Review

1. These molecules have an amino group, $-NH_2$, and an acid group, $-COOH$.

2.
$$\begin{array}{c} H \\ | \\ H_2N-C-COOH \\ | \\ R \end{array}$$

3. A peptide bond links carbon and nitrogen atoms of different amino acids and is formed when amino acids condense in the synthesis of polypeptides. Proteolytic enzymes break down proteins through hydrolysis.

4. **a.** Answers may include cysteine, glutamic acid, glycine, valine, asparagine, and histidine.

b. The structures are found in **Table 2.**

c. Answers may include: cysteine covalently bonds to other cysteine units, glutamic acid is acidic, glycine acts as a spacer, valine is nonpolar, asparagine is polar, and histidine is basic.

5. the substitution of one amino acid in one of the polypeptide chains of hemoglobin by a "wrong" amino acid

6. The *secondary structure* refers to regions in which the polypeptide has a regular shape, such as a helix or a pleated sheet.

Answers continued on p. 745A

Nucleic Acids

KEY TERMS
- nucleic acid
- DNA
- gene
- DNA fingerprint
- clone
- recombinant DNA

OBJECTIVES

1 **Relate** the structure of nucleic acids to their function as carriers of genetic information.

2 **Describe** how DNA uses the genetic code to control the synthesis of proteins.

3 **Describe** important gene technologies and their significance.

Nucleic Acids and Information Storage

You are probably like one or both of your parents in personality or physical features. Some traits may be due to the environment you grew up in, but many traits you inherited from your parents. Before you were born, you began as a single cell that had equal amounts of information from your mother and father about *their* hereditary characteristics. As that cell divided and redivided, that information was duplicated and now resides in every cell of your body.

Hereditary information is not just about the shape and color of your eyes, but also about the very fact that you have eyes—and that you are a human and not a snail or a cabbage. All that information, including the "construction plans" for building your body, is stored chemically in compounds called **nucleic acids.**

Nucleic-Acid Structure

Like polysaccharides and polypeptides, nucleic acids are biological polymers. Nucleic acids are formed from equal numbers of three chemical units: a sugar, a phosphate group, and one of several nitrogenous bases. The "backbone" of the nucleic acid is a -sugar-phosphate-sugar-phosphate-chain, with various nitrogenous bases connected to the sugar units. **Figure 9** shows the structures of the four most common nitrogenous bases.

nucleic acid

an organic compound, either RNA or DNA, whose molecules are made up of one or two chains of nucleotides and carry genetic information

Thymine   Cytosine   Adenine   Guanine

Figure 9
There are four common nitrogenous bases of nucleic acids. Thymine and cytosine bases have a single six-membered ring. Adenine and guanine bases have connected six- and five-membered rings.

725

Chapter Resource File
- Lesson Plan

Focus

Overview
Before beginning this section, review with your students the Objectives listed in the Student Edition. This section introduces students to the structure and function of nucleic acids. Students will learn about DNA, the genetic code, and gene technologies such as DNA fingerprinting, cloning, and recombinant DNA.

Bellringer
Write the following questions on the board. Are children similar in appearance to their parents? What is the cause of these similarities?

Motivate

Identifying Preconceptions — GENERAL

1. How do chemical bonds differ from intermolecular forces? **Ans.** Ionic and covalent bonds link atoms within a compound, whereas intermolecular forces can act between separate molecules. Also, ionic and covalent bonding are stronger than intermolecular forces.

2. Can a limited number of symbols be used to represent an infinite number of combinations? List several examples. **Ans.** Yes, the alphabet, counting numbers, and Morse code are a few common examples.
LS Verbal

Using the Figure — BASIC
Figure 9 Have students examine the structures of the nitrogenous bases. What do they all have in common? How do they differ? Point out to students that *nitrogenous* means "nitrogen containing" and that all of the bases have nitrogen atoms in their rings. Adenine and guanine have two connected rings, or fused rings. Have students copy the skeletal structures and add the carbon atoms where they belong. **LS** Visual

Discussion ——— ADVANCED

Many diseases are caused by malfunctioning genes, resulting in the lack of a needed protein or the production of a harmful protein. Ask students how comparing the DNA of healthy people with the DNA of people that have a specific disease can offer valuable information to researchers looking for a cure to the disease. Ask students what technologies would be necessary to sample and compare the DNA of many different people. Ask students to research topics such as the human genome project and gene chips and report their findings to the class. **LS** Verbal

Teach

Using the Figure

Figure 10 Point out to students that the T-A and C-G nitrogenous base pairings each involve the hydrogen bonding of a single-ring base (T or C) with a double-ring base (A or G) on the opposing strand.

Group Activity ——— BASIC

Have pairs of students take 12 index cards. Each card should have a question or an incomplete statement about the four common nitrogenous bases of nucleic acids on one side and the answer on the other side. Sample statements: A stands for ___; A pairs with ___; A forms ___ hydrogen bonds. Students should deal the cards with the questions showing in two rows. Alternating turns, each player attempts to answer one of his or her cards. The player that answers the most questions correctly wins. **LS** Interpersonal

DNA

deoxyribo**n**ucleic **a**cid, the material that contains the information that determines inherited characteristics

internet connect

www.scilinks.org
Topic: DNA
SciLinks code: HW4042

SCiLINKS Maintained by the National Science Teachers Association

Figure 10
The three-dimensional structure of DNA is made stable by hydrogen bonding between base pairs.

726

Deoxyribonucleic Acid, or DNA

Deoxyribonucleic acid is the full name of the most famous nucleic acid, which is usually known by the abbreviation **DNA.** DNA acts as the biochemical storehouse of genetic information in the cells of all living things.

The sugar in DNA is *deoxyribose,* which has a ring in which four of the atoms are carbon and the fifth atom is oxygen. The phosphate group comes from phosphoric acid, $(HO)_3PO$. Two of the $-OH$ groups from the phosphoric acid condense with the $-OH$ groups on two different sugar molecules, linking all three together as shown below.

The nitrogenous bases connect to the sugar units in the backbone. There is one base per sugar unit. Any one of the four bases—adenine, guanine, thymine, and cytosine—is connected along the strand at the sugar units. All genetic information is encoded in the sequence of the four bases, which are abbreviated to A, G, T, and C. Just as history is written in books using a 26-letter alphabet, heredity is written in DNA using a 4-letter alphabet.

Living things vary in the size and number of DNA molecules in their cells. Cells may have just one or many molecules of DNA. Some bacteria cells have a single molecule of DNA that has about 8 million bases. Human cells have 46 molecules of DNA that have a total of about 6 billion bases.

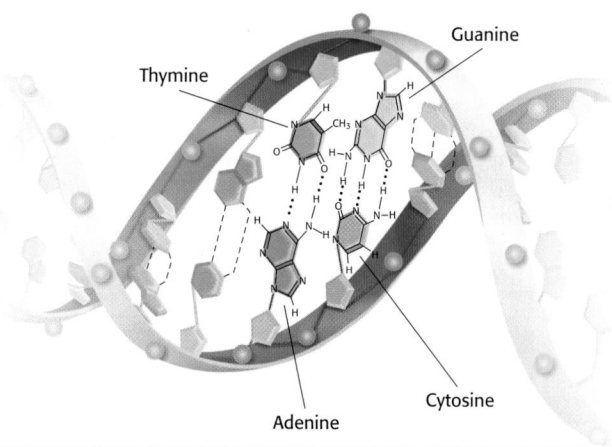

did you know?

The heaviest known elements have molar masses that are considerably less than 300 g/mol. The DNA molecule, which is comprised of several different elements, can have a molar mass as high as several billion grams per mole!

Transparencies

TT Structure of DNA

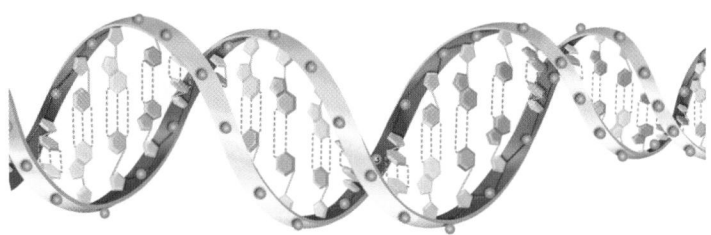

DNA's Three-Dimensional Structure

There are single strands of DNA, but the biological polymer is mostly found as a double helix in which two DNA strands spiral around each other as shown in **Figure 11**. The two strands are not duplicates of each other. Instead, they are complementary. This means that where an adenine (A) is found in one strand, thymine (T) is found in the other. Likewise, a guanine (G) in one strand is matched with a cytosine (C) in the other.

The reason for the complementary nature of DNA can be seen in **Figure 10**. When A and T are lined up opposite each other, the two bases are ideally placed for forming two hydrogen bonds, which bond the two strands together. Likewise, G and C can easily form three hydrogen bonds between themselves. No other pairing can form the right hydrogen bonds to keep the strands together. Thus, the three-dimensional configuration of DNA looks like a twisted ladder or spiral staircase, with A–T and G–C base pairs providing the rungs or steps.

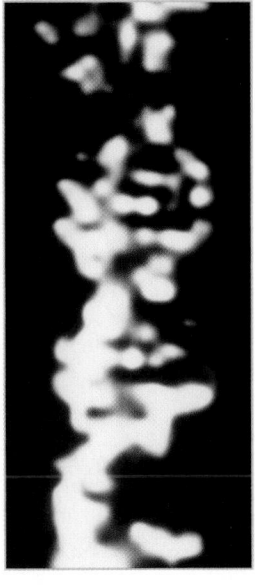

Figure 11
The double helix of DNA can be seen by scanning-tunneling microscopy (above) or shown as a molecular model (above left).

Quick LAB

Isolation of Onion DNA

SAFETY PRECAUTIONS

PROCEDURE

1. Place **5 mL of onion extract** in a **test tube.** The extract was taken from whole onions that were processed in a laboratory.

2. Hold the test tube at a 45° angle. Use a **pipet** to add **5 mL of ice-cold ethanol** to the tube one drop at a time. Note: Allow the ethanol to run slowly down the side of the tube so that it forms a distinct layer.

3. Let the test tube stand for 2–3 min.

4. Insert a **glass stirring rod** into the boundary between the onion extract and ethanol. Gently twirl the stirring rod by rolling the handle between your thumb and finger.

5. Remove the stirring rod from the liquids, and examine any material that has stuck to it. You are looking at onion DNA. Touch the

DNA to the lip of the test tube, and observe how it acts as you try to remove it.

ANALYSIS

1. Why do you think the DNA is now visible?

2. How has the DNA changed from when it was undisturbed in the onion's cells?

727

HISTORY

CONNECTION — **ADVANCED**

Writing The history of genetics reaches back more than one hundred years and includes the work of many different scientists. Provide the following key timeline events to the class: 1866, Gregor Mendel's work with pea plants; 1868, Friedrich Miescher extracts DNA from a cell nucleus; 1910, Walter Sutton and Theodor Boveri suggest inheritance is linked to chromosomes; 1928, Frederick Griffith's mouse experiments; 1944, Oswald Avery suggests DNA is likely the genetic material of life; 1952, Maurice Wilkins and Rosalind Franklin use X rays to photograph DNA; 1952, Alfred Hershey and Martha Chase confirm DNA's function as genetic material; 1953, James Watson and Francis Crick model the DNA molecule. Have students research and write reports on events given in the timeline or on developments that occurred after 1953. **LS** Verbal

Figure 12
DNA replicates by building
complementary strands on
the single strands that form
as the original helix unwinds.

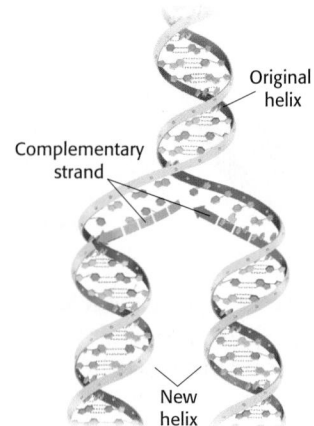

Original
helix

Complementary
strand

New
helix

Teach, continued

Teaching Tip

Protein Synthesis Use figures from a biology text to give students a better understanding of protein synthesis. RNA molecules are responsible for transmitting the genetic information stored in a cell's DNA to the cell's ribosomes, cell structures where protein synthesis occurs. First, a special form of RNA, called *messenger RNA* (mRNA), is produced in the cell nucleus. The mRNA is built from a specific section (gene) of the DNA and contains three-base segments called *codons*.

The mRNA then moves out into the cell's cytoplasm. Floating in the cytoplasm are ribosomes, amino acids, and tRNA (another form of RNA). The tRNA binds with a special amino acid and moves to the ribosome where it attaches to the "start" codon of the mRNA. As the ribosome moves along the mRNA, amino acids are added and a polypeptide chain forms. When the mRNA "stop" codon is reached, both the ribosome and the completed protein are released from the mRNA.

Teaching Tip

DNA Replication The conversion of a double strand into two single strands is often likened to the unzipping of a zipper. This analogy is imperfect, however, because the replication of human DNA actually starts from multiple sites along the helix, rather than from one of its ends.

internet connect

www.scilinks.org
Topic: DNA Replication
SciLinks code: HW4044

SciLINKS. Maintained by the National Science Teachers Association

gene

a segment of DNA that is located in a chromosome and that codes for a specific hereditary trait

internet connect

www.scilinks.org
Topic: Protein Synthesis
SciLinks code: HW4103

SciLINKS. Maintained by the National Science Teachers Association

728

DNA Replication

There is a copy of your DNA in each cell in your body, because DNA is able to replicate itself efficiently. To begin replication, a part of the double helix unwinds, providing two strands. Each strand acts as a template for the making of a new strand. New nucleic acid units made by the cell meet up one by one with their complementary bases on the template. Hydrogen bonds form between the correct base pairs: A to T, T to A, C to G, and G to C. As the nucleic acid units line up on the template strand, covalent bonds form between the sugars and phosphate groups of neighboring units or the complementary strand, as shown in **Figure 12.** Eventually, the original double helix is replaced by two perfect copies.

RNA and Protein Synthesis

Our proteins determine what our cells do. However, our DNA determines what these proteins are made of. A **gene** is a segment of DNA that has the code for the amino acid sequence to build a polypeptide. The way that the gene is translated into an amino-acid sequence is elaborate. It uses many proteins and another nucleic acid, *ribonucleic acid,* or RNA.

Protein synthesis begins with the cell making an RNA strand that codes for a specific protein. The DNA double helix unwinds and RNA units match up with the DNA bases. The process is similar to DNA replication. However, instead of using DNA units, the cell uses RNA units, which differ from DNA by an oxygen on the sugar unit and in one of the bases. RNA has the base *uracil,* shown in **Figure 13,** instead of thymine. The uracil bases hydrogen-bond with the adenine on the DNA strand, as in the following base sequence.

DNA strand: C C C C A C C C T A C G G T G
RNA strand: G G G G U G G G A U G C C A C

The cell then uses the RNA strand as instructions for building a protein. Amino acids line up according to the sequence of bases in the RNA. The polypeptide chain grows as bonds form between the amino acids.

Transparencies

TT DNA Replication

did you know? ———— **ADVANCED**

In 1950, before the structure of DNA was known, Erwin Chargaff analyzed the amounts of nitrogenous bases in cells from different organisms. He was surprised to find that the amount of adenine matched the amount of thymine and that the amount of cytosine matched the amount of guanine. Have students explain Chargaff's discovery. **Ans.** In double-stranded DNA, each adenine base is paired with a thymine base and each cytosine is paired with a guanine. What Chargaff found is what would be expected if the nitrogenous bases were always paired G-C and A-T in DNA. **LS Logical**

The Genetic Code

There are 20 different amino acids but only four RNA bases. Thus, a single base cannot specify a single amino acid. In fact, a group of three, or a *triplet* of bases in RNA indicates a particular amino acid. For example, the sequence of bases GUC causes valine to be added to a growing polypeptide. The complete *genetic code* lists the RNA triplets and their corresponding amino acids. You can use **Skills Toolkit 1** to decode RNA sequences to their corresponding amino acid sequences, as shown below.

RNA strand:	GGG	GUG	GGA	UGC	CAC
amino acid:	glycine	valine	glycine	cysteine	histidine

Because there are $4^3 = 64$ triplet combinations of the four bases, most of the 20 amino acids are encoded by more than one triplet. Almost all living things use the same code to translate their proteins.

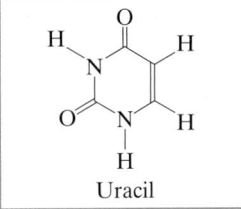

Figure 13
Uracil is a nitrogenous base that is unique to RNA. Uracil pairs with adenine.

SKILLS Toolkit 1

Using the Genetic Code

This table shows the triplet codes of RNA that specify each of the 20 amino acids. The triplets UAA, UAG, UGA, and AUG signal the end of the gene and the start of the next gene.

1. Find the first base of the RNA triplet along the left side of the table.
2. Follow that row to the right until you are beneath the second base of the triplet.
3. Move up or down in that section until you are even, on the right side of the chart, with the third base of the triplet.

The Genetic Code

First base	Second base				Third base
	U	C	A	G	
U	UUU UUC Phenylalanine — UUA UUG Leucine	UCU UCC UCA UCG Serine	UAU UAC Tyrosine — UAA Stop UAG	UGU UGC Cysteine — UGA—Stop UGG—Tryptophan	U C A G
C	CUU CUC CUA CUG Leucine	CCU CCC CCA CCG Proline	CAU CAC Histidine — CAA CAG Glutamine	CGU CGC CGA CGG Arginine	U C A G
A	AUU AUC Isoleucine AUA — AUG—Start	ACU ACC ACA ACG Threonine	AAU AAC Asparagine — AAA AAG Lysine	AGU AGC Serine — AGA AGG Arginine	U C A G
G	GUU GUC GUA GUG Valine	GCU GCC GCA GCG Alanine	GAU GAC Aspartic acid — GAA GAG Glutamic acid	GGU GGC GGA GGG Glycine	U C A G

729

Demonstration
DNA Strings

1. Cut a 1-meter long piece of thick string and tie its ends together in order to form a large loop. Tell students the string represents the sugar-phosphate "backbone" of DNA.

2. Hang the loop from a doorknob and then insert toothpicks into opposite sides of the loop at 5-centimeter intervals. The toothpicks represent the base pairs that link the two complementary strands together. For added detail, the base pair toothpicks may be color coded and assembled in the proper sequence.

3. Twist the bottom end of the loop around the axis of the DNA model and have students observe the resulting characteristic double helix shape.

Homework — GENERAL

1. A segment of a DNA strand has the base sequence CGACCTTTT.
   a. What is the corresponding RNA sequence? Ans. GCUGGAAAA
   b. What is the corresponding amino acid sequence? Ans. alanine-glycine-lysine
   c. What is the sequence in a complementary strand of DNA? Ans. GCTGGAAAA

2. There are 4^3, or 64, triplet bases comprised from the four A, T, G, and C bases. Suppose that instead of a triplet of bases, amino acids were encoded by a quartet of bases. How many quartets are possible? Ans. 4^4, or 256 possibilities

3. How many hydrogen bonds does an A-T pairing form? Ans. 2

Transparencies

TT Using the Genetic Code

Using the Figure

The kitten shown in **Figure 15** is a calico cat. Having a calico fur pattern is a genetic trait. However, the specific pattern of colors that a calico cat's fur has is not determined by genes. Thus, though this kitten is genetically identical to its parent, it does not look identical. Have interested students discuss the role of genes in determining the characteristics of an organism.

internet connect

www.scilinks.org
Topic: DNA Fingerprinting
SciLinks code: HW4043

SCi
LINKS   Maintained by the
National Science
Teachers Association

DNA fingerprint

the pattern of bands that results when an individual's DNA sample is fragmented, replicated, and separated

Figure 14
Scientists study images called autoradiographs, which show the pattern of nitrogenous bases in the DNA of an organism.

Gene Technology

After learning the role that DNA plays in life, biological chemists have gone on to research ways of using DNA that differ from natural processes. These efforts have many benefits and promise many more to come. But at the same time, gene technology has raised fears about the possibilities of misuse or mistake, as well as ethical issues about the uniqueness and sanctity of life.

Mapping and Identifying DNA

There are thought to be about 30 000 genes in human DNA. However, genes are only a tiny part of our DNA. There are large parts of our DNA that either have no function or have functions that have not been found yet.

Both the coding and noncoding base sequences differ from person to person. Unless you have an identical twin, the chance that someone else shares your DNA pattern is next to zero. Because no one else has the same DNA as you, your DNA pattern gives a unique "fingerprint" of you and your cells. Scientists use a technique called **DNA fingerprinting** to identify where a sample of DNA comes from. In DNA fingerprinting, scientists compare *autoradiographs* of DNA samples, such as those shown in **Figure 14.** Autoradiographs are images that show the DNA's pattern of nitogenous bases.

You may have heard that DNA fingerprinting is used in forensics to prove whether a suspect can be linked to a crime. There are other applications. Two people who are closely related to each other have DNA patterns that are more similar than the DNA of two unrelated people, so DNA is useful in identifying a person's family members and tracing heredity. Likewise, because species that share a common extinct ancestor have similar DNA patterns, scientists can track presumed evolutionary links.

Identifying DNA from Small Samples

It takes a lot of DNA to make a DNA fingerprint. However, forensic applications of DNA fingerprinting can make use of a single hair, or the smallest trace of blood. Scientists can use small samples of DNA because they can rapidly copy, or "amplify," DNA strands. By making many copies of a tiny sample of DNA, a scientist can make enough DNA to see the pattern of bases.

Scientists use a method called *polymerase chain reaction,* or PCR, which replicates a short "targeted" sequence of double-stranded DNA. Large amounts of the four monomeric components of DNA are added to a solution that has the DNA, an enzyme, and *primers.* A primer is a short length of single-stranded DNA that has the complementary sequence of the first few bases of the target. The solution is then subjected to a number of heating-cooling cycles. Heating denatures the DNA and separates the double strands. Cooling causes the primer to connect to the end of the target. The enzyme then replicates the DNA using the primer as a starting point. In this way, the amount of DNA is doubled during each cooling. After 20 cycles, the amount of DNA increases by a factor of 2^{20}, or more than 1 million.

FORENSICS
CONNECTION   **ADVANCED**

Writing **DNA Fingerprinting** First developed in England in 1985, DNA testing has revolutionized the field of forensic science. In 1992 the National Research Council released an official opinion stating that DNA testing was a reliable method for identifying criminal suspects. After this opinion, DNA testing quickly became a common tool of criminal prosecution. Have students research the use of DNA testing in local and state level courts. Who conducts the DNA testing? Does a government body regulate the testing? What techniques are commonly used and what types of evidence are submitted during the course of a trial? What potential problems can occur with DNA fingerprinting? Ask students to select one or several areas to investigate and to submit a written report based on their findings. **LS Verbal**

Figure 15
a Each of these identical twins has the same genetic information as her sister.

b Growers can produce many orchids by artificial cloning of the meristem tissue of a single orchid plant.

c The kitten at left is an artificial clone of an adult calico cat.

Cloning

Identical twins arise from the chance splitting of a group of embryonic cells early in the growth of a human baby. Each cell of a very young embryo can grow into a complete organism, but this ability is lost as an embryo grows larger and its cells become more specialized.

Undifferentiated cells are cells that have not yet specialized to become part of a specific tissue in the body. These cells include *stem cells* in animals and *meristem* cells in plants, which may be cultured artificially so they grow into complete organisms. These organisms are genetically identical to the organisms from which the cells were harvested and are **clones** of their "parent." Cloning a mammal is a difficult task. However, it was accomplished in 1997 by Scottish scientist Ian Wilmut. His work produced a sheep named Dolly. Dolly's genes were taken from the mammary cell of one sheep and placed in the enucleated, or empty, egg cell of another sheep. Dolly's embryo was then raised in the uterus of a third sheep. Scientists have artificially cloned many other living things—not only sheep, but plants, such as orchids, and other animals, such as the kitten shown in **Figure 15.**

clone

an organism that is produced by asexual reproduction and that is genetically identical to its parent; to make a genetic duplicate

Close

1. What forms the backbone of a nucleic acid? Ans. alternating sugar and phosphate units

2. In what way are polysaccharides, polypeptides, and polynucleotides similar? Ans. They are all biological polymers that contain carbon and oxygen.

3. In recombinant DNA technology, what is inserted into the host DNA strand? Ans. a foreign gene from another organism

4. Name two types of nucleic acid that all human cells contain. Ans. DNA and RNA

5. Which intermolecular force links complementary strands of DNA? Ans. hydrogen bonding

LS Verbal

Reteaching ———————— BASIC

Have students write down several base sequences that are at least twelve bases in length. Then use **Skills Toolkit 1** from the text to determine the amino acid sequence represented and to determine the sequence of the complementary DNA strand. LS Logical

SKILL BUILDER — GENERAL

Vocabulary Ask students to define *inhibitor*. In the case of HIV treatment, the protease inhibitor interferes with reverse transcriptase, thus blocking the formation of new viruses and preventing the spread of the infection. LS Verbal

---

Chapter Resource File

• Concept Review
• Quiz

---

recombinant DNA

DNA molecules that are artificially created by combining DNA from different sources

---

Recombinant DNA

The greatest advances in gene technology have come from *recombinant DNA technology*. Making use of proteins that cut and reconnect DNA molecules, scientists have learned to insert genes from one species into the DNA of another. When this **recombinant DNA** is placed in a cell, the cell is able to make the protein coded by the foreign gene.

The earliest success was in redesigning the DNA of bacteria to make human insulin, a protein that people with diabetes lack. Many proteins can be made in this way, and drug companies are rapidly finding ways to cure diseases and make life-saving drugs using recombinant DNA.

Bacteria are not the only living things that have been treated with recombinant DNA. Plants have been made more resistant to insects and frost damage. Spiders do not make large quantities of spider silk proteins, which may be used as strong building materials, so genetically changed goats with spider genes make milk that has these potentially useful proteins. This very active scientific field has grown much since the late 1900s.

Though genetically changed organisms offer new solutions to many difficult problems, many people worry about the drawbacks of using such technologies. For example, a genetically changed organism may thrive so well in an ecosystem that natural organisms cannot compete and are wiped out. Also, some people object to products that come from recombinant DNA because of ethical issues about the creation of new life forms for human use.

③ Section Review

UNDERSTANDING KEY IDEAS

1. From what three components is DNA made?

2. Describe the three-dimensional shape of DNA.

3. Describe how DNA uses the genetic code to control the synthesis of proteins.

4. Why is a very small trace of blood enough for DNA fingerprinting?

5. What was the first protein to be made commercially by recombinant DNA technology?

PRACTICE PROBLEMS

6. For what sequence of amino acids does the RNA base sequence AUGAAGUUUG-GCUAA code?

7. A segment of a DNA strand has the base sequence ACGTTGGCT.
   a. What is the base sequence in a complementary strand of RNA?
   b. What is the corresponding amino acid sequence?
   c. What is the base sequence in a complementary strand of DNA?

CRITICAL THINKING

8. Why might identical twins be called clones?

9. What features of the four base pairs make them ideal for holding DNA strands together?

10. Is it possible to specify the 20 amino acids using only two base pairs as the code? Explain.

Answers to Section Review

1. the sugar deoxyribose, phosphoric acid, and nitrogenous bases: adenine, guanine, thymine, and cytosine

2. Two sugar-phosphate backbones form intertwined spirals; hydrogen-bonded base pairs attach the backbones like the rungs of a ladder.

3. A sequence of triplets in DNA is transcribed into a sequence of RNA triplets. RNA triplets code for specific amino acids. Proteins are synthesized using RNA according to the sequence of amino acids dictated by the coding of the gene on the DNA molecule.

4. A small amount of DNA can be copied many times by the PCR method.

5. human insulin

6. (start)lysine-phenylalanine-glycine(stop)

7. a. UGC AAC CGA
   b. cysteine-asparagine-arginine
   c. TGC AAC CGA

8. Like artificial clones, identical twins have identical DNA. However, identical twins are not artificial like organisms more commonly known as clones.

Answers continued on p. 745A

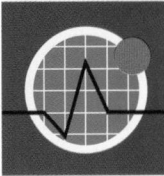

Protease Inhibitors

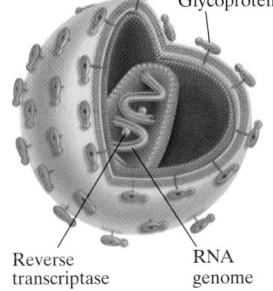

Glycoprotein

Reverse transcriptase

RNA genome

HIV Virus

HIV, or human immunodeficiency virus, is the virus that causes AIDS by severely weakening the human immune system. Since the discovery of HIV in 1983, scientists have searched for drugs that will combat the growth of the virus in human cells. Protease inhibitors are one of the newest classes of drugs to be developed.

Viruses are not living cells. They are bits of genetic material (RNA or DNA) combined with protein molecules. Viruses enter (infect) cells and release their genetic material. The cell uses this genetic material as a code to make more viruses.

The HIV virus is a *retrovirus*, a virus that contains RNA, which it carries into the cell along with an enzyme called *reverse transcriptase*. The HIV virus uses the reverse transcriptase enzyme to make a DNA copy of the RNA genetic pattern. The DNA segment enters the cell's nucleus, where it becomes a part of the cell's genes. There, it causes the cell to make all of the parts needed to make new viruses. The new viruses assemble and leave the cell to infect new cells. The cell is usually destroyed in the process.

Inhibiting Viral Reproduction

Most of the drugs that have been used to treat HIV infections are compounds that inhibit the reverse transcriptase enzyme, in turn preventing the RNA from forming a DNA copy. The new drugs, protease inhibitors, do their work after the parts of the virus have been made. The polypeptides that are needed to put together new viruses must be cut apart into the individual proteins. Protease is an enzyme that breaks the polypeptides in the right places. Inhibiting protease keeps many of the new viruses from forming.

Questions

1. Research to find out more about HIV. Identify the kind of cells the virus attacks, and describe how the viral infection leads to AIDS.
2. Find out more about other retroviruses. How are drugs used to combat infections caused by these retroviruses?

CAREER APPLICATION

Nurse Practitioner

A nurse practitioner does all of the things that registered nurses in hospitals or physicians' offices do. In fact, most nurse practitioners (NPs) begin as nurses and, after a few years of experience, study to become a nurse practitioner. Nurse practitioners have some of the same responsibilities as physicians. NPs can do extensive diagnoses of disease, carry out medical tests, counsel families, and in some cases, prescribe medicine. They often have specialties, such as pediatrics, mental health, or geriatrics. For some families, the NP is the primary health care provider.

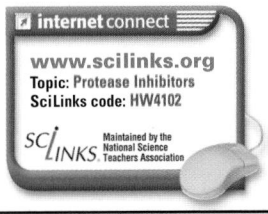

internet connect

www.scilinks.org
Topic: Protease Inhibitors
SciLinks code: HW4102

SCLINKS Maintained by the National Science Teachers Association

733

Protease Inhibitors Review the characteristics of retroviruses and the workings of the human immune system.

- Retroviruses contain RNA instead of DNA.
- Retroviruses have the enzyme reverse transcriptase, which is used in the production of viral DNA.
- During the normal process of transcription, DNA is a template for RNA. Reverse transcriptase reverses the normal process, and RNA is a template for DNA.
- HIV evolves rapidly from protein mutations, making vaccine development difficult.

Answers to Feature Questions

1. HIV attacks white blood cells called *helper T-cells*. The helper T-cells help regulate the activity of many other blood cells that destroy invading pathogens. This response to pathogens is called the *immune response*; it enables humans to fight off bacterial and viral diseases. Destruction of many T-cells weakens immune defenses so much that the body fails to ward off common infections. This condition results in AIDS.

2. Answers may include tumor viruses, such as leukemia.

CAREER APPLICATION

GENERAL

Interview Have students make a list of questions to interview a nurse practitioner about his or her job. Before the interview, have them carefully plan out what they will ask. Then have them arrange an interview with a nurse practitioner in their community. Students should write an article about what they have learned. **LS Verbal**

---

did you know? ———— **ADVANCED**

Treating AIDS AIDS treatment currently requires a combination of drugs to overcome the development of resistance by the virus. Have students research the types of drugs that AIDS patients must take. How many different types of medication are prescribed? What are the side effects of some of the medications? What is the approximate total cost for this type of treatment? Students may research AIDS treatment on the Internet, or you may

wish to provide students with books and brochures from local clinics. Student answers may include discussions of drug types such as protease inhibitors, nucleoside/nucleotide reverse transcriptase inhibitors, and non-nucleoside reverse transcriptase inhibitors and side effects such as nausea, diarrhea, fatigue, anemia, pain in the hands and feet, and a change in body weight. **LS Verbal**

Overview

Before beginning this section, review with your students the Objectives listed in the Student Edition. This section introduces students to energy in living systems. Students will learn how plants use energy through photosynthesis and how animals harvest energy through respiration.

 **Bellringer**

Write the following questions on the board. Where do we get the energy we need to fuel our growth and maintain our biological processes? Where did the energy necessary to produce the food we eat come from?

Motivate

Identifying
Preconceptions — GENERAL

1. Photosynthesis is a reaction that occurs in the leaves of plants in the presence of sunlight. Is photosynthesis exothermic or endothermic? **Ans.** endothermic; Students may confuse the two reaction types, both of which play important roles in the generation and use of energy in plants and animals.

2. What is the primary function of our lungs? **Ans.** to get oxygen into our bodies and to get rid of carbon dioxide; Many students will mention oxygen intake, but forget the equally important job of getting rid of carbon dioxide.

LS Verbal

Energy in Living Systems

KEY TERMS
- photosynthesis
- respiration
- ATP

www.scilinks.org
Topic: Biochemical Processes
SciLinks code: HW4022

OBJECTIVES

1 **Explain** how plants use photosynthesis to gather energy.

2 **Explain** how plants and animals use energy from respiration to carry out biological functions.

Obtaining Energy

Moving our muscles is one way in which we use energy, but many other ways that we use energy are harder to see. We use energy in digesting our food, in pumping our blood, in keeping warm, and in making the many compounds that our bodies need to function and grow. Energy is needed for every action of every organ in our bodies. All living things need energy to build and repair themselves and to fuel their activities. With rare exceptions, all forms of life on Earth draw energy ultimately from sunlight. Green plants get energy directly from the sun's rays through the process of **photosynthesis.** Other living things rely on plants, directly or indirectly, as their source of energy.

The flow of energy throughout an ecosystem is related to the carbon cycle. The carbon cycle follows carbon atoms as they become part of one compound and then another. The reactions that involve these carbon compounds, shown in **Figure 16,** give plants and animals the energy that they need.

photosynthesis

the process by which plants, algae, and some bacteria use sunlight, carbon dioxide, and water to produce carbohydrates and oxygen

Figure 16
Plants use carbon dioxide, water, and sunlight to produce oxygen and glucose. Glucose is used by plants and animals to produce chemical energy in the form of a substance called *ATP.*

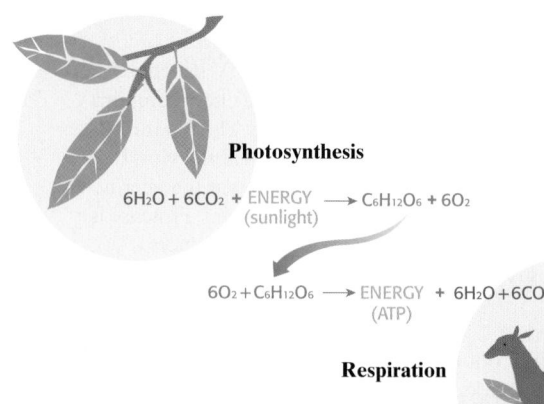

Photosynthesis

$$6H_2O + 6CO_2 + \text{ENERGY} \longrightarrow C_6H_{12}O_6 + 6O_2$$
(sunlight)

$$6O_2 + C_6H_{12}O_6 \longrightarrow \text{ENERGY} + 6H_2O + 6CO_2$$
(ATP)

Respiration

734

MISCONCEPTION
///ALERT\\\

Many students fail to understand the connection between plants as producers and humans as consumers that cannot obtain energy on their own. Humans and all other animals eat food in the form of plants, or other animals that consumed plants, in order to obtain a supply of carbon-containing compounds (mainly glucose). Animals harvest the energy contained in the chemical bonds of these carbon-containing compounds through cellular respiration. The energy generated through cellular respiration is used to power cellular functions in the animal. Go over the carbon cycle and reinforce the concept that ultimately all the energy used in living organisms comes from the sun, and that humans (consumers) rely on plants to convert solar energy into a form they can use (glucose).

Plants Use Photosynthesis as a Source of Carbohydrates

Look at the diagram of the carbon cycle shown in **Figure 16**. Notice that the reactants needed for the second equation—glucose and oxygen—are produced in the first equation, photosynthesis. And the reactants and conditions needed for the first equation—carbon dioxide, water, and energy—are produced in the second equation, although the energy is in a different form.

Most plants use *chlorophyll*, a magnesium-containing organic molecule, to capture the energy of sunlight. The light absorbed by the chlorophyll is mostly from the red and the blue regions of the visible spectrum. What is reflected is green light, from the central region of the spectrum, which accounts for the color of most plants.

The overall chemistry of photosynthesis, which takes place in green plants and many other living things, such as the algae in **Figure 17**, is described by the following endothermic equation.

$$6 \; O=C=O \; + \; 6 \; H_2O \xrightarrow{\text{light}} \; \text{glucose} \; + \; 6 \; O=O$$

carbon dioxide   water   glucose   oxygen

Animals Consume Carbohydrates as a Source of Energy

The carbohydrates that plants make by photosynthesis are used as a source of energy, not only by plants themselves, but also by animals.

Both plants and animals need carbohydrates for energy, and both plants and animals store simple carbohydrates by making them into larger carbohydrate polymers, such as starch and glycogen. Because animals cannot make carbohydrates directly from the sun's energy as plants do, animals eat plants or other animals to obtain the carbohydrates that plants have made. **Figure 18** shows one way that we get plant carbohydrates. Once an animal eats a plant, it breaks the plant's larger carbohydrates down into simpler carbohydrates, such as glucose. Glucose, which is soluble in blood, can be carried to the rest of the body for energy use.

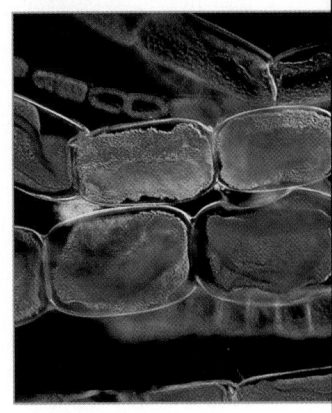

Figure 17
Green plants and algae have chlorophyll, a multiringed compound that contains magnesium.

Figure 18
Carbohydrates, such as the starch found in this baked potato, are the main energy source for most humans.

Activity ———— ADVANCED

Divide the class into small groups. Have each group construct a large-sized drawing of the carbon cycle. Ask each group to consider how changes in Earth's environment and atmosphere might alter the carbon cycle. Have them consider factors such as deforestation, global warming, ozone depletion, and acid precipitation. Each group should then pick one of the given changes and explain to the class, by use of their diagram, how they think the change will affect the carbon cycle. Support conflicting conclusions to stimulate discussion. Discuss how scientists often disagree on the possible environmental impact of various factors. **LS Visual**

Teach

SKILL BUILDER — GENERAL

Math Skills Write the balanced equation for the photosynthesis reaction on the board. Discuss how the coefficients show the relative number of moles of each substance involved in the reaction. Then have students calculate the moles of glucose formed from photosynthesis of 132 g of carbon dioxide. **Ans.** 0.5 mole **LS Logical**

Chapter Resource File
• Lesson Plan

Transparencies
TT Photosynthesis and Respiration

REAL-WORLD
CONNECTION ——— GENERAL

Exploring Fermentation Alcoholic fermentation and lactic acid fermentation both generate energy without using oxygen. Lactic acid fermentation occurs in the muscles of organisms when the energy supplied through cellular respiration is inadequate. Alcoholic fermentation is commonly used in baking to make breads rise and in brewing to convert sugars into alcohol. Each of these two fermentation processes produces only two ATP molecules per molecule of glucose fermented.

Alcoholic fermentation produces ethanol and carbon dioxide. Have students investigate the formation of carbon dioxide by letting them design their own simple experiment using balloons, yeast, warm water, beakers, scales, funnels, and various sweeteners as needed. Explain the fermentation reaction and tell students they are to design an experiment to determine the amount of fermentable sugars contained in various sweeteners such as table sugar, corn syrup, honey, fruit juice, aspartame, and saccharin. Have students predict which sweetener will produce the greatest volume of carbon dioxide gas. **LS Logical**

Teach, *continued*

Activity ──────── BASIC

Have students write the complete reaction equations (including energy) for photosynthesis and cellular respiration. Ask them to draw a colored box around each of the following items in the equations: Energy, $6O_2$, $6CO_2$, $6H_2O$, and $C_6H_{12}O_6$. Have them use a different color for each of the five items. Use their drawings to reinforce the fact that the products of the photosynthesis reaction are the reactants of the cellular respiration reaction and vice versa. **LS** Visual

Group Activity ──────── GENERAL

Have students use models of atoms to help visualize the reaction that takes place during respiration. Have each group make a model of each of the reactant molecules present in the equation. If needed, students should refer to their texts for constructing the glucose molecule. Have students break the reactant molecules as needed so they can be reformed into the required product molecules. Ask students if energy is absorbed or liberated in the reaction. **LS** Visual

Teaching Tip ──────── ADVANCED

Reaction Spotlight The oxidation of one mole of glucose by cellular respiration releases 2803 kJ of energy. Ask students to calculate the amount of energy released when 35 g of glucose is oxidized through cellular respiration.

Ans. (35 g)(1 mole/180 g) (2803 kJ/mole) = 545 kJ **LS** Logical

Figure 19
Muscular activity leads to an increase in respiration rate.

respiration

the process by which cells produce energy from carbohydrates; atmospheric oxygen combines with glucose to form water and carbon dioxide

internet connect

www.scilinks.org
Topic: Respiration
SciLinks code: HW4111

*SCi*LINKS. Maintained by the National Science Teachers Association

736

Using Energy

Glucose itself is changed into a more readily available source of energy through *respiration*. The equation for chemical respiration is shown in **Figure 16**. In everyday speech, *respiration* means getting gases into and out of the lungs. In biological chemistry, **respiration** refers to the entire process of getting oxygen into body tissues and allowing it to react with glucose to generate energy.

Respiration Requires Oxygen and Glucose

You may have noticed that you breathe more heavily when you exercise, as does the runner in **Figure 19**. This is because you need to get more oxygen into your system and you need to remove carbon dioxide more rapidly from your system.

The lungs move oxygen from the air into the blood as oxygen-carrying hemoglobin. The lungs also move carbon dioxide out of the blood—where it is present as $HCO_3^-(aq)$ and $H_2CO_3(aq)$—and into the air. The bloodstream carries oxygen and glucose to all the cells of your body for respiration. The bloodstream must also remove the products of respiration. That is, it takes carbon dioxide to the lungs, and it takes water to the kidneys.

Chemical respiration, or *cellular respiration*, takes place in the cells of a plant or animal and is fueled by glucose and oxygen. The overall process is the opposite of the photosynthesis reaction, as shown in the following equation.

$$\text{glucose} \qquad \text{oxygen} \qquad \text{carbon dioxide} \qquad \text{water}$$

For every molecule of glucose that is broken down by respiration, six molecules of oxygen, O_2, are consumed. The overall process produces six molecules of carbon dioxide, CO_2, and six molecules of water.

Respiration Is Exothermic

While photosynthesis takes in energy, respiration gives off energy. The thermodynamic values for the equation below show that the reaction is very exothermic ($\Delta H = -2803$ kJ) and highly spontaneous ($\Delta G = -2880$ kJ).

$$C_6H_{12}O_6(s) + 6O_2(g) \longrightarrow 6CO_2(g) + 6H_2O(l)$$

However, the goal of cellular respiration is not to liberate energy as heat or light but to produce chemical energy in the form of special polyatomic ions, as discussed next.

Adenosine triphosphate (ATP)   **Adenosine diphosphate (ADP)**

Adenosine Triphosphate and Adenosine Diphosphate

Adenosine triphosphate, **ATP,** and adenosine diphosphate, ADP, are the high-energy and low-energy forms of a chemical that acts as energy "cash" in biological systems. The structures of ATP and ADP are shown in **Figure 20.** The main structural difference between them is that ATP has an extra phosphate group, $-PO_3^-$.

The hydrolysis of ATP to ADP is exothermic ($\Delta H = -20$ kJ) and spontaneous, as the following equation shows.

$$\text{ATP}(aq) + H_2O(l) \longrightarrow \text{ADP}(aq) + H_2PO_4^-(aq) \qquad \Delta G = -31 \text{ kJ}$$

Many reactions in a cell would not take place spontaneously if left alone. These reactions can "use" the spontaneity of ATP hydrolysis to take place by coupling with the ATP $\longrightarrow$ ADP reaction. ATP hydrolysis thus allows these other nonspontaneous reactions to take place.

The Two Stages of Cellular Respiration

Cellular respiration has two stages. Both stages produce ATP. The first stage of cellular respiration includes *glycolysis.* The name means "glucose-splitting," which makes sense because the six-carbon glucose is split into two molecules of pyruvic acid, $CH_3COCOOH$ or $C_3H_4O_3$. The glycolysis reaction has about a dozen steps. Other products react further to make more ATP. The net gain of eight ATP is shown in the following equation.

$$C_6H_{12}O_6 + O_2 + 8\text{ADP} + 8H_2PO_4^- \longrightarrow 2C_3H_4O_3 + 8\text{ATP} + 10H_2O$$

The second stage of cellular respiration, called the *Krebs cycle,* also has several steps. The overall result is the oxidation of pyruvic acid to form CO_2, as shown in the following equation.

$$2C_3H_4O_3 + 5O_2 + 30\text{ADP} + 30H_2PO_4^- \longrightarrow 6CO_2 + 30\text{ATP} + 34H_2O$$

The two stages together produce 38 ATP ions per glucose molecule. The reaction shown for glucose has an enthalpy change of -2803 kJ. Thus, $(38 \times -20 \text{ kJ})/(-2803 \text{ kJ})$ or 27% of the energy of glucose can be stored as ATP. The remaining energy helps to keep the body warm.

Figure 20
The hydrolysis of ATP produces ADP and releases energy.

ATP
adenosine **trip**hosphate, an organic molecule that acts as the main energy source for cell processes; composed of a nitrogenous base, a sugar, and three phosphate groups

internet connect

www.scilinks.org
Topic: ATP
SciLinks code: HW4018

SCILINKS Maintained by the National Science Teachers Association

Teaching Tips
Structure of ATP Point out to students that the structure of ATP is similar to that of the adenine nitrogenous base unit of DNA.

Reaction Spotlight Point out to students that the breakdown of ATP into ADP requires water, and that each mole of ATP hydrolyzed back to ADP releases 30.5 kJ of energy.

Homework ——— BASIC

Graphic Organizer Have students create a graphic organizer or concept map that shows the relationship between the following key ideas: *energy storage, conversion of ATP to ADP, energy production, cellular respiration, glucose,* and *glycogen.* LS Visual

READING SKILL BUILDER — BASIC

Paired Reading Have students read this page and the next with a partner. They should take turns summarizing in their own words the concepts in each paragraph. LS Verbal

737

Graphing Have students use the data in **Table 4** to create a bar graph showing the amount of energy required in kJ to do each of the following activities for 30 min: running, swimming, bicycling, and walking. **LS** Logical

Close

Quiz ——————— GENERAL

1. What are the three types of ATP-powered work done within cells? Ans. synthetic, mechanical, and transport

2. Given carbon dioxide, sunlight, water, and oxygen, which is not required in order for photosynthesis to occur? Ans. oxygen

3. What chemical substance do plants use to capture energy from sunlight? Ans. chlorophyll

4. Which step of the two-step cellular respiration process produces more energy in the form of ATP? Ans. The second stage produces more ATP.
LS Logical

Reteaching ——————— BASIC

Have students draw a diagram of the carbon cycle and label the flow of energy through the system. Have them list the reactants and the products of photosynthesis and cellular respiration and have them explain why plants are called *producers* and animals are called *consumers*. **LS** Visual

---

Table 4 Approximate "Cost" of Daily Activities

Activity (for 30 min)	Energy required (kJ)	ATP required (mol)
Running	1120	56
Swimming	840	42
Bicycling	1400	70
Walking	560	28

ATP Is Energy Currency

The conversion of ATP to ADP gives the energy needed for many cellular activities. So, ATP represents energy that is immediately available in the cell. Also, ATP is continuously resynthesized by cellular respiration as long as an organism is alive.

On the molecular level, there are three kinds of work that a cell does, and ATP gives the energy needed for them all. ATP gives the energy needed for *synthetic work,* making compounds that do not form spontaneously because they are accompanied by a positive ΔG. By coupling the reaction to the ATP $\longrightarrow$ ADP conversion, the overall process becomes spontaneous. ATP also gives the energy needed for *mechanical work.* The ATP $\longrightarrow$ ADP conversion changes the shape of muscle cells, which allows muscles to flex and move. Finally, the ATP $\longrightarrow$ ADP conversion fuels *transport work,* carrying solutes across a membrane. Again, the ATP $\longrightarrow$ ADP conversion is harnessed to allow specific proteins in the membrane to pump ions into or out of the cell. **Table 4** shows just how much ATP is needed for some daily activities.

④ Section Review

UNDERSTANDING KEY IDEAS

1. What are two reactions that involve carbon and together give plants energy?

2. Write the chemical reaction representing the photosynthesis of glucose.

3. What role does the bloodstream play in respiration?

4. Write the net equation for the reaction that makes ATP and pyruvic acid from glucose during cellular respiration.

5. Briefly, what biological role is played by ATP?

CRITICAL THINKING

6. To some small extent, plants make some ATP during photosynthesis. Why can't plants use photosynthesis as an energy source all the time instead of making carbohydrates?

7. In what sense is it true to say that sunlight fills the energy needs of a cheetah?

8. Explain the roles of glycogen, glucose, and ATP as energy sources in animals.

9. For chemical reactions, Gibbs energy is a more important quantity than enthalpy. Show that the efficiency of the glucose to ATP conversion in terms of ΔG is only 41%.

10. Explain how nonspontaneous biochemical reactions can take place with the help of ATP.

Answers to Section Review

1. photosynthesis and respiration

2. $6CO_2 + 6H_2O + \text{sunlight} \rightarrow C_6H_{12}O_6 + 6O_2$

3. It carries oxygen and glucose to the cells and removes carbon dioxide and excess water.

4. $C_6H_{12}O_6 + O_2 + 8ADP + 8H_2PO_4^- \rightarrow$
$2C_3H_4O_3 + 8ATP + 10H_2O$

5. It provides cells with the energy to perform three kinds of work: synthetic, mechanical, and transport.

6. They need access to energy when the sun's energy is not available for photosynthesis, such as during the night.

7. The cheetah gets its energy from the flesh of an antelope. Antelopes get their energy from plants. Plants get their energy from sunlight.

8. Glycogen is used for long-term energy storage. Glucose is the medium by which energy is distributed to cells. ATP is the energy source available inside cells.

9. $(38 \times -31 \text{ kJ})/(-2880 \text{ kJ}) = 0.41 = 41\%$

10. By coupling the nonspontaneous reaction with the very spontaneous ATP $\rightarrow$ ADP reaction, the overall process becomes spontaneous.

Magnesium

Element Spotlight

Mg
Magnesium
24.3050
[Ne]3s²
12

Where is Mg?
Earth's crust
2.5% by mass
Sea water
0.13% by mass

Magnesium: An Unlimited Resource

Extracting magnesium from sea water is an efficient and economical process. Sea water is mixed with lime, CaO, from oyster shells to form insoluble magnesium hydroxide, Mg(OH)₂, which can be easily filtered out. Hydrochloric acid is added to the solid to form magnesium chloride. The electrolysis of molten magnesium chloride will produce pure magnesium metal.

Industrial Uses

- Magnesium oxide, MgO, is used in paper manufacturing, as well as in fertilizers, medicine, and household cleaners.
- Aqueous magnesium hydroxide, Mg(OH)₂, is known as *milk of magnesia,* an antacid.
- Magnesium alloys are used in aircraft fuselages, engine parts, missiles, luggage, optical and photo equipment, lawn mowers, and portable tools.

Spinach is a good source of dietary magnesium. Magnesium is the central atom in the green plant pigment chlorophyll.

Real-World Connection If 90 million metric tons of magnesium were extracted per year for 1 million years, the magnesium content of the oceans would drop by 0.01%.

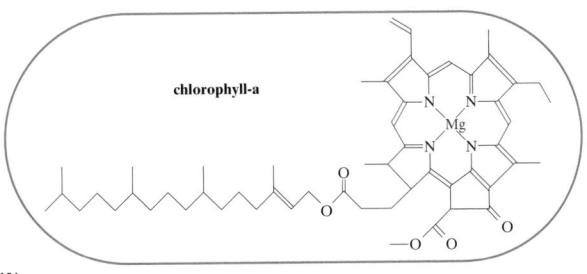

chlorophyll-a

A Brief History

1808: Humphry Davy discovers that the compound magnesia alba is the oxide of a new metal.

1828: A.A.B. Bussy obtains the first pure magnesium metal.

1944: L. M. Pidgeon discovers how to extract magnesium from its ore, dolomite.

1700 — 1800 — 1900

1833: Michael Faraday makes magnesium metal through the electrolysis of molten magnesium chloride.

1852: Robert Bunsen designs an electrolytic cell that allows molten Mg to be collected without burning when it makes contact with the air.

Questions

1. Find out more about chlorophyll. How is chlorophyll's structure important to its role in photosynthesis?
2. Magnesium is used to make fireworks. Find out what property makes this substance useful in fireworks.

☑ internet connect
www.scilinks.org
Topic: Magnesium
SciLinks code: HW4077
SCiLINKS Maintained by the National Science Teachers Association

739

Element Spotlight

Magnesium Magnesium is the least dense metal used in industrial construction. Its density is less than two-thirds that of aluminum. Magnesium is quite active and is used to reduce other metals from their salts. Magnesium is an important alloying agent, although it is too reactive to be used as a structural component itself. Many reactions, including organic condensation, reduction, addition, and dehalogenation reactions, can use magnesium as a catalyst.

Chapter Resource File
- Concept Review
- Quiz

Answers to Feature Questions
1. The structure of chlorophyll contains a complex of many connected rings with double-bonds, which allows magnesium electrons to absorb the energy of a photon of visible light.
2. Magnesium burns brightly in air.

20 CHAPTER HIGHLIGHTS

Alternative Assessments

SKILL BUILDER
• Pages 715, 715, 720

READING SKILL BUILDER

• Page 719

Homework
• Pages 721, 737

Group Activity
• Page 726

Reteaching
• Page 738

CHAPTER REVIEW
• Items 51, 52

Portfolio Assessments

Homework
• Page 714

HEALTH CONNECTION
• Page 718

SKILL BUILDER
• Page 720

REAL-WORLD CONNECTION
• Page 735

KEY IDEAS

SECTION ONE Carbohydrates and Lipids

• Carbohydrates are compounds of carbon, hydrogen, and oxygen made by living things for energy storage and support. They can be ringed and have many –OH groups.

• Carbohydrates are classified into monosaccharides, disaccharides, or polysaccharides according to the number of rings present. The smaller carbohydrates are called *sugars*.

• Sugars combine by condensation, a reaction in which a water molecule is formed. The reverse reaction, hydrolysis, breaks down polysaccharides into smaller carbohydrate units.

• Lipids are nonpolar molecules that include fats, phospholipids, steroids, and waxes.

SECTION TWO Proteins

• The 20 amino acids from which proteins are formed all have the formula $H_2N\text{-CHR-COOH}$. They differ in the identity of *R,* which stands for different side chains.

• Proteins are formed by condensation of amino acids.

• The form and function of a protein depends on its three-dimensional shape, which itself depends on the amino acid sequence in the polypeptide chain.

SECTION THREE Nucleic Acids

• Nucleic acids are made of -phosphate-sugar-phosphate-sugar- chains with nitrogenous bases connected to the sugar units.

• DNA uses four bases and forms a double helix by specific A–T and G–C pairing. Replication can take place only when the helix splits apart.

• The arrangement of base triplets on DNA encodes genetic information by dictating the synthesis of proteins.

• Gene technologies involve working with DNA and include DNA fingerprinting, cloning, and recombinant DNA.

SECTION FOUR Energy in Living Systems

• Green plants use solar energy, carbon dioxide, and water to synthesize glucose during photosynthesis.

• The reverse of photosynthesis is respiration, in which glucose is broken down into carbon dioxide and water. Energy is harvested in the process by the production of about 38 ATP ions per glucose molecule.

• Through the release of energy during the breaking of its third phosphate bond, ATP fuels life's processes: motion, synthesis, and transport.

740

KEY TERMS

carbohydrate
monosaccharide
disaccharide
polysaccharide
condensation reaction
hydrolysis
lipid

protein
amino acid
polypeptide
peptide bond
enzyme
denature

nucleic acid
DNA
gene
DNA fingerprint
clone
recombinant DNA

photosynthesis
respiration
ATP

KEY SKILLS

Using the Genetic Code
Skills Toolkit 1 p. 729

Chapter Resource File

• **Consumer Lab** All Fats Are Not Equal

USING KEY TERMS

1. What do all carbohydrates have in common?

2. How are carbohydrates classified?

3. What type of reaction changes sugars into polysaccharides?

4. Where are peptide bonds found?

5. Describe and name the four different levels of protein structure.

6. What is an enzyme, and what is a proteolytic enzyme?

7. Felicia adds vinegar (acetic acid) to milk, which causes casein (a protein in the milk) to curdle. The casein has the same amino acid sequence before and after the vinegar is added, so Felicia knows that no new substances have formed. How might she explain what has happened to the casein?

8. Contrast the terms nucleic acid, DNA, and gene.

9. Name three examples of gene technologies.

10. Describe how green plants use sunlight.

11. What does *respiration* mean in everyday language, and what larger meaning does it have in biological chemistry?

12. Describe the role of ATP and what the name *ATP* stands for.

UNDERSTANDING KEY IDEAS

Carbohydrates and Lipids

13. Name two examples from each of the following classes of carbohydrate: monosaccharide, disaccharide, and polysaccharide.

14. What different roles do the polysaccharides starch and cellulose play in plant systems?

15. Identify the reaction that changes sugars into polysaccharides and the reaction that changes polysaccharides into sugars.

16. Both cholesterol and oleic acid are lipids. What property do they have in common?

Proteins

17. List the four groups attached to the central carbon of an amino acid.

18. What are the products of protein synthesis?

19. Explain the cause of genetic disease sickle cell anemia.

20. Describe five properties that particular amino-acid side-chain groups give to the polypeptide formed by their condensation.

21. How is a disulfide bridge formed?

22. What is the *lock-and-key* model of enzyme action?

Nucleic Acids

23. Describe the structure of a DNA molecule and what the name *DNA* stands for.

24. In DNA replication, why is a G on the original strand partnered by a C on the complementary strand, and not by an A, a T, or a G?

25. What is the genetic code? Give an example of how it is used.

26. What is recombinant DNA technology?

27. What was significant about the sheep Dolly?

28. Describe the procedure for DNA amplification by polymerase chain reaction (PCR).

741

Assignment Guide

Section	Questions
1	1–4, 13–16, 41, 52
2	5–7, 17–22, 44, 45, 54–59
3	8, 9, 23–28, 35–38, 40, 42, 43, 47–49, 51, 53, 60
4	10–12, 29–34, 39, 46, 50

REVIEW ANSWERS

Using Key Terms

1. They are compounds of carbon, hydrogen, and oxygen produced by living things for energy storage and support. They can possess rings and many –OH groups.

2. They are classified into monosaccharides, disaccharides, or polysaccharides according to the number of rings present. The smaller carbohydrates are called *sugars*.

3. Sugars combine by condensation, a reaction in which a water molecule is formed.

4. Peptide bonds are found in polypeptides and proteins.

5. The primary structure describes the order of amino acids in the polypeptide chain. The secondary structure is how the side chains interact with one another to form α-helices or β-pleated sheets. The tertiary structure describes how different sections of the protein with helical and pleated-sheet secondary structures can fold in different directions due to intermolecular forces, giving the protein a distinct three-dimensional shape. The quaternary structure describes how different polypeptide chains fit together to form the larger protein.

6. An enzyme is a protein catalyst. A proteolytic enzyme is an enzyme that catalyzes the cleavage of a peptide bond.

7. The casein has been denatured because of the change in pH. Denaturing a protein changes its tertiary and quaternary structures, but it does not affect the protein's primary structure.

8. A gene is a section of a DNA molecule that codes for a specific protein. DNA is a type of nucleic acid that is responsible for coding hereditary information. Nucleic acids are biological polymers with basic units that consist of a sugar, a phosphate group, and a nitrogenous base.

9. Examples of gene technologies include DNA fingerprinting, cloning, and recombinant DNA.

10. Green plants use sunlight to convert carbon dioxide and water into glucose and oxygen through photosynthesis.

11. Respiration refers to the process of breathing (gas exchange). In biological chemistry the term refers to the overall process in which oxygen and glucose are used to provide energy throughout a living system.

12. ATP is the molecule that provides the immediate energy to fuel all of life's processes; it stands for adenosine triphosphate.

Understanding Key Ideas

13. possible answers—monosaccharides: glucose, fructose, ribose; disaccharides: sucrose, maltose, lactose; polysaccharides: starch, cellulose, glycogen, chitin

14. Starch is present for long-term energy storage. Cellulose is for support and structure.

15. Sugars combine by condensation, a reaction in which a water molecule is formed. The reverse reaction, hydrolysis, breaks down polysaccharides into smaller units.

16. Neither cholesterol nor oleic acid is soluble in water.

17. The groups attached to the central carbon include a hydrogen, an amino group, a carboxylic acid group, and a side chain.

18. polypeptides and water

19. A substitution error in the amino-acid sequence causes a change in the structure of the hemoglobin protein.

Energy in Living Systems

29. Identify the specialized molecule that absorbs light in photosynthesis.

30. Write the balanced chemical equation that describes the overall process in photosynthesis.

31. Explain why plants are generally green.

32. What is glycolysis?

33. How is ATP produced?

34. How are living things able to respond immediately to energy-demanding situations?

PRACTICE PROBLEMS

Skills Toolkit 1 Using the Genetic Code

35. What sequence of amino acids do the following RNA base sequences code?
   a. AAG AUU GGA CAC
   b. AUG UCU UCG AGU UCA UAG
   c. AAA GGG CCC UUU

36. A segment of a DNA strand has the base sequence TACACACGTTGGATT.
   a. What is the base sequence in a complementary strand of RNA?
   b. What is the corresponding amino acid sequence?
   c. What is the base sequence in a complementary strand of DNA?

37. a. Write one possible RNA sequence that codes for the following amino acids: aspartic acid-glutamine-tryptophan.
   b. What is the sequence in a complementary strand of DNA?

MIXED REVIEW

38. Imagine that you have created a very short polypeptide from the following RNA sequence: GACGAAGGAGAG.
   a. What is the amino acid sequence of the polypeptide?
   b. What property does the polypeptide have?

39. Write balanced chemical equations to describe the following metabolic processes: (a) starch $\longrightarrow$ glucose; (b) glucose $\longrightarrow$ carbon dioxide; (c) ATP $\longrightarrow$ ADP.

40. A nucleic acid has the following sequence: UCAUCGUGGAACUUG.
   a. Is the nucleic acid RNA or DNA?
   b. How can you tell?
   c. For what amino acid sequence does this nucleic acid code?

41. Hydrolysis of the disaccharide lactose produces glucose and galactose. What type of carbohydrate is galactose?

42. Identify each of the following structures as a carbohydrate, an amino acid, or a nitrogenous base.

   a.

   b.

   c.

CRITICAL THINKING

43. Explain how a similar reaction forms three kinds of biological polymers: polysaccharides, polypeptides, and nucleic acids.

44. How is it possible to denature a protein without breaking the polypeptide chain?

45. Why is a special molecule, hemoglobin, needed to move oxygen *from* the lungs, while no molecule is needed to move carbon dioxide *to* the lungs?

46. Study the image above. Identify whether respiration or photosynthesis take place in organisms A and B. Explain your answer.

47. Compare the advantages and disadvantages of DNA fingerprinting compared with literal fingerprints as a forensic tool.

48. A lab technician sweeps his hair back while he prepares a sample for polymerase chain reaction (PCR). Later, the DNA fingerprinting tests of the sample indicate that the lab technician was at the scene of the crime. What other explanation is there for the results of the DNA test?

49. How can spider silk proteins be present in the milk of goats?

50. Explain how all of the following statements can be true: "Many plants use *starch* to provide energy"; "Energy is supplied by *glucose* in both plants and animals"; and "*ATP* is the energy source in all living cells."

ALTERNATIVE ASSESSMENT

51. News reports about gene technology are sometimes one-sided, stressing the advantages but ignoring the dangers, or vice versa. Find such a report and write "the other side of the story."

52. Research to find out more about the structure of phospholipids and the properties that make them ideal for the construction of cell membranes.

CONCEPT MAPPING

53. Use the following terms to complete the concept map below: *DNA, polypeptides, amino acids, nucleic acids,* and *carbohydrates.*

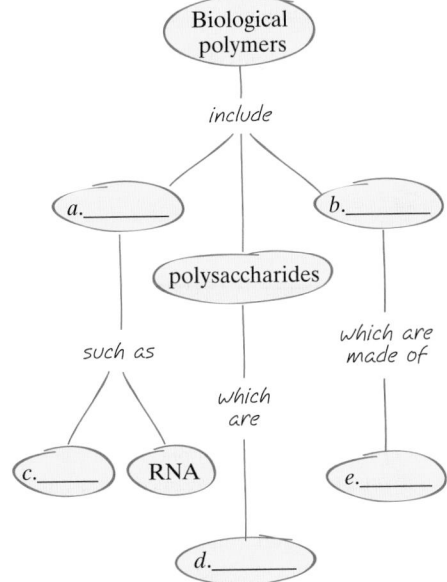

REVIEW ANSWERS
continued

20. Nonpolar organic groups lead to hydrophobicity. The –S–H group gives the ability to form a disulfide bridge. The –O–H and –N–H groups give hydrogen-bonding sites. Several nitrogen groups lead to basicity. The –COOH group leads to acidity.

21. The side chains of two cysteines at different sites on a polypeptide form a $-CH_2-S-S-CH_2-$ covalent bond.

22. The model states that the shape of an active site on the enzyme's surface is complementary to the shape of the substrate.

23. DNA consists of sugar (deoxyribose) units linked by $-O-(POOH)-O-$ bridges. One of four bases, adenine, cytosine, guanine, or thymine, is connected to each deoxyribose unit. DNA stands for deoxyribonucleic acid.

24. G bases are ideally structured to form three hydrogen bonds with a C base. There is no equivalent opportunity for hydrogen bonding with A, G, or T bases.

25. The genetic code is a list of the 64 triplet sequences in RNA and the amino acid specified by each combination. It is used to determine the amino acid coded by a triplet, for example, the triplet CGA codes for arginine.

26. Recombinant DNA technology is the splicing of a gene from a "guest" species into the DNA of a host species and thereby enabling the host to produce proteins typical of the guest.

27. She was the first artificial mammalian clone.

28. A DNA sample is added to a solution containing an enzyme, a supply of monomers, and primers (short lengths of single-stranded DNA with the complementary bases of the beginning of the DNA segment they are desired to amplify). A number of heating and cooling cycles follow, and the DNA is replicated many times.

29. Chlorophyll is the molecule that absorbs light in photosynthesis.

30. $6CO_2 + 6H_2O + sunlight \rightarrow$
$C_6H_{12}O_6 + 6O_2$

31. The plant pigment chlorophyll absorbs light from the red and blue ends of the light spectrum, leaving green to be reflected.

32. Glycolysis is the first stage in cellular respiration, in which the glucose molecule is split into two pyruvic acid molecules.

33. ATP is produced from ADP and phosphate ions in reactions coupled to the oxidation of glucose.

34. by having a supply of ATP constantly available in every cell, ready for instantaneous use

Practice Problems

35. a. lysine-isoleucine-glycine-histidine

b. (start)-serine-serine-serine-serine-(stop)

c. lysine-glycine-proline-phenylalanine

36. a. AUG UGU GCA ACC UAA

b. (start)-cysteine-alanine-threonine-(stop)

c. ATGTGTGCAACCTAA

37. a. Sample answer: GAU CAA UGG

b. Sample answer: CTA GTT ACC

Mixed Review

38. a. aspartic acid-glutamic acid-glycine-glutamic acid

b. The polypeptide is acidic, because it has three acidic side chains.

39. a. $C_{6n}H_{10n}O_{5n} + nH_2O \rightarrow nC_6H_{12}O_6$;

b. $C_6H_{12}O_6 + 6O_2 \rightarrow 6CO_2 + 6H_2O$;

c. $ATP(aq) + H_2O(l) \rightarrow ADP(aq) + H_2PO_4^-(aq)$.

Answers continued on p. 745A

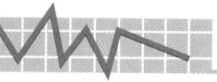

FOCUS ON GRAPHING

Study the graph below, and answer the questions that follow.
For help in interpreting graphs, see Appendix B, "Study Skills for Chemistry."

54. What characteristic of the two proteins in this bar graph is being compared?

55. The two proteins compared are α-keratin in wool and fibroin in spider silk. Which color represents the protein found in wool?

56. According to the graph, what is significant about spider fibroin protein?

57. What are the mole percentages of alanine in α-keratin and fibroin?

58. Why do you think the mole percentages of all of the amino acids are not shown?

59. Spider fibroin protein is a much stronger material than α-keratin in wool. Violet would like to create a strong protein for manufacturing fishing line. What amino acids might she decide to use to build the protein? Use the graph to support your answer.

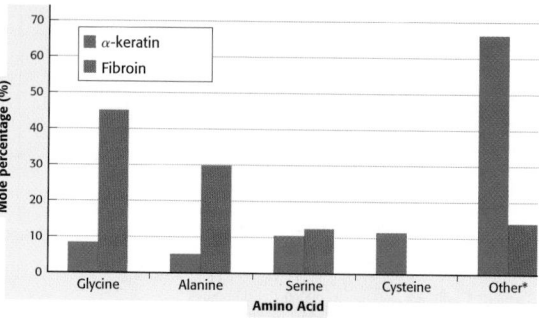

Amino Acid Composition of Proteins in Wool and Silk

TECHNOLOGY AND LEARNING

60. Graphing Calculator

Polypeptides and Amino Acids

Go to Appendix C. If you are using a TI-83 Plus, you can download the program **PEPTIDE** and run the application as directed. If you are using another calculator, your teacher will provide you with keystrokes to use. There are 20 amino acids that occur in proteins found in nature. The program will prompt you to input a number of amino acids. After you do, press **ENTER**. The program will respond with the number of different straight-chain polypeptides possible given that number of amino acid units.

a. Aspartame is an artificial sweetener that is a dipeptide, a protein made of two amino acids. How many possible dipeptides are there?

b. Enkephalins produced in the brain serve to help the body deal with pain. Several of them are pentapeptides. That is, they are polypeptides made of five amino acids. How many different pentapeptides are there?

c. The calculator uses the following equation:

number of polypeptides = $20^{(\text{number of amino acids})}$

This equation can also be expressed as:

number of polypeptides =
$(2^{(\text{number of amino acids})})(10^{(\text{number of amino acids})})$

Given this equation, estimate how many possible polypeptides there are that are made of 100 amino acids? (Hint: The answer is too large for your calculator. However, you can use the graphing calculator to find the value of $2^{(\text{number of amino acids})}$.)

744

Chapter Resource File

• Chapter Test

Answers

1. d
2. b
3. a
4. c
5. a
6. c
7. c
8. a
9. b
10. d
11. c
12. a

1. _____ are compounds used by plants for energy storage and structural support.
   a. Proteins
   b. Disaccharides
   c. Polypeptides
   d. Polysaccharides

2. The mechanism of enzyme catalysis requires the substrate to bind
   a. to DNA.
   b. to an active site.
   c. through a disulfide bridge.
   d. permanently.

3. In forming a DNA double helix, _____ links with thymine.
   a. adenine
   b. cytosine
   c. guanine
   d. thymine

4. In the following list, _____ is *not* an amino acid.
   a. glutamic acid
   b. histidine
   c. nucleic acid
   d. cysteine

5. The _____ group is present in all amino acids.
   a. –COOH
   b. –OH
   c. –CO—NH—
   d. –SH

6. The _____ bond is never employed in establishing the tertiary structure of proteins.
   a. covalent
   b. ionic
   c. metallic
   d. hydrogen

7. Glycolysis produces
   a. glucose.
   b. ethanol.
   c. pyruvic acid.
   d. carbon dioxide.

8. In the genetic code, groups of _____ bases specify an amino acid.
   a. 3
   b. 4
   c. 20
   d. 38

9. The products of photosynthesis that are the reactants of respiration are
   a. carbon dioxide and water.
   b. glucose and oxygen.
   c. oxygen and water.
   d. carbon dioxide and pyruvic acid.

10. In forensic work, polymerase chain reaction is used to
    a. compare the DNA of a suspect with a crime-scene sample.
    b. test if a DNA sample is of human origin.
    c. establish the innocence of the victim.
    d. increase the amount of DNA available.

11. _____ is not a protein.
    a. Keratin
    b. Trypsin
    c. Chitin
    d. Fibroin

12. _____ is a carbohydrate and a sugar, but not a disaccharide.
    a. Fructose
    b. Lactose
    c. Maltose
    d. Sucrose

Continuation of Answers

Answers to Section 1 Review, *continued*

12. Like starch in plants, glycogen is the polysaccharide energy-storage molecule in animals. Glycogen is also structurally similar to starch.

13. a. hydrolysis of a disaccharide

  b. Maltose and water are the reactants. Two glucose molecules are the products.

Answers to Section 2 Review, *continued*

 7. An enzyme is denatured when it loses its tertiary or quaternary structure. Changes in solvent, temperature, salinity, or pH could lead to denaturing.

 8. The enzyme has an active site with which the substrate associates. The bond is then broken and the products leave.

 9. They both produce water molecules.

10. A polypeptide is simply a string of amino acids and refers to the primary structure of a protein—the polymer chain with a specific amino-acid sequence. However, the properties of proteins depend also on the secondary, tertiary, and quaternary structures—the way that one or more polypeptides form an active molecule with a specific shape.

11. (1) An ionic bond can form between an acidic side chain on one segment and a basic side chain on the other. (2) Disulfide bridges (covalent bonds) can form between cysteine side chains. (3) Hydrogen bonds can form between some side chains. (4) Nonpolar side chains can form nonpolar environments.

12. valine and glycine

Answers to Section 3 Review, *continued*

 9. The atoms of thymine are ideally positioned to form H⋯N and O⋯H hydrogen bonds with adenine. The atoms of guanine are ideally positioned to form O⋯H, H⋯N, and H⋯O hydrogen bonds with cytosine. Also, the two pairs form "rungs" of the same length between the DNA backbones.

10. No, the number of twofold combinations of four bases is $4^2 = 16$, less than 20.

Answers to Chapter Review, *continued*

40. a. RNA

  b. It contains uracil, a base unique to RNA.

  c. serine-serine-tryptophan-asparagine-leucine

41. a monosaccharide

42. a. an amino acid (cysteine)

  b. a carbohydrate (maltose)

  c. a nitrogenous base (cytosine)

Critical Thinking

43. The same reaction is used to link the units in all three compounds: condensation with creation of a water molecule.

44. The properties of a protein relate to its secondary and tertiary structure, which can be changed without affecting the primary structure.

45. Because CO_2 is much more soluble in aqueous solutions than O_2 is.

46. Organism A is a plant, which can use respiration and photosynthesis. Organism B is an animal, so it can use only respiration.

47. Sample answer: advantages are surer identification, less delicate samples, and greater difficulty of concealment. Disadvantages include the need for actual body tissue and the ease of (intentional or unintentional) contamination.

48. The lab technician's hair was accidentally introduced into the sample when he touched his hair, so the results show his DNA fingerprint.

49. By recombinant DNA technology, the gene used by spiders to code for silk is spliced into the DNA of goats, permitting them to synthesize it.

50. Starch is for long-term energy storage, glucose is the vehicle for energy distribution, and the need for immediate energy is met by ATP.

Alternative Assessment

51. Answers may vary.

52. Answers should include discussion of the hydrophobic tails and hydrophilic head of phospholipids. This property allows them to form a bilipid layer.

Concept Mapping

53. a. nucleic acids

    b. polypeptides

    c. DNA

    d. carbohydrates

    e. amino acids

Focus on Graphing

54. The graph compares the proteins' amino acid compositions by mole percentage.

55. blue

56. Spider fibroin protein contains larger percentages of glycine and alanine amino acids.

57. The mole percentage of alanine in a-keratin is 5% and in fibroin is 29.4%.

58. The other amino acids most likely did not show significant differences between the two proteins.

59. Because spider fibroin protein contains larger percentages of glycine and alanine amino acids, Violet might use those amino acids to build a strong protein.

Technology and Learning

60. a. 400

    b. 3 200 000

    c. $(2^{100})(10^{100}) = 1.27 \times 10^{130}$

LABORATORY PROGRAM

Skills Practice Labs and Inquiry Labs

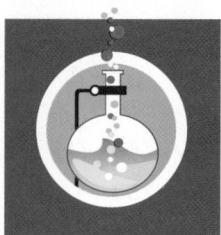

Working in the World of a Chemist

Meeting Today's Challenges

Even though you have already taken science classes with lab work, you will find the two types of laboratory experiments in this book organized differently from those you have done before. The first type of lab is called a Skills Practice Lab. Each Skills Practice Lab helps you gain skills in lab techniques that you will use to solve a real problem presented in the second type of lab, which is called an Investigation. The Skills Practice Lab serves as a Technique Builder, and the Investigation is presented as an exercise in Problem Solving.

Both types of labs refer to you as an employee of a professional company, and your teacher has the role of supervisor. Lab situations are given for real-life circumstances to show how chemistry fits into the world outside of the classroom. This will give you valuable practice with skills that you can use in chemistry and in other careers, such as creating a plan with available resources, developing and following a budget, and writing business letters.

As you work in these labs, you will better understand how the concepts you studied in the chapters are used by chemists to solve problems that affect life for everyone.

Skills Practice Labs

The Skills Practice Labs provide step-by-step procedures for you to follow, encouraging you to make careful observations and interpretations as you progress through the lab session. Each Skills Practice Lab gives you an opportunity to practice and perfect a specific lab technique or concept that will be needed later in an Investigation.

Skills Practice Lab

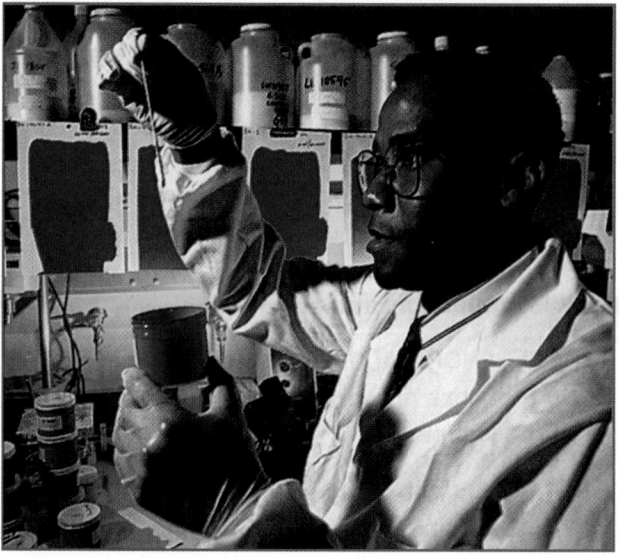

What Should You Do Before a Skills Practice Lab?

Preparation will help you work safely and efficiently. The evening before a lab, be sure to do the following:

◆ Read the lab procedure to make sure you understand what you will do.

◆ Read the safety information that begins on page 751, as well as any safety information provided in the lab procedure itself.

◆ Write down any questions you have in your lab notebook so that you can ask your teacher about them before the lab begins.

◆ Prepare all necessary data tables so that you will be able to concentrate on your work when you are in the lab.

What Should You Do After a Skills Practice Lab?

Most teachers require a lab report as a way of making sure that you understand what you are doing. Your teacher will give you specific details about how to organize your lab reports, but most lab reports will include the following:

◆ title of the lab

- summary paragraph(s) describing the purpose and procedure

- data tables and observations that are organized and comprehensive

- worked-out calculations with proper units

- answers that are, boxed, circled, or highlighted for items in the Analysis and Interpretation, Conclusion, and Extensions sections

Inquiry Labs

Inquiry Labs
Design Your Own Experiment

The Inquiry Labs differ from Skills Practice Labs because they do not provide step-by-step instructions. In each Inquiry Lab, you are required to develop your own procedure to solve a problem presented to your company by a client. You must decide how much money to spend on the project and what equipment to use. Although this may seem difficult. Inquiry Labs contain a number of clues about how to successfully solve the problem.

What Should You Do Before an Inquiry Lab?

Before you will be allowed to work on the lab, you must turn in a preliminary report. Usually, you must describe in detail the procedure you plan to use, provide complete data tables for the data and observations you will collect, and list exactly what equipment you will need and the costs. Only after your teacher, acting as your supervisor, approves your plans are you allowed to proceed. Before you begin writing a preliminary report, follow these steps.

- Read the Inquiry Lab thoroughly, and search for clues.

- Jot down notes in your lab notebook as you find clues.

- Consider what you must measure or observe to solve the problem.

- Think about Skills Practice Labs you have done that used a similar technique or reaction.

- Imagine working through a procedure, keeping track of each step, and determining what equipment you need.

- Carefully consider whether your approach is the best, most efficient one.

What Should You Do After an Inquiry Lab?

After you finish, organize a report of your data as described in the Memorandum. This is usually in the form of a one- or two-page letter to the client. Your teacher may have additional requirements for your report. Carefully consider how to convey the information the client needs to know. In some cases, a graph or diagram can communicate information better than words can.

If you need help with graphing or with using significant figures, ask your teacher.

Materials List for Inquiry Labs

Refer to the Equipment and Chemical lists below when planning your procedure for the Inquiry Labs. Include in your budget only the items you will need to solve the problem presented to your company by the client. Remember, you must always include the cost of lab space and the standard disposal fee in your budget.

Equipment	Equipment (continued)
Aluminum foil	Plastic bags
Balance	Ring stand/ring/wiregauze or pipestem triangle
Beaker, 250 mL	Ring stand with buretclamp
Beaker, 400 mL	Rubber policeman
Beaker tongs	Spatula
Büchner funnel	Spectroscope
Bunsen burner/related equipment	Standard disposal fee
Buret	Stopwatch
Cobalt glass plate	6 test tubes/holder/rack
Crucible and cover	Thermistor probe
Crucible tongs	Thermometer
Desiccator	Wash bottle
Drying oven	Watch glass
Erlenmeyer flask, 250 mL	Weighing paper
Evaporating dish	
Filter flask with sink attachment	**Reagents and Additional Materials**
Filter paper	Ring stand with buret clamp
Flame-test wire	Rubber policeman
Glass funnel	Spatula
Glass plate	Spectroscope
Glass stirring rod	Standard disposal fee
Graduated cylinder,100 mL	Stopwatch
Hot plate	Ring stand with buretclamp
Index card (3 in. x 5 in.)	Rubber policeman
Lab space/fume hood/utilities	Spatula
Litmus paper	6 test tubes/holder/rack
Magnetic stirrer	Thermistor probe
Mortar and pestle	Thermometer
Paper clips	Wash bottle
pH meter	Watch glass
	Weighing paper

Safety in the Chemistry Laboratory

Any chemical can be dangerous if it is misused. Always follow the instructions for the experiment. Pay close attention to the safety notes. Do not do anything differently unless you are instructed to do so by your teacher.

Chemicals, even water, can cause harm. The challenge is to know how to use chemicals correctly. If you follow the rules stated below, pay attention to your teacher's directions, and follow the precautions on chemical labels and in the experiments, then you will be using chemicals correctly.

These Safety Precautions Always Apply in the Lab

1. Always wear a lab apron and safety goggles.

Laboratories contain chemicals that can damage your clothing even if you aren't working on an experiment at the time. Keep the apron strings tied.

Some chemicals can cause eye damage and even blindness. If your safety goggles are uncomfortable or if they cloud up, ask your teacher for help. Try lengthening the strap, washing the goggles with soap and warm water, or using an anti-fog spray.

2. Do not wear contact lenses in the lab.

Even if you wear safety goggles, chemicals can get between contact lenses and your eyes and cause irreparable eye damage. If your

doctor requires you to wear contact lenses instead of glasses, then you should wear eye-cup safety goggles in the lab. Ask your doctor or your teacher how to use this very important and special eye protection.

3. NEVER WORK ALONE IN THE LABORATORY.

Do lab work only under the supervision of your teacher.

4. Wear the right clothing for lab work.

Necklaces, neckties, dangling jewelry, long hair, and loose clothing can knock things over or catch on fire. Tuck in neckties, or take them off. Do not wear a necklace or other dangling jewelry, including hanging earrings. It also might be a good idea to remove your wristwatch so that it is not damaged by a chemical splash.

Pull back long hair, and tie it in place. Wear cotton clothing if you can. Nylon and polyester fabrics burn and melt more readily than cotton does. It's best to wear fitted garments, but if your clothing is loose or baggy, tuck it in or tie it back so that it does not get in the way or catch on fire. It is also important to wear pants, not shorts or skirts.

Wear shoes that will protect your feet from chemical spills. Do not wear open-toed shoes or sandals or shoes with woven leather straps. Shoes made of solid leather or polymer are preferred over shoes made of cloth.

5. Only books and notebooks needed for the experiment should be in the lab.

Do not bring textbooks, purses, bookbags, backpacks, or other items into the lab; keep these things in your desk or locker.

6. Read the entire experiment before entering the lab.

Memorize the safety precautions. Be familiar with the instructions for the experiment. Only materials and equipment authorized by your teacher should be used. When you do your lab work, follow the instructions and safety precautions described in the experiment.

7. Read chemical labels.

Follow the instructions and safety precautions stated on the labels.

8. Walk with care in the lab.

Sometimes you will have to carry chemicals from the supply station to your lab station. Avoid bumping into other students and spilling the chemicals. Stay at your lab station at other times.

9. Food, beverages, chewing gum, cosmetics, and smoking are NEVER allowed in the lab.

(You should already know this.)

10. NEVER taste chemicals or touch them with your bare hands.

Keep your hands away from your face and mouth while working, even if you are wearing gloves.

11. Use a sparker to light a Bunsen burner.

Do not use matches. Be sure that all gas valves are turned off and that all hot plates are turned off and unplugged when you leave the lab.

12. Be careful with hot plates, Bunsen burners, and other heat sources.

Keep your body and clothing away from flames. Do not touch a hot plate after it has just been turned off because it is probably still hot. The same is true of glassware, crucibles, and other things that have been removed from the flame of a Bunsen burner or from a drying oven.

13. Do not use electrical equipment with frayed or twisted wires.

14. Be sure your hands are dry before you use electrical equipment.

Before plugging an electrical cord into a socket, be sure the equipment is turned off. When you are finished with the equipment, turn it off. Before you leave the lab, unplug the equipment, but be sure to turn it off FIRST.

15. Do not let electrical cords dangle from work stations.

Dangling cords can cause tripping or electrical shocks. The area under and around electrical equipment should be dry, and cords should not lie in puddles of spilled liquid.

16. Know fire-drill procedures and the locations of exits.

17. Know the location and operation of safety showers and eyewash stations.

18. If your clothes catch on fire, walk to the safety shower, stand under it, and turn it on.

19. If you get a chemical in your eyes, walk immediately to the eyewash station, turn it on, and lower your head so that your eyes are in the running water.

Hold your eyelids open with your thumbs and fingers, and roll your eyeballs around. Flush your eyes continuously for at least 15 minutes. Call out to your teacher as you do this.

20. If you spill anything on the floor or lab bench, call your teacher rather than trying to clean it up by yourself.

Your teacher will tell you if it is OK for you to do the cleanup; if not, your teacher will know how the spill should be cleaned up safely.

21. If you spill a chemical on your skin, wash the chemical off at the sink and call your teacher.

If you spill a solid chemical on your clothing, brush it off carefully without scattering it onto somebody else, and call your teacher. If you get liquid on your clothing, wash it off right away using the faucet at the sink, and call your teacher. If the spill is on your pants or somewhere else that will not fit under the sink faucet, use the safety shower. Remove the pants or other affected clothing while you are under the shower, and call your teacher. (It may be temporarily embarrassing to remove pants or other clothing in front of your class, but failing to flush that chemical off your skin could cause permanent damage.)

22. The best way to prevent an accident is to stop it before it happens.

If you have a close call, tell your teacher so that you and your teacher can find a way to prevent it from happening again. Otherwise, the next time, it could be a harmful accident instead of just a close call. If you get a headache, feel sick to your stomach, or feel dizzy, tell your teacher immediately.

23. All accidents, no matter how minor, should be reported to your teacher.

24. For all chemicals, take only what you need.

If you take too much and have some left over, DO NOT put it back in the bottle. If a chemical is accidently put into the wrong bottle, the next person to use it will have a contaminated sample. Ask your teacher what to do with leftover chemicals.

25. NEVER take any chemicals out of the lab.

26. Horseplay and fooling around in the lab are very dangerous.

NEVER be a clown in the laboratory.

27. Keep your work area clean and tidy.

After your work is done, clean your work area and all equipment.

28. Always wash your hands with soap and water before you leave the lab.

29. All of these rules apply all of the time you are in the lab.

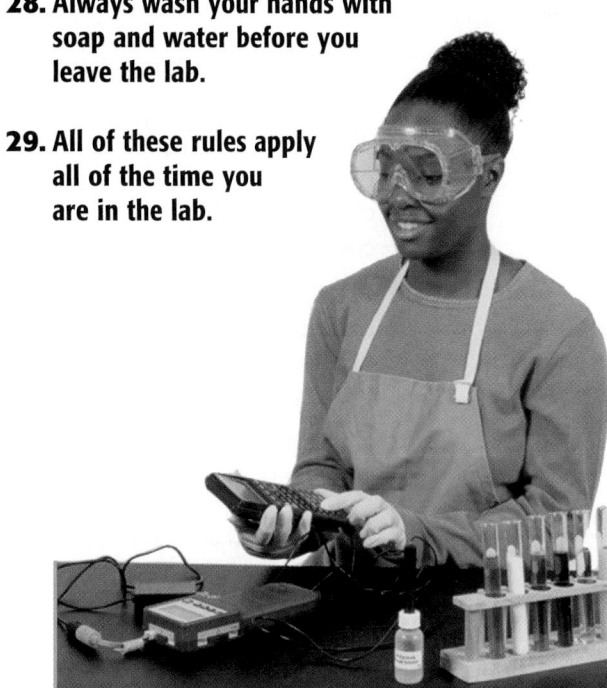

Safety in the Chemistry Laboratory **753**

CLOTHING PROTECTION

◆ Wear laboratory aprons in the laboratory. Keep the apron strings tied so that they do not dangle.

EYE SAFETY

◆ Wear safety goggles in the laboratory at all times. Know how to use the eyewash station.

CLEAN UP

◆ Keep your hands away from your face and mouth.

◆ Always wash your hands before leaving the laboratory.

CHEMICAL SAFETY

◆ Never taste, eat, or swallow any chemicals in the laboratory. Do not eat or drink any food from laboratory containers. Beakers are not cups, and evaporating dishes are not bowls.

◆ Never return unused chemicals to their original containers.

◆ It helps to label the beakers and test tubes containing chemicals. (This is not a new rule, just a good idea.)

◆ Never transfer substances by sucking on a pipet or straw; use a suction bulb.

WASTE DISPOSAL

◆ Some chemicals are harmful to our environment. You can help protect the environment by following the instructions for proper disposal.

GLASSWARE SAFETY

◆ Never place glassware, containers of chemicals, or anything else near the edges of a lab bench or table.

HAND SAFETY

◆ If a chemical gets on your skin or clothing or in your eyes, rinse it immediately, and alert your teacher.

CAUSTIC SAFETY

◆ If a chemical is spilled on the floor or lab bench, tell your teacher, but do not clean it up yourself unless your teacher says it is OK to do so.

HEATING SAFETY

◆ When heating a chemical in a test tube, always point the open end of the test tube away from yourself and other people.

Safety Quiz

Refer to the list of rules on p. 751–753, and identify whether a specific rule applies or whether the rule presented is a new rule.

1. Tie back long hair, and confine loose clothing. (Rule ? applies)

2. Never reach across an open flame. (Rule ? applies)

3. Use proper procedures when lighting Bunsen burners. Turn off hot plates, Bunsen burners, and other heat sources when they are not in use. (Rule ? applies)

4. Heat flasks or beakers on a ring stand with wire gauze between the glass and the flame. (Rule ? applies)

5. Use tongs when heating containers. Never hold or touch containers while heating them. Always allow heated materials to cool before handling them. (Rule ? applies)

6. Turn off gas valves when they are not in use. (Rule ? applies)

7. Use flammable liquids only in small amounts. (Rule ? applies)

8. When working with flammable liquids, be sure that no one else is using a lit Bunsen burner or plans to use one. (Rule ? applies)

9. What additional rules apply to every lab? (Rule ? applies)

10. Check the condition of glassware before and after using it. Inform your teacher of any broken, chipped, or cracked glassware because it should not be used. (Rule ? applies)

11. Do not pick up broken glass with your bare hands. Place broken glass in a specially designated disposal container. (Rule ? applies)

12. Never force glass tubing into rubber tubing, rubber stoppers,, or wooden corks. To pro-tect your hands, wear heavy cloth gloves or wrap toweling around the glass and the tubing, stopper, or cork, and gently push in the glass. (Rule ? applies)

13. Do not inhale fumes directly. When instructed to smell a substance, use your hand to wave the fumes toward your nose, and inhale gently. (Rule ? applies)

14. Keep your hands away from your face and mouth. (Rule ? applies)

15. Always wash your hands before leaving the laboratory.(Rule ? applies)

Finally, if you are wondering how to answer the question that asks what additional rules apply to every lab, here is the correct answer.

Any time you see any of the safety symbols, you should remember that all 29 of the numbered laboratory rules apply.

LABORATORY TECHNIQUES

Teacher's Notes

Time Required
1 lab period

Ratings

	1	2	3	4
	EASY			HARD

TEACHER PREPARATION	2
STUDENT SETUP	1
CONCEPT LEVEL	1
CLEANUP	1

Skills Acquired
• Collecting Data
• Communicating
• Interpreting
• Measuring
• Organizing and Analyzing Data

The Scientific Method
In this lab, students will:
• Make Observations
• Analyze the Results
• Draw Conclusions

Skills Practice Lab

OBJECTIVES

◆ Demonstrate proficiency in using a Bunsen burner, a balance, and a graduated cylinder.

◆ Demonstrate proficiency in handling solid and liquid chemicals.

◆ Develop proper safety techniques for all lab work.

◆ Use neat and organized data-collecting techniques.

◆ Use graphing techniques to plot data.

MATERIALS
◆ balance
◆ beakers, 250 mL (2)
◆ Bunsen burner and related equipment
◆ copper wire
▸ crucible tongs
◆ evaporating dish
◆ graduated cylinder, 100 mL
▸ heat-resistant mat
◆ NaCl
◆ spatula
◆ test tube
◆ wax paper or weighing paper

1 Laboratory Techniques

Introduction

You have applied to work at a company that does research, development, and analysis work. Although the company does not require employees to have extensive chemical experience, all applicants are tested for their ability to follow directions, heed safety precautions, perform simple laboratory procedures, clearly and concisely communicate results, and make logical inferences.

The company will consider your performance on the test in deciding whether to hire you and determining what your initial salary will be. Pay close attention to the procedures and safety precautions because you will continue to use them throughout your work if you are hired by this company. In addition, you will need to pay attention to what is happening around you, make careful observations, and keep a clear and legible record of these observations in your lab notebook.

This laboratory orientation session will teach you some of the following techniques:
• how to use a Bunsen burner
• how to handle solids and liquids
• how to use a balance
• how to practice basic safety techniques in lab work

756

Safety Cautions
• Read all safety precautions, and discuss them with your students.
• Safety goggles and a lab apron must be worn at all times.
• Long hair and loose clothing must be tied back.
• If students find a gas valve that has not been turned off properly, the room must be thoroughly ventilated—using the fume hood and windows—in order to prevent a fire hazard.

Disposal Information
There are no special disposal guidelines for this lab. The sodium chloride can be reused by each class if you designate a special container for students to dispose of the NaCl after each part. Keep the used NaCl separate from the NaCl you use as a reagent. The copper wire can also be reused.

Safety Procedures

- Wear safety goggles when working around chemicals, acids, bases, flames, or heating devices. Contents under pressure may become projectiles and cause serious injury.
- Never look directly at the sun through any optical device or use direct sunlight to illuminate a microscope.
- Avoid wearing contact lenses in the lab.
- If any substance gets in your eyes, notify your instructor immediately and flush your eyes with running water for at least 15 min.

- Secure loose clothing, and remove dangling jewelry. Don't wear open-toed shoes or sandals in the lab.
- Wear an apron or lab coat to protect your clothing when working with chemicals.
- If a spill gets on your clothing, rinse it off immediately with water for at least 5 min while notifying your instructor.

- Always use caution when working with chemicals.
- Never mix chemicals unless specifically directed to do so.
- Never taste, touch, or smell chemicals unless specifically directed to do so.
- Add acid or base to water; never do the opposite.
- Never return unused chemicals to the original container.
- Never transfer substances by sucking on a pipette or straw; use a suction bulb.
- Follow instructions for proper disposal.

- Avoid wearing hair spray or hair gel on lab days.
- Whenever possible, use an electric hot plate as a heat source instead of an open flame.
- When heating materials in a test tube, always angle the test tube away from yourself and others.
- Glass containers used for heating should be made of heat-resistant glass.
- Know your school's fire-evacuation routes.

- Clean and decontaminate all work surfaces and personal protective equipment as directed by your instructor.
- Dispose of all sharps (broken glass and other contaminated sharp objects) and other contaminated materials (biological and chemical) in special containers as directed by your instructor.

Data Table 1

Material	Mass (g) step 11	Mass (g) step 12
empty beaker		
beaker and 50 mL of water		
50 mL of water		
beaker and 100 mL of water		
100 mL of water		
beaker and 150 mL of water		
150 mL of water		

Procedure

1. Copy Data Tables 1 and 2 in your lab notebook. Be sure that you have plenty of room for observations about each test.

Data Table 2

Material	Mass (g)
weighing paper	
weighing paper and NaCl	

2. Record in your lab notebook the location and use of the following emergency items: lab shower, eyewash station, and emergency telephone numbers.

3. Check to be certain that the gas valve at your lab station and at the neighboring lab stations are turned off. Notify your teacher immediately if a valve is on, because the fumes must be cleared before any work continues.

757

Tips and Tricks

Thoroughly discuss all safety precautions outlined in this laboratory and in the safety section. Because this laboratory will be one of the first ones of the year, be sure to make safety a priority.

Point out the location and demonstrate the operation of all of the safety equipment, especially the lab shower, eyewash station, and fire blankets. Be sure students know what steps to follow in an emergency. Demonstrate the proper use of safety goggles and a lab apron. If students think that the lab equipment and chemicals used in this Exploration are not particularly dangerous, remind them that other items in the lab may be dangerous, so safety precautions must be taken. Explain the need to tie back loose clothing and long hair when working in the lab. Emphasize the need to wear appropriate clothing, as outlined in the safety section.

Have students make the data tables before they enter the lab.

Remind students to make very careful observations, especially for the portion of the lab using the copper wire and the Bunsen burner.

Techniques to Demonstrate

Bunsen burners tend to vary in design; therefore, show students exactly how to light the Bunsen burners used in your lab. Tell students that the burner flame is a chemical reaction in which methane and oxygen react to form carbon dioxide and water vapor.

Lighting Bunsen burners with a match can be dangerous, especially if the matches are not safety matches. It is best to use a striker.

Be sure to demonstrate the proper technique for measuring mass with a balance. Make sure students know how to help keep the lab equipment in good shape. Be certain that neither chemicals nor hot objects are put directly on the balance pan. Students may also have difficulty understanding how to use the Vernier scale on the triple-beam balance unless they can see a demonstration. If you are using another kind of balance, explain how it works.

Show students how to read volume from the meniscus. This can be demonstrated with a quick sketch on the board or on an overhead projector. Remind students to keep their eyes at the same level as the top of the fluid when reading volume.

4. Compare the Bunsen burner in **Figure A** with your burner. Construction may vary, but the air and methane gas, CH_4, always mix in the barrel, the vertical tube in the center of the burner.

Figure A

5. Partially close the air ports at the base of the barrel, turn the gas on full, hold the sparker about 5 cm above the top of the barrel, and proceed to light. Adjust the gas valve until the flame extends about 8 cm above the barrel. Adjust the air supply until you have a quiet, steady flame with a sharply defined, light-blue inner cone. If an internal flame develops, turn off the gas valve, and let the burner cool down. Otherwise, the metal of the burner can get hot enough to set fire to anything nearby that is flammable. Before you relight the burner, partially close the air ports.

6. Using crucible tongs, hold a 10 cm piece of copper wire for 2–3 s in the part of the flame labeled "a" in **Figure B**. Repeat this step for the parts of the flame labeled "b" and "c." Record your observations in your lab notebook.

7. Experiment with the flame by completely closing the air ports at the base of the burner. Observe and record the color of the flame and the sounds made by the burner. Using crucible tongs, hold an evaporating dish in the tip of the flame for about 3 min. Place the dish on a heat-

resistant mat, and shut off the burner. After the dish cools, examine its underside, and record your observations.

8. Before using the balance, make sure that it is level and showing a mass of zero. If necessary, adjust the calibration knob. To avoid discrepancies, use the same balance for all measurements during a lab activity. Never put chemicals directly on the balance pan.

9. Place a piece of weighing paper on the balance pan. Determine the mass of the paper, and record this mass to the nearest 0.01 g in your data table. Put a small quantity of NaCl on a separate piece of weighing paper. Then, transfer 13 g of the NaCl to the weighing paper on the balance pan. Record the exact mass to the nearest 0.01 g in your data table.

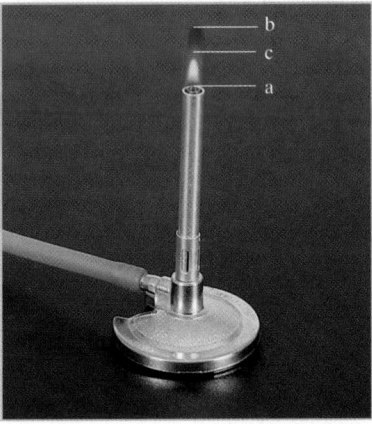

Figure B

10. Remove the weighing paper and NaCl from the balance pan. Lay the test tube flat on the table, and transfer the NaCl into the tube by rolling the weighing paper and sliding it into the test tube. As you lift the test tube to a vertical position, tap the paper gently, and the solid will slip into the test tube, as shown in **Figure C.**

11. Measure the mass of a dry 250 mL beaker, and record the mass in your data table. Add water up to the 50 mL mark, determine the new mass,

758

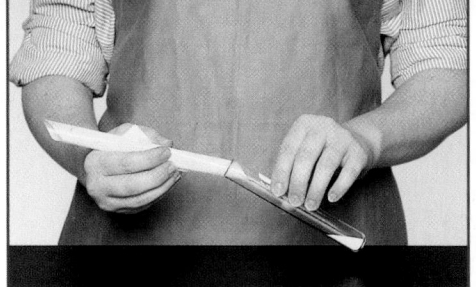

Figure C

and record the new mass in your data table. Repeat the procedure by filling the beaker to the 100 mL mark and then to the 150 mL mark, and record the mass each time. Subtract the mass of the empty beaker from the other measurements to determine the masses of the water.

12. Repeat step 11 with a second dry 250 mL beaker, but use a graduated cylinder to measure the volumes of water to the nearest 0.1 mL before pouring the water into the beaker. Read the volumes by using the bottom of the meniscus, the curve formed by the water's surface.

13. Clean all apparatus and your lab station. Put the wire, NaCl, and weighing paper in the containers designated by your teacher. Pour the water from the beakers into the sink. Scrub the cooled evaporating dish with soap, water, and a scrub brush. Be certain that the gas valves at your lab station and the nearest lab station are turned off. Be sure lab equipment is completely cool before storing it. Always wash your hands thoroughly after all lab work is finished and before you leave the lab.

Analysis

1. **Analyzing data** Based on your observations, which type of flame is hotter: the flame formed when the air ports are open or the flame formed when they are closed? What is the hottest part of the flame? (Hint: The melting point of copper is 1083°C.)

2. **Examining data** Which of the following measurements could have been made by your balance: 3.42 g of glass, 5.666 72 g of aspirin, or 0.000 017 g of paper?

3. **Constructing graphs** Make a graph of mass versus volume for data from steps 11 and 12. The mass of water (g) should be graphed along the y-axis as the dependent variable, and the volume of water (mL) should be graphed along the x-axis as the independent variable.

Conclusions

4. **Interpreting information** When methane is burned, it usually produces carbon dioxide and water. If there is a shortage of oxygen, the flame is not as hot and black carbon solid is formed. Which steps in the lab demonstrate these flames?

5. **Applying conclusions** Which is the most accurate method for measuring volumes of liquids, a beaker or a graduated cylinder? Explain why.

6. **Evaluating data** In Mandeville High School, Jarrold got only partway through step 7 of this experiment when he had to put everything away. Soon after Jarrold left, his lab drawer caught on fire. How did this happen?

7. **Drawing conclusions** The density of water is equal to its mass divided by its volume. Calculate the density of water by using your data from step 11. Then, calculate the density of water by using your data from step 12.

Extensions

8. **Designing experiments** You have been asked to design an experiment to find the density of sand. The density of sand is equal to its mass divided by its volume. Describe how you could measure the density of sand by using the equipment from this lab.

9. **Research and communications** Scientists use a number of different instruments to measure the mass of an object. Find information on different types of balances, and make a poster comparing at least three different kinds of balances. The poster should show the smallest amount of mass that could be measured on the balance and identify something appropriate to measure on the balance.

759

Answers to Analysis

1. The hottest flame is formed when the air ports are open. The hottest part of the flame should be at the top of the light-blue inner core of the Bunsen burner flame, which corresponds to c in Figure B. (Students should have observed the copper melting.)

2. Only the 3.42 g measurement could have been made using the laboratory balance. All of the other measurements indicate precision that is not possible with the laboratory balance.

3. Accept all graphs that are labeled correctly and represent data accurately.

Answers to Conclusions

4. Step 5 demonstrated the quiet, steady light-blue flame that produced carbon dioxide and water vapor. Step 7 demonstrated the loud, yellow flame with the shortage of oxygen that produced black carbon solid.

5. Using a graduated cylinder is the best method for measuring volumes of liquids because it is marked in increments of 1 mL instead of increments of 50 mL.

6. The fire may have started because Jarrold did not allow the burner or the evaporating dish to cool completely. When he put them into the lab drawer, they may have been hot enough to ignite some paper.

7. Densities for step 11
   1.10 g/mL (50 mL H_2O)
   1.01 g/mL (100 mL H_2O)
   1.00 g/mL (150 mL H_2O)
   Densities for step 12
   1.010 g/mL (50 mL H_2O)
   0.997 g/mL (100 mL H_2O)
   1.005 g/mL (150 mL H_2O)

Answers to Extensions

8. First, a 250 mL beaker is weighed. Then, 50 mL of sand is measured with the graduated cylinder and then weighed in a beaker. After the weight of the beaker is subtracted, the mass of the sand is found. The mass is then divided by the volume. The result is the density of sand.

9. Students should easily find three different types of balances. Triple beam balances usually measure to the nearest 0.01 gram, while analytical balances may measure to the nearest 0.0001 gram.

CONSERVATION OF MASS: PERCENTAGE OF WATER IN POPCORN

Teacher's Notes

Time Required
2 lab periods

Ratings

```
        1    2    3    4
Ratings EASY       HARD
```

TEACHER PREPARATION	1
STUDENT SETUP	2
CONCEPT LEVEL	2
CLEANUP	1

Skills Acquired
- Collecting Data
- Communicating
- Designing Experiments
- Experimenting
- Inferring
- Interpreting
- Measuring
- Organizing and Analyzing Data

The Scientific Method
In this lab, students will:
- Make Observations
- Ask Questions
- Form a Hypothesis
- Test the Hypothesis
- Analyze the Results
- Draw Conclusions
- Communicate the Results

Inquiry LAB
Design Your Own Experiment

1 Conservation of Mass
Percentage of Water in Popcorn

THE PROBLEM

January 9, 2004

Director of Research
CheMystery Labs, Inc.
52 Fulton Street
Springfield, VA 22150

Dear Director of Research:

Juliette Brand Foods is preparing to enter the rapidly expanding popcorn market with a new popcorn product. As you may know, the key to making popcorn pop is the amount of water contained within the kernel.

As of today, the product development division has created three different production techniques for the popcorn, each of which creates popcorn that contains differing amounts of water. We need an independent lab such as yours to measure the percentage of water contained in each sample and to determine which technique produces the best-popping popcorn.

I have enclosed samples from each of the three techniques, labeled "technique beta," "technique gamma," and "technique delta." Please send us the bill when the work is complete.

Sincerely,

Mary Biedenbecker

Mary Biedenbecker, Director
Product Development Division

References

Popcorn pops because of the natural moisture inside each kernel. When the internal water is heated above 100°C, the kernel expands rapidly and the liquid water changes to a gas, which takes up much more space than the liquid.

The percentage of water in popcorn can be determined by the following equation.

$$\frac{\text{initial mass} - \text{final mass}}{\text{initial mass}} \times 100 = \text{percent } H_2O$$

The popping process works best when the kernels are first coated with a small amount of vegetable oil. Make sure you account for the presence of this oil when measuring masses.

760

Materials
- aluminum foil (1 sheet)
- beaker, 250 mL
- Bunsen burner with gas tubing and striker
- kernels of popcorn for each of the three brands (80)
- oil (to coat the bottom of the beaker)
- ring stand, iron ring, and wire gauze

Safety Cautions
- Safety goggles and a lab apron must be worn at all times.
- Long hair and loose clothing must be tied back.
- Remind students to use beaker tongs to handle heated glassware; heated glass does not always look hot.
- Remind students that the popcorn is for testing only and must not be eaten.
- The popped popcorn should be disposed of in the designated waste container.

THE PLAN

CheMystery Labs, Inc. 52 Fulton Street, Springfield, VA 22150

CheMystery Labs, Inc.
52 Fulton Street
Springfield, VA 22150

Memorandum

Date: January 11, 2004
To: Leon Fuller
From: Martha Li-Hsien

Your team needs to design a procedure for determining the percentage of water in three samples of popcorn. Some of the popcorn was damaged in the mail, so each team will have only 80 kernels of popcorn per technique. Make sure to use your samples carefully!

Before you begin the lab work, I must approve your procedure. Give the following items to me ASAP:

- a detailed one-page plan for your procedure, including any necessary data tables
- a detailed list of the equipment and materials you will need

When you finish your experiment, prepare a report in the form of a two-page letter to Mary Biedenbecker that includes the following:

- a paragraph summarizing how you analyzed the samples
- your findings about the percentage of water in each sample, including calculations and a discussion of the multiple trials
- a detailed and organized data table
- a graph comparing your findings with those of the other teams
- suggestions for improving the analysis procedure

Required Precautions

- Wear safety goggles when working around chemicals, acids, bases, flames, or heating devices. Contents under pressure may become projectiles and cause serious injury.
- Avoid wearing contact lenses in the lab.
- If any substance gets in your eyes, notify your instructor immediately and flush your eyes with running water for at least 15 min.
- Secure loose clothing, and remove dangling jewelry. Don't wear open-toed shoes or sandals in the lab.
- Wear an apron or lab coat to pro-

tect your clothing when working with chemicals.
- If a spill gets on your clothing, rinse it off immediately with water for at least 5 min while notifying your instructor.
- Always use caution when working with chemicals.
- Never taste, touch, or smell chemicals unless specifically directed to do so.
- Follow instructions for proper disposal.
- Whenever possible, use an electric hot plate as a heat source instead of an open flame.

- When heating materials in a test tube, always angle the test tube away from yourself and others.
- Know your school's fire-evacuation routes.
- Clean and decontaminate all work surfaces and personal protective equipment as directed by your instructor.
- Dispose of all sharps (broken glass and other contaminated sharp objects) and other contaminated materials (biological and chemical) in special containers as directed by your instructor.

761

Checkpoint 1: By the end of the first class period, students should turn in a detailed one-page plan/procedure for approval. This should be compared to the sample procedure found in the answer section of the wrap. The student procedure should be evaluated according to organization and accuracy and how well it follows the steps of the scientific method.

Checkpoint 2: During the second class period, students should revise procedures according to the teacher's approval and begin the procedure. By the end of the second class period, students should be able to determine the percentage of water in popcorn. Students should be evaluated on lab technique, quality and clarity of observations, and the explanation of observations/conclusions.

Sample Answer to Procedure

Cover the beaker in which the popcorn is being popped with foil. Put holes in the foil to allow the water vapor to escape.

- Measure the mass of the unpopped kernels. (Twenty kernels work well.)
- Measure the mass of the beaker.
- Measure the mass of the beaker with the uncooked popcorn and vegetable oil coating.

Heat the popcorn until the majority of the kernels have popped. The popcorn pops more efficiently if the beaker is held firmly with tongs and gently shaken side to side on the wire gauze.

Allow the beaker to cool for 5 minutes. Measure the mass of the beaker, popped corn, and vegetable oil. Subtract the "before" and "after" masses of the popcorn to determine the mass of water in the kernels.

Sample Data

	Technique beta	Technique gamma	Technique delta
20 kernels	2.47 g	2.93 g	3.76 g
250 mL beaker	101.40 g	101.40 g	101.40 g
Beaker + oil	101.50 g	101.50 g	101.50 g
Beaker + oil + 20 kernels (before)	103.97 g	104.43 g	105.26 g
Beaker + oil + 20 kernels (after)	103.50 g	103.50 g	103.80 g
Mass of water in 20 kernels	0.47 g	0.93 g	1.46 g
% water in popcorn	19%	32%	39%

SEPARATION OF MIXTURES

Teacher's Notes

Time Required

2 lab periods if filtering is done once (includes overnight drying time)

Ratings

$$\underset{\substack{\text{EASY}}}{\overset{1}{\rule{0pt}{0pt}}} \quad \overset{2}{\rule{0pt}{0pt}} \quad \overset{3}{\rule{0pt}{0pt}} \quad \underset{\substack{\text{HARD}}}{\overset{4}{\rule{0pt}{0pt}}}$$

TEACHER PREPARATION	3
STUDENT SETUP	2
CONCEPT LEVEL	2
CLEANUP	2

Skills Acquired

• Collecting Data
• Communicating
• Experimenting
• Interpreting
• Measuring
• Organizing and Analyzing Data

The Scientific Method

In this lab, students will:
• Make Observations
• Test the Hypothesis
• Analyze the Results
• Draw Conclusions

Skills Practice Lab

OBJECTIVES

◆ **Recognize** how the solubility of a salt varies with temperature.

◆ **Demonstrate** proficiency in fractional crystallization and in filtration.

◆ **Solve** the percentage of two salts recovered by fractional crystallization.

MATERIALS

◆ balance
◆ beaker tongs or hot mitt
◆ beakers, 150 mL (4)
◆ Bunsen burner or hot plate
◆ filter paper
◆ graduated cylinder, 100 mL
◆ ice and rock salt
◆ NaCl–KNO₃ solution (50 mL)
◆ nonmercury thermometer
◆ ring stand set up
◆ rubber policeman
◆ spatula
◆ stirring rod, glass
◆ tray, tub, or pneumatic trough
◆ vacuum filtration setup or gravity-filtration setup

OPTIONAL EQUIPMENT

◆ CBL unit
◆ graphing calculator with cable
◆ Vernier temperature probe

2 Separation of Mixtures

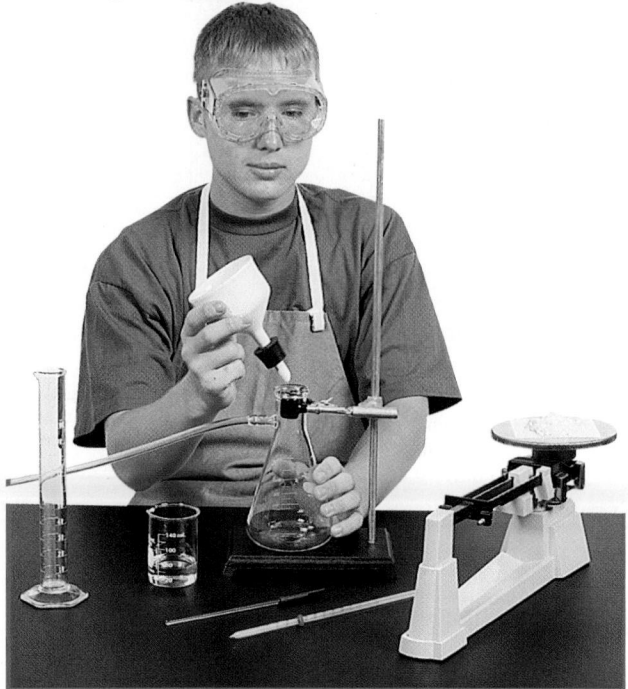

Introduction

Your company has been contacted by a fireworks factory. A mislabeled container of sodium chloride, NaCl, was accidentally mixed with potassium nitrate, KNO_3. KNO_3 is used as an oxidizer in fireworks to ensure that the fireworks burn thoroughly. The fireworks company wants your company to investigate ways they could separate the two compounds. They have provided an aqueous solution of the mixture for you to work with.

The substances in a mixture can be separated by physical means. For example, if one substance dissolves in a liquid solvent but another does not, the mixture can be filtered. The substance that dissolved will be carried through the filter by the solvent, but the other substance will not.

Because both NaCl and KNO_3 dissolve in water, filtering alone cannot separate them. However, there are differences in the way they dissolve. The graph in **Figure A** shows the same amount of NaCl dissolving in water regardless of the temperature of the water. On the other hand, KNO_3 is very soluble in warm water but much less soluble at 0°C.

Safety Procedures

- Wear safety goggles when working around chemicals, acids, bases, flames, or heating devices. Contents under pressure may become projectiles and cause serious injury.
- Never look directly at the sun through any optical device or use direct sunlight to illuminate a microscope.
- Avoid wearing contact lenses in the lab. If any substance gets in your eyes, notify your instructor immediately and flush your eyes with running water for at least 15 min.

- Secure loose clothing, and remove dangling jewelry. Don't wear open-toed shoes or sandals in the lab.
- Wear an apron or lab coat to protect your clothing when working with chemicals.
- If a spill gets on your clothing, rinse it off immediately with water for at least 5 min while notifying your instructor.

- Always use caution when working with chemicals.
- Never mix chemicals unless specifically directed to do so.
- Never taste, touch, or smell chemicals unless specifically directed to do so.

- Add acid or base to water; never do the opposite.
- Never return unused chemicals to the original container.
- Never transfer substances by sucking on a pipette or straw; use a suction bulb.
- Follow instructions for proper disposal.

- Avoid wearing hair spray or hair gel on lab days.
- Whenever possible, use an electric hot plate as a heat source instead of an open flame.
- When heating materials in a test tube, always angle the test tube away from yourself and others.
- Glass containers used for heating should be made of heat-resistant glass.
- Know your school's fire-evacuation routes.

- Clean and decontaminate all work surfaces and personal protective equipment as directed by your instructor.
- Dispose of all sharps (broken glass and other contaminated sharp objects) and other contaminated materials (biological and chemical) in special containers as directed by your instructor.

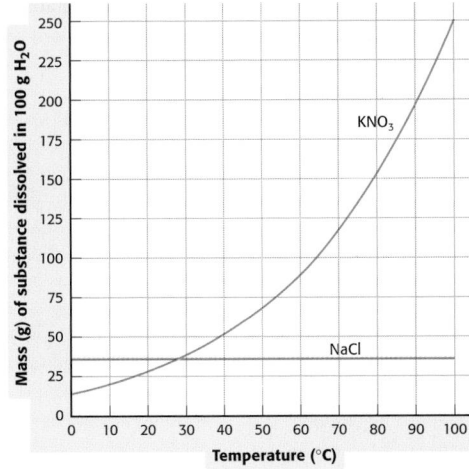

Solubility vs. Temperature for Two Salts

Figure A
This graph shows the relationship between temperature and the solubility of $NaCl$ and KNO_3.

You will make use of the differences in solubility to separate the two salts. This technique is known as fractional crystallization. If the water solution of $NaCl$ and KNO_3 is cooled from room temperature to a temperature near 0°C, some KNO_3 will crystallize. This KNO_3 residue can then be separated from the $NaCl$ solution by filtration. The $NaCl$ can be isolated from the filtrate by evaporation of the water. To determine whether this method is efficient, you will measure the mass of each of the recovered substances. Then, your client can decide whether this method is cost-effective.

Materials
Vacuum Filtration Option
- aspirator for spigot
- Büchner funnel (either ceramic or plastic)
- filter paper
- one-hole rubber stopper or sleeve
- vacuum flask (sidearm flask) and tubing

Optional Equipment
- CBL unit
- graphing calculator with link cable
- Vernier temperature probe

Safety Cautions
- Safety goggles and a lab apron must be worn at all times.
- Read all safety precautions, and discuss them with your students.
- Remind students that when equipment has been heated, it should be handled with tongs or a hot mitt; heated glassware does not always look hot.

Disposal Information
Potassium nitrate, KNO_3, cannot be disposed of in any manner because it is a strong oxidizer. However, it can be reused the next time you do this lab. Dry the crystals in a separate disposal container, and then store the dried crystals in sealed, labeled glass bottles completely away from the bottles of reducing substances, especially those that can burn. Reuse the crystals the following year instead of ordering more. Because this is a separation exercise, it does not matter if impurities are present.

763

Filtration-Technique Option

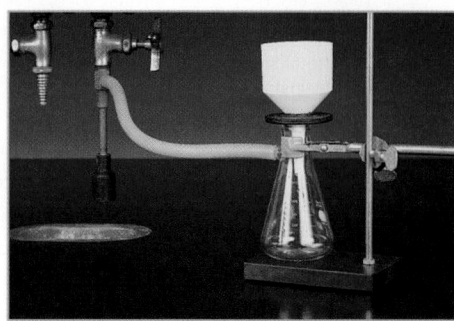

Figure B
Vacuum filtration

Figure C
Gravity filtration

Vacuum-Filtration Setup

1. To set up a vacuum filtration, screw an aspirator nozzle onto the faucet. Attach the other end of the plastic tubing to the side arm of the filter flask.

2. Place a one-hole rubber stopper on the stem of the funnel, and fit the stopper snugly in the neck of the filter flask, as shown in **Figure B.**

3. Place a piece of filter paper on the bottom of the funnel so that it is flat and covers all of the holes in the funnel.

4. When you are ready, turn on the water at the faucet that has the aspirator nozzle attached. This action creates a vacuum, which helps the filtering process go much faster. If the suction is working properly, the filter paper should be pulled against the bottom of the funnel, which results in covering all of the holes. If the filter paper appears to have bubbles of air under it or is not centered well, turn the water off, reposition the filter paper, and begin again.

Gravity-Filtration Setup

1. Set up a ring stand with a ring. Gently rest a glass funnel inside the ring, and place a beaker under the glass funnel, as shown in **Figure C.**

2. Fold a piece of filter paper in half along its diameter, and then fold it again to form a quadrant, as shown in **Figure D.** Separate the folds of the filter paper so that three thicknesses are on one side and one thickness is on the other.

3. Fit the filter paper in the funnel, and wet it with a little water so that it will adhere to the sides of the funnel. Gently but firmly press the paper against the sides of the funnel so that no air is between the funnel and the filter paper. Be certain that the filter paper does not extend above the sides of the funnel.

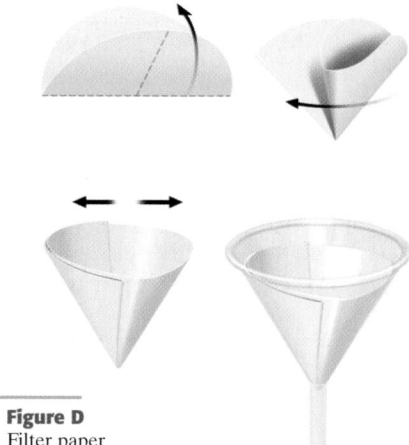

Figure D
Filter paper

764

Procedure

Advance Preparation

1. Copy the data table below in your lab note-book. Be sure that you have plenty of room for observations about each test.

Data Table 1
Mass of beaker 1
Volume of NaCl–KNO$_3$ solution added to beaker 1
Temperature of mixture before cooling
Mass of filter paper
Mass of beaker 4
Mass of beaker 4 with NaCl
Mass of beaker 1 with filter paper and KNO$_3$
Temperature of mixture after cooling

2. Obtain four clean, dry 150 mL beakers, and label them 1, 2, 3, and 4.

Thermometer procedure continues on page 767.

CBL and Sensors

3. Connect the CBL to the graphing calculator with the unit-to-unit link cable using the I/O ports located on each unit. Connect the temperature probe to the CH1 port. Turn on the CBL and the graphing calculator. Start the program CHEMBIO on the graphing calculator.

   a. Select option *SET UP PROBES* from the MAIN MENU. Enter 1 for the number of probes. Select the temperature probe from the list. Enter 1 for the channel number.

   b. Select the *COLLECT DATA* option from the MAIN MENU. Select the *TRIGGER* option from the DATA COLLECTION menu.

4. Set up your filtering apparatus. If you are using a Büchner funnel for vacuum filtration or a glass funnel for gravity filtration, follow the setup procedure under "Filtration-Technique Option."

5. Measure the mass of beaker 1 to the nearest 0.01 g, and record the mass in your data table.

6. Measure about 50 mL of the NaCl–KNO$_3$ solution into a graduated cylinder. Record the exact volume in your data table. Pour this mixture into beaker 1.

7. Using the temperature probe, measure the temperature of the mixture. Press TRIGGER on the CBL to collect the temperature reading of the mixture. Record this temperature in your data table. Select *CONTINUE* from the TRIGGER menu on the graphing calculator.

8. Measure the mass of a piece of filter paper to the nearest 0.01 g, and record the mass in your data table.

9. Make an ice bath by filling a tray, tub, or trough half-full with ice. Add a handful of rock salt. The salt lowers the freezing point of water so that the ice bath can cool to a lower temperature. Fill the ice bath with water until it is three-quarters full.

10. Using a fresh supply of ice and distilled water, fill beaker 2 half-full with ice, and add water. Do not add rock salt to this ice-water mixture. You will use this water to wash your purified salt.

First Filtration

11. Put beaker 1 with your NaCl–KNO$_3$ solution into the ice bath. Place the temperature probe in the solution to monitor the temperature. Stir the solution with a stirring rod while it cools. (Do not stir the solution with the temperature probe.) The lower the temperature of the mixture is, the more KNO$_3$ that will crystallize out of solution. When the temperature nears 4°C, press TRIGGER on the CBL to collect the temperature reading of the mixture. Record this temperature in your data table. Select *STOP* from the TRIGGER menu on the graphing calculator. Proceed with step 11a if you are using the Büchner funnel or step 11b if you are using a glass funnel.

Techniques to Demonstrate

Be sure to review the vocabulary associated with filtration, especially the difference between the filtrate and the residue.

Review the vacuum filtration setup if your students are using this technique. This method speeds up the lab considerably. (Plastic Büchner funnels are far less prone to breakage than glass funnels or ceramic Büchner funnels.)

Review the gravity filtration, especially folding the filter paper, if your students are using this technique.

You will also need to show students how to use a rubber policeman to collect crystals. Remind students to evaporate the water gradually to avoid violent bubbling and loss of product.

Misconception Alert

Students get very frustrated when their results do not match calculated ideal values. Students should realize when they answer Conclusions item 9 that it is impossible to achieve a perfect separation using this technique.

This exploration provides an opportunity for discussions about the uncertain nature of science and the constant need to improve on lab techniques at all levels of science.

a. Vacuum filtration

Prepare the filtering apparatus by pouring approximately 50 mL of ice-cold distilled water from beaker 2 through the filter paper. After the water has gone through the funnel, empty the filter flask into the sink. Reconnect the filter flask, and pour the salt-and-water mixture in beaker 1 into the funnel. Use the rubber policeman to transfer all of the cooled mixture into the funnel, especially any crystals that are visible. It may be helpful to add small amounts of ice-cold water from beaker 2 to beaker 1 to wash any crystals onto the filter paper. After all of the solution has passed through the funnel, wash the KNO_3 residue by pouring a very small amount of ice-cold water from beaker 2 over it. When this water has passed through the filter paper, turn off the faucet and carefully remove the tubing from the aspirator. Empty the filtrate, which has passed through the filter paper and is now in the filter flask, into beaker 3. When finished, continue with step 12.

b. Gravity filtration

Place beaker 3 under the glass funnel. Prepare the filtering apparatus by pouring approximately 50 mL of ice-cold water from beaker 2 through the filter paper. The water will pass through the filter paper and drip into beaker 3. When the dripping stops, empty beaker 3 into the sink. Place beaker 3 back under the glass funnel so that it will collect the filtrate from the funnel. Pour the salt-water mixture into the funnel. Use the rubber policeman to transfer all of the cooled mixture into the funnel, especially any visible crystals. It may be helpful to add small amounts of ice-cold water from beaker 2 to beaker 1 to wash any crystals onto the filter paper. After all of the solution has passed through the funnel, wash the KNO_3 by pouring a very small amount of ice-cold water from beaker 2 over it.

12. After you have finished filtering, use either a hot plate or a Bunsen burner, ring stand, ring, and wire gauze to heat beaker 3. When the liquid in beaker 3 begins to boil, continue heating gently

until enough water has vaporized to decrease the volume to approximately 25–30 mL. Be sure to use beaker tongs. Remember that hot glassware does not always look hot.

Second Filtration

13. Allow the solution in beaker 3 to cool. Then set it in the ice bath and stir until the temperature is approximately 4°C.

14. Measure the mass of beaker 4, and record the mass in your data table.

15. Repeat step 11a or step 11b, pouring the solution from beaker 3 onto the filter paper and using beaker 4 to collect the filtrate that passes through the filter.

16. Wash and dry beaker 1. Carefully remove the filter paper with the KNO_3 from the funnel, and put it in the beaker. Avoid spilling the crystals. Place the beaker in a drying oven overnight.

Figure E
Use beaker tongs to move a beaker that has been heated, even if you believe that the beaker is cool.

Recovery of NaCl

17. Heat beaker 4 with a hot plate or Bunsen burner until the water begins to boil. Continue to heat the beaker gently until all of the water has vaporized and the salt appears dry. Turn off the hot plate or burner, and allow the beaker to cool. Use beaker tongs to move the beaker, as shown in **Figure E.** Measure the mass of beaker

4 with the NaCl to the nearest 0.01 g, and record the mass in your data table.

18. The next day, use beaker tongs to remove beaker 1 with the filter paper and KNO_3 from the drying oven. Allow the beaker to cool. Measure the mass using the same balance you used to measure the mass of the empty beaker. Record the new mass in your data table. Be sure to use beaker tongs. Remember that hot glassware does not always look hot.

19. Clean all apparatus and your lab station. Once the mass of the NaCl has been determined, add water to dissolve the NaCl, and rinse the solution down the drain. Do not wash KNO_3 down the drain. Dispose of the KNO_3 in the waste container designated by your teacher. Wash your hands thoroughly after all lab work is finished and before you leave the lab.

Thermometer

3. Set up your filtering apparatus. If you are using a Büchner funnel for vacuum filtration or a glass funnel for gravity filtration, follow the setup procedure under "Filtration-Technique Option."

4. Measure the mass of beaker 1 to the nearest 0.01 g, and record the mass in your data table.

5. Measure about 50 mL of the NaCl–KNO_3 solution into a graduated cylinder. Record the exact volume in your data table. Pour this mixture into beaker 1.

6. Using a thermometer, measure the temperature of the mixture. Record this temperature in your data table.

7. Measure the mass of a piece of filter paper to the nearest 0.01 g, and record the mass in your data table.

8. Make an ice bath by filling a tray, tub, or trough half-full with ice. Add a handful of rock salt. The salt lowers the freezing point of water so that the ice bath can cool to a lower temperature. Fill the ice bath with water until it is three-quarters full.

9. Using a fresh supply of ice and distilled water, fill beaker 2 half-full with ice, and add water. Do not add rock salt to this ice-water mixture. You will use this water to wash your purified salt.

First Filtration

10. Put beaker 1 with your NaCl–KNO_3 solution into the ice bath. Place a thermometer in the solution to monitor the temperature. Stir the solution with a stirring rod while it cools. The lower the temperature of the mixture is, the more KNO_3 that will crystallize out of solution. When the temperature nears 4°C, record the temperature in your data table. Proceed with step 10a if you are using the Büchner funnel or step 10b if you are using a glass funnel. Never stir a solution with a thermometer; the bulb is very fragile.

   a. Vacuum filtration
   Prepare the filtering apparatus by pouring approximately 50 mL of ice-cold distilled water from beaker 2 through the filter paper. After the water has gone through the funnel, empty the filter flask into the sink. Reconnect the filter flask, and pour the salt-and-water mixture in beaker 1 into the funnel. Use the rubber policeman to transfer all of the cooled mixture into the funnel, especially any crystals that are visible. It may be helpful to add small amounts of ice-cold water from beaker 2 to beaker 1 to wash any crystals onto the filter paper. After all of the solution has passed through the funnel, wash the KNO_3 residue by pouring a very small amount of ice-cold water from beaker 2 over it. When this water has passed through the filter paper, turn off the faucet and carefully remove the tubing from the aspirator. Empty the filtrate, which has passed through the filter paper and is now in the filter flask, into beaker 3. When finished, continue with step 11.

   b. Gravity filtration
   Place beaker 3 under the glass funnel. Prepare the filtering apparatus by pouring approximately 50 mL of ice-cold water from beaker 2 through the filter paper. The water

Tips and Tricks

Thoroughly discuss the procedure used in this lab. Students need to work quickly and efficiently in order to complete this lab in one lab period. If students work in a step-by-step process, the lab may take too long. Instead, encourage them to perform multiple tasks simultaneously through teamwork. While one partner is cooling the solution, the other lab partner could be cleaning, drying, and measuring the mass of the other beakers.

Students may have difficulty reading the solubility graph. It is not necessary for students to understand all aspects of solutions and solubility, but an understanding of this graph is important to understanding the concept of fractional crystallization; therefore, explain the graph thoroughly. Practice taking several readings from the graph, asking students how many grams of each salt would dissolve in 100 g of water at a given temperature. Relate the results to the cycles of cooling and heating in the procedure.

Results are greatly improved if a second filtration step is performed after a little water has evaporated, but this takes more time.

For every liter of solution, add about 140 g of NaCl and 320 g of KNO_3. It is necessary to stir and gently heat the solution to completely dissolve the salts. A hot-plate stirrer is a valuable tool.

Students should use only non-mercury thermometers. If a mercury thermometer breaks, the droplets that are not cleaned up will quickly evaporate, creating toxic mercury vapors.

Centigram balances will give best results, but less precise balances are acceptable.

Answers to Analysis

1. Answers will vary.
2. Answers will vary.
3. Answers will vary.

Answers to Conclusions

4. Answers will vary.
5. At room temperature, 25°C, about 32 g of KNO_3 and 35 g of NaCl will dissolve in 100 g of water. At –2°C, 15 g of KNO_3 and 33 g of NaCl will dissolve in 100 g of water.
6. Answers will vary.
7. Answers will vary.
8. As much as 7 g of KNO_3 could still be contaminating the NaCl if the solution was filtered only once.
9. Because the solubility of KNO_3 is never 0.0 g, it is always possible that some of the KNO_3 is dissolved in the solution after filtration.

will pass through the filter paper and drip into beaker 3. When the dripping stops, empty beaker 3 into the sink. Place beaker 3 back under the glass funnel so that it will collect the filtrate from the funnel. Pour the salt-water mixture into the funnel. Use the rubber policeman to transfer all of the cooled mixture into the funnel, especially any visible crystals. It may be helpful to add small amounts of ice-cold water from beaker 2 to beaker 1 to wash any crystals onto the filter paper. After all of the solution has passed through the funnel, wash the KNO_3 by pouring a very small amount of ice-cold water from beaker 2 over it.

11. After you have finished filtering, use either a hot plate or a Bunsen burner, ring stand, ring, and wire gauze to heat beaker 3. When the liquid in beaker 3 begins to boil, continue heating gently until enough water has vaporized to decrease the volume to approximately 25–30 mL. Be sure to use beaker tongs. Remember that hot glassware does not always look hot.

Second Filtration

12. Allow the solution in beaker 3 to cool. Then set it in the ice bath and stir until the temperature is approximately 4°C.

13. Measure the mass of beaker 4, and record the mass in your data table.

14. Repeat step 10a or step 10b, pouring the solution from beaker 3 onto the filter paper and using beaker 4 to collect the filtrate that passes through the filter.

15. Wash and dry beaker 1. Carefully remove the filter paper with the KNO_3 from the funnel, and put it in the beaker. Avoid spilling the crystals. Place the beaker in a drying oven overnight.

Recovery of NaCl

16. Heat beaker 4 with a hot plate or Bunsen burner until the water begins to boil. Continue to heat the beaker gently until all of the water has vaporized and the salt appears dry. Turn off the hot plate or burner, and allow the beaker to cool. Use beaker tongs to move the beaker, as shown in Figure E. Measure the mass of beaker 4 with the NaCl to the nearest 0.01 g, and record the mass in your data table.

17. The next day, use beaker tongs to remove beaker 1 with the filter paper and KNO_3 from the drying oven. Allow the beaker to cool. Measure the mass using the same balance you used to measure the mass of the empty beaker. Record the new mass in your data table. Be sure to use beaker tongs. Remember that hot glassware does not always look hot.

18. Clean all apparatus and your lab station. Once the mass of the NaCl has been determined, add water to dissolve the NaCl, and rinse the solution down the drain. Do not wash KNO_3 down the drain. Dispose of the KNO_3 in the waste container designated by your teacher. Wash your hands thoroughly after all lab work is finished and before you leave the lab.

Analysis

1. **Analyzing results** Find the mass of NaCl in your 50 mL sample by subtracting the mass of the empty beaker 4 from the mass of beaker 4 with NaCl.

2. **Analyzing data** Find the mass of KNO_3 in your 50 mL sample by subtracting the mass of beaker 1 and the mass of the filter paper from the mass of beaker 1 with the filter paper and KNO_3.

3. **Analyzing data** Determine the total mass of the two salts.

Conclusions

4. Applying conclusions How many grams of KNO_3 and NaCl would be found in a 1.0 L sample of the solution? (Hint: For each substance, make a conversion factor by using the mass of the compound and the volume of the solution.)

5. Analyzing graphs Use the graph at the beginning of this exploration to determine how much of each compound would dissolve in 100 g of water at room temperature and at the temperature of your ice-water bath.

6. Drawing conclusions Calculate the percentage by mass of NaCl in the salt mixture. Calculate the percentage by mass of KNO_3 in the salt mixture. Assume that the density of your 50 mL solution is 1.0 g/mL.

7. Applying conclusions The fireworks company has another 55 L of the salt mixture dissolved in water just like the sample you worked with. How many kilograms of each compound can the company expect to recover from this sample? (Hint: Use your answer from item 4 to help you answer this question.)

8. Evaluating methods Use the graph shown at the beginning of this lab to estimate how much KNO_3 could still be contaminating the NaCl you recovered.

9. Relating ideas Use the graph shown at the beginning of this lab to explain why it is impossible to completely separate the two compounds by fractional crystallization.

10. Evaluating methods Why was it important to use ice-cold water to wash the KNO_3 after filtration?

11. Evaluating methods If it was important to use very cold water to wash the KNO_3, why was the salt-and-ice-water mixture from the bath not used? After all, it had a lower temperature than the ice and distilled water from beaker 2 did. (Hint: Consider what is contained in rock salt.)

12. Evaluating methods Why was it important to keep the amount of cold water used to wash the KNO_3 as small as possible?

13. Interpreting graphics Using the graph shown at the beginning of this lab, determine the minimum mass of water necessary to dissolve the amounts of each compound from Analysis items 1 and 2. Calculate the mass dissolved at room temperature and at 4°C. What volumes of water would be necessary? (Hint: The density of water is about 1.0 g/mL.)

Extensions

1. Designing experiments Describe how you could use the properties of the compounds to test the purity of your recovered samples. If your teacher approves your plan, use it to check your separation of the mixtures. (Hint: Check a chemical handbook for more information about the properties of NaCl and KNO_3.)

2. Designing experiments How could you improve the yield or the purity of the compounds you recovered? If you can think of ways to modify the procedure, ask your teacher to approve your plan and run the procedure again.

10. Because the water was ice cold, the KNO_3 did not dissolve very well in it. But if there was any NaCl mixed with the KNO_3 crystals, the NaCl would have dissolved in the ice-cold water and passed through the filter paper.

11. Although the rock salt and ice mixture was cooler, it contained NaCl and would have contaminated the KNO_3 crystals if it had been used in the rinsing step.

12. By using a small amount of water to wash the crystals, students kept the amount of KNO_3 that redissolved in the water to a minimum.

13. 17.5 g H_2O to dissolve 6.12 g NaCl at 25°C

38.8 g H_2O to dissolve 12.42 g KNO_3 at 25°C

18.5 g H_2O to dissolve 6.12 g NaCl at 2°C

82.8 g H_2O to dissolve 12.42 g KNO_3 at 2°C

Answers to Extensions

1. Suggestions for determining purity will vary but could include measuring the density or testing the melting point. The latter would require special equipment.

2. Suggestions for improving purity will vary. Be sure answers are safe and include carefully planned procedures.

769

SEPARATION OF MIXTURES: MINING CONTRACT

Teacher's Notes

Time Required
2 lab periods

Ratings
$$\overset{\substack{1 \quad 2 \quad 3 \quad 4}}{\underset{\text{EASY} \qquad \text{HARD}}{\rule{3cm}{0.4pt}}}$$

TEACHER PREPARATION	2
STUDENT SETUP	2
CONCEPT LEVEL	2
CLEANUP	1

Skills Acquired
- Collecting Data
- Communicating
- Designing Experiments
- Experimenting
- Identifying/Recognizing Patterns
- Inferring
- Interpreting
- Measuring
- Organizing and Analyzing Data
- Predicting

The Scientific Method
In this lab, students will:
- Make Observations
- Ask Questions
- Form a Hypothesis
- Test the Hypothesis
- Analyze the Results
- Draw Conclusions
- Communicate the Results

Inquiry LAB
Design Your Own Experiment

2 Separation of Mixtures
Mining Contract

THE PROBLEM

January 20, 2004

George Taylor
Director of Analytical Services
CheMystery Labs, Inc.
52 Fulton Street
Springfield, VA 22150

Dear George:

I thought of your new company when a problem came up here at Goldstake. I think I have some work for your company. While performing exploratory drilling for natural gas near Afton in western Wyoming, our engineers encountered a new subterranean, geothermal aquifer. We estimate the size of the aquifer to be 1×10^{12} L.

The Bureau of Land Management advised us to alert the Environmental Protection Agency. Preliminary qualitative tests of the water identified two dissolved salts: potassium nitrate and copper nitrate.

The EPA is concerned that a full-scale mining operation may harm the environment if the salts are present in large quantities. They are requiring us to halt all operations while we obtain more information for an environmental impact statement. We need your firm to separate the sample, purify the sample, and make a determination of the amounts of the two salts in the Afton Aquifer.

Sincerely,

Lynn L. Brown

Lynn L. Brown
Director of Operations
Goldstake Mining Corporation

References
The procedure for this Investigation is similar to one your team recently completed involving the separation of sodium chloride, NaCl, and potassium nitrate, KNO_3.

770

Materials
- balance
- beakers, 250 mL (2)
- Celsius thermometer, nonmercury type, with a range from −10°C to 120°C
- $Cu(NO_3)_2 \times KNO_3$ solution (50 mL)
- graduated cylinder, 100 mL
- hot plate or Bunsen burner with gas tubing and striker
- ice, 1.5 mL
- plastic bags (2)
- ring stand, ring, and wire gauze (nonasbestos)
- rock salt (20 g)
- rubber policeman and glass stirring rod
- spatula
- tray, tub, or pneumatic trough for ice bath

CheMystery Labs, Inc. 52 Fulton Street, Springfield, VA 22150

CheMystery Labs, Inc.
52 Fulton Street
Springfield, VA 22150

Memorandum

Date: January 23, 2004
To: Andre Kalaviencz
From: George Taylor

Because this is our first mining-industry contract, we need to plan carefully to get good results at minimum cost. Each research team will receive only a 50.0 mL sample of the aquifer water.

I need the following information from each team before the work begins.
- a detailed, one-page plan for the procedure that you will use to accomplish the analysis, including all necessary data tables
- a list of the materials and supplies you will need

When you have completed your labwork, present the following information to Goldstake in a two page report:
- the mass of potassium nitrate, KNO_3, and copper nitrate, $Cu(NO_3)_2$, in the 50.0 mL sample
- the extrapolated mass of KNO_3 and $Cu(NO_3)_2$ in the Afton Aquifer
- a short paragraph that summarizes and describes the procedures you used
- detailed and organized data and analysis section that shows your calculations and explanations of any possible sources of error

Required Precautions

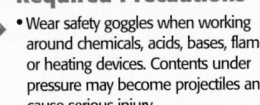 • Wear safety goggles when working around chemicals, acids, bases, flames, or heating devices. Contents under pressure may become projectiles and cause serious injury.
- Avoid wearing contact lenses in the lab.
- If any substance gets in your eyes, notify your instructor immediately and flush your eyes with running water for at least 15 min.
- Secure loose clothing, and remove dangling jewelry. Don't wear open-toed shoes or sandals in the lab.
 • Wear an apron or lab coat to pro-tect your clothing when working with chemicals.
- If a spill gets on your clothing, rinse it off immediately with water for at least 5 min while notifying your instructor.
- Always use caution when working with chemicals.
- Never taste, touch, or smell chemicals unless specifically directed to do so.
- Follow instructions for proper disposal.
- Whenever possible, use an electric hot plate as a heat source instead of an open flame.
- When heating materials in a test tube, always angle the test tube away from yourself and others.
- Know your school's fire-evacuation routes.
- Clean and decontaminate all work surfaces and personal protective equipment as directed by your instructor.
- Dispose of all sharps (broken glass and other contaminated sharp objects) and other contaminated materials (biological and chemical) in special containers as directed by your instructor.

771

Sample Data

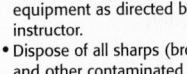

Volume of solution	46.50 mL
Mass of beaker 1	65.80 g
Mass of filter paper	0.40 g
Mass of beaker 1 with filter paper and KNO_3 (Day 2)	78.22 g

Mass of beaker 2	66.25 g
Mass of beaker 2 with $Cu(NO_3)_2$	72.37 g
Temperature before cooling	25°C
Temperature after cooling	-2°C

Most students will recover approximately 12 g of KNO_3 and 7 g of $Cu(NO_3)_2$ from 50 mL of solution. Given this data, the aquifer would contain 2.4×10^{14} g of KNO_3 and 1.4×10^{14} g of $Cu(NO_3)_2$.

Safety Precautions
- Safety goggles and a lab apron must be worn at all times.
- Read all safety precautions, and discuss them with your students.
- Remind students to use beaker tongs or a hot mitt when handling glassware that has been heated.
- If a hot plate is used, the precautions listed under "Safety in the Chemistry Laboratory" must be followed to prevent electric shock.

Checkpoints
Checkpoint 1: By the end of the first class period, students should turn in a detailed one-page plan/procedure for approval. This should be compared to the sample procedure found in the answer section of the wrap. The student procedure should be evaluated according to organization and accuracy and how well it follows the steps of the scientific method.

Checkpoint 2: During the second class period, students should revise procedures according to the teacher's approval and begin the procedure. By the end of the second class period, students should be able to draw a reasonable conclusion and determine the mass of potassium nitrate and copper nitrate from the Afton Aquifer sample. Students should be evaluated on lab technique, quality and clarity of observations, and the explanation of observations/conclusions.

Sample Answer to Procedure
Measure the mass of the two beakers and a piece of filter paper. Pour the solution in a plastic bag and cool it in an ice-water bath so the KNO_3 will crystallize. Filter the cold mixture. Rinse the crystals on the filter paper with a small amount of ice-cold distilled water. Heat the filtrate to evaporate some of the water. Cool the solution again. Filter the cold mixture. Rinse the crystals again. Dry the filter paper in one of the beakers. Pour the filtrate into the other beaker, and evaporate the water. Measure the mass of the beaker.

Teacher's Notes

Time Required
1–2 lab periods

Ratings

```
        1   2   3   4
Ratings EASY      HARD
```

TEACHER PREPARATION	3
STUDENT SETUP	2
CONCEPT LEVEL	1
CLEANUP	2

Skills Acquired
- Classifying
- Collecting Data
- Experimenting
- Identifying/Recognizing Patterns
- Interpreting
- Organizing and Analyzing Data

The Scientific Method
In this lab, students will:
- Make Observations
- Ask Questions
- Analyze the Results
- Draw Conclusions.

Skills Practice Lab

OBJECTIVES

- Identify a set of flame-test color standards for selected metal ions.
- Relate the colors of a flame test to the behavior of excited electrons in a metal ion.
- Draw conclusions and identify an unknown metal ion by using a flame test.

USING SCIENTIFIC **METHODS**

- Demonstrate proficiency in performing a flame test and in using a spectroscope.

MATERIALS

- beaker, 250 mL
- Bunsen burner
- $CaCl_2$ solution
- cobalt glass plates
- crucible tongs
- distilled water
- flame-test wire
- glass test plate
- HCl solution (1.0 M)
- K_2SO_4 solution
- Li_2SO_4 solution
- Na_2SO_4 solution
- NaCl crystals
- NaCl solution
- spectroscope
- $SrCl_2$ solution
- unknown solution

772

3 Flame Tests

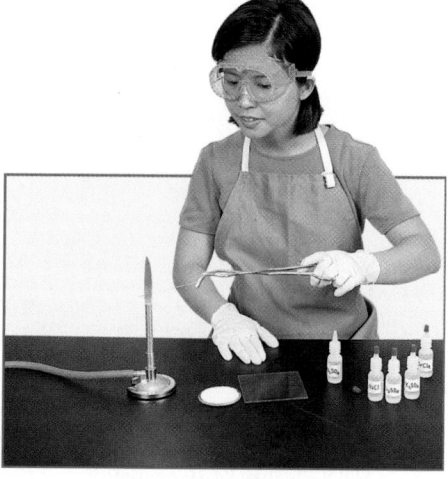

Introduction

Your company has been contacted by Julius and Annette Benetti. They are worried about some abandoned, rusted barrels of chemicals that their daughter found while playing in the vacant lot behind their home. The barrels have begun to leak a colored liquid that flows through their property before emptying into a local sewer. The Benettis want your company to identify the compound in the liquid. Earlier work indicates that it is a dissolved metal compound. Many metals, such as lead, have been determined to be hazardous to our health. Many compounds of these metals are often soluble in water and are therefore easily absorbed into the body.

Electrons in atoms jump from their ground state to excited states by absorbing energy. Eventually these electrons fall back to their ground state, re-emitting the absorbed energy in the form of light. Because each atom has a unique structure and arrangement of electrons, each atom emits a unique spectrum of light. This characteristic light is the basis for the chemical test known as a flame test. In this test the atoms are excited by being placed within a flame. As they re-emit the absorbed energy in the form of light, the color of the flame changes. For most metals, these changes are easily visible. However, even the presence of a tiny speck of another substance can interfere with the identification of the true color of a particular type of atom.

To determine what metal is contained in the barrels behind the Benettis' house, you must first perform flame tests with a variety of standard solutions of different metal compounds. Then you will perform a flame test with the unknown sample from the site to see if it matches any of the solutions you've used as standards. Be sure to keep your equipment very clean, and perform multiple trials to check your work.

Safety Procedures

- Wear safety goggles when working around chemicals, acids, bases, flames, or heating devices. Contents under pressure may become projectiles and cause serious injury.
- Never look directly at the sun through any optical device or use direct sunlight to illuminate a microscope.
- Avoid wearing contact lenses in the lab.
- If any substance gets in your eyes, notify your instructor immediately and flush your eyes with running water for at least 15 min.

- Secure loose clothing, and remove dangling jewelry. Don't wear open-toed shoes or sandals in the lab.
- Wear an apron or lab coat to protect your clothing when working with chemicals.
- If a spill gets on your clothing, rinse it off immediately with water for at least 5 min while notifying your instructor.

- If a chemical gets on your skin or clothing or in your eyes, rinse it immediately, and alert your instructor.
- If a chemical is spilled on the floor or lab bench, alert your instructor, but do not clean it up yourself unless your teacher says it is OK to do so.

- Always use caution when working with chemicals.

- Never mix chemicals unless specifically directed to do so.
- Never taste, touch, or smell chemicals unless specifically directed to do so.
- Add acid or base to water; never do the opposite.
- Never return unused chemicals to the original container.
- Never transfer substances by sucking on a pipette or straw; use a suction bulb.
- Follow instructions for proper disposal.

- Avoid wearing hair spray or hair gel on lab days.
- Whenever possible, use an electric hot plate as a heat source instead of an open flame.
- When heating materials in a test tube, always angle the test tube away from yourself and others.
- Glass containers used for heating should be made of heat-resistant glass.
- Know your school's fire-evacuation routes.

- Clean and decontaminate all work surfaces and personal protective equipment as directed by your instructor.
- Dispose of all sharps (broken glass and other contaminated sharp objects) and other contaminated materials (biological and chemical) in special containers as directed by your instructor.

Data Table 1

Metal Compound	Color of flame	Wavelengths (nm)
$CaCl_2$ solution		
K_2SO_4 solution		
Li_2SO_4 solution		
Na_2SO_4 solution		
$SrCl_2$ solution		
Na_2SO_4 (cobalt glass)		
K_2SO_4 (cobalt glass)		
Na_2SO_4 and K_2SO_4		
Na_2SO_4 and K_2SO_4 (cobalt glass)		
NaCl solution		
NaCl crystals		
Unknown solution		

Materials
Optional Equipment
- spectroscope
- wooden splints

1. To prepare 1.0 M HCl, observe the required precautions. Add 83 mL of concentrated HCl to enough distilled water to make 1.00 L of solution. Add the acid slowly while stirring to avoid overheating.

2. To prepare 0.5 M $CaCl_2$, dissolve 55 g of $CaCl_2$ in enough water to make 1.00 L of solution.

3. To prepare 0.5 M K_2SO_4, dissolve 87 g of K_2SO_4 in enough water to make 1.00 L of solution.

4. To prepare 0.5 M Li_2SO_4, dissolve 64 g of $Li_2SO_4 \cdot H_2O$ in enough water to make 1.00 L of solution.

5. To prepare 0.5 M Na_2SO_4, dissolve 71 g of Na_2SO_4 in enough water to make 1.00 L of solution.

6. To prepare 0.5 M NaCl, dissolve 29 g of NaCl in enough water to make 1.00 L of solution.

7. To prepare 0.5 M $SrCl_2$, dissolve 133.3 g of $SrCl_2 \cdot 6H_2O$ in enough water to make 1.00 L of solution.

8. For the unknown solution, any of the above solutions can be used. Mixtures may prove to be too complicated for students to analyze. Excess solutions can be stored and used next year.

9. For flame-test wire, use either number 24 platinum wire or nichrome wire. Some teachers prefer to use wooden splints for the flame tests.

Safety Cautions

- Read all safety precautions, and discuss them with your students.

- Safety goggles and a lab apron must be worn at all times.

- Tie back long hair and loose clothing when working in the lab.

- Wear goggles, a face shield, impermeable gloves, and a lab apron when you prepare the HCl. Work in a hood known to be in good operating condition, and have another person standing by to call for help in case of an emergency. Work within a 30 second walk of a safety shower and eyewash station.

- Students should not handle concentrated acid solutions. In case of an acid spill, dilute the spill with water. Then mop up the spill with wet cloths or a wet cloth mop designated for spill cleanup. Wear disposable plastic gloves while cleaning spills.

Disposal Information

Set out a disposal container for the students. After all of the waste beakers have been emptied, neutralize the resulting solution with 0.1 M NaOH. When the solution's pH is between 5 and 9, pour it down the drain. If you use wooden splints, collect them after they have been labeled, and reuse them next year for the same compounds.

Procedure

1. Copy the Data Table 1 in your lab notebook. Be sure that you have plenty of room for observations about each test.

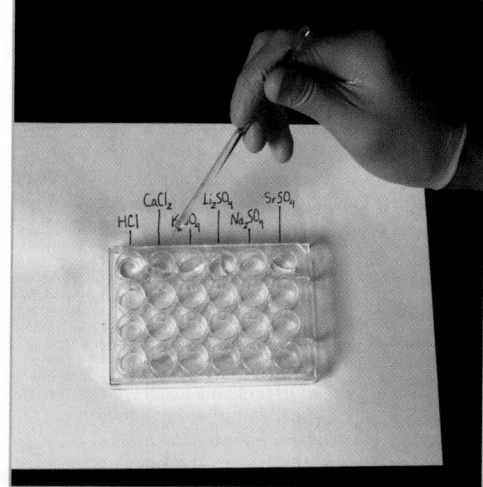

Figure A
Be sure that you record the position of the various metal ion solutions in each well of the well strip.

2. Label a beaker "Waste." Thoroughly clean and dry a well strip. Fill the first well one-fourth full with 1.0 M HCl. Clean the test wire by first dipping it in the HCl and then holding it in the flame of the Bunsen burner. Repeat this procedure until the flame is not colored by the wire. When the wire is ready, rinse the well with distilled water, and collect the rinse water in the waste beaker.

3. Put 10 drops of each metal ion solution listed in the materials list except NaCl in a row in each well of the well strip. Put a row of 1.0 M HCl drops on a glass plate across from the metal ion solutions. Record the position of all of the chemicals placed in the wells. The wire will need to be cleaned thoroughly with HCl between each test solution to avoid contamination from the previous test.

4. Dip the wire into the $CaCl_2$ solution, as shown in **Figure A,** and then hold it in the Bunsen burner flame. Observe the color of the flame, and record it in the data table. Repeat the procedure again, but this time look through the spectroscope to view the results. Record the wavelengths you see from the flame. Perform each test three times. Clean the wire with the HCl as you did in step 2.

5. Repeat step 4 with the K_2SO_4 and with each of the remaining solutions in the well strip. For each solution that you test, record the color of each flame and the wavelength observed with the spectroscope. After the solutions are tested, clean the wire thoroughly, rinse the well strip with distilled water, and collect the rinse water in the waste beaker.

6. Test another drop of Na_2SO_4, but this time view the flame through two pieces of cobalt glass. Clean the wire, and repeat the test by using the K_2SO_4. View the flame through the cobalt glass. Record in your data table the colors and wavelengths of the flames. Clean the wire and the well strip, and rinse the well strip with distilled water. Pour the rinse water into the waste beaker.

7. Put a drop of K_2SO_4 in a clean well. Add a drop of Na_2SO_4. Flame-test the mixture. Observe the flame without the cobalt glass. Repeat the test, this time observing the flame through the cobalt glass. Record the colors and wavelengths of the flames in the data table. Clean the wire, and rinse the well strip with distilled water. Pour the rinse water into the waste beaker.

8. Test a drop of the NaCl solution in the flame, and then view it through the spectroscope. (Do not use the cobalt glass.) Record your observations. Clean the wire, and rinse the well strip with distilled water. Pour the rinse water into the waste beaker. Place a few crystals of NaCl in a clean well, dip the wire in the crystals, and do the flame test once more. Record the color of the flame test. Clean the wire, and rinse the well strip with distilled water. Pour the rinse water into the waste beaker.

Figure B
Flame test

9. Dip the wire into the unknown solution; then hold it in the Bunsen burner flame, as shown in **Figure B.** Perform flame tests for the wire, both with and without the cobalt glass. Record your observations. Clean the wire, and rinse the well strip with distilled water. Pour the rinse water into the waste beaker.

10. Clean all apparatus and your lab station. Dispose of the contents of the waste beaker into the container designated by your teacher. Wash your hands thoroughly after cleaning up the lab area and equipment.

Analysis

1. **Organizing data** Examine your data table, and create a summary of the flame test for each metal ion.

2. **Analyzing data** Account for any differences in the individual trials for the flame tests for the metal ions.

3. **Explaining events** Explain how viewing the flame through cobalt glass can make analyzing the ions being tested easier.

4. **Explaining events** Explain how the lines seen in the spectroscope relate to the position of electrons in the metal atom.

5. **Identifying patterns** For three of the metal ions tested, explain how the flame color you saw relates to the lines of color you saw when you looked through the spectroscope.

Conclusions

6. **Evaluating results** What metal ions are in the unknown solution from the barrels on the vacant lot?

7. **Evaluating methods** How would you characterize the flame test with respect to its sensitivity? What difficulties could occur when identifying ions by the flame test?

8. **Evaluating methods** Explain how you can use a spectroscope to identify the components of solutions containing several different metal ions.

9. **Applying ideas** Some stores sell jars of "fireplace crystals." When sprinkled on a log, these crystals make the flames blue, red, green, and violet. Explain how these crystals can change the flame's color. What ingredients would you expect the crystals to contain?

Extensions

10. **Designing experiments** A student performed flame tests on several unknown substances and observed that they all were shades of red. What could the student do to correctly identify these substances? Explain your answer.

11. **Designing experiments** During a flood, the labels from three bottles of chemicals were lost. The three unlabeled bottles of white solids were known to contain the following substances: strontium nitrate, ammonium carbonate, and potassium sulfate. Explain how you could easily test the substances and relabel the three bottles. (Hint: Ammonium ions do not provide a distinctive flame color.)

775

Answers

(Note: Assign only items 1–3, 6, and 7 if spectroscopes are unavailable.)

1. See the sample data table.

2. Answers will vary. Some students may have had difficulty cleaning the wire properly, so the first test of a new compound may have had traces of the previous compound.

3. The flame color of potassium is purple, but it is so weak that it can be overpowered by the yellow sodium light if a mixture is tested. The cobalt glass screens out the yellow sodium light.

4. Each line in the spectroscope represented the energy emitted as excited electrons moved from a specific high-energy orbital back to their ground state.

5. Answers will vary, but students should realize that the colors seen by the eye were the result of combining the colors of light seen in the line spectra.

6. Answers will vary.

7. The flame test is fairly specific because it can show an easily detectable signal with a very small amount of material. Possible difficulties include problems with contamination and the fact that some metals have similar colors when flame tested.

8. The flame test of the mixture can be examined with the spectroscope. By comparing the lines in the spectra with those for other metals, one can determine which lines are due to which metals.

9. The student should compare the red shades with those of the known samples. Information about spectral lines would also help determine which metal is the unknown.

10. Strontium nitrate will change the flame color to red; potassium sulfate will change the flame color to purple; and ammonium carbonate will not change the flame color.

11. The crystals contain a mixture of metal salts. When sprinkled on a fire, the crystals cause the flame to glow, just as if several flame tests were being performed.

Techniques to Demonstrate

Demonstrate the flame-test technique, including how to clean the flame-test wire. Point out that because the color lasts briefly, several trials may be necessary. If you have spectroscopes, each pair of students can use one to identify the specific lines in the spectra of the light emitted in the flame test. One student can view through the spectroscope while the other performs the flame test. Be sure you explain how to use the spectroscope.

Tips and Tricks

Remind students that it is important to clean the wire between tests to prevent interference.

SPECTROSCOPY AND FLAME TESTS: IDENTIFYING MATERIALS

Teacher's Notes

Time Required
2 lab periods

Ratings
EASY 1 2 3 4 HARD

TEACHER PREPARATION	2
STUDENT SETUP	2
CONCEPT LEVEL	2
CLEANUP	2

Skills Acquired
- Classifying
- Collecting Data
- Communicating
- Designing Experiments
- Experimenting
- Identifying/Recognizing Patterns
- Inferring
- Interpreting
- Organizing and Analyzing Data
- Predicting

The Scientific Method
In this lab, students will:
- Make Observations
- Ask Questions
- Form a Hypothesis
- Test the Hypothesis
- Analyze the Results
- Draw Conclusions
- Communicate the Results

Inquiry LAB
Design Your Own Experiment

3 Spectroscopy and Flame Tests
Identifying Materials

THE PROBLEM

January 27, 2004

Director of Investigations
CheMystery Labs, Inc.
52 Fulton Street
Springfield, VA 22150

Dear Director:

As you may have seen in news reports, one of our freelance pilots, David Matthews, was killed in a crash of an experimental airplane.

The reports did not mention that Matthews's airplane was a recently perfected design that he had developed for us. The notes he left behind indicate that the coating on the nose cone was the key to the plane's speed and maneuverability. Unfortunately, he did not reveal what substances he used, and we were able to recover only flakes of material from the nose cone after the accident.

We have sent you samples of these flakes dissolved in a solution. Please identify the material Matthews used so that we can duplicate his prototype. We will pay $200,000 for this work, provided that you can identify the material within three days.

Sincerely,

Jared MacLaren

Jared MacLaren
Experimental Testing Agency

References
Review information about spectroscopic analysis. The procedure is similar to one your team recently completed to identify an unknown metal in a solution. As before, use small amounts of metal, and clean equipment carefully to avoid contamination. Perform multiple trials for each sample.

The following information is the bright-line emission data (in nm) for the four possible metals.

- Lithium: 670, 612, 498, 462
- Potassium: 700, 695, 408, 405
- Strontium: 710, 685, 665, 500, 490, 485, 460, 420, 405
- Calcium: 650, 645, 610, 485, 460, 445, 420

776

Materials
- beaker, 250 mL
- Bunsen burner, gas tubing, and striker
- cobalt glass plate
- crucible tongs
- flame-test wire, 5 cm
- glass plate (a 7 cm × 15 cm plate or a microchemistry well plate)
- HCl solution, 1.0 M (5 mL)
- unknown solution (0.5 M Li_2SO_4 is recommended)

Optional Equipment
- spectroscope
- wooden splints

THE PLAN

CheMystery Labs, Inc. 52 Fulton Street, Springfield, VA 22150

CheMystery Labs, Inc.
52 Fulton Street
Springfield, VA 22150

Memorandum

Date: January 28, 2004
To: Edwin Thien
From: Marissa Bellinghausen

We have narrowed down the material used to four possibilities. It is a compound of either lithium, potassium, strontium, or calcium. Using flame tests and the wavelengths of spectroscopic analysis, you should be able to identify which of these is in the sample.

Because our contract depends on timeliness, give me a preliminary report that includes the following as soon as possible:
- a detailed, one-page summary of your plan for the procedure
- an itemized list of equipment

After you complete your analysis, prepare a report in the form of a two-page letter to MacLaren. The report must include the following:
- the identity of the metal in the sample
- a summary of your procedure
- a detailed and organized analysis and data sections showing tests and results

Required Precautions

- Wear safety goggles when working around chemicals, acids, bases, flames, or heating devices. Contents under pressure may become projectiles and cause serious injury.
- Avoid wearing contact lenses in the lab.
- If any substance gets in your eyes, notify your instructor immediately and flush your eyes with running water for at least 15 min.
- Secure loose clothing, and remove dangling jewelry. Don't wear open-toed shoes or sandals in the lab.
- Wear an apron or lab coat to pro-

tect your clothing when working with chemicals.
- If a spill gets on your clothing, rinse it off immediately with water for at least 5 min while notifying your instructor.
- Always use caution when working with chemicals.
- Never taste, touch, or smell chemicals unless specifically directed to do so.
- Follow instructions for proper disposal.
- Whenever possible, use an electric hot plate as a heat source instead of an open flame.

- When heating materials in a test tube, always angle the test tube away from yourself and others.
- Know your school's fire-evacuation routes.
- Clean and decontaminate all work surfaces and personal protective equipment as directed by your instructor.
- Dispose of all sharps (broken glass and other contaminated sharp objects) and other contaminated materials (biological and chemical) in special containers as directed by your instructor.

777

Sample Data

Substance	UNKNOWN
Flame color	RED (CARMINE)
Wavelengths (nm)	462, 498, 612, 670

Student data should have multiple trials. Data is shown for Li_2SO_4. Data will vary if a different unknown is chosen.

Checkpoints
Checkpoint 1: By the end of the first class period, students should turn in a detailed one-page plan/procedure for approval. This should be compared to the sample procedure found in the answer section of the wrap. The student procedure should be evaluated according to organization and accuracy and how well it follows the steps of the scientific method.

Checkpoint 2: During the second class period, students should revise procedures according to the teacher's approval and begin the procedure. By the end of the second class period, students should be able to draw a reasonable conclusion and determine the feasibility of identifying the metal in the samples. Students should be evaluated on lab technique, quality and clarity of observations, and the explanation of observations/conclusions.

Sample Answer to Procedure
Place three drops of the unknown solution and several drops of 1.0 M HCl on a glass plate. Dip the flame-test wire in a drop of HCl to clean it. Dip the flame-test wire into the first drop of unknown solution. Hold the wire in a Bunsen burner flame, and record the color. (Record the wavelength as well if a spectroscope is available.) Clean the wire in another drop of HCl, and repeat the process. When finished, rinse the plate with distilled water, and collect the rinse water in a waste beaker.

THE MENDELEEV LAB OF 1869

Teacher's Notes

Time Required

1 lab period if alternate procedure is used

Ratings

EASY 1 2 3 4 HARD

TEACHER PREPARATION	1
STUDENT SETUP	1
CONCEPT LEVEL	2
CLEANUP	1

Skills Acquired

• Classifying
• Constructing Models
• Inferring
• Organizing and Analyzing Data
• Predicting

The Scientific Method

In this lab, students will:
• Make Observations
• Analyze the Results

Materials

Li, He, Na, Ca, Ne, Br$_2$, K, I$_2$, Si, Ge, and Mg can be added if available. See the Chapter Resource File for the "Periodic Table" chapter for element cards.

1. When you are ready to begin the lab, place the element cards of the knowns and unknowns around the classroom. Give the students blank note cards so that they can write down the properties of each element. Elements that are present in samples require students to make observations about their physical state or color. These observations should be written on

Skills Practice Lab

OBJECTIVES

◆ Observe the physical properties of common elements.

◆ Observe the properties and trends in the elements on the periodic table.

◆ Draw conclusions and identify unknown elements based on observed trends in properties.

USING SCIENTIFIC **METHODS**

MATERIALS

◆ blank periodic table

◆ elemental samples of Ar, C, Cu, Sn, and Pb

◆ note cards, 3 × 5

◆ periodic table

4 The Mendeleev Lab of 1869

Introduction

Russian chemist Dmitri Mendeleev is generally credited as being the first chemist to observe that patterns emerge when the elements are arranged according to their properties. Mendeleev's arrangement of the elements was unique because he left blank spaces for elements that he claimed were undiscovered as of 1869. Mendeleev was so confident that he even predicted the properties of these undiscovered elements. His predictions were eventually proven to be quite accurate, and these new elements fill the spaces that originally were blank in his table.

Use your knowledge of the periodic table to determine the identity of each of the nine unknown elements in this activity. The unknown elements are from the groups in the periodic table that are listed below. Each group listed below contains at least one unknown element.

1   2   11   13   14   17   18

None of the known elements serves as one of the nine unknown elements.

No radioactive elements are used during this experiment. The relevant radioactive elements include Fr, Ra, At, and Rn. You may not use your textbook or other reference materials. You have been provided with enough information to determine each of the unknown elements.

the students' note cards. The other information can be copied directly from the element cards.

2. When all of the element cards have been copied, the students should organize the known elements. The cards can be organized in the form of a periodic table. Photocopy this model for each group and have the groups use it as a guide for arranging the elements. For example, the cards for lithium, sodium, and potassium should be arranged in a vertical column so students can observe the trends in their properties. While this organizational strategy seems effective, the students may wish to create

their own methods. This is fine as long as students make progress toward identifying the unknowns. Students should note that the elements highlighted in red are radioactive.

3. To save time, an alternate procedure may be used. Instead of placing element cards around the classroom, photocopy the pages showing the element cards and distribute one copy to each group. Designate one person from each group to cut the pages into individual element cards. When all of the element cards have been cut out, the students should organize the known elements in the form of a periodic table showing elements with similar properties in the same group.

Safety Procedures

- Wear safety goggles when working around chemicals, acids, bases, flames, or heating devices. Contents under pressure may become projectiles and cause serious injury.
- Never look directly at the sun through any optical device or use direct sunlight to illuminate a microscope.
- Avoid wearing contact lenses in the lab.

- If any substance gets in your eyes, notify your instructor immediately and flush your eyes with running water for at least 15 min.
- Secure loose clothing, and remove dangling jewelry. Don't wear open-toed shoes or sandals in the lab.
- Wear an apron or lab coat to protect your clothing when working with chemicals.
- If a spill gets on your clothing, rinse it off immediately with water for at least 5 minutes while notifying your instructor.

Safety Cautions
- Read all safety precautions, and discuss them with your students.
- Safety goggles, gloves, and a lab apron must be worn at all times.
- Students should wash their hands immediately and avoid touching their eyes if they accidentally touch the tin or lead samples. These elements pose a health risk if they enter a student's body.

Disposal Information
None. The samples and the set of element cards can be kept for future use.

Answers to Conclusion

Unknown #	Element
1	Si
2	F
3	Rb
4	Kr
5	Au
6	Sr
7	Ge
8	Mg
9	Tl

Data Table 1

Unknown	Element
1	
2	
3	
4	

Procedure

1. Copy the data table in your lab notebook. Be sure that you have plenty of room for observations about each test.

2. Use the note cards to copy the information listed on each of the sample cards in the worksheets that your teacher has given you. If the word *observe* is listed, you will need to visually inspect the sample and then write the observation in the appropriate space.

3. Arrange the note cards of the known elements in a crude representation of the periodic table. In other words, all of the known elements from Group 1 should be arranged in the appropriate order. Arrange all of the other cards accordingly.

4. Once the cards of the known elements are in place, inspect the properties of the unknowns to see where their properties would best "fit" the trends of the elements of each group.

5. Assign the proper element name to each of the unknowns. Add the symbol for each one of the unknown elements to your data table.

6. Clean up your lab station, and return the leftover note cards and samples of the elements to your teacher. Do not pour any of the samples down the drain or in the trash unless your teacher directs you to do so. Wash your hands thoroughly before you leave the lab and after all your work is finished.

Conclusions

1. **Interpreting information** Summarize your group's reasoning for the assignment of each unknown. Explain in a few sentences exactly how you predicted the identity of the nine unknown elements.

779

Tips and Tricks

Discuss the work of Dmitri Mendeleev and the concept of periodicity with your students before starting this activity. In your discussion, emphasize the trends among the groups of elements in the periodic table, but do not explain the chemical basis for these trends, because doing so might distract the students from observing the various patterns and trends in the properties.

This experiment can seem overwhelming to students because of the large number of elements and properties. Students should work in groups of 3–4 students, where they can verbalize their reasoning as they attempt to identify the unknowns. Encourage students to divide the experiment into smaller tasks such as data collection, organization of the known elements, and identification of the unknowns. Minimize some of the initial frustration by directing students' attention to the trends in the properties among the known alkali metals. Seeing the trends in the properties of lithium, sodium, and potassium should help students as they try to identify the nine unknown elements.

PERCENT COMPOSITION OF HYDRATES

Teacher's Notes

Time Required

1 lab period

Ratings

EASY 1 2 3 4 HARD

TEACHER PREPARATION	2
STUDENT SETUP	1
CONCEPT LEVEL	2
CLEANUP	2

Skills Acquired

- Collecting Data
- Identifying/Recognizing Patterns
- Measuring
- Organizing and Analyzing Data

The Scientific Method

In this lab, students will:
- Make Observations
- Analyze the Results
- Draw Conclusions
- Communicate Results

Materials

If you do not have a desiccator, you can use a jar with a lid or a similar container. To make a resting place for a crucible within the jar, use a pipe-stem triangle. Bend the ends of the pipe-stem triangle down to form legs. Fill the bottom of the jar with granular anhydrous $CaCl_2$.

Skills Practice Lab

OBJECTIVES

- Demonstrate proficiency in using the balance and the Bunsen burner.
- Determine that all the water has been driven from a hydrate by heating your sample to a constant mass.
- Relate results to the law of conservation of mass and the law of multiple proportions. *USING SCIENTIFIC* **METHODS**
- Perform calculations by using the molar mass.
- Analyze the results and determine the empirical formula of the hydrate and its percentage by mass of water.

MATERIALS

- balance
- Bunsen burner
- crucible and cover
- crucible tongs
- $CuSO_4$, hydrated crystals
- desiccator
- distilled water
- dropper or micropipet
- ring and pipe-stem triangle
- ring stand
- spatula
- stirring rod, glass
- weighing paper

780

7 Percent Composition of Hydrates

Introduction

You are a research chemist working for a company that is developing a new chemical moisture absorber and indicator. The company plans to seal the moisture absorber into a transparent, porous pouch attached to a cellophane window on the

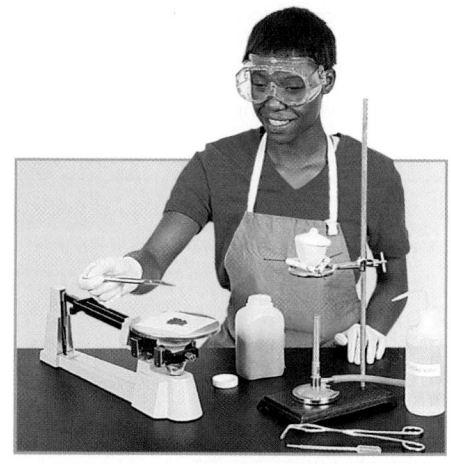

inside of packages for compact disc players. This way, moisture within the packages will be absorbed, and any package that has too much moisture can be quickly detected and dried out. Your company's efforts have focused on copper(II) sulfate, $CuSO_4$, which can absorb water to become a hydrate that shows a distinctive color change.

When many ionic compounds are crystallized from a water solution, they include individual water molecules as part of their crystalline structure. If the substances are heated, this water of crystallization may be driven off and leave behind the pure anhydrous form of the compound. Because the law of multiple proportions also applies to crystalline hydrates, the number of moles of water driven off per mole of the anhydrous compound should be a simple whole-number ratio. You can use this information to help you determine the formula of the hydrate.

To help your company decide whether $CuSO_4$ is the right substance for the moisture absorber and indicator, you will need to examine the hydrated and anhydrous forms of the compound and determine the following:
- the empirical formula of the hydrate, including its water of crystallization
- if the compound is useful as an indicator when it changes from the hydrated to the anhydrous form
- the mass of water absorbed by the 25 g of anhydrous compound, which the company proposes to use

Even if you can guess what the formula for the hydrate should be, carefully perform this lab so that you know how well your company's supply of $CuSO_4$ absorbs moisture.

Safety Cautions

- Read all safety precautions, and discuss them with your students.
- Safety goggles and a lab apron must be worn at all times.
- Remind students to confine long hair and loose clothing before lighting a burner.
- Remind students that heated objects can be hot enough to burn even if they look cool. Students should always use crucible tongs to handle crucibles and crucible covers.

Disposal Information

Provide a labeled container for the disposal of the rehydrated and anhydrous copper sulfate and any excess of the original compound. Later, redissolve the contents of the container in distilled water. Let the solution evaporate, and then recover the crystals for reuse next year. Do not dispose of copper sulfate in a landfill or an incinerator, and do not pour it down a drain.

Safety Procedures

- Wear safety goggles when working around chemicals, acids, bases, flames, or heating devices. Contents under pressure may become projectiles and cause serious injury.
- Never look directly at the sun through any optical device or use direct sunlight to illuminate a microscope.
- Avoid wearing contact lenses in the lab.
- If any substance gets in your eyes, notify your instructor immediately and flush your eyes with running water for at least 15 min.

- Secure loose clothing, and remove dangling jewelry. Don't wear open-toed shoes or sandals in the lab.
- Wear an apron or lab coat to protect your clothing when working with chemicals.
- If a spill gets on your clothing, rinse it off immediately with water for at least 5 min while notifying your instructor.

- If a chemical gets on your skin or clothing or in your eyes, rinse it immediately, and alert your instructor.
- If a chemical is spilled on the floor or lab bench, alert your instructor, but do not clean it up yourself unless your teacher says it is OK to do so.

- Always use caution when working with chemicals.
- Never mix chemicals unless specifically directed to do so.
- Never taste, touch, or smell chemicals unless specifically directed to do so.

- Add acid or base to water; never do the opposite.
- Never return unused chemicals to the original container.
- Never transfer substances by sucking on a pipette or straw; use a suction bulb.
- Follow instructions for proper disposal.

- Avoid wearing hair spray or hair gel on lab days.
- Whenever possible, use an electric hot plate as a heat source instead of an open flame.
- When heating materials in a test tube, always angle the test tube away from yourself and others.
  Glass containers used for heating should be made of heat-resistant glass.
- Know your school's fire-evacuation routes.

- Check the condition of glassware before and after using it. Inform your teacher of any broken, chipped, or cracked glassware, because it should not be used.
- Do not pick up broken glass with your bare hands. Place broken glass in a specially designated disposal container.

- Clean and decontaminate all work surfaces and personal protective equipment as directed by your instructor.
- Dispose of all sharps (broken glass and other contaminated sharp objects) and other contaminated materials (biological and chemical) in special containers as directed by your instructor.

Tips and Tricks

Students should be familiar with the concepts of empirical formula and percentage by mass as well as the process of mass-mole conversions. Remind students that the symbol in a hydrate is not a multiplication symbol.

To review the law of conservation of mass and the law of multiple proportions, write the following reaction on the board.

$$CuSO_4 \cdot nH_2O \rightarrow CuSO_4 + nH_2O$$

According to the law of conservation of mass, the mass will be the same on both sides of the equation. This allows the mass of water to be calculated by comparing the mass of the substance before and after the reaction. According to the law of multiple proportions, n, the number of water molecules per formula unit of $CuSO_4$, should be a small whole number.

Data Table 1

Mass of empty crucible and cover	
Initial mass of sample, crucible, and cover	
Mass of sample, crucible, and cover after first heating	
Mass of sample, crucible, and cover after second heating	
Constant mass of sample, crucible, and cover	

781

Techniques to Demonstrate

Students will need to be given detailed instructions on how to use a crucible; how to set it on the pipe-stem triangle; and how to place the crucible lid on top of the crucible, leaving a small opening. Be certain they realize that they should always use crucible tongs to touch the crucible, even when the crucible is cool. If they touch the crucible, oils from their hands can contribute to errors in mass measurements. Remind students to gently break up the crystals with a stirring rod after placing them in the crucible. This will prevent the crucible from cracking and breaking as it is heated.

Students should use the same balance throughout this lab. Remind students not to put hot crucibles on balances.

Students should reheat the crucible, cover, and anhydrous copper sulfate until successive masses are within 0.02 g.

Emphasize the need to heat the copper sulfate very slowly. If it is heated too rapidly, it may splatter or decompose. As noted in the procedure, any yellowish color that persists after the cooling is a sign of decomposition of the sulfate due to overheating.

The crystals may turn slightly yellow when heated, but this color should not remain after cooling. A wing top on the burner will produce a steady, warm flame that can help prevent decomposition.

Procedure

1. Copy Data Table 1 in your lab notebook. Be sure that you have plenty of room for observations about each test.

2. Make sure that your equipment is very clean so that you will get the best possible results. Once you have heated the crucible and cover, do not touch them with your bare hands. Remember that you will need to cool the heated crucible in the desiccator before you measure its mass. Never put a hot crucible on a balance; it will damage the balance.

3. Place the crucible and cover on the triangle with the lid slightly tipped, as shown in **Figure A.** The small opening will allow gases to escape. Heat the crucible and cover until the crucible glows slightly red. Use the tongs to transfer the crucible and cover to the desiccator, and allow them to cool for 5 min. Determine the mass of the crucible and cover to the nearest 0.01 g, and record the mass in your data table.

Figure A

4. Using a spatula, add approximately 5 g of copper sulfate hydrate crystals to the crucible. Break up any large crystals before placing them in the crucible. Determine the mass of the covered crucible and crystals to the nearest 0.01 g, and record the mass in your data table.

5. Place the crucible with the copper sulfate hydrate on the triangle, and again position the cover so there is only a small opening. If the opening is too large, the crystals may spatter as they are heated. Heat the crucible very gently on a low flame to avoid spattering. Increase the temperature gradually for 2 or 3 min, and then heat until the crucible glows red for at least 5 min. Be very careful not to raise the temperature of the crucible and its contents too suddenly. You will observe a color change, which is normal, but if the substance remains yellow after cooling, it was overheated and has begun to decompose. Allow the crucible, cover, and contents to cool for 5 min in the desiccator, and then measure their mass. Record the mass in your data table.

6. Heat the covered crucible and contents to redness again for 5 min. Allow the crucible, cover, and contents to cool in the desiccator, and then determine their mass and record it in the data table. If the two mass measurements differ by no more than 0.01 g, you may assume that all of the water has been driven off. Otherwise, repeat the process until the mass no longer changes, which indicates that all of the water has evaporated. Record this constant mass in your data table.

7. After recording the constant mass, set aside a part of your sample on a piece of weighing paper. Using the dropper or pipet, as shown in **Figure B,** put a few drops of water onto this sample to rehydrate the crystals. Record your observations in your lab notebook.

Figure B

8. Clean all apparatus and your lab station. Make sure to completely shut off the gas valve before leaving the laboratory. Remember to wash your hands thoroughly. Place the rehydrated and anhydrous chemicals in the disposal containers designated by your teacher.

Analysis

1. Explaining events
Why do you need to heat the clean crucible before using it in this lab? Why do the tongs used throughout this lab need to be especially clean?

2. Explaining events
Why do you need to use a cover for the crucible? Could you leave the cover off each time you measure the mass of the crucible and its contents and still get accurate results? Explain your answer.

3. Examining data
Calculate the mass of anhydrous copper sulfate (the residue that remains after heating to constant mass) by subtracting the mass of the empty crucible and cover from the mass of the crucible, cover, and heated $CuSO_4$. Use the molar mass for $CuSO_4$, determined from the periodic table, to calculate the number of moles present.

4. Analyzing data
Calculate the mass and moles of water originally present in the hydrate by using the molar mass determined from the periodic table.

Conclusions

5. Interpreting information Explain why the mass of the sample decreased after it was heated, despite the law of conservation of mass.

6. Drawing conclusions Using your answers from items 3 and 4, determine the empirical formula for the copper sulfate hydrate.

7. Analyzing results What is the percentage by mass of water in the original hydrated compound?

8. Applying conclusions How much water could 25 g of anhydrous $CuSO_4$ absorb?

9. Applying conclusions When you rehydrated the small amount of anhydrous copper sulfate, what were your observations? Explain whether this substance would make a good indicator of moisture.

10. Applying conclusions Some cracker tins include a glass vial of drying material in the lid. This is often a mixture of magnesium sulfate and cobalt chloride. As the mixture absorbs moisture to form hydrated compounds, the cobalt chloride changes from blue-violet $CoCl_2 \cdot 2H_2O$ to pink $CoCl_2 \cdot 6H_2O$. When this hydrated mixture becomes totally pink, it can be restored to the dihydrate form by being heated in the oven. Write equations for the reactions that occur when this mixture is heated.

11. Drawing conclusions Three pairs of students obtained the results in the table below when they heated a solid. In each case, the students observed that when they began to heat the solid, drops of a liquid formed on the sides of the test tube.
   a. Could the solid be a hydrate? Explain how you could find out.
   b. If the solid has a molar mass of 208 g/mol after being heated, how many formula units of water are there in one formula unit of the unheated compound?

Data Table 2

Sample number	Mass before heating (g)	Constant mass after heating (g)
1	1.92	1.26
2	2.14	1.40
3	2.68	1.78

Extensions

12. Designing experiments Some electronic equipment is packaged for shipping with a small packet of drying material. You are interested in finding out whether the electronic equipment was exposed to moisture during shipping. How could you determine this?

783

Answers to Analysis

1. The crucible was heated to be certain it was completely dry. The tongs had to be clean so that they did not transfer any dirt or debris to the crucible. The presence of any water, dirt, or debris causes error in the mass measurements.

2. The cover was used to keep debris out of the crystals. It also helped prevent water from condensing on the crystals as they cooled. Without the cover, each mass measurement would have been slightly larger because the crystals would have absorbed water.

3. 3.24g $CuSO_4$; 2.03×10^{-2} mol $CuSO_4$

4. 1.76g H_2O; 0.0977 mol H_2O

Answers to Conclusions

5. As the sample was heated, water evaporated from the crystal to become water vapor. The mass of the water vapor and the crystals was the same as the mass of the hydrated crystals, so the law of conservation of mass was not violated.

6. $CuSO_4 \cdot 5H_2O$

7. 35.2%

8. 13.6g

9. Anhydrous copper sulfate is a white powder. When water is added, copper sulfate turns blue. This dramatic color change makes it a good indicator of moisture.

10. $CoCl_2 \cdot 6H_2O \rightarrow CoCl_2 \cdot 2H_2O + 4H_2O$

11. a. The solid could be a hydrate, because it has less mass after heating. If the ratio of the amount of the water apparently lost to the amount of the apparently anhydrous compound is a small, whole-number ratio that remains constant with different samples, the solid is a hydrate.

b. 6

Answers to Extension

1. The packet of drying material could be weighed and then heated in a drying oven. After cooling, the packet of drying material would be re-weighed. If the second mass is less than the first mass, the electronic equipment was exposed to moisture.

2. Drying materials are used in many foods. Amorphous silica is used in powdered foods such as table salt and powdered hot chocolate.

HYDRATES: GYPSUM AND PLASTER OF PARIS

Teacher's Notes

Time Required

2 lab periods

Ratings

1 — 2 — 3 — 4
EASY — HARD

TEACHER PREPARATION	2
STUDENT SETUP	2
CONCEPT LEVEL	2
CLEANUP	1

Skills Acquired

- Collecting Data
- Communicating
- Designing Experiments
- Experimenting
- Interpreting
- Measuring
- Organizing and Analyzing Data

The Scientific Method

In this lab, students will:
- Make Observations
- Ask Questions
- Form a Hypothesis
- Test the Hypothesis
- Analyze the Results
- Draw Conclusions
- Communicate the Results

Materials

- balance (0.01 g accuracy)
- Bunsen burner
- crucible tongs (not beaker tongs)
- crucibles and covers (2)
- desiccator
- gypsum sample (5 g)
- mortar and pestle
- plaster of Paris sample (5 g)
- ring stand, ring, pipe-stem triangle
- spatula

7 Hydrates
Gypsum and Plaster of Paris

Director of Research
CheMystery Labs, Inc.
52 Fulton Street
Springfield, VA 22150

Dear Director:

Lost Art Gypsum Mine previously sold its raw gypsum to a manufacturing company that used the gypsum to make anhydrous calcium sulfate, $CaSO_4$ (a desiccant), and plaster of Paris. That company has now gone out of business, and we are currently negotiating the purchase of the firm's equipment to process our own gypsum into $CaSO_4$ and plaster of Paris.

Your company has been recommended to plan the large-scale industrial process for our new plant. We will need a detailed report on the development of the process and formulas for these products. This report will be presented to the bank handling our loan for the new plant. As we discussed on the telephone today, we are willing to pay you $250,000 for the work, and the contract papers will arrive under separate cover today.

Sincerely,

Alex Farros

Alex Farros
Vice President
Lost Art Gypsum Mine

References

Review information about hydrates and water of crystallization. Gypsum and plaster of Paris are hydrated forms of calcium sulfate, $CaSO_4$. One of the largest gypsum mines in the world is located outside Paris, France. Plaster of Paris contains less water of crystallization than gypsum. Plaster of Paris is commonly used in plaster walls and art sculptures.

784

Safety Cautions

- Read all safety precautions, and discuss them with your students.

- Safety goggles and a lab apron must be worn at all times.

- Remind students to confine long hair and loose clothing before lighting a burner.

- Remind students that heated objects can be hot enough to burn even if they look cool. Students should always use crucible tongs to handle crucibles and crucible covers.

THE PLAN

CheMystery Labs, Inc. 52 Fulton Street, Springfield, VA 22150

CheMystery Labs, Inc.
52 Fulton Street
Springfield, VA 22150

Memorandum

Date: February 10, 2004
To: Kenesha Smith
From: Martha Li-Hsien

Your team needs to develop a procedure to experimentally determine the correct empirical formulas for both hydrates of this anhydrous compound. You will use gypsum samples from the mine and samples of the plaster of Paris product.

As soon as possible, I need a preliminary report from you that includes the following:
- a detailed one-page summary of your plan for the procedure, including all necessary data tables
- an itemized list of equipment

After you complete the analysis, prepare a two-page report that includes the following information:
- formulas for anhydrous calcium sulfate, plaster of Paris, and gypsum
- a summary of your procedure
- detailed and organized data and analysis sections that show calculations, along with estimates and explanations of any possible sources of error

Required Precautions

- Wear safety goggles when working around chemicals, acids, bases, flames, or heating devices. Contents under pressure may become projectiles and cause serious injury.
- Avoid wearing contact lenses in the lab.
- If any substance gets in your eyes, notify your instructor immediately and flush your eyes with running water for at least 15 min.
- Secure loose clothing, and remove dangling jewelry. Don't wear open-toed shoes or sandals in the lab.

- Wear an apron or lab coat to pro-

tect your clothing when working with chemicals.
- If a spill gets on your clothing, rinse it off immediately with water for at least 5 min while notifying your instructor.

- Always use caution when working with chemicals.
- Never taste, touch, or smell chemicals unless specifically directed to do so.
- Follow instructions for proper disposal.
- Whenever possible, use an electric hot plate as a heat source instead of an open flame.

- When heating materials in a test tube, always angle the test tube away from yourself and others.
- Know your school's fire-evacuation routes.
- Clean and decontaminate all work surfaces and personal protective equipment as directed by your instructor.
- Dispose of all sharps (broken glass and other contaminated sharp objects) and other contaminated materials (biological and chemical) in special containers as directed by your instructor.

785

Checkpoints

Checkpoint 1: By the end of the first class period, students should turn in a detailed one-page plan/procedure for approval. This should be compared to the sample procedure found in the answer section of the wrap. The student procedure should be evaluated according to organization and accuracy and how well it follows the steps of the scientific method.

Checkpoint 2: During the second class period, students should revise procedures according to the teacher's approval and begin the procedure. By the end of the second class period, students should be able to draw a reasonable conclusion and determine formulas for anhydrous calcium sulfate, plaster of Paris, and gypsum. Students should be evaluated on lab technique, quality and clarity of observations, and the explanation of observations/conclusions.

Sample Answer to Procedure

Heat a crucible and lid, cool them, and measure their mass. Add some crushed crystals of gypsum. Measure the mass, and then heat gently. Cool and measure the new mass. Keep reheating until the mass is fairly constant. Repeat for the plaster of Paris.

Sample Data

	Gypsum	Plaster of Paris
Mass of empty crucible and lid	32.18 g	32.18 g
Mass of crucible, lid, and compound	37.18 g	37.18 g
Mass of crucible, lid, and compound after first heating	36.10 g	36.91 g
Mass of crucible, lid, and compound after second heating	36.08 g	36.89 g

STOICHIOMETRY AND GRAVIMETRIC: ANALYSIS

Teacher's Notes

Time Required

2 lab periods (includes overnight drying time)

Ratings

1 2 3 4
EASY   HARD

TEACHER PREPARATION	2
STUDENT SETUP	2
CONCEPT LEVEL	3
CLEANUP	2

Skills Acquired

- Collecting Data
- Experimenting
- Interpreting
- Measuring
- Organizing and Analyzing Data

The Scientific Method

In this lab, students will:
- Make Observations
- Analyze the Results
- Draw Conclusions

Materials

1. To prepare 0.3 M $SrCl_2$, dissolve 80.0 g of $SrCl_2 \cdot 6H_2O$ in enough H_2O to make 1.00 L of solution.

2. For the unknown, 0.5 M Na_2CO_3 is recommended. To prepare 0.5 M Na_2CO_3, dissolve 53.0 g of Na_2CO_3 in enough H_2O to make 1.00 L of solution.

Skills Practice Lab

OBJECTIVES

- Observe the reaction between strontium chloride and sodium carbonate, and write a balanced equation for the reaction.

- Demonstrate proficiency with gravimetric methods.

- Measure the mass of insoluble precipitate formed.  USING SCIENTIFIC METHODS

- Relate the mass of precipitate formed to the mass of reactants before the reaction.

- Calculate the mass of sodium carbonate in a solution of unknown concentration.

MATERIALS

- balance
- beaker tongs
- beakers, 250 mL (3)
- distilled water
- drying oven
- filter paper
- glass funnel or Büchner funnel
- glass stirring rod
- graduated cylinder, 100 mL
- Na_2CO_3 solution (15 mL)
- ring and ring stand
- rubber policeman
- spatula
- $SrCl_2$ solution, 0.30 M (45 mL)
- water bottle

---

9 Stoichiometry and Gravimetric Analysis

Introduction

You are working for a company that makes water-softening agents for homes with hard water. Recently, there was a mix-up on the factory floor, and sodium carbonate solution was mistakenly mixed in a vat with an unknown quantity of distilled water. You must determine the amount of Na_2CO_3 in the vat in order to predict the percentage yield of the water-softening product.

When chemists are faced with problems that require them to determine the quantity of a substance by mass, they often use a technique called gravimetric analysis. In this technique, a small sample of the material undergoes a reaction with an excess of another reactant. The chosen reaction is one that almost always provides a yield near 100%. If the mass of the product is carefully measured, you can use stoichiometry calculations to determine how much of the reactant of unknown amount was involved in the reaction. Then by comparing the size of the analysis sample with the size of the original material, you can determine exactly how much of the substance is present.

This procedure involves a double-displacement reaction between strontium chloride, $SrCl_2$, and sodium carbonate, Na_2CO_3. In general, this reaction can be used to determine the amount of any carbonate compound in a solution.

You will react an unknown amount of sodium carbonate with an excess of strontium chloride. After purifying the product, you will determine the following:

- how much product is present
- how much Na_2CO_3 must have been present to produce that amount of product
- how much Na_2CO_3 is contained in the 575 L of solution

Safety Procedures

- Wear safety goggles when working around chemicals, acids, bases, flames, or heating devices. Contents under pressure may become projectiles and cause serious injury.
- Never look directly at the sun through any optical device or use direct sunlight to illuminate a microscope.
- Avoid wearing contact lenses in the lab.
- If any substance gets in your eyes, notify your instructor immediately and flush your eyes with running water for at least 15 min.

- Secure loose clothing, and remove dangling jewelry. Do not wear open-toed shoes or sandals in the lab.
- Wear an apron or lab coat to protect your clothing when working with chemicals.
- If a spill gets on your clothing, rinse it off immediately with water for at least 5 min while notifying your instructor.

- Always use caution when working with chemicals.
- Never mix chemicals unless specifically directed to do so.
- Never taste, touch, or smell chemicals unless specifically directed to do so.

- Add acid or base to water; never do the opposite.
- Never return unused chemicals to the original container.
- Never transfer substances by sucking on a pipette or straw; use a suction bulb.
- Follow instructions for proper disposal.

- Avoid wearing hair spray or hair gel on lab days.
- Whenever possible, use an electric hot plate as a heat source instead of an open flame.
- When heating materials in a test tube, always angle the test tube away from yourself and others.
- Glass containers used for heating should be made of heat-resistant glass.
- Know your school's fire-evacuation routes.

- Clean and decontaminate all work surfaces and personal protective equipment as directed by your instructor.
- Dispose of all sharps (broken glass and other contaminated sharp objects) and other contaminated materials (biological and chemical) in special containers as directed by your instructor.

Safety Cautions

- Read all safety precautions, and discuss them with your students.
- Safety goggles and a lab apron must be worn at all times.
- Remind students that heated objects can be hot enough to burn even if they look cool. Students should always use beaker tongs to place samples in a drying oven.

Disposal Information

Be sure to set out two disposal containers, one for solids and another for liquids. The solids can be disposed of in the trash. The liquids can be washed down the drain with plenty of water.

Data Table 1

Volume of Na_2CO_3 solution added
Volume of $SrCl_2$ solution added
Mass of dry filter paper
Mass of beaker with paper towel
Mass of beaker with paper towel, filter paper, and precipitate
Mass of precipitate

Procedure

1. Organizing Data

Copy the data table in your lab notebook. Be sure that you have plenty of room for observations about each test.

2. Clean all of the necessary lab equipment with soap and water. Rinse each piece of equipment with distilled water.

3. Measure the mass of a piece of filter paper to the nearest 0.01 g, and record this value in your data table.

4. Refer to page 764 to set up a filtering apparatus, either a Büchner funnel or a gravity filtration, depending on what equipment is available.

5. Label a paper towel with your name, your class, and the date. Place the towel in a clean, dry 250 mL beaker, and measure and record the mass of the towel and beaker to the nearest 0.01 g.

787

Review the procedures for the filtration technique your students will be using. Remind students of the importance of using clean glassware and avoiding loss of product.

Tips and Tricks

Thoroughly discuss mass-mass stoichiometry and its application in the laboratory. It would be useful to perform calculations with sample data before performing the lab.

6. Measure about 15 mL of the Na_2CO_3 solution into the graduated cylinder. Record this volume to the nearest 0.5 mL in your data table. Pour the Na_2CO_3 solution into a clean, empty 250 mL beaker. Carefully wash the graduated cylinder, and rinse it with distilled water.

7. Measure about 25 mL of the 0.30 M $SrCl_2$ solution into the graduated cylinder. Record this volume to the nearest 0.5 mL in your data table. Pour the $SrCl_2$ solution into the beaker with the Na_2CO_3 solution, as shown in **Figure A.** Gently stir the solution and precipitate with a glass stirring rod.

Figure A
Graduated cylinder pouring solution into beaker.

8. Carefully measure another 10 mL of $SrCl_2$ into the graduated cylinder. Record the volume to the nearest 0.5 mL in your data table. Slowly add it to the beaker. Repeat this step until no more precipitate forms.

9. Once the precipitate has settled, slowly pour the mixture into the funnel. Be careful not to overfill the funnel because some of the precipitate could be lost between the filter paper and the funnel. Use the rubber policeman to transfer as much of the precipitate into the funnel as possible.

10. Rinse the rubber policeman into the beaker with a small amount of distilled water, and pour this solution into the funnel. Rinse the beaker several more times with small amounts of distilled water, as shown in **Figure B.** Pour the rinse water into the funnel each time.

Figure B
Washing a beaker with water bottle

11. After all of the solution and rinses have drained through the funnel, slowly rinse the precipitate on the filter paper in the funnel with distilled water to remove any soluble impurities.

12. Carefully remove the filter paper from the funnel, and place it on the paper towel that you have labeled with your name. Unfold the filter paper, and place the paper towel, filter paper, and precipitate in the rinsed beaker. Then place the beaker in the drying oven. For best results, allow the precipitate to dry overnight.

13. Using beaker tongs, remove your sample from the drying oven, and allow it to cool. Measure and record the mass of the beaker with paper towel, filter paper, and precipitate to the nearest 0.01 g.

14. Dispose of the precipitate in a designated waste container. Pour the filtrate in the other 250 mL beaker into the designated waste container. Clean up the lab and all equipment after use, and dispose of substances according to your teacher's instructions. Wash your hands thoroughly after all lab work is finished and before you leave the lab.

788

Analysis

1. Organizing Data
Write a balanced equation for the reaction. What is the precipitate? Write its empirical formula. (Hint: It was a double-displacement reaction.)

2. Examining Data
Calculate the mass of the dry precipitate. Calculate the number of moles of precipitate produced in the reaction. (Hint: Use the results from step 13.)

3. Examining Data
How many moles of Na_2CO_3 were present in the 15 mL sample?

Conclusions

4. Evaluating Methods
There was 0.30 mol of $SrCl_2$ in every liter of solution. Calculate the number of moles of $SrCl_2$ that were added. Determine whether $SrCl_2$ or Na_2CO_3 was the limiting reactant. Would this experiment have worked if the other reactant had been chosen as the limiting reactant? Explain why or why not.

5. Evaluating Methods
Why was the precipitate rinsed in step 11? What soluble impurities could have been on the filter paper along with the precipitate? How would the calculated results vary if the precipitate had not been completely dry? Explain your answer.

6. Applying Conclusions
How many grams of Na_2CO_3 were present in the 15 mL sample?

7. Applying Conclusions
How many grams of Na_2CO_3 are present in the 575 L? (Hint: Create a conversion factor to convert from the sample, with a volume of 15 mL, to the entire solution, with a volume of 575 L.)

8. Evaluating Methods
Ask your teacher for the theoretical mass of Na_2CO_3 in the sample, and calculate your percentage error.

Extensions

1. Designing Experiments
What possible sources of error can you identify with your procedure? If you can think of ways to eliminate them, ask your teacher to approve your plan, and run the procedure again.

Answers to Analysis

1. $SrCl_2(aq) + Na_2CO_3(aq) \rightarrow 2NaCl(aq) + SrCO_3(s)$; the precipitate is strontium carbonate, $SrCO_3$.

2. Answers will vary for mass of the dry precipitate.

3. Answers will vary.

4. Sodium carbonate is the limiting reactant. Only sodium carbonate could be used as the limiting reactant, because the purpose of the experiment was to use gravimetric analysis to determine the amount of sodium carbonate in solution. If another reactant had been used as the limiting reactant, the mass of Na_2CO_3 in solution could not have been determined before the reaction went to completion.

5. The precipitate was rinsed to remove any NaCl impurities that may have remained on the $SrCO_3$.

Answers to Conclusions

6. Answers will vary.

7. Answers will vary.

8. If students were given a 0.5 M solution, the correct mass of Na_2CO_3 is 0.795 g for every 15.0 mL.

Answers to Extensions

1. Suggestions for improving the procedure will vary. Students may suggest using larger amounts of each reactant or running multiple trials. Be sure experiments are safe and include carefully planned procedures.

Teacher's Notes

Time Required

2–3 lab periods (with overnight drying time)

Ratings

$$\overset{\displaystyle 1 \quad\quad 2 \quad\quad 3 \quad\quad 4}{\underset{\text{EASY} \qquad\qquad\qquad \text{HARD}}{\rule{7cm}{0.4pt}}}$$

TEACHER PREPARATION	2
STUDENT SETUP	2
CONCEPT LEVEL	3
CLEANUP	1

Skills Acquired

• Collecting Data
• Designing Experiments
• Experimenting
• Measuring
• Organizing and Analyzing Data

The Scientific Method

In this lab, students will:
• Make Observations
• Ask Questions
• Form a Hypothesis
• Test the Hypothesis
• Analyze the Results
• Draw Conclusions
• Communicate the Results

Materials

• balance
• beaker tongs
• beakers, 250 mL (2)
• $CaCO_3$ solution (20 mL)
• drying oven
• graduated cylinder, 100 mL
• Na_2CO_3, 0.5 M (75 mL or less)
• pipe-stem triangle
• ring and ring stand

Exploration 10B has additional information on materials for filtration.

THE PROBLEM

March 3, 2004

George Taylor, Director of Analysis
CheMystery Labs, Inc.
52 Fulton Street
Springfield, VA 22150

Dear Mr. Taylor:

The city's Public Works Department is investigating new sources of water. One proposal involves drilling wells into a nearby aquifer that is protected from brackish water by a unique geological formation. Unfortunately, this formation is made of calcium minerals. If the concentration of calcium ions in the water is too high, the water will be "hard," and treating it to meet local water standards would be too expensive for us.

Water containing more than 120 mg of calcium per liter is considered hard. I have enclosed a sample of water that has been distilled from 1.0 L to its present volume. Please determine whether the water is of suitable quality.

We are seeking a firm to be our consultant for the entire testing process. Interested firms will be evaluated based on this water analysis. We look forward to receiving your report.

Sincerely,

Dana Rubio

Dana Rubio
City Manager

References

Review the "Stoichiometry" chapter for information about mass-mass stoichiometry. In this investigation, you will use a double-displacement reaction, but Na_2CO_3 will be used as a reagent to identify how much calcium is present in a sample. Like strontium and other Group 2 metals, calcium salts react with carbonate-containing salts to produce an insoluble precipitate.

790

Safety Cautions

• Read all safety precautions, and discuss them with your students.

• Safety goggles and a lab apron must be worn at all times.

• Remind students that heated objects can be hot enough to burn even if they look cool. Students should always use beaker tongs to place samples in a drying oven.

THE PLAN

CheMystery Labs, Inc. 52 Fulton Street, Springfield, VA 22150

CheMystery Labs, Inc.
52 Fulton Street
Springfield, VA 22150

Memorandum

Date: March 4, 2004
To: Shane Thompson
From: George Taylor

We can solve the city's problem by doing some careful gravimetric analysis, because calcium salts and carbonate compounds undergo double-displacement reactions to yield insoluble calcium carbonate as a precipitate.

Before you begin your work, I will need the following information from you so that I can create our bid:

- a detailed one-page summary of your plan for the procedure as well as all necessary data tables
- a description of necessary calculations
- an itemized list of equipment

After you complete the analysis, prepare a two-page report for Dana Rubio. Make sure to include the following items:

- a calculation of calcium concentration in mg/L for the water from the aquifer
- an explanation of how you determined the amount of calcium in the sample, including measurements and calculations
- a balanced chemical equation for the reaction
- explanations and estimations for any possible sources of error

Required Precautions

- Wear safety goggles when working around chemicals, acids, bases, flames, or heating devices. Contents under pressure may become projectiles and cause serious injury.
- Avoid wearing contact lenses in the lab.
- If any substance gets in your eyes, notify your instructor immediately and flush your eyes with running water for at least 15 min.
- Secure loose clothing, and remove dangling jewelry. Don't wear open-toed shoes or sandals in the lab.
- Wear an apron or lab coat to pro-

tect your clothing when working with chemicals.
- If a spill gets on your clothing, rinse it off immediately with water for at least 5 min while notifying your instructor.
- Always use caution when working with chemicals.
- Never taste, touch, or smell chemicals unless specifically directed to do so.
- Follow instructions for proper disposal.
- Whenever possible, use an electric hot plate as a heat source instead of an open flame.

- When heating materials in a test tube, always angle the test tube away from yourself and others.
- Know your school's fire-evacuation routes.
- Clean and decontaminate all work surfaces and personal protective equipment as directed by your instructor.
- Dispose of all sharps (broken glass and other contaminated sharp objects) and other contaminated materials (biological and chemical) in special containers as directed by your instructor.

791

Checkpoint 1: By the end of the first class period, students should turn in a detailed one-page plan/procedure for approval. This should be compared to the sample procedure found in the answer section of the wrap. The student procedure should be evaluated according to organization and accuracy and how well it follows the steps of the scientific method.

Checkpoint 2: During the second class period, students should revise procedures according to the teacher's approval and begin the procedure. By the end of the second class period, students should be able to draw a reasonable conclusion and determine calculation of calcium concentration in mg/L for the Afton Aquifer water. Students should be evaluated on lab technique, quality and clarity of observations, and the explanation of observations/conclusions.

Sample Answer to Procedure

Measure the mass of a piece of filter paper and a beaker. Add an excess of Na_2CO_3 solution of known concentration to a carefully measured volume of a solution containing calcium ions. Filter and dry the $CaCO_3$ precipitate that forms. Measure its mass.

Sample Data

Volume of water sample	20.00 mL
Volume of 0.5 M Na_2CO_3	40.00 mL
Mass of 150 mL beaker	65.80 mL
Mass of filter paper	0.31 g
Mass of beaker, filter paper, and Na_2CO_3	67.07 g

CALORIMETRY AND HESS'S LAW

Teacher's Notes

Time Required
1 lab period

Ratings
EASY 1 2 3 4 HARD

TEACHER PREPARATION	2
STUDENT SETUP	1
CONCEPT LEVEL	2
CLEANUP	2

Skills Acquired
• Collecting Data
• Experimenting
• Identifying/Recognizing Patterns
• Interpreting
• Measuring
• Organizing and Analyzing Data

The Scientific Method
In this lab, students will:
• Make Observations
• Analyze the Results
• Test the Hypothesis
• Draw Conclusions

Skills Practice Lab

OBJECTIVES

♦ **Demonstrate** proficiency in the use of calorimeters and related equipment.

♦ **Relate** temperature changes to enthalpy changes.

♦ **Determine** the heat of reaction for several reactions.

♦ **Demonstrate** that the heat of reaction can be additive.

MATERIALS

♦ balance
♦ distilled water
♦ glass stirring rod
♦ graduated cylinder, 100 mL
♦ HCl solution, 0.50 M (100 mL)
♦ HCl solution, 1.0 M (50 mL)
♦ NaOH pellets (4 g)
♦ NaOH solution, 1.0 M (50 mL)
♦ plastic-foam cups (or calorimeters)
♦ spatula
♦ thermometer
♦ watch glass

OPTIONAL EQUIPMENT

♦ CBL unit
♦ graphing calculator with link cable
♦ Vernier temperature probe

792

10 Calorimetry and Hess's Law

Introduction

A man working for a cleaning firm was told by his employer to pour some old cleaning supplies into a glass container for disposal. Some of the supplies included muriatic (hydrochloric) acid, $HCl(aq)$, and a drain cleaner containing lye, $NaOH(s)$. When the substances were mixed, the container shattered, spilling the contents onto the worker's arms and legs. The worker claims that the hot spill

caused burns, and he is therefore suing his employer. The employer claims that the worker is lying because the solutions were at room temperature before they were mixed. The employer says that a chemical burn is unlikely because tests after the accident revealed that the mixture had a neutral pH, indicating that the HCl and NaOH were neutralized. The court has asked you to evaluate whether the worker's story is supported by scientific evidence.

Chemicals can be dangerous because of their special storage needs. Chemicals that are mixed and react are even more dangerous because many reactions release large amounts of heat. Glass is heat-sensitive and can shatter if there is a sudden change in temperature due to a reaction. Some glassware, such as Pyrex, is heat-conditioned but can still fracture under extreme heat conditions, especially if scratched.

You will measure the amount of heat released by mixing the chemicals in two ways. First you will break the reaction into steps and measure the heat change of each step. Then you will measure the heat change of the reaction when it takes place all at once. When you are finished, you will be able to use the calorimetry equation from the chapter "Causes of Change" to determine the following:
• the amount of heat evolved during the overall reaction
• the amount of heat for each step
• the amount of heat for the reaction in kilojoules per mole
• whether this heat could have raised the temperature of the water in the solution high enough to cause a burn

Materials

1. To prepare 0.50 M HCl, observe the required precautions. Add 42 mL of concentrated HCl to enough distilled water to make 1.00 L of solution. Add the acid slowly while stirring to avoid overheating.

2. To prepare 1.0 M HCl, add 83 mL of concentrated HCl to enough distilled water to make 1.00 L of solution. Add the acid slowly while stirring to avoid overheating.

3. To prepare 1.00 L of 1.0 M NaOH, dissolve 40.0 g of NaOH in enough water to make 1.00 L of solution.

4. To prevent the hygroscopic NaOH pellets from absorbing too much water, keep them in a reagent bottle with a stopper. Instruct students to replace the stopper after they have obtained what they need.

Safety Procedures

- Wear safety goggles when working around chemicals, acids, bases, flames, or heating devices. Contents under pressure may become projectiles and cause serious injury.
- Never look directly at the sun through any optical device or use direct sunlight to illuminate a microscope.
- Avoid wearing contact lenses in the lab.
- If any substance gets in your eyes, notify your instructor immediately and flush your eyes with running water for at least 15 min.

- Secure loose clothing, and remove dangling jewelry. Don't wear open-toed shoes or sandals in the lab.
- Wear an apron or lab coat to protect your clothing when working with chemicals.
- If a spill gets on your clothing, rinse it off immediately with water for at least 5 min while notifying your instructor.

- If a chemical gets on your skin or clothing or in your eyes, rinse it immediately, and alert your instructor.
  If a chemical is spilled on the floor or lab bench, alert your instructor, but do not clean it up yourself unless your teacher says it is OK to do so.

- Always use caution when working with chemicals.
- Never mix chemicals unless specifically directed to do so.
- Never taste, touch, or smell chemicals unless specifically directed to do so.
- Add acid or base to water; never do the opposite.
- Never return unused chemicals to the original container.
- Never transfer substances by sucking on a pipette or straw; use a suction bulb.
- Follow instructions for proper disposal.

- Check the condition of glassware before and after using it. Inform your teacher of any broken, chipped, or cracked glassware, because it should not be used.
- Do not pick up broken glass with your bare hands. Place broken glass in a specially designated disposal container.

- Clean and decontaminate all work surfaces and personal protective equipment as directed by your instructor.
- Dispose of all sharps (broken glass and other contaminated sharp objects) and other contaminated materials (biological and chemical) in special containers as directed by your instructor.

Procedure

1. Copy the data table below in your lab notebook. Reactions 1 and 3 will each require two additional spaces to record the mass of the empty watch glass and the mass of the watch glass with NaOH.

2. If you are not using a plastic-foam cup as a calorimeter, ask your teacher for instructions on using the calorimeter. At various points in the procedure, you will need to measure the temperature of the solution within the calorimeter.

Thermometer procedure continues on page 796.

Data Table 1

	Reaction 1	Reaction 2	Reaction 3
Total volumes of liquid(s)			
Initial temperature			
Final temperature			
Mass of empty watch glass			
Mass of watch glass with NaOH			

Safety Cautions

- Read all safety precautions, and discuss them with your students.
- Safety goggles and a lab apron must be worn at all times.
- Wear safety goggles, a face shield, impermeable gloves, and a lab apron while preparing the HCl. Work in a hood known to be in good operating condition, and have another person standing by to call for help in case of an emergency.
- Be sure you are within a 30 second walk of a safety shower and eyewash station.
- Students should not handle concentrated acid solutions. In case of an acid spill or base spill, dilute the spill with water. Then mop up the spill with wet cloths or a wet cloth mop designated for spill cleanup. Wear disposable plastic gloves while cleaning spills.

Disposal Information

Set out four disposal containers. Designate one for acidic liquids, one for basic liquids, one for neutral liquids, and one for excess NaOH pellets. When students are finished, slowly combine the liquid contents of the containers one at a time. If there are any excess NaOH pellets, add them a few at a time to the mixture, stirring constantly to be sure the pellets dissolve. Then check the pH. Add 1.0 M acid or base until the pH is within the range of 5–9, and then pour the solution down the drain.

CBL and Sensors

3. Connect the CBL to the graphing calculator with the unit-to-unit link cable using the I/O ports located on each unit. Connect the temperature probe to the CH1 port. Turn on the CBL and the graphing calculator. Start the program CHEMBIO on the graphing calculator.

   a. Select option *SET UP PROBES* from the MAIN MENU. Enter 1 for the number of probes. Select the temperature probe from the list. Enter 1 for the channel number. Select *USE STORED* from the CALIBRA-TION menu.

   b. Select the *COLLECT DATA* option from the MAIN MENU. Select the *TRIGGER* option from the DATA COLLECTION menu.

4. Measure the temperature by gently inserting the Vernier temperature probe into the hole in the calorimeter lid.

Reaction 1: Dissolving NaOH

5. Pour about 100 mL of distilled water into a graduated cylinder. Measure and record the volume of the water to the nearest 0.1 mL. Pour the water into your calorimeter.

6. Using the temperature probe, measure the temperature of the water. Press TRIGGER on the CBL to collect the temperature reading. Record this temperature in your data table. Select *STOP* from the TRIGGER menu on the graphing calculator. Leave the probe in the calorimeter.

7. Select the *COLLECT DATA* option from the MAIN MENU. Select the *TIME GRAPH* option from the DATA COLLECTION menu. Enter 6 for the time between samples, in sec-

onds. Enter 99 for the number of samples (the CBL will collect data for 9.9 min). Press ENTER. Select *USE TIME SETUP* to continue. If you want to change the number of samples or the time between samples, select *MODIFY SETUP.* Enter 0 for *Ymin,* enter 100 for *Ymax,* and enter 5 for *Yscl.*

8. Determine and record the mass of a clean and dry watch glass to the nearest 0.01 g. Remove the watch glass from the balance. While wearing gloves, obtain about 2 g of NaOH pellets, and put them on the watch glass. Use forceps when handling NaOH pellets. Measure and record the mass of the watch glass and the pellets to the nearest 0.01 g. **It is important that this step be done quickly because NaOH is hygroscopic. It absorbs moisture from the air, and its mass increases as long as it remains exposed to the air.**

9. Press ENTER on the graphing calculator to begin collecting the temperature readings for the water in the calorimeter.

10. Immediately place the NaOH pellets in the calorimeter cup, and gently stir the solution with a stirring rod. Place the lid on the calorimeter.

11. When the CBL displays DONE, use the arrow keys to trace the graph. Time in seconds in graphed on the *x*-axis, and the temperature readings are graphed on the *y*-axis. Record the highest temperature reading from the CBL in your data table.

12. When the reaction is finished, pour the solution into the container designated by your teacher for disposal of basic solutions.

13. Be sure to clean all equipment and rinse it with distilled water before continuing with the next procedure.

Reaction 2: NaOH and HCl in solution

14. Pour about 50 mL of 1.0 M HCl into a graduated cylinder. Measure and record the volume of the HCl solution to the nearest 0.1 mL. Pour the HCl solution into your calorimeter.

15. Select the *COLLECT DATA* option from the MAIN MENU. Select the *TRIGGER* option from the DATA COLLECTION menu. Using the temperature probe, measure the temperature of the HCl solution. Press TRIGGER on the CBL to collect the temperature reading. Record this temperature in your data table.

16. Pour about 50 mL of 1.0 M NaOH into a graduated cylinder. Measure and record the volume of the NaOH solution to the nearest 0.1 mL. **For this step only, rinse the temperature probe in distilled water.** Using the temperature probe, measure the temperature of the NaOH solution. Press TRIGGER on the CBL to collect the temperature reading. Record this temperature in your data table. Select *STOP* from the TRIGGER menu on the graphing calculator. Put the probe in the calorimeter.

17. Select the *COLLECT DATA* option from the MAIN MENU. Select the *TIME GRAPH* option from the DATA COLLECTION menu. Enter 6 for the time between samples, in seconds. Enter 99 for the number of samples. Press ENTER. Select *USE TIME SETUP* to continue. If you want to change the number of samples or the time between samples, select *MODIFY SETUP*. Enter 0 for *Ymin*, enter 100 for *Ymax*, and enter 5 for *Yscl*. Press ENTER on the calculator to begin collecting temperature readings.

18. Pour the NaOH solution into the calorimeter cup, and stir gently. Place the lid on the calorimeter.

19. When the CBL displays DONE, use the arrow keys to trace the graph. Time in seconds in

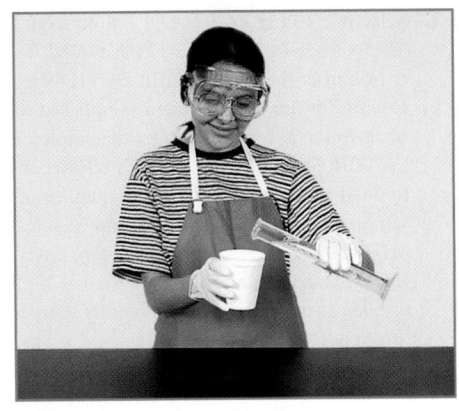

Figure A

graphed on the *x*-axis, and the temperature readings are graphed on the *y*-axis. Record the highest temperature reading from the CBL in your data table.

20. Pour the solution into the container designated by your teacher for disposal of mostly neutral solutions. Clean and rinse all equipment before continuing with the next procedure.

Reaction 3: Solid NaOH and HCl in solution

21. Pour about 100 mL of 0.50 M HCl into a graduated cylinder. Measure and record the volume to the nearest 0.1 mL. Pour the HCl solution into your calorimeter, as shown in **Figure A.**

22. Select the *COLLECT DATA* option from the MAIN MENU. Select the *TRIGGER* option from the DATA COLLECTION menu. Using the temperature probe, measure the temperature of the HCl solution. Press TRIGGER on the CBL to collect the temperature reading. Record this temperature in your data table. Select *STOP* from the TRIGGER menu on the graphing calculator.

795

23. Select the *COLLECT DATA* option from the MAIN MENU. Select the *TIME GRAPH* option from the DATA COLLECTION menu. Enter 6 for the time between samples, in seconds. Enter 99 for the number of samples. Press ENTER. Select *USE TIME SETUP* to continue. If you want to change the number of samples or the time between samples, select *MODIFY SETUP.* Enter 0 for *Ymin,* enter 100 for *Ymax,* and enter 5 for *Yscl.* Press ENTER on the calculator to begin collecting temperature readings.

24. Measure the mass of a clean and dry watch glass, and record it in your data table. Obtain approximately 2 g of NaOH. Place it on the watch glass, and record the total mass to the nearest 0.01 g. **It is important that this step be done quickly because NaOH is hygroscopic.**

25. Press ENTER on the graphing calculator to begin collecting the temperature readings for the water in the calorimeter.

26. Immediately place the NaOH pellets in the calorimeter, and gently stir the solution. Place the lid on the calorimeter.

27. When the CBL displays DONE, use the arrow keys to trace the graph. Time in seconds in graphed on the *x*-axis, and the temperature readings are graphed on the *y*-axis. Record the highest temperature reading from the CBL in your data table.

28. When the reaction is finished, pour the solution into the container designated by your teacher for disposal of basic solutions.

29. Clean all apparatus and your lab station. Check with your teacher for the proper disposal procedures. Any excess NaOH pellets should be disposed of in the designated container. Always wash your hands thoroughly after cleaning up the lab area and equipment.

Thermometer

3. Measure the temperature by gently inserting the thermometer into the hole in the calorimeter lid, as shown in **Figure B.** The thermometer takes time to reach the same temperature as the solution inside the calorimeter, so wait to be sure you have an accurate reading. **Thermometers break easily, so be careful with them, and do not use them to stir a solution.**

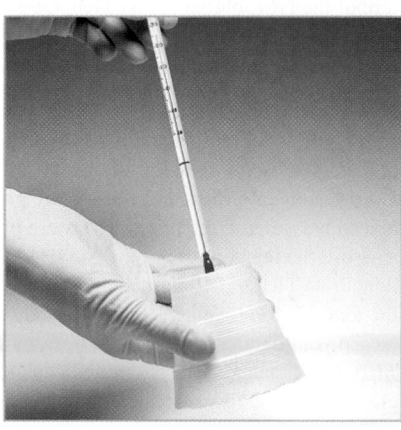

Figure B

Reaction 1: Dissolving NaOH

4. Pour about 100 mL of distilled water into a graduated cylinder. Measure and record the volume of the water to the nearest 0.1 mL. Pour the water into your calorimeter. Measure and record the water temperature to the nearest 0.1°C.

5. Determine and record the mass of a clean and dry watch glass to the nearest 0.01 g. Remove the watch glass from the balance. While wearing gloves, obtain about 2 g of NaOH pellets, and put them on the watch glass. Use forceps when handling NaOH pellets. Measure and record the mass of the watch glass and the pellets to the nearest 0.01 g. **It is important that this step be done quickly because NaOH is hygroscopic. It absorbs moisture from the air, and increases its mass as long as it remains exposed to the air.**

6. Immediately place the NaOH pellets in the calorimeter cup, and gently stir the solution with a stirring rod. **Do not stir with a thermometer.** Place the lid on the calorimeter. Watch the thermometer, and record the highest temperature in the data table. When the reaction is finished, pour the solution into the container designated by your teacher for disposal of basic solutions.

7. Be sure to clean all equipment and rinse it with distilled water before continuing with the next procedure.

Reaction 2: NaOH and HCl in solution

8. Pour about 50 mL of 1.0 M HCl into a graduated cylinder. Measure and record the volume of the HCl solution to the nearest 0.1 mL. Pour the HCl solution into your calorimeter. Measure and record the temperature of the HCl solution to the nearest 0.1°C.

9. Pour about 50 mL of 1.0 M NaOH into a graduated cylinder. Measure and record the volume of the NaOH solution to the nearest 0.1 mL. **For this step only, rinse the thermometer in distilled water, and measure the temperature of the NaOH solution in the graduated cylinder to the nearest 0.1°C. Record the temperature in your data table, and then replace the thermometer in the calorimeter.**

10. Pour the NaOH solution into the calorimeter cup, and stir gently. Place the lid on the calorimeter. Watch the thermometer, and record the highest temperature in the data table. When finished with this reaction, pour the solution into the container designated by your teacher for disposal of mostly neutral solutions.

11. Clean and rinse all equipment before continuing with the next procedure.

Reaction 3: Solid NaOH and HCl in solution

12. Pour about 100 mL of 0.50 M HCl into a graduated cylinder. Measure and record the volume to the nearest 0.1 mL. Pour the HCl solution into your calorimeter, as shown in **Figure C.** Measure and record the temperature of the HCl solution to the nearest 0.1°C.

Figure C

13. Measure the mass of a clean and dry watch glass, and record it in your data table. Obtain approximately 2 g of NaOH. Place it on the watch glass, and record the total mass to the nearest 0.01 g. **It is important that this step be done quickly because NaOH is hygroscopic.**

14. Immediately place the NaOH pellets in the calorimeter, and gently stir the solution. Place the lid on the calorimeter. Watch the thermometer, and record the highest temperature in the data table. When finished with this reaction, pour the solution into the container designated by your teacher for disposal of mostly neutral solutions.

15. Clean all apparatus and your lab station. Check with your teacher for the proper disposal procedures. Any excess NaOH pellets should be disposed of in the designated container. Always wash your hands thoroughly after cleaning up the lab area and equipment.

Answers to Analysis

1. $NaOH(s) \rightarrow NaOH(aq)$
$NaOH(aq) + HCl(aq) \rightarrow$
$H_2O(l) + NaCl(aq)$
$NaOH(s) + HCl(aq) \rightarrow H_2O(l) +$
$NaCl(aq)$

2. Adding the first two equations in the answer to item 1 will yield the third equation.

3. A good calorimeter must insulate; any heat created by the reaction should be absorbed by the water instead of the surroundings. Plastic-foam cups insulate better than paper cups.

4. Answers will vary.

5. Answers will vary.

6. Answers will vary.

7. Answers will vary.

8. Answers will vary.

9. The sum of the heats of the first two reactions should equal the heat of the third reaction.

10. Reaction 1 involved heat of solution. Reaction 2 involved heat of reaction. Reaction 3 involved heat of solution and heat of reaction.

Answers to Conclusions

11. Answers will vary.

12. Answers will vary.

13. HCl is the limiting reactant. There was 0.05 mol NaOH left.

14. The chemists use an ice bath because the heat of solution for NaOH pellets is high enough to make the solution dangerously hot.

15. The same procedure could be used for endothermic reactions. However, the temperature of the water will decrease, and the enthalpy change for the reaction will have a positive value.

16. NaOH solution is more stable than solid NaOH. The products of an exothermic reaction are more stable than the reactants.

Analysis

1. **Organizing Data**
Write a balanced chemical equation for each of the three reactions that you performed. (Hint: Be sure to include states of matter for all substances in each equation.)

2. **Analyzing Results**
Find a way to get the equation for the total reaction by adding two of the equations from Analysis and Interpretation item 1 and then canceling out substances that appear in the same form on both sides of the new equation. (Hint: Start with the equation whose product is a reactant in a second equation. Add those two equations together.)

3. **Explaining Events**
Explain why a plastic-foam cup makes a better calorimeter than a paper cup does.

4. **Organizing Data**
Calculate the change in temperature (Δt) for each of the reactions.

5. **Organizing Data**
Assuming that the density of the water and the solutions is 1.00 g/mL, calculate the mass, m, of liquid present for each of the reactions.

6. **Analyzing Results**
Using the calorimeter equation, calculate the heat released by each reaction. (Hint: Use the specific heat capacity of water in your calculations; $c_{p,H_2O} = 4.180$ J/g•°C.)

$$\text{Heat} = m \times \Delta t \times c_{p,H_2O}$$

7. **Organizing Data**
Calculate the moles of NaOH used in each of the reactions. (Hint: To find the number of moles in a solution, multiply the volume in liters by the molar concentration.)

8. **Analyzing Results**
Calculate the ΔH value in terms of kilojoules per mole of NaOH for each of the three reactions.

9. **Analyzing Results**
Using your answer to Analysis and Interpretation item 2 and your knowledge of Hess's law from the chapter "Causes of Change," explain how the enthalpies for the three reactions should be mathematically related.

10. **Analyzing Results**
Which of the following types of heat of reaction apply to the enthalpies calculated in Analysis and Interpretation item 8: heat of combustion, heat of solution, heat of reaction, heat of fusion, heat of vaporization, and heat of formation?

Conclusions

11. **Evaluating Methods**
Use your answers from Analysis and Interpretation items 7 and 8 to determine the ΔH value for the reaction of solid NaOH with HCl solution by direct measurement and by indirect calculation.

12. **Drawing Conclusions**
Third-degree burns can occur if skin comes into contact for more than 4 s with water that is hotter than 60°C (140°F). Suppose someone accidentally poured hydrochloric acid into a glass-disposal container that already contained the drain cleaner NaOH and the container shattered. The solution in the container was approximately 55 g of NaOH and 450 mL of hydrochloric acid solution containing 1.35 mol of HCl (a 3.0 M HCl solution). If the initial temperature of the solutions was 25°C, could a mixture hot enough to cause burns have resulted?

798

13. Applying Conclusions

For the reaction between the drain cleaner and HCl described in item 12, which chemical is the limiting reactant? How many moles of the other reactant remained unreacted?

14. Evaluating Results

When chemists make solutions from NaOH pellets, they often keep the solution in an ice bath. Explain why.

15. Evaluating Methods

You have worked with heats of solution for exothermic reactions. Could the same type of procedure be used to determine the temperature changes for endothermic reactions? How would the procedure stay the same? What would change about the procedure and the data?

16. Drawing Conclusions

Which is more stable, solid NaOH or NaOH solution? Explain your answer.

Extensions

1. Designing Experiments

You have worked with heats of solution for exothermic reactions. Could the same type of procedure be used to determine the temperature changes for endothermic reactions? How would the procedure stay the same? What would change about the procedure and the data?

2. Designing Experiments

A chemical supply company is going to ship NaOH pellets to a very humid place, and you have been asked to give advice on packaging. Design a package for the NaOH pellets. Explain the advantages of your package's design and materials. (Hint: Remember that the reaction in which NaOH absorbs moisture from the air is exothermic and that NaOH reacts exothermically with other compounds as well.)

Answers to Extensions

1. If acid or base spills are neutralized instead of diluted, the heat of reaction for the neutralization could cause a heat burn. In addition, the acid or base could cause a chemical burn.

2. Suggestions for package design will vary. Be sure each design addresses the dangers of moisture, breakage, and spills.

Techniques to Demonstrate

Make sure that students understand how to handle NaOH pellets. Students should not use their fingers to pick up the pellets; they should use forceps instead. NaOH pellets are hygroscopic, so the mass measurements must be taken quickly, before the pellets absorb moisture from the air. Make sure students use a watch glass instead of weighing paper for measuring the mass of the NaOH.

Tips and Tricks

Thoroughly discuss the calorimetry equation. Consider working through some sample data with the class. At first students may have difficulty understanding that reactions 1 and 2 are the equivalent of reaction 3. A thorough discussion of Hess's law and combining equations should clear up students' confusion.

PAPER CHROMATOGRAPHY OF COLORED MARKERS

Teacher's Notes

Time Required
1 lab period

Ratings

Ratings $\overset{1}{\underset{\text{EASY}}{\vdash}}\quad\overset{2}{\vdash}\quad\overset{3}{\vdash}\quad\overset{4}{\underset{\text{HARD}}{\dashv}}$

TEACHER PREPARATION	1
STUDENT SETUP	1
CONCEPT LEVEL	2
CLEANUP	1

Skills Acquired
- Classifying
- Collecting Data
- Experimenting
- Identifying/Recognizing Patterns
- Interpreting
- Predicting

The Scientific Method
In this lab, students will:
- Make Observations
- Analyze the Results
- Draw Conclusions

Materials
1. To prepare the NaCl solution, dissolve 10 g of NaCl in enough water to make 1.00 L of solution.

Skills Practice Lab

OBJECTIVES

- Conduct a paper chromatography experiment with three different water-soluble colored markers.

- Design a successful method to ensure that the chromatography paper remains vertical throughout the experiment.

- Observe the dye components of three different water-soluble markers.

MATERIALS

- beaker, 250 mL
- chromatography paper
- developing solution: NaCl solution, 0.1% by mass
- graduated cylinder, 10 mL
- hot plate
- markers
- paper clips
- pencils
- ruler
- scissors

---

13 Paper Chromatography of Colored Markers

Introduction

There is a wide variety of marker products on the market today ranging in color and function. All of these markers contain different dye components that are responsible for their color.

Paper chromatography is an analytical technique that uses paper as a medium to separate the different dye components dissolved in a mixture. In this process, the mixture to be separated is placed on a piece of chromatography paper. A solvent is then allowed to soak up into the paper. As the solvent travels across the paper, some of the components of the mixture are carried with it. Particles of the same component group together. The components that are most soluble and least attracted to the paper travel farther than others. A color band is created and the different components can be seen separated on the paper. The success of chromatography hinges on the slight difference in the physical properties of the individual components.

In this activity you will use a paper chromatography to determine the components of the dyes found in water-soluble markers. Your goal is to use paper chromatography to determine the dye components of three different water-soluble markers. You will also need to design a simple method that will keep the chromatography paper vertical while it is in the developing solution.

Safety Procedures

- Wear safety goggles when working around chemicals, acids, bases, flames, or heating devices. Contents under pressure may become projectiles and cause serious injury.
- Never look directly at the sun through any optical device or use direct sunlight to illuminate a microscope.
- Avoid wearing contact lenses in the lab.
- If any substance gets in your eyes, notify your instructor immediately and flush your eyes with running water for at least 15 min.

- Secure loose clothing, and remove dangling jewelry. Don't wear open-toed shoes or sandals in the lab.
- Wear an apron or lab coat to protect your clothing when working with chemicals.
- If a spill gets on your clothing, rinse it off immediately with water for at least 5 min while notifying your instructor.

- Always use caution when working with chemicals.
- Never mix chemicals unless specifically directed to do so.
- Never taste, touch, or smell chemicals unless specifically directed to do so.
- Add acid or base to water; never do the opposite.
- Never return unused chemicals to the original container.
- Never transfer substances by sucking on a pipette or straw; use a suction bulb.
- Follow instructions for proper disposal.

- Clean and decontaminate all work surfaces and personal protective equipment as directed by your instructor.
- Dispose of all sharps (broken glass and other contaminated sharp objects) and other contaminated materials (biological and chemical) in special containers as directed by your instructor.

Procedure

1. Copy the data table below in your lab notebook.

Data Table 1

Marker Color	Dye Components

2. Obtain a clean 250 mL beaker and a 7.0×2.5 cm piece of chromatography paper.

3. Choose three different markers for this activity. Write the color of each marker in your data table.

4. Using a ruler, draw a horizontal line in pencil approximately 1.0 cm from one of the ends of the paper. Mark three small dots on this line, using a different marker for each dot.

Safety Cautions

- Read all safety precautions, and discuss them with your students.

- Always wear safety goggles and a lab apron to protect your eyes and clothing. The developing solution could irritate the eyes and other exposed skin.

- If a hot plate is used, it should be equipped with a three-prong plug. Each electrical socket in the laboratory must have three holes, with a GFI (ground-fault interrupter) circuit. Check the polarity of the circuit with a polarity tester from an electronics supply store. Repair any incorrectly wired sockets.

- Be certain electrical equipment is turned off before plugging it in. Turn it off again before unplugging it. Wiring hookups should be made or altered only when the apparatus is disconnected from the power source and the power switch is in the "off" position.

- Do not let electrical cords dangle from work stations; dangling cords are a tripping and shock hazard.

- Do not use electrical equipment with frayed or kinked cords.

- Be certain the area under and around electrical equipment is dry.

- Cords should not lie in puddles or spilled liquids. Students should have dry hands when they are using electrical equipment.

- Turn off and unplug all electrical equipment before leaving the laboratory.

- Remind students that heated objects can be hot enough to burn even though they look cool. Students should always use tongs when handling heated equipment.

Disposal Information

Students can pour the remaining developing solution into the sink.

Techniques to Demonstrate

You may wish to hold a piece of chromatography paper in a beaker to show students the basic idea of this experiment. However, avoid recommending specific methods to keep the chromatography paper vertical inside the developing chamber.

Tips and Tricks

Before beginning the lab, students should review basic information about chromatography and look at a developed chromatogram. Direct the students' attention to the fact that they will have to design a method for the chromatography paper to remain vertical as it develops. Encourage the students to create their own method to achieve this task.

Stress that the initial spots on the chromatography paper must remain above the liquid in the beaker. Some students may have the misconception that their experiment was unsuccessful if an orange marker does not reveal a mixture of yellow and red dyes. One often overlooked explanation is that the orange marker is composed only of orange dyes. The lack of color separation could be attributed to the presence of a single dye.

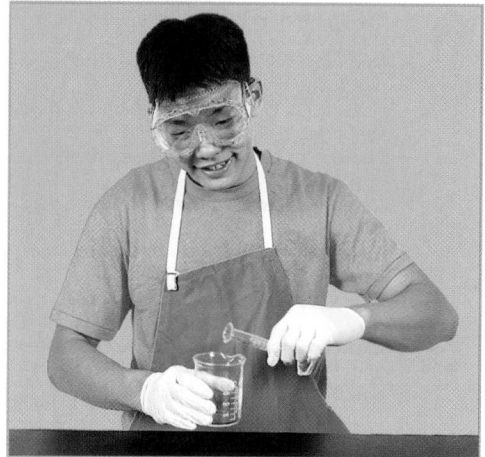

Figure A

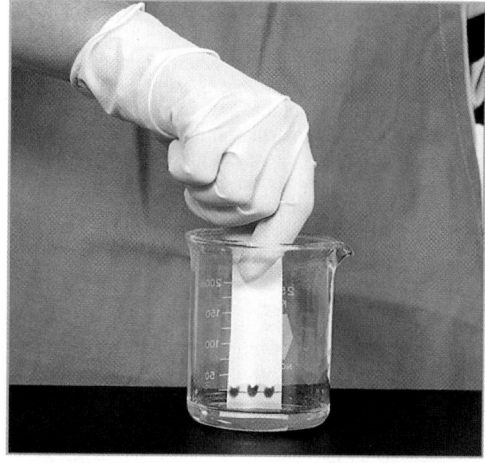

Figure B

5. Using a pencil, label each of the dots on the chromatography paper according to the color of the markers.

6. Measure out 7.0 mL of the developing solution in a 10 mL graduated cylinder.

7. Pour the 7.0 mL of solution in a 250 mL beaker, as shown in **Figure A.** Make sure the bottom of the beaker is completely covered. The level of the liquid must be below the marks on your chromatography paper.

8. You will need to design an experimental technique to ensure that your paper sample does not slide into the developing solution. The chromatography paper must remain vertical as the developing solution rises into the paper.

9. Carefully place your paper (with the dots at the

bottom) into the liquid, as shown in **Figure B.**

10. When the level of the liquid has advanced through most of the paper, remove the paper from the developing solution. Hold up the paper and observe the colors.

11. The chromatography samples can be carefully dried on a hot plate.

12. You may repeat this process using overwrite or color-change markers.

13. Clean all apparatus and your lab station. Return equipment to its proper place. Dispose of chemicals and solutions in the containers designated by your teacher. Do not pour any chemicals down the drain or in the trash unless your teacher directs you to do so.

802

Analysis

1. Describing Events
What was the purpose of this experiment?

2. Explaining Events
Why were only water-soluble markers used in this experiment? Could permanent markers be used?

3. Explaining Events
Why must the spotted marks remain above the level of the liquid in the beaker?

Conclusions

4. Applying Conclusions
Why shouldn't you use a ballpoint pen when marking the initial line and spots on the chromatography paper? Explain.

5. Evaluating Results
Make observations about the dye components (colors) of each marker based on your results.

6. Applying Conclusions
Explain how law enforcement officials could use paper chromatography to identify a pen that was used in a ransom note.

7. Applying Conclusions
List some other applications for using paper chromatography.

8. Evaluating Methods
Compare your results with those of another lab group. Were the dye components found in other markers different from those found in yours?

Extensions

1. Research and Communications
Gasoline is a mixture of many different chemicals. Chemists can identify the different components of the mixture using chromatography. Research what gasoline is composed of and make a chart of the common components.

Answers to Analysis

1. to use paper chromatography to separate the dyes in a water soluble marker

2. Permanent markers should not be used because they are insoluble in water. We would have to use a developing solution that would dissolve the permanent markers.

3. The dyes would simply dissolve in the developing solution rather than move up the chromatography paper.

Answers to Conclusions

4. The ink could be separated into its different dyes. This could interfere with the separation of the dyes of the markers.

5. Answers will vary.

6. Black ink usually consists of a mixture of different colors of ink. Each brand of pen generally has a unique set of components. Therefore, two similar black pens could be analyzed by paper chromatography to reveal the dye components. These mixtures could be compared with a chromatogram taken from the ransom note.

7. Answers may vary but might include determining whether an original document has been altered.

Answers to Extension

1. Answers will vary.

DRIP-DROP ACID-BASE EXPERIMENT

Teacher's Notes

Time Required
1 lab period

Ratings

EASY 1 — 2 — 3 — 4 HARD

TEACHER PREPARATION	2
STUDENT SETUP	1
CONCEPT LEVEL	2
CLEANUP	1

Skills Acquired
- Collecting Data
- Identifying/Recognizing Patterns
- Interpreting
- Measuring
- Organizing and Analyzing Data

The Scientific Method
In this lab, students will:
- Make Observations
- Analyze the Results
- Draw Conclusions

Skills Practice Lab

OBJECTIVES

- Translate word equations into chemical formulas.
- Count the number of drops of sodium hydroxide needed to completely react with different acid samples.
- Calculate the average number of drops of sodium hydroxide needed for each acid.
- Relate the number of drops to the coefficients in the balanced chemical equations.

MATERIALS

- buret clamps
- burets (2)
- H_2SO_4, 0.1 M
- HCl, 0.1 M
- NaOH, 0.3 M
- phenolphthalein indicator
- pipets
- ring stands
- test tubes
- test-tube rack

15A Drip-Drop Acid-Base Experiment

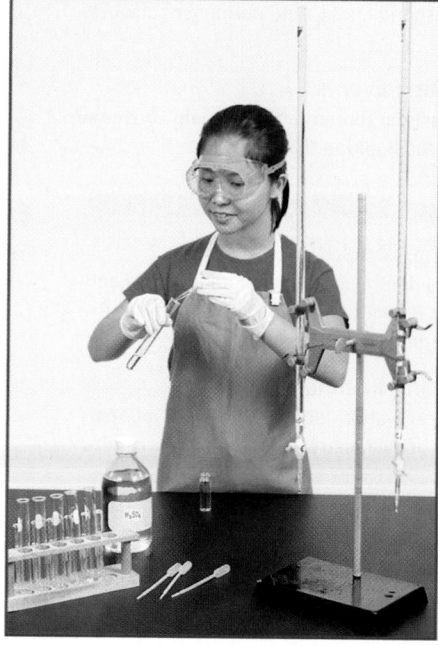

Introduction

The purpose of this lab is to investigate the simple reaction between two different acids and a base. We will be counting the number of drops of sodium hydroxide that are needed to react completely with all of the acid. The starting acid and base solutions are colorless and clear, and the final products are colorless and clear.

To monitor the progress of the chemical reaction, the acid-base indicator phenolphthalein will be used. Phenolphthalein is colorless when acidic and pink in color when neutral or basic. In this activity, we will know that all of the acid has been consumed by the base when the test-tube solution starts to turn pink. We can monitor the progress of the reaction so that a single drop of the base results in a sudden change from colorless to pink. At that point, we will know that all of the acid has reacted with the base.

You will need to count the number of drops of sodium hydroxide that are necessary to neutralize two different acids. Find the relationship between the sodium hydroxide drops necessary and the coefficients in the balanced chemical equation.

Materials

1. To prepare 0.10 M HCl, dissolve 8.3 mL of concentrated HCl in enough water to make 1.00 L of solution.

2. To prepare 0.10 M H_2SO_4, dissolve 5.6 mL of concentrated H_2SO_4 in enough water to make 1.00 L of solution.

3. To prepare 0.30 M NaOH, dissolve 12 g of solid NaOH in enough water to make 1.00 L of solution.

Safety Procedures

- Wear safety goggles when working around chemicals, acids, bases, flames, or heating devices. Contents under pressure may become projectiles and cause serious injury.
- Never look directly at the sun through any optical device or use direct sunlight to illuminate a microscope.
- Avoid wearing contact lenses in the lab.
- If any substance gets in your eyes, notify your instructor immediately and flush your eyes with running water for at least 15 min.

- Secure loose clothing, and remove dangling jewelry. Don't wear open-toed shoes or sandals in the lab.
- Wear an apron or lab coat to protect your clothing when working with chemicals.
- If a spill gets on your clothing, rinse it off immediately with water for at least 5 min while notifying your instructor.

- Always use caution when working with chemicals.
- Never mix chemicals unless specifically directed to do so.

- Never taste, touch, or smell chemicals unless specifically directed to do so.
- Add acid or base to water; never do the opposite.
- Never return unused chemicals to the original container.
- Never transfer substances by sucking on a pipette or straw; use a suction bulb.
- Follow instructions for proper disposal.

- Check the condition of glassware before and after using it. Inform your teacher of any broken, chipped, or cracked glassware, because it should not be used.
- Do not pick up broken glass with your bare hands. Place broken glass in a specially designated disposal container.

- Clean and decontaminate all work surfaces and personal protective equipment as directed by your instructor.
- Dispose of all sharps (broken glass and other contaminated sharp objects) and other contaminated materials (biological and chemical) in special containers as directed by your instructor.

Procedure

1. Translate each of the word equations shown below into chemical equations.

   hydrochloric acid + sodium hydroxide ⟶
   sodium chloride + water

   sulfuric acid + sodium hydroxide ⟶
   sodium sulfate + water

2. Copy Data Tables 1 and 2 in your lab notebook. Be sure that you have plenty of room for observations about each test.

3. Clean six test tubes, and rinse them with distilled water. They do not need to be dry.

4. Obtain approximately 10 mL of sodium hydroxide solution in a small beaker.

Data Table 1

HCl volume (mL)	NaOH (drops)
2.00	
2.00	
4.00	
4.00	

Data Table 2

H_2SO_4 volume (mL)	NaOH (drops)
2.00	
2.00	

Safety Cautions

- Read all safety precautions, and discuss them with your students.
- Safety goggles and lab aprons should be worn at all times.
- Wash hands with water if the chemicals in this experiment come in contact with your skin.
- Be sure to completely close the stopcock valve on the burets.

Disposal Information

The contents of the test tubes can be poured down the sink with plenty of water. The stock solutions of HCl, H_2SO_4, and NaOH can be kept for other experiments.

Tips and Tricks

Make sure you have discussed the nomenclature of ionic compounds with your students before allowing them to perform this experiment. Have each student translate the word equations shown in the Preparation section. Before the students begin the experiment, check their chemical formulas.

If you do not have enough burets for each group, simply set up the acids from four different "buret stations" placed around the room. Each buret station should consist of two labeled burets on a ring stand. Fill one buret with hydrochloric acid, and fill the other with sulfuric acid. Using four buret stations helps to avoid congestion as students measure their acids directly into their test tubes. With four small dropper bottles of the phenolphthalein indicator solution available, the students progress quickly through the lab.

805

Techniques to Demonstrate

Demonstrate how to use burets to ensure the accuracy of the volume measurements. It may be a good time to remind students about the importance of reading the meniscus and measuring the height of the liquids at eye level. Using colored water generally improves the effectiveness of this demonstration because the students can easily observe the buret's contents and see that turning the stopcock changes the contents.

Also demonstrate the pink color of the end point. A highly effective technique is to demonstrate how the addition of one drop of NaOH can produce the desired end point. Also demonstrate that additional drops of base do not change the color of the indicator.

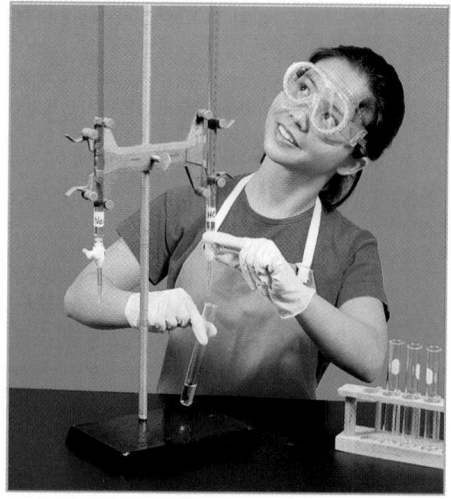

Figure A

Part I

5. Use a buret to put exactly 2.00 mL of hydrochloric acid directly into your test tube, as shown in **Figure A.**

6. Add two drops of phenolphthalein indicator solution to the test tube.

7. Use a pipet to add the sodium hydroxide solution dropwise to the test tube. Count the number of drops of sodium hydroxide as you add them. Gently shake the test tube from side to side after adding each drop. **Continue adding drops until the color just changes from colorless to pink.**

8. Record in your data table the total number of drops of sodium hydroxide needed to reach the color change. To obtain consistent results, repeat this trial.

Part II

9. Use a buret to add exactly 4.00 mL of hydrochloric acid directly into a clean test tube.

10. Add two drops of phenolphthalein indicator solution to the test tube.

11. Using a pipet, add one drop of sodium hydroxide solution at a time to the test tube. Count the number of drops of sodium hydroxide as you add them. Gently swirl the test tube after adding each drop. **Continue adding drops until the color just changes from colorless to a pink.**

12. Record in your data table the total number of drops of sodium hydroxide needed to reach the color change. Repeat this trial.

Part III Sulfuric Acid

13. Use a buret to add exactly 2.00 mL of sulfuric acid directly into your test tube.

14. Add two drops of phenolphthalein indicator solution to the test tube.

15. Using a pipet, add one drop of sodium hydroxide solution at a time to the test tube. Count the number of drops of sodium hydroxide as you add them. Gently swirl the test tube after adding each drop. **Continue adding drops until the color just changes from colorless to pink.**

16. Record in your data table the total number of drops of sodium hydroxide needed to reach the color change. Repeat this trial.

17. Clean all apparatus and your lab station. Return equipment to its proper place. Dispose of chemicals and solutions in the containers designated by your teacher. Do not pour any chemicals down the drain or in the trash unless your teacher directs you to do so. Wash your hands thoroughly after all work is finished and before you leave the lab.

Analysis

1. Examining Data
What was the average number of drops of sodium hydroxide required to consume 2.00 mL of HCl? Show your work. 4.00 mL of HCl? Show your work.

2. Examining Data
What was the average number of drops of sodium hydroxide required to consume 2.00 mL of H_2SO_4? Show your work.

3. Analyzing Results
Compare your responses to Analysis item 1. Is there a difference in the average number of drops? What is the ratio between these two numbers? Is it 1:1, 1:2, 2:1, or 1:3? Explain the "chemistry" behind this ratio.

4. Analyzing Results
Now compare your responses to Analysis and Interpretation items 1 and 2. Is there a difference in the average number of drops? What is the ratio between these two numbers? Is it 1:1, 1:2, 1:3, etc? Explain the "chemistry" behind this ratio.

Conclusions

5. Applying Conclusions
Based on your observed results, how many drops of sodium hydroxide would be needed to react completely with a 2.00 mL sample of HNO_3?

$$HNO_3 + NaOH \longrightarrow NaNO_3 + H_2O$$

Answers to Analysis
1. Answers will vary.
   Answers will vary.
2. Answers will vary.
3. The number of drops of sodium hydroxide needed for 2.00 mL of HCl and 4.00 mL of HCl doubled. Therefore, the ratio is 1:2. Because the volume of HCl was doubled, the number of drops of NaOH should double as well.
4. The ratio is 1:2. The coefficients in the balanced chemical equation explains the 1:2 ratio. The same volumes of acid were used, but different amounts of base were required.

Answers to Conclusion
5. This chemical equation is balanced as written. Therefore, a 1:1 ratio exists between the reactants. One would expect a number of drops of NaOH equal to 2.00 mL to be required to react completely with the HNO_3.

807

ACID-BASE TITRATION OF AN EGGSHELL

Teacher's Notes

Time Required
1–2 lab periods

Ratings
EASY 1 2 3 4 HARD

TEACHER PREPARATION	2
STUDENT SETUP	2
CONCEPT LEVEL	2
CLEANUP	2

Skills Acquired
- Collecting Data
- Experimenting
- Inferring
- Interpreting
- Measuring
- Organizing and Analyzing Data

The Scientific Method
In this lab, students will:
- Make Observations
- Analyze the Results
- Draw Conclusions

Skills Practice Lab

OBJECTIVES

- Determine the amount of calcium carbonate present in an eggshell.

- Relate experimental titration measurements to a balanced chemical equation.

- Infer a conclusion from experimental data.

- Apply reaction-stoichiometry concepts.

MATERIALS

- balance
- beaker, 100 mL
- bottle, 50 mL or small Erlenmeyer flask
- desiccator (optional)
- distilled water
- drying oven
- eggshell
- forceps
- graduated cylinder, 10 mL
- HCl, 1.00 M
- medicine droppers or thin-stemmed pipets (3)
- mortar and pestle
- NaOH, 1.00 M
- phenolphthalein solution
- weighing paper
- white paper or white background

808

15B Acid-Base Titration of an Eggshell

Introduction

You are a scientist working with the Department of Agriculture. A farmer has brought a problem to you. In the past 10 years, his hens' eggs have become increasingly fragile. So many of them have been breaking that he is beginning to lose money. The farmer believes his problems are linked to a landfill upstream, which is being investigated for illegal dumping of PCBs and other hazardous chemicals. Your job is to find out if the PCBs are the cause of the hens' fragile eggs.

Birds have evolved a chemical process that allows them to rapidly produce the calcium carbonate, $CaCO_3$, required for eggshell formation. Research has shown that some chemicals, like DDT and PCBs, can decrease the amount of calcium carbonate in the eggshell, resulting in shells that are thin and fragile.

You need to determine how much calcium carbonate is in sample eggshells from chickens that were not exposed to PCBs. The farmer's eggshells contain about 78% calcium carbonate. The calcium carbonate content of eggshells can easily be determined by means of an acid-base back-titration. A carefully measured excess of a strong acid will react with the calcium carbonate. Because the acid is in excess, there will be some left over at the end of the reaction. The resulting solution will be titrated with a strong base to determine how much acid remained unreacted. Phenolphthalein will be used as an indicator to signal the endpoint of the titration. From this measurement, you can determine the following:

- the amount of excess acid that reacted with the eggshell
- the amount of calcium carbonate that was present to react with this acid

Safety Procedures

- Wear safety goggles when working around chemicals, acids, bases, flames, or heating devices. Contents under pressure may become projectiles and cause serious injury.
- Never look directly at the sun through any optical device or use direct sunlight to illuminate a microscope.
- Avoid wearing contact lenses in the lab.
- If any substance gets in your eyes, notify your instructor immediately and flush your eyes with running water for at least 15 minutes.

- Secure loose clothing, and remove dangling jewelry. Don't wear open-toed shoes or sandals in the lab.
- Wear an apron or lab coat to protect your clothing when working with chemicals.
- If a spill gets on your clothing, rinse it off immediately with water for at least 5 min while notifying your instructor.

- If a chemical gets on your skin or clothing or in your eyes, rinse it immediately, and alert your instructor.
- If a chemical is spilled on the floor or lab bench, alert your instructor, but do not clean it up yourself unless your teacher says it is OK to do so.

- Always use caution when working with chemicals.
- Never mix chemicals unless specifically directed to do so.
- Never taste, touch, or smell chemicals unless specifically directed to do so.
- Add acid or base to water; never do the opposite.
- Never return unused chemicals to the original container.
- Never transfer substances by sucking on a pipette or straw; use a suction bulb.
- Follow instructions for proper disposal.

- Whenever possible, use an electric hot plate as a heat source instead of an open flame.
- When heating materials in a test tube, always angle the test tube away from yourself and others.
- Glass containers used for heating should be made of heat-resistant glass.
- Know your school's fire-evacuation routes.

- Clean and decontaminate all work surfaces and personal protective equipment as directed by your instructor.
- Dispose of all sharps (broken glass and other contaminated sharp objects) and other contaminated materials (biological and chemical) in special containers as directed by your instructor.

Procedure

1. Make data and calculation tables like the following tables.

Data Table 1

Total volume of acid drops	
Average volume of each drop	
Total volume of base drops	
Average volume of each drop	

Data Table 2

Titration Steps	
Mass of entire eggshell	
Mass of ground eggshell sample	
Number of drops of 1.00 M HCl added	150
Volume of 1.00 M HCl added	
Number of drops of 1.00 M NaOH added	
Volume of 1.00 M NaOH added	
Volume of HCl reacting with NaOH	
Volume of HCl reacting with eggshell	
Number of moles of HCl reacting with eggshell	
Number of moles of $CaCO_3$ reacting with HCl	
Mass of $CaCO_3$	
Percentage of $CaCO_3$ in eggshell sample	

Materials

1. Desiccators are optional. Less equipment will be required if students share.
2. If droppers or pipets are labeled as "acid," "base," and "indicator," these can be reused from class to class. For drops of equal size, place the pipet or dropper bulb in a tubing screw-clamp. Each turn of the screw will force out a drop of equal size.
3. To prepare 1.00 M HCl, follow the required precautions. Add 82.6 mL of concentrated HCl to enough distilled water to make 1.00 L of solution. Add the acid slowly while stirring to avoid overheating.
4. To prepare 1.00 M NaOH, dissolve 40.0 g of NaOH pellets in enough distilled water to make 1.00 L of solution. Add a few pellets at a time while stirring to avoid overheating.
5. To prepare phenolphthalein solution, dissolve 1.00 g phenolphthalein in 50.0 mL of denatured alcohol, and add 50.0 mL of distilled water.
6. A Vernier pH probe can be used to monitor the titration and its end point.

Small-Scale Option
- 15 mL of 1.00 M NaOH
- 15 mL of 1.00 M HCl
- medicine droppers or thin-stemmed pipets (3)

Full-Scale Option
- 100 mL of 1.00 M NaOH
- 100 mL of 1.00 M HCl
- burets

Optional Equipment
- CBL unit
- graphing calculator with link cable
- Vernier pH probe with amplifier

Safety Cautions

- Read all safety precautions, and discuss them with your students.
- Safety goggles and a lab apron must be worn at all times.
- Wear goggles, a face shield, impermeable gloves, and a lab apron when you prepare the HCl and NaOH. Work in a hood known to be in good operating condition, and have another person standing by to call for help in case of an emergency. Be sure you are within a 30 second walk of a safety shower and eyewash station.
- Students should not handle concentrated acid solutions.
- In case of an acid or base spill, dilute the spill with water. Then mop up the spill with wet cloths or a wet cloth mop designated for spill cleanup. Wear disposable plastic gloves while cleaning up spills.
- Remind students to use tongs to remove beakers from the oven because the beakers will be hot.

Trial	Initial acid pipet	Final acid pipet	Initial base pipet	Final base pipet
1				
2				
3				

Figure A

Figure B

Figure C

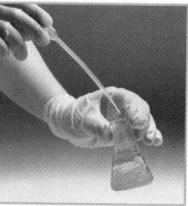
Figure D

Disposal Information

Set out three disposal containers: one for unused acid solutions, one for unused base solutions, and one for partially neutralized substances and the contents of the waste beaker. While stirring, slowly combine the solutions one at a time. Adjust the pH of the final waste liquid with 1.0 M acid or base until the pH is between 5 and 9. Pour the neutralized liquid down the drain.

Techniques to Demonstrate

Demonstrate titration procedures, showing the proper way to add drops from burets or droppers and showing the proper way to swirl the flask after adding each drop. Show the end point of the titration. Remind students that the pink color that first appears often disappears when the flask is swirled.

Tips and Tricks

Be sure to explain the nature of a back-titration. Students may not realize that the amount of $CaCO_3$ is being measured indirectly.

2. Remove the white and the yolk from the egg, as shown in **Figure A**. Dispose of these according to your teacher's directions. Wash the shell with distilled water, and carefully peel all the membranes from the inside of the shell. Discard the membranes. Place ALL of the shell in a pre-massed beaker, and dry the shell in the drying oven at 110°C for about 15 min.

3. Put exactly 5.0 mL of water in the 10.0 mL graduated cylinder. Record this volume in the data table in your lab notebook. Fill the first dropper or pipet with water. This dropper should be labeled "Acid." **Do not use this dropper for the base solution.** Holding the dropper vertical, add 20 drops of water to the cylinder. For the best results, keep the sizes of the drops as even as possible throughout this investigation. Record the new volume of water in the first data table as Trial 1.

4. Without emptying the graduated cylinder, add an additional 20 drops from the dropper, as you did in step 3, and record the new volume as the final volume for Trial 2. Repeat this procedure once more for Trial 3.

5. Repeat steps 3 and 4 for the second thin-stemmed dropper. Label this dropper "Base." **Do not use this dropper for the acid solution.**

6. Make sure that the three trials produce data that are similar to each other. If one is greatly different from the others, perform steps 3–5 over again. If you're still waiting for the eggshell in the drying oven, calculate and record in the first data table the total volume of the drops and the average volume per drop.

7. Remove the eggshell and beaker from the oven. Cool them in a desiccator. Record the mass of the entire eggshell in the second data table. Place half the shell in a clean mortar, and grind it to a very fine powder, as shown in **Figure B**. This will save time when dissolving the eggshell. (If time permits, dry the crushed eggshell again, and cool it in the desiccator.)

8. Measure the mass of a piece of weighing paper. Transfer about 0.1 g of ground eggshell to a piece of weighing paper, and measure the eggshell's mass as accurately as possible. Record the mass in the second data table. Place this eggshell sample in a clean 50 mL bottle or Erlenmeyer flask. A flask will make it easier to swirl the mixture when needed.

9. Fill the acid dropper with the 1.00 M HCl acid solution, and then empty the dropper into an extra 100 mL beaker. Label the beaker "Waste." Fill the base dropper with the 1.00 M NaOH base solution, and then empty the dropper into the 100 mL beaker.

810

Procedural Changes for Full-Scale Option

- Prepare burets for titrations.
- Delete the first data table and the "number of drops" entries in the second data table.
- Omit steps 3–6 for calibrating droppers.
- Use a sample size near 1.00 g instead of 0.10 g in step 8.
- Rinse burets in the same way that droppers are rinsed in step 9.
- Instead of 150 drops, use 50.0 mL of HCl from the acid buret to react with the $CaCO_3$ in step 10.
- Titrate the mixture with NaOH from the base buret until the phenolphthalein changes color in step 11.
- Omit item 3 in the Analysis and Interpretation section.

10. Fill the acid dropper once more with 1.00 M HCl. Using the acid dropper, add exactly 150 drops of 1.00 M HCl to the bottle (or flask) with the eggshell, as shown in **Figure C.** Swirl gently for 3–4 minutes. Rinse the sides of the flask with about 10 mL of distilled water. Using a third dropper, add two drops of phenolphthalein solution. Record the number of drops of HCl used in the second data table.

11. Fill the base dropper with the 1.00 M NaOH. Slowly add NaOH from the base dropper into the bottle or flask with the eggshell mixture until a faint pink color persists, even after it is swirled gently, as shown in **Figure D.** It may help to use a white piece of paper as a background so you will be able to see the color as soon as possible. **Be sure to add the base drop by drop, and be certain the drops end up in the reaction mixture and not on the side of the bottle or flask. Keep a careful count of the number of drops used.** Record the number of drops of base used in the second data table.

12. Clean all apparatus and your lab station. Return the equipment to its proper place. Dispose of chemicals and solutions in the containers designated by your teacher. Do not pour any chemicals down the drain or in the trash unless your teacher directs you to do so. Wash your hands thoroughly before you leave the lab and after all work is finished.

Analysis

1. Explaining Events
The calcium carbonate in the eggshell sample undergoes a double-replacement reaction with the hydrochloric acid in step 10. Then the carbonic acid that was formed decomposes. Write a balanced chemical equation for these reactions. (Hint: The gas observed was carbon dioxide.)

2. Explaining Events
Write the balanced chemical equation for the acid-base neutralization of the excess unreacted HCl with the NaOH.

3. Organizing Data
Make the necessary calculations from the first data table to find the number of milliliters in each drop. Using this milliliter/drop ratio, convert the number of drops of each solution in the second data table to volumes in milliliters.

4. Analyzing Results
Using the relationship between the molarity and volume of acid and the molarity and volume of base needed to neutralize the acid, calculate the volume of the HCl solution that was neutralized by the NaOH. Then subtract this amount from the initial volume of HCl to determine how much HCl reacted with $CaCO_3$.

Conclusions

5. Evaluating Data
Use the stoichiometry of the reaction in Analysis and Interpretation item 1 to calculate the number of moles of $CaCO_3$ that reacted with the HCl, and record this number in your table.

6. Evaluating Data
Workers in a lab in another city have also tested eggs, and they found that a normal eggshell is about 97% $CaCO_3$. Calculate the percent error for your measurement.

Extensions

1. Building Models
Calculate an estimate of the mass of $CaCO_3$ present in the entire eggshell, based on your results. (Hint: Apply the percent composition of your sample to the mass of the entire eggshell.)

2. Designing Experiments
What possible sources of error can you identify in this procedure? If you can think of ways to eliminate them, ask your teacher to approve your plan, and run the procedure again.

Time Required
2 lab periods

Ratings

	1	2	3	4
	EASY			HARD

TEACHER PREPARATION	2
STUDENT SETUP	2
CONCEPT LEVEL	3
CLEANUP	2

Skills Acquired
- Classifying
- Collecting Data
- Communicating
- Designing Experiments
- Experimenting
- Identifying/Recognizing Patterns
- Inferring
- Interpreting
- Measuring
- Organizing and Analyzing Data

The Scientific Method
In this lab, students will:
- Make Observations
- Ask Questions
- Form a Hypothesis
- Test the Hypothesis
- Analyze the Results
- Draw Conclusions
- Communicate the Results

Inquiry LAB
Design Your Own Experiment

15B Acid-Base Titration

THE PROBLEM

DELIVER BY OVERNIGHT COURIER

Date: April 21, 2004
To: EPA National Headquarters
From: Anthony Wong, Plant Supervisor
Re: Vacaville Bleachex Corp. Plant Spill

As a result of last night's earthquake, the Bleachex plant in the industrial park south of Vacaville was severely damaged. The safety control measures failed because of the magnitude of the earthquake.

Bleachex manufactures a variety of products using concentrated acids and bases. Plant officials noticed a large quantity of liquid, which was believed to be either sodium hydroxide or hydrochloric acid solution, flowing through the loading bay doors. An Emergency Toxic Spill Response Team attempted to determine the source and identity of the unknown liquid. A series of explosions and the presence of chlorine gas forced the team to abandon its efforts. The unknown liquid continues to flow into the nearly full containment ponds.

We are sending a sample of the liquid to you by overnight courier, and we hope that you can quickly and accurately identify the liquid and notify us of the proper method for cleanup and disposal. We need your answer as soon as possible.

Sincerely,

Anthony Wong

Anthony Wong

Materials
- beaker, 250 mL
- buret tubes (2)
- double buret clamp
- Erlenmeyer flask, 250 mL
- index card
- NaOH, 0.5 M (50 mL)
- pH paper
- phenolphthalein solution (1 mL)
- ring stand
- unknown solution (0.20 M HCl), 50 mL
- wash bottle

Optional Equipment
- CBL unit
- graphing calculator with link cable
- Vernier pH probe with amplifier
- stopwatch

THE PLAN

CheMystery Labs, Inc. 52 Fulton Street, Springfield, VA 22150

Memorandum

Date: April 22, 2004
To: Cicely Jackson
From: Marissa Bellinghausen

This project is a high priority. First we must determine the pH of the unknown so that we know whether it is an acid or a base. Then titrate the unknown using a standard solution to determine its concentration so that we can advise Bleachex on the amount of neutralizing agents that will be needed for the three containment ponds.

I need the following items:
- a detailed one-page plan for your procedure that includes all necessary data tables (include multiple trials)
- a detailed list of the equipment and materials you will need

When you have completed your experiment, prepare a report in the form of a two-page letter that we can fax to Anthony Wong. The letter must include the following:
- the identity of the unknown and its concentration
- the pH of the unknown and an explanation of how you determined the pH
- paragraph summarizing how you titrated the sample to determine its concentration
- a detailed and organized data table
- a detailed analysis section, including calculations, a discussion of the multiple trials, and a statistical analysis of your precision
- your proposed method for cleanup and disposal, including the amount of neutralizing agents that will be needed

Required Precautions

- Wear safety goggles when working around chemicals, acids, bases, flames, or heating devices. Contents under pressure may become projectiles and cause serious injury.
- Avoid wearing contact lenses in the lab.
- If any substance gets in your eyes, notify your instructor immediately and flush your eyes with running water for at least 15 min.
- Secure loose clothing, and remove dangling jewelry. Don't wear open-toed shoes or sandals in the lab.
- Wear an apron or lab coat to pro-

tect your clothing when working with chemicals.
- If a spill gets on your clothing, rinse it off immediately with water for at least 5 min while notifying your instructor.
- Always use caution when working with chemicals.
- Never taste, touch, or smell chemicals unless specifically directed to do so.
- Follow instructions for proper disposal.
- Whenever possible, use an electric hot plate as a heat source instead of an open flame.

- When heating materials in a test tube, always angle the test tube away from yourself and others.
- Know your school's fire-evacuation routes.
- Clean and decontaminate all work surfaces and personal protective equipment as directed by your instructor.
- Dispose of all sharps (broken glass and other contaminated sharp objects) and other contaminated materials (biological and chemical) in special containers as directed by your instructor.

813

- Read all safety precautions, and discuss them with your students.
- Safety goggles and a lab apron must be worn.
- Students should not handle concentrated acids.
- In case of an acid or base spill, dilute the spill with water. Then mop up the spill with wet cloths or a wet cloth mop designated for spill cleanup. Wear disposable plastic gloves while cleaning up spills.

Checkpoints

Checkpoint 1: By the end of the first class period, students should turn in a detailed one-page plan/procedure for approval. This should be compared to the sample procedure found in the answer section of the wrap. The student procedure should be evaluated according to organization and accuracy and how well it follows the steps of the scientific method.

Checkpoint 2: During the second class period, students should revise procedures according to the teacher's approval and begin the procedure. By the end of the second class period, students should be able to draw a reasonable conclusion and determine identity of the unknown and its concentration. Students should be evaluated on lab technique, quality and clarity of observations, and the explanation of observations/conclusions.

Sample Answer to Procedure

The unknown can be identified as an acid with phenolphthalein, pH paper, or a pH probe. Place 25.0 mL of unknown in a flask. After cleaning and rinsing a buret with 0.50 M NaOH, fill the buret with NaOH. Transfer small amounts of NaOH from the buret to the flask, swirling the flask after each addition, until the end point is reached. Repeat the procedure with 25.0 mL of unknown.

Sample Data

Solution	Trial 1	Trial 2
Unknown (HCl)	25.0 mL	25.0 mL
0.5 M NaOH	9.9 mL	10.0 mL

REACTION RATES

Teacher's Notes

Time Required
1 lab period

Ratings

Easy 1 — 2 — 3 — 4 Hard

TEACHER PREPARATION	2
STUDENT SETUP	1
CONCEPT LEVEL	3
CLEANUP	1

Skills Acquired
- Classifying
- Collecting Data
- Experimenting
- Identifying/Recognizing Patterns
- Interpreting
- Measuring
- Organizing and Analyzing Data
- Predicting

The Scientific Method
In this lab, students will:
- Make Observations
- Analyze the Results
- Draw Conclusions

Skills Practice Lab

OBJECTIVES

- Prepare and observe several different reaction mixtures.

- Demonstrate proficiency in measuring reaction rates.

- Analyze the results and relate experimental results to a rate law that you can use to predict the results of various combinations of reactants.

USING SCIENTIFIC **METHODS**

MATERIALS

- 8-well microscale reaction strips (2)
- distilled or deionized water
- fine-tipped dropper bulbs or small micro-tip pipets (3)
- solution A
- solution B
- stopwatch or clock with second hand

16 Reaction Rates

Introduction

Executive "toys" are a big business. Your company has been contacted by a toy company that wants technical assistance in designing a new executive desk gadget. The company wants to investigate a reaction that turns a distinctive color in a specific amount of time. Although it will not be easy to determine the precise combination of chemicals that will work, the profit the company stands to make would be worthwhile in the end.

In this experiment you will determine the rate of an *oxidation-reduction*, or *redox*, reaction. Reactions of this type are discussed in the chapter "Electrochemistry." The net equation for the reaction you will study is as follows:

$$3Na_2S_2O_5(aq) + 2KIO_3(aq) + 3H_2O(l) \xrightarrow{H^+}$$
$$2KI(aq) + 6NaHSO_4(aq)$$

One way to study the rate of this reaction is to observe how fast $Na_2S_2O_5$ is used up. After all the $Na_2S_2O_5$ solution has reacted, the concentration of iodine, I_2, an intermediate in the reaction, builds up. A starch indicator solution added to the reaction mixture will signal when this happens. The colorless starch will change to a blue-black color in the presence of I_2.

In the experiment, the concentrations of the reactants are given in terms of drops of Solution A and drops of Solution B. Solution A contains $Na_2S_2O_5$, the starch-indicator solution, and dilute sulfuric acid to supply the hydrogen ions needed to catalyze the reaction. Solution B contains KIO_3. You will run the reaction with several different concentrations of the reactants and record the time it takes for the blue-black color to appear.

To determine the best conditions and concentrations for the reaction, you will determine the following:
- how changes in reactant concentrations affect the reaction outcome
- how much time elapses for each reaction
- a rate law for the reaction that will allow you to predict the results with other combinations

814

Safety Procedures

- Wear safety goggles when working around chemicals, acids, bases, flames, or heating devices. Contents under pressure may become projectiles and cause serious injury.
- Never look directly at the sun through any optical device or use direct sunlight to illuminate a microscope.
- Avoid wearing contact lenses in the lab.
- If any substance gets in your eyes, notify your instructor immediately and flush your eyes with running water for at least 15 min.

- Secure loose clothing, and remove dangling jewelry. Don't wear open-toed shoes or sandals in the lab.
- Wear an apron or lab coat to protect your clothing when working with chemicals.
- If a spill gets on your clothing, rinse it off immediately with water for at least 5 min while notifying your instructor.

- If a chemical gets on your skin or clothing or in your eyes, rinse it immediately, and alert your instructor.
- If a chemical is spilled on the floor or lab bench, alert your instructor, but do not clean it up yourself unless your teacher says it is OK to do so.

- Always use caution when working with chemicals.
- Never mix chemicals unless specifically directed to do so.
- Never taste, touch, or smell chemicals unless specifically directed to do so.
- Add acid or base to water; never do the opposite.
- Never return unused chemicals to the original container.
- Never transfer substances by sucking on a pipette or straw; use a suction bulb.
- Follow instructions for proper disposal.

- Clean and decontaminate all work surfaces and personal protective equipment as directed by your instructor.
- Dispose of all sharps (broken glass and other contaminated sharp objects) and other contaminated materials (biological and chemical) in special containers as directed by your instructor.

Procedure

1. Copy the data table below in your lab notebook.

Data Table 1

	Well 1	Well 2	Well 3	Well 4	Well 5
Time reaction began					
Time reaction stopped					
Drops of A					
Drops of B					
Drops of H_2O					

Materials

1. To prepare clock solution A, make a paste of 1 g of water-soluble starch and about 10 mL of water. Add 225 mL of boiling distilled water. Reheat and boil the solution for a few minutes. After the solution has cooled, add 0.05 g of sodium metabisulfate, $Na_2S_2O_5$, and 1.3 mL of 1.0 M H_2SO_4. Dilute with enough distilled water to make 250 mL of solution. (Note: This solution should be prepared within months of its use.)

2. To prepare 10 mL of 1.0 M H_2SO_4, observe the required precautions. Add 0.6 mL of concentrated H_2SO_4 to 9.4 mL of distilled water. Add the acid slowly while stirring to avoid overheating.

3. To prepare clock solution B, dissolve 1.07 g KIO_3 in enough distilled water to make 250 mL of solution.

4. If the reaction suggested here is too slow, add a few milligrams of $Na_2S_2O_5$ to solution A. If it is too fast, add some distilled water to clock solution B to dilute the KIO_3 solution.

5. Test the reaction before students begin. Make certain you are familiar with proper disposal methods for the reactants and products.

6. Fine-tipped dropper bulbs can be made out of thin-stem pipets. Stretch the pipets by hand, narrowing the tube near the bulb, and then cut off the excess tube.

Safety Cautions

- Read all safety precautions, and discuss them with your students.
- Safety goggles and a lab apron must be worn at all times.
- Students should not handle concentrated acid solutions.
- In case of an acid or base spill, dilute the spill with water. Then mop up the spill with wet cloths or a wet cloth mop designated for spill cleanup. Wear disposable plastic gloves while cleaning up spills.
- Wear safety goggles, a face shield, impermeable gloves, and a lab apron when you prepare the HCl. Work in a fume hood known to be in good operating condition, and have another person standing by to call for help in case of an emergency. Be sure you are within a 30 second walk of a safety shower and eyewash station.

Disposal Information

Set out one container for disposal. Treat the waste with 1.0 M $Na_2S_2O_3$ solution to be certain all iodine is reduced to iodide. Neutralize the solution with 1.0 M acid or base, and pour it down the drain.

Techniques to Demonstrate

Point out that holding the dropper bottles vertically will help ensure consistently sized drops.

Tips and Tricks

Thoroughly discuss the concept of rate. Work through a calculation with sample data, relating time elapsed to rate and the resulting rate data to a rate expression that involves concentration.

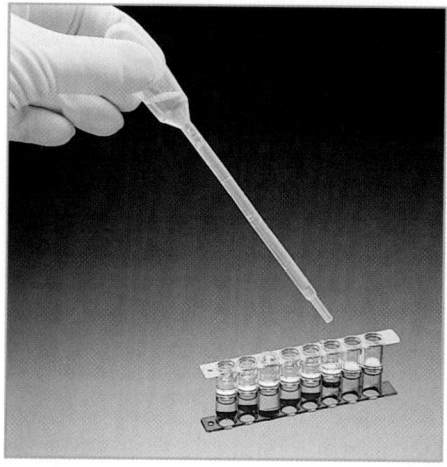

Figure A

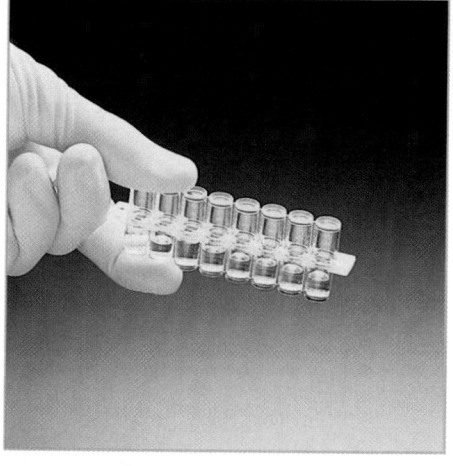

Figure B

2. Obtain three dropper bulbs or small microtip pipets, and label them "A," "B," and "H_2O."

3. Fill bulb or pipet A with solution A, fill bulb or pipet B with solution B, and fill the bulb or pipet for H_2O with distilled water.

4. Using the first 8-well strip, place five drops of Solution A into each of the first five wells, as shown in **Figure A.** (Disregard the remaining three wells.) Record the number of drops in the appropriate places in your data table. For best results, try to make all of the drops about the same size.

5. In the second 8-well reaction strip, place one drop of Solution B in the first well, two drops in the second well, three drops in the third well, four drops in the fourth well, and five drops in the fifth well. Record the number of drops in the appropriate places in your data table.

6. In the second 8-well strip that contains drops of Solution B, add four drops of water to the first well, three drops to the second well, two drops to the third well, and one drop to the fourth well. Do not add any water to the fifth well.

7. Carefully invert the second strip. The surface tension should keep the solutions from falling out of the wells. Place the second strip well-to-well on top of the first strip, as shown in **Figure B.**

8. Holding the strips tightly together, record the exact time or set the stopwatch as you shake the strips. This procedure should effectively mix the upper solutions with each of the corresponding lower ones.

9. Observe the lower wells. Note the sequence in which the solutions react, and record the number of seconds it takes for each solution to turn a blue-black color.

10. Dispose of the solutions in the container designated by your teacher. Wash your hands thoroughly after cleaning up the area and equipment.

Analysis

1. Organizing Data

Calculate the time elapsed for the complete reaction of each combination of Solution A and Solution B.

2. Constructing Graphs

Make a graph of your results. Label the x-axis "Number of drops of Solution B." Label the y-axis "Time elapsed." Make a similar graph for drops of Solution B versus rate (1/time elapsed).

3. Analyzing Data

Which mixture reacted the fastest? Which mixture reacted the slowest?

4. Explaining Events

Why was it important to add the drops of water to the wells that contained fewer than five drops of Solution B? (Hint: Figure out the total number of drops in each of the reaction wells.)

Conclusions

5. Evaluating Methods

How can you be sure that each of the chemical reactions began at about the same time? Why is this important?

6. Evaluating Results

Of the following variables that can affect the rate of a reaction, which is tested in this experiment: temperature, catalyst, concentration, surface area, or nature of reactants? Explain your answer.

7. Analyzing Graphs

Use your data and graphs to determine the relationship between the concentration of Solution B and the rate of the reaction. Describe this relationship in terms of a rate law.

8. Evaluating Data

Share your data with other lab groups, and calculate a class average for the rate of the reaction for each concentration of B. Compare the results from other groups with your results. Explain why there are differences in the results.

9. Evaluating Methods

What are some possible sources of error in this

procedure? If you can think of ways to eliminate them, ask your teacher to approve your plan and run your procedure again.

10. Making Predictions

How would your data be different if the experiment were repeated but Solution A was diluted with one part solution for every seven parts distilled water?

Extensions

1. Designing Experiments

What combination of drops of Solutions A and B would you use if you wanted the reaction to last exactly 2.5 min? Design an experiment to test your answer. If your teacher approves your plan, perform the experiment, and record these results. Make another graph that includes both the old and new data.

2. Designing Experiments

How would you determine the smallest interval of time during which you could distinguish a clock reaction? Design an experiment to find out. If your teacher approves your plan, perform your experiment.

3. Designing Experiments

How would the results of this experiment be affected if the reaction took place in a cold environment? Design an experiment to test your answer using materials available. If your teacher approves your plan, perform your experiment and record the results. Make another graph, and compare it with your old data.

4. Designing Experiments

Devise a plan to determine the effect of Solution A on the rate law. If your teacher approves your plan, perform your experiment, and determine the rate law for this reaction.

5. Building Models

If Solution B contains 0.02 M KIO_3, calculate the value for the constant, k, in the expression below. (Hint: Remember that Solution B is diluted when it is added to Solution A.)

$$rate = k[KIO_3]$$

817

Answers to Analysis

1. Students will need to convert from minutes and seconds to just seconds.

2. Accept all graphs that are labeled correctly and represent data accurately.

3. The most concentrated solution (5 drops of B) had the fastest reaction time. The least concentrated solution (1 drop of B and 4 drops of water) had the slowest reaction time.

4. The total number of drops for each well should be the same to make valid comparisons of different concentrations.

Answers to Conclusions

5. The reactions began at the same time because the shake-start mixes all of the solutions at once. In this way, all the reactions take place at the same time and under the same conditions.

6. The effect of reactant concentration (for Solution B) on reaction rate is being measured in this experiment.

7. The rate of the reaction is directly proportional to the concentration of Solution B. R = k[B] or R α [B]

8. Answers will vary because of variations in drop size.

9. Suggestions for improving the procedure will vary. Possibilities may include minimizing error by performing multiple trials or keeping the drop size standard. Be sure answers include safe, carefully planned procedures.

10. Suggestions will vary. Be sure answers include safe, carefully planned procedures. Students should find that the rate is seven times slower.

Answers to Extensions

1. Answers will vary depending on the conditions in your lab. Students may need to use a different reaction vessel if this is too much for the micro-well strips. Be sure answers include safe, carefully planned procedures.

2. Answers may vary. Determine the difference in time for the smallest difference in the amounts of the reactants. Be sure answers include safe, carefully planned procedures.

3. Answers will vary. Be sure students' suggestions are safe and include carefully planned procedures. The rate is slower at lower temperatures.

4. Students may suggest using drops of water and Solution A and keeping the amount of Solution B constant. Another linear relationship is involved.
R = k[A][B]

5. Answers should be in the range of 3.5 to 4.0.

REDOX TITRATION

Teacher's Notes

Time Required
1 lab period

Ratings

TEACHER PREPARATION	3
STUDENT SETUP	3
CONCEPT LEVEL	3
CLEANUP	2

Skills Acquired
• Collecting Data
• Experimenting
• Interpreting
• Measuring
• Organizing and Analyzing Data

The Scientific Method
In this lab, students will:
• Make Observations
• Analyze the Results
• Draw Conclusions
• Communicate the Results

Skills Practice Lab

OBJECTIVES

♦ Demonstrate proficiency in performing redox titrations and recognizing the end point of a redox reaction.

♦ Determine the concentration of a solution using stoichiometry and volume data from a titration.

MATERIALS

♦ beaker 250 mL (2)
♦ beaker 400 mL
♦ burets (2)
♦ distilled water
♦ double buret clamp
♦ Erlenmeyer flask, 125 mL (4)
♦ $FeSO_4$ solution
♦ graduated cylinder, 100 mL
♦ H_2SO_4, 1.0 M
♦ $KMnO_4$, 0.0200 M
♦ ring stand
♦ wash bottle

17 Redox Titration

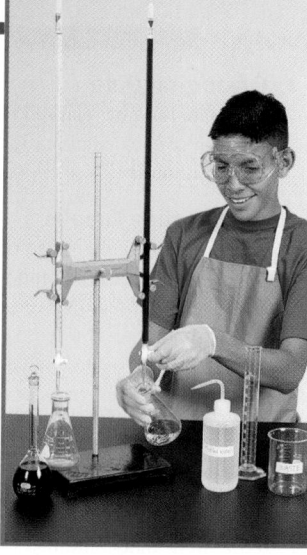

Introduction

You are a chemist working for a chemical analysis firm. A large pharmaceutical company has hired you to help salvage some products that were damaged by a small fire in their warehouse. Although there was only minimal smoke and fire damage to the warehouse and products, the sprinkler system ruined the labeling on many of the pharmaceuticals. The firm's best-selling products are iron tonics used to treat low-level anemia. The tonics are produced from hydrated iron(II) sulfate, $FeSO_4 \cdot 7H_2O$. The different types of tonics contain different concentrations of $FeSO_4$. You have been hired to help the pharmaceutical company figure out the proper label for each bottle of tonic.

In the chapter "Acids and Bases" you studied acid-base titrations in which an unknown amount of acid is titrated with a carefully measured amount of base. In this procedure a similar approach called a redox titration is used. In a redox titration, the reducing agent, Fe^{2+}, is oxidized to Fe^{3+} by the oxidizing agent, MnO_4^-. When this process occurs, the Mn in MnO_4^- changes from a +7 to a +2 oxidation state and has a noticeably different color. You can use this color change in the same way that you used the color change of phenolphthalein in acid-base titrations—to signify a redox reaction end point. When the reaction is complete, any excess MnO_4^- added to the reaction mixture will give the solution a pink or purple color. The volume data from the titration, the known molarity of the $KMnO_4$ solution, and the mole ratio from the following balanced redox equation will give you the information you need to calculate the molarity of the $FeSO_4$ solution.

$$5Fe^{2+}(aq) + MnO_4^-(aq) + 8H^+(aq) \longrightarrow$$
$$5Fe^{3+}(aq) + Mn^{2+}(aq) + 4H_2O(l)$$

To determine how to label the bottles, you must determine the concentration of iron(II) ions in the sample from an unlabeled bottle from the warehouse by answering the following questions:

• How can the volume data obtained from the titration and the mole ratios from the balanced redox reaction be used to determine the concentration of the sample?

• Which tonic is in the sample, given information about the concentration of each tonic?

Materials

1. To prepare 0.0200 M $KMnO_4$, dissolve 3.16 g of $KMnO_4$ in enough distilled water to make 1.00 L of solution. For best results, this solution must be prepared shortly before the lab.

2. To prepare 0.15 M $FeSO_4$, dissolve 41.70 g of $FeSO_4 \cdot 7H_2O$ in enough distilled water to make 1.00 L of solution.

3. To prepare 1.00 L of 1.0 M H_2SO_4, observe the required precautions. Slowly add 56 mL of concentrated H_2SO_4 to enough distilled water to make 1.00 L of solution. Add the acid slowly while stirring to avoid overheating.

Safety Procedures

- Wear safety goggles when working around chemicals, acids, bases, flames, or heating devices. Contents under pressure may become projectiles and cause serious injury.
- Never look directly at the sun through any optical device or use direct sunlight to illuminate a microscope.
- Avoid wearing contact lenses in the lab.
- If any substance gets in your eyes, notify your instructor immediately and flush your eyes with running water for at least 15 min.

- Secure loose clothing, and remove dangling jewelry. Don't wear open-toed shoes or sandals in the lab.
- Wear an apron or lab coat to protect your clothing when working with chemicals.
- If a spill gets on your clothing, rinse it off immediately with water for at least 5 min while notifying your instructor.

- If a chemical gets on your skin or clothing or in your eyes, rinse it immediately, and alert your instructor.
- If a chemical is spilled on the floor or lab bench, alert your instructor, but do not clean it up yourself unless your teacher says it is OK to do so.

- Always use caution when working with chemicals.
- Never mix chemicals unless specifically directed to do so.
- Never taste, touch, or smell chemicals unless specifically directed to do so.
- Add acid or base to water; never do the opposite.
- Never return unused chemicals to the original container.
- Never transfer substances by sucking on a pipette or straw; use a suction bulb.
- Follow instructions for proper disposal.

- Check the condition of glassware before and after using it. Inform your teacher of any broken, chipped, or cracked glassware, because it should not be used.
- Do not pick up broken glass with your bare hands. Place broken glass in a specially designated disposal container.

- Clean and decontaminate all work surfaces and personal protective equipment as directed by your instructor.
- Dispose of all sharps (broken glass and other contaminated sharp objects) and other contaminated materials (biological and chemical) in special containers as directed by your instructor.

Safety Cautions

- Read all safety precautions, and discuss them with your students.
- Safety goggles and a lab apron must be worn at all times.
- Students should not handle concentrated acid solutions. In case of an acid spill, dilute the spill with water. Then mop up the spill with wet cloths or a wet cloth mop designated for spill cleanup. Wear disposable plastic gloves while cleaning up spills.
- Wear safety goggles, a face shield, impermeable gloves, and a lab apron when you prepare the HCl. Work in a fume hood known to be in good operating condition, and have another person standing by to call for help in case of an emergency. Be sure you are within a 30 second walk of a safety shower and eyewash station.

Disposal Information

Set out one container for disposal. The mixture that results should be acidic. If the mixture is purple, add $FeSO_4$ solution slowly while stirring until the color is gone. Precipitate the iron by adding 1.0 M NaOH. Filter the iron, and put the precipitate in the trash. Neutralize the filtrate with 1.0 M acid or base until its pH is between 5 and 9, and pour it down the drain.

Data Table 1

Trial	Initial $KMnO_4$ volume	Final $KMnO_4$ volume	Initial $FeSO_4$ volume	Final $FeSO_4$ volume
1				
2				
3				

Procedure

1. Organizing Data
Copy the data table above in your lab notebook.

2. Clean two 50 mL burets with a buret brush and distilled water. Rinse each buret at least three times with distilled water to remove any contaminants.

3. Label two 250 mL beakers "0.0200 M $KMnO_4$," and "$FeSO_4$ solution." Label three of the flasks 1, 2, and 3. Label the 400 mL beaker "Waste." Label one buret "$KMnO_4$" and the other "$FeSO_4$."

4. Measure approximately 75 mL of 0.0200 M $KMnO_4$, and pour it into the appropriately labeled beaker. Obtain approximately 75 mL of $FeSO_4$ solution, and pour it into the appropriately labeled beaker.

5. Rinse one buret three times with a few milliliters of 0.0200 M $KMnO_4$ from the appropriately labeled beaker. Collect these rinses in the

819

Techniques to Demonstrate

At this point, students should have had plenty of practice in operating burets with acid-base titrations. Remind students to create the end-point standard in step 7.

Tips and Tricks

Thoroughly discuss the concept of oxidation-reduction and how this type of reaction can result in a color change. Review the concept of molar ratios and balancing redox equations.

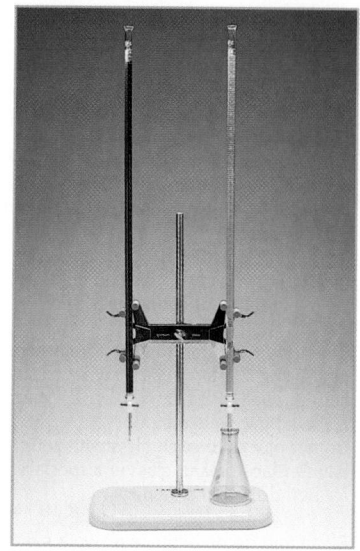

Figure A

Figure B

waste beaker. Rinse the other buret three times with small amounts of $FeSO_4$ solution from the appropriately labeled beaker. Collect these rinses in the waste beaker.

6. Set up the burets as shown in **Figure A.** Fill one buret with approximately 50 mL of the 0.0200 M $KMnO_4$ from the beaker, and fill the other buret with approximately 50 mL of the FeSO4 solution from the other beaker.

7. With the waste beaker underneath its tip, open the $KMnO_4$ buret long enough to be sure the buret tip is filled. Repeat for the $FeSO_4$ buret.

8. Add 50 mL of distilled water to one of the 125 mL Erlenmeyer flasks, and add one drop of 0.0200 M $KMnO_4$ to the flask. Set this aside to use as a color standard, as shown in **Figure B,** to compare with the titration and to determine the end point.

9. Record the initial buret readings for both solutions in your data table. Add 10.0 mL of the

hydrated iron(II) sulfate, $FeSO_4 \cdot 7H_2O$, solution to flask 1. Add 5 mL of 1.0 M H_2SO_4 to the $FeSO_4$ solution in this flask. The acid will help keep the Fe^{2+} ions in the reduced state, allowing you time to titrate.

10. Slowly add $KMnO_4$ from the buret to the $FeSO_4$ in the flask while swirling the flask, as shown in **Figure C.** When the color of the solution matches the color standard you prepared in step 8, record the final readings of the burets in your data table.

11. Empty the titration flask into the waste beaker. Repeat the titration procedure in steps 9 and 10 with flasks 2 and 3.

12. Always clean up the lab and all equipment after use. Dispose of the contents of the waste beaker in the container designated by your teacher. Also pour the contents of the color-standard flask into this container. Wash your hands thoroughly after cleaning up the area and equipment.

820

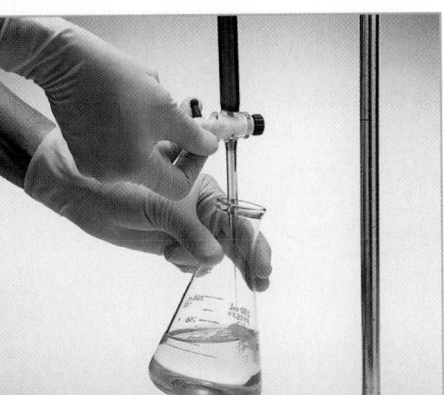

Figure C

Analysis

1. Analyzing Data
Calculate the number of moles of MnO_4^- reduced in each trial.

2. Analyzing Data
Calculate the number of moles of Fe^{2+} oxidized in each trial.

3. Analyzing Data
Calculate the average concentration (molarity) of the iron tonic.

4. Explaining Events
Explain why it was important to rinse the burets with $KMnO_4$ or $FeSO_4$ before adding the solutions. (Hint: Consider what would happen to the concentration of each solution if it were added to a buret that had been rinsed only with distilled water.)

Conclusions

5. Evaluating Data
The company makes three different types of iron tonics: Feravide A, with a concentration of 0.145 M $FeSO_4$; Feravide Extra-Strength, with 0.225 M $FeSO_4$; and Feravide Jr., with 0.120 M $FeSO_4$. Which tonic is your sample?

6. Evaluating Methods
What possible sources of error can you identify with this procedure? If you can think of ways to eliminate them, ask your teacher to approve your plan, and run the procedure again.

Extensions

1. Research and Communication
Blueprints are based on a photochemical reaction. The paper is treated with a solution of iron(III) ammonium citrate and potassium hexacyanoferrate(III) and dried in the dark. When a tracing-paper drawing is placed on the blueprint paper and exposed to light, Fe^{3+} ions are reduced to Fe^{2+} ions, which react with hexacyanoferrate(III) ions in the moist paper to form the blue color on the paper. The lines of the drawing block the light and prevent the reduction of Fe^{3+} ions, resulting in white lines. Find out how sepia prints are made, and report on this information.

2. Building Models
Electrochemical cells are based on the process of electron flow in a system with varying potential differences. Batteries are composed of such systems and contain different chemicals for different purposes and price ranges. You can make simple experimental batteries using metal wires and items such as lemons, apples, and potatoes. What are some other "homemade" battery sources, and what is the role of these food items in producing electrical energy that can be measured as battery power? Explain your answers.

821

Answers to Analysis

1. Answers will vary.
2. Answers will vary.
3. Answers will vary.
4. The concentrations of each solution is very important. If you rinse the buret only with distilled water, the solutions might be diluted if there was any water left in the buret. By using the solution being measured to rinse the buret, you can be sure any solution left behind will not dilute the solution added.

Answers to Conclusion

5. The sample tonic is most likely Feravide A, 0.145 M $FeSO_4$.
6. Suggestions for improving the procedure will vary. Possible suggestions include performing repeated trials or using larger volumes for the titration. Be sure answers include safe, carefully planned procedures.

Answers to Extensions

1. Sepia printing is similar to blueprint printing. In sepia printing, the potassium hexacyanoferrate(III) oxidizes the crystals of silver in the print so that they become silver ions again. Then the print is redeveloped in sodium sulfide.
2. Answers will vary. The food is the conductor that completes the circuit between the two metals, each of which has a different reduction potential. In terms of a traditional cell, the food acts as both the porous barrier that separates the metal and the solution of aqueous ions that completes the circuit.

REDOX TITRATION: MINING FEASIBILITY STUDY

Teacher's Notes

Time Required
2 lab periods

Ratings
$$\underset{\text{EASY} \qquad\qquad \text{HARD}}{\overset{1 \quad 2 \quad 3 \quad 4}{\rule{3cm}{0.4pt}}}$$

TEACHER PREPARATION   3
STUDENT SETUP   3
CONCEPT LEVEL   3
CLEANUP   2

Skills Acquired
• Collecting Data
• Communicating
• Designing Experiments
• Experimenting
• Identifying/Recognizing Patterns
• Inferring
• Interpreting
• Measuring
• Organizing and Analyzing Data
• Predicting

The Scientific Method
In this lab, students will:
• Make Observations
• Ask Questions
• Form a Hypothesis
• Test the Hypothesis
• Analyze the Results
• Draw Conclusions
• Communicate the Results

Inquiry LAB
Design Your Own Experiment

17 Redox Titration
Mining Feasibility Study

THE PROBLEM

May 11, 2004

George Taylor
Director of Analytical Services
CheMystery Labs, Inc.
52 Fulton Street
Springfield, VA 22150

Dear Mr. Taylor:

Because of the high quality of your firm's work in the past, Goldstake is again asking that you submit a bid for a mining feasibility study. A study site in New Mexico has yielded some promising iron ore deposits, and we are evaluating the potential yield.

Your bid should include the cost of evaluating the sample we are sending with this letter and the fees for 20 additional analyses to be completed over the next year. The sample is a slurry extracted from the mine using a special process that converts the iron ore into iron(II) sulfate, $FeSO_4$, dissolved in water. The mine could produce up to 1.0×10^5 L of this slurry daily, but we need to know how much iron is in that amount of slurry before we proceed.

The contract for the other analyses will be awarded based on the accuracy of this analysis and the quality of the report. Your report will be used for two purposes: to evaluate the site for quantity of iron and to determine who our analytical consultant will be if the site is developed into a mining operation. I look forward to reviewing your bid proposal.

Sincerely,

Lynn L. Brown

Lynn L. Brown
Director of Operations
Goldstake Mining Company

References
Review more information on redox reactions. Remember to add a small amount of sulfuric acid, H_2SO_4, so the iron will stay in the Fe^{2+} form. Calculate your disposal costs based on the mass of potassium permanganate, $KMnO_4$, and $FeSO_4$ in your solutions, as well as the mass of the H_2SO_4 solution.

822

Materials
• beaker, 400 mL
• beakers, 250 mL (2)
• burets (2)
• distilled water
• double buret clamp
• Erlenmeyer flasks, 125 mL (4)
• graduated cylinder, 100 mL
• H_2SO_4, 1.0 M (5 mL)
• $KMnO_4$, 0.0200 M (100 mL)
• ring stand
• unknown solution, 50 mL (0.25 M $FeSO_4$)
• wash bottle

THE PLAN

CheMystery Labs, Inc. 52 Fulton Street, Springfield, VA 22150

Memorandum

Date: May 12, 2004
To: Crystal Sievers
From: George Taylor

Good news! The quality of our work has earned us a repeat customer, Goldstake Mining Company. This analysis could turn into a long-term contract. Perform the analysis more than one time so that we can be confident of our accuracy. Before you begin your analysis, send Ms. Brown the following items:

• a detailed, one-page plan for the procedure and all necessary data tables
• a detailed sheet that lists all of the equipment and materials you plan to use

When you have completed the laboratory work, please prepare a report in the form of a two-page letter to Ms. Brown. Include the following information:

• moles and grams of $FeSO_4$ in 10 mL of sample
• moles, grams, and percentage of iron(II) in 10 mL of the sample
• the number of kilograms of iron that the company could extract from the mine each year, assuming that 1.0×10^5 L of slurry could be mined per day, year round
• a balanced equation for the redox equation
• a detailed and organized data and analysis section showing calculations of how you determined the moles, grams, and percentage of iron(II) in the sample (include calculations of the mean, or average, of the multiple trials)

Required Precautions

• Wear safety goggles when working around chemicals, acids, bases, flames, or heating devices. Contents under pressure may become projectiles and cause serious injury.
• Avoid wearing contact lenses in the lab.
• If any substance gets in your eyes, notify your instructor immediately and flush your eyes with running water for at least 15 min.
• Secure loose clothing, and remove dangling jewelry. Don't wear open-toed shoes or sandals in the lab.
• Wear an apron or lab coat to pro-

tect your clothing when working with chemicals.
• If a spill gets on your clothing, rinse it off immediately with water for at least 5 min while notifying your instructor.

• Always use caution when working with chemicals.
• Never taste, touch, or smell chemicals unless specifically directed to do so.
• Follow instructions for proper disposal.

• Whenever possible, use an electric hot plate as a heat source instead of an open flame.

• When heating materials in a test tube, always angle the test tube away from yourself and others.
• Know your school's fire-evacuation routes.
• Clean and decontaminate all work surfaces and personal protective equipment as directed by your instructor.

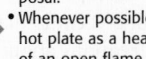

• Dispose of all sharps (broken glass and other contaminated sharp objects) and other contaminated materials (biological and chemical) in special containers as directed by your instructor.

823

Sample Data

Trial	Initial FeSO₄ volume (mL)	Final FeSO₄ volume (mL)	Initial KMnO₄ volume (mL)	Final KMnO₄ volume (mL)
1	50.0	40.0	50.0	25.0
2	40.0	30.0	50.0	25.2
3	30.0	20.0	50.0	24.8

Safety Cautions

• Safety goggles and a lab apron must be worn at all times.
• Read all safety precautions, and discuss them with your students.
• Students should not handle concentrated acids. In case of an acid spill, dilute the spill with water. Then mop up the spill with wet cloths or a wet cloth mop designated for spill cleanup. Wear disposable plastic gloves while cleaning up spills.

Checkpoints

Checkpoint 1: By the end of the first class period, students should turn in a detailed one-page plan/procedure for approval. This should be compared to the sample procedure found in the answer section of the wrap. The student procedure should be evaluated according to organization and accuracy and how well it follows the steps of the scientific method.

Checkpoint 2: During the second class period, students should revise procedures according to the teacher's approval and begin the procedure. By the end of the second class period, students should be able to draw a reasonable conclusion and determine the feasibility of evaluating ore samples. Students should be evaluated on lab technique, quality and clarity of observations, and the explanation of observations/conclusions.

Sample Answer to Procedure

Prepare the burets for the titration using $KMnO_4$ in one buret and $FeSO_4$ in the other. Put a drop of $KMnO_4$ in one Erlenmeyer flask to serve as a color standard to recognize the end point.

Put 10.0 mL of the $FeSO_4$ solution in an Erlenmeyer flask. Add 5 mL of 1.0 M H_2SO_4. Slowly add $KMnO_4$ from the buret. Swirl after each drop is added. When the color matches the standard, record the final volumes of the burets. Repeat the titration procedure twice.

POLYMERS AND TOY BALLS

Teacher's Notes

Time Required
1 lab period

Ratings
EASY 1 2 3 4 HARD

TEACHER PREPARATION	2
STUDENT SETUP	1
CONCEPT LEVEL	1
CLEANUP	2

Skills Acquired
• Collecting Data
• Experimenting
• Interpreting
• Measuring
• Organizing and Analyzing Data

The Scientific Method
In this lab, students will:
• Make Observations
• Analyze the Results
• Draw Conclusions

Skills Practice Lab

OBJECTIVES

♦ Synthesize two different polymers.

♦ Prepare a small toy ball from each polymer.

♦ Observe the similarities and differences between the two types of balls.

♦ Measure the density of each polymer.

♦ Compare the bounce height of the two balls.

MATERIALS

♦ acetic acid solution (vinegar), 5% (10 mL)
♦ beaker, 2 L, or plastic bucket or tub
♦ distilled water
♦ ethanol solution, 50% (3 mL)
♦ gloves
♦ graduated cylinder, 10 mL
♦ graduated cylinder, 25 mL
♦ liquid latex (10 mL)
♦ meterstick
♦ paper cups, 5 oz (2)
♦ paper towels
♦ sodium silicate solution (12 mL)
♦ wooden stick

19 Polymers and Toy Balls

Introduction

Your company has been contacted by a toy company that specializes in toy balls made from vulcanized rubber. Recent legislation has increased the cost of disposing of the sulfur and other chemical byproducts of the manufacturing process for this type of rubber. The toy company wants you to research some other materials.

Rubber is a polymer of covalently bonded atoms. When rubber is vulcanized, it is heated with sulfur. The sulfur atoms form bonds between adjacent molecules of rubber, which increases its strength and making it more elastic.

Latex rubber is a colloidal suspension that can be made synthetically or found naturally in plants. Latex is composed of approximately 60% water, 35% hydrocarbon monomers, 2% proteins, and some sugars and inorganic salts.

The polymer formed from ethanol, C_2H_5OH, and a solution of sodium silicate, mostly in the form of $Na_2Si_3O_7$, also has covalent bonds. When the polymer is formed, water is also a product.

Latex rubber and the ethanol–sodium silicate polymer are the two materials you will become familiar with as you do the following:
• Synthesize each polymer.
• Make a ball 2–3 cm in diameter from each polymer.
• Make observations about the physical properties of each polymer.
• Measure how well each ball bounces.

Safety Procedures

- Wear safety goggles when working around chemicals, acids, bases, flames, or heating devices. Contents under pressure may become projectiles and cause serious injury.
- Never look directly at the sun through any optical device or use direct sunlight to illuminate a microscope.
- Avoid wearing contact lenses in the lab.
- If any substance gets in your eyes, notify your instructor immediately and flush your eyes with running water for at least 15 min.

- Secure loose clothing, and remove dangling jewelry. Don't wear open-toed shoes or sandals in the lab.
- Wear an apron or lab coat to protect your clothing when working with chemicals.
- If a spill gets on your clothing, rinse it off immediately with water for at least 5 min while notifying your instructor.

- If a chemical gets on your skin or clothing or in your eyes, rinse it immediately, and alert your instructor.

- If a chemical is spilled on the floor or lab bench, alert your instructor, but do not clean it up yourself unless your teacher says it is OK to do so.

- Always use caution when working with chemicals.
- Never mix chemicals unless specifically directed to do so.
- Never taste, touch, or smell chemicals unless specifically directed to do so.
- Add acid or base to water; never do the opposite.
- Never return unused chemicals to the original container.
- Never transfer substances by sucking on a pipette or straw; use a suction bulb.
- Follow instructions for proper disposal.

- Use flammable liquids only in small amounts.
- When working with flammable liquids, be sure that no one else in the lab is using a lit Bunsen burner or plans to use one. Make sure there are no other heat sources present.

Data Table 1

Trial	Height (cm)	Mass (g)	Diameter (cm)
1			
2			
3			

Procedure

1. Copy Data Table 1 above in your lab notebook. Be sure that you have plenty of room for observations about each test.

Organizing Data

2. Fill the 2 L beaker, bucket, or tub about half full with distilled water.

3. Using a clean 25 mL graduated cylinder, measure 10 mL of liquid latex and pour it into one of the paper cups.

4. Thoroughly clean the 25 mL graduated cylinder with soap and water, and then rinse it with distilled water.

5. Measure 10 mL of distilled water. Pour it into the paper cup with the latex.

6. Measure 10 mL of the 5% acetic acid solution, and pour it into the paper cup with the latex and water.

7. Immediately stir the mixture by using the wooden stick.

8. As you continue stirring, a polymer "lump" will form around the wooden stick. Pull the stick with the polymer lump from the paper cup, and immerse the lump in the 2 L beaker, bucket, or tub.

825

Materials

1. For 5% acetic acid, use white vinegar. Do not dilute glacial acetic acid.

2. Liquid latex may be purchased from Flinn Scientific, Inc., in Batavia, IL.

3. Sodium silicate solution, also known as water glass, can be purchased ready to use from many scientific supply companies.

4. Keep the bottles of latex and acetic acid solution in an operating hood because the vapors are unpleasant. Keep the bottle of ethanol in an operating fume hood because the vapors are flammable. Remind students to keep all three containers closed when not in use.

5. If balls need to be kept overnight, place them in plastic bags. If the ethanol-silicate polymer ball crumbles, add a few drops of water to it.

Safety Cautions

- Wear safety goggles and a lab apron during the lab.

- Read all safety precautions, and discuss them with your students.

- Promptly clean up all spills with paper towels.

- Ethanol is flammable. Ensure that there are no flames anywhere in the room when an alcohol is present. Keep the alcohol in a hood, use a container with a lid, and restrict the amount kept in the hood to the minimum needed by the students.

- Do not allow students to take any ethanol-silicate polymer from the laboratory.

Disposal Information

Paper cups, paper towels, disposable gloves, latex, and ethanol-silicate polymer balls and fragments should be thrown in the trash can. Waste liquids from this lab can be poured down the drain.

Techniques to Demonstrate

Show students how to roll the latex and the ethanol-silicate polymer into a ball. Students will have difficulty making a perfect sphere, but they should try to make it as regular as possible. The more irregular the shape is, the more difficulty students will have calculating the volume and determining the bounce height because the ball will not bounce straight up.

To determine the volume of the balls, students should use mathematical calculations and measurements of the diameters. Instruct students not to measure the volumes by water displacement. The ethanol-silicate polymer will dissolve in water.

Remind students to be patient with the ethanol-silicate polymer, which tends to crumble. If it crumbles too much, a few drops of water will rehydrate it so it can be reshaped into a ball.

Tips and Tricks

Review the material on polymers in the "Carbon and Organic Compounds" chapter. Be certain students understand the differences between a monomer and a polymer. Be sure to relate the properties of these polymers, such as strength, flexibility, and elasticity, to the nature of the covalent bonds holding them together.

9. While wearing gloves, gently pull the lump from the wooden stick. Be sure to keep the lump immersed in the water, as shown in **Figure A.**

10. Keep the latex rubber underwater, and use your gloved hands to mold the lump into a ball, as shown in **Figure B.** Squeeze the lump several times to remove any unused chemicals. You may remove the latex rubber from the water as you roll it in your hands to smooth the ball.

11. Set aside the latex-rubber ball to dry. While it is drying, begin to make a ball from the ethanol and sodium silicate solutions.

12. In a clean 25 mL graduated cylinder, measure 12 mL of sodium silicate solution and pour it into the other paper cup.

13. In a clean 10 mL graduated cylinder, measure 3 mL of 50% ethanol. Pour the ethanol into the paper cup with the sodium silicate, and mix with the wooden stick until a solid substance is formed.

14. While wearing gloves, remove the polymer that forms and place it in the palm of one hand, as shown in **Figure C.** Gently press it with the palms of both your hands until a ball that does not crumble is formed. This process takes a little time and patience. The liquid that comes out of the ball is a combination of ethanol and water. Occasionally moisten the ball by letting a small amount of water from a faucet run over it. When the ball no longer crumbles, you are ready to go on to the next step.

15. Observe as many physical properties of the balls as possible, and record your observations in your lab notebook.

16. Drop each ball several times, and record your observations.

17. Drop each ball from a height of 1 m, and measure its bounce. Perform three trials for each ball, and record the values in your data table.

18. Measure the diameter and the mass of each ball, and record the values in your data table.

19. Clean all apparatus and your lab station. Dispose of any extra solutions in the containers indicated by your teacher. Clean up your lab area. Remember to wash your hands thoroughly when your lab work is finished.

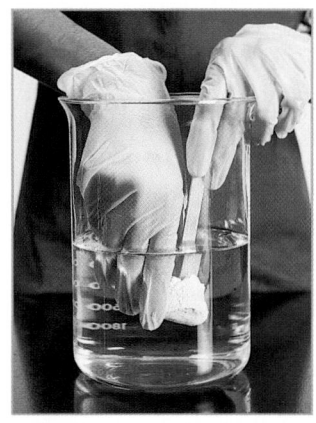

Figure A

Figure B

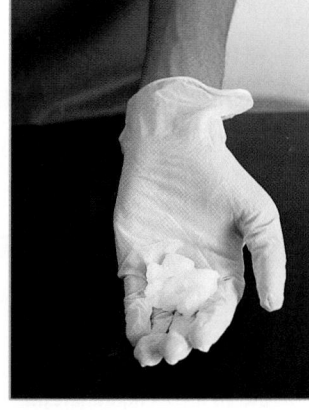

Figure C

826

Analysis

1. **Analyzing data** Give the chemical formula for the latex (isoprene) monomer and the ethanol–sodium silicate polymer.

2. **Analyzing data** List at least three observations you made of the properties of the two different balls.

3. **Explaining events** Explain how your observations in item 2 indicate that the polymers in each ball are not ionically bonded.

4. **Organizing data** Calculate the average height of the bounce for each type of ball.

5. **Organizing data** Calculate the volume for each ball. Even though the balls may not be perfectly spherical, assume that they are. (Hint: The volume of a sphere is equal to $\frac{4}{3} \times \pi \times r^3$, where r is the radius of the sphere, which is one-half of the diameter.)

6. **Organizing data** Using your measurements for the volumes from item 5 and the recorded mass, calculate the density of each ball.

Conclusions

7. **Evaluating data** Which polymer would you recommend for the toy company's new toy balls? Explain your reasoning.

8. **Evaluating results** Using the table shown below, calculate the unit cost, that is, the amount of money it costs to make a single ball. (Hint: Calculate how much of each reagent is needed to make a single ball.)

Data Table 2

Reagent	Price (dollars per liter)
Acetic acid solution	1.50
Ethanol solution	9.00
Latex solution	20.00
Sodium silicate solution	10.00

9. **Evaluating results** What are some other possible practical applications for each of the polymers you made?

10. **Making predictions** When a ball bounces up, kinetic energy of motion is converted into potential energy. With this in mind, explain which will bounce higher, a perfectly symmetrical, round sphere or an oblong shape that vibrates after it bounces.

11. **Evaluating methods** Explain why you didn't measure the volume of the balls by submerging them in water.

Extensions

1. **Research and communication** Polymers are used in our daily lives. Describe or list the polymers you come into contact with during a one-day period in your life.

2. **Designing experiments** Design a mold for a polymer ball that will make it symmetrical and smooth. If your teacher approves of your design, try the procedure again with the mold.

827

Answers

1. isoprene monomer: C_5H_8; ethanol-silicate monomer: $SiC_4H_{10}O_3$

2. Answers will vary but could include the following: The latex ball is more opaque, less smooth, and less crumbly than the ethanol-silicate polymer ball. The ethanol-silicate polymer ball breaks down after a period of time. Both balls bounce.

3. Ionic substances are not crumbly nor do they tend to bounce.

4. Answers will vary.

5. Answers will vary.

6. Answers will vary.

Answers to Conclusions

7. Answers will vary but should be based on the properties of the ball. Students may argue that because the ethanol-silicate polymer ball crumbles in time, it would be a poor candidate for a toy.

8. To make a latex ball you need 0.010 L of liquid latex and 0.010 L of acetic acid. The cost per liter of these combined substances is $21.50. The unit cost for a latex ball is $0.215. To make an ethanol-silicate polymer ball you need 0.012 L of sodium silicate solution and 0.003 L of 50% ethanol. The cost per liter of sodium silicate is $0.12. The cost per liter of 50% ethanol is $0.027. The unit cost for an ethanol-silicate polymer ball is $0.147.

9. Answers will vary but should be based on the properties of each substance.

10. A ball with an oblong shape will not bounce as high because some of the kinetic energy will be transferred into the energy of vibration, leaving less to be transferred into gravitational potential energy. A sphere that bounces will transfer nearly all of its kinetic energy into gravitational potential energy.

11. The ethanol-silicate polymer ball would break down in the water, so density had to be calculated from measurements of diameter.

Answers to Extensions

1. Answers will vary. Some possible polymers include rubber, proteins, cellulose, nylon, polyester, polyurethane, etc.

2. Suggestions for a mold will vary. Be sure students choose a material that will not dissolve in the solvents used in this process.

Appendix A

A CHEMICAL REFERENCE HANDBOOK

TABLE A-1   SI MEASUREMENT

Prefix	Symbol	Factor of Base Unit	Prefix	Symbol	Factor of Base Unit
giga	G	1 000 000 000	*centi*	c	0.01
mega	M	1 000 000	*milli*	m	0.001
kilo	k	1 000	*micro*	μ	0.000 001
hecto	h	100	*nano*	n	0.000 000 001
deka	da	10	*pico*	p	0.000 000 000 001
deci	d	0.1			

TABLE A-2   UNIT ABBREVIATIONS

amu	=	atomic mass unit (mass)	**mol**	=	mole (quantity)
atm	=	atmosphere (pressure, non-SI)	**M**	=	molarity (concentration)
Bq	=	becquerel (nuclear activity)	**N**	=	newton (force)
°C	=	degree Celsius (temperature)	**Pa**	=	pascal (pressure)
J	=	joule (energy)	**s**	=	second (time)
K	=	kelvin (temperature, thermo-dynamic)	**V**	=	volt (electric potential difference)

TABLE A-3   SYMBOLS

Symbol		Meaning	Symbol		Meaning
α	=	helium nucleus (also ^4_2He) emission from radioactive materials	ΔG^0	=	standard free energy of reaction
β	=	electron (also $^0_{-1}e$) emission from radioactive materials	ΔG^0_f	=	standard molar free energy of formation
γ	=	high-energy photon emission from radioactive materials	H	=	enthalpy
			ΔH^0	=	standard enthalpy of reaction
Δ	=	change in a given quantity (e.g., ΔH for change in enthalpy)	ΔH^0_f	=	standard molar enthalpy of formation
c	=	speed of light in vacuum	K_a	=	ionization constant (acid)
c_p	=	specific heat capacity (at constant pressure)	K_b	=	dissociation constant (base)
			K_{eq}	=	equilibrium constant
D	=	density	K_{sp}	=	solubility-product constant
E_a	=	activation energy	KE	=	kinetic energy
E^0	=	standard electrode potential	m	=	mass
E^0_{cell}	=	standard potential of an electrochemical cell	N_A	=	Avogadro's number
			n	=	number of moles
G	=	Gibbs free energy	P	=	pressure

828

TABLE A-3 CONTINUED

Symbol		Meaning
pH	=	measure of acidity $(-\log[H_3O^+])$
R	=	ideal gas law constant
S	=	entropy
S^0	=	standard molar entropy

Symbol		Meaning
T	=	temperature (thermodynamic, in kelvins)
t	=	temperature (in degrees Celsius)
V	=	volume
v	=	velocity or speed

TABLE A-4   PHYSICAL CONSTANTS

Quantity	Symbol	Value
Atomic mass unit	amu	$1.660\ 5402 \times 10^{-27}$ kg
Avogadro's number	N_A	$6.022\ 137 \times 10^{23}$/mol
Electron rest mass	m_e	$9.109\ 3897 \times 10^{-31}$ kg 5.4858×10^{-4} amu
Ideal gas law constant	R	8.314 L•kPa/mol•K 0.0821 L•atm/mol•K
Molar volume of ideal gas at STP	V_M	22.414 10 L/mol
Neutron rest mass	m_n	$1.674\ 9286 \times 10^{-27}$ kg 1.008 665 amu
Normal boiling point of water	T_b	373.15 K = 100.0°C
Normal freezing point of water	T_f	273.15 K = 0.00°C
Planck's constant	h	$6.626\ 076 \times 10^{-34}$ J•s
Proton rest mass	m_p	$1.672\ 6231 \times 10^{-27}$ kg 1.007 276 amu
Speed of light in a vacuum	c	$2.997\ 924\ 58 \times 10^8$ m/s
Temperature of triple point of water		273.16 K = 0.01°C

TABLE A-5   PROPERTIES OF COMMON ELEMENTS

Name	Form/color	Density (g/cm³)	Melting point (°C)	Boiling point (°C)	Common oxidation states
Aluminum	silver metal	2.702	660.37	2467	3+
Arsenic	gray metalloid	5.72714	817 (28 atm)	613 (sublimes)	3–, 3+, 5+
Barium	bluish white metal	3.51	725	1640	2+
Bromine	red-brown liquid	3.119	27.2	58.78	1–, 1+, 3+, 5+, 7+
Calcium	silver metal	1.54	839 ± 2	1484	2+
Carbon	diamond	3.51	3500 (63.5 atm)	3930	2+, 4+
	graphite	2.25	3652 (sublimes)	—	
Chlorine	green-yellow gas	3.214*	2100.98	234.6	1–, 1+, 3+, 5+, 7+
Chromium	gray metal	7.2028	1857 ± 20	2672	2+, 3+, 6+

continued on next page

TABLE A-5 CONTINUED

Name	Form/color	Density (g/cm³)	Melting point (°C)	Boiling point (°C)	Common oxidation states
Cobalt	gray metal	8.9	1495	2870	2+, 3+
Copper	red metal	8.92	1083.4 ± 0.2	2567	1+, 2+
Fluorine	yellow gas	1.69‡	2219.62	2188.14	1−
Germanium	gray metalloid	5.32325	937.4	2830	4+
Gold	yellow metal	19.31	1064.43	2808 ± 2	1+, 3+
Helium	colorless gas	0.1785*	2272.2 (26 atm)	2268.9	0
Hydrogen	colorless gas	0.0899*	2259.34	2252.8	1−, 1+
Iodine	blue-black solid	4.93	113.5	184.35	1−, 1+, 3+, 5+, 7+
Iron	silver metal	7.86	1535	2750	2+, 3+
Lead	bluish white metal	11.343716	327.502	1740	2+, 4+
Lithium	silver metal	0.534	180.54	1342	1+
Magnesium	silver metal	1.745	648.8	1107	2+
Manganese	gray-white metal	7.20	1244 ± 3	1962	2+, 3+, 4+, 6+, 7+
Mercury	silver liquid metal	13.5462	238.87	356.58	1+, 2+
Neon	colorless gas	0.9002*	2248.67	2245.9	0
Nickel	silver metal	8.90	1455	2730	2+, 3+
Nitrogen	colorless gas	1.2506*	2209.86	2195.8	3−, 3+, 5+
Oxygen	colorless gas	1.429*	2218.4	2182.962	2−
Phosphorus	yellow solid	1.82	44.1	280	3−, 3+, 5+
Platinum	silver metal	21.45	1772	3827 ± 100	2+, 4+
Potassium	silver metal	0.86	63.25	760	1+
Silicon	gray metalloid	2.33 ± 0.01	1410	2355	2+, 4+
Silver	white metal	10.5	961.93	2212	1+
Sodium	silver metal	0.97	97.8	882.9	1+
Strontium	silver metal	2.6	769	1384	2+
Sulfur	yellow solid	1.96	119.0	444.674	2−, 4+, 6+
Tin	white metal	7.28	231.88	2260	2+, 4+
Titanium	white metal	4.5	1660 ± 10	3287	2+, 3+, 4+
Uranium	silver metal	19.05 ± 0.0225	1132.3 ± 0.8	3818	3+, 4+, 6+
Zinc	blue-white metal	7.14	419.58	907	2+

* Densities of gases given in g/L at STP
† Densities obtained at 20°C unless otherwise noted (superscript)
‡ Density of fluorine given in g/L at 1 atm and 15°C

TABLE A-6   KEY OF ATOM COLORS USED IN MOLECULAR MODELS IN *HOLT CHEMISTRY*

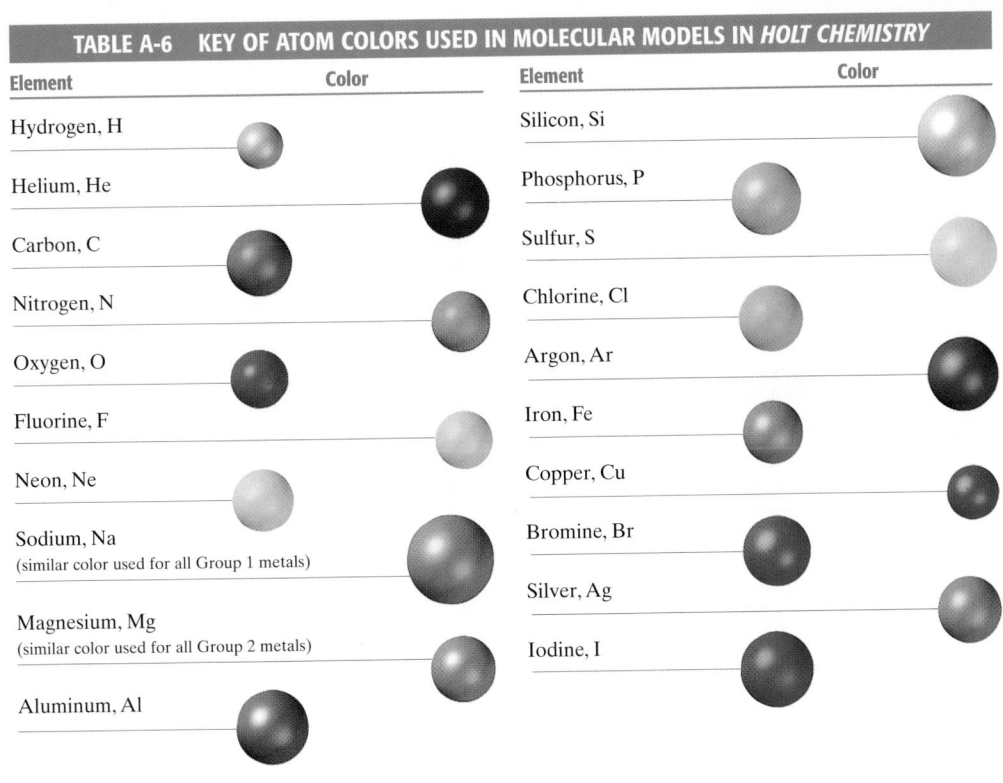

Element	Color		Element	Color
Hydrogen, H			Silicon, Si	
Helium, He			Phosphorus, P	
Carbon, C			Sulfur, S	
Nitrogen, N			Chlorine, Cl	
Oxygen, O			Argon, Ar	
Fluorine, F			Iron, Fe	
Neon, Ne			Copper, Cu	
Sodium, Na (similar color used for all Group 1 metals)			Bromine, Br	
Magnesium, Mg (similar color used for all Group 2 metals)			Silver, Ag	
Aluminum, Al			Iodine, I	

TABLE A-7   COMMON IONS

Cation	Symbol	Cation	Symbol	Anion	Symbol	Anion	Symbol
Aluminum	Al^{3+}	Lead(II)	Pb^{2+}	Acetate	CH_3COO^-	Hydrogen sulfate	HSO_4^-
Ammonium	NH_4^+	Magnesium	Mg^{2+}	Bromide	Br^-	Hydroxide	OH^-
Arsenic(III)	As^{3+}	Mercury(I)	Hg_2^{2+}	Carbonate	CO_3^{2-}	Hypochlorite	ClO^-
Barium	Ba^{2+}	Mercury(II)	Hg^{2+}	Chlorate	ClO_3^-	Iodide	I^-
Calcium	Ca^{2+}	Nickel(II)	Ni^{2+}	Chloride	Cl^-	Nitrate	NO_3^-
Chromium(II)	Cr^{2+}	Potassium	K^+	Chlorite	ClO_2^-	Nitrite	NO_2^-
Chromium(III)	Cr^{3+}	Silver	Ag^+	Chromate	CrO_4^{2-}	Oxide	O_2^-
Cobalt(II)	Co^{2+}	Sodium	Na^+	Cyanide	CN^-	Perchlorate	ClO_4^-
Cobalt(III)	Co^{3+}	Strontium	Sr^{2+}	Dichromate	$Cr_2O_7^{2-}$	Permanganate	MnO_4^-
Copper(I)	Cu^+	Tin(II)	Sn^{2+}	Fluoride	F^-	Peroxide	O_2^{2-}
Copper(II)	Cu^{2+}	Tin(IV)	Sn^{4+}	Hexacyanoferrate(II)	$Fe(CN)_6^{4-}$	Phosphate	PO_4^{3-}
Hydronium	H_3O^+	Titanium(III)	Ti^{3+}	Hexacyanoferrate(III)	$Fe(CN)_6^{3-}$	Sulfate	SO_4^{2-}
Iron(II)	Fe^{2+}	Titanium(IV)	Ti^{4+}	Hydride	H^-	Sulfide	S^{2-}
Iron(III)	Fe^{3+}	Zinc	Zn^{2+}	Hydrogen carbonate	HCO_3^-	Sulfite	SO_3^{2-}

TABLE A-8 PREFIXES FOR NAMING COVALENT COMPOUNDS

Prefix	Number of Atoms	Example	Name	Prefix	Number of Atoms	Example	Name
mono-	1	CO	carbon monoxide	hexa-	6	CeB_6	cerium hexaboride
di-	2	SiO_2	silicon dioxide	hepta-	7	IF_7	iodine heptafluoride
tri-	3	SO_3	sulfur trioxide	octa-	8	Np_3O_8	trineptunium octoxide
tetra-	4	SCl_4	sulfur tetrachloride	nona-	9	I_4O_9	tetraiodine nonoxide
penta-	5	$SbCl_5$	antimony pentachloride	deca-	10	S_2F_{10}	disulfur decafluoride

TABLE A-9 ACTIVITY SERIES OF THE ELEMENTS

Activity of Metals		Activity of Halogens
Li	react with cold H_2O and acids, replacing hydrogen; react with oxygen, forming oxides	F_2
Rb		Cl_2
K		Br_2
Ca		I_2
Ba		
Sr		
Ca		
Na		
Mg	react with steam (but not cold water) and acids; replacing hydrogen; react with oxygen, forming oxides	
Al		
Mn		
Zn		
Cr		
Fe		
Cd		
Co	do not react with water; react with acids, replacing hydrogen; react with oxygen, forming oxides	
Ni		
Sn		
Pb		
H_2	react with oxygen, forming oxides	
Sb		
Bi		
Cu		
Hg		
Ag	fairly unreactive, forming oxides only indirectly.	
Pt		
Au		

832

TABLE A-10   STATE SYMBOLS AND REACTION CONDITIONS

Symbol	Meaning
$(s), (l), (g)$	substance in the solid, liquid, or gaseous state
(aq)	substance in aqueous solution (dissolved in water)
$\longrightarrow$	"produces" or "yields," indicating the result of a reaction
$\rightleftharpoons$	reversible reaction in which products can reform into reactants; final result is a mixture of products and reactants
$\xrightarrow{\Delta}$ or $\xrightarrow{heat}$	reactants are heated; temperature is not specified
$\xrightarrow{Pd}$	name or chemical formula of a catalyst, added to speed a reaction
$(c), \downarrow$	product is a solid precipitate
$\uparrow$	product is a gas

TABLE A-11   THERMODYNAMIC DATA

Substance	ΔH_f^0 (kJ/mol)	S^0 (J/mol·K)	ΔG_f^0 (kJ/mol)	Substance	ΔH_f^0 (kJ/mol)	S^0 (J/mol·K)	ΔG_f^0 (kJ/mol)
$Ag(s)$	0.0	42.7	0.0	$C_6H_{14}(g, \text{n-hexane})$	−167.1	388.4	0.0
$AgCl(s)$	−127.1	96.2	−109.8	$C_7H_{16}(g, \text{n-heptane})$	−187.7	427.9	8.0
$AgNO_3(s)$	−124.4	140.9	−33.5	$C_8H_{18}(g, \text{n-octane})$	−208.6	466.7	16.3
$Al(s)$	0.0	28.3	0.0	$C_8H_{18}(g, \text{iso-octane})$	−224.0	423.2	12.6
$AlCl_3(s)$	−705.6	110.7	−628.9	$CaCO_3(s)$	−1206.9	92.9	−1128.8
$Al_2O_3(s, \text{corundum})$	−1676.0	51.0	−1582.4	$CaCl_2(s)$	−795.8	108.4	−748.1
$Br_2(l)$	0.0	152.2	0.0	$Ca(OH)_2(s)$	−986.1	83.4	−898.6
$Br_2(g)$	30.9	245.5	30.9	$Ca(s)$	0.0	41.6	0.0
$C \ (s, \text{diamond})$	1.9	2.4	2.90	$CaO(s)$	−634.9	38.2	−604.04
$C \ (s, \text{graphite})$	0.0	5.7	0.0	$Cl_2(g)$	0.0	223.1	0.0
$CCl_4(l)$	−132.8	216.2	−65.3	$Cu(s)$	0.0	33.2	0.0
$CCl_4(g)$	−95.8	309.9	−60.2	$CuCl_2(s)$	−220.1	108.1	−175.7
$CH_3OH(l)$	−239.1	127.2	−166.4	$CuSO_4(s)$	−770.0	109.3	−661.9
$CH_4(g)$	−74.9	186.3	−50.8	$F_2(g)$	0.0	202.8	0.0
$CO(g)$	−110.5	197.6	−137.2	$Fe(s)$	0.0	27.3	0.0
$CO_2(g)$	−393.5	213.8	−394.4	$FeCl_3(s)$	−399.4	142.3	−334.05
$CS_2(g)$	117.1	237.8	67.2	$Fe_2O_3(s, \text{hematite})$	−824.8	87.4	−742.2
$C_2H_6(g)$	−83.8	229.1	32.9	$Fe_3O_4(s, \text{magnetite})$	−1120.9	145.3	−1015.5
$C_2H_4(g)$	52.5	219.3	68.1	$H_2(g)$	0.0	130.7	0.0
$C_2H_5OH(l)$	−277.0	161.0	−174.9	$HBr(g)$	−36.4	198.6	−53.4
$C_3H_8(g)$	−104.7	270.2	−24.3	$HCl(g)$	−92.3	186.8	−95.3
$C_4H_{10}(g, \text{n-butane})$	−125.6	310.1	−16.7	$HCN(g)$	135.1	201.7	124.7
$C_4H_{10}(g, \text{isobutane})$	−134.2	294.6	−20.9	$HCOOH(l)$	−425.1	129.0	−361.4

continued on next page

833

TABLE A-11 CONTINUED

Substance	ΔH_f^0 (kJ/mol)	S^0 (J/mol·K)	ΔG_f^0 (kJ/mol)	Substance	ΔH_f^0 (kJ/mol)	S^0 (J/mol·K)	ΔG_f^0 (kJ/mol)
HF(g)	−272.5	173.8	−273.2	NO₂(g)	33.1	240.0	51.3
HNO₃(g)	−134.3	266.4	−74.8	N₂O(g)	82.4	220.0	104.2
H₂O(g)	−241.8	188.7	−228.6	N₂O₄(g)	9.1	304.4	97.8
H₂O(l)	−285.8	70.0	−237.2	Na(s)	0.0	51.5	0.0
H₂O₂(l)	−187.8	109.6	−120.4	NaCl(s)	−411.2	72.1	−384.2
H₂S(g)	−20.5	205.7	−33.6	NaOH(s)	−425.9	64.4	−379.5
H₂SO₄(l)	−814.0	156.9	−690.1	O₂(g)	0.0	205.0	0.0
K(s)	0.0	64.7	0.0	O₃(g)	142.7	238.9	163.2
KCl(s)	−436.7	82.6	−409.2	Pb(s)	0.0	64.8	0.0
KNO₃(s)	−494.6	133.0	−394.9	PbCl₂(s)	−359.4	136.2	−317.9
KOH(s)	−424.7	78.9	−379.1	PbO(s, red)	−219.0	66.3	−188.95
Li(s)	0.0	29.1	0.0	S(s)	0.0	32.1	0.0
LiCl(s)	−408.6	59.3	−384.4	SO₂(g)	−296.8	248.1	−300.2
LiOH(s)	−484.9	42.8	−439.0	SO₃(g)	−395.8	256.8	−371.1
Mg(s)	0.0	32.7	0.0	Si(s)	0.0	18.8	0.0
MgCl₂(s)	−641.6	89.6	−591.8	SiCl₄(g)	−657.0	330.9	−617.0
Hg(l)	0.0	76.0	0.0	SiO₂(s, quartz)	−910.9	41.5	−856.7
Hg₂Cl₂(s)	−264.2	192.5	−210.8	Sn(s, white)	0.0	51.6	0.0
HgO(s, red)	−90.8	70.3	−55.6	Sn(s, gray)	−2.1	44.1	0.13
N₂(g)	0.0	191.6	0.0	SnCl₄(l)	−511.3	258.6	−440.2
NH₃(g)	−45.9	192.8	−16.5	Zn(s)	0.0	41.6	0.0
NH₄Cl(s)	−314.4	94.6	−203.0	ZnCl₂(s)	−415.0	111.5	−369.4
NO(g)	90.3	210.8	86.6	ZnO(s)	−348.3	43.6	−318.32

TABLE A-12   HEAT OF COMBUSTION

Formula	ΔH_c (kJ/mol)	Formula	ΔH_c (kJ/mol)	Formula	ΔH_c (kJ/mol)
H₂(g)	−285.8	C₆H₁₄(l)	−4163.2	C₁₀H₈(s)	−5156.3
C(s, graphite)	−393.5	C₇H₁₆(l)	−4817.0	C₁₄H₁₀(s)	−7076.5
CO(g)	−283.0	C₈H₁₈(l)	−5470.5	CH₃OH(l)	−726.1
CH₄(g)	−890.8	C₂H₄(g)	−1411.2	C₂H₅OH(l)	−1366.8
C₂H₆(g)	−1560.7	C₃H₆(g)	−2058.0	(C₂H₅)₂O(l)	−2751.1
C₃H₈(g)	−2219.2	C₂H₂(g)	−1301.1	CH₂O(g)	−570.7
C₄H₁₀(g)	−2877.6	C₆H₆(l)	−3267.6	C₆H₁₂O₆(s)	−2803.0
C₅H₁₂(g)	−3535.6	C₇H₈(l)	−3910.3	C₁₂H₂₂O₁₁(s)	−5640.9

TABLE A-13 WATER-VAPOR PRESSURE

Temperature (°C)	Pressure (mm Hg)	Pressure (kPa)	Temperature (°C)	Pressure (mm Hg)	Pressure (kPa)
0.0	4.6	0.61	23.0	21.1	2.81
5.0	6.5	0.87	23.5	21.7	2.90
10.0	9.2	1.23	24.0	22.4	2.98
15.0	12.8	1.71	24.5	23.1	3.10
15.5	13.2	1.76	25.0	23.8	3.17
16.0	13.6	1.82	26.0	25.2	3.36
16.5	14.1	1.88	27.0	26.7	3.57
17.0	14.5	1.94	28.0	28.3	3.78
17.5	15.0	2.00	29.0	30.0	4.01
18.0	15.5	2.06	30.0	31.8	4.25
18.5	16.0	2.13	35.0	42.2	5.63
19.0	16.5	2.19	40.0	55.3	7.38
19.5	17.0	2.27	50.0	92.5	12.34
20.0	17.5	2.34	60.0	149.4	19.93
20.5	18.1	2.41	70.0	233.7	31.18
21.0	18.6	2.49	80.0	355.1	47.37
21.5	19.2	2.57	90.0	525.8	70.12
22.0	19.8	2.64	95.0	633.9	84.53
22.5	20.4	2.72	100.0	760.0	101.32

TABLE A-14 DENSITIES OF GASES AT STP

Gas	Density (g/L)	Gas	Density (g/L)
Air, dry	1.293	Hydrogen	0.0899
Ammonia	0.771	Hydrogen chloride	1.639
Carbon dioxide	1.997	Hydrogen sulfide	1.539
Carbon monoxide	1.250	Methane	0.7168
Chlorine	3.214	Nitrogen	1.2506
Dinitrogen monoxide	1.977	Nitrogen monoxide (at 10°C)	1.340
Ethyne (acetylene)	1.165	Oxygen	1.429
Helium	0.1785	Sulfur dioxide	2.927

835

TABLE A-15   DENSITY OF LIQUID WATER

Temperature (°C)	Density (g/cm³)	Temperature (°C)	Density (g/cm³)
0	0.999 84	25	0.997 05
2	0.999 94	30	0.995 65
3.98 (maximum)	0.999 973	40	0.992 22
4	0.999 97	50	0.988 04
6	0.999 94	60	0.983 20
8	0.999 85	70	0.977 77
10	0.999 70	80	0.971 79
14	0.999 24	90	0.965 31
16	0.998 94	100	0.958 36
20	0.998 20		

TABLE A-16   MEASURES OF CONCENTRATION

Name	Symbol	Units	Areas of application
Molarity	M	$\dfrac{\text{mol solute}}{\text{L solution}}$	in solution stoichiometry calculations
Molality	m	$\dfrac{\text{mol solute}}{\text{kg solvent}}$	boiling-point elevation and freezing-point depression calculations
Mole fraction	X	$\dfrac{\text{mol solute}}{\text{total mol solution}}$	in solution thermodynamics
Volume percent	% V/V	$\dfrac{\text{volume solute}}{\text{volume solution}} \times 100$	with liquid-liquid mixtures
Mass or weight percent	% or %w/w	$\dfrac{\text{g solute}}{\text{g solution}} \times 100$	in biological research
Parts per million	ppm	$\dfrac{\text{g solute}}{1\ 000\ 000\ \text{g solution}}$	to express small concentrations
Parts per billion	ppb	$\dfrac{\text{g solute}}{1\ 000\ 000\ 000\ \text{g solution}}$	to express very small concentrations, as in pollutants or contaminants

TABLE A-17 SOLUBILITIES OF GASES IN WATER*

Gas	0°C	10°C	20°C	60°C
Air	0.029 18	0.022 84	0.018 68	0.012 16
Ammonia	1130	870	680	200
Carbon dioxide	1.713	1.194	0.878	0.359
Carbon monoxide	0.035 37	0.028 16	0.023 19	0.014 88
Chlorine	—	3.148	2.299	1.023
Hydrogen	0.021 48	0.019 55	0.018 19	0.016 00
Hydrogen chloride	512	475	442	339
Hydrogen sulfide	4.670	3.399	2.582	1.190
Methane	0.055 63	0.041 77	0.033 08	0.019 54
Nitrogen†	0.023 54	0.018 61	0.015 45	0.010 23
Nitrogen monoxide	0.073 81	0.057 09	0.047 06	0.029 54
Oxygen	0.048 89	0.038 02	0.031 02	0.019 46
Sulfur dioxide	79.789	56.647	39.374	—

* Volume of gas (in liters) at STP that can be dissolved in 1 L of water at the temperature (°C) indicated.
† Atmospheric nitrogen: 98.815% N_2, 1.185% inert gases

TABLE A-18 SOLUBILITY OF COMPOUNDS*

Formula	0°C	20°C	60°C	100°C
$Al_2(SO_4)_3$	31.2	36.4	59.2	89.0
NH_4Cl	29.4	37.2	55.3	77.3
NH_4NO_3	118	192	421	871
$(NH_4)_2SO_4$	70.6	75.4	88	103
$BaCl_2 \cdot 2H_2O$	31.2	35.8	46.2	59.4
$Ba(OH)_2$	1.67	3.89	20.94	$101.40^{80°}$
$Ba(NO_3)_2$	4.95	9.02	20.4	34.4
$Ca(HCO_3)_2$	16.15	16.60	17.50	18.40
$Ca(OH)_2$	0.189	0.173	0.121	0.076
$CuCl_2$	68.6	73.0	96.5	120
$CuSO_4 \cdot 5H_2O$	23.1	32.0	61.8	114
$PbCl_2$	0.67	1.00	1.94	3.20
$Pb(NO_3)_2$	37.5	54.3	91.6	133
$LiCl$	69.2	83.5	98.4	128
Li_2SO_4	36.1	34.8	32.6	$30.9^{90°}$
$MgSO_4$	22.0	33.7	54.6	68.3
$HgCl_2$	3.63	6.57	16.3	61.3

continued on next page

TABLE A-18 CONTINUED

Formula	0°C	20°C	60°C	100°C
KBr	53.6	65.3	85.5	104
KClO$_3$	3.3	7.3	23.8	56.3
KCl	28.0	34.2	45.8	56.3
K$_2$CrO$_4$	56.3	63.7	70.1	74.5$^{90°}$
KI	128	144	176	206
KNO$_3$	13.9	31.6	106	245
K$_2$SO$_4$	7.4	11.1	18.2	24.1
AgC$_2$H$_3$O$_2$	0.73	1.05	1.93	2.59$^{80°}$
AgNO$_3$	122	216	440	733
NaC$_2$H$_3$O$_2$	36.2	46.4	139	170
NaClO$_3$	79.6	95.9	137	204
NaCl	35.7	35.9	37.1	39.2
NaNO$_3$	73.0	87.6	122	180
C$_{12}$H$_{22}$O$_{11}$	179.2	203.9	287.3	487.2

* Solubilities are given in grams of solute that can be dissolved in 100 g of water at the temperature (°C) indicated.

TABLE A-19   EXAMPLES OF COMPLEX IONS

Complex cation	Color	Complex cation	Color	Complex anion	Color
[Co(NH$_3$)$_6$]$^{3+}$	yellow-orange	[Fe(H$_2$O)$_5$SCN]$^{2+}$	deep red	[Co(CN)$_6$]$^{3-}$	pale yellow
[Co(NH$_3$)$_5$(H$_2$O)]$^{3+}$	bright red	[Ni(NH$_3$)$_6$]$^{2+}$	blue-violet	[CoCl$_4$]$^{2-}$	blue
[Co(NH$_3$)$_5$Cl]$^{2+}$	violet	[Ni(H$_2$O)$_6$]$^{2+}$	green	[Cu$_2$Cl$_6$]$^{2-}$	red
[Co(H$_2$O)$_6$]$^{2+}$	pink	[Zn(NH$_3$)$_4$]$^{2+}$	colorless	[Fe(CN)$_6$]$^{3-}$	red
[Cu(NH$_3$)$_4$]$^{2+}$	blue-purple	[Zn(NH$_3$)$_6$]$^{2+}$	colorless	[Fe(CN)$_6$]$^{4-}$	yellow
[Cu(H$_2$O)$_4$]$^{2+}$	light blue			[Fe(C$_2$O$_4$)$_3$]$^{3-}$	green

TABLE A-20 EQUILIBRIUM CONSTANTS

Equation	K_{eq} expression	Values	
$N_2(g) + 3H_2(g) \rightleftarrows 2NH_3(g)$	$\dfrac{[NH_3]^2}{[N_2][H_2]^3}$	3.3×10^8 (25°C)	4.2×10^1 (327°C)
$NH_3(aq) + H_2O(l) \rightleftarrows NH_4^+(aq) + OH^-(aq)$	$\dfrac{[NH_4^+][OH^-]}{[NH_3]}$	1.8×10^{-5} (25°C)	
$2NO_2(g) \rightleftarrows N_2O_4(g)$	$\dfrac{[N_2O_4]}{[NO_2]^2}$	1.25×10^3 (0°C) 1.65×10^2 (25°C) 2.0×10^1 (100°C)	
$N_2(g) + O_2(g) \rightleftarrows 2NO(g)$	$\dfrac{[NO]^2}{[N_2][O_2]}$	4.5×10^{-31} (25°C)	6.7×10^{-10} (627°C)
$CO_2(g) + H_2(g) \rightleftarrows CO(g) + H_2O(g)$	$\dfrac{[CO][H_2O]}{[CO_2][H_2]}$	2.2 (1400°C)	4.6 (2000°C)
$H_2CO_3(aq) + H_2O(l) \rightleftarrows H_3O^+(aq) + HCO_3^-(aq)$	$\dfrac{[H_3O^+][HCO_3^-]}{[H_2CO_3]}$	4.3×10^{-7} (25°C)	
$HCO_3^-(aq) + H_2O(l) \rightleftarrows CO_3^{2-}(aq) + H_3O^+(aq)$	$\dfrac{[CO_3^{2-}][H_3O^+]}{[HCO_3^-]}$	4.7×10^{-11} (25°C)	
$H_2(g) + I_2(g) \rightleftarrows 2HI(g)$	$\dfrac{[HI]^2}{[H_2][I_2]}$	1.13×10^2 (250°C)	1.8×10^1 (1127°C)
$Hg^{2+}(aq) + Hg(l) \rightleftarrows Hg_2^{2+}(aq)$	$\dfrac{[Hg_2^{2+}]}{[Hg^{2+}]}$	8.1×10^1 (25°C)	

TABLE A-21 APPROXIMATE CONCENTRATION OF IONS IN OCEAN WATER

Ion	Concentration (mol/L)	Ion	Concentration (mol/L)
Cl^-	0.554	K^+	0.010
Na^+	0.470	Ca^{2+}	0.009
Mg^{2+}	0.047	CO_3^{2-}	0.002
SO_4^{2-}	0.015	Br^-	0.001

839

TABLE A-22   STANDARD ELECTRODE POTENTIALS

Electrode reaction	$E°$ (V)
$Li^+(aq) + e^- \rightleftharpoons Li(s)$	−3.0401
$K^+(aq) + e^- \rightleftharpoons K(s)$	−2.931
$Ca^{2+}(aq) + 2e^- \rightleftharpoons Ca(s)$	−2.868
$Na^+(aq) + e^- \rightleftharpoons Na(s)$	−2.71
$Mg^{2+}(aq) + 2e^- \rightleftharpoons Mg(s)$	−2.372
$Al^{3+}(aq) + 3e^- \rightleftharpoons Al(s)$	−1.662
$Zn(OH)_2(s) + 2e^- \rightleftharpoons Zn(s) + 2OH^-(aq)$	−1.249
$2H_2O(l) + 2e^- \rightleftharpoons H_2(g) + 2OH^-(aq)$	−0.828
$Zn^{2+}(aq) + 2e^- \rightleftharpoons Zn(s)$	−0.7618
$Fe^{2+}(aq) + 2e^- \rightleftharpoons Fe(s)$	−0.447
$PbSO_4(s) + H_3O^+(aq) + 2e^- \rightleftharpoons Pb(s) + HSO_4^-(aq) + H_2O(l)$	−0.42
$Cd^{2+}(aq) + 2e^- \rightleftharpoons Cd(s)$	−0.4030
$Pb^{2+}(aq) + 2e^- \rightleftharpoons Pb(s)$	−0.1262
$Fe^{3+}(aq) + 3e^- \rightleftharpoons Fe(s)$	−0.037
$2H_3O^+(aq) + 2e^- \rightleftharpoons H_2(g) + 2H_2O(l)$	0.000
$AgCl(s) + e^- \rightleftharpoons Ag(s) + Cl^-(aq)$	+0.222
$Cu^{2+}(aq) + 2e^- \rightleftharpoons Cu(s)$	+0.3419
$O_2(g) + 2H_2O(l) + 4e^- \rightleftharpoons 4OH^-(aq)$	+0.401
$I_2(s) + 2e^- \rightleftharpoons 2I^-(aq)$	+0.5355
$Fe^{3+}(aq) + e^- \rightleftharpoons Fe^{2+}(aq)$	+0.771
$Hg_2^{2+}(aq) + 2e^- \rightleftharpoons 2Hg(l)$	+0.7973
$Ag^+(aq) + e^- \rightleftharpoons Ag(s)$	+0.7996
$Br_2(l) + 2e^- \rightleftharpoons 2Br^-(aq)$	+1.066
$MnO_2(s) + 4H_3O^+(aq) + 2e^- \rightleftharpoons Mn^{2+}(aq) + 6H_2O(l)$	+1.224
$O_2(g) + 4H_3O^+(aq) + 4e^- \rightleftharpoons 6H_2O$	+1.229
$Cl_2(g) + 2e^- \rightleftharpoons 2Cl^-(aq)$	+1.358
$PbO_2(s) + 4H_3O^+(aq) + 2e^- \rightleftharpoons Pb^{2+}(aq) + 6H_2O(l)$	+1.455
$MnO_4^-(aq) + 8H_3O^+(aq) + 5e^- \rightleftharpoons Mn^{2+}(aq) + 12H_2O(l)$	+1.507
$PbO_2(s) + HSO_4^-(aq) + 3H_3O^+(aq) + 2e^- \rightleftharpoons PbSO_4(s) + 5H_2O(l)$	+1.691
$Ce^{4+}(aq) + e^- \rightleftharpoons Ce^{3+}(aq)$	+1.72
$Ag_2O_2(s) + 4H^+(aq) + e^- \rightleftharpoons 2Ag(s) + 2H_2O(l)$	+1.802
$F_2(g) + 2e^- \rightleftharpoons 2F^-(aq)$	+2.866

TABLE A-23 SOME CLASSES OF ORGANIC COMPOUNDS

Class	Functional group	Example	Use
Alcohol	—OH	2-propanol	disinfectant
Aldehyde	$\overset{O}{\underset{\|}{-C-H}}$	benzaldehyde	almond flavor
Halide	—F, Cl, Br, I	trichlorofluoromethane (Freon-11)	refrigerant
Amide	$\overset{O}{\underset{\|}{-C-NH_2}}$	niacinamide (nicotinamide)	nutrient
Amine	$-\overset{\|}{\underset{\|}{N}}$	caffeine	beverage ingredient
Carboxylic acid	$\overset{O}{\underset{\|}{-C-OH}}$	tetradecanoic acid (myristic acid)	soap-making ingredient
Ester	$\overset{O}{\underset{\|}{-C-O-}}$	ethyl butanoate	perfume ingredient
Ether	—O—	methyl phenyl ether (anisole)	perfume ingredient
Ketone	$\overset{O}{\underset{\|}{-C-}}$	propanone (acetone)	solvent in nail-polish remover

841

TABLE A-24   SOLUBILITY PRODUCT CONSTANTS AT 25°C

Salt	K_{sp}	Salt	K_{sp}
Ag_2CO_3	8.4×10^{-12}	$FeCO_3$	3.1×10^{-11}
$AgCl$	1.8×10^{-10}	$Fe(OH)_2$	4.9×10^{-17}
Ag_2CrO_4	1.1×10^{-12}	$Fe(OH)_3$	2.6×10^{-39}
Ag_2S	1.1×10^{-49}	FeS	1.6×10^{-19}
$AgBr$	5.4×10^{-13}	$MgCO_3$	6.8×10^{-6}
AgI	8.5×10^{-17}	$Mg(OH)_2$	5.6×10^{-12}
$AlPO_4$	9.8×10^{-21}	$Mg_3(PO_4)_2$	9.9×10^{-25}
$BaSO_4$	1.1×10^{-10}	$MnCO_3$	2.2×10^{-11}
$CaCO_3$	5.0×10^{-9}	$Pb(OH)_2$	1.4×10^{-20}
$Ca(OH)_2$	4.7×10^{-7}	PbS	9.0×10^{-29}
$Ca_3(PO_4)_2$	2.1×10^{-33}	$PbSO_4$	1.8×10^{-8}
$CaSO_4$	7.1×10^{-5}	$SrSO_4$	3.4×10^{-7}
CuS	1.3×10^{-36}	$ZnCO_3$	1.2×10^{-10}
		ZnS	2.9×10^{-25}

STUDY SKILLS FOR CHEMISTRY
TABLE OF CONTENTS

Succeeding in Your Chemistry Class

Your success in this course will depend on your ability to apply some basic study skills to learning the material. Studying chemistry can be difficult, but you can make it easier using simple strategies for dealing with the concepts and problems. Becoming skilled in using these strategies will be your keys to success in this and many other courses.

Reading the Text

- **Read the assigned material before class** so that the class lecture makes sense. Use a dictionary to help you interpret vocabulary. Remember while reading to figure out what information is important.

  Working together with others using Paired Reading and Discussion strategies can help you decide what is important and clarify the material. (For more discussion, see Other Reading Strategies on page 853.)

- **Select a quiet setting** away from distractions so that you can concentrate on what you are reading.

- **Have a pencil and paper nearby to jot down notes and questions** you may have. Be sure to get these questions answered in class. Power Notes (see page 849) can help you organize the notes you take and prepare you for class.

- **Use the Objectives in the beginning of each section as a list of what you need to know** from the section. Teachers generally make their tests based on the text objectives or their own objectives. Using the objectives to focus your reading can make your learning more efficient. Using the K/W/L strategy (see page 851) can help you relate new material to what you already know and what you need to learn.

Taking Notes in Class

- **Be prepared to take notes during class.** Have your materials organized in a notebook. Separate sheets of paper can be easily lost.

- **Don't write down everything your teacher says.** Try to tell which parts of the lecture are important and which are not. Reading the text before class will help in this. You will not be able to write down everything, so you must try to write down only the important things.

- **Recopying notes later is a waste of time** and does not help you learn material for a test. Do it right the first time. Organize your notes as you are writing them down so that you can make sense of your notes when you review them without needing to recopy them.

Reviewing Class Notes

- **Review your notes as soon as possible after class.** Write down any questions you may have about the material covered that day. Be sure to get these questions answered during the next class. You can work with friends to use strategies such as Paired Summarizing and L.I.N.K. (See page 853.)

- **Do not wait until the test to review.** By then you will have forgotten a good portion of the material.

- **Be selective about what you memorize.** You cannot memorize everything in a chapter. First of all, it is too time consuming. Second, memorizing and understanding are not the same thing. Memorizing topics as they appear in your notes or text does not guarantee that you will be able to correctly answer questions that require understanding of those topics. You should only memorize material that you understand. Concept Maps and other Reading Organizers, Sequencing/Pattern Puzzles, and Prediction Guides can help you understand key ideas and major concepts. (See pages 846, 852, and 854.)

844

Working Problems

In addition to understanding the concepts, the ability to solve problems will be a key to your success in chemistry. You will probably spend a lot of time working problems in class and at home. The ability to solve chemistry problems is a skill, and like any skill, it requires practice.

- **Always review the Sample Problems in the chapter.** The Sample Problems in the text provide road maps for solving certain types of problems. Cover the solution while trying to work the problem yourself.

- **The problems in the Chapter Review are similar to the Sample Problems.** If you can relate an assigned problem to one of the Sample Problems in the chapter, it shows that you understand the material.

- **The four steps: Gather information, Plan your work, Calculate, and Verify should be the steps you go through when working assigned problems.** These steps will allow you to organize your thoughts and help you develop your problem-solving skills.

- **Never spend more than 15 minutes trying to solve a problem.** If you have not been able to come up with a plan for the solution after 15 minutes, additional time spent will only cause you to become frustrated. What do you do? Get help! See your teacher or a classmate. Find out what it is that you do not understand.

- **Do not try to memorize the Sample Problems; spend your time trying to understand how the solution develops.** Memorizing a particular sample problem will not ensure that you understand it well enough to solve a similar problem.

- **Always look at your answer and ask yourself if it is reasonable and makes sense.** Check to be sure you have the correct units and numbers of significant figures.

Completing Homework

Your teacher will probably assign questions and problems from the Section Reviews and Chapter Reviews or assign Concept Review worksheets. The purpose of these assignments is to review what you have covered in class and to see if you can use the information to answer questions or solve problems. As in reviewing class notes, do your homework as soon after class as possible while the topics are still fresh in your mind. Do not wait until late at night, when you are more likely to be tired and to become frustrated.

Reviewing for an exam

- **Don't panic and don't cram!** It takes longer to learn if you are under pressure. If you have followed the strategies listed here and reviewed along the way, studying for the exam should be less stressful.

- When looking over your notes and concept maps, **recite ideas out loud.** There are two reasons for reciting:

   1. You are hearing the information, which is effective in helping you learn.
   2. If you cannot recite the ideas, it should be a clue that you do not understand the material, and you should begin rereading or reviewing the material again.

- **Studying with a friend provides a good opportunity for recitation.** If you can explain ideas to your study partner, you know the material.

Taking an exam

- **Get plenty of rest before the exam** so that you can think clearly. If you have been awake all night studying, you are less likely to succeed than if you had gotten a full night of rest.

- **Start with the questions you know.** If you get stuck on a question, save it for later. As time passes and you work through the exam, you may recall the information you need to answer a difficult question or solve a difficult problem.

Good luck!

845

Making Concept Maps

MAP **A**

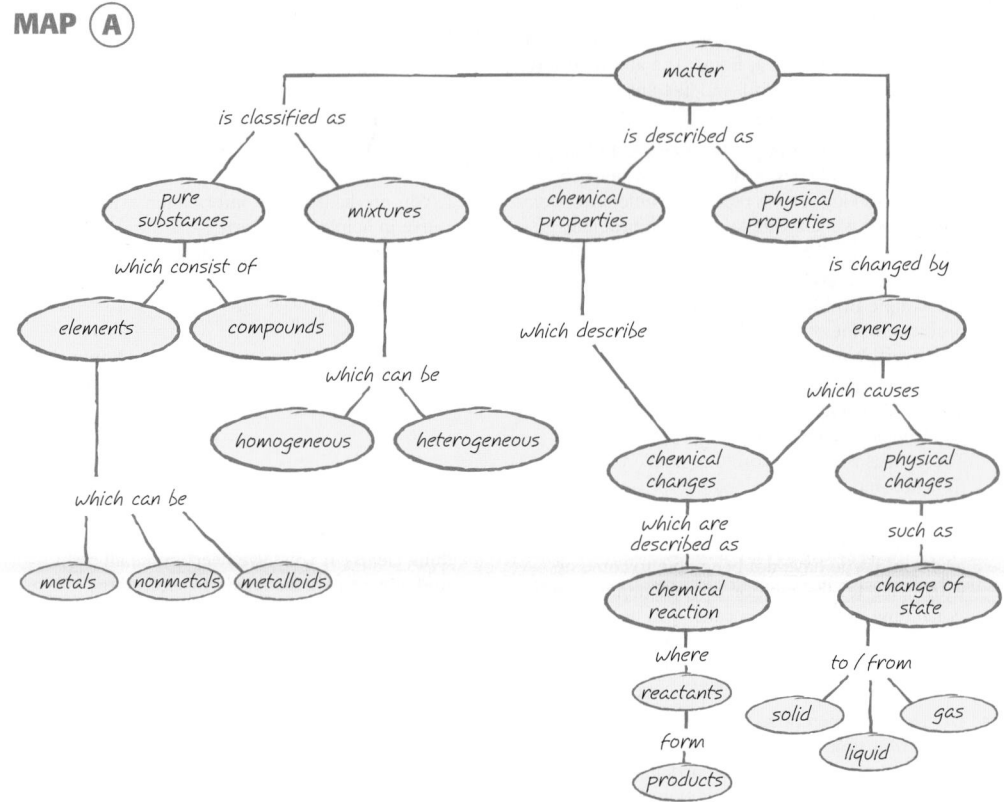

Making concept maps can help you decide what material in a chapter is important and how to efficiently learn that material. A concept map presents key ideas, meanings, and relationships for the concepts being studied. It can be thought of as a visual road map of the chapter. Learning happens efficiently when you use concept maps because you work with only the key ideas and how they fit together.

The concept map shown as **Map A** was made from vocabulary terms from the first few chapters of the book. Vocabulary terms are generally labels for concepts, and concepts are generally nouns. In a concept map, linking words are used to form propositions that connect concepts and give them meaning in context. For example, on the map above, "matter is described by physical properties" is a proposition.

Studies show that people are better able to remember materials presented visually. A concept map is better than an outline because you can see the relationships among many ideas. Because outlines are linear, there is no way of linking the ideas from various sections of the outline. Read through the map to become familiar with the information presented. Then look at the map in relation to all of the text pages in the first few chapters; which gives a better picture of the important concepts—the map or the full chapters?

To Make a Concept Map

1. List all the important concepts.

We'll use some of the boldfaced and italicized terms from the chapter "Matter and Energy."

matter	mixture
compound	pure substance
element	heterogeneous mixture
homogeneous mixture	

• From this list, group similar concepts together. For example, one way to group these concepts would be into two groups—one that is related to mixtures and one that is related to pure substances.

mixture	pure substance
heterogeneous mixture	compound
homogeneous mixture	element

2. Select a main concept for the map.

We will use matter as the main concept for this map.

3. Build the map by placing the concepts according to their importance under the main concept. For this map the main concept is *matter*.

One way of arranging the concepts is shown in **Map B.**

MAP (B)

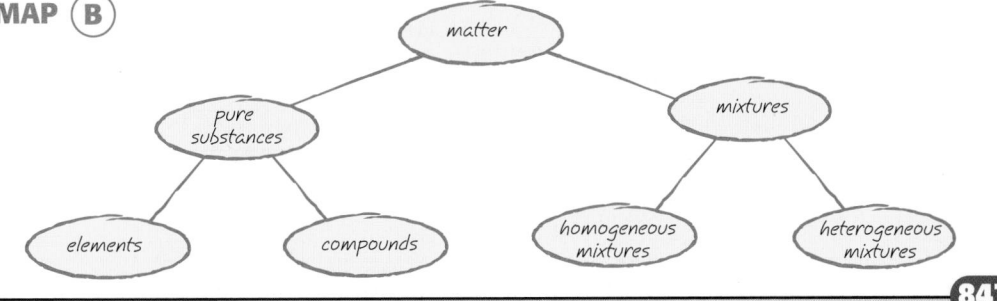

847

Appendix B

Answers to Practice

1. **a.** concept
   b. linking words
   c. linking word
   d. linking words
   e. concept
   f. linking words
   g. concept
   h. linking word

2. Answers may vary. Sample answer: Matter is changed by energy; matter is classified as pure substances and mixtures; matter is described by chemical properties, which describe chemical changes.

3. Answers may vary. Sample answer: Matter can be pure substances or mixtures.

MAP C

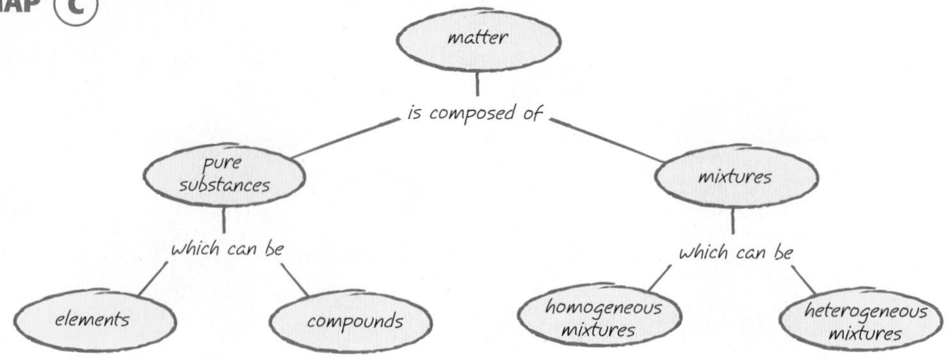

matter

is composed of

pure substances

mixtures

which can be

which can be

elements

compounds

homogeneous mixtures

heterogeneous mixtures

4. **Add linking words to give meaning to the arrangement of concepts.**

When adding the links, be sure that each proposition makes sense. To distinguish concepts from links, place your concepts in circles, ovals, or rectangles, as shown in the maps. Then make cross-links. Cross-links are made of propositions and lines connecting concepts across the map. Links that apply in only one direction are indicated with an arrowhead. **Map C** is a finished map covering the main ideas listed in Step 1.

Making maps might seem difficult at first, but the process forces you to think about the meanings and relationships among the concepts. If you do not understand those relationships, you can get help early on.

Practice mapping by making concept maps about topics you know. For example, if you know a lot about a particular sport, such as basketball, or if you have a particular hobby, such as playing a musical instrument, you can use that topic to make a practice map. By perfecting your skills with information that you know very well, you will begin to feel more confident about making maps from the information in a chapter.

Remember, the time you devote to mapping will pay off when it is time to review for an exam.

PRACTICE

1. Classify each of the following as either a concept or linking word(s).

   a. classification _____

   b. is classified as _____

   c. forms _____

   d. is described by _____

   e. reaction _____

   f. reacts with _____

   g. metal _____

   h. defines _____

2. Write three propositions from the information in **Map A.**

   _____
   _____
   _____

3. List two cross-links shown on **Map C.**

   _____
   _____
   _____

848

Making Power Notes

Power notes help you organize the chemical concepts you are studying by distinguishing main ideas from details. Similar to outlines, power notes are linear in form and provide you with a framework of important concepts. Power notes are easier to use than outlines because their structure is simpler. Using the power notes numbering system you assign a 1 to each main idea and a 2, 3, or 4 to each detail.

Power notes are an invaluable asset to the learning process, and they can be used frequently throughout your chemistry course. You can use power notes to organize ideas while reading your text or to restructure your class notes for studying purposes.

To learn to make power notes, practice first by using single-word concepts and a subject you are especially interested in, such as animals, sports, or movies. As you become comfortable with structuring power notes, integrate their use into your study of chemistry. For an easier transition, start with a few boldfaced or italicized terms. Later you can strengthen your notes by expanding these single-word concepts into more-detailed phrases and sentences. Use the following general format to help you structure your power notes.

Power 1: Main idea
 Power 2: Detail or support for power 1
  Power 3: Detail or support for power 2
   Power 4: Detail or support for power 3

1. **Pick a Power 1 word from the text.**

The text you choose does not have to come straight from your chemistry textbook. You may be making power notes from your lecture notes or from an outside source. We'll use the term atom found in the chapter entitled "Atoms and Moles" in your textbook.

Power 1: Atom

2. **Using the text, select some Power 2 words to support your Power 1 word.**

We'll use the terms nucleus and electrons, which are two parts of an atom.

Power 1: Atom
 Power 2: Nucleus
 Power 2: Electrons

3. **Select some Power 3 words to support your Power 2 words.** We'll use the terms positively charged and negatively charged, two terms that describe the Power 2 words.

Power 1: Atom
 Power 2: Nucleus
  Power 3: Positively charged
 Power 2: Electrons
  Power 3: Negatively charged

4. **Continue to add powers to support and detail the main idea as necessary.**

There are no restrictions on how many power numbers you can use in your notes. If you have a main idea that requires a lot of support, add more powers to help you extend and organize your ideas. Be sure that words having the same power number have a similar relationship to the power above. Power 1 terms do not have to be related to each other. You can use power notes to organize the material in an entire section or chapter of your text. Doing so will provide you with an invaluable study guide for your classroom quizzes and tests.

Power 1: Atom
 Power 2: Nucleus
  Power 3: Positively charged
  Power 3: Protons
   Power 4: Positively charged
  Power 3: Neutrons
   Power 4: No charge
 Power 2: Electrons
  Power 3: Negatively charged

Practice

1. **Use a periodic table and the power notes structure below to organize the following terms: alkaline-earth metals, nonmetals, calcium, sodium, halogens, metals, alkali metals, chlorine, barium, and iodine.**

1 _____
 2 _____
  3 _____
 2 _____
  3 _____
  3 _____
1 _____
 2 _____
  3 _____
  3 _____

Answers to Practice

1 metals
 2 alkali metals
  3 sodium
 2 alkaline-earth metals
  3 calcium
  3 barium
1 nonmetals
 2 halogens
  3 chlorine
  3 iodine

Making Two-Column Notes

Two-column notes can be used to learn and review definitions of vocabulary terms, examples of multiple-step processes, or details of specific concepts. The two-column-note strategy is simple: write the term, main idea, step-by-step process, or concept in the left-hand column, and the definition, example, or detail on the right.

One strategy for using two-column notes is to organize main ideas and their details. The main ideas from your reading are written in the left-hand column of your paper and can be written as questions, key words, or a combination of both. Details describing these main ideas are then written in the right-hand column of your paper.

1. **Identify the main ideas.** The main ideas for a chapter are listed in the section objectives. However, you decide which ideas to include in your notes. For example, here are some main ideas from the objectives in Section 4-2.

• Describe the locations in the periodic table and the general properties of the alkali metals, alkaline-earth metals, the halogens, and the noble gases.

2. **Divide a blank sheet of paper into two columns and write the main ideas in the left-hand column.** Summarize your ideas using quick phrases that are easy for you to understand and remember. Decide how many details you need for each main idea, and write that number in parentheses under the main idea.

3. **Write the detail notes in the right-hand column.** Be sure you list as many details as you designated in the main-idea column. Here are some main ideas and details about some of the groups in the Periodic Table.

The two-column method of review is perfect whether you use it to study for a short quiz or for a test on the material in an entire chapter. Just cover the information in the right-hand column with a sheet of paper, and after reciting what you know, uncover the notes to check your answers. Then ask yourself what else you know about that topic. Linking ideas in this way will help you to gain a more complete picture of chemistry.

Main Idea	Detail Notes
• Alkali metals (4 properties)	• Group 1 • highly reactive • ns^1 electron configuration • soft, silvery
• Alkaliine-earth metals (4 properties)	• Group 2 • reactive • ns^2 electron configuration • harder than alkali metal
• Halogens (3 properties)	• Group 17 • reactive • nonmetallic
• Noble gases	• Group 18 • low reactivity • stable ns^2-np^6 configuration

Using the K/W/L/ Strategy

The K/W/L strategy stands for "what I **K**now—what I **W**ant to know—what I **L**earned." You start by brainstorming about the subject matter before reading the assigned material. Relating new ideas and concepts to those you have learned previously will help you better understand and apply the new knowledge you obtain. The section objectives throughout your textbook are ideal for using the K/W/L strategy.

1. **Read the section objectives.** You may also want to scan headings, boldfaced terms, and illustrations before reading. Here are two of the objectives from Section 1-2 to use as an example.

   • Explain the gas, liquid, and solid states in terms of particles.
   • Distinguish between a mixture and a pure substance.

2. **Divide a sheet of paper into three columns, and label the columns "What I Know," "What I Want to Know," and "What I Learned."**

3. **Brainstorm about what you know about the information in the objectives, and write these ideas in the first column.** Because this chart is designed primarily to help you integrate your own knowledge with new information, it is not necessary to write complete sentences.

4. **Think about what you want to know about the information in the objectives, and write these ideas in the second column.** Include information from both the section objectives and any other objectives your teacher has given you.

5. **While reading the section or afterwards, use the third column to write down the information you learned.** While reading, pay close attention to any information about the topics you wrote in the "What I Want to Know" column. If you do not find all of the answers you are looking for, you may need to reread the section or reference a second source. Be sure to ask your teacher if you still cannot find the information after reading the section a second time.

It is also important to review your brainstormed ideas when you have completed reading the section. Compare your ideas in the first column with the information you wrote down in the third column. If you find that some of your brainstormed ideas are incorrect, cross them out. It is extremely important to identify and correct any misconceptions you had prior to reading before you begin studying for your test.

What I Know	What I want to Know	What I Learned
• gas has no definite shape or volume	• how gas, liquid, and solid states are related to particles	• molecules in solid and liquid states are close together, but are far apart in gas state
• liquid has no definite shape, but has definite volume	• how mixtures and pure substances are different	• molecules in solid state have fixed positions, but molecules in liquid and gas states can flow
• solid has definite shape and volume		• mixtures are combinations of pure substances
• mixture is a combination of substances		• pure substances have fixed compositions and definite properties
• pure substance has only one component		

851

Appendix B

Appendix B

Using Sequencing/Pattern Puzzles

You can use pattern puzzles to help you remember sequential information. Pattern puzzles are not just a tool for memorization. They also promote a greater understanding of a variety of chemical processes, from the steps in solving a mass-mass stoichiometry problem to the procedure for making a solution of specified molarity.

Here's a step-by-step example showing how to make a pattern puzzle, and how to use it to help you study. For other topics that require remembering information in a particular order, just follow the same steps.

1. Write down the steps of a process in your own words.

For an example, we will use the process for converting the amount of a substance in moles to mass in grams. (See Sample Problem D in the chapter on "Atoms and Moles.")

On a sheet of notebook paper, write down one step per line, and do not number the steps.

Do not copy the process straight from your textbook. Writing the steps in your own words promotes a more thorough understanding of the process.

You may want to divide longer steps into two or three shorter steps.

- Gather information about what you know and what you don't know.
- Write a set-up that shows what is given and what is desired.
- Look at the periodic table to determine the molar mass of the substance.
- Write the correct conversion factor that has mass in the numerator and amount in moles in the denominator.
- Multiply the amount of substance by the conversion factor.
- Solve the equation and verify that your answer is reasonable.

2. Cut the sheet of paper into strips with only one step per strip of paper. Shuffle the strips of paper so that they are out of sequence.

- Look at the periodic table to determine the molar mass of the substance.
- Solve the equation and verify that your answer is reasonable.
- Write the correct conversion factor that has mass in the numerator and amount in moles in the denominator.
- Write a set-up that shows what is given and what is desired.
- Gather information about what you know and what you don't know.
- Multiply the amount of substance by the conversion factor.

3. Place the strips in their proper sequence. Confirm the order of the process by checking your text or your class notes.

Pattern puzzles are especially helpful when you are studying for your chemistry tests. Before tests, use your puzzles to practice sequencing and to review the steps of chemistry processes. You and a classmate can also take turns creating your own pattern puzzles of different chemical processes and putting each other's puzzles in the correct sequence. Studying with a classmate in this manner will help make studying fun and will enable you to help each other.

- Gather information about what you know and what you don't know.
- Write a set-up that shows what is given and what is desired.
- Look at the periodic table to determine the molar mass of the substance.
- Write the correct conversion factor that has mass in the numerator and amount in moles in the denominator.
- Multiply the amount of substance by the conversion factor.
- Solve the equation and verify that your answer is reasonable.

852

Other Reading Strategies

Brainstorming

Brainstorming is a strategy that helps you recognize and evaluate the knowledge you already have before you start reading. It works well individually or in groups. When you brainstorm, you start with a central term or idea, then quickly list all the words, phrases, and other ideas that you think are related to it.

Because there are no "right" or "wrong" answers, you can use the list as a basis for classifying terms, developing a general explanation, or speculating about new relationships. For example, you might brainstorm a list of terms related to the word element before you read about elements early in the textbook. The list might include gold, metals, chemicals, silver, carbon, oxygen, and water. As you read the textbook, you might decide that some of the terms you listed are not elements. Later, you might use that information to help you distinguish between elements and compounds.

Building/Interpreting Vocabulary

Using a dictionary to look up the meanings of prefixes and suffixes as well as word origins and meanings helps you build your vocabulary and interpret what you read. If you know the meaning of prefixes like kilo- (one thousand) and milli- (one thousandth), you have a good idea what kilograms, kilometers, milligrams, and millimeters are and how they are different. (See Appendix A for a list of SI Prefixes.)

Knowledge of prefixes, suffixes, and word origins can help you understand the meaning of new words. For example, if you know the prefix *–poly* comes from the word meaning *many,* it will help you understand what polysaccharides and polymers are.

Reading Hints

Reading hints help you identify and bookmark important charts, tables, and illustrations for easy reference. For example, you may want to use a self-adhesive note to bookmark the periodic table in the chapter describing it or on the inside back cover of your book so you can easily locate it and use it for reference as you study different aspects of chemistry and solve problems involving elements and compounds.

Interpreting Graphic Sources of Information

Charts, tables, photographs, diagrams, and other illustrations are graphic, or visual, sources of information. The labels and captions, together with the illustrations help you make connections between the words and the ideas presented in the text.

Reading Response Logs

Keeping a reading response log helps you interpret what you read and gives you a chance to express your reactions and opinions about what you have read. Draw a vertical line down the center of a piece of paper. In the left-hand column, write down or make notes about passages you read to which you have reactions, thoughts, feelings, questions, or associations. In the right-hand column, write what those reactions, thoughts, feelings, questions, or associations are. For example, you might keep a reading response log when studying about Nuclear Energy.

853

Graphing Skills

Line Graphs

In laboratory experiments, you will usually be controlling one variable and seeing how it affects another variable. Line graphs can show these relations clearly. For example, you might perform an experiment in which you measure the growth of a plant over time to determine the rate of the plant's growth. In this experiment, you are controlling the time intervals at which the plant height is measured. Therefore, time is called the *independent variable*. The height of the plant is the *dependent variable*. **Table 1** gives some sample data for an experiment to measure the rate of plant growth.

The independent variable is plotted on the *x*-axis. This axis will be labeled *Time (days)*, and will have a range from 0 days to 35 days. Be sure to properly label your axis including the units on the values.

The dependent variable is plotted on the *y*-axis. This axis will be labeled *Plant Height (cm)* and will have a range from 0 cm to 5 cm.

Think of your graph as a grid with lines running horizontally from the y-axis, and vertically from the *x*-axis. To plot a point, find the *x* (in this example time) value on the *x*-axis. Follow the vertical line from the *x*-axis until it intersects

the horizontal line from the *y*-axis at the corresponding *y* (in this case height) value. At the intersection of these two lines, place your point. **Figure 1** shows what a line graph of the data in **Table 1** might look like.

Figure 1

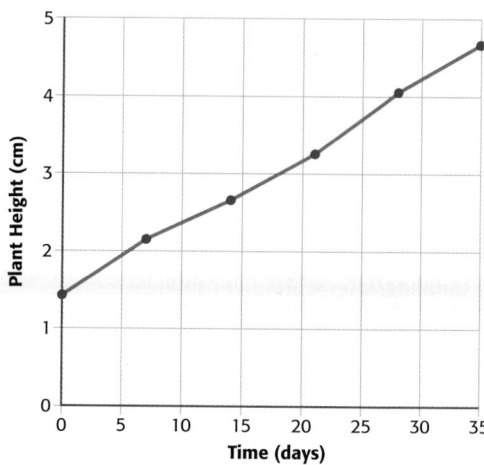

Table 1 Experimental Data for Plant Growth versus Time

Time (days)	Plant height (cm)
0	1.43
7	2.16
14	2.67
21	3.25
28	4.04
35	4.67

Practice

1. What does the line in **Figure 1** show, and what can you conclude about the plants used in the experiment?

2. Create a line graph of the following data.

Number of Days	Plant height (cm)
0	1.46
7	2.67
14	3.89
21	4.82

3. Compare the graph you made with **Figure 1**. What can you conclude about the two different groups of plants?

854

Scatter Plots

Some experiments or groups of data are best represented in a graph that is similar to a line graph and that is called a scatter plot. As in a line graph, the data points are plotted on the graph by using values on an *x*-axis and a *y*-axis. Scatter plots are often used to find trends in data. Instead of connecting the data points with a line, a trend can be represented by a best-fit line. A best-fit line is a line that represents all of the data points without necessarily going through all of them. To find a best-fit line, pick a line that is equidistant from as many data points as possible. Examine the graph below.

Figure 2

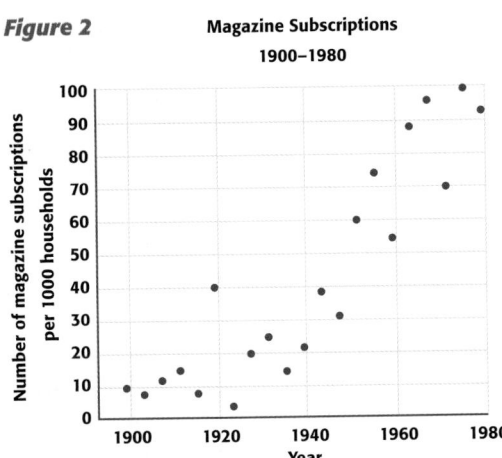

Magazine Subscriptions
1900–1980

If we connected all of the data points with lines, the lines would create a zigzag pattern that would not tell us much about our data. But if we find a best-fit line, we can see a trend more clearly. Furthermore, if we pick two points on the best-fit line, we can estimate its slope. Examine the dotted lines on *Figure 3.*

The points can be estimated as 18 magazine subscriptions per 100 households in 1920, and 42 magazine subscriptions per 100 households in 1940. If we subtract 1920 from 1940, and 18 sub-

scriptions from 42 subscriptions (using the point slope formula), we see that the line shows a trend of an increase of 24 subscriptions per 1000 households acres every 20 years. Scatter plots can also be used when there are two or more trends within one group of data or when there is no distinct trend at all.

Figure 3

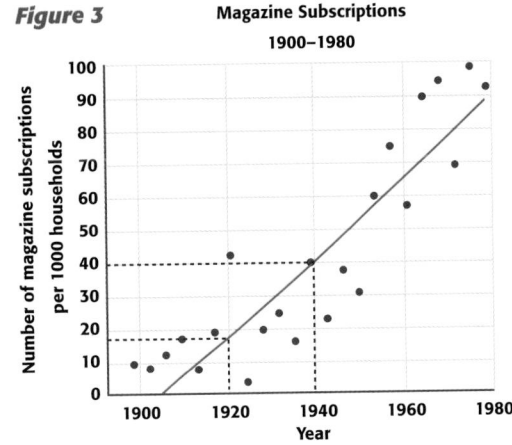

Magazine Subscriptions
1900–1980

Practice

1. Copy the graph below, and draw a best-fit line.

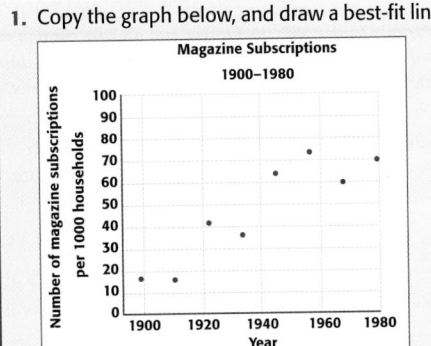

Magazine Subscriptions
1900–1980

2. What does that line represent?
3. If these were the data from a different city than the data in *Figure 3,* what conclusions could you draw about the two cities?

855

GRAPHING CALCULATOR TECHNOLOGY

Downloading

To solve the Graphing Calculator problems in Technology & Learning sections of the Chapter Reviews, you will need to download the programs and the datasets. These files can be found at the website go.hrw.com.

internet connect

www.scilinks.org
Topic: go.hrw.com
SciLinks code: HW4 TI83

SCiLINKS. Maintained by the National Science Teachers Association

This Web site contains links for downloading programs and applications you will need for Technology and Learning exercises.

Note: In order to transfer programs and applications from a computer to your calculator, you will need a TI-Graph Link cable. Programs can also be transferred directly between calculators using a unit-to-unit cable. Refer to the TI Web site or to your calculator's user's manual for instructions.

If your computer does not already have TI-Graph Link software installed, click Step 1: TI-Graph Link Software and follow the links for downloading and installing TI-Graph Link from the TI Web site.

Click Step 2: HChmProg. This will load the file HCHMPROG.ZIP onto your computer. Once the file is downloaded, double-click the icon and the file will be extracted into a file called hchmprog.8xg.

Click Step 3: Getting Started and follow the instructions for your TI-Graph Link to load hchmprog.8xg onto your TI calculator. When the file is sent to the calculator, it should expand into 17 programs. These programs should appear in the PRGM menu.

Download the CHEMAPPS application from go.hrw.com.

Using the Data Sets

Once you have loaded the application, you can use it via the "apps" function on the calculator. Press the blue [APPS] button in the upper left portion of the keypad. A menu will be displayed of the applications in your calculator's memory. Look for an application titled **CHEMAPPS**. Select it either by using the arrow keys and pressing **ENTER** or by using the number keys. Do this each time you choose to open the application. A title screen will appear for a few seconds, and then a menu will be displayed, listing all of the available data sets. The sets are listed by chapter and can be selected with either the arrow keys and [ENTER] or with the number keys.

The instructions are essentially the same for every question that uses the datasets. You can select the lists to be loaded by placing the cursor beside them and pressing the [ENTER] key or, if you would like to load all of the lists from that chapter at one time, you can use the arrow keys to go to the **All** menu and choose the first option **All+** and then press [ENTER]. When you have chosen all of the data sets that you need, use the arrow keys to go to the **Load** menu. The **SetUpEditor** allows you to decide where you would like the lists to be stored. 1: Add to Editor adds the lists to the end of the List editor. By choosing [STAT] **and 1: Edit…,** you can see these lists behind L_6.
2: Exchange Lists replaces the data values in those with the sets from the application. They will be replaced in order, so if you have only one list, L_1 will be replaced, and so on. **3: No change** will not add the list to the List editor; however, by going to the List menu via [2nd] [LIST], you will see the names of the data sets listed under L_6. After you have finished, choose **Load,** and the sets will be loaded into the desired location.

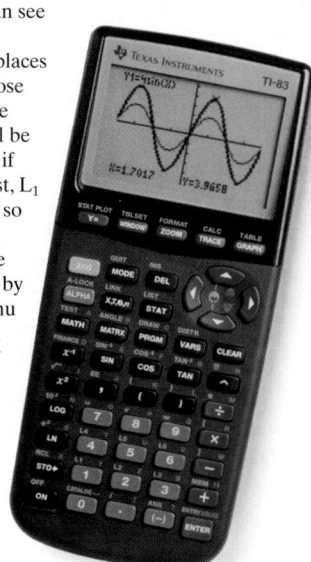

856

Troubleshooting

- Calculator instructions in the Holt Chemistry program are written for the TI-83 Plus. You may use other graphing calculators, but some of the programs and instructions may not work exactly as described.

- If you have problems loading programs or applications onto your calculator, you may need to clear programs or other data from your calculator's memory.

- Always make sure that you are downloading correct versions of the software. TI-Graph Link has different versions for Windows and for Macintosh as well as different versions for different calculators.

- If you need additional help, Texas Instruments can provide technical support at 1-800 TI-CARES.

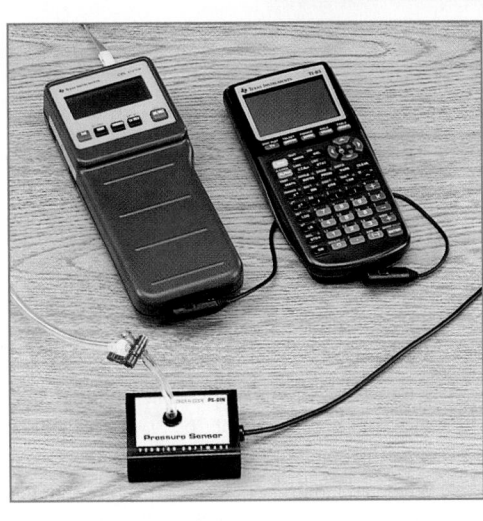

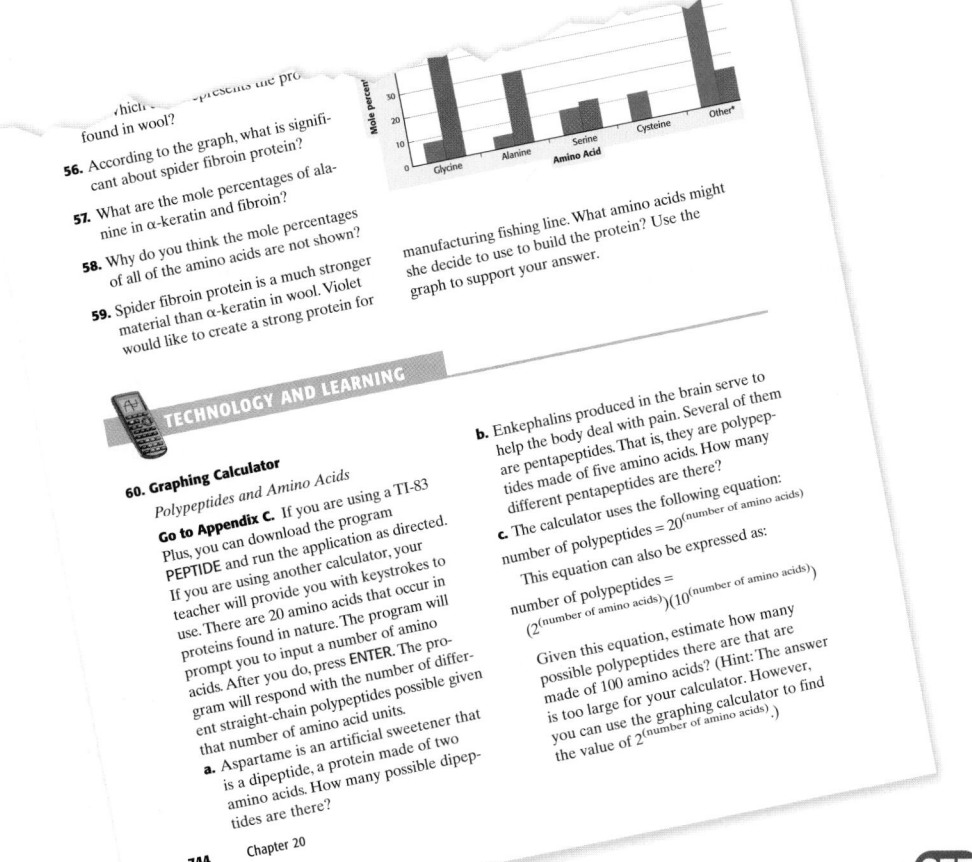

which ... -presents the pro... found in wool?

56. According to the graph, what is significant about spider fibroin protein?

57. What are the mole percentages of alanine in α-keratin and fibroin?

58. Why do you think the mole percentages of all of the amino acids are not shown?

59. Spider fibroin protein is a much stronger material than α-keratin in wool. Violet would like to create a strong protein for manufacturing fishing line. What amino acids might she decide to use to build the protein? Use the graph to support your answer.

TECHNOLOGY AND LEARNING

60. Graphing Calculator

Polypeptides and Amino Acids

Go to Appendix C. If you are using a TI-83 Plus, you can download the program PEPTIDE and run the application as directed. If you are using another calculator, your teacher will provide you with keystrokes to use. There are 20 amino acids that occur in proteins found in nature. The program will prompt you to input a number of amino acids. After you do, press **ENTER.** The program will respond with the number of different straight-chain polypeptides possible given that number of amino acid units.

a. Aspartame is an artificial sweetener that is a dipeptide, a protein made of two amino acids. How many possible dipeptides are there?

b. Enkephalins produced in the brain serve to help the body deal with pain. Several of them are pentapeptides. That is, they are polypeptides made of five amino acids. How many different pentapeptides are there?

c. The calculator uses the following equation:

number of polypeptides $= 20^{(\text{number of amino acids})}$

This equation can also be expressed as:

number of polypeptides $=$
$(2^{(\text{number of amino acids})})(10^{(\text{number of amino acids})})$

Given this equation, estimate how many possible polypeptides there are that are made of 100 amino acids? (Hint: The answer is too large for your calculator. However, you can use the graphing calculator to find the value of $2^{(\text{number of amino acids})}$.)

744 Chapter 20

857

Appendix D

PROBLEM BANK

Answers to Problem Bank
The Science of Chemistry

1. 2.8 g/cm^3
2. 1.14 g
3. 5.60 mL
4. 3.65 g/cm^3
5. 111 cm^3
6. 1645 cm
7. 0.01645 km
8. 1.4 × 10^{-5} g
9. 0.0105 kg
10. 1570 m
11. 3.54 × 10^{-6} g
12. 3.5 × 10^6 μmol
13. 1200 mL
14. 0.000358 m^3
15. 548.6 cm^3
16. 1.71 g/cm^3
17. 4.54 cm^3
18. 86400 s
19. 625000 cg
20. 440 g
21. 33.7 cm^3
22. 4.35 g

The Science of Chemistry

Section: Describing Matter

1. What is the density of a block of marble that occupies 310 cm^3 and has a mass of 853 g?

2. Diamond has a density of 3.26 g/cm^3. What is the mass of a diamond that has a volume of 0.350 cm^3?

3. What is the volume of a sample of liquid mercury that has a mass of 76.2 g, given that the density of mercury is 13.6 g/mL?

4. What is the density of a sample of ore that has a mass of 74.0 g and occupies 20.3 cm^3?

5. Find the volume of a sample of wood that has a mass of 95.1 g and a density of 0.857 g/cm^3?

6. Express a length of 16.45 m in centimeters.

7. Express a length of 16.45 m in kilometers.

8. Express a mass of 0.014 mg in grams.

9. Complete the following conversion:
   10.5 g = _____ kg

10. Complete the following conversion:
    1.57 km = _____ m

11. Complete the following conversion:
    3.54 μg = _____ g

12. Complete the following conversion:
    3.5 mol = _____ μmol

13. Complete the following conversion:
    1.2 L = _____ mL

14. Complete the following conversion:
    358 cm^3 = _____ m^3

15. Complete the following conversion:
    548.6 mL = _____ cm^3

16. What is the density of an 84.7 g sample of an unknown substance if the sample occupies 49.6 cm^3?

17. What volume would be occupied by 7.75 g of a substance with a density of 1.70766 g/cm^3?

18. Express a time period of exactly 1 day in terms of seconds. Try to write out all the equalities needed to solve this problem.

19. How many centigrams are there in 6.25 kg?

20. Polycarbonate plastic has a density of 1.2 g/cm^3. A photo frame is constructed from two 3.0 mm sheets of polycarbonate. Each sheet measures 28 cm by 22 cm. What is the mass of the photo frame?

21. Find the volume of a cube that is 3.23 cm on each edge.

22. Calculate the density of a 17.982 g object that occupies 4.13 cm^3.

Matter and Energy

Section: Measurements and Calculations in Chemistry

1. Determine the specific heat of a material if a 35.0 g sample absorbed 48.0 J as it was heated from 293 K to 313 K.

2. A piece of copper alloy with a mass of 85.0 g is heated from 30.0°C to 45.0°C. In the process, it absorbs 523 J of energy as heat.
   a. What is the specific heat of this copper alloy?
   b. How much energy will the same sample lose if it is cooled to 25.0°C?

3. The temperature of a 74.0 g sample of material increases from 15.0°C to 45.0°C when it absorbs 2.00 kJ of energy as heat. What is the specific heat of this material?

4. How much energy is needed to raise the temperature of 5.00 g of gold (c_p = 0.129 J/(g•K)) by 25.0°C?

5. Energy in the amount of 420 J is added to a 35.0 g sample of water (c_p = 4.18 J/(g•K)) at a temperature of 10.0°C. What will be the final temperature of the water?

6. What mass of liquid water (c_p = 4.18 J/(g•K)) at room temperature (25°C) can be raised to its boiling point with the addition of 24.0 kJ of energy?

7. How much energy would be absorbed as heat by 75 g of iron ($c_p = 0.449$ J/(g•K)) when heated from 295 K to 301 K?

Atoms and Moles

Section: Structure of Atoms

1. How many protons, electrons, and neutrons are in an atom of bromine-80?

2. Write the nuclear symbol for carbon-13.

3. Write the hyphen notation for the element that contains 15 electrons and 15 neutrons.

4. How many protons, electrons, and neutrons are in an atom of carbon-13?

5. Write the nuclear symbol for oxygen-16.

6. Write the hyphen notation for the element whose atoms contains 7 electrons and 9 neutrons.

Section: Counting Atoms

7. What is the mass in grams of 2.25 mol of the element iron, Fe?

8. What is the mass in grams of 0.375 mol of the element potassium, K?

9. What is the mass in grams of 0.0135 mol of the element sodium, Na?

10. What is the mass in grams of 16.3 mol of the element nickel, Ni?

11. What is the mass in grams of 3.6 mol of the element carbon, C?

12. What is the mass in grams of 0.733 mol of the element chlorine, Cl?

13. How many moles of calcium, Ca, are in 5 g of calcium?

14. How many moles of gold, Au, are in 3.6×10^{-10} g of gold?

15. How many moles of copper, Cu, are in 3.22 g of copper?

16. How many moles of lithium, Li, are in 2.72×10^{-4} g of lithium?

17. How many moles of lead, Pb, are in 1.5×10^{12} atoms of lead?

18. How many moles of tin, Sn, are in 2500 atoms of tin?

19. How many atoms of aluminum, Al, are in 2.75 mol of aluminum?

20. How many moles of carbon, C, are in 2.25×10^{22} atoms of carbon?

21. How many moles of oxygen, O, are in 2×10^6 atoms of oxygen?

22. How many atoms of sodium, Na, are in 3.8 mol of sodium?

23. What is the mass in grams of 7.5×10^{15} atoms of nickel, Ni?

24. How many atoms of sulfur, S, are in 4 g of sulfur?

25. What mass of gold, Au, contains the same number of atoms as 9.00 g of aluminum, Al?

26. What is the mass in grams of 5×10^9 atoms of neon, Ne?

27. How many atoms of carbon, C, are in 0.02 g of carbon?

28. What mass of silver, Ag, contains the same number of atoms as 10 g of boron, B?

29. How many moles of atoms are there in 3.25×10^5 g Pb?

30. How many moles of atoms are there in 150 g S?

The Mole and Chemical Composition

Section: Avogadro's Number and Mole Conversions

1. What is the mass in grams of 3.25 mol $Fe_2(SO_4)_3$?

2. How many moles of molecules are there in 250 g of hydrogen nitrate, HNO_3?

3. How many molecules of aspirin, $C_9H_8O_4$, are there in a 100 mg tablet of aspirin?

4. What is the mass in grams of 3.04 mol of ammonia vapor, NH_3?

5. Calculate the mass of 0.257 mol of calcium nitrate, $Ca(NO_3)_2$.

6. How many moles are there in 6.60 g of $(NH_4)_2SO_4$?

7. How many moles are there in 4.50 kg of $Ca(OH)_2$?

Appendix D

Matter and Energy
1. 0.069 J/(g•K)
2. a. 0.410 J/(g•K)
   b. 697 J = 700J
3. 0.901 J/(g•K)
4. 16.1 J
5. 12.9°C
6. 77 g
7. 200 J

Appendix D

Atoms and Moles

1. 35 protons, 35 electrons, 45 neutrons
2. $^{13}_{6}C$
3. phosphorus-30
4. 6 protons, 6 electrons, 7 neutrons
5. $^{16}_{8}O$
6. nitrogen-16
7. 126 g
8. 14.7 g
9. 0.3104 g
10. 957 g
11. 43 g
12. 26.0 g
13. 0.1 mol
14. 1.8×10^{-12} mol
15. 0.051 mol
16. 3.92×10^{-5} mol
17. 2.49087×10^{-12} mol
18. 4.1×10^{-21} mol
19. 1.66×10^{24} atoms
20. 0.037 mol
21. 3×10^{-18} mol
22. 2.3×10^{24} atoms
23. 7.3×10^{-7} g
24. 7×10^{22} atoms
25. 65.7 g
26. 2×10^{-13} g
27. 1×10^{21} atoms
28. 100 g
29. 1.5×10^{3} mol
30. 4.7 mol

8. How many molecules are there in 25.0 g of H_2SO_4?
9. How many molecules are there in 125 g of sugar, $C_{12}H_{22}O_{11}$?
10. What is the mass in grams of 6.25 mol of copper(II) nitrate?
11. How many moles are there in 3.82 g of SO_2?
12. How many moles are there in 4.15×10^{-3} g of $C_6H_{12}O_6$?
13. How many moles are there in 77.1 g of Cl_2?
14. How many molecules are there in 3.82 g of SO_2?
15. How many molecules are there in 4.15×10^{-3} g of $C_6H_{12}O_6$?
16. Determine the number of moles in 4.50 g of H_2O?
17. Determine the number of moles in 471.6 g of $Ba(OH)_2$?
18. Determine the number of moles in 129.68 g of $Fe_3(PO_4)_2$?
19. What is the mass in grams of 1.00 mol NaCl?
20. What is the mass in grams of 2.00 mol H_2O?
21. What is the mass in grams of 3.500 mol $Ca(OH)_2$?
22. What is the mass in grams of 0.6250 mol $Ba(NO_3)_2$?

Section: Relative Atomic Mass and Chemical Formulas

23. Find the formula mass for the following: H_2SO_4.
24. Find the formula mass for the following: $Ca(NO_3)_2$.
25. Find the formula mass for the following: PO_4^{3-}.
26. Find the formula mass for the following: $MgCl_2$.
27. Find the formula mass for the following: Na_2SO_3.
28. Find the formula mass for the following: $HClO_3$.

29. Find the formula mass for the following: MnO_4^-.
30. Find the formula mass for the following: C_2H_6O.
31. Find the molar mass for the following: Al_2S_3.
32. Find the molar mass for the following: $NaNO_3$.
33. Find the molar mass for the following: $Ba(OH)_2$.
34. Find the molar mass for the following: K_2SO_4.
35. Find the molar mass for the following: $(NH_4)_2CrO_4$.
36. Determine the formula mass of the following: calcium acetate, $Ca(CH_3COO)_2$.
37. Determine the molar mass of the following: glucose, $C_6H_{12}O_6$.
38. Determine the molar mass of the following: calcium acetate, $Ca(CH_3COO)_2$.
39. Determine the formula mass of the following: the ammonium ion, NH_4^+.
40. Determine the molar mass of the following: the chlorate ion, ClO_3^-.
41. Determine the molar mass of XeF_4.
42. Determine the molar mass of $C_{12}H_{42}O_6$.
43. Determine the molar mass of Hg_2I_2.
44. Determine the molar mass of CuCN.
45. Determine the formula mass of the following: the ammonium ion, NH_4^+.
46. Determine the formula mass of the following: the chlorate ion, ClO_3^-.

Section: Formulas and Percentage Composition

47. Calculate the percentage composition of lead (II) chloride, $PbCl_2$.
48. Calculate the percentage composition of barium nitrate, $Ba(NO_3)_2$.
49. Find the mass percentage of water in $ZnSO_4 \cdot 7H_2O$.

50. a. Magnesium hydroxide is 54.87% oxygen by mass. How many grams of oxygen are in 175 g of the compound?

   b. How many moles of oxygen is this?

51. Calculate the percent composition of sodium nitrate, $NaNO_3$.

52. Calculate the percent composition of silver sulfate, Ag_2SO_4.

53. What is the mass percentage of water in the hydrate $CuSO_4 \cdot 5H_2O$?

54. a. Zinc chloride, $ZnCl_2$ is 52.02% chlorine by mass. What mass of chlorine is contained in 80.3 g of $ZnCl_2$?

   b. How many moles of Cl is this?

55. A compound is found to contain 63.52% iron and 36.48% sulfur. Find its empirical formula.

56. Determine the formula mass and molar mass of ammonium carbonate, $(NH_4)_2CO_3$.

   a. The formula mass is _____ ?

   b. The molar mass is _____ ?

57. Calculate the percent composition of $(NH_4)_2CO_3$.

58. A compound is found to contain 36.48% Na, 25.41% S, and 38.11% O. Find its empirical formula.

59. Calculate the percent composition of sodium chloride, NaCl.

60. Calculate the percent composition of silver nitrate, $AgNO_3$.

61. Calculate the percent composition of magnesium hydroxide, $Mg(OH)_2$.

62. What is the mass percentage of water in the hydrate $CuSO_4 \cdot 5H_2O$?

63. Determine the percent composition for NaClO.

64. Determine the percent composition for H_2SO_3.

65. Determine the percent composition for C_2H_5COOH.

66. Determine the percent composition for $BeCl_2$.

67. a. A compound is found to contain 54.5% carbon, 9.1% hydrogen, and 36.4% oxygen Determine the simplest formula.

   b. The molar mass of a compound is 88.1 g. What is the molecular formula if the simplest formula is C_2H_4O?

68. Find the empirical formula of a compound found to contain 26.56% potassium, 35.41% chromium, and the remainder oxygen.

69. Analysis of 20.0 g of a compound containing only calcium and bromine indicates that 4.00 g of calcium are present. What is the empirical formula of the compound formed?

70. A compound is analyzed and found to contain 36.70% potassium, 33.27% chlorine, and 30.03% oxygen. What is the empirical formula of the compound?

71. Determine the empirical formula of the compound that contains 17.15% carbon, 1.44% hydrogen, and 81.41% fluorine.

72. A 60.00 g sample of tetraethyl-lead, a gasoline additive, is found to contain 38.43 g lead, 17.83 g carbon, and 3.74 g hydrogen. Find its empirical formula.

73. A 100.00 g sample of an unidentified compound contains 29.84 g sodium, 67.49 g chromium, and 72.67 g oxygen. What is the compound's empirical formula?

74. A compound is found to contain 53.70% iron and 46.30% sulfur. Find its empirical formula.

75. Analysis of a compound indicates that it contains 1.04 g K, 0.70 g Cr, and 0.86 g O. Find its empirical formula.

76. If 4.04 g of N combine with 11.46 g of O to produce a compound with the formula mass of 108.0 amu, what is the molecular formula of this compound?

77. Determine the empirical formula of a compound containing 63.50% silver, 8.25% nitrogen, and the remainder oxygen.

78. Determine the empirical formula of a compound found to contain 52.11% carbon, 13.14% hydrogen, and 34.75% oxygen.

The Mole and Chemical Composition

1. 1300 g
2. 4.0 mol
3. 3×10^{20} molecules
4. 51.8 g
5. 42.1 g
6. 0.05 mol
7. 61 mol
8. 1.5×10^{23} molecules
9. 2.2×10^{23} molecules
10. 1170 g
11. 0.0596 mol
12. 2.30×10^{-5} mol
13. 1.09 mol
14. 3.59×10^{22} molecules
15. 1.39×10^{19} molecules
16. 0.25 mol
17. 2.752 mol
18. 0.3627 mol
19. 58.4 g
20. 36.0 g
21. 259.3 g
22. 163.3 g
23. 98.09
24. 164.10
25. 94.97
26. 95.21
27. 126.05
28. 84.46
29. 118.94
30. 46.08

Appendix D

31. 150.17
32. 85.00
33. 171.35
34. 174.27
35. 152.10
36. 158.18 amu
37. 180.18 g/mol
38. 158.18 g/mol
39. 18.05 g/mol
40. 83.45 g/mol
41. 207.29 g/mol
42. 264.36 g/mol
43. 654.98 g/mol
44. 89.57 g/mol
45. 18.05 amu
46. 83.45 amu
47. 74.51% Pb, 25.49% Cl
48. 52.55% Ba, 10.72% N, 36.73% O
49. 43.85%
50. a. 96.0 g
b. 6.00 mol
51. 27.05% Na, 16.48% N, 56.47% O
52. 69.19% Ag, 10.29% S, 20.53% O
53. 36.08%
54. a. 41.8 g
b. 1.18 mol
55. FeS
56. a. 96.11 amu
b. 96.11 g/mol
57. N: 29.15%, H: 8.407%, C: 12.50%, O: 49.94%
58. Na_2SO_3
59. 39.34% Na, 60.66% Cl
60. 63.50% Ag, 8.25% N, 28.26% O

79. Chemical analysis of citric acid shows that it contains 37.51% C, 4.20% H, 58.29% O. What is its empirical formula?

80. A 175.0 g sample of a compound contains 56.15 g C, 9.43 g H, 74.81 g O, 13.11 g N, and 21.49 g Na. What is its empirical formula?

81. In the laboratory, a sample of pure nickel was placed in a clean, dry, weighed crucible. The crucible was heated so that the nickel would react with the oxygen in the air. After the reaction appeared complete, the crucible was allowed to cool and the mass was determined. The crucible was reheated and allowed to cool. Its mass was then determined again to be certain that the reaction was complete. The following data were collected:
Mass of crucible = 30.02 g
Mass of nickel and crucible = 31.07 g
Mass of nickel oxide and crucible = 31.36 g
Determine the following information based on the data given above:
Mass of nickel =
Mass of nickel oxide =
Mass of oxygen =
Based on your calculations, what is the empirical formula for the nickel oxide?

82. Determine the molecular formula of the compound with an empirical formula of CH and a formula mass of 78.110 amu.

83. A sample compound with a formula mass of 34.00 amu is found to consist of 0.44 g H and 6.92 g O. Find its molecular formula.

84. The empirical formula for trichloroisocyanuric acid, the active ingredient in many household bleaches, is OCNCl. The molar mass of this compound is 232.41 g/mol. What is the molecular formula of trichloroisocyanuric acid?

85. Determine the molecular formula of a compound with an empirical formula of NH_2 and a formula mass of 32.06 amu.

86. The molar mass of a compound is 92 g/mol. Analysis of a sample of the compound indicates that it contains 0.606 g N and 1.390 g O. Find its molecular formula.

87. What is the molecular formula of the molecule that has an empirical formula of CH_2O and a molar mass of 120.12 g/mol?

88. A compound with a formula mass of 42.08 amu is found to be 85.64% carbon and 14.36% hydrogen by mass. Find its molecular formula.

Stoichiometry

Section: Calculating Quantities in Reactions

1. A seashell, composed largely of calcium carbonate, is placed in a solution of HCl. As a result, 1500 mL of dry CO_2 gas at STP is produced. The other products are $CaCl_2$ and H_2O.

   Based on this information, how many grams of $CaCO_3$ are consumed in the reaction?

2. *Acid precipitation* is the term generally used to describe rain or snow that is more acidic than normal. One cause of acid precipitation is the formation of sulfuric and nitric acids from various sulfur and nitrogen oxides produced in volcanic eruptions, forest fires, and thunderstorms. In a typical volcanic eruption, for example, 3.50×10^8 kg of SO_2 may be produced. If this amount of SO_2 were converted to H_2SO_4 according to the two-step process given below, how many kilograms of H_2SO_4 would be produced from such an eruption?

$$SO_2 + \frac{1}{2}O_2 \longrightarrow SO_3$$

$$SO_3 + H_2O \longrightarrow H_2SO_4$$

3. Solid iron(III) hydroxide decomposes to produce iron(III) oxide and water vapor. If 0.750 L of water vapor is produced at STP,

   a. How many grams of iron(III) hydroxide were used?

   b. How many grams of iron(III) oxide are produced?

4. Balance the following chemical equation.

$$Mg(s) + O_2(g) \longrightarrow MgO(s)$$

   Then, based on the quantity of reactant or product given, determine the corresponding quantities of the specified reactants or products, assuming that the system is at STP.

a. How many moles of MgO are produced from 22.4 L O_2?

b. How many moles of MgO are produced from 11.2 L O_2?

c. How many moles of MgO are produced from 1.40 L O_2?

5. Assume that 8.50 L of I_2 are produced using the following reaction that takes place at STP:

$KI(aq) + Cl_2(g) \longrightarrow KCl(aq) + I_2(g)$

Balance the equation before beginning your calculations.

a. How many moles of I_2 are produced?

b. How many moles of KI were used?

c. How many grams of KI were used?

6. Suppose that 650 mL of hydrogen gas are produced through a replacement reaction involving solid iron and sulfuric acid, H_2SO_4, at STP. How many grams of iron(II) sulfate are also produced?

Section: Limiting Reactants and Percentage Yield

7. Methanol, CH_3OH, is made by causing carbon monoxide and hydrogen gases to react at high temperature and pressure. If 450 mL of CO and 825 mL of H_2 are mixed,

a. Which reactant is present in excess?

b. How much of that reactant remains after the reaction?

c. What volume of CH_3OH is produced, assuming the same pressure?

Causes of Change

Section: Using Enthalpy

1. When 1 mol of methane is burned at constant pressure, –890 kJ/mol of energy is released as heat. If a 3.2 g sample of methane is burned at constant pressure, what will be the value of ΔH? (Hint: Convert the grams of methane to moles. Also make sure your answer has the correct sign for an exothermic process.)

2. How much energy is needed to raise the temperature of a 55 g sample of aluminum from 22.4°C to 94.6°C? The specific heat of aluminum is 0.897 J/(g•K).

3. 3500 J of energy are added to a 28.2 g sample of iron at 20°C. What is the final temperature of the iron in kelvins? The specific heat of iron is 0.449 J/(g•K).

Section: Changes in Enthalpy During Reactions

4. The combustion of methane gas, CH_4, forms $CO_2(g) + H_2O(l)$. Calculate the energy as heat produced by burning 1 mol of the methane gas.

5. Calculate ΔH for the following reaction:

$2N_2(g) + 5O_2(g) \longrightarrow 2N_2O_5(g)$

Use the following data in your calculation:

$H_2(g) + \frac{1}{2}O_2(g) \longrightarrow H_2O(l)$

$\Delta H_f^0 = -285.8$ kJ/mol

$N_2O_5(g) + H_2O(l) \longrightarrow 2NHO_3(l)$
$\Delta H = -76.6$ kJ/mol

$\frac{1}{2}N_2(g) + \frac{3}{2}O_2(g) + \frac{1}{2}H_2(g) \longrightarrow HNO_3(l)$

$\Delta H_f^0 = -174.1$ kJ/mol

6. Calculate the enthalpy of formation of butane, C_4H_{10}, using the following balanced chemical equation and information. Write out the solution according to Hess's law.

$C(s) + O_2(g) \longrightarrow CO_2(g)$
$\Delta H_f^0 = -393.5$ kJ/mol

$H_2(g) + \frac{1}{2}O_2(g) \longrightarrow H_2O(l)$
$\Delta H_f^0 = -285.8$ kJ/mol

$4CO_2(g) + 5H_2O(l) \longrightarrow C_4H_{10}(g) + \frac{13}{2}O_2(g)$
$\Delta H = 2877.6$ kJ/mol

7. Calculate the enthalpy of combustion of 1 mol of nitrogen, N_2, to form NO_2.

8. Calculate the enthalpy of formation for 1 mol sulfur dioxide, SO_2, from its elements, sulfur and oxygen. Use the balanced chemical equation and the following information.

$S(s) + \frac{3}{2}O_2(g) \longrightarrow SO_3(g)$
$\Delta H = -395.2$ kJ/mol

$2SO_2(g) + O_2(g) \longrightarrow 2SO_3(g)$
$\Delta H = -198.2$ kJ/mol

Appendix D

61. 41.67% Mg, 54.87% O, 3.46% H
62. 36.08%
63. 30.88% Na, 47.62% Cl, 21.49% O
64. 2.46% H, 39.07% S, 58.47% O
65. 48.63% C, 8.18% H, 43.19% O
66. 11.28% Be, 88.72% Cl
67. a. C_2H_4O
    b. $C_4H_8O_2$
68. $K_2Cr_2O_7$
69. $CaBr_2$
70. $KClO_2$
71. CHF_3
72. PbC_8H_{20}
73. $Na_2Cr_2O_7$
74. Fe_2S_3
75. K_2CrO_4
76. N_2O_5
77. $AgNO_3$
78. C_2H_6O
79. $C_6H_8O_7$
80. $C_5H_{10}O_5NNa$
81. NiO
82. C_6H_6
83. H_2O_2
84. $O_3C_3N_3Cl_3$
85. N_2H_4
86. N_2O_4
87. $C_4H_8O_4$
88. C_3H_6

863

Appendix D

Stoichiometry

1. 6.7 g

2. 5.36×10^8 kg

3. a. 2.38 g

  b. 1.78 g

4. a. 2 mol

  b. 1 mol

  c. 0.125 mol

5. a. 0.379 mol

  b. 0.758 mol

  c. 126 g

6. 4.41 g

7. a. CO

  b. 38 mL

  c. 412 mL

9. Use enthalpy data given after the question to calculate the enthalpy of reaction for each of the following. Solve each by combining the known thermochemical equations.

a. $CaCO_3(s) \longrightarrow CaO(s) + CO_2(g)$

  Given:

  $2CaCO_3(s) \longrightarrow 2Ca(s) + 2C(s) + 3O_2(g)$

  $\Delta H = 2413.8$ kJ/mol

  $2Ca(s) + O_2(g) \longrightarrow 2CaO(s)$

  $\Delta H = -1269.8$ kJ/mol

  $C(s) + O_2(g) \longrightarrow CO_2(g)$

  $\Delta H = -393.51$ kJ/mol

b. $Ca(OH)_2(s) \longrightarrow CaO(s) + H_2O(g)$

  Given:

  $Ca(OH)_2(s) \longrightarrow Ca(s) + O_2(g) + H_2(g)$

  $\Delta H = 983.2$ kJ/mol

  $2Ca(s) + O_2(g) \longrightarrow 2CaO(s)$

  $\Delta H = -1269.8$ kJ/mol

  $2H_2(g) + O_2(g) \longrightarrow 2H_2O(g)$

  $\Delta H = -483.6$ kJ/mol

c. $Fe_2O_3(s) + 3CO(g) \longrightarrow 2Fe(s) + 3CO_2(g)$

  Given:

  $2Fe_2O_3(s) \longrightarrow 4Fe(s) + 3O_2(g)$

  $\Delta H = 1651$ kJ/mol

  $CO(g) \longrightarrow C(s) + \frac{1}{2}O_2(g)$

  $\Delta H = 110.5$ kJ/mol

  $C(s) + O_2(g) \longrightarrow CO_2(g)$

  $\Delta H = -393.5$ kJ/mol

10. Calculate the enthalpies for reactions in which ethane, C_6H_6, are the respective reactants and $CO_2(g)$ and $H_2O(l)$ are the products in each. Solve each by combining the known thermochemical equations using the ΔH values given below. Verify the result by using the general equation for finding enthalpies of reaction from enthalpies of formation.

a. $C_2H_6(g) + O_2(g) \longrightarrow$

  Given:

  $C_2H_6(g) \longrightarrow 2C(s) + 3H_2(g)$

  $\Delta H = 83.8$ kJ/mol

  $C(s) + O_2(g) \longrightarrow CO_2(g)$

  $\Delta H = -393.5$ kJ/mol

  $H_2(g) + \frac{1}{2}O_2(g) \longrightarrow H_2O(l)$

  $\Delta H = -285.8$ kJ/mol

b. $C_6H_6(g) + O_2(g) \longrightarrow$

  Given:

  $C_6H_6(l) \longrightarrow 6C(s) + 3H_2(g)$

  $\Delta H = -49.08$ kJ/mol

  $C(s) + O_2(g) \longrightarrow CO_2(g)$

  $\Delta H = -393.5$ kJ/mol

  $H_2(g) + \frac{1}{2}O_2(g) \longrightarrow H_2O(l)$

  $\Delta H = -285.8$ kJ/mol

11. The enthalpy of formation of ethanol, C_2H_5OH, is -277 kJ/mol at 298.15 K. Calculate the enthalpy of combustion of one mole of ethanol from the information given below, assuming that the products are $CO_2(g)$ and $H_2O(l)$.

  $C_2H_5OH(l) \longrightarrow 2C(s) + 3H_2(g) + \frac{1}{2}O_2(g)$

  $\Delta H = -(-277$ kJ/mol$)$

  $C(s) + O_2(g) \longrightarrow CO_2(g)$

  $\Delta H = -393.5$ kJ/mol

  $H_2(g) + \frac{1}{2}O_2(g) \longrightarrow H_2O(l)$

  $\Delta H = -285.8$ kJ/mol

12. The enthalpy of formation for sulfur dioxide gas is -0.2968 kJ/(mol•K). Calculate the amount of energy given off in kJ when 30 g of $SO_2(g)$ is formed from its elements.

Section: Order and Spontaneity

13. Predict the sign of ΔS for each of the following reactions:

a. the thermal decomposition of solid calcium carbonate

  $CaCO_3(s) \longrightarrow CaO(s) + CO_2(g)$

b. the oxidation of SO_2 in air

  $2SO_2(g) + O_2(g) \longrightarrow 2SO_3(g)$

14. Calculate the value of ΔG for the reaction below, given the values of ΔH and ΔS. The temperature is 298 K.

  $Cu_2S(s) + S(s) \longrightarrow 2CuS(s)$

  $\Delta H = -26.7$ kJ/mol

  $\Delta S = -0.0197$ kJ/(mol•K)

15. Will the reaction in item 14 be spontaneous at 298 K?

Appendix D

16. Predict whether the value of ΔS for each of the following reactions will be greater than, less than, or equal to zero.
   a. $3H_2(g) + N_2(g) \longrightarrow 2NH_3(g)$
   b. $2Mg(s) + O_2(g) \longrightarrow 2MgO(s)$
   c. $C_6H_{12}O_6(s) + 6O_2(g) \longrightarrow$
   $$6CO_2(g) + 6H_2O(g)$$
   d. $KNO_3(s) \longrightarrow K^+(aq) + NO_3^-(aq)$

17. Based on the following values, compute ΔG values for each reaction and predict whether the reaction will occur spontaneously.
   a. $\Delta H = 125$ kJ/mol, $T = 293$ K,
   $\Delta S = 0.035$ kJ/(mol•K)

   b. $\Delta H = -85.2$ kJ/mol, $T = 400$ K,
   $\Delta S = 0.125$ kJ/(mol•K)

   c. $\Delta H = -275$ kJ/mol, $T = 773$ K,
   $\Delta S = 0.45$ kJ/(mol•K)

18. The ΔS for the reaction shown, at 298.15 K, is 0.003 kJ/(mol•K). Calculate the ΔG for this reaction, and determine whether it will occur spontaneously at 298.15 K.
   $C(s) + O_2(g) \longrightarrow CO_2(g) + 393.51$ kJ/mol

19. When graphite reacts with hydrogen at 300K, ΔH is -74.8 kJ/mol and ΔS is -0.0809 kJ/(mol•K). Will this reaction occur spontaneously?

20. The thermite reaction used in some welding applications has the following enthalpy and entropy changes at 298.15 K. Assuming ΔS and ΔH are constant, calculate ΔG at 448 K.
   $Fe_2O_3(s) + 2Al(s) \longrightarrow 2Fe(s) + Al_2O_3(s)$
   $\Delta H = -851.5$ kJ/mol
   $\Delta S = -0.0385$ kJ/(mol•K)

21. Calculate the change in enthalpy for the following reaction.
   $4FeO(s) + O_2(g) \longrightarrow 2Fe_2O_3(s)$
   Use the following enthalpy data.
   $FeO(s) \longrightarrow Fe(s) + \frac{1}{2}O_2(g)$

   $\Delta H = 272$ kJ/mol

   $2Fe(s) + \frac{3}{2}O_2(g) \longrightarrow Fe_2O_3(g)$

   $\Delta H = -824.2$ kJ/mol

States of Matter and Intermolecular Forces

Section: Intermolecular Forces

1. a. Which contains more molecules of water, 5.00 cm^3 of ice at 0.0°C or 5.00 cm^3 of liquid water at 0.0°C?

   b. How many more molecules?

GAS

Section: Characteristics of Gases

1. Convert a pressure of 1.75 atm to kPa.

2. Convert the pressure of 1.75 atm to mm Hg.

3. Convert a pressure of 570 torr to atmospheres.

4. Convert a pressure of 570 torr to kPa.

5. A weather report gives a current atmospheric pressure reading of 745.8 mm Hg. Express this reading in atmospheres.

6. Convert the pressure of 745.8 mm Hg to torrs.

7. Convert the pressure of 745.8 mm Hg to kilopascals.

8. Convert a pressure of 151.98 kPa to pressure in standard atmospheres.

9. Convert a pressure of 456 torr to pressure in standard atmospheres.

10. Convert a pressure of 912 mm Hg to pressure in standard atmospheres.

Section: The Gas Laws

11. A balloon filled with helium gas has a volume of 500 mL at a pressure of 1 atm. The balloon is released and reaches an altitude of 6.5 km, where the pressure is 0.5 atm. Assuming that the temperature has remained the same, what volume does the gas occupy at this height?

12. A gas has a pressure of 1.26 atm and occupies a volume of 7.4 L. If the gas is compressed to a volume of 2.93 L, what will its pressure be, assuming constant temperature?

13. Divers know that the pressure exerted by the water increases about 100 kPa with every 10.2 m of depth. This means that at 10.2 m below

865

Appendix D

Causes of Change
1. -180 kJ
2. 3600 J
3. 570 K
4. 890.2 kJ
5. 28.4 kJ/mol
6. -125.4 kJ/mol
7. 66.4 kJ
8. 296.1 kJ
9. a. 178.49 kJ/mol
   b. 106.5 kJ/mol
   c. -23.5 kJ/mol
10. a. -1560 kJ/mol
   b. -3267 kJ/mol
11. -1367.4 kJ/mol
12. 0.14 kJ
13. a. $+$
   b. $-$
14. -20.8 kJ/mol
15. Yes
16. a. $\Delta S < 0$
   b. $\Delta S < 0$
   c. $\Delta S > 0$
   d. $\Delta S > 0$
17. a. No, not spontaneous
   b. Yes, spontaneous
   c. Yes, spontaneous
18. a. -394.404 kJ/mol $= 400$ kJ/mol
   b. Yes, spontaneous
19. Yes, it will occur spontaneously
20. -834.252 kJ/mol
21. -560 kJ/mol

Appendix D

States of Matter and Intermolecular Forces

1. a. the liquid water

   b. 1.38686×10^{22} molecules

---

the surface, the pressure is 201 kPa; at 20.4 m below the surface, the pressure is 301 kPa; and so forth. Given the volume of a balloon is 3.5 L at STP and that the temperature of the water remains the same, what is the volume 51 m below the water's surface?

14. The piston of an internal combustion engine compresses 450 mL of gas. The final pressure is 15 times greater than the initial pressure. What is the final volume of the gas, assuming constant temperature?

15. A helium-filled balloon contains 125 mL of gas at a pressure of 0.974 atm. What volume will the gas occupy at standard pressure?

16. A weather balloon with a volume of 1.375 L is released from Earth's surface at sea level. What volume will the balloon occupy at an altitude of 20.0 km, where the air pressure is 10 kPa?

17. A sample of helium gas has a volume of 200 mL at 0.96 atm. What pressure, in atm, is needed to reduce the volume at constant temperature to 50 mL?

18. A certain mass of oxygen was collected over water when potassium chlorate was decomposed by heating. The volume of the oxygen sample collected was 720 mL at 25°C and a barometric pressure of 755 torr. What would the volume of the oxygen be at STP? (Hint: First calculate the partial pressure of the oxygen. Then use the combined gas law.)

19. Use Boyle's law to solve for the missing value in the following:
$P_1 = 350$ torr, $V_1 = 200$ mL, $P_2 = 700$ torr, $V_2 = ?$

20. Use Boyle's law to solve for the missing value in the following:
$P_1 = 0.75$ atm, $V_2 = 435$ mL, $P_2 = 0.48$ atm, $V_1 = ?$

21. Use Boyle's law to solve for the missing value in the following:
$V_1 = 2.4 \times 10^5$ mL, $P_2 = 180$ mm Hg, $V_2 = 1.8 \times 10^3$ mL, $P_1 = ?$

22. The pressure exerted on a 240 mL sample of hydrogen gas at constant temperature is increased from 0.428 atm to 0.724 atm. What will the final volume of the sample be?

23. A flask containing 155 cm^3 of hydrogen was collected under a pressure of 22.5 kPa. What pressure would have been required for the volume of the gas to have been 90 cm^3, assuming the same temperature?

24. A gas has a volume of $V_1 = 450$ mL. If the temperature is held constant, what volume would the gas occupy if the pressure $P_2 = 2P_1$?

25. What volume would the gas occupy if the pressure $P_2 = 0.25P_1$?

26. A sample of oxygen that occupies 1.00×10^6 mL at 575 mm Hg is subjected to a pressure of 1.25 atm. What will the final volume of the sample be if the temperature is held constant?

27. A helium-filled balloon has a volume of 2.75 L at 20°C. The volume of the balloon decreases to 2.46 L after it is placed outside on a cold day.

   a. What is the outside temperature in K?

   b. What is the outside temperature in °C?

28. A gas at 65°C occupies 4.22 L. At what Celsius temperature will the volume be 3.87 L, assuming the same pressure?

29. A certain quantity of gas has a volume of 0.75 L at 298 K. At what temperature, in degrees Celsius, would this quantity of gas be reduced to 0.50 L, assuming constant pressure?

30. A balloon filled with oxygen gas occupies a volume of 5.5 L at 25°C. What volume will the gas occupy at 100°C?

31. A sample of nitrogen gas is contained in a piston with a freely moving cylinder. At 0°C, the volume of the gas is 375 mL. To what temperature must the gas be heated to occupy a volume of 500 mL?

32. Use Charles's law to solve for the missing value in the following:
$V_1 = 80$ mL, $T_1 = 27°C$, $T_2 = 77°C$, $V_2 = ?$

33. Use Charles's law to solve for the missing value in the following:
$V_1 = 125$ L, $V_2 = 85$ L, $T_2 = 127°C$, $T_1 = ?$

34. Use Charles's law to solve for the missing value in the following:
$T_1 = -33°C$, $V_2 = 54$ mL, $T_2 = 160°C$, $V_1 = ?$

35. A sample of air has a volume of 140 mL at 67°C. At what temperature will its volume be 50 mL at constant pressure?

36. At standard temperature, a gas has a volume of 275 mL. The temperature is then increased to 130°C, and the pressure is held constant. What is the new volume?

37. An aerosol can contains gases under a pressure of 4.5 atm at 20°C. If the can is left on a hot sandy beach, the pressure of the gases increases to 4.8 atm. What is the Celsius temperature on the beach?

38. Before a trip from New York to Boston, the pressure in an automobile tire is 1.8 atm at 293 K. At the end of the trip, the pressure gauge reads 1.9 atm. What is the new Celsius temperature of the air inside the tire? (Assume tires with constant volume.)

39. At 120°C, the pressure of a sample of nitrogen is 1.07 atm. What will the pressure be at 205°C, assuming constant volume?

40. A sample of helium gas has a pressure of 1.2 atm at 22°C. At what Celsius temperature will the helium reach a pressure of 2 atm?

41. An empty aerosol-spray can at room temperature (20°C) is thrown into an incinerator where the temperature reaches 500°C. If the gas inside the empty container was initially at a pressure of 1.0 atm, what pressure did it reach inside the incinerator? Assume the gas was at constant volume and the can did not explode.

42. The temperature within an automobile tire at the beginning of a long trip is 25°C. At the conclusion of the trip, the tire has a pressure of 1.8 atm. What is the final Celsius temperature within the tire if its original pressure was 1.75 atm?

43. A sample of gas in a closed container at a temperature of 100°C and a pressure of 3.0 atm is heated to 300°C. What pressure does the gas exert at the higher temperature?

44. A sample of hydrogen at 47°C exerts a pressure of 0.329 atm. The gas is heated to 77°C at constant volume. What will its new pressure be?

45. To what temperature must a sample of nitrogen at 27°C and 0.625 atm be taken so that its pressure becomes 1.125 atm at constant volume?

46. The pressure on a gas at −73°C is doubled, but its volume is held constant. What will the final temperature be in degrees Celsius?

Section: Molecular Composition of Gases

47. Quantitatively compare the rates of effusion for the following pairs of gases at the same temperature and pressure.
a. Hydrogen and nitrogen
b. Fluorine and chlorine

48. Some hydrogen gas is collected over water at 20°C. The levels of water inside and outside the gas-collection bottle are the same. The partial pressure of hydrogen is 742.5 torr. What is the barometric pressure at the time the gas is collected?

49. What is the volume, in liters, of 0.100 g of $C_2H_2F_4$ vapor at 0.928 atm and 22.3°C?

50. What is the molar mass of a 1.25 g sample of gas that occupies a volume of 1.00 L at a pressure of 0.961 atm and a temperature of 27.0°C?

51. What pressure, in atmospheres, is exerted by 0.325 mol of hydrogen gas in a 4.08 L container at 35°C?

52. A gas sample occupies 8.77 L at 20.0°C. What is the pressure, in atmospheres, given that there are 1.45 mol of gas in the sample?

53. A 2.07 L cylinder contains 2.88 mol of helium gas at 22°C. What is the pressure in atmospheres of the gas in the cylinder?

54. A tank of hydrogen gas has a volume of 22.9 L and holds 14.0 mol of the gas at 12°C. What is the reading on the pressure gauge in atmospheres?

55. A sample that contains 4.38 mol of a gas at 250.0 K has a pressure of 0.857 atm. What is the volume?

867

Gases
1. 177 kPa
2. 1330 mm Hg
3. 0.75 atm
4. 76 kPa
5. 0.9813 atm
6. 745.8 torr
7. 99.43 kPa
8. 1.4999 atm
9. 0.600 atm
10. 1.20 atm
11. 1 L
12. 3.2 atm
13. 0.59 L
14. 30 mL
15. 122 mL
16. 14 L
17. 3.8 atm
18. 630 mL
19. 100 mL
20. 280 mL
21. 1.4 mm Hg
22. 140 mL
23. 40 kPa
24. 225 mL
25. 1800 mL
26. 6.05×10^5 mL
27. **a.** 260 K
    b. −11
28. 37°C
29. −74°C
30. 6.9 L
31. 91°C
32. 93 mL
33. 315°C
34. 30 mL
35. −150°C
36. 406 mL
37. 40°C
38. 36°C
39. 1.3 atm
40. 220°C
41. 2.6 atm
42. 37°C
43. 4.6 atm
44. 0.360 atm
45. 267°C
46. 127°C

Appendix D

47. a. 3.72
  b. 1.4
48. 760.0 torr
49. 0.0256 L
50. 32.0 g/mol
51. 2.01 atm
52. 3.98 atm
53. 33.7 atm
54. 14.3 atm
55. 105 L
56. 33.0 L
57. 240 mL
58. 247 mL
59. 81.9 g
60. 111 g
61. 0.90 atm
62. 74 g
63. 60.3 g
64. 83.8 g/mol
65. 0.572 g/L
66. 33 g/mol
67. 1.19 g/L
68. 72.7 g/mol
69. 29.0 g/mol
70. 544 L
71. a. 14.2 atm
  b. 4.46 atm
  c. 77.7 atm
72. a. 39.4 L
  b. 14.9 L
  c. 3.81 L
73. a. 0.0645 mol
  b. 0.0300 mol
  c. 0.0377 mol
74. a. 15 g
  b. 2.22 g
  c. 0.364 g
75. a. 11.7 g/mol
  b. 13.5 g/mol
  c. 11.5 g/mol

56. How many liters are occupied by 0.909 mol of nitrogen at 125°C and 0.901 atm pressure?

57. A reaction yields 0.00856 mol of O_2 gas. What volume in mL will the gas occupy if it is collected at 43.0°C and 0.926 atm pressure?

58. A researcher collects 0.00909 mol of an unknown gas by water displacement at a temperature of 16.0°C and 0.873 atm pressure (after the partial pressure of water vapor has been subtracted). What volume of gas in mL does the researcher have?

59. How many grams of carbon dioxide gas are there in a 45.1 L container at 34.0°C and 1.04 atm?

60. What is the mass, in grams, of oxygen gas in a 12.5 L container at 45.0°C and 7.22 atm?

61. A sample of carbon dioxide with a mass of 0.30 g was placed in a 250 mL container at 400.0 K. What is the pressure exerted by the gas?

62. What mass of ethene gas, C_2H_4, is contained in a 15.0 L tank that has a pressure of 4.40 atm at a temperature of 305 K?

63. NH_3 gas is pumped into the reservoir of a refrigeration unit at a pressure of 4.45 atm. The capacity of the reservoir is 19.4 L. The temperature is 24.0°C. What is the mass of the gas in g?

64. What is the molar mass of a gas if 0.427 g of the gas occupies a volume of 125 mL at 20.0°C and 0.980 atm?

65. What is the density of a sample of ammonia gas, NH_3, if the pressure is 0.928 atm and the temperature is 63.0°C?

66. The density of a gas was found to be 2.0 g/L at 1.50 atm and 27.0°C. What is the molar mass of the gas?

67. What is the density of argon gas, Ar, at a pressure of 551 torr and a temperature of 25.0°C?

68. A chemist determines the mass of a sample of gas to be 3.17 g. Its volume is 942 mL at a temperature of 14.0°C and a pressure of 1.09 atm. What is the molar mass of the gas?

69. The density of dry air at sea level (1 atm) is 1.225 g/L at 15.0°C. What is the average molar mass of the air?

70. How many liters of gaseous carbon monoxide at 27.0°C and 0.247 atm can be produced from the burning of 65.5 g of carbon according to the following equation?

$$2C(s) + O_2(g) \longrightarrow 2CO(g)$$

71. Calculate the pressure, in atmospheres, exerted by each of the following.
  a. 2.50 L of HF containing 1.35 mol at 320 K
  b. 4.75 L of NO_2 containing 0.860 mol at 300 K
  c. 750 mL of CO_2 containing 2.15 mol at 57.0°C

72. Calculate the volume, in liters, occupied by each of the following.
  a. 2.00 mol of H_2 at 300.0 K and 1.25 atm
  b. 0.425 mol of NH_3 at 37.0°C and 0.724 atm
  c. 4.00 g of O_2 at 57.0°C and 0.888 atm

73. Determine the number of moles of gas contained in each of the following.
  a. 1.25 L at 250 K and 1.06 atm
  b. 0.800 L at 27.0°C and 0.925 atm
  c. 750 mL at –50.0°C and 0.921 atm

74. Find the mass of each of the following.
  a. 5.60 L of O_2 at 1.75 atm and 250 K
  b. 3.50 L of NH_3 at 0.921 atm and 27.0°C
  c. 0.125 mL of SO_2 at 0.822 atm and –53.0°C

75. Find the molar mass of each gas measured at the specified conditions.
  a. 0.650 g occupying 1.12 L at 280 K and 1.14 atm
  b. 1.05 g occupying 2.35 L at 37.0°C and 0.840 atm
  c. 0.432 g occupying 750 mL at –23.0°C and 1.03 atm

76. If the density of an unknown gas is 3.20 g/L at –18.0°C and 2.17 atm, what is the molar mass of this gas?

77. One method of estimation the temperature of the center of the sun is based on the assumption that the center consists of gases that have an average molar mass of 2.00 g/mol. If the density of the center of the sun is 1.40 g/cm^3 at a pressure of 1.30×10^9 atm, calculate the temperature in degrees Celsius.

78. Three of the primary components of air are carbon dioxide, nitrogen, and oxygen. In a sample containing a mixture of only these gases at exactly one atmosphere pressure, the partial pressures of carbon dioxide and nitrogen are given as $P_{CO_2} = 0.285$ torr and $P_{N_2} = 593.525$ torr. What is the partial pressure of oxygen?

79. Determine the partial pressure of oxygen collected by water displacement if the water temperature is 20.0°C and the total pressure of the gases in the collection bottle is 730 torr. P_{H_2O} at 20.0°C is equal to 17.5 torr.

80. A sample of hydrogen effuses through a porous container about 9.00 times faster than an unknown gas. Estimate the molar mass of the unknown gas.

81. Compare the rate of effusion of carbon dioxide with that if hydrogen chloride at the same temperature and pressure.

82. If a molecule of neon gas travels at an average of 400 m/s at a given temperature, estimate the average speed of a molecule of butane gas, C_4H_{10}, at the same temperature.

83. Nitrogen effused through a pinhole 1.7 times as fast as another gaseous element at the same conditions. Estimate the other element's molar mass.

84. Determine the molecular mass ratio of two gases whose rates of diffusion have a ratio of 16.0:1.

85. Estimate the molar mass of a gas that effuses at 1.60 times the effusion rate of carbon dioxide.

86. List the following gases in order of increasing average molecular velocity at 25°C: H_2O, He, HCl, BrF, and NO_2.

87. What is the ratio of the average velocity of hydrogen molecules to that of neon atoms at the same temperature and pressure?

88. At a certain temperature and pressure, chlorine molecules have an average velocity of 0.0380 m/s. What is the average velocity of sulfur dioxide molecules under the same conditions?

89. A sample of helium effuses through a porous container 6.50 times faster than does unknown gas X. What is the molar mass of the unknown gas?

90. How many liters of H_2 gas at STP can be produced by the reaction of 4.60 g of Na and excess water, according to the following equation?
$$2Na(s) + 2H_2O(l) \longrightarrow H_2(g) + 2NaOH(aq)$$

91. How many grams of Na are needed to react with H_2O to liberate 400 mL H_2 gas at STP?

92. What volume of oxygen gas in liters can be collected at 0.987 atm pressure and 25.0°C when 30.6 g of $KClO_3$ decompose by heating, according to the following equation?
$$2KClO_3(s) \xrightarrow[\text{MnO}_2]{\Delta} 2KCl(s) + 3O_2(g)$$

93. What mass of sulfur must be used to produce 12.6 L of gaseous sulfur dioxide at STP according to the following equation?
$$S_8(s) + 8O_2(g) \longrightarrow 8SO_2(g)$$

94. How many grams of water can be produced from the complete reaction of 3.44 L of oxygen gas, at STP, with hydrogen gas?

95. Aluminum granules are a component of some drain cleaners because they react with sodium hydroxide to release both energy and gas bubbles, which help clear the drain clog. The reaction is
$$2NaOH(aq) + 2Al(s) + 6H_2O(l) \longrightarrow$$
$$2NaAl(OH)_4(aq) + 3H_2(g)$$

What mass of aluminum would be needed to produce 4.00 L of hydrogen gas at STP?

96. What volume of chlorine gas at 38.0°C and 1.63 atm is needed to react completely with 10.4 g of sodium to form NaCl?

97. Air bags in cars are inflated by the sudden decomposition of sodium azide, NaN_3, by the following reaction.
$$2NaN_3(s) \longrightarrow 3N_2(g) + 2Na(s)$$
What volume of N_2 gas, measured at 1.30 atm and 87.0°C, would be produced by the reaction of 70.0 g of NaN_3?

98. Assume that 5.60 L of H_2 at STP react with CuO according to the following equation:
$$CuO(s) + H_2(g) \longrightarrow Cu(s) + H_2O(g)$$
How many moles of H_2 react?

76. 30.9 g/mol
77. 2.26×10^7 °C
78. 166.19 torr
79. 713 torr
80. 162 g/mol
81. 0.91
82. 235 m/s
83. 81 g/mol
84. 256
85. 17.2 g/mol
86. Rates of effusion: BrF < NO_2 < HCl < H_2O < He
87. 3.16
88. 0.0400 m/s
89. 169 g/mol
90. 2.24 L
91. 0.8 g
92. 9.29 L
93. 18.0 g
94. 5.53 g
95. 3.21 g
96. 3.54 L
97. 36.7 L
98. 0.250 mol
99. 190 L
100. 358 L
101. 1.18×10^8 L
102. 2.35×10^8 L
103. 1240 mL
104. **a.** 0.50 mol
　　b. 0.75 mol
　　c. 17 L

Appendix D

Solutions

1. 0.1249 M
2. 1.1 M
3. 0.413 mol
4. 0.571 M
5. 0.0146 M
6. **a.** 40.00 g/mol
   b. 0.167 M
7. 0.953 M
8. 343 g
9. 1140 g
10. 0.143 mol
11. **a.** 132.2 g
    b. 4.003 M
12. **a.** 169.88 g/mol
    b. 1.0×10^2 mL
13. **a.** 142.0 g
    b. 0.113 mol
14. **a.** 192 g
    b. 0.000391 M
15. **a.** 75 g
    b. 160 g
16. **a.** 40.00 g
    b. 0.435 M
17. d, a, b, c

99. A modified Haber process for making ammonia is conducted at 550°C and 250 atm. If 10.0 kg of nitrogen (the limiting reactant) is used and the process goes to completion, what volume of ammonia is produced?

100. When liquid nitroglycerin, $C_3H_5(NO_3)_3$, explodes, the products are carbon dioxide, nitrogen, oxygen, and water vapor. If 500.0 g of nitroglycerin explode at STP, what is the total volume, at STP, for all the gases produced?

101. The principal source of sulfur on Earth is deposits of free sulfur occurring mainly in volcanically active regions. The sulfur was initially formed by the reaction between the two volcanic vapors SO_2 and H_2S to form $H_2O(l)$ and $S_8(s)$. What volume of SO_2, at 0.961 atm and 22.0°C, was needed to form a sulfur deposit of 4.50×10^5 kg on the slopes of a volcano in Hawaii?

102. What volume of H_2S, at 0.961 atm and 22.0°C, was needed to form a sulfur deposit of 4.50×10^5 kg on the slopes of a volcano in Hawaii?

103. A 3.25 g sample of solid calcium carbide, CaC_2, reacted with water to produce acetylene gas, C_2H_2, and aqueous calcium hydroxide. If the acetylene was collected over water at 17.0°C and 0.974 atm, how many milliliters of acetylene were produced?

104. Assume that 13.5 g of Al react with HCl according to the following equation, at STP:
    $Al(s) + HCl(aq) \longrightarrow AlCl_3(aq) + H_2(g)$
    Remember to balance the equation first.
    a. How many moles of Al react?
    b. How many moles of H_2 are produced?
    c. How many liters of H_2 at STP are produced?

Solutions

Section: Concentration and Molarity

1. What is the molarity of a 2.000 L solution that is made from 14.60 g of NaCl?

2. What is the molarity of a HCl solution that contains 10.0 g of HCl in 250 mL of solution?

3. How many moles of NaCl are in 1.25 L of 0.330 M NaCl?

4. What is the molarity of a solution composed of 6.250 g of HCl in 0.3000 L of solution?

5. 5.00 grams of sugar, $C_{12}H_{22}O_{11}$, are dissolved in water to make 1.00 L of solution. What is the concentration of this solution expressed as molarity?

6. Supppose you wanted to dissolve 40.0 g NaOH in enough H_2O to make 6.00 L of solution. You want to calculate the molarity, M, of the resulting solution.
   a. What is the molar mass of NaOH?
   b. What is the molarity of this solution?

7. What is the molarity of a solution of 14.0 g of NH_4Br in enough H_2O to make 150 mL of solution?

8. Suppose you wanted to produce 1.00 L of a 3.50 M solution of H_2SO_4. How many grams of solute are needed to make this solution?

9. How many grams of solute are needed to make 2.50 L of a 1.75 M solution of $Ba(NO_3)_2$?

10. How many moles of NaOH are contained in 65.0 mL of a 2.20 M solution of NaOH in H_2O?

11. A solution is made by dissolving 26.42 g of $(NH_4)_2SO_4$ in enough H_2O to make 50.00 mL of solution.
    a. What is the molar mass of $(NH_4)_2SO_4$?
    b. What is the molarity of this solution?

12. Suppose you wanted to find out how many milliliters of 1.0 M $AgNO_3$ are needed to provide 168.88 of pure $AgNO_3$.
    a. What is the molar mass of $AgNO_3$?
    b. How many mL of solution are needed?

13. Na_2SO_4 is dissolved in water to make 450 mL of a 0.250 M solution.
    a. What is the molar mass of Na_2SO_4?
    b. How many moles of Na_2SO_4 are needed?

14. Citric acid is one component of some soft drinks. Suppose that a 2.00 L solution is made from 150 mg of citric acid, $C_6H_8O_7$.
    a. What is the molar mass of $C_6H_8O_7$?
    b. What is the molarity of citric acid in the solution?

870

15. Suppose you wanted to know how many grams of KCl would be left if 350 mL of a 6.0 M KCl solution were evaporated to dryness.
    a. What is the molar mass of KCl?
    b. How many grams of KCl would remain?

16. Sodium metal reacts violently with water to form NaOH and release hydrogen gas. Suppose that 10.0 g of Na reacts completely with 1.00 L of water, and the final volume of the system is 1.00 L.

    $$2Na(s) + 2H_2O(l) \longrightarrow 2NaOH(aq) + H_2(g)$$

    a. What is the molar mass of NaOH?
    b. What is the molarity, M, of the NaOH solution formed by the reaction?

Section: Physical Properties of Solutions

17. Given 0.01 m aqueous solutions of each of the following, arrange the solutions in order of increasing change in the freezing point of the solution.
    a. NaI
    b. $CaCl_2$
    c. K_3PO_4
    d. $C_6H_{12}O_6$ (glucose)

Chemical Equilibrium

Section: Systems at Equilibrium

1. At equilibrium a mixture of N_2, H_2, NH_3 gas at 500°C is determined to consist of 0.602 mol/L of N_2, 0.420 mol/L of H_2, and 0.113 mol/L of NH_3. What is the equilibrium constant for the reaction $N_2(g) + 3H_2(g) \rightleftarrows 2NH_3(g)$ at this temperature?

2. The reaction $AB_2C(g) \rightleftarrows B_2(g) + AC(g)$ reached equilibrium at 900 K in a 5.00 L vessel. At equilibrium 0.0840 mol of AB_2C, 0.0350 mol of B_2, and 0.0590 mol of AC were detected. What is the equilibrium constant at this temperature for this system?

3. At equilibrium at 1.0 L vessel contains 20.00 mol of H_2, 18.00 mol of CO_2, 12.00 mol of H_2O, and 5.900 mol of CO at 427°C. What is the value of K_{eq} at this temperature for the following reaction?

    $$CO_2(g) + H_2(g) \rightleftarrows CO(g) + H_2O(g)$$

4. A reaction between gaseous sulfur dioxide and oxygen gas to produce gaseous sulfur trioxide takes place at 600°C. At that temperature, the concentration of SO_2 is found to be 1.50 mol/L, the concentration of O_2 is 1.25 mol/L, and the concentration of SO_3 is 3.50 mol/L. Using the balanced chemical equation, calculate the equilibrium constant for this system.

5. At equilibrium at 2500 K, [HCl] = 0.0625 and $[H_2] = [Cl_2] = 0.00450$ for the reaction

    $$H_2(g) + Cl_2(g) \rightleftarrows 2HCl(g).$$

    Find the value of K_{eq}.

6. An equilibrium mixture at 435°C is found to consist of 0.00183 mol/L of H_2, 0.00313 mol/L of I_2, and 0.0177 mol/L of HI. Calculate the equilibrium constant, K_{eq}, for the reaction $H_2(g) + I_2(g) \rightleftarrows 2HI(g)$.

7. For the reaction

    $$H_2(g) + I_2(g) \rightleftarrows 2HI(g)$$

    at 425°C, calculate [HI], given $[H_2] = [I_2] = 0.000479$ and $K_{eq} = 54.3$.

8. At 25°C, an equilibrium mixture of gases contains 0.00640 mol/L PCl_3, 0.0250 mol/L Cl_2, and 0.00400 mol/L PCl_5. What is the equilibrium constant for the following reaction?

    $$PCl_5(g) \rightleftarrows PCl_3(g) + Cl_2(g)$$

9. At equilibrium a 2 L vessel contains 0.360 mol of H_2, 0.110 mol of Br, and 37.0 mol of HBr. What is the equilibrium constant for the reaction at this temperature?

    $$H_2(g) + Br_2(g) \rightleftarrows 2HBr(g)$$

10. Calculate the solubility-product constant, K_{sp}, of lead(II) chloride, $PbCl_2$, which has a solubility of 1.00 g/100.0 g H_2O at a temperature other than 25°C.

11. 5.00 g of Ag_2SO_4 will dissolve in 1.00 L of water. Calculate the solubility product constant for this salt.

Chemical Equilibrium

1. 0.286
2. 0.00492
3. 0.1967
4. 4.36
5. 193
6. 54.7
7. 0.00353 mol/L
8. 0.0400
9. 34600
10. 0.000186
11. 1.65×10^{-5}
12. 1.2×10^{-25}
13. 2.8×10^{-13}
14. 8.9×10^{-14} mol/L
15. 5.7×10^{-4} mol/L
16. 5.0×10^{-7} mol/L
17. 1.3×10^{-12} mol/L
18. 2.08×10^{-5}
19. 1.3×10^{-26} mol/L
20. 0.056×10^{-9} mol/L
21. 6.1×10^{-8} mol/L

871

Appendix D

Acids and Bases

1. acidic
2. basic
3. acidic
4. basic
5. 0.35 M
6. **a.** 1×10^{-8} M
   b. 10×10^{-7} M
   c. 5×10^{-7} M
7. 3
8. 5
9. 10
10. 12
11. **a.** 1×10^{-4} M
    b. 5×10^{-5} M
12. 1×10^{-4} M
13. 1×10^{-10} M
14. 1×10^{-3} M
15. 1×10^{-11} M
16. 3×10^{-2} M
17. 3.33×10^{-13} M
18. **a.** 2×10^{-4} M
    b. 5×10^{-11} M
19. **a.** 0.01 M
    b. 1×10^{-12} M
20. **a.** 1×10^{-3} M
    b. 5×10^{-4} M
21. 0.0067 M

12. What is the value of K_{sp} for tin(II) sulfide, given that its solubility is 5.2×10^{-12} g/100.0 g water?

13. Calculate the solubility product constant for calcium carbonate, given that it has a solubility of 5.3×10^{-5} g/L of water.

14. Calculate the solubility of cadmium sulfide, CdS, in mol/L, given the K_{sp} value as 8.0×10^{-27}.

15. Determine the concentration of strontium ions in saturated solution of strontium sulfate, $SrSO_4$, if the K_{sp} for K_{sp} is 3.2×10^{-7}.

16. What is the solubility in mol/L of manganese(II) sulfide, MnS, given that its K_{sp} value is 2.5×10^{-13}?

17. Calculate the concentration of Zn^{2+} in saturated solution of zinc sulfide, ZnS, given that K_{sp} of zinc sulfide equals 1.6×10^{-24}.

18. What is the value of K_{sp} for Ag_2SO_4 if 5.40 g is soluble in 1.00 L of water?

19. Calculate the concentration of Hg^{2+} ions in a saturated solution of $HgS(s)$. K_{sp} is 1.6×10^{-52}.

20. At 25°C, the value of K_{eq} is 1.7×10^{-13} for the following reaction.

$$N_2O(g) + \frac{1}{2}O_2(g) \rightleftarrows 2NO(g)$$

It is determined that $[N_2O] = 0.0035$ mol/L and $[O_2] = 0.0027$ mol/L. Using this information, what is the concentration of $NO(g)$ at equilibrium?

21. Tooth enamel is composed of the mineral hydroxyapatite, $Ca_5(PO_4)_3OH$, which has a K_{sp} of 6.8×10^{-37}. The molar solubility of hydroxyapatite is 2.7×10^{-5} mol/L. When hydroxyapatite is reacted with fluoride, the OH^- is replaced with the F^- ion on the mineral, forming fluorapatite, $Ca_5(PO_4)_3F$. (The latter is harder and less susceptible to caries.) The K_{sp} of fluorapatite is 1×10^{-60}. Calculate the solubility of fluorapatite in water. Given your calculations, can you support the fluoridation of drinking water? Your answer must be within ± 0.5%.

Acids and Bases

Section: Acidity, Basicity and pH

1. Identify the following as being true of acidic or basic solutions at 25°C:
   $[H_3O^+] = 1.0 \times 10^{-3}$ M

2. Identify the following as being true of acidic or basic solutions at 25°C:
   $[OH^-] = 1.0 \times 10^{-4}$ M

3. Identify the following as being true of acidic or basic solutions at 25°C:
   pH = 5

4. Identify the following as being true of acidic or basic solutions at 25°C:
   pH = 8

5. The pH of a hydrochloric acid solution for cleaning tile is 0.45. What is the $[H_3O^+]$ in the solution?

6. A $Ca(OH)_2$ solution has a pH of 8.
   a. Determine $[H_3O^+]$ for the solution.
   b. Determine $[OH^-]$.
   c. Determine $[Ca(OH)_2]$.

7. Determine the pH of the following solution:
   1×10^{-3} M HCl.

8. Determine the pH of the following solution:
   1×10^{-5} M HNO_3.

9. Determine the pH of the following solution:
   1×10^{-4} M NaOH.

10. Determine the pH of the following solution:
    1×10^{-2} M KOH.

11. The pH of a solution is 10.
    a. What is the concentration of hydroxide ions in the solution?
    b. If the solution is $Sr(OH)_2(aq)$, what is its molarity?

12. Determine the hydronium ion concentration in a solution that is 1×10^{-4} M HCl.

13. Determine the hydroxide ion concentration in a solution that is 1×10^{-4} M HCl.

14. Determine the hydronium ion concentration in a solution that is 1×10^{-3} M HNO_3.

15. Determine the hydroxide ion concentration in a solution that is 1×10^{-3} M HNO_3.

16. Determine the hydroxide ion concentration in a solution that is 3×10^{-2} M NaOH.

17. Determine the hydronium ion concentration in a solution that is 3.00×10^{-2} M NaOH.

18. a. Determine the hydroxide ion concentration in a solution that is 1×10^{-4} M $Ca(OH)_2$.
    b. Determine the hydronium ion concentration in a solution that is 1×10^{-4} M $Ca(OH)_2$.

19. a. Determine the $[H_3O^+]$ in a 0.01 M solution of $HClO_4$.
    b. Determine the $[OH^-]$ in a 0.01 M solution of $HClO_4$.

20. An aqueous solution of $Ba(OH)_2$ has a $[H_3O^+]$ of 1×10^{-11} M.
    a. What is the $[OH^-]$?
    b. What is the molarity of $Ba(OH)_2$ in the solution?

Section: Neutralization and Titration

21. If 20 mL of 0.01 M aqueous HCl is required to neutralize 30 mL of an aqueous solution of NaOH, determine the molarity of the NaOH solution.

Reaction Rates

Section: How Can Reaction Rates Be Explained?

1. A reaction involving reactants A and B is found to occur in the one-step mechanism: $2A + B \longrightarrow A_2B$. Write the rate law for this reaction, and predict the effect of doubling the concentration of either reactant on the overall reaction rate.

2. A chemical reaction is expressed by the balanced chemical equation $A + 2B \longrightarrow C$. Using the data below, determine the rate law for the reaction.
   Experiment # 1. initial [A] = 0.2 M
   initial [B] = 0.2 M
   initial rate of formation of C = 0.0002 M/min

   Experiment # 2. initial [A] = 0.2 M
   initial [B] = 0.4 M
   initial rate of formation of C = 0.0008 M/min

   Experiment # 3. initial [A] = 0.4 M
   initial [B] = 0.4 M/min
   initial rate of formation of C = 0.0016 M

3. A particular reaction is found to have the following rate law.
   $R = k[A][B]^2$
   How is the rate affected if

   a. the initial concentration of A is cut in half?

   b. the initial concentration of B is tripled?

   c. the initial concentration of A is doubled, but the concentration of B is cut in half?

Oxidation, Reduction, and Electrochemistry

Section: Oxidation-Reduction Reactions

1. Name the following acid: HNO_2

2. Assign oxidation numbers to each atom in H_2SO_3.

3. Assign oxidation numbers to each atom in H_2CO_3.

4. Assign oxidation numbers to each atom in HI.

5. Assign oxidation numbers to each atom in CO_2.

6. Assign oxidation numbers to each atom in NH_4^+.

7. Assign oxidation numbers to each atom in MnO_4^-.

8. Assign oxidation numbers to each atom in $S_2O_3^{2-}$.

9. Assign oxidation numbers to each atom in H_2O_2.

10. Assign oxidation numbers to each atom in P_4O_{10}.

11. Assign oxidation numbers to each atom in OF_2.

12. Assign oxidation numbers to each atom in SO_3.

13. Determine the oxidation state of the metal in CdS.

14. Determine the oxidation state of the metal in ZnS.

Appendix D

Reaction Rates

1. a. $R = k[A]^2[B]$
   b. [A] doubled $\Rightarrow$ the rate will increase fourfold;
      [B] doubled $\Rightarrow$ the rate will double

2. $R = k[A][B]^2$

3. a. rate is reduced by $\frac{1}{2}$
   b. rate is increased by a factor of 9
   c. rate is reduced by $\frac{1}{2}$

Appendix D

Oxidation, Reduction, and Electrochemistry

1. $H = +1$, $N = +3$, $O = -2$
2. $H = +1$, $S = +4$, $O = -2$
3. $H = +1$, $C = +4$, $O = -2$
4. $H = +1$, $I = -1$
5. $C = +4$, $O = -2$
6. $N = -3$, $H = +1$
7. $Mn = +7$, $O = -2$
8. $S = +2$, $O = -2$
9. $H = +1$, $O = -1$
10. $P = +5$, $O = -2$
11. $O = +2$, $F = -1$
12. $S = +6$, $O = -2$
13. $Cd = +2$
14. $Zn = +2$
15. $Pb = +2$, $Cr = +6$
16. $Fe = +3$
17. $Mn = +7$
18. $Co = +2$
19. $Cu = +2$
20. $N = +3$
21. $N = +5$
22. **a.** yes
    b. yes
    c. no
    d. yes
    e. no
23. **a.** nonredox
    b. redox
    c. nonredox

---

15. Determine the oxidation state of the metals in $PbCrO_4$.

16. Determine the oxidation state of the metal in $Fe(SCN)^{2+}$.

17. Determine the oxidation state of the metal in MnO_4^-.

18. Determine the oxidation state of the metals in $CoCl_2$.

19. Determine the oxidation state of the metal in $[Cu(NH_3)_4](OH)_2$.

20. Determine the oxidation state of the nitrogen in N_2O_3.

21. Determine the oxidation state of the nitrogen in N_2O_5.

22. Which of the following equations represent redox reactions?

    a. $2KNO_3(s) \longrightarrow 2KNO_2(s) + O_2(g)$

    b. $H_2(g) + CuO(s) \longrightarrow Cu(s) + H_2O(l)$

    c. $NaOH(aq) + HCl(aq) \longrightarrow NaCl(aq) + H_2O(l)$

    d. $H_2(g) + Cl_2(g) \longrightarrow 2HCl(g)$

    e. $SO_3(g) + H_2O(l) \longrightarrow H_2SO_4(aq)$

23. Identify if the following reactions are redox or nonredox:

    a. $2NH_4Cl(aq) + Ca(OH)_2(aq) \longrightarrow$
    $2NH_3(aq) + 2H_2O(l) + CaCl_2(aq)$

    b. $2HNO_3(aq) + 3H_2S(g) \longrightarrow 2NO(g) + 4H_2O(l) + 3S(s)$

    c. $[Be(H_2O)_4]^{2+}(aq) + H_2O(l) \longrightarrow$
    $H_3O^+(aq) + [Be(H_2O)_3OH]^+(aq)$

Nuclear Chemistry

Section: Atomic Nuclei and Nuclear Stability

1. The mass of a $^{20}_{10}Ne$ atom is 19.992 44 amu. Calculate the mass defect.

2. The mass of a $^{7}_{3}Li$ atom is 7.016 00 amu. Calculate its mass defect.

3. Calculate the nuclear binding energy of one lithium-6 atom. The measured atomic mass of lithium-6 is 6.015 amu.

4. Calculate the nuclear binding energy of the nucleus $^{35}_{19}K$. The measured atomic mass of $^{35}_{19}K$ is 34.988011 amu.

5. Calculate the nuclear binding energy of the nucleus $^{23}_{11}Na$. The measured atomic mass of $^{23}_{11}Na$ is 22.989 767 amu.

6. The nuclear binding energy of $^{35}_{19}K$ is 4.47×10^{-11} J. Calculate the binding energy per nucleon for $^{35}_{19}K$.

7. The nuclear binding energy of $^{23}_{11}Na$ is 2.99×10^{-11} J. Calculate the binding energy per nucleon for $^{23}_{11}Na$.

8. Calculate the binding energy per nucleon of $^{238}_{92}U$ in joules. The atomic mass of $^{238}_{92}U$ is 238.050 784 amu.

9. The energy released by the formation of a nucleus of $^{56}_{26}Fe$ is 7.89×10^{-11} J. Use Einstein's equation, $E = mc^2$, to determine how much mass is lost (in kilograms) in this process.

10. Calculate the nuclear binding energy of one mole of deuterium atoms. The measured mass of deuterium is 2.0140 amu.

Section: Nuclear Charge

11. Balance the nuclear equation:
    $^{43}_{19}K \longrightarrow ^{43}_{20}Ca + \underline{\quad ? \quad}$

12. Balance the nuclear equation:
    $^{233}_{92}U \longrightarrow ^{229}_{90}Th + \underline{\quad ? \quad}$

13. Balance the nuclear equation:
    $^{11}_{6}C + \underline{\quad ? \quad} \longrightarrow ^{11}_{5}B$

14. Balance the nuclear equation:
    $^{13}_{7}N \longrightarrow ^{0}_{-1}e + \underline{\quad ? \quad}$

15. Write the nuclear equation for the release of an alpha particle by $^{210}_{84}Po$.

16. Write the nuclear equation for the release of an alpha particle by $^{210}_{82}Pb$.

17. Balance the nuclear equation:
    $^{239}_{93}Np \longrightarrow ^{0}_{-1}e + \underline{\quad ? \quad}$

18. Balance the nuclear equation:
    $^{9}_{4}Be + ^{4}_{2}He \longrightarrow \underline{\quad ? \quad}$

19. Balance the nuclear equation:
$$^{32}_{15}P + \underline{\quad?\quad} \longrightarrow ^{33}_{15}P$$

20. Balance the nuclear equation:
$$^{236}_{92}U \longrightarrow ^{94}_{36}Kr + \underline{\quad?\quad} + 3^{1}_{0}n$$

Section: Uses of Nuclear Chemistry

21. The Environmental Protection Agency and health officials nationwide are concerned about the levels of radon gas in homes. The half-life of the radon-222 isotope is 3.8 days. If a sample of gas taken from a basement contains 4.38 μg of radon-222, how much radon will remain in the sample after 15.2 days?

22. Uranium-238 decays through alpha decay with a half-life of 4.46×10^9 years. How long would it take for 7/8 of a sample of uranium-238 to decay?

23. The half-life of carbon-14 is 5715 years. How long will it be until only half of the carbon-14 in a sample remains?

24. The half-life of iodine-131 is 8.040 days. What percentage of an iodine-131 sample will remain after 40.2 days?

25. The half-life of plutonium-239 is 24 110 years. Of an original mass of 100 g, how much remains after 96 440 years?

26. The half-life of thorium-227 is 18.72 days. How many days are required for three-fourths of a given amount to decay?

27. The half-life of protactinium-234 is 6.69 hours. What fraction of a given amount remains after 26.76 hours?

28. How many milligrams remain of a 15 mg sample of radium-226 after 6396 years? The half life of radium-226 is 1599 years.

29. After 4797 years, how much of an original 0.25 g of radium-226 remains? Its half-life is 1599 years.

30. The half-life of radium-224 is 3.66 days. What was the original mass of radium-224 if 0.05 g remains after 7.32 days?

Appendix D

Nuclear Chemistry

1. 0.172 46 amu
2. 0.042 13 amu
3. 5.2×10^{-12} J
4. 4.47×10^{-11} J
5. 2.99×10^{-11} J
6. 1.28×10^{-12} J
7. 1.30×10^{-12} J
8. 1.21×10^{-12} J/nucleon
9. 8.77×10^{-28} kg
10. 2.24×10^{11} J/mol
11. $? = ^{0}_{-1}e$
12. $? = ^{4}_{2}He$
13. $? = ^{0}_{-1}e$
14. $? = ^{13}_{6}C$
15. $^{210}_{84}Po \rightarrow ^{4}_{2}He + ^{206}_{82}Pb$
16. $^{210}_{82}Pb \rightarrow ^{0}_{-1}\beta + ^{210}_{83}Bi$
17. $? = ^{239}_{94}Pu$
18. $? = ^{13}_{6}C$
19. $? = ^{1}_{0}n$
20. $? = ^{139}_{56}Ba$
21. 0.274 μg
22. 1.34×10^{10} years
23. 5715 years
24. 3.13%
25. 6.25 g
26. 37.44 days
27. 1/16
28. 0.94 mg
29. 0.031 g
30. 0.2 g

Appendix E

SELECTED ANSWERS

The Science of Chemistry

Practice Problems A

1. **a.** 0.000 765 kg
   b. 1340 mg
   c. 0.0342 g
   d. 23 745 000 000 mg
3. shortest: 0.0128 km; longest: 17 931 mm

Section 2 Review

9. **a.** 17 300 ms
   b. 0.000 002 56 km
   c. 5.67 g
   d. 0.005 13 km
11. about 1081 beans

Chapter Review

31. **a.** 0.357 L
    b. 2.5×10^7 mg
    c. 35 L
    d. 2460 cm^3
    e. 2.5×10^{-4} g
    f. 2.5×10^{-7} kg
    g. 1500 ms
    h. 10 500 mmol
37. 151 g

Matter and Energy

Section 1 Review

7. **a.** 373 K
   b. 1058 K
   c. 273 K
   d. 236 K

Practice Problems A

1. **a.** 1273 mL
   b. 98.5 cm^2
   c. 8.2 g
3. 4593 kJ/min

Practice Problems B

1. 0.069 J/g•K
3. 329 K

Section 3 Review

7. 0.30 J/g•K
9. 5.2×10^3 s

Chapter Review

25. 6.411 g
27. 2.79 m^2
29. 8.82×10^{-4} g
31. **a.** 9.225×10^{-2} km
    b. 9.225×10^3 cm
33. 3.6×10^3 J
35. 13°C
37. **a.** 6.730×10^{-4}
    b. $5.000 00 \times 10^4$
    c. 3.010×10^{-6}
39. 8.57×10^8 m^2
41. 4

Atoms and Moles

Practice Problems A

1. 11 protons and 11 electrons
3. 80

Practice Problems B

1. Both isotopes have 17 protons and 17 electrons. Cl-35 has 18 neutrons and Cl-37 has 20 neutrons.

Section 2 Review

5. **a.** 35 electrons, 35 protons, 45 neutrons
   b. 46 electrons, 46 protons, 60 neutrons
   c. 55 electrons, 55 protons, 78 neutrons

Practice Problems C

1. $1s^2 2s^2 2p^4$ or $[He]2s^2 2p^4$

Section 3 Review

5. $1s^2 2s^2 2p^6 3s^2 3p^1$ or $[Ne]3s^2 3p^1$
7. 5

Practice Problems D

1. 238 g
3. 0.84 mol; 0.86 g

Practice Problems E

1. 4.2×10^{23} atoms
3. 0.58 mol

Section 4 Review

7. 2.4×10^{24} atoms
9. 1.3 mol

Chapter Review

39. 54
41. 9
43. **a.** $^{234}_{92}U$
    b. $^{235}_{92}U$
    c. $^{238}_{92}U$
45. $^{12}_6C$ and $^{13}_6C$
47. $1s^2 2s^2 2p^6 3s^2 3p^6 3d^8 4s^2$
49. 6
51. **a.** 0.500 mol
    b. 4.7 mol
    c. 0.100 mol
53. 0.005 g
55. **a.** 1.2×10^{24} atoms
    b. 6.02×10^{23} atoms
    c. 2.7×10^{24} atoms
59. 99.8g
61. $1s^2 2s^2 2p^6 3s^2 3p^6 3d^{10} 4s^2$
63. **a.** 42 g
    b. 1.7×10^{-13} g
    c. 0.449 g
69. 0.307 kg
71. 0.39 mol; 2.3×10^{23} atoms
73. 5
79. $1s^2 2s^2 2p^6 3s^2 3p^3$

Ionic Compounds

Practice Problems A

1. **a.** Ca(CN)$_2$
   b. Rb$_2$S$_2$O$_3$
   c. Ca(CH$_3$COO)$_2$
   d. (NH$_4$)$_2$SO$_4$

Chapter Review

27. **a.** Li$_2$SO$_4$
    b. Sr(NO$_3$)$_2$
    c. NH$_4$CH$_3$COO
    d. Ti$_2$(SO$_4$)$_3$

876

29. barium; Cl^-; chromium(III); fluoride; Mn^{2+}; O^{2-}; **a.** $MnCl_2$; **b.** CrF_3; **c.** BaO

31. a. peroxide
    b. chromate
    c. ammonium
    d. carbonate

33. a. CN^-
    b. SO_4^{2-}
    c. NO_3^-
    d. MnO_4^-

35. a. 3
    b. 1
    c. 4
    d. 7

Covalent Compounds

Practice Problems A

1.

Practice Problems B

1.

Practice Problems C

1. $:\ddot{O}=C=\ddot{O}:$

   $:C\equiv O:$

Practice Problems D

1. trigonal pyramidal
3. trigonal planar

Section 3 Review

7. a. bent
    b. trigonal pyramidal
    c. trigonal pyramidal
    d. tetrahedral

Chapter Review

33. a.

35. a. tetrahedral
    b. bent

36. a.

b.

41. a. silicon tetrachloride, tetrahedral

    b. boron trichloride, trigonal planar

    c. nitrogen tribromide, trigonal pyramidal

The Mole and Chemical Composition

Practice Problems A

1. 1.13×10^{23} ions Na$^+$
3. 2.544×10^{24} molecules $C_2H_4O_2$

Practice Problems B

1. 0.940 mol Xe
3. 4.5×10^{-7} mol termites
5. a. 1.050×10^{-2} mol O
    b. 5.249×10^{-3} mol C
    c. 3.690 mol O
    d. 8.841×10^{-8} mol K$^+$
    e. 3.321×10^{-10} mol Cl$^-$
    f. 6.64×10^{-10} mol N
    g. 6.63×10^2 mol Cl$^-$

Practice Problems C

1. 223 g Cu
3. 1063 g CH_4

Practice Problems D

1. 2.25×10^{24} atoms Cu
3. 9.33×10^{25} atoms As

Section 1 Review

7. a. 3.61×10^{24} Na$^+$ ions
    b. 7.23×10^{24} Na$^+$ ions
    c. 3.08×10^{24} Na$^+$ ions
9. a. 2.86×10^{-7} g He
    b. 15.22 g CH_4
    c. 200.5 g Ca^{2+}
11. 206.3 g ibuprofen
13. a. 26.7 g Ca
    b. 50. g boron-11
    c. 7.032×10^{-4} g Na$^+$

Practice Problems E

1. 69.73 amu

Practice Problems F

1. a. 259.80 g/mol
    b. 136.06 g/mol
    c. 342.34 g/mol
    d. 253.80 g/mol
    e. 60.06 g/mol
    f. 262.84 g/mol
3. a. 92.15 g/mol
    b. 0.0815 mol $C_6H_5CH_3$

Section 2 Review

9. 10.80 amu; boron
11. a. SrS, 119.69 g/mol, 1.76×10^{-2} mol SrS

877

Appendix E

b. PF_3, 87.97 g/mol, 2.40×10^{-2} mol PF_3

c. $Zn(C_2H_3O_2)_2$, 183.49 g/mol, 1.15×10^{-2} mol $Zn(C_2H_3O_2)_2$

d. $Hg(BrO_3)_2$, 456.39 g/mol, 4.62×10^{-3} mol $Hg(BrO_3)_2$

e. $Ca(NO_3)_2$, 164.10 g/mol, 1.29×10^{-2} mol $Ca(NO_3)_2$

Practice Problems G

1. Mn_2O_3

3. Fe_3O_4

Practice Problems H

1. C_6H_6

3. NO_2

Practice Problems I

1. 93.311% Fe, 6.689% C

3. 35.00% N, 5.05% H, 59.96% O

5. a. Both are 39.99% C, 6.73% H, and 53.28% O because, if you combine the hydrogen atoms in acetic acid, the empirical formulas are the same.

b. The percentage composition of the empirical formula is the same as the percentage compositions of the molecular formulas.

Section 3 Review

5. SO_2

7. a. 64.62% Ag, 14.39% C, 1.82% H, and 19.17% O

b. 55.39% Pb, 18.95% Cl, and 25.66% O

c. 27.93% Fe, 24.06% S, and 48.01% O

d. 39.81% Cu, 20.09% S, and 40.10% O

Chapter Review

23. 1.20×10^{24} molecules $C_{12}H_{22}O_{11}$

25. 7.53×10^{21} atoms Hg

27. 5.815×10^{24} ions Ni^{2+}

29. 41.5 mol MgO

31. 12.5 mol C_6H_6

33. 6.82×10^{-2} mol Na^+

35. 0.006192 g P

37. 1.46×10^{-8} g CO_2

39. 81.6 g O_2

41. 160 g NaCl

43. a. 58.44 g NaCl

b. 36.04 g H_2O

c. 260 g $Ca(OH)_2$

d. 163 g $Ba(NO_3)_2$

45. 1.337×10^{24} formula units $ZnCl_2$

47. 2.79×10^{24} atoms Al

49. a. 4.99×10^{-2} mol $(NH_4)_2SO_4$

b. 61 mol $Ca(OH)_2$

c. 7.49×10^{-2} mol H_2SO_4

51. 7.25 mol $NaNO_2$

53. 57.8 mol C_3H_8

55. 79.90 amu

57. 55.84 amu

59. 94.97 g/mol

61. 75.08 g/mol

63. P_2O_3

65. $C_9H_{18}N_6$

67. $Co_2C_8O_8$

69. $C_4H_4N_2$

71. 37.56% NH_4^+

73. 22.57% N, 6.51% H, 19.35% C, 51.56% O

75. a. 0.00152 mol Na^+

b. 0.0072 mol Ca^{2+}

77. 40 mol H

79. 28.09 amu

81. a. 180.18 g/mol

b. 180.18 g/mol

83. $AlPO_4$ is 22.12% Al, while $AlCl_3$ is 20.23% Al, so aluminum phosphate has more aluminum per gram.

Stoichiometry

Practice Problems A

1. a. 0.670 mol O_2

b. 1.34 mol H_2O

Practice Problems B

1. 45.6 g Al

3. 679 g Fe_2O_3

Practice Problems C

1. 315 mL C_5H_8

3. 113 mL C_5H_{12}

Practice Problems D

1. 2.89×10^{24} molecules BrF_5

Section 1 Review

5. a. 1.42 mol CO_2

b. 47.2 mL CO_2

Practice Problems E

1. PCl_3 is excess, H_2O is limiting, theoretical yield is 109 g HCl

3. PCl_3 is excess, H_2O is limiting, theoretical yield is 101 g HCl

Practice Problems F

1. N_2 is limiting, 85.3%

3. Br_2 is limiting, 90.9%

Practice Problems G

1. 1.04×10^3 g NH_3

3. 439 g BrCl

Section 2 Review

7. a. $P_4O_{10} + 6H_2O \longrightarrow 4H_3PO_4$

b. 138.1 g H_3PO_4

c. 91.4%

9. 3.70 g Cu

11. a. $Mg + 2H_2O \longrightarrow Mg(OH)_2 + H_2$

b. 86.8%

c. 55 g $Mg(OH)_2$

Practice Problems H

1. 33 g Na

3. 121 g $NaHCO_3$

Practice Problems I

1. 2.17 cycles, so after 3 full cycles all of the 1.00 mL of isooctane will have reacted

3. $2CH_3OH + 3O_2 \longrightarrow 2CO_2 + 4H_2O$; 2.2×10^2 L air

Practice Problems J

1. 4.01 g CO_2

Section 3 Review

5. 10.7 g Na_2O

Chapter Review

27. a. 6.6 mol H_2

b. 3.36 mol O_2

c. 8.12 mol H_2

29. a. 1.08 mol O_2

b. 2.62 mol Al_2O_3

c. 1.99 mol Al_2O_3

31. a. 49.0 g O_2

b. 748 g $KClO_3$

c. 12.7 g KCl

33. 107 g O_2
35. a. 113 L O_2
   b. 25 L O_2
   c. 7.41 L O_2
37. a. 4.61×10^{23} molecules H_2
   b. 4.31×10^{23} atoms Na
   c. 3.30×10^{20} molecules H_2
39. a. excess, H_2O; limiting, CaC_2
   b. 26 g C_2H_2
   c. 74 g $Ca(OH)_2$
41. 75.6%
43. 72.5%
45. 4.7 g Al
47. 46.6 L CO_2
49. $\dfrac{25\ \text{mol}\ O_2}{2\ \text{mol}\ C_8H_{18}}$, or 25:2
51. 1.71×10^3 L O_2
53. 2.16×10^3 g CO_2; 88.0%
55. a. $\dfrac{2\ \text{mol}\ NaN_3}{2\ \text{mol}\ Na}$, $\dfrac{2\ \text{mol}\ NaN_3}{3\ \text{mol}\ N_2}$,

$\dfrac{2\ \text{mol}\ Na}{2\ \text{mol}\ NaN_3}$,

$\dfrac{2\ \text{mol}\ Na}{3\ \text{mol}\ N_2}$, $\dfrac{3\ \text{mol}\ N_2}{2\ \text{mol}\ NaN_3}$,

$\dfrac{3\ \text{mol}\ N_2}{2\ \text{mol}\ Na}$

   b. 4.0 mol N_2
   c. 11.0 g Na
   d. 4.86×10^{23} molecules N_2
57. 1.44 L N_2
59. 1.74×10^3 g CO_2

Causes of Change

Practice Problems A
1. 97 J
3. 220 J

Section 1 Review
9. 29.2 kJ
11. 0.52 mol
13. 301 K
15. 0.864 J/K•mol

Practice Problems B
1. 2.60×10^3 J/mol
3. 3.6×10^2 J/mol

Practice Problems C
1. -2.56×10^3 J/mol
3. -2.8×10^2 J/mol

Section 2 Review
5. 120 J/mol
7. 3690 J/mol
9. 42.8 J/K•mol

Practice Problems D
1. −57.2 kJ

Practice Problems E
1. −1428.6 kJ; exothermic

Section 3 Review
5. −818.6 kJ

Practice Problems F
1. −332.2 J/K
3. −95 J/K

Practice Problems G
1. −41 kJ, spontaneous
3. −1.2 kJ, spontaneous

Practice Problems H
1. −394.4 kJ, spontaneous

Section 4 Review
7. −146.5 J/K
9. 182.1 kJ, no
11. 60 kJ; The result is half the result of problem 10.

Chapter Review
27. 35.2 J/K•mol
29. 7040 J
31. −5030 J
33. −510 kJ
35. 117.6 J
37. −663 J/K
39. −345 kJ   The reaction is spontaneous.
41. −227.9 kJ   The reaction is spontaneous.
45. −57 kJ
51. −184.614 kJ; yes
53. Reaction 1: ΔG = 115 kJ, not spontaneous; Reaction 2: ΔG = −101 kJ, spontaneous; Reaction 3: ΔG = −310 kJ, spontaneous
55. 108.7 J/K

States of Matter and Intermolecular Forces

Practice Problems A
1. T_{mp} = 159 K, T_{bp} = 351 K
3. T_{mp} = 195 K, T_{bp} = 240 K

Section 3 Review
7. 266 K
9. 233 K

Practice Problems B
1. a. Phase diagram for sulfur dioxide, SO_2

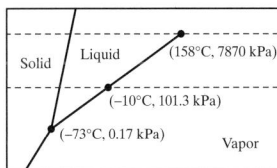

   b. solid
   c. vapor
   d. The sulfur dioxide changes from a liquid to a vapor.
   e. The sulfur dioxide remains a liquid.

Section 4 Review
7. a. Phase diagram for benzene, C_6H_6

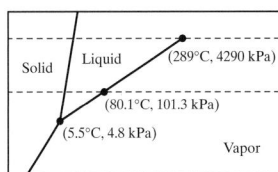

   b. liquid
   c. vapor
   d. The benzene changes from a solid to a liquid.
   e. The benzene changes from a liquid to a vapor.

Chapter Review
53. 290 K
55. 273 K
57. 391 K
59. 334 K

Appendix E

61. Phase diagram for ammonia, NH_3

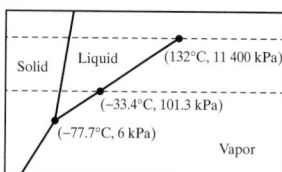

63. Phase diagram for iodine, I_2

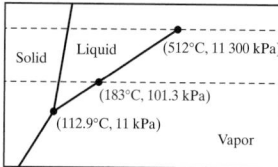

65. 3130 K

Gases

Practice Problems A
1. 7.37×10^6 Pa
3. 0.9869 atm

Section 1 Review
9. 610.5 Pa

Practice Problems B
1. 142 mL
3. 7.9×10^5 L

Practice Problems C
1. 0.67 L
3. $-11.0°C$

Practice Problems D
1. 1.29 atm
3. 491 K, or 218°C

Section 2 Review
5. 31.0 mL
7. 114 kPa
9. 5.00 L

Practice Problems E
1. 7.97×10^{-2} mol
3. 1500 kPa

Practice Problems F
1. N_2 has a higher speed; 1.069 times faster
3. 48.6 g/mol

Practice Problems G
1. 11.4 L
3. 3.87 g Na

Section 3 Review
7. 0.781 mol
9. 5.3×10^{-3} mol SO_2
11. 15.0 L

Chapter Review
31. 101 325 newtons
33. 290 kPa
35. 113 mL
37. 1100 mL
39. 66.3 mL
41. 93.3 mL
43. 0.540 L
45. 3.1 L
47. 152 kPa
49. 26 kPa
51. 8.4 atm
53. 0.781 mol
55. 266 kPa
57. 4.0×10^3 L
59. $M = 64$ g/mol. It is SO_2.
61. 1.91×10^3 m/s
63. 10.4 L
65. 3.56×10^{-2} g C_8H_{18}
67. 2.28×10^3 L
69. 1.80 g CO_2; 1.80 g/L CO_2; 1.31 g/L O_2
71. 8.1×10^3 g LiOH
73. 5.71 g/L
75. 171 kPa
77. 12.5 g O_2
79. 2.64 L H_2
81. 171°C
83. 44 g He
91. 0.65 g O_2

Solutions

Practice Problems A
1. 15 ppm
3. 4250 ppm
5. 63 ppm
7. 26 ppm

Practice Problems B
1. 0.83 M acetic acid
3. 0.816 M sulfuric acid
5. 0.2501 M Ba(OH)$_2$
7. 11 g NaCl

Practice Problems C
1. 109 g HCl
3. 451 g CdS

Section 2 Review
5. 438 ppm Cd
7. 4.00 g NaOH
9. 0.838 M NaOCl
11. 5.8×10^3 g Ca$_3$(PO$_4$)$_2$ and 2.0×10^3 g H_2O

Chapter Review
43. 1.7 ppm
45. 5×10^{-2} g Cl_2
47. 0.7776 M NaOH
49. 2.0 mol AgNO$_3$
51. 1.61 M CuCl$_2$
53. 0.123 M H_3PO_4
55. 5.85 g NaOH
57. 5.02 M HCl
59. 0.259 g LiF
61. 0.74 M $C_6H_{12}O_6$
63. 0.50 M Ba$_3$(PO$_4$)$_2$
65. 0.0309 M AgNO$_3$
67. 440 g H_2SO_4
69. 23 g K_2CrO_4
71. 0.111 M $C_6H_{12}O_6$
73. 0.0230 M PbBr$_2$
75. 1.9 M NaCl
77. 1.93 M NaCl

Chemical Equilibrium

Practice Problems A
1. 2.0

Practice Problems B
1. 3.5×10^{-5}

Practice Problems C
1. 6.2×10^{-9}
3. 1.80×10^{-10}

Practice Problems D
1. 2.6×10^{-4}
3. 1.43×10^{-2}

Section 2 Review

5. 3.53×10^{-3}

7. $C(s) + CO_2(g) \rightleftarrows 2CO(g)$; 2.5×10^{-2}

Chapter Review

33. a. 0.67

b. 0.52

c. 311

35. 0.048 mol/ L

37. 0.046 mol/ L

39. 2.9×10^{-26}

41. 8.22×10^{-96}

43. $1.7 \times 10^{-14}, 1.3 \times 10^{-7}, 2.7 \times 10^{-7}, 1.6 \times 10^{-4}$

45. 4.0×10^{-5} M

57. 2.8×10^{-3} mol/ L

Acids and Bases

Practice Problems A

1. $[H_3O^+] = 1.38 \times 10^{-11}$ M

3. $[H_3O^+] = 2.67 \times 10^{-13}$ M

5. $[OH^-] = 2.4 \times 10^{-4}$ M; $[H_3O^+] = 4.2 \times 10^{-11}$ M

Practice Problems B

1. 2.3

3. 11.3

Practice Problems C

1. $[H_3O^+] = 5.0 \times 10^{-4}$ M

3. $[H_3O^+] = 7.9 \times 10^{-9}$ M; $[OH^-] = 1.3 \times 10^{-6}$ M

Section 2 Review

7. $[OH^-] = 3.16 \times 10^{-12}$ M; pH = 2.50

9. $[OH^-] = 0.088$ M; $[H_3O^+] = 1.1 \times 10^{-13}$ M; pH = 12.95

11. 1.00×10^{-2} mol HBr

Practice Problems D

1. 6.9×10^{-3} M

3. 4.674×10^{-3} moles

Section 3 Review

9. 13.5 mL

11. 0.18 M

Practice Problems E

1. $[H_3O^+] = 1.62 \times 10^{-3}$ M

3. $K_a = 6.4 \times 10^{-5}$

Section 4 Review

7. $K_a = 3.97 \times 10^{-8}$

9. $K_a = 6.8 \times 10^{-4}$

13. 1.0×10^{-4}

15. $K_{eq} = 9.19 \times 10^3$

Chapter Review

41. 2.74×10^{-14} M

43. 5.35×10^{-12} M

45. 1.41×10^{-7} M

47. 12.13

49. 12.91

51. a. 2.3

b. 1.3

c. 0.3

d. −0.7

53. 13.48

55. 0.17

57. 5.72

59. $[H_3O^+] = 3.2 \times 10^{-10}$ M; $[OH^-] = 3.2 \times 10^{-5}$ M

61. 3×10^{-5} mol

63. $[OH^-] = 5.25 \times 10^{-6}$ M

65. $[H_3O^+] = 7.9 \times 10^{-11}$ M; $[OH^-] = 1.3 \times 10^{-4}$ M

67. $[OH^-] = 2.0 \times 10^{-10}$ M

69. $[H_3O^+] = 1.0 \times 10^{-3}$ M

71. $[H_3O^+] = 5.0 \times 10^{-14}$ M

75. 55.0 mL

77. 0.5260 M

79. 0.798 M

81. 0.1544 M

83. 1.8×10^{-4}

85. 1.6×10^{-5}

87. 2.5×10^{-3}

89. 1.3

91. $[H_3O^+] = 0.400$ M; pH = 0.40; $[OH^-] = 2.5 \times 10^{-14}$ M

93. 1.7×10^{-7}

95. $[H_3O^+] = 1.1 \times 10^{-3}$ M; $[C_6H_5COOH] = 0.020$ M; $[C_6H_5COO^-] = 1.2 \times 10^{-3}$ M

99. 1.24 g

101. 2.42

Reaction Rates

Practice Problems A

1. $rate = \dfrac{-\Delta[Br^-]}{2\Delta t} = \dfrac{-\Delta[H_2O_2]}{\Delta t}$

3. 9.0×10^{-5} M/s

Practice Problems B

1. 1

3. a factor of 3.2

Section 2 Review

7. 8.1×10^{-5} M/s

Chapter Review

25. 1.4×10^{-6} M/s

27. 1.8×10^{-6} M/s

29. $rate = k[NO]^2[Br_2]$

31. double

Oxidation, Reduction, and Electrochemistry

Practice Problems A

1. a. N = −3; H = +1

c. H = +1; O = −2

e. H = 0

g. K = +1; Cl = +5; O = −2

i. Ca = +2; O = −2; H = +1

k. H = +1; P = +5; O = −2

Practice Problems B

1. half-reactions: $4Fe(s) \longrightarrow 4Fe^{3+}(aq) + 12e^-$ and $3O_2(aq) + 12H_3O^+(aq) + 12e^- \longrightarrow 18H_2O(l)$; overall: $4Fe(s) + 3O_2(aq) + 12H_3O^+(aq) \longrightarrow 4Fe^{3+}(aq) + 18H_2O(l)$

3. half-reactions: $2Br^-(aq) \longrightarrow Br_2(aq) + 2e^-$ and $H_2O_2(aq) + 2H_3O^+(aq) + 2e^- \longrightarrow 4H_2O(l)$; overall: $2Br^-(aq) + H_2O_2(aq) + 2H_3O^+(aq) \longrightarrow Br_2(aq) + 4H_2O(l)$

Section 1 Review

7. a. C = −4; H = +1

b. H = +1; S = +4; O = −2

c. Na = +1; H = +1; C = +4; O = −2

d. Na = +1; Bi = +5; O = −2

Practice Problems C

1. $(+0.401 \text{ V}) - (-0.037 \text{ V}) = +0.438$ V

3. $(+0.7996 \text{ V}) - (+0.3419 \text{ V}) = +0.4577$ V

Section 3 Review

5. The cell voltage is $(+1.358 \text{ V}) - (+0.222 \text{ V}) = +1.136$ V. The silver electrode is the anode.

Chapter Review

39. C = +4; O = −2

41. Ba = +2; Cl = −1

43. S = −2

45. C = −4; H = +1

47. Ca = +2; C = +4; O = −2

49. C = +4; O = −2; Cl = −1

51. $Fe(s) \longrightarrow Fe^{2+}(aq) + 2e^-$

53. $Fe(s) + Cl_2(g) \longrightarrow$
$$Fe^{2+}(aq) + 2Cl^-(aq)$$

55. $6H_2O(l) \longrightarrow O_2(g) +$
$$4H_3O^+(aq) + 4e^-$$

57. $O_2(g) + 4H_3O^+(aq) + 4e^- \longrightarrow$
$$6H_2O(l)$$

59. $4H_2O(l) + 2SO_2(aq) +$
$O_2(g) \longrightarrow 2HSO_4^-(aq) + 2H_3O^+(aq)$

61. $(+1.30 \text{ V}) - (+0.45 \text{ V}) = +0.85 \text{ V}$

63. $(0.0000 \text{ V}) - (+0.771 \text{ V}) =$
$$-0.771 \text{ V}$$

65. $(+0.401 \text{ V}) - (-0.828 \text{ V}) =$
$$+1.229 \text{ V}$$

67. $(+1.691 \text{ V}) - (-0.42 \text{ V}) =$
$$+2.11 \text{ V}$$

75. 0

77. $Fe^{3+}(aq) + e^- \longrightarrow Fe^{2+}(aq)$;
reduction

79. $(-0.4030 \text{ V}) - (+1.358 \text{ V}) =$
$$-1.761 \text{ V}$$

Nuclear Chemistry

Practice Problems A

1. $^{214}_{82}Pb$

3. $^{256}_{101}Md$

Section 2 Review

5. a. $^{233}_{92}U \longrightarrow ^{229}_{90}Th + ^4_2He$

b. $^{66}_{29}Cu \longrightarrow ^{66}_{30}Zn + ^0_{-1}e$

c. $^9_4Be \longrightarrow ^4_2He \longrightarrow ^{13}_6C$
$^{13}_6C \longrightarrow ^{12}_6C + ^1_0n$

d. $^{238}_{92}U + ^1_0n \longrightarrow ^{239}_{92}U$
$^{239}_{92}U \longrightarrow ^{239}_{93}Np + ^0_{-1}e$
$^{239}_{93}Np \longrightarrow ^{239}_{94}Pu + ^0_{-1}e$

Practice Problems B

1. 6396 y

3. 12 min

Practice Problem C

1. 0.25 mg

3. 0.32 g

Section 3 Review

5. 1/8

7. 1/16

Chapter Review

33. $^{235}_{92}U \longrightarrow ^{231}_{90}Th + ^4_2He$

$^{231}_{90}Th \longrightarrow ^{231}_{91}Pa + ^0_{-1}e$

$^{231}_{91}Pa \longrightarrow ^{227}_{89}Ac + ^4_2He$

$^{227}_{89}Ac \longrightarrow ^{227}_{90}Th + ^0_{-1}e$

$^{227}_{90}Th \longrightarrow ^{223}_{88}Ra + ^4_2He$

$^{223}_{88}Ra \longrightarrow ^{219}_{86}Rn + ^4_2He$

$^{219}_{86}Rn \longrightarrow ^{215}_{84}Po + ^4_2He$

$^{215}_{84}Po \longrightarrow ^{211}_{82}Pb + ^4_2He$

$^{211}_{82}Pb \longrightarrow ^{211}_{83}Bi + ^0_{-1}e$

$^{211}_{83}Bi \longrightarrow ^{211}_{84}Po + ^0_{-1}e$

$^{211}_{84}Po \longrightarrow ^{207}_{82}Pb + ^4_2He$

35. a. ^2_1H

b. $^{12}_6C$

c. $^0_{-1}e$

37. 12.8 h

39. 5715 y

41. 0.274 g

43. 0.056 g

45. a. 1.55:1; outside

b. 1:1; within

c. 1.15:1; within

d. 1.6:1; outside

47. 8.77×10^{-28} kg

51. for example,

$^{238}_{92}U + ^1_0n \longrightarrow ^{239}_{92}U$

$^{239}_{92}U \longrightarrow ^{239}_{93}Np + ^0_{-1}e$

$^{239}_{93}Np \longrightarrow ^{239}_{94}Pu + ^0_{-1}e$

53. 0.200 g

55. 0.04049 amu per atom

57. 1.34×10^{-10} y

59. 3%

61. 4.9×10^{-12} J

63. 10.5 days

65. $^{217}_{89}Ac \longrightarrow ^{213}_{87}Fr + ^4_2He$

67. The total mass of this nucleus—56 amu—will be greater than 55.847 amu. The mass is not equal to 55.847 amu because that value is an average of the masses of several different isotopes.

69. 29.1 y

71. 64.4 min

73. 22 860 y

75. $^{130}_{52}Te \longrightarrow ^{130}_{53}I + ^0_{-1}e$

Carbon and Organic Compounds

Practice Problems A

1. a. 2,2,4-trimethylpentane

b. 1-pentyne

c. 2,3,4-trimethylnonane

d. 2-methyl-3-hexene

Practice Problems B

1. a. ethanol

b. 2-pentanone

c. butanoic acid

d. 3-hexanol

Section 2 Review

7. methylbenzene

Chapter Review

41. a. iodomethane

b. 1, 3-dichloroethane

47. a. 3-hexanone

b. 3-bromo-1,1-dichlorobutane

c. cyclohexanol

882

GLOSSARY

accuracy
a description of how close a measurement is to the true value of the quantity measured (p. 55)

acid-ionization constant, K_a
the equilibrium constant for a reaction in which an acid donates a proton to water (p. 559)

actinide
any of the elements of the actinide series, which have atomic numbers from 89 (actinium, Ac) through 103 (lawrencium, Lr) (p. 130)

activated complex
a molecule in an unstable state intermediate to the reactants and the products in the chemical reaction (p. 590)

activation energy
the minimum amount of energy required to start a chemical reaction (p. 590)

activity series
a series of elements that have similar properties and that are arranged in descending order of chemical activity; examples of activity series include metals and halogens (p. 280)

actual yield
the measured amount of a product of a reaction (p. 316)

addition reaction
a reaction in which an atom or molecule is added to an unsaturated molecule (p. 694)

alkali metal
one of the elements of Group 1 of the periodic table (lithium, sodium, potassium, rubidium, cesium, and francium) (p. 125)

alkaline-earth metal
one of the elements of Group 2 of the periodic table (beryllium, magnesium, calcium, strontium, barium, and radium) (p. 126)

alkane
a hydrocarbon characterized by a straight or branched carbon chain that contains only single bonds (p. 681)

alkene
a hydrocarbon that contains one or more double bonds (p. 681)

alkyne
a hydrocarbon that contains one or more triple bonds (p. 681)

alloy
a solid or liquid mixture of two or more metals (p. 130)

amino acid
any one of 20 different organic molecules that contain a carboxyl and an amino group and that combine to form proteins (p. 717)

amphoteric
describes a substance, such as water, that has the properties of an acid and the properties of a base (p. 538)

anion
an ion that has a negative charge (p. 161)

anode
the electrode on whose surface oxidation takes place; anions migrate toward the anode, and electrons leave the system from the anode (p. 614)

aromatic hydrocarbon
a hydrocarbon that contains six-carbon rings and is usually very reactive (p. 682)

atom
the smallest unit of an element that maintains the properties of that element (p. 21)

atomic mass
the mass of an atom expressed in atomic mass units (p. 100)

atomic number
the number of protons in the nucleus of an atom; the atomic number is the same for all atoms of an element (p. 84)

ATP
adenosine triphosphate, an organic molecule that acts as the main energy source for cell processes; composed of a nitrogenous base, a sugar, and three phosphate groups (p. 737)

Aufbau principle
the principle that states that the structure of each successive element is obtained by adding one proton to the nucleus of the atom and one electron to the lowest-energy orbital that is available (p. 97)

average atomic mass
the weighted average of the masses of all naturally occurring isotopes of an element (p. 235)

Avogadro's law
the law that states that equal volumes of gases at the same temperature and pressure contain equal numbers of molecules (p. 432)

Avogadro's number
6.02×10^{23}, the number of atoms or molecules in 1 mol (p. 101, p. 224)

beta particle
a charged electron emitted during certain types of radioactive decay, such as beta decay (p. 649)

boiling point
the temperature and pressure at which a liquid and a gas are in equilibrium (p. 382)

bond energy
the energy required to break the bonds in 1 mol of a chemical compound (p. 192)
bond length
the distance between two bonded atoms

at their minimum potential energy; the average distance between the nuclei of two bonded atoms (p. 192)

bond radius
half the distance from center to center of two like atoms that are bonded together (p. 135)

Boyle's law
the law that states that for a fixed amount of gas at a constant temperature, the volume of the gas increases as the pressure of the gas decreases and the volume of the gas decreases as the pressure of the gas increases (p. 424)

Brønsted-Lowry acid
a substance that donates a proton to another substance (p. 535)

Brønsted-Lowry base
a substance that accepts a proton (p. 536)

buffer
a solution made from a weak acid and its conjugate base that neutralizes small amounts of acids or bases added to it (p. 561)

C

calorimeter
a device used to measure the heat absorbed or released in a chemical or physical change (p. 351)

calorimetry
the measurement of heat-related constants, such as specific heat or latent heat (p. 351)

carbohydrate
any organic compound that is made of carbon, hydrogen, and oxygen and that provides nutrients to the cells of living things (p. 712)

catalysis
the acceleration of a chemical reaction by a catalyst (p. 593)

catalyst
a substance that changes the rate of a chemical reaction without being consumed or changed significantly (p. 593)

cathode
the electrode on whose surface reduction takes place (p. 613)

cation
an ion that has a positive charge (p. 161)

chain reaction
a reaction in which a change in a single molecule makes many molecules change until a stable compound forms (p. 654)

Charles's law
the law that states that for a fixed amount of gas at a constant pressure, the volume of the gas increases as the temperature of the gas increases and the volume of the gas decreases as the temperature of the gas decreases (p. 426)

chemical
any substance that has a defined composition (p. 4)

chemical change
a change that occurs when one or more substances change into entirely new substances with different properties (p. 39)

chemical equation
a representation of a chemical reaction that uses symbols to show the relationship between the reactants and the products (p. 263)

chemical equilibrium
a state of balance in which the rate of a forward reaction equals the rate of the reverse reaction and the concentrations of products and reactants remain unchanged (p. 497)

chemical kinetics
the area of chemistry that is the study of reaction rates and reaction mechanisms (p. 576)

chemical property
a property of matter that describes a substance's ability to participate in chemical reactions (p. 18)

chemical reaction
the process by which one or more substances change to produce one or more different substances (p. 5, p. 260)

clone
an organism that is produced by asexual reproduction and that is genetically identical to its parent; to make a genetic duplicate (p. 731)

coefficient
a small whole number that appears as a factor in front of a formula in a chemical equation (p. 268)

colligative property
a property that is determined by the number of particles present in a system but that is independent of the properties of the particles themselves (p. 482)

colloid
a mixture consisting of tiny particles that are intermediate in size between those in solutions and those in suspensions and that are suspended in a liquid, solid, or gas (p. 456)

combustion reaction
the oxidation reaction of an organic compound, in which heat is released (p. 276)

common-ion effect
the phenomenon in which the addition of an ion common to two solutes brings about precipitation or reduces ionization (p. 517)

compound
a substance made up of atoms of two or more different elements joined by chemical bonds (p. 24)

concentration
the amount of a particular substance in a given quantity of a mixture, solution, or ore (p. 460)

condensation
the change of state from a gas to a liquid (p. 382)

condensation reaction
a chemical reaction in which two or more molecules combine to produce water or another simple molecule (p. 699, p. 715)

conductivity
the ability to conduct an electric current (p. 478)

conjugate acid
an acid that forms when a base gains a proton (p. 537)

conjugate base
a base that forms when an acid loses a proton (p. 537)

conversion factor
a ratio that is derived from the equality of two different units and that can be used to convert from one unit to the other (p. 13)

corrosion
the gradual destruction of a metal or alloy as a result of chemical processes such as oxidation or the action of a chemical agent (p. 620)

covalent bond
a bond formed when atoms share one or more pairs of electrons (p. 191)

critical mass
the minimum mass of a fissionable isotope that provides the number of neutrons needed to sustain a chain reaction (p. 654)

critical point
the temperature and pressure at which the gas and liquid states of a substance become identical and form one phase (p. 402)

crystal lattice
the regular pattern in which a crystal is arranged (p. 174)

D

Dalton's law of partial pressures
the law that states that the total pressure of a mixture of gases is equal to the sum of the partial pressures of the component gases (p. 439)

decomposition reaction
a reaction in which a single compound breaks down to form two or more simpler substances (p. 278)

density
the ratio of the mass of a substance to the volume of the substance; often expressed as grams per cubic centimeter for solids and liquids and as grams per liter for gases (p. 16)

denature
to change irreversibly the structure or shape—and thus the solubility and other properties—of a protein by heating, shaking, or treating the protein with acid, alkali, or other species (p. 723)

detergent
a water-soluble cleaner that can emulsify dirt and oil (p. 484)

diffusion
the movement of particles from regions of higher density to regions of lower density (p. 436)

dipole
a molecule or a part of a molecule that contains both positively and negatively charged regions (p. 195)

dipole-dipole forces
interactions between polar molecules (p. 386)

disaccharide
a sugar formed from two monosaccharides (p. 712)

dissociation
the separating of a molecule into simpler molecules, atoms, radicals, or ions (p. 472)

DNA
deoxyribonucleic acid, the material that contains the information that determines inherited characteristics (p. 726)

DNA fingerprint
the pattern of bands that results when an individual's DNA sample is fragmented, replicated, and separated (p. 730)

double bond
a covalent bond in which two atoms share two pairs of electrons (p. 204)

double-displacement reaction
a reaction in which a gas, a solid precipitate, or a molecular compound forms from the apparent exchange of atoms or ions between two compounds (p. 283)

effusion
the passage of a gas under pressure through a tiny opening (p. 437)

electrochemical cell
a system that contains two electrodes separated by an electrolyte phase (p. 613)

electrochemistry
the branch of chemistry that is the study of the relationship between electric forces and chemical reactions (p. 612)

electrode
a conductor used to establish electrical contact with a nonmetallic part of a circuit, such as an electrolyte (p. 613)

electrolysis
the process in which an electric current is used to produce a chemical reaction, such as the decomposition of water (p. 627)

electrolyte
a substance that dissolves in water to give a solution that conducts an electric current (p. 478)

electrolytic cell
an electrochemical device in which electrolysis takes place when an electric current is in the device (p. 627)

electromagnetic spectrum
all of the frequencies or wavelengths of electromagnetic radiation (p. 92)

electron
a subatomic particle that has a negative charge (p. 80)

electron shielding
the reduction of the attractive force between a positively charged nucleus and its outermost electrons due to the cancellation of some of the positive charge by the negative charges of the inner electrons (p. 133)

electron configuration
the arrangement of electrons in an atom (p. 96)

electronegativity
a measure of the ability of an atom in a chemical compound to attract electrons (p. 137)

electroplating
the electrolytic process of plating or coating an object with a metal (p. 630)

element
a substance that cannot be separated or broken down into simpler substances by chemical means; all atoms of an element have the same atomic number (p. 22)

elimination reaction
a reaction in which a simple molecule, such as water or ammonia, is removed and a new compound is produced (p. 699)

empirical formula
a chemical formula that shows the composition of a compound in terms of the relative numbers and kinds of atoms in the simplest ratio (p. 242)

emulsion
any mixture of two or more immiscible liquids in which one liquid is dispersed in the other (p. 484)

endothermic
describes a process in which heat is absorbed from the environment (p. 40)

end point
the point in a titration at which a marked color change takes place (p. 554)

energy
the capacity to do work (p. 38)

enthalpy
the sum of the internal energy of a system plus the product of the system's volume multiplied by the pressure that the system exerts on its surroundings (p. 340)

entropy
a measure of the randomness or disorder of a system (p. 358)

enzyme
a type of protein that speeds up metabolic reactions in plant and animals without being permanently changed or destroyed (p. 595, p. 722)

equilibrium
in chemistry, the state in which a chemical process and the reverse chemical process occur at the same rate such that the con-centrations of reactants and products do not change (p. 400)

equilibrium constant
a number that relates the concentrations of starting materials and products of a reversible chemical reaction to one another at a given temperature (p. 503)

evaporation
the change of a substance from a liquid to a gas (p. 39, p. 382)

excess reactant
the substance that is not used up completely in a reaction (p. 313)

excited state
a state in which an atom has more energy than it does at its ground state (p. 94)

exothermic
describes a process in which a system releases heat into the environment (p. 40)

freezing
the change of state in which a liquid becomes a solid as heat is removed (p. 383)

freezing point
the temperature at which a solid and liquid are in equilibrium at 1 atm pressure; the temperature at which a liquid substance freezes (p. 383)

functional group
the portion of a molecule that is active in a chemical reaction and that determines the properties of many organic compounds (p. 683)

gamma ray
the high-energy photon emitted by a nucleus during fission and radioactive decay (p. 649)

Gay-Lussac's law
the law that states that the pressure of a gas at a constant volume is directly proportional to the absolute temperature (p. 430)

Gay-Lussac's law of combining volumes of gases
the law that states that the volumes of gases involved in a chemical change can be represented by the ratio of small whole numbers (p. 439)

gene
a segment of DNA that is located in a chromosome and that codes for a specific hereditary trait (p. 728)

Gibbs energy
the energy in a system that is available for work (p. 362)

Graham's law of diffusion
the law that states that the rate of diffusion of a gas is inversely proportional to the square root of the gas's density (p. 437)

ground state
the lowest energy state of a quantized system (p. 94)

group
a vertical column of elements in the periodic table; elements in a group share chemical properties (p. 119)

half-life
the time required for half of a sample of a radioactive substance to disintegrate by radioactive decay or by natural processes (p. 658)

half-reaction
the part of a reaction that involves only oxidation or reduction (p. 608)

halogen
one of the elements of Group 17 (fluorine, chlorine, bromine, iodine, and astatine); halogens combine with most metals to form salts (p. 126)

heat
the energy transferred between objects that are at different temperatures; energy is always transferred from higher-temperature objects to lower-temperature objects until thermal equilibrium is reached (p. 41, p. 338)

Henry's law
the law that states that at constant temperature, the solubility of a gas in a liquid is directly proportional to the partial pressure of the gas on the surface of the liquid (p. 477)

Hess's law
the law that states that the amount of heat released or absorbed in a chemical reaction does not depend on the number of steps in the reaction (p. 353)

heterogeneous
composed of dissimilar components (p. 26)

homogeneous
describes something that has a uniform structure or composition throughout (p. 26)

Hund's rule
the rule that states that for an atom in the ground state, the number of unpaired electrons is the maximum possible and these unpaired electrons have the same spin (p. 98)

hydration
the strong affinity of water molecules for particles of dissolved or suspended substances that causes electrolytic dissociation (p. 472)

hydrocarbon
an organic compound composed only of carbon and hydrogen (p. 680)

hydrogen bond
the intermolecular force occurring when a hydrogen atom that is bonded to a highly electronegative atom of one molecule is attracted to two unshared electrons of another molecule (p. 387)

GLOSSARY

hydrolysis
a chemical reaction between water and another substance to form two or more new substances; a reaction between water and a salt to create an acid or a base (p. 716)

hydronium ion
an ion consisting of a proton combined with a molecule of water; H_3O^+ (p. 480)

hypothesis
a theory or explanation that is based on observations and that can be tested (p. 50)

I

ideal gas
an imaginary gas whose particles are infinitely small and do not interact with each other (p. 433)

ideal gas law
the law that states the mathematical relationship of pressure (P), volume (V), temperature (T), the gas constant (R), and the number of moles of a gas (n); $PV = nRT$ (p. 434)

immiscible
describes two or more liquids that do not mix with each other (p. 470)

indicator
a compound that can reversibly change color depending on the pH of the solution or other chemical change (p. 546)

intermediate
a substance that forms in a middle stage of a chemical reaction and is considered a stepping stone between the parent substance and the final product (p. 589)

intermolecular forces
the forces of attraction between molecules (p. 386)

ion
an atom, radical, or molecule that has gained or lost one or more electrons and has a negative or positive charge (p. 161)

isomer
one of two or more compounds that have the same chemical composition but different structures (p. 686)

isotope
an atom that has the same number of protons (atomic number) as other atoms of the same element do but that has a different number of neutrons (atomic mass) (p. 88)

K

kinetic energy
the energy of an object that is due to the object's motion (p. 42)

kinetic-molecular theory
a theory that explains that the behavior of physical systems depends on the combined actions of the molecules constituting the system (p. 421)

L

lanthanide
a member of the rare-earth series of elements, whose atomic numbers range from 58 (cerium) to 71 (lutetium) (p. 130)

lattice energy
the energy associated with constructing a crystal lattice relative to the energy of all constituent atoms separated by infinite distances (p. 168)

law
a summary of many experimental results and observations; a law tells how things work (p. 52)

law of conservation of energy
the law that states that energy cannot be created or destroyed but can be changed from one form to another (p. 40)

law of conservation of mass
the law that states that mass cannot be created or destroyed in ordinary chemical and physical changes (p. 52, p. 76)

law of definite proportions
the law that states that a chemical compound always contains the same elements in exactly the same proportions by weight or mass (p. 75)

law of multiple proportions
the law that states that when two elements combine to form two or more compounds, the mass of one element that combines with a given mass of the other is in the ratio of small whole numbers (p. 77)

Le Châtelier's principle
the principle that states that a system in equilibrium will oppose a change in a way that helps eliminate the change (p. 512)

Lewis structure
a structural formula in which electrons are represented by dots; dot pairs or dashes between two atomic symbols represent pairs in covalent bonds (p. 199)

limiting reactant
the substance that controls the quantity of product that can form in a chemical reaction (p. 313)

London dispersion force
the intermolecular attraction resulting from the uneven distribution of electrons and the creation of temporary dipoles (p. 390)

M

main-group element
an element in the s-block or p-block of the periodic table (p. 124)

mass
a measure of the amount of matter in an object; a fundamental property of an object that is not affected by the forces that act on the object, such as the gravitational force (p. 10)

mass defect
the difference between the mass of an atom and the sum of the masses of the atom's protons, neutrons, and electrons (p. 644)

mass number
the sum of the numbers of protons and neutrons in the nucleus of an atom (p. 85)

matter
anything that has mass and takes up space (p. 10)

melting
the change of state in which a solid becomes a liquid by adding heat or changing pressure (p. 383)

melting point
the temperature and pressure at which a solid becomes a liquid (p. 383)

miscible
describes two or more liquids that can dissolve into each other in various proportions (p. 470)

mixture
a combination of two or more substances that are not chemically combined (p. 25)

molarity
a concentration unit of a solution expressed as moles of solute dissolved per liter of solution (p. 462)

molar mass
the mass in grams of 1 mol of a substance (p. 101, p. 230)

mole
the SI base unit used to measure the amount of a substance whose number of particles is the same as the number of atoms of carbon in exactly 12 g of carbon-12 (p. 101, p. 224)

molecular formula
a chemical formula that shows the number and kinds of atoms in a molecule, but not the arrangement of the atoms (p. 244)

molecular orbital
the region of high probability that is occupied by an individual electron as it travels with a wavelike motion in the three-dimensional space around one of two or more associated nuclei (p. 191)

molecule
the smallest unit of a substance that keeps all of the physical and chemical properties of that substance; it can consist of one atom or two or more atoms bonded together (p. 23)

monosaccharide
a simple sugar that is the basic subunit of a carbohydrate (p. 712)

N

neutral
describes an aqueous solution that contains equal concentrations of hydronium ions and hydroxide ions (p. 542)

neutralization reaction
the reaction of the ions that characterize acids (hydronium ions) and the ions that characterize bases (hydroxide ions) to form water molecules and a salt (p. 548)

neutron
a subatomic particle that has no charge and that is found in the nucleus of an atom (p. 82)

newton
the SI unit for force; the force that will increase the speed of a 1 kg mass by 1 m/s each second that the force is applied (abbreviation, N) (p. 419)

noble gas
an unreactive element of Group 18 of the periodic table; the nobles gases are helium, neon, argon, krypton, xenon, or radon (p. 127)

nonelectrolyte
a liquid or solid substance or mixture that does not allow an electric current (p. 479)

nonpolar covalent bond
a covalent bond in which the bonding electrons are equally attracted to both bonded atoms (p. 194)

nuclear fission
the splitting of the nucleus of a large atom into two or more fragments; releases additional neutrons and energy (p. 654)

nuclear fusion
the combination of the nuclei of small atoms to form a larger nucleus; releases energy (p. 656)

nuclear reaction
a reaction that affects the nucleus of an atom (p. 143)

nucleic acid
an organic compound, either RNA or DNA, whose molecules are made up of one or two chains of nucleotides and carry genetic information (p. 725)

nucleon
a proton or neutron (p. 642)

nucleus
in physical science, an atom's central region, which is made up of protons and neutrons (p. 81)

nuclide
an atom that is identified by the number of protons and neutrons in its nucleus (p. 642)

octet rule
a concept of chemical bonding theory that is based on the assumption that atoms tend to have either empty valence shells or full valence shells of eight electrons (p. 159)

orbital
a region in an atom where there is a high probability of finding electrons (p. 91)

order
in chemistry, a classification of chemical reactions that depends on the number of molecules that appear to enter into the reaction (p. 586)

oxidation
a reaction that removes one or more elec-

trons from a substance such that the substance's valence or oxidation state increases (p. 604)

oxidation number
the number of electrons that must be added to or removed from an atom in a combined state to convert the atom into the elemental form (p. 606)

oxidation-reduction reaction
any chemical change in which one species is oxidized (loses electrons) and another species is reduced (gains electrons); also called redox reaction (p. 605)

oxidizing agent
the substance that gains electrons in an oxidation-reduction reaction and that is reduced (p. 611)

partial pressure
the pressure of each gas in a mixture (p. 439)

pascal
the SI unit of pressure; equal to the force of 1 N exerted over an area of 1 m^2 (abbreviation, Pa) (p. 419)

Pauli exclusion principle
the principle that states that two particles of a certain class cannot be in exactly the same energy state (p. 96)

peptide bond
the chemical bond that forms between the carboxyl group of one amino acid and the amino group of another amino acid (p. 718)

percentage composition
the percentage by mass of each element in a compound (p. 241)

period
in chemistry, a horizontal row of elements in the periodic table (p. 122)

periodic law
the law that states that the repeating chemical and physical properties of elements change periodically with the atomic numbers of the elements (p. 119)

pH
a value that is used to express the acidity or alkalinity (basicity) of a system; each whole number on the scale indicates a tenfold change in acidity; a pH of 7 is neutral, a pH of less than 7 is acidic, and a pH of greater than 7 is basic (p. 542)

phase
in chemistry, a part of matter that is uniform (p. 399)

phase diagram
a graph of the relationship between the physical state of a substance and the temperature and pressure of the substance (p. 402)

photosynthesis
the process by which plants, algae, and some bacteria use sunlight, carbon dioxide, and water to produce carbohydrates and oxygen (p. 734)

physical change
a change of matter from one form to another without a change in chemical properties (p. 39)

physical property
a characteristic of a substance that does not involve a chemical change, such as density, color, or hardness (p. 15)

polar covalent bond
a covalent bond in which a pair of electrons shared by two atoms is held more closely by one atom (p. 194)

polyatomic ion
an ion made of two or more atoms (p. 178)

polymer
a large molecule that is formed by more than five monomers, or small units (p. 696)

polypeptide
a long chain of several amino acids (p. 718)

polysaccharide
one of the carbohydrates made up of long chains of simple sugars; polysaccharides include starch, cellulose, and glycogen (p. 712)

precision
the exactness of a measurement (p. 55)

pressure
the amount of force exerted per unit area of a surface (p. 419)

product
a substance that forms in a chemical reaction (p. 8)

protein
an organic compound that is made of one or more chains of amino acids and that is a principal component of all cells (p. 717)

proton
a subatomic particle that has a positive charge and that is found in the nucleus of an atom; the number of protons of the nucleus is the atomic number, which determines the identity of an element (p. 82)

pure substance
a sample of matter, either a single element or a single compound, that has definite chemical and physical properties (p. 22)

quantity
something that has magnitude, size, or amount (p. 12)

quantum number
a number that specifies the properties of electrons (p. 95)

radioactivity
the process by which an unstable nucleus emits one or more particles or energy in the form of electromagnetic radiation (p. 648)

rate-determining step
in a multistep chemical reaction, the step that has the lowest velocity, which determines the rate of the overall reaction (p. 589)

rate law
the expression that shows how the rate of formation of product depends on the concentration of all species other than the solvent that take part in a reaction (p. 586)

reactant
a substance or molecule that participates in a chemical reaction (p. 8)

reaction mechanism
the way in which a chemical reaction takes place; expressed in a series of chemical equations (p. 586)

reaction rate
the rate at which a chemical reaction takes place; measured by the rate of formation of the product or the rate of disappearance of the reactants (p. 578)

recombinant DNA
DNA molecules that are artificially created by combining DNA from different sources (p. 732)

reducing agent
a substance that has the potential to reduce another substance (p. 611)

reduction
a chemical change in which electrons are gained, either by the removal of oxygen, the addition of hydrogen, or the addition of electrons (p. 605)

resonance structure
in chemistry, any one of two or more possible configurations of the same compound that have identical geometry but different arrangements of electrons (p. 206)

respiration
in chemistry, the process by which cells produce energy from carbohydrates; atmospheric oxygen combines with glucose to form water and carbon dioxide (p. 736)

reversible reaction
a chemical reaction in which the products re-form the original reactants (p. 497)

salt
an ionic compound that forms when a metal atom or a positive radical replaces the hydrogen of an acid (p. 167)

saturated hydrocarbon
an organic compound formed only by carbon and hydrogen linked by single bonds (p. 688)

saturated solution
a solution that cannot dissolve any more solute under the given conditions (p. 474)

scientific method
a series of steps followed to solve problems, including collecting data, formulating a hypothesis, testing the hypothesis, and stating conclusions (p. 46)

self-ionization constant of water, K_w
the product of the concentrations of the two ions that are in equilibrium with water; $[H_3O^+][OH^-]$ (p. 540)

significant figure
a prescribed decimal place that determines the amount of rounding off to be done based on the precision of the measurement (p. 56)

single bond
a covalent bond in which two atoms share one pair of electrons (p. 200)

soap
a substance that is used as a cleaner and that dissolves in water (p. 484)

solubility
the ability of one substance to dissolve in another at a given temperature and pressure; expressed in terms of the amount of solute that will dissolve in a given amount of solvent to produce a saturated solution (p. 468)

solubility equilibrium
the physical state in which the opposing processes of dissolution and crystallization of a solute occur at equal rates (p. 476)

solubility product constant
the equilibrium constant for a solid that is in equilibrium with the solid's dissolved ions (p. 507)

solute
in a solution, the substance that dissolves in the solvent (p. 455)

solution
a homogeneous mixture of two or more substances uniformly dispersed throughout a single phase (p. 454)

solvent
in a solution, the substance in which the solute dissolves (p. 455)

specific heat
the quantity of heat required to raise a unit mass of homogeneous material 1 K or 1°C in a specified way given constant pressure and volume (p. 45)

spectator ions
ions that are present in a solution in which a reaction is taking place but that do not participate in the reaction (p. 286)

standard electrode potential
the potential developed by a metal or other material immersed in an electrolyte solution relative to the potential of the hydrogen electrode, which is set at zero (p. 622)

standard solution
a solution of known concentration (p. 550)

standard temperature and pressure
for a gas, the temperature of 0°C and the pressure 1.00 atm (p. 420)

states of matter
the physical forms of matter, which are solid, liquid, gas, and plasma (p. 6)

stoichiometry
the proportional relationships between two or more substances during a chemical reaction (p. 303)

strong acid
an acid that ionizes completely in a solvent (p. 532)

strong base
a base that ionizes completely in a solvent (p. 534)

strong force
the interaction that binds nucleons together in a nucleus (p. 643)

sublimation
the process in which a solid changes directly into a gas (The term is sometimes also used for the reverse process.) (p. 383)

substitution reaction
a reaction in which one or more atoms replace another atom or group of atoms in a molecule (p. 696)

superheavy element
an element whose atomic number is greater than 106 (p. 147)

supersaturated solution
a solution that holds more dissolved solute than is required to reach equilibrium at a given temperature (p. 475)

surface tension
the force that acts on the surface of a liquid and that tends to minimize the area of the surface (p. 380)

surfactant
a compound that concentrates at the boundary surface between two immiscible phases, solid-liquid, liquid-liquid, or liquid-gas (p. 484)

suspension
a mixture in which particles of a material are more or less evenly dispersed throughout a liquid or gas (p. 454)

synthesis reaction
a reaction in which two or more substances combine to form a new compound (p. 277)

T

temperature
a measure of how hot (or cold) something is; specifically, a measure of the average kinetic energy of the particles in an object (p. 43, p. 339)

thermodynamics
the branch of science concerned with the energy changes that accompany chemical and physical changes (p. 348)

titrant
a solution of known concentration that is used to titrate a solution of unknown concentration (p. 550)

titration
a method to determine the concentration of a substance in solution by adding a

solution of known volume and concentration until the reaction is completed, which is usually indicated by a change in color (p. 550)

transition range
the pH range over which a variation in a chemical indicator can be observed (p. 554)

transition metal
one of the metals that can use the inner shell before using the outer shell to bond (p. 129)

triple bond
a covalent bond in which two atoms share three pairs of electrons (p. 205)

triple point
the temperature and pressure conditions at which the solid, liquid, and gaseous phases of a substance coexist at equilibrium (p. 402)

unit
a quantity adopted as a standard of measurement (p. 12)

unit cell
the smallest portion of a crystal lattice that shows the three-dimensional pattern of the entire lattice (p. 175)

unsaturated hydrocarbon
a hydrocarbon that has available valence bonds, usually from double or triple bonds with carbon (p. 688)

unsaturated solution
a solution that contains less solute than a saturated solution does and that is able to dissolve additional solute (p. 474)

unshared pair
a nonbonding pair of electrons in the valence shell of an atom; also called lone pair (p. 200)

valence electron
an electron that is found in the outermost shell of an atom and that determines the atom's chemical properties (p. 119, p. 199)

vapor pressure
the partial pressure exerted by a vapor that is in equilibrium with its liquid state at a given temperature (p. 400)

voltage
the potential difference or electromotive force, measured in volts; it represents the amount of work that moving an electric charge between two points would take (p. 613)

volume
a measure of the size of a body or region in three-dimensional space (p. 10)

VSEPR theory
a theory that predicts some molecular shapes based on the idea that pairs of valence electrons surrounding an atom repel each other (p. 209)

weak acid
an acid that releases few hydrogen ions in aqueous solution (p. 532)

weak base
a base that releases few hydroxide ions in aqueous solution (p. 534)

weight
a measure of the gravitational force exerted on an object; its value can change with the location of the object in the universe (p. 10)

GLOSARIO

A

accuracy/exactitud
término que describe qué tanto se aproxima una medida al valor verdadero de la cantidad medida (p. 55)

acid-ionization constant/constante de ionización ácida, el término K_a
la constante de equilibrio para una reacción en la cual un ácido dona un protón al agua (p. 559)

actinide/actínido
cualquiera de los elementos de la serie de los actínidos, los cuales tienen números atómicos del 89 (actinio, Ac) al 103 (laurencio, Lr) (p. 130)

activated complex/complejo activado
una molécula que está en un estado inestable, intermedio entre los reactivos y los productos en una reacción química (p. 590)

activation energy/energía de activación
la cantidad mínima de energía que se requiere para iniciar una reacción química (p. 590)

activity series/serie de actividad
una serie de elementos que tienen propiedades similares y que están ordenados en orden descendiente respecto a su actividad química; algunos ejemplos de series de actividad incluyen a los metales y los halógenos (p. 280)

actual yield/rendimiento real
la cantidad medida del producto de una reacción (p. 316)

addition reaction/reacción de adición
una reacción en la que se añade un átomo o una molécula a una molécula insaturada (p. 696)

alkali metal/metal alcalino
uno de los elementos del Grupo 1 de la tabla periódica (litio, sodio, potasio, rubidio, cesio y francio) (p. 125)

alkaline-earth metal/metal alcalinotérreo
uno de los elementos del Grupo 2 de la tabla periódica (berilio, magnesio, calcio, estroncio, bario y radio) (p. 126)

alkane/alcano
un hidrocarburo formado por una cadena simple o ramificada de carbonos que únicamente contiene enlaces sencillos (p. 681)

alkene/alqueno
un hidrocarburo que contiene uno o más enlaces dobles (p. 681)

alkyne/alquino
un hidrocarburo que contiene uno o más enlaces triples (p. 681)

alloy/aleación
una mezcla sólida o líquida de dos o más metales (p. 130)

amino acid/aminoácido
cualquiera de las 20 distintas moléculas orgánicas que contienen un grupo carboxilo y un grupo amino y que se combinan para formar proteínas (p. 717)

amphoteric/anfotérico
término que describe una substancia, como el agua, que tiene propiedades tanto de ácido como de base (p. 538)

anion/anión
un ion que tiene carga negativa (p. 161)

anode/ánodo
el electrodo en cuya superficie ocurre la oxidación; los aniones migran hacia el ánodo y los electrones se alejan del sistema por el ánodo (p. 614)

aromatic hydrocarbon/hidrocarburo aromático
un hidrocarburo que tiene anillos de seis carbonos y que normalmente es muy reactivo (p. 682)

atom/átomo
la unidad más pequeña de un elemento que conserva las propiedades de ese elemento (p. 21)

atomic mass/masa atómica
la masa de un átomo, expresada en unidades de masa atómica (p. 100)

atomic number/número atómico
el número de protones en el núcleo de un átomo; el número atómico es el mismo para todos los átomos de un elemento (p. 84)

ATP/ATP
adenosín trifosfato; molécula orgánica que funciona como la fuente principal de energía para los procesos celulares; formada por una base nitrogenada, un azúcar y tres grupos fosfato (p. 737)

Aufbau principle/principio de Aufbau
el principio que establece que la estructura de cada elemento sucesivo se obtiene añadiendo un protón al núcleo del átomo y un electrón a un orbital de menor energía que se encuentre disponible (p. 245)

average atomic mass/masa atómica promedio
el promedio ponderado de las masas de todos los isótopos de un elemento que se encuentran en la naturaleza (p. 235)

Avogadro's law/ley de Avogadro
la ley que establece que volúmenes iguales de gases a la misma temperatura y presión contienen el mismo número de moléculas (p. 432)

Avogadro's number/número de Avogadro
6.02×10^{23}, el número de átomos o moléculas que hay en 1 mol (p. 101, p. 224)

B

beta particle/partícula beta
un electrón con carga, emitido durante ciertos tipos de desintegración radiactiva, como por ejemplo, durante la desintegración beta (p. 649)

boiling point/punto de ebullición
la temperatura y presión a la que un líquido y un gas están en equilibrio (p. 382)

bond energy/energía de enlace
la energía que se requiere para romper los enlaces de 1 mol de un compuesto químico (p. 192)

bond length/longitud de enlace
la distancia entre dos átomos que están enlazados en el punto en que su energía potencial es mínima; la distancia promedio entre los núcleos de dos átomos enlazados (p. 192)

bond radius/radio de enlace
la distancia mitad del centro al centro de dos como los átomos que se pegan juntos (p. 135)

Boyle's law/ley de Boyle
la ley que establece que para una cantidad fija de gas a una temperatura constante, el volumen del gas aumenta a medida que su presión disminuye y el volumen del gas disminuye a medida que su presión aumenta (p. 424)

Brønsted-Lowry acid/ácido de Brønsted-Lowry
una substancia que le dona un protón a otra substancia (p. 535)

Brønsted-Lowry base/base de Brønsted-Lowry
una substancia que acepta un protón (p. 536)

buffer/búfer
una solución que contiene un ácido débil y su base conjugada y que neutraliza pequeñas cantidades de ácidos y bases que se le añaden (p. 561)

C

calorimeter/calorímetro
un aparato que se usa para medir la cantidad de calor absorbido o liberado en un cambio físico o químico (p. 351)

calorimetry/calorimetría
la medida de las constantes relacionadas con el calor, tales como el calor específico o el calor latente (p. 351)

carbohydrate/carbohidrato
cualquier compuesto orgánico que está hecho de carbono, hidrógeno y oxígeno y que proporciona nutrientes a las células de los seres vivos (p. 712)

catalysis/catálisis
la aceleración de una reacción química por un catalizador (p. 593)

catalyst/catalizador
una substancia que cambia la tasa de una reacción química sin ser consumida ni cambiar significativamente (p. 593)

cathode/cátodo
el electrodo en cuya superficie ocurre la reducción (p. 613)

cation/catión
un ion que tiene carga positiva (p. 161)

chain reaction/reacción en cadena
una reacción en la que un cambio en una sola molécula hace que muchas moléculas cambien, hasta que se forma un compuesto estable (p. 654)

Charles's law/ley de Charles
la ley que establece que para una cantidad fija de gas a una presión constante, el volumen del gas aumenta a medida que su temperatura aumenta y el volumen del gas disminuye a medida que su temperatura disminuye (p. 426)

chemical/substancia química
cualquier substancia que tiene una composición definida (p. 4)

chemical change/cambio químico
un cambio que ocurre cuando una o más substancias se transforman en substancias totalmente nuevas con propiedades diferentes (p. 39)

chemical equation/ecuación química
una representación de una reacción química que usa símbolos para mostrar la relación entre los reactivos y los productos (p. 263)

chemical equilibrium/equilibrio químico
un estado de equilibrio en el que la tasa de la reacción directa es igual a la tasa de la reacción inversa y las concentraciones de los productos y reactivos no sufren cambios (p. 497)

chemical kinetics/cinética química
el área de la química que se ocupa del estudio de las tasas de reacción y de los mecanismos de reacción (p. 576)

chemical property/propiedad química
una propiedad de la materia que describe la capacidad de una substancia de participar en reacciones químicas (p. 18)

chemical reaction/reacción química
el proceso por medio del cual una o más substancias cambian para producir una o más substancias distintas (p. 5, p. 260)

clone/clon
un organismo producido por reproducción asexual que es genéticamente idéntico a su progenitor; clonar significa hacer un duplicado genético (p. 731)

coefficient/coeficiente
un número entero pequeño que aparece como un factor frente a una fórmula en una ecuación química (p. 268)

colligative property/propiedad coligativa
una propiedad que se determina por el número de partículas presentes en un sistema, pero que es independiente de las propiedades de las partículas mismas (p. 482)

colloid/coloide
una mezcla formada por partículas diminutas que son de tamaño intermedio entre las partículas de las soluciones y las de las suspensiones y que se encuentran suspendidas en un líquido, sólido o gas (p. 456)

combustion reaction/reacción de combustión
la reacción de oxidación de un compuesto orgánico, durante la cual se libera calor (p. 276)

common-ion effect/efecto del ion común
el fenómeno en el que la adición de un ion común a dos solutos produce precipitación o reduce la ionización (p. 517)

compound/compuesto
una substancia formada por átomos de dos o más elementos diferentes unidos por enlaces químicos (p. 24)

concentration/concentración
la cantidad de una cierta substancia en una cantidad determinada de mezcla, solución o mena (p. 460)

condensation/condensación
el cambio de estado de gas a líquido (p. 382)

condensation reaction/reacción de condensación
una reacción química en la que dos o más moléculas se combinan para producir agua u otra molécula simple (p. 699, p. 715)

conductivity/conductividad
la capacidad de conducir una corriente eléctrica (p. 478)

conjugate acid/ácido conjugado
un ácido que se forma cuando una base gana un protón (p. 537)

conjugate base/base conjugada
una base que se forma cuando un ácido pierde un protón (p. 537)

conversion factor/factor de conversió
una razón que se deriva de la igualdad entre dos unidades diferentes y que se puede usar para convertir una unidad en otra (p. 13)

corrosion/corrosión
la destrucción gradual de un metal o de una aleación como resultado de procesos químicos tales como la oxidación o la acción de un agente químico (p. 620)

covalent bond/enlace covalente
un enlace formado cuando los átomos comparten uno más pares de electrones (p. 191)

critical mass/masa crítica
la cantidad mínima de masa de un isótopo fisionable que proporciona el número de neutrones que se requieren para sostener una reacción en cadena (p. 654)

critical point/punto crítico
la temperatura y presión a la que los estados líquido y gaseoso de una substancia se vuelven idénticos para formar una fase (p. 402)

crystal lattice/red cristalina
el patrón regular en el que un cristal está ordenado (p. 174)

Dalton's Law of Partial Pressures/ley de Dalton de las presiones parciales
la ley que establece que la presión total de una mezcla de gases es igual a la suma de las presiones parciales de los gases componentes (p. 439)

decomposition reaction/reacción de descomposición
una reacción en la que un solo compuesto se descompone para formar dos o más substancias más simples (p. 278)

density/densidad
la relación entre la masa de una substancia y su volumen; comúnmente se expresa en gramos por centímetro cúbico para los sólidos y líquidos, y como gramos por litro para los gases (p. 16)

denature/desnaturalice
para hacer una proteína perder sus estructuras terciarias y cuaternarios (p. 723)

detergent/detergente
un limpiador no jabonoso, soluble en agua, que emulsiona la suciedad y el aceite (p. 484)

diffusion/difusión
el movimiento de partículas de regiones de mayor densidad a regiones de menor densidad (p. 436)

dipole/dipolo
una molécula o parte de una molécula que contiene regiones cargadas tanto positiva como negativamente (p. 195)

dipole-dipole forces/fuerzas dipolo-dipolo
interacciones entre moléculas polares (p. 386)

disaccharide/disacárido
un azúcar formada a partir de dos monosacáridos (p. 712)

dissociation/disociación
la separación de una molécula en moléculas más simples, átomos, radicales o iones (p. 472)

DNA/ADN
ácido desoxirribonucleico, el material que contiene la información que determina las características que se heredan (p. 726)

DNA fingerprint/huella de ADN
el patrón de bandas que se obtiene cuando los fragmentos de ADN de un individuo se separan (p. 730)

double bond/doble enlace
un enlace covalente en el que dos átomos comparten dos pares de electrones (p. 204)

double-displacement reaction/reacción de doble desplazamiento
una reacción en la que un gas, un precipitado sólido o un compuesto molecular se forma a partir del intercambio aparente de iones entre dos compuestos (p. 283)

GLOSARIO

E

effusion/efusión
el paso de un gas bajo presión a través de una abertura diminuta (p. 437)

electrochemical cell/celda electroquímica
un sistema que contiene dos electrodos separados por una fase electrolítica (p. 615)

electrochemistry/electroquímica
la rama de la química que se ocupa del estudio de la relación entre las fuerzas eléctricas y las reacciones químicas (p. 614)

electrode/electrodo
un conductor que se usa para establecer contacto eléctrico con una parte no metálica de un circuito, tal como un electrolito (p. 615)

electrolysis/electrólisis
el proceso por medio del cual se utiliza una corriente eléctrica para producir una reacción química, como por ejemplo, la descomposición del agua (p. 629)

electrolyte/electrolito
una substancia que se disuelve en agua y crea una solución que conduce la corriente eléctrica (p. 478)

electrolytic cell/celda electrolítica
un aparato electroquímico en el que se da lugar la electrólisis cuando hay una corriente eléctrica en el aparato (p. 629)

electromagnetic spectrum/espectro electromagnético
todas las frecuencias o longitudes de onda de la radiación electromagnética (p. 92)

electron/electrón
una partícula subatómica que tiene carga negativa (p. 80)

electron shielding/blindaje de los electrónes
la reducción de la fuerza atractiva entre un núcleo positivamente cargado y sus electrones exteriores debido a la cancelación de algo de la carga positiva por las cargas negativas de los electrones internos (p. 133)

electron configuration/configuración electrónica
el ordenamiento de los electrones en un átomo (p. 96)

electronegativity/electronegatividad
una medida de la capacidad de un átomo de un compuesto químico de atraer electrones (p. 137)

electroplating/electrochapado
el proceso de recubrir o aplicar una capa de un metal a un objeto (p. 630)

element/elemento
una substancia que no se puede separar o descomponer en substancias más simples por medio de métodos químicos; todos los átomos de un elemento tienen el mismo número atómico (p. 22)

elimination reaction/reacción de eliminación
una reacción en la que se remueve una molécula simple, como el agua o el amoníaco, y se produce un nuevo compuesto (p. 699)

empirical formula/fórmula empírica
la composición de un compuesto en función del número relativo y el tipo de átomos que hay en la proporción más simple (p. 242)

emulsion/emulsión
cualquier mezcla de dos o más líquidos inmiscibles en la que un líquido se encuentra disperso en el otro (p. 484)

endothermic/endotérmico
término que describe un proceso en que se absorbe calor del ambiente (p. 40)

end point/punto de equivalencia
el punto en una titulación en el que ocurre un cambio marcado de color (p. 554)

energy/energía
la capacidad de realizar un trabajo (p. 38)

enthalpy/entalpía
la suma de la energía interna de un sistema más el producto del volumen del sistema multiplicado por la presión que el sistema ejerce en su ambiente (p. 340)

entropy/entropía
una medida del grado de aleatoriedad o desorden de un sistema (p. 358)

enzyme/enzima
un tipo de proteína que acelera las reacciones metabólicas en las plantas y animales, sin ser modificada permanentemente ni ser destruida (p. 595, p. 722)

equilibrium/equilibrio
en química, el estado en el que un proceso químico y el proceso químico inverso ocurren a la misma tasa, de modo que las concentraciones de los reactivos y los productos no cambian (p. 400)

equilibrium constant/constante de equilibrio
un número que relaciona las concentraciones de los materiales de inicio y los productos de una reacción química reversible a una temperatura dada (p. 503)

evaporation/evaporación
el cambio de una substancia de líquido a gas (p. 39, p. 382)

excess reactant/reactivo en exceso
la substancia que no se usa por completo en una reacción (p. 313)

excited state/estado de excitación
un estado en el que un átomo tiene más energía que en su estado fundamental (p. 94)

exothermic/exotérmico
término que describe un proceso en el que un sistema libera calor al ambiente (p. 40)

F

freezing/congelamiento
el cambio de estado de líquido a sólido al eliminar calor del líquido (p. 383)

freezing point/punto de congelación
la temperatura a la que un sólido y un líquido están en equilibrio a 1 atm de presión; la temperatura a la que una substancia en estado líquido se congela (p. 383)

functional group/grupo funcional
la porción de una molécula que está involucrada en una reacción química y que determina las propiedades de muchos compuestos orgánicos (p. 683)

G

gamma ray/rayo gamma
el fotón de alta energía emitido por un núcleo durante la fisión y la desintegración radiactiva (p. 649)

Gay-Lussac's law/ley de Gay-Lussac
la ley que establece que la presión por un gas a volumen constante es directamente proporcional a la temperatura absoluta (p. 430)

Gay-Lussac's law of combining volumes of gases/ley de combinación de los volúmenes de los gases de Gay-Lussac
la ley que establece que los volúmenes de los gases que participan en un cambio químico se pueden representar por razones de números pequeños enteros (p. 439)

gene/gene
un segmento de ADN ubicado en un cromosoma, que codifica para un carácter hereditario específico (p. 728)

Gibbs energy/energia Gibbs
la energía de un sistema disponible para realizar un trabajo (p. 362)

Graham's law of diffusion/ley de efusión de Graham
la ley que establece que la tasa de difusión de un gas es inversamente proporcional a la raíz cuadrada de su densidad (p. 437)

ground state/estado fundamental
el estado de energía más bajo de un sistema cuantificado (p. 94)

group/grupo
una columna vertical de elementos de la tabla periódica; los elementos de un grupo comparten propiedades químicas (p. 119)

H

half-life/vida media
el tiempo que tarda la mitad de una muestra de una substancia radiactiva en desintegrarse por desintegración radiactiva o por procesos naturales (p. 658)

half-reaction/media reacción
la parte de una reacción que sólo involucra oxidación o reducción (p. 608)

halogen/halógeno
uno de los elementos del Grupo 17 (flúor, cloro, bromo, yodo y ástato); se combinan con la mayoría de los metales para formar sales (p. 126)

heat/calor
la transferencia de energía entre objetos que están a temperaturas diferentes; la energía siempre se transfiere de los objetos que están a la temperatura más alta a los objetos que están a una temperatura más baja, hasta que se llega a un equilibrio térmico (p. 41, p. 338)

Henry's law/ley de Henry
la ley que establece que a una temperatura constante, la solubilidad de un gas en un líquido es directamente proporcional a la presión parcial de un gas en la superficie del líquido (p. 477)

Hess's law/ley de Hess
la ley que establece que la cantidad de calor liberada o absorbida en una reacción química no depende del número de pasos que tenga la reacción (p. 353)

heterogeneous/heterogéneo
compuesto de componentes que no son iguales (p. 26)

homogeneous/homogéneo
término que describe a algo que tiene una estructura o composición global uniforme (p. 26)

Hund's rule/regla de Hund
la regla que establece que para un átomo en estado fundamental, el número de electrones no apareados es el máximo posible y que estos electrones no apareados tienen el mismo espín (p. 98)

hydration/hidratación
la fuerte afinidad de las moléculas del agua a partículas de substancias disueltas o suspendidas que causan disociación electrolítica (p. 472)

hydrocarbon/hidrocarburo
un compuesto orgánico compuesto únicamente por carbono e hidrogeno (p. 680)

hydrogen bond/enlace de hidrógeno
la fuerza intermolecular producida por un átomo de hidrógeno que está unido a un átomo muy electronegativo de una molécula y que experimenta atracción a dos electrones no compartidos de otra molécula (p. 387)

hydrolysis/hidrólisis
una reacción química entre el agua y otras substancias para formar dos o más substancias nuevas; una reacción entre el agua y una sal para crear un ácido o una base (p. 716)

hydronium ion/ion hidronio
un ion formado por un protón combinado con una molécula de agua; H_3O^+ (p. 480)

hypothesis/hipótesis
una teoría o explicación basada en observaciones y que se puede probar (p. 50)

ideal gas/gas ideal
un gas imaginario con partículas que son infinitamente pequeñas y que no interactúan unas con otras (p. 433)

ideal gas law/ley de los gases ideales
la ley que establece la relación matemática entre la presión (P), volumen (V), temperatura (T), la constante de los gases (R) y el número de moles de un gas (n); $PV = nRT$ (p. 434)

immiscible/inmiscible
término que describe dos o más líquidos que no se mezclan uno con otro (p. 470)

indicator/indicador
un compuesto que puede cambiar de color de forma reversible dependiendo del pH de la solución o de otro cambio químico (p. 546)

intermediate/intermediario
una substancia que se forma en un estado medio de una reacción química y que se considera un paso importante entre la substancia original y el producto final (p. 589)

intermolecular forces/fuerzas intermoleculares
las fuerzas de atracción entre moléculas (p. 386)

ion/ion
un átomo, radical o molécula que ha ganado o perdido uno o más electrones y que tiene una carga negativa o positiva (p. 161)

isomer/isómero
uno de dos o más compuestos que tienen la misma composición química pero diferentes estructuras (p. 686)

isotope/isótopo
un átomo que tiene el mismo número de protones (número atómico) que otros átomos del mismo elemento, pero que tiene un número diferente de neutrones (masa atómica) (p. 88)

kinetic energy/energía cinética
la energía de un objeto debido al movimiento del objeto (p. 42)

kinetic-molecular theory/teoría cinética molecular
una teoría que explica que el comportamiento de los sistemas físicos depende de las acciones combinadas de las moléculas que constituyen el sistema (p. 421)

lanthanide/lantánido
la serie de elementos de tierras raras, cuyos números atómicos van del 58 (cerio) al 71 (lutecio) (p. 130)

lattice energy/energía de la red cristalina
la energía asociada con la construcción de una red cristalina en relación con la energía de todos los átomos que la constituyen cuando éstos están separados por distancias infinitas (p. 168)

law/ley
un resumen de muchos resultados y observaciones experimentales; una ley dice cómo funcionan las cosas (p. 52)

law of conservation of energy/ley de la conservación de la energía
la ley que establece que la energía ni se crea ni se destruye, sólo se transforma de una forma a otra (p. 40)

law of conservation of mass/ley de la conservación de la masa
la ley que establece que la masa no se crea ni se destruye por cambios químicos o físicos comunes (p. 52, p. 76)

law of definite proportions/ley de las proporciones definidas
la ley que establece que un compuesto químico siempre contiene los mismos elementos en exactamente las mismas proporciones de peso o masa (p. 75)

law of multiple proportions/ley de las proporciones múltiples
la ley que establece que cuando dos elementos se combinan para formar dos o más compuestos, la masa de un elemento que se combina con una cantidad determinada de masa de otro elemento es en la proporción de número enteros pequeños (p. 77)

Le Chatelier's principle/principio de Le Chatelier
el principio que establece que un sistema en equilibrio se opondrá a un cambio de modo tal que ayude a eliminar el cambio (p. 512)

Lewis structure/estructura de Lewis
una fórmula estructural en la que los electrones se representan por medio de puntos; pares de puntos o líneas entre dos símbolos atómicos representan pares en los enlaces covalentes (p. 199)

limiting reactant/reactivo limitante
la substancia que controla la cantidad de producto que se puede formar en una reacción química (p. 313)

London dispersion force/fuerza de dispersión de London
la atracción intermolecular que se produce como resultado de la distribución desigual de los electrones y la creación de dipolos temporales (p. 390)

main-group element/elemento de grupo principal
un elemento que está en el bloque *s-* o *p-* de la tabla periódica (p. 124)

mass/masa
una medida de la cantidad de materia que tiene un objeto; una propiedad fundamental de un objeto que no está afectada por las fuerzas que actúan sobre el objeto, como por ejemplo, la fuerza gravitacional (p. 10)

mass defect/defecto de masa
la diferencia entre la masa de un átomo y la suma de la masa de los protones, neutrones y electrones del átomo (p. 644)

mass number/número de masa
la suma de los números de protones y neutrones que hay en el núcleo de un átomo (p. 85)

matter/materia
cualquier cosa que tiene masa y ocupa un lugar en el espacio (p. 10)

melting/fusión
el cambio de estado en el que un sólido se convierte en líquido al añadir calor o al cambiar la presión (p. 383)

melting point/punto de fusión
la temperatura y presión a la cual un sólido se convierte en líquido (p. 383)

miscible/miscible
término que describe a dos o más líquidos que son capaces de disolverse uno en el otro en varias proporciones (p. 470)

mixture/mezcla
una combinación de dos o más substancias que no están combinadas químicamente (p. 25)

molarity/molaridad
una unidad de concentración de una solución, expresada en moles de soluto disuelto por litro de solución (p. 462)

molar mass/masa molar
la masa en gramos de 1 mol de una substancia (p. 101, p. 230)

mole/mol
la unidad fundamental del sistema internacional de unidades que se usa para medir la cantidad de una substancia cuyo número de partículas es el mismo que el número de átomos de carbono en exactamente 12 g de carbono-12 (p. 101, p. 224)

molecular formula/fórmula molecular
una fórmula química que muestra el número y los tipos de átomos que hay en una molécula, pero que no muestra cómo están distribuidos (p. 244)

molecular orbital/orbitál molecular
una región entre dos núcleos donde hay grande probabilidad de tener un electron que mueve como una onda (p. 191)

molecule/molécula
la unidad más pequeña de una substancia que conserva todas las propiedades físicas y químicas de esa substancia; puede estar formada por un átomo o por dos o más átomos enlazados uno con el otro (p. 23)

monosaccharide/monosacárido
un azúcar simple que es una subunidad fundamental de los carbohidratos (p. 712)

neutral/neutro
describe una solución acuosa que contenga concentraciones iguales de los iones del hydronium y de los iones del hidróxido (p. 542)

neutralization reaction/reacción de neutralización
la reacción de los iones que caracterizan a los ácidos (iones hidronio) y de los iones que caracterizan a las bases (iones hidróxido) para formar moléculas de agua y una sal (p. 548)

neutron/neutrón
una partícula subatómica que no tiene carga y que se encuentra en el núcleo de un átomo (p. 82)

newton/newton
la unidad de fuerza del sistema internacional de unidades; la fuerza que aumentará la rapidez de un kg de masa en 1 m/s cada segundo que se aplique la fuerza (abreviatura: N) (p. 419)

noble gas/gas noble
un elemento no reactivo del Grupo 18 de la tabla periódica; los gases nobles son: helio, neón, argón, criptón, xenón o radón (p. 127)

nonelectrolyte/no-electrolito
una substancia o una mezcla líquida o sólida que no permite el flujo de una corriente eléctrica (p. 479)

nonpolar covalent bond/enlace covalente no polar
un enlace covalente en el que los electrones de enlace tienen la misma atracción por los dos átomos enlazados (p. 194)

nuclear fission/fisión nuclear
la partición del núcleo de un átomo grande en dos o más fragmentos; libera neutrones y energía adicionales (p. 654)

nuclear fusion/fusión nuclear
combinación de los núcleos de átomos pequeños para formar un núcleo más grande; libera energía (p. 656)

nuclear reaction/reacción nuclear
una reacción que afecta el núcleo de un átomo (p. 143)

nucleic acid/ácido nucleico
un compuesto orgánico, ya sea ARN o ADN, cuyas moléculas están formadas por una o más cadenas de nucleótidos y que contiene información genética (p. 725)

nucleon/nucleón
un protón o neutrón (p. 642)

nucleus/núcleo
en ciencias físicas, la región central de un átomo, la cual está constituida por protones y neutrones (p. 81)

nuclide/nucleido
un átomo que se identifica por el número de protones y neutrones que hay en su núcleo (p. 642)

octet rule/regla del octeto
un concepto de la teoría de formación de enlaces químicos que se basa en la suposición de que los átomos tienden a tener orbitales de valencia vacíos u orbitales de valencia llenos de ocho electrones (p. 159)

orbital/orbital
una región en un átomo donde hay una alta probabilidad de encontrar electrones (p. 91)

order/orden
en la química, una clasificación de reacciones químicas que depende del número de las moléculas que aparecen entrar en la reacción (p. 586)

oxidation/oxidación
una reacción en la que uno o más electrones son removidos de una substancia, aumentado su valencia o estado de oxidación (p. 604)

oxidation number/número de oxidación
el número de electrones que se deben añadir o remover de un átomo en estado de combinación para convertirlo a su forma elemental (p. 606)

oxidation-reduction reaction/reacción de óxido-reducción
cualquier cambio químico en el que una especie se oxida (pierde electrones) y otra especie se reduce (gana electrones); también se denomina reacción redox (p. 605)

oxidizing agent/agente oxidante
la substancia que gana electrones en una reacción de óxido-reducción y que es reducida (p. 611)

partial pressure/presión parcial
la presión de cada gas en una mezcla (p. 439)

pascal/pascal
la unidad de presión del sistema internacional de unidades; es igual a la fuerza de 1 N ejercida sobre un área de 1 m^2 (abreviatura: Pa) (p. 419)

Pauli exclusion principle/principio de exclusión de Pauli
el principio que establece que dos partículas de una cierta clase no pueden estar en exactamente el mismo estado de energía (p. 96)

peptide bond/enlace peptídico
el enlace químico que se forma entre el grupo carboxilo de un aminoácido y el grupo amino de otro aminoácido (p. 718)

percentage composition/composición porcentual
el porcentaje en masa de cada elemento que forma un compuesto (p. 241)

period/período
en química, una hilera horizontal de elementos en la tabla periódica (p. 122)

periodic law/ley periódica
la ley que establece que las propiedades químicas y físicas repetitivas de un elemento cambian periódicamente en función del número atómico de los elementos (p. 119)

pH/pH
un valor que expresa la acidez o la alcalinidad (basicidad) de un sistema;

cada número entero de la escala indica un cambio de 10 veces en la acidez; un pH de 7 es neutro, un pH de menos de 7 es ácido y un pH de más de 7 es básico (p. 542)

phase/fase
en química, una parte de la materia que es uniforme (p. 399)

phase diagram/diagrama de fases
una gráfica de la relación entre el estado físico de una substancia y la temperatura y presión de la substancia (p. 402)

photosynthesis/fotosíntesis
el proceso por medio del cual las plantas, algas y algunas bacterias utilizan la luz solar, dióxido de carbono y agua para producir carbohidratos y oxígeno (p. 734)

physical change/cambio físico
un cambio de materia de una forma a otra sin que ocurra un cambio en sus propiedades químicas (p. 39)

physical property/propiedad física
una característica de una substancia que no implica un cambio químico, tal como la densidad, el color o la dureza (p. 15)

polar covalent bond/enlace covalente polar
un enlace en el que un par de electrones que está siendo compartido por dos átomos se mantiene más unido a uno de los átomos (p. 194)

polyatomic ion/ion poliatómico
un ion formado por dos o más átomos (p. 178)

polymer/polímero
una molécula grande que está formada por más de cinco monómeros, o unidades pequeñas (p. 698)

polypeptide/polipéptido
una cadena larga de varios aminoácidos (p. 718)

polysaccharide/polisacárido
uno de los carbohidratos formados por cadenas largas de azúcares simples; algunos ejemplos de polisacáridos incluyen al almidón, celulosa y glucógeno (p. 712)

precision/precisión
la exactitud de una medición (p. 55)

pressure/presión
la cantidad de fuerza ejercida en una superficie por unidad de área (p. 419)

product/producto
una substancia que se forma en una reacción química (p. 8)

protein/proteína
un compuesto orgánico que está hecho de una o más cadenas de aminoácidos y que es el principal componente de todas las células (p. 717)

proton/protón
una partícula subatómica que tiene una carga positiva y que se encuentra en el núcleo de un átomo; el número de protones que hay en el núcleo es el número atómico, y éste determina la identidad del elemento (p. 82)

pure substance/substancia pura
una muestra de materia, ya sea un solo elemento o un solo compuesto, que tiene propiedades químicas y físicas definidas (p. 22)

Q

quantity/cantidad
algo que tiene magnitud o tamaño (p. 12)

quantum number/número cuántico
un número que especifica las propiedades de los electrones (p. 95)

R

radioactivity/radiactividad
el proceso por medio del cual un núcleo inestable emite una o más partículas o energía en forma de radiación electro-magnética (p. 648)

rate-determining step/paso determinante de la tasa
en una reacción química de varios pasos, el paso que tiene la velocidad más baja, el cual determina la tasa global de la reacción (p. 589)

rate law/ley de la tasa
la expresión que muestra la manera en que la tasa de formación de producto depende de la concentración de todas las especies que participan en una reacción, excepto del solvente (p. 586)

reactant/reactivo
una substancia o molécula que participa en una reacción química (p. 8)

reaction mechanism/mecanismo de reacción
la manera en la que ocurre una reacción química; se expresa por medio de una serie de ecuaciones químicas (p. 586)

reaction rate/tasa de reacción
la tasa a la que ocurre una reacción química; se mide por la tasa de formación del producto o por la tasa de desaparición de los reactivos (p. 578)

recombinant DNA/ADN recombinante
moléculas de ADN que son creadas artificialmente al combinar ADN de diferentes fuentes (p. 732)

reducing agent/agente reductor
una substancia que tiene el potencial de reducir otra substancia (p. 611)

reduction/reducción
un cambio químico en el que se ganan electrones, ya sea por la remoción de oxígeno, la adición de hidrógeno o la adición de electrones (p. 605)

resonance structure/estructura de resonancia
en la química, una de dos o más configuraciones posibles del mismo compuesto que tienen geometría idéntica pero diversos arreglos de electrones (p. 206)

respiration/respiración
en química, el proceso por medio del cual las células producen energía a partir de los carbohidratos; el oxígeno atmosférico se combina con la glucosa para formar agua y dióxido de carbono (p. 736)

reversible reaction/reacción inversa
una reacción química en la que los productos vuelven a formar los reactivos originales (p. 497)

S

salt/sal
un compuesto iónico que se forma cuando el átomo de un metal o un radical positivo reemplaza el hidrógeno de un ácido (p. 167)

saturated hydrocarbon/hidrocarburo saturado
un compuesto orgánico formado sólo por carbono e hidrógeno unidos por enlaces simples (p. 688)

saturated solution/solución saturada
una solución que no puede disolver más soluto bajo las condiciones dadas (p. 474)

scientific method/método científico
una serie de pasos que se siguen para solucionar problemas, los cuales incluyen recopilar información, formular una hipótesis, comprobar la hipótesis y sacar conclusiones (p. 46)

self-ionization constant of water, K_w constante de la auto-ionización de agua
el producto de las concentraciones de los dos iones que están en equilibrio con agua (p. 540)

significant figure/cifra significativa
un lugar decimal prescrito que determina la cantidad de redondeo que se hará con base en la precisión de la medición (p. 56)

single bond/enlace simple
un enlace covalente en el que dos átomos comparten un par de electrones (p. 200)

soap/jabón
una sustancia que se usa como limpiador y que se disuelve en el agua (p. 484)

solubility/solubilidad
la capacidad de una sustancia de disolverse en otra a una temperatura y presión dadas; se expresa en términos de la cantidad de soluto que se disolverá en una cantidad determinada de solvente para producir una solución saturada (p. 468)

solubility equilibrium/equilibrio de solubilidad
el estado físico en el que los procesos opuestos de disolución y cristalización de un soluto ocurren a la misma tasa (p. 476)

solubility product constant/constante del producto de solubilidad
la constante de equilibrio de un sólido que está en equilibrio con los iones disueltos del sólido (p. 507)

solute/soluto
en una solución, la sustancia que se disuelve en el solvente (p. 455)

solution/solución
una mezcla homogénea de dos o más sustancias dispersas de manera uniforme en una sola fase (p. 454)

solvent/solvente
en una solución, la sustancia en la que se disuelve el soluto (p. 455)

specific heat/calor específico
la cantidad de calor que se requiere para aumentar una unidad de masa de un material homogéneo 1 K ó 1°C de una manera especificada, dados un volumen y una presión constantes (p. 45)

spectator ions/iones espectadores
iones que están presentes en una solución en la que está ocurriendo una reacción, pero que no participan en la reacción (p. 286)

standard electrode potential/potencial estándar del electrodo
el potencial que desarrolla un metal u otro material que se encuentre sumergido en una solución de electrolitos respecto al potencial del electrodo de hidrógeno, al cual se le da un valor de cero (p. 622)

standard solution/solución estándar
una solución de concentración conocida (p. 550)

standard temperature and pressure/ temperatura y presión estándar
para un gas, la temperatura de 0°C y la presión de 1.00 atm (p. 420)

states of matter/estados de la materia
las formas físicas de la materia, que son sólida, líquida, gaseosa y plasma (p. 6)

stoichiometry/estequiometría
las relaciones proporcionales entre dos o más substancias durante una reacción química (p. 303)

strong acid/ácido fuerte
un ácido que se ioniza completamente en un solvente (p. 532)

strong base/base fuerte
un base que se ioniza completamente en un solvente (p. 534)

strong force/fuerza fuerte
la interacción que mantiene unidos a los nucleones en un núcleo (p. 643)

sublimation/sublimación
el proceso por medio del cual un sólido se transforma directamente en un gas o un gas se transforma directamente en un sólido (p. 383)

substitution reaction/reacción de sustitución
una reacción en la cual uno o más átomos reemplazan otro átomo o grupo de átomos en una molécula (p. 696)

superheavy element/elemento superheavy
un elemento que número atómico es mayor de 106 (p. 147)

supersaturated solution/solución sobresaturada
una solución que contiene más soluto disuelto que el que se requiere para llegar al equilibro a una temperatura dada (p. 475)

surface tension/tensión superficial
la fuerza de atracción entre las moléculas que están debajo de la superficie de un líquido, la cual crea una fuerza hacia adentro que tiende a evitar que el líquido fluya (p. 380)

surfactant/surfactant
un compuesto que se concentra en la superficie del límite entre dos fases, solid-liquid, líquido-líquidos, o líquido-gases inmiscibles (p. 484)

suspension/suspensión
una mezcla en la que las partículas de un material se encuentran dispersas de manera más o menos uniforme a través de un líquido o de un gas (p. 454)

synthesis reaction/reacción de síntesis
una reacción en la que dos o más sustancias se combinan para formar un compuesto nuevo (p. 277)

temperature/temperatura
una medida de qué tan caliente (o frío) está algo; específicamente, una medida de la energía cinética promedio de las partículas de un objeto (p. 43, p. 339)

thermodynamics/termodinámica
la ramificación de la ciencia referida a los cambios de la energía que acompañan cambios del producto químico y de la comprobación (p. 348)

titrant/titrant
una solución de la concentración sabida que se utiliza para titular una solución de la concentración desconocida (p. 550)

titration/titulación
un método para determinar la concentración de una sustancia en una solución al añadir una solución de volumen y concentración conocidos hasta que se completa la reacción, lo cual normalmente es indicado por un cambio de color (p. 550)

transition range/intervalo de transición
el rango de pH en el cual se puede observar una variación en un indicador químico (p. 554)

transition metal/metal de transición
uno de los metales que tienen la capacidad de usar su orbital interno antes de usar su orbital externo para formar un enlace (p. 129)

triple bond/enlace triple
un enlace covalente en el que dos átomos comparten tres pares de electrones (p. 205)

triple point/punto triple
las condiciones de temperatura y presión en las que las fases sólida, líquida y gaseosa de una sustancia coexisten en equilibrio (p. 402)

unit/unidad
una cantidad adoptada como un estándar de medición (p. 12)

unit cell/celda unitaria
la porción más pequeña de una red cristalina, la cual muestra el patrón tridimensional de la red completa (p. 175)

unsaturated hydrocarbon/hidrocarburo no saturado
un hidrocarburo que tiene enlaces de valencia disponibles, normalmente de enlaces dobles o triples con carbono (p. 688)

unsaturated solution/solución no saturada
una solución que contiene menos soluto que una solución saturada, y que tiene la capacidad de disolver más soluto (p. 474)

unshared pair/par compartido
un par de electrones que no están enlazados en el orbital de valencia de un átomo; también se llama par solitario (p. 200)

valence electron/electrón de valencia
un electrón que se encuentra en el orbital más externo de un átomo y que determina las propiedades químicas del átomo (p. 119, p. 199)

vapor pressure/presión de vapor
la presión parcial ejercida por el vapor, la cual está en equilibrio con su estado líquido a una temperatura dada (p. 400)

voltage/voltaje
la diferencia de potencial o fuerza electromotriz, medida en voltios; representa la cantidad de trabajo que tomaría mover una carga eléctrica entre dos puntos (p. 613)

volume/volumen
una medida del tamaño de un cuerpo o región en un espacio de tres dimensiones (p. 10)

VSEPR theory/teoría VSEPR
una teoría que predice algunas formas moleculares con base en la idea de que los pares de electrones de valencia que rodean un átomo se repelen unos a otros (p. 209)

weak acid/ácido débil
un ácido que libera pocos iones de hidrógeno en una solución acuosa (p. 532)

weak base/base débil
un base que libera pocos iones de hidroxido en una solución acuosa (p. 534)

weight/peso
una medida de la fuerza gravitacional ejercida sobre un objeto; su valor puede cambiar en función de la ubicación del objeto en el universo (p. 10)

INDEX

chlorofluorocarbons (CFCs), 52

chromatography, 458, 458f

cloning, 731

coefficients, 268, 303

cohesion, 379

colligative properties, 481–483, 481f, 482f, 483f

colloids, 456, 456f

color, 15

color change, 8, 9f. *See also* **indicators**

combining volumes, Gay-Lussac's law of, 439, 439f

combustion, 276, 276f
calorimetry of, 351–352, 351f
fire extinguishers and, 290
of gasoline, 323–324, 323f, 325
rate of, 582

common ion effect, 517, 517f, 561–562

completion reactions, 302, 496, 496f

complex ions, 500, 500f, 500t
equilibria with, 500–501, 513, 513f, 515, 515f

composition, percentage, 241–243, 241f, 246–248, 246f

compounds, 24–25, 24f. *See also* **ionic compounds; molecular compounds**
atomic theory and, 74–78, 75f, 76f, 77f, 77t
versus mixtures, 27

compressibility, of gases, 417, 417f, 421

computer chips, 214

concentration, 460–461, 460t. *See also* **solutions**
defined, 460
from equilibrium constant, 506
Le Châtelier's principle and, 512, 513, 513f, 517, 561–562
molarity, 460t, 462–467, 462f
parts per million, 460, 460t, 461
reaction rate and, 582, 582f, 586–588, 588f
from solubility product constant, 510
in stoichiometry problems, 308
from titration data, 555–556

condensation, 381, 381f, 382, 382f
energy and, 396, 396f

condensation reactions, 699–700, 699f, 700f, 715, 718

conductivity, 478. *See also* **electrical conductivity**

conjugate acid, 537, 537t, 558, 559, 559t

conjugate base, 537, 537t, 558, 559, 559t
in buffer, 561–563

conservation of energy, 40–41, 41f

conservation of mass, 52, **76,** 76f, 77
balanced equation and, 263, 267, 267t, 270

controlled experiment, 51

conversion factors, 13–14, 226–227, 230
significant figures and, 58

coordination compounds, 500–501, 500f, 500t

corrosion, 620, 620f
prevention of, 131, 621, 621f, 631

Coulomb's law, 83

counting units, 101, 225, 225t

count value, 58

covalent bonds, 190–198. *See also* **Lewis structures**
defined, 191
electronegativity and, 194–197, 194f, 195f, 196t
energy and, 192–193, 192f, 193t
polar, 194–196, 195f, 197, 212–213, 212f
shared electrons in, 190–191, 191f, 194, 198, 200

covalent compounds. *See also* **molecular compounds**
as electrolytes, 479
formulas of, 236
naming of, 206–207, 207t
properties of, 197t, 198

critical mass, 654

critical point, 402, 402f

CRT (cathode-ray tube), 79–80, 79f, 80f

crystal lattice, 173, **174–175,** 174f
energy of, 168–169, 169f

crystallization, 476, 476f

cyclotron, 145

D

Dalton (unit of atomic mass), 100

Dalton, John, 77–78, 439, 643

Dalton's law of partial pressures, 439

Daniell cell, 616, 616f, 622f

dating objects, 658–662, 658f, 661f

decanting, 457f

decay series, 651, 651f

decomposition reaction, 278–279, 278f

definite proportions, law of, 75, 75f, 77

denaturing, 723

density, 16–17, 16f, 17f, 17t
calculation of, 17, 57
defined, 16
of gases, 417, 417f
QuickLAB experiment on, 18
in stoichiometry problems, 308, 309

deoxyribonucleic acid (DNA), 726–728, 726f, 727f, 728f
HIV treatment and, 738
hydrogen bonding in, 388, 388f, 726f, 727, 728
QuickLAB isolation of, 727
recombinant, 731

deposition, 381, 381f, 384

derived units, 15, 15f

detergent, 484, 484f, 485–486, 486f

diamond, 679, 679f

diatomic elements, 23, 23f
molar masses of, 237

diffraction
of electrons, 91
of X rays, 175

diffusion
of gases, 436–438, 436f
in solution, 358f, 359

dipole, 195

dipole-dipole forces, 386–389, 387t, 388f, 389f. *See also* **hydrogen bonds**
dispersion forces and, 391, 392

disaccharides, 712, 713, 714, 714f, 715. *See also* **sucrose**

disorder, 358, 359, 363

displacement reactions, 280–282, 280f, 281t
double, 283, 283f
net ionic equation for, 287

dissociation, 472

distillation, 459

disulfide bridge, 719, 719f

DNA (deoxyribonucleic acid), 726–728, 726f, 727f, 728f
HIV treatment and, 738
hydrogen bonding in, 388, 388f, 726f, 727, 728
QuickLAB isolation of, 727
recombinant, 731

DNA fingerprinting, 730–731

***d* orbitals,** 95, 96, 96f, 97, 97f
periodic table and, 122, 129

double bonds
defined, 204
in Lewis structures, 204, 205, 206, 206f
in organic compounds, 681, 682, 688, 688f, 689–691

double-displacement reaction, 283, 283f

Downs cell, 628, 628f

dry cells, 617, 617f

ductility, 129

dyes, synthetic, 49

dynamic equilibrium, 499, 499f

E

effective nuclear charge, 135, 136, 137, 138, 139

effusion, 437–438, 437f

Einstein, Albert, 62, 91, 143

electrical conductivity
of acid solutions, 480, 531, 531f
bond type and, 197, 197t
of metals, 129, 197
of salts, 172, 172f
of solutions, 478–481, 480f
of superconductors, 148

electric current, 479, 481, 531, 613

CREDITS

STAFF CREDITS

Executive Editor,
High School Physical Sciences
Mark Grayson

Managing Editor
Debbie Starr

Editorial Development Team
John A. Benner
Molly Frohlich
Michael Mazza
Micah Newman
Kharissia Pettus

Copyeditors
Dawn Marie Spinozza, Copyediting Manager
Anne-Marie De Witt
Jane A. Kirschman
Kira J. Watkins

Editorial Support Staff
Jeanne Graham
Mary Helbling
Shannon Oehler
Stephanie S. Sanchez
Tanu'e White

Editorial Interns
Kristina Bigelow
Erica Garza
Sarah Ray
Kenneth G. Raymond
Kyle Stock
Audra Teinert

Online Products
Bob Tucek, Executive Editor
Wesley M. Bain
Catherine Gallagher
Douglas P. Rutley

Production
Eddie Dawson, Production Manager
Suzanne Brooks, Production Coordinator

Design
Book Design
Joe Melomo, Design Director
Lori Male, Senior Designer
Ed Diaz, Design Associate
Alicia Sullivan, Designer, Teacher Edition
Sally Bess, Designer, Teacher Edition
Charlie Taliaferro, Design Associate, Teacher Edition

Image Acquistions
Curtis Riker, Director
Jeannie Taylor, Photo Research Supervisor
Elaine Tate, Art Buyer Supervisor

Design New Media
Ed Blake, Design Director
Kimberly Cammerata, Design Manager
Michael Rinella, Senior Designer

Media Design
Richard Metzger, Design Director
Chris Smith, Senior Designer

Graphic Services
Kristen Darby, Director
Jeff Robinson, Senior Ancillary Designer
Jane Dixon, Image Designer

Cover Design
Joe Melomo, Design Director

Design Implementation and Page Production
AARTPACK, Inc.

Electronic Publishing
EP Manager
Robert Franklin

EP Team Leaders
Juan Baquera
Sally Dewhirst
Christopher Lucas
Nanda Patel
JoAnn Stringer

Senior Production Artists
Katrina Gnader
Lana Kaupp
Kim Orne
Production Artists
Sara Buller
Ellen Kennedy
Patty Zepeda
Quality Control
Barry Bishop
Becky Golden-Harrell
Angela Priddy
Ellen Rees

New Media
Armin Gutzmer, Director of Development
Melanie Baccus, New Media Coordinator
Lydia Doty, Senior Project Manager
Marsh Flournoy, Quality Assurance Analyst
Cathy Kuhles, Technical Assistant
Tara F. Ross, Senior Project Manager

Ancillary Development and Production
GTS, Inc., Boston, Massachusetts

PHOTO & ART CREDITS

Art

Abbreviated as follows: (t) top; (b) bottom; (l) left; (r) right; (c) center.

All art unless otherwise noted by Holt, Rinehart, and Winston.

Chapter 1: Page 6-8, Kristy Sprott; 16 (br), David Chapman; 22-26, Kristy Sprott; 28 (flow chart), David Chapman, 28 (molecules), Kristy Sprott; 32, Kristy Sprott; 34, David Chapman. **Chapter 2:** Page 39,40, Kristy Sprott; 41-46, David Chapman; 48-53, Kristy Sprott; 70, David Chapman. **Chapter 3:** Page 74 (br), Kristy Sprott; 74 (bl), David Chapman; 75 (cr), Kristy Sprott; 75 (br), David Chapman; 76-80, Kristy Sprott; 81 (bl), Uhl Studios, Inc.; 83-91, Kristy Sprott; 92, Uhl Studios, Inc.; 93, David Chapman; 96, J/B Woolsey Associates; 97, David Chapman. **Chapter 4:** Page 117, Jack Scott; 119-130, David Chapman; 133, Kristy Sprott; 134, David Chapman; 135 (t), Kristy Sprott; 135 (b), David Chapman; 136-140 David Chapman; 143-144, Kristy Sprott; 146, 151, David Chapman; 153, Kristy Sprott; 154, David Chapman.**Chapter 5:** Page 160, David Chapman; 161, Kristy Sprott; 162-163, David Chapman; 169 (flow chart), David Chapman, (molecules), Kristy Sprott; 172, 174 (zooms), 183, Kristy Sprott; 186, David Chapman. **Chapter 6:** Page 190-192, Kristy Sprott; 194, David Chapman; 195-212, Kristy Sprott. **Chapter 7:** Page 236-238, Kristy Sprott; 241, David Chapman; 242, Kristy Sprott; 246-256, David Chapman. **Chapter 8:** Page 261-297, Kristy Sprott. **Chapter 9:** Page 315, (zooms) Kristy Sprott, (realia) Stephen Durke/Washington Artists; 321, 323, Stephen Durke/Washington Artists 326, Kristy Sprott; 334, David Chapman. **Chapter 10:** Page 339, Leslie Kell; 341, David Chapman; 348, Kristy Sprott; 351, Uhl Studios, Inc.; 353, Stephen Durke/Washington Artists; 363-366, Kristy Sprott; 373,374, David Chapman; **Chapter 11:** Page 379, 380, Kristy Sprott; 381, Stephen Durke/Washington Artists; 384-390, Kristy Sprott; 394, David Chapman; 399, 400 (tc), Kristy Sprott; 400 (bl), Stephen Durke/Washington Artists; 401-412, David Chapman. **Chapter 12:** Page 417 Kristy Sprott; 418 (t,b),419, Stephen Durke/Washington Artists; 421, Kristy Sprott; 422, David Chapman; 424 (t), Stephen Durke/Washington Artists; 424 (br), 427, David Chapman; 429 (t), Stephen Durke/Washington Artists; 429 (b), David Chapman; 431, Kristy Sprott; 434, David Chapman; 437, Kristy Sprott; 439-450, Stephen Durke/Washington Artists. **Chapter 13:** Page 455, 456, 462, 464, 471, Kristy Sprott; 472, David Chapman; 473, Kristy Sprott; 474, David Chapman; 476, 479, 480, 482, Kristy Sprott; 483, David Chapman; 484, David Chapman/Kristy Sprott; 485, Kristy Sprott. **Chapter 14:** Page 496, 497, Kristy Sprott; 498, 505, David Chapman; 507, 513, Kristy Sprott; 514, David Chapman; 515-524, Kristy Sprott. **Chapter 15:** Page 531-539, Kristy Sprott; 540, David Chapman; 549, Kristy Sprott; 551, David Chapman; 557-572, Kristy Sprott.

Chapter 16: Page 578, Kristy Sprott; 580, David Chapman; 588 (clocks) David Chapman, (zooms) Kristy Sprott; 591 (chart) David Chapman, (molecules) Kristy Sprott; 592, Kristy Sprott; 594, David Chapman. **Chapter 17:** Page 605, 608, Kristy Sprott; 613 (bl), Stephen Durke/Washington Artists; 613 (br), Kristy Sprott; 614 (art), Stephen Durke/Washington Artists; 614 (zoom), Kristy Sprott; 615 (tr), Stephen Durke/Washington Artists; 616 (art), Stephen Durke/Washington Artists; 616 (zooms), Kristy Sprott; 616, Kristy Sprott; 617-620 (art), Stephen Durke/Washington Artists; 620 (zooms), Kristy Sprott; 626 (art), Stephen Durke/Washington Artists; 626 (zooms), Kristy Sprott; 628-635, Stephen Durke/Washington Artists; 638, David Chapman. **Chapter 18:** Page 643-644, Kristy Sprott; 645-648, David Chapman; 649,650, Kristy Sprott; 651, David Chapman; 655, Uhl Studios, Inc.; 661, 663, David Chapman. **Chapter 19:** Page 680-700 Kristy Sprott. **Chapter 20:** Page 717-720, Stephen Durke/Washington Artists; 722, Christy Krames; 726-728, 733, Morgan-Cain & Assoc.; 743, John White/The Neis Group; 744, David Chapman. **Labs:** Page 758, Uhl Studios, Inc.; 763, David Chapman; 764, 788, Uhl Studios, Inc. **Appendix:** Page 831, Kristy Sprott.

Photo

Abbreviated as follows: (t) top; (b) bottom; (l) left; (r) right; (c) center.

All photos unless otherwise noted by Holt, Rinehart, and Winston.

Cover and page i: Charlie Winters; **Table of Contents:** Page vi, © Russell Johnson (t); © Kraft/Explorer/Photo Researchers, Inc. (b); vii, IBM Research, Almaden Research Center (t); © Science Pictures Limited/CORBIS (b); viii, © CORBIS (t); © Adastra/Getty Images (b); ix, © Ted Mahieu/Corbis Stock Market (t); © Roland Birke/Peter Arnold (b); x, © David Nardini /Getty Images (t); Peter Van Steen/HRW (b); xi, © Charles O'Rear/ CORBIS (t); Brand X (b); xii, © H. Abernathy/H. Armstrong Roberts Stock Photography (t); © Jerry Shulman/SuperStock (b); **Chapter 1:** Page 2, © Russell Johnson; 4, Victoria Smith/HRW; 6, Peter Van Steen/HRW; 7, Victoria Smith/HRW; 8, Charlie Winter/HRW; 9 (all), Sergio Purtell/Foca/HRW; 10, Victoria Smith/HRW; 10 (inset), Victoria Smith/HRW; 11, Charlie Winters/HRW; 12, Andrew Brookes Creative/Corbis Stock Market; 15, Victoria Smith/HRW; 17, Denis Fagan/HRW; 19, Charlie Winter/HRW; 20, © S. Feld/H. Armstrong Roberts Stock Photography; 21, © Larry Lefever/Grant Heilman Photography; 22 (tl), © Leonard Lessin/Peter Arnold; 22 (tm), © Tom Pantages Photography; 22 (tr) © Charles D. Winters/Photo Researchers, Inc.; 23 (l), Victoria Smith/HRW; 23 (r), © Steve Vidler/SuperStock; 24, © Francisco Cruz/SuperStock; 25 (l), © Neal Mishler/FPG International/Getty Images; 25 (r), Sergio Purtell/Foca/HRW; 26 (both), Charlie Winter/HRW; 29, © A. Ramey/Photo Edit. **Chapter 2:** Page 36, © Kraft/Explorer/Photo Researchers, Inc.; 38, Charlie Winters/HRW; 39, Ice Sculpture by Duncan Hamilton. Photograph by Mike Venables; 40, Charlie Winters/HRW; 42, © Bettmann/CORBIS; 43, © Ana/The Image Works; 48, © Andre Jenny/South Stock/PictureQuest; 49, Photo Researchers, Inc.; 50, HRW Photo; 51, © Photo by Joe Raedle/Newsmakers/Getty Images ; 52, © Newsmakers/Getty Images; 54, © Kristen Brochmann/Fundamental Photography; 56, Charlie Winters/HRW; 57, Victoria Smith/HRW; 64, Scoones/SIPA Press. **Chapter 3:** Page 72, IBM Research, Almaden Research Center; 75, Peter Van Steen/HRW; 79, © Kelvin Murray/Stone/Getty Images; 80, Charlie Winters/HRW; 82 (l), Sergio Purtell/Foca; 82 (r), © Steve Bronstein/Getty Images; 100, Sam Dudgeon/HRW; 104, John Macfie/ImageState; 105 (t), E.R. Degginger/Color-Pic, Inc.; 105 (b), © Carl Frank/Photo Researchers, Inc. **Chapter 4:** Page 114, © Phil Schermeister/CORBIS; 116 (all), Charlie Winters/HRW; 123, © Linda S. Nye/Phototake; 125, 126, Charlie Winters/HRW; 127 (t), Charlie Winters/HRW; 127 (b), © Bob Burch/Index Stock Imagery; 128, © Tom Pantages; 129, © Leonard Lessin/Peter Arnold; 130, © Roger Ressmeyer/CORBIS; 131, © Fritz Henle/Photo Researchers, Inc.; 132 (all), Charlie Winters, 142, NASA; 143, © Roger Russmeyer/ CORBIS; 148 (l), NASA; 148 (r), © Gabe Palmer/Corbis Stock Market. **Chapter 5:** Page 156, © Science Pictures Limited/CORBIS; 158 (both), 159, 164 (both), Charlie Winters/ HRW; 166, © Eric Simmons/Stock Boston/PictureQuest; 170, © David Young-Wolff/Photo Edit; 172 (all), Charlie Winters/HRW;174 (l), © Edward R. Degginger/ Bruce Coleman Inc./PictureQuest; 174 (r), © Anthony Mercieca/SuperStock; 181, Rick Lance/ Phototake; 181 (inset), Sam Dudgeon/HRW. **Chapter 6:** Page 188, Peter Van Steen/HRW; 190, © T. McCarthy/Custom Medical Stock Photo; 191, 193 (both), 195, Victoria Smith/HRW; 197 (all), Charlie Winters/HRW; 199, Courtesy of Edgar Fahs Smith Collection/University of Pennsylvania Library; 200, Sam Dudgeon/HRW; 203, 204 (both), Sergio Purtell/Foca; 206, © IFA/Peter Arnold, Inc.; 208, Victoria Smith/ HRW; 214, © Jim Karageorge/FPG International/Getty Images. **Chapter 7:** Page 222, © CORBIS; 224, Charlie Winters/HRW; 225, Victoria Smith/HRW; 234, 236 (all), 241 (both), Charlie Winters/HRW; 244, Sam Dudgeon/HRW; 249, © Jim Karageorge/FPG International/Getty Images. **Chapter 8:** Page 258, © Adastra/Getty Images; 260, © Phil Schermeister/CORBIS; 261 (both), ©1988 Richard Megna Fundamental Photographs; 262 (l), Sam Dudgeon/HRW; 262 (r), Charlie Winters/HRW; 263, 264 (both), Sam Dudgeon/HRW; 272 (both), 275,

Credits **909**

Periodic Table of the Elements

Key:

6 — Atomic number
C — Symbol
Carbon — Name
12.0107 — Average atomic mass
[He]$2s^2 2p^2$ — Electron configuration

Group 1	Group 2

Period

1 **H** Hydrogen 1.007 94 $1s^1$								
3 **Li** Lithium 6.941 [He]$2s^1$	4 **Be** Beryllium 9.012 182 [He]$2s^2$							
11 **Na** Sodium 22.989 770 [Ne]$3s^1$	12 **Mg** Magnesium 24.3050 [Ne]$3s^2$							

		Group 3	Group 4	Group 5	Group 6	Group 7	Group 8	Group 9
19 **K** Potassium 39.0983 [Ar]$4s^1$	20 **Ca** Calcium 40.078 [Ar]$4s^2$	21 **Sc** Scandium 44.955 910 [Ar]$3d^1 4s^2$	22 **Ti** Titanium 47.867 [Ar]$3d^2 4s^2$	23 **V** Vanadium 50.9415 [Ar]$3d^3 4s^2$	24 **Cr** Chromium 51.9961 [Ar]$3d^5 4s^1$	25 **Mn** Manganese 54.938 049 [Ar]$3d^5 4s^2$	26 **Fe** Iron 55.845 [Ar]$3d^6 4s^2$	27 **Co** Cobalt 58.933 200 [Ar]$3d^7 4s^2$
37 **Rb** Rubidium 85.4678 [Kr]$5s^1$	38 **Sr** Strontium 87.62 [Kr]$5s^2$	39 **Y** Yttrium 88.905 85 [Kr]$4d^1 5s^2$	40 **Zr** Zirconium 91.224 [Kr]$4d^2 5s^2$	41 **Nb** Niobium 92.906 38 [Kr]$4d^4 5s^1$	42 **Mo** Molybdenum 95.94 [Kr]$4d^5 5s^1$	43 **Tc** Technetium (98) [Kr]$4d^6 5s^1$	44 **Ru** Ruthenium 101.07 [Kr]$4d^7 5s^1$	45 **Rh** Rhodium 102.905 50 [Kr]$4d^8 5s^1$
55 **Cs** Cesium 132.905 45 [Xe]$6s^1$	56 **Ba** Barium 137.327 [Xe]$6s^2$	57 **La** Lanthanum 138.9055 [Xe]$5d^1 6s^2$	72 **Hf** Hafnium 178.49 [Xe]$4f^{14} 5d^2 6s^2$	73 **Ta** Tantalum 180.9479 [Xe]$4f^{14} 5d^3 6s^2$	74 **W** Tungsten 183.84 [Xe]$4f^{14} 5d^4 6s^2$	75 **Re** Rhenium 186.207 [Xe]$4f^{14} 5d^5 6s^2$	76 **Os** Osmium 190.23 [Xe]$4f^{14} 5d^6 6s^2$	77 **Ir** Iridium 192.217 [Xe]$4f^{14} 5d^7 6s^2$
87 **Fr** Francium (223) [Rn]$7s^1$	88 **Ra** Radium (226) [Rn]$7s^2$	89 **Ac** Actinium (227) [Rn]$6d^1 7s^2$	104 **Rf** Rutherfordium (261) [Rn]$5f^{14} 6d^2 7s^2$	105 **Db** Dubnium (262) [Rn]$5f^{14} 6d^3 7s^2$	106 **Sg** Seaborgium (263) [Rn]$5f^{14} 6d^4 7s^2$	107 **Bh** Bohrium (264) [Rn]$5f^{14} 6d^5 7s^2$	108 **Hs** Hassium (265) [Rn]$5f^{14} 6d^6 7s^2$	109 **Mt** Meitnerium (268)† [Rn]$5f^{14} 6d^7 7s^2$

† Estimated from currently available IUPAC data.

* The systematic names and symbols for elements greater than 109 will be used until the approval of trivial names by IUPAC.

58 **Ce** Cerium 140.116 [Xe]$4f^1 5d^1 6s^2$	59 **Pr** Praseodymium 140.907 65 [Xe]$4f^3 6s^2$	60 **Nd** Neodymium 144.24 [Xe]$4f^4 6s^2$	61 **Pm** Promethium (145) [Xe]$4f^5 6s^2$	62 **Sm** Samarium 150.36 [Xe]$4f^6 6s^2$
90 **Th** Thorium 232.0381 [Rn]$6d^2 7s^2$	91 **Pa** Protactinium 231.035 88 [Rn]$5f^2 6d^1 7s^2$	92 **U** Uranium 238.0289 [Rn]$5f^3 6d^1 7s^2$	93 **Np** Neptunium (237) [Rn]$5f^4 6d^1 7s^2$	94 **Pu** Plutonium (244) [Rn]$5f^6 7s^2$

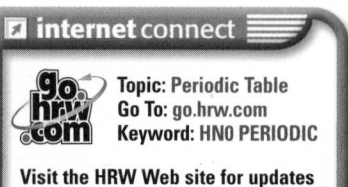

internet connect

go.hrw.com

Topic: Periodic Table
Go To: go.hrw.com
Keyword: HN0 PERIODIC

Visit the HRW Web site for updates on the periodic table.